The Geology of Japan

The Geological Society of London
Books Editorial Committee

Chief Editor

Rick Law (USA)

Society Books Editors

Jim Griffiths (UK)
Dave Hodgson (UK)
Phil Leat (UK)
Nick Richardson (UK)
Daniela Schmidt (UK)
Randell Stephenson (UK)
Rob Strachan (UK)
Mark Whiteman (UK)

Society Books Advisors

Ghulam Bhat (India)
Marie-Francoise Brunet (France)
Martin Hand (Australia)
Jasper Knight (South Africa)
Mario Parise (Italy)
Satish-Kumar (Japan)
Marco Vecoli (Saudi Arabia)
Gonzalo Veiga (Argentina)

Geological Society books refereeing procedures

The Society makes every effort to ensure that the scientific and production quality of its books matches that of its journals. Since 1997, all book proposals have been refereed by specialist reviewers as well as by the Society's Books Editorial Committee. If the referees identify weaknesses in the proposal, these must be addressed before the proposal is accepted.

Once the book is accepted, the Society Book Editors ensure that the volume editors follow strict guidelines on refereeing and quality control. We insist that individual papers can only be accepted after satisfactory review by two independent referees. The questions on the review forms are similar to those for *Journal of the Geological Society*. The referees' forms and comments must be available to the Society's Book Editors on request.

Although many of the books result from meetings, the editors are expected to commission papers that were not presented at the meeting to ensure that the book provides a balanced coverage of the subject. Being accepted for presentation at the meeting does not guarantee inclusion in the book.

More information about submitting a proposal and producing a book for the Society can be found on its website: www.geolsoc.org.uk.

It is recommended that reference to all or part of this book should be made in one of the following ways:

Moreno, T., Wallis, S., Kojima, T. & Gibbons, W. (eds) 2016. *The Geology of Japan*. Geological Society, London.

Saito, Y., Ikehara, K. & Tamura, T. 2016. Coastal geology and oceanography. *In*: Moreno, T., Wallis, S., Kojima, T. & Gibbons, W. (eds) *The Geology of Japan*. Geological Society, London, 413–434.

The Geology of Japan

EDITED BY

TERESA MORENO

Institute of Environmental Assessment and Water Research (IDÆA),
Spanish National Research Council (CSIC), Barcelona, Spain

SIMON WALLIS

Department of Earth and Planetary Sciences, Graduate School of
Environmental Studies, Nagoya University, Japan

TOMOKO KOJIMA

Department of Earth and Environmental Sciences, Graduate School of
Science and Technology, Kumamoto University, Japan

and

WES GIBBONS

WPS, 08870 Sitges, Barcelona, Spain

2016
Published by
The Geological Society
London

THE GEOLOGICAL SOCIETY

The Geological Society of London (GSL) was founded in 1807. It is the oldest national geological society in the world and the largest in Europe. It was incorporated under Royal Charter in 1825 and is Registered Charity 210161.

The Society is the UK national learned and professional society for geology with a worldwide Fellowship (FGS) of over 10 000. The Society has the power to confer Chartered status on suitably qualified Fellows, and about 2000 of the Fellowship carry the title (CGeol). Chartered Geologists may also obtain the equivalent European title, European Geologist (EurGeol). One-fifth of the Society's fellowship resides outside the UK. To find out more about the Society, log on to www.geolsoc.org.uk.

The Geological Society Publishing House (Bath, UK) produces the Society's international journals and books, and acts as European distributor for selected publications of the American Association of Petroleum Geologists (AAPG), the Indonesian Petroleum Association (IPA), the Geological Society of America (GSA), the Society for Sedimentary Geology (SEPM) and the Geologists' Association (GA). Joint marketing agreements ensure that GSL Fellows may purchase these societies' publications at a discount. The Society's online bookshop (accessible from www.geolsoc.org.uk) offers secure book purchasing with your credit or debit card.

To find out about joining the Society and benefiting from substantial discounts on publications of GSL and other societies worldwide, consult www.geolsoc.org.uk, or contact the Fellowship Department at: The Geological Society, Burlington House, Piccadilly, London W1J 0BG: Tel. +44 (0)20 7434 9944; Fax +44 (0)20 7439 8975; E-mail: enquiries@geolsoc.org.uk.

For information about the Society's meetings, consult *Events* on www.geolsoc.org.uk. To find out more about the Society's Corporate Affiliates Scheme, write to enquiries@geolsoc.org.uk.

Published by The Geological Society from:
The Geological Society Publishing House, Unit 7, Brassmill Enterprise Centre, Brassmill Lane, Bath BA1 3JN, UK

The Lyell Collection: www.lyellcollection.org
Online bookshop: www.geolsoc.org.uk/bookshop
Orders: Tel. +44 (0)1225 445046, Fax +44 (0)1225 442836

British Library Cataloguing in Publication Data

A catalogue record for this book is available from the British Library.
ISBN 978-1-86239-742-2 (hardback)
ISBN 978-1-86239-743-9 (softback)

Distributors

For details of international agents and distributors see:
www.geolsoc.org.uk/agentsdistributors

Typeset by Techset Composition India (P) Ltd, Bangalore and Chennai, India
Printed by Henry Ling Ltd at the Dorset Press, Dorchester, UK

Contents

Preface

The idea of editing a third 'Geology of' book developed during an extended stay in Japan in 2011, financial support for which was generously provided by the Generalitat de Catalunya and by the Japan Society for the Promotion of Science (Project No. S-11019). Wes Gibbons and I visited first in the spring and again in the autumn, working with our colleague Tomoko Kojima at Kumamoto University on the physicochemical characterization of atmospheric aerosols arriving in Japanese airspace from mainland Asia. It became apparent to us that this time the editorial team would be greatly enhanced by adding geological editors living locally in Japan, fluent in Japanese and with a good understanding of the local Geoscience community. Simon Wallis and Tomoko Kojima kindly agreed to join the project, consent was reached with the Geological Society of London and letters sent to potential senior authors in late 2011. As is usually the case with major book projects, progress was slower than initially planned: our senior authors are all research-active and most have additional heavy commitments to teaching and administration. Furthermore, as anyone who has ever tried it will know, writing book chapters in one's 'spare time' and in a foreign language is never easy. Two chapters (Earthquakes and Hokkaido) threatened delays that were becoming terminal and needed drastic action. In the case of Earthquakes, we are forever indebted to Professor Hiroyuki Tsutsumi of Kyoto University for suggesting Professor Shinji Toda at Tohoku University who saved the day. We are similarly grateful to Professor Hayato Ueda of Niigata University who joined us at a very late stage in order to produce the Hokkaido chapter. Another special contribution to the project was made by Professor Yoshiki Saito (senior author of Chapter 9) who proposed that the book would benefit by the addition of a chapter on Neogene and Quaternary sedimentary successions, and in this context we greatly appreciate the efforts of Professor Makoto Ito of Chiba University who agreed at a relatively late stage to join our list of senior authors.

Arriving in Osaka Kansai airport just two days after the terrible earthquake and tsunami on the 11 March 2011 was a heart-breaking and horrifying reminder of the relevance of geology to the people of Japan. The whole country was in shock and mourning, made worse by subsequent events at Fukushima, but the deep sense of social solidarity in the struggle to overcome natural disaster was also very obvious. Japanese society has always had to live with unusually frequent extreme events resulting from living on the most complex and active plate boundary system on Earth. This propensity is further exacerbated by the effects on young mountainous and volcanic landforms of a climate prone to typhoons and heavy thunderstorms. It is this same tectonic setting and geomorphology, of course, that makes Japan one of the most interesting places in the world for a geologist to visit, and I hope that the final chapter in this book will encourage many readers to see it for themselves: the country is truly amazing.

On behalf of the editorial team, I wish to express my deep thanks to all contributors and reviewers and for the enormous collective effort they have made. Once again Angharad Hills at the Geological Society of London has from the beginning been consistently helpful, encouraging and supportive, and thanks are also due to Helen Floyd-Walker, and Rick Law, our Series Editor. I would also like thank my 'task force' team in Japan, Simon Wallis and Tomoko Kojima, who have indeed more than proved themselves essential to the success of the project, frequently encouraging authors on the telephone when my emails became too persistent to be effective. Tomoko Kojima produced, to my knowledge for the first time, the lists of Kanji and Hiragana symbols for geological names. Finding agreement on the naming of geological units, always controversial in any country, has eluded us at times and we have been forced to take a 'state-of-the-art' approach that reflects the ongoing situation in the research community. Finally, I wish to thank my husband Wes Gibbons for his support and the hours he has put in, applying his by-now considerable experience of improving English text written by authors writing in what is for them a foreign language. It was he, back in 2011, who first suggested the interesting challenge of organizing a Geology of Japan book and convinced me that 'yes we can' or, as we said in atrocious English here in Spain during the 2010 Football World Cup, 'impossible is nothing'.

TERESA MORENO

Barcelona, May 2015

The contents of the *Geology of Japan* reflect the considerable advances in our understanding in Earth Sciences of the Japanese islands and surrounding seafloor that have been made over the last 20 years. This is not the first book to review the geology of Japan; however, our new book has a wider scope covering not only regional geology but also many related topics such as earthquakes, resource geology and engineering geology. We hope this will serve as an entry point for those wishing to study the geology of Japan more closely and those who are looking for a reliable and up-to-date summary.

In preparing the text for this book, one of the problems we faced is the lack of a clear consensus on how to transcribe Japanese using the Roman alphabet. One particular issue is representing the length of vowels: long vowels may be indicated by use of overbars known as macrons, additional vowels or inserting the letter 'h'. In other cases, no distinction is made and the correct pronunciation requires additional information such as knowledge of the characters used to write the name. Here, we employ a simple system that does not clutter the text with bars, extra vowels or extraneous h letters that would, in any case, be an incomplete solution. A second related problem is that the pronunciation of the phonetic symbol 'ん' may be closer to the English 'n' or 'm' depending on the word. Some try to reflect this by changing the romanization accordingly. In the face of strong but contradictory opinions from several contributing authors, we soon relinquished the idea of standardizing the representation of 'ん'. As a result of this and other similar issues, the same geological term may be spelt slightly differently in different chapters. We hope that, with a little forbearance, this should not be too irritating or cause too much confusion. Japanese phonetic syllabary is unambiguous and, as an aid to overcoming these problems, we have included an extensive list of geological and geographical names with the corresponding Japanese script. If necessary, we hope that visitors will take the opportunity to ask Japanese speakers the correct pronunciation or even try their hand at mastering some first steps in written Japanese.

I am grateful to all the authors and reviewers who have contributed to this book. The reward to all these workers is to see the results before others and the satisfaction of knowing they have made a lasting contribution to the geological studies of Japan. I have also received considerable support and encouragement for this project from the Geological Society of Japan and in particular the former and present Presidents, Akira Ishiwatari and Yasufumi Iryu. I am also pleased to acknowledge my co-editors, Teresa, Wes and Tomoko. Teresa and Wes had the foresight to recognize the need for an up-to-date English language overview of the geology of Japan. Teresa has maintained a firm but diplomatic hand on the editorial rudder, while Wes has provided both wisdom and good humour to keep things moving in the right direction. Tomoko has provided much-needed logistical support and compiled information on Japanese geological terms, museums, etc. that help to make this book a unique contribution. It has been a long journey – longer than we expected! The good companionship of my fellow editors helped to make it a pleasure, but it is also good to reach the destination.

Simon Wallis

Nagoya, May 2015

The editors gratefully acknowledge the work of the referees who gave so much of their time to reviewing the initial manuscripts (listed below in alphabetical order):

Bernard BONIN	Universite de Paris-Sud, France
Peter BURGESS	Royal Holloway, University of London, UK
Timothy BYRNE	University of Connecticut, USA
Moonsup CHO	Seoul National University, Korea
Anthony EDWARD	Centre de Recherche et d'Enseignement de Géosciences, France
John C EICHELBERGER	University of Alaska Fairbanks, USA
Masaki ENAMI	Nagoya University, Japan
Bogdan ENESCU	University of Tsukuba, Japan
Olivier FABBRI	Université de Franche-Comté, France
Jose Ignacio GIL IBARGUCHI	Universidad del Pais Vasco, Spain
Kathy GILLIS	University of Victoria, Canada
Richard GOLDFARB	United States Geological Survey, USA
George HELFFRICH	University of Bristol, UK
Takeshi IKEDA	Kyushu University, Japan
Shunso ISHIHARA	Geological Survey of Japan (AIST), Japan
Laurent JOLIVET	Université d'Orléans, France
Koen de JONG	Seoul National University, Korea
Kyuichi KANAGAWA	Chiba University, Japan
Shuichi KODAIRA	Japan Agency for Marine-Earth Science and Technology, Japan
José Luis MACÍAS	Universidad Nacional Autónoma de México, Mexico
Mauri McSAVENEY	GNS Science, New Zealand
Yujiro OGAWA	University of Tsukuba, Japan
Shigeru OTOH	University of Toyama, Japan
Dave PETLEY	Durham University, UK
Greg POWER	University of Portsmouth, UK
Martin REYNERS	Institute of Geological and Nuclear Sciences, New Zealand
Takeshi SAGIYA	Nagoya University, Japan
Masayuki SAKAKIBARA	Ehime University, Japan
Robert J. STERN	University of Texas at Dallas, USA
Makoto TAKEUCHI	Nagoya University, Japan
Michael UNDERWOOD	University of Missouri, USA
Tetsuro URABE	The University of Tokyo, Japan
Joe WHALEN	Natural Resources Canada, Canada
Atsushi YAMAJI	Kyoto University, Japan

Contributors

Natsue ABE
Research and Development Center for Ocean Drilling Science, Japan Agency for Marine-Earth Science and Technology (JAMSTEC), Japan. abenatsu@jamstec.go.jp.

Shoji ARAI
Department of Earth Sciences, Kanazawa University, Japan. ultrasa@staff.kanazawa-u.ac.jp.

Masahiro CHIGIRA
Disaster Prevention Research Institute, Kyoto University, Kyoto, Japan. chigira@slope.dpri.kyoto-u.ac.jp.

Masayuki EHIRO
The Tohoku University Museum, Sendai, Japan. ehiro@m.tohoku.ac.jp.

Wes GIBBONS
WPS, 08870 Sitges, Spain. wes.punto.es@gmail.com.

Akira HASEGAWA
Research Center for Prediction of Earthquakes and Volcanic Eruptions, Tohoku University, Sendai, Japan. hasegawa@aob.geophys.tohoku.ac.jp.

Yoshitaka HASHIMOTO
Department of Applied Science, Faculty of Science, Kochi University, Kochi, Japan.

Yasutaka HAYASAKA
Department of Earth and Planetary Systems Science, Graduate School of Science, Hiroshima University, Japan. hayasaka@hiroshima-u.ac.jp.

Yoshikuni HIROI
Department of Earth Sciences, Chiba University, Japan. yhiroi@earth.s.chiba-u.ac.jp.

Yoshihisa HIROKI
Faculty of Education, Osaka Kyoiku University, Japan. hiroki@cc.osaka-kyoiku.ac.jp.

Ken IKEHARA
Geological Survey of Japan, National Institute of Advanced Industrial Science and Technology (AIST), Tsukuba, Japan. k-ikehara@aist.go.jp.

Teruyoshi IMAOKA
Department of Geology, Yamaguchi University, Japan.

Yasuhumi IRYU
Department of Earth Science, Tohoku University, Japan. iryu@m.tohoku.ac.jp

Satoko ISHIMARU
Department of Earth and Environmental Sciences, Kumamoto University, Japan. ishimaru@sci.kumamoto-u.ac.jp.

Akira ISHIWATARI
Nuclear Regulation Authority, Tokyo, Japan. stepstone@tulip.ocn.ne.jp.

Osamu ISHIZUKA
Geological Survey of Japan, National Institute of Advanced Industrial Science and Technology (AIST), Tsukuba, Japan. o-ishizuka@aist.go.jp.

Makoto ITO
Department of Earth Sciences, Graduate School of Science, Chiba University, Japan. mito@faculty.chiba-u.jp.

Koji KAMEO
Department of Earth Sciences, Graduate School of Science, Chiba University, Japan. kameo@faculty.chiba-u.jp.

Yuji KANAORI
Department of Geosphere Sciences, Graduate School of Science and Engineering, Yamaguchi University, Japan. kanaori@yamaguchi-u.ac.jp.

Nobuyuki KANEKO
Institute for Geo-Resources and Environment, National Institute of Advanced Industrial Science and Technology (AIST), Tsukuba, Japan.

Gaku KIMURA
Department of Earth and Planetary Science, University of Tokyo, Tokyo, Japan. gaku@eps.s.u-tokyo.ac.jp.

Yujin KITAMURA
Department of Earth and Environmental Sciences, Kagoshima University, Kagoshima, Japan. yujin@sci.kagoshima-u.ac.jp.

Satoru KOJIMA
Department of Civil Engineering, Gifu University, Japan. skojima@gifu-u.ac.jp.

Tomoko KOJIMA
Department of Earth and Environmental Sciences, Kumamoto University, Japan. tkojima@sci.kumamoto-u.ac.jp.

Fukashi MAENO
Earthquake Research Institute, University of Tokyo, Japan. fmaeno@eri.u-tokyo.ac.jp.

Fujio MASUDA
Doshisha University, Kyoto, Japan. masuda@mail.doshisha.ac.jp.

Atsushi MATSUOKA
Department of Geology, Niigata University, Japan. amatsuoka@geo.sc.niigata-u.ac.jp.

Yuichiro MIYATA
Department of Geosphere Sciences, Yamaguchi University, Japan. miyata@mail.sci.yamaguchi-u.ac.jp.

Kazuhiro MIYAZAKI
Geological Survey of Japan, National Institute of Advanced Industrial Science and Technology (AIST), Tsukuba, Japan. kazu-miyazaki@aist.go.jp.

Teresa MORENO
Institute of Environmental Assessment & Water Research (IDÆA-CSIC), Barcelona, Spain. teresa.moreno@idaea.csic.es.

Setsuya NAKADA
Earthquake Research Institute, University of Tokyo, Japan. nakada@eri.u-tokyo.ac.jp.

Junichi NAKAJIMA
Department of Earth and Planetary Sciences, Tokyo Institute of Technology, Tokyo, Japan. nakajima@geo.titech.ac.jp.

Takashi NAKAJIMA
Geological Survey of Japan, National Institute of Advanced Industrial Science and Technology (AIST), Geological Museum, Tsukuba, Japan. tngeoch@aist.go.jp.

Takeshi NAKAJIMA
Geological Survey of Japan, National Institute of Advanced Industrial Science and Technology (AIST), Geological Museum, Tsukuba, Japan. takeshi.nakajima@aist.go.jp.

Alexander R. L. NICHOLS
Japan Agency for Marine-Earth Sciences and Technology (JAMSTEC), Japan.

Manchuk NURUMKHAAN
Mongolian University of Science and Technology, Ulaanbaatar, Mongolia. manchukn@gmail.com.

Yasuhiko OHARA
Hydrographic and Oceanographic Department of Japan & Japan Agency for Marine-Earth Science and Technology, Japan. ohara@jodc.go.jp.

Takamoto OKUDAIRA
Department of Geosciences, Faculty of Science, Osaka City University, Japan. oku@sci.osaka-cu.ac.jp.

Masaki OZAKI
Geological Survey of Japan, National Institute of Advanced Industrial Science and Technology (AIST), Tsukuba, Japan. masa-ozaki@aist.go.jp.

Kazuhito OZAWA
Department of Earth and Planetary Science, University of Tokyo, Japan. ozawa@eps.s.u-tokyo.ac.jp.

Makoto SAITO
Geological Survey of Japan, National Institute of Advanced Industrial Science and Technology (AIST), Tsukuba, Japan. saitomkt@aist.go.jp.

Yoshiki SAITO
Geological Survey of Japan, National Institute of Advanced Industrial Science and Technology (AIST), Tsukuba, Japan. yoshiki.saito@aist.go.jp.

Hiroyoshi SANO
Department of Earth & Planetary Sciences, Faculty of Science, Kyushu University, Japan. sano@geo.kyushu-u.ac.jp.

Yasufumi SATOGUCHI
Lake Biwa Museum, Shiga, Japan. satoguti@lbm.go.jp.

Toshiaki SHIMURA
Department of Geology, Yamaguchi University, Japan. smr@yamaguchi-u.ac.jp.

Yoshiaki SUGAMORI
Department of Regional Environment, Faculty of Regional Sciences, Tottori University, Japan. sugamori@rs.tottori-u.ac.jp.

Noritoshi SUZUKI
Department of Earth Science, Graduate School of Science, Tohoku University, Sendai, Japan. norinori@m.tohoku.ac.jp.

Noriyuki SUZUKI
Hokkaido University, Japan. suzu@sci.hokudai.ac.jp.

Yuichiro SUZUKI
Institute for Geo-Resources and Environment, National Institute of Advanced Industrial Science and Technology (AIST), Tsukuba, Japan.

Asahiko TAIRA
Japan Agency for Marine-Earth Science and Technology (JAMSTEC), Japan. ataira@jamstec.go.jp.

Tetsuichi TAKAGI
Institute for Geo-Resources and Environment, National Institute of Advanced Industrial Science and Technology (AIST), Tsukuba, Japan.

Masaki TAKAHASHI
Department of Earth System Science, Nihon University, Tokyo, Japan.

Narumi TAKAHASHI
Japan Agency for Marine-Earth Science and Technology (JAMSTEC), Japan. narumi@jamstec.go.jp.

Osamu TAKANO
Japan Petroleum Exploration Co., Ltd, Japan. osamu.takano@japex.co.jp.

Shizuo TAKEMURA
Hyogo University of Teacher Education, Japan. takesizu@hyogo-u.ac.jp.

Miyuki TAKEUCHI
Department of Earth Sciences, Kanazawa University, Japan. takeuchi-miyuki@stu.kanazawa-u.ac.jp.

Toru TAMURA
Geological Survey of Japan, National Institute of Advanced Industrial Science and Technology (AIST), Tsukuba, Japan. toru.tamura@aist.go.jp.

Yoshihiko TAMURA
Japan Agency for Marine-Earth Science and Technology (JAMSTEC), Japan. tamuray@jamstec.go.jp.

Ken-Ichiro TANI
National Museum of Nature and Science, Tokyo, Japan. kentani@kahaku.go.jp.

Yoshiyuki TATSUMI
Kobe University, Japan. tatsumi@diamond.kobe-u.ac.jp.

Shinji TODA
International Research Institute of Disaster Science, Tohoku University, Sendai, Japan. toda@irides.tohoku.ac.jp.

Seiichi TOSHIMITSU
Geological Survey of Japan, National Institute of Advanced Industrial Science and Technology (AIST), Tsukuba, Japan. stoshimitsu@aist.go.jp.

Tatsuki TSUJIMORI
Center for Northeast Asian Studies, Tohoku University, Sendai, Japan. tatsukix@m.tohoku.ac.jp.

Kazuhiro TSUKADA
Nagoya University Museum, Nagoya, Japan. tsukada@num.nagoya-u.ac.jp.

Takayuki UCHINO
Institute of Geology and Geoinformation, Geological Survey of Japan, National Institute of Advanced Industrial Science and Technology (AIST), Tsukuba, Japan. t-uchino@aist.go.jp.

Hayato UEDA
Department of Geology, Faculty of Science, Niigata University, Japan. ueta@geo.sc.niigata-u.ac.jp.

Kohtaro UJIIE
Graduate School of Life and Environmental Sciences, University of Tsukuba, Japan. kujiie@geol.tsukuba.ac.jp.

Yasuhiko WAKIZAKA
Public Works Research Institute, Tsukuba, Japan. wakizaka@pwri.go.jp.

Simon WALLIS
Department of Earth and Planetary Sciences, Graduate School of Environmental Studies, Nagoya University, Japan. swallis@eps.nagoya-u.ac.jp.

Yasushi WATANABE
Faculty of International Resource Sciences, Akita University, Japan. y-watanabe@gipc.akita-u.ac.jp.

Asuka YAMAGUCHI
Atmosphere and Ocean Research Institute, University of Tokyo, Chiba, Japan. asuka@aori.u-tokyo.ac.jp.

Takahiro YAMAMOTO
Geological Survey of Japan, National Institute of Advanced Industrial Science and Technology (AIST), Tsukuba, Japan. t-yamamoto@aist.go.jp.

Hidekazu YOSHIDA
Nagoya University Museum, Nagoya, Japan. dora@num.nagoya-u.ac.jp.

Dapeng ZHAO
Research Center for Prediction of Earthquakes and Volcanic Eruptions, Tohoku University, Sendai, Japan. zhao@aob.gp.tohoku.ac.jp.

Foreword

The publication of *The Geology of Japan* by the Geological Society of London in collaboration with the Geological Society of Japan is a milestone in the history of our Society. As the present and previous presidents of the Geological Society of Japan, we would like to express our heartfelt congratulations to the four editors and all the authors on the publication of this important book. This will surely become a valuable reference work for geologists working in education, scientific research and industry in Japan, and we sincerely hope the book will also be widely read by researchers, students and engineers from all over the globe.

The Japanese islands are located in one of the most tectonically active regions of the world where four plates converge. The geology of this region stretches back into the Precambrian and reflects the long interaction of the area with numerous different plates, many of which have ceased to exist. This complex geological history is reflected in the great variety of geology seen in Japan and the complex geological structure. This variety can be encapsulated by a sentence commonly found in the Earth Science section of elementary school science textbooks: 'the Japanese islands are a mineral display cabinet'. Over the generations, Japanese people have made use of different mineral resources taken from this mineral cabinet. One of the most notable is silver; approximately one-third of the silver circulated worldwide in the sixteenth century is thought to have been produced in Japan, most of which came from Iwami Ginzan Silver Mine.

Japan's geological setting is also reflected in the constant threat of large-scale geohazards such as earthquakes, tsunamis, volcanic eruptions and landslides. In 1792, one of several lava domes of Mt Unzen in Kyushu collapsed causing a mega-tsunami that killed about 15 000 people; this was Japan's worst-ever volcanic-related disaster. It is also still fresh in our minds that nearly 20 000 people were killed by the 2011 Tohoku Earthquake and its associated tsunami off the Pacific coast. This disaster was a stark reminder to Japanese people of the importance of understanding the Earth with a view both to the ancient record and ongoing geological processes. There is renewed concern in the importance of disaster prevention and mitigation from the central government down to the grassroots levels. This publication reviews a wide range of different aspects of the geology of Japan, and we hope this will help readers to appreciate the close relationships between the Japanese people and the geology of this region, including natural resources and geohazards. These relationships have helped form both the physical and cultural landscapes of Japan and underpin its present economic prosperity.

The Geological Societies of London and Japan signed a Memorandum of Understanding to promote exchange and collaboration between our societies in 2013. This year also marked the 400th anniversary of the first official relations between the UK and Japan. We are delighted that publication of *The Geology of Japan* will be complete in time for another anniversary: the 125th year since the founding of our Society in 1893. This publication fills a 20 year gap since the last English-language book with a similar title. During this time, our knowledge about the development and evolution of the Japanese islands has greatly expanded. In particular, advances in technology have made it possible to survey the ocean floor in unprecedented detail, and there has been a leap in our understanding of the marine geology around Japan. The contents of this new book summarize insights into the geology of Japan gained over the last quarter-century and have been expertly crafted to present the latest understanding of the geological development and evolution of the Japanese islands and their environs.

We gratefully acknowledge the support of the present and former Presidents of the Geological Society of London, Mr David Shilston and Professor David Manning, and the Executive Secretary, Mr Edmund Nickless, for their support in promoting the publication of this book and collaboration between our societies. We also express our deep appreciation to Professor Simon Wallis for his devoted management and liaising between the Geological Societies of London and Japan. Finally, we thank the chief editor Dr Teresa Moreno who, along with Dr Tomoko Kojima, Professor Simon Wallis and Dr Wes Gibbons, worked tirelessly to bring this project to a successful conclusion.

Yasaufumi Iryu
President of the Geological Society of Japan
(June 2014–present)

Akira Ishiwatari
President of the Geological Society of Japan
(June 2012–May 2014)

1 Geological evolution of Japan: an overview

A. TAIRA, Y. OHARA, S. R. WALLIS, A. ISHIWATARI & Y. IRYU

In this chapter we present an overview of the geology of the Japanese archipelago and the surrounding ocean basins. To set the scene we begin by examining the main geological features of the present-day Japanese arc system. We then move back in time to the oldest rocks known in Japan and show how these origins have developed into the modern island arc. The early geological history of Japan is represented by Palaeozoic rocks, which were derived from both continental and oceanic domains. It is unclear if this early continental margin was passive or the site of plate convergence. After the Jurassic the geological record preserves abundant evidence for oceanic plate subduction beneath Japan. This convergence has continued until the present with numerous plates interacting with and leaving their mark on the geology of Japan. At *c.* 25 Ma the Japan Sea began to open causing the separation of Japan from the East Asia margin to form the present island arc.

The present contribution builds on previously published overviews of the geology of Japan (e.g. Taira 2001). This new review incorporates the results of ongoing geological mapping that has contributed to a better understanding of the distribution of different units. The last decade has also seen a major increase in the amount of data available from seismic, geodetic and age dating and other studies. One of the biggest advances has been an improved documentation and understanding of the submarine geology around Japan. The main impetus for this advance has been the scientific results from Japan's legal continental shelf survey project in the context of the United Nations Convention on the Law of the Sea (UNCLOS), and in this chapter we incorporate many of the new results from this work.

Overview of the present Japanese arc system

The Japanese island arc system can be divided into four distinct domains: (1) the western Kuril–eastern Hokkaido Arc; (2) the western Hokkaido–Honshu Arc; (3) the Ryukyu Arc; and (4) the Izu–Ogasawara (Bonin) Arc. These arcs were formed as a result of subduction of the Pacific and Philippine Sea plates (Taira 2001; Figs 1.1 & 1.2). Important information about the arcs can be gleaned from on-land exposures but the arcs are broad domains stretching from the deformation front at the trenches to the back-arc basins, most of which lie beneath the sea (Fig. 1.3). In order to develop a good overall understanding of the geotectonic and palaeogeographic evolution of the Japanese arc system, it is therefore important to incorporate information both from land- and marine-based studies into a single scheme. The present-day configuration of the Japanese arc system was mainly established during the Neogene (Fig. 1.4). The development of each of these arcs is summarized below.

The Kuril Arc stretches *c.* 1000 km between eastern Hokkaido and the Kamchatka Peninsula. The volcanic front produced by subduction of the Pacific Plate extends from the Kuril Arc to the central part of Hokkaido (Figs 1.1–1.3). The Kuril back-arc basin formed during the Neogene with kinematics analogous to the opening of a fan (Kimura 1986). Arc volcanism in Hokkaido initiated at *c.* 20 Ma with a major change in geochemistry at *c.* 14 Ma: the older volcanic rocks show relatively high alkalinity and Ti–Nb contents probably related to plume activity during the opening of the Japan Sea, whereas the younger volcanic rocks display a more typical arc-like geochemistry (Hirose & Nakagawa 1999). Oblique subduction of the Pacific Plate beneath the Kuril Arc and the associated strike-parallel shear stresses have caused westwards migration of the Kuril forearc, or Kuril forearc sliver (Kimura 1986), resulting in the formation of a late Palaeogene–early Neogene collision zone in central Hokkaido known as the Hidaka Collision Zone (Komatsu *et al.* 1989). The Hidaka mountains now expose geological units dipping to the NE and expose a remarkably complete cross-section through arc crust including a high-T/P metamorphic domain locally with granulite domains that represent the lower crust and a kilometre-scale body derived from the underlying mantle. The crustal thickness beneath the Hidaka Mountains is 30–50 km (Miyauchi & Moriya 1984; Ogawa *et al.* 1994), and is probably the thickest in Hokkaido. Arita *et al.* (1998) proposed that collision tectonics resulted in crustal delamination in the southern Hidaka Mountains. A more complex interleaving of the mantle and lower crust is suggested by the seismic studies of Kita *et al.* (2012).

To the south, the Honshu Arc can be divided into the northeast (NE) Honshu and southwest (SW) Honshu arcs (Figs 1.1–1.3), which are separated by the Izu Collision Zone (ICZ). The ICZ was formed by the NW-directed motion of the Izu–Ogasawara Arc and convergence with the Honshu Arc that ultimately caused collision-related tectonics. We summarize the evolution of these two arc domains separately.

The NE Honshu Arc extends over a distance of *c.* 700 km, and displays two prominent parallel arrays of Quaternary volcanoes (Tatsumi *et al.* 1983) (Fig. 1.2). Overall, both western Hokkaido and NE Honshu show the development of north–south-trending reverse fault systems, indicating that these areas are under an east–west-aligned compressional stress regime. The forearc of the NE Honshu Arc has a relatively smooth topography with a sharp trench-slope break at a water depth of *c.* 2000 m (Figs 1.3 & 1.5a). In the Japan Trench, the relatively rapid subduction of the Pacific Plate (*c.* 8–9 cm a^{-1}; DeMets *et al.* 1994) is associated with very active seismicity. This is the same region where the 11 March 2011 M_w 9.0 mega-earthquake and mega-tsunami occurred (Fujiwara *et al.* 2011; Sato *et al.* 2011). At the seaward end of the continental plate, seismic studies have revealed the presence of a distinct wedge-shaped sedimentary unit with relatively low P-wave velocities (2–3 km s^{-1}) that underthrusts the overriding plate. The low seismic velocities and wide extent of this unit (Fig. 1.5a) suggest it is an important fluid reservoir in the region (Tsuru *et al.* 2002).

From: Moreno, T., Wallis, S., Kojima, T. & Gibbons, W. (eds) 2016. *The Geology of Japan*. Geological Society, London, 1–24.

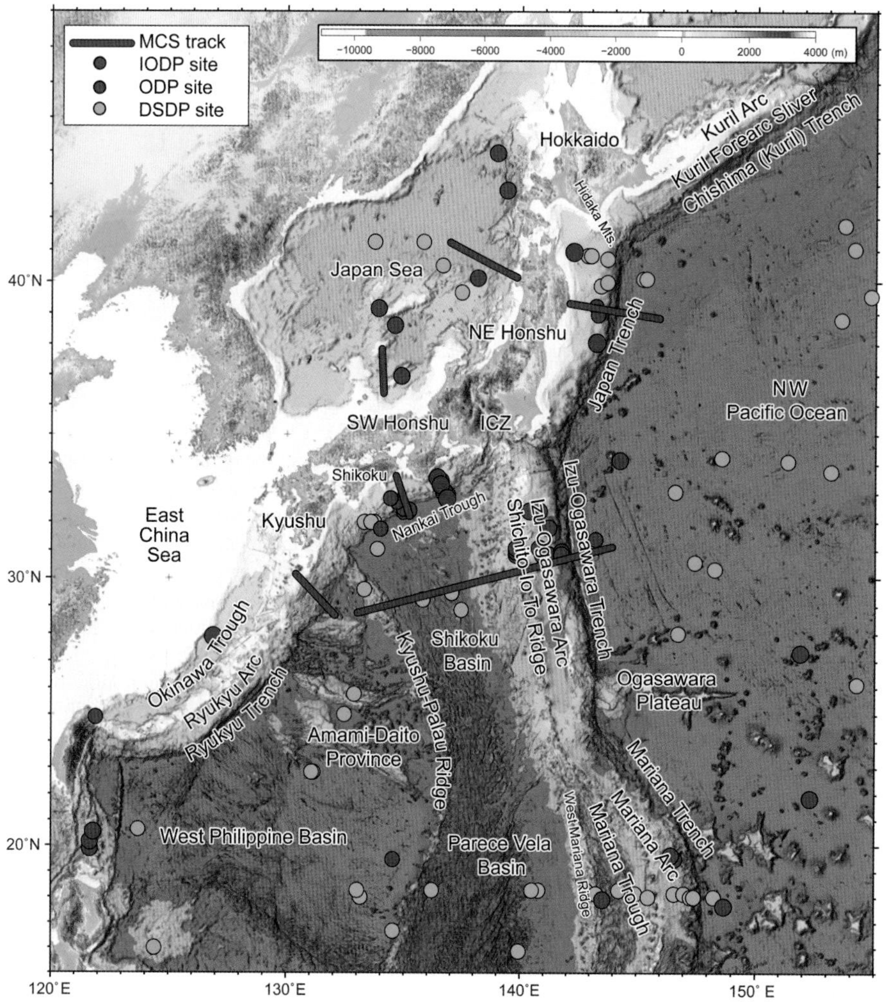

Fig. 1.1. Index map for the Japanese island arc system. Blue lines indicate the location of multi-channel seismic profiles (Fig. 1.5). Past ocean drilling sites (those from DSDP, ODP and IODP) are also shown.

The section of the NE Honshu Arc that lies beneath the Japan Sea exhibits a distinct 'basin and range' morphology (Fig. 1.5a; Okamura *et al.* 1995). This formed as the result of neotectonic east–west-directed compression that caused inversion of a sequence of normal faults and thrusting of associated sedimentary sequences (Okamura *et al.* 1995) (Fig. 1.4). The normal faulting formed as a result of rifting of Japan away from the east Asian margin associated with the opening of the Japan Sea. The most recent phase of compressional tectonics may represent incipient subduction of the oceanic lithosphere of the Japan Sea (Nakamura 1983). The present compressional tectonic phase is also shown in the pattern of intra-arc seismicity, which helps define the east boundary of the eastwards-moving Amur Plate (Taira 2001; Figs 1.2 & 1.6). The basement geology of the NE Honshu Arc is composed of a complex arrangement of Mesozoic–Palaeozoic subduction-related accretionary prisms and tectonic slices of a variety of rocks, including ophiolites and continental margin sedimentary sections (e.g. Ozawa 1988 and later in this chapter in the section on 'Precambrian–Palaeozoic geology of Japan').

The SW Honshu Arc extends *c.* 600 km and includes the large island of Shikoku, which is interpreted as a forearc non-volcanic ridge (Fig. 1.1). Central Honshu is characterized by the presence of high mountain ranges. This topography developed as a result of the collision with the Izu peninsula – the Izu Collisional Zone – and the convergence between the SW Honshu and NE Honshu Arcs. The boundary between these two arc domains is commonly considered to be the boundary between the Amur and Okhotsk Plates (Figs 1.2, 1.4 & 1.6).

The basement geology of the SW Honshu Arc shows a similar rock assemblage to that seen in NE Honshu but exhibits a more linear arrangement of accretionary complexes evolving southeastwards from Palaeozoic to Cenozoic time, including the presently active Nankai accretionary prism (Taira *et al.* 1982, 1997). The Nankai Trough is the surface manifestation of the subduction of the Shikoku Basin that is a part of the Philippine Sea Plate. The forearc side of Shikoku is characterized by a fold–thrust structure on the landwards slope of the Nankai Trough and by a series of forearc basins (Fig. 1.3). A buried circular seamount subducting beneath the prism is observed in a reflection seismic profile, causing deformation of the accretionary wedge (Park *et al.* 1999; Fig. 1.5b). The topography within the Nankai forearc sliver is the result of the active accretion of a trench wedge and the internal deformation of the accretionary prism as a result of the oblique subduction of the Philippine Sea Plate. The Quaternary volcanoes in SW Honshu are located along the Japan Sea coast and mainly show alkaline geochemistry, in contrast to the calc-alkaline suite of the NE Japan Arc.

Palaeomagnetic data (Otofuji & Matsuda 1983) demonstrate that the Japan Sea back-arc basin formed by a combined anticlockwise rotation of NE Honshu and clockwise rotation of SW Honshu, analogous to the opening of a double door (Tamaki *et al.* 1992). Seismic surveys suggest that the northernmost part of the Japan Sea is

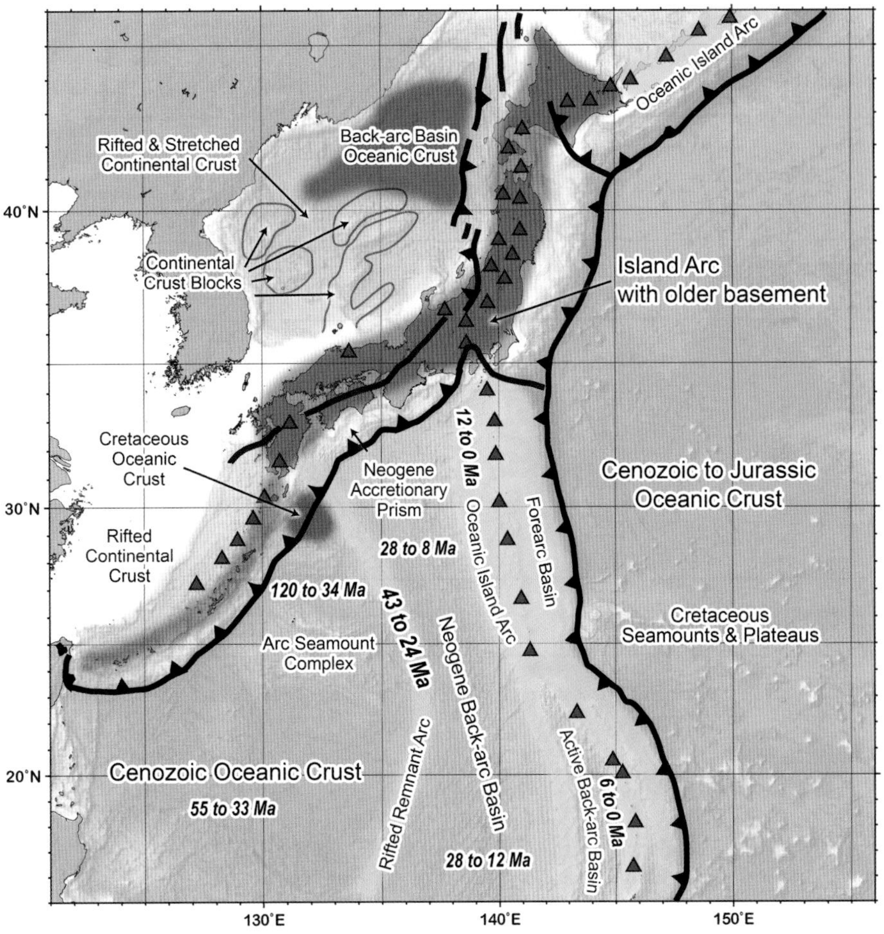

Fig. 1.2. Tectonic summary map for the Japanese island arc system. Small red triangles indicate the locations of active volcanoes. Representative ages for each of the main tectonic elements of the Philippine Sea Plate are shown.

underlain by oceanic crust and rifted and fragmented continental crust elsewhere (Figs 1.2 & 1.5b). The middle Miocene opening of the Japan Sea (Tamaki *et al.* 1992) was accompanied by substantial submarine magmatism, resulting in accumulation of voluminous 'Green Tuff' deposits associated with the formation of massive sulphide ore deposits (Kuroko).

To the south of the Honshu Arc the Ryukyu Arc extends from Kyushu to the islands of Okinawa, forming a connection with the northern tip of Taiwan (Fig. 1.1). The arc basement in the NE part of this Arc is an extension of the SW Honshu Arc geology, including Mesozoic and Cenozoic accretionary prisms. However, the SW section shows similarities with Taiwan, being mainly composed of continental margin sedimentary rocks with slices of metamorphic rocks (Park *et al.* 1998). The volcanic chains trending from southern Kyushu connect to submerged volcanic activity in the Okinawa Trough back-arc rift (Fig. 1.2; Shinjo *et al.* 2000). The Okinawa Trough

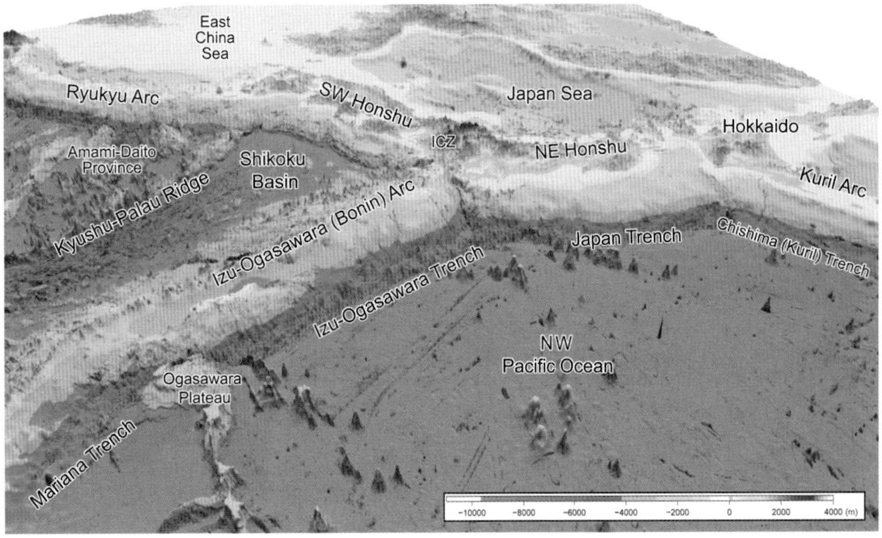

Fig. 1.3. Overview of the 3D morphology of the Japanese island arc system, looking from the NE.

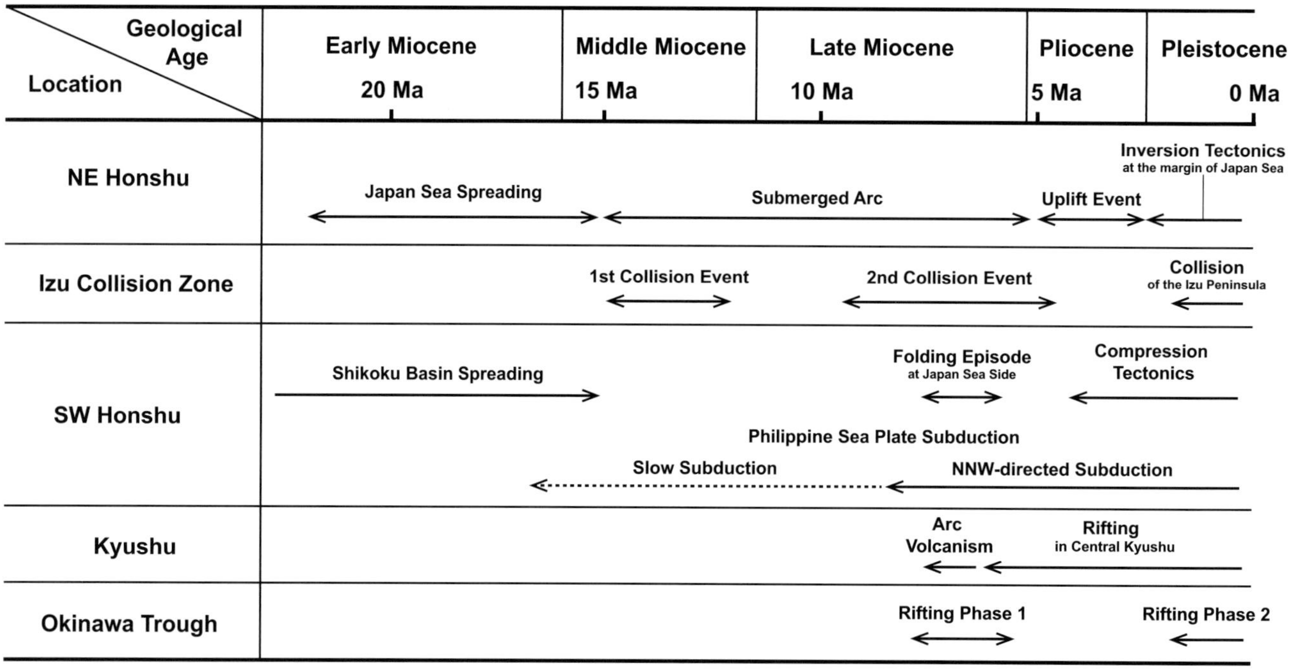

Fig. 1.4. Summary of tectonic events during the Cenozoic evolution of East Asia (modified after Taira 2001).

formed in two distinct stages with a period of quiescence in between; the first began in the Miocene and ended in the Pliocene, and the second phase began at *c.* 2 Ma in the Quaternary (Fig. 1.4). The Ryukyu Trench is the surface manifestation of the subduction of the West Philippine Basin and Amami-Daito Province that is a part of the Philippine Sea Plate. The Amami-Daito Province is composed of a rather heterogeneous collage of oceanic plateaus, remnant arcs and intervening basins (Figs 1.1–1.3 & 1.5c).

The final active arc of Japan is the Izu–Ogasawara Arc, which is a largely submerged oceanic island arc extending from the ICZ with the Honshu Arc to the northern end of the Mariana Trough back-arc basin (Figs 1.1 & 1.2). Other prominent morphological features of the Philippine Sea Plate include the Kyushu–Palau Ridge, which was once part of the proto Izu–Ogasawara Arc that became separated by intra-arc spreading. The oceanic area that formed as a result of this spreading is represented by the Shikoku and Parece Vela basins (Fig. 1.1; Karig 1971, 1975; Hall 2002; Ishizuka *et al.* 2011*b*). The Shikoku and Parece Vela basins display well-defined linear magnetic anomalies, suggesting three distinct stages of spreading (Okino *et al.* 1994, 1999). The Izu–Ogasawara Arc grew during the Neogene after the formation of the Shikoku and Parece Vela basins into a *c.* 20 km thick crustal structure, including a 6 km mid-crustal layer with a P-wave velocity of 6 km s^{-1} that has continental affinities (Fig. 1.5c; Suyehiro *et al.* 1996; Kodaira *et al.* 2007). The Izu–Ogasawara Trench is underlain by some of the oldest parts of the Pacific Plate, dating back to the early Cretaceous, and is characterized by numerous seamounts and plateaus (Fig. 1.2). One of the larger of these bathymetric highs, the Ogasawara Plateau, is in the process of colliding with the forearc high domain of the Izu–Ogasawara Arc, producing a wedge of off-scraped oceanic rocks. Minor seamount chains are also recognized in the rear-arc domain, and show *en echelon* alignments with respect to the main arc (Ishizuka *et al.* 1998; Hochstaedter *et al.* 2000, 2001). The Izu–Ogasawara Arc has been interpreted as an archetypal young island arc domain built on ocean crust, but recent findings of various ancient rocks show this domain has a longer and more complex evolution than previously recognized (Tani *et al.* 2012).

Geological evolution of Japan from the Precambrian to Cenozoic

Having set the scene by reviewing our knowledge of the present geology of the Japanese island arcs and surrounding oceanic basins, we now examine the older record in more detail to provide our understanding of how the geology of the Japanese island arcs developed. A summary of the main geological units and terranes discussed in the following sections is given in Figure 1.6.

Precambrian–Palaeozoic geology of Japan

The oldest known rocks of Japan are granite and gneiss boulders included in the conglomerate beds at Kamiaso, Hichiso Town, Gifu Prefecture, southwest Japan (central Honshu) (Shibata & Adachi 1974; Suzuki *et al.* 1991; Shimizu *et al.* 1996). An example of these boulders is displayed in the Hichiso Precambrian Museum (or 'The-Oldest-Rocks-in-Japan Museum') that was built near the conglomerate outcrop. The area belongs to the Mino Belt, and the conglomerate occurs as a part of the Jurassic accretionary complex that formed along the ancient subduction trench. Whole-rock Rb–Sr, K–Ar and zircon and monazite chemical Th–U–Total Pb ages mostly fall within the range 2000–1500 Ma. The oldest zircon grain yields an age of 3040 ± 180 Ma. Precambrian zircon grains are widespread in sandstone of the Japanese Jurassic accretionary complexes and shelf sediments. Excluding those younger than 200 Ma, the age distribution shows three major peaks at 300–200, 500–400 and 2500–1000 Ma with a minor peak at 900–700 Ma (Isozaki *et al.* 2010). These zircon U–Pb ages (and bulk-rock Nd Depleted MORB-type Mantle (DMM) model ages) suggest that their main provenance is located in the North China craton (Sino-Korean continental block) continental block where 1700 Ma and older rocks are widely exposed, although the presence of 900–700 Ma zircons suggests additional inputs from the South China craton (Yangtze continental block) where late Proterozoic basement is pervasive (Kagami *et al.* 2006; Nutman *et al.* 2006; Isozaki *et al.* 2010). Precambrian ages are also reported from the granites and gneisses of the Hida

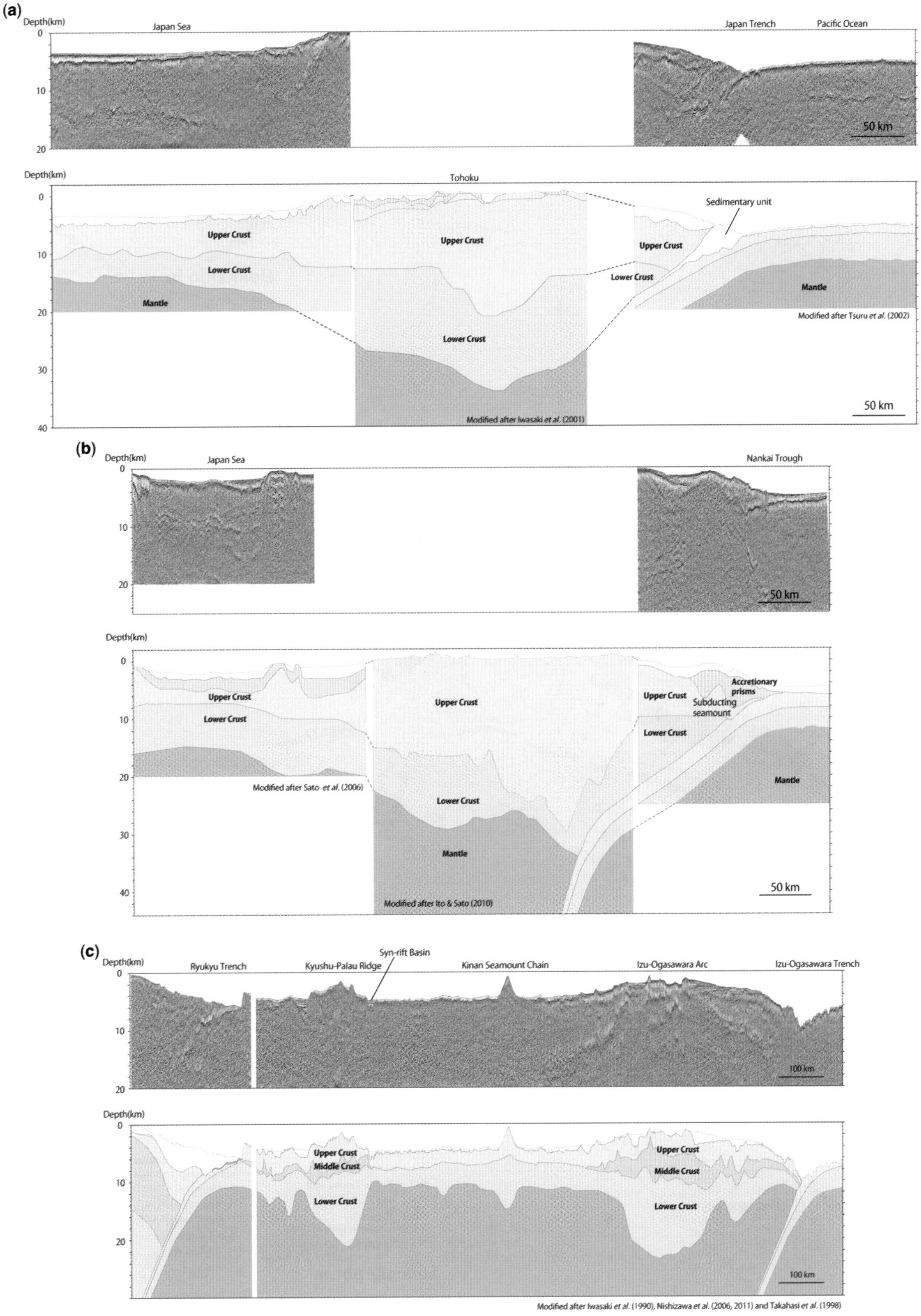

Fig. 1.5. (**a–c**) Reflection seismic profiles and interpreted crustal structure of the Japanese island arc system (after Iwasaki *et al.* 1990, 2001; Takahashi *et al.* 1998; Tsuru *et al.* 2002; Nishizawa *et al.* 2006, 2011; Sato *et al.* 2006; Ito & Sato 2010). The locations of the profiles are shown in Figure 1.1.

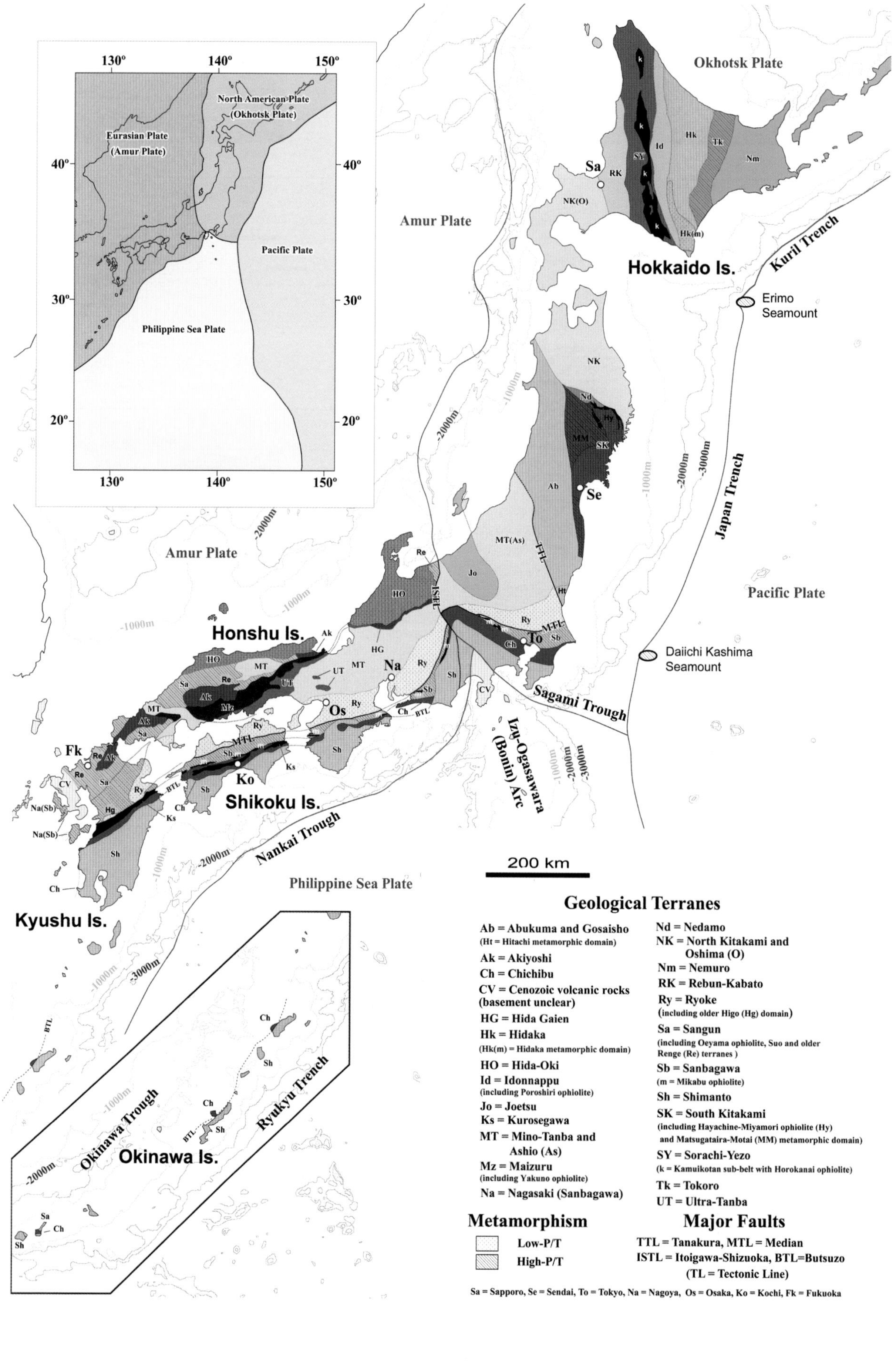

130° 140° 150°
40° 30° 20°
North American Plate (Okhotsk Plate)
Eurasian Plate (Amur Plate)
Pacific Plate
Philippine Sea Plate
Okhotsk Plate
Amur Plate
Hokkaido Is.
Kuril Trench
Erimo Seamount
Japan Trench
Pacific Plate
Daiichi Kashima Seamount
Sagami Trough
Izu-Ogasawara (Bonin) Arc
Honshu Is.
Shikoku Is.
Kyushu Is.
Nankai Trough
Philippine Sea Plate
Okinawa Trough
Ryukyu Trench
Okinawa Is.
200 km
Geological Terranes
Ab = Abukuma and Gosaisho
(Ht = Hitachi metamorphic domain)
Ak = Akiyoshi
Ch = Chichibu
CV = Cenozoic volcanic rocks (basement unclear)
HG = Hida Gaien
Hk = Hidaka
(Hk(m) = Hidaka metamorphic domain)
HO = Hida-Oki
Id = Idonnappu
(including Poroshiri ophiolite)
Jo = Joetsu
Ks = Kurosegawa
MT = Mino-Tanba and Ashio (As)
Mz = Maizuru
(including Yakuno ophiolite)
Na = Nagasaki (Sanbagawa)
Nd = Nedamo
NK = North Kitakami and Oshima (O)
Nm = Nemuro
RK = Rebun-Kabato
Ry = Ryoke
(including older Higo (Hg) domain)
Sa = Sangun
(including Oeyama ophiolite, Suo and older Renge (Re) terranes)
Sb = Sanbagawa
(m = Mikabu ophiolite)
Sh = Shimanto
SK = South Kitakami
(including Hayachine-Miyamori ophiolite (Hy) and Matsugataira-Motai (MM) metamorphic domain)
SY = Sorachi-Yezo
(k = Kamuikotan sub-belt with Horokanai ophiolite)
Tk = Tokoro
UT = Ultra-Tanba
Metamorphism
Low-P/T
High-P/T
Major Faults
TTL = Tanakura, MTL = Median
ISTL = Itoigawa-Shizuoka, BTL=Butsuzo
(TL = Tectonic Line)
Sa = Sapporo, Se = Sendai, To = Tokyo, Na = Nagoya, Os = Osaka, Ko = Kochi, Fk = Fukuoka

Mountains and Oki Island, including the oldest ion microprobe U–Pb zircon age in Japan (3400 Ma from the Hida gneiss; Sano *et al.* 2000). However, these basement rocks are thought to be derived mostly from Palaeozoic protoliths (a Carboniferous age has been locally confirmed by relict fossils; Hiroi *et al.* 1979) and metamorphosed at *c.* 250 Ma (Sano *et al.* 2000).

Palaeozoic sedimentary rocks in Japan, together with the associated igneous and metamorphic rocks, are sporadically distributed in various parts of all four main islands of Japan (Fig. 1.7), and are divided into two main groups based on their characteristic occurrences and lithologies: so-called 'shelf facies' and 'oceanic facies' (Fig. 1.7). This grouping is frequently used in Japanese texts such as the *Regional Geology Series* (edited by the Geological Society of Japan and published by Asakura Shoten in 2006–2016) to describe the Palaeozoic and Mesozoic rocks in Japan.

The shelf facies sediments are mainly composed of a stratified succession of limestone and clastic rocks with some tuff beds but are almost entirely lacking chert, suggesting their formation in shallow water off an active continental margin or an island arc. In Japan, they occur in three distinct areas: the South Kitakami Belt in NE Honshu, the Hida Gaien Belt (also referred to as the Hida Marginal Belt) in the Inner Zone of SW Japan (central Honshu) and the Kurosegawa Belt in the Outer Zone of SW Japan (mainly in Shikoku and Kyushu). The shelf facies beds constitute a relatively continuous stratigraphic succession (but of course with several unconformities and faults) ranging from Ordovician (or Silurian) to Permian, and are associated with analogous shelf facies beds of Mesozoic age. It should be noted that these Palaeozoic sedimentary rocks are closely associated with the early Palaeozoic supra-subduction zone (SSZ) -type ophiolites (Oeyama ophiolite in SW Japan and Miyamori–Hayachine ophiolite in NE Japan; see Chapter 3). The oldest fossils such as late Ordovician ostracodes (Adachi & Igo 1980; Igo *et al.* 1980) and conodonts (Tsukada 1997; Tsukada & Koike 1997) have only been reported from the Hida Gaien Belt, but Silurian and younger fossils are present in all three areas. The metagabbro and jadeitite blocks in the Oeyama and Miyamori–Hayachine ophiolites show Cambrian and Ordovician whole-rock K–Ar and zircon U–Pb ages. Metamorphic rocks (basic and pelitic rocks of the greenschist and amphibolite facies with some blueschist) and granitic rocks (Hikami Granite) constitute the pre-Silurian basement in the South Kitakami Belt. The close association with ophiolitic rocks and abundance of acidic tuff beds indicate a circum-Pacific type (or Andean type) continental margin environment during early Palaeozoic time. However, some workers prefer a passive continental margin setting for this period.

The shelf facies Palaeozoic stratigraphy in Japan is best developed in the South Kitakami Belt (Minato *et al.* 1965; Mori *et al.* 1992), and its main features are briefly introduced here. The Silurian Kawauchi Formation is distributed in the Hikoroichi area near coastal cities of Rikuzen-Takata and Ofunato (Iwate Prefecture) and coeval formations are more widely developed along the southern foot of Mt Hayachine in the northern inland area (Iwate Prefecture). The Kawauchi Formation mainly consists of limestone with *Favosites*, and unconformably covers the Hikami Granite in the Hikoroichi area. The Okuhinotsuchi Formation in the Setamai area yields *Halysites*. These macrofossils and conodonts suggest a Silurian age. The Ono Formation consists of basal brecciated limestone with latest Silurian fossils and overlying Devonian clastic and pyroclastic rocks. The Devonian Nakazato Formation conformably overlies the Ono Formation, and consists of basaltic or andesitic tuff, black shale and sandstone and some limestone lenses. Its upper part is mostly composed of shale and sandstone. Abundant brachiopod, bryozoan, pelecypod and crinoid fossils are found from black shale. The Carboniferous series is divided into several formations such as the Hikoroichi, Onimaru and Nagaiwa formations in ascending order. The lower part (Hikoroichi) is rich in basaltic tuff, acidic tuff and tuffaceous sandstone, but the upper part (Onimaru and Nagaiwa) is mostly composed of limestone. Abundant coral, brachiopod, trilobite and foraminifera fossils occur. The Permian series is divided into three stages represented by the Sakamotozawa, Kanokura and Toyoma formations in ascending order. The Sakamotozawa Formation consists of conglomerate, sandstone and shale with some bedded limestone in its lower part, and is dominated by limestone with abundant fusulinid fossils in its upper part. The Kanokura Formation consists of green sandstone in its lower part and limestone in its upper part. The Toyoma Formation is almost wholly composed of black shale with some limestone lenses, which contain some Fusulinid, Ammonoid, Pelecypod and Gastropod fossils. 'Toyoma' is the old village name for part of the present Tome City, Miyagi Prefecture. The absence of chert, scarcity of flysch-type turbidites and abundance of fossiliferous bedded limestone and tuff beds are prominent feature of the Palaeozoic formations in the South Kitakami Belt, and these features are mostly shared with those in the Hida Gaien and Kurosegawa belts. Geochronological population analysis of detrital zircon grains in the shelf facies sediments in the South Kitakami Belt suggests that the belt evolved as a Gondwana continental margin until Silurian–Devonian time, then was rifted and drifted in the ocean as an island arc through the Permian–Early Jurassic, and finally amalgamated with the North China continental margin in the Middle Jurassic (Okawa *et al.* 2013).

Metamorphosed Cambrian rocks have recently been reported from the Hitachi area, southern Abukuma Belt near the Pacific coast of NE Japan (Fig. 1.7). Rhyolitic lava, tuff and clastic sedimentary rocks are the main constituents. This Cambrian formation is covered unconformably by a Carboniferous formation, and Tagiri *et al.* (2011) suggest a North China (Sino-Korean) affinity for this sequence: a Silurian–Devonian 'great hiatus' is a widespread feature of the Sino-Korean block (e.g. Ishiwatari & Tsujimori 2003). As described above, the Palaeozoic stratigraphy in the other areas of Japan shows closer affinities with the Yangtze type of South China that is characterized by continuous sedimentation through Silurian–Devonian time, and the Palaeozoic sequence in the Hitachi area is in strong contrast to them. It should be noted, however, that the Cambrian and Ordovician series in the Sino-Korean block is dominated by limestone, but no limestone of this age is present in the Hitachi area. The Palaeozoic palaeogeography of the region is disputed, but one proposal is summarized in Figure 1.8.

The late Palaeozoic, SSZ-type Yakuno ophiolite and the associated Permian sedimentary rocks in the Maizuru Belt of the Inner Zone of SW Japan are another important constituent of the shelf

Fig. 1.6. Modern plate boundaries and main geological divisions of the Japanese island arc basement. The geological basement units are commonly referred to as belts. However, this term does not accurately describe the distributions of many of these units and the term 'terrane' is preferred by some workers. This figure does not include most of the superficial basin deposits younger than Cretaceous or the younger volcanic deposits. However, we have distinguished the forearc basin deposits that make up the Yezo group of Hokkaido because of its widespread development and the importance in the Yezo–Sorachi Terrane. The outline of the land territories was plotted using GMT software and the depth contours for the ocean were drawn using the NOAA dataset ETOPO1. The distribution of the geological basement units is adapted from Isozaki (1996) with modifications incorporating the results of more recent geological surveys, including the observations of the co-authors of this book. In particular, information provided by K. Miyazaki and H. Ueda was used in preparing the Kyushu and Hokkaido areas, respectively. Two seamounts that have influenced the geometry of the plate boundary along the Japan Trench are also indicated.

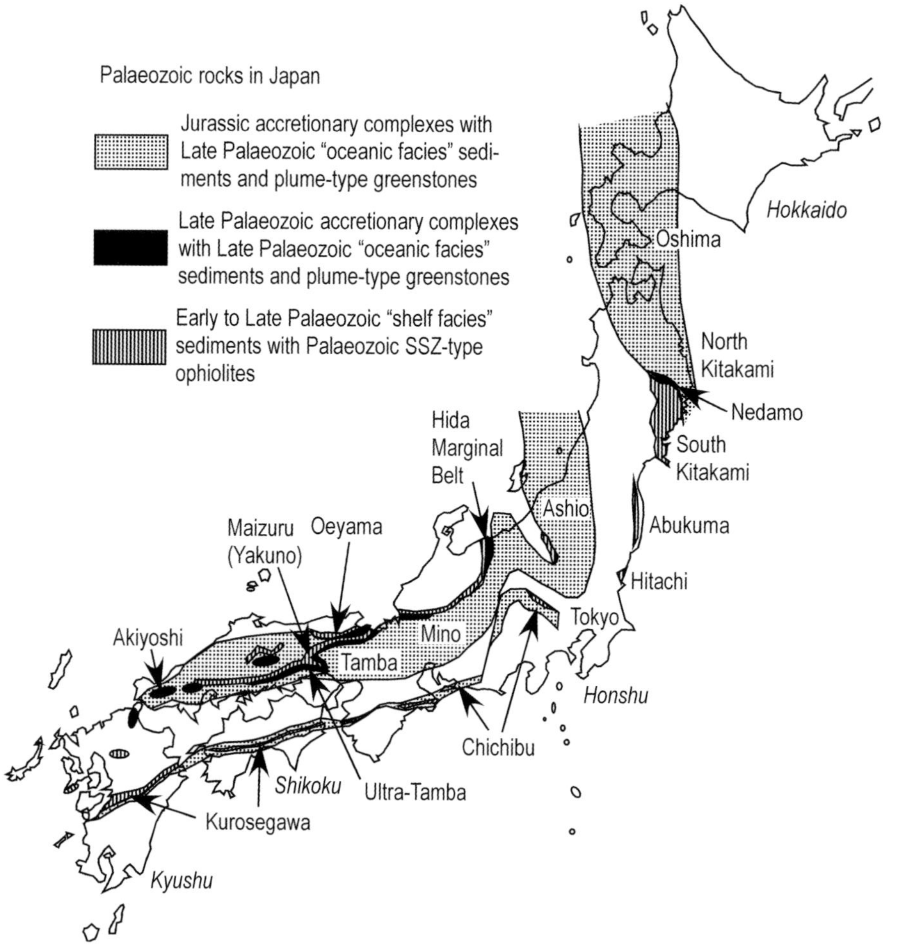

Fig. 1.7. Distribution of Palaeozoic rocks in Japan. Shelf facies rocks occur in the South Kitakami, Hida Gaien and Kurosegawa belts. The other indicated areas are accretionary complexes that contain Palaeozoic oceanic facies rocks.

facies Palaeozoic rocks (see Chapter 3). The black shale associated with the basaltic rocks of the Yakuno ophiolite is of Permian age, and thick late Permian and Triassic clastic rocks cover the ophiolite. The clastic rocks are almost free of chert, associated with acidic tuff and bear abundant brachiopod and shell fossils. Analogous Permo-Triassic formations can be found in the Vladivostok–Nakhodka area in Primorye, Russia and the Yanji Belt in North Korea and Jilin Province, China (Ishiwatari & Tsujimori 2012).

The oceanic facies consists of the late Palaeozoic chert, reef limestone and greenstones (altered basaltic rocks) that originated in seamounts, oceanic plateaus and mid-ocean ridges (i.e. normal ocean floor) as well as trench-fill sediments (shale and sandstone), constituting the 'ocean plate stratigraphy' (Isozaki *et al.* 1990; Isozaki 1997) (Fig. 1.8). They comprise the late Palaeozoic Akiyoshi, Ultra-Tamba (Tanba) and Nedamo accretionary complexes, but late Palaeozoic chert, limestone and greenstone also occur in the Jurassic Tamba, Mino, Ashio, Chichibu, North Kitakami and Oshima (South Hokkaido) accretionary complexes as thick slab-like greenstone-limestone (and/or chert) body or as blocks in the mudstone mélanges. The limestone plateau with greenstone basement is especially well developed in western Honshu such as Akiyoshi, Oga and Taishaku and in northern Kyushu such as Zomeki and Hirao. Clastic sedimentary rocks forming the matrix of the late Palaeozoic accretionary complexes such as the Nedamo Belt in the Kitakami Mountains and the Akiyoshi and Ultra-Tamba Belts in the Inner Zone of SW Japan are volumetrically the most important Palaeozoic rocks in Japan. More fragmental greenstone-limestone-chert bodies occur in the Jurassic accretionary complexes such as the Tamba–Mino–Ashio Belt, Chichibu Belt and North Kitakami–Oshima Belt as tectonic blocks of a few metres to several kilometres in size. Examples of prominent limestone bodies of this type are the late Palaeozoic Ibuki-yama, Funafuse-yama (SW Japan) and Kuzuu body (to the north of Tokyo) and the Triassic Akka body (NE Japan).

The 'oceanic facies' rocks in the Palaeozoic are best developed and studied in the Akiyoshi area in the westernmost Honshu (Kanmera *et al.* 1990; Sano & Kanmera 1991). The Permian Akiyoshi accretionary complex is divided into two units, which show similar ages but contrasting dominant lithologies: a limestone-rich unit consisting of several large limestone plateaus; and a chert-rich unit consisting of limestone-poor oceanic sediments. Both units formed by continuous sedimentation through the same period, but the chert-dominated unit has a thickness of only 150 m compared to the limestone-dominated unit with a thickness of 800–1600 m. The chert and limestone are both associated with greenstone in their lower parts and are both covered by thick shale and sandstone sequences of middle–late Permian age. The rock sequence therefore represents a typical ocean plate stratigraphy, and can be interpreted in terms of the movement of an oceanic plate from its generation in the mid-ocean ridge to its destruction in the subduction zone (Isozaki *et al.* 1990; Isozaki 1997). The Akiyoshi Limestone plateau comprises a continuous stratigraphic sequence ranging from early Carboniferous to middle Permian age. This sequence may form a recumbent anticline with a widespread distribution of the reversed stratigraphy in its lower flank (Ozawa 1925; Morinaga & Ota 1971), or an assemblage of numerous 100 m sized limestone blocks

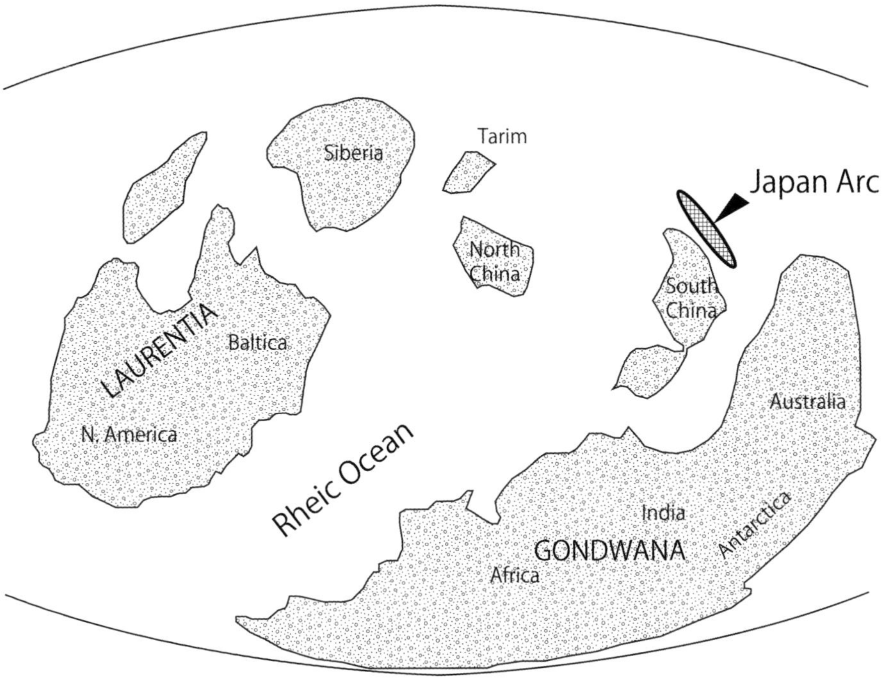

Fig. 1.8. Proposed palaeogeographic setting for the Palaeozoic of Japan (after Williams *et al.* 2014), based on fossil assemblages that show clear links to the Yangtze (South China) Craton and less well-developed associations with Australia and the Sino-Korean (North China) Craton.

of various ages that are set in a matrix of limestone breccia, possibly fortuitously showing a consistent reversed 'stratigraphy' when viewed on the large scale (Sano & Kanmera 1991) (Fig. 1.10). The latter concept is based on the subduction–destruction processes of a large seamount that are currently taking place at the Japan Trench off eastern Honshu, namely the Daiichi–Kashima Seamount. In Japan, coal (carbon) of the Carboniferous age is essentially absent, although a small amount of coal of Permian age is reported from the Funafuse-yama limestone body of the Mino Belt. The coal may have formed upon a seamount in a lagoon environment surrounded by coral reef (Suzuki 1997).

The Nedamo accretionary complex on the north of the South Kitakami Belt (Fig. 1.7) was recently recognized as the oldest accretionary complex (Carboniferous) in Japan (Uchino & Kawamura 2010). The trench-fill sediments of this complex include clastic particles of ophiolite and blueschist that were exposed in the forearc area at that time.

The geochemistry and mineralogy of oceanic greenstones (metabasalts) suggest their origin in seamounts (hot spots), oceanic plateaus (superplumes) and normal ocean floor (mid-oceanic ridges) (Tatsumi *et al.* 2000; Koizumi & Ishiwatari 2006; Ichiyama *et al.* 2008). The greenstones of seamount and oceanic plateau origin comprise the plume-type ophiolite (Dilek & Furnes 2011).

The Palaeozoic sequences in Japan show many features in common with those in the other circum-Pacific areas. Palaeozoic ophiolites, blueschists and accreted oceanic sediments are present in Russian Primorye on the other side of the Japan Sea (Ishiwatari & Tsujimori 2003), and its northern extension can be found in the Taigonos Peninsula, Ganichalan area and Koryak Mountains in NE Russia, where abundant Palaeozoic and Mesozoic ophiolites are present (Ishiwatari *et al.* 2003). An analogous ophiolitic nappe pile is found in the Klamath Mountains in California and Oregon (USA), where the early Palaeozoic Trinity Ophiolite tectonically overlies late Palaeozoic–Triassic ophiolites and blueschist, which in turn tectonically overly the Jurassic Josephine Ophiolite and the Cretaceous Franciscan Complex (Ishiwatari 1991). This forms a mirror image of the multiple ophiolitic nappe pile in SW Japan (see Chapter 3).

Palaeozoic ophiolites are also found in New Zealand (Dun Mountain Ophiolite; late Palaeozoic; Kimbrough *et al.* 1992) and eastern Australia (Great Serpentine Belt and Tasmania; early Palaeozoic; Spaggiari *et al.* 2003).

In the nearby continental areas, the early Palaeozoic ophiolite and blueschist are well developed in the Qilian Shan in NW China. The Palaeozoic greenstones, chert and limestone comprising typical accretionary complexes are present in the Central Asian Orogenic Belt in Mongolia (Erdenesaihan *et al.* 2013). This orogenic belt bears late Proterozoic, early Palaeozoic and late Palaeozoic ophiolite complexes.

In summary, Palaeozoic rocks in Japan consist of two types: (1) early Palaeozoic ophiolites and its sedimentary cover that was deposited in an active continental margin (or island arc) environment; and (2) late Palaeozoic oceanic crust and its sedimentary cover that was deposited on the deep ocean floor or around the oceanic islands, and now occurs in the late Palaeozoic and Jurassic accretionary complexes. These deposits share many features in common with other circum-Pacific areas and the Central Asian (Mongol–Okhotsk) Orogenic Belt.

Mesozoic–Palaeogene geology of Japan

During the Mesozoic period (252–65 Ma) Japan was situated along the eastern margin of Asia. The major geological event at the beginning of the Mesozoic was the continental collision between the North and South China cratons resulting in the Dabie–Sulu Orogeny with widespread ultrahigh-pressure metamorphism at *c.* 240–220 Ma (e.g. Hacker *et al.* 2006). From this time onwards, the eastern margin of the Asian continent and hence of the Japanese islands has been dominated by oceanic convergent tectonics associated with the formation of accretionary complexes. Detritus from the Dabie–Sulu Orogeny is the probable source for much of the Jurassic clastic rocks in Japan. There were no major continental collisions with the Paleozoic basement domains of Japan, but its Mesozoic geological record does show evidence for accretion of ocean islands plateaus and ridges. Several distinct phases of

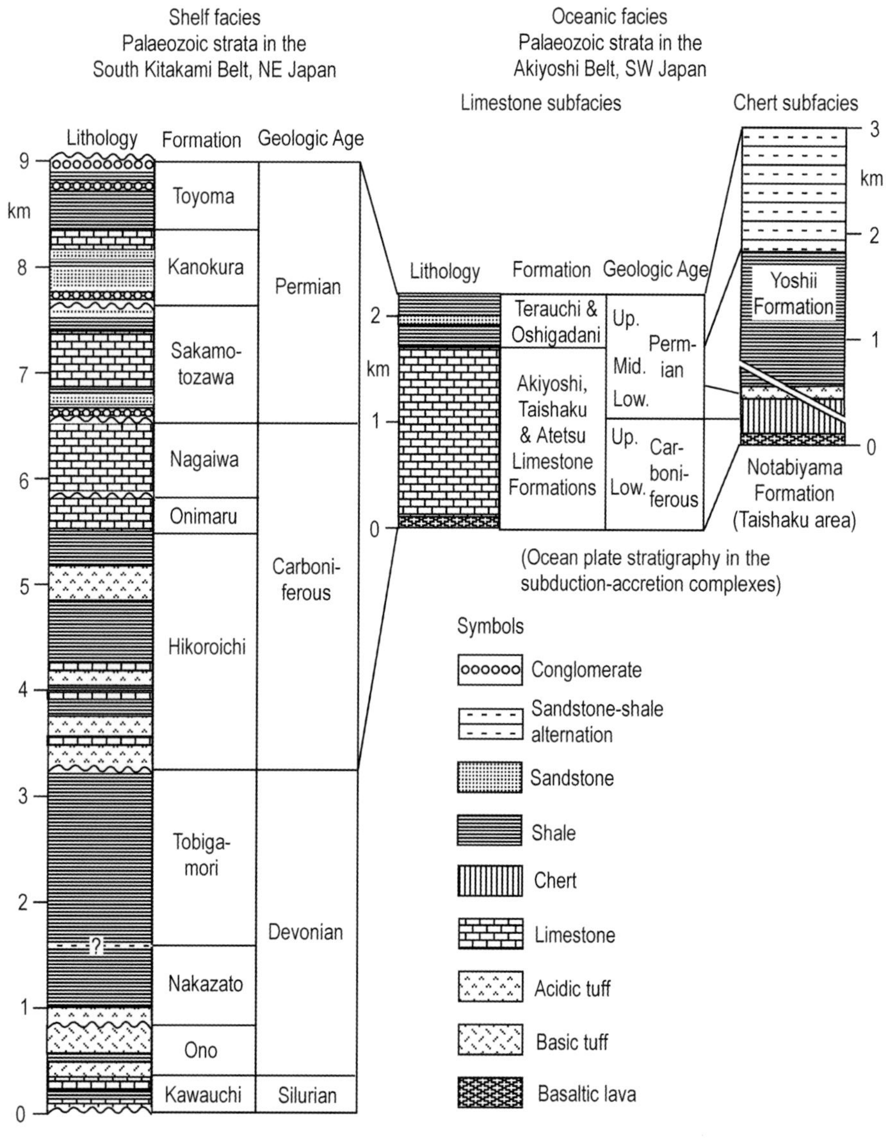

Fig. 1.9. Stratigraphy of the typical shelf facies and oceanic facies Palaeozoic rocks in Japan. Modified after Minato *et al.* (1965), Mori *et al.* (1992), Kanmera *et al.* (1990) and *Regional Geology Series* of the Geological Society of Japan (Asakura Shoten).

approach and subduction of active spreading ridges have been proposed, and these are likely to have left their mark on the geological development (e.g. Osozawa 1994; Isozaki 1996; Maruyama *et al.* 1997; Aoya *et al.* 2003; Wallis *et al.* 2009) (Fig. 1.11).

Three major oceanic plates can be identified that subducted beneath the Japanese arc during the Mesozoic: the Izanagi, Kula and Pacific plates (Engerbretson *et al.* 1985). The Kula Plate probably formed during the late Cretaceous by splitting from the Pacific Plate (Woods & Davies 1982). An additional plate, the Philippine Sea Plate, began to develop in the Cretaceous but is mainly younger than 50 Ma (see section on 'Cenozoic evolution of the Izu–Ogasawara Arc and Phippine Sea Plate' below). The best data for recognizing ocean plates and their palaeo-motion comes from palaeomagnetic studies of the ocean floor. In addition, on-land geological information (in particular kinematic data) can help constrain which plate was subducting beneath the Japanese island at a given time and hence define the location of plate boundaries (e.g. Maruyama *et al.* 1997; Wallis *et al.* 2009).

The geological units that formed in relation to subduction throughout the Mesozoic can be divided into: accreted units which are locally metamorphosed; igneous intrusions cross-cutting the accreted units; and sedimentary basin deposits overlying both the igneous and accretionary units. There is a general progression from older units on the Japan Sea or continental side to younger units on the Pacific Ocean side. The only major exception is seen in SW Japan where the Kurosegawa and associated units are surrounded by younger rocks (Fig. 1.6). We describe these various units in more detail below.

Jurassic

Jurassic plate convergence in east Asia is witnessed both by a belt of calc-alkaline plutons found throughout the Korean Peninsula and SE China and a contemporaneous accretionary complex that is developed throughout Japan. The main Jurassic accretionary terranes are the Mino–Tanba Belt in SW Japan, the Northern Kitakami Belt in NE Japan and the Oshima Belt in Hokkaido (e.g. Matsuda & Isozaki 1991). The best exposure is in SW Japan due to the recent uplift and erosion of recent sedimentary cover in this region. The equivalent rocks in NE Japan are covered by thick sequences of volcaniclastic material and are less well studied. The southern part of the Mino–Tanba Belt has undergone high-T/P Cretaceous metamorphism and is referred to as the Ryoke Belt (Nakajima 1994; Suzuki & Adachi 1998; Chapter 2c). The Kitakami–Oshima Belt of NE

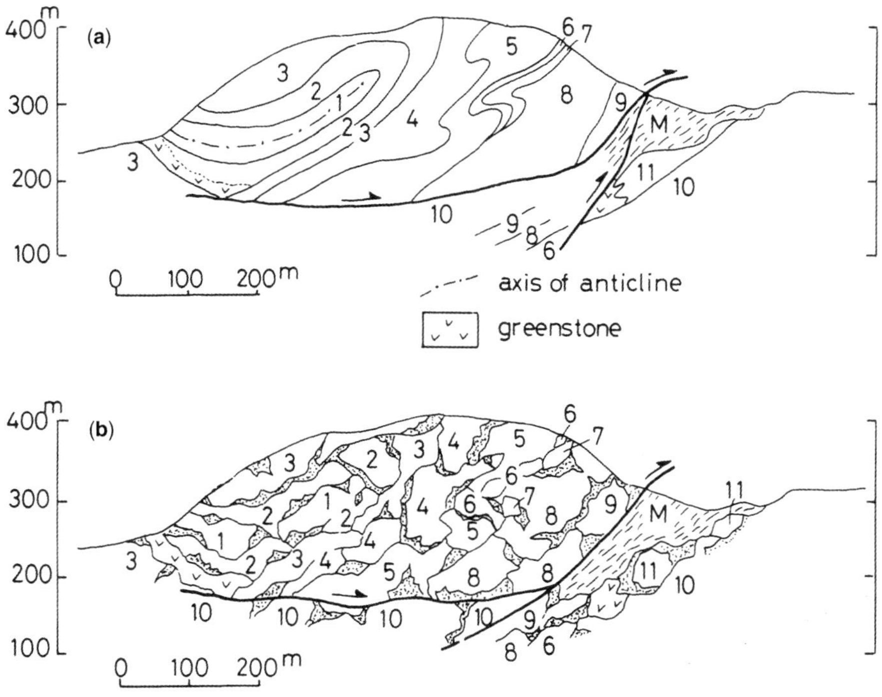

Fig. 1.10. Two alternative interpretations of the folded and inverted structure of the Akiyoshi Limestone. Numbers 1–11 indicate coral and fusulinid zones by Morinaga & Ota (1971) in younging order from Lower Carboniferous to Middle Permian. (**a**) A coherent recumbent anticlinal fold suggested by Morinaga & Ota (1971); while (**b**) Sano & Kanmera (1991) views it as an aggregate of collapsed limestone blocks of various ages.

Japan may show a similar relationship with the Abakuma Gozaisho high-T/P metamorphic units (Fig. 1.11).

The main rock types of the Jurassic accretionary complexes are a combination of early Triassic–Middle Jurassic deep-sea units including chert and lesser amounts of mafic rocks derived from ocean floor basalt. These units are overlain by clastic material including both mudstone and sandstone, suggesting proximity to the continental margin. Large limestone blocks significantly older than the surrounding matrix are also widespread. The presence of common 300–200 Ma detrital zircon ages suggests that much of the detrital material for the sandstone was derived from the Sulu–Dabie Collision Belt. The Jurassic units are commonly divided into coherent types and mixed or mélange types (Isozaki 1996) and are thought to exist as large thrust sheets separated by low-angle faults. Detailed structural analysis has been used to infer the palaeo-convergence direction of the associated oceanic plate (Kimura & Hori 1993). The Triassic–Jurassic boundary is locally well exposed and detailed recent studies have revealed evidence for a large meteorite impact, possibly contributing to the mass extinction at this time (Sato *et al.* 2013).

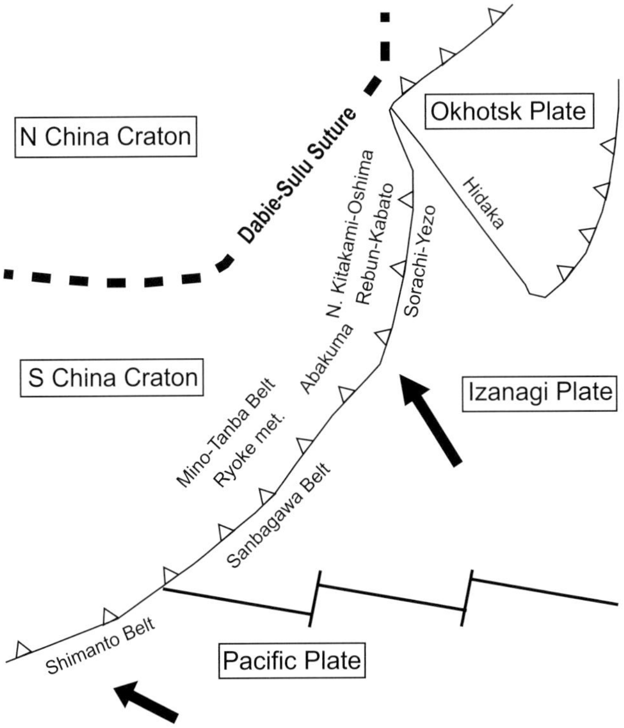

Fig. 1.11. Proposed main features of the Mesozoic palaeogeography of Japan in the middle Cretaceous showing general relationships between the main terranes and one proposed ridge subduction event (adapted after Maruyama *et al.* 1997). The Kula Plate formed by splitting of the northern part of the Pacific Plate (Woods & Davies 1982). The Izanagi and Pacific plates were moving at rates of around 20 and 10 cm a^{-1}, respectively.

The Sanbosan Belt of SW Japan is separated from the main body of the Mino–Tanba Belt by younger rocks, but lithological and geochronological comparisons suggest the two can be correlated and the Sanbosan Belt represents a klippe of the Mino–Tanba sequence (Isozaki *et al.* 1992). Despite its relative proximity to the contemporaneous high-grade Ryoke Belt, the Sanbosan Belt does not show any significant metamorphism. The reasons for this are unclear, but there may have been significant strike-slip motion between the two domains.

The Jurassic accretionary units are commonly intruded by granitic plutonic rocks of Cretaceous and younger ages (see chapter 4 'Granitic Rocks'). The older plutons show evidence of strong ductile deformation. The accretionary and intrusive units are commonly unconformably overlain by Jurassic–Cretaceous sedimentary basin deposits. Similar basins also develop above the older basement units such as the Hida gneiss. The oldest basin sediments are present in the Kuruma Formation (*c.* 200–160 Ma), which correlates with the lower parts of the more widespread Tetori Formation (*c.* 160–100 Ma). Many of the best Japanese dinosaur fossils have been reported from the Tetori Formation. Conglomerates of the Tetori Formation locally contain chert clasts derived from the Mino–Tanba Belt

(Kunugiza *et al.* 2002). Sedimentary basins of similar age but of marine associations are found in the outer zone of Japan. These are associated with the distinct marine Ryoseki faunal assemblages.

Cretaceous–Palaeogene

Cretaceous high P/T metamorphism of the Sanbagawa Belt and equivalent parts of the Nagasaki metamorphics shows that subduction continued beneath the Asian eastern margin after the Jurassic. The protoliths for most of the Sanbagawa Belt are Jurassic to Cretaceous in age and consist of mafic, quartz and pelitic schists with smaller amounts of ultramafic rocks, marble and metagabbro. The Sanbagawa Belt is also associated with once commercially important Cu deposits probably formed as hydrothermal ocean floor deposits. The contemporaneous high-T/P Ryoke metamorphic belt – probably including the Higo and Abukuma belts – lies immediately to the north of the Sanbagawa Belt. The Sanbagawa and Ryoke belts are one of the best-known examples of paired metamorphism, characteristic of convergent margins (e.g. Miyashiro 1961, 1973). Both belts show strong orogeny-oblique movements, implying that the associated plate convergence was also oblique. This is compatible with sinistral strike-slip faulting documented along the Median Tectonic Line (MTL) and other parts of east Asia (Okada & Sakai 2000). The only contemporaneous plate that has been proposed in this region with a compatible palaeo-convergence vector is the Izanagi Plate. Wallis *et al.* (2009) argue that this oblique subduction continued until interaction with the Izanagi–Kula Ridge at *c.* 85 Ma. There is a general consensus among Japan-based geologists that ridge approach and subduction played an important role in the formation of both the Sanbagawa and Ryoke belts (see Chapter 2c for more details).

The Ryoke and Sanbagawa belts are separated by the MTL, a major strike-slip fault. This tectonic boundary has a long movement history that began in the Cretaceous and is in part contemporaneous with the Ryoke and Sanbagawa metamorphic stages (e.g. Takagi 1986). A series of sedimentary basins, the Onogawa–Izumi sediments, is developed along the northern side of the MTL unconformably overlying the Ryoke Belt. The clastic sediments of the Onogawa–Izumi basins have depositional ages of Turoninan (90–94 Ma) in the east and Maastrichtian (72–66 Ma) in the west and are generally interpreted to have formed in pull-apart basins formed by sinistral movement along the MTL (Noda & Toshimitsu 2009). The Sanbagawa Belt is unconformably overlain by the Kuma Formation, which has a basal age of *c.* 55 Ma (Narita *et al.* 1999), and this gives a clear younger age to constrain the duration of the Sanbagawa Orogeny.

To the south, the Sanbagawa Belt is in contact with the Triassic–Jurassic Chichibu accretionary belt. Further south lies the enigmatic Kurosegawa Belt, consisting of a wide variety of blocks in a serpentinite matrix, and the Triassic–Jurassic Sanbosan Belt. These three belts form a relatively old outlier within the overall oceanwards-younging sequence of SW Japan. The Kurosegawa Belt is overlain by narrow and relatively thick Cretaceous sedimentary basin deposits. A well-studied example is the Ryoseki Formation. These shallow- to deep-marine facies include thick sequences of turbidites and are distributed from Kyushu to Kanto Mountains. The depositional ages range from Valanginian (140–133 Ma) to Cenomanian (100–94 Ma).

Cretaceous igneous activity is recognized in the Ryoke Belt (Suzuki & Adachi 1998) and younger Sanyo Belt to the north of the MTL (Nakajima 1994). The intrusions show a range of different compositions generally from tonalite to two-mica granite. There is also a large region of Cretaceous volcaniclastic deposits, known collectively as the Nohi Rhyolite, consisting of welded tuff, dacite and rhyolite. This is an important marker for relative dating of the Ryoke granitoids, because some of the earlier granites are overlain by the Nohi Rhyolite and others intrude it (Suzuki & Adachi 1998). These observations are significant because they show that granites were intruded to relatively shallow levels in the crust and there has been no more than a few kilometres of erosion since the Cretaceous.

At the same time that the Sanbagawa and Ryoke metamorphic events were occurring, there was a major phase of accretion represented by the Shimanto Belt. The Shimanto Belt is the largest Cretaceous unit in Japan (Fig. 1.11) with a northern boundary defined by the Butsuzo Tectonic Line (BTL), which generally separates it from the Chichibu Belt. However, in the Kii Peninsula, the Shimanto Belt locally appears to be in direct contact with the Sanbagawa Belt (Isozaki *et al.* 1992), lending support to the idea that the intervening Chichibu and Kurosegawa belts form a klippe. The Shimanto Belt has been widely studied as a type example of an accretionary prism. The stratigraphy has been firmly established on the basis of microfossils (in particular radiolarians) and it can be clearly divided into a Cretaceous northern sub-belt and a younger southern sub-belt dominantly of Eocene–Miocene age. There are very few rocks of upper Cretaceous (Maastrichtian) and earliest Palaeogene (Paleocene), suggesting reduced accretion during this time or subsequent subduction erosion. The Shimanto Belt stretches from the Ryukyu Arc through Kyushu and Shikoku islands and through central Japan, with the final outcrops occurring in the Kanto Mountains.

The dominant rock types of the Shimanto Belt are mudstone and sandstone with lesser amounts of bedded chert and locally ocean floor mafic rocks. The accreted units are commonly strongly deformed and there are significant domains of tectonic mélange. The strata dip mainly north and locally show clear evidence of folding and thrusting (Taira *et al.* 1988). Documented changes in the kinematics of deformation may be related to subduction of different plates and the passage of a spreading ridge (Lewis & Byrne 2001; Tokiwa 2009); likely candidates are the Izanagi–Pacific and Kula–Pacific ridges. The high thermal gradients recognized in the Shimanto Belt (Underwood *et al.* 1993) are compatible with the passage of a spreading ridge. The sedimentary facies of the Shimanto Belt include forearc basins, slope and trench deposits. Unusual igneous intrusions of a similar age to the accretion are present in the Palaeogene–Neogene Shimanto of Shikoku (e.g. Muroto gabbro and Ashizurimisaki granite). These occurrences are also compatible with the idea of a relatively hot accretionary complex. To the south, the Shimanto Belt is in contact with the younger Nankai accretionary complex, which is the active accretionary margin related to the present subduction of the Philippine Sea Plate.

Hokkaido has a distinct Cretaceous to Paleocene geological history from the rest of Japan. In the early Cretaceous, westwards subduction of a remnant of the Izanagi or Kula plates, known as the Okhotsk Plate, formed a volcanic arc represented by the subaqueous calcalkaline and alkaline volcano-sedimentary sequences of the Rebun–Kabato Belt (Niida & Kito 1986; Kiyokawa 1992; Niida 1992) and an accretionary complex represented by the Sorachi–Yezo Group (Fig. 1.11). Detritus derived from the Rebun–Kabato Arc was deposited in a series of forearc basins, continental slope and trench environments. The Yezo Group represents a well-developed forearc basin fill that is in unconformable contact with the Sorachi accreted units. The Yezo Group has yielded some spectacular examples of large Cretaceous ammonites. Limestone blocks are locally present within the lower part of the Yezo Group. The lowermost part of the Sorachi Group is represented by a prominent sliver of basaltic oceanic crust (Kiyokawa 1992). Structurally underlying this is an accretionary domain associated with serpentinite that has undergone high-pressure metamorphism referred to as the Kamuikotan Belt (Ueda *et al.* 2000). The Idonnappu Belt crops out further to the east and lies between the Sorachi–Yezo and Hidaka belts. The Idonnappu Belt is a strongly deformed accretionary complex commonly associated with mélange (Kiyokawa 1992; Ueda *et al.* 2000) and is thought to be a shallow-level equivalent to the Kamuikotan Belt.

The youngest unit formed by westward subduction in central Hokkaido is the Hidaka Belt. This dominantly Palaeogene accretion belt is separated from the Idonnappu Belt by a major east dipping thrust. The Hidaka Belt mainly consists of turbidites locally with mélange containing blocks of basalt, chert and limestone.

On the opposite side of a relatively small Sorachi Ocean lay the Okhotsk Landmass. In the late Cretaceous subduction also began under this block forming the Tokoro accretionary complex and the associated Nemuro volcanic arc complex. The intervening Izanagi or Kula Plate was gradually consumed leading to the late Miocene collision between NE-Japan and Kuril Arc systems. The collision caused widespread deformation affecting the Sorachi-Ezo, Hidaka and Tokoro belts. The Hidaka Belt is strongly affected by high T/P metamorphism including some of the youngest known granulite domains (*c.* 19 Ma; Kemp *et al.* 2007) underlain by the Horoman peridotite, which represents an unusually pristine slab of upper mantle. Heat transport for metamorphism is aided by multiple intrusions of gabbroic magma.

Summary of the Mesozoic–Palaeogene evolution of Japan

The Mesozoic history of Japan is essentially one of accretion now represented mainly by the large accretionary complexes of the Mino–Tanba, Shimanto, North Kitakami–Oshima and Sorachi–Yezo belts. These all dominantly consist of siliciclastic deposits formed in the associated ocean trenches. Cretaceous subduction is associated with both subduction-type high-P/T metamorphic domains and arc-related high-T/P metamorphism. In SW Japan the main domain of high-P/T metamorphism is the Sanbagawa Belt; in Hokkaido it is the Kamuikotan Belt. The main belt of Cretaceous high-T/P metamorphism is the Ryoke Belt, which is associated with numerous granitoid intrusions. The Jurassic–middle Cretaceous accretionary units and associated metamorphic and igneous domains are overlain by a series of sedimentary basins ranging from dominantly continental facies in the inner arc region, typically represented by the Tetori Formation, to dominantly marine facies close to the Median Tectonic Line to the Izumi and Ryoseki formations in the outer arc region.

Cenozoic geology of Japan

In this section we focus primarily on the sedimentary record of the Cenozoic Era of Japan. Palaeogene deposits in Japan are composed mainly of two different types of strata: (1) accretionary prism deposits forming the Shimanto Belt of the Outer Zone of SW Japan, which continued until the Miocene; and (2) terrestrial, brackish and shallow-marine deposits, which accumulated along continental or island arc margins and are currently distributed chiefly in Hokkaido, southern Tohoku, western Honshu and northern Kyushu. These deposits yielded significant amounts of coal and are referred to as the Yubari, Joban, Ube and Chikuho coalfields, respectively. Eocene–Oligocene non-marine and marine deposits are distributed in Okayama and Hyogo prefectures, western Honshu (Tsubamoto *et al.* 2007; Matsubara 2013). Palaeogene volcanic rocks (andesite to rhyolite) and granites are widespread throughout Japan, for example at Jodogahama (Kitakami, NE Japan), Eocene Akashima and Oligocene Monzen (Oga Peninsula, NE Japan), Futomiyama (central Japan), etc. Palaeogene deposits also occur in the South Ryukyus (Ryukyu Islands) and the Ogasawara (Bonin) islands. The deposits in the Ryukyus consist of volcanic and volcaniclastic rocks (Nosoko Formation) overlying shallow-water carbonates (Miyara Formation) rich in coralline algae, both of which are Eocene in age (Nakagawa *et al.* 1982). Eocene volcanic and volcaniclastic deposits occur, associated with Eocene and Oligocene limestones, in the Ogasawara islands which formed by arc volcanism caused by subduction of the Pacific Plate beneath the Philippine Sea Plate (Ishizuka *et al.* 2011*a*).

The Neogene sedimentary succession in the Japan Sea coastal areas of northern Honshu formed in association with the rifting and rotation of NE and SW Japan that led to the formation of the Japan Sea; the main movement occurred prior to 15–10 Ma. The pre-rifting deposits include non-marine sandstones/conglomerates and volcanic/volcaniclastic rocks (Daijima Formation) resting on volcanic rocks (Monzen Formation). However, recent studies show that the volcanic rocks of the Monzen Formation and the underlying Akashima Formation date back to the Eocene (Kobayashi *et al.* 2008) and the Cretaceous (Kano *et al.* 2012), respectively. Highly alkalic, typical continental rift zone magma erupted at 25–20 Ma in northern central Japan to produce 'moonstone rhyolite' ignimbrites (Ayalew & Ishiwatari 2011). The pre-rifting deposits are overlain by altered volcanic and volcaniclastic rocks called the 'Green Tuff' (Nishikurosawa Formation), which formed by active volcanism that occurred at the early stage of the opening of the Japan Sea (early–middle Miocene). The metal deposits formed by this volcanism and associated hydrothermal activity are called 'Kuroko', yielding copper, lead and zinc. Subsequent sedimentation is represented by thick marine siliciclastics; the lower part is dominated by hard shale (middle–upper Miocene Onnagawa Formation) and the upper part is dominated by mudstone (upper Miocene–Pleistocene Funakawa Formation). This succession has been regarded as a 'standard succession', based on which a local chronostratigraphy was established (Hujioka 1959). The correlative deposits are distributed along the central mountains or backbone range to the Japan Sea side of mainland Japan extending from Central and West Hokkaido to Chugoku.

A major sedimentary basin is defined by a large-scale graben called the Fossa Magna, which separates domains of NE and SW Japan. This graben initiated as a rift in the east Asian continental margin and expanded when opposed rotations of the SW and NE Japan blocks caused stretching between them. The resulting graben structure formed a deep sedimentary basin that accommodated a thick marine succession (total thickness of up to 6000 m) dominantly consisting of volcanic and volcaniclastic rocks. The southern part of the Fossa Magna (the area around Mt Fuji) was formed by the collision and accretion of micro-continental blocks (or volcanic island arc blocks) due to the northwards movement of the Philippine Sea Plate. This domain is referred to as the Izu Collision Zone (ICZ). The collided micro-continents or continental fragments are the Koma (13–12 Ma), Misaka (12–11 Ma), Tanzawa (*c.* 6 Ma) and Izu (*c.* 1 Ma) massifs. The uplifted hinterlands (the previously collided micro-continents) supplied large amounts of detritus to the forearc trough (Amano & Ito 1990). After the initial phase of spreading the Fossa Magna area began to undergo east–west shortening, which can be correlated to the convergence between the Okhotsk and Amur plates from *c.* 4 Ma (Ohtake *et al.* 2002). This shortening has resulted in rapid uplift of the Hida Mountains, exposing young mid-crustal rocks including remarkable Quaternary granitic rocks, the youngest in the world (Harayama *et al.* 2003; Ito *et al.* 2013).

To the north in East Hokkaido, a Neogene marine succession developed in the boundary region between the Kuril and NE Japan arcs that overlie the coal-bearing Palaeogene deposits. In the

Kanto district a thick Miocene–Pleistocene succession developed throughout the Miura and Boso peninsulas in the southern Kanto district (Ogawa & Taniguchi 1988; Saito 1992). The succession has abundant turbidites and is considered to have been deposited in a forearc slope-to-basin setting.

In the Outer Zone of SW Japan a prominent sequence of marine deposits (e.g. the Miyazaki Group with a total thickness up to 3000 m) developed above accretionary prism deposits of the Shimanto Belt. The Neogene strata of the Ryukyus comprise both shallow-marine deposits formed on the continental margin (Yaeyama Group) and those in shelf–slope basins (Shimajiri Group). The Miocene Yaeyama Group, composed mainly of sandstone associated with shale, limestone conglomerate and coal beds, is exposed in the South Ryukyus, and has been found beneath the shelf off Miyako-jima. The Shimajiri Group consists chiefly of mudstone and sandstone accompanied by tuff and volcanic and volcaniclastic rocks, and ranges in age from late Miocene–early Pleistocene (Ujiié & Kaneko 2006; Imai *et al.* 2013). The on-land distribution of the group is limited to the eastern periphery of the Ryukyu Arc. However, sediments correlative to this group extend to the Okinawa Trough and the eastern margin of the East China Sea Shelf as well as the region around the Ryukyu Islands (Aiba & Sekiya 1979).

Thick Quaternary sediments occur beneath coastal plains, such as seen in the Sendai, Kanto, Niigata, Nobi and Osaka plains. Quaternary sediments also occur covering coastal and river terraces or filling local basins (e.g. between Oga Peninsula and Akita Plain). The formation of coral reefs is considered to have been initiated in the Ryukyus after the opening of the Okinawa Trough at *c.* 1.71–1.39 Ma and the subsequent influx of the Kuroshio current into the back-arc basin. Most islands in the Central and South Ryukyus are therefore capped or rimmed by the Pleistocene reef complex (coral reef and associated fore-reef and shallow lagoon) limestones accompanied by marine and non-marine siliciclastic deposits called the Ryukyu Group, and fringed by present-day coral reefs.

A recent discovery of young volcanoes on the Pacific Plate close to Japan has been shown to represent a previously unrecognized type of igneous activity (Hirano *et al.* 2006). The volcanoes dominantly consist of strongly alkaline mafic rocks erupted since *c.* 8.5 Ma and some may be as young as 50 ka. The eruption of these rocks on the ocean floor is probably related to stresses associated with the formation of the fore-bulge as the Pacific Plate approaches the subduction zone. The discovery of these rocks provides a new window from which to view the deeper parts of old oceanic plates and lends support to the idea that the asthenosphere contains small amounts of melt (Hirano 2011).

Summary of the on-land Cenozoic geological evolution of Japan

Subduction of the Pacific Plate, which started in the Mesozoic, continued in the Cenozoic. In the Neogene a major change in tectonic configuration occurred with the onset of subduction of the Philippine Plate and the opening of the Japan Sea as a back-arc basin between 21 and 14 Ma. The resulting complex local tectonic movements were reflected in the widespread development of extensional basins and sedimentary fill, such as seen in the Fossa Magna, at the same time that accretionary sedimentary units continued to develop. The opening of the Japan Sea is also associated with the widespread deposition of the Green Tuff and associated Kuroko mineralization. Collision of the Izu Arc with central Japan caused significant distortion of the pre-existing geological structure and is associated with major surface uplift. Convergence between the Okhotsk and Amur plates from *c.* 4 Ma has caused rapid uplift. Associated high erosion rates have exposed Quaternary granites formed in the mid-crust. The Ryukyu region is associated with the development of major carbonate deposits during the Cenozoic.

Cenozoic evolution of the Izu–Ogasawara Arc and Philippine Sea Plate

The Philippine Sea Plate and Izu–Ogasawara Arc are two of the main geological features of the Cenozoic development of the Japanese arc system. Major new surveys of the Philippine Sea and adjacent areas in the Pacific Ocean have been carried out as part of Japan's preparations for an application to have an extended continental shelf recognized as defined in the United Nations Convention on the Law of the Sea (UNCLOS) (Ohara *et al.* 2007). The extensive new datasets mean that the geology of the Philippine Sea Plate is now one of the best documented of any ocean region in the world. In this section, we review the main scientific results from Japan's legal continental shelf survey project and how these have helped develop our understanding of the Cenozoic evolutionary history of this area.

Japan's legal continental shelf survey project

Japan's legal continental shelf survey project (CSSP) consists of two major phases. The first phase started in 1983 with swath-mapping of the northern Philippine Sea and adjacent areas in the NW Pacific Ocean by the Hydrographic and Oceanographic Department of Japan (HODJ). The first phase greatly increased our understanding of the Philippine Sea Plate, allowing reconstruction of the first detailed evolutionary history of the Shikoku and Parece Vela basins (Okino *et al.* 1994, 1998, 1999). Along with HODJ's swath-mapping project, Japan Oil, Gas and Metals National Corporation (JOGMEC) started the national project 'Deep Sea Survey Technologies for Natural Resources in Japan' in 1998. The project's main objective was to gather information about the natural resources in the Philippine Sea and adjacent areas in the Pacific Ocean, and to develop survey skills for studies of deep-sea areas. The project acquired a large amount of state-of-the-art multichannel seismic reflection profiles and cored bottom materials with a wire-line rock drill, Deep-Sea Boring Machine System (BMS) in the Philippine Sea and adjacent areas in the Pacific Ocean.

After 20 years of HODJ's continued efforts, in 2003 the Japanese government started the second phase of the CSSP. Along with HODJ and JOGMEC, the Japan Agency for Marine-Earth Science and Technology (JAMSTEC) and the Geological Survey of Japan (GSJ-AIST) also took joint responsibility for the project in the second phase.

Bathymetry

The bathymetric map of the Philippine Sea obtained through Japan's CSSP is shown in Figure 1.12. The Amami Plateau, Daito Ridge, Oki–Daito Ridge and Oki–Daito Rise are the oldest (Cretaceous–early Eocene) components of the Philippine Sea Plate (Karig 1971, 1975; Mizuno *et al.* 1978; Hickey-Vargas 2005). Together with the northern part of the Kyushu–Palau Ridge, these oldest components comprise a larger bathymetric complex: the Amami–Daito Province. The West Philippine Basin occupies the western part of the Philippine Sea, and extends from 5° to 23 °N. It is the largest and deepest inactive basin on the Philippine Sea Plate. The water depth of the basin ranges from *c.* 5000 to *c.* 6000 m; however, it reaches *c.* 7900 m locally at the extinct spreading centre known as the CBF Rift (CBF as 'Central Basin Fault') (Fujioka *et al.* 1999). A wide and rugged bathymetric high, the Urdaneta Plateau, is located in the NW part of the West Philippine Basin. This area typically displays complex sinuous seafloor-spreading fabrics that were

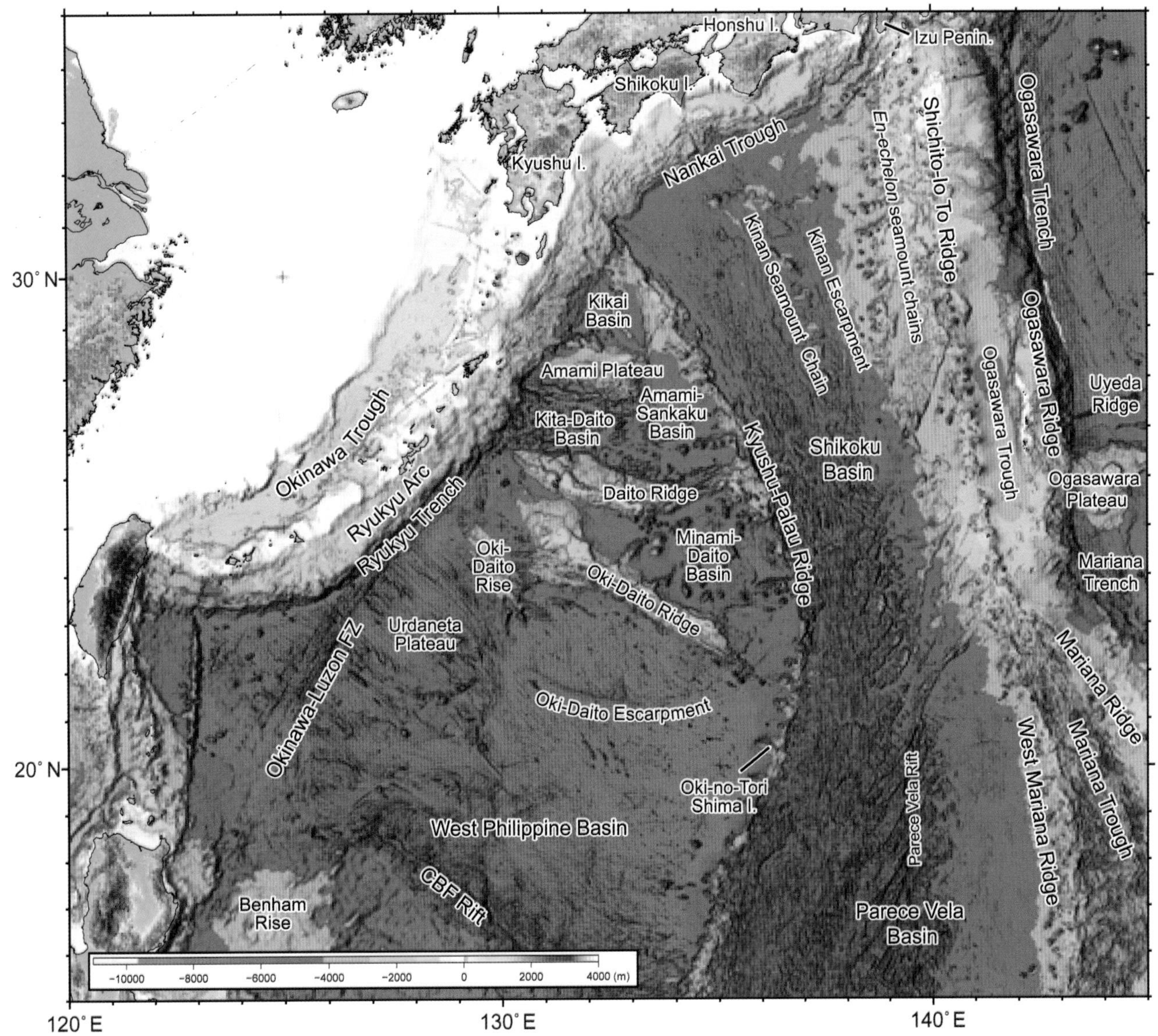

Fig. 1.12. Index map of the Philippine Sea, showing prominent bathymetric features (bathymetric data after Japan's CSSP).

due to robust activity of ocean island basalt (OIB) -like magma (Ishizuka *et al.* 2013).

The Kyushu–Palau Ridge is a prominent and continuous topographic feature that can be traced over *c.* 3000 km all the way from Kyushu Island in the north to the Palau islands in the south. The northern part of the Kyushu–Palau Ridge is composed of an elongated ridge morphology overlain by seamounts. Typically, these seamounts demonstrate a NE–SW-aligned trend, oblique to the ridge axis. The seamounts are relatively common in the northern part and scarcer and more isolated in the south.

The Shikoku Basin is an approximately north–south-elongated fan-shaped basin bounded by the Kyushu–Palau Ridge on its west and by the Kinan Escarpment on its east. The northern end of the Basin is defined by the Nankai Trough. To the south, the Shikoku Basin is contiguous with the Parece Vela Basin with the boundary located at *c.* 24 °N. The water depth of the basin gradually increases towards the SW, and the maximum depth of >5000 m is attained in the western margin close to the Kyushu–Palau Ridge. The NNE part of the basin is generally flat-bottomed, indicating that the basin floor is covered with relatively thick sediments (Fig. 1.5c). In contrast, the SW part of the basin is rugged and displays linear, approximately NNW–SSE-trending abyssal hill fabrics. The extinct spreading axis of the basin is marked by a chain of seamounts, the Kinan Seamount Chain, with an overall NW–SE-aligned trend (Ishizuka *et al.* 2009). The eastern margin of the Shikoku Basin is marked by the Kinan Escarpment, which extends north–south for *c.* 500 km along the eastern Shikoku Basin. The Kinan Escarpment could be a large normal fault formed by post-spreading deformation of Shikoku Basin lithosphere along the mechanical boundary between the active buoyant island arc (i.e. the Izu–Ogasawara Arc) and the inactive isostatically sinking back-arc basin (i.e. the Shikoku Basin) (Kasuga & Ohara 1997; Ohara *et al.* 1997).

From east to west, the topographic highs of the Izu–Ogasawara Arc can be divided into the Ogasawara Ridge, the Shichito–Ioto Ridge (also commonly known as the Shichito–Iwojima Ridge), and the *en echelon* seamount chains oblique to the overall trend of the arc (Fig. 1.12). The Ogasawara Ridge is one of the oldest components of the Philippine Sea Plate, consisting mainly of Eocene volcanic rocks. This ridge is located in the outer-arc of the Izu–Ogasawara Arc adjacent to the Izu–Ogasawara Trench; it extends for *c.* 400 km in a north–south direction and has a maximum width of *c.* 110 km. The Ogasawara Ridge is associated with a chain of islands, the Ogasawara islands, the type locality of boninite. The Shichito–Ioto Ridge is the present active volcanic front of the arc, which is associated with multiple volcanic islands and seamounts. In the northern part, between 30° and 32 °N, the four sets of rear-arc en echelon seamount chains are oblique to the Shichito–Ioto Ridge. Each seamount in the chain rises to 1000–2400 m above the surrounding seafloor and has a basal diameter of 10–30 km.

Gravity signatures

The Bouguer gravity anomaly map of the Philippine Sea obtained through Japan's CSSP is shown in Figure 1.13. The northern Philippine Sea is characterized by low Bouguer anomalies over the bathymetric highs and high anomalies over the basins, indicating the presence of thicker crust beneath the bathymetric highs and typical oceanic crust beneath the basins.

The Amami Plateau and the Daito and Oki–Daito ridges are associated with Bouguer anomalies of 200–300 mGal, *c.* 100 mGal less than those of the surrounding basins. These observations indicate that these ridges have crust domains thicker than the surrounding basins. Bouguer anomalies of >400 mGal occur in the central part of the Kita–Daito Basin and in the West Philippine Basin, suggesting these basins have thin crust. The Kyushu–Palau Ridge is associated with lower Bouguer anomalies of 200–300 mGal. Bouguer anomalies vary rapidly at the boundary between the Kyushu–Palau Ridge and the Shikoku Basin, suggesting that the crustal thickness changes very rapidly there. Bouguer anomalies over the northern Izu–Ogasawara Arc decrease northwards. There is a general westwards increase in the anomalies from the Izu–Ogasawara Arc to the Shikoku Basin. The Shikoku and Parece Vela basins generally show Bouguer anomalies of 350–400 mGal. However, areas with Bouguer gravity anomalies less than *c.* 350 mGal do occur around the Kinan Seamount Chain.

Three-dimensional modelling of gravity data for the northern part of the Philippine Sea shows that the Kyushu–Palau and

Fig. 1.13. Bouguer gravity anomaly of the Philippine Sea (gravity data after Japan's CSSP).

Oki–Daito Ridges are characterized by crust of thickness in the range *c.* 10–15 km (Ishihara & Koda 2007).

Magnetic signatures

The total magnetic field anomaly map of the Philippine Sea obtained through Japan's CSSP is shown in Figure 1.14. In the western area of the Philippine Sea Plate, east–west-trending magnetic anomaly zones are predominant, that is, there exist anomaly belts along the east–west-trending Amami Plateau, the Daito and the Oki–Daito ridges . These magnetic anomaly belts are therefore readily interpreted as the result of a linearly aligned series of volcanic entities on the plateau and ridges, each of which gives a pair of negative and positive anomaly values in the north and the south sides.

The amplitude of the magnetic anomalies in the West Philippine Basin is within ±300 nT and several clear lineations can be recognized. This is in striking contrast to the weak and ambiguous magnetic anomaly pattern of the Parece Vela Basin, located at approximately the same latitude.

A clear magnetic anomaly pattern is recognized on the seafloor of the Shikoku Basin. Okino *et al.* (1994) identify the presence of anomalies 7 to 5B (26–15 Ma) in the basin. In the western part of the basin, the NNW–SSE-aligned lineations are identified as anomalies 7 to 6B (26–23 Ma). The linear anomalies abut the Kyushu–Palau Ridge. The approximately north–south-aligned lineations are identified as 6B to 6 (23–20 Ma). The magnetic lineations in the central part of the basin are interpreted as anomalies 5E to 5B. The anomalies are symmetric about the Kinan Seamount Chain, suggesting that the seamount chains are located on the site of the former spreading axis of the Shikoku Basin and the basin opened symmetrically (Okino *et al.* 1994). However, the magnetic anomaly isochrons older than 6 are very faint in the eastern margin of the Shikoku Basin, indicating the easternmost Shikoku Basin floor is now under the inner-arc slope of the Izu–Ogasawara Arc. Magnetic dipolar anomalies that are not associated with any particular topographic highs can be seen along the middle part of the inner-arc slope of the Izu–Ogasawara Arc. These dipolar anomalies are interpreted to result from magmatic intrusion along the faults when rifting occurred (Okino *et al.* 1994). The conjugate

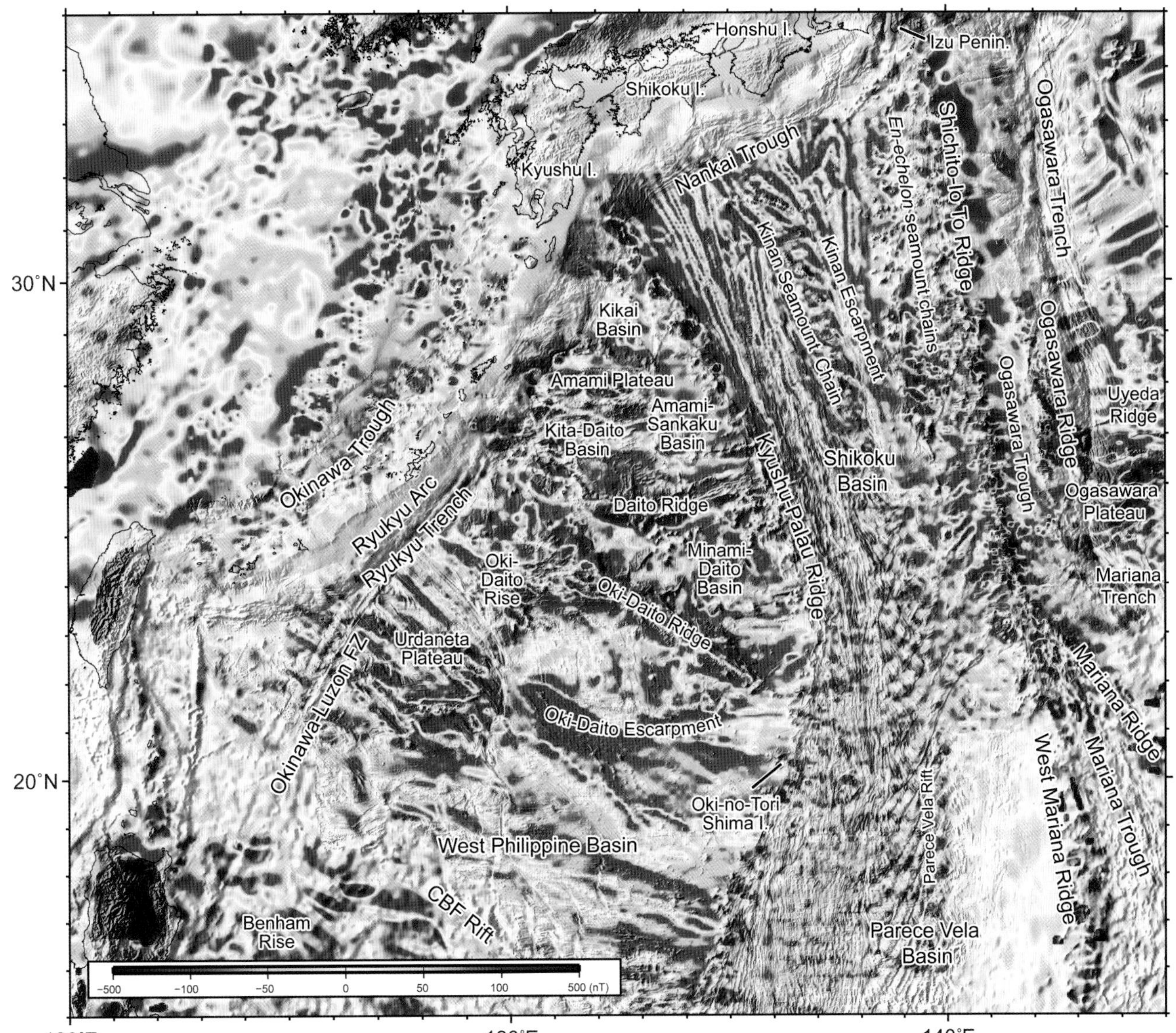

Fig. 1.14. Total magnetic anomaly map of the Philippine Sea (magnetic data after Japan's CSSP).

dipolar anomalies can be seen along the eastern flank of the Kyushu–Palau Ridge.

The amplitude of the anomalies in the Parece Vela Basin is generally very low and does not exceed ± 100 nT except adjacent to the Kyushu–Palau Ridge. It is difficult to recognize a magnetic lineation pattern associated with the overall north–south-trending fabric in the Parece Vela Basin, possibly due to the east–west spreading having occurred near the magnetic equator (Okino *et al.* 1998).

In the eastern area, three prominent magnetic anomalies trending north–south correspond to the Ogasawara and Shichito–Ioto ridges and rear-arc en echelon seamount chains. Each volcanic complex along the central axis of the Shichito–Ioto Ridge is characterized by either isolated high-amplitude anomalies or a pair of positive and negative magnetic anomalies. This is a typical anomaly pattern in the northern middle latitudes.

Crustal structure

The maximum crustal thickness beneath the Kyushu–Palau Ridge is *c.* 20 km (Fig. 1.5c; Nishizawa *et al.* 2007). The crustal structure of the Daito Ridge is similar to that of the Kyushu–Palau Ridge, with maximum crustal thickness *c.* 20 km (Nishizawa *et al.* 2014). The crustal structure of the Shikoku Basin is directly comparable with that of characteristic normal oceanic crust (Fig. 1.5c; Nishizawa *et al.* 2011). The crust of the Kyushu–Palau Ridge is, therefore, significantly thicker than the oceanic crust. The crust of the northern Izu–Ogasawara Arc is *c.* 20 km thick in between the volcanic front and the rear-arc *en echelon* seamount chains (Fig. 1.5c; Suyehiro *et al.* 1996). The Izu–Ogasawara and Mariana arcs are characterized by the presence of a thick (>5 km) middle crust with a P-wave velocity of *c.* 6 km s^{-1} (Suyehiro *et al.* 1996; Takahashi *et al.* 1998, 2007; Kodaira *et al.* 2007), interpreted to be of granitic composition. The 6 km s^{-1} middle crust persists beneath the arc from the outer-arc to the western edge of the rear-arc en echelon seamount chains, and attains a maximum thickness of *c.* 8 km (Fig. 1.5c).

The sediment layer of the western flank of the Kyushu–Palau Ridge is continuous with the flank of the ridge, and this layer is interpreted as a package of volcanogenic sediments associated with the volcanic-arc activity of the proto-Izu–Ogasawara Arc (Fig. 1.5c). The eastern boundary of the Kyushu–Palau Ridge can be correlated with a series of rift-related structures. In some places the east-dipping scarps show a series of sharp steps with sediment-filled depressions between them, interpreted to be synrift basins (Fig. 1.5c).

The inner-arc slope of the Izu–Ogasawara Arc is characterized by a sedimentary sequence that formed as a result of arc volcanism (Fig. 1.5c). Distinct changes are recognized in the thickness of the sedimentary sequence and the structure of the top of acoustic basement near the boundary between the Kinan Escarpment and the Kinan Seamount Chain. A thick sedimentary sequence, with seismic travel times of >1 s, characterizes the inner-arc slope from the rear-arc *en echelon* seamount chains to the Kinan Escarpment. This sedimentary sequence is commonly bounded or dammed by a small elevation of acoustic basement related to the Kinan Escarpment (Fig. 1.5c). The sedimentary section is continuous from the slope on the island arc side, and is interpreted to be volcanogenic sediments supplied by volcanism along the Izu–Ogasawara Arc. West of the Kinan Seamount Chain, the sedimentary section in the Shikoku Basin has a maximum thickness of *c.* 100 m (Fig. 1.5c).

Seafloor geology

Numerous bottom samplings have been carried out in the Izu–Ogasawara Arc and Philippine Sea Plate, including the DSDP (Deep Sea Drilling Project), ODP (Ocean Drilling Program) and IODP (International Ocean Discovery Program) drillings (Fig. 1.1). In addition to geochemistry, many of these samples have been dated using high-precision Ar–Ar geochronology and the formation ages have been well documented (Fig. 1.2).

The northernmost West Philippine Basin floor drilled at DSDP Site 294/295 yields lavas with typical OIB-like signatures (Hickey-Vargas 1998). The volcanic rocks drilled at other locations in the West Philippine Basin (i.e. at DSDP Site 447 and ODP Site 1201D) show N-MORB (normal mid-ocean ridge basalt) to E-MORB (enriched mid-ocean ridge basalt) -like signatures, however (Hickey-Vargas 1998; Savov *et al.* 2006). The volcanic rocks from the Oki–Daito Ridge and Urdaneta Plateau are enriched both in large-ion lithophile elements (LILE) and high-field-strength elements (HFSE) compared to island arc rocks, and show characteristics more similar to OIB. DSDP Site 446 drilled the western part of the Minami–Daito Basin, yielding lavas with clear OIB-like signatures with ages of *c.* 51–43 Ma (Hickey-Vargas 1998).

Ishizuka *et al.* (2013) further demonstrated that the volcanic rocks collected by the CSSP from the Urdaneta Plateau, Oki–Daito Rise, Oki–Daito Ridge and Minami–Daito Basin are in general OIB-like. Precise Ar–Ar ages of these basalts lie within the range *c.* 51–36 Ma. Ishizuka *et al.* (2013) argued that this region evolved with the interaction of a mantle plume (Oki–Daito mantle plume), contemporaneous with the spreading of the West Philippine Basin.

The volcanic rocks from the Amami Plateau include basalts and tonalites, which have typical arc signatures (Hickey-Vargas 2005). The volcanic rocks from the Daito Ridge also have typical arc signatures, yielding very old ages of *c.* 119 and *c.* 117 Ma. These ages are comparable to new hornblende Ar–Ar ages for the igneous rocks from the Amami Plateau (*c.* 114 Ma; Hickey-Vargas 2005). These results confirm that the Amami Plateau is one of the oldest components within the Philippine Sea Plate with a history stretching back to the early Cretaceous, although younger volcanic rocks (*c.* 33 and *c.* 37 Ma) also exist in the same area.

The volcanic rocks collected by the CSSP from the Kyushu–Palau Ridge are basaltic to andesitic, and have geochemical signatures typical of medium-K series. This contrasts with the low-K series suites from the Eocene high-Mg andesite and boninite on Chichijima Island (Ishizuka *et al.* 2011*b*). The volcanic rocks from the Kyushu–Palau Ridge show a clear arc signature (i.e. depletion in HFSE and enrichment in LILE). Ar–Ar ages of rocks from this ridge lie within the range *c.* 48–25 Ma, mostly falling within the range *c.* 28–25 Ma (Ishizuka *et al.* 2011*b*). These results imply that the onset of seafloor spreading in the Shikoku and Parece Vela basins in this region is no older than *c.* 25 Ma. Extensive exposures of tonalites are known from Komahashi–Daini and Minami–Koho seamounts on the Kyushu–Palau Ridge. These tonalites have clear arc signatures (Haraguchi *et al.* 2003) and ages of *c.* 37–38 Ma (Shibata *et al.* 1977).

Several DSDP and ODP sites sampled the Shikoku Basin (Fig. 1.1). Volcanic rocks recovered from drilling of the Shikoku Basin are basalts, which occur both as extrusive pillow lavas and intrusive sills. The basalts show a N-MORB-like signature (e.g. Dick *et al.* 1980). The volcanic rocks from the Kinan Seamount Chain have E-MORB to OIB-like signatures. The basaltic sills drilled at DSDP Site 444 also share these characteristics (Hickey-Vargas 1998). Volcanism of the Kinan Seamount Chain was active for *c.* 15–8 Ma, therefore occurring for at least 7 million years after spreading in the Shikoku Basin ceased (Ishizuka *et al.* 2009). This phase of volcanism therefore coincided with the earliest stage of the current volcanism in the Izu–Ogasawara and Mariana arcs. Ishizuka *et al.* (2009) concluded that the Kinan Seamount Chain basalts do not have the signature suggestive of significant slab input, and argued that the origin of the Kinan Seamount Chain is due to the

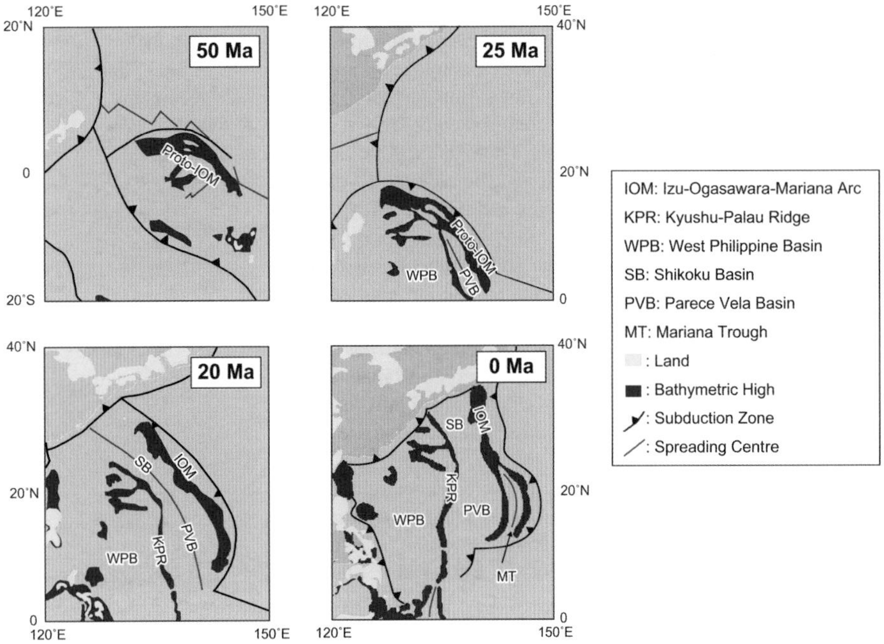

Fig. 1.15. Tectonic reconstruction of the Philippine Sea from 50 Ma to present (modified after Hall 2002).

contribution of enriched component in the asthenosphere after the cessation of Shikoku Basin spreading at *c.* 15 Ma. Post-spreading-enriched magmatism may be a common process during the last stage and after the cessation of back-arc spreading (Sato *et al.* 2002; Ishizuka *et al.* 2009).

The Shichito–Ioto Ridge is the active volcanic front of the Izu–Ogasawara Arc, bearing many Quaternary volcanoes from Izu–Oshima Island in the north to Minami–Ioto Island in the south. The modern magmatism at the Shichito–Ioto Ridge is bimodal with basalt and rhyolite predominating (Tamura & Tatsumi 2002), as evidenced by the presence of submarine calderas (Fiske *et al.* 2001) as well as the tephras obtained by ODP drilling (Fujioka *et al.* 1992). The volcanic rocks of the rear-arc *en echelon* seamount chains have clear arc signatures, showing andesitic as well as basaltic and dacitic compositions (Ishizuka *et al.* 1998; Hochstaedter *et al.* 2000, 2001). Volcanism forming the rear-arc *en echelon* seamount chains was active from *c.* 17 to 3 Ma (Ishizuka *et al.* 1998). The volcanic rocks from the Kinan Escarpment include basaltic lavas and dolerite. Major element compositions of the volcanic rocks from the Kinan Escarpment are similar to those of volcanic rocks from the rear-arc *en echelon* seamount chains (Ishizuka *et al.* 2009). The volcanic rocks from the Kinan Escarpment yield ages of *c.* 14–12 Ma, and constitute the oldest ages for the volcanic rocks from the rear-arc *en echelon* seamount chains, suggesting that arc volcanism has migrated to the east with time.

The Ogasawara Ridge is the Eocene component of the Izu–Ogasawara Arc, yielding the basalt that erupted during subduction initiation at *c.* 52 Ma (i.e. forearc basalt or FAB; Ishizuka *et al.* 2011*a*), and boninites and high-Mg andesites that erupted during the early stages of the Philippine Sea evolution (i.e. *c.* 48–46 Ma; Ishizuka *et al.* 2006).

Evolutionary history of the Philippine Sea Plate

The following evolutionary history for the region is proposed based on the compilation of the available data (Fig. 1.15; Okino *et al.* 1994, 1999; Hall 2002; Ishizuka *et al.* 2011*b*, 2013).

The Amami Plateau, Daito Ridge, Oki–Daito Ridge and Oki–Daito Rise are Cretaceous–early Eocene in age and represent the oldest components of the Philippine Sea Plate (Karig 1971, 1975; Mizuno *et al.* 1978). The West Philippine Basin is the oldest back-arc basin in the Philippine Sea, and developed by spreading along the now-extinct spreading ridge, the CBF Rift. Rifting started at *c.* 55 Ma and spreading ended at *c.* 30 Ma (Deschamps & Lallemand 2002). OIB-like volcanic activity occurred during the early evolutionary stage of the West Philippine Basin (Ishizuka *et al.* 2013), resulting in the formation of the Oki–Daito Ridge and Urdaneta Plateau, as well as the volcanic terrain in the Minami–Daito Basin and the bathymetric terrace of the Kyushu–Palau Ridge at *c.* 15 °N. The Benham Rise, located in the southern part of the Philippine Basin, is considered to be a counterpart of the Urdaneta Plateau (Hilde & Lee 1984). The proto-Izu–Ogasawara and Mariana arcs were formed in association with subduction along the east and north of the proto-West Philippine Basin. Rifting of the proto-Izu–Ogasawara and Mariana arcs during the Oligocene (*c.* 28–25 Ma) split the arcs into the remnant Kyushu–Palau Ridge and the active Izu–Ogasawara and Mariana arcs. This rifting evolved to the seafloor spreading that formed the Shikoku Basin from *c.* 25 to *c.* 15 Ma and the Parece Vela Basin from *c.* 25 to *c.* 7 Ma. Ar–Ar radiometric ages of the volcanic rocks from the main body of the Kyushu–Palau Ridge mostly fall within the range *c.* 28–25 Ma, further confirming that seafloor spreading of the Shikoku and Parece Vela basins initiated at or after *c.* 25 Ma (Ishizuka *et al.* 2011*b*). The Kinan Seamount Chain was mainly formed by post-spreading volcanism (Okino *et al.* 1994; Sato *et al.* 2002; Ishizuka *et al.* 2009) that started shortly before the end of the spreading in the Shikoku Basin. The current active Izu–Ogasawara and Mariana arcs, including that of the rear-arc *en echelon* seamount chains, were formed during the last stages of the Philippine Sea evolution.

Concluding remarks

The geological history of Japan starts in the early Palaeozoic as part of a continental margin with hints of the earlier history in the presence of older mineral grains and clasts. The location of the Palaeozoic margin with respect to the North China, South China and Australian cratons remains disputed. From the Mesozoic onwards, the eastern margin of the newly amalgamated Asia continent, including the future Japanese islands, became the site of oceanic plate convergence. This convergence resulted in the accretion of several distinct domains including the Akiyoshi Limestone and the Yakuno Ophiolite. A series of major accretionary complexes were subsequently added to these accreted terranes, the largest being the Mino–Tanba and Shimanto belts. Accretion is commonly associated with strong deformation and metamorphism and key examples are the subduction-type Sanbagawa Belt and the arc-type Ryoke Belt. Subduction of the Izanagi, Kula, Pacific and most recently the Philippine Sea plates is closely related to the Mesozoic and Cenozoic evolution of Japan, and the approach and subduction of the ridge boundaries between these plates is thought to strongly influence the geological history by many workers. The recent rapid increase in our knowledge of the modern seas surrounding Japan has revealed the complex nature of the oceanic domains around Japan, and this information will help to further our understanding of how interaction between oceanic and island arc domains are reflected in the geological record.

Some key questions remain to be answered concerning the geological evolution of Japan: (1) the original location of Japan and how it links with the cratons of East Asia and Australia, including the much-disputed eastward extension of the Sulu–Dabie Suture; (2) the role of large-scale strike-slip movements in the Mesozoic history of Japan (both left- and right-lateral displacements are proposed by some, while others emphasize the importance of strike-normal movements and downplay the importance of strike-slip tectonics); (3) the relationship between metamorphism and plate movements, in particular ridge subduction and the formation, deformation and exhumation of paired metamorphic belts; (4) the causes of back-arc spreading including the rapid opening of the Japan Sea; and (5) the origin of the oldest domain of the Philippine Sea Plate, the Amami–Daito Province.

We are grateful to the following people for their assistance in gathering data and preparing figures for this chapter: Tsuyoshi Yoshida, Kentaro Kaneda, Narumi Takahashi, Yuka Kaiho, Kiyoyuki Kisimoto and Hiroshi Mori. This chapter was improved by the comments of an anonymous referee and the editors.

Appendix

English to Kanji and Hiragana translations for geological and place names

Abukuma	阿武隈	あぶくま
Akashima	赤島	あかしま
Akita	秋田	あきた
Akiyoshi	秋吉	あきよし
Akka	安家	あっか
Amami	奄美	あまみ
Ashio	足尾	あしお
Ashizurimisaki	足摺岬	あしずりみさき
Bonin	ボニン	ぼにん
Butsuzo	仏像	ぶつぞう
Chichi-jima	父島	ちちじま
Chichibu	秩父	ちちぶ
Chikuho	筑豊	ちくほう
Chugoku	中国	ちゅうごく
Daiichi-Kashima	第一鹿島	だいいちかしま
Daijima	台島	だいじま
Daito	大東	だいとう
Fuji	富士	ふじ
Funafuse-yama	舟伏山	ふなふせやま
Funakawa	船川	ふなかわ

(*Continued*)

English to Kanji and Hiragana translations for geological and place names (Continued)

Futomiyama	太美山	ふとみやま
Gifu	岐阜	ぎふ
Gozaisho	御在所	ございしょ
Hayachine	早池峰	はやちね
Hichiso	七宗	ひちそう
Hida	飛騨	ひだ
Hida Gaien	飛騨外縁	ひだがいえん
Hidaka	日高	ひだか
Hikami	氷上	ひかみ
Hikoroichi	日頃市	ひころいち
Hirao	平尾	ひらお
Hitachi	常陸	ひたち
Hokkaido	北海道	ほっかいどう
Honshu	本州	ほんしゅう
Horoman	幌満	ほろまん
Hyogo	兵庫	ひょうご
Ibuki-yama	伊吹山	いぶきやま
Idonnappu	イドンナップ	いどんなっぷ
Ioto（Iwojima）	硫黄島	いおうとう
Iwate	岩手	いわて
Izanagi	イザナギ	いざなぎ
Izu	伊豆	いず
Izu-Oshima	伊豆大島	いずおおしま
Izumi	和泉	いずみ
Joban	常磐	じょうばん
Jodogahama	浄土ヶ浜	じょうどがはま
Kabato	樺戸	かばと
Kamiaso	上麻生	かみあそう
Kamuikotan	神居古潭	かむいこたん
Kanokura	叶倉	かのくら
Kanto	関東	かんとう
Kawauchi	川内	かわうち
Kii	紀伊	きい
Kinan	紀南	きなん
Kita-Daito	北大東	きただいとう
Kitakami	北上	きたかみ
Koma	巨摩	こま
Komahashi-Daini	駒橋第二	こまはしだいに
Kuroko	黒鉱	くろこう
Kurosegawa	黒瀬川	くろせがわ
Kuroshio	黒潮	くろしお
Kuruma	来馬	くるま
Kuzuu	葛生	くずう
Kyushu	九州	きゅうしゅう
Maizuru	舞鶴	まいづる
Minami-Daito	南大東	みなみだいとう
Minami-Ioto	南硫黄島	みなみいおうとう
Minami-Koho	南高鵬	みなみこうほう
Mino	美濃	みの
Misaka	御坂	みさか
Miyagi	宮城	みやぎ
Miyako-jima	宮古島	みやこじま
Miyamori	宮守	みやもり
Miyara	宮良	みやら
Miyazaki	宮崎	みやざき
Monzen	門前	もんぜん
Muroto	室戸	むろと
Nagaiwa	長岩	ながいわ
Nakazato	中里	なかざと
Nankai	南海	なんかい
Nedamo	根田茂	ねだも
Nemuro	根室	ねむろ

(*Continued*)

English to Kanji and Hiragana translations for geological and place names (Continued)

Niigata	新潟	にいがた
Nishikurosawa	西黒沢	にしくろさわ
Nobi	濃尾	のうび
Nohi	濃飛	のうひ
Nosoko	野底	のそこ
Oeyama	大江山	おおえやま
Ofunato	大船渡	おおふなと
Oga	男鹿	おが
Ogasawara	小笠原	おがさわら
Okayama	岡山	おかやま
Oki	隠岐	おき
Oki-Daito	沖大東	おきだいとう
Okinawa	沖縄	おきなわ
Okuhinotsuchi	奥火の土	おくひのつち
Onimaru	鬼丸	おにまる
Onnagawa	女川	おんながわ
Ono	大野	おおの
Onogawa	大野川	おおのがわ
Osaka	大阪	おおさか
Oshima	渡島	おしま
Rebun	礼文	れぶん
Rikuzen-Takata	陸前高田	りくぜんたかた
Ryoke	領家	りょうけ
Ryoseki	領石	りょうせき
Ryukyu	琉球	りゅうきゅう
Sakamotozawa	坂本沢	さかもとざわ
Sanbosan (Sambosan)	三宝山	さんぼうさん
Sanyo	山陽	さんよう
Sendai	仙台	せんだい
Setamai	世田米	せたまい
Shichito	七島	しちとう
Shikoku	四国	しこく
Shimajiri	島尻	しまじり
Shimanto	四万十	しまんと
Sorachi	空知	そらち
Taishaku	帝釈	たいしゃく
Tanba (Tamba)	丹波	たんば
Tanzawa	丹沢	たんざわ
Tetori	手取	てとり
Tohoku	東北	とうほく
Tokyo	東京	とうきょう
Tome	登米	とめ
Toyoma	登米	とよま
Ube	宇部	うべ
Yaeyama	八重山	やえやま
Yakuno	夜久野	やくの
Yezo	蝦夷	えぞ
Yubari	夕張	ゆうばり
Zomeki	蔵目喜	ぞうめき

References

Adachi, S. & Igo, H. 1980. A new Ordovician leperditiid ostracode from Japan. *Proceedings of the Japan Academy Series B*, **56**, 504–507.

Aiba, J. & Sekiya, E. 1979. Distribution and characteristics of the Neogene sedimentary basins around the Nansei-Shoto (Ryukyu Islands). *Journal of the Japanese Association of Petroleum Technologists*, **44**, 97–108.

Amano, K. & Ito, K. 1990. Construction of the South Fossa Magna – Collision and accretion tectonics of arcs deduced from the sedimentary sequences. *Memoirs of the Geological Society of Japan*, **34**, 45–56 [in Japanese with English abstract].

Aoya, M., Uehara, S-I., Matsumoto, M., Wallis, S. R. & Enami, M. 2003. Subduction-stage pressure-temperature path of eclogite from the Sambagawa belt: prophetic record for oceanic-ridge subduction. *Geology*, **31**, 1045–1048.

Arita, K., Ikawa, T. *et al.* 1998. Crustal structure and tectonics of the Hidaka collision zone, Hokkaido (Japan), revealed by vibroseis seismic reflection and gravity surveys. *Tectonophysics*, **290**, 197–210.

Ayalew, D. & Ishiwatari, A. 2011. Comparison of rhyolites from continental rift, continental arc and oceanic island arc: implication for the mechanism of silicic magma generation. *Island Arc*, **20**, 78–93.

DeMets, C., Gordon, R. G., Argus, D. F. & Stein, S. 1994. Effect of recent revisions to the geomagnetic reversal time scale on estimates of current plate motions. *Geophysical Research Letters*, **21**, 2191–2194.

Deschamps, A. & Lallemand, S. 2002. The West Philippine Basin: an Eocene to early Oligocene back arc basin opened between two opposed subduction zones. *Journal of Geophysical Research*, **107**, 2322, http://doi.org/10.1029/2001JB001706

Dick, H. J. B., Marsh, N. G. & Bullen, T. D. 1980. Deep Sea Drilling Project Leg 58: abyssal basalts from the Shikoku Basin, their petrology and major element geochemistry. *In*: deVries Klein, G., Kobayashi, K. *et al.* (eds) *Initial Report DSDP*. Washington, DC, US Government Printing Office, **58**, 843–872.

Dilek, Y. & Furnes, H. 2011. Ophiolite genesis and global tectonics: geochemical and tectonic fingerprinting of ancient oceanic lithosphere. *Geological Society of America Bulletin*, **123**, 387–411.

Engerbretson, D. C., Cox, A. & Gordon, R. G. 1985. Relative motions between oceanic plates and continental plates in the Pacific Basin. *Geological Society of America Special Publication*, **206**, 59.

Erdenesaihan, G., Ishiwatari, A., Orolmaa, D., Arai, S. & Tamura, A. 2013. Middle Paleozoic greenstones in the Hangay region, central Mongolia: remnants of an accreted oceanic plateau and forearc magmatism. *Journal of Mineralogical and Petrological Sciences*, **108**, 303–325.

Fiske, R. S., Naka, J., Iizasa, K., Yuasa, M. & Klaus, A. 2001. Submarine silicic caldera at the front of the Izu-Bonin arc, Japan: voluminous seafloor eruptions of rhyolite pumice. *Geological Society of America Bulletin*, **113**, 813–824.

Fujioka, K., Nishimura, A., Matsuo, Y. & Rodolfo, K. S. 1992. Correlation of Quaternary tephras throughout the Izu-Bonin areas. *In*: Taylor, B., Fujioka, K. *et al.* (eds) *Proceedings of ODP, Scientific Results*. College Station, TX, Ocean Drilling Program, **126**, 23–45.

Fujioka, K., Okino, K., Kanamatsu, T., Ohara, Y., Ishizuka, O., Haraguchi, S. & Ishii, T. 1999. An enigmatic extinct spreading center in the West Philippine backarc basin unveiled. *Geology*, **27**, 1135–1138.

Fujiwara, T., Kodaira, S., No, T., Kaiho, Y., Takahashi, N. & Kaneda, Y. 2011. The 2011 Tohoku-Oki Earthquake: displacement reaching the trench axis. *Science*, **334**, 1240.

Hacker, B. R., Wallis, S. R., Ratschbacher, L., Grove, M. & Gehrels, G. 2006. High-temperature geochronology constraints on the tectonic history and architecture of the ultrahigh-pressure Dabie-Sulu Orogen. *Tectonics*, **25**, http://doi.org/10.1029/2005TC001937

Hall, R. 2002. Cenozoic geological and plate tectonic evolution of SE Asia and the SW Pacific: computer-based reconstructions, model and animations. *Journal of Asian Earth Sciences*, **20**, 353–431.

Haraguchi, S., Ishii, T., Kimura, J. & Ohara, Y. 2003. Formation of tonalite from basaltic magma at the Komahashi-Daini Seamount, northern Kyushu-Palau Ridge in the Philippine Sea, and growth of Izu-Ogasawara (Bonin)-Mariana arc crust. *Contributions to Mineralogy and Petrology*, **145**, 151–168.

Harayama, S., Ohyabu, K., Miyama, Y., Adachi, H. & Shukuwa, R. 2003. Eastward tilting and uplifting after the late Early Pleistocene in the eastern-half area of the Hida Mountain Range. *Quaternary Research*, **42**, 127–140.

Hickey-Vargas, R. 1998. Origin of the Indian Ocean-type isotopic signature in basalts from Philippine Sea plate spreading centers: an assessment of local v. large-scale processes. *Journal of Geophysical Research*, **103**, 20 963–20 979.

Hickey-Vargas, R. 2005. Basalt and tonalite from the Amami Plateau, Northern West Philippine Basin: new early Cretaceous age and geochemical results and their petrologic and tectonic implications. *Island Arc*, **14**, 653–665.

Hilde, T. W. C. & Lee, C. S. 1984. Origin and evolution of the West Philippine Basin: a new interpretation. *Tectonophysics*, **102**, 85–104.

Hirano, N. 2011. Petit-spot volcanism: a new type of volcanic zone discovered near a trench. *Geochemical Journal*, **45**, 157–167.

Hirano, N., Takahashi, E. *et al.* 2006. Volcanism in response to plate flexure. *Science*, **313**, 1426–1428.

Hiroi, Y., Fuji, N. & Okimura, Y. 1979. New Fossil discovery from the Hida metamorphic rocks in the Unazuki area, central Japan. *Proceedings of the Japan Academy, Series B*, **54**, 268–271.

Hirose, W. & Nakagawa, M. 1999. Neogene volcanism in central-eastern Hokkaido: beginning and evolution of arc volcanism inferred from volcanological parameters and geochemistry. *Journal of Geological Society of Japan*, **105**, 247–265 [in Japanese with English abstract].

Hochstaedter, A. G., Gill, J. B., Taylor, B., Ishizuka, O., Yuasa, M. & Morita, S. 2000. Across-arc geochemical trends in the Izu-Bonin arc: constraints on source composition and mantle melting. *Journal of Geophysical Research*, **105**, 495–512.

Hochstaedter, A. G., Gill, J. B., Peters, R., Broughton, P., Holden, P. & Taylor, B. 2001. Across-arc geochemical trends in the Izu-Bonin arc: contributions from the subducting slab. *Geochemistry, Geophysics, Geosystems*, **2**, 1019, http://doi.org/10.1029/2000GC000105

Huchon, P. & Kitazato, H. 1984. Collision of the Izu block with central Japan during the Quaternary and geological evolution of the Ashigara area. *Tectonophysics*, **110**, 201–210.

Hujioka, K. 1959. *Explanatory Text of the Geological Map of Japan, Scale 1:50,000, Toga & Funakawa, Akita Nos. 1, 2*. Geological Survey of Japan, Kawasaki [in Japanese with English abstract].

Ichiyama, Y., Ishiwatari, A. & Koizumi, K. 2008. Petrogenesis of greenstones from the Mino-Tamba belt, SW Japan: evidence for an accreted Permian oceanic plateau. *Lithos*, **100**, 127–146.

Igo, H., Adachi, S., Furutani, H. & Nishiyama, H. 1980. Ordovician fossils first discovered in Japan. *Proceedings of the Japan Academy, Series B*, **56**, 499–503.

Imai, R., Sato, T. & Iryu, Y. 2013. Chronological and paleoceanographic constraints of Miocene to Pliocene 'mud sea' in the Ryukyu Islands (southwestern Japan) based on calcareous nannofossil assemblages. *Island Arc*, **22**, 522–537.

Ishihara, T. & Koda, K. 2007. Variation of crustal thickness in the Philippine Sea deduced from three-dimensional gravity modeling. *Island Arc*, **16**, 322–337.

Ishiwatari, A. 1991. Ophiolites in the Japanese islands: typical segment of the circum-Pacific multiple ophiolite belts. *Episodes*, **14**, 274–279.

Ishiwatari, A. & Tsujimori, T. 2003. Paleozoic ophiolites and blueschists in Japan and Russian Primorye in the tectonic framework of East Asia: a synthesis. *Island Arc*, **12**, 190–206.

Ishiwatari, A. & Tsujimori, T. 2012. Japan and the 250 Ma continental collision belt in East Asia: Yaeyama promontory hypothesis revisited. *Journal of Geography*, **121**, 460–470 [in Japanese with English abstract].

Ishiwatari, A., Sokolov, S. D. & Vysotskiy, S. V. 2003. Petrological diversity and origin of ophiolites in Japan and Far East Russia with emphasis on depleted harzburgite. *In*: Dilek, Y. & Robinson, P. T. (eds) *Ophiolites in Earth History*. Geological Society, London, Special Publications, **218**, 597–618.

Ishizuka, O., Uto, K., Yuasa, M. & Hochstaedter, A. G. 1998. K–Ar ages from the seamount chains in the back-arc region of the Izu-Ogasawara arc. *Island Arc*, **7**, 408–421.

Ishizuka, O., Kimura, J.-I. *et al.* 2006. Early stages in the evolution of Izu-Bonin arc volcanism: new age, chemical and isotopic constraints. *Earth and Planetary Science Letters*, **250**, 385–401.

Ishizuka, O., Yuasa, M., Taylor, R. N. & Sakamoto, I. 2009. Two contrasting magmatic types coexist after the cessation of back-arc spreading. *Chemical Geology*, **266**, 283–305.

Ishizuka, O., Tani, K. *et al.* 2011*a*. The timescales of subduction initiation and subsequent evolution of an oceanic island arc. *Earth and Planetary Science Letters*, **306**, 229–240.

Ishizuka, O., Taylor, R. N., Yuasa, M. & Ohara, Y. 2011*b*. Making and breaking an island arc: a new perspective from the Oligocene Kyushu-Palau arc, Philippine Sea. *Geochemistry, Geophysics, Geosystems*, **12**, Q05005, http://doi.org/10.1029/2010GC003440

Ishizuka, O., Taylor, R. N., Ohara, Y. & Yuasa, M. 2013. Upwelling, rifting, and age-progressive magmatism from the Oki-Daito mantle plume. *Geology*, **41**(9), 1011–1014.

Isozaki, Y. 1996. Anatomy and genesis of a subduction related orogeny: a new view of geotectonic subdivision and evolution of the Japanese Islands. *Island Arc*, **5**, 289–320.

Isozaki, Y. 1997. Jurassic accretion tectonics of Japan. *Island Arc*, **6**, 25–51.

Isozaki, Y., Maruyama, S. & Furuoka, F. 1990. Accreted oceanic materials in Japan. *Tectonophysics*, **181**, 179–205.

Isozaki, Y., Hashimoto, T. & Itaya, T. 1992. The Kurosegawa klippe: an examination. *Journal of Geological Society of Japan*, **98**, 917–941 [in Japanese with English abstract].

Isozaki, Y., Aoki, K., Nakama, T. & Yanai, S. 2010. New insight into a subduction-related orogen: a reappraisal of the geotectonic framework and evolution of the Japanese Islands. *Gondwana Research*, **18**, 82–105.

Ito, H., Yamada, R., Tamura, A., Arai, S., Horie, K & Hokada, T. 2013. *Earth's Youngest Exposed Granite and Its Tectonic Implications: The 10–0.8 Ma Kurobegawa Granite*. Scientific Reports 3, Article number: 1306, http://doi.org/10.1038/srep01306

Ito, T. & Sato, H. 2010. Crustal structure of the trench-island arc-backarc sea system from the Nankai Trough to the northern margin of the Yamato Basin, Southwest Japan. *Journal of Geography*, **119**, 235–244 [in Japanese with English abstract].

Iwasaki, T., Hirata, N. *et al.* 1990. Crustal and upper mantle structure in the Ryukyu Island Arc deduced from deep seismic sounding. *Geophysical Journal International*, **102**, 631–651.

Iwasaki, T., Kato, W. *et al.* 2001. Extensional structure in Northern Honshu Arc as inferred from seismic refraction/wide-angle reflection profiling. *Geophysical Research Letters*, **28**, 2329–2332.

Kagami, H., Kawano, Y. *et al.* 2006. Provenance of Paleozoic-Mesozoic sedimentary rocks in the Inner Zone of Southwest Japan: an evaluation based on Nd model ages. *Gondwana Research*, **9**, 142–151.

Kanaori, Y., Kawakami, S. & Yairi, K. 1992. The block structures and Quaternary strike-slip block rotation of central Japan. *Tectonics*, **11**, 47–56.

Kanmera, K., Sano, H. & Isozaki, Y. 1990. Akiyoshi terrane. *In*: Ichikawa, K., Mizutani, S. *et al.* (eds) *Pre-Cretaceous Terranes of Japan*. International Geological Correlation Program, Osaka, Project No. **224**, 49–62.

Kano, K., Tani, K., Iwano, H., Danbara, T., Ishizuka, O., Ohguchi, T. & Dunkleu, D. J. 2012. Radiometric ages of the Akashima Formation, Oga Peninsula, NE Japan. *Journal of the Geological Society of Japan*, **118**, 351–364 [in Japanese with English abstract].

Karig, D. E. 1971. Origin and development of marginal basins in the Western Pacific. *Journal of Geophysical Research*, **76**, 2542–2561.

Karig, D. E. 1975. Basin genesis in the Philippine Sea. *In*: Karig, D. E., Ingle, J. C. Jr. *et al.* (eds) *Initial Report of DSDP*. Washington, DC, US Government Printing Office, **31**, 857–879.

Kasuga, S. & Ohara, Y. 1997. Evolution of the escarpments located in the backarc basins in the southern waters of Japan: mechanical boundary between sinking backarc basins and buoyant active island arc. *Report of Hydrographic Researches*, **33**, 39–51 [in Japanese with English abstract].

Kemp, A. I. S., Shimura, T. & Hawkesworth, C. J. 2007. Linking granulites, silicic magmatism, and crustal growth in arcs: ion microprobe (zircon) U–Pb ages from the Hidaka metamorphic belt, Japan. *Geology*, **5**, 807–810.

Kimbrough, D. L., Mattinson, J. M., Coombs, D. S., Landis, C. A. & Johnston, M. R. 1992. Uranium-lead ages from the Dun Mountain ophiolite belt and Brooks Street terrane, South Island, New Zealand. *Geological Society of America Bulletin*, **104**, 429–443.

Kimura, G. 1986. Oblique subduction and collision: forearc tectonics of the Kuril arc. *Geology*, **14**, 404–407.

Kimura, K. & Hori, R. 1993. Offscraping accretion of Jurassic chert clastic complexes in the Mino-Tamba Belt, central Japan. *Journal of Structural Geology*, **15**, 145–161.

Kita, S., Hasegawa, A., Nakajima, J., Okada, T., Matsuzawa, T. & Katsumata, K. 2012. High-resolution seismic velocity structure beneath the Hokkaido corner, northern Japan: arc-arc collision and origins of the 1970 M 6.7 Hidaka and 1982 M 7.1 Urakawa-oki earthquakes. *Journal of Geophysical Research*, **117**, B12301, http://doi.org/10.1029/2012JB009356

Kiyokawa, S. 1992. Geology of the Idonnappu Belt, Hokkaido, Japan: evolution of a Cretaceous accretionary complex. *Tectonics*, **11**, 1180–1026.

Kobayashi, N., Ohguchi, T. & Kano, K. 2008. Stratigraphic revision of the Monzen Formation, Oga Peninsula, NE Japan. *Bulletin of the Geological Survey of Japan*, **59**, 211–224 [in Japanese with English abstract].

Kodaira, S., Sato, T., Takahashi, N., Ito, A., Tamura, Y., Tatsumi, Y. & Kaneda, Y. 2007. Seismological evidence for variable growth of crust along the Izu intraoceanic arc. *Journal of Geophysical Research*, **112**, B05104, http://doi.org/10.1029/2006JB004593

Koizumi, K. & Ishiwatari, A. 2006. Oceanic plateau accretion inferred from Late Paleozoic greenstones in the Jurassic Tamba accretionary complex, southwest Japan. *Island Arc*, **15**, 58–83.

Komatsu, M., Osanai, Y., Toyoshima, T. & Miyashita, S. 1989. Evolution of the Hidaka metamorphic belt, northern Japan. *In*: Daly, J. S., Cliff, R. A. & Yardley, B. W. D. (eds) *Evolution of Metamorphic Belts*. Geological Society, London, Special Publications, **43**, 487–493.

Kunugiza, K., Goto, A. & Yokoyama, K. 2002. *Tectonics and Paleogeography of the Tetori Group. Geological Report of the Mesozoic Tetori Formation in the Tetori River area*. Ishikawa Prefecture [in Japanese].

Lewis, J. C. & Byrne, T. B. 2001. Fault kinematics and past plate motions at a convergent plate boundary: tertiary Shimanto Belt, southwest Japan. *Tectonics*, **20**, 1066–1067.

Maruyama, S., Isozaki, Y., Kimura, G. & Terabayashi, M. 1997. Paleogeographic maps of the Japanese Islands: plate tectonic synthesis from 750 Ma to the present. *Island Arc*, **6**, 121–142.

Matsubara, T. 2013. Molluscan fauna of the 'Miocene' Namigata Formation in the Namigata area, Okayama Prefecture, southwest Japan. *Journal of the Geological Society of Japan*, **119**, 249–266.

Matsuda, T. & Isozaki, Y. 1991. Well-documented travel history of Mesozoic Pelagic Chert in Japan – from Remote Ocean to Subduction Zone. *Tectonics*, **10**, 475–499.

Minato, M., Gorai, M. & Hunahashi, M. (eds) 1965. *The Geologic Development of the Japanese Islands*. Tsukiji Shokan, Tokyo.

Miyashiro, A. 1961. Evolution of metamorphic belts. *Journal of Petrology*, **2**, 277–311.

Miyashiro, A. 1973. Paired and unpaired metamorphic belts. *Tectonophysics*, **17**, 241–254.

Miyauchi, H. & Moriya, T. 1984. Velocity structure beneath Hidaka Mountains in Hokkaido, Japan. *Journal of Physics of the Earth*, **32**, 13–42.

Mizuno, A., Okuda, Y., Nagumo, S., Kagami, H. & Nasu, N. 1978. Subsidence of the Daito Ridge and associated basins, North Philippine Sea. *In*: Watkins, J. S., Montadert, L. & Dickerson, P. W. (eds) *Geological and Geophysical Investigations of Continental Margins*. American Association of Petroleum Geologists Memoir, Tulsa, Oklahoma, **29**, 239–243.

Mori, K., Okami, K. & Ehiro, M. 1992. Paleozoic and Mesozoic sequences in the Kitakami Mountains. 29th IGC Field Trip Guide Book, Nagoya University (A05), **1**, 81–114.

Morinaga, Y. & Ota, M. 1971. Subsurface geology of the Akiyoshi limestone group in the Maki and Kyoei area, Shuho Town, Southwest Japan. *Bulletin of Akiyoshi-dai Museum of Natural History*, **7**, 25–56.

Nakagawa, H., Doi, N., Shirao, M. & Araki, Y. 1982. Geology of Ishigakijima and Iriomote-jima. Yaeyama Gunto, Ryukyu Islands. *Contribution from the Institute of Geology and Paleontology, Tohoku University*, **84**, 1–22 [in Japanese with English abstract].

Nakajima, T. 1994. The Ryoke Plutonometamorphic Belt – crustal section of the Cretaceous Eurasian Continental-Margin. *Lithos*, **33**, 51–66.

Nakamura, K. 1983. Possible nascent trench along the eastern Japan Sea as the convergent boundary between Eurasian and North American plates. *Bulletin of Earthquake Research Institute, University of Tokyo*, **58**, 711–722.

Narita, K., Yamaji, T., Tagami, T., Kurita, H., Obuse, A. & Matsuoka, K. 1999. The sedimentary age of the Tertiary Kuma Group in Shikoku and its significance. *Journal of the Geological Society of Japan*, **105**, 305–308.

Niida, K. 1992. Basalts and dolerites in the Sorachi Yezo Group, Central Hokkaido, Japan. *Journal of Faculty of Science, Hokkaido University, Series IV*, **23**, 301–319.

Niida, K. & Kito, N. 1986. Cretaceous arc-trench systems in Hokkaido. *Monograph of the Association for Geological Collaboration in Japan*, **31**, 379–402 [in Japanese with English abstract].

Nishizawa, A., Kaneda, K., Nakanishi, A., Takahashi, N. & Kodaira, S. 2006. Crustal structure of the ocean-island arc transition at the mid Izu-Ogasawara (Bonin) arc margin. *Earth Planets Space*, **58**, e33–e36.

Nishizawa, A., Kaneda, K., Katagiri, Y. & Kasahara, J. 2007. Variation in crustal structure along the Kyushu-Palau Ridge at 15–21°N on the Philippine Sea Plate based on seismic refraction profiles. *Earth Planets Space*, **59**, e17–e20.

Nishizawa, A., Kaneda, K. & Oikawa, M. 2011. Backarc basin oceanic crust and uppermost mantle seismic velocity structure of the Shikoku Basin, south of Japan. *Earth Planets Space*, **63**, 151–155.

Nishizawa, A., Kaneda, K., Katagiri, Y. & Oikawa, M. 2014. Wide-angle refraction experiments in the Daito Ridges region at the northwestern end of the Philippine Sea plate. *Earth Planets and Space*, **66**, 25, http://doi.org/10.1186/1880-5981-66-25

Noda, A. & Toshimitsu, S. 2009. Backward stacking of submarine channel-fan successions controlled by strike-slip faulting: the Izumi Group (Cretaceous), southwest Japan. *Lithosphere*, **1**, 41–59.

Nutman, A. P., Sano, Y., Terada, K. & Hidaka, H. 2006. 743+/−17 Ma granite clast from Jurassic conglomerate, Kamiaso, Mino Terrane, Japan: the case for South China Craton provenance (Korean Gyeonggi Block?). *Journal of Asian Earth Sciences*, **26**, 99–104.

Ogawa, Y. & Taniguchi, H. 1988. Geology and tectonics of the Miura-Boso Peninsulas and the adjacent area. *Modern Geology*, **12**, 147–168.

Ogawa, Y., Nishida, Y. & Makino, M. 1994. A collision boundary imaged by magnetotellurics, Hidaka Mountains, central Hokkaido, Japan. *Journal of Geophysical Research*, **99**, 22 373–22 388.

Ohara, Y., Kasuga, S., Okino, K. & Kato, Y. 1997. Survey maps Philippine Sea structure. *EOS Transactions, American Geophysical Union*, **78**, 555.

Ohara, Y., Tokuyama, H. & Stern, R. J. 2007. Thematic section: geology and geophysics of the Philippine Sea and adjacent areas in the Pacific Ocean. *Island Arc*, **16**, 319–321.

Ohtake, M., Taira, A. & Ota, Y. 2002. *Active Faults and Seismotectonics along the Eastern Margin of the Japan Sea*. University of Tokyo Press, Tokyo.

Okada, H. & Sakai, T. 2000. The Cretaceous System of the Japanese Island and its physical environments. *In*: Okada, H. & Mateer, N. J. (eds) *Cretaceous Environments of Asia*. Elsevier, Amsterdam, 113–144.

Okamura, Y., Watanabe, M., Morijiri, R. & Satoh, M. 1995. Rifting and basin inversion in the eastern margin of the Japan Sea. *The Island Arc*, **4**, 166–181.

Okawa, H., Shimojo, M. et al. 2013. Detrital zircon geochronology of the Silurian-Lower Cretaceous continuous succession of the South Kitakami Belt, Northeast Japan. *Memoir of the Fukui Prefectural Dinosaur Museum*, **12**, 35–78.

Okino, K., Shimakawa, Y. & Nagaoka, S. 1994. Evolution of the Shikoku Basin. *Journal of Geomagnetism and Geoelectricity*, **46**, 463–479.

Okino, K., Kasuga, S. & Ohara, Y. 1998. A new scenario of the Parece Vela Basin genesis. *Marine Geophysical Researches*, **20**, 21–40.

Okino, K., Ohara, Y., Kasuga, S. & Kato, Y. 1999. The Philippine Sea: new survey results reveal the structure and the history of the marginal basins. *Geophysical Research Letters*, **26**, 2287–2290.

Osozawa, S. 1994. Plate reconstruction based upon age data of Japanese accretionary complexes. *Geology*, **22**, 1135–1138.

Otofuji, Y. & Matsuda, T. 1983. Paleomagnetic evidence for the clockwise rotation of Southwest Japan. *Earth and Planetary Science Letters*, **62**, 349–359.

Ozawa, K. 1988. Ultramafic tectonite of the Miyamori ophiolitic complex in the Kitakami Mountains, Northeast Japan: hydrous upper mantle in an island arc. *Contributions to Mineralogy and Petrology*, **99**, 159–175.

Ozawa, Y. 1925. Paleontological and stratigraphical studies on the Permo-Carboniferous limestone of Nagato, Pt. II. Paleontology. *The Journal of the College of Science, Imperial University of Tokyo, Japan*, **45**, 1–90.

Park, J.-O., Tokuyama, H., Shinohara, M., Suyehiro, K. & Taira, A. 1998. Seismic record of tectonic evolution and backarc rifting in the southern Ryukyu island arc system. *Tectonophysics*, **294**, 21–42.

Park, J.-O., Tsuru, T., Kaneda, Y., Kono, Y., Kodaira, S., Takahashi, N. & Kinoshita, H. 1999. A subducting seamount beneath the Nankai accretionary prism off Shikoku, southwestern Japan. *Geophysical Research Letters*, **26**, 931–934.

Saito, S. 1992. Stratigraphy of Cenozoic strata in the southern terminus area of Boso Peninsula, Central Japan. *Contribution from the Institute of Geology and Paleontology, Tohoku University*, **93**, 1–37 [in Japanese with English abstract].

Sano, H. & Kanmera, K. 1991. Collapse of ancient oceanic reef complex: what happened during collision of Akiyoshi reef complex? *Journal of Geological Society of Japan*, **97**, 297–309.

Sano, Y., Hidaka, H., Terada, K., Shimizu, H. & Suzuki, M. 2000. Ion microprobe U-Pb zircon geochronology of the Hida gneiss: finding of the oldest minerals in Japan. *Geochemical Journal*, **34**, 135–153.

Sato, H., Machida, S., Kanayama, S., Taniguchi, H. & Ishii, T. 2002. Geochemical and isotopic characteristics of the Kinan Seamount Chain in the Shikoku Basin. *Geochemical Journal*, **36**, 519–526.

Sato, H., Onoue, T., Nozaki, T. & Suzuki, K. 2013. Osmium isotope evidence for a large Late Triassic impact event. *Nature Communications*, **4**, http://doi.org/10.1038/ncomms3455

Sato, M., Ishikawa, T., Ujihara, N., Yoshida, S., Fujita, M., Mochizuki, M. & Asada, A. 2011. Displacement above the hypocenter of the 2011 Tohoku-oki earthquake. *Science*, **332**, 1395.

Sato, T., Takahashi, N., Miura, S., Fujie, G., Kang, D-H., Kodaira, S. & Kaneda, Y. 2006. Last stage of the Japan Sea back-arc opening deduced from the seismic velocity structure using wide-angle data. *Geochemistry, Geophysics, Geosystems*, **7**, Q06004, http://doi.org/10.1029/2005 GC001135

Savov, I. P., Hickey-Vargas, R., D'Antonio, M., Ryan, J. G. & Spadea, P. 2006. Petrology and geochemistry of West Philippine Basin basalts and early Palau-Kyushu arc volcanic clasts from ODP leg 195, Site 1201D: implications for the early history of the Izu-Bonin-Mariana Arc. *Journal of Petrology*, **47**, 277–299.

Shibata, K. & Adachi, M. 1974. Rb–Sr whole-rock ages of Precambrian metamorphic rocks in the Kamiaso conglomerate from central Japan. *Earth and Planetary Science Letters*, **21**, 277–287.

Shibata, K., Mizuno, A., Yuasa, M., Uchiumi, S. & Nakagawa, T. 1977. Further K-Ar dating of tonalite dredged from the Komahashi-Daini Seamount. *Bulletin of Geological Survey of Japan*, **28**, 1–4.

Shimizu, H., Lee, S.-G., Masuda, A. & Adachi, M. 1996. Geochemistry of Nd and Ce isotopes and REE abundances in Precambrian orthogneiss clasts from the Kamiaso conglomerate, central Japan. *Geochemical Journal*, **30**, 57–69.

Shinjo, R., Woodhead, J. D. & Hergt, J. M. 2000. Geochemical variation within the northern Ryukyu Arc: magma source compositions and geodynamic implications. *Contributions to Mineralogy and Petrology*, **140**, 263–282.

Spaggiari, C. V., Gray, D. R. & Foster, D. A. 2003. Tethyan- and Cordilleran-type ophiolites of eastern Australia: implications for the evolution of Tasmanides. *In*: Dilek, Y. & Robinson, P. T. (eds) *Ophiolites in Earth History*. Geological Society, London, Special Publications, **218**, 517–540.

Suyehiro, K., Takahashi, N. *et al.* 1996. Continental crust, crustal underplating, and low-Q upper mantle beneath an oceanic island arc. *Science*, **272**, 390–392.

Suzuki, K. & Adachi, M. 1998. Denudation history of the high T/P Ryoke metamorphic belt, southwest Japan: constraints from CHIME monazite ages of gneisses and granitoids. *Journal of Metamorphic Geology*, **16**, 23–37.

Suzuki, K., Adachi, M. & Tanaka, T. 1991. Middle Precambrian provenance of Jurassic sandstone in the Mino Terrane, central Japan: Th-U-total Pb evidence from an electron microprobe monazite study. *Sedimentary Geology*, **75**, 141–147.

Suzuki, S. 1997. Petrology of anthracite in the Permian Funabuseyama Limestone, Mino Belt, central Japan. *Journal of Geological Society of Japan*, **103**, 869–879 (in Japanese with English abstract).

Tagiri, M., Dunkley, D. J., Adachi, T., Hiroi, Y. & Fanning, C. M. 2011. SHRIMP dating of magmatism in the Hitachi metamorphic terrane, Abukuma Belt, Japan: evidence for a Cambrian volcanic arc. *Island Arc*, **20**, 259–279.

Taira, A. 2001. Tectonic evolution of the Japanese Island arc system. *Annual Review of Earth and Planetary Sciences*, **29**, 109–134.

Taira, A., Okada, H., Whitaker, J. H. & Smith, A. J. 1982. The Shimanto Belt of Japan: Cretaceous-lower Miocene active-margin sedimentation. *In*: Leggett, J. K. (ed.) *Trench-Forearc Geology: Sedimentation and Tectonics on Modern and Ancient Active Plate Margins*. Geological Society, London, Special Publications, Blackwell Scientific Publications, Osney Mead, Oxford, **10**, 5–26.

Taira, A., Katto, J., Tashiro, M., Okamura, M. & Kodama, K. 1988. The Shimanto Belt in Shikoku Japan – evolution of Cretaceous to Miocene Accretionary Prism. *Modern Geology*, **12**, 5–46.

Taira, A., Kiyokawa, S., Aoike, K. & Saito, S. 1997. Accretion tectonics of the Japanese islands and evolution of continental crust. *Comptes Rendus de l'Academie des Sciences Serie II Fascicule A: Sciences de la Terre et des Planetes*, **325**, 467–478.

Takagi, H. 1986. Implications of mylonitic microstructures for the geotectonic evolution of the Median Tectonic Line, Central Japan. *Journal of Structural Geology*, **8**, 3–14.

Takahashi, N., Suyehiro, K. & Shinohara, M. 1998. Implications from the seismic crustal structure of the northern Izu-Bonin arc. *Island Arc*, **7**, 383–394.

Takahashi, N., Kodaira, S., Klemperer, S. L., Tatsumi, Y., Kaneda, Y. & Suyehiro, K. 2007. Crustal structure and evolution of the Mariana intra-oceanic island arc. *Geology*, **35**, 203–206.

Tamaki, K., Suyehiro, K., Allan, J., Ingle, J. C. & Pisciotto, K. A. 1992. Tectonic synthesis and implications of Japan Sea ODP drilling. *In*: Tamaki, K., Suyehiro, K. *et al.* (eds) *Proceedings of ODP, Scientifc Results*. Pt. 2, College Station, TX, Ocean Drilling Program, **127/128**, 1333–1348.

Tamura, Y. & Tatsumi, Y. 2002. Remelting of an andesitic crust as a possible origin for rhyolitic magma in oceanic arcs: an example from the Izu-Bonin arc. *Journal of Petrology*, **43**, 1029–1047.

Tani, K., Ishizuka, O. *et al.* 2012. Izu-Bonin Arc: intra-oceanic from the beginning? Unraveling the crustal structure of the Mesozoic proto-Philippine Sea Plate. *AGU 2012 Fall Meeting*, Abstract T11B-2569.

Tatsumi, Y., Sakuyama, M., Fukuyama, H. & Kushiro, I. 1983. Generation of arc basalt magmas and thermal structure of the mantle wedge in subduction zones. *Journal of Geophysical Research*, **88**, 5815–5825.

Tatsumi, Y., Kani, T., Ishizuka, H., Maruyama, S. & Nishimura, Y. 2000. Activation of Pacific mantle plumes during the Carboniferous: evidence from accretionary complexes in southwest Japan. *Geology*, **28**, 580–582.

Tokiwa, T. 2009. Timing of dextral oblique subduction along the eastern margin of the Asian continent in the Late Cretaceous: evidence from the accretionary complex of the Shimanto Belt in the Kii Peninsula, Southwest Japan. *Island Arc*, **18**, 306–319.

Tsubamoto, T., Matsubara, T., Tanaka, S. & Saegusa, H. 2007. Geological age of the Yokawa Formation of the Kobe Group (Japan) on the basis of terrestrial mammalian fossils. *Island Arc*, **16**, 479–492.

Tsukada, K. 1997. Stratigraphy and structure of Paleozoic rocks in the Hitoegane area, Kamitakara Village, Gifu Prefecture. *Journal of Geological Society of Japan*, **103**, 658–668 [in Japanese with English abstract].

Tsukada, K. & Koike, T. 1997. Ordovician conodonts from the Hitoegane area, Kamitakara Village, Gifu Prefecture. *Journal of Geological Society of Japan*, **103**, 171–174 [in Japanese with English abstract].

Tsuru, T., Park, J.-O., Miura, S., Kodaira, S., Kido, Y. & Hayashi, T. 2002. Along-arc structural variation of the plate boundary at the Japan Trench margin: implication of interplate coupling. *Journal of Geophysical Research*, **107**, 2357, http://doi.org/10.1029/2001JB001664

Uchino, T. & Kawamura, M. 2010. Tectonics of an Early Carboniferous forearc inferred from a high-P/T schist-bearing conglomerate in the Nedamo Terrane, Northeast Japan. *Island Arc*, **19**, 177–191.

Ueda, H., Kawamura, M. & Niida, K. 2000. Accretion and tectonic erosion processes revealed by the mode of occurrence and geochemistry of greenstones in the Cretaceous accretionary complexes of the Idonnappu Zone, southern central Hokkaido, Japan. *The Island Arc*, **9**, 237–257.

Ujiié, H. & Kaneko, N. 2006. *Geology of the Naha and Okinawashi-Nambu District. Quadrangle Series, 1:50,000.* Geological Survey of Japan, Tsukuba [in Japanese with English abstract].

Underwood, M., Hibbard, J. P. & DiTullio, L. 1993. Geological summary and conceptual framework for the study of thermal maturity within the Eocene-Miocene Shimanto belt, Shikoku Japan. *In*: Underwood, M. B. (ed.) *Thermal Evolution of the Tertiary Shimanto Belt, Southwest Japan: An Example of Ridge-Trench Interaction*. Geological Society of America Special Paper, **273**, 1–24.

Wallis, S. R., Anczkiewicz, R., Endo, S., Aoya, M., Platt, J. P., Thirlwall, M. & Hirata, T. 2009. Plate movements, ductile deformation and geochronology of the Sanbagawa belt, SW Japan: tectonic significance of 89–88 Ma Lu-Hf eclogite ages. *Journal of Metamorphic Geology*, **27**, 93–105.

Williams, M., Wallis, S., Oji, T. & Lane, P. D. 2014. Ambiguous biogeographical patterns mask a more complete understanding of the Ordovician to Devonian evolution of Japan. *Island Arc*, **23**, 76–101.

Woods, M. T. & Davies, G. F. 1982. Late Cretaceous genesis of the Kula plate. *Earth and Planetary Science Letters*, **58**, 161–166.

2a Palaeozoic basement and associated cover

MASAYUKI EHIRO (COORDINATOR), TATSUKI TSUJIMORI, KAZUHIRO TSUKADA & MANCHUK NURAMKHAAN

Pre-Cenozoic rocks of the Japanese islands are largely composed of latest Palaeozoic to Cretaceous accretionary complexes and Cretaceous granitic intrusives. Exposures of older rocks are restricted to a limited number of narrow terranes, notably the Hida, Oeyama and Hida Gaien belts (Inner Zone of SW Japan), the Kurosegawa Belt (Outer Zone of SW Japan) and the South Kitakami Belt (NE Japan). In these belts, early Palaeozoic basement rocks are typically overlain by a cover of middle Palaeozoic to Mesozoic shelf facies strata. This chapter describes these basement inliers and their cover, grouping them under four subheadings: Hida, Oeyama, Hida Gaien and South Kitakami/Kurosegawa belts. Although opinions are varied among authors whether the Unazuki Schist should be placed in the Hida Belt (TT) or in the Hida Gaien Belt (KT & NM) sections, this chapter will describe the Unazuki Schist in the Hida Belt section.

Hida Belt (TT)

The overall structure of the Japanese archipelago, particularly well displayed in SW Japan, comprises a stack of NW-rooting, subhorizontal nappes, with older sheets normally occupying upper structural positions. The Hida Belt, situated along the back-arc (northern) side of SW Japan from the Hida Mountains to Oki Island, is essentially a remnant fragment of moderately deep-level continental crust that was once a part of the Asian continental margin prior to the opening of the Japanese Sea in Miocene times. The continental orogenic history recorded in the Hida Belt is therefore markedly different from other geotectonic units that instead record oceanwards growth and landwards erosion during Pacific (Cordilleran) -type orogenesis. Komatsu (1990) postulated that the Hida Belt has been thrust southwards as a large-scale nappe onto the Hida Gaien (Hida Marginal) Belt.

Polymetamorphosed Permo-Triassic granite-gneiss complexes with migmatite, impure marble, amphibolite and minor high-aluminous pelitic schist occur as a basement terrane in the Hida Mountains in central western Japan (Fig. 2a.1), and have been referred to informally as the 'Hida Gneiss' since the 1930s (Kobayashi 1938). Most rocks in this Hida Gneiss Complex are characterized by amphibolite facies mineralogy, and a few small inliers of similar rocks in Oki Island and the northern Chugoku Mountains of SW Japan have been regarded as western extensions of the same metamorphic unit. Miyashiro (1961) postulated the concept of 'paired metamorphic belts', and considered that the Hida low-pressure/high-temperature metamorphic belt was paired with the 'Sangun' high-pressure/low-temperature metamorphic belt (Fig. 2a.2). However, geochronological data for these metamorphic rocks have revealed that the timing of metamorphism of the Hida Belt was not coeval with that in the 'Sangun' Belt (e.g. Isozaki 1997; Isozaki *et al.* 2010; Wakita 2013). The eastern margin of the Hida Gneiss Complex is separated by a mylonite zone from Barrovian-type, medium-pressure, pelitic schists ('Unazuki Schist') which crop out as a north–south-aligned elongated, narrow subunit 2–3 km wide and 17 km long (e.g. Kano 1990; Takagi & Hara 1994). Another important tectonic boundary within the Hida Belt is the 'Funatsu Shear Zone', which comprises dextrally sheared, mostly metagranitoid mylonitic rocks (Komatsu *et al.* 1993).

The presence in the Hida Gneiss Complex of high-aluminous metapelites, metamorphosed acidic volcanic rocks and abundant impure siliceous marble associated with orthogneiss suggest a passive-margin lithology for the protoliths, probably as continental shelf sediments and basement rock on a rifted continental margin (e.g. Sohma & Kunugiza 1993; Isozaki 1996, 1997; Wakita 2013). Finally, the granite-gneiss complexes of the Hida Belt are unconformably overlain by a cover sequence of Lower Jurassic–Lower Cretaceous shallow marine and non-marine sedimentary rocks with rare dinosaur fossils, and by a thick Cenozoic volcaniclastic succession.

Hida Gneiss Complex and related granitic rocks

In the Hida Mountains, gneissose rocks show north–south- to NNW–SSE-striking and west-dipping foliations with mineral lineations plunging gently southwards (e.g. Kano 1980; Arakawa 1982; Sohma & Akiyama 1984). The metamorphic lithologies exposed are mainly of calcareous gneiss, quartzofeldspathic gneiss, marble, amphibolite (hornblende gneiss), granitic gneiss and minor pelitic gneiss (e.g. Naito 1993). These rocks are associated with granitic plutons and migmatite showing multiple stages of anatexis, deformation and intrusion (Figs 2a.3 & 2a.4).

The Hida Gneiss Complex has been subdivided into 'inner' lower-temperature and 'outer' higher-temperature regions, based on the mineralogy of pelitic gneiss (Suzuki *et al.* 1989). Rare staurolite, cordierite and andalusite have been found in sillimanite-bearing pelitic gneiss of the inner region (e.g. Asami & Adachi 1976). Mafic gneiss and amphibolite in this inner region contain hornblende + biotite + plagioclase + K-feldspar + clinopyroxene and scapolite occurs in calcareous gneiss (Jin & Ishiwatari 1997). Garnet-biotite Fe–Mg exchange geothermometry on pelitic gneiss suggests re-equilibration at a temperature of *c.* 550–650°C and a pressure of 0.4–0.5 GPa (e.g. Suzuki *et al.* 1989; Jin & Ishiwatari 1997). Overall, the metamorphic rocks of the inner region record amphibolite facies conditions and contrast with the outer region where granulite facies orthopyroxene- and/or spinel-bearing mineral assemblages have been described from mafic gneiss and sillimanite-bearing pelitic gneiss (e.g. Sohma *et al.* 1986). Outer zone granulite facies felsic gneiss contains rare corundum + K-feldspar mineral assemblage (Suzuki & Kojima 1970), and Fe-rich mafic gneiss (up to 26.3 wt% total FeO) contains garnet +

From: Moreno, T., Wallis, S., Kojima, T. & Gibbons, W. (eds) 2016. *The Geology of Japan*. Geological Society, London, 25–60.

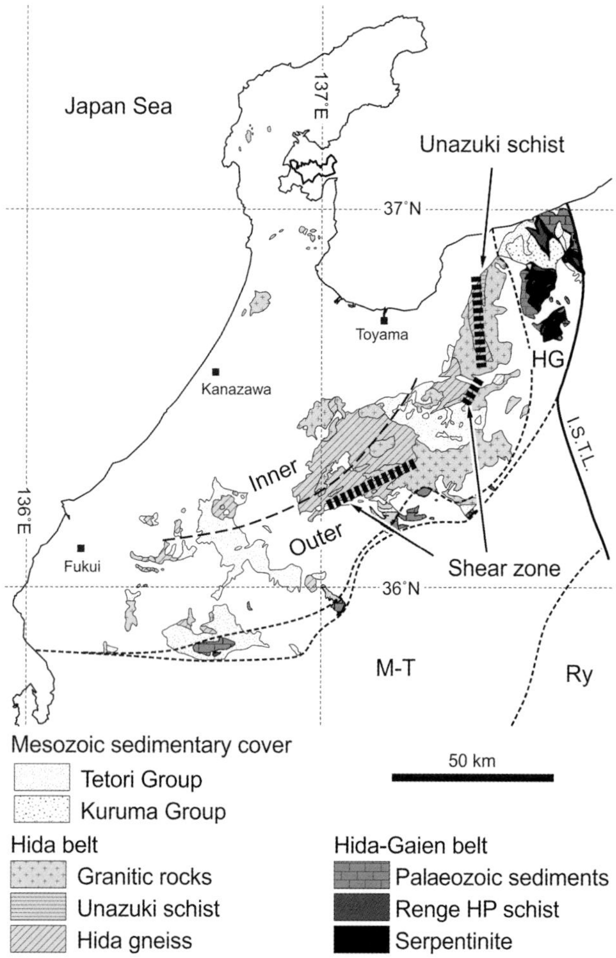

Fig. 2a.1. Simplified geological map of the Hida Mountains showing exposures of the gneiss-granite complexes of the Hida Belt and Unazuki Schist (modified after Sohma & Kunugiza 1993). Exposures of Palaeozoic rocks of the Hida Gaien Belt and Mesozoic sedimentary rocks are also shown. Dashed line is a boundary between inner and outer metamorphic regions proposed by Suzuki *et al.* (1989). Dotted lines are boundaries of geotectonic units (HG, Hida Gaien Belt; M-T, Mino–Tanba Belt; Ry, Ryoke Belt).

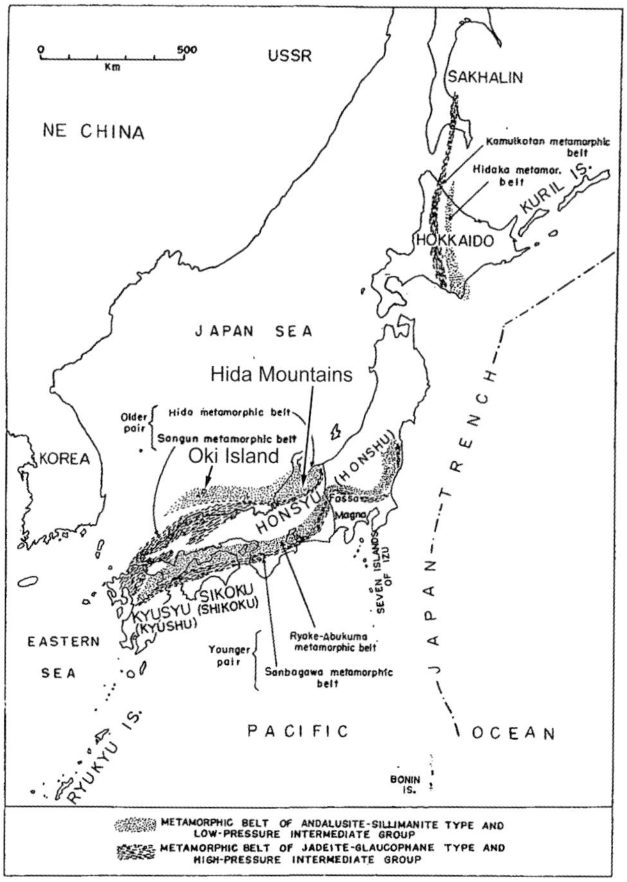

Fig. 2a.2. A conceptual map of 'paired metamorphic belts' proposed by Miyashiro (1961) [Copyright© by the Oxford University Press] showing the Hida Belt.

clinopyroxene (augite) (Suzuki 1973). In this outer region, garnet in pelitic gneiss is characterized by a higher pyrope component (up to 28 mol%) than that of the inner region, and the garnet-biotite geothermometry data indicate apparent temperatures of *c.* 700°C at a pressure of *c.* 0.4–0.7 GPa (Suzuki *et al.* 1989). Calcareous gneiss contains clinopyroxene and scapolite, and marbles contain dolomite and rare olivine (Naito 1993). Some of the clinopyroxene-rich gneiss has been interpreted as a reaction product with marble and amphibolite (Kunugiza & Goto 2006). Calcite-graphite carbon isotope geothermometry for marble of the outer zone gives a temperature of >750°C (Wada 1988). In the granulite facies marble, a millimetre-scale oxygen isotope anomaly (depleted $\delta^{18}O$ value) along calcite-calcite grain boundaries and a systematic correlation of oxygen isotope with Mn and Sr suggest a cooling rate of >1°C myr^{-1} after amphibolite facies conditions (Wada 1988; Graham *et al.* 1998). In some places Zn–Pb ore deposits are associated with skarns and migmatitic gneiss (e.g. Kano 1991; Kunugiza 1999).

Based on deformation and cross-cutting and deformational relationships in the field, 'granitic rocks' (*sensu lato*) associated with the Hida Gneiss Complex have been grouped into at least two types: older and younger 'granites'. The older granites display a wide range of rocks from gabbro (diorite) to granite and are commonly associated with migmatitic gneiss and skarns, with the gabbroic/dioritic rocks occurring as dykes (e.g. Kano & Watanabe 1995; Arakawa *et al.* 2000), and are characterized by high Sr/Y ratios and $\delta^{18}O$ values >9.8‰ (Ishihara 2005). In contrast, the younger granites have $\delta^{18}O$ values <9.1‰, cross-cut older granites, Hida Gneiss and Unazuki Schist, and can be further subdivided into pre- and post-mylonitization intrusions (Ishihara 2005). The pre-mylonitic younger granites are characterized by initial $^{87}Sr/^{86}Sr$ ratios <0.705, and are clearly distinguished from other granitic rocks in the Japanese archipelago (Arakawa 1990; Arakawa & Shinmura 1995; Jahn 2010).

Zircon ion microprobe U–Pb geochronology for pelitic gneiss has given a concordant detrital age of *c.* 1.84 Ga, whereas metamorphic zircons record 250 Ma (and more rarely *c.* 285 Ma) with discordant detrital zircons suggesting upper intercept ages of 3.42 and 2.56 Ga (Sano *et al.* 2000). Zircons from the older granites and the pre-mylonitic younger granites yielded ion microprobe U–Pb ages of 248–245 and 193 Ma, respectively; zircons from a felsic gneiss yielded *c.* 242 Ma and rare 330 Ma (Zhao *et al.* 2013). Euhedral zircons in migmatitic gneiss hosting Zn–Pb skarn deposits yield a U–Pb age of 234 Ma for a regional metamorphic event (Sakoda *et al.* 2006). Zircons in mylonitized granites in the Funatsu Shear Zone show an ion microprobe U–Pb age of 250–240 Ma and zircons from a post-mylonitization granitic intrusion yield 191 Ma, constraining the timing of mylonitization to Triassic–Early Jurassic times (Takahashi *et al.* 2010).

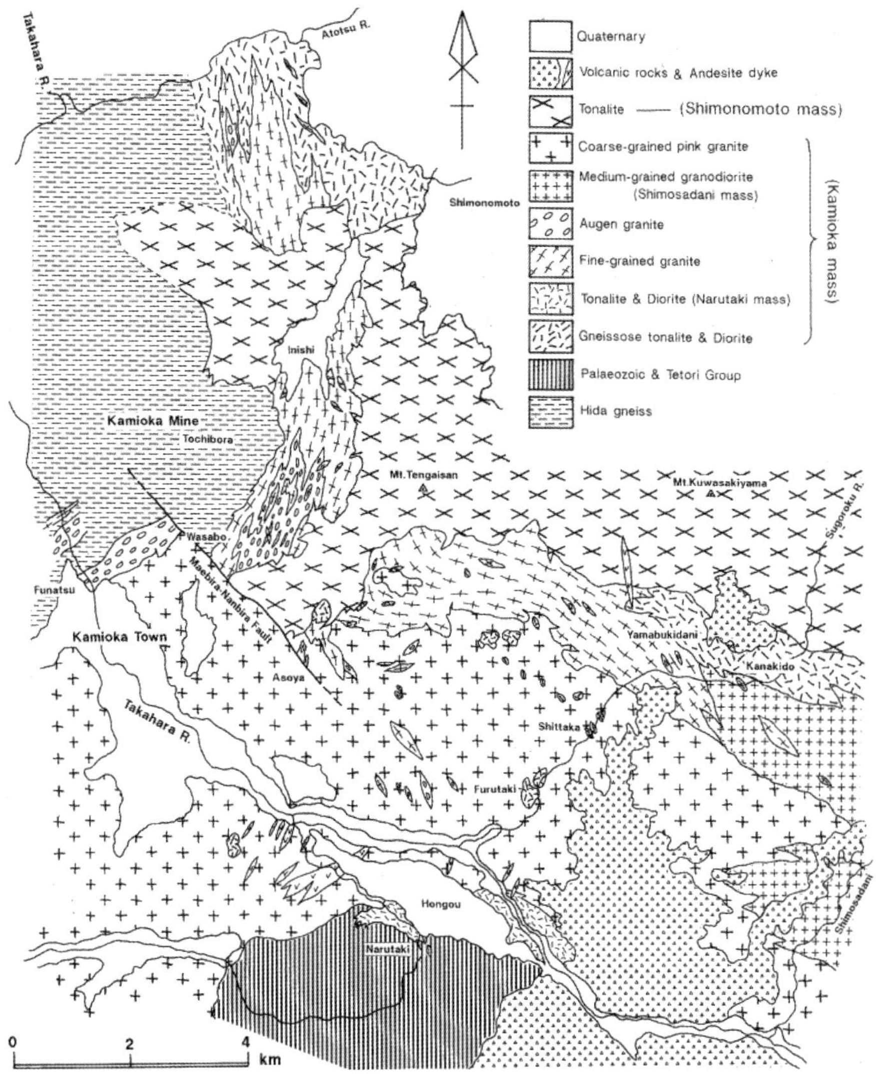

Fig. 2a.3. Geological map of the Hida gneiss-granite complex in the Kamioka area by Kano & Watanabe (1995) [Copyright© by the Geology Society of Japan]. The map shows multiple stages of deformation and igneous activities.

Electron microprobe Th–U–total Pb chemical ages of zircon in the pelitic gneiss have yielded 250–230 Ma for sillimanite-grade amphibolites facies metamorphism (Suzuki & Adachi 1994).

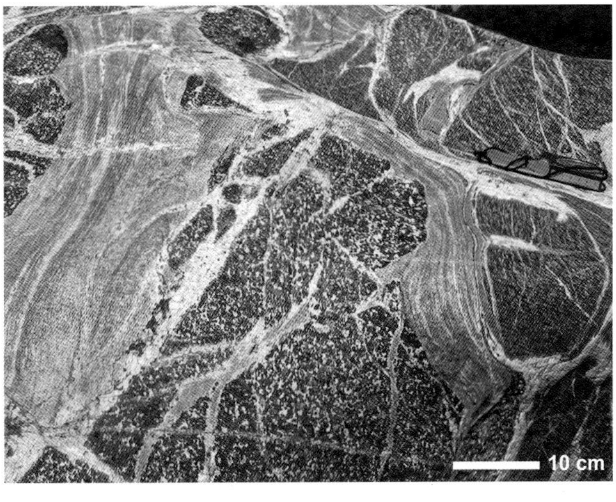

Fig. 2a.4. Amphibolite migmatite of the Hida Belt showing different stages of melting, segregation and deformation process.

Sm–Nd whole-rock–mineral isochron ages of amphibolite and pelitic gneiss are 413 $\pm$ 60 Ma and 274 $\pm$ 13 Ma, respectively (Asano *et al.* 1990). Finally, K–Ar hornblende or biotite areas in both metamorphic rocks and granitic intrusions cluster at *c.* 180 Ma (cf. Ohta & Itaya 1989).

Unazuki Schist

This noth–south-aligned elongated subunit of the Hida Belt is composed of medium-pressure-type metamorphic rocks and gabbro–diorite and granite plutons. The metamorphic rocks include mafic schist, quartzofeldspathic schist (metamorphosed rhyolite and acidic tuff, which have been described as 'leptite'), high-aluminous pelitic schists, metamorphosed impure limestone and rare conglomerate schist (Suwa 1966; Hiroi 1978). Schistose rocks show north–south- to NNW–SSE-striking and west-dipping foliation, and the mineral lineation plunges gently to the south. The high-aluminous pelitic schists contain abundant staurolite porphyroblasts (Fig. 2a.5), with the discovery of staurolite- and kyanite-bearing pelitic schist by Ishioka (1949) being the first recognition of medium-pressure regional metamorphism in Japan. Bulk-rock compositions of the Unazuki pelitic schist are high in Al_2O_3 but low in K_2O. The compositional trend on Thompson (1957)'s AFM ternary diagram (A–Al_2O_3–K_2O, F–FeO, M–MgO, projected

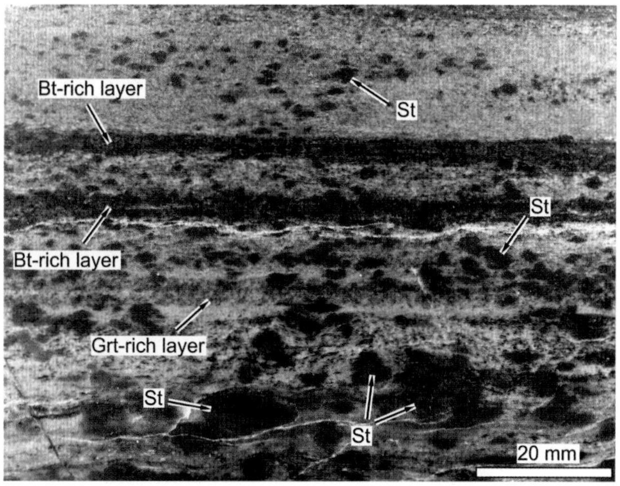

Fig. 2a.5. Polished slab of staurolite-bearing Unazuki Schist of Hiroi (1983)'s zone III (St, staurolite; Bt, biotite; Grt, garnet). The specimen shows compositional layering that might represent an original sedimentary alternation.

from ideal muscovite) is very different from that of metamorphosed trench-fill semi-pelagic sediments such as Sanbagawa pelitic schist (Fig. 2a.6), and it is noteworthy that pelitic gneiss in the Hida Belt is also typically peraluminous.

The metamorphic grade of the Unazuki Schist increases steadily southwards and shows systematic changes based on the appearance of characteristically zoned Fe–Mg silicate and aluminosilicate minerals in high-aluminous metapelite. Hiroi (1983) therefore identified four zones in order of increasing metamorphic grade: I – chloritoid + quartz; II – staurolite + chlorite + muscovite; III – kyanite + biotite; and IV – sillimanite + muscovite. These Barrovian-type mineral isograds lie obliquely over the lithological

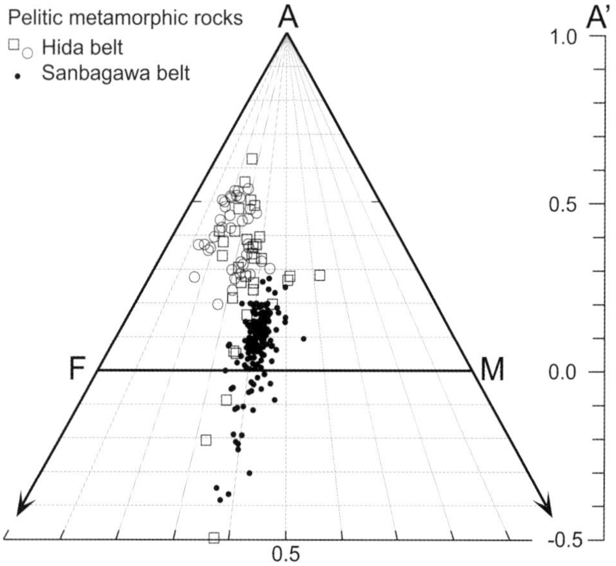

Fig. 2a.6. Thompson (1957)'s AFM diagram (A–Al_2O_3-K_2O, F–FeO, M–MgO, projected from ideal muscovite) showing bulk-rock compositions of metapelites from the Hida Belt. Unazuki Schist, open circle (Hiroi 1984); Hida Gneiss, open square (Asami & Adachi 1976; Suzuki 1977; Sohma *et al.* 1986; Naito 1993; Jin & Ishiwatari 1997) and Sanbagawa Belt (Goto *et al.* 1996). A' is Al-index in Thompson AFM projection, defined as A'= moles Al_2O_3 – 3 × moles K_2O.

structure of the schist. Bedded impure limestone of zone I contains rare fossils of Late Carboniferous bryozoa and foraminifera (Hiroi *et al.* 1978). In addition, the Unazuki Schist has been partly overprinted by contact metamorphism near younger granites, with indialite (a high-temperature polymorph of cordierite) having been found as veins in contact metamorphosed metapelite (Kitamura & Hiroi 1982) and vesuvianite + wollastonite in contact metamorphosed limestone (Okui 1985).

Zircon ion microprobe U–Pb geochronology for granitic rocks yields a concordant detrital age of 229 and 256 Ma, and inherited domains show 3.75–3.55 Ga and 1.94 Ga (Horie *et al.* 2010). The Permo-Triassic ages are consistent with Rb–Sr isochron age of the Unazuki Schist (Ishizaka & Yamaguchi 1969), whereas Rb–Sr whole-rock–muscovite and K–Ar biotite yield cooling ages of 214 and 175 Ma, respectively (Shibata *et al.* 1970).

A clast of staurolite-bearing high-aluminous pelitic schist was found in the Upper Jurassic conglomerate of the Tetori Group overlying the Hida Belt (Tsujimori 1995), and detrital chloritoid has been found in Lower Jurassic shallow-marine sandstone of the Kuruma Group, which lies on the Hida Gaien Belt (Kamikubo & Takeuchi 2010). Considering the limited number of occurrences of staurolite- and/or chloritoid-bearing metamorphic rocks in Japan, these sedimentary records suggest that the Unazuki Schist (or its equivalent) had already been exposed at the surface by Jurassic times.

Mesozoic sedimentary cover sequences

The Hida Gneiss Complex and related granitic rocks are unconformably overlain by Middle Jurassic–Lower Cretaceous Tetori Group marine and non-marine deposits. Various macrofossils, including plants, molluscs, fishes, reptiles, turtles and dinosaurs, have been found in the Tetori Group. The group is divided into Kuzuryu, Itoshiro and Akaiwa subgroups in ascending order (e.g. Maeda 1961). Illite crystallinity indicates that the Kuzuryu Subgroup exhibits a higher degree of diagenesis than the overlying Itoshiro and Akaiwa subgroups (Kim *et al.* 2007). The marine Kuzuryu Subgroup is unconformably covered by the non-marine Itoshiro Subgroup, which in turn is overlain by the non-marine Akaiwa Subgroup. Electron microprobe Th–U–total Pb geochronology shows that detrital monazites are mostly *c.* 1.74–1.25 Ga with a small peak of *c.* 250 Ma (Obayashi 1995). Recent laser ablation inductively coupled plasma mass spectrometry (LA-ICP-MS) U–Pb zircon geochronology for acidic tuffs limits minimum depositional ages to 130 Ma for Kuzuryu Subgroup and 118 Ma for Itoshiro Subgroup (Kusuhashi *et al.* 2006). In conglomerate beds of the Tetori Group most abundant clasts are granitic rocks and orthoquartzites, which are likely derived from a basement rock in the East Asian continental block. However, rare radiolarian cherts derived from a Jurassic accretionary complex occur in the Akaiwa Subgroup; this suggests a change of lithology of provenance area during Early Cretaceous times (Takeuchi *et al.* 1991). Palaeomagnetic analyses show that the Tetori Basin was located at *c.* 24° N until the Late Jurassic, but during the Early Cretaceous epoch it moved northwards to 40° N (Hirooka *et al.* 2002).

Oki Gneiss Complex

In the Oki–Dogo islands, a small exposure (6 × 8 km) of gneiss-granite has been considered to mark a western extension of the Hida Gneiss Complex of the Hida Mountains. This Oki Gneiss Complex consists mainly of felsic gneiss with migmatite, granitic intrusions and minor amphibolite (Hoshino 1979). Amphibolite contains orthopyroxene-bearing mineral assemblage, and two-pyroxene geothermometry gives a granulite facies temperature of *c.* 800°C. A zircon Pb–Pb age of *c.* 1.9 Ga from the Oki Gneiss Complex has been obtained using isotope dilution thermal ionization mass spectrometry (Yamashita & Yanagi 1994), a result consistent with a Sm–Nd whole-rock isochron age 1.98 ± 0.18 Ga from amphibolite (Tanaka & Hoshino, 1987). Zircon ion microprobe U–Pb geochronology for granitic gneisses yields concordant ages of 1.87 and 1.88 Ga for the crystallization age of the granitic protolith (Cho *et al.* 2012). Moreover, recrystallized zircon rims give a concordant age of *c.* 236 Ma (Tsutsumi *et al.* 2006; Cho *et al.* 2012). Electron microprobe Th–U–total Pb chemical dating of monazite in paragneisses shows *c.* 250 Ma for regional metamorphism and rare zoned monazite preserves older ages of 1.69 ± 0.23 Ga at the core and 440 ± 30 Ma further out, with the 250 ± 20 Ma event recorded at the rims (Suzuki & Adachi 1994). Biotite K–Ar and muscovite $^{40}Ar/^{39}Ar$ plateau ages suggest a timing of cooling through closure temperature of micas at *c.* 170 Ma (Shibata & Nozawa 1966; Dallmeyer & Takasu 1998). A small outcrop of granitic gneisses found near the Daisen Volcano, interpreted as an inlier of the Hida Gneiss, has yielded a Rb–Sr whole-rock–mineral isochron (mafic gneiss) of 186 ± 6 Ma (Ishiga *et al.* 1989) and a U–Pb age of 198 ± 3 Ma (Ishihara *et al.* 2012).

Geotectonic correlation with East Asian continental margin

How does the Hida Belt correlate with petrotectonic units in NE China and the Korean Peninsula (Fig. 2a.7)? Hiroi (1981) first correlated the late Permian geotectonic unit of Unazuki Schist with the Ogcheon Belt in the Korean Peninsula, based on the occurrence of staurolite-bearing high-aluminous metapelites. Using protolith lithology and metamorphic character (Barrovian-type medium-pressure metamorphism) of the Unazuki Schist, Isozaki & Maruyama (1991) postulated that the Unazuki Schist represented an eastern extension of a Permo-Triassic collisional unit between the Shino-Korea (North China) and Yangtze (South China) blocks, and that gneiss-granite complexes in the Hida Mountains and Oki Island could be correlated with North China and South China blocks, respectively (cf. Isozaki 1996, 1997). More recently, Isozaki *et al.* (2010) suggested that the medium-pressure-type metamorphic rocks with passive margin protoliths in the Higo, Unazuki and Hitachi units represented an eastern extension of the Permo-Triassic continent–continental collision zone in east-central China.

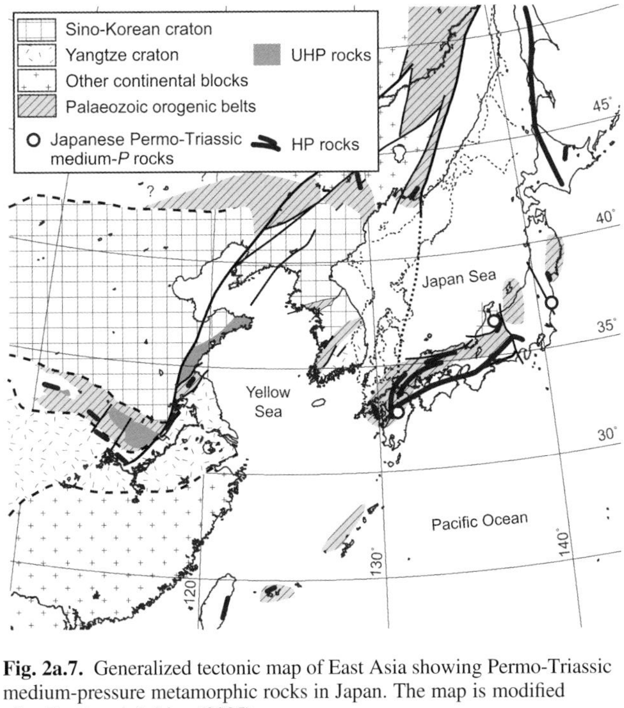

Fig. 2a.7. Generalized tectonic map of East Asia showing Permo-Triassic medium-pressure metamorphic rocks in Japan. The map is modified after Tsujimori & Liou (2005).

Ishiwatari & Tsujimori (2003) proposed the Yaeyama promontory hypothesis. They essentially supported the idea of the Unazuki Schist correlating with Permo-Triassic petrotectonic units in NE China and the Korean Peninsula, but considered that both Permo-Triassic collision-type orogeny and Pacific-type orogeny had occurred along the same plate boundary. Ernst *et al.* (2007) proposed an amalgamated suture zone 'Tongbai–Dabie–Sulu (east-central China)–Imjingang–Gyeonggi (central Korea)–Renge–Suo (SW Japan)–Sikhote-Alin Orogen' that reflects collision between the Sino–Korean and Yangtze blocks on the SW portion, and accretion of outboard oceanic arcs ± sialic fragments against the NE margin.

In the Hida Gneiss and Unazuki granites, the presence of inherited Archean and Palaeoproterozoic zircon ages and the lack of Neoproterozoic zircon suggest that this region is genetically related to the Sino–Korean Block (Sano *et al.* 2000; Horie *et al.* 2010). Based on the deformational style, zircon U–Pb geochronology and palaeogeographical location of granitic rocks, Takahashi *et al.* (2010) correlated the Funatsu Shear Zone of the Hida Mountains to the Cheongsan Shear Zone of the south-central Ogcheon Belt in the Korean Peninsula. Sr–Nd–Pb isotope geochemical study of Permo-Triassic granitic rocks in the Korean Peninsula confirmed that granitic plutons of the Gyeongsang Basin, characterized by low initial $^{87}Sr/^{86}Sr$ ratios (0.704–0.705), correlated with the pre-mylonitic younger granites of the Hida Mountains (Cheong *et al.* 2002). Jahn *et al.* (2000) linked the Hida Belt to the Central Asian Orogenic Belt, based on isotope geochemical characteristics of granitic rocks. Recently, this idea was supported by Chang & Zhao (2012)'s model in which the Permo-Triassic collisional zone was terminated by the Yellow Sea Transform Fault and did not continue into the Korean Peninsula or Hida Belt.

Because of considerable modification of the Japanese archipelago in Cenozoic times, there is still debate regarding the possible correlations between the Hida Belt and other Permo-Triassic geotectonic units along the East Asian continental margin. To further our understanding of the geotectonic configuration of this margin, a more detailed and integrated approach to geology, petrology and geochronology of metamorphic and associated granitic rocks than has so far been available is required.

Oeyama Belt (TT)

Kilometre-scale Alpine-type ultramafic bodies are sporadically exposed in the Chugoku Mountains of SW Japan, running parallel to the orogenic trend west from Oeyama to Tari–Misaka in the central Chugoku Mountains. The largest body (Sekinomiya) covers an area of *c.* 20 × 5 km (Fig. 2a.8) and the area with the most numerous outcrops is concentrated in the Central Chugoku Mountains. Arai (1980) first pointed out that these ultramafic bodies had originally constituted the lowest part of an ophiolitic suite subsequently emplaced as dismembered fragments, with the 'ophiolitic succession' being best developed in the Oeyama body (Kurokawa 1985). Overall compositional trends of Cr-spinel overlap with either forearc or abyssal peridotite. Since these ultramafic bodies have different petrological, geochemical and geochronological features from the mafic-ultramafic bodies of the Yakuno Ophiolite, Ishiwatari (1991) grouped these ultramafic bodies into the 'Oeyama Ophiolite', a unit also referred to as the 'Oeyama Belt' in geotectonic models (e.g. Isozaki & Maruyama 1991; Isozaki *et al.* 2010). The ultramafic bodies of the Oeyama Belt occupy the structurally highest position in the

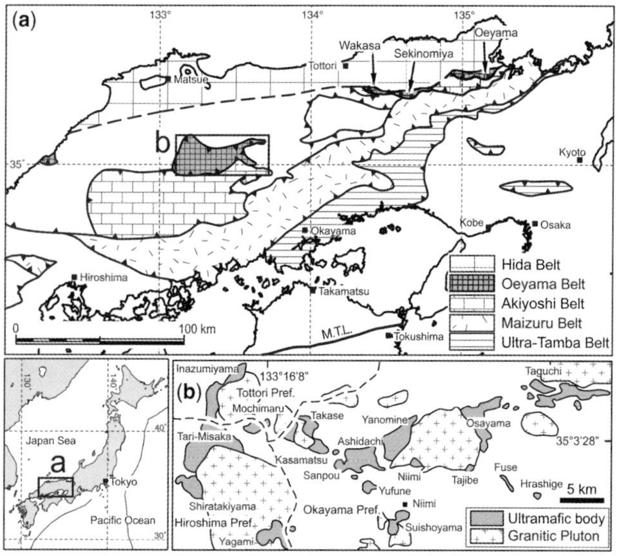

Fig. 2a.8. (**a**) Tectonic framework of the Chugoku Mountains, showing the Oeyama Belt and other Palaeozoic geotectonic units (after Tsujimori & Liou 2004). (**b**) Distributions of ultramafic bodies in the central Chugoku Mountains (after Matsumoto *et al.* 1995).

Chugoku Mountains, above the Renge, Akiyoshi and Maizuru belts (e.g. Uemura *et al.* 1979; Tsujimori & Itaya 1999), and have been broadly correlated with similar ultramafic bodies in the Hida Mountains and northern Kyushu. In the geological history of the Japanese archipelago since Early Palaeozoic time, the Oeyama Belt has acted as a basement unit for Early Palaeozoic forearc basin sediments and as a forearc mantle wedge involved with Palaeozoic subduction zone metamorphic rocks.

The ultramafic bodies of the Oeyama Belt are composed mainly of serpentinized harzburgite (more lherzolitic in eastern bodies) and dunite with minor podiform chromitite and mafic intrusions (e.g. Arai 1980). The chromitites enclosed in dunite are best developed in the Central Chugoku Mountains (Arai 1980; Hirano *et al.* 1987; Matsumoto *et al.* 1997) and more rarely in the Hida Mountains (Yamane *et al.* 1988; Tsujimori 2004). Amphibolites (metacumulate and gneissose metagabbro) occur as tectonic blocks only in the Eastern Chugoku Mountains (Kurokawa 1985; Nishimura & Shibata 1989; Tsujimori & Liou 2004), and associated jadeitite (jade) has been recorded in some places (cf. Chihara 1989).

Due to intrusions of Late Cretaceous granitic plutons, most bodies have been partly overprinted by contact metamorphism (e.g. Arai 1975; Kurokawa 1985; Matsumoto *et al.* 1995) although some ultramafic bodies, particularly in the Hida Mountains, had already undergone a regional metamorphism before contact heating (Uda 1984; Tsujimori 2004; Nozaka 2005; Khedr & Arai 2010; Machi & Ishiwatari 2010; Nozaka & Ito 2011). Sm–Nd isochron ages of *c.* 560 Ma for the gabbroic intrusions suggest a Cambrian or late Neoproterozoic igneous age (Hayasaka *et al.* 1995). Jadeitite dykes or blocks yield hydrothermal zircon U–Pb ages of *c.* 520–470 Ma (Tsujimori *et al.* 2005; Kunugiza & Goto 2010), whereas amphibolites and gneissose metagabbro give hornblende K–Ar ages of *c.* 440–400 Ma (Nishimura & Shibata 1989; Tsujimori *et al.* 2000).

Characteristics of primary peridotites

The ultramafic bodies of the Oeyama Belt in the Chugoku Mountains preserve a kilometre-scale original lithological structure of the crust–mantle transition zone or the uppermost oceanic mantle. They consist mainly of serpentinized harzburgite and dunite with minor podiform chromitite and gabbroic intrusions (e.g. Arai 1980; MITI 1993, 1994; Matsumoto *et al.* 1997). The largest chromitite pod in the Wakamatsu Mine of the Tari–Misaka body has dimensions of $40 \times 25 \times 210$ m. Serpentinized harzburgite is characterized by occurrence of vermicular-like intergrowth of Cr-spinal and orthopyroxene, whereas Cr-spinels in dunite are euhedral to subhedral (Fig. 2a.9). The overall layered structure is not well developed in those bodies. Podiform chromitites are enclosed by dunite envelope (Fig. 2a.10), and the occurrences of relatively large chromitite pods are limited in dunite-dominant bodies. The lithological relations and mineralogical features of chromitites suggest a melt–mantle interaction and related melt mixing to precipitate Cr-spinel. The chromitite-precipitated melt was a mixture of secondary Si-rich melts formed by this interaction and the primitive magmas in the upper mantle (e.g. Arai & Yurimoto 1995). The Cr# (Cr/Cr + Al) of Cr-spinel in harzburgite, dunite and chromitite varies from 0.4 to 0.6, Cr-spinels in dunite and chromitite have slightly higher TiO_2 and the latter can contain hydrous mineral inclusions such as pargasite and Na-phlogopite (Arai 1980; Matsumoto *et al.* 1995, 1997).

In the Hida Mountains, several ultramafic bodies of the Oeyama Belt are exposed along the Hida–Gaien Belt. Although they are highly serpentinized or recrystallized, chemical compositions of relict Cr-spinel and bulk-rock compositions of serpentinite suggest mostly harzburgite and subordinate dunite protolith (e.g. Khedr & Arai 2010, 2011; Machi & Ishiwatari 2010). Based on petrological features, the ultramafic rocks have been grouped into ‘serpentinized peridotite’ and ‘metamorphosed peridotite’. The serpentinized peridotites are subdivided into high-Al and high-Cr groups, each with Cr-spinel Cr# values of 0.3–0.4 and 0.5–0.6, respectively (Machi & Ishiwatari 2010; Khedr & Arai 2011). In contrast, relict primary Cr-spinels in metamorphosed peridotite are characterized by very high Cr# of 0.7–0.9 (Tsujimori 2004; Khedr & Arai 2011). Rare Cr-spinel in chromitite preserves primary pargasite (Tsujimori 2004).

Metamorphosed peridotites in the Hida Mountains have been subjected to hydration and metamorphism. Due to the regional deformation, highly deformed metamorphosed peridotites often show a penetrative schistosity defined by preferred orientation of antigorite and tremolite and with a trend similar to that of high-pressure/low-temperature schists in the Renge Belt (Yamazaki

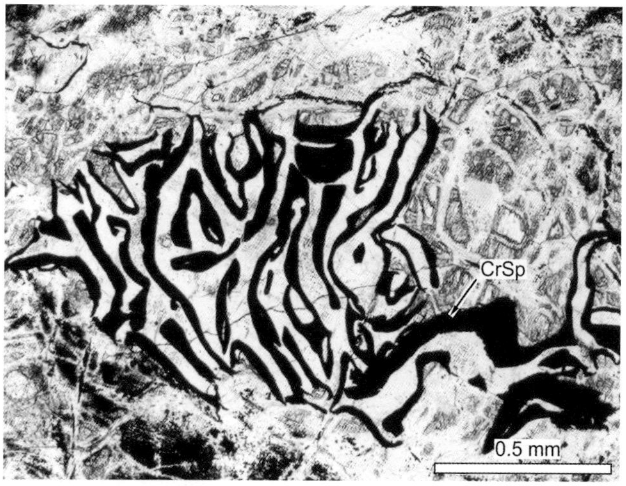

Fig. 2a.9. Photomicrograph of Cr-spinel (CrSp) in serpentinized harzburgite exhibiting a vermicular-like texture.

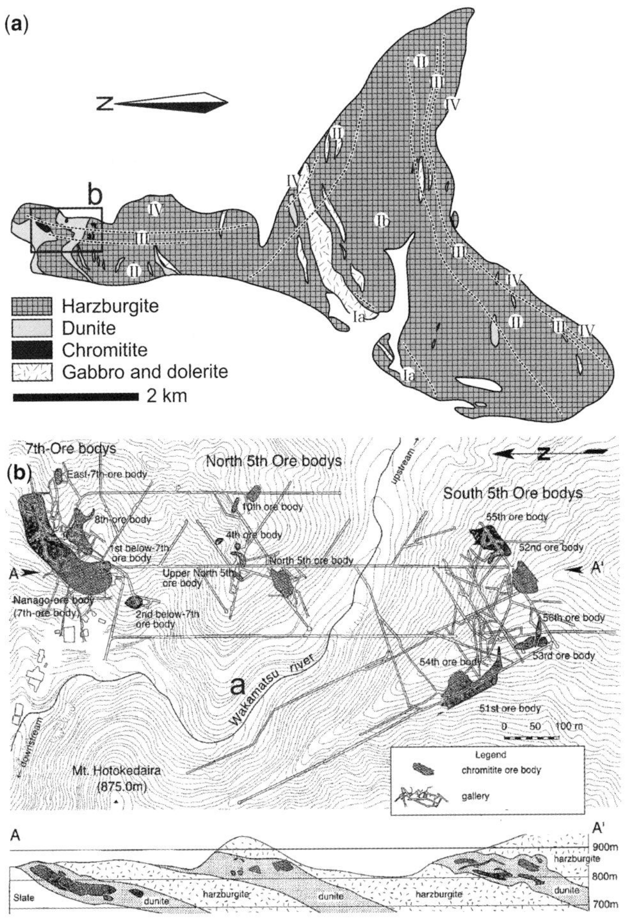

Fig. 2a.10. (**a**) Lithological map of the Tari–Misaka ultramafic body (after Matsumoto *et al.* 1997). Mineral zones of contact metamorphism by Matsumoto *et al.* (1995) are also shown. (**b**) Distribution of chromitite pods of the Wakamatsu Mine of the Tari–Misaka ultramafic body (Matsumoto *et al.* 2002 [Copyright© by the Society of Resource Geology]).

1981; Nakamizu *et al.* 1989). In the Happo peridotite body (8 × 5 km), Nakamizu *et al.* (1989) and Nozaka (2005) identified three mineral zones: the diopside zone with olivine (relic) + antigorite + diopside; the tremolite zone with olivine + tremolite + orthopyroxene; and the talc zone with olivine + talc + tremolite. The talc zone metamorphosed peridotites (tremolite–chlorite peridotites) contain a mineral assemblage of olivine + low-Al orthopyroxene + tremolite + chlorite + high-Ti-Cr# Cr-spinel, suggesting metamorphic conditions of $T = 650$–$750°C$ and $P = 1.6$–2.0 GPa (Khedr & Arai 2010). Metamorphic olivine is intergrown with antigorite and exhibits a 'cleavable' texture (Kuroda & Shimoda 1967). Tremolite contains significant Na (*c.* 0.4 wt% Na_2O), and rare richterite and edenite are associated with tremolite (Khedr & Arai 2010). In the Kotaki area, Machi & Ishiwatari (2010) have also identified deformed meta-peridotite with similar petrological characteristics to the Happo peridotite.

In the Chugoku Mountains, the eastern ultramafic bodies (Oeyama, Izushi, Sekinomiya and Wakasa) often contain a characteristic mineral assemblage of olivine intergrown with antigorite showing a 'cleavable' texture (Uemura *et al.* 1979; Uda 1984; Kurokawa 1985; Nozaka & Ito 2011). The distribution of this 'cleavable' olivine is not related to aureoles of contact metamorphism by younger granitic intrusion (Nozaka & Ito 2011), but rather its ubiquitous presence in these ultramafic bodies indicates regional metamorphism of the forearc wedge-mantle in a subduction zone. However, the timing of this regional metamorphism of ultramafic rocks is still under debate.

In the topographically highest portion of the Oeyama body are exposures of metamorphosed mafic-ultramafic cumulate and amphibolite (4.5 × 1.5 km) (Fig. 2a.11) known as the 'Fuko Pass metacumulate' (Kurokawa 1975, 1985; Tsujimori 1999; Tsujimori & Ishiwatari 2002; Tsujimori & Liou 2004). This metacumulate body has been interpreted as a mafic-ultramafic cumulate member of an ophiolitic succession (Kurokawa 1985). Mafic metacumulate is subdivided three lithologic types: foliated epidote-amphibolite; leucocratic metagabbro; and melanocratic metagabbro. The foliated epidote-amphibolite and leucocratic metagabbro contain low-variance epidote-amphibolite facies mineral assemblage with kyanite, staurolite, and paragonite (Kuroda *et al.* 1976; Kurokawa 1975; Tsujimori & Liou 2004), whereas the melanocratic metagabbros preserve relict granulite facies minerals (Tsujimori & Ishiwatari 2002). Microtextural relationships and mineral chemistry define three metamorphic stages: relict granulite facies metamorphism; high-pressure epidote–amphibolite facies metamorphism; and retrogression. Relict Al-rich diopside (up to 8.5 wt% Al_2O_3) and pseudomorphs after spinel and plagioclase suggest medium-P granulite facies conditions ($P = 0.8$–1.3 GPa at $T > 850°C$) (Tsujimori & Ishiwatari 2002). An unusually low-variance assemblage (hornblende + clinozoisite + kyanite ± staurolite + paragonite + rutile ± albite ± corundum) constrains metamorphic P/T conditions to *c.* 1.1–1.9 GPa at 550–800°C for the high-pressure metamorphism. The breakdown of kyanite to produce a retrograde margarite-bearing assemblage at $P < 0.5$ GPa and $T = 450$–$500°C$ indicates a greenschist facies overprint during decompression (Tsujimori & Liou 2004). Foliated epidote-amphibolite yields hornblende K–Ar ages of *c.* 400–440 Ma (Tsujimori *et al.* 2000); similarly, Early Palaeozoic gneissose epidote amphibolites (but without kyanite) occur in the Wakasa ultramafic body *c.* 20 km to the west of the Oeyama area (Nishimura & Shibata 1989). The presence of Early Palaeozoic high-pressure metamorphic rocks in the Kurosegawa and Oeyama belts provides a petrotectonic constraint for the earliest subduction event in the Japanese Orogen (Tsujimori 2010). The regional metamorphism of the Oeyama Belt occurred in an Early Palaeozoic subduction zone with geothermal gradient of the order 15°C km^{-1}, a relatively high value that might explain the

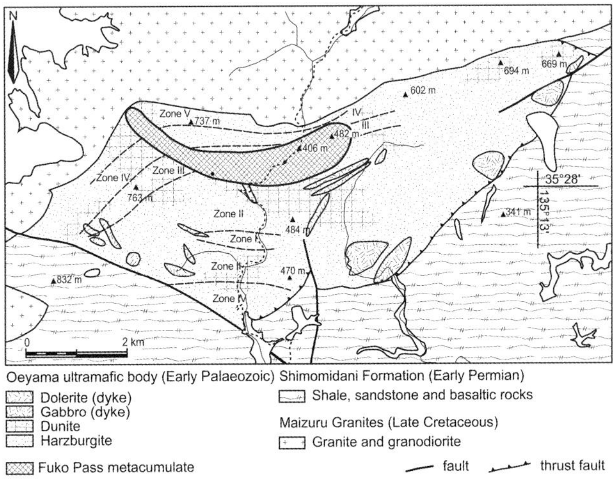

Fig. 2a.11. Geological map of the Oeyama area (after Kurokawa 1985; Tsujimori *et al.* 2000).

epidote-amphibolite facies hydrous recrystallization of the mafic cumulates and ultramafic rocks.

Jadeitite

Jadeitite is a plate tectonic gemstone that correlates forearc mantle wedge and high-pressure and low-temperature metamorphism within a supra-subduction zone at relatively shallow depths (<100 km) (Stern *et al.* 2013; Harlow *et al.* 2015). Since Kawano (1939) first identified jadeitite as boulders in the Kotaki River of the Hida Mountains, numerous jadeitite boulders have been found from at least six ultramafic bodies (Kotaki, Omi, Happo, Sekinomiya, Wakasa and Osayama) within the Oeyama Belt (e.g. Masutomi 1966; Chihara 1971; Tazaki & Ishiuchi 1976; Kobayashi *et al.* 1987) (Fig. 2a.12). The jadeitite localities are characterized by serpentinite mélange and accompanied by Late Palaeozoic high-pressure and low-temperature metamorphic rocks (see 'Renge rocks in the Hida Mountains' section in Chapter 2b). It is noteworthy that the age of jadeitite formation is significantly older than the blueschist facies metamorphism of the Renge rocks in the same mélange (Tsujimori & Harlow 2012).

Most jadeitites are principally fluid precipitates (P-type), but a few formed by metasomatic replacement (R-type) and hence preserve relict minerals and protolith textures (Tsujimori & Harlow 2012). Rutile and zircon are common accessory minerals in the jadeitites, with the rutile-bearing jadeitite having formed at higher T and P than blueschist facies jadeitite. In the Osayama jadeitite, oscillatory-zoned rutile- and jadeite-bearing zircon yielded ion-microprobe U–Pb ages scattering over the range 521–451 Ma (weighted mean age 472 $\pm$ 8.5 Ma). Inherited igneous cores of zoned zircons suggest possible protoliths of gabbro, norite or plagiogranite and therefore an oceanic crustal origin (Fu *et al.* 2010), with ages in the range 523–488 Ma (Tsujimori *et al.* 2005). Oxygen isotope compositions of zircon formed during jadeitite formation is lighter ($\delta^{18}O = 3.6 \pm 0.6$) than that of the inherited igneous core formed in equilibrium with mantle compositions ($\delta^{18}O = 5.0 \pm 0.4$) (Fu *et al.* 2010). In the Itoigawa–Omi area, zircons in jadeitites interpreted as fluid precipitates on the basis of their rhythmic zoning, inclusion suites and rare earth element (REE) patterns, yield ion-microprobe U–Pb ages of 519 $\pm$ 17 and 512 $\pm$ 7 Ma (Kunugiza & Goto 2010). The jadeitite formation is likely to be coeval with the Early Palaeozoic epidote-amphibolite facies metamorphism of the Oeyama Belt.

Contact metamorphism around granitic plutons

Mineral zones of contact metamorphism have been mapped in the ultramafic bodies of the Central Chugoku Mountains (Arai 1975; MITI 1993, 1994; Matsumoto *et al.* 1995). Arai (1975) described four zones in order of increasing metamorphic grade: I – antigorite; II – olivine + talc; III – olivine + 'anthophyllite'; and IV – olivine + orthopyroxene. Zone I was not subjected to thermal metamorphism and can be subdivided into chrysotile-lizardite and antigorite subzones (Matsumoto *et al.* 1995). In the highest grade of contact aureole, porphyroblastic spinifex-like olivine and radial aggregates or spherulitic shapes of orthopyroxene occur (Arai 1975). 'Anthophyllite' in the Tari–Misaka body has *Pnma* crystal structure and is mineralogically classified as protoanthophyllite; some protoanthophyllite contains lamellae of anthophyllite with $C2/m$ structure (Konishi *et al.* 2002, 2003).

Hida Gaien Belt (KT & MN)

This belt, variously referred to as the Hida marginal belt, Circum-Hida tectonic belt and so on, is defined here as the Hida Gaien Belt after Kojima *et al.* (2004). The rocks of the Hida Gaien Belt are restricted to narrow outcrops between the Hida and Sangun–Renge–Akiyoshi–Maizuru–Ultra-Tanba–Mino belts in central Japan (Figs 2a.13 & 2a.14). The Carboniferous rocks in the Unazuki area, referred to as the Unazuki Schist, is included in the Hida Gaien Belt in some studies based on its lithostratigraphy and palaeontology (e.g. Yamakita & Otoh 1987; Tsukada *et al.* 2004); however, it is described in the Hida Belt section here. The Nagato tectonic zone dividing the Sangun–Renge and Akiyoshi belts, SW Japan (Kabashima *et al.* 1993), is considered to be a western extension of this belt (Isozaki & Tamura 1989).

This belt was firstly defined by Kamei (1955*a*) as a complex zone dividing the continental massif of the Hida Belt and the 'Palaeozoic geosynclinal facies' rocks of the Mino Belt, and has been subsequently regarded as a suture zone between the continental mass

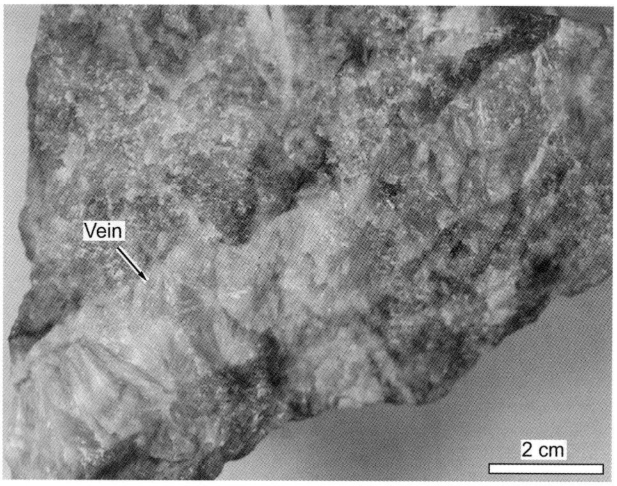

Fig. 2a.12. Early Palaeozoic jadeitite from the Osayama area, showing a texture of fluid precipitates. Radial aggregates of coarse-grained jadeite crystals were precipitated directly from a Na–Al–Si-rich aqueous fluid in a vein within pre-existing jadeitite (cf. Tsujimori & Harlow 2012).

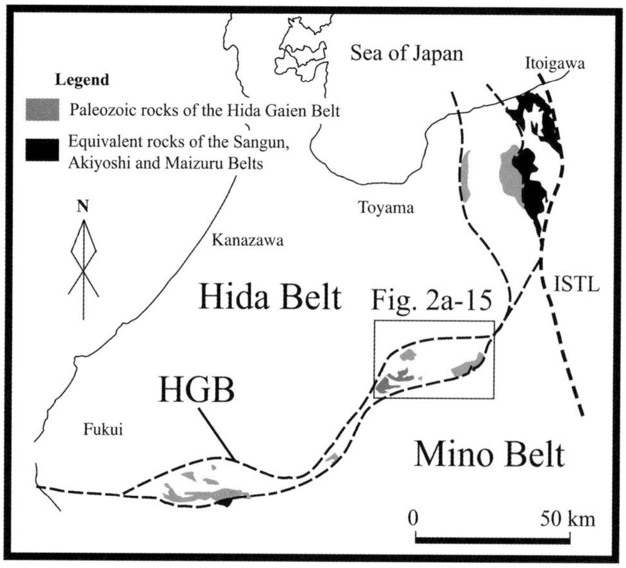

Fig. 2a.13. Index map of the Hida Gaien Belt. ISTL, Itoigawa–Shizuoka Tectonic Line; HGB, Hida Gaien Belt.

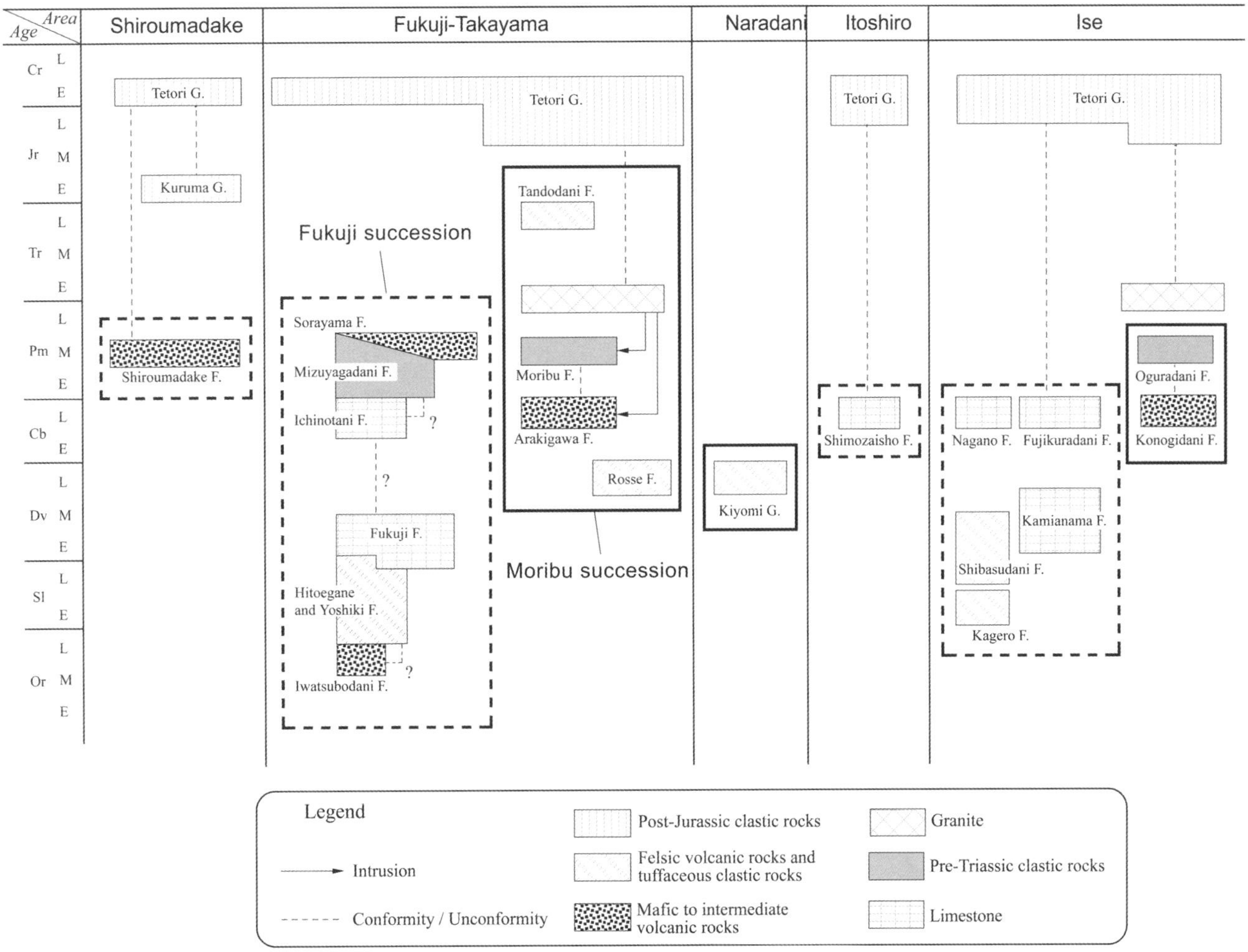

Fig. 2a.14. Stratigraphic relationship among the rocks of the Hida Gaien Belt. Cr, Cretaceous; Jr, Jurassic; Tr, Triassic; Pm, Permian; Cb, Carboniferous; Dv, Devonian; Sl, Silurian; Or, Ordovician; E, Early; M, Middle; L, Late; F., Formation; G., Group.

and middle Palaeozoic oceanic crust (Horikoshi 1972), and as a serpentinite mélange zone caused by 'Hida Nappe' emplacement (Chihara & Komatsu 1982; Komatsu *et al.* 1985; Sohma & Kunugiza 1993) onto the Sangun, Akiyoshi, Maizuru and Mino belts. In a more recent study, the belt has been viewed as a tectonic zone caused by Jurassic (dextral) and Cretaceous (sinistral) shearing along the continental margin (Tsukada 2003; Matsumoto 2012).

The Hida Gaien Belt is composed of fault-bounded blocks of Palaeozoic–Mesozoic shelf facies rocks which can be divided into Moribu and Fukuji types based on their differing lithostratigraphy (Figs 2a.14 & 2a.15): (1) the Moribu succession mainly comprises Upper Devonian felsic tuffaceous rocks, Carboniferous volcanic rocks and Middle Permian clastic rocks in ascending order; and (2) the Fukuji succession comprises mainly Ordovician (?) mafic volcanic rocks, Ordovician–Devonian (?) felsic tuffaceous rocks, Lower–Middle (?) Devonian limestone, Carboniferous limestone and Lower–Middle Permian clastic and pyroclastic rocks in ascending order. Whereas both the Moribu and Fukuji successions record similar Devonian palaeoenvironments, they diverged in Carboniferous times when the Moribu rocks record abundant volcanism. In contrast, coeval Fukuji rocks reflect quiet tropical lagoonal conditions until Permian times when volcanism suddenly began, whereas in the Moribu succession volcanism was replaced by clastic sediment deposition (Fig. 2a.15).

The rocks exposed at the Fukuji–Takayama area, which is the type locality of this belt (Tsukada *et al.* 2004), are briefly described in this section and we examine a model that might explain its tectonic development.

Moribu stratigraphy

Devonian–Triassic rocks of the Moribu succession include the Upper Devonian Rosse (mainly felsic tuffaceous rocks), Carboniferous Arakigawa (mafic volcanic rocks with felsic tuffaceous rocks and limestone), Middle Permian Moribu (clastic rocks with felsic tuffaceous rocks and limestone) and Upper Triassic Tandodani (mainly felsic tuffaceous rocks) formations (Fig. 2a.14).

The Rosse Formation, which is in fault contact with the other formations, contains Upper Devonian fossils such as brachiopods, crinoids and plants, with a Lower Devonian limestone block yielding corals (e.g. Tazawa *et al.* 1997, 2000*b*). The Arakigawa Formation yields Upper Visean–Late Kasimovian (or Gzhelian?) fossils such as corals, goniatites, foraminifers, fusulinoideans, brachiopods and trilobites (Isomi & Nozawa 1957; Fujimoto *et al.* 1962; Igo 1964; Kobayashi & Hamada 1987; Tazawa *et al.* 2000*a*). The Moribu Formation, unconformably overlying the Arakigawa Formation, yields Middle Permian brachiopods and fusulinoideans with recycled Lower Permian fusulinoideans in its lowermost horizon

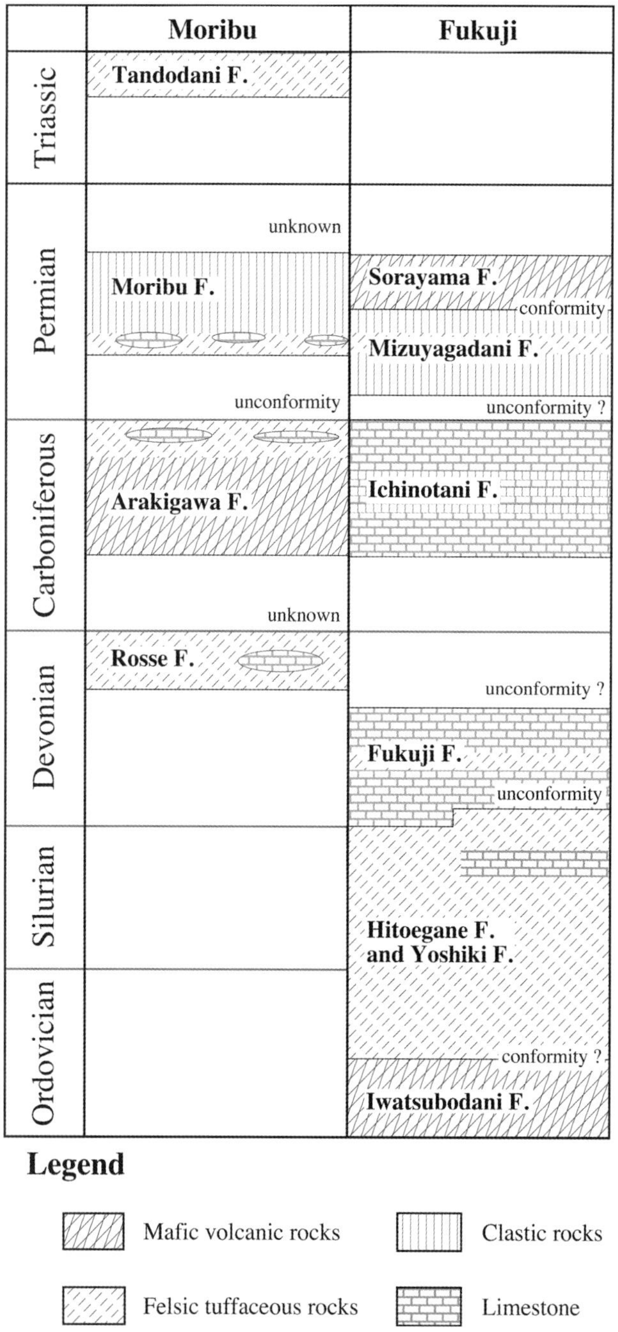

Fig. 2a.15. A columnar section of the Moribu and Fukuji successions showing their major lithologies and characteristic layers and blocks. F., Formation.

(Fujimoto *et al.* 1962; Horikoshi *et al.* 1987; Tazawa *et al.* 1993; Niwa *et al.* 2004). The Tandodani Formation yielding Carnian–Norian conodonts is in fault contact with the Arakigawa Formation (Tsukada *et al.* 1997; Tsukada & Niwa 2005; Fig. 2a.15).

The Moribu Formation is partly metamorphosed into biotite hornfels by intrusion of the 250–240 Ma granite (Funatsu Granite), which is overlain unconformably by Middle Jurassic beds of the Tetori Group (Fig. 2a.14). All pre-Cretaceous rocks are unconformably overlain by uppermost Cretaceous volcanic rocks (Fig. 2a.16).

The equivalent rocks are exposed as the Carboniferous Konogidani and Permian Oguradani formations in the Ise area and as the Devonian Kiyomi Group in the Naradani area (e.g. Tsukada *et al.* 2004; Figs 2a.13 & 2a.14).

Fukuji stratigraphy

The Fukuji succession includes a number of formations in ascending order: (1) Ordovician (?) Iwatsubodani Formation (mafic volcanic rocks); (2) Ordovician–Devonian (?) Hitoegane Formation (felsic tuffaceous rocks); (3) Upper Silurian Yoshiki Formation (felsic tuffaceous rocks); (4) Lower–Middle (?) Devonian Fukuji Formation (mainly limestone); (5) Carboniferous Ichinotani Formation (mainly limestone); (6) Lower–Middle Permian Mizuyagadani Formation (clastic rocks with tuffaceous rocks); and (7) Middle Permian Sorayama Formation (mainly mafic volcanic rocks) (e.g. Igo 1990; Tsukada & Takahashi 2000; Kurihara 2004; Manchuk *et al.* 2013*a*; Fig. 2a.15).

The Iwatsubodani Formation is completely made up of basaltic rocks, whereas the overlying Hitoegane Formation consists mostly of felsic tuffaceous rocks without any sign of basaltic volcanism except in its lower horizons (Tsukada 1997). The lithological contrast between these formations indicates a rapid change from mafic to felsic magmatism in Ordovician times. The Hitoegane Formation ranges from Middle or Late Ordovician to probably Early Devonian age based on fossil records of conodonts, trilobites, plants and radiolarians (Kobayashi & Hamada 1987; Igo 1990; Tazawa & Kaneko 1991; Tsukada & Koike 1997; Kurihara 2007).

The Yoshiki Formation, yielding uppermost Silurian radiolarians and coeval zircons with SHRIMP concordant ages of *c.* 420 Ma (Kurihara 2007; Manchuk *et al.* 2013*a*), can be regarded as an equivalent of the Upper Member of the Hitoegane Formation based on its lithological and palaeontological similarities (Manchuk *et al.* 2013*b*). The Yoshiki Formation is largely in fault contact with the surrounding geological units, although it is partly overlain by the Fukuji Formation (Igo 1990). The Fukuji Formation consists mainly of limestone with corals and stromatoporoids as the main allochemical constituents, along with minor felsic tuff layers (Kamei 1962; Tsukada 2005). Early–Middle (?) Devonian fossils, including corals, conodonts, trilobites and brachiopods, have been obtained from this formation (e.g. Kamei 1952, 1955*b*; Kamei & Igo 1955; Kobayashi & Igo 1956; Hamada 1959*a*, *b*; Koizumi & Kakegawa 1970; Research Group for the Palaeozoic of Fukuji 1973; Okazaki 1974; Ohno 1977; Niikawa 1980; Kuwano 1986, 1987). The Ichinotani Formation, which possibly overlies the Fukuji Formation (Tsukada *et al.* 1999), has yielded many Visean–Gzhelian fossils such as fusulinoideans and smaller foraminifers that indicate an origin in a tropical to subtropical environment (Igo 1956; Kato 1959; Igo & Adachi 1981; Adachi 1985). Remarkable red shale layers in this formation suggest that a pale soil formed by chemical weathering of limestone was deposited in a freshwater lake (Igo 1960). The Mizuyagadani Formation comprises mainly clastic rocks with felsic tuffaceous layers in its lower horizons, and yields Lower Permian corals and Middle Permian radiolarians (Igo 1959; Niko *et al.* 1987; Umeda & Ezaki 1997). The Sorayama Formation yielding autochthonous Middle Permian fusulinoidean fauna conformably overlies the Mizuyagadani Formation (Tsukada *et al.* 1999; Tsukada & Takahashi 2000).

The equivalent rocks of this succession are exposed as the Permian Shiroumadake Formation in the Shiroumadake area, as the Carboniferous Shimozaisho Formation in the Itoshiro area, and as the Silurian Kagero, Silurian–Devonian Shibasudani, Devonian Kamianama and Carboniferous Nagano and Fujikuradani formations in the Ise area (e.g. Tsukada *et al.* 2004; Figs 2a.13 & 2a.14).

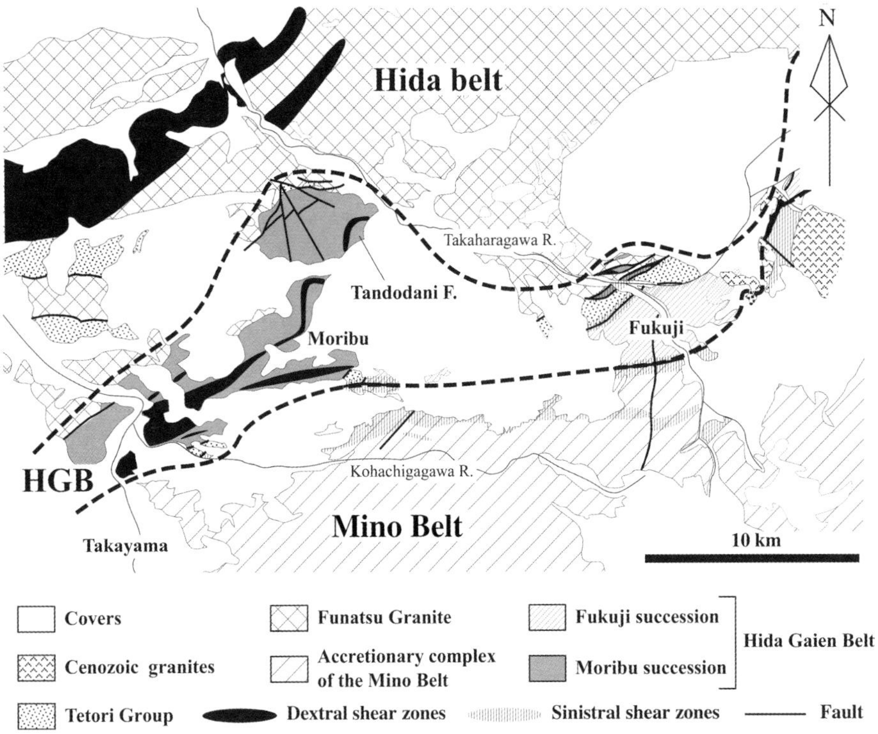

Fig. 2a.16. Simplified geological map of the Fukuji–Takayama area. F., Formation.

Tectonic history of the Hida Gaien Belt

The fault-bounded blocks of the Moribu and Fukuji successions, along with Lower Cretaceous beds belonging to the Middle Jurassic–Lower Cretaceous Tetori Group, are distributed along a narrow zone between the Hida and Mino belts in the Fukuji–Takayama area. In this area, the 250–240 Ma granite intrudes the Moribu succession (Figs 2a.14 & 2a.16) as well as the Hida Gneiss Complex, demonstrating that the Moribu rocks must have been juxtaposed against the Hida Belt by Triassic times (Fig. 2a.17). The Middle to Upper Jurassic Tetori Group overlies the rocks of the Fukuji succession and the Hida Belt, therefore the timing of the juxtaposition of the Fukuji succession against the Hida Belt needs to be before Middle–Late Jurassic times (Fig. 2a.17).

Dextral ductile shear zones, correlated with the Honam Shear Zone in Korea, cut the Carboniferous (?)–Triassic granite to form mylonitic augen gneisses (Yanai *et al.* 1985; Tsukada 2003; Fig. 2a.16). Clasts of similar augen gneisses occur in Middle Jurassic beds within the Tetori Group, which unconformably overlies rocks of the Hida Belt. The timing of deformation has been further constrained by a 191 Ma zircon age on a post-dextral shearing intrusion (Takahashi *et al.* 2010). Taken together, these facts suggest that the shearing lasted until Late Triassic times (timing of deposition of the Tandodani Formation) or later and ended by 191 Ma (Fig. 2a.17). These Mesozoic dextral movements, produced during northwards drift of the continental blocks of East Asia (Otoh *et al.* 1999), were responsible for the fragmentation and dispersal of the Moribu succession and produced the essentially fault-collage disrupted structure of the Hida Gaien Belt, that is, formation of the 'proto-Hida Gaien Belt' (Fig. 2a.17). The equivalent rocks of the Sangun–Renge Belt are in contact with the Moribu rocks across a dextral shear zone in the Ise area (Otoh *et al.* 1999). This fact may suggest that the juxtaposition of the rocks of the Sangun–Renge–Akiyoshi–Maizuru belts against those of the Moribu and Fukuji is attributable to dextral shearing.

A younger deformation phase in the area is recorded by sinistral brittle shear zones which cross the southern part of the Hida Gaien Belt and the northern part of the Mino Belt. One of these sinistral shear zones forms the boundary between the Lower Cretaceous beds of the Tetori Group and rocks of the Fukuji succession (Fig. 2a.15), and is unconformably covered by undeformed Maastrichtian volcanic rocks (Kasahara 1979; Tsukada 2003). The sheared rocks are intruded by undeformed granitoids dated as *c.* 64 Ma in age (Harayama 1990), so it can be deduced that sinistral shearing lasted until Early Cretaceous times or later but had finished by *c.* 64 Ma (Fig. 2a.17). This phase of sinistral shearing was presumably linked to that registered along the eastern margin of Asia in Cretaceous times (e.g. Ozawa 1987; Tashiro 1994; Otoh & Yanai 1996). It restructured the 'proto-Hida Gaien Belt' during sinistral oblique collision with the Mino Belt to complete the present features of the Hida Gaien Belt (Fig. 2a.16). The rocks of the Sangun–Renge–Akiyoshi–Maizuru–Ultra-Tanba belts which had been widely distributed between the proto-Hida Gaien Belt and the Mino Belt might have been crushed by the sinistral movement, to be narrowly scattered in the outer margin of the Hida Gaien Belt (Umeda *et al.* 1996; Otoh *et al.* 1999; Niwa *et al.* 2002; Tsukada *et al.* 2004).

South Kitakami and Kurosegawa belts (ME)

The South Kitakami Belt of NE Japan occupies the southern half of the Kitakami Massif and the eastern marginal part of the Abukuma Massif (Fig. 2a.18), and is composed of older basement rocks with a cover of shallow-marine Ordovician–Lower Cretaceous strata. The cover sequence shows a basically coherent lithostratigraphy, although affected by faulting and folding and containing several stratigraphic breaks represented by unconformities. The belt is bounded to the west by a left-lateral strike-slip fault (the Hatakawa Tectonic Line) which separates it from the Abukuma Belt, which

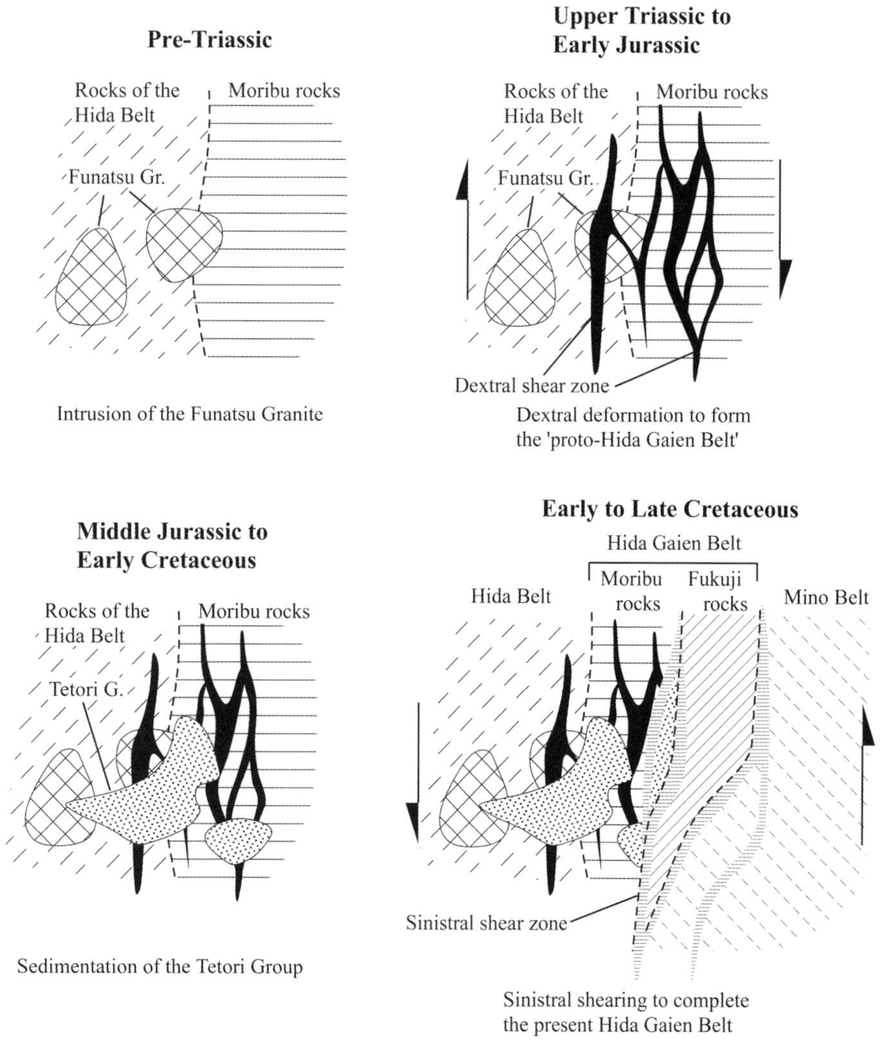

Fig. 2a.17. Schematic model for tectonic development of the Hida Gaien Belt.

comprises metamorphosed Jurassic accretionary complexes and Early Cretaceous (130–110 Ma, 100–90 Ma) granitic rocks (Fig. 2a.18). To the north it is in fault contact with Late Palaeozoic (Nedamo Belt) and Jurassic (North Kitakami Belt) accretionary complexes. The southernmost part of the Abukuma Massif, where the so-called 'Hitachi Palaeozoic' metamorphic rocks crop out (see Chapter 2b), is probably a southern extension of the South Kitakami Belt. The lower part of this Hitachi sequence is considered to be Cambrian in age (Tagiri *et al.* 2010, 2011), whereas the lithofacies and ages of the upper part are similar to the Upper Palaeozoic rocks of the South Kitakami Belt.

The Kurosegawa Belt in the Outer Zone of SW Japan is a southern extension of the South Kitakami Belt. It occupies an east–west-trending narrow zone from Kanto to Kyushu, and is in fault contact on both sides with Jurassic accretionary complexes. There are differing opinions on the tectonic division of SW Japan and tectonic setting of the Kurosegawa Belt (e.g. Matsuoka *et al.* 1998; Yao 2000; Yamakita & Otoh 2000; Isozaki *et al.* 2010). The Kurosegawa Belt includes the Kurosegawa Tectonic Zone (Ichikawa *et al.* 1956), a Late Palaeozoic accretionary complex and its metamorphic equivalent, and well-bedded clastic sequences of Late Palaeozoic, Triassic and Jurassic ages (Yoshikura *et al.* 1990; Hada *et al.* 2000). The Kurosegawa Tectonic Zone is characterized by the pre-Silurian basement, Siluro-Devonian covering strata, and serpentinite and tectonic blocks included therein. The Late Triassic shallow-marine strata unconformably rest not only on the Permian shallow-marine strata, but also on Late Palaeozoic accretionary complexes. The Kurosegawa rocks are well known in Kyushu and Shikoku (especially central Shikoku).

Basement rocks and Early–Middle Palaeozoic cover in the South Kitakami Belt

In the South Kitakami Belt the basement and Middle Palaeozoic strata are distributed in three separate areas, each showing different lithostratigraphic successions: the Nagasaka–Soma district in the west; the Miyamori–Hayachine–Kamaishi district in the north; and the Hikoroichi–Setamai district in the eastern-central part of the belt (Fig. 2a.19).

Nagasaka–Soma district

The Soma district in the eastern margin of the Abukuma Massif is isolated *c.* 150 km to the south of the Nagasaka district of the Southern Kitakami Massif, due to Early Cretaceous left-lateral strike-slip faulting (Otsuki & Ehiro 1992). The basement exposed in the Nagasaka-Soma district comprises the Matsugadaira–Motai Metamorphic Complex and the Shoboji Diorite (Fig. 2a.20).

The Matsugadaira–Motai Metamorphic Complex is of high-pressure/low-temperature type, containing alkali-amphiboles and pumpellyite (Kanisawa 1964; Maekawa 1988), and individual

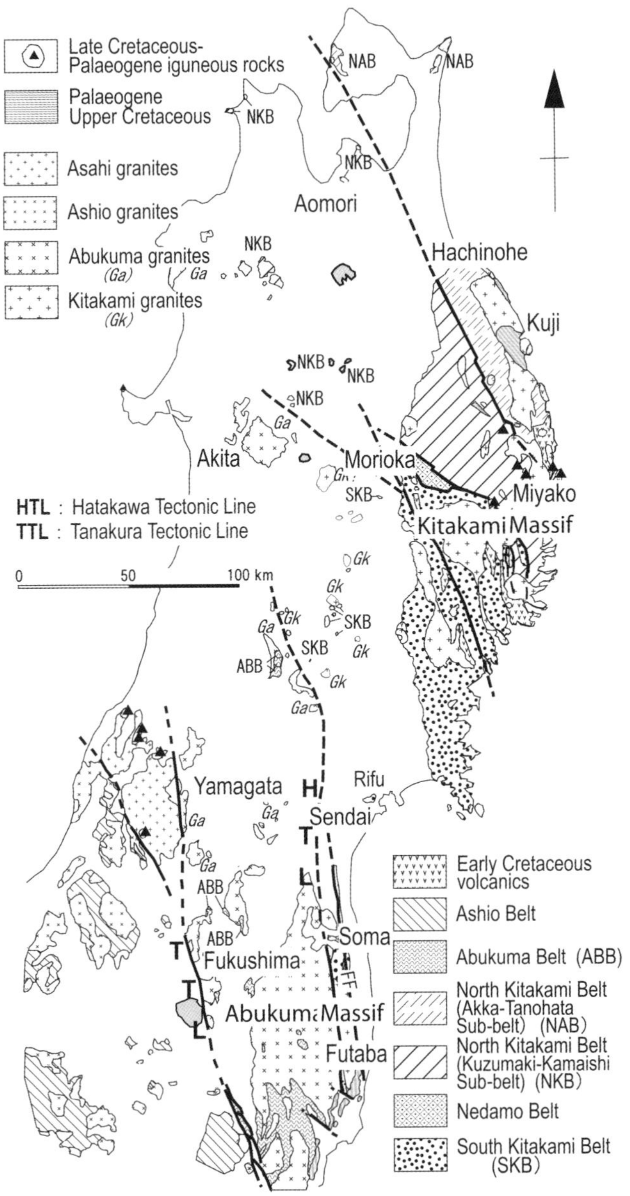

Fig. 2a.18. Geotectonic division map of the pre-Neogene of NE Japan.

metamorphic units have been given various local names such as Motai and Unoki in the Nagasaka district and Matsugadaira, Suketsune, Yamagami and Wariyama in the Soma district (Figs 2a.18 & 2a.19). These metamorphic outcrops comprise amphibolites (basalt and gabbro origin), greenschist, pelitic schist and serpentinized ultrabasic rocks, with subordinate amounts of siliceous and psammitic schists. The chemical composition of the basaltic rocks is similar to that of MORB (mid-oceanic ridge basalts) (Tanaka 1975; Kawabe *et al.* 1979) and the siliceous schist is considered to be derived from chert. The metamorphic lithologies are a mixture of rocks of both continental and oceanic origin and some parts show block-in-matrix mélange structure (Maekawa 1981). The metamorphics are therefore considered to be of accretionary complex origin (Umemura & Hara 1985; Ehiro & Okami 1991). The Matsugadaira Unit is unconformably overlain by the Upper Devonian Ainosawa Formation (Ehiro & Okami 1990). A pre-Late Devonian age is also indicated in the Nagasaka district by the common occurrence of metamorphic clasts in the Upper Devonian Tobigamori Formation (Ehiro & Okami 1991). Sasaki *et al.* (1997) reported an unconformity outcrop between Motai ultrabasic rocks and covering middle Palaeozoic sediments of the Tobigamori Formation (see below) at Natsuyama. Finally, K–Ar hornblende metamorphic ages of *c.* 500 Ma have been obtained from banded amphibolites (Kanisawa *et al.* 1992), indicating the basement rocks to be at least Cambrian in age.

The Shoboji Diorite is sporadically distributed as small bodies near and to the east of Shoboji, Nagasaka district, and is in fault contact with Motai metamorphic rocks and the lower part of the Tobigamori Formation. It is composed of nearly massive diorite to gabbro, and not thought to have been overprinted by the Matsugadaira–Motai high-pressure/low-temperature metamorphism. N-MORB normalized incompatible trace element patterns show that these plutonic rocks were derived from subduction zone magma (Kanisawa & Ehiro 1997), with an intrusion age considered to be latest Ordovician (K–Ar hornblende ages: *c.* 440 Ma; Kanisawa & Ehiro 1997) or latest Cambrian (U–Pb zircon age: 493 Ma; Isozaki *et al.* 2015).

The overlying Middle Palaeozoic strata comprise the Upper Devonian (to lowermost Carboniferous) Tobigamori (Nagasaka district) and Ainosawa (Soma district) formations, meaning that Silurian–Middle Devonian rocks are missing. The Tobigamori Formation consists mainly of mudstone and thin alternating beds of sandstone and mudstone, with purple-coloured tuff and tuff breccia interbedded in the middle part. Metabasite clasts similar to the Motai amphibolites, along with minor amounts of schist clasts, are abundant in the breccia (Fig. 2a.21). The Ainosawa Formation consists of tuffaceous mudstone and mudstone above a basal pebbly sandstone that contains schist and granitic clasts and unconformably covers the underlying metamorphic Matsugadaira Unit (Ehiro & Okami 1990).

Yabe & Noda (1933) reported the Late Devonian (Famennian) brachiopod *Spirifer verneuili* (*Cyrtospirifer tobigamoriensis*; Noda & Tachibana 1959) from the Tobigamori Formation, these being the first Devonian fossils discovered in Japan. The middle part of the formation yielded the land plant *Leptophloeum rhombicum* (Tachibana 1950), whereas in the uppermost part Ehiro & Takaizumi (1992) reported two Late Devonian (and an earliest Carboniferous) ammonoids. The main part of the formation is therefore assumed to be Late Devonian in age, ranging up to earliest Carboniferous. The Ainosawa Formation has also yielded *Cyrtospirifer* (Hayasaka & Minato 1954) and *Leptophloeum* (Koseki & Hamada 1988; Tazawa *et al.* 2006).

Miyamori–Hayachine–Kamaishi district

In this district the Hayachine Complex (Ehiro *et al.* 1988) and its equivalents are widely distributed and comprise the Nakadake Serpentinite, Kagura Unit and Koguro Formation in the eastern part of the Mt Hayachine district (Fig. 2a.20), with similar rocks also cropping out in the Kamaishi and Miyamori districts (Fig. 2a.19). The Nakadake Serpentinite consists of serpentinized peridotite, pyroxenite and hornblendite with subordinate amount of gabbro. The lower part of the Kagura Unit consists of schistose gabbro, peridotite and small amounts of dolerite, whereas the upper part is composed of a dolerite and trondhjemite sheeted dyke complex (Fig. 2a.22) sometimes including gabbro. The Koguro Formation consists mainly of dolerite and basalt, with intercalated thin beds of tuffaceous sandstone and mudstone, and reddish hematite-quartz rock in the upper part. The Hayachine Complex was once inferred to have been formed in a rift zone (Osawa 1983; Ehiro *et al.* 1988) but N-MORB normalized patterns indicate island arc basalt (Mori *et al.* 1992), an origin also inferred by the presence of abundant hornblende in the mafic and ultramafic rocks (Ozawa 1984). Since the overlying Palaeozoic strata yield Silurian fossils, the Hayachine

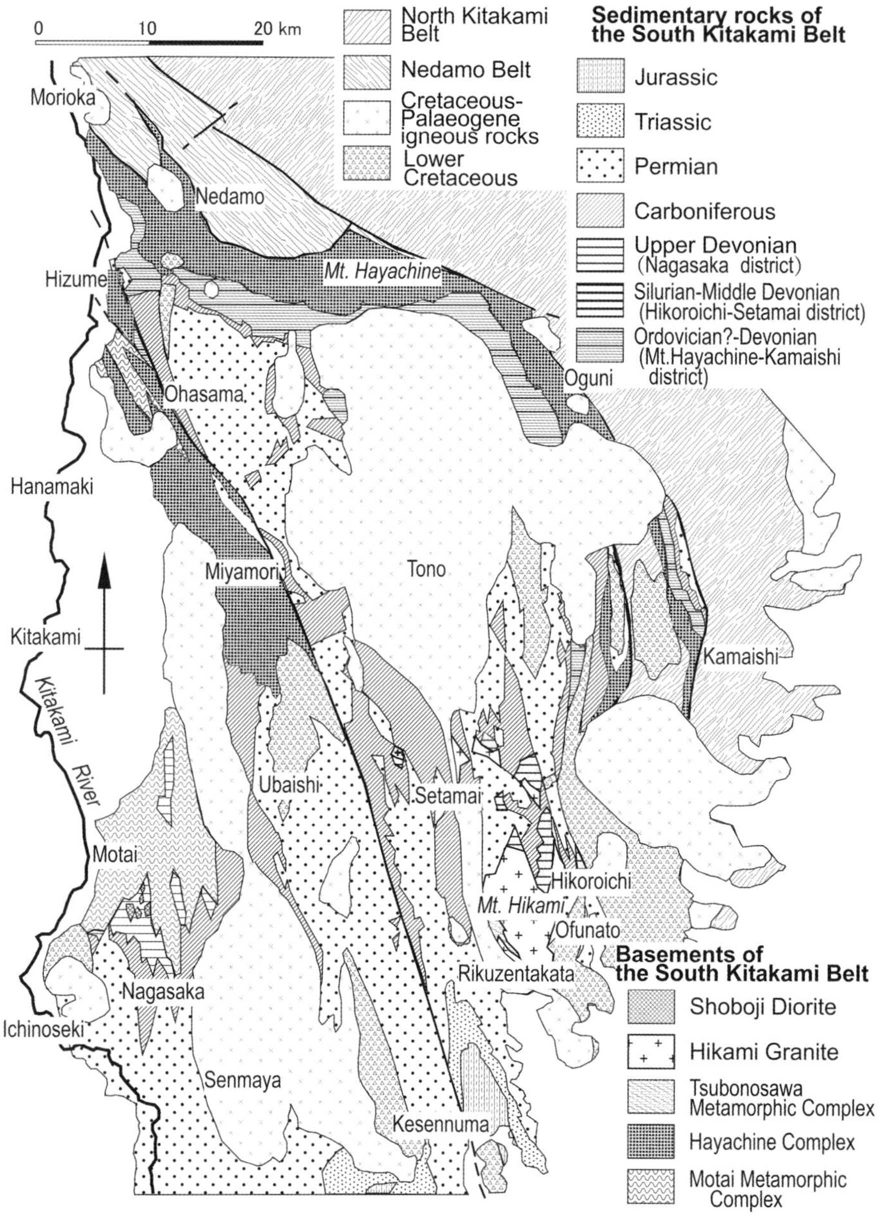

Fig. 2a.19. Geological map of the South Kitakami Belt in the northern half of the Southern Kitakami Massif, NE Japan.

Complex has been considered to be of Ordovician age (Okami *et al.* 1986; Ehiro *et al.* 1988). K–Ar ages of the complex show a wide range of 244–484 Ma (Ozawa *et al.* 1988; Shibata & Ozawa 1992), probably due to the thermal effect of the Early Cretaceous granites. Shimojo *et al.* (2010) reported Middle and Late Ordovician zircon LA-ICP-MS U–Pb ages from the upper part of the complex (466 $\pm$ 5 Ma) and the lowermost part of the overlying Yakushigawa Formation (457 $\pm$ 10 Ma).

The Lower–Middle Palaeozoic cover strata in the eastern Mt Hayachine district comprise the Yakushigawa and Odagoe formations in ascending order. The Yakushigawa Formation conformably overlies the Koguro Formation and consists of mudstone and sandstone in association with tuff and basalt in the lower part. The Odagoe Formation conformably covers the Yakushigawa Formation and is composed of mudstone and siliceous mudstone with impure limestone in the main part and basalt, tuff and limestone in the upper part. A Silurian brachiopod was reported from the lower part (Ehiro *et al.* 1986). In the western Mt Hayachine district, the cover strata comprise the fault-bounded Nameirizawa and Orikabetoge formations and undivided Devonian (in ascending order). The Nameirizawa Formation is lithologically similar to the Yakushigawa Formation, whereas the Orikabetoge Formation is composed of conglomerate, arkose and mudstone with subordinate amounts of limestone and tuff, and a rich fauna of Silurian corals and trilobites (Okami *et al.* 1986). The conglomerate contains many granitic clasts lithologically similar to those exposed in the Hikami granitic complex of the Hikoroichi–Setamai district (see following section). Finally, undivided Devonian strata in this district comprise tuff, mudstone, sandstone and conglomerate.

In the Kamaishi district the uppermost Silurian–Upper Devonian Senjyogataki Formation rests on the Kentosan Unit (equivalent of the Kagura Unit) of the Hayachine Complex, although their stratigraphic relationship is unclear. The Senjyogataki Formation consists, from base to top, of basalt and dolerite, tuff and alternating beds of tuffaceous and siliceous mudstones, and mudstone (Okami *et al.* 1987). The tuffaceous-siliceous mudstone yielded latest Silurian

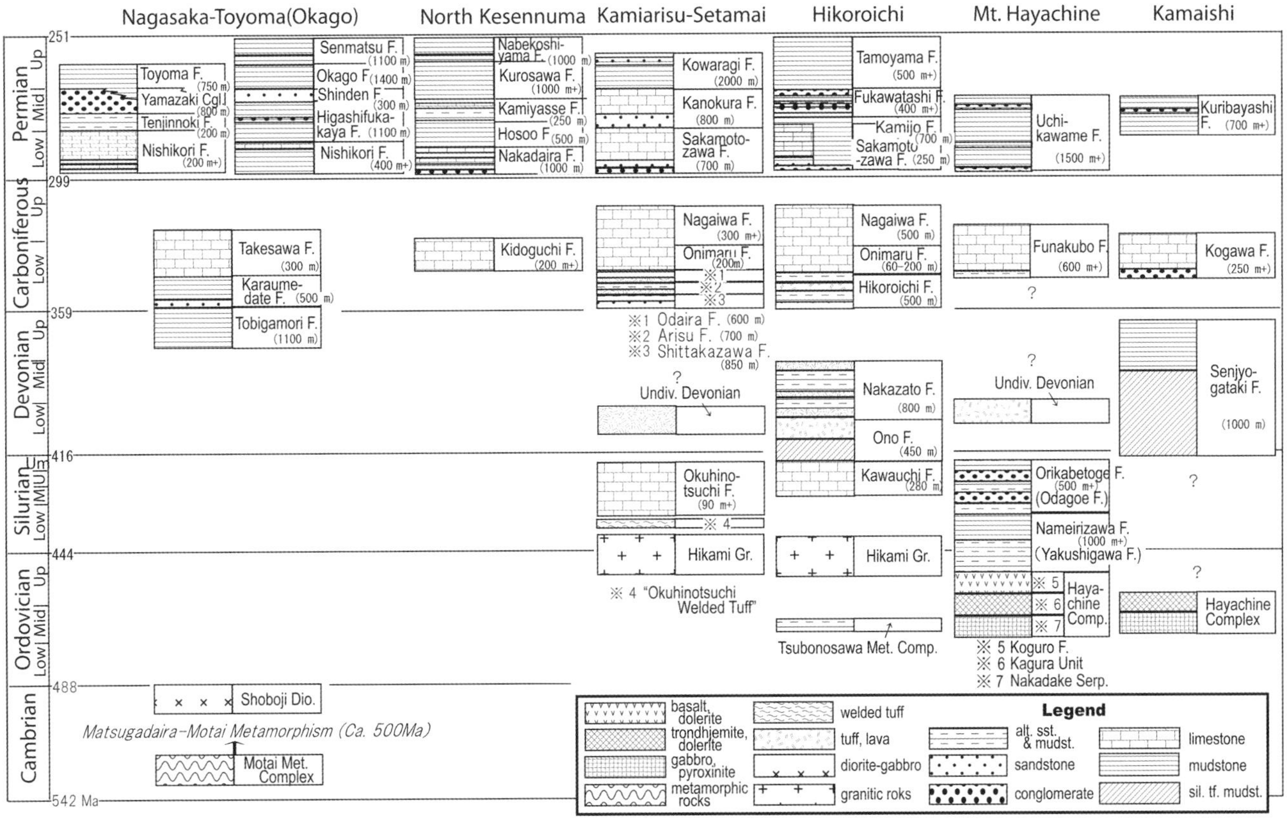

Fig. 2a.20. Stratigraphy of the Palaeozoic of the South Kitakami Belt, NE Japan. The Thicknesses of the strata are given in parentheses. alt., alternating; Comp., Complex; Dio., Diorite; F., Formation; GR., Granite; mudst., mudstone; Met., Metamorphic; Serp., Serpentinite; sil., siliceous; sst., sandstone; tf., tuff.

(Pridolian) (Suzuki *et al.* 1996) to Early Devonian radiolarians (Umeda 1998*a*), and the uppermost part of the formation contains the Late Devonian (Famennian) plant *Leptophloeum* (Okami *et al.* 1987).

Hikoroichi–Setamai district

The Hikami Granite (Murata *et al.* 1974) and Tsubonosawa Metamorphic Complex (Ishii *et al.* 1956) form the basement in this district. The largest body of the Hikami Granite, which consists of massive or schistose granite, granodiorite and tonalite, crops out around Mt Hikami (Fig. 2a.19), with other bodies forming rather small and scattered outcrops in the Komatsu-Pass, Shiraishi-Pass, Yokamachi, Ezo, Hirasawa and Okuhinotsuchi areas. All of these exposures, except for that at Hirasawa, are associated with Middle Palaeozoic sedimentary rocks of Silurian–Middle Devonian in age.

Fig. 2a.21. Polished specimen of tuff breccia from the middle part of the Upper Devonian Tobigamori Formation, showing the common occurrence of amphibolite clasts (am).

Ishii *et al.* (1956) divided the Hikami outcrop into 'Hikamisan-type' and 'Ono-type', while Asakawa *et al.* (1999) later divided it into 'D-type', most of which show schistosity, and massive 'S-type', with roughly coincides with the Ono-type. Kobayashi & Takagi (2000) considered that the massive Ono-type granodiorite intruded after the schistose granites had been mylonitized. The radiometric ages obtained from the Hikami Granite are 442 Ma (Watanabe *et al.* 1995; SHRIMP U–PB age) and 440 Ma (Asakawa *et al.* 1999; Rb–Sr whole-rock isochron age). Some younger ages (Devonian–Permian) had been reported (e.g. CHIME ages: Suzuki & Adachi 1991; Suzuki *et al.* 1992; Adachi *et al.* 1994; LA-ICP-MS U–Pb age: Shimojo *et al.* 2010) but these are in conflict with the fact that the Hikami plutons are unconformably overlain by Silurian strata (see the following paragraph). Moreover, a granitic clast included in the lower part of the Ono Formation (Pridolian) has been dated at *c.* 250 Ma by Suzuki *et al.* (1992), so it seems likely that the post-Silurian ages obtained from the Hikami Granite may be related to later thermal events. Finally, the Tsubonosawa Metamorphic Complex comprises contact metamorphic lithologies derived from mudstone and sandstone protoliths of probable Cambro-Ordovician age, and included in the Hikami Granite (Hikamisan-type) as xenoblocks of various sizes. Detrital zircon grains from these metasediments have ages in the range

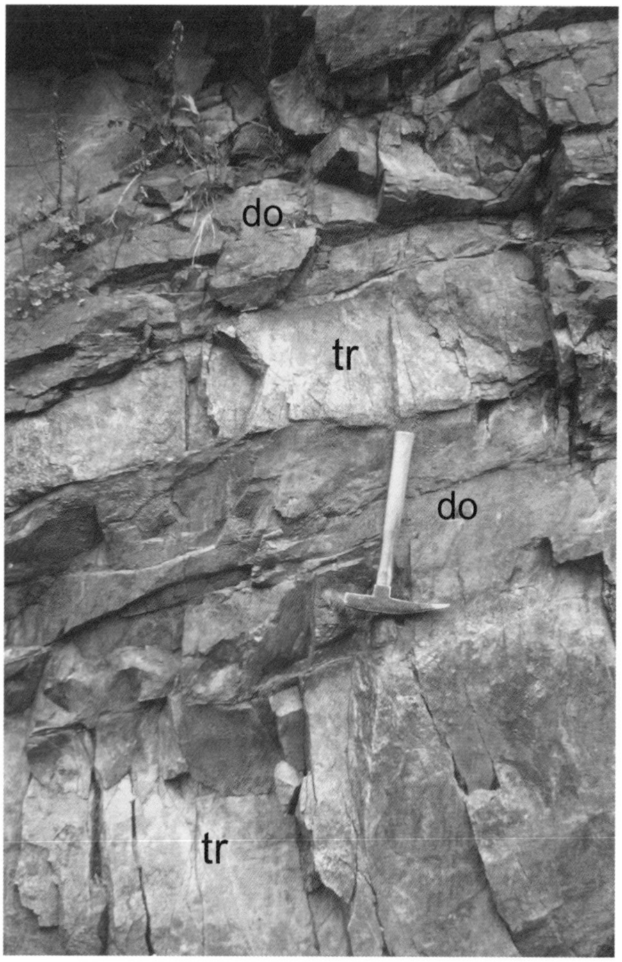

Fig. 2a.22. Photograph of an outcrop showing the sheeted dyke complex of the Kagura Unit of the Hayachine Complex. tr, trondhjemite; do, dolerite.

3800–500 Ma and euhedral zircon grains yield ages of *c.* 440 Ma (CHIME ages: Suzuki & Adachi 1991; SHRIMP U–Pb ages: Watanabe *et al.* 1995).

The Middle Palaeozoic strata in this district comprise the Silurian Kawauchi Formation (and its equivalents), the latest Silurian–Early Devonian Ono Formation and the Middle Devonian Nakazato Formation; Upper Devonian rocks are absent (Fig. 2a.20). Onuki (1937) first reported Silurian fossils from this district, this being the first record of Silurian fossils in Japan. The Kawauchi Formation consists mainly of limestone with basal arkose, and is associated with calcareous mudstone in the upper part. It has yielded many corals (*Schedohalysites*, *Falsicatenipora*, *Favosites*, *Heliolites*, etc.) and trilobites (*Encrinurus*), and is considered to be Middle–Late Silurian in age (Kato *et al.* 1980). The basal arkose unconformably covers the Hikami plutonic basement (Murata *et al.* 1974, 1982; Kitakami Paleozoic Research Group 1982). In the Okuhinotsuchi district, the Early–Late Silurian Okuhinotsuchi Formation (Kawamura 1980) rests unconformably on a welded tuff (the Okuhinotsuchi Welded Tuff; Murata *et al.* 1982), which in turn unconformably covers the Hikami Granite.

The Ono Formation is divided into the Oh1, Oh2 and Oh3 members (Minato *et al.* 1979). The Oh1 Member, which has yielded Pridolian radiolarians (Umeda 1996*b*), is a slump bed which incorporates variously sized clasts of granite, arkose and limestone in a tuffaceous and siliceous mudstone. These clasts are lithologically similar to the underlying Hikami Granite and basal arkose and limestone of the Kawauchi Formation, respectively. The Oh2 Member is composed of acidic tuff and alternating beds of acidic tuff and tuffaceous-siliceous mudstone, and the Oh3 Member similarly consists mainly of tuff with subordinate amount of tuffaceous sandstone and mudstone. The Nakazato Formation conformably overlies the Ono Formation and is divided into the N1, N2, N3 and N4 members (Minato *et al.* 1979). The N1 Member consists mainly of basic tuff and lapilli tuff in association with mudstone; the N2 Member is composed of alternating beds of acidic tuff and mudstone; and the N3 Member comprises alternating beds of sandstone and mudstone, a part of which is tuffaceous. The lower part of the N4 Member consists of sandstone and pebbly sandstone, and the upper part of it consists of sandstone and mudstone with subordinate amounts of tuff. The N3 Member has yielded Middle Devonian trilobites (Kobayashi & Hamada 1977; Minato *et al.* 1979) and Middle Devonian (Givetian) radiolarians (Umeda 1996*b*).

Basement rocks and Middle Palaeozoic cover in the Kurosegawa Belt

Basement rocks

The basement rocks of the Kurosegawa Belt in Shikoku are composed of the Terano Metamorphic Complex (including the Miyagadani Metamorphic Complex) and Mitaki Igneous Complex (Ichikawa *et al.* 1956). The Terano Metamorphic Complex comprises medium-pressure type (Karakida 1981) biotite gneiss and amphibolites with radiometric ages ranging from Ordovician to Jurassic (mostly clustering around 400 Ma; Yoshikura *et al.* 1990), and is in fault contact with the Mitaki Igneous Complex. U–Pb ages of detrital zircons from the Terano Metamorphic Complex show peaks at 450–500 and *c.* 600 Ma (Yoshimoto *et al.* 2013). The garnet-clinopyroxene granulite and amphibolite of the Kurosegawa Belt in Kyushu District have Sm–Nd ages of *c.* 420, and 490 and 540 Ma, respectively (Osanai *et al.* 2000). The Mitaki lithologies – which include the Yokokurayama Granite (Yasui 1984) or Gomi Granite (Yoshikura 1985) – are typically medium- to coarse-grained, commonly sheared, calc-alkaline granitic rocks ranging from granite to diorite (mostly granodiorite) with U–Pb zircon ages of *c.* 440–442 Ma (Hada *et al.* 2000).

Middle Palaeozoic covering strata

Middle Palaeozoic strata in central Shikoku (Fig. 2a.23) are represented by the Siluro-Devonian Yokokurayama Group which is

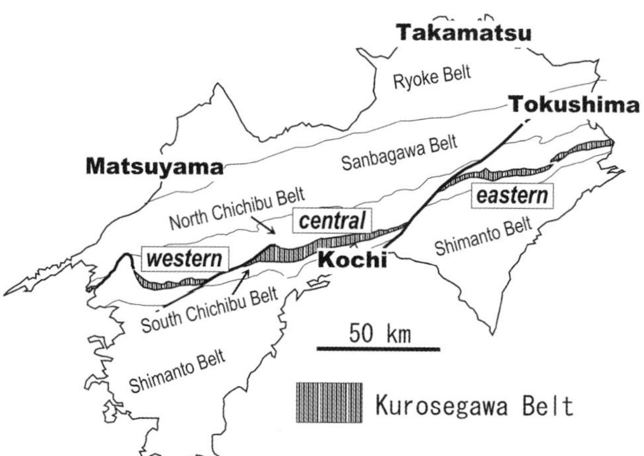

Fig. 2a.23. Map showing the distribution of the Kurosegawa Belt in Shikoku, SW Japan.

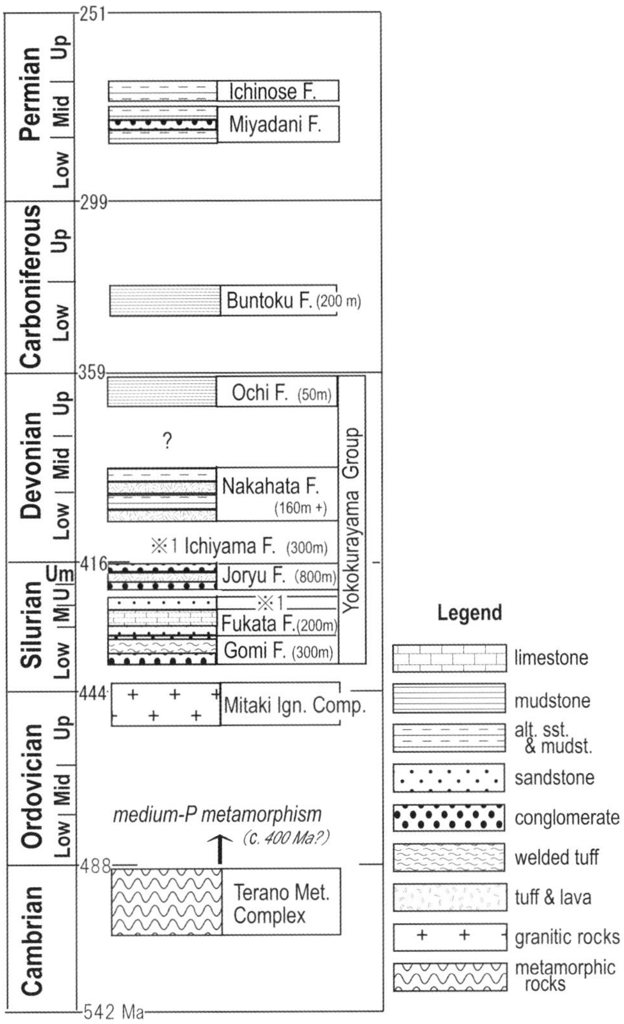

Fig. 2a.24. Stratigraphy of the Palaeozoic of the Kurosegawa Belt in the central part of Shikoku, Southwest Japan. Thicknesses of the strata are given in parentheses. alt., alternating; Comp., Complex; Ign., Igneous; F., Formation; mudst., mudstone; Met., Metamorphic; sst., sandstone.

divided into the Gomi, Fukata, Ichiyama, Joryu, Nakahata and Ochi formations, in ascending order (Umeda 1998*b*; Fig. 2a.24). The stratigraphic relationships among the first three formations are conformable, whereas there are hiatuses between the other formations, based on the ages of radiolarians (Umeda 1998*b*, *c*). The Okanaro Group (Ichikawa *et al.* 1956; Umeda 1994) in western Shikoku, the Suberidani Group (Hirayama *et al.* 1956) in eastern Shikoku and the Gionyama Formation (Hamada 1959*c*) in Kyushu are equivalent strata, but exclude the equivalent of the Ochi Formation.

The lower part of the Gomi Formation consists of basal conglomerate, which unconformably covers the Yokokurayama Granite (Yasui 1984), overlain by conglomeratic sandstone and sandstone. The main part is composed of welded tuff and intercalated acidic tuff, tuffaceous sandstone, sandstone and mudstone (Yoshikura & Sato 1976). A Wenlockian U–Pb SHRIMP age (427 Ma) has been reported from the welded tuff (Aitchison *et al.* 1996). The Fukata Formation (Umeda 1998*b*) is subdivided into a Lower Member, comprising alternating beds of sandstone and mudstone, and mudstone, with calcareous sandstone and conglomerate in its lower and upper parts, and an Upper Member comprising mainly limestone with limestone conglomerate, tuffaceous sandstone and mudstone. This formation yields late Wenlockian–early Ludlovian corals, trilobites (Hamada 1959*c*; Kobayashi & Hamada 1985), conodonts (Niko *et al.* 1989) and radiolarians (Wakamatsu *et al.* 1990). The corals are rich in halysitids such as *Halysites*, *Schedohalysites* and *Falsicatenipora*, and favositids. The Ichiyama Formation consists of conglomerate, tuffaceous sandstone and mudstone, and acidic tuff. Based on the radiolarians, this formation dated as early–middle Ludlovian (Umeda 1998*b*). The Joryu Formation consists of conglomerate in the lower part and tuffaceous sandstone, acidic tuff and conglomerate in the upper part. The acidic tuff contains rich Pridolian radiolarians (Umeda 1998*b*, *c*). The Nakahata Formation is dominated by acidic tuff and tuffaceous sandstone, including basal conglomerate-sandstone with abundant granite, rhyolite and acidic tuff clasts. This formation yields an abundant radiolarian fauna, ranging in age from middle Early Devonian to early Middle Devonian (Umeda 1998*b*, 1998*c*).

The Ochi Formation (Hirata 1966; Yoshikura 1982) mainly comprises alternating beds of sandstone and mudstone, with intercalated conglomerate, sandstone, mudstone and acidic tuff. It is in fault contact with the other formations (Yoshikura 1985), although Yasui (1984) has stated that there is a possibility that this formation unconformably rests on the underlying acidic-tuff-dominated Devonian. The Late Devonian plant *Leptophloeum rhombicum* has been reported (Hirata 1966), and a similar *Leptophloeum*-bearing clastic facies is known from Upper Devonian rocks in eastern Shikoku where conglomerate, sandstone and mudstone cover Bed G4 of the Suberidani Group in probable unconformity (Yasui & Okitsu 2007). In Kyushu, the Naidaijin Formation has also yielded *Leptophloeum* (Miyamoto & Tanimoto 1993; Saito *et al.* 2003).

Upper Palaeozoic strata of the South Kitakami Belt

Carboniferous and (especially) Permian shallow-marine strata are widely distributed in the South Kitakami Belt (Fig. 2a.19).

Carboniferous

The Carboniferous succession ranges in age from Tournaisian to Moscovian, lacks Late Carboniferous rocks, and can be divided into two lithologically distinct lower and upper sequences. The lower (Tournaisian–upper Visean) consists mainly of volcaniclastics with minor amounts of limestone, whereas the upper (upper Visean–Moscovian) is dominated by limestone and partly intercalated with tuff (Kawamura & Kawamura 1989*a*). The volcanics in these strata, as in the underlying Devonian sequence, show bimodal SiO_2 contents and belong to a volcanic-arc to back-arc tectonic setting (Kawamura & Kawamura 1989*b*).

The lower sequence comprises the Hikoroichi Formation (Hikoroichi district), the Shittakasawa, Arisu and Odaira formations (Setamai–Yokota district), the Karaumedate Formation (Nagasaka district), Mano Formation (Soma district) and their equivalents in other areas (Fig. 2a.20). They differ in their lithological component ratio and thickness, and are considered to have been deposited in different basins (Kawamura & Kawamura 1989*a*, *b*). Those in the Hikoroichi and Setamai–Yokota districts are dominated by volcanic rocks and are very thick, whereas those in the Nagasaka–Soma district are mainly composed of clastic rocks with minor amounts of tuff. In contrast, the Mano Formation in the Soma district is very thin. In most districts in the South Kitakami Belt, the Lower Carboniferous strata rest unconformably above Middle Devonian strata. In the Nagasaka–Soma district, however, the Tobigamori Formation ranges from Upper Devonian to lowermost Carboniferous and is covered conformably by the Lower Carboniferous Karaumedate Formation. The Hikoroichi Formation and its equivalents yield a rich fauna of corals, brachiopods and trilobites, and its lower–middle part and upper part are dated as Tournaisian–lower Visean and upper Visean, respectively (Mori & Tazawa 1980; Kawamura 1983; Tazawa 1984).

The upper sequence comprises the Onimaru and Nagaiwa formations in the Hikoroichi, Setamai and Yokota districts, and their equivalents in other areas. The Onimaru Formation is composed of bedded limestone and muddy limestone, and the Nagaiwa Formation consists of massive limestone or limestone with siliceous nodules and sometimes intercalated pyroclastic rocks. The Onimaru Formation is rich in corals belonging to the *Kueichouphyllum* fauna. It is biostratigraphically divided into the *Kueichouphyllum glacial–Actinocyathus japonicus* Zone, *Saccaminopsis* Zone, *Arachnolasma–Palaeosmilia regia* Zone and Barren Zone, and dated as upper Visean (Niikawa 1983*a*, 1983*b*). The Nagaiwa Formation is divided into the *Millerella* and *Profusulinella* zones (fusulinoideans; Kobayashi 1973) or *Declinognathodus noduliferus* and *Idiognathoides sulcatus* zones (conodont; Minato *et al.* 1979), and dated as Serpukhovian–Moscovian.

Permian

The Permian succession in the South Kitakami Belt consists mainly of shallow-marine clastic sediments and limestone, almost entirely lacks pyroclastic rocks (unlike the Devonian and Carboniferous sequences), and has been divided into the Sakamotozawan, Kanokuran and Toyoman series (Fig. 2a.20) in ascending order (e.g. Minato *et al.* 1978, 1979).

The Sakamotozawa Formation (the type of the Sakamotozawan Series) in the Hikoroichi district consists of basal conglomerate, sandstone and alternating beds of sandstone and mudstone in the lower part, massive limestone with mudstone in the main part and sandstone and mudstone in the uppermost part (Kanmera & Mikami 1965*a*). Equivalent strata of the Sakamotozawa Formation in other areas also comprise basal conglomerate or sandstone, sandstone, mudstone and limestone, although the amounts of these different components varies from place to place (Ehiro 1989, 2016). The formation unconformably overlies the Carboniferous Nagaiwa Formation or its equivalents but without any notable structural break, except in the Setamai district where it progressively oversteps various horizons of the Carboniferous from the uppermost Shittakasawa Formation to the Nagaiwa Formation (Saito 1968). The Sakamotozawa Formation and its equivalents yield a rich fauna of fusulinoideans and corals. Kanmera & Mikami (1965*b*) established five fusulinoidean zones – the *Zellia nunosei*, *Monodiexodina langsonensis*, *Pseudofusulina vulgaris*, *P. fusiformis* and *P. ambigua* zones – and dated the formation as Sakmarian–Artinskian. Ueno *et al.* (2009) extended the uppermost part into the Kungurian stage, based on the occurrence of *Misellina* sp. The Sakamotozawan Nishikori Formation in the Toyoma district yielded plant fossils belonging to the Cathaysian flora such as *Cathaysiopteris*, *Sphenopteris*, *Odontopteris*, *Pecopteris*, *Taeniopteris* and *Cordites* (Asama 1956, 1967).

The Kanokura Formation (the type of the Kanokuran Series) in the Setamai district rests conformably on the Sakamotozawa Formation, and comprises a lower sequence of conglomeratic sandstone, sandstone and mudstone overlain by muddy limestone and lenticular limestones. The upper part of this formation consists mainly of massive limestone, but changes laterally to calcareous sandstone or mudstone (Ehiro 1989, 2015). Thick conglomerate beds, called the Usuginu-type Conglomerate, are sometimes intercalated in the Kanokuran Series (especially in the upper part) and the overlying Toyoman Series. The conglomerate contains well-rounded pebbles to boulders rich in granitic clasts in a muddy or sandy matrix. Permian–Triassic radiometric ages have been known from these granitic clasts (K–Ar ages of 276–225 Ma, Shibata 1973; CHIME ages of 257–244 Ma, Takeuchi & Suzuki 2000). Various fossils such as fusulinoideans, corals, brachiopods and molluscs have been reported from the Kanokura Formation (e.g. Minato *et al.* 1978), allowing Choi (1973) to establish three fusulinoidean zones: the *Monodiexodina matsubaishi* Zone (lower part, which has also yielded *Cancellina*), *Colania kotsuboensis* Zone (uppermost part of the lower) and *Lepidolina multiseptata* Zone (upper part). Based on these fusulinoideans, the Kanokura Formation is dated as Roadian–Capitanian.

The Permian in the northern Kesennuma district comprises the Nakadaira, Hosoo, Kamiyasse, Kurosawa and Nabekoshiyama formations in ascending order. The Hosoo Formation is dominated by mudstone, and its upper and uppermost parts yield Roadian and Wordian ammonoids, respectively (Ehiro & Misaki 2005). The Kamiyasse Formation consists of calcareous sandstone and mudstone with limestone, and contains Wordian–Capitanian fusulinoideans and ammonoids (Ehiro & Misaki 2005) with a rich fauna of brachiopods. Finally, the lower part of the Kurosawa Formation, in which mudstone is dominant, contains Capitanian ammonoids (Ehiro & Araki 1997) along with the agassizodontid shark *Helicoprion* sp. (Araki 1980).

The Toyoman Series includes the Toyoma Formation (Toyoma, Ogatsu and southern Kesennuma districts), Okago and Senmatsu formations (Okago district), the main part of the Suenosaki and the Tanoura formations (Utatsu district), the upper part of the Kurosawa and the Nabekoshiyama formations (northern Kesennuma district) and the main part of the Kowaragi Formation (Karakuwa district), as well as various unnamed beds in the Ofunato district and other places. Compared to the Lower–Middle Permian successions of the South Kitakami Belt, these younger formations are lithologically monotonous and dominated by mudstone with minor sandstone and rare limestone and conglomerate. The mudstones of these strata typically show strong slaty cleavage, except for those in the Okago district, and so have been used as roofing slate. The Toyoma and equivalent formations yield Wuchiapingian ammonoids (Murata & Bando 1975; Ehiro & Bando 1985; Ehiro 2010), whereas the Senmatsu, Tanoura and Nabekoshiyama formations, as well as unnamed beds in the Ofunato district, contain Changhsingian fusulinoideans, smaller foraminifers and ammonoids (Ishii *et al.* 1975; Tazawa 1975; Murata & Shimoyama 1979; Ehiro 1996; Kobayashi 2002).

Upper Palaeozoic covering strata in the Kurosegawa Belt

Carboniferous

In the Kurosegawa Belt in Shikoku Carboniferous rocks are represented by the shallow-marine Buntoku Formation (Nakai 1980), fragments of which are dispersed within a fault zone in Yokokurayama district, central Shikoku. It consists mainly of mudstone intercalated with basic tuff containing limestone lenses which have yielded Late Visean corals such as *Lithostrotion, Palaeosmilia, Diphyphyllum* and *Carcinophyllum*. Equivalent strata also crop out in Kyushu (Yuzuruha Formation, Miyamoto & Tanimoto 1993; Kakisako Formation, Kanmera 1952) where they are dominated by limestone with a late Visean *Kueichouphyllum*-coral fauna (Kanmera 1952; Miyamoto & Tanimoto 1993; Kido *et al.* 2007).

Permian

Permian strata in central Shikoku are represented by the shallow-marine Miyadani and Ichinose formations (Hada *et al.* 1992), these having equivalents in western Shikoku as the Miyanaro and Doi formations, respectively. They are all in fault contact with other strata including Jurassic accretionary complexes. The Miyadani and Miyanaro formations mainly consist of sandstone, mudstone and alternating beds of sandstone and mudstone, in association with conglomerate and rare acidic tuff. The latter also includes small limestone lenses with early–middle Middle Permian fusulinoideans (Hada 1974) and siliceous mudstones with late Early Permian radiolarians (Hada *et al.* 1992). The Ichinose and Doi formations are composed of sandstone, conglomerate and mudstone, with subordinate amount of acidic tuff and limestone lenses. Limestone lenses yield the late Middle Permian fusulinoid *Lepidolina*, and radiolarians from these formations are also late Middle Permian in age (Hada *et al.* 1992). Similarly, the middle–late Middle Permian ammonoid *Cibolites* has been reported from the Katsura Sandstone of the Ichinose Formation (Group) (Koizumi *et al.* 1994). The Haigyu Group (Hirayama *et al.* 1956) in eastern Shikoku is also Middle Permian in age, whereas to the west in Kyushu coeval siliciclastic strata comprise the Early Permian Tsurukoba Formation (Miyamoto *et al.* 1997), Middle Permian Kozaki Formation (Kanmera 1961, 1963) and the late Middle–Late Permian Kuma Formation (Kanmera 1953, 1954).

Late Palaeozoic–Mesozoic metamorphic rocks and serpentinite in the Kurosegawa Belt

Various kinds of metamorphic rocks (excluding the Terano Metamorphic Complex) and serpentinites occur in the Kurosegawa Belt, although their relationships with the basements and their covering strata are not clear. The metamorphic rocks were known as the Ino Formation, which consists of low-grade metamorphic rocks and tectonic blocks of high-grade metamorphic rocks in serpentinite and Ino Formation. Wakita *et al.* (2007) referred to the former as the Younger Ino Metamorphic Complex and the latter as the Older Ino Metamorphic Complex.

High-grade metamorphic rocks

The Older Ino Metamorphic Complex occurs as lenticular small bodies in and around the Younger Ino Metamorphic Complex and is composed mainly of mafic schist with minor amount of pelitic schist. It records an epidote-glaucophane schist subfacies to albite-epidote-amphibolite subfacies metamorphism (Wakita *et al.* 2007). The metamorphic age is estimated to be Carboniferous based on the K–Ar ages of phengite from mafic schists (327–317 Ma: Ueda *et al.* 1980; 350 Ma: Wakita *et al.* 2007). High-grade metamorphic rocks in serpentinite consist of basic, pelitic and psammitic schists and show glaucophene schist to pumpellyite-actinolite facies metamorphism (Nakajima & Maruyama 1978); a basic and siliceous schist block preserving jadeite-glaucophene facies metamorphism is also known. K–Ar ages of phengite from the schist are 240 and 208 Ma (Maruyama *et al.* 1978).

Younger Ino Metamorphic Complex

The Younger Ino Metamorphic Complex (Ino Formation) is rather widely distributed in the Kurosegawa Belt of central Shikoku and composed of basic, pelitic, psammitic and siliceous schists. It records pumpellyite-actinolite subfacies metamorphism and K–Ar ages of phengite in the range 185–148 Ma (Wakita *et al.* 2007). The unmetamorphosed part of the complex shows an Oceanic Plate Stratigraphy including the Upper Carboniferous–Upper Triassic bedded chert, Upper Triassic–Lower Jurassic siliceous mudstone and Lower Jurassic mudstone (Hori & Wakita 2004).

Serpentinite

The serpentinite is distributed along the fault zones as sheet-like to lenticular bodies up to several kilometres wide and their protoliths are considered to be dunite and harzburgite with lesser amount of lherzolite, based on relict minerals and pseudomorphic textures (Hada *et al.* 2001). Their chemistry is similar to subduction-related ultramafic rocks from the Mariana and Tonga trenches (Yoshikura & Miyaji 1988).

Triassic–Lower Cretaceous strata in the South Kitakami Belt

Triassic to lowest Cretaceous strata in the South Kitakami Belt were deposited in a shallow marine or alluvial environment and are mainly composed of clastic rocks in association with rare limestone and tuff. They are distributed in the southern part of the Southern Kitakami Massif (Figs 2a.18 & 2a.25) and in the eastern margin of the Abukuma Massif (Soma district), forming three large, south-plunging synclinal structures (Figs 2a.25 & 2a.26): the Shizugawa (Shizukawa)–Hashiura Sub-belt (Western Sub-belt); Karakuwa–Oshika Sub-belt (Central Sub-belt); and the Ofunato Sub-belt (Eastern Sub-belt) (Takizawa 1977, 1985). The stratigraphy, lithology and thickness of the strata are different in the three synclines.

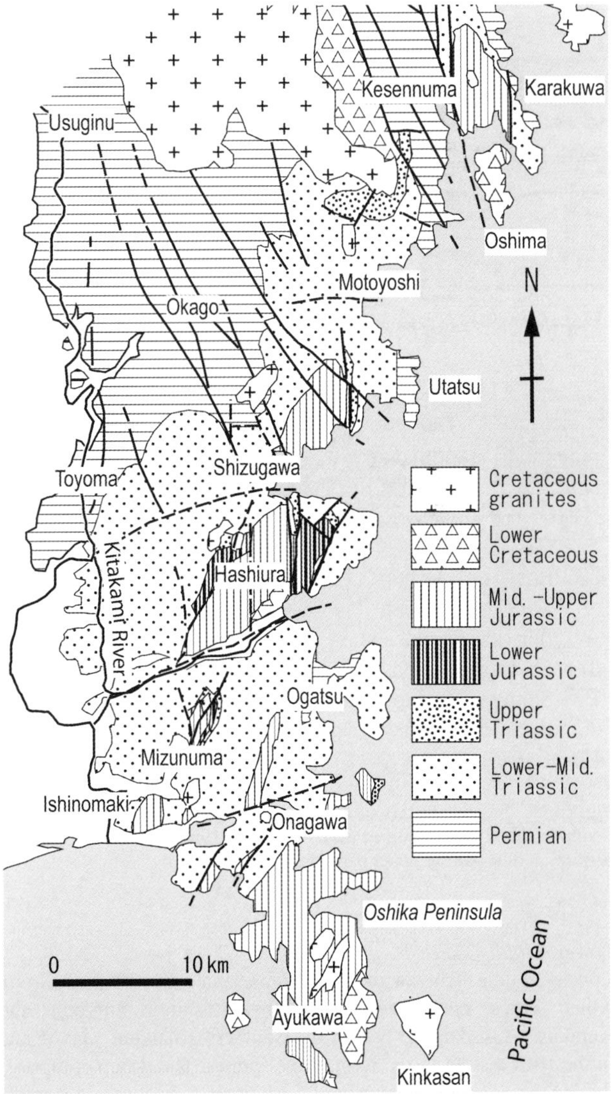

Fig. 2a.25. Geological map of the southern part of the Southern Kitakami Massif, showing the distribution of the Mesozoic strata of the South Kitakami Belt.

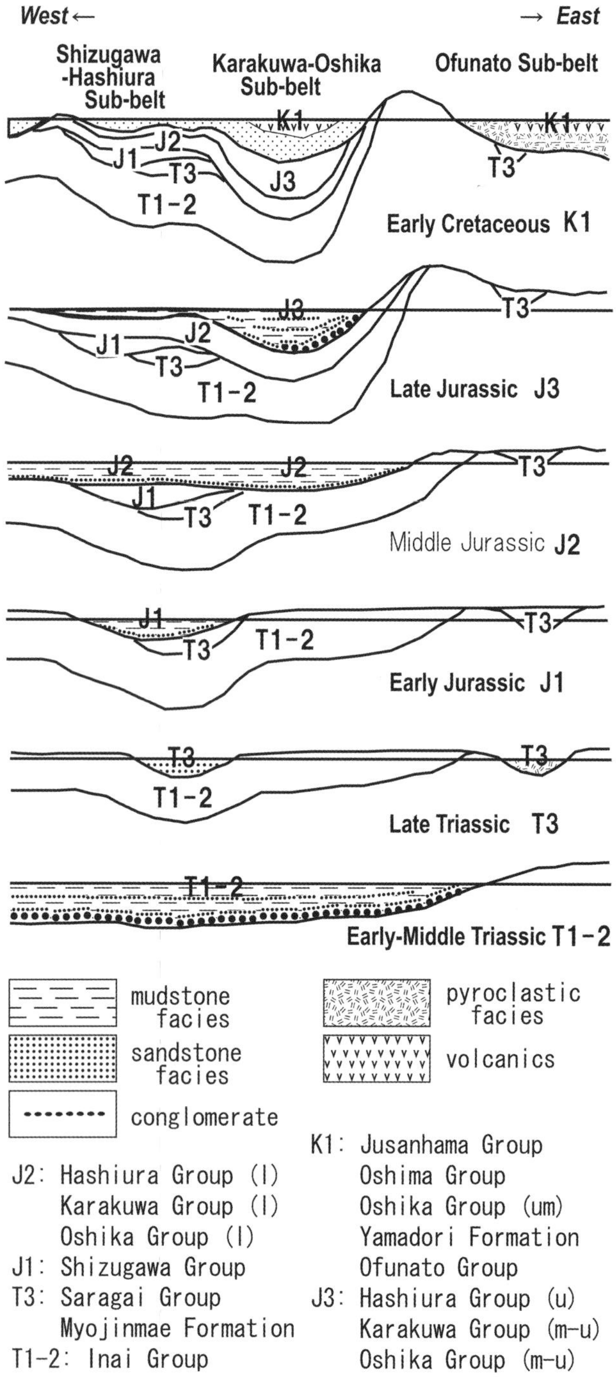

Fig. 2a.26. Tectonic history of the Mesozoic sedimentary basin in the South Kitakami Belt (modified from Yamashita 1957). l, lower part; m, middle part; u, upper part; um, uppermost part.

Lower–Middle Triassic

Lower–Middle Triassic rocks are represented by the Inai Group which crops out in the Shizugawa–Hashiura Sub-belt and Karakuwa–Oshika Sub-belt of the Southern Kitakami Massif and in the Rifu district to the west of the former. The group comprises the Hiraiso, Osawa, Fukkoshi and Isatomae formations in conformable ascending order (Onuki & Bando 1959). The Hiraiso Formation rests unconformably on the Upper Permian formations and consists of a basal conglomerate overlain by bedded calcareous sandstones followed by alternating beds of calcareous sandstone and mudstone in the upper part. The Osawa Formation consists mainly of calcareous laminated or massive mudstone, but sometimes contains lenticular slump beds of sandstones and conglomerates in the middle part (e.g. Kamada 1980, 1983). The Fukkoshi Formation is made up of sandstones and mudstones, and the thick Isatomae Formation (>1500 m) consists of laminated sandy mudstone to muddy sandstone, intercalated with thick sandstone beds.

The Osawa Formation has yielded the Ichthyosaurian fossil *Utatsusaurus hataii* in association with abundant ammonoids belonging to the *Columbites–Subcolumbites* fauna, and is dated as late Olenekian (Spathian) (Bando & Shimoyama 1974; Bando & Ehiro 1982). Bando (1970) described an early Induan (Griesbachian) ammonoid *Glyptophiceras* cf. *gracile* from the Hiraiso Formation, but it was probably derived from the Osawa Formation and the specific assignment needs to be re-examined (Ehiro 2002). The Hiraiso Formation rests unconformably on the various horizons of the Upper Permian (Wuchiapingian–Changhsingian), and a stratigraphic interval from the lower Induan (uppermost Changhsingian?) to lower Olenekian is missing (Ehiro 2002). The Fukkoshi and Isatomae formations yield Anisian ammonoids (Mojsisovics 1888; Diener 1916; Shimizu 1930; Onuki & Bando 1959; Bando 1964), with the work of Mojsisovics (1888) providing the first description of ammonoid fossils from Japan.

Upper Triassic and lowest Middle Jurassic

Upper Triassic rocks are represented by the Saragai Group which is limited to narrow outcrops within the Shizugawa–Hashiura and Ofunato sub-belts. In contrast, Lower to lower Middle Jurassic rocks are placed within the Shizugawa Group which is found only in the Shizugawa–Hashiura Sub-belt, and Upper Triassic–Lower Jurassic sequences are totally absent from the Karakuwa–Oshika Sub-belt (Fig. 2a.26).

The Upper Triassic Saragai Group comprises the Shindate and Chonomori formations (Shizugawa area; Onuki & Bando 1958) and Uchinohara Formation (Hashiura–Mizunuma area; Takahashi & Onuki 1959) in the Shizugawa–Hashiura Sub-belt, and the Myojinmae Formation in the Ofunato Sub-belt (Kanagawa & Ando 1983). The Shindate Formation, resting unconformably on the Isatomae Formation, is composed of thick sandstone beds with subordinate amounts of mudstone and rare coaly mudstone. The Chonomori Formation conformably covers the Shindate Formation and consists of alternating beds of sandstone and mudstone. Naumann (1881) reported an occurrence of bivalve fossil *Monotis* from the Chonomori Formation, which is the first report of the Triassic fossils from Japan. The Chonomori Formation yields a rich *Monotis* fauna in association with ammonoids *Placites* and *Arcestes* (Shimizu & Mabuti 1932; Onuki & Bando 1958; Nakazawa 1964; Ando 1987), and is dated as being of late Carnian–Norian age. The Uchinohara Formation, unconformably overlying the Isatomae Formation, consists mainly of massive arkose and lacks age-diagnostic fossils, although some authors (e.g. Takizawa *et al.* 1984, 1990) have treated this formation as of Early Jurassic age. The Myojinmae Formation, lying unconformably on Permian strata, consists of pyroclastic rocks, volcanic conglomerate and tuffaceous sandstone, with an occurrence of *Monotis ochotica* having been reported by Kanagawa & Ando (1983).

Lowest Middle Jurassic rocks are represented by the Shizugawa (Shizukawa) Group which comprises the Niranohama Formation and overlying Hosoura Formation. The Niranohama Formation rests unconformably on the Saragai Group and consists of sandstone and sandy mudstone deposited in a brackish–littoral environment (Hayami 1961*a*) and yielded abundant bivalves and middle–upper Hettangian ammonoids (Matsumoto 1956; Hayami 1961*b*; Sato

1962; Takahashi 1969; Sato & Westermann 1991). A diverse belemnite fauna has been reported from this formation (Iba *et al.* 2012). The Hosoura Formation conformably overlies the Niranohama Formation and consists of laminated and massive mudstone, and thin alternating beds of sandstone and mudstone. Many ammonoid species ranging from Sinemurian to Aalenian in age have been obtained from this formation (Sato 1962; Takahashi 1969; Sato & Westermann 1991).

Middle Jurassic–Lower Cretaceous

The Middle Jurassic–Lower Cretaceous sequence in the Shizugawa–Hashiura Sub-belt comprises the Hashiura Group and the overlying Jusanhama Group. The Hashiura Group in the Shizugawa area is, in ascending order, divided into the Aratozaki, Arato and Sodenohama formations. The Aratozaki Formation correlates in other areas with the Nakahara Formation (Hashiura area) and the Kojima Formation (Mizunuma area), and the Arato Formation correlates with the Nagao Formation (Hashiura area) and the Owada Formation (Mizunuma area). The Aratozaki Formation and its equivalents rest unconformably on the Hosoura Formation or the Isatomae Formation, consist mainly of arkose in association with conglomerate and alternating beds of sandstone and mudstone, and yield bivalve fossils such as *Inoceramus*, *Trigonia* and *Vaugonia* (Hayami 1961*b*). The Arato Formation and its equivalents conformably rest on the Aratozaki and equivalent formations and comprise bedded mudstone with alternating beds of sandstone and mudstone. The Arato Formation yielded upper Bajocian ammonoids (Sato 1962; Takahashi 1969). The Nagao Formation yields middle Bajocian–Lower Cretaceous ammonoids (Takahashi 1969; Kase 1979), whereas the lower and middle–upper parts of the Owada Formation contain Bajocian (Suzuki *et al.* 1998) and Oxfordian (Takahashi 1969) ammonoids, respectively. The Sodenohama Formation conformably overlies the Arato Formation, consists of sandstone and laminated mudstone, and contains ammonoid fossils of Oxfordian–Kimmeridgian age (Takahashi 1969).

The Jusanhama Group crops out only in the Hashiura area (Fig. 2a.25) where it has been divided into the Tsukihama and overlying Tategami formations (Takahashi 1961, but see Kase 1979; Takizawa *et al.* 1990 for alternative lithostratigraphic divisions). According to Takahashi (1961), the Tsukihama Formation unconformably covers the Nagao Formation and is mainly composed of massive or bedded, quartzose or arkosic sandstones with intercalated mudstones. The Tategami Formation rests conformably on the Tsukihama Formation, consists of alternating beds of quartzose sandstone and mudstone, and has yielded Late Jurassic–Early Cretaceous brackish water molluscs (Hayami 1960). Given its stratigraphic position above the Nagao Formation, the Jusanhama Group is considered to be Early Cretaceous in age.

In the Karakuwa district of the Karakuwa–Oshika Sub-belt, Mesozoic rocks are represented by the Karakuwa Group and the Oshima Group within the wide, south-plunging Tsunakizaka Syncline (Fig. 2a.24). The Middle Jurassic–lowermost Cretaceous Karakuwa Group is divided into the Kosaba, Tsunakizaka, Ishiwaritoge, Mone, Kogoshio and Isokusa formations in conformably ascending order (Shiida 1940; Ehiro 1974; Takizawa 1985). The Kosaba Formation consists of basal conglomeratic sandstone resting unconformably on the Isatomae Formation, then sandstone intercalated with mudstone that yielded Bajocian molluscs (Hayami 1961*a*). The Tsunakizaka Formation consists of mudstone in its main part and alternating beds of sandstone and mudstone in the uppermost part. It is about 400 m in thickness at the NE part of its outcrop, thins to about 200 m towards the west and south, and has yielded a rich fauna of Bajocian ammonoids (Sato 1962, 1972). The overlying Ishiwaritoge Formation (20–150 m) is composed of conglomerate with alternating beds of sandstone and mudstone deposited in a fluvial–alluvial environment (Takizawa 1985), and also thins towards the west and south (Ehiro 1974). The conglomerate is dominated by granitic pebbles to boulders, thought to be derived from the pre-Silurian Hikami Granite (Kano 1959). The Mone Formation consists of bedded mudstone overlain by alternating beds of arkosic sandstone and mudstone, which is in turn capped by bedded sandstone. It has yielded the bivalve *Myophorella* (*Haidaia*) (Hayami 1961*c*) and ammonoid *Perisphinctes* (Kato *et al.* 1977). The Kogoshio Formation is rich in sandstone and mudstone with rare conglomerate. The sandstone is arkosic, and those in the lower part are thick, coarse-grained and quartzose. This formation has yielded shallow-marine bivalves (Hayami 1961*a*), although the middle part is thought to be an alluvial deposit (Takizawa 1985). The Isokusa (Nagasaki) Formation crops out on Oshima Island in Kesennuma Bay. It consists mainly of sandy mudstone and has yielded Early Cretaceous (Berriasian–Valanginian) ammonoids (Sato 1958; Takahashi 1973; Nara *et al.* 1994). The Oshima Group mostly crops out on Oshima Island and comprises the Kanaegaura and overlying Yokonuma formations (Onuki 1969). The Kanaegaura Formation is very thick (1200 m) and consists mainly of andesite lavas, andesitic pyroclastic rocks and volcanic conglomerates. The covering Yokonuma Formation is composed of sandstone, mudstone and alternating beds of sandstone and mudstone with rare limestone and tuffaceous sandstone; since it has yielded the ammonoid *Crioceratites* (*C.*) *ishiwarai* (Yabe & Shimizu 1925) along with corals and bivalves, it is dated as Hauterivian–Barremian (Obata 1988).

The Jurassic–lowest Cretaceous Oshika Group and overlying Lower Cretaceous Yamadori Formation, widely distributed across the Oshika Peninsula, belong to the Karakuwa–Oshika Sub-belt. The Oshika Group comprises the Tsukinoura, Oginohama and Ayukawa formations in ascending order (Takizawa *et al.* 1974). The Tsukinoura Formation unconformably overlies the Isatomae Formation, and consists of conglomerate and sandstone in the lower part and monotonous mudstone in the upper part. The lower part yields Bajocian ammonoids (Sato 1972) in association with rich bivalves (Hayami 1959, 1961*a*). The Oginohama Formation, conformably covering the Tsukinoura Formation, is composed of sandstone and alternating beds of sandstone and mudstone (Fig. 2a.27) with intercalated conglomerate, and has been divided into the Kitsunezaki, Makinohama, Kozumi and Fukiura members (Takizawa *et al.*

Fig. 2a.27. Photograph showing the folded alternating beds of sandstone and mudstone of the Oginohama Formation (Oshika Group).

1974). The Oginohama Formation yields Oxfordian–Tithonian ammonoids (Inai & Takahashi 1940; Sato 1962; Takahashi 1969; Takizawa *et al.* 1974) and abundant plant fossils belonging to the Ryoseki Flora (Kimura & Ohana 1989*a*, *b*). The Ayukawa Formation conformably covers the Oginohama Formation, consists of arkose and mudstone with rare conglomerate, and is divided into the Kiyosaki, Kobitawatashi, Futawatashi and Domeki members in ascending order (Takizawa *et al.* 1974). The lower part of the Kobitawatashi Member has yielded the Berriasian ammonoid *Berriasella* (Takizawa 1970), whereas the upper part (and Futawatashi Member) yields Valanginian ammonoids (Takizawa 1970; Obata 1988). The Kiyosaki and Domeki members are considered to be continental deposits because they have no marine fossils and contain rich plant fragments and coaly mudstones. The Yamadori Formation unconformably overlies the Ayukawa Formation and consists of andesitic to dacitic pyroclastic rocks in the lower part and basalt and basaltic pyroclastic rocks in the upper part (Takizawa *et al.* 1974).

The Somanakamura Group, distributed in the Soma district of the Abukuma Massif, also belongs to the Karakuwa–Oshika Sub-belt. It comprises the Awazu, Yamagami, Tochikubo, Nakanosawa, Tomizawa and Koyamada formations in ascending order (Yanagisawa *et al.* 1996). These units are dominated by clastic rocks, except for the Nakanosawa Formation which includes thick limestone beds. Based on their ammonoid fauna, the Awazu, Nakanosawa and Koyamada formations have been given ages of Bajocian–Bathonian, Kimmeridgian–Tithonian and Tithonian–Valanginian, respectively (Sato 1962; Mori 1963; Sato *et al.* 2005; Sato & Taketani 2008). The Tochikubo Formation is of particular interest in yielding a rich flora of Ryoseki-type plants (Kimura & Ohana 1989*a*, *b*; Takimoto *et al.* 2008), and the Tomizawa Formation has been interpreted as an alluvial deposit (Takizawa 1985).

In the Ofunato Sub-belt the Lower Cretaceous Ofunato Group is widely distributed, rests unconformably on the Permian strata, and has been divided into the Hakoneyama, Funagawara, Hijochi, Kobosoura and Takonoura formations in conformable ascending order (Onuki & Mori 1961). (Note however that Kanagawa & Ando (1983) interpreted the Hakoneyama Formation, which consists of volcanic conglomerate, as a southern extension of the Upper Triassic Myojinmae Formation.) The Funagawara Formation is composed of conglomerate, sandstone, mudstone and tuffaceous sandstone, and has yielded late Hauterivian ammonoids (Obata & Matsumoto 1977; Matsumoto *et al.* 1982) and late Hauterivian–early Barremian, brackish to shallow-marine bivalves (Kozai & Tashiro 1993). The Hijochi Formation consists of alternating beds of sandstone and mudstone in association with conglomerate and tuff breccias, with the Barremian ammonoid *Holcodiscus* having been reported from its upper part (Obata & Matsumoto 1977). The Kobosoura Formation consists of pyroclastic rocks, tuffaceous sandstone and conglomerate with intercalated sandstone and mudstone, and contains the bivalve *Pterotrigonia* (*Pterotrigonia*) dated as Hauterivian–Aptian (Tashiro & Kozai 1989). The Takonoura Formation consists mainly of alternating beds of sandstone and mudstone with conglomerate and tuff.

Lower Cretaceous volcanic-dominated strata are widely distributed throughout the Kitakami Massif, resting on Palaeozoic formations in the Southern Kitakami Massif and on Jurassic accretionary complexes in the Northern Kitakami Massif. The volcanic rocks are subduction-related, range from basalt to rhyolite, and include a volcano-plutonic complex with Early Cretaceous (120–110 Ma) granitic rocks (Kanisawa 1974). The abundant presence of Early Cretaceous adakitic volcanic and granitic rocks implies the subduction of young, hot oceanic plate (Tsuchiya & Kanisawa 1994; Tsuchiya *et al.* 2005).

Mesozoic covering strata in the Kurosegawa Belt

In central Shikoku, a shallow-marine Mesozoic cover sequence is represented by the Middle Triassic Zohoin Formation, Upper Triassic Kochigatani Group, Triassic–Jurassic Naruo Formation and the Jurassic Keta Formation (Hada *et al.* 1992).

The Zohoin Formation consists of mudstone with sandstone and rare tuffaceous mudstone, is in fault contact with other strata, and yields bivalves and ammonoids of Ladinian age (Bando 1964). It is therefore coeval with the *Daonella*-bearing mudstone-dominated Ladinian Usugatani Formation which crops out in eastern Shikoku (Hirayama *et al.* 1956). The Kochigatani Group, which is also found in western Shikoku, is of Carnian–Norian age (Bando 1964) and divided into lower and upper subgroups. The lower subgroup is composed of sandstone, mudstone and alternating beds of sandstone and mudstone, with conglomeratic sandstone, and has yielded the bivalves *Halobia*, *Tosapecten* and the ammonoid *Paratrachyceras*. The upper subgroup is composed of mudstone and sandstone and contains the bivalve *Monotis* and ammonoid *Sirenites* (Tsujino *et al.* 2013). The Sabudani Formation in eastern Shikoku is an equivalent of this group, which rests unconformably on the Permian accretionary complex and Devonian shallow-marine strata (Ichikawa *et al.* 1956; Yoshikura *et al.* 1990).

The Naruo and its equivalent Nakanose (western Shikoku) formations are composed of clastic rocks and contain latest Triassic–Early Jurassic radiolarians (Hada *et al.* 1992). The Keta Formation, and its equivalent Kagio Formation in western Shikoku, are divided into the lower (sandstones and conglomerates) and upper (mudstone) members. The Kagio Formation unconformably covers the Mitaki Igneous Complex (Hada 1974; Tominaga *et al.* 1979). Radiolarians from these formations are of early Middle Jurassic age (Matsuoka 1985; Hada *et al.* 1992).

Lower Cretaceous Miyako Group, Upper Cretaceous and Palaeogene strata

The Aptian–Albian Miyako Group is distributed along the Pacific coast of the Kitakami Massif, and Upper Cretaceous and associated Palaeogene strata are distributed in the Kuji district (Kuji and Noda groups) and some narrow areas in the northern Kitakami Massif, and the Futaba district (Futaba and Shiramizu groups) are distrubted in the SE Abukuma Massif (Fig. 2a.18). Although these strata were deposited after the Early Cretaceous amalgamation of the South Kitakami and North Kitakami belts and are therefore not strictly part of the South Kitakami Belt, they are briefly described to understand the geological development of the Kitakami Massif.

The Miyako Group, which rests on Jurassic accretionary complexes and Early Cretaceous volcanic and granitic rocks with remarkable unconformity (see Fig. 2a.28), comprises mainly storm-dominated shallow-marine sequences (Mochizuki & Ando 2003). The group has yielded abundant fossils such as ammonoids, molluscs and corals, and is dated as late Aptian–early Albian based on ammonoids (Matsumoto 1953, 1963; Hanai *et al.* 1968).

The Upper Cretaceous Kuji Group (Shimazu & Teraoka 1962) in the Kuji district was deposited in an alluvial-shallow marine basin (Minoura & Yamauchi 1989; Terui & Nagahama 1995), and consists of clastic rocks with minor amounts of tuff. Based on the ammonoids and inoceramids, Matsumoto *et al.* (1982) dated the group as late Coniacian–Campanian, whereas Futakami *et al.* (1987) considered the group to be Santonian–early Campanian. The Futaba Group (Matsumoto 1943) in the Futaba district is composed mainly of clastic rocks deposited during three sedimentary cycles, each of which begins with fluvial conditions and finishes with a lagoonal or shallow-marine facies (Ando *et al.* 1995). It

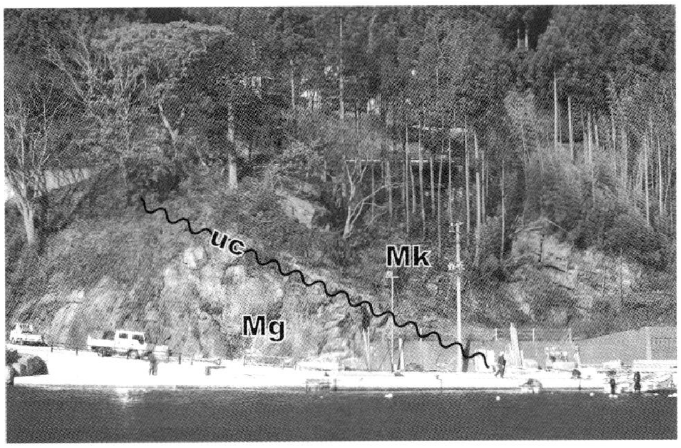

Fig. 2a.28. Photograph showing the unconformity between the Jurassic accretionary complex of the North Kitakami Belt and Lower Cretaceous Miyako Group at Raga, Tanohata, Iwate Prefecture. Mg, sandstone of the Jurassic accretionary complex (Magisawa Unit), dipping vertically; Mk, alternating beds of sandstone and mudstone of the Miyako Group, dipping gently eastwards.

has yielded a rich fauna of ammonoids and inoceramids of Coniacian–early Santonian age (Obata & Suzuki 1969; Matsumoto *et al.* 1990; Kubo *et al.* 2002), including the plesiosaur *Futabasaurus suzukii* (Sato *et al.* 2006).

The Palaeogene Noda Group in the Kuji district unconformably covers the Upper Cretaceous succession and consists of clastic rocks deposited in an alluvial environment showing four sedimentary cycles, each starting with conglomerate and ending with coaly mudstone rich in land plants (Shimazu & Teraoka 1962). Shimazu & Teraoka (1962) correlated the whole group with the Oligocene, whereas Horiuchi & Kimura (1986) & Uemura (1997) considered that the lower part of the group is Paleocene in age. The upper Eocene–lower Oligocene Shiramizu Group (Kubo *et al.* 2002) in the Futaba district overlies unconformably both metamorphic rocks of the Abukuma Belt and Upper Cretaceous strata, and comprises clastic rocks deposited in a fluvial to shallow-marine environment with a coal bed in the lower part (Kubo *et al.* 2002; Suto *et al.* 2005).

Geological development and palaeogeography of the South Kitakami–Kurosegawa Belt

Although fragmental and scattered in fault-bounded small outcrops, the rocks of the Kurosegawa Belt closely resemble those of the South Kitakami Belt; both include Cambro-Ordovician granitic and metamorphic basement, Early Silurian welded tuff, halysitid-bearing Silurian limestone, Siluro-Devonian volcaniclastics, *Leptophloeum*-bearing Late Devonian and late Early Carboniferous *Kueichouphyllum*-bearing limestone (Figs 2a.19 & 2a.23). The Kurosegawa Belt is therefore interpreted to have been a part of the South Kitakami Belt ('South Kitakami Palaeoland'; Ehiro & Kanisawa 1999) during Palaeozoic times (Umeda 1996*a*; Ehiro 2000).

Early–Middle Palaeozoic

The South Kitakami Cambro-Ordovician basement comprises high-*P* (or medium-*P*) metamorphic rocks of accretionary complex origin and subduction-related intrusives. These rocks represent the continental crust of a 'South Kitakami Palaeoland' created along the active continental margin of an ancient continent (Ehiro & Kanisawa 1999), a tectonic setting that persisted through Devonian and Carboniferous times (Kawamura & Kawamura 1989*b*). Silurian palaeomagnetic data from the Kurosegawa Belt indicate low latitudes (5–15°; Shibuya *et al.* 1983). Since no valid palaeomagnetic data have been obtained from the Pre-Cretaceous rocks of the South Kitakami Belt due to the thermal effect by the Early Cretaceous granitic intrusives, the palaeogeography of the South Kitakami Palaeoland has been reconstructed mainly by using palaeobiogeographic data. Kato (1990) argued that the Silurian–Devonian coral faunas from the South Kitakami Belt and Kurosegawa Belt bear a close resemblance to those from eastern Australia and South China, especially to the former. On the other hand, Kido & Sugiyama (2011) stressed that the Silurian corals from the Kurosegawa Belt are more closely comparable with those from South China, often down to individual species level, than to those from Australia. It is therefore highly probable that the South Kitakami Palaeoland was established and developed in a marginal part of South China or nearby continental area during Early–Middle Palaeozoic times.

Late Palaeozoic

In contrast to the Carboniferous strata, Permian outcrops almost completely lack pyroclastic sediments, a fact leading Minoura (1985) to deduce that the South Kitakami Belt was a passive environment from Permian until earliest Cretaceous times when arc volcanism resumed. The Middle Permian Usuginu-type conglomerates contain abundant granitic clasts dated as 257–244 Ma (CHIME ages: Takeuchi & Suzuki 2000), but there is no outcrop of the source granitic rocks in the South Kitakami Belt. Kano (1971) and Iwai & Ishizaki (1966) considered that the granitic clasts were derived from the dissected, lost uplift zones in the belt, but Yoshida *et al.* (1994) and Yoshida & Machiyama (1998) concluded that the hinterland was a magmatic arc situated to the west of the belt. Based on the CHIME ages of the granitic clasts, Takeuchi & Suzuki (2000) speculated that the source granitic rocks were intruded just before the deposition of the Usuginu-type conglomerate, uplifted rapidly, and eroded and transported in the sedimentary basin.

The Australian affinity of the corals of the South Kitakami Palaeoland persisted until early Visean times when the coral faunas changed drastically to have affinity only with those of the South China area (Kato 1990; Niikawa 1994). The Early Carboniferous flora also belongs to the Cathaysian Floral Province (Asama *et al.* 1985; Kato *et al.* 1989). The southwards drift of Australia (Eastern Gondwana) took place during Late Devonian or Early Carboniferous times (Powell & Li 1994), and the Palaeotethys Sea which separated the Cathaysian and Gondwana continents already existed in the Devonian Period (e.g. Metcalfe 1996). This palaeobiogeographic change is therefore considered to be due to the separation of South China (along with South Kitakami) from the Gondwana continent during Devonian times.

The Permian flora of the South Kitakami Belt similarly belongs to the Cathaysian Floral Province (Asama 1956, 1967, 1974; Asama & Murata 1974), and so also differs from that of the Australia area (Gondwanan Province). It is noteworthy that in North China the species of the genus *Gigantopteris*, one of the typical constituents of Cathaysian flora and widespread in South China, have only been reported from the southernmost areas, namely in Xu-Huai-Yu subprovince (Mei *et al.* 1996) or the concurrent *Gigantonoclea–Gigantopteris* region (Zhang *et al.* 1999), as well as the South Kitakami Belt. The Permian bivalve (Fang 1985; Nakazawa 1991; Fang & Yin 1995) and ammonoid faunas (Ehiro 1997; Ehiro *et al.* 2005) are closely allied to those of South China, as are Middle–Late Permian fusulinoideans (Ozawa 1987). In the same way, according to Wang *et al.* (2006), Early–Middle Permian coral faunas closely resemble those of South China (sometimes down to species level), and Kawamura & Machiyama (1995) observed that the biotic composition of the reefal Middle Permian Iwaizaki Limestone is typical tropical and closely allied to those of South China and Indochina.

Although Tazawa (1991, 2002) and Shi *et al.* (1995) used Middle Permian brachiopod fauna to argue that the South Kitakami Belt was located near the NE margin of North China, the majority of palaeobiogeographic data indicate that the 'South Kitakami Palaeoland' was located at or near the eastern margin of South China during Late Palaeozoic time (Ehiro 2001).

Mesozoic

Following marine regression at the Permian–Triassic boundary, the sea re-entered the South Kitakami Belt in late Early Triassic times (Ehiro 2002), spreading uniformly over the Shizugawa–Hashiura and Karakuwa–Oshika sub-belts by Middle Triassic times so that there are only minor differences in stratal lithofacies and thicknesses. After a regressive phase at the Middle–Late Triassic boundary, subsequent transgression was restricted only to the narrow area represented by the Shizugawa–Hashiura and Ofunato sub-belts. The deposition of Lower–lower Middle Jurassic strata was restricted only to the Shizugawa–Hashiura Sub-belt, and Upper Triassic–Lower Jurassic rocks are totally absent from the Karakuwa–Oshika Sub-belt. Whereas the thicknesses of the Middle Jurassic strata of the Shizugawa–Hashiura and Karakuwa–Oshika sub-belts are similar, Upper Jurassic–Lower Cretaceous strata are very much thicker in the Karakuwa–Oshika Sub-belt. We may therefore deduce that the sedimentary basin depocentre lay initially in the Shizugawa–Hashiura Sub-belt (Late Triassic–Early Jurassic) but shifted east to the Karakuwa–Oshika Sub-belt in Late Jurassic times (Fig. 2a.26; Yamashita 1957; Takizawa 1977, 1985).

The South Kitakami Belt continued to be located at or near the eastern margin of South China until Early Cretaceous times, as indicated by Mesozoic faunas and floras in the South Kitakami Belt (Ehiro & Kanisawa 1999; Ehiro 2001). Early Triassic ammonoids in the Osawa Formation belong to the lower latitudinal *Columbites–Subcolumbites* fauna (Ehiro 1997; Brayard *et al.* 2009), although Nakazawa (1991) considered that the Middle Triassic *Daonella* fauna has some affinity with that of Siberia and Ando (1987) stressed that the Late Triassic *Monotis* fauna from the Saragai Group show similarity to that of the Boreal province. Although the Jurassic ammonoid fauna similarly includes Boreal elements (the genus Kepplerites; Takizawa 1977), most Jurassic ammonoids in the South Kitakami Belt are of Tethys-Pacific type (Bando *et al.* 1987). Finally, Late Jurassic–Early Cretaceous flora in East Asia can be divided into the Ryoseki Floral Province, which typically developed in the Outer Zone of SW Japan and South China, and the Tetori Floral Province, distributed at higher latitudes and so now found in the Inner Zone of SW Japan and the NE China–east Siberia area (Kimura 1987). The Late Jurassic–earliest Cretaceous floral assemblages from the South Kitakami and Kurosegawa belts belong to the Ryoseki Floristic Province (Kimura 1987; Kimura & Ohana 1989*a*, *b*). Palaeomagnetic study of the Mesozoic sedimentary rocks of the Kurosegawa Belt in Kyushu District shows palaeolatitudes for the Late Triassic and Early Cretaceous as $3.5 \pm 3.8°$ N and $18.4 \pm 2.5°$ N, respectively, which coincide with those of South China (Uno *et al.* 2011).

The Palaeozoic–Lower Cretaceous (Barremian) strata in the Kitakami and Abukuma massifs, including the Jurassic accretionary complexes, were strongly folded and faulted during the Early Cretaceous Oshima Orogeny (Kobayashi 1941) and then intruded by Early Cretaceous granitic rocks (Fig. 2a.27). NNW-trending left-lateral strike-slip faults, with displacements of several tens to several hundreds of kilometres, were still active during the early phase of the granitic activity, and the granites (120–110 Ma) along the faults consequently display strong mylonitic fabrics (e.g. Otsuki & Ehiro 1978, 1992). In contrast, the Aptian–Albian Miyako Group and Upper Cretaceous strata are only weakly deformed and rest on the pre-Miyakoan folded strata and granites with pronounced unconformity (Fig. 2a.28). The distribution area of these strata on the land surface is rather narrow, due to the shift of the sedimentary depocentre to the east after the Oshima Orogeny. Thick Cretaceous–Palaeogene strata are widely distributed beneath the continental shelf off the coasts of Kitakami (Osawa *et al.* 2002) and Abukuma (Iwata *et al.* 2002), however.

Appendix

English to Kanji and Hiragana translations for geological and place names

Abukuma	阿武隈	あぶくま
Ainosawa	合ノ沢	あいのさわ
Akaiwa	赤岩	あかいわ
Akiyoshi	秋吉	あきよし
Arakigawa	荒城川	あらきがわ
Arato	荒砥 (荒戸)	あらと
Aratozaki	荒砥崎	あらとざき
Arisu	有住	ありす
Awazu	粟津	あわづ
Ayukawa	鮎川	あゆかわ
Buntoku	文徳	ぶんとく
Chonomori	長の森	ちょうのもり
Chugoku	中国	ちゅうごく
Dogo	島後	どうご
Doi	土居	どい
Domeki	百目木	どうめき
Ezo	恵蘇	えぞ
Fujikuradani	藤倉谷	ふじくらだに
Fukata	深田	ふかた
Fukkiura	福貴浦	ふっきうら
Fukkoshi	風越	ふっこし
Fuko	普甲	ふこう
Fukuji	福地	ふくじ
Funagawara	船河原	ふながわら
Funatsu	船津	ふなつ
Futaba	双葉	ふたば
Futawatashi	長渡	ふたわたし
Gionyama	祇園山	ぎおんやま
Gomi	五味	ごみ
Haigyu	拝宮	はいぎゅう
Hakoneyama	箱根山	はこねやま
Happo (Happou)	八方	はっぽう
Hashiura	橋浦	はしうら
Hatakawa	畑川	はたかわ
Hayachine	早池峰	はやちね
Hida	飛騨	ひだ
Hida Gaien	飛騨外縁	ひだがいえん
Higo	肥後	ひご
Hijochi	飛定地	ひじょうち
Hikami	氷上	ひかみ
Hikamisan	氷上山	ひかみさん
Hikoroichi	日頃市	ひころいち
Hiraiso	平磯	ひらいそ
Hirasawa	平沢	ひらさわ
Hitachi	日立	ひたち
Hitoegane	一重ヶ根	ひとえがね
Hosoo	細尾	ほそお
Hosoura	細浦	ほそうら
Ichinose	市ノ瀬	いちのせ
Ichinotani	一の谷	いちのたに
Ichiyama	市山	いちやま

(*Continued*)

English to Kanji and Hiragana translations for geological and place names (Continued)

Inai	稲井	いない
Ino	伊野	いの
Isatomae	伊里前	いさとまえ
Ise	伊勢	いせ
Ishiwaritoge	石割峠	いしわりとうげ
Isokusa	磯草	いそくさ
Itoigawa	糸魚川	いといがわ
Itoshiro	石徹白	いとしろ
Iwaizaki	岩井崎	いわいざき
Iwatsubodani	岩坪谷	いわつぼだに
Izushi	出石	いずし
Joryu	上流	じょうりゅう
Kagero	影路	かげろ
Kagio	嘉義尾	かぎお
Kagura	神楽	かぐら
Kakisako	柿迫	かきさこ
Kamaishi	釜石	かまいし
Kamianama	上穴馬	かみあなうま
Kamioka	神岡	かみおか
Kamiyasse	上八瀬	かみやっせ
Kanaegaura	鼎ヶ浦	かなえがうら
Kanokura	叶倉	かのくら
Karakuwa	唐桑	からくわ
Karaumedate	唐梅舘	からうめだて
Katsura	桂	かつら
Kawauchi	川内	かわうち
Kentosan	犬頭山	けんとうさん
Kesennuma	気仙沼	けせんぬま
Keta	毛田	けた
Kitakami	北上	きたかみ
Kitsunezaki	狐崎	きつねざき
Kiyomi	清見	きよみ
Kiyosaki	清崎	きよさき
Kobitawatashi	小長渡	こびたわたし
Kobosoura	小細浦	こぼそうら
Kochigatani	川内ヶ谷	こうちがたに
Kogoshio	小々汐	こごしお
Koguro	小黒	こぐろ
Kojima	小島	こじま
Komatsu	小松	こまつ
Konogidani	此木谷	このぎだに
Kosaba	小鯖	こさば
Kotaki	小滝	こたき
Kowaragi	小原木	こわらぎ
Koyamada	小山田	こやまだ
Kozaki	小崎	こざき
Kozumi	小積	こづみ
Kuji	久慈	くじ
Kuma	球磨	くま
Kurosawa	黒沢	くろさわ
Kurosegawa	黒瀬川	くろせがわ
Kuruma	来馬	くるま
Kuzuryu	九頭竜	くずりゅう
Maizuru	舞鶴	まいづる
Makinohama	牧の浜	まきのはま
Mano	真野	まの
Matsugadaira	松ヶ平	まつがだいら
Mino	美濃	みの
Misaka	三坂	みさか
Mitaki	三滝	みたき
Miyadani	宮谷	みやだに
Miyagadani	宮ヶ谷	みやがだに

(Continued)

English to Kanji and Hiragana translations for geological and place names (Continued)

Miyako	宮古	みやこ
Miyamori	宮守	みやもり
Miyanaro	宮成	みやなろ
Mizunuma	水沼	みずぬま
Mizuyagadani	水屋ヶ谷	みずやがだに
Mone	舞根	もうね
Moribu	森部	もりぶ
Motai	母体	もたい
Myojinmae	明神前	みょうじんまえ
Nabekoshiyama	鍋越山	なべこしやま
Nagaiwa	長岩	ながいわ
Nagao	長尾	ながお
Nagasaka	長坂	ながさか
Nagasaki	長崎	ながさき
Naidaijin	内大臣	ないだいじん
Nakadaira	中平	なかだいら
Nakadake	中岳	なかだけ
Nakahara	中原	なかはら
Nakahata	中畑	なかはた
Nakanosawa	中ノ沢	なかのさわ
Nakanose	中之瀬	なかのせ
Nakazato	中里	なかざと
Nameirizawa	名目入沢	なめいりざわ
Naradani	楢谷	ならだに
Naruo	成穂	なるほ
Natsuyama	夏山	なつやま
Nedamo	根田茂	ねだも
Niranohama	韮の浜	にらのはま
Nishikori	錦織	にしこり
Noda	野田	のだ
Ochi	越智	おち
Odagoe	小田越	おだごえ
Odaira	大平	おおだいら
Oeyama	大江山	おおえやま
Ofunato	大船渡	おおふなと
Ogatsu	雄勝	おがつ
Oginohama	荻の浜	おぎのはま
Oguradani	小椋谷	おぐらだに
Okago	大籠	おおかご
Okanaro	岡成	おかなろ
Oki	隠岐	おき
Okuhinotsuchi	奥火の土	おくひのつち
Omi	青海	おうみ
Onimaru	鬼丸	おにまる
Ono	大野	おおの
Orikabetoge	折壁峠	おりかべとうげ
Osawa	大沢	おおさわ
Osayama	大佐山	おおさやま
Oshika	牡鹿	おしか
Oshima	大島	おおしま
Owada	大和田	おおわだ
Raga	羅賀	らが
Renge	蓮華	れんげ
Rifu	利府	りふ
Rosse	呂瀬	ろっせ
Ryoke	領家	りょうけ
Ryoseki	領石	りょうせき
Sabudani	寒谷	さぶだに
Sakamotozawa	坂本沢	さかもとざわ
Sanbagawa (Sambagawa)	三波川	さんばがわ
Sangun	三郡	さんぐん
Saragai	皿貝	さらがい

(Continued)

English to Kanji and Hiragana translations for geological and place names (Continued)

Sekinomiya	関宮	せきのみや
Senjogataki	千丈ヶ滝	せんじょうがたき
Senmatsu	千松	せんまつ
Setamai	世田米	せたまい
Shibasudani	子馬巣谷	しばすだに
Shimozaisho	下在所	しもざいしょ
Shindate	新舘	しんだて
Shiraishi	白石	しらいし
Shiramizu	白水	しらみず
Shiroumadake	白馬岳	しろうまだけ
Shittakasawa	尻高沢	しったかさわ
Shizugawa (Shizukawa)	志津川	しづがわ
Shoboji	正法寺	しょうぼうじ
Sodenohama	袖の浜	そでのはま
Soma	相馬	そうま
Somanakamura	相馬中村	そうまなかむら
Sorayama	空山	そらやま
Suberidani	辷谷	すべりだに
Suenosaki	末の崎	すえのさき
Suketsune	助常	すけつね
Takayama	高山	たかやま
Takonoura	蛸浦	たこのうら
Tanba (Tamba)	丹波	たんば
Tandodani	谷土谷	たんどだに
Tanohata	田野畑	たのはた
Tanoura	田の浦	たのうら
Tari	多里	たり
Tategami	立神	たてがみ
Terano	寺野	てらの
Tetori (Tedori)	手取	てとり
Tobigamori	鳶ヶ森	とびがもり
Tochikubo	栃窪	とちくぼ
Tomizawa	富沢	とみざわ
Toyoma	登米	とよま
Tsubonosawa	壺の沢	つぼのさわ
Tsukihama	月浜	つきはま
Tsukinoura	月の浦	つきのうら
Tsunakizaka	綱木坂	つなきざか
Tsurukoba (Tsurunokoba)	鶴木場	つるこば
Uchinohara	内の原	うちのはら
Unazuki	宇奈月	うなづき
Unoki	鵜ノ木	うのき
Usugatani	臼ヶ谷	うすがたに
Usuginu	薄衣	うすぎぬ
Wakamatsu	若松	わかまつ
Wakasa	若桜	わかさ
Wariyama	割山	わりやま
Yaeyama	八重山	やえやま
Yakuno	夜久野	やくの
Yakushigawa	薬師川	やくしがわ
Yamagami	山上	やまがみ
Yokamachi	八日町	ようかまち
Yokokurayama	横倉山	よこくらやま
Yokonuma	横沼	よこぬま
Yokota	横田	よこた
Yoshiki	吉城	よしき
Yuzuruha	湯鶴葉	ゆづるは
Zohoin	蔵法院	ぞうほういん

References

ADACHI, M., SUZUKI, K., YOGO, S. & YOSHIDA, S. 1994. The Okuhinotsuchi granitic mass in the South Kitakami terrane: pre-Silurian basement or Permian intrusives. *Journal of Mineralogy, Petrology and Economic Geology*, **89**, 21–26.

ADACHI, S. 1985. Smaller foraminifers of the Ichinotani Formation (Carboniferous – Permian), Hida Massif, Central Japan. *Science Reports of the Institute of Geoscience, University of Tsukuba, B*, **6**, 59–139.

AITCHISON, J. C., HADA, S., IRELAND, T. & YOSHIKURA, S. 1996. Ages of Silurian radiolarians from the Kurosegawa terrane, southwest Japan constrained by U/Pb SHRIMP data. *Journal of Southeastern Asian Earth Sciences*, **14**, 53–70.

ANDO, H. 1987. Evolution and paleobiogeography of Late Triassic bivalve *Monotis* from Japan. *In*: MCKENZIE, K. G. (ed.) *Shallow Tethys 2*. AA Balkema, Rotterdam, 233–246.

ANDO, H., SEISHI, M., OSHIMA, M. & MATSUMARU, T. 1995. Fluvial-shallow marine depositional systems of the Futaba Group (Upper Cretaceous) – depositional facies and sequences. *Journal of Geography*, **104**, 284–303.

ARAI, S. 1975. Contact metamorphosed dunite-harzburgite complex in the Chugoku district, western Japan. *Contributions to Mineralogy and Petrology*, **52**, 1–16.

ARAI, S. 1980. Dunite-harzburgite-chromitite complexes in refractory residue in the Sangun-Yamaguchi zone, western Japan. *Journal of Petrology*, **21**, 141–165.

ARAI, S. & YURIMOTO, H. 1995. Possible subarc origin of podiform chromitites. *Island Arc*, **4**, 104–111.

ARAKAWA, Y. 1982. Deformational history of the Hida metamorphic rocks in the northern part of Gifu Prefecture, central Japan. *Journal of the Geological Society of Japan*, **88**, 753–767 [in Japanese with English abstract].

ARAKAWA, Y. 1990. Two types of granitic intrusions in the Hida belt, Japan: Sr isotopic and chemical characteristics of the Mesozoic Funatsu granitic rocks. *Chemical Geology*, **85**, 101–117.

ARAKAWA, Y. & SHINMURA, T. 1995. Nd-Sr isotopic and geochemical characteristics of two contrasting types of calc-alkaline plutons in the Hida Belt, Japan. *Chemical Geology*, **124**, 217–232.

ARAKAWA, Y., SAITO, Y. & AMAKAWA, H. 2000. Crustal development of the Hida belt, Japan: evidence from Nd-Sr isotopic and chemical characteristics of igneous and metamorphic rocks. *Tectonophysics*, **328**, 183–204.

ARAKI, H. 1980. Discovery of *Helicoprion* a Chondrichthyes from Kesennuma City, Miyagi Prefecture, Japan. *Journal of the Geological Society of Japan*, **86**, 135–137 [in Japanese].

ASAKAWA, Y., MARUYAMA, T. & YAMAMOTO, M. 1999. Rb-Sr whole-rock isochron ages of the Hikami Granitic Body in the South Kitakami Belt, Northeast Japan. *Memoirs of the Geological Society of Japan*, **53**, 221–234 [in Japanese with English abstract].

ASAMA, K. 1956. Permian plants from Maiya in northern Honshu, Japan (Preliminary note). *Proceedings of Japan Academy*, **32**, 469–471.

ASAMA, K. 1967. Permian plants from Maiya, Japan 1, *Cathaysiopteris* and *Psygmophyllum*. *Bulletin of the National Science Museum*, **13**, 291–317.

ASAMA, K. 1974. Permian plants from Takakurayama, Japan. *Bulletin of the National Science Museum*, **17**, 239–248.

ASAMA, K. & MURATA, M. 1974. Permian plants from Setamai, Japan. *Bulletin of the National Science Museum*, **17**, 251–256.

ASAMA, K., ASANO, T., SATO, E. & YAMADA, Y. 1985. Early Carboniferous plants from the Hikoroichi Formation, Southern Kitakami Massif, Northeast Japan (Preliminary report). *Journal of the Geological Society of Japan*, **91**, 425–426 [in Japanese].

ASAMI, M. & ADACHI, M. 1976. Staurolite-bearing cordierite-sillimanite gneiss from the Toga area in the Hida metamorphic terrane, central Japan. *Journal of the Geological Society of Japan*, **82**, 259–271.

ASANO, M., TANAKA, T. & SUWA, K. 1990. Sm-Nd and Rb-Sr ages of the Hida metamorphic rocks in the Wada-gawa area, Toyama Prefecture. *Journal of the Geological Society of Japan*, **96**, 957–966 [in Japanese with English abstract].

BANDO, Y. 1964. The Triassic stratigraphy and ammonite fauna of Japan. *Science Reports of the Tohoku University, Second Series*, **36**, 1–137.

BANDO, Y. 1970. Lower Triassic ammonodids from the Kitikami Massif. *Transactions and Proceeedings of the Palaeontological Society of Japan, New Series*, **79**, 337–354.

BANDO, Y. & EHIRO, M. 1982. On some Lower Triassic ammonites from the Osawa Formation at Asadanuki, Towa-cho, Tome-gun, Miyagi Prefecture, Northeast Japan. *Proceeedings of the Palaeontological Society of Japan, New Series*, **128**, 375–385.

Bando, Y. & Shimoyama, S. 1974. Late Schythian ammonoids from the Kitakami Massif. *Transactions and Proceeedings of the Palaeontological Society of Japan, New Series*, **94**, 293–312.

Bando, Y., Sato, T. & Matsumoto, T. 1987. Palaeobiogeography of the Mesozoic Ammonoidea, with special reference to Asia and the Pacific. *In*: Taira, A. & Tashiro, M. (eds) *Historical Biogeography and Plate Tectonic Evolution of Japan and Eastern Asia*. Terra Scientific Publishing Company, Tokyo, 65–95.

Brayard, A., Escarguel, G., Bucher, H. & Bruhwiler, T. 2009. Smithian and Spathian (Early Triassic) ammonoid assemblages from terranes: paleoceanographic and paleogeographic implications. *Journal of Asian Earth Scences*, **36**, 420–433.

Chang, K. H. & Zhao, X. X. 2012. North and South China suturing in the east end: what happened in Korean Peninsula? *Gondwana Research*, **22**, 493–506.

Cheong, C. S., Kwon, S. T. & Sagong, H. 2002. Geochemical and Sr-Nd-Pb isotopic investigation of Triassic granitoids and basement rocks in the northern Gyeongsang Basin, Korea: implications for the young basement in the East Asian continental margin. *Island Arc*, **11**, 25–44.

Chihara, K. 1971. Mineralogy and paragenesis of jadeites from the Omi-Kotaki area, Central Japan. *Mineralogical Society of Japan, Special Paper*, **1**, 147–156.

Chihara, K. 1989. Tectonic significance of jadeitite in Hida marginal belt and Sangun metamorphic belt. *Memoir of the Geological Society of Japan*, **33**, 37–51.

Chihara, K. & Komatsu, M. 1982. The recent study in the Hida marginal zone, especially the Omi-Renge & the Jo-etsu Belts – a review. *Memoir of the Geological Society of Japan*, **21**, 101–116 [in Japanese with English abstract].

Cho, D.-L., Takahashi, Y., Yi, K. & Lee, S. R. 2012. SHRIMP U-Pb zircon ages of granite gneiss and paragneiss from Oki-Dogo Island, southwest Japan, and their tectonic implications. *Abstract of the EGU General Assembly 2012*, Vienna, Austria, 1720.

Choi, D. R. 1973. Permian fusulinids from the Setamai-Yahagi district, Southern Kitakami Mountains, N. E. Japan. *Journal of the Faculty of Sciences, Hokkaido University, Series 4*, **16**, 1–132.

Dallmeyer, R. D. & Takasu, A. 1998. $^{40}Ar/^{39}Ar$ mineral ages from the Oki metamorphic complex, Oki-Dogo, southwest Japan: implications for regional correlations. *Journal of Asian Earth Sciences*, **16**, 437–448.

Diener, C. 1916. Japanische Triasfaunen. *Denkschriften der Akademie der Wissenschaften in Wien*, **92**, 1–30.

Ehiro, M. 1974. Geological and structural studies of the area along the Hizume-Kesennuma Tectonic Line, in Southern Kitakami Massif. *Journal of the Geological Society of Japan*, **80**, 457–474 [in Japanese with English abstract].

Ehiro, M. 1989. 2.1(2)-5 the Permian. *In*: Editorial committee of ‘geology of japan 2 tohoku district’ (ed.) *Geology of Japan 2 Tohoku District*. Kyoritsu Shuppan, Tokyo, 43–47 [in Japanese].

Ehiro, M. 1996. Latest Permian ammonoid Paratirolites from the Ofunato district, Southern Kitakami Massif, Northeast Japan. *Transactions and Proceedings of the Palaeontological Society of Japan, New Series*, **184**, 592–596.

Ehiro, M. 1997. Ammonoid palaeobiogeography of the South Kitakami Palaeoland and palaeogeography of eastern Asia during Permian to Triassic time. *In*: Jin, Y.-G. & Dineley, D. (eds) *Proceedings of the 30th International Geological Congress*, VSP Press, Utrecht, the Netherlands, **12**, 18–28.

Ehiro, M. 2000. Relationships in tectonic framework among the South Kitakami and Hayachine Tectonic Belts, Kurosegawa Belt, and ‘Paleo-Ryoke Belt’. *Memoirs of the Geological Society of Japan*, **56**, 53–64 [in Japanese with English abstract].

Ehiro, M. 2001. Origins and drift histories of some microcontinents distributed in the eastern margin of Asian Continent. *Earth Science*, **55**, 71–81.

Ehiro, M. 2002. Time-gap at the Permian-Triassic boundary in the South Kitakami Belt, Northeast Japan: an examination based on the ammonoid fossils. *Saito Ho-on Kai Museum of Natural History, Reseach Bulletin*, **68**, 1–12.

Ehiro, M. 2010. Permian ammonoids of Japan: their stratigraphic and paleobiogeographic significance. *In*: Tanabe, K., Shigeta, Y., Sasaki, T. & Hirano, H. (eds) *Cephalopods – Present and Past*. Tokai University Press, Tokyo, 233–241.

Ehiro, M. 2016. 4.2.4 the Permian. *In*: The geological society of Japan (ed.) *Regional Geology of Japan, 2 Tohoku Rejion*. Asakura Publishing Co., Tokyo [in Japanese] (in press).

Ehiro, M. & Araki, H. 1997. Permian cephalopods of Kurosawa, Kesennuma City in the Southern Kitakami Massif, Northeast Japan. *Palaeontological Research*, **1**, 55–66.

Ehiro, M. & Bando, Y. 1985. Late Permian ammonoids from the Southern Kitakami Massif, Northeast Japan. *Transactions and Proceeedings of the Palaeontological Society of Japan, New Series*, **137**, 25–49.

Ehiro, M. & Kanisawa, S. 1999. Origin and evolution of the South Kitakami Microcontinent during the Early-Middle Palaeozoic. *In*: Metcalfe, I. (ed.) *Gondwana Dispersion and Asian Accretion: IGCP 321 Final Results Volume*. A. A. Balkema, Rotterdam, 283–295.

Ehiro, M. & Misaki, A. 2005. Middle Permian ammonoids from the Kamiyasse-Imo district in the Southern Kitakami Massif, Northeast Japan. *Paleontological Research*, **9**, 1–14.

Ehiro, M. & Okami, K. 1990. Stratigraphic relation between the Matsugadaira metamorphic rocks and the Upper Devonian Ainosawa Formation in the eastern marginal part of the Abukuma Massif, Northeast Japan. *Journal of the Geological Society of Japan*, **96**, 537–547 [in Japanese with English abstract].

Ehiro, M. & Okami, K. 1991. Geologic relation between the ‘Matsugadaira-Motai’ and South Kitakami Belts, with special reference to the Palaeozoic tectonic evolution of the Northeast Japan. *Essays in Geology–Professor Hisao Nakagawa Commemorative Volume*. Professor Hisao Nakagawa Taikan Kinenjigyo-kai, Sendai, Japan, 23–29 [in Japanese with English abstract].

Ehiro, M. & Takaizumi, Y. 1992. Late Devonian and Early Carboniferous ammonoids from the Tobigamori Formation in the Southern Kitakami Massif, Northeast Japan and their stratigraphic significance. *Journal of the Geological Society of Japan*, **98**, 197–204.

Ehiro, M., Tazawa, J., Oishi, M. & Okami, K. 1986. Discovery of *Trimerella* (Silurian Brachiopoda) from the Odagoe Formation, south of Mt. Hayachine in the Kitakami Massif, Northeast Japan and its significance. *Journal of the Geological Society of Japan*, **92**, 753–756 [in Japanese].

Ehiro, M., Okami, K. & Kanisawa, S. 1988. Recent progress and further subject in studies on the ‘Hayachine Tectonic Belt’ in the Kitakami Massif, Northeast Japan. *Earth Science*, **42**, 317–335 [in Japanese with English abstract].

Ehiro, M., Hasegawa, H. & Misaki, A. 2005. Permian ammonoids *Prostacheoceras* and *Perrinites* from the Southern Kitakami Massif, Northeast Japan. *Journal of Paleontology*, **79**, 1222–1228.

Ernst, W. G., Tsujimori, T., Zhang, R. Y. & Liou, J. G. 2007. Permo-Triassic collision, subduction-zone metamorphism, and tectonic exhumation along the East Asian continental margin. *Annual Review of Earth and Planetary Sciences*, **35**, 73–110.

Fang, Z. 1985. A preliminary study of the Cathaysia faunal province. *Acta Paleontologia Sinica*, **24**, 344–349 [in Chinese with English abstract].

Fang, Z. & Yin, D. 1995. Discovery of fossil bivalves from Early Permian of Dongfang, Hainan Island with a review on glaciomarine origin of Nanlong diamictites. *Acta Palaeontologia Sinica*, **34**, 301–315 [in Chinese with English abstract].

Fu, B., Valley, J. W. et al. 2010. Multiple origins of zircons in jadeitite. *Contributions to Mineralogy and Petrology*, **159**, 769–780.

Fujimoto, H., Kanuma, M. & Igo, H. 1962. Upper Paleozoic rocks of the Hida Mountainlands. *In*: Fujimoto, H. (ed.) *Studies of Geology in the Hida Mountainlands*. Research Group of Geology of the Hida Mountainlands, Tokyo, 44–70 [in Japanese].

Futakami, M., Kawakami, T. & Obata, I. 1987. Santonian texanitine ammonites from the Kuji Group, Northeast Japan. *Bulletin of the Iwate Prefectural Museum*, **5**, 103–115.

Goto, A., Higashino, T. & Sakai, C. 1996. XFR analyses of Sanbagawa pelitic schists in central Shikoku, Japan. *Memoirs of the Faculty of Science, Kyoto University, Series of Geology and Mineralogy*, **58**, 1–19.

Graham, C. M., Valley, J. W. & Eiler, J. M. 1998. Timescales and mechanisms of fluid infiltration in a marble: an ion microprobe study. *Contributions to Mineralogy and Petrology*, **132**, 371–389.

Hada, S. 1974. Construction and evolution of the intrageosynclinal tectonic lands in the Chichibu Belt of western Shikoku, Japan. *Journal of Geoscience, Osaka City University*, **17**, 1–52.

HADA, S., SATO, E., TAKESHIMA, H. & KAWAKAMI, A. 1992. Age of the covering strata in the Kurosegawa Terrane: dismembered continental fragment in southwest Japan. *Palaeogeography, Palaeoclimatology, Palaeoecology*, **96**, 59–69.

HADA, S., YOSHIKURA, S. & GABITES, J. E. 2000. U–Pb zircon ages for the Mitaki iguneous rocks, Siluro-Devonian tuff and granitic boulders in the Kurosegawa Terrane, Southwest Japan. *Memoirs of the Geological Society of Japan*, **56**, 183–198.

HADA, S., ISHII, K., LANDIS, C. A., AITCHISON, J. & YOSHIKURA, S. 2001. Kurosegawa Terrane in Southwest Japan: disrupted remnant of a Gondwana-derived terrane. *Gondwana Research*, **4**, 27–38.

HAMADA, T. 1959*a*. Discovery of a Devonian Ostracoda in the Fukuji area, Gifu Prefecture, Central Japan. *Japanese Journal of Geology and Geography*, **30**, 39–51.

HAMADA, T. 1959*b*. On the taxonomic position of *Favosites hidensis* and its age. *Japanese Journal of Geology and Geography*, **30**, 201–213.

HAMADA, T. 1959*c*. Gotlandian stratigraphy of the Outer Zone of Southwest Japan. *Journal of the Geological Society of Japan*, **65**, 688–700.

HANAI, T., OBATA, I. & HAYAMI, I. 1968. Notes on the Cretaceous Miyako Group. *Memoirs of the National Science Museum–Natural History of Rikuchu Province, Northeastern Japan*, **1**, 20–28.

HARAYAMA, S. 1990. *Geology of the Kamikochi District. With Geological Sheet Map at 1: 50,000*. Geological Survey of Japan, Tsukuba [in Japanese with English abstract].

HARLOW, G. E., TSUJIMORI, T. & SORENSEN, S. S. 2015. Jadeitites and plate tectonics. *Annual Review of Earth and Planetaly Sciences*, **43**, 105–138.

HAYAMI, I. 1959. Some pelecypods from the Tsukinoura Formation in Miyagi Prefecture. *Transactions and Proceeedings of the Palaeontological Society of Japan, New Series*, **35**, 133–137.

HAYAMI, I. 1960. Pelecypods of the Jusanhama Group (Purbeckian or Wealden) in Hasiura area, northeast Japan. *Japanese Journal of Geology and Geography*, **31**, 13–21.

HAYAMI, I. 1961*a*. Successions of the Kitakami Jurassic. Jurassic stratigraphy of South Kitakami, Japan I. *Japanese Journal of Geology and Geography*, **32**, 159–177.

HAYAMI, I. 1961*b*. Sediments and correlation of the Kitakami Jurassic, Jurassic stratigraphy of South Kitakami, Japan II. *Japanese Journal of Geology and Geography*, **32**, 179–190.

HAYAMI, I. 1961*c*. Geologic history recorded in the Kitakami Jurassic. Jurassic stratigraphy of South Kitakami, Japan III. *Japanese Journal of Geology and Geography*, **32**, 191–204.

HAYASAKA, I. & MINATO, M. 1954. A *Sinospirifer*-faunule from the Abukuma Plateau, northeast Japan, in composition with the so-called Upper Devonian brachiopod faunule of the Kitakami Mountains. *Transactions and Proceeedings of the Palaeontological Society of Japan, New Series*, **16**, 201–211.

HAYASAKA, Y., SUGIMOTO, T. & KANO, T. 1995. Ophiolitic complex and metamorphic rocks in the Niimi-Katsuyama area, Okayama Prefecture. *Excursion Guidebook of 102th Annual Meeting of the Geological Society of Japan*, Hiroshima, 71–87 [in Japanese].

HIRANO, H., HIGASHIMOTO, S. & KAMITANI, M. 1987. Geology and chromite deposits of the Tari district, Tottori Prefecture. *Bulletin of the Geological Survey of Japan*, **29**, 61–71 [in Japanese with English abstract].

HIRATA, S. 1966. Mt. Yokokurayama in Kochi Prefecture, Japan. *Chigaku-Kenkyu*, **17**, 258–273 [in Japanese].

HIRAYAMA, K., YAMASHIYA, N., SUYARI, K. & NAKAGAWA, C. 1956. *Geological Map of Tokushima, 1: 75,000 Tsurugisan with Explanatory Text*. Tokushima Prefecture [in Japanese].

HIROI, Y. 1978. Geology of the Unazuki district in the Hida metamorphic terrane, central Japan. *Journal of the Geological Society of Japan*, **84**, 521–530 [in Japanese with English abstract].

HIROI, Y. 1981. Subdivision of the Hida metamorphic complex, central Japan, and its bearing on the geology of the Far East in pre-Sea of Japan time. *Tectonophysics*, **76**, 317–333.

HIROI, Y. 1983. Progressive metamorphism of the Unazuki pelitic schists in the Hida terrane, central Japan. *Contributions to Mineralogy and Petrology*, **82**, 334–350.

HIROI, Y. 1984. Petrography of Unazuki pelitic schists, Hida terrane, central Japan: part 2. *Bulletin of the Faculty of Education, Kanazawa University, Natural Science*, **33**, 69–78 [in Japanese with English abstract].

HIROI, Y., FUJI, N. & OKIMURA, Y. 1978. New fossil discovery from the Hida metamorphic rocks in the Unazuki area, central Japan. *Proceedings of the Japan Academy, Series B*, **54**, 268–271.

HIROOKA, K., KATO, M., MORISADA, T. & AZUMA, Y. 2002. Paleomagnetic study on the dinosaur-bearing strata of the Tetori Group, central Japan. *Memoir of the Fukui Prefectural Dinosaur Museum*, **1**, 54–62.

HORI, N. & WAKITA, K. 2004. Reconstructed oceanic plate stratigraphy of the Ino Formation in the Ino district, Kochi prefecture, central Shikoku, Japan. *Journal of Asian Earth Sciences*, **24**, 185–197.

HORIE, K., YAMASHITA, M. *ET AL*. 2010. Eoarchean–Paleoproterozoic zircon inheritance in Japanese Permo-Triassic granites (Unazuki area, Hida Metamorphic Complex): unearthing more old crust and identifying source terranes. *Precambrian Research*, **183**, 145–157.

HORIKOSHI, E. 1972. Orogenic belt of Japan and Plate. *Kagaku (Science)*, **42**, 665–673 [in Japanese].

HORIKOSHI, E., TAZAWA, J., NAITO, N. & KANEDA, J. 1987. Permian brachiopods from Moribu, north of Takayama City, Hida Mountains, central Japan. *Journal of the Geological Society of Japan*, **93**, 141–143 [in Japanese with English abstract].

HORIUCHI, J. & KIMURA, T. 1986. *Ginkgo tzagajanica* Samylina from the Palaeogene Noda Group, Northeast Japan, with special reference to its external morphology and cuticular features. *Transactions and Proceeedings of the Palaeontological Society of Japan, New Series*, **142**, 341–353.

HOSHINO, M. 1979. Two pyroxene amphibolite in Dogo, Oki Islands, Shimane-ken, Japan. *Journal of the Japanese Association of Mineralogists, Petrogists and Economic Geologists*, **74**, 87–99.

IBA, Y., SANO, S., MUTTERLOSE, J. & KONDO, Y. 2012. Belemnites originated in the Triassic – a new look at an old group. *Geology*, **40**, 911–914.

ICHIKAWA, K., ISHII, K., NAKAGAWA, C., SUYARI, K. & YAMASHITA, N. 1956. Die Kurosegawa Zone. *Journal of the Geological Society of Japan*, **62**, 82–103 [in Japanese with German abstract].

IGO, H. 1956. On the Carboniferous and Permian of the Fukuji district, Hida Massif, with special reference to the fusulinid zones of the Ichinotani Group. *Journal of the Geological Society of Japan*, **62**, 217–240 [in Japanese with English abstract].

IGO, H. 1959. Notes on some Permian corals from Fukuji, Hida Massif, Central Japan. *Transactions and Proceedings of the Palaeontological Society of Japan, New Series*, **34**, 79–85.

IGO, H. 1960. First discovery of non-marine sediments in the Japanese Carboniferous. *Proceedings of the Japan Academy*, **36**, 498–502.

IGO, H. 1964. On the occurrence of *Goniatites* (s.s.) from the Hida Massif, central Japan. *Transactions and Proceedings of the Palaeontological Society of Japan, New Series*, **54**, 234–238.

IGO, H. 1990. Paleozoic strata in the Hida 'Gaien' Belt. *In*: ICHIKAWA, K., MIZUTANI, S., HARA, I., HADA, S. & YAO, A. (eds) *Pre-Cretaceous Terranes of Japan*. Nippon Insatsu shuppan, Tokyo, 41–48.

IGO, H. & ADACHI, S. 1981. Foraminiferal biostratigraphy of the Ichinotani Formation (Carboniferous-Permian), Hida Massif, Central Japan Part 1-Some foraminifers from the upper part of the Lower Member of the Ichinotani Formation. *Science Reports of the Institute of Geoscience, University of Tsukuba*, **2**, 101–118.

INAI, Y. & TAKAHASHI, T. 1940. On the geology of the southernmost part of the Kitakami Massif. *Contributions from the Institute of Geology and Paleontology, Tohoku University*, **34**, 1–40 [in Japanese].

ISHIGA, H., SUZUKI, M., IIZUMI, S., NISHIMURA, K., KAGAMI, H. & TANAKA, S. 1989. Western extension of Hida terrane: with special reference to gneisses and mylonites discovered in Mizoguchi-cho, western part of Daisen, Tottori Prefecture, Southwest Japan. *Journal of the Geological Society of Japan*, **95**, 129–132 [in Japanese].

ISHIHARA, S. 2005. Source diversity of the older and early Mesozoic granitoids in the Hida Belt, central Japan. *Bulletin of the Geological Survey of Japan*, **56**, 117–126 [in Japanese with English abstract].

ISHIHARA, S., HIRANO, H. & TANI, K. 2012. Jurassic granitoids intruding into the Hida and Sangun metamorphic rocks in the central Sanin District, Japan. *Bulletin of the Geological Survey of Japan*, **63**, 227–231 [in Japanese with English abstract].

ISHII, K., SENDO, T., UEDA, Y. & SHIMAZU, M. 1956. *Explanatory text for the Geology of Iwate Prefecture II, Iguneous Rocks of Iwate Prefecture*. Iwate Prefecture [in Japanese].

ISHII, K., OKIMURA, Y. & NAKAZAWA, K. 1975. On the genus *Colaniella* and its biostratigraphic significance. *Journal of Geoscience, Osaka City University*, **19**, 107–138.

ISHIOKA, K. 1949. Staurolite and kyanite near Unazuki in the lower Kurobegawa area. *Journal of the Geological Society of Japan*, **55**, 156 [in Japanese].

Ishiwatari, A. 1991. Time-space distribution and petrologic diversity of Japanese ophiolite. *In*: Peters, Tj., Nicolas, A. & Coleman, R. G. (eds) *Ophiolite Genesis and Evolution of the Oceanic Lithosphere*, Kluwer Academic Publisher, Dordrecht, the Netherlands, 723–743.

Ishiwatari, A. & Tsujimori, T. 2003. Paleozoic ophiolites and blueschists in Japan and Russian Primorye in the tectonic framework of East Asia: a synthesis. *Island Arc*, **12**, 190–206.

Ishizaka, K. & Yamaguchi, M. 1969. U–Th–Pb ages of sphene and zircon from the Hida metamorphic terrain, Japan. *Earth Planetary Science Letters*, **6**, 179–185.

Isomi, H. & Nozawa, T. 1957. *Geological Sheet Map at 1:5000, Funatsu. With Geological Sheet Map at 1:50,000*. Geological Survey of Japan, Tsukuba [in Japanese with English abstract].

Isozaki, Y. 1996. Anatomy and genesis of a subduction-related orogen: a new view of geotectonic subdivision and evolution of the Japanese Islands. *Island Arc*, **5**, 289–320.

Isozaki, Y. 1997. Contrasting two types of orogen in Permo-Triassic Japan: accretionary v. collisional. *Island Arc*, **6**, 2–24.

Isozaki, Y. & Maruyama, S. 1991. Studies on orogeny based on plate tectonics in Japan and new geotectonic subdivision of the Japanese Island. *Journal of Geography*, **100**, 697–761.

Isozaki, Y. & Tamura, H. 1989. Late Carboniferous and Early Permian radiolarians from the Nagato Tectonic Zone and their implication to geologic structure of the Inner Zone, Southwest Japan. *Memoirs of the Geological Society of Japan*, **33**, 167–176.

Isozaki, Y., Aoki, K., Nakama, T. & Yanai, S. 2010. New insight into a subduction related orogen: a reappraisal of the geotectonic framework and evolution of the Japanese Islands. *Gondwana Research*, **18**, 82–105.

Isozaki, Y., Ehiro, M., Nakahata, H., Aoki, K., Sakata, S. & Hirata, T. 2015. Cambrian plutonism in Northeast Japan and its significance for the earliest arc-trench system of proto-Japan: new U-Pb zircon ages of the oldest granitoids in the Kitakami and Ou Mountains. *Journal of Asian Earth Sciences*, **108**, 136–149.

Iwai, J. & Ishizaki, K. 1966. A preliminary study on the Usuginu type conglomerate–with special reference to its paleogeographical and structural significance. *Contributions from the Institute of Geology and Paleontology, Tohoku University*, **62**, 35–53 [in Japanese with English abstract].

Iwata, T., Hirai, A., Inaba, T. & Hirano, M. 2002. Petroleum system in the offshore Joban Basin, northeast Japan. *Journal of the Japanese Association for Petroleum Technology*, **67**, 62–71 [in Japanese with English abstract].

Jahn, B. M. 2010. Accretionary orogen and evolution of the Japanese islands – implications from a Sr-Nd isotopic study of the Phanerozoic granitoids from SW Japan. *American Journal of Science*, **310**, 1210–1249.

Jahn, B. M., Wu, F., Hu, A. & Chen, B. 2000. Granitoids of the Central Asian Orogenic Belt and Continental Growth in the Phanerozoic. *Transactions of the Royal Society of Edinburgh: Earth Sciences* (Special Issue: Fourth Hutton Symposium on the Origin of Granites and Related Rocks), **91**, 181–193.

Jin, F. & Ishiwatari, A. 1997. Petrological and geochemical study on Hida gneisses in the upper reach area of Tetori River: comparative study on the pelitic metamorphic rocks with the other areas of Hida belt, Sino-Korean block and Yangtze block. *Journal of Mineralogy, Petrology, and Economic Geology*, **92**, 213–230 [in Japanese with English abstract].

Kabashima, T., Isozaki, Y. & Nishimura, Y. 1993. Find of boundary thrust between 300 Ma high-P/T type schists and weakly metamorphosed Permian accretionary complex in the Nagano tectonic zone, Southwest Japan. *Journal of the Geological Society of Japan*, **99**, 877–880 [in Japanese].

Kamada, K. 1980. The Triassic Inai Group in the Karakuwa Area, Southern Kitakami Mountains, Japan (Part 2) – On the 'intraformational disturbances' in the Lower Triassic Osawa Formation. *Journal of the Geological Society of Japan*, **86**, 713–726 [in Japanese with English abstract].

Kamada, K. 1983. Triassic Inai Group in the Toyoma area in the southern Kitakami Mountains, Japan with special reference to the submarine sliding deosits in the Triassic Osawa Formation. *Earth Science*, **37**, 147–161 [in Japanese with English abstract].

Kamei, T. 1952. The stratigraphy of the Fukuji district, southern part of Hida Mountainland (Study on Paleozoic rocks of Hida 1). *Journal of Shinshu University*, **2**, 43–74.

Kamei, T. 1955*a*. Geology of the Hida Marginal belt. *Journal of the Geological Society of Japan*, **61**, 414 [in Japanese].

Kamei, T. 1955*b*. Classification of the Fukuji formation (Silurian) on the basis of *Favosites* with description of some *Favosites*. *Journal of Shinshu University*, **5**, 39–63.

Kamei, T. 1962. On the Devonian System in the Hida Mountainlands. *In*: Fujimoto, H. (ed.) *Studies of Geology in the Hida Mountainlands*. Research Group of Geology of Hida Mountainlands, Tokyo, 33–43 [in Japanese].

Kamei, T. & Igo, H. 1955. On the discovery of *Cheirurus sterenbergi* BOEK from the Gotlandian Fukuji Group. *Journal of the Geological Society of Japan*, **61**, 457 [in Japanese with English abstract].

Kamikubo, H. & Takeuchi, M. 2010. Detrital heavy minerals from Lower Jurassic clastic rocks in the Joetsu area, central Japan: Paleo-Mesozoic tectonics in the East Asian continental margin constrained by limited chloritoid occurrences in Japan. *Island Arc*, **20**, 221–247.

Kanagawa, K. & Ando, H. 1983. Discovery of *Monotis* in the Ofunato area, southern Kitakami Mountains and its significance. *Journal of the Geological Society of Japan*, **89**, 187–190.

Kanisawa, S. 1964. Metamorphic rocks of the southwestern part of the Kitakami Mountainland, Japan. *Science Report of the Tohoku University, Series 3*, **9**, 155–198.

Kanisawa, S. 1974. Granitic rocks closely associated with the Lower Cretaceous volcanic rocks in the Kitakami Mountains, Northeast Japan. *Journal of the Geological Society of Japan*, **80**, 355–367.

Kanisawa, S. & Ehiro, M. 1997. Pre-Devonian Shoboji Diorite distributed in the western border of the South Kitakami Belt: its bearing on the characteristics of petrology and K–Ar age. *Journal of Mineralogy, Petrology and Economic Geology*, **92**, 195–204 [in Japanese with English abstract].

Kanisawa, S., Ehiro, M. & Okami, K. 1992. K–Ar ages of amphibolites from the Matsugadaira-Motai Metamorphics and their significance. *Japanese Journal of Mineralogists, Petrologists and Economic Geologists*, **87**, 412–419 [in Japanese with English abstract].

Kanmera, K. 1952. The Lower Carboniferous Kakisako Formation of southern Kyushu, with a description of some corals and fusulinids. *Memoirs of the Faculty of Science, Kyushu University, Series D*, **3**, 157–177.

Kanmera, K. 1953. The Kuma Formation with special reference to the Upper Permian in Japan. *Journal of the Geological Society of Japan*, **59**, 449–468 [in Japanese with English abstract].

Kanmera, K. 1954. Fusulinids from the Upper Permian Kuma Formation, Southern Kyushu, Japan- With special reference to the fusulinid zone in the Upper Permian of Japan. *Memoirs of the Faculty of Science, Kyushu University, Series D, Geology*, **4**, 1–38.

Kanmera, K. 1961. Middle Permian Kozaki formation. *Science Report of the Faculty of Science, Kyushu University, Geology*, **5**, 196–215.

Kanmera, K. 1963. Fusulines of the Middle Permian Kozaki Formation of Southern Kyushu. *Memoirs of the Faculty of Science, Kyushu University, Series D, Geology*, **14**, 79–141.

Kanmera, K. & Mikami, T. 1965*a*. Succession and sedimentary features of the Lower Permian Sakamotozawa Formation. *Memoirs of the Faculty of Science, Kyushu University, Series D, Geology*, **16**, 265–274.

Kanmera, K. & Mikami, T. 1965*b*. Fusuline zonation of the Lower Permian Sakamotozawa Series. *Memoirs of the Faculty of Science, Kyushu University, Series D, Geology*, **16**, 275–320.

Kano, H. 1959. On the granite-pebbles from the Shishiori Formation (Upper Jurassic) and their origin. *Journal of the Geological Society of Japan*, **65**, 750–759 [in Japanese with English abstract].

Kano, H. 1971. Studies on the Usuginu conglomerates in the Kitakami Mountains – Studies on the granite-bearing conglomerates in Japan, No.22. *Journal of the Geological Society of Japan*, **77**, 415–440 [in Japanese with English abstract].

Kano, T. 1980. Geological study of northern-half of western part of Hida metamorphic region, central Japan. *Journal of the Geological Society of Japan*, **86**, 687–704 [in Japanese with English abstract].

Kano, T. 1990. Intrusive relation of the Okumayama granitic mass (Shimonomoto type) into the Iori granitic mass (Funatsu type) in the Hayatsukigawa area: re-examination of the sub-division for early Mesozoic granites (Funatsu granites) in the Hida region. *Journal of the Geological Society of Japan*, **96**, 379–388 [in Japanese with English abstract].

Kano, T. 1991. Metasomatic origin of augen gneisses and related mylonitic rocks in the Hida metamorphic complex, central Japan. *Mineralogy and Petrology*, **45**, 29–45.

KANO, T. & WATANABE, T. 1995. Geology and structure of the early Mesozoic granitoids in the east of the Kamioka mining area, southern Hida metamorphic region, central Japan. *Journal of the Geological Society of Japan*, **101**, 499–514 [in Japanese with English abstract].

KARAKIDA, Y. 1981. Metamorphic conditions of high tempereture metamorphic rocks in the Kurosegawa tectonic zone. *In*: HARA, I. (ed.) *Tectonics of Paired Metamorphic Belt.* Tanishi Print Kikaku, Hiroshima, 165–169.

KASAHARA, Y. 1979. Geology of the Oamamiyama Group –Latest Cretaceous acid volcanism on the Hida Marginal Belt, Central Japan. *Memoir of the Geological Society of Japan*, **17**, 177–186 [in Japanese with English abstract].

KASE, T. 1979. Stratigraphy of the Mesozoic formatins in the shiura area, Southern Kitakami Mountainland, northern Japan. *Journal of the Geological Society of Japan*, **85**, 111–122 [in Japanese with English abstract].

KATO, M. 1959. Some Carboniferous rugose corals from Ichinotani formation, Japan. *Journal of the Faculty of Science, Hokkaido University, Series 4*, **10**, 263–287.

KATO, M. 1990. Paleozoic corals. *In*: ICHIKAWA, K., MIZUTANI, S., HARA, I., HADA, S. & YAO, A. (eds) *Pre-Cretaceous Terranes of Japan*. Nippon Insatsu shuppan, Tokyo, 307–312.

KATO, M., KUMANO, S., MINOURA, N., KAMADA, K. & KOSHIMIZU, S. 1977. Perisphinctes Ozikaensi from the Karakuwa Peninsula, Northeast Japan. *Journal of the Geological Society of Japan*, **83**, 305–306 [in Japanese].

KATO, M., MINATO, M., NIIKAWA, I., KAWAMURA, M., NAKAI, H. & HAGA, S. 1980. Silurian and Devonian corals of Japan. *Acta Palaeontologica Polonica*, **25**, 557–566.

KATO, M., KAWAMURA, M., KAWAMURA, T., TAZAWA, J., NIIKAWA, I. & NAKAMURA, T. 1989. Recent knowledge on the Carboniferous of the Kitakami Mountains. *XI Congress of International Stratigraphy and Geology of Carboniferous*, Beijing, 1987, *Compte Rendu 2*, 64–73.

KAWABE, I., SUGISAKI, R. & TANAKA, T. 1979. Petrochemistry and tectonic setting of Paleozoic-Early Mesozoic geosynclenal volcanics in the Japanese Islands. *Journal of the Geological Society of Japan*, **85**, 339–354.

KAWAMURA, M. 1980. Silurian halysitids from the Shimoarisu District, Iwate Prefecture, Northeast Japan. *Journal of the Faculty of Science, Hokkaido University, Series 4*, **19**, 273–303.

KAWAMURA, T. 1983. The Lower Carboniferous formations in the Hikoroichi region, southern Kitakami Mountains, northeast Japan (Part 1). Stratigraphy of the Hikoroichi Formation. *Journal of the Geological Society of Japan*, **89**, 707–722 [in Japanese with English abstract].

KAWAMURA, T. & KAWAMURA, M. 1989*a*. The Carboniferous System of the South Kitakami Terrane, northeast Japan (Part): Summary of the stratigraphy. *Earth Science*, **43**, 84–97 [in Japanese with English abstract].

KAWAMURA, T. & KAWAMURA, M. 1989*b*. The Carboniferous System of the South Kitakami Terrane, northeast Japan (Part 2): Sedimentary and tectonic environment. *Earth Science*, **43**, 157–167 [in Japanese with English abstract].

KAWAMURA, T. & MACHIYAMA, H. 1995. A Late Permian coral reef complex, South Kitakami Terrane, northeast Japan. *Sedimentary Geology*, **99**, 135–150.

KAWANO, Y. 1939. A new occurrence of jade (jadeite) in Japan and its chemical properties. *Journal of the Japanese Association of Mineralogy, Petrology and Economic Geology*, **22**, 195–201.

KHEDR, M. Z. & ARAI, S. 2010. Hydrous peridotites with Ti-rich chromian spinel as a low-temperature forearc mantle facies: evidence from the Happo-O'ne metaperidotites (Japan). *Contributions to Mineralogy and Petrology*, **159**, 137–157.

KHEDR, M. Z. & ARAI, S. 2011. Petrology and geochemistry of chromian spinel-bearing serpentinite in the Hida marginal belt (Ise area, Japan): characteristics of their protoliths. *Journal of Mineralogical and Petrological Sciences*, **106**, 255–260.

KIDO, E. & SUGIYAMA, T. 2011. Silurian rugose corals from the Kurosegawa Terrane, Southwest Japan, and their paleobiogeographic implication. *Bulletin of Geosciences*, **86**, 49–61.

KIDO, E., SUGIYAMA, T. & MUKAI, T. 2007. *Early Carboniferous Corals Found in the Limestone Exposed on the Southern Slope of Mt. Chunobori-dake, Gokase-cho, Miyazaki Prefecture, Southwest Japan.* Fukuoka University Science Reports **37**, 79–91.

KIM, J. C., LEE, Y. I. & HISADA, K. I. 2007. Depositional and compositional controls on sandstone diagenesis, the Tetori Group (Middle Jurassic–Early Cretaceous), central Japan. *Sedimentary Geology*, **195**, 183–202.

KIMURA, T. 1987. Geographical distribution of Palaeozoic and Mesozoic plants in east and southeast Asia. *In*: TAIRA, A. & TASHIRO, M. (eds) *Historical Biogeography and Plate Tectonic Evolution of Japan and Eastern Asia*. Terra Science Publishing Company, Tokyo, 135–200.

KIMURA, T. & OHANA, T. 1989*a*. Late Jurassic plants from the Oginohama Formation, Oshika Group in the Outer Zone of Northeast Japan (I). *Bulletin of National Science Museum, Tokyo, Series C*, **15**, 1–24.

KIMURA, T. & OHANA, T. 1989*b*. Late Jurassic plants from the Oginohama Formation, Oshika Group in the Outer Zone of Northeast Japan (II). *Bulletin of National Science Museum, Tokyo, Series*, **15**, 53–70.

KITAKAMI PALEOZOIC RESEARCH GROUP 1982. Pre-Silurian basement rocks of the Southern Kitakami Belt. *Memoirs of the Geological Society of Japan*, **21**, 261–281.

KITAMURA, M. & HIROI, Y. 1982. Indialite from Unazuki schist, Japan, and its transition texture to cordierite. *Contributions to Mineralogy and Petrology*, **80**, 110–116.

KOBAYASHI, F. 1973. On the Middle Carboniferous Nagaiwa Formation. *Journal of the Geological Society of Japan*, **79**, 69–78 [in Japanese with English abstract].

KOBAYASHI, F. 2002. Lithology and foraminiferal fauna of allochthonous limestones (Changhsingian) in the upper part of the Toyoma Formation in the South Kitakami Belt, Northeast Japan. *Paleontological Research*, **6**, 331–342.

KOBAYASHI, S., MIYAKE, H. & SHOJI, T. 1987. A jadeite rock from Oosa-cho, Okayama Prefecture, Southwestern Japan. *Mineralogical Journal*, **13**, 314–327.

KOBAYASHI, T. 1938. A tectonic view on the Oga Decke in the inner zone of western Japan. *Journal of the Geological Society of Japan*, **14**, 121–124.

KOBAYASHI, T. 1941. The Sakawa orogenic cycle and its bearing on the origin of the Japanese Islands. *Journal of the Faculty of Science, Imperial University of Tokyo, Section 2*, **5**, 219–578.

KOBAYASHI, T. & HAMADA, T. 1977. Devonian trilobites of Japan in comparison with Asian, Pacific and other faunas. *Palaeontological Society of Japan, Special Paper*, **23**, 1–132.

KOBAYASHI, T. & HAMADA, T. 1985. Additional Silurian Trilobites to the Yokokura-Yama Fauna from Shikoku, Japan. *Transactions and Proceeedings of the Palaeontological Society of Japan, New Series*, **139**, 206–217.

KOBAYASHI, T. & HAMADA, T. 1987. A new Carboniferous Trilibite from the Hida Plateau, West Japan. *Procedings of the Japan Academy, Series B*, **63**, 115–118.

KOBAYASHI, T. & IGO, H. 1956. On the occurrence of *Crotalocephalus*, Devonian trilobites in Hida, west Japan. *Japanese Journal of Geology and Geography*, **27**, 143–155.

KOBAYASHI, Y. & TAKAGI, H. 2000. Lithology, structure and petrochemistry of the Hikami Granitic Rocks in the South Kitakami Belt (Geotectonic evolution of the Paleo-Ryoke and Kurosegawa Terranes). *Memoirs of the Geological Society of Japan*, **56**, 103–122 [in Japanese with English abstract].

KOIZUMI, H. & KAKEGAWA, S. 1970. New occurrence of Devonian trilobites from Fukuji, Gifu Prefecture, Central Japan. *Chikyu Kagaku (Earth Science)*, **24**, 182–187 [in Japanese].

KOIZUMI, H., MIMOTO, K. & YOSHIHARA, N. 1994. Permian ammo-noid from Katsura, Sakawa, Kouchi Prefecture, Southwest Japan. *Chigaku Kenkyu*, **43**, 29–33 [in Japanese].

KOJIMA, S., TAKEUCHI, M. & TSUKADA, K. 2004. On the English expression of the Hida Gaien belt. *Journal of the Geological Society of Japan*, **110**, 565–566 [in Japanese with English abstract].

KOMATSU, M. 1990. Hida 'Gaien' belt and Joetsu belt. *In*: ICHIKAWA, K., MIZUTANI, S., HARA, I., HADA, S. & YAO, A. (eds) *Pre-Cretaceous Terranes of Japan*. Nippon Insatsu shuppan, Tokyo, 25–40.

KOMATSU, M., UJIHARA, M. & CHIHARA, K. 1985. Pre-Tertiary basement structure in the Inner zone of Honshu and the North Fossa Magna region. *Science Report of the Niigata University, Series E, Geology and Mineralogy*, **5**, 133–148.

KOMATSU, M., NAGASE, M., NAITO, K., KANNO, T., UJIHARA, M. & TOYOSHIMA, T. 1993. Structure and tectonics of the Hida massif, central Japan. *Memoirs of the Geological Society of Japan*, **42**, 39–62 [in Japanese with English abstract].

KONISHI, H., DÓDONY, I. & BUSECK, P. R. 2002. Protoanthophyllite from three metamorphosed serpentinites. *American Mineralogist*, **87**, 1096–1103.

KONISHI, H., GROY, T. L., DÓDONY, I., MIYAWAKI, R., MATSUBARA, S. & BUSECK, P. R. 2003. Crystal structure protoanthophyllite: a new mineral from the Takase ultramafic complex. *American Mineralogist*, **88**, 1718–1723.

KOSEKI, O. & HAMADA, T. 1988. On Leptophloeum discovered from the Upper Devonian Ainosawa Formation in the Abukuma Massif, northeastern Fukushima Prefecture. *Abstracts of the 137th Regular Meetings of the Palaeontological Society of Japan*, 1–1 [in Japanese].

KOZAI, T. & TASHIRO, M. 1993. Bivalve fauna from the Lower Cretaceous Funagawara Formation, Northeast Japan. *Memoirs of the Faculty of Science, Kochi University, Geology*, **14**, 25–43.

KUBO, K., YANAGISAWA, Y., TOSHIMITSU, S., BANNO, Y., KANEKO, N., YOSHIOKA, T. & TAKAGI, T. 2002. *Geology of the Kawamae and Ide District, Quadrangle Series, 1:50,000*. Geological Survey of Japan, AIST, Tsukuba [in Japanese with English abstract].

KUNUGIZA, K. 1999. Incipient stage of ore formation process of the Kamioka Zn–Pb ore deposit in the Hida metamorphic belt, central Japan: leaching and precipitation of clinopyroxene. *Resource Geology*, **49**, 199–212.

KUNUGIZA, K. & GOTO, A. 2006. 1.5 Clinopyroxene-bearing granitic rock (Inishi-type) and grey granite. *In*: Geological society of japan (ed.) *Regional Geology of Japan, 4, Chubu Region*. Asakura Publishing Co., Tokyo, 148–149 [in Japanese].

KUNUGIZA, K. & GOTO, A. 2010. Juvenile Japan: hydrothermal activity of the Hida-Gaien belt indicating initiation of subduction of proto-pacific plate in *c*. 520 Ma. *Journal of Geography*, **119**, 279–293.

KURIHARA, T. 2004. Silurian and Devonian radiolarian biostratigraphy of the Hida Gaien belt, central Japan. *Journal of the Geological Society of Japan*, **110**, 620–639 [in Japanese with English abstract].

KURIHARA, T. 2007. Uppermost Silurian to Lower Devonian radiolarians from the Hitoegane area of the Hida-gaien terrane, central Japan. *Micropaleontolgy*, **53**, 221–237.

KURODA, Y. & SHIMODA, S. 1967. Olivine with well-developed cleavages: its geological and mineralogical meanings. *Journal of the Geological Society of Japan*, **73**, 377–388.

KURODA, Y., KUROKAWA, K., URUNO, K., KINUGAWA, T., KANO, H. & YAMADA, T. 1976. Staurolite and kyanite from epidote hornblende rocks in the Oeyama (Komori) ultramafic mass, Kyoto Prefecture, Japan. *Earth Sciences (Chikyu Kagaku)*, **30**, 331–333.

KUROKAWA, K. 1975. Discovery of kyanite from epidote amphibolite in the Oeyama ultramafic mass, inner zone of southwestern Japan. *Journal of the Geological Society of Japan*, **81**, 273–274.

KUROKAWA, K. 1985. Petrology of the Oeyama ophiolitic complex in the Inner Zone of Southwest Japan. *Science Report of Niigata University (Series E)*, **6**, 37–113.

KUSUHASHI, N., MATSUMOTO, A. ET AL. 2006. Zircon U–Pb ages from tuff beds of the upper Mesozoic Tetori Group in the Shokawa district, Gifu Prefecture, central Japan. *Island Arc*, **15**, 378–390.

KUWANO, Y. 1986. Geologicalal age of the Fukuji Formation, Central Japan. *Memoirs of the National Science Museum*, **19**, 67–70 [in Japanese].

KUWANO, Y. 1987. Early Devonian conodonts and Ostracodes from central Japan. *Bulletin of the National Science Museum, Tokyo, Series C*, **13**, 77–105 [in Japanese].

MACHI, S. & ISHIWATARI, A. 2010. Ultramafic rocks in the Kotaki area, Hida Marginal Belt, central Japan: peridotites of the Oeyama ophiolite and their metamorphism. *Journal of the Geological Society of Japan*, **116**, 293–308 [in Japanese with English abstract].

MAEDA, S. 1961. On the geological history of the Mesozoic Tetori Group in Japan. *Journal of College of Arts and Science, Chiba University*, **3**, 369–426.

MAEKAWA, H. 1981. Geology of the Motai Group in the southwestern part of the Kitakami Mountains. *Journal of the Geological Society of Japan*, **87**, 543–554 [in Japanese with English abstract].

MAEKAWA, H. 1988. High P/T metamorphic rocks in Northeast Japan: Motai-Matsugadaira zone. *Earth Science*, **42**, 212–219 [in Japanese with English abstract].

MANCHUK, N., HORIE, K. & TSUKADA, K. 2013*a*. SHRIMP U-Pb age of the radiolarian-bearing Yoshiki Formation in Japan. *Bulletin of Geosciences*, **88**, 223–240.

MANCHUK, N., KURIHARA, T. ET AL. 2013*b*. U–Pb zircon age from the radiolarian-bearing Hitoegane Formation in the Hida Gaien Belt, Japan. *Island Arc*, **22**, 494–507.

MARUYAMA, S., UEDA, Y. & BANNO, S. 1978. 208–240 m.y. old Jedeite-Glaucophene schists in the Kurosegawa Tectonic Zone near Kochi City, Shikoku. *Journal of the Japanese Association of Mineralogy, Petrology and Economic Geology*, **73**, 300–310.

MASUTOMI, J. 1966. The discovery of Jadeite from the Sangun metamorphic belt, Tottori. *Chigaku-Kenkyu*, **17**, 83 [in Japanese].

MATSUMOTO, I., ARAI, S. & HARADA, T. 1995. Hydrous mineral inclusions in chromian spinel from the Yanomine ultramafic complex of the Sangun zone, Southwest Japan. *Journal of Mineralogy, Petrology and Economic Geology*, **90**, 333–338 [in Japanese with English abstract].

MATSUMOTO, I., ARAI, S. & YAMAUCHI, H. 1997. High-Al podiform chromitites in dunite-harzburgite complexes of the Sangun zone, central Chugoku district, Southwest Japan. *Journal of Asian Earth Sciences*, **15**, 295–302.

MATSUMOTO, I., ARAI, S. & YAMANE, T. 2002. Significance of magma/peridotite reaction for size of chromitite: example for Wakamatsu chromite mine of the Tari-Misaka ultramafic complex, southwestern Japan. *Resource Geology*, **52**, 135–146 [in Japanese with English abstract].

MATSUMOTO, TAKAYUKI. 2012. Geology of the Hida Gaien Belt in the upper Kuzuryu-gawa River Area in Ono City, Fukui Prefecture, Central Japan. *Resource Geology*, **62**, 384–407.

MATSUMOTO, TATSURO. 1943. Fundamentals in the Cretaceous stratigraphy of Japan. Parts II & III. *Memoirs of the Faculty of Science, Kyushu Imperial University, Series D*, **2**, 97–237.

MATSUMOTO, TATSURO. 1953. *The Cretaceous System in the Japanese Islands*. Japanese Society for Promotion of Science, Tokyo.

MATSUMOTO, TATSURO. 1956. *Yebisites*, a New Lower Jurassic Ammonite from Japan. *Transactions and Proceeedings of the Palaeontological Society of Japan, New Series*, **23**, 205–212.

MATSUMOTO, TATSURO. 1963. The Cretaceous. *In*: TAKAI, F., MATSUMOTO, T. & TORIYAMA, R. (eds) *Geology of Japan*. University of Tokyo Press, Tokyo, 99–128.

MATSUMOTO, TATSURO, OBATA, I., TASHIRO, M., OHTA, Y., TAMURA, M., MATSUKAWA, M. & TANAKA, H. 1982. Correlation of marine and non-marine formations in the Cretaceous of Japan. *Fossils*, **31**, 1–26 [in Japanese with English abstract].

MATSUMOTO, TATSURO, NEMOTO, M. & SUZUKI, C. 1990. Gigantic ammonites from the Cretaceous Futaba Group of Fukushima Prefecture. *Transactions and Proceeedings of the Palaeontological Society of Japan, New Series*, **157**, 366–381.

MATSUOKA, A. 1985. Middle Jurassic Keta Formation of the southern part of the Middle Chichibu Terrane in the Sakawa area, Kochi Prefecture, Southwest Japan. *Journal of the Geological Society of Japan*, **91**, 411–420 [in Japanese with English abstract].

MATSUOKA, A., YAMAKITA, S., SAKAKIBARA, M. & HISADA, K. 1998. Unit division for the Chichibu Composite Belt from a view point of accretionary tectonics and geology of western Shikoku, Japan. *Journal of the Geological Society of Japan*, **104**, 634–653 [in Japanese with English abstract].

MEI, M., HUANG, Q. C., DU, M. & DILCHER, D. L. 1996. The Xu-Huai-Yu Subprovince of the Cathaysian Floral Province. *Review of Palaeobotany and Palynology*, **90**, 63–77.

METCALFE, I. 1996. Pre-Cretaceous evolution of SE Asian terranes. *In*: HALL, R. & BLUNDELL, D. (eds) *Tectonic Evolution of Southeast Asia*. Geological Society, London, 97–122.

MINATO, M., KATO, M., NAKAMURA, K., HASEGAWA, Y., CHOI, D. R. & TAZAWA, J. 1978. Biostratigraphy and correlation of the Permian of Japan. *Journal of the Faculty of Science, Hokkaido University, Series 4*, **18**, 11–47.

MINATO, M., HUNAHASHI, M., WATANABE, J. & KATO, M. (eds) 1979. *The Abean Orogeny, Variscan Geohistory of Northern Japan*. Tokai University Press, Tokyo.

MINOURA, K. 1985. Where did the Kitakami and Abukuma massifs come from? *Kagaku*, **55**, 14–23 [in Japanese].

MINOURA, K. & YAMAUCHI, H. 1989. Upper Cretaceous-Paleogene Kuji Basin of Northeast Japan: tectonic controls on strike-slip basin. *In*: TAIRA, A. & MASUDA, F. (eds) *Sedimentary Facies in the Active Margin*. Terra Science Publishing Company, Tokyo, 633–658.

MITI (JAPANESE MINISTRY OF INTERNATIONAL TRADE AND INDUSTRY) 1993. Report on the rare-metal survey of the Dogoyama district (1992) [in Japanese].

MITI (JAPANESE MINISTRY OF INTERNATIONAL TRADE AND INDUSTRY) 1994. Report on the rare-metal survey of the Dogoyama district (1993) [in Japanese].

MIYAMOTO, M. & TANIMOTO, Y. 1993. Formation in the Chichibu Belt of South Kyushu, Southwest Japan. *News of Osaka Micropaleontologist*, **9**, 19–33.

MIYAMOTO, T., KUWAZURU, J. & OKIMURA, Y. 1997. The Lower Permian formation discovered from the Kurosegawa Terrane, Kyushu. *News of Osaka Micropaleontologists (NOM), Special Volume 10, Proceedings of the Fifth Radiolarian Synposium*, 33–40 [in Japanese].

MIYASHIRO, A. 1961. Evolution of metamorphic belts. *Journal of Petrology*, **2**, 277–311.

MOCHIZUKI, K. & ANDO, H. 2003. Molluscan fossil beds in the storm-dominated shallow-marine sequences of the Lower Cretaceous Miyako Group. *Fossils*, **74**, 1–2 [in Japanese].

MOJSISOVICS, E. von. 1888. Über einige japanische Triasfossilien. *Beiträge zur Paläontologie Österreich-Ungarns und des Orients*, **7**, 163–178.

MORI, K. 1963. Geology and paleontology of the Jurassic Somanakamura Group, Fukushima Prefecture, Japan. *Science Report of the Tohoku University, Series 2*, **35**, 33–65.

MORI, K. & TAZAWA, J. 1980. Discovery and significance of Visean rugose corals and brachiopods from the type locality of Lower Carboniferous Hikoroichi Formation. *Journal of the Geological Society of Japan*, **86**, 143–146 [in Japanese].

MORI, K., OKAMI, K. & EHIRO, M. 1992. Paleozoic and Mesozoic sequences in the Kitakami Mountains (29th IGC Field Trip A05). *In*: ADACHI, M. & SUZUKI, K. (eds) *29th IGC Field Trip Guide Book Vol. 1, Paleozoic and Mesozoic Terranes: Basement of the Japanese Islands Arcs*. Nagoya University, Japan, 81–114.

MURATA, M. & BANDO, Y. 1975. Discovery of Late Permian *Arax-oceras* from the Toyoma Formation in the Kitakami Massif, Northeast Japan. *Transactions and Proceedings of the Palaeon-tological Society of Japan, New Series*, **97**, 22–31.

MURATA, M. & SHIMOYAMA, S. 1979. Stratigraphy near the Permian Triassic Boundary and the Pre Triassic Unconformity in the Kitakami Massif, Northeast Japan. *Kumamoto Journal of Science (Earth Science)*, **11**, 11–31 [in Japanese with English abstract].

MURATA, M., KANISAWA, S., UEDA, Y. & TAKEDA, N. 1974. Base of the Silurian System and the Pre-Silurian Granites in the Kitakami Massif, Northeast Japan. *Journal of the Geological Society of Japan*, **80**, 475–486 [in Japanese with English abstract].

MURATA, M., OKAMI, K., KANISAWA, S. & EHIRO, M. 1982. Additional evidence for the Pre-Silurian Basement in the Kitakami Massif, Northeast Honshu, Japan. *Memoirs of the Geological Society of Japan*, **21**, 245–259.

NAITO, K. 1993. Geochemical investigation of the Hida metamorphic rock. *Memoir of the Geological Society of Japan*, **42**, 21–37 [in Japanese with English abstract].

NAKAI, H. 1980. New occurrence of Lower Carboniferous in Shikoku with description of a new aulate rugosa. *Earth Science*, **34**, 138–143.

NAKAJIMA, T. & MARUYAMA, S. 1978. Barroisite-bearing schist blocks in serpenitinite of the Kurosegawa Tectonic Zone, west of Kochi City, central Shikoku. *Journal of the Geological Society of Japan*, **84**, 231–242.

NAKAMIZU, M., OKADA, M., YAMAZAKI, T. & KOMATSU, M. 1989. Metamorphic rocks in the Omi-Renge serpentinite melange, Hida Marginal Tectonic Belt, Central Japan. *Memoirs of the Geological Society of Japan*, **33**, 21–35 [in Japanese with English abstract].

NAKAZAWA, K. 1964. On the Upper Triassic *Monotis* Beds, especially, on the *Monotis typica* Zone. *Journal of the Geological Society of Japan*, **70**, 523–535 [in Japanese with English abstract].

NAKAZAWA, K. 1991. Mutual relation of Tethys and Japan during Permian and Triassic time viewed from bivalve fossils. *In*: KOTAKA, T. J. M., MCKENZIE, K. G., MORI, K., OGASAWARA, K. & STANLEY, G. D. JR (eds) *Shallow Tethys*. Saito Ho-on Kai, Sendai, **3**, 3–20.

NARA, Y., TAKETANI, Y. & MINOURA, K. 1994. Jurassic-Cretaceous stratigraphy in the Kesennuma and Karakuwa areas of Southern Kitakami Mountains, Northeast Japan. *Bulletin of the Fukushima Museum*, **8**, 29–63 [in Japanese with English abstract].

NAUMANN, E. 1881. Über das Vorkommen von Triasbildungen im nördichen Japan. *Jahrbuch der Geologischen Reichsanstalt, Wien*, **31**, 519–628.

NIIKAWA, I. 1980. Geology and biostratigraphy of the Fukuji district, Gifu Prefecture, Central Japan. *Journal of the Geological Society of Japan*, **86**, 25–36 [in Japanese with English abstract].

NIIKAWA, I. 1983*a*. Biostratigraphy and correlation of the Onimaru Formation in the southern Kitakami Mountains, Part 1 Geology and biostratigraphy. *Journal of the Geological Society of Japan*, **89**, 347–357 [in Japanese with English abstract].

NIIKAWA, I. 1983*b*. Biostratigraphy and correlation of the Onimaru Formation in the southern Kitakami Mountains, Part 2 Correlation and conclusion. *Journal of the Geological Society of Japan*, **89**, 549–557 [in Japanese with English abstract].

NIIKAWA, I. 1994. The palaeobiogeography of *Kueichouphyllum*. *Courier Forschungsinstitut Senckenberg*, **172**, 43–50.

NIKO, S., YAMAKITA, S., OTOH, S., YANAI, S. & HAMADA, T. 1987. Permian radiolarians from the Mizuyagadani Formation in Fukuji area, Hida Marginal Belt and their significance. *Journal of the Geological Society of Japan*, **93**, 431–433 [in Japanese].

NIKO, S., HAMADA, T. & YASUI, T. 1989. Silurian Orthocerataceae (Mollusca: Cephalopoda) from the Yokokurayama Formation, Kurosegawa Terrane. *Transactions and Proceedings of the Palaeontological Society of Japan, New Series*, **154**, 59–67.

NISHIMURA, Y. & SHIBATA, K. 1989. Modes of occurrence and K–Ar ages of metagabbroic rocks in the 'Sangun metamorphic belt', Southwest Japan. *Memoir of Geological Society of Japan*, **33**, 343–357 [in Japanese with English abstract].

NIWA, M., TSUKADA, K. & KOJIMA, S. 2002. Permian clastic formation in the Yokoo area, Nyukawa Village, Gifu Prefecture, central Japan. *Journal of the Geological Society of Japan*, **108**, 75–87 [in Japanese with English abstract].

NIWA, M., HOTTA, K. & TSUKADA, K. 2004. Middle Permian fusulinoideans from the Moribu Formation in the Hida-gaien Tectonic Zone, Nyukawa Village, Gifu Prefecture, central Japan. *Journal of the Geological Society of Japan*, **110**, 384–387 [in Japanese with English abstract].

NODA, M. & TACHIBANA, K. 1959. Some Upper Devonian cyrtospiriferids from the Nagasaka district, Kitakami Mountainland. *Science Bulletin of the Faculty of Liberal Arts and Education, Nagasaki University*, **10**, 15–21.

NOZAKA, T. 2005. Metamorphic history of serpentinite mylonites from the Happo ultramafic complex, central Japan. *Journal of Metamorphic Geology*, **23**, 711–723.

NOZAKA, T. & ITO, Y. 2011. Cleavable olivine in serpentinite mylonites from the Oeyama ophiolite. *Journal of Mineralogical and Petrological Sciences*, **106**, 36–50.

OBATA, I. 1988. Cretaceous formations in Northeast Japan. *Earth Science*, **42**, 385–395 [in Japanese with English abstract].

OBATA, I. & MATSUMOTO, T. 1977. *Correlation of the Lower Cretaceous formations in Japan*. Science Reports, Department of Geology, Kyushu University **12**, 165–179 [in Japanese with English abstract].

OBATA, I. & SUZUKI, T. 1969. Additional note on the upper limit of the Cretaceous Futaba Group. *Journal of the Geological Society of Japan*, **75**, 443–445 [in Japanese with English abstract].

OBAYASHI, T. 1995. Provenance nature of the Tetori Group in the Shiramine area, central Japan, based on the chemical composition of detrital garnets. *Journal of the Geological Society of Japan*, **101**, 235–248 [in Japanese with English abstract].

OHNO, T. 1977. Lower Devonian brachiopods from the Fukuji Formation, central Japan. *Memoirs of the Faculty of Science, Kyoto University (Geology and Mineralogy)*, **44**, 79–126.

OHTA, K. & ITAYA, T. 1989. Radiometric ages of granitic and metamorphic rocks in the Hida metamorphic belt, central Japan. *Bulletin of Hiruzen Research Institute, Okayama University of Science*, **15**, 1–12.

OKAMI, K., EHIRO, M. & OISHI, M. 1986. Geology of the Lower-Middle Palaeozoic around the northern marginal part of the Southern Kitakami Massif with reference to the geologic development of the 'Hayachine Tectonic Belt'. *Essays in Geology*, Professor Nobu Kitamura Commemorative Volume, Sendai, Japan, 313–330 [in Japanese with English abstract].

OKAMI, K., EHIRO, M., KURIYAGAWA, H. & ASANUMA, A. 1987. *Leptophloeum* bearing formation in the 'Hayachine Tectonic Belt', Kitakami Massif, Northeast Japan. *Journal of the Geological Society of Japan*, **93**, 321–327 [in Japanese with English abstract].

OKAZAKI, Y. 1974. Devonian trilobites from the Fukuji formation in the Hida Massif, central Japan. *Memoirs of the Faculty of Science, Kyoto University (Geology and Mineralogy)*, **40**, 84–94.

OKUI, A. 1985. Polymetamorphism in the Hida metamorphic rocks of upper Katakai river area, Toyama Prefecture, central Japan, with special reference to the effect of intrusion of the Funatsu granitic rocks. *Journal of Mineralogy, Petrology, and Economic Geology*, **80**, 387–397 [in Japanese with English abstract].

ONUKI, Y. 1937. Discovery of the Gotlandian formation and the stratigraphy of the Paleozoic in the Kesen district, Iwate prefecture, Kitakami Massif (Preliminary report). *Journal of the Geological Society of Japan*, **44**, 600–604 [in Japanese].

ONUKI, Y. 1969. Geology of the Kitakami Massif, Northeast Japan. *Contributions from the Institute of Geology and Paleontology, Tohoku University*, **69**, 1–239 [in Japanese with English abstract].

ONUKI, Y. & BANDO, Y. 1958. On the Saragai Group of the Upper Triassic System (Stratigraphical and paleontological studies of the Triassic System in the Kitakami Masif, Northeast Japan: 1). *Journal of the Geological Society of Japan*, **64**, 481–493 [in Japanese with English abstract].

ONUKI, Y. & BANDO, Y. 1959. On the Inai Group of the Lower and Middle Triasic System (Stratigraphical and paleontological studies of the Triassic System in the Kitakami Masif, Northeast Japan: 3). *Contributions from the Institute of Geology and Paleontology, Tohoku University*, **50**, 1–69 [in Japanese with English abstract].

ONUKI, Y. & MORI, K. 1961. Geology of the Ofunato district, Iwate Prefecture, southern part of theKitakami Massif, Japan. *Journal of the Geological Society of Japan*, **67**, 641–654 [in Japanese with English abstract].

OSANAI, Y., HAMAMOTO, T., KAGAMI, H., OWADA, M., DOYAMA, D. & ANDO, T. K. 2000. Protolith and Sm-Nd geochronology of garnet-clinopyroxine granulite and garnet amphibolite from the Kurosegawa Belt in Kyushu, southwest Japan. *Memoir of the Geological Society of Japan*, **56**, 199–212 [in Japanese with English abstract].

OSAWA, M. 1983. Geological studies on the 'Hatachine Tectonic Belt'. *Contributions from the Institute of Geology and Paleontology, Tohoku University*, **85**, 1–30 [in Japanese with English abstract].

OSAWA, M., NAKANISHI, S., TANAHASHI, M. & ODA, M. 2002. Structure, tectonic evolution and gas exploration potential of offshore Sanriku and Hidaka provinces, Pacific Ocean, off northern Honshu and Hokkaido, Japan. *Journal of the Japanese Association for Petroleum Technology*, **67**, 38–51 [in Japanese with English abstract].

OTOH, S. & YANAI, S. 1996. Mesozoic inversive wrench tectonics in far east Asia: examples from Korea and Japan. *In*: YIN, A. & HARRISON, M. (eds) *The Tectonic Evolution of Asia*. Cambridge University Press, Cambridge, 401–419.

OTOH, S., TSUKADA, K., SANO, K., NOMURA, R., JWA, Y. & YANAI, S. 1999. Triassic to Jurassic dextral ductile shearing along the eastern margin of Asia: a Synthesis. *In*: METCALFE, I., JISHUN, R., CHARVET, J. & HADA, S. (eds) *Gondwana Dispersion and Asian Accretion IGCP 321 Final Results Volume*. A. A. Balkema, Rotterdam, 89–113.

OTSUKI, K. & EHIRO, M. 1978. Major strike-slip faults and their bearing on spreading in the Japan Sea. *Journal of Physical Earth*, **26**, 537–555.

OTSUKI, K. & EHIRO, M. 1992. Cretaceous left-lateral faulting in Northeast Japan and its bearing on the origin of geologic structure of Japan. *Journal of the Geological Society of Japan*, **98**, 1097–1112 [in Japanese with English abstract].

OZAWA, K. 1984. Geology of the Miyamori ultramafic complex in the Kitakami Mountains, Northeast Japan. *Journal of the Geological Society of Japan*, **90**, 697–716.

OZAWA, K., SHIBATA, K. & UCHIUMI, S. 1988. K–Ar ages of hornblende in gabbroic rocks from the Miyamori ultramafic complex of the Kitakami Mountains. *Journal of Mineralogy, Petrology and Economic Geology*, **83**, 150–159 [in Japanese with English abstract].

OZAWA, T. 1987. Permian fusulinacean biogeographic provinces in Asia and their tectonic implications. *In*: TAIRA, A. & TASHIRO, M. (eds) *Historical Biogeography and Plate Tectonic Evolution of Japan and Eastern Asia*. Terra Science Publishing Company, Tokyo, 45–63.

POWELL, C. M. & LI, Z. 1994. Reconctruction of the Panthalassan margin of Gondwanaland. *Geological Society of America, Memoir*, **184**, 5–9.

RESEARCH GROUP FOR THE PALEOZOIC OF FUKUJI 1973. On the occurrence of *Rhizophyllum* (Rugosa) from the Fukuji formation, Central Japan. *Journal of the Geological Society of Japan*, **79**, 423–424 [in Japanese].

SAITO, M., SAIKI, K. & TOSHIMITSU, S. 2003. Late Devonian *Leptophloeum* from the coherent strata in the Kurosegawa Belt, Tomochi district, central Kyusyu. *Journal of the Geological Society of Japan*, **109**, 293–298 [in Japanese with English abstract].

SAITO, Y. 1968. Geology of the younger Paleozoic System of the southern Kitakami Massif, Iwate Prefecture Japan. *Science Report of the Tohoku University, Series 2*, **40**, 79–139.

SAKODA, M., KANO, T., FANNING, C. M. & SAKAGUCHI, T. 2006. SHRIMP U-Pb zircon age of the Inishi migmatite around the Kamioka mining area, Hida metamorphic complex, central Japan. *Resource Geology*, **56**, 17–26.

SANO, Y., HIDAKA, H., TERADA, K., SHIMIZU, H. & SUZUKI, M. 2000. Ion microprobe U-Pb zircon geochronology of the Hida gneiss: finding of the oldest minerals in Japan. *Geochemical Journal*, **34**, 135–153.

SASAKI, M., TSUKADA, K. & OTOH, S. 1997. An outcrop of unconformity at the base of the Upper Devonian Tobigamori Formtion, Southern Kitakami Mountains. *Journal of the Geological Society of Japan*, **103**, 647–655 [in Japanese with English abstract].

SATO, Tadashi. 1958. Presence du Berriasian dans La stratigraphie de plateau de Kitakami (Japon septentrional). *Bulletin de la Société Géologique de France, Série 6*, **8**, 585–599.

SATO, Tadashi. 1962. Etudes biostratigraphiques des ammonites du Jurassique du Japon. *Mémoires de la Société Géologique de France. Nouvelle série*, **41**, 1–122.

SATO, Tadashi. 1972. Some Bajosian ammonites from Kitakami, Northeast Japan. *Transactions and Proceedings of the Palaeontological Society of Japan, New Series*, **85**, 280–292.

SATO, Tadashi & TAKETANI, Y. 2008. Late Jurassic to Early Cretaceous ammonite fauna from the Somanakamura Group in Northeast Japan. *Paleontological Research*, **12**, 261–282.

SATO, Tadashi & WESTERMANN, G. E. G. 1991. Jurassic taxa ranges and correlation charts for the Circum Pacific, 4. Japan and South-East Asia. *Newsletter on Stratigraphy*, **24**, 81–108.

SATO, Tadashi. TAKETANI, Y. *ET AL.* 2005. Newly collected ammonites from the Jurassic-Cretaceous Somanakamura Group. *Bulletin of the Fukushima Museum*, **19**, 1–14 [in Japanese with English abstract].

SATO, Tamaki, HASEGAWA, Y. & MANABE, M. 2006. A new elasmosaurid plesiosaur from the Upper Cretaceous of Fukushima, Japan. *Palaeontology*, **49**, 467–484.

SHI, G. R., ARCHBOLD, N. W. & ZHAN, L. 1995. Distribution and characteristics of mixed (transitional) mid-Permian (Late Artinskian-Ufimian) marine faunas in Asia and their palaeogeographical implications. *Palaeogeography, Palaeoclimatology, Palaeoecology*, **114**, 241–271.

SHIBATA, K. 1973. K–Ar ages of the Hikami granites and the Usuginu granitic clasts. *Journal of the Geological Society of Japan*, **79**, 705–707 [in Japanese with English abstract].

SHIBATA, K. & NOZAWA, T. 1966. K–Ar Ages of Hida Metamorphic Rocks, Amo-Tsunokawa Area and Oki Area, Japan. *Bulletin of the Geological Survey of Japan*, **17**, 410–416.

SHIBATA, K. & OZAWA, K. 1992. Ordovician arc ophiolite, the Hayachine and Miyamori complexes, Kitakami Mountains, Northeast Japan. *Geochemical Journal*, **26**, 85–97.

SHIBATA, K., NOZAWA, T. & WANLESS, R. K. 1970. Rb-Sr geochronology of the Hida metamorphic belt, Japan. *Canadian Journal of Earth Sciences*, **7**, 1383–1401.

SHIBUYA, H., SASAJIMA, S. & YOSHIKURA, S. 1983. A paleomagnetic study on the Silurian acidic tuffs in the Yokokurayama lenticular body of the Kurosegawa tectonic zone, Kochi Prefecture, Japan. *Journal of the Geological Society of Japan*, **89**, 307–309 [in Japanese with English abstract].

SHIIDA, I. 1940. On the geology of the vicinity of Kesennuma, Miyagi Prefecture. *Contributions from the Institute of Geology and Paleontology, Tohoku University*, **33**, 1–72 [in Japanese].

SHIMAZU, M. & TERAOKA, Y. 1962. *Geological Map of 'Rikuchu-Noda' in the Scale 1:50,000 and its Explanatory Text*. Geological Survey of Japan, Kawasaki, Japan [in Japanese].

SHIMIZU, S. 1930. On some Anisic ammonites from the *Hollandites* beds of the Kitakami Mountainland. *Science Report of the Tohoku Imperial University, Series 2*, **14**, 63–74.

SHIMIZU, S. & MABUTI, S. 1932. Upper Triassic strata in the Kitakami Massif. *Journal of the Geological Society of Japan*, **39**, 313–317 [in Japanese].

SHIMOJO, M., OTOH, S., YANAI, S., HIRATA, T. & MARUYAMA, S. 2010. LA-ICP-MS U-Pb age of some older rocks of the South Kitakami Belt, Northeast Japan. *Journal of Geography*, **119**, 257–269 [in Japanese with English abstract].

SOHMA, T. & AKIYAMA, S. 1984. Geological structure and lithofacies in the central part of the Hida metamorphic belt. *Journal of the Geological Society of Japan*, **90**, 609–628 [in Japanese with English abstract].

SOHMA, T. & KUNUGIZA, K. 1993. The formation of the Hida nappe and the tectonics of Mesozoic sediments: the tectonic evolution of the Hida region, central Japan. *Memoir of the Geological Society of Japan*, **42**, 1–20 [in Japanese with English abstract].

SOHMA, T., KONAGAI, K., SAWADA, N. & MARUYAMA, S. 1986. Petrological Study of the Northern Part of the Hida Central Metamorphic Complex in Central Honshu, Japan. *Memoirs of the Faculty of Education, Toyama University, Series B*, **34**, 9–24 [in Japanese].

STERN, R. J., TSUJIMORI, T., HARLOW, G. E. & GROAT, L. A. 2013. Plate tectonic gemstones. *Geology*, **41**, 723–726.

SUTO, I., YANAGISAWA, Y. & OGASAWARA, K. 2005. Tertiary geology and chronostratigraphy of the Joban area and its environs, northeastern Japan. *Bulletin of the Geological Survey of Japan*, **56**, 375–409 [in Japanese with English abstract].

SUWA, K. 1966. Finding of conglomerate schist from the Upper Katalao River area, Toyama Prefecture, central Japan. *Journal of the Geological Society of Japan*, **72**, 585–591.

SUZUKI, K. & ADACHI, M. 1991. Precambrian provenance and Silurian metamorphism of the Tsubonosawa paragneiss in the South Kitakami terrane, Northeast Japan, revealed by the chemical Th-U-total Pb isochron ages of monazite, zircon and xenotime. *Geochemical Journal*, **25**, 357–376.

SUZUKI, K. & ADACHI, M. 1994. Middle Precambrian detrital monazite and zircon from the Hida gneiss on Oki-Dogo Island, Japan: their origin and implications for the correlation of basement gneiss of Southwest Japan and Korea. *Tectonophysics*, **235**, 277–292.

SUZUKI, K., ADACHI, M., SANGO, K. & CHIBA, H. 1992. Chemical Th-U-total Pb isochron ages of manazites and zircons from the Hikami Granite and 'Siluro-Devonian' clastic rocks in the South Kitakami terrane. *Journal of Mineralogy, Petrology and Economic Geology*, **87**, 30–349 [in Japanese with English abstract].

SUZUKI, M. 1973. An occurrence of 'eclogitic rock' in the Hida metamorphic belt. *Journal of the Japanese Association of Mineralogists, Petrologists and Economic Geologists*, **12**, 372–382.

SUZUKI, M. 1977. Polymetamorphism in the Hida metamorphic belt, central Japan. *Journal of Science of the Hiroshima University, Series C*, **7**, 217–296.

SUZUKI, M. & KOJIMA, G. 1970. On the association of potassium feldspar and corundum found in the Hida metamorphic belt. *Journal of the Japanese Association of Mineralogists, Petrologists and Economic Geologists*, **63**, 266–274.

SUZUKI, M., NAKAZAWA, S. & OSAKABE, T. 1989. Tectonic development of the Hida Belt: with special reference to its metamorphic history and late Carboniferous to Triassic orogenies. *Memoir of the Geological Society of Japan*, **33**, 1–10 [in Japanese with English abstract].

SUZUKI, N., TAKAHASHI, D. & KAWAMURA, T. 1996. Late Silurian an Early Devonian Policystine (Radiolaria) from the Middle Paleozoic deposits in the Kamaishi area, northeast Japan. *Journal of the Geological Society of Japan*, **102**, 824–827 [in Japanese with English abstract].

SUZUKI, Y., EHIRO, M. & MORI, K. 1998. Middle Jurassic ammonoids from the Owada Formation in the Mizunuma district, Southern Kitakami Massif, Northeast Japan. *Journal of the Geological Society of Japan*, **104**, 268–271 [in Japanese with English abstract].

TACHIBANA, K. 1950. Devonian plant first discovered in Japan. *Proceedings of the Japan Academy*, **26**, 54–60.

TAGIRI, M., MORIMOTO, M., MOCHIZUKI, R., YOKOSUKA, A., DUNKLEY, D. J. &. ADACHI, T. 2010. Hitachi metamorphic rocks–Occurrence and geology of metagranitic rocks with Cambrian SHRIMP zircon age. *Journal of Geography*, **119**, 245–256 [in Japanese with English abstract].

TAGIRI, M., DUNKLEY, J. D., ADACHI, T., HIROI, Y. & FANNING, C. M. 2011. SHRIMP dating of magmatism in the Hitachi metamoephic terrane, Abukuma Belt, Japan: evidence for a Cambrian volcanic arc. *Island Arc*, **20**, 259–279.

TAKAGI, H. & HARA, T. 1994. Kinematic pictures of the ductile shear zones in the Hida terrane and their tectonic implication. *Journal of the Geological Society of Japan*, **100**, 931–950 [in Japanese with English abstract].

TAKAHASHI, H. 1961. Mesozoic stratigraphy of the Hashiura-Jusanhama area, Southern Kitakami Mountains. *Bulletin of the Faculty of Arts and Sciences, Ibaraki University (Natural Science)*, **12**, 145–159 [in Japanese with English abstract].

TAKAHASHI, H. 1969. Stratigraphy and ammonite fauna of the Jurassic System of the Southern Kitakami Massif, northeast Honshu, Japan. *Science Report of the Tohoku University, Series 2*, **41**, 1–93.

TAKAHASHI, H. 1973. The Isokusa Formation and its late Upper Jurassic and early Lower Cretaceous ammonite fauna. *Science Report of the Tohoku University, Series 2*, Special Volume **6** (Hatai Memorial Volume) 319–336.

TAKAHASHI, H. & ONUKI, Y. 1959. On the Jurassic System of the Mizunuma-Owada areas in the Southern Kitakami Mountains. *Journal of the Geological Society of Japan*, **65**, 766 [in Japanese].

TAKAHASHI, Y., CHO, D. L. & KEE, W. S. 2010. Timing of mylonitization in the Funatsu Shear Zone within Hida Belt of southwest Japan: implications for correlation with the shear zones around the Ogcheon Belt in the Korean Peninsula. *Gondwana Research*, **17**, 102–115.

TAKEUCHI, M. & SUZUKI, K. 2000. Permian CHIME ages of leucocratic tonalite clasts from Middle Permian Usuginu-type conglomerate in the South Kitakami Terrane, northeast Japan. *Journal of the Geological Society of Japan*, **106**, 812–815.

TAKEUCHI, M., SAITO, M. & TAKIZAWA, F. 1991. Radiolarian fossils obtained from conglomerate of the Tetori Group in the upper reaches of the Kurobegawa River, and its geologic significance. *Journal of Geological Society of Japan*, **97**, 345–356 [in Japanese with English abstract].

TAKIMOTO, H., OHANA, T. & KIMURA, T. 2008. New fossil plants from the Upper Jurassic Tochikubo and Tomizawa formations, Somanakamura Group, Fukushima Prefecture, Northeast Japan. *Paleontological Reseach*, **12**, 129–144.

TAKIZAWA, F. 1970. Ayukawa Formation of the Ojika Peninsula, Miyagi Prefecture, northeast Japan. *Bulletin of the Geological Survey of Japan*, **21**, 567–578.

TAKIZAWA, F. 1977. Some Aspects of the Mesozoic Sedimentary Basins in the South Kitakami Belt, Northeast Japan. *Monograph, Association for the Geological Collaboration in Japan*, **20**, 61–73 [in Japanese with English abstract].

TAKIZAWA, F. 1985. Jurassi sedimentation in the South Kitakami Belt, Northeast Japan. *Bulletin of the Geological Survey of Japan*, **36**, 203–320.

TAKIZAWA, F., ISSHIKI, N. & KATADA, M. 1974. *Geology of the Kinkasan District, Quadrangle Series, Scale 1:50,000*. Geological Survey of Japan, Kawasaki, Japan [in Japanese with English abstract].

TAKIZAWA, F., KAMBE, N., KUBO, K., HATA, M., SANGAWA, A. & KATADA, M. 1984. *Geology of the Ishinomaki District, Quadrangle Series, Scale 1:50,000*. Geological Survey of Japan, Yatabe, Japan [in Japanese with English abstract].

TAKIZAWA, F., KAMADA, K., SAKAI, A. & KUBO, K. 1990. *Geology of the Toyoma District, Quadrangle Series, Scale 1:50,000*. Geological Survey of Japan, Tsukuba, Japan [in Japanese with English abstract].

TANAKA, T. 1975. Geological significance of rare earth elements in Japanese geosynclinal basalts. *Contributions to Mineralogy and Petrology*, **52**, 233–246.

TANAKA, T. & HOSHINO, M. 1987. Sm-Nd ages of Oki metamorphic rocks and their geological significance. *Abstract of 94th Annual Meeting of the Geological Society of Japan*, Osaka City University, Osaka, 492.

TASHIRO, M. 1994. Cretaceous tectonic evolution of southwest Japan from the bivalve faunal view-points. *Research Reports of the Kochi University*, **43**, 43–54.

TASHIRO, M. & KOZAI, T. 1989. Bivalve faunal correlation of the Cretaceous System of Northeast Japan with that of Southwest Japan. *Earth Science*, **43**, 129–139 [in Japanese with English abstract].

TAZAKI, K. & ISHIUCHI, K. 1976. Coexisting jadeite and paragonite in albitite. *Journal of the Mineralogical Society of Japan*, **12**, 184–194.

TAZAWA, J. 1975. Uppermost Permian fossils from the Southern Kitakami Mountains, Northeast Japan. *Journal of the Geological Society of Japan*, **81**, 629–640.

TAZAWA, J. 1984. Early Carboniferous (Visean) Brachiopods from the Hikoroichi Formation of the Kitakami Mountains, Northeast Japan. *Transactions and Proceedings of the Palaeontological Society of Japan, New Series*, **133**, 300–312.

TAZAWA, J. 1991. Middle Permian brachiopod biogeography of Japan and adjacent regions in East Asia. *In*: ISHII, K., LIU, X., ICHIKAWA, K. & HUANG, B. (eds) *Pre-Jurassic geology of Inner Mongolia, China. Report of China-Japan Cooperative Research Group, 1987–1989*. Matsuya Insatsu, Osaka, 213–230.

TAZAWA, J. 2002. Late Paleozoic brachiopod faunas of the South Kitakami Belt, northeast Japan, and their paleobiogeographic and tectonic implications. *Island Arc*, **11**, 287–301.

TAZAWA, J. & KANEKO, A. 1991. *Encrinurus (Silurian trilobite) from tuff in Hitoegane of the Fukuji District, Hida Mountain, Central Japan and its significance. Earth Science (Chikyu Kagaku)* **45**, 61–64 [in Japanese with English abstract].

TAZAWA, J., TSUSHIMA, K. & HASEGAWA, Y. 1993. Discovery of *Monodiexodina* from the Permian Moribu Formation in the Hida Gaien Belt,

Central Japan. *Chikyu-Kagaku (Earth Science)*, **47**, 345–348 [in Japanese].

Tazawa, J., Niikawa, I., Furuichi, K., Miyake, Y., Ohkura, M., Furutani, H. & Kaneko, N. 1997. Discovery of Devonian tabulate corals and crinoids from the Moribu district, Hida Gaien Belt, central Japan. *Journal of the Geological Society of Japan*, **103**, 399–401 [in Japanese].

Tazawa, J., Hasegawa, Y. & Yoshida, K. 2000*a*. Schwagerina (Fusulinacean) and Choristites (Brachiopoda) from the Carboniferous Arakigawa Formation in the Hida Gaien Belt, Central Japan. *Earth Science (Chikyu Kagaku)*, **54**, 196–199 [in Japanese].

Tazawa, J., Yang, W. & Miyake, Y. 2000*b*. Cyrtospirifer and Leptophloeum from the Devonian Rosse Formation, Hida Gaien Belt, central Japan. *Journal of the Geological Society of Japan*, **106**, 727–735 [in Japanese with English abstract].

Tazawa, J., Saiki, K. & Yokota, A. 2006. Leptophloeum from the Ainosawa Formation of the Soma area, Fukushima Prefecture, northeast Japan, and the tectono-sedimentary setting of the Leptophloeum-bearing Upper Devonian in Japan. *Earth Science*, **60**, 69–72 [in Japanese with English abstract].

Terui, K. & Nagahama, H. 1995. Depositional facies and sequences of the Upper Cretaceous Kuji Group, Northeast Japan. *Memoir of the Geological Society of Japan*, **45**, 238–249 [in Japanese with English abstract].

Thompson, J. B. 1957. The graphical analysis of mineral assemblages in pelitic schists. *American Mineralogists*, **42**, 842–858.

Tominaga, R., Hara, I. & Kuwano, Y. 1979. Geologic structure of the northern margin of the Kurosegawa Tectonic Zone in the Mt. Mitaki area, Ehime Prefecture. *Studies on Late Mesozoic Tectonism in Japan*, **1**, 31–38 [in Japanese].

Tsuchiya, N. & Kanisawa, S. 1994. Early Cretaceous Sr-rich silicic magmatism by slab melting in the Kitakami Mountains, northeast Japan. *Journal of Geophysical Research*, **99**, 22205–22220.

Tsuchiya, N., Suzuki, S., Kimura, J. & Kagami, H. 2005. Evidence for slab melt/mantle reaction: petrogenesis of Early Cretaceous and Eocene high-Mg andesites from the Kitakami Mountains, Japan. *Lithos*, **79**, 179–206.

Tsujimori, T. 1995. Staurolite-bearing sillimanite schist cobble from the Upper Jurassic Tetori Group in the Kuzuryu area, Hida Mountains, central Japan. *Journal of the Geological Society of Japan*, **101**, 971–977.

Tsujimori, T. 1999. Petrogenesis of the Fuko Pass high-pressure metacumulate from the Oeyama peridotite body, southwestern Japan: evidence for Early Paleozoic subduction metamorphism. *Memoirs of Geological Society of Japan*, **52**, 287–302.

Tsujimori, T. 2004. Origin of serpentinites in the Omi serpentinite melange (Hida Mountains, Japan) deduced from zoned chromian spinel. *Journal of the Geological Society of Japan*, **110**, 591–597 [in Japanese with English abstract].

Tsujimori, T. 2010. Paleozoic subduction-related metamorphism in Japan: new insights and perspectives. *Journal of Geography*, **119**, 294–312 [in Japanese with English abstract].

Tsujimori, T. & Harlow, G. E. 2012. Petrogenetic relationships between jadeitite and associated high-pressure and low-temperature metamorphic rocks in worldwide jadeitite localities: a review. *European Journal of Mineralogy*, **24**, 371–390.

Tsujimori, T. & Ishiwatari, A. 2002. Granulite facies relics in the early Paleozoic kyanite-bearing ultrabasic metacumulate in the Oeyama belt, the Inner Zone of southwestern Japan. *Gondwana Research*, **5**, 823–835.

Tsujimori, T. & Itaya, T. 1999. Blueschist-facies metamorphism during Paleozoic orogeny in southwestern Japan: phengite K–Ar ages of blueschist-facies tectonic blocks in a serpentinite melange beneath early Paleozoic Oeyama ophiolite. *Island Arc*, **8**, 190–205.

Tsujimori, T. & Liou, J. G. 2004. Metamorphic evolution of kyanite-staurolite-bearing epidote-amphibolite from the Early Paleozoic Oeyama belt, SW Japan. *Journal of Metamorphic Geology*, **22**, 301–313.

Tsujimori, T. & Liou, J. G. 2005. Eclogite-facies mineral inclusions in clinozoisite from Paleozoic blueschist, central Chugoku Mountains, Southwest Japan: evidence of regional eclogite-facies metamorphism. *International Geology Review*, **47**, 215–232.

Tsujimori, T., Nishina, K., Ishiwatari, A. & Itaya, T. 2000. 403–443 Ma kyanite-bearing epidote amphibolite from the Fuko Pass metacumulate in Oeyama, the Inner Zone of southwestern Japan. *Journal of the Geological Society of Japan*, **106**, 646–649 [in Japanese with English abstract].

Tsujimori, T., Liou, J. G., Wooden, J. & Miyamoto, T. 2005. U–Pb dating of large zircons in low-temperature jadeitite from the Osayama serpentinite melange, Southwest Japan: insights into the timing of serpentinization. *International Geology Review*, **47**, 1048–1057.

Tsujino, Y., Shigeta, Y., Maeda, H., Komatsu, T. & Kusuhashi, N. 2013. Late Triassic ammonoid *Sirenites* from the Sabudani Formation in Tokushima, Southwest Japan, and its biostratigraphic and paleobiogeographic implications. *Island Arc*, **22**, 549–561.

Tsutsumi, Y., Yokoyama, K., Horie, K., Terada, K. & Hidaka, H. 2006. SHRIMP U-Pb dating of detrital zircons in paragneiss from Oki-Dogo Island, western Japan. *Journal of Mineralogical and Petrological Sciences*, **101**, 289–298.

Tsukada, K. 1997. Stratigraphy and structure of Paleozoic rocks in the Hitoegane area, Kamitakara Village, Gifu Prefecture. *Journal of the Geological Society of Japan*, **103**, 658–668 [in Japanese with English abstract].

Tsukada, K. 2003. Jurassic dextral and Cretaceous sinistral movements along the Hida marginal belt. *Gondwana Research*, **6**, 687–698.

Tsukada, K. 2005. Tabulata corals from the Devonian Fukuji Formation, Hida gaien belt, central Japan - Part 1. *Bulletin of the Nagoya University Museum*, **21**, 57–125.

Tsukada, K. & Koike, T. 1997. Ordovician conodonts from the Hitoegane area, Kamitakara Village, Gifu Prefecture. *Journal of the Geological Society of Japan*, **103**, 171–174 [in Japanese].

Tsukada, K. & Niwa, K. 2005. The Triassic Tandodani Formation in the Hongo area, Hida Gaien belt, central Japan. *Journal of the Earth and Planetary Science, Nagoya University*, **52**, 1–10.

Tsukada, K. & Takahashi, Y. 2000. Redefinition of the Permian strata in the Hida-gaien Tectonic Zone, Fukuji area, Gifu Prefecture, Central Japan. *Journal of the Earth and Planetary Science, Nagoya University*, **47**, 1–35.

Tsukada, K., Yamakita, S. & Koike, T. 1997. Late Triassic conodonts from the Hongo area in the Hida Marginal Belt. *Journal of the Geological Society of Japan*, **103**, 1175–1178 [in Japanese].

Tsukada, K., Takahashi, Y. & Ozawa, T. 1999. Stratigraphic relationship between the Mizuyagadani and Sorayama Formations, and age of the Sorayama Formation, in the Hida-gaien Tectonic Zone, Kamitakara Village, Gifu Prefecture, central Japan. *Journal of the Geological Society of Japan*, **105**, 496–507 [in Japanese with English abstract].

Tsukada, K., Takeuchi, M. & Kojima, S. 2004. Redefinition of the Hida Gaien belt. *Journal of the Geological Society of Japan*, **110**, 640–658 [in Japanese with English abstract].

Uda, S. 1984. The contact metamorphism of the Oeyama ultrabasic mass and the genesis of the cleavable olivine. *Journal of the Geological Society of Japan*, **90**, 393–410 [in Japanese with English Abstract].

Ueda, Y., Nakajima, T., Matsuoka, K. & Maruyama, S. 1980. K–Ar ages of muscovite from greenstone in the Ino Formation and schists blocks associated with the Kurosegawa Tectonic Zone near Kochi City, central Shikoku. *Journal of Japanese Association of Mineralogy, Petrology and Economic Geology*, **75**, 230–233 [in Japanese with English abstract].

Uemura, F., Sakamoto, T. & Yamada, N. 1979. *Geology of the Wakasa District, Quadrangle Series, Scale 1:50,000*. Geological Survey of Japan, Kawasaki, Japan.

Uemura, K. 1997. Ceniozoic history of *Ginkgo* in East Asia. *In*: Hori, T., Ridge, R. W., Tulecke, R., Tredici, P. D., Treouillaux-Guiller, J. & Tobe, H. (eds) *Ginkgo Biloba-A Global Treasure*. Springer-Verlag, Tokyo, 207–221.

Ueno, K., Shintani, T. & Tazawa, J. 2009. *Fusuline foraminifera from the upper part of the Sakamotozawa Formation, South Kitakami Belt, Northeast Japan*. Science Report of the Niigata University (Geology) **24**, 27–61.

Umeda, Masaki. 1994. Mesozoic and Paleozoic radiolarians from the Kurosegawa Terrane, southwestern Ehime Prefecture, Japan. *Journal of the Geological Society of Japan*, **100**, 513–515 [in Japanese].

Umeda, Masaki. 1996*a*. Correlation among the Middle Paleozoic formations in the Kurosegawa, Hida-Gaien and South Kitakami belts, based on the radiolarian fossils. *Earth Monthly*, **18**, 718–723 [in Japanese].

Umeda, Masaki. 1996*b*. Radiolarian fossils from the Ono and Nakazato Formations in the Southern Kitakami Massif, Northeast Japan. *Earth Science*, **50**, 331–336 [in Japanese].

Umeda, Masaki. 1998*a*. Devonian radiolarians from the Senjyogataki Formation of the Southern Kitakami Terrane in the Kamaishi area, Northeast

Japan. *Journal of the Geological Society of Japan*, **104**, 276–279 [in Japanese].

UMEDA, MASAKI. 1998*b*. The Siluro-Devonian Yokokurayama Group in the Yokokurayama area, Kochi, Southwest Japan. *Journal of the Geological Society of Japan*, **104**, 365–376 [in Japanese with English abstract].

UMEDA, MASAKI. 1998*c*. Upper Silurian to Middle Devonian radiolarian zones of the Yokokurayama and Konomoriareas in the Kurosegawa Belt, Southwest Japan. *Island Arc*, **7**, 637–646.

UMEDA, MASAKI & EZAKI, Y. 1997. Middle Permian radiolarian fossils from the acidic tuffs of the Kanayama and Fukuji areas in the Hida 'Gaien' Terrane, central Japan. *Fossils*, **62**, 37–44 [in Japanese with English abstract].

UMEDA, MIYUKI, TAGA, H. & HATTORI, I. 1996. Discovery and its geologic significance of Permian radiolarians from clastic rocks at the northern margin of the Nanjo Massif, Fukui Prefecture, Central Japan. *Journal of the Geological Society of Japan*, **102**, 635–638 [in Japanese with English abstract].

UMEMURA, H. & HARA, I. 1985. Tectonics in the Abukuma metamorphic belt. *Memoir of the Geological Society of Japan*, **25**, 127–136 [in Japanese with English abstract].

UNO, K., FURUKAWA, K. & HADA, S. 2011. Margin-parallel translation in the western Pacific: Paleomagnetic evidence from an allochthonous terrane in Japan. *Earth and Planetary Science Letters*, **303**, 153–161.

WADA, H. 1988. Microscale isotopic zoning in calcite and graphite crystals in marble. *Nature*, **331**, 61–63.

WAKAMATSU, H., SUGIYAMA, K. & FURUTANI, H. 1990. Silurian and Devonian radiolarians from the Kurosegawa Tectonic Zone, Suthwest Japan. *Journal of Earth Sciences, Nagoya University*, **37**, 157–192.

WAKITA, K. 2013. Geology and tectonics of Japanese Islands: a review – The key to understanding the geology of Asia. *Journal of Asian Earth Sciences*, **72**, 75–87.

WAKITA, K., MIYAZAKI, K., TOSHIMITSU, S., YOKOYAMA, S. & NAKAGAWA, M. 2007. *Geology of the Ino District, Quadrangle Series, 1:50,000*. Geological Survey of Japan, AIST, Tsukuba [in Japanese with English abstract].

WANG, X. D., SUGIYAMA, T., KIDO, E. & WANG, X. J. 2006. Permian rugose coral faunas of Inner Mngolia–Northeast China and Japan: paleobiogeographical implications. *Journal of Asian Earth Sciences*, **26**, 369–379.

WATANABE, T., FANNING, M., URUNO, K. & KANO, H. 1995. Pre-Middle Silurian granitic magmatism and associated metamorphism in northern Japan: SHRIMP U–Pb zircon chronology. *Geological Journal*, **30**, 273–280.

YABE, H. & NODA, M. 1933. On the discovery of *Spirifer verneuili* Murchison in Japan. *Proceedings of the Imperial Academy of Tokyo*, **9**, 521–522.

YABE, H. & SHIMIZU, S. 1925. A new Cretaceous ammonite, *Crioceras ishiwarai*, from Oshima, province of Rikuzen. *Japanese Journal of Geology and Geography*, **4**, 85–87.

YAMAKITA, S. & OTOH, S. 1987. Symmetrical occurrence of pre-Jurassic rocks and strata in Southwest Japan, and its tectonic significances. *Journal of the Tectonic Research Group of, Japan*, **32**, 87–101 [in Japanese with English abstract].

YAMAKITA, S. & OTOH, S. 2000. Cretaceous rearrangement processes of pre-Cretaceous geologic units of the Japanese Islands by MTL-Kurosegawa left-lateral strike-slip fault system. *Memoirs of the Geological Society of Japan*, **56**, 23–38 [in Japanese with English abstract].

YAMANE, M., BAMBA, M. & BAMBA, T. 1988. The first finding of orbicular chromite ore in Japan. *Mining Geology*, **38**, 501–508.

YAMASHITA, K. & YANAGI, T. 1994. U-Pb and Rb-Sr dating of the Oki metamorphic rocks, the Oki Island, Southwest Japan. *Geochemical Journal*, **28**, 333–339.

YAMASHITA, N. 1957. *Geoscience Series 10. The Mesozoic I, II*. Association for Geological Collaboration of Japan, Tokyo [in Japanese].

YAMAZAKI, T. 1981. Metamorphism of metamorphic rocks and ultramafic rocks in the Happo-O'ne area. *In*: CHIHARA, K. (ed.) *Report of Scientific Project, 'Hida-Gaien Belt', Grant-in-Aid for Scientific Research (A) X00050–434045*, **1**, Department of Geology, Niigata University, Niigata, Japan, 31–37 [in Japanese].

YANAGISAWA, Y., YAMAMOTO, T., BANNO, Y., TAZAWA, J., YOSHIOKA, T., KUBO, K. & TAKIZAWA, F. 1996. *Geology of the Somanakamura District, Quadrangle Series, Scale 1:50,000*. Geological Survey of Japan, Tsukuba, Japan [in Japanese with English abstract].

YANAI, S., PARK, B. S. & OTOH, S. 1985. The Honan shear zone: deformation and tectonic implication in the Far-East. *Science Papers of the Colleage of Arts and Sciences, University of Tokyo*, **35**, 181–209.

YAO, A. 2000. Terrane arrangement of Southwest Japan in view of the Paleozoic-Mesozoic tectonics of East Asia. *Monograph, Association for the Geological Collaboration in Japan*, **49**, 145–155 [in Japanese with English abstract].

YASUI, T. 1984. On the Pre-Silurian Basement in the Yokokurayama Lenticular Body of the Kurosegawa Tectonic Zone. *Earth Science*, **38**, 89–101 [in Japanese with English abstract].

YASUI, T. & OKITSU, N. 2007. Discovery of Late Devonian plant *Leptophloeum* from the Suberidani area in the Kurosegawa Belt, Tokushima Prefecture, southwest Japan. *Journal of the Geological Society of Japan*, **113**, 15–18 [in Japanese with English abstract].

YOSHIDA, K. & MACHIYAMA, H. 1998. Middle Permian coarse clastics in the western marginal areaofthe South Kitakami Terrane, northeast Japan. *Journal of the Geological Society of Japan*, **104**, 71–89 [in Japanese with English abstract].

YOSHIDA, K., KAWAMURA, M. & MACHIYAMA, H. 1994. Transition in the composition of the Permian clastic rocks in the South Kitakami Terrane, Northeast Japan. *Journal of the Geological Society of Japan*, **100**, 744–761 [in Japanese with English abstract].

YOSHIKURA, S. 1982. The geology of the Yokokurayama Lenticular Body and the Usuginu Conglomerate. *Memoirs of the Geological Society of Japan*, **21**, 213–229 [in Japanese with English abstract].

YOSHIKURA, S. 1985. Igneous and high-grade metamorphic rocks in the Kurosegawa Tectonic Zone and its tectonic significance. *Journal of Geoscience, Osaka City University*, **28**, 45–83.

YOSHIKURA, S. & MIYAJI, K. 1988. Igneous petrology of ultramafic rocks from the Kurosegawa tectonic zone in central Shikoku, Japan. *Abstract of Third International Symposium IGCP Project-224*, Beijing, **141**.

YOSHIKURA, S. & SATO, K. 1976. A few evidences on the Kurosegawa Tectonic Zone near Yokokurayama, Kochi Prefecture. *Island-arc Basement (Tokokiban)*, **3**, 53–56 [in Japanese].

YOSHIKURA, S., HADA, S. & ISOZAKI, Y. 1990. Kurosegawa Terrane. *In*: ICHIKAWA, K., MIZUTANI, S., HARA, I., HADA, S. & YAO, A. (eds) *Pre-Cretaceous Terranes of Japan*. Nippon Insatsu Shuppan Co. Ltd., Osaka, 185–201.

YOSHIMOTO, A., OSANAI, Y., NAKANO, N., ADACHI, T., YONEMURA, K. & ISHIZUKA, H. 2013. U-Pb detrital zircon dating of pelitic and quartzite from the Kurosegawa Tectonic Zone, Southwest Japan. *Journal of Mineralogical and Petrological Sciences*, **108**, 178–183.

ZHANG, H., WANG, Y., SHEN, G., HE, Z. & WANG, J. 1999. Palaeophytogeography and palaeoclimatic implications of Permian Gigantopterids on the North China Plate. *In*: YIN, H. & TONG, J. (eds) *Proceedings of the International Conference on Pangea and the Paleozoic-Mesozoic-Transition*, March 9–11 1999, China Univ Geosci, Wuhan, China, 167–168.

ZHAO, X., MAO, J., YE, H., LIU, K. & TAKAHASHI, Y. 2013. New SHRIMP U-Pb zircon ages of granitic rocks in the Hida Belt, Japan: implications for tectonic correlation with Jiamushi massif. *Island Arc*, **22**, 508–521.

2b Pre-Cretaceous accretionary complexes

SATORU KOJIMA, YASUTAKA HAYASAKA, YOSHIKUNI HIROI, ATSUSHI MATSUOKA, HIROYOSHI SANO, YOSHIAKI SUGAMORI, NORITOSHI SUZUKI, SHIZUO TAKEMURA, TATSUKI TSUJIMORI & TAKAYUKI UCHINO

Accretionary complexes (AC) form at convergent plate margins by the subduction of oceanic plate underneath the continental plate (Fig. 2b.1). The oceanic plate is created at the mid-oceanic ridge, and moves to the trench while accumulating pelagic sediments. After arriving at the trench, where the pelagic sediments are covered by continent-derived clastic materials, the plate is subducted and part of the sediments accrete to the continental plate, producing fault stacking and several types of mélanges (Fig. 2b.1). The characteristic AC succession reflects the ocean plate stratigraphy (OPS) (Matsuda & Isozaki 1991) starting with basaltic basement covered by radiolarian ribbon chert, then siliceous mudstone and finally coarse clastic rocks.

A major part of the pre-Cretaceous basement rocks of Honshu and Shikoku consists of ACs and their metamorphic equivalents (Fig. 2b.2). The fundamental process underlying the origin of these rocks has been the subduction of oceanic plates from the SE (present direction), so that the ACs become younger from NW to SE (from upper to lower structural plate), although fold structures with a NW-dipping enveloping surface partly disturb the regularity. The accretionary process has been somewhat episodic, producing mainly Permian (Akiyoshi, Maizuru and Ultra-Tamba–Tamba belts and part of the Northern Chichibu Sub-belt) and Jurassic (Mino–Tamba–Ashio, Northern and Southern Chichibu, and North Kitakami belts) ACs and their metamorphic equivalents, although fragments of Carboniferous AC, known as the Nedamo Belt, also occur.

In this chapter the lithologies, stratigraphy, geological structure and noteworthy characteristics of these older ACs are documented in order to further elucidate the geological development of the Japanese islands described in Chapter 1. Although parts of the Maizuru and Abukuma belts are not strictly composed of ACs, they are included in this chapter for convenience. The geology of the Ryoke and Sanbagawa belts, the protoliths of which are ACs, is described in Chapter 2e.

Renge high-*P*/*T* rocks (TT)

Late Palaeozoic high-pressure and low-temperature (high-*P*/*T*) metamorphic rocks known as the Renge metamorphic rocks were originally grouped within the so-called 'Sangun' Belt proposed by Kobayashi (1941) to describe high-*P*/*T* schists widely scattered across the Inner Zone of SW Japan (central-western Honshu, northern Kyushu and the Ryukyu islands). Before the late 1980s, the 'Sangun' belt was considered to define a single coherent high-*P*/*T* metamorphic belt of pre-Jurassic age, paired with the low-pressure and high-temperature (high-*T*/*P*) Hida belt (e.g. Miyashiro 1961; Fig. 2b.2). However, with the accumulation of geochronological data of mainly phengite K–Ar ages in the 1990s (e.g. Shibata & Nishimura 1989; Nishimura 1998; Tsujimori & Itaya 1999), the Sangun Belt was subdivided into two discrete geotectonic units: (1) an Early Mesozoic 'Suo' Belt and (2) an older group of Late Palaeozoic 'Renge' rocks associated with the Oeyama Belt (Fig. 2b.3). This subdivision has been generally accepted in modern geotectonic interpretations of the Japanese archipelago (e.g. Isozaki *et al.* 2010; Tsujimori 2010; Wakita 2013). The limited outcrop of the Renge rocks owes much to the overprint of voluminous Mesozoic batholith belts and an unconformable cover of Cenozoic volcaniclastic rocks in the back-arc region of Honshu. However, based on the lithological, structural, metamorphic and geochronological similarities, the Renge rocks are considered to have been constituents of a Late Palaeozoic regional high-*P*/*T* metamorphic belt.

The exposure of Renge rocks is limited to several relatively small areas in comparison with the high-*P*/*T* rocks of the Suo Belt (Fig. 2b.4). Many localities are distributed along a NE–SW-directed line from the Itoigawa–Shizuoka Tectonic Line, and occur as tectonic sheets and/or blocks in serpentinite mélange associated with the Oeyama Belt (Fig. 2b.4). In the Chugoku Mountains, the Renge rocks occur within serpentinite mélange units that lie beneath the ultra\mafic bodies of the Oeyama Belt; the serpentinite mélange units have been emplaced upon rocks of the Akiyoshi and/or Suo belts (Uemura *et al.* 1979; Kabashima *et al.* 1993; Tsujimori 1998; Tsujimori & Liou 2007). In the Hida Mountains, the Renge rocks are tectonically mixed with the ultramafic rocks of the Oeyama Belt (Nakamizu *et al.* 1989; Kunugiza *et al.* 2004).

Most of the metamorphism recorded by the Renge rocks took place in the epidote-blueschist facies and/or greenschist/blueschist transitional facies to epidote-amphibolite facies, although lawsonite-blueschist and glaucophane-bearing eclogite locally occur (cf. Tsujimori 2010). The presence of Middle–Late Palaeozoic Renge lawsonite-blueschist and glaucophane-eclogite provides evidence of a cold geotherm in the palaeosubduction zone. The protoliths of the Renge rocks were pelagic and semi-pelagic siliceous-clayey deposits, trench-fill turbidites, basaltic oceanic crusts and rare mantle wedge materials. The presence of high-*P*/*T* metamorphosed forearc ophiolitic materials (fragments of the Oeyama Belt) suggests that significant landwards subduction erosion has occurred since Early Palaeozoic time.

Renge rocks in the Hida Mountains

In the Hida Mountains, fragments of Renge rocks occur in serpentinite mélange units of the Hida-Gaien Belt exposed in areas such as Itoigawa–Omi, Renge–Shirouma, Happo-O'ne, Gamata, Naradani

From: Moreno, T., Wallis, S., Kojima, T. & Gibbons, W. (eds) 2016. *The Geology of Japan*. Geological Society, London, 61–100.

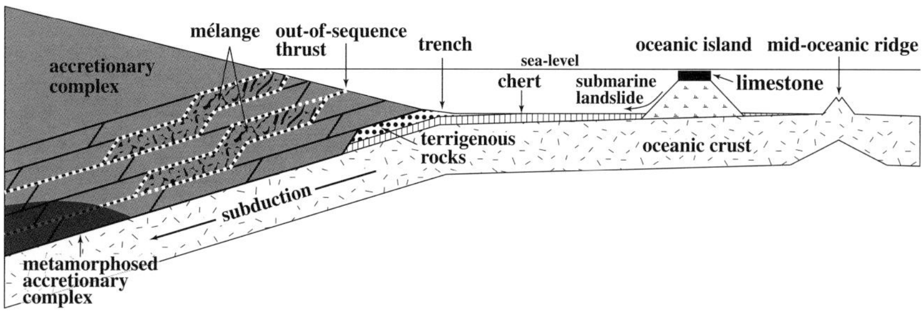

Fig. 2b.1. Formation process of accretionary complex (AC).

and Kuzuryu. These serpentinite mélanges with Renge rocks are unconformably overlain by Mesozoic sedimentary rocks of the Lower Jurassic Kuruma Group and the Middle Jurassic–Lower Cretaceous Tetori Group. The Renge rocks in the Hida Mountains record mainly greenschist/blueschist transitional facies to epidote-amphibolite facies metamorphism and locally preserve blueschist-eclogite facies metamorphism (e.g. Tsujimori 2002; Kunugiza & Maruyama 2011; Matsumoto *et al.* 2011). The mineral assemblages of these subgroups, and consequently the compositional trends of amphiboles in mafic rocks, vary from one group to another (Fig. 2b.5). The greenschist/blueschist transitional facies and epidote-amphibolite facies groups include medium- to coarse-grained garnet-amphibolites as mafic layers and lenses within garnet- and biotite-bearing pelitic schists, and are commonly characterized by the epidote-amphibolite facies mineral assemblage garnet + hornblende + plagioclase ± clinozoisite ± biotite + rutile ± ilmenite + quartz (e.g. Nakamizu *et al.* 1989; Matsumoto *et al.* 2011), with rare paragonite occurring within porphyroblastic garnet and plagioclase (Tsujimori & Matsumoto 2006). In contrast, the blueschist and eclogite facies groups are characterized by glaucophane-bearing mineral assemblages (Banno 1958; Nakamizu *et al.* 1989), with medium- to coarse-grained eclogite and garnet-blueschist occurring as mafic layers within paragonite-bearing pelitic schists, with the assemblage garnet + omphacite + glaucophane + clinozoisite + rutile + quartz ± phengite (Tsujimori 2002). Rare lawsonite and pumpellyite occur in both pelitic and mafic schists exposed in the western Hida Mountains (Miyakawa 1982; Sohma *et al.* 1983), and piemontite occurs in metachert. Phengitic white micas from these Renge rocks have yielded K–Ar and $^{40}Ar/^{39}Ar$ ages of *c.* 360–280 Ma, and zircon U–Pb geochronology constrains the timing of peak metamorphism to *c.* 360 Ma (cf. Tsujimori 2010).

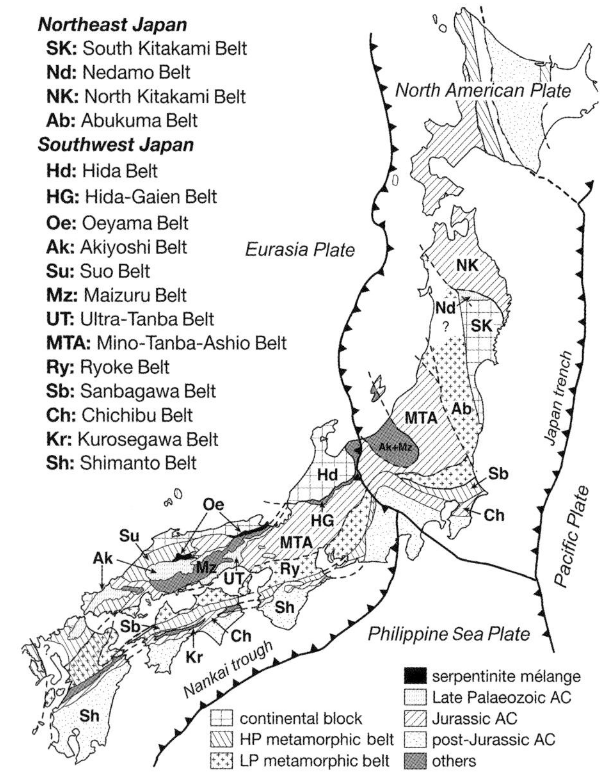

Fig. 2b.2. Tectonic map of Japan. Base map after Isozaki & Maruyama (1991).

Renge rocks in the Chugoku Mountains

In the Chugoku Mountains in western Honshu (Fig. 2b.4), two different types of high-pressure metamorphic rocks are associated with the Oeyama Belt as tectonic blocks: epidote-amphibolite facies gabbroic rocks with 470–400 Ma hornblende K–Ar ages (e.g. Nishimura & Shibata 1989; Tsujimori & Liou 2004); and blueschist facies pelitic and mafic schists with phengite K–Ar ages of *c.* 350–280 Ma (Shibata & Nishimura 1989; Nishimura 1998; Tsujimori & Itaya 1999). Of these, the younger high-P/T rocks in particular have been regarded as slices of the Late Palaeozoic Renge metamorphic belt, tectonically underlying the Oeyama Belt (e.g. Uemura *et al.* 1979; Tsujimori 1998). Renge blueschists have been described from at least five localities. A blueschist-bearing serpentinite mélange in the Osayama area (the 'Osayama serpentinite mélange'; Tsujimori 1998; Tsujimori & Itaya 1999; Fig. 2b.6) is of special interest. The matrix of the serpentinite mélange consists of schistose, friable, fine-grained serpentinite with pebble- to boulder-size fragments of serpentinized peridotite. Chrysotile and lizardite are the most dominant serpentine minerals, with rare winchitic amphibole. Blueschist blocks in this Osayama serpentinite mélange consist mainly of metasediments (pelitic, psammitic and siliceous schists), metabasites and minor marble. The blueschist

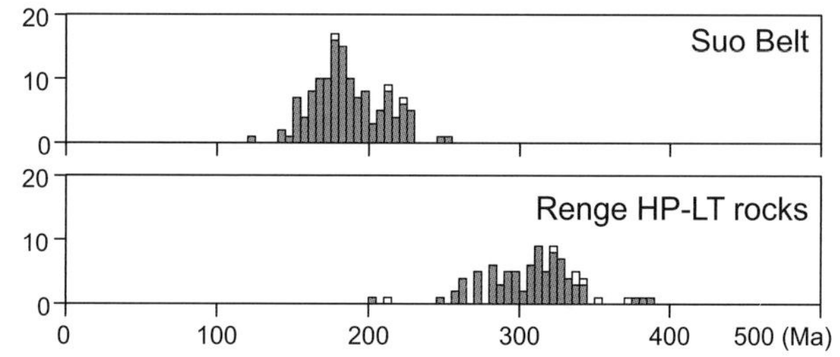

Fig. 2b.3. K–Ar and $^{40}Ar/^{39}Ar$ age histograms showing difference between the Suo and Renge high-P/T metamorphic belts; the diagram is modified after Itaya *et al.* (2011). Grey bars, phengite ages of pelitic schists; white bars, whole-rock, paragonite and amphibole ages from psammitic schists, mafic schists, amphibolites and eclogites. HP-LT: high pressure-low temperature.

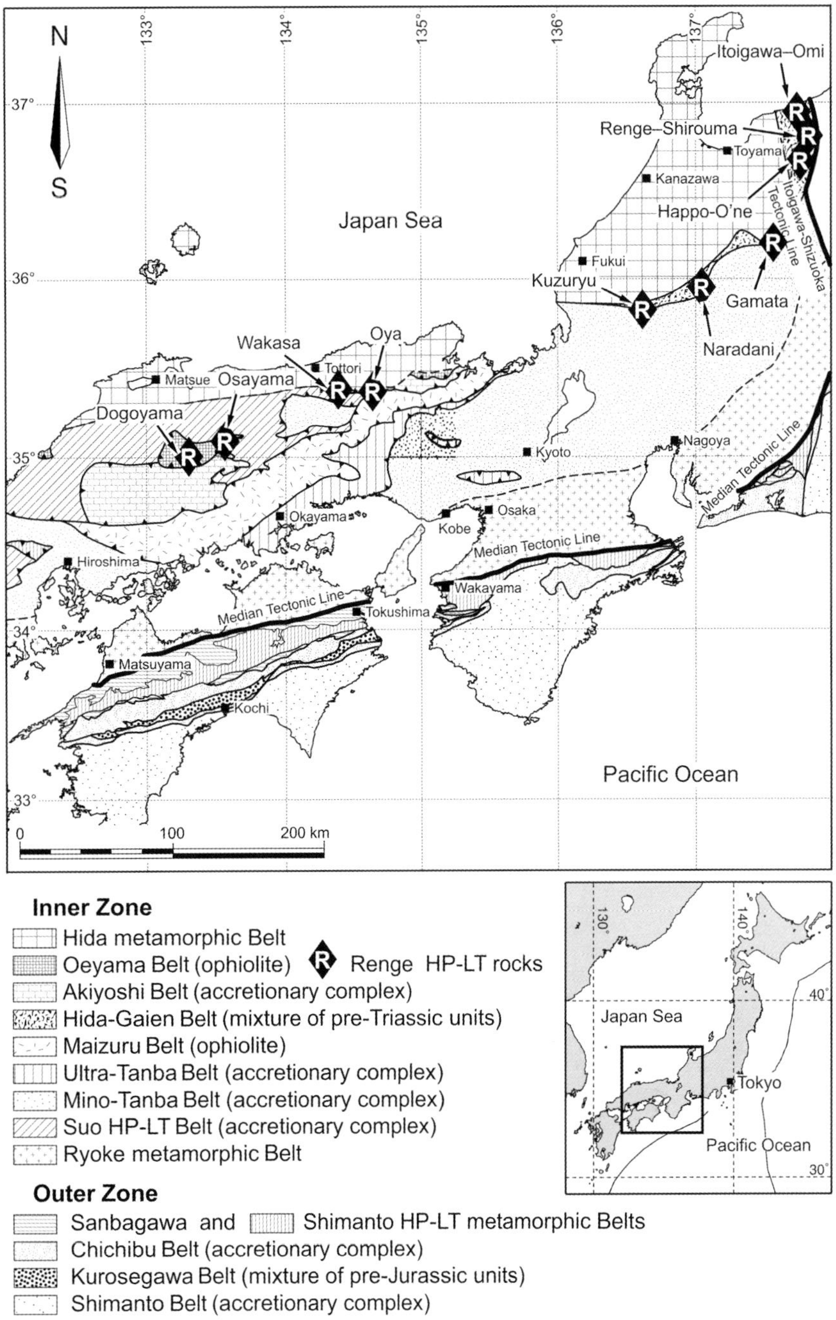

Fig. 2b.4. Tectonic framework of SW Japan, showing the localities of the Renge high-P/T schists. The map is modified after Tsujimori (2002). HP-LT: high pressure-low temperature.

facies mafic mineralogy has allowed subdivision into lawsonite-pumpellyite grade (the most abundant type) and epidote grade (sometimes accompanied by garnet in pelitic schists). Gabbro and dolerite blocks within serpentinite mélange, interpreted as fragments of the ophiolitic succession of the Oeyama Belt, also contain blueschist facies assemblages of lawsonite-pumpellyite grade. Phengite K–Ar ages of the blueschist blocks are in the range 327–273 Ma, regardless of metamorphic grade (Tsujimori & Itaya 1999). A lawsonite-blueschist-bearing small serpentinite mélange unit has also been described in the Oya area, located *c.* 150 km east of the Osayama area (Tsujimori & Liou 2007). Although there are no occurrences here of high-grade blueschists, lawsonite blueschists contains abundant metamorphic Ca–Na clinopyroxene and the rock preserves rare flattened pillow structures.

Metamorphic evolution of Renge eclogites and high-grade blueschists

The highest-grade Renge rocks of the Hida Mountains are glaucophane-bearing low-temperature eclogites of the Omi serpentinite mélange (Tsujimori *et al.* 2000; Tsujimori 2002). These eclogite facies rocks and related high-grade blueschists have mid-ocean ridge basalt (MORB)-like bulk-rock composition and contain glaucophane and epidote as both prograde and retrograde phases. Prograde-zoned porphyroblastic garnet of eclogites preserves mineral inclusions showing a transition from epidote-blueschist to eclogite facies (Fig. 2b.7). Moreover, the matrix of eclogite has been subjected to a blueschist facies overprinting after the peak eclogite facies metamorphism (at $T \approx 550$–$600°C$ and $P > 1.8$ GPa).

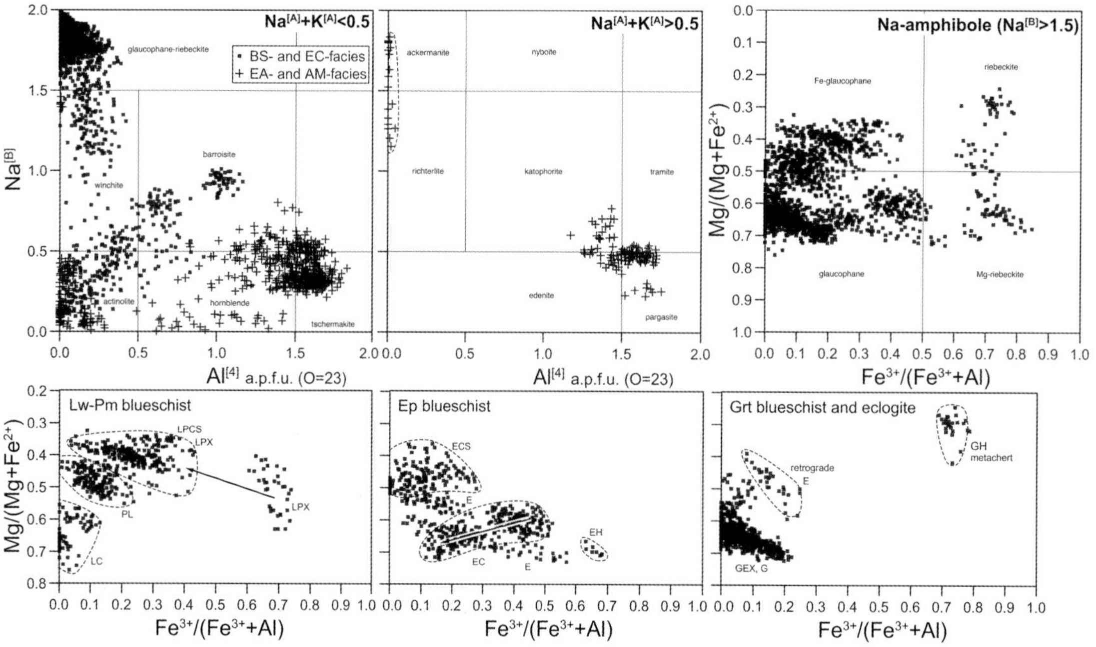

Fig. 2b.5. Compositional trend of amphiboles from Renge metamorphic rocks (Tsujimori 2010). Arrows show chemical zoning from core to rim. BS, blueschist; EC, eclogite; EA, epidote-amphibolite; AM, amphibolite; C, chlorite; S, stilpnomelane; L, lawsonite; P, pumpellyite; E, epidote; G, garnet; X, clinopyroxene; H, hematite.

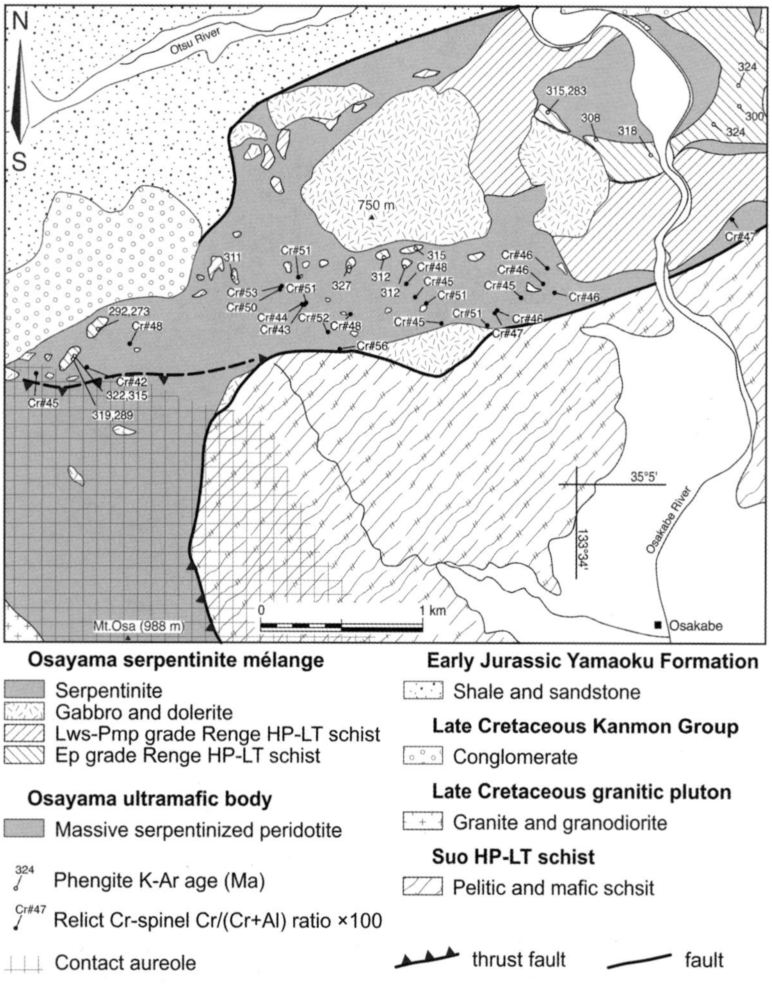

Fig. 2b.6. Geological map of the Osayama serpentinite mélange. The map is modified after Tsujimori (1998). HP-LT: high pressure-low temperature.

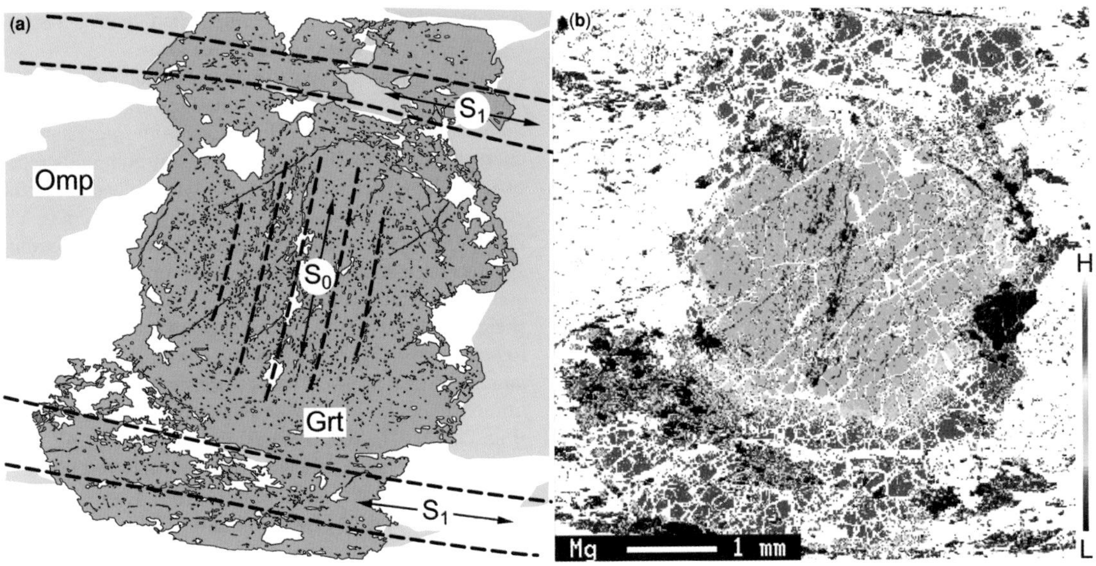

Fig. 2b.7. Representative texture of prograde-zoned garnet preserving a transition from blueschist to eclogite (Tsujimori 2002). (**a**) Textural sketch of garnet (Grt) and omphacite (Omp). Dashed lines represents schistosities S_0 (blueschist stage) and S_1 (eclogite stages). (**b**) Mg distribution of garnet of (a). A gradual colour filling bar represents concentration of Mg from low (L) to high (H).

Eclogite facies mineral assemblages also occur in a garnet-bearing blueschist block of the Osayama serpentinite mélange in the Chugoku Mountains. The blueschist block contains a retrograde lawsonite-pumpellyite grade mineral assemblage (Tsujimori 1998; Tsujimori & Liou 2005), suggesting that the rocks were 'refrigerated' during exhumation.

Detrital records of the Renge rocks

Detrital clasts derived from the Renge rocks provide important geological constraints on a 'missing' Late Palaeozoic metamorphic belt. The presence of detrital fragments of glaucophane-lawsonite and lawsonite-pumpellyite aggregates in Early Triassic sandstones of the Shidaka Group of the Maizuru Belt (Adachi 1990) indicates surface exposure of Renge rocks by earliest Mesozoic time. Furthermore, clasts of Renge rocks have been found widely in Mesozoic and Cenozoic shelf deposits and (more rarely) in back-arc and trench-fill deposits in Japan. For instance, lawsonite-bearing Late Palaeozoic schist clasts occur in the Lower Cretaceous Sasayama Group overlying the Mino–Tamba–Ashio Belt in the Eastern Chugoku Mountains (Kobayashi & Goto 2008). Similarly, Miocene Aizawa Formation conglomerate of the Joetsu Mountains (central Japan) contains abundant epidote-blueschist clasts. Furthermore, detrital phengitic micas yielding K–Ar ages of *c.* 320 Ma occur in the Eocene sandstone of the Shimanto Belt in southern Kyushu (Teraoka *et al.* 1994). We may therefore conclude that Late Palaeozoic Renge rocks were widely exposed across palaeo-Japan as a metamorphic belt, but that most of the belt has been eroded during the maturation of the Japanese Orogen.

Akiyoshi Belt (HS)

The Akiyoshi Belt is defined as a Permian (late Capitanian–early Wuchiapingian) AC that consists of an unmetamorphosed Carboniferous–Permian oceanic assemblage containing remnants of a middle Visean–Capitanian oceanic atoll(s) on basaltic seamount(s) with a Panthalassic affinity. The accreted rocks were uplifted by Late Triassic times, because they are overlain unconformably by post-orogenic Upper Triassic fluvio-lacustrine sediments. The Akiyoshi atoll stratigraphy records long-term (70–80 Ma) climatic events and sea-level changes related to the growth and retreat of the Late Palaeozoic Gondwana ice sheet. This provides an important record of Mississippian–Capitanian environmental change in the Panthalassa Ocean and tectonic events at its western margin (Fig. 2b.8). The main exposures of the Akiyoshi Belt occur in the Omi, Atetsu, Taishaku, Akiyoshi and Hirao areas in SW Japan (Fig. 2b.9). This section summarizes the lithostratigraphy and age of the Akiyoshi accreted rocks, overviews their tectonostratigraphic

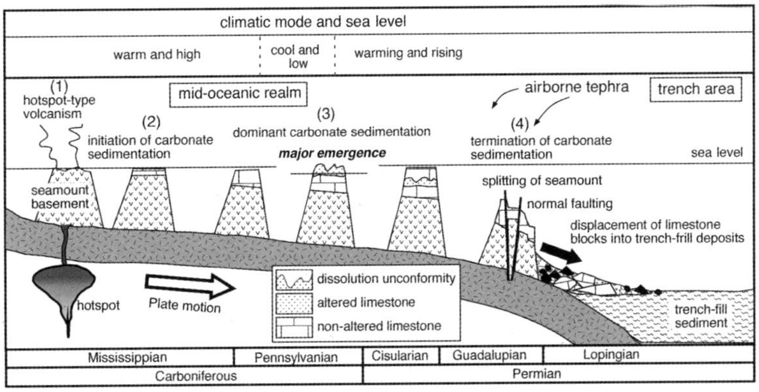

Fig. 2b.8. Birth to demise of the atoll-capped Akiyoshi seamount, modified after Sano (2006). (1) Middle Mississippian generation of an oceanic island in an equatorial zone of the Panthalassa Ocean. (2) Middle Mississippian initiation of shallow-marine carbonate sedimentation. (3) Mississippian–Guadalupian dominant accumulation of shallow-marine carbonates and Late Mississippian prominent emergence events. (4) Late Guadalupian termination of the atoll sedimentation, immediately followed by the encroachment to a trench area and the normal fault-induced tectonic collapse at an outer bulge.

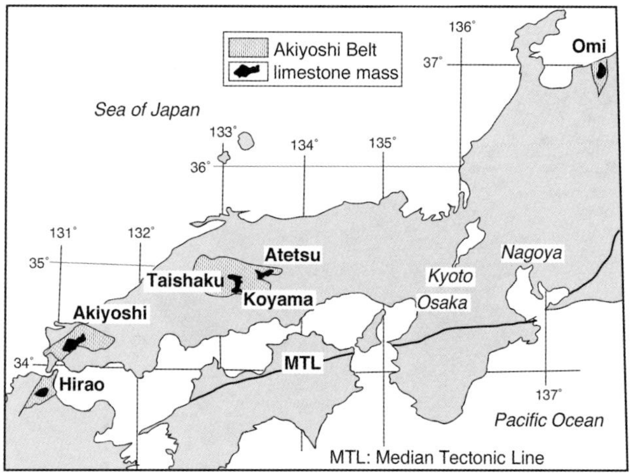

Fig. 2b.9. Approximate distribution of the Akiyoshi belt and its representative Mississippian–Permian limestone masses. Modified after Nakashima & Sano (2007).

classification and briefly introduces recent approaches to palaeoenvironmental analyses of the Akiyoshi atoll carbonates.

Stratigraphy and age

The major lithological components of the Akiyoshi belt are the Tournasian (?) to Visean basaltic rocks, middle Visean–Capitanian shallow-marine limestones and deep-water cherts, and upper Capitanian–lower Wuchiapingian siliceous tuff and terrigenous rocks (Fig. 2b.10). On the basis of the regional geology, Sano & Kanmera (1988) grouped the Carboniferous–Permian rocks into five lithologic units and reconstructed their depositional setting (A1, A2, B1, B2 and C in Fig. 2b.10).

Units A1 and A2 typically show a basaltic basement overlain by a thick pile of massive shallow-marine limestone and a succession dominated by redeposited limestone with intercalations of spicular chert (Fig. 2b.10). The basaltic rocks of both units represent relicts of seamount(s) formed by hotspot-type volcanism in the equatorial zone of the Panthalassa Ocean (S. Sano *et al.* 2000; Y. Sano *et al.* 2000). The shallow-marine limestone of unit A1 commonly forms spectacular plateaus exhibiting karst topography (Fig. 2b.11a) and consists predominantly of shallow subtidal facies rich in diverse skeletal debris, locally with metazoan reefal facies constructed by rugose corals, *Chaetetes* and algae with stromatolitic facies (Sugiyama & Nagai 1994). The occurrence of metazoan reefal facies is limited to rocks of middle Mississippian–late Pennsylvanian age (Fig. 2b.10). The redeposited limestone of unit A2 consists chiefly of limestone talus deposits, debrites and turbidites, which bear shallow-marine skeletal debris derived from the shallow-marine limestone of unit A1. Units A1 and A2 therefore represent an atoll facies upon a seamount and its upper slope facies, respectively.

Both units B1 and B2 also have basaltic rocks at the base, but these are overlain by upper Visean–upper Capitanian cherts (Fig. 2b.10). While the chert of unit B1 is characterized by abundant siliceous sponge spicules with scarce or no radiolarians, the chert of unit B2 contains radiolarians with some sponge spicules. The spicular chert of unit B1 is also distinguished by containing redeposited limestone beds and pods from radiolarian-bearing chert of unit B2. The redeposited limestone in the spicular chert dominantly yields smaller foraminifers and fusulinids of Visean–Bashkirian and Artinskian–Wordian ages, as well as conodonts. Calcitic debris of the redeposited limestone are inferred to have been shed from the atoll margins by sediment-gravity flows (Fig. 2b.10; Nakashima & Sano 2007). Both of the B1 and B2 chert successions are overlain

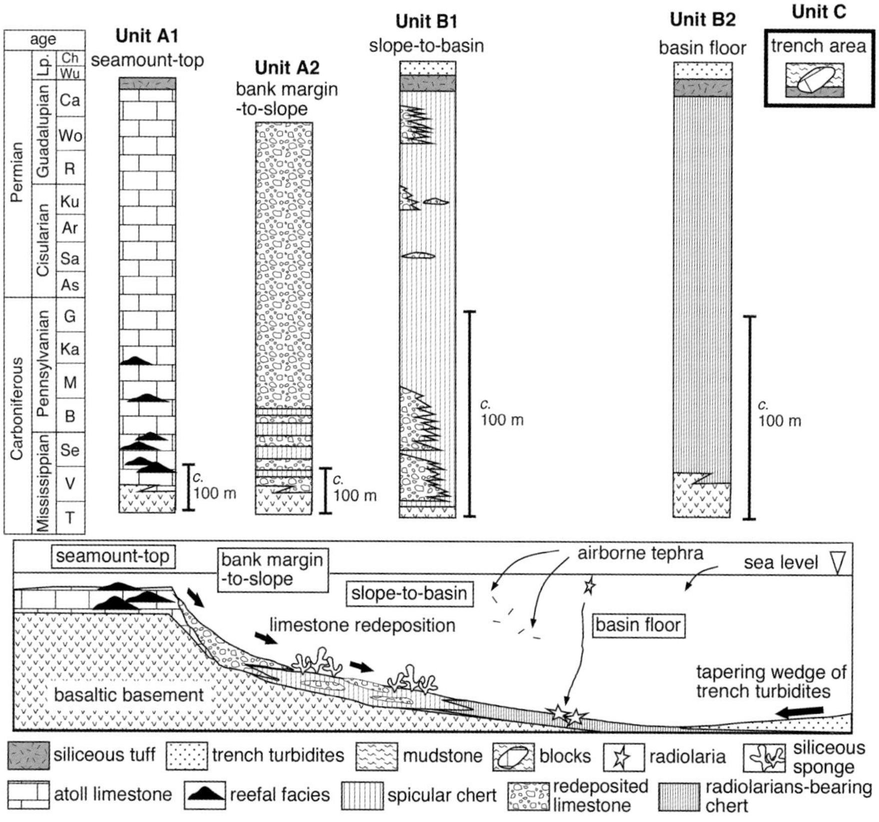

Fig. 2b.10. Sedimentary profile of the middle Lower Carboniferous–upper Middle Permian oceanic rocks on and around a mid-Panthalassic seamount, with simplified lithologic columns. The mudstone with siliceous tuff boxed at the upper right margin represents the matrix of the upper Middle–lower Upper Permian mélange unit containing exotic blocks of the oceanic rocks. Modified from Sano & Kanmera (1988) and Nakashima & Sano (2007). T, Tournaisian; V, Visean; Se, Serpukhovian; B, Bashkirian; G, Gzhelian; As, Asselian; Sa, Sakmarian; Ar, Artinskian; Ku, Kungurian; R, Roadian; Wo, Wordian; Ca, Capitanian; Wu, Wuchiapingian; Ch, Changhsingian.

Fig. 2b.11. Exposure views of key facies of the Akiyoshi atoll carbonates and their structural relation to the upper Capitanian–lower Wuchiapingian scaly mudstone. (**a**) Numerous pinnacles of the completely massive, Cisularian–Guadalupian light-grey limestone on gently rolling hills with karstic topography. Akiyoshi-dai Plateau. (**b**) Isolated limestone block (ls) embedded in the upper Capitanian scaly mudstone. Approximate width of view is 10 m. Western margin of the Akiyoshi-dai Plateau. (**c**) Limestone fragments (ls) with clusters of crinoid debris in the upper Capitanian scaly mudstone. Western margin of the Akiyoshi-dai Plateau. (**d**) Complicated contacts between the Pennsylvanian–lower Guadalupian limestone (ls) and the upper Capitanian scaly mudstone (m) containing blocks of the limestone breccia (br). Approximately 300 m wide view in the working quarry at the northwestern margin of the Akiyoshi-dai Plateau. (e) *Microcodium* structure (mc) composed of aggregates of dark-grey to brown calcite crystals replacing the host rock (h) of the middle Kasimovian light-grey skeletal limestone. Akiyoshi-dai Plateau.

by upper Capitanian–lower Wuchiapingian siliceous tuff and alternations of turbiditic sandstone and mudstone. Sano & Kanmera (1988) considered the spicular chert and the radiolarian-bearing chert as representing deep-marine sediments on the lower slope of a seamount and the surrounding ocean floor, respectively (Fig. 2b.10). The siliceous tuff originated from airborne tephra, and the alternation of sandstone and mudstone represents the deep-marine tapering slope wedge of trench-fill turbidites. The successions of units B1 and B2 therefore both record the deep-marine OPS formed during the Mississippian–Lopingian migration of their depositional sites from a mid-oceanic setting to a trench area (Fig. 2b.10).

Unit C is characterized by the dominance of upper Capitanian–lower Wuchiapingian scaly-cleaved mudstone with subordinate siliceous tuff and sandstone (Fig. 2b.10). The scaly mudstone is strongly contorted and chaotically contains various-sized blocks of massive shallow-marine limestone and limestone breccia, thin graded beds of limestone conglomerate and sandstone and lithic fragments of limestone, with isolated individuals of fusulinids and crinoids (Fig. 2b.11b, c). These blocks and related calcareous debris occur most frequently in the scaly mudstone near its boundary with mappable-scale masses of the shallow-marine limestone of unit A (Figs 2b.11d & 2b.12; Kanmera & Nishi 1983; Sano & Kanmera 1991*a*). The scaly mudstone of unit C is interpreted as originally deposited as argillaceous sediments within a trench-fill turbidite sequence.

Tectonostratigraphically, these five units are organized into two coherent units and one mélange unit. The two coherent units correspond to units B1 and B2, and are commonly characterized by an orderly stratigraphic succession comprising basaltic rocks, spicular chert or radiolarians-bearing cherts, siliceous tuff and turbiditic terrigenous rocks, in ascending order. Each of these successions forms fault-bounded slabs, which are stacked to form a highly complicated imbricate structure (Kanmera & Nishi 1983). The mélange unit is characterized by chaotic mixing of various-sized blocks (mainly of the shallow-marine limestone dominantly derived from unit A1) and polymictic limestone breccias with an argillaceous matrix derived from the scaly mudstone of unit C (Sano & Kanmera 1991*b*, *c*). Units A1 and A2 are therefore tectonostratigraphically defined as blocks within the mélange unit. This is exemplified by the structural relation of unit A1 (Akiyoshi Limestone) to unit C (Tsunemori Formation) in the Akiyoshi-dai Plateau area, where the former structurally rests upon the latter with highly complicated contacts (Figs 2b.11d & 2b.12; Sano & Kanmera 1991*a*). The age of the scaly mudstone of unit C constrains the timing of mixing to have been Capitanian–early Wuchiapingian.

Accretion of atoll-capped Akiyoshi seamount

The structural relation of the Akiyoshi Limestone and the Tsunemori Formation, along with their ages and the lithological character of related limestone breccia rocks, led Sano & Kanmera (1991*d*)

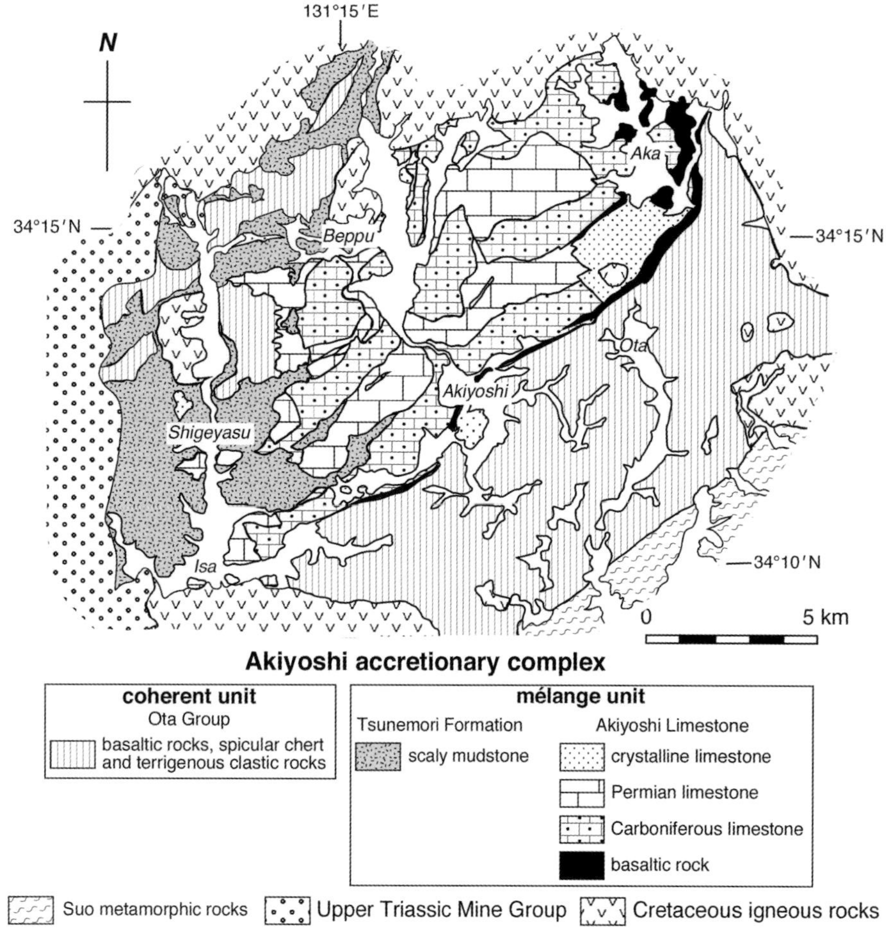

Fig. 2b.12. Simplified geological map of the Akiyoshi-dai Plateau area chiefly underlain by the Mississippian–Permian Akiyoshi accreted rocks. After Kanmera & Nishi (1983). See text for the tectonostratigraphic classification of the Akiyoshi accretionary rocks. Mapped area is shown in Figure 2b.9.

to interpret the accretionary events of the atoll-capped Akiyoshi seamount in terms of normal faulting in an outer trench area (Fig. 2b.8). Their hypothesis invokes normal fault-induced tectonic collapse of the seamount in an outer bulge, fragmenting the atoll carbonates into various-sized blocks. These blocks were gravitationally displaced down toward a trench area and mingled with argillaceous trench-fill sediments to form the chaotic rocks of unit C during late Capitanian–early Wuchiapingian times. The intermingling of smaller limestone debris with the trench-fill sediments formed the polymictic limestone breccia with its argillaceous matrix. Finally, the chaotic mixture of blocks and finer debris derived from the shallow-marine limestone of unit A1 mixed with trench-fill sediments were incorporated into the accretionary prism.

Unlocking palaeoenvironmental changes

Since the pioneering study by Ozawa (1925), a great deal of stratigraphic, sedimentological and palaeontological work has been carried out, dominantly in the shallow-marine limestone of unit A1 (e.g. fusulinoidean biostratigraphy: Toriyama 1958; Watanabe 1991; facies analysis of organic reef complex: Ota 1968; Sugiyama & Nagai 1994). In addition, plate-tectonics-based studies have revealed the tectonic and depositional settings of the Akiyoshi seamount (Kanmera & Nishi 1983; Sano & Kanmera 1988), so there has been much recent progress in our understanding of Panthalassic Mississipiian–Permian palaeoenvironments as recorded by Akiyoshi atoll stratigraphy. This section outlines these recent studies on palaeoclimatic and sea-level changes preserved in the Akiyoshi rocks.

Climatic episodes and sea-level changes recorded in the Akiyoshi atoll stratigraphy

In the course of its long journey (70–80 m.y.) across the Panthalassa Ocean, the Akiyoshi atoll recorded climate and sea-level changes that occurred during Mississippian–Permian time. Recent studies have shown that the environmental changes had pronounced impacts on carbonate sedimentation and biotic assemblages on the Akiyoshi atoll (Nakazawa & Ueno 2004; Sano *et al.* 2004; Sano 2006; Nakazawa *et al.* 2009).

Chiefly on the basis of literature review, Sano (2006) summarized the temporal changes of the dominant lithofacies and biotas of the entire Akiyoshi Limestone, the most thoroughly studied atoll unit within the Akiyoshi Belt. Its succession begins with upper Visean–Bashkirian reefal facies constructed by warm-adapted metazoan reef-builders (corals, *Chaetetes*) and oolitic-crinoidal limestone facies, which indicates deposition in a warm climate with elevated sea levels. The Visean–Bashkirian succession is followed by a Moscovian–Kasimovian muddy limestone-dominant association which records the demise of the metazoan reef builders and frequent emergence events affected by cool climatic conditions under generally lowered sea levels (Fig. 2b.12). A subsequent Gzhelian–Capitanian, muddy limestone-skeletal grainstone association is characterized by an upwards-increase in calcimicrobes, calcareous algae and calcisponges, and records less-frequent emergence events than the preceding Moscovian–Kasimovian succession. These biotic and facies characteristics are interpreted as reflecting a warming climate and sea-level rise during Gzhelian–Capitanian time.

The climate episodes and sea-level changes interpreted from Akiyoshi atoll stratigraphy generally correspond to the long-term global trend of Mississippian–Permian climatic changes and sea-level fluctuation related to Gondwana glaciation (Fig. 2b.8). These include Early–Middle Mississippian pre-glacial warm climate and elevated sea levels, Middle–Late Pennsylvanian cooling and related glacio-eustatic sea-level fluctuation at the culmination of Gondwana glaciation, and Early–Middle Permian global warming and sea-level rise due to the retreat of Gondwana ice sheets.

Following the results of detailed facies analysis using drilled core samples and sequence stratigraphic interpretation, Nakazawa & Ueno (2004) recognized the sequence boundary between the karstified skeletal grainstone of shoal facies and the overlying lime-mudstone and dolomitic micrite of transgressive peritidal facies in the upper Guadalupian section of the Akiyoshi Limestone. These authors suggested that a sea-level fall resulted in a biotic turnover, as exemplified by the coincidence of the fusulinid biostratigraphic boundary with the sequence boundary.

Also based on the drill core data, Nakazawa *et al.* (2009) recognized a long-term sea-level change comprising an late Cisularian (Kungurian) to late Guadalupain (Capitanian) gentle sea-level fall, punctuated by a short-lived middle Guadalupian (Wordian) sea-level rise event. On the basis of the correlation of the sea-level changes with sea-level curves proposed from shelf regions (e.g. Ross & Ross 1987), the Kungurian–Capitanian sea-level fluctuation is considered to have been eustatically driven. Furthermore, Nakazawa *et al.* (2009) have concluded that this long-term eustatic sea-level change controlled the development of parasequences in the Kungurian–Capitanian succession of the Akiyoshi Limestone. One such parasequence, for example, comprises shoal to lagoonal skeletal grainstone that accumulated under conditions of low accommodation during the late Wordian–Capitanian second phase of sea-level fall.

Palaeo-monsoon

Although the Akiyoshi atoll carbonates have been viewed as lacking terrigenous grains (Sano & Kanmera 1988), Soreghan *et al.* (2011) recently found atmospheric dust from the Pennsylvanian section of unit A1 in the Akiyoshi area. This finding is the first-known record of atmospheric dust from the Palaeozoic ocean. The dust is dominated by the clay fraction, but also contains medium-sand-sized grains mainly of rock fragments, quartz and plagioclase, indicating a continental source (at least in part). On the basis of the sedimentological data and climate modelling, Soreghan *et al.* (2011) have preliminarily concluded that East Pangea is the most likely provenance for the sand-sized grains, implying the operation of a palaeo-Tethyan monsoon. This conclusion implies that the Akiyoshi atoll carbonates offer us a potentially valuable record of strong and geologically frequent atmospheric convection events during Late Palaeozoic times.

Suo Belt (TT)

The Suo Belt is an Early Mesozoic unit of the so-called 'Sangun' Belt (see previous description of Renge rocks and Fig. 2b.3; Nishimura 1998). Although the belt overlaps the San'in and Sanyo batholith belts, the spatial distribution of the Suo schist can be traced in the Inner Zone of SW Japan for more than 500 km from the Chugoku Mountains to Kyushu. The Triassic high-*P*/*T* schists of the Tomuru Formation in the Ishigaki–Iriomote islands (Nishimura *et al.* 1983) have been considered a SW extension of the Suo Belt (Nishimura *et al.* 1983; Nuong *et al.* 2008), and other exposures of the Suo schists crop out in the Kurosegawa Sub-belt in Shikoku and central Kyushu (e.g. Isozaki & Itaya 1991). The Suo Belt is essentially a metamorphosed Permo–Triassic accretionary complex, consisting of pelagic/hemi-pelagic siliceous-clayey deposits, trench-fill turbidites, basaltic oceanic crusts and subordinate amounts of limestone, gabbro, serpentinite and serpentinized peridotite (Fig. 2b.13). The Ochiai–Hokubo ultramafic body in the central Chugoku Mountains, which is quite different in petrological features from the ultramafic bodies of the Oeyama Belt (Arai *et al.* 1988), is also considered to belong to the Suo Belt in view of the radiometric ages obtained from deformed metagabbro (245–237 Ma: Nishimura & Shibata 1989). Pumpellyite-actinolite facies metapelite contains rare Middle Permian radiolarian fossils (Takeshita *et al.* 1987). Bedded chert is more dominant as a protolith of the schist in the eastern exposures, where Triassic conodont and Jurassic radiolarian fossils have been described from the Hatto Formation (Hayasaka 1987). Because of these fossils, we cannot rule out a possibility that the eastern exposure represents a metamorphosed equivalent of the Mino–Tamba–Ashio Belt. Overall, the Suo schists exhibit multiple phases (up to five) of deformation (e.g. Oho 1990), typically with isoclinal folds associated with a penetrative schistosity overprinted more than twice by open to tight folds with crenulation cleavages.

The Suo Belt is structurally overlain by Palaeozoic geotectonic units (Oeyama and Akiyoshi). In the Nishiki area, for instance, clastic rocks of the Nishiki Group (Akiyoshi Belt) are thrust over the Suo high-*P*/*T* schists of the Tsuno Group (Fig. 2b.14; Nishimura 1971). The Suo belt is in fault contact with the Yakuno Ophiolite of the Maizuru Belt, and occupies a structurally higher part of the Mino–Tamba–Ashio Belt (Hayasaka 1987). The boundary between the Suo and Hida belts is still unclear, although cataclastic fault rocks separating a Jurassic granitic pluton from Suo schist have been described. In the Nomo Peninsula, the Suo schists are underlain by the Cretaceous high P/T schists of the Sanbagawa Belt (Nishimura *et al.* 2004).

Phengitic micas from the Suo schists have yielded K–Ar ages of *c.* 220 Ma in the east and *c.* 190 Ma in the west, whereas some Jurassic or much younger K–Ar ages were considered as a thermal effect of the Mesozoic and Cenozoic igneous activity (Shibata & Nishimura 1989; Nishimura 1998). Metagabbroic rocks have yielded hornblende K–Ar ages of 253–245 Ma (Nishimura & Shibata 1989), whereas the zircon U–Pb geochronology of psammitic schists suggests *c.* 2.0–1.9 Ga detrital ages for the metasediments and *c.* 230 Ma for the subduction-zone metamorphism (Miyamoto & Yanagi 1996; Tsutsumi *et al.* 2003).

Characteristics of metamorphism and metamorphic zonation

The metamorphic mineral assemblages in the Suo schists cover a wide range of metamorphic facies that include the pumpellyite-actinolite, epidote-blueschist, greenschist facies and epidote-amphibolite facies. Amphibolite facies rocks are also locally found in metagabbroic rocks, but these are considered to be relics from an event unrelated to the Suo metamorphism (Nishimura & Shibata 1989). Overall, metamorphic rocks of the pumpellyite-actinolite facies and blueschist/greenschist transitional facies predominate, with the presence of pumpellyite-actinolite facies being a characteristic feature of most exposures (e.g. Hashimoto 1968; Nishimura 1971; Hayasaka 1987; Watanabe *et al.* 1989; Nishimura *et al.* 2004). Barroisitic Ca–Na amphibole (with abundant porphyroblastic albite) has been described in mafic schists that have been considered to record the highest-grade reached in the Suo Belt (e.g. Hashimoto 1972). Also, it is noteworthy that the occurrence of lawsonite in the pumpellyite-actinolite facies metapelites has been described from at least four localities of the Suo Belt (Nishimura & Okamoto 1976; Watanabe *et al.* 1983; Hayasaka 1987; Karakida *et al.* 1989). A lawsonite + pumpellyite + Na-amphibole mineral assemblage only

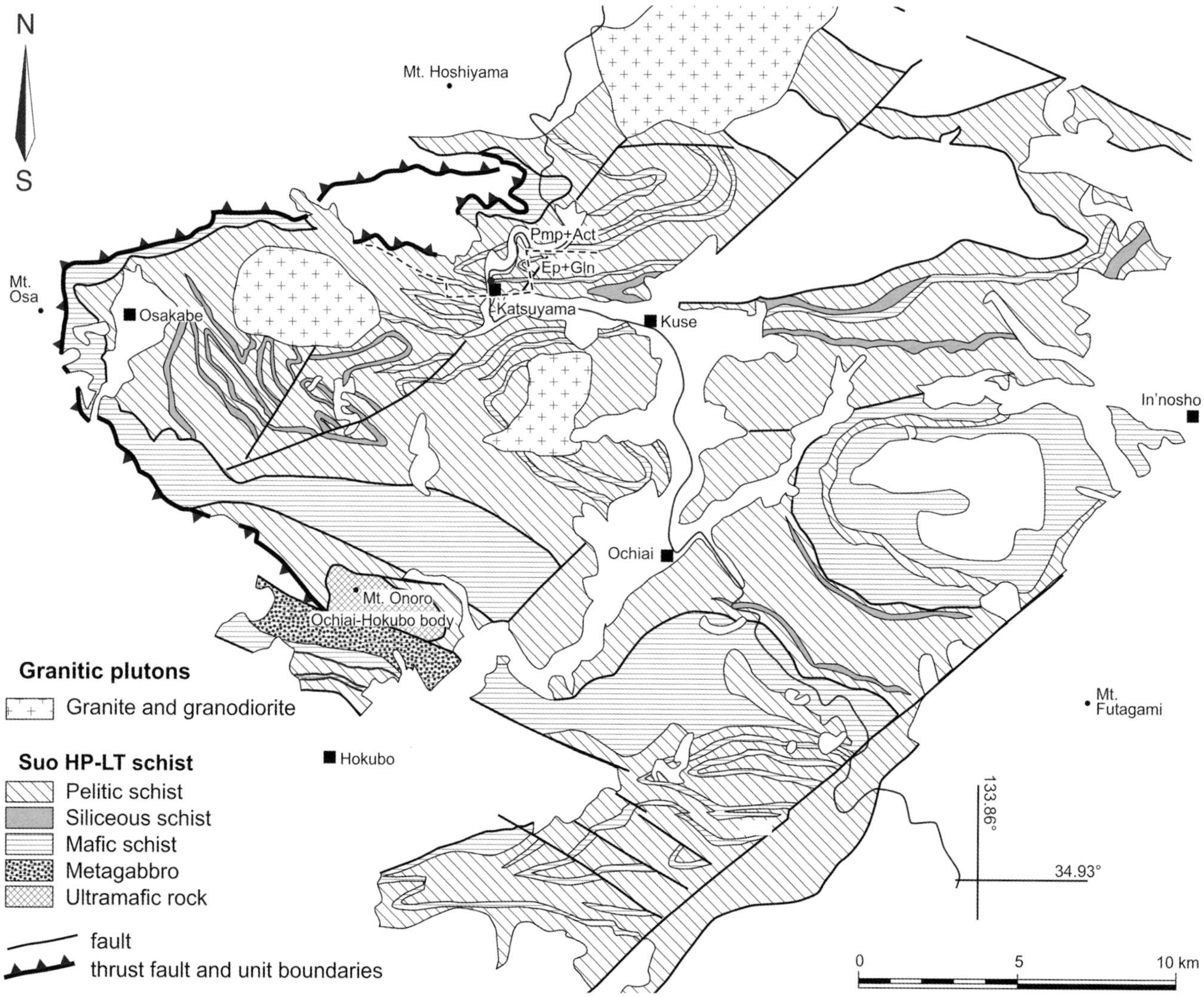

Fig. 2b.13. Simplified geological map of the Katsuyama–Asahi area, showing lithological variations and tectonic boundaries between the overlying Palaeozoic geotectonic units. The map was modified after Teraoka *et al.* (1996).

occurs in the Tomuru Formation in the Ishigaki–Iriomote islands (Nishimura *et al.* 1983), with metamorphic aragonite having been found in the lawsonite-bearing schists (Ishizuka & Imaizumi 1988). Chemical compositions of Na-amphibole of the Suo blueschists tend to have significant amounts of riebeckite and/or actinolite components, suggesting a lower pressure/temperature ratio than that of the Renge blueschists (Tsujimori 1998). High-Al glaucophane, which is common in the Renge blueschist, is relatively rare in the Suo Belt, whereas so-called 'crossite' and/or winchite are rather common.

Based on the appearance of characteristic Fe–Mg silicate minerals in mafic rocks, Nishimura (1971) identified three mineral zones in the Nishiki–Yamaguchi area: (A) pumpellyite + chlorite; (B) pumpellyite + actinolite; and (C) epidote + Na-amphibole, in ascending order of metamorphic grade. The B and C zones correspond to the Suo schists whereas the A zone belongs to the Akiyoshi Belt. Hashimoto (1972) proposed the existence of pumpellyite-actinolite and epidote-Na-amphibole zones for the Suo schists in the Katsuyama area (Hashimoto 1968). Relict igneous Ti-rich augite is commonly found in mafic schists, and metamorphosed picritic basalt contains pseudomorphs after olivine. In the Tomuru Formation of the Ishigaki Island, Nishimura *et al.* (1983) identified pumpellyite-glaucophane, epidote-glaucophane and epidote-barroisite zones.

The Suo schists exposed along the back-arc side of the Chugoku Mountains were overprinted by a thermal metamorphism from granitic plutons of the San'in batholith belt (Miyakawa 1961; Shibata & Nishimura 1989). The appearance of biotite in the pelitic schist defines the contact aureole, with hornblende and augitic clinopyroxene occurring in mafic hornfels.

K–Ar-age–temperature relationships

Two contrasting patterns of phengitic mica K–Ar ages from progressive metamorphic sequences have been described from the Nishiki area and Ishigaki Island (Shibata & Nishimura 1989; Nuong *et al.* 2008) (Fig. 2b.15). The Nishiki metamorphic sequence displays younger ages in higher-grade metamorphic rocks, and a thermal structure in which the higher-grade zone is in the lower part of the apparent stratigraphic succession. In contrast, the Ishigaki metamorphic sequence indicates that age becomes progressively older with increasing metamorphic temperature and the thermal structure is inverted, such that the highest-grade zone occurs in the uppermost parts of the apparent stratigraphic succession. Consideration of the

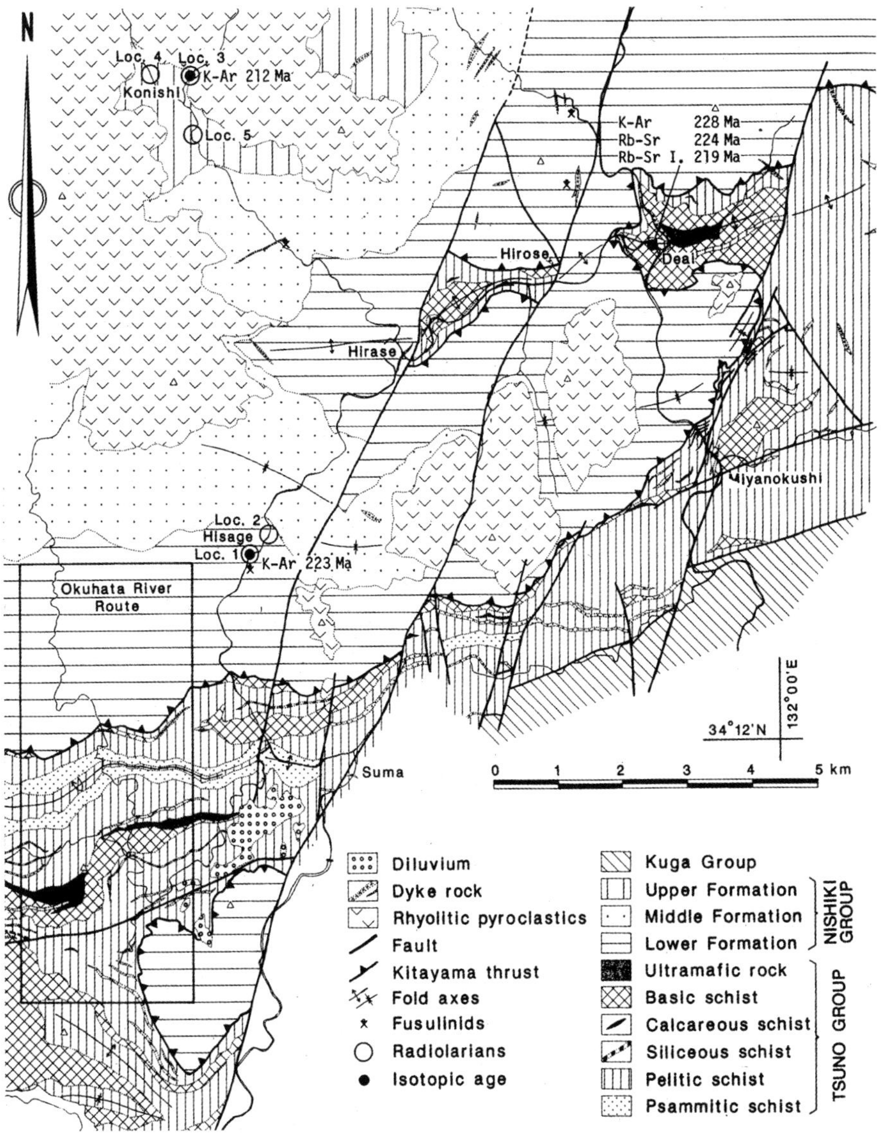

Fig. 2b.14. Geological map of the Nishiki area by Nishimura (1971) cited in Shibata & Nishimura (1989).

K–Ar-age-temperature relationships trajectory of the rocks from these two areas suggests that a shorter period of deformation could explain the Nishiki pattern, whereas a more prolonged duration of deformation could produce the Ishigaki pattern (Itaya *et al.* 2011; Fig. 2b.16).

Maizuru Belt (YHa)

The Maizuru Belt is a narrow (10–30 km wide) but long (380 km) zone extending from the Oshima Peninsula (on the Sea of Japan side at the western end of Fukui Prefecture) through Kyoto, Hyogo, Okayama and Hiroshima prefectures to the SW end of Shimane Prefecture (Fig. 2b.17). Exposures of this belt are rather sporadic in the western areas because of intrusions of Late Cretaceous granites and extensive cover by coeval ignimbrite. Nevertheless, its coherence as one terrane is justified by a close association of Permian clastic formations with metabasites and sheared granites. A similar association is known as the Katashina Belt exposed in central Japan, where it is considered to be the eastern extension of the Maizuru Belt although details are yet unknown. The Maizuru Belt has been subdivided into Southern, Central and Northern zones (Kano *et al.* 1959; Fig. 2b.18) as described in the following sections. Overall, the belt can be summarized as a collided arc–back-arc system formed during Late Palaeozoic (mainly Permian) time (Hayasaka *et al.* 1996), with the three zones representing (from south to north): a collided intra-oceanic island arc; back-arc basin deposits; and continental crust.

Southern Zone

The Southern Zone consists mainly of the Yakuno Ophiolite which represents the juvenile crust of an intra-oceanic island arc originating on Late Palaeozoic oceanic crust (Ishiwatari 1985; Hayasaka 1990; Ishiwatari *et al.* 1990; Suda 2004). This igneous complex records at least two stages of activity, with the older showing T-type MORB or oceanic plateau affinity and the younger of island-arc origin. Sano (1992) reported Sm–Nd whole-rock/mineral internal isochron ages of 410–430 Ma, disregarding the above classification. However, zircon U–Pb ages show *c.* 340–320 Ma for the oceanic crust and *c.* 290–280 Ma for the island-arc crust (Herzig *et al.* 1997). The geology of the Yakuno Ophiolite is described in more detail in Chapter 3.

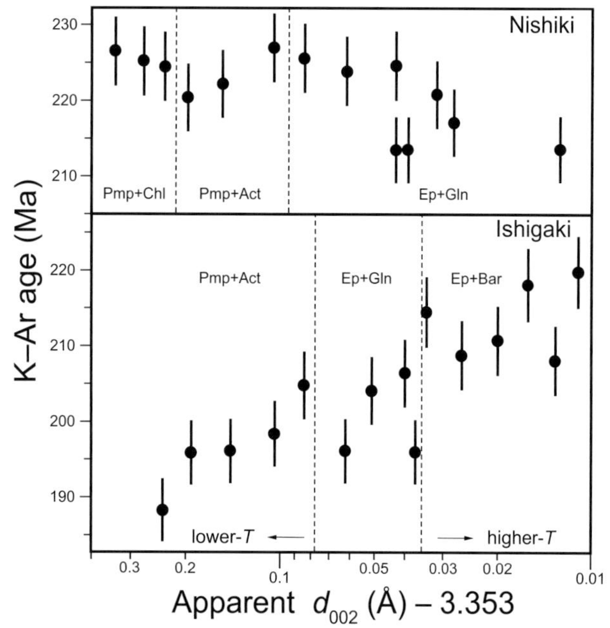

Fig. 2b.15. Age–temperature relations of the Ishigaki and Nishiki metamorphic sequences of the Suo high-pressure schist belt (Nuong *et al.* 2008).

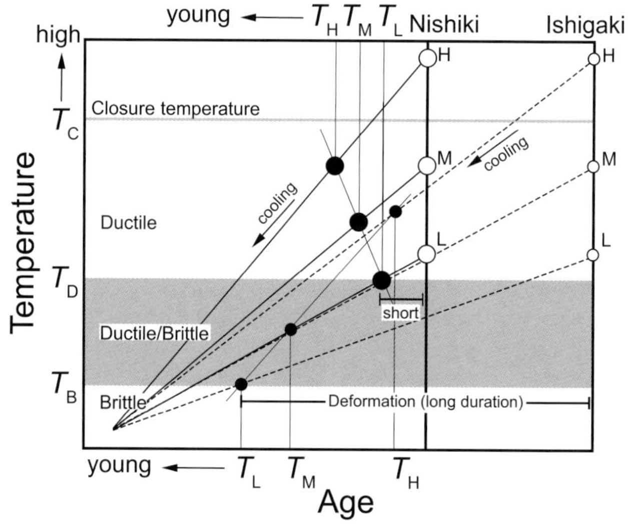

Fig. 2b.16. Generalized temperature–age diagram showing a possible mechanism to form negative and positive K–Ar-age–temperature (T) relationships in the Suo belt (Itaya *et al.* 2011). T_L, K–Ar age of low-T portion; T_M, K–Ar age of moderate-T portion; T_H, K–Ar age of high-T portion; T_C, closure T of phengite; T_D, maximum T of ductile/brittle boundary; T_B, minimum T of ductile/brittle boundary.

Central Zone

The Central Zone consists mainly of the Permian Maizuru Group and unconformably overlying Triassic formations (Nakazawa 1958; Shimizu 1962; Suzuki 1987).

Permian Maizuru Group

The Maizuru Group comprises three formations (Lower, Middle and Upper), with the Lower Formation comprising metabasalt and basaltic tuff with subordinate amounts of metagabbro, metadolerite and siliceous reddish brown claystone. Based on their chemistry these metabasites can be considered as a dismembered basement complex

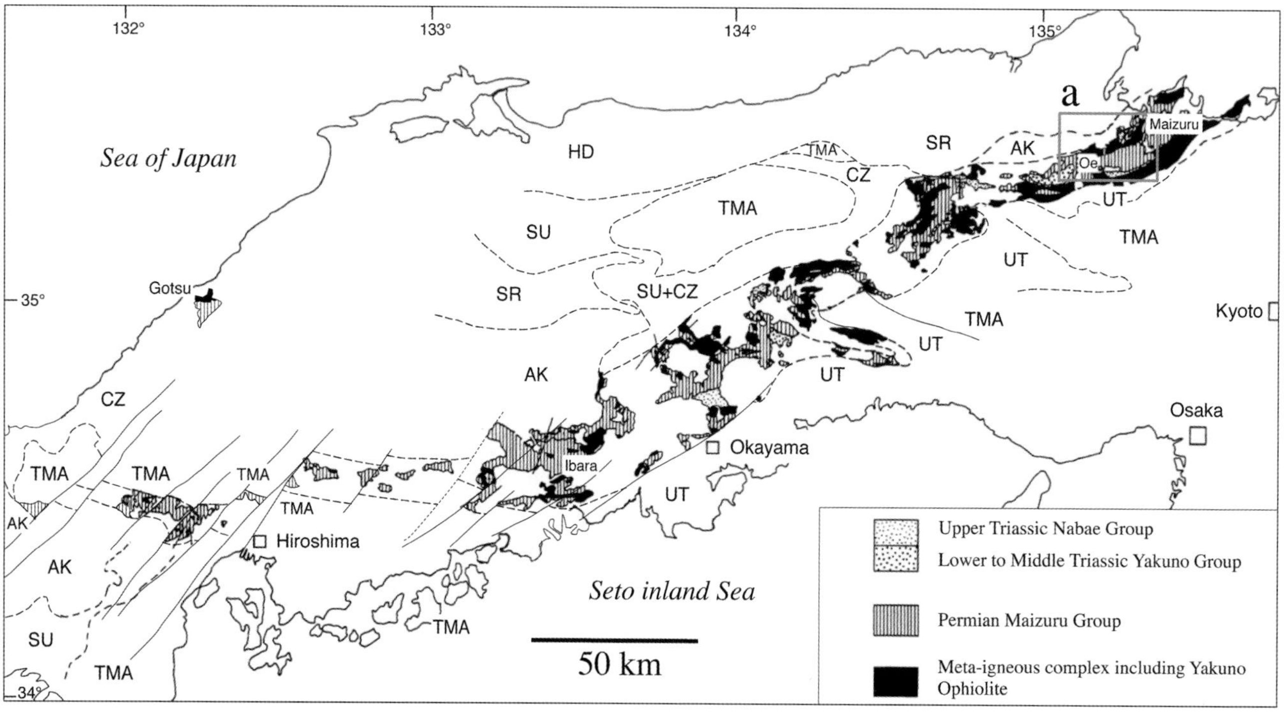

Fig. 2b.17. Geological map of the Maizuru Belt modified after Hayasaka (1990). HD, Hida–Oki Belt; SR, Sangun–Renge Belt; AK, Akiyoshi Belt; UT, Ultra-Tamba Belt; SU, Suo metamorphic belt; CZ, Chizu Belt; TMA, Tamba–Mino–Ashio Belt. Classification and name of each belt are adopted from those of *Pre-Cretaceous Terranes of Japan* compiled by Ichikawa (1990). The box 'a' shows the area of Figure 2b.18.

of a back-arc basin (Koide 1986), and have yielded Rb–Sr whole-rock isochron ages of 290 ± 26 Ma and 281 ± 8 Ma (Koide *et al.* 1987). These ages coincide with the formation of younger-stage rocks of the Yakuno Ophiolite, indicating coeval igneous activity of Early Permian 'Yakuno island arc' and the opening of the 'Maizuru back-arc basin'. The preservation of hemipelagic claystone indicates a rather wide spreading of the back-arc basin, analogous to the modern Shikoku Basin on the Philippine Sea Plate.

The Middle Formation is dominated by massive mudstone with subordinate amounts of sandstone and thin alternating beds of sandstone and mudstone representing a distal turbiditic sequence, mixed with occasional intercalations of siliceous tuffs. The upper half of the Middle Formation yields late Middle Permian fusulinids (Working Group on the Permian–Triassic Systems 1975) and late Middle–early Late Permian radiolarians (Ishiga 1984; Ishiga *et al.* 1988).

The Upper Formation is further subdivided into two members. The Lower Member consists mostly of sandstone with subordinate amounts of conglomerate and alternating beds of sandstone and mudstone representing proximal turbidite facies. The Upper Member is dominated by mudstone, and is characterized by intercalations of displaced limestone lenses. The Upper Member of the Upper Formation contains Late Permian fusulinids and smaller foraminifers (Ishii *et al.* 1975). An uppermost unit known as the Gujyo Formation consists mainly of greywacke-type sandstone with subordinate conglomerate, mudstone and siltstone, is exposed only in the Oe district (Fig. 2b.18), and has yielded latest Permian molluscs and brachiopods (Suzuki 1987). Osozawa *et al.* (2004) considered the lithologies comprising the Gujyo Formation to be forearc basin deposits.

The upwards coarsening observed from the Middle to Upper formations, combined with their complex south-vergent fold–thrust structure, indicate that the sedimentary basin approached the convergent margin to form a collision-accretion complex together with the Yakuno Ophiolite (Hayasaka *et al.* 1996; Osozawa *et al.* 2004). The Maizuru back-arc basin was closed and disappeared due to the collision of a Yakuno island arc by the end of Permian time.

Triassic rocks

Triassic formations are distributed sporadically in the Maizuru Belt, unconformably overlying the Maizuru Group. They are formally classified into the Lower–lower Middle Triassic Yakuno Group and Upper Triassic Nabae Group (Nakazawa 1958). Obvious Ladinian strata are not known in the Maizuru Belt and the basin tectonics of the above two groups are thought to have been quite different. The Yakuno Group (up to 700 m thick) is exposed in the Oe region and other areas further southwest (Figs 2b.17 & 2b.18). Marked sedimentary and biofacies changes indicate the change from deltaic through nearshore to offshore environments from north to south (Nakazawa 1958). This strongly suggests the presence of an uplifting landmass just to the north of the Central Zone of the Maizuru Belt in Early Triassic time. The Yakuno Group can therefore be defined in terms of foreland basin deposits, recording one broad cycle of transgression and regression. The Yakuno Group yields abundant molluscan fossils, indicating an earliest Triassic–early Middle Triassic age.

The Upper Triassic Nabae Group is exposed as a narrow tectonic basin bounded by high-angle faults and lying between the Central and Southern zones of the Maizuru Belt (Fig. 2b.18). This group

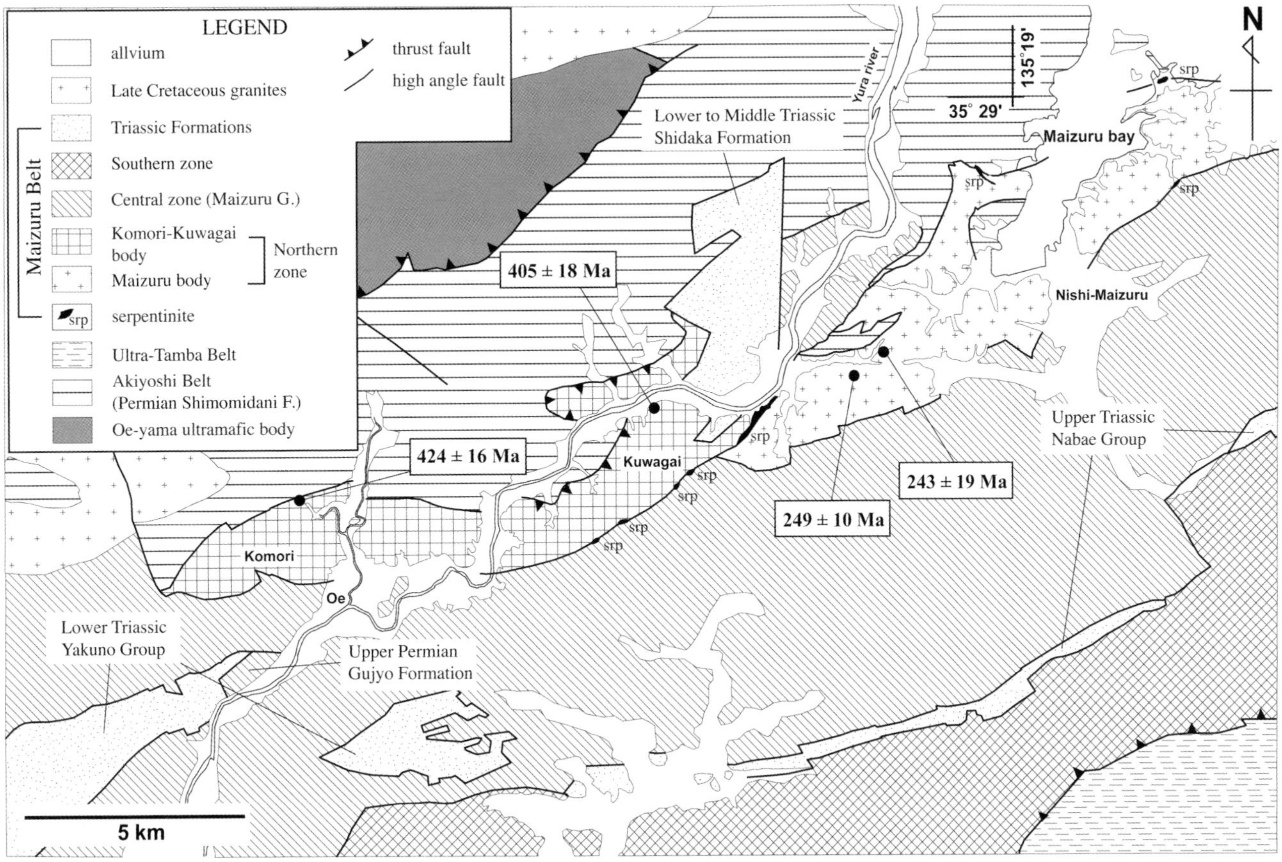

Fig. 2b.18. Geological map of the Maizuru-Oe district partly modified after Fujii *et al.* (2008). Original data are from Igi *et al.* (1961), Igi & Kuroda (1965), Suzuki (1987) and Ikeda & Hayasaka (1994). Location of the mapped area is shown in Figure 2b.17.

(up to 940 m thick) consists mainly of sandstone, mudstone, alternating beds of sandstone and mudstone, and conglomerate in order of abundance, and is characterized by intercalations of thin coal seams (Nakazawa 1958). Lithofacies and biofacies indicate that the Nabae Group was deposited in a neritic- to brackish-water environment, and it records one cycle of transgression and regression. Adachi & Suzuki (1992) reported EPMA (electron probe microanalyser) U–Th–total-Pb ages of detrital zircon and monazite from sandstones in the Nabae Group consisting of age clusters around 260 and 380–520 Ma. These ages demonstrate a provenance from the Northern Zone of the Maizuru Belt, as described in the following section.

Northern Zone

The Northern Zone is exposed only in the Maizuru–Oe district, northwestern Kyoto Prefecture, and is subdivided into the Komori–Kuwagai Complex in the west and the Maizuru Complex in the east (Fig. 2b.18). These two bodies are bounded by a high-angle fault associated with serpentinite lenses. Moreover, they differ from each other in their rock associations and ages, as well as in the degree of deformation. The Komori–Kuwagai Complex consists mainly of moderately to highly deformed granites with subordinate amounts of metagabbros, metadolerite, amphibolite and garnet-biotite gneiss. On the other hand, the Maizuru Complex consists exclusively of weakly deformed or undeformed granites (referred to informally as the 'Maizuru Granite'). SHRIMP zircon U–Pb ages of 424 ± 16 Ma and 405 ± 18 Ma for the granites from the Komori–Kuwagai Complex and 249 ± 10 Ma and 243 ± 19 Ma for the Maizuru Granite have been reported (Fujii *et al.* 2008; Fig. 2b.19). The granites of the Komori–Kuwagai body contain much older component of zircon (*c.* 580 and 765 Ma), which are likely inherited from host rock. They also contain monazite grains dated at *c.* 470 and 450 Ma, and their $^{87}Sr/^{86}Sr$ initial ratio is 0.70645 (Ikeda & Hayasaka 1994).

The Komori–Kuwagai Complex is thrust over the Permian Shimomidani Formation of the Akiyoshi belt to the north, whereas to the south it is brought into contact with the Permian Maizuru Group of the Central Zone by high-angle faults. The Lower–Middle Triassic Shidaka Formation unconformably overlies the Komori–Kuwagai body, which is composed mainly of fluvial fan sandstone and conglomerate with intercalated mudstone. From lithofacies, ages and geological structure, the Komori–Kuwagai body is postulated to be derived from the Khanka Massif of Prymorie in far eastern Russia, having been displaced by dextral fault movement during Late Triassic–Early Cretaceous time (Fujii *et al.* 2008). The Maizuru body can be age-correlated with granites in the Hida–Oki Belt.

Development of the Maizuru tectonic belt as a large-scale dextral shear zone

The above-mentioned history of the Maizuru Belt is summarized in Figure 2b.19 in the postulated context of a collided arc–back-arc system. The distribution of the Maizuru Belt shows a marked zonal arrangement cutting across the subhorizontal piled nappe structure of other belts present in the Inner Zone of SW Japan (Fig. 2b.17). This zonal arrangement becomes clear by removing the effect of Late Cretaceous NE–SW and NW–SE transverse fault movements. The northern boundary of the Maizuru Belt represents a high-angle dextral fault, whereas its southern boundary is a low-angle fault thrusting over the Ultra-Tamba belt (Hayasaka 1987). High-angle dextral faults are also ubiquitous within the Central Zone, forming a tectonic mélange zone that differs from those of more typical accretionary complexes. Many lenticular blocks of metagabbro, metadolerite as well as some granite lie in a longitudinally subparallel alignment. Moreover, kilometre-long exotic duplexes displaced from the Suo and Akiyoshi belts have been brought into the Central Zone. Because of this structural complexity imposed by major dextral shearing deformation, the Maizuru Belt has often been referred

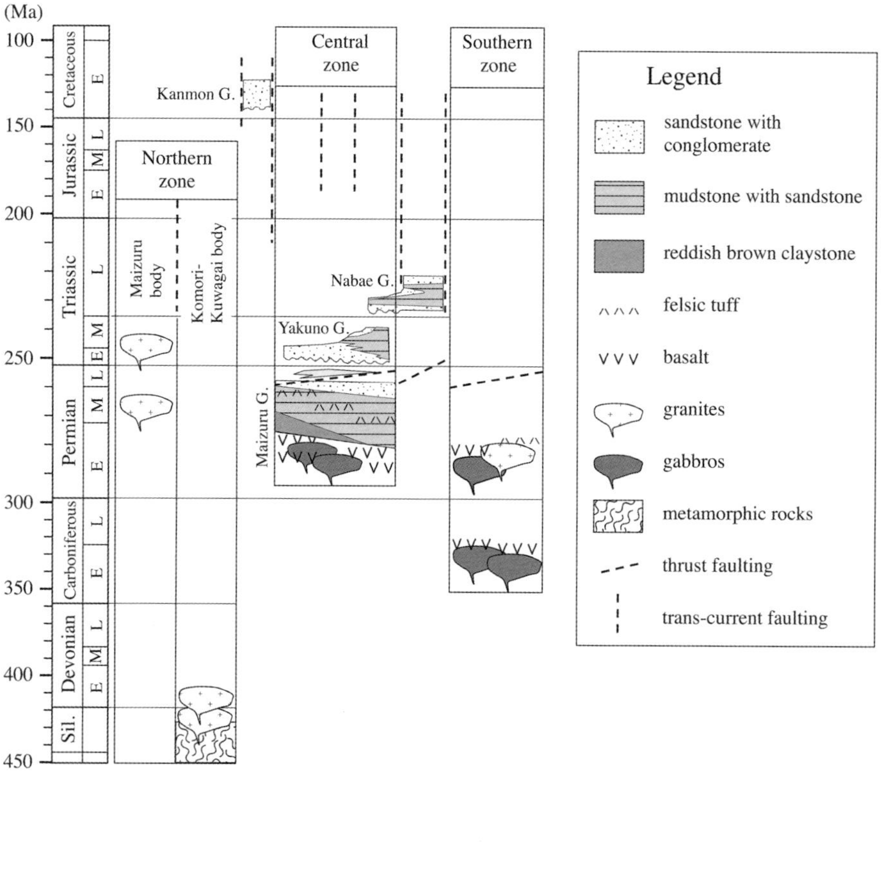

Fig. 2b.19. Generalized geo-historical column for the Maizuru Belt arranged from Hayasaka (1990), Hayasaka *et al.* (1996) and Fujii *et al.* (2008) with new data. Time scale is relatively expanded to 167% between 300 and 200 Ma.

to as the 'Maizuru tectonic belt'. An isolated and remote exposure of probable Maizuru Belt is known in the Gotsu area, northwestern Chugoku Province (Fig. 2b.17), and may have been separated from its main body by large-scale transcurrent displacement related to the dextral shearing.

Ultra-Tamba Belt (ST & YS)

The Ultra-Tamba Belt was first defined by Caridroit *et al.* (1985) and Ishiga (1986), although this definition was subsequently modified by Ishiga (1990). The belt extends from western Fukui Prefecture to western Okayama Prefecture, trending ENE–WSW except for where it has been subjected to intense post-Jurassic deformation (Fig. 2b.2). Correlatives of the belt include those found in the Tohoku region (Nakae & Kurihara 2011) and in the Sikhote–Alin Mountains of the Russian Far East (Kojima *et al.* 2000). The Ultra-Tamba belt is situated between the Maizuru and Tamba belts, and is tectonically divided into three sub-belts (UT-3, UT-2 and UT-1), each separated by a thrust fault (Fig. 2b.20).

Lithology and ages of sub-belts

The UT-3 sub-belt, which occupies the structurally uppermost part of the Ultra-Tamba belt, includes the Kozuki Formation (SW Hyogo Prefecture), the Kunisaki Complex (SE Hyogo Prefecture; Fig. 2b.20) and the Hiehara Formation (western Okayama Prefecture). These units consist mainly of mudstone, sandstone, felsic tuff and greenstones, along with minor chert and limestone which display coherent, broken and mixed facies. Mudstone, felsic tuff and chert yield early Middle–early Late Permian, Carboniferous and early Early Permian radiolarians, respectively (e.g. Pillai & Ishiga 1987; Takemura *et al.* 1993). Limestone associated with greenstones contains Carboniferous corals and Carboniferous and Permian fusulinids (e.g. Igi 1969).

The UT-2 sub-belt, which is structurally located between UT-3 and UT-1, contains strata named the Oi Formation, correlated with the Tokura Formation in northern Kyoto Prefecture, the Takatsuki Formation in northern Osaka Prefecture and the Inagawa Complex in southeastern Hyogo Prefecture (Fig. 2b.20). The sub-belt mainly consists of sandstone, laminated mudstone, chert, siliceous mudstone and felsic tuff, with minor amounts of 'greenstone' metabasalt. The chert yields late Middle–early Late Permian radiolarians, and the mudstone, siliceous mudstone and felsic tuff contain early–middle Late Permian radiolarians (e.g. Sugamori 2009*a*).

UT-1 occupies the structurally lowermost part of the Ultra-Tamba belt. Strata of this sub-belt are generally called the Hikami Formation, which is correlative with the Kuchikanbayashi Formation in northern Kyoto Prefecture and part of the Yamasaki Formation in southwestern Hyogo Prefecture. These formations consist of sandstone and mudstone with minor chert and greenstones. Mudstone contains poorly preserved radiolarians that appear to indicate a Middle–Late Permian age (Kurimoto 1986). The age of the formation of the UT-1 sub-belt remains unknown. Although some formations of the Ultra-Tamba belt were previously believed to contain Mesozoic forearc sediments, they have since yielded Late Permian radiolarians (e.g. Sugamori 2011) and are now considered to be part of the UT-2 sub-belt.

Tectonics of the Ultra-Tamba Belt

The following points are important in the context of the tectonics of the Ultra-Tamba belt: (1) some formations display broken or mixed facies; (2) the terrigenous clastic rocks are the same age as, or younger than, cherts that occur in the same unit; (3) typical chert–clastic sequences are exposed locally; and (4) shallow-water limestone containing corals and fusulinids is associated with metabasaltic 'greenstones'. These lines of evidence indicate that the Ultra-Tamba belt is an AC that formed in a subduction zone during Middle–Late Permian times.

The Maizuru, Ultra-Tamba and Tamba belts occur as a pile of nappes (Fig. 2b.21) although the nature of the rock assemblages, combined with their ages, indicate that these belts formed at different times and in different tectonic settings. The present distribution of three belts was probably formed after Late Jurassic times as the Tamba Belt contains Late Jurassic rocks. Deformation structures (e.g. slaty cleavage and folding) indicating southwards vergence are penetratively developed in the Ultra-Tamba Belt, and the deformation and metamorphism increase in grade towards structurally underlying units (Kimura 1988; Takemura & Suzuki 1996). These deformation structures probably formed when the Ultra-Tamba Belt was thrust over the Tamba Belt. The Ultra-Tamba and Tamba belts consist mainly of AC materials that formed at Permian and late Triassic–Jurassic times, respectively. Triassic tectonics in the region are poorly understood, however. Middle Triassic radiolarians have been reported from the Kamitaki Formation (mid-eastern Hyogo Prefecture) in the UT-1 sub-belt (Sugamori 2009*b*). Further study is required to understand the original relationship between the Ultra-Tamba and Tamba belts, and to accurately reconstruct Triassic tectonics.

Mino–Tamba–Ashio Belt (SK)

The Mino–Tamba–Ashio Belt is located geographically south of, and structurally beneath, the Palaeozoic belts, always with fault boundaries, and gradually changes into the Ryoke metamorphic belt to the south. Rocks of this belt comprise Jurassic ACs, and they can be traced from the northeastern Ashio area to the southwestern Iwakuni area in Honshu (Figs 2b.2 & 2b.22). The Jurassic ACs extend north to the Sikhote–Alin Range in far-eastern Russia and Nadanhada Mountains in NE China (Kojima 1989; Kojima *et al.* 2000). The lithologies common to these areas are Permian basalt and limestone, Permian–Jurassic radiolarian ribbon chert, and Jurassic clastic rocks and, as such, the succession represents a typical OPS (Matsuda & Isozaki 1991). In this section the ACs in the Mino area of central Japan, one of the best-studied areas, are described in detail and the more important findings emphasized.

Subdivision

The ACs in the Mino area have been subdivided into six tectonostratigraphic units (Sakamoto-toge, Samondake, Funafuseyama, Nabi, Kamiaso and Kanayama) on the basis of their lithology, geological structure and age of accretion (Wakita 1988). Accretion ages of these Mino units become younger from north to south with the structurally lower units occurring in the southern area, although the surface distribution is disturbed by a fold structure with a gently north-dipping enveloping surface (Fig. 2b.23). The Sakamoto-toge unit is the oldest AC in the Mino area, accreted during Early Jurassic times, and mostly composed of mélanges with blocks of sandstone, chert, basalt and limestone within a weakly sheared shale matrix. The Samondake unit is characterized by Middle Jurassic massive sandstone and alternating sandstone and mudstone, whereas the Funafuseyama unit consists of mélanges including large slabs and blocks of Permian limestone and chert. The accretionary age of both the Samondake and Funafuseyama units is Middle Jurassic. Much of the Nabi unit is also stratigraphically chaotic, varying from highly deformed mélanges with Triassic chert slabs to weakly broken Jurassic turbidite, but the timing of

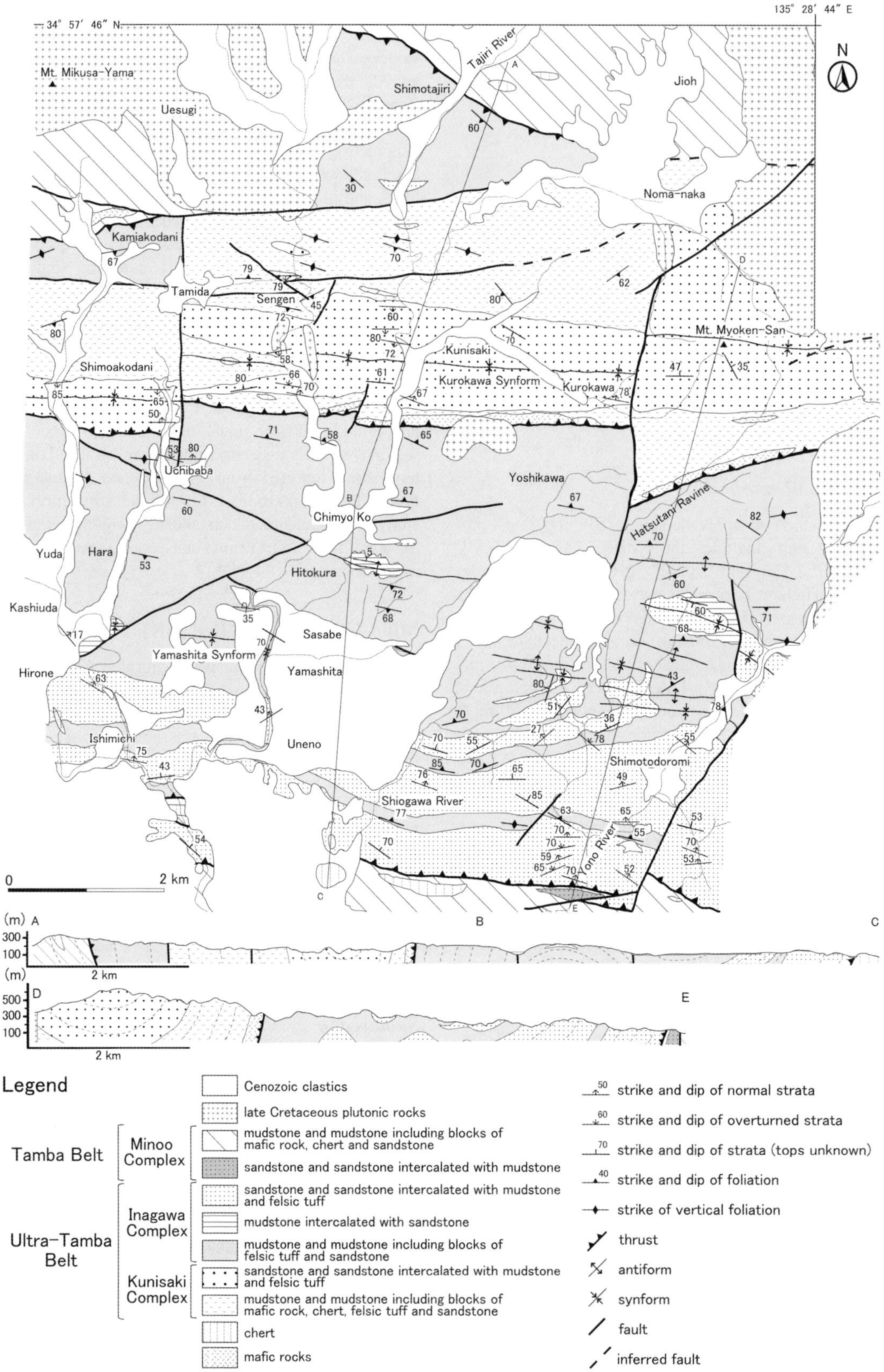

Fig. 2b.20. Geological map and geological cross-section of the Kawanishi–Inagawa area, southeastern part of Hyogo Prefecture and northwestern part of Osaka Prefecture (modified after Sugamori 2009*a*). The Ultra-Tamba Belt in the area is thrust over the Tamba Belt (Jurassic AC) and divided into the Kunisaki and Inagawa complexes (in structurally descending order). These complexes form a pile of nappes that in the Inagawa Complex includes clastic rocks younger than those in the Kunisaki Complex, and are interpreted as being formed by successive accretionary growth. The age of east–west-trending upright folds developing into the Ultra-Tamba and Tamba belts is probably early Cretaceous, because of late Cretaceous plutonic rocks intruded.

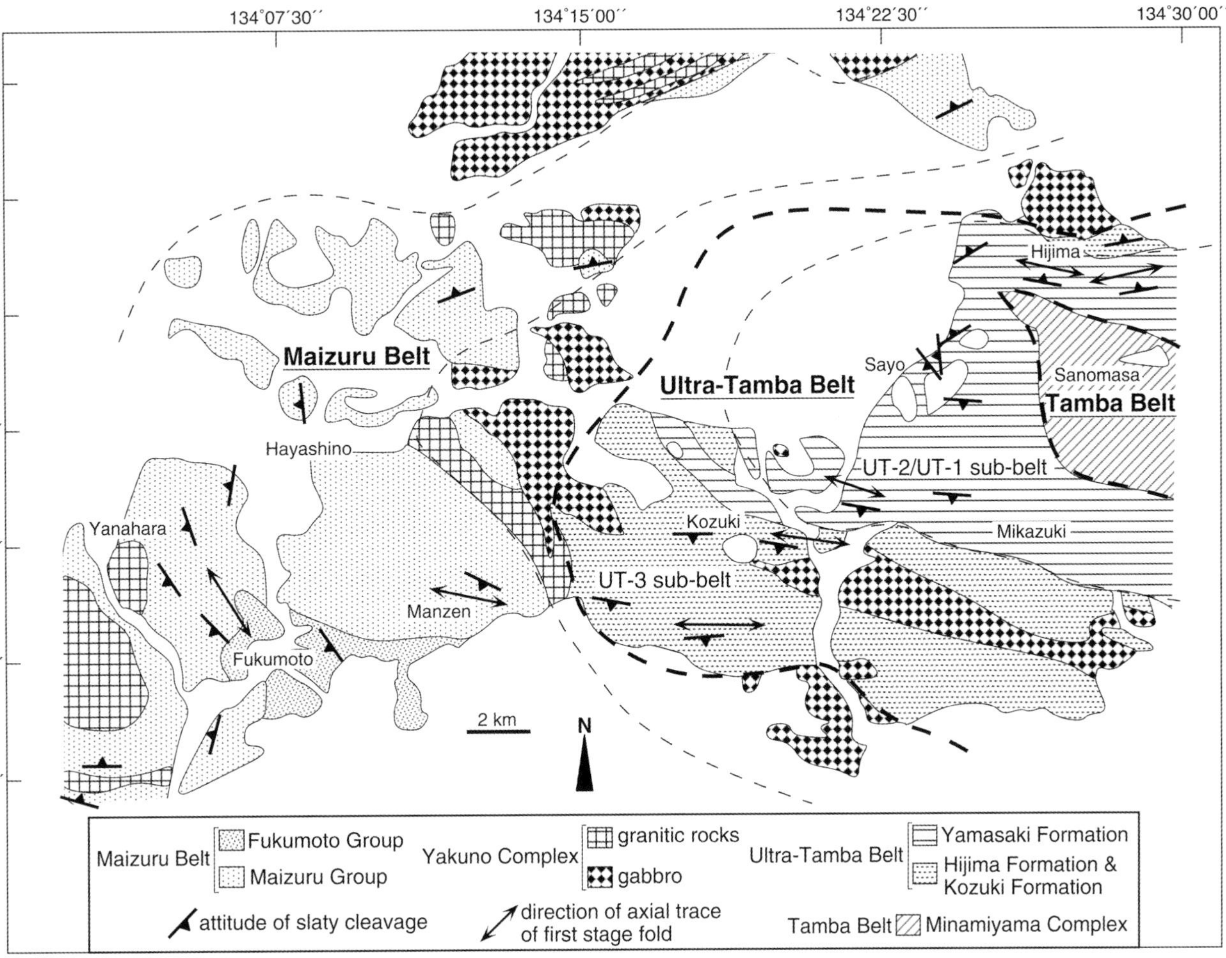

Fig. 2b.21. Regional structure in western Hyogo and eastern Okayama prefectures (modified after Takemura & Suzuki 1996). The Maizuru, Ultra-Tamba and Tamba belts occur as a pile of nappes and form an antiform in this area.

accretion was early Late Jurassic. The Kamiaso unit shows a stratigraphically coherent upper oceanic plate succession, comprising Lower Triassic siliceous claystone, Middle Triassic–Early Jurassic radiolarian ribbon chert and Middle Jurassic siliceous mudstone conformably covered by turbidite with rare conglomerate interbeds. Finally, the Kanayama unit consists of mélanges and is the youngest of the Mino ACs, yielding earliest Cretaceous radiolarians from mudstones.

Lithology

Although varying from unit to unit, the main lithologies of the ACs in the Mino area are summarized in Figure 2b.24.

Basalt

The basaltic rocks are massive, pillowed, brecciated (hyaloclastite), heavily altered and weakly metamorphosed to prehnite-pumpellyite

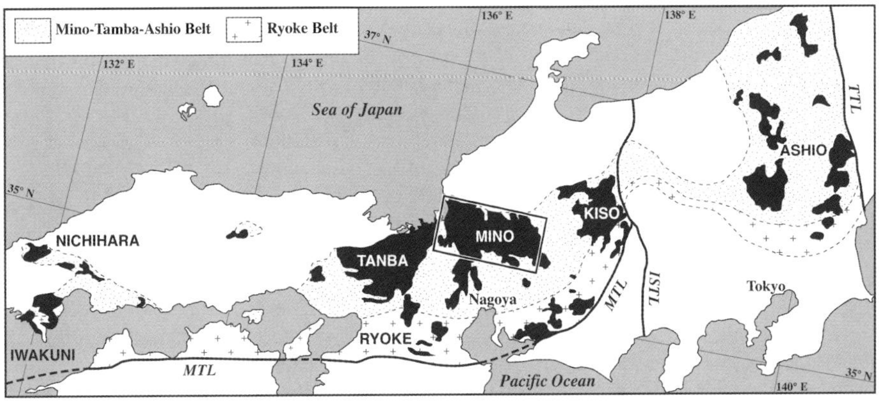

Fig. 2b.22. Map showing the location of the Mino–Tamba–Ashio Belt and the Ryoke Belt (metamorphic equivalent of the former). Black areas indicate the actual distribution of the rocks of these belts (modified from Nakae 2000). Open box indicates the map area of Figure 2b.23. MTL, Median Tectonic Line; ISTL, Itoigawa–Shizuoka Tectonic Line; TTL, Tanakura Tectonic Line.

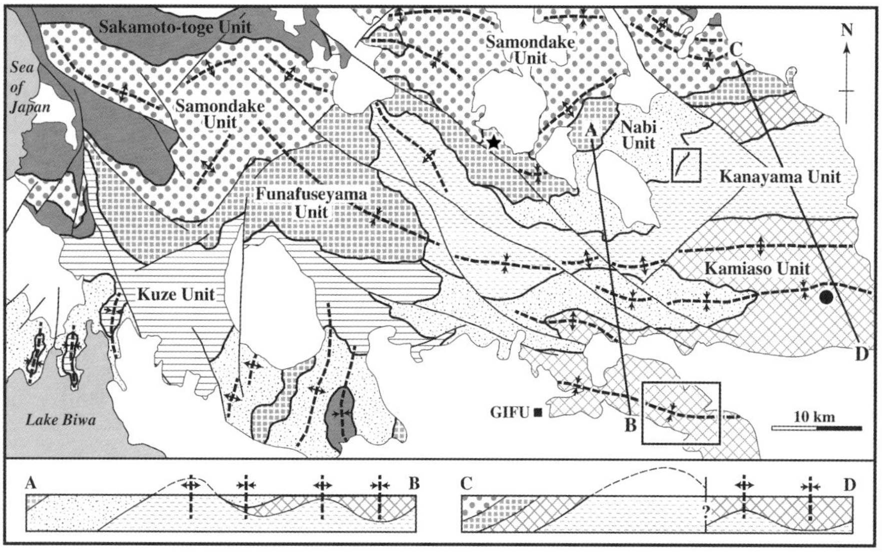

Fig. 2b.23. Map showing the tectonostratigraphic subdivision of the ACs in the Mino area and geological cross-sections along the lines A–B and C–D (modified from Nakae 2000). Map area is indicated in Figure 2b.22. Larger and smaller open boxes indicate the map areas of Figures 2b.25 and 2b.27, respectively. Black star is the locality of Permo-Triassic boundary section reported by Sano *et al.* (2010), and black circle is the locality of Kamiaso conglomerate including *c.* 2000 Ma gneiss clast (Adachi 1971).

facies. The radiometric ages are difficult to measure because of the alteration, but conformable relationships with overlying limestone and chert formations suggest that the basalts are Early Permian or older. Triassic basalts have also been reported from a few localities on the basis of their intimate occurrence with Triassic radiolarian chert (e.g. Wakita 1984). Chemical compositions indicate that the basalts formed in a hot-spot oceanic island setting near mid-ocean ridges (Jones *et al.* 1993).

Limestone

The majority of the limestones in the Mino area are Early–Middle Permian cap carbonates resting on the basaltic oceanic island basement, although Carboniferous limestone blocks in mélanges and Upper Triassic deep-marine micritic limestones are known at several localities (Sano & Kojima 2000). The Permian limestone yields fossils such as fusulinaceans, bivalves, brachiopods, corals and gastropods. No Triassic shallow-marine limestone has been reported from the Mino Belt, but Triassic conodonts have been recovered from cave-filling clastic carbonates in the Permian limestone (Sano & Kojima 2000). Further, Middle Triassic submarine landslide deposits occur embedded within deep-marine chert (Kojima & Sano 2011), implying that shallow carbonate accumulation continued into Triassic time (Fig. 2b.24). The Upper Triassic deep-marine limestones (5–10 cm in thickness) are interbedded with chert, and characteristically include thin shells of planktonic bivalves and radiolarian tests.

Chert

This lithology is mostly radiolarian ribbon chert composed of alternating beds of several-centimetres-thick chert and several-millimetres-thick siliceous shale. The chert consists of micro- to crypto-crystalline quartz and radiolarian remains, and is lacking in coarse clastic materials. The chert and overlying siliceous mudstone have been biostratigraphically studied by using radiolarian and conodont fossils. The cherts rest conformably on the basalt with interbeds of hydrothermal chert and resedimented dolomite at the boundary horizon (Sano 1988). The Upper Permian chert is covered by Permo-Triassic boundary carbonaceous black shale, followed by Early Triassic siliceous lithologies comprising alternating light grey siliceous shale and carbonaceous black shale. This Early Triassic siliceous unit, referred to as siliceous claystone in this chapter, grades upwards into radiolarian chert. These Middle Triassic cherts contain rare coarse clastic horizons containing chert, siliceous shale, basic volcanic rock, apatite, dolomite, glauconite, polycrystalline quartz and Permian and Triassic radiolarian and conodont remains (Kojima *et al.* 1999). The clastic materials are considered to have been transported by submarine landslides from nearby oceanic islands (Kojima & Sano 2011). The chert sequence gradually changes to overlying Early–Late Jurassic siliceous mudstone, deposited as the oceanic plate approached the trench (Fig. 2b.24). Manganese-carbonate nodules often found near the chert-mudstone boundary contain extraordinarily well-preserved radiolarians and provide important age controls.

Clastic sediments

The siliceous mudstones mentioned above are locally interbedded with acidic tuff derived from volcanic arc on the continental margin, and are overlain by trench-fill turbidites. Although the palaeocurrent directions analysed by using sole markings of the turbidites are not consistent throughout the Mino area, north–south-aligned across-trench and west–east-aligned trench-parallel transportation of clastic materials are dominant (Adachi & Mizutani 1971). Mineralological studies of sandstones and conglomerates in this clastic succession indicate a metamorphic provenance (e.g. Mizutani 1959) because of the occurrence of pyrope-rich garnet indicating a granulite facies protolith (Adachi & Kojima 1983; Takeuchi 2000). Rb–Sr whole-rock isochron data from gneiss clasts in the Kamiaso conglomerate (see Fig. 2b.23 for the locality) yielded an age of *c.* 2000 Ma (Adachi 1971; Shibata & Adachi 1974; Fig. 2b.23), and CHIME and SHRIMP ages of zircons and monazites in sandstone and clasts of conglomerate range from 3420 to 161 Ma (Suzuki *et al.* 1991; Y. Sano *et al.* 2000; Hidaka *et al.* 2002; Nutman *et al.* 2006). Taken together, these lines of evidence described above demonstrate that the clastic materials of the Mino ACs were derived from a Precambrian continent exposing high-grade metamorphic rocks.

Geological structure

The ACs in the Mino–Tamba–Ashio Belt have two types of characteristic geological structures: fault-bounded stack of chert-clastic sequence; and mélange.

Fault-bounded stack of chert-clastic sequence

Rocks of the Kamiaso unit are characterized by the repeated occurrence of chert and clastic rocks, with one of the best examples being observed in the Unuma area (Fig. 2b.25). Early Triassic siliceous

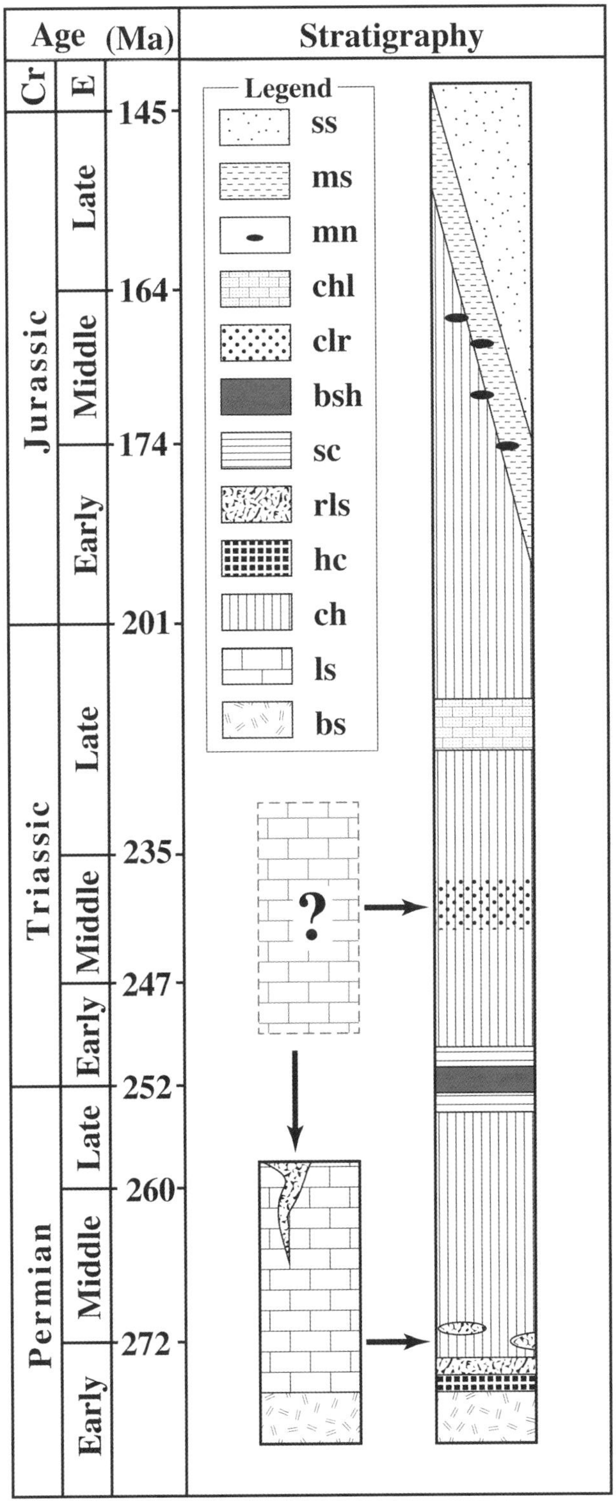

Fig. 2b.24. Schematic diagram illustrating stratigraphy of the Permian–earliest Cretaceous shallow-marine (left) and deep-marine (right) oceanic rock assemblages of the Mino–Tamba–Ashio Belt (after Sano & Kojima 2000). Presence of the Triassic shallow-marine carbonate is estimated on the basis of the Triassic fossils found in the cave deposits in Permian limestone and of the submarine landslide deposits in the Triassic deep-marine chert derived from shallow-marine oceanic rocks. ss, sandstone; ms, mudstone; mn, manganese carbonate nodule; chl, alternating chert and limestone; clr, submarine landslide deposit; bsh, black carbonaceous shale; sc, siliceous claystone; rls, resedimented limestone and dolomite; hc, hydrothermal chert; ch, chert; ls, limestone; bs, basalt.

claystone, Middle Triassic–Early Jurassic radiolarian ribbon chert and Middle Jurassic siliceous mudstone covered by coarse clastic rocks together define a synformal structure with a westwards-plunging fold axis (Kondo & Adachi 1975; Kimura & Hori 1993; Yoshida & Wakita 1999). Detailed radiolarian and conodont biostratigraphic studies of the rocks exposed along the Kiso River reveal that the geological package, apparently composed of many layers of chert and clastic rocks, actually comprises thin sheets of one chert-clastic formation repeated by thrust faults (Fig. 2b.26; Yao *et al.* 1980; Matsuoka *et al.* 1994). The stacking structure is considered to have formed during the off-scraping process of subduction-accretion (Fig. 2b.2), with the Permo/Trias boundary black shale deforming more easily than the underlying and overlying cherts and so acting as a décollement surface during accretion (Matsuda & Isozaki 1991; Kimura & Hori 1993).

Mélange

Although the occurrence of mélange is common throughout the Mino–Tamba–Ashio Belt, only one example (Kanayama unit) is described here. The Kariyasu area near the boundary between the Kanayama and Nabi units is underlain by mélanges composed of blocks and sheets of variable size, shape and lithology within a sheared shale matrix (Wakita 1984, 1995). As well as the larger blocks (depicted in Fig. 2b.27a), the mélange also includes many smaller blocks such as metre- to centimetre-scale blocks of chert, siliceous mudstone and sandstone as seen in exposures along the Nagara River near Kariyasu (Fig. 2b.27b). The larger sheets preserve, in part, the original OPS, with Early Triassic siliceous claystone, Middle Triassic–Early Jurassic chert and Middle–Late Jurassic siliceous mudstone forming a coherent mass in the mélange (Fig. 2b.27a). The original OPS can be reconstructed even in the more chaotic mélanges as shown in Figure 2b.27b, indicating that the southeastern part of the mélange formed during the Early–Late Triassic part of the OPS and the northwestern part comprises the Middle Triassic–Jurassic part. Most of the mélanges in the Mino–Tamba–Ashio Belt are thought to have formed along out-of-sequence thrusts during structural thickening within the AC (Fig. 2b.1), although some might have formed by sedimentary and diapiric processes (Wakita 2000).

Recent discoveries

Low-latitude and Southern Hemisphere origin of the Middle Triassic chert

The exact origins of the Mino ACs are difficult to pinpoint because plate configurations during Palaeozoic–Jurassic time are still unclear. However, Ando *et al.* (2001) performed palaeomagnetic and micropalaeontological analyses on the Middle Triassic–earliest Jurassic chert in the Unuma area (Fig. 2b.25). By comparing the magnetic reversal patterns with those of the coeval European sections, they concluded that the chert was deposited in an equatorial region with a palaeolatitude between 10° N and 10° S and that the lower–middle Anisian (Middle Triassic) chert depositional basin was situated in the Southern Hemisphere.

Continuous Permo-Triassic boundary section

The lithological contrast between the Permo-Triassic boundary black shale and the underlying and overlying cherts is too great to have allowed accretion of the succession as a whole without some degree of thrust décollement. While Permian and Triassic chert formations occur ubiquitously not only in the Mino–Tamba–Ashio Belt but also in the Northern and Southern Chichibu belts, completely continuous sections across the system boundary are very rare. However, Sano *et al.* (2010) discovered one such continuous section in the Funafuseyama unit of the Mino area (see Fig. 2b.23 for the

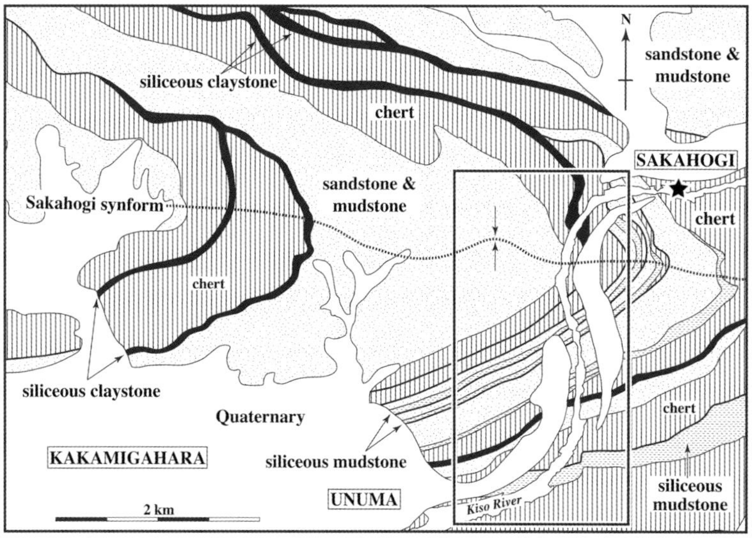

Fig. 2b.25. Geological map of the Unuma area (modified from Yoshida & Wakita 1999). Map area is shown in Figure 2b.23. Star indicates the locality of the Middle Norian (Upper Triassic) impact ejecta reported by Onoue *et al.* (2012). Open box indicates the map areas of Figure 2b.26.

locality). The section comprises uppermost Permian black chert including radiolarians of *Neoalbaillella optima* Zone and lowermost Triassic black claystone with thin black chert beds yielding conodonts of the *Hindeodus parvus* Zone. Sano *et al.* (2012) discussed palaeoenvironmental changes in the Panthalassic Ocean on the basis of the geochemical data obtained from this section.

Upper Triassic impact ejecta preserved in chert

The slow accumulation of sediments such as the radiolarian chert of the Mino–Tamba–Ashio Belt, with apparent average sedimentation rate of several metres per million years (Matsuda & Isozaki 1991), provides the potential opportunity to identify very rare events such as giant meteorite impacts. Onoue *et al.* (2012) reported evidence for such an impact event (based on a platinum group elements anomaly along with nickel-rich magnetite and microspherules) from the middle Norian (Upper Triassic) chert at Sakahogi (see Fig. 2b.25 for the locality). They were able to precisely determine the age by using the microfossils, and suggested the probable correlation of the event with the 215.5 Ma Manicouagan impact crater in Canada.

Chichibu Belt (AM)

The Chichibu Belt is one of several elements in the Outer Zone of SW Japan characterized by a zonal arrangement of basement rocks outcropping broadly parallel to the Japanese islands. The geological units of the belt are composed mainly of Late Palaeozoic–middle Mesozoic ACs and are in fault contact with the Sanbagawa metamorphic rocks to the north and with Cretaceous ACs of the Shimanto Belt to the south. The belt is distributed over a distance of 1500 km from the Kanto Mountains in the NE through Shikoku and Kyushu to the Ryukyu islands in the SW. The Chichibu Belt is divided into three sub-belts based on characteristic features of their components and geological structures. From north to south these are the Northern Chichibu, Kurosegawa (Middle Chichibu) and Southern Chichibu sub-belts, the three being juxtaposed with each other in most areas from the Kanto Mountains to Kyushu.

This section begins with a brief overview of research history of the Chichibu Belt followed by descriptions of the stratigraphy and age. The OPS reconstructed from components of the Chichibu Belt is summarized in Figure 2b.28. Finally, formative processes are discussed on the basis of comparisons of OPS between the Northern and Southern Chichibu sub-belts.

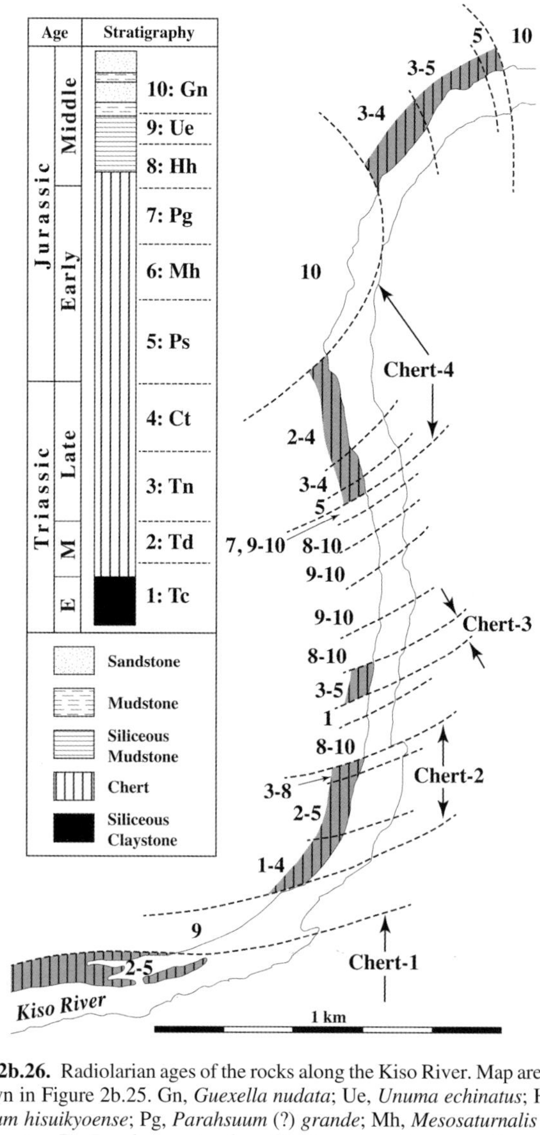

Fig. 2b.26. Radiolarian ages of the rocks along the Kiso River. Map area is shown in Figure 2b.25. Gn, *Guexella nudata*; Ue, *Unuma echinatus*; Hh, *Hsuum hisuikyoense*; Pg, *Parahsuum* (?) *grande*; Mh, *Mesosaturnalis hexagonus*; Ps, *Parahsuum simplum*; Ct, *Canoptum triassicum*; Tn, *Triassocampe nova*; Td, *Triassocampe deweveri*; Tc, *Triassocampe coronata* (after Matsuoka *et al.* 1994).

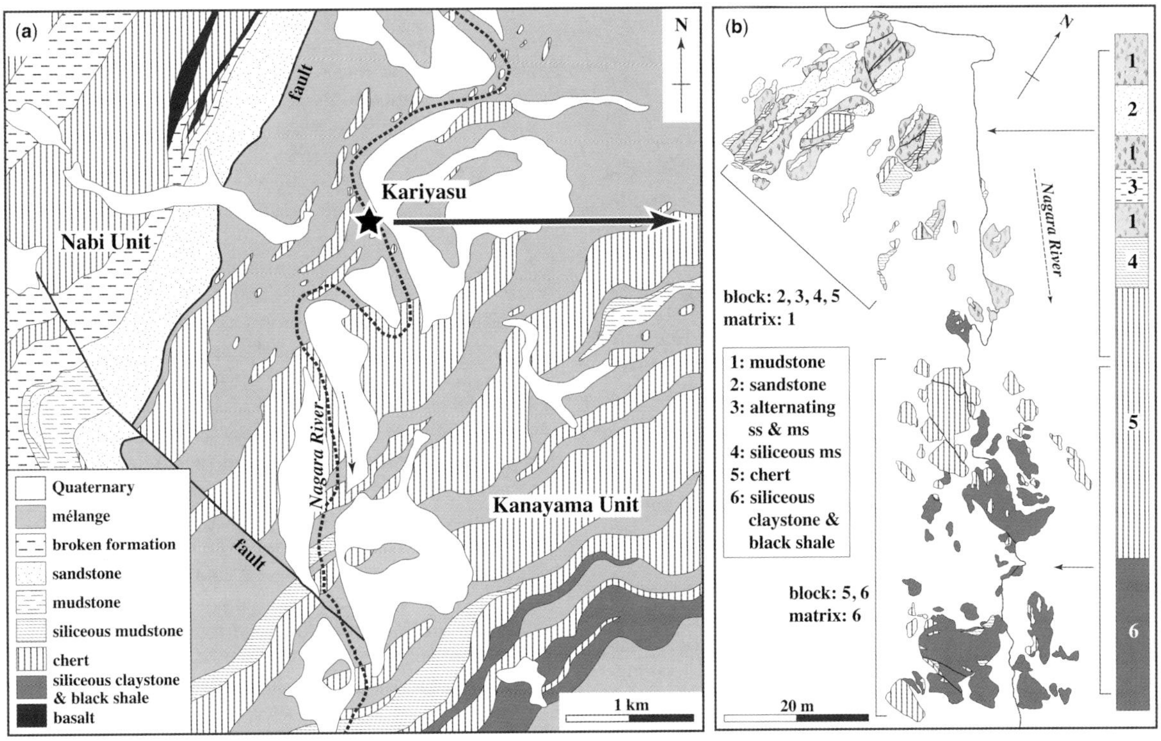

Fig. 2b.27. (**a**) Geological map of the Kariyasu area (modified from Wakita 1984, 1995). Map area is shown in Figure 2b.23. (**b**) Geological map along the right bank of the Nagara River (modified from Wakita 2000). Map area is indicated as black star in (a).

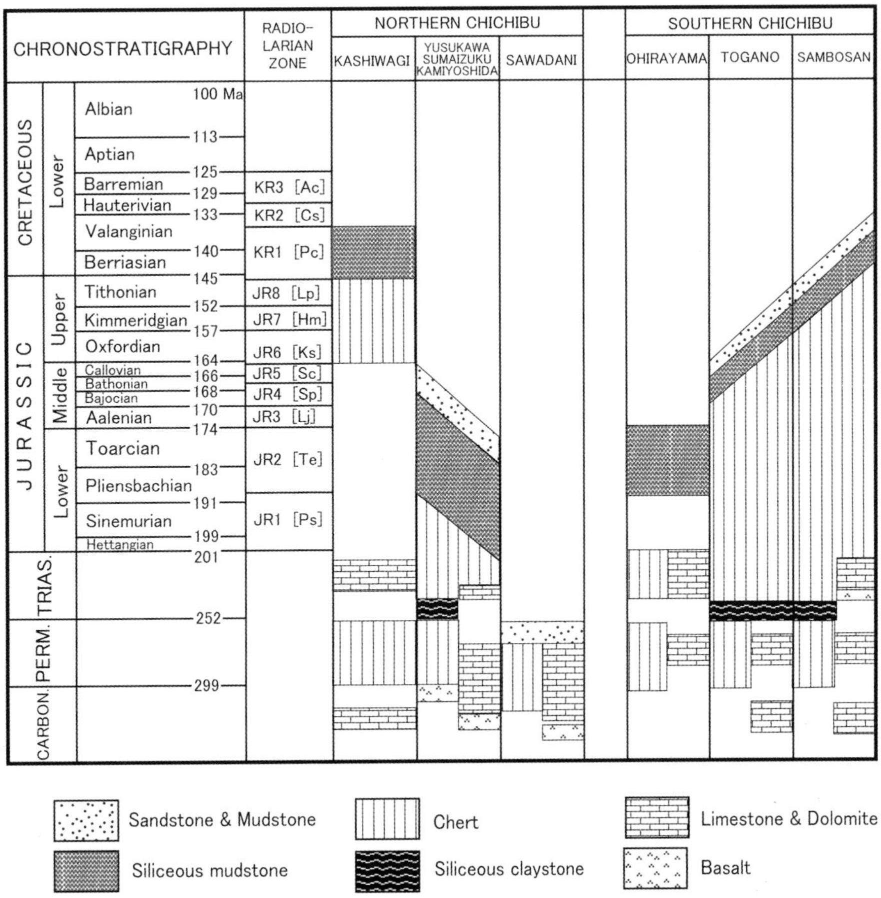

Fig. 2b.28. Stratigraphy and age of units in the Chichibu Belt. Adopted from Matsuoka *et al.* (1998). Radiolarian zonation for the Jurassic and Lower Cretaceous is updated from Matsuoka (1995).

Historical review

Geological investigations of the Chichibu Belt were initiated at the end of the 19th century, and included early reconnaissance work by Naumann (1885). A comprehensive study was completed by Kobayashi (1941) who wrote an article entitled 'Sakawa Orogenic Cycle and its bearing on the origin of the Japanese Islands'. The Sakawa area is located in the central part of Shikoku and has become regarded as a key area for clarifying the stratigraphy and age of the geological units in the Chichibu Belt (e.g. Yehara 1927; Kobayashi 1941). In the 1950s extensive field research was carried out mainly in Shikoku, detailed geological maps were produced and the threefold subdivision was introduced (Northern, Middle and Southern sub-belts). The Middle Chichibu Sub-belt, which is characterized by serpentinites, granitic rocks of *c.* 400 Ma, metamorphic rocks, and marine fossil-bearing beds ranging from middle Palaeozoic to Cretaceous in age, has come to be called the Kurosegawa Sub-belt (Ichikawa *et al.* 1956). Since the 1970s Triassic conodonts and Triassic–Early Cretaceous radiolarians have been reported from the Chichibu Belt and, with the introduction of plate tectonic theory, the sedimentary rocks of the Chichibu Belt have been interpreted as representing slices of ACs. Previous publications are summarized in Hada & Kurimoto (1990) for the Northern Chichibu Sub-belt and Matsuoka & Yao (1990) for the Southern Chichibu Sub-belt. Matsuoka *et al.* (1998) proposed a division of the ACs into units applicable for the entire Chichibu Belt and well as the Kanto Mountains and Shikoku, where sufficient geological and micropalaeontological data were available.

Stratigraphy and age

The stratigraphy and age of each of the Chichibu units defined by Matsuoka *et al.* (1998) are illustrated on Figure 2b.28. These ACs essentially consist of pelagic sequences and terrigenous sediments, with the former being further divided into deep-water chert successions and a shallow-water limestone-basalt association. The terrigenous sediments consist chiefly of sandstone-dominated alternating sandstone/mudstone beds of trench-fill origin. Characteristic facies of the ACs include coherent chert-clastic sequences and mélange, with the former preserving a stratigraphic succession from pelagic chert through hemi-pelagic siliceous mudstone to terrigenous coarse clastic rocks and typically represented by the Togano Group (Matsuoka 1984, 1992) in the Sakawa area, central Shikoku (Fig. 2b.29). On the other hand, the mélanges comprise mixtures of pelagic rocks (including chert) and limestone-basalt lithologies in a mudstone matrix, as represented by the Sambosan Group in central Shikoku. Rock successions in Figure 2b.28 are based either on the real stratigraphy of chert-clastic sequence or reconstructed sequences based on the relationship between rock types and their microfossil (radiolarian, conodont, foraminifer) ages.

Northern Chichibu Sub-belt

The rocks of the Northern Chichibu Sub-belt are composed mainly of Mesozoic ACs subdivided into the Early–Middle Jurassic Kamiyoshida, Sumaizuku and Yusukawa units and earliest Cretaceous Kashiwagi unit. In addition there are subordinate amounts of Late Palaeozoic AC rocks, known as the Sawadani unit. The spatial distributions of these ACs are well documented in the Kanto Mountains and Shikoku, and the boundary between the Northern Chichibu Sub-belt and the Kurosegawa Sub-belt has been discussed by Yamakita (1998). A younger age limit for the Chichibu rocks is provided by an unconformable cover of Lower Cretaceous shallow-marine and brackish sediments, collectively known as the Ryoseki–Monobegawa groups.

The Kashiwagi unit rests directly on the Mikabu metabasalts, which are sandwiched between the Chichibu ACs and the Sanbagawa metamorphic rocks. The type area of the Kashiwagi unit is located in the Kanto Mountains, and consists mainly of weakly metamorphosed oceanic sequences such as chert, limestone and metabasalt. It is not well dated by microfossils, although the youngest bed is siliceous mudstone and known to be of earliest Cretaceous in age.

The Kamiyoshida, Sumaizuku and Yusukawa units are Lower–Middle Jurassic ACs. The Kamiyoshida and Sumaizuku units have their type areas in the Kanto Mountains, and mainly comprise disrupted chert-clastic sequences associated with basalt. In contrast, the Yusukawa unit, which crops out only in the southern part of the Northern Chichibu Sub-belt, has a type section in western Shikoku and is characterized by terrigenous clastics-dominated mélange facies. A reconstructed stratigraphy for these three units is shown together in Figure 2b.28 as age-diagnostic microfossil data are too sporadic to allow more specific detail. However, it is known that clastic rocks of the Kamiyoshida unit are younger (and tectonostratigraphically lower) than those of the Sumaizuku unit. Finally, the oldest rocks of the sub-belt, represented by the Sawadani unit, crop out mainly in Shikoku and consist of Carboniferous–Permian limestone-basalt associations and Late Permian clastic rocks. This unit occupies the uppermost part of the ACs exposed in the Northern Chichibu Sub-belt.

The geological units in the Northern Chichibu Sub-belt exhibit a classic thrust-stacked geometry. The structurally lowest but geochronologically youngest Kashiwagi unit (lowest Cretaceous) is overlain by the Lower–Middle Jurassic Kamiyoshida–Sumaizuku–Yusukawa units which themselves lie beneath the oldest (Upper Permian) and highest Sawadani unit. This tectonic stacking was formed by successive accretion from Late Permian to earliest Cretaceous times. The notable apparent absence of Triassic and Upper Jurassic AC rocks in the Northern Chichibu Sub-belt may be related to the effects of tectonic erosion or strike-slip movement along the plate boundary.

Southern Chichibu Sub-belt

The rocks of the Southern Chichibu Sub-belt mainly comprise Jurassic and earliest Cretaceous ACs, and have been subdivided into the Ohirayama, Togano and Sambosan units from north to south. No Palaeozoic ACs have been reported so far. The accreted units are associated with Jurassic–Lower Cretaceous shallow-marine deposits distributed from the Kanto Mountains to Kyushu and collectively known as the Torinosu Group. These shallow marine deposits (which characteristically include 'Torinosu-type' reef limestone) are now intercalated with the ACs, but probably unconformably rested upon them originally.

All of the type localities of the Ohirayama, Togano and Sambosan units are in central Shikoku (Fig. 2b.29), but equivalent geological entities are traceable from the Kanto Mountains to the Ryukyu islands. The Ohirayama unit has been studied in detail only in its type locality, the Sakawa area, central Shikoku, where it consists mainly of mélange facies containing Permian limestone and Triassic chert (Triassic limestone blocks are also reported from western Shikoku).

The Togano unit is composed predominantly of Triassic–Jurassic chert-clastic sequences associated with subordinate mélange. The chert-clastic sequences show an upwards-coarsening succession which reflects landwards drift of the seafloor from a pelagic realm towards a trench (Fig. 2b.1). The unit is characterized by a south-vergent imbricate structure and south-younging polarity of the chert-clastic sequences.

The Sambosan unit consists mainly of Upper Jurassic–Lower Cretaceous mélange sequences which include abundant

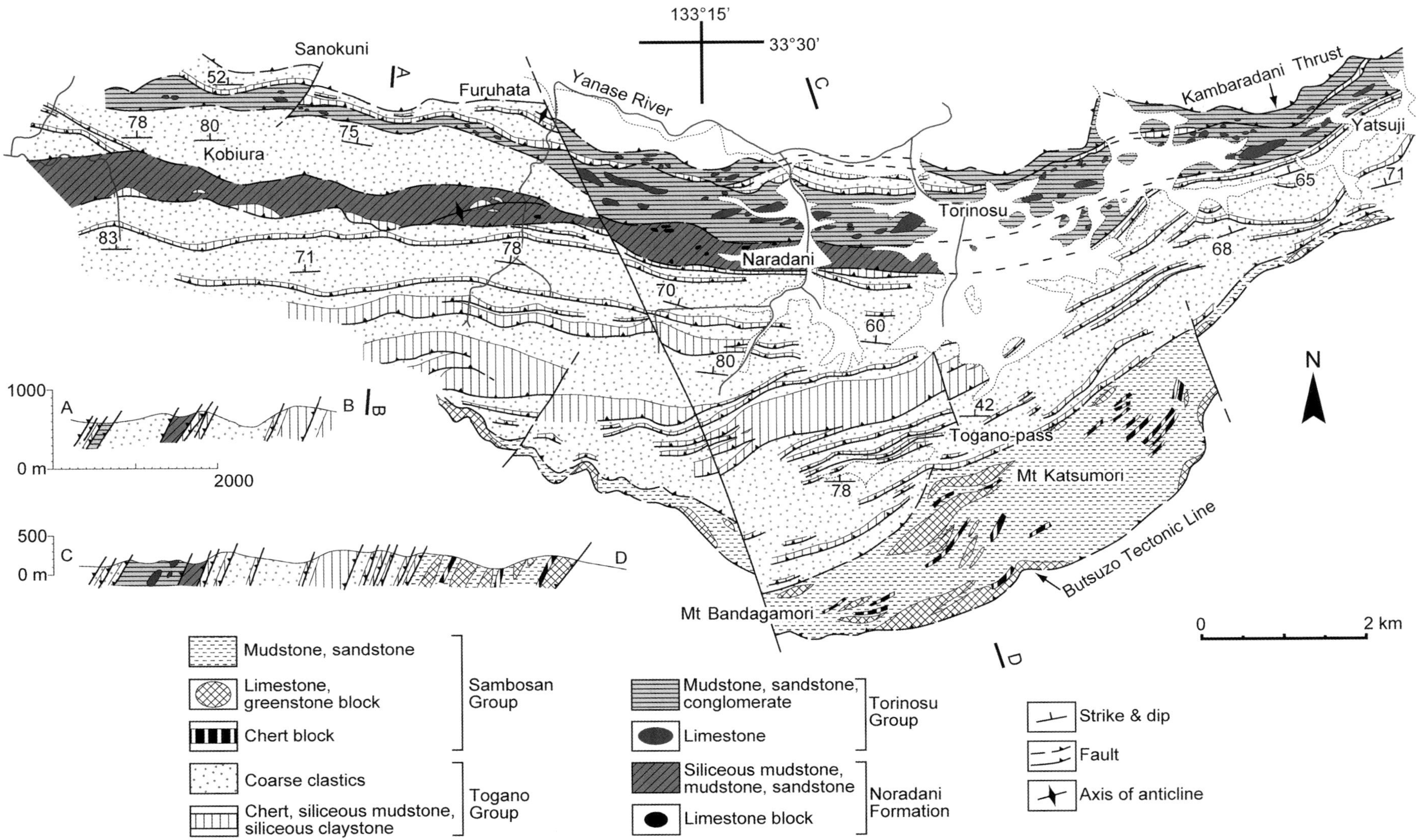

Fig. 2b.29. Geological map of the Southern Chichibu Belt in the Sakawa area, central Shikoku. The Togano and Sambosan groups are representative of the Togano and the Sambosan units, respectively. The Togano Group is characterized by alternating occurrence of chert and coarse clastic rocks. This is due to tectonic repetition of chert-clastic sequences. The Sambosan Group is distributed south of the Togano Group and is characterized by chert and limestone-greenstone blocks in a matrix composed mainly of mudstone. The Naradani Formation and the Torinosu Group are shallow-marine sequences which originally covered the Togano Group with an unconformity. Adopted from Matsuoka (1992).

Triassic–Jurassic chert blocks and Triassic limestone blocks accompanied by basalts of seamount origin. Late Triassic megalodont (bivalve) -bearing limestones have been found from many localities in the Sambosan unit, indicating that the limestones were formed in the Tethyan Province (Tamura 1987). Terrigenous rocks in the Sambosan unit are always younger than those in the Togano unit juxtaposed to the north.

Finally, two major stages of tectonic development have been detected in the Southern Chichibu Sub-belt (Fig. 2b.30). The Togano unit was formed by successive offscrape-accretion mainly during Middle–Late Jurassic times, whereas the Sambosan unit was constructed by Late Jurassic–Early Cretaceous collision-accretion of seamounts. This difference in tectonic setting is considered to be related to the topography of the subducting oceanic plate: an oceanic plate with an abyssal plain for the Togano unit and an oceanic plate with seamounts for the Sambosan unit.

Formative processes of the Chichibu Belt

Diverse scenarios have been proposed for the origins of the Chichibu belt, with Yao (2000) categorizing these into five major models used to explain the tectonic evolution of SW Japan. When limited to the Chichibu Belt, such models can be divided into two types – nappe model and translation model – with the differences between these depending largely on how the Kurosegawa Sub-belt is interpreted. The nappe model requires that the Kurosegawa rocks rest on the Jurassic ACs as a nappe, with the Northern and Southern Chichibu sub-belts being directly connected to each other beneath the nappe (Isozaki & Itaya 1991; Hara *et al.* 1992; Isozaki 1996; Yao 2000). The translation model assumes that the Kurosegawa rocks are sandwiched between the ACs of the Northern and Southern Chichibu sub-belts (Taira & Tashiro 1987). These models can be tested by comparing the ACs in the Northern and Southern sub-belts (Fig. 2b.28). In this context, it is clear that there is an overall similarity between the two sub-belts, both being dominated by Jurassic–earliest Cretaceous ACs with Lower Triassic siliceous claystone, Middle Triassic–Jurassic chert and younger terrigenous sequences. Furthermore, limestone blocks within mélange facies have similar Carboniferous–Triassic ages. On the other hand, considerable differences are recognizable in the stratigraphy of the two sub-belts, the most critical being the duration of hemi-pelagic siliceous mudstone sedimentation. The deposition of siliceous mudstone in the Northern Chichibu Sub-belt lasted two to three times longer than that in the Southern Chichibu Sub-belt. In addition, whereas Upper Jurassic ACs are common in the Southern Chichibu Sub-belt, they are missing from the Northern Chichibu Sub-belt. To satisfy both these similarities and differences at the same time, we view the translation of the model as being more suitable as an explanation for the structural evolution of the Chichibu Belt (Fig. 2b.31). Such a model allows for the ACs of both sub-belts to have been formed along a single convergent plate margin but apart from each other, with lateral variations in oceanic plate stratigraphy. Following their accretion, rocks from the two areas were subsequently juxtaposed with each other during northwards translation of the Southern Chichibu Sub-belt. Such movements, probably linked to left-lateral faulting during Early Cretaceous time along the eastern margin of Asia, have resulted in the Kurosegawa Sub-belt representing the suture zone between the Northern Chichibu and Southern Chichibu sub-belts.

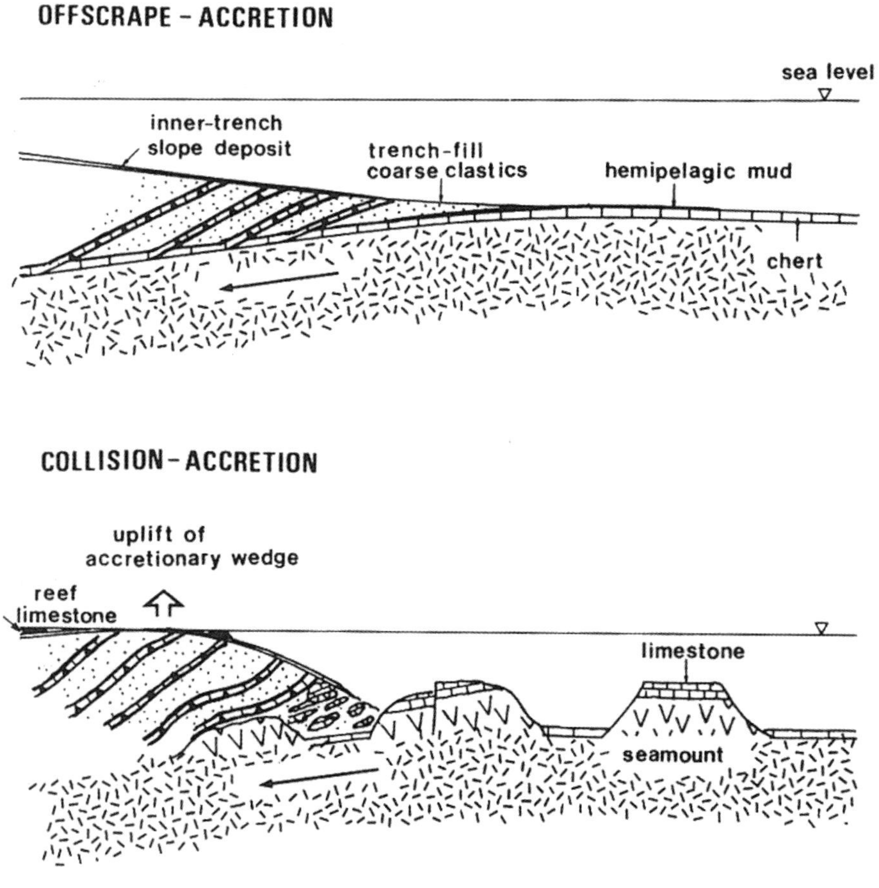

Fig. 2b.30. Schematic diagram showing two major tectonic stages of the Southern Chichibu Belt. Adopted from Matsuoka (1992).

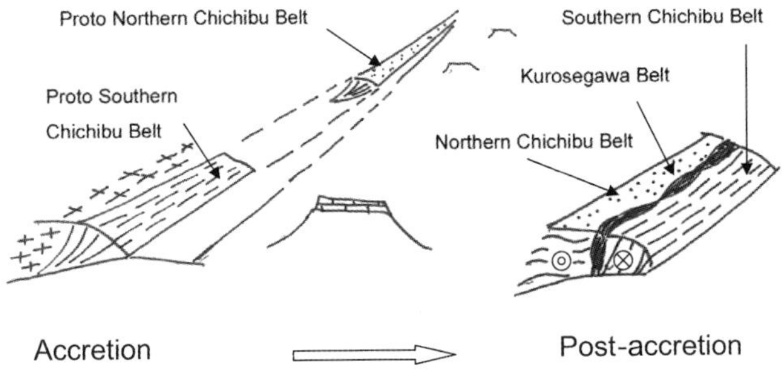

Fig. 2b.31. Formative processes of the Chichibu Belt, characterized by an accretion event along a single subduction zone and a post-accretion translation event.

Nedamo Belt (TU)

The Nedamo Belt (previously referred to as part of the 'Hayachine Tectonic Belt') represents an Early Carboniferous AC named the Nedamo Complex, and crops out across an area 40 km long and 10 km wide. It is a fault-bounded belt sandwiched between the South Kitakami Belt (mainly Palaeozoic shallow-marine deposits; see Chapter 2a) and the North Kitakami Belt (Jurassic AC; see 'North Kitakami Belt' section below) in the Kitakami Massif, Tohoku district, NE Japan (Figs 2b.2 & 2b.32) The boundary faults are generally high-angle and associated with serpentinite, and the tectonic relationships between the Nedamo Belt and the North/South Kitakami belts remain unclear. Generally, both bedding and cleavage strike NW–SE and are steeply SW-dipping, with closed folds (wavelengths of several metres to several hundred metres) being commonplace.

Lithostratigraphic and fossil age constraints

The Nedamo Complex comprises mafic rock, chert, alternations of mudstone and felsic tuff, sandstone and conglomerate, most of them affected by low-grade regional metamorphism and shearing deformation. On the map scale of 1:50 000, the Nedamo Complex indicates large-scale mélange facies disruption with a matrix of alternating mudstone and felsic tuff containing more competent blocks such as basalt and chert. On the outcrop scale, however, exposures display smaller-scale disruption, producing the so-called broken or dismembered facies of Raymond (1984).

Mafic rocks are common in the belt, and consist primarily of basalt (volcaniclastic rock and lava, locally pillowed) with minor amounts of dolerite. Chemical signatures of the mafic rocks show affinities with oceanic island basalt (alkali basalt and within-plate tholeiite) and MORB (Hamano *et al.* 2002; Uchino & Kawamura 2009). The mafic rocks have been affected by metamorphism at prehnite-pumpellyite facies (Moriya 1972), pumpellyite-actinolite facies (Onuki *et al.* 1988), greenschist facies (Uchino & Kawamura 2010*a*) and blueschist facies (epidote-blueschist subfacies; Uchino & Kawamura 2010*a*).

Chert provides a minor component of the belt, occurring as both red hydrothermal massive types with basalt as well as grey varieties. The former often contains iron-manganese layers several centimetres in thickness and microfossils of radiolarians, conodonts and sponge spicules.

Alternations of mudstone and felsic tuff (Fig. 2b.33a, b) are common in the belt, and characterize the Nedamo Complex because they

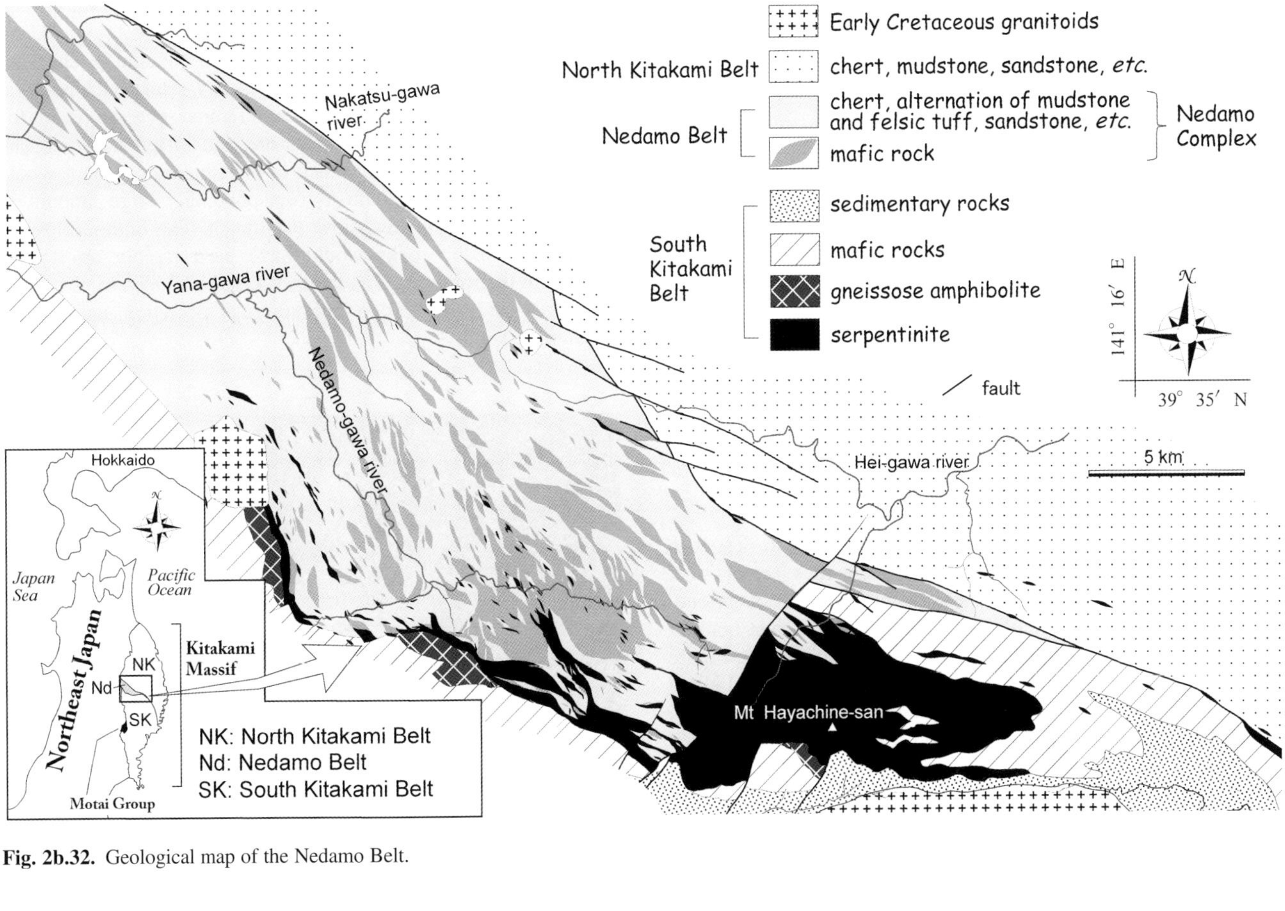

Fig. 2b.32. Geological map of the Nedamo Belt.

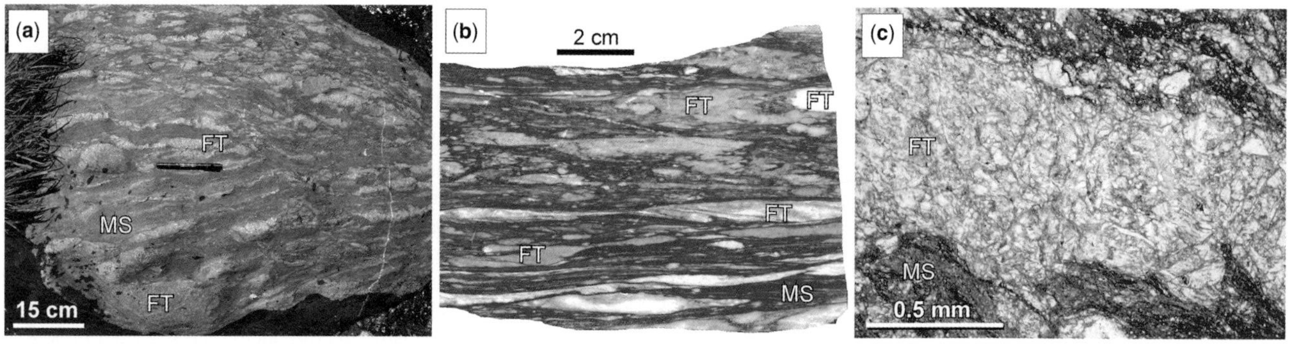

Fig. 2b.33. Outcrop (**a**) and specimen (**b**) of alternation of mudstone and felsic tuff, and photomicrograph (**c**) of the felsic tuff which shows the vitroclastic texture in the alternation (after Kawamura *et al.* 2013). FT, felsic tuff; MS, mudstone.

are few in ACs of other belts in the Japanese islands. Although the pale green felsic tuff looks like chert on outcrop, the vitroclastic texture is recognizable under the microscope (Fig. 2b.33c). The mudstone and felsic tuff beds range in thickness from several millimetres to centimetres, but some felsic tuffs occur as a layer of several metres in thickness within the couplet.

Sandstone occurs sporadically in the belt and is classified as lithic wacke, with the ratio of rock fragments occasionally reaching >75%. Conglomerate, rare in this belt, contains a wide variety of clasts that include dacite-rhyodacite, felsic tuff, basalt, sandstone, chert, limestone and schist.

Limestone clasts in the conglomerate have yielded Chaetetids of uncertain age (Kawamura *et al.* 2013). However, the hydrothermal massive chert associated with MORB includes Late Devonian conodonts (*Palmatolepis glabra prima* and *Palmatolepis* cf. *minuta minuta*) (Hamano *et al.* 2002) and Middle–Late Devonian radiolarians (*Trilonche* sp.) (Kawamura *et al.* 2013), and mudstones have yielded Early Devonian–Early Carboniferous radiolarians (*Palaeoscenidium cladophorum*). The age of trench-fill sediments (i.e. accretion age) is therefore considered to be Early Carboniferous (Uchino *et al.* 2005).

Despite the intense shearing deformation that disturbs the original stratigraphy of the Nedamo Complex, the OPS can be reconstructed (in ascending order) as MORB, hydrothermal massive chert, thin-bedded chert, alternations of mudstone and felsic tuff, and sandstone. Such an oceanic stratigraphy is in marked contrast to that of other Jurassic ACs, which are characterized by limestone, thick-bedded chert and rare felsic tuff. The duration of the OPS of the Nedamo Complex appears to have occupied a relatively short range lasting from Late Devonian (massive chert with MORB) to Early Carboniferous (mudstone) times.

Tectonostratigraphic constraints, regional correlations and geotectonic model

In some places within the Nedamo Belt, there are small (less than several metres in size) exotic masses of ultramafic rock (serpentinite, hornblendite and pyroxenite), hornblende gabbro, granitoids (tonalite and quartz diorite), gneissose amphibolite and schist (garnet-bearing pelitic schist and glaucophane-bearing mafic schist; Figure 2b.34). They are considered to be tectonic blocks displaced along faults during post-accretionary tectonic movements (e.g. Kawamura *et al.* 2007). Phengite in the pelitic and mafic schists has yielded *c.* 380 Ma $^{40}Ar/^{39}Ar$ radiometric ages (Kawamura *et al.* 2007), which is older than the accretion age of the Nedamo Belt. The amphibolite and igneous rocks are considered to be fragments of Ordovician island-arc basement rocks of the South Kitakami Belt (Kawamura *et al.* 2013), and the schist is considered to be part of the high-P/T schist fragmentary distributed in the Abukuma Massif.

No geological body in the Japanese islands can be fully correlated with the Early Carboniferous AC of the Nedamo Belt. However, the Motai Group (Onuki *et al.* 1962) (Fig. 2b.32) located in the western part of the Kitakami Massif could be lined up as a possible candidate, judging from its geographical location before Early Cretaceous sinistral strike-slip faulting (e.g. Hizume-Kesennuma fault; Ehiro 1977) and its lithologic similarity to the Nedamo Complex (Kawamura & Kitakami Paleozoic Research Group 1988; Uchino & Kawamura 2010*a*). Protoliths of the Renge schists (Carboniferous high-P/T metamorphic complex; see 'Renge high-P/T rocks' section) in SW Japan could be another correlative unit, given the metamorphic ages of the Renge schists.

Uchino & Kawamura (2010*b*) provide a geotectonic model (Fig. 2b.35) depicting the Early Carboniferous forearc of the eastern palaeo-Asian arc–trench system, based on sedimentological and petrological studies from the conglomerate containing clasts of ultramafic rock and on 347–317 Ma high-P/T schist (Uchino *et al.* 2008). Given that the Lower Carboniferous successions in the South Kitakami and Nedamo belts are rich in igneous rock and felsic tuff, respectively, volcanism in the island-arc system would have been very active, with a correspondingly large amount of volcanic ash brought down into the trench. The high-P/T schist and the ultramafic rocks would have been rapidly uplifted to crop out in the forearc, given the constraints provided by the schist clast ages and the Early Carboniferous accretion age of the Nedamo Belt.

Fig. 2b.34. 380 Ma glaucophane-bearing mafic schist (tectonic block). Ab, albite; Ep, epidote; Phn, phengite; Qtz, quartz.

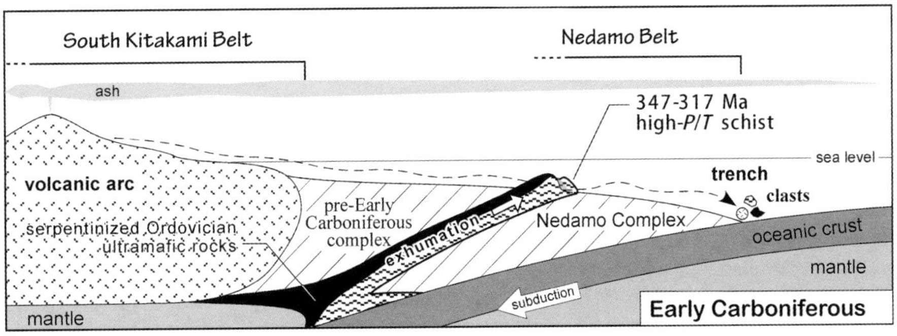

Fig. 2b.35. Geotectonic model of the eastern palaeo-Asian arc-trench system during the Early Carboniferous. The high-P/T schist with ultramafic rock uplifted and cropped out in the forearc within 30 m.y. (after Uchino & Kawamura 2010*b*).

Abukuma Belt (YHi)

The Abukuma Plateau is located in the southernmost part of the pre-Palaeogene outcrop in NE Japan, and is underlain mostly by Early–middle Cretaceous plutonic rocks. Metamorphic rocks occur rather sporadically in this region, and have been divided into four groups based on spatial distribution, metamorphic ages and lithological characteristics (Fig. 2b.36).

The first of these groups is limited to sparse exposures along the northeastern margin of the Abukuma Plateau and collectively named the Matsugadaira–Motai metamorphic rocks. They are characterized by middle Palaeozoic high-P/T metamorphism and belong to the South Kitakami Belt. The second group occurs as isolated roof remnants on plutonic rocks in the northern Abukuma Plateau and consists of thermally metamorphosed limestones, pelites, cherts and basic-ultrabasic rocks (Ehiro *et al.* 1989), similar in lithology to those constituting oceanic islands. Neither fossil nor radiometric age data have been obtained from these rocks. The third group is found in the south-central part of the plateau (the Gosaisyo–Takanuki district) and has been referred to as the Gosaisyo–Takanuki metamorphic rocks. More specifically, however, they are divided into the Gosaisyo Complex of oceanic-crust-type which is thrust over

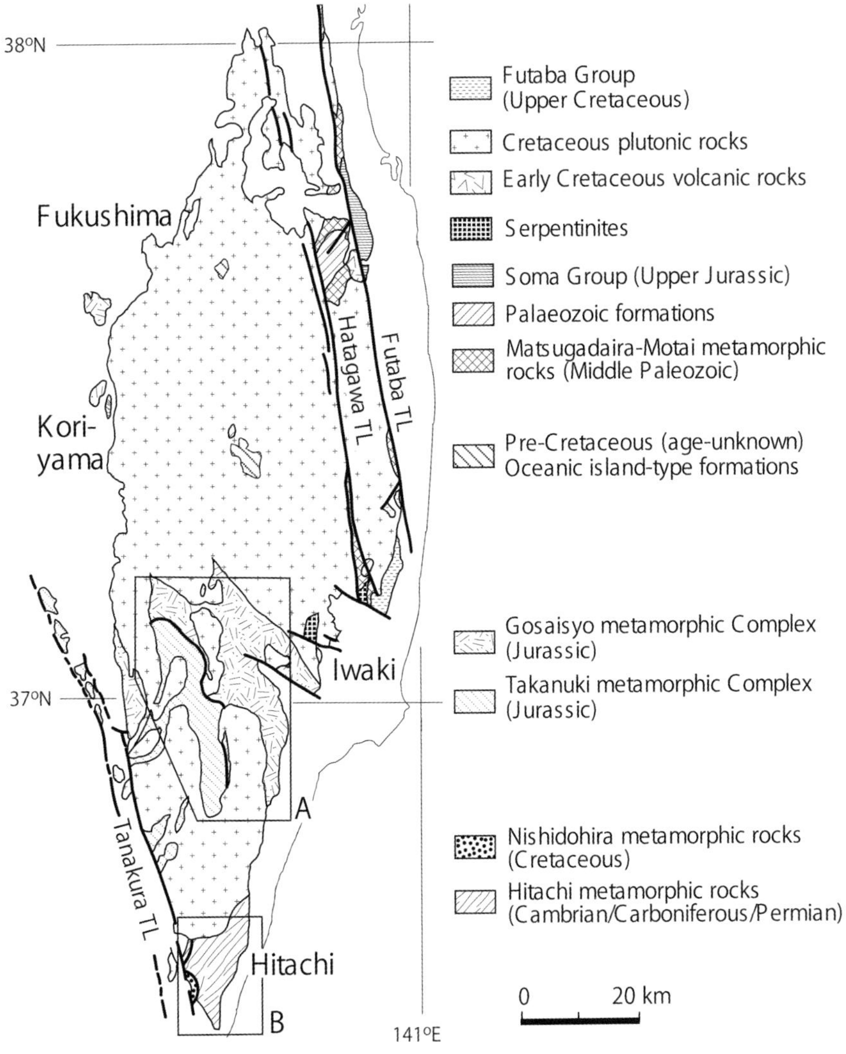

Fig. 2b.36. Geological map of Abukuma Plateau, modified after Tagiri *et al.* (2011). A, Gosaisho–Takanuki district; B, Hitachi district; TL, tectonic line. Note that in the text the Gosaisho and Takanuki metamorphic rocks are referred to as 'complexes'.

the Takanuki Complex of terrigenous origin. Finally, the fourth metamorphic group crops out in the Hitachi district of the southernmost part of the Abukuma Plateau, comprising the Hitachi and Nishidohira metamorphic rocks. The protoliths of the Hitachi metamorphic rocks are Early and Late Palaeozoic sedimentary, volcanic and related intrusive rocks, including Late Cambrian sediments which are the oldest rocks in Japan. The Nishidohira metamorphic rocks originate from Cretaceous sedimentary rocks, and are in fault contact with the Hitachi lithologies.

The exact relationships between these four metamorphic groups are not yet established. The metamorphic rocks in the Gosaisyo–Takanuki and Hitachi districts will be described below as those constituting the Abukuma Belt, mainly after Hiroi *et al.* (1987, 1998) and Tagiri *et al.* (2010, 2011). The northwestern extension of the belt is obscured by an extensive cover formed by Cenozoic volcanic activity.

Gosaisyo–Takanuki district

The metamorphic rocks of the Gosaisyo and Takanuki complexes are extensively intruded and thermally metamorphosed by Cretaceous plutonic complexes. Several of these plutons were emplaced along or close to the boundary between the two complexes, along with sheets and lenses of ultrabasic rocks enclosed within metamorphic rocks (Fig. 2b.37).

The Gosaisyo Complex is composed mainly of basic and siliceous rocks with subordinate pelitic and calcareous rocks. Chemical compositions of homogeneous and massive basic rocks (probably derived from lava) are similar to those of T-type MORB. Some siliceous rocks preserve Early Jurassic radiolarian fossils and are intercalated with iron ore (magnetite and hematite) layers similar to banded iron formations (Hiroi *et al.* 1987). Zircon grains from a pelitic-siliceous rock yield a broad age clustering at *c.* 450 Ma (Fig. 2b.38b). However, the SHRIMP zircon ages range to *c.* 520 Ma and form a continuum on the Tera–Wasserburg concordia plot. These analyses do not reflect the age of sedimentation, but rather the provenance. It is noteworthy that a fairly uniform provenance age is recorded, in marked contrast with the case of the Takanuki pelitic rocks (Fig. 2b.38e, f). It is significant that a small number of andesitic to rhyolitic dykes belonging to the calc-alkalic rock series occur and have experienced deformation and metamorphic events that are nearly identical to those

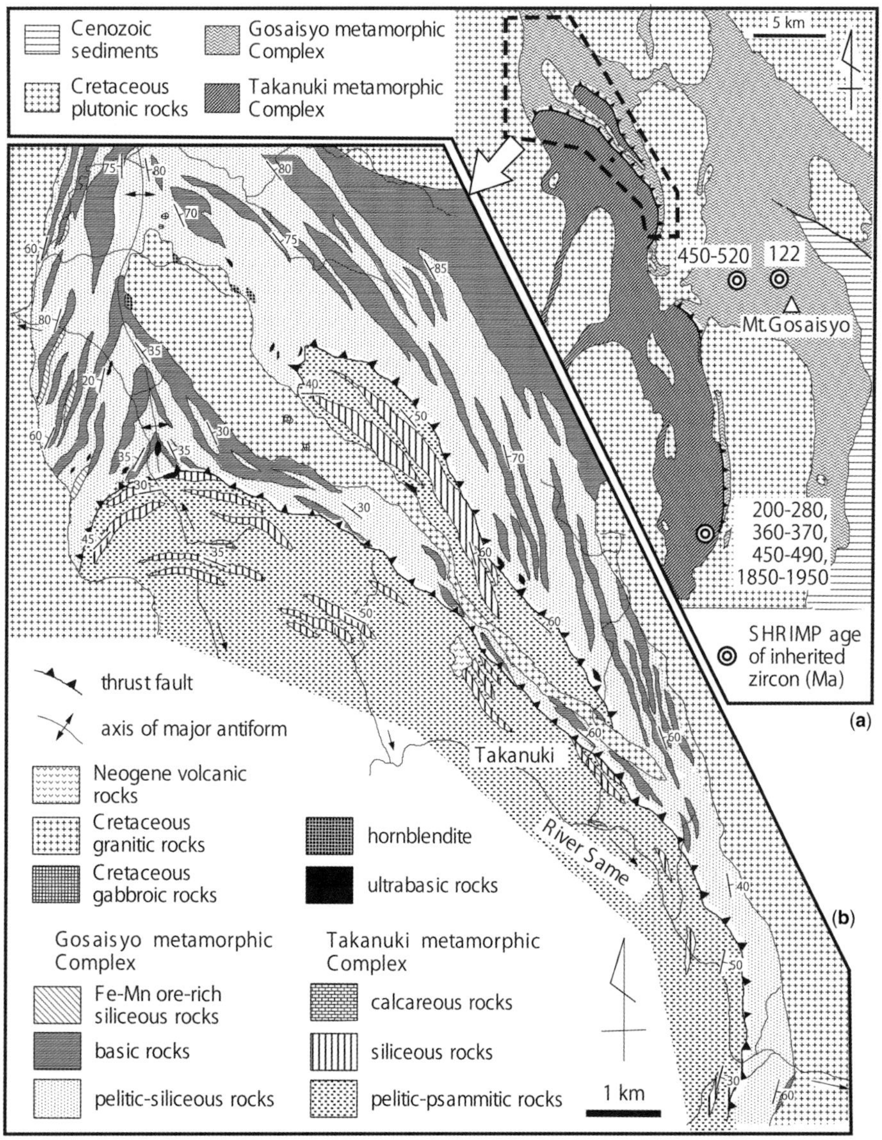

Fig. 2b.37. Geological map of Gosaisyo–Takanuki district, modified after Goto (1991) and figure 2 of Ishikawa *et al.* (1996). Note that in the text the Gosaisho and Takanuki metamorphic rocks are referred to as 'complexes'.

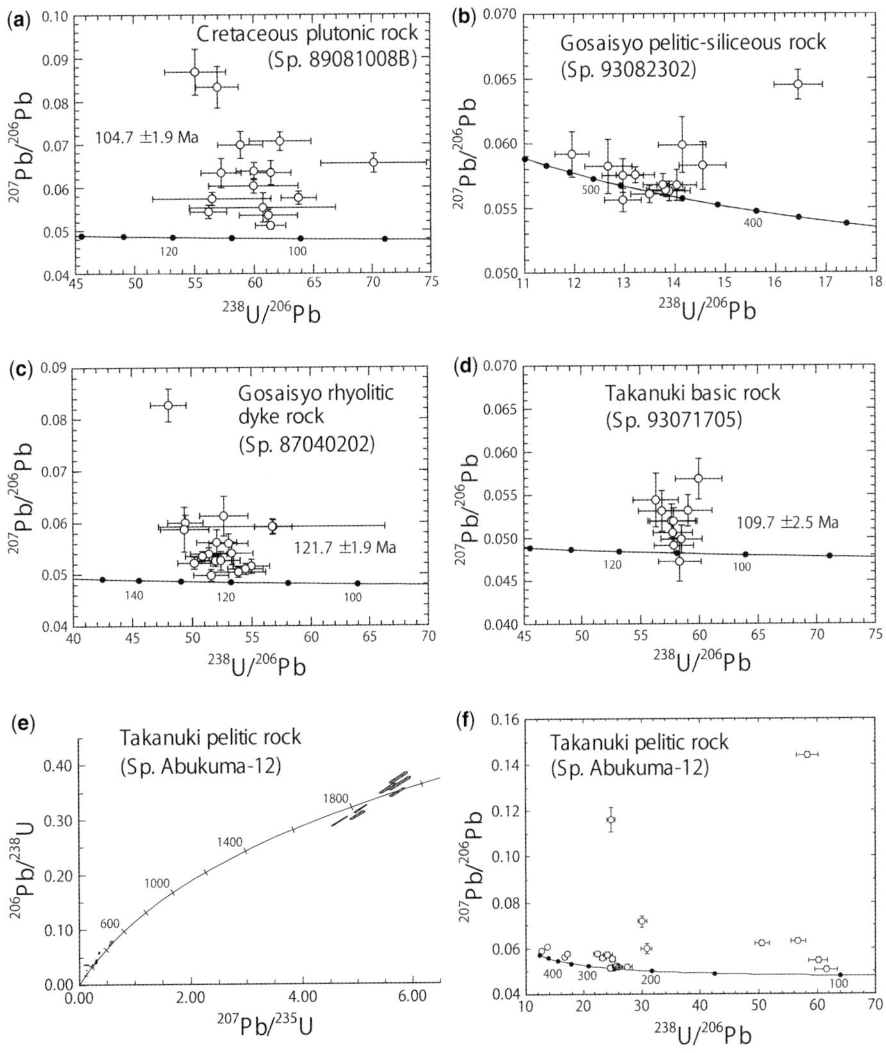

Fig. 2b.38. Tera–Wasserburg and U–Pb concordia plots for zircons from plutonic and metamorphic rocks in the Gosaisyo–Takanuki district.

of the host rocks. Zircons in one of these rhyolitic dykes yielded an age population of igneous zircon grains. A weighted mean of U–Pb ages for most analyses gives an age of 121.7 ± 1.9 Ma (Fig. 2b.38c), which is interpreted to reflect the time of igneous crystallization. Some analyses suggest a loss of radiogenic Pb during the 110 Ma metamorphic overprint. Three to four deformation events have been distinguished in the Gosaisyo Complex on the basis of four different types of fold: F1, isoclinal flow folds; F2, predominant buckle folds with a subvertical axial plane and subhorizontal fold axis nearly parallel to mineral lineation; F3, drag folds associated with axial plane cleavage; and F4, kink folds (Ishikawa & Otsuki 1990). These authors also showed that the regional thermal structure is discordant with the geological structure, and suggested a close link between the left-lateral ductile shearing (F3) within the Gosaisyo Complex and the activities of the Tanakura, Hatakawa and Futaba tectonic lines (shear zones) shown in Figure 2b.36.

The Takanuki Complex consists dominantly of pelitic-psammitic rocks with small amounts of calcareous, lateritic, siliceous and basic rocks. Pelitic-psammitic rocks are commonly migmatitic, indicating that partial melting has taken place during high-grade metamorphism. Although the Takanuki siliceous rocks are far more coarse-grained than those in the Gosaisyo Complex, they may have originated from cherts. Silica-poor lateritic rocks are found as lenses completely enclosed by coarse-grained marbles. A weighted mean of U–Pb ages of zircons from a minor basic rock gives an age of 109.7 ± 2.5 Ma (Fig. 2b.38d), and zircon morphology and age determinations indicate that this rock went through a single high-grade metamorphic event at *c.* 110 Ma. Cores of zoned zircon grains in a pelitic rock give concordant to nearly concordant ages of 200–280 Ma while rims yield a concordant age of 110 Ma (Fig. 2b.38e, f). Scattered Proterozoic (1850–1950 Ma) and Ordovician–Devonian ages (360–370 and 450–490 Ma) were also obtained mainly in the inherited cores of zircons which may be of detrital origin. The depositional age of the Takanuki sedimentary protoliths may therefore have been between 110 and 200 Ma, with new zircon growth and overgrowth taking place at about 110 Ma during the regional and almost contemporaneous contact metamorphic events. The Takanuki Complex as a whole shows gentle dome structures around cores of Cretaceous plutonic masses (Fig. 2b.37).

Although the dominant mineralogy indicates a classic andalusite-sillimanite type progressive metamorphism as defined by Miyashiro (1961), there are several lines of evidence suggesting that the Takanuki and Gosaisyo complexes underwent rapid high-temperature (>700°C) loading (up to more than 10 kbar in the kyanite stability field), followed by subsequent unloading during a single Cretaceous metamorphic event (Hiroi & Kishi 1989; Hiroi *et al.* 1998). Such an event may have resulted from ridge–trench interactions (Brown 1998), such as the obduction of oceanic crust. Similar high-temperature loading has been inferred from several Cretaceous regional metamorphic terrains around the Pacific Ocean (e.g. the Fiordland granulites in New Zealand; Bradshaw 1989).

Hitachi district

The Hitachi metamorphic rocks predominantly dip moderately–steeply SE, metamorphic grade increases westwards, and they are intruded and locally thermally metamorphosed by the Cretaceous Irishiken granodiorite mass in the north. SHRIMP U–Pb zircon ages have revealed these rocks to have originated from Early and Late Palaeozoic formations with a ‘great hiatus’ between Late Cambrian and Early Carboniferous times (Sakashima *et al.* 2003; Tagiri *et al.* 2010, 2011; Fig. 2b.39). The Early Palaeozoic rocks are divided into the Akazawa and Tamadare units, which are bounded by serpentinite shear zones and are almost contemporary in age. The Akazawa Unit consists mainly of volcanic rocks with subordinate amounts of clastic sediments. The unit is subdivided into upper and lower subunits based on the stratigraphic position and metamorphic grade, with the upper subunit being intruded by metagranitoids which have been grouped into northern and southern bodies by Tagiri *et al.* (2011). Sakashima *et al.* (2003) first reported the age of 491 Ma for zircons from a metagranitoid of the northern body. Granitic porphyry also occurs sporadically as dykes in the Akazawa Unit and the metagranitoids. The Tamadare Unit is composed mainly of biotite-hornblende gneiss originating from diorite, along with minor amounts of mica schist. Tagiri *et al.* (2011) reported the Late Cambrian ages for zircons from various rocks as follows: andesitic lava of the Lower Akazawa Subunit, diorite of the Tamadare Unit and shallow-level quartz porphyry dyke intruding into the Upper Akazawa Subunit.

The Upper Palaeozoic deposits are subdivided into the Daioin, Ayukawa and Omika units. The Daioin Unit has a basal conglomerate containing granitic boulders within a fine-grained quartzo-feldspathic to micaceous matrix, and is overlain by the Ayukawa Unit that consists mainly of slate and sandstone. Fossiliferous

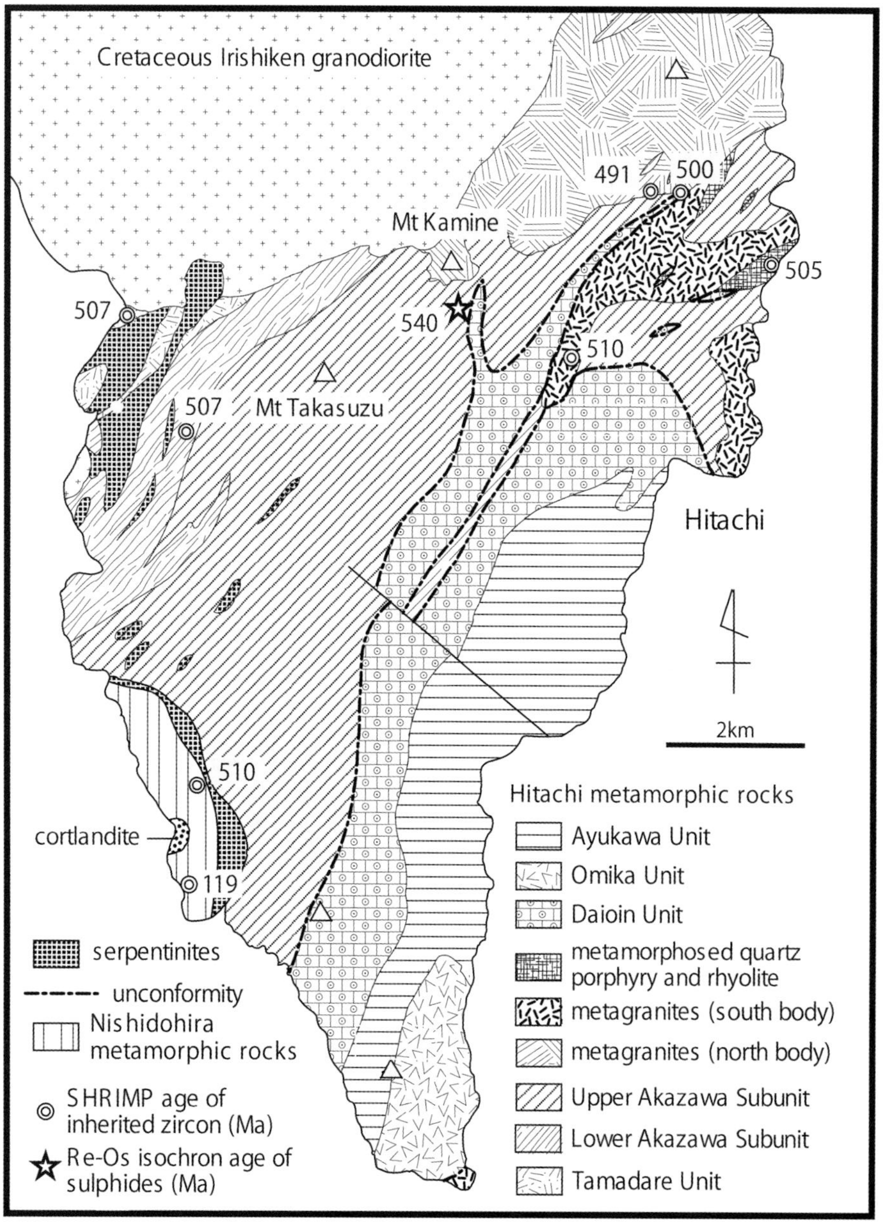

Fig. 2b.39. Geological map of Hitachi district, modified after Tagiri *et al.* (2011). Re-Os isochron age of sulfides is after Nozaki *et al.* (2012).

limestones occur in both the Daioin and Ayukawa units, yielding Early Carboniferous and Early Permian ages, respectively. The Omika Unit crops out further to the SE and is composed mainly of greenstones with a small amount of fossiliferous limestone. This unit also contains blocks of metagranitoids similar to the possibly correlative Daioin Unit, although its depositional age remains unknown.

Finally, the previously mentioned Cretaceous Nishidohira metamorphic rocks in the Hitachi district are composed mainly of arkosic sandstone and mudstone with subordinate volcaniclastic sediments and granitic conglomerate, metamorphosed into micaceous and felsic schists, fine-grained gneisses and amphibolites. They dip gently (*c.* 30°) to ENE, and the total thickness is estimated to be *c.* 300 m. Sedimentary structures preserved in coarser-grained mica schist beds and alternating layers of micaceous and felsic schists demonstrate that some beds are overturned. Metamorphic grade increases up-sequence from lower to upper amphibolite facies, and cortlandite and gneissose granite are intruded into the upper levels, inducing thermal effects on surrounding rocks. The Nishidohira metamorphic rocks were once considered to be a member of the Early Palaeozoic units because of the occurrence of *c.* 510 Ma detrital zircons (Tagiri *et al.* 2011), but they have been revealed by additional dating to be Cretaceous in age (Kanamitsu *et al.* 2011; M. Tagiri pers. comm. 2012) (Fig. 2b.39).

North Kitakami Belt (NS)

The North Kitakami Belt occupies a wide area in the northern part of the Kitakami Massif, Iwate Prefecture, and occurs as small remnants in Aomori Prefecture. The rocks in this belt consist of: Upper Carboniferous–Upper Jurassic chert; Middle–Upper Jurassic hemipelagic siliceous mudstone and terrigenous mudstone; Middle Jurassic–lowest Cretaceous sandstone; Upper Carboniferous–Lower Permian or Middle–Upper Triassic sequences of basaltic rocks, limestone and ribbon chert; and small lenses of Lower–Middle Permian limestone. Upper Jurassic limestone with corals is also patchily found, and can be correlated with the Torinosu-type limestone of the Chichibu Belt. These rocks are, in part, thermally metamorphosed by Early Cretaceous plutonic activities (Suzuki *et al.* 2007*a*).

The North Kitakami Belt is tectonically a southern extension of the Oshima Belt in SW Hokkaido as well as a northern extension of the Chichibu Belt (Ehiro *et al.* 2008). The lithologic and tectonic units of the North Kitakami belt show NW–SE-trending distributions. The western and southwestern boundary with the Nedamo Belt (Carboniferous AC) and the South Kitakami Belt (Ordovician–Jurassic isolated continental fragment) is defined by the SE–NW-trending Hayachine Eastern Marginal Fault (Fig. 2b.40), whereas the eastern boundary of this belt is submerged beneath the Pacific Ocean (Ehiro & Suzuki 2003). The North Kitakami Belt is divided by the Seki–Odaira Fault (or Iwaizumi Tectonic Line) into two sub-belts: Kuzumaki–Kamaishi Sub-belt in the SW and the Akka–Tanohata Sub-belt in the NE (Fig. 2b.40).

The Kuzumaki–Kamaishi Sub-belt is different from the Akka–Tanohata Sub-belt in the presence of Palaeozoic sequences and plagioclase-rich sandstone in the former as opposed to K-feldspathic sandstones in the latter (Okami *et al.* 1992), although the change is gradual (Takahashi *et al.* 2006). Another distinctive feature of the Kuzumaki–Kamaishi Sub-belt is the presence along its western

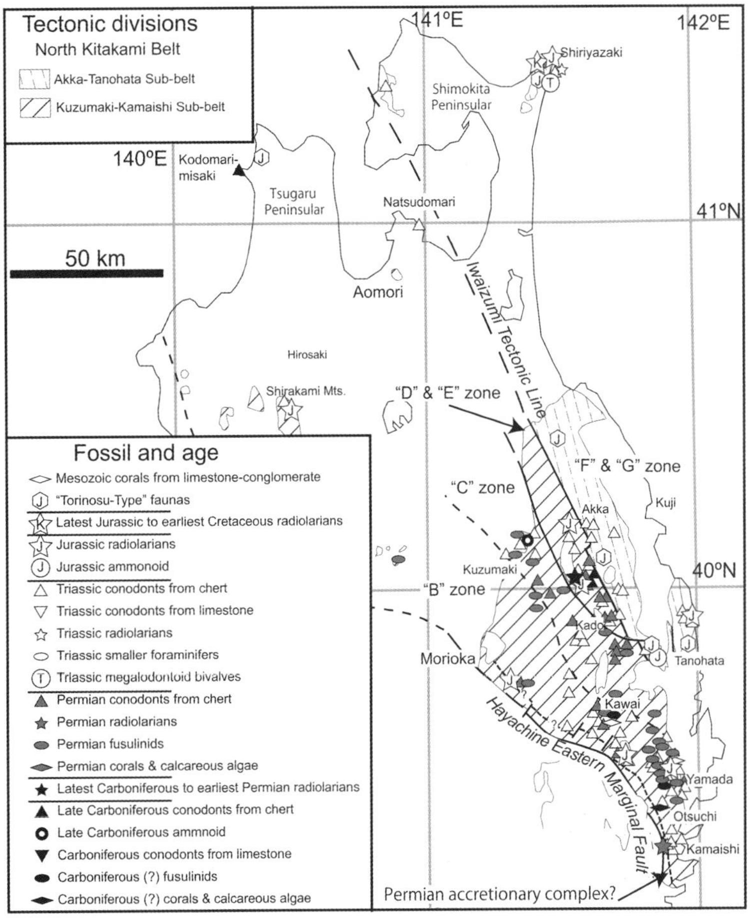

Fig. 2b.40. Schematic tectonic division for the North Kitakami Belt in Tohoku, NE Japan, with all fossil localities. Map based on Suzuki *et al.* (2007*a*); tectonic divisions in the North Kitakami Belt are mainly referred to Otoh & Sasaki (2003), Kanisawa *et al.* (2006) and Nakae & Kurihara (2011). Note that Carboniferous and Permian fossils are limited in the Kuzumaki–Kamaishi Sub-belt. This figure was the first to distinguish the Permian accretionary complex in the sense of Nakae & Kurihara (2011) from the North Kitakami Belt. The precise distribution of the Permian accretionary complex is still unknown.

margin of a unit known as the Kirinai Formation. This unit consists of black phyllitic mudstone (yielding Late Permian radiolarians) and greenish-grey massive sandstone in association with a minor amount of basic tuff. These rocks have been interpreted to be a Permian AC correlative to the Ultra-Tamba Belt (Nakae & Kurihara 2011), despite the occurrences of Triassic conodonts in chert and Jurassic-type radiolarians in chert and siliceous mudstone known from the western part of the Kirinai outcrop (Ehiro & Suzuki 2003).

The typical lithologies and tectonic structures of the North Kitakami Belt are best documented in the Kawai area for the Kuzumaki–Kamaishi Sub-belt (Suzuki *et al.* 2007*a*), in the Akka area for the Akka–Tanohata Sub-belt (Sugimoto 1974), and in the southern part of the Kuzumaki–Kamaishi Sub-belt for the Permian AC (Nakae & Kurihara 2011) (Fig. 2b.40). The geological structure of the Kuzumaki–Kamaishi Sub-belt is different between outcrop and map scales because bedding planes in exposed outcrops dip subvertically–vertically, but commonly dip shallowly to the SSW on the map. This difference is caused by the development of high-angle intrafolial folding with a shallow-dipping enveloping surface. In contrast, the Akka–Tanohata Sub-belt on the map scale is characterized by NNW–SSE-trending anticlines and synclines with wavelengths of a few kilometres in the northern North Kitakami Massif (Sugimoto 1974).

The Kuzumaki–Kamaishi and Akka–Tanohata Sub-belts have been further subdivided into three 'zones' comprising ten units and a single 'zone' with five units, respectively, based on detailed field mapping (Fig. 2b.41) (e.g. Otoh & Sasaki 2003; Takahashi *et al.* 2006; Suzuki *et al.* 2007*a*; Ehiro *et al.* 2008; Kawamura 2010). Although the tectonostratigraphic correlation between the North Kitakami and Chichibu belts is still in dispute (e.g. Otoh & Sasaki 2003; Suzuki *et al.* 2007*a*), the 'C' zone, 'D' and 'E' zones, and 'F' and 'G' zones of the North Kitakami Belt are correlated with the Ohirayama, Togano and Sambosan units of the Southern Chichibu Belt, respectively. No unit correlating with the 'B' zone of the North Kitakami Belt has been detected in the Chichibu Belt.

Differences and similarities between the Chichibu and the Mino–Tamba–Ashio belts

The North Kitakami Belt is considered to have been originally connected with the Chichibu Belt, but there are several differences between these two belts. The width of the former reaches up to

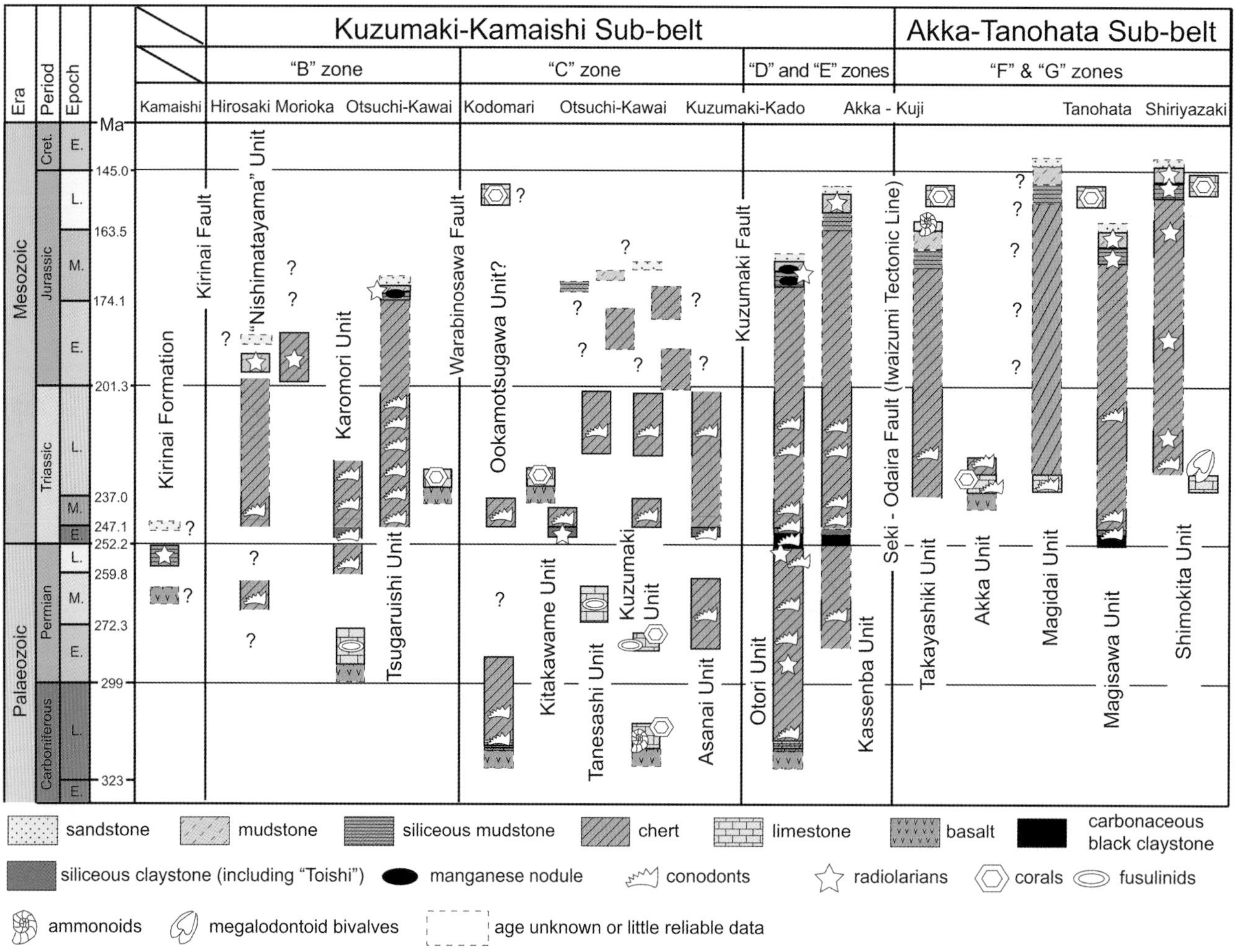

Fig. 2b.41. Schematic compilation of age and lithostratigraphic diagram of the North Kitakami Belt and the newly recognized Permian accretionary complex. As the data source exceeds more than 45 references, the original source should be referred to Suzuki *et al.* (2007*a*) and Ehiro *et al.* (2008). The tectonostratigraphic divisions, formation names and assigned numerical ages are partly updated from both Suzuki *et al.* (2007*a*) and Ehiro *et al.* (2008). The age of the siliceous mudstone and mudstone tends to become younger from the 'Nishimatayama' Unit of 'B' Zone in the Kuzumaki–Kamaishi Sub-belt to the Shimokita Unit in the Akka–Tanohata Sub-belt.

150 km, whereas that of the latter is 10 km at maximum. The Upper Triassic limestones with megalodontoid bivalves typically found in the Chichibu and North Kitakami belts have not been reported from the Mino–Tamba–Ashio Belt (Sano *et al.* 2009), and they are much larger in the North Kitakami Belt exemplified by the Akka Limestone body than those in the Chichibu Belt. Furthermore, large Permian limestone bodies common in the Chichibu and Mino–Tamba–Ashio belts have never been found in the North Kitakami Belt. On the other hand, the Upper Carboniferous sequence from basalt through alternating beds of dolomite and red ribbon chert, to red ribbon chert in the Otori Unit of the 'D' and 'E' zones of the Akka–Tanohata Sub-belt, is equivalent to the red ribbon chert sequence of the Kamiyoshida Unit of the Northern Chichibu Belt.

The broad width of the North Kitakami Belt is similar to that of the Mino–Tamba–Ashio Belt. Radiolarian-bearing manganese carbonate nodules of Middle Jurassic age are stratigraphically embedded in siliceous mudstone and mudstone of the Tsugaruishigawa Unit of the 'B' zone and the Otori Unit of 'D' and 'E' zones in the North Kitakami Belt. Similar Middle Jurassic manganese nodules in siliceous mudstone or mudstone are only found in the Mino–Tamba–Ashio Belt based on the compiled occurrence data of radiolarian-bearing carbonate nodules by Yamakita & Hori (2009) and Suzuki *et al.* (2007*b*). The distribution and tectonic relationship between the Ultra-Tamba Belt (Permian AC) and Mino–Tamba–Ashio Belt (Jurassic AC) is similar to those of the Kirinai Formation (Permian AC) and the North Kitakami Belt (Jurassic AC).

Tectonic events recorded in the North Kitakami Belt

The juxtaposition of the North and South Kitakami belts was completed after the deposition of the Jurassic Torinosu-type limestone, the youngest sediments of the North Kitakami Belt, but before the Oshima Orogeny which has affected both belts. The deformation and thermal metamorphism induced by the Oshima Orogeny occurred between Barremian and probably Aptian times, and no later than the deposition of the undeformed Miyako Group of Aptian–Albian age (Kobayashi 1941; Onuki 1981; Kanisawa & Ehiro 1989). More extensive information, including the main references for the North Kitakami Belt, are summarized in Suzuki *et al.* (2007*a*) and Nakae & Kurihara (2011), including a historical overview of changing ideas on how these rocks have been interpreted.

Relationship between the older AC and mainland Asia (SK)

The older ACs described in this chapter had been developed in relation to mainland Asia as the Sea of Japan opened around 15 Ma. Kojima (1989) and Kojima *et al.* (2000, 2008) reconstructed and illustrated the northern extension of the Maizuru, Ultra-Tamba, Mino–Tamba–Ashio, Southern Chichibu and North Kitakami belts into NE Asia before the opening of the Sea of Japan (Fig. 2b.42). Different views of the reconstruction are provided by Niitsuma *et al.* (1985) and Matsuda *et al.* (1998). The complicated arrangement of the older ACs, high- and low-P/T metamorphic complexes and fragments of continent and island arc (Figs 2b.2 & 2b.42) formed through the processes of subduction-accretion, nappe formation, ophiolite emplacement and large-scale strike-slip faulting before 15 Ma is described in Chapter 1.

The kinematics of each geological entity have been analysed by several different methods including palaeomagnetic and palaeobiogeographic methods, and provenance and structural analyses. One of the results of the palaeomagnetic analyses, namely the low-latitude and Southern Hemisphere origin of the Middle Triassic chert in the Mino–Tamba–Ashio Belt, is explained in this chapter.

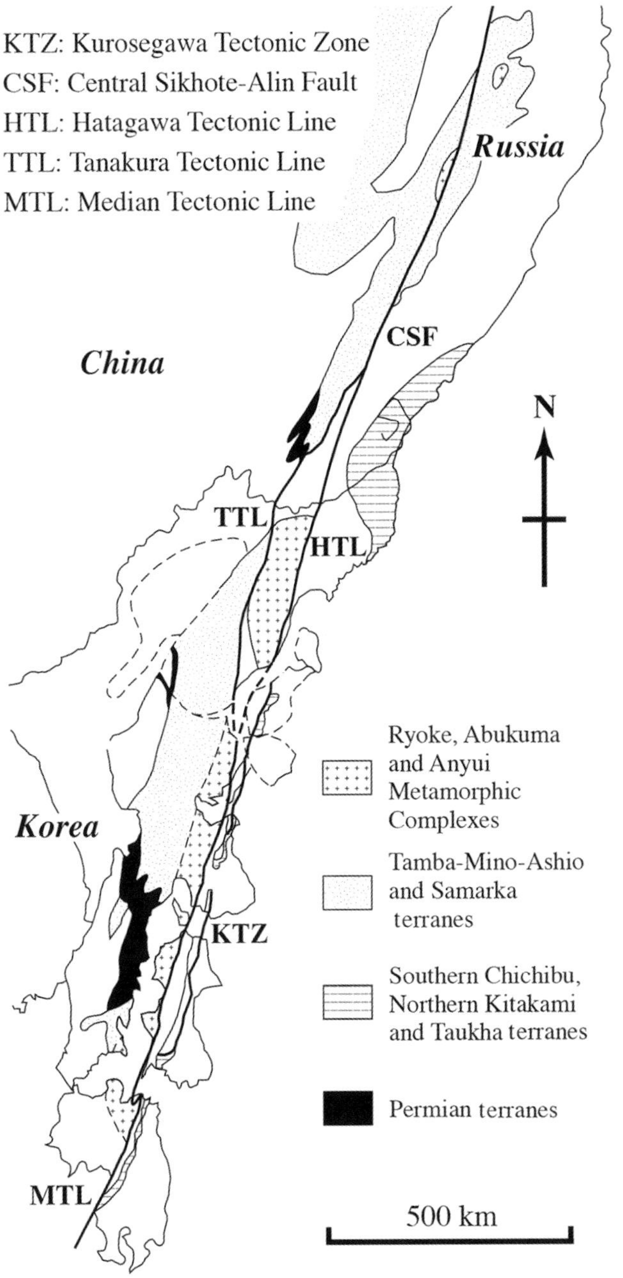

Fig. 2b.42. Map showing northern extension of the Japanese older AC belts into NE Asia before the opening of the Sea of Japan (after Kojima *et al.* 2008). The Permian terranes in this figure include Maizuru and Ultra-Tamba belts of this book.

Palaeobiogeographic analyses using endemic faunas and floras are also useful to estimate the palaeo-position of sedimentary basins; for example, Tazawa (1993) reconstructed the original arrangement of the Palaeozoic–Mesozoic belts on the basis of Permian brachiopod fauna, and Ehiro (1997) indicated close relationships between Permian ammonoid faunas among the South Kitakami, South China and Khanka massifs. The relationships between ACs and the continental margins where they accreted have been analysed by using clastic materials in trench-fill turbidites, as demonstrated in this chapter (e.g. Mino–Tamba–Ashio Belt). Recent progress in detrital zircon chronology by using laser ablation inductively coupled plasma mass spectrometry has accumulated a vast amount of age data (e.g. Otoh *et al.* 2013) which should reveal the nature of the continental provenance.

Y. Hiroi is most grateful to M. Tagiri for the detailed and new information about the Hitachi and Nishidohira metamorphic rocks and M. Fanning for the zircon SHRIMP dating. H. S. gives special thanks to Gerilyn S. Soreghan (University of Oklahoma) for her review on the early version of the manuscript.

Appendix

English to Kanji and Hiragana translations for geological and place names

Abukuma	阿武隈	あぶくま
Aizawa	会沢	あいざわ
Akazawa	赤沢	あかざわ
Akiyoshi	秋吉	あきよし
Akiyoshi-dai	秋吉台	あきよしだい
Akka	安家	あっか
Aomori	青森	あおもり
Asahi	旭	あさひ
Ashio	足尾	あしお
Atetsu	阿哲	あてつ
Ayukawa	鮎川	あゆかわ
Butsuzo	仏像	ぶつぞう
Chihibu	秩父	ちちぶ
Chizu	智頭	ちず
Chugoku	中国	ちゅうごく
Daioin	大雄院	だいおういん
Dogoyama	道後山	どうごやま
Fukui	福井	ふくい
Fukumoto	福本	ふくもと
Funafuseyama	舟伏山	ふなふせやま
Futaba	双葉	ふたば
Gamata	蒲田	がまた
Gosaisho	御斎所	ごさいしょ
Gotsu	江津	ごうつ
Gujyo	公庄	ぐじょう
Happo-one	八方尾根	はっぽうおね
Hatakawa	畑川	はたかわ
(Hatagawa)		
Hatto	八東	はっとう
Hayachine	早池峰	はやちね
Hida	飛騨	ひだ
Hida-Gaien	飛騨外縁	ひだがいえん
Hiehara	稗原	ひえはら
Hikami	氷上	ひかみ
Hirao	平尾	ひらお
Hiroshima	広島	ひろしま
Hitachi	日立	ひたち
Hizume	日詰	ひづめ
Hokubo	北房	ほくぼう
Honshu	本州	ほんしゅう
Hyogo	兵庫	ひょうご
Inagawa	猪名川	いながわ
Iriomote	西表	いりおもて
Irishiken	入四間	おりしけん
Ishigaki	石垣	いしがき
Itoigawa	糸魚川	いといがわ
Iwaizumi	岩泉	いわいずみ
Iwakuni	岩国	いわくに
Joetsu	上越	じょうえつ
Kamaishi	釜石	かまいし
Kamiaso	上麻生	かみあそう

(*Continued*)

English to Kanji and Hiragana translations for geological and place names (Continued)

Kamitaki	上滝	かみたき
Kamiyoshida	上吉田	かみよしだ
Kanayama	金山	かなやま
Kanto	関東	かんとう
Kariyasu	刈安	かりやす
Kashiwagi	柏木	かしわぎ
Katashina	片品	かたしな
Katsuyama	勝山	かつやま
Kawai	川井	かわい
Kawanishi	川西	かわにし
Kesennuma	気仙沼	けせんぬま
Kirinai	桐内	きりない
Kiso	木曽	きそ
Kitakami	北上	きたかみ
Komori	河守	こうもり
Kozuki	上月	こうづき
Kuchikanbayashi	口上林	くちかんばやし
Kuga	玖珂	くが
Kunisaki	国崎	くにさき
Kurosegawa	黒瀬川	くろせがわ
Kuruma	来馬	くるま
Kuwagai	桑飼	くわがい
Kuzumaki	葛巻	くずまき
Kuzuryu	九頭竜	くずりゅう
Kyoto	京都	きょうと
Kyushu	九州	きゅうしゅう
Maizuru	舞鶴	まいづる
Matsugadaira	松ヶ平	まつがだいら
Miharaiyama	御祓山	みはらいやま
Mikabu	御荷鉾	みかぶ
Mikata	三方	みかた
Mino	美濃	みの
Miyako	宮古	みやこ
Monobegawa	物部川	ものべがわ
Motai	母体	もたい
Nabae	難波江	なばえ
Nabi	那比	なび
Nagara	長良	ながら
Naradani	楢谷	ならだに
Nedamo	根田茂	ねだも
Nishidohira	西堂平	にしどうひら
Nishiki	錦	にしき
Nishimatayama	西股山	にしまたやま
Nomo	野母	のも
Ochiai	落合	おちあい
Odaira	大平	おおだいら
Oe	大江	おおえ
Oeyama	大江山	おおえやま
Ohirayama	大平山	おおひらやま
Oi	大飯	おおい
Okayama	岡山	おかやま
Oki	隠岐	おき
Omi	青海	おうみ
Omika	大甕	おおみか
Osayama	大佐山	おおさやま
Oshima*	渡島	おしま
Oshima†	大島	おおしま
Otori	大鳥	おおとり
Otsuchi	大槌	おおつち
Oya	大屋	おおや
Renge	蓮華	れんげ
Ryoke	領家	りょうけ

(*Continued*)

English to Kanji and Hiragana translations for geological and place names (Continued)

Ryoseki	領石	りょうせき
Ryukyu	琉球	りゅうきゅう
Sakahogi	坂祝	さかほぎ
Sakamoto-toge	坂本峠	さかもととうげ
Sakawa	佐川	さかわ
Samondake	左門岳	さもんだけ
San'in	山陰	さんいん
Sanbagawa (Sambagawa)	三波川	さんばがわ
Sanbosan (Sambosan)	三宝山	さんぼうさん
Sangun	三郡	さんぐん
Sanyo	山陽	さんよう
Sasayama	篠山	ささやま
Sawadani	沢谷	さわだに
Seki	関	せき
Shidaka	志高	しだか
Shikoku	四国	しこく
Shimane	島根	しまね
Shimanto	四万十	しまんと
Shimokita	下北	しもきた
Shimomidani	下見谷	しもみだに
Shirouma	白馬	しろうま
Shizuoka	静岡	しずおか
Sumaizuku	住居附	すまいづく
Suo	周防	すおう
Taishaku	帝釈	たいしゃく
Takanuki	竹貫	たかぬき
Takatsuki	高槻	たかつき
Tamadare	玉簾	たまだれ
Tanakura	棚倉	たなくら
Tanba (Tamba)	丹波	たんば
Tanesashi	種差	たねさし
Tanohata	田野畑	たのはた
Tanzawa	丹沢	たんざわ
Tetori (Tedori)	手取	てとり
Togano	斗賀野	とがの
Tohoku	東北	とうほく
Tokura	十倉	とくら
Tomuru	トムル	とむる
Torinosu	鳥の巣	とりのす
Toyogadake	豊ヶ岳	とよがだけ
Tsugaruishigawa	津軽石川	つがるいしがわ
Tsunemori	常森	つねもり
Tsuno	都濃	つの
Unuma	鵜沼	うぬま
Wakasa	若桜	わかさ
Yakuno	夜久野	やくの
Yamada	山田	やまだ
Yamaguchi	山口	やまぐち
Yamasaki	山崎	やまさき
Yusukawa	遊子川	ゆすかわ

*For 'Oshima belt'.
†For 'Oshima Peninsula' and 'Oshima Orogeny'.

References

ADACHI, M. 1971. Permian intraformational conglomerate at Kamiaso, Gifu Prefecture, central Japan. *Journal of Geological Sociey of Japan*, **77**, 471–482.

ADACHI, M. 1990. Heavy mineral assemblage of the Triassic Shidaka Group in the Maizuru terrane. Abstract of the Annual Meeting of the Geological Society of Japan, Tokyo, Japan, 278 [in Japanese].

ADACHI, M. & KOJIMA, S. 1983. Geology of the Mt. Hikagedaira area, east of Takayama, Gifu Prefecture, central Japan. *Journal of Earth Sciences, Nagoya University*, **31**, 37–67.

ADACHI, M. & MIZUTANI, S. 1971. Sole markings and paleocurrent system in the Paleozoic group of the Mino terrain, central Japan. *Memoir of Geological Society of Japan*, **6**, 39–48.

ADACHI, M. & SUZUKI, K. 1992. A preliminary note on the age of detrital monazites and zircons from sandstones in the Upper Triassic Nabae Group, Maizuru terrane. *Memoirs of Geological Society of Japan*, **38**, 111–120 [in Japanese with English abstract].

ANDO, A., KODAMA, K. & KOJIMA, S. 2001. Low-latitude and Southern Hemisphere origin of Anisian (Triassic) bedded chert in the Inuyama area, Mino terrane, central Japan. *Journal of Geophysical Research*, **106**, 1973–1986.

ARAI, S., INOUE, T. & OYAMA, T. 1988. Igneous petrology of the Ochiai-Hokubou ultramafic complex, the Sangun zone, western Japan: a preliminary report. *Journal of Geological Society of Japan*, **94**, 91–102 [in Japanese with English abstract].

BANNO, S. 1958. Glaucophane schists and associated rocks in the Omi district, Niigata Prefecture, Japan. *Japanese Journal of Geology and Geography*, **29**, 29–44.

BRADSHAW, J. Y. 1989. Origin and metamorphic history of an Early Cretaceous polybaric granulite terrain, Fiordland, southwest New Zealand. *Contributions to Mineralogy and Petrology*, **103**, 346–360.

BROWN, M. 1998. Ridge-trench interactions and high-T-low-P metamorphism, with particular reference to the Cretaceous evolution of the Japanese Islands. *In*: TRELOAR, P. J. & O'BRIEN, P. L. (eds) *What Drives Metamorphism and Metamorphic Reactions?* Geological Society, London, Special Publications, **138**, 137–169.

CARIDROIT, M., ICHIKAWA, K. & CHARVET, J. 1985. The Ultra-Tamba Zone, a new unit in the Inner Zone of Southwest Japan: its importance in the nappe structure after the example of the Maizuru area. *Earth Science (Chikyu-kagaku)*, **39**, 210–219.

EHIRO, M. 1977. The Hizume-Kesennuma Fault –with special reference to its character and significance on the geologic development. *Contributions from the Institute of Geology and Paleontology, Tohoku University*, **77**, 1–37 [in Japanese with English abstract].

EHIRO, M. 1997. Ammonoid palaeobiogeography of the South Kitakami Palaeoland and palaeogeography of eastern Asia during Permian to Triassic time. *In*: JIN, Y. G. & DINELEY, D. (eds) *Proceedings of 30th International Geological Congress*. VSP, Utrecht, **12**, 18–28.

EHIRO, M. & SUZUKI, N. 2003. Re-definition of the Hayachine Tectonic Belt of Northeast Japan and a proposal of a new tectonic unit, the Nedamo Belt. *Japanese Journal of Structural Geology*, **47**, 13–21 [in Japanese with English abstract].

EHIRO, M., KANISAWA, S. & TAKETANI, Y. 1989. Pre-tertiary Takine Group in the central Abukuma massif. *Bulletin of Fukushima Museum*, **3**, 21–37 [in Japanese with English abstract].

EHIRO, M., YAMAKITA, S., TAKAHASHI, S. & SUZUKI, N. 2008. Jurassic accretionary complexes of the North Kitakami Belt in the Akka-Kuji area, Northeast Japan. *Journal of Geological Society of Japan*, **114** (supplement), 121–139 [in Japanese with English abstract].

FUJII, M., HAYASAKA, Y. & TERADA, K. 2008. SHRIMP zircon and EPMA monazite dating of granitic rocks from the Maizuru terrane, southwest Japan: correlation with East Asian Paleozoic terranes and geological implications. *Island Arc*, **17**, 322–341.

GOTO, J. 1991. *Geological and Petrological Study of the Boundary Area Between the Gosaisyo and Takanuki Metamorphic Rocks in the Abukuma Metamorphic Terrane*. MSc thesis, Chiba University, Japan.

HADA, S. & KURIMOTO, C. 1990. Northern Chichibu Terrane. *In*: ICHIKAWA, K., MIZUTANI, S., HARA, I., HADA, S. & YAO, A. (eds) *Pre-Cretaceous Terranes of Japan*. Department of Geosciences, Osaka City University, Osaka, 165–183.

HAMANO, K., IWATA, K., KAWAMURA, M. & KITAKAMI PALEOZOIC RESEARCH GROUP 2002. Late Devonian conodont age of red chert intercalated in greenstones of the Hayachine Belt, Northeast Japan. *Journal of Geological Society of Japan*, **108**, 114–122 [in Japanese with English abstract].

HARA, I., SHIOTA, T. *ET AL*. 1992. Tectonic evolution of the Sambagawa schists and its implications in convergent margin processes. *Journal of Science of Hiroshima University, Series C*, **9**, 495–595.

HASHIMOTO, M. 1968. Glaucophanitic metamorphism of the Katsuyama district, Okayama Prefecture, Japan. *Journal of the Faculty of Sciences, University of Tokyo, Section II*, **17**, 99–162.

Hashimoto, M. 1972. Mineral facies of the Sangun metamorphic rocks of the Chugoku Provinces. *Bulletin of Japan National Science Museum*, **15**, 767–775.

Hayasaka, Y. 1987. Study on the late Paleozoic-early Mesozoic tectonic development of western half of the Inner zone of Southwest Japan. *Geological Report of Hiroshima University*, **27**, 119–204 [in Japanese with English abstract].

Hayasaka, Y. 1990. Maizuru Terrane. *In*: Ichikawa, K., Mizutani, S., Hara, I., Hada, S. & Yao, A. (eds) *Pre-Cretaceous Terranes of Japan*. Department of Geosciences, Osaka City University, Osaka, 81–95.

Hayasaka, Y., Ikeda, K., Shishido, T. & Ishizuka, M. 1996. Geological reconstruction of the Maizuru Terrane as an arc-back arc system. *In*: Shimamoto, T., Hayasaka, Y., Shiota, T., Oda, M., Takeshita, T., Yokoyama, S. & Ohtomo, Y. (eds) *Tectonics and Metamorphism (The Hara Volume)*. SOUBUN, Tokyo, 134–144 [in Japanese with English abstract].

Herzig, C. T., Kimbrough, D. L. & Hayasaka, Y. 1997. Early Permian zircon uranium–lead ages for plagiogranites in the Yakuno ophiolite, Asago district, Southwest Japan. *Island Arc*, **6**, 396–403.

Hidaka, H., Shimizu, H. & Adachi, M. 2002. U-Pb geochronology and REE geochemistry of zircons from Palaeoproterozoic paragneiss clasts in the Mesozoic Kamiaso conglomerate, central Japan: evidence for an Archean provenance. *Chemical Geology*, **187**, 279–293.

Hiroi, Y. & Kishi, S. 1989. P-T evolution of Abukuma metamorphic rocks in north-east Japan: metamorphic evidence for oceanic crust obduction. *In*: Daly, J. S., Cliff, R. A. & Yardley, B. D. D. (eds) *Evolution of Metamorphic Belts*. Geological Society, London, Special Publications, **43**, 481–486.

Hiroi, Y., Yokose, M., Oba, T., Kishi, S., Nohara, T. & Yao, A. 1987. Discovery of Jurassic radiolaria from acmite-rhodonite-nearing metachert of the Gosaisyo metamorphic rocks in the Abukuma terrane, northeast Japan. *Journal of Geological Society of Japan*, **93**, 445–448.

Hiroi, Y., Kishi, S., Nohara, T., Sato, K. & Goto, J. 1998. Cretaceous high-temperature rapid loading and unloading in the Abukuma metamorphic terrane, Japan. *Journal of Metamorphic Geology*, **16**, 67–81.

Ichikawa, K. 1990. Pre-Cretaceous Terranes of Japan. *In*: Ichikawa, K., Mizutani, S., Hara, I., Hada, S. & Yao, A. (eds) *Pre-Cretaceous Terranes of Japan*. Department of Geosciences, Osaka City University, Osaka, 1–12.

Ichikawa, K., Ishii, K., Nakagawa, C., Suyari, K. & Yamashita, N. 1956. Die Kurosegawa Zone. *Journal of Geological Society of Japan*, **62**, 82–103.

Igi, S. 1969. Findings of the Carboniferous corals from the so-called Kamigori Structural Belt, Japan. *Bulletin of Geological Survey of Japan*, **20**, 77–78.

Igi, S. & Kuroda, K. 1965. *Explanatory Text of the Geological Map of Japan, Scale 1:50000, Oeyama*. Geological Survey of Japan, Kawasaki [in Japanese with English abstract].

Igi, S., Kuroda, K. & Hattori, H. 1961. *Explanatory Text of the Geological Map of Japan, Scale 1:50000, Maizuru*. Geological Survey of Japan, Kawasaki [in Japanese with English abstract].

Ikeda, K. & Hayasaka, Y. 1994. Rb–Sr ages of the Yakuno rocks from the Northen subzone of the Maizuru Terrane, Kyoto Prefecture, Southwest Japan. *Journal of Mineralogy, Petrology and Economic Geology*, **89**, 443–54.

Ishiga, H. 1984. Follicucullus (Permian Radiolaria) from the Maizuru Group in the Maizuru Belt, Southwest Japan. *Earth Science (Chikyu Kagaku)*, **38**, 427–434.

Ishiga, H. 1986. Ultra-Tamba Zone of Southwest Japan. *Journal of Geosciences, Osaka City University*, **29**, 89–100.

Ishiga, H. 1990. Ultra-Tamba Terrane. *In*: Ichikawa, K., Mizutani, S., Hara, I., Hada, S. & Yao, A. (eds) *Pre-Cretaceous Terranes of Japan*. Department of Geosciences, Osaka City University, Osaka, 97–107.

Ishiga, H., Takamatsu, M., Takigawa, T., Nishimura, K. & Tokuoka, T. 1988. Radiolarian biostratigraphy of Maizuru Group in the northwest of Ibara and Kanagawa areas, Okayama Prefecture. *Geological Report of Shimane University*, **7**, 39–48 [in Japanese with English abstract].

Ishii, K., Okimura, K. & Nakazawa, K. 1975. On the genus *Colaniella* and its biostratigraphic significance. *Journal of Geosciences, Osaka City University*, **19**, 107–138.

Ishikawa, M. & Otsuki, K. 1990. Fold structures and left-lateral ductile shear in the Gosaisho metamorphic belt, Northeast Japan. *Journal of Geological Society of Japan*, **96**, 719–730 [in Japanese with English abstract].

Ishikawa, M., Hiroi, Y. & Tagiri, M. 1996. Metamorphic rocks and geologic structures of the Takanuki-Gosaisho metamorphic belt. *In*: Mori, K. & Nagahama, H. (eds) *Excursion Guidbook, 103rd Annual Meeting of Geological Society of Japan*. Tokyo, Japan, 155–176.

Ishiwatari, A. 1985. Granulite-facies metacumulates of the Yakuno ophiolite, Japan: evidence for unusually thick oceanic crust. *Journal of Petrology*, **26**, 1–30.

Ishiwatari, A., Ikeda, Y. & Koide, Y. 1990. The Yakuno ophiolite, Japan: fragments of Permian island arc and marginal basin crust with a hotspot. *In*: Malpas, J. G., Moores, E. M., Panayiotou, A. & Xenophontos, C. (eds) *Ophiolites: Oceanic Crustal Analogues (Proceedings of the Troodos '87 Symposium)*. Geological Survey Department, Ministry of Agriculture and Natural Resources, Strovolos, Cyprus, 479–506.

Ishizuka, H. & Imaizumi, M. 1988. Metamorphic aragonite from the Yaeyama metamorphic rocks on Ishigaki-jima, southwest Ryukyu Islands. *Journal of Geological Society of Japan*, **94**, 719–722.

Isozaki, Y. 1996. Anatomy and genesis of a subduction-related orogen: a new view of geotectonic subdivision and evolution of the Japanese Islands. *Island Arc*, **5**, 289–320.

Isozaki, Y. & Itaya, T. 1991. Pre-Jurassic Klippe in northern Chichibu Belt in west-central Shikoku, Southwest Japan. Kurosegawa Terrane as a tectonic outlier of the pre-Jurassic rocks of the Inner Zone. *Journal of Geological Society of Japan*, **97**, 431–450 [in Japanese with English abstract].

Isozaki, Y. & Maruyama, S. 1991. Studies on orogeny based on plate tectonics in Japan and new geotectonic subdivision of the Japanese Islads. *Journal of Geography (Tokyo)*, **100**, 697–761 [in Japanese with English abstract].

Isozaki, Y., Aoki, K., Nakama, T. & Yanai, S. 2010. New insight into a subduction related orogen: a reappraisal of the geotectonic framework and evolution of the Japanese Islands. *Gondwana Research*, **18**, 82–105.

Itaya, T., Tsujimori, T. & Liou, J. G. 2011. Evolution of the Sanbagawa and Shimanto high-pressure belts in SW Japan: insights from K–Ar (Ar–Ar) geochronology. *Journal of Asian Earth Sciences*, **42**, 1075–1090.

Jones, G., Valsami-Jones, E. & Sano, H. 1993. Nature and tectonic setting of accreted basalts from the Mino terrane, central Japan. *Journal of the Geological Society of London*, **150**, 1167–1181.

Kabashima, T., Isozaki, Y. & Nishimura, Y. 1993. Find of boundary thrust between 300 Ma high-P/T type schists and weakly metamorphosed Permian accretionary complex in the Nagato tectonic zone, Southwest Japan. *Journal of the Geological Society of Japan*, **101**, 397–400 [in Japanese].

Kanamitsu, G., Otoh, S., Shimojo, M., Hirata, T. & Yokoyama, T. 2011. New detrital-zircon ages and geological setting of the Hitachi area, Ibaraki Prefecture. *Abstracts of 2011 Joint Annual Meeting of Japan Association of Mineralogical Sciences and the Geological Society of Japan*, 10 September, 2011, Mito, Japan, T10-O-6 [in Japanese].

Kanisawa, S. & Ehiro, M. 1989. The Oshima Orogeny and the tectonic development from the Cretaceous to the Paleogene. *In*: Editorial Committee of Tohoku, Part 2 of Regional Geology of Japan (ed.) *Regional Geology of Japan, Part 2, Tohoku*, Kyoritsu Shuppan, Tokyo, 244–246 [in Japanese].

Kanisawa, S., Ehiro, M. & Otsuki, K. 2006. Chapter 3. The Geology of Paleozoic, Mesozoic and Paleogene in Tohoku. *In*: Tohoku Construction Association (ed.) *Geology of Tohoku Region for Construction Engineers*, Tohoku Construction Association, Sendai, 151–193 [in Japanese].

Kanmera, K. & Nishi, H. 1983. Accreted oceanic reef complex in southwest Japan. *In*: Hashimoto, M. & Ueda, S. (eds) *Accretion Tectonics in the Circum-Pacific Regions*. Terra Scientific Publishing Company, Tokyo, 195–206.

Kano, H., Nakazawa, K., Igi, S. & Shiki, T. 1959. On the high-grade metamorphic rocks associated with the Yakuno intrusive rocks of the Maizuru zone. *Journal of Geological Society of Japan*, **65**, 267–271.

Karakida, Y., Yamamoto, H. & Hayama, Y. 1989. The Manotani metamorphic rocks, Kumamoto Prefecture and its reversion. *Memoir of Geological Society of Japan*, **33**, 199–215 [in Japanese with English abstract].

Kawamura, M. & Kitakami Paleozoic Research Group 1988. On the geology of Hayachine Tectonic Belt. *Earth Science (Chikyu-Kagaku)*, **42**, 371–384 [in Japanese with English abstract].

Kawamura, M., Uchino, T., Gouzu, C. & Hyodo, H. 2007. 380 Ma $^{40}Ar/^{39}Ar$ ages of the high-P/T schists obtained from the Nedamo

Terrane, Northeast Japan. *Journal of Geological Society of Japan*, **113**, 492–499 [in Japanese with English abstract].

Kawamura, N. 2010. The Oshima Belt, a Jurassic accretionary complex. *In*: Geological Society of Japan (eds) *Geology of Hokkaido*. Asakura, Tokyo, 19–25 [in Japanese].

Kawamura, T., Uchino, T., Kawamura, M., Yoshida, K., Nakagawa, M. & Nagata, H. 2013. *Geology of the Hayachine San district, Quadrangle series, 1:50,000*. Geological Survey of Japan, AIST, Tsukuba [in Japanese with English abstract].

Kimura, K. 1988. Geology and tectonic setting of the Ultra-Tamba Belt in the western part of Ayabe City, Kyoto Prefecture, Southwest Japan. *Journal of Geological Society of Japan*, **94**, 361–379.

Kimura, K. & Hori, R. 1993. Offscraping accretion of Jurassic chert-clastic complexes in the Mino-Tamba Belt, central Japan. *Journal of Structural Geology*, **15**, 145–161.

Kobayashi, F. & Goto, A. 2008. Stratigraphy of the Lower Formation of the Sasayama Group (Lower Cretaceous) in the Kamitaki-Shimotaki area, Tamba City, Hyogo Prefecture, Japan and the K-Ar age of a schist cobble contained in the conglomerate of the formation. *Journal of Geological Society of Japan*, **114**, 577–586 [in Japanese with English abstract].

Kobayashi, T. 1941. The Sakawa orogenic cycle and its bearing on the origin of the Japanese Islands. *Journal of Faculty of Science, Imperial University of Tokyo, Section II*, **5**, 219–578.

Koide, Y. 1986. Origin of the Ibara metabasalt from the Maizuru Tectonic Belt, Southwest Japan. *Journal of Geological Society of Japan*, **92**, 329–348.

Koide, Y., Tazaki, K. & Kagami, H. 1987. Sr isotopic study of Ibara dismembered ophiolite from Maizuru Tectonic Belt, Southwest Japan. *Journal of Mineralogy, Petrology and Economic Geology*, **82**, 1–15.

Kojima, S. 1989. Mesozoic terrane accretion in Northeast China, Sikhote-Alin and Japan regions. *Palaeogeography, Palaeoclimatology, Palaeoecology*, **69**, 213–232.

Kojima, S. & Sano, H. 2011. Permian and Triassic submarine landslide deposits in a Jurassic accretionary complex in central Japan. *In*: Yamada, Y., Kawamura, K., Ikehara, K., Ogawa, Y., Urgeles, R., Mosher, D., Chaytor, J. & Strasser, M. (eds) *Submarine Mass Movements and Their Consequences*. Springer, Dordrecht, Advances in Natural and Technological Hazards Research **31**, 639–648.

Kojima, S., Ando, H., Kida, M., Mizutani, S., Sakata, Y., Sugiyama, K. & Tsukada, H. 1999. Clastic rocks in Triassic bedded chert in the Mino terrane, central Japan: their petrographic properties and radiolarian ages. *Journal of Geological Sociey of Japan*, **105**, 421–434 [in Japanese with English abstract].

Kojima, S., Kemkin, I. V., Kametaka, M. & Ando, A. 2000. A correlation of accretionary complexes between southern Sikhote-Alin of Russia and Inner Zone of Southwest Japan. *Geosciences Journal*, **4**, 175–185.

Kojima, S., Tsukada, K. et al. 2008. Geological relationship between Anyui Metamorphic Complex and Samarka terrane, Far East Russia. *Island Arc*, **17**, 502–516.

Kondo, N. & Adachi, M. 1975. Mesozoic strata of the area north of Inuyama, with special reference to the Sakahogi conglomerate. *Journal of Geological Sociey of Japan*, **81**, 373–386 [in Japanese with English abstract].

Kunugiza, K. & Maruyama, S. 2011. Geotectonic evolution of the Hida marginal belt, central Japan: reconstruction of the oldest Pacific-type orogeny of Japan. *Journal of Geography (Tolyo)*, **120**, 960–980 [in Japanese with English abstract].

Kunugiza, K., Goto, A., Itaya, T. & Yokoyama, K. 2004. Geological development of the Hida Gaien belt: constraints from K-Ar ages of high P/T metamorphic rocks and U-Th-Pb EMP ages of granitic rocks affecting contact metamorphism of serpentinite. *Journal of Geological Society of Japan*, **110**, 580–590.

Kurimoto, C. 1986. Ultra-Tamba Zore in the Fukuchiyama area, Kyoto Prefecture: its constituent rocks and distribution. *Earth Science (Chikyu-kagaku)*, **40**, 64–67.

Matsuda, T. & Isozaki, Y. 1991. Well-documented travel history of Mesozoic pelagic chert in Japan: from remote ocean to subduction zone. *Tectonics*, **10**, 475–499.

Matsuda, T., Sakiyama, T. et al. 1998. Fission-track ages and magnetic susceptibility of Cretaceous to Paleogene volcanic rocks in southeastern Sikhote Alin, Far East Russia. *Resource Geology*, **48**, 285–290.

Matsumoto, K., Sugimura, K., Tokita, I., Kunugiza, K. & Maruyama, S. 2011. Geology and metamorphism of the Itoigawa-Omi area of the Hida Gaien belt, central Japan: reconstruction of the oldest Pacific-type high P/T type metamorphism and hydration metamorphism during exhumation. *Journal of Geography (Tokyo)*, **120**, 4–29 [in Japanese with English abstract].

Matsuoka, A. 1984. Togano Group of the Southern Chichibu Terrane in the western part of Kochi Prefecture, southwest Japan. *Journal of Geological Society of Japan*, **90**, 455–477 [in Japanese with English abstract].

Matsuoka, A. 1992. Jurassic-Early Cretaceous tectonic evolution of the Southern Chichibu Terrane, Southwest Japan. *Palaeogeography, Palaeoclimatology, Palaeoecology*, **96**, 71–88.

Matsuoka, A. 1995. Jurassic and Lower Cretaceous radiolarian zonation in Japan and in the western Pacific. *Island Arc*, **4**, 140–153.

Matsuoka, A. & Yao, A. 1990. Southern Chichibu terrane. *In*: Ichikawa, K., Mizutani, S., Hara, I., Hada, S. & Yao, A. (eds) *Pre-Cretaceous Terranes of Japan*. Department of Geosciences, Osaka City University, Osaka, 203–216.

Matsuoka, A., Hori, R., Kuwahara, K., Hiraishi, M., Yao, A. & Ezaki, Y. 1994. Triassic-Jurassic radiolarian-bearing sequences in the Mino terrane, central Japan. *In*: Organizing Committee of Interrad VII (eds) *Guide Book for INTERRAD VII Field Excursion*. Department of Geosciences, Osaka City University, Osaka, 19–61.

Matsuoka, A., Yamakita, S., Sakakibara, M. & Hisada, K. 1998. Unit division for the Chichibu Composite Belt from a view point of accretionary tectonics and geology of western Shikoku, Japan. *Journal of Geological Society of Japan*, **104**, 634–653 [in Japanese with English abstract].

Miyakawa, K. 1961. General considerations on the Sangun metamorphic rocks on the basis of their petrographical features observed in the San-in provinces, Japan. *Journal of Earth Sciences, Nagoya University*, **9**, 345–393.

Miyakawa, K. 1982. Low-grade metamorphic rocks of the Hida marginal belt in the upper Kuzuryu river area, central Japan. *Journal of Japanese Association of Mineralogist, Petrologist and Economic Geologist*, **77**, 256–265 [in Japanese with English abstract].

Miyamoto, T. & Yanagi, T. 1996. U-Pb dating of detrital zircons from the Sangun metamorphic rocks, Kyushu, Southwest Japan: an evidence for 1.9–2.0 Ga granite emplacement in the provenance. *Geochemical Journal*, **30**, 261–271.

Miyashiro, A. 1961. Evolution of metamorphic belts. *Journal of Petrology*, **2**, 277–311.

Mizutani, S. 1959. Clastic plagioclase in Permian greywacke from the Mugi area, Gifu Prefecture, central Japan. *Journal of Earth Sciences, Nagoya University*, **7**, 108–136.

Moriya, S. 1972. Low-grade metamorphic rocks of the northern Kitakami Mountainland. *Science Reports of Tohoku University, Series 3*, **11**, 239–282.

Nakae, S. 2000. Regional correlation of the Jurassic accretionary complex in the Inner Zone of Southwest Japan. *Memoir of Geological Society of Japan*, **55**, 73–98 [in Japanese with English abstract].

Nakae, S. & Kurihara, T. 2011. Direct age for Upper Permian accretionary complex (Kirinai Formation), Kitakami Mountains, Northeast Japan. *Palaeoworld*, **20**, 146–157.

Nakamizu, M., Okada, M., Yamazaki, T. & Komatsu, M. 1989. Metamorphic rocks in the Omi-Renge serpentinite melange, Hida Marginal Tectonic Belt, Central Japan. *Memoirs of Geological Society of Japan*, **33**, 21–35 [in Japanese with English abstract].

Nakashima, K. & Sano, H. 2007. Palaeoenvironmental implication of resedimented limestones shed from Mississippian–Permian mid-oceanic atoll-type buildup into slope-to-basin facies, Akiyoshi, Japan. *Palaeogeography, Palaeoclimatology, Palaeoecology*, **247**, 329–356.

Nakazawa, K. 1958. The Triassic system in the Maizuru zone, Southwest Japan. *Memoirs of College of Science, University of Kyoto, Series B*, **24**, 285–313.

Nakazawa, T. & Ueno, K. 2004. Sequence boundary and related sedimentary and diagenetic facies formed on Middle Permian mid-oceanic carbonate platform: core observation of Akiyoshi Limestone, Southwest Japan. *Facies*, **50**, 301–311.

Nakazawa, T., Ueno, K., Kawahata, H., Fujikawa, M. & Kashiwagi, K. 2009. Facies stacking patterns in high-frequency sequences influenced by long-term sea-level change on a Permian Panthalassan oceanic atoll: an example from the Akiyoshi Limestone, SW Japan. *Sedimentary Geology*, **214**, 35–48.

Naumann, E. 1885. *Uber den Bau und die Entstehung der japanischen Inseln*. R. Friedlaender und Sohn, Berlin.

NIITSUMA, N., TAIRA, A. & SAITO, Y. 1985. Japanese Islands before the opening of the Sea of Japan. *Science (Kagaku)*, **55**, 744–747 [in Japanese].

NISHIMURA, Y. 1971. Regional metamorphism of the Nishiki-cho district Southwest japan. *Journal of Science, Hiroshima University, Series C*, **6**, 203–268.

NISHIMURA, Y. 1998. Geotectonic subdivision and areal extent of the Sangun belt, Inner Zone of Southwest Japan. *Journal of Metamorphic Geology*, **16**, 129–140.

NISHIMURA, Y. & OKAMOTO, T. 1976. Lawsonite-albite schist from the Masuda district, southwest Japan. *Jubilee Publication in the Commemoration of Professor G. Kojima, Sixtieth Birthday*. Hiroshima University Press, Hiroshima, Japan, 144–152 [in Japanese].

NISHIMURA, Y. & SHIBATA, K. 1989. Modes of occurrence and K-Ar ages of metagabbroic rocks in the Sangun metamorphic belt, Southwest Japan. *Memoir of Geological Society of Japan*, **33**, 343–357 [in Japanese with English abstract].

NISHIMURA, Y., MATSUBARA, Y. & NAKAMURA, E. 1983. Zonation and K-Ar ages of the Yaeyama metamorphic rocks, Ryukyu Islands. *Memoir of Geological Society of Japan*, **22**, 27–37 [in Japanese with English abstract].

NISHIMURA, Y., HIROTA, Y., SHIOZAKI, D., NAKAHARA, N. & ITAYA, T. 2004. The Nagasaki metamorphic rocks and their geotectonics in Mogi area, Nagasaki Prefecture, Southwest Japan: juxtaposition of the Suo belt with the Sanbagawa belt. *Journal of Geological Society of Japan*, **110**, 372–383 [in Japanese with English abstract].

NOZAKI, T., KATO, Y., SUZUKI, K. & KASE, K. 2012. Re-Os geochronology of the Hitachi VMS deposit, Ibaraki Prefecture: the oldest sulfide deposit in the Japanese Islands. *Abstracts of Japan Geoscience Union 2012*, 23 May, 2012, Makuhari, Japan, SGL43-05.

NUONG, N. D., ITAYA, T. & NISHIMURA, Y. 2008. Age (K–Ar phengite)–temperature–structure relations: a case study from the Ishigaki high-pressure schist belt, southern Ryukyu Arc, Japan. *Geological Magazine*, **145**, 677–684.

NUTMAN, A. P., SANO, Y., TERADA, K. & HIDAKA, H. 2006. 743 ± 17 Ma granite clast from Jurassic conglomerate, Kamiaso, Mino Terrane, Japan: the case for South China Craton provenance (Korean Gyeonggi Block?). *Journal of Asian Earth Sciences*, **26**, 99–104.

OHO, Y. 1990. Superimposed folds of the Sangun metamorphic rocks in the Ochiai-Asahi area, Okayama Prefecture, Southwest Japan. *Journal of Geological Society of Japan*, **95**, 541–551 [in Japanese with English abstract].

OKAMI, K., KOSHIYA, S. & EHIRO, M. 1992. Compositions of the Paleozoic and Mesozoic sandstones distributed in the Kitakami and Abukuma Mountains, Northeast Japan. *Memoir of Geological Society of Japan*, **38**, 43–57 [in Japanese with English abstract].

ONOUE, T., SATO, H. ET AL. 2012. Deep-sea record of impact apparently unrelated to mass extinction in the Late Triassic. *Proceedings of National Academy of Sciences, USA*, **109**(47), 19134–19139, http://www.pnas.org/cgi/doi/10.1073/pnas.1209486109

ONUKI, H., SHIBA, M., KAGAWA, H. & HORI, H. 1988. Low-temperature regional metamorphic rocks in the northern Kitakami Mountains. I. Kuzakai-Morioka area. *Journal of Mineralogy, Petrology and Economic Geology*, **83**, 495–506 [in Japanese with English abstract].

ONUKI, Y. 1981. Part 1. The Kitakami Mountains. *In*: HASE CHICHITSU CHOSA JIMUSHO (ed.) *Explanation of the Geological Map of the Kitakami River Area, Kitakami Mountains, at a Scale of 1:200,000*. Hase Chichitsu Chosa Jimusho, Sendai, 3–223 [in Japanese].

ONUKI, Y., TAKAHASHI, K. & ABE, T. 1962. On the Motai Group of the Kitakami Massif, Japan. *Journal of Geological Society of Japan*, **68**, 629–639 [in Japanese with English abstract].

OSOZAWA, S., TAKEUCHI, H. & KOITABASHI, T. 2004. Formation of the Yakuno ophiolite; accretionary subduction under medium-pressure-type metamorphic conditions. *Tectonophysics*, **393**, 197–219.

OTA, M. 1968. The Akiyoshi Limestone Group: a geosynclinal organic reef complex. *Akiyoshi-dai Science Museum Bulletin*, **5**, 1–44.

OTOH, S. & SASAKI, M. 2003. Tectonostratigraphic divisions and regional correlation of the sedimentary complex of the North Kitakami Belt. *Journal of Geography (Tokyo)*, **112**, 406–410 [in Japanese with English abstract].

OTOH, S., OBARA, H. ET AL. 2013. Provenance of pre-Aptian sandstones of Japan viewed from detrital zircon geochronology. *Japan Geoscience Union Meeting 2013*, 19–24 May, 2013, Makuhari, Japan, Abstract.

OZAWA, Y. 1925. Paleontological and stratigraphical studies on the Permo-Carboniferous limestone of Nagato, Pt. II. Paleontology. *Journal of College of Science, Imperial University of Tokyo*, **45**, 1–90.

PILLAI, D. & ISHIGA, H. 1987. Discovery of Late Permian radiolarians from Kozuki Formation, Kozuki-Tatsuno Belt, Southwest Japan. *Journal of Geological Society of Japan*, **93**, 847–850.

RAYMOND, L. A. 1984. Classification of mélanges. *Geological Society of America Specical Papers*, **198**, 7–20.

ROSS, C. A. & ROSS, J. R. P. 1987. Late Paleozoic sea levels and depositional sequences. *In*: ROSS, C. A. & HAMAN, D. (eds) *Timing and Depositional History of Eustatic Sequences: Constraints on Seismic Stratigraphy*. Cushman Foundation for Foraminiferal Research, Special Publication, **24**, 137–149.

SAKASHIMA, T., TERADA, K., TAKESHITA, T. & SANO, Y. 2003. Large-scale displacement along the Median Tectonic Line, Japan: evidence from SHRIMP zircon U-Pb dating of granites and gneisses from the South Kitakami and paleo-Ryoke belts. *Journal of Asian Earth Sciences*, **21**, 1019–1039.

SANO, H. 1988. Permian oceanic-rocks of Mino terrane, central Japan. Part I Chert facies. *Journal of Geological Society of Japan*, **94**, 697–709.

SANO, H. 2006. Impact of long-term climate change and sea-level fluctuation on Mississippian to Permian mid-oceanic atoll sedimentation (Akiyoshi Limestone Group, Japan). *Palaeogeography, Plaeoclimatology, Paleoecology*, **236**, 169–189.

SANO, H. & KANMERA, K. 1988. Paleogeographic reconstruction of accreted oceanic rocks, Akiyoshi, southwest Japan. *Geology*, **16**, 600–603.

SANO, H. & KANMERA, K. 1991*a*. Collapse of ancient reef complex. What happened during collision of Akiyoshi reef complex? Geologic setting and age of Akiyoshi terrane rocks on western Akiyoshi-dai plateau. *Journal of Geological Society of Japan*, **97**, 113–133.

SANO, H. & KANMERA, K. 1991*b*. Collapse of ancient reef complex. What happened during collision of Akiyoshi reef complex? Broken limestone as collapse products. *Journal of Geological Society of Japan*, **97**, 217–229.

SANO, H. & KANMERA, K. 1991*c*. Collapse of ancient reef complex. What happened during collision of Akiyoshi reef complex? Limestone breccias, redeposited limestone debris and mudstone injection. *Journal of Geological Society of Japan*, **97**, 297–309.

SANO, H. & KANMERA, K. 1991*d*. Collapse of ancient oceanic reef complex. What happened during collision of Akiyoshi reef complex? Sequence of Collisional collapse and generation of collapse products. *Journal of Geological Society of Japan*, **97**, 631–644.

SANO, H. & KOJIMA, S. 2000. Carboniferous to Jurassic oceanic rocks of Mino-Tamba-Ashio terrane, southwest Japan. *Memoir of Geological Society of Japan*, **55**, 123–144 [in Japanese with English abstract].

SANO, H., FUJII, S. & MATSUURA, F. 2004. Response of Carboniferous–Permian mid-oceanic seamount-capping buildup to global cooling and sea-level change: Akiyoshi, Japan. *Palaeogeography, Palaeoclimatology, Palaeoecology*, **213**, 187–206.

SANO, H., KUWAHARA, K., YAO, A. & AGEMATSU, S. 2010. Panthalassan seamount associated Permian-Triassic boundary siliceous rocks, Mino terrane, central Japan. *Paleontological Research*, **14**, 293–314.

SANO, H., WADA, T. & NARAOKA, H. 2012. Late Permian to Early Triassic environmental changes in the Panthalassic Ocean: record from the seamount-associated deep-marine siliceous rocks, central Japan. *Palaeogeography, Palaeoclimatology, Palaeoecology*, **363–364**, 1–10.

SANO, S. 1992. Neodymium isotopic compositions of Silurian Yakuno metagabbros. *Journal of Mineralogy, Petrology and Economic Geology*, **87**, 272–282 [in Japanese with English abstract].

SANO, S., HAYASAKA, Y. & TAZAKI, K. 2000. Geochemical characteristics of Carboniferous greenstones in the inner zone of Southwest Japan. *Island Arc*, **9**, 81–96.

SANO, S., SUGISAKA, N. & SHIMAGUCHI, T. 2009. Discovery of megalodontid bivalves in the Shiriya area, northern Honshu, Northeast Japan, and its geological implications. *Memoir of Fukui Prefectural Dinosaur Museum*, **8**, 51–57 [in Japanese with English abstract].

SANO, Y., HIDAKA, H., TERADA, K., SHIMIZU, H. & SUZUKI, M. 2000. Ion microprobe U-Pb zircon geochronology of the Hida gneiss: finding of the oldest minerals in Japan. *Geochemical Journal*, **34**, 135–153.

SHIBATA, K. & ADACHI, M. 1974. Rb-Sr whole-rock ages of Precambrian metamorphic rocks in the Kamiaso conglomerate from central Japan. *Earth and Planetary Science Letters*, **21**, 277–287.

Shibata, K. & Nishimura, Y. 1989. Isotopic ages of the Sangun crystalline schists, Southwest Japan. *Memoir of Geological Society of Japan*, **33**, 317–341.

Shimizu, D. 1962. The Permian Maizuru Group, its stratigraphy and syntectonic faunal succession through the latest Paleozoic Orogeny. *Memoirs of College of Science, University of Kyoto, Series B*, **28**, 571–609.

Sohma, T., Maruyama, S., Matsushita, S., Yamamoto, M. & Matsumoto, K. 1983. Olistoststrome in western area of the Hida-Gaien belt and its geotectonic bearings. *Journal of Faculty of Education, Toyama University*, **31**, 13–23 [in Japanese].

Soreghan, G. S., Heavens, N., Patterson, E. P., Sano, H., Mahowald, N., Davydov, V. & Soreghan, M. J. 2011. Giant grains from Pennsylvanian Dust of the Panthalassic Ocean: evidence for Extreme Winds and a Paleo-Tethyan Monsoon. *AGU Annual Meeting Abstract*, 5–9 December, 2011, San Francisco, USA, PP22D-07.

Suda, Y. 2004. Crustal anatexis and evolution of granitoid magma in Permian intra-oceanic island arc, the Asago body of the Yakuno ophiolite, Southwest Japan. *Journal of Mineralogical and Petrological Sciences*, **99**, 339–356.

Sugamori, Y. 2009*a*. Ultra-Tamba Terrane in the Kawanishi-Inagawa area, southwestern part of Hyogo Prefecture, Southwest Japan. *Journal of Geological Society of Japan*, **115**, 80–95 [in Japanese with English abstract].

Sugamori, Y. 2009*b*. Middle Triassic Kamitaki Formation between the Ultra-Tamba and Tamba terranes, Southwest Japan and its geological implication for tectonic evolution of the eastern margin of East Asia. *In*: Luo, H., Aitchison, J. C. et al. (eds) *Abstracts of InterRad 12*, 14–17 September, 2009. Nanjing, China, 162–163.

Sugamori, Y. 2011. Late Permian radiolarians from the Ajima Formation of the Ultra-Tamba Terrane in the Sasayama area, southwest Japan. *Palaeoworld*, **20**, 158–165.

Sugimoto, M. 1974. Stratigraphical study in the Outer Belt of the Kitakami Massif, Northeast Japan. *Contributions from Institute of Geology and Paleontology, Tohoku University*, **74**, 1–48 [in Japanese with English abstract].

Sugiyama, T. & Nagai, K. 1994. Reef facies and paleoecology of reef-building corals in the lower part of the Akiyoshi Limestone Group (Carboniferous), Southwest Japan. *Courier Forschungsinstitus Senckenberg*, **172**, 140–231.

Suzuki, K., Adachi, M. & Tanaka, T. 1991. Middle Precambrian provenance of Jurassic sandstone in the Mino Terrane, central Japan: Th-U-total Pb evidence from an electron microprobe monazite study. *Sedimentary Geology*, **75**, 141–147.

Suzuki, N., Ehiro, M., Yoshihara, K., Kimura, Y., Kawashima, G., Yoshimoto, H. & Nogi, T. 2007*a*. Geology of the Kuzumaki-Kamaishi Subbelt of the North Kitakami Belt (a Jurassic accretionary complex), Northeast Japan: case study of the Kawai-Yamada area, eastern Iwate Prefecture. *Bulletin of Tohoku University Museum*, **6**, 103–174.

Suzuki, N., Yamakita, S., Takahashi, S. & Ehiro, M. 2007*b*. Middle Jurassic radiolarians from carbonate manganese nodules in the Otori Formation in the eastern part of the Kuzumaki-Kamaishi Subbelt, the North Kitakami Belt, Northeast Japan. *Journal of Geological Society of Japan*, **113**, 274–277 [in Japanese with English abstract].

Suzuki, S. 1987. Sedimentary and tectonic history of the eastern part of Maizuru zone, Southwest Japan. *Geological Peport of Hiroshima University*, **27**, 1–54 [in Japanese with English abstract].

Tagiri, M., Morimoto, M., Mochizuki, R., Yokosuka, A., Dunkley, D. J. & Adachi, T. 2010. Hitachi metamorphic rocks: Occurrence and geology of metagranitic rocks with Cambrian SHRIMP zircon age. *Journal of Geography (Tokyo)*, **119**, 245–56 [in Japanese with English abstract].

Tagiri, M., Horie, K., Adachi, T. & Hiroi, Y. 2011. SHRIMP dating of magmatism in the Hitachi metamorphic terrane, Abukuma Belt, Japan: evidence for a Cambrian volcanic arc. *Island Arc*, **20**, 259–279.

Taira, A. & Tashiro, M. 1987. Late Paleozoic and Mesozoic accretion tectonics in Japan and Eastern Asia. *In*: Taira, A. & Tashiro, M. (eds) *Historical Biogeography and Plate Tectonic Evolution of Japan and Eastern Asia*. Terra Scientific Publishing Company, Tokyo, 1–43.

Takahashi, S., Ehiro, M. & Suzuki, N. 2006. Preliminary report on the geology of the North Kitakami Belt, a Jurassic accretionary complex in the west Akka area, Iwaizumi-Town, Iwate Prefecture, northeast Japan. *Geology of Iwate*, **35/36**, 66–71 [in Japanese].

Takemura, S. & Suzuki, S. 1996. The geology and tectonics of the Ultra-Tamba Zone, western Hyogo Prefecture, Southwest Japan. *Journal of Geological Society of Japan*, **102**, 1–12 [in Japanese with English abstract].

Takemura, S., Suzuki, S. & Ishiga, H. 1993. Stratigraphy of the Kozuki Formation, Kamigori Zone, Southwest Hyogo Prefecture, Japan, reconsideration with the discvery of *Albaillella asymmtrica* (radiorarian fossil) and the structural analysis. *Journal of Geological Society of Japan*, **99**, 675–678 [in Japanese with English abstract].

Takeshita, H., Watanabe, T. & Ishiga, H. 1987. Discovery of Permian radiolarians from the Tanoharagawa Formation (the Sangun metamrophic rocks), Gotsu, Shimane Prefecture, Japan. *Journal of Geological Society of Japan*, **93**, 435–438 [in Japanese].

Takeuchi, M. 2000. Origin of Jurassic coarse clastic sediments in the Mino-Tamba Belt. *Memoir of Geological Society of Japan*, **55**, 107–121 [in Japanese with English abstract].

Tamura, M. 1987. Distribution of Japanese Triassic bivalve faunas and sedimentary environment of megalodont limestone in Japan. *In*: Taira, A. & Tashiro, M. (eds) *Historical Biogeography and Plate Tectonic Evolution of Japan and Eastern Asia*. Terra Scientific Publishing Company, Tokyo, 97–110.

Tazawa, J. 1993. Pre-Neogene tectonics of the Japanese Islands from the viewpoint of paleobiogeography. *Journal of Geological Society of Japan*, **99**, 525–543 [in Japanese with English abstract].

Teraoka, Y., Shibata, K., Okumura, K. & Uchiumi, S. 1994. K-Ar ages of detrial K-feldspars and muscovites from the Shimanto Supergroup in east Kyushu and west Shikoku, Southwest Japan. *Journal of Geological Society of Japan*, **100**, 477–485 [in Japanese with English abstract].

Teraoka, Y., Matsuura, H. et al. 1996. Geological Map of Japan 1:200000 NI-53-26 Takahashi. *Geological Survey of Japan*, Tsukuba.

Toriyama, R. 1958. Geology of Akiyoshi: part III. Fusulinids of Akiyoshi. *Memoir of Faculty of Science, Kyushu University, D*, **VII**, 1–48.

Tsujimori, T. 1998. Geology of the Osayama serpentinite melange in the central Chugoku Mountains, southwestern Japan: 320 Ma blueschist-bearing serpentinite melange beneath the Oeyama ophiolite. *Journal of Geological Society of Japan*, **104**, 213–231 [in Japanese with English abstract].

Tsujimori, T. 2002. Prograde and retrograde P-T paths of the late Paleozoic glaucophane eclogite from the Renge metamorphic belt, Hida Mountains, Southwest Japan. *International Geology Review*, **44**, 797–818.

Tsujimori, T. 2010. Paleozoic subduction-related metamorphism in Japan: new insights and perspectives. *Journal of Geography (Tokyo)*, **119**, 294–312 [in Japanese with English abstract].

Tsujimori, T. & Itaya, T. 1999. Blueschist-facies metamorphism during Paleozoic orogeny in southwestern Japan: phengite K-Ar ages of blueschist-facies tectonic blocks in a serpentinite melange beneath early Paleozoic Oeyama ophiolite. *Island Arc*, **8**, 190–205.

Tsujimori, T. & Liou, J. G. 2004. Metamorphic evolution of kyanite-staurolite-bearing epidote-amphibolite from the Early Paleozoic Oeyama belt, SW Japan. *Journal of Metamorphic Geology*, **22**, 301–313.

Tsujimori, T. & Liou, J. G. 2005. Eclogite-facies mineral inclusions in clinozoisite from Paleozoic blueschist, central Chugoku Mountains, Southwest Japan: evidence of regional eclogite-facies metamorphism. *International Geology Review*, **47**, 215–232.

Tsujimori, T. & Liou, J. G. 2007. Significance of the Ca-Na pyroxene-lawsonite-chlorite assemblage in blueschist-facies metabasalts: an example from the Renge metamorphic rocks, SW Japan. *International Geology Review*, **49**, 415–430.

Tsujimori, T. & Matsumoto, K. 2006. P-T pseudosection of a glaucophane-epidote eclogite from Omi serpentinite melange, SW Japan: a preliminary report. *Journal of Geological Society of Japan*, **112**, 407–414 [in Japanese with English abstract].

Tsujimori, T., Ishiwatari, A. & Banno, S. 2000. Eclogitic glaucophane schist from the Yunotani valley in Omi Town, the Renge metamorphic belt, the Inner Zone of southwestern Japan. *Journal of Geological Society of Japan*, **106**, 353–362 [in Japanese with English abstract].

Tsutsumi, Y., Yokoyama, K., Terada, K. & Sano, Y. 2003. SHRIMP U-Pb dating of zircons in metamorphic rocks from northern Kyushu, western Japan. *Journal of Mineralogical and Petrological Sciences*, **98**, 181–193.

Uchino, T. & Kawamura, M. 2009. Chemical composition of the green rocks in the Nedamo Terrane, Northeast Japan. *Journal of Geological Society of Japan*, **115**, 242–247 [in Japanese with English abstract].

Uchino, T. & Kawamura, M. 2010*a*. Glaucophane found from meta-basalt in the Nedamo Terrane, Northeast Japan, and its geologic significance. *Bulletin of Geological Survey of Japan*, **61**, 445–452 [in Japanese with English abstract].

Uchino, T. & Kawamura, M. 2010*b*. Tectonics of an Early Carboniferous forearc inferred from a high-P/T schist-bearing conglomerate in the Nedamo Terrane, Northeast Japan. *Island Arc*, **19**, 177–199.

Uchino, T., Kawamura, M. & Kurihara, T. 2005. Early Carboniferous radiolarians discovered from the Hayachine Terrane, Northeast Japan: the oldest fossil age for clastic rocks of accretionary complex in Japan. *Journal of Geological Society of Japan*, **111**, 249–252 [in Japanese with English abstract].

Uchino, T., Kawamura, M., Gouzu, C. & Hyodo, H. 2008. Phengite $^{40}Ar/^{39}Ar$ age of garnet-bearing pelitic schist pebble obtained from conglomerate in the Nedamo Terrane, Northeast Japan. *Journal of Geological Society of Japan*, **114**, 314–317 [in Japanese with English abstract].

Uemura, F., Sakamoto, T. & Yamada, N. 1979. *Geology of the Wakasa District, with Geological Sheet Map at 1:50,000.* Geological Survey of Japan, Tsukuba [in Japanese with English abstract].

Wakita, K. 1984. *Geology of the Hachiman District, with Geological Sheet Map at 1:50,000.* Geological Survey of Japan, Tsukuba [in Japanese with English abstract].

Wakita, K. 1988. Origin of chaotically mixed rock bodies in the Early Jurassic to Early Cretaceous sedimentary complex of the Mino terrane, central Japan. *Bulletin of Geological Survey of Japan*, **39**, 675–757.

Wakita, K. 1995. *Geology of the Mino District, with Geological Sheet Map at 1:50,000.* Geological Survey of Japan, Tsukuba [in Japanese with English abstract].

Wakita, K. 2000. Melanges of the Mino terrane. *Memoir of Geological Society of Japan*, **55**, 145–163 [in Japanese with English abstract].

Wakita, K. 2013. Geology and tectonics of Japanese islands: a review. The key to understanding the geology of Asia. *Journal of Asian Earth Sciences*, **72**, 75–87.

Watanabe, K. 1991. Fusuline biostratigraphy of the Upper Carboniferous and Lower Permian of Japan, with special reference to the Carboniferous-Permian boundary. *Paleontological Society of Japan, Special Paper*, **32**, 1–150.

Watanabe, T., Kobayashi, H. & Sengan, H. 1983. Lawsonite from quartzfeldspathic schist in the Sangun metamorphic belt, Shikuma, Shimane Prefecture. *Memoirs of Faculty of Science, Shimane University*, **17**, 81–86.

Watanabe, T., Sengan, H. & Kobayashi, H. 1989. The Hazumi-minami tectonic melange in the eastern Goutsu City, Shimane Prefecture: geology, petrology and structure. *Memoir of Geological Society of Japan*, **33**, 107–124 [in Japanese with English abstract].

Working Group on the Permian-Triassic Systems 1975. Stratigraphy near the Permian-Triassic boundary in Japan and its correlation. *Journal of Geological Society of Japan*, **81**, 165–184 [in Japanese with English abstract].

Yamakita, S. 1998. What belongs to the Northern Chichibu Belt? Tectonic division between the Northern Chichibu Belt and the Kurosegawa Belt. *Journal of Geological Society of Japan*, **104**, 623–633.

Yamakita, S. & Hori, R. S. 2009. Early Jurassic radiolarians from a carbonate nodule in the Northern Chichibu Belt in western central Shikoku, Southwest Japan. *News of Osaka Micropaleontologists, Special Volume*, **14**, 497–505 [in Japanese with English abstract].

Yao, A. 2000. Tectonic arrangement of Southwest Japan in view of the Paleozoic-Mesozoic tectonics of East Asia. *In*: Ishiga, H., Takazu, A. & Hayasaka, Y. (eds) *Tectonic Division and Correlation of Terranes of Southwest Japan.* Association for Geological Collaboration in Japan, Monograph, Tokyo, Japan, **49**, 145–155.

Yao, A., Matsuda, T. & Isozaki, Y. 1980. Triassic and Jurassic radiolarians from the Inuyama area, central Japan. *Journal of Geosciences, Osaka City University*, **23**, 135–154.

Yehara, S. 1927. Faunal and stratigraphical study of the Sakawa Basin, Shikoku. *Japanese Journal of Geology and Geography*, **5**, 1–40.

Yoshida, S. & Wakita, K. 1999. *Geology of the Gifu District, with geological sheet map at 1:50,000.* Geological Survey of Japan, Tsukuba [in Japanese with English abstract].

2c Paired metamorphic belts of SW Japan: the geology of the Sanbagawa and Ryoke metamorphic belts and the Median Tectonic Line

SIMON RICHARD WALLIS & TAKAMOTO OKUDAIRA

The idea of paired metamorphic belts proposed by Miyashiro (1961, 1973) links the presence of parallel belts of high and low heat flow along convergent margins with paired regions of metamorphism below the surface that form under contrasting *P–T* conditions. Quantitative modelling of convergent plate margins by Oxburgh & Turcotte (1971) showed that the belt of low heat flow can be accounted for by the inflow of cold lithosphere into the mantle along oceanic trenches (Fig. 2c.1), and the belt of high heat flow can be accounted for by the rise of hot magma and fluids in volcanic arcs. The idea of paired metamorphic belts provided a new dynamic way of looking at regional metamorphism, linking the physical movements of plates to the resulting movement of heat. Researchers in the two fields of structural geology and metamorphic petrology now had a common purpose: establishing a link between their observations of regional metamorphic belts and plate-scale tectonic processes. This new insight still provides the framework and motive for many studies of orogenic belts. The Sanbagawa and Ryoke belts are both the type locality of paired metamorphism and the starting point for this new field of endeavour in basement geology. In this chapter we summarize the present state of knowledge of these belts and consider how their geological development reflects plate movements and the development of the Mesozoic East Asian subduction zone.

The Sanbagawa Belt

Distribution and lithology

The Sanbagawa Belt was first named after the region of flat-lying pelitic, mafic and quartz schists exposed in the Sanbagawa River area in the Kanto Mountains about 80 km to the NW of Tokyo. In particular, the rocks exposed in the Nagatoro and Mt Mikabo areas are type localities for the Sanbagawa schists and Mikabu Greenrock (Koto 1888; Fig. 2c.1). To the west of this region similar rocks appear within the Tenryu region. In contrast to the Sanbagawa Belt of the Kanto Mountains, the rocks in the Tenryu area are steeply dipping and the trend follows the line of the collision between the Izu Peninsula and the Japanese mainland. The continuation of the Sanbagawa Belt can then be traced through central Japan and the Kii Peninsula to Shikoku. The main exposures are found in central Shikoku, and geological highlights of this region include: the world-class Besshi Cu deposit mine, which was active until 1973 (Aoya *et al.* 2013*b*; Chapter 10); large exposures of strongly deformed eclogite (Aoya & Wallis 1999, 2003; Ota *et al.* 2004; Miyagi & Takasu 2005; Endo 2010); and the only known kilometre-scale exposure of garnet peridotite from an oceanic subduction setting (Mizukami & Wallis 2005). The exposure of the Sanbagawa Belt thins as it approaches the western edge of Shikoku and only a small region is found on the facing eastern shores of Kyushu Island. Probable correlative rocks are also present around the town of Nagasaki (Nagasaki metamorphics; see Chapter 2e). In total, the Sanbagawa Belt has a length of >1000 km, a maximum width of 30 km and is the most extensive of all the Japanese metamorphic belts.

A closer discussion of the Sanbagawa Belt requires a more complete geological definition. Here, we use the Sanbagawa Belt to mean the area comprising high-strain ductilely deformed relatively high-*P–T* metamorphic rocks that lie between the Median Tectonic Line (MTL) in the north – the largest on-land fault of the Japanese islands – and a prominent kilometre-scale suite of dominantly

From: Moreno, T., Wallis, S., Kojima, T. & Gibbons, W. (eds) 2016. *The Geology of Japan*. Geological Society, London, 101–124.

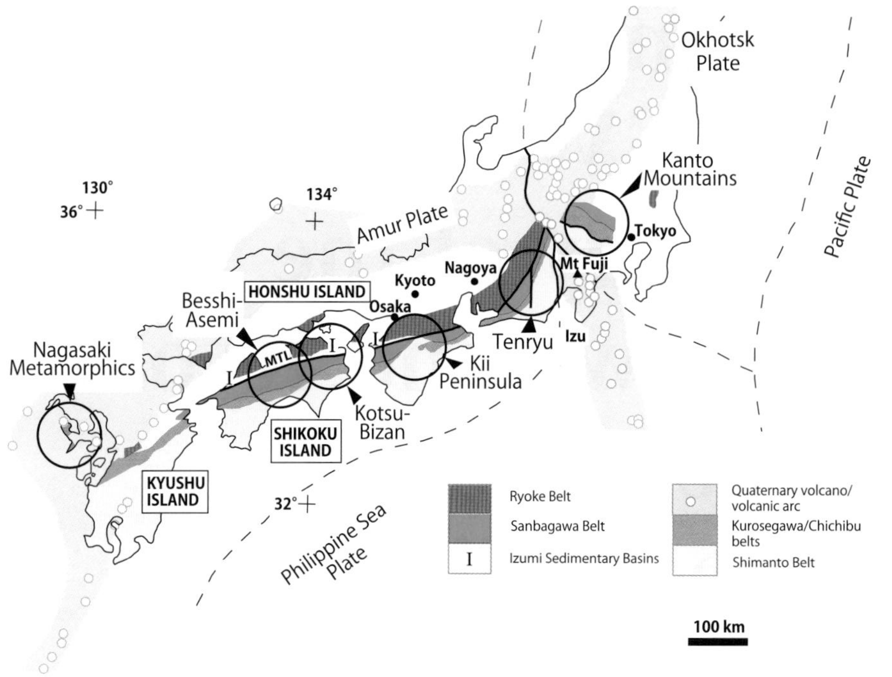

Fig. 2c.1. Distribution of the Sanbagawa and Ryoke belts in SW Japan. Recent plate configuration, volcanic arcs and the main locations for the geographical localities mentioned in the description of the Sanbagawa Belt are also shown. The localities for the Ryoke Belt are summarized in Figure 2c.5.

mafic to ultramafic low-grade metamorphic rocks in the south that are known as the Mikabu Belt (Fig. 2c.2). Between the Mikabu Belt and MTL the main rock types of the Sanbagawa Belt are graphite-bearing pelitic schist, psammitic schist (locally containing conglomeratic layers), mafic schist (for historical reasons referred to as basic schist in most papers on the region) and siliceous schist (commonly associated with Mn-rich minerals). Mantle-derived ultramafic rocks are found scattered throughout the belt with outcrop sizes up to several kilometres. Significant domains of marble, metagabbro and Cu ore deposits are present in the Besshi area of Shikoku.

Many workers treat the Mikabu Belt as distinct from the Sanbagawa Belt. However, the same type of relatively high-*P–T* metamorphism is recognized in both lithological domains and dating suggest the metamorphism was of a similar age (see summary in Suzuki & Ishizuka 1998). For these reasons, we consider the Mikabu rocks to represent a prominent domain of mafic to ultramafic lithologies that lies within the Sanbagawa Belt. To the south, the Mikabu Belt is in contact with the Chichibu Belt, which is an accretionary complex consisting mainly of oceanic sedimentary rocks. The relationship between the Chichibu and Sanbagawa belts is not well defined, but there is also evidence of broadly contemporaneous high-*P–T* metamorphism in the northern part of this belt. Similar evidence for high-*P* metamorphism is not seen in the southern part of the Chichibu Belt, and it has become common to distinguish the northern and southern parts as distinct tectonic units (Suzuki & Ishizuka 1998).

Most of SW Japan shows a very similar north–south-trending distribution of geological belts to that described above. The main exception is the Kii Peninsula where the Mikabu Belt is missing, as is the entire section of not only the Chichibu Belt but also the more southerly Kurosegawa and Sanbosan belts (also referred to as the Kurosegawa Sub-belt and Sambosan Group or unit, e.g.

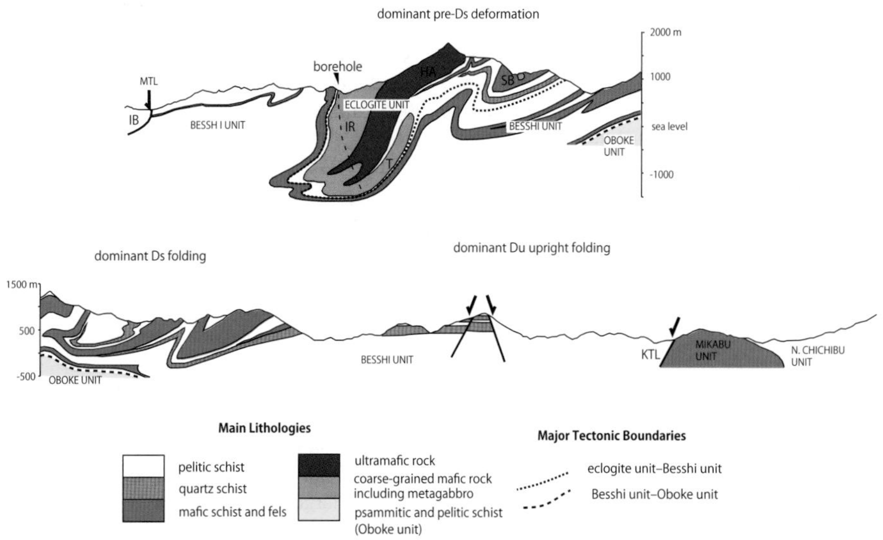

Fig. 2c.2. Cross-sections through the Sanbagawa Belt showing folded structure and main tectonic divisions of the Sanbagawa Belt in central Shikoku (based on Aoya & Yokoyama 2009 and Aoya *et al.* 2013*b*). IR, Iratsu; T, Tonaru; HA, Higashiakaishi; SB, Sebadani; KTL, Kiyomizu Tectonic Line; IB, Izumi Basin; MTL, Median Tectonic Line.

Chapter 2b). This leaves the Sanbagawa Belt in direct contact with the Shimanto Belt (Sasaki & Isozaki 1992). Without the intervening geological units, it is difficult to establish clear criteria to distinguish the Sanbagawa Belt from metamorphosed Shimanto Belt and some workers have proposed definitions that expand the domain of metamorphosed Shimanto belt and consequently greatly reduce the generally accepted distribution of the Sanbagawa Belt (Aoki *et al.* 2011).

In addition to the above rock types, several Miocene igneous bodies intrude the Sanbagawa Belt. It is not significant to our chapter, but it is worth noting that the diamond-bearing lamprophyre dyke reported by Mizukami *et al.* (2008) lies within the Sanbagawa Belt; it is part of the much younger Miocene event, however, and there is no reason to link this unusual occurrence of forearc diamond to the Sanbagawa metamorphism.

Protoliths and interpretation as a convergent margin

The Sanbagawa Belt is considered to consist primarily of oceanic crust and trench deposits. One clear example of deep-sea deposits is quartz schist, which is commonly associated with Mn-rich minerals and relict radiolarian fossils (e.g. Toriumi 1975). Chert and limestone of the low-grade regions locally contains identifiable radiolarians and conodonts (see summary in Isozaki & Itaya 1990) mostly of Triassic–Jurassic age although some are older. The presence of oceanic mafic rocks including pillow lava is further evidence that the Sanbagawa Belt contains oceanic crustal protoliths. Graphitic pelitic schist makes up a significant proportion of the Sanbagawa rocks. This is in common with the presence of Cretaceous black shales throughout the world in relatively shallow oceanic regions (e.g. Wignall 1994). There are also locally psammitic rocks and some conglomerate horizons (Shinjoe & Tagami 1994; Moriyama & Wallis 2002), possibly deposited as part of a deep-sea fan. These clasts include large (up to 50 cm diameter) felsic volcanic rocks (Suzuki 1927), showing the relative proximity of a continental margin. The suite of meta-siliciclastic rocks can be interpreted as continental-derived material that accumulated in an oceanic trench. This type of tectonic and sedimentary mixing of oceanic and continental material is one of the characteristics of convergent margins.

Geochemistry has been used to study the origin of the widespread mafic units. There is evidence for mid-ocean ridge basalt (MORB) chemistry for many of the mafic schist units in the main part of the Sanbagawa Belt (Okamoto *et al.* 2000; Nozaki *et al.* 2006). The mafic rocks of the Mikabu greenstone belt also show dominantly MORB geochemical affinities with minor amounts of rocks classified as oceanic island-arc basalt or hawaiite (Hashimoto 1989; Agata 1998; Tatsumi *et al.* 1998). The Iratsu body has been interpreted as having an origin as a seamount (Kugimiya & Takasu 2002; Aoya *et al.* 2006), an oceanic plateau (Terabayashi *et al.* 2005) and an intra-oceanic arc (Utsunomiya *et al.* 2011).

In addition to various rocks derived from the crust, mantle-derived serpentinite and other ultramafic rock types are also present as a minor but widely distributed component of the Sanbagawa Belt. These occur mainly as blocks in contact with pelitic schist and some metamorphic reaction is commonly observed along the boundary, typically forming actinolite or tremolite rock with a variably developed schistosity. The largest ultramafic body in the Sanbagawa Belt is the Higashiakaishi Peridotite. The origin of this body is contentious; both a hanging-wall mantle wedge origin (Hattori *et al.* 2010) and a footwall subducting plate origin (Terabayashi *et al.* 2005) have been proposed. Banno (2004) proposes that the smaller ultramafic bodies are a mixture of wedge mantle and subducted oceanic origins. A key study to resolve this issue is provided by the mapping of the occurrences of all known ultramafic rocks in an area of $23 \times 30\ km^2$ in central Shikoku (Aoya *et al.* 2013*a*). This work clearly shows mantle-derived rocks are only present in the higher-grade (garnet-grade and above) higher-pressure parts of the Sanbagawa Belt. There is no record of mantle-derived rocks in the lower-grade material that has been less deeply subducted. These observations imply there was a minimum depth requirement for subducted sediments to come into contact with the preserved mantle rocks. This is only compatible with a hanging-wall origin for all the observed mantle rocks.

Metamorphism

Subduction metamorphism characterized by high-*P–T* metamorphic conditions is typically associated with an assemblage of jadeitic pyroxene and quartz, and with the development of blueschist and eclogite rock types. In many cases the metamorphic minerals aragonite and lawsonite are also present. These petrological features are all present in the Sanbagawa Belt. However, the most widely developed mineral assemblages are broadly compatible with metamorphism characterized by moderate-*P–T* conditions more commonly associated with continental metamorphism. This stage of the Sanbagawa metamorphism has been termed Barrovian-type metamorphism by some workers. However, the Sanbagawa metapelite shows distinct mineral parageneses to the Barrovian type locality and the field metamorphic gradient is also clearly distinct. The appellation 'Barrovian' is strictly speaking inappropriate, but it is widely used in the literature and in the interest of simplicity we maintain this use here. Recrystallization in this secondary Barrovian stage shows that the Sanbagawa Belt represents an unusually warm subduction complex.

The difference between the relatively high-*T* subduction metamorphism of the Sanbagawa Belt and lower-*T* subduction metamorphism recorded in, for instance, the Franciscan Belt of the western USA was recognized early in the history of the research, and Miyashiro (1961) suggested the Sanbagawa Belt metamorphism should be described as intermediate–high-*P*/*T*. A good modern analogue for the Sanbagawa Belt is the present SW Japan subduction zone, where the young warm Philippine Sea Plate is being subducted.

Microstructural studies in the high-grade zones show the dominant Barrovian Sanbagawa metamorphism commonly overprints an earlier and less well-preserved phase of eclogite facies metamorphism (Fig. 2c.3). The two stages of metamorphism can be distinguished microstructurally and petrologically, and their development reflects a metamorphic history with two distinct temperature peaks. Both phases are interpreted as separate manifestations of an essentially continuous metamorphism in the same Sanbagawa convergent margin.

The grade of the later Barrovian phase metamorphism is generally discussed in terms of key Fe–Mg minerals developed in pelitic schist and separated into the chlorite, garnet, albite biotite and oligoclase biotite zones in order of increasing metamorphic grade (Higashino 1975; Enami 1982). Each zone formed over a range of metamorphic conditions with estimates as follows: 300–360°C, 5.5–6.5 kbar (chlorite zone); 425–495°C, 7.0–10.0 kbar (garnet zone); 470–590°C, 8.0–11.0 kbar (albite biotite zone); and 585–635°C, 9.0–11.0 kbar (oligoclase biotite zone) (Enami *et al.* 1994).

The chlorite zone can be locally subdivided into high- and low-*T* domains by the presence or absence of pumpellyite (Nakajima *et al.* 1977). This suite of mineral zones corresponds to the greenschist to epidote amphibolite facies series (Enami *et al.* 1994). Definition of these mineral zones has made a major contribution to documenting the thermal structure of the Sanbagawa Belt. However, most of the Sanbagawa Belt comprises the low-*T* chlorite zone, and the thermal structure of this extensive domain remains largely unknown. The metamorphic grade based on these minerals zones and other

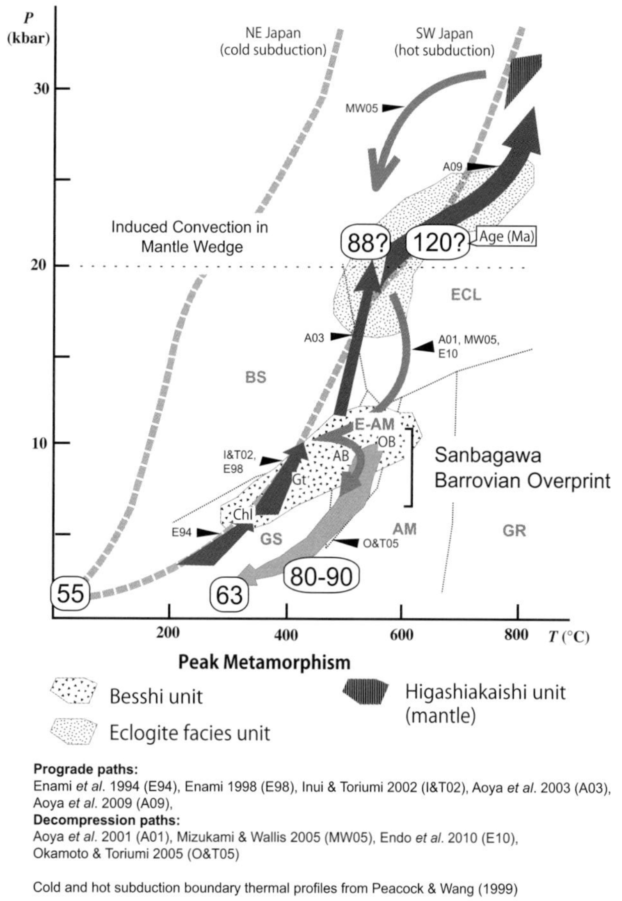

Fig. 2c.3. Summary of metamorphic information and chronological constraints in the Sanbagawa Belt of central Shikoku. For simplicity this figure does not include the earliest recorded phase of metamorphism in the granulite and amphibolite facies. Metamorphic facies abbreviations: ECL, eclogite facies; AM, amphibolite; BS, blueschist; GS, greenschist; E-AM, epidote amphibolite; GR, granulite. Chl, Gt, AB, OB refer to the chlorite, garnet, albite-biotite and oligoclase biotite zones of the secondary 'Barrovian' stage of the Sanbagawa metamorphism. The numbers associated with the *P–T* paths represent ages in units of Ma.

qualitative indications of metamorphic *T*, such as the degree of recrystallization of original features (sedimentary clasts, radiolarian tests, igneous minerals etc.), generally increases from south to north across the belt. However, in the well-studied Besshi and Asemi areas, the highest-grade material occurs within the structural pile and is bounded both to the north and south by lower-grade material (Banno & Sakai 1989; Higashino 1990; Wallis & Aoya 2000; Ota *et al.* 2004; Mori & Wallis 2010; Aoya *et al.* 2013*b*).

Eclogitic rocks are mainly found associated with but pre-dating the highest-grade part of the late metamorphic sequence. The clearest eclogites are found in coarse-grained mafic lithologies of the western and eastern Iratsu bodies (Takasu 1989; Ota *et al.* 2004; Endo *et al.* 2012). Careful field and microstructural studies have also revealed evidence for eclogite metamorphism affecting finer-grained mafic schist (Aoya & Wallis 1999; Wallis & Aoya 2000) and rarely pelitic schist (Matsumoto *et al.* 2003; Kouketsu & Enami 2010; Kouketsu *et al.* 2010). However, the rarity of characteristic minerals and the effects of the Barrovian overprint mean that the true extent of the eclogite facies metamorphism – particularly in the pelitic schists – is unclear. Recent work using Raman quartz residual pressure has had some success in recognizing eclogitic pressures in pelitic rocks (e.g. Mouri & Enami 2008; Fig. 2c.3). Attempts have been made to define boundaries of the eclogite unit based on the distribution of exotic rocks such as serpentinite (e.g. Wallis & Aoya 2000), but these have only had limited success.

Estimation of peak conditions

To discuss the peak conditions, we first distinguish between eclogitic and non-eclogitic domains. These two domains are thought by many workers to represent distinct structural units (Wallis & Aoya 2000). The non-eclogitic domain can be further subdivided into the Oboke and Besshi (or Shirataki) units, and a major tectonic boundary is thought to exist between these two (Takasu & Dallmeyer 1990; Aoki *et al.* 2011; Aoya *et al.* 2013*b*). The Mikabu Unit may be structurally linked to the lithologically distinct Oboke Unit (Wallis 1998), but here we consider estimates for metamorphic conditions of the Mikabu, Oboke, Besshi and Eclogite units separately.

Mikabu Belt. There are few quantitative estimates of *P* and *T* from this unit. The presence of metamorphic aragonite (Suzuki 1995) and sodic pyroxene and blue amphibole in mafic rocks (Suzuki & Ishizuka 1998) suggests that metamorphism was of the high-P/T type. The preservation of aragonite suggests temperatures less than *c.* 300°C, which is compatible with the restricted recrystallization and the common preservation of original minerals such as pyroxene. Estimates of metamorphic pressure based on the Na content of pyroxene is hampered by the small amounts of quartz in the mafic rocks, but metamorphic parageneses suggest metamorphism at the high-pressure pumpellyite actinolite facies (Suzuki & Ishizuka 1998), implying pressures of 400–600 MPa.

Oboke Unit. Aoki *et al.* (2008) use phase equilibria and amphibole compositions in mafic lithologies to estimate peak *P–T* conditions of 400–450 MPa and 240–270°C. Relatively low *T* (less than *c.* 300°C) accompanying deformation is indicated by dominance of solution transfer processes over crystal plastic and the presence of original clastic grains of feldspar and quartz. Most of this unit consists of metapelite and metapsammite that generally lack minerals suitable for *P–T* estimates. However, careful observation has revealed the presence of low-variance mineral assemblages of the pumpellyite-actinolite facies (Sakaguchi & Ishizuka 2008), and suggests the composition of epidote can be used as an indication of temperature. This temperature indicator shows a gradual decrease of *T* with increasing structural depth in the Oboke Unit.

Besshi Unit (also the Shirataki Unit of Aoya et al. *(2013b)).* The peak metamorphic conditions of the later-stage metamorphism have been derived from numerous methods including garnet-chlorite thermometry and Na-pyroxene-albite-quartz barometry (Banno & Sakai 1989; Enami *et al.* 1994). Barometry suggests a greater pressure of metamorphism at the same temperature to the north of the belt close to the MTL (Enami *et al.* 1994). This implies that the original isotherms also dipped to the north, which is in agreement with the idea of metamorphism in a subduction zone that dipped towards the MTL.

Eclogite Unit including the Higashiakaishi garnet peridotite. Eclogite facies rocks are found mainly in the Besshi area (Takasu 1989; Aoya 2001; Ota *et al.* 2004; Miyamoto *et al.* 2007; Endo 2010; Kabir & Takasu 2010). Eclogite facies metamorphism has also been reported in the Kotsu (Matsumoto *et al.* 2003; Tsuchiya & Hirajima 2013), Funaokayama (Endo *et al.* 2013) and Asemi river (Aoki *et al.* 2009) areas.

The peak *P–T* conditions of the eclogite facies metamorphism generally fall between 500 and 750°C and between 1.5 and 2.5 GPa. Ultra-high-pressure conditions (>3.0 GPa) are recorded in the Higashiakaishi mantle unit (Enami *et al.* 2004; Mizukami & Wallis 2005). There is an apparent gap in peak pressures suggesting the eclogite facies rocks can be divided into at least two tectonically distinct subunits, and this is supported by microstructural studies (Mizukami & Wallis 2005).

Early metamorphism

In addition to the eclogite and later metamorphic stages described in the previous sections, the eclogitic rocks also locally preserve evidence for earlier metamorphism in the granulite or high-*T* amphibolite facies (Yokoyama 1980; Endo *et al.* 2009, 2012). Endo *et al.* (2009) use a combination of petrological studies and Lu–Hf dating of garnet to propose that this early phase took place at *c.* 117 Ma. Other early phases of eclogite facies metamorphism have also proposed by Kabir & Takasu (2010). The nature of early phases of metamorphism is of considerable significance for the tectonic models that can be proposed for the region. However, at present it is unclear if a single phase or multiple phases should be considered and the interpretation in terms of the development of the Sanbagawa subduction zone remains speculative.

Subduction P–T paths: low P

One of the reasons for studying convergent-margin-type metamorphism is to reveal the physical conditions in the deeper and inaccessible parts of subduction zones. The peak temperatures and associated pressures summarized above are important, but to understand how these conditions were achieved and how they relate to the processes of subduction and exhumation, it is essential to determine both burial and exhumation parts of the *P–T* paths. *P–T* paths have been studied using chemical zoning in metamorphic minerals including Na–Px, Gt and amphibole (Enami *et al.* 1994; Enami 1998; Inui & Toriumi 2002; Aoya *et al.* 2003; Mizukami & Wallis 2005; Okamoto & Toriumi 2005) and pseudosection modelling (Kabir & Takasu 2010; Endo *et al.* 2012, 2013). Suitable minerals for similar approaches to determining *P–T* paths are rare in low-grade rocks and most of the information is from the Besshi and Eclogite units.

Compilations (Aoya & Wallis 2003; Ota *et al.* 2004) show that the prograde *P–T* paths generally describe a semi-continuous curve in *P–T* space characterized by major increases in *P–T* gradients with increasing *P* (Fig. 2c.3). Comparisons with the results of thermal modelling show that this type of curved *P–T* path is characteristic of subduction of a young slab, when a spreading ridge lies close to an active subduction zone (Peacock & Wang 1999; Uehara & Aoya 2005). The recorded prograde *P–T* paths provide a powerful argument that development of the Sanbagawa metamorphism was associated with subduction of a young slab and with the approach of an active spreading ridge. One of the underlying assumptions of this work is that the recorded prograde metamorphism is broadly synchronous, something that has yet to be conclusively demonstrated.

Subduction P–T paths: high P

A distinct but continuous section of the high-*P* prograde *P–T* path can also be defined at the highest recorded pressures. The contrast between the lower-*P* and high-*P* sections is revealed by plotting *P–T* re-equilibration conditions for the highest-grade units in the Iratsu area (Aoya *et al.* 2009). These conditions show a wide range of temperatures associated with a relatively small change in pressure from 2.0 to 2.5 GPa describing a flat-lying section of the *P–T* path. The presence of this change in curvature in the *P–T* path requires a major change in thermal structure in the original subduction zone at *c.* 65 km depth. Comparison with thermal models of subduction zones reveals a likely explanation for this two-tier structure to the prograde Sanbagawa *P–T* paths. Heat flow data from modern convergent margins show that there is a change from shallow depths where the subducting slab is essentially decoupled from the overlying mantle to depths greater than *c.* 80 km where the wedge mantle and slab become strongly coupled (Wada & Wang 2009). This coupling induces flow in the mantle wedge. The depth where the slab and mantle become coupled is associated with a rapid increase in *T* along the subduction surface due to the influx of hot mantle towards the cool slab. The flat *P–T* path recorded in the highest-grade part of the Sanbagawa Belt (Fig. 2c.3) provides strong petrological evidence for the presence of induced flow in the mantle wedge that starts at a depth of 65 km. This is a good example of how a combination of thermal modelling and observation of natural examples can improve our understanding of subduction zone processes.

Subduction P–T paths: anticlockwise paths

The Higashiakaishi Unit preserves a distinct *P–T* path from the above crustal rocks that shows an increase in *P* associated with a decrease in *T*: an anticlockwise *P–T* path (Mizukami & Wallis 2005). Endo *et al.* (2012) also recognize an early anticlockwise path in the western Iratsu body but at a lower pressure than in the Higashiakaishi Unit. A decrease in temperature with progressive subduction is characteristic of the earliest stage of convergence shortly after initiation of the subduction zone, and this is the interpretation placed on these paths by Mizukami & Wallis (2005) and Endo *et al.* (2012). Lu–Hf dating (Endo *et al.* 2009; Wallis *et al.* 2009) supports the idea that the anticlockwise *P–T* paths record a distinct early part of the Sanbagawa metamorphism.

Exhumation P–T paths

The decompression or exhumation *P–T* paths are potentially useful in discussions of exhumation rates and recognizing temporal changes in the thermal structure in the Sanbagawa subduction zone. Aoya (2001) and Endo (2010) present detailed examinations of the exhumation *P–T* path of the Eclogite Unit. These studies suggest that decompression was initially associated with a small but significant temperature rise before the Eclogite Unit became juxtaposed with the Besshi Unit at lower temperatures.

The next stage of the *P–T* evolution is a static phase of recrystallization associated with heating before further decompression (Enami *et al.* 1994; Aoya 2001; Mizukami & Wallis 2005). The proposed decompression *P–T* path with a double temperature peak (Fig. 2c.3) can explain why the later phase of recrystallization cuts across the boundaries of the eclogite metamorphism and the common presence of a phase of static intertectonic porphyroblast growth. The *P–T* conditions of the final stages of decompression have been studied using amphibole zonation (Okamoto & Toriumi 2005) and show decompression under relatively high thermal gradients of around 35°C km^{-1}.

Detailed thermal modelling of the exhumation *P–T* paths with double temperature peaks has not been carried out, but in qualitative terms this can be explained by the heating associated with approach of a spreading ridge to the subduction zone as the rocks are exhumed (Wallis *et al.* 2009). A simpler decompression *P–T* path with no double temperature peak has also been proposed (Ota *et al.* 2005), but this is based on less petrological information and does not adequately account for the microstructural development.

Complex P–T paths

Some workers propose more complex cycling *P–T* paths (Kabir & Takasu 2010) than those discussed in the previous sections. If correct, these would imply material is repeatedly subducted and exhumed during a single phase of convergence. The microstructural basis for these P–T paths is the evidence for repeated breakdown and formation of omphacitic pyroxene in eclogitic material. Other approaches to deriving P–T paths do not recognize similar cycling paths (Aoya 2001; Ota *et al.* 2004; Endo *et al.* 2012) and it is possible that simpler explanations are possible for the complex microstructural observations.

Structure and deformation

The concept of paired tectonic belts has focused our attention on the way that thermal conditions within convergent plate margins are related to large-scale tectonic movements. The Sanbagawa Belt represents the subduction side of such a system where inflow of relatively cold lithosphere produces cooling and metamorphism under high P/T conditions. The movement of lithosphere also causes deformation, which is recorded in the geological structures. This mechanical record of plate interaction is the subject of the present chapter.

The Sanbagawa Belt has undergone high-strain ductile deformation. The dominantly ductile nature of the deformation is reflected in the good lateral continuity of lithological units on the kilometre scale and the wide development of strong folding (Fig. 2c.2). This is in contrast to cooler subduction-type metamorphic belts such as the Franciscan Belt where deformation of some lithologies, such as chert, is dominantly brittle.

Labelling of deformation phases

The Sanbagawa Belt preserves multiple phases of deformation related to different metamorphic conditions and with distinct tectonic significance. Some nomenclature is needed to differentiate these distinct phases. In tectonic studies it is usual to denote different stages of deformation as D1, D2 etc. in order of sequence. This approach has been used in the Sanbagawa Belt (e.g. Kojima & Suzuki 1958; Faure 1985) where D1 refers to the most widespread high-strain deformation. However, D1 is not the earliest phase of ductile deformation and in many outcrops is it not the dominant phase. Various alternative schemes have been proposed for different areas and tectonics units. In this contribution, rather than adding to the complexity by proposing a new scheme, we select a combination of those already published that we consider the most convenient to describe the main features of deformation in the different tectonic units.

The dominant phase of ductile deformation in much of the Sanbagawa Belt (in particular the Besshi Unit) is referred to as Ds (Wallis 1990) with an associated schistosity Ss. This phase can be generally recognized as the dominant phase of ductile deformation that post-dates a well-defined phase of largely static intertectonic growth of albite porphyroblasts (Wallis 1998; Aoya 2002; Fukunari & Wallis 2007; Mori & Wallis 2010). Later phases of deformation are labelled Dt, Du etc. In contrast, in the main regions of eclogite metamorphism, penetrative Ds deformation is commonly weak and early tectonic histories are well preserved. Endo *et al.* (2012) proposes a deformation scheme for these regions where D_{E1} is the earliest recognized phase of deformation in the Eclogite Unit. Deformation phases in the Higashiakaishi Unit are also distinct, and here we use the scheme proposed by Mizukami & Wallis (2005) consisting of D1 to D4. Ds is equivalent to D_{E3} in the Eclogite Unit and D4 in the Higashiakaishi Unit.

Deformation history of the Sanbagawa Belt

Higashiakaishi Unit. The earliest deformation fabrics recorded in the Sanbagawa Belt are from the Higashiakaishi Unit (Mizukami & Wallis 2005). D1 is represented solely by the presence of aligned olivine, and the tectonic significance of this deformation phase is unclear. Dynamic recrystallization of D1 fabrics leads to D2 fabrics (Mizukami *et al.* 2004). The associated chemical zoning in orthopyroxene (opx) suggests D2 occurred during subduction at relatively high temperatures. The same deformation occurred during antigorite formation at the later stages, which places an important constraint on the P–T path (see earlier section). The D2 planer fabric, S2, is subsequently folded and the development of a new D3 foliation defined by aligned antigorite. D3 structures are cross-cut by new static mineral growth (mainly tremolite) related to heating before the Barrovian facies metamorphism, and this stage is a clear marker event that can be used in correlating deformation. Before D3 the Higashiakaishi Unit is at high pressure ($\geq$3.0 GPa), whereas it is at moderate pressures (*c.* 1 GPa) after D3. This change suggests the D3 deformation was related to a major phase of exhumation including amalgamation with the Eclogite Unit. A final phase of ductile deformation D4 forms a series of open folds. This phase can be correlated with Ds.

Eclogite Unit deformation. Structural analyses by Aoya (2002) and Endo *et al.* (2012) on separate parts of the Eclogite Unit give very similar results despite being established in two separate bodies. This agreement lends support to the idea that such deformation schemes have regional significance. Grain-shape-preferred orientation of barroisitic amphibole, omphacite and paragonite in metabasite defines the dominant D_{E2} fabric with both linear and planer fabrics. In contrast to the roughly east–west-aligned (or strike-parallel) stretching direction Ls of Ds that dominates in most of the Sanbagawa Belt (e.g. summary in Wallis *et al.* 2009), the orientation of L_{E2} shows considerable variation and on average is oriented approximately north–south (or dip-parallel) (Aoya & Wallis 1999; Kugimiya & Takasu 2002; Endo *et al.* 2012). The variation in the L_{E2} direction can be explained as the result of overprinting during the D_{E3} phase; open D_{E3} folding at a scale of centimetres to tens of metres is commonly observed. The D_{E3} deformation becomes strong in regions thought to be adjacent to the boundary with the non-eclogitic Besshi Unit. In these areas, tight to isoclinal D_{E3} folding is common and the S_{E2} schistosity is transposed to a new S_{E3}. Before this deformation the unit is at eclogite facies pressures of *c.* 2.5 GPa; after this deformation it is at pressures of *c.* 1 GPa. This phase of deformation therefore represents a stage of major decompression and exhumation. It is also the stage at which the Eclogite Unit becomes conjoined with the Besshi Unit.

The earliest deformation phase D_{E1} is recognized in the fabrics preserved in the hinges of isoclinal D_{E2} folds. A foliation S_{E1} is also recognized as alternating layers of granoblastic bimineralic eclogite and barroisite-rich eclogite. The tectonic significance of this fabric is unclear.

Besshi Unit and low-grade units. The Besshi Unit has undergone penetrative ductile deformation associated with the formation of a well-developed foliation. A variably well-developed stretching lineation and mineral lineation is oriented east–west to NW–SE (summary in Wallis *et al.* 2009). This lineation is oblique to the overall trend of the Sanbagawa Belt; in many publications this is described as orogen-parallel, but a more accurate description is orogen-oblique.

Large-sale folding is a characteristic of the Ds deformation. These folds are recognized by direct mapping of lithologies and changes in fold vergence (Aoya 2002; Mizukami & Wallis 2005; Mori & Wallis 2010; Aoya *et al.* 2013*b*) and confirmed in boring core (Hara *et al.* 1992; Aoya *et al.* 2013*b*). Ds folds locally develop a sheath geometry (Wallis 1990). There are also two phases of folding that post-date Ds: a locally well-developed Dt folding with relatively flat-lying axial planes; and a more ubiquitous phase of generally upright folds, Du (Wallis 1990; Aoya *et al.* 2013*b*). Du folds have axes lying with an en echelon relationship with the

MTL (Hara *et al.* 1992). These folds are also recognized in the Ryoke Belt (Okudaira *et al.* 2009). The meso- and microstructures of Du suggest that it represents a regional phase of deformation associated with subhorizontal shortening and left lateral shearing. There is also a pre-Ds deformation, Dr, recognized mainly in the hinges of Ds folds, which is associated with subduction and burial (Wallis *et al.* 1992). An indication of the degree of folding during the different stages is given in the cross-section of Figure 2c.2

Role of discontinuities

Differences in the peak metamorphic conditions and early deformation histories imply the presence of major tectonic discontinuities between the Besshi and Oboke units and between the Besshi and Eclogite units (Wallis 1998). However, defining the exact location of these discontinuities is rarely easy. Within the Besshi Unit, lithological layers commonly pinchout on metre to kilometre scales (Hara *et al.* 1990). This suggests the presence of discontinuities at a low angle to bedding but in many cases the stage at which these discontinuities developed is unclear; many are likely to have formed early in the evolution of the Sanbagawa Belt before metamorphism. There are also brittle to semi-brittle shear zones that occurred associated with low-grade metamorphism. These can be ascribed to a late stage of Ds or some independent phase (Takeshita & Yagi 2004; Fukunari & Wallis 2007; Fukunari *et al.* 2011).

A zone of strong upright foliation in the south of the Sanbagawa Belt is commonly referred to as the Kiyomizu Tectonic Line, and Aoya *et al.* (2013*b*) interpret this as a Du shear zone. The main brittle faults that affect the Sanbagawa Belt occur adjacent to the MTL, where both normal and strike-slip phases are recognized (Fukunari & Wallis 2007).

Finite strain and the kinematics of deformation

The regionally dominant phase of deformation is Ds and a considerable amount of information is available on the kinematics of this deformation. Despite the widespread presence of a strong foliation and stretching lineation (i.e. S–L tectonites) it has been suggested that the Sanbagawa Belt is characterized by uniaxial extensional strains (Toriumi 1985; Iwamori 2003). This conclusion is supported by finite strain analysis of deformed radiolarians in metachert. However, finite strain analysis in higher-grade zones of deformed conglomerate shows flattening strains (Moriyama & Wallis 2002) and these workers show that a more likely explanation for the uniaxial extensional strains is polyphase deformation with a flat-lying Ds overprinted by the horizontal shortening due to Du. The net result is uniaxial extension; this is the result of two independent tectonic phases however and, in this case, attempts to model the result in terms of a single tectonic event are unlikely to yield meaningful results. A similar result is reported by Okudaira & Beppu (2008).

In addition to the finite strain axes, a second key aspect of the kinematics of deformation – and one that is important to relate regional deformation to plate movements – is the sense of shear. Numerous studies have shown that Ds is dominantly associated with a top-to-the-west or -NW sense of shear (summary in Wallis *et al.* 2009). However, in some regions of low metamorphic grade at low structural levels, the opposite top-to-the-east sense of shear is well developed (Wallis 1995; Abe *et al.* 2001; Takeshita & Yagi 2004).

The top-to-the-west sense of shear and orogen-oblique stretching is generally interpreted to reflect subduction of the Izanagi Plate beneath the Sanbagawa Belt (Wallis 1998; Takeshita & Yagi 2004; Wallis *et al.* 2009). More detailed kinematic analyses of the Ds structures suggest that it is associated with deformation between simple and pure shear causing significant ductile thinning and contributing to the exhumation of the belt (Wallis 1992, 1995).

The top-to-the-east sense of shear shown by the low-grade low structural parts of the Sanbagawa Belt may be related to the movement of the Kula or Pacific plates that were in contact with Japan after the Izanagi subduction and which converge with the opposite sense of obliquity (Wallis *et al.* 2009). Top-to-the-east sense of shear reported in the Besshi area (Okamoto 1998) may be related to this phase. The obliquity of the Du fold axes suggests an overall dextral shear as the last major phase of ductile deformation in the Sanbagawa Belt. Masago *et al.* (2005) report a change in the sense of shear between the lower (top-to-the-SE) and upper (top-to-the-NW) boundaries of the Sanbagawa Belt. These workers propose the opposed senses of shear developed during the same tectonic event related to extrusion of a thin sheet. A mechanically more likely explanation is that the observed structures developed as the result of two separate deformation phases. Kinematic analysis of the fault zone at the top of the Sanbagawa Belt in the Kanto Mountains shows east–west shearing overprinted by top-to-the-north displacement (Wallis *et al.* 1990; Kobayashi 1996).

The Nagasaki metamorphic rocks show similar age and type of metamorphism to the Sanbagawa Belt and the two are commonly correlated. The main difference is the orientation of the stretching lineation: the lineation in the Nagasaki area is roughly north–south compared to the dominant east–west to SE–NW direction in much of the rest of the belt. This major change in stretching direction may reflect original along-strike tectonic changes in the Sanbagawa subduction belt. However, the recent tectonics of Kyushu are complex, including rotations probably related to the subduction of an aseismic ridge (Wallace *et al.* 2009). The difference may therefore be due to this localized and recent large-angle rigid body rotation.

Large-scale structure and correlation of tectonic units

Large-scale unit structure

The tectonic division of the Sanbagawa Belt into the Eclogite, Besshi and Oboke units given here is largely based on work in Shikoku. The Eclogite Unit extends to the Kii Peninsula (Endo *et al.* 2013) and in general this domain of high-grade rocks overlies lower-grade lithologies of the Besshi Unit. The ultramafic rocks recognized in Shikoku – which are derived from the mantle wedge (Aoya *et al.* 2013*a*) – are also seen at the highest structural levels in the Kii Peninsula (Hirota 1991). There is some dispute as to whether the Eclogite Unit should be treated as a coherent unit (Wallis & Aoya 2000) or a collection of independent blocks (Takasu 1989). The wide variety of different lithologies that have suffered eclogite facies metamorphism shows that at least parts of this unit have early distinct and discrete histories. However, most metamorphic studies show largely consistent subduction and exhumation *P–T* paths (Aoya 2001; Ota *et al.* 2004; Endo 2010) suggesting the separate parts of the Eclogite Unit amalgamated at depth before being exhumed together. Ideas that the eclogitic domain is a tectonic mélange with a serpentinite matrix (Takasu 1989) are not supported by detailed mapping (Aoya *et al.* 2013*a*, *b*) which shows that, although widespread, serpentinite is a very minor component of the Sanbagawa Belt.

In Shikoku, the Eclogite Unit is underlain by the Besshi Unit although the precise contact is not well defined. The two units have been strongly folded after their juxtaposition. Below the Besshi Unit lie the Oboke and Mikabu units. These are lower temperature and have suffered less ductile deformation than the Besshi Unit. Ar ages of phengite are younger in the Oboke Unit than the Besshi Unit. Further to the east in central Japan, the Sanbagawa Belt can be divided into two units separated by a fault and with distinctly higher metamorphic grade and older cooling ages close to the MTL. These distinct domains can be correlated with the Besshi and Oboke units (Nuong *et al.* 2011).

The Mikabu Unit is present intermittently along the southern margin of the Sanbagawa Belt and forms a distinct lithological marker. The relationship between the Mikabu Unit and the rest of the Sanbagawa Belt is contentious. In the Kii Peninsula and Kanto Mountains the Mikabu Unit structurally overlies the rest of the Sanbagawa Belt (Shimizu 1988; Abe *et al.* 2001). However, in Shikoku most evidence suggests the Mikabu Belt is structurally overlain by the Besshi Unit (Sakakibara *et al.* 1998; Aoya & Yokoyama 2009; Murata & Maekawa 2011). Wallis (1998) suggests that despite their different lithologies, the Mikabu Unit structurally correlates with the Oboke Unit.

Correlation between the Oboke Unit and Shimanto Belt

The Shimanto Belt is an accretionary complex that started to form in the Cretaceous and, like the Sanbagawa Belt, is a major geological constituent of SW Japan. The age of metamorphism in the Sanbagawa Belt overlaps with the formation of the Shimanto Belt. It is therefore clear that in some sense the two can be correlated. The Shimanto Belt contains a much higher proportion of coarse clastic sediments than the Sanbagawa Belt. This probably reflects the relatively shallow formation depths for the Shimanto Belt consisting mainly of offscraped trench-fill sediments, in contrast to the more deeply buried rocks of the Sanbagawa Belt with a greater proportion of rocks derived from oceanic basement. The two belts could therefore represent different levels of the same subduction zone. However, for much of their distribution, the Sanbagawa and Shimanto belts are separated by a significantly older domain of rocks: the Kurosegawa Belt. The Kurosegawa Belt has been interpreted as an arc-collision zone (Maruyama *et al.* 1984; Charvet 2013), a major strike-slip zone (Taira *et al.* 1983) and a superficial klippe (e.g. Isozaki *et al.* 1992). If the Kurosegawa Belt has accommodated major strike-slip movements or is the site of an arc collision, then the Shimanto and Sanbagawa belts represent contemporaneous belts that did not form adjacent to one another and cannot be simply correlated.

The most widely accepted view is that the Kurosegawa Belt represents a klippe and that there is structural continuity between the units to the north and south (Isozaki *et al.* 1992). In support of this hypothesis, several workers have emphasized similarities between the Oboke Unit and northern Shimanto Belt (e.g. Aoki *et al.* 2007) and propose the two can be directly correlated. This proposed correlation implies a large-scale flat-lying geometry of geological units in SW Japan and relegates the role of strike-slip faulting to a secondary event. The main lines of evidence for the proposed Oboke–Shimanto link are: (1) geological mapping in Kii Peninsula indicating that the Sanbagawa Belt rocks are locally in direct contact with the Shimanto Belt (Sasaki & Isozaki 1992); (2) similarities between bulk-rock compositions (Kiminami *et al.* 1999); and (3) consistent ages of metamorphism (Aoki *et al.* 2007). A seismic profile across SW Japan shows a series of north-dipping reflectors associated with large-scale folds (Ito *et al.* 2009). Some workers use this as evidence in support of a flat-lying thin nappe structure for the Sanbagawa Belt, although the details of the structure in the critical upper 10 km of crust are not well defined and other geophysical cross-sections suggest more steeply dipping contacts (Kato *et al.* 2010).

The results of geological mapping in several key areas (Isozaki *et al.* 1992) show steeply dipping contacts between the southern part of the Sanbagawa Belt and the proposed klippe. Other studies of the southern part of the Sanbagawa Belt in Shikoku also reveal steeply dipping contacts (Wallis 1998; Aoya *et al.* 2009; Murata & Maekawa 2011). The nappe hypothesis requires these units to be cut at a high angle by a major flat-lying discontinuity just below the present surface, and it is perhaps surprising there is no surface expression of this structure. However, Isozaki *et al.* (1992) proposes a structure where the steep dips are the result of tight folding of an originally flat-lying klippe.

Thickness of the Sanbagawa Belt

The thickness of tectonic units is a key factor when trying to relate *P–T* paths to exhumation rates and discussing mechanical models of exhumation: thick units are associated with a long time scale for cooling and greater buoyancy forces compared to the traction on the boundaries. There is an oft-repeated mantra that the total thickness of the Sanbagawa Belt is less than 2 km; this is clearly incorrect. The Besshi Cu mine in the Sanbagawa Belt of central Shikoku reached depths of 950 m below sea level and the highest mountain in the same region is higher than 1700 m. The belt as a whole dips to the north, implying a thickness of the Sanbagawa Belt in this region in excess of 7 km (e.g. Aoya *et al.* 2013*b*; Fig. 2c.2). There is also good evidence for significant ductile thinning of the belt during Ds deformation (Wallis 1995). This suggests the original thickness of the belt was even greater and claims that the whole Sanbagawa Belt exists as a sheet of no more than 2 km thickness should be disregarded.

Chronology

Two key pieces of evidence constraining the age of the Sanbagawa metamorphism are: (1) sparse fossil evidence yields ages for the low-grade part of the Sanbagawa Belt from Carboniferous to Jurassic times (see summary in Isozaki & Itaya 1990; Sakakibara *et al.* 1993); and (2) the local presence of *c.* 55 Ma continental deposits unconformably overlying the Sanbagawa schists that shows the main phase of Sanbagawa Orogenesis was complete by the early Palaeogene (Narita *et al.* 1999). In addition, mica Ar and garnet Lu–Hf radiometric dating shows the main metamorphism is Cretaceous (Monié *et al.* 1987; Faure *et al.* 1988; Itaya & Takasugi 1988; Takasu & Dallmeyer 1990; De Jong *et al.* 1992; Hirajima *et al.* 1992; Endo *et al.* 2009; Wallis *et al.* 2009) and U–Pb detrital zircon ages (Aoki *et al.* 2007; Tsutsumi *et al.* 2009; Otoh *et al.* 2010) suggest that some of the metasediments were deposited after middle Cretaceous times. Based on these results, there is good consensus that the tectono-metamorphic history of the Sanbagawa Belt mainly reflects plate movements in the East Asian margin during the middle–latest Cretaceous (Fig. 2c.4).

The main area of dispute in this field is the age of the peak of metamorphism (Fig. 2c.4). This is important because the difference between peak and cooling ages constrains the time scales for orogeny. Wallis *et al.* (2009) propose the Sanbagawa Orogeny was a short-lived event largely complete within a few million years. The more traditional view is that the peak metamorphism is older and the Sanbagawa Orogeny represents a process lasting several tens of millions of years (Isozaki & Itaya 1990; Aoki *et al.* 2008). The interpretation of metamorphic ages also plays a central role in proposed correlations between the Sanbagawa and Shimanto belts and divisions within the Sanbagawa Belt, both of which have major implications for the geological architecture of SW Japan. Here we summarize the main points of dispute.

There is good evidence for mineral growth in central Shikoku at *c.* 115–120 Ma recorded by both zircon U–Pb dating (Okamoto *et al.* 2004) and garnet Lu–Hf dating (Endo *et al.* 2009) in the Iratsu Unit. Some workers consider this age representative of the age of peak metamorphism throughout the Sanbagawa Belt. In contrast, Wallis *et al.* (2009) interpret 88 Ma Lu–Hf ages of garnet and omphacite as dating the peak of eclogite metamorphism. These workers use petrological and structural data to correlate this phase of mineral growth with most of the high-grade metamorphism implying the peak of

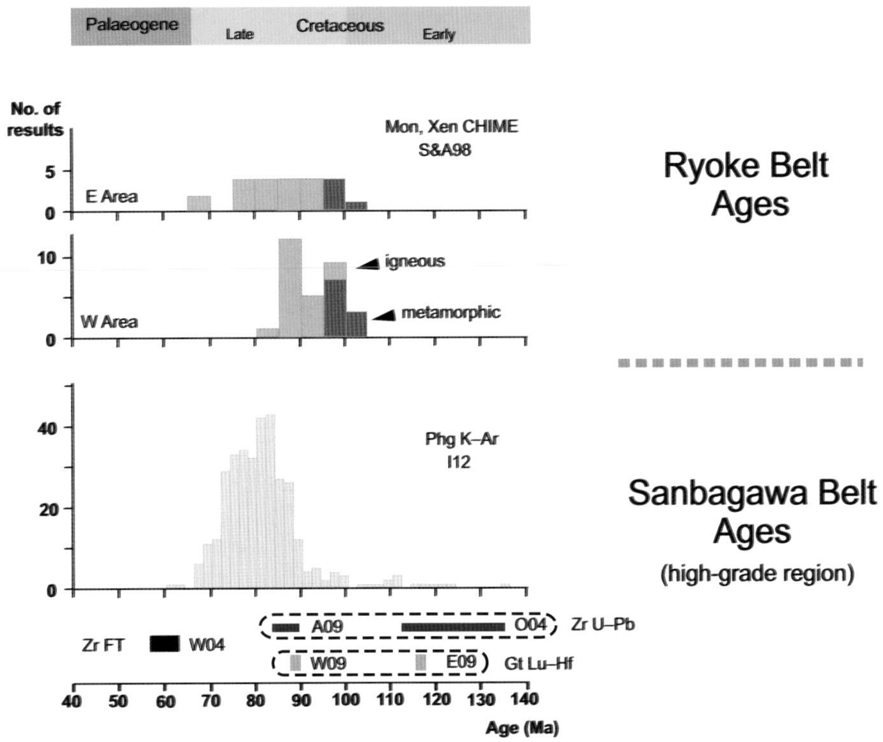

Fig. 2c.4. Summary of the radiometric dates of metamorphism in the Sanbagawa Belt of Shikoku showing the Sanbagawa Orogeny is largely restricted to the Late Cretaceous. S&A98, Suzuki & Adachi (1998); I12, Itaya *et al.* (2011); W04, Wallis *et al.* (2004); A09, Aoki *et al.* (2009); W09, Wallis *et al.* (2009); O04, Okamoto *et al.* (2004); E09, Endo *et al.* (2009).

metamorphism for most high-grade material is *c.* 88 Ma. A peak age at *c.* 88 Ma is supported by U–Pb dating of zircon rims in the Besshi Unit of *c.* 85.6 Ma (Aoki *et al.* 2009). However, this zircon age cannot be unequivocally linked to the metamorphic history and Aoki *et al.* (2009) interpret the age as recording growth during retrograde recrystallization. The Lu–Hf dating of undisputed metamorphic minerals allows a clear link to be made between the age data and metamorphic history. However, there were clearly several stages of distinct mineral growth during the Sanbagawa metamorphism. Settling the dispute about the age of peak metamorphism will require more well-defined age data that can be clearly linked to the metamorphic history. More details of this debate are presented in recent publications (e.g. Endo *et al.* 2009; Wallis *et al.* 2009; Aoki *et al.* 2010, 2011; Wallis & Endo 2010; Itaya *et al.* 2011; Aoya *et al.* 2013*b* pp. 67–68).

U–Pb ages of detrital zircon derived from siliciclastic rocks have made a significant contribution to our understanding of the chronology of the Sanbagawa Belt. Much of the detrital material is younger than the fossil ages of the oceanic basement recorded in the low-grade zones. The youngest zircon grain recorded in the Oboke Unit of central Shikoku yields an age of 82 ± 11 Ma (Aoki *et al.* 2007). If the age of peak metamorphism of the high-grade zones is taken to be 120 Ma, then the Oboke zircon ages imply two distinct metamorphic periods separated by several tens of millions of years. However, if the age of the peak of high-grade metamorphism is taken to be 88 Ma, then the two age datasets are compatible and only a single metamorphic phase is required. These considerations underlie much of the discussion about whether the Oboke and related regions should be considered part of the Shimanto Belt and distinct from a more narrowly defined Sanbagawa Belt.

Another string of the argument concerning possible correlations between the Sanbagawa and Shimanto belts is based on estimates of the age of peak metamorphism in the Oboke Unit. Aoki *et al.* (2008) interpret K–Ar ages of *c.* 65 Ma as the peak age. If correct, this would show that metamorphism in the Oboke Unit is indeed significantly younger than the overlying higher-grade Besshi Unit which shows cooling to *c.* 250°C at the same time (Wallis *et al.* 2004). However, the zircon fission track study in the Oboke Unit demonstrates the presence of a secondary phase of heating in the area (Wallis *et al.* 2004). Wallis *et al.* (2004) suggest the re-heating may be correlated to the widespread middle Miocene phase of igneous activity in SW Japan. Such reheating may have caused degassing of Ar, and the Ar ages may therefore be underestimating the true age of the peak metamorphism.

Detrital zircon in the Kii Peninsula and the Kanto Mountains yield U–Pb age populations with the youngest ages of 71 ± 9.2 Ma (Otoh *et al.* 2010) and 78.8 ± 1.3 Ma (Tsutsumi *et al.* 2009), respectively. Both of these age ranges are incompatible even with the younger 88 Ma age estimate of the peak of the high-grade metamorphism of the Besshi Unit. This can be interpreted as evidence for two separate metamorphic phases within the Sanbagawa Belt. However, there is no independent constraint on peak ages of high-grade material in these other areas, and the young ages may reflect diachronous metamorphism along the orogen.

Ar ages of the low-grade units have also been reported throughout the Sanbagawa Belt. In central Shikoku Ar–Ar and K–Ar ages (Itaya & Takasugi 1988; Takasu & Dallmeyer 1990) give ages in the range of 71–63 Ma. Aoki *et al.* (2008) suggest some of this spread is due to inclusion of some detrital material and give an estimate for the age of metamorphic mica of 65–61 Ma. These authors argue that peak temperature of metamorphism in the Oboke Unit is close to the closure temperature for these methods, implying that these ages reflect the peak of metamorphism. Ar dating in central Japan (Nuong *et al.* 2011) shows a separation into two different units with the younger one further from the MTL. K–Ar dating in the Kanto Mountains (Hirajima *et al.* 1992) shows older ages at high structural levels and low metamorphic grade and younger ages at low structural levels and high metamorphic grade (chlorite zone=84–72 Ma; garnet zone=78–66 Ma; biotite zone=67–53 Ma). The younging of ages with increasing structural depth is the same as that observed in the Sanbagawa

Belt of Shikoku. Takasu *et al.* (1996) suggest there is a progressive younging in Ar cooling ages from west to east along the Sanbagawa Belt.

Plate movements and the formation of the Sanbagawa Belt

Records of plate movements

There have been several attempts to link the tectonic history recorded in the Sanbagawa Belt with the reconstruction of plate movements in the Pacific realm. It is generally accepted that there are four plates at this time that need to be taken into consideration: the Farallon, Izanagi, Kula and Pacific plates. This four-fold division follows Engebretson *et al.* (1985). There has been some revision of the plate constructions that allow for relative movement of hot spots (Doubrovine *et al.* 2012). However, the general features of the plate divisions and their associated movement directions have not changed, and here we refer to the Engebretson *et al.* (1985) model.

Plate reconstructions using magnetic anomalies and bathymetry allow the direction of movement between the Farallon, Izanagi, Kula and Pacific plates and the eastern Asian margin to be estimated. Such reconstructions cannot however locate the plate boundaries. One way to constrain which plate was responsible for the Sanbagawa metamorphism is to consider the on-land deformation. Normal convergence will be associated with movements roughly perpendicular to the orogen. In contrast, highly oblique plate convergence will be associated with movements that have an orogen-parallel component. The sense of shear may also be used to determine the sense of obliquity of convergence. Several studies relate the formation of the Sanbagawa Belt with the subduction of the Kula Plate. However, the roughly normal movement direction of this plate during middle Cretaceous times is not compatible with the orogeny-oblique lineations of the Sanbagawa Belt. Wallis *et al.* (2009) use such information in combination with sense-of-shear information to infer that subduction of the Izanagi Plate caused the Sanbagawa metamorphism.

Interaction with a spreading ridge

Since the early days of plate tectonics there has been a recognition that multiple spreading ridges existed in the Mesozoic Pacific realm and that at least one major phase of ridge subduction has taken place beneath Japan (Larson & Chase 1972). In addition, there have for a long time been suggestions that some aspects of the geology of SW Japan are related to the approach and passing of a spreading ridge (Miyashiro 1973; Maruyama *et al.* 1997; Aoya *et al.* 2003; Okudaira & Yoshitake 2004; Wallis *et al.* 2009).

In the Sanbagawa Belt, the metamorphic *P–T* paths provide some of the best evidence for the involvement of a spreading ridge in orogenesis. The strongly curved prograde *P–T* paths are a clear indication of subduction of a young slab, implying that an active spreading ridge was located close by (Aoya *et al.* 2003; Uehara & Aoya 2005). The subsequent approach of the ridge can also account for the dual temperature peak recorded by the metamorphism. Plate reconstructions suggest the Izanagi Plate subducted rapidly and resulted in the approach of the spreading ridge at the Pacific Plate (Engebretson *et al.* 1985). The arrival of a ridge at a subduction zone is likely to result in a major rearrangement of plate movements such as that recognized at *c.* 110–80 Ma (Engebretson *et al.* 1985). This is in good agreement with the Lu–Hf ages of 88 Ma for the peak of metamorphism and some of the U–Pb zircon ages (see 'Chronology' section). However, it should be borne in mind that this period corresponds to the Cretaceous normal superchron where there was an exceptionally long period of stability in the polarity of the Earth's magnetic field. This limits the amount of information that can be gleaned from magnetic anomalies for this time.

Ridge approach can also be related to exhumation processes. One proposal is that the forearc mantle is sufficiently rigid to act as a cantilever and, when the buoyant ridge enters the subduction zone, high-*P* metamorphic rocks are squeezed back to the Earth's surface (Maruyama *et al.* 1996). The strength of the mantle is highly dependent on temperature. It is therefore likely that the mantle close to the subducting slab is cold enough to be relatively strong. However, thermal modelling predicts that the temperature increases rapidly towards the centre of the wedge. The presence of melts and fluids means that there is very little long-term strength in these regions, and the wedge as a whole cannot be considered a rigid block on geological time scales. An additional problem with this proposed mechanism is that it implies orogeny-perpendicular movements throughout the exhumation process. This is not compatible with the observed transport direction.

Wallis *et al.* (2009) suggest that exhumation occurs in at least two distinct phases, initially at rates of more than 1 cm a^{-1} followed by much slower rates of millimetres per annum. The initial stage of exhumation is triggered by the rise in temperature associated with ridge approach. The increase in *T* will cause a major reduction in the strength of the subducted rocks. These include metapelite and amphibole-rich metabasite that are more buoyant than the mantle. The combination of buoyancy forces and weakening allows the subducted rocks to detach from the down-going slab and rise to lower crustal levels (Endo *et al.* 2012). This model requires a stiff but not necessarily rigid leading edge to the overlying wedge mantle. Wallis *et al.* (2004) suggest that the later stages of exhumation are dominated by ductile extension and erosion. A summary of the large-scale processes is given in Figure 2c.5.

There are proposals for alternative models of the Sanbagawa Orogenesis that involve collision of continental blocks or arcs rather than interaction with a ridge (e.g. Maruyama *et al.* 1984; Charvet 2013). Charvet (2013) considers that the high rate of exhumation, presence of nappes and clockwise *P–T* paths in the Sanbagawa Belt are all incompatible with oceanic subduction. However, as we have outlined above, we consider these features are adequately explained by processes of oceanic subduction. Geochemical evidence reveals the incorporation of old crustal material, which is compatible with many different origins including a subducted arc. The present location of the putative arc remains unclear.

Kinematics of deformation and large-scale tectonics

Oblique convergence can account for the obliquity of the Ds stretching lineation in the Sanbagawa Belt, but in the reconstruction suggested in the previous section the finite stretching direction is not parallel to the predicted movement direction. This can be accounted for by a combination of oblique convergence and gravity-driven movements perpendicular to the wedge. The action of gravity-driven deformation is supported by the relatively low degree of noncoaxiality of Ds deformation.

The orientation of the pre-Ds stretching is clearly distinct from Ds and in most locations is roughly orogen-perpendicular, although this direction has been strongly affected by Ds deformation. One interpretation of this change in lineation is to link them temporally and suggest they represent different parts of a cyclic flow in the subduction wedge (e.g. Toriumi & Kohsaka 1995). However, as discussed in the section 'Structure and deformation', deformation in the Sanbagawa Belt can be separated into distinct phases with clear overprinting relationships; lumping these together leads to erroneous inferences. The orogeny-perpendicular lineations shown in the eclogite units formed during exhumation at a time when the Besshi Unit was still subducting. The most reasonable

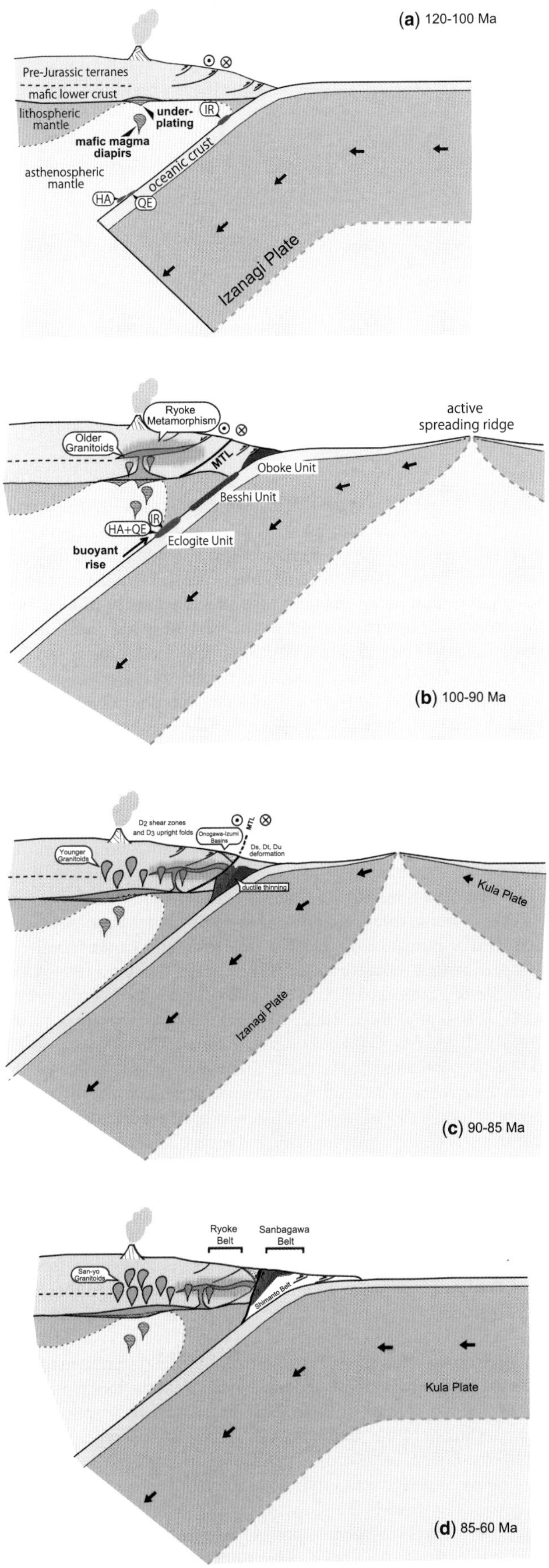

Fig. 2c.5. Plausible tectonic model in the Cretaceous SW Japan. (**a**) 120–100 Ma: During this period the oldest fragments of the Sanbagawa metamorphic domain were formed including the Iratsu and quartz eclogite blocks. The early prograde Ryoke metamorphism occurred associated with emplacement of I-type Older Ryoke granitoids formed by partial melting of the lower crustal mafic rocks. This early history may be related to the onset of a subduction cycle and before cooling to steady state. (**b**) 100–90 Ma: Large-scale dehydration melting producing I-type magma that can be accounted for by approximately 20 vol.% melt of amphibolites that is likely to occur at *c.* 950°C (e.g. Petford *et al.* 2000). Deep-seated horizontal shear zones associated with magma intrusion may have formed a crustal-scale detachment in the forearc region. The main units of the Sanbagawa Belt were subducted at this time. There was some buoyant return flow along the subduction boundary by the deepest subducted units: the quartz eclogite and the Higashiakaishi blocks. (**c**) 90–85 Ma: This is the period when the Izanagi–Kula plate ridge boundary was approaching Japan. This is the main period of exhumation in the Sanbagawa Belt. The exhumation from mantle to lower crustal depths is driven by buoyancy triggered by a decrease in the strength of the rocks cause by an increase in temperature due to the ridge approach. At crustal levels, tectonic thinning, both ductile and brittle, combines with erosion to bring the high-*P* rocks to the Earth's surface. Deformation in this period includes phases of both horizontal extension and horizontal shortening. The shortening is shown by the presence of upright folding in both the Sanbagawa (Du) and Ryoke (D_3) belts. Large-scale strike-slip faulting along the MTL is associated with the formation of the Onogawa–Izumi sedimentary basins and excision of part of the forearc. As a consequence there is a relative shift in the locus of the volcanic activity. (**d**) 85–60 Ma: There was an increase in the age of the oceanic lithosphere after the subduction of an oceanic ridge. The associated igneous activity moved continent-wards to form the San-yo granitoids. At the same time, left-lateral pull-apart forearc basins continued to form, and these were the site for deposition of the Onogawa–Izumi Group.

explanation for this set of lineations is the operation of buoyancy forces acting in the direction of dip, up the subduction boundary (Endo *et al.* 2012).

After the Izanagi–Kula Ridge arrived at the Sanbagawa convergent margin, it is likely that left-lateral subduction of the Izanagi Plate would be followed by right lateral subduction of the Kula Plate, and this may be seen in the top-to-the-east sense of shear shown by the lower-grade regions (see 'Finite strain and the kinematics of deformation' above). This shift from the tectonics related to the Izanagi to the Kula plates may also be recorded in the Shimanto Belt (Tokiwa 2009).

The Ryoke Belt

Distribution and lithology

The Ryoke Belt consists of the Cretaceous I-type/ilmenite-series granitoids and low-*P–T* metamorphic rocks that are derived mainly from the Jurassic accretionary complex of the Mino–Tamba Terrane (e.g. Banno & Nakajima 1992; Kagami *et al.* 1992; Nakajima 1994, 1996). The Ryoke Belt is exposed throughout SW Japan from Kyushu to the Kanto Region, extending for nearly 1000 km in length. This belt has a width of 30–50 km (Fig. 2c.6) and is similar in size to that of the Sierra Nevada and Peninsular ranges batholiths in western North America. The MTL represents the southern boundary fault to the Ryoke Belt (e.g. Hara *et al.* 1980; Ichikawa 1980). The Ryoke Belt is a part of the circum-Pacific Cordilleran-type orogenic belt formed at an ancient arc-trench system or active continental margin during Cretaceous–Paleocene times (e.g. Ishihara 1977; Nakajima 1994, 1996).

The Ryoke metamorphic rocks have been considered to be a typical example of low-*P* facies series or andalusite-sillimanite type of Miyashiro (1961). However, metamorphic rocks occupy a mere 20–30% of the total area of the belt: the remainder comprises granitoids. Granitoids in the Ryoke Belt are divided into two groups: Younger

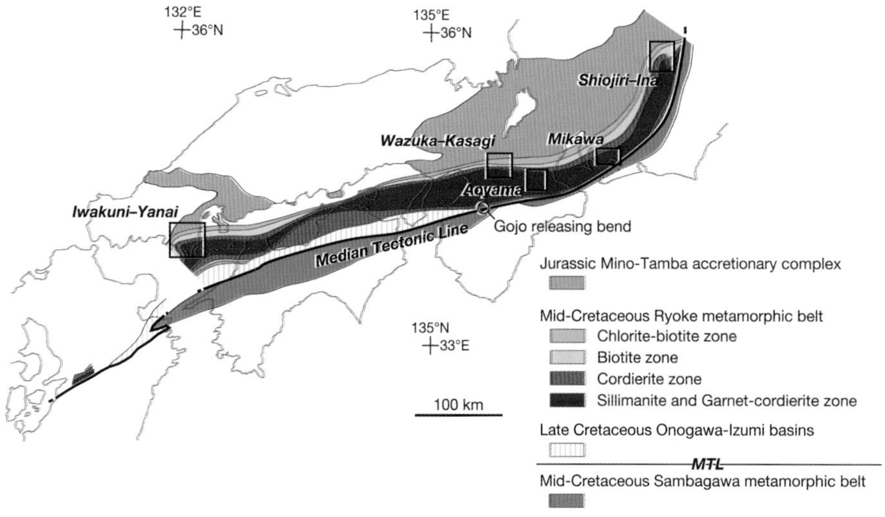

Fig. 2c.6. Distribution of the Ryoke Belt, SW Japan (modified from Okudaira *et al.* 2009).

Ryoke granitoids and Older Ryoke granitoids (e.g. Koide 1958). The Younger Ryoke granitoids are nearly unfoliated and were intruded as relatively small stock-like bodies cross-cutting the general structure of the weakly metamorphosed host rocks. In contrast, the Older Ryoke granitoids are foliated and occur as sheet-like bodies intruding high-grade gneisses concordant with their gneissose structure. The Cretaceous San-yo granitoids that are located north of the Younger Ryoke granitoids are unfoliated and occur as large batholithic bodies. The Older Ryoke granitoids are therefore synkinematic and the Younger Ryoke and San-yo granitoids are post-kinematic. Details of the Cretaceous–Paleocene granitoids are described in Chapter 4; here we only present a summary of the main petrogenetic features of the Ryoke granitoid rocks.

Geochemical characteristics of the granitoids are similar to other I-type granitoids in the world, implying that the major source of these granitoids is mafic magmatic rocks or its metamorphosed equivalents (Kagami *et al.* 1992; Nakajima 1996). Kutsukake (2002) indicated that the I-type granitoids of the Ryoke Belt could be generated by the dehydration melting of amphibolite or hydrous melting of tholeiite at pressures of $\geq$1 GPa, based on their rare earth element (REE) patterns. Furthermore, minor element chemistry shows these granitoids are best designated as arc-type granitoids (Kagami *et al.* 1992; Kutsukake 1993, 2002; Nakajima 1996). On the initial $^{87}Sr/^{86}Sr$ (SrI) and ϵNd diagram the Ryoke granitoids plot in high SrI and low ϵNd region (Yuhara *et al.* 2000), and any contribution of upper crustal recycled materials, that is, sediments of an accretionary complex, seems to be small (Kagami *et al.* 1992). Geochemical characteristics of gabbroic and dioritic rocks within the Ryoke Belt suggest that they are closely related to the granitoids (Iizumi *et al.* 2000; Nakajima *et al.* 2004).

Metamorphism

Metamorphic zonation

Metamorphic rocks have been subdivided into a number of mineral zones (Fig. 2c.6), with the highest-grade zone being lower granulite facies (800°C at 600 MPa; Ikeda 2002, 2004). The metamorphic field gradient (apparent geothermal gradient) is up to *c.* 50°C km^{-1}, suggesting that the belt represents mid-crustal material that formed beneath a volcanic front.

The main metamorphism occurred during middle Cretaceous times (*c.* 100 Ma; Suzuki & Adachi 1998). The protolith of most of the metamorphic rocks is considered to be a Jurassic accretionary complex of the Mino–Tamba Terrane (e.g. Ichikawa 1990; Banno & Nakajima 1992). The non- to weakly metamorphosed sedimentary rocks of the Mino–Tamba Terrane show a gradual transition into the higher-grade Ryoke metamorphic rocks, with the boundary being rather arbitrarily defined as the first appearance of metamorphic biotite within pelitic rocks. The highest-grade rocks of the Mino–Tamba Terrane are chlorite-bearing metapelites, and the lowest-grade rocks of the Ryoke metamorphic belt are biotite–chlorite metapelites (Fig. 2c.6). The Ryoke metamorphic rocks represent that part of the Mino–Tamba accretionary complex, which was regionally metamorphosed due to the intrusion of the sheet-like Older Ryoke granitoids (Okudaira *et al.* 1993, 2001; Brown 1998; Ikeda 1998; Miyazaki 2007, 2010).

Using pelitic mineral assemblages in several districts (Iwakuni–Yanai, Wazuka–Kasagi, Aoyama, Mikawa and Shiojiri–Ina), the progressive metamorphism can be described as a sequence of six zones (from the lowest to the highest): chlorite–biotite (Chl–Bt); biotite (Bt); muscovite–cordierite (Ms–Crd); K-feldspar–cordierite (Kfs–Crd); sillimanite–K-feldspar (Sil–Kfs); and garnet–cordierite (Grt–Crd) zones (Okudaira *et al.* 1993, 2001; Nakajima 1994; Okudaira 1996*b*; Hokada 1998; Ikeda 1998, 2002, 2004; Ozaki *et al.* 2000; Kawakami 2001, 2002; Miyazaki 2007). The areas in the Chl–Bt, Bt and Ms–Crd zones locally overlap with contact aureoles of the post-kinematic granitoids. Even though the extent of the contact aureoles is difficult to discern, cordierite with trilling and randomly oriented muscovite are taken as evidence for contact metamorphism. The following description for each mineral zone is based on the study of the Iwakuni–Yanai area by Ikeda (1998). In the Higo district (Kyushu) and the Tsukuba district (Kanto Region), the low-*P*–*T* metamorphic rocks equivalent to rocks of the Ryoke Belt have been observed (e.g. Osanai *et al.* 1998; Miyazaki 1999, 2004). Because the distribution of these rocks is completely isolated from the Ryoke Belt, we do not describe the rocks from these districts in detail.

Chl–Bt zone. This zone is characterized by the occurrence of biotite coexisting with chlorite and muscovite. Near the biotite isograd, biotite is fine-grained (<50 μm long) and occurs only sporadically as a constituent of the schistosity. The dominant assemblage in pelitic rocks is Chl + Bt + Ms + plagioclase (Pl). K-feldspar occurs only rarely in pelitic rocks, but is common in siliceous rocks.

Bt zone. The disappearance of chlorite from most pelitic rocks defines the start of the biotite zone. Both biotite and muscovite are coarse-grained (0.1 mm long) compared with those in the chlorite–biotite zone. In this zone, the dominant mineral assemblage is Bt + Ms + Pl.

Ms–Crd zone. The coexistence of cordierite and muscovite defines the zone. Cordierite is minor, and now occurs as fine-grained pinite in the assemblage of Crd + Bt + Ms + Pl. However, this assemblage is uncommon and the most frequent assemblage in the pelitic rocks is the same as that of the biotite zone. The grain sizes of muscovite and biotite are similar to those in the biotite zone and increase up to 0.3 mm.

Kfs–Crd zone. The first appearance of cordierite that coexists with K-feldspar defines the start of this zone. There is a marked reduction in the preferred orientation of muscovite and biotite, and the development of compositional banding composed of quartzo-feldspathic and micaceous layers occurs in this zone. The first appearance of andalusite coincides approximately with this isograd. In the high-grade part of this zone, andalusite (And) is transformed to sillimanite. Most pelitic and siliceous rocks contain muscovite: the muscovite in this zone and at higher metamorphic grades is thought to be of retrograde origin. However, there is a possibility that some of the muscovite was stable at the peak metamorphism. The representative mineral assemblages in pelitic rocks are And and/or Sil + Crd + Bt + Kfs + Pl and Crd or Grt + Bt + Kfs + Pl. A tourmaline-out isograd that is defined by the disappearance of tourmaline is delineated within this zone in the Iwakuni–Yanai district (Kawakami & Ikeda 2003), but within the Grt–Crd zone in Aoyama area (Kawakami 2001).

Sil–Kfs zone. The sillimanite–K-feldspar zone is defined by the occurrence of sillimanite and K-feldspar in the matrix assemblage. Sillimanite occurs as fibrous or columnar grains and is commonly associated with biotite. The assemblages of Sil + Bt + Kfs + Pl and Sil + Grt + Bt + Kfs + Pl are dominant in pelitic and siliceous rocks. Cordierite is less common in this zone compared to the Kfs–Crd zone, coexisting with sillimanite, biotite and K-feldspar while no coexisting garnet is found.

Grt–Crd zone. The coexistence of garnet and cordierite is characteristic of this zone. The dominant mineral assemblages in pelitic rocks are Crd and/or Grt + Bt + Kfs + Pl. The Grt + Bt + Kfs + Pl assemblage is dominant in both pelitic and siliceous rocks. Cordierite locally includes fibrous or rounded sillimanite in the centre of a single grain. Spinel is observed only in this zone and occurs rarely along grain boundaries within cordierite clusters. The spinel coexists with quartz, biotite and K-feldspar at the peak metamorphism. Orthopyroxene occurs only in mafic rocks of this zone and coexists with hydrous minerals such as biotite, hornblende (Ikeda 2002). In the Aoyama area, metatexite and inhomogeneous diatexite developed in the Grt–Crd zone (Kawakami 2001).

P–T *conditions*

As shown in Figure 2c.7, based on internally consistent *P–T* estimation, Ikeda (2004) estimates the *P–T* conditions of each zone in the Iwakuni–Yanai district to be: *c.* 470°C at 20–50 MPa (Chl–Bt zone); 450°C at 90–120 MPa (Bt zone); 520–550°C at 40–100 MPa (Ms–Crd zone); 600–680°C at 120–370 MPa (Kfs–Crd zone); 700–760°C at 500–640 MPa (Sil–Kfs zone); and 790–860°C at 460–760 MPa (Grt–Crd zone). In the Aoyama district, *P–T* conditions are estimated to be 300–400 MPa and 620–670°C for Sil–Kfs zone and 450–600 MPa and 650–800°C for Grt–Crd zone (Kawakami 2002). In the Mikawa district, *P–T* conditions estimated for Bt, Sil–Kfs and Grt–Crd zones are 510–590°C at 290–370 MPa, 570–710°C at 370–430 MPa and 720–800°C at 430–570 MPa,

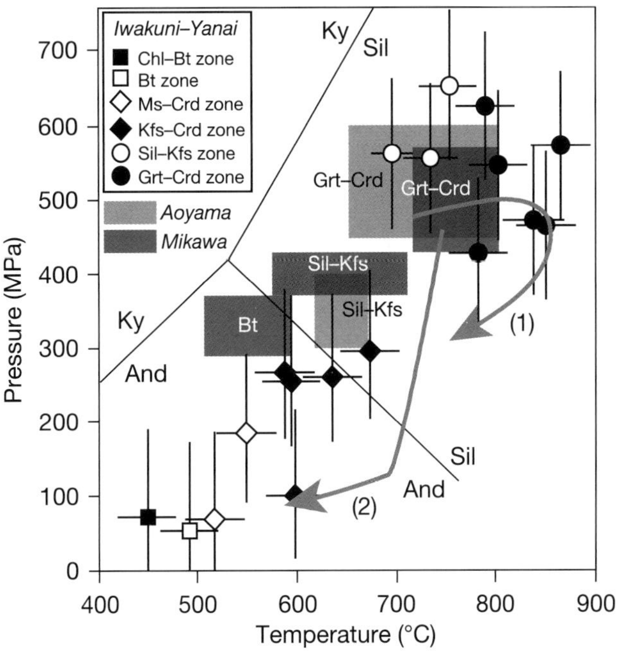

Fig. 2c.7. Temperature and pressure conditions for rocks in the Iwakuni–Yanai district (Ikeda 2004), the Aoyama district (Kawakami 2002) and the Mikawa district (Miyazaki 2010). *P–T* paths of the Grt–Crd zone in the Iwakuni–Yanai district (1: Brown 1998) and Aoyama district (2: Kawakami 2002) are also shown.

respectively (Miyazaki 2010). The inferred metamorphic field gradient of the metamorphic belt is 40–50°C km^{-1}.

P–T *paths*

Metamorphic *P–T* paths for the rocks in the Iwakuni–Yanai district have been estimated petrologically (*P–T* path (1) in Fig. 2c.7; Brown 1998); the prograde and retrograde *P–T* paths for the Kfs–Crd zone and Grt–Crd zone are nearly isobaric, suggesting the possibility that the heat source for the low-*P–T* metamorphism is that supplied by the intrusion of the synkinematic granitoids. However, in the Aoyama district, the crystallization of magmatic andalusite from anatectic melt in the low-temperature part of the Grt–Crd zone after the peak *T* of metamorphism shows that the *P–T* path includes a period of near-isothermal decompression after the thermal peak (*P–T* path (2) of Fig. 2c.7; Kawakami 2002): a clockwise loop rather than a hairpin shape. Kamitomo *et al.* (2008) also suggested near-isothermal decompression after the thermal peak, based on the reaction microstructure of garnet + quartz → orthopyroxene + plagioclase observed in garnet-bearing mafic granulite xenoliths in the synkinematic granitoids of the Iwakuni–Yanai district. Okudaira *et al.* (2001) investigated the thermal evolution of the Ryoke metamorphic belt, based on a simple one-dimension (1D) thermal model incorporating the exhumation history. This study shows retrograde *P–T* paths for the high-grade zones are expected to show an initial phase of isobaric cooling followed by near-isothermal decompression. The apparently contradictory isobaric *P–T* paths of Brown (1998) and near-isothermal paths of Kawakami (2002) and Kamitomo *et al.* (2008) may therefore represent the early and late stages of the retrograde *P–T* path, respectively.

Structure and deformation

Deformation events

Three generations of ductile deformation structures are recognized within the Ryoke Belt (D_1, D_2 and D_3; Okudaira *et al.* 1993,

1995, 2001; Okudaira 1996*b*). A schistosity oriented parallel or subparallel to bedding – a so-called bedding-parallel schistosity – is well developed in metapelites; the constituent minerals of the schistosity crystallized during the main low *P–T* metamorphism. The schistosity-forming deformation is referred to as D_1. In some metapelites, there are many kinds of asymmetric deformation structures such as: intrafolial folds with axial planes parallel to the schistosity; slightly rotated boudins of quartzo-feldspathic veins and thin layers of metachert; extensional crenulation cleavages; rotated porphyroblasts; and melt-filled R_1 fractures. The intrusion of the Older granitoids also occurred during D_1.

A phase of localized deformation that formed granite mylonites near the MTL is assigned to D_2. In some districts, large-scale overturned folds and their associated parasitic folds formed during D_2. D_3 resulted in a series of the upright folds with east–west-trending axes, which occurred during and/or after D_2 mylonitization (Shimada *et al.* 1998). The Younger granitoid rocks intruded during and after the formation of the D_3-upright folds (Nishiwaki & Okudaira 2007).

A broad D_2 mylonite zone (<2 km wide), referred to as the Kashio mylonite zone or the Ryoke southern marginal shear zone, formed within the Ryoke Belt during sinistral top-to-the-west shearing along the MTL (Hara *et al.* 1980; Takagi 1985; Ohtomo 1993; Yamamoto 1994; Sakakibara 1995, 1996; Shimada *et al.* 1998; Okudaira *et al.* 2009). According to Takagi & Shibata (2000), the onset of a significant strike-slip component of displacement along the MTL occurred from middle Cretaceous times (*c.* 95 Ma) onwards.

Minor shear zones also developed within the Ryoke granitoids *c.* 10–50 km away from the MTL. This deformation has been referred to as the Ryoke inner shear zone (Hara *et al.* 1980; Takagi *et al.* 1988; Sakakibara 1995; Sakamaki *et al.* 2006). It has been suggested that mylonites formed in a zone with a width of *c.* 1–2 km along the MTL at high temperatures of *c.* 450–500°C, as inferred from the temperature of deformation-controlled myrmekite replacement (Michibayashi & Masuda 1993; Sakakibara 1996). The grain size of dynamically recrystallized quartz decreases systematically towards the MTL, possibly indicating the shear stress applied on rocks and/or the finite strain of deformed rocks increases toward the MTL (Hara *et al.* 1980; Takagi 1985; Sakakibara 1996; Shimada *et al.* 1998).

The mylonites were subsequently localized in a narrow zone (*c.* 500 m wide) next to the MTL at lower temperatures of *c.* 300–400°C (Michibayashi & Masuda 1993), as inferred from a thermochronological analysis of biotite that recrystallized during mylonitization (Dallmeyer & Takasu 1991). These observations suggest that strain localization occurred during the cooling and exhumation of the Ryoke Belt. Pseudotachylytes are locally present in the inner and southern marginal shear zones (Shimada *et al.* 2001; Sakamaki *et al.* 2006; Shigematsu *et al.* 2012). The mylonites that developed along the MTL have been considered to be rocks of the exhumed deep levels of the Cretaceous MTL (e.g. Okudaira & Shigematsu 2012; Shigematsu *et al.* 2012).

Kinematics of deformation

Strain analysis shows D_1 is dominantly associated with plane to flattening finite strain (Okudaira & Beppu 2008; Okudaira *et al.* 2009). Prolate strain is locally observed in rocks where outcrop-scale D_3 folds are present, and the finite strain in these areas likely represents the combined tectonic strain of D_1 and D_3. In these cases, the orientations of the stretching lineation of D_1 and hinge lines of mesoscopic D_3 folds are parallel to each other.

Tightly folded metachert is commonly observed. The geometry of the folds is not simply related to the geometry of strain ellipsoids inferred from the radiolarian fossils preserved within metachert layers, which were deformed mainly during D_1. This suggests that many of the folds in the metachert were formed by pre-D_1 deformation (Okudaira & Beppu 2008). Because the metapelite may have been derived from rocks of the Mino–Tamba accretionary complex with a mélange fabric, pre-D_1 deformation may be related to shallow subduction-related processes such as the offscraping or underplating of subducted material.

In Figure 2c.8, we summarize the orientations of measured D_1 stretching lineations throughout the belt, as defined by stretched radiolarian fossils within metacherts of the Shiojiri–Ina district (Toriumi 1985; Toriumi & Kuwahara 1988), Wazuka–Kasagi district (Okudaira *et al.* 2009) and Iwakuni–Yanai district (Beppu & Okudaira 2008; Okudaira & Beppu 2008), and by stretched psammitic patches within a pelitic matrix or elongate aggregates of biotite in the Mikawa district (Adachi & Wallis 2008). According to Toriumi (1985) and Toriumi & Kuwahara (1988) samples with prolate strain characterize the high-grade zone, whereas samples with oblate strain occur in the low-grade zone. However, these studies did not distinguish between the D_1 and D_3 strains. Since upright folds are common in the high-grade zone, samples representing prolate strain are likely to have been strongly affected by D_3. In the Shiojiri–Ina

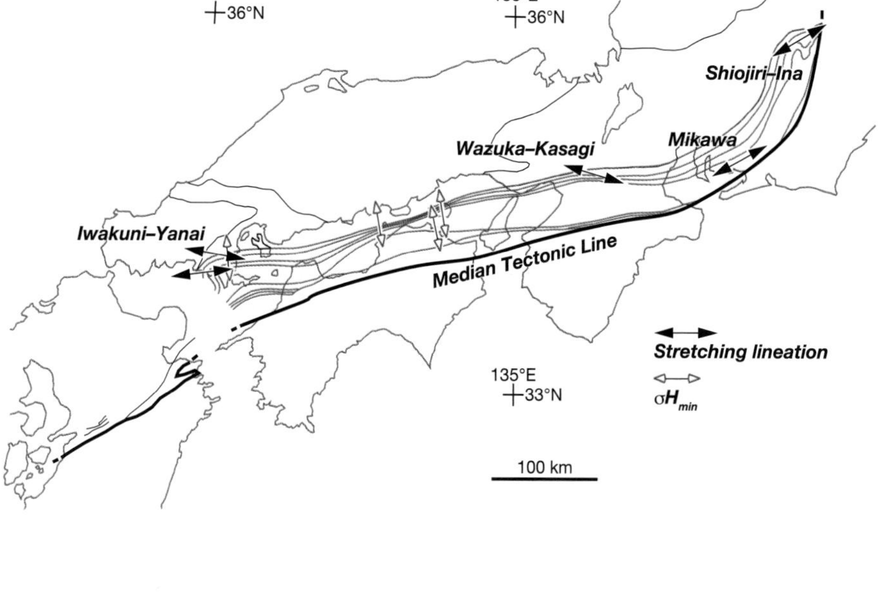

Fig. 2c.8. Maximum stretching direction of the schistosity-forming deformation (D_1) measured at different sites throughout the Ryoke Belt (Okudaira *et al.* 2009). Data sources: Shiojiri–Ina, Toriumi (1985) and Toriumi & Kuwahara (1988); Mikawa, Adachi & Wallis (2008); Wazuka–Kasagi, Okudaira *et al.* (2009); Iwakuni–Yanai, Beppu & Okudaira (2008) and Okudaira & Beppu (2008). Orientations of horizontal minimum stress axis (σH_{min}) during D_1 are from Yokoyama (1984) and Okudaira *et al.* (1995).

district we therefore selected the data in the low-grade zone as representative of the D_1 strain geometry. In the Wazuka–Kasagi and Iwakuni–Yanai districts, the stretching lineation is also defined by stretched sandstone clasts within a pelitic matrix (Okudaira & Beppu 2008; Okudaira *et al.* 2009). The orientations of measured D_1 stretching lineations summarized in Figure 2c.8 show a dominant east–west trend, approximately parallel to the trend of the Ryoke Belt.

The boundaries of mineral zones are parallel/subparallel to the schistosity (i.e. *XY*-plane of the strain ellipsoid of D_1) throughout the belt (Okudaira *et al.* 1993; Ikeda 1998, 2004), suggesting that the schistosity developed horizontally at the time of the main metamorphism. Since a general combination of simple shear and shortening normal to the shear plane leads to a general flattening of the strain ellipsoid, the strain geometry of D_1 deformation indicates the likely formation of a horizontal shear zone with a shear direction parallel to the arc length; the shear zone would have developed at the depth at which the Ryoke Belt formed (*c.* 15 km).

Asymmetric deformation structures related to shear deformation are rarely observed; consequently, we are unable to estimate the shear sense during D_1 deformation. However, Miyashita (1996) and Adachi & Wallis (2008) documented some asymmetric structures with a consistent top-to-the-west shear during D_1. These structures indicate that arc-parallel shearing occurred during D_1, although the coaxial deformation structures dominate. If the deformation is considered as consisting of coaxial and non-coaxial components, the coaxial component has an increasingly large effect on fabric stability with increasing strain; the fabrics are dominated by the coaxial component (e.g. Teyssier & Tikoff 1999).

It has been proposed that some deformation structures indicate arc-normal extension during D_1 (Fig. 2c.8). In the northern part of the Iwakuni–Yanai district, for example, the biotite zone metapelites contain kinematic indicators (e.g. asymmetric extensional crenulation cleavage) showing north-side-down displacement at a low angle to the D_1 foliation, indicating north–south extension during D_1 (Okudaira *et al.* 1995). Yokoyama (1984) reported that the horizontal minimum stress axis (σH_{min}) inferred from dyke orientations in SW Japan was oriented north–south during the early stages (D_1 of the present study), indicating that D_1 was dominated by arc-normal extension. These observations suggest that simultaneous arc-parallel stretching and arc-normal extension (i.e. transtension) occurred within the SW Japan arc during the middle Cretaceous.

Chronology

There are many chronological data of various isotopic systems (K–Ar, Rb–Sr, Sm–Nd and U–Pb systems) for the Ryoke granitoid intrusions (e.g. Herzig *et al.* 1998; Suzuki & Adachi 1998; Yuhara *et al.* 2000). Ages of the peak or near-peak metamorphism have been estimated only by the chemical Th–U–total Pb isochron method (CHIME) monazite ages (Suzuki & Adachi 1998; Kawakami & Suzuki 2011). Although there are also K–Ar and Rb–Sr biotite ages of metamorphic rocks (e.g. Nakajima 1994), they represent certain retrograde stages but not the peak metamorphism. Figure 2c.9 shows the summary of monazite (and xenotime) ages for the high-grade metamorphic rocks and the Ryoke and San-yo granitoids. CHIME monazite dating of high-grade metamorphic rocks yields ages of 102–98 Ma both in the west, central and east of the belt, whereas ages of the Older granitoid intrusions are 95–90 Ma. CHIME monazite ages of metasediments represent the timing of prograde monazite growth at lower amphibolite facies conditions, whereas those of the granitoid rocks represent the timing of their crystallization in cooling magmas (Suzuki & Adachi 1998). These results therefore indicate that the high-grade metamorphism occurred simultaneously with the formation (or intrusion) of the Older granitoid rocks. Activity of the Ryoke and San-yo granitoid intrusions continued from *c.* 95 to *c.* 68 Ma in the eastern part, whereas it continued from *c.* 95 to *c.* 85 Ma in the western part.

Since the K–Ar ages and Rb–Sr mineral isochron ages of granitoid and metamorphic rocks showed along-arc, eastwards younging, the Cretaceous granitic magmatism in SW Japan has been ascribed to the interaction of a mid-oceanic ridge to the trench of the Eurasian continental margin (Kinoshita & Ito 1986; Nakajima *et al.* 1990; Nakajima 1994, 1996; Kinoshita 1995). Based on the CHIME monazite age data, Suzuki & Adachi (1998) argue that the eastwards younging of the Ryoke granitoids so far documented reflects the differential uplift of the Ryoke Belt rather than lateral migration of magmatism. On the other hand, based on Rb–Sr whole-rock isochron ages, Yuhara *et al.* (2000) consider the outbreak of magmatism all over SW Japan to be *c.* 120 Ma.

Exhumation histories of the Ryoke Belt have been proposed by thermochronological analyses (e.g. Suzuki & Adachi 1998; Okudaira *et al.* 2001; Kawakami & Suzuki 2011). Thermochronological analysis for a variety of different isotopic systems, combined with thermal modelling for one of the Older granitoid intrusions in the Iwakuni–Yanai district, reveal two distinctive cooling stages: (1) a rapid cooling stage (>40°C myr^{-1}) for a period of *c.* 7 myr after

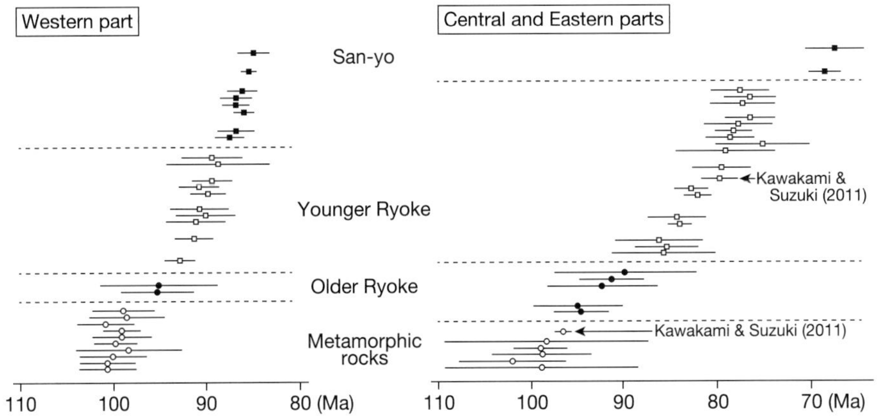

Fig. 2c.9. Summary of CHIME monazite and xenotime ages for the high-grade metamorphic rocks and the Ryoke and San-yo granitoids. This figure is modified from Suzuki & Adachi (1998), with additional data from Kawakami & Suzuki (2011).

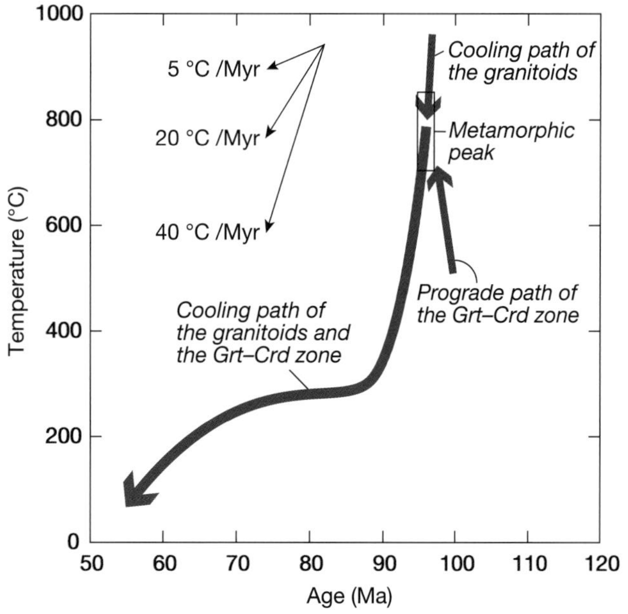

Fig. 2c.10. Cooling history of the Older granitoids in the Iwakuni–Yanai district (modified from Okudaira *et al.* 2001). Two distinct cooling stages can be recognized as: (1) a rapid cooling (>40°C myr^{-1}) for a period (lasting *c.* 7 myr) soon after the thermal peak (at *c.* 98 Ma) and subsequently (2) the slow cooling stage (*c.* 5°C myr^{-1}) after *c.* 88 Ma.

the peak metamorphism (*c.* 95 Ma); followed by (2) a slow cooling stage (*c.* 5°C myr^{-1}) after *c.* 88 Ma (Fig. 2c.10; Okudaira *et al.* 2001). The first rapid cooling stage corresponds to thermal relaxation of the intruded granodiorite magma and its associated metamorphic rocks during D_1, and to the exhumation related to displacement along low-angle D_2 faults that initiated soon after the intrusion of the magma. Formation of large-scale D_3 upright folds resulted in a crustal thickening and subsequent isostatic uplift which, combined with an increased surface gradient and corresponding increase in erosion rates, may have contributed to exhumation during the first stage. The average exhumation rate of the first stage is >2 mm yr^{-1}. During the second stage, the rocks were not accompanied by ductile deformation and were exhumed at a much slower rate of 0.1–0.2 mm yr^{-1}.

Tectonic interpretation

Heat source for low-P–T metamorphism

The low *P–T* metamorphism shows a strong spatial and temporal association with the emplacement of the Older Ryoke granitoid intrusions. This relationship is demonstrated by three lines of geological evidence: (1) Older granitoid intrusion into the high-grade metamorphic rocks is synchronous with D_1; (2) throughout the high-grade metamorphic zones, distinct contact aureoles caused by the intrusion of the granodiorite are lacking; and (3) intrusion ages (*c.* 95 Ma) of the granodiorite are similar to the ages of the thermal peak (*c.* 98 Ma). Although not well constrained, based on the Al-in-hornblende barometer, the depth of emplacement of the Older Ryoke granitoid rocks was >400 MPa and that of the Younger Ryoke and San-yo granitoid rocks was <400 MPa (Takahashi 1993; Nishiwaki & Okudaira 2007). Pressure estimation of associated metamorphic rocks gives similar results to those of the granitoids (Ikeda 2004). The prograde *P–T* paths for the Kfs–Crd and Grt–Crd zones are nearly isobaric (Brown 1998; Ikeda 1998). All these observations point towards the intrusion (or transport) and solidification of granitic melts as a significant heat source for the low *P–T* metamorphism (Okudaira *et al.* 1993; Okudaira 1996*b*; Miyazaki 2007, 2010). The viability of this idea has been tested quantitatively using numerical modelling that combines reconstruction of chemical zoning in metamorphic garnet with the known *T–t* history (Okudaira 1996*a*; Tirone & Ganguly 2010).

Tectonic evolution

In order to account for the prolate strain of the Ryoke metamorphic rocks, a tectonic model for the Ryoke Belt has been suggested that emphasizes 3D corner flow in the forearc wedge (Toriumi 1985; Iwamori 2003). However, as described in the section ‘Kinematics of deformation’, plane strain to general flattening strain dominates in the Ryoke metamorphic rocks (Okudaira & Beppu 2008; Okudaira *et al.* 2009). Previous studies (Toriumi 1985; Toriumi & Kuwahara 1988) that reported general prolate strain as a dominant strain in the metamorphic rocks did not take into account the polyphase nature of deformation. Prolate strain is only observed in rocks with the mesoscopic D_3 folds, suggesting this strain type represents the combined tectonic strain of D_1 and D_3. A similar interpretation has been reported in the Sanbagawa Belt by Moriyama & Wallis (2002).

Metapelite was mainly deformed during the formation of the D_1 schistosity, and the effect of D_3 is relatively small. Assuming D_3 to be plane strain ($X > Y = 1 > Z$) for cases where the principal axes of the D_1 and D_3 finite strain ellipsoids were parallel, the strain magnitude of D_3 should be $\leq$ 0.3–0.4. Furthermore, since prograde metamorphism up to the peak metamorphism has occurred during the subduction stage in the 3D corner flow model, the *X*-axis of the strain ellipsoid of rocks formed at the peak metamorphism would tend to be parallel to the subduction direction. However, the *X*-axis of the strain ellipsoid of rocks formed at the peak metamorphism (D_1) is oriented parallel to the length of the Ryoke Belt, that is likely to be parallel to the volcanic arc. The 3D corner flow models of the forearc wedge proposed by Toriumi (1985) and Iwamori (2003) are therefore considered to be unsuitable for explaining the geodynamic process in the forearc region of the Late Cretaceous SW Japan.

The southern marginal part of the Ryoke Belt is overlain by the Late Cretaceous left-lateral pull-apart forearc basin deposits of the Onogawa–Izumi Group that are distributed on the north side of the MTL (e.g. Teraoka *et al.* 1998; Yamakita & Otoh 2000; see Fig. 2c.6). In Kyushu, the Onogawa forearc basin formed during Cenomanian–Santonian time (Teraoka *et al.* 1998). In Shikoku and Kinki, the Izumi forearc basins existed during Campanian–Maastrichtian time (e.g. Taira *et al.* 1983; Miyata 1990). They are now considered to be a continuous pull-apart forearc sedimentary basin formed during Cenomanian–Maastrichtian time (Yamakita & Otoh 2000). These basins were filled with sediment supplied mostly from the north, which comprised mainly clastic material derived from felsic to intermediate volcanic rocks with an additional granitic component (Teraoka *et al.* 1998). No detritus from the Sambagawa metamorphic belt has been identified in the Onogawa–Izumi Group (Ichikawa 1980; Teraoka *et al.* 1998). The depocentres in the Onogawa–Izumi sedimentary basins migrated eastwards (Taira *et al.* 1983; Miyata 1990; Yamakita & Otoh 2000). The formation of the Izumi sedimentary basin has been explained as stepwise migration of the centre of subsidence due to tilting by secondary antithetic normal faults along the Gojo releasing bend (see Fig. 2c.6) associated with sinistral movements on the high- to moderate-angle master fault (i.e. MTL) at that time (Miyata 1990).

The sinistral movement associated with a high- to moderate-angle MTL is generally thought to have started at the Cenomanian stage (Teraoka *et al.* 1998). If so, then this would imply that relatively shallow movements along the MTL were taking place at the same time as the middle Cretaceous horizontal shear was developing in the deeper parts of the forearc region. It is likely that these adjacent and contemporaneous structures are linked to form a crustal-scale detachment (Fig. 2c.5a, b). It is noteworthy that the distribution of the southern boundary of the Ryoke Belt corresponds with the distribution of the highest-grade zone (Fig. 2c.6), suggesting that the horizontal shear strain was localized into the highest-grade zone to form the detachment. Furthermore, a downward extension of the present-day MTL dips gently northwards as revealed by recent seismic reflection studies (Ito *et al.* 1996, 2009).

In a case of oblique plate convergence decoupled into a component of convergence normal to the plate boundary and a shear component taken up by strike-slip faulting on a transcurrent fault in the interior of the over-riding plate, intra-arc strike-slip faults at the surface tend to develop close to the axes of chains of active volcanoes where the geothermal gradient is high (e.g. Fitch 1972; McCaffrey 1992). Fitch (1972) suggested that strike-slip faulting forming the MTL might have developed close to the high-geotherm region within which the Ryoke Belt formed. Palaeomagnetic studies (Otofuji & Matsuda 1987) show the Cretaceous SW Japan arc was oriented NE–NNE, although the orientation of the plate boundary (trench) between the continental Eurasian plate and oceanic plate (Izanagi and Pacific plates) is not well known. In middle Cretaceous times (*c.* 100 Ma), the rapid northwards movement of the Izanagi Plate continued with highly oblique subduction beneath the Eurasian continent (e.g. Taira *et al.* 1989; Maruyama *et al.* 1997; Wallis *et al.* 2009). This plate configuration possibly resulted in the development of sinistral strike-slip tectonics within the forearc region of the SW Japan arc (Shimada *et al.* 1998; Takagi 1999; Takagi & Shibata 2000).

D_2 mylonitization developed as a continuous mylonite zone in the Older Ryoke granitoid rocks along the MTL. Almost all of the sense-of-shear indicators within the mylonites record top-to-the-west movement (e.g. Hara *et al.* 1980; Takagi 1986; Yamamoto 1994; Sakakibara 1995; Shimada *et al.* 1998). Shimada *et al.* (1998) proposed that D_3 folding occurred during and/or after D_2 mylonitization. Simultaneous upright folding and horizontal mylonitization can be explained as the partitioning of compression in the over-riding unit and sinistral transcurrent movement along the north-dipping detachment fault (i.e. MTL), respectively under a transpression regime during the Late Cretaceous epoch (Shimada *et al.* 1998; Takagi 1999; Fig. 2c.5c). The north-dipping detachment fault may be formed at D_1 under transtension and was reactivated during D_2–D_3 to form the D_2 mylonites and D_3 folds. During and/or after the formation of D_3 folds, the Younger Ryoke and San-yo granitoid intrusions were emplaced at middle to shallow crustal depths (Fig. 2c.5c, d). Development of the Onogawa–Izumi forearc basin along the MTL may be associated with the Late Cretaceous transpression.

Transition from the D_1 to D_2–D_3 was probably related to the increase in the arc-normal compressional stress component due to decrease in the obliquity of the oblique convergence. According to Maruyama *et al.* (1997), during the Late Cretaceous the subduction direction of the oceanic plate changed from NNW to WNW with respect to a reference point at 35° N and 135° E. If the plate boundary was parallel to the SW Japan arc, the obliquity between the two plates decreased during the Late Cretaceous and so the arc-normal compressional stress component in the forearc region may have increased. Alternatively, the subduction of a spreading ridge or very young oceanic lithosphere, proposed as an explanation of granitic magmatism and the formation of the Ryoke and Sanbagawa belts, may have controlled the stress field in the forearc region at this time (e.g. Okudaira & Suda 2011).

MTL: link between the Sanbagawa and Ryoke belts

Distribution and age

The MTL is a major crustal-scale fault with a strike length of >1000 km and a displacement history from the middle Cretaceous to the present. Seismic reflection profiles across the MTL indicate that it dips to the north (Ito *et al.* 1996, 2009). The MTL also defines the prominent tectonic boundary between the Ryoke and Sanbagawa belts. Rocks of both the Ryoke and Sanbagawa belts show progressive increases in deformation towards this contact. The MTL can be traced from Kyushu to the Kanto Mountain regions, but it is most clearly recognized in the central region including Shikoku and central Japan where it is commonly associated with distinct breaks in topography. This link with topography indicates that recent movements have affected this boundary, as confirmed by GPS measurements across the fault (Tabei *et al.* 2003). The association of the MTL with mylonites that have been exhumed from mid-crustal levels shows that this boundary has been the focus for deformation for a long period of time. Correlation with synmetamorphic deformation in the Ryoke and Sanbagawa belts suggests that the ductile deformation extends back to the Cretaceous Period. This inference is supported by the presence of synsedimentary deformation in the Izumi basins deposited along the MTL (Noda & Toshimitsu 2009). Takagi & Shibata (1992) report K–Ar ages of fault gouge along the MTL. There is considerable variation, but ages up to 63 Ma are compatible with a long history of movement.

Movement history and kinematics

The MTL is generally considered to be a long-lived strike-slip fault, although the details of the movement history are more complex (e.g. Ichikawa 1980). The main evidence for an early strike-slip origin is the shape and nature of the Izumi sedimentary basin with the age of deposition progressively changing along the length of the belt. This suggests the sedimentation occurred in a pull-apart basin (e.g. Noda & Toshimitsu 2009). Mineral lineations formed by ductile deformation in both the Ryoke and Sanbagawa belts close to the MTL are dominantly subhorizontal and this also implies the strong influence of strike-slip movements. However, there are reports of significant amounts of normal movement along the MTL in several regions (Takagi *et al.* 1992; Fukunari & Wallis 2007; Kubota & Takeshita 2008; Shigematsu *et al.* 2012) compatible with stress fields reported by Kanai *et al.* (2014). This normal faulting can help account for the exhumation of the Sanbagawa Belt. The Tobe thrust is one area which has been interpreted as a thrust fault that has subsequently been reworked as a normal fault (Takagi *et al.* 1992; Fukunari & Wallis 2007).

The sense of shear for the oldest strike-slip movements is consistently sinistral. However, the most recent movements in the Shikoku and western Kii Peninsula regions show the opposite dextral sense; the present sense of shear can be related to the oblique subduction of the Philippine Sea Plate.

The full displacement on the fault is a matter for debate. Some workers propose that the strike-slip component is minor and the MTL is a dominantly low-angle fault with normal and thrust stages. If the deep elongate geometry of the Izumi basin deposits is accepted as sufficient evidence for formation in a strike-slip-associated domain, then the dimensions of this basin suggest that larger-scale strike-slip should be considered. Large-scale sinistral strike-slip movement is compatible with the overall highly oblique convergence at the time of formation.

The Sanbagawa–Ryoke metamorphic pair and the problem of the missing forearc

The Sanbagawa and Ryoke metamorphic belts formed during roughly the same time interval, they have compatible tectonic histories showing clear evidence for highly oblique convergence being key to their deformation, and they both reveal strong evidence for metamorphism being related to the approach and subduction of an active spreading ridge or very young oceanic lithosphere. In addition to their close spatial association, these geological features strongly support the idea that a common model is needed to account for their formation. The original idea of Miyashiro, that this pair of belts in some way represents the two thermal regimes identified through the surface heat flow data, still stands. There is however a significant problem that needs to be addressed: how to account for the missing 100–200 km of forearc that is normally seen in convergent margins. The following five ideas attempt to do just this.

Thrusting

Thrusting does not seem at first a suitable tectonic setting for removing material from between two domains. However, a flat-lying detachment coupled with significant erosion is a geometrically feasible explanation for the juxtaposition of the Ryoke and Sanbagawa belts. Isozaki (1996) suggests this type of tectonic movement began at *c.* 20 Ma associated with the opening of the Japan Sea. There are two significant problems with this idea. To remove sufficient amounts of material in 20×10^6 years to account for the Ryoke–Sanbagawa juxtaposition requires long-term erosion rates of 5–10 mm a^{-1}. This rate of erosion is only likely to be achieved in highly glaciated domains, incompatible with the relatively low latitude of Japan. Secondly, as shown unequivocally by the opening of the Japan Sea, the post-20 Ma period is one of extensional stress in the region and this is mechanically difficult to reconcile with large-scale thrusting.

Tectonic erosion

Another process that has recently been invoked to account for the formation of some aspects of the Ryoke–Sanbagawa metamorphic pair is that of tectonic erosion by subduction: the base of the forearc domain is progressively scraped away and dragged down into the mantle where it is recycled (Aoki *et al.* 2012). This type of tectonic erosion is a recognized process in some convergent margins and significant removal of the forearc is a possibility. As tectonic erosion occurs, the associated volcanic arc should retreat and the original high-*P–T* belt should be removed. The preservation of the Sanbagawa Belt is therefore strong evidence that the proposed subsequent significant tectonic erosion of the Cretaceous forearc has not taken place.

Forearc magmatism

Iwamori (2000) uses the results of thermal modelling to suggest that the Ryoke Belt may have been formed by forearc magmatism related to a ridge approach. A significant problem with this model is that it predicts the Ryoke Belt should form seaboard of the Sanbagawa Belt; the inverse of that is actually observed. An alternative idea for forearc magmatism relates large-scale dehydration of serpentinite and flux melting of the forearc and mantle wedge to the formation of the Ryoke Belt (Aoya *et al.* 2009). The cause of the heating is proposed to be the approach of the Izanagi–Kula Ridge. However, there are also significant problems with this model. (1) The Ryoke magmatism and metamorphism should be dominantly at the very end of the Sanbagawa metamorphism. However, prograde Ryoke metamorphism is thought to have occurred at *c.* 100 Ma, which is long before the ridge is predicted to have arrived in the subduction zone. (2) The identity of the original arc is not made clear in this contribution, but should lie within the East Asian margin. There is no clear corresponding domain of high *T* and magmatic activity.

Strike-slip faulting along the MTL

The forearc originally present between the Ryoke arc region and the Sanbagawa subduction domain may have been removed by large-scale strike-slip faulting (e.g. Miyashiro 1972; Brown 1998). Major displacement along the MTL could account for this excision. This requires that the original locations of the Ryoke and Sanbagawa belts have been displaced relatively by an amount equal to or greater than their full length of 800 km. The evidence for such large displacements is not compelling, but the results of recent thermal modelling gives evidence in support of displacements of the order hundreds of kilometres (Mori 2014). It is possible that the large displacements were accommodated by a combination of both left- and right-lateral displacements, limiting the relative translation (Fig. 2c.5c, d).

Vertical rise and normal faulting

Vertical rise of the Sanbagawa Belt through the forearc mantle accommodated by normal faulting and erosion (Fig. 2c.5c, d) can also account for the removal of material that once separated the two metamorphic belts. This model could account for the presence of pods of serpentinite along the MTL (Mori 2014).

With the present state of knowledge, we consider that a combination of strike-slip faulting that acted to excise part of the forearc region and normal faulting that accommodated vertical movement of the high-*P* units is the most likely explanation for the present-day juxtaposition of the Sanbagawa and Ryoke belts.

Summary and conclusions

The Ryoke and Sanbagawa belts are contemporaneous metamorphic belts with contrasting thermal histories. Their history reflects the Cretaceous tectonics of eastern Asia and studies of these belts are important in providing key case studies of ancient convergent tectonics.

There is very good evidence that the formation of both belts was related to interaction with an active spreading ridge or a very young oceanic lithosphere. The somewhat older ages recorded in the Ryoke Belt compared to the Sanbagawa Belt are difficult to reconcile with a simple model of progressive warming of a convergent margin associated with the approach of a single spreading ridge. One possible explanation is that the two belts represent different parts of a single convergent margin with different types of ridge interaction. The present juxtaposition of these two belts suggests there was large-scale strike-slip and normal faulting along the MTL. Testing these and other hypotheses requires several key disputes to be resolved, including the age of peak metamorphism in the Sanbagawa Belt and the amount of displacement on the MTL.

We would like to acknowledge numerous co-workers who have helped form many of the ideas presented here, in particular M. Aoya, M. Enami, S. Endo, I. Hara, Y. Hayasaka, T. Ikeda, H. Mori, T. Mizukami, K. Suzuki, T. Takeshita, Y. Ohtomo and H. Takagi. This work was partially supported by grant JSPS grant 30263065 awarded to SW.

Appendix

English to Kanji and Hiragana translations for geological and place names

Aoyama	青山	あおやま
Asemi	汗見	あせみ
Besshi	別子	べっし
Chichibu	秩父	ちちぶ
Funaokayama	船岡山	ふなおかやま
Gojo	五條	ごじょう
Higashiakaishi	東赤石	ひがしあかいし
Higo	肥後	ひご
Ina	伊那	いな
Iratsu (Irazu)	五良津	いらつ
Iwakuni	岩国	いわくに
Izu	伊豆	いず
Izumi	和泉	いずみ
Kanto	関東	かんとう
Kasagi	笠置	かさぎ
Kashio	鹿塩	かしお
Kii	紀伊	きい
Kinki	近畿	きんき
Kiyomizu	清水	きよみず
Kotsu	高越	こうつ
Kurosegawa	黒瀬川	くろせがわ
Kyushu	九州	きゅうしゅう
Mikabo	御荷鉾	みかぼ
Mikabu	御荷鉾	みかぶ
Mikawa	三河	みかわ
Mino-Tanba (Tamba)	美濃-丹波	みの–たんば
Nagasaki	長崎	ながさき
Nagatoro	長瀞	ながとろ
Oboke	大歩危	おおぼけ
Onogawa	大野川	おおのがわ
Ryoke	領家	りょうけ
San-yo	山陽	さんよう
Sanbagawa (Sambagawa)	三波川	さんばがわ
Sanbosan	三宝山	さんぼうさん
Sebadani	瀬場谷	せばだに
Shikoku	四国	しこく
Shimanto	四万十	しまんと
Shiojiri	塩尻	しおじり
Shirataki	白滝	しらたき
Tenryu	天竜	てんりゅう
Tobe	砥部	とべ
Tonaru	東平	とうなる
Tsukuba	筑波	つくば
Wazuka	和束	わづか
Yanai	柳井	やない

References

ABE, T., TAKAGI, H., SHIMADA, K., KIMURA, S., IKEYAMA, K. & MIYASHITA, A. 2001. Ductile shear deformation of the Sambagawa metamorphic rocks in the Kanto Mountains. *Journal of the Geological Society of Japan*, **107**, 337–353.

ADACHI, Y. & WALLIS, S. 2008. Ductile deformation and development of andalusite microstructures in the Hongusan area: constraints on the metamorphism and tectonics of the Ryoke Belt. *Island Arc*, **17**, 41–56.

AGATA, T. 1998. Geochemistry of ilmenite from the Asama ultramafic-mafic layered igneous complex, Mikabu greenstone belt, Sambagawa metamorphic terrane, central Japan. *Geochemical Journal*, **32**, 231–241.

AOKI, K., IIZUKA, T., HIRATA, T., MARUYAMA, S. & TERABAYASHI, M. 2007. Tectonic boundary between the Sanbagawa belt and the Shimanto belt in central Shikoku. *Journal of the Geological Society of Japan*, **113**, 171–183.

AOKI, K., ITAYA, T. *ET AL*. 2008. The youngest blueschist belt in SW Japan: implication for the exhumation of the Cretaceous Sanbagawa high-P/T metamorphic belt. *Journal of Metamorphic Geology*, **26**, 583–602, http://doi.org/10.1111/J.1525-1314.2008.00777.X

AOKI, K., KITAJIMA, K. *ET AL*. 2009. Metamorphic P–T-time history of the Sanbagawa belt in central Shikoku, Japan and implications for retrograde metamorphism during exhumation. *Lithos*, **113**, 393–407, http://doi.org/10.1016/j.lithos.2009.04.033

AOKI, K., KITAJIMA, K. *ET AL*. 2010. Reply to 'Comment on "Metamorphic *P–T*-time history of the Sanbagawa belt in central Shikoku, Japan and implications for retrograde metamorphism during exhumation"' by S. R. Wallis and S. Endo. *Lithos*, **116**, 197–199, http://doi.org/10.1016/j.lithos.2010.01.015

AOKI, K., MARUYAMA, S., ISOZAKI, Y., OTOH, S. & YANAI, S. 2011. Recognition of the Shimanto HP metamorphic belt within the traditional Sanbagawa HP metamorphic belt: new perspectives of the Cretaceous-Paleogene tectonics in Japan. *Journal of Asian Earth Sciences*, **42**, 355–369, http://doi.org/10.1016/J.Jseaes.2011.05.001

AOKI, K., ISOZAKI, Y., YAMAMOTO, S., MAKI, K., YOKOYAMA, T. & HIRATA, T. 2012. Tectonic erosion in a Pacific-type orogen: detrital zircon response to Cretaceous tectonics in Japan. *Geology*, **40**, 1087–1090.

AOYA, M. 2001. *P–T*-D path of eclogite from the Sanbagawa belt deduced from combination of petrological and microstructural analyses. *Journal of Petrology*, **42**, 1225–1248.

AOYA, M. 2002. Structural position of the Seba eclogite unit in the Sambagawa belt: supporting evidence for an eclogite nappe. *Island Arc*, **11**, 91–110.

AOYA, M. & WALLIS, S. R. 1999. Structural and microstructural constraints on the mechanism of eclogite formation in the Sambagawa belt, SW Japan. *Journal of Structural Geology*, **21**, 1561–1573, http://doi.org/10.1016/S0191-8141(99)00114-5

AOYA, M. & WALLIS, S. R. 2003. Role of nappe boundaries in subduction-related regional deformation: spatial variation of meso- and microstructures in the Seba eclogite unit, the Sambagawa belt, SW Japan. *Journal of Structural Geology*, **25**, 1097–1106, http://doi.org/10.1016/S0191-8141(02)00147-5

AOYA, M. & YOKOYAMA, S. 2009. *Geology of the Hibihara District. Quadrangle Series 1:50, 000*. Geological Survey of Japan, Tsukuba [in Japanese with English abstract].

AOYA, M., UEHARA, S., MATSUMOTO, M., WALLIS, S. R. & ENAMI, M. 2003. Subduction-stage pressure-temperature path of eclogite from the Sambagawa belt: prophetic record for oceanic-ridge subduction. *Geology*, **31**, 1045–1048, http://doi.org/10.1130/G19927.1

AOYA, M., TSUBOI, M. & WALLIS, S. R. 2006. Origin of eclogitic metagabbro mass in the Sambagawa belt: geological and geochemical constraints. *Lithos*, **89**, 107–134, http://doi.org/10.1016/j.lithos.2005.10.001

AOYA, M., MIZUKAMI, T., UEHARA, S. & WALLIS, S. R. 2009. High-P metamorphism, pattern of induced flow in the mantle wedge, and the link with plutonism in paired metamorphic belts. *Terra Nova*, **21**, 67–73, http://doi.org/10.1111/J.1365-3121.2008.00860.X

AOYA, M., ENDO, S., MIZUKAMI, T. & WALLIS, S. R. 2013*a*. Paleo-mantle wedge preserved in the Sambagawa high-pressure metamorphic belt and the thickness of forearc continental crust. *Geology*, **41**, 451–454, http://doi.org/10.1130/G33834.1

AOYA, M., NODA, A. *ET AL*. 2013*b*. *Geology of the Niihama District. Quadrangle Series, 1:50, 000*. Geological Survey of Japan, Tsukuba.

BANNO, S. 2004. Brief history of petrotectonic research on the Sanbagawa Belt, Japan. *Island Arc*, **13**, 475–483, http://doi.org/10.1111/J.1440-1738.2004.00439.X

BANNO, S. & NAKAJIMA, T. 1992. Metamorphic belts of Japanese Islands. *Annual Review of Earth and Planetary Sciences*, **20**, 159–179.

BANNO, S. & SAKAI, C. 1989. Geology and metamorphic evolution of the Sanbagawa metamorphic belt, Japan. *In*: DALY, J. S., CLIFF, R. A., YARDLEY, B. W. D. (eds) *The Evolution of Metamorphic Belts*. Geological Society, London, Special Publications, **43**, 519–532.

BEPPU, Y. & OKUDAIRA, T. 2008. Strain analysis of rocks within a metamorphosed accretionary complex in the Iwakuni–Yanai district, SW Japan. *Journal of Geosciences, Osaka City University*, **51**, 1–8.

BROWN, M. 1998. Unpairing metamorphic belts: P–T paths and a tectonic model for the Ryoke Belt, southwest Japan. *Journal of Metamorphic Geology*, **16**, 3–22.

Charvet, J. 2013. Late Paleozoic-Mesozoic tectonic evolution of SW Japan: a review – reappraisal of the accretionary orogeny and revalidation of the collisional model. *Journal of Asian Earth Sciences*, **72**, 88–101, http://doi.org/10.1016/j.jseaes.2012.04.023

Dallmeyer, R. D. & Takasu, A. 1991. Middle Paleocene terrane juxtaposition along the Median Tectonic Line, southwest Japan: evidence from $^{40}Ar/^{39}Ar$ mineral ages. *Tectonophysics*, **200**, 281–297.

de Jong, K., Wijbrans, J. R. & Feraud, G. 1992. Repeated thermal resetting of phengites in the Mulhacen Complex (Betic Zone, southeastern Spain) shown by $40^{Ar}/39^{Ar}$ step heating and single grain laser probe dating. *Earth and Planetary Science Letters*, **110**, 173–191.

Doubrovine, P. V., Steinberger, B. & Torsvik, T. H. 2012. Absolute plate motions in a reference frame defined by moving hot spots in the Pacific, Atlantic, and Indian oceans. *Journal of Geophysical Research-Solid Earth*, **117**, Article no. B09101, http://doi.org/10.1029/2011jb009072

Enami, M. 1982. Oligoclase-biotite zone of the Sanbagawa metamorphic terrain in the Bessi district, central Shikoku, Japan. *Journal of Geological Society of Japan*, **88**, 887–900 [in Japanese with English abstract].

Enami, M. 1998. Pressure-temperature path of Sanbagawa prograde metamorphism deduced from grossular zoning of garnet. *Journal of Metamorphic Geology*, **16**, 97–106, http://doi.org/10.1111/j.1525-1314.1998.00058.x

Enami, M., Wallis, S. R. & Banno, Y. 1994. Paragenesis of sodic pyroxene-bearing quartz schists – implications for the *P–T* history of the Sambagawa Belt. *Contributions to Mineralogy and Petrology*, **116**, 182–198, http://doi.org/10.1007/Bf00310699

Enami, M., Mizukami, T. & Yokoyama, K. 2004. Metamorphic evolution of garnet-bearing ultramafic rocks from the Gongen area, Sanbagawa belt, Japan. *Journal of Metamorphic Geology*, **22**, 1–15.

Endo, S. 2010. Pressure-temperature history of titanite-bearing eclogite from the Western Iratsu body, Sanbagawa Metamorphic Belt, Japan. *Island Arc*, **19**, 313–335, http://doi.org/10.1111/j.1440-1738.2010.00708.x

Endo, S., Wallis, S., Hirata, T., Anczkiewicz, R., Platt, J., Thirlwall, M. & Asahara, Y. 2009. Age and early metamorphic history of the Sanbagawa belt: Lu-Hf and *P–T* constraints from the Western Iratsu eclogite. *Journal of Metamorphic Geology*, **27**, 371–384, http://doi.org/10.1111/j.1525-1314.2009.00821.x

Endo, S., Wallis, S. R., Tsuboi, M., Aoya, M. & Uehara, S. 2012. Slow subduction and buoyant exhumation of the Sanbagawa eclogite. *Lithos*, **146**, 183–201, http://doi.org/10.1016/j.lithos.2012.05.010

Endo, S., Nowak, I. & Wallis, S. R. 2013. High-pressure garnet amphibolite from the Funaokayama unit, western Kii Peninsula and the extent of eclogite facies metamorphism in the Sanbagawa belt. *Journal of Mineralogical and Petrological Sciences*, **108**, 189–200.

Engebretson, D. C., Cox, A. & Gordon, G. 1985. Relative plate motions between oceanic and continental plates in the Pacific basin. *Geological Society of America Special Paper*, **206**, 159.

Faure, M. 1985. Microtectonic evidence for eastward ductile shear in the Jurassic orogen of SW Japan. *Journal of Structural Geology*, **7**, 175–186.

Faure, M., Fabbri, O. & Monie, P. 1988. The Miocene bending of south west Japan: new Ar^{39}/Ar^{40} and microtectonic constraints from the Nagasaki schists (western Kyushu) an extension of the Sanbagawa high pressure belt. *Earth and Planetary Science Letters*, **91**, 105–116.

Fitch, T. J. 1972. Plate convergence, transcurrent faults and internal plate deformation adjacent to southeast Asia and western Pacific. *Journal of Geophysical Research*, **77**, 4432–4460.

Fukunari, T. & Wallis, S. R. 2007. Structural evidence for large-scale top-to-the-north normal displacement along the Median Tectonic Line in southwest Japan. *Island Arc*, **16**, 243–261, http://doi.org/10.1111/j.1440-1738.2007.00570.x

Fukunari, T., Wallis, S. R. & Tsunogae, T. 2011. Fluid inclusion microthermometry for *P–T* constraints on normal displacement along the Median Tectonic Line in Northern Besshi area, Southwest Japan. *Island Arc*, **20**, 426–438, http://doi.org/10.1111/j.1440-1738.2011.00778.x

Hara, I., Shoji, K., Sakurai, Y., Yokoyama, S. & Hide, K. 1980. Origin of the Median Tectonic Line and its initial shape. *Memoirs of the Geological Society of Japan*, **18**, 27–49.

Hara, I., Shiota, T., Hide, K., Okamoto, K., Takeda, K., Hayasaka, Y. & Sakurai, Y. 1990. Nappe structure of the Sambagawa Belt. *Journal of Metamorphic Geology*, **8**, 441–456.

Hara, I., Shiota, T. *et al.* 1992. Tectonic evolution of the Sambagawa schists and its implications in convergent margin processes. *Journal of Science, Hiroshima University, Series C*, **9**, 495–595.

Hashimoto, M. 1989. On the Mikabu green rocks. *Journal of the Geological Society of Japan*, **95**, 789–798.

Hattori, K., Wallis, S., Enami, M. & Mizukami, T. 2010. Subduction of mantle wedge peridotites: evidence from the Higashi-akaishi ultramafic body in the Sanbagawa metamorphic belt. *Island Arc*, **19**, 192–207, http://doi.org/10.1111/j.1440-1738.2009.00696.x

Herzig, C. T., Kimbrough, D. L., Tainosho, Y., Kagami, H., Iizumi, S. & Hayasaka, Y. 1998. Late Cretaceous U/Pb zircon ages and Precambrian crustal inheritance in Ryoke granitoids, Kinki and Yanai districts, Japan. *Geochemical Journal*, **32**, 21–31.

Higashino, T. 1975. Biotite zone of Sanbagawa metamorphic terrain in the Siragayama area, central Sikoku, Japan. *Journal of Geological Society of Japan*, **81**, 653–670 [in Japanese with English abstract].

Higashino, T. 1990. The higher-grade metamorphic zonation of the Sambagawa metamorphic belt in central Shikoku, Japan. *Journal of Metamorphic Geology*, **8**, 413–423.

Hirajima, T., Isono, T. & Itaya, T. 1992. K-Ar age and chemistry of white mica in the Sanbagawa metamorphic rocks in the Kanto Mountains, central Japan. *Journal of the Geological Society Japan*, **98**, 445–455.

Hirota, Y. 1991. Geology of the Sanbagawa metamorphic belt in western Kii Peninsula, Japan. *Memoirs of the Faculty of Sciences, Shimane University*, **25**, 131–142 [in Japanese].

Hokada, T. 1998. Phase relations and P–T conditions of the Ryoke pelitic and psammitic metamorphic rocks in the Ina district, Central Japan. *Island Arc*, **7**, 609–620.

Ichikawa, K. 1980. Geohistory of the Median Tectonic Line of southwest Japan. *Memoirs of the Geological Society of Japan*, **18**, 187–212.

Ichikawa, K. 1990. Pre-Cretaceous terranes of Japan. *In*: Ichikawa, K., Mizutani, S., Hara, I., Hada, S. & Yao, A. (eds) *Pre-Cretaceous Terranes of Japan*. Publication of IGCP Project No. 224, Nippon Insatsu Shuppan Co. Ltd., Osaka, 1–12.

Iizumi, S., Imaoka, T. & Kagami, H. 2000. Sr-Nd ratios of gabbroic and dioritic rocks in a Cretaceous–Paleocene granitic terrane, Southwest Japan. *Island Arc*, **9**, 113–127.

Ikeda, T. 1998. Progressive sequence of reactions of the Ryoke metamorphism in the Yanai district, southwest Japan: the formation of cordierite. *Journal of Metamorphic Geology*, **16**, 39–52.

Ikeda, T. 2002. Regional occurrence of orthopyroxene-bearing basic rocks in the Yanai district, SW Japan: evidence for granulite-facies Ryoke metamorphism. *Island Arc*, **11**, 185–192.

Ikeda, T. 2004. Pressure-temperature conditions of the Ryoke metamorphic rocks in Yanai district, SW Japan. *Contributions to Mineralogy and Petrology*, **146**, 577–589.

Inui, M. & Toriumi, M. 2002. Prograde pressure temperature paths in the pelitic schists of the Sanbagawa metamorphic belt, SW Japan. *Journal of Metamorphic Geology*, **20**, 563–580.

Ishihara, S. 1977. The magnetite-series and ilmenite-series granitic rocks. *Mining Geology*, **27**, 293–305.

Isozaki, Y. 1996. Anatomy and genesis of a subduction-related orogen: a new view of geotectonic subdivision and evolution of the Japanese Islands. *Island Arc*, **5**, 289–320.

Isozaki, Y. & Itaya, T. 1990. Chronology of Sambagawa metamorphism. *Journal of Metamorphic Geology*, **8**, 401–411.

Isozaki, Y., Hashiguchi, T. & Itaya, T. 1992. The Kurosegawa klippe: an examination. *Journal of the Geological Society of Japan*, **98**, 917–941.

Itaya, T. & Takasugi, H. 1988. Muscovite K-Ar ages of the Sanbagawa schists, Japan and argon depletion during cooling and deformation. *Contributions to Mineralogy and Petrology*, **100**, 281–290.

Itaya, T., Tsujimori, T. & Liou, J. G. 2011. Evolution of the Sanbagawa and Shimanto high-pressure belts in SW Japan: insights from K–Ar (Ar–Ar) geochronology. *Journal of Asian Earth Sciences*, **42**, 1075–1090, http://doi.org/10.1016/j.jseaes.2011.06.012

Ito, T., Ikawa, T., Yamakita, S. & Maeda, T. 1996. Gently north-dipping Median Tectonic Line (MTL) revealed by recent seismic reflection studies, southwest Japan. *Tectonophysics*, **264**, 51–63.

Ito, T., Kojima, Y. *et al.* 2009. Crustal structure of southwest Japan, revealed by the integrated seismic experiment Southwest Japan 2002. *Tectonophysics*, **472**, 124–134, http://doi.org/10.1016/j.tecto.2008.05.013

Iwamori, H. 2000. Thermal effects of ridge subduction and its implications for the origin of granitic batholith and paired metamorphic belts. *Earth and Planetary Science Letters*, **181**, 131–144.

Iwamori, H. 2003. Viscous flow and deformation of regional metamorphic belts at convergent plate boundaries. *Journal of Geophysical*

Research-Solid Earth, **108**, article no. 2321, http://doi.org/10.1029/2002jb001808

Kabir, M. F. & Takasu, A. 2010. Evidence for multiple burial-partial exhumation cycles from the Onodani eclogites in the Sambagawa metamorphic belt, central Shikoku, Japan. *Journal of Metamorphic Geology*, **28**, 873–893, http://doi.org/10.1111/j.1525-1314.2010.00898.x

Kagami, H., Iizumi, S., Tainosho, Y. & Owada, M. 1992. Spatial variations of Sr and Nd isotope ratios of Cretaceous–Paleogene granitoid rocks, Southwest Japan Arc. *Contributions to Mineralogy and Petrology*, **112**, 165–177.

Kamitomo, I., Imaoka, T. & Owada, M. 2008. Decompressional microstructure from garnet-bearing mafic granulite in Yanai district, Ryoke belt Southwest Japan. *Journal of the Geological Society of Japan*, **114**, 88–91 [in Japanese with English abstract].

Kanai, T., Yamaji, A. & Takagi, H. 2014. Paeostress analyses by means of mixed Bingham distributions of healed microcracks in the Ryoke granites, central Japan. *Journal of the Geological Society of Japan*, **120**, 23–35 [in Japanese with English abstract].

Kato, A., Iidaka, T. *et al.* 2010. Variations of fluid pressure within the subducting oceanic crust and slow earthquakes. *Geophysical Research Letters*, **37**, L14310, http://doi.org/10.1029/2010GL043723

Kawakami, T. 2001. Tourmaline breakdown in the migmatite zone of the Ryoke metamorphic belt, SW Japan. *Journal of Metamorphic Geology*, **39**, 61–75.

Kawakami, T. 2002. Magmatic andalusite from the migmatite zone of the Aoyama area, Ryoke metamorphic belt, SW Japan, and its importance in constructing the P–T path. *Journal of Mineralogy, Petrology and Economic Geology*, **97**, 241–253.

Kawakami, T. & Ikeda, T. 2003. Boron in metapelites controlled by the breakdown of tourmaline and retrograde formation of borosilicates in the Yanai area, Ryoke metamorphic belt, SW Japan. *Contributions to Mineralogy and Petrology*, **145**, 131–150.

Kawakami, T. & Suzuki, K. 2011. CHIME monazite dating as a tool to detect polymetamorphism in high-temperature metamorphic terrane: example from the Aoyama area, Ryoke metamorphic belt, Southwest Japan. *Island Arc*, **20**, 439–453.

Kiminami, K., Hamasaki, A. & Matsuura, T. 1999. Geochemical contrast between the Sanbagawa psammitic schists (Oboke unit) and the Cretaceous Shimanto sandstones in Shikoku, Southwest Japan and its geologic significance. *Island Arc*, **8**, 373–382, http://doi.org/10.1046/J.1440-1738.1999.00248.X

Kinoshita, O. 1995. Migration of igneous activity related to ridge subduction in southwest Japan and the East Asian continental margin from the Mesozoic to the Paleogene. *Tectonophysics*, **245**, 25–35.

Kinoshita, O. & Ito, H. 1986. Migration of Cretaceous igneous activity in southwest Japan related to ridge subduction. *Journal of the Geological Society of Japan*, **92**, 723–735.

Kobayashi, K. 1996. Rotation of slip direction of the Atokura Nappe viewed from microstructural analyses of brittle shear zones in the Sambagawa belt, Southwest Japan. *Journal of Structural Geology*, **18**, 563–571, http://doi.org/10.1016/S0191-8141(96)80024-1

Koide, H. 1958. *Dando Granodioritic Intrusives and their Associated Metamorphic Complex*. Maruzen, Tokyo.

Kojima, G. & Suzuki, T. 1958. Rock structure and quartz fabric in a thrusting shear zone: the Kiyomizu tectonic zone. *Journal of Science of Hiroshima University, Series C*, **2**, 173–193.

Koto, B. 1888. On the so-called crystalline schists of Chichibu. *Journal of the College of Science, Imperial University of Tokyo*, **2**, 77–141.

Kouketsu, Y. & Enami, M. 2010. Aragonite and omphacite-bearing metapelite from Besshi region, Sambagawa belt in central Shikoku, Japan and its implication. *Island Arc*, **19**, 165–176, http://doi.org/10.1111/J.1440-1738.2009.00690.X

Kouketsu, Y., Enami, M. & Mizukami, T. 2010. Omphacite-bearing metapelite from the Besshi region, Sambagawa metamorphic belt, Japan: prograde eclogite facies metamorphism recorded in metasediment. *Journal of Mineralogical and Petrological Sciences*, **105**, 9–19, http://doi.org/10.2465/Jmps.090217

Kubota, Y. & Takeshita, T. 2008. Paleocene large-scale normal faulting along the Median Tectonic Line, western Shikoku, Japan. *Island Arc*, **17**, 129–151.

Kugimiya, Y. & Takasu, A. 2002. Geology of the western Iratsu mass within the tectonic melange zone in the Sambagawa metamorphic belt, Besshi district, central Shikoku, Japan. *Journal of the Geological Society of Japan*, **108**, 644–662.

Kutsukake, T. 1993. An initial continental margin plutonism–Cretaceous Older Ryoke granitoids, southwest Japan. *Geological Magazine*, **130**, 15–28.

Kutsukake, T. 2002. Geochemical characteristics and variations of the Ryoke granitoids, southwest Japan: petrogenetic implications for the plutonic rocks of a magmatic arc. *Gondwana Research*, **5**, 355–372.

Larson, R. L. & Chase, C. G. 1972. Late Mesozoic evolution of western Pacific Ocean. *Geological Society of America Bulletin*, **83**, 3627–3643, http://doi.org/10.1130/0016-7606(1972)83[3627:Lmeotw]2.0.Co;2

Maruyama, S., Banno, S., Matsuda, T. & Nakajima, T. 1984. Kurosegawa Zone and Its Bearing on the Development of the Japanese Islands. *Tectonophysics*, **110**, 47–60, http://doi.org/10.1016/0040-1951(84)90057-X

Maruyama, S., Luiou, J. G. & Terabayashi, M. 1996. Blueschists and eclogites of the world and their exhumation. *International Geology Review*, **38**, 384–594.

Maruyama, S., Isozaki, Y., Kimura, G. & Terabayashi, M. 1997. Paleogeographic maps of the Japanese Islands: plate tectonic synthesis from 750 Ma to the present. *Island Arc*, **6**, 121–142.

Masago, H., Okamoto, K. & Terabayashi, M. 2005. Exhumation of the Sanbagawa high-pressure metamorphic belt, SW Japan-constraints from the upper and lower boundary faults. *International Geology Review*, **47**, 1194–1206.

Matsumoto, M., Wallis, S., Aoya, M., Enami, M., Kawano, J., Seto, Y. & Shimobayashi, N. 2003. Petrological constraints on the formation conditions and retrograde P–T path of the Kotsu eclogite unit, central Shikoku. *Journal of Metamorphic Geology*, **21**, 363–376.

McCaffrey, R. 1992. Oblique plate convergence, slip vectors, and forearc deformation. *Journal of Geophysical Research*, **97**, 8905–8915.

Michibayashi, K. & Masuda, T. 1993. Shearing during progressive retrogression in granitoids: abrupt grain size reduction of quartz at the plastic-brittle transition for feldspar. *Journal of Structural Geology*, **15**, 1421–1432.

Miyagi, Y. & Takasu, A. 2005. Prograde eclogites from the Tonaru epidote amphibolite mass in the Sambagawa Metamorphic Belt, central Shikoku, southwest Japan. *Island Arc*, **14**, 215–235, http://doi.org/10.1111/J.1440-1738.2005.00468.X

Miyamoto, A., Enami, M., Tsuboi, M. & Yokoyama, K. 2007. Peak conditions of kyanite-bearing quartz eclogites in the Sanbagawa metamorphic belt, central Shikoku, Japan. *Journal of Mineralogical and Petrological Sciences*, **102**, 352–367, http://doi.org/10.2465/Jmps.070120

Miyashiro, A. 1961. Evolution of metamorphic belts. *Journal of Petrology*, **2**, 277–311.

Miyashiro, A. 1972. Metamorphism and related magmatism in plate tectonics. *American Journal of Science*, **272**, 629–656.

Miyashiro, A. 1973. Paired and unpaired metamorphic belts. *Tectonophysics*, **17**, 241–254, http://doi.org/10.1016/0040-1951(73)90005-X

Miyashita, Y. 1996. Garnet porphyroblast growth in relation to the deformation in the Ryoke metamorphic belt in the southern Yanai area, southwest Japan. *Journal of the Geological Society of Japan*, **102**, 84–104 [in Japanese with English abstract].

Miyata, T. 1990. Slump strain indicative of paleoslope in Cretaceous Izumi sedimentary basin along Median Tectonic Line, southwest Japan. *Geology*, **18**, 392–394.

Miyazaki, K. 1999. Thermobarometry of the Tsukuba metamorphic rocks. *Bulletin of the Geological Survey of Japan*, **50**, 525–525 [in Japanese with English abstract].

Miyazaki, K. 2004. Low-P–high-T metamorphism and the role of heat transport by melt migration in the Higo Metamorphic Complex, Kyushu, Japan. *Journal of Metamorphic Geology*, **22**, 793–809.

Miyazaki, K. 2007. Formation of a high-temperature metamorphic complex due to pervasive melt migration in the hot crust. *Island Arc*, **16**, 69–82.

Miyazaki, K. 2010. Development of migmatites and the role of viscous segregation in high-*T* metamorphic complexes: example from the Ryoke metamorphic Complex, Mikawa Plateau, Central Japan. *Lithos*, **116**, 287–299.

Mizukami, T. & Wallis, S. R. 2005. Structural and petrological constraints on the tectonic evolution of the garnet-lherzolite facies Higashi-akaishi peridotite body, Sanbagawa belt, SW Japan. *Tectonics*, **24**, TC6012, http://doi.org/10.1029/2004TC001733

Mizukami, T., Wallis, S. R. & Yamamoto, J. 2004. Natural examples of olivine lattice preferred orientation patterns with a flow-normal a-axis maximum. *Nature*, **427**, 432–436

Mizukami, T., Wallis, S., Enami, M. & Kagi, H. 2008. Forearc diamond from Japan. *Geology*, **36**, 219–222, http://doi.org/10.1130/G24350a.1

Monié, P., Faure, M. & Maluski, H. 1987. Premières datations 39^{Ar}-40^{Ar} du métamorphisme mésozoïque de haute-pression de Sanbagawa (SW Japon). *Comptes Rendus de l'Académie des Sciences, Paris, Série II*, **20**, 1221–1224.

Mori, H. 2014. *Origin of the Thermal Structure Recorded in Metamorphic Rocks of SW Japan: Shear Heating Along the MTL and Large-Scale Folding in the Sanbagawa Belt*. PhD thesis, Nagoya University.

Mori, H. & Wallis, S. 2010. Large-scale folding in the Asemi-gawa region of the Sanbagawa Belt, southwest Japan. *Island Arc*, **19**, 357–370, http://doi.org/10.1111/J.1440-1738.2010.00713.X

Moriyama, Y. & Wallis, S. 2002. Three-dimensional finite strain analysis in the high-grade part of the Sanbagawa Belt using deformed metaconglomerate. *Island Arc*, **11**, 111–121.

Mouri, T. & Enami, M. 2008. Areal extent of eclogite facies metamorphism in the Sanbagawa belt, Japan: new evidence from a Raman microprobe study of quartz residual pressure. *Geology*, **36**, 503–506, http://doi.org/10.1130/G24599a.1

Murata, A. & Maekawa, H. 2011. Geological structures of the Mikabu greenstones of the Uchiko-Oda area, west Shikoku. *Research Reports of the Tokushima Socio-Arts and Science Faculty*, **25**, 29–38.

Nakajima, T. 1994. The Ryoke plutonometamorphic belt: crustal section of the Cretaceous Eurasian continental margin. *Lithos*, **33**, 51–66.

Nakajima, T. 1996. Cretaceous granitoids in SW Japan and their bearing on the crust-forming processes in the eastern Eurasian margin. *Transactions of the Royal Society of Edinburgh: Earth Science*, **87**, 183–191.

Nakajima, T., Banno, S. & Suzuki, T. 1977. Reactions leading to the disappearance of pumpellyite in low-grade metamorphic rocks of the Sanbagawa metamorphic belt in central Shikoku, Japan. *Journal of Petrology*, **18**, 263–284.

Nakajima, T., Shirahase, T. & Shibata, T. 1990. Along-arc lateral variation of Rb–Sr and K–Ar ages of Cretaceous granitic rocks in Southwest Japan. *Contributions to Mineralogy and Petrology*, **104**, 381–389.

Nakajima, T., Kamiyama, H., Williams, I. S. & Tani, K. 2004. Mafic rocks from the Ryoke Belt, southwest Japan: implication for Cretaceous Ryoke/San-yo granitic magma genesis. *Transactions of the Royal Society of Edinburgh: Earth Sciences*, **95**, 249–263.

Narita, K., Yamaji, T., Tagami, T., Kurita, H., Obuse, A. & Matsuoka, K. 1999. The sedimentary age of the Tertiary Kuma Group in Shikoku and its significance. *Journal of the Geological Society of Japan*, **105**, 305–308.

Nishiwaki, H. & Okudaira, T. 2007. Emplacement process of the Hatsuse plutonic complex, central Kinki Province, SW Japan. *Journal of the Geological Society of Japan*, **113**, 249–265 [in Japanese with English abstract].

Noda, A. & Toshimitsu, S. 2009. Backward stacking of submarine channel-fan successions controlled by strike-slip faulting: the Izumi Group (Cretaceous), southwest Japan. *Lithosphere*, **1**, 41–59.

Nozaki, T., Nakamura, K., Awaji, S. & Kato, Y. 2006. Whole-rock geochemistry of basic schists from the Besshi area, central Shikoku: implications for the tectonic setting of the Besshi sulfide deposit. *Resource Geology*, **56**, 423–432, http://doi.org/10.1111/J.1751-3928.2006.Tb00295.X

Nuong, N. D., Thanh, N. X., Gouzu, C. & Itaya, T. 2011. Phengite geochronology of crystalline schists in the Sakuma-Tenryu district, central Japan. *Island Arc*, **20**, 401–410, http://doi.org/10.1111/J.1440-1738.2011.00773.X

Ohtomo, Y. 1993. Origin of the Median Tectonic Line. *Journal of Science of the Hiroshima University, Series C*, **9**, 611–669.

Okamoto, A. & Toriumi, M. 2005. Progress of actinolite-forming reactions in mafic schists during retrograde metamorphism: an example from the Sanbagawa metamorphic belt in central Shikoku, Japan. *Journal of Metamorphic Geology*, **23**, 335–356, http://doi.org/10.1111/J.1525-1314.2005.00580.X

Okamoto, K. 1998. Inclusion-trail geometry of albite porphyroblasts in a fold structure in the Sambagawa Belt, central Shikoku, Japan. *Island Arc*, **7**, 283–294, http://doi.org/10.1046/J.1440-1738.1998.00176.X

Okamoto, K., Maruyama, S. & Isozaki, Y. 2000. Accretionary complex origin of the Sanbagawa, high P/T metamorphic rocks, Central Shikoku, Japan – Layer-parallel shortening sructure and greenstone chemistry. *Journal of the Geological Society of Japan*, **106**, 70–86.

Okamoto, K., Shinjoe, H., Katayama, I., Terada, K., Sano, Y. & Johnson, S. 2004. SHRIMP U-Pb zircon dating of quartz-bearing eclogite from the Sanbagawa Belt, south west Japan: implications for metamorphic evolution of subducted protolith. *Terra Nova*, **16**, 81–89.

Okudaira, T. 1996*a*. Temperature–time path for low-pressure Ryoke metamorphism, Japan, based on chemical zoning in garnet. *Journal of Metamorphic Geology*, **14**, 427–440.

Okudaira, T. 1996*b*. Thermal evolution of the Ryoke metamorphic belt, southwestern Japan: tectonic and numerical modeling. *Island Arc*, **5**, 373–385.

Okudaira, T. & Beppu, Y. 2008. Inhomogeneous deformation of metamorphic tectonites of contrasting lithologies: strain analysis of metapelite and metachert from the Ryoke metamorphic belt, SW Japan. *Journal of Structural Geology*, **30**, 39–49.

Okudaira, T. & Shigematsu, N. 2012. Estimates of stress and strain rate in mylonites based on the boundary between the fields of grain-size sensitive and insensitive creep. *Journal of Geophysical Research*, **117**, B03210, http://doi.org/10.1029/2011JB008799

Okudaira, T. & Suda, Y. 2011. Cretaceous events at the eastern margin of the East Asia recorded in rocks of the Ryoke belt, SW Japan. *Journal of Geography*, **120**, 452–465 [in Japanese with English abstract].

Okudaira, T. & Yoshitake, Y. 2004. Thermal consequences of the formation of a slab window beneath the Mid-Cretaceous southwest Japan arc: a 2-D numerical analysis. *Island Arc*, **13**, 520–532.

Okudaira, T., Hara, I., Sakurai, Y. & Hayasaka, Y. 1993. Tectono-metamorphic processes of the Ryoke belt in the Iwakuni-Yanai district, southwest Japan. *Memoirs of the Geological Society of Japan*, **42**, 91–120.

Okudaira, T., Takeshita, T. & Hara, I. 1995. Emplacement mechanism of the Older Ryoke granites in the Yanai district, southwest Japan, with special reference to extensional deformation in the Ryoke metamorphic belt. *Journal of Science of the Hiroshima University, Series C*, **10**, 357–366.

Okudaira, T., Hayasaka, Y., Himeno, O., Watanabe, K., Sakurai, Y. & Ohtomo, Y. 2001. Cooling and inferred exhumation history of the Ryoke metamorphic belt in the Yanai district, south-west Japan: constraints from Rb–Sr and fission-track ages of gneissose granitoid and numerical modeling. *Island Arc*, **10**, 98–115.

Okudaira, T., Beppu, Y., Yano, R., Tsuyama, M. & Ishii, K. 2009. Mid-crustal horizontal shear zone in the forearc region of the mid-Cretaceous SW Japan arc, inferred from strain analysis of rocks within the Ryoke metamorphic belt. *Journal of Asian Earth Sciences*, **35**, 34–44.

Osanai, Y., Hamamoto, T., Maishima, O. & Kagami, H. 1998. Sapphirine-bearing granulites and related high-temperature metamorphic rocks from the Higo metamorphic terrane, west-central Kyushu, Japan. *Journal of Metamorphic Geology*, **16**, 53–66.

Ota, T., Terabayashi, M. & Katayama, I. 2004. Thermobaric structure and metamorphic evolution of the Iratsu eclogite body in the Sanbagawa belt, central Shikoku, Japan. *Lithos*, **73**, 95–126, http://doi.org/10.1016/J.Lithos.2004.01.001

Otofuji, Y. & Matsuda, T. 1987. Amount of clockwise rotation of Southwest Japan–fan shape opening of the southwestern part of the Japan Sea. *Earth and Planetary Science Letters*, **85**, 289–301.

Otoh, S., Shimojo, M., Aoki, K., Nakayama, T., Maruyama, S. & Yanai, S. 2010. Age distribution of detrital zircons in the psammitic schist of the Sanbagawa belt, southwest Japan. *Journal of Geography*, **119**, 333–346 [in Japanese with English abstract].

Oxburgh, E. R. & Turcotte, D. L. 1971. Origin of paired metamorphic belts and crustal dilation in island arc regions. *Journal of Geophysical Research*, **76**, 1315–1327, http://doi.org/10.1029/Jb076i005p01315

Ozaki, M., Sangawa, A., Miyazaki, K., Nishioka, Y., Miyachi, Y., Takeuchi, K. & Tagutschi, Y. 2000. *Geology of the Nara District. Quadrangle Series, Scale 1:50,000*. Geological Survey of Japan, Tsukuba [in Japanese with English abstract].

Peacock, S. M. & Wang, K. 1999. Seismic consequences of warm v. cool subduction metamorphism: examples from southwest and northeast Japan. *Science*, **286**, 937–939.

Petford, N., Cruden, A. R., McCaffrey, K. J. W. & Vigneresse, J.-L. 2000. Granite magma formation, transport and emplacement in the Earth's crust. *Nature*, **408**, 669–673.

Sakaguchi, M. & Ishizuka, H. 2008. Subdivision of the Sanbagawa pumpellyite-actinolite facies region in central Shikoku, southwest

Japan. *Island Arc*, **17**, 305–321, http://doi.org/10.1111/J.1440-1738.2008.00620.X

Sakakibara, M., Hori, R. S. & Murakami, T. 1993. Evidence from radiolarian chert xenoliths for post-Early Jurassic volcanism of the Mikabu Greenrocks, Okuki area, western Shikoku, Japan. *Journal of the Geological Society of Japan*, **99**, 831–834.

Sakakibara, M., Oyama, Y., Umeki, M., Sakakihara, H., Shono, H. & Goto, S. 1998. Geotectonic division and regional metamorphism of Northern Chichibu belt in western Shikoku. *Journal of the Geological Society of Japan*, **104**, 604–622.

Sakakibara, N. 1995. Structural evolution of multiple ductile shear zone system in the Ryoke belt, Kinki Province. *Journal of Science of the Hiroshima University, Series C*, **10**, 267–332.

Sakakibara, N. 1996. Qualitative estimation of deformation temperature and strain rate from microstructure and lattice preferred orientation in plastically deformed quartz aggregates. *Journal of the Geological Society of Japan*, **102**, 199–210 [in Japanese with English abstract].

Sakamaki, H., Shimada, K. & Takagi, H. 2006. Selective generation surface of pseudotachylyte – example from the Asuke Shear Zone, SW Japan. *Journal of the Geological Society of Japan*, **112**, 519–530 [in Japanese with English abstract].

Sasaki, H. & Isozaki, Y. 1992. Low-angle thrust between the Sanbagawa and Shimanto belts in central Kii Peninsula, southwest Japan. *Journal of the Geological Society of Japan*, **98**, 57–60.

Shigematsu, N., Fujimoto, K., Tanaka, N., Mori, H. & Wallis, S. 2012. Internal structure of the Median Tectonic Line fault zone, SW Japan, revealed by borehole analysis. *Tectonophysics*, **532–535**, 103–118.

Shimada, K., Takagi, H. & Osawa, H. 1998. Geotectonic evolution in transpressional regime: time and space relationships between mylonitization and folding in the southern Ryoke belt, eastern Kii Peninsula, southwest Japan. *Journal of the Geological Society of Japan*, **104**, 825–844 [in Japanese with English abstract].

Shimada, K., Kobari, Y., Okamoto, T., Takagi, H. & Saka, Y. 2001. Pseudotachylyte veins associated with granitic cataclasite along the Median Tectonic Line, eastern Kii Peninsula, Southwest Japan. *Journal of the Geological Society of Japan*, **107**, 117–128.

Shimizu, I. 1988. Ductile deformation in the low-grade part of the Sambagawa metamorphic belt in the northern Kanto Moutnains, central Japan. *Journal of the Geological Society of Japan*, **94**, 609–628.

Shinjoe, H. & Tagami, T. 1994. Cooling History of the Sambagawa Metamorphic Belt Inferred from Fission-Track Zircon Ages. *Tectonophysics*, **239**, 73–79, http://doi.org/10.1016/0040-1951(94)90108-2

Suzuki, J. 1927. On a conglomerate schist from Iya valley in Shikoku. *Proceedings of the Imperial Academy*, **III**, 675–678.

Suzuki, S. 1995. Metamorphic aragonite from the Mikabu and Northern Chichibu belts in central Shikoku SW Japan: identification by Micro-area X-ray, diffraction analysis. *Journal of the Geological Society of Japan*, **101**, 1003–1006.

Suzuki, K. & Adachi, M. 1998. Denudation history of the high T/P Ryoke metamorphic belt, southwest Japan: constraints from CHIME monazite ages of gneisses and granitoids. *Journal of Metamorphic Geology*, **16**, 23–38.

Suzuki, S. & Ishizuka, H. 1998. Low-grade metamorphism of the Mikabu and northern Chichibu belts in central Shikoku, SW Japan: implications for the areal extent of the Sanbagawa low-grade metamorphism. *Journal of Metamorphic Geology*, **16**, 107–116, http://doi.org/10.1111/J.1525-1314.1998.00066.X

Tabei, T., Hashimoto, M., Miyazaki, S. & Ohta, Y. 2003. Present-day deformation across the southwest Japan arc: oblique subduction of the Philippine Sea plate and lateral slip of the Nankai forearc. *Earth Planets and Space*, **55**, 643–647.

Taira, A., Saito, Y. & Hashimoto, M. 1983. The role of oblique subduction and strike-slip tectonics in the evolution Japan. *In*: Hilde, T. W. C. & Uyeda, S. (eds) *Geodynamics of the Western Pacific-Indonesian Regions*. Geodynamics Series, **11**, American Geophysical Union, Washington DC, 303–316.

Taira, A., Tokuyama, H. & Soh, W. 1989. Accretion tectonics and evolution of Japan. *In*: Ben-Avraham, A. (ed.) *The Evolution of the Pacific Ocean Margins*. Oxford University Press, Oxford, 100–123.

Takagi, H. 1985. Mylonitic rocks of the Ryoke belt in the Kayumi area, eastern part of Kii Peninsula. *Journal of the Geological Society of Japan*, **91**, 637–651 [in Japanese with English abstract].

Takagi, H. 1986. Implications of mylonitic microstructures for the geotectonic evolution of the Median Tectonic Line, central Japan. *Journal of Structural Geology*, **8**, 3–14.

Takagi, H. 1999. Partitioning tectonics: kinematic partitioning of strike-slip and thrust faulting (or folding) at transpression zones. *Structural Geology*, **43**, 21–31 [in Japanese with English abstract].

Takagi, H. & Shibata, K. 1992. K–Ar dating of fault gouge – examples along the Median Tectonic Line. *Memoirs of the Geological Society of Japan*, **40**, 31–38 [in Japanese with English abstract].

Takagi, H. & Shibata, K. 2000. Constituents of the Paleo-Ryoke Belt and restoration of the Paleo-Ryoke and Kurosegawa Terranes. *Memoirs of the Geological Society of Japan*, **56**, 1–12 [in Japanese with English abstract].

Takagi, H., Mizutani, T. & Hirooka, K. 1988. Deformation of quartz in an inner shear zone of the Ryoke belt – an example in the Kishiwada area, Osaka Prefecture. *Journal of the Geological Society of Japan*, **94**, 869–886 [in Japanese with English abstract].

Takagi, H., Takeshita, T., Shibata, K., Uchiumi, S. & Inoue, M. 1992. Middle Miocene normal faulting along the Tobe Thrust in western Shikoku. *Journal of the Geological Society of Japan*, **98**, 1069–1072 [in Japanese].

Takahashi, Y. 1993. Al in hornblende as a potential geobarometer for granitoids: a review. *Bulletin of Geological Survey of Japan*, **44**, 597–608 [in Japanese with English abstract].

Takasu, A. 1989. P–T histories of peridotite and amphibolite tectonic blocks in the Sanbagawa metamorphic belt, Japan. *In*: Daly, J. S., Cliff, R. A. & Yardley, B. W. D. (eds) *The Evolution of Metamorphic Belts*. Geological Society, London, Special Publications, **43**, 533–538.

Takasu, A. & Dallmeyer, R. D. 1990. ^{40}Ar–^{39}Ar mineral age constraints for the tectonothermal evolution of the Sambagawa metamorphic belt, central Shikoku, Japan: a Cretaceous accretionary prism. *Tectonophysics*, **185**, 111–139.

Takasu, A., Dallmeyer, R. D. & Hirota, Y. 1996. $40^{Ar}/39^{Ar}$ muscovite ages of the Sanbagawa schists in the Iimori district, Kii peninsula, Japan: implications for orogen-parallel diachronism. *Journal of the Geological Society of Japan*, **102**, 406–418.

Takeshita, T. & Yagi, K. 2004. Flow patterns during exhumation of the Sanbagawa metamorphic rocks, SW Japan, caused by brittle-ductile, arc-parallel extension. *In*: Grocott, J., McCaffrey, K. J. W., Taylor, G. & Tikoff, B. (eds) *Vertical Coupling and Decoupling in the Lithosphere*. Geological Society, London, Special Publications, **227**, 279–296.

Tatsumi, Y., Shinjoe, H., Ishizuka, H., Sager, W. W. & Klaus, A. 1998. Geochemical evidence for a mid-Cretaceous superplume. *Geology*, **26**, 151–154, http://doi.org/10.1130/0091-7613(1998)026<0151:Gefamc>2.3.Co;2

Terabayashi, M., Okamoto, K. *et al.* 2005. Accretionary complex origin of the mafic-ultramafic bodies of the Sanbagawa belt, Central Shikoku, Japan. *International Geology Review*, **47**, 1058–1073.

Teraoka, Y., Suzuki, M. & Kawakami, K. 1998. Provenance of Cretaceous and Paleogene sediments in the Median Zone of Southwest Japan. *Bulletin of the Geological Survey of Japan*, **49**, 395–411 [in Japanese with English abstract].

Teyssier, C. & Tikoff, B. 1999. Fabric stability in oblique convergence and divergence. *Journal of Structural Geology*, **21**, 969–974.

Tirone, M. & Ganguly, J. 2010. Garnet composition as recorders of P–T–t history of metamorphic rocks. *Gondwana Research*, **18**, 138–146.

Tokiwa, T. 2009. Timing of dextral oblique subduction along the eastern margin of the Asian continent in the Late Cretaceous: evidence from the accretionary complex of the Shimanto Belt in the Kii Peninsula, Southwest Japan. *Island Arc*, **18**, 306–319, http://doi.org/10.1111/J.1440-1738.2009.00665.X

Toriumi, M. 1985. Two types of ductile deformation regional metamorphic belt. *Tectonophysics*, **113**, 307–326, http://doi.org/10.1016/0040-1951(85)90203-3

Toriumi, M. & Kohsaka, Y. 1995. Cyclic P–T path and plastic deformation of eclogite mass in the Sanbagawa metamorphic belt. *Journal of the Faculty of Science University of Tokyo, Section II*, **22**, 211–231.

Toriumi, M. & Kuwahara, H. 1988. Inhomogeneous progressive deformation during low P/T type Ryoke regional metamorphism in central Japan. *Lithos*, **21**, 109–116.

Tsuchiya, S. & Hirajima, T. 2013. Evdence of the lawsonite-eclogite facies metamorphism from an epidote-glaucophane eclogite in the Kotsu area of the Sanbagawa belt, Japan. *Journal of Mineralogical and Petrological Sciences*, **108**, 166–171.

Tsutsumi, Y., Miyashita, A., Terada, K. & Hidaka, H. 2009. SHRIMP U-Pb dating of detrital zircons from the Sanbagawa Belt, Kanto Mountains, Japan: need to revise the framework of the belt. *Journal of Mineralogical and Petrological Sciences*, **104**, 12–24, http://doi.org/10.2465/Jmps.080416

Uehara, S. & Aoya, M. 2005. Thermal model for approach of a spreading ridge to subduction zones and its implications for high-P/high-T metamorphism: importance of subduction v. ridge approach ratio. *Tectonics*, **24**, Article no. Tc4007, http://doi.org/10.1029/2004tc001715

Utsunomiya, A., Jahn, B. M., Okamoto, K., Ota, T. & Shinjoe, H. 2011. Intra-oceanic island arc origin for Iratsu eclogites of the Sanbagawa belt, central Shikoku, southwest Japan. *Chemical Geology*, **280**, 97–114, http://doi.org/10.1016/J.Chemgeo.2010.11.001

Wada, I. & Wang, K. L. 2009. Common depth of slab-mantle decoupling: reconciling diversity and uniformity of subduction zones. *Geochemistry, Geophysics, Geosystems*, **10**, Article no. Q10009, http://doi.org/10.1029/2009gc002570

Wallace, L. M., Ellis, S., Miyao, K., Miura, S., Beavan, J. & Goto, J. 2009. Enigmatic, highly active left-lateral shear zone in southwest Japan explained by aseismic ridge collision. *Geology*, **37**, 143–146, http://doi.org/10.1130/G25221a.1

Wallis, S. R. 1990. The timing of folding and stretching in the Sanbagawa belt, the Asemigawa region, central Shikoku. *Journal of the Geological Society of Japan*, **96**, 345–352.

Wallis, S. R. 1992. Vorticity analysis in a metachert from the Sanbagawa belt SW Japan. *Journal of Structural Geology*, **14**, 271–280.

Wallis, S. R. 1995. Vorticity analysis and recognition of ductile extension in the Sanbagawa belt, SW Japan. *Journal of Structural Geology*, **17**, 1077–1093.

Wallis, S. R. 1998. Exhuming the Sanbagawa metamorphic belt; the importance of tectonic discontinuities. *Journal of Metamorphic Geology*, **16**, 83–95.

Wallis, S. R. & Aoya, M. 2000. A re-evaluation of eclogite facies metamorphism in SW Japan: proposal for an eclogite nappe. *Journal of Metamorphic Geology*, **18**, 653–664.

Wallis, S. R. & Endo, S. 2010. Comment on 'Metamorphic P–T-time history of the Sanbagawa belt in central Shikoku, Japan and implications for retrograde metamorphism during exhumation' by K. Aoki, K. Kitajima, H. Masago, M. Nishizawa, M. Terabayashi, S. Omori, T. Yokoyama, N. Takahata, Y. Sano, S. Maruyama [Lithos 113 (2009) 393–407]. *Lithos*, **116**, 195–196, http://doi.org/10.1016/J.Lithos.2010.01.016

Wallis, S. R., Hirajima, T. & Yanai, S. 1990. Sense and direction of movement along the Atokura fault at Shimonita, Kanto Mountains, central Japan. *Journal of the Geological Society of Japan*, **96**, 977–980.

Wallis, S. R., Banno, S. & Radvanec, M. 1992. Kinematics, structure and relationship to metamorphism of the east-west flow in the Sanbagawa belt, southwest Japan. *Island Arc*, **1**, 176–185.

Wallis, S., Moriyama, Y. & Tagami, T. 2004. Exhumation rates and age of metamorphism in the Sanbagawa belt: new constraints from zircon fission track analysis. *Journal of Metamorphic Geology*, **22**, 17–24, http://doi.org/10.1111/J.1525-1314.2003.00493.X

Wallis, S. R., Anczkiewicz, R., Endo, S., Aoya, M., Platt, J. P., Thirlwall, M. & Hirata, T. 2009. Plate movements, ductile deformation and geochronology of the Sanbagawa belt, SW Japan: tectonic significance of 89–88 Ma Lu-Hf eclogite ages. *Journal of Metamorphic Geology*, **27**, 93–105, http://doi.org/10.1111/J.1525-1314.2008.00806.X

Wignall, P. B. 1994. *Black Shales*. Clarendon Press, Oxford.

Yamakita, S. & Otoh, S. 2000. Estimation of the amount of Late Cretaceous left-lateral strike-slip displacement along the Median Tectonic Line and its implications in the juxtaposition of some geologic belts of Southwest Japan. *Monograph of the Association of Geological Collaboration of Japan*, **49**, 93–104 [in Japanese with English abstract].

Yamamoto, H. 1994. Kinematics of mylonitic rocks along the Median Tectonic Line, Akashi Range, central Japan. *Journal of Structural Geology*, **16**, 61–70.

Yokoyama, K. 1980. Nikubuchi peridotite body in the Sanbagawa metamorphic belt; Thermal history of the 'Al-pyroxene-rich suite' peridotite body in high pressure metamorphic terrain. *Contributions to Mineralogy and Petrology*, **73**, 1–13.

Yokoyama, S. 1984. Geological and petrological studies of Late Mesozoic dyke swarms in the Inner Zone of Southwest Japan. *Geological Report of the Hiroshima University*, **24**, 1–63 [in Japanese with English abstract].

Yuhara, M., Kagami, H. & Nagao, K. 2000. Geochronological characterization and petrogenesis of granitoids in the Ryoke belt, Southwest Japan: constraints from K–Ar, Rb–Sr and Sm–Nd systematics. *Island Arc*, **9**, 64–80.

2d Cretaceous–Neogene accretionary units: Shimanto Belt

GAKU KIMURA, YOSHITAKA HASHIMOTO, ASUKA YAMAGUCHI, YUJIN KITAMURA & KOHTARO UJIIE

The Shimanto Belt is one of the most-studied ancient accretionary complexes in the world and yields an opportunity to investigate deep plate boundary processes including seismogenesis in a subduction zone. It is extensively exposed south-westwards from central Japan through the Kii Peninsula, Shikoku, Kyushu and out to the Ryukyu islands (Fig. 2d.1). In this chapter, we overview recent research progress on this classic Cretaceous–Neogene accretionary unit, emphasizing how its tectonic history informs us about the ongoing processes in modern subduction zones.

Historical background and overview

The Shimanto Belt was the name given to the poorly known but prominent geological belt distributed along the Pacific side of Kyushu, Shikoku and the Kii Peninsula in 1900 (Explanation note for 1:1000 000 geological map of Japan; Geological Survey of Japan 1900). The first geological map, edited by Naumann (1885), had classified the belt as Palaeozoic (e.g. Yamashita 1990). Subsequent to this, Takuji Ogawa (famous for being the father of the first Japanese Nobel laureate Hideki Yukawa) described the belt at a typical exposure along the Shimanto River in southwestern Shikoku (Ogawa, T. in Explanation note for 1:1000 000 Geological map of Japan, 1900). Based on this description, Kobayashi (1941) interpreted the Shimanto rocks as representing a Mesozoic mio-geosyncline or flysch basin associated with the Sakawa Orogeny. As elsewhere in the world, geosynclinal theory was widely applied in Japan (e.g. Knopf 1948) and, in the 1960s to early 1970s, the belt was regarded as the product of the Shimanto geosyncline (e.g. Minato *et al.* 1965; Kishu Shimanto Research Group 1975). The geosyncline model was challenged by plate tectonic theory in the late 1970s (e.g. Kanmera & Sakai 1975; Suzuki & Hada 1979) when there was an intense debate in Japan on 'fixism' (geosynclines) v. 'mobilism' (plate tectonics). Taira and his co-workers in Kochi University finally resolved this controversy by dating the Shimanto Belt using radiolarian microfossils in the early 1980s (Taira *et al.* 1980, 1988), a geological breakthrough now known as the 'radiolarian revolution' (Ishigaki & Yao 1982). Since that time, the Shimanto Belt has been regarded as a fossilized accretionary complex along the Asian continental margin produced in Cretaceous–early Miocene times (e.g. Taira *et al.* 1988; Kimura 1997).

The Shimanto rocks have been divided into northern and southern units on the basis of age and lithofacies (e.g. Taira *et al.* 1980, 1988). The Geological Survey of Japan, AIST (2012) later refined this approach on their 1:200 000 digital geological map of Japan by defining three units, namely Early and Late Cretaceous units in the northern Shimanto Belt and a mainly Eocene–early Miocene unit in the southern Shimanto Belt. The two northern Shimanto units are separated from the southern unit by the Aki Tectonic Line in Shikoku, the Gobo–Hagi Tectonic Line in the Kii Peninsula and by the Nobeoka Thrust in Kyushu.

The youngest part of the Sanbagawa metamorphic belt in Shikoku has recently been recognized to be a metamorphosed equivalent of the northern Shimanto Belt on the basis of U–Pb dating on zircon grains in metamorphosed sandstone (Isozaki *et al.* 2010; Aoki *et al.* 2011), a conclusion already reached in the Kii Peninsula on the basis of radiolarian biostratigraphy studies in the early 1980s (Kurimoto 1982; Yamato-Omine Research Group 1994). Further NE along the Japanese archipelago Cretaceous–Palaeogene accretionary complexes are also recognized in Hokkaido (Fig. 2d.1) and extend to the northern Sakhalin Island, although the setting changes to an intra-continental orogenic belt due to an Eocene to Oligo-Miocene collision between the Asian continental margin and the ancient Okhotsk Plate (e.g. Kimura 1994). In these areas the northern extension of the Shimanto Belt might be buried beneath the modern forearc (Fig. 2d.1), as suggested by geophysical observation as well as the Palaeogene–Neogene tectonic history (e.g. Taira *et al.* 1988; Kimura 1994; Maruyama *et al.* 1997).

General geology and large-scale structure

The general geology and structure of the Shimanto Belt in SW Japan shows a southwards-verging fold-and-thrust belt with southwards and downwards younging age polarity; first pointed out by Kanmera & Sakai (1975) this was later justified by dating, mainly using radiolarian microfossils (e.g. Taira *et al.* 1980, 1988). While this overall structural geometry is broadly accepted, along-strike structural and compositional variations and the possible presence of large-scale orogen-parallel strike-slip fault zones add further complications to the overall model. Yanai (1986), Kano *et al.* (1990) and Sugiyama (1994) interpreted east–west structural variations along the Shimanto Belt as a result of Quaternary along-orogen compression caused by the contraction between NE and SW Japan in the central mountainous region of the country (Japan Alps); this interpretation is based to some extent on an early tectonic model presented by Hujita (1968). In this context the present topography of the outer zone of SW Japan includes north–south-trending mountains with prominent capes at their southern tips (e.g. Cape Shiono in the Kii Peninsula, capes Muroto and Ashizuri in Shikoku, Fig. 2d.2).

From: MORENO, T., WALLIS, S., KOJIMA, T. & GIBBONS, W. (eds) 2016. *The Geology of Japan*. Geological Society, London, 125–137.

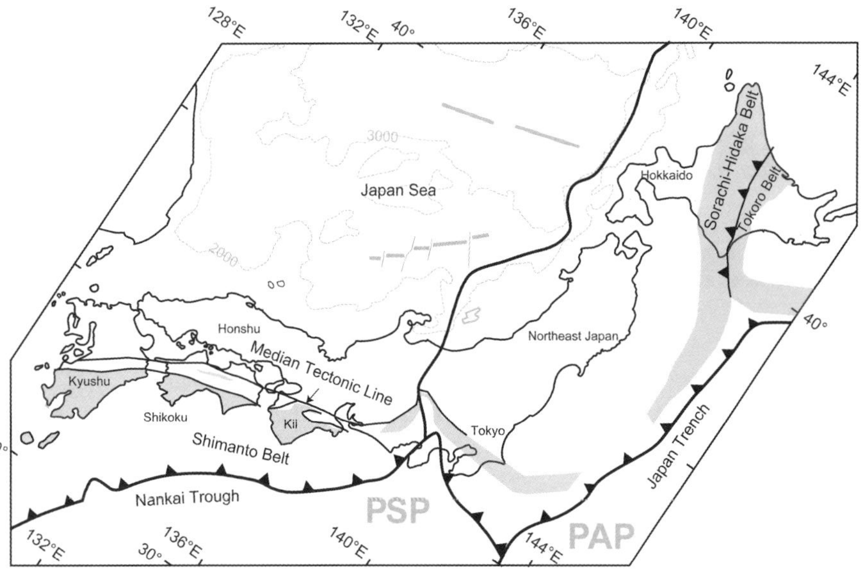

Fig. 2d.1. Index map of the Cretaceous–Neogene accretionary complexes in Japan. The Shimanto Belt is in SW Japan and the Sorachi–Hidaka and Tokoro belts are in Hokkaido. The Cretaceous–Neogene accretionary complexes off NE Japan are inferred beneath the forearc near the Japan Trench and/or truncated by the trench.

The geological structures of the Shimanto Belt, especially those of the Neogene southern belt, show northwards-plunging antiforms (Fig. 2d.2), the presence of which has been explained as a 'mega-kink structure' formed by a recent along-strike contraction (Yanai 1986; Kano *et al.* 1990; Sugiyama 1994). This interpretation argues that the Shimanto Belt was firstly formed in the simple prism of an accretionary fold-and-thrust belt with east–west-directed strike and was modified due to a later deformation. However, Kimura *et al.* (2014) have recently offered an alternative interpretation for these large-scale antiforms, comparing them with modern examples of indentation structure due to seamount and/or arc–arc collision.

A large embayment structure in the modern Nankai accretionary prism is observed to the SE of Cape Muroto (Fig. 2d.2). The northern margin of this embayment is uplifted and forms a shallow bank called Tosa-Bae, known as a good fishing ground. This bank is underlain by a large seamount subducted at the Nankai Trough (Kodaira *et al.* 2000, 2006). The seamount is the northern extension of the Kinan Seamount Chain of the Shikoku Basin in the Philippine Sea Plate (Fig. 2d.2). This seamount chain was formed along the spreading ridge of the Shikoku Basin, which opened at *c.* 15 Ma although post-spreading intermittent leaky volcanism continued until *c.* 7 Ma (Okino *et al.* 1999). The width of the embayment formed by the seamount collision is *c.* 50 km. The splay fault that

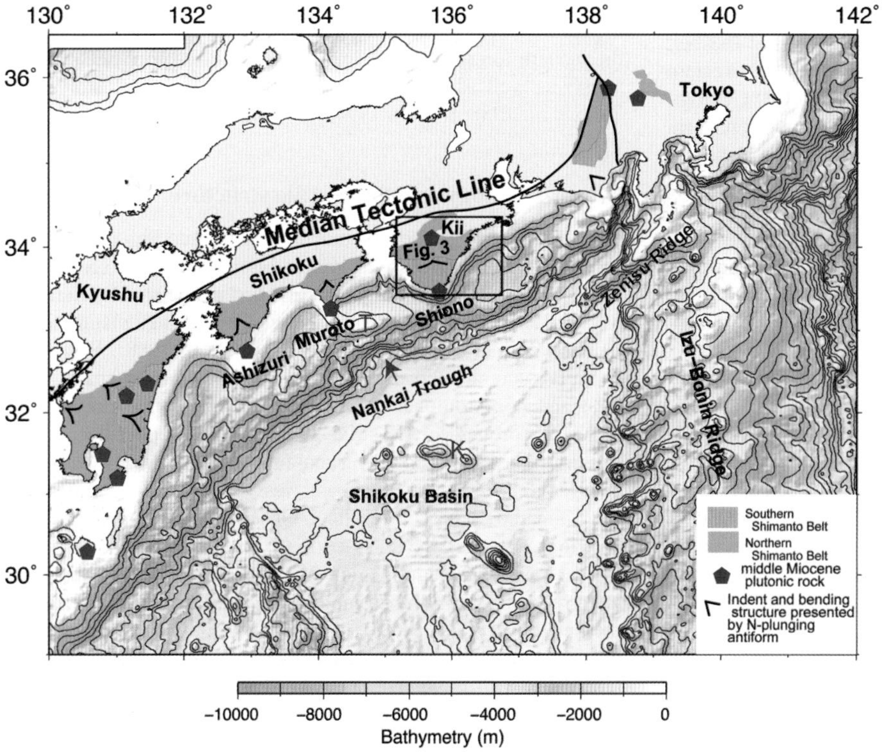

Fig. 2d.2. Location of the indentation structure represented by northwards-plunging antiforms in the Shimanto Belt and middle Miocene plutonic rocks at Capes Ashizuri, Muroto and Shiono. A red T shows the Tosa-bae formed by seamount collision and embayment presented by red arrow. K, Kinan Seamount chain.

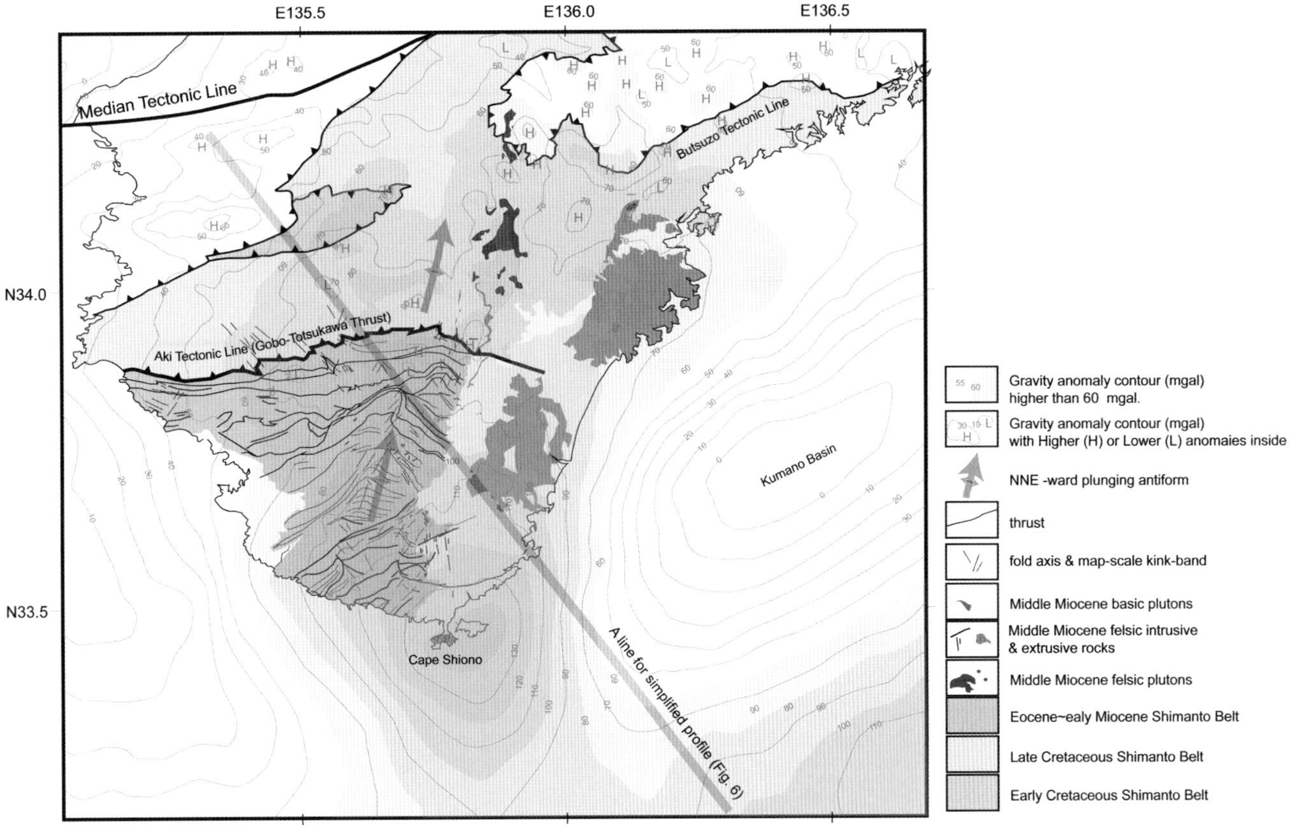

Fig. 2d.3. Simplified geological structure of the Shimanto Belt in the Kii Peninsula north to Cape Shiono with Bouger gravity anomaly assuming density of 2.0 g cm^{-3} After Kimura *et al.* (2014), compiled from Nakaya (2012) and Seamless Geological map series by Geological Survey of Japan, AIST (2012). Note the NNE-plunging antiform and felsic plutons. The antiform and igneous rock distribution is associated with positive gravity anomaly and is traced to the offshore continental shelf. The gravity anomaly around Cape Shiono suggests a large pluton below.

developed in the modern Nankai accretionary prism is also bent, similar to the bending of the Gobo–Hagi Tectonic Line in the Shimanto Belt in the Kii Peninsula (Fig. 2d.3) and the Aki Tectonic Line in Shikoku. Sugiyama (1994) originally explained this embayment structure as a result of thrusting caused by east–west-aligned compression, but Kodaira *et al.* (2006) clearly documented the seamount collision-related indentation on the basis of seismic studies. The indentation structure is also obvious to the north of the Izu collision zone in central Japan. These modern indentation structures are similar to the north-plunging antiforms observed in the Shimanto Belt on land. Kimura *et al.* (2014) suggested that the indentation structure is a result of collision of the proto-Izu arc in middle Miocene times. They also suggested that the collision swiftly migrated from the west to the east, revived the classic reconstruction model of Marshak & Karig (1977), and supported the tectonic model by Hall (1996, 2002).

The Shimanto Belt comprises a Cretaceous–early Miocene accretionary complex, and a common view has been that the same accretionary rocks continue downwards from the surface to the Moho (Taira 2001; Isozaki *et al.* 2010). However, Kimura *et al.* (2014) have suggested that the Shimanto Belt in the eastern Kii Peninsula to the NE of Cape Shiono, north of Cape Muroto and Cape Ashizuri in Shikoku, and in Kyushu have all been intruded by granitic and/or gabbroic plutonic rocks during middle Miocene times (*c.* 17–12 Ma; Fig. 2d.2) and these have thickened the crust to more than twice the volume of the accretionary complex. This estimation is based on the geophysical observation of gravity anomalies suggestive of a mass beneath those areas, the presence of a high resistive mass revealed by magnetotelluric observation in the Kii Peninsula and the velocity structure beneath the Kii peninsula (Fig. 2d.3) (Kodaira *et al.* 2006; Tsuji *et al.* 2013) and offshore Cape Muroto (Kodaira *et al.* 2006). Volcano-geological research on the Kumano acidic volcanic rocks (Miura & Wada 2007) also suggests the presence of a large volume of felsic plutons beneath the eastern Kii Peninsula (Fig. 2d.4). Furthermore, granitic dykes are distributed along the crest of northwards-plunging antiforms (Fig. 2d.3). These facts are consistent with the collision and indentation of an active proto-Izu volcanic arc during middle Miocene times (Kimura *et al.* 2014). Following Hasegawa & Ohno (2006) and Kodaira *et al.* (2006), Kimura *et al.* (2014) suggested that the distribution of these plutonic rocks controls the boundary of the rupture area of the historical large earthquakes in the Nankai Trough (Fig. 2d.4; Ando 1975).

Kula–Pacific ridge subduction hypothesis

Uyeda & Miyashiro (1974) proposed an idea that the interactions between the Kula–Pacific spreading ridge and trench in the Late Cretaceous western Pacific margin controlled several major tectonic events, back-arc spreading of the Japan Sea and both igneous activity and tensional basin formation along the continental margin. This paper was a great milestone and stimulated many studies on the possible effects of ridge subduction during the earlier stages of Japanese plate tectonic evolution. The opening of the Japan Sea was, however, subsequently documented not as Late Cretaceous but emphasized more as an event during Oligo-Miocene times (e.g. Otofuji *et al.* 1985; Tamaki 1985). Any causal relationship between ridge

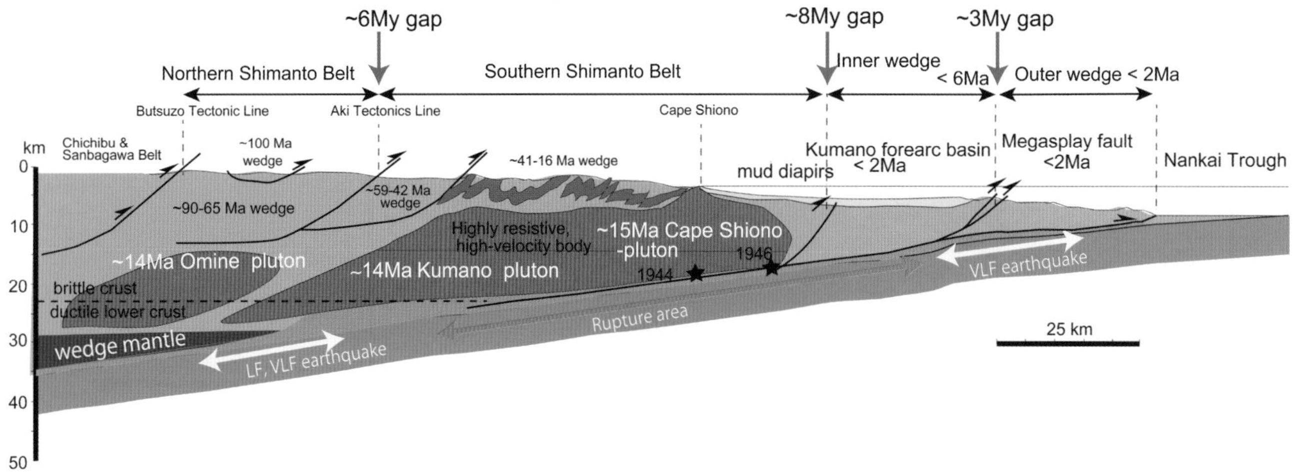

Fig. 2d.4. A simplified profile across the Kii Peninsula region (Kimura *et al.* 2014). A large plutonic rock might have intruded during the middle Miocene and built up the upper plate of island arc crust. Profile location is shown in Figure 2d.3. Depth ranges for the 1944 rupture area, deep low-frequency (LF) and very-low-frequency (VLF) and shallow VLF earthquakes are also depicted. Intermittent growth of accretionary wedge of the Shimanto Belt and the Nankai prism since *c.* 100 Ma is after Suzuki (2012). Note that large gaps of several million years are clearly recorded. The modern Nankai prism might have been initiated at *c.* 6 Ma.

subduction and back-arc opening ceased to be the focus of much discussion, although the link between granitic batholiths and ridge subduction continued to be supported by many geologists (Takahashi 1983; Kinoshita 1995; Maruyama *et al.* 1997).

The Late Cretaceous Shimanto Belt was regarded as an accretionary complex developed, and paired with igneous activity, along the Asian continental margin (e.g. Kanmera & Sakai 1975; Taira *et al.* 1980, 1988). Taira and his co-workers gathered large amounts of radiolarian and nannoplankton data (Taira *et al.* 1980, 1988) and dated the ages of various sedimentary rocks constituting the accretionary complex of the Shimanto Belt in Shikoku. They reconstructed a Cretaceous accretion history along the East Asian continental margin on the basis of this dating, and suggested that the main phase of accretion in the northern Shimanto Belt was Late Cretaceous in age. Age differences between the accreted oceanic material and terrigenous sediments were ascribed mainly to the age of the subducting plate (Taira *et al.* 1988).

Kiminami *et al.* (1992) reported that the accreted basalt and terrigenous shale in the southernmost part of the northern Shimanto Belt are *in situ*, meaning that the basalts were erupted near or in the trench, and supported the Late Cretaceous ridge subduction hypothesis previously proposed by Uyeda & Miyashiro (1974). They speculated that the ridge subduction migrated from west to east on the basis of roughly reconstructed 'eruptive' ages. Maruyama *et al.* (1997) also invoked the Kula–Pacific ridge subduction at *c.* 80 Ma in order to explain the exhumation of the Sanbagawa metamorphic rocks and granitic activity in SW Japan. The age suggested by Kiminami *et al.* (1992) is younger than that of Maruyama *et al.* (1997) and Isozaki *et al.* (2010).

Onishi & Kimura (1995) investigated the *in situ* basalt outcrops described by Kiminami *et al.* (1992), and pointed out that the boundary between the chilled margin of basalt and the baked margin of mélange host rocks supposed by Kiminami *et al.* (1992) is in fact a sheared boundary altered by thermal fluid flow. They emphasized that the basaltic blocks are not *in situ* but exotic and tectonically mixed with terrigenous sediments within a tectonic mélange. The outcrops of this 'Mugi mélange' have subsequently been investigated by many workers, and the unit is now recognized to represent an underplated mélange formed at 150–220°C and several kilometres depth within a latest Cretaceous–Neogene accretionary complex (Matsumura *et al.* 2003; Ikesawa *et al.* 2005; Kitamura *et al.* 2005; Kawabata *et al.* 2007; Ujiie *et al.* 2007*a*, *b*; Shibata *et al.* 2008; Hashimoto *et al.* 2009; Kimura *et al.* 2012; Yamaguchi *et al.* 2012; Ujiie & Kimura 2014). All these papers document the tectonic relationship occasionally formed by seismogenic faulting as reviewed by Kimura *et al.* (2012) and Ujiie & Kimura (2014) (Figs 2d.5 & 2d.6). The young oceanic plate subduction and the consequently relatively warm conditions of the subduction zone are well documented by various geothermometers applied to the Mugi mélange (e.g. Matsumura *et al.* 2003). The setting is almost the same as that of the modern-day Nankai Trough (Spinelli & Harris 2011). Based on the fact that nobody views the Nankai Trough as the site of ongoing active ridge subduction of the Shikoku Basin, the Kula–Pacific ridge subduction hypothesis for the Shimanto Belt is not applicable.

Plate boundary processes

Subduction plate boundaries with thick trench-fill sediments, such as the Nankai, Barbados, Cascadia, Aleutian, Sumatran and southern Chile margins, initiate as an incipient décollement within a sedimentary layer (e.g. Moore *et al.* 1988) and evolve to include oceanic basement at depth (Kimura & Ludden 1995). Hence, plate boundary megathrusts similarly evolve from the deformation of initially unlithified sediments to deformation of lithified rocks at depth. This process is directly related to the aseismic to seismic plate boundary processes in the classic sense (Byrne *et al.* 1988). However, recent findings of very-low-frequency earthquakes (VLFE) in the shallow 'aseismic' portion of the subduction zone (Ito & Obara 2006*a*, *b*; Obana & Kodaira 2009; Sugioka *et al.* 2012) require a revision of faulting processes at plate boundaries. What faulting and deformation mechanisms generate VLFE and how are they different from normal earthquake faulting? How do these two different regimes change at the traditional updip limit of the seismogenic zone? And how do underthrust sediments affect the deformation mechanisms? We review and discuss these points in this section, including the effects of underthrust sediments on the deformation and fault mechanisms involved and the relationships between the tectonic mélange formation and plate boundary processes, taking account of recent progress in research within the Shimanto Belt.

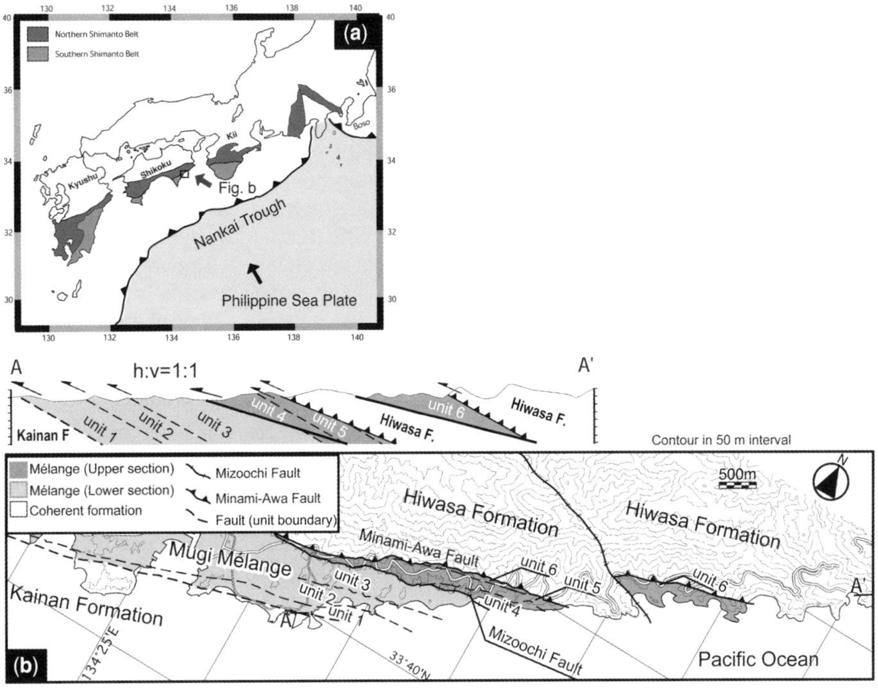

Fig. 2d.5. (a) Location and (b) geological map with profile of an area in eastern Shikoku. Note that the mélange units are stacked by thrusts. Geological map is modified after Shibata *et al.* (2008) and Kimura *et al.* (2012).

Effect of underthrust sediments on deformation mechanisms

The lithology and stratigraphy of sediments on the subducting plate at the trench may have a strong influence on plate boundary processes at depth because the sequence beneath the décollement, which is initially formed within sediments, is underthrust together with oceanic basement. The trench sequence typically comprises basaltic oceanic basement, pelagic to hemipelagic sediments, distal terrigenous sediments and trench-fill coarse turbiditic sediments in ascending order (Fig. 2d.6) (Piper *et al.* 1973; Underwood 2007), a sequence known as oceanic plate stratigraphy in Japan (Matsuda & Isozaki 1991). This original trench sequence is recognizable in the mélange units such as the Mugi mélange even though the primary stratigraphy is mostly broken and disrupted (Taira *et al.* 1988; Kano *et al.* 1991; Kimura & Mukai 1991; Kimura & Ludden 1995; Onishi *et al.* 2001; Ikesawa *et al.* 2005). Volcanic ash layers are occasionally intercalated within the sediments, especially in the middle latitude area of the western Pacific region due to the eastwards atmospheric flow of the westerlies. The upper part of this trench sequence starts to be scraped off and accreted at the deformation front (Fig. 2d.6). The mechanical behaviour of all these underthrust sequences is thought to influence plate boundary processes which change progressively from shallow to deep levels as the décollement undergoes deformation hardening via tectonic lithification (Moore & Byrne 1987; Kimura *et al.* 2007).

Deformation of terrigenous sediments

The deformation mechanisms of sandstone-dominated mélange are dominated by cataclasis resulting in breakage of sandstone and

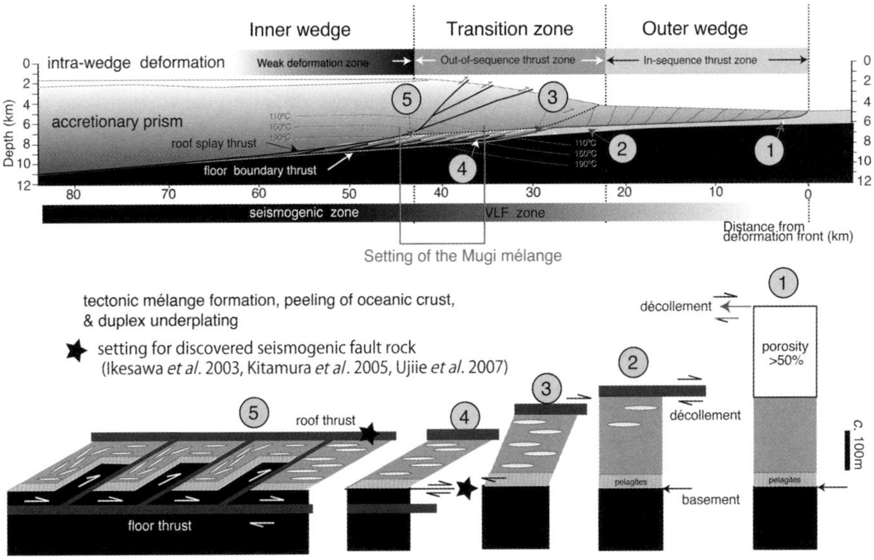

Fig. 2d.6. A cartoon explaining underthrust, tectonic mélange formation, oceanic crust peeling and underplating. Isotherm lines are inferred from modern Nankai Trough and thermal structure of the Mugi mélange (Kimura *et al.* 2012). Setting of inner wedge, transition zone and outer wedge is from Kimura *et al.* (2007). VLFE (very-low-frequency earthquake) zone is after Obana & Kodaira (2009) and Sugioka *et al.* (2012). 1, Deformation front. 2, 3, Underthrust sections at depths from a few to several kilometres where the shear propagates downwards due to deformation hardening. 4, Where tectonic mélange is formed, accompanied by cataclastic breakage of competent sandstone layers and plastic flow of shale matrix. 5, Oceanic crust is peeled and underplated together with underthrust mélange.

extensional cracking. This cataclastic comminution is especially pronounced near the necks and tails of boudinaged sandstone blocks in mélange. The fractal dimension of the cataclasite is about 1.5 (in two dimensions), which is a critical value for fragmented granular material (Sammis *et al.* 1987; Monzawa & Otsuki 2003) because the grains are compacted by a maximum of more than 80% volume fraction. This means that sandstone with an original porosity of more than 50% on the ocean floor would be compacted (due to grain crushing and cementation) to reach a porosity less than 20% (Hashimoto *et al.* 2006). Monzawa & Otsuki (2003) pointed out that grains in cataclasites with such a fractal dimension are not able to move independently, while intergranular slip is prevented because stresses are concentrated at the grain–grain contacts and grains are resistant to slip without grain breakage. In the case of larger fractal dimensions (>1.6), more pore space is necessary so that cataclasites could become fluidized and induce slip weakening (Monzawa & Otsuki 2003).

Mélange in the Shimanto Belt records a change in shape of sandstone boudins from elongate and oblate to a more spherical shape, simultaneously with an increase in the degree of diagenesis and metamorphic grade at seismogenic depth (Kitamura & Kimura 2012). A comparable change is seen in the anisotropic magnetic susceptibility (AMS) ellipsoids of the shale matrix (Kitamura & Kimura 2012). The internal deformation mechanisms of these sandstone blocks in mélanges hotter (250–350°C) than the Mugi mélange (150–250°C) are not yet understood, but the texture of cataclasite should be compared to the heating temperature signals. If the fractal dimension is higher than 1.6 (2D) and the evidence for fluidization is clear, the mechanism is seismic-related slip (Monzawa & Otsuki 2003; Ujiie *et al.* 2007*a*; Ujiie & Kimura 2014) and the process of mélange formation could be strongly related to seismogenic plate boundary processes. In contrast to sandstone boudins, the dominant deformation mechanism in the illite-dominated shale matrix is the development of a preferred orientation via plastic deformation, accommodated by pressure solution and crenulation folding. This type of deformation may be promoted through the smectite–illite transition with dehydration and creeping plate boundary process during the interseismic period or microseismic processes.

Recent friction experiments on fault zone materials at slip rates of several hundreds of microns per second (Ikari *et al.* 2011; Takahashi *et al.* 2014) indicate that velocity-weakening behaviour occurs when the clay content is small. However, fault zone materials always show slip-weakening behaviour regardless of mineral composition at slip rates of *c.* 1 m s^{-1} (Ujiie & Tsutsumi 2010).

Deformation of hemipelagic to pelagic sediments

Hemipelagic to pelagic sediments commonly cover the oceanic basement rocks. These sediments are composed of smectite–illite–chlorite clay minerals that originated from terrigenous sediments and altered oceanic basement. Beneath the hemipelagic sediments, pelagic radiolarian or diatomaceous siliceous ooze and/or calcareous ooze are commonly recognized. These hemipelagic and pelagic sediments undergo progressive diagenesis and become lithified to cherts or micritic limestones. As the oceanic plate becomes older, thicker pelagic sediments (mainly planktonic tests of nannofossils and radiolarians and/or diatoms) are thickly deposited. Older oceanic plates are relatively cold, as indicated by heat-flow measurements (Wada *et al.* 2008). In an old oceanic plate, the heat flow is insufficient to promote the chemical diagenesis of pelagic sediments (e.g. the transition of siliceous ooze made up of opal aggregates to opal-CT (Cristobalite and/or Tridymite) and radiolarian and/or diatomaceous 'chert' and the transition of nannofossil ooze to micritic limestones). To progress the diagenesis of siliceous rocks, temperatures higher than 20–100°C are required (Moore & Saffer 2001). After the accumulation of pelagic sediments, such high temperatures do not prevail except for regions where intraplate magmatism and related hydrothermal activity take place. The main phase of diagenesis may therefore occur when the sediments are underthrust in the subduction zone, together with the oceanic basement.

Radiolarian cherts are commonly observed in ancient orogens (e.g. Taira *et al.* 1988; Matsuda & Isozaki 1991). Such cherts are generally folded at the outcrop scale and the folding pattern is consistent with a map-scale fold-and-thrust belt or accretionary prism (e.g. Kimura & Hori 1993; Kameda *et al.* 2012). Such deformation proceeds in a ductile fashion on an outcrop scale without the plastic flow of quartz during the diagenesis of siliceous ooze in a subduction zone (Kimura & Hori 1993). The deformation of hemipelagic–pelagic sediments is therefore only possible in the temperature range (i.e. temperatures less than *c.* 150°C) experienced after underthrusting but before entering the seismogenic zone. Such sediments release a large amount of fluid during diagenesis at temperatures of 50–150°C (e.g. Moore & Saffer 2001; Saffer *et al.* 2008), resulting in anomalous pore pressures and a decrease in effective shear stress if it is enclosed in the impermeable layers. Dehydrated and transformed minerals within sediments show changes in their frictional properties. Clay-rich hemipelagic sediments show velocity-strengthening behaviour, whereas quartz or carbonate-rich pelagic sediments show velocity-weakening behaviour in low-velocity experiments (Moore *et al.* 2007). The Mugi mélange contains hemipelagic red shale layers several metres thick, as well as Mn-rich umber deposits (Fujinaga *et al.* 1999) and hydrothermal red cherts (Onishi & Kimura 1995; Fujinaga *et al.* 1999). These rocks are tectonically mixed with terrigenous black shales as blocks, and show internal flow structures (Onishi & Kimura 1995; Kitamura *et al.* 2005). This example from the Mugi mélange suggests the subduction of a young oceanic plate, as indicated by the age difference between the hemipelagic shale and terrigenous sandstone (Shibata *et al.* 2008). The hemipelagites were hard enough to be preserved as blocks at the time of formation of the sandstone block-in-matrix structure.

Earthquake-related rocks in mélange

Although pseudotachylite has been considered to be the only definitive geological evidence for seismogenic slip (Cowan 1999), rather enigmatically it is nevertheless rarely observed in tectonic mélange despite strenuous searching. In the Mugi and Okitsu mélanges, pseudotachylites have been found in an upper boundary thrust of the mélange and/or a duplex roof thrust, both representing the appropriate setting for plate boundary décollement (Ikesawa *et al.* 2003; Kitamura *et al.* 2005; Ujiie *et al.* 2007*b*). Ikesawa *et al.* (2003) argued that the roof thrust of the mélange might be saturated by dehydrated fluid due to the presence of an impermeable hanging wall, so that effective strength was small and slip propagating from depth reached the thrust and induced friction melting. As subduction zones contain large amounts of water as interstitial fluids and/or hydrated minerals, the main dynamic weakening mechanisms may not only be melt lubrication but also thermal pressurization, fluidization and silica-gel lubrication, as observed from the tops of the basaltic layers in the Mugi mélange (Ujiie *et al.* 2007*a*, *b*; Ujiie & Tsutsumi 2010; Ujiie & Kimura 2014). In this mélange the maximum temperature is much less than that requiring melting, although recent studies indicate that localized temperature rises of at least a few hundred degrees can occur even in shallow portions of faults in accretionary prisms (Ishikawa *et al.* 2008; Sakaguchi *et al.* 2011; Yamaguchi *et al.* 2011*b*).

Mineralogical and geochemical analyses are important in detecting thermal signals and determining the fault mechanisms that cause earthquakes. In addition to pseudotachylites and ultracataclasites, mélange itself, especially when involving cataclastic breakage of sandstone blocks, is a possible candidate for seismic fault rock. The cataclastic breakage of sandstone boudinage may be induced by the aftershocks of large earthquakes or VLFE. Such earthquakes have recently been detected along the basal décollement beneath the imbricate thrust zone shallower than the classic seismogenic zone (Sugioka *et al.* 2012). Recently, Fagereng & Sibson (2010) pointed out that sandstone-dominated mélange may play an important role in earthquake mechanisms because sandstone is relatively competent and promotes shear localization. In the Mugi mélange, shear localization is recognized along the Riedel shear surfaces, which cut the sandstone boudins (e.g. Kimura *et al.* 2012). The displacement along each Riedel shear surface is less than a few decimetres. These deformational features suggest that Riedel shear surfaces are not generated during major earthquakes with large displacement or during diffusive creeping slip, but are more related to aftershocks or slow slip processes such as VLFE (Saito *et al.* 2013; Ujiie & Kimura 2014). Except for this localized shear, all the other deformation mechanisms in the mélange matrix indicate the dominance of very slow diffusive mass-transfer processes driven by pressure-solution and clay-mineral plastic deformation, in contrast with seismogenic and VLFE processes. More detailed studies are necessary to document this hypothesis.

Comparison of the tectonic mélange with modern subduction plate boundary

The stepping down of the plate boundary décollement in the Nankai Trough has been proposed based on interpretations of seismic profiles (Park *et al.* 1999, 2002; Bangs *et al.* 2004). Kimura *et al.* (2007) noted that the stepping-down model is consistent with the geological interpretation of the Mugi mélange, including the recognition of oceanic slabs along the lowest portion of the thrust sheet. Moore *et al.* (2009) questioned the validity of the step-down model, based on a 3D seismic survey in the Kumano region off the Kii Peninsula, although the evidence is somewhat unclear. The temperature range recorded from the Mugi mélange is the same as that for the stepping down of décollement in the Nankai Trough. It is therefore inferred that underplating of the oceanic slab took place at the same portion of the step-down (Kimura *et al.* 2007, 2012), as proposed by Park *et al.* (2002). The ubiquitous nature of shearing in the Mugi mélange suggests the downwards propagation of shear or simultaneous shearing of the entire mélange sequence. In this case, the 'stepping-down model' is not appropriate because it indicates that the localized roof thrusts switch to the floor thrust through a ramp thrust in a duplex. In the case of the downwards growth of a shear zone or bulk shear, the shear is distributed and thrust-sheet stacking is a continuous process (Fig. 2d.6). The obscure nature of seismic profiles across the Kumano region of the Nankai trough appears to be due to imaging difficulties resulting from geological complexity. Future drilling into this portion by a Riser mission of the NanTroSEIZE project, the multi-stage drilling program of Integrated Ocean Drilling Program target in the Nankai Trough (Tobin & Kinoshita 2006), may provide insights into the nature of underplating such as those derived from analyses of on-land analogues such as the Mugi mélange.

Criteria for tectonic mélange of the subduction megathrust
One of the long-standing debates regarding mélange is the 'sheared olistostrome syndrome', which suggests that primary block-in-matrix texture was formed as mass wasting around the trench but obliterated by shear along the plate boundary megathrust (e.g. Cowan & Page 1975; Aalto 1989; Festa *et al.* 2010). The olistostrome of the mass wasting can include exotic blocks (olistoliths) such as highly metamorphosed rocks and/or older accretionary prism materials as it slips down from the inner trench slope to the trench. In such cases, it is rather easy to define the 'olistolith' in terms of differences in metamorphic grade (e.g. Orange & Underwood 1995; Wakabayashi 2011) or in age (e.g. Osozawa *et al.* 2009) between the blocks and the matrix of the trench-filling sediments. As the envisaged mass wasting is a superficial phenomenon, the primary deformation mechanism of the matrix might not present deep signals such as pressure solution and/or other fluid–rock interaction under the P/T conditions of subduction zone megathrust. Even in the case of a thoroughly overprinted olistostrome, careful geological observation on various scales might reveal the primary processes. Under P/T conditions such as those represented by the shallow portion of the Mugi mélange (130–150°C, several kilometres depth), strain exhibited by the mélange is not so large and the primary texture is well preserved. Recently developed dating methods and analytical estimation of P/T should be capable of resolving the problem of this 'sheared olistostrome syndrome'.

Kimura *et al.* (2012) proposed that four criteria are key to distinguish a true tectonic mélange generated in subduction megathrusts from other settings of the tectonic mélange (e.g. Festa *et al.* 2010) and other mélange origins. These are as follows: (1) lithological mixing of terrigenous trench-fill sediments with oceanic basement rocks, even though original stratigraphy has been disrupted (e.g. Taira *et al.* 1988); (2) systematic younging of pelagic sediments to the trench-fill which represents the original stratigraphy (e.g. Taira *et al.* 1988; Matsuda & Isozaki 1991); (3) the presence of a systematic shearing-related fabric, consistent with the relative palaeo-convergence (e.g. Kano *et al.* 1991; Kimura & Mukai 1991; Onishi & Kimura 1995; Hashimoto & Kimura 1999); and (4) evidence for pressure–temperature (P/T) conditions indicative of the onset of mélange formation in the depth range expected in subduction zones.

Tectonic boundary thrusts

As mentioned in 'Historical background and overview', the Shimanto Belt has been subdivided into northern and southern units separated by a tectonic boundary fault which has been given different names in different parts of the outcrop (Fig. 2d.7a). In Japan such major tectonic boundary faults are called 'tectonic lines', a term introduced by the first geological professor of the University of Tokyo, Edmund Naumann (Naumann 1885); a prime example is the Median Tectonic Line (MTL), separating the Inner and Outer zones of SW Japan (Figs 2d.1 & 2d.2). In the case of the Shimanto Belt on a map scale the tectonic boundary faults are thrust zones, although detailed kinematic descriptions have so far only been provided for the Nobeoka Thrust in Kyushu.

The Nobeoka Thrust

Kondo *et al.* (2005) described in detail the deformation and thermal structure of the Nobeoka Thrust and suggested a displacement of 8–14 km at a depth of several to 11 km beneath the seafloor. The hanging-wall rock of the thrust is composed of Eocene Kitagawa Group shales and sandstones. In contrast, the footwall strata belong to the Eocene–early Oligocene Hyuga Group and are composed of a mélange of shale matrix with sandstone and basaltic blocks deformed in a brittle manner (Kondo *et al.* 2005; Kimura *et al.* 2013). Geothermometry using vitrinite reflectance indicates that the hanging-wall Kitagawa and the footwall Hyuga Groups experienced heating up to a maximum temperature of *c.* 320 and 250°C, respectively (Kondo *et al.* 2005). Chlorite-quartz-H_2O geothermometry (Kameda *et al.* 2011), illite crystallinity and fission-track analysis (Hara & Kimura 2008), and a re-examination of fluid inclusion

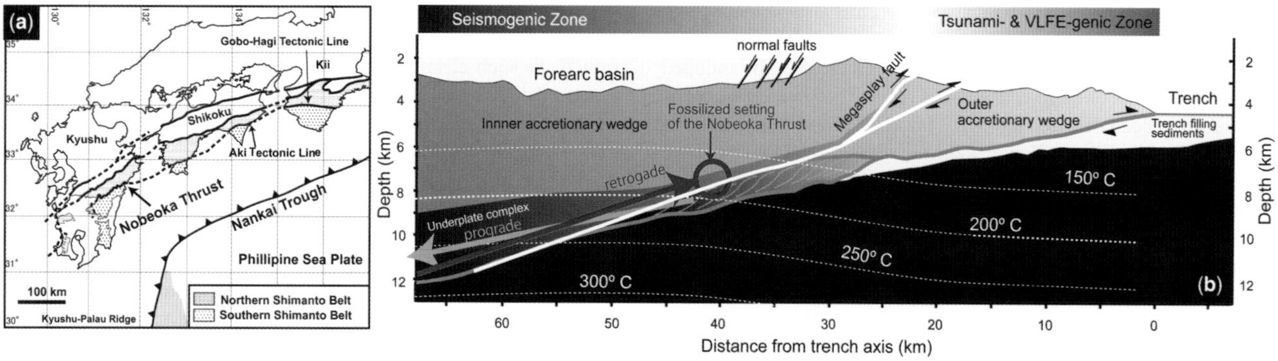

Fig. 2d.7. (**a**) Tectonic setting of the Nobeoka Thrust. (**b**) The tectonic setting of the Nobeoka Thrust as a megasplay in subduction zone (Kimura *et al.* 2013). Oceanward limit of the sesimogenic zone is after Park *et al.* (2002), while the tsunamigenic-slow slip zone is after Ito & Obara (2006*a*, *b*), Obana & Kodaira (2009), Sugioka *et al.* (2012) and Sakaguchi *et al.* (2011).

data (Raimbourg *et al.* 2014) also point to the same temperature range (Fig. 2d.7b).

The Nobeoka Thrust is characterized by a cataclastic fault core *c.* 20 cm thick bordered by brittle damage zones several tens of metres thick for the hanging wall and *c.* 100 m thick for the footwall (Kondo *et al.* 2005; Yamaguchi *et al.* 2011*a*). Ubiquitous shear zones of gouges of a few millimetres to several centimetres thickness, orientated almost parallel to the fault core, are well developed in the damage zone. Subsidiary fractures and cataclastic composite-planar fabrics defined as Y, R, P and T (Logan *et al.* 1981) are observed in the footwall damage zone (Kondo *et al.* 2005; Yamaguchi *et al.* 2011*a*). Kondo *et al.* (2005) also suggested that the Nobeoka Thrust was located at the seismogenic depth from the viewpoint of thermal models (Hyndman *et al.* 1997; Oleskevich *et al.* 1999) and comparable to the modern splay fault in the Nankai Trough (Park *et al.* 2002). Okamoto *et al.* (2006) found a pseudotachylite with implosion breccia in the hanging wall in the vicinity of the fault core. Geothermobarometry using H_2O-CH_4 fluid inclusions in tension crack-filling quartz veins in the fault core and the footwall shear zone indicates that thermal fluid *c.* 50°C hotter than the host rocks flowed into the zone (Kondo *et al.* 2005). Yamaguchi *et al.* (2011*a*) analysed major and trace elements of such veins, revealing that the fluid was oxidized during the pre-seismic period but changed to reduced conditions immediate after the slip, and suggested exotic fluid inflow into the fault zone from depth.

Raimbourg *et al.* (2009) reported that the hanging-wall Kitagawa Group experienced three stages of deformation: (1) horizontal shortening represented by outcrop-scale folding and thrusting; (2) vertical loading documented by the development of horizontal slaty cleavages; and (3) deformation by simple shear, limited to the vicinity of the Nobeoka Thrust. These observations led these authors to suggest that the accretionary prism was stable under the normal fault-type stress field in the deep setting of the inner wedge before upthrusting. Kameda *et al.* (2011) examined the clay minerals of the cleavages of the Kitagawa Group in detail and defined anchizone–epizone-grade metamorphism characterizing reactions near the downdip limit of the seismogenic zone. On the basis of K–Ar and fission track dating, Hara & Kimura (2008) suggested that the timing of the thrusting was at least *c.* 48–40 Ma, but it might be much younger because the thrust overlies the early Oligocene (28.4–33.9 Ma) Hyuga Group (Murata 1991).

Kimura *et al.* (2013) analysed the Nobeoka Thrust hanging wall in detail and suggested that it experienced four stages of deformation: initial horizontal shortening documented by folding and thrusting; then vertical loading shown by horizontal slaty cleavages; followed by cleavage-parallel shear only along the vicinity of the Nobeoka Thrust; and finally brittle fracturing associated with pseudotachylite-bearing faults. Taking account of the dynamic changes in stress field suggested by Wang & Hu (2006) and Raimbourg *et al.* (2009), Kimura *et al.* (2013) interpreted the structural evolution as follows. During the interseismic period vertical loading (σ_1) was dominant, as recorded by diffusive deformation involving cleavage formation, while a smaller horizontal compression (σ_2) was related to the locking of the megathrust; together, these maintained a stable taper.

Elastic strain energy in the hanging wall of the inner wedge was co-seismically released during slip on the megathrust associated with rheological weakening of the fault plane. This sudden release of elastic strain caused brittle cracking, with σ_1 oriented at a high angle to the shear surface of the Nobeoka Thrust. According to analyses of crack-filling veins and brittle fractures parallel to the fault core, most deformation associated with the displacement suggested from the difference in metamorphic grade between the hanging- and footwalls has been localized in the footwall. The hanging wall appears to have gravitationally collapsed during the co-seismic period, but remained stable and at an extensionally subcritical condition during the interseismic period.

Hamahashi *et al.* (2013) analysed the physical properties of this exhumed and fossilized subduction-zone megasplay fault using geophysical logging data. The footwall cataclasite shows higher averages of neutron porosity (*c.* 7.6%) and lower values of electric resistivity (*c.* 232 Ωm) and P-wave velocity (*c.* 4.2 km s^{-1}) compared to the hanging-wall phyllite (*c.* 4.8%, *c.* 453 Ωm, *c.* 4.8 km s^{-1}). Despite this contrast on the mesoscopic scale, the resistivity and porosity data from both the hanging wall and footwall can fit into a single curve of Archie's law, suggesting a similarity in microstructure and mineralogy at this low-porosity range. Bounding the main fault core of the Nobeoka Thrust, the brittle damage zone in the hanging wall yields the pseudotachylite evidence for seismogenic slip and does not follow Archie's law. A similar hanging-wall damage zone is also observed in the modern shallow splay fault in the Nankai Trough, but is much thicker than 50 m. The footwall damage zone is more extensive than that of the hanging wall in the Nobeoka Thrust. Splay faults may exhibit strong deformation in the hanging wall in the early stage of their development, but as fault rocks become buried deeper and strain accumulates along the fault the damage effects may eventually increase in the footwall.

Origin of the tectonic boundary thrust

The origin of major tectonic boundary thrusts in these accretionary sequences is an important question, one which is still not fully understood in the case of SW Japan. One idea has been that it represents a tectonic boundary between the colliding and accreted terranes, but it is difficult to know how to define the colliding exotic terrane and discriminate it from *in situ* terranes. The terrane 'school' (e.g. Jones *et al.* 1982) challenged the orogenic model of Dewey & Bird (1970) by introducing the possibility of large-scale strike-slip movements along major boundary faults. This model however failed in general to clarify the origin of such tectonic boundaries, not least because the geological units comprising accretionary prisms in subduction zones and collisional fold-and-thrust belts present very complex geological features and behaviours (e.g. McClay 1992).

Another interpretation for the tectonic boundary fault in SW Japan is that of the 'out-of-sequence thrust' in the accretionary prism (Ohmori *et al.* 1997; Kimura 1998; Kondo *et al.* 2005; Hara & Kimura 2008). This model assumes that the supply of terrigenous sediments and the convergence of relative plates are constant, and the prism grows constantly (Suppe 1981; Dahlen 1990). The prism grows by in-sequence thrusting at the deformation front (Morley 1988), with out-of-sequence thrusting occurring at the inboard of the prism and maintaining the critical taper. The resulting accretionary prism can therefore thicken until it obtains the thickness of continental crust of *c.* 30 km and more (Suppe 1981; Dahlen 1990).

If we attempt to apply this out-of-sequence thrust model to tectonic boundary faults in the Shimanto Belt, it implies that the northern Shimanto Belt was continuously followed by the growth of the southern Shimanto Belt without any serious break since Late Cretaceous–early Miocene times. However, the tectonic history of SW Japan presents several complications such as: global change in relative convergence (Uyeda & Miyashiro 1974; Onishi & Kimura 1995); Kula–Pacific ridge subduction (Uyeda & Miyashiro 1974; Maruyama *et al.* 1997); intermittent sediment supply to the trench in Late Cretaceous and Palaeogene times (Maruyama *et al.* 1997); and tectonic erosion (Isozaki *et al.* 2010). Given that one or some combination of these tectonic scenarios may have controlled the onset and evolution of the main tectonic boundaries within the Shimanto Belt, at this stage it would seem wise to avoid generic interpretations. Onishi & Kimura (1995) and Onishi *et al.* (2001) have discussed the possible relationship between the tectonic gap between the northern and southern parts of the Shimanto Belt and global changes in relative plate convergence. How such interpretations can be tested and perhaps applied to other circum-Pacific regions remains to be seen.

Accretion v. tectonic erosion

Accretionary and erosive margins provide tectonic end-members within our understanding of the subduction zone setting (von Huene & Scholl 1991; Clift & Vannucchi 2004); how these tectonic processes might be recorded and recognizable in ancient subduction complexes remains a challenging issue, however. Tectonic erosion includes sediment subduction and basal erosion along the plate boundary megathrust which drags down the crust of the upper plate into the mantle. This process means that the 'evidence' for such erosion is commonly based on lost geological tectonostratigraphic data, that is, gaps in the record, and must be speculated from indirect phenomena such as subsidence of the forearc slopes (von Huene & Lallemand 1990). A topographically rough surface to the subducting oceanic plate, such as that provided by seamounts, has been suggested to work like an erosive saw carving into the upper plate (von Huene & Lallemand 1990). Another mechanism of basal erosion has been suggested as hydrofracturing of upper plate materials due to dehydration-induced fluid pressures, resulting in entrainment of upper plate materials into the basal décollement (von Huene & Ranero 2003). Considering the interaction between the *c.* 30 km thick crust of the upper plate and subducting oceanic plate, a subduction dip angle of *c.* 15° and convergent rate of *c.* 10 cm a^{-1}, at least *c.* 1 Ma of continuous basal erosion is necessary to induce clear subsidence of the forearc because the length of plate interface between the upper crustal and subducting plates is *c.* 115 km (30/cos15°). In several examples of subduction zones, for example the Japan Trench and the Middle America Trench off Costa Rica, the subsidence of a few thousand metres of the forearc combined with a lack of accretionary prism over a period of several million years suggest that the erosive condition needs to be maintained for several to tens of millions of years (von Huene & Lallemand 1990; Vannucchi *et al.* 2001).

Such an age gap in the ancient accretionary complex would be one of the signals for tectonic erosion in the past. Isozaki *et al.* (2010) have recently proposed the hypothesis that a tremendous amount of tectonic erosion took place during Early Cretaceous and middle–late Miocene times, based on age gaps in the accretionary complex. However, such age gaps in the accretionary complex do not automatically imply that tectonic erosion has taken place; other interpretations such as no accretion, cessation of subduction and/or later tectonic modification are also possible. In the case of the middle–late Miocene period, for example, a drastic tectonic change after the opening of the Japan Sea and clockwise rotation of SW Japan may be linked to ridge subduction (Seno & Maruyama 1984) or a switch in subduction from the Pacific to Philippine Sea Plate (Hall 2002). Recent drilling in the forearc of the Nankai Trough (Saffer *et al.* 2009) suggests that the accretion was renewed at *c.* 6 Ma after igneous activity intruding the early Miocene accretionary prism (Kimura *et al.* 2014). Kimura *et al.* (2014) interpreted that the subduction ceased between *c.* 12 and 8 Ma due to the transference of subduction from the Pacific Plate to the Philippine Sea Plate, as opposed to the 'continuous subduction with subduction erosion' viewpoint (Isozaki *et al.* 2010). These different scenarios need to be tested in the future.

Shibata *et al.* (2008) dated the Mugi mélange located in the southernmost part of the northern Shimanto Belt and revealed that it can be separated into a cooler-temperature (*c.* 150–170°C) lower part and a higher-temperature (*c.* 220°C) upper part. Furthermore, zircon ages from tuffs and sandstones reveal a clear age gap of several millions of years between these two parts of the mélange. Shibata *et al.* (2008) explained this age gap in terms of either later tectonic overthrusting, or no underplating and/or basal tectonic erosion. A gap of several million years along an ancient plate boundary fault implies that there is no geological record for the subduction of more than several hundred kilometres of oceanic plate. In the Inner Zone of SW Japan a large amount of felsic igneous rocks are exposed and have been extensively dated, although comparisons with ages from the Shimanto accretionary complex using relatively modern methods such as detrital Zircon U–Pb dating are still at an early stage. Such data will likely help clarify the tectonic evolution of the Shimanto Belt, one of the world's best-preserved accretionary prisms, and improve our understanding of platc boundary processes.

This research was financially supported by a Grant-in-Aid for Scientific Research from MEXT of Japan (#21107005). We acknowledge A. Mukai, T. C. Onishi, A. Sakaguchi, K. Omori, H. Yamaguchi, J. Kameda, H. Kondo, E. Ikesawa, K. Kawabata, M. Ito, K. Sato, T. Shibata, S. Okamoto, H. Raimbourg, M. Hojo, Y. Kusaba, M. Hamahashi, R. Kawasaki and A. Koge for their fieldwork in the Shimanto Belt, many discussions and suggestions. The Seamless Digital Geological Map of Japan (https://gbank.gsj.jp/seamless/) was used for the basemap of our figures.

Appendix

English to Kanji and Hiragana translations for geological and place names

Aki	安芸	あき
Ashizuri	足摺	あしずり
Gobo	御坊	ごぼう
Hagi	萩	はぎ
Hidaka	日高	ひだか
Hokkaido	北海道	ほっかいどう
Hyuga	日向	ひゅうが
Izu	伊豆	いず
Kii	紀伊	きい
Kinan	紀南	きなん
Kitagawa	北川	きたがわ
Kumano	熊野	くまの
Kyushu	九州	きゅうしゅう
Mugi	牟岐	むぎ
Muroto	室戸	むろと
Nankai	南海	なんかい
Nobeoka	延岡	のべおか
Okitsu	興津	おきつ
Ryukyu	琉球	りゅうきゅう
Sakawa	佐川	さかわ
Sanbagawa (Sambagawa)	三波川	さんばがわ
Shikoku	四国	しこく
Shimanto	四万十	しまんと
Shiono	潮	しおの
Sorachi	空知	そらち
Tokoro	常呂	ところ
Tosa-Bae	土佐バエ	とさばえ

References

Aalto, K. 1989. Franciscan complex olistostrome at Crescent City, northern California. *Sedimentology*, **36**, 471–495.

Ando, M. 1975. Source mechanisms and tectonic significance of historical earthquakes along the Nankai Trough, Japan. *Tectonophysics*, **27**, 119–140.

Aoki, K., Maruyama, S., Isozaki, Y., Otoh, S. & Yanai, S. 2011. Recognition of the Shimanto HP metamorphic belt within the traditional Sanbagawa HP metamorphic belt: new perspectives of the Cretaceous–Paleogene tectonics in Japan. *Journal of Asian Earth Sciences*, **42**, 355–369, http://doi.org/10.1016/j.jseaes.2011.05.001

Bangs, N. L., Shipley, T. H., Gulick, S. P., Moore, G. F., Kuramoto, S. & Nakamura, Y. 2004. Evolution of the Nankai Trough decollement from the trench into the seismogenic zone: inferences from three-dimensional seismic reflection imaging. *Geology*, **32**, 273–276.

Byrne, D. E., Davis, D. M. & Sykes, L. R. 1988. Loci and maximum size of thrust earthquakes and the mechanics of the shallow region of subduction zones. *Tectonics*, **7**, 833–857.

Clift, P. & Vannucchi, P. 2004. Controls on tectonic accretion v. erosion in subduction zones: implications for the origin and recycling of the continental crust. *Reviews of Geophysics*, **42**, RG2001, http://doi.org/10.1029/2003RG000127

Cowan, D. S. 1999. Do faults preserve a record of seismic slip? A field geologist's opinion. *Journal of Structural Geology*, **21**, 995–1001.

Cowan, D. S. & Page, B. M. 1975. Recycled Franciscan material in Franciscan melange west of Paso Robles, California. *Geological Society of America Bulletin*, **86**, 1089–1095.

Dahlen, F. 1990. Critical taper model of fold-and-thrust belts and accretionary wedges. *Annual Review of Earth and Planetary Sciences*, **18**, 55–99, http://doi.org/10.1146/annurev.ea.18.050190.000415

Dewey, J. F. & Bird, J. M. 1970. Mountain belts and the new global tectonics. *Journal of Geophysical Research*, **75**, 2625–2647.

Fagereng, Å. & Sibson, R. H. 2010. Melange rheology and seismic style. *Geology*, **38**, 751–754.

Festa, A., Pini, G. A., Dilek, Y. & Codegone, G. 2010. Mélanges and mélange-forming processes: a historical overview and new concepts. *International Geology Review*, **52**, 1040–1105.

Fujinaga, K., Kato, S., Kiminami, K., Miura, K. & Nakamura, K. 1999. Geochemistry of red shale from Mugi Formation in the Upper Cretaceous Shimanto Belt, Shikoku Japan. *Memoir of the Geological Society of Japan*, **52**, 205–216 [In Japanese with English abstract].

Geological Survey of Japan (ed.) 1900. *1:1 000 000, Geological Map of Japan*. Ministry of Agriculture and Economy, Tokyo.

Geological Survey of Japan, AIST (ed.) 2012. *Seamless Digital Geological Map of Japan 1: 200 000. Jul 3, 2012 Version*. Research Information Database DB084, Geological Survey of Japan, National Institute of Advanced Industrial Science and Technology.

Hall, R. 1996. Reconstructing Cenozoic SE Asia. *In*: Hall, R. & Blundell, D. (eds) *Tectonic Evolution of Southeast Asia*. Geological Society, London, Special Publications, **106**, 153–184.

Hall, R. 2002. Cenozoic geological and plate tectonic evolution of SE Asia and the SW Pacific: computer-based reconstructions, model and animations. *Journal of Asian Earth Sciences*, **20**, 353–431.

Hamahashi, M., Saito, S. *et al.* 2013. Contrasts in physical properties between the hanging wall and footwall of an exhumed seismogenic megasplay fault in a subduction zone – An example from the Nobeoka Thrust Drilling Project. *Geochemistry, Geophysics, Geosystems*, **14**, 5354–5370.

Hara, H. & Kimura, K. 2008. Metamorphic and cooling history of the Shimanto accretionary complex, Kyushu, Southwest Japan: implications for the timing of out-of-sequence thrusting. *Island Arc*, **17**, 546–559.

Hasegawa, S. & Ohno, H. 2006. Middle miocene igenous rocks controlling the distribution active faults. *Chikyu Monthly Symposium*, Tokyo, **54**, 50–58 [in Japanese].

Hashimoto, Y. & Kimura, G. 1999. Underplating process from melange formation to duplexing: example from the Cretaceous Shimanto Belt, Kii Peninsula, southwest Japan. *Tectonics*, **18**, 92–107.

Hashimoto, Y., Nakaya, T., Ito, M. & Kimura, G. 2006. Tectonolithification of sandstone prior to the onset of seismogenic subduction zone: evidence from tectonic mélange of the Shimanto Belt, Japan. *Geochemistry, Geophysics, Geosystems*, **7**, Q06013.

Hashimoto, Y., Nikaizo, A. & Kimura, G. 2009. A geochemical estimation of fluid flux and permeability for a fault zone in Mugi mélange, the Cretaceous Shimanto Belt, SW Japan. *Journal of Structural Geology*, **31**, 208–214.

Hujita, K. 1968. Rokko movement. *Quaternary Research*, **7**, 248–260 [in Japanese].

Hyndman, R. D., Yamano, M. & Oleskevich, D. A. 1997. The seismogenic zone of subduction thrust faults. *Island Arc*, **6**, 244–260.

Ikari, M. J., Marone, C. & Saffer, D. M. 2011. On the relation between fault strength and frictional stability. *Geology*, **39**, 83–086.

Ikesawa, E., Sakaguchi, A. & Kimura, G. 2003. Pseudotachylyte from an ancient accretionary complex: evidence for melt generation during seismic slip along a master décollement? *Geology*, **31**, 637–640.

Ikesawa, E., Kimura, G. *et al.* 2005. Tectonic incorporation of the upper part of oceanic crust to overriding plate of a convergent margin: an example from the Cretaceous–early Tertiary Mugi Mélange, the Shimanto Belt, Japan. *Tectonophysics*, **401**, 217–230.

Ishigaki, S. & Yao, A. 1982. Radiolarian revolution: conversation between the micropaleontologist and school teacher. *Geological Education and Scientific Movement*, **11**, 93–102 [in Japanese].

Ishikawa, T., Tanimizu, M. *et al.* 2008. Coseismic fluid–rock interactions at high temperatures in the Chelungpu fault. *Nature Geoscience*, **1**, 679–683.

Isozaki, Y., Aoki, K., Nakama, T. & Yanai, S. 2010. New insight into a subduction-related orogen: a reappraisal of the geotectonic framework and evolution of the Japanese Islands. *Gondwana Research*, **18**, 82–105.

Ito, Y. & Obara, K. 2006*a*. Dynamic deformation of the accretionary prism excites very low frequency earthquakes. *Geophysical Research Letters*, **33**, L02311, http://doi.org/10.1029/2005GL025270

Ito, Y. & Obara, K. 2006*b*. Very low frequency earthquakes within accretionary prisms are very low stress-drop earthquakes. *Geophysical Research Letters*, **33**, L09302, http://doi.org/10.1029/2006GL025883

Jones, D. L., Silberling, N. J., Gilbert, W. & Coney, P. 1982. Character, distribution, and tectonic significance of accretionary terranes in the Central Alaska Range. *Journal of Geophysical Research*, **87**, 3709–3717, http://doi.org/10.1029/JB087iB05p03709

Kameda, J., Raimbourg, H., Kogure, T. & Kimura, G. 2011. Low-grade metamorphism around the down-dip limit of seismogenic subduction zones: example from an ancient accretionary complex in the Shimanto Belt, Japan. *Tectonophysics*, **502**, 383–392.

Kameda, J., Hina, S. *et al.* 2012. Silica diagenesis and its effect on interplate seismicity in cold subduction zones. *Earth and Planetary Science Letters*, **317**, 136–144.

Kanmera, K. & Sakai, T. 1975. Where is the modern setting for the Shimanto Group? *GDP Newsletter*, **II-I-(1)**, I, 55–64 [in Japanese].

Kano, K., Kosaka, K., Murata, A. & Yanai, S. 1990. Intra-arc deformations with vertical rotation axes: the case of the pre-Middle Miocene terranes of Southwest Japan. *Tectonophysics*, **176**, 333–354.

Kano, K., Nakaji, M. & Takeuchi, S. 1991. Asymmetrical mélange fabrics as possible indicators of the convergent direction of plates: a case study from the Shimanto Belt of the Akaishi Mountains, central Japan. *Tectonophysics*, **185**, 375–388.

Kawabata, K., Tanaka, H. & Kimura, G. 2007. Mass transfer and pressure solution in deformed shale of accretionary complex: examples from the Shimanto Belt, southwestern Japan. *Journal of Structural Geology*, **29**, 697–711.

Kiminami, K., Kashiwagi, N. & Miyashita, S. 1992. Occurrence and significance of in-situ greenstones from the Mugi formation in the upper Cretaceous Shimanto supergroup, eastern Shikoku, Japan. *The Journal of the Geological Society of Japan*, **98**, 867–883.

Kimura, G. 1994. The latest Cretaceous-early Paleogene rapid growth of accretionary complex and exhumation of high pressure series metamorphic rocks in northwestern Pacific margin. *Journal of Geophysical Research*, **99**, 22147–22164, http://doi.org/10.1029/94jb00959

Kimura, G. 1997. Cretaceous episodic growth of the Japanese Islands. *Island Arc*, **6**, 52–68.

Kimura, G. & Ludden, J. 1995. Peeling oceanic crust in subduction zones. *Geology*, **23**, 217–220.

Kimura, G. & Mukai, A. 1991. Underplated units in an accretionary complex: melange of the Shimanto Belt of eastern Shikoku, southwest Japan. *Tectonics*, **10**, 31–50.

Kimura, G., Kitamura, Y., Hashimoto, Y., Yamaguchi, A., Shibata, T., Ujiie, K. & Okamoto, S. 2007. Transition of accretionary wedge structures around the up-dip limit of the seismogenic subduction zone. *Earth and Planetary Science Letters*, **255**, 471–484.

Kimura, G., Yamaguchi, A. *et al.* 2012. Tectonic mélange as fault rock of subduction plate boundary. *Tectonophysics*, **568**, 25–38.

Kimura, G., Hamahashi, M. *et al.* 2013. Hanging wall deformation of a seismogenic megasplay fault in an accretionary prism: the Nobeoka Thrust in southwestern Japan. *Journal of Structural Geology*, **52**, 136–147, http://doi.org/10.1016/j.jsg.2013.03.015

Kimura, G., Hashimoto, Y., Kitamura, Y., Yamaguchi, A. & Koge, S. 2014. Middle Miocene swift migration of the TTT triple junction and rapid crustal growth in southwest Japan: a review. *Tectonics*, **33**, http://doi.org/10.1002/2014TC003531

Kimura, K. 1998. Out-of-sequence thrust of accretionary complex. *Memoir of the Geological Society of Japan*, **50**, 131–146.

Kimura, K. & Hori, R. 1993. Offscraping accretion of Jurassic chert-clastic complexes in the Mino-Tamba Belt, central Japan. *Journal of Structural Geology*, **15**, 145–161.

Kinoshita, O. 1995. Migration of igneous activities related to ridge subduction in Southwest Japan and the East Asian continental margin from the Mesozoic to the Paleogene. *Tectonophysics*, **245**, 25–35.

Kishu Shimanto Research Group 1975. Geological Development of Shimanto Geosyncline. *Monograph, the Association for the Geological Collaboration in Japan*, **19**, 1453–1156 [in Japanese].

Kitamura, Y. & Kimura, G. 2012. Dynamic role of tectonic mélange during interseismic process of plate boundary mega earthquakes. *Tectonophysics*, **568**, 39–52.

Kitamura, Y., Sato, K. *et al.* 2005. Mélange and its seismogenic roof décollement: a plate boundary fault rock in the subduction zone: An example from the Shimanto Belt, Japan. *Tectonics*, **24**, TC5012, http://doi.org/10.1029/2004TC001635

Knopf, A. 1948. The geosynclinal theory. *Geological Society of America Bulletin*, **59**, 649–670.

Kobayashi, T. 1941. The Sakawa orogenic cycle and its bearing on the origin of the Japanese Islands: dedicated to Professor Matajiro Yokoyama. *Journal of the Faculty of Science, Imperial University of Tokyo, Section II*, **5**, 219–578.

Kodaira, S., Takahashi, N., Nakanishi, A., Miura, S. & Kaneda, Y. 2000. Subducted seamount imaged in the rupture zone of the 1946 Nankaido earthquake. *Science*, **289**, 104–106.

Kodaira, S., Hori, T. *et al.* 2006. A cause of rupture segmentation and synchronization in the Nankai trough revealed by seismic imaging and numerical simulation. *Journal of Geophysical Research*, **111**, B09301, http://doi.org/10.1029/2005JB004030

Kondo, H., Kimura, G. *et al.* 2005. Deformation and fluid flow of a major out-of-sequence thrust located at seismogenic depth in an accretionary complex: Nobeoka Thrust in the Shimanto Belt, Kyushu, Japan. *Tectonics*, **24**, TC6008.

Kurimoto, C. 1982. 'Chichibu System' in the area southwest of Koyasan, Wakayama Prefecture-Upper Cretaceous Hanazono Formation. *Journal of the Geological Society of Japan*, **88**, 901–914.

Logan, J. M., Higgs, N. G. & Friedman, M. 1981. Laboratory studies on natural gouge from the U.S. Geological Survey Dry Lake Valley No. 1 well, San Andreas fault zone. *In*: Carter, N. L., Friedman, M., Logan, J. M. & Stearns, D. W. (eds) *Mechanical Behavior of Crustal Rocks: The Handin Volume*. American Geophysical Union, Geophysical Monograph **24**, 121–134.

Marshak, R. S. & Karig, D. E. 1977. Triple junctions as a cause for anomalously near-trench igneous activity between the trench and volcanic arc. *Geology*, **5**, 233–236.

Maruyama, S., Isozaki, Y., Kimura, G. & Terabayashi, M. 1997. Paleogeographic maps of the Japanese Islands: plate tectonic synthesis from 750 Ma to the present. *Island Arc*, **6**, 121–142.

Matsuda, T. & Isozaki, Y. 1991. Well-documented travel history of Mesozoic pelagic chert in Japan: from remote ocean to subduction zone. *Tectonics*, **10**, 475–499.

Matsumura, M., Hashimoto, Y., Kimura, G., Ohmori-Ikehara, K., Enjohji, M. & Ikesawa, E. 2003. Depth of oceanic-crust underplating in a subduction zone: inferences from fluid-inclusion analyses of crack-seal veins. *Geology*, **31**, 1005–1008.

McClay, K. R. 1992. Glossary of thrust tectonics terms. *In*: McClay, K. R. (ed.) *Thrust Tectonics*. Chapman & Hall, London, 419–433.

Minato, M., Gorai, M. & Hunahashi, M. (eds) 1965. *The Geologic Developments of the Japanese Islands*. Tsukiji Shokan Co., Tokyo.

Miura, D. & Wada, Y. 2007. Effects of stress in the evolution of large silicic magmatic systems: an example from the Miocene felsic volcanic field at Kii Peninsula, SW Honshu, Japan. *Journal of Volcanology and Geothermal Research*, **167**, 300–319.

Monzawa, N. & Otsuki, K. 2003. Comminution and fluidization of granular fault materials: implications for fault slip behavior. *Tectonophysics*, **367**, 127–143.

Moore, G. F., Park, J.-O. *et al.* 2009. Structural and seismic stratigraphic framework of the NanTroSEIZE State 1 transect. *In*: *NanTroSEIZE Stage 1: Investigations of Seismogenesis, Nankai Trough, Japan*, Proceedings of the Integrated Ocean Drilling Program, **314/315/316**, http://doi.org/10.2204/iodp.proc.314315316.102.2009

Moore, J. C. & Byrne, T. 1987. Thickening of fault zones: a mechanism of melange formation in accreting sediments. *Geology*, **15**, 1040–1043.

Moore, J. C. & Saffer, D. 2001. Updip limit of the seismogenic zone beneath the accretionary prism of southwest Japan: an effect of diagenetic to low-grade metamorphic processes and increasing effective stress. *Geology*, **29**, 183–186.

Moore, J. C., Mascle, A. *et al.* 1988. Tectonics and hydrogeology of the northern Barbados Ridge: results from Ocean Drilling Program leg 110. *Geological Society of America Bulletin*, **100**, 1578–1593.

Moore, J. C., Rowe, C. C. & Meneghini, F. 2007. How accretionarty prisms eucidate seismogeninesis in subduction zone. *In*: Dixon, H. & Moore, J. C. (eds) *The Sesimogeniczone of Subduction Thrust Faults*. Columbia University Press, New York, 288–300.

Morley, O. K. 1988. Out-of-sequence thrusts. *Tectonics*, **7**, 539–561.

Murata, A. 1991. Duplex structures of the Uchinohae Formation in the Shimanto Terrane, Southwest Japan. *Journal of Geological Society of Japan*, **97**, 39–52 [in Japanese with English abstract].

Nakaya, S. 2012. Bending structure of the Shimanto accretionary prism in the Kii Peninsula. *Association for the Geological Collaboration in Japan, Monograph*, **59**, 119–128 [in Japanese with English abstract].

Naumann, E. 1885. Ueber den Bau und die Entstehung der japanischen Inseln: Begleitworte zu den von der geologischen Aufnahme von Japan für den. *Internationalen Geologen-Congress in Berlin bearbeiteten topographischen und geologischen Karten*. R. Friedländer & Sohn, Berlin.

Obana, K. & Kodaira, S. 2009. Low-frequency tremors associated with reverse faults in a shallow accretionary prism. *Earth and Planetary Science Letters*, **287**, 168–174.

Ohmori, K., Taira, A., Tokuyama, H., Sakaguchi, A., Okamura, M. & Aihara, A. 1997. Paleothermal structure of the Shimanto accretionary prism, Shikoku, Japan: role of an out-of-sequence thrust. *Geology*, **25**, 327–330.

Okamoto, S., Kimura, G., Takizawa, S. & Yamaguchi, H. 2006. Earthquake fault rock indicating a coupled lubrication mechanism. *Earth Discussions*, **1**, 135–149.

Okino, K., Ohara, Y., Kasuga, S. & Kato, Y. 1999. The Philippine Sea: new survey results reveal the structure and the history of the marginal basins. *Geophysical Research Letters*, **26**, 2287–2290.

Oleskevich, D. A., Hyndman, R. D. & Wang, K. 1999. The updip and downdip limits to great subduction earthquakes: thermal and structural models of Cascadia, south Alaska, SW Japan, and Chile. *Journal of Geophysical Research*, **104**, 14965–14991.

Onishi, C. T. & Kimura, G. 1995. Change in fabric of melange in the Shimanto Belt, Japan: change in relative convergence? *Tectonics*, **14**, 1273–1289.

Onishi, C. T., Kimura, G., Hashimoto, Y., Ikehara-Ohmori, K. & Watanabe, T. 2001. Deformation history of tectonic melange and its relationship to the underplating process and relative plate motion: an example from the deeply buried Shimanto Belt, SW Japan. *Tectonics*, **20**, 376–393.

Orange, D. L. & Underwood, M. B. 1995. Patterns of thermal maturity as diagnostic criteria for interpretation of melanges. *Geology*, **23**, 1144–1148.

Osozawa, S., Morimoto, J. & Flower, M. F. J. 2009. 'Block-in-matrix' fabrics that lack shearing but possess composite cleavage planes: a sedimentary mélange origin for the Yuwan accretionary complex in the Ryukyu island arc, Japan. *Geological Society of America Bulletin*, **121**, 1190–1203.

Otofuji, Y.-I., Matsuda, T. & Nohda, S. 1985. Opening mode of the Japan Sea inferred from the palaeomagnetism of the Japan Arc. *Nature*, **317**, 603–604.

Park, J.-O., Tsuru, T., Kaneda, Y., Kono, Y., Kodaira, S., Takahashi, N. & Kinoshita, H. 1999. A subducting seamount beneath the Nankai accretionary prism off Shikoku, southwestern Japan. *Geophysical Research Letters*, **26**, 931–934.

Park, J.-O., Tsuru, T., Kodaira, S., Cummins, P. R. & Kaneda, Y. 2002. Splay fault branching along the Nankai subduction zone. *Science*, **297**, 1157–1160.

Piper, D. J. W., von Huene, R. & Duncan, J. R. 1973. Late Quaternary sedimentation in the active eastern Aleutian Trench. *Geology*, **1**, 19–22.

Raimbourg, H., Shibata, T., Yamaguchi, A., Yamaguchi, H. & Kimura, G. 2009. Horizontal shortening v. vertical loading in accretionary prisms. *Geochemistry, Geophysics, Geosystems*, **10**, Q04007, http://doi.org/10.1029/2008GC002279

Raimbourg, H., Thiéry, R. *et al.* 2014. A new method of reconstituting the P–T conditions of fluid circulation in an accretionary prism (Shimanto, Japan) from microthermometry of methane-bearing aqueous inclusions. *Geochimica et Cosmochimica Acta*, **125**, 96–109.

Saffer, D. M., Underwood, M. B. & McKiernan, A. W. 2008. Evaluation of factors controlling smectite transformation and fluid production in subduction zones: application to the Nankai Trough. *Island Arc*, **17**, 208–230.

Saffer, D. M., McNiell, L., Araki, E., Byrne, T., Eguchi, N., Toczko, S. & Takahashi, K. & The Expedition 319 Scientists 2009. NanTroSEIZE Stage 2: NanTroSEIZE riser/riserless observatory. *In*: *IODP Preliminary Report*. International Ocean Drilling Program, **319**, 82, http://doi.org/10.2204/iodp.pr.319.2009

Saito, T., Ujiie, K., Tsutsumi, A., Kameda, J. & Shibazaki, B. 2013. Geological and frictional aspects of very-low-frequency earthquakes in an accretionary prism. *Geophysical Research Letters*, **40**, 703–708, http://doi.org/10.1002/grl/50175

Sakaguchi, A., Chester, F. *et al.* 2011. Seismic slip propagation to the updip end of plate boundary subduction interface faults: vitrinite reflectance geothermometry on Integrated Ocean Drilling Program NanTro SEIZE cores. *Geology*, **39**, 395–398.

Sammis, C., King, G. & Biegel, R. 1987. The kinematics of gouge deformation. *Pure and Applied Geophysics*, **125**, 777–812.

Seno, T. & Maruyama, S. 1984. Paleogeographic reconstruction and origin of the Philippine Sea. *Tectonophysics*, **102**, 53–84.

Shibata, T., Orihashi, Y., Kimura, G. & Hashimoto, Y. 2008. Underplating of mélange evidenced by the depositional ages: U–Pb dating of zircons from the Shimanto accretionary complex, southwest Japan. *Island Arc*, **17**, 376–393.

Spinelli, G. A. & Harris, R. N. 2011. Thermal effects of hydrothermal circulation and seamount subduction: temperatures in the Nankai Trough Seismogenic Zone Experiment transect, Japan. *Geochemistry, Geophysics, Geosystems*, **12**, Q0AD21, http://doi.org/10.1029/2011GC003727

Sugioka, H., Okamoto, T. *et al.* 2012. Tsunamigenic potential of the shallow subduction plate boundary inferred from slow seismic slip. *Nature Geoscience*, **5**, 414–418.

Sugiyama, Y. 1994. Neotectonics of southwest Japan due to the right-oblique subduction of the Philippine Sea plate. *Geofisica Internacional*, **33**, 53–76.

Suppe, J. 1981. Mechanics of mountain building and metamorphism in Taiwan. *Memoir of the Geological Society of China*, **4**, 67–89.

Suzuki, H. 2012. Structural features of the Shimanto Accretionary Prism in the Kii Peninsula. *The Association for the Geological Collaboration in Japan, Monograph*, **59**, 101–110 [in Japanese with English abstract].

Suzuki, T. & Hada, S. 1979. Cretaceous tectonic melange of the Shimanto belt in Shikoku, Japan. *Journal of the Geological Society of Japan*, **85**, 467–479.

Taira, A. 2001. Tectonic evolution of the Japanese island arc system. *Annual Review of Earth and Planetary Sciences*, **29**, 109–134.

Taira, A., Okamura, M. *et al.* 1980. Lithofacies and geologic age relationship within mélange zone of northern Shimanto Belt (Cretaceous), Kochi Prefecture, Japan. *In*: Taira, A. & Tashiro, M. (eds) *Geology and Paleontology of the Shimanto Belt*. Rinya-Kosaikai Press, Kochi, 179–214.

Taira, A., Katto, J., Tashiro, M., Okamura, M. & Kodama, K. 1988. The Shimanto Belt in Shikoku, Japan: Evolution of Cretaceous to Miocene accretionary prism. *Modem Geology*, **12**, 5–46.

Takahashi, M. 1983. Space-time distribution of late Mesozoic to early Cenozoic magmatism in East Asia and its tectonic implications. *In*: Hashimoto, M. & Uyeda, S. (eds) *Tectonics in the Circum-Pacific Regions*. TERAPUB, Tokyo, 69–88.

Takahashi, M., Azuma, S., Ito, H., Kanagawa, K. & Inoue, A. 2014. Frictional properties of the shallow Nankai Trough accretionary sediments dependent on the content of clay minerals. *Earth, Planets and Space*, **66**, 75.

Tamaki, K. 1985. Two modes of back-arc spreading. *Geology*, **13**, 475–478.

Tobin, H. J. & Kinoshita, M. 2006. NanTroSEIZE: the IODP Nankai Trough seismogenic zone experiment. *Scientific Drilling*, **2**, 23–27.

Tsuji, T., Kodaira, S., Ashi, J. & Park, J.-O. 2013. Widely distributed thrust and strike-slip faults within subducting oceanic crust in the Nankai Trough off the Kii Peninsula, Japan. *Tectonophysics*, **600**, 52–62, http://doi.org/10.1016/j.tecto.2013.03.014

Ujiie, K. & Kimura, G. 2014. Earthquake faulting in subduction zones: insights from fault rocks in accretionary prisms. *Progress in Earth and Planetary Science*, **1**, 7, http://doi.org/10.1186/2197-4284-1-7

Ujiie, K. & Tsutsumi, A. 2010. High-velocity frictional properties of clay-rich fault gouge in a megasplay fault zone, Nankai subduction zone. *Geophysical Research Letters*, **37**, L24310, http://doi.org/10.1029/2010GL046002

Ujiie, K., Yamaguchi, A., Kimura, G. & Toh, S. 2007*a*. Fluidization of granular material in a subduction thrust at seismogenic depths. *Earth and Planetary Science Letters*, **259**, 307–318.

Ujiie, K., Yamaguchi, H., Sakaguchi, A. & Toh, S. 2007*b*. Pseudotachylytes in an ancient accretionary complex and implications for melt lubrication during subduction zone earthquakes. *Journal of Structural Geology*, **29**, 599–613.

Underwood, M. 2007. Sediment inputs to subduction zones: why lithostratigraphy and clay mineralogy matter. *In*: Dixon, T. & Moore, J. C. (eds) *The Seismogenic Zone of Subduction Thrust Faults*. Colombia University Press, New York, 42–85.

Uyeda, S. & Miyashiro, A. 1974. Plate tectonics and the Japanese Islands: a synthesis. *Geological Society of America Bulletin*, **85**, 1159–1170.

Vannucchi, P., Scholl, D. W., Meschede, M. & McDougall-Reid, K. 2001. Tectonic erosion and consequent collapse of the Pacific margin of Costa

Rica: combined implications from ODP Leg 170, seismic offshore data, and regional geology of the Nicoya Peninsula. *Tectonics*, **20**(5), 649–668.

von Huene, R. & Lallemand, S. 1990. Tectonic erosion along the Japan and Peru convergent margins. *Geological Society of America Bulletin*, **102**, 704–720.

von Huene, R. & Ranero, C. R. 2003. Subduction erosion and basal friction along the sediment-starved convergent margin off Antofagasta, Chile. *Journal of Geophysical Research*, **108**, 2079.

von Huene, R. & Scholl, D. W. 1991. Observations at convergent margins concerning sediment subduction, subduction erosion, and the growth of continental crust. *Reviews of Geophysics*, **29**, 279–316.

Wada, I., Wang, K., He, J. & Hyndman, R. D. 2008. Weakening of the subduction interface and its effects on surface heat flow, slab dehydration, and mantle wedge serpentinization. *Journal of Geophysical Research*, **113**, B04402, http://doi.org/10.1029/2007JB005190

Wakabayashi, J. 2011. Mélanges of the Franciscan Complex, California: diverse structural settings, evidence for sedimentary mixing, and their connection to subduction processes. *In*: Wakabayashi, J. & Dilek, Y. (eds) *Mélanges: Processes and Societal Significance*. Geological Society of America Special Papers, **480**, 117–141.

Wang, K. & Hu, Y. 2006. Accretionary prisms in subduction earthquake cycles: the theory of dynamic Coulomb wedge. *Journal of Geophysical Research*, **111**, B06410, http://doi.org/10.1029/2005JB004094

Yamaguchi, A., Cox, S. F., Kimura, G. & Okamoto, S. 2011*a*. Dynamic changes in fluid redox state associated with episodic fault rupture along a megasplay fault in a subduction zone. *Earth and Planetary Science Letters*, **302**, 369–377, http://doi.org/10.1016/j.epsl.2010.12.029

Yamaguchi, A., Sakaguchi, A. *et al.* 2011*b*. Progressive illitization in fault gouge caused by seismic slip propagation along a megasplay fault in the Nankai Trough. *Geology*, **39**, 995–998.

Yamaguchi, A., Ujiie, K., Nakai, S. & Kimura, G. 2012. Sources and physicochemical characteristics of fluids along a subduction-zone megathrust: a geochemical approach using syn-tectonic mineral veins in the Mugi mélange, Shimanto accretionary complex. *Geochemistry, Geophysics, Geosystems*, **13**, Q0AD24, http://doi.org/10.1029/2012GC004137

Yamashita, N. 1990. Nauman's studies on volcanoes and volcanic rocks – Naumann's contribution to the geology of Japan, (1). *Journal of the Geological Society of Japan*, **96**, 479–491 [in Japanese with English abstract].

Yamato-Omine Research Group 1994. Paleozoic-Mesozoic System in the central mountain of the Kii Peninsula (part 4). *Earth Science (Chikyu-Kagaku)*, **46**, 185–198 [in Japanese with English abstract].

Yanai, S. 1986. Megakink bands and Miocene regional stress field in outer southwestern Japan. *Scientific Papers of the College of Arts and Sciences, University of Tokyo*, **36**, 55–79.

2e The Kyushu–Ryukyu Arc

KAZUHIRO MIYAZAKI, MASAKI OZAKI, MAKOTO SAITO & SEIICHI TOSHIMITSU

The Kyushu–Ryukyu Arc extends for over 1000 km south from the large island of Kyushu down the Ryukyu Island chain towards Taiwan. It has been produced by subduction of oceanic plates beneath the Eurasian continent and exposes a wide range of Palaeozoic, Mesozoic and Cenozoic rocks formed in an active continental margin setting (Maruyama *et al.* 1997; Isozaki *et al.* 2010). In this chapter we describe the geology of the arc from the oldest (Palaeozoic and Triassic) to youngest (Neogene and Quaternary) rocks, and discuss the geodynamic evolution of the area.

Pre-Jurassic and Jurassic (KM, MS, ST)

Pre-Jurassic and Jurassic geological units occur as isolated outcrops on the northern side of the Usuki–Yatsushiro Tectonic Line (UYTL), in the zone between the UYTL and Butsuzo Tectonic Line (BTL) and in the Ryukyu islands (Figs 2e.1 & 2e.2). These units consist of Palaeozoic–early Mesozoic, mostly accretionary complex materials that include high-pressure–low-temperature (HP-LT) metamorphic rocks, mélanges and low-grade metasedimentary rocks. In this section we first describe the outcrops of these rocks in northern Kyushu, where they include the Renge and Suo metamorphic complexes as well as several Permian and Jurassic accretionary units (Fig. 2e.1). This is followed by a focus on pre-Jurassic and Jurassic rocks outcropping in the complex zone between the UYTL and BTL in central Kyushu, especially those in the well-studied Tomochi district. Further south, outcrops of pre-Cretaceous rocks are limited to minor exposures in several of the Ryukyu islands (Fig. 2e.1).

North of the Usuki–Yatsushiro Tectonic Line

Renge metamorphic complex

The Renge metamorphic complex ('metamorphic rocks in the Renge Belt' of Nishimura & Shibata 1989; Nishimura 1998) comprises a range of Carboniferous–Permian high-pressure metamorphic rocks. The southeastern margin of the complex is in fault contact with Permian accretionary complex and the Suo metamorphic complex (Fig. 2e.1). Because the Renge metamorphic complex has been extensively intruded by Cretaceous granitic rocks, their original HP metamorphic signature has been largely modified, with the exception of some areas such as the Sasaguri district (outcrop on NW side of the Sasaguri metagabbro in Fig. 2e.1). The following descriptions are based on the review of Yamamoto (1992).

The Renge metamorphic complex in northern Kyushu consists of ultramafic rocks, metagabbro, mafic, calcareous, siliceous, pelitic and psammitic schists. These rocks are gently folded by NE–SW-trending large-scle antiforms and synforms in the Sasaguri district. The apparently lower structural levels of the complex occur in the NW part of the district from where metamorphic grade increases south-eastwards towards apparently upper structural levels. Metamorphic mineral assemblages belong to the transitional zone between epidote-blueschist subfacies and greenschist facies, with pelitic schists being similar to those in the chlorite and garnet zones (Higashino 1990) of the Cretaceous Sanbagawa metamorphic complex in central Shikoku. Phengite K–Ar ages from pelitic schists are 259 ± 6 and 272 ± 8 Ma which, along with a phengite-whole-rock Rb–Sr isochron age of 298 ± 10 Ma and a whole-rock Rb–Sr isochron age of 308.4 ± 19.0 Ma (Shibata & Nishimura 1989), indicate Carboniferous–Permian metamorphic ages for the Renge metamorphic complex. Detrital zircon U–Pb ages from psammitic schists exposed in the Sasaguri and Nokonoshima areas (Tsutsumi *et al.* 2011*b*) show age clusters around 540–430 Ma and 470–390 Ma, respectively, these being similar to those from the complex in the Sefuri Mountains (Adachi *et al.* 2012). Hornblende K–Ar data (Nishimura & Shibata 1989) from amphibolite associated with schists in the Sasaguri district (the Sasaguri metagabbro in Fig. 2e.1) have yielded a date of 430 ± 15 Ma, suggesting a Silurian age for ocean-floor metamorphism before the high-pressure subduction event (Nishimura & Shibata 1989).

Further to the south, the Kiyama metamorphic unit in central Kyushu crops out across a small area SW of the Quaternary Aso volcanic caldera (Fig. 2e.1) and has been attributed to the Renge metamorphic complex (Nishimura 1998). The southern margin of the Kiyama exposures is in fault contact with serpentinite, and the southeastern margin is in fault contact with the Upper Cretaceous Mifune Group. The Kiyama metamorphic unit consists of mafic, pelitic and psammitic schists, with a fault separating epidote-blueschist subfacies rocks in the east from albite-epidote amphibolite subfacies rocks in the west (Kabashima *et al.* 1995). Phengite K–Ar isotopic data (Kabashima *et al.* 1995) yield ages of 306–290 Ma, and detrital zircon U–Pb data (Tsutsumi *et al.* 2003) show age clusters around 590–380 Ma with a peak around 420 Ma.

Permian accretionary complex

Permian accretionary complexes correlated with the Akiyoshi Belt in nearby western Honshu crops out in NE Kyushu (Fig. 2e.1). The northwestern margin of the complex is in contact with the Renge metamorphic complex across a NE–SW- to NNE–SSW-trending fault, and the southeastern margin is in fault contact with the underlying Suo metamorphic complex (Matsushita *et al.* 1969). The following descriptions are largely based on the recent review of Sano (2010*a*). These rocks are characterized by the occurrence of a large body of limestone, along with chert, sandstone and mudstone. Mapping reveals both stratigraphically coherent sequences (such as the Kiku Unit) and mélanges (such as the Yobuno Unit). The Kiku Unit consists of basaltic volcaniclastic rocks overlain progressively by bedded chert, siliceous mudstone, felsic tuff, alternations of sandstone and mudstone, and sandstone. Middle Permian radiolarian fossils have been reported from chert and felsic tuff, and

From: Moreno, T., Wallis, S., Kojima, T. & Gibbons, W. (eds) 2016. *The Geology of Japan*. Geological Society, London, 139–174.

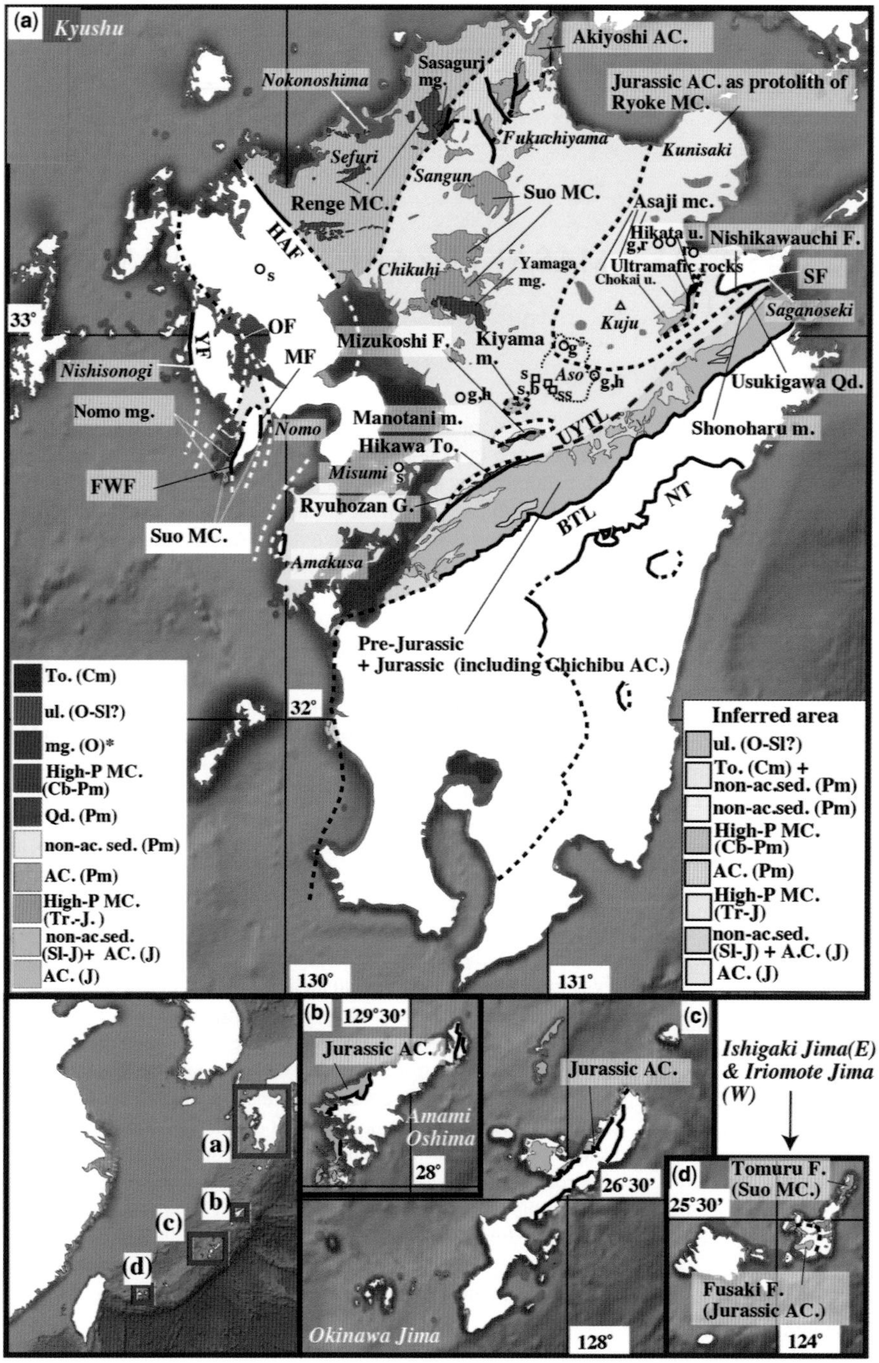

Fig. 2e.1. Distribution of Pre-Jurassic and Jurassic rocks in the Kyushu–Ryukyu Arc, based on the seamless digital geological map of Japan (1:200 000) (Geological Survey of Japan; https://gbank.gsj.jp/seamless/index_en.html). AC, Accretionary Complex; MC, Metamorphic Complex; mg, metagabbro; m, metamorphic unit; non-ac.sed, non-accretionary sedimentary rocks; ul, ultramafic rocks; G, Group; F, Formation; u, unit; Qd, Quartzdiorite; UYTL, Usuki–Yatsushiro Tectonic Line; BTL, Butsuzo Tectonic Line; NT, Nobeoka Thrust; SF, Sashu Fault; HAF, Hatashima–Ariake Fault; OF, Omurawan–Amakusa Fault; YF, Yobikonoseto Fault; FWF, Fukabori–Wakimisaki Fault; MF, Mogi Fault; Cm, Cambrian; O, Ordovician; Sl, Silurian; D, Devonian; Cb, Carboniferous; Pm, Permian; Tr, Triassic; J, Jurassic. Grey areas represent distribution of the Ryoke MC as basement rocks. Small open circles represent xenoliths and small open squares show borehole sites, with symbols: s, schist; h, hornfels; g, granite; b, metagabbro; ss, sandstone; r, metamorphic rocks of the Ryoke MC.

Middle–Late Permian radiolarian fossils from mudstone. The mélanges of the Yobuno Unit consist of a matrix of highly sheared mudstone and alternations of mudstone and sandstone surrounding blocks of basalt, limestone, chert and siliceous mudstone. Large (kilometre-scale) bodies of limestone are distributed across the Hiraodai, Kawaradake and Funaoyama areas. The mélanges are inferred to have been formed in latest Middle Permian–latest Late Permian times.

The Hikata unit of the Asaji metamorphic complex in east-central Kyushu forms another outcrop of Permian accretionary rocks (Fig. 2e.1). It comprises low-grade metamorphic rocks derived from mudstone, sandstone and siliceous rock (Hayasaka *et al.* 1989; Teraoka *et al.* 1992), with Permian radiolarian fossils reported from mudstone (Toyohara & Murata 2003). The unit is thought to correlate with the Nishiki Formation, a Permian accretionary complex in the Akiyoshi Belt of the Chugoku region of western Honshu (Fujii *et al.* 2008). As will be described in the section on 'Cretaceous sedimentary basin', the Hikata unit is in fault contact with or partly unconformably covered by the Onogawa Group (Fujii *et al.* 2008).

Suo metamorphic complex

The Suo metamorphic complex (metamorphic rocks in the Suo Belt of Nishimura 1998) comprises a Triassic–Jurassic high-pressure unit intruded by Cretaceous granitic rocks and subsequently widely overprinted by high-temperature metamorphism. Despite this thermal overprint, high-pressure metamorphic minerals are preserved south

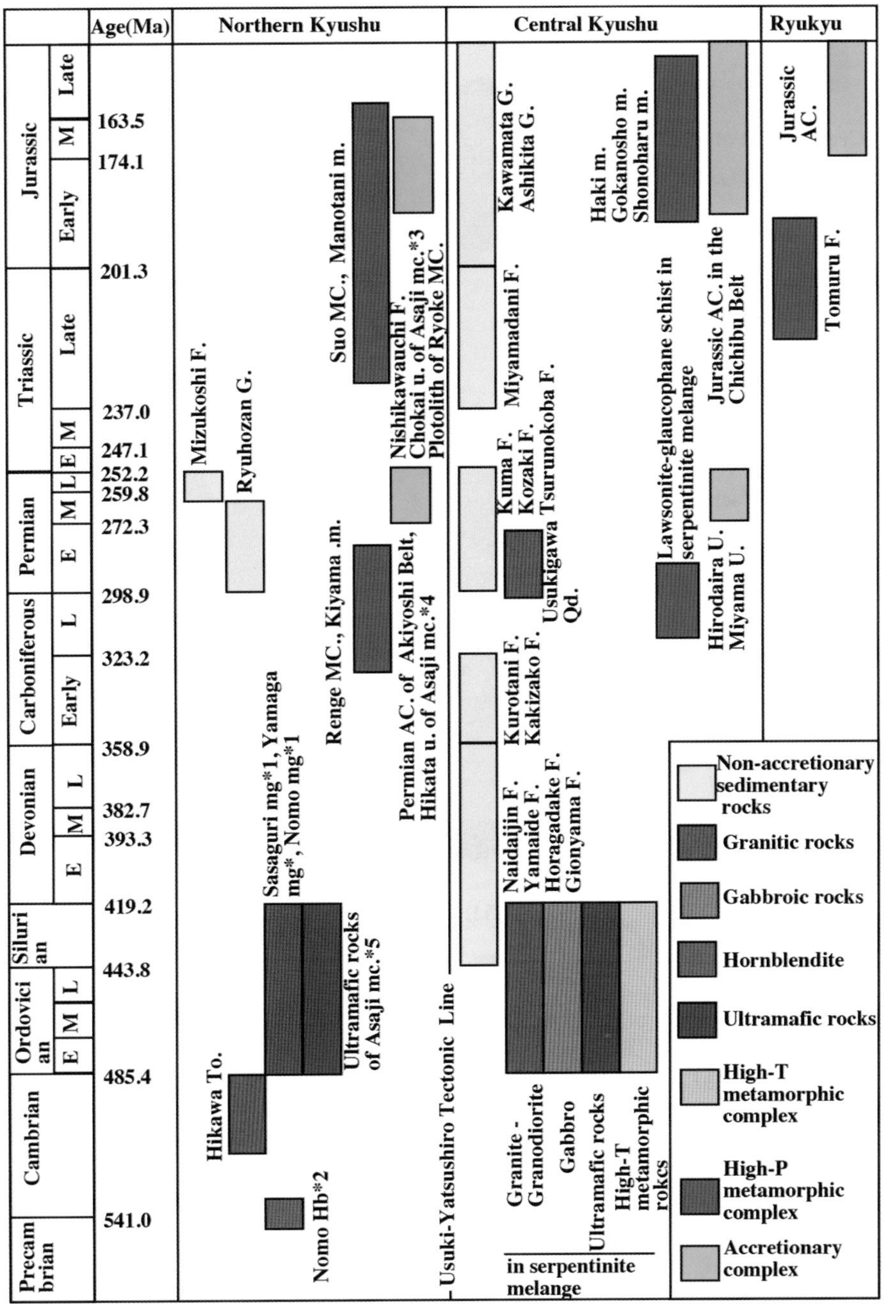

Fig. 2e.2. Stratigraphic succession for pre-Jurassic and Jurassic strata in the Kyushu–Ryukyu Arc. Abbreviations as for Figure 2e.1. *1, Formational age before high-pressure metamorphism; *2, formational age of hornblendite included in metagabbro; *3, protolith age; there is no fossil evidence and there is a possibility that the Chokai unit is a Permian accretionary complex; *4, protolith age; and *5, there are no age data and there is a possibility that formational age of ultramafic rocks of the Asaji metamorphic complex is younger or older than Ordovician–Silurian.

of Fukuoka in the Kurume–Yame–Yamaga area (Fig. 2e.3). The Suo metamorphic complex in this area consists of pelitic, psammitic, mafic and siliceous schists and metagabbro with small amounts of calcareous schist and serpentinite. These rocks are gently folded by east–west-trending antiforms and synforms, and are overlain tectonically by the large Yamaga Metagabbro (Figs 2e.1 & 2e.3). The metamorphic grade of the schists and metagabbro is pumpellyite-actinolite subfacies to epidote-blueschist subfacies and the metamorphic age is estimated to be Triassic–Jurassic, with phengite K–Ar isotopic data from pelitic schists yielding dates of 220–180 Ma (Shibata & Nishimura 1989; Nishimura 1998). Detrital zircon U–Pb ages (Tsutsumi *et al.* 2003) from psammitic schists cluster around 260 Ma. Hornblende K–Ar data from metagabbro and amphibolite and zircon U–Pb data from plagiogranite of the Yamaga Metagabbro yield ages of 306 $\pm$ 19 and 477 $\pm$ 11 Ma (Nishimura & Shibata 1989) and 489 Ma (Miyamoto & Enokibara 2009), respectively. The last two ages presumably relate to Ordovician ocean-floor metamorphism and igneous activity of the Yamaga Metagabbro prior to subduction (Nishimura & Shibata 1989; Miyamoto & Enokibara 2009).

Exposures of the Suo metamorphic complex in the Omuta–Tamana area experienced Cretaceous high-temperature polymetamorphism (Fig. 2e.3) (Tomita *et al.* 2008). This thermal overprint has produced the Konoha high-temperature metamorphic unit which, with detrital zircon U–Pb ages from psammites clustering around 310–175 Ma with a peak at 250 Ma, are likely derived from a Suo metamorphic complex protolith (Tsutsumi *et al.* 2003).

Another likely correlative with the Suo metamorphic complex is provided by the Manotani metamorphic unit (Manotani metamorphic rocks) which crops out on the northern side of the UYTL

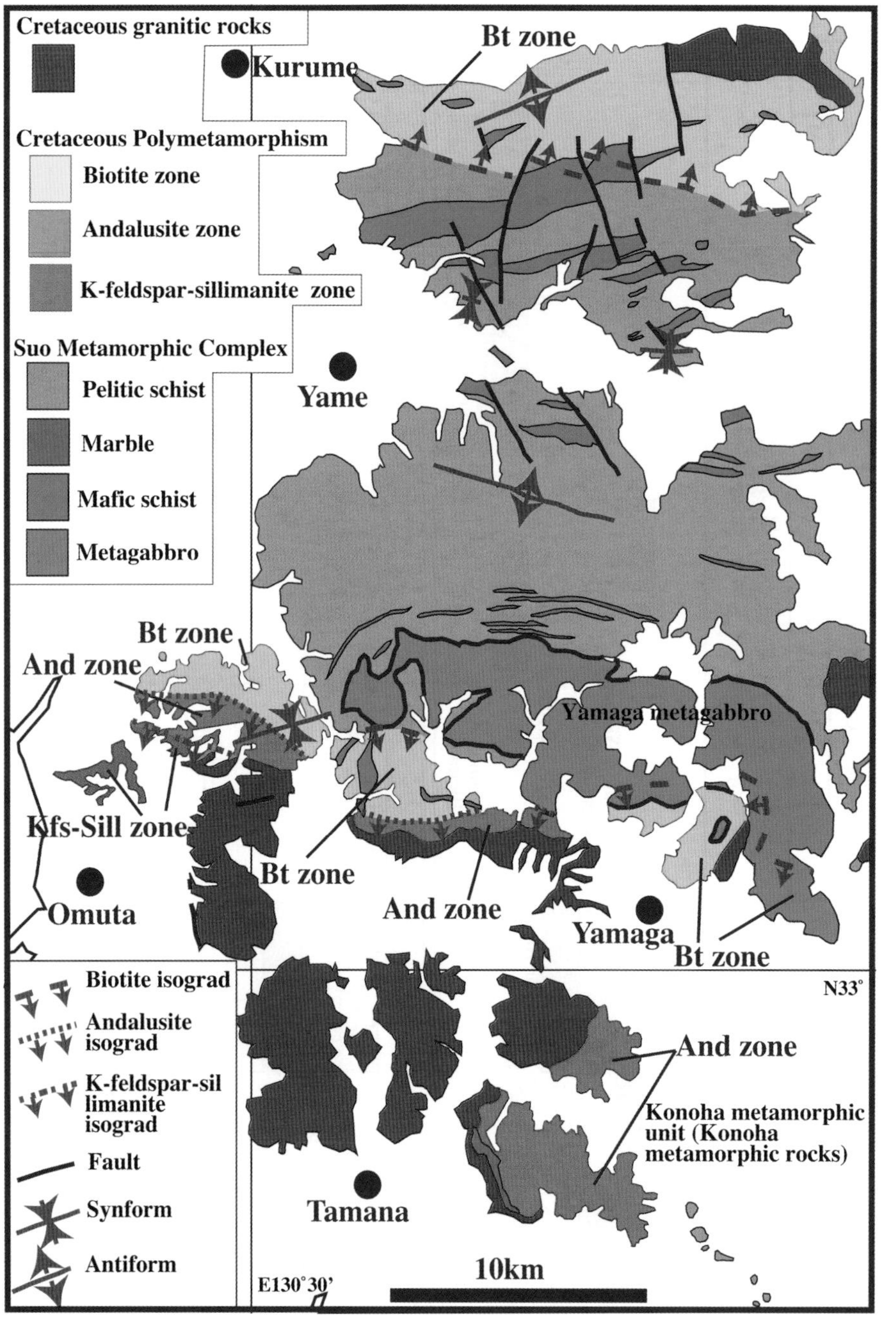

Fig. 2e.3. The Suo metamorphic complex and its zonation due to Cretaceous polymetamorphism in the Kurume-Yame-Yamaga-Omuta-Tamana district. Bt, biotite; And, andalusite; Kfs, K-feldspar; Sill, sillimanite.

(Fig. 2e.1; see also Figs 2e.6 & 2e.16). These rocks consist of mafic, pelitic and siliceous schists metamorphosed under pumpellyite-actinolite to epidote-blueschist subfacies conditions, although overprinted by Cretaceous high-temperature metamorphism along the southern margin of their outcrop (Fig. 2e.1; see also Figs 2e.6 & 2e.16). Phengite K–Ar ages of pelitic schists from the northern part of the metamorphic rocks lie in the range 214–176 Ma (Nagakawa *et al.* 1997), with detrital zircon U–Pb data on psammitic schists yielding an age of 294.2 ± 5.0 Ma (Maki *et al.* 2011), similar to those from the Suo metamorphic complex.

The Higo metamorphic complex was produced by the Cretaceous high-temperature metamorphism and tectonically underlies the Manotani Metamorphic Rocks (Fig. 2e.1; see also Figs 2e.6 & 2e.16). The protolith of the complex is similar to the Suo metamorphic complex (except that there is a relatively large amount of limestone). Detrital zircon U–Pb ages (Sakashima *et al.* 2003) from pelitic gneiss show age clusters around 2155, 1353, 421 and 350–184 Ma, which are similar to those of the Suo metamorphic complex, suggesting the possibility of a Suo protolith for the Higo metamorphic complex.

Basement rocks cropping out below the Cenozoic strata in the area between the Aso Caldera–Kuju Mountains and the UYTL are restricted, except for exposure of the Higo metamorphic complex. However, metamorphic rocks have been reported from borehole samples (New Energy and Industrial Technology Development Organization 1995), xenoliths in the Cenozoic volcanic rocks (Karakida *et al.* 1969) and clasts in conglomeratic Cretaceous non-accretionary sedimentary rocks (Teraoka *et al.* 1998). Pelitic and

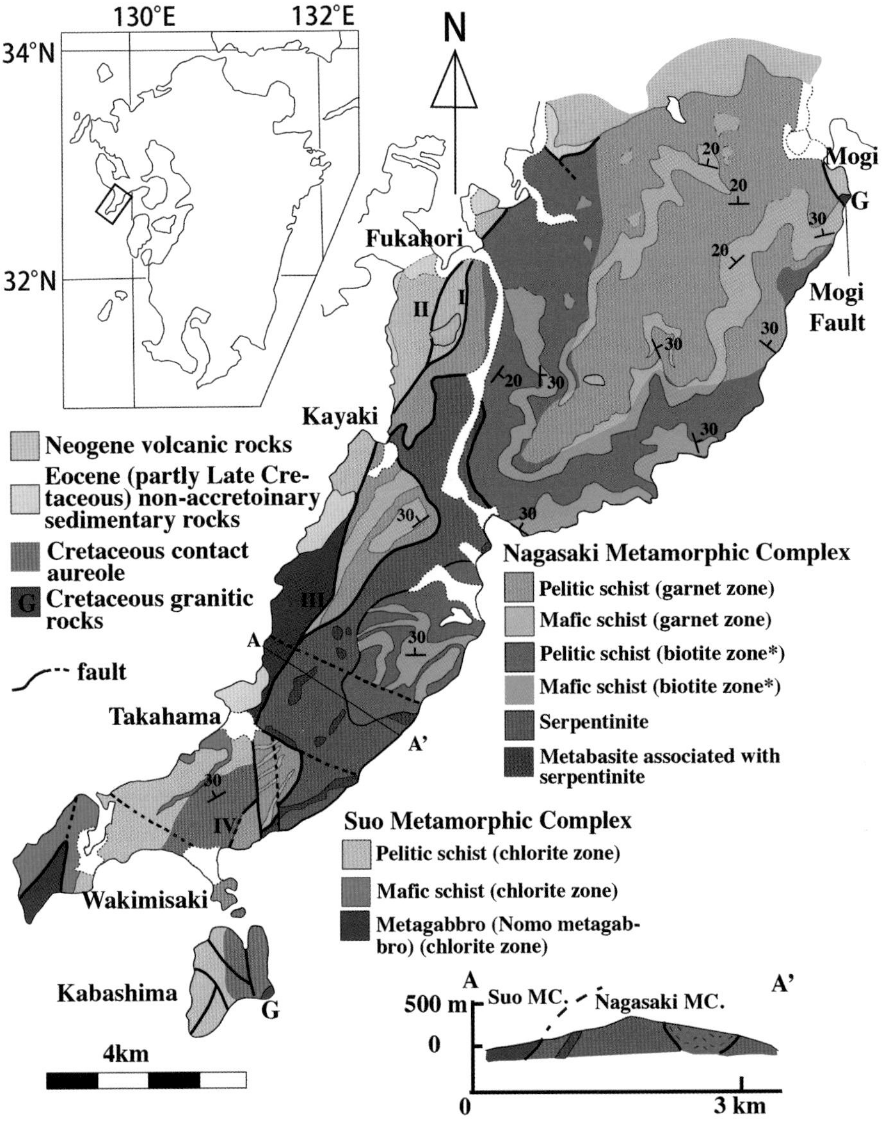

Fig. 2e.4. Geological map of the Suo and Nagasaki metamorphic complexes in the Nomo Peninsula (modified from Miyazaki & Nishiyama 1989). Biotite zone* includes the lower garnet zone of Miyazaki & Nishiyama (1989), of which the peak temperature is the same as that of the biotite zone. Fault I is the Joyama Thrust. Fault I-III-IV is the Fukabori–Wakimisaki Fault.

mafic schists occur as xenoliths in Cenozoic volcanic rocks in the Misumi Peninsula. Similarly, xenoliths of granite, ultramafic rocks and high-temperature metamorphic rocks have been found in lava of the Nekodake Volcano in the Aso Caldera, and pelitic schists and metagabbro were reported from borehole data on the western side of the caldera. These associations of metamorphic and igneous rocks are similar to those of the Suo metamorphic complex and their Cretaceous high-temperature polymetamorphic rocks on the northern side of the UYTL. In addition, clasts of granitic rocks, schists, hornfels and high-temperature metamorphic rocks are found in conglomerates of the Cretaceous Goshonoura, Mifune and Onogawa groups. Phengite K–Ar data from a large angular boulder in the Onogawa Group have yielded ages of 199–182 Ma (Isozaki & Itaya 1989) and imply a source from nearby exposures of the Suo metamorphic complex. These lines of evidence suggest that the Suo metamorphic complex and high-temperature metamorphic rocks derived from it are widely present in subcrop on the northern side of the UYTL (Fig. 2e.1).

The Suo metamorphic complex also crops out on the Nomo Peninsula on the western side of the Omurawan–Amakusa Fault (OF on Fig. 2e.1). More specifically, the complex is distributed on the western side of the Fukahori–Wakimisaki Fault and the eastern side of the Mogi Fault, on either side of the underlying Cretaceous high-pressure Nagasaki metamorphic complex (Fig. 2e.4). The Suo metamorphic complex exposed on the peninsula comprises pelitic, psammitic, mafic, siliceous and calcareous schists and metagabbro (Nomo Metagabbro), and their metamorphic grade is pumpellyite-actinolite subfacies with sub-calcic amphiboles in mafic schist (Miyazaki & Nishiyama 1989). Phengite K–Ar ages of pelitic schist range from 248 to 150 Ma (mostly 200–150 Ma; Nishimura 1998; Nishimura *et al.* 2004). Detrital zircon U–Pb ages of psammitic schist (Tsutsumi *et al.* 2003) cluster in the range 330–240 Ma. These radiometric ages are comparable with those from elsewhere in the Suo metamorphic complex. Hornblende K–Ar ages of metagabbro yield 480–457 Ma (Hattori & Shibata 1982). Hornblende K–Ar data from hornblendite enclaves in the metagabbro show an age of 592–589 Ma (Hattori & Shibata 1982). Both metagabbro and hornblendite have been overprinted by the high-pressure metamorphism seen in the surrounding schists. The hornblende K–Ar ages of the metagabbro and hornblendite are therefore interpreted as their crystallization date prior to the high-pressure metamorphism. A similar metagabbro also crops out in the marine area to the west of the Amakusa Peninisula (Fig. 2e.1) and is found as clasts in conglomerate of the Upper Cretaceous Himenoura Group (Hattori & Isomi 1976).

Other pre-Jurassic units north of the UYTL in north-central Kyushu
In addition to the main pre-Jurassic inliers previously mentioned, more minor exposures of older basement rocks include the Mizukoshi Formation, Ryuhozan Group, Hikawa Tonalite and Asaji ultramafic rocks, all of which crop out on the northern side of the UYTL of east-central Kyushu.

Mizukoshi Formation. This stratigraphic unit comprises Permian shelf sediments that are unconformably covered by the Upper Cretaceous Mifune Group to the north, and in high-angle fault contact to the south with the Manotani metamorphic rocks. The following descriptions are mainly based on the review of Miyamoto (2010). The Mizukoshi Formation consists of mudstone, conglomerate and sandstone with limestone lenses. A lower member consists mainly of sandstone and alternations of sandstone and mudstone, whereas an upper member lacks the argillaceous component and consists mainly of conglomerate and sandstone. Bedding strikes ENE–WSW and dips north, and sedimentary structures show upwards-younging to the north. Late Permian *fusulina* fossils have been reported from limestone (Yanagida 1958), and Middle–Late Permian brachiopod fossils have also been described. The age of sedimentation is estimated to be Late Permian on the basis of radiolarian fossils from mudstone. Hornblende K–Ar ages of granitic clasts from conglomerate yield 260 $\pm$ 13 Ma (Tobe *et al.* 2000) which, along with chemical compositions of detrital garnet, indicates that the Mizukoshi Formation is correlative with Permian non-accretionary sedimentary rocks cropping out on the southern side of the UYTL.

Ryuhozan Group. This group is in fault contact with the Hikawa Tonalite on the northern side of the UYTL (Fig. 2e.1; see also Figs 2e.6 & 2e.16). Its northern margin is intruded by the Cretaceous Higo plutonic rocks and its southern margin is bounded by the UYTL (Fig. 2e.1; see also Figs 2e.6 & 2e.16). The group is composed of Permian shelf limestone, sandstone, mudstone, and felsic and mafic volcaniclastic rocks. Early–Middle Permian *fusulina* fossils have been reported from the limestone (Murata *et al.* 1981). The group records an Early Cretaceous 'Higo' metamorphic event that increases in grade northwards from greenschist to amphibolite facies. Chloritoid-bearing pelitic schist derived from laterite has been reported (Sakashima *et al.* 1995), and the metamorphic age is estimated to be *c.* 100 Ma based on muscovite and biotite K–Ar isotopic data (Sakashima *et al.* 1999).

Hikawa Tonalite. The Hikawa Tonalite distributed with the Ryuhozan Group along the UYTL (Figs 2e.1; see also Figs 2e.6 & 2e.16) consists of hornblende tonalite to granodiorite. A zircon U–Pb age of 502.5 $\pm$ 9.6 Ma from the tonalite (Sakashima *et al.* 2003) represents the crystallization age of the pluton. The tonalite subsequently underwent intense mylonitization in Cretaceous times, as recorded by K–Ar ages of 112.7 $\pm$ 5.6 and 79.0 $\pm$ 4.0 Ma obtained from fine-grained hornblende formed during dynamic crystallization (Sakashima *et al.* 1999).

Asaji ultramafic rocks. Ultramafic rocks crop out between the Hikata and Chokai units of the Asaji metamorphic complex in eastern Kyushu (Fig. 2e.1). This ultramafic unit, which is bounded above and below by low-angle faults, consists of serpentinite and clinopyroxenite with small amounts of metasomatic quartz-magnesite rock, metagabbro and amphibolite. The lithology and mineral composition of these ultramafic rocks are similar to those cropping out on the southern side of the UYTL (Soda & Takagi 2004).

Jurassic accretionary complex
Remnants of Jurassic accretionary complexes are exposed in several places across Kyushu north of the UYTL. The geology of these areas is described in the three following subsections.

Nishikawauchi Formation. This unit crops out in eastern Kyushu as a small lenticular body along the Sashu Fault which separates the Upper Cretaceous Onogawa Group from the Sanbagawa metamorphic complex (Figs 2e.1 & 2e.5). The formation consists of conglomerate, sandstone and mudstone, from which Jurassic radiolarians have been reported (Saito *et al.* 1993).

Chokai Unit. The Chokai Unit is found within the structurally lower part of the Asaji metamorphic complex outcrop in NE Kyushu (Fig. 2e.1). The unit consists of metamorphic lithologies derived from chert, mudstone, sandstone and mafic rocks. The precursor of the Chokai Unit is inferred to be Jurassic accretionary complex material on the basis of the lithological similarities between the two (Teraoka *et al.* 1992), although as yet there is no fossil evidence to confirm this. There is a possibility that the unit belongs to the Permian accretionary complex of the Ultra-Tamba Belt in SW Japan, also on the basis of the lithological similarities (Fujii *et al.* 2008).

Protolith of Ryoke metamorphic complex. Rocks attributed to the Ryoke metamorphic complex form the basement to Cenozoic volcanic rocks from the Kunisaki Peninsula in NE Kyushu, through Mt Kuju to the Aso Caldera (Fig. 2e.1; Sasada 1987). In SW Japan, the metamorphic grade can be shown to increase continuously from a non-metamorphosed Jurassic accretionary complex into the Ryoke metamorphic complex without any major tectonic discontinuity (e.g. Ikeda 1998).

Zone between the Usuki–Yatsushiro and Butsuzo tectonic lines

Pre-Jurassic units and Jurassic non-accretionary sedimentary rocks
Pre-Jurassic and Jurassic rocks crop out in the structurally complex zone between the UYTL and BTL (Fig. 2e.1). The publication of

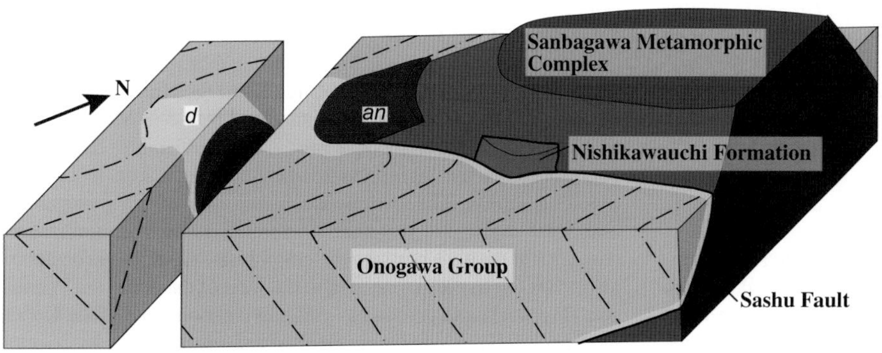

Fig. 2e.5. Relation among the Nishikawauchi Formation, Sanbagawa metamorphic complex and Onogawa Group in the Saganoseki area of east Kyushu (Teraoka 1970). an, andesite (Miocene); d, disrupted zone of the Onogawa Group due to activation of the Sashu Fault.

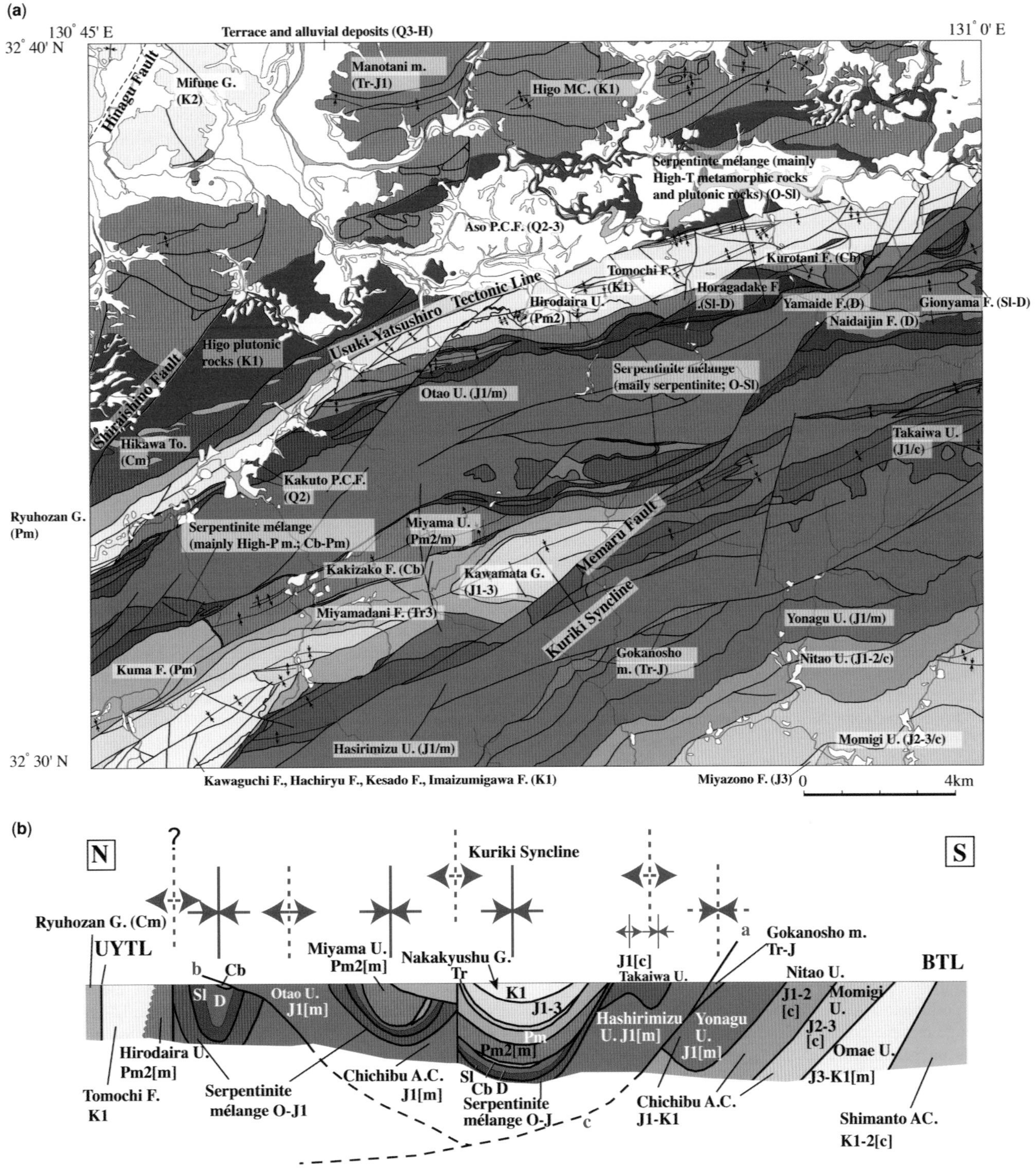

Fig. 2e.6. Geological map and cross-section of the Tomochi district of west-central Kyushu (modified from Saito *et al.* 2005). (**a**) Geological map. (**b**) Schematic cross-section from the southern side of the Usuki–Yatsushiro Tectonic Line to the northern area of the Shiiba district that is contiguous to the southeast. Pm2, Middle Permian; J1, Early Jurassic; J2, Middle Jurassic; J3, Late Jurassic; K, Cretaceous; K1, Early Cretaceous; K2, Late Cretaceous; Q2, Middle Pleistocene; Q3, Late Pleistocene; H, Holocene; /m or [m], mélange unit; /c or [c], coherent unit; Aso P.C.F., Aso pyroclastic flow deposit; Kakuto P.C.F., Kakuto pyroclastic flow deposit. Other abbreviations as for Figure 2e.1.

the 1:50 000 geological map sheet of the Tomochi district (Saito *et al.* 2005) has revealed many details regarding the geological structure of the zone and forms the basis for the following section. In the Tomochi district, pre-Jurassic geological units and Jurassic non-accretionary sedimentary rocks tectonically overlie rocks belonging to the Jurassic accretionary complex (so-called 'Jurassic accretionary complex in the Chichibu Belt'; Fig. 2e.6). This primary structure has been modified by east–west-trending to

ENE–WSW-trending antiforms, synforms and high-angle faults formed by north–south compression and by NE–SW-trending high-angle left-lateral faults (Fig. 2e.6). The rocks in this zone comprise a sequence of serpentinite mélange overlain progressively by Silurian–Devonian shelf sediments, Carboniferous shelf sediments, Permian accretionary complex, Permian shelf sediments, Triassic shelf sediments and Jurassic shelf sediments. Cretaceous sediments tectonically or unconformably cover these older geological units (Fig. 2e.6).

The serpentinite mélange (Fig. 2e.6) consists of sheared serpentinite matrix and tectonic blocks of plutonic and metamorphic rocks analogous to the so-called 'plutonic and metamorphic rocks in the Kurosegawa Belt' seen in neighbouring Shikoku. However, the presence of serpentinite mélange in its strict sense is rare and confined to faults, with tectonic blocks being dominant over the sheared serpentinite matrix. Detailed geological mapping of the Ino district in central Shikoku (Wakita *et al.* 2007) has shown that tectonic blocks in 'the Kurosegawa Belt' represent stacked slices of Early Jurassic metamorphic rocks overlain progressively by Carboniferous high-pressure metamorphic rocks, ultramafic rocks and Ordovician–Silurian plutonic and high-temperature metamorphic rocks from bottom to top. It is likely that the serpentinite mélange in the Tomochi district in Kyushu may have had a similar geological structure prior to later tectonic modification.

Plutonic rocks in the serpentinite mélange consist of granite-granodiorite, gabbro and clinopyroxenite, with high-temperature metamorphic rocks comprising amphibolite to granulite facies metabasite and amphibolite facies metapelite. Hornblende K–Ar data from granodiorite in the Kuraoka area located to the east of the Tomochi district has yielded an age of 450 $\pm$ 12 Ma (Umeda *et al.* 1986). These plutonic rocks and high-temperature metamorphic rocks are correlated with the Ordovician–Silurian Mitaki (igneous) and Terano (metamorphic) units described in Shikoku (Hada *et al.* 2000).

High-pressure metamorphic rocks in the serpentinite mélanges have been divided into lawsonite-blueschist subfacies and pumpellyite-actinolite subfacies to epidote-blueschist subfacies lithologies. Lawsonite-glaucophane schist (Karakida *et al.* 1977), glaucophane-jadeite-metagabbro (Saito & Miyazaki 2006) and omphacite schist (Miyazoe *et al.* 2009) have been recognized in the lawsonite-blueschist subfacies metamorphic rocks. Phengite K–Ar data on glaucophane-alkali pyroxene-lawsonite-bearing pelitic schist have yielded ages of 299.0 $\pm$ 6.1 and 280.2 $\pm$ 5.8 Ma (Kamihara *et al.* 2012). Detrital zircon U–Pb ages from glaucophane-phengite schist in the Itsuki district cluster around 530–420 Ma (Yoshimoto *et al.* 2013), which is a range similar to that of the Renge metamorphic complex in northern Kyushu. Phengite K–Ar data on pelitic schist of the Gokanosho metamorphic unit yield an age range of 182–144 Ma for the Gokanosho metamorphic unit. The reason why there is such a large discrepancy in these ages is not well understood, but is probably due to later deformation because the Gokanosho metamorphic unit is distributed within narrow belts sandwiched by faults. Argon depletion of phengite by low-temperature deformation (Itaya & Takasugi 1988) may contribute to the age discrepancy, so that the older age should be closer to the metamorphic ages of the Gokanosho metamorphic unit. In terms of regional correlations, the Gokanosho metamorphic unit outcrops are thought to be a NE extension of the Haki metamorphic unit in the Hinagu district south of Kumamoto, which also exhibits pumpellyite-actinolite subfacies metamorphism and has yielded phengite K–Ar pelitic schist ages of 182–187 Ma (Nishizono *et al.* 1996). These radiometric ages suggest that the Gokanosho and Haki metamorphic units can be correlated with the Suo metamorphic complex of northern Kyushu.

The Silurian–Devonian non-accretionary sedimentary unit in the Tomochi district comprises the Gionyama, Horagadake, Yamaide and Naidaijin formations (Fig. 2e.6), which together tectonically overlie the serpentinite mélange. The Gionyama and Horagadake formations consist mainly of felsic tuff and limestone with sandstone and alternations of sandstone and mudstone. Late Silurian–Early Devonian radiolarian fossils have been reported from felsic tuff in the Horagatake Formation (Kurihara 2004). The Naidaijin Formation conformably overlies the Gionyama Formation and yielded Late Devonian *leptophloeum* and *brachiopoda. Coralite* and *trilobite* fossils were reported from limestones. On the basis of these fossils, the sedimentation age of the Naidaijin Formation is estimated to be Late Devonian (partly Middle Devonian). The Yamaide Formation conformably overlies the Horagadake Formation, consists of mudstone with small amounts of sandstone, conglomerate and limestone, and yielded the Late Devonian fossil *leptophloeum* (Saito *et al.* 2003).

Carboniferous non-accretionary sedimentary rocks have been divided into the Kakizako and Kurotani formations (Fig. 2e.6), both of which tectonically underlie Permian accretionary complex rocks. The Kakizako Formation consists of fossiliferous (Early Carboniferous) limestone, mudstone, sandstone and conglomerate. The Yuzuruha Formation in the Kuraoka area located to the east of the Tomochi district is correlative with the Kakizako Formation (Murata 1992).

Permian non-accretionary sedimentary rocks are represented by the Kuma Formation, the base of which is in fault contact with serpentinite or serpentinite mélange and the top of which is in fault contact with non-accretionary Triassic sedimentary rocks. The formation consists mainly of sandstone with conglomerate, mudstone, alternations of mudstone and sandstone, and limestone. Late Permian radiolarian fossils have been reported from mudstone and Permian *fusulina* fossils from limestone. The lithologies and sedimentation age of the Kuma Formation resemble those of the Mizukoshi Formation on the northern side of the UYTL. Other Permian non-accretionary sedimentary rocks in central Kyushu comprise the Early Permian Tsurunokoba Formation and Middle Permian Kozaki Formation (Matsumoto & Kanmera 1964; Fig. 2e.2).

Triassic non-accretionary sedimentary rocks are represented by the Miyamadani Formation which consists of calcareous clastic rocks such as sandstone, mudstone and alternations of sandstone and mudstone. Many *Monotis* fossils have been reported from this formation, the sedimentation age of which is estimated to be Late Triassic.

Jurassic non-accretionary sedimentary rocks are represented by the Kawamata Group, which has yielded Early–Late Jurassic radiolaria (Saito *et al.* 2005). In the Tomochi district the group has been subdivided into the Nishinoiwa, Bisho and Ikenohara formations, and consists mainly of sandstone and mudstone with felsic tuff. Formations that are correlative with the Kawamata Group are distributed in other areas of the zone between the UYTL and BTL from eastern to western Kyushu. The Ashikita Group in the Hinagu area for example is characterized by Early–Late Jurassic turbidites intercalated with debris-flow deposits that include blocks of Middle Permian–Late Jurassic limestone, limestone conglomerate, siliceous mudstone, sandstone and conglomerate (Sano 2010*b*).

Permian accretionary complex outcrops in the Tomochi district comprise the Miyama and Hirodaira units, both of which tectonically overlie Silurian–Devonian and Carboniferous non-accretionary sedimentary rocks with a fault contact. The top of both units is either unconformably covered by the Tomochi Formation of the Cretaceous Monobegawa Group, or is in fault contact with Permian non-accretionary sedimentary rocks of the Kuma Formation. These units consist of mélange with a mudstone matrix including blocks of basalt, chert, felsic tuff, mudstone and sandstone. Late Permian radiolarian fossils were reported from mudstone (Saito & Toshimitsu 2003). The Kamiwashidani Formation (Sakai *et al.* 1993) in eastern Kyushu is also a coeval Permian accretionary complex.

Jurassic accretionary complex

The Jurassic accretionary complex crops out in the Tomochi district and includes both mélanges and coherent units, with the latter preserving an oceanic plate stratigraphy in which a chert–clastic rock sequence is repeated as thrust sheets (Fig. 2e.7). These rocks have been subdivided into seven tectonostratigraphic units which, from structurally highest to lowest, comprise: Early Jurassic coherent strata (Takadake Unit); Early Jurassic mélange (Otao, Hashirimizu and Yonagu units); latest Early–early Middle Jurassic coherent strata (Nitao Unit); Middle–early Late Jurassic coherent strata (Momigi Unit); and Late Jurassic mélange (Omae Unit) (Fig. 2e.8). Similar Jurassic complex materials are also found in eastern Kyushu where they have been divided into the Youra (coherent), Tsukumi (mélange) and Konose (mélange) units (Sano & Onoue 2010). The Omae Unit is distinctive in including Triassic limestone and basalt originating from an oceanic seamount (Fig. 2e.8), as well as mudstones that are much younger (Late Jurassic–Early Cretaceous) than the oceanic sediments of the other units (Fig. 2e.8). These younger sedimentary rocks may represent a later cover to the accretionary complex.

The overall thickness of chert–clastic sedimentary rocks in the Middle–early Late Jurassic coherent unit (Momigi Unit) in the Tomochi district is clearly thinner than that of coeval rocks of the Kamiaso Unit in the Mino-Tanba Belt of the Inner Zone of SW Japan (Fig. 2e.9). It is not clear whether this is due to original differences in stratigraphic thickness or to a greater amount of structural thickening due to folding and faulting in the Inner Zone of SW Japan.

Late Jurassic slope basin deposits (Ishida 2006, 2009) and sedimentary rocks consisting of alternations of sandstone and mudstone, which include slumped blocks of Late Jurassic limestone (Nishi 1994), together form a cover to the Jurassic accretionary complex. In addition, equivalent strata lacking intense deformation have been recognized in some areas within the Jurassic accretionary complex (Saito *et al.* 2010).

Geological structure in the Tomochi district

It is clear that the Jurassic accretionary complex tectonically underlies both pre-Jurassic units and Jurassic non-accretionary sedimentary rocks (Fig. 2e.6). This basic relationship has been modified by later deformation (Saito *et al.* 2005) so that, for example, the Hashirimizu Unit (Early Jurassic mélange) has been thrust over the Gokanosho metamorphic unit (fault *a* in Fig. 2e.6). Removing the effect of this faulting reveals an antiform and synform with a 2–3 km wavelength. These folds are thought to have been initiated during pre-Cretaceous north–south compression, but were reactivated after sedimentation of the Late Cretaceous Nakakyushu Group which shows similarly oriented but gentler folding. The timing of thrusting (fault *a* in Fig. 2e.6) is considered to be post-Late Cretaceous but pre-Middle Miocene, when an extensional tectonic setting was producing normal faults cutting granitic rocks in the SE of Kyushu (Fabbri *et al.* 1997; Saito *et al.* 2005).

It is inferred that the Silurian–Devonian non-accretionary sedimentary rocks unconformably covered the serpentinite mélange, or were at least deposited close by. Carboniferous strata overlie these Silurian–Devonian sedimentary rocks, and underlie the Permian accretionary complex. These stratigraphic relationships are seen throughout the Tomochi district. As shown in the cross-section on Figure 2e.6, alongside the UYTL the Permian accretionary complex is unconformably covered by the Lower Cretaceous Tomochi Formation of the Monobegawa Group. In the centre of the zone between the UYTL and BTL, the Permian accretionary complex within the Kuriki syncline tectonically underlies Permian–Jurassic non-accretionary sedimentary rocks and the Early Cretaceous Nakakyushu Group. The latter group contains clasts eroded from the

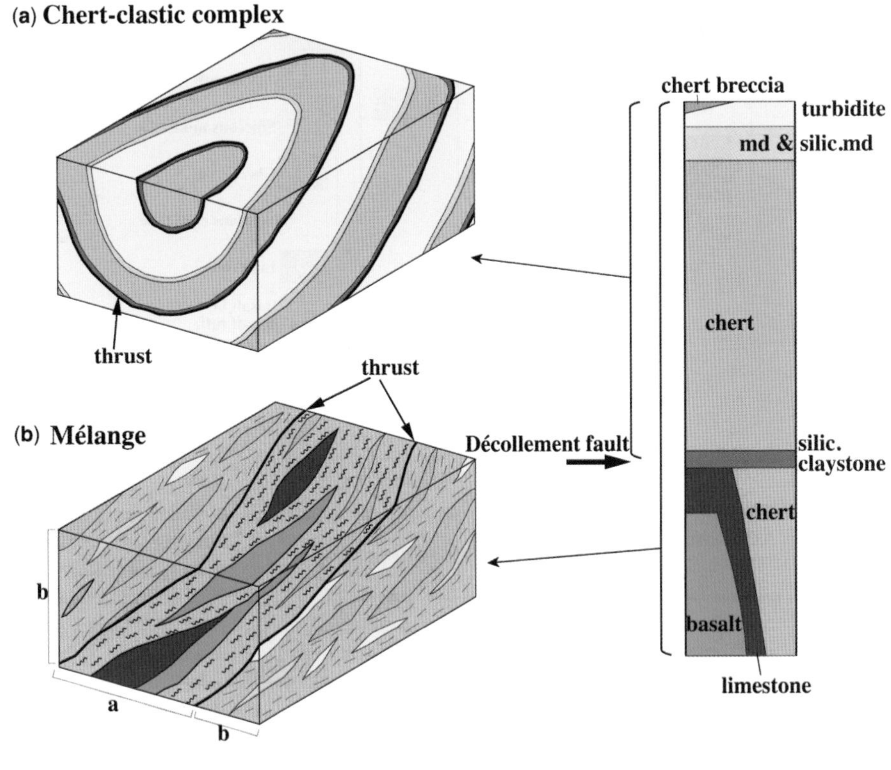

Fig. 2e.7. Block diagram and original stratigraphy of the Jurassic accretionary complex for (**a**) coherent unit (chert-clastic complex) and (**b**) mélange unit (Saito 2008). a, mélange with basalt; b, mélange without basalt. silic., siliceous; md, mudstone.

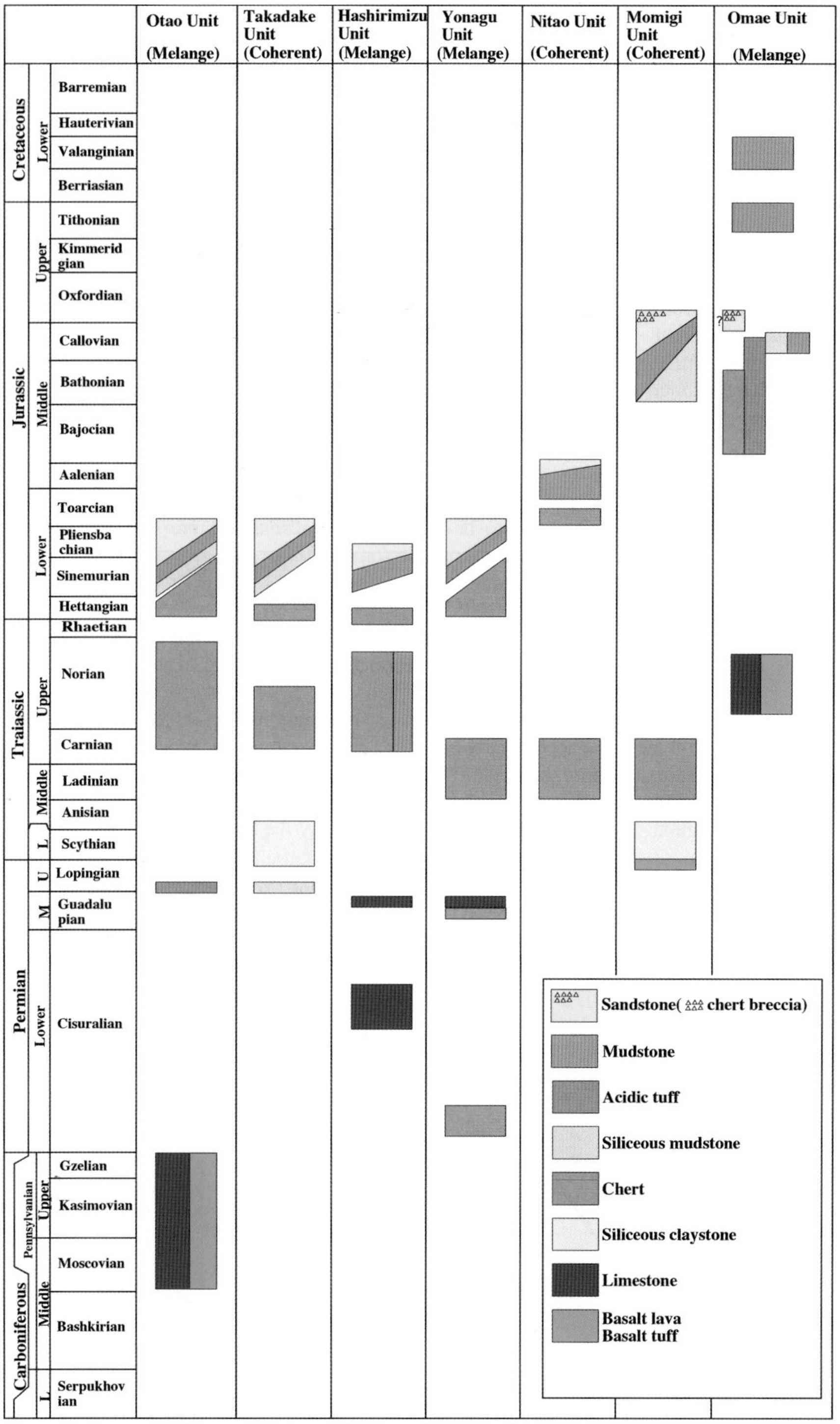

Fig. 2e.8. Stratigraphic succession of Jurassic accretionary complex of the Chichibu Belt in the Tomochi and Shiibamura districts (modified from Saito *et al.* 2005).

Jurassic accretionary complex, which was therefore presumably exposed by Albian times.

Other plutonic and metamorphic units outside the Tomochi district
The Usukigawa Quartzdiorite and the Shonoharu metamorphic unit form minor outcrops on the southern side of the UYTL in eastern Kyushu (Fig. 2e.1). The northern margin of the Usukigawa Quartzdiorite is bounded by the UYTL which juxtaposes it against the Upper Cretaceous Onogawa Group, and the southern margin is bounded by a high-angle fault with the Upper Cretaceous Tano Group. The quartzdiorite, which is locally mylonitized and contains hornblendite enclaves, has yielded hornblende K–Ar ages of 272–252 Ma, whole-rock Rb–Sr ages of 275 ± 56 Ma (Takagi *et al.* 1997) and zircon U–Pb ages of 292 ± 12.4 Ma (Sakashima *et al.* 2003). These radiometric ages suggest that the Usukigawa Quartzdiorite may be correlated with the Permian Kinshozan Quartzdiorite in the Kanto Mountains of the Outer Zone of SW Japan (Takagi *et al.* 1997).

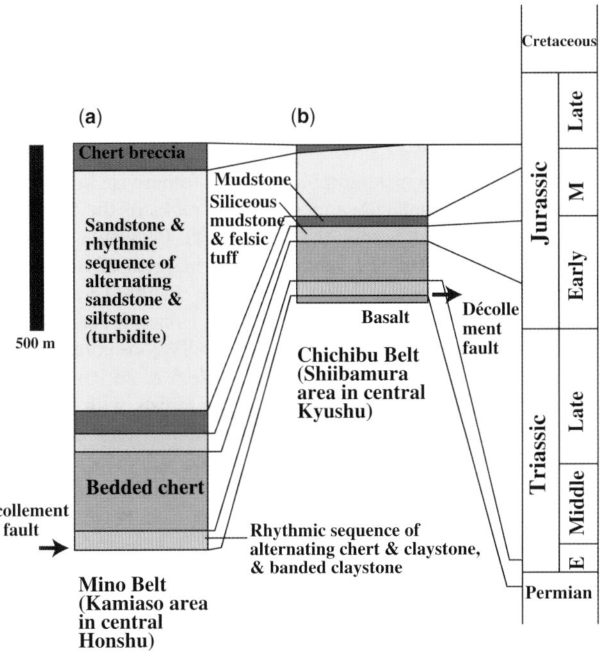

Fig. 2e.9. Comparison of the estimated thickness of the reconstructed stratigraphy of Jurassic accretionary complex between the Mino Belt (**A**) and Chichibu Belt (**B**). (Saito 2008). (**a**) Reconstructed stratigraphy in the Kamiaso area of the Mino Belt, central Honshu, Japan. (**b**) Reconstructed stratigraphy in the Shiibamura area of the Chichibu Belt, central Kyushu, Japan.

Amphibolite is locally distributed along the UYTL at the northern margin of the Usukigawa Quartzdiorite, and pelitic gneiss occurs similarly along the high-angle fault defining the southern margin. The pelitic gneiss becomes foliated cataclasite near the fault. Hornblende K–Ar isotopic data from the amphibolite yield an age of 124 $\pm$ 6 Ma (Takagi *et al.* 1997), which is similar to that of the Higo metamorphic complex.

Finally, pelitic schist and phyllite of the Shonoharu metamorphic unit known as the Shonoharu metamorphic rocks are distributed along the southern side of the UYTL in eastern Kyushu and have yielded a phengite K–Ar age 190 $\pm$ 10 Ma (Teraoka *et al.* 1992), similar to that of the Suo metamorphic complex.

Ryukyu islands

The Tomuru Formation crops out on Ishigaki-jima and Iriomote-jima in the Ryukyu islands, and is correlated with the Suo metamorphic complex (Fig. 2e.1; Faure *et al.* 1988*b*; Nishimura 1998). The Tomuru Formation tectonically overlies the Jurassic accretionary complex of the Fusaki Formation (Isozaki & Nishimura 1989), and consists of ultramafic rocks, metagabbro, mafic, siliceous and pelitic schists. The metamorphic grade ranges from lawsonite-blueschist facies through epidote-blueschist subfacies to albite-epidote amphibolite subfacies, with metamorphic grade increasing upwards towards apparent upper structural levels (Nakae *et al.* 2009). The mineral assemblages of pelitic schists are correlative with those of chlorite and garnet zones of the Sanbagawa metamorphic complex in central Shikoku. Phengite K–Ar data from pelitic schists yield ages of 220–190 Ma (Nishimura 1998) and detrital zircon U–Pb data from psammitic schist (Ishizuka *et al.* 2011) show age clusters around 303–234 Ma.

Jurassic accretionary complex rocks are exposed from Amami Oshima to Okinawa-jima and Ishigaki-jima (Fig. 2e.1). The following descriptions are based on the reviews of Takeuchi (2010) and Takami *et al.* (1999). The Jurassic accretionary complex in the Ryukyu islands yields radiolarian fossils and comprises Carboniferous–Early Cretaceous oceanic rocks, such as limestone and mafic volcanic and siliceous rocks, and Early Jurassic–Early Cretaceous clastic rocks. The complex can be divided into coherent sequences (Ie Unit) and mélanges (Fusaki, Iheya, Yuwan and Motobu units). The Fusaki Unit on Ishigaki-jima can be correlated with the Jurassic accretionary complex in the Mino-Tanba Belt of the Inner Zone of SW Japan. The other units are correlative with the Jurassic accretionary complex in the Chichibu Belt of the Outer Zone of SW Japan. The Motobu Unit on Okinawa-jima has undergone sub-greenschist facies (pumpellyite-actinolite subfacies) metamorphism, and has a metamorphic age of *c.* 70 Ma (Kojima *et al.* 1999).

Discussion

The spatial distribution of pre-Jurassic and Jurassic rocks in the Kyushu–Ryukyu Arc shows that the Jurassic accretionary complex always underlies the pre-Jurassic units. As demonstrated in Figure 2e.1, the distribution of the Suo metamorphic complex in east-central Kyushu surrounds the Jurassic accretionary complex due to the presence of the large-scale antiform developed on the northern side of the UYTL. Further north (Fig. 2e.1), the Permian accretionary complex structurally overlies the Suo metamorphic complex, whereas to the south the large-scale structure becomes increasingly complex closer to the UYTL. Small bodies of pre-Suo units, such as the Renge metamorphic complex, Permian non-accretionary sedimentary rocks and Cambrian tonalite, are locally distributed on the northern side of the UYTL. Minor outcrops of the Permian accretionary complex and ultramafic rocks are also distributed in the Asaji area of eastern Kyushu (Fig. 2e.1). The local distribution of many of these geological units may be explained by modification due to activation of east–west-trending faults and folds similar to those developing in the zone between the UYTL and BTL.

As shown in the previous section, the geological structure in the zone between the UYTL and BTL is even more complicated than that on the northern side of the UYTL, and there are distinct differences between the two areas. The apparent thicknesses of geological units correlative with the Renge and Suo metamorphic complexes and the Permian accretionary complex are clearly thinner than those in northern Kyushu. Furthermore, pre-Jurassic geological units near to and within this central zone between the UYTL and BTL show more variety in terms of their formational ages and environments than those to the north of the UYTL. Another notable feature of this area is that the primary geological structures have been modified by east–west-trending faults and folds. As outlined in the Cretaceous section below ('Cretaceous accretionary and metamorphic complexes and plutonic rocks'), a strike-slip basin later developed on the northern side of the UYTL; this strike-slip deformation may contribute to modification of original geological structure near the UYTL and between UYTL and BTL. In addition, post-Cretaceous north–south compression and strike-slip movement also contributed to further modification of geological structures near UYTL and between the UYTL and BTL.

Cretaceous accretionary and metamorphic complexes and plutonic rocks (KM & MS)

Accretionary complex

The Cretaceous accretionary complex materials in Kyushu are represented by the Morotsuka Group which is part of the Shimanto Cretaceous accretionary complex of SW Japan. Similar accretionary complex rocks crop out on Amami Oshima, Tokunoshima, Okinoerabu-jima, Yoronto and Okinawa-jima in the Ryukyu islands (Figs 2e.10 & 2e.11).

The northwestern margin of the Morotsuka Group is bounded by the NW-dipping Butsuzo Tectonic Line (BTL), and underlies the Jurassic accretionary complex of the Chichibu Belt. The southeastern margin of the group is bounded by the Nobeoka Thrust (NT), and overlies the Palaeogene accretionary complex of the Hyuga Group (Fig. 2e.11; see also Fig. 2e.23a). The Morotsuka Group can be divided into the Saiki Subgroup and structurally underlying Kamae Subgroup. The subgroups are separated by the NW-dipping Tsukabaru Thrust and have bedding gently dipping northwards. On the basis of sedimentation ages of terrigenous clastic rocks of turbidites, the strata of the Saiki Subgroup were accreted during Aptian–Turonian times, whereas strata of the Kamae Subgroup have Cenomanian–Campanian accretion ages (Fig. 2e.11).

The Saiki Subgroup consists mainly of sandstone with turbidite and mudstone, and in central Kyushu has been subdivided into the thrust-stacked Choshigasa (top), Fudono and Kamishiiba (base) units (Figs 2e.12 & 2e.13). The structurally lowest unit (Kamishiiba) has the youngest accretion age (Fig. 2e.13). Basalt, red mudstone and chert occasionally occur on the upper side of the boundary thrusts between the units (or along thrusts within the units), and remnants of oceanic plate stratigraphy involving basalt, chert and clastic rocks can be recognized in places. However, the sequences near the thrusts are disrupted into mélange. Mudstone can be seen to conformably overlie basalt, suggesting that basaltic volcanism occurred just before accretion. Compared to the Saiki Subgroup, the underlying Kamae Subgroup is more schistose; the latter has been subdivided from top to base into the sandstone-dominated Yato Formation and Sampodake Unit and the mudstone-basalt-dominated Makimine Formation.

Provenance studies of sediments within the Morotsuka Group have demonstrated that while sandstones of the Saiki Subgroup are mainly derived from granitic rocks, those from the Kamae Subgroup are mainly sourced from volcanic rocks. This suggests the onset of a major phase of volcanic activity in SW Japan during early Late Cretaceous times (Teraoka & Okumura 1992).

The Morotsuka Group has undergone low-grade metamorphism increasing downwards from prehnite-actinolite subfacies in the Saiki Subgroup and the upper part of the Kamae Subgroup to greenschist and transitional greenschist-amphibolite facies in the lower part of the Kamae Subgroup (Fig. 2e.14; Toriumi & Teruya 1988; Nagae & Miyashita 1999; Miyazaki & Okumura 2002). Prehnite-actinolite subfacies metamorphism took place due to subduction of a young and hot plate (Miyazaki & Okumura 2002). Metamorphic ages of the Morotsuka Group are estimated to be 60–50 Ma based on muscovite K–Ar data from meta-mudstone (Hara & Kimura 2008) and K-feldspar K–Ar data from sandstones (Teraoka *et al.* 1994) of the Kamae Subgroup.

Finally, in Okinawa-jima, the Nago Formation can also be correlated with the Cretaceous accretionary complex in the Shimanto Belt. The upper part of this formation underwent metamorphism up to greenschist facies and yielded metamorphic ages of 77–60 Ma (Kojima *et al.* 1999). These ages are similar to those of the Sanbagawa metamorphic complex of SW Japan; however, in Okinawa-jima there are no typical high-pressure minerals present. Biotite occurs in pelitic schist without garnet, implying that metamorphic pressure is clearly lower than that of the Sanbagawa rocks, and that metamorphism may have been of low-pressure type rather than high-temperature type. The lower part of the Nago Formation underwent sub-greenschist facies metamorphism at 55–40 Ma (Kojima *et al.* 1999).

Metamorphic complexes

Cretaceous metamorphic rocks in the Kyushu–Ryukyu Arc can be divided into high-pressure (Sanbagawa and Nagasaki) and high-temperature (Ryoke, Asaji, Higo) metamorphic complexes, each of which is described below.

The Sanbagawa metamorphic complex (otherwise known in the literature as the Sanbagawa metamorphic rocks of the Sanbagawa Belt) is distributed across the Saganoseki Peninsula in eastern Kyushu (Fig. 2e.10). The south and SW margins of the complex are in tectonic contact across the Sashu Fault (Teraoka 1970; Yamakita *et al.* 1995) with the overlying Upper Cretaceous Onogawa Group cropping out around a ENE–SWS-trending antiform in the central part of the peninsula (Figs 2e.5 & 2e.10). The complex consists mainly of pelitic and mafic schists with psammitic, siliceous and calcareous schists, serpentinite and metagabbro (Miyazaki & Yoshioka 1994). The metamorphic facies is pumpellyite-actinolite subfacies to epidote-blueschist subfacies, and pelitic schists show a chlorite zone mineral assemblage (Higashino 1990) comparable to that of the Sanbagawa metamorphic complex in central Shikoku.

The Nagasaki metamorphic complex is the western extension of the Sanbagawa metamorphic complex (Faure *et al.* 1988*a*; Nishimura 1998; Nishimura *et al.* 2004) and crops out on the Nishisonogi and Nomo peninsulas and on Amakusa Shimoshima (Figs 2e.4, 2e.10 & 2e.15), with gently dipping foliations typifying all three areas. The Triassic–Jurassic high-pressure metamorphic rocks exposed on the Nomo Peninsula are assigned to the Suo metamorphic complex (Nishimura 1998; Nishimura *et al.* 2004) and overlie the Nagasaki metamorphic complex. In the Nishisonogi Peninsula, the Nagasaki metamorphic complex has been brought into contact with Palaeogene non-accretionary sedimentary rocks by the north–south-trending Yobikonoseto Fault (Fig. 2e.10), whereas in the Nomo Peninsula the Nagasaki metamorphic complex is in contact with the Suo metamorphic complex along the Fukabori–Wakimisaki and Mogi faults (Fig. 2e.4). In the Amakusa Shimoshima area, the Nagasaki metamorphic complex underlies the Upper Cretaceous Himenoura Group with low-angle fault contacts (Fig. 2e.15).

The Nagasaki metamorphic complex consists of pelitic, mafic, siliceous and psammitic schists and serpentinite. The complex in the Nishisonogi Peninsula and Amakusa Shimoshima consists mainly of pelitic schist, whereas exposures in the Nomo Peninsula additionally include substantial amounts of mafic schists and serpentinite. The metamorphic grade in the Nishisonogi and Amakusa Shimoshima areas belongs to the epidote-blueschist subfacies, and the mineral assemblage of pelitic schist is equivalent to the garnet zone of Higashino (1990). Metamorphic mineral assemblages in the Nomo Peninsula record albite-epidote amphibolite subfacies conditions, with metamorphic grade increasing structurally downwards (Miyazaki & Nishiyama 1989) from the garnet zone to the albite-biotite zone of Higashino (1990).

In the Nishisonogi Peninsula, quartz-bearing jadeite rock (Shigeno *et al.* 2005) and garnet-bearing amphibole-clinopyroxenite (Nishiyama 1989) have been reported as forming anomalously high-grade tectonic blocks in serpentinite. In the Amakusa Shimoshima area, a mylonitized unit consisting of kyanite-bearing amphibole gneiss, garnet amphibolite and felsic gneiss tectonically overlies the more typical schists of the Nagasaki metamorphic complex. The metamorphic facies of the mylonitized unit is kyanite-amphibolite subfacies to high-pressure granulite, recording much higher temperatures than the underlying schist (Ikeda *et al.* 2005; Miyazaki *et al.* 2013). In the Nomo Peninsula, mylonitized quartz-diorite (Oshima 1964) and high-temperature metamorphic rocks (Takeda *et al.* 2002) occur under the Joyama Thrust (Fig. 2e.4).

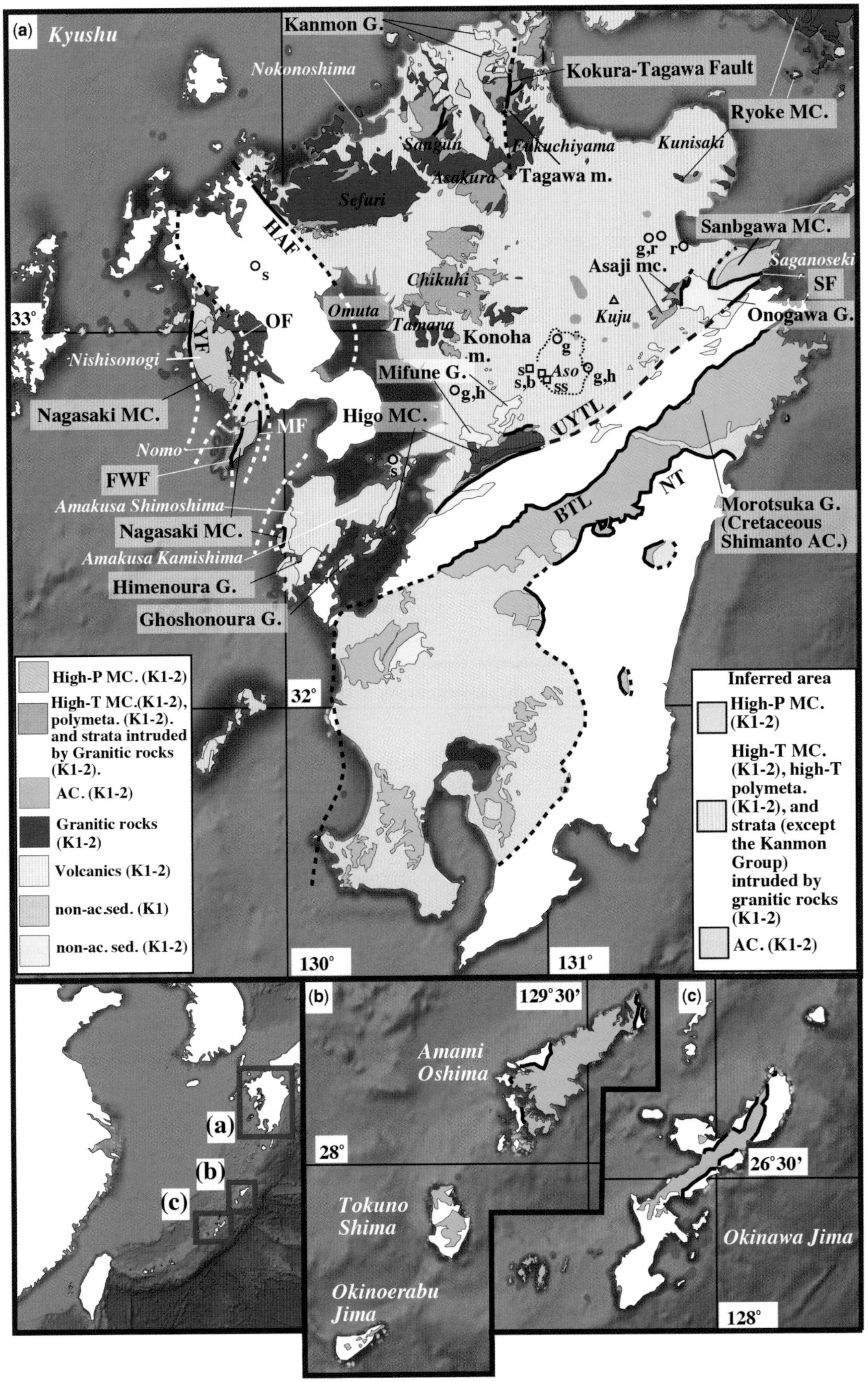

Fig. 2e.10. Distribution of Cretaceous accretionary complex, metamorphic complex, plutonic rocks, volcanic rocks and non-accretionary sedimentary rocks in the Kyushu–Ryukyu Arc, based on the seamless digital geological map of Japan (1:200 000) Geological Survey of Japan. https://gbank.gsj.jp/seamless/index_en.html. Abbreviations as for Figure 2e.1. polymeta., polymetamorphic rocks; K1–2, Early–Late Cretaceous; K1, Early Cretaceous.

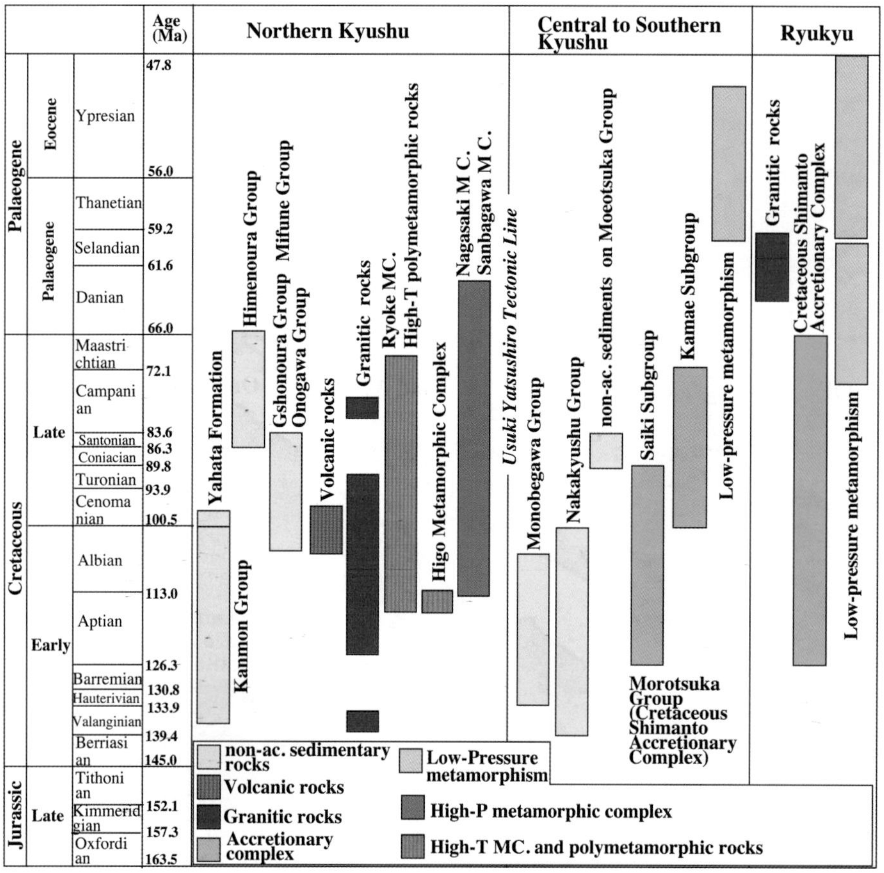

Fig. 2e.11. Stratigraphic succession of Cretaceous accretionary complex, metamorphic complex, plutonic rocks, volcanic rocks and non-accretionary sedimentary rocks in the Kyushu–Ryukyu Arc. Abbreviations as for Figure 2e.1.

Detrital zircon U–Pb ages (Kouchi *et al.* 2011; Tsutsumi *et al.* 2011*a*) from psammitic schists cluster around 2500–1500, 270–160 and 100–90 Ma, whereas phengite K–Ar and Ar–Ar data yield ages between 60 and 90 Ma (Hattori & Shibata 1982; Faure *et al.* 1988*a*; Nishimura *et al.* 2004; Takagi *et al.* 2009; Tsutsumi *et al.* 2011*a*). The protolith age of the jadeite rock in the Nishisonogi Peninsula is estimated to be 126 ± 6 Ma, and the metamorphic age has been estimated to be 84 ± 6 Ma (Mori *et al.* 2011). The metamorphic age of the high-pressure granulite in Amakusa Shimoshima is estimated to be 113.7 ± 1.6 Ma (Miyazaki *et al.* 2013).

The Ryoke metamorphic complex crops out on the Kunisaki Peninsula in eastern Kyushu (Fig. 2e.10). The complex consists mainly of pelitic gneiss with siliceous gneiss and amphibolite (Ishizuka *et al.* 2005). The metamorphic facies is sillimanite-amphibolite subfacies and the mineral assemblage is comparable with the K-feldspar-sillimanite zone (sillimanite-K-feldspar zone of Ikeda 1998) of the Ryoke metamorphic complex in the Yanai area of the Chugoku region.

The Chokai unit of the Asaji metamorphic complex represents the western extension of the Ryoke metamorphic complex (Fig. 2e.10). It consists of metamorphic rocks derived from mafic volcanic rocks, chert, mudstone and sandstone. The highest-grade part of the Chokai unit is up to amphibolite facies and metamorphic grade increases towards the Nioki Granite, which is considered to have been one of the heat sources driving the Cretaceous high-temperature metamorphism.

The Higo metamorphic complex is distributed across west-central Kyushu from the Tomochi district to the east coast of the Amakusa Kamishima area (Fig. 2e.16), and ranges in metamorphic grade from amphibolite to granulite facies. The Permian Ryuhozan Group, Cambrian Hikawa Tonalite and the Triassic–Jurassic Manotani metamorphic unit have also been overprinted by the high-temperature Higo metamorphism. The complex is disrupted into many small blocks by ENE–WSE- and NNE–SSE-aligned high-angle faults (Fig. 2e.16) and consists of metamorphic lithologies derived from sandstone, mudstone, mafic volcanic rocks, limestone, siliceous rock and ultramafic rocks. Meta-mudstone varies from pelitic schist through pelitic gneiss to pelitic migmatite with increasing metamorphic grade, allowing subdivision of the complex into biotite, K-feldspar-sillimanite, garnet-cordierite-I, garnet-cordierite-II and locally orthopyroxene zones with increasing metamorphic grade (Obata *et al.* 1994; Miyazaki 2004; Maki *et al.* 2009). Sapphirine granulite occurs with metamorphosed ultramafic rocks (Osanai *et al.* 1998). Metamorphic grade increases towards apparent lower structural levels in each block, and is generally higher in the southern blocks than in the northern blocks (Fig. 2e.16). Integrated analysis of Higo lithostratigraphy has demonstrated how, prior to fault disruption, the complex had a coherent structure with temperature and pressure increasing downwards through a sequence *c.* 10 km thick (Miyazaki 2004). Metamorphic zircon U–Pb ages and monazite U–Th–Pb ages (Sakashima *et al.* 2003; Dunkley *et al.* 2008; Maki *et al.* 2011) lie in the range 110–120 Ma, so that the peak metamorphism of the Higo metamorphic complex is estimated to have taken place in Early Cretaceous times. Older ages of between 290 and 200 Ma have been obtained from Higo metamorphic rocks (Osanai *et al.* 1993; Osanai *et al.* 1998; Hamamoto *et al.* 1999), although their meaning remains uncertain (Dunkley *et al.* 2008).

Cretaceous high-temperature polymetamorphic rocks crop out in northern Kyushu (Fig. 2e.10) as granulite in the Sefuri Mountains (Owada *et al.* 2005) and migmatites around the Omuta district (Miyazaki & Matsuura 2005; Tomita *et al.* 2008), as well as in the Tamana (Konoha metamorphic rocks: Hashimoto & Fujimoto 1962),

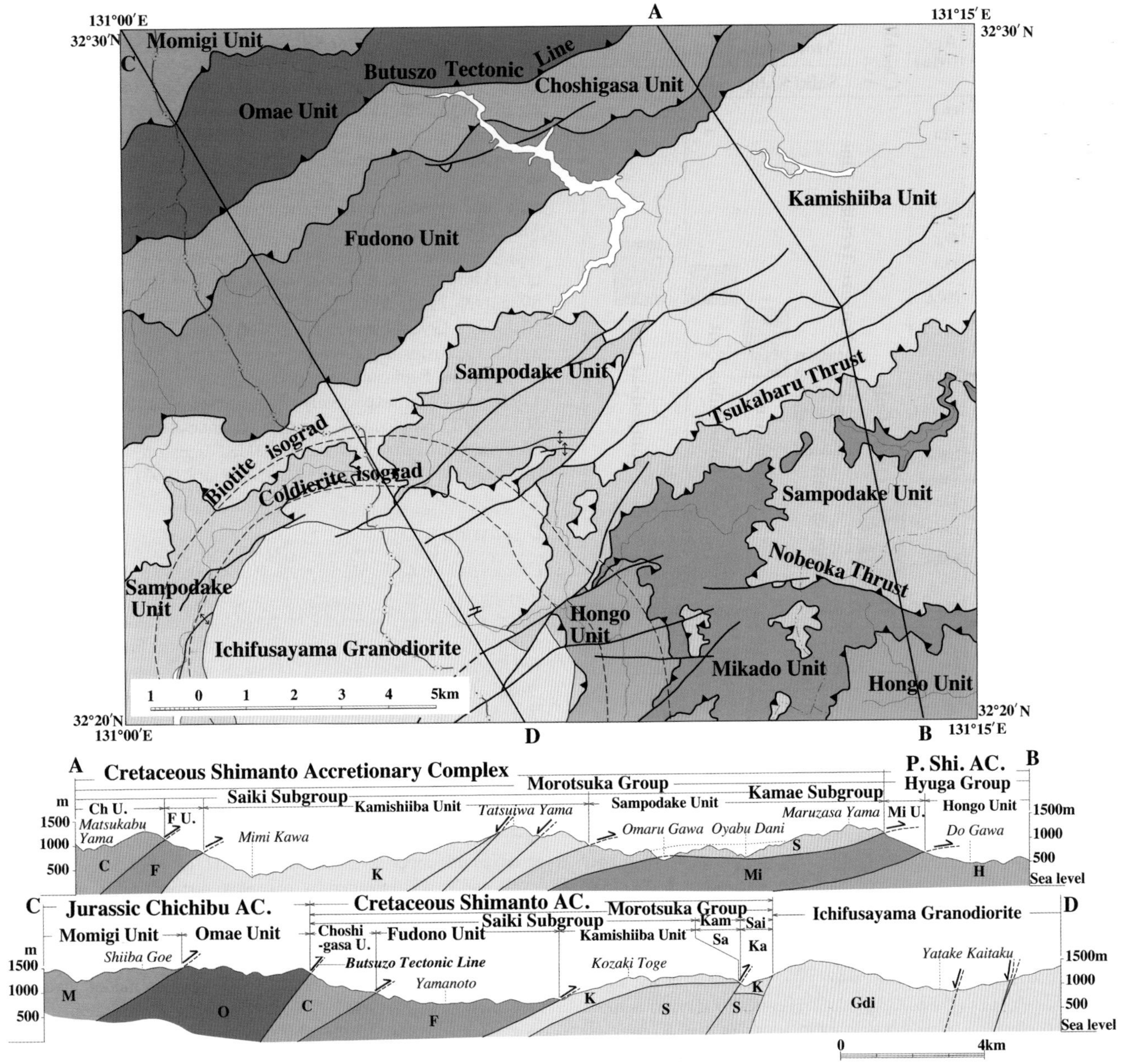

Fig. 2e.12. Geological map of the Shiibamura district (modified from Saito *et al.* 1996). Ch, Choshigasa; Mi, Mikado; Sa, Saiki; Ka, Kamae; M, Momigi Unit; O, Omae Unit; C, Choshigasa Unit; F, Fudono Unit; K, Kamishiiba Unit; S, Sampodake Unit; Mi, Mikado Unit; H, Hongo Unit; Gdi, granodiorite.

Hiraodai (Tagawa metamorphic rocks: Fukuyama *et al.* 2004) and Asakura (Kitano & Ikeda 2012) areas. The protolith of the granulite in the Sefuri Mountains is the Renge metamorphic complex, whereas that for the other metamorphic units is the Suo metamorphic complex. In the Omuta district, the Suo metamorphic complex demonstrates a gradual and continuous change to high-temperature metamorphic rocks across biotite, andalusite and K-feldspar-sillimanite zones with increasing metamorphic grade (Fig. 2e.3; Tomita *et al.* 2008). Biotite K–Ar ages of pelitic migmatites are *c.* 97 Ma (Tomita *et al.* 2008).

Cretaceous high-temperature metamorphic rocks crop out across various small islands westwards from the Nishisonogi Peninsula in NW Kyushu (Fig. 2e.10). The protolith for these rocks is the Lower Cretaceous Kanmon Group (Tachibana 1962; Katada & Matsui 1973), and whole-rock K–Ar dating of hornfels yield ages of 83 ± 3 Ma (Shibata 1968). Further along the coast, andalusite-bearing hornfels formed in response to intrusion of Cretaceous granitic plutons occur in the Kabashima area in the southern part of the Nomo Peninsula (Fig. 2e.4; Iwasaki 1954). The protolith of the hornfels is the Suo metamorphic complex, and biotite and muscovite K–Ar dating from the hornfels yield ages of 86–71 Ma (Nishimura *et al.* 2004).

Plutonic rocks

Cretaceous plutonic rocks in Kyushu are distributed on the northern side of the UYTL (Fig. 2e.10) and comprise granite, granodiorite and tonalite with small amounts of gabbro and quartz diorite. They intruded the Renge, Suo, Ryoke and Higo metamorphic complexes and the Cretaceous Kanmon Group.

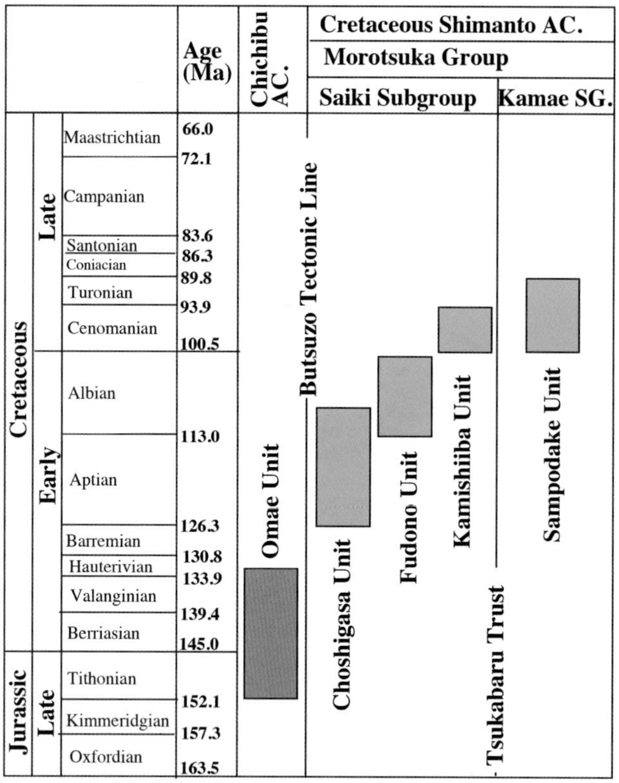

Fig. 2e.13. Stratigraphic succession of the Cretaceous Shimanto accretionary complex (Saito *et al.* 1996). AC, accretionary complex; SG, subgroup.

The Cretaceous granitic rocks on the western side of the Kokura–Tagawa Fault in northern Kyushu (Fig. 2e.10) have generally higher Sr content and magnetic susceptibility than those elsewhere in SW Japan (Izawa *et al.* 1989), with magnetite-series intrusions to the west contrasting with ilmenite-series plutons east of the fault (Ishihara *et al.* 1979). The initial Sr ratio of the Cretaceous granitic rocks in Kyushu is generally lower than elsewhere in the Inner Zone of SW Japan (Owada *et al.* 1999; Kamei 2002), and is more similar to that of Cretaceous granitic rocks of the Kyeongsan Basin in SE Korea (Owada *et al.* 1999). The Shiraishino Granodiorite (Fig. 2e.16), intruded into the Higo metamorphic complex, has an adakaitic composition (Kamei 2004). Mafic plutonic rocks with high-Mg andesitic composition have been reported from the Kitataku area in the southern margin of the Sefuri Mountains, the Shikanoshima area near Fukuoka and on the Kunisaki Peninsula (Kamei *et al.* 2004; Tiepolo *et al.* 2012).

Radiometric ages of the Cretaceous plutonic rocks in Kyushu range from 135 to 70 Ma; most are of middle Cretaceous age (120–90 Ma), broadly overlapping in age with volcanic activity in northern Kyushu as recorded for example by the Shimonoseki Subgroup. Rb–Sr isochron and K–Ar ages for plutonic rocks in northern Kyushu are 116–75 Ma (Owada *et al.* 1999). Zircon U–Pb and monazite U–Th–Pb dating of granitic rocks and mafic plutonic rocks in and near the Sefuri Mountains show ages of 104–102 Ma (Adachi *et al.* 2012; Tiepolo *et al.* 2012). Hornblende and biotite K–Ar ages from granitic rocks from the Chikushi Mountains through the Aso–Kuju–Asaji area to the Kunisaki Peninsula range from 112 to 71 Ma (Kawano & Ueda 1966; Matsumoto & Narishige 1985; Sasada 1987; Takagi *et al.* 2000; Fujii *et al.* 2008; Miyoshi *et al.* 2011), and zircon U–Pb and monazite U–Th-Pb data from the area yield ages between 135 and 75 Ma (Takagi *et al.* 2001; Tobe 2006; Takagi *et al.* 2007). Rb–Sr whole-rock isochron data of granitic rocks from the Kunisaki Peninsula and the Asaji area (Osanai *et al.* 1993) yielded ages of 142.2 ± 2.1 and 164.0 ± 14.7 Ma, clearly older than zircon U–Pb and monazite U–Th–Pb ages from the area, and as yet of uncertain meaning. Hornblende and biotite K–Ar ages of granitic rocks in the Higo area are 110–104 Ma (Nakajima *et al.* 1995; Kamei *et al.* 2000), and zircon U–Pb ages of granitic and gabbroic rocks in the Higo area are similarly 113–110 Ma (Sakashima *et al.* 2003). Biotite K–Ar ages from granitic rocks on the western side of the Yobikonoseto Fault in the Nishisonogi Peninsula and at Kabashima on the Nomo peninsula yield 93–89 and 89–79 Ma (Kawano & Ueda 1966; Hattori & Shibata 1982; Nishimura 1998; Nishimura *et al.* 2004), respectively. Zircon U–Pb ages in the Nishisonogi area and on the eastern side of the Mogi Fault in the Nomo Peninsula have produced ages of 99.8 ± 4.5 Ma and 117.0 ± 0.4 Ma, respectively (Tsutsumi *et al.* 2010; Kouchi *et al.* 2011), and hornblende K–Ar data on andesite in the Wakino and Shimonoseki subgroups have yielded ages of 109.5 ± 2.4 Ma (Owada *et al.* 1999) and 103 Ma (Matsuura 1998), respectively.

Discussion

Cretaceous metamorphism, plutonism and subduction zone thermal structure

In this section we discuss the origin of the Cretaceous metamorphic complexes in the context of subduction zone thermal structure. We focus on three main units: (1) the Higo metamorphic complex; (2) the high-temperature polymetamorphic rocks of 'Sangun metamorphic rocks' (Renge and Suo metamorphic complexes); and (3) the Nagasaki metamorphic complex.

Peak pressure–temperature (P/T) conditions of the Higo metamorphic complex increase down towards lower structural levels, that is, southwards. Correlation between metamorphic thermal gradient (dT/dP) and metamorphic rocks shows that high dT/dP characterizes the upper levels whereas low dT/dP appears at the lower levels and is accompanied by pervasive distribution of migmatite. This observation suggests that effective heat transport by melt migration operated in the lower part of the Higo metamorphic complex (Miyazaki 2004). The P/T conditions and volume fraction of solidified melt can be explained by a thermal model involving melt migration (Miyazaki 2004, 2007, 2010). Given the metamorphic ages of 116–110 Ma obtained from the migmatites (Maki *et al.* 2011), the likely tectonic setting for the Higo metamorphic complex was middle–lower crust beneath the volcanic arc in late Early Cretaceous times (120–110 Ma).

High-T polymetamorphic rocks with high-P 'Sangun'-type protoliths occur in the Omuta district of northern Kyushu (Tomita *et al.* 2008). The highest metamorphic conditions are up to 0.5 GPa and 700°C (Miyazaki & Matsuura 2005), with biotite K–Ar ages around 97–96 Ma (Tomita *et al.* 2008). These data suggest that the polymetamorphic rocks were formed in the middle crust of the volcanic arc slightly later (middle Cretaceous; *c.* 100 Ma) than those of the Higo metamorphic complex. Similar high-T polymetamorphic rocks are also distributed around Hiraodai (Fukuyama *et al.* 2004) and Asakura (Kitano & Ikeda 2012) in northern Kyushu.

As previously mentioned (in 'Metamorphic complexes' section above), the Nagasaki metamorphic complex in Amakusa is subdivided into lower and upper units (Ikeda *et al.* 2005). The lower unit consists largely of garnet-bearing pelitic schist (phengite Ar–Ar ages: *c.* 90 Ma, Takagi *et al.* 2009) with minor amounts of glaucophane-bearing mafic schists, siliceous schists and marble, whereas the upper unit mainly consists of mafic and felsic gneisses with mafic to felsic intrusions. The metamorphic facies of the upper

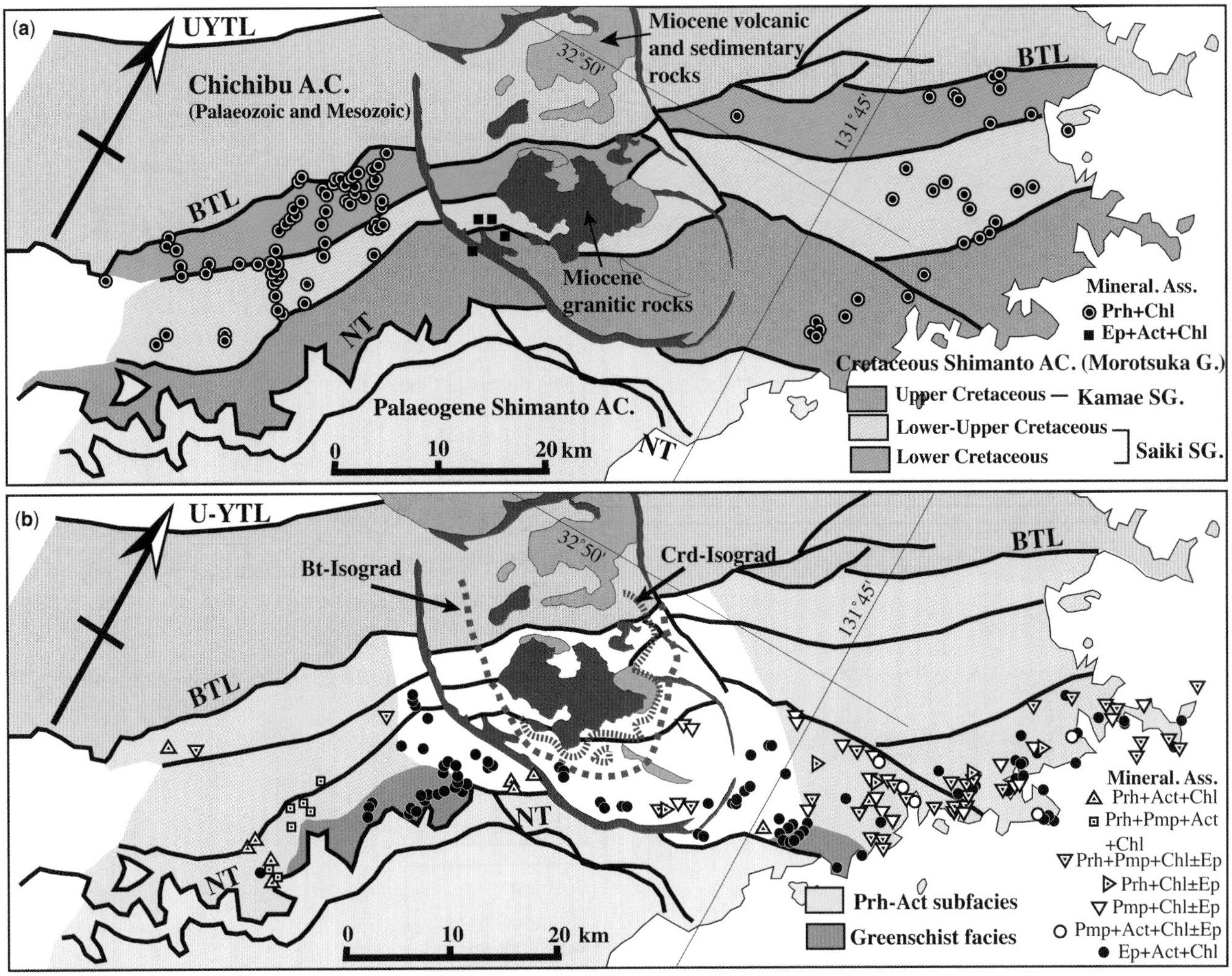

Fig. 2e.14. Low-pressure metamorphism of the Cretaceous Shimanto accretionary complex (modified from Miyazaki & Okumura 2002). (**a**) Spatial distribution of mineral assemblages in greywackes. (**b**) Mineral assemblages of mafic volcanic rocks and metamorphic facies. All assemblages in greywackes contain additional qtz + ab + ms + cc + titanite + opaque minerals. All assemblages in mafic volcanic rocks contain additional ab + titanite + opaque minerals ± qtz ± cc. Bt-Isograd (biotite isograd) and Crd-Isograd (cordierite-isograd) are due to Miocene contact metamorphism. Prp, prehnite; Act, actinolite; Chl, chlorite; Ep, epidote; Pmp, pumpellyite; qtz, quartz; ab, albite; cc, calcite. Other abbreviations as for Figure 2e.1.

part of the upper unit reaches high-*P* granulite grade ($P = 1.1$ GPa, Ikeda *et al.* 2005; Miyazaki *et al.* 2013) in rocks yielding metamorphic zircon ages of 120–110 Ma (Miyazaki *et al.* 2013), and are extensively mylonitized (Arima *et al.* 2011). These P/T conditions (Fig. 2e.17) suggest that the upper unit is a slice of the volcanic arc lower crust built at the Eurasian plate margin and horizontally transported in late Early Cretaceous times. Such a tectonic model is consistent with thermal modelling of the subduction zone (Miyazaki *et al.* 2013), with material transport from volcanic arc towards the trench taking place at lower crustal depths due to asthenospheric corner flow under the high heat flow anomaly present beneath the volcanic arc (Fig. 2e.18).

Modification by Cenozoic tectonics

Cenozoic tectonic movements have modified the spatial distribution of the Cretaceous metamorphic complexes as demonstrated by the following two key points. Firstly, the location of the Nagasaki metamorphic complex in the Nishisonogi and Nomo peninsulas is too far north relative to its predicted outcrop along-strike from the Sanbagawa metamorphic complex on the Saganoseki Peninsula and the Nagasaki metamorphic complex in the Amakusa Shimoshima area (Fig. 2e.10). This can be explained by the opening of the Sea of Japan in Middle Miocene times. A negative Bouguer anomaly (Kubodera *et al.* 1976) is observed between the Hatashima–Ariakekai and Omurawan–Amakusa faults. Palaeogene–Neogene non-accretionary sedimentary rocks are distributed in this area, and there is a possibility that a right-lateral fault operating during the opening of the Sea of Japan is hidden underneath the area of the negative Bouguer anomaly. However, the detailed geology of the basement in the area is not well known.

Secondly, the Cretaceous high-pressure metamorphic units, such as the Sanbagawa and Nagasaki metamorphic complexes, underlie Cretaceous high-temperature metamorphic rocks formed during the intrusion of arc-related Cretaceous plutons (Fig. 2e.10). This low-angle juxtaposition of rocks, formed under such contrasted geothermal gradients, is abnormal for the tectonic setting of the high-pressure metamorphic complex when there has been no post-Cretaceous tectonic movements. The current boundary between the high-pressure metamorphic complex and overlying hornfels formed by intrusion of Cretaceous granitic rock in the Nomo Peninsula of western Kyushu is a fault with cataclasite. In east Kyushu, the Sashu Fault lies between the Upper Cretaceous Onogawa Group and

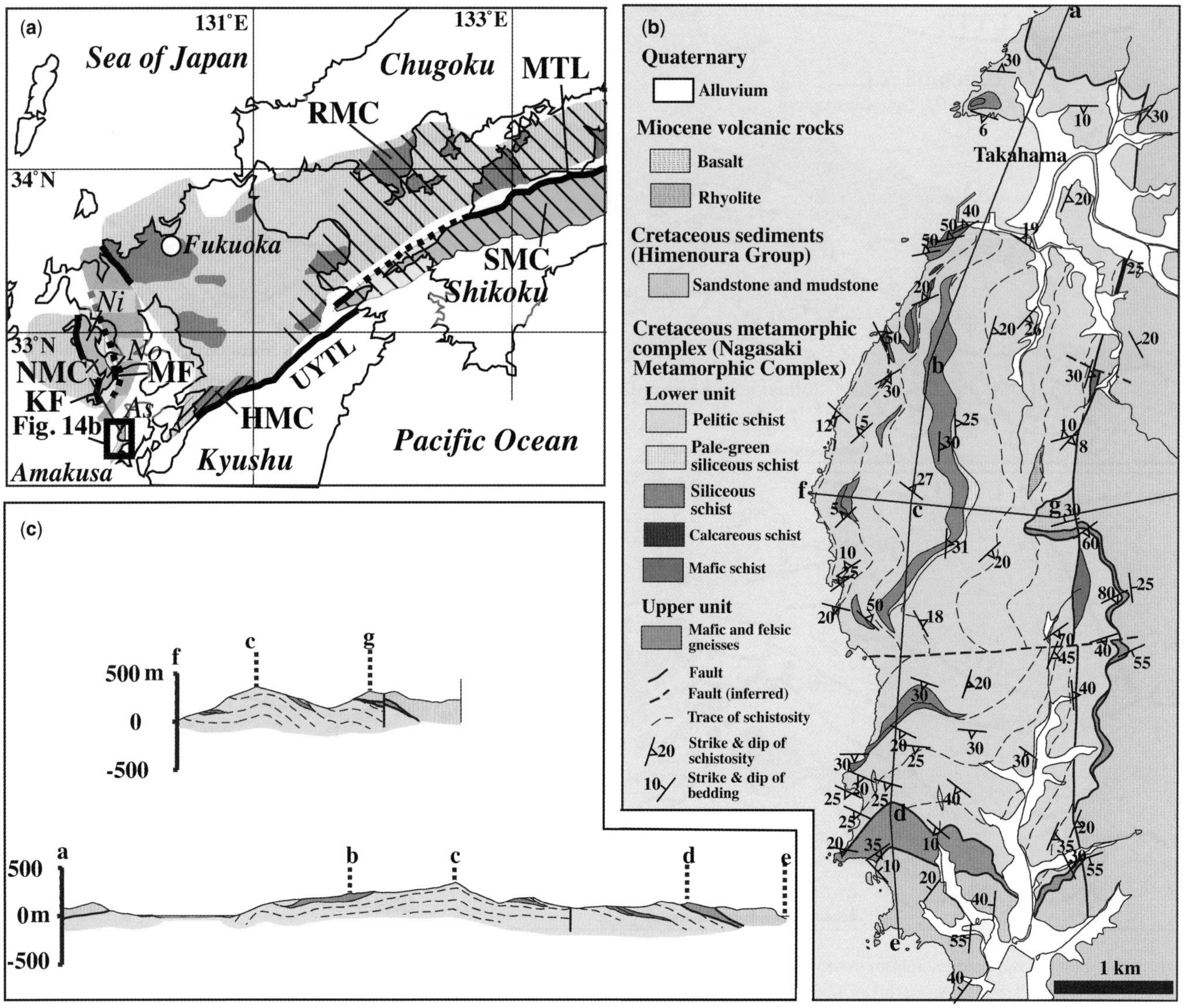

Fig. 2e.15. Geological map of the Nagasaki metamorphic complex in Amakusa Shimoshima (Miyazaki *et al.* 2013). (**a**) Distribution of Cretaceous metamorphic complexes in SW Japan. Cretaceous plutonic and high-*T* metamorphic complexes (pink) and their inferred area (light pink), Cretaceous high-*P* metamorphic complexes (blue) and their inferred area (light blue) and the north–south-trending area of low Bouguer anomalies (grey; Kubodera *et al.* 1976). The Sanbagawa metamorphic complex (SMC), Ryoke metamorphic complex and plutonic rocks (RMC) and Higo metamorphic complex (HMC) are hatched areas. KF and MF are the Fukabori–Wakimisaki Fault and Mogi Fault, respectively. Ni, No and As represent the Nishisonogi Peninsula, Nomo Peninsula and Amakusa Shimoshima, respectively. (**b**) Geological map of the Nagasaki metamorphic complex (NMC). (**c**) Geological cross-sections of the NMC.

the Sanbagawa metamorphic complex. As discussed in the section 'Cretaceous sedimentary basins', the Onogawa Group was deposited on a basement intruded by Cretaceous plutonic rocks and near Cretaceous volcanic centres. The shear sense of the Sashu fault zone shows left-lateral slip with top-to-the-south movement sense (Yamakita *et al.* 1995). This suggests oblique subduction of the high-pressure metamorphic complex under the metamorphic and plutonic rocks formed in or near the Cretaceous volcanic arc. K–Ar dating of the fault gouge of the Sashu Fault yields an Early Oligocene age (Yamakita *et al.* 1995); however, the initial activation age of this fault may be older.

Cretaceous sedimentary basins (KM, ST & SM)

Cretaceous non-accretionary sedimentary rocks in Kyushu crop out in the north, centre and south of the island. Each of these three areas is described below.

Northern Kyushu

Cretaceous sedimentary rocks in northern Kyushu are subdivided into the Lower Cretaceous Kanmon Group and the Upper Cretaceous Yahata Formation. The following descriptions are largely based on the reviews of Nakae (2010) and Ota & Yabumoto (1992). The continental Kanmon Group unconformably covers the Renge metamorphic complex and Permian accretionary complex in the Akiyoshi Belt, and is unconformably covered by the Upper Cretaceous Yahata Formation. The group is subdivided into the Wakino and Shimonoseki subgroups. The Wakino Subgroup is composed mainly of Valanginian–Hauterivian lake deposits which yielded the remains of large dinosaurs and comprise conglomerate, sandstone and mudstone with abundant freshwater molluscan fossils. The younger Shimonoseki Subgroup consists of felsic or intermediate lava and volcaniclastic rocks and is considered to be of Albian age. A zircon fission track age on dacite welded tuff is 101 Ma (Ueda & Nishimura 1982) and hornblende K–Ar data

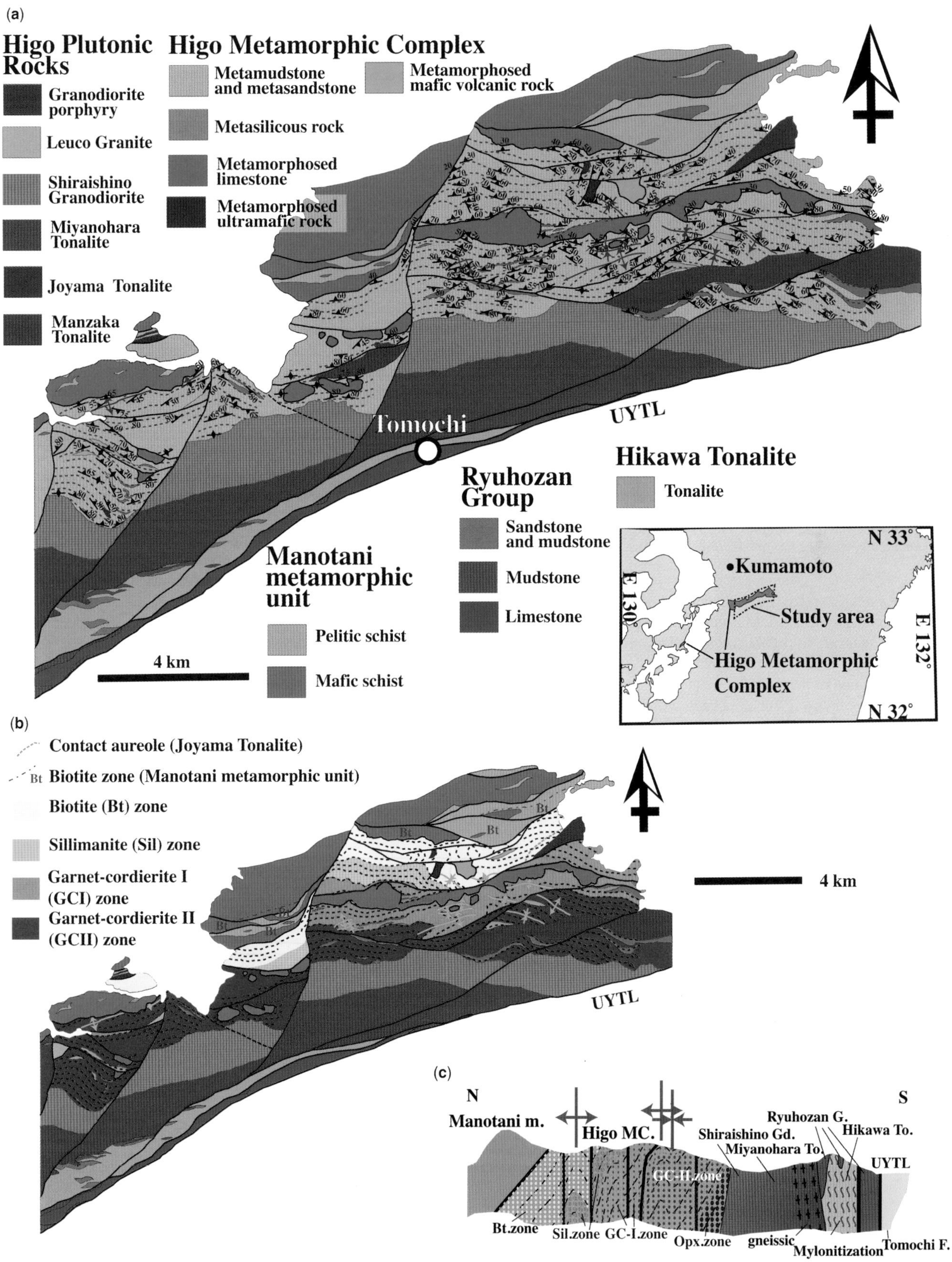

Fig. 2e.16. Geological map and metamorphic zonation of the Higo metamorphic complex in the Tomochi district, west central Kyushu (Miyazaki 2004). (**a**) Geological map (the Pleistocene Aso pyroclastic flow deposit in this district is omitted). (**b**) Metamorphic zonation. (**c**) Schematic cross-section.

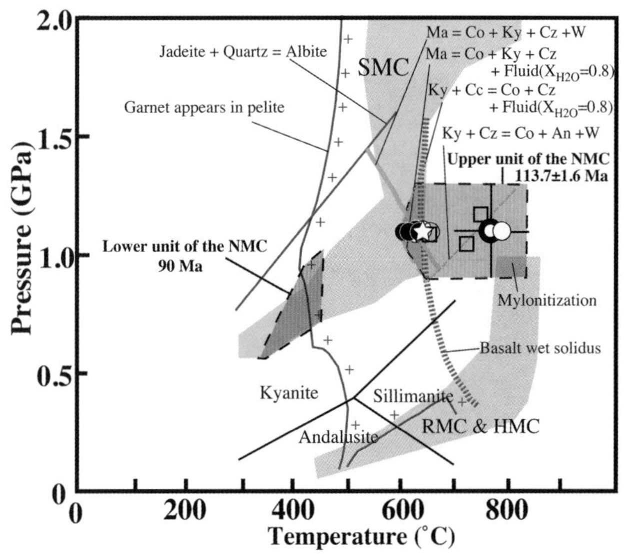

Fig. 2e.17. *P*/*T* conditions of the Nagasaki metamorphic complex (NMC) (Miyazaki *et al.* 2013). The upper unit is the pale red area with a broken line, and the lower unit is the pale blue area with a broken line. Grey areas show *P*/*T* conditions of the Sanbagawa metamorphic complex (SMC), metamorphic rocks of the Ryoke plutono-metamorphic Complex (RMC), and the Higo metamorphic complex (HMC). Ma, margarite; Co, corundum; Ky, kyanite; Cz, clinozoisite; An, anorthite; W, water.

from andesite lava and dacite welded tuff yield similar dates of 107–103 Ma (Imaoka *et al.* 1993; Matsuura 1998). The Yahata Formation consists of andesitic to rhyolitic volcaniclastic rocks and lavas, and sandstone.

Sedimentation of the Wakino Subgroup is thought to have taken place in a lake environment associated with fault activity that was caused by tectonically rapid subsidence and uplift. In contrast, volcanic activity dominated the depositional setting of the Shimonoseki Subgroup. It has been suggested that the Kanmon Group was deposited in a half-graben basin formed by left-lateral strike-slip fault system occurring along the Early Cretaceous eastern margin of the Asian continent (Fig. 2e.19; Okada 2000). Given this tectonic setting, the Kanmon Group is likely to be correlative with the Kyongsang Group in the southeastern part of the Korean Peninsula (Chough & Sohn 2010), and the two stratigraphic units would originally have lain much closer to each other prior to the opening of the Sea of Japan (Fig. 2e.19).

North-Central Kyushu

The Cretaceous non-accretional sedimentary rocks cropping out on the northern side of the UYTL in central Kyushu are composed of the Goshonoura, Mifune, Onogawa and Himenoura groups (Fig. 2e.20; Teraoka *et al.* 1998). The following descriptions are largely based on the reviews of Teraoka *et al.* (1998) and Kondo & Kikuchi (2010). The Goshonoura Group is composed of shallow-marine to non-marine delta deposits and has yielded dinosaur remains as well as abundant brackish molluscan fossils. Angular clasts of the Higo metamorphic complex are included in the lower part of the group, indicating an erosive unconformity between the two originally. The group consists of conglomerate, sandstone and mudstone, and is estimated to be late Albian–late Cenomanian in age. The Mifune Group is similarly shallow-marine to non-marine deltaic in origin, with its lower part yielding abundant marine molluscan fossils and the upper part yielding abundant brackish molluscs as well as dinosaur fossils in some places. The group, which unconformably covers the Permian Mizukoshi Formation and the Manotani metamorphic unit, comprises conglomerate, sandstone, mudstone and purple red mudstone with felsic tuff. The lower part of the group yields Cenomanian ammonite fossils and a fission track age of 83.6 $\pm$ 3.1 Ma was reported from the upper part of the group, indicating a sedimentation age for the group ranging from Cenomanian to Coniacian. The Onogawa Group (Teraoka 1970) unconformably covers the Asaji metamorphic complex (Fujii *et al.* 2008) and comprises marine conglomerate, sandstone and mudstone. The group has yielded abundant invertebrate fossils, such as ammonites, bivalves, gastropods and echinoids, and the sedimentation age is estimated to be Cenomanian–Santonian. The Himenoura Group and its equivalent unconformably cover the Higo metamorphic complex, Mifune and Goshonoura groups and

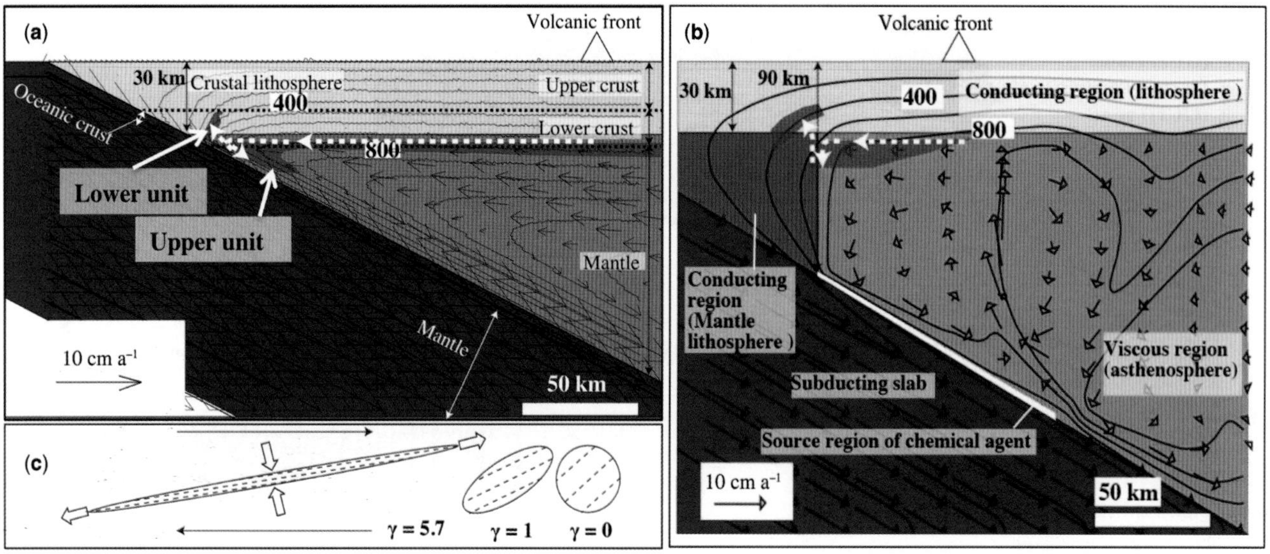

Fig. 2e.18. Thermal-advection models of a subduction zone (Miyazaki *et al.* 2013). (**a**) Steady-state model. (**b**) Non-steady-state model. Blue and red areas show *P*/*T* conditions of the Nagasaki metamorphic complex. White arrows represent possible trajectories of the upper unit. (**c**) Schematic diagram for thinning due to simple shear. Broken lines represent thermal contours. γ represents shear strain.

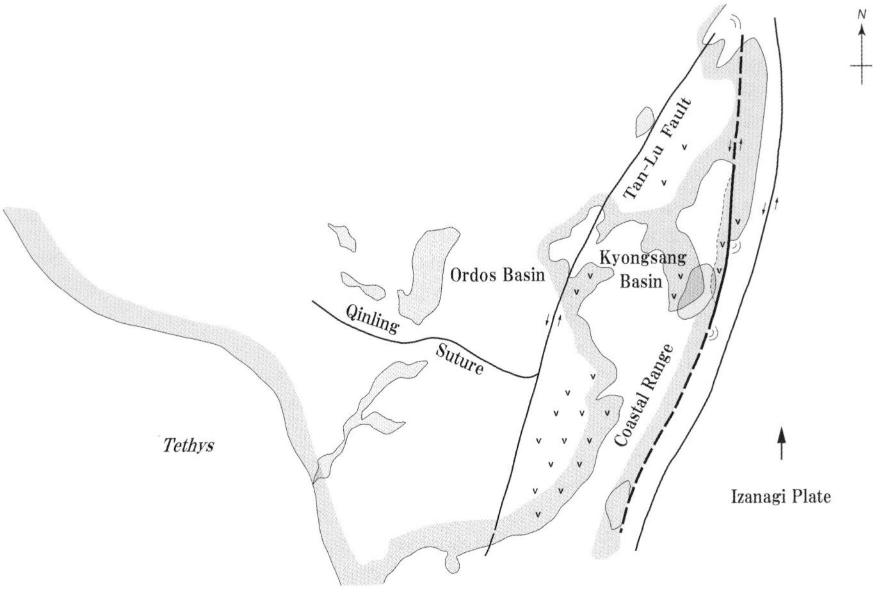

Fig. 2e.19. Palaeogeography of East Asia during the Early Cretaceous (Okada 2000). Light yellow, basins and rivers; v, igneous activity; double wave mark, point source of clastics. Light brown parts indicate ancient and modern coasts.

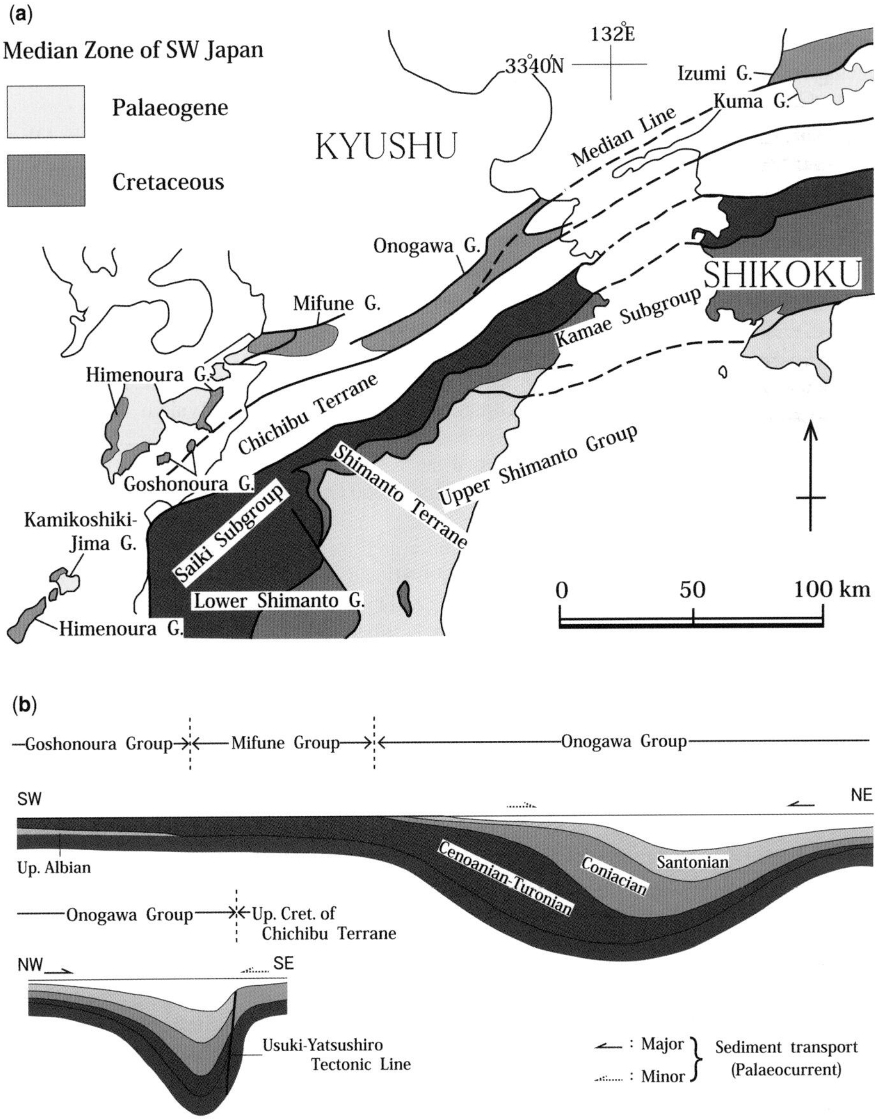

Fig. 2e.20. (**a**) Geological sketch map of middle Kyushu and west Shikoku showing the distribution of the Cretaceous and Palaeogene (from Teraoka *et al.* 1998). (**b**) Schematic cross-sections of the Goshonoura–Mifune–Onogawa sedimentary basin, showing the mode of sedimentation of the Cretaceous deposits in the Median Zone of SW Japan (from Teraoka 1970; Teraoka *et al.* 1998).

are dominated by marine turbidite, although non-marine deposits also occur. The group consists of sandstone, mudstone and alternations of sandstone and mudstone with felsic tuff, and has yielded ammonites, bivalves, gastropods, echinoids and marine and non-marine reptiles. On the basis of the fossils, the sedimentation age is estimated to be Santonian–Maastrichian. Given that a fission track age of 60.3 ± 2.9 Ma has been reported from felsic tuff, there is a possibility that the sedimentation age of the top of the group might extend up into the Paleocene epoch.

It is thought that the Goshonoura, Mifune and Onogawa groups were deposited in the same sedimentary basin, with Late Cretaceous sedimentation starting in the west (Fig. 2e.20). The Goshonoura and Mifune groups together represent shallow-marine to non-marine sediments deposited in the western part of the basin, where subsidence of the basement was relatively small. The Onogawa Group comprises marine sedimentary rocks deposited in the more rapidly subsiding eastern part of the basin. The basin has a half-graben structure that was formed by left-lateral strike-slip and down-to-dip-slip movements along basement faults (Teraoka *et al.* 1998). Conglomerate clasts in the Goshonoura, Mifune and Onogawa groups consist mainly of felsic to intermediate volcanic rocks and granitic rocks with sedimentary rocks, schist and gneiss. The proportion of volcanic rocks to granitic rocks increases over time, implying that development of the basin was associated with nearby volcanic activity. Metre-sized boulders of schist occur in the upper and middle part of the Onogawa Group but, rather than sourcing from the Sanbagawa metamorphic complex (currently cropping out adjacent to the Onogawa Group in the Saganoseki Peninsula), their phengite K–Ar ages imply derivation from the Suo metamorphic complex. This interpretation is supported by the fact that the middle part of the Onogawa Group is in fault contact with the Shonoharu metamorphic unit (correlated with the Suo metamorphic complex) along the UYTL. The existence of these coarse sedimentary deposits in the Onogawa Group strongly suggests that the Suo metamorphic complex formed a basement beneath the Onogawa Group. In addition, the compositions of detrital garnets from the Mifune and Onogawa groups implicate the Higo metamorphic complex as a sediment source. Overall, these lines of evidence and unconformable relationships suggest that the Higo and Suo metamorphic complexes (and thermal or high-temperature polymetamorphic equivalents of the Suo metamorphic complex) are likely to have formed the basement to the sedimentary basins in which the Goshonoura, Mifune and Onogawa groups were deposited.

Elsewhere in central Kyushu, Cretaceous non-accretionary sedimentary rocks crop out between the UYTL and BTL and comprise the Nakakyushu and Monobegawa groups (Fig. 2e.11). The Nakakyushu Group crops out in the west of this area, is subdivided into the Kawaguchi, Hachiryuzan, Kesado, Imaizumigawa and Yatsushiro formations, and consists of sandstone, mudstone, alternation of sandstone and mudstone, and conglomerate. These rocks are in fault contact with non-accretionary sedimentary rocks of the Kuma Formation (Permian) and the Ikenohara Formation (Jurassic). In the Hinagu district, the group unconformably covers Jurassic non-accretionary marine sediments of the Kurosaki Formation (Tamura & Murakami 1986; Tashiro *et al.* 1994). Bivalve fossils found within the group are similar to those from the Goshonoura and Mifune groups (Tanaka 2010), and the sedimentation age is estimated to be Valanginian–Middle Albian (Takahashi *et al.* 2003).

The Monobegawa Group in the Hinagu–Tomochi area in west Kyushu is composed of the Kohara, Mitsumineyama, Miyaji and Tomochi formations. The Kohara Formation forms the lower part of the Monobegawa Group in this district, but is in fault contact with surrounding strata. In contrast, the Tomochi Formation unconformably covers the Permian accretionary complex of the Hirodaira Unit (Saito & Toshimitsu 2003). The group consists mainly of conglomerate, sandstone and mudstone, and yields abundant molluscan fossils including ammonites which indicate (1) a Barremian sedimentation age for the Mitsumineyama Formation; and (2) a late Aptian–early Albian age for the Tomochi Formation. The sedimentation age of the group is estimated to be Hauterivian–early Albian.

Southern Kyushu

The Cretaceous non-accretionary sedimentary rocks on the southern side of the BTL tectonically overlie the Cretaceous accretionary complex (the Morotsuka Group) of the Shimanto Belt (Figs 2e.10 & 2e.11). On the Satsuma Peninsula, the Late Cretaceous Chiran Formation lies above the Early Cretaceous–early Late Cretaceous Takasakiyama Formation on a low-angle fault (Ministry of International Trade and Industry, Agency for Natural Resources and Energy 1985). The Chiran Formation consists only of clastic rocks (such as mudstone and sandstone), contains fossils (including ammonites: Matsumoto *et al.* 1973) and, unlike the Takasakiyama Formation, does not include basalt. These lines of evidence suggest that the Chiran Formation represents sediments that overlay the accretionary complex of the Takasakiyama Formation (Kawanabe *et al.* 2004), similar to the fossiliferous clastic sedimentary rocks of the Kukino Formation in the Hokusatsu area of SW Kyushu (Hayasaka 1999). Overall, these Cretaceous non-accretionary sedimentary rocks are correlative with the Uwajima Group in western Shikoku (Kawanabe *et al.* 2004; Saito *et al.* 2010).

Eocene–Early Oligocene sedimentary basin, accretionary complex and volcanic activity (MS & MO)

This section deals with the Palaeogene accretionary complex, forearc basin and volcanic arc system developed due to subduction of the Pacific Plate under the Eurasian Plate along the present Kyushu–Ryukyu Arc from Eocene to Early Oligocene times (Figs 2e.21 & 2e.22a). The volcanic front at that time was distributed from the San'in area of SW Honshu to offshore northern Kyushu (Figs 2e.22a & 2e.23a). Towards the west, the volcanic front in the Ryukyu Arc was shifted several hundred kilometres to the south (Fig. 2e.22a). This dislocation is due to extension of the Okinawa Trough on the SW side of Kyushu (latest Miocene–present day).

Middle Eocene–Early Oligocene sedimentary rocks, such as those of the Hyuga and Kitagawa Groups, crop out in the Shimanto Belt in southern Kyushu.

The Hyuga Group forms a Palaeogene accretionary complex tectonically underlying the Cretaceous Shimanto accretionary complex across the Nobeoka Thrust in SE Kyushu. The accretionary age of the Hyuga Group is estimated to be Middle Eocene–Early Oligocene (Nishi 1988). On the basis of detailed research of planktonic foraminifera fossils and geological mapping, the north part of the Hyuga Group is characterized by an imbricate structure where strata display remnants of ocean plate stratigraphy which is repeated by thrust faults (Nishi 1988; Kimura *et al.* 1991; Saito 2008; Hara *et al.* 2009). The thrust décollement surface lies within oceanic red mudstone (Fig. 2e.24), as is also seen in the thrust geometry of the Jurassic accretionary complex in Kyushu. The accretionary complex represented by the Hyuga Group formed over a very short duration in Middle Eocene times (Saito 2008), possibly in response to a change of subduction direction of the Pacific Plate (Fig. 2e.25) or a change of subducting plate from the Izanagi to the Pacific Plate. Surficial mudstones and limestones covering the Hyuga Group are also involved in the fault imbricate structure (Saito *et al.* 2007; Saito 2008). On the basis of microfossils Early Oligocene fossils were reported from the group in central Kyushu (Nishi 1988),

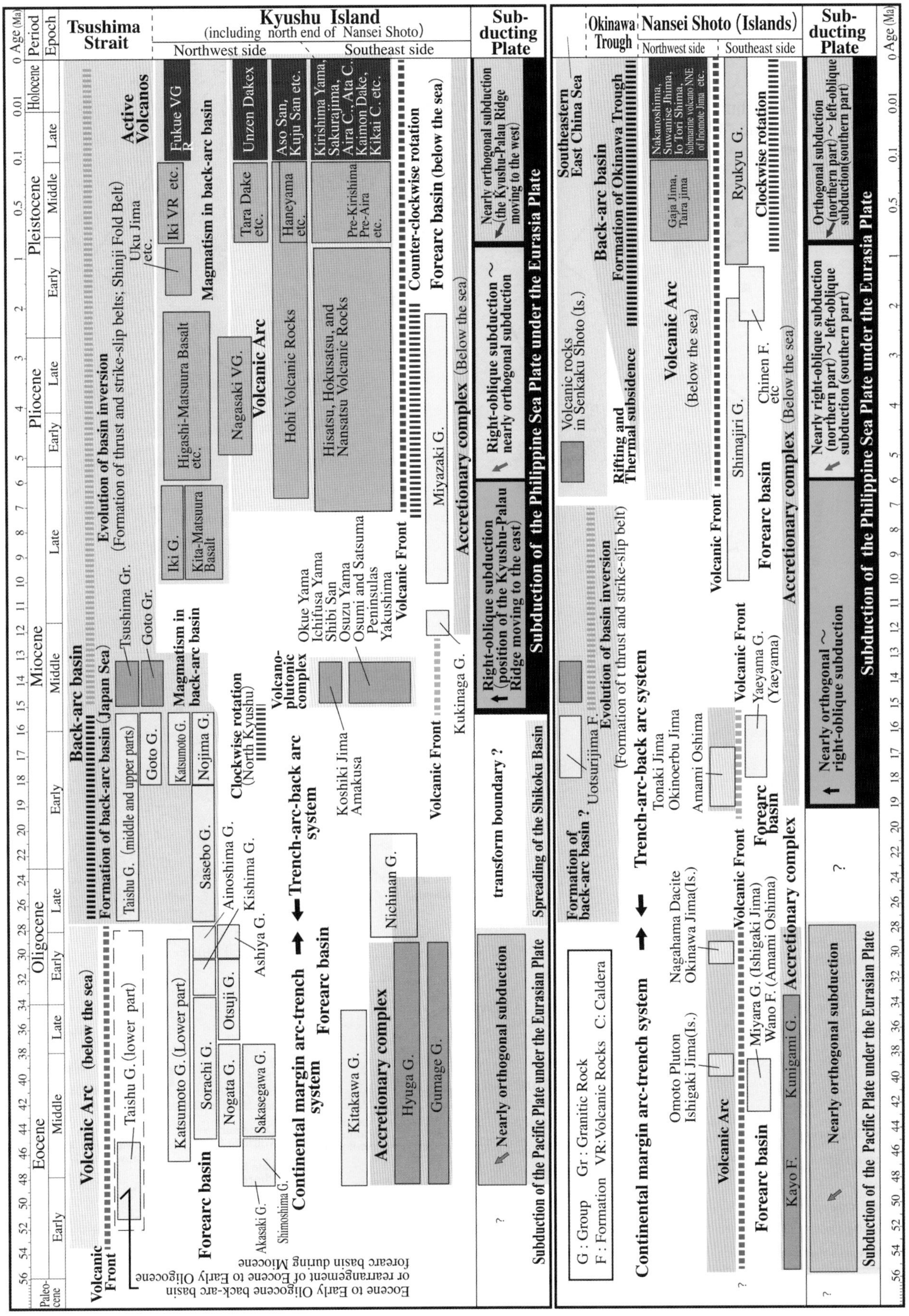

Fig. 2e.21. Summary of the geology of the Kyushu–Ryukyu Arc during Eocene–Quaternary time.

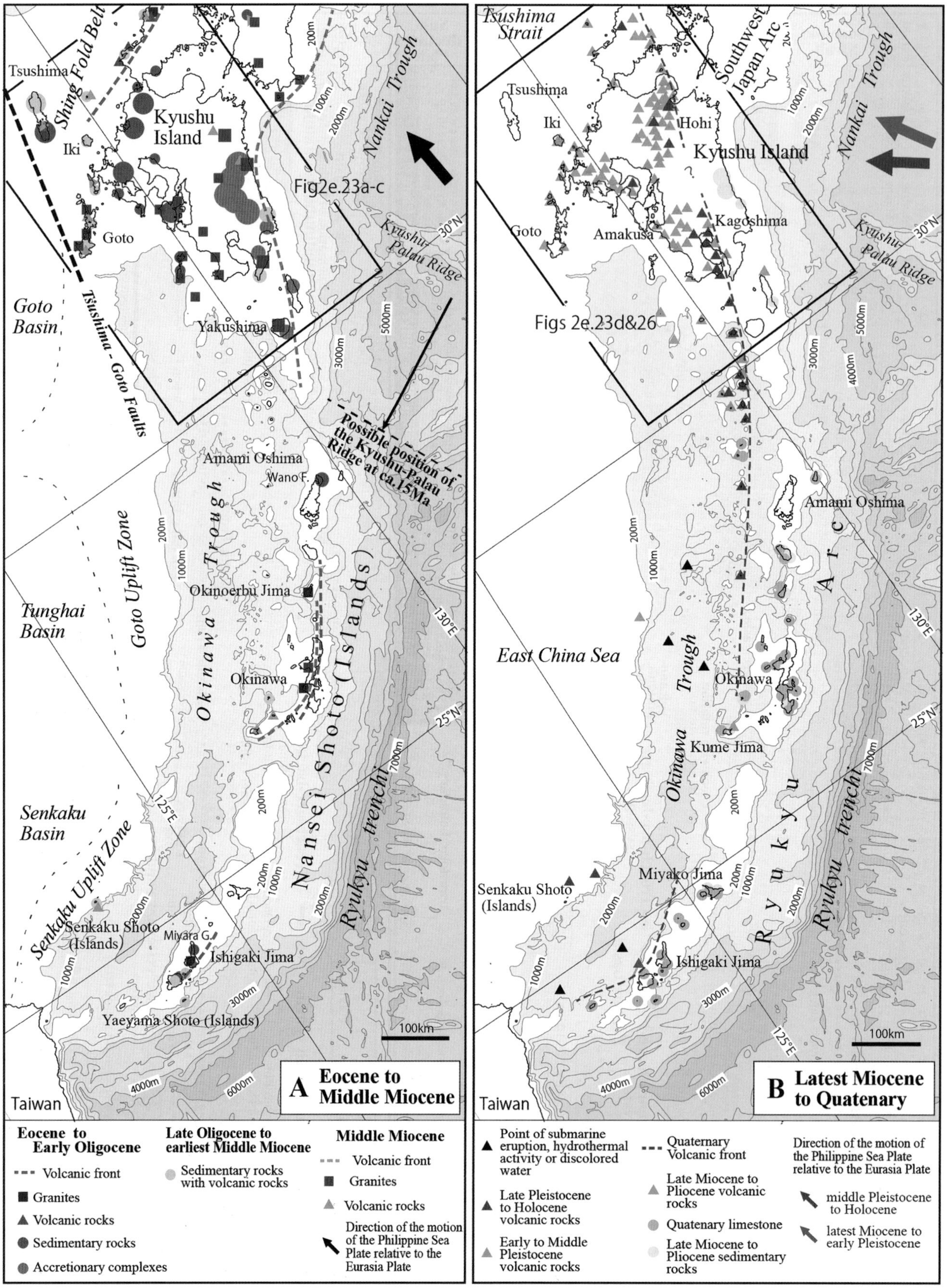

Fig. 2e.22. Distribution of Eocene–Quaternary units in the Kyushu–Ryukyu Arc. Based on the seamless digital geological map of Japan (1:200 000) (Geological Survey of Japan; https://gbank.gsj.jp/seamless/index_en.html) and *Volcanoes of Japan* (Nakano *et al.* 2013).

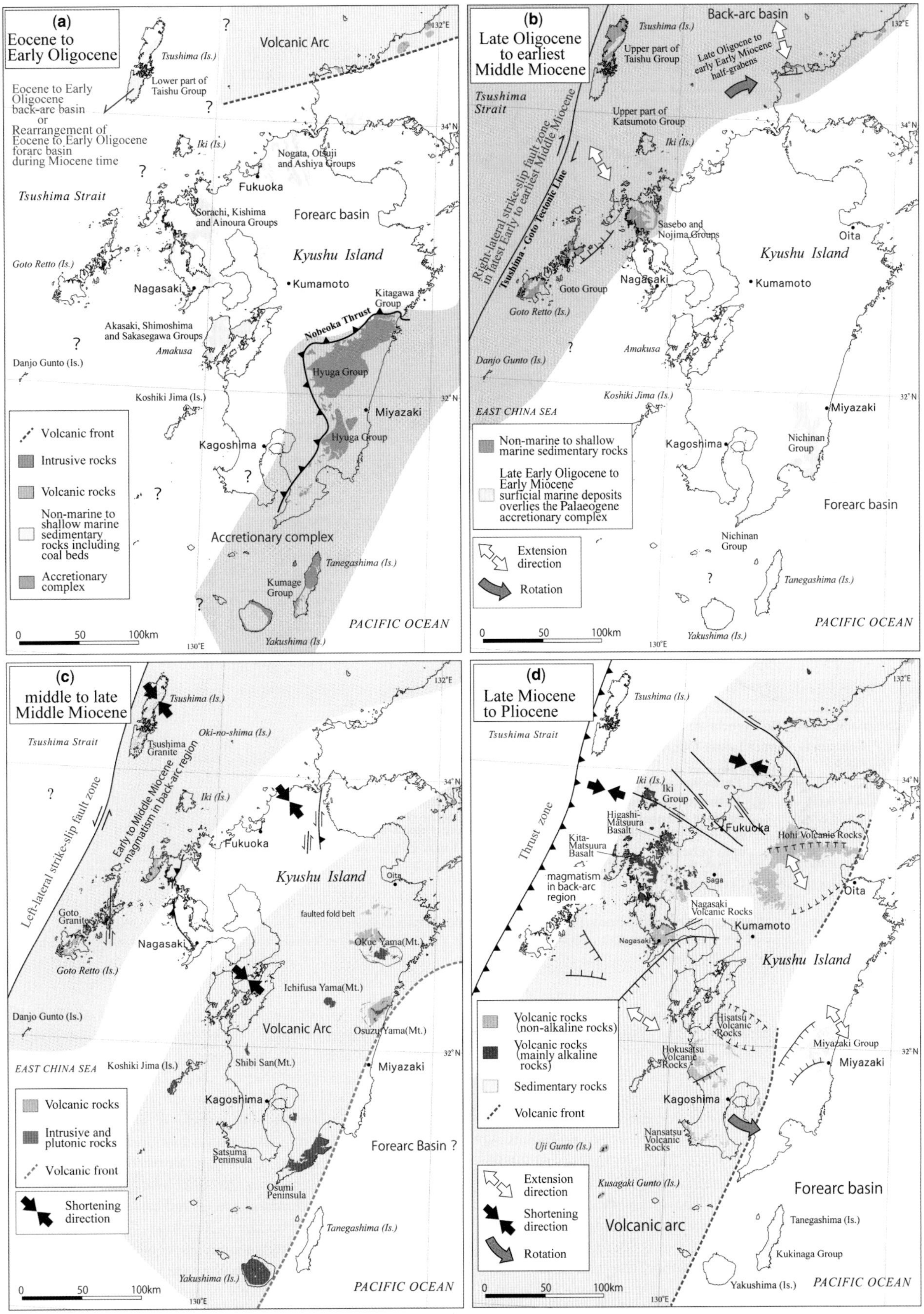

Fig. 2e.23. Distribution and tectonic setting of the Eocene–Pliocene units in Kyushu Island and surrounding area. Based on the seamless digital geological map of Japan (1:200 000) (Geological Survey of Japan; https://gbank.gsj.jp/seamless/index_en.html) and *Volcanoes of Japan* (Nakano *et al.* 2013).

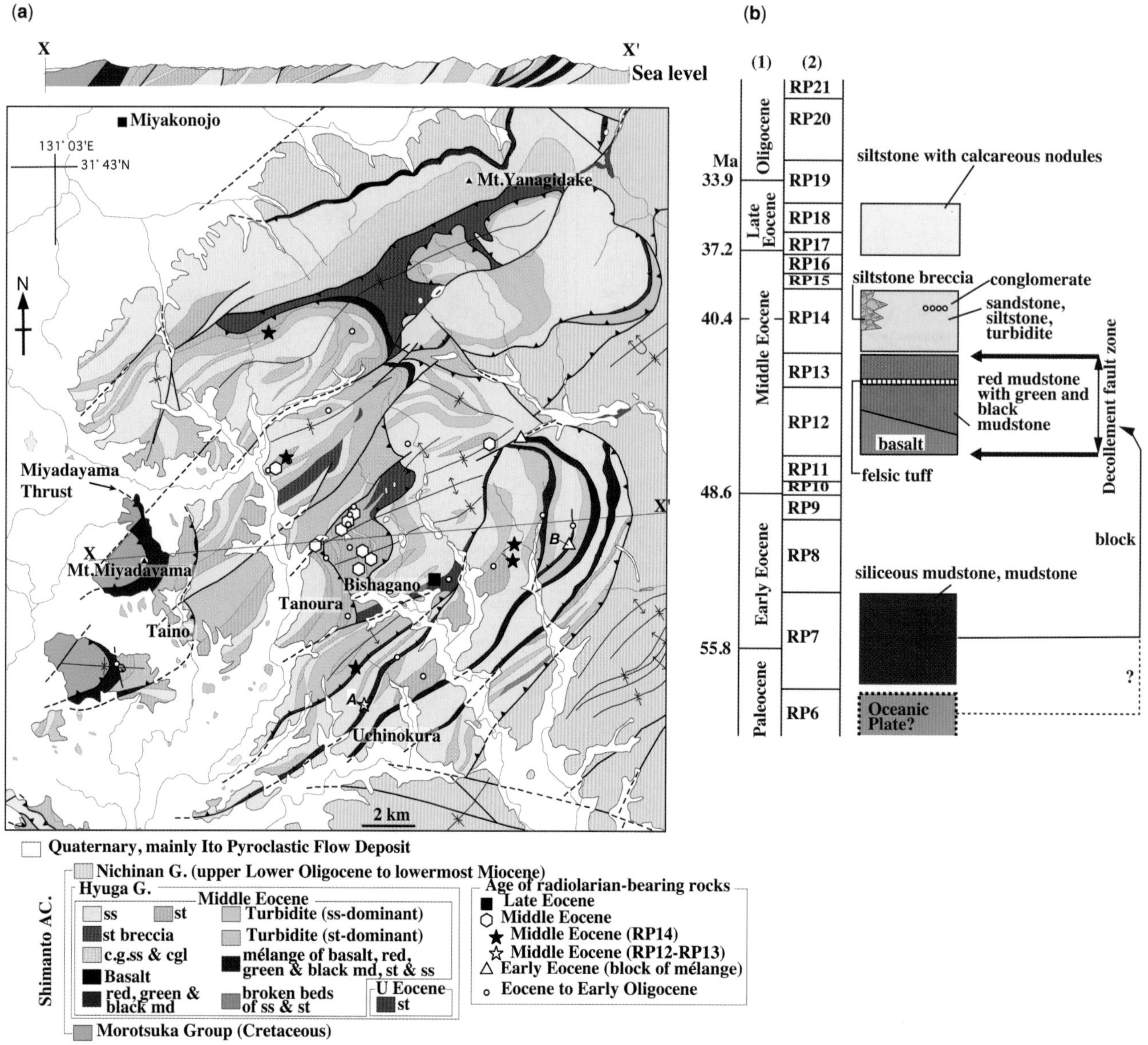

Fig. 2e.24. (**a**) Geological map and cross-section of Eocene accretionary complex (Hyuga Group) in southeastern Kyushu (Saito 2008). (**b**) Reconstructed stratigraphy of the Hyuga Group (Saito 2008). (1) Time scale, (2) radiolarian zones from Sanfilippo & Nigrini (1998). ss, sandstone; st, siltstone; c.g., coarse grained; cgl, conglomerate.

although the accretion age of the Hyuga Group in SE Kyushu has been estimated to be Middle Eocene (Saito *et al.* 1994; Saito 2008). The relationship between the Early Oligocene strata and the Middle Eocene accretionary complex is obscure. The Hyuga Group outcrop extends to the islands of Yakushima and Tanegashima. Towards the south, the Kayo Formation in Okinawa-jima has yielded large Eocene foraminifera and possibly represents an Eocene accretionary complex (Ujiie 2002; Nakae *et al.* 2010), although detailed correlation with the Palaeogene accretionary complex in Kyushu is still required. The group in central Kyushu has undergone sub-greenschist facies metamorphism (Imai *et al.* 1971), although details are lacking.

The Kitagawa Group is sandwiched between (and in fault contact with) the underlying Hyuga Group and the overlying Morotsuka Group (Murata 1998). The Nobeoka Thrust forms a boundary fault between the Kitagawa and Hyuga groups in SE Kyushu, with slickensides on the fault surface indicating a top-to-the-SE (Saito *et al.* 1996) or a top-to-the-SSE (Kondo *et al.* 2005; Kimura *et al.* 2013) slip direction. Based on K–Ar ages of illite, zircon fission-track ages and the youngest depositional age, Hara & Kimura (2008) suggest that the thrusting occurred during Middle Eocene times (48–40 Ma); Eocene radiolarian fossils have been reported from the group (Ogawauchi *et al.* 1984). Although the group has been interpreted as non-accretionary sediments covering the accretionary complex (Geological Survey of Japan 1992), details of its exact geological setting are not known. The group experienced greenschist facies metamorphism, as indicated by temperatures estimated from vitrinite reflectance (Kondo *et al.* 2005).

Palaeogene forearc sedimentary rocks crop out in central-west to north Kyushu and comprise shallow-marine to fluvial deposits with coal beds that were extensively mined until the 1960s. These strata unconformably overlie the Cretaceous Himenoura Group (partly Paleocene; Yoshida *et al.* 1985), with deposition beginning in Middle Eocene times (in the Amakusa area) and probably linked to the

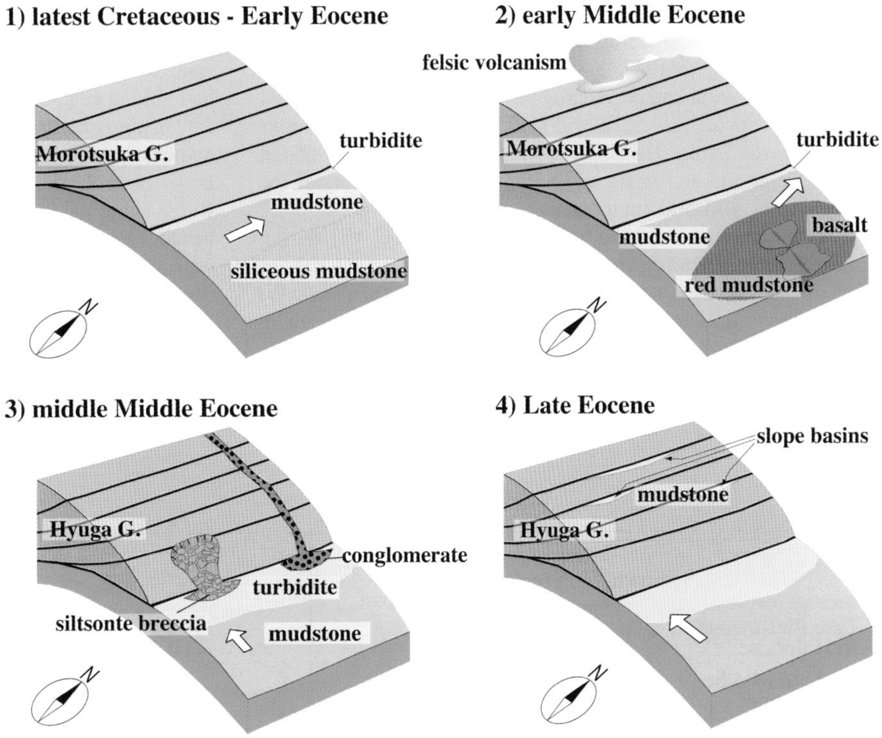

Fig. 2e.25. A model of the evolution of the Hyuga Group (Saito 2008).

development of the broadly coeval Hyuga Group (accretionary complex) in southern Kyushu.

Stratal equivalents to the NW Kyushu forearc sedimentary rocks are found on Ishigaki-jima and Amami Oshima in the southern Ryukyu Arc. In contrast, almost all of the sedimentary rocks along the eastern side of Kyushu and in the eastern forearc basin region of Kyushu (SW Japan Arc) were eroded out. This is likely due to differences in tectonic setting, with the SW Japan Arc having undergone rapid uplift resulting from the subduction of the young, hot and buoyant crust of the Shikoku Basin after *c.* 15 Ma. In addition, Eocene sedimentary rocks accumulated in a back-arc environment along the Okinawa Trough in the southeastern East China Sea, and crop out in the Goto, Tunghai and Senkaku basins (Kizaki 1986; Fig. 2e.22a).

Palaeogene sedimentary rocks also crop out in the Tsushima archipelago between Kyushu and the Korean Peninsula, where they are known as the Taishu Group and are estimated to be Early–early Middle Eocene in age based on radiolarian (Nakajo & Funakawa 1996) and radiometric ages (Sakai & Yuasa 1998). In contrast, the upper part of this group has yielded Miocene foraminifera (Ibaragi 1994) and Early Miocene molluscan fossils. This sedimentation gap implies that the Early Miocene back-arc basin overlapped the Eocene basin, with the Taishu Group being deposited in the intra-arc basin (Sakai 1993). However, given that strike-slip faulting between the Tsushima archipelago and Kyushu also occurred during Miocene time, the possibility that the Eocene Tsushima basin was linked continuously to the Eocene forearc basin in northern Kyushu Island cannot be discarded.

Late Oligocene–Middle Miocene sedimentary basin, accretionary complex and magmatism (MO & MS)

The Sea of Japan was formed in Late Oligocene–middle Middle Miocene times (*c.* 27–15 Ma; e.g. Tamaki *et al.* 1992; Figs 2e.21 & 2e.23b). Opening of the Shikoku Basin (the northeastern part of the Philippine Sea Plate) took place during the same period (Okino *et al.* 1994, 1999). Subsequently, the subduction of the Shikoku Basin beneath SW Japan started at *c.* 15 Ma. At the same time, the tectonic regime in the island-arc and back-arc basin switched to compressional (e.g. Yamamoto 1991), and igneous activity in the Pacific Ocean coastal (previously forearc) areas became dramatically active (Fig. 2e.23c).

Upper Oligocene–uppermost Middle Miocene sediments were continuously deposited in back-arc basins such as in northwestern Kyushu, Tsushima and the Goto islands (Fig. 2e.23b). These basins are considered to have formed during transtensional right-lateral strike-slip movements along a major NNE–SSW-trending fault system (Tsushima–Goto Tectonic Line) running between the Tsushima archipelago and the Korean Peninsula until *c.* 15 Ma (e.g. Yoon & Chough 1995; Fabbri *et al.* 1996). Marine sediments, such as those represented by the Nojima Group (Komatsubara *et al.* 2005), were deposited widely in this region from late Early Miocene to earliest Middle Miocene times due to globally high sea level and rapid rifting. These transgressive deposits in the forearc basin also crop out at Yaeyama islands in the southern Ryukyu Arc.

In SE Kyushu, facing the open Pacific Ocean, there are extensive outcrops of Early Oligocene–earliest Middle Miocene sedimentary rocks belonging to the Nichinan Group (e.g. Sakai *et al.* 1987). Low-angle thrust faulting has been recognized between the Hyuga and Nichinan groups (Saito *et al.* 1997), with the latter group forming the thrust footwall. The Nichinan Group near the thrust consists of mixed rock or broken formation of alternating sandstone and mudstone, and may be viewed as an accretionary complex. In other places, the group contains many large blocks up to a few kilometres in size with fold axial directions in each block bearing no relation to each other, suggesting important block rotations. Some of these mixed blocks include Eocene rocks correlated with the Hyuga Group. The Nichinan Group yields molluscan fossils which can be correlated with the Oligocene Asia Fauna (Shuto 1962; Saito *et al.* 1994). It is not clear whether these sediments are accretionary complex in origin or not, and they have been referred to as 'the Misaki Olistostrome' (Kanmera 1977). The Nichinan Group is therefore interpreted as composed mainly of submarine landslide deposits (Sakai *et al.* 1987; Sakai 1988*b*), mixing blocks derived from diverse sedimentary environments ranging from shallow marine to deep-sea fan (Sakai 1988*a*). Similar sedimentary rocks are recognized in the Nabae Group in the southern parts of the Muroto Peninsula and strata exposed in the southern part of the Kii Peninsula, 300–500 km to the east.

It is thought that SW Japan rotated about 40° clockwise during *c.* 17–15 Ma (e.g. Otofuji *et al.* 1991). The Palaeogene outcrops of northern and western Kyushu Island similarly rotated clockwise (36°) relative to eastern Asia, whereas the Miocene exposures in the Goto and Tsushima islands are known to have rotated slightly anticlockwise (Ishikawa 1997). This implies that the Tsushima and Goto islands were not involved in the regional rotations induced by the opening of the Sea of Japan.

Igneous activity occurred in the forearc region of SW Japan after *c.* 15 Ma. In comparison with the position of Eocene–Early Oligocene volcanism, the Kyushu–Ryukyu volcanic front at 15–12 Ma had shifted *c.* 250 km towards the Pacific Ocean (Figs 2e.22a & 2e.23d). It is believed that this shift in position was caused by subduction of the young, hot Philippine Sea Plate beneath the Shikoku Basin (e.g. Kimura *et al.* 2005). The present-day Kyushu–Palau Ridge along the western edge of the Shikoku Basin is subducting beneath the southeastern part of Kyushu. Because a Middle Miocene granite is recognized at Yakushima (Fig. 2e.23c), and the Kyushu–Palau ridge has moved eastwards due to subduction of the Philippine Sea Plate towards the north during 15–7 Ma, the ridge is estimated to have been located on the west side of Yakushima at 15 Ma (Fig. 2e.22a).

Intrusions of granitic rocks occurred in the back-arc of this subduction system around 15 Ma, as recorded by the plutons exposed in

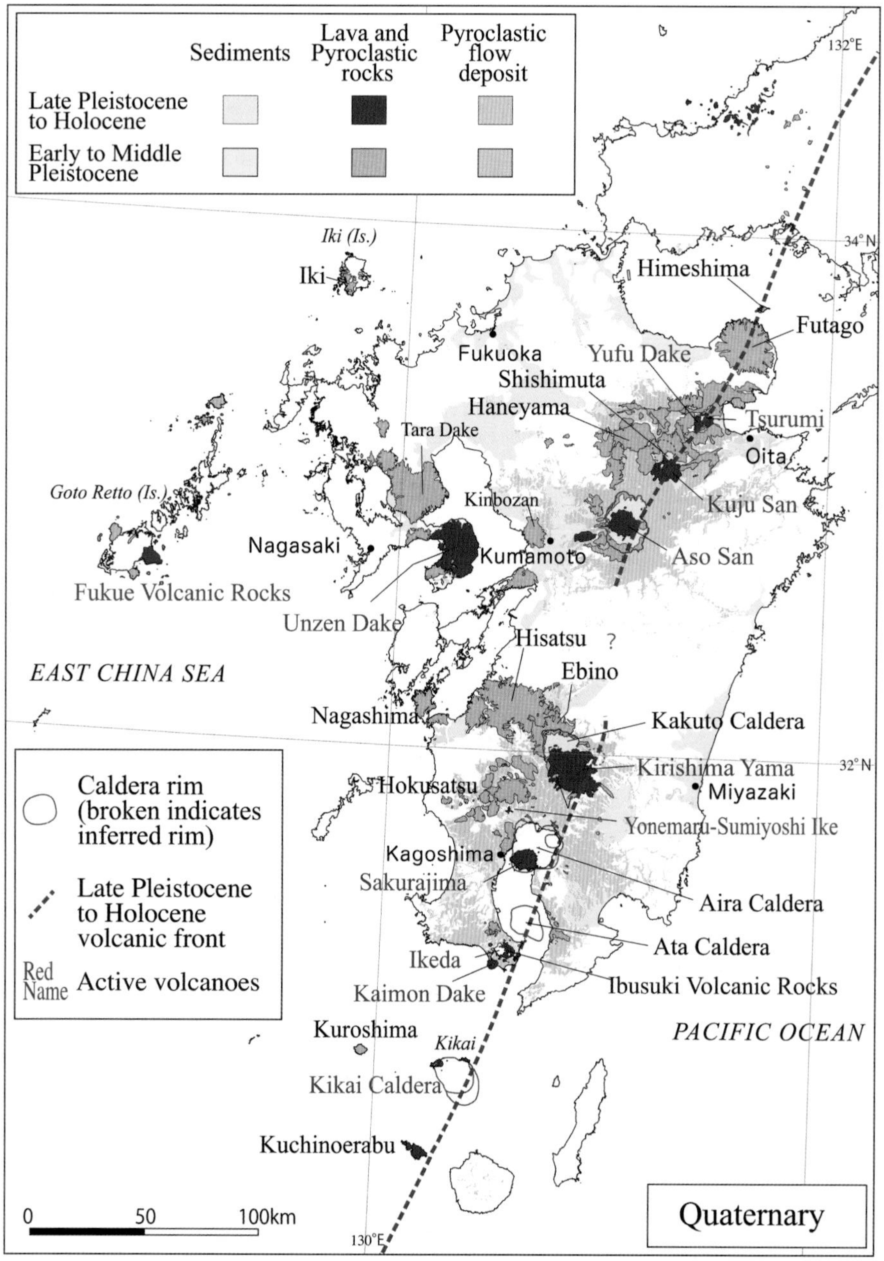

Fig. 2e.26. Distribution and tectonic setting of Quaternary units in Kyushu Island and surrounding area. Based on the seamless digital geological map of Japan (1:200 000) (Geological Survey of Japan; https://gbank.gsj.jp/seamless/index_en.html) and *Volcanoes of Japan* (Nakano et al. 2013).

the Tsushima and Goto islands and NW Kyushu. Similar igneous activity involving intrusions of granitic rocks is recognized across a wide area of the southern margin of the Sea of Japan (Figs 2e.22a & 2e.23c). This activity coincides with the development of fold and strike-slip belts along the southern margin of the back-arc region, such as in the offshore area of San'in, between Kyushu and Tsushima, around the Goto islands and along the continental shelf in the northwestern margin of the Okinawa Trough. These belts are referred to as the Shinji Fold Belt (Otsuka 1939), Goto Uplift Zone and Senkaku Uplift Zone (Kizaki 1986). Finally, Pliocene sediments cover the Palaeogene strata on the continental shelf at the NW margin of the Okinawa Trough, with Miocene sediments absent from many places. This basin-inversion in northern Kyushu, Tsushima and Goto, caused by compression due to the subduction of the Philippine Sea plate, started *c.* 15 Ma ago and continues through to the present day.

Late Miocene–Quaternary sedimentary basin and volcanic activity (MO)

After the cessation of Middle Miocene igneous activity along the Pacific coastal margin, Late Miocene–Pleistocene forearc sediments of mainly shallow marine origin (such as the Kukinaga, Miyazaki and Shimajiri groups) began to be deposited (e.g. Kizaki 1986; Figs 2e.22b & 2e.23d).

The Okinawa Trough and the Ryukyu Arc started to form in Late Miocene times, probably with rifting inducing volcanism in central and SW Kyushu (Hohi and western Kagoshima areas) and along the eastern margin of the Okinawa Trough in the Ryukyu Arc at the same time (Kamata 1989; Nagao *et al.* 1999). Late Miocene–Pliocene volcanic rocks in the Senkaku islands are also thought to record this initial volcanism associated with the formation of the back-arc basin.

Voluminous alkali and tholeitic basalt volcanism are widely recognized across NW Kyushu, the Goto islands and Iki Island during Late Miocene–Quaternary times. This magmatism is assumed to have derived from mantle diapiric upwelling in the back-arc basin in the extensional back-arc related to the formation of the Okinawa Trough (e.g. Nakada *et al.* 1997; Figs 2e.23d & 2e.26).

Since Late Miocene times, southern Kyushu has been subjected to north–south extension so that normal faults have developed here and further south on the neighbouring island of Tanegashima. Under this new tectonic regime, the Miyazaki Plain–Miyakonojo Basin was formed as a graben, and the sedimentary depocentre of the Miyazaki Group shifted northwards into the Miyazaki Plain–Miyakonojo Basin. In contrast, in northern Kyushu, NW–SE to east–west compression has continued, producing left-lateral strike-slip and reverse faults from Late Miocene times to the present day.

The subduction direction of the Philippine Sea Plate changed from north to NW at 6 Ma and from NNW to WNW at 2–1 Ma (e.g. Seno & Maruyama 1984; Yamazaki & Okamura 1989; Kamata & Kodama 1999; Fig. 2e.22b). The opening of the Okinawa Trough has accelerated as a result, with major Quaternary subsidence. In addition, anticlockwise rotation (*c.* 27°) in southern Kyushu and the northernmost Ryukyu Arc (Kamata & Kodama 1994) and clockwise rotation (*c.* 19°) in the southern Ryukyu Arc (Miki *et al.* 1990) have taken place, along with strike-slip faulting throughout the southern Kyushu and Ryukyu arcs.

Since latest Miocene–Quaternary times, the location of the volcanic front has not varied greatly (Figs 2e.22b, 2e.23d & 2e.26). The western Ryukyu Arc, where the left-oblique subduction of the Philippine Sea Plate is clear, is the most tectonically active area in the Okinawa Trough where thinning of continental crust is taking place.

Quaternary volcanoes in central Kyushu and the northern Ryukyu arc are characterized by large-scale calderas forming huge pyroclastic eruptions, for example Aso, Aira, Ata, Kikai and Shishimuta (Fig. 2e.26). This intense volcanic activity is the result of anomalously high heat flow during the formation of rhomboidal basins in central Kyushu by dextral motion along the Median Tectonic Line, combined with crustal rotation and incipient rifting in south Kyushu and the northern Okinawa Trough (Kamata & Kodama 1999). The Unzen volcano is a special case, being offset west of the main volcanic front located at the northeastern end of the Okinawa Trough. Seismic tomographic modelling in Kyushu has revealed that the Philippine Sea Plate slab has reached a position beneath Unzen so that hot upwelling materials related both to the opening of the Okinawa Trough and the dehydration processes of subducting Philippine Sea Plate slab are currently combining to induce the volcanism (Wang & Zhao 2006).

Since latest Early Pleistocene times coral-reef limestone (the Ryukyu Group) has formed on islands of the southern and central Ryukyu Arc, as the supply of clastic materials decreased due to the Ryukyu Arc separating from the continent as the Okinawa Trough formed (e.g. Nakagawa 1977).

Finally, not much is known about the development of accretionary complex materials during Late Miocene–Quaternary times, because there is no exposure of them on land. The distance between trench and forearc in the Ryukyu Arc is shorter than that in the SW Japan Arc due to the high-angle of the Philippine Sea Plate subduction, producing a generally slower uplift rate of the Ryukyu Arc (except for some areas such as on Kikai-jima).

We thank H. Hoshizumi and Y. Kawanabe for their constructive suggestions.

Appendix

English to Kanji and Hiragana translations for geological and place names

Aira	姶良	あいら
Akiyoshi	秋吉	あきよし
Amakusa	天草	あまくさ
Amakusa Kamishima	天草上島	あまくさかみしま
Amakusa Shimoshima	天草下島	あまくさしもしま
Amami Oshima	奄美大島	あまみおおしま
Ariake	有明	ありあけ
Asaji	朝地	あさじ
Asakura	朝倉	あさくら
Ashikita	芦北	あしきた
Aso	阿蘇	あそ
Ata	阿多	あた
Bisho	美生	びしょう
Butsuzo	仏像	ぶつぞう
Chichibu	秩父	ちちぶ
Chikushi	筑紫	ちくし
Chiran	知覧	ちらん
Chokai	朝海	ちょうかい
Choshigasa	銚子笠	ちょうしがさ
Chugoku	中国	ちゅうごく
Fudono	不土野	ふどの
Fukahori (Fukabori)	深堀	ふかほり
Fukuoka	福岡	ふくおか
Funaoyama	船尾山	ふなおやま
Fusaki	富崎	ふさき
Gionyama	祇園山	ぎおんやま
Gokanosho	五家荘	ごかのしょう

(Continued)

English to Kanji and Hiragana translations for geological and place names (Continued)

Goshonoura	御所浦	ごしょのうら	
Goto	五島	ごとう	
Hachiryuzan	八竜山	はちりゅうざん	
Haki	葉木	はき	
Hashirimizu	走水	はしりみず	
Hatashima	畑島	はたしま	
Higo	肥後	ひご	
Hikata	日方	ひかた	
Hikawa	氷川	ひかわ	
Himenoura	姫浦	ひめのうら	
Hinagu	日奈久	ひなぐ	
Hiraodai	平尾台	ひらおだい	
Hirodaira	広平	ひろだいら	
Hohi	豊肥	ほうひ	
Hokusatsu	北薩	ほくさつ	
Hongo	本郷	ほんごう	
Honshu	本州	ほんしゅう	
Horagadake	洞が岳	ほらがだけ	
Hyuga	日向	ひゅうが	
Ie	伊江	いえ	
Iheya	伊平屋	いへや	
Ikenohara	池の原	いけのはら	
Iki	壱岐	いき	
Imaizumigawa	今泉川	いまいずみがわ	
Ino	伊野	いの	
Iriomote-jima	西表島	いりおもてじま	
Ishigaki-jima	石垣島	いしがきじま	
Itsuki	五木	いつき	
Kabashima	樺島	かばしま	
Kagoshima	鹿児島	かごしま	
Kakisako	柿迫	かきさこ	
Kakuto	加久藤	かくとう	
Kamae	蒲江	かまえ	
Kamiaso	上麻生	かみあそう	
Kamishiiba	上椎葉	かみしいば	
Kamiwashidani	上鷲谷	かみわしだに	
Kanmon	関門	かんもん	
Kanto	関東	かんとう	
Kawaguchi	川口	かわぐち	
Kawamata	河俣	かわまた	
Kawaradake	香春岳	かわらだけ	
Kayo	嘉陽	かよう	
Kesado	袈裟堂	けさどう	
Kii	紀伊	きい	
Kikai	鬼界	きかい	
Kikai-jima	喜界島	きかいじま	
Kiku	企救	きく	
Kinshozan	金勝山	きんしょうざん	
Kitagawa	北川	きたがわ	
Kitataku	北多久	きたたく	
Kiyama	木山	きやま	
Kohara	小原	こはら	
Kokura	小倉	こくら	
Konoha	木の葉	このは	
Konose	神瀬	こうのせ	
Kozaki	小崎	こざき	
Kuju	九重	くじゅう	For 'Kuju Mountains and Kuju area'
Kuju	久住	くじゅう	For 'Mt. Kuju'
Kukinaga	茎永	くきなが	
Kukino	久木野	くきの	
Kuma	球磨	くま	
Kunisaki	国東	くにさき	
Kuraoka	鞍岡	くらおか	
Kuriki	栗木	くりき	
Kurosaki	黒崎	くろさき	

(*Continued*)

English to Kanji and Hiragana translations for geological and place names (Continued)

Kurosegawa	黒瀬川	くろせがわ
Kurotani	黒谷	くろたに
Kurume	久留米	くるめ
Kyushu	九州	きゅうしゅう
Makimine	槙峰	まきみね
Manotani	間の谷	まのたに
Mifune	御船	みふね
Mikado	神門	みかど
Mino	美濃	みの
Misaki	岬	みさき
Misumi	三角	みすみ
Mitaki	三滝	みたき
Mitsumineyama	三峰山	みつみねやま
Miyaji	宮地	みやじ
Miyakonojo	都城	みやこのじょう
Miyama	深山	みやま
Miyamadani	深山谷	みやまだに
Miyazaki	宮崎	みやざき
Mizukoshi	水越	みずこし
Mogi	茂木	もぎ
Momigi	樅木	もみぎ
Monobegawa	物部川	ものべがわ
Morotsuka	諸塚	もろつか
Motobu	本部	もとぶ
Muroto	室戸	むろと
Nabae	難波江	なばえ
Nagasaki	長崎	ながさき
Nago	名護	なご
Naidaijin	内大臣	ないだいじん
Nakakyushuu	中九州	なかきゅうしゅう
Nekodake	根子岳	ねこだけ
Nichinan	日南	にちなん
Nioki	荷尾杵	におき
Nishikawauchi	西川内	にしかわうち
Nishiki	錦	にしき
Nishinoiwa	西の岩	にしのいわ
Nishisonogi	西彼杵	にしそのぎ
Nitao	仁田尾	にたお
Nobeoka	延岡	のべおか
Nojima	野島	のじま
Nokonoshima	能古島	のこのしま
Nomo	野母	のも
Okinawa	沖縄	おきなわ
Okinawa-jima	沖縄島	おきなわじま
Okinoerabu-jima	沖永良部島	おきのえらぶじま
Omae	尾前	おまえ
Omurawan	大村湾	おおむらわん
Omuta	大牟田	おおむた
Onogawa	大野川	おおのがわ
Otao	小田尾	おたお
Renge	蓮華	れんげ
Ryoke	領家	りょうけ
Ryuhozan	竜峰山	りゅうほうざん
Ryukyu	琉球	りゅうきゅう
Saganoseki	佐賀関	さがのせき
Saiki	佐伯	さいき
San'in	山陰	さんいん
Sanbagawa (Sambagawa)	三波川	さんばがわ
Sangun	三郡	さんぐん
Sanpodake (Sampodake)	三方岳	さんぼうだけ
Sasaguri	篠栗	ささぐり
Sashu	佐志生	さしう
Satsuma	薩摩	さつま
Sefuri	脊振	せふり
Senkaku	尖閣	せんかく
Shiiba	椎葉	しいば
Shiibamura	椎葉村	しいばむら

(*Continued*)

English to Kanji and Hiragana translations for geological and place names (Continued)

Shikanoshima	志賀島	しかのしま
Shikoku	四国	しこく
Shimajiri	島尻	しまじり
Shimanto	四万十	しまんと
Shimonoseki	下関	しものせき
Shinji	宍道	しんじ
Shiraishino	白石野	しらいしの
Shiroyama	城山	しろやま
Shishimuta	猪牟田	ししむた
Shonoharu	庄ノ原	しょうのはる
Suo	周防	すおう
Tagawa	田川	たがわ
Taishu	対州	たいしゅう
Takadake	高岳	たかだけ
Takasakiyama	高崎山	たかさきやま
Tamana	玉名	たまな
Tanba (Tamba)	丹波	たんば
Tanegashima	種子島	たねがしま
Tano	田野	たの
Terano	寺野	てらの
Tokunoshima	徳之島	とくのしま
Tomochi	砥用	ともち
Tomuru	トムル	とむる
Tsukabaru	塚原	つかばる
Tsukumi	津久見	つくみ
Tsurukoba (Tsurunokoba)	鶴木場	つるこば
Tsushima	対馬	つしま
Unzen	雲仙	うんぜん
Usuki	臼杵	うすき
Usukigawa	臼杵川	うすきがわ
Uwajima	宇和島	うわじま
Wakimisaki	脇岬	わきみさき
Wakino	脇野	わきの
Yaeyama-shoto	八重山諸島	やえやましょとう
Yahata	八幡	やはた
Yakushima	屋久島	やくしま
Yamaga	山鹿	やまが
Yamaide	山出	やまいで
Yame	八女	やめ
Yanai	柳井	やない
Yato	八戸	やと
Yatsushiro	八代	やつしろ
Yobikonoseto	呼子ノ瀬戸	よびこのせと
Yobuno	呼野	よぶの
Yonagu	与奈久	よなぐ
Yoronto	与論島	よろんとう
Youra	四浦	ようら
Yuwan	湯湾	ゆわん
Yuzuruha	湯鶴葉	ゆづるは

References

Adachi, T., Osanai, Y., Nakano, N. & Owada, M. 2012. LA-ICP-MS zircon and FE-EPMA U-Th-Pb monazite dating of pelitic granulites from the Mt. Ukidake area, Sefuri Mountains, northern Kyushu. *Journal of the Geological Society of Japan*, **118**, 39–52.

Arima, K., Ikeda, T. & Miyazaki, K. 2011. Reaction microstructures in corundum and kyanite bearing mafic mylonites from the Takahama metamorphic rocks, western Kyushu, Southwest Japan. *Island Arc*, **20**, 248–258.

Chough, S. K. & Sohn, Y. K. 2010. Tectonic and sedimentary evolution of a Createceous continental arc-backarc system in the Korean peninsula: new view. *Earth-Science Reviews*, **101**, 225–249.

Dunkley, D. J., Suzuki, K., Hokada, T. & Kusiak, M. A. 2008. Contrasting ages between isotopic chronometers in granulites: monazite dating and metamorphism in the Higo Complex, Japan. *Gondwana Research*, **14**, 624–643.

Fabbri, O., Charvet, J. & Fournier, M. 1996. Alternate senses of displacement along the Tsushima fault system during the Neogene based on fracture analyses near the western margin of the Japan Sea. *Tectonophysics*, **257**, 257–295.

Fabbri, O., Tokushige, H. & Hayamizu, M. 1997. Normal faulting in the Middle Miocene Osumi granodioritic pluton, southern Kyushu, and its significance. *Journal of the Geological Society of Japan*, **103**, 141–151.

Faure, M., Fabbri, O. & Monie, P. 1988*a*. The Miocene bending of southwest Japan: new $^{39}Ar/^{40}Ar$ and microtectonic constraints from the Nagasaki schist (western Kyushu), an extension of the Sanbagawa high-pressure belt. *Earth and Planetary Science Letters*, **91**, 13–31.

Faure, M., Monine, P. & Fabbri, O. 1988*b*. Microtectonics and $^{40}Ar/^{39}Ar$ datings of high pressure metamorphic rocks in the south Ryukyu Arc and their bearings on the pre-Eocene geodynamic evolution of Eastern Asia. *Tectonohysics*, **156**, 133–143.

Fujii, M., Hayasaka, Y. & Horie, K. 2008. Metamorphism and timing of the nappe movement in the Asaji metamorphic area, eastern Kyushu. *Journal of the Geological Society of Japan*, **114**, 127–140 [in Japanese with English abstract].

Fukuyama, M., Urata, K. & Nishiyama, T. 2004. Geology and petrology of the Hirao limestone and the Tagawa metamorphic rocks – with special reference to the contact metamorphism by Cretaceous granodiorite. *Journal of Mineralogical and Petrological Sciences*, **99**, 25–41.

Geological Survey of Japan 1992. 1:1 000 000 Geological Map of Japan (3rd edn). Geological Survey of Japan, Tokyo.

Hada, S., Yoshikura, S. & Gabites, J. E. 2000. U–Pb zircon ages for the Mitaki igneous rocks, Siluro-Devonian tuff, and granitic boulders in the Kurosegawa Terrane, Southwest Japan. *Memoirs of the Geological Society of Japan*, **56**, 183–198.

Hamamoto, T., Osanai, Y. & Kagami, H. 1999. Sm–Nd, Rb–Sr and K–Ar geochronology of the Higo metamorphic terrane, west-central Kyushu, Japan. *Island Arc*, **8**, 323–334.

Hara, H. & Kimura, K. 2008. Metamorphic and cooling history of the Shimanto accretionary complex, Kyushu, Southwest Japan: implications for the timing of out-of-sequence thrusting. *Island Arc*, **17**, 546–559.

Hara, H., Kimura, K. & Naito, K. 2009. Geology of the Murasho Distict. Quadrangle Serise, 1:50 000, Geological Survey of Japan, AIST [in Japanese with English abstract].

Hashimoto, M. & Fujimoto, M. 1962. The konoha metamorphic rocks, Kyushu. *Bulletin of the National Museum of Nature and Science*, **6**, 17–35.

Hattori, H. & Isomi, H. 1976. Nagasaki metamorphic rocks in northwest Kyushu, with particular emphasis on the geology of westernmost Amakusa-shimoshima. *Bulletin of the Geological Survey of Japan*, **27**, 11–28.

Hattori, H. & Shibata, K. 1982. Radiometric dating of pre-Neogene granitic and metamorphic rocks in northwest Kyushu, Japan, with emphasis on geotectonics of the Nishisonogi zone. *Bulletin of the Geological Survey of Japan*, **33**, 57–84.

Hayasaka, S. 1999. Inoceramus fossils from the Shimanto Belt in south Kyushu. *Shizen-Aigo*, **25**, 2 [in Japanese].

Hayasaka, Y., Hara, I. & Yoshigai, T. 1989. Nappe structure of the Asaji metamorphic rocks, with special reference to geological structure of the basement complexes in Kyushu. *Memoirs of the Geological Society of Japan*, **33**, 177–186 [in Japanese with English abstract].

Higashino, T. 1990. The higher grade metamorphic zonation of the Sambagawa metamorphic belt in central Shikoku, Japan. *Journal of Metamorphic Geology*, **8**, 413–423.

Ibaragi, M. 1994. Age and paleoenvironment of the Tertiary system in the northwestern Kyushu on the basis of foraminiferous. *Monthly Earth*, **16**, 150–153 [in Japanese].

Ikeda, T. 1998. Progressive sequence of reactions of the Ryoke metamorphism in the Yanai district, southwest Japan: the formation of cordierite. *Journal of Metamorphic Geology*, **16**, 39–52.

Ikeda, T., Yoshida, H., Arima, K., Nishiyama, T., Yanagi, T. & Miyazaki, K. 2005. Garnet-clinopyroxene amphibolite from the Takahama metamorphic rocks, western Kyushu, SW Japan: evidence for high-pressure granulite facies metamorphism. *Journal of Mineralogical and Petrological Sciences*, **100**, 104–105.

Imai, I., Teraoka, Y. & Okumura, K. 1971. Geologic structure and metamorphic zonation of the northeastern part of the Shimanto terrane in

Kyushu, Japan. *Journal of Geological Society of Japan*, **77**, 207–220 [in Japanese with English abstract].

Imaoka, I., Nakajima, T. & Itaya, T. 1993. K–Ar ages of hornblendes in andesite and dacite from the Cretaceous Kanmon Group, Southwest Japan. *Journal of Mineralogy, Petrology and Economic Geology*, **88**, 265–271.

Ishida, N. 2006. Sedimentary evolution of the Torinosu-type limestone bearing strata in the Southern Chichibu terrane: a case study of the Upper Jurassic Ebirase Formation in the middle stream of the Kuma River, Kumamoto Prefecture. *Kumamoto Journal of Science, Earth Sciences*, **18**, 69–87 [in Japanese with English abstract].

Ishida, N. 2009. Jurassic to Early Cretaceous accretionary complexes and Upper Jurassic trench-slope basin deposits of the Southern Chichibu Terrane in the Itsuki-Gokanosho area, western Kyushu. *News of Osaka Micropaleontologists, Special Volume*, **14**, 375–403 [in Japanese with English abstract].

Ishihara, S., Karakida, Y. & Sato, K. 1979. Distribution of the magnetite-series and ilmenite-series granitoids in the northern Kyushu-western Chugoku district. With emphasis on re-evaluation of the Kokura-Tagawa fault zone. *Journal of the Geological Society of Japan*, **85**, 47–50 [in Japanese with English abstract].

Ishikawa, N. 1997. Differential rotations of north Kyushu Island related to middle Miocene clockwise rotation of SW Japan. *Journal of Geophysical Research*, **102**, 17 729–17 745.

Ishizuka, H., Ichigi, K., Osanai, Y., Nakano, N., Adachi, T. & Yoshimoto, A. 2011. Detritus zircon age and formation tectonics of the Yaeyama metamorphic rocks. *Abstract of 2011 Joint Annual Meeting of Japan Association of Mineralogical Sciences and the Geological Society of Japan*, 9–11 September 2011, Mito, **9** [in Japanese].

Ishizuka, Y., Mizuno, K., Matsuura, H. & Hoshizumi, H. 2005. Geology of the Bungo-Kituski district. Quadrangle Series, 1:50 000, Geological Survey of Japan, AIST [in Japanese with English abstract].

Isozaki, Y. & Itaya, T. 1989. Origin of schist clasts of Upper Cretaceous Onogawa Group, southwest Japan. *Journal of the Geological Society of Japan*, **95**, 361–368.

Isozaki, Y. & Nishimura, Y. 1989. Fusaki Formation, Jurassic subduction-accretion complex on Ishigaki Island, southern Ryukyus and its geologic implication to Late Mesozoic convergent margin of East Asia. *Memoirs of the Geological Society of Japan*, **33**, 259–257 [in Japanese with English abstract].

Isozaki, Y., Aoki, K., Nakama, T. & Yanai, S. 2010. New insight into a subduction-related orogen: a reappraisal of the geotectonic framework and evolution of the Japanese Island. *Gondwana Research*, **18**, 82–105.

Itaya, T. & Takasugi, H. 1988. Muscovite K–Ar ages of the Sanbagawa schist, Japan and argon depletion during cooling and deformation. *Contributions to Mineralogy and Petrology*, **100**, 281–290.

Iwasaki, M. 1954. Contact metamorphic rocks in the Kaba-shima, Nagasaki Prefecture. *Journal of Gakugei, Tokushima University, Natural Science*, **4**, 97–102 [in Japanese with English abstract].

Izawa, E., Karakida, Y., Shimada, N. & Takahashi, M. 1989. High-strontium granites in northern Kyushu, Japan. *Structural Development of the Japanese Islands, DELP Publication*, **28**, 62–67.

Kabashima, T., Isozaki, Y., Nishimura, Y. & Itaya, T. 1995. Re-examination on K-Ar ages of the Kiyama high-P/T schist in central Kyushu. *Journal of the Geological Society of Japan*, **101**, 297–400 [in Japanese with English abstract].

Kamata, H. 1989. Volcanic and structural history of the Hohi volcanic zone, central Kyushu, Japan. *Bulletin of Volcanology*, **51**, 315–332.

Kamata, H. & Kodama, K. 1994. Tectonics of an arc-arc junction: an example from Kyushu Island at the junction of the southwest Japan Arc and Ryukyu Arc. *Tectonophysics*, **233**, 69–81.

Kamata, H. & Kodama, K. 1999. Volcanic history and tectonics of the Southwest Japan Arc and the Ryukyu Arc since 2 Ma. *Island Arc*, **8**, 393–403.

Kamei, A. 2002. Petrogenesis of Cretaceous peraluminous granite suites with low initial Sr isotopic ratios, Kyushu Island, southwest Japan Arc. *Gondowana Research*, **5**, 813–822.

Kamei, A. 2004. An adakitic pluton on Kyushu Island, southwest Japan arc. *Journal of Asian Earth Sciences*, **24**, 43–58.

Kamei, A., Owada, M., Hamamoto, T., Osanai, Y., Yuhara, M. & Kagami, H. 2000. Isotopic equilibration ages for the Miyanohara tonalite from the Higo metamorphic belt in central Kyushu, Southwest Japan: implications for the tectonic setting during the Triassic. *Island Arc*, **9**, 97–112.

Kamei, A., Owada, M., Nagao, T. & Shiraki, K. 2004. High-Mg diorites derived from sanukitic HMA magmas, Kyushu Island, southwest Japan arc: evidence from clinopyroxene and whole rock compositions. *Lithos*, **75**, 359–371.

Kamihara, K., Fujimoto, Y. & Hirajima, T. 2012. Phengite K-Ar ages of high-P type metamorphic rocks of the Hakoishi sub-unit of the Kurosegawa belt in Yatsushiro area, Kyushu. *Abstract of the 2012 Annual Meeting of the Japan Association of Mineralogical Sciences*, **254**, 19–21 September 2012, Kyoto [in Japanese].

Kanmera, K. 1977. General aspects and recognition of olistostromes in the geosynclinal sequence. *Monograph, Association for the Geological Collaboration in Japan*, **20**, 145–159 [in Japanese with English abstract].

Karakida, Y., Yamamoto, H., Miyachi, S., Oshima, T. & Inoue, T. 1969. Characteristics and geological situations of metamorphic rocks in Kyushu. *Memoirs of the Geological Society of Japan*, **4**, 3–21 [in Japanese with English abstract].

Karakida, T., Oshima, T. & Miyachi, S. 1977. The Kurosegawa Tectonic Zone and the Chichibu Belt in Kyushu. *In*: Hide, K. (ed.) *The Sambagawa Belt*. Geological Society of Japan, Tokyo, 165–175 [in Japanese].

Katada, M. & Matsui, K. 1973. *Metamorphic Rocks of the Ainoshima Belt in Northwestern Kyushu, Japan*. Report no. 246, Geological Survey of Japan [in Japanese with English abstract].

Kawanabe, Y., Sakaguchi, K., Saito, M., Komazawa, M. & Yamazaki, T. 2004. Geological Map of Japan 1:200 000, Kaimon Dake and a part of Kuro Shima. Geological Survey of Japan, Tsukuba, AIST [in Japanese with English abstract].

Kawano, Y. & Ueda, Y. 1966. K-Ar dating on the igneous rocks in Japan (V) – Granitic rocks in southwestern Japan. *Journal of the Japanese Association of Mineralogists, Petrologists and Economic Geologists*, **56**, 191–211 [in Japanese with English abstract].

Kimura, G., Hamahashi, M. *et al.* 2013. Hanging wall deformation of a seismogenic megasplay fault in an accretionary prism: the Nobeoka Trust in southwestern Japan. *Journal of Structural Geology*, **52**, 136–147.

Kimura, J., Stern, R. J. & Yoshida, T. 2005. Reinitiation of subduction and magmatic responses in SW Japan during Neogene time. *Geological Society of America Bulletin*, **117**, 969–986.

Kimura, K., Iwaya, T., Mimura, K., Sato, Y., Suzuki, Y. & Sakamaki, Y. 1991. Geology of the Osuzuyama District. Quadrangle Serise, 1:50 000, Geological Survey of Japan [in Japanese with English abstract].

Kitano, I. & Ikeda, T. 2012. Pressure and temperature conditions of contact metamorphism of the Suo metamorphic rocks in Asakura area, Fukuoka Prefecture, Japan: constrains on uplifting process. *Journal of the Geological Society of Japan*, **118**, 801–809.

Kizaki, K. 1986. Geology and tectonics of the Ryukyu Islands. *Tectonophysics*, **125**, 193–207.

Kojima, T., Nishimura, Y., Takami, M. & Itaya, T. 1999. K–Ar ages and Metamorphism o the Nago Formation, Okinawa Island. *Abstract of Kansai branch and Southwest branch of the Geological Society of Japan*, 21 November 1998, Matsuyama, no. 126 and no. 113, 27 [in Japanese].

Komatsubara, J., Ugaji, H. *et al.* 2005. Fission-track age and inferred subsidence rate of the Lower to Middle Miocene Nojima Group in NW Kyushu. *Journal of the Geological Society of Japan*, **111**, 350–360.

Kondo, H., Kimura, G. *et al.* 2005. Deformation and fluid flow of a major out-of-sequence thrust located at seismogenic depth in an accretionary complex: Nobeoka Thrust in the Shimanto Belt, Kyushu, Japan. *Tetonics*, **24**, TC6008.

Kondo, Y. & Kikuchi, N. 2010. Cretaceous in central belt of Kyushu. *In*: Geological Society of Japan (ed.) *Nihon-Chiho-Chishitsushi, Kyushu-Okinawa-Chiho*. Asakura-Shoten, Tokyo, 229–240 [in Japanese].

Kouchi, Y., Orihashi, Y. *et al.* 2011. Discovery of Shimanto High-P/T metamorphic rocks from the western margin of Kyushu. *Journal of Geography*, **120**, 30–39 [in Japanese with English abstract].

Kubodera, A., Mitsunami, T., Satomura, M. & Inoue, M. 1976, Compilation of gravity values and drawing the detailed gravity (Bouguer anomaly) map in the central part of Kyushu. *Shizen Saigai Kagaku Shiryo Kaiseki Kenkyu*, **3**, 45–53.

Kurihara, T. 2004. Late Silurian and Early Devonian radiolarians from the Tomochi district in the Kurosegawa Terrane, central Kyushu. *Abstract, the 153rd Regular Meeting, The Palaeontological Society of Japan*, 12–13 June 2010, Tsukuba, **44** [in Japanese].

MAKI, K., FUKUNAGA, T., NISHIYAMA, T. & MORI, T. 2009. Prograde P-T path of medium-pressure granulite facies calc-silicate rocks, Higo metamorphic terrane, central Kyushu, Japan. *Journal of Metamorphic Geology*, **27**, 107–124.

MAKI, K., FUKUYAMA, M., MIYAZAKI, K., YUI, T.-F. & GROVE, M. 2011. Zircon SHRIMP U-Pb ages and trace element analyses from the Higo, Abukuma and Ryoke metamorphic terranes. *Abstracts (Section B) of 2011 Joint Annual Meeting of Japan Association of Mineralogical Sciences and the Geological Society of Japan*, 9–11 September 2011, Mito, **9** [in Japanese].

MARUYAMA, S., ISOZAKI, Y., KIMURA, G. & TERABAYASHI, M. 1997. Paleogeographic maps of the Japanese Islands: plate tectonic synthesis from 750 Ma to the present. *Island Arc*, **6**, 121–142.

MATSUMOTO, H. & NARISHIGE, K. 1985. Volcanic geology of Kunisaki Peninsula, Oita Prefecture. *Memoirs of the Faculty of Education, Kumamoto University, Natural Science*, **20**, 61–76 [in Japanese with English abstract].

MATSUMOTO, T. & KANMERA, K. 1964. Geology of the Hinagu district. Quadrangle Series, 1:50 000, Geological Survey of Japan [in Japanese with English abstract].

MATSUMOTO, T., OTSUKA, H. & OKI, K. 1973. Cretaceous fossils from the Shimanto Belt of Kagoshima Prefecture. *Journal of the Geological Society of Japan*, **79**, 703–704 [in Japanese with English abstract].

MATSUSHITA, H., NAGAI, T. & KANEKO, N. 1969. The geological structure of Hiraodai and its neighborhood, Fukuoka Prefecture (The geological structures of northern Kyushu, Part 1). *The Science Reports of the Faculty of Science, Kyushu University, Geology*, **9**, 113–119.

MATSUURA, H. 1998. K–Ar ages of the Early Cretaceous Shimonoseki Subgroup and the Kawara Granodiorite, Southwest Japan. *Journal of Mineralogy, Petrology and Economic Geology*, **93**, 307–312 [in Japanese with English abstract].

MIKI, M., MATSUDA, T. & OTOFUJI, Y. 1990. Opening mode of the Okinawa trough: paleomagnetic evidence from the South Ryukyu arc. *Tectonophysics*, **175**, 335–347.

MINISTRY OF INTERNATIONAL TRADE AND INDUSTRY, AGENCY FOR NATURAL RESOURCES AND ENERGY 1985. S.59 Regional research report. Nansatsu district [in Japanese].

MIYAMOTO, T. 2010. Permian mizukoshi formation. *In*: GEOLOGICAL SOCIETY OF JAPAN (ed.) *Nihon-Chiho-Chishitsushi, Kyushu-Okinawa-Chiho*. Asakura-Shoten, Tokyo, 227–229 [in Japanese].

MIYAMOTO, T. & ENOKIBARA, A. 2009. Occurrence and U-Pb age of Yamaga meta-gabbro complex. *Abstract of 2009 Annual Meeting of Japan Association of Mineralogical Sciences*, 8–10 September 2009, Sapporo, **42** [in Japanese].

MIYAZAKI, K. 2004. Low-P – high-T metamorphism and the role of heat transport by melt migration in the Higo Metamorphic Complex, Kyushu, Japan. *Journal of Metamorphic Geology*, **22**, 793–809.

MIYAZAKI, K. 2007. Formation of a high-temperature metamorphic complex due to pervasive melt migration in the hot crust. *Island Arc*, **16**, 69–82.

MIYAZAKI, K. 2010. Development of migmatites and the role of viscous segregation in high-T metamorphic complexes: example from the Ryoke Metamorphic Complex, Mikawa Plateau, Central Japan. *Lithos*, **116**, 287–299.

MIYAZAKI, K. & MATSUURA, H. 2005. K-feldspar-garnet-sillimanite-biotite gneiss and clinopyroxene-garnet-quartz gneiss from Omuta district, northern Kyushu, Japan. *Abstract of 112th Annual Meeting of the Geological Society of Japan*, 18–20 September 2005, Kyoto, **295** [in Japanese].

MIYAZAKI, K. & NISHIYAMA, T. 1989. Petrological study of the Nagasaki metamorphic rocks in the Nomo Peninsula with special reference to biotite zone. *Memoirs of the Geological Society of Japan*, **33**, 217–236 [Japanese with English abstract].

MIYAZAKI, K. & OKUMURA, M. 2002. Thermal modeling in shallow subduction: an implication to low P/T metamorphism of the Cretaceous Shimanto Accretionary Complex, Japan. *Journal of Metamorphic Geology*, **20**, 441–452.

MIYAZAKI, K. & YOSHIOKA, T. 1994. Geology of the Saganoseki district. With Geological Sheet Map at 1:50 000. Geological Survey of Japan [in Japanese with English abstract].

MIYAZAKI, K., IKEDA, T., FUKUYAMA, M., MAKI, K., YUI, T.-F. & GROVE, M. 2013. Pressure-temperature structure of a mylonitized metamorphic pile, and the role of advection of the lower crust, Nagasaki Metamorphic Complex, Kyushu, Japan. *Lithos*, **162–163**, 14–26.

MIYAZOE, T., NISHIYAMA, T., UYETA, K., MIYAZAKI, K. & MORI, Y. 2009. Coexistence of pyroxenes jadeite, omphacite, and diopside/hedenbergite in an albite-omphacite rock from a serpentinite mélange in the Kurosegawa Zone of Central Kyushu, Japan. *American Mineralogist*, **94**, 34–40.

MIYOSHI, M., YUGUCHI, T., SHINMURA, T., MORI, Y., ARAKAWA, Y. & TOYOHAMA, F. 2011. Petrological characteristics and K-Ar age of borehole core samples of basement rocks from the northwestern caldera floor of Aso, central Japan. *Journal of the Geological Society of Japan*, **114**, 585–590 [in Japanese with English abstract].

MORI, Y., ORIHASHI, Y., MIYAMOTO, T., SHIMADA, K., SHIGENO, M. & NISHIYAMA, T. 2011. Origin of zircon in jadeitite from the Nishisonogi metamorphic rocks, Kyushu, Japan. *Journal of Metamorphic Geology*, **29**, 673–684.

MURATA, A. 1998. Duplexes and low-angle nappe structures of the Shimanto Terrane, southwest Japan. *Memoirs of the Geological Society of Japan*, **50**, 147–158.

MURATA, M. 1992. Chichibu belt. *In*: KARAKIDA, Y., HAYASAKA, S. & HASE, Y. (eds) *Regional Geology, Part 9*. Kyoritsu Shuppan, Tokyo, Kyushu, 47–57 [in Japanese].

MURATA, M., NISHIZONO, Y. & ITOYAMA, T. 1981. Ryohozan Group and Usuki-Yatsushiro Tectonic Line. *Abstract of 88th Annual Meeting of the Geological Society of Japan*, 1–3 April 1981, **184**, Tokyo [in Japanese].

NAGAE, S. & MIYASHITA, S. 1999. Low-P/T type metamorphism and deformation of the northern Shimanto Belt, Kyushu, southwest Japan. *Memoirs of the Geological Society of Japan*, **52**, 225–272 [in Japanese with English abstract].

NAGAKAWA, N., OBATA, M. & ITAYA, T. 1997. K-Ar ages of the Higo metamorphic belt. *Journal of the Geological Society of Japan*, **103**, 943–952.

NAGAO, T., HASE, Y., NAGAMINA, S., KAKUBUCHI, S. & SAKAGUCHI, K. 1999. Late Miocene to middle Pleistocene Hisatsu Volcanic Rocks generated from heterogeneous magma sources: evidence from temporal-spatial variation of distribution and chemistry of the rocks. *Journal of Mineralogy, Petrology and Economic Geology*, **94**, 461–481.

NAKADA, M., YANAGI, T. & MAEDA, S. 1997. Lower crustal erosion induced by mantle diapiric upwelling: constraints from sedimentary basin formation followed by voluminous basalt volcanism in northwest Kyushu, Japan. *Earth and Planetary Science Letters*, **146**, 415–429.

NAKAE, S. 2010. The kanmon group. *In*: GEOLOGICAL SOCIETY OF JAPAN (ed.) *Nihon-Chiho-Chishitsushi, Kyushu-Okinawa-Chiho*. Asakura-Shoten, Tokyo, 222–227 [in Japanese].

NAKAE, S., NAGAMORI, H., MIYAZAKI, K. & KOMAZAWA, M. 2009. Geological map of Japan 1:200 000, Ishigaki Jima. Geological Survey of Japan, AIST [in Japanese with English abstract].

NAKAE, S., KANEKO, N., MIYAZAKI, K., OHNO, T. & KOMAZAWA, M. 2010. Geological Map of Japan 1:200 000, Yoron Jima and Naha. Geological Survey of Japan, AIST [in Japanese with English abstract].

NAKAGAWA, H. 1977. Cenozoic history of the Ryukyu Island–some problems. *Geological Studies of the Ryukyu Islands*, **2**, 1–9.

NAKAJIMA, T., NAGAKAWA, K., OBATA, M. & UCHIUMI, S. 1995. Rb-Sr and K-Ar ages of the Higo metamorphic rocks and related granitic rocks, Southwest Japan. *Journal of the Geological Society of Japan*, **101**, 615–620 [in Japanese with English abstract].

NAKAJO, T. & FUNAKAWA, S. 1996. Eocene radiolarians from the Lower Formation of the Taishu Group, Tsushima islands, Nagasaki Prefecture, Japan. *Journal of the Geological Society of Japan*, **102**, 751–754.

NAKANO, S., NISHIKI, K. ET AL. 2013. *Volcanoes of Japan* (3rd edn). 1:2 000 000, no. 11. Geological Survey of Japan, Tsukuba.

NEW ENERGY AND INDUSTRIAL TECHNOLOGY DEVELOPMENT ORGANIZATION 1995. *Report for development facilitation of geothermal energy*, no. 28 west area of Aso mountain, 1508 [in Japanese].

NISHI, H. 1988. Structural analysis of part of the Shimanto accretionary complex, Kyushu, Japan, based on planktonic foraminiferal zonation. *Modern Geology*, **12**, 47–69.

NISHI, T. 1994. Geology and tectonics of the Sambosan Terrane in eastern Kyushu, southwest Japan. Stratigraphy, sedimentological features and depositional setting of the Shakumasan Group. *Journal of the Geological Society of Japan*, **100**, 199–215.

NISHIMURA, Y. 1998. Geotectonic subdivision and areal extent of the Sangun belt, Inner Zone of Southwest Japan. *Journal of Metamorphic Geology*, **16**, 129–140.

Nishimura, Y. & Shibata, K. 1989. Modes of occurrence and K-Ar ages of metagabbroic rocks in the 'Sangun metamorphic belt', Southwest Japan. *Memoirs of the Geological Society of Japan*, **33**, 343–357 [in Japanese with English abstract].

Nishimura, Y., Hirota, Y., Shiokai, D., Nakahara, N. & Itaya, T. 2004. The Nagasaki metamorphic rocks and their geotectonics in Mogi area, Nagasaki Prefecture, Southwest Japan – juxtaposition of the Suo belt with the Sanbagawa belt. *Journal of Geological Society of Japan*, **110**, 372–383 [in Japanese with English abstract].

Nishiyama, T. 1989. Petrological study of the Nagasaki metamorphic rocks in the Nishisonogi Peninsula – with special reference to the greenrock complex and the reaction-enhanced ductility. *Memoirs of the Geological Society of Japan*, **33**, 237–257 [Japanese with English abstract].

Nishizono, Y., Nishimura, Y., Itaya, T. & Murata, M. 1996. K-Ar ages of weakly metamorphosed rocks from the border between the Kurosegawa and the Southern Chichibu Terranes in western Kyushu, Southwest Japan. *Abstract of the 103th Annual Meeting of the Geological Society of Japan*, 1–3 April 1996, Sendai, **132** [in Japanese].

Obata, M., Yoshimura, Y., Nagakawa, K., Odawara, S. & Osanai, Y. 1994. Crustal anatexis and melt migrations in the Higo metamorphic terrane, west-central Kyushu, Kumamoto, Japan. *Lithos*, **32**, 135–147.

Ogawauchi, Y., Iwamatsu, A. & Tanabe, A. 1984. Stratigraphy and geologic structures of the Shimanto Supergroup in the northeastern part of Nobeoka City, Miyazaki Prefecture, Japan. *Reports of the Faculty of Science, Kagoshima University (Earth Science & Biology)*, **17**, 67–88 [in Japanese with English abstract].

Okada, H. 2000. Nature and development of Cretaceous sedimentary basins in East Asia: a review. *Geoscience Journal*, **4**, 271–282.

Okino, K., Shimakawa, Y. & Nagaoka, S. 1994. Evolution of the Shikoku Basin. *Journal of Geomagnetism and Geoelectricity*, **46**, 463–479.

Okino, K., Ohara, Y., Kasuga, S. & Kato, Y. 1999. The Philippine Sea: new survey results reveal the structure and the history of the marginal basins. *Geophysical Research Letters*, **26**, 2287–2290.

Osanai, Y., Masao, S. & Kagami, M. 1993. Rb-Sr whole rock isochron ages of granitic rocks from the central Kyushu, Japan. *Memoirs of the Geological Society of Japan*, **42**, 135–150 [in Japanese with English abstract].

Osanai, Y., Hamanoto, T., Maishima, O. & Kagami, H. 1998. Sapphirine-bearing granulites and related high-temperature metamorphic rocks from the Higo metamorphic terrane, west-central Kyushu, Japan. *Journal of Metamorphic Geology*, **16**, 53–66.

Oshima, T. 1964. Crystalline schists in the Nomo peninsula, Nagasaki Prefecture. *Memoirs of the Faculty of Science, Kyushu University, Geology*, **7**, 39–45.

Ota, Y. & Yabumoto, Y. 1992. The Kanmon Group. *Geology of Japan, Part 9*. Kyoritsu Shuppan, Tokyo, Kyushu, 19–22 [in Japanese].

Otofuji, Y., Itaya, T. & Matsuda, T. 1991. Rapid rotation of southwest Japan – palaeomagnetism and K-Ar ages of Miocene volcanic rocks of southwest Japan. *Geophysical Journal International*, **105**, 397–405.

Otsuka, Y. 1939. Tertiary crustal deformations in Japan (with short remarks on Tertiary paleogeography). *Jubilee Publication in the Commemoration of Professor H. Yabe, M.I.A. 60th Birthday*, **1**, 481–519.

Owada, M., Kamei, A., Yamamoto, K., Osanai, Y. & Kagami, H. 1999. Spatial–temporal variations and origin of granitic rocks from central to northern part of Kyushu, southwest Japan. *Memoirs of the Geological Society of Japan*, **53**, 349–363 [in Japanese with English abstract].

Owada, M., Hada, R., Yada, J., Nakamura, M. & Osanai, Y. 2005. Newly found granulites from the western part of Sefuri Mountains, North Kyushu. *Journal of Geological Society of Japan*, **111**, 50–53 [in Japanese with English abstract].

Saito, M. 2008. Rapid evolution of the Eocene accretionary complex (Hyuga Group) of the Shimanto terrane in southeastern Kyushu, southwestern Japan. *Island Arc*, **17**, 242–260.

Saito, M. & Miyazaki, K. 2006. Jadeite-bearing metagabbro in serpentinite mélange of the 'Kurosegawa Belt' in Izumi Town, Yatsushiro City, Kumamoto Prefecture, central Kyushu. *Bulletin of the Geological Survey of Japan*, **57**, 196–176.

Saito, M. & Toshimitsu, S. 2003. Permian radiolarian fossils from the basement of the Lower Cretaceous Tomochi Formation in the Tomochi area, central Kyushu. *Journal of the Geological Society of Japan*, **109**, 71–74 [in Japanese with English abstract].

Saito, M., Teraoka, Y., Miyazaki, K. & Toshimitsu, S. 1993. Radiolarian fossils from the Nishikawauchi Formation in the Onogawa Basin, east Kyushu, and their geological significance. *Journal of the Geological Society of Japan*, **99**, 479–482 [in Japanese].

Saito, M., Sato, Y. & Yokoyama, S. 1994. Geology of Sueyoshi district, with Geological Sheet Map at 1:50 000. Geological Survey of Japan [in Japanese with English abstract].

Saito, M., Kimura, K., Naito, K. & Sakai, A. 1996. Geology of Shiibamura district, with Geological Sheet Map at 1:50 000. Geological Survey of Japan [in Japanese with English abstract].

Saito, M., Sakaguchi, K. & Komazawa, M. 1997. Geological Map of Japan 1:200 000, Miyazaki. Geological Survey of Japan, Tsukuba.

Saito, M., Saiki, K. & Toshimitsu, S. 2003. Late Devonian Leptophloeum from the coherent strata in the Kurosegawa Belt, Tomochi district, central Kyushu. *Journal of the Geological Society of Japan*, **109**, 293–298 [in Japanese with English abstract].

Saito, M., Miyazaki, K., Toshimitsu, S. & Hoshizumi, H. 2005. Geology of the Tomochi district. Quadrangle Series, 1:50 000, Geological Survey of Japan, AIST [in Japanese with English abstract].

Saito, M., Ogasawara, M., Nagamori, H., Geshi, N. & Komazawa, M. 2007. Geological Map of Japan 1:200 000, Yaku Shima. Geological Survey of Japan, AIST [in Japanese with English abstract].

Saito, M., Takarada, S. *et al.* 2010. Geological Map of Japan 1:200 000, Yatsushiro and a part of Nomo Zaki. Geological Survey of Japan, AIST [in Japanese with English abstract].

Sakai, A., Teraoka, Y., Miyazaki, K., Hoshizumi, H. & Sakamaki, Y. 1993. Geology of the Miemachi district, with geological sheet map at 1:50 000. Geological Survey of Japan, Tsukuba [in Japanese with English abstract].

Sakai, H. 1988*a*. Toi-misaki olistostrome of Southern Belt of the Shimanto Terrane, South Kyushu – I. Reconstruction of depositional environments and stratigraphy before the collapse. *Journal of the Geological Society of Japan*, **94**, 733–747 [in Japanese with English abstract].

Sakai, H. 1988*b*. Toi-misaki olistostrome of the Southern Belt of the Shimanto Terrane, South Kyushu – II. Deformation structures of huge submarine slides and their processes of formation. *Journal of the Geological Society of Japan*, **94**, 837–853 [in Japanese with English abstract].

Sakai, H. 1993. Tectonics and sedimentation of the Tertiary sedimentary basin in the northern Kyushu. *Memoirs of the Geological Society of Japan*, **42**, 183–201.

Sakai, H. & Yuasa, T. 1998. K-Ar ages of the Mogi and Ugetsuiwa subaqueous pyroclastic flow deposits in the Taishu Group, Tsushima Islands. *Memoirs of the National Museum of Nature and Science*, **31**, 23–28.

Sakai, T., Kusaba, T., Nishi, H., Komori, M. & Watanabe, M. 1987. Olistostrome of the Shimanto Terrane in the Nichinan area, southern part of the Miyazaki Prefecture, South Kyushu. With reference to deformation and mechanism of emplacement of olistolith. *Science Reports, Department of Geology, Kyushu University*, **15**, 167–199 [in Japanese with English abstract].

Sakashima, T., Yamamoto, H., Iwamatsu, A., Yokota, S., Takeshita, T. & Hayasaka, Y. 1995. Finding of chloritoid from the Ryohozan zone in the northeastern part of the Yatsushiro City, Kumamoto Prefecture. *Journal of the Geological Society of Japan*, **101**, 999–1002 [in Japanese with English Abstract].

Sakashima, T., Takeshita, T., Itaya, T. & Hayasaka, Y. 1999. Stratigraphy, geologic structures, and K-Ar ages of the Ryuhozan metamorphic rocks in western Kyushu, Japan. *Journal of the Geological Society of Japan*, **105**, 161–180 [in Japanese with English abstract].

Sakashima, T., Terada, K., Takeshita, T. & Sano, Y. 2003. Large-scale displacement along the Median Tectonic Line, Japan: evidence from SHRIMP zircon U-Pb dating of granites and gneisses from the South Kitakami and paleo-Ryoke belts. *Journal of Asian Earth Sciences*, **21**, 1019–1039.

Sanfilippo, A. & Nigrini, C. 1998. Code numbers for Cenozoic low-latitude radiolarian biostratigraphic zones and GPTS conversion tables. *Marine Micropaleontology*, **33**, 109–1056.

Sano, Y. 2010*a*. Permian accretionary complex in the Akiyoshi Belt. *In*: Geological Society of Japan (ed.) *Nihon-Chiho-Chishitsushi, Kyushu-Okinawa-Chiho*. Asakura-Shoten, Tokyo, 178–182 [in Japanese].

Sano, Y. 2010*b*. Jurassic. *In*: Geological Society of Japan (ed.) *Nihon-Chiho-Chishitsushi, Kyushu-Okinawa-Chiho*. Asakura-Shoten, Tokyo, 248–251 [in Japanese].

Sano, Y. & Onoue, T. 2010. Jurassic accretionary complex in the Chichibu and Sambosan Belts. *In*: Geological Society of Japan (ed.) *Nihon-Chiho-Chishitsushi, Kyushu-Okinawa-Chiho*. Asakura-Shoten, 190–201 [in Japanese].

Sasada, M. 1987. Pre-Tertiary basement rocks of Hohi area, central Kyushu. *Bulletin of the Geological Survey of Japan*, **38**, 385–422 [in Japanese with English abstract].

Seno, T. & Maruyama, S. 1984. Paleogeographic reconstruction and origin of the Philippine Sea. *Tectonophysics*, **102**, 53–84.

Shibata, K. 1968. *K-Ar Age Determinations on Granitic and Metamorphic Rocks in Japan*. Report, no. 277. Geological Survey of Japan.

Shibata, K. & Nishimura, Y. 1989. Isotopic ages of the Sangun crystalline schists, Southwest Japan. *Memoirs of the Geological Society of Japan*, **33**, 317–341 [in Japanese with English abstract].

Shigeno, M., Mori, Y. & Nishiyama, T. 2005. Reaction microtextures of jadeitite form the Nishisonogi metamorphic rocks, Kyushu, Japan. *Journal of Mineralogical and Petrological Science*, **100**, 237–246.

Shuto, T. 1962. A revision of the Pleistocene stratigraphy in Kyushu for the basis of interprovincial correlation (Notes on the Pleistocene history in Kyushu Island-2). *Journal of the Geological Society of Japan*, **68**, 301–312 [in Japanese with English abstract].

Soda, Y. & Takagi, H. 2004. Mineral chemistry of ultramafic rocks associated with the Asaji metamorphic rocks, eastern Kyushu. *Journal of the Geological Society of Japan*, **110**, 698–714 [in Japanese with English abstract].

Tachibana, K. 1962. On the Pre-Tertiary Enoshima formation, Ainoshima thermally metamorphosed rocks and the Cretaceous Nishisonogi granitic rocks exposed on the sea between the Goto Islands and the Nishisonogi Peninsula – in special reference to the granitic rocks of Nagasaki Prefecture. *Bulletin of Faculty of Liberal Art, Nagasaki University*, **3**, 24–43.

Takagi, H., Shibata, K., Suzuki, K., Tanaka, T. & Ueda, K. 1997. Isotopic ages of the Usukigawa quartz diorite along the Usuki-Yatsushiro Tectonic Line in eastern Kyushu and their geological significance. *Journal of the Geological Society of Japan*, **103**, 368–376 [in Japanese with English abstract].

Takagi, H., Soda, Y. & Yoshimura, J. 2000. K-Ar ages of the granitic clasts from the Onogawa Group, eastern Kyushu. *Memoirs of the Geological Society of Japan*, **56**, 213–220 [in Japanese with English abstract].

Takagi, H., Tobe, E., Sakashima, T. & Terada, K. 2001. Western extension of the Median Tectonic Line in Kyushu. *Abstract of the 108th Annual Meeting of the Geological Society of Japan*, 21–23 September 2001, Kanazawa, **118** [in Japanese].

Takagi, H., Ishii, T., Tobe, E., Soda, Y., Suzuki, K., Iwano, H. & Danhara, T. 2007. Petrology and radiogenic age of accidental clasts of granitic mylonite from the Aso-4 pyroclastic flow deposit and their correlation to the Nioki Granite. *Journal of the Geological Society of Japan*, **113**, 1–14 [in Japanese with English abstract].

Takagi, H., Hunt, J., Gouzu, C. & Hyodo, H. 2009. $^{40}Ar/^{39}Ar$ ages of the Takahama metamorphic rocks in the Amakusa Island. *Abstract of 116th Annual Meeting of the Geological Society of Japan*, 4–6 September 2009, Okayama, **142** [in Japanese].

Takahashi, T., Tanaka, H., Sakamoto, D., Nagata, Y. & Nakamoto, E. 2003. Cretaceous formations and their bivalve faunas in Mukujima Island, Tsukumi City, Oita Prefecture (Part 1). *Bulletin of Goshonoura Cretaceous Museum*, **4**, 1–10 [in Japanese with English abstract].

Takami, M., Takemura, R., Nishimura, Y. & Kojima, T. 1999. Reconstruction of oceanic plate stratigraphies and unit division of Jurassic-Early Cretaceous accretionary complexes in Okinawa Islands, central Ryukyu Islands. *Journal of the Geological Society of Japan*, **105**, 866–880 [in Japanese with English abstract].

Takeda, K., Nishimura, Y., Itaya, T. & Hayasaka, Y. 2002. High-grade metamorphic rocks from the Nomo Peninsula and Amakusa-shimoshima, Western Kyushu. *Abstract of 109th Annual Meeting of the Geological Society of Japan*, 14–16 September 2002, Niigata, **170**.

Takeuchi, M. 2010. Jurassic accretionary complex of Chichibu belt and Sanpozan belt in Ryukyu arc. *In*: Geological Society of Japan (ed.) *Nihon-Chiho-Chishitsushi, Kyushu-Okinawa-Chiho*. Asakura-Shoten, Tokyo, 201–207 [in Japanese].

Tamaki, K., Suyehiro, K., Allan, J., Ingle, J. C., Jr. & Piscitto, K. A. 1992. Tectonic synthesis and implications of Japan Sea ODP drilling. *In*: Tamaki, K., Suyehiro, K. *et al.* (eds) *Proceedings of the Ocean Drilling Program – Scientific Results*. Ocean Drilling Program, College Station, TX, **127/128**(Pt. 2), 1333–1348.

Tamura, M. & Murakami, K. 1986. Upper Jurassic Kurosaki Formation discovered at Kurosaki, Tanoura Town, Kumamoto Prefecture, Japan. *Memoirs of Faculty of Education of the Kumamoto University, Natural Science*, **35**, 47–55.

Tanaka, H. 2010. Cretaceous normal sediments in Kurosegawa zone. *In*: Geological Society of Japan (ed.) *Nihon-Chiho-Chishitsushi, Kyushu-Okinawa-Chiho*. Asakura-Shoten, Tokyo, 251–255 [in Japanese].

Tashiro, M., Tanaka, H., Sakamoto, T. & Takahashi, T. 1994. Cretaceous system at Tanoura and Hinagu areas, southwest Kyushu. *Research Reports of Kochi University, Natural Science*, **43**, 69–78.

Teraoka, Y. 1970. *Cretaceous Formations in the Onogawa Basin and its Vicinity, Kyushu, Southwest Japan*. Reports of Geological Survey of Japan, no. 237 1–87 [in Japanese with English abstract].

Teraoka, Y. & Okumura, K. 1992. Tectonic division and Cretaceous sandstone compositions of the Northern Belt of the Shimanto Terrane, Southwest Japan. *Memoirs of the Geological Society of Japan*, **38**, 261–270 [in Japanese with English abstract].

Teraoka, Y., Miyazaki, K., Hoshizumi, H., Yoshioka, T., Sakai, A. & Ono, K. 1992. Geology of the Inukai district. Quadrangle Series, 1:50 000, Geological Survey of Japan, Tsukuba [in Japanese with English abstract].

Teraoka, Y., Shibata, K., Okumura, K. & Uchiumi, S. 1994. K-Ar ages of detrital K-feldspars and muscovites from the Shimanto Supergroup in east Kyushu and west Shikoku, Southwest Japan. *Journal of the Geological Society of Japan*, **100**, 477–485 [in Japanese with English abstract].

Teraoka, Y., Suzuki, M. & Kawakami, K. 1998. Provenance of Cretaceous and Paleogene sediments in the Median Zone of Southwest Japan. *Bulletin of the Geological Survey of Japan*, **49**, 395–411.

Tiepolo, M., Langone, A., Morishita, T. & Yuhara, M. 2012. On the recycling of amphibole-rich ultramafic intrusive rocks in the arc crust: evidence from Shikanoshima Island (Kyushu, Japan). *Journal of Petrology*, **53**, 1255–1285.

Tobe, E. 2006. *Implications of isotopic ages of granites for the geotectonic evolution of middle Kyushu*. Doctoral thesis, Division of Geology, Mineral Resources Engineering and Materials Science and Engineering Major, Graduate School of Science and Engineering, Waseda University.

Tobe, E., Takagi, H. & Shibata, K. 2000. K-Ar ages of the granitic clasts from the Mizukoshi Formation in the Higo Belt, western Kyushu. *Memoirs of the Geological Society of Japan*, **56**, 221–228 [in Japanese with English abstract].

Tomita, S., Shimoyama, S., Matsuura, H., Miyazaki, K., Ishibasi, K. & Miki, T. 2008. Geology of the Omuta district. Quadrangle Series, 1:50 000, Geological Survey of Japan, AIST [in Japanese with English abstract].

Toriumi, M. & Teruya, J. 1988. Tectono-metamorphism of the Shimanto Belt. *Modern Geology*, **12**, 303–324.

Toyohara, F. & Murata, M. 2003. Age of strata in the Ryoke zone in Kyushu. *Abstracts of the 110th Annual Meeting of the Geological Society of Japan*, 19–21 September 2003, Shizuoka, **80** [in Japanese].

Tsutsumi, Y., Yokoyama, K., Terada, K. & Sano, Y. 2003. SHRIMP U-Pb dating of detrital zircons in metamorphic rocks from northern Kyushu, western Japan. *Journal of Mineralogical and Petrological Sciences*, **98**, 181–193.

Tsutsumi, Y., Horie, K., Shiraishi, K. & Yokoyama, K. 2010. Zircon U-Pb age of tectonic block of granitic rock from eastern area of the Nomo Peninsula, Nagasaki Prefecture. *Abstract of Annual Meeting of the Geochemical Society of Japan*, 7–9 September 2010, Kumagaya, **1A27** [in Japanese].

Tsutsumi, Y., Horie, K., Miyashita, A. & Shiraishi, K. 2011*a*. Timings of deposition and metamorphism of the Nagasaki Metamorphic Rocks, southwest Japan; what is their attribution? *Abstract of Joint Annual Meeting of Japan Association of Mineralogical Sciences and the Geological Society of Japan*, 9–11 September 2011, Mito, **84** [in Japanese].

Tsutsumi, Y., Yokoyama, K., Terada, K. & Hidaka, H. 2011*b*. SHRIMP dating of detrital zircon from the Sangun-Renge Belt of Sangun Metamorphic Rocks, Northern Kyushu, Southwest Japan. *Bulletin of the National Museum of Nature and Science*, **37**, 5–16 [in Japanese with English abstract].

Ueda, K. & Nishimura, S. 1982. Fission-Track ages of Late Cretaceous volcanic rocks in the Agawa-Umoto and Aoshima areas. *Abstracts of the 89th Annual Meeting of the Geological Society of Japan*, 1–3 May 1982, Niigata, **382**.

UJIIE, K. 2002. Evolution and kinematics of an ancient decollement zone, mélange in the Shimanto accretionary complex of Okinawa, Ryukyu arc. *Journal of Structural Geology*, **24**, 937–952.

UMEDA, M., SHIBATA, K. & IGI, S. 1986. K-Ar ages of hornblende from the granodiorite in the Kuraoka igneous rocks around Mt. Gionyama in the Kurosegawa structural belt, central Kyushu. *Journal of the Geological Society of Japan*, **92**, 1565–1158 [in Japanese].

WAKITA, K., MIYAZAKI, K., TOSHIMITSU, S., YOKOYAMA, S. & NAKAGAWA, M. 2007. Geology of the Ino district. Quadrangle Series, 1:50 000. Geological Survey of Japan, AIST [in Japanese with English abstract].

WANG, Z. & ZHAO, D. 2006. Vp and Vs tomography of Kyushu, Japan: new insight into arc magmatism and forearc seismotectonics. *Physics of the Earth and Planetary Interiors*, **157**, 269–285.

YAMAKITA, S., ITO, T., TANAKA, H. & WATANABE, H. 1995. Early Oligocene top-to-the-west motion along the Sashu fault, a low-angle oblique thrust of the Paleo-Median Tectonic Line, east Kyushu, Japan. *Journal of the Geological Society of Japan*, **101**, 978–988.

YAMAMOTO, H. 1992. Sangun mountains. *In*: KARAKIDA, Y., HAYASAKA, S. & HASE, Y. (eds) *Regional Geology of Japan Part 9*. Kyushu, Kyoritsu Shuppan, Tokyo, 8–10 [in Japanese].

YAMAMOTO, T. 1991. Late Cenozoic dike swarms and tectonic stress field in Japan. *Bulletin of the Geological Survey of Japan*, **42**, 131–148.

YAMAZAKI, T. & OKAMURA, Y. 1989. Subducting seamounts and deformation of overriding forearc wedges around Japan. *Tectonophysics*, **160**, 207–229.

YANAGIDA, J. 1958. The Upper Permian Mizukoshi Formation. *Journal of the Geological Society of Japan*, **64**, 222–231 [in Japanese].

YOON, S. H. & CHOUGH, S. K. 1995. Regional strike slip in the eastern continental margin of Korea and its tectonic implications for the evolution of Ulleung Basin, East Sea (Sea of Japan). *Geological Society of America Bulletin*, **107**, 83–97.

YOSHIDA, S., TASHIRO, M., OTSUKA, M. & NAKAZATO, H. 1985. Re-examination of geology of the Upper Himenoura Subgroup in Amakusa-Shimojima Island, Kyushu, Japan. *Fossils (Paleontological Society of Japan)*, 17–22 [in Japanese with English abstract].

YOSHIMOTO, A., OSANAI, Y., NAKANO, N., ADACHI, T., YONEMURA, K. & ISHIZUKA, T. 2013. U-Pb detrital dating of pelitic schists and quartzite from the Kurosegawa tectonic zone, southwest Japan. *Journal of Mineralogical and Petrological Sciences*, **108**, 178–183.

2f Izu–Bonin Arc

YOSHIYUKI TATSUMI, YOSHIHIKO TAMURA, ALEXANDER R. L. NICHOLS, OSAMU ISHIZUKA, NARUMI TAKAHASHI & KEN-ICHIRO TANI

The Izu–Bonin–Mariana (IBM) arc system extends 2800 km from the Izu Peninsula to Guam Island (Fig. 2f.1) and provides an excellent example of an intra-oceanic convergent margin where the effects of crustal anatexis and assimilation are considered to be minimal (Stern *et al.* 2003; Tatsumi & Stern 2006). The current IBM activity is caused by subduction of the Pacific Plate beneath the Philippine Sea Plate, which dips at 35° at the northern tip of this system and ia nearly vertical at Mariana. Evolution of the IBM Arc since 50 Ma has strongly influenced the present architecture of the Japanese archipelago. One important event during this arc system's evolution is back-arc rifting from 15–25 Ma that separated the IBM Arc system from the remnant Kyushu–Palau Ridge by *c.* 500 km and created the young oceanic lithosphere of the Shikoku Basin that is now being subducted beneath the SW Japan Arc. It should be further stressed that recent geophysical and geological survey results suggest this intra-oceanic arc is an active site of both creation and growth of the continental crust. This chapter will highlight the tectonic and structural evolution of the IBM arc system.

Oceanic arcs and the birth of continental crust (Y Tatsumi)

The Earth: a planet with a rough surface

Planet Earth shows a bimodal height distribution at the surface, in contrast to the topography of other terrestrial planets (Fig. 2f.2). Although Mars also exhibits an apparent crustal dichotomy, that is, the lower northern hemisphere v. the higher southern hemisphere, the distribution of topography on Mars becomes unimodal if the effect of the discrepancy between the geometric centre of figure and the centre of mass is removed (Smith *et al.* 1999; Aharonson *et al.* 2001). In contrast, the unique 'lowland and highland' topography of the Earth reflects the distribution of oceanic and continental crust, each with different density and thickness of on average 2900 v. 2700 kg m^{-3} and 6 v. 40 km, respectively. This density difference is due to the contrast in their average composition, that is, mafic (*c.* 50 wt% SiO_2) for oceanic crust v. intermediate (*c.* 60 wt% SiO_2) for continental crust (e.g. Christensen & Mooney 1995; Kelemen 1995; Rudnick 1995; Taylor 1995).

It is well established that mafic oceanic crust is manufactured at oceanic ridges by decompression melting of mantle materials, whereas intermediate igneous rocks produced at convergent plate boundaries implicate arcs as central to continental crust formation (McLennan & Taylor 1982). It should be stressed, however, that intermediate (andesitic) magmatism is dominant in continental arcs which already have mature continental crust, but is rare in juvenile intra-oceanic arcs which lack continental basement (e.g. Tamura & Tatsumi 2002). It is clearly absurd that the existence of continental crust is necessary for the creation of further new continental crust, yet the means by which the initial arc crust, composed of mantle-derived mafic magma, evolves into intermediate continental crust remains largely unknown. In order to explain the intermediate composition of the average continental crust, two processes are postulated: first, differentiated magmas of intermediate composition must be generated, either by fractionation of mafic melts or by anatexis of mafic crust; and second a significant fraction of mafic–ultramafic components must be removed by delamination or lower-crust foundering (e.g. Rudnick 1995; Tatsumi 2005). In addition, mantle-derived, primary andesites such as boninites, which typify magmatism at the initial stage of IBM arc system evolution, could comprise an important part of the continental crust (e.g. Shirey & Hanson 1984).

Intra-oceanic arcs are better places to examine the process of continental crust formation than continental arcs, as the effects of pre-existing crust, which should contribute to the crustal growth processes via complex interaction, are minimized because the upper plate is initially mafic. High-resolution seismic profiles along and across the IBM Arc crust (Suyehiro *et al.* 1996; Kodaira *et al.* 2007*a*, *b*; Takahashi *et al.* 2008) reveal a middle crust with P-wave velocity (V_p) of 6.0–6.8 km s^{-1}, a value equivalent to the average V_p of continental crust and typical of plutonic rocks having intermediate compositions (Christensen & Mooney 1995). This characteristic layer has therefore been interpreted as intermediate in composition (Kitamura *et al.* 2003; Takahashi *et al.* 2008; Tatsumi *et al.* 2008). This interpretation is supported by the presence of tonalitic xenoliths in lavas and as exposures along intra-oceanic arcs such as the IBM and Kyushu–Palau systems (e.g. Sakamoto *et al.* 1999), and by the exposure of tonalitic plutons in the Tanzawa Mountains of central Honshu, where the IBM arc system collides with Honshu Island (Kawate & Arima 1998; Tani *et al.* 2010). Furthermore, the presence of a middle crust layer exhibiting such a characteristic V_p has been documented for other arcs in oceanic regions such as Tonga (Crawford *et al.* 2003) and Kurile (Nakanishi *et al.* 2009). These lines of evidence provide a compelling foundation to support the challenging and paradoxical hypothesis that continents have been born in the ocean via manufacturing processes in a subduction 'factory'.

Initiation of plate subduction

The creation of two types of crust on Earth is due to magmatism at divergent and convergent plate boundaries. Heat from the deep interior of the Earth is removed predominantly by creation and recycling of the mobile oceanic plate, that is, via plate tectonics and mantle convection. It is interesting that of all the current terrestrial planets in the solar system plate tectonics only operates on Earth, although other planets may also have the liquid, and hence high-temperature core, that transfers heat via the mantle towards the surface. Since mantle viscosity depends strongly on temperature, a very viscous cold boundary layer will move so slowly that the cooling due to the surface recycling becomes far less effective than the conductive heat transport through the boundary layer, a situation referred to as the stagnant lid regime (e.g. Solomatov & Moresi 1996). If this is the case, then why are the Earth's plates or lithosphere weak enough to move? An intuitive answer to this question is that the presence of liquid water at the surface weakens the plate, allowing fault

From: Moreno, T., Wallis, S., Kojima, T. & Gibbons, W. (eds) 2016. *The Geology of Japan*. Geological Society, London, 175–199.

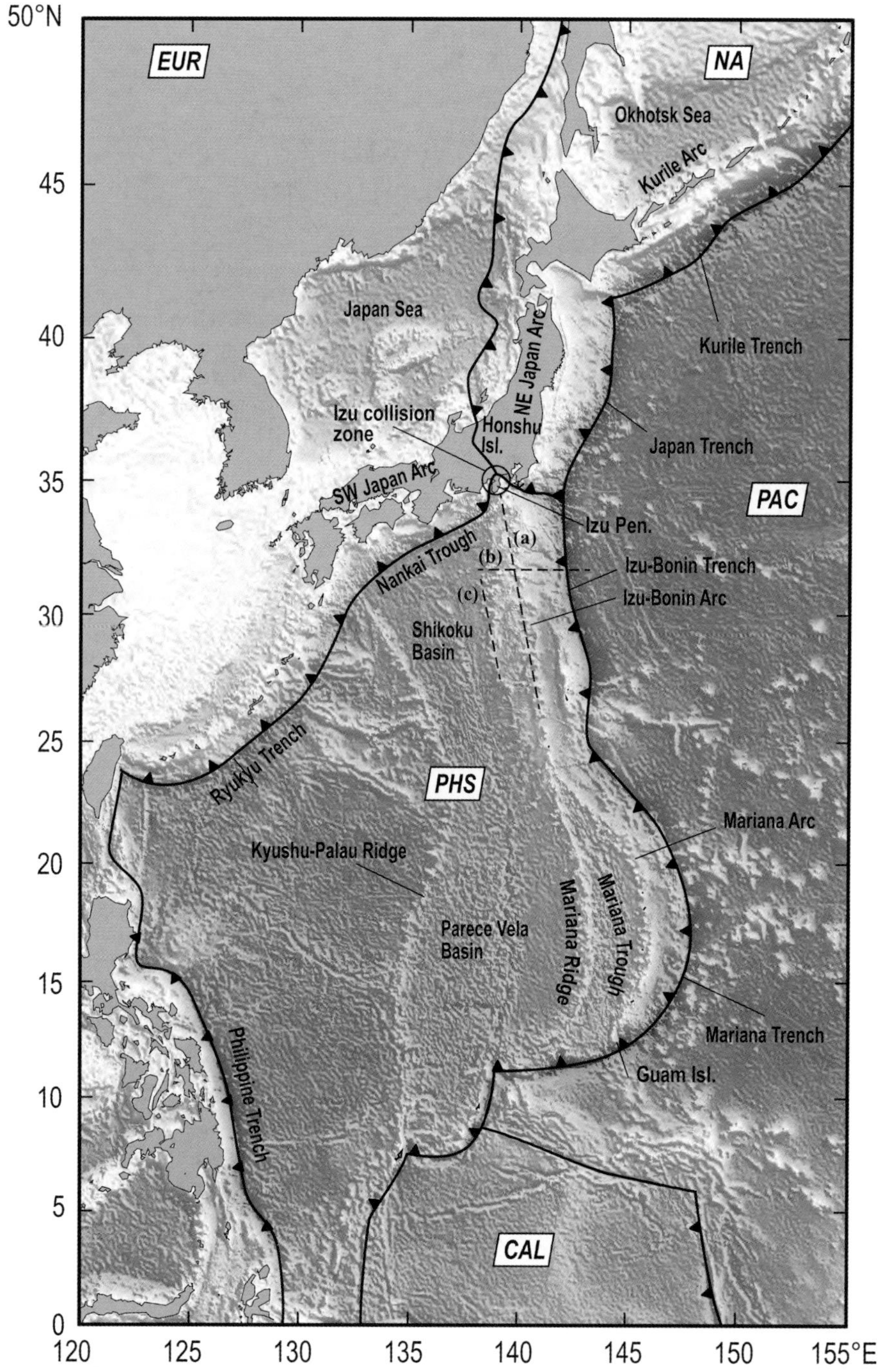

Fig. 2f.1. Tectonic map of the Izu–Bonin–Mariana (IBM) arc system and its surrounding plates. CAL, Caroline Plate; EUR, Eurasia Plate; NA, North America Plate; PAC, Pacific Plate; PHS, Philippine Sea Plate. Old seafloor (180–135 Ma) of the western Pacific Plate subducts beneath the active IBM Arc along the Izu–Bonin–Mariana trenches. Spreading centres are active in the Mariana Trough (7–0 Ma) and relic in the Shikoku and Parece Vela basins (30–15 Ma) and West Philippine Basin (50–35 Ma). The Ogasawara Plateau, Amami Plateau, Daito and Oki-Daito ridges are Cretaceous–Eocene features. The Kyushu–Palau Ridge (KPR) marks the rifted western edge of the initial IBM arc system (50–30 Ma), subsequently separated by back-arc spreading into the Shikoku and Parece Vela basins. The black dashed lines show the locations of the wide-angle seismic profiles, shown in Figure 2f.3, along the present-day volcanic front (Kodaira *et al.* 2007*a*, *b*), across the arc (Suyehiro *et al.* 1996) and along the back-arc *c.* 150 km west of the volcanic front (Kodaira *et al.* 2008).

boundaries to develop within the rigid plate and finally initiating collapse or subduction of denser plates (e.g. Kohlstedt *et al.* 1995). Comprehending the process of subduction initiation is therefore of primary importance for understanding Earth evolution.

Two general mechanisms have been advanced for subduction initiation: induced and spontaneous. The former results from continued convergence resulting from slab pull along-strike of a given system despite local jamming of the subduction zone by buoyant continental or thickened oceanic lithosphere (Gurnis *et al.* 2004). On the other hand, the IBM arc system is proposed to be an example of spontaneous subduction initiation wherein subsidence of relatively old, hence dense, lithosphere commenced along transform faults and/or fracture zones adjacent to relatively young and buoyant lithosphere (Stern 2004). This idea has developed from an analysis of IBM forearc volcanic stratigraphy, which provides a basis for determining the petrogenetic evolution of the magmas that immediately post-date subduction initiation.

Volcano-tectonic evolution (OI)

The evolution of the IBM arc system is among the best known of any convergent margin (e.g. Stern *et al.* 2003). Extensive sampling utilizing dredging, submersible dives and ocean drilling revealed the outline of the volcano-tectonic history of this arc system (e.g. Taylor 1992; Stern *et al.* 2003). Recent efforts to obtain reliable age data on volcanic rocks have provided time constraints on this history (Cosca

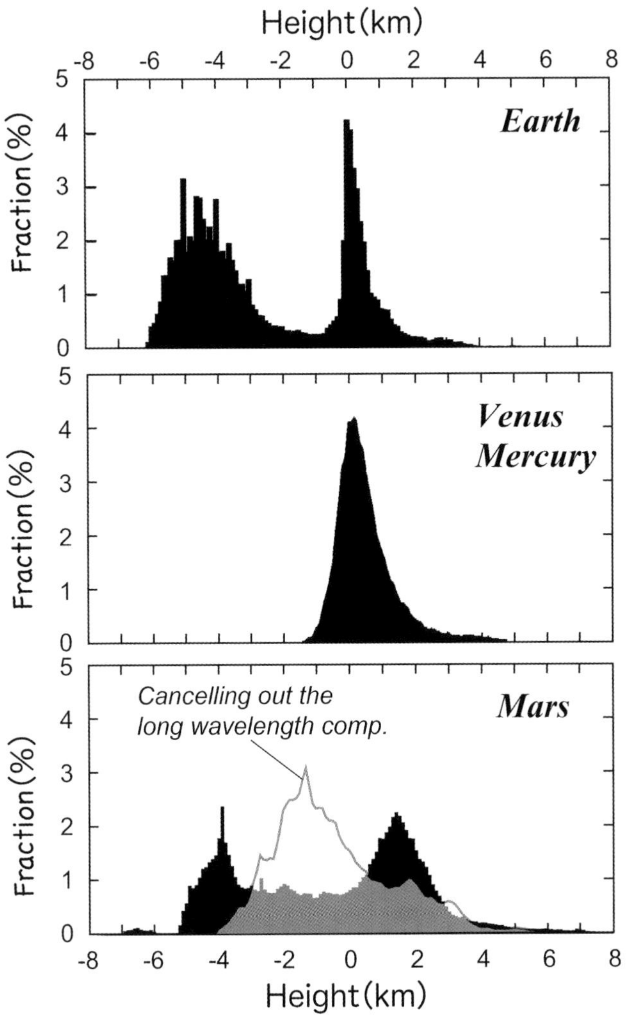

Fig. 2f.2. Surface topography of the terrestrial planets (Sharpton & Head 1985; Smith *et al.* 1999; Aharonson *et al.* 2001).

et al. 1998; Ishizuka *et al.* 1998, 2002*a*, *b*, 2003*a*, 2006*a*, *b*, 2007, 2009, 2010, 2011*a*, *b*). The volcano-tectonic history of the IBM Arc system can be divided into six stages (Fig. 2f.3):

(1) subduction initiation stage, 52–48 Ma;
(2) first arc volcanism (boninite, high-Mg andesite), 48–44 Ma;
(3) second arc volcanism (tholeiitic, calc-alkaline), 44–25 Ma;
(4) first spreading of back-arc basin (the Shikoku Basin), 25–15 Ma;
(5) third arc volcanism (tholeiitic, calc-alkaline), 17–3 Ma; and
(6) rifting in the back-arc and tholeiitic volcanism along the present volcanic front, 2.8–0 Ma.

Subduction initiation volcanism, 52–48 Ma

The IBM forearc is an excellent location for investigating the record of subduction initiation and subsequent arc evolution (e.g. Meijer 1980; Stern & Bloomer 1992; Bloomer *et al.* 1995), partly because of the exposure of early-arc lavas on the forearc islands (e.g. Reagan & Meijer 1984; Umino 1985). Stratigraphy of the forearc crust in the Bonin Ridge area comprises, from bottom to top (Ishizuka *et al.* 2011*a*): (1) gabbroic rocks; (2) a sheeted dyke complex; (3) basaltic lava flows; (4) volcanic breccia and conglomerate containing boninitic clasts; (5) boninite and tholeiitic andesite lava flows and dykes; and (6) tholeiitic and calc-alkaline arc volcanic rocks (with the latter two divisions exposed on the Bonin islands). Basaltic rocks recovered in the IBM forearc from stratigraphic levels below boninite have chemical compositions that are similar to mid-ocean ridge basalt (MORB) and the term 'forearc basalt' (FAB) has been proposed by Reagan *et al.* (2010). Most of the reliable $^{40}Ar/^{39}Ar$ ages for FAB from the submarine Bonin Ridge are identical within error and indicate that FAB magmatism occurred at *c.* 50–52 Ma, preceding boninite eruption by 2–3 myr. (Ishizuka *et al.* 2011*a*). U–Pb zircon ages from gabbro below the FAB indicate that these are contemporaneous (Ishizuka *et al.* 2011*a*) and probably comagmatic. Forearc basalt and related gabbro represent the first magmas produced as the IBM subduction zone began to form. One important characteristic of the common forearc stratigraphy in the IBM forearc is the association of sheeted dykes with basaltic pillow lavas, which strongly implies that the eruption of FAB was associated with seafloor spreading. This is supported by the seismic velocity structure of the Bonin Ridge area (Kodaira *et al.* 2010), showing it to have a thin ocean-ridge-like crust (<10 km). It appears that the FAB was produced by seafloor spreading associated with subduction initiation along the length of the IBM forearc, as suggested in the conceptual model of Stern & Bloomer (1992) and the numerical model of Hall *et al.* (2003).

Published numerical models of subduction initiation require at least 100 km of convergence before a subduction zone nucleates, with self-sustaining subduction occurring shortly thereafter (Hall *et al.* 2003). The 52–51 Ma age for the onset of FAB magmatism can be considered as the age of initiation of slab sinking followed by self-sustaining subduction. This age nearly coincides with the best estimate of the change in motion of the Pacific Plate deduced from the age of the Hawaiian–Emperor bend (*c.* 50 Ma: Sharp & Clague 2006). Because the FAB volcanism appears to be nearly synchronous with (at least not later than) the change in plate motion, it appears that it was the onset of subduction that changed the plate motion.

The potential location of subduction nucleation is along the Mesozoic-aged crust that is now found at the margins of the West Philippine Basin (e.g. Amami Plateau: Hickey-Vargas 2005; Daito Ridge: Ishizuka *et al.* 2011*a*, *b*; Huatung Basin: Deschamps *et al.* 2000; Hickey-Vargas *et al.* 2008). One possible scenario for subduction initiation at the IBM Arc was that it was induced by overthrusting of the Mesozoic arc and back-arc or forearc terranes bounding the east side of the Asian Plate over the Pacific Plate, followed by failure of the Pacific Plate lithosphere and subduction initiation. Alternatively, subduction could have begun spontaneously, facilitated by the density contrast between the arc-bearing Mesozoic Asian crust and the old oceanic Pacific lithosphere to its west.

First arc volcanism (boninite, high-Mg andesite), 48–44 Ma

Boninite volcanism occurred as the nascent subduction zone evolved and followed FAB volcanism. The type locality of boninite is in the Bonin islands, an uplifted segment of the IBM forearc. Exposures of boninite and other early arc lavas are better exposed on the Bonin islands than anywhere else in the world. $^{40}Ar/^{39}Ar$ dating indicates that boninitic volcanism on Chichijima Island occurred briefly during Eocene time, between 48 and 46 Ma. Boninite has been found in the same stratigraphic position at several locations along the entire *c.* 2500 km length of the IBM forearc, suggesting that boninite volcanism occurred contemporaneously along the entire length of the IBM arc system beginning at *c.* 48 Ma (Ishizuka *et al.* 2006*a*).

Boninites from the Bonin islands are characterized by high MgO at given SiO_2, low abundances of high-field strength elements

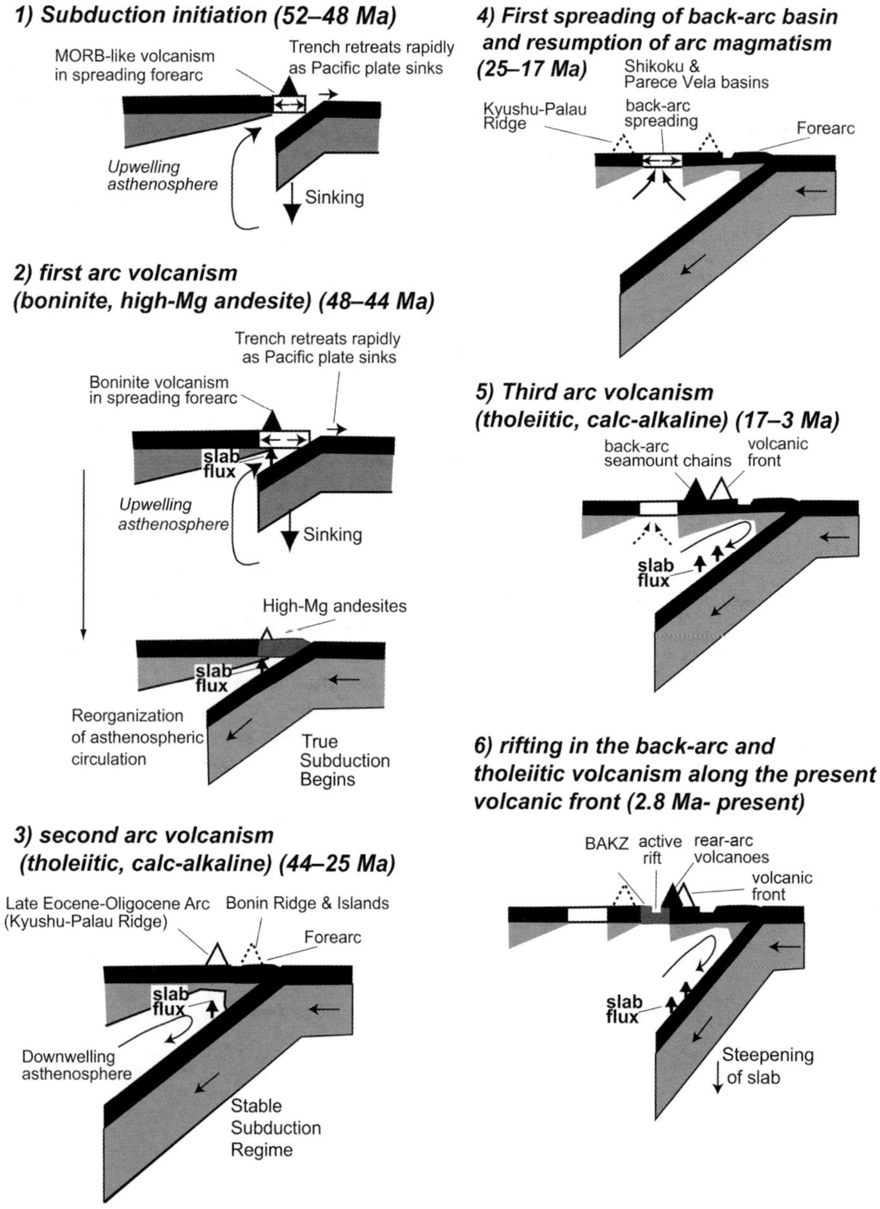

Fig. 2f.3. Schematic diagram of volcano-tectonic evolution of the Izu–Bonin section of the IBM Arc, modified after Ishizuka *et al.* (2006*a*, 2008).

(HFSE), low Sm/Zr, low abundances of rare Earth elements (REE) and U-shaped REE patterns. These 'low-Ca boninites' (Crawford *et al.* 1989) can be explained by low-pressure melting of a depleted harzburgite that was massively affected by a slab flux (e.g. Taylor *et al.* 1994; Ishizuka *et al.* 2006*a*). This is in contrast to FAB, which was much less affected by subduction-related fluids or melts.

Following boninite magmatism, transitional high-Mg andesite (Taylor *et al.* 1994; Ishizuka *et al.* 2006*a*; Reagan *et al.* 2008) started to erupt after *c.* 45 Ma. Boninite series lavas from Guam with compositions that are similar to high-Mg andesite lavas from the Bonin Ridge erupted at *c.* 44 Ma. This implies that eruption of high-Mg andesites with higher REE and TiO_2 concentrations than the Bonin Ridge boninites are also contemporaneous along the length of the arc.

In summary, shallow melting of depleted mantle with the aid of hydrous fluids from the newly subducted slab produced boninitic volcanism nearly simultaneously along the entire length of the IBM arc system.

Second arc volcanism (tholeiitic, calc-alkaline), 44–25 Ma

After *c.* 44 Ma, tholeiitic to calc-alkaline andesites from the Bonin Ridge show strong characteristics of arc magmas: they are relatively depleted in Nb and enriched in fluid-mobile elements such as Sr, Ba, U and Pb. These characteristics indicate that by 45 Ma, near-normal configurations of mantle flow and melting, as well as subduction-related fluid formation and metasomatism, were established in the IBM Arc system. These andesites therefore represent a transitional stage from forearc spreading (represented by forearc basalt and perhaps boninites) to the stable, mature arc that developed in Late Eocene–Early Oligocene time. These OPX-bearing high-Mg tholeiitic to calc-alkaline andesites were erupted along the western escarpment of the Bonin Ridge, as the arc magmatic axis localized and retreated from the trench. Overall, after *c.* 44 Ma a continued influx of hydrous fluids into increasingly fertile lherzolitic mantle generated tholeiitic and calc-alkaline magma (Ishizuka *et al.* 2006*a*), marking the point at which a mature, stable arc magmatic system was finally established.

The post-44 Ma arc magmatic system that developed in Late Eocene–Oligocene time is well exposed on the Kyushu–Palau Ridge (KPR), which is a remnant arc separated from the IBM arc system by Shikoku–Parece Vela Basin back-arc spreading from 25 to 15 Ma. Radiometric ages indicate that samples collected from the northern to central KPR range in age between 43 and 25 Ma (mostly between 27 and 25 Ma), with the youngest defining the time at which arc volcanism ceased on the KPR. In other words, back-arc rifting (or spreading) of the Shikoku and Parece Vela basins initiated at *c.* 25 Ma (Ishizuka *et al.* 2011*b*). This interpretation is generally consistent with the timing of rifting and spreading of the Shikoku Basin based on magnetic anomaly data and seafloor fabric observations. Okino *et al.* (1994) identified a magnetic lineation corresponding to Anomaly 7 in the western part of the basin, and interpreted this as recording that the Shikoku Basin spreading started at 26 Ma. The lack of a systematic age variation of volcanic rocks along the KPR indicates that rifting was initiated almost contemporaneously along the entire KPR between 30 and 11°N. By extension, this means that initiation of the Shikoku and Parece Vela basins occurred at about the same time.

An exact location for the Oligocene volcanic front, which is the conjugate pair of the KPR, is not clear. Recent dating results demonstrate that the forearc volcanic edifices are all Eocene (ODP Sites 786, 792 and 793: Cosca *et al.* 1998; Ishizuka *et al.* 2011*a*, *b*). This implies that the Oligocene volcanic front was situated west of the Eocene volcanism, near the modern volcanic front. However, no evidence of the Oligocene volcanism is exposed on the surface, presumably due to accumulation of younger volcanic rocks and volcaniclastics. One exception could be the Omachi Seamount (31.79 and 37.29 Ma: Ishizuka *et al.* 2011*b*), which could represent an Eocene–Oligocene volcano situated slightly towards the back of the arc. This seamount seems to have been isolated from the volcanic front by *c.* 31 Ma by rifting in the Ogasawara Trough and rifting after 25 Ma seems to have taken place to the west of this seamount, leaving it stranded in the frontal arc. Basaltic to andesitic lavas from KPR are enriched relative to MORB in large-ion lithophile elements (LILE) and depleted in HFSE such as Nb, confirming its arc affinities and the involvement of depleted MORB-type mantle. Sr–Pb–Nd isotopic relations confirm that KPR lavas have end-member components in their sources that are similar to those in post-45 Ma Eocene forearc lavas, which are distinct from Chichi-jima boninite sources.

First spreading of back-arc basin and resumption of arc magmatism, 25–15 Ma

IBM Arc volcanism was quiescent between 22 and 17 Ma during the spreading of the Shikoku and Parece Vela back-arc basins. Recent efforts to date the magmatic events during the last stage of the back-arc spreading revealed that they were more prolonged than previously thought. The Kinan Seamount Chain, which overlaps with the extinct spreading centre of the Shikoku Basin, gave $^{40}Ar/^{39}Ar$ ages of 15.5–7.7 Ma (Ishizuka *et al.* 2009). Volcanic rocks from the Kinan Seamount Chain show a distinct systematic temporal variation in their geochemical characteristics (Ishizuka *et al.* 2009). Abundances of the incompatible elements are significantly higher in the younger samples, such that the lavas progress from a medium-K through high-K to shoshonitic LREE/HREE, and more-incompatible/less-incompatible element ratios all progressively increase with time as their Nd isotopic composition becomes less radiogenic. This change could be explained by weaker, fading mantle convection beneath the Shikoku Basin. After Shikoku Basin spreading stopped, the upward flow of asthenosphere decreased; decompression mantle melting would consequently have been suppressed, so that as a result only enriched blobs were preferentially melted.

Tani *et al.* (2011*a*, *b*) recently reported a much younger episode of magmatism (as young as 7.9 Ma) from an oceanic core complex adjacent to the extinct spreading centre of the Parece Vela Basin than the previous estimate (*c.* 12 Ma) which was based on interpretations from magnetic anomaly data. These new age constraints on the final magmatic event of the back-arc basins make the period between this spreading event and the next rifting/spreading event much shorter or almost non-existent.

Third arc volcanism (tholeiitic, calc-alkaline), 17–3 Ma

Toward the end of the Shikoku–Parece Vela basin spreading, IBM Arc volcanism resumed at *c.* 17 Ma (Taylor 1992; Ishizuka *et al.* 2003*a*). Activity at the volcanic front was characterized by bimodal eruptions of low-K tholeiitic basalts and rhyolites (Taylor 1992). From Late Oligocene–Neogene times the Izu–Bonin Arc source became depleted in incompatible trace elements during back-arc basin formation, and re-enriched after spreading stopped at *c.* 17 Ma in the basin (e.g. Gill *et al.* 1994; Hochstaedter *et al.* 2001; Straub 2003; Tollstrup *et al.* 2010). Neogene volcanism behind the volcanic front (but not associated with back-arc rifting: a setting sometimes termed 'rear-arc' e.g. Stern 2010) is recorded in seamount chains crossing the rear regions of the Izu–Bonin Arc. After 15 Ma, low-K tholeiitic volcanism at the volcanic front and tholeiitic-calc-alkaline volcanism with more enriched chemical characteristics in the back-arc region are continuously dominant and only subtle temporal differences are evident (Ishizuka *et al.* 1998; Taylor & Nesbitt 1998; Hochstaedter *et al.* 2000, 2001; Bryant *et al.* 2003; Ishizuka *et al.* 2003*a*, *b*; Straub 2003; Machida *et al.* 2008). There is a consistent difference in chemistry between magmas erupted at the front and rear of the arc throughout Neogene times.

The Neogene seamount chains form a reasonable age progression, with older chains furthest from the current Izu–Bonin volcanic front (Ishizuka *et al.* 2003*a*, 2009, 2010). Ishizuka *et al.* (2003*a*) interpreted this trend as an eastwards retreat of the area of active seamount volcanism, perhaps due to the steepening of Pacific Plate subduction with time (Ishizuka *et al.* 2006*b*). Provided that dehydration and melting of the subducting slab takes place across a limited pressure and temperature range, slab steepening would narrow the lateral extent of fluid release. This in turn would reduce the extent of volcanism and the distance between the volcanism and the trench with time. Steepening of the slab seems to be consistent with initiation of rifting after 2.8 Ma in the Izu–Bonin Arc (Taylor 1992; Ishizuka *et al.* 2002*a*) as slab steepening could cause extension of the overriding plate (e.g. Carlson & Mortera-Gutierrez 1990). Another possibility is that the mantle temperature and/or convection decreased with time. This may have changed the magmatic focus in at least two ways: lower temperatures may narrow the region of mantle melting, while diminished mantle convection might affect the decompression melting as well as the thermal structure of the wedge. In addition, a change in slab surface temperature could affect the depth of dehydration and/or melting of the slab, which in turn would influence the extent and position of magma production.

Rifting in the back-arc and tholeiitic volcanism along the present volcanic front, 2.8–0 Ma

Back-arc rifting initiated just behind the volcanic front in the central part of the Izu–Bonin Arc after 2.8 Ma (Taylor 1992; Ishizuka *et al.* 2002*a*). The back-arc knolls zone (BAKZ), lying west of the active rift zone, consists of north–south-trending ridges and small volcanoes (Ishizuka *et al.* 1998, 2002*a*). Using detailed bathymetry and sidescan sonar data, Morita *et al.* (1999) and Ishizuka *et al.* (2002*a*) demonstrated that extension affected the 50 km width of the BAKZ. Basaltic volcanism formed north–south-trending ridges at 2.8–2.5 Ma and is presumed to be the earliest rift-related volcanism (Ishizuka *et al.* 2002*a*). Subsequent volcanism was widely distributed in the whole BAKZ, where it continued until at least 1 Ma. These volcanic rocks differ in composition from those of the Neogene seamount chains which predate them. Volcanism post-3 Ma behind the volcanic front has been 'rift-type' which is bimodal in silica content and has trace element and isotopic compositions that can be distinguished from both the volcanic front and the seamount chains in the rear part of the arc. BAKZ basalts show a stronger arc-like signature compared to the Sumisu Rift lavas. This implies that rifting in the Izu–Bonin Arc had an initial stage when the lavas with a stronger arc signature erupted, prior to the eruption of back-arc basin basalts (Fryer *et al.* 1990). However, these rocks are not intermediate in composition and so are not typical arc rocks (Hochstaedter *et al.* 2001; Ishizuka *et al.* 2003*a*). The differences have been attributed to some combination of a transition from flux to decompression mantle melting as arc rifting commences, a change in the character of the slab-derived flux or a change in the mantle (Hochstaedter *et al.* 1990*a*, *b*, 2001; Ishizuka *et al.* 2003*a*, *b*, 2006*a*, *b*). Two different magmatic suites can therefore be differentiated behind the Izu–Bonin Arc since Early Miocene times: 'seamount-type' from 17 to 3 Ma and 'rift-type' from 3 to 0 Ma. Neither one of these formed in a typical spreading back-arc basin.

The active rift zone is located 10–15 km behind the Izu volcanic front and includes the Aogashima, Myojin and Sumisu rifts from north to south. Holocene basalt lavas (<0.1 Ma) occur along the central axis of the Sumisu Rift, whereas older lavas (<1.4 Ma) crop out along the rift walls (Hochstaedter *et al.* 1990*a*). Lavas from the active rifts are compositionally bimodal (Hochstaedter *et al.* 1990*a*). These active rift lavas show much weaker arc-like signatures relative to the volcanic front, back-arc seamount chains and BAKZ, and are more similar to MORB, that is, back-arc basin basalt. Sumisu Rift, where rifting is currently taking place, erupts lavas with lower Sr and Pb isotopic ratios and weaker enrichment in fluid-mobile elements than the volcanic front. This is consistent with slab-derived fluid contributing less to the Sumisu Rift basalts than to those at the volcanic front (Hochstaedter *et al.* 1990*b*; Taylor & Nesbitt 1998).

Rifting and associated basin formation have not been recognized in the Izu–Bonin Arc north of 34° N. However, the location of active volcanism also seems to have changed in this area in Late Pliocene times (Ishizuka *et al.* 2006*b*). Quaternary volcanism is limited to within 40 km of the volcanic front and appears to have narrowed since, whereas during the Miocene and Pliocene epochs active volcanism occurred in the back-arc seamounts and ridges as far as 120 km west of the current volcanic front.

In the southernmost part of the Izu–Bonin Arc, the northern tip of the Mariana Trough has been propagating northwards. This propagation has been recorded by the northwards migration of the eruption of unusually enriched magmas at 4.3 cm a^{-1} for the last 6 myr (Ishizuka *et al.* 2010). This process still appears to be operating, and at the present day the northern tip of the Mariana Trough is located around Io-to Island.

Geological and geophysical structure of the IBM arc system (Y Tamura & AN)

Recent studies demonstrate that most of the submarine exposures in the IBM forearc are Eocene–Oligocene in age (Ishizuka *et al.* 2006*a*, *b*). Moreover, most of these rocks have differentiated from mantle melts to varying degrees, implying that plutonic rocks from this era are likely to make up most of the crust. The abundance of Eocene–Oligocene volcaniclastic deposits drilled in the forearc (ODP sites 787, 792 and 793) further emphasizes the important role that Eocene and Oligocene magmatism played in the evolution of the Izu–Bonin Arc (Hiscott & Gill 1992; Gill *et al.* 1994). The *c.* 50 myr old IBM arc system therefore consists mostly of Eocene–Oligocene middle and lower crust that underlies the upper crust; these units are in turn covered by Quaternary volcanic rocks. However, the recent discovery of Mesozoic basement in the Izu–Bonin forearc (Ishizuka *et al.* 2011*a*, *b*) illustrates that pre-existing crust might have played an important role in the early development of the IBM arc system.

Magnetic anomalies and crustal structure

Yamazaki & Yuasa (1998) recognized three conspicuous north–south-trending rows of long-wavelength magnetic anomalies along the Izu–Bonin Arc, oriented slightly oblique to the present volcanic front. The eastern anomalies correlate with the frontal arc bathymetric highs such as the Omachi Seamount; the western anomalies coincide with the Kyushu–Palau Ridge (the remnant arc); and the middle anomalies lying near 139°E cross linear arrays of Miocene volcanoes. They attributed all three magnetic anomalies to loci of Oligocene magmatic centres and suggested that the magnetic anomalies were caused by induced magnetization associated with mafic Palaeogene plutonic bodies constituting the middle–lower crust. Observed crustal velocity profiles are consistent with these magnetic anomalies.

Full-crustal velocity profiles for the IBM Arc system, which resulted from several active-source wide-angle seismic studies in the northern Izu–Bonin Arc (Suyehiro *et al.* 1996; Kodaira *et al.* 2007*a*, *b*, 2008) and extend for 1050 km along the volcanic front and 500 km along the rear part of the arc *c.* 150 km west of the volcanic front, define continuous layers extending *c.* 200 km across the arc (Fig. 2f.4). Kodaira *et al.* (2007*a*, *b*) documented systematic crustal variations beneath the volcanic front (on wavelengths of 80–100 km), defined on the basis of average crustal velocities and the thickness of felsic- to intermediate-composition middle crust (P-wave velocity V_p of 6.0–6.8 km s^{-1}; Fig. 2f.4a). Kodaira *et al.* (2008) also showed undulating crustal thicknesses crossing Miocene volcanic cross-chains in the rear part of the Izu–Bonin Arc, *c.* 150 km west of the present volcanic front (Fig. 2f.4c). Crustal thicknesses are as great as 25–30 km beneath the rear part of the arc and have along-arc wavelengths of about 100 km. Interestingly the crust underlying this part of the arc thickens and thins, but these variations are not related to the surface volcanoes in contrast to the situation along the volcanic front (Kodaira *et al.* 2007*a*; Tamura *et al.* 2009). It is therefore suggested that most of the thick crust in the rear arc was created in Eocene–Oligocene times, before the Shikoku Basin opened. Moreover, the undulating pattern of total- and middle-crust thicknesses and the variations of average seismic velocity reflecting bulk composition of crust, form patterns similar to those along the present volcanic front. Three discrete thick crustal segments (20–25 km thick) in the rear side of the arc, and possible counterparts beneath the volcanic front (Kodaira *et al.* 2008), are shown in Figure 2f.4. The crust beneath the volcanic front is thicker than that to the rear, possibly because of Quaternary magmatism (Kodaira *et al.* 2007*a*, *b*; Tatsumi *et al.* 2008; Tamura *et al.* 2009) and/or a greater magmatic production rate beneath the Eocene–Oligocene volcanic front. Taylor (1992) interpreted the frontal arc highs (*c.* 50 km east of the present volcanic front) to mark the Oligocene volcanic front. However, the similarities in

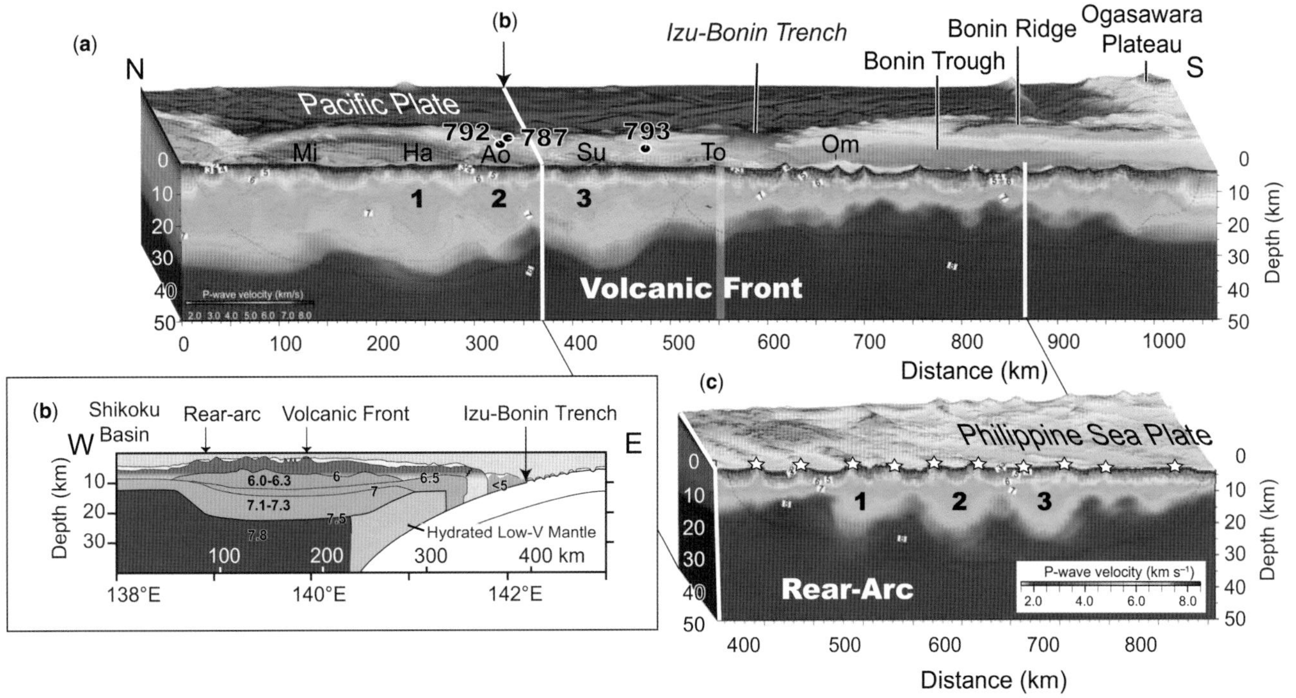

Fig. 2f.4. Three-dimensional block diagrams (bounded by the dashed lines in Fig. 2f.1) show seismic profiles of (**a**) the volcanic front, (**b**) across-arc and (**c**) the back-arc *c.* 150 km west of the volcanic front (after Suyehiro *et al.* 1996; Kodaira *et al.* 2008). Numbered circles indicate sites drilled during ODP Legs 125 and 126, which recovered Oligocene and Neogene turbidites. Abbreviations show basalt-dominant Quaternary volcanoes (Mi, Miyake; Ha, Hachijo; Ao, Aogashima; Su, Sumisu; To, Torishima) on the volcanic front and the andesitic Oligocene volcano (Om, Omachi seamount) east of the front. The stars on the back-arc profile indicate Mio-Pliocene volcanoes. Three discrete thick crustal segments (20–25 km thick) in the back-arc and their possible counterparts below the volcanic front (Kodaira *et al.* 2008) are numbered from 1 to 3. Schematic across-arc profile (P-wave velocity) of the Izu–Bonin Arc is shown in (b) after Suyehiro *et al.* (1996). The 6.0–6.3, 7.1–7.3 and 7.8 km s^{-1} layers correspond to parts of the middle crust, lower crust and upper mantle, respectively.

crustal variation patterns beneath the volcanic front and the possible palaeo-arc crust (preserved at the rear) suggest that many parts of the crust and its undulating pattern beneath the volcanic front might have been created before the Miocene opening of the Shikoku Basin. The seismic profile (Fig. 2f.4d) lies 20–50 km west of the magnetic anomaly along the rear part of the arc recognized by Yamazaki & Yuasa (1998). Kodaira *et al.* (2008) found good correlations between the seismic velocity image and the arrangement of magnetic highs; the strong magnetic highs lie immediately east of the thick crustal segments. The undulating crustal structure coincides with magnetic anomalies along the Izu–Bonin Arc, suggesting the possible existence and volumetric importance of Eocene–Oligocene crust.

Quaternary volcanoes

Bimodal volcanism of basalts and rhyolites

Volume-weighted histograms of rock types from the Quaternary Izu–Bonin Arc (30.5–35° N) indicate that basalt and basaltic andesite ($<$57 wt% SiO_2) are the predominant eruptive products, but dacite and/or rhyolite also form a major mode (Fig. 2f.5a, Tamura & Tatsumi 2002). About half of the edifices at the Quaternary volcanic front are calderas dominated by rhyolite (i.e. volcanic rocks with $>$70 wt% SiO_2) (Yuasa & Kano 2003; Tani *et al.* 2008). Turbidites drilled during ODP Leg 126 in the Izu–Bonin Arc, which have an age range of 31–0.1 Ma, are similarly bimodal (Gill *et al.* 1994). Rhyolites were also abundant in the Oligocene and Eocene (Reagan *et al.* 2008; Straub 2008) so that the IBM intra-oceanic arc system has included abundant rhyolite from its earliest stage to the present. This is not unique to the Izu–Bonin Arc. The 30–36.5° S sector of the Kermadec Arc (Fig. 2f.5b) is another example of an intra-oceanic arc with abundant Quaternary rhyolite (Smith *et al.* 2003, 2006; Wright *et al.* 2006). The crustal evolution of intra-oceanic arcs must therefore explain this bimodal volcanism of basalts and rhyolites.

Alternating basalt-dominant island volcanoes and rhyolite-dominant submarine volcanoes

Large basalt-dominated volcanoes are spaced at *c.* 100 km intervals along the Quaternary volcanic front and rhyolite-dominated calderas lie between the basaltic volcanoes north of 31° N, with gaps of 50–75 km where there are no volcanic edifices (Fig. 2f.6). Kodaira *et al.* (2007*a*) conducted an active-source wide-angle seismic study of this area and Figure 2f.6 shows the location of their seismic profile. A total of 103 ocean bottom seismographs (OBSs) were deployed at *c.* 5 km intervals along the profile. Figure 2f.7 shows the average SiO_2 content (in wt%) of Quaternary volcanic rocks and the thickness of the middle crust along the profile based on Kodaira *et al.* (2007*a*) and Tamura *et al.* (2009). Kodaira *et al.* (2007*a*) showed: (1) a remarkable along-arc periodic structural variation with a wavelength of 80–100 km in the average crustal velocities and in the thickness of the felsic to intermediate composition middle crust with a P-wave velocity (V_p) of 6.0–6.8 km s^{-1}; and (2) the structural variations correlate well with the average chemical compositions of overlying arc volcanoes. That is, the thickest total crust and the thickest intermediate-composition middle crust occur predominantly below the basaltic volcanoes, while the intermediate-composition middle crust tends to be thinner beneath the dacitic to rhyolitic volcanoes.

It is amazing that the velocity structure of this part of the Izu–Bonin Arc crust, which has a complex 50-million-year history, is related to the chemical composition of Quaternary volcanoes. Moreover, why should basaltic volcanoes stand on thicker average-velocity (more continental-like) crust than silicic volcanoes?

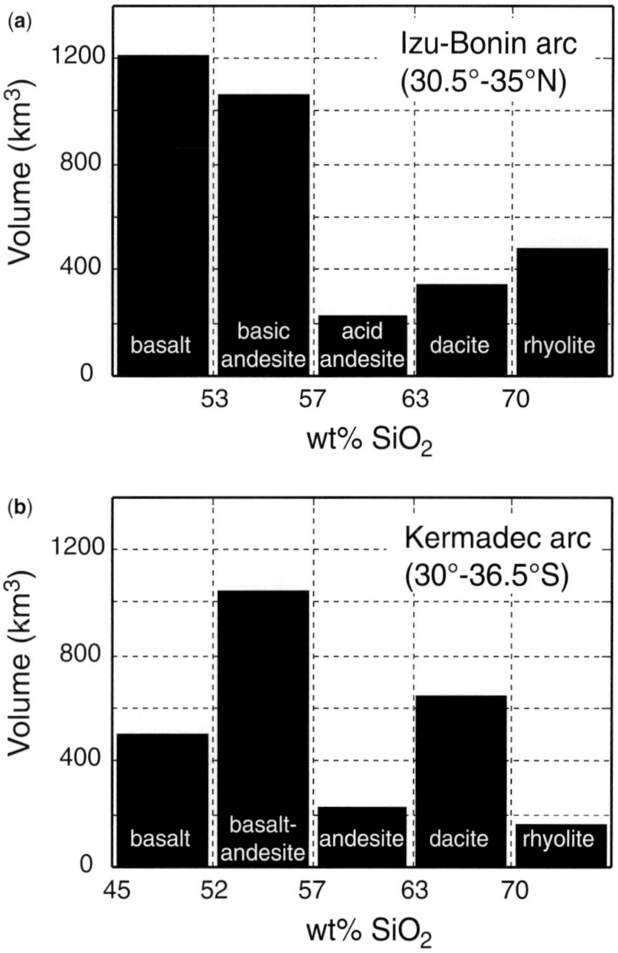

Fig. 2f.5. Volume-weighted histograms of rock types from (**a**) the Quaternary Izu–Bonin Arc (30.5–35° N) (Tamura & Tatsumi 2002) and (**b**) the 30–36.5° S sector of the Kermadec Arc (Wright *et al.* 2006).

Rhyolite genesis and crustal evolution

Tamura *et al.* (2009) suggest that three chemical types of Quaternary rhyolites exist in the Izu–Bonin Arc front, and there is a close relationship between volcano type and crustal structure: rhyolites from basalt-dominant volcanoes (R1), rhyolites from rhyolite-dominant volcanoes (R2) and rift-type rhyolites (R3). R1 rhyolites have the highest CaO/Al_2O_3, which is reflected in higher An content in their plagioclase phenocrysts. R1 rhyolites have lower Zr/Y and LREE-depleted patterns whereas the other two types have higher Zr/Y and flat to slightly enriched REE patterns. Whereas R1 rhyolites have Sr–Nd–Pb isotopes similar to Quaternary Izu–Bonin frontal basalts, the other two rhyolite types have significantly lower Sr–Pb and higher Nd isotope ratios. All of these characteristics are akin to rhyolites from the Oligocene arc and the Plio-Pleistocene back-arc extensional zone. Tamura *et al.* (2009) conclude that the crustal source materials of these three types of rhyolites are different. It is not well understood, however, why and how these crustal sources differ systematically and alternately along and behind the Izu–Bonin Arc. A new petrological model for the general evolution of arc crust and sub-arc mantle in the Izu–Bonin–Mariana arc-trench system was presented recently (Tatsumi *et al.* 2008). However, the correlation between smaller-scale along-strike crustal variations with a wavelength of *c.* 80 km and their correlation with Quaternary volcano chemistry (Kodaira *et al.* 2007*a*) are the primary concerns of this model.

It is possible that locally developed hot regions within the mantle wedge (hot fingers) could be responsible for the large basaltic volcanoes (Tamura *et al.* 2002). If these mantle sources remain stationary for millions of years, then basalt-dominant volcanoes will eventually overlie thicker crust (Kodaira *et al.* 2007*a*). In addition, this would suggest that rhyolite volcanoes have no mantle roots beneath the crust. Instead, dykes from the basalt volcanoes may provide the heat source to partially melt the crust (Tamura *et al.* 2009).

Can basaltic magma be transported laterally for long distances (10–50 km) from basalt volcanoes?

Recently, long-distance lateral magma transport within the crust has been recognized between Izu–Bonin Arc volcanoes (e.g. Nishimura *et al.* 2001; Geshi *et al.* 2002; Toda *et al.* 2002; Ishizuka *et al.* 2008). When Miyakejima Volcano erupted in 2000, seismic activity began beneath the volcano and migrated about 30 km north-westwards for 6 days, accompanied by ground deformation (Nishimura *et al.* 2001). These events are interpreted to reflect long-distance dyke injection from Miyakejima Volcano (Geshi *et al.* 2002; Toda *et al.* 2002). Another example is from Hachijojima Volcano. Ishizuka *et al.* (2008) showed geochemically that magma was transported laterally for $>$20 km at a depth of 10–20 km (middle–lower crust) from Hachijojima Volcano, which formed a 20 km long submarine volcanic chain (Hachijo NW chain). Moreover, seismic reflection profiles of the Izu–Bonin Arc crust support the existence of sills under rhyolite volcanoes. Kodaira *et al.* (2007*a*) found laterally continuous subhorizontal crustal reflectors within the arc crust along the volcanic front of the Izu–Bonin Arc. Many of these reflectors indicate major crustal layer boundaries that have an abrupt change of velocity gradient across them. However, other reflectors that crossed isovelocity contours are interpreted to be floating reflectors, representing features such as laterally intruded sills up to 50 km long. These lines of evidence suggest that magma can be transported laterally for long distances (10–50 km) from basalt volcanoes in the Izu–Bonin Arc. A forearc Pliocene basalt sill drilled at ODP Site 792 also indicates that magma intruded up to 50 km from a basalt-dominated volcano, Aogashima (Taylor *et al.* 1995).

Across-arc variation

The most obvious features of the Izu–Bonin Arc are the several *c.* 50 km long en echelon back-arc chains (Kan'ei, Manji, Enpo and Genroku chains) of large seamounts striking N60° E (Fig. 2f.6). The eastern end of the chains lie above the middle row of magnetic anomalies (Yamazaki & Yuasa 1998), whereas the western end lies on Shikoku Basin crust. In some cases (e.g. Manji and Genroku), the seamount chains align with large volcanoes on the volcanic front (e.g. Aoga-shima and Sumisu, respectively) and with areas of thickened middle and total crust, but the association is imperfect. Basalts to dacites of age 17–3 Ma have been dredged from many Neogene seamounts. Volcanism along these chains occurred sporadically along their total length, but lavas dredged from the top of the seamounts in the western part of the chains are generally older than those to the east (Ishizuka *et al.* 1998). Neogene volcanism along these rear seamount chains and at adjacent isolated seamounts therefore began at *c.* 17 Ma, slightly before the Shikoku Basin ceased spreading, and continued until *c.* 3 Ma (Ishizuka *et al.* 1998; 2003*b*). This type of volcanism ceased altogether at the initiation of rifting behind the volcanic front at *c.* 2.8 Ma (Ishizuka *et al.* 2002*a*, *b*).

Several explanations for the seamount chains have been proposed. They might be related to compression caused by collision between Honshu and the Izu–Bonin Arc associated with the Japan Sea opening (Karig & Moore 1975; Bandy & Hilde 1983). They may overlie Shikoku Basin transform faults (Yamazaki & Yuasa

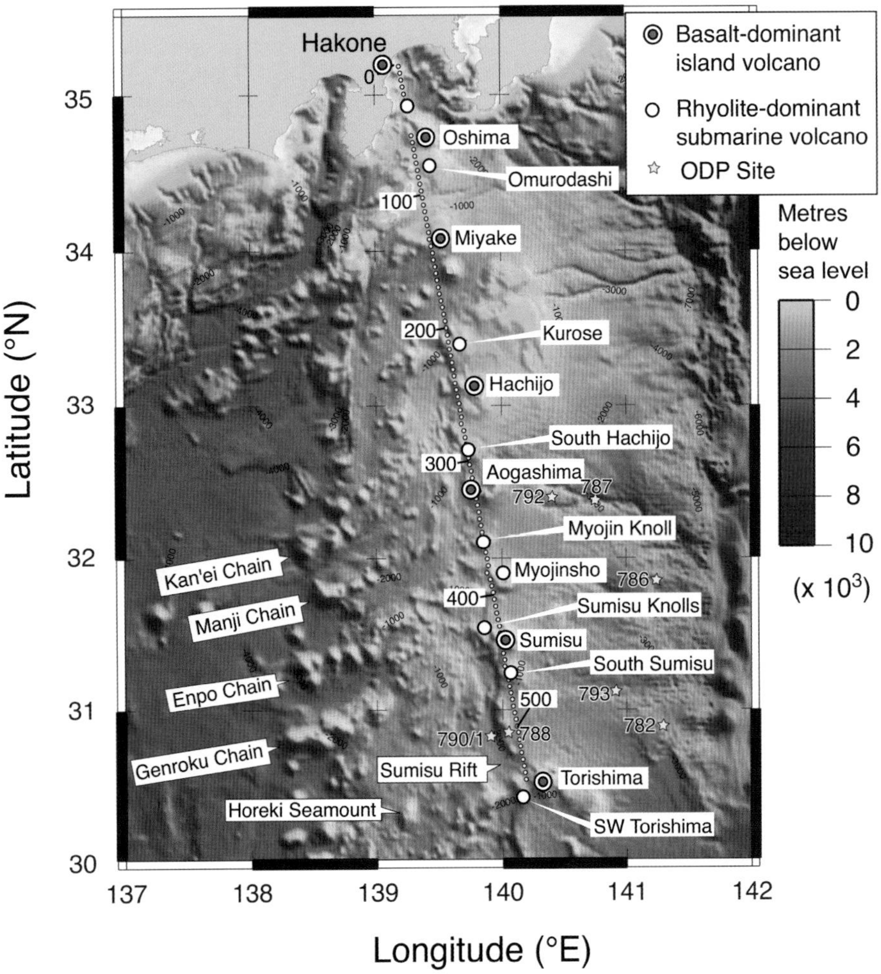

Fig. 2f.6. Map showing the 16 Quaternary volcanoes of the Izu Arc and the location of the wide-angle seismic profile after Tamura *et al.* (2009). Dots on the profile indicate a total of 103 ocean bottom seismographs (OBSs) which were deployed at *c.* 5 km intervals, and the numbers on the profile show the distance from its northern end (Kodaira *et al.* 2007*a*). Numbered stars indicate sites drilled on the Philippine Sea Plate in the Izu–Bonin region during ODP Legs 125 and 126. We refer to four tectonic settings of magmatism: the volcanic front; active rifts, which are located just behind and between the volcanic front volcanoes; a 100 km wide extensional zone that extends westwards from the active rifts; and back-arc seamount chains including Genroku, Enpo, Manji and Kan'ei that start in the Shikoku (back-arc) Basin west of the arc and continue into the extensional zone. We refer to magmatism in the active rifts and extensional zone as 'rift-type' and magmatism in the back-arc seamount chains as 'back-arc type'.

1998) or diapirs in the mantle wedge, similar to the 'hot fingers' proposed for NE Japan (Tamura *et al.* 2002; Honda *et al.* 2007).

A striking characteristic of subduction zone magmatism within many volcanic arcs of modest width is the consistent increase of their incompatible element concentrations, notably K_2O, away from the volcanic front (Gill 1981). Arc-front lavas are markedly dissimilar from those erupted in the rear of the arc in terms of major and trace element abundances, REE patterns and isotopic ratios (Gill 1981; Tatsumi & Eggins 1995). Such cross-arc asymmetry was recognized before the evolution of plate tectonic concepts (e.g. Kuno 1959; Dickinson & Hatherton 1967).

Figure 2f.8 shows the variation between K_2O and SiO_2 for lavas from frontal and rear-arc volcanoes of the northern part of the Izu–Bonin Arc. Basalts and andesites along the Izu–Bonin volcanic front contain significantly less K_2O than those from the rear arc (Fig. 2f.8). Lavas from the frontal volcanoes form a low-K suite as defined by Gill (1981), whereas rear-arc lavas are medium- and high-K. This across-arc variation is almost the same as that found in the NE Japan Arc (e.g. Tamura 2003; Kimura & Yoshida 2006). Figure 2f.9 shows a C1 chondrite-normalized REE plot for the Izu–Bonin basalts and andesites. All basalts from arc-front volcanoes are strongly depleted in the more incompatible light rare Earth elements (LREE) compared with the middle and heavy REE (MREE and HREE). In contrast, rear-arc basalts and andesites are enriched in the LREE and MREE compared with the HREE. In summary, basalt and andesite magmas at the front of the Izu–Bonin Arc are so depleted in K_2O and LREE that they are dissimilar to the 'average continental crust', whereas compositions of lavas erupted in the rear parts of the arc are closer approximations to the average continental crust of Rudnick & Gao (2003).

Figure 2f.10 shows along-arc Sr–Nd–Pb isotopic variations of lavas from volcanoes in the front and rear of the arc, the Sumisu Rift and Knolls (Fig. 2f.6). Rear-arc basalts and andesites contain less radiogenic Sr, Nd and Pb isotopes compared to those at the volcanic front (Hochstaedter *et al.* 2000; Ishizuka *et al.* 2003*a*, *b*; Tollstrup *et al.* 2010). Both rear and frontal volcanoes have a limited range of isotopic compositions compared with lavas from the rift. Moreover, the along-arc $^{206}Pb/^{204}Pb$ variations are coherent between the volcanic front and rear (Ishizuka *et al.* 2003*a*, *b*). Tollstrup *et al.* (2010) and Kimura *et al.* (2010) proposed that across-arc differences in the geochemistry of the Izu–Bonin Arc basalts are controlled by the addition of aqueous slab fluids to the volcanic front and hydrous partial melt of the slab to the more posterior parts of the arc.

Isotopic variation along the IBM subduction zone and slow anomalies of shear wave speed in the mantle wedge

Along the IBM arc system, Isse *et al.* (2009) detected three separate slow anomalies of shear wave speed in the mantle wedge at depths shallower than 100 km beneath the arc. Moreover, these three anomalies have a close relationship with the three groups of frontal and rear-arc volcanoes distinguished by their $^{87}Sr/^{86}Sr$, $^{143}Nd/^{144}Nd$ and $^{206}Pb/^{204}Pb$ along the arc from Oshima (34.72° N) to Chaife (14.66° N) (Volpe *et al.* 1987, 1990; Woodhead 1989; Hochstaedter *et al.* 1990*a*, *b*, 2001; Lin *et al.* 1990; Stern *et al.* 1993, 2003, 2006;

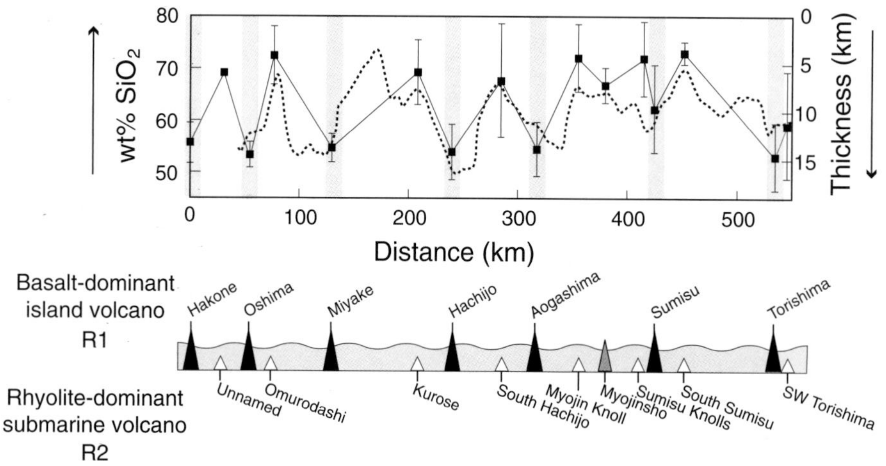

Fig. 2f.7. Along-arc crustal structure (dotted line: thickness of middle crust with V_p of 6.0–6.8 kms^{-1} at depths between 5 and 20 km) and average SiO_2 content (in wt%) of volcanic rocks (solid squares) sampled and dredged from the 16 Quaternary volcanoes of the Izu–Bonin Arc shown in Figure 2f.6. The basalt-dominant island volcanoes produce small volumes of rhyolites (R1). The rhyolite-dominant submarine volcanoes erupt mostly rhyolite (R2) (after Tamura *et al.* 2009).

Gribble *et al.* 1996, 1998; Elliott *et al.* 1997; Taylor & Nesbitt 1998; Sun & Stern 2001; Kuritani *et al.* 2003; Yokoyama *et al.* 2003; Tamura *et al.* 2005; Wade *et al.* 2005; Kohut *et al.* 2006; Ishizuka *et al.* 2007, 2010).

$^{87}Sr/^{86}Sr$ decreases gradually from north to south along the Izu–Bonin Arc between 35° N and 27.5° N. However, the isotopic trend is reversed with $^{87}Sr/^{86}Sr$ increasing south of 27.5° N (Taylor & Nesbitt 1998; Ishizuka *et al.* 2007). The maximum $^{87}Sr/^{86}Sr$ appears in arc volcanoes between 21° N and 25° N in the shoshonitic province in the southern Izu–Bonin Arc (Sun & Stern 2001) and northern seamount province (NSP) of the Mariana Arc. Volcanoes in the central island province (CIP) and southern seamount province (SSP) of the Mariana Arc have lower and more constant $^{87}Sr/^{86}Sr$, which are similar to those from the northern Izu–Bonin Arc (Fig. 2f.11a). $^{143}Nd/^{144}Nd$ variations along the arc mostly mirror variations seen for Sr isotopes, with the shoshonitic province of the southern Izu–Bonin Arc and NSP of the Mariana Arc having less radiogenic Nd than the rest of the arc (Stern *et al.* 2003). $^{143}Nd/^{144}Nd$ of Mariana Arc CIP and SSP lavas are lower than in lavas of the northern Izu–Bonin Arc (Fig. 2f.11b). $^{206}Pb/^{204}Pb$ increases continuously southwards in the Izu–Bonin Arc, but increases sharply south of 27.7° N to reach a maximum of *c.* 19.6 at Kita-io-to island (Taylor & Nesbitt 1998; Ishizuka *et al.* 2007). Ishizuka *et al.* (2007) showed that trace element compositions also vary systematically along-arc

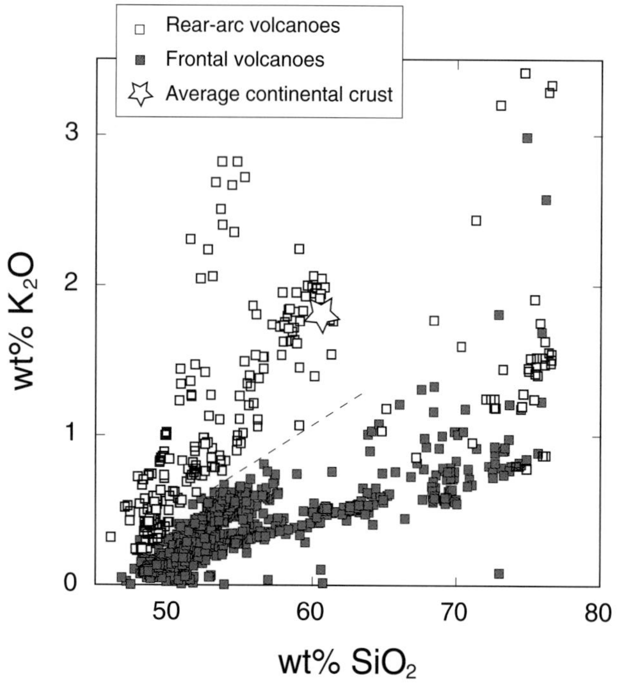

Fig. 2f.8. K_2O v. SiO_2 (wt%) in lavas from the volcanic front (Oshima, Miyakejima, Mikurajima, Hachijojima, Aogashima, Myojin Knoll, Sumisu and Torishima), the back-arc (Kan'ei, Manji, Enpo, Genroku, Horeki) and average continental crust (Rudnick & Gao 2003). Data are from Tamura & Tatsumi (2002), Machida & Ishii (2003), Tamura *et al.* (2005, 2007, 2009) and Machida (unpublished data). For reasons of clarity, no field is shown for rift-type magmas (<3 Ma).

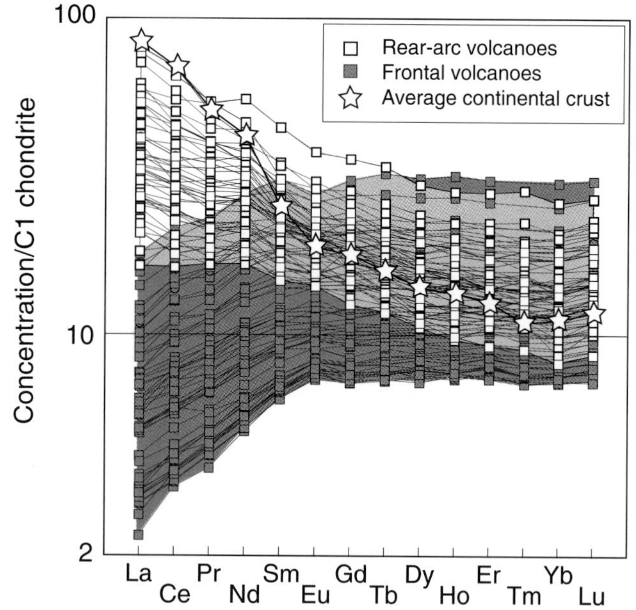

Fig. 2f.9. Chondrite-normalized rare Earth element (REE) abundances in the volcanic front and the back-arc basalts and andesites compared to average continental crust (Rudnick & Gao, 2003). Data from the volcanic front (Oshima, Miyakejima, Hachijojima, Aogashima, Sumisu and Torishima) from Taylor & Nesbitt (1998) and Tamura *et al.* (2005, 2007). Back-arc data (Kan'ei, Manji, Enpo, Genroku and Horeki) from Ishizuka *et al.* (2003*a*, *b*), Hochstaedter *et al.* (2001), Machida & Ishii (2003) and Ishizuka (unpublished data). Back-arc patterns deviate only slightly from average continental crust.

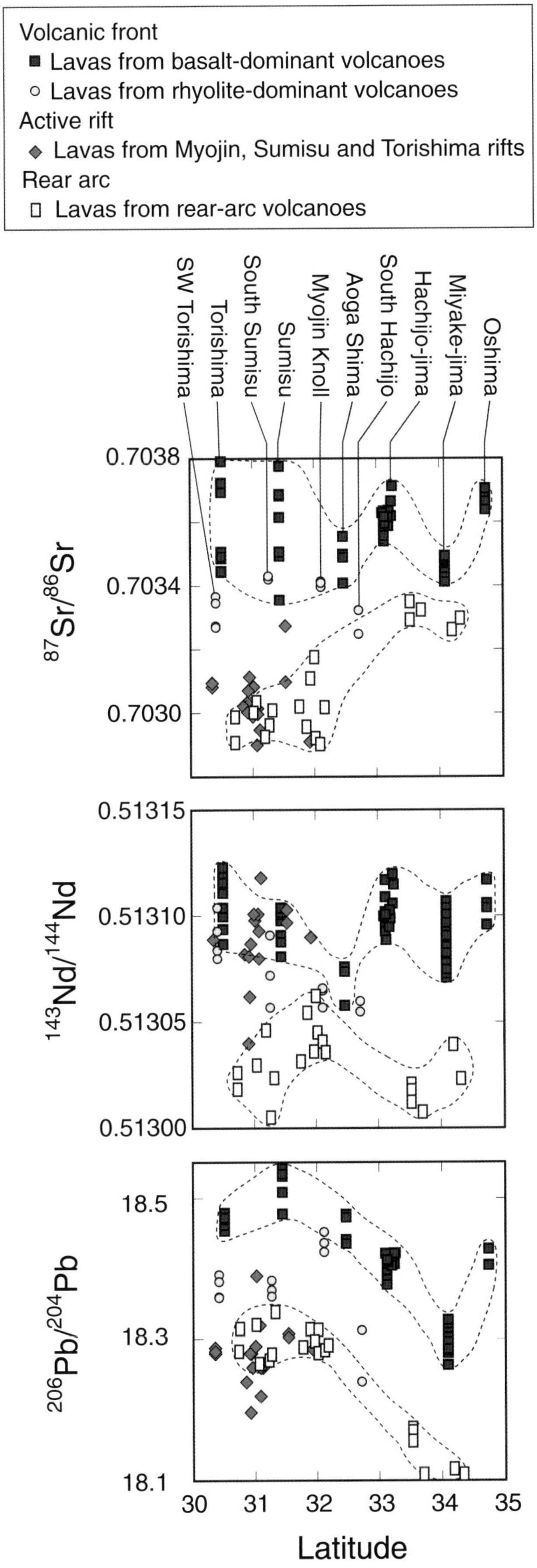

from north (35° N) to south (24.5° N) along the arc. $^{206}Pb/^{204}Pb$ values of the Mariana Arc (CIP and SSP) are higher than those of the northern Izu–Bonin Arc (Fig. 2f.11c).

The IBM Arc system can be divided into three geographic segments based on the isotopic values and the degree of isotopic variability. The Izu–Bonin segment north of 27.5° N is characterized by high $^{143}Nd/^{144}Nd$ (0.513093 on average) and low $^{206}Pb/^{204}Pb$ (18.47 on average). The volcanoes in the southern Izu–Bonin Arc (the shoshonitic province) and the NSP of the Mariana Arc between 27.5° N and 21° N have the highest $^{87}Sr/^{86}Sr$ and $^{206}Pb/^{204}Pb$, and the lowest $^{143}Nd/^{144}Nd$. Isotopic variability within this segment is characteristically large. The Mariana segment including the CIP and SSP, south of 21° N, is distinguished by lower $^{87}Sr/^{86}Sr$ (0.703443 on average) and $^{143}Nd/^{144}Nd$ (0.513013 on average), and higher $^{206}Pb/^{204}Pb$ (18.7741 on average) than the northern Izu–Bonin Arc. Ishizuka *et al.* (2007) suggested that the volcaniclastic sediments originating from HIMU (high-μ) oceanic islands on the subducting Pacific plate are possible candidates that could have introduced a component with high $^{206}Pb/^{204}Pb$ and low $^{143}Nd/^{144}Nd$ into the mantle wedge between 25° N and 27.5° N.

Nevertheless, the segmentation seen in the isotope ratios matches the segmentation in the shear wave speed anomalies (Fig. 2f.11d). The isotopic signatures of the back-arc volcanoes are less influenced by fluid-mobile recycled slab components, so the differences between each segment suggests that the areas of mantle being imaged in the northern and southern shear wave speed anomalies are isotopically different and independent of current subduction. Isse *et al.* (2009) suggest that each of the anomalies is a site of large-scale flow of upper mantle into the mantle wedge, and that each already contains a component from the adjacent subducting slab. How do the variations of Isse *et al.* (2009) correlate with these isotopic variations?

Seismic structure of the Izu–Bonin Arc (NT)

The Japan Agency for Marine-Earth Science and Technology (JAMSTEC) has been carrying out seismic surveys around the Izu–Bonin intra-oceanic arc area since 2002. These include refraction surveys using ocean bottom seismographs (OBSs) and multichannel reflection system (MCS) using a 6 km long streamer. Airgun arrays, each with a total capacity of 7800 cubic inches, were adopted as the seismic sources. Because the main objective of these surveys has been to understand the seismic structure of the whole system, the seismic lines have been distributed across the entire arc (Fig. 2f.12). The specification for the data acquisition is unified for all surveys to enable the structures obtained in each line to be compared. The OBSs, with three component velocity sensors, were deployed along the seismic lines at intervals of 5 km. For the refraction surveys we usually shot the airgun array every 200 m, while for the reflection survey the interval was 50 m.

Fig. 2f.10. Along-arc Sr–Nd–Pb isotopic variations in lavas from frontal-arc volcanoes (basalt-dominant and rhyolite-dominant volcanoes), the Myojin–Sumisu–Torishima rifts and back-arc volcanoes, Izu–Bonin Arc. Data from basalt-dominant volcanoes, those of rhyolite-dominant volcanoes, rifts and back-arc volcanoes are shown as black squares, grey squares, dark grey diamonds and white rectangles, respectively. Data from Tamura *et al.* (2009) (SW Torishima, Torishima, South Sumisu, Myojin Knoll, South Hachijo, Torishima Rift, Sumisu Rift and Myojin Rift), Taylor & Nesbitt (1998), Ishizuka *et al.* (2003*a*) (Quaternary frontal volcanoes, back-arc volcanoes and back-arc knolls), Ishizuka *et al.* (2008) (Hachijojima submarine parts), Yokoyama *et al.* (2003) (Miyakejima) and Hochstaedter *et al.* (1990*b*) (Sumisu Rift).

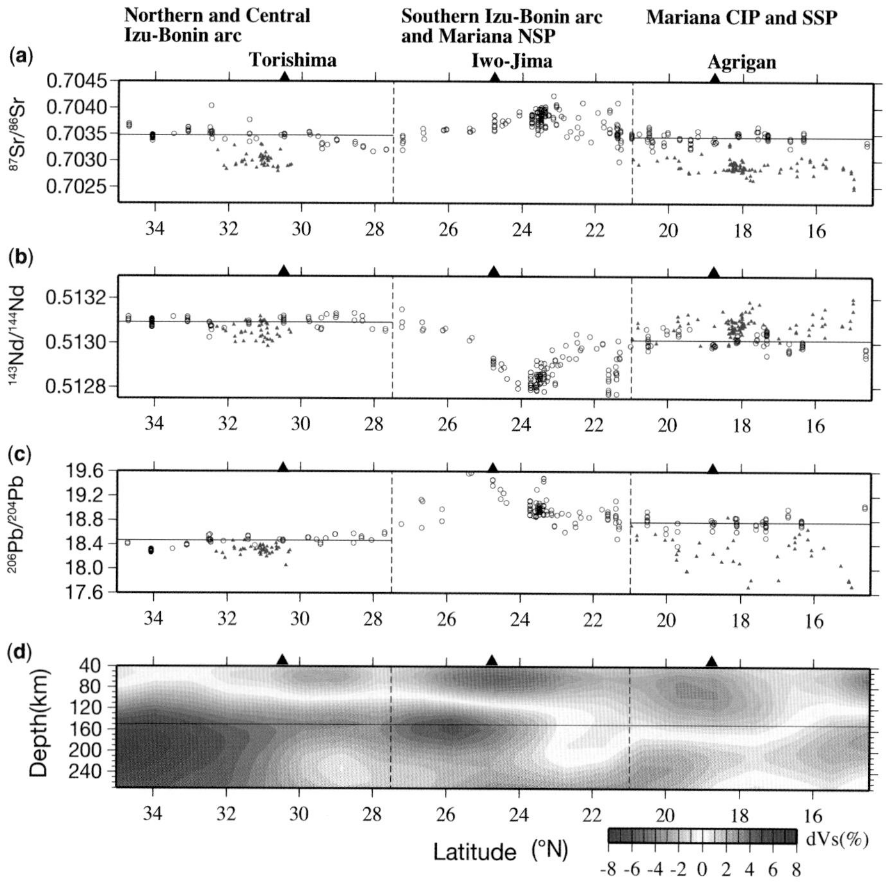

Fig. 2f.11. Along-arc variation of (**a**) $^{87}Sr/^{86}Sr$, (**b**) $^{143}Nd/^{144}Nd$ and (**c**) $^{206}Pb/^{204}Pb$ values from frontal volcanoes (circles) and back-arc volcanoes of the northern Izu and Mariana Trough (blue triangles). (**d**) Along-arc shear wave speed profile. Solid black line shows the approximate location of slab upper surface (after Isse *et al.* 2009).

The analysis methods also needed to be unified, and the tomographic inversion technique (Zhang *et al.* 1998) was adopted. The most important advantage of using this inversion is its objectivity. It is expected that the seismic structure of the arc is complicated due to the presence of many active volcanoes and intrusions. Since the analysis method had to be simple, the inversion technique using the first arrivals of the refractions was adopted. The methods depend strongly on the initial model, however. A complicated initial model results in a worse final model with many local minima and artificial noise. The initial model used here is homogeneous and constructed using the shallowest velocity structure obtained by the MCS data. Firstly, the shallow structure was estimated by inversion using first arrivals within a 20 km offset from each OBS. Using this result, a second inversion was performed using first arrivals within a 60 km offset. Finally, an image was obtained using all the offset arrivals and the results of the second inversion. The resolution of the final model was confirmed by a checkerboard test.

Suyehiro *et al.* (1996) and Takahashi *et al.* (1998) showed that the Izu–Bonin Arc has a middle crustal layer with a seismic velocity of 6 km s^{-1}. Beneath is a thick lower crust with a velocity of >7 km s^{-1}, which makes up approximately 50% of the whole crust, and above is a thin upper crust layer with a velocity of 6.5–7.0 km s^{-1}. The lower crust thins at the arc-back-arc transition zone where its thickness is similar to that of typical oceanic crust; however, the velocity is still very high, reaching 7.3–7.4 km s^{-1}. Kodaira *et al.* (2007*a*) and Obana *et al.* (2010) obtained the mantle velocity from controlled source seismic survey data and tomography using natural earthquake observation data, and showed that the mantle velocity beneath the volcanic front is slower than 8.0 km s^{-1}.

There are both similarities and differences between the across-arc crustal structures along the Izu–Bonin Arc. Figure 2f.13 shows tomographic images across the whole Izu–Bonin Arc. Kodaira *et al.* (2007*a*) showed that the northern Izu–Bonin Arc has a thick crust, with a maximum thickness of 25–35 km and a large variation in the thickness of the 6.0–6.8 km s^{-1} velocity layer. The crust thins in the central Izu–Bonin Arc to the south of Torishima, reaching only *c.* 10–15 km (Kodaira *et al.* 2007*b*). The variations in crustal thickness can be seen in Figure 2f.13. The middle crust, with a velocity of 6.0–6.5 km s^{-1}, is continuous across the entire arc region. South of Io-To, the crust thickens again to a thickness similar to that of the northern Izu–Bonin Arc (Sato *et al.* 2009*a*, *b*). There is a maximum in crustal thickness at the junction of the Izu–Bonin Arc, the Mariana Arc and the West Mariana Ridge. Each crustal section has two or three local maxima below the volcanic front, forearc and more posterior parts of the arc. The West Mariana Ridge is a remnant arc of the Mariana Arc and its crust has similar characteristics to the Izu–Bonin Arc (Takahashi *et al.* 2007, 2008) and the Kyushu Palau Ridge (Nishizawa *et al.* 2007). Here we divide the Izu–Bonin Arc into three regions according to the structural variations: the northern part, north of Torishima; the central part, from Torishima to Nishinoshima; and the southern part, south of Nishinoshima. Their structural characteristics are summarized in the following sections.

Northern Izu–Bonin Arc

Figure 2f.14 shows the cross-section along Line IBr6. Thick crust is detected beneath the forearc region and the thickening is largely caused by the 6.0–6.8 km s^{-1} layer. The relationship between a thicker 6.0–6.8 km s^{-1} layer and thicker overall crust occurs over the entire Izu–Bonin Arc. Figure 2f.15 shows reflection profiles beneath the forearc region along Lines KT5, IBr5, IBr6 and

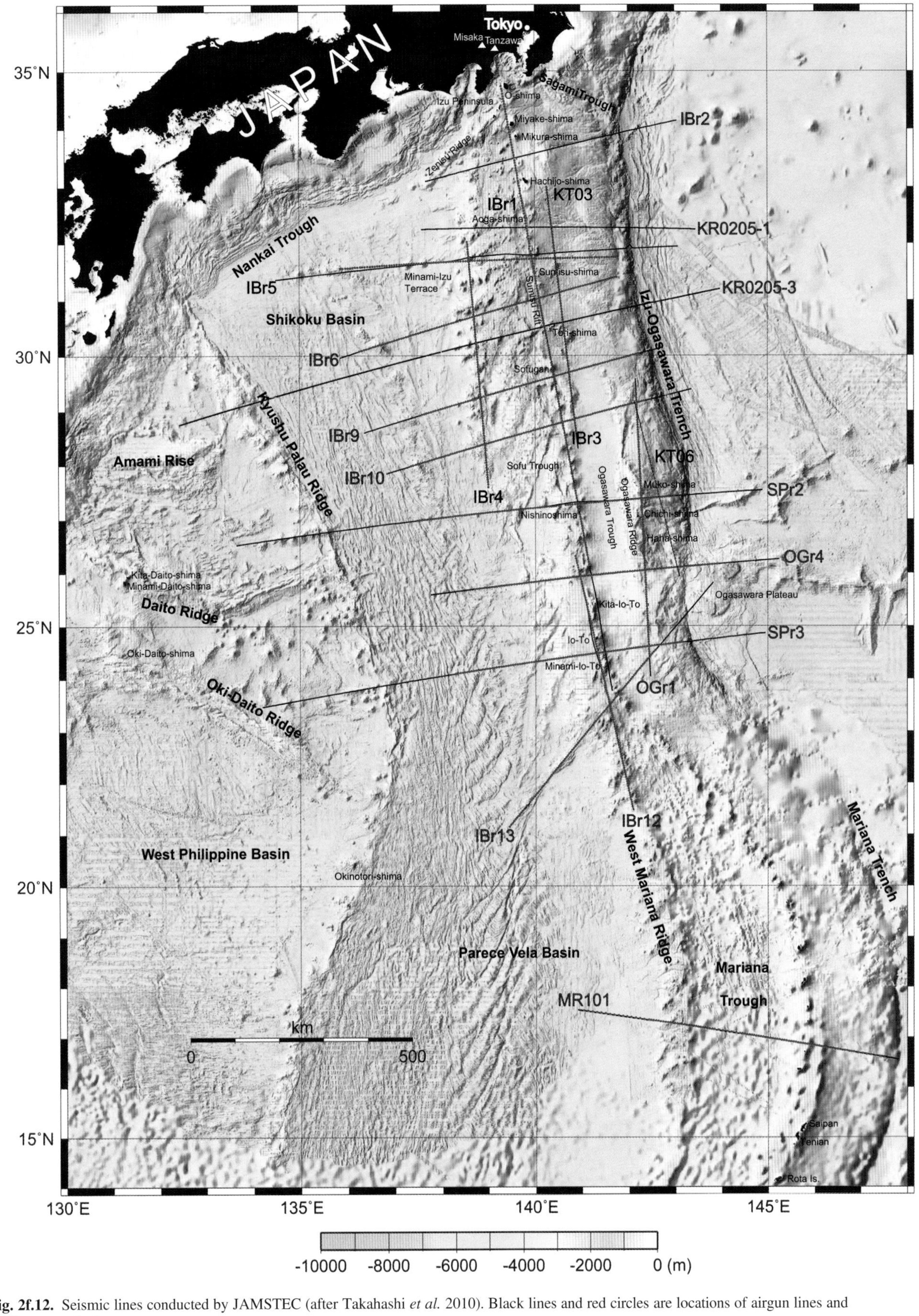

Fig. 2f.12. Seismic lines conducted by JAMSTEC (after Takahashi *et al.* 2010). Black lines and red circles are locations of airgun lines and OBSs, respectively.

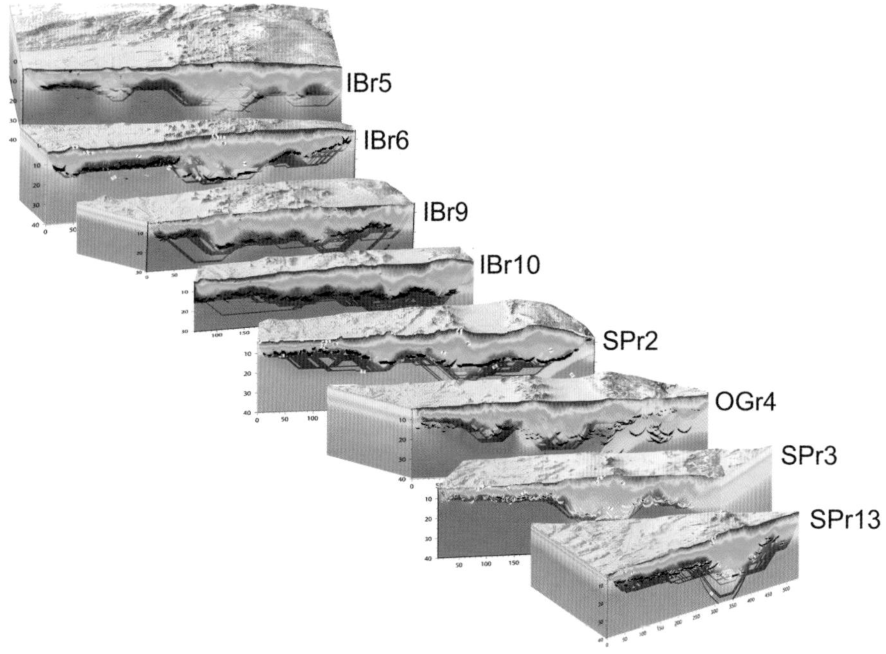

Fig. 2f.13. Tomographic images of crustal structures across the Izu–Bonin Arc (after Takahashi *et al.* 2010). Black lines indicate locations of the Moho. Horizontal and vertical axes are distance from the western end (km) and the depth (km), respectively.

KR0205–3. Most sections show two or three basement highs beneath the forearc, which correspond to the two regions with thick crust detected in the refraction data. JAMSTEC also carried out refraction surveys in north–south directions in the forearc basin, and Takahashi *et al.* (2011*a*, *b*) found that there are also large heterogeneities in crustal thickness here, similar to that of the volcanic front (Fig. 2f.16). Thick total crust and a thick 6.0–6.8 km s^{-1} layer occur beneath Sumisu Spur, Higashitorishima Seamount No. 2 and Omachi Seamount, and are similar to the crust beneath the basaltic volcanoes at the volcanic front. The MCS survey found half-graben structures developed in the shallow sediments (Yamashita *et al.* 2009).

Kodaira *et al.* (2008) obtained a velocity image of the crustal structure of the rear region of the arc and showed that crustal thickness varies in a north–south direction in a similar way to that at the volcanic front, suggesting that the frontal and rear parts of the arc separated during rifting. The trace of this rifting can be seen in the velocity structure along Line 6 in Figure 2f.14; the middle crust, with a velocity of 6.0–6.5 km s^{-1}, becomes thinner than at the volcanic front and a higher velocity is detected in the lower crust. Just west of the volcanic front, the Sumisu Rift with narrow basins can be identified on the topography map. The velocity structure along Line 6 suggests that the thickest crust is beneath the Sumisu Rift, just west of the volcanic front. The high velocity of >7.0 km s^{-1} suggests that the dense magmatic materials have been accreted to the crust underlying the Sumisu Rift.

Magnetic lineations have been detected in the Shikoku Basin, a back-arc basin behind the IBM arc system (Okino *et al.* 1994). The Kinan Seamount Chain runs through the centre of the basin, with the Kinan Escarpment to the east of the chain. The seafloor becomes slightly shallower east of the escarpment, rising to a topographic terrace known as the Minami–Izu Terrace. The eastern Shikoku Basin is not underlain by typical oceanic crust (Fig. 2f.14), being *c.* 8–10 km thick, slightly thicker than the typical 7.2 km of oceanic crust (White *et al.* 1992). The velocity of the lower crust is also higher than that of typical oceanic crust (6.7–7.0 km s^{-1}), reaching 7.0–7.5 km s^{-1} beneath the Minami–Izu Terrace and the Kinan Escarpment.

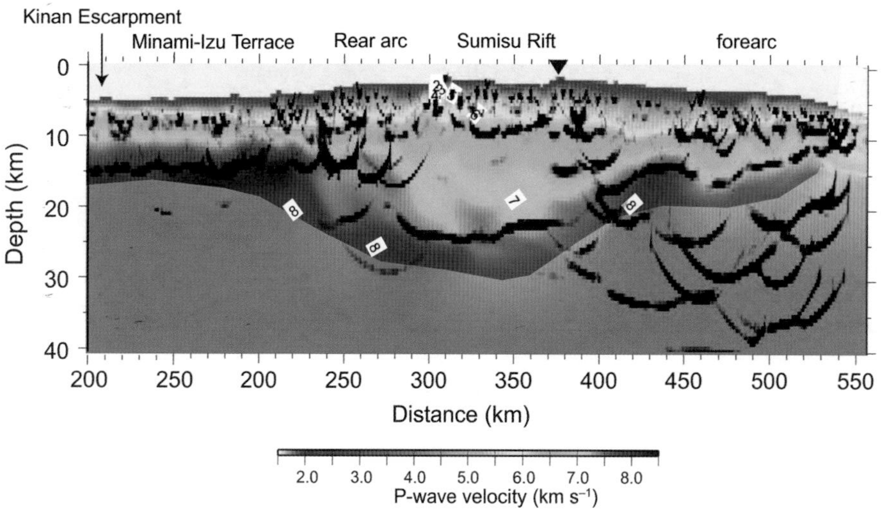

Fig. 2f.14. Close-up of the arc region from the tomographic image of Line IBr6 (after Takahashi *et al.* 2010). Black lines indicate reflector-mapped travel-times of picked reflections on the velocity image. The black triangle indicates the location of the volcanic front.

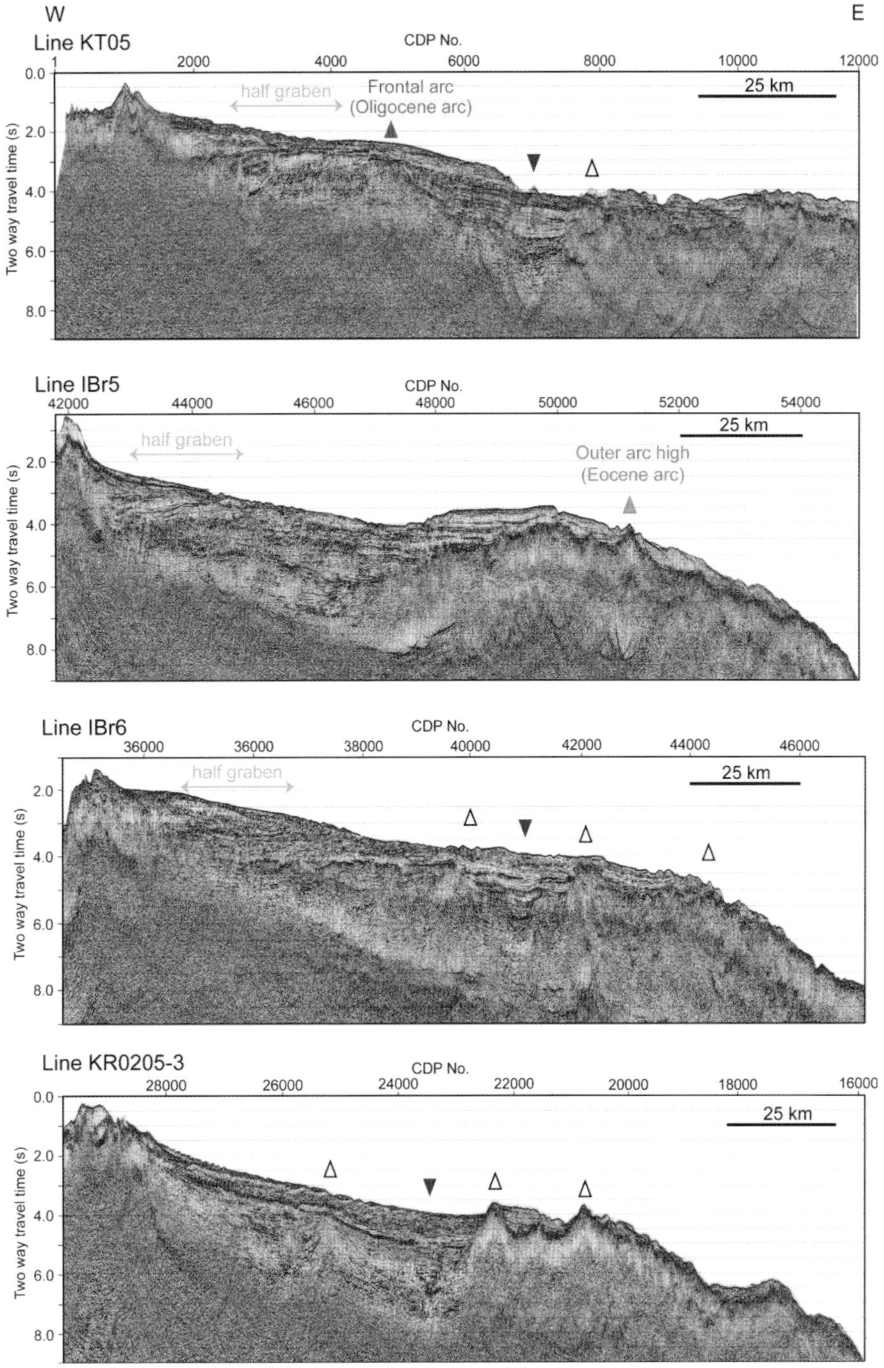

Fig. 2f.15. Comparison of migrated reflection profiles (Yamashita *et al.* 2009). Basement highs and lows are shown by open white triangles and solid blue triangles, respectively. Purple and green triangles indicate ODP sites where basement had already been dated.

Central Izu–Bonin Arc

The structure of the central part of the Izu–Bonin Arc from Torishima to Nishinoshima is characterized by: (1) immature arc crust along the volcanic front; (2) crustal thickening in the forearc region; (3) rifted crust in the rear arc; and (4) back-arc basin crust (Fig. 2f.13). Kodaira *et al.* (2007*b*) indicated that the crust beneath the volcanic front also varies in the same way as that in the northern part, but is relatively thin. The crust is *c.* 10–20 km thick and the middle crust, with a velocity of 6.0–6.5 km s^{-1}, occurs throughout the region. Takahashi *et al.* (2011*a*, *b*) found two velocity structures along Lines IBr9 and IBr10 using refraction data (Fig. 2f.17). Forearc crust has a maximum thickness of *c.* 15 km, but crust beneath the northern tip of the Ogasawara Ridge has a greater thickness of *c.* 20–25 km thick.

The rear area of the central Izu–Bonin Arc has thinner crust of *c.* 15 km except below An'ei Seamount where it reaches 20 km (Fig. 2f.17). The reflection sections show many normal faults in this region (Takahashi *et al.* 2011*a*, *b*). In particular, the western half of the rear arc has twice as much normal faulting as the eastern half, suggesting rifting in a tensional field. The back-arc region has similar structural characteristics to that in the northern part. The crust is thin, reaching a minimum of *c.* 7 km, and gradually thins westwards from the rear arc to the Kinan Escarpment. The lower crust includes high-velocity materials in excess of 7.0 km s^{-1}.

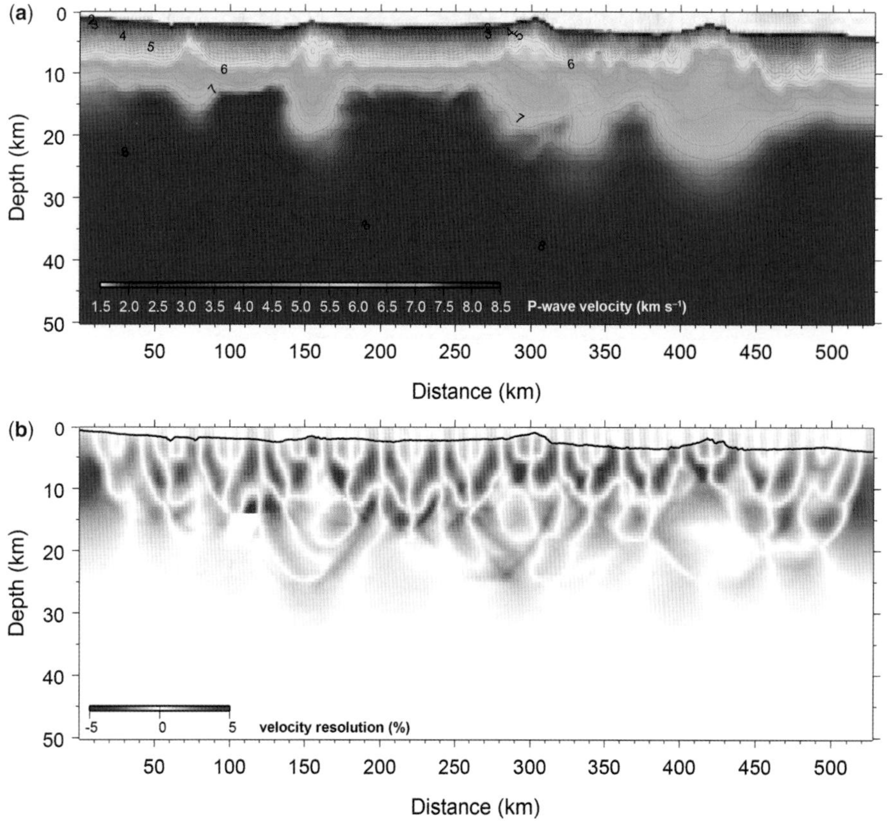

Fig. 2f.16. Crustal structure along Line KT03 (after Takahashi *et al.* 2011*b*). (**a**) Tomographic image of the crustal structure. (**b**) Checkerboard test to understand resolution of the obtained image. Resolved area distributes shallower part from 15 to 20 km.

Southern Izu–Bonin Arc

The structural characteristics of the southern Izu–Bonin Arc are: (1) mature arc crust along the volcanic front; (2) Ogasawara Ridge crust in the forearc region; (3) transition structure between the volcanic front and the back-arc region; (4) thin back-arc basin crust; and (5) anomalously thick crust at the junction of the Izu–Bonin Ridge, Mariana Arc and the West Mariana Ridge.

Sato *et al.* (2009*a*, *b*) obtained a crustal velocity image along the volcanic front of the southern part, which is similar to the across-arc profiles of OGr4 and SPr3 (Fig. 2f.13). The crust thickens beneath this region, reaching *c.* 20–25 km, and its characteristics are similar to that of the northern part. The forearc region has a large bathymetric high, the Ogasawara Ridge, which has a complex velocity structure. The velocity structure along Line SPr2 indicates that the Ogasawara Ridge crust is *c.* 20 km thick (Takahashi *et al.* 2009),

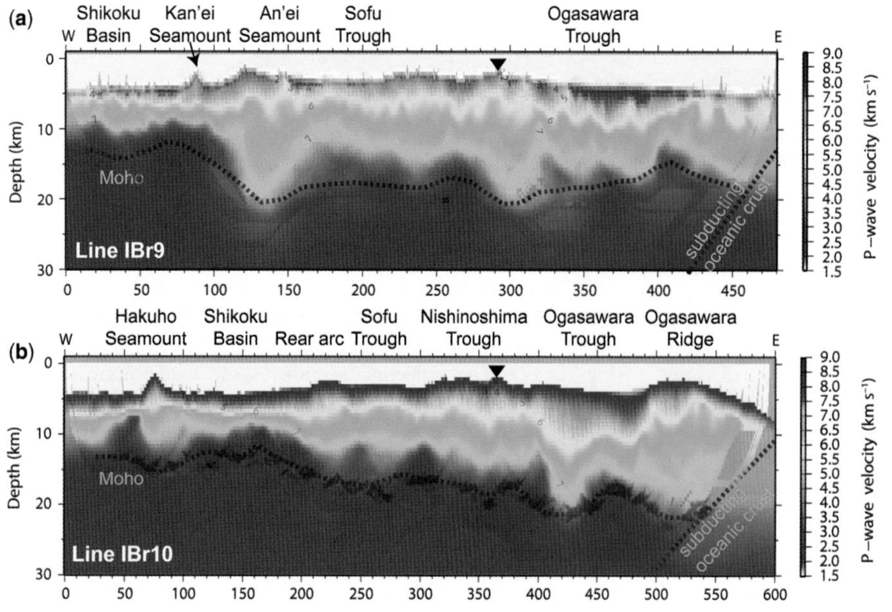

Fig. 2f.17. Comparison of crustal structures along Lines 9 (a) and 10 (b) (after Takahashi *et al.* 2011*a*). Black triangles indicate location of volcanic front. Dotted lines define the Moho. Masked areas have no resolution without ray paths.

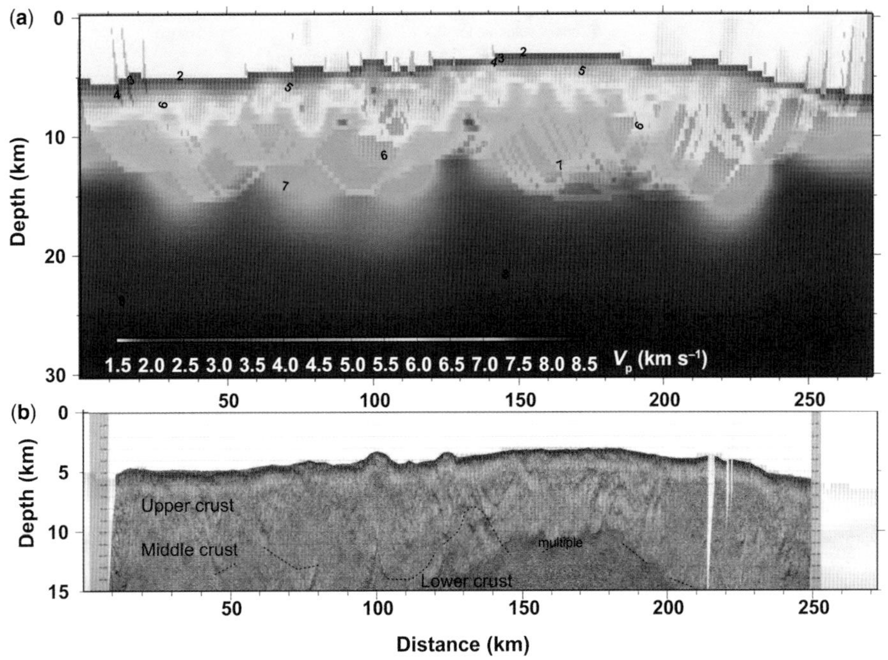

Fig. 2f.18. Comparison of crustal structures along Line KT06 (after Takahashi *et al.* 2011*b*). (**a**) Tomographic image. Masked areas have no resolution without ray paths. (**b**) Migrated reflection profile on (a).

and seems to have a different composition compared to the volcanic front. The velocity of the middle crust is 6.4–6.6 km s^{-1} and that of the lower crust is 6.8–7.4 km s^{-1}, both of which are slightly faster than middle and lower crust beneath the volcanic front. Kodaira *et al.* (2010) obtained the velocity image along the Ogasawara Ridge showing a minimum thickness of *c.* 7 km, similar to the oceanic crust, and a maximum thickness of *c.* 20 km, similar to the arc crust. The thin crustal region is narrow (<10 km). Recently JAMSTEC carried out further seismic surveys near the trench-slope break to the east of the Ogasawara Ridge. It was revealed that this is not simple oceanic crust but strongly heterogeneous, suggesting that it is a small crustal block (Fig. 2f.18).

The back-arc region of the southern Izu–Bonin Arc is located at the junction of Shikoku and Parece Vela basins, where the Kinan Escarpment becomes less well defined. The crust is *c.* 7 km thick, thinner than that of the Shikoku back-arc basin to the north. However, there are high-velocity materials in the lower crust and intrusive materials can also be identified. The structural characteristics are similar to that of the northern part, but the volume of the high-velocity materials is small.

Finally, as mentioned above, the thickest crust is not at the volcanic front but in the initial rifted areas, such as in the Sumisu Rift. A similar situation occurs at the junction of the Izu–Bonin Arc, the Mariana Arc and the West Mariana Ridge. Line SPr13 runs from the forearc through the northernmost Mariana Arc, the Mariana Trough and the West Mariana Ridge as far as the western end of the Minami-Io-To Spur. The maximum crustal thickness along this line is >30 km beneath the Mariana Trough, which is the initial rifting area (Fig. 2f.13). The 6.0–6.8 km s^{-1} layer has two blocks with a thickness of 10 and 15 km beneath the Mariana Arc and the West Mariana Ridge, respectively. Just beneath the Mariana Trough it is the lower crust that is thick and not the 6.0–6.8 km s^{-1} layer. Moreover, V_p of the lower crust is high, reaching 7.0 km s^{-1}. This crustal structure is similar to that below the Sumisu Rift.

Collision of the Ogasawara Plateau

From its formation in the Eocene epoch until *c.* 15 Ma, the Izu–Bonin Arc crust developed without any interaction with continental crust. However, since *c.* 15 Ma it has been colliding with the Japan islands. Arai *et al.* (2009) and Tamura *et al.* (2010*b*) show that the Philippine Sea Plate delaminates here, with the upper crust accreted to Japan and the lower crust subducted beneath it. Such delamination has been detected elsewhere, for example where the Ontong Java Plateau collides with the Solomon islands (Miura *et al.* 2004).

Takahashi *et al.* (2010) obtained a velocity image along Line OGr4, which runs from the Ogasawara Plateau to the Parece Vela Basin via the Ogasawara Trough (Fig. 2f.19). On the OBS records, clear reflections from the Moho and the top of the subducting Pacific plate can be identified. They show the tomographic image of the crustal structure, and project these reflections on it using the pre-stack depth migration technique (Fujie *et al.* 2006). The continuous Moho and reflections from the top and within the incoming plate can be identified. The image suggests that the upper crust of the Ogasawara Plateau is colliding into the Izu–Bonin Arc but the lower crust is subducting, resulting in delamination of the Ogasawara Plateau; the Philippine Sea Plate is acting in a similar manner at the northern end of the Izu–Bonin Arc.

Izu collision zone (KT)

The Izu collision zone (ICZ) is located in central Japan (Fig. 2f.1), where the northern end of the Izu–Bonin Arc has been colliding with the Honshu Arc since middle Miocene times (e.g. Taira *et al.* 1989). This collision is contemporaneous with the opening of the Japan Sea, when the eastern margin of the Eurasian continent rifted off to form the Japanese archipelago and the southern half of this rotated clockwise to form the SW Honshu Arc (Otofuji & Matsuda 1983). Since then, blocks of the Izu–Bonin Arc have been successively accreted onto Honshu as the collision boundary migrated south. Each block is separated by a large thrust fault and thick piles of clastic sediments; these are interpreted to be the abandoned plate boundary and trough-fill deposits, respectively (Amano 1991). The collision caused northwards bending of the Honshu Arc forming a syntaxis, where Jurassic (Chichibu Belt) and Cretaceous–Paleocene (Shimanto Belt) accretionary complexes of weakly metamorphosed terrigenous clastic sediments are exposed (Fig. 2f.20). As a result of the collision the arc crust beneath the

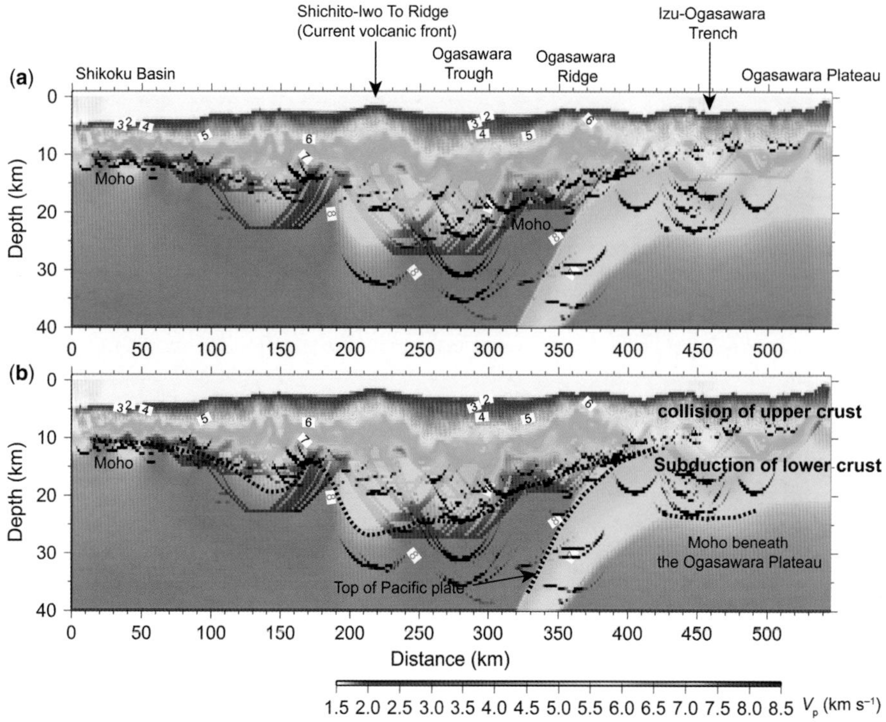

Fig. 2f.19. Tomographic image of Line OGr4 and reflection mapping. (**a**) Reflection mapping on the tomographic image. Masked areas have no resolution without ray paths. (**b**) Interpretation of the reflection mapping.

ICZ is thickened to *c.* 40 km (Asano *et al.* 1985), with voluminous granitic intrusions now exposed within the collision zone (here referred to as the ICZ granitoids, Fig. 2f.20).

The ICZ is presently the only place on Earth where an active intra-oceanic arc is currently colliding with a mature, continental arc, providing a unique opportunity to study the geological processes that occur during arc–continent collision. Successive coalescence of juvenile island arcs has been proposed as a fundamental process for Phanerozoic continental growth (e.g. McLennan & Taylor 1982). Studying the tectonic and magmatic processes observed in the ICZ may therefore provide insights into the genesis of the continental crust.

The ongoing ICZ collision is also important to modern society as this collision is responsible for one of the most disastrous earthquakes in the 20th century: the magnitude 7.9 Great Kanto earthquake of 1923 which killed more than 100 000 people. Furthermore, Fuji and Hakone volcanoes, located *c.* 20–40 km north of the Izu Peninsula, are two large Quaternary active volcanoes within the ICZ (Fig. 2f.21); and historical eruptions of these volcanoes, especially of Fuji Volcano, have posed volcanic hazards to the surrounding region.

Tectonic framework

Amano (1991) identified four accreted thrust-bounded blocks within the ICZ (Kushigatayama, Misaka, Tanzawa and Izu blocks; Fig. 2f.20), mostly composed of Miocene submarine volcanic rocks and associated volcaniclastic and gravity-flow deposits such as

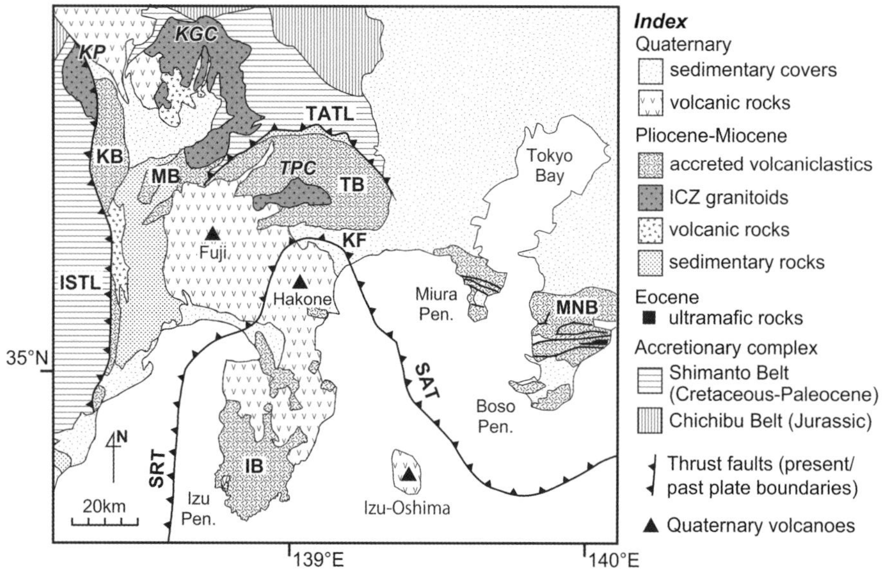

Fig. 2f.20. Generalized geological map of the ICZ (modified after Aoike 1999; Saito *et al.* 2007*a*). IB, Izu Block; ISTL, Itoigawa–Shizuoka Tectonic Line; KB, Koma Block; KF, Kannawa Fault; KGC, Kofu Granitic Complex; KP, Kaikoma Pluton; MB, Misaka Block; MNB, Mineoka Belt; SAT, Sagami Trough; SRT, Suruga Trough; TB, Tanzawa Block; TPC, Tanzawa Plutonic Complex; TATL, Tonoki–Aikawa Tectonic Line.

turbidite and debrite (Aoike 1999), which correspond to the upper-crustal sequence of the Izu–Bonin Arc. Amano (1991) also proposed a sequential southwards migration of the collision boundary, with the first collision occurring at *c.* 13 Ma between the Kushigatayama Block and the Honshu Arc. The collision then migrated to the Misaka Block at *c.* 9 Ma followed by a collision of the Tanzawa Block at *c.* 5 Ma, and the Izu Block has been currently colliding at the Kannawa Fault (Fig. 2f.20) from *c.* 1 Ma onwards (Amano 1991). However, there are uncertainties and controversies about the timing when these collisions occurred (Aoike 1999). For example, recent palaeomagnetic and biostratigraphic study in the southern Boso Peninsula has shown that clockwise deformation related to the Tanzawa Block collision began at 6.8 Ma (Yamamoto & Kawakami 2005). During the migration of collision from the Tanzawa Block to the Izu Block, the Ashigara Basin (a trough basin formed at the northern margin of the Izu Block along the Kannawa Fault) abruptly subsided at a maximum rate of 4.6 mm a^{-1} and contemporaneously filled with turbidites derived from the uplifted Tanzawa Mountains (Soh *et al.* 1998).

Miocene volcanic rocks and volcaniclastics in the ICZ are mostly of tholeiitic to calc-alkaline magma-series types showing bimodal compositions of basalt to basaltic andesite and dacite to rhyolite (Aoike 1999), similar to those of the Quaternary volcanoes in the northern Izu–Bonin volcanic front (Tamura & Tatsumi 2002). However, tectonic settings (i.e. volcanic front, rear arc, back-arc basin, etc.) where the accreted ICZ blocks were originally formed in the Izu–Bonin Arc are largely unknown, chiefly due to the tectonic complexities within the collision zone. Whole-rock compositions of the Miocene volcanic rocks in the ICZ generally show arc-signature, such as depletion in Nb and enrichment of large-ion lithophile elements (LILE) against high-field-strength elements (HFSE), typical of Izu–Bonin volcanic rocks (Aoike 1999). However, there are suggestions of volcanic rocks of different tectonic origins, such as the basaltic lavas and pillows with back-arc basin basalt-like geochemistry in the Misaka Block (Aoike 1999) and the high-magnesian andesite clasts in the volcaniclastic deposit in Tanzawa Block (Arima *et al.* 1999). Furthermore, many of these sequences in the ICZ have been weathered and diagenetically altered, making original geochemical and geochronological features difficult to decipher.

A recent zircon U–Pb geochronological study utilizing the sensitive high-resolution ion microprobe (SHRIMP) of the widespread submarine volcanic and intrusive rocks exposed in the central-southern Izu Peninsula showed that these rocks formed during Late Miocene–Pleistocene times (Tani *et al.* 2011*a*, *b*) and do not, as previously believed, form part of a regional Tertiary basement. Furthermore, whole-rock compositions of the volcanic rocks in the Izu Peninsula are enriched in LILE and HFSE similar to those erupted from volcanoes in the rear part of the Izu–Bonin Arc, indicating that the southernmost Izu Block is an accreted terrane of Izu–Bonin rear-arc seamount chains active through Miocene–Pleistocene times (Tani *et al.* 2011*a*, *b*).

With these new insights suggesting that Late Miocene–Pleistocene Izu Peninsula rocks have an origin in the Izu–Bonin rear arc, the tectonic framework of Quaternary volcanism in the ICZ also needs to be reinterpreted. The Hakone Volcano dominantly erupts low-K tholeiitic basalt similar to those of the Izu–Bonin volcanic front volcanoes. In contrast, modern ejecta from the Fuji Volcano, situated *c.* 30 km NW of Hakone volcano, have a distinct affinity with the Izu–Bonin rear-arc lavas, characterized by enrichments in LILE and HFSE (Tani *et al.* 2011*a*, *b*). The geochemical variations observed within the Hakone and Fuji volcanoes are therefore likely to represent Izu–Bonin across-arc variations, still undisturbed at an earlier phase of collision. An estimation of the isodepth of the subducting slabs of the Pacific Plate and Philippine Sea Plate in the northern Izu–Bonin Arc and ICZ (Nakajima *et al.* 2009) show that the volcanic front of the northern Izu–Bonin Arc lies 120 km above the Pacific Plate. North of Izu–Oshima volcano and into the ICZ the volcanic front shifts to the west; beneath Hakone Volcano the Pacific Plate is *c.* 20 km deeper. This is due to the deformation of the overriding Philippine Sea Plate as a result of the NW-directed collision of the buoyant Izu–Bonin Arc crust against the Honshu Arc (Nakajima *et al.* 2009).

Another important feature of the ICZ is the presence of accreted ultramafic rocks to the east. The Mineoka Belt (Fig. 2f.20) is located in the southern Boso Peninsula, where the forearcs of Izu–Bonin and NE Honshu arcs collide. The Mineoka Belt includes a suite of rocks that characterize the ophiolite sequence, mainly mafic to ultramafic rocks with pelagic to terrigenous sediments (Ogawa 1983). A wide variety of igneous rocks occur as tectonic blocks within the terrigenous sediments or serpentine bodies, such as alkali-basalt, tholeiitic basalt, gabbro and diorite. Eocene–Miocene Ar–Ar and K–Ar ages have been reported from these rocks (Hirano *et al.* 2003). The precise timing of when the Mineoka Belt accreted to the Honshu Arc is not well known, but it is inferred to have begun from *c.* 16 Ma (Saito 1996), contemporaneous with the onset of Izu collision. Ogawa & Taniguchi (1988) proposed that Tertiary rocks that comprise the Mineoka Belt are remnants of the ‘Mineoka Plate’ that existed NE of the present Izu–Bonin Arc, which has been subducted beneath the Honshu Arc. However, recent research cruises to the landward slopes of the Izu–Bonin Trench reveal exposures of *in situ* ophiolitic rocks of Eocene age, as well as subsequent post-Eocene volcanic rocks formed during early Izu–Bonin Arc magmatism (Ishizuka *et al.* 2011*a*, *b*). The lithologies and ages of the rocks in the Izu–Bonin Trench slope are identical to those that comprise the Mineoka Belt, suggesting that the Mineoka Belt may also be part of the ICZ, potentially representing the accreted Izu–Bonin forearc region.

Syncollisional magmatism and the formation of Izu collision zone granitoids

Voluminous Neogene granitic plutons are exposed within the ICZ, except for the southernmost Izu Block (Fig. 2f.20). Three large plutonic bodies are recognized: the Kaikomagatake Pluton; the Kofu Granitic Complex (KGC); and the Tanzawa Plutonic Complex (TPC). These plutons have been exhumed by tectonic uplift during arc–arc collision, and are composed of various types of granitic rocks ranging from tonalite to granite (Takahashi 1989). Even though the Izu Block lacks surficial exposure of granitic plutons, the existence of silicic plutons beneath the Izu Peninsula is also inferred from tonalitic xenoliths within the volcaniclastic deposits of the Quaternary volcanoes and small porphyritic dacite intrusives found in the western part of the peninsula (Sakamoto *et al.* 1999).

The Kaikomagatake Pluton is located in the NW corner of the ICZ, intruding the Shimanto Belt. The pluton is truncated in the west by the Itoigawa–Shizuoka Tectonic Line (ISTL), which is the western boundary between the Honshu Arc and the accreted blocks of the Izu–Bonin Arc. The pluton has been divided into internal (Hoo type) and external (Kaikoma type) zones based on petrography (Fujimoto *et al.* 1965). The Hoo type consists mainly of hornblende-biotite granodiorite, whereas the Kaikoma type consists of biotite granite. The rocks along the ISTL are strongly deformed and locally show mylonitization (Shimamoto *et al.* 1991). Recent SHRIMP zircon ages obtained from the Kaikomagatake pluton give a narrow range of ages of 13.9–12.7 Ma (Saito *et al.* 2012), indicating that emplacement and solidification of the pluton occurred early during ICZ collision.

The KGC, the largest plutonic complex in the ICZ, is located at the northern end of the ICZ and is composed of various types of granitic rocks ranging from biotite granite, hornblende-biotite granodiorite and hornblende tonalite (Saito *et al.* 2007*a*). The northern and central KGC intrudes the Shimanto Belt whereas the southernmost part is emplaced into the Miocene Nishiyatsushiro Group, corresponding to the accreted submarine volcanic sequence of the Izu–Bonin Arc. SHRIMP zircon ages for the KGC range from 16.8 to 10.6 Ma (Saito *et al.* 2007*a*), indicating that emplacement of the KGC was simultaneous to that of the Kaikomagatake Pluton during the earlier phase of collision.

The TPC is located in the central part of the ICZ, and is predominantly composed of hornblende tonalite plutons and associated gabbroic bodies (Takita 1980). The TPC intrudes the Miocene Tanzawa Group, resulting in amphibolite facies contact metamorphism (Seki *et al.* 1969). The emplacement age of the TPC is controversial. Previously reported hornblende and biotite K–Ar ages from the TPC vary widely over the range 16–4 Ma (e.g. Sato *et al.* 1986). Taking the older K–Ar ages, TPC emplacement has been considered to predate the onset of Tanzawa Block collision at 6.8 Ma (Yamamoto & Kawakami 2005). However, recent SHRIMP zircon ages show that most of the TPC was emplaced at *c.* 5–4 Ma, simultaneously to the collision (Tani *et al.* 2010). Furthermore, a thermogeochronological study revealed a very high post-emplacement cooling rate of up to *c.* 660°C myr^{-1} (Tani *et al.* 2010), making the TPC one of the most rapidly cooled granitic plutons on Earth. This suggests that the Tanzawa region experienced very rapid uplift in association with the emplacement of granitic magma during the collision.

In addition to the three large plutonic complexes, there are numerous small granitic intrusive bodies within the ICZ. These bodies are mostly exposed along the eastern margin of the ISTL, intruding Miocene submarine volcaniclastics and sedimentary rocks. They occur as small complexes of north–south-trending intrusive bodies, parallel to the direction of the ISTL. Most are hornblende or hornblende-biotite tonalitic rocks, geochemically and petrographically similar to those of the TPC and KGC (e.g. Yajima 1970; Ozaki *et al.* 2002). However, the Kokkaibashi Pluton, located adjacent to the northeastern part of the Kaikomagatake Pluton, consists of a hornblende gabbro body that is in fault contact with an ultramafic cortlandite body (Tsunoda & Shimizu 1983).

Petrogenesis of the ICZ granitoids and the timing of ICZ granitic magma formation have been studied in detail (e.g. Kawate & Arima 1998; Saito *et al.* 2007*a*, 2012; Tamura *et al.* 2010*a*; Tani *et al.* 2010) since these granitic plutons may represent the juvenile granitic crust formed in the intra-oceanic arc environment. *In situ* seismic velocity measurements suggest that ranges of V_p in the seismically defined Izu–Bonin middle crust layer ($V_p = 6.0–6.5$ km s^{-1}) can be explained if the middle crust layer is composed of tonalitic rocks with intermediate composition ($SiO_2 \approx 60$ wt%) similar to those exposed in the TPC (Kitamura *et al.* 2003). However, recent zircon U–Pb geochronological studies of the ICZ granitoids show that these granitic plutons were generated during the collision (Saito *et al.* 2007*a*; Tani *et al.* 2011*a*, *b*), and rather represent the exposed deep crustal section of the accreted Izu–Bonin Arc (Taira *et al.* 1998). Furthermore, whole-rock geochemistry of the ICZ granitoids shows evidence for the assimilation of Shimanto sediment during granitic magma generation, such as elevated whole-rock strontium isotopic compositions in the KGC and Kaikomagatake Pluton (Saito *et al.* 2007*b*, 2012) and elevated whole-rock and zircon Th–Nb ratios in the TPC rocks (Tani *et al.* 2010). Incorporation of terrigenous Shimanto sediment is considered to play a fundamental role in the ICZ granitoids attaining geochemical characteristics with continental affinities, such as enrichments in incompatible elements and genesis of peraluminous granitic magma (Saito *et al.* 2007*b*, 2012). These pieces of evidence suggest that voluminous syncollisional granitic crust was generated by arc–arc collision, perhaps accompanying compositional modification of the juvenile Izu–Bonin Arc crust.

In addition to the voluminous granitic magmatism in the ICZ, there is growing evidence for explosive silicic volcanism associated with the collision. Tamura *et al.* (2010*a*) showed that Pleistocene (*c.* 2.5 Ma) garnet-bearing pumiceous tephra, widely distributed in the southern Kanto area, is sourced from a volcano that existed in the southern Tanzawa Mountains, where a compositionally and petrologically similar garnet-bearing rhyolite dyke with concordant K–Ar age of 2.45 Ma has been reported (Arima *et al.* 1990). The garnet phenocrysts in the tephra and in the rhyolite dyke are manganese-rich (Arima *et al.* 1990; Tamura *et al.* 2010*a*), suggesting that these garnet phenocrysts are the shallow crystallizing phase from a peraluminous granitic magma, potentially formed by partial melting of pelitic sediments (Arima *et al.* 1990). The volume of tephra from this eruption is estimated to be at least 2.8 km^3, similar in volume to the large explosive Horeki eruption of the Fuji Volcano which occurred in 1707 (Tamura *et al.* 2010*a*).

The Higashi–Yamanashi Volcano-Plutonic Complex, located to the north of Kofu Basin, is a caldera that is 25 km across north–south and 6 km wide east–west, and is composed of rhyolitic to dacitic densely welded tuffs that are intruded by a co-genetic granodiorite pluton (Kanamaru & Takahashi 2009). K–Ar ages of the granodiorite are 4.38–4.27 Ma (Shibata *et al.* 1984), contemporaneous with the formation of TPC. The proximal thickness of the welded tuff is more than 1.3 km, suggesting that voluminous pyroclastic eruptions occurred during the formation of the caldera (Kanamaru & Takahashi 2009). Granodioritic rocks in the Higashi–Yamanashi Volcano-Plutonic Complex show elevated whole-rock and zircon Th–Nb ratios, indicating that they are related to the syncollisional magmatism and that the assimilation of Shimanto sediments was involved in their magma genesis (Tani *et al.* 2010).

The recent recognition of voluminous syncollisional granitic crust formation and associated silicic volcanism in the ICZ indicates vigorous tectono-magmatic activity during arc–continent collision. Compositional modifications resulting from syncollisional magmatism led to a continental geochemical signature, indicating that ICZ igneous rocks provide valuable analogues for collisional magmatism in ancient suture zones.

Appendix

English to Kanji and Hiragana translations for geological and place names

Amami	奄美	あまみ
An'ei	安永	あんえい
Aogashima	青ヶ島	あおがしま
Ashigara	足柄	あしがわ
Bonin	ボニン	ぼにん
Boso	房総	ぼうそう
Chichibu	秩父	ちちぶ
Chichijima	父島	ちちじま
Daito	大東	だいとう
Enpo	延宝	えんぽう
Fuji	富士	ふじ
Genroku	元禄	げんろく
Hachijo	八丈	はちじょう
Hachijojima	八丈島	はちじょうじま
Hakone	箱根	はこね

(*Continued*)

English to Kanji and Hiragana translations for geological and place names (Continued)

Higashi-Yamanashi	東山梨	ひがしやまなし
Honshu	本州	ほんしゅう
Hoo	鳳凰	ほうおう
Horeki	宝暦	ほうれき
Io-to	硫黄島	いおうとう
Itoigawa	糸魚川	いといがわ
Izu	伊豆	いず
Izu-Oshima	伊豆大島	いずおおしま
Kaikoma	甲斐駒	かいこま
Kaikomagatake	甲斐駒ケ岳	かいこまがたけ
Kan'ei	寛永	かんえい
Kannawa	神縄	かんなわ
Kanto	関東	かんとう
Kinan	紀南	きなん
Kita-Ioto	北硫黄島	きたいおうとう
Kofu	甲府	こうふ
Kokkaibashi	国界橋	こっかいばし
Kushigatayama	櫛形山	くしがたやま
Kyushu	九州	きゅうしゅう
Manji	万治	まんじ
Mariana	マリアナ	まりあな
Mikurajima	御蔵島	みくらじま
Minami-Ioto	南硫黄島	みなみいおうとう
Minami-Izu	南伊豆	みなみいず
Mineoka	嶺岡	みねおか
Misaka	御坂	みさか
Miyakejima	三宅島	みやけじま
Myojin	明神	みょうじん
Nishinoshima	西之島	にしのしま
Nishiyatsushiro	西八代	にしやつしろ
Ogasawara	小笠原	おがさわら
Oki-Daito	沖大東	おきだいとう
Omachi	大町	おおまち
Oshima	大島	おおしま
Shikoku	四国	しこく
Shimanto	四万十	しまんと
Shizuoka	静岡	しずおか
Sumisu	須美寿	すみす
Tanzawa	丹沢	たんざわ
Torishima	鳥島	とりしま

References

AHARONSON, O., ZUBER, M. T. & ROTHMAN, D. H. 2001. Statistics of Mars' topography from the Mars orbiter laser altimeter: slopes, correlations, and physical models. *Journal of Geophysical Research*, **106**, 23 723–23 735.

AMANO, K. 1991. Multiple collision tectonics of the South Fossa Magna in central Japan. *Modern Geology*, **15**, 315–329.

AOIKE, K. 1999. Tectonic evolution of the Izu collision zone. *Research Report of the Kanagawa Prefectural Museum, Natural History*, **9**, 111–151 [in Japanese with English abstract].

ARAI, R., IWASAKI, T., SATO, H., ABE, S. & HIRATA, N. 2009. Collision and subduction structure of the Izu-Bonin Arc, central Japan, revealed by refraction/wide-angle reflection analysis. *Tectonophysics*, **475**, 438–453, http://doi.org/10.1016/j.tecto.2009.05.023

ARIMA, M., SUEKANE, T., KADOTA, M., KATO, H. & YAMASHITA, H. 1990. Garnet-bearing rhyolite discovered in the Tanzawa Mountains. *Kanagawa Chigaku*, **70/71**, 1–6 [in Japanese].

ARIMA, M., AOIKE, K. & KAWATE, S. 1999. Tectonic evolution of the Tanzawa Mountain land. *Research Report of the Kanagawa Prefectural Museum, Natural History*, **9**, 57–77 [in Japanese with English abstract].

ASANO, N., WADA, K. ET AL. 1985. Crustal structure in the northern part of the Philippine Sea Plate as derived from seismic observations of Hatoyama-off Izu Peninsula explosion. *Journal of Physics of the Earth*, **33**, 173–189.

BANDY, W. L. & HILDE, T. W. C. 1983. Structural features of the Bonin arc: implications for its tectonic history. *Tectonophysics*, **99**, 331–353.

BLOOMER, S. H., TAYLOR, B., MACLEOD, C. J., STERN, R. J., FRYER, P., HAWKINS, J. W. & JOHNSON, L. 1995. Early Arc volcanism and the Ophiolite problem: a perspective from drilling in the Western Pacific. *In*: TAYLOR, B. & NATLAND, J. (eds) *Active Margins and Marginal Basins of the Western Pacific*. American Geophysical Union, Washington DC, 67–96.

BRYANT, C. J., ARCULUS, R. J. & EGGINS, S. M. 2003. The geochemical evolution of the Izu-Bonin Arc system: a perspective from tephras recovered by deep-sea drilling. *Geochemistry, Geophysics, Geosystems*, **4**, 1094, http://doi.org/10.1029/2002GC000427

CARLSON, R. L. & MORTERA-GUTIERREZ, C. A. 1990. Subduction hinge migration along the Izu-Bonin-Mariana Arc. *Tectonophysics*, **181**, 331–344.

CHRISTENSEN, N. I. & MOONEY, W. D. 1995. Seismic velocity structure and composition of the continental crust: a global view. *Journal of Geophysical Research*, **100**, 9761–9788.

COSCA, M. A., ARCULUS, R. J., PEARCE, J. A. & MITCHELL, J. G. 1998. $^{40}Ar/^{39}Ar$ and K-Ar geochronological age constraints for the inception and early evolution of the Izu-Bonin- Mariana Arc system. *Island Arc*, **7**, 579–595.

CRAWFORD, A. J., FALLOON, T. J. & GREEN, D. H. 1989. Classification, petrogenesis and tectonic setting of boninites. *In*: CRAWFORD, A. J. (ed.) *Boninites*. Unwin-Hyman, London, 1–49.

CRAWFORD, W. C., HILDEBRAND, J. A., DORMAN, L. M., WEBB, S. C. & WIENS, D. A. 2003. Tonga Ridge and Lau Basin crustal structure from seismic refraction data. *Journal of Geophysical Research*, **108**, 2195, http://doi.org/10.1029/2001JB001435

DESCHAMPS, A., MONIE, P., LALLEMAND, S., HSU, S.-K. & YEH, K. Y. 2000. Evidence for Early Cretaceous oceanic crust trapped in the Philippine Sea Plate. *Earth and Planetary Science Letters*, **179**, 503–516.

DICKINSON, W. R. & HATHERTON, T. 1967. Andesite volcanism and seismicity around the Pacific. *Science*, **157**, 801–803.

ELLIOTT, T., PLANK, T., ZINDLER, A., WHITE, W. & BOURDON, B. 1997. Element transport from slab to volcanic front at the Mariana Arc. *Journal of Geophysical Research*, **102**, 14 991–15 019.

FRYER, P., TAYLOR, B., LANGMUIR, C. H. & HOCHSTAEDTER, A. G. 1990. Petrology and geochemistry of lavas from the Sumisu and Torishima backarc rifts. *Earth and Planetary Science Letters*, **100**, 161–178, http://doi.org/10.1016/0012-821X(90)90183-X

FUJIE, G., ITO, A., KODAIRA, S., TAKAHASHI, N. & KANEDA, Y. 2006. Confirming shark bending of the Pacific plate in the northern Japan trench subduction zone by applying a traveltime mapping method. *Physics of Earth and Planetary Interiors*, **157**, 72–85.

FUJIMOTO, U., ICHIKI, K. ET AL. 1965. On the granitic rocks and the Itoigawa-Shizuoka Tectonic Line of the Northern Akaishi Massif – Geology of the Northern Akaishi Massif, Part 2. *Chikyukagaku (Earth Science)*, **76**, 15–24 [in Japanese with English abstract].

GESHI, N., SHIMANO, T., CHIBA, T. & NAKADA, S. 2002. Caldera collapse during the 2000 eruption of Miyakejima Volcano, Japan. *Bulletin of Volcanology*, **64**, 55–68.

GILL, J. B. 1981. *Orogenic Andesites and Plate Tectonics*. Berlin, Springer-Verlag.

GILL, J. B., HISCOTT, R. N. & VIDAL, Ph. 1994. Turbidite geochemistry and evolution of the Izu-Bonin Arc and continents. *Lithos*, **33**, 135–168.

GRIBBLE, R. F., STERN, R. J., BLOOMER, S. H., STUBEN, D., O'HEARN, T. & NEWMAN, S. 1996. MORB mantle and subduction components interact to generate basalts in the southern Mariana Trough backarc basin. *Geochimica et Cosmochimica Acta*, **60**, 2153–2166.

GRIBBLE, R. F., STERN, R. J., NEWMAN, S., BLOOMER, S. H. & O'HEARN, T. 1998. Chemical and isotopic composition of lavas from the Northern Mariana Trough: implications for magmagenesis in backarc basins. *Journal of Petrology*, **39**, 125–154.

GURNIS, M., HALL, L. C. & LAVIER, L. 2004. Evolving force balance during incipient subduction. *Geochemistry, Geophysics, Geosystems*, **5**, http://doi.org/10.1029/2003GC000681

HALL, C. E., GURNIS, M., SDROLIAS, M., LAVIER, L. L. & MUELLAR, R. D. 2003. Catastrophic initiation of subduction following forced convergence across fracture zones. *Earth and Planetary Science Letters*, **212**, 15–30.

HICKEY-VARGAS, R. 2005. Basalt and tonalite from the Amami Plateau, northern West Philippine Basin: new Early Cretaceous ages and geochemical results, and their petrologic and tectonic implications. *Island Arc*, **14**, 653–665.

Hickey-Vargas, R., Bizimis, M. & Deschamps, A. 2008. Onset of the Indian Ocean isotopic signature in the Philippine Sea Plate: hf and Pb isotope evidence from Early Cretaceous terranes. *Earth and Planetary Science Letters*, **268**, 255–267.

Hirano, N., Ogawa, Y., Saito, K., Yoshida, T., Sato, H. & Taniguchi, H. 2003. Multi-stage evolution of the Tertiary Mineoka ophiolite, Japan: new geochemical and age constraints. *In*: Dilek, Y. & Robinson, R. T. (eds) *Ophiolites in Earth History*. Geological Society, London, Special Publications, **218**, 279–298.

Hiscott, R. N. & Gill, J. B. 1992. Major and trace element geochemistry of Oligocene to Quaternary volcaniclastic sands and sandstones from the Izu-Bonin Arc. *In*: Taylor, B., Fujioka, K. *et al.* (eds) *Proceedings of the Ocean Drilling Program, Scientific Results*, Ocean Drilling Program, College Station, TX, **126**, 467–485.

Hochstaedter, A. G., Gill, J. B., Kusakabe, M., Newman, S., Pringle, M., Taylor, B. & Fryer, P. 1990*a*. Volcanism in the Sumisu rift, I. Major element, volatile, and stable isotope geochemistry. *Earth and Planetary Science Letters*, **100**, 179–194.

Hochstaedter, A. G., Gill, J. B. & Morris, J. D. 1990*b*. Volcanism in the Sumisu rift, II. Subduction and non-subduction related components. *Earth and Planetary Science Letters*, **100**, 195–209.

Hochstaedter, A. G., Gill, J. B., Taylor, B., Ishizuka, O., Yuasa, M. & Morita, S. 2000. Across-arc geochemical trends in the Izu-Bonin Arc: constraints on source composition and mantle melting. *Journal of Geophysical Research*, **105**, 495–512.

Hochstaedter, A., Gill, J., Peters, R., Broughton, P. & Holden, P. 2001. Across-arc geochemical trends in the Izu-Bonin Arc: contributions from the subducting slab. *Geochemistry, Geophysics, Geosystems*, **2**, 2000GC000105.

Honda, S., Yoshida, T. & Aoike, K. 2007. Spatial and temporal evolution of arc volcanism in the northeast Honshu and Izu-Bonin Arcs: evidence of small-scale convection under the island arc? *Island Arc*, **16**, 214–223.

Ishizuka, O., Uto, K., Yuasa, M. & Hochstaedter, A. G. 1998. K-Ar ages from seamount chains in the backarc region of the Izu-Ogasawara arc. *Island Arc*, **7**, 408–421.

Ishizuka, O., Uto, K., Yuasa, M. & Hochstaedter, A. G. 2002*a*. Volcanism in the earliest stage of backarc rifting in the Izu-Bonin Arc revealed by laser-heating $^{40}Ar/^{39}Ar$ dating. *Journal of Volcanology and Geothermal Research*, **120**, 71–85.

Ishizuka, O., Yuasa, M. & Uto, K. 2002*b*. Evidence of porphyry copper-type hydrothermal activity from a submerged remnant backarc volcano of the Izu-Bonin Arc: implications for the volcanotectonic history of backarc seamounts. *Earth and Planetary Science Letters*, **198**, 381–399.

Ishizuka, O., Taylor, R. N., Milton, J. A. & Nesbitt, R. W. 2003*a*. Fluid-mantle interaction in an intra-oceanic arc: constraints from high-precision Pb isotopes. *Earth and Planetary Science Letters*, **211**, 221–236.

Ishizuka, O., Uto, K. & Yuasa, M. 2003*b*. Volcanic history of the backarc region of the Izu-Bonin (Ogasawara) Arc. *In*: Larter, R. D. & Leat, P. H. (eds) *Intra-Oceanic Subduction Systems: Tectonic and Magmatic Processes*. Geological Society, London, Special Publications, **219**, 187–205.

Ishizuka, O., Kimura, J.-I. *et al.* 2006*a*. Early stages in the evolution of Izu-Bonin Arc volcanism: new age, chemical, and isotopic constraints. *Earth and Planetary Science Letters*, **250**, 385–401.

Ishizuka, O., Taylor, R. N., Milton, J. A., Nesbitt, R. W., Yuasa, M. & Sakamoto, I. 2006*b*. Variation in the source mantle of the northern Izu arc with time and space: Constraints from high-precision Pb isotopes. *Journal of Volcanology and Geothermal Research*, **156**, 266–290.

Ishizuka, O., Taylor, R. N., Milton, J. A., Nesbitt, R. W., Yuasa, M. & Sakamoto, I. 2007. Processes controlling along-arc isotopic variation of the southern Izu-Bonin Arc. *Geochemistry, Geophysics, Geosystems*, **8**(6), Q06008, http://doi.org/10.1029/2006GC001475

Ishizuka, O., Geshi, N., Itoh, J., Kawanabe, Y. & TuZino, T. 2008. The magmatic plumbing of the submarine Hachijo NW volcanic chain, Hachijo-jima, Japan: long-distance magma transport? *Journal of Geophysical Research*, **113**, http://doi.org/10.1029/2007JB005325

Ishizuka, O., Yuasa, M., Taylor, R. N. & Sakamoto, I. 2009. Two contrasting magmatic types coexist after the cessation of backarc spreading. *Chemical Geology*, **266**, 283–305.

Ishizuka, O., Yuasa, M. *et al.* 2010. Migrating shoshonitic magmatism tracks Izu-Bonin-Mariana intra-oceanic arc rift propagation. *Earth and Planetary Science Letters*, **294**, 111–122.

Ishizuka, O., Tani, K. *et al.* 2011*a*. The timescales of subduction initiation and subsequent evolution of an oceanic island arc. *Earth and Planetary Science Letters*, **306**, 229–240, http://doi.org/10.1016/j.epsl.2011.04.006

Ishizuka, O., Taylor, R. N., Yuasa, M. & Ohara, Y. 2011*b*. Making and breaking an Island arc: a new perspective from the Oligocene Kyushu-Palau arc, Philippine Sea. *Geochemistry, Geophysics, Geosystems*, **12**, Q05005, http://doi.org/10.1029/2010GC003440

Isse, T., Shiobara, H. *et al.* 2009. Seismic structure of the upper mantle beneath the Philippine Sea from seafloor and land observation: implications for mantle convection and magma genesis in the Izu–Bonin–Mariana subduction zone. *Earth and Planetary Science Letters*, **278**, 107–119.

Kanamaru, T. & Takahashi, M. 2009. Geology and structure of pyroclastic rocks in the Higashi-Yamanashi volcano-plutonic complex and its implications for the process of cauldron formation. *Proceedings of the Institute of Natural Sciences, Nihon University*, **44**, 121–138 [in Japanese with English abstract].

Karig, D. E. & Moore, G. F. 1975. Tectonic complexities in the Bonin arc system. *Tectonophysics*, **27**, 97–118.

Kawate, S. & Arima, M. 1998. Petrogenesis of the Tanzawa plutonic complex, central Japan: exposed felsic middle crust of the Izu-Bonin-Mariana Arc system. *Island Arc*, **7**, 342–358.

Kelemen, P. B. 1995. Genesis of high Mg-andesites and the continental crust. *Contributions to Mineralogy and Petrology*, **120**, 1–19.

Kimura, J.-I. & Yoshida, T. 2006. Contributions of slab fluid, mantle wedge and crust to the origin of Quaternary lavas in the NE Japan arc. *Journal of Petrology*, **47**, 2185–2232.

Kimura, J.-I., Kent, A. J. R. *et al.* 2010. Origin of cross-chain geochemical variation in Quaternary lavas from the northern Izu arc: using a quantitative mass balance approach to identify mantle sources and mantle wedge processes. *Geochemistry, Geophysics, Geosystems*, **11**, Q10011, http://doi.org/10.1029/2010GC003050

Kitamura, K., Ishikawa, M. & Arima, M. 2003. Petrological model of the northern Izu–Bonin–Mariana Arc crust: constraints from high-pressure measurements of elastic wave velocities of the Tanzawa plutonic rocks, central Japan. *Tectonophysics*, **371**, 213–221.

Kodaira, S., Sato, T., Takahashi, N., Ito, A., Tamura, Y., Tatsumi, Y. & Kaneda, Y. 2007*a*. Seismological evidence for variable growth of crust along the Izu intraoceanic arc. *Journal of Geophysical Research*, **112**, B05104, http://doi.org/10.1029/2006JB004593

Kodaira, S., Sato, T., Takahashi, N., Miura, S., Tamura, Y., Tatsumi, Y. & Kaneda, Y. 2007*b*. New seismological constraints on growth of continental crust in the Izu-Bonin intra-oceanic arc. *Geology*, **35**(11), 1031–1034, http://doi.org/10.1130/G23901A.1

Kodaira, S., Sato, T., Takahashi, N., Yamashita, M., NoT. & Kaneda, Y. 2008. Seismic imaging of a possible paleo-arc in the Izu-Bonin intra-oceanic arc and its implications for arc evolution processes. *Geochemistry, Geophysics, Geosystem*, **9**, Q10X01, http://doi.org/10.1029/2008GC002073

Kodaira, S., Noguchi, N., Takahashi, N., Ishizuka, O. & Kaneda, Y. 2010. Evolution from forearc oceanic crust to island arc crust: a seismic study along the Izu-Bonin forearc. *Journal of Geophysical Research*, **115**, B09102, http://doi.org/10.1029/2009JB006968

Kohlstedt, D. L., Evans, B. & Mackwell, S. J. 1995. Strength of the lithosphere: constraints imposed by laboratory experiments. *Journal of Geophysical Research*, **100**, 17 587–17 602.

Kohut, E. J., Stern, R. J., Kent, A. J. R., Nielsen, R. L., Bloomer, S. H. & Leybourne, M. 2006. Evidence for adiabatic decompression melting in the Southern Mariana Arc from high-Mg lavas and melt inclusions. *Contributions to Mineralogy and Petrology*, **152**, 201–221.

Kuno, H. 1959. Origin of Cenozoic petrographic provinces of Japan and surrounding areas. *Bulletin of Volcanology*, **32**, 141–176.

Kuritani, T., Yokoyama, T., Kobayashi, K. & Nakamura, E. 2003. Shift and rotation of composition trends by magma mixing: 1983 eruption at Miyake-jima volcano, Japan. *Journal of Petrology*, **44**, 1895–1916.

Lin, P. N., Stern, R. J., Morris, J. & Bloomer, S. H. 1990. Nd- and Sr-isotopic compositions of lavas from the northern Mariana and southern volcano arcs; implications for the origin of island arc melts. *Contributions to Mineralogy and Petrology*, **105**, 381–392.

MACHIDA, S. & ISHII, T. 2003. Backarc volcanism along the en echelon seamounts: the Empo seamount chain in the northern Izu-Ogasawara arc. *Geochemistry, Geophysics, Geosystems*, **4**, 9006, http://doi.org/10.1029/2003GC000554

MACHIDA, S., ISHII, T., KIMURA, J.-I., AWAJI, S. & KATO, Y. 2008. Petrology and geochemistry of cross-chains in the Izu-Bonin back arc: three mantle components with contributions of hydrous liquids from a deeply subducted slab. *Geochemistry, Geophysics, Geosystems*, **9**, Q05002, http://doi.org/10.1029/2007GC001641

MCLENNAN, S. M. & TAYLOR, S. R. 1982. Geochemical constraints on the growth of the continental crust. *Journal of Geology*, **90**, 347–361.

MEIJER, A. 1980. Primitive arc volcanism and a boninite series: examples from western Pacific island arcs. *In*: HAYES, D. E. (ed.) *The Tectonic Evolution of Southeast Asian Seas and Islands*. American Geophysical Union, Washington, DC, Geophysical Monograph **23**, 269–282.

MIURA, S., SUYEHIRO, K., SHINOHARA, M., TAKAHASHI, N., ARAKI, E. & TAIRA, A. 2004. Seismological structure and implications of collision between the Ontong Java Plateau and Solomon arc from ocean bottom seismometer-airgun data. *Techtonophysics*, **389**, 191–220, http://doi.org/10.1016/j.tecto.2003.09.029

MORITA, S., ISHIZUKA, O., HOCHSTAEDTER, A. G., ISHII, T., YAMAMOTO, F., TOKUYAMA, H. & TAIRA, A. 1999. Volcanic and structural evolution of the northern Izu-Bonin Arc. *Chikyu Monthly*, **23**, 79–88 [in Japanese].

NAKAJIMA, J., HIROSE, F. & HASEGAWA, A. 2009. Seismotectonics beneath the Tokyo metropolitan area, Japan: effect of slab–slab contact and overlap on seismicity. *Journal of Geophysical Research*, **114**, B08309, http://doi.org/10.1029/2008JB006101

NAKANISHI, A., KURASHIMO, E. *ET AL*. 2009. Crustal evolution of the southwestern Kuril Arc deduced from seismic velocity and geochemical structure. *Tectonophysics*, **472**, 105–123.

NISHIMURA, T., OZAWA, S., MURAKAMI, M., SAGIYA, T., TADA, T., KAIZU, M. & UKAWA, M. 2001. Crustal deformation caused by magma migration in the northern Izu Islands, Japan. *Geophysical Research Letters*, **28**, 3745–3748.

NISHIZAWA, A., KANEDA, K., KATAGIRI, Y. & KASAHARA, J. 2007. Variation in crustal structure along the Kyushu-Palau Ridge at 15–21N on the Philippine Sea plate based on seismic refractions profiles. *Earth, Planets and Space*, **59**, e17–e20.

OBANA, K., KAMIYA, S., KODAIRA, S., SUETSUGU, D., TAKAHASHI, N., TAKAHASHI, T. & TAMURA, Y. 2010. Along-arc variation in seismic velocity structure related to variable growth of arc crust in northern Izu-Bonin intraoceanic arc. *Geochemistry, Geophysics, Geosystem*, **11**, Q08012, http://doi.org/10.1029/2010GC003146

OGAWA, Y. 1983. Mineoka ophiolite belt in the Izu forearc area – Neogene accretion of oceanic and island arc assemblages in the northeastern corner of the Philippine Sea plate. *In*: HASHIMOTO, M. & UYEDA, S. (eds) *Accretion Tectonics in the Circum-Pacific Region*. Terra Scientific Publishing Company (Terrapub), Tokyo, 245–260.

OGAWA, Y. & TANIGUCHI, H. 1988. Geology and tectonics of the Miura-Boso Peninsulas and the adjacent area. *Modern Geology*, **12**, 147–168.

OKINO, K., SHIMAKAWA, Y. & NAGAOKA, S. 1994. Evolution of the Shikoku basin. *Journal of Geomagnetism and Geoelectronics*, **46**, 463–479.

OTOFUJI, Y. & MATSUDA, T. 1983. Paleomagnetic evidence for the clockwise rotation of Southwest Japan: *Earth and Planetary Science Letters*, **62**, 349–359.

OZAKI, M., MAKIMOTO, H. *ET AL*. 2002. *Geological map of Japan 1:200,000*, Kofu, Geological Survey of Japan/AIST [in Japanese with English abstract].

REAGAN, M. K. & MEIJER, A. 1984. Geology and geochemistry of early arc volcanic rocks from Guam. *Geological Society of America Bulletin*, **95**, 701–713.

REAGAN, M. K., HANAN, B. B., HEIZLER, M. T., HARTMAN, B. S. & HICKEY-VARGAS, R. 2008. Petrogenesis of volcanic rocks from Saipan and Rota, Mariana, Islands, and implications for the evolution of nascent island arcs. *Journal of Petrology*, **49**, 441–464.

REAGAN, M. K., ISHIZUKA, O. *ET AL*. 2010. Forearc basalts and subduction initiation in the Izu-Bonin-Mariana system. *Geochemistry, Geophysics, Geosystems*, **11**, Q03X12, http://doi.org/10.1029/2009GC002871

RUDNICK, R. L. 1995. Making continental crust. *Nature*, **378**, 571–578.

RUDNICK, R. L. & GAO, S. 2003. Composition of the continental crust. *In*: RUDNICK, R. L. (ed.) *Treatise on Geochemistry*. The Crust Elsevier, Oxford, **3**, 1–64.

SAITO, S. 1996. Stratigraphy of Cenozoic strata in the southern terminus area of Boso Peninsula, Central Japan. *Contributions from the Institute of Geology and Paleontology, Tohoku University*, **93**, 1–37.

SAITO, S., ARIMA, M., NAKAJIMA, T., MISAWA, K. & KIMURA, J. I. 2007*a*. Formation of distinct granitic magma batches by partial melting of hybrid lower crust in the Izu arc Collision Zone, central Japan. *Journal of Petrology*, **48**, 1761–1791.

SAITO, S., ARIMA, M. & NAKAJIMA, T. 2007*b*. Hybridization of a shallow 'I-type' granitoid pluton and its host migmatite by magma-chamber wall collapse: the Tokuwa Pluton, central Japan. *Journal of Petrology*, **48**, 79–111.

SAITO, S., ARIMA, M. *ET AL*. 2012. Petrogenesis of the Kaikomagatake granitoid pluton in the Izu Collision Zone, central Japan: implications for transformation of juvenile oceanic arc into mature continental crust. *Contributions to Mineralogy and Petrology*, **163**, 611–629, http://doi.org/10.1007/s00410-011-0689-1

SAKAMOTO, I., HIRATA, D. & FUJIOKA, K. 1999. Description of basement rocks from the Izu-Bonin Arc. *Research Report of Kanagawa Prefectural Museum*, **9**, 21–39 [in Japanese with English abstract].

SATO, K., SHIBATA, K. & UCHIUMI, S. 1986. Discordant K-Ar ages between hornblende and biotite from the tonalitic pluton in the south Fossa Magna, central Japan. *Journal of Geological Society of Japan*, **92**, 439–446 [in Japanese with English abstract].

SATO, T., KODAIRA, S., TAKAHASHI, N., TATSUMI, Y. & KANEDA, Y. 2009*a*. Amplitude modeling of the seismic reflectors in the crust-mantle transition layer beneath the volcanic front along the northern Izu-Bonin island arc. *Geochemistry Geophysics Geosystem*, **10**, Q02X04, http://doi.org/10.1029/2008GC001990

SATO, T., KODAIRA, S., TAKAHASHI, N., KAIHO, Y. & KANEDA, Y. 2009*b*. Crustal image along the southern Izu-Bonin Arc and the northern tip of the West Mariana Ridge deduced from a seismic refraction/reflection data. *Eos, Transactions of the American Geophysical Union*, **90**, Fall Meeting Suppl., Abstract T23A-1870.

SEKI, Y., OKI, Y., MATSUDA, T., MIKAMI, K. & OKUMURA, K. 1969. Metamorphism in the Tanzawa Mountains, central Japan. *Journal of the Japanese Association of Mineralogists, Petrologists and Economic Geologists*, **61**, 50–75.

SHARP, W. D. & CLAGUE, D. A. 2006. 50-Ma initiation of Hawaiian-Emperor Bend records major change in Pacific Plate motion. *Science*, **313**, 1281–1284.

SHARPTON, V. L. & HEAD, J. W. III. 1985. Analysis of regional slope characteristics on Venus and Earth. *Journal of Geophysical Research*, **90**, 3733–3740.

SHIBATA, K., KATO, Y. & MIMURA, K. 1984. K-Ar ages of granites and related rocks from the northern Kofu area. *Bulletin of the Geological Survey of Japan*, **35**, 19–24 [in Japanese with English abstract].

SHIMAMOTO, T., KANAORI, Y. & ASAI, K.-I. 1991. Cathodoluminescence observations on low-temperature mylonites: potential for detection of solution-precipitation microstructures. *Journal of Structural Geology*, **13**, 967–973.

SHIREY, S. B. & HANSON, G. N. 1984. Mantle-derived Archaean monzodiorites and trachyandesites. *Nature*, **310**, 222–224.

SMITH, D. E., ZUBER, M. T. *ET AL*. 1999. The global topography of Mars and implications for surface evolution. *Science*, **284**, 1495–1503.

SMITH, I. E. M., STEWART, R. B. & PRICE, R. C. 2003. The petrology of a large intra-oceanic silicic eruption: the Sandy Bay Tephra, Kermadec Arc, Southwest Pacific. *Journal of Volcanology and Geothermal Research*, **124**, 173–194.

SMITH, I. E. M., WORTHINGTON, T. J., PRICE, R. C., STEWART, R. B. & MAAS, R. 2006. Petrogenesis of dacites in an oceanic subduction environment: raoul Island, Kermadec arc. *Journal of Volcanology and Geothermal Research*, **156**, 252–265.

SOH, W., NAKAYAMA, K. & KIMURA, T. 1998. Arc-arc collision in the Izu collision zone, central Japan, deduced from the Ashigara Basin and adjacent Tanzawa Mountains. *Island Arc*, **7**, 330–341.

SOLOMATOV, V. S. & MORESI, L.-N. 1996. Stagnant lid convection on Venus. *Journal of Geophysical Research*, **101**, 4737–4753.

STERN, R. J. 2004. Subduction initiation: spontaneous and induced. *Earth and Planetary Science Letters*, **226**, 275–292.

STERN, R. J. 2010. The anatomy and ontogeny of modern intra-oceanic arc systems. *In*: KUSKY, T. M., ZHAI, M.-G. & XIAO, W. (eds) *The Evolving Continents: Understanding Processes of Continental Growth*. Geological Society, London, Special Publications, **338**, 7–34.

Stern, R. J. & Bloomer, S. H. 1992. Subduction zone infancy: examples from the Eocene Izu-Bonin-Mariana and Jurassic California. *Geological Society of America Bulletin*, **104**, 1621–1636.

Stern, R. J., Jackson, M. C., Fryer, P. & Ito, E. 1993. O, Sr, Nd and Pb isotopic composition of the Kasuga Cross-Chain in the Mariana Arc; a new perspective on the K-h relationship. *Earth and Planetary Science Letters*, **119**, 459–475.

Stern, R. J., Fouch, M. J. & Klemperer, S. L. 2003. An overview of the Izu-Bonin-Mariana subduction factory. *In*: Eiler, J. (ed.) *Inside the Subduction Factory*. American Geophysical Union, Washington, DC, Geophysical Monograph **138**, 175–222.

Stern, R. J., Kohut, E., Bloomer, S. H., Leybourne, M., Fouch, M. & Vervoort, J. 2006. Subduction factory processes beneath the Guguan cross-chain, Mariana Arc: no role for sediments, are serpentinites important? *Contributions to Mineralogy and Petrology*, **151**, 202–221.

Straub, S. M. 2003. The evolution of the Izu Bonin - Mariana volcanic arcs (NW Pacific) in terms of major elements. *Geochemistry, Geophysics, Geosystems*, **4**, 1018, http://doi.org/10.1029/2002GC000357

Straub, S. M. 2008. Uniform processes of melt differentiation in the central Izu Bonin volcanic arc (NW Pacific) *In*: Annen, C. & Zellmer, G. F. (eds) *Dynamics of Crustal Magma Transfer, Storage and Differentiation*. Geological Society, London, Special Publications, **304**, 261–283.

Sun, C. H. & Stern, R. J. 2001. Genesis of Mariana shoshonites: contribution of the subduction component. *Journal of Geophysical Research*, **106**, 589–608.

Suyehiro, K., Takahashi, N. *et al.* 1996. Continental crust, crustal underplating, and low-Q upper mantle beneath an oceanic island arc. *Science*, **272**, 390–392.

Taira, A., Tokuyama, H. & Soh, W. 1989. Accretion tectonics and evolution of the Pacific Ocean Margins. *In*: Ben-Avraham, Z. (ed.) *The Evolution of the Pacific Ocean Margins*. New York, Oxford University Press, 100–123.

Taira, A., Saito, S. *et al.* 1998. Nature and growth rate of the Northern Izu-Bonin (Ogasawara) arc crust and their implications for continental crust formation. *Island Arc*, **7**, 395–407.

Takahashi, M. 1989. Neogene granitic magmatism in the South Fossa Magna collision zone, central Japan. *Modern Geology*, **14**, 127–143.

Takahashi, N., Suyehiro, K. & Shinohara, M. 1998. Implications from the seismic crustal structure of the northern Izu-Ogasawara arc. *Island Arc*, **7**, 383–394.

Takahashi, N., Kodaira, S., Klemperer, S. L., Tatsumi, Y., Kaneda, Y. & Suyehiro, K. 2007. Crustal structure and evolution of the Mariana intra-oceanic island arc. *Geology*, **35**, 203–206.

Takahashi, N., Kodaira, S., Tatsumi, Y., Kaneda, Y. & Suyehiro, K. 2008. Structure and growth of the Izu-Bonin-Mariana Arc crust: i. Seismic constraint on crust and mantle structure of the Mariana Arc - backarc system. *Journal of Geophysical Research*, **113**, http://doi.org/10.1029/2007JB005120

Takahashi, N., Kodaira, S. *et al.* 2009. Structural variations of arc crusts and rifted margins in southern Izu-Ogasawara arc-backarc system. *Geochemistry, Geophysics, Geosystems*, **10**, Q09X08, http://doi.org/10.1029/2008GC002146

Takahashi, N., Yamashita, M. *et al.* 2010. Delamination of the Ogasawara Plateau. *Abstract SCG086-16 presented at Japan Geoscience Union Meeting*, Chiba, 23–28 May.

Takahashi, N., Yamashita, M., Kodaira, S., Miura, S., Sato, T., No, T. & Tatsumi, Y. 2011*a*. Crustal growth of the Izu-Ogasawara arc estimated from structural characteristics of Oligocene arc. *In*: *Proceedings of 2011 Fall AGU Meeting*, AGU, San Francisco, California, 5–9 December. Abstract T53A-2498.

Takahashi, N., Yamashita, M. *et al.* 2011*b*. Rifting structure of central Izu-Ogasawara (Bonin) arc crust: results of seismic crustal imaging. *In*: Ogawa, Y., Annma, R. & Dilek, Y. (eds) *Accretionary Prisms and Convergent Margin Tectonics in the Northwest Pacific Margin*. Springer, Netherlands, 75–95.

Takita, R. 1980. Petrologic study on the gabbroic rocks in the Tanzawa Mountains, central Japan: with special reference to the genetic relation between the gabbroic rocks and the tonalites. *Journal of the Geological Society of Japan*, **86**, 369–387 [in Japanese with English abstract].

Tamura, I., Takagi, H. & Yamazaki, H. 2010*a*. The Tanzawa-garnet pumice: widely distributed 2.5 Ma tephra in the southern Kanto area, Japan. *Journal of the Geological Society of Japan*, **116**, 360–373 [in Japanese with English abstract].

Tamura, Y. 2003. Some geochemical constraints on hot fingers in the mantle wedge: evidence from NE Japan. *In*: Larter, R. D. & Leat, P. T. (eds) *Intra-Oceanic Subduction Systems: Tectonic and Magmatic Processes*. Geological Society, London, Special Publications, **219**, 221–237.

Tamura, Y. & Tatsumi, Y. 2002. Remelting of an andesitic crust as a possible origin for rhyolitic magma in oceanic arcs: an example from the Izu-Bonin Arc. *Journal of Petrology*, **43**, 1029–1047.

Tamura, Y., Tatsumi, Y., Zhao, D., Kido, Y. & Shukuno, H. 2002. Hot fingers in the mantle wedge: new insights into magma genesis in subduction zones. *Earth and Planetary Science Letters*, **197**, 107–118.

Tamura, Y., Tani, K., Ishizuka, O., Chang, Q., Shukuno, H. & Fiske, R. S. 2005. Are arc basalts dry, wet, or both? Evidence from the Sumisu caldera volcano, Izu-Bonin Arc, Japan. *Journal of Petrology*, **46**, 1769–1803.

Tamura, Y., Tani, K., Chang, Q., Shukuno, H., Kawabata, H., Ishizuka, O. & Fiske, R. S. 2007. Wet and dry basalt magma evolution at Torishima volcano, Izu-Bonin Arc, Japan: the possible role of phengite in the downgoing slab. *Journal of Petrology*, **48**, 1999–2031.

Tamura, Y., Gill, J. B. *et al.* 2009. Silicic magmas in the Izu-Bonin oceanic arc and implications for crustal evolution. *Journal of Petrology*, **50**, 685–723.

Tamura, Y., Ishizuka, O. *et al.* 2010*b*. Missing oligocene crust of the Izu-Bonin Arc: consumed or rejuvenated during collision? *Journal of Petrology*, **51**, 823, http://doi.org/10.1093/petrology/egq002

Tani, K., Fiske, R. S., Tamura, Y., Kido, Y., Naka, J., Shukuno, H. & Takeuchi, R. 2008. Sumisu volcano, Izu-Bonin Arc, Japan: site of a silicic caldera-forming eruption from a small open-ocean island. *Bulletin of Volcanology*, **70**, 547–562.

Tani, K., Dunkley, D. J., Kimura, J. I., Wysoczanski, R. J., Yamada, K. & Tatsumi, Y. 2010. Syncollisional rapid granitic magma formation in an arc-arc collision zone: evidence from the Tanzawa plutonic complex, Japan. *Geology*, **38**, 215–218.

Tani, K., Dunkley, D. J. & Ohara, Y. 2011*a*. Termination of backarc spreading: zircon dating of a giant oceanic core complex. *Geology*, **39**, 47–50.

Tani, K., Fiske, R. S., Dunkley, D. J., Ishizuka, O., Oikawa, T., Isobe, I. & Tatsumi, Y. 2011*b*. The Izu Peninsula, Japan: zircon geochronology reveals a record of intra-oceanic backarc magmatism in an accreted block of Izu–Bonin upper crust. *Earth and Planetary Science Letters*, **303**, 225–239, http://doi.org/10.1016/j.epsl.2010.12.052

Tatsumi, Y. 2005. The subduction factory: how it operates in the evolving Earth. *GSA Today*, **15**, 4–10.

Tatsumi, Y. & Eggins, S. M. 1995. *Subduction Zone Magmatism*. Blackwell, Cambridge, MA.

Tatsumi, Y. & Stern, R. J. 2006. Manufacturing continental crust in the subduction factory. *Oceanography*, **19**, 104–112.

Tatsumi, Y., Shukuno, H., Tani, K., Takahashi, N., Kodaira, S. & Kogiso, T. 2008. Structure and growth of the Izu-Bonin-Mariana arc crust: 2. Role of crust-mantle transformation and the transparent Moho in arc crust evolution. *Journal of Geophysical Research*, **113**, B02203, http://doi.org/10.1029/2007JB005121

Taylor, B. 1992. Rifting and the volcanic-tectonic evolution of the Izu-Bonin-Mariana Arc system. *In*: Taylor, B., Fujioka, K. *et al.* (eds) *Proceedings of Ocean Drilling Program, Scientific Results*. Texas A&M University, College Station, **126**, 627–650.

Taylor, R. N. 1995. The geochemical evolution of the continental crust. *Reviews in Geophysics*, **33**, 241–265.

Taylor, R. N. & Nesbitt, R. W. 1998. Isotopic characteristics of subduction fluids in an intra-oceanic setting, Izu-Bonin Arc, Japan. *Earth and Planetary Science Letters*, **164**, 79–98.

Taylor, R. N., Nesbitt, R. W., Vidal, P., Harmon, R. S., Auvray, B. & Croudace, I. W. 1994. Mineralogy, chemistry, and genesis of the Boninite Series Volcanics, Chichijima, Bonin Islands, Japan. *Journal of Petrology*, **35**, 577–617.

Taylor, R. N., Narlow, M. S., Johnson, L. E., Taylor, B., Bloomer, S. H. & Mitchell, J. G. 1995. Intrusive volcanic rocks in western Pacific Forearcs. *In*: Taylor, B. & Natland, J. (eds) *Active Margins and Marginal Basins of the Western Pacific*. American Geophysical Union, Washington, DC, Geophysical Monograph, **88**, 31–43.

Toda, S., Stein, R. S. & Sagiya, T. 2002. Evidence from the AD 2000 Izu islands earthquake swarm and stressing rate governs seismicity. *Nature*, **419**, 58–61.

Tollstrup, D., Gill, J., Kent, A., Prinkey, D., Williams, R., Tamura, Y. & Ishizuka, O. 2010. Across-arc geochemical trends in the Izu-Bonin

Arc: contributions from the subducting slab, revisited. *Geochemistry, Geophysics, Geosystems*, **11**, http://doi.org/10.1029/2009GC002847

Tsunoda, K. & Shimizu, M. 1983. Ultrabasic rocks from the Northern part of the Kaikomagatake Granitic rocks, Yamanashi Prefecture, Japan. *Memoirs of the Faculty of Liberal Arts & Education, Part II, Mathematics & Natural Sciences, Yamanashi University*, **34**, 114–120 [in Japanese with English abstract].

Umino, S. 1985. Volcanic geology of Chichijima, the Bonin Islands (Ogasawara Islands). *Journal of the Geological Society of Japan*, **91**, 505–523.

Volpe, A. M., MacDougall, J. D. & Hawkins, J. W. 1987. Mariana Trough Basalts (MTB) trace-element and Sr-Nd isotopic evidence for mixing between MORB-like and arc-like melts. *Earth and Planetary Science Letters*, **82**, 241–254.

Volpe, A. M., MacDougall, J. D., Lugmair, G. W., Hawkins, J. W. & Lonsdale, P. 1990. Fine-scale isotopic variation in Mariana Trough basalts: Evidence for heterogeneity and a recycled component in backarc basin mantle. *Earth and Planetary Science Letters*, **100**, 241–254.

Wade, J. A., Plank, T. *et al.* 2005. The May 2003 eruption of Anatahan volcano, Mariana Islands: geochemical evolution of a silicic island-arc volcano. *Journal of Volcanology and Geothermal Research*, **146**, 139–170.

White, R. S., McKenzie, D. & O'Nions, R. K. 1992. Oceanic crustal thickness from seismic measurements and rate earth element inversions. *Journal of Geophysical Research*, **97**, 19 683–19 715.

Woodhead, J. D. 1989. Geochemistry of the Mariana Arc (western Pacific); source composition and processes. *Chemical Geology*, **76**, 1–24.

Wright, I. C., Worthington, T. J. & Gamble, J. A. 2006. New multibeam mapping and geochemistry of the 30°–35° S sector, and overview of southern Kermadec arc volcanism. *Journal of Volcanology and Geothermal Research*, **149**, 263–296.

Yajima, T. 1970. The Sanogawa gabbro-diorite complex. *Science Reports of the Saitama University, Series B*, **5**, 199–230 [in Japanese with English abstract].

Yamamoto, Y. & Kawakami, S. 2005. Rapid tectonics of the Late Miocene Boso accretionary prism related to the Izu–Bonin arc collision. *Island Arc*, **14**, 178–198.

Yamashita, M., Kodaira, S., Takahashi, N., Tatsumi, Y. & Kaneda, Y. 2009. The forearc crustal evolution of Izu-Bonin (Ogasawara) region obtained by seismic reflection and refraction surveys. *Eos, Transactions, American Geophysical Union*, **90**, Fall Meeting Suppl., Abstract T32A-03.

Yamazaki, T. & Yuasa, M. 1998. Possible Miocene rifting of the Izu-Ogasawara (Bonin) arc deduced from magnetic anomalies. *Island Arc*, **7**, 374–382.

Yokoyama, T., Kobayashi, K., Kuritani, T. & Nakamura, E. 2003. Mantle metasomatism and rapid ascent of slab components beneath island arcs: evidence from ^{238}U-^{230}Th-^{226}Ra disequilibria of Miyakejima volcano, Izu arc, Japan. *Journal of Geophysical Research*, **108**, http://doi.org/10.1029/2002JB002103

Yuasa, M. & Kano, K. 2003. Submarine silicic calderas on the northern Shichito-Iwojima ridge, Izu-Ogasawara (Bonin) arc, western Pacific. *In*: White, J. D. L., Smellie, J. L. & Clague, D. A. (eds) *Explosive Subaqueous Volcanism*. American Geophysical Union, Washington, DC, Geophysical Monograph, **140**, 231–243.

Zhang, J., Ten Brink, U. S. & Toksoz, M. N. 1998. Nonlinear refraction and reflection travel time tomography. *Journal of Geophysical Research*, **103**, 29 763–29 757.

2g Hokkaido

HAYATO UEDA

Hokkaido is the northernmost of the four main islands of Japan and lies between Honshu Island to the SW, the neighbouring far-east Russian territories such as Sakhalin Island to the north, Primorye (Sikhote–Alin) to the west and the Kuril islands to the east. It has close geological relations with all these adjacent areas. The island also faces three ocean basins: the Pacific to the south, the Kuril Basin of the Sea of Okhotsk to the NE, and the Sea of Japan to the west. The latter two are back-arc basins opened in the Miocene, so that most of the basement rocks were generated before their opening and related to Jurassic–Palaeogene arc-trench systems involving subduction of oceanic plates in the Pacific.

However, there are several unique characteristics in Hokkaido. First of all, about one-third of the island (eastern Hokkaido) belongs not to the Honshu Arc but to the Kuril Arc. This tectonic setting applies back to the Cretaceous Period when its basement geology was constructed, as well as to the present-day situation. As a result, pre-Neogene rocks there cannot be correlated with anything in the Honshu Arc, but rather to the Sakhalin and Kuril islands. The eastern part of Hokkaido, together with southwestern Kurils, must have been attached to the eastern parts of Sakhalin before the Kuril Basin opened. The rest of the island, that is, central and SW parts of Hokkaido, was presumably attached to the coastal Primorye (Sikhote–Alin) of SE Russia. Central and southwestern Hokkaido represent the northern continuation of the Honshu Arc, and might have been part of the Eurasian active continental margin before the opening of the Japan Sea. However, the presence of abundant Mesozoic ophiolites, extensive forearc basin deposits and an Early Cretaceous volcanic chain, all characteristic of Hokkaido, inhibit the simple application of the same tectonic divisions as seen in SW Japan. This chapter considers this question of correlation between the two areas, in addition to reviewing major basement components and the tectono-stratigraphic framework of Hokkaido.

Hokkaido Island belongs to the North America (Okhotsk) Plate, facing the Eurasia (Amur) Plate to the west. The present-day plate boundary, which became active in early Quaternary times, runs offshore along the western coast of the island (i.e. eastern margin of the Sea of Japan). However, in the Neogene period the major plate boundary is thought to have crossed through central Hokkaido. Hokkaido has also been a collision zone between the Kuril and Honshu arcs at least since the Miocene epoch. This collision coincided with, and was closely related to, opening of the back-arc basins corresponding to both arcs. During this collision, pre-Neogene basement rocks were significantly modified in their structure by dextral shear zones and west-verging thrust faults, and partly underwent intense metamorphism. Those in eastern Hokkaido were rotated over 90° and central Hokkaido forms a complicated fold-and-thrust belt, with thick Neogene foreland basin deposits and the exhumation of Miocene granulites at its core. These young events are also key to any reconstruction of the original basement structure and tectonic histories, as well as providing in themselves a case study of recent arc–arc collision.

Overview of Hokkaido geology

The distribution of pre-Neogene basement rocks is essentially classified, from west to east, as the Oshima, Sorachi–Yezo, Hidaka, Tokoro and Nemuro belts (Fig. 2g.1) as summarized in Kiminami *et al.* (1986*a*, 1992*c*); there have been some differences in divisions, nomenclature and definitions in later publications by subsequent researchers, however. These tectonostratigraphic belts are generally north–south aligned, except the Nemuro Belt which has an east–west trend in parts. The Oshima, Sorachi–Yezo and at least parts of the Hidaka belts are attributed to the Honshu (NE Japan) arc-trench system (Fig. 2g.2). Accretionary complexes show general eastwards-younging trends when their complicated structures are refolded back to their orientation during Tertiary collision tectonics. The Tokoro and Nemuro belts (and some units here assigned to the Hidaka Belt) belong to the Kuril Arc. Clastic rocks are richer in basic to intermediate volcanic components, with only a minor contribution of continent-derived sialic materials (Kiminami & Kontani 1983*a*). These two arc-trench systems have opposite polarities: the former verging to the east and the latter to the west. The boundary of the two arc-trench systems facing each other is however not distinct.

NE Japan Arc Terrane

Oshima Belt

The Oshima Belt in the SW is characterized by the occurrence of Jurassic accretionary complexes and the intrusion of Early Cretaceous granitoid rocks. These Mesozoic rocks are exposed as separate horsts and/or windows among Neogene volcanic and sedimentary cover sequences. The accretionary complexes consist of terrigenous clastic rocks (mainly sandstone and mudstone) and pelagic chert, with lesser amounts of limestone and basalt. They occur as structurally stacked thrust sheets of chert-clastic sequences or mélange facies (Kawamura *et al.* 1986, 2000). Based on radiolarians and conodonts, the age of chert ranges from Late Carboniferous or Early Permian to Early Jurassic and the age of mudstone ranges from Middle Jurassic to Late Jurassic or even possibly to earliest Cretaceous (Tajika *et al.* 1984; Ishiga & Ishiyama 1987; Fig. 2g.3). Late Carboniferous, Permian and Late Triassic ages have been assigned for limestones based on fossils such as corals, algae, fusulinids and conodonts (e.g. Minato & Rowett 1967; Sakagami *et al.* 1969; Yoshida & Aoki 1972; Minoura & Kato 1978). Basaltic lava and breccia, closely associated with limestones, are of oceanic-island origins (Kawamura *et al.* 1986). According to Kawamura *et al.* (2000), the Oshima Belt can be sub-divided into a western region where Carboniferous–Permian limestone and chert occur, and an eastern region where Palaeozoic outcrop is lacking and Triassic limestone characteristically occurs.

Sandstones in the Oshima Belt are arkosic, being rich in quartz and feldspars and poor in lithic fragments, which consist mainly of granite, felsic volcanic and sedimentary rocks (Tajika *et al.* 1984;

From: Moreno, T., Wallis, S., Kojima, T. & Gibbons, W. (eds) 2016. *The Geology of Japan*. Geological Society, London, 201–221.

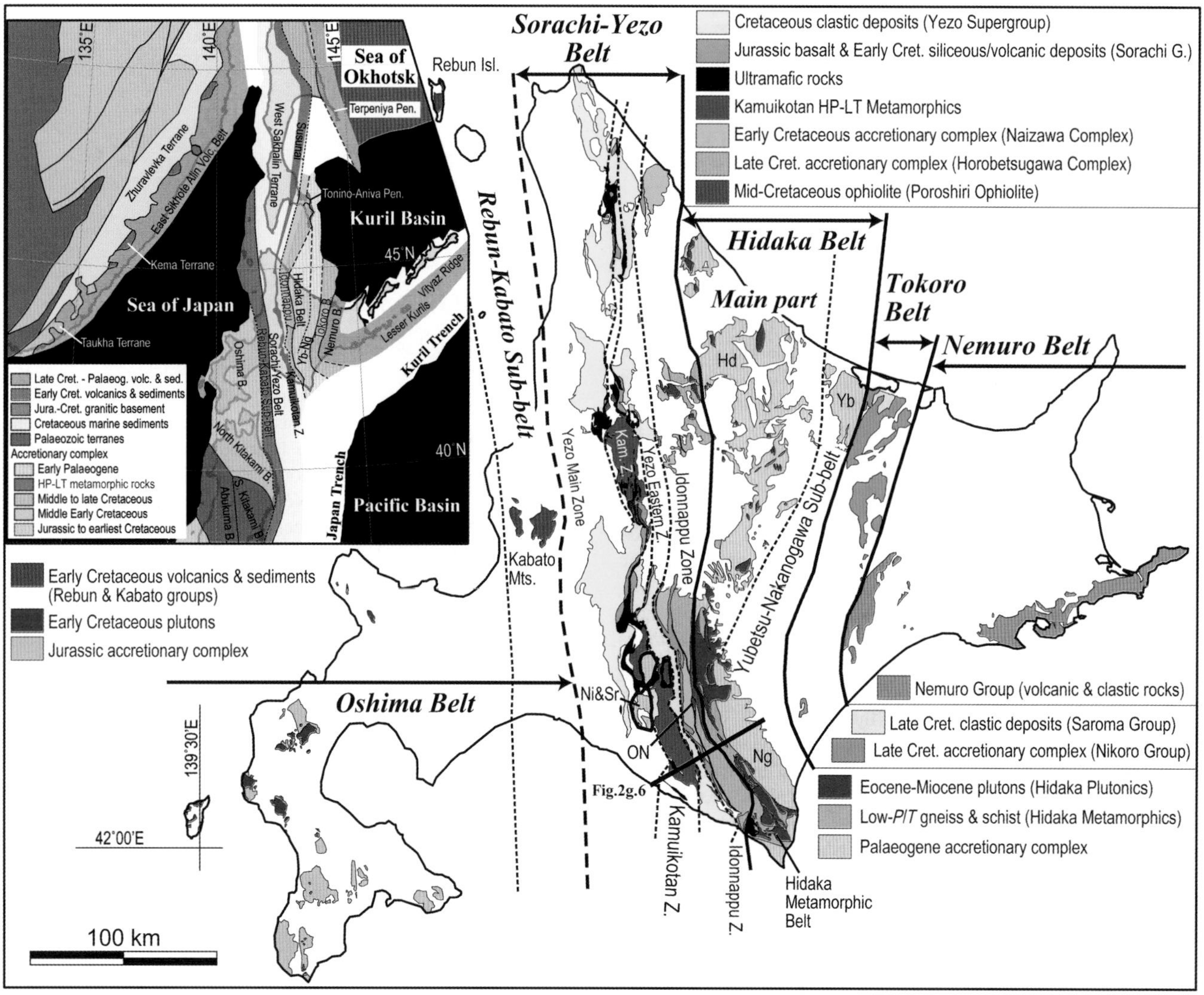

Fig. 2g.1. Tectonostratigraphic zonation of Hokkaido and its inferred continuation to surrounding areas. Hd, Hidaka Supergroup; Ni&Sr, Niseu and Sarugawa formations; Ng, Nakanogawa Group; ON, Okuniikappu Complex; and Yb, Yubetsu Group.

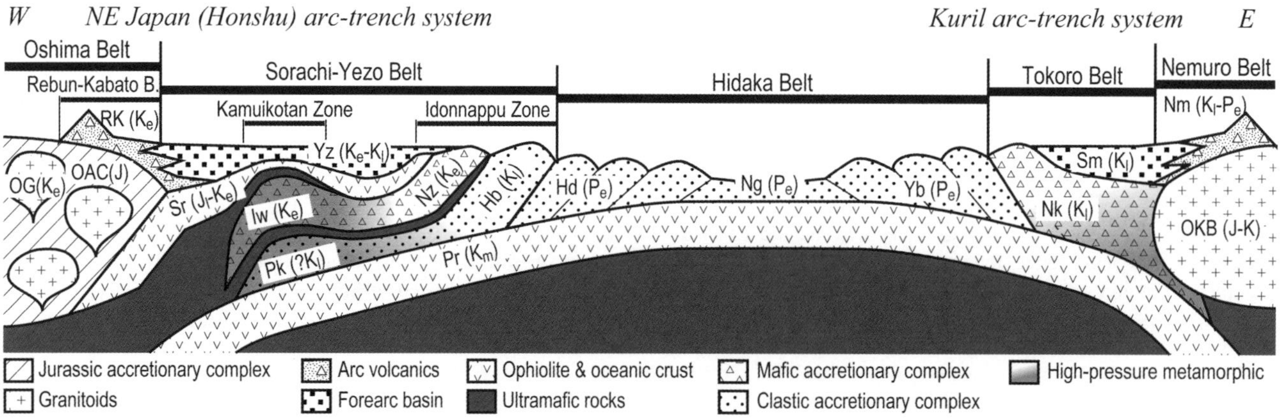

Fig. 2g.2. Schematic cross-section of inferable pre-collision architecture of two arc-trench systems which comprise basement rocks of Hokkaido. Abbreviations for geological units: Hb, Horobetsugawa Complex; Hd, Hidaka Supergroup; Iw, Iwashimizu Complex of Kamuikotan metamorphic rocks; Ng, Nakanogawa Group; Nk, Nikoro Group; Nm, Nemuro Group; Nz, Naizawa Complex; OAC, Oshima Belt accretionary complexes; OG, Oshima Belt granites; OKB, Granitic basement of the Sea of Okhotsk; Pk, Pankehoronai Unit of Kamuikotan metamorphic rocks; Pr, Poroshiri Ophiolite; Sm, Saroma Group; Sr, Sorachi Group (Horokanai Ophiolite); Yb, Yubetsu Group; and Yz, Yezo Group. Age abbreviations: J, Jurassic; K, Cretaceous; and P, Palaeogene (with subscripts e, early; m, middle, and l, late).

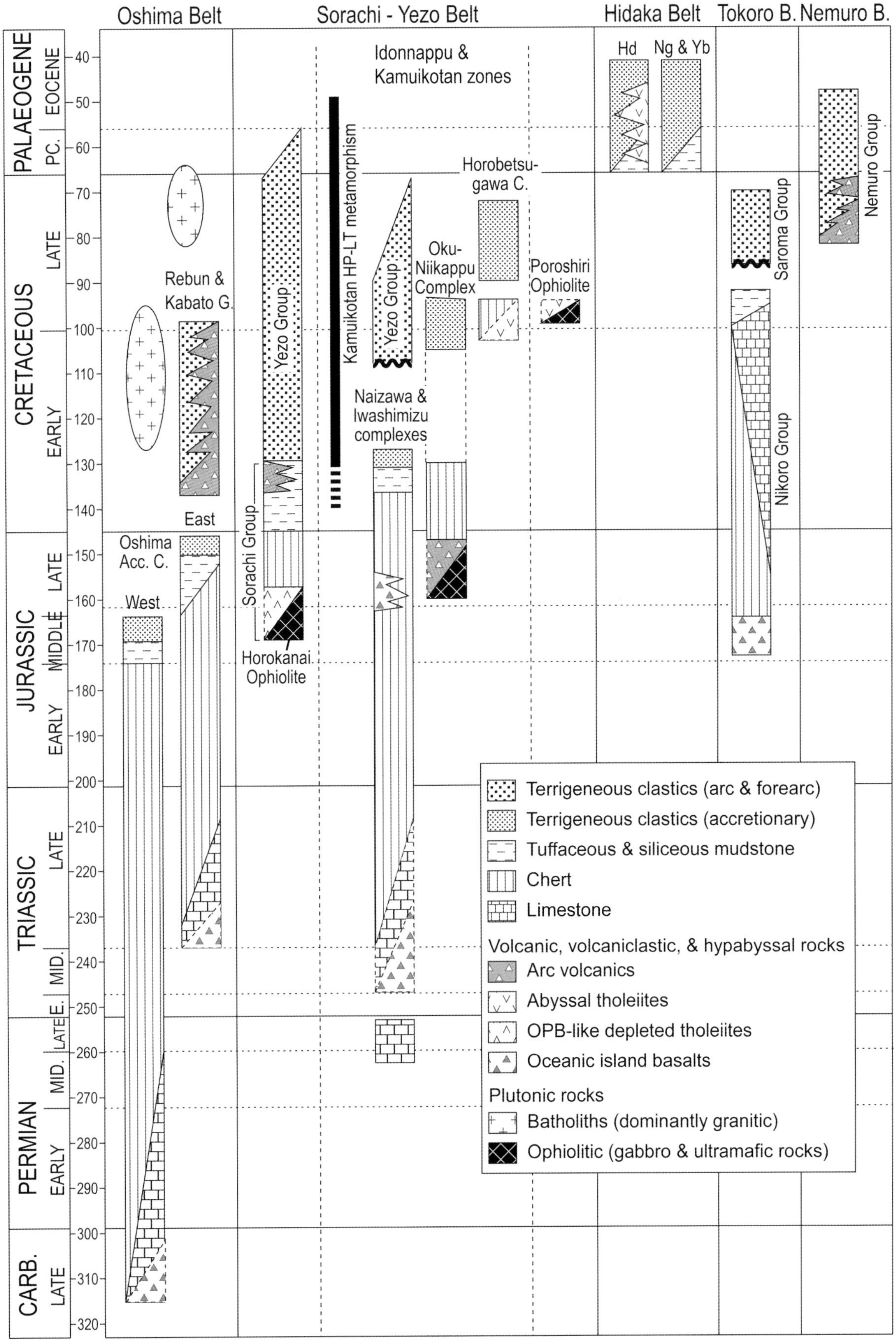

Fig. 2g.3. Summary of geological ages of the basement rocks in Hokkaido. For data sources, see the text.

Kusunoki 1996; Kawamura *et al.* 2000). According to Kawamura *et al.* (2000) and Roser *et al.* (2003), their highly silicic whole-rock composition suggests a felsic continental source. Kawamura *et al.* (2000) also reported pyrope-rich detrital garnet, presumably of granulite origins, and SHRIMP U–Pb ages of *c.* 1800 and 2500 Ma detrital zircons from a sandstone sample. Based on these data, they considered their source to have been the North China Craton. Kusunoki (1996) and Kusunoki & Musashino (2001) compared the modal and chemical compositions of the Oshima Belt sandstones with the Taukha Terrane of the south Sikhote–Alin, far-east Russia, and suggested they share the same sedimentary provenances.

Cretaceous plutonic rocks intruded into the Jurassic accretionary complexes are summarized by Tsuchiya *et al.* (1986) and Tsuchiya (2010). According to them, the plutonic rocks are dominated by calc-alkaline tonalite to granodiorite batholiths both of ilmenite and magnetite series. Alkaline (shoshonitic) series rocks of gabbro, ultramafic cumulate and monzo-gabbro/diorite are also locally present. In addition to these intrusives, Early Cretaceous rhyolitic volcaniclastic rocks including welded tuff also occur in the Okushiri Island off SW Hokkaido. Radiometric ages (mainly by K–Ar method) from Oshima Belt granites range from 130 to 65 Ma with a cluster at 115–100 Ma. Plutonic rocks in the Oshima Belt form part of the Cretaceous granite belt exposed in the NE Honshu Arc. Although some radiometric ages of the Oshima Belt granites are coeval with those in the Northern and Southern Kitakami belts, adakite granites known in the latter are lacking in the Oshima Belt.

Rebun–Kabato Belt

Along the eastern marginal parts of the Oshima Belt, Early Cretaceous arc volcanic rocks and related sedimentary rocks are exposed in the Kabato Mountains (Kumaneshiri Group) and on Rebun Island (Rebun Group), and comprise a north–south-trending volcanic chain called the Rebun–Kabato Belt. The basement of these volcanogenic sequences is inferred to belong to the Oshima Belt accretionary complex, and the Rebun–Kabato Belt is therefore defined as a sub-belt of volcanic cover in the Oshima Belt (Kiminami *et al.* 1986*a*). Its southern extension is traced by an array of marine aeromagnetic anomalies (Finn 1994) connecting the belt to coastal NE Honshu (e.g. Harachiyama Formation in the Northern Kitakami Belt), where Early Cretaceous volcanogenic deposits unconformably overlie the Jurassic accretionary complex. It is notable that this volcanic zone obliquely crosses the basement architecture: it overlaps Jurassic accretionary terranes of the Oshima and Northern Kitakami belts in the central area, and the Palaeozoic terrane of the Southern Kitakami Belt in the south.

The Kumaneshiri Group in the Kabato Mountains near Sapporo consists mainly of mudstone, volcaniclastic sandstone and tuff, with formations dominated by volcanic conglomerate, tuff breccia, pillow lava and dykes (Nagata *et al.* 1986; Kondo 1991). Volcanic rocks are high-K basalt to andesite, including both tholeiitic and calc-alkaline series, with low-TiO_2 island-arc characteristics (Nagata *et al.* 1986; Kondo 1993). Early Cretaceous radiolarians from mudstones (Nagata *et al.* 1986; Kondo 1991) and a 101.3 ± 2.7 Ma Ar–Ar age from a volcanic rock clast (Kondo 1991) have been reported. Clasts of Permian fusulina limestone (Hashimoto *et al.* 1960) and of siliceous sedimentary rocks with Jurassic radiolarians (Kondo 1991) in conglomerates suggest that parts of the sedimentary clasts were supplied from Jurassic accretionary complex materials in addition to arc volcanoes.

The Rebun Group in Rebun Island consists dominantly of volcanic and tuff breccia, associated with volcaniclastic conglomerate and sandstone, mudstone and dykes of dolerite-gabbro and diorite (Nagao *et al.* 1963; Ikeda & Komatsu 1986). Volcanic rocks are TiO_2-poor basalt to andesite of tholeiitic series (Ikeda & Komatsu 1986). The lowermost formation has yielded Barremian ammonoids (Nagao *et al.* 1963) and Early Cretaceous (Valanginian–Aptian) radiolarians (Iwata *et al.* 1997).

Sorachi–Yezo Belt

The Sorachi–Yezo Belt is characterized by Cretaceous forearc basin clastic sediments (Yezo Group) which overlie an ophiolite and siliceous sedimentary sequence (Horokanai Ophiolite and Sorachi Group) ranging from Late Jurassic to Early Cretaceous in age, and the Cretaceous high-pressure/low-temperature (HP–LT) Kamuikotan metamorphic rocks (Kiminami & Kontani 1983*a*). This belt is regarded as a part of an extensive (up to 1400 km long) and continuous Cretaceous forearc basin along the Eurasian continental margin (Fig. 2g.1), running from Sakhalin Island to offshore NE Honshu (Ando 2003). The Kamuikotan HPLT metamorphic rocks and related serpentinite mélange structurally underlie the ophiolite to forearc basin sequence (Fig. 2g.4), and are exposed as tectonic windows in the cores of anticlines. These exposures of the HP–LT metamorphic rocks have been referred to as the Kamuikotan Zone (or Kamuikotan Metamorphic Belt), a sub-belt of the Sorachi–Yezo Belt. Another component of this underlying Cretaceous accretionary complex is expressed as the Idonnappu Zone (or Idonnappu Belt) which crops out along the eastern margin of the forearc basin. The Idonnappu zone has been included in the Hidaka Belt (Kiminami *et al.* 1986*b*), although more recently it has come to be viewed as the unmetamorphosed equivalent of the Kamuikotan metamorphic rocks (Watanabe *et al.* 1994; Iwasaki *et al.* 1995). Ueda *et al.* (2000, 2001) therefore regarded it as the eastern marginal zone of the Sorachi–Yezo Belt, and distinguished it from Palaeogene accretionary complexes in the Hidaka Belt which have no direct structural or sedimentary relations with the forearc basin. Structurally below the Idonnappu Zone accretionary complex, another metamorphosed ophiolite (Poroshiri Ophiolite) occurs along the suture of the Hidaka Collision Zone. The Poroshiri Ophiolite has been included in the Hidaka Belt (as the western zone of the Hidaka Metamorphic Belt); however, Komatsu *et al.* (1992) and Ueda (2006) included it into the Sorachi–Yezo Belt. Figure 2g.4 summarizes the tectono-stratigraphy of the Sorachi–Yezo Belt compared to the crustal section of the Hidaka Metamorphic Belt. Following these ideas, this chapter regards the Sorachi–Yezo Belt in a somewhat wider sense, representing both the Cretaceous forearc basin and the underlying subduction–accretionary units, where exposed.

Horokanai Ophiolite

The Horokanai Ophiolite (Asahina & Komatsu 1979; Ishizuka 1980, 1981, 1985, 1987) typically occurs in the Horokanai area in the north-central part of the Sorachi–Yezo Belt. Although significantly disrupted, it shows an ophiolite succession from serpentinized peridotite via gabbro, amphibolite and MORB-like basalt, to cherty (or tuffaceous) siliceous sedimentary rocks which are conformably overlain by clastic rocks of the Yezo Group. The base of the peridotite is in fault contact with crystalline schists of the Kamuikotan metamorphic rocks. Among these, basalt and siliceous sedimentary rocks are correlated with the Sorachi Group. The peridotite consists mainly of harzburgite and dunite (both with highly depleted chemical characteristics: Katoh & Nakagawa 1986) and serpentinites, and is intruded by numerous gabbro and diorite dykes with chilled margins (Igarashi *et al.* 1985). Basalt and gabbro underwent low-pressure metamorphism from zeolite to granulite facies, and associated amphibolite can be distinguished from the adjacent high-pressure amphibolites of the Kamuikotan metamorphic rocks (Ishizuka *et al.* 1983*a*; Ishizuka 1985). Chert overlying the basalt

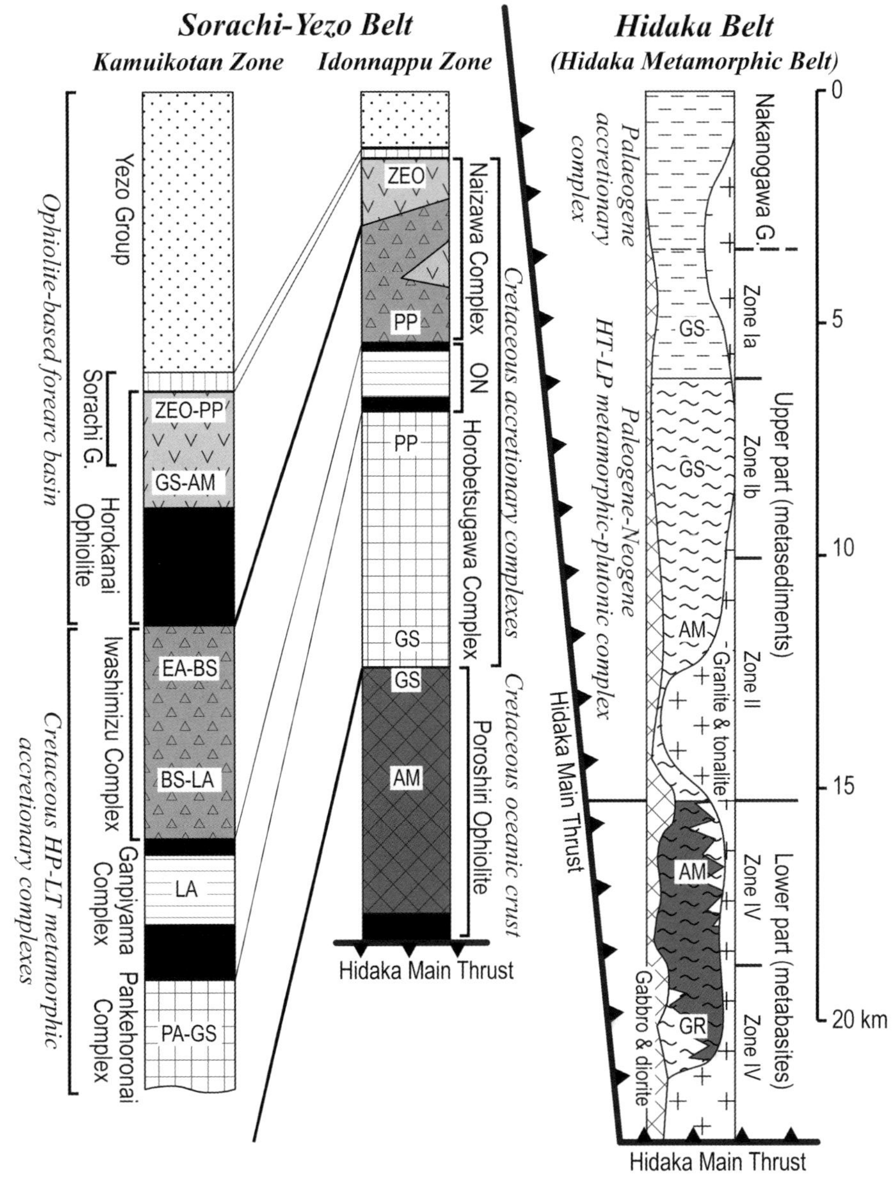

Fig. 2g.4. Reconstructed crustal sections of the Sorachi–Yezo Belt (modified from Ueda 2006) and the Hidaka Metamorphic Belt (simplified from Komatsu *et al.* 1994; Shimura *et al.* 2004). AM, amphibolite facies; BS, blueschist facies; EA, epidote-albite amphibolite facies; GR, granulite facies; GS, greenschist facies; LA, lawsonite-albite facies; PA, pumpellyite-actinolite facies; PP, prehnite-pumpellyite facies; ZEO, zeolite facies; and ON, Okuniikappu Complex.

layer yields Late Jurassic radiolarians (Ishizuka *et al.* 1983*b*; Kawabata 1988; Kiminami *et al.* 1992*a*).

Sorachi Group

The name Sorachi Group has traditionally been applied to geological units dominated by basalt and chert in the Sorachi–Yezo Belt. However, stratigraphic and geochemical studies have revealed that they comprise a mixture of differing components with contrasting origins. Some of these components are now regarded as belonging to the accretionary complex of the Kamuikotan Zone, whereas others represent allochthonous units (klippe) derived from the Hidaka Belt and emplaced during Neogene collision. Excluding these exotic components, coherent sequences typical of the stratigraphy summarized by Kito *et al.* (1986) are here described as the Sorachi Group (*sensu stricto*).

The Sorachi Group is generally divided into a basaltic lower part and a volcaniclastic upper part (Kito *et al.* 1986). The lower part consists almost entirely of pillow and massive basalt and their volcaniclastc rocks (=the basaltic section of the Horokanai Ophiolite). These basalts are characterized by non-vesicular and aphyric tholeiites, associated with picrite in places. Basalts show chemical compositions similar to N-MORB, and were therefore regarded as an obducted or trapped ocean floor (Ishizuka 1981; Kiminami *et al.* 1985*a*, *b*; Niida 1992). Nagahashi & Miyashita (2002) however pointed to the more extensive melting involved in the production of the lower Sorachi basalts as compared to typical N-MORB, and correlated them with oceanic plateau basalts (OPB). Ichiyama *et al.* (2012, 2014) showed that picrites of the Sorachi Group originated from extremely high-temperature mantle plume, which they attributed to an oceanic plateau, whereas Takashima *et al.* (2002) considered the depleted nature of the lower Sorachi basalts to have resulted from back-arc basin spreading along the Eurasian continental margin.

The upper Sorachi Group is characterized by siliceous and volcanogenic sedimentary rocks such as chert, siliceous mudstone, felsic tuff and volcanic sandstone and conglomerate, with occasionally intercalated basaltic pillow lavas, dolerite sills (Kito 1987; Niida 1992), felsic tuff breccia and quartz diorite dykes (Girard *et al.* 1991).

Siliceous sedimentary rocks yield Late Jurassic–middle Early Cretaceous radiolarians (e.g. Kanie *et al.* 1981; Kiminami *et al.* 1985*a*, *b*; Kito 1987, 1995, 1997; Takashima *et al.* 2001). Lavas and volcanic rock clasts of the upper Sorachi Group show chemical characteristics of an arc setting (Girard *et al.* 1991; Niida 1992; Takashima *et al.* 2002).

Yezo Group

The Yezo Group is characterized by thick terrigeneous clastic sequences ranging from middle Early Cretaceous (Barremian) to Late Cretaceous (Campanian or locally to Paleocene), and is also known to yield many fossils (especially ammonoids and inoceramids) along with well-preserved sedimentological and palaeoenvironmental records (e.g. Ando 2003; Takashima *et al.* 2004). It has been classically treated as a supergroup divided into Lower Yezo, Middle Yezo, Upper Yezo and Hakobuchi groups (e.g. Ando 2003), although some researchers regard the sequence as a single group (e.g. Takashima *et al.* 2004). The lower part of the Yezo Group conformably overlies the Sorachi Group, and its base is defined as the onset of terrigenous clastic supply such as siliciclastic sandstone or black mudstone deposited upon the volcanogenic upper Sorachi deposits (Kanie *et al.* 1981; Kito 1987). Based on this stratigraphic relation, the Yezo Group is regarded to be a residual forearc basin (Kiminami *et al.* 1985a, *b*; Niida & Kito 1986) after the classification by Dickinson & Seely (1979), generated upon oceanic crust connected to a continental margin. Although there are significant variations in local depositional environments and influences related to sea-level change, the Yezo Group shows a general shallowing-upwards sequence presumably reflecting infill of the basin. The lower Yezo Group (*c.* Barremian–Aptian) consists of deep-sea turbidites devoid of calcareous fossils except for slide blocks of shallow-marine Orbitolina limestone. The middle Yezo Group (Albian–Turonian) is dominanted by turbidites with calcareous micro- and mega-fossils. Littoral deposits reflecting eustatic sea-level changes are also found in several sections (Ando 1990*a*, *b*). The upper Yezo Group (Coniacian–Campanian) is represented by severely bioturbated shelf mudstone and sandstone with abundant occurrences of molluscan fossils. The uppermost parts (Campanian–Maastrichtian Hakobuchi Group) become sandy and gravelly, reflecting fluvial and/or littoral deposition.

Clastic compositions of sandstone vary from quartzo-feldspathic in the lower parts to more lithic with volcanogenic clasts in the middle and upper parts (Okada 1971). Beside these modal variations, the Yezo Group sandstones show chemical characters ranging from continental or dissected arc to evolved island arc (Kiminami *et al.* 1992*b*; Roser *et al.* 1999; Kawamura & Fujimoto 2000).

From the upper parts of the lower Yezo Group to the basal parts of the middle Yezo Group (Aptian–Albian), there are several kinds of characteristic deposits suggestive of regional tectonic events. In the middle and northern areas of the basin, deep-sea deposits containing slide blocks of shallow-water, *Orbitolina* limestones occur (Sano 1995; Iba & Sano 2007; Iba *et al.* 2011). In the middle–north areas of the basin, detrital serpentinites occur as blocks in slumps (Nagata *et al.* 1987) or clasts in debris-flow conglomerate and sandstone associated with calcareous bioclasts and ooids (Yoshida *et al.* 2003, 2010), suggesting exhumation and uplift of mantle materials within the basin. In the southern areas, submarine or locally fluvial unconformity can be recognized at several localities, where the middle Yezo Group overlies metabasites and metasediments of the Kamuikotan Zone rocks. The basal conglomerate contains pebbles and blocks of HP–LT metabasites including blueschist and epidote amphibolite correlative to the basement rocks, indicating local seafloor and subaerial exposures of the exhumed HP–LT subduction complex (Kawamura *et al.* 1999; Ueda *et al.* 2002). Yoshida *et al.* (2010) also showed detrital glaucophane particles in the serpentinite-bearing conglomerate in the northern part of the basin.

Kamuikotan Zone

The Kamuikotan Zone is characterized by the occurrence of rocks recording HP–LT metamorphism (Kamuikotan metamorphism) and serpentinites, exposed in the cores of anticlines within the Yezo forearc basin. Although the geological structure of these exposures is complicated, the HP–LT rocks occur mainly either as coherent masses of crystalline schist and lower-grade accretionary complexes, or as blocks in serpentinite mélanges (Sakakibara & Ota 1994). In addition to these HP–LT rocks, fault-bounded sheets and blocks of ophiolitic rocks such as serpentinized peridotite, metagabbro and amphibolite of low-pressure series correlative with the Horokanai Ophiolite also occur (Ishizuka *et al.* 1983*a*; Nakagawa & Toda 1987; Kawamura *et al.* 1998).

The Kamuikotan metamorphic rocks can be grouped into at least three types in terms of protolith, metamorphic condition and ages. The oldest rocks so far known comprise epidote-albite amphibolite with minor garnet amphibolite and garnet-quartz schist, and occur as isolated blocks in serpentinite mélange or in lower-grade schists (Bikerman *et al.* 1971; Katoh *et al.* 1979; Ishizuka & Imaizumi 1980; Nakagawa & Toda 1987). This type of metamorphism may have occurred prior to the cooling age of 125–130 Ma (white mica K–Ar and Ar–Ar ages: Bikerman *et al.* 1971; Imaizumi & Ueda 1981; Ota *et al.* 1993). These lithologies commonly display the effects of retrograde blueschist facies metamorphism to varying degrees, represented for example by sodic amphiboles rimming hornblende.

The second group (Types I and II by Sakakibara & Ota 1994) consists dominantly of metabasites and chert associated with pelitic and psammitic rocks, and occurs mainly as coherent masses (Iwashimizu Complex by Kawamura *et al.* 1998; Ueda 2005). These rocks underwent blueschist facies and lower-grade (lawsonite-albite facies) metamorphism characterized by common occurrences of aragonite, lawsonite, sodic amphiboles and jadeitic pyroxene, as well as epidote and pumpellyite. Lower grade parts of them are very weakly deformed, occasionally preserving original sedimentary and volcanic structures. In the least recrystallized parts of the Iwashimizu Complex, chert yields Late Triassic–earliest Cretaceous radiolarians, and mudstone contains middle Early Cretaceous radiolarians (Hori & Sakakibara 1994). MidCretaceous (Albian–Cenomanian) radiolarians were also reported from another aragonite-bearing pelitic unit (Ganpiyama Complex: Kawamura *et al.* 2001) isolated within serpentinites. Metabasites of the Iwashimizu Complex show chemical characteristics dominantly of oceanic-island basalts (Nakano & Komatsu 1979; Kimura *et al.* 1994; Skakibara *et al.* 1999). These fossil ages and the basalt geochemistry of the Iwashimizu Complex are the same as those of the Naizawa Complex in the Idonnappu Zone, and Watanabe *et al.* (1994) and Iwasaki *et al.* (1995) correspondingly proposed that the former was the deeply subducted equivalent of the latter.

The third group is characterized by highly foliated and folded pelitic schist with minor siliceous, psammitic and basic lithologies (type III, i.e. Pankehoronai unit of Sakakibara & Ota 1994). It is also characterized by a pumpellyite-actinolite facies metamorphism with calcite instead of aragonite, and actinolite instead of glaucophane, indicating a higher thermal gradient than that of the Iwashimizu Complex.

Idonnappu Zone

The Idonnappu Zone is a zone of accretionary complexes between the Sorachi–Yezo forearc basin deposits in the west and the Hidaka metamorphic rocks (including Poroshiri Ophiolite) in the east (Jolivet & Miyashita 1985). This zone extends northwards to where the Hidaka metamorphic rocks are not exposed, so that its eastern boundary with the adjacent Hidaka Belt is not well defined. The zone is a composite of accretionary complexes of different ages and lithologies. Most exposures belong either to western Early Cretaceous units containing abundant oceanic rocks, or to eastern Late Cretaceous clastic-dominant units (Kiyokawa 1992; Ueda *et al.* 2000, 2001). Following the subdivisions proposed by Ueda *et al.* (2000, 2001), these are here described as the Naizawa and Horobetsugawa complexes, respectively. Between them, Ueda & Miyashita (2005) identified the ophiolitic Okuniikappu Complex which includes mid- Cretaceous clastic sedimentary rocks. Although sedimentary structures generally face west, these three units overall display an eastwards (downwards) -younging arrangement, a structure that is compatible with generally accepted west-dipping subduction models (e.g. Niida & Kito 1986).

The Naizawa Complex is dominated by oceanic rocks composed of basalt, Permian–Triassic limestone (Igo *et al.* 1974; Hashimoto *et al.* 1975; Ishizaki 1979; Sakagami & Sakai 1979; Kato *et al.* 1986) and Triassic–earliest Cretaceous chert (Kato *et al.* 1986; Igo *et al.* 1987; Kato & Iwata 1989; Kiyokawa 1992; Sakakibara *et al.* 1997*a*, *b*; Ueda *et al.* 2001), associated with hemipelagic mudstone and terrigenous clastic rocks of middle Early Cretaceous ages (Kato *et al.* 1986; Igo *et al.* 1987; Tumanda & Sashida 1988; Kato & Iwata 1989; Tumanda 1989; Kiyokawa 1992; Ueda *et al.* 2001). Basaltic rocks consist of two distinct chemical types: depleted tholeiite identical to the lower Sorachi Group; and oceanic-island basalts. Ueda *et al.* (2000) interpreted the complex as having been formed by tectonic mixing of subducted seamounts with the upper plate ophiolite (Sorachi Group) forming the inner trench slope.

The Okuniikappu Complex is an ophiolitic mélange consisting of slices and blocks of serpentinized peridotite, gabbro-diabase, altered volcanic rocks, earliest Cretaceous chert and midCretaceous (Albian–Cenomanian: Kiyokawa 1992) clastic sedimentary rocks separated by sheared serpentinites. All the igneous rocks show island-arc chemical characteristics. Andesitic volcanic rocks with minor boninite are overlain by earliest Cretaceous pelagic chert intercalated with volcanogenic debris beds. Ueda & Miyashita (2005) attributed these arc rocks and andesite-chert sequences to a remnant arc behind a back-arc basin which was subducted along the Eurasian continental margin in mid-Cretaceous times, analogous to the present-day Kyushu–Palau Ridge.

The Horobetsugawa Complex is dominated by Late Cretaceous (Campanian) clastic sedimentary rocks with minor tectonic sheets of basalt, and chert and pelagic red mudstone of mid-Cretaceous (Albian–Cenomanian) age (Kiyokawa 1992; Ueda *et al.* 2001). The basalts are intensely sheared and show MORB chemistry (Ueda *et al.* 2000). Repetition of these lithologies suggests an imbricate structure. The complex also locally contains sedimentary mélange containing blocks of presumably Naizawa Complex origins, and a Paleocene clastic sequence presumed to represent trench-slope basin deposits. The structurally lowermost parts of the complex underwent greenschist facies metamorphism, and are overthrust by the Poroshiri Ophiolite and Hidaka metamorphic rocks.

Sandstones of accretionary units in the Idonnappu Zone have similar clastic and chemical compositions to equivalent horizons of the Yezo Group: relatively quartzo-feldspathic in Lower Cretaceous times, and felsic volcano-lithic in Upper Cretaceous times (Sakai & Kanie 1986; Roser *et al.* 1999), suggesting that the accretionary complex and the forearc basin shared the same continental-arc sedimentary provenance.

Poroshiri Ophiolite

On the east (structurally lower part) of the Idonnappu Zone, a north–south-trending zone exposes the Poroshiri Ophiolite (Miyashita & Yoshida 1988; Miyashita *et al.* 1994) and is bounded by a thrust fault (Hidaka Western Boundary Thrust or WBT). It underwent greenschist to amphibolite (or locally granulite) facies metamorphism, and was once regarded as a western part (the Western Zone) of the Hidaka Metamorphic Belt (Osanai *et al.* 1986; Komatsu *et al.* 1989). However, its oceanic rock character strongly contrasts with the continental crustal section typifying the main zone of the Hidaka Metamorphic Belt. The Poroshiri Ophiolite is here included in the Sorachi–Yezo Belt, taking into account its genetic relation with the Idonnappu Zone accretionary complexes (Komatsu *et al.* 1992; Ueda 2006).

The ophiolite consists mainly of amphibole schists, metagabbro and metacumulates of the amphibolite facies, with peridotite and metasediments (represented by viridine-muscovite-quartz schist: Suzuki *et al.* 1965). These rocks are locally intruded by dykes, which also underwent metamorphism (Miyashita & Niida 1981), and the original ophiolite sequence has been rearranged by thrusts and folds (Miyashita 1983) with synmetamorphic deformation occurring during dextral transpression (Jolivet & Miyashita 1985; Arita *et al.* 1986; Arai & Miyashita 1994). The Hidaka metamorphic rocks to the east are partly involved in folds and thrusts within the ophiolite (Arai & Miyashita 1994; Arai *et al.* 1995).

The geochemistry of basaltic rocks and the mineralogy of metacumulate and peridotite all suggest that the ophiolite, as well as the later dykes, originated from normal oceanic crust consisting solely of mid-oceanic ridge magmatic rocks (Miyashita & Niida 1981; Miyashita & Yoshida 1988; Miyashita *et al.* 1994, 2007). Kizaki (2000) reported a zircon U–Pb age of 96.7 $\pm$ 2.6 Ma from a quartz amphibolite presumably of plagiogranite origin, suggesting that the Poroshiri Ophiolite represents mid-Cretaceous oceanic crust. Because this age is contemporaneous with pelagic sediments in the Late Cretaceous unit of the Idonnappu Zone, Ueda (2006) considered that the Poroshiri Ophiolite is a relic of the subducted oceanic slab on which the Idonnappu Zone accretionary complex was formed.

Metamorphic grade generally increases from west to east (Miyashita 1981, 1983; Osanai *et al.* 1986; Miyashita *et al.* 1994; Tanaka *et al.* 2012). Metabasites along the western margins (WBT) of the ophiolite show greenschist facies mineralogy (zone A) and these grade into the amphibolite facies (zone C) via the epidote amphibolite facies (zone B), across several-hundred-metre intervals showing apparent geothermal gradients up to *c.* 200°C km^{-1}. Central and eastern parts (zone C) of the ophiolite show relatively homogeneous metamorphic temperatures, except for the easternmost parts (zone D) which grade through upper amphibolite to granulite facies. Metamorphic amphiboles commonly show prograde compositional zoning. This is attributed to early ocean-floor metamorphism followed by heating from the east as the ophiolite was juxtaposed with hot granulites of the Hidaka Metamorphic Belt (Osanai *et al.* 1986). A white mica K–Ar age of 17.1 $\pm$ 0.4 Ma (Shibata *et al.* 1984) suggests Miocene exhumation in conjunction with the Hidaka Metamorphic Belt.

Crustal section of the Sorachi–Yezo Belt

Ueda (2006) summarized the tectonostratigraphy in the Sorachi–Yezo Belt, and argued for a pre-collisional crustal structure reaching down as deep as 20 km within the NE Japan forearc (Fig. 2g.4). It is characterized by a pile of originally flat-lying, fault-bounded units (i.e. nappes) with an overall structurally downwards-younging polarity. In and around the Idonnappu Zone, metamorphic grade increases downwards from the Jurassic Sorachi Group and overlying

Cretaceous Yezo Groups (zeolite facies) via the Early Cretaceous Naizawa accretionary complex (zeolite to prehnite-pumpellyite facies), the Late Cretaceous Horobetsugawa accretionary complex (prehnite-pumpellyite to greenschist facies) and the Poroshiri Ophiolite (greenschist to amphibolite facies). Downwards younging with increasing metamorphic grade may record crustal thickening by underthrusting and tectonic accretion along the subduction margin. In and around the Kamuikotan Zone, the Sorachi group (Horokanai Ophiolite) is underlain by mafic-dominant (Iwashimizu Complex) and then by clastic-dominant (Pankehoronai unit) HP metamorphic rocks, correlative with the Naizawa and Horobetsugawa accretionary complexes of the Idonnappu Zone, respectively, with decreasing metamorphic ages structurally downwards. However, metamorphic pressure also decreases downwards. Exhumation coeval with subduction and tectonic underplating might be necessary to produce such a structure.

Hidaka Belt

The Hidaka Belt is characterized by outcrops of Palaeogene accretionary complexes in central Hokkaido. There is some uncertainty as to the exact definition and extent of this belt, these varying with different authors, but in this chapter we describe those units presumed to result from west-dipping subduction along the NE Japan margin. The Yubetsu and Nakanogawa groups, sometimes included in the Hidaka Belt, are described in the section dealing with the Kuril arc-trench system ('Kuril Arc Terrane').

The Hidaka Belt accretionary complexes (Hidaka Supergroup) are widely distributed in northern and central parts of the belt but reduce in extent southwards, presumably as a result of Neogene collision of the Kuril Arc. Northern parts of the belt comprise zones with differing structural trends (north–south v. NE–SW: Tajika 1989, 1992) bounded by faults and shear zones (Watanabe & Kimura 1987). In addition to the dominant occurrence of mudstone and turbidite sandstones, lesser amounts of green and red mudstones, basalts, conglomerate and chaotic mélange facies including blocks of chert and limestone also occur. Except for Cretaceous deposits in the western marginal zones, here assigned to the Idonnappu Zone, Paleocene–Eocene radiolarians have been reported from hemipelagic and terrigenous mudstones in central to eastern parts of the belt (Watanabe & Iwata 1985; Kiminami *et al.* 1990*a*; Tajika & Iwata 1990). Radiolarians from mudstones originally thought to be Late Cretaceous in age (Tajika & Iwata 1983; Watanabe & Iwata 1987; Iwata & Tajika 1989) have been re-assigned to the early Paleocene by Tajika (1992), reflecting subsequent progress in early Palaeogene biostratigraphy. Permian fusulinids (Endo & Hashimoto 1955) and Triassic holothurians sclerites (Iwata *et al.* 1983*a*) have been identified in limestone clasts and blocks, and the age of chert blocks ranges from Late Triassic to mid-Cretaceous (Tajika & Iwata 1983, 1990; Watanabe & Iwata 1987; Iwata & Tajika 1989). These older rocks, including Late Cretaceous mudstone, are considered as sedimentary blocks derived from the Idonnappu Zone in the west (Tajika & Iwata 1990).

The Hidaka Belt accretionary complexes are occasionally accompanied by basalt and dolerite. They show field relationships suggestive of synsedimentary *in situ* eruption and intrusion into unconsolidated terrigenous mudstone and sandstone (Mariko 1984; Miyashita & Katsushima 1986; Kiminami *et al.* 1999; Nakayama 2003), as represented by inter-pillow mudstone, silicified baked margins of mudstones in contact with dolerite and mudstone fragments involved in dolerites. Geochemical studies indicate a mid-oceanic ridge basalt origin (Miyashita & Katsushima 1986; Mariko & Kato 1994; Miyashita & Yoshida 1994; Miyashita & Kiminami 1999). Based on these characteristics, a seafloor spreading environment adjacent to a continent, either associated with continental margin rifting (Mariko 1984) or oceanic ridge subduction (Miyashita & Katsushima 1986), has been assumed for the origin of the basalt-sediment complex.

The Hidaka Belt is also characterized by intrusions of granitoid rocks and gabbro (Maeda *et al.* 1986). The zone of these intrusions within the non-metamorphic Hidaka Supergroup extends to the Hidaka Metamorphic Belt in the south (Hidaka Mountains), where deep crustal sections are exposed. The Hidaka Metamorphic Belt is described later in 'Hidaka Collision Zone'.

Continuation to the Honshu Arc

The Oshima Belt is clearly a continuation of the Northern Kitakami Belt of NE Honshu Island (Fig. 2g.1), as both are merely separated by the narrow Tsugaru Straight. Both the belts share a common distribution of geological ages: accretionary units containing Carboniferous–Permian chert and limestone in the west, and units without Palaeozoic rocks in the east (cf. Suzuki *et al.* 2007). Yamakita & Otoh (2002) reviewed the subsurface geology of the Kanto sedimentary basin near Tokyo, where the NE and SW Japan arcs are connected. They proposed that both the arcs are not structurally distinct but continuous, although they are significantly bent beneath the Kanto Basin. According to their correlation, the Oshima and Northern Kitakami belts are the northern continuation of the Southern Chichibu Belt in the Outer Zone of SW Japan. This correlation is supported by the occurrences of Triassic limestone blocks containing megalodont bivalves in the Oshima and Northern Kitakami belts (Sano *et al.* 2009), because these bivalve fossils exclusively occur in the Outer Zone (Chichibu Belt) in SW Japan (Tamura 1990).

The correlation made above suggests that other NE Japan arc components in Hokkaido may also be continuations of the Outer Zone of SW Japan (Fig. 2g.5). However, there are many discrepancies in Cretaceous and even younger geology, obstructing a simple connection between the two areas. The first problem is that the Early Cretaceous arc magmatism common in NE Japan (including Hokkaido) is absent from the Outer Zone of SW Japan. The Early Cretaceous volcanic front found in the Rebun–Kabato Belt of Hokkaido is clearly traceable to the south along an offshore aeromagnetic anomaly (Finn 1994) extending into Early Cretaceous arc-volcanic rocks along the Sanriku coast (e.g. Harachiyama Formation) and further to offshore Joban (Fukushima), both in NE Japan. Early Cretaceous intrusions (mainly granitic) in Hokkaido and NE Japan also run parallel to the volcanic front in the east. This Early Cretaceous magmatic arc obliquely crosses pre-Cretaceous basement structures; it overprints Palaeozoic coherent sequences of the Southern Kitakami Belt in the south and Jurassic accretionary complexes of the Oshima and Northern Kitakami belts in the north. Although the significance of this obliquity has not so far been well discussed, it implies that the arc-trench system was reorganized in the Early Cretaceous epoch.

The second major difference is the existence of the ophiolite-based large forearc basin of the Sorachi–Yezo Belt in Hokkaido. Although coeval Cretaceous forearc deposits sporadically occur in SW Japan, they overlie non-ophiolitic units such as Jurassic accretionary complexes (Chichibu Belt), Palaeozoic sequences (Kurosegawa and Hida Gaien belts) or sialic metamorphic units (Ryoke and Hida belts). It would seem therefore that the emplacement of the Jurassic ophiolite of the lower Sorachi Group (including the Horokanai Ophiolite), whose origin is still controversial, is critically responsible for this difference.

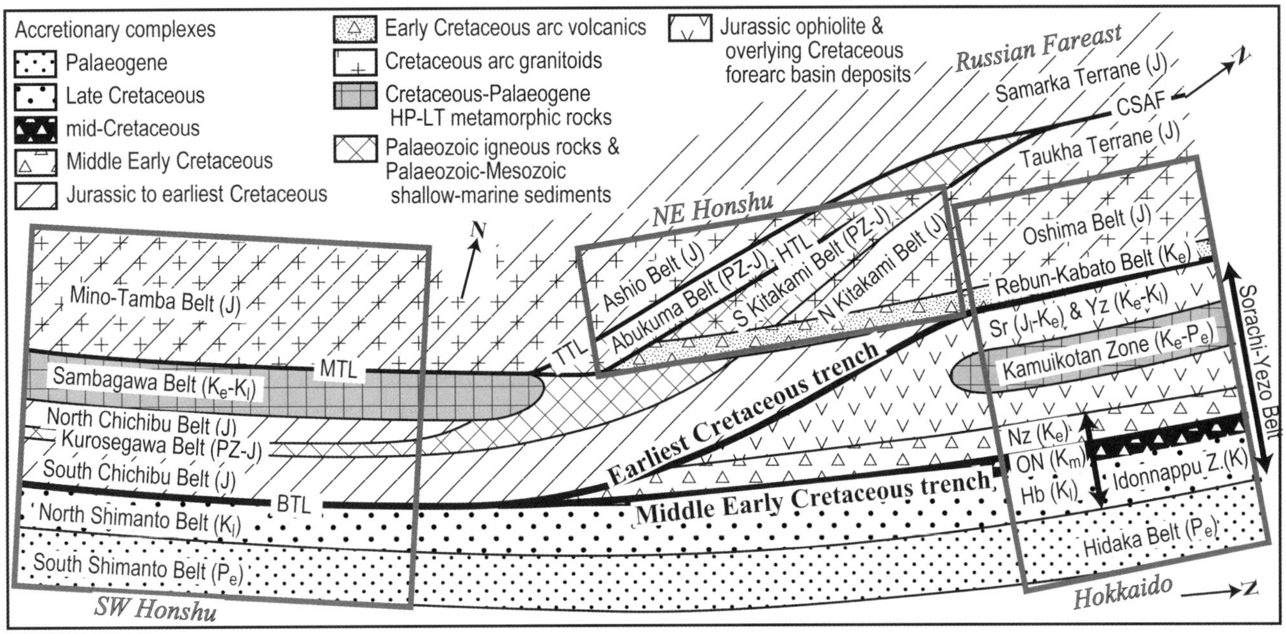

Fig. 2g.5. A hypothetical correlation of basement rocks of central-SW Hokkaido with SW and NE parts of the Honshu Arc (not to scale). Abbreviations for geological units and ages as for Figure 2g.2 with additional ON, Okuniikappu Complex; PZ, Palaeozoic; BTL, Butsuzo Tectonic Line; MTL, Median Tectonic Line; TTL, Tanakura Tectonic Line; and HTL, Hatagawa Tectonic Line. Unlike the SW Honshu arc, tectonic features younger than middle Early Cretaceous (volcanic arc, granite belt, HP–LT metamorphic belt and trench) obliquely cross the older structures in NE Honshu, along with insertion of the Sorachi Group ophiolite in Hokkaido, presumably responsible for structural discrepancies between SW Honshu arc and Hokkaido.

Late Cretaceous accretionary units (Horobetsugawa Complex) in the Idonnappu Zone and Palaeogene accretionary units in the Hidaka Belt are both correlative with the northern and southern Shimanto Belt in SW Japan, respectively, in terms of age and clastic-dominant lithology, although those in Hokkaido tend to be more argillaceous than in SW Japan. In SW Japan, the Shimanto Belt is contiguous with the Chichibu Belt, although bounded by a fault (the Butsuzo Tectonic Line). In Hokkaido, by contrast, the Idonnappu and Hidaka belts crop out at some distance from the Oshima Belt, with a gap of *c.* 70 km being occupied by the Sorachi–Yezo forearc basin. This gap may have been much wider, taking into account the shortening effect of intense Tertiary folding in the forearc basin. It would therefore appear that the ophiolite of the Sorachi Group can be regarded as an extra component unique to Hokkaido in Cretaceous Japan.

Emplacement of this ophiolite may also have affected differences in the position of the Cretaceous HP–LT metamorphic belts: the Kamuikotan Zone in Hokkaido and the Sanbagawa Belt in SW Japan. The Sanbagawa Belt lies among Jurassic accretionary complexes (Chichibu and Ryoke belts), partly with older units. Parts of the Jurassic accretionary complex in the adjacent northern Chichibu Belt also underwent weak HP–LT metamorphism, suggesting a close genetic relationship between the two since the timing of metamorphism. In contrast, the Kamuikotan Zone is situated inside the Sorachi–Yezo Belt and its metamorphic rocks are exposed as tectonic windows among the hanging-wall ophiolite and/or the forearc basin deposits. It lies more than 30 km east of the eastern tip (Rebun–Kabato Sub-belt) of the Oshima Belt, and no direct structural relationship with the Jurassic accretionary complexes is apparent.

The differences in the position of the volcanic front and HP–LT metamorphism between SW Japan and Hokkaido can be understood if we consider the trench was episodically shifted as much as 100 km eastward as the consequence of emplacement of the Jurassic ophiolite (Sorachi Group) into the Hokkaido forearc during Early Cretaceous times.

Early Cretaceous (Valanginian–Barremian) accretionary complexes in the Idonnappu and Kamuikotan zones are unique to Hokkaido, with no equivalent units being found in SW Japan. These units contain Triassic–Jurassic oceanic rocks such as chert, limestone and associated seamount volcanic lithologies. Such characteristics are more similar to the Chichibu Belt rather than to the Shimanto Belt of SW Japan. However, both the Idonnappu and Kamuikotan zones lie remote from the Oshima Belt which is correlated with the Chichibu Belt, and so seem different. Because these Early Cretaceous units structurally underlie the Sorachi Group ophiolites and/or the Yezo forearc basin deposits, they might have formed in a trench on the east of the ophiolite in contrast to the setting of the Oshima Belt, as depicted by Niida & Kito (1986).

Mid-Cretaceous accretionary complex lithologies (Albian–Cenomanian Okuniikappu Complex of Idonnappu Zone) are also characteristic of Hokkaido, though their volume is relatively small. Such rocks comprise a narrow ophiolitic zone with fragments of intraoceanic-arc magmatic rocks as well as overlying chert and deep-sea turbidites.

Continuation into Russian Far East

Geological continuity with the Russian Far East and Hokkaido (Fig. 2g.1) has been suggested by various authors (e.g. Kiminami 1986; Zonenshain *et al.* 1990; Nokleberg *et al.* 2000; Zharov 2005). The Oshima Belt, together with the Northern Kitakami Belt, is considered to be continuous with the Jurassic–Early Cretaceous accretionary complex of the Taukha Terrane in Sikhote-Alin of Primorye (Yamakita & Otoh 1999; Kojima *et al.* 2000; Nokleberg *et al.* 2000). Cretaceous volcanic-arc rocks cored from the subsurface basement of Moneron Island, offshore SW Sakhalin (Piskunov & Khvedchuk 1976; Simanenko *et al.* 2011), are regarded as the northern continuation of the Rebun–Kabato Belt. The volcano-sedimentary sequence of the Kema Terrane in eastern Sikhote–Alin is also considered to have been a continuation (or attached to the back-arc side of the volcanic front) of the Rebun–Kabato Belt (Nokleberg *et al.* 2000; Malinovsky *et al.* 2006). The Early Cretaceous sedimentary basin of the Zhuravlevka Terrane in eastern Sikhote–Alin is located on the back-arc side of the Kema volcanic chain, and seems to have no counterpart in Hokkaido. Components of the Sorachi–Yezo Belt are clearly correlative with geological units of southern Sakhalin Island. The Yezo forearc basin evidently continues into the West Sakhalin Terrane (Ando 2003), HP–LT metamorphic rocks of the Kamuikotan Zone correlate with the Susunai Terrane (Sakakibara *et al.* 1997*a*, *b*), the Early Cretaceous accretionary complex of the Aniva Terrane is viewed as an extension of the Idonnappu Zone, and the clastic accretionary complex of the Hidaka Belt correlates with the Tonin–Aniva Complex (Zharov 2005).

Kuril Arc Terrane

In contrast to the NE Japan Arc, pre-Neogene rocks of westernmost parts of the Kuril Arc exposed on eastern Hokkaido show an almost opposite polarity (Fig. 2g.2); arc-volcanic rocks and related sediments form the Nemuro Belt in the east, and the Cretaceous–Palaeogene accretionary complex forms the Tokoro Belt in the west. Clastic sedimentary rocks in both belts are much more dominated by mafic to intermediate volcanogenic clasts than those in NE Japan arc rocks. It is therefore considered that rocks of these belts were formed in distinct arc-trench systems away from the Eurasian continental margin, and the Mesozoic crust of the Sea of Okhotsk is assumed to have been their major source area. It is also notable that rocks older than Middle Jurassic have so far not been found in these belts (Fig. 2g.3), in contrast to NE Japan arc sequences which occasionally include Palaeozoic and Triassic rocks.

Eastern parts of the Nemuro Belt show an east–west structural trend whereas the Tokoro Belt and western parts of the Nemuro Belt trend north–south, parallel to the NE Japan arc components (Fig. 2g.1). However, palaeomagnetic studies (Nanayama *et al.* 1993; Fujiwara *et al.* 1995) have shown that Cretaceous–Late Eocene rocks in the latter areas rotated clockwise 70–140°. This rotation occurred until latest Eocene–Early Oligocene times, after which sediments do not record significant rotation (Fujiwara & Kanamatsu 1994). Before the rotation, the structural trends of these belts restore to an east–west strike almost parallel to the present-day Kuril outer arc.

The Nemuro and Tokoro belts must have been positioned on the east side of central Sakhalin Island before Miocene back-arc opening of the Kuril back-arc basin in the southern parts of the Okhotsk Sea.

Nemuro Belt

The Nemuro Belt is characterized by the Late Cretaceous (Campanian) to early Palaeogene (Paleocene and locally Early Eocene) deposits of the Nemuro Group. The group consists of clastic sedimentary rocks with volcanic and intrusive rocks (Kiminami 1983; Okada *et al.* 1987; Kiminami *et al.* 1992*c*). The lowermost formation (Campanian Nokkamappu Formation) contains lava and tuff breccia as well as coarse clastic rocks. The volcanic rocks consist of basalt, andesite and dacite with both tholeiitic and calc-alkaline chemical affinities (Kiminami *et al.* 1992*c*). Other upper sequences consist of muddy, sandy and conglomeratic formations, represented by turbidite, bottom current and slump deposits (Kiminami 1975, 1983; Naruse 2003; Naruse & Otsubo 2011). Cretaceous sections occasionally yield megafossils such as ammonoids and inoceramids (e.g. Naruse *et al.* 2000). A carbonaceous bed assigned to the Cretaceous–Tertiary (K/T) boundary layer has been found by Kaiho & Saito (1986) and Saito *et al.* (1986) in the western parts of the belt (Shiranuka Hills). The Nemuro Group is also characterized by occasional intrusions of mafic sills and associated pillow lavas within Maastrichtian–Paleocene horizons. These consist of highly potassic alkali dolerite (shoshonite) and monzonite (Yagi 1969; Ishikawa *et al.* 1971; Kiminami *et al.* 1992*c*; Simura & Ozawa 2006).

The Nemuro Group has been regarded as representing forearc basin deposits based on their sedimentary facies (Kiminami 1983), or as a frontal arc sequence based on the occurrences of effusive and intrusive rocks (Kimura & Tamaki 1985). In these interpretations, basalt and andesite in the lowermost formation are regarded as parts of arc volcanoes. Alkali intrusions are also regarded as varieties of the arc magmatism. The Nemuro Belt extends to the present-day non-volcanic outer arc of the Lesser Kuril islands, and further on to the submarine Vityaz Ridge in the east where Upper Cretaceous and lower Palaeogene volcanic rocks are more abundant (e.g. Tsvetkov *et al.* 1986; Lelikov & Emelyanova 2011). Cretaceous–early Palaeogene volcanic-clastic sequences potentially correlative with the Nemuro Group (Kiminami 1986) are exposed on the Terpeniya Peninsula of eastern Sakhalin Island (Terekhov *et al.* 2010).

The basement of the Nemuro Group is not exposed on land. Goto *et al.* (1991) dated float stones of granitoids found on the east coast of the Shiretoko Peninsula, a Neogene–Quaternary volcanic field of the Nemuro Belt, and obtained K–Ar ages of 192 Ma from an adamellite and 124 and 115 Ma from mylonitic rocks (quartz monzonite and quartz diorite, respectively). They considered that these float stones originated from xenoliths of Tertiary volcanic rocks rather than ice-rafted pebbles because of the rather limited location of their occurrences, and that they represent the basement crustal rock of the belt.

Tokoro Belt

The Tokoro Belt is represented by abundant occurrences of metabasalts associated with chert and volcaniclastic rocks of the Nikoro Group. It is associated with the overlying clastic rocks of the Saroma Group. They are regarded as having formed along the palaeo-Kuril arc-trench system (or the Okhotsk block margins in other literature). In addition to these diagnostic units, parts of Palaeogene clastic-dominant accretionary complexes, the Yubetsu and Nakanogawa groups, have also been included in this belt by several authors (Kiminami *et al.* 1986*a*, 1992*b*) who have emphasized differences in sedimentary provenances and therefore of the arc-trench system to which the basin belonged. However, as emphasized by Nanayama *et al.* (1993), clastic composition and sedimentary basins are not distinct but transitional between the NE Japan and Kuril trenches in the Palaeogene period. It is therefore still difficult to define any exact boundary between these two clastic-dominant accretionary complexes of apparently different arc-trench systems.

Nikoro Group

The Nikoro Group is characterized by abundant occurrences of volcanic and volcanogenic clastic rocks associated with chert and limestone. The volcanic rocks consist mainly of basaltic pillow and massive lava, with minor amounts of dolerite and felsic rocks. Minor and small bodies of troctolite, gabbro, wehrlite and tonalite also occur (Bamba 1981, 1984; Niida 1981). Clastic rocks are dominantly green-coloured sandstone and conglomerate with abundant volcaniclastic materials, and have also been described as sandy or epiclastic hyaloclastites (Bamba 1984; Sakakibara *et al.* 1986). These sandstones are polymictic consisting dominantly of basaltic volcanic clasts occasionally associated with rock fragments of andesite, granitoid rocks and crystalline schists. Chert occurs overlying basaltic rocks and has been dated as Late Jurassic–Early Cretaceous in age using radiolarian fossils (Kiminami *et al.* 1983; Okada *et al.* 1989; Iwata *et al.* 1990). Early Cretaceous (Aptian) planktonic foraminifers were reported from pelagic limestone conformably overlying Late Jurassic chert (Okada *et al.* 1989). Precise ages from the clastic rocks have not been successfully obtained, but they are inferred to be Late Cretaceous based on radiolarian evidence (Sakakibara *et al.* 1993).

The Nikoro Group was once considered to represent island-arc rocks (Komatsu 1981; Kiminami & Kontani 1983*a*) although, based on whole-rock chemistry, Bamba (1984) identified both tholeiitic and alkaline basalts. He considered the former to be of abyssal basalt origin, comprising an ophiolitic assemblage together with gabbro and troctolite. Alkali basalts are considered as accreted fragments of subducted seamounts (Niida 1981; Sakakibara *et al.* 1986).

The majority of the Nikoro Group underwent zeolite to prehnite-pumpellyite facies metamorphism. However, in the northeastern parts of the belt, high-pressure–low-temperature (HP–LT) metamorphic minerals represented by aragonite, lawsonite, sodic pyroxene and sodic amphibole occur in basaltic rocks (Sakakibara 1986*a*, *b*, 1991). These parts of the Tokoro Belt therefore comprise a HP–LT metamorphic subduction complex. In these rocks, epidote and lawsonite co-occur with pumpellyite and sodic pyroxene and not with sodic amphibole, showing slightly lower pressure than typical blueschists in which epidote or lawsonite + sodic amphibole assemblages are common. Metamorphic conditions of 200–300°C at 0.5–0.6 GPa have been estimated by Sakakibara (1991).

Saroma Group

The Saroma Group comprises coherent sedimentary sequences of conglomerate, sandstone and mudstone, and occurs in synclines unconformably overlying the Nikoro Group. The basal conglomerates contain poorly sorted and rounded boulders and cobbles, and show a generally finings-upward succession grading into turbidites. The Saroma Group is therefore considered to represent deposits of a forearc basin developed over a Nikoro Group accretionary complex (Kiminami & Kontani 1983*a*; Kiminami *et al.* 1985*a*, *b*). Late Cretaceous ages have been assigned using bivalves (Sakakibara & Tanaka 1986; Obata *et al.* 1993) and radiolarians (Iwata *et al.* 1983*b*; Kanamatsu *et al.* 1992). Sandstones and conglomerates within the sequence are highly volcanogenic. Sandstones are lithic wacke consisting dominantly of mafic to intermediate volcanic rock fragments and plagioclase (Kiminami & Kontani 1983*b*), with abundant clinopyroxene clasts with island-arc-type chemistry (Kanamatsu *et al.* 1992). Conglomerates are dominated by pyroxene- or hornblende-andesite clasts, with lesser amounts of basalt, rhyolite, alkali plutonic rock and sedimentary rock clasts (Research Group of the Tokoro Belt 1984). Clasts of alkali plutonic rocks such as monzonite and syenite suggest close genetic links with the Nemuro Group.

Yubetsu and Nakanogawa groups in the Hidaka Belt

The Yubetsu and Nakanogawa groups in the northeastern and southeastern parts of the Hidaka Belt, respectively, are considered to have been deposited within, or adjacent to, the Kuril arc-trench system. They consist mainly of flysch turbidites, and Kiminami & Kontani (1983*a*, *b*) considered them as forearc basin deposits. However, later researchers came to regard them as accretionary complexes (Tajika 1988; Nanayama 1992*a*, *b*).

Yubetsu Group

The Yubetsu Group occurring near the Okhotsk coast of northeastern Hokkaido is characterized by turbidite sandstone and mudstone accompanied by conglomerate and red mudstone. Tajika (1988) identified west-verging imbricate structures of thrust sheets, each of which comprises coarsing-upwards sedimentary sequences from basal red and green pelagic and hemipelagic mudstone to turbidite sandstone. Red and green mudstones yield radiolarians fossils which were initially assigned a Late Cretaceous age (Iwata & Tajika 1986; Iwata & Kato 1986). However, they contain species later known to be diagnostic of Paleocene times, and were therefore re-examined and re-assigned as early Paleocene by Iwata & Tajika (1992). A Late Paleocene K–Ar age of a tuff bed (Kiminami *et al.* 1990*b*) and Early Eocene dinoflagellate cysts from a mudstone (Kurita & Tajika 1998) suggest that the turbidites are also early Palaeogene in age.

Sandstones of the Yubetsu Group are lithic, containing abundant rock fragments of mafic to intermediate volcanics (Kontani & Kiminami 1980). Whole-rock geochemistry of the sandstones suggests sedimentary provenances of an immature island arc (Kumon & Kiminami 1994).

Nakanogawa Group

The Nakanogawa Group occurs on the eastern foothills of the Hidaka Mountains in south-central Hokkaido. It is characterized by turbidites accompanied by minor red and green hemipelagic mudstone, with local occurrences of mélange containing blocks of oceanic rocks such as basalt, limestone and chert (Kontani 1978; Nanayama 1992*b*). Red and green mudstones and the mélange matrix yield Paleocene radiolarians (Nanayama 1992*b*), and felsic tuff beds have been dated as Early Eocene in age using fission track (Nanayama & Ganzawa 1997).

Nanayama (1992*a*) and Nanayama *et al.* (1993) divided the Nakanogawa Group into petroprovince zones I to III based on palaeo-current and sandstone clastic compositions. Zone I in the south is characterized by lithic sandstone richer in andesite clasts derived from the SSE and east, and the petrographic characteristics resemble those of the Nemuro and Saroma groups of the palaeo-Kuril arc. Zone III sandstones are richer in quartz and felsic volcanic clasts, in common with Yezo Group and Hidaka Supergroup sediments belonging to the NE Japan Arc. Zone III and transitional Zone II rocks were transported from the north. Nanayama *et al.* (1993) also reported palaeomagnetic data which suggests clockwise rotation after sedimentation. They proposed that the Nakanogawa Group was deposited in sediment-flooded trenches at the junction of the Kuril and NE Japan arcs: Zone III was originally in the east adjacent to the Kuril Arc; Zone I was closer to the NE Japan Arc in the west; and Zone II lay somewhere between zones I and III.

The Nakanogawa Group is also known as the major protolith of the upper crustal parts of the Hidaka Metamorphic Belt. Western parts of the Nakanogawa Group are intruded by granites and metamorphosed to biotite hornfels, which grade into biotite schist of the Hidaka Metamorphic Belt westwards. Bedding planes and fold axes are obliquely overprinted by the metamorphic foliation.

The Sea of Okhotsk

Before the opening of the Kuril Basin, the Nemuro and Tokoro belts and Lesser Kuril islands are thought to have lain along the southwest margin of the Okhotsk Sea block, as part of a somewhat thin but sialic crust 20–25 km thick (Burk & Gnibidenko 1977; Maeda 1990). From structural highs and slopes where the acoustic basement is exposed, granitic and basic to acidic volcanic rocks of arc character occur and are associated with minor amounts of schists based on dredging results summarized by Burk & Gnibidenko (1977), Gnibidenko (1985), Gnibidenko *et al.* (1995) and Lelikov & Emelyanova (2007). Using K–Ar and Ar–Ar ages, Early–Late Cretaceous granitoid and volcanic rocks appear to be the most dominant and widespread. Lesser amounts of Jurassic granitoid rocks and Eocene volcanic rocks are also present. It is generally considered that the Sea of Okhotsk is a microcontinental terrane distinct from the Eurasian margin (e.g. Parfenov & Natal'in 1986; Zonenshain *et al.* 1990). This idea can be supported by the 1000 km wide submerged continental crust of the Okhotsk Sea offshore the Cretaceous active continental margin of the Okhotsk–Chukotka Belt. The likely microcontinental origins of the Sea of Okhotsk have also been generally accepted in interpretations of Hokkaido geology. However, because no Precambrian and/or Palaeozoic rocks have been found in the sea, Kiminami (1986) and Ueda & Miyashita (2005) suggested the possibility that the Sea of Okhotsk crust is in fact an amalgam of Cretaceous intraoceanic remnant arcs, as found in the western Philippine Sea.

Hidaka Collision Zone

The Hidaka Collision Zone is the boundary zone between the two arc-trench systems of the NE Japan and Kuril arcs (Kimura 1986; Arita *et al.* 1998), these belonging to the Eurasian and Okhotsk margins, respectively. The zone overlaps the Sorachi–Yezo, Hidaka and Tokoro belts, and is characterized by west-verging fold-and-thrust structures with dextral shears on the surface, and also by exhumation of rocks as deep as lower crustal levels along the suture (Hidaka Main Thrust or HMT). In contrast to Alpine and Himalayan collision zones, where a passive continental margin was subducted beneath the upper plate of active margin, the Hidaka Collision Zone is a boundary between two active margins comprising an arc–arc junction. Absence of slab-pull may be responsible for the absence of collision-related HP rocks.

Two stages have been assumed for the collision of the two arcs (Kimura *et al.* 1983; Kimura 1996). In the first stage, the Okhotsk Block collided with the Eurasian margin along central–north Sakhalin in late Eocene–Oligocene times, before the back-arc opening of the Kuril Basin and the Sea of Japan. At that time, the Kuril Arc including the Nemuro and Tokoro belts was connected to the east of central Sakhalin (adjacent to the Terpeniye Peninsula). Central Hokkaido was not yet the site of collision, but was still facing the open Pacific to the east. As a result of the Kuril Basin opening, presumably in Early Miocene times, the Kuril Arc drifted southwards and its western tip came to reach the position of eastern Hokkaido aided by dextral strike-slip movement (Jolivet & Miyashita 1985; Jolivet & Huchon 1989). This dextral movement is also considered to be responsible for the pull-apart-style back-arc opening of the Sea of Japan (Lallemand & Jolivet 1985; Kimura & Tamaki 1986; Jolivet *et al.* 1990; Jolivet & Tamaki 1992), which coincided with Kurile back-arc opening. The Miocene strike-slip movement is attributed to oblique convergence of Eurasia (Amur) and North America (Okhotsk) plates (Kimura *et al.* 1983), or relative northwards migration of Eurasia due to collision with India (Kimura & Tamaki 1986; Jolivet *et al.* 1990). Once the back-arc openings had been completed, the Kuril forearc was subsequently thrust westwards onto the NE Japan Arc, dragged by oblique subduction of the Pacific Plate (Kimura 1986).

The Hidaka Collision Zone verges to the west, generally consisting (from east to west) of a metamorphic core (Hidaka Metamorphic Belt or HMB, and Poroshiri Ophiolite) and a frontal fold-and-thrust belt of pre-Neogene basement rocks and younger deposits (Fig. 2g.6). The HMT is the tectonic boundary between the hanging-wall Kuril forearc crustal section (HMB) on the east and the footwall NE Japan forearc (Poroshiri Ophiolite and Idonnappu Zone accretionary complex) on the west. The boundary between the Poroshiri Ophiolite and sedimentary rocks of the Idonnappu Zone defines another major fault (Hidaka Western Boundary Thrust or WBT) with a significant metamorphic gap.

Hidaka Metamorphic Belt

The HMB consists of high-temperature and low-pressure (HT-LP) metamorphic rocks up to granulite facies associated with abundant mafic to felsic plutonic rocks. The belt defines a NNW-trending arc *c.* 120 km long and 10–20 km wide, with foliations generally steeply dipping towards the east except in southernmost areas with flat-lying structures. Metamorphic grade increases westwards and is cut by the east-dipping HMT along the western (basal) margin, whereas the eastern low-grade parts of the complex grade into non-metamorphic accretionary sediments of the Nakanogawa Group lying to the east. The whole metamorphic sequence, together with the non-metamorphic sediments, is regarded as an arc crustal section down to 22–23 km (Komatsu *et al.* 1989, 1994; Fig. 2g.4). The HMB is divided into four metamorphic zones (Osanai *et al.* 1986, 1991, 1992; Komatsu *et al.* 1989, 1992, 1994): Zone I is characterized by Ms + Chl (greenschist facies); Zone II comprises Bt + Ms; Zone III consists of Grt + Sill + Kf (both amphibolite faceis); and Zone IV is characterized by Grt + Crd + Kf ± Opx (granulite facies) with anatectites. The upper parts (consisting of zones I and II) are dominated by pelitic rocks, and are also characterized by calc-alkaline plutonic intrusions typically of granite. The lower part (zones III and IV) is dominated by mafic rocks (amphibolites) with lesser amounts of pelitic gneiss. Amphibolites show chemical affinities with MORB (Kawanami *et al.* 2006). Tholeiitic gabbro intrusions (Maeda & Kagami 1996) and peridotite bodies (represented by the Horoman peridotite) also occur in the lower parts. Inclusions of higher-grade rocks (Grt–Opx granulite) in tonalites suggest the existence of the deeper lowermost crust that has not been exhumed in the Hidaka Belt (Shimura *et al.* 2004). Tonalite intrusions, occasionally mylonitized, are concentrated at the middle horizons separating the upper and lower parts, and are also common at the base of the lower parts along the HMT. According to structural analysis by Toyoshima *et al.* (1994), metamorphic rocks first underwent flattening (D1) and subsequent subhorizontal shear with a top-to-south sense accompanied by duplexing (D2), followed by listric thrusting and tilting with dextral shearing during exhumation (D3). Pseudotachylite veins were also formed as metamorphic rocks arrived at shallow crustal levels (Toyoshima 1990; Toyoshima *et al.* 2004).

Recent U–Pb geochronology (Usuki *et al.* 2006; Kemp *et al.* 2007) suggests that the major granulite facies metamorphic event occurred in late Early Miocene times (*c.* 19 Ma), representing one of the youngest exposed granulite complexes in the world. Hornblende and biotite K–Ar ages of the high-grade rocks cluster around 19–16 Ma, suggesting rapid exhumation (Arita *et al.* 1993). Palaeogene Rb–Sr and K–Ar ages for plutonic and metamorphic rocks in middle–upper horizons have also been obtained (Owada *et al.* 1991, 1997; Saheki *et al.* 1995); these were explained by Kemp *et al.* (2007) as recording Eocene sub-arc magmatism and metamorphism later overprinted by Early Miocene granulite metamorphism associated with MORB-like gabbro

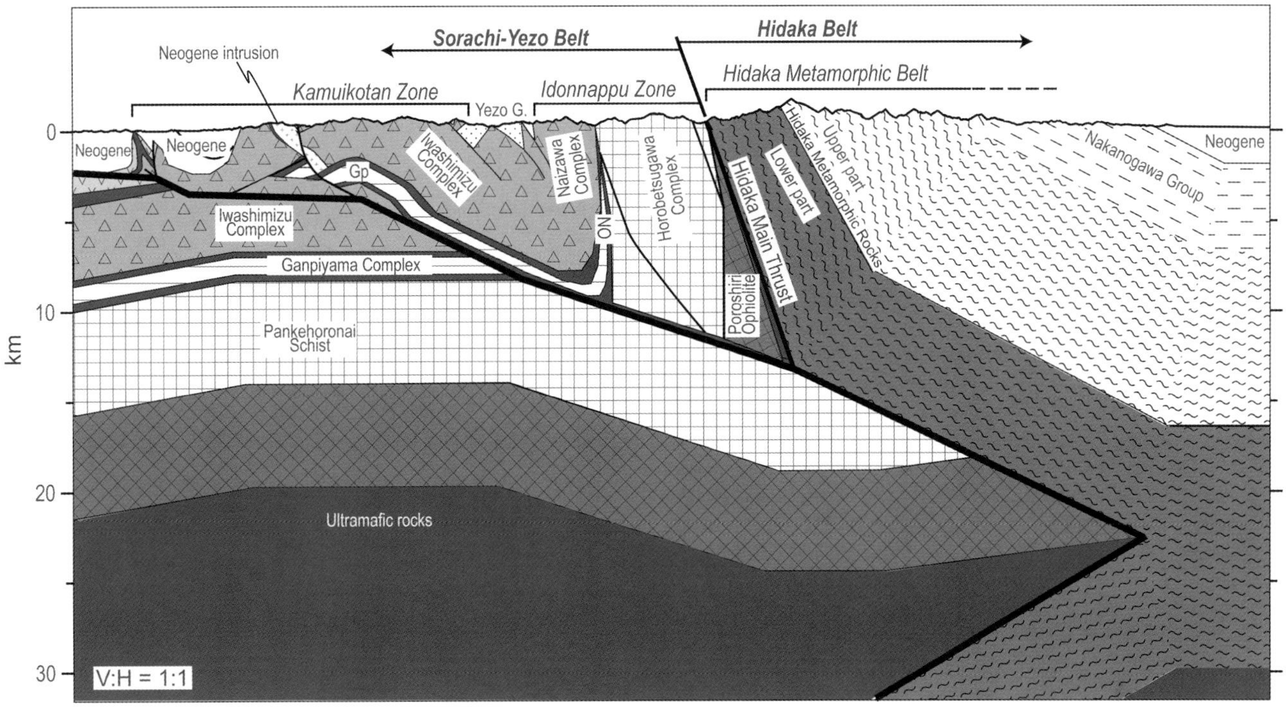

Fig. 2g.6. An interpretative deep geological cross-section of the Hidaka Collision Zone across the Hidaka Mountains (revised from Ueda 2006). Section line is shown in Figure 2g.1. Crustal sections of the Hidaka and Sorachi–Yezo belts follow Shimura *et al.* (2004) and Ueda (2006), respectively, as summarized in Figure 2g.4. Delamination structure of the lowermost crust beneath the Hidaka Metamorphic Belt follows Ito (2000). *Abbreviations*: Gp, Ganpiyama Complex; and ON, Oku-Niikappu Complex.

intrusion related to back-arc opening of the Sea of Japan and Kuril Basin.

Deep seismic reflection surveys were conducted across the Hidaka Collision Zone in the 1990s, and the results displayed deep crustal to uppermost mantle structures (Arita *et al.* 1998; Tsumura *et al.* 1999; Ito 2000, 2002). According to them, lower crustal lamination of the Kuril arc crust shallower than two-way time (TWT) 7.8 s (*c.* 23 km) dips to the east and extends westwards to the surface exposures of the HMB, whereas deeper lamination dips to the west. This structure was interpreted as a delamination at a lower crustal level: the Kuril arc crust shallower than 23 km is detached and thrust onto the NE Japan arc crust, with the remaining deeper portions (lowermost crust and uppermost mantle) being subducted beneath the NE Japan arc crust (Fig. 2g.6). This delamination-wedge structure, also imaged by seismic tomography (Murai *et al.* 2003), is quite consistent with both surface geology and petrology of the HMB, where a near-complete crustal section shallower than 22–23 km is exposed dipping to the east (Komatsu *et al.* 1994; Shimura *et al.* 2004); it also provides a good explanation of the whereabouts of the detached deeper rocks.

Frontal fold-and-thrust belt

The basement fold-and-thrust belt comprises rocks of the Sorachi–Yezo Belt including the Idonnappu and Kamuikotan zones (Fig. 2g.6). It shows complicated structures with numerous folds and faults with both thrust and strike-slip displacements, presumably as a result of collision contraction overprinting pre-collision structures by trench accretion and blueschist exhumation. Major folds and faults generally verge to the west, with common occurrences of steeply dipping and overturned deposits facing west. The west-verging structures also deform and cross-cut Miocene cover deposits. Deep seismic reflection (Arita *et al.* 1998; Ito 2002) imaged several east-dipping reflectors not obviously present in the surface geology. Ueda (2006) deduced a deep crustal section of the frontal zone based on the surface geology of the Sorachi–Yezo Belt, which he assumed to have comprised a flat-lying stack of accretionary units before the collisional faulting and thrusting. This model emphasized geological asymmetry across each of the two major anticlines (HMT as a thrust anticline, and the Kamuikotan Zone), and regarded them as corresponding to major ramps of the basal detachment fault.

In addition to the rather high-angle collision structures, nappes and klippes of the Cretaceous Yezo Group, overlying Palaeogene carbonaceous deposits (Ishikari Group) are known both from surface and well data (summarized by Ito 2000; Kazuka *et al.* 2002*b*). Nappe structures in basement rocks, where klippes from eastern units (Idonnappu Zone) overlie the autochthon of the Kamuikotan Zone, have also been proposed by Jolivet & Cadet (1984). Although not all the units proposed by them are currently accepted as klippe (e.g. Kawamura *et al.* 2001), two flat-lying geological units (Sarugawa and Niseu formations; Fig. 2g.1) overlying the Kamuikotan Zone rocks are considered as allochthons. The Sarugawa Formation consists mostly of pillow basalts of MORB chemistry (Katoh *et al.* 2000), with intercalating mudstones yielding Eocene radiolarians (Kazuka *et al.* 2002*a*). The Niseu Formation, structurally overlying the Sarugawa Formation, is a deformed clastic sedimentary sequence presumably of accretionary origin, and is also dated as Eocene age by radiolarians and fission track (Kumagai *et al.* 1995). Because no Eocene but only Cretaceous accretionary units occur elsewhere in the Sorachi–Yezo Belt, these formations are considered to have originated from the Palaeogene Hidaka Belt accretionary complex to the east of the HMT (Kumagai *et al.* 1995), and were subsequently displaced more than 50 km by nappe tectonics. These klippe units might have been emplaced before they were unconformably overlain by lower Middle Miocene deposits.

Collision-related sediments

The frontal parts of the collision zone comprised a Neogene marine sedimentary basin, whose sediments are now incorporated within the fold-and thrust belt. This basin is interpreted as a foredeep resulting from flexure of the lower plate (NE Japan forearc) due to the overburden of the thrusting Kuril Arc (Hoyanagi *et al.* 1986; Kimura & Miyashita 1986; Kawakami *et al.* 2004). The oldest deposits are local conglomeratic gravity-flow deposits and associated turbidites of the Late Oligocene Erimo Formation (dated by Kurita & Kusunoki 1997), located at the southern end of the Hidaka Mountains (Cape Erimo). They contain boulders of granites with biotite K–Ar ages of 30–33 Ma (Arita *et al.* 2001), and partly underwent intense dextral shear deformation (Kusunoki & Kimura 1998) suggesting incipient uplift and erosion of the uppermost HMB crust before the main collision. Thick gravelly turbidites of the foredeep (Kawabata Formation and equivalents) were developed in Middle Miocene times, suggesting the climax of the mountain uplift (Kawakami *et al.* 2004). This timing overlaps the cooling age range of the high-grade metamorphic rocks (Arita *et al.* 1993; Ono 2002), given that only upper crustal rocks had been exposed and were supplied as clasts at that time (Kawakami *et al.* 2004, 2006). Clasts of amphibolites- and granulite-grade HMB rocks began to be supplied in late Miocene and Pliocene times, respectively (Miyasaka *et al.* 1986), at which time the exhumation rate of the HMB had already decreased (Arita *et al.* 2001; Ono 2002).

Summary

Hokkaido is a collision zone between the two distinct arc-trench systems of NE Japan (Honshu) and Kuril. The NE Japan arc components (Oshima Belt, Sorachi–Yezo Belt and the main part of the Hidaka Belt) occupy western and central parts of Hokkaido. They are generally regarded as a continuation of NE Japan and the Outer Zone of SW Japan. However, the existence of the ophiolite-based large forearc basin introduces marked differences in geological configuration of the volcanic front, accretionary units and exhumation site of HP–LT metamorphic rocks from those of SW Japan. The Kuril arc components (Nemuro Belt, Tokoro Belt and parts of the Hidaka Belt) in east Hokkaido probably developed along with the Sea of Okhotsk, and have no direct genetic relation to any other parts of Japan. The two arc-trench systems of opposite polarities have collided since Miocene times, resulting in rapid exhumation of lower crustal metamorphic rocks in the Hidaka Metamorphic Belt along the boundary and in the development of a fold-and-thrust belt and foredeep basins in the frontal zones.

Hokkaido is one of the places on Earth where the development and amalgamation of continental and island arcs are well recorded. The character of the arc–arc junction and the occurrence of diverse types of ophiolites offer keys to help reconstruct the palaeogeography and plate configuration of the NW Pacific area. The collisional orogeny exhumed deep sections of magmatic arc crust (Hidaka Metamorphic Belt) and forearc crust (Sorachi–Yezo Belt), from which we can learn much about the arc-forearc architecture and evolution. Although the accumulation of geological data on Hokkaido has so far been less comprehensive compared to other areas of Japan, the unique tectonics and diverse geology of this island have great potential to improve our understanding of convergent margin evolution.

Appendix

English to Kanji and Hiragana translations for geological and place names

Butsuzo	仏像	ぶつぞう
Chichibu	秩父	ちちぶ
Erimo	襟裳	えりも
Fukushima	福島	ふくしま
Ganpiyama	雁皮山	がんぴやま
Hakobuchi	函淵	はこぶち
Harachiyama	原地山	はらちやま
Hatagawa	畑川	はたがわ
Hida	飛騨	ひだ
Hida Gaien	飛騨外縁	ひだがいえん
Hidaka	日高	ひだか
Hokkaido	北海道	ほっかいどう
Honshu	本州	ほんしゅう
Horobetsugawa	幌別川	ほろべつがわ
Horokanai	幌加内	ほろかない
Horoman	幌満	ほろまん
Idonnappu	イドンナップ	いどんなっぷ
Ishikari	石狩	いしかり
Iwashimizu	岩清水	いわしみず
Joban	常磐	じょうばん
Kabato	樺戸	かばと
Kamuikotan	神居古潭	かむいこたん
Kanto	関東	かんとう
Kawabata	川端	かわばた
Kitakami	北上	きたかみ
Kumaneshiri	隈根尻	くまねしり
Kurosegawa	黒瀬川	くろせがわ
Kyushu	九州	きゅうしゅう
Naizawa	ナイ沢	ないざわ
Nakanogawa	中の川	なかのがわ
Nemuro	根室	ねむろ
Nikoro	仁頃	にころ
Niseu	ニセウ（仁世宇）	にせう
Nokkamappu	ノッカマップ	のっかまっぷ
Okuniikappu	奥新冠	おくにいかっぷ
Okushiri	奥尻	おくしり
Oshima	渡島	おしま
Pankehoronai	パンケ幌内（班渓幌内）	ぱんけほろない
Poroshiri	ポロシリ（幌尻）	ぽろしり
Rebun	礼文	れぶん
Ryoke	領家	りょうけ
Sanbagawa	三波川	さんばがわ
Sanriku	三陸	さんりく
Sapporo	札幌	さっぽろ
Saroma	佐呂間	さろま
Sarugawa	沙流川	さるがわ
Shimanto	四万十	しまんと
Shiranuka	白糠	しらぬか
Shiretoko	知床	しれとこ
Sorachi	空知	そらち
Tanakura	棚倉	たなくら
Tokoro	常呂	ところ
Tokyo	東京	とうきょう
Tsugaru	津軽	つがる
Yezo	エゾ（蝦夷）	えぞ
Yubetsu	湧別	ゆうべつ

References

Ando, H. 1990*a*. Stratigraphy and shallow marine sedimentary facies of the Mikasa Formation Middle Yezo Group (Upper Cretaceous). *Journal of the Geological Society of Japan*, **96**, 279–295 [in Japanese with English abstract].

Ando, H. 1990*b*. Shallow-marine sedimentary facies distribution and progradational sequences of the Mikasa Formation Middle Yezo Group (Upper Cretaceous). *Journal of the Geological Society of Japan*, **96**, 453–469 [in Japanese with English abstract].

Ando, H. 2003. Stratigraphic correlation of Upper Cretaceous to Paleocene forearc basin sediments in northeast Japan: cyclic sedimentation and basin evolution. *Journal of Asian Earth Science*, **21**, 921–935.

Arai, T. & Miyashita, S. 1994. Shear deformation and metamorphism of the Poroshiri Ophiolite in the Shunbetsu River region, the Hidaka belt, Hokkaido, Japan. *Journal of the Geological Society of Japan*, **100**, 162–176 [in Japanese with English abstract].

Arai, T., Miyashita, S. & Shimura, T. 1995. Exotic slices derived from the Hidaka metamorphic belt in the Poroshiri ophiolite, Hokkaido, Japan. *Journal of Mineralogy, Petrology, and Economic Geology*, **90**, 388–402 [in Japanese with English abstract].

Arita, K., Toyoshima, T., Owada, M., Miyashita, S. & Jolivet, L. 1986. Tectonic movements of the Hidaka metamorphic belt, Hokkaido, Japan. *Memoirs of the Association for Geological Collaboration in Japan*, **31**, 247–263 [in Japanese with English abstract].

Arita, K., Shingu, H. & Itaya, T. 1993. K-Ar geochronological constraints on tectonics and exhumation of the Hidaka metamorphic belt, Hokkaido, northern Japan. *Journal of Mineralogy, Petrology, and Economic Geology*, **88**, 101–113.

Arita, K., Ikawa, T. *et al.* 1998. Crustal structure and tectonics of the Hidaka Collision Zone, Hokkaido (Japan), revealed by vibroseis seismic reflection and gravity surveys. *Tectonophysics*, **290**, 197–210.

Arita, K., Ganzawa, T. & Itaya, T. 2001. Tectonics and uplift process of the Hidaka Mountains, Hokkaido, japan inferred from thermochronology. *Bulletin of the Earthquake Research Institute, University of Tokyo*, **76**, 93–104 [in Japanese with English abstract].

Asahina, T. & Komatsu, M. 1979. The Horokanai ophiolitic complex in the Kamuikotan Tectonic Belt, Hokkaido, Japan. *Journal of the Geological Society of Japan*, **85**, 317–330.

Bamba, T. 1981. Troctolite and gabbro from the Tokoro Belt, central axial zone of Hokkaido, Japan. *Journal of the Japanese Association of Mineralogists, Petrologists and Economic Geologists*, **76**, 386–394.

Bamba, T. 1984. The Tokoro Belt, a tectonic unit of the central axial zone of Hokkaido. *Journal of Faculty of Science, Hokkido University Series IV*, **21**, 21–75.

Bikerman, M., Minato, M. & Hunahashi, M. 1971. K-Ar age of the garnet amphibolite of the Mitsuishi district, Hidaka province, Hokkaido, Japan. *Earth Science (Chikyu-Kagaku)*, **25**, 27–29.

Burk, C. A. & Gnibidenko, H. S. 1977. The structure and age of acoustic basement in the Okhotsk Sea. *In*: Talwani, M. & Pitman, W. C., III (eds) *Island Arcs, Deep Sea Trenches and Back-Arc Basins*. American Geophysical Union, Washington, 451–461.

Dickinson, W. R. & Seely, D. R. 1979. Structure and stratigraphy of forearc regions. *AAPG Bulletin*, **63**, 2–31.

Endo, R. & Hashimoto, W. 1955. Unquestionably Paleozoic (Permian) fossils found in Hokkaido, Japan. *Proceedings of the Japan Academy*, **31**, 704–708.

Finn, C. 1994. Aeromagnetic evidence for a buried Early Cretaceous magmatic arc, northeast Japan. *Journal of Geophysical Research, Solid Earth*, **99**, 22 165–22 185.

Fujiwara, Y. & Kanamatsu, T. 1994. On the age of bending structure in eastern Hokkaido. *Chishitsu News*, **478**, 45–48 [in Japanese].

Fujiwara, Y., Nanayama, F. & Kanamatsu, T. 1995. Tectonic evolution of two paleo arc-trench systems in Hokkaido, northern Japan. *Geofísica Internacional*, **34**, 283–291.

Girard, M., Jolivet, L., Nakagawa, M., Aguirre, L. & Niida, K. 1991. Acidic volcanic products in lower Cretaceous deposits of the Sorachi-Yezo Belt, Hokkaido, Northeast Japan. *Journal of the Geological Society of Japan*, **97**, 1–14.

Gnibidenko, H. S. 1985. The Sea of Okhotsk – Kuril Island Ridge and Kuril-Kamchatka Trench. *In*: Narin, A. E. M., Stehli, F. G. & Uyeda, S. (eds) *The Ocean Basins and Margins, Vol. 7A, The Pacific Ocean*. Plenum Press, New York, 377–418.

Gnibidenko, H. S., Hilde, T. W. C., Gretskaya, E. V. & Andreyev, A. A. 1995. Kuril (South Okhotsk) backarc basin. *In*: Taylor, B. (ed.) *Backarc Basins: Tectonics and Magmatism*. Plenum Press, New York, 421–449.

Goto, Y., Gouchi, N. & Itaya, T. 1991. Finding of 192 Ma granitic cobbles from the Cape Shiretoko area, eastern Hokkaido, Japan. *Journal of Mineralogy, Petrology, and Economic Geology*, **86**, 381–388.

Hashimoto, W., Igo, H., Asakura,K., Tateno, T. & Nagase, K. 1960. Discovery of the Fusulinids from the Kabato Mountainland, Ishikarino-kuni, Hokkaido. *Journal of the Geological Society of Japan*, **66**, 361 [in Japanese].

Hashimoto, W., Koike, T. & Hasegawa, T. 1975. First confirmation of the Permian system in the central part of Hokkaido. *Proceedings of the Japan Academy*, **51**, 34–37.

Hori, R. & Sakakibara, M. 1994. A chert-clastic sequence spanning the late Triassic – Early Cretacoeus period of the Kamuikotan Complex in the Shizunai area, south-central Hokkaido, Japan. *Journal of the Geological Society of Japan*, **100**, 575–583.

Hoyanagi, K., Miyasaka, S., Watanabe, Y., Kimura, G. & Matui, M. 1986. Depositions of turbidites in the Miocene collision zone, central Hokkaido. *Monographs of the Association for Geological Collaboration in Japan*, **31**, 265–284 [in Japanese with English abstract].

Iba, Y. & Sano, S. 2007. Mid-Cretaceous step-wise demise of the carbonate platform biota in the Nothwest Pacific and establishment of the North Pacific biotic province. *Paleogeography, Paleoclimatology, Paleoecology*, **245**, 262–482.

Iba, Y., Sano, S. & Miura, T. 2011. Orbitolinid foraminifers in the Northwest Pacific: their taxonomy and stratigraphy. *Micropaleontology*, **57**, 163–171.

Ichiyama, Y., Ishiwatari, A., Kimura, J., Senda, R., Kawabata, H. & Tatsumi, Y. 2012. Picrites in central Hokkaido: evidence of extremely high temperature magmatism in the Late Jurassic ocean recorded in and accreted oceanic plateau. *Geology*, **40**, 411–414.

Ichiyama, Y., Ishiwatari, A., Kimura, J.-I., Senda, R. & Miyamoto, T. 2014. Jurassic plume-origin ophiolites in Japan: accreted fragments of oceanic plateaus. *Contributions to Mineralogy and Petrology*, **168**, http://doi.org/10.1007/s00410-014-1019-1

Igarashi, T., Katoh, T. & Niida, K. 1985. The Takadomari seroentinites in the Kamuikotan ophiolite belt, Hokkaido, Japan. *Journal of the Faculty of Science, Hokkaido University, Series IV*, **21**, 305–319.

Igo, H., Koike, T., Igo, H. & Kinoshita, T. 1974. On the occurrence of Triassic conodonts from the Sorachi Group in the Hidaka Mountains, Hokkaido. *Journal of the Geological Society of Japan*, **80**, 135–136 [in Japanese].

Igo, H., Sashida, K. & Ueno, H. 1987. Early Cretaceous radiolarians from the Esashi Mountains, northern Hokkaido. *Annual Report of the Institute of Geoscience, the University of Tsukuba*, **13**, 105–109.

Ikeda, I. & Komatsu, M. 1986. Early Cretaceous volcanic rocks of Rebun Island, north Hokkaido, Japan. *Memoirs of the Association for Geological Collaboration in Japan*, **31**, 51–62 [in Japanese with English abstract].

Imaizumi, M. & Ueda, Y. 1981. On the K-Ar ages of the rocks of two kinds existed in the Kamuikotan metamorphic rocks located in the Horokanai district, Hokkaido. *Journal of Mineralogy, Petrology, and Economic Geology*, **76**, 88–92 [in Japanese with English abstract].

Ishiga, H. & Ishiyama, D. 1987. Jurassic accretionary complex in Kaminokuni Terrane, southwestern Hokakido, Japan. *Mining Geology*, **37**, 381–394.

Ishikawa, H., Berman, S. & Yagi, K. 1971. Geochemical study of trace elements in the alkali rocks of Nemuro Peninsula, Hokkaido, Japan. *Geochemical Journal*, **5**, 187–206.

Ishizaki, S. 1979. Find of Triassic Bryozoans from the Pre-Yezo Group in the Esashi Mountains, Hokkaido. *Earth Science (Journal of Association for the geological collaboration in Japan)*, **33**, 355–359 [in Japanese].

Ishizuka, H. 1980. Geology of the Horokanai Ophiolite in the Kamuikotan Tectonic Belt, Hokkaido, Japan. *Journal of the Geological Society of Japan*, **86**, 119–134 [in Japanese with English abstract].

Ishizuka, H. 1981. Geochemistry of the Horokanai Ophiolite in the Kamuikotan Tectonic Belt, Hokkaido, Japan. *Journal of the Geological Society of Japan*, **87**, 17–34.

Ishizuka, H. 1985. Prograde metamorphism of the Horokanai Ophiolite in the Kamuikotan Zone, Hokkaido, Japan. *Journal of Petrology*, **26**, 391–417.

ISHIZUKA, H. 1987. Igneous and metamorphic petrology of the Horokanai Ophiolite in the Kamuikotan Zone, Hokkaido, Japan: a synthetic thesis. *Memoirs of the Faculty of Science, Kochi University, Series E, Geology*, **8**, 1–70.

ISHIZUKA, H. & IMAIZUMI, M. 1980. Epidote-garnet amphibolite from the Horokanai Pass area in the Kamuikotan Tectonic Belt, Hokkaido. *Journal of the Geological Society of Japan*, **86**, 15–24.

ISHIZUKA, H., IMAIZUMI, M., GOUCHI, N. & BANNO, S. 1983*a*. The Kamuikotan zone in Hokkaido, Japan: tectonic mixing of high-pressure and low-pressure metamorphic rocks. *Journal of Metamorphic Geology*, **1**, 263–275.

ISHIZUKA, H., OKAMURA, M. & SAITO, Y. 1983*b*. Latest Jurassic radiolarians from the Horokanai Ophiolite in the Kamuikotan Zone, Hokkaido, Japan. *Journal of the Geological Society of Japan*, **89**, 731–732.

ITO, T. 2000. Crustal structure of the Hidaka collision zone and its foreland fold-and-thrust belt, Hokkaido, Japan. *Journal of the Japanese Association for Petroleum Technology*, **65**, 103–109 [in Japanese with English abstract].

ITO, T. 2002. Active faulting, lower crustal delamination and ongoing Hidaka arc-arc collision, Hokkaido, Japan. *In*: FUJINAWA, Y. & YOSHIDA, A. (eds) *Seismotectonics in Convergent Plate Boundary*. Terrapub, Tokyo, 219–224.

IWASAKI, I., WATANABE, T., ITAYA, T., YAMAZAKI, M. & TAKIGAMI, Y. 1995. Paleogene K-Ar ages from the Kamuikotan metamorphic rocks, southern area of the Kamuikotan Gorge, central Hokkaido, northern Japan. *Geological Journal*, **30**, 281–295.

IWATA, K. & KATO, Y. 1986. Upper Cretaceous radiolarians of the Yubetsu Group and the Hidaka Supergroup in the northern Hidaka Belt. *News of Osaka Micropaleontologists*, Special Volume, **7**, 75–86 [in Japanese with English abstract].

IWATA, K. & TAJIKA, J. 1986. Late Cretaceous radiolarians of the Yubetsu Group, Tokoro Belt, northeast Hokkaido. *Journal of the Faculty of Science, Hokkaido University Series IV*, **21**, 619–644.

IWATA, K. & TAJIKA, J. 1989. Jurassic and Cretaceous radiolarians from the pre-Tertiary system in the Hidaka Belt, Maruseppu Region, northeast Hokkaido. *Journal of the Faculty of Science, Hokkaido University, Series IV*, **22**, 453–466.

IWATA, K. & TAJIKA, J. 1992. *Early Paleogene Radiolarians from Green and Red Mudstones in the Yubetsu Group & Reconsideration of the Age of their Sedimentation*. Report of the Geological Survey of Hokkaido, Sapporo, Japan, no. 63, http://www.hro.or.jp/list/environmental/research/gsh/publication/report/report0202/index.html#no63

IWATA, K., UOZUMI, S., NAKAMURA, K. & TAJIKA, J. 1983*a*. Discovery of radiolarians and holothurians sclerites from the pre-Tertiary system around Nishiokoppe, northeast Hokkaido (preliminary report). *Journal of the Geological Society of Japan*, **89**, 55–56.

IWATA, K., WATANABE, M., NAKAMURA, K. & UOZUMI, S. 1983*b*. Occurrence of Jurassic and Cretaceous radiolarians from the pre-Tertiary systems around Lake Saroma, northeast Hokkaido. *Earth Science ('Chikyu Kagaku')*, **37**, 225–228 [in Japanese].

IWATA, K., HARIYA, Y., CHOI, J. H., YAGI, E. & MIURA, T. 1990. Radiolarian age of the manganese deposits of the Tokoro Belt, northeast Hokkaido. *Journal of Faculty of Science, Hokkaido University Series IV*, **22**, 565–576.

IWATA, K., OBUT, O. T. & TAJIKA, J. 1997. Lower Cretaceous radiolaria from the Rebun Group, northern Hokkaido. *News of Osaka Micropaleontologists*, Special Volume, **10**, 211–215.

JOLIVET, L. & CADET, J. P. 1984. The Iwanai nappe in the Kamuikotan tectonic belt, southern Hokkaido, Japan. *Journal of the Faculty of Science, Hokkaido University, Series IV*, **21**, 293–304.

JOLIVET, L. & HUCHON, P. 1989. Crustal-scale strike-slip deformation in Hokkaido, northern Japan. *Journal of Structural Geology*, **5**, 509–522.

JOLIVET, L. & MIYASHITA, S. 1985. The Hidaka shear zone (Hokkaido, Japan): genesis during a right-lateral strike-slip movement. *Tectonics*, **4**, 289–302.

JOLIVET, L. & TAMAKI, K. 1992. Neogene kinematics in the Japan Sea region and volcanic activity of the northeast Japan arc. *In*: TAMAKI, K., SUYEHIRO, L., ALLAN, J., MCWILLIAMS, M. ET AL. (eds) *Proceedings of the Ocean Drilling Program vol. 127/128 Scientific Results Part 2*. Ocean Drilling Program, College Station, TX, 1311–1331.

JOLIVET, L., DAVY, P. & COBBOLD, P. 1990. Right-lateral shear along the northwest Pacific margin and the India-Eurasia collision. *Tectonics*, **9**, 1409–1419.

KAIHO, K. & SAITO, T. 1986. Terminal Cretaceous sedimentary sequence recognized in the northernmost Japan based on planktonic froeaminiferal evidence. *Proceedings of Japan Academy*, **62**, 145–148.

KANAMATSU, T., NANAYAMA, F., IWATA, K. & FUJIWARA, T. 1992. Pre-Tertiary Systems on the wesstern side of the Abashiri Tectonic Line in the Shiranuka area, eastern Hokkaido, Japan: implications to the tectonic relationship between the Nemuro and Tokoro Belts. *Journal of the Geological Society of Japan*, **98**, 1113–1128 [in Japanese with English abstract].

KANIE, Y., TAKETANI, Y., SAKAI, A. & MIYATA, Y. 1981. Lower Cretaceous deposits beneath the Yezo Group in the Urakawa area, Hokkaido. *Journal of the Geological Society of Japan*, **87**, 527–533 [in Japanese with English abstract].

KATO, Y. & IWATA, K. 1989. Radiolarian biostratigraphic study of the pre-Tertiary system around the Kamikawa basin, central Hokkaido, Japan. *Journal of the Faculty of Science, Hokkaido University, Series IV*, **22**, 425–452.

KATO, Y., IWATA, K., UOZUMI, S. & NAKAMURA, K. 1986. Re-examination on the Pre-Tertiary System distributed in the Toma-Kaimei district of central Hokkaido. *Journal of the Geological Society of Japan*, **92**, 239–242 [in Japanese with English abstract].

KATOH, T. & NAKAGAWA, M. 1986. Tectogenesis of ultramafic rocks in the Kamuikotan tectonic belt, Hokkaido, Japan. *Memoirs of the Association for Geological Collaboration in Japan*, **31**, 119–135.

KATOH, T., NIIDA, K. & WATANABE, T. 1979. Serpentinite Melange around Mt. Shirikomadake in the Kamuikotan Structural Belt, Hokkaido. *Journal of the Geological Society of Japan*, **85**, 279–285 [in Japanese with English abstract].

KATOH, T., UEDA, H., GANZAWA, Y., KIZAKI, K., KAWAMURA, M., ONO, M. & NAKAGAWA, M. 2000. Geology and tectonics of the Sorachi-Yezo Belt around Hidaka-Town, Hokkaido, Japan: Age and igneous petrology of the Sarugawa Formation. *Abstract, Annual Meeting of the Geological Society of Japan*, 29th September–1st October 2001, Matsue, Japan, **107**, 262 [in Japanese].

KAWABATA, K. 1988. New species of latest Jurassic and earliest Cretaceous radiolarians from the Sorachi Group in Hokkaido, Japan. *Bulletin of the Osaka Museum of Natural History*, **43**, 1–13.

KAWAKAMI, G., ARITA, K., OKADA, T. & ITAYA, T. 2004. Early exhumation of the collisional orogen and concurrent infill of foredeep basins in the Miocene Eurasian-Okhotsk Plate boundary, central Hokkaido, Japan: inferences from K-Ar dating of granitoids clasts. *Island Arc*, **13**, 359–369.

KAWAKAMI, G., OHIRA, H., ARITA, K., ITAYA, T. & KAWAMURA, M. 2006. Uplift history of the Hidaka Mountains, Hokkaido, Japan: a thermochronologic view. *Journal of the Geological Society of Japan*, **112**, 684–698 [in Japanese with English abstract].

KAWAMURA, M. & FUJIMOTO, S. 2000. Chemical composition of the sandstones from the Cretaceous Yezo Supergroup, south central Hokkaido. *Memoirs of the Geological Society of Japan*, **57**, 119–132 [in Japanese with English abstract].

KAWAMURA, M., TAJIKA, J., KAWAMURA, T. & KATO, Y. 1986. Constitution and occurrences of the Paleozoic and Mesozoic formations in S.W. Hokkaido, northern Japan. *Monographs of the Association for Geological Collaboration in Japan*, **31**, 17–32.

KAWAMURA, M., NAKAGWA, M. ET AL. 1998. Geologic structure of the metamorphosed Mesozoic accretionary complex in the Sorachi-Yezo Belt, central Hokkaido, Japan: Unit-classification of the Biei Complex and proposal of Oichan Nappe. *Earth Science (Chikty Kagaku)*, **52**, 433–452 [in Japanese with English abstract].

KAWAMURA, M., UEDA, H. & NARUSHIMA, T. 1999. Unconformities and slump bodies recognized in forearc sediments-stratigraphic phenomena at the basal part of the Middle Yezo Group, Sorachi-Yezo Belt, central Hokkaido. *Memoirs of the Geological Society of Japan*, **52**, 37–52 [in Japanese with English abstract].

KAWAMURA, M., YASUDA, N., WATANABE, T., FANNING, M. & TERADA, T. 2000. Composition and provenance of the Jurassic quartzofeldspathic sandstones of the Oshima Accretionary Belt, SW Hokkaido, Japan. *Memoirs of the Geological Society of Japan*, **57**, 63–72 [in Japanese with English abstract].

KAWAMURA, M., UEDA, H., NAKAGAWA, M., KATOH, T. & Research Group of Hidaka Convergent Zone 2001. High P/T metamorphic minerals found from the 'undivided Hidaka Supergroup' in the Sorachi-Yezo Belt, Hokkaido. *Journal of the Geological Society of Japan*, **107**, 237–240 [in Japanese with English abstract].

KAWANAMI, S., NAKANO, N., OSANAI, Y., KAGAMI, H. & OWADA, M. 2006. Protoliths of the high-grade amphibolites from the Main Zone of the Hidaka metamorphic belt in Hokkaido, northern Japan and comparison with greenstones in the northern Hidaka belt. *Journal of the Geological Society of Japan*, **112**, 639–653 [in Japanese with English abstract].

KAZUKA, T., ITO, T. & AITA, Y. 2002*a*. Eocene radiolarian fossils from the Sarugawa Formation, the Hidaka foreland fold-and-thrust belt, Hokkaido, Japan. *Journal of the Geological Society of Japan*, **108**, 474–477 [in Japanese with English abstract].

KAZUKA, T., KIKUCHI, S. & ITO, T. 2002*b*. Structure of the foreland fold-and-thrust belt, Hidaka Collision Zone, Hokkaido, Japan: reprocessing and re-interpretation of the JNOC seismic reflection profiles 'Hidaka' (H91–2 and H91–3). *Bulletin of the Earthquake Research Institute, University of Tokyo*, **77**, 97–109.

KEMP, A. I. S., SHIMURA, T., HAWKESWORTH, C. J. & EIMF 2007. Linking granulites, silicic magmatism, and crustal growth in arcs: ion microprobe (zircon) U–Pb ages from the Hidaka metamorphic belt, Japan. *Geology*, **35**, 807–810.

KIMINAMI, K. 1975. Sedimentology of the Nemuro Group (Part I). *Journal of the Geological Society of Japan*, **81**, 215–232.

KIMINAMI, K. 1983. Sedimentary history of the Late Cretaceous-Paleocene Nemuro Group, Hokkaido, Japan: a forearc basin of the paleo-Kuril arc-trench system. *Journal of the Geological Society of Japan*, **89**, 607–624.

KIMINAMI, K. 1986. Cretaceous tectonics of Hokkaido and environs of the Okhotsk Sea. *Monographs of the Association for Geological Collaboration of Japan*, **31**, 403–418 [in Japanese with English abstract].

KIMINAMI, K. & KONTANI, Y. 1983*a*. Mesozoic arc-trench systems in Hokkaido, Japan. *In*: HASHIMOTO, M. & UYEDA, S. (eds) *Accretion Tectonics in the Circum-Pacific Regions*. Terrapub, Tokyo, 107–122.

KIMINAMI, K. & KONTANI, Y. 1983*b*. Sedimentology of the Saroma Group, northern Tokoro Belt, Hokkaido. *Earth Science ('Chikyu Kagaku')*, **37**, 38–47 [in Japanese.

KIMINAMI, K., SUIZU, M. & KONTANI, Y. 1983. Discovery and significance of Cretaceous radiolarians from the Mesozoic in the Tokoro Belt, eastern Hokkaido, Japan. *Journal of the Association for the Geological Collaboration in Japan*, **37**, 48–52 [in Japanese].

KIMINAMI, K., KITO, N. & TAJIKA, J. 1985*a*. Mesozoic Group in Hokkaido – stratigraphy and age, and their significance. *Earth Science ('Chikyu Kagaku')*, **39**, 1–17 [in Japanese with English abstract].

KIMINAMI, K., KONTANI, T. & MIYASHITA, S. 1985*b*. Lower Cretaceous strata covering the abyssal tholeiite (the Hidaka Western Greenstone Belt) in the Chiroro area, central Hokkaido, Japan. *Journal of the Geological Society of Japan*, **91**, 27–42.

KIMINAMI, K., KOMATSU, M., NIIDA, K. & KITO, N. 1986*a*. Tectonic divisions and stratigraphy of the Mesozoic rocks of Hokkaido, Japan. *Monographs of the Association for Geological Collaboration in Japan*, **31**, 1–15.

KIMINAMI, K., MIYASHITA, S. ET AL. 1986*b*. Mesozoic rocks in the Hidaka Belt – Hidaka Supergroup. *Monographs of the Association for Geological Collaboration in Japan*, **31**, 137–155 [in Japanese with English abstract].

KIMINAMI, K., KAWABATA, K. & MIYASHITA, S. 1990*a*. Discovery of Paleogene radiolarians from the Hidaka Supergroup and its significance with special reference to ridge subduction. *Journal of the Geological Society of Japan*, **96**, 323–326 [in Japanese with English abstract].

KIMINAMI, K., SHIBATA, K. & UCHIUMI, S. 1990*b*. K-Ar age of a tuff from the Yubetsu Group in the Tokoro Belt, Hokkaido, Japan. *Journal of the Geological Society of Japan*, **96**, 77–80 [in Japanese].

KIMINAMI, K., KOMATSU, M. & KAWABATA, K. 1992*a*. Composition of clastics of the Sorachi Group and the Lower Yezo Group in the Inushibetsugawa area, Hokkaido, Japan, and their significance. *Memoirs of the Geological Society of Japan*, **38**, 1–11 [in Japanese with English abstract].

KIMINAMI, K., KUMON, F., NISHIMURA, T. & SHIKI, T. 1992*b*. Chemical composition of sandstones derives from magmatic arcs. *Memoirs of the Geological Society of Japan*, **38**, 361–372.

KIMINAMI, K., NIIDA, K. ET AL. 1992*c*. Cretaceous-Paleogene arc-trench systems in Hokkaido. *In*: ADACHI, M. & SUZUKI, K. (eds) *29th IGC Field Trip Guide Book. Vol. 1: Paleozoic and Mesozoic Terranes: Basement of the Japanese Island Arcs*. Nagoya University, 1–43.

KIMINAMI, K., MIYASHITA, S. & KAWABATA, K. 1999. Occurrence and significance of *in-situ* basaltic rocks from the Rurochi Formation in the Hidaka Supergroup, northern Hokkaido, Japan. *Memoirs of the Geological Society of Japan*, **52**, 103–112 [in Japanese with English abstract].

KIMURA, G. 1986. Oblique subduction and collision: forearc tectonics of the Kuril arc. *Geology*, **43**, 404–407.

KIMURA, G. 1996. Collision orogeny at arc-arc junctions in the Japanese Islands. *Island Arc*, **5**, 262–275.

KIMURA, G. & MIYASHITA, S. 1986. Triple junction of trench-trench-oblique collision zone and origin of the Hidaka metamorphic belt. *Monographs of the Association for Geological Collaboration in Japan*, **31**, 451–458 [in Japanese with English abstract].

KIMURA, G. & TAMAKI, K. 1985. Tectonic framework of the Kuril arc since its initiation. *In*: NASU, N., UYEDA, S., KOBAYASHI, K., KUSHIRO, I. & KAGAMI, H. (eds) *Formation of Active Ocean Margins*. Terrapub, Tokyo, 641–676.

KIMURA, G. & TAMAKI, K. 1986. Collision, rotation, and back-arc spreading in the ergion of the Okhotsk and Japan seas. *Tectonics*, **5**, 389–401.

KIMURA, G., MIYASHITA, S. & MIYASAKA, S. 1983. Collision tectonics in Hokkaido and Sakhalin. *In*: HASHIMOTO, M., & UYEDA, S. (eds) *Accretion Tectonics in the Circum Pacific Regions*. Terrapub, Tokyo, 123–134.

KIMURA, G., SAKAKIBARA, M. & OKAMURA, M. 1994. Plumes in central Panthalassa? Deductions from accreted oceanic fragments in Japan. *Tectonics*, **13**, 905–916.

KITO, N. 1987. Stratigraphic relation between greenstones and clastic sedimentary rocks in the Kamuikotan Belt, Hokkaido, Japan. *Journal of the Geological Society of Japan*, **93**, 21–35 [in Japanese with English abstract].

KITO, N. 1995. Upper Jurassic to Lower Cretaceous stratigraphy of Hokkaido, Japan. *Mémoires de Géologie (Lausanne)*, **23**, 923–935.

KITO, N. 1997. Early Cretaceous radiolarian biostratigraphy of Hokkaido Central Belt by means of unitary association method. *In*: KAWAMURA, M., OKA, T. & KONDO, T. (eds) *Commemorative Volume for Professor Makoto Kato*. Nakanishi Printing, Sapporo, 29–40 [in Japanese with English abstract].

KITO, N., KIMINAMI, K., NIIDA, K., KANIE, Y., WATANABE, T. & KAWAGUCHI, M. 1986. The Sorachi Group and the Yezo Supergroup: late Mesozoic ophiolites and forearc basin sediments in the axial zone of Hokkaido. *Monographs of the Association for Geological Collabroration in Japan*, **31**, 81–96.

KIYOKAWA, S. 1992. Geology of the Idonnappu Belt, Central Hokkaido, Japan: evolution of a Cretaceous accretionary complex. *Tectonics*, **11**, 1180–1206.

KIZAKI, K. 2000. *Age of the Formation and the Metamorphic Process of the Poroshiri Ophiolite, Hokkaido, Japan*. PhD thesis, Hokkaido University.

KOJIMA, S., KEMUKIN, T. V., KAMETAKA, M. & ANDO, A. 2000. A correlation of accretionary complexes of southern Sikhote-Alin of Russia and the Inner Zone of Southwest Japan. *Geoscience Journal*, **4**, 175–185.

KOMATSU, M. 1981. Tectonics of Hokkaido, with special reference to the Hidaka metamorphic belt. *In*: HARA, I. (ed.) *Tectonics of Paired Metamorphic Belts*. Tanishi Print Kikaku, Hiroshima, 55–59.

KOMATSU, M., MIYASHITA, S., MAEDA, J., OSANAI, Y. & TOYOSHIMA, T. 1989. Evolution of the Hidaka metamorphic belt, Hokkaido, northern Japan. *In*: DALY, J. S., CLIFF, R. A. & YARDLEY, B. W. D. (eds) *Evolusion of Metamorphic Belts*. Geological Society, London, Special Publications, **43**, 487–493.

KOMATSU, M., SHIBAKUSA, H., MIYASHITA, S., ISHIZUKA, H., OSANAI, Y. & SAKAKIBARA, M. 1992. Subduction and collision related high and low P/T metamorphic belts in Hokkaido. *In*: KATO, H. & NORO, H. (eds) *29th IGC Field Trip Guide Book Vol. 5: Metamorphic Belts and Related Plutonism in the Japanese Islands*. Geological Society of Japan, 1–61.

KOMATSU, M., TOYOSHIMA, T., OSANAI, Y. & ARAI, M. 1994. Prograde and anatectic reactions in the deep arc crust exposed in the Hidaka metamorphic belt, Hokkaido, Japan. *Lithos*, **33**, 31–49.

KONDO, H. 1991. Stratigraphy and geological structure of the Kumaneshiri Group in the Kabato Mountains, Hokkaido, Japan. *Journal of the Geological Society of Japan*, **97**, 357–376 [in Japanese with English abstract].

KONDO, H. 1993. Igneous rocks of the Kumaneshiri Group in the Kabato Mountains, Hokkaido, Japan: Characterization of the alkali rich igneous rocks in the Cretaceous volcanic zone of Northeast Japan. *Journal of the Geological Society of Japan*, **99**, 347–364 [in Japanese with English abstract].

KONTANI, Y. 1978. Geological study of the Hidaka Supergroup distributed on the east side of the Hidaka metamorphic belt (part 1): Stratigraphy and geological structure. *Journal of the Geological Society of Japan*, **84**, 1–14 [in Japanese with English abstract].

KONTANI, Y. & KIMINAMI, K. 1980. Petrological study of the sandstones in the Pre-Cretaceous Yubetsu Group, northeastern Hidaka Belt, Hokkaido, Japan. *Earth Science (Journal of the Association for Geological Collaboration in Japan)*, **33**, 307–319.

KUMAGAI, T., KITO, N. & GANZAWA, Y. 1995. Occurrence of Eocene radiolarians and fission track dating from the Niseu Formation, Hokkaido Central Belt, Northern Japan. *Journal of the Geological Society of Japan*, **101**, 965–969 [in Japanese with English abstract].

KUMON, F. & KIMINAMI, K. 1994. Modal and chemical compositions of the representative sandstones from the Japanese Islands and their tectonic implications. *In*: KUMON, F. & YU, K. M. (eds) *Proceedings of the 29th International Geological Congress Part A: Sandstone Petrology in Relation to Tectonics*. VSP, Utrecht, 135–151.

KURITA, H. & KUSUNOKI, K. 1997. A Late Oligocene age of dinoflagellate cysts from the Erimo Formation, southern central Hokkaido, Japan, and its implications for tectonic history. *Journal of the Geological Society of Japan*, **103**, 1179–1182 [in Japanese with English abstract].

KURITA, H. & TAJIKA, J. 1998. Tectonic implications of Early Eocene dinoflagellate cysts from the Mukai-engaru Formation of the Yubetsu Group, Tokoro Belt, eastern Hokkaido. *Journal of the Geological Society of Japan*, **104**, 808–811 [in Japanese with English abstract].

KUSUNOKI, K. & KIMURA, G. 1998. Collision and extrusion at the Kuril-Japan arc junction. *Tectonics*, **17**, 843–858.

KUSUNOKI, T. 1996. The similarity in the Jurassic – early Cretaceous sandstone compositions between Dalnegorsk region in Southern Sikhote-Alin, Russia and the Oshima Belt in Hokkaido, Japan. *Earth Science ('Chikyu Kagaku')*, **50**, 414–418.

KUSUNOKI, T. & MUSASHINO, M. 2001. Comparison of the Jurassic to earliest Cretaceous sandstones from the Japanese Islands and South Sikhote-Alin. *Earth Science ('Chikyu Kagaku')*, **55**, 293–306.

LALLEMAND, S. E. & JOLIVET, L. 1985. Japan Sea: a pull-apart basin? *Earth and Planetary Science Letters*, **76**, 375–389.

LELIKOV, E. P. & EMELYANOVA, T. A. 2007. Volcanogenic complexes of the Sea of Okhotsk and the Sea of Japan (Comparative Analysis). *Oceanology*, **47**, 294–303.

LELIKOV, E. P. & EMELYANOVA, T. A. 2011. Geology and volcanism of the underwater Vityaz Ridge (Pacific slope of the Kuril island arc). *Oceanology*, **51**, 315–328.

MAEDA, J. 1990. Opening of the Kuril Basin deduced from the magmatic history of central Hokkaido, North Japan. *Tectonophysics*, **174**, 235–255.

MAEDA, J. & KAGAMI, H. 1996. Interaction of a spreading ridge and an accretionary prism: implications from MORB magmatism in the Hidaka magmatic zone, Hokkaido, Japan. *Geology*, **24**, 31–34.

MAEDA, J., SUETAKE, S., IKEDA, Y., TOMURA, S., MOTOYOSHI, Y. & OKAMOTO, Y. 1986. Tertiary plutonic rocks in the axial zone of Hokkaido: distribution, age, major element chemistry, and tectonics. *Monographs of the Association for Geological Collaboration of Japan*, **31**, 223–246 [in Japanese with English abstract].

MALINOVSKY, A. I., GOLOZOUBOV, V. V. & SIMANENKO, V. P. 2006. The Kema Island-Arc Terrane, Eastern Sikhote Alin: formation Settings and Geodynamics. *Doklady Earth Sciences*, **410**, 1026–1029.

MARIKO, T. 1984. Sub-sea hydrothermal alteration of basalt, diabase and sedimentary rocks in the Shimokawa copper mining area, Hokkaido, Japan. *Mining Geology*, **34**, 307–321.

MARIKO, T. & KATO, Y. 1994. Host rock geochemistry and tectonic setting of some volcanogenic massive sulfide deposits in Japan: examples of the Shimokawa and the Hitachi ore deposits. *Resource Geology*, **44**, 353–367.

MINATO, M. & ROWETT, C. L. 1967. New Paleozoic fossils from southern Hokkaido, Japan. *Journal of the Faculty of Science, Hokkaido University, Series IV*, **13**, 321–332.

MINOURA, N. & KATO, M. 1978. Permian calcareous algae found in the Matsumae Group, Matsumae Peninsula, southwestern Hokkaido. *Journal of the Faculty of Science, Hokkaido University, Series IV*, **18**, 377–383.

MIYASAKA, S., HOYANAIGI, K., WATANABE, Y. & MATUI, M. 1986. Late Cenozoic mountain-building history in central Hokkaido deduced from the composition of conglomerate. *Monographs of the Association for Geological Collaboration in Japan*, **31**, 285–294 [in Japanese with English abstract].

MIYASHITA, S. 1981. Ophiolite succession and metamorphism of the western zone of the Hidaka Metamorphic Belt, Hokkaido. *In*: HARA, I. (ed.) *Tectonics of Paired Metamorphic Belts*. Tanishi Print, Hiroshima, 25–30.

MIYASHITA, S. 1983. Reconstruction of the ophiolite succession in the western zone of the Hidaka Metamorphic Belt, Hokkaido. *Journal of the Geological Society of Japan*, **89**, 69–86 [in Japanese with English abstract].

MIYASHITA, S. & KATSUSHIMA, T. 1986. The Tomuraushi greenstone complex of the central Hidaka zone: contemporaneous occurrence of abyssal tholeiite and terrigeneous sediments. *Journal of the Geological Society of Japan*, **92**, 535–557.

MIYASHITA, S. & KIMINAMI, K. 1999. Petrology of the greenstones in the Rurochi Formation in the northern Hidaka Belt, Hokkaido. *Memoirs of the Geological Society of Japan*, **52**, 113–124 [in Japanese with English abstract].

MIYASHITA, S. & NIIDA, K. 1981. Metamorphosed dolerite intrusive from the western zone of the Hidaka Metamorphic Belt, Hokkaido. *Journal of the Faculty of Science, Hokkaido University Series 4*, **20**, 113–133.

MIYASHITA, S. & YOSHIDA, A. 1988. Pre-Cretaceous and Cretaceous ophiolites in Hokkaido, Japan. *Bulletin de la Societe Geologique de France*, **IV**, 251–260.

MIYASHITA, S. & YOSHIDA, A. 1994. Geology and petrology of the Shimokawa ophiolite (Hokkaido, Japan): ophiolite possibly generated near R-T-T triple junction. *In*: ISHIWATARI, A., MALPES, J. & ISHIZUKA, H. (eds) *Proceedings of the 29th International Geological Congress Part D: Circum-Pacific Ophiolites*. VSP, Netherlands, 163–182.

MIYASHITA, S., KIZAKI, K., ARAI, T. & TOYOSHIMA, T. 1994. *Field Guide to the Poroshiri Ophiolite of the Hidaka Zone, Central Hokkaido*, Japan (IGCP Project 283, 321 & 359 Field Trip Guidebook). Kyobunsha Co. Ltd., Sapporo.

MIYASHITA, S., ADACHI, Y., TANAKA, S., NAKAGAWA, M. & KIMURA, J. 2007. Genesis of the Poroshiri ophiolite, Hokkaido, Japan; Inference from geochemical evidence. *Journal of the Geological Society of Japan*, **113**, 212–221 [in Japanese with English abstract].

MURAI, Y., AKIYAMA, S. *ET AL.* 2003. Delamination structure imaged in the source area of the 1982 Urakawa-oki earthquake. *Geophysics Research Letters*, **30**, 1490, http://doi.org/10.1029/2002GL016459

NAGAO, S., AKIBA, C. & OMORI, T. 1963. Explanatory text of the Geological Map of Japan (Scale 1:50 000), Rebunto. Hokkaido Development Agency, Sappro, Japan [in Japanese with English abstract].

NAGAHASHI, T. & MIYASHITA, S. 2002. Petrology of the greenstones of the Lower Sorachi Group in the Sorachi-Yezo Belt, central Hokkaido, Japan, with special reference to discrimination between oceanic plateau basalts and mid-oceanic ridge basalts. *Island Arc*, **11**, 122–141.

NAGATA, H., KITO, N. & NAKAGAWA, M. 1987. Serpentinite gravels found in the Yezo Supergroup, Hokkaido, Japan. *Earth Science (Chikyu Kagaku)*, **41**, 57–60 [in Japanese].

NAGATA, M., KITO, N. & NIIDA, K. 1986. The Kumaneshiri Group in the Kabato Mountains: the age and nature as an Early Cretaceous volcanic arc. *Monographs of the Association for Geological Collaboration in Japan*, **36**, 63–79 [in Japanese with English abstract].

NAKAGAWA, M. & TODA, H. 1987. Geology and petrology of Yubari-dake Serpentinite Melange in the Kamuikotan Tectonic Belt, central Hokkaido, Japan. *Journal of the Geological Society of Japan*, **93**, 733–748.

NAKANO, N. & KOMATSU, M. 1979. Kaersutite-aegirine alkali diabase in the Shizunai-Mitsuishi area in the Kamuikotan greenrock zone,Hokkaido. *Journal of the Geological Society of Japan*, **85**, 367–376 [in Japanese with English abstract].

NAKAYAMA, K. 2003. Internal structure of the Shimokawa greenstone-argillaceous sediment complex. Reconstruction of the paleo-sedimented spreading center. *Shigen-Chishitsu*, **53**, 81–94 [in Japanese with English abstract].

NANAYAMA, F. 1992*a*. Three petroprovinces identified in the Nakanogawa Group, Hidaka Belt, central Hokkaido, Japan, and their geotectonic significance. *Memoirs of the Geological Society of Japan*, **38**, 27–42.

NANAYAMA, F. 1992*b*. Stratigraphy and facies of the Paleocene Nakanogawa Group in the southern part of central Hokkaido, Japan. *Journal of the Geological Society of Japan*, **98**, 1041–1059 [in Japanese with English abstract].

NANAYAMA, F. & GANZAWA, Y. 1997. Sedimentary stratigraphy, environment and age of the northern unit of Nakanogawa Group in the Hidaka belt, central Japan. *Memoirs of Geological Society of Japan*, **47**, 279–293 [in Japanese with English abstract].

NANAYAMA, F., KANAMATSU, T. & FUJIWARA, Y. 1993. Sedimentary petrology and paleotectonic analysis of the arc-arc junction: the Paleocene Nakanogawa Group in the Hidaka Belt, central Hokkaido, Japan. *Paleogeography, Paleoclimatology, Paleoecology*, **105**, 53–69.

NARUSE, H. 2003. Cretaceous to Paleocene depositional history of North-Pacific subduction zone: reconstruction from the Nemuro Group, eastern Hokkaido, northern Japan. *Cretaceous Research*, **24**, 55–71.

NARUSE, H. & OTSUBO, M. 2011. Heterogeneity of internal structures in a mass-transport deposit, Upper Cretaceous to Paleocene Akkeshi Formation, Hokkaido Island, northern Japan. *In*: SHIPP, R. C., WEIMER, P. & POSAMENTIER, H. W. (eds) *Mass-Transport Deposits in Deepwater Settings*. SEPM, Tulsa, Special Publications, **96**, 279–290.

NARUSE, H., MAEDA, H. & SHIGETA, Y. 2000. Newly discovered Late Cretaceous molluscan fossils and inferred K/T boundary in the Nemuro Group, eastern Hokkaido, northern Japan. *Journal of the Geological Society of Japan*, **106**, 161–164 [in Japanese with English abstract].

NIIDA, K. 1981. Geology of the Tokoro Belt, Hokkaido. *In*: HARA, I. (ed.) *Tectonics of Paired Metamorphic Belts*. Tanishi Print Kikaku, Hiroshima, 49–54.

NIIDA, K. 1992. Basalts and dolerites in the Sorachi-Yezo Belt, central Hokkaido, Japan. *Journal of the Faculty of Science, Hokkaido University, Series IV*, **23**, 301–319.

NIIDA, K. & KITO, N. 1986. Cretaceous arc-trench system in Hokkaido. *Monographs of the Association for Geological Collaboration in Japan*, **31**, 379–402.

NOKLEBERG, W. J., PARFENOV, L. M. ET AL. 2000. Phanerozoic tectonic evolution of the circum-North Pacific. USGS Professional Paper 1626.

OBATA, I., HAYAMI, I., MATSUKAWA, M., TERAOKA, Y. & TAKETANI, Y. 1993. A problematic bivalve from the Saroma Group of northeastern Hokkaido and its geological significance. *Memoirs of National Science Musium, Tokyo*, **26**, 31–37.

OKADA, H. 1971. Clastic sediments of the Cretaceous Yezo geosynclines. *Memoirs of the Geological Society of Japan*, **6**, 61–74.

OKADA, H., YAMADA, M., MATSUOKA, H., MUROTA, T. & INOSE, T. 1987. Calcareous nannofossils and biostratigraphy of the Upper Cretaceous and Lower Paleogene Nemuro Group, eastern Hokkaido, Japan. *Journal of the Geological Society of Japan*, **93**, 329–348.

OKADA, H., TARDUNO, J. A., NAKASEKO, K., NISHIMURA, A., SLITER, W. V. & OKADA, H. 1989. Microfossil assemblages from the Late Jurassic to Early Cretaceous Nikoro pelagic sediments, Tokoro Belt, Hokkaido, Japan. *Memoirs of Faculty of Science, Kyushu University Series D, Geology*, **26**, 193–214.

ONO, M. 2002. Uplift history of the Hidaka Mountains: fission-track analysis. *Bulletin of the Earthquake Institute, University of Tokyo*, **77**, 123–130 [in Japanese with English abstract].

OSANAI, Y., MIYASHITA, S., ARITA, K. & BAMBA, M. 1986. The metamorphism and thermal structure of the collisional terrain of a continental and oceanic crusts: a case of the Hidaka metamorphic belt, Hokkaido, Japan. *Monograph of the Association for the Geological Collaboration in Japan*, **31**, 205–222 [in Japanese with English abstract].

OSANAI, Y., KOMATSU, M. & OWADA, M. 1991. Metamorphism and granite genesis in the Hidaka Metamorphic Belt, Hokkaido, Japan. *Journal of Metamorphic Geology*, **9**, 111–124.

OSANAI, Y., OWADA, M. & KAWASAKI, T. 1992. Tertiary deep crustal ultrametamorphism in the Hidaka metamorphic belt, northern Japan. *Journal of Metamorphic Geology*, **10**, 401–414.

OTA, T., SAKAKIBARA, M. & ITAYA, T. 1993. K-Ar ages of the Kamuikotan metamorphic rocks in Hokkaido, Japan. *Journal of the Geological Society of Japan*, **99**, 335–345 [in Japanese with English abstract].

OWADA, T., OSANAI, Y. & KAGAMI, H. 1991. Timing of anatexis in the Hidaka metamorphic belt, Hokkaido, Japan. *Journal of the Geological Society of Japan*, **97**, 751–754.

OWADA, T., OSANAI, Y. & KAGAMI, H. 1997. Rb-Sr isochron ages for hornblende tonalite from the southeastern part of the Hidaka metamorphic belt, Hokkaido, Japan: implication for timing of peak metamorphism. *Memoirs of the Geological Society of Japan*, **47**, 21–17.

PARFENOV, L. M. & NATAL'IN, B. A. 1986. Mesozoic tectonic evolution of northeastern Asia. *Tectonophysics*, **127**, 291–304.

PISKUNOV, B. N. & KHVEDCHUK, I. I. 1976. New compositional and age data on the deposits of Moneron Island, northern part of the Sea of Japan. *Doklay Akademii Nauk SSSR*, **226**, 647–650.

RESEARCH GROUP OF THE TOKORO BELT 1984. Petrographic constitution of the Nikoro Group and the significance of unconformity at the base of the Saroma Group, Tokoro Belt, Hokkaido. *Earth Science ('Chikyu Kagakui')*, **38**, 408–419.

ROSER, B., UEDA, H. & MAEHARA, K. 1999. Major and trace element compositions of Cretaceous-Paleocene sandstones and mudrocks from the Idonnappu Zone and Yezo Supergroup, Urakawa, Hokkaido. *Geoscience Report of the Shimane University*, **18**, 11–24.

ROSER, B. P., UEDA, H. & KAWAMURA, M. 2003. Whole-rock analyses of sandstones and mudrocks from the Oshima Belt, Oshima Peninsula, SW Hokkaido. *Geoscience Report of the Shimane University*, **22**, 93–105.

SAHEKI, K., SHIBA, M., ITAYA, T. & ONUKI, H. 1995. K-Ar ages of the metamorphic and plutonic rocks in the southern part of the Hidaka belt, Hokkaido and their implications. *Journal of Mineralogy, Petrology, and Economic Geology*, **90**, 297–309.

SAITO, T., YAMANOI, T. & KAIHO, K. 1986. End-Cretaceous devastation of terrestrial flora in the boreal Far East. *Nature*, **323**, 253–255.

SAKAGAMI, S. & SAKAI, A. 1979. Triassic bryozoans from the Hidaka Group in Hokkaido, Japan. *Transactions and Proceedings of the Palaeontological Society of Japan*, **114**, 77–86.

SAKAGAMI, S., MINAMIKAWA, S. & KAWASHIMA, M. 1969. Conodonts from the Kamiiso limestone and consideration of its geological age. *Journal of Geography*, **78**, 38–43 [in Japanese with English abstract].

SAKAI, A. & KANIE, Y. 1986. 1:50000 quadrangle series: Geology of the Nishicha district. Geological Survey of Japan [in Japanese with English abstract].

SAKAKIBARA, M. 1986*a*. A newly discovered high-pressure terrane in eastern Hokkaido, Japan. *Journal of Metamorphic Geology*, **4**, 401–408.

SAKAKIBARA, M. 1986*b*. Aragonite from the Nikoro greenstone in the Tokoro Belt, eastern Hokkaido, Japan. *Journal of Japanese Association of Mineralogy, Petrology, and Economic Geology*, **81**, 446–453.

SAKAKIBARA, M. 1991. Metamorphic petrology of the northern Tokoro metabasites, eastern Hokkaido, Japan. *Journal of Petrology*, **32**, 333–364.

SAKAKIBARA, M. & OTA, T. 1994. Metamorphic evolution of the Kamuikotan high-pressure and low-temperature metamorphic rocks in central Hokkaido, Japan. *Journal of Geophysical Research*, **99**, 22 221–22 235.

SAKAKIBARA, M. & TANAKA, K. 1986. Discovery of Inoceramus from the Saroma Group, the Tokoro Belt, northeast Hokkaido. *Earth Science ('Chikyu Kagaku')*, **40**, 205–206 [in Japanese].

SAKAKIBARA, M., NIIDA, K. ET AL. 1986. Nature and tectonic history of the Tokoro belt. *Monograph of the Association of Geological Collaboration in Japan*, **31**, 173–187.

SAKAKIBARA, M., ISOZAKI, T., NANAYAMA, F. & NARUI, E. 1993. Radiolarian age of greenrock-chert-limestone sequence and its accretionary process of the Nikoro Group in the Tokoro belt, eastern Hokkaido, Japan. *Journal of the Geological Society of Japan*, **99**, 615–627 [in Japanese with English abstract].

SAKAKIBARA, M., HORI, R. S., IKEDA, M. & UMEKI, M. 1997*a*. Petrologic characteristics and geologic age of green rocks including chert xenoliths in the Pippu area, central Hokkaido, Japan. *Journal of the Geological Society of Japan*, **103**, 953–963 [in Japanese with English abstract].

SAKAKIBARA, M., OFUKA, H., KIMURA, G., ISHIZUKA, H., MIYASHITA, S., OKAMURA, M. & MELINIKOV, O. A. 1997*b*. Metamorphic evolution of the Susunai metabasites in southern Sakhalin, Russian Republic. *Journal of Metamorphic Geology*, **15**, 565–580.

SAKAKIBARA, M., HORI, R. S., KIMURA, G., IKEDA, M., KOUMOTO, T. & KATO, H. 1999. The age of magmatism and petrochemical characteristics of the Sorachi plateau reconstructed in Cretaceous accretionary complex, central Hokkaido, Japan. *Memoirs of the Geological Society of Japan*, **52**, 1–15 [in Japanese with English abstract].

SANO, S. 1995. Litho- and biofacies of Early Cretaceous rudist-bearing carbonate sediments in northeastern Japan. *Sedimentary Geology*, **99**, 179–189.

SANO, S., SUGISAWA, T. & SHIMAGUCHI, T. 2009. Discovery of megalodontoid bivalves in the Shiriya area, northern Honshu, Northeast Japan, and its geological implications. *Memoirs of the Fukui Prefectural Dinosaur Museum*, **8**, 51–57 [in Japanese with English abstract].

SHIBATA, K., UCHIUMI, S., UTO, K. & NAKAGAWA, T. 1984. K-Ar age results-2: new data from the Geological Survey of Japan. *Bulletin of Geological Survey of Japan*, **35**, 331–340 [in Japanese with English abstract].

SHIMURA, T., OWADA, M., OSANAI, Y., KOMATSU, M. & KAGAMI, H. 2004. Variety and genesis of the pyroxene-bearing S- and I-type granitoids from the Hidaka Metamorphic Belt, Hokkaido, northern Japan. *Transaction of the Royal Society of Edinburgh: Earth Sciences*, **95**, 161–179.

SIMANENKO, V. P., RASSKAZOV, S. V., YASNYGINA, T. A., SIMANENKO, L. F. & CHASHCHIN, A. A. 2011. Cretaceous complexes of the frontal zone of the Moneron-Samarga island arc: geochemical data on the basalts from the deep borehole on Moneron Island, the Sea of Japan. *Russian Journal of Pacific Geology*, **5**, 26–46.

SIMURA, R. & OZAWA, K. 2006. Mechanism of crystal redistribution in a sheet-like magma body: constraints from the Nosappumisaki and other shoshonite intrusions in the Nemuro Peninsula, northern Japan. *Journal of Petrology*, **47**, 1809–1851.

SUZUKI, J., BAMBA, T. & SUZUKI, Y. 1965. On the viridine quartz schist from the Chiei area, Hidaka province, Hokkaido, Japan. *Proceedings of the Japan Academy*, **41**, 722–727.

SUZUKI, N., EHIRO, M., YOSHIHARA, K., KIMURA, Y., KAWASHIMA, G., YOSHIMOTO, H. & NOGI, T. 2007. Geology of the Kuzumaki-Kamaishi Subbelt of the North Kitakami Belt (a Jurassic accretionary complex), Northeast Japan: case study of the Kawai-Yamada area, eastern Iwate Prefecture. *Bulletin of the Tohoku University Museum*, **6**, 103–174.

TAJIKA, J. 1988. Stratigraphy and structure of the Upper Cretaceous Yubetsu Group, Tokoro Belt, eastern Hokkaido – an application of trench accretion model. *Journal of the Geological Society of Japan*, **94**, 817–836 [in Japanese with English abstract].

TAJIKA, J. 1989. Cretaceous accretionary complex in the northern Hidaka Belt. *Monthly Chikyu*, **11**, 323–327 [in Japanese].

TAJIKA, J. 1992. Modal and chemical compositions of the 'Paleogene' sandstones from the northern Hidaka belt, Hokkaido. *Memoirs of the Geological Society of Japan*, **38**, 13–26 [in Japanese with English abstract].

TAJIKA, J. & IWATA, K. 1983. Occurrence of Cretaceous radiolarians from the Hidaka Supergroup around Maruseppu, northeast Hokkaido. *Journal of the Geological Society of Japan*, **89**, 535–538 [in Japanese].

TAJIKA, J. & IWATA, K. 1990. Paleogene melange of the northern Hidaka Belt – Geology and radiolarian age of the Kamiokoppe Formation. *Journal of Hokkai-Gakuen University*, **66**, 35–55.

TAJIKA, J., IWATA, K. & KUROSAWA, K. 1984. Geology of the Mesozoic System around Mt. Obira, Shimamaki, southwest Hokkaido. *Earth Science ('Chikyu-Kagaku')*, **38**, 397–407 [in Japanese with English abstract].

TAKASHIMA, R., YOSHIDA, T. & NISHI, H. 2001. Stratigraphy and sedimentary environments of the Sorachi and Yezo Groups in the Yubari-Ashibetsu area, Hokkaido, Japan. *Journal of the Geological Society of Japan*, **107**, 359–378 [in Japanese with English abstract].

TAKASHIMA, R., NISHI, H. & YOSHIDA, T. 2002. Geology, petrology, and tectonic setting of the Late Jurassic ophiolite in Hokkaido, Japan. *Journal of Asian Earth Sciences*, **21**, 197–215.

TAKASHIMA, R., KAWABE, F., NISHI, H., MORIYA, K., WANI, R. & ANDO, H. 2004. Geology and stratigraphy of forearc basin sediments in Hokkaido, Japan: cretaceous environmental events on the north-west Pacific margin. *Cretaceous Research*, **25**, 365–390.

TAMURA, M. 1990. The distribution of Japanese Triassic bivalve faunas with special reference to parallel distribution of inner Arcto-Pacific fauna and outer Tethyan fauna in Upper Triassic. *In*: ICHIKAWA, K., MIZUTANI, S., HARA, I., HADA, S. & YAO, A. (eds) *Pre-Cretaceous Terranes of Japan, IGCP Project No.224: Pre-Jurassic Evolution of Eastern Asia*. Nippon Insatu Shuppan C. Ltd, Osaka, 347–359.

TANAKA, S., KIZAKI, K. & MIYASHITA, S. 2012. Geology and metamorphism in the northern Poroshiri ophiolite, Hokkaido, Japan. *Journal of the Geological Society of Japan*, **118**, 723–740 [in Japanese with English abstract].

TEREKHOV, E. P., MOZHEROVSKY, A. V., GOROVAYA, M. T., TSOY, I. B. & VASHCHENKOVA, N. G. 2010. Composition of the rocks of the Kotikovo Group and the main stages in the Late Cretaceous-Paleogene evolution of the Terpeniya Peninsula, Sakhalin Island. *Russian Journal of Pacific Geology*, **4**, 260–273.

TOYOSHIMA, T. 1990. Pseudotachylite from the Main Zone of the Hidaka metamorphic belt, Hokkaido, northern Japan. *Journal of Metamorphic Geology*, **8**, 507–523.

TOYOSHIMA, T., KOMATSU, M. & SHIMURA, T. 1994. Tectonic evolution of lower crustal rocks in an exposed magmatic arc section in the Hidaka metamorphic belt, Hokkaido, northern Japan. *Island Arc*, **3**, 182–198.

TOYOSHIMA, T., OBARA, T. ET AL. 2004. Pseudotachylytes, related fault rocks, asperities, and crustal structures in the Hidaka metamorphic belt, Hokkaido, northern Japan. *Earth Planets Space*, **56**, 1209–1215.

TSUCHIYA, N. 2010. Cretaceous igneous rocks in the Oshima Belt. *In*: The Geological Society of Japan (ed.) *Regional Geology of Japan 1, Hokkaido District*. Asakura Publishing, Tokyo, 87–92.

TSUCHIYA, N., MIKI, J., NISHIKAWA, J. & HASHIMOTO, M. 1986. Cretaceous plutonic rocks in southwestrn Hokkaido – continental amrgin type magmatism related to the subduction in Cretaceous time. *Monographs of the Association for Geological Collaboration in Japan*, **31**, 33–50 [in Japanese with English abstract].

TSUMURA, N., IKAWA, H. ET AL. 1999. Delamination wedge structure beneath the Hidaka Collision Zone, Central Hokkaido, Japan, inferred from seismic reflection profiling. *Geophysical Research Letters*, **26**, 1057–1060.

TSVETKOV, A. A., GOVOROV, G. I., TSVETKOVA, M. V. & ARAKELYANTS, M. M. 1986. The evolution of magmatism in the Lesser Kuril Ridge of the Kuril island-arc system. *International Geology Review*, **28**, 180–196.

TUMANDA, F. P. 1989. Cretaceous radiolarian biostratigraphy in the Esashi Mountain area, Northern Hokkaido, Japan. *Science Report of the Institute of Geoscience, University of Tsukuba, Sec. B*, **10**, 1–44.

TUMANDA, F. & SASHIDA, K. 1988. On the occurrence of inoceramid bivalves and radiolarians from the Manokawa Formation, Esashi Mountain area, northern Hokkaido. *Annual Report of the Institute of Geoscience, University of Tsukuba*, **14**, 56–63.

UEDA, H. 2005. Accretion and exhumation structures formed by deeply subducted seamounts in the Kamuikotan high-P/T Zone, Hokkaido, Japan. *Tectonics*, **24**, TC2007.

UEDA, H. 2006. Geologic structure of Cretaceous accretionary complexes in the frontal Hidaka collision zone, Hokkaido, Japan. *Journal of the Geological Society of Japan*, **112**, 699–717 [in Japanese with English abstract].

UEDA, H. & MIYASHITA, S. 2005. Tectonic accretion of a subducted intraoceanic remnant arc in Cretaceous Hokkaido, Japan, and implications for evolution of the Pacific northwest. *Island Arc*, **14**, 582–598.

UEDA, H., KAWAMURA, M. & NIIDA, K. 2000. Accretion and tectonic erosion processes revealed by the mode of occurrence and geochemistry of greenstones in the Cretaceous accretionary complexes of the Idonnappu Zone, southern central Hokkaido, Japan. *Island Arc*, **9**, 237–257.

UEDA, H., KAWAMURA, M. & IWATA, K. 2001. Tectonic evolution of Cretaceous accretionary complex in the Idonnappu Zone, Urakawa area, central Hokkaido, Northern Japan: with reference to radiolarian ages and thermal structure. *Journal of the Geological Society of Japan*, **107**, 81–98.

UEDA, H., KAWAMURA, M. & YOSHIDA, K. 2002. Blueschist-bearing fluvial conglomerate and unconformity in the Cretaceous forearc sequence, south central Hokkaido, northern Japan: rapid exhumation of a high-P/T metamorphic accretionary complex. *Journal of the Geological Society of Japan*, **108**, 133–152.

USUKI, T., KAIDEN, H., MISAWA, K. & SHIRAISHI, K. 2006. Sensitive high-resolution ion microprobe U–Pb ages of the Latest Oligocene-Early Miocene rift-related Hidaka high-temperature metamorphism in Hokkaido, northern Japan. *Island Arc*, **15**, 503–516.

WATANABE, T., IWASAKI, I., UEDA, H. & KOITABASHI, S. 1994. K-Ar age and structure of Mesozoic accretionary prism in NW Pacific rim – a case study of the Kamuikotan metamorphic and Idonnappu belts. *In*: WU, H., TIAN, B. & LIU, Y. (eds) *IGCP Project 294 International Symposium, Very Low Grade Metamorphism: Mechanisms and Geological Application*. Seismological Press, Baijing, 145–155.

WATANABE, Y. & IWATA, K. 1985. Discovery of Paleogene radiolarians from the Yuyanbetsu Formation, central Hokkaido, and its geological significance. *Earth Science (Chikyu Kagaku)*, **39**, 446–452.

WATANABE, Y. & IWATA, K. 1987. The Hidaka Supergroup in the Tomuraushi region, Hidaka Belt, Hokkaido, Japan. *Earth Science (Journal of the Association for the Geological Collaboration in Japan)*, **41**, 35–47.

WATANABE, Y. & KIMURA, G. 1987. Strike-slip fault (Nayorogawa Fault) in northen Hokkaido. *Journal of the Geological Society of Japan*, **93**, 1–10.

YAGI, K. 1969. Geochemistry of the alkali dolerites of the Nemuro Peninsula, Japan. *Bulletin Volcanologique*, **33**, 1101–1117.

YAMAKITA, S. & OTOH, S. 1999. Reconstruction of the geological continuity between Primorye and Japan before the opening of the Sea of Japan.

Institute for Northeast Asian Studies Research Annual, **24**, 1–16 [in Japanese].

Yamakita, S. & Otoh, S. 2002. Intra-arc deformation of Southwest and Northeast Japan Arcs due to the collision of Izu-Bonin Arc and reconstruction of their connection before the opening of the Japan Sea. *Bulletin of the Earthquake Research Institute, University of Tokyo*, **77**, 249–266 [in Japanese with English abstract].

Yoshida, K., Taki, S., Iba, Y., Sugawara, M. & Hikida, Y. 2003. Discovery of serpentinite bearing conglomerate in the Lower Yezo Group, Hokkaido, northern Japan and its tectonic significance. *Journal of the Geological Society of Japan*, **109**, 336–344 [in Japanese with English abstract].

Yoshida, K., Iba, Y., Taki, S., Sugawara, M., Tsugane, T. & Hikida, Y. 2010. Deposition of serpentinites-bearing conglomerate and its implications for Early Cretaceous tectonics in northern Japan. *Sedimentary Geology*, **232**, 1–14.

Yoshida, T. & Aoki, C. 1972. Paleozoic formation in the Matsumae Peninsula, Hokkaido, and the occurrence of conodonts from the southern part of the Oshima Peninsula, Hokkaido. *Bulletin of the Geological Survey of Japan*, **23**, 635–646 [in Japanese with English abstract].

Zharov, A. E. 2005. South Sakhalin tectonics and geodynamics: a model for the Cretaceous-Paleogene accretion of the East Asian continental margin. *Russian Journal of Earth Sciences*, **7**, ES5002, http://doi.org/10.2205/2005ES000190

Zonenshain, L. P., Kuzmin, M. I. & Natapov, L. M. 1990. *Geology of the USSR: A Plate-Tectonic Synthesis*. American Geophysical Union, Washington, D.C.

3 Ophiolites and ultramafic rocks

AKIRA ISHIWATARI, KAZUHITO OZAWA, SHOJI ARAI, SATOKO ISHIMARU, NATSUE ABE & MIYUKI TAKEUCHI

This chapter describes the geological occurrences, mineralogical and geochemical diversity and petrogenetic importance of the mafic and ultramafic igneous (mainly plutonic) rocks in Japan. This chapter is divided into five parts: (1) time–space distribution of ophiolitic rocks in Japan (AI); (2) ophiolitic rocks in SW Japan (AI); (3) ophiolitic rocks in Hokkaido (AI); (4) ophiolitic rocks in NE Honshu (KO); and (5) ultramafic xenoliths from Japan (SA, SI, NA and MT).

Time–space distribution of ophiolitic rocks in Japan (AI)

Ophiolites are on-land remnants of the ancient oceanic lithosphere (oceanic crust and upper mantle) which were tectonically emplaced on the continental crust or into the orogenic belt. The most typical ophiolite on Earth is exposed in Oman, eastern Arabian Peninsula, where it measures 500 km long, 100 km wide, 15 km thick and comprises three major layers: a lower layer of peridotite or serpentinite (upper mantle); a middle layer of gabbro (lower crust); and an upper layer of basalt-dolerite occurring as pillow lava and dykes (upper crust). The close association and genetic kinship of the three rock layers was first recognized in the Alps by Steinmann (1927). Bailey & McCallien (1960) stressed their association with radiolarian chert (deep-sea sediments), and designated the assemblage of serpentinite, pillow lava and radiolarian chert as the Steinmann Trinity that characterizes deep-sea magmatic activity. The layers of the original Steinmann Trinity, that is, peridotite, gabbro and basalt, are now recognized as representing the seismically distinguished strata of the oceanic lithosphere, namely the upper mantle (Layer 4), lower crust (Layer 3) and upper crust (Layer 2), respectively. The overlying sediments (Layer 1) are typically represented by radiolarian chert that was deposited on the deep ocean floor below the carbonate compensation depth. The ophiolite members may form small bodies isolated either by tectonic destruction or erosion, such a body being referred to as a dismembered ophiolite. In some cases, numerous dismembered ophiolite bodies occur in the matrix of serpentinite or mudstone, and such a mixture is called ophiolite (serpentinite) mélange ('mixture' in French, a descriptive term without implication on its origin). The mixture is referred to as ophiolite olistostrome (or sedimentary mélange) if it originated in submarine landslide, or tectonic mélange if it formed by fragmentation along faults and shear zones. Isolated mantle peridotite bodies with fertile composition (lherzolite) have been referred to as Alpine-type peridotite, and are thought to represent a fragment of the upwelling mantle diapir. Ultramafic igneous rocks also occur as lavas, dykes and sills (picrite, komatiite, meimechite, etc.), Alaskan-type intrusions (dunite, wehrlite, clinopyroxenite, etc.) and the lower parts of continental layered intrusions, although the latter two types do not occur in Japan.

Ophiolites and ultramafic rocks are distributed all over Japan (Fig. 3.1), with the best-developed examples occurring as large thrust slices (nappes) at Yakuno (SW Japan), Horokanai and Poroshiri (Hokkaido). Dismembered ophiolite nappes (mostly peridotite/serpentinite) include those of Oeyama (SW Japan) and Hayachine–Miyamori (NE Honshu). Ophiolite mélanges are also common (Omi–Renge, Kurosegawa, Mikabu, Setogawa–Mineoka (SW Japan), Motai (NE Honshu), Kamuikotan and Tokoro (Hokkaido)), and Alpine-type (Horoman) lherzolite is present in Hokkaido. Picritic lavas are abundant in the Sorachi, Kamuikotan and Mikabu ophiolites as well as within greenstone bodies in the Mino–Tamba–Ashio–Chichibu Jurassic accretionary complexes of SW Japan.

Ophiolites in SW Japan form distinct belts running subparallel to the general elongation of the islands, and their ages generally become younger towards the Pacific Ocean. The oldest (Early Palaeozoic) Oeyama Ophiolite is distributed along the Japan Sea coast, followed oceanwards by the Late Palaeozoic Yakuno Ophiolite, the Jurassic Mikabu Ophiolite and finally the Cenozoic Mineoka Ophiolite, which is exposed in the Miura and Boso peninsulas to the south of Tokyo. Structurally, the oldest ophiolite occupies the highest position with the youngest ophiolite lowest (Fig. 3.2), obeying the general rule of tectonic superposition (underplating of younger units) of accreted sediments in subduction zones.

The overall geological structure in NE Japan is the same as that in SW Japan, with an Early Palaeozoic (Miyamori–Hayachine) ophiolite forming the basement of a nappe (South Kitakami) which occupies the highest structural position, and this ophiolite is underlain successively by younger accretionary complexes such as the Carboniferous Nedamo Belt, the Jurassic North Kitakami–SW Hokkaido (Oshima) Belt and the Cretaceous Hidaka Belt in eastern Hokkaido. The Horokanai Ophiolite and Sorachi metabasalts (greenstones) in central Hokkaido occupy the position between Jurassic and Cretaceous accretionary complexes. The Hidaka Mountains in south-central Hokkaido mark the site of a relatively minor Miocene orogenic movement caused by collision of the Honshu and Chishima (Kuril) arcs, and westwards thrusting of the Kuril Arc locally disturbs the generally eastwards-thrusting nappe pile typical of NE Japan. The Alpine-type Horoman peridotite body was emplaced as a sheet-like solid intrusion along the west-vergent Hidaka Main Thrust. This very fresh peridotite (lherzolite) body provides an excellent manifestation of the magmatic and tectonic processes in the mantle, as well as attractive scenery in the Mt Apoi Geopark with a characteristic flora growing on the ultramafic soil. In the Tohoku–Hokkaido area of NE Japan there is also a nappe pile of accretionary complexes and ophiolites with downwards (oceanwards) -younging polarity. Although there are some irregularities, the rule of downwards (oceanwards) -younging is similarly valid in other circum-Pacific orogenic belts such as those in eastern Russia (Koryak Mountains) and western USA (Klamath Mountains) (Irwin 1981; Ishiwatari 1991; Ishiwatari *et al.* 2003). These

From: Moreno, T., Wallis, S., Kojima, T. & Gibbons, W. (eds) 2016. *The Geology of Japan*. Geological Society, London, 223–250.

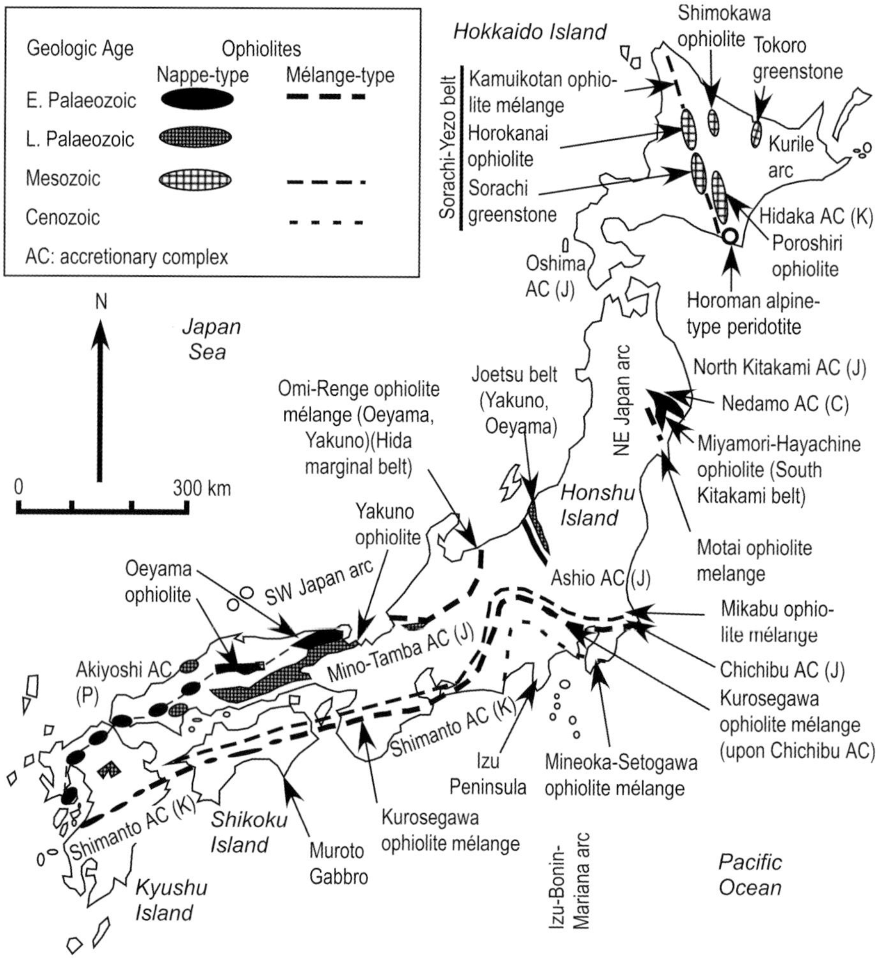

Fig. 3.1. Distribution of ophiolitic complexes in Japan (modified after Ishiwatari & Tsujimori 2003). Ages of accretionary complexes are indicated as follows: C, Carboniferous; P, Permian; J, Jurassic; K, Cretaceous–Cenozoic.

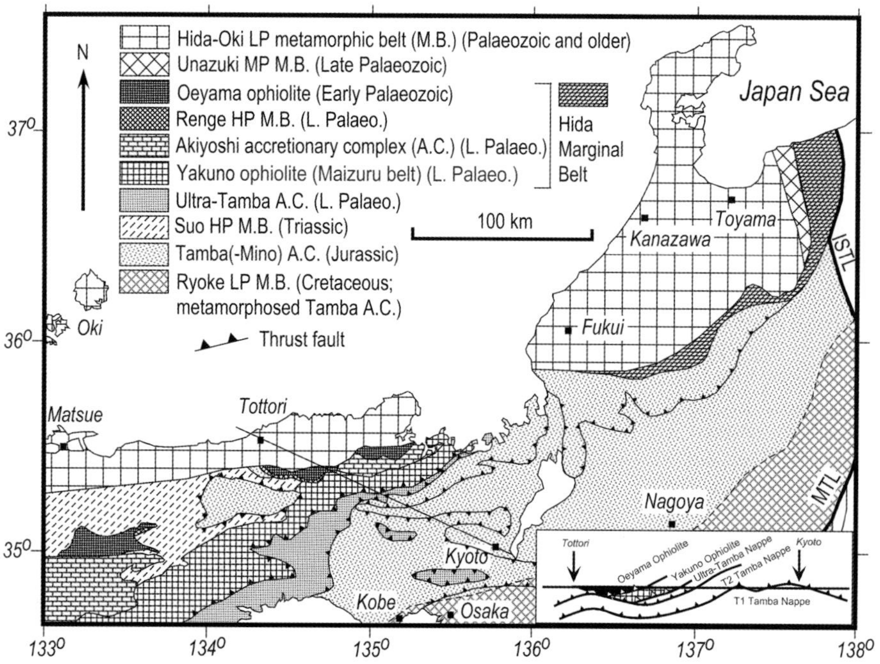

Fig. 3.2. Tectonic map and schematic cross-section of the Inner Zone of SW Japan around Osaka, Kyoto and Nagoya, showing superposition of gently folded nappes of ophiolites, high-pressure metamorphic rocks and accretionary complexes with downwards-younging polarity.

circum-Pacific multiple ophiolite belts (Ishiwatari 1991, 1994) can be distinguished from the intra-continental ophiolite belts (Appalachian, Caledonian, Urals, Alps, Mediterranean, etc.) where ophiolite ages tend to be single and uniform over a wide area.

Dilek & Furnes (2011) have proposed a new ophiolite classification based on the tectonic setting of their original birthplace and define: (1) continental rift-passive margin type (e.g. Alps); (2) mid-ocean ridge (MOR) type (e.g. Maquary Island); (3) plume type (e.g. Gorgona Island, Mino–Tamba greenstone); (4) supra-subduction zone (SSZ) type (e.g. Oman, Troodos); and (5) active continental margin type (e.g. Sierra Nevada). The former three types originate in the subduction-free environment, but the latter two types are formed by the subduction zone magmatism. Japanese ophiolites are diverse, and sometimes still controversial, in their likely origin. For example, the Yakuno Ophiolite in SW Japan includes both transitional MORB-type and island-arc-type segments (Ishiwatari *et al.* 2003), and ophiolites and greenstones in the Sorachi–Yezo Belt, central Hokkaido, include the Horokanai Ophiolite which shows MORB-type basalt chemistry (Ishizuka 1985, 1987), metabasalts and picrites of oceanic superplume origin (Tatsumi *et al.* 1998; Ichiyama *et al.* 2012) and a boninite-bearing forearc origin (Ueda & Miyashita 2005).

Ophiolitic rocks in SW Japan (AI)

Inner Zone

The geology of SW Japan is divided into the Inner and Outer zones by the Median Tectonic Line (MTL), a 1000 km long active fault that marks the boundary between the Ryoke low-*P* metamorphic belt to the north and the Sanbagawa high-*P* metamorphic belt to the south, both of Cretaceous age. Ophiolites in the Inner Zone are mostly of Palaeozoic age, whereas those in the Outer Zone yield mostly Mesozoic and Cenozoic ages. Ophiolites in SW Japan form a typical circum-Pacific Phanerozoic multiple ophiolite belt with general oceanwards (structurally downwards)-younging of the ophiolite ages from Early Palaeozoic to Cenozoic times (Ishiwatari 1991) (Figs 3.1 & 3.2).

Oeyama Ophiolite

This unit is of Early Palaeozoic age and occupies the innermost part of SW Japan at the structurally highest position. It is mostly composed of mantle peridotite that is serpentinized to various degrees and thermally metamorphosed by Cretaceous granite intrusions (Arai 1973, 1974, 1975, 1980; Uda 1984; Kurokawa 1985). Outcrops are mostly distributed in the eastern Chugoku Mountains (such as Tari-Misaka, Ashidachi, Wakasa, Sekinomiya, Izushi and Oeyama), but they extend to the west as far as northern Kyushu and to the east in the Hida marginal and Joetsu belts (Fig. 3.1).

The ultramafic rocks of the Oeyama body are mostly serpentinized, but their protoliths are sometimes preserved. The dominant rock type is the mantle peridotitc such as lherzolite, harzburgite and dunite (Fig. 3.3). The ophiolite includes metamorphosed cumulate rocks in the Fuko Pass area of the easternmost Oeyama body (Kurokawa 1985; Tsujimori & Ishiwatari 2002). Although the peridotites were intruded by many doleritic (diabasic) plutons, volcanic components of this ophiolite have not so far been recorded. It should be noted, however, that a mafic rock complex exposed at Saijo, eastern Hiroshima Prefecture, comprises metabasalt, metadolerite and metagabbros with some chert and tuff, and might belong to the Oeyama Ophiolite due to its structural position overlying the Permian Akiyoshi accretionary complex and its position adjacent to the Tari–Misaka serpentinite body of the Oeyama Ophiolite.

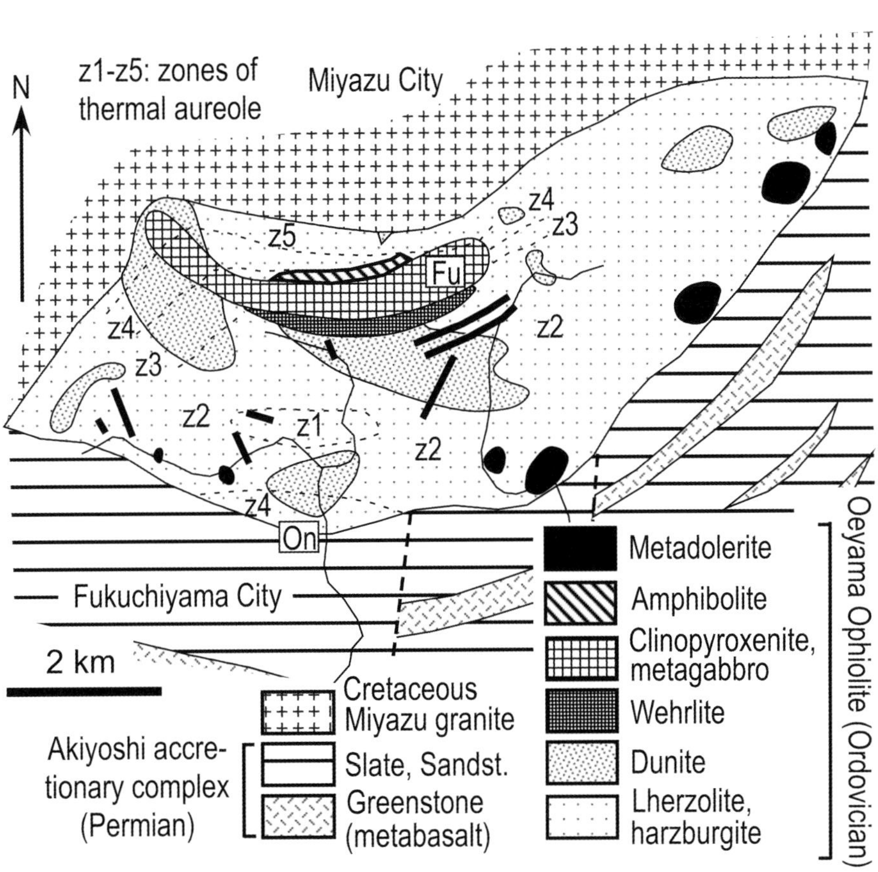

Fig. 3.3. Geological map of the Oeyama ophiolitic body in northern Kyoto Prefecture, Japan (after Uda 1984). Fu, Fuko Pass; On, Onigajaya. Zone 1 (z1) to Zone 5 (z5) are metamorphic mineral zones with increasing temperature, forming the contact aureole of the Cretaceous granite.

Residual mantle peridotite of the Oeyama Ophiolite is relatively fertile (i.e. little basaltic partial melt has been extracted from it) and is grouped into harzburgite (spinel Cr#50) and lherzolite (spinel Cr#30; where Cr# = 100Cr/(Al + Cr)). The lherzolite comprises only the Oeyama and Izushi bodies, while all other bodies are dominated by harzburgite. The harzburgite and lherzolite occur together in the Kotaki area, Hida Gaien Belt (Machi & Ishiwatari 2010), with both being characterized by unique occurrences of spinel in the form of vermicular microscopic aggregates with pyroxene.

The serpentinite shows various degrees of recrystallization due to thermal metamorphism by Cretaceous granite. As the serpentinite contains abundant H_2O, mineral recrystallization is well developed by dehydration. In general, metamorphic temperature decreases away from the granite (from 800°C to less than 200°C), producing a concentric zonation. The zonal mineral assemblages occur as: Zone 1, serpentine without recrystallized olivine; Zone 2, antigorite-olivine-clinopyroxene; Zone 3, antigorite-olivine-tremolite; Zone 4, olivine-talc-tremolite; and Zone 5, olivine-orthopyroxene-tremolite (or hornblende) (Fig. 3.3). Anthophyllite or cummingtonite may occur between zones 4 and 5 in some cases. The olivine in zones 2 and 3 characteristically shows partings in three perpendicular directions (called 'cleavable olivine').

An Early Palaeozoic (Cambro Ordovician) age for the Oeyama Ophiolite has been deduced from K–Ar data on gabbroic mafic intrusions (*c.* 450 Ma) and zircon U–Pb data on jadeitic metamorphic rocks (530 Ma) occurring as blocks (or 'knockers') in the associated mélange. Metagabbros in the Nomo Peninsula, Kyushu are also >500 Ma. In the circum-Pacific orogenic belts, ophiolites with similar ages are found in northeastern Russia (Sergeevka massif, Primorye and Ust'-Belaya and Ganychalan ophiolites, Koryak Mountains), western USA (Trinity ophiolite) and eastern Australia and Tasmania (Ishiwatari 1991; Ishiwatari *et al.* 2003). In continental settings, Early Palaeozoic ophiolites are abundant in the Appalachian–Caledonian–Urals belts, the central Asian Orogenic Belt and in Qilianshan, China.

The Oeyama Ophiolite is underlain by serpentinite mélange with blocks of Renge high-pressure metamorphic rocks (blueschist, metagabbro, garnet amphibolite, jadeitite, etc.) yielding Carboniferous ages of *c.* 300 Ma. The Osayama Mélange in the Chugoku Mountains (Okayama Prefecture) as mapped by Tsujimori (1998) (Fig. 3.4) is a typical example of this type of unit, and analogous mélanges are reported from the Oya area (Hyogo Prefecture) and Omi–Kotaki area (Niigata and Nagano prefectures).

Yakuno Ophiolite

The Yakuno Ophiolite preserves a complete sequence of mantle peridotite, cumulate ultramafic rocks, gabbroic rocks, doleritic and basaltic rocks and associated black mudstone (Fig. 3.5). The mantle peridotite is mostly composed of foliated, coarse-grained harzburgite banded with dunite and orthopyroxenite. The harzburgite, having 2–3 volume per cent of clinopyroxene, a Fo_{90-91} olivine and a Cr#70 spinel, represents a moderately depleted residue of partially melted mantle. The ultramafic cumulates are dunite, wehrlite and clinopyroxenite, showing clinopyroxene-type crystallization sequence. The boundary between the ultramafic cumulate rocks and the overlying gabbro is gradual, and alternating layers of gabbro, pyroxenite and peridotite are well exposed. This part of the ophiolite succession corresponds to the Moho discontinuity that marks the crust–mantle boundary. The rocks at the Moho level of the Yakuno Ophiolite are characterized by the reaction relationship between olivine and plagioclase as well as the occurrence of aluminous (green) spinel and very aluminous pyroxene, suggesting that the Moho lay as deep as 20–25 km. The upper part of the gabbro is devoid of ultramafic rocks, and is cut by basalt/dolerite

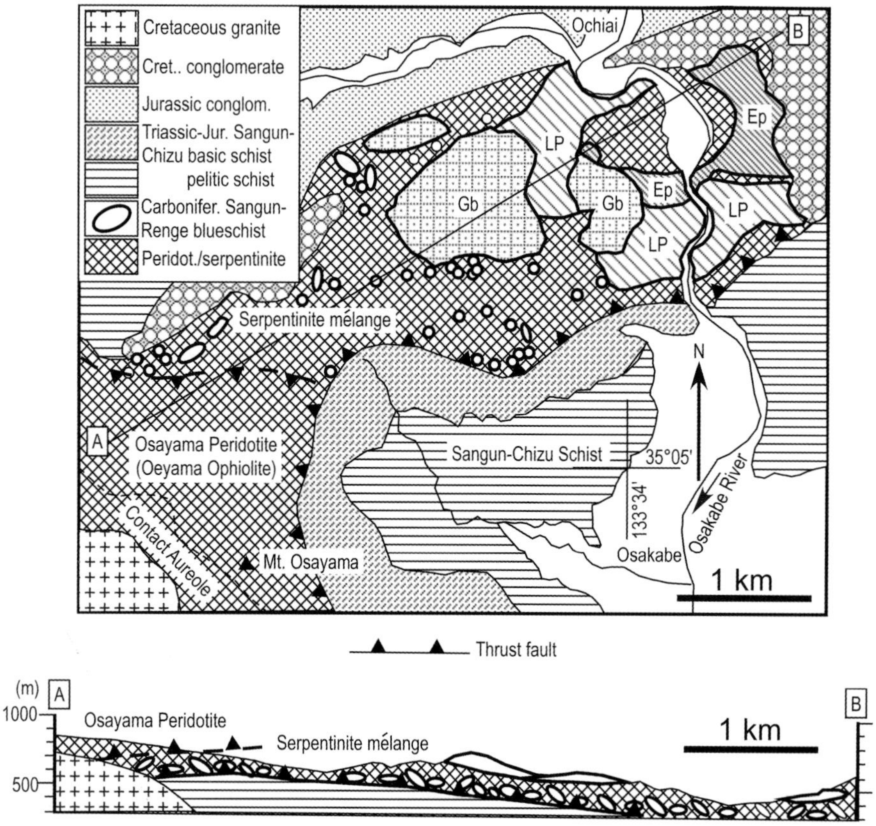

Fig. 3.4. Geological map of the serpentinite mélange associated with the Oeyama Ophiolite at Osayama, Okayama Prefecture, Japan (Tsujimori 1998).

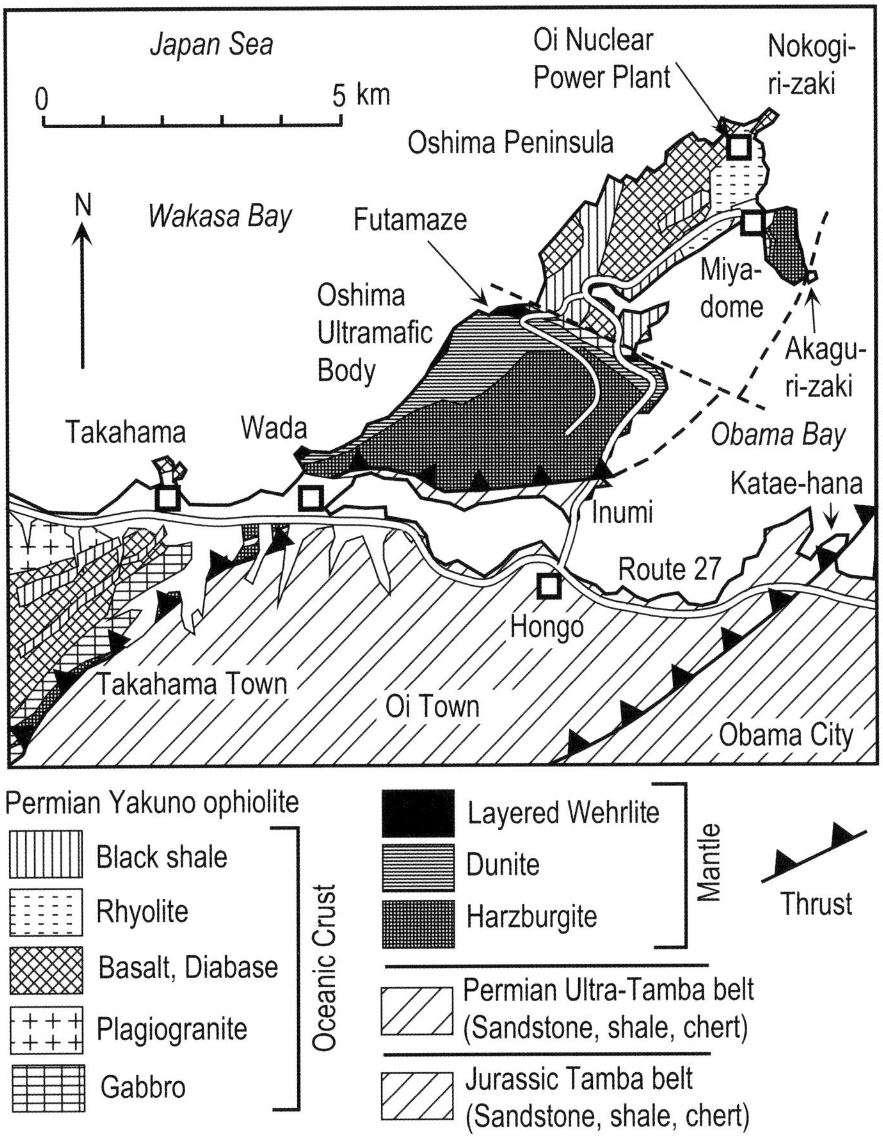

Fig. 3.5. Geological map of the Yakuno ophiolite in Fukui Prefecture, central Japan (after Ishiwatari 1985*a*, *b*; Ishiwatari & Hayasaka 1992).

dykes that are metamorphosed to amphibolite. The gabbro is also partly recrystallized to metagabbroic amphibolite. Intrusive bodies of plagiogranite of various sizes are abundant at or near the gabbro/basalt boundary. The basalt occurs mostly as massive lava flow but minor pillow lava also occurs. These basaltic lavas apparently flowed on or into the unconsolidated black mudstone and form basalt-mud mixtures. The basalt is mostly aphyric, but in some cases minor olivine (altered), plagioclase (altered), clinopyroxene (fresh) and spinel (fresh) occur as phenocrysts. Some rhyolite bodies are present in the basalt but intermediate rock (andesite) is rare, indicating essentially bimodal volcanism. Radiolarian fossils obtained from the black mudstone indicate a Middle Permian age, and radiometric data (zircon U–Pb and hornblende K–Ar) of the granitic and gabbroic rocks have yielded Early–Middle Permian ages. The association with black mudstone indicates that the basaltic lavas erupted near a land mass, and the basalt chemistry closely resembles that of marginal (back-arc) basin basalt (Ichiyama & Ishiwatari 2004). Gabbroic rocks in the Kamigori and Ohara areas, about 200 km to the SW, may have formed from the island-arc magma in view of the high An content of plagioclase and low Mg# of coexisting clinopyroxene. The Yakuno Ophiolite as a whole may therefore represent a Permian back-arc system subsequently thrust over the Late Permian accretionary complex represented by the Ultra-Tamba nappe (Fig. 3.5), and is itself tectonically overlain by the Renge metamorphic rocks (Carboniferous) and the Oeyama Ophiolite (Early Palaeozoic) (Fig. 3.2) (Ishiwatari *et al.* 1999).

Outer Zone

Mikabu Ophiolite Mélange

This mélange crops out as a narrow (maximum 5 km, generally 1–2 km width) but distinctive belt that extends for *c.* 800 km along the Pacific coast from the Kanto Mountains near Tokyo through the Hamamatsu and Toba (Kii Peninsula) areas to the western end of Shikoku Island (Fig. 3.1). This mélange is tectonically underlain by the Sanbagawa high-pressure metamorphic belt to the north, and is tectonically overlain by the Chichibu accretionary complex to the south. This mélange was first identified by Ernst (1972) as 'Permian oceanic crust', but later radiolarian and isotopic geochronology indicated its Jurassic age (Ichiyama *et al.* 2014). No mantle peridotite has been so far discovered from this mélange, but cumulate mafic-ultramafic rocks such as gabbro, troctolite, dunite, wehrlite, clinopyroxenite and their hornblende-bearing varieties are abundant (Nakamura 1971). These plutonic rocks contain minor orthopyroxene (Mizukami 2002) and rare inverted pigeonite (Nakayama *et al.* 1973), suggesting their crystallization from tholeiitic magma. Basalts include both tholeiitic and alkaline affinities, and ultramafic volcanic rocks (picrite) of komatiitic composition are also abundant (Ichiyama *et al.* 2014). The most prominent geological feature of the Mikabu unit is its highly fragmented occurrence (Fig. 3.6). The ultramafic rocks, gabbro and pillow lavas occur as blocks 1–1000 m in size within a matrix of so-called 'gabbro sandstone' or 'gabbro conglomerate' (Inomata 1978; Iwasaki 1984). The gabbro sandstone consists of mafic clastic grains with abundant pyroxene, often showing ophitic (doleritic) or granular (gabbroic) texture. Schistosity is developed in some places and alkali pyroxene and sodic amphibole occur as metamorphic minerals, suggesting that this ophiolite has been overprinted by high-*P* (Sanbagawa?) metamorphism. The abundance of ultramafic volcanic rocks and their komatiitic chemistry suggest an origin in accreted oceanic plateau, with a similar age and origin as the much less fragmented Sorachi Belt in Hokkaido (Ichiyama *et al.* 2014). The mechanism responsible for the intense fragmentation of the Mikabu ophiolitic rocks remains enigmatic.

Kurosegawa Ophiolitic Mélange

This unit occurs in the Chichibu accretionary complex of Jurassic accretion age, and extends for 1000 km from the Kanto Mountains in a narrow, discontinuous outcrop through Honshu and Shikoku to Kyushu, where it is most widely developed. This is essentially a serpentinite mélange (Maruyama 1981), although serpentinite is far less abundant than in other, more typical serpentinite mélanges. The allochthonous blocks included in this mélange include: (1) fossil-bearing limestone and other sedimentary rocks (from Silurian to Cretaceous age without any major age gaps or metamorphism); and (2) a diverse array of Mitaki granitic rocks (Palaeozoic), Terano metamorphic rocks (garnet-biotite gneiss, amphibolite, rare granulite or charnockite), crystalline schists (with high-*P* mineral assemblages of Palaeozoic–Mesozoic ages) as well as greenstones, gabbros and peridotite all enveloped by films of serpentinite. The peridotite bears cleavable olivine and vermicular spinel as seen in the Oeyama Ophiolite, and radiometric ages of the metamorphic rocks and granites range from 450 to 180 Ma. The extensive Kurosegawa outcrop in Kyushu is composed of three parts: (1) serpentinite mélange with 400 Ma igneous rocks (granites), metamorphic rocks (of amphibolite and granulite facies) and high-*P* schists (Miyazoe *et al.* 2012 also report peculiar metasomatic rock (basic schist) with epidote including 17 wt% SrO); (2) Silurian–Cretacous shallow-marine sediments with abundant limestone with fossils; and (3) Permian accretionary complex (Hara *et al.* 2013 recently

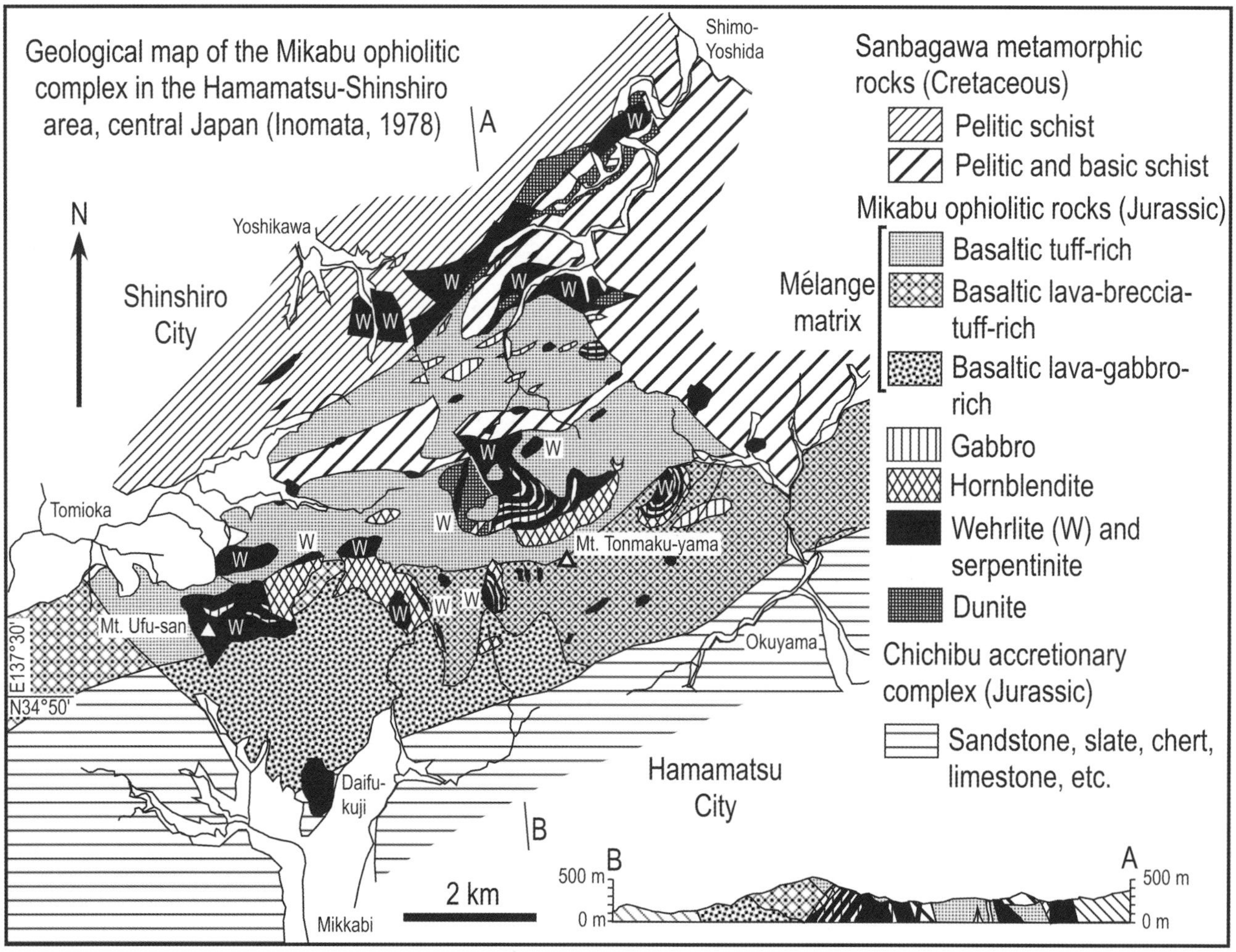

Fig. 3.6. Geological map of the Mikabu ophiolitic complex in the Hamamatsu–Shinshiro area, central Japan (after Inomata 1978).

found Early Jurassic accretionary complex from the Kurosegawa mélange in eastern Shikoku). All these units tectonically overlie the Jurassic accretionary complex of the Chichibu Belt (Saito *et al.* 2005). The Kurosegawa unit is therefore far more complex than the Mikabu Ophiolitic Mélange, and contains various rocks of diverse lithology and age. Maruyama *et al.* (1984) proposed that the Kurosegawa mélange represents an exotic terrain that collided with Japan in Cretaceous times, whereas Isozaki (1996) considered the mélange as a klippe (tectonic outlier) of the Palaeozoic–Mesozoic units that were overthrust from the Inner Zone of SW Japan. Another school stresses the importance of strike-slip tectonics (Taira *et al.* 1983). Yoshimoto *et al.* (2013) suggest a South China origin for detrital zircons in the pelitic schists from the Kurosegawa mélange. These zircons show distinct age peaks at 450–500 and 600 Ma and closely resemble those from the metamorphic belts in the Korean Peninsula, which mostly belongs to the North China Block.

Mineoka–Setogawa Ophiolitic Mélange

This unit has also been referred to as the ‘circum-Izu serpentinite belt’ (Arai 1991), because it forms an arc-like belt in front of the Izu Arc that is colliding with Honshu. This mélange is situated in the Shimanto accretionary complex of Cretaceous–Cenozoic age, and forms the youngest ophiolite belt in Japan. From west to east, the mélange appears in eastern Shizuoka City, curving north through the Kobotoke Mountains and ends in the Miura and Boso peninsulas to the south of Tokyo. The Mineoka–Setogawa Ophiolite Mélange includes abundant serpentinite, gabbro, basaltic rocks and a considerable amount of ultramafic volcanic rocks (picrite, meimechite and boninite) (Sakamoto *et al.* 1993; Hirano *et al.* 2003) in association with serpentine sandstone and conglomerate (Arai *et al.* 1983). Fresh mantle peridotites have been described from the Miura and Boso peninsulas, and they contain highly calcic plagioclase (anorthite). Some greenschist and amphibolite are also associated as blocks, and the overall highly fragmental occurrence of this ophiolite may be related to its emplacement mechanism at a trench–trench–trench triple junction (Mori & Ogawa 2005). Radiometric ages of the MORB-type basalts range from 80 to 50 Ma, those of arc-type volcanic and gabbroic rocks range from 45 to 32 Ma, those of oceanic-island-type alkali basalt and picrite range from 20 to 16 Ma, and the final emplacement of this mélange may be slightly later (Hirano *et al.* 2003). The formation mechanism and Cenozoic development history of this mélange in the forearc of the Izu Arc and later in the collision zone between the Izu and Honshu arcs are discussed by Mori *et al.* (2011).

Accretionary complex metabasalts

Metabasaltic rocks (or ‘greenstones’) crop out widely across SW Japan where oceanic sediments transported into the trench have been accreted in thrust duplex structures that preserve slices of oceanic plate stratigraphy (Isozaki 1996). These typically comprise basaltic pillow lava, limestone or chert, siliceous mudstone, mudstone, sandstone or turbidite in ascending (younging) order. The stratigraphy may be preserved as an intact, undisturbed sequence in some cases, but is more commonly fragmented and highly disturbed by fault movement along the subduction zone; in more extreme cases, it forms a mélange with a mudstone or serpentinite matrix. The basaltic rocks occur in two modes, either: (1) as a sheet-like body a few kilometres long and a few 100 m thick, forming the basal part of a thrust sheet; or (2) as blocks in a fragmented geological unit or mélange with a mudstone matrix. The basaltic rocks are usually overprinted by low-grade metamorphism in the zeolite, prehnite-pumpellyite, pumpellyite-actinolite or greenschist facies, and therefore transformed into ‘greenstones’.

These metabasites can form the major portion of an accreted slice occurring as pillow lava, massive lava, hyaloclastite and tuff breccias, as well as dykes and sills and, in some cases, associated with ultramafic volcanic rocks, dolerite and gabbro. Prominent limestone masses tend to be associated with the larger greenstone bodies, suggesting that this pairing represent fragments of reef limestone and the underlying volcanic edifice that constituted oceanic islands (Jones *et al.* 1993).

The basal greenstone shows chemical characteristics more enriched in incompatible elements than the normal mid-ocean ridge basalt, which is consistent with a seamount or oceanic plateau origin. The greenstone is mostly composed of low-Ti series basalts, but high-Ti series basalt and picrite (or meimechite) occurs as later (upper) lavas and intrusive rocks (dykes and sills). The greenstone blocks in mélanges are extremely variable in chemistry, ranging from normal mid-ocean ridge basalt to highly enriched oceanic island tholeiite and alkali basalt (Ichiyama & Ishiwatari 2005; Ichiyama *et al.* 2006, 2007, 2008; Koizumi & Ishiwatari 2006).

In the Late Palaeozoic (Permian) accretionary complex in the Inner Zone of SW Japan, major limestone plateaus such as Oga, Atetsu, Taishaku, Akiyoshi, Zomeki and Hirao crop out from western Honshu to northern Kyushu and are associated with greenstone bodies originated in the superplume magmatism of Carboniferous age (Tatsumi *et al.* 2000).

In the Jurassic accretionary complex of the Mino–Tamba–Ashio Belt, the greenstone bodies are also associated with large limestone bodies in some places (Funafuseyama, Ibukiyama, Ryozen, Kuzuu, etc.) but are associated with chert in other parts. In the Outer Zone, the Jurassic accretionary complex in the Chichibu Belt also bears abundant greenstones that are associated with chert and limestone. These greenstones are mostly of Carboniferous and Permian age, but those in the lower nappes tend to be younger (Triassic). The greenstones include some ultramafic volcanic rocks such as picrite, meimechite, ferropicrite and limbergite. The Chichibu Belt also includes the Kurosegawa Ophiolitic (or serpentinite) Mélange as described earlier.

The Shimanto Belt also bears some minor greenstones of MORB and alkali basalt chemistry. The eastern part of the Shimanto Belt in the southern Kanto area (near Tokyo) surrounding the Izu Peninsula includes the Mineoka–Setogawa Ophiolite as described earlier. The youngest (southernmost) part of the Shimanto Belt in eastern Shikoku is exposed around Cape Muroto, where the accretionary complex contains a post-accretionary gabbroic intrusion up to 200 m thick with a 10 m thick hornfelsic contact aureole and a doleritic chilled margin with MORB-like chemistry (Miyake 1985) rather than the island-arc basalt chemistry expected from its tectonic position in the accretionary wedge above a subduction zone. The central part of the gabbro body is coarse-grained and partly pegmatitic, but olivine accumulation has occurred near the bottom and roof (Hoshide *et al.* 2006). This gabbro intrusion may represent a ridge subduction event or forearc MORB magmatism as documented from the Eocene Izu forearc area. This gabbro body is one of the more spectacular attractions of the Muroto Geopark.

Ophiolitic rocks obtained from the adjacent seafloor

Peridotites, gabbros and amphibolites along with basaltic volcanic rocks have been recovered from the Philippine Seafloor to the south of Japan along spreading centres of the Parece Vela (dormant) and Mariana Trough (active) back-arc basins. The Parece Vela Basin spreading centre exhibits characteristics indicating a magma-starved, slow spreading setting with core complexes such as Godzilla Mullion (Harigane *et al.* 2011) exposing uplifted mantle and lower crustal rocks (Ohara & Ishii 1998). The Izu–Bonin–Mariana (IBM) forearc area has yielded an abundance of ophiolitic rocks in many dredge and drill sites. For example, the Hahajima Seamount along the Ogasawara Trench has mantle harzburgite, cumulate ultramafic rocks (dunite, wehrlite, pyroxenite), gabbros, dolerite and various volcanic rocks such as MORB-like basalt, island-arc tholeiite, calc-alkali andesite, boninite, adakite, dacite and rhyolite as well as minor sedimentary and metamorphic rocks (Ishiwatari *et al.* 2006; Murata *et al.* 2009; Li *et al.* 2013). The boninite is an ultramafic volcanic rock with >52 wt% SiO_2 and >8 wt% (up to 25 wt%) MgO, and is a high-magnesian andesite. Adakite is also a high-magnesian andesite (or dacite) with high Sr content (>600 ppm) and high Sr/Y ratio. Boninite, dacite and island-arc tholeiite are also exposed in the Ogasawara (Bonin) islands as pillow lavas and dykes (especially at Chichijima and Hahajima: Umino 1986). The crust–mantle section exposed on the seafloor of the IBM forearc is now widely accepted as the best analogue of the on-land ophiolite (Ishizuka *et al.* 2014). An association with high-pressure metamorphic rocks (blueschist, eclogite, etc.) is not only widespread among on-land ophiolites but also reported from the ocean floor of the IBM forearc, as demonstrated by blueschist blocks found from the Conical Seamount in the Mariana forearc (Maekawa *et al.* 1993) and eclogite outcrops reported from the Omachi Seamont of the Izu Arc (Ueda *et al.* 2004).

Ophiolitic rocks in Hokkaido (AI)

The geological structure of Hokkaido was essentially produced by the Miocene collision between the NE Japan and Kuril arcs. Deep crustal and mantle rocks representing the suture zone between these two units are exposed as a result of this small-scale orogeny in the Hidaka mountain range in central Hokkaido. The Kuril Arc was thrust over the NE Japan Arc from east to west (Fig. 3.7), and west-verging recumbent folds are well developed in the coal-bearing Palaeogene and Miocene sedimentary rocks in the foreland around the Ishikari plain on which Sapporo City is situated. Mantle peridotites such as the Horoman Complex and deep crustal metamorphic rocks (granulites) are exposed along the Hidaka Main Thrust that runs through the western slope of the Hidaka mountain range, where the metamorphic grade decreases eastwards through amphibolite and greenschist facies to almost unmetamorphosed sediments of the Hidaka accretionary complex in the eastern foothills. This transect provides a complete crustal cross-section through an island arc (Komatsu *et al.* 1983). The protoliths of these Hidaka metamorphic rocks are mostly sediments of the Palaeogene accretionary complex, but intrusions of I-type granite and gabbro are abundant in the high-grade part. The metamorphic facies series belongs to the typical low-pressure andalusite-sillimanite type.

Although westwards thrusting is prominent in the Hidaka mountain range and its eastern foreland, the overall geological structure of Hokkaido comprises a nappe pile of accretionary complexes with eastwards (oceanwards) thrusting and younging polarity similar to that seen in SW Japan. Southwestern Hokkaido (Oshima Peninsula) is a continuation of the North Kitakami Belt exposed in northern Honshu Island, where Jurassic accretionary complex rocks are widely developed. The eastern margin of the North Kitakami–Oshima Belt is marked by the Rebun–Kabato Cretaceous volcanic arc. Further east, the Sorachi–Yezo–Kamuikotan Belt forms a 30 km wide ophiolitic zone running north–south through the central axis of Hokkaido (Fig. 3.1). This belt may owe its origin to the accretion of an oceanic plateau with oceanic island-arc complexes of Jurassic–Cretaceous age. The Sorachi Group is mostly composed of Late Jurassic basaltic rocks that in some places form an ophiolite sequence with radiolarian chert, gabbro and ultramafic rocks (e.g. Horokanai Ophiolite), and grades upwards into the Yezo Group that is characterized by Cretaceous forearc basin sediments. The

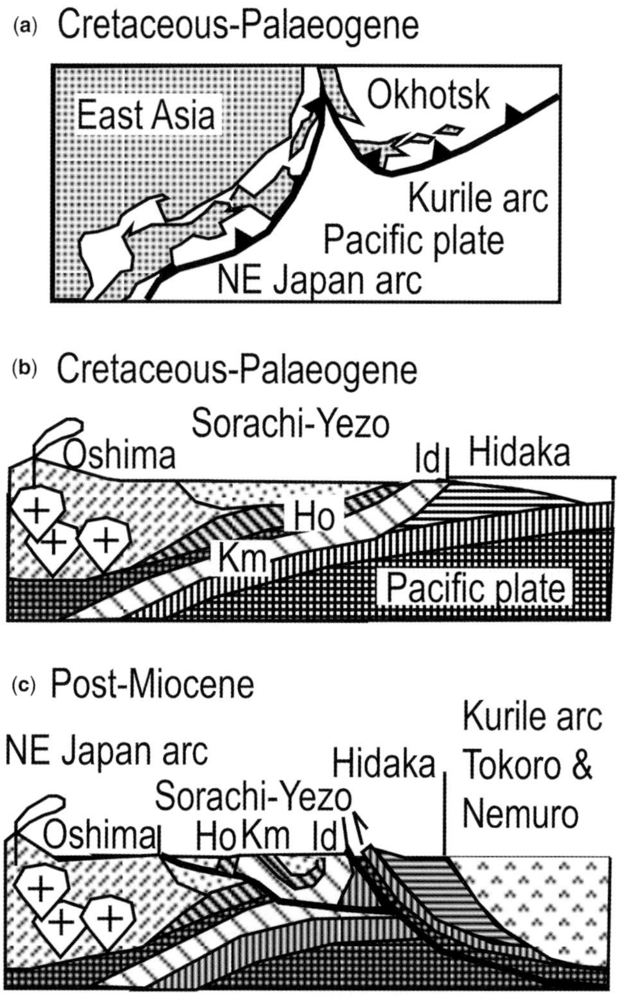

Fig. 3.7. (**a**) Possible palaeotectonic configuration of the western and eastern Hokkaido before Miocene collision of the NE Japan Arc and the Kurile Arc. (**b**) East–west cross-section showing geological structure of the Cretaceous–Palaeogene accretionary complexes of the NE Japan Arc before the Miocene collision. Eastwards thrusting dominated. (**c**) Schematic east–west cross-section of Central Hokkaido after the Miocene collision. Westwards thrusting dominates in this stage (after Ueda 2005).

Sorachi–Yezo sequence is not affected by regional metamorphism except for the so-called ocean-floor metamorphism, but the tectonically underlying Kamuikotan Belt is strongly affected by high-pressure metamorphism with glaucophane, sodic pyroxene, lawsonite, jadeite (coexisting with quartz) and aragonite having been reported. The Kamuikotan metamorphic rocks occur as several thrust slices underlying the Sorachi–Yezo Belt, with each slice showing downwards younging of metamorphic age and a downwards decrease of metamorphic grade (Sakakibara & Ota 1994).

Between the Sorachi–Yezo–Kamuikotan Belt and the Hidaka Metamorphic Belt, there exists the Idonnappu Belt which is essentially composed of accretionary complexes, ophiolites and oceanic island-arc complexes of Cretaceous age. In its eastern margin, adjacent to the Hidaka Metamorphic Belt, the Poroshiri Ophiolite Complex occurs directly beneath the Hidaka Main Thrust.

From west to east, the Oshima (Jurassic) accretionary complex, the Sorachi–Yezo Belt (Jurassic–Cretaceous), the Idonnappu accretionary complex (Cretaceous) and the Hidaka accretionary complex (Palaeogene) are progressively crossed, forming an eastwards (oceanwards, downwards)-younging pile of accretionary complexes, although partly disrupted by the Kuril Arc collision along the Hidaka mountain range in south-central Hokkaido.

Further details on the main ophiolites and ultramafic bodies cropping out in Hokkaido are provided in the following sections.

Horokanai Ophiolite

This unit crops out in the northern part of the Sorachi–Yezo Belt as a nappe overlying the Kamuikotan metamorphic rocks, and is exposed along the Uryu River along the axis of a north–south-trending antiform (Fig. 3.8). The basal mantle peridotite of this ophiolite appears on both flanks of the antiform, but the overlying members (ultramafic cumulate, gabbro and basaltic rock) occur only on the eastern flank (Ishizuka 1985, 1987). The mantle peridotite comprises strongly depleted harzburgite, and the ultramafic cumulate is characterized by layered orthopyroxenite and dunite. The gabbro member and lower part of the basalt member were metamorphosed under hornblende granulite to amphibolite facies conditions, and partly transformed into foliated metagabbro and amphibolite. The metamorphic grade of the upper part of the basalt member is very low (zeolite facies). Pillow structure is well preserved in the upper part, which is covered by younger Jurassic chert with radiolarian fossils (Ishizuka 1985, 1987). This ophiolite is characterized by the very depleted nature of the mantle peridotite, peculiar orthopyroxene-rich cumulates, intense 'ocean-floor' metamorphism varying from zeolite to hornblende granulite facies downwards through the ophiolite stratigraphy, and MORB-like basal chemistry.

The ophiolitic mantle peridotite of the Sorachi Belt is highly depleted harzburgite (spinel Cr#>80) in northern Hokkaido, but is less depleted in southern Hokkaido (e.g. Iwanai-dake), and becomes lherzolitic (spinel Cr#<30) in the southernmost Shizunai–Mitsuishi area. Strongly depleted harzburgite is also present further to the north in Sakhalin (Ishiwatari *et al.* 2003).

The ophiolitic volcanic rocks are distributed across wide areas in central Hokkaido around Furano City. They are mostly composed of tholeiitic basalt, but large amounts of picritic rocks also occur (Takashima *et al.* 2002; Ichiyama *et al.* 2012). Olivine phenocrysts in the picritic rocks are as magnesian as Fo_{94} and the NiO-Fo variation trend closely follows that of mantle olivine, suggesting a very high magma temperature as in the case of komatiites (Tatsumi *et al.* 1998; Ichiyama *et al.* 2012). The abundant occurrence of very-high-temperature picrites and basalt geochemistry with high Nb/Zr and Nb/Y ratios suggests that the Sorachi basaltic rocks originated in an oceanic plateau formed by the activity of a large mantle plume (superplume). On the other hand, however, the association with acidic tuffs in some places suggests that the magmatism occurred near a volcanic arc (Takashima *et al.* 2002). It is possible that the oceanic plateau magmatism of the Sorachi Belt took place near a subduction zone and oceanic island arc.

Poroshiri Ophiolite

This unit is situated in the eastern margin of the Idonnappu Belt, adjacent to the Hidaka Metamorphic Belt (Fig. 3.9), and is of

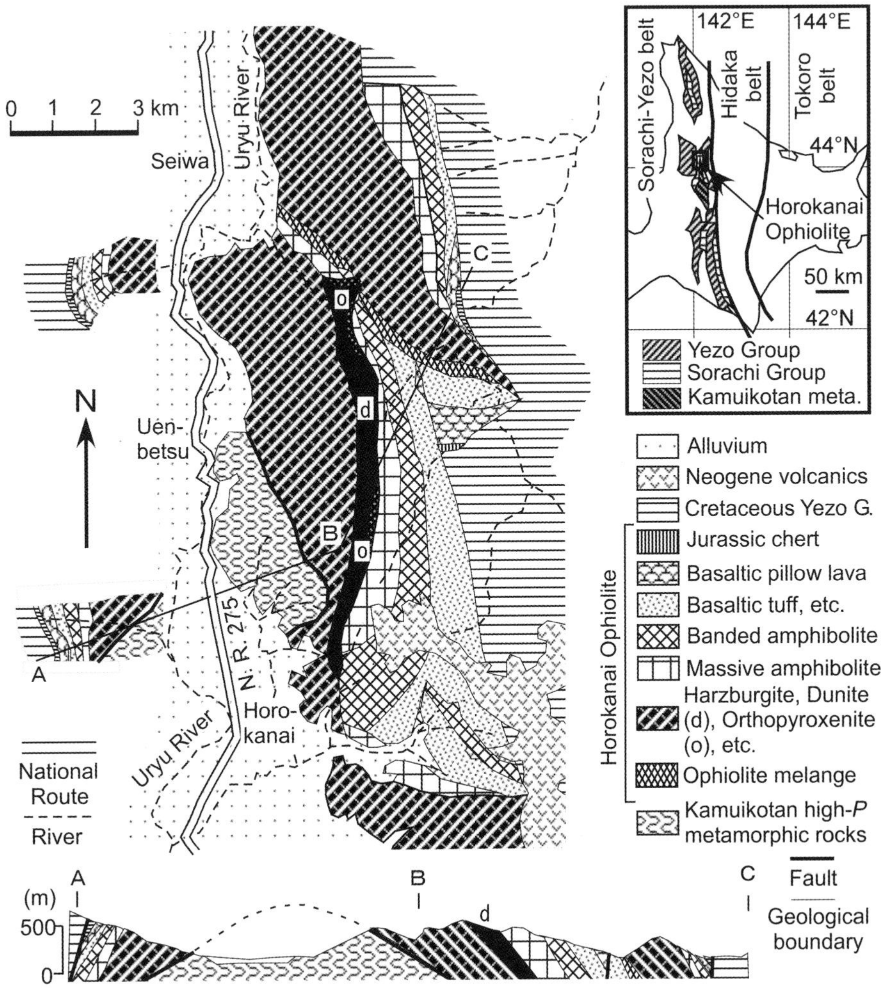

Fig. 3.8. Geological map and cross-section of the Horokanai Ophiolite and Kamuikotan metamorphic rocks in the Horokanai area (after Ishizuka 1985, 1987).

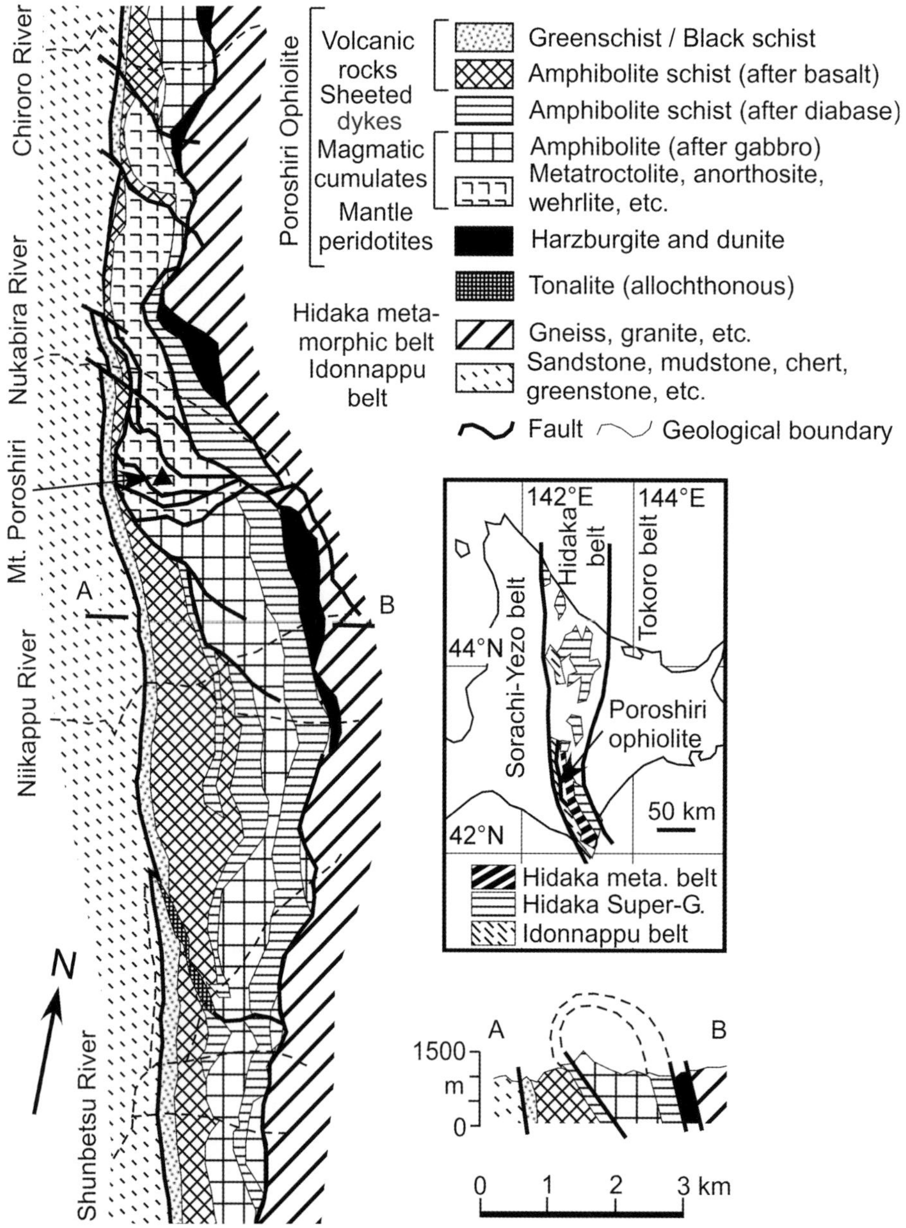

Fig. 3.9. Geological map and cross-section of the Poroshiri Ophiolite in the Hidaka area (after Miyashita *et al.* 1994).

uncertain age although generally assumed to be Cretaceous in view of the absence of Jurassic chert and the close association of basalts and clastic sediments (Miyashita & Yoshida 1988). The adjacent peridotite body to the SE of Mt Poroshiri is thought to be the mantle portion of this ophiolite. The body is composed of a western dunite part and an eastern harzburgite part, both with Fo_{90-93} (average 91.4) olivine and Cr#35–65 (average 55) spinel (Suzuki & Niida 1997). The cumulate member of this ophiolite is well developed in the Mt Poroshiri area (Fig. 3.9), and is mainly composed of a layered troctolite-anorthosite sequence. Olivine is Fo_{84-89} and plagioclase is An_{82-88} (Miyashita & Hashimoto 1975; Miyashita *et al.* 1994). Olivine gabbro is distributed in the eastern half of the ophiolite body. The sheeted dykes and volcanic rocks of this ophiolite are thoroughly metamorphosed under amphibolite and greenschist facies, and the original volcanic structure has been largely destroyed. The protoliths of the associated metasediments are thought to be mudstone and sandstone with a minor amount of Mn-rich hydrothermal deposits, which has been recrystallized into viridine (Mn-andalusite)-bearing schist.

The overall structure of this ophiolite appears to be a tight anticline with the axial plane dipping to the east (Fig. 3.9). Amphibolites which originated as dykes and lavas are distributed along the margins, and gabbros and layered troctolites occupy the axial zone of the anticline. There are prominent curved thrust faults cutting this ophiolite in a NW direction, suggesting right-lateral movement between the Sorachi–Yezo and Hidaka belts.

The metabasaltic rocks of the Poroshiri Ophiolite show MORB-like geochemistry with a considerable amount of ferrobasaltic varieties. However, the Okuniikappu volcanic rock complex of the Idonnappu Belt that is situated only 2 km to the west of the Poroshiri Ophiolite is composed of low-K calc-alkaline andesite and boninite, indicating definite island-arc affinity (Ueda & Miyashita 2005). It is interesting that the Poroshiri Ophiolite does not bear chert (nor metachert), whereas in contrast the Okuniikappu volcanic complex has abundant chert with Early Cretaceous radiolarian fossils (Ueda & Miyashita 2005).

Horoman peridotite complex

A number of mantle peridotite bodies are distributed along the Hidaka Main Thrust and can be divided into harzburgitic and lherzolitic types. The harzburgite-type complexes are interpreted to be the mantle section of the Poroshiri Ophiolite (see the previous section), whereas well-layered, lherzolite-type complexes such as the Horoman peridotite complex are interpreted to be the tectonically emplaced, sub-arc mantle of the Hidaka Belt. The latter is classified as Alpine-type peridotite in view of the absence of an associated thick basalt-gabbro sequence (oceanic crust) and very fertile peridotite chemistry (close to primitive mantle) with mineralogical features suggestive of a deep-mantle origin (e.g. the common occurrence of spinel-pyroxene symplecite after garnet).

The Horoman peridotite complex forms a well-layered, subhorizontal body with its lower part dominated by spinel lherzolite and harzburgite and its upper part dominated by plagioclase lherzolite (Fig. 3.10) (Niida 1984). Deformation texture also shows contrasts between the upper and lower part: porphyroclastic with basal shear zone in the lower part and equigranular in the upper part (Sawaguchi 2004). These main lithologies of peridotites are interlayered with lesser amounts of dunite, wehrlite, pyroxenite and mafic rocks. Takahashi (1991) divided the Horoman Complex into three series: the main harzburgite-lherzolite series (MHL); a spinel-rich dunite-wehrlite series (SDW); and a rare banded dunite-harzburgite series (BDH). The MHL harzburgite contains Cr#45–65 spinel, the MHL lherzolite bears Cr#20–45 spinel and plagioclase lherzolite has even more aluminous spinel (Cr#5–15). The SDW is interpreted to be a magma conduit that was filled by the cumulate minerals, and SDW olivines are slightly richer in iron (Fo_{89-90}) than MHL olivines (Fo_{90-92}). The BDH is rich in Mg (Fo_{91-94}), bears highly chromian spinel (Cr#80–90) and is interpreted to be a cumulate from a high-Mg magma. The spinel lherzolite in some places bears purple-coloured, elongated and flattened aggregates of fine-grained two-pyroxenes and spinel, up to a few centimetres in length and usually with a core of two-pyroxene spinel symplectite. This symplectite and the surrounding aggregates are interpreted to be the decomposition product of garnet (Tazaki *et al.* 1972; Takahashi & Arai 1989; Morishita & Arai 2003; Odashima *et al.* 2008). Ozawa & Takahashi (1995) showed that even plagioclase lherzolites were derived from a garnet lherzolite and that the Horoman Complex as a whole was exhumed adiabatically from a depth of >60 km. Mafic layers rarely bear corundum, which also suggests a deep origin (>30 km Morishita & Arai 2001). Takazawa *et al.* (1999, 2000) argued that the variations in bulk-rock composition of peridotite and mafic layers are consistent with partial melting processes. Ozawa (2004) showed that the Horoman Complex underwent a heating event during its decompression, and proposed asthenospheric upwelling to heat the Horoman lithospheric mantle (peridotite).

Other greenstones in Hokkaido

The Jurassic Oshima accretionary complex bears greenstones that are strongly folded and altered. They may represent accreted seamount basalts (alkali basalt, tholeiite, picrite, etc.) of Late Palaeozoic–Triassic ages similar to those in the North Kitakami Belt, although they have not yet been well studied. The Hidaka accretionary complex to the east of the Hidaka Metamorphic Belt bears so-called '*in situ* greenstones' that intruded into or erupted over the adjacent sediments during sedimentation or subduction-accretion, the largest example being the Shimokawa greenstone complex (20 × 2 km) which has basalts of MORB affinity (Miyashita & Yoshida 1988, 1994). This may represent magmatic activity in the subduction-accretion zone and would have needed abnormal thermal input in this area, perhaps that induced by subduction of an oceanic ridge for example. The Tokoro Belt to the south of Lake Saroma is thought to belong to the Kuril Arc, and bears abundant greenstones that are associated with Late Jurassic–Early

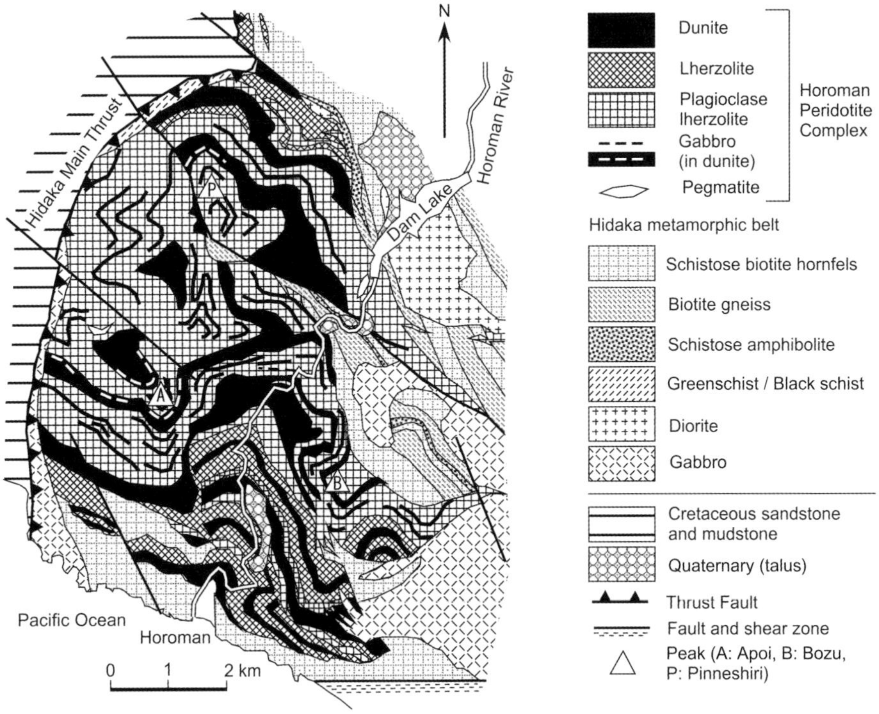

Fig. 3.10. Geological map of the Horoman peridotite complex in Samani Town, Hokkaido (Niida 1984). The area is included in the Mt Apoi Geopark. The layering structure is subhorizontal, and sinuous layers on the map actually follow topographic contour lines.

Cretaceous chert. The geochemistry indicates a seamount origin for these metabasalts, which have experienced high-pressure, low-temperature metamorphism in the subduction zone (Niida *et al.* 1981; Sakakibara 1986, 1991).

Ophiolitic rocks in NE Honshu (KO)

The Hayachine–Miyamori Ophiolite of NE Honshu was formed and emplaced during the Ordovician ophiolite pulse (Abbate *et al.* 1985; Ishiwatari 1994; Yakubchuk *et al.* 1994), which is characterized by a predominance of supra-subduction zone ophiolites (Dilek & Furnes 2011). This ophiolite is shown to have formed in an arc environment on the basis of many lines of geological, petrological and geochemical evidence (e.g. Ozawa 1988).

Geological setting

Pre-Cretaceous geological units of the Kitakami Mountains in NE Japan are divided into the North and South Kitakami massifs (Onuki 1969). The North Kitakami Massif is a Jurassic accretionary complex (Ehiro *et al.* 2008) and bordered to the SW by a Carboniferous accretionary complex known as the Nedamo Belt (Hamano *et al.* 2002; Ehiro & Suzuki 2003; Uchino *et al.* 2005, 2008; Fig. 3.11). The South Kitakami Massif consists mainly of Silurian–Jurassic shallow-marine–subaerial strata coherently stratified in an arc environment without major tectonic breaks (Onuki 1969; Oide *et al.* 1989; Yoshida & Machiyama 2004). Silurian strata including welded tuff and other acidic volcaniclastic rocks are distributed in the South Kitakami Massif (Onuki 1937; Murata *et al.* 1982), indicating the presence of mature continental crust by Silurian times (Fig. 3.11). Granitic intrusions (Hikami Granite) accompany these Silurian strata are granodiorite-tonalite with minor hornblende gabbro (Hoe 1976; Kobayashi *et al.* 2000). The whole-rock Rb–Sr isotope systems (Asakawa *et al.* 1999) and zircon U–Pb ages (Watanabe *et al.* 1995) of the Hikami Granite clearly indicates that parts of the granite formed in Late Ordovician–Early Silurian times.

Ordovician ultramafic rocks with subordinate amounts of mafic rocks are distributed along the boundary of the two massifs to the south of the Nedamo Belt (the Hayachine ultramafic complex) and along the Hizume–Kesennuma Fault (Miyamori ultramafic complex) (Seki 1951, 1952; Onuki 1963, 1964, 1965; Ozawa 1983, 1984, 1988, 1990, 1994, 2001; Fujimaki & Yomogida 1986; Ozawa *et al.* 1988; Shibata & Ozawa 1992; Ozawa & Shimizu 1995; Ehiro & Suzuki 2003; Yoshikawa & Ozawa 2007; Yoshikawa *et al.* 2012; Fig. 3.11). The Hayachine Complex is bordered to the south by mafic volcanic and intrusive rocks (Nekozoko mafic complex) and further still to the south by Silurian–Devonian strata of the South Kitakami Massif (Fig. 3.11; Ehiro *et al.* 1988; Tazawa 1988; Ehiro & Suzuki 2003; Shimojo *et al.* 2010), suggesting that the Nekozoko rocks are pre-Silurian and probably Ordovician (Ehiro & Suzuki 2003). The Hizume–Kesennuma Fault is a NW–SE- to NNW–SSE-trending sinistral transcurrent structure with *c.* 30 km displacement during Early Cretaceous times (Ehiro 1977). The Miyamori Complex was transported from the north to the current site along this fault as a thrust sheet (Ozawa 1984). Before the displacement, the Miyamori Complex is inferred to have been located to the west of the Hayachine Complex. This and their petrological, geochemical and geochronological similarity indicate that the two complexes represent a geological unit constituting an Orodovician arc system (Ozawa 1984), so that these ultramafic complexes and the Nekozoko mafic complex as a whole are therefore referred to as the Hayachine–Miyamori Ophiolite (Ozawa & Shimizu 1995).

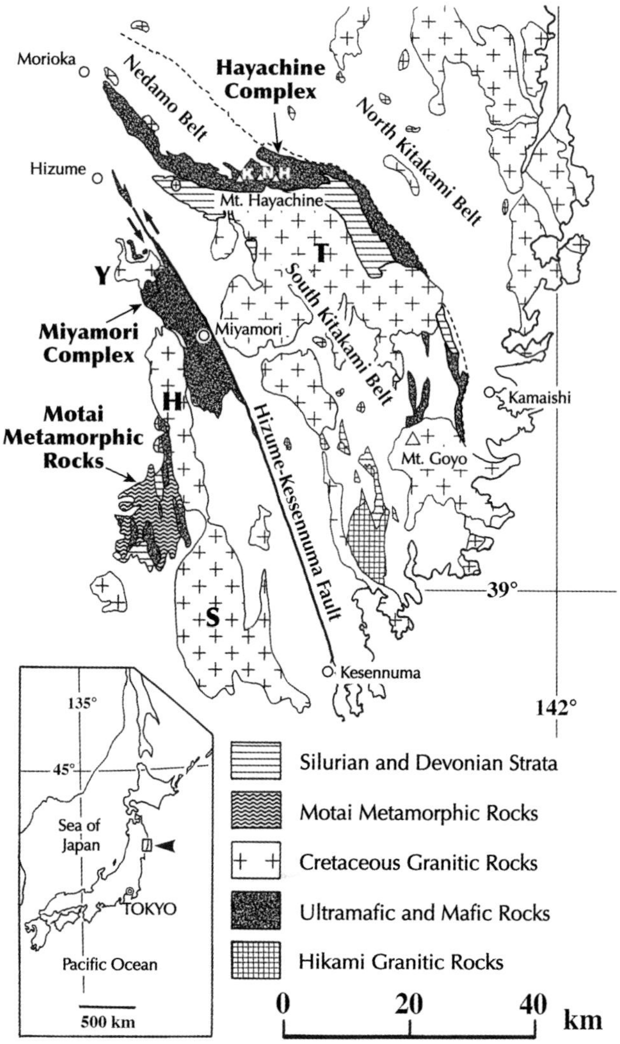

Fig. 3.11. Geological map and tectonic subdivision of the southern Kitakami Mountains (modified from Ehiro & Suzuki 2003; Kawamura *et al.* 2013). The location of Silurian–Devonian units along the boundary of the South and North Kitakami belts is after Ehiro (2000), and the location of the Nedamo Belt is after Kawamura *et al.* (2013). A dashed line bordering the Nedamo Belt and the southeastern part of the Hayachine Complex indicates the southern limit of the North Kitakami Belt (Yoshida & Katada 1964; Ehiro *et al.* 1988). Cretaceous granites referred to within the main text: T, Tono granite; S, Senmaya granite; H, Hitokabe granite; Y, Yamaya granite (in black). Three peaks of Mt Hayachine: H, Hayachine; N, Nakadake; and K, Keitosan (in white).

Hayachine–Miyamori Ophiolite

This ophiolite consists mainly of serpentinized peridotites with hornblende-rich mafic-ultramafic rocks (Fig. 3.12; Ozawa 1984; Shibata & Ozawa 1992). The Hayachine and Miyamori complexes are lithologically contrasting: lherzolite dominates in the Hayachine Complex, whereas harzburgite and dunite are more abundant in the Miyamori Complex (Fig. 3.13b; Ozawa 1984, 1988). Smaller peridotite bodies found in the western part of the Hayachine Complex were not affected by the Cretaceous granite intrusions; they are, without exception, lherzolite with abundant clinopyroxene and Al-rich spinel (Fig. 3.13b). The main part of the Hayachine Complex exposed along the ridges of Keitosan–Nakadake–Hayachine (Fig. 3.11), however, was overprinted by contact metamorphism

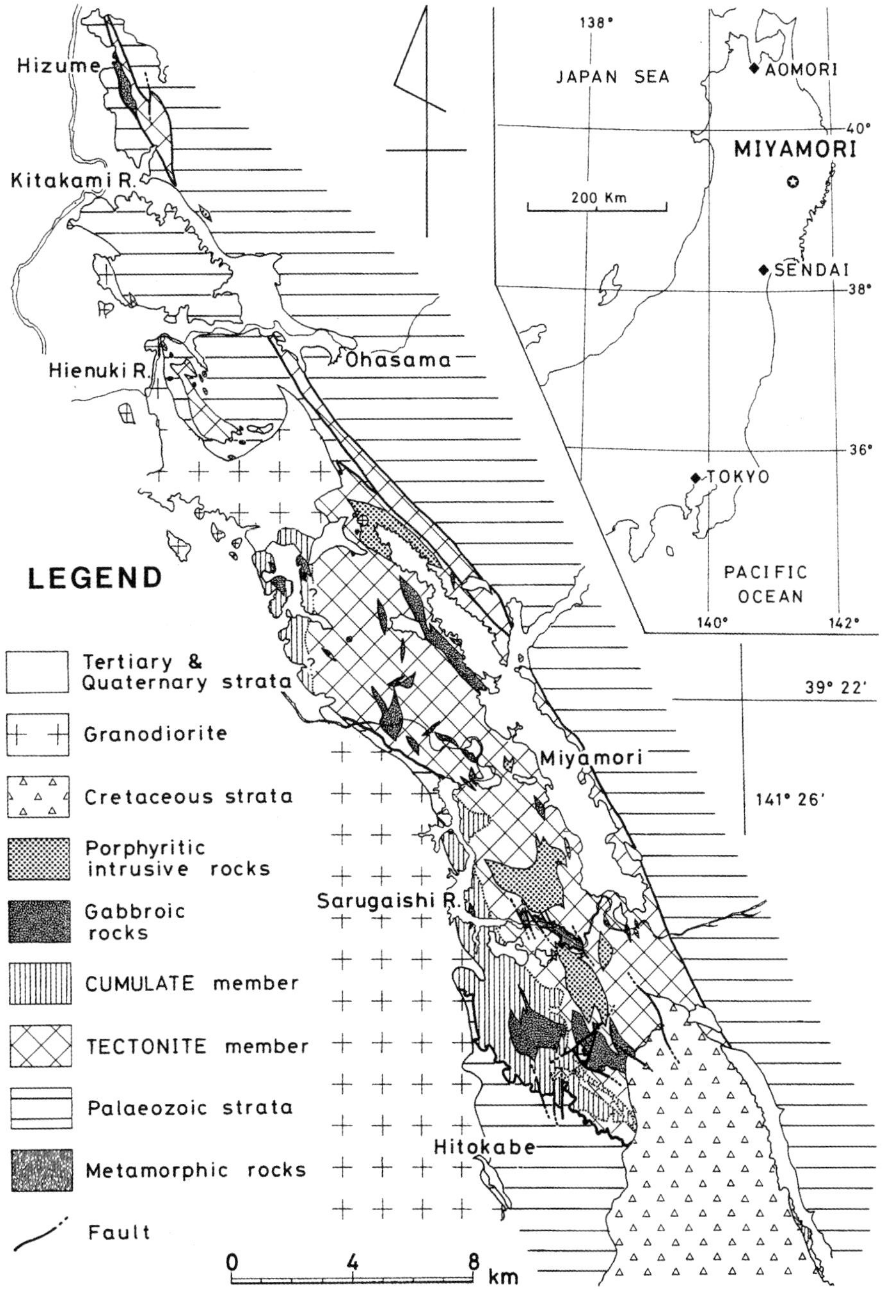

Fig. 3.12. Geological map of the Miyamori ultramafic complex (after Ozawa 1984).

of the Cretaceous Tono Granite, obscuring the original petrographic features (Fujimaki & Yomogida 1986). The original lithology of the peridotites can be identified as lherzolite by the presence of chromian diopside and rare relicts of pale brown spinel, if the metamorphic grade is as low as olivine + antigorite ± diopside. Even for peridotites extensively recrystallized at much higher metamorphic conditions, the abundance of relict of spinel lamellae, now replaced by fine-grained magnetite, and the predominance of tremolite aggregates, forming pseudomorphs after clinopyroxene, suggest that the dominant protolithic rock type of the Hayachine Complex was lherzolite (Ozawa *et al.* 2013). The Miyamori Complex was also overprinted by contact metamorphism of Cretaceous granites, and the highest-grade mineral assemblage (orthopyroxene and olivine) showing polygonal aggregates of metamorphic olivine occurs near the Hitokabe and Yamaya granite intrusions (Seki 1951; Ozawa *et al.* 1988). Anthophyllite and cummingtonite-bearing assemblages are also well developed (Seki 1951). In the eastern parts of the Miyamori Complex, which Seki (1951) considered not to have been metamorphosed, rocks with antigorite + olivine ± diopside assemblage are heterogeneously and irregularly distributed in the predominant serpentinized peridotites with chrysotile/lizardite ± brucite (± relict olivine) assemblage. Such antigorite-bearing peridotites include 'cleavable olivine' (Kuroda & Shimoda 1967; Velinsky & Pinus 1969), as do similar mineral assemblages of antigorite + olivine ± diopside in the Hayachine Complex (Ozawa *et al.* 2013). 'Cleavable olivine' is common among seafloor peridotites of the Izu–Bonin–Mariana forearc (Ohara & Ishii 1998; Murata *et al.* 2009) where granite intrusion is scarce, suggesting its formation is linked to processes with the hydrated lithosphere in the supra-subduction zone.

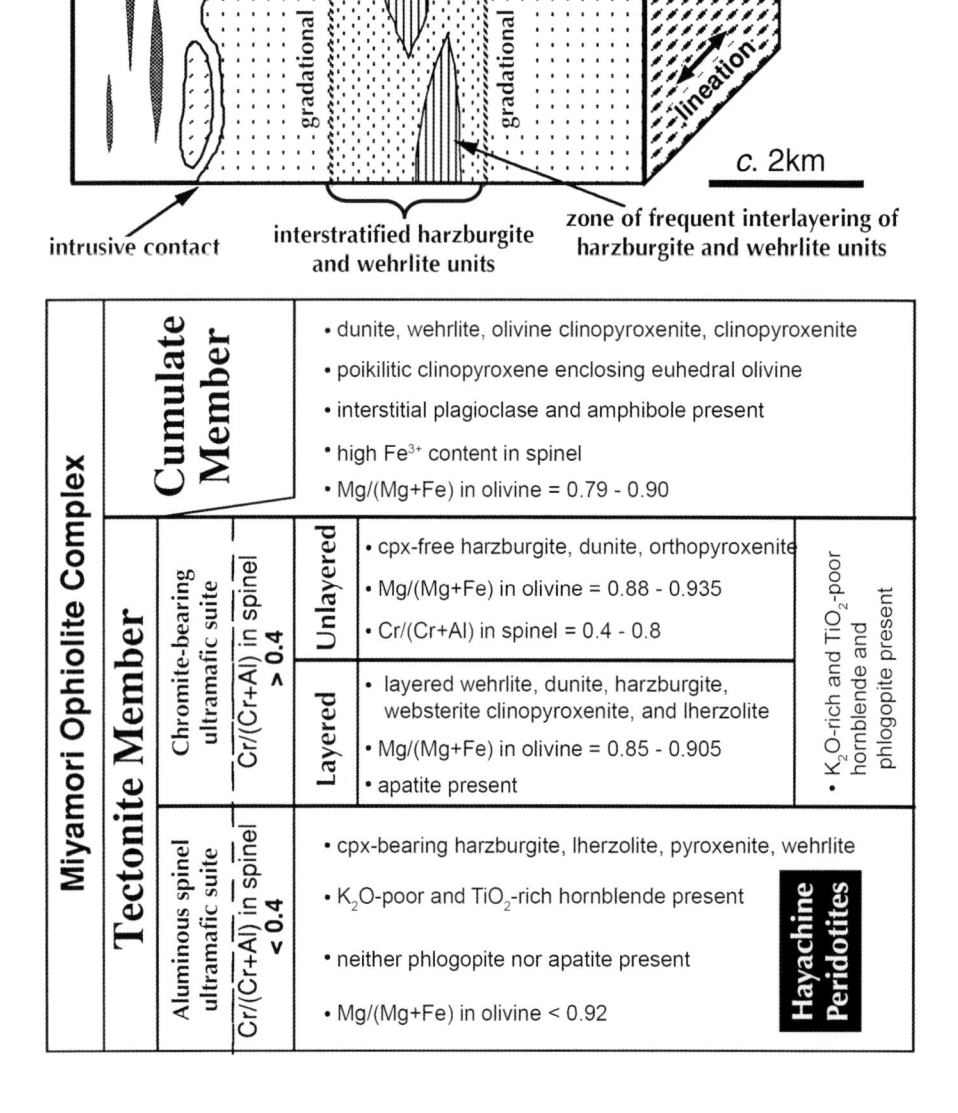

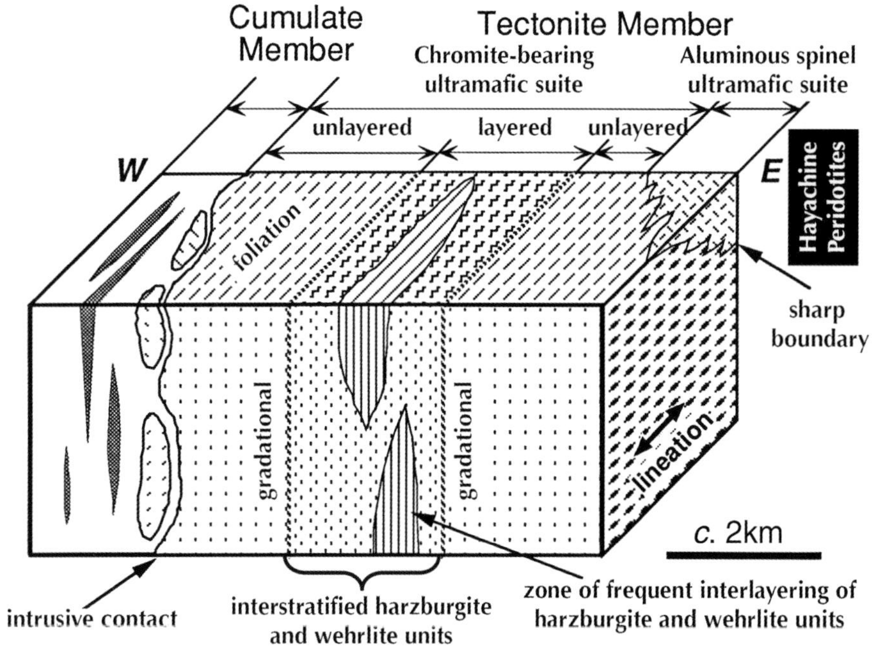

Complex	Member	Suite	Unit	Features	
Miyamori Ophiolite Complex	Cumulate Member			• dunite, wehrlite, olivine clinopyroxenite, clinopyroxenite • poikilitic clinopyroxene enclosing euhedral olivine • interstitial plagioclase and amphibole present • high Fe^{3+} content in spinel • Mg/(Mg+Fe) in olivine = 0.79 - 0.90	
	Tectonite Member	Chromite-bearing ultramafic suite Cr/(Cr+Al) in spinel > 0.4	Unlayered	• cpx-free harzburgite, dunite, orthopyroxenite • Mg/(Mg+Fe) in olivine = 0.88 - 0.935 • Cr/(Cr+Al) in spinel = 0.4 - 0.8	• K_2O-rich and TiO_2-poor hornblende and phlogopite present
			Layered	• layered wehrlite, dunite, harzburgite, websterite clinopyroxenite, and lherzolite • Mg/(Mg+Fe) in olivine = 0.85 - 0.905 • apatite present	
		Aluminous spinel ultramafic suite Cr/(Cr+Al) in spinel < 0.4		• cpx-bearing harzburgite, lherzolite, pyroxenite, wehrlite • K_2O-poor and TiO_2-rich hornblende present • neither phlogopite nor apatite present • Mg/(Mg+Fe) in olivine < 0.92	Hayachine Peridotites

Fig. 3.13. Subdivision of the ultramafic section of the Miyamori ultramafic complex (after Ozawa 1994). This block diagram indicates the basic internal structure of the complex and contact relationships between subdivisions (members, suits and units). The lower diagram indicates the principle features of the divisions, and ultramafic rocks of the Hayachine Complex form part of the aluminous spinel ultramafic suite within the Tectonite Member, identified using white letters within a black box. Hornblende-rich mafic and ultramafic late intrusive rocks are omitted for simplicity, though they are genetically related to the peridotites and pyroxenites.

Internal structure of the Miyamori Complex

A schematic illustration of the internal structure of the Miyamori Complex, the best preserved part of the Hayachine–Miyamori Ophiolite, is shown in Figure 3.13a after Ozawa (1983, 1984, 1988, 1994). The complex can be divided into tectonites, which show penetrative plastic deformation texture, and cumulates, which show cumulus texture without evidence for solid-state deformation.

The cumulate ultramafic rocks show a lithologic diversity reflected by a large variation in the ratio of clinopyroxene and olivine, and are mostly dunite and wehrlite with lesser amounts of olivine clinopyroxenite and clinopyroxenite (Fig. 3.13b). Plagioclase, which is completely replaced by secondary minerals such as grossular, is rarely present in these rocks, filling the interstitial part of euhedral-subhedral clinopyroxene and olivine. The olivine-plagioclase assemblage indicates that the cumulate formation occurred in the plagioclase-peridotite field at a magmatic temperature (<1 GPa considering the uncertainty in plagioclase composition; Ozawa 1986). There is always a reaction zone between plagioclase and olivine, which is composed of pargasite-spinel symplectite on the plagioclase side and orthopyroxene on the olivine side. This reaction texture indicates that pargasite was crystallized near solidus or at subsolidus conditions and therefore that the upper pressure limit of the cumulate formation was 0.7 GPa (Ozawa 1986). The Fo content of cumulate olivines (100 × Mg/(Mg + Fe)) ranges from 79 to 90 with an average value of *c.* 85 (Ozawa 1988).

Ultramafic tectonites show strong lineations, a less well-developed foliation and lattice-preferred orientation of olivine indicating (010)[100] slip (Fukudome 1978; Ozawa 1989). The lithology of this member is diverse, but the most dominant rock types are harzburgite and dunite with subordinate amounts of lherzolite, orthopyroxenite and websterite (Fig. 3.13b). Wehrlite, olivine clinopyroxenite and clinopyroxenite, which are common in the cumulates, also occur in the tectonites, although with penetrative plastic deformation. The Fo content of olivine in ultramafic rocks of the tectonites shows a wide variation over the range 75–95 but is mostly 90–92, which is greater than that of the cumulates (Ozawa 1988). Tectonites always contain hornblende (usually more than a few percentage by volume), which is one of their peculiar features not common to the other ophiolites. Some tectonite websterites contain very Al-rich green spinel, which indicates that these deformed rocks were subjected to pressures >0.5 GPa, as estimated from the two pyroxene thermometry of 800°C and orthopyroxene–clinopyroxene–spinel assemblages (Ozawa 1986).

Along the boundary between these ultramafic cumulates and tectonites, cumulate wehrlite, dunite and pyroxenite enclose angular fragments of tectonite lithology such as harzburgite and dunite (Ozawa 1983, 1984). Such harzburgitic fragments are of cobble to boulder size and exhibit a fine-grained polygonal texture of orthopyroxene aggregates derived from porphyroclasts, indicating static annealing of plastic deformation microstructures by intruding magmas, and enrichment of Fe from the centre to the contact with the host wehrlite, indicating its entrapment in Fe-rich magma forming wehrlite (Ozawa 1983). The boundary is very irregular on various scales, and harzburgite blocks of centimetre to kilometre size are recognized (Fig. 3.12). These facts suggest that a magma forming the cumulates intruded into already deformed tectonites. From the pressure constraints and the appearance of 1 km sized blocks of harzburgite in wehrlite, it is inferred that the tectonite unit was overlain by the cumulates at a depth of *c.* 18 ± 3 km (*c.* 0.6 ± 0.1 GPa), so that the cumulates represent a solidified magma body formed in the uppermost mantle (Fig. 3.13). The *c.* 30 km horizontal extension of the cumulate (as shown in Fig. 3.12) suggests that this magma body formed along the Moho discontinuity, where neutral buoyancy of mafic magmas is to be expected. The crustal thickness of the ophiolite would therefore have been *c.* 15 km, which is much thicker than the present-day oceanic crust of 6–8 km.

The tectonite lithologies can be divided into an Aluminous Spinel Ultramafic Suite (ASUS) and a Chromite-bearing Ultramafic Suite (CRUS) according to Cr/(Cr + Al) values of spinel (Fig. 3.13b; Ozawa 1988). The Miyamori Complex is mostly composed of CRUS which contains 1–2 km sized blocks of relatively refractory ASUS (Fig. 3.12; Ozawa 1987). The boundary between the CRUS and ASUS has not been observed, but discontinuity in the Cr/(Cr + Al) value of spinel indicates that it is fairly sharp (Ozawa 1988). The Hayachine Complex is composed of ASUS-type lherzolite (Fig. 3.13).

The dominant rock type of the Miyamori Complex, CRUS, is subdivided into layered and unlayered units (Ozawa 1994). The unlayered unit consists mainly of clinopyroxene-free harzburgite and dunite, whereas the layered unit contains wehrlite and clinopyroxenite in addition to harzburgite and dunite (Ozawa 1984, 1994). Clinopyroxenite occurs as sill-like layers which can be traced into dykes highly oblique to the layering (Ozawa 1984). The lineation and foliation, exhibited by elongate/flattened shapes of spinel, pyroxenes and hornblende aggregates, are concordant with layering. Clinopyroxene crystals in dykes elongate highly oblique to the dyke wall but parallel to the lineation. On the geological map, the layered rocks are distributed as elongate, NW–SE-trending bodies, characterized by lower Fo content in olivine, and show gradational boundaries with the unlayered lithologies. Phlogopite, present in both layered and unlayered rocks, commonly occurs along the margin of hornblende crystals as small euhedral grains or as isolated grains showing preferred shape orientation concordant with the foliation. Wehrlite and dunite in the layered rocks rarely contain apatite.

The CRUS tectonites contain hornblende-rich mafic-ultramafic rocks, such as hornblende gabbro, clinopyroxene hornblende gabbro, hornblendite and pyroxene hornblendite. They appear as tabular bodies in the peridotite host trending NW–SE with thickness varying from a few tens of centimetres up to 2 km (Fig. 3.12; Ozawa 1984). They locally cross-cut foliation or contain blocks of host harzburgite on the outcrop scale, indicating that they formed from magma-filling fractures or melt-rich permeable conduits (Seki 1952; Ozawa 1984; Ozawa *et al.* 2013).

Evidence for arc origin

The Hayachine–Miyamori Ophiolite is mostly composed of a mantle section, but its petrologic and geochemical features clearly show that it represents the upper mantle of an Ordovician arc–back-arc system (Ozawa 1984, 1988, 1990, 1994, 2001; Ozawa & Shimizu 1995; Yoshikawa & Ozawa 2007; Yoshikawa *et al.* 2012). Limited geochemical and mineralogical data from volcanic and shallow intrusive rocks further support an arc origin (Shibata & Ozawa 1992; Shimojo *et al.* 2010).

The most peculiar feature of the Hayachine–Miyamori Ophiolite is that all mantle peridotites contain significant amounts of hornblende and, more rarely, of phlogopite and apatite (Ozawa 1984). It is also peculiar in that the hornblende gabbro and hornblendite commonly occur in the mantle peridotites. As discussed in Ozawa (1984, 1988), the hornblende in peridotites is not a metasomatic reaction product of subsolidus fluid introduction, but is magmatic in origin. The presence of hydrous magmatic minerals therefore strongly suggests that the tectonic setting of the Hayachine–Miyamori was supra-subduction ophiolite, and geochemical data strengthen this model.

The TiO_2 and K_2O contents of hornblende Hayachine–Miyamori peridotites are different between the ASUS and CRUS (Fig. 3.13b; Ozawa 1988). The whole-rock chemistry of ASUS is rich in Al and Ca due to the high abundance of clinopyroxene and aluminous spinel but the hornblende is poor in K_2O, whereas the whole-rock chemistry of CRUS is poor in Al and Ca but the hornblende is rich in K_2O. The TiO_2 content in hornblende of CRUS is comparable to or lower than that of ASUS which has a higher TiO_2/K_2O ratio (>2 with an average of *c.* 10 as compared to <1 and *c.* 0.5 for CRUS). Comparison with the TiO_2/K_2O ratios of primitive magmas produced in various tectonic settings shows that the data for CRUS lie within the range of high-magnesian andesite (HMA) or island-arc basalt (IAB), whereas those for ASUS plot within the range of normal and enriched mid-ocean ridge basalt (N-MORB and E-MORB) or back-arc basin basalt (BAB) (Ozawa 1988).

Ozawa & Shimizu (1995) measured trace element contents in clinopyroxene and hornblende in CRUS and ASUS and revealed systematic variations of trace element patterns. The most fertile lherzolites of ASUS from the Hayachine Complex contain clinopyroxene with heavy rare Earth element (HREE) contents close to those of the primitive mantle (McDonough & Sun 1995), but are depleted in light rare Earth elements (LREE), and the CI-chondrite-normalized REE pattern is similar to that of MORB. Clinopyroxene in ASUS rocks within the Miyamori Complex is poorer in HREE and less depleted in LREE than that of the Hayachine lherzolites, showing a flat or V-shaped CI-chondrite-normalized REE pattern. Clinopyroxene in CRUS of the Miyamori Complex is more depleted in HREE but is enriched in middle rare Earth elements (MREE) and LREE relative to HREE, showing a LREE-enriched CI-chondrite-normalized pattern with a maximum in MREE (Nd and Sm). High-field-strength elements (HFSE) such as Ti and Zr are slightly fractionated from REE for clinopyroxene in the Hayachine ASUS but are markedly depleted relative to REE for clinopyroxene in the Miyamori CRUS, showing sawtooth trace element patterns.

These data and the variation of Fo content of olivine and Cr/(Cr+Al) of spinel suggest that ASUS underwent minor partial melting/melt segregation during the influx of a small amount of slab-derived H_2O-rich fluid into N-MORB source, whereas the CRUS underwent extensive partial melting/melt segregation under an influx of significant amounts of slab-derived fluid (Ozawa 1988, 1990). Yoshikawa & Ozawa (2007) examined Rb–Sr and Sm–Nd isotope systems and trace element patterns on carefully separated clinopyroxene in ASUS and CRUS peridotites and strengthened the flux melting model by revealing: (1) higher $^{143}Nd/^{144}Nd$ and $^{147}Sm/^{144}Nd$ ratio for the Hayachine ASUS, which indicates a lower degree of partial melting than in the Miyamori CRUS; and (2) more enriched Sr and Nd isotope initial ratios than those of Ordovician MORB. They also recognized several distinct slab-derived components in clinopyroxene trace element patterns on the basis of the variable fractionation of Sr and Zr relative to REE, from which the depths of derivation of involved fluid/melt for the Hayachine ASUS, Miyamori ASUS and Miyamori CRUS are estimated. They inferred most importantly that the Hayachine ASUS underwent melting in the back-arc region (e.g. back-arc basin) and that the Miyamori ASUS and CRUS underwent melting in the forearc region.

Geotectonic evolution of the South Kitakami Massif

High-*P* and low-*T* metamorphic rocks (Motai unit) are distributed in the southwestern part of the South Kitakami Massif (Maekawa 1981, 1988; Fig. 3.11). The age of the protoliths for these rocks is broadly Ordovician–Devonian, that is, younger than the emplacement of the Hayachine–Miyamori Ophiolite, although the metamorphic ages remain controversial. East and northeast of the Motai outcrops is a series of linear Cretaceous granites (the Senmaya, Hitokabe and Yamaya plutons; Fig. 3.11). The shape of these plutons, along with the distribution patterns of Triassic strata to the south of the Senmaya intrusion, suggest a sinistral strike-slip fault with *c.* 10 km displacement since Jurassic times, referred to as the Hitokabe–Iriya Fault (Ehiro 1977). The Motai metamorphic rocks were positioned very closely to the west of the Hayachine–Miyamori Ophiolite prior to this fault movement. It is inferred that the subduction in NE Honshu during Ordovician–Devonian time was from the west to the east. This subduction polarity is based on the fact that the Motai metamorphic rocks are situated to the west of the Hayachine–Miyamori arc ophiolite and that the trace element patterns of the most fertile ASUS of the Hayachine Complex underwent a small degree of partial melting in a back-arc basin with a limited influx of fluid derived from a very deep part of the subducting slab (Yoshikawa & Ozawa 2007), placing Miyamori in the forearc and Hayachine in the back-arc. Kawamura *et al.* (2007) reported $^{40}Ar/^{39}Ar$ ages of 380 Ma from separated muscovite in the high-*P/T* metamorphic rocks (Tateishi Schists), which occur as a tectonic block in the Carboniferous accretionary complex (Nedamo Belt). Its mode of occurrence as a tectonic block (Uchino & Kawamura 2006), along with the presence of Late Devonian greenstones with MORB or within-plate tholeiite compositions in the Nedamo Belt (Hamano *et al.* 2002), indicate that SW-dipping subduction of oceanic lithosphere had already started in Late Devonian times. A large change in subduction polarity from eastwards- to westwards-dipping might have taken place during Late Devonian times. The polarity change may have been related to a major tectonic event, such as the break-up of a Gondwana supercontinent.

Ultramafic xenoliths from Japan: an insight into sub-arc mantle (SA, SI, NA, MT)

Ultramafic xenoliths from island arcs are generally less well known than those from oceanic hotspots and continental rifts (e.g. Nixon 1987). The Japan Arc area however is one of the classical examples that have provided us with direct petrological information on sub-arc mantle through the study of peridotite xenoliths (e.g. Kuno 1967; Takahashi 1978, 1980, 1986; Aoki 1987; Abe *et al.* 1998; Arai *et al.* 2000). The ultramafic xenoliths are mainly from the rear-arc side of the Japan Arc. Our understanding of the sub-arc mantle (mantle wedge) has been recently widened by an abundance of data from peridotite xenoliths from the Kamchatka Arc, especially from Avacha volcano (e.g. Ishimaru *et al.* 2007; Ishimaru & Arai 2008), which may be representative of the mantle materials beneath a

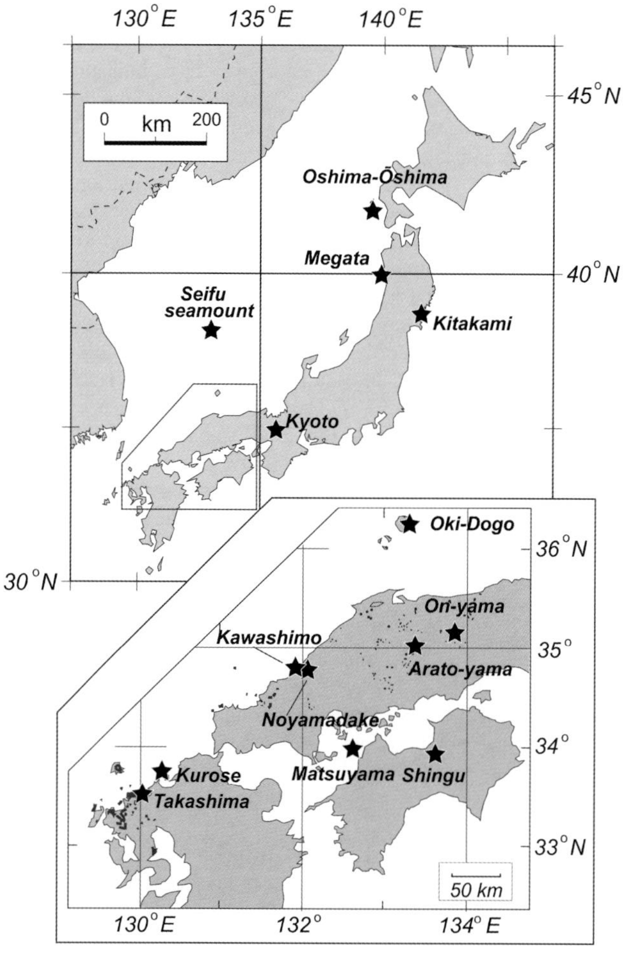

Fig. 3.14. Distribution of main localities of ultramafic xenoliths in the Japan Arc and the Sea of Japan. Cenozoic basalts related to the opening of the Sea of Japan are shown in dark grey on the SW Japan Arc (Takamura 1973). Black star, main xenolith locality.

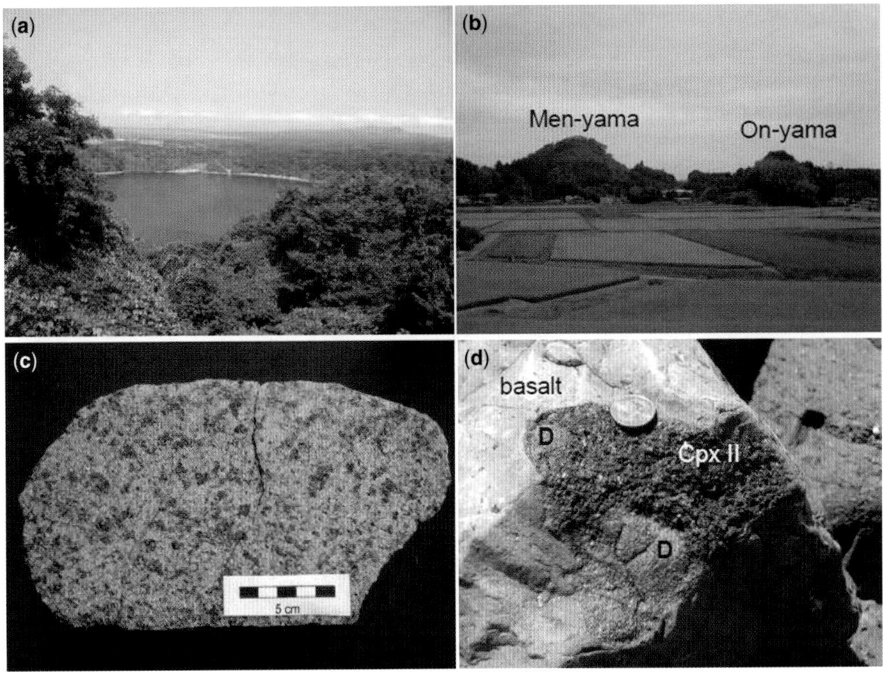

Fig. 3.15. Photographs of xenolith-bearing monogenetic volcanoes and xenoliths. (**a**) Ichinomegata crater of Megata Volcano, the NE Japan Arc (Fig. 3.1). (**b**) Monogenetic volcanoes providing deep-seated xenoliths from the SW Japan Arc. On-yama and Men-yama of Tsuyama basin cluster (Fig. 3.1). (**c**) Sawed surface of one of the largest peridotite (harzburgite) xenoliths from Ichinomegata. It occurred as a rounded bomb in andesitic to dacitic pyroclastics. The dark part is orthopyroxene. (**d**) Composite xenolith of black coarse-grained Group II clinopyroxenite (Cpx II) containing dunite clasts (D) from Takashima, the SW Japan Arc (Fig. 3.1).

volcanic front, complimentary to the data from the Japan Arc xenoliths. Xenoliths from the Sea of Japan (e.g. Seifu seamount and Oki-Dogo; Fig. 3.14) provide information on deep parts of the back-arc basin (e.g. Arai & Ninomiya 2006; Ninomiya *et al.* 2007). Peridotite xenoliths from the Japan Arc have been compared with other xenoliths from the western Pacific rim (Arai *et al.* 2007) and with sub-arc peridotite xenoliths from around the world (Arai & Ishimaru 2008). Virtually all of the mantle peridotite xenoliths are of chromian spinel-bearing type, with plagioclase lherzolites being found only from Megata Volcano, and no garnet-bearing peridotites have been recognized from the Japan Arc (e.g. Abe & Arai 2001). However, Mizukami *et al.* (2008) found microdiamond inclusions in clinopyroxenes in pyroxenites from the SW Japan Arc. Other ultramafic xenoliths (Fig. 3.14) have been reported from Lower Cretaceous high-Mg andesites at Kitakami (mainly pyroxenites: Tsuchiya *et al.* 2005) and from the Setouchi high-Mg andesites of Miocene age at Matsuyama, Shikoku (dunite and pyroxenites: Tsuchiya *et al.* 2005). Finally, Ishimaru *et al.* (2009) described hydrous peridotite xenoliths from a spessartite dyke (108 Ma) in Kyoto (Fig. 3.14).

We adopt the standard nomenclature for rock names except for lherzolite and harzburgite, the boundary between which is more meaningfully placed at a clinopyroxene/(orthopyroxene + clinopyroxene) volume ratio of 0.1 (see Arai 1994) instead of at a clinopyroxene 5 vol% (Arai 1994). We distinguish between Group I (with metamorphic textures and green clinopyroxene) and Group II (with igneous textures and black clinopyroxene)-type xenoliths (Frey & Prinz 1978). Group I xenoliths are widespread across Japan, whereas Group II xenoliths are present only in the SW Japan Arc.

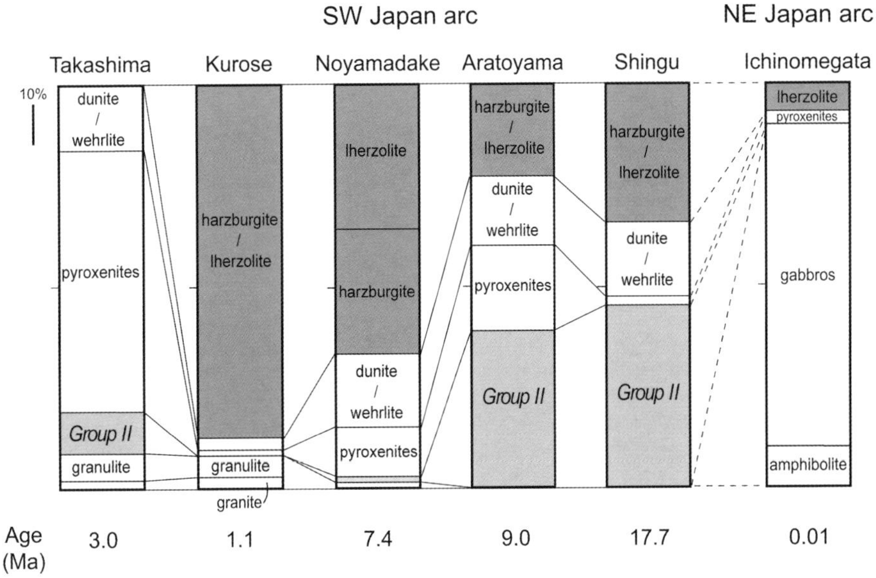

Fig. 3.16. Relative abundances of rock species in xenolith suites from the main localities from the Japan Arc. Data for the SW Japan Arc and for Ichinomegata (NE Japan Arc) are from Arai *et al.* (2005) and Takahashi (1986), respectively. Note that harzburgite is less abundant than lherzolite in the Ichinomegata xenolith suite.

Geological background

Only two of the five arcs (Kuril, NE Japan, Izu–Ogasawara, SW Japan and Ryukyu) forming the Japanese islands provide us with mantle-derived ultramafic xenoliths (i.e. the NE and SW Japan arcs: Fig. 3.14). Host rocks for xenoliths in the SW Japan Arc are mostly Cenozoic alkaline basalts related to mantle diapirism responsible for the opening of the Sea of Japan. In contrast, in NE Japan xenolith host rocks are mainly andesites and basalts of arc type.

Only two localities of mantle xenoliths (Megata and Oshima-Ōshima) have been reported from the NE Japan Arc (Arai *et al.* 2007; Arai & Ishimaru 2008; Fig. 3.14). Magata Volcano on the Oga Peninsula has three craters: Ichinomegata, Ninomegata and Sannomegata (Fig. 3.15). Volcanic rocks from these three craters provide ultramafic to mafic xenoliths in pyroclastics of *c.* 0.01 Ma in age, although only the xenoliths from Ichinomegata are important in abundance. The host rock is calc-alkaline dacite to andesite (Ichinomegata and Ninomegata) and high-Al basalt (Sannomegata) (Katsui *et al.* 1979; Aoki 1987). Oshima-Ōshima Volcano, which is active, is located in the Sea of Japan *c.* 50 km off Hokkaido, and is the most continent-wards volcano in the NE Japan Arc (Fig. 3.14). The calc-alkaline andesite from Oshima-Ōshima Volcano contains ultramafic xenoliths (Ninomiya & Arai 1992).

Many localities yielding ultramafic xenoliths are scattered throughout the Japan Sea side of western Japan (Fig. 3.14). Their host alkaline basalts form small monogenetic volcanoes (Fig. 3.15) composed of basal pyroclastics and overlying lavas (Hirai 1983; Arai *et al.* 2005), with ages in the range 12–1 Ma (Uto 1990). An exception is provided by alkali basalt (or lamprophyre) dykes within Sanbagawa high-*P* schists, exposed at Shingu, Shikoku Island (Goto & Arai 1987). The Shingu dykes are located on the Pacific-side (in the Outer Zone of SW Japan) of all mantle-xenolith-bearing volcanics in Japan. They are older (18 Ma; Uto *et al.* 1987) than the other xenolith-bearing alkaline basalts in the Japan-Sea side (the Inner Zone) of SW Japan (Fig. 3.14). It is noteworthy that all of the ultramafic xenoliths from the Japan Arc (except those from Shingu) were produced after the establishment of the current arc system following the opening of the Sea of Japan. They are therefore representative of deep-seated materials of the current Japan Arc (Figs 3.14 & 3.16). Finally, some of the xenolith localities are very easy to access, especially Oki-Dogo and Takashima (Figs 3.14 & 3.15d), where alkaline basalts containing abundant deep-seated xenoliths can be observed on beach boulders and outcrops.

Petrographic descriptions

NE Japan Arc

Peridotite xenoliths from Megata Volcano vary from lherzolite to harzburgite (Aoki 1987; Abe 1997) (Figs 3.15 & 3.17), with the former being more abundant and finer grained than the latter (Takahashi 1980; Abe 1997). Textures vary from slightly fine-grained protogranular in lherzolites to coarser-grained prophyroclastic in harzburgites (Abe *et al.* 1999), and some of the Megata lherzolites are finer grained than ordinary continental mantle peridotites with similar textures (Mercier & Nicolas 1975). These fertile lherzolites contain plagioclase which has been converted, partially or completely, to spinel-pyroxene symplectite through reaction with olivine during cooling (Takahashi 1986). These symplectites, sometimes globular in shape, were once considered to be garnet pseudomorphs (Kuno 1967; Kuno & Aoki 1970). Some of the Ichinomegata peridotites are hydrated, producing pargasite and, rarely, phlogopite (Aoki & Shiba 1973; Arai 1986; Abe *et al.* 1995). Pargasites sometimes cut pyroxenes as veinlets and replace the symplectite, indicating a

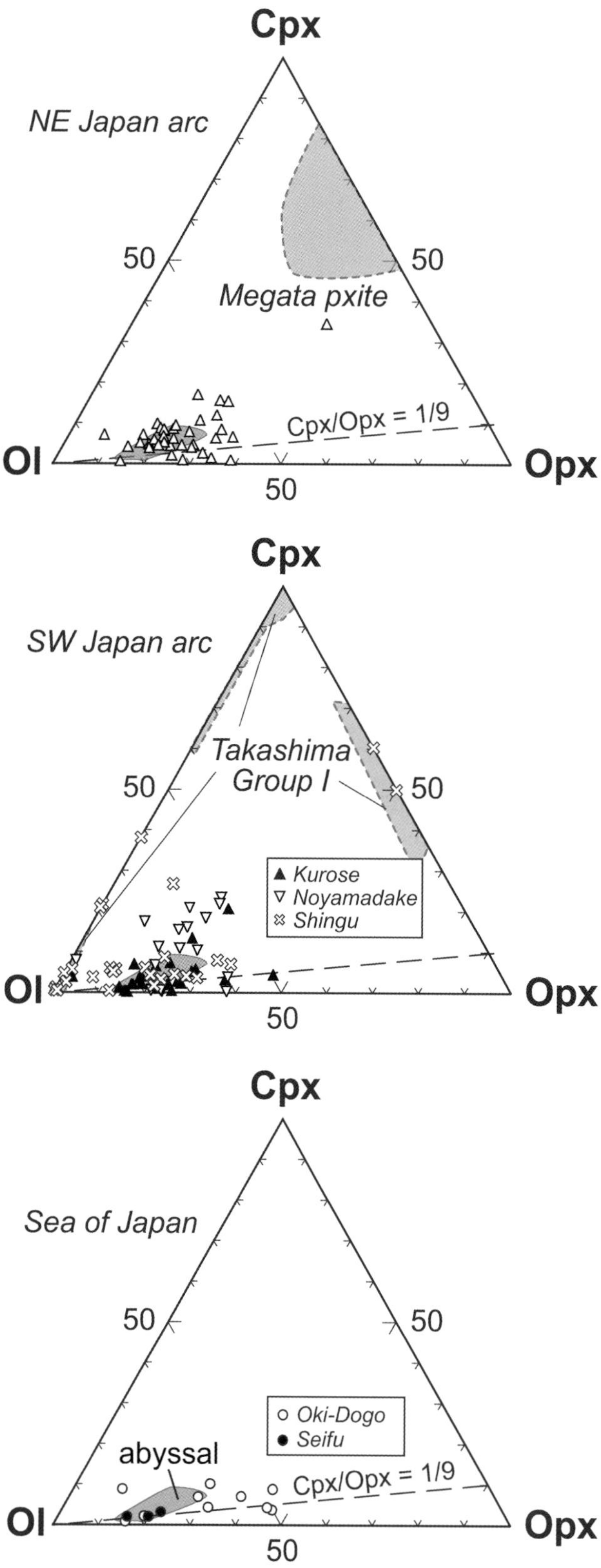

Fig. 3.17. Modal compositions of peridotite xenoliths from the main localities from Japan. Ol, olivine; Opx, orthopyroxene; Cpx, clinopyroxene. Some pyroxenites (pxite) are shown for comparison. The field for abyssal peridotites (Dick *et al.* 1984) is indicated by shaded area. Broken line denotes Cpx/(Opx + Cpx) = 0.1. Data are from Arai (1986), Hirai (1986), Goto & Arai (1987), Abe *et al.* (1992, 1995, 2003), Abe (1997) and Ninomiya *et al.* (2007).

subsolidus metasomatic product (Arai 1986; Takahashi 1986). Websterites and clinopyroxenites, which are hydrated to various degrees, are commonly found from Ichinomegata (Aoki & Shiba 1973; Takahashi 1986) (Fig. 3.17). Gabbroic rocks (which are mostly rich in hornblende) and amphibolites are abundant as xenoliths from Megata (Aoki 1971; Takahashi 1986) (Fig. 3.16). Arai & Saeki (1980) argued for an almost anhydrous nature of mafic and ultramafic xenoliths from Sannomegata, in contrast to the Ichinomegata xenoliths which are more or less hydrous. Curiously, xenoliths of dunite, one of the key mantle rocks (Quick 1981), are nevertheless rather rare from Megata Volcano (Fig. 3.16).

Andesite lavas from Oshima-Ōshima Volcano contain a large amount of ultramafic rocks (mainly dunites and wehrlites and clinopyroxenites) as xenoliths (Yamamoto 1983; Ninomiya & Arai 1992). Most of them form composite xenoliths with gabbroic rocks (Ninomiya & Arai 1992; Arai & Ninomiya 2006). Harzburgite (weakly porphyroclastic) has been very rarely found as clasts within the gabbroic xenoliths from this volcano (Ninomiya & Arai 1992; Arai & Ninomiya 2006) (Fig. 3.16).

SW Japan Arc

The Cenozoic volcanoes (Takamura 1973) related to the opening of the Sea of Japan have a common age and lava chemistry and in several cases yield ultramafic xenoliths (Uto 1990; Iwamori 1991) (Fig. 3.14). Some of the mantle xenolith-bearing basalts have erupted from a solitary monogenetic volcano, for example Kurose and Noyamadake (Arai *et al.* 2000) (Figs 3.14 & 3.15), whereas others emanate from volcanic clusters. The frequency of deep-seated xenoliths in terms of rock species is highly variable from one locality to another (e.g. Arai *et al.* 2000) (Fig. 3.16), but Group II xenoliths (mostly pyroxenites) are abundant in almost all localities, especially in monogenetic volcanoes from a volcano cluster (Arai *et al.* 2000) (Fig. 3.16). In contrast, they are almost absent or scarce from the solitary volcanoes such as Kurose and Noyamadake (Arai *et al.* 2000) (Fig. 3.16).

Mantle peridotites intermediate between lherzolite and harzburgite in composition are especially abundant in the xenolith suite collected from Kurose, northern Kyushu, and occur along with fine-grained Fe-rich lherzolites, spinel websterites and granulites (Arai *et al.* 2000; Arai & Abe 2008) (Figs 3.16 & 3.17). Xenolith compositions similarly intermediate between lherzolite and harzburgite are also found at Shingu (Goto & Arai 1987; Arai *et al.* 2000) (Fig. 3.17), where minute diamond inclusions have been found in clinopyroxene from clinopyroxenites (Mizukami *et al.* 2008). Lherzolites are predominant in a xenolith suite from On-yama (Tsuyama Basin cluster) (Arai & Muraoka 1992) (Figs 3.14 & 3.15); in other cases xenolith suite compositions are highly variable and range from typical harzburgite to lherzolite, such as at Noyamadake and Arato-yama (Kibi Plateau cluster) (Fig. 3.17; Hirai 1983, 1986; Abe *et al.* 1999; Arai *et al.* 2000). Xenoliths of Group I dunites and pyroxenites are commonly associated with the lherzolite and harzburgite xenoliths (Fig. 3.16; Arai *et al.* 2000, 2005). Takashima (Higashi–Matsuura cluster), northern Kyushu, is very famous for large amounts of ultramafic xenoliths, some of which are more than 30 cm across (Figs 3.16 & 3.17), and includes both Group I (dunites, wehrlites, clinopyroxenites, websterites) and younger Group II (pyroxenites and pyroxene megacrysts) lithologies (Figs 3.15 & 3.16).

The textures of mantle peridotites in SW Japan are highly variable (Abe *et al.* 1999). Peridotites from Kurose, On-yama, Arato-yama and Shingu show protogranular to porphyroclastic textures as seen in the Megata peridotites (Abe *et al.* 1999). The Noyamadake peridotites, protogranular to porphyroclastic in texture, are very poor in modal chromian spinel and their pyroxene grains show rounded shapes (Hirai 1986; Abe *et al.* 1999) due to high equilibrium temperatures as described below.

Sea of Japan

The Oki islands in the Sea of Japan off Honshu Island form a remnant of continental margin trapped by back-arc-basin lithosphere during the opening of the Japan Sea. Xenoliths from these islands (Fig. 3.14) are highly variable, ranging from mantle peridotites (mainly lherzolite) to gabbroic rocks through various-textured pyroxenites and peridotites (Groups I, II and intermediate type) (Takahashi 1978; Abe *et al.* 2003). Peridotite xenoliths have been obtained from a small alkali basalt (*c.* 8 Ma) seamount near Takeshima (Seifu seamount, 'Takeshima seamount' in Ninomiya *et al.* 2007) (Fig. 3.14). Most of them are slightly fertile (clinopyroxene-rich) harzburgites to depleted lherzolite (Ninomiya *et al.* 2007). The Oshima-Ōshima Volcano is considered to be located on the western rim of the NE Japan Arc rim (e.g. Tamaki *et al.* 1992) (Fig. 3.14), and some of the xenoliths from this volcano may represent the deep part of lithosphere of the Sea of Japan (Arai & Ninomiya 2006).

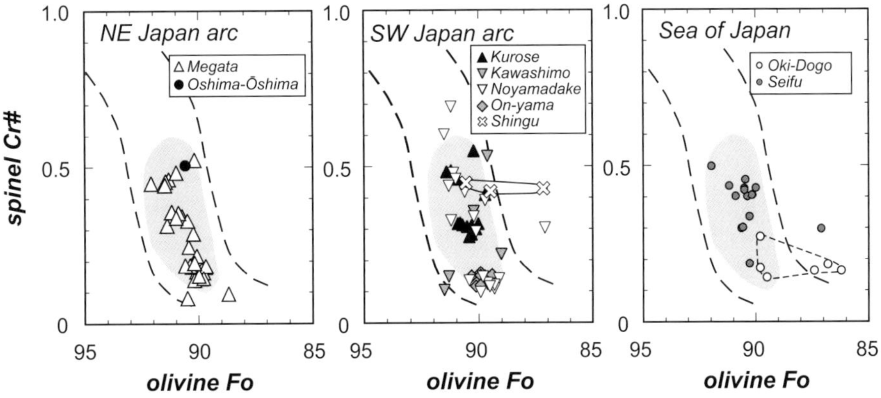

Fig. 3.18. Relations between the Fo content of olivine and the Cr/(Cr + Al) ratio in mantle peridotite xenoliths from Japan. Cr# = Cr/(Cr + Al) atomic ratio. OSMA, olivine-spinel mantle array denoting residual peridotite trend in spinel peridotite stability field (Arai 1994). Shaded area indicates the field for abyssal peridotites (Arai 1994). Peridotites off the OSMA are possibly metasomatized peridotites (cf. Arai *et al.* 2000). Note that almost all peridotite xenoliths from Japan are similar in Fo (olivine)–Cr# (spinel) relation to abyssal peridotites. Data are from Arai (1986), Hirai (1986), Goto & Arai (1987), Abe *et al.* (1992, 1995, 2003), Arai & Muraoka (1992), Ninomiya & Arai (1992), Abe (1997) and Ninomiya *et al.* (2007).

Mineral chemistry

NE Japan Arc

The mantle peridotites from the NE Japan Arc are similar in major-element mineral chemistry to abyssal peridotites recovered from the current ocean floor (Arai & Ishimaru 2008) (Figs 3.18 & 3.19). The Fo content (=100 Mg/(Mg + Fe^{2+}) atomic ratio) of olivine, from 89 to 92, shows a weak positive correlation with the Cr# (=Cr/(Cr + Al) atomic ratio) of chromian spinel, from *c.* 0.5 to 0.1 (Fig. 3.18; Arai & Ishimaru 2008). The harzburgite from Oshima-Ōshima is more depleted in its major-element mineral chemistry and has a slightly higher Cr# of chromian spinel than peridotites from Megata (Fig. 3.18). Hydrous peridotites from Ichinomegata that contain >10 vol% hornblende (pargasite) show clearly lower Fo contents (82–87) of olivine than anhydrous and slightly hydrous peridotites (Abe *et al.* 1995). The Y_{Fe} (=Fe^{3+}/(Cr + Al + Fe^{3+})) atomic ratio of chromian spinel is higher in the NE Japan peridotites, especially in the hydrated Ichinomegata perdiotites, than in abyssal peridotites (Arai & Ishimaru 2008). Clinopyroxenes are characterized by low Na contents compared with peridotite with the same Cr# of spinel in peridotite xenoliths from the NE Japan Arc, as also seen in abyssal peridotites (Fig. 3.19). It is noteworthy that low-Na clinopyroxene can even be found in anhydrous peridotites (Abe *et al.* 1995). Plagioclase in the Megata lherzolite is An_{72} in composition (Takahashi 1986).

Finally, pargasites and phlogopites in the Megata peridotites are low in TiO_2 relative to those in subcontinental peridotite xenoliths (Arai & Ishimaru 2008), with most of the Megata pargasites also being very low in K_2O (<0.1 wt%) (Abe *et al.* 1992; Abe 1997).

SW Japan Arc

The mineral chemistry of the peridotite xenoliths from the SW Japan Arc are complicated compared with those of the NE Japan Arc. This is mainly due to local differences in metasomatic overprints from magmas involved in the formation of Group II rocks (Arai *et al.* 2000). Peridotites veined by the Group II melts were chemically modified with or without a modal change, so that olivines and spinels became variously enriched in Fe, and Fe^{3+} and Ti, respectively (Goto & Arai 1987; Arai *et al.* 2000). Macroscopically, peridotite xenoliths from monogenetic volcanoes (e.g. Arato-yama) from a volcano cluster are clearly more metasomatized in chemistry than those from the solitary monogenetic volcanoes of Noyamadake and Kurose (Fig. 3.15) (Arai *et al.* 2000).

The SW Japan Arc peridotites (with the exception of the metasomatized peridotites) as a whole show a weak positive correlation between the Fo content of olivine (89–91) and the Cr# of spinel (0.7–0.1), preserving a residual character (Fig. 3.18). The mantle peridotites are least affected by the metasomatism so mostly have a similar mineral chemistry to those in NE Japan, that is, to abyssal peridotites (Arai & Ishimaru 2008; Fig. 3.18). A notable exception to this is provided by the harzburgite from Noyamadake which is more refractory than abyssal harzburgite; it contains chromian spinel with high Cr#s (higher than 0.6–0.7) and more sodic clinopyroxene (Hirai 1986; Arai *et al.* 2000; Arai & Ishimaru 2008; Figs 3.18 & 3.19).

Sea of Japan

Lherzolites are predominant in the mantle peridotite xenoliths from Oki–Dogo (Takahashi 1978; Abe *et al.* 2003), with a Cr# of chromian spinel ranging from 0.1 to 0.3 and olivine Fo_{90} characterizing unmetasomatized examples, decreasing to Fo_{86-87} where metasomatized by the Group II melt (Abe *et al.* 2003). In the case of the Seifu Seamount, the Fo content of olivine and the Cr# of associated spinel vary from 90 to 92 and 0.5 to 0.2, respectively, in harzburgites and lherzolites (Fig. 3.18; Ninomiya *et al.* 2007). Peridotite xenoliths from the Sea of Japan area show Cr# (spinel)-Na_2O (clinopyroxene) compositions intermediate between abyssal peridotites and subcontinental peridotites (Fig. 3.19).

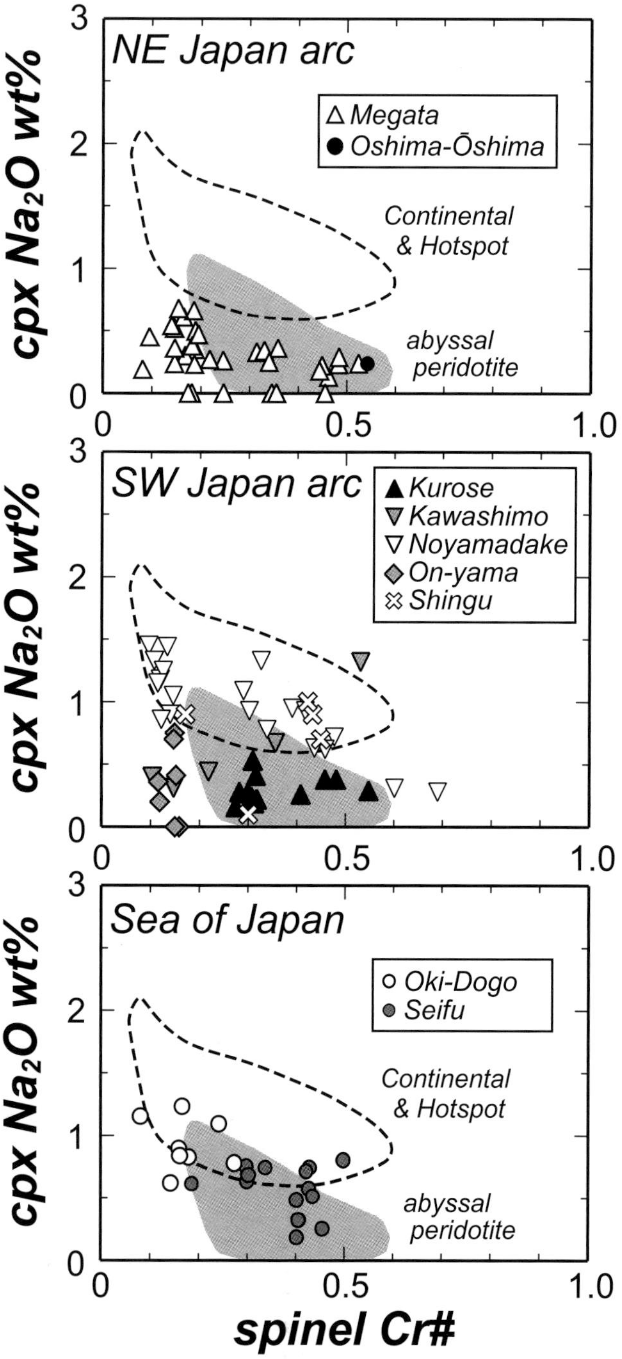

Fig. 3.19. Relationships between the Na_2O content of clinopyroxene and the Cr#(Cr/(Cr + Al) atomic ratio) of chromian spinel in peridotite xenoliths from the Japan Arc. Fields for continental and hotspot peridotites and abyssal peridotite are after Arai (1991). Data are from Arai (1986), Hirai (1986), Goto & Arai (1987), Abe *et al.* (1992, 1995, 2003), Arai & Muraoka (1992), Ninomiya & Arai (1992), Abe (1997) and Ninomiya *et al.* (2007).

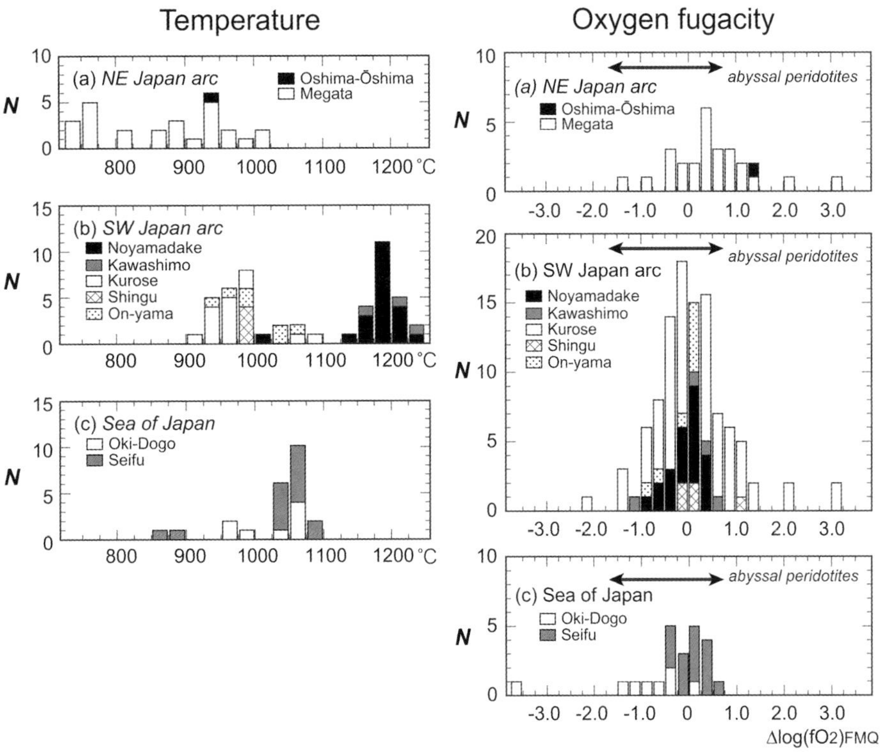

Fig. 3.20. Histograms of the calculated equilibrium temperature (left) and those of $\Delta\log(fO_2)_{FMQ}$ in peridotite xenoliths from the Japan Arc. The two-pyroxene thermometer of Wells (1977) was used for the temperature calculation. $\Delta\log(fO_2)_{FMQ}$ (=oxygen fugacity relative to the FMQ value) was calculated after Ballhaus *et al.* (1990).

Trace-element geochemistry

The trace-element geochemistry of clinopyroxenes in mantle xenoliths indicates that the Japan Arc peridotites are intermediate between abyssal and subcontinental compositions in terms of clinopyroxene geochemistry (Abe *et al.* 1998, 1999; Abe & Arai 2005). Some of the lherzolitic peridotites from Japan, which are mostly protogranular in texture, are similar to abyssal lherzolites, whereas harzburgitic peridotites, some of which are strongly deformed or sheared, show enrichment in light to middle rare Earth elements (Abe *et al.* 1998). In the case of the Megata peridotites in NE Japan, it can be shown that clinopyroxene geochemical characteristics are independent of the presence or absence of hydration (pargasite formation) (Abe *et al.* 1998; Abe & Arai 2005).

Thermobarometry

As stated above, almost all the ultramafic xenoliths documented from the Japan arcs probably equilibrated in the spinel-peridotite (to plagioclase-peridotite) facies, although high-pressure relics can be locally preserved (e.g. Mizukami *et al.* 2008).

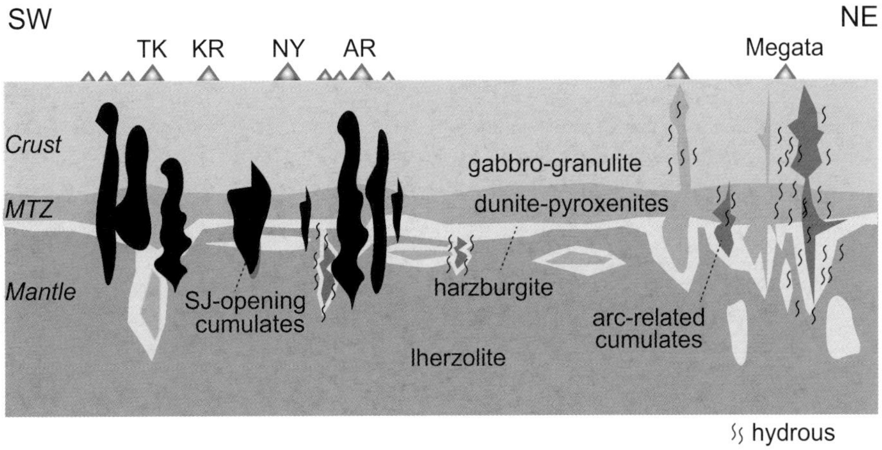

Fig. 3.21. Petrologic model of the deep part of the Japan Sea side of the NE Japan Arc to the SW Japan Arc. The across-arc profile was provided by Arai & Ishimaru (2008). The basic structure of the Japan Arc had been established at the eastern rim of the Eurasian continent before the opening of the Sea of Japan. Cumulates from arc-related magmas formed, and related metasomatic precipitation of hydrous minerals proceeded more prominently beneath the NE Japan Arc than beneath the SW Japan Arc. Cumulates (mainly pyroxenites) from basaltic magmas related to the opening of the Sea of Japan ('SJ-opening cumulates') are widely distributed in the deep part of the SW Japan Arc. For more detailed petrologic structures beneath the Japan Arc and the Sea of Japan, see Takahashi (1978), Aoki (1987), Arai *et al.* (2000) and Ninomiya *et al.* (2007). MTZ, Moho transition zone; AR, Arato-yama; NY, Noyamadake; KR, Kurose; TK, Takashima.

The equilibrium temperatures of mantle peridotites based on the two-pyroxene thermometry of Wells (1977) are highly variable (<800°C to >1200°C; Fig. 3.20). Peridotites from the NE Japan Arc show temperatures mostly lower than 1000°C (Fig. 3.20), whereas the Noyamadake peridotites from SW Japan Arc give temperatures of *c.* 1200°C (Fig. 3.20), consistent with their peculiar textures with rounded pyroxene grains (Abe & Arai 2005) as well as their high Mg# of spinel at a given Cr# (Arai *et al.* 2000; Arai & Ishimaru 2008). Most of the other mantle peridotite xenoliths from the NE Japan Arc yield intermediate values (900–1000°C: Fig. 3.23), although those from the Sea of Japan area show slightly high temperatures of 1000–1100°C (Fig. 3.20).

The redox state of the upper mantle recorded by the Japan Arc peridotite xenoliths can be examined by using the oxygen barometer $\Delta\log(fO_2)_{FMQ}$ (=oxygen fugacity relative to the Fayalite–Magnetite–Quartz buffer) value of Ballhaus *et al.* (1990, 1991) (Fig. 3.20). Such values range from −1 to +2 (mostly 0 to +1) in peridotite xenoliths from the NE Japan Arc, which is slightly higher than that for abyssal peridotites (−2 to +1: Ballhaus *et al.* 1991) (Fig. 3.20). By comparison, the range of such values in peridotites from the SW Japan Arc is mostly from −1.5 to +1.5 in peridotites, almost similar to or slightly larger than the abyssal peridotite value, and the Japan Sea peridotites are almost similar to abyssal peridotites.

Petrologic model of the sub-arc mantle

The xenoliths described above are representative of the residual mantle beneath the Japan Arc (especially the back-arc side) and the Sea of Japan (Kuno & Aoki 1970; Takahashi 1978; Abe *et al.* 1998; Arai *et al.* 2000, 2007; Arai & Ishimaru 2008). Most of them were derived from the uppermost mantle in the spinel (or plagioclase) stability field, although the depth of their derivation is variable. The Noyamadake peridotites, which have the highest equilibrium temperature, possibly represent the deepest material. Dunites, wehrlites and pyroxenites of Group I may form a Moho transition zone (or cumulus mantle; Takahashi 1978) between the residual mantle and the lower crust (Fig. 3.21). Observations on individual xenoliths indicate that these lithologies also form bands and dykes within the residual mantle peridotite. They are mostly cumulates, but some of the dunites may be a reaction product between harzburgite and melt (Arai & Abe 1994). Spinel-bearing websterites were possibly converted from troctolitic rocks through a subsolidus reaction between olivine and plagioclase (Arai & Abe 2008).

The variation of the mantle peridotite from lherzolite to harzburgite was made by the difference in degree of partial melting (e.g. Arai 1994). The melts involved were basically arc-related magmas, produced by melting of the mantle wedge assisted by slab-derived fluids (see Abe *et al.* 1998).

Pargasitic amphiboles were commonly produced in the residual mantle peridotite beneath the NE Japan Arc (Aoki & Shiba 1973; Arai 1986). Associated with Al-rich spinels, they form around the Moho transition zone as a result of the reaction between olivine, plagioclase and water supplied from the slab, obscuring the initial Moho (i.e. the original physical contact between the plagioclase-rich lower crust and olivine-rich upper mantle) (Arai & Abe 2008).

In the SW Japan Arc, Group II rocks (dunites, wehrlites and pyroxenites) were widely produced in the upper mantle to lower crust beneath the Cenozoic monogenetic volcano clusters (Arai *et al.* 2000) (Fig. 3.21). In contrast, such rocks are rare or absent from the Sea of Japan area where there are few such volcanoes (Arai *et al.* 2000), and only minimally developed in the NE Japan Arc (Fig. 3.21). This difference reflects the fact that in the NE Japan Arc a mantle diapir likely upwelled along a simple rift to form the Yamato Basin (oceanic lithosphere) in the northern part of the Sea of Japan (e.g. Tamaki & Honza 1991; Tamaki *et al.* 1992), and the roots of the arc have been more modified by crystal accumulation and metasomatism from recent arc magma. This is consistent with the wide distribution of thinned and rifted continental crust (lithosphere) in the southern part of the Sea of Japan (Tamaki *et al.* 1992; Ninomiya *et al.* 2007). Hornblendite dykes with metasomatic aureole within the Megata peridotites (Abe *et al.* 1992) provide a good example of such modification by arc-related magmatism. Hydration of the lower crust to upper mantle related to arc magmatism may have occurred at the eastern active margin of the Eurasian continent even before the opening of the Sea of Japan (e.g. Ishimaru *et al.* 2009) (Fig. 3.21).

Overviewing all of the evidence, it is clear that the upper mantle beneath the Japan Arc as revealed by deep-seated xenoliths is inhomogeneous and complicated in lithology, reflecting continual magmatic activity which modifies pre-existing rocks and varies in its spatial extent along the Western Pacific rim. Harzburgites more depleted than abyssal harzburgite are present in the sub-arc mantle (e.g. Noyamadake) and may be more common in frontal parts of the Japan Arc, although this is not verified due to the lack of xenolith localities (Ishimaru *et al.* 2007; Arai & Ishimaru 2008). Fluids released from the slab or solidifying arc magmas hydrated deep-seated rocks, especially in the NE Japan Arc, and mantle peridotites from the Japan Arc exhibit a relatively high oxidation state as predicted by Ballhaus *et al.* (1991).

The authors of this section thank H. Hirai, M. Fujiwara, K. Goto, C. Tanaka and A. Ninomiya for their collaboration. Comments by an anonymous reviewer and A. Ishiwatari were very helpful in revision of the manuscript.

Appendix

English to Kanji and Hiragana translations for geological and place names

Akiyoshi	秋吉	あきよし
Apoi	アポイ	あぽい
Arato-yama	荒戸山	あらとやま
Ashidachi	足立	あしだち
Ashio	足尾	あしお
Atetsu	阿哲	あてつ
Boso	房総	ぼうそう
Chichibu	秩父	ちちぶ
Chichijima	父島	ちちじま
Chishima	千島	ちしま
Chugoku	中国	ちゅうごく
Fuko	普甲	ふこう
Fukui	福井	ふくい
Funafuseyama	舟伏山	ふなふせやま
Furano	富良野	ふらの
Hahajima	母島	ははじま
Hamamatsu	浜松	はままつ
Hayachine	早池峰	はやちね
Hida	飛騨	ひだ
Hida Gaien	飛騨外縁	ひだがいえん
Hidaka	日高	ひだか
Higashi-Matsuura	東松浦	ひがしまつうら

(*Continued*)

English to Kanji and Hiragana translations for geological and place names (Continued)

Hikami	氷上	ひかみ
Hirao	平尾	ひらお
Hiroshima	広島	ひろしま
Hitokabe	人首	ひとかべ
Hizume	日詰	ひづめ
Hokkaido	北海道	ほっかいどう
Honshu	本州	ほんしゅう
Horokanai	幌加内	ほろかない
Horoman	幌満	ほろまん
Hyogo	兵庫	ひょうご
Ibukiyama	伊吹山	いぶきやま
Ichinomegata	一ノ目潟	いちのめがた
Idonnappu	イドンナップ	いどんなっぷ
Iriya	入谷	いりや
Ishikari	石狩	いしかり
Iwanai-dake	岩内岳	いわないだけ
Izu	伊豆	いず
Izushi	出石	いずし
Joetsu	上越	じょうえつ
Kabato	樺戸	かばと
Kamigori	上郡	かみごおり
Kamuikotan	神居古潭	かむいこたん
Kanto	関東	かんとう
Keitosan	鶏頭山	けいとうさん
Kesennuma	気仙沼	けせんぬま
Kibi	吉備	きび
Kii	紀伊	きい
Kitakami	北上	きたかみ
Kobotoke	小仏	こぼとけ
Kotaki	小滝	こたき
Kurose	黒瀬	くろせ
Kurosegawa	黒瀬川	くろせがわ
Kuzuu	葛生	くずう
Kyoto	京都	きょうと
Kyushu	九州	きゅうしゅう
Matsuyama	松山	まつやま
Megata	目潟	めがた
Men-yama	女山	めんやま
Mikabu	御荷鉾	みかぶ
Mineoka	嶺岡	みねおか
Mino	美濃	みの
Misaka	三坂	みさか
Mitaki	三滝	みたき
Mitsuishi	三石	みついし
Miura	三浦	みうら
Miyamori	宮守	みやもり
Motai	母体	もたい
Muroto	室戸	むろと
Nagano	長野	ながの
Nagoya	名古屋	なごや
Nakadake	中岳	なかだけ
Nedamo	根田茂	ねだも
Nekozoko	猫底	ねこぞこ
Niigata	新潟	にいがた
Ninomegata	二ノ目潟	にのめがた
Nomo	野母	のも
Noyamadake	野山岳	のやまだけ
Oeyama	大江山	おおえやま
Oga*	大賀	おおが
Oga[†]	男鹿	おが
Ogasawara	小笠原	おがさわら
Ohara	大原	おおはら

(Continued)

English to Kanji and Hiragana translations for geological and place names (Continued)

Okayama	岡山	おかやま
Oki	隠岐	おき
Oki-Dogo	隠岐島後	おきどうご
Okuniikappu	奥新冠	おくにいかっぷ
Omachi	大町	おおまち
Omi	青海	おうみ
On-yama	男山	おんやま
Onigajaya	鬼ヶ茶屋	おにがぢゃや
Osaka	大阪	おおさか
Osayama	大佐山	おおさやま
Oshima	渡島	おしま
Oshima–Ōshima	渡島大島	おしまおおしま
Oya	大屋	おおや
Poroshiri	幌尻	ぽろしり
Rebun	礼文	れぶん
Renge	蓮華	れんげ
Ryoke	領家	りょうけ
Ryozen	霊仙	りょうぜん
Ryukyu	琉球	りゅうきゅう
Saijo	西条	さいじょう
Samani	様似	さまに
Sanbagawa (Sambagawa)	三波川	さんばがわ
Sannomegata	三ノ目潟	さんのめがた
Sapporo	札幌	さっぽろ
Saroma	サロマ	さろま
Seifu	清風	せいふう
Sekinomiya	関宮	せきのみや
Senmaya	千厩	せんまや
Setogawa	瀬戸川	せとがわ
Setouchi	瀬戸内	せとうち
Shikoku	四国	しこく
Shimanto	四万十	しまんと
Shimokawa	下川	しもかわ
Shingu	新宮	しんぐう
Shinshiro	新城	しんしろ
Shizunai	静内	しずない
Shizuoka	静岡	しずおか
Sorachi	空知	そらち
Taishaku	帝釈	たいしゃく
Takashima	高島	たかしま
Takeshima	竹島	たけしま
Tamba (Tanba)	丹波	たんば
Tari	多里	たり
Tateishi	建石	たていし
Terano	寺野	てらの
Toba	鳥羽	とば
Tohoku	東北	とうほく
Tokoro	常呂	ところ
Tokyo	東京	とうきょう
Tono	遠野	とおの
Tsuyama	津山	つやま
Uryu	雨竜	うりゅう
Wakasa	若桜	わかさ
Yakuno	夜久野	やくの
Yamato	大和	やまと
Yamaya	山屋	やまや
Yezo	エゾ（蝦夷）	えぞ
Zomeki	蔵目喜	ぞうめき

*For the limestone plateau in Chugoku region.

[†]For 'Oga Peninsula'.

References

Abbate, E., Bortolotti, V., Passerini, P. & Principi, G. 1985. The rhythm of Phanerozoic ophiolites. *Ofioliti*, **10**, 109–138.

Abe, N. 1997. *Petrology of Mantle Xenoliths from the Arcs: Implications for the Petrochemical Evolution of the Wedge Mantle.* PhD thesis, Kanazawa University, Japan.

Abe, N. & Arai, S. 2001. Comments on ‘Garnet-bearing spinel harzburgite xenolith from Arato-yama alkali basalt, southwest Japan’ by Yamamoto *et al. Japanese Magazine of Mineralogical and Petrological Sciences*, **30**, 190–193 [in Japanese with English abstract].

Abe, N. & Arai, S. 2005. Petrography and geochemistry of the mantle xenoliths: implications for lithospheric mantle beneath the Japan arcs. *Japanese Magazine of Mineralogical and Petrological Sciences*, **34**, 143–158 [in Japanese with English abstract].

Abe, N., Arai, S. & Saeki, Y. 1992. Hydration processes in the arc mantle; petrology of the Megata peridotite xenoliths, the Northeast Japan arc. *Journal of Mineralogy, Petrology and Economic Geology*, **87**, 305–317 [in Japanese with English abstract].

Abe, N., Arai, S. & Ninomiya, A. 1995. Peridotite xenoliths and essential ejecta from the Ninomegata crater, the Northeastern Japan arc. *Journal of Mineralogy, Petrology and Economic Geology*, **90**, 41–49 [in Japanese with English abstract].

Abe, N., Arai, S. & Yurimoto, H. 1998. Geochemical characteristics of the uppermost mantle beneath the Japan island arcs: implications for upper mantle evolution. *Physics of Earth and Planetary Interiors*, **107**, 233–247.

Abe, N., Arai, S. & Yurimoto, H. 1999. Texture-dependent geochemical variations of sub-arc mantle peridotite from Japan island arcs. *Proceedings of VIIth International Kimberlite Conference*, 11–17 April 1998, J.B. Dawson Vol., Cape Town, South Africa, 13–22.

Abe, N., Takami, M. & Arai, S. 2003. Petrological feature of spinel lherzolite xenoliths from Oki-Dogo Island: an implication for variety of the upper mantle peridotite beneath southwest Japan. *Island Arc*, **12**, 219–232.

Aoki, K. 1971. Petrology of mafic inclusions from Itinome-gata, Japan. *Contributions to Mineralogy and Petrology*, **30**, 314–331.

Aoki, K. 1987. Japanese island arc: xenoliths in alkali basalts, high-alumina basalts, and calc-alkaline andesites and dacites. *In*: Nixon, P. H. (ed.) *Mantle Xenoliths*. John Wiley, New York, 319–333.

Aoki, K. & Shiba, I. 1973. Pargasite in lherzolite and websterite inclusions from Itinome-gata, Japan. *Journal of Japanese Association of Mineralogists, Petrologists and Economic Geologists*, **68**, 303–310.

Arai, S. 1973. Compositional variation of olivine in rocks of the Tari-Misaka ultramafic complex and its interpretation. *Proceedings of the Japan Academy*, **49**, 649–653.

Arai, S. 1974. ‘Non-calciferous’ orthopyroxene and its bearing on the petrogenesis of ultramafic rocks in Sangun and Joetsu zones. *Journal of the Japanese Association of Mineralogists, Petrologists and Economic Geologists*, **69**, 343–353.

Arai, S. 1975. Contact metamorphosed dunite-harzburgite complex in the Chugoku district, western Japan. *Contributions to Mineralogy and Petrology*, **52**, 1–16.

Arai, S. 1980. Dunite-harzburgite-chromitite complexes as refractory residue in the Sangun-Yamaguchi zone, western Japan. *Journal of Petrology*, **21**, 141–165.

Arai, S. 1986. K/Na variation in phlogopite and amphibole of upper mantle peridotites due to fractionation of the metasomatic fluids. *Journal of Geology*, **94**, 436–444.

Arai, S. 1991. The circum-Izu massif peridotite, central Japan, as back-arc mantle fragments of the Izu-Bonin arc system. *In*: Peters, T. J., Nicolas, A. *et al.* (eds) *Ophiolite Genesis and Evolution of the Oceanic Lithosphere (Proceedings of Oman ’90 Ophiolite Conference)*. Ministry of Petroleum & Minerals, Sultanate of Oman, Kluwer Academic Publishing, Dordrecht, 801–816.

Arai, S. 1994. Characterization of spinel peridotites by olivine-spinel compositional relationships: review and interpretation. *Chemical Geology*, **113**, 191–204.

Arai, S. & Abe, N. 1994. Podiform chromitite in the arc mantle: chromitite xenoliths from the Takashima alkali basalt, southwest Japan arc. *Mineralium Deposita*, **29**, 434–438.

Arai, S. & Abe, N. 2008. Investigation of the petrologic nature of the Moho toward the Mohole. *Journal of Geography*, **117**, 110–123 [in Japanese with English abstract].

Arai, S. & Ishimaru, S. 2008. Insights into petrological characteristics of the lithosphere of mantle wedge beneath arcs through peridotite xenoliths: a review. *Journal of Petrology*, **49**, 665–695.

Arai, S. & Muraoka, H. 1992. Peridotite xenoliths in alkali basalts from On-yama, the Chugoku district, as a suite of fertile upper mantle peridotites beneath the Southwest Japan arc. *Journal of Petrology, Mineralogy and Economic Geology*, **87**, 240–251 [in Japanese with English abstract].

Arai, S. & Ninomiya, C. 2006. What is the upper mantle peridotite of back-arc basin? *Journal of the Geological Society of Thailand*, No. 1, 1–8.

Arai, S. & Saeki, Y. 1980. Ultramafic-mafic inclusions from Sannomegata crater, Oga Peninsula, Japan, with special reference to the petrographical difference from the Ichinomegata inclusions. *Journal of the Geological Society of Japan*, **86**, 705–708.

Arai, S., Ito, T. & Ozawa, K. 1983. Ultramafic-mafic clastic rocks from the Mineoka Belt, central Japan. *Journal of the Geological Society of Japan*, **89**, 287–297 [in Japanese with English abstract].

Arai, S., Hirai, H. & Uto, K. 2000. Mantle peridotite xenoliths from the Southwest Japan arc and a model for the sub-arc upper mantle structure and composition of the Western Pacific rim. *Journal of Mineralogical and Petrological Sciences*, **95**, 9–23.

Arai, S., Hirai, H. & Abe, N. 2005. Geological aspects of peridotite and related xenoliths in volcanic rocks: an example from the Japan arcs. *Japanese Magazine of Mineralogical and Petrological Sciences*, **34**, 133–142 [in Japanese with English abstract].

Arai, S., Abe, N. & Ishimaru, S. 2007. Mantle peridotites from the Western Pacific. *Gondwana Research*, **11**, 180–199.

Asakawa, Y., Maruyama, T. & Yamamoto, M. 1999. Rb-Sr whole-rock isochron ages of the Hikami Granitic Body in the South Kitakami Belt, Northeast Japan. *Memoirs of the Geological Society of Japan*, **53**, 221–234 [in Japanese with English abstract].

Bailey, E. B. & McCallien, W. J. 1960. Some aspect of the Steinmann trinity, mainly chemical. *Quaterly Journal of the Geological Society of London*, **116**, 365–395.

Ballhaus, C., Berry, R. F. & Green, D. H. 1990. Oxygen fugacity controls in the Earth’s upper mantle. *Nature*, **348**, 437–440.

Ballhaus, C., Berry, R. F. & Green, D. H. 1991. High pressure experimental calibration of the olivine-orthopyroxene-spinel oxygen geobarometer: implications for the oxidation state of the upper mantle. *Contributions to Mineralogy and Petrology*, **107**, 27–40.

Dick, H. J. B., Fisher, R. L. & Bryan, W. B. 1984. Mineralogic variability of the uppermost mantle along mid-ocean ridge. *Earth and Planetary Science Letters*, **69**, 88–106.

Dilek, Y. & Furnes, H. 2011. Ophiolite genesis and global tectonics: geochemical and tectonic fingerprinting of ancient oceanic lithosphere. *Geological Society of America Bulletin*, **123**, 387–411.

Ehiro, M. 1977. The Hizume-Kesennuma fault–with special reference to its character and significance on the geologic development. *Contributions from the Institute of Geology and Paleontology of Tohoku University*, **77**, 1–37 [in Japanese with English abstract].

Ehiro, M. 2000. Relationships in tectonic framework among the South Kitakami and Hayachine Tectonic Belts, Kurosegawa Belt, and ‘Paleo-RyokeBelt’. *Memoirs of the Geological Society of Japan*, **56**, 53–64 [in Japanese with English abstract].

Ehiro, M. & Suzuki, N. 2003. Re-definition of the Hayachine Tectonic Belt of Northeast Japan and a proposal of a new tectonic unit, the Nedamo Belt. Japan. *Journal of Structural Geology*, **47**, 13–21 [in Japanese with English abstract].

Ehiro, M., Okami, K. & Kanisawa, S. 1988. Recent progress and further subjects in studies on the ‘Hayachine Tectonic Belt’ in the Kitakami Massif, Northeast Japan. *Chikyu-Kagaku (Earth Science)*, **42**, 317–335 [in Japanese with English abstract].

Ehiro, M., Yamakita, S., Takahashi, S. & Suzuki, N. 2008. Jurassic accretionary complexes of the North Kitakami Belt in the Akka-Kuji area, Northeast Japan. *Journal of the Geological Society of Japan*, **114** (Suppl.), 121–139 [in Japanese with English abstract].

Ernst, W. G. 1972. Possible Permian oceanic crust and plate junction in central Shikoku, Japan. *Tectonophysics*, **15**, 233–239.

Frey, F. A. & Prinz, M. 1978. Ultramafic inclusions from San Carlos, Arizona: petrologic and geochemical data bearing on their petrogenesis. *Earth and Planetary Science Letters*, **38**, 129–176.

Fujimaki, H. & Yomogida, K. 1986. Petrology of Hayachine ultramafic complex in contact aureole, NE Japan. (I). primary and metamorphic minerals, (II) metamorphism and origin of the complex. *Journal of the*

Japanese Association of Mineralogists, Petrologists and Economic Geologists, **81**, 1–11, 59–66.

Fukudome, T. 1978. Emplacement mechanism of the Miyamori ultramafic rock body, Kitakami mountainland, Northeast Japan. *Contributions from the Institute of Geology and Paleontology of Tohoku University*, **79**, 1–32.

Goto, K. & Arai, S. 1987. Petrology of peridotite xenoliths in lamprophyre from Shingu, Southwestern Japan: implications for origin of Fe-rich mantle peridotite. *Mineralogy and Petrology*, **37**, 137–155.

Hamano, K., Iwata, K., Kawamura, M. & Kitakami Paleozoic Research Group 2002. Late Devonian conodont age of red chert intercalated in greenstone of the Hayachine Belt, Northeast Japan. *Journal of the Geological Society of Japan*, **108**, 114–122.

Hara, H., Kurihara, T. & Mori, H. 2013. Tectono-stratigraphy and low-grade metamorphism of Late Permian and Early Jurassic accretionary complexes within the Kurosegawa belt, Southwest Japan: implications for mechanisms of crustal displacement within active continental margin. *Tectonophysics*, **76**, 635–647.

Harigane, Y., Michibayashi, K. & Ohara, Y. 2011. Relicts of deformed lithospheric mantle within serpentinites and weathered peridotites from the Godzilla Megamullion, Parece Vela back-arc basin, Philippine Sea. *Island Arc*, **20**, 174–187.

Hirai, H. 1983. Mode of occurrence of alkali basaltic volcanic products and their inclusions of Noyamadake, Shimane Prefecture, southwestern Japan. *Journal of the Japanese Association for Mineralogists, Petrologists and Economic Geologists*, **78**, 211–220 [in Japanese with English abstract].

Hirai, H. 1986. *Petrology of Ultramafic Xenoliths from Noyamadake and Kurose, Southwestern Japan*. PhD thesis, University of Tsukuba, Japan.

Hirano, N., Ogawa, Y., Saito, K., Yoshida, T., Sato, H. & Taniguchi, H. 2003. Multi-stage evolution of the Tertiary Mineoka ophiolite, Japan: new geochemical and age constraints. *In*: Dilek, Y. & Robinson, P. T. (eds) *Ophiolites in Earth History*. Geological Society, London, Special Publications, **218**, 279–298.

Hoe, S. 1976. The Hikami granitic complex. *Chikyu-kagaku (Earth Science)*, **30**, 39–53.

Hoshide, T., Obata, M. & Akatsuka, T. 2006. Magmatic differentiation by means of segregation and diapiric ascent of anorthositic crystal mush – the Murotomisaki Gabbroic Complex, Shikoku, Japan. *Journal of Mineralogical and Petrological Sciences*, **101**, 334–339.

Ichiyama, Y. & Ishiwatari, A. 2004. Petrochemical evidence for off-ridge magmatism in a back-arc setting from the Yakuno ophiolite, Japan. *Island Arc*, **13**, 157–177.

Ichiyama, Y. & Ishiwatari, A. 2005. HFSE-rich picritic rocks from the Mino accretionary complex, southwestern Japan. *Contributions to Mineralogy and Petrology*, **149**, 373–387.

Ichiyama, Y., Ishiwatari, A., Hirahara, Y. & Shuto, K. 2006. Geochemical and isotopic constraints on the genesis of the Permian ferropicritic rocks from the Mino-Tamba belt, SW Japan. *Lithos*, **89**, 47–65.

Ichiyama, Y., Ishiwatari, A., Koizumi, K., Ishida, Y. & Machi, S. 2007. Olivine-spinifex basalt from the Tamba Belt, southwest Japan: evidence for Fe- and high field strength element-rich ultramafic volcanism in Permian Ocean. *Island Arc*, **16**, 493–503.

Ichiyama, Y., Ishiwatari, A. & Koizumi, K. 2008. Petrogenesis of greenstones from the Mino-Tamba belt, SW Japan: evidence for an accreted Permian oceanic plateau. *Lithos*, **100**, 127–146.

Ichiyama, Y., Ishiwatari, A., Kimura, J.-I., Senda, R., Kawabata, H. & Tatsumi, Y. 2012. Picrites in central Hokkaido: evidence of extremely high temperature magmatism in the Late Jurassic ocean recorded in an accreted oceanic plateau. *Geology*, **40**, 411–414.

Ichiyama, Y., Ishiwatari, A., Kimura, J.-I., Senda, R. & Miyamoto, T. 2014. Jurassic plume-origin ophiolites in Japan: accreted fragments of oceanic plateaus. *Contributions to Mineralogy and Petrology*, **167**, http://doi.org/10.1007/s00410-014-1019-1

Inomata, M. 1978. Geology of the Mikabu belt and ultrabasic complexes in Mt. Ufu-san and Mt. Tonmaku-yama area to the north of Lake Hamana, Central Japan. *Chikyu Kagaku (Earth Science)*, **32**, 336–344 [in Japanese with English abstract].

Irwin, W. P. 1981. Tectonic accretion of the Klamath Mountains. *In*: Ernst, W. G. (ed.) *The Geotectonic Development of California* (Rubey Volume 1), Princeton Hall, Jersey City, 29–49.

Ishimaru, S. & Arai, S. 2008. Nickel enrichment in mantle olivine beneath a volcanic front. *Contributions to Mineralogy and Petrology*, **156**, 119–131.

Ishimaru, S., Arai, S., Ishida, Y., Shirasaka, M. & Okrugin, V. M. 2007. Melting and multi-stage metasomatism in the mantle wedge beneath a frontal arc inferred from highly depleted peridotite xenoliths from the Avacha volcano, southern Kamchatka. *Journal of Petrology*, **48**, 395–433.

Ishimaru, S., Arai, S., Tamura, A., Takeuchi, M. & Kiji, M. 2009. Subarc magmatic and hydration processes inferred from a hornblende peridotite xenolith in spessartite from Kyoto, Japan. *Journal of Mineralogical and Petrological Sciences*, **114**, 97–104.

Ishiwatari, A. 1985*a*. Granulite-facies metacumulates of the Yakuno ophiolite, Japan: evidence for unusually thick oceanic crust. *Journal of Petrology*, **26**, 1–30.

Ishiwatari, A. 1985*b*. Igneous petrogenesis of the Yakuno ophiolite (Japan) in the context of the diversity of ophiolites. *Contributions to Mineralogy and Petrology*, **89**, 155–167.

Ishiwatari, A. 1991. Ophiolites in the Japanese islands: typical segment of the circum-Pacific multiple ophiolite belts. *Episodes*, **14**, 274–279.

Ishiwatari, A. 1994. Circum-Pacific Phanerozoic multiple ophiolite belts. *In*: Ishiwatari, A., Malpas, J. & Ishizuka, H. (eds) *Circum-Pacific Ophiolites (Proceedings of the 29th IGC (Kyoto)*, Part D. VSP Publishers, Netherlands, 7–28.

Ishiwatari, A. & Hayasaka, Y. 1992. Ophiolite nappes and blue-schists of the Inner Zone of Southwest Japan. *In*: Kato, H. & Noro, H. (eds) *29th IGC Field Trip Guidebook*. Geological Survey of Japan, Tsukuba, **5**, 285–325.

Ishiwatari, A. & Tsujimori, T. 2003. Paleozoic ophiolites and blueschists in Japan and Russian Primorye in the tectonic framework of East Asia: a synthesis. *Island Arc*, **12**, 190–206.

Ishiwatari, A., Tsujimori, T., Hayasaka, Y., Sugimoto, T. & Ishiga, H. 1999. Nappe-bounding thrust faults in the Paleozoic-Mesozoic accretionary orogen in the Inner Zone of southwestern Japan. *Journal of the Geological Society of Japan*, **105**, III–IV [in Japanese with English explanation].

Ishiwatari, A., Sokolov, S. D. & Vysotskiy, S. V. 2003. Petrological diversity and origin of ophiolites in Japan and Far East Russia with emphasis on depleted harzburgite. *In*: Dilek, Y. & Robinson, P. T. (eds) *Ophiolites in Earth History*. Geological Society, London, Special Publications, **218**, 597–618.

Ishiwatari, A., Yanagida, Y., Li, Y.-B., Ishii, T., Haraguchi, S., Koizumi, K., Ichiyama, Y. & Umeka, M. 2006. Dredge petrology of the boninite- and adakite-bearing Hahajima Seamount of the Ogasawara (Bonin) forearc: an ophiolite or a serpentinite seamount? *Island Arc*, **15**, 102–118.

Ishizuka, H. 1985. Prograde metamorphism of the Horokanai ophiolite in the Kamuikotan Zone, Hokkaido, Japan. *Journal of Petrology*, **26**, 391–417.

Ishizuka, H. 1987. Igneous and metamorphic petrology of the Horokanai ophiolite in the Kamuikotan zone, Hokkaido, Japan: a synthesis. *Memoirs of the Faculty of Science, Kochi University, Series E, Geology*, **8**, 1–70.

Ishizuka, O., Tani, K. & Reagan, M. K. 2014. Izu-Bonin-Mariana forearc crust as a modern ophiolite analogue. *Elements*, **10**, 115–120.

Isozaki, Y. 1996. Anatomy and genesis of a subduction-related orogen: a new view of geotectonic subdivision and evolution of the Japanese Islands. *Island Arc*, **5**, 289–320.

Iwamori, H. 1991. Zonal structure of Cenozoic basalts related to mantle upwelling in southwest Japan. *Journal of Geophysical Research*, **96**, 6157–6170.

Iwasaki, M. 1984. Sequence of igneous events and ocean-floor metamorphism in the greenstone (ophiolitic detritus deposit) from Eastern Shikoku, Japan. *Ofioliti*, **9**, 443–462.

Jones, G., Sano, H. & Valsami-Jones, E. 1993. Nature and tectonic setting of accreted basalts from the Mino terrane, central Japan. *Journal of the Geological Society of London*, **150**, 1167–1181.

Katsui, Y., Yamamoto, M., Nemoto, S. & Niida, K. 1979. Genesis of calc-alkalic andesites from Oshima-Oshima and Ichinomegata volcanoes, north Japan. *Journal of Faculty of Science, Hokkaido University, Series IV*, **19**, 157–168.

Kawamura, M., Uchino, T., Gouzu, C. & Hyodo, H. 2007. 380 Ma $^{40}Ar/^{39}Ar$ ages of the high-P/T schists obtained from the Nedamo

Terrane, Northeast Japan. *Journal of the Geological Society of Japan*, **113**, 492–499.

Kawamura, T., Uchino, T., Kawamura, M., Yoshida, K., Nakagawa, M. & Nagata, H. 2013. Geology of Hayachine San District. *Quadrangle Series, 1:50 000*. Geological Survey of Japan, AIST, Tsukuba, 101 [in Japanese with English abstract].

Kobayashi, Y., Takagi, H., Katoh, K., Sango, K. & Shibata, K. 2000. Petrochemistry and correlation of Paleozoic granitic rocks in Japan. *Memoirs of the Geological Society of Japan*, **56**, 65–88 [in Japanese with English abstract].

Koizumi, K. & Ishiwatari, A. 2006. Oceanic plateau accretion inferred from Late Paleozoic greenstones in the Jurassic Tamba accretionary complex, southwest Japan. *Island Arc*, **15**, 58–83.

Komatsu, M., Miyashita, S., Maeda, J., Osanai, Y. & Toyoshima, T. 1983. Disclosing of a deepest section of continental-type crust up-thrust as the final event of collision of arcs in Hokkaido, North Japan. *In*: Hashimoto, M. & Uyeda, S. (eds) *Accretion Tectonics in the Circum-Pacific Regions*, Terra Science Publishing, Tokyo, 149–165.

Kuno, H. 1967. Mafic and ultramafic nodules from Itinome-gata, Japan. *In*: Wyllie, P. J. (ed.) *Ultramafic and Related Rocks*. John Wiley & Sons, New York, 337–346.

Kuno, H. & Aoki, K. 1970. Chemistry of ultramafic nodules and their bearing on the origin of basaltic magmas. *Physics of Earth and Planetary Interiors*, **3**, 273–301.

Kuroda, Y. & Shimoda, S. 1967. Olivine with developed cleavages–its geological and mineralogical meanings. *Journal of the Geological Society of Japan*, **73**, 377–388.

Kurokawa, K. 1985. *Petrology of the Oeyama Ophiolitic Complex in the Inner Zone of Southwest Japan*. Science Reports of Niigata University, Series E, No. 6

Li, Y.-B., Kimura, J.-I. *et al.* 2013. High-Mg adakite and low-Ca boninite from a Bonin fore-arc seamount: implications for the reaction between slab melts and depleted mantle. *Journal of Petrology*, **54**, 1149–1175.

Machi, S. & Ishiwatari, A. 2010. Ultramafic rocks in the Kotaki area, Hida Marginal Belt, central Japan: peridotites of the Oeyama ophiolite and their metamorphism. *Journal of the Geological Society of Japan*, **116**, 293–308 [in Japanese with English abstract].

Maekawa, H. 1981. Geology of the Motai Group in the southwestern part of the Kitakami Mountain. *Journal of the Geological Society of Japan*, **87**, 543–554 [in Japanese with English abstract].

Maekawa, H. 1988. High P/T metamorphic rocks in Northeast Japan–Motai-Matsugadaira zone. *Chikyu Kagaku (Earth Science)*, **42**, 212–219 [in Japanese with English abstract].

Maekawa, H., Shozui, M., Ishii, T., Fryer, P. & Pearce, J. A. 1993. Blueschist metamorphism in an active subduction zone. *Nature*, **364**, 520–523.

Maruyama, S. 1981. The Kurosegawa melange zone in the Ino district to the north of Kochi City, central Shikoku. *Journal of the Geological Society of Japan*, **87**, 569–583.

Maruyama, S., Banno, S., Matsuda, T. & Nakajima, T. 1984. Kurosegawa Zone and its bearing on the development of the Japanese Islands. *Tectonophysics*, **110**, 47–60.

McDonough, W. F. & Sun, S. S. 1995. The composition of the Earth. *Chemical Geology*, **120**, 223–253.

Mercier, J. C. & Nicolas, A. 1975. Textures and fabrics of upper-mantle peridotites as illustrated by xenoliths from basalts. *Journal of Petrology*, **16**, 454–487.

Miyake, Y. 1985. MORB-like tholeiites formed within the Miocene forearc basin, Southwest Japan. *Lithos*, **18**, 23–34.

Miyashita, S. & Hashimoto, S. 1975. The layered basic complex of Mt. Poroshiri, Hokkaido, Japan. *Journal of the Faculty of Science*, Hokkaido University, Ser. IV, **16**, 421–452.

Miyashita, S. & Yoshida, A. 1988. Pre-Cretaceous and Cretaceous ophiolites in Hokkaido, Japan. *Bulletin de la Societe Géologique de France*, **1988**, 251–260.

Miyashita, S. & Yoshida, A. 1994. Geology and petrology of the Shimokawa ophiolite (Hokkaido, Japan): ophiolite possibly generated near R-T-T triple junction. *In*: Ishiwatari, A., Malpas, J. *et al.* (eds) *Circum-Pacific Ophiolites*. Proceedings of the 29th International Geology Congress Part D, VSP, Utrecht, 163–182.

Miyashita, S., Kizaki, K., Arai, T. & Toyoshima, T. 1994. The Poroshiri ophiolite of the Hidaka zone, Central Hokkaido, Japan. *In*: *IGCP Project 283, 321 & 359 in Japan*, Field Trip Guidebook, 20–23 September 1994, Kyobunsha, Sapporo.

Miyazoe, T., Enami, M., Nishiyama, T. & Mori, Y. 2012. Retrograde strontium metasomatism in serpentinite mélange of the Kurosegawa Zone in central Kyushu, Japan. *Mineralogical Magazine*, **76**, 635–647.

Mizukami, T. 2002. Orthopyroxene and compositional layering in the Sugashima Ultramafic Mass, Mikabu belt. *Japanese Magazine of Mineralogical and Petrological Sciences*, **31**, 87–96 [in Japanese with English abstract].

Mizukami, T., Wallis, S., Enami, M. & Kagi, H. 2008. Forearc diamond from Japan. *Geology*, **36**, 219–222.

Mori, R. & Ogawa, Y. 2005. Transpressional tectonics of the Mineoka ophiolite belt in a trench-trench-trench-type triple junction, Boso Peninsula, Japan. *Island Arc*, **14**, 571–581.

Mori, R., Ogawa, Y., Hirano, N., Tsunogae, T., Kurosawa, M. & Chiba, T. 2011. Role of plutonic and metamorphic block exhumation in a forearc ophiolite melange belt: an example from the Mineoka belt, Japan. *In*: Wakabayashi, J. & Dilek, Y. (eds) *Melanges: Processes of Formation and Societal Significance*. GSA Special Paper, **480**, 95–115.

Morishita, T. & Arai, S. 2001. Petrogenesis of corundum-bearing mafic rock in the Horoman Peridotite Complex, Japan. *Journal of Petrology*, **42**, 1279–1299.

Morishita, T. & Arai, S. 2003. Evolution of spinel-pyroxene symplectite in spinel-lherzolites from the Horoman Complex, Japan. *Contributions to Mineralogy and Petrology*, **144**, 509–522.

Murata, M., Okami, K., Kanisawa, S. & Ehiro, M. 1982. Additional evidence for the pre-Silurian basement in the Kitakami massif, northeast Honshu, Japan. *Memoirs of the Geological Society of Japan*, **21**, 245–259.

Murata, K., Maekawa, H., Yokose, H., Yamamoto, K., Fujioka, K., Ishii, T., Chiba, H. & Wada, Y. 2009. Significance of serpentinization of wedge mantle peridotites beneath Mariana forearc, western Pacific. *Geosphere*, **5**, 90–104.

Nakamura, Y. 1971. Petrology of the Toba ultrabasic complex, Mie Prefecture, Central Japan. *Journal of the Faculty of Science, University of Tokyo, Section II*, **18**, 1–51.

Nakayama, I., Kaji, A., Shioda, T. & Iwasaki, M. 1973. Finding of inverted pigeonite in the Mikabu zone at Kamiyama. *Memoirs of the Faculty of Science, Kyoto University*, **40**, 27–33.

Niida, K. 1984. Petrology of the Horoman ultramafic rocks in the Hidaka metamorphic belt, Hokkaido, Japan. *Journal of the Faculty of Science, Hokkaido University, Series IV*, **21**, 197–250.

Niida, K. & Research Group of the Tokoro Belt 1981. Geology of the Tokoro Belt, Hokkaido. *In*: Hara, I. (ed.) *Tectonics of Paired Metamorphic Belts*. Tanishi Print, Hiroshima, 49–54.

Ninomiya, A. & Arai, S. 1992. Harzburgite fragment in a composite xenolith from an Oshima-Oshima andesite, the Northeast Japan. *Arc. Bulletin of the Volcanological Society of Japan*, **37**, 269–273 [in Japanese with English abstract].

Ninomiya, C., Arai, S. & Ishii, T. 2007. Peridotite xenoliths from the Takeshima seamount, Japan: an insight into the upper mantle beneath the Sea of Japan. *Japanese Magazine of Mineralogical and Petrological Sciences*, **36**, 1–14 [in Japanese with English abstract].

Nixon, P. H. (ed.) 1987. *Mantle Xenoliths*. John Wiley, New York.

Odashima, N., Morishita, T., Ozawa, K., Nagahara, H., Tsuchiyama, A. & Nagashima, R. 2008. Formation and deformation mechanisms of two-pyroxene spinel symplectite in an ascending mantle, the Horoman peridotite complex, Japan: an EBSD (electon backscatter diffraction) study. *Journal of Mineralogical and Petrological Sciences*, **103**, 1–15.

Ohara, Y. & Ishii, T. 1998. Peridotites from the southern Mariana forearc: heterogeneous fluid supply in mantle wedge. *Island Arc*, **7**, 541–558.

Oide, K., Nakagawa, N. & Kanisawa, S. 1989. *Regional Geology of Japan, Part 2, TOHOKU*. Kyoritsu Shuppan, Tokyo.

Onuki, H. 1963. Petrology of the Hayachine ultramafic complex in the Kitakami Mountainland, northern Japan. *Science Reports of Tohoku University, Series III*, **8**, 241–295.

Onuki, H. 1964. Hornblendes from ultramafic intrusives in the Kitakami Mountainland. *Journal of the Japanese Association of Mineralogists, Petrologists and Economic Geologists*, **51**, 210–222.

Onuki, H. 1965. Petrochemical research on the Horoman and Miyamori ultramafic intrusives, northern Japan. *Science Reports of Tohoku University, Series III*, **9**, 217–276.

Onuki, Y. 1937. New discovery of Gotlandian deposits from the Kesen-gun, Iwate Prefecture, Kitakami mountains, and Paleozoic stratigraphy

(Preliminary report). *Journal of the Geological Society of Japan*, **44**, 600–604 [in Japanese with English abstract].

ONUKI, Y. 1969. Geology of the Kitakami massif, Northeast Japan. *Contributions from the Institute of Geology and Paleontology of Tohoku University*, **69**, 1–239 [in Japanese with English abstract].

OZAWA, K. 1983. Relationships between tectonite and cumulate in ophiolites: the Miyamori ultramafic complex, Kitakami Mountains, northeast Japan. *Lithos*, **16**, 1–16.

OZAWA, K. 1984. Geology of the Miyamori ultramafic complex in the Kitakami Mountains, northeast Japan. *Journal of Geological Society of Japan*, **90**, 697–716.

OZAWA, K. 1986. Partitioning of elements between constituent minerals in peridotites from the Miyamori ultramafic complex, Kitakami Mountains, northeast Japan: estimation of P-T condition and igneous composition of minerals. *Journal of the Faculty of Science, University of Tokyo, Section II*, **21**, 115–137.

OZAWA, K. 1987. Petrology of aluminous spinel peridotites and pyroxenites of the Miyamori ultramafic complex, northeast Japan. *Journal of the Faculty of Science, University of Tokyo, Section II*, **21**, 309–332.

OZAWA, K. 1988. Ultramafic tectonite of the Miyamori ophiolitic complex in the Kitakami Mountains, Northeast Japan: hydrous upper mantle in an island arc. *Contributions to Mineralogy and Petrology*, **99**, 159–175.

OZAWA, K. 1989. Stress-induced Al-Cr zoning of spinel in deformed peridotites. *Nature*, **338**, 141–144.

OZAWA, K. 1990. Origin of the Miyamori ophiolitic complex, northeast Japan: TiO_2/K_2O of amphibole and TiO_2/Na_2O of clinopyroxene as discriminants for the tectonic setting of ophiolites. *Proceedings of the Troodos 1987 Ophiolite Conference*, 4–10 October 1987, Cyprus, 485–495.

OZAWA, K. 1994. Melting and melt segregation in the mantle wedge above a subduction zone: evidence from the chromite-bearing peridotites of the Miyamori ophiolite complex, northeastern. *Journal of Petrology*, **35**, 647–678.

OZAWA, K. 2001. Mass balance equations for open magmatic systems: trace element behavior and its application to open system melting in the upper mantle. *Journal of Geophysical Research*, **106**, 13 407–13 434.

OZAWA, K. 2004. Thermal history of the Horoman peridotite complex: a record of thermal perturbation in the lithospheric mantle. *Journal of Petrology*, **45**, 253–273.

OZAWA, K. & SHIMIZU, N. 1995. Open-system melting model in the upper mantle: constraints from the Hayachine-Miyamori ophiolite, northeastern Japan. *Journal of Geophysical Research*, **100**, 22 315–22 335.

OZAWA, K. & TAKAHASHI, N. 1995. P-T history of a mantle diaper: the Horoman peridotite complex, Hokkaido, northern Japan. *Contributions to Mineralogy and Petrology*, **120**, 223–248.

OZAWA, K., SHIBATA, K. & UCHIUMI, S. 1988. K-Ar ages of hornblende in gabbroic rocks from the Miyamori ultramafic complex of the Kitakami Mountains. *Journal of Mineralogy, Petrology, and Economic Geology*, **83**, 150–159 [in Japanese with English abstract].

OZAWA, K., MAEKAWA, H. & ISHIWATARI, A. 2013. Reconstruction of the structure of Ordovician-Devonian arc system and its evolution processes: hayachine – Miyamori ophiolite and Motai high-pressure metamorphic rocks in Iwate Prefecture. *Journal of the Geological Society of Japan*, **119**, 134–153 [in Japanese with English abstract].

QUICK, J. E. 1981. The origin and significance of large, tabular dunite bodies in the Trinity peridotite, northern California. *Contributions to Mineralogy and Petrology*, **78**, 413–422.

SAITO, M., MIYAZAKI, K., TOSHIMITSU, S. & HOSHIZUMI, H. 2005. Geology of the Tomochi district. Quadrangle Series, 1:50,000, Geological Survey of Japan, AIST [in Japanese with English abstract].

SAKAKIBARA, M. 1986. A newly discovered high-pressure terrane in eastern Hokkaido, Japan. *Journal of Metamorphic Geology*, **4**, 401–408.

SAKAKIBARA, M. 1991. Metamorphic petrology of the Northern Tokoro metabasites, eastern Hokkaido, Japan. *Journal of Petrology*, **32**, 333–364.

SAKAKIBARA, M. & OTA, T. 1994. Metamorphic evolution of the Kamuikotan high-pressure and low-temperature metamorphic rocks in central Hokkaido, Japan. *Journal of Geophysical Research*, **99**, 22 221–22 235.

SAKAMOTO, T., OGAWA, Y. & NAKADA, S. 1993. Origin of the greenstones in the Setogawa accretionary complex and their tectonic significance. *Journal of the Geological Society of Japan*, **99**, 9–28 [in Japanese with English abstract].

SAWAGUCHI, T. 2004. Deformation history and exhumation process of the Horoman Peridotite Complex, Hokkaido, Japan. *Tectonophysics*, **379**, 109–126.

SEKI, Y. 1951. Metamorphism of ultrabasic rocks caused by intrusion of granodiorite in Miyamori district, Iwate Prefecture–metamorphism of serpentine rock. *Journal of the Geological Society of Japan*, **57**, 35–43 [in Japanese with English abstract].

SEKI, Y. 1952. The studies on Miyamori ultrabasic mass, Iwate Pref., N-E Japan (No. 4)–on the structural studies. *Journal of the Geological Society of Japan*, **58**, 505–516 [in Japanese with English abstract].

SHIBATA, K. & OZAWA, K. 1992. Ordovician arc ophiolite, the Hayachine and Miyamori ultramafic complexes, Kitakami Mountains, Northeast Japan: isotopic ages and geochemistry. *Geochemical Journal*, **26**, 85–97.

SHIMOJO, M., OHTO, S., YANAI, S., HIRATA, T. & MARUYAMA, S. 2010. LA-ICP-MS U-Pb age of some older rocks of the South Kitakami Belt, Northeast Japan. *Journal of Geography (Chigaku Zasshi)*, **119**, 257–269 [in Japanese with English abstract].

STEINMANN, G. 1927. Die ophiolitischen Zonen in den Mediterranen Kettengebirgen. *Proceedings of the 14th IGC (Madrid)*, **2**, 638–667. [English translation by Bernoulli, D. & Friedman, G.M. 2003. Geological Society of America, Special Papers, 373, 77–92].

SUZUKI, M. & NIIDA, K. 1997. The Oku-Niikappu peridotites: dunite in ophiolitic uppermost mantle. *Memoirs of the Geological Society of Japan*, **47**, 219–229 [in Japanese with English abstract].

TAIRA, A., SAITO, Y. & HASHIMOTO, M. 1983. The role of oblique subduction and strike-slip tectonics in the evolution of Japan. *In*: *Geodynamics of Western Pacific-Indonesian Region*. American Geophysical Union, Washington, Geodynamics Series 11, 303–316.

TAKAHASHI, E. 1978. Petrological model of the crust and upper mantle of the Japanese island arcs. *Bulletin Volcanologique*, **41**, 529–547.

TAKAHASHI, E. 1980. Thermal history of lherzolite xenoliths – petrology of lherzolite xenoliths from the Ichinomegata crater, Oga peninsula, northeast Japan, Part I. *Geochimica et Cosmochimica Acta*, **44**, 1643–1658.

TAKAHASHI, E. 1986. Genesis of calc-alkali andesite magma in a hydrous mantle-crust boundary: petrology of lherzolite xenoliths from the Ichinomegata crater, Oga Peninsula, Northeast Japan, part II. *Journal of Volcanology and Geothermal Research*, **29**, 355–395.

TAKAHASHI, N. 1991. Evolutional history of the uppermost mantle of an arc system: petrology of the Horoman peridotite massif, Japan. *In*: PETERS, T. J., NICOLAS, A. ET AL. (eds) *Ophiolite Genesis and Evolution of the Oceanic Lithosphere (Proceedings of Oman'90 Ophiolite Conference)*. Ministry of Petroleum & Minerals, Sultanate of Oman. Kluwer Academic Publishing, Dordrecht, 195–206.

TAKAHASHI, N. & ARAI, S. 1989. Textural and chemical features of chromian spinel pyroxene symplectites in the Horoman peridotites, Hokkaido, Japan. *Science Reports of the Institute of Geoscience, University of Tsukuba, Section B*, **10**, 45–55.

TAKAMURA, H. 1973. Petrographical and petrochemical studies of the Cenozoic basaltic rocks on Chugoku Province. *Geological Reports of Hiroshima University*, **18**, 1–167 [in Japanese with English abstract].

TAKASHIMA, R., NISHI, H. & YOSHIDA, T. 2002. Geology, petrology and tectonic setting of the Late Jurassic ophiolite in Hokkaido, Japan. *Journal of Asian Earth Sciences*, **21**, 197–215.

TAKAZAWA, E., FREY, F. A., SHIMIZU, N., SAAL, A. & OBATA, M. 1999. Polybaric petrogenesis of mafic layers in the Horoman peridotite complex, Japan. *Journal of Petrology*, **40**, 1827–1851.

TAKAZAWA, E., FREY, F. A., SHIMIZU, N. & OBATA, M. 2000. Whole rock compositional variations in an upper mantle peridotite (Horoman, Hokkaido, Japan): are they consistent with a partial melting process? *Geochimica et Cosmochimica Acta*, **64**, 695–716.

TAMAKI, K. & HONZA, E. 1991. Global tectonics and formation of marginal basins: role of the western Pacific. *Episodes*, **14**, 224–230.

TAMAKI, K., SUYEHIRO, K., ALLAN, J., INGL, J. C. & PISCIOTTO, K. A. 1992. Tectonic synthesis and implications of Japan Sea ODP drilling. *Proceedings of ODP, Scientific Results*, **127/128**, 1333–1348.

TATSUMI, Y., SHINJOE, H., ISHIZUKA, H., SAGER, W. W. & KLAUS, A. 1998. Geochemical evidence for a mid-Cretaceous superplume. *Geology*, **26**, 151–154.

TATSUMI, Y., KANI, T., ISHIZUKA, H., MARUYAMA, S. & NISHIMURA, Y. 2000. Activation of Pacific mantle plumes during the Carboniferous: evidence from accretionary complexes in southwest Japan. *Geology*, **28**, 580–582.

TAZAKI, K., ITO, E. & KOMATSU, M. 1972. Experimental study on a pyroxene-spinel symplectite at high pressures and temperatures. *Journal of the Geological Society of Japan*, **78**, 347–354.

TAZAWA, J. 1988. Palaeozoic-Mesozoic stratigraphy and tectonics of the Kitakami Mountains, northeast Japan. *Chikyu Kagaku (Earth Science)*, **42**, 165–178 [in Japanese with English abstract].

Tsuchiya, N., Suzuki, S., Kimura, J.-I. & Kagami, H. 2005. Evidence for slab melt/mantle reaction: petrogenesis of Early Cretaceous and Eocene high-Mg andesites from the Kitakami Mountains, Japan. *Lithos*, **79**, 179–206.

Tsujimori, T. 1998. Geology of the Osayama serpentinite melange in the central Chugoku Mountains, southwestern Japan: 320 Ma blueschist-bearing serpentinite melange beneath the Oeyama ophiolite. *Journal of the Geological Society of Japan*, **104**, 213–231.

Tsujimori, T. & Ishiwatari, A. 2002. Granulite facies relics in the Early Paleozoic kyanite-bearing ultrabasic metacumulate in the Oeyama belt, the Inner Zone of southwestern Japan. *Gondwana Research*, **5**, 823–835.

Uchino, T. & Kawamura, M. 2006. Glaucophane-bearing mafic schist discovered from the Nedamo Terrane (ex-"Hayachine Terrane"), Northeast Japan, and its geologic implications. *Journal of the Geological Society of Japan*, **112**, 478–481 [in Japanese with English abstract].

Uchino, T., Kurihara, T. & Kawamura, M. 2005. Early Carboniferous radiolarians discovered from the Hayachine Terrane, Northeast Japan: the oldest fossil age for clastic rocks of accretionary complex in Japan. *Journal of the Geological Society of Japan*, **111**, 249–252 [in Japanese with English abstract].

Uchino, T., Kawamura, M. & Kawamura, T. 2008. Lithology of the Nedamo Terrane, an Early Carboniferous accretionary complex, and its southern boundary with the South Kitakami Terrane. *Journal of the Geological Society of Japan*, **114**, 141–157 [in Japanese with English abstract].

Uda, S. 1984. The contact metamorphism of the Oeyama ultrabasic mass and the genesis of the 'cleavable olivine'. *Journal of the Geological Society of Japan*, **90**, 393–410 [in Japanese with English abstract].

Ueda, H. 2005. Accretion and exhumation structures formed by deeply subducted seamounts in the Kamuikotan high-pressure/temperature zone, Hokkaido, Japan. *Tectonics*, **24**, TC2007, http://doi.org/10.1029/2004TC001690

Ueda, H. & Miyashita, S. 2005. Tectonic accretion of a subducted intraoceanic remnant arc in Cretaceous Hokkaido, Japan, and implications for evolution of the Pacific northwest. *Island Arc*, **14**, 582–568.

Ueda, H., Usuki, T. & Kuramoto, Y. 2004. Intraoceanic unroofing of eclogite facies rocks in the Omachi Seamount, Izu-Bonin frontal arc. *Geology*, **32**, 849–852.

Umino, S. 1986. Magma mixing in Boninite sequence of Chichijima, Bonin Islands. *Journal of Volcanology and Geothermal Research*, **29**, 125–157.

Uto, K. 1990. *Neogene Volcanism of Southwest Japan: its Time and Space on K-Ar Dating*. PhD thesis, University of Tokyo, Japan.

Uto, K., Hirai, H., Goto, K. & Arai, S. 1987. K-Ar ages of carbonate- and mantle nodule-bearing lamprophyre dikes from Shingu, central Shikoku, Southwest Japan. *Geochemical Journal*, **21**, 283–290.

Velinsky, V.V. & Pinus, G.V. 1969. Olivines with perfect cleavages in ultrabasites of Chukotok. *Doklady of the Academy of Sciences of the USSR Earth Science Sections* (English translation edition), **185**, 99–101.

Watanabe, T., Fanning, C. M., Uruno, K. & Kano, H. 1995. Pre-middle Silurian granitic magmatism and associated metamorphism in northern Japan: SHRIMP U-Pb zircon chronology. *Geological Journal*, **30**, 273–280.

Wells, P. R. A. 1977. Pyroxene thermometry in simple and complex systems. *Contributions to Mineralogy and Petrology*, **62**, 129–139.

Yamamoto, M. 1983. Spinels in basaltic lavas and ultramafic inclusions of Oshima–Oshima volcano, north Japan. *Journal of Faculty of Science, Hokkaido University*, **20**, 135–143.

Yakubchuk, A. S., Nikishin, A. M. & Ishiwatari, A. 1994. Late Proterozoic ophiolite pulse. *In*: Ishiwatari, A. et al. (eds) *Circum-Pacific Ophiolites: Proceedings of the 29th IGC Ophiolite Symposium*. VSP Publishers, Netherlands, 273–286.

Yoshida, K. & Machiyama, H. 2004. Provenance of Permian sandstones, South Kitakami Terrane, Northeast Japan: implications for Permian arc evolution. *Sedimentary Geology*, **166**, 185–207.

Yoshida, T. & Katada, M. 1964. *1:50000 Geologic Map of Otsuchi and Karodake and its Explanatory Text*. Geological Survey of Japan, Kawasaki City.

Yoshikawa, M. & Ozawa, K. 2007. Rb-Sr and Sm-Nd isotopic systematics of the Hayachine-Miyamori ophiolitic complex: melt generation process in the mantle wedge beneath an Ordovician island arc. *Gondwana Research*, **11**, 234–246.

Yoshikawa, M., Suzuki, K., Shibata, T. & Ozawa, K. 2012. Geochemical and Os isotopic characteristics of a fresh harzburgite in the Hayachine-Miyamori ophiolite: evidence for melting under influx of carbonate-rich silicate melt in an infant arc environment. *Journal of Mineralogical and Petrological Sciences*, **107**, 250–255.

Yoshimoto, A., Osanai, Y., Nakano, N., Adachi, T., Yonemura, K. & Ishizuka, H. 2013. U-Pb detrital zircon dating of pelitic schists and quartzite from the Kurosegawa Tectonic Zone, Southwest Japan. *Journal of Mineralogical and Petrological Sciences*, **108**, 178–183.

4 Granitic rocks

TAKASHI NAKAJIMA, MASAKI TAKAHASHI,
TERUYOSHI IMAOKA & TOSHIAKI SHIMURA

General occurrences and settings (TN)

Granitic rocks occupy nearly 30% of the ground surface of Japan (Geological Survey of Japan 1982) and are mainly Mesozoic and Cenozoic in age (Fig. 4.1). Palaeozoic granites do exist, but outcrops are sparse and mostly confined to geological fragments rather than large granitic provinces. Many Japanese granites belong within the Late Mesozoic granitic superprovince of the NW Pacific rim, continuing northeastwards from Japan to Sikhote Alin and Chukotka and southwestwards to the Korean Peninsula and Southern Coastal China (Fig. 4.1). As part of the 'Late Mesozoic Circum-Pacific Ring of Fire', these Asian granites provide a mirror image of the Canadian Cordillera, Sierra Nevada and Peninsular Range granitic provinces of western North America.

Japanese granitoids were mostly (>90%) produced when Japan belonged within the Asian continent, and in this sense it can be said that most Japanese granitoids are Asian granitoids. In this context the Japanese islands had been the growing front of the collage continent Asia as a site of continental-margin-type orogeny, as evidenced by the widespread occurrence of Permo-Carboniferous to modern accretionary complexes beneath which granitic continental crust was formed. In addition, however, Japanese Cenozoic granitoids also include examples as young as Quaternary in age, demonstrating the continuing reality of the ongoing orogeny.

Japanese granitoids occur in two modes, being closely associated either with low- to medium-pressure-type metamorphic rocks forming plutono-metamorphic belts, or with coeval silicic volcanic rocks forming volcano-plutonic complexes. In this chapter, the characteristic nature and occurrence of Japanese granitoids are described and discussed, especially from the viewpoint of magmatism in relation to tectonics.

Spatio-temporal distribution (TN, MT, TI)

Figure 4.2 provides a histogram of surface exposure areas of granitoids in the Japanese islands at 10 Ma intervals, on the basis of isotope ages reported so far. It shows that granitic magmatism in the Japanese islands is highly episodic. There is a strong age concentration (80%) within the range of 130–40 Ma, and another smaller but distinct cluster at 20–10 Ma. Very young granitoids (10–0 Ma), including those of Quaternary age mentioned above, do not clearly appear on this histogram because of their limited exposure. Pre-Cretaceous granitoids define small and obscure clusters at 190–160 Ma and 250–200 Ma.

Before 1990, Japanese granitoids had been dated mostly using K–Ar and Rb–Sr methods. U–Pb age dating was not available until ion microprobe (IMP, e.g. SHRIMP) ages were documented in the 1990s (e.g. Watanabe *et al.* 1995; Nakajima 1996). Instead of IMP, U–Th–Pb age determinations using EMP (Electron Micro-Probe), called CHIME ages (Suzuki & Adachi 1991), were undertaken in Japan. After laser ablation inductively coupled plasma mass spectrometry (LA-ICP-MS) and SHRIMP were installed at several institutes and universities recently, U–Pb ages have been released gradually although many of them are still unpublished.

These new U–Th–Pb ages for granitoids have confirmed old K–Ar and Rb–Sr age data in some cases but produced different results in others, opening up new interpretations of Japanese geological history. A granitoid map with recent U–Th–Pb age data is shown in Figure 4.3a, b.

Pre-Cretaceous

Palaeozoic granitoids occur as small, fault-bounded isolated blocks in tectonized zones rather than as batholithic belts. The oldest age from granitic rocks so far documented in Japan is around 500 Ma, confirming that the Japanese islands are underlain by continental crust formed by the Phanerozoic orogeny with no evidence of Precambrian basement. These oldest rocks have been documented from two localities: (1) a tonalite from central Kyushu (Sakashima *et al.* 2003); and (2) 'deformed granite' and 'granitic porphyry' documented by Tagiri *et al.* (2011) and tonalite by Sakashima *et al.* (2003), both from the Hitachi area in NE Japan.

Granitic rocks with Siluro-Devonian ages are exposed in two modes. One of these is as small (<2 km) tectonic blocks of weakly foliated to unfoliated hornblende biotite granodiorites in serpentinite mélange, long described from the Kurosegawa Tectonic Zone and its equivalents in central Shikoku, western Chugoku, Kii Peninsula and other related areas (U–Pb ages by Hada *et al.* 2000; Osanai *et al.* 2014). The other type is as several kilometre-sized bodies, which include the Hikami Granite (U–Pb ages by Watanabe *et al.* 1995; Shimojo *et al.* 2010) in northern NE Japan and the Maizuru Granite (U–Pb ages by Fujii *et al.* 2008; Tsutsumi *et al.* 2014) in central Chugoku, SW Japan. Both types of granitoid lithology occur surrounded by younger rocks, either Jurassic accretionary complexes or Cretaceous ignimbrites, and are interpreted as fault-bounded.

Carboniferous–Permian granitic rocks yielding U–Pb zircon ages of *c.* 300 Ma have been documented in eastern Kyushu (Sakashima *et al.* 2003) and along the eastern margin of the Abukuma Plateau. Plagiogranite from the Yakuno ophiolite has yielded Early Permian zircon U–Pb ages of 285–280 Ma (Herzig *et al.* 1997).

It is difficult to identify whether these Palaeozoic granitoids (500–280 Ma) are of volcano-plutonic or plutono-metamorphic type, because they occur as isolated blocks without accompanying coeval silicic volcanic rocks or metamorphic rocks. Their chemical affinity is mostly of arc-type granite.

Permian–Jurassic ages have been reported from granitic rocks within the Hida Belt in central Japan. These represent the oldest plutono-metamorphic belt in Japan, located in the most rear-arc position within the archipelago. U–Pb ages from these Hida granitic rocks range from 260 to 190 Ma, i.e. from end-Permian–Triassic to some Jurassic examples, with clusters at *c.* 250 Ma (foliated biotite granites) and 200 Ma (unfoliated hornblende biotite granodiorites) (Horie *et al.* 2010; Takahashi *et al.* 2010).

From: Moreno, T., Wallis, S., Kojima, T. & Gibbons, W. (eds) 2016. *The Geology of Japan*. Geological Society, London, 251–272.

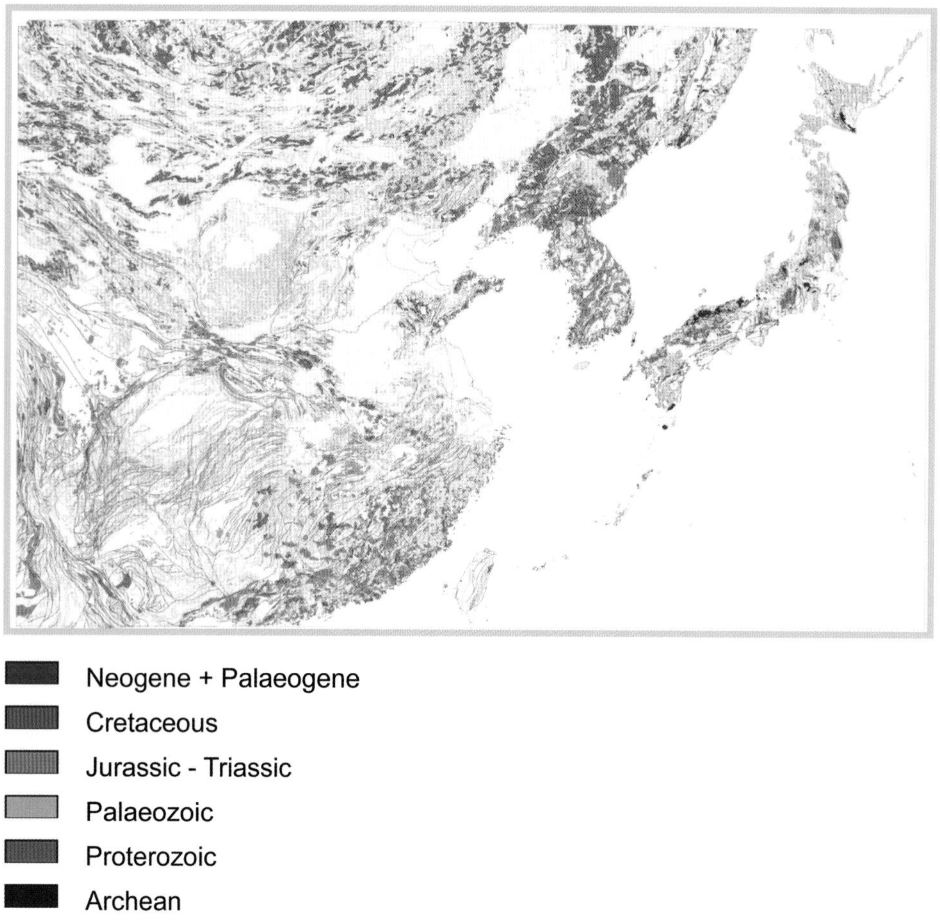

Fig. 4.1. Granite map of eastern Asia (reproduced from Teraoka & Okumura 2003).

Cretaceous–Early Palaeogene

SW Japan

The Cretaceous–Early Palaeogene granitic province in SW Japan is the largest in the Japanese islands. It extends from central Japan (U–Pb ages by Nakajima 1996) to Kyushu Island (U–Pb ages by Sakashima *et al.* 2003; Tiepolo *et al.* 2012) on the northern side of the Median Tectonic Line (MTL), and is comparable in size to the Sierra Nevada or Peninsular Ranges in the western US.

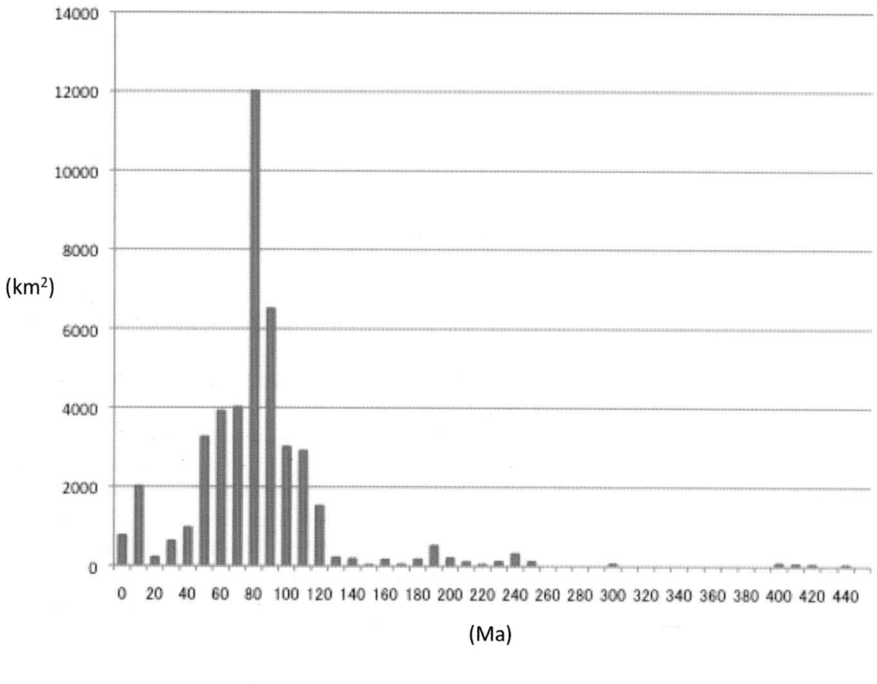

Fig. 4.2. A histogram of surface exposure areas of granitoids in Japanese islands in terms of 10 Ma duration. Data from 1:200 000 Seamless Geological Map of Japan (GSJ 2009) and Takagi (2004).

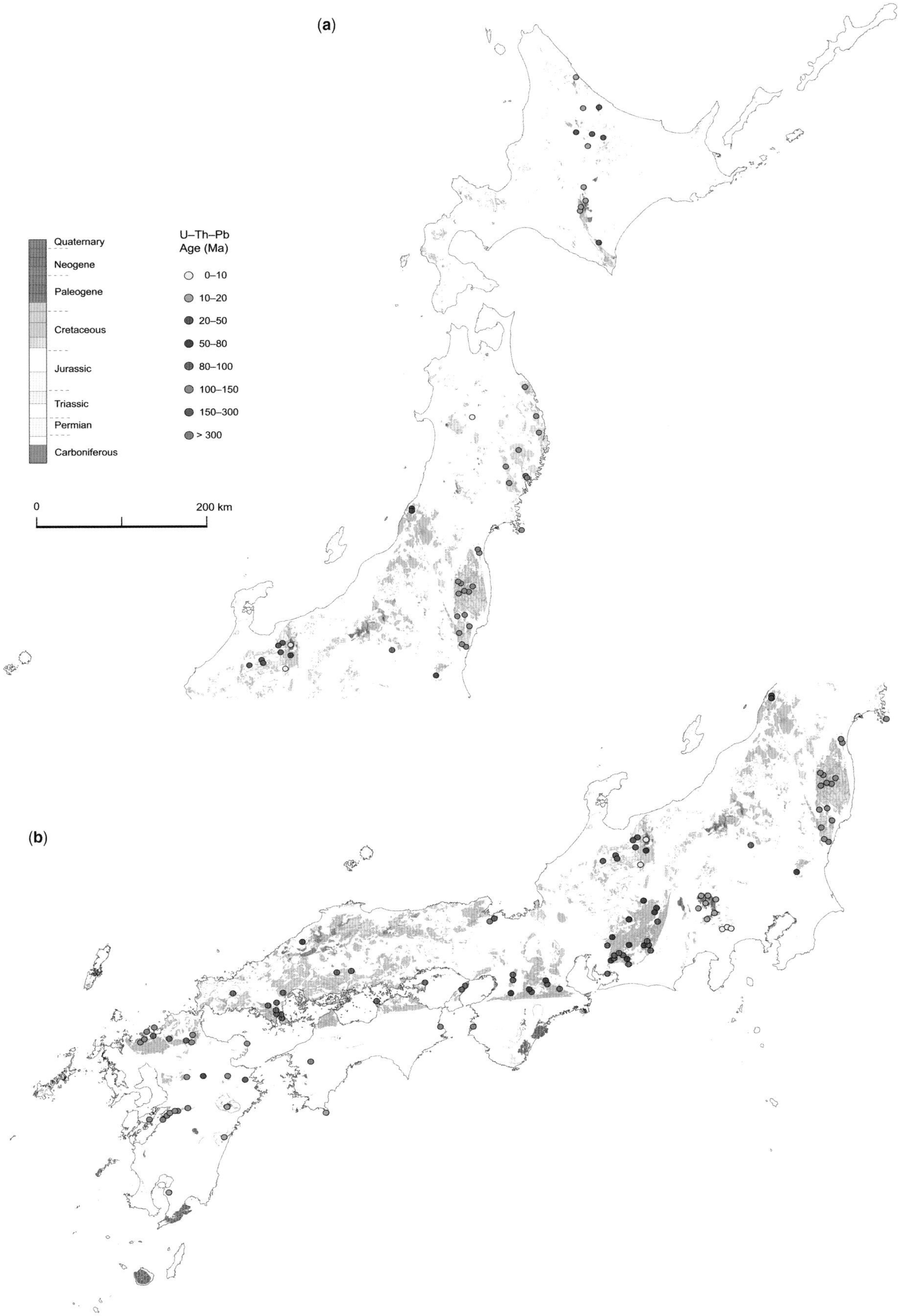

Fig. 4.3. Granitoid map of Japan with U–Th–Pb age data shown in coloured circles (**a**) NE Japan and (**b**) SWJapan. The granitoid distribution map is based on the Seamless Geological Map of Japan (GSJ 2009). The small patches shown in blue are predominantly intermediate to mafic composition.

Table 4.1. *Cretaceous–Early Palaeogene granitoid zones in SW Japan*

Zone	Occurrences	Age (Ma)	I/S types	Series
SW Japan				
Ryoke	Plutono-metamorphic	115–65	I-type	Ilmenite
San'yo	Volcano-plutonic	110–70	I-type	Ilmenite
San'in	Volcano-plutonic	110–30	I-type	Magnetite
NE Japan				
Abukuma	Plutono-metamorphic	115–100	I-type	Ilmenite
Kitakami	Volcano-plutonic	130–115	I-type	Magnetite

The province is divided into three arc-parallel zones which, from forearc to back-arc position, are called Ryoke, San'yo and San'in zones (Table 4.1: note that these 'zones' define Cretaceous granitic subprovinces and are independent of tectonic divisions shown in Fig. 1.6). The San'yo and San'in zones are composed of volcano-plutonic complexes, while the Ryoke zone is a plutono-metamorphic belt with accompanying migmatites. Granitic rocks in both Ryoke and San'yo zones are entirely Cretaceous (115–65 Ma), while those in the San'in zone are of Middle Cretaceous–Early Palaeogene age (110–55 Ma). Granitoids of the three zones in Kyushu Island, located at the western end of SW Japan, have ages restricted to 115–100 Ma, which makes them distinctly older than those from other areas in SW Japan. For example, granitoids in the San'in zone excluding those from Kyushu Island are mostly younger than 80 Ma.

An across-arc transect from San'yo zone to Ryoke zone is regarded as a surface-exposed crustal cross-section of the Cretaceous Eurasian continental margin (Nakajima 1994). Comparing Th–Pb monazite age data using EMP (CHIME ages) on granitic rocks from eastern and western parts of SW Japan (Suzuki & Adachi 1998), ages of 70–100 Ma granitoids in the San'yo zone appear to have an eastwards-younging polarity. This age polarity of the granitoids seems to prevail all over SW Japan from Kyushu Island, which is located at the western end, to the eastern part of the Chubu district. Enormous quantities of volcanic rocks, variously referred to as Takada, Abu-Hikimi and Nohi rhyolites (Sonehara & Harayama 2007), are associated with granitoids in the San'yo zone and also seem to show the same eastwards-younging polarity.

NE Japan

In NE Japan, Cretaceous granitic rocks occur in the Kitakami and Abukuma zones (Table 4.1). Kitakami granitoids are of volcano-plutonic type, whereas Abukuma granitoids are of plutono-metamorphic type. Ages of Kitakami and Abukuma zone granitoids are *c.* 127–113 Ma and *c.* 115–100 Ma, respectively (Kon & Takagi 2012; Tsuchiya *et al.* 2012; Takahashi *et al.* 2014), making them slightly older than the Cretaceous granitoids exposed in SW Japan. Both of these zones extend towards the N-NNW along the NE Japan arc. In these northern areas, however, Cretaceous igneous rocks are poorly exposed due to a cover of overlying Cenozoic sediments and younger volcanic rocks.

Late Palaeogene

In addition to Late Cretaceous–Early Palaeogene granitoids, a number of Late Palaeogene calderas are exposed in the San'in zone of the Chugoku district. U–Pb ages of these granitic and volcanic rocks have unfortunately not yet been published, except one age of 33 Ma (Iwano *et al.* 2012), but they are presumed to be Late Eocene–Early Oligocene (30–45 Ma) in age on the basis of available K–Ar and Rb–Sr ages (Imaoka *et al.* 2011). This Late Palaeogene caldera-forming granitic magmatism is included in the San'in zone in Table 4.1. It has been noted that there appears to have been a transient break of magmatic activity lasting for *c.* 10 Ma between Late Palaeogene magmatism and Early Palaeogene granitoids (Imaoka *et al.* 2011; Iida *et al.* 2013).

Another group of Late Palaeogene (46–37 Ma) granitic rocks occurs in central Hokkaido (Jahn *et al.* 2014). These occur sporadically in relatively small-sized bodies, some of which are exposed in the Hidaka plutono-metamorphic belt and are described in 'Hidaka Belt' below.

Neogene

A small but sharp peak at 20–10 Ma in the age histogram (Fig. 4.2) marks activity in two granitic provinces, namely in central Hokkaido, including the Hidaka plutono-metamorphic belt, and a short-lived burst of magmatism in Middle Miocene times along the forearc region of SW Japan. The Hidaka Belt shows a crustal section that spans unmetamorphosed accretionary complexes with high-level granitoid plutons to high-grade metamorphic complexes of granulite grade with deep-seated granitoids. Both the deepest-seated and high-level granitoids are *c.* 20 Ma in age (Kamiyama *et al.* 2007; Kemp *et al.* 2007; Jahn *et al.* 2014).

Middle Miocene granitoids in SW Japan are exposed sporadically but on a much larger scale, mostly defining volcano-plutonic complexes or high-level granitoid stocks. All K–Ar ages documented so far are concentrated around 15–14 Ma (Sumii *et al.* 1998), indicating a highly transient phase of magmatism.

Late Neogene granitoids 9–4 Ma in age are exposed in the Tanzawa Mountains in the Izu Collision Zone, central Japan. These are discussed later, particularly in relation to the collision history of the Izu volcanic arc with its tonalitic middle crust (Tani *et al.* 2010).

Quaternary

Quaternary granitoids are exposed at two sites in central Japan, where the uplift rate has been exceptionally high (2–3 mm a^{-1}). These are Takidani Granite in Central Highlands (Harayama 1992) and Kurobegawa Granite in the Hida Mountains (Ito *et al.* 2013*a*, *b*). The Takidani Granite has yielded ages of 2–1 Ma (K–Ar and Rb–Sr: Harayama 1992) and Kurobegawa Granite has yielded a U–Pb age of 0.6–0.5 Ma (Ito *et al.* 2013*a*, *b*). An even younger granite intrusion of effectively 0 Ma is known in the Kakkonda geothermal area of NE Japan, where a borehole penetrated a still hot subsurface granitic body at 2800 m below ground level (Sasaki *et al.* 2003).

Chemical characters and typology (MT, TN)

Whole-rock major-element chemistry

An average value for major-element chemical composition of granitic rocks in Japanese islands is shown in Table 4.2. The most common whole-rock SiO_2 content is 70–72 wt%, as demonstrated by a frequency distribution histogram (Aramaki *et al.* 1972; Aramaki & Nozawa 1978). Siliceous granitoids of $SiO_2 > 64$ wt% ($SiO_2 = 64$ wt% corresponds roughly to quartz diorite) comprise *c.* 75% of the total, an observation similar to granitoids in South Korea, the Malaya Peninsula, Sikhote Alin, Central Alaska and Sierra Nevada. In contrast, granitoids from southern California, Boulder Batholith, central and southern Peru are less siliceous, with more than 35% having $SiO_2 < 64$ wt% (Aramaki *et al.* 1972).

As would be expected, average SiO_2 content varies among the different granitic provinces. In SW Japan, the most voluminous Cretaceous–Palaeogene granitoids show an across-arc variation. The

Table 4.2. *Average major-element chemical composition of granitic rocks from (a) Japan (Hattori* et al. *1960), (b) Lachlan Fold Belt, Australia, I-type and (c) Lachlan Fold Belt, S-type (Chappell & White 1992)*

	Japan ($n = 440$)	Lachlan I-type ($n = 1074$)	Lachlan S-type ($n = 704$)
SiO_2	69.17	69.50	70.91
TiO_2	0.39	0.41	0.44
Al_2O_3	15.00	14.21	14.00
Fe_2O_3	1.05	1.01	0.52
FeO	2.48	2.22	2.59
MnO	0.10	0.07	0.06
MgO	1.15	1.38	1.24
CaO	3.15	3.07	1.88
Na_2O	3.45	3.16	2.51
K_2O	3.01	3.48	4.09
P_2O_5	0.13	0.11	0.15
H_2O (+)	0.74		
H_2O (−)	0.3		
Total	100.12	98.62	98.39

proportion of $SiO_2 > 64$ wt% granitoids is lowest in the Ryoke Zone (around 75%) on the forearc side, increasing to around 85% through the San'yo Zone to reach a maximum within the San'in Zone (as high as > 90%) in the more back-arc position. By contrast, granitoid rocks in the Cretaceous Abukuma and Kitakami zones in NE Japan are less silicic with $SiO_2 > 64$ wt% granitoids forming only 50–60% of the total, whereas more than 90% of Miocene granitoids in SW Japan have $SiO_2 > 64$ wt%.

K_2O contents of most Japanese granitoids correspond to the medium-K to high-K series of Peccerillo & Taylor (1976) (Fig. 4.4), but there is no distinct variation of K_2O contents among the granitic provinces. Cretaceous–Palaeogene granitic provinces have a statistic peak frequency at *c.* 3.0 wt% for $SiO_2 = 70$ wt%. The systematic across-arc variation of K_2O content, as seen in the Sierra Nevada (Bateman & Dodge 1970), is not found either along the traverse through the Ryoke, San'yo to San'in zones in SW Japan or from the Kitakami to Abukuma zones in NE Japan. In contrast, the Miocene granitic province in SW Japan shows an across-arc gradient of K_2O content from 2.5 wt% on the back-arc side to 3.2 wt% on the forearc side for $SiO_2 = 65\%$ (Nakada & Takahashi 1979).

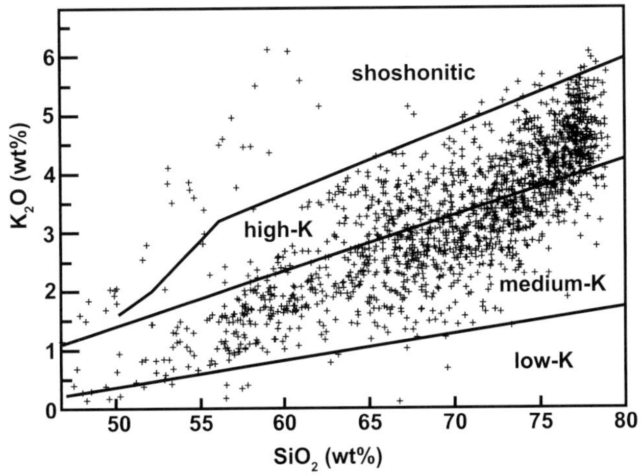

Fig. 4.4. Chemical compositions of the granitoids of Japan on the SiO_2-K_2O diagram by Peccerillo & Taylor (1976).

K_2O/Na_2O ratios vary across Cretaceous–Palaeogene granitic provinces with, for example, the San'in Zone being slightly lower in K_2O/Na_2O ratio than Ryoke and San'yo zones. In particular, it is clear that Late Palaeogene granitoids in the coastal area of the Sea of Japan are lower in their K_2O/Na_2O ratios than other granitoids in SW Japan. Granitoids of the Kitakami and Abukuma zones are slightly lower in K_2O/Na_2O ratio than those of Cretaceous to Early Palaeogene granitoids in SW Japan (Takahashi *et al.* 1984), although some shoshonitic granitoids occur in the Kitakami zone, NE Japan (Katada *et al.* 1991). There is also spatial variation in $Fe^{3+}/(Fe^{2+}+Fe^{3+})$ ratios of the granitoids, these being higher in the Kitakami and San'in zones but lower in Abukuma, Ryoke and San'yo zones.

Trace-element geochemistry

Trace-element geochemistry is popular because it shows the nature of magma more clearly and with a greater variation range than major elements. Another reason is that it allows us to discuss petrogenesis or magmatic evolution without detailed petrological knowledge and data. The development and widespread use of X-ray fluorescence (XRF) and ICP-MS techniques in recent decades has resulted in a huge amount of chemical analyses data, stimulating the use of many trace-element discrimination diagrams and facilitating petrogenetic interpretations.

In MORB-normalized 'spidergrams', depletion of high-field-strength elements (HFSE) such as Nb and Hf is commonly observed in arc magmas, and most Japanese granitoids do indeed show Nb depletion as seen in other arcs throughout the world (as shown in Fig. 4.5). The (Y + Nb)–Rb diagram introduced by Pearce *et al.* (1984) is often used for the discrimination of granitoids into syncollision granitoid (syn-COLG), within-plate granitoid (WPG), oceanic-ridge granitoid (ORG) and volcanic-arc granitoid (VAG). The granitoids in arc settings, such as in the Cretaceous Circum-Pacific plutonic belts that include most of the Japanese examples, plot in the area of volcanic-arc type (Fig. 4.6).

Chondrite-normalized rare earth element (REE) patterns are often employed to discuss petrogenesis and magmatic settings, and here once again Cretaceous–Palaeogene granitoids from SW Japan show the slightly enriched light-REE and nearly flat heavy-

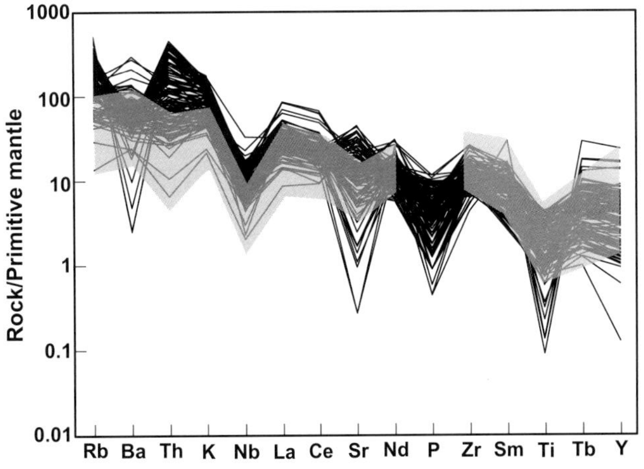

Fig. 4.5. Spidergram of representative granitoids in Japan. The shaded bands show the compositional range of felsic to intermediate rocks from representative arcs in the world including Aleutian, Mariana, Solomon, Sangihe, South Sandwich, Kermadec and Lesser Antile arcs (after Saito 2014). Normalizing values are from Sun & McDonough (1989).

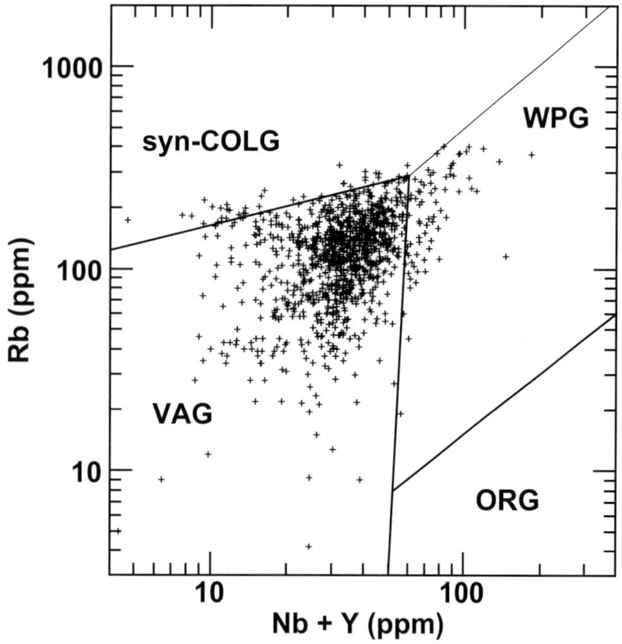

Fig. 4.6. Cretaceous–Palaeogene granitoids of SW Japan shown on the (Y + Nb)–Rb diagram by Pearce *et al.* (1984). They are mostly plotted in the area of VAG (volcanic arc granite).

REE pattern (Fig. 4.7) seen in other arc-related granitoids around the world.

Adakitic lithologies occur ubiquitously in the Early–Middle Cretaceous granitoids (120–90 Ma) of Japan. Adakites are characterized by high light-REE and Sr and low heavy-REE and Y contents, reflecting magma generation leaving a garnet residue. They occur sporadically in the San'yo and Ryoke zones in SW Japan in Kyushu, Kinki and Chubu districts (Kiji *et al.* 2000; Kamei 2004; Imaoka *et al.* 2013), extend into the San'yo Zone in NE Japan (Takahashi *et al.* 2005), and are the most abundant in the Kitakami Zone (Tsuchiya & Kanisawa 1994; Tsuchiya *et al.* 2007).

Although adakite first drew attention as a possible modern analogue of Archean tonalite-trondhjemite-granodiorite (TTG) suites associated with magma generation from slab melting (e.g. Defant & Drummond 1990; Martin *et al.* 2005), various other genetic explanations have been proposed. In the case of the Japanese adakitic granitoids, petrogenetic models include: (1) melting of subducted oceanic basaltic crust and subsequent magmatic differentiation; (2) melting of wedge mantle metasomatized by silicic melts derived from the fused subducted oceanic basaltic crust and/or sediments and subsequent magmatic differentiation; (3) melting of the lower crust and subsequent magmatic differentiation; and (4) magmatic differentiation of wet basaltic magma (Takahashi *et al.* 2005; Tsuchiya *et al.* 2007; Kamei *et al.* 2009).

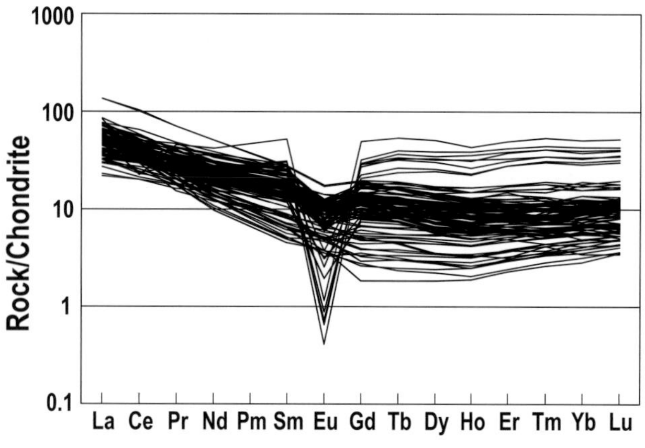

Fig. 4.7. Chondrite-normalized REE patterns of Japanese granitoids.

Granite types: I/S and others

The classification of I/S type granitoids was first proposed by Chappell & White (1974), with igneous protoliths providing the origin of I-type and sedimentary protoliths the origin of S-type in the Kosciusko Batholith of SE Australia, with which both types are intimately associated (Hine *et al.* 1978). Mantle-sourced M-type (White 1979) and alkalic and anhydrous A-type (Loiselle & Wones 1979; Eby 1990) were later added to the classification although these two types are small in amount compared with I- and S-type intrusions, which together represent most of the granitoids in the world. Unfortunately, however, the discrimination between I- and S-types has been confused because this classification has been adopted by researchers using different criteria such as whole-rock chemistry, mineralogy, isotopic character and so on. In this chapter, we adopt the strictest definition of S-type, that is, based on the presence of magmatic cordierite. Figure 4.8 shows the positive correlation between alumina saturation index (ASI: $Al_2O_3/(K_2O + Na_2O + 2CaO)$ in moles) and SiO_2 contents of Cretaceous–Palaeogene granitoids in SW Japan. The positive correlation of SiO_2 with ASI is characteristically seen in I-type granitoids, reflecting magmatic fractionation of clinopyroxene and/or hornblende (Zen 1986; Nakajima 1996). This positive correlation is commonly seen in other I-type granitoids in the world (e.g. Harrison & Piercy 1990; John & Wooden 1990).

Granitoids in the Japanese islands are therefore mostly of I-type, with other types such as S-type being relatively small in amount and both M-type and A-type being rare (Takahashi *et al.* 1980). The protoliths of I-type granitoids are deduced to be various igneous/meta-igneous rocks. Typical S-type granitoids characterized by

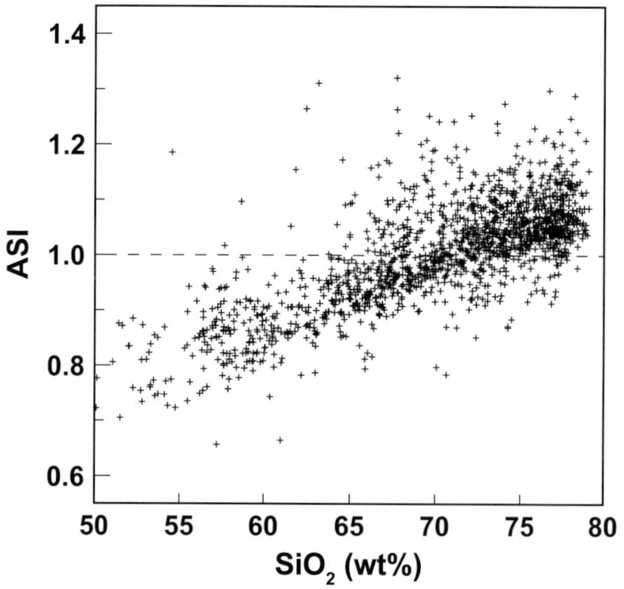

Fig. 4.8. Cretaceous–Palaeogene granitoids from SW Japan on the SiO_2-ASI diagram. Note that they show a trend of positive correlation.

magmatic cordierite occur only on the oceanic side of the SW Japan magmatic belt (Nakada & Takahashi 1979) and in the Hidaka plutono-metamorphic belt in Hokkaido (Shimura *et al.* 1992), both of these being Middle Miocene in age. Cretaceous S-type-like cordierite-free peraluminous muscovite granites (e.g. the Busetsu granite in central Japan) are exposed in the Ryoke Zone, and considerable involvement of pelitic rocks is presumed in their generation.

M-type granitoids in Japan occur mainly in the Izu collisional region, for example Tanzawa tonalite (Kawate & Arima 1998; Tani *et al.* 2010) and Ashigawa tonalite (Saito *et al.* 2004). These M-type lithologies comprise tonalite to trondhjemite with low K_2O/Na_2O ratios, the protoliths of which are igneous rocks depleted in K_2O and enriched in Na_2O and which constitute the crust of the Izu–Bonin oceanic volcanic arc (Takahashi 1989; Sato 1991; Saito *et al.* 2007).

Granitoids with A-type affinity are scarce in Japan. They can be found in the Middle Miocene Ashizuri–Misaki plutonic complex on the Pacific coast of Shikoku Island, SW Japan (Murakami *et al.* 1983; Ishihara *et al.* 1990; Shinjoe *et al.* 2010) and in the alkali element-enriched Late Cretaceous Aoki granitic complex in central Japan (Ueki & Harayama 2012).

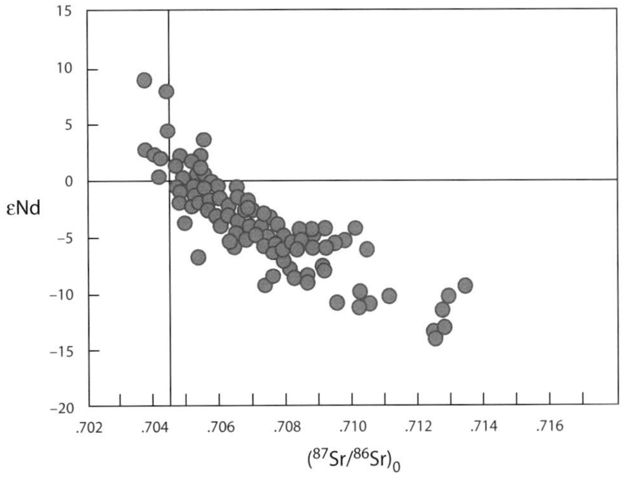

Fig. 4.9. Representative Japanese granitoids in the $^{87}Sr/^{86}Sr$ – εNd systematics, showing their negative correlation and flattening trend in the range of isotopically enriched granitoids.

Magnetite and ilmenite series

In another well-known classification, Japanese granitoids have been subdivided into magnetite-series and ilmenite-series (Ishihara 1977) based on their magnetic susceptibility, which reflects the presence of magnetite in the rock. Magnetite-series and ilmenite-series granitoids are generated under oxidized and reduced environments, respectively. The reduced condition of the ilmenite series is considered to be caused by involvement of pelitic sedimentary rocks with organic carbonaceous material. Magnetite-series granitoids include I- and M-types, whereas ilmenite-series granitoids include I-, S- and A-types (Takahashi *et al.* 1980). Cretaceous–Early Palaeogene granitoids in the Kitakami and San'in zones belong to the magnetite series, whereas those of the Abukuma, Ryoke and San'yo zones are of ilmenite series. The Triassic–Jurassic granitoids of the Hida Zone, as well as the late Palaeogene granitoids along the coastal region of Sea of Japan fall within the magnetite series. Miocene granitoids of SW Japan belong to the ilmenite series, whereas Miocene–Pliocene granitoids in the Izu Collisional Zone include both ilmenite-series and magnetite-series examples. The Palaeogene and Neogene granitoids of the Hidaka plutono-metamorphic belt belong to the ilmenite series.

Thus, I-type ilmenite-series granitoids predominate in the Japanese islands. Indeed, the proportion of ilmenite-series granitoids is higher than within the granitoid batholiths of Sierra Nevada and Peninsular ranges.

Isotopic composition (TN, MT)

A large amount of isotopic data on Japanese granitic rocks has been produced since the 1970s, with discussion mainly focusing on the origin of the granitic magma or the unexposed basement structure to the archipelago. Most of these data are derived from Sr and Nd isotopes with some oxygen and sulphur isotopes, whereas Pb and other isotopic data on the granitic rocks are scarce.

Initial $^{87}Sr/^{86}Sr$ ratio of granitoids

Initial $^{143}Nd/^{144}Nd$ ratios, denoted as εNd, are negatively correlated with initial $^{87}Sr/^{86}Sr$ ratios in arc granites (e.g. Kay *et al.* 1991; Kagami *et al.* 1992) as shown in Figure 4.9. Sr isotope ratios are more sensitive than are Nd isotopes for revealing the involvement of crustal components in the petrogenesis of granitic magmas. For this reason, we will mainly focus on Sr isotopes in the following discussions.

Initial $^{87}Sr/^{86}Sr$ ratios of Japanese granitoids show remarkable variations in the range of 0.704–0.712 throughout geological time (Shibata & Ishihara 1979; Kagami *et al.* 1992, 2000; Jahn 2010). Given that these values are far below initial $^{87}Sr/^{86}Sr$ ratios of granitoids in Precambrian cratons in the continents, it is generally accepted that Japanese granitoids were formed mainly from mantle-derived juvenile source materials without involvement of ancient cratons. This is consistent with the basement geology of Japanese islands being composed of Late Palaeozoic–Cenozoic accretionary complexes and their metamorphic equivalents. These sedimentary/metasedimentary rocks are not as isotopically enriched as Precambrian cratons, although they have higher initial $^{87}Sr/^{86}Sr$ ratios than mantle or MORB.

Sr isotopes in the volcano-plutonic and plutono-metamorphic complexes

Initial $^{87}Sr/^{86}Sr$ ratios of granite-related volcanic rocks are not well known, partly because their U–Pb ages have not been measured. K–Ar ages are too sensitive to secondary alteration and easily lose their magmatic ages. Rb–Sr whole-rock isochron ages are also problematic because the calculation assumes isotopic equilibration even in an unequilibrated geological unit, as clearly shown by Tsuboi & Suzuki (2003). Furthermore, we have to be careful in the case of very high Rb/Sr granitoids because age corrections induce large extrapolation errors in the calculation of initial $^{87}Sr/^{86}Sr$ ratios. To avoid such errors, the $^{87}Sr/^{86}Sr$ ratio of apatite can be useful (Tsuboi & Suzuki 2003).

Looking into the raw data carefully, however, such as selecting $^{87}Sr/^{86}Sr$ data from low Rb/Sr rocks, it appears that initial $^{87}Sr/^{86}Sr$ ratios of volcanic rocks are largely similar to associated granitoids in volcano-plutonic complexes. On the other hand, in plutono-metamorphic belts, initial $^{87}Sr/^{86}Sr$ ratios of granitoids are not close to those of associated metamorphic rocks. This appears consistent with the fact that Japanese granitoids are mostly I-type, even if they occur closely associated with migmatitic rocks or display slightly peraluminous minerals.

Spatio-temporal distribution of initial $^{87}Sr/^{86}Sr$ ratio of granitoids

Cretaceous–Palaeogene

Initial $^{87}Sr/^{86}Sr$ ratios of Cretaceous–Palaeogene granitoids, which represent the most widespread felsic magmatism in the Japanese Archipelago, lie in the 0.704–0.712 range, thus covering nearly the whole range exhibited by all Japanese granitoids. Palaeogene granitoids, which occur mostly in the back-arc side of SW Japan (San'in Zone), as well as associated mafic rocks, have relatively low initial $^{87}Sr/^{86}Sr$ ratios of 0.7040–0.7055 in the western part of SW Japan (Imaoka *et al.* 2011); they can be as high as 0.7077–0.710 in the central to eastern part, however (Shibata & Ishihara 1979; Terakado & Nohda 1993). Cretaceous granitoids in SW Japan (San'yo and Ryoke zones) display a range of initial $^{87}Sr/^{86}Sr$ ratios (0.706–0.712), which is wider than most of the Palaeogene granitoids. A particularly high ratio 'hot spot' occurs in the middle to eastern part of SW Japan (initial $^{87}Sr/^{86}Sr$ ratios of 0.709–0.712). This 'hot spot' appears to cover the uppermost Cretaceous and the earliest Palaeogene granitoids in the eastern San'yo and San'in zones (Shibata & Ishihara 1979; Arakawa & Takahashi 1989; Tainosho *et al.* 1999).

Terakado & Nakamura (1984) and Kagami *et al.* (1992) proposed a segmented structure for the crust to explain regional differences in isotopic compositions of granitoids. But the data quality and density are not sufficient to support their discontinuous domain boundaries. Kagami *et al.* (1992) and his group introduced transitional zones with overlapping values between their domains, which means that their domains are not discontinuous against their own scheme. Although there is a remarkable spatial Sr isotopic heterogeneity in the Cretaceous source area, and isotopic gradients exist both locally and regionally, it is not appropriate to define discontinuous isotopic domains. Moreover, a distribution map of initial $^{87}Sr/^{86}Sr$ ratios of granitoids so far documented (e.g. Kagami *et al.* 1992) does not show a snapshot of the contemporaneous crustal isotopic structure because the 'initial $^{87}Sr/^{86}Sr$ ratios' are obtained as the $^{87}Sr/^{86}Sr$ ratios at their respective emplacement ages which differ from body to body. Caution therefore needs to be applied to interpret isotopic distribution maps in terms of isotopic signature of the granitoid source. Figure 4.10 is an initial $^{87}Sr/^{86}Sr$ distribution map showing the snapshot at *c.* 80 Ma, which can be used for the discussion on the heterogeneity of the source region.

It is noteworthy that initial $^{87}Sr/^{86}Sr$ ratios of mafic rocks associated with Cretaceous granitoids in SW Japan are 0.707–0.708, similar to the low $^{87}Sr/^{86}Sr$ members of the granite suite (Kagami *et al.* 2000; Nakajima *et al.* 2004), suggesting similar source materials and parental magmas. This may imply a somewhat enriched mantle beneath SW Japan, specifically the Cretaceous eastern Asian margin. It may well be regarded as subduction-related lithospheric mantle, as documented from Sierra Nevada (e.g. Mukhopadhyay & Manton 1994; Ducea & Saleeby 1998).

Finally, Early–Middle Cretaceous granitoids from NE Japan, of both volcano-plutonic and plutono-metamorphic types, have relatively low initial $^{87}Sr/^{86}Sr$ ratios of 0.7042–0.7055.

Miocene granitoids

Middle Miocene granitic rocks on the forearc side in SW Japan have initial $^{87}Sr/^{86}Sr$ ratios increasing systematically oceanwards from 0.7065 near the MTL to 0.7085 at the Pacific coast (Shibata & Ishihara 1979). This is coincident with an across-arc petrographical change from I-type to S-type character (Nakada & Takahashi 1979). Mafic rocks of the same ages occur ubiquitously and have low initial $^{87}Sr/^{86}Sr$ ratios of 0.7031–0.7038.

Middle Miocene–Pleistocene granitoids exposed around the Izu Collision Zone display initial $^{87}Sr/^{86}Sr$ ratios of 0.7033–0.7062, clustering mainly at 0.7040–0.7045 in the 14–10 Ma bodies (Saito *et al.* 2007, 2012) and 0.7033–7037 in the 8–4 Ma bodies (Ishizaka & Yanagi 1977; Shibata & Ishihara 1979; Terakado *et al.* 1988).

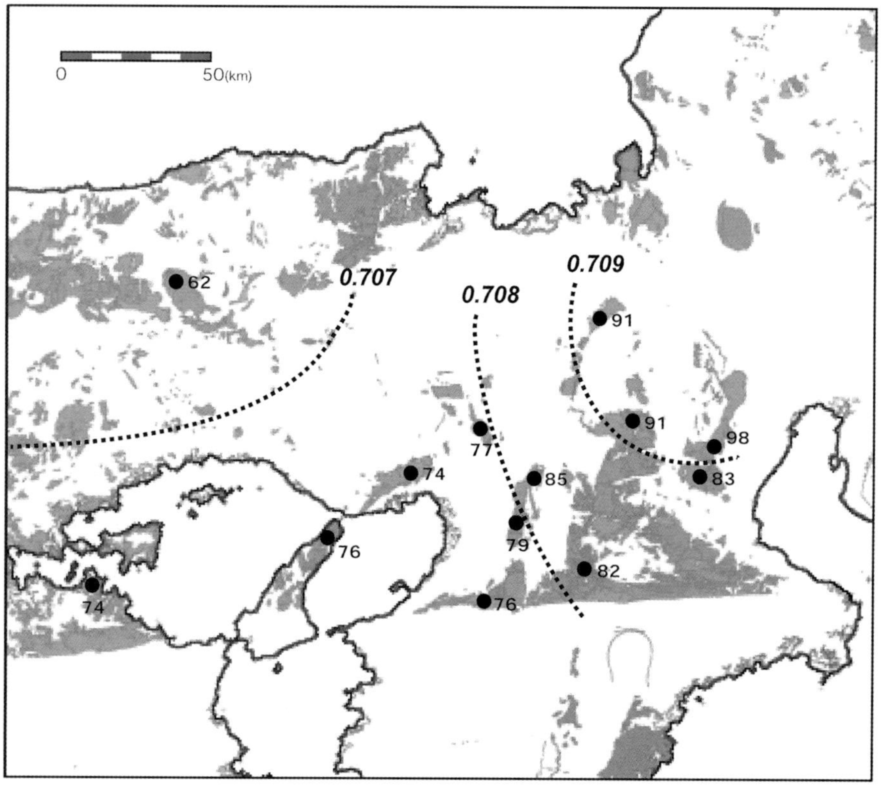

Fig. 4.10. $^{87}Sr/^{86}Sr$ distribution map of the *c.* 80 Ma granitoids in the Kinki and eastern Chugoku districts of the Japanese islands. The two-digit numbers besides the measured sample points (solid circle) were calculated via ($^{87}Sr/^{86}Sr$–0.7000) × 10 000.

They are interpreted as recording syncollisional granitic magmatism contemporaneous with the collision of the Izu and Honshu arcs (see Chapter 2f).

Early Miocene Hidaka granitoids in Hokkaido have initial $^{87}Sr/^{86}Sr$ ratios of 0.7041–0.7042, while mafic rocks have 0.7025 (Maeda & Kagami 1996). Kamiyama *et al.* (2007) envisaged an isotopically evolving process of magma from low-level juvenile mantle-related gabbroic rock with $^{87}Sr/^{86}Sr = 0.7025$ to isotopically equilibrated granite of $^{87}Sr/^{86}Sr = 0.7041$ at the top of a single plutonic body in Hidaka Belt at *c.* 20 Ma.

Quaternary granitoids

Quaternary granitoids that occur in the central highlands (Harayama 1992; Wada *et al.* 2004; Ito *et al.* 2013*a*, *b*) possess an initial $^{87}Sr/^{86}Sr$ ratio of 0.7076 (Harayama 1994), similar to nearby Quaternary volcanic rocks (Notsu 1983; Nakamura *et al.* 2008). These data imply a relatively enriched source beneath present-day central Japan.

Pre-Cretaceous granitoids

Triassic–Jurassic granitoids in the Hida Mountains can be divided into two groups on the basis of their isotopic character, one having initial $^{87}Sr/^{86}Sr$ ratios of 0.7044–7053 and the other having a higher range of values of 0.7057–0.7105 (Arakawa 1990). Granitoids belonging to the first group are exposed on the forearc side of the Hida Terrane while the second group crops out on the inner side of the terrane (Arakawa 1990). These isotopic data are not well associated with reliable age data. Granitoids of the first group seem to be mostly Early Jurassic in age, while the second group appears to include Early Triassic examples.

Other fragmental pre-Cretaceous granitoid blocks, mainly of alleged Permo-Triassic age (*c.* 250 Ma), give relatively low initial $^{87}Sr/^{86}Sr$ ratios of 0.7038–0.7042 (Shibata & Takagi 1989). Granitoids with Devonian–Silurian ages (*c.* 450–400 Ma) have similar initial $^{87}Sr/^{86}Sr$ ratios to those of the Cretaceous granitoids of SW Japan.

Stable isotopes

Oxygen and sulphur isotopes reflect the degree of upper crustal contamination in the granitic magma. In this context, the Cretaceous and Palaeogene granitoids of SW Japan show a zonal variation with $\delta^{18}O$ increasing from the back-arc region (San'in Zone) towards the MTL (Ryoke Zone), with a rise from 6‰ to over 11‰ at a normalized values of $SiO_2 = 70$ wt% with a corresponding decrease in magnetic susceptibility (Ishihara & Matsuhisa 2002). This $\delta^{18}O$ range is comparable to that found in granitoids of the Peninsular Range Batholith, western US (Taylor & Silver 1978). A similar across-arc zonal variation of $\delta^{18}O$ is seen in Middle Miocene granitoids on the Pacific side of SW Japan (Ishihara & Matsuhisa 1999). Late Neogene granitoids with the lowest initial $^{87}Sr/^{86}Sr$ ratio in the Izu Collision Zone also have the lowest $\delta^{18}O$ values of *c.* 5.0–5.5‰ (Ishihara & Matsuhisa 2005).

Sulphur isotope ratios correlate with magnetic susceptibility. Broadly speaking, granitoids belonging to the magnetite series show positive $\delta^{34}S$ values while ilmenite-series granitoids display negative $\delta^{34}S$ (Sasaki & Ishihara 1979; Ishihara & Sasaki 2002).

Isotopic character and typology

Sr isotopic signatures reflect the source material of plutonic intrusions. Given that Japanese granitoids are not derived from ancient cratons but rather from Phanerozoic arc-related sources, variation in initial Sr isotopic ratios is limited although some differences can be identified still between granite types and series.

As concluded previously, most Japanese granitoids are of I-type. Their initial $^{87}Sr/^{86}Sr$ ratios range from 0.705 to 0.712 with a modal peak at 0.706–0.708. S-type granitoids in the strict sense (having magmatic cordierite) are rare and show initial $^{87}Sr/^{86}Sr$ ratios of 0.708–0.710 in contrast to M-type granitoids, which display values in the 0.7035–0.7045 range. Magnetite-series granitoids have relatively low initial $^{87}Sr/^{86}Sr$ ratios of 0.7035–0.7070 in the western part of SW Japan and in NE Japan, and of 0.707–0.709 in the central to eastern part of SW Japan. Ilmenite-series granitoids lie within the range 0.706–0.712.

Oxygen isotope patterns correlate with Sr isotopes. Magnetite-series granitoids have mostly $SiO_2 = 70$ wt%-normalized $\delta^{18}O$ of less than +8.0‰, whereas ilmenite-series granitoids are higher than +8.0‰ for Cretaceous–Palaeogene granitoids in SW Japan (Ishihara & Matsuhisa 2002). Similarly, Middle Miocene granitoids of S-type in the broad sense have higher $\delta^{18}O$ than I-type granitoids (Ishihara & Matsuhisa 1999).

Volcano-plutonic complexes (MT, TI, TN)

High-level granitic rocks often occur associated with (and often intruding) coeval felsic volcanic rocks, being exposed together as volcano-plutonic complexes. Volcanic rocks include voluminous densely welded silicic ash-flow tuffs and associated lavas, sometimes organized into large-scale cauldrons and volcano-tectonic depressions.

Geochronological data on these volcanic rocks still depend mainly on K–Ar ages, since U–Th–Pb ages are quite limited. K–Ar ages are problematic however, due to more excess argon and secondary release of radiogenic argon than within plutonic rocks. In fact, some K–Ar age data have been replaced by recent U–Pb ages (e.g. Sawada & Itaya 1993 v. Sato *et al.* 2013). The following spatio-temporal descriptions therefore depend mainly on K–Ar ages until U–Th–Pb ages are available.

Cretaceous–Palaeogene

Large-scale volcano-plutonic activity with batholithic granitoids that developed in Cretaceous–Early Palaeogene times has been traditionally classified into stages I (130–100 Ma), II (100–80 Ma), III (80–68 Ma), IV (68–50 Ma) and V (43–30 Ma) based on K–Ar and Rb–Sr ages (Imaoka *et al.* 2011). This magmatism was temporally continuous, except for a break between stages IV and V. The southern limit of volcano-plutonic complexes is regarded as approximating the volcanic front of each stage in SW Japan (Fig. 4.11).

Gigantic Cretaceous volcano-plutonic complexes prevail in the San'yo Zone of SW Japan. Although some of them must have been eroded away, a huge amount of granitic and felsic volcanic rocks still remain on the present surface. These volcanic rocks constitute several geological units, for example Nohi Rhyolites in Chubu district (Sonehara & Harayama 2007), Koto Rhyolites in Kinki district (Sawada & Itaya 1993) and Aioi, Takada and Abu-Hikimi rhyolites in Chugoku district (Yoshida 1961; Imaoka *et al.* 1988; Yamamoto 2003). All of them are intruded by high-level granitoids. Of these various complexes, the volcano-stratigraphic sequences of the Nohi Rhyolites is the best established. Nohi Rhyolites crop out over an area of *c.* 120 × 40 km, giving a total volume of 5000–7000 km^3 of welded tuff (Sonehara & Harayama 2007). The volume is far beyond any present-day large-scale volcanic field in the world, such as Yellowstone or Taupo (Christiansen 1983; Wilson *et al.* 1984). The Nohi Rhyolites cover an area elongated in a NNW–SSE direction and extending from the northern margin of the Ryoke Belt to the Hida Belt. Other units such as Takada and Abu-Hikimi Rhyolites in Chugoku district, western

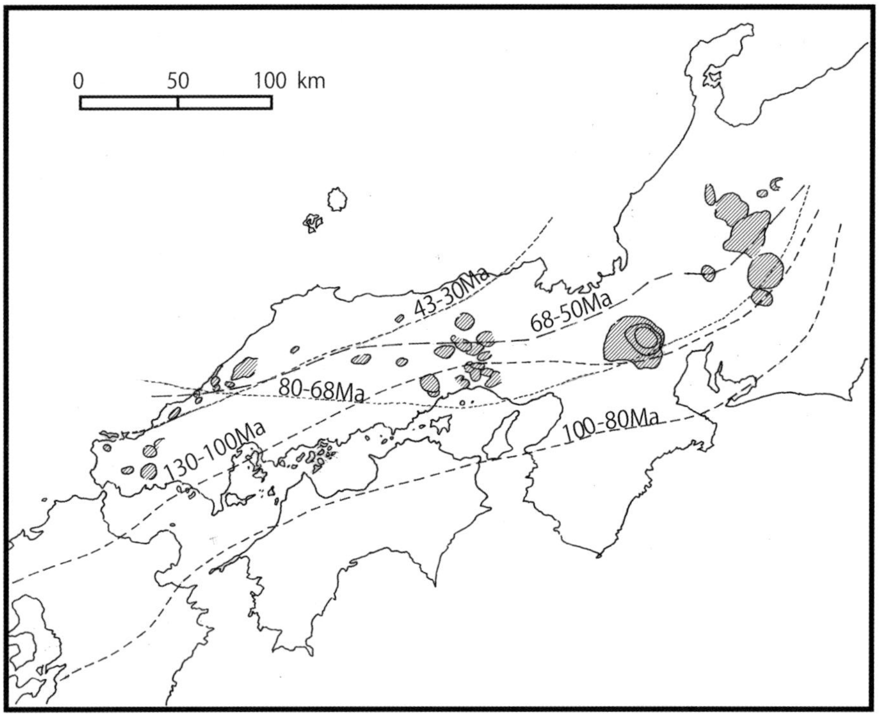

Fig. 4.11. Temporal migration of the volcano-plutonic field during the Palaeogene–Cretaceous and cauldrons in SW Japan.

SW Japan, have lithological associations, sequences and volumes largely similar to Nohi Rhyolites, with an estimated maximum total thickness of *c.* 3000 m and magmatic volumes reaching several thousands of cubic kilometres (Imaoka *et al.* 1988).

Volcanic rocks are predominantly rhyolites, rhyodacites and dacites with minor amounts of andesites, and are mostly strongly welded crystal-rich ash-flow tuffs with subordinate amounts of lava. Their chemical character exhibits arc magmatism affinity and is similar to the associated granitoids in the San'yo zone, which are high-K, I-type and ilmenite series.

Although age data on the early stage of this magmatic event are not abundant, it seems plausible that the Cretaceous–Palaeogene magmatism started in the Kitakami Zone, NE Japan, at *c.* 130 Ma, according to recently reported U–Pb ages (Tsuchiya *et al.* 2012) which represents the onset of Stage I. This magmatic activity is characterized by high-Mg andesite and basalt and associated with calc-alkaline, adakitic and shoshonitic granitoids (Tsuchiya & Kanisawa 1994; Tsuchiya *et al.* 2007). In SW Japan, it commenced with Middle Cretaceous andesite to dacite in northern Kyushu and western Chugoku, accompanied by 110–100 Ma granitoids (U–Pb uraninite ages by Yokoyama *et al.* 2010; zircon ages by Adachi *et al.* 2012).

The climax of this magmatism seems to have been reached at *c.* 90–70 Ma, during stages II and III. Late Cretaceous U–Pb zircon ages have been documented from the Koto Rhyolites in the eastern Kinki region (75–70 Ma: Sato *et al.* 2013), the Arima-Aioi Rhyolites in western Kinki region (86–81 Ma: Sato *et al.* 2013) and the Nohi-Rhyolite-related high-level granitoids (71 Ma: Nakajima 1996). The oldest CHIME (Th–Pb) monazite age of granitoids intruding the Nohi Rhyolites reported so far is 67 Ma (Suzuki & Adachi 1998).

The duration of volcanic activity is quite an important problem, but not yet well constrained due to a lack of reliable age data. Volcanic rocks are intruded by granitoids and never unconformably overlie them. Field observations of intruding relationships, including mafic pillows (MME; mafic microgranular enclaves) in granitic matrix, have been re-examined recently from the viewpoint of magma chamber processes (e.g. Nakajima *et al.* 2004).

Late Palaeogene volcano-plutonic complexes with caldera clusters are well preserved in the coastal area of the San'in Zone (Imaoka *et al.* 2011); they were formed at *c.* 45–30 Ma during Stage V, and are all of I-type and magnetite series.

Neogene and Quaternary

Middle Miocene granitoids associated with large-sized volcano-plutonic complexes include Osuzu-yama (Nakada 1983) and Okue-yama (Takahashi 1986) in Kyushu, Ishizuchi (Yoshida 1984) in Shikoku, Kumano (Aramaki 1965; Miura 1999; Kawakami *et al.* 2007) in the Kinki district, and stocks and batholiths in the Pacific coastal area of SW Japan. They record a transient period of magmatism at 15–14 Ma (Sumii *et al.* 1998; Orihashi *et al.* 2013), with most of them exhibiting the clear outlines of volcanic caldera structures and some of them being composite. They are exposed in the region of the Palaeogene Shimanto Supergroup (Chapter 2d) and some of them have chemical and petrographic features of S-type granitoids, implying magmatic interaction with sedimentary host rocks (Nakada & Takahashi 1979).

Late Miocene–Early Pliocene volcano-plutonic activity has been documented in the Izu Collisional Zone, where the Higashi–Yamanashi complex forms a graben-like volcano-tectonic depression (Takahashi 1989).

The Quaternary Takidani and Kurobegawa granites exposed in the Hida mountain range of central Japan (Harayama 1992; Wada *et al.* 2004) have yielded U–Pb ages of 1.36 ± 0.23 Ma (Sano *et al.* 2002) and 0.5–0.6 Ma (Ito *et al.* 2013*a*, *b*), respectively, which are nearly identical to their K–Ar biotite ages (Harayama 1992, 1994). Both granite bodies intrude Quaternary felsic volcanic rocks and are interpreted as tilted caldera (Harayama 1994).

In the geothermal field of Kakkonda in NE Japan, a subsurface granite body was detected by drilling 2800 m below ground level (Sasaki *et al.* 2003). The lithology is granodiorite-tonalite of

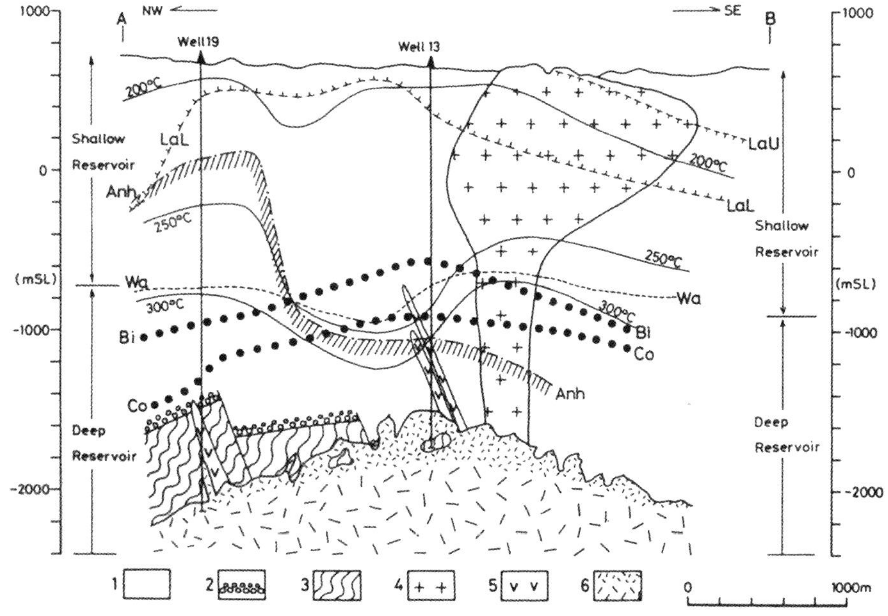

Fig. 4.12. Cross-section of the Kakkonda geothermal field, NE Japan (Kato *et al.* 1993). Legend: 1, Neogene andesite to dacite tuffs; 2, Conglomerate; 3, pre-Paleogene formation; 4, dacite intrusion; 5, old intrusion; 6, Kakkonda granite. LaU, upper limit of laumontite; LaL, lower limit of laumontite; Wa, lower limit of wairakite; Bi, biotite isograd in the contact aureole; Co, cordierite isograd.

I-type and magnetite-series affinity. The top of the body is still hot at 300°C, rising to >400°C at depths of 900 m (Fig. 4.12). The Kakkonda granite intrudes Neogene andesite to dacite tuffs and tuffaceous sediments, overlain by Pleistocene–Holocene rhyolite tuffs that are argued to be genetically related to the Kakkonda granite magma (Kanisawa *et al.* 1994).

Deep-seated granites (TN, TS)

Deep-seated granites are usually accompanied by migmatites, these together forming plutono-metamorphic belts such as those documented in other areas including the Appalachians (DeYoreo *et al.* 1989; Solar *et al.* 1998), the Armorican Saint-Malo zone (Brown

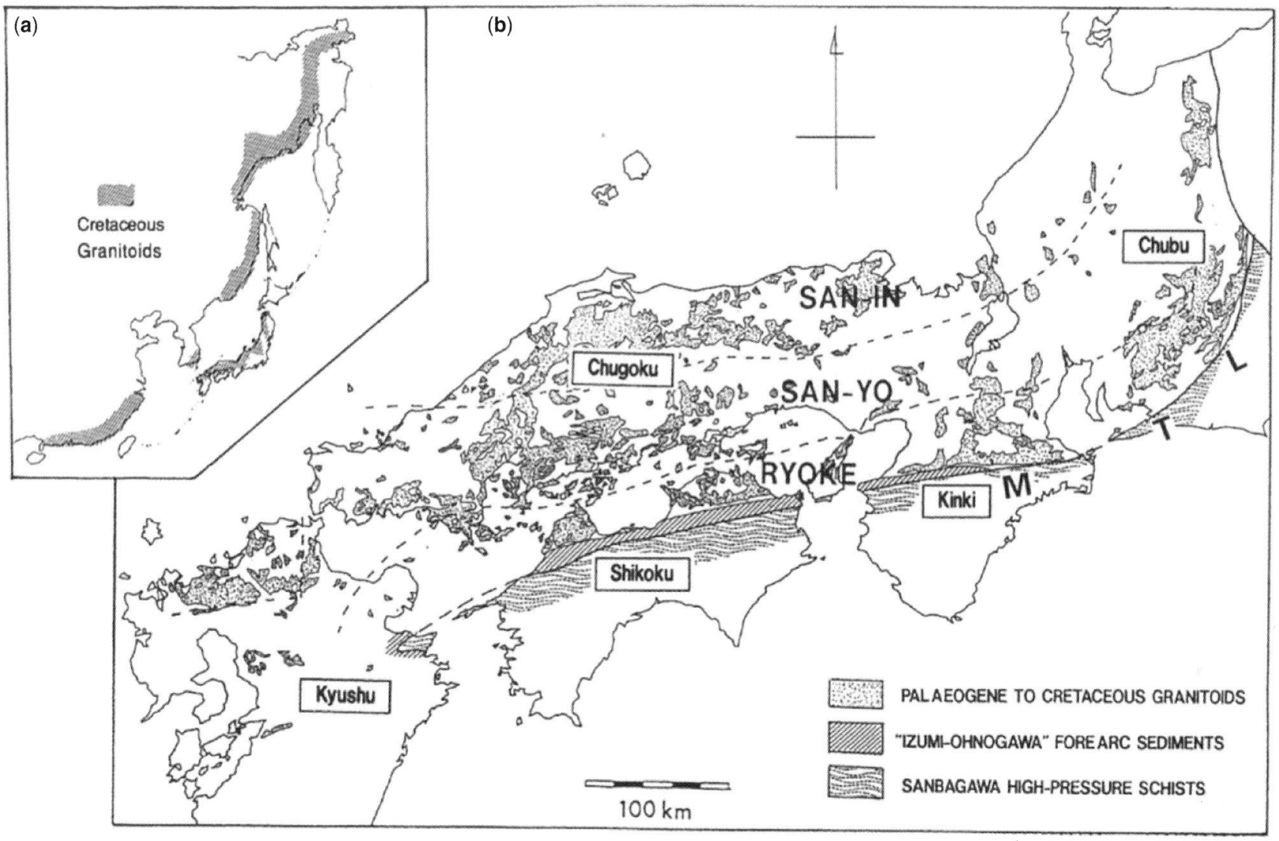

Fig. 4.13. Granitic provinces in SW Japan, modified after Nakajima (1996).

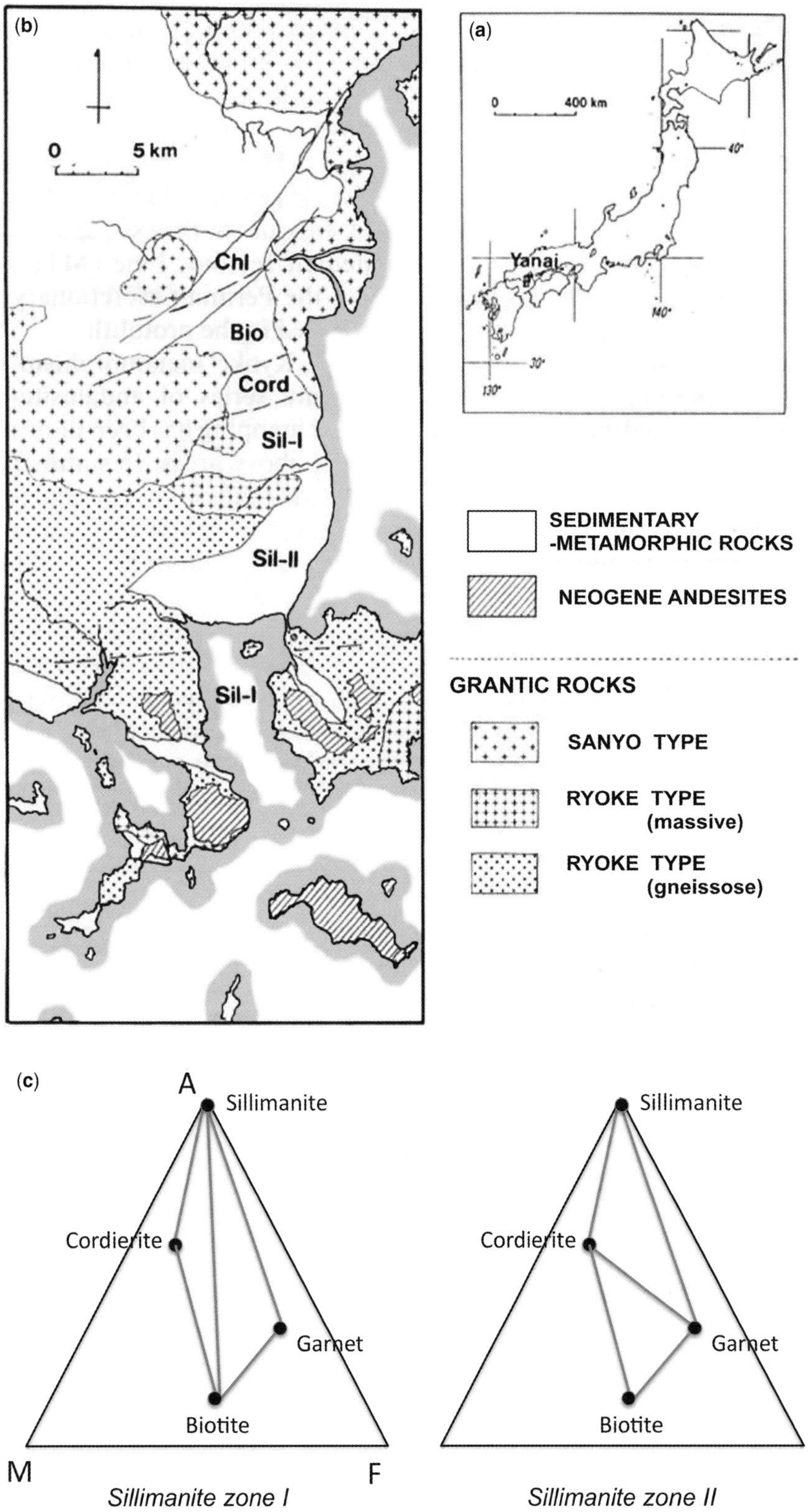

Fig. 4.14. Occurrence of granitic rocks and metamorphic mineral zones in the San'yo–Ryoke transect in the Yanai area, SW Japan, modified after Nakajima (1994). (**a**) Locality of the Yanai area. (**b**) Map showing the metamorphic mineral zones and granitoids. Chl, chlorite zone; Bio, biotite zone; Cord, cordierite zone; Sil-I, Sillimanite zone I; Sil-II, Sillimanite zone II. (**c**) Phase equilibria with excess quartz, K-feldspar and plagioclase, expressing higher-grade equilibration in Sillimanite zone II than Sillimanite zone I, described by the equation biotite + sillimanite <Sil-I> = garnet + cordierite (+K feldspar + SiO_2 + H_2O) <Sil-II>.

& D'Lemos 1991; Milord *et al.* 2001) and the Ivrea-Verbano zone (Quick *et al.* 1995). They represent the surface-exposed deep crust of orogens. In the Japanese islands, deep-seated granites occur in several plutono-metamorphic belts, the Ryoke Belt in SW Japan, the Hidaka Belt in Hokkaido Island, the Hida Belt in the most back-arc position of Japan and the Abukuma Belt in NE Japan.

Ryoke Belt

The Ryoke Belt is well known as a low-pressure-type metamorphic belt in the classical Miyashiro's 'paired metamorphic belt' theory (see Chapter 2c). It comprises pre-Cretaceous accretionary complexes metamorphosed up to migmatite grade and is associated

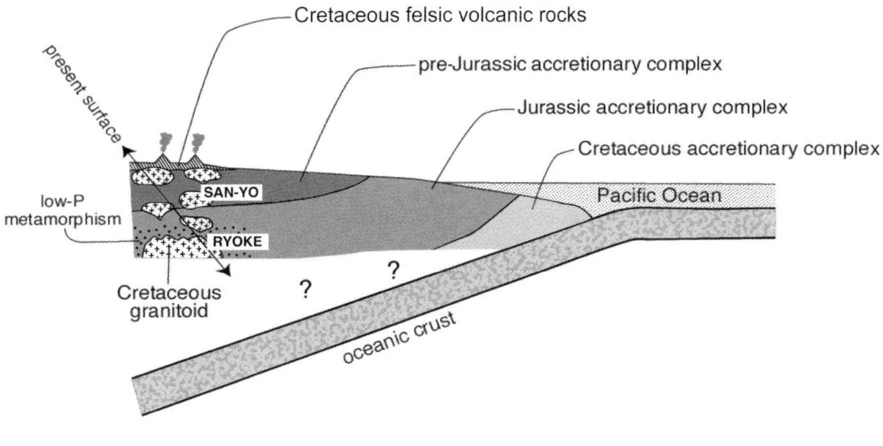

Fig. 4.15. Schematic crustal section of the Cretaceous orogen at the Eurasian continental margin, now exposed in SW Japan as the Ryoke and San'yo zones. Not to scale. Modified after Nakajima (1994, 1996). 'Cretaceous felsic volcanic rocks' denotes Nohi–Takada–Abu Rhyolites.

with Cretaceous granitoids. The exposure area of the granitoids is several times larger than that of the metamorphic rocks.

Deep-seated granitic rocks in SW Japan are exposed as plutono-metamorphic belts only in the Ryoke Zone on the forearc side (Fig. 4.13). No Palaeogene deep-seated granitoids are exposed in SW Japan. Cretaceous Ryoke granitic rocks are exposed close to the MTL, in association with high-grade gneisses and migmatites. The granitoids are mostly deformed and intrusive contacts are concordant with the gneissosity of metamorphic host rocks, partly showing 'lit-par-lit' (layer-by-layer) intrusion features. This emplacement mode, along with their outcrop patterns, implies sheet-like intrusion of granitic magma at a relatively deep crustal level. The main lithology is biotite hornblende granodiorite to tonalite, with SiO_2 contents of 65–70 wt%. Magmatic garnet, cordierite and sillimanite are not present and, though muscovite sometimes occurs, it is mostly a secondary mineral. The ASI correlates positively with SiO_2 contents, which indicates an I-type chemical character (Nakajima 1996).

These deep-seated gneissose granitoids have classically been called 'Older Ryoke granites', in contrast to unfoliated 'Younger Ryoke granites'. The deep-seated Older Ryoke granites were interpreted to have provided the heat source for Ryoke regional metamorphism (Okudaira 1996; Miyazaki 2004), although this view is contradicted by the fact that the age data of granitoids so far documented are younger than Ryoke metamorphic rocks (Suzuki & Adachi 1998).

Migmatites are mostly metatexite, though diatexite also occurs. The metamorphic grade of the migmatites is typically upper amphibolite facies as indicated by their mineral assemblage of garnet-cordierite-K feldspar-biotite-plagioclase-quartz (Nakajima 1994; Ikeda 2004). Granulites rarely occur. Geothermobarometry indicates pressure–temperature (P/T) conditions of the migmatite zone in the Yanai area to have been 0.45–0.65 GPa and 700–850°C (Ikeda 2004), which is close to the solidus of deep-seated granites. Based on such data, the deepest level of surface-exposed deep-seated Ryoke granitoids is estimated to have been 15–20 km. Figure 4.14 shows the granite-featuring geological map with metamorphic mineral zones of the Ryoke–San'yo belts in the Yanai area, SW Japan. It demonstrates that high-level granites of the San'yo Belt intrude unmetamorphosed to weakly metamorphosed rocks, while deep-seated granites of Ryoke Belt intrude high-grade metamorphic rocks. The granitoids in the San'yo Belt also intrude comagmatic felsic volcanic rocks overlying sedimentary rocks and represent volcano-plutonic complexes. In this San'yo–Ryoke transect, the metamorphic grade of wall rocks increases continuously from a low-grade zone, where chlorite is the only metamorphic mineral, to a high-grade zone where garnet-cordierite-K-feldspar stably coexist (Fig. 4.14b, c). This transect is therefore interpreted to represent a surface-exposed cross-section through upper to middle crust of a Cretaceous orogen. Figure 4.15 provides a schematic crustal section of the Cretaceous Eurasian continental margin before the opening of the Japan Sea (Nakajima 1994).

Migmatite has drawn attention as a possible source for granitic magma. It has been speculated that anatectic melts are extracted and segregated to grow and coalesce to form a pluton-sized magma bodies (e.g. D'Lemos *et al.* 1992; Bons *et al.* 2004). Likewise, Obata *et al.* (1994) described the development of leucosomes in Ryoke migmatites of the Higo area, Kyushu Island, with increasing metamorphic grade and discussed potential growth processes able to produce a granitic body. However, Nakajima *et al.* (2008) examined the U–Pb zircon ages of migmatites and associated granitic bodies and argued that leucocratic melts do not represent the origin of granitic plutons.

Hidaka Belt

The Hidaka Belt on Hokkaido Island is a young plutono-metamorphic complex, which comprises low P/T -type metamorphic rocks up to granulite grade and intrusive rocks of various compositions from granites to gabbros emplaced at various depths (Komatsu *et al.* 1989; Osanai *et al.* 1991). High-level granitoids occur also, but are not accompanied by cogenetic volcanic rocks.

The metamorphic complex consists of steeply eastwards-dipping thrust sheets of metamorphosed Palaeogene accretionary complexes composed of pelitic-psammitic, intermediate, mafic and ultramafic rocks and igneous rocks (gabbros, diorites and granitoids) (Fig. 4.16). The Hidaka plutono-metamorphic belt is interpreted as a crustal section of an island arc (Komatsu *et al.* 1989), uplifted since the Miocene collision of the Honshu and Kuril arcs (Kimura 1986). An idealized crustal section through the Hidaka plutono-metamorphic belt is shown in Figure 4.17.

Metamorphic rocks are divided into four zones, from I at the top (east) to IV at the bottom (west) of the thrust sheets (Osanai *et al.* 1991). Metamorphic conditions of zones I, II, III and IV are subgreenschist-greenschist facies, greenschist-amphibolite facies, amphibolite facies and granulite facies, respectively. Peak metamorphic P/T conditions of zone IV were 870°C and 7.3 kbar (0.73 GPa) (Osanai *et al.* 1991), and the age of the granulite facies metamorphism has been dated as *c.* 20–18 Ma (Usuki *et al.* 2006; Kemp *et al.* 2007). As such, the granulites of the Hidaka Belt represent some of the youngest exposures of lower crust on Earth.

Granitic rocks are widely exposed in the Hidaka Belt, consisting mainly of tonalites and granodiorites and including both paraluminous and metaluminous types, both of which belong to the ilmenite series. Some peraluminous tonalites are S-type, as they contain

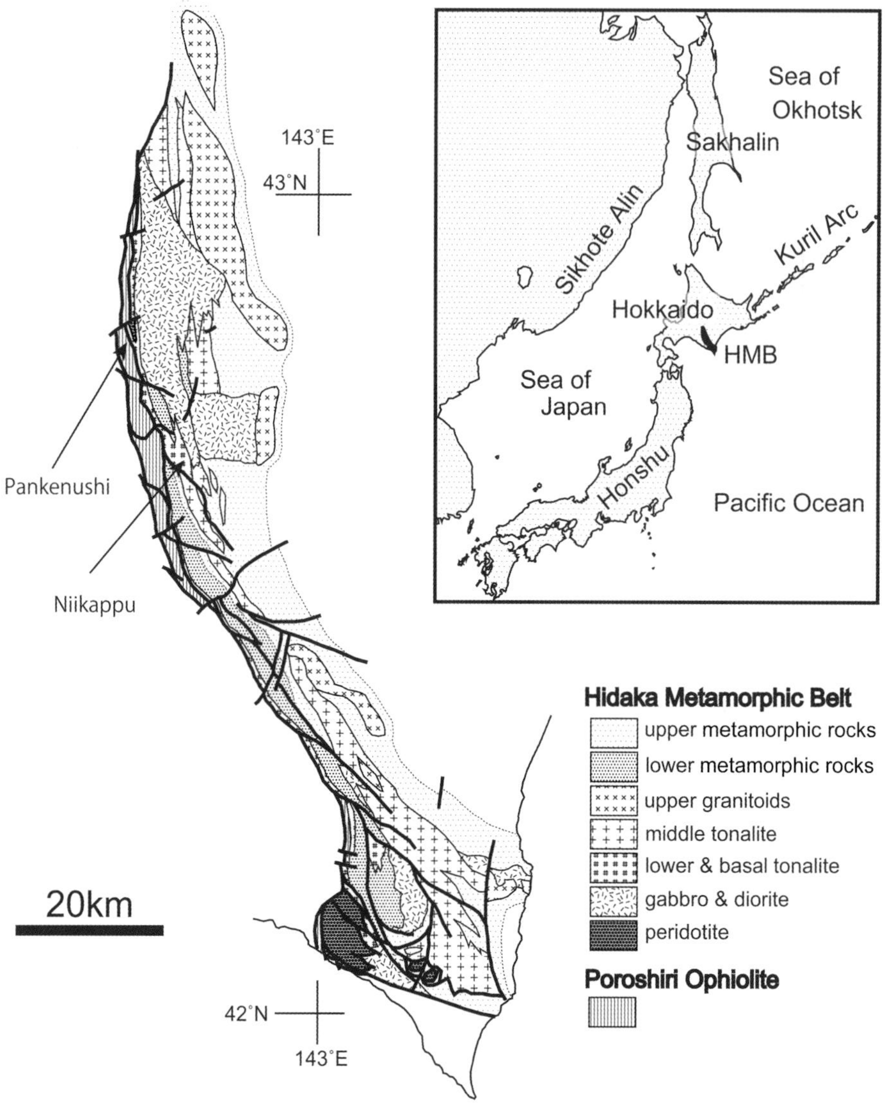

Fig. 4.16. Simplified geological map of the Hidaka Metamorphic Belt, modified after Komatsu *et al.* (1986).

magmatic cordierite (Shimura *et al.* 1992). They have also garnet and orthopyroxene.

The granitic suites were emplaced into the various metamorphic layers (zones I–IV), and can be classified into four depth levels: upper, middle, lower and basal intrusions (Table 4.3). U–Pb zircon ages show two thermal pulses at 37 and 19 Ma (Kemp *et al.* 2007; Jahn *et al.* 2014). The oldest age was given by 'middle' metaluminous tonalite emplaced in the amphibolite facies layers, whereas the youngest is from 'basal' peraluminous tonalite intruded into granulite facies rocks. These two age clusters are consistent with previously reported K–Ar mica age groupings of 40–35 and 20–18 Ma obtained from plutonic and metamorphic rocks in the southernmost

Table 4.3. *Characteristics of the granitic suite of the Hidaka Metamorphic Belt*

			Peraluminous			Metaluminous		
				Approx. age (Ma)			Approx. age (Ma)	
Depth type	Rock type	Intrusive level	Characteristic mineral assemblage	37	19	Characteristic mineral assemblage	37	19
Upper	Granite, granodiorite	Chl–Ms metasediment unit	Bt + Pl + Qz + Kfs		X	Bt + Hbl + Pl + Qz + Kfs	X	X
Middle	Tonalite	Bt–Ms gneiss and schist unit	Bt ± Ms ± Crd + Pl + Qz ± Kfs	X		Bt + Hbl + Pl + Qz ± Kfs	X	
Lower	Tonalite	Amphibolite unit	Bt + Opx ± Crd + Pl + Qz ± Kfs Bt + Opx + Grt ± Crd + Pl + Qz ± Kfs		X	Bt + Hbl + Opx ± Cpx + Pl + Qz	X	
Basal	Tonalite	Granulite unit	Bt + Grt ± Crd + Pl + Qz ± Kfs		X	Bt + Hbl + Opx ± Cpx + Pl + Qz	X	

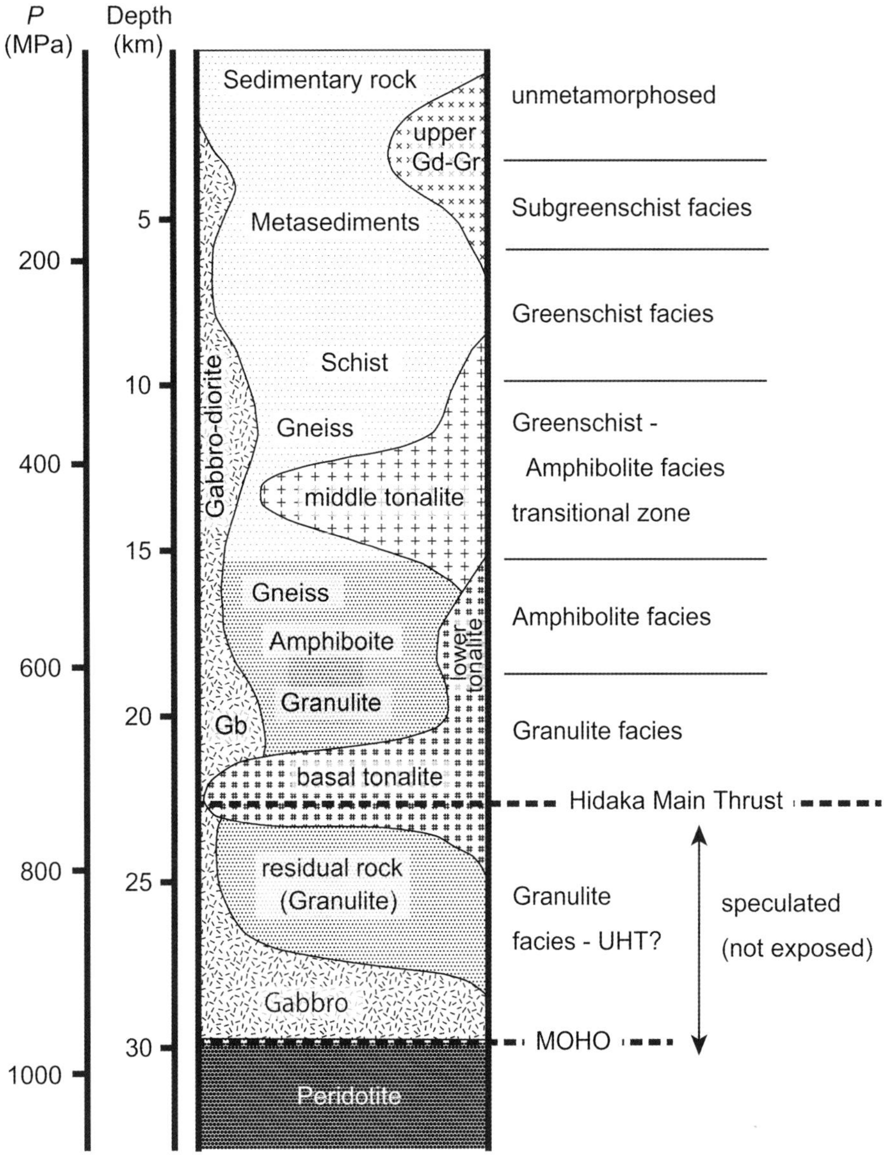

Fig. 4.17. Idealized crustal succession of the Hidaka Metamorphic Belt, modified after Komatsu *et al.* (1986) and Shimura *et al.* (2004). Deeper levels below the Hidaka Main Thrust (lowermost crust) are not exposed.

area of the Hidaka Belt (Saeki *et al.* 1995). Typical 19 Ma basal-lower tonalites enclose a large number of metamorphic and igneous enclaves, such as mafic and pelitic/psammitic granulites, gabbros, diorites and also fragments of 37 Ma metaluminous tonalite. The 37 Ma magmatic activity is interpreted as arc magmatism (Kemp *et al.* 2007), and the 19 Ma peraluminous tonalite could have been produced from 37 Ma metaluminous tonalite by crustal assimilation (Takahashi 1992). In addition, 20–19 Ma metaluminous granitoids occur ubiquitously in the upper levels (e.g. Kamiyama *et al.* 2007).

Hida Belt

The Hida Belt, located in a back-arc position in central Honshu, is the oldest metamorphic belt in the Japanese islands. Pre-Cretaceous age clusters in Figure 4.2 include those from Hida granitoids and other isolated small bodies and fragments of ancient rocks. The Hida Belt is composed of high-grade gneisses and migmatites with deformed and undeformed granitoids. Low-grade to unmetamorphosed rocks and volcanic rocks are absent.

The ages of the granitoids documented so far are spread across the range 300–180 Ma and are not yet fully interpreted. Takahashi *et al.* (2010) documented SHRIMP U–Pb zircon ages and detected two groups at 200–190 and 250–240 Ma, as seen in Figure 4.3. The 250–240 Ma granitoids are highly deformed, together with metamorphic rocks. The main lithology is gneissose biotite granite, while locally a distinctive strongly sheared K-feldspar-phyric rock occurs, which was once called ‘augen gneiss’. These rocks are interpreted as deep-seated with an emplacement depth equivalent to upper amphibolite-granulite facies metamorphism, assuming that the granitoids intruded during peak metamorphism.

Cpx-bearing monzonitic to trondjemitic migmatites, called ‘Inishi-type rock’, occur and have attracted attention since the early studies of Japanese geology. They are interpreted to have formed by interaction between limestone and granitic magma (Kunugiza & Goto 2006). If so, they represent another type of migmatite not derived from anatexis of metamorphic rock.

Abukuma Belt

The Abukuma Belt represents a Cretaceous plutono-metamorphic belt in NE Japan. It was once argued to be an eastern extension of the Ryoke Belt, but the ages of Abukuma granitoids are distinctly older than those of their supposed Ryoke equivalents nearby (Kon & Takagi 2012). The Abukuma plutono-metamorphic belt, as for the Hidaka and Hida belts, also lacks coeval felsic volcanic rocks. Some granitoids are gneissose and intruded concordantly with the gneissosity of high-grade gneisses.

Tectonics and granitic magmatism (TN, MT, TI, TS)

The Japanese islands have been located along the eastern margin of the Eurasian continent and the convergent plate boundary since Late Palaeozoic times. Granitic magmatism at this convergent plate boundary is essentially controlled by the nature of the subducting plate, especially its age and convergence rate, and upwelling of mantle plumes or collision events nearby.

The tectonic settings in which Japanese granitoids formed can be briefly summarized as: (1) Triassic–Jurassic normal plate subduction, with low convergent rate along the continental margin; (2) Cretaceous–Early Palaeogene abnormal plate subduction, with high convergent rate or ridge subduction; (3) Late Palaeogene upwelling of a mantle plume along the Eurasian continental margin; (4) Middle Miocene upwelling of a mantle plume causing opening of the Japan Sea basin and Kuril back-arc basin; (5) Miocene–Pliocene collision of the Izu–Bonin (Ogasawara) Arc with Honshu island; and (6) Quaternary normal subduction of oceanic plates, with relatively high (NE Japan) and moderate (SW Japan) convergent rates. We summarize the main Japanese granitoid-forming events within the context of this plate tectonic history in the following sections.

Cretaceous–Early Palaeogene: subduction of oceanic plate with extremely high convergent rate and/or active mid-oceanic ridge subduction

The most voluminous granitic and felsic volcanic rocks in the Japanese islands were formed during Cretaceous–Palaeogene times, when the subduction rate of the Pacific plate beneath the Japanese islands was extremely high (Mueller *et al.* 2008). As they constitute a part of the 'Late Mesozoic Circum-Pacific ring of fire', their generation appears to be related to the now classical concept of a 'Cretaceous pulse' (Pitman 1978), during which global relative plate motions were unusually rapid. In addition, in the Japan area the Cretaceous–Palaeogene period was when the Kula–Pacific ridge collided with the Eurasian margin and a ridge–trench–trench (RTT) junction migrated along the eastern margin of the Eurasian plate (Engebretson *et al.* 1985). Nakajima *et al.* (1990) and Kinoshita & Ito (1986) documented eastwards-younging isotopic ages in Cretaceous granitoids of SW Japan and interpreted them as tracking the along-arc migration of the RTT junction. This ridge subduction model was later also applied to Palaeogene–Neogene granitoid plutons in coastal Alaska, where systematic along-arc eastwards-younging ages were documented (Bradley *et al.* 1993; see also Sisson *et al.* 2010) although without citing the Japanese papers. They featured a similar RTT junction migration along the Alaskan coastal trench and Farallon–Pacific plate ridge, which is a mirror image of SW Japan (Nakajima *et al.* 1990).

Recently, Tani *et al.* (2014) have proposed that three magmatic pulses occurred in Cretaceous–Palaeogene times (95–80, 70–55, 40–30 Ma), based on a number of new SHRIMP U–Pb zircon ages on granitoids from the Kinki and Chugoku districts, and ascribed them to global pulses demonstrated by Mueller *et al.* (2008). This new interpretation presents a radical challenge to the traditional view of Cretaceous–Palaeogene continuous granite magmatism in SW Japan shown in Figure 4.2

It is well established that throughout Earth history large-scale granitic magmatism has taken place episodically, not only in Phanerozoic but also in Proterozoic and Archean times (e.g. Condie 1982; Windley 1984; Rino *et al.* 2004). Global-scale events are presumably controlled by global dynamics, such as pulses of mantle activation. Ridge-subduction-induced granitic magmatism may be a regional-scaled close-up view of such a global pulse, and the subject will doubtlessly require much further data and discussion.

Late Palaeogene: upwelling of mantle plume in the eastern margin of the Eurasian continent

Late Palaeogene magmatism that produced clusters of volcano-plutonic calderas in the coastal region of the Japan Sea in western Japan occurred during an extensional phase 15–10 Ma after the gigantic Cretaceous–Palaeogene magmatic pulse. It is presumed to be a result of mantle upwelling at the eastern margin of Eurasian continent, which may have been a precursor to back-arc rifting (Imaoka *et al.* 2011).

Middle Miocene: opening of marginal sea basins

In Middle Miocene times, magmatism caused by back-arc opening took place at two different sites with different tectonic settings within the Japanese islands, namely the Pacific coastal area of SW Japan and central Hokkaido.

Pacific coastal Miocene magmatism was caused by the opening of the Japan Sea at *c.* 15 Ma. This back-arc opening resulted in obduction of a SW Japan forearc sliver over the newly born Shikoku Basin. This magmatism was very short-lived (15–14 Ma) and the resulting intrusions are all high-level granitoids associated with volcano-plutonic complexes.

In Hokkaido, Middle Miocene magmatic activity and granulite facies metamorphism are thought to have been caused by upwelling of MORB-source mantle (Maeda & Kagami 1996) related to opening of the Japan and Kuril basins (Usuki *et al.* 2006; Kemp *et al.* 2007). Opening of the Kuril Basin caused uplift and denudation of a crustal cross-section through the arc down to 20 km depth. S-type tonalite in the basal level evolved by crustal melting during Middle Miocene granulite facies metamorphism, and formed the Hidaka plutono-metamorphic belt.

Late Miocene–Pliocene: collision of Izu and Honshu arcs

Granitoids yielding 16–4 Ma SHRIMP U–Pb zircon ages occur in the Izu Collision Zone, intruding Miocene trench-fill sediments. They were originally interpreted to be on-land accreted equivalents of the tonalitic middle crust of Izu Arc accreted to Honshu Arc, based on an along-arc V_p profile of the Izu–Bonin–Mariana (IBM) Arc (Suyehiro *et al.* 1996). Based on new geochronological data however, they are now interpreted as syncollisional granitoids (Tani *et al.* 2010).

Evolution of granitic crust in Japanese islands (TN)

Temporal change in isotope ratios

The Japanese islands are composed of multiple granitic provinces, which partly overlap and cross each other. In such settings, we can sometimes obtain information on temporal change or evolution of crustal material by elucidating the isotopic character of granitoids and related rocks. A significant difference in initial $^{87}Sr/^{86}Sr$ ratios between contiguous plutonic bodies at nearly the same locality with intrusive relations and distinctly different ages, for example, allows us to speculate that an event occurred in which the source region of granitic magma was renewed or re-organized. Possible examples include intrusive contacts of Cretaceous (*c.* 80 Ma) and Palaeogene (*c.* 40 Ma) granitic bodies to the north of Hiroshima in western SW Japan (Shibata & Ishihara 1974), and Plio-Pleistocene granitic plutons intruding Late Palaeogene, Cretaceous and Jurassic granitoids in the Japan Alps of central Japan (Harayama *et al.* 2010).

Possible crustal re-organization

Initial $^{87}Sr/^{86}Sr$ ratios of the granitic rocks in the Japanese islands are mostly 0.704–0.706 with the exception of some Cretaceous and Middle Miocene granitoids in SW Japan, which may be ascribed to unusually extensive magmatism affecting the crust in this area. Isotopic re-organization of the crust can occur due to various factors such as: (1) infiltration of slab-derived fluid into the mantle wedge; (2) episodic underplating of mafic magma onto the lower crust; (3) partial anatexis of upper crustal materials; or (4) rift-related asthenospheric upwelling.

Judging from their isotopic character, the main source of granitoids in the Japanese islands is re-melting of lower crustal mafic rocks originated in underplated basaltic input from mantle. Upper crustal material is involved to some extent, producing a variety of granitic intrusions with a wide range of $^{87}Sr/^{86}Sr$. This interpretation is consistent with the fact that S-type granitoids are rare. The mantle source was slightly enriched ($^{87}Sr/^{86}Sr$ ratios up to 0.706–0.708) at certain times in SW Japan, probably due to accumulation of subduction fluids from the downgoing slab.

The difference in isotopic signature between the Cretaceous granitoids of SW and NE Japan is problematic. It may be ascribed to local heterogeneity of the source mantle, but their original relative palaeogeographic positions and later displacements are not yet clear. The discrepancy of timing between Cretaceous granitic magmatism in SW v. NE Japan is also not yet well interpreted or understood. Despite these uncertainties, however, it is apparent that the upper crust of the Japanese islands was re-organized or generated at least three times, wholly or partly. These three events relate to Middle–Late Cretaceous subduction-related orogeny, Palaeogene–Neogene back-arc rift-related asthenospheric input, and Middle Miocene short-ranged magmatism driven by forceful subduction of the young Philippine Sea Plate beneath the SW Japan forearc sliver caused by the opening of Sea of Japan. Of these three, Middle–Late Cretaceous magmatism is likely to have been the most extensive and of the largest scale from the viewpoint of total volume of magma produced. This important event is viewed as having thoroughly re-organized the whole crust beneath present SW Japan and probably also NE Japan. Palaeogene–Neogene magmatism would have partially rejuvenated the crust. In the Middle Miocene re--organization, the nature of previously existing crust is not understood clearly; it has been a site of the growth front of the accretionary orogen and pre-Neogene upper crustal components are not exposed in the Middle Miocene granitic province.

Generation and survival of granitoids and granitic crust

It has been suggested that the occurrence and age frequency of granitic rocks have been strongly affected by tectonic erosion in the orogenic belt (e.g. von Huene *et al.* 2004; Yamamoto *et al.* 2009). Granitic crust does not always stay relatively close to the Earth's surface, and can instead be brought down to mantle depths with the subducting slab. This is evidenced by ultra-high-pressure metamorphic rocks with granitic protolith (e.g. Chopin 1987) which were metamorphosed under mantle depth conditions.

Nakama *et al.* (2010) argued that there have been five extensive crust-forming granitic magmatic events at 520–400, 280–210, 190–160, 110–90 and 80–60 Ma in the Japanese islands, on the basis of U–Pb age clusters of detrital zircons in accretionary complexes from SW Japan, and that the granitic crusts formed by the first four events have been lost through tectonic erosion.

From surface-exposed granitic rocks in Japanese islands, however, granitic magmatism continued from 120 to 60 Ma with a climax at 80–90 Ma (Fig. 4.2), which contradicts Nakama *et al.* (2010)'s hypothetical magmatic break. Granitic rocks in the Hida Belt show U–Pb ages of 200–190 and 260–230 Ma, which may correspond to their second and third pulses. It is well known that Jurassic granitic plutons widely occur in the Korean Peninsula (e.g. Chough *et al.* 2000). Whereas 500–400 Ma granitic rocks only occur in the Japanese islands sporadically as small tectonic blocks and boulders (e.g. Sakashima *et al.* 2003; Osanai *et al.* 2014), in contrast they form a large granitic province in southern China (Teraoka & Okumura 2003). Their eastern extension to the on-land geology of the Japanese islands is not known because they lie beneath the East China Sea.

Consequently, we may conclude that the granitic crust exposed in Japan has been generated episodically since Early Palaeozoic times, along the eastern margin of the Asian continent in and around the site of what would later become the Japanese islands. Some of this granitic crust may have disappeared from the surface due to tectonic erosion, although enough has survived to form a granitic province that crops out both in the Japanese islands as well as within mainland Asia. In some cases, evidence for this province and its long geological history is only preserved in small isolated blocks or as clastic particles such as detrital zircons in younger sediments.

Drs K. Tani (National Museum of Science) and K. Iida (TITech) showed unpublished data and participated in fruitful discussions. Dr K. Okumura (GSJ) produced the Asian Granite Map modifying the GSJ Geological Map of Asia, which his group has published. Dr S. Takarada (GSJ) produced the Granite Map of Japan for us from GSJ Seamless Geologic Map. Dr K. Miyazaki (GSJ) provided up-to-date information on the geology of Kyushu district, including his compilation of U–Pb age data on the granitoids. We extend our sincere gratitude to all for their kind assistance.

Appendix

English to Kanji and Hiragana translations for geological and place names

Abu	阿武	あぶ
Abukuma	阿武隈	あぶくま
Aioi	相生	あいおい
Arima	有馬	ありま
Ashigawa	芦川	あしがわ
Ashizuri-misaki	足摺岬	あしずりみさき
Bonin	ボニン	ぼにん
Busetsu	武節	ぶせつ
Chubu	中部	ちゅうぶ
Chugoku	中国	ちゅうごく
Hida	飛騨	ひだ
Hidaka	日高	ひだか
Higashi-Yamanashi	東山梨	ひがしやまなし
Higo	肥後	ひご
Hikami	氷上	ひかみ
Hikimi	匹見	ひきみ
Hiroshima	広島	ひろしま
Hitachi	日立	ひたち

(*Continued*)

English to Kanji and Hiragana translations for geological and place names (Continued)

Hokkaido	北海道	ほっかいどう
Honshu	本州	ほんしゅう
Inishi	伊西	いにし
Ishizuchi	石鎚	いしづち
Izu	伊豆	いず
Kakkonda	葛根田	かっこんだ
Kii	紀伊	きい
Kinki	近畿	きんき
Kitakami	北上	きたかみ
Koto	湖東	ことう
Kumano	熊野	くまの
Kurobegawa	黒部川	くろべがわ
Kurosegawa	黒瀬川	くろせがわ
Kyushu	九州	きゅうしゅう
Maizuru	舞鶴	まいづる
Nohi	濃飛	のうひ
Ogasawara	小笠原	おがさわら
Okue-yama	大崩山	おおくえやま
Osuzu-yama	尾鈴山	おすずやま
Ryoke	領家	りょうけ
San'in	山陰	さんいん
San'yo	山陽	さんよう
Shikoku	四国	しこく
Shimanto	四万十	しまんと
Takada	高田	たかだ
Takidani	滝谷	たきだに
Tanzawa	丹沢	たんざわ
Yakuno	夜久野	やくの
Yanai	柳井	やない

References

ADACHI, T., OSANAI, Y., NAKANO, N. & OWADA, M. 2012. LA-ICP-MS U-Pb zircon and FE-EPMA U-Th-Pb monazite datings on pelitic granulites from the Mt. Ukidake area, Sefuri Mountains, northern Kyushu. *Journal of the Geological Society of Japan*, **118**, 39–52.

ARAKAWA, Y. 1990. Two types of granitic intrusions in the Hida belt, Japan: Sr isotopic and chemical characteristics of the Mesozoic Funatsu granitic rocks. *Chemical Geology*, **85**, 101–117.

ARAKAWA, Y. & TAKAHASHI, Y. 1989. Strontium isotopic and chemical variations of the granitic rocks in the Tsukuba district, Japan. *Contributions to Mineralogy and Petrology*, **101**, 46–56.

ARAMAKI, S. 1965. Mode of emplacement of acid igneous complex (Kumano Acidic Rocks) in Southeastern Kii peninsula. *Journal of the Geological Society of Japan*, **71**, 525–540 [in Japanese with English abstract].

ARAMAKI, S. & NOZAWA, T. 1978. *A Reference Book of Chemical Data for Japanese Granitoids*. Contribution from Geodynamic Project of Japan, **78–1**.

ARAMAKI, S., HIRAYAMA, K. & NOZAWA, T. 1972. Chemical composition of Japanese granites, Part 2. Variation trends and average composition of 1200 analyses. *Journal of the Geological Society of Japan*, **78**, 39–49.

BATEMAN, P. C. & DODGE, F. C. W. 1970. Variations of major chemical constituents across the Central Sierra Nevada Batholith. *Geological Society of America Bulletin*, **81**, 409–420.

BONS, P. D., ARNOLD, J., ELBURG, M. A., KALDA, J., SOESOO, A. & VAN MILLIGAN, B. P. 2004. Melt extraction and accumulation from partially molten rocks. *Lithos*, **78**, 25–42.

BRADLEY, D. C., HAEUSSLER, P. & KUSKY, T. M. 1993. Timing of Early Tertiary ridge subduction in southern Alaska. *U. S. Geological Survey Bulletin*, **2068**, 163–177.

BROWN, M. & D'LEMOS, R. S. 1991. The Cadomian granites of Mancellia, northeast Armorican Massif of France: relationship to the St. Malo migmatite belt, petrogenesis and tectonic setting. *Precambrian Research*, **51**, 393–427.

CHAPPELL, B. W. & WHITE, A. J. R. 1974. Two contrasting granite types. *Pacific Geology*, **8**, 173–174.

CHAPPELL, B. W. & WHITE, A. J. R. 1992. I- and S-type granites in the Lachlan Fold Belt. *Transactions of the Royal Society of Edinburgh, Earth Sciences*, **83**, 1–26.

CHOPIN, C. 1987. Very high-pressure metamorphism in the western Alps: implication for subduction of the continental crust. *Philosophical Transaction of the Royal Society of London, Series A*, **321**, 183–197.

CHOUGH, S. K., KWON, S.-T., REE, J.-H & CHOI, D. K. 2000. Tectonic and sedimentary evolution of the Korean Peninsula: a review and new view. *Earth-Science Reviews*, **52**, 175–235.

CHRISTIANSEN, R. L. 1983. Yellowstone magmatic evolution: its bearing on understanding large-volume explosive volcanism. *In*: *Explosive Volcanism, Inception, Evolution, and Hazard*. National Academy Press, Washington DC, 84–95.

CONDIE, K. C. 1982. *Plate Tectonics and Crustal Evolution*. 2nd edn. Pargamon Press, New York.

DEFANT, M. J. & DRUMMOND, M. S. 1990. Derivation of some modern arc magmas by melting of young subducted lithosphere. *Nature*, **347**, 662–665.

DEYOREO, J. J., LUX, D. R., GUIDOTTI, C. V., DECKER, E. R. & OSBERG, P. H. 1989. The Acadian thermal history of western Maine. *Journal of Metamorphic Geology*, **7**, 169–190.

D'LEMOS, R. S., BROWN, M. & STRACHAN, R. A. 1992. Granite magma generation, ascent and emplacement within a transpressional orogen. *Journal of the Geological Society, London*, **149**, 487–490.

DUCEA, M. N. & SALEEBY, J. B. 1998. The age and origin of a thick mafic-ultramafic keel from beneath the Sierra Nevada batholith. *Contributions to Mineralogy and Petrology*, **133**, 169–185.

EBY, G. N. 1990. The A-type granitoids: a review of their occurrence and chemical characteristics and speculations on their petrogenesis. *Lithos*, **26**, 115–134.

ENGEBRETSON, D. C., COX, A. & GORDON, R. G. 1985. *Relative Motions between Oceanic and Continental Plates in the Pacific Basin*. Geological Society of America Special Paper, **206**.

FUJII, M., HAYASAKA, Y. & TERADA, K. 2008. SHRIMP zircon and EPMA monazite dating of granitic rocks from the Maizuru terrane, southwest Japan: correlation with East Asian Paleozoic terranes and geological implications. *Island Arc*, **17**, 322–341.

GEOLOGICAL SURVEY OF JAPAN 1982. *Geological Atlas of Japan*. Geological Survey of Japan, Tsukuba.

GEOLOGICAL SURVEY OF JAPAN 2009. *Seamless Geological Map of Japan at a Scale of 1:200,000, DVD Edition*. Geological Survey of Japan, AIST, Tsukuba.

HADA, S., YOSHIKURA, S. & GABITES, E. J. 2000. U-Pb zircon ages for the Mitaki igneous rocks, Siluro-Devonian tuff, and granitic boulders in the Kurosegawa Terrane, Southwest Japan. *Memoirs of the Geological Society of Japan*, **56**, 183–198.

HARAYAMA, S. 1992. Youngest exposed granitoid pluton on Erath: cooling and uplift of the Pliocene-Quaternary Takidani granodiorite in the Japan Alps, central Japan. *Geology*, **20**, 657–660.

HARAYAMA, S. 1994. Cooling history of the youngest exposed pluton in the world – The Plio-Pleistocene Takidani Granodiorite (Japan Alps, central Japan). *Memoirs of the Geological Society of Japan*, **43**, 87–97 [in Japanese with English abstract].

HARAYAMA, S., TAKAHASHI, M., SHUKUWA, R., ITAYA, T. & YAGI, K. 2010. High-temperature hot springs and Quaternary Kurobegawa Granite along the Kurobegawa River. *Journal of the Geological Society of Japan*, **116** (Supplement), 63–81 [in Japanese].

HARRISON, S. M. & PIERCY, B. A. 1990. The evolution of the Antarctic Peninsular magmatic arc; evidence from northwestern Palmer Land. *In*: KAY, S. M. & RAPELA, C. W. (eds) *Plutonism from Antarctica to Alaska*. Geological Society of America Special Paper, **241**, 9–25.

HATTORI, H., NOZAWA, T. & SAITO, M. 1960. On chemical composition of granitic rocks of Japan. *In*: BARTH, T. F. W. & SORENSEN, H. (eds) *Report of the International Geological Congress, XXI Session, Norden, Part XIV, The Granite-Gneiss Problem,* Copenhagen, 40–46.

HERZIG, C. T., KIMBROUGH, D. L. & HAYASAKA, Y. 1997. Early Permian zircon uranium-lead ages for plagiogranites in the Yakuno ophiolite, Asago district, Southwest Japan. *Island Arc*, **6**, 396–403.

HINE, R., WILLIAMS, I. S., CHAPPELL, B. W. & WHITE, A. J. R. 1978. Contrasts between I- and S-type granitoids of the Kosciusko Batholith. *Journal of the Geological Society of Australia*, **25**, 219–234.

Horie, K., Yamashita, M. *et al.* 2010. Eoarchean-Paleoproterozoic zircon inheritance in Japanese Permo-Triassic granites (Unazuki area, Hida Metamorphic Complex): unearthing more old crust and identifying source terranes. *Precambrian Research*, **183**, 145–157.

Iida, K., Iwamori, H. *et al.* 2013. Chronology and isotope study for Cretaceous and Paleogene granitic rocks, SW Japan. *Abstract, Japan Geoscience Union Meeting*, Makuhari, 2013, SCG61-03.

Ikeda, T. 2004. Pressure-temperature conditions of the Ryoke metamorphic rocks in the Yanai district, SW Japan. *Contributions to Mineralogy and Petrology*, **146**, 577–589.

Imaoka, T., Murakami, N., Matsumoto, T. & Yamasaki, H. 1988. Paleogene cauldrons in the western San-in district, Southwest Japan. *Journal of Faculty of Liberal Arts, Yamaguchi University*, **22**, 41–75.

Imaoka, T., Kiminami, K. *et al.* 2011. K–Ar age and geochemistry of the SW Japan Paleogene cauldron cluster: implications for Eocene-Oligocene thermo-tectonic reactivation. *Journal of Asian Earth Sciences*, **40**, 509–533.

Imaoka, T., Nakashima, K. *et al.* 2013. Episodic magmatism at 105 Ma in the Kinki district, SW Japan: petrogenesis of Nb-rich lamprophyres and adakites, and geodynamic implications. *Lithos*, **184–187**, 105–131.

Ishihara, S. 1977. The magnetite-series and ilmenite-series granitic rocks. *Mining Geology*, **27**, 293–305.

Ishihara, S. & Matsuhisa, Y. 1999. Oxygen isotopic constraints on the geneses of the Miocene Outer Zone granitoids in Japan. *Lithos*, **46**, 523–534.

Ishihara, S. & Matsuhisa, Y. 2002. Oxygen isotopic constraints on the geneses of the Cretaceous-Paleogene granitoids in the Inner Zone of Southwest Japan. *Bulletin of the Geological Survey of Japan*, **53**, 421–438 [in Japanese with English abstract].

Ishihara, S. & Matsuhisa, Y. 2005. Oxygen isotopic constraints on the geneses of the Late Cenozoic plutonic rocks of the Green Tuff Belt, Northeast Japan. *Bulletin of the Geological Survey of Japan*, **56**, 315–324.

Ishihara, S. & Sasaki, A. 2002. Paired sulfur isotopic belts: late Cretaceous-Paleogene ore deposits of Southwest Japan. *Bulletin of the Geological Survey of Japan*, **53**, 461–477.

Ishihara, S., Tanaka, T., Terashima, S., Togashi, S., Murao, S. & Kamioka, H. 1990. Peralkaline rhyolite dikes at the Cape Ashizuri: a new type of REE and rare metal mineral resources. *Mining Geology*, **40**, 107–115.

Ishizaka, K. & Yanagi, T. 1977. K, Rb and Sr abundances and Sr isotopic compositions of the Tanzawa granitic and associated gabbroic rocks, Japan: low-potash island arc plutonic complex. *Earth and Planetary Science Letters*, **33**, 345–352.

Ito, H., Yamada, R., Tamura, A., Arai, S., Horie, K. & Hokada, T. 2013*a*. Earth's youngest exposed granite and its tectonic implications: the 1.0–0.8 Ma Kurobegawa Granite. *Nature Science Report*, **3**, 1306.

Ito, H., Tamura, A., Morishita, T., Arai, S., Arai, F. & Kato, O. 2013*b*. Quaternary plutonic magma activities in the southern Hachimantai geothermal area (Japan) inferred from zircon LA-ICP-MS U–Th–Pb dating method. *Journal of Volcanology and Geothermal Research*, **265**, 1–8.

Iwano, H., Orihashi, Y., Danhara, T., Hirata, T. & Ogasawara, M. 2012. Evaluation of fission track and U-Pb double dating methods for detrital zircon grains: using homogeneous zircon grains in Kawamoto Granodiorite in Shimane prefecture, Japan. *Journal of the Geological Society of Japan*, **118**, 365–375 [in Japanese with English abstract].

Jahn, B.-M. 2010. Accretionary orogen and evolution of the Japanese Islands – Implications from a Sr–Nd isotopic study of the Phanerozoic granitoids from SW Japan. *American Journal of Science*, **310**, 1210–1249.

Jahn, B.-M., Usuki, M., Usuki, T. & Chung, S.-L. 2014. Generation of Cenozoic granitoids in Hokkaido (Japan): constraints from zircon geochronology, Sr–Nd–Hf isotopic and geochemical analyses, and implications for crustal growth. *American Journal of Science*, **314**, 704–750.

John, B. E. & Wooden, J. 1990. Petrology and geochemistry of the metaluminous to peraluminous Chemehuevi Mountains Plutonic Suite, southeastern California. *Geological Society of America Memoir*, **174**, 71–98.

Kagami, H., Iizumi, S., Tainosho, Y. & Owada, M. 1992. Spatial variations of Sr and Nd isotope ratios of Cretaceous-Paleogene granitoid rocks, Southwest Japan arc. *Contributions to Mineralogy and Petrology*, **112**, 165–177.

Kagami, H., Yuhara, M. *et al.* 2000. Continental basalts in the accretionary complexes of the Southwest Japan Arc: constraints from geochemical and Sr and Nd isotopic data of metadiabase. *Island Arc*, **9**, 3–20.

Kamei, A. 2004. An adakitic pluton on Kyushu Island, southwest Japan arc. *Journal of Asian Earth Sciences*, **24**, 43–58.

Kamei, A., Miyake, Y., Owada, M. & Kimura, J.-I. 2009. A pseudo adakite derived from partial melting of tonalitic to granodioritic crust, Kyushu, southwest Japan arc. *Lithos*, **112**, 615–625.

Kamiyama, H., Nakajima, T. & Kamioka, H. 2007. Magmatic stratigraphy of the tilted Tottabetsu Plutonic Complex, Hokkaido, North Japan: magma chamber dynamics and pluton construction. *Journal of Geology*, **115**, 295–314.

Kanisawa, S., Doi, N., Kato, O. & Ishikawa, K. 1994. Quaternary Kakkonda Granite underlying the Kakkonda Geothermal Field, Northwest Japan. *Journal of Mineralogy, Petrology and Economic Geology*, **89**, 390–407 [in Japanese with English abstract].

Katada, M., Kanaya, H. & Onuki, H. 1991. Magmatic differentiation of the Himekami pluton in the northwest Kitakami Mountains. *Journal of Petrology and Mineralogy*, **86**, 100–111 [in Japanese with English abstract]

Kato, O., Doi, N. & Muramatsu, Y. 1993. Neo-granite pluton and geothermal reservoir at the Kakkonda geothermal field, Iwate Prefecture, Japan. *Journal of the Geothermal Research Society of Japan*, **15**, 41–57 [in Japanese with English abstract].

Kawakami, Y., Hoshi, H. & Yamaguchi, Y. 2007. Mechanism of caldera collapse and resurgence: observations from the northern part of the Kumano acidic rocks, Kii peninsula, southwest Japan. *Journal of Volcanology and Geothermal Research*, **167**, 263–281.

Kawate, S. & Arima, M. 1998. Tanzawa plutonic complex, central Japan: exposed felsic miidle crust of Izu-Bonin-Mariana arc. *Island Arc*, **7**, 342–358.

Kay, S. M., Mpodozis, C., Ramos, V. A. & Munizaga, F. 1991. Magma space variations for mid-late Tertiary magmatic rocks associated with a shallowing subduction zone and a thickening crust in the central Andes (28–33°S). *In*: Harmon, R. S. & Rapela, C. W. (eds) *Andean Magmatism and its Tectonic Setting*. Geological Society of America Special Paper, **265**, 113–137.

Kemp, A. I. S., Shimura, T. & Hawkesworth, C. J., EIMF. 2007. Linking granulites, silicic magmatism, and crustal growth in arcs: Ion microprobe (zircon) U-Pb ages from the Hidaka metamorphic belt, Japan. *Geology*, **35**, 807–810.

Kiji, M., Ozawa, H. & Murata, M. 2000. Cretaceous adakitic Tamba granitoid in northern Kyoto, San'yo belt, Southwest Japan. *Japanese Magazine of Mineralogical and Petrological Sciences*, **29**, 136–149.

Kimura, G. 1986. Oblique subduction and collision: forearc tectonics of the Kuril arc. *Geology*, **14**, 404–407.

Kinoshita, O. & Ito, H. 1986. Migration of Cretaceous igneous activity in southwest Japan related to ridge subduction. *Journal of the Geological Society of Japan*, **92**, 723–735 [in Japanese with English abstract].

Komatsu, M., Miyashita, S. & Arita, K. 1986. Composition and structure of the Hidaka metamorphic belt, Hokkaido – historical review and present status. *Monograph of the Association for the Geological Collaboration in Japan*, **31**, 189–203.

Komatsu, M., Osanai, Y., Toyoshima, T. & Miyashita, S. 1989. Evolution of the Hidaka metamorphic belt, Northern Japan. *In*: Daly, J. S., Cliff, R. A. & Yardley, B. W. D. (eds) *Evolution of Metamorphic Belts*. Geological Society, London, Special Publications, **43**, 487–493.

Kon, Y. & Takagi, T. 2012. U–Pb zircon ages of Abukuma granitic rocks in the western Abukuma plateau, northeastern Japan Arc. *Journal of Mineralogical and Petrological Sciences*, **107**, 183–191.

Kunugiza, K. & Goto, A. 2006. Cpx-granite, Inishi-type rock and grey granite. *In*: Regional Geology of Japan (ed.) *Chubu Region, Asakura*. Geological Society of Japan Publishing Co. Ltd., **4**, 148–149 [in Japanese].

Loiselle, M. C. & Wones, D. R. 1979. Characteristics and origin of anorogenic granites. *1979 Annual Meeting of Geological Society of America at San Diego, Abstracts with Program*, **11**, 468.

Maeda, J. & Kagami, H. 1996. Interaction of a spreading ridge and an accretionary prism: implications from MORB magmatism in the Hidaka metamorphic zone, Hokkaido, Japan. *Geology*, **24**, 31–34.

Martin, H., Smithies, R. H., Rapp, R., Moyen, J.-F. & Champion, D. 2005. An overview of adakite, tonalite-trondhjemite-granodiorite (TTG), and sanukitoid: relationships and some implications for crustal evolution. *Lithos*, **79**, 1–24.

Milord, I., Sawyer, E. W. & Brown, M. 2001. Formation of diatexite migmatite and granite magma during anatexis of semi-pelitic metasedimentary rocks: an example from St. Malo, France. *Journal of Petrology*, **42**, 487–505.

Miura, D. 1999. Arcuate pyroclastic conduits, ring faults, and coherent floor at Kumano caldera, southwest Honshu, Japan. *Journal of Volcanology and Geothermal Research*, **92**, 271–294.

Miyazaki, K. 2004. Low-P-high-T metamorphism and the role of heat transport by melt migration in the Higo Metamorphic Complex, Kyushu, Japan. *Journal of Metamorphic Geology*, **22**, 793–809.

Mueller, R. D., Sdrolias, M., Gaina, C., Steinberger, B. & Heine, C. 2008. Long-term sea-level fluctuations driven by ocean basin dynamics. *Science*, **319**, 1357–1362.

Mukhopadhyay, B. & Manton, W. I. 1994. Upper-mantle fragments from beneath the Sierra Nevada batholith: partial fusion, fractional crystallization, and metasomatism in a subduction-related ancient lithosphere. *Journal of Petrology*, **35**, 1417–1450.

Murakami, N., Kanisawa, S. & Ishikawa, K. 1983. High fluorine content of Tertiary igneous rocks from the Cape Ashizuri, Hochi Prefecture, Southwest Japan. *Journal of the Japanese Association of Mineralogy, Petrology and Economic Geology*, **78**, 497–504.

Nakada, S. 1983. Zoned magma chamber of the Osuzuyama Acid Rocks, Southwest Japan. *Journal of Petrology*, **24**, 471–494.

Nakada, S. & Takahashi, M. 1979. Regional variation in chemistry of the Miocene intermediate to felsic magmas in the Outer Zone and Setouchi Province of Southwest Japan. *Journal of the Geological Society of Japan*, **85**, 571–582 [in Japanese with English abstract].

Nakajima, T. 1994. The Ryoke plutonometamorphic belt: cretaceous crustal section of the Eurasian continental margin. *Lithos*, **33**, 51–66.

Nakajima, T. 1996. Cretaceous granitoids in SW Japan and their bearing on the crust-forming process in the eastern Eurasian margin. *Transactions of the Royal Society of Edinburgh, Earth Sciences*, **87**, 183–191.

Nakajima, T., Shirahase, T. & Shibata, K. 1990. Along-arc variation of Rb-Sr and K-Ar ages of Cretaceous granitic rocks in Southwest Japan. *Contributions to Mineralogy and Petrology*, **104**, 381–389.

Nakajima, T., Kamiyama, H., Williams, I. S. & Tani, K. 2004. Mafic rocks from the Ryoke Belt, Southwest Japan: implications for Cretaceous Ryoke/San-yo granitic magma genesis. *Transactions of the Royal Society of Edinburgh, Earth Sciences*, **95**, 249–263.

Nakajima, T., Orihashi, Y., Miyazaki, K. & Danhara, T. 2008. From migmatites to plutons: the origin of granitic magma, U–Pb zirconological approach. *Abstract, 33rd International Geological Congress*, Oslo.

Nakama, T., Hirata, T., Otoh, S., Aoki, K., Yanai, S. & Maruyama, S. 2010. Paleogeography of the Japanese Islands: age spectra of detrital zircon and provenance history of the orogen. *Journal of Geography*, **119**, 1161–1172 [in Japanese with English abstract].

Nakamura, H., Iwamori, H. & Kimura, J. I. 2008. Geochemical evidence for enhanced fluid flux due to overlapping subducting plates. *Nature Geoscience*, **1**, 380–384.

Notsu, K. 1983. Strontium isotope composition in volcanic rocks from the Northeast Japan arc. *Journal of Volcanology and Geothermal Research*, **18**, 531–548.

Obata, M., Yoshimura, Y., Nagakawa, K., Odawara, S. & Osanai, Y. 1994. Crustal anatexis and melt migrations in the Higo metamorphic terrane, west-central Kyushu, Kumamoto, Japan. *Lithos*, **32**, 135–147.

Okudaira, T. 1996. Thermal evolution of the Ryoke metamorphic belt, southwestern Japan: tectonic and numerical modeling. *Island Arc*, **5**, 373–385.

Orihashi, Y., Shinjoe, H. & Anma, R. 2013. Newly revealed NNW shift of granitic magmatism during Mid-Miocene period, Kyushu, Japan. *Abstract volume, Goldschmidt Conference 2013, Mineralogical Magazine*, Florence, **77–5**, 1896.

Osanai, Y., Komatsu, M. & Owada, M. 1991. Metamorphism and granite genesis of the Hidaka Metamorphic belt, Hokkaido, Japan. *Journal of Metamorphic Geology*, **9**, 111–124.

Osanai, Y., Yoshimoto, A. *et al.* 2014. LA-ICP-MS zircon U-Pb geochronology of Paleozoic granitic rocks and related igneous rocks from the Kurosegawa tectonic belt in Kyushu, Southwest Japan. *Journal of Mineralogical and Petrological Sciences*, **43**, 71–99 [in Japanese with English abstract]

Pearce, J. A., Harris, N. V.W. & Tindle, A. G. 1984. Trace element discrimination diagrams for the tectonic interpretation of granitic rocks. *Journal of Petrology*, **25**, 956–983.

Peccerillo, A. & Taylor, S. R. 1976. Geochemistry of the Eocene calc-alkaline volcanic rocks from Kastamonu area, northern Turkey. *Contributions to Mineralogy and Petrology*, **58**, 63–81.

Pitman, W. C., III 1978. Relationship between eustasy and stratigraphic sequences of passive margins. *Geological Society of America Bulletin*, **89**, 1389–1403.

Quick, J., Sinigoi, S. & Mayer, A. 1995. Emplacement of mantle peridotite in the lower continental crust, Ivrea-Verbano zone, northwest Italy. *Geology*, **23**, 739–742.

Rino, S., Komiya, T., Windley, B. F., Katayama, I., Motoki, A. & Hirata, T. 2004. Major episodic increases of continental crustal growth determined from zircon ages of river sands; implications for mantle overturns in the Early Precambrian. *Physics of the Earth and Planetary Interiors*, **146**, 360–394.

Saeki, K., Shiba, M., Itaya, T. & Onuki, H. 1995. K–Ar ages of the metamorphic and plutonic rocks in the southern part of the Hidaka belt, Hokkaido, and their implications. *Journal of Mineralogy, Petrology and Economic Geology*, **90**, 297–309 [in Japanese with English abstract].

Saito, S. 2014. Geochemical diversity of Neogene granitoid plutons in the Izu Collision Zone; Implications for transformation of juvenile oceanic arc into mature continental crust. *Journal of Mineralogical and Petrological Sciences*, **43**, 115–130 [in Japanese with English abstract].

Saito, S., Arima, M., Nakajima, T. & Kimura, J.-I. 2004. Petrogenesis of Ashigawa and Tonogi granitic intrusions, southern part of the Miocene Kofu Granitic Complex central Japan: M-type granite in the Izu arc collision zone. *Journal of Mineralogical and Petrological Sciences*, **99**, 104–117.

Saito, S., Arima, M., Nakajima, T., Misawa, K. & Kimura, J. 2007. Formation of distinct granitic magma chamber batches by partial melting of hybrid lower crust in the Izu arc collision zone, central Japan. *Journal of Petrology*, **48**, 1761–1791.

Saito, S., Arima, M. *et al.* 2012. Petrogenesis of the Kaikomagatake granitoid pluton in the Izu Collision Zone, central Japan: implications for transformation of juvenile oceanic arc into mature continental crust. *Contributions to Mineralogy and Petrology*, **163**, 611–629.

Sakashima, T., Terada, K., Takeshita, T. & Sano, Y. 2003. Large-scale displacement along the Median Tectonic Line, Japan: evidence from SHRIMP zircon U–Pb dating of granites and gneisses from the South Kitakami and paleo-Ryoke belts. *Journal of Asian Earth Sciences*, **21**, 1019–1039.

Sano, Y., Tsutsumi, Y., Terada, K. & Kaneoka, I. 2002. Ion microprobe U–Pb dating of Quaternary zircon: implication for magma cooling and residence time. *Journal of Volcanology and Geothermal Research*, **117**, 285–296.

Sasaki, A. & Ishihara, S. 1979. Sulfur isotopic composition of the magnetite-series and ilmenite-series granitoids in Japan. *Contributions to Mineralogy and Petrology*, **68**, 107–115.

Sasaki, M., Fujimoto, K. *et al.* 2003. Petrographic features of a high-temperature granite just newly solidified magma at the Kakkonda geothermal field, Japan. *Journal of Volcanology and Geothermal Research*, **121**, 247–269.

Sato, K. 1991. Miocene granitoid magmatism at the island-arc junction, central Japan. *Modern Geology*, **15**, 367–399.

Sato, D., Matsuura, H., Danhara, T., Iwano, H. & Hirata, T. 2013. U–Pb zircon ages of the Late Cretaceous Aioi Group caldera cluster, and Arima Group and Koto Rhyolites with calderas, SW Japan. *Abstract, 2013 Annual Meeting of the Geological Society of Japan*, Sendai, R3-O6 [in Japanese].

Sawada, Y. & Itaya, T. 1993. K–Ar ages of Late Cretaceous granitic ring complex around Lake Biwa. *Journal of the Geological Society of Japan*, **99**, 975–990 [in Japanese with English abstract].

Shibata, K. & Ishihara, S. 1974. K–Ar ages of biotites across the central part of the Hiroshima granites. *Journal of the Geological Society of Japan*, **80**, 431–433 [in Japanese].

Shibata, K. & Ishihara, S. 1979. Initial $^{87}Sr/^{86}Sr$ ratios of plutonic rocks from Japan. *Contributions to Mineralogy and Petrology*, **70**, 381–390.

Shibata, K. & Takagi, H. 1989. Tectonic relationship between the Median Tectonic Line and the Tanakura Tectonic Line viewed from isotopic ages and Sr isotopes of granitic rocks in the northern Kanto Mountains. *Journal of the Geological Society of Japan*, **95**, 687–700 [in Japanese with English abstract].

Shimojo, M., Ohto, S., Yanai, S., Hirata, T. & Maruyama, S. 2010. LA-ICP-MS U–Pb ages of some older rocks of the South Kitakami

Belt, Northeast Japan. *Journal of Geography*, **119**, 257–269 [in Japanese with English abstract].

Shimura, T., Komatsu, M. & Iiyama, J. T. 1992. Genesis of the lower crustal Grt-Opx tonalite (S-type) in the Hidaka Metamorphic Belt, northern Japan. *Transactions of the Royal Society of Edinburgh, Earth Sciences*, **83**, 259–268.

Shimura, T., Owada, M., Osanai, Y., Komatsu, M. & Kagami, H. 2004. Variety and genesis of the pyroxene-bearing S- and I-type granitoids from the Hidaka Metamorphic Belt, Hokkaido, northern Japan. *Transactions of the Royal Society of Edinburgh, Earth Sciences*, **95**, 161–179.

Shinjoe, H., Orihashi, Y. & Sumii, T. 2010. U–Pb zircon ages of syenitic and granitic rocks in the Ashizuri igneous complex, southwestern Shikoku: constraint for the origin of forearc alkaline magmatism. *Geochemical Journal*, **44**, 275–283.

Sisson, V. B., Roeske, S. M. & Pavlis, T. L. (eds) 2010. *Geology of a Transpressional Orogen Developed during Ridge-trench Interaction along the North Pacific Margin.* Geological Society of America Special Paper, **371**.

Solar, G. S., Pressley, R. A., Brown, M. & Tucker, R. D. 1998. Granite ascent in convergent orogenic belts: testing a model. *Geology*, **26**, 711–714.

Sonehara, T. & Harayama, S. 2007. Petrology of the Nohi Rhyolite and its related granitoids: a Late Cretaceous large silicic igneous field in central Japan. *Journal of Volcanology and Geothermal Research*, **167**, 57–80.

Sumii, T., Uchiumi, S., Shinjoe, H. & Shimoda, H. 1998. Re-examination on K-Ar age of the Kumano Acidic Rocks in Kii Peninsula, Southwest Japan. *Journal of the Geological Society of Japan*, **104**, 387–394 [in Japanese with English abstract].

Sun, S.-S. & McDonough, W. F. 1989. Chemical and isotopic systematics of oceanic basalt: implications for mantle composition and processes. *In*: Sanders, A. D. & Norry, M. J. (eds) *Magmatism in the Ocean Basins*. Geological Society, London, Special Publications, **42**, 313–345.

Suyehiro, K., Takahashi, N. *et al.* 1996. Continental crust, crustal underplating, and low-Q upper mantle beneath an oceanic island arc. *Science*, **272**, 390–392.

Suzuki, K. & Adachi, M. 1991. The chemical Th-U-total Pb isochron ages of zircon and monazite from the Grey granite of the Hida terrane, Japan. *Journal of Earth Science of Nagoya University*, **38**, 11–37.

Suzuki, K. & Adachi, M. 1998. Denudation history of the high T/P Ryoke metamorphic belt, southwest Japan: constraints from CHIME monazite ages of gneisses and granitoids. *Journal of Metamorphic Geology*, **16**, 23–37.

Tagiri, M., Dunkley, D. J., Adachi, T., Hiroi, Y. & Fanning, M. 2011. SHRIMP dating of magmatism in the Hitachi metamorphic terrane, Abukuma Belt, Japan: evidence for a Cambrian volcanic arc. *Island Arc*, **20**, 259–279.

Tainosho, Y., Kagami, H., Yuhara, M., Nakano, S., Sawada, K. & Morioka, K. 1999. High initial Sr isotopic ratios of Cretaceous to Early Paleogene granitic rocks in Kinki district, Southwest Japan. *Memoirs of the Geological Society of Japan*, **53**, 309–321 [in Japanese with English abstract].

Takagi, T. 2004. Origin of magnetite- and ilmenite-series granitic rocks in the Japan arc. *American Journal of Science*, **304**, 169–202.

Takahashi, M. 1986. Anatomy of a middle Miocene Valles-type caldera cluster: geology of the Okueyama volcano-plutonic complex, Southwest Japan. *Journal of Volcanology and Geothermal Research*, **29**, 33–70.

Takahashi, M. 1989. Neogene granitic magmatism in the South Fossa Magna collision zone, central Japan. *Modern Geology*, **14**, 127–143.

Takahashi, M., Aramaki, S. & Ishihara, S. 1980. Magnetite-series/Ilmenite-series v. I-type/S-type granitoids. *Mining Geology*, Special Issue **8**, 13–28.

Takahashi, M., Aramaki, S. & Tsusue, A. 1984. Segmented structure in the Cretaceous to Early Paleogene granitic batholiths oh the Japanese Islands based on statistics of major element chemistry: a reconnaissance study. *Mining Geology*, **34**, 373–384.

Takahashi, Y. 1992. Petrological study of tonalitic rocks in the upper reaches of Satsunai River, Main Zone of the Hidaka Metamorphic Belt. – coexistent relation of S-type with I-type granite. *Journal of the Geological Society of Japan*, **98**, 295–308 [in Japanese with English abstract].

Takahashi, Y., Kagashima, S. & Mikoshiba, U. M. 2005. Geochemistry of adakitic quartz diorite in the Yamizo Mountains, central Japan – implications for Early Cretaceous adakitic magmatism in the Inner Zone of Southwest Japan. *Island Arc*, **14**, 150–164.

Takahashi, Y., Cho, D.-L. & Kee, W.-S. 2010. Timing of mylonitization in the Funatsu Shear Zone within Hida Belt of southwest Japan: implications for correlation with the shear zones around the Ogcheon Belt in the Korean Peninsula. *Gondwana Research*, **17**, 102–115.

Takahashi, Y., Mikoshiba, M., Kubo, K., Danhara, T., Iwano, H. & Hirata, T. 2014. U-Pb ages of zircon in plutonic rocks within the southern Abukuma Mountains. *Abstract, Japan Geoscience Union Meeting 2014*, Yokohama, SGL43-03.

Tani, K., Dunkley, D. J., Kimura, J.-I., Wysoczanski, R. G., Yamada, K. & Tatsumi, Y. 2010. Syn-collisional rapid granitic magma formation in an arc-arc collision zone: evidence from Tanzawa plutonic complex, Japan. *Geology*, **38**, 215–218.

Tani, K., Horie, K., Dunkley, D. & Ishihara, S. 2014. Pulsed granitic crust formation revealed by comprehensive SHRIMP zircon dating of the SW Japan granitoids. *Abstract, Japan Geoscience Union Meeting 2014*, Yokohama.

Taylor, H. P., Jr & Silver, L. T. 1978. Oxygen isotope relationships in plutonic igneous rocks of the Peninsular Ranges Batholith, southern and Baja California. *U. S. Geological Survey Open File Report*, **78–701**, 423–426.

Terakado, Y. & Nakamura, N. 1984. Nd and Sr isotopic variations in acidic rocks from Japan: significance of upper mantle heterogeneity. *Contributions to Mineralogy and Petrology*, **87**, 407–417.

Terakado, Y. & Nohda, S. 1993. Rb-Sr dating of acidic rocks from the middle part of the Inner Zone of Southwest Japan: tectonic implications for the migration of the Cretaceous to Paleogene igneous activity. *Chemical Geology*, **109**, 69–87.

Terakado, Y., Shimizu, H. & Masuda, A. 1988. Nd and Sr isotopic variations in acidic rocks formed under a peculiar tectonic environment in Miocene Southwest Japan. *Contributions to Mineralogy and Petrology*, **99**, 1–10.

Teraoka, Y. & Okumura, K. 2003. *Geological map of East Asia 1:3,000,000*. Geological Survey of Japan, AIST.

Tiepolo, M., Langone, A., Morishita, T. & Yuhara, M. 2012. On the recycling of amphibole-rich ultramafic intrusive rocks in the arc crust: evidence from Shikanoshima Island (Kyushu, Japan). *Journal of Petrology*, **53**, 1255–1285.

Tsuboi, M. & Suzuki, K. 2003. Heterogeneity of initial $^{87}Sr/^{86}Sr$ ratios within a single pluton: evidence from apatite strontium isotopic study. *Chemical Geology*, **199**, 189–197.

Tsuchiya, N. & Kanisawa, S. 1994. Early Cretaceous Sr-rich silicic magmatism by slab melting in the Kitakami Mountains, Northeast Japan. *Journal of Geophysical Research*, **99**, 22 205–22 220.

Tsuchiya, N., Kimura, J. & Kagami, H. 2007. Petrogenesis of Early Cretaceous adakitic granites from the Kitakami Mountains, Japan. *Journal of Volcanology and Geothermal Research*, **167**, 134–159.

Tsuchiya, N., Takeda, T., Tani, K., Adachi, T., Nakano, N. & Osanai, Y. 2012. Zircon U-Pb geochronology and petrochemistry of Early Cretaceous adakitic granites in the Kitakami Mountains, Japan. *Abstract, 2012 Meeting of Japan Association of Mineralogical Sciences*, Kyoto, R6–08 [in Japanese].

Tsutsumi, Y., Yokoyama, K., Kasatkin, S. A. & Golozubov, V. V. 2014. Zircon U-Pb age of granitoids in the Maizuru Belt, southwest Japan and the southernmost Khanka Massif, Far East Russia. *Journal of Mineralogical and Petrological Sciences*, **109**, 97–102.

Ueki, T. & Harayama, S. 2012. Late Cretaceous hot and dry felsic magmatism in the Nishina Mountains, Northern Japan Alps. *Journal of the Geological Society of Japan*, **118**, 207–219 [in Japanese with English abstract].

Usuki, T., Kaiden, H., Misawa, K. & Shiraishi, K. 2006. Sensitive high-resolution ion microprobe U-Pb ages of the Latest Oligocene – Early Miocene rift-related Hidaka high-temperature metamorphism in Hokkaido, northern Japan. *Island Arc*, **15**, 503–516.

von Huene, R., Ranero, C. R. & Vannucci, P. 2004. Generic model of subduction erosion. *Geology*, **32**, 913–916.

Wada, H., Harayama, S. & Yamaguchi, Y. 2004. Mafic enclaves floating through a vertical fractionating felsic magma chamber: the Kurobegawa Granitic Pluton, Hida Mountain Range, central Japan. *Geological Society of America Bulletin*, **116**, 788–801.

Watanabe, T., Fanning, C. M., Uruno, K. & Kano, H. 1995. Pre-middle Silurian granitic magmatism and associated metamorphism in northern Japan: SHRIMP U-Pb zircon chronology. *Geological Journal*, **30**, 273–280.

White, A. J. R. 1979. Source of granite magmas. *1979 Annual Meeting of Geological Society of America at San Diego, Abstract with Program*, **11**, 539.

Wilson, C. J. N., Rogan, A. M., Smith, I. E. M., Northey, D. J., Nairn, I. A. & Houghton, B. F. 1984. Caldera volcanoes of Taupo volcanic zone, New Zealand. *Journal of Geophysical Research*, **89B**, 8463–8484.

Windley, B. F. 1984. *The Evolving Continents*. 2nd edn. John Wiley & Sons.

Yamamoto, S., Senshu, H., Rino, S., Omori, S. & Maruyama, S. 2009. Granitic subduction: arc subduction, tectonic erosion and sediment subduction. *Gondwana Research*, **15**, 443–453.

Yamamoto, T. 2003. Lithofacies and eruption ages of Late Cretaceous caldera volcanoes in the Himeji-Yamasaki district, southwest Japan: implications for ancient large-scale felsic arc volcanism. *Island Arc*, **12**, 294–309.

Yokoyama, K., Shigeoka, M., Goto, A., Terada, K., Hidaka, H. & Tsutsumi, Y. 2010. U–Th–total Pb ages of uraninite and thorite from granitic rocks in the Japanese Islands. *Bulletin of National Museum, Natural Science, Series C*, **36**, 7–18.

Yoshida, H. 1961. The Late Mesozoic igneous activities in the middle Chugoku province. *Science Report of Geology, Hiroshima University*, **8**, 1–39 [in Japanese with English abstract].

Yoshida, T. 1984. Tertiary Ishizuchi cauldron, Southwest Japan arc: formation by ring fracture. *Journal of Geophysical Research*, **89B**, 8502–8510.

Zen, E. 1986. Aluminum enrichment in silicate melts by fractional crystallization: some mineralogic and petrographic constraints. *Journal of Petrology*, **27**, 1095–1117.

5 Miocene–Holocene volcanism

SETSUYA NAKADA, TAKAHIRO YAMAMOTO & FUKASHI MAENO

Post-Palaeogene volcanism in Japan is closely related to the evolution of the Japanese island-arc system that was initiated with the Japan Sea opening (JSO) in Early–Middle Miocene times, and is reflected in the distribution of volcanoes and geochemical variations of volcanic rocks presently observed. In the NE Japan Arc at the beginning of the JSO, the volcanic field migrated towards the oceanic side and the isotopic composition of magma temporally changed in the back-arc side of the system. This was likely caused by injection of hot asthenosphere into the mantle wedge and damming of the subducting slab. During the JSO, rift volcanism in the back-arc region was characterized by numerous grabens that filled with voluminous volcanic rocks derived from bimodal low-K tholeiitic basalt and silicic magmas. In the SW Japan Arc, a period of Middle Miocene forearc volcanism at 14–12 Ma was fed by high-magnesium andesite and caldera-forming silicic magma, and is interpreted as a result of subduction initiation associated with the oceanwards migration of the SW Japan Arc and underthrusting of the hot Shikoku Basin.

In more recent times, over the last two million years volcanism has focused on the present volcanic front to produce the modern east and west Japan volcanic belts. Volcanic rocks in both belts can be divided into different rock series, these being tholeiitic in the forearc, alkaline in the back-arc and calc-alkaline in the middle. In addition a systematic variation of hydrous minerals in volcanic rocks can be recognized in the across-arc direction, especially in NE Japan. Such geochemical differences may reflect original variations in the primary basalt and the process of fluid transport in the mantle wedge. The enrichment in incompatible elements and water towards the back-arc side in Central Japan, where the Pacific and Philippine Sea plates overlap, may be explained by the chemical differences between fluid fluxes from different subducting slabs. Caldera-forming silicic magmatism also reflects the plate tectonic system and the crust-mantle thermochemical structures of the mantle wedge.

The frequencies of occurrence of large-scale caldera-forming events are very low, but they have occurred repeatedly since the JSO. The youngest examples are currently concentrated along the volcanic front of Hokkaido Island, the northern end of Honshu Island and the central and southern parts of Kyushu Island, represented by the eruptions of Kutcharo, Aso, Aira and Kikai calderas. During the last few decades volcanic eruptions in Japan have occurred at a limited number of volcanoes, notably at Sakurajima but also at Usu in 1977–1978 and 2000, Izu-Oshima in 1986–1989, Miyakejima in 1983 and 2000, Unzen in 1990–1995 and Shinmoedake (Kirishima) in 2011. In contrast, the basaltic Mt Fuji has remained quiescent over the last 300 years (with the exception of swarms of low-frequency earthquakes during 2000–2001), although it remains one of the representative active volcanoes of Japan. At present there are 110 active volcanoes in Japan, these having been classified into three ranks (A, B and C) based on the record of eruption activity over the last ten thousand years, and 47 of them are currently monitored by the Japan Meteorological Agency.

Neogene volcanism

The Japanese island-arc system was formed in Early–Middle Miocene times resulting from the opening of the Japan Sea, a typical back-arc basin lying behind both the NE and SW Japan arcs. A number of palaeomagnetic results have shown that the SW Japan Arc rotated clockwise (e.g. Otofuji *et al.* 1985; Otofuji & Matsuda 1987), while the NE Japan Arc rotated anticlockwise (Tosha & Hamano 1988; Otofuji *et al.* 1994; Takahashi *et al.* 1999), producing the curved shape of the present Japanese archipelago. During the Miocene epoch, the NE and SW Japan arcs detached from the Sikhote Alin region of the Eurasian continent margin and moved towards the Pacific and Philippine Sea plates, respectively (Fig. 5.1). Retreat of the trench at the western margin of the Pacific Plate reached up to 500 km from the pre-opening position. Although the Japan Basin (which is made up of oceanic crust) appeared in the northern gap between the Sikhote Alin and the NE Japan Arc, there are abundant topographic highs of deformed continental fragments in the southern part of the Japan Sea (Tamaki 1988). The Ocean Drilling Program (ODP) drilling in the Japan Sea revealed that voluminous basaltic rocks erupted at 21–18 Ma (Kaneoka *et al.* 1992). The Shikoku and Chishima basins (Fig. 5.1) simultaneously opened with the Japan Sea behind the Izu–Mariana and Chishima arcs, respectively (Kimura & Tamaki 1986; Okino *et al.* 1998).

At the beginning of the JSO, two important changes took place in Early–Middle Miocene volcanism in the Japanese islands as follows: (1) migration of the volcanic field towards the oceanic side at 21 Ma (Ohki *et al.* 1993; Yoshida *et al.* 1995; Fig. 5.2); and (2) a temporal change of basalt Nd–Sr isotopic signatures from enriched to depleted in the back-arc side of the NE Japan Arc (Nohda *et al.* 1988). Tatsumi *et al.* (1989) proposed that these phenomena were caused by injection of hot asthenosphere into the mantle wedge resulting from a large-scale eastwards asthenospheric flow and a consequent blocking of the subducting slab in terms of both material and thermal input. The enriched compositions of Miocene back-arc rocks in the NE Japan Arc may represent the characteristics of subcontinental lithospheric mantle material, and the isotopic change could then be interpreted as the result of thinning of subcontinental lithosphere by the injection of asthenosphere with Nd–Sr isotopic compositions similar to MORB source (Nohda *et al.* 1988). Cousens *et al.* (1994) have shown that Japan Sea Miocene basalts cover a wide range of isotopic compositions (e.g. $^{87}Sr/^{86}Sr = 0.7037–0.7050$; $^{206}Pb/^{204}Pb = 17.65–18.36$) and define a mixing trend between depleted and enriched mantle components. Their combined

From: Moreno, T., Wallis, S., Kojima, T. & Gibbons, W. (eds) 2016. *The Geology of Japan*. Geological Society, London, 273–308.

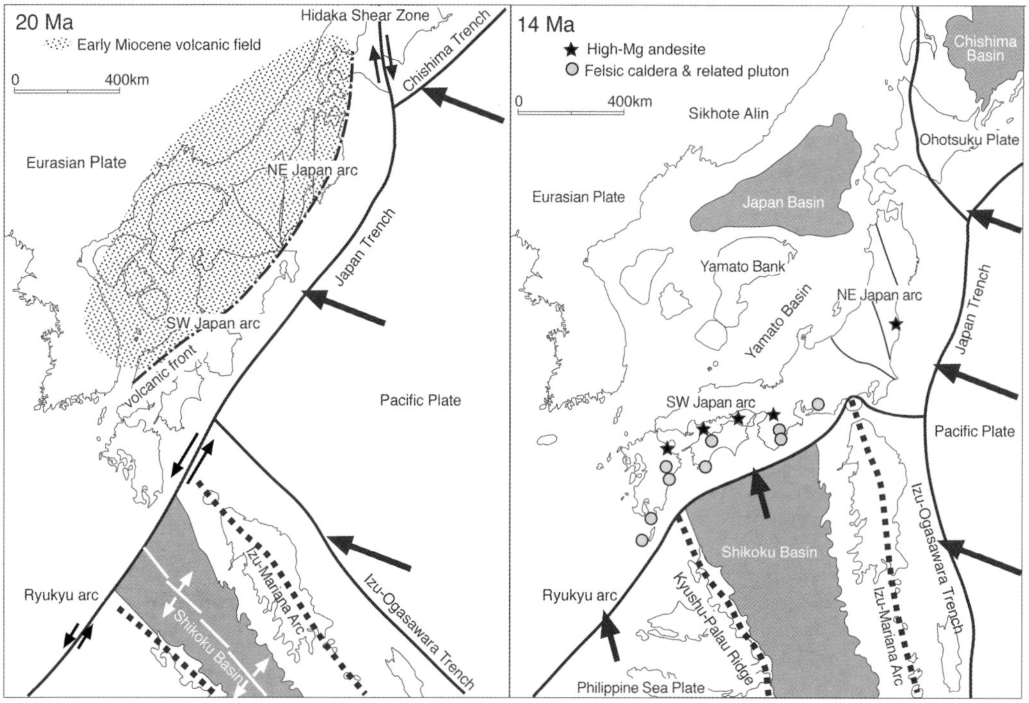

Fig. 5.1. Tectonic reconstruction around the Japan islands during Early–Middle Miocene times. Left: initial opening stage. Volcanic fields expanded toward the forearc regions. The plate boundary between the Eurasian and the Philippine Sea, including the Izu–Mariana Arc and the Kyushu–Palau Ridge, is a transform fault with left-lateral slip movement. Arrows show directions of lateral movement for plates. The Shikoku Basin within the Philippine Sea Plate had been spreading from 30 to 15 Ma. Right: post-opening stage. The opening of the Japan Sea, Shikoku and Chishima basins stopped at 15 Ma. The NE and SW Japan arcs grew above the subducting Pacific and Philippine Sea ocean plates, respectively. High-Mg andesite and felsic caldera-forming eruptions occurred in the Setouchi and Outer Zone of the SW Japan Arc (Fig. 5.3). A small amount of high-Mg andesite simultaneously erupted in the forearc region of the NE Japan Arc. After Yamamoto & Hoang (2009).

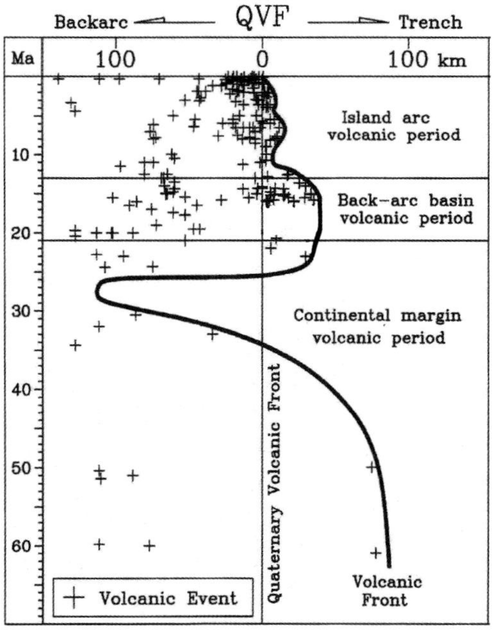

Fig. 5.2. Migration of the volcanic front of the NE Japan arc over the past 60 Ma. The ages of each volcanic event are plotted against distance of the eruption centre from the Quaternary volcanic front (QVF). During the continental margin volcanic period, the volcanic front moved far to the back-arc side. In contrast, the Early Miocene volcanic field spread towards the trench side beyond QVF during 25–21 Ma. Modified from Yoshida *et al.* (1995).

isotope and trace-element systematics support a model of two-component mixing between Indian MORB-like mantle and subduction-modified Pacific sediments, which are isotopically enriched but chemically depleted.

During the JSO, many graben appeared in the back-arc region and were filled by voluminous volcanic rocks associated with marine formations (Yamaji 1990). This rift volcanism was characterized by bimodal eruptions of low-K tholeiitic basalt and more-enriched felsic magma suggesting a lower crustal origin (Shuto *et al.* 2006). In contrast, the 21–15 Ma forearc volcanism in the NE Japan Arc was dominated by low-K tholeiitic basalt with minor amounts of adakitic dacite. The geochemistry of this adakite suggests that incursion of the hot asthenosphere above the Pacific slab caused melting to occur, although the downgoing slab was Cretaceous in age and therefore too cold to melt under normal conditions (Yamamoto & Hoang 2009). Furthermore, the unusual 14–12 Ma forearc volcanism in the SW Japan Arc, which is characterized by high-magnesium andesite (HMA) and caldera-forming felsic magma, is interpreted as the result of subduction nucleation between the oceanwards-migrating SW Japan Arc and the underthrusting hot Shikoku Basin (Furukawa & Tatsumi 1999; Kimura *et al.* 2005; Fig. 5.3). The HMA presumably resulted from the interaction of melts from the subducted Shikoku Basin plate with the overlying mantle (Shimoda *et al.* 1998; Tatsumi 2001). Felsic magmas in the forearc may have been generated by melting of Shimanto accretionary complex sediments caused by intrusion of HMA magmas (Shinjoe 1997).

After the cessation of the JSO, the main part of the NE Japan Arc became submerged due to thermal subsidence (Sato & Amano

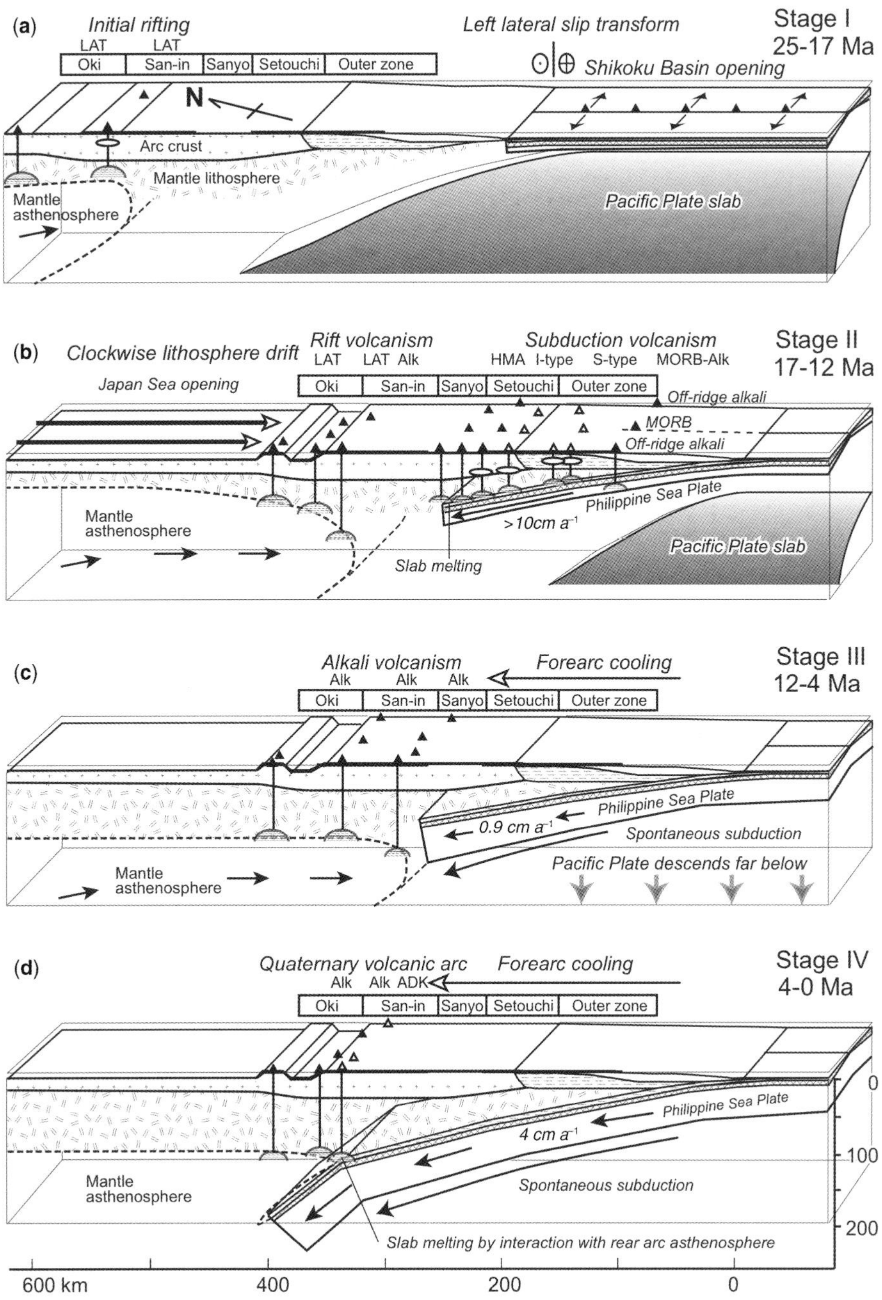

Fig. 5.3. Diagrams showing Neogene development of the upper mantle structure and volcanism beneath the SW Japan Arc. (**a**) Stage I. Prior to opening of the Japan Sea, the SW Japan Arc was situated along the margin of the Eurasia continent, and initial rifting volcanism took place in the back-arc region. The Shikoku Basin was spreading during this stage, but the plate boundary between the Shikoku Basin and SW Japan acted as a transform fault. (**b**) Stage II. The plate boundary along the SW Japan Arc accommodated opening of the Japan Sea and permitted clockwise rotation of the arc; subduction must therefore have been underway by 17 Ma. Initial descent of the Philippine Sea slab was marked by an episode of high-Mg andesite and felsic-caldera-forming volcanism in the forearc region. (**c**) Stage III. After the Japan Sea opened, monogenetic alkali basalt volcanism dominated in the SW Japan Arc, but progressive penetration of the Philippine Sea slab terminated this activity in the Sanyo zone by *c.* 4 Ma. Cessation of volcanism in the forearc region and narrowing of the volcanic field indicate forearc cooling due to the subduction. (**d**) Stage IV. Subduction of the Philippine Sea slab accelerated at *c.* 5 Ma, to become comparable to the present subduction rate. The slab surface is *c.* 100 km beneath adakite volcanic centres, which is consistent with the idea that slab melting occurs by dehydration from amphibolite transforming to eclogite. ADK, adakite; Alk, alkali basalt; HMA, high-magnesium andesite; LAT, low-alkali tholeiite; MORB, mid-oceanic-ridge basalt. After Kimura *et al.* (2005).

1991). It started uplifting, accompanied by caldera-forming silicic volcanism and development of intra-arc highlands and basins, at *c.* 10 Ma however (Sato 1994; Yamamoto 2009). Late Miocene–Quaternary caldera volcanoes are distributed along the NE Japan Arc in clusters 40–80 km apart (Fig. 5.4). Each cluster comprises 3 to >10 calderas with most calderas being >10 km in diameter, although some clusters include smaller calderas. These calderas have erupted abundant ignimbrite sheets and generated voluminous volcaniclastic sediments that fill the intra-arc and back-arc basins (Yamamoto 2009). Seismic tomography and pre-eruption uplifting of the caldera cluster suggest that a substantial amount of mantle-derived magma was stored within the lower crust, and the geochemistry of ignimbrites also suggests that silicic magmas were generated by the melting of variably assimilated mafic intrusions (Yamamoto 2011).

In the SW Japan Arc, monogenetic alkali basalt volcanism continues up to the present due to mantle upwelling, whereas forearc volcanism ceased at 12 Ma (Iwamori 1992; Hoang & Uto 2003). This volcanic field narrowed towards the back-arc side with time as the Philippine Sea slab descended and slowly cooled (Kimura *et al.* 2005; Fig. 5.3). Adakitic dacites erupted in the San'in region after 1.7 Ma above the 100 km depth contour of the subducted Philippine Sea Plate, suggesting that melting resulted from interaction of the slab with upwelling asthenosphere (Morris 1995). This change from alkali basalt to adakite volcanism suggests penetration of the slab into the asthenosphere (Fig. 5.3). In contrast to this volcanic system, arc-type sub-alkali volcanism with graben formation has become prominent behind the Ryukyu Arc since *c.* 6 Ma (Kamata 1989; Shinjo *et al.* 2000). This flare-up is presumably related to opening of the back-arc Okinawa Trough (Sibuet *et al.* 1995; Shinjo *et al.* 1999).

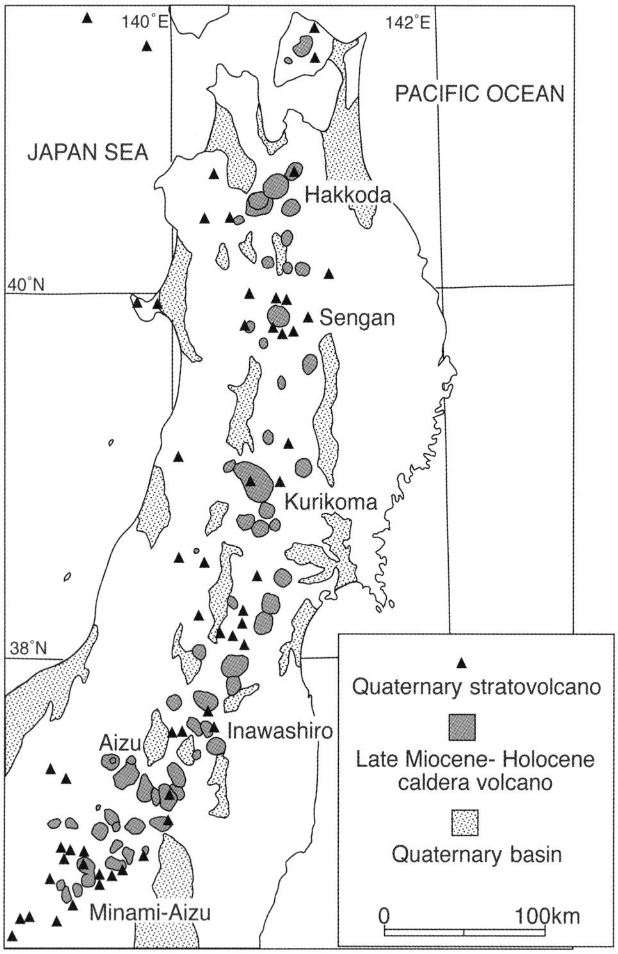

Fig. 5.4. Distribution of Late Miocene–Quaternary caldera volcanoes in the NE Japan Arc. Quaternary stratovolcanoes overlap the caldera volcanoes in volcano clusters, forming high terrain along the arc. After Yamamoto (2009).

Modern volcanic setting

Volcanic activity of the Japan Arc has been focused on the present volcanic front since around 2 million years ago, producing the east and west volcanic belts (Fig. 5.5) The NE Japan volcanic belt runs down from the Kurile islands to the Mariana Arc through Hokkaido Island and the NE and central parts of Honshu Island, bending around Asama Volcano. The west Japan volcanic belt starts from the northwestern coast of Honshu and connects to the Ryukyu Arc through Kyushu Island, bending around Aso Volcano. The eruption rate is generally highest on the volcanic front and, in the case of the east Japan volcanic belt, it decreases towards the back-arc side. Large-scale felsic magmatic activity is concentrated on the volcanic front of Hokkaido Island, the northern end of Honshu Island and the central and southern parts of Kyushu Island, where caldera-forming eruptions have occurred. Nakagawa *et al.* (1986) pointed out the localization of small Quaternary volcanoes in a narrow zone on the Pacific side of the volcanic front. They also emphasized the higher eruption rate and magmatic temperature, and the relative scarcity of hornblende phenocryst-bearing rocks in the volcanic front as opposed to volcanoes on either side (back-arc and Pacific side).

The volcanic fronts run in general parallel to the trenches and the troughs. However, the details of the alignment along the arcs are not uniform. For example, in the NE and SW Japan arcs and the Izu–Ogasawara Arc extending to the south from Izu–Oshima (Fig. 5.5), the alignment parallel to the trench and the trough is recognizable along the volcanic front. In central Japan, however, this parallelism is violated as the volcanic chain significantly deflects towards the back-arc side. In central Japan, in addition to subduction of the Pacific Plate from the east, the overlapping subduction of the Philippine Sea Plate occurs. Iwamori (2007) use numerical modelling to demonstrate that most of the H_2O is subducted to a depth where serpentine and chlorite break down in the serpentinite layer, which is formed just above the subducting slab in the wedge. This depth depends on the thermal structure of the slab and the aqueous fluid and melt are not, in general, supplied straight upwards to the volcanic front. Due to the cold environment caused by the double subducting of plates in central Japan, the geotherm is low compared with NE and SW Japan, causing the volcanic region in central Japan to deflect towards the back-arc side. Tamura *et al.* (2002) showed that volcanoes in NE Japan are grouped into clusters traversing the arc elongation with an average width of *c.* 50 km at intervals of 30–75 km. Since the clusters correspond to the areas with negative gravity anomaly and negative perturbation in the seismic wave velocity, they proposed a hot finger model for the mantle wedge of the subduction zone where finger-like hot mantle materials flow from below the back-arc towards the volcanic front.

Rocks of back-arc volcanoes are generally enriched in alkaline and other trace elements such as Rb, Sr, Ba, Th, U and rare Earth elements (REE). Kuno (1966, 1969) pointed out the alkaline enrichment of basaltic rocks in the NE Japan back-arc, with tholeiitic series in the forearc region and high-alumina series in the central part of the arc. Such a distribution is not clear in SW Japan however, where no tholeiitic basalt series has erupted. Miyashiro (1974) showed a similar variation for andesitic rocks in Japan, where tholeiitic, calc-alkalic and alkalic series rocks were newly classified. Sakuyama (1977) described the zonal variation of phenocryst assemblages in volcanic rocks along the NE Japan Arc, with a lack of hydrous minerals on the forearc side, hornblende-bearing rocks without biotite in the middle and biotite-bearing rocks in the back-arc. He proposed an increase in H_2O in magma towards the back-arc side based on this variation. Nakamura *et al.* (2008) confirmed a chemical distinction between fluid fluxes incoming from different subducting slabs in central Japan, under which the Pacific and Philippine Sea plates overlap. Recently it was suggested that primary magma formed in the mantle wedge beneath the volcanic front contained 1–2 wt% H_2O, based on melt-inclusion studies and the presence of Ca-rich plagioclase phenocrysts (e.g. Hamada & Fujii 2008).

Large-scale eruptions in Japan

Large-scale caldera-forming eruptions (generally classified by the volcanic explosivity index of Newhall & Self 1982, VEI ≥ 7) are catastrophic volcanic events that impose devastating impacts on the natural environment and human activity at a local to global scale. The latest event with VEI 7 in Japan occurred at Kikai Caldera, southern Kyushu at 7.3 ka, erupting a magma volume of >170 km^3 with a serious effect on human activity in the Kyushu area. The frequencies of such large-scale eruptions are very low, but they have occurred repeatedly on the Japanese islands throughout the Neogene and Quaternary periods.

These young calderas are located along or behind the present volcanic front. More specifically, the distribution of the late Quaternary large calderas with a diameter >10 km and an erupted tephra volume >50 km^3 (produced by greater- or similar-scale eruptions as the historically largest volcanic event at Tambora, Indonesia in 1815) is limited to NE and SW Japan (Kutcharo, Akan, Shikotsu and Toya

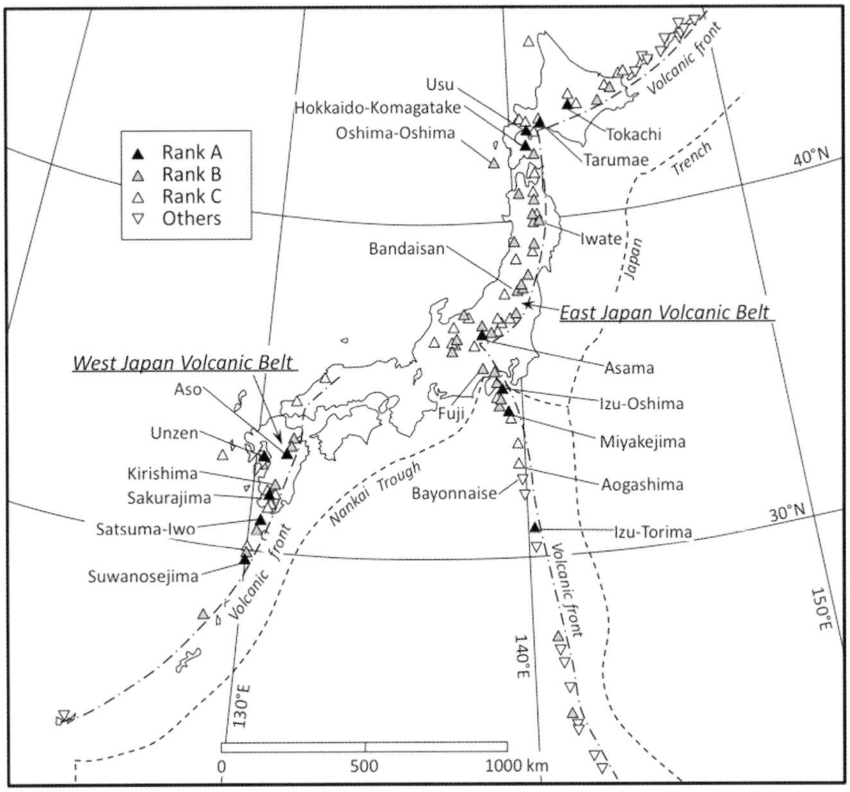

Fig. 5.5. Distribution of active volcanoes (<10 ka) in Japan. Ranks A to C are explained in the text (Fig. 5.23).

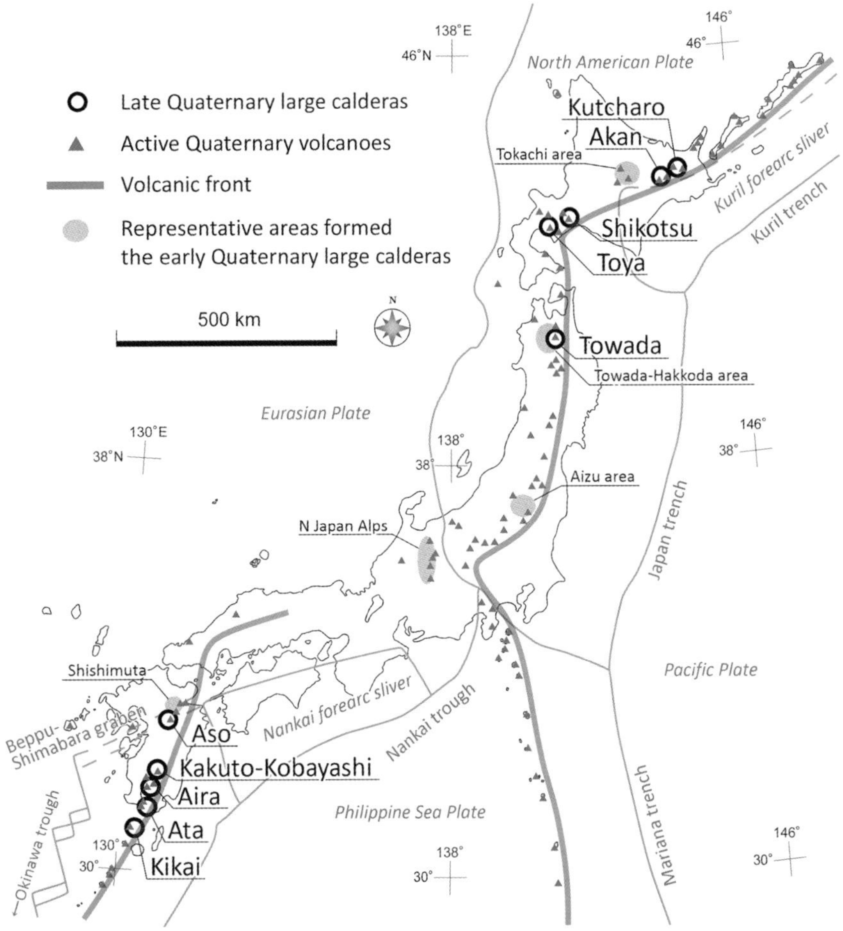

Fig. 5.6. Distribution of late Quaternary large calderas, active Quaternary volcanoes and the volcanic front. Shaded areas indicate selected early Quaternary large calderas. Plate tectonic framework around the present-day Japanese islands is from Isozaki (1996).

calderas in Hokkaido; Towada Caldera in northern Tohoku; and Aso, Kakuto, Kobayashi, Aira, Ata and Kikai calderas in Kyushu) (Fig. 5.6). In the Neogene and early Quaternary periods (*c.* 1 Ma), large calderas also formed in Hokkaido (Tokachi area etc.), Tohoku (Aizu area, Towada–Hakkoda area etc.), Central Japan (North Japan Alps, etc.) and NE Kyushu (Shishimuta area, etc.). The varying scales of the younger caldera-forming eruptions have been determined based on the distribution of ignimbrites and widespread ash-fall layers in and around the Japanese islands (Machida & Arai 2003); those for older periods are less well constrained however, not only because tephra deposits are poorly preserved but also because the original caldera geometry and structure is usually eroded and altered by younger volcanic activity.

A noticeable feature of caldera-forming eruptions is their cyclic behaviour. For example, the large Aso Caldera has had four major caldera-forming events since 0.3 Ma, demonstrating the cyclic behaviour of a large silicic magmatic system (Fig. 5.7). The frequencies of the eruptions for each caldera have been estimated at a few ten thousand to a hundred thousand years based on tephrochronological studies (Machida & Arai 2003). Although only a single gigantic eruption has been identified at Shikotsu, Toya and Aira calderas, for many of the other calderas they have formed by multiple caldera-forming events that have created their present shape and structure.

The eruption chronology of a single caldera-forming eruption is generally characterized by an initial Plinian phase followed by the most intense phase producing large-scale pyroclastic flows,

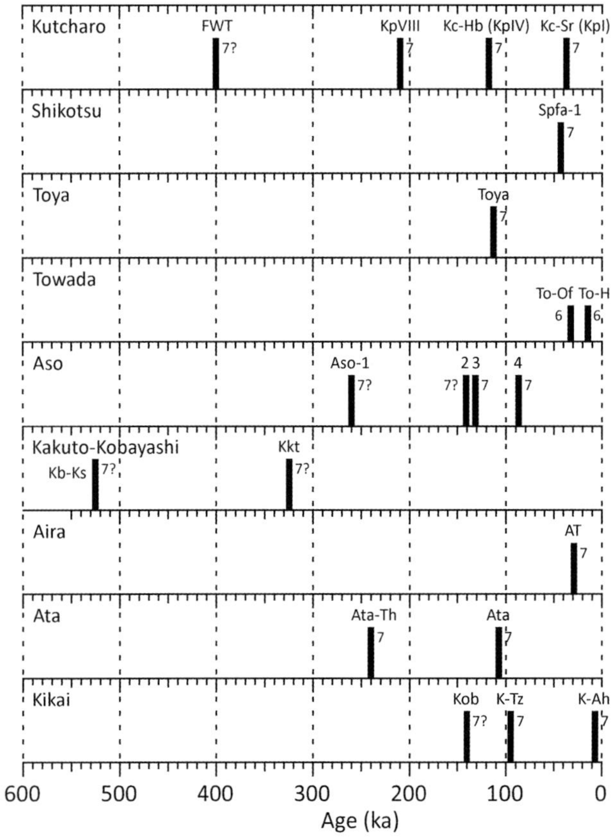

Fig. 5.7. Time distribution of selected large-scale caldera-forming eruptions in the late Quaternary period. Locations of calderas are shown in Figure 5.6. The tephra layer name from each eruption and VEI are also indicated near bars. The original data are from Machida & Arai (2003), but some eruption ages are modified following recent studies.

co-ignimbrite ash-falls and coincidental caldera collapse, as recorded in products from, for example, Spfa-1, AT, Ata and K-Ah eruptions (for the names, see Fig. 5.7). However, some examples (Aso-4, K-Tz, etc.) lack the initial Plinian phase. The eruptions resulted in huge ash-falls that covered very wide areas of the Japanese islands (Fig. 5.8) as shown for example by Aso-4 tephra, which deposited an ash layer still >10 cm thick even in the Hokkaido area. These large-scale eruptions have provided valuable markers, or isochrones, that allow reconstruction of the palaeoenvironmental history. The release of an enormous quantity of volatiles dissolved in the silicic magmas must have caused a long-lasting global effect due to the injection of sulphuric gases into the stratosphere and aerosol formation, as evidenced in the ice cores in the polar region (e.g. Zielinski *et al.* 1996).

Caldera-forming eruptions show that large magmatic systems, storing a vast amount of silicic magmas (generally, orders of 10–100 km^3 in dense rock equivalent), have been produced and evolved in the arc crust. The development of a large silicic magmatic system along a volcanic front reflects the evolution of the plate tectonic system and the crust-mantle thermochemical structures of the mantle wedge, where mantle-derived basalts fuel crustal magmatism, creating evolved silicic magmas in the upper or lower crust (e.g. Yoshida *et al.* 1995; Kimura *et al.* 2005; Yamamoto 2011). In fact, the crustal structure would have been modified by intensive basaltic intrusions to form large silicic magma storages beneath the calderas (e.g. Yoshida *et al.* 2005). For some calderas, the roles of basaltic injection to develop into silicic magmas and their generation, the magma chamber evolution and the eruption processes have been widely discussed from points of view as diverse as the origin of hybrid magmas erupted (e.g. at Aso Caldera, Kaneko *et al.* 2007) or bimodal magmatism in post-caldera activity (e.g. at Kikai Caldera, Saito *et al.* 2002). Cyclic eruptions at the same caldera may be attributed to magma fluxing, magma differentiation and episodic generation of silicic magma chambers. Local tectonic and crustal parameters such as age, composition, thickness of crust, convergence rate of the subducting slab and change of the stress field might also control the magmatic system and the occurrence of caldera-forming eruptions (e.g. Hughes & Mahood 2008).

In Kyushu, a number of large calderas have formed in the past million years. The two major areas of caldera formation appear to coincide with areas of extensional normal faulting, namely the Kagoshima volcano-tectonic graben (KVTG, Fig. 5.9) (for Kakuto–Kobayashi, Aira, Ata and Kikai calderas) and the Beppu–Shimabara graben (for Aso Caldera). The KVTG is a trough-shaped depression *c.* 30 km wide and 120 km long and trending NNE–SSW across south Kyushu Island. Quaternary volcanism in this area is confined to the width of the KVTG, which represents the volcanic front at the northern part of the Ryukyu Arc. The graben has been subject to continuous extension and downfaulting since late Pliocene times (Shiono *et al.* 1980; Aramaki 1984). It can be demonstrated by the tectonics that a large part of southern Kyushu has rotated anticlockwise with respect to northern Kyushu and the Eurasia Plate during the past two million years (Kodama *et al.* 1995). There are six major calderas (Kakuto, Kobayashi, Aira, North Ata, South Ata and Kikai), all of which include post-caldera cones that have formed inside the graben during the late Quaternary period. The relationship between calderas and grabens indicates that the calderas may have formed in periods of local extension within their respective grabens; otherwise intense magmatism associated with calderas and underplating weakened the upper plate to induce extension (Mahony *et al.* 2011).

In the NE Japan Arc, several gigantic caldera-forming eruptions have also occurred during the late Quaternary period. In eastern Hokkaido, the Kutcharo and Akan calderas have formed by

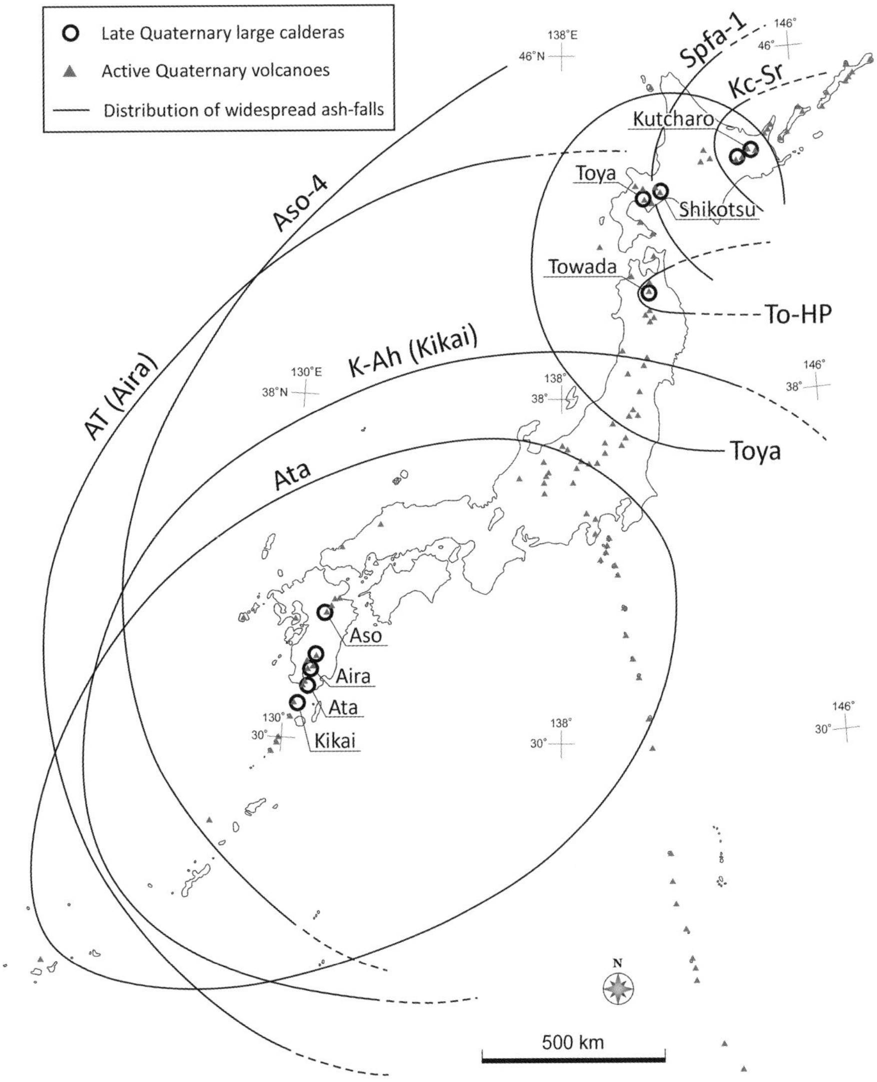

Fig. 5.8. Distribution of widespread ash-falls from selected late Quaternary caldera-forming eruptions. Modified from Machida & Arai (2003).

explosive volcanism since at least 1.5 Ma (Hasegawa & Nakagawa 2007). The volcanoes were developed under the oblique subduction of the Pacific Plate with strike-slip motion, or uplift and anticline formation of the Kuril forearc sliver (Kimura 1996; Goto *et al.* 2000). In NE Honshu and its northern extension large silicic calderas (Towada etc.) formed during the Neogene and Quaternary periods, corresponding to the evolution of an intra-crustal stress regime and crust–mantle thermal structures controlled mainly by the motion of the subducting Pacific Plate (Yoshida *et al.* 2005). Although a number of silicic calderas were created in the Tohoku area during the late Miocene and Pliocene epoch (Yoshida *et al.* 1995; Yamamoto 2011), their activity declined by the Quaternary period probably because the regional tectonic stress field changed from neutral, which is more suitable to develop large-scale magma chambers, to compressional at *c.* 3.5 Ma ago (Sato 1994). The rate of erupted magma volume from caldera-forming events in northern Tohoku, including Towada Caldera, has also declined with time (e.g. Kudo *et al.* 2007).

Kutcharo Group

Four Quaternary calderas, Akan, Kutcharo, Atosanupuri and Mashu, are clustered within a 50 km^2 area of eastern Hokkaido at the southern end of the Kurile Arc (Fig. 5.10). The diameter of the Akan and Kutcharo calderas is larger than 20 km, whereas Atosanupuri and Mashu calderas are smaller and located within and on the rim of the Kutcharo Caldera. In eastern Hokkaido, the concentration of large-scale explosive eruptions (VEI >5) has migrated from Akan to Mashu during the Quaternary period (Hasegawa *et al.* 2009*b*, 2012).

Akan Caldera is a rectangular-shaped structure (24 × 13 km) with a complex history of explosive eruptions. The pyroclastic deposits from this caldera are grouped into 17 eruptive units classified as Ak1 to Ak17 in descending stratigraphic order (Fig. 5.11). The explosive activity of Akan Caldera occurred over a period of more than 1 million years, from >1.5 Ma to *c.* 150 ka. On the basis of the lithic component analysis of ejecta and negative gravity anomalies, the Akan Caldera is considered to be a composite one having multiple vent areas (Hasegawa & Nakagawa 2007; Hasegawa *et al.* 2009*a*). Within this caldera, there are four post-caldera stratovolcanoes, including the active Meakan-dake and Oakan-dake volcanoes.

Kutcharo Caldera (26 × 20 km) is the largest caldera in Japan. The caldera-forming eruption began with a large-scale pyroclastic flow dubbed the Furume Welded Tuff (FWT) at 400 ka. Subsequent caldera-forming eruptive units, the Kutcharo Pumice Flows VIII to I (KpVIII to KpI; Figs 5.11 & 5.12) in ascending order, occurred during 210–40 ka every 20–40 thousand years (Hasegawa *et al.* 2012).

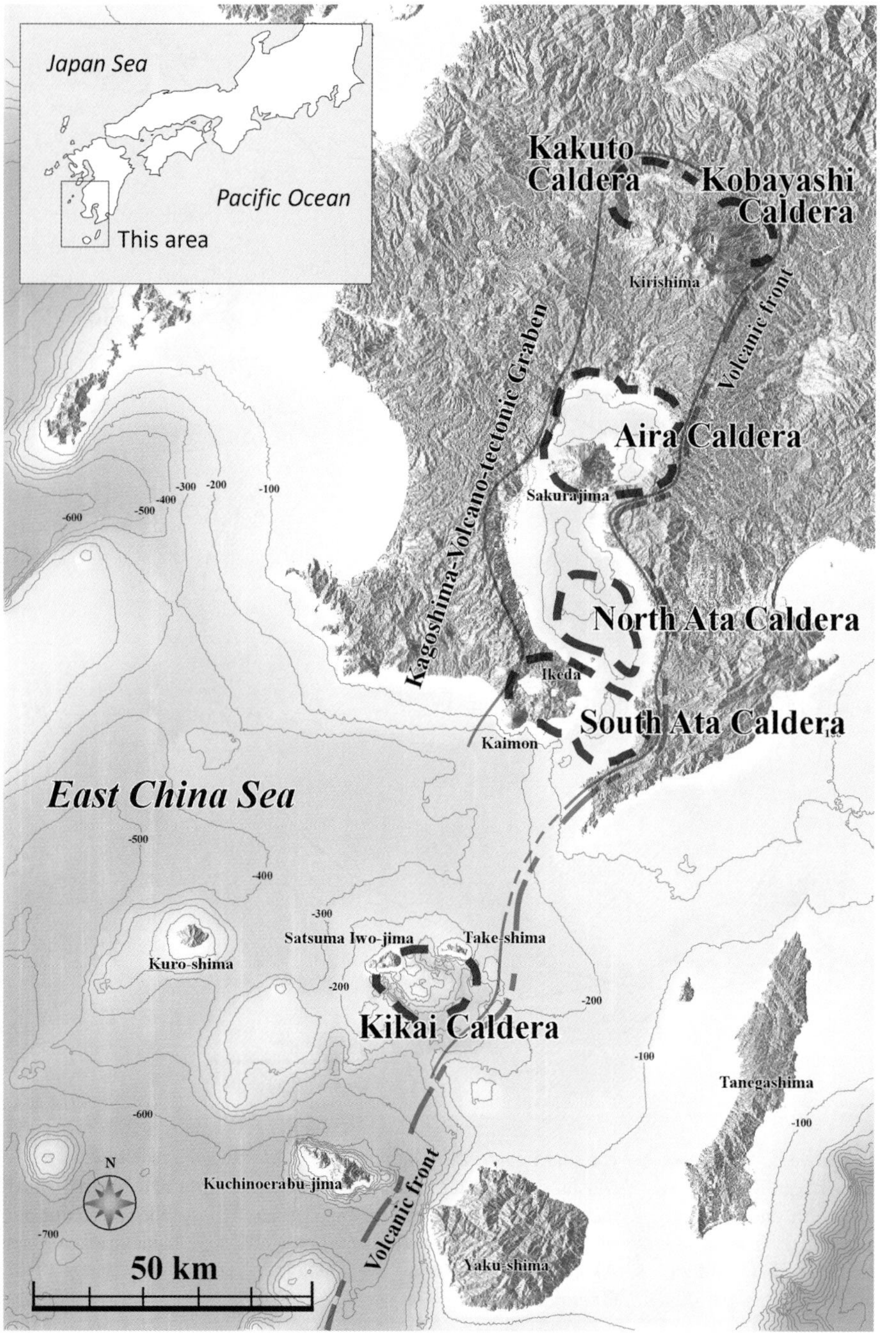

Fig. 5.9. Distribution of large calderas in southern Kyushu (modified from Maeno & Taniguchi 2007).

KpIII and KpII are different flow units within the single eruptive event. KpVI is the most voluminous pyroclastic flow, with 175 km^3 in apparent volume that covered the whole of the eastern part of Hokkaido. Post-caldera vents of this caldera make a circular alignment, suggesting a ring-dyke system.

Atosanupuri Caldera, a relatively small structure, has an approximate diameter of 6 km and formed within Kutcharo Caldera as a post-caldera volcano. Large-scale pyroclastic eruptions with the Atosanupuri Caldera occurred from *c.* 30 ka to 24 ka (Hasegawa *et al.* 2009*b*). Subsequent lava domes extruded within the caldera, and the youngest domes appeared between 1.5 and 1.0 ka.

Mashu Caldera (7.5 × 5.5 km) was formed on the eastern rim of the Kutcharo Caldera and includes the Kamuinupuri crater in its eastern part. The present caldera and crater were formed by large-scale explosive eruptions since 14 ka, immediately after the stratovolcano-formation stage (Kishimoto *et al.* 2009). The Mashu main caldera-forming eruption occurred at 7.5 ka, producing the Ma-j and Ma-f units (Fig. 5.11) with a total apparent tephra volume of 18.6 km^3. The 1 ka Ma-b eruption resulted in the formation of the Kamuinupuri crater and ejected 4.6 km^3 of tephra. Between Ma-l and KpI, at least 15 large-eruptive units occurred from this caldera volcano (Hasegawa *et al.* 2009*b*).

Although the main essential ejecta by these calderas consist of similar orthopyroxene- clinopyroxene andesite to dacite, they are chemically distinct (Fig. 5.13). The Kutcharo ejecta are medium-K in composition while the Mashu ejecta are low-K. Despite these contrasting bulk-rock compositions of the two calderas, observed medium-K series were presumably generated by the mixing of

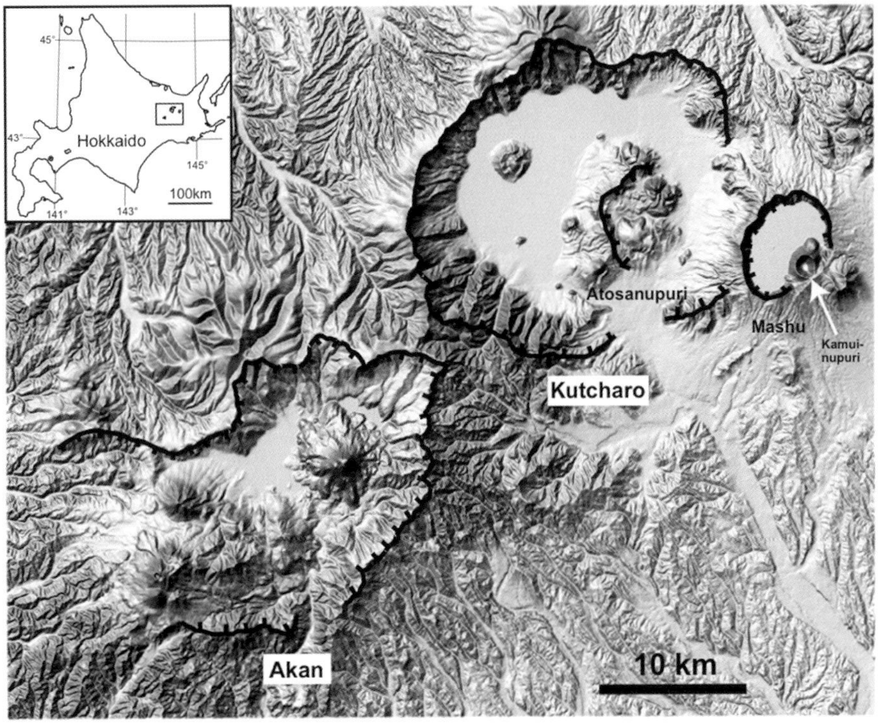

Fig. 5.10. Shaded relief map of Akan, Kutcharo, Atosanupri and Mashu calderas. After Hasegawa *et al.* (2012).

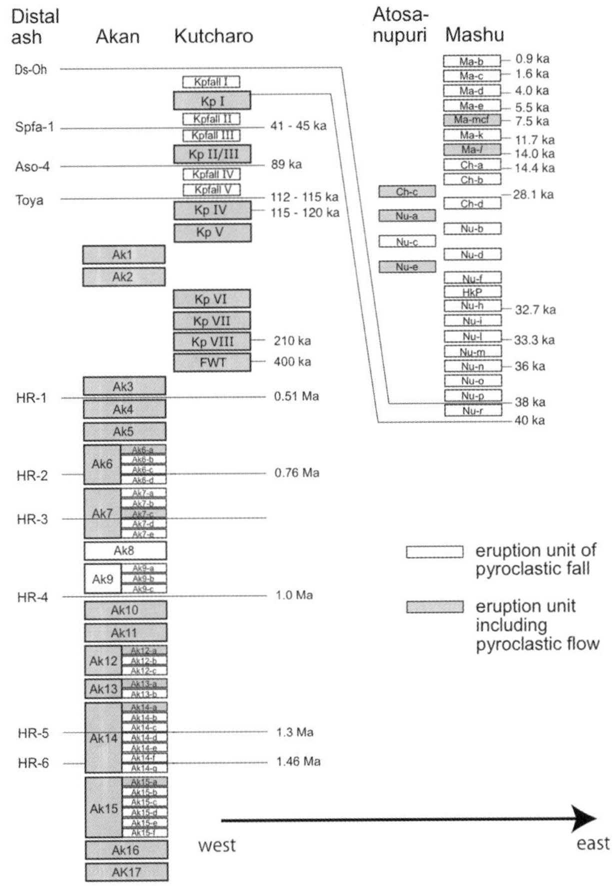

Fig. 5.11. The Quaternary tephrostratigraphy in eastern Hokkaido. Note that the eruption centre has migrated eastwards with time. Modified from Hasegawa *et al.* (2012). Radiocarbon ages are calibrated to calendar ages.

low-K basalt and medium-K rhyolite derived from solidified low-K basalt (Miyagi *et al.* 2012). In addition, the similar Sr, Nd and Pb isotopic compositions of the basalt and rhyolite of both calderas, combined with the progressive enrichment of trace elements from basalt to rhyolite, support the interpretation that the two rock types are genetically related.

Aso Caldera

Aso Caldera in central Kyushu has the dimension of 18 km × 25 km (Fig. 5.14) and is a compound caldera formed by four cycles of caldera-forming eruptions during *c.* 270–90 ka and named Aso-1 to -4 (Fig. 5.15). The total eruption volumes of Aso-1 (270 ka), Aso-2 (140 ka), Aso-3 (120 ka) and Aso-4 (90 ka) are approximately 50, 50, >150 and >600 km^3 respectively (reviewed by Kaneko *et al.* 2007). The volume of pyroclastic flow deposits that has remained into modern times is about 175 km^3, with their original estimated volume ranging from >25 km^3 for Aso-2 (smallest) to >80 km^3 for Aso-4 (largest) (Ono & Watanabe 1983). Volcanic ash emitted during the eruption of Aso-4 (90 ka) is the most voluminous among the four cycles. As previously mentioned, it covered the Japanese islands, with a layer over 10 cm deposited in Hokkaido some 1600 km from the source. Aso-4 pyroclastic flow deposits crop out in the western end of Honshu about 150 km from the source, implying that pyroclastic flows moved over the sea. Soon after the Aso-4 caldera formation (90 ka), eruptions of magma ranging chemically from basalt to rhyolite began, resulting in the formation of the present central cones (Fig. 5.14). Abundant pyroclastic deposits or lavas were observed between products of the two successive caldera-forming eruptions. It is presumed that the central cones also formed between the two cycles (Ono & Watanabe 1985). Although it is located east of the present central cone and inside the present caldera, Nekodake stratocone formed before the Aso-3 eruption.

The circular caldera rim consists of repeated embayments which were formed by landslides from the unstable caldera wall after the eruption. Drilling data on the caldera floor and gravity surveys

Fig. 5.12. Pyroclastic flow deposits from Akan and Kutcharo calderas. See Figure 5.11 for the stratigraphic position of these units. Kitami City, Hokkaido.

show that the original caldera rim lies a short distance within the present shape (Ono *et al.* 1993; Komazawa 1995) (Fig. 5.16). Although the present caldera shape was completed basically by the time of the Aso-4 eruption, the original shape had evolved through the three previous cycles. Lack of an accompanying Plinian pumice fall in the Aso-4 eruption suggests gushing out of a large amount of magma-gas mixture through ring fractures rather than a single vent during the climactic eruption, but no direct evidence of ring fractures has been found. The collapse scenario during the caldera eruption has not been clearly defined. The subsurface structure of the Aso Caldera suggests a flat-bottomed basin rather than a funnel-shaped depression (Ono *et al.* 1993). Wells of the geothermal exploration project (NEDO 1994) that reached down to 1.8 km depth revealed horizontal deposition of the Aso-4 pyroclastic flows in the deeper parts (Hoshizumi *et al.* 1997). Gravity anomaly data support the model of a flat-bottom caldera and showed the existence of multiple basement grabens within the caldera (Fig. 5.16). The Aso-4 pyroclastic flow deposits consist of multiple flow units with slightly different lithological and mineralogical characteristics and with different distribution patterns (e.g. Kaneko *et al.* 2007). This implies that not all the flows issued from the same vent area. Each gravitational graben may mark the source from which different flow units issued or may be remnants of older calderas buried by the Aso-4 pyroclastic flow deposits.

Rocks of caldera-forming eruptions range in composition from basalt to dacite; for example, mafic scoria in Aso-4 ranges between 50 and 57 wt% SiO_2 and pumice ranges from 66 to 70 wt% SiO_2. Aso-4 can be distinguished chemically from Aso-1 to -3 by its high K_2O content and the presence of amphibole phenocrysts. Kaneko *et al.* (2007) proposed that the magma chamber of Aso-4 was chemically zoned with more silicic magma above, and that they were mixed during the eruption.

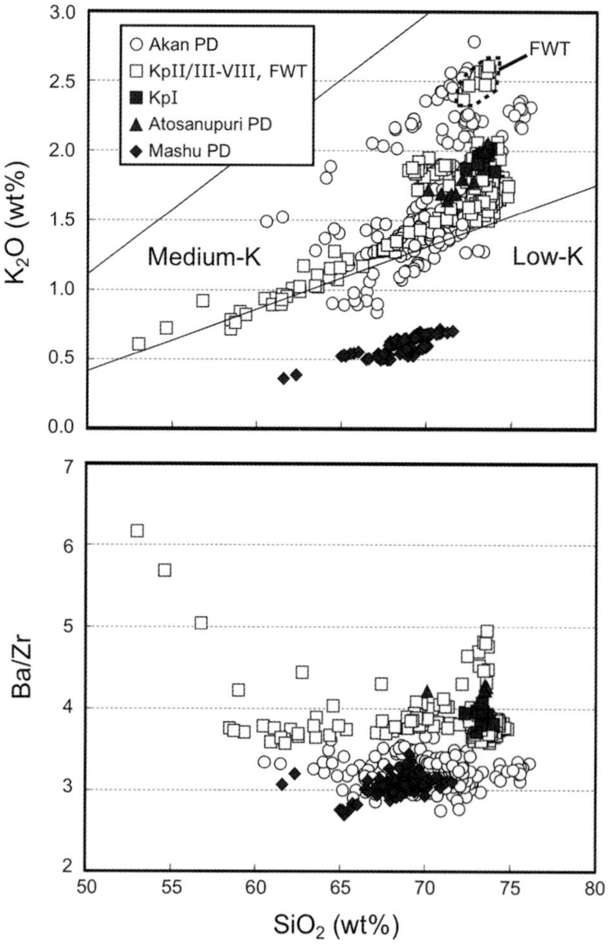

Fig. 5.13. Variations of SiO_2 with K_2O (upper) and Ba/Zr (lower) in the bulk-rock compositions of essential ejecta from Akan, Kutcharo, Atosanupri and Mashu calderas. After Hasegawa *et al.* (2012).

Aira C`aldera

Aira Caldera occupies the northern end of Kagoshima Bay, in Kyushu, and is controlled by regional fault systems that make it

Fig. 5.14. Southerly view of Aso caldera. The mountains inside the caldera are the central cones. Nekodake, a stratocone older than the Aso-3 caldera-forming eruption (150 ka), is middle right. The flat caldera floor, extending between the central cone to the caldera rim as a broad moat, developed during at least three periods of lake formation. Photo by Yasuo Miyabuchi.

rectangular in shape (Fig. 5.17). In its northeastern portion is the 10 km diameter submarine caldera known as the Wakamiko Caldera, which has a flat bottom located at 200 m below sea level. Pre-caldera volcanism erupting basalt, andesite and rhyolite took place from 2.9 Ma until 29 ka (Sudo *et al.* 2000, 2001). Nagaoka (1988) and Nagaoka *et al.* (2001) observed that at least seven pyroclastic eruptions have taken place during the last 100 kyr, occurring every 14 kyr years on average from the Aira Caldera, resulting in a complex topography in its eastern portion.

A large pyroclastic eruption occurred within the Aira Caldera at 29 ka (Okuno 2002) (Aramaki 1984). It started with a Plinian eruption, forming the Osumi pumice fall (98 km^3), and was followed by

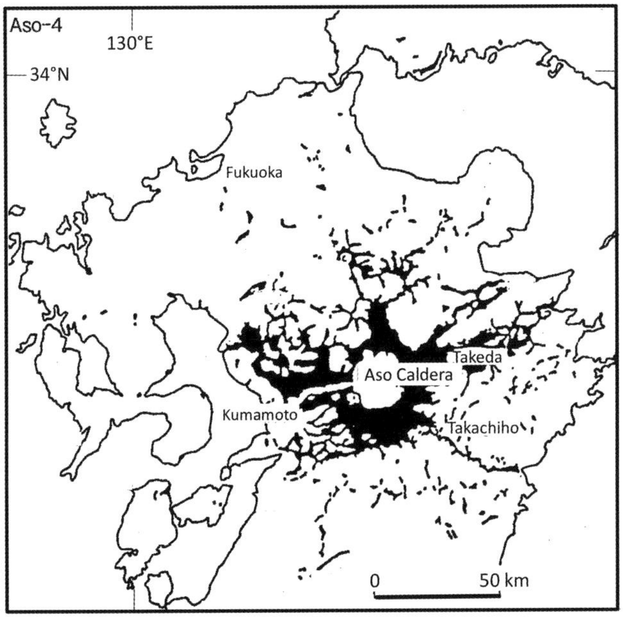

Fig. 5.15. Distribution of Aso-4 pyroclastic flow deposits. After Ono & Watanabe (1983).

the fine-grained Tsumaya pyroclastic flow (13 km^3). Both erupted from a vent located near the presently active Sakurajima Volcano which rises on the southern rim of the Aira Caldera. The subsequent violent explosion ejected basement rock fragments and pumiceous rhyolite from the central vent, this gradually developing into a huge eruptive column that rapidly collapsed, emplacing the Ito pyroclastic flow (*c.* 300 km^3 in volume). The earliest phase of this eruption produced the Kamewarizaka lag breccias, which is up to 30 m thick, was emplaced at the caldera rim and is rich in basement lithics up to 2 m across (Ueno 2001; Fig. 5.18). The lag breccia is a near-vent variety of the bottom concentration zone of lithics found in the Ito pyroclastic flow deposit. Textural features and its monotonous petrologic character indicate that the main part of the Ito pyroclastic flow was emplaced by a simple, short-lived eruptive mechanism. The Osumi and Ito pumices are orthopyroxene rhyolites with crystal contents varying from 2 to 5% in volume and from 75 to 76 wt% in SiO_2. This large-scale eruption resulted in the formation of a vast pyroclastic plateau covering southern Kyushu. The most distal deposits are found 90 km to the north of the caldera centre (Fig. 5.19; Yokoyama 2000). The Aira-Tn ash, a fine-grained counterpart of the Ito pyroclastic flow, extended for more than 1400 km from the vent and covered more than 500 000 km^2 (Machida & Arai 2003). The total volume of magma erupted during the Ito eruption is estimated to have been *c.* 110 km^3 in dense-rock equivalent (DRE) (Aramaki 1984).

Medium- or small-volume pyroclastic deposits younger than the Ito pyroclastic flow are found not only within the Aira Caldera but also on its eastern periphery. The 11 ka Shinjima pumice, a submarine pyroclastic flow, was deposited within the caldera (Kano *et al.* 1996), whereas the 16 ka Takano base surge was deposited on land (Okuno 2002). These deposits are intercalated within tephra layers of Sakurajima Volcano (Fig. 5.19). Although they occupy different stratigraphic positions, the units have similar pumice compositions (ferrohypersthene rhyolite), which are quite different from those of Sakurajima Volcano. This means that new silicic pyroclastic eruptions occurred within the Aira Caldera coevally with the formation and growth of the Sakurajima Volcano, although their exact vent positions are as yet unknown. Three submarine peaks, probably post-caldera volcanoes, are aligned along the

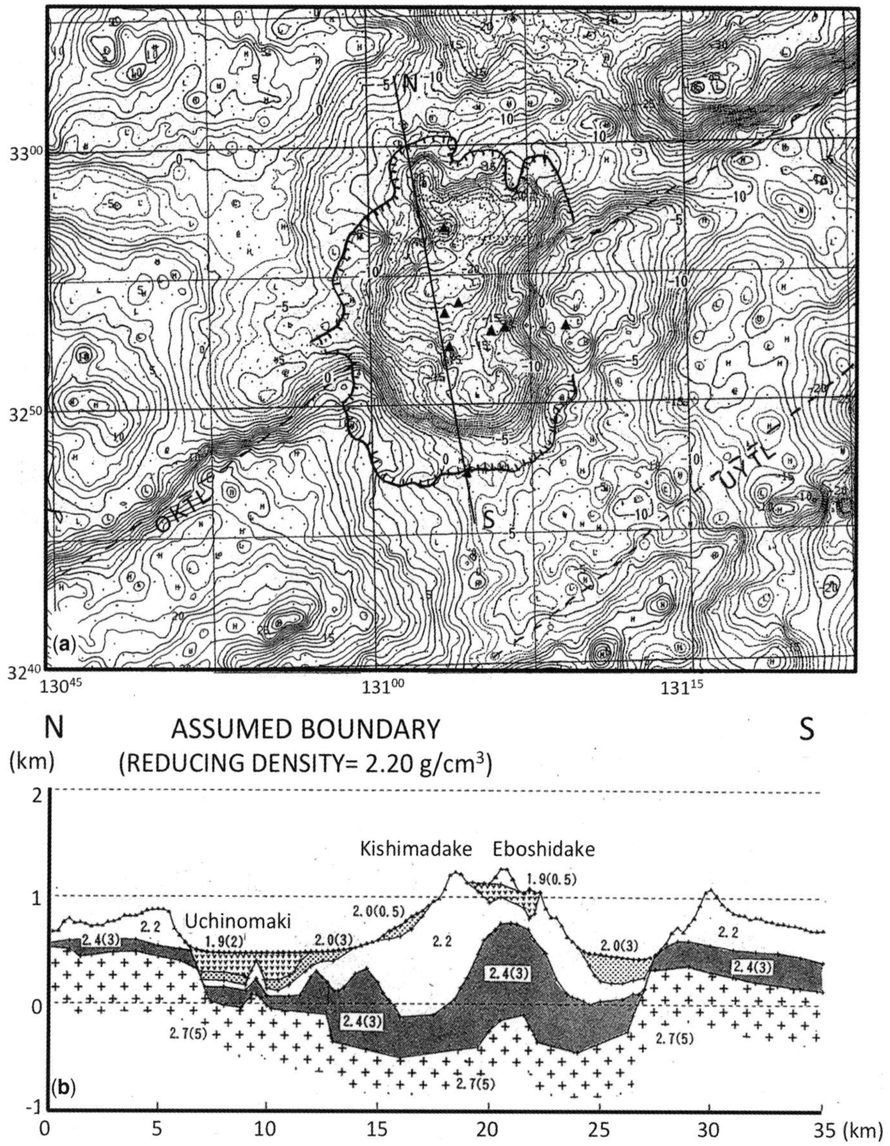

Fig. 5.16. Bouguer anomalies of the Aso caldera (**a**) and cross-section of the subsurface structure of Aso caldera (**b**). After Komazawa (1995). Contour interval in (a) is 1 mGal. KOTL, Kumamoto–Oita tectonic line; UYTL Usuki–Yatsushiro tectonic lines.

eastern margin of the Aira Caldera. Vigorous submarine fumarole activity, referred to as 'Tagiri', occurs on the floor of the Wakamiko Caldera.

Kikai Caldera

Kikai Caldera, 17 km wide and 20 km long, is located in the East China Sea and on the southern extension of the Kagoshima volcano-tectonic graben (KVTG) (Fig. 5.9). Most of the Kikai Caldera is presently submerged. Subaerial parts form the two main islands on the northern caldera rim, Satsuma Iwojima and Takeshima. The rhyolitic Iwo-dake and basaltic Inamura-dake volcanoes on Satsuma Iwojima are post-caldera stratocones (Ono *et al.* 1982). The caldera floor is 300–500 m below sea level and some submarine cones have grown after caldera formation (Fig. 5.20). The central part of the caldera, thought to be a resurgent dome, grew after the 7.3 ka caldera-forming eruption.

Volcanic activity associated with the Kikai Caldera began at *c.* 700 ka with bimodal magmatism (Ono *et al.* 1982). After this activity, multiple caldera-forming eruptions have occurred over the last 200 ka. Deposits from these eruptions are distributed on islands around the Kikai Caldera. On the caldera rim, three sheets of ignimbrite are exposed: the 140 ka Koabi ignimbrite; the 95 ka Nagase ignimbrite; and the latest 7.3 ka Koya–Takeshima ignimbrite (Ono *et al.* 1982). The Komoriko tephra group (13–8 ka), represented by andesitic activity preceeding the 7.3 ka eruption, is exposed on both Satsuma Iwojima and Takeshima (Okuno *et al.* 2000).

The scale of the 7.3 ka eruption was VEI 7, making it the most recent large-scale caldera-forming event in Japan, and the event produced four major volcano-stratigraphic units. The lowermost unit consists of Plinian pumice-fall deposits, overlain by the second unit which comprises intra-Plinian flow deposits showing various types of stratification and degrees of welding, and observed only in proximal islands. The third unit is a voluminous ignimbrite (named the Koya ignimbrite in distal areas or the Takeshima ignimbrite for proximal areas; Ui 1973; Ono *et al.* 1982; Walker *et al.* 1984). This eruptive unit is traceable up to 100 km away from the source and has been interpreted as a low-aspect ratio ignimbrite (Ui *et al.* 1984; Walker *et al.* 1984). The topmost unit is a

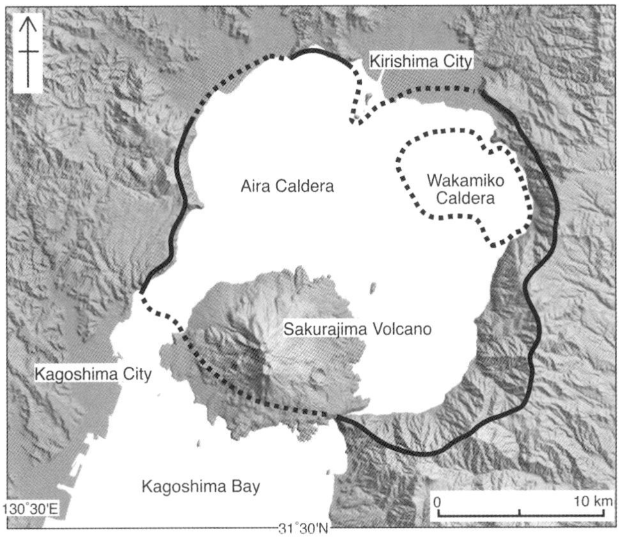

Fig. 5.17. Shaded relief map of Aira calderas. Sakurajima volcano and Wakamiko caldera formed after Aira caldera.

co-ignimbrite ash-fall deposit (K-Ah, or Akahoya ash). This tephra was dispersed over a wide area of Japan, more than 1000 km distant from the Kikai Caldera (Fig. 5.21), with a total tephra volume of >170 km^3 (Machida & Arai 2003). The main products were two-pyroxene rhyolite with 72–74 wt% in SiO_2 and 7–10 vol% phenocrysts. The eruption represents the evacuation of a large silicic magma chamber with an estimated magma volume of 70–80 km^3 in DRE (Maeno & Taniguchi 2007). The magma chamber depth of the 7.3 ka eruption is estimated at 3–7 km, based on the gas-saturation pressure of melt inclusions (Saito *et al.* 2001). Banded pumices with andesitic composition (56–58 wt% in SiO_2) in the ignimbrite indicate that mafic magma co-existed with or was intruded into a magma chamber deeper than 7 km. This eruption caused a major environmental and cultural impact in western Japan. Discovery of tsunami deposits on the northeast coast of Oita, mainland Kyushu (Fujiwara *et al.* 2010), suggests that huge tsunamis were generated and propagated through the Pacific Ocean from this eruption. Numerical simulations showed that a likely mechanism of tsunami generation is rapid caldera collapse rather than the entrance of a large pyroclastic flow into the sea (Maeno *et al.* 2006; Maeno & Imamura 2007).

Post-caldera volcanic activity started from 5.2 ka (Okuno *et al.* 2000), with growth of the old Iwo-dake Volcano on the northwestern side of the caldera. Activity of the Inamura-dake volcano started from 3.6 ka (Okuno *et al.* 2000) and was characterized by basaltic lava flows and tephras, producing a small stratocone. This basaltic activity was sandwiched between the rhyolitic activity of the old and young Iwo-dake volcanoes (Fig. 5.21). In 1934–1935, submarine volcanic activity produced a silicic lava dome and the island of Showa Iwojima, which represents the latest magmatic eruption in the Kikai Caldera (Maeno & Taniguchi 2006). Geochemical and petrographical studies of the rhyolites and mafic inclusions from the volcanoes indicate that a stratified magma chamber exists beneath the caldera, comprising a lower basaltic layer, an upper rhyolitic layer and a thin middle andesitic layer during the post-caldera stage (Saito *et al.* 2002).

Most post-caldera eruptions seem to have occurred along a trend within the northwestern side of the caldera. The distribution of the 7.3 ka deposits and post-caldera volcanoes indicate that the sources of both rhyolite and basalt have co-existed beneath the northwestern side of the caldera for over 7 kyr (Saito *et al.* 2002; Maeno & Taniguchi 2007). It is also noteworthy that the distribution of vents is consistent with the direction of the southwestern edge of the KVTG (Fig. 5.9). These lines of evidence indicate that the Kikai Caldera is still very active, probably because of its position along the structurally weak, extensionally faulted southwestern edge of the volcanic graben (Maeno & Taniguchi 2007).

Recent eruptions in Japan

The Japan Arc is similar in scale to the Indonesian Sumatra–Java Arc but has recorded fewer large volcanic eruptions. Figure 5.22

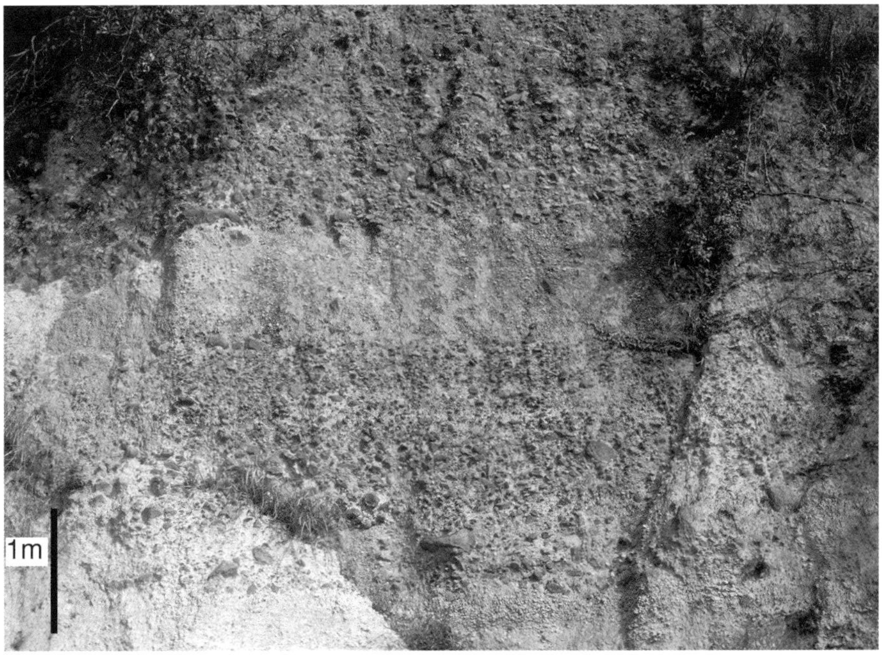

Fig. 5.18. Outcrop of the Kamewarizaka breccia containing abundant lithic fragments. This facies corresponds to near-vent lag breccia of the Ito pyroclastic flow deposit. There are two normal-graded breccia units interbedded with massive pumice flow deposits. Scale bar is 1 m. Kokubu–Shikine, Kagishima Prefecture.

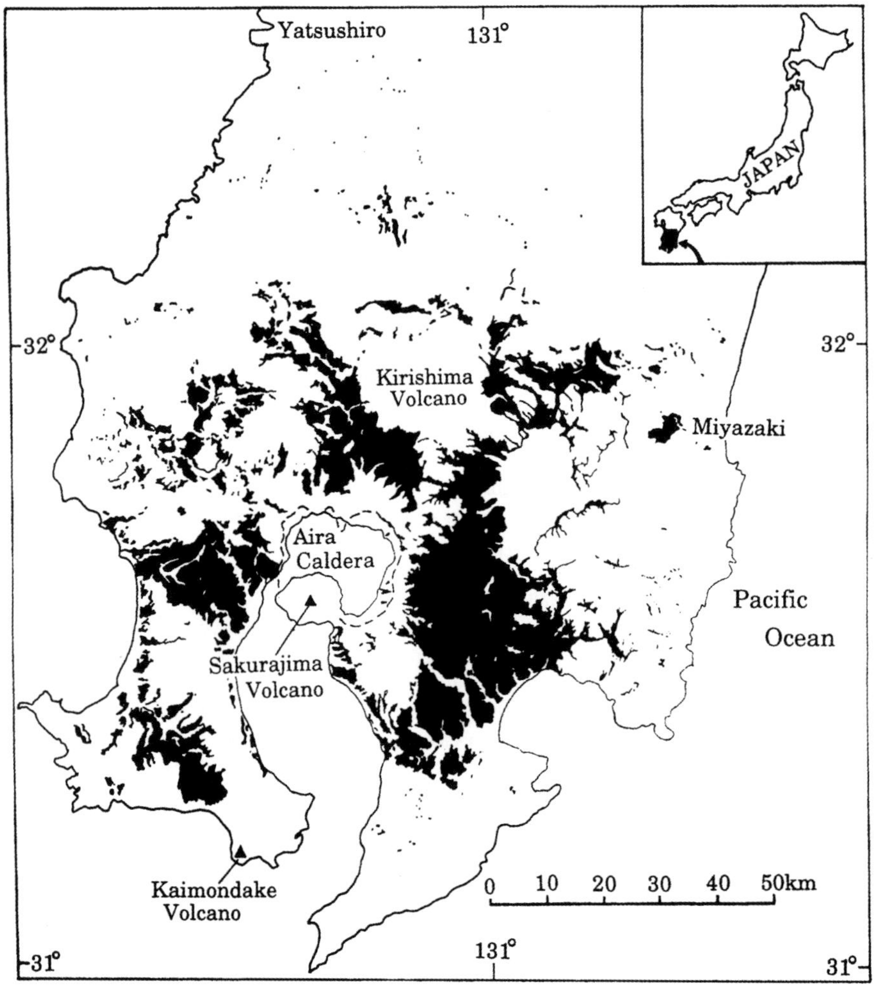

Fig. 5.19. The distribution of the Ito pyroclastic flow deposit (black part). After Yokoyama (2000).

compares the two volcanic areas over the last four centuries, beginning when documentation of volcanic eruptions started in Indonesia. It is immediately evident from this figure that eruptions of VEI 4 or greater have not occurred in Japan since the eruption of Tarumae Volcano in 1739 (VEI 5), and no eruptions larger than VEI 3 have occurred after the eruptions of Sakurajima in 1914 and Hokkaido–Komagatake in 1929.

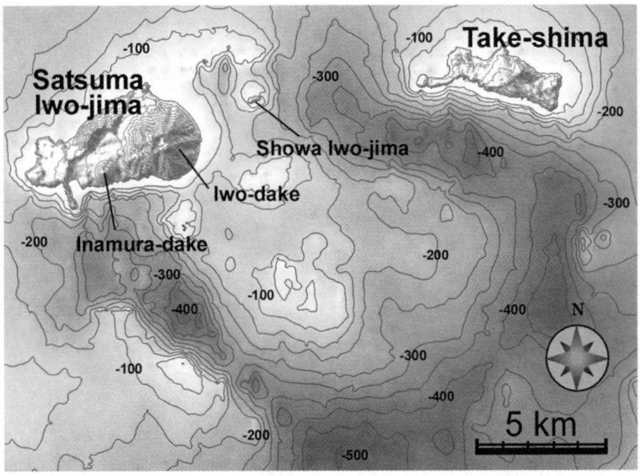

Fig. 5.20. Topography and bathymetry map of the Kikai caldera (modified from Maeno & Taniguchi 2007).

Volcanic hazard events in Japan during the last 300 years are shown in Table 5.1. Eruptions have occurred at a limited number of volcanoes, most frequently at Sakurajima and Suwanosejima. Recent major eruptions include Usu in 1977–1978 and 2000, Izu–Oshima in 1986–1989, Miyakejima in 1983 and 2000, Unzen in 1990–1995 and Shinmoedake (Kirishima) in 2011. Although Asama and Aso volcanoes have been active over historic time, only small-scale eruptions have occurred during recent decades. A seismic crisis accompanied by inflation was recorded without any eruption at Iwate Volcano in 1998. Similarly, swarms of low-frequency earthquakes were observed at Mount Fuji in 2000 and 2001, although no surface manifestation of volcanic activity has been observed.

The Tohoku–Oki Earthquake (Mw 9.0) involved extensive deformation in the eastern half of the Japan Arc (Ozawa *et al.* 2011). Following this event, 20 volcanoes in the NE and central Japan, including Fuji Volcano, showed anomalous seismicity for as long as a few months. At Fuji Volcano, a M 6.4 earthquake took place on 15 March 2011 (four days after the M 9.0 earthquake) about 9 km beneath the summit, and was followed by swarms of smaller magnitude earthquakes. Such seismic triggering of volcanic activity was demonstrated by the 1707 eruption of Fuji Volcano (VEI 5), which occurred 49 days after a M 8.7 earthquake in the Nankai Trough.

Fig. 5.21. Outcrop showing the eruptive history in Satsuma Iwojima after the 7.3 ka caldera-forming eruption. The Takeshima ignimbrite is a product of the climactic phase of the 7.3 ka caldera-forming eruption. OIo1a–2b, In1–3 and YIo1–4a are derived from the rhyolitic Old Iwo-dake Volcano, the basaltic Inamura-dake Volcano, and the rhyolitic Young Iwo-dake Volcano, respectively. Modified from Maeno & Taniguchi (2005).

Ranking and monitoring systems of active volcanoes in Japan

The National Coordination Committee for Volcanic Eruption Prediction has classified active volcanoes into three ranks – A, B and C – based on their record of eruption activity during periods of 100 years and 10 kyr (JMA 2005) (Fig. 5.23). Rank A includes volcanoes with a 100 year activity index of >5 and 10 kyr activity index of >10. Rank B covers volcanoes with a 100 year activity index of >1 and 10 kyr year activity index of >7 except rank A volcanoes. Finally, all other volcanoes are classified as Rank C.

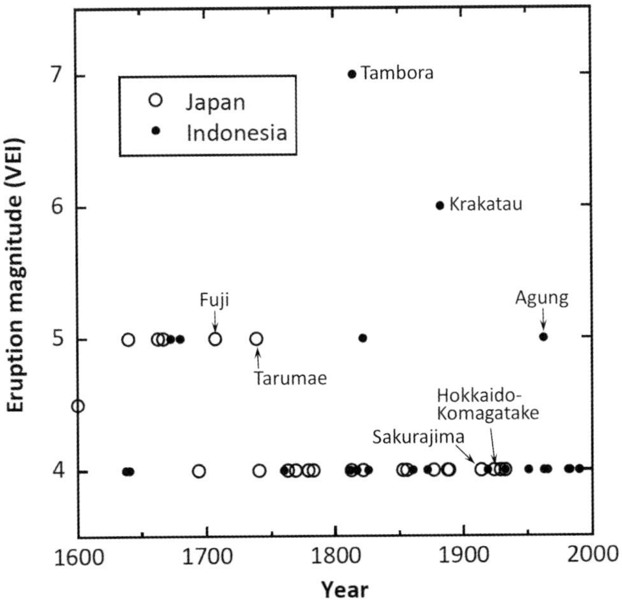

Fig. 5.22. Eruption magnitudes of volcanic eruptions between Japan and Indonesia during 1600–2000. Data from Siebelt *et al.* (2010).

Evaluating the potential of volcanic eruption

Generally, the eruption activity index (AI) can be expressed by frequency, scale and style of eruptions as follows:

$$\mathrm{AI} = (\text{frequency}) \times (\text{scale}) \times (\text{mode})$$

As the frequency and scale vary over orders of magnitude depending on eruptions, the factors should be expressed in log units:

$$\mathrm{AI} = \log_{10}(\text{frequency}) + \log_{10}(\text{scale}) + \log_{10}(\text{mode})$$

The 10 kyr activity index is calculated as the sum of three factors:

$$10\ \text{kyrAI} = \text{frequency factor} + \text{scale factor} + \text{mode factor},$$

where frequency factor is increased by 0.5 when eruptions occurred in the latest 300 years, 1000 years and 10 kyr, respectively; scale factor is expressed as VEI; and the mode factor is represented by the characteristic style of eruptions experienced. Depending on speed of movement, impact area, temperature and intensity of volcanic products, the number assigned is 3 for pyroclastic flow, pyroclastic surge or debris avalanche; 2 for lahar or phreatomagmatic explosion; and 1 for lava flow. To add weight to eruption events over the last 1000 years, the points for events before 1000 years ago are multiplied by 0.5. The calculated mode factor is represented by the highest point among these events. For example, in a given volcano the last pyroclastic-flow(s) occurred 2000 years ago, while phreatomagmatic explosion(s) plus lahar(s) occurred 600 and 300 years ago, respectively. In this case, mode factor is 2. The frequency of eruptions is ignored here.

The 100 year activity index is calculated by summing three factors to evaluate the eruption potential based on the observation for these 100 years:

$$\begin{aligned}100\ \text{year activity index} = {} & 100\ \text{year frequency factor}\\ & + 30\ \text{year frequency factor}\\ & + 100\ \text{year scale factor},\end{aligned}$$

where 100 year frequency factor is the log of number of years when eruptions or anomalies were observed for these 100 years, plus 1 and the 30 year frequency factor is the log of the number of years when eruptions or volcanic anomalies were observed for these 30 years,

Table 5.1. *Volcanic hazards with >10 casualties in Japan since the eighteenth century*

Year	Volcano	Casualties	Hazards
1721	Asama	15	Volcanic bomb
1741	Oshima–Oshima	1467	Debris avalanche and tsunami
1779	Sakurajima	>150	Volcanic bomb and lava flow
1781	Sakurajima	15	Tsunami by submarine eruption
1783	Asama	1151	Pyroclastic flow, debris flow and flood
1785	Aogashima	130–140	Leaving the island
1792	Unzen	15 000	Debris avalanche and tsunami
1822	Usu	50–103	Pyroclastic flow
1856	Hokkaido–Komagatake	21–29	Pumice fall and pyroclastic flow
1888	Bandai	461 (or 477)	Debris avalanche
1900	Adatara	72	Pyroclastic surge
1902	Izu–Torishima	125	Isolation in the island
1914	Sakurajima	58	Lava flow and earthquake
1926	Tokachi	144	Mud flow
1940	Miyakejima	11	Volcanic bomb and lava flow
1952	Bayonnaise	31	Submarine eruption
1958	Aso	12	Volcanic bomb
1991	Unzen	43	Pyroclastic flow

According to JMA (2005).

plus 1. When fumarolic activity is observed, add 1 to the number of years before calculating the logarithm. Finally, the 100 year scale factor is the logarithm of the total volume of eruption products (in units of 10^4 m^3) over a period of 100 years.

A total of 110 active volcanoes exist in Japan, including the four islands claimed by Russia (Fig. 5.5). Of these, 47 volcanoes were selected by the National Coordinating Committee of for Prediction of Volcanic Eruption as volcanoes for which continuous monitoring and observation are needed to minimize the risk of volcanic disasters. These volcanoes are monitored around the clock by the Japan Meteorological Agency (JMA) with seismometers, tiltmeters, microphones (infrasound monitors), GPS instruments and cameras. Monitoring is further supported by universities, local governments and institutions for disaster protection. When it is expected that volcanic activity will affect residential areas, a volcanic alert is issued by the JMA showing the danger area.

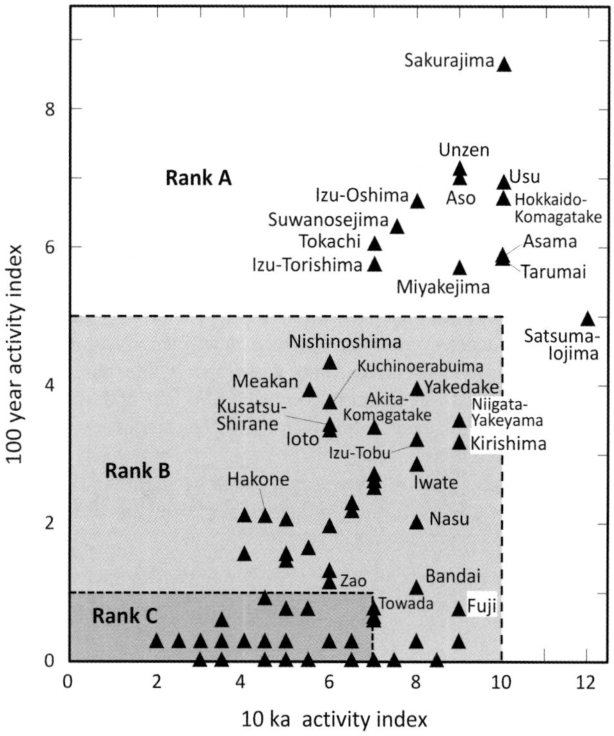

Fig. 5.23. Ranking of eruption potential for active volcanoes (JMA 2005).

Usu Volcano

Usu is a stratovolcano of basalt to andesite composition, growing on the southern rim of the Toya Caldera (*c.* 110 ka). The present summit has a somma with a diameter of *c.* 1.8 km, filled by several lava domes of dacite to rhyolite. It is considered that the somma formed by a sector collapse at *c.* 7–8 ka, when a debris avalanche flowed to the south leaving a hummocky surface which extends across the southern flank of the volcano (Soya *et al.* 2007).

Usu is one of the most active volcanoes in Japan. The present eruptive period started in 1663 with a Plinian outbreak and pyroclastic surges (eruptive volume of 2.5 km^3). Since then, five summit eruptions with Plinian columns occurred at the end of seventeenth century and in 1769, 1822, 1853, 1880–1889 and 1977–1978, and three flank eruptions with phreatic to phreatomagmatic events occurred in 1910, 1943–1945 and 2000. The magma for historical eruptions became less evolved with time, decreasing from 73 wt% SiO_2 in 1663 to 67 wt% SiO_2 in 2000 (Oba & Katsui 1983; Soya *et al.* 2007). Recent modelling of the petrological evolution of the magma chamber has been presented by Tomiya & Takahashi (1995, 2005) and Matsumoto & Nakagawa (2010). Tomiya & Takahashi (1995; 2005) proposed mixing of a chemically stratified dacite magma with a deep-seated rhyolite magma, and the involvement of new dacite magma since 1977. Matsumoto & Nakagawa (2010) identified three cycles of the magma plumbing system, each of which injected mafic magma into a chemically zoned, shallow magma chamber.

Irrespective of eruption locations, recent eruptions always ended with the formation of lava domes or cryptodomes. For example,

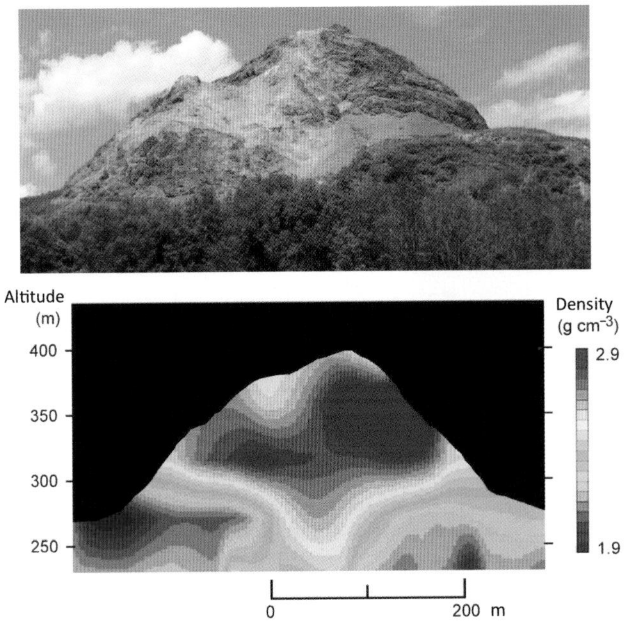

Fig. 5.24. Southwesterly view of Showa–Shinzan lava dome, Usu Volcano (top) and its cosmic-ray muon image showing the average density distribution. After Tanaka *et al.* (2007).

Showa–Shinzan (new mountain in the Showa era) was formed during the 1943–1945 eruption. The growing process of Showa–Shinzan was recorded in detail by a postmaster of the local post office, Mr Masao Mimatsu, his record becoming famous as the Mimatsu diagram (Minakami *et al.* 1951). The top of Showa–Shinzan is covered by sedimentary rocks deposited before the eruption, part of which are baked like brick, so that it is an upheaved plug rather than a cryptodome or a lava dome (Blake 1990). Cosmic-ray muon imaging of the Showa–Shinzan dome was carried out by Tanaka *et al.* (2007) who demonstrated that the dome root becomes as narrow as *c.* 100 m (Fig. 5.24).

1977–78 and 2000 eruptions

Plinian eruption of dacite magma took place at the summit crater on 7 August 1977, following earthquakes that were felt for 32 hours (Yokoyama *et al.* 1981). The eruption column reached *c.* 12 km above the crater one hour later (Fig. 5.25), and the eruption activity continued until the early morning of 14 August. The volume of erupted material was *c.* 8.3×10^6 m^3. Following this first eruptive phase, the dacite magma formed an intrusion as a cryptodome within the summit crater, uplifting the summit area 100–250 m over the first 10 weeks or so. On 16 November, the second eruptive phase began with a small phreatic explosion. There were repeated phreatomagmatic explosions from July to September 1978, this phase finally ending on 27 October. Heavy rains generated lahars on the slopes around the volcano in October 1978, causing damage to houses and the deaths of two people with one person missing. The deformation of the summit area continued until January 1980, by which time the summit area had been uplifted about 170 m and the northeastern summit part had been pushed out 160 m laterally (Watanabe 1984).

A phreatomagmatic eruption occurred on 31 March 2000 following earthquakes felt for three days (Nakada 2001; Ui *et al.* 2002; Oshima & Ui 2003), a volcanic alert having been issued by the JMA on 29 March. New craters were opened at the western flank of the main volcano (the western flank of Nishiyama). The plume reached about 3.5 km above the crater and the eruptive activity continued for about 6 hours, with a small pyroclastic surge being generated. The volume of the erupted material by 31 March was more than 2.3×10^5 m^3 (Endo *et al.* 2001). The next day a new crater opened with a phreatic explosion in the hillside (Kompirayama) just behind the resort area, about 500 m NE of the first crater (Fig. 5.26a, b). Phreatic explosions then occurred in both areas, creating 65 new craters by middle April. From early April, the western flank of Nishiyama had been uplifted by shallow magma intrusion and the height reached about 80 m (uplift volume of 4×10^7 m^3), generating

Fig. 5.25. Plinian eruption at Usu volcano on 7 August 1977. (**a**) Easterly view of eruption column at the onset of the explosion (09:12 local time). Courtesy of Toya Caldera–Usu Volcano Global Geopark. (**b**) Eruption cloud growing over Usu volcano, taken from the north at 09h51 m by Hokkaido Shinbun (Katsui *et al.* 1978).

Fig. 5.26. The 2000 eruption at Usu volcano. (**a**) Phreatic explosions from craters formed just behind the hot-spring resort village, Toyako–Onsen. (**b**) House roofs were penetrated, appearing like a honeycomb, by ballistics of the explosions in Toyako–Onsen. (**c**) Step faults developed on slopes of the area under which new magma intruded.

many step-faults (Fig. 5.26c). The phreatic stage continued until September 2000.

Izu–Oshima Volcano

Izu–Oshima volcano is an active basaltic stratovolcano that forms the northernmost island (15 × 9 km) of the Izu–Mariana Arc. The elevation of this island is 764 m a.s.l., but the volcanic edifice height is over 1000 m from the sea floor. The island consists of highly dissected remnants of three old (Pliocene or Early Pleistocene?) volcanoes which are exposed on the northern and western coasts, as well as in the products of Izu–Oshima volcano proper. These products cover the older edifices and occupy most of the subaerial portion of the island (Fig. 5.27).

Izu–Oshima volcano is composed of lava flows and volcaniclastic rocks of low-K, arc-type tholeiitic olivine basalt and

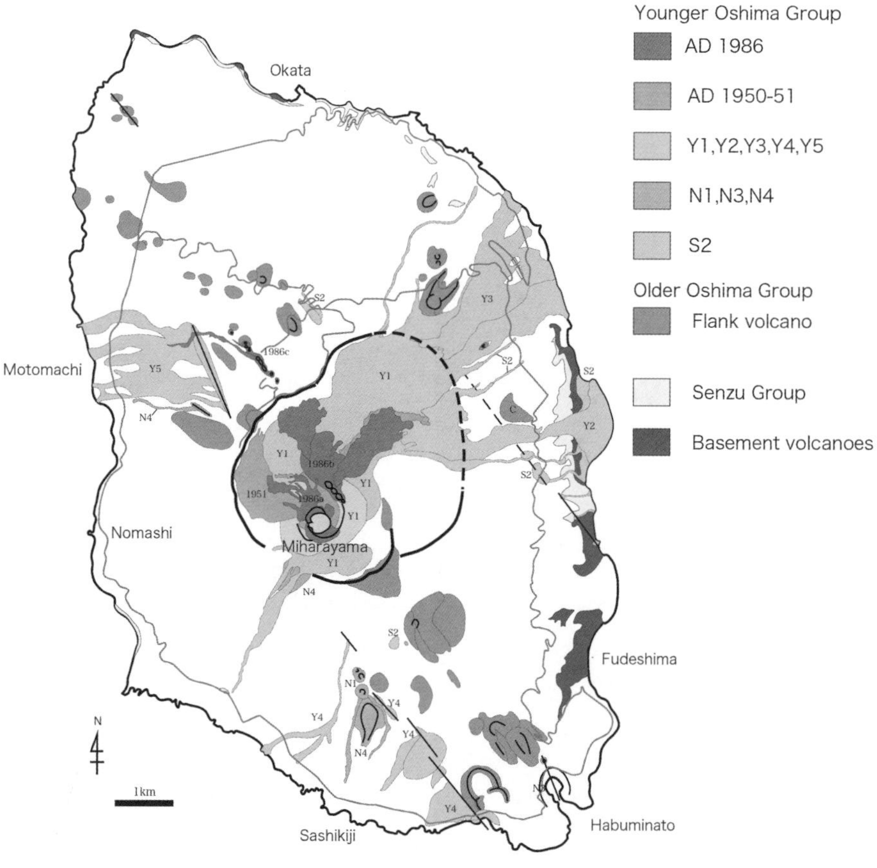

Fig. 5.27. Simplified geological map for Izu–Oshima Volcano. Modified from Kawanabe (1998).

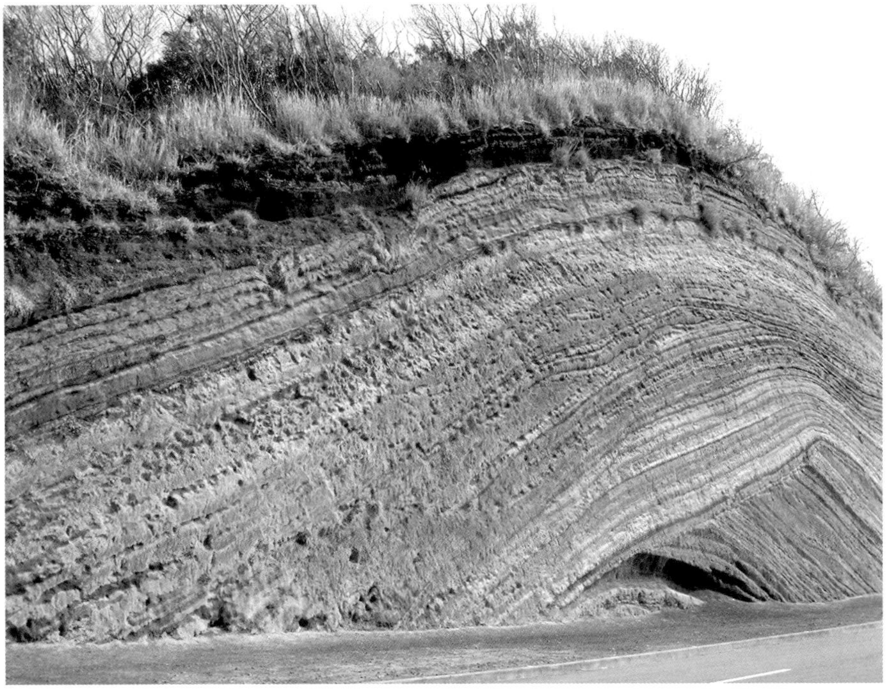

Fig. 5.28. Multi-layered pyroclastic fall deposits of Izu–Oshima volcano, *c.* 10 kyr old, along the western coast on the ring road.

pyroxene-olivine basalt (Kawanabe 1991). The older edifice of the volcano is called the Senzu Group (Nakamura 1964) and is made up of phreatomagmatic coarse ejecta and lahar deposits with a small amount of lava flows. Volcanic activity started at *c.* 40–50 ka, and represents the stage of emergence of an island. The younger edifice of the volcano consists of normal subaerial alternations of lava flows and scoria and ash falls. This activity began at *c.* 20 ka and marks the stage of continuous growth of a stratovolcano above sea level. The upper unit is subdivided into pre-caldera and syn- and post-caldera deposits (Nakamura 1964). The pre-caldera deposits are made up of about 100 layers of pyroclastic deposits (Fig. 5.28), with the intervals between the layers being 150 years on average. There are many flank volcanoes forming fissures, scoria cones and tuff cones, with their vents being elongated in a NNW–SSE direction, reflecting the regional stress field.

After a scoria eruption from the summit and several flank eruptions had occurred at *c.* 1.66 ka, a large phreatic explosion took place in the summit area and a high-speed pyroclastic density current covered almost the entire island (Yamamoto 2006; S_2 in Fig. 5.29). The present shape of the summit caldera is thought to be formed in this syn-caldera stage (Nakamura 1964). The post-caldera products resulted from ten large eruptions and several minor eruptions (Koyama & Hayakawa 1996; Kawanabe 2012). The eruptive volume of these large eruptions is up to a cubic kilometre and the average recurrent interval of the large eruptions is 100–150 years (Fig. 5.29). The most recent large eruption occurred in 1778–1778 (Y_1). Large eruptions usually began with scoria fall deposition followed by effusion of lava flows. In some events, flank eruptions took place and caused violent phreatomagmatic explosions near the coast. The emission of phreatic ash emission activities from the central cone, Mt Mihara, followed for several years after the early magmatic activity. After the Y_1 eruption, many medium to small eruptions occurred. The 1876–1877, 1912–1914, 1950–1951 and 1986–1987 eruptions were relatively large and erupted several tens of millions of cubic metres of magma.

The 1986 event (Fig. 5.30) began with a Strombolian eruption and lava effusion in Mt Mihara on 15 November. On 21 November, after a short repose of activity, a fissure eruption occurred in the northwestern caldera floor, extending beyond the caldera and causing a lava flow which moved towards the largest town of Motomachi (resulting in the evacuation of residents and visitors from the island). The eruption itself ceased in the morning of 22 November, but the evacuations continued for about one month. On 16 November 1987, accompanied by loud explosions, the lava filling the old pit crater in Mt Mihara exploded and collapsed. Several collapses were accompanied by small eruptions afterwards, recreating the pit crater in Mt Mihara. Since a small eruption on 4 October 1990 there has been only fumarolic activity; earthquake and volcanic tremors are sometimes observed and a slow inflation of the volcano continues.

The magma of this volcano consists of two types (Nakano & Yamamoto 1991). One is 'plagioclase-controlled' and the other is 'differentiated' magma (multimineral-controlled); the bulk chemistry of the former is controlled by plagioclase addition or removal, while that of the latter is controlled by fractionation of plagioclase, orthopyroxene, clinopyroxene and titanomagnetite (Fig. 5.31). Summit eruptions of this volcano tap only plagioclase-controlled magmas, while flank eruptions are supplied by both magma types. It is considered unlikely that both magma types would coexist in the same magma chamber based on the petrology. In the case of the 1986 eruption, the flank magma was isolated from the summit magma chamber or central conduit and formed small magma pockets, where further differentiation occurred due to relatively rapid cooling. In a period of quiescence prior to the 1986 eruption, new magma was supplied to the summit magma chamber and the summit eruption began. Dyke intrusion or fracturing around the small magma pockets triggered the flank eruption of the differentiated magma.

Miyakejima Volcano

Miyakejima Volcano, located about 180 km south of Tokyo, is one of Japan's most active volcanoes with repeated eruptions of basaltic to basaltic andesite magmas having occurred every

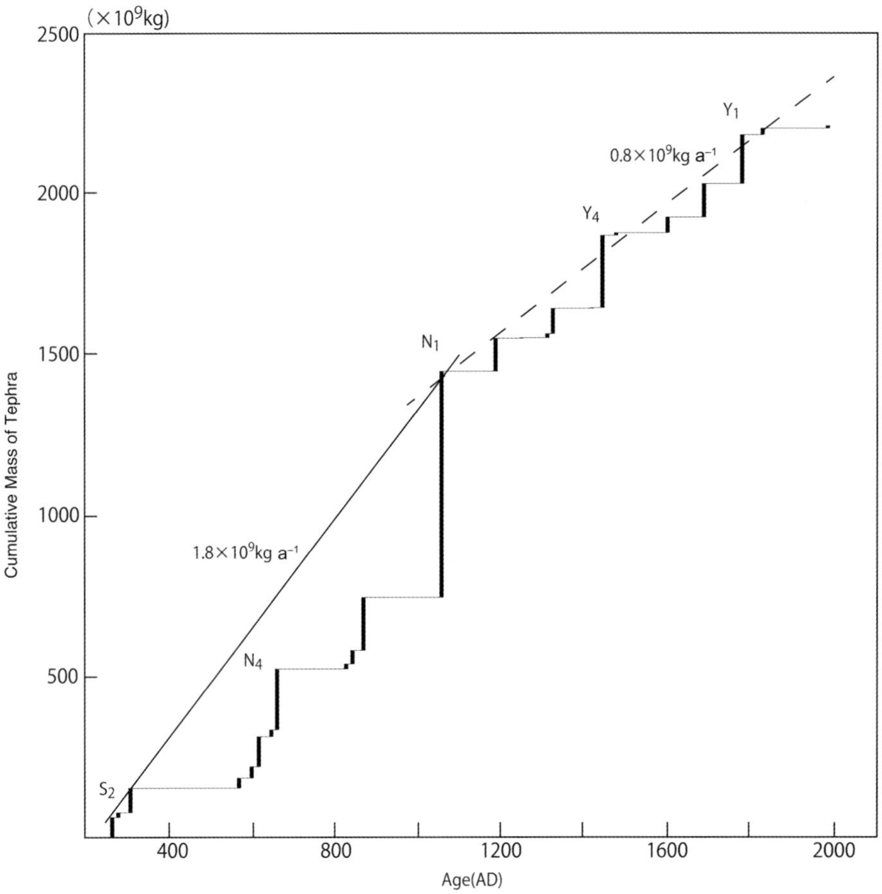

Fig. 5.29. Cumulative mass of tephra with time. After Kawanabe (2012).

20 years over the last three centuries (Tsukui *et al.* 2001). Miyakejima is nested by two calderas, the outer 4 km wide Kuwanokitaira Caldera (*c.* 10 ka) and the inner 1.6 km wide Hatchodaira Caldera. The inner caldera was formed at *c.* 2.5–2.2 ka and was filled completely by eruption products until the nineth century; a central cone, Mt Oyama, was formed by the recent eruptions. During the 2000 eruption the new Oyama Caldera formed at the same locality as the inner caldera.

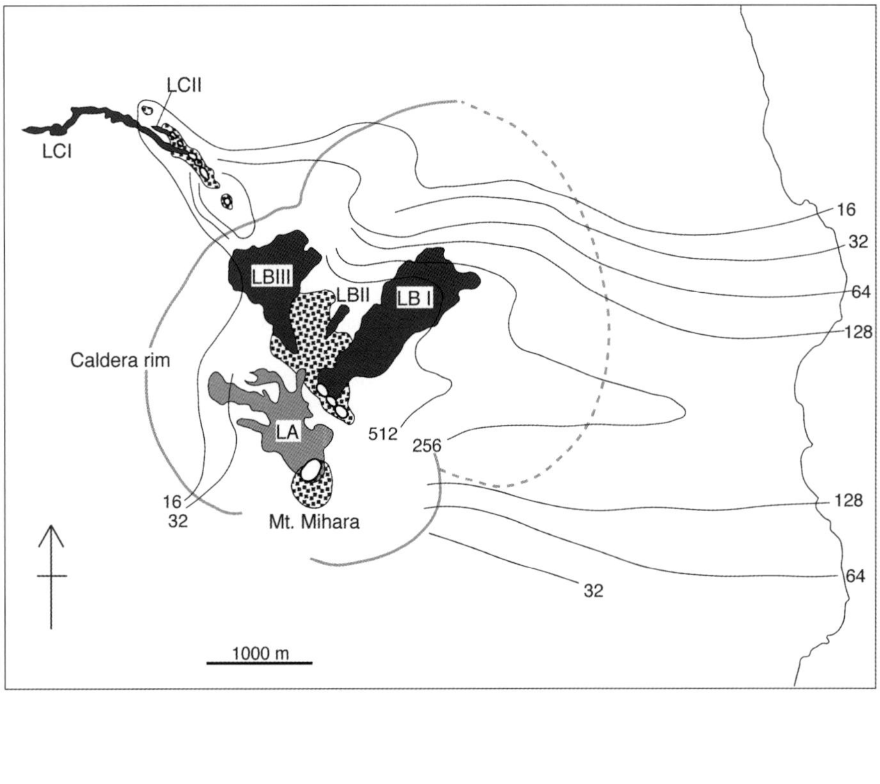

Fig. 5.30. Distribution of the ejecta of the 1986 eruption. Numerals of each isopach are the thickness of the scoria fall deposits in millimetres. LA, LBI, LBII, LBIII, LCI and LCII are names of lava flows. Shaded areas are scoria cones. Mt Mihara is the central cone within the summit caldera. Modified from Soya *et al.* (1987).

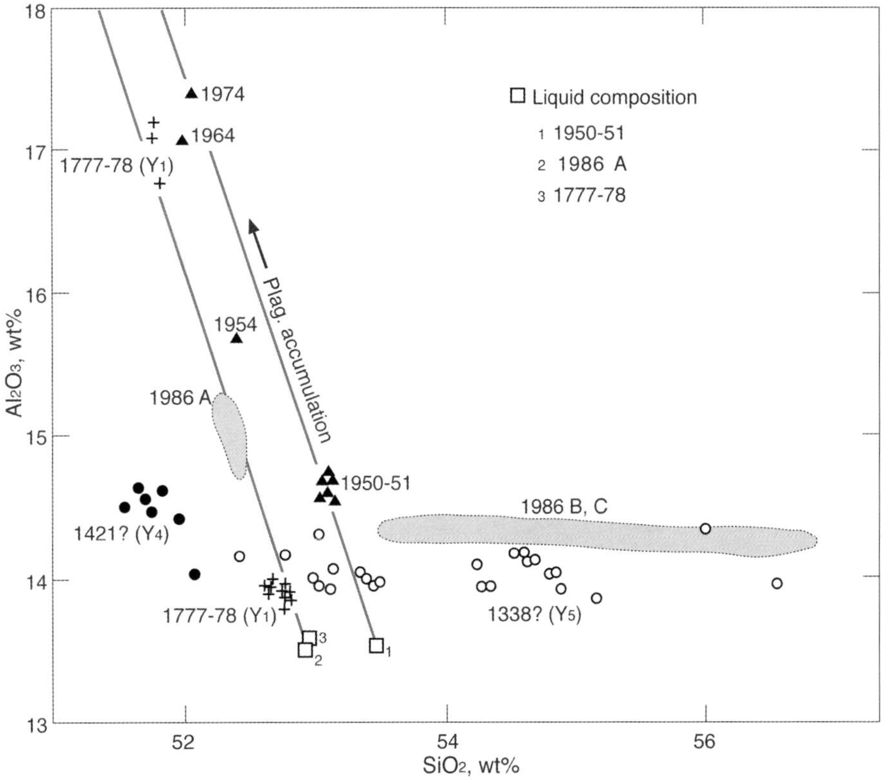

Fig. 5.31. Al_2O_3 v. SiO_2 diagram of products of the 1338, 1421, 1777–1778, 1950–1974 and 1986 eruptions; 1986A designates the 1986 summit eruption and 1986B, C the 1986 flank eruption. Three open squares indicate the calculated liquid compositions. Two lines denote respective plagioclase-control lines connected with plagioclase. After Nakano & Yamamoto (1991).

Historically documented eruptions have taken place in 1085, 1154, 1469, 1535, 1595, 1643, 1712, 1763, 1811, 1835, 1874, 1940, 1962 and 1983, with the latest being in 2000 (e.g. Miyazaki 1984). Precursory seismic events were usually short, lasting a few hours prior to the eruptions. Once commenced, most eruptions ended within several hours to a couple of weeks. Typical volumes of eruptive products of these eruptions range from 0.001 km^3 to 0.1 km^3 in dense-rock equivalent (DRE; Tsukui *et al.* 2001).

The 1983 and 2000 eruptions

Small volcanic earthquakes were felt for only 90 min prior to the 1983 eruption, which began on the afternoon of 3 October (Aramaki & Hayakawa 1984). During the first 2 h, fire curtains of 100–500 m height were observed throughout the 4.5 km long flank fissure (Fig. 5.32). Scoria fall extensively covered the eastern part of the fissure. Successive lava flows moved to the west (Ako village) and southwest, destroying about 340 houses. Violent phreatomagmatic explosions occurred near the southern end of the fissure, and craters with tuff rings formed on the coast. The eruption ended early the next morning, following which an instrumental monitoring network including seismometers, tiltmeters, GPS and electromagnetic instrument was established. Traditional level surveys indicated renewed inflation of the volcano started after the 1983 eruption, with the pressure source under the southwestern part of the island, and continued to eventually approach pre-1983 eruption levels by the beginning of 2001.

Although fissure eruptions of basaltic magma have characterized the recent eruptions of this island (Nakamura 1984) (Fig. 5.32), the 2000 eruption was different because it involved the collapse of the volcano summit accompanied by phreatic to phreatomagmatic explosions (Nakada *et al.* 2005*a*). An earthquake swarm took place on the evening of 26 June 2000, clustered under the southern part of the summit. The hypocentres moved to the northwest during the night and submarine eruptions began early the next morning (Uhira *et al.* 2005). Felt earthquakes occurred under the sea up to 40 km NW of the island, while small earthquakes still continued within the island. The former hypocentres showed the form of a large dyke running in a NW–SE direction. Small phreatic explosions occurred at the summit on the evening of 8 July, associated with 500 m deep and 800 m wide subsidence of the summit area (Fig. 5.33). The size of the depression increased gradually by middle August and reached the size of the existing caldera, 1.8 km across (Geshi *et al.* 2002; Nakada *et al.* 2005*a*) (Fig. 5.33). The subsidence was accompanied by intermittent phreatic to phreatomagmatic explosions and repeated very-long-period (VLP) earthquake events. Small earthquakes increased in number and scale before each VLP event. The latter occurred every time subsidence (deflation) of the summit area was interrupted, according to tiltmeter survey (Ukawa *et al.* 2000). This was interpreted as due to the caldera block sinking as a piston within a cylinder into the magma chamber. Magma was pushed by the sudden release of friction between the piston and cylinder, generating the VLP event (Kumagai *et al.* 2001).

On the evening of 18 August, an intense Vulcanian to sub-Plinian eruption took place ejecting abundant juvenile projectiles, and the whole island was covered by volcanic ash (Nakada *et al.* 2005*a*). The eruption column of this eruption reached *c.* 16 km in altitude. On the morning of 29 August, phreatic explosions continued for hours and an ash cloud spilled out from the summit depression, generating low-temperature ash-cloud flows (dust flows) which slowly moved down to the northern sea through the Kamitsuki district. The weather station recorded a temperature increase of about 4 degrees when it was engulfed by the ash cloud. The major eruptive events ended with this explosion.

By the end of August, SO_2 emissions from the newly formed caldera (Oyama Caldera) became strong. The daily SO_2 emission was 40 000–50 000 tons/day during the period from October 2000 to January 2001 (Kazahaya *et al.* 2004). This rate

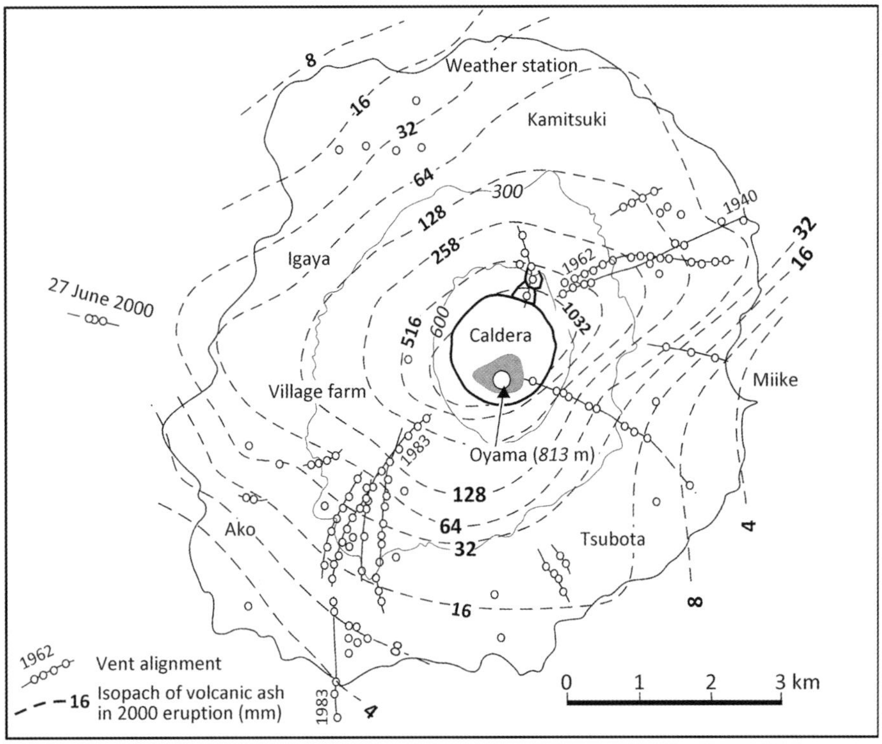

Fig. 5.32. Map showing the alignment of vents of recent eruptions with caldera and tephra thickness of the 2000 eruption at Miyakejima Volcano. Vent alignments show radial fissures developed from the centre of the volcano. The previous summit area including Mt Oyama subsided due to lateral migration of magma from its chamber to the NW in the 2000 eruption, resulting in the formation of new caldera, Oyama Caldera. A pyroclastic cone (shaded) was formed in the southern end of the caldera floor, from which a huge amount of SO_2 was emitted.

peaked at the beginning of September and then decreased with time to 2000 tons/day at the end of 2006 and just below 1000 tons/day in 2012.

The total volume of tephra of the 2000 eruption was estimated as 9.3×10^6 m^3 in DRE, which is much smaller than the volume of the Oyama Caldera (6×10^8 m^3; Nakada *et al.* 2005*a*). The GPS-measured distance between Kozushima and Niijima islands, which are located on either side of a large dyke about 30 km NW of Miyakejima, had increased since the end of June. The elongation is concordant with the deflation of Miyakejima Island, which continued during the summit subsidence. This was interpreted as due to magma withdrawal from the magma chamber beneath Miyakejima Volcano and intrusion to the NW of the volcano along with additional magma from the lower crust or mantle (Yamaoka *et al.* 2005). This elongation almost ceased on 18 August when a large explosion occurred at the summit, while the shrinkage of Miyakejima Island continued. This was probably caused by the continuous emission of a large amount of gas from the Miyakejima magma chamber, as indicated by the exceptional rate of gas emission that started in the end of August (Kazahaya *et al.* 2004).

Fig. 5.33. Photographs showing the progress of the summit subsidence (Oyama Caldera) during the 2000 eruption. Left: northerly view of the summit area of Miyakejima soon after the first explosions, taken on 9 July 2000. Right: northeasterly view of the summit area after a series of major eruption, taken on 5 September 2000. Smoking vent is located at the base of the southern caldera wall.

The magma that first erupted under the sea in 2000 was basaltic andesite (54 wt% SiO_2), while the magma erupted as volcanic bombs from the summit after the caldera collapse was basalt (52 wt% SiO_2) (Amma-Miyasaka *et al.* 2005; Kaneko *et al.* 2005; Saito *et al.* 2005). These latter authors proposed a magma plumbing system wherein a basaltic andesite magma chamber is located above a basalt magma chamber and the former was emptied by the lateral migration of the magma to the NW.

Fuji Volcano

Fuji Volcano (3776 m a.s.l.) is the largest stratovolcano in Japan (Fig. 5.34), with a volume of 400–500 km^3. This volcano has grown on the boundary between the Eurasian and Philippine Sea plates. Its magmatic activity is caused by the subduction of the Pacific Plate beneath the Eurasian Plate. Although surface volcanic activity of Fuji volcano has been quiet since its last eruption in 1707, deep low-frequency earthquakes increased greatly in the period from October 2000 to May 2001 (Ukawa 2005). The hypocentres of these deep earthquakes were located some 2–4 km NE from the summit. Growth of this volcano started at *c.* 100 ka, with only basaltic ejecta and lava from the beginning. The eruptive history of Fuji Volcano is outlined in the following sections (Miyaji 1988; Yamamoto *et al.* 2005*a*).

100–17 ka (15 cal ka BC)

The activity of this stage, known as the Older Fuji volcano (Tsuya 1968) or Hoshiyama stage (Yamamoto *et al.* 2007), was characterized by sub-Plinian explosive eruptions of basaltic magma. More than 140 scoria fall deposits are found on the eastern flank of the volcano. Lahar deposits are common, creating thick volcanic fans on the distal flanks of the volcano. The main part of Older Fuji volcano collapsed towards the SW at *c.* 18 000 BC, emplacing the Tanukiko debris avalanche deposit (Yamamoto *et al.* 2007).

15–6 cal ka BC

The volcanic edifice started to build again constructing the Younger Fuji Volcano (Tsuya 1968). Abundant voluminous basalt lava flows erupted from the summit and flanks during this period (Fujinomiya stage; Yamamoto *et al.* 2007). Erupted volumes of several lava flows are more than 1 km^3 each, and the longest lava flow reached more than 40 km from the summit. This period was apparently the highest in terms of eruption rate (Fig. 5.35; Miyaji 1988, 2007).

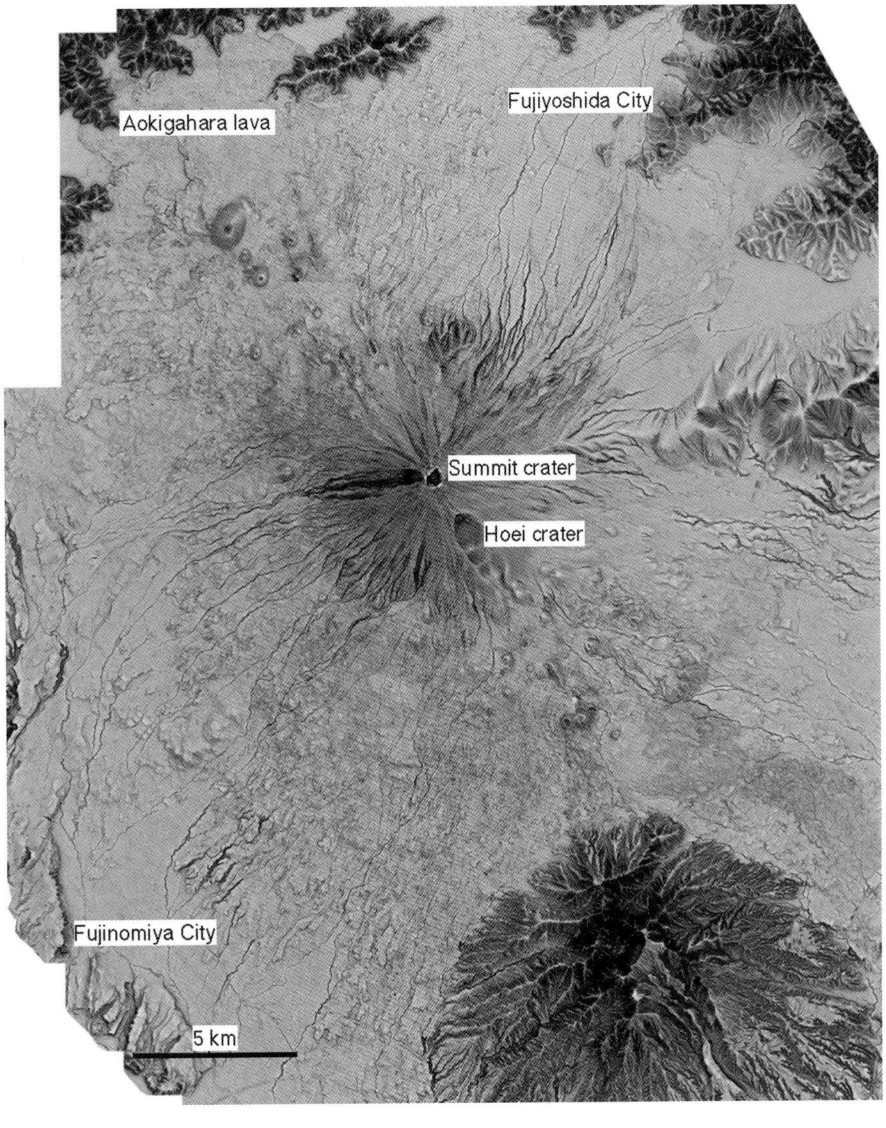

Fig. 5.34. Relief map of Fuji Volcano (Fuji Sabo Office, Ministry of Land, Infrastructure, Transport and Tourism). The Aokigahara lava flow extruded from the northern foot of the volcano in AD 864–866. The Hoei crater was formed by the 1707 eruption.

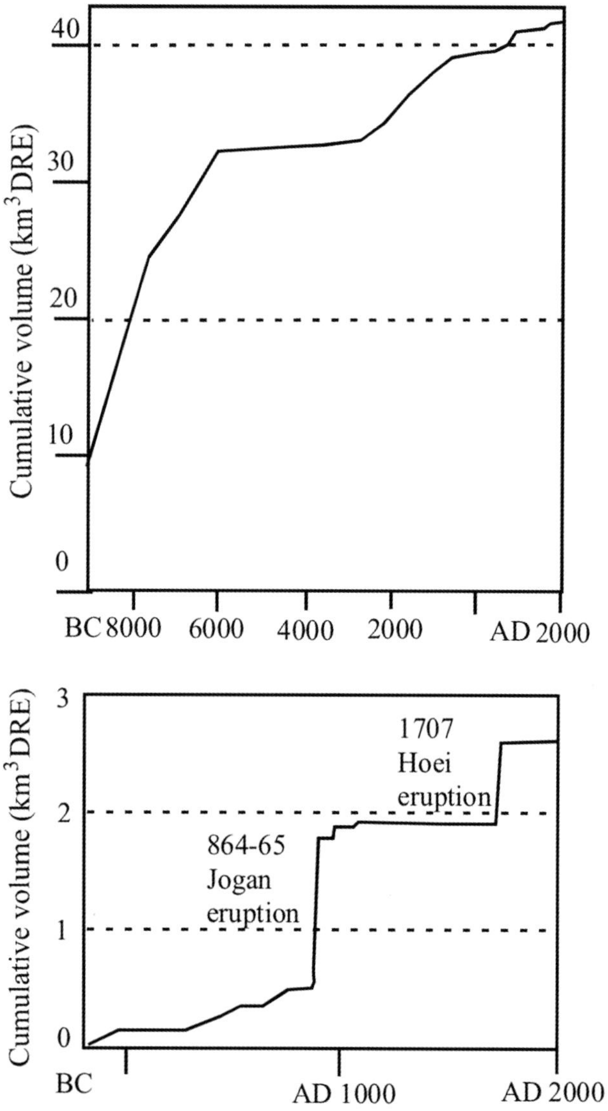

Fig. 5.35. Cumulative volume for total products of Fuji volcano during the last 11 kyr (upper) and 2200 years (lower). After Miyaji (2007).

6–3.6 cal ka BC
The volcanic activity became low during this period, and Fuji Black Soil (Machida 1964) is accumulated on the foot of the volcano.

36–17 cal a BC
Eruptive activity increased again to build the present volcanic cone, which is made up of many thin basaltic lava flows. The summit region was formed by 2100 cal a BC. Flank lava extrusions also occurred during this period.

1700–300 cal a BC
Explosive sub-Plinian eruptions from the summit and flanks were dominant during this period. These eruptive products cover the summit region as agglutinate sheets. Some fall deposits emplaced on the steep western upper flank generated pyroclastic avalanches towards the west (Yamamoto *et al.* 2005*b*). Other fall deposits emplaced on the upper flank transformed into secondary lava flows. The final sub-Plinian eruption from the summit took place in 300 cal a BC.

300 cal a BC–1707 AD
During this period, eruptions only occurred from the flanks. The volcanic activity on the eastern and northeastern flank was relatively higher than that on the other flanks before 700 AD. The frequency of flank eruptions increased during 700–1000 AD in every direction from the summit (Takada *et al.* 2007). Fissure eruption sites were restricted within 13.5 km from the summit. The Jogan eruption, one of the huge eruptions in Fuji Volcano, occurred on the northwestern foot in 864–866 AD, producing the Aokigahara lava flow (Fig. 5.34). The total erupted volume is 1.4 km^3 (Takahashi *et al.* 2007). Volcanic activity became low during 1100–1700 AD. Some historical documents reported the existence of smoking at the summit during this period.

1707 AD–present day
The last eruption, the 1707 AD Hoei eruption, was Plinian and occurred at the three craters on the southeastern flank (Fig. 5.34). This eruption is subdivided into four stages and its erupted volume was estimated at 0.7 km^3 in DRE (Miyaji & Koyama 2007; Miyaji *et al.* 2011). The eruption started on 6 December 1707 and finished on 1 January 1708. Fallout tephra spread eastwards and reached Edo, present metropolitan Tokyo (Fig. 5.36). Miyaji *et al.* (2011) have divided this eruption into three stages on the basis of the patterns of the eruptive pulses. Stage I had two energetic eruptive pulses forming a column at least 20 km high, each pulse showing an intense initial outburst followed by a decrease in intensity. The eruption sequence indicates ruptures of highly overpressured dacite and andesite magma chambers (Fig. 5.37). The initial silicic eruption was followed by basaltic magma withdrawal from a deep and voluminous magma chamber. Stage II consisted of discrete sub-Plinian pulses of relatively degassed basaltic magma. Stage III was mainly characterized by sustained column activity without any clear pauses. During this stage, the column height appears to have been more than 16 km. The Cu-rich vesicular scoria and continuous eruption in stage III indicate a stable supply of volatile-rich magma from depth. Some villages of the eastern foot were buried by voluminous pumice and scoria. The Hoei eruption was followed by many lahars, and the passage of about 100 years was required to recover from the disaster.

Fuji Volcano has issued mostly basaltic magmas during its early 100 kyr. A temporal shift in incompatible elements in the magma was caused by original differences in primary magma (Togashi & Takahashi 2007). The Fuji basalts have evolved with an FeO*/MgO ratio larger than 1.6, and show a large variation in

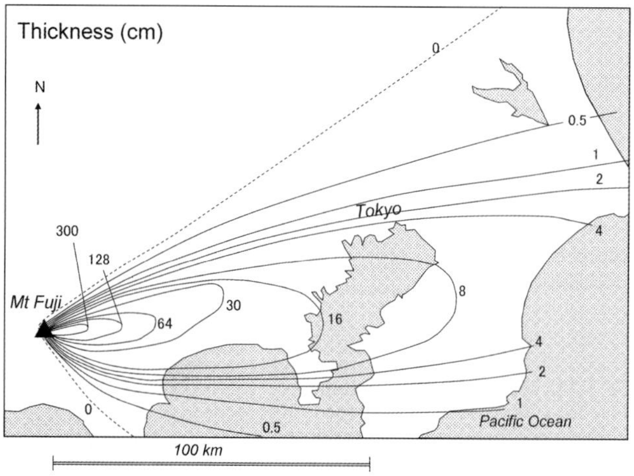

Fig. 5.36. Distribution of the Fuji 1707 tephra. After Miyaji & Koyama (2007).

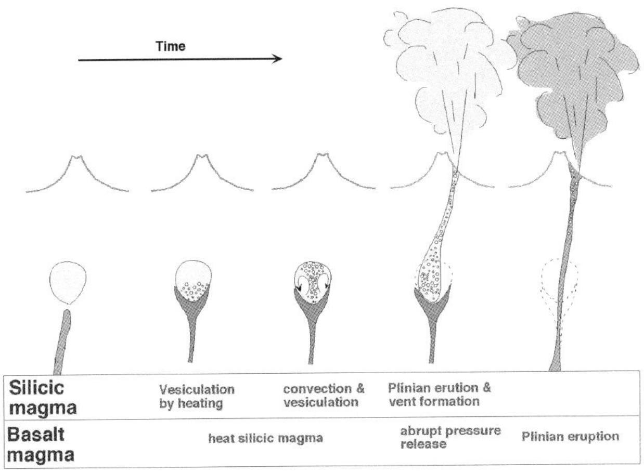

Fig. 5.37. Cartoons showing the successive development of magma reservoir and the eruption sequence of 1707 eruption. After Fujii (2007).

incompatible elements concentration independent of silica content (Fujii 2007; Togashi & Takahashi 2007). This trend is quite different from the other volcanoes of the Izu Arc. The magma reservoir beneath Fuji Volcano may be deeper than those of the other Izu volcanoes, which is consistent with geophysical observations (Fujii 2007). The magma chamber may have been formed in the granitic middle crust of the Philippine Sea Plate beneath the Eurasian Plate.

Unzen Volcano

Unzen is a composite andesite to dacite volcano built during the past 500 myr within an active volcanic graben in the Shimabara Peninsula, Kyushu (Hoshizumi *et al.* 1999). Although the initial activity of Unzen was explosive, most eruptions throughout its history are characterized by less explosive events with the formation of thick lava flows and domes and associated collapsed materials (Watanabe & Hoshizumi 1995). The summit area is occupied by a group of lava domes and its flanks and apron are composed of pyroclastic flows, lahars and debris avalanche deposits (Fig. 5.38). The centre of volcanic activity at Unzen has migrated to the east with time, whereas the depression of the volcanic graben has continued. As a result, the topographical embayment of Tachibana Bay has formed in the western part of the Shimabara Peninsula.

Historical eruptions at Unzen volcano occurred in 1663, 1792 and 1990–1995. In 1663, andesite lava (Furuyake lava flow) with a volume of *c.* 2×10^6 m^3 flowed down from a crater near the summit of Mt Fugen, Hatoana and descended on the northern slope. About 60 people living in the eastern foot of this volcano were killed by lahars generated during the explosion at the Kujukushima Crater. In 1792, dacite lava (Shin-yake lava flow) issued on the upper northern slope following phreatic explosions at the Jigokuato Crater, which in turn was preceded by felt earthquakes in the western part of the peninsula. The lava effusion continued for about 2 months and the volume of eruption reached *c.* 3×10^7 m^3. Felt earthquakes continued even after the halt of lava effusion, migrating to the east until a large earthquake on 21 May 1792 triggered a flank failure of an old lava dome, Mayuyama, in the eastern part of the peninsula. About one-third of Mayuyama collapsed. The resultant debris avalanche swept over the ancient city of Shimabara and rushed into the Ariake Sea, generating tsunamis which inundated the opposite shore in the Higo district (now Kumamoto) and returned to inundate the whole coast of the Shimabara Peninsula. In total 15 000 people were killed by this event, remembered as the Shimabara Catastrophe.

1990–1995 eruption

About one year before the start of phreatic explosions at the Jigokuato and Kujukushima craters on 17 November 1990, swarms of volcanic earthquakes began under Tachibana Bay and migrated towards the summit. Extrusion of lava occurred at Jigokuato Crater on 20 May 1991 (Fig. 5.39) (Nakada & Fujii 1993; Nakada *et al.* 1999; Umakoshi *et al.* 2001). Lava effusion had been expected because juvenile clasts were found in ashes from the phreatic explosions during February–May 1991 and the rate of shallow earthquakes had increased, along with inflation and demagnetization of the summit area. Indeed, a lava dome started growing at the Jigokuato Crater; the lava overflowed the crater and flowed down the eastern slope of the summit. The lava front suffered partial or

Fig. 5.38. Easterly view of Unzen Volcano. Shimabara city is in foreground, adjacent to the Ariake Sea. The mountain rising behind Shimabara city is a lava dome, Mayuyama, which collapsed in 1792. Small hills in the city and small islets are remnants of this flank failure.

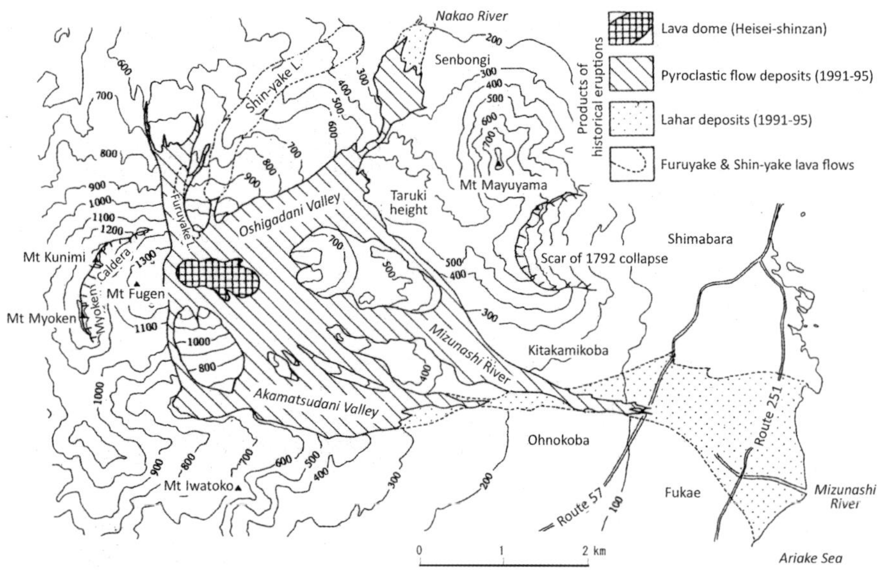

Fig. 5.39. Distribution of products of the 1991–1995 eruption at Unzen Volcano.

large collapses, with the generation of pyroclastic flows descending along the small gullies. The deposits extensively covered the eastern slope of the volcano (Fig. 5.40). On 3 June 1991, a gravitational collapse of the eastern part of the summit dome produced pyroclastic flows that engulfed and killed 43 people who were watching the eruption at an observation site along the Mizunashi River. Most of the casualties were journalists who were reporting the state of the eruption within the evacuation-recommended zone (Nakada 2000). Three volcanologists were also killed, Katia and Maurice Krafft and Harry Glicken. On 23 May 1991, there had already been one casualty when a pyroclastic flow descended to the NE (Fig. 5.40). The largest pyroclastic flows occurred on 15 September 1991, reaching a runout of *c.* 5.5 km from the source.

The lava effusion rate was highest in the early stage (about 5×10^5 m^3/day) and decreased with time, showing two pulses of magma supply. The lava dome grew exogenously during the first half of the eruption, extending and forming multiple lava lobes with different directions, while the lava dome grew endogenously during the second half of the eruption, increasing its thickness (Nakada *et al.* 1999). Over 9000 pyroclastic flows were seismologically monitored. A lava dome had grown by early 1995, becoming *c.* 1.2 km long, 0.8 km wide and 0.25–0.4 km high. The total volume of the erupted magma reached *c.* 0.21 km^3, about half of which remained at the summit as a lava dome complex.

Dome lavas during the 1991–1995 eruption were porphyritic dacite (66–64 wt% SiO_2), more silicic when the effusion rate was high (Nakada & Motomura 1999). The rocks showed a mineralogical disequilibrium indicating mixing of magmas with different compositions. Nakamura (1995) suggested mixing between a felsic crystal mush and an aphyric mafic magma at a few months prior to the eruption, based on the diffusion time calculated in zoned profiles of magnetite phenocrysts. Nakada & Motomura (1999) proposed a

Fig. 5.40. Pyroclastic flow inundating the Senbongi district of Shimabara city at Unzen Volcano, taken on the early morning of 24 May 1993. The volcano summit is covered by the new lava dome (Heisei Shinzan) and the eastern and northeastern slopes are covered by pyroclastic flow deposits.

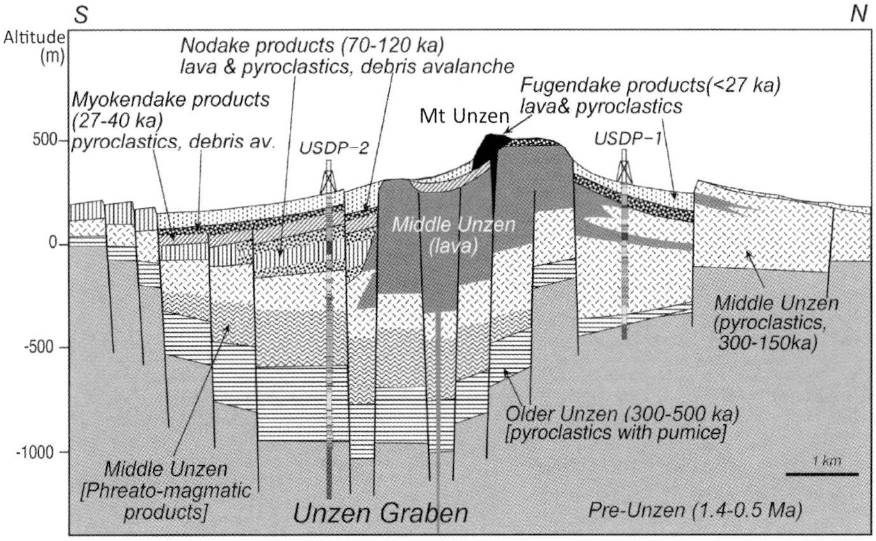

Fig. 5.41. Cross-section of Unzen Volcano based on the results of Unzen Scientific Drilling Project (USDP). After Takarada *et al.* (2007).

different model in which aphyric rhyodacite magma was mixed with xenocrysts crystallizing in the marginal zone of the magma chamber.

Scientific drilling project

The scientific drilling project carried out at Unzen Volcano between 1999 and 2004 was aimed at obtaining a better understanding of the geological history and eruption mechanism of this lava-dome-dominant volcano (Nakada & Wilson 2008). Three wells were drilled on the volcano flank to investigate the geological structure and history of the volcano, and one slant directionally drilled borehole penetrated the conduit of the last eruption from the northern middle slope of Mt Unzen. Drilled cores revealed that the tilted, half-graben structure where Unzen is sited (Matsumoto *et al.* 2012) is filled with the volcanic edifice, which can be classified into old, middle and young stages (Takarada *et al.* 2007) (Fig. 5.41). The conduit of the last eruption was penetrated *c.* 1200 m below and 500 m west of the summit crater. Although the rock collected at the conduit was cooled (<200°C) and highly hydrothermally altered, its correlation with the summit lava dome was demonstrated based on unique mineral, chemical and isotopic compositions (Nakada *et al.* 2005*b*). The conduit of the last eruption was located in the southern end of the conduit zone, which is an east–west-trending composite dyke enclosed in volcanic breccia and is as much as 350 m thick (Goto *et al.* 2008). Porosity of the conduit zone at the depth of penetration was, as a whole, too low for the rising magma to degas efficiently into the conduit wall. The effusive nature of the eruption in this volcano, despite substantial volatile content of the magma, is concluded to have occurred by efficient degassing at shallower levels. Despite its low porosity, the conduit zone cooled quickly (9 years from the end of eruption to sampling by drilling) due to hydrothermal fluid circulation within the edifice.

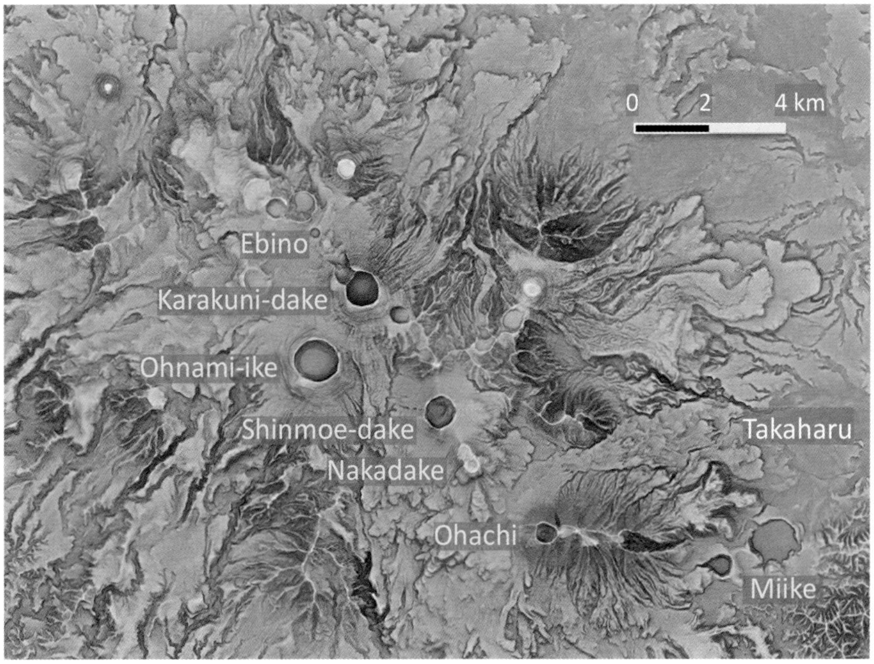

Fig. 5.42. Relief map of the Kishirima volcano group (Asia Air Survey Co. Ltd.; JMA 2005).

Fig. 5.43. The first sub-Plinian explosion during the 2011 eruption at Shinmoedake (Kirishima). Taken from the north by Kazuo Shimousuki on the evening of 26 January 2011.

Shinmoedake (Kirishima) Volcano

Kirishima is a volcano group consisting of more than 25 volcanic cones (Fig. 5.42). Shinmoedake is one of the presently active cones in this Kirishima group, and its latest eruption occurred in 2011. The volcanic activity of the Kirishima group started at *c.* 600 ka with caldera-forming eruptions, and changed at *c.* 330 ka to growth of stratovolcanoes with Plinian, Vulcanian, Strombolian and phreatomgamatic eruptions (e.g. Nagaoka & Okuno 2011). Migration of eruption centres during these 330 thousand years has produced many stratocones and pyroclastic cones. The most recent eruptive activity has occurred at Ohachi and Shinmoedake craters. The activity at Shinmoedake between the 1716–1717 and 2000 eruptions has been characterized by repeated phreatic explosions (e.g. Imura & Kobayashi 1991).

According to Imura & Kobayashi (1991), the main phase of the 1716–1717 eruption continued for about three months with multiple sub-Plinian and Vulcanian events. This series was preceded by a phreatic event 8 months before, and followed by a large sub-Plinian event 7 months after. During sub-Plinian events, small-scale pyroclastic flows were issued and travelled *c.* 2 km from the crater. The total volume of tephra is estimated to be *c.* 0.07 km^3 in DRE (Imura & Kobayashi 1991).

The 2011 eruption

The main phase of the 2011 eruption was characterized by three successive sub-Plinian explosions (Fig. 5.43) which were preceded by phreatic to phreatomagmatic eruptions beginning in 2008 (Ozawa & Kozono 2013; Nakada *et al.* 2013). The level of seismicity

Fig. 5.44. Lava dome formed within the summit crater of Shinmoedake (Kirishima) on 31 January 2011. Taken from the east by Tetsuo Kobayashi.

Fig. 5.45. Vulcanian explosion from the Showa crater of Sakurajima Volcano on 3 November 2009. Photo by Nobuo Geshi.

around the volcano became elevated one year before the onset of the eruptions. Inflation of the volcanic body was clearly recognized from the end of 2009 when the GPS network surrounding Shinmoedake volcano showed extension (Nakao *et al.* 2013). The calculated source of the inflation was located *c.* 6–7 km NW of Shinmoedake at *c.* 9 km depth. Inflation stopped at the beginning of the first sub-Plinian explosion on 26 January 2011 and deflation of the similar source was observed during 26–31 January (Nakao *et al.* 2013). The inflation resumed in early February and continued to the end of 2011.

Volcanic ash first contained juvenile materials (up to 8%) in the phreatomagmatic eruption on 19 January 2011 (Suzuki *et al.* 2013*a*). After this eruption, volcanic tremors continued up to the climactic eruption when eruption columns of three sub-Plinian explosions

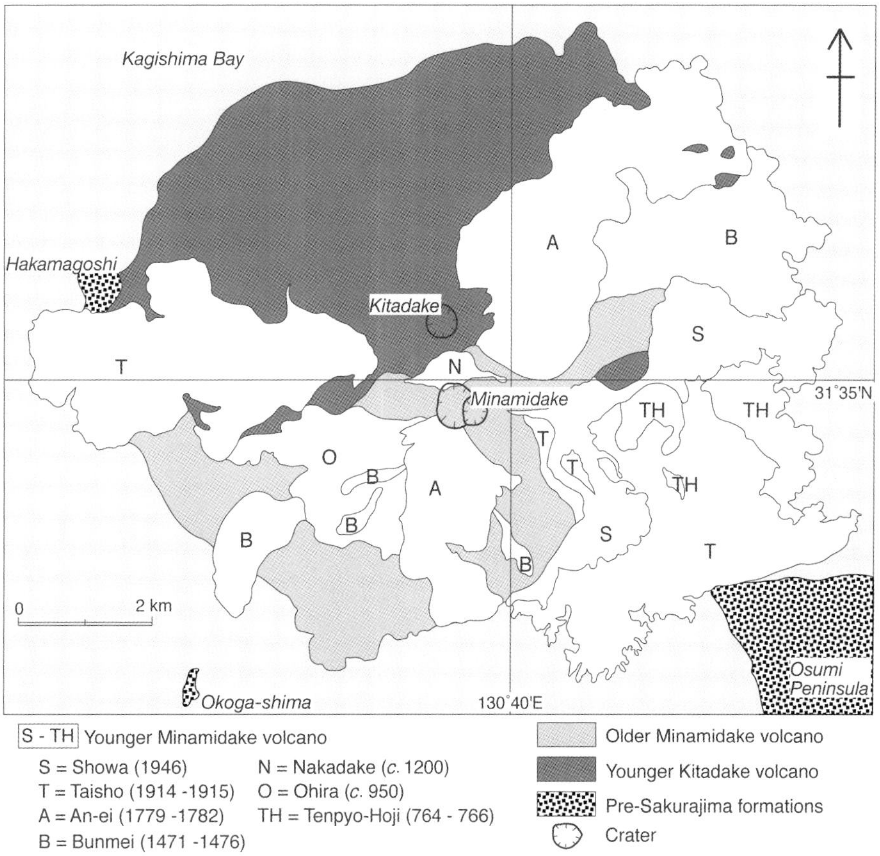

Fig. 5.46. Simplified geological map for Sakurajima Volcano. Modified from Kobayashi *et al.* (2013).

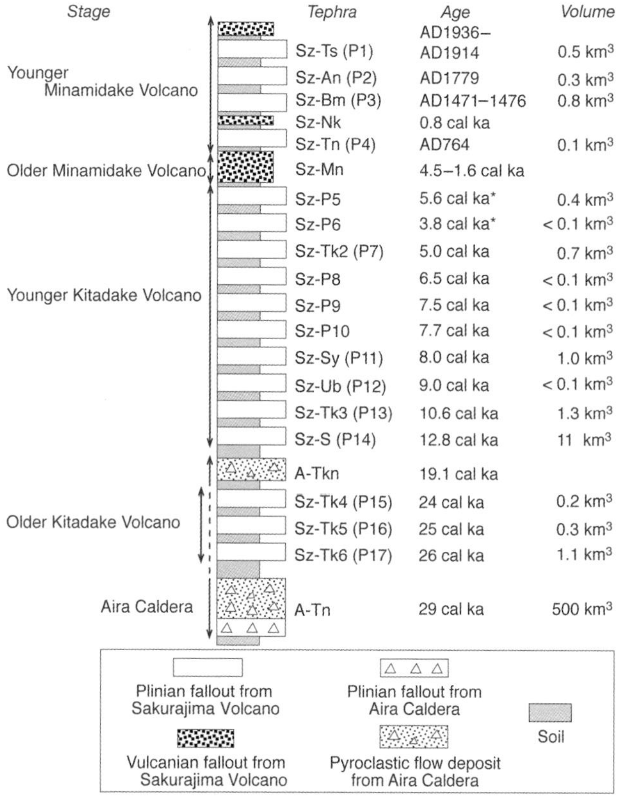

Fig. 5.47. Stratigraphy, ages and volumes of the tephra from Sakurajima volcano. Eruption ages are taken from Okuno *et al.* (1998) and Okuno (2002) (*, inconsistent with stratigraphy). Volume is apparent volume for tephra. Modified from Kobayashi *et al.* (2013).

during 26 and 27 January 2011 rose *c.* 7 km above the crater. The third sub-Plinian column was preceded by a Vulcanian explosion. During the intermittent Vulcanian explosions, lava issued from the summit crater and grew as a lava dome until 31 January. The next day, with the lava dome completely covering the crater floor and sealing the vent (Fig. 5.44), extensive Vulcanian explosions began. Vulcanian explosions repeated until April 2011, being initially more frequent but with explosion intervals increasing with time. The eruptive activity ended with phreatic to phreatomagmatic events during the summer of 2011. Ground deformation continued to the end of that year as noted above.

The total eruption volume of the products has been estimated at 2.1–2.7 × 10^7 m^3 in DRE (0.7–1.2 × 10^7 m^3 for tephra and 1.5 × 10^7 m^3 for the lava dome; Maeno *et al.* 2014; Ozawa & Kozono 2013). The pumice erupted during the sub-Plinian explosions was grey to white in colour, some with a mingling texture, and with bulk composition of 57–63 wt% SiO_2. The dome lava is 58 wt% SiO_2. Suzuki *et al.* (2013*b*) proposed that a magma-mixing event between silicic andesite and basalt magmas took place before and during the eruption.

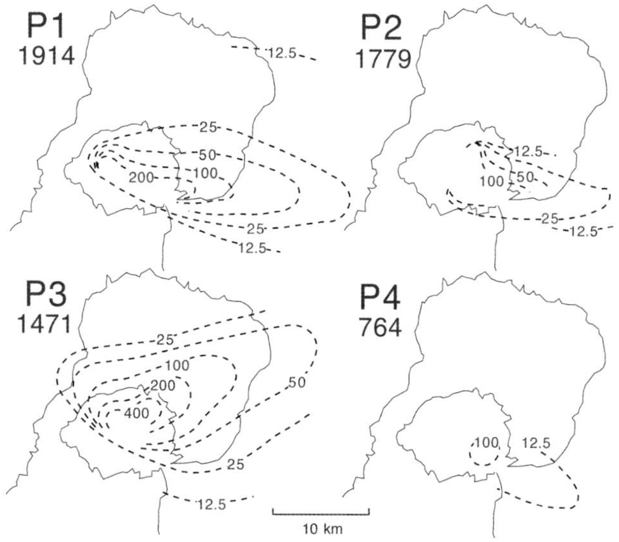

Fig. 5.48. Distributions of Plinian pumice fallouts (P1–P4) for historical large-scale eruptions. Taken from Kobayashi & Tameike (2002). Numerals are thickness in centimetres.

Sakurajima Volcano

Sakurajima Volcano (Fig. 5.45), located in southern Kyushu, is one of the most active volcanoes in the world. This is a post-caldera volcano of the Aira Caldera (see above), formed by the 29 ka catastrophic pyroclastic AT eruption. Sakurajima Volcano, which started its growth at least 26 ka ago on the southern rim of the caldera (Fig. 5.45), is composed of two adjacent stratovolcanoes: the Kitadake and the overlying Minamidake volcanoes (Fukuyama 1978; Kobayashi *et al.* 2013; Fig. 5.46). Kitadake volcano ejected 13 Plinian tephra units (P17 to P5; Fig. 5.47) and thick dacitic lava flows. The largest eruption (P14) of this volcano occurred at 12.8 ka, marking the beginning of the Younger Kitadake stage after a 10 kyr quiescence period. Minamidake Volcano was born at 4.5 ka, and constructed an Older Minamidake edifice and ejected Sakurajima–Minamidake Volcanic Sand (Sz–Mn; Fig. 5.47) during violent Vulcanian eruptions from the summit crater.

The activity of the Younger Minamidake volcano includes four large historic eruption events, namely the Tenpyo–Hoji (P4; 764–766 AD), Bunmei (P3; 1471–1476 AD), An'ei (P2; 1779–1782 AD) and Taisho (P1; 1914–1915 AD) eruptions. At each large event voluminous pumiceous tephra was ejected in the initial-stage Plinian eruption (Fig. 5.48) followed by lava effusion from lateral vents. The volumes of ejected magma in each large event are estimated at 0.3–2.0 km^3. The An'ei eruption started as a subaerial Plinian eruption with lava extrusion in 1779, followed by submarine eruptions issued NE off the island from 1779 to 1782 (Kobayashi 2009). The Taisho eruption, which was the largest

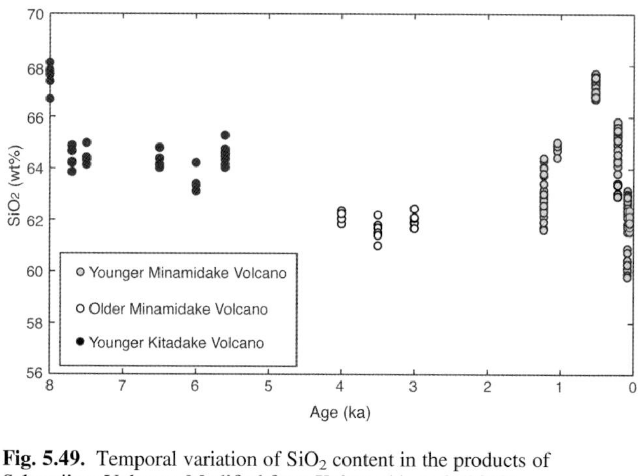

Fig. 5.49. Temporal variation of SiO_2 content in the products of Sakurajima Volcano. Modified from Kobayashi *et al.* (2013).

eruption of Japan during the 20th century, started on 12 January 1914 and consisted of three stages (Yasui *et al.* 2007). After an initial vigorous Plinian eruption that lasted about 36 hours (Stage 1), extrusion of lava associated with intermittent ash-emitting eruptions lasted for 20 days on the eastern and western flanks (Stage 2), followed by the extrusion of lava for more than 1.5 years on the eastern vents (Stage 3). The resulting Taisho Lava buried a strait between the island of Sakurajima and the Osumi Peninsula, making the island into a peninsula. During the 1946 eruption the Showa Lava, 0.18 km^3 in volume, flowed out without ejection of pumice. Since 1955, Vulcanian explosions from the summit crater of Minamidake have continued, although with some pauses. The total number of explosions from 1955 to the end of 2002 exceeds 7600 with 474 explosions occurring during 1985 alone, the most active year of Minamidake. In 2006, Vulcanian explosions resumed at Showa crater (Fig. 5.45).

All products of these two volcanoes are composed of olivine-bearing and/or olivine-free pyroxene andesite and pyroxene dacite, and show a relatively narrow compositional range of 62–67 wt% SiO_2. However, there have been temporal variations in the chemical composition of Sakurajima rocks: SiO_2 contents increased from 4.5 ka to the Bunmei eruption time then decreased to the present (Fig. 5.49). Furthermore, the Younger Minamidake magma has higher Ti, P, Zr and Y contents than the Older Minamidake and Younger Kitadake products, and the Zr/Th ratio has increased gradually over the past 4500 years (Uto *et al.* 2005). These authors have concluded that mantle-derived mafic magma has been supplied to the magma chamber beneath the Aira Caldera, and has mixed with crust-derived felsic magma. It is suggested that the supply of mafic magma has increased with time over the past 4500 years.

The authors are grateful for review comments by John C. Eichelberger and José Luis Macias. Yoshihisa Kawanabe is thanked for providing the geologic map of Izu-Oshima volcano. Tetsuo Kobayashi, Yasuo Miyabuchi, Kazuo Shimousuki and Toya Caldera-Usu Volcano Global Geopark kindly provided photographs.

Appendix

English to Kanji and Hiragana translations for geological and place names

Aira	姶良	あいら
Aizu	会津	あいづ
Akahoya	アカホヤ	あかほや
Akan	阿寒	あかん
Ako	阿古	あこ
An'ei	安永	あんえい
Aokigahara	青木ヶ原	あおきがはら
Ariake Sea	有明海	ありあけかい
Asama	浅間	あさま
Aso	阿蘇	あそ
Ata	阿多	あた
Atosanupuri	アトサヌプリ	あとさぬぷり
Beppu	別府	べっぷ
Bunmei	文明	ぶんめい
Chishima	千島	ちしま
Edo	江戸	えど
Fugen	普賢	ふげん
Fuji	富士	ふじ
Fujinomiya	富士宮	ふじのみや
Furume	古梅	ふるめ

(Continued)

English to Kanji and Hiragana translations for geological and place names (Continued)

Furuyake	古焼	ふるやけ
Hakkoda	八甲田	はっこうだ
Hatchodaira	八丁平	はっちょうだいら
Hatoana	鳩穴	はとあな
Heisei Shinzan	平成新山	へいせいしんざん
Higo	肥後	ひご
Hoei	宝永	ほうえい
Hokkaido	北海道	ほっかいどう
Hokkaido-Komagatake	北海道駒ヶ岳	ほっかいどうこまがたけ
Honshu	本州	ほんしゅう
Hoshiyama	星山	ほしやま
Inamura-dake	稲村岳	いなむらだけ
Ito	入戸	いと
Iwate	岩手	いわて
Iwo-dake	硫黄岳	いおうだけ
Izu-Oshima	伊豆大島	いずおおしま
Jigokuato	地獄跡	じごくあと
Jogan	貞観	じょうがん
Kagoshima	鹿児島	かごしま
Kakuto	加久藤	かくとう
Kamewarizaka	亀割坂	かめわりざか
Kamitsuki	神着	かみつき
Kamuinupuri	カムイヌプリ	かむいぬぷり
Kikai	鬼界	きかい
Kirishima	霧島	きりしま
Kitadake	北岳	きただけ
Kitami	北見	きたみ
Koabi	小アビ	こあび
Kobayashi	小林	こばやし
Kokubu-Shikine	国分敷根	こくぶしきね
Komoriko	籠港	こもりこう
Kompirayama	金比羅山	こんぴらやま
Koya	幸屋	こうや
Kozushima	神津島	こうづしま
Kujukushima	九十九島	くじゅうくしま
Kumamoto	熊本	くまもと
Kutcharo (Kussharo)	屈斜路	くっちゃろ
Kuwanokitaira	桑の木平	くわのきたいら
Kyushu	九州	きゅうしゅう
Mashu	摩周	ましゅう
Mayuyama	眉山	まゆやま
Meakan-dake	雌阿寒岳	めあかんだけ
Mihara	三原	みはら
Minamidake	南岳	みなみだけ
Miyakejima	三宅島	みやけじま
Mizunashi River	水無川	みずなしがわ
Motomachi	元町	もとまち
Nagase	長瀬	ながせ
Nankai	南海	なんかい
Nekodake	根子岳	ねこだけ
Niijima	新島	にいじま
Nishiyama	西山	にしやま
Oakan-dake	雄阿寒岳	おあかんだけ
Ogasawara	小笠原	おがさわら
Ohachi	御鉢	おはち
Oita	大分	おおいた
Okinawa	沖縄	おきなわ
Osumi	大隅	おおすみ
Oyama	雄山	おやま
Ryukyu	琉球	りゅうきゅう
Sakurajima	桜島	さくらじま
San'in	山陰	さんいん

(Continued)

English to Kanji and Hiragana translations for geological and place names (Continued)

Sanyo	山陽	さんよう
Satsuma Iwojima (Satsuma Io-jima)	薩摩硫黄島	さつまいおうじま
Senbongi	千本木	せんぼんぎ
Senzu	泉津	せんづ
Shikoku	四国	しこく
Shikotsu	支笏	しこつ
Shimabara	島原	しまばら
Shimanto	四万十	しまんと
Shin-yake	新焼	しんやけ
Shinjima	新島	しんじま
Shinmoedake	新燃岳	しんもえだけ
Shishimuta	猪牟田	ししむた
Showa	昭和	しょうわ
Showa-Shinzan	昭和新山	しょうわしんざん
Suwanosejima	諏訪之瀬島	すわのせじま
Tachibana	橘	たちばな
Taisho	大正	たいしょう
Takano	高野	たかの
Takeshima	竹島	たけしま
Tanukiko	田貫湖	たぬきこ
Tarumae (Tarumai)	樽前	たるまえ
Tenpyo-Hoji	天平宝字	てんぴょうほうじ
Tohoku	東北	とうほく
Tokachi	十勝	とかち
Tokyo	東京	とうきょう
Towada	十和田	とわだ
Toya	洞爺	とうや
Toyako-onsen	洞爺湖温泉	とうやこおんせん
Tsumaya	妻屋	つまや
Unzen	雲仙	うんぜん
Usu	有珠	うす
Wakamiko	若尊	わかみこ

References

Amma-Miyasaka, M., Nakagawa, M. & Nakada, S. 2005. Magma plumbing system of the 2000 eruption of Miyakejima Volcano, Japan. *Bulletin of Volcanology*, **67**, 254–267.

Aramaki, S. 1984. Formation of the Aira Caldera, southern Kyushu, ~22 000 years ago. *Journal of Geophysical Research*, **89**, 8485–8501.

Aramaki, S. & Hayakawa, Y. 1984. Sequence and mode of eruption of the October 3–4, 1983 eruption of Miyakejima. *Bulletin of Volcanological Society of Japan*, **29**, S24–S35 [in Japanese with English abstract].

Blake, S. 1990. Viscoplastic models of lava domes. *In*: Fink, J. H. (ed.) *Lava Flows and Domes. Emplacement Mechanism and Hazard Implications*, IAVCEI Proceedings in Volcanology, Springer-Verlag, Berlin, **2**, 88–126.

Cousens, B. L., Allan, J. F. & Gorton, M. P. 1994. Subduction-modified pelagic sediments as the enriched component in back-arc basalts from the Japan Sea: Ocean Drilling Program Sites 797 and 794. *Contributions to Mineralogy and Petrology*, **117**, 421–434.

Endo, K., Ohno, M. *et al.* 2001. Phreatomagmatic explosions of the 2000 eruption of Usu Volcano. *Proceedings of Institute of Natural Science, Nihon University*, **36**, 66–73 [in Japanese with English abstract].

Fujii, T. 2007. Magmatology of Fuji Volcano. *In*: Aramaki, S., Fujii, T., Nakada, S. & Miyaji, N. (eds) *Fuji Volcano*. Yamanashi Institute of Environmental Sciences, Fuji-Yoshida, 233–244 [in Japanese with English abstract].

Fujiwara, O., Machida, H. & Shiochi, J. 2010. Tsunami deposit from the 7300 cal BP Akahoya eruption preserved in the Yokoo midden, north Kyushu, West Japan. *The Quaternary Research (Daiyonki-Kenkyu)*, **46**, 293–302 [in Japanese with English abstract].

Fukuyama, H. 1978. Geology of Sakurajima volcano, southern Kyushu. *Bulletin of the Geological Society of Japan*, **84**, 309–316 [in Japanese with English abstract].

Furukawa, Y. & Tatsumi, Y. 1999. Melting of a subducting slab and production of high-Mg andesite magmas: unusual magmatism in SW Japan at 13~15 Ma. *Geophysical Research Letters*, **26**, 2271–2274.

Geshi, N., Shimano, T., Chiba, T. & Nakada, S. 2002. Caldera collapse during the 2000 eruption of Miyakejima volcano. *Bulletin of Volcanology*, **64**, 55–68.

Goto, Y., Funayama, A., Gouchi, N. & Itaya, T. 2000. K-Ar ageas of the Akan-Shiretoko volcanic chain lying oblique to the Kurile trench: implications for tectonic control of volcanism. *Island Arc*, **9**, 203–218.

Goto, Y., Nakada, S. *et al.* 2008. Character and origin of lithofacies in the conduit of Unzen volcano, Japan. *Journal of Volcanology and Geothermal Research*, **175**, 45–59.

Hamada, M. & Fujii, T. 2008. Experimental constraints on the effects of pressure and H_2O on the fractional crystallization of high-Mg island arc basalt. *Contribution to Mineralogy and Petrology*, **155**, 767–790.

Hasegawa, T. & Nakagawa, M. 2007. Stratigraphy of Early to Middle Pleistocene pyroclastic deposits around Akan caldera, eastern Hokkaido, Japan. *Journal of Geological Society of Japan*, **113**, 53–72 [in Japanese with English abstract].

Hasegawa, T., Yamamoto, A., Kamiyama, H. & Nakagawa, M. 2009*a*. Gravity structure of Akan composite caldera, eastern Hokkaido, Japan: application of lake water corrections. *Earth Planets Space*, **61**, 933–938.

Hasegawa, T., Kishimoto, H., Nakagawa, M., Itoh, J. & Yamamoto, T. 2009*b*. Eruptive history of post-caldera volcanoes of Kutcharo caldera, eastern Hokkaido, Japan, as inferred from tephrostratigraphy in the Konsen and Shari areas for the period 35–12 ka. *Journal of Geological Society of Japan*, **115**, 369–390 [in Japanese with English abstract].

Hasegawa, T., Nakagawa, M. & Kishimoto, H. 2012. The eruption history and silicic magma systems of caldera-forming eruptions in eastern Hokkaido, Japan. *Journal of Mineralogical and Petrological Sciences*, **107**, 39–43.

Hoang, N. & Uto, K. 2003. Geochemistry of Cenozoic basalts in the Fukuoka district (northern Kyushu, Japan): implications for asthenosphere and lithospheric mantle interaction. *Chemical Geology*, **198**, 249–268.

Hoshizumi, H., Watanabe, K., Sakaguchi, K., Uto, K., Ono, K. & Nakamura, T. 1997. The Aso-4 pyroclastic flow deposit confirmed from the deep drill holes inside the Aso Caldera. *In*: *Programme and Abstracts of the Volcanological Society of Japan 1997*, Matsumoto, **2**, 5. [in Japanese].

Hoshizumi, H., Uto, K. & Watanabe, K. 1999. Geology and eruptive history of Unzen Volcano, Shimabara Peninsula, Kyushu, SW Japan. *Journal of Volcanology and Geothermal Research*, **89**, 81–94.

Hughes, G. R. & Mahood, G. A. 2008. Silicic calderas in arc settings: characteristics, distribution, and tectonic controls. *Geological Society of America Bulletin*, **123**, 1577–1595.

Imura, R. & Kobayashi, T. 1991. Eruptions of Shinmoedake Volcano, Kirishima Volcano Group, in the last 300 years. *Bulletin of Volcanological Society of Japan*, **36**, 135–148 [in Japanese with English abstract].

Isozaki, Y. 1996. Anatomy and genesis of a subduction-related orogeny: a new view of geotectonic subdivision and evolution of the Japanese Islands. *Island Arc*, **5**, 289–320.

Iwamori, H. 1992. Degree of melting and source composition of Cenozoic basalts in southwest Japan: evidence for mantle upwelling by flux melting. *Journal of Geophysical Research*, **97**, 10 983–10 995.

Iwamori, H. 2007. Transportation of H_2O beneath the Japan arcs and its implications for global water circulation. *Chemical Geology*, **239**, 182–198.

JMA (Japan Meteorological Agency) 2005. *National Catalogue of the Active Volcanoes in Japan*, 3rd ed. Japan Meteorological Agency, Tokyo [in Japanese].

Kamata, H. 1989. Volcanic and structural history of the Hohi volcanic zone, central Kyushu, Japan. *Bulletin of Volcanology*, **51**, 315–332.

Kaneko, K., Kamata, H., Koyaguchi, T., Yoshikawa, M. & Furukawa, K. 2007. Repeated large-scale eruptions from a single compositionally stratified magma chamber: an example from Aso volcano, Southwest Japan. *Journal of Volcanology and Geothermal Research*, **167**, 160–180.

Kaneko, T., Yasuda, A. *et al.* 2005. Submarine flank eruption preceding caldera subsidence during the 2000 eruption of Miyakejima Volcano, Japan. *Bulletin of Volcanology*, **67**, 243–253.

KANEOKA, I., TAKIGAMI, Y., TAKAOKA, N., YAMASHITA, S. & TAMAKI, K. 1992. ^{40}Ar-^{39}Ar analyses of volcanic rocks recovered from the Japan Sea floor: constrains on the age of formation of the Japan Sea. *Proceedings of Ocean Drilling Program, Scientific Results*, **127/128**, 819–836.

KANO, K., YAMAMOTO, T. & ONO, K. 1996. Subaqueous eruption and emplacement of the Shinjima Pumice, Shinjima (Moeshima) Island, Kagoshima Bay, SW Japan. *Journal of Volcanology and Geothermal Research*, **71**, 187–206.

KATSUI, Y., OBA, Y. ET AL. 1978. Preliminary report of the 1977 eruption of Usu Volcano. *Journal of Faculty of Sciences, Hokkaido University, Series IV*, **18**, 385–408.

KAWANABE, Y. 1991. Petrological evolution of Izu Oshima volcano. *Bulletin of Volcanological Society of Japan*, **36**, 297–310 [in Japanese with English abstract].

KAWANABE, Y. 1998. *Geological map of Izu-Oshima volcano*. Geological Survey of Japan, Tsukuba.

KAWANABE, Y. 2012. New ^{14}C ages of the Younger Oshima Group, Izu-Oshima volcano, Izu-Ogawara arc Japan. *Bulletin of Geological Survey of Japan*, **63**, 283–289 [in Japanese with English abstract].

KAZAHAYA, K., SHINOHARA, H. ET AL. 2004. Gigantic SO_2 emission from Miyakejima volcano, Japan, caused by caldera collapse. *Geology*, **32**, 425–428.

KIMURA, G. 1996. Collision orogeny at arc-arc junctions in the Japanese Islands. *Island Arc*, **5**, 262–275.

KIMURA, G. & TAMAKI, K. 1986. Collision, rotation and back arc spreading: the case of the Okhotsk and Japan Seas. *Tectonics*, **5**, 389–401.

KIMURA, J.-I., STERN, R. J. & YOSHIDA, T. 2005. Reinitiation of subduction and magmatic responses in SW Japan during Neogene time. *Geological Society of America Bulletin*, **117**, 969–986.

KISHIMOTO, H., HASEGAWA, T., NAKAGAWA, M. & WADA, K. 2009. Tephrostratigraphy and eruption style of Mashu volcano, during the last 14 000 years, eastern Hokkaido, Japan. *Bulletin of Volcanological Society of Japan*, **54**, 15–36 [in Japanese with English abstract].

KOBAYASHI, T. 2009. Origin of new islets (An-ei islets) formed during the An-ei eruption (1779–1782) of Sakurajima volcano, southern Kyushu, Japan. *Bulletin of Volcanological Society of Japan*, **54**, 1–13 [in Japanese with English abstract].

KOBAYASHI, T. & TAMEIKE, T. 2002. History of eruptions and volcanic damage from Sakurajima volcano, Southern Kyushu, Japan. *The Quaternary Research (Daiyonki-Kenkyu)*, **41**, 269–278 [in Japanese with English abstract].

KOBAYASHI, T., MIKI, D., SASAKI, H., IGUCHI, M., YAMAMOTO, T. & UTO, K. 2013. *Geological map of Sakurajima volcano*, Second edition. Geological Survey of Japan, Tsukuba.

KODAMA, K., TASHIRO, H. & TAKEUCHI, T. 1995. Quaternary counterclockwise rotation of southern Kyushu, southwest Japan. *Geology*, **23**, 823–826.

KOMAZAWA, M. 1995. Gravimetric analysis of Aso Volcano and interpretation. *Journal of the Geodetic Society of Japan*, **41**, 17–45.

KOYAMA, M. & HAYAKAWA, Y. 1996. Syn- and post-caldera eruption history of Izu Oshima volcano based on tephra and loess stratigraphy. *Journal of Geography (Chigakuzasshi)*, **105**, 133–162 [in Japanese with English abstract].

KUDO, T., SASAKI, M., UCHIYAMA, Y., NOZAWA, A., SASAKI, H., TOKIZAWA, T. & TAKARADA, S. 2007. Petrological variation of large-volume felsic magmas from Hokkaido-Towada caldera cluster: implication for the origin of high-K felsic magmas in the Northeastern Japan Arc. *Island Arc*, **16**, 133–155.

KUMAGAI, I., OHMINATO, T., NAKANO, M., OOI, M., KUBO, A., INOUE, H. & OIKAWA, J. 2001. Very-long-period seismic signals and caldera formation at Miyake island, Japan. *Science*, **293**, 687–690.

KUNO, H. 1966. Lateral variation of basalt magma type across continental margins and island arcs. *Bulletin of Volcanology*, **29**, 195–222.

KUNO, H. 1969. Origin of andesite and its bearing on the island structure. *Bulletin of Volcanolgy*, **32**, 141–176.

MACHIDA, H. 1964. Tephrochronological study of Volcano Fuji and adjacent areas. *Journal of Geography (Chigakuzasshi)*, **73**, 337–350 [in Japanese with English abstract].

MACHIDA, H. & ARAI, F. 2003. *Atlas of Tephra in and Around Japan (Revised edition)*. University of Tokyo Press, Tokyo [in Japanese].

MAENO, F. & IMAMURA, F. 2007. Numerical investigations of tsunamis generated by pyroclastic flows from the Kikai caldera, Japan. *Geophysical Research Letters*, **34**, L23303, http://doi.org/10.1029/2007GL031222

MAENO, F. & TANIGUCHI, H. 2005. Eruptive history of Satsuma Iwo-jima Island, Kikai, caldera, after a 6.5 ka caldera-forming eruption. *Bulletin of the Volcanological Society of Japan, Series 2*, **50**, 71–85 [in Japanese with English abstract].

MAENO, F. & TANIGUCHI, H. 2006. Silicic lava dome growth in the 1934–1935 Showa Iwo-jima eruption, Kikai caldera, south of Kyushu, Japan. *Bulletin of Volcanology*, **68**, 673–688.

MAENO, F. & TANIGUCHI, H. 2007. Spatiotemporal evolution of a marine caldera-forming eruption, generating a low-aspect ratio pyroclastic flow, 7.3 ka, Kikai caldera, Japan: implication from near-vent eruptive deposits. *Journal of Volcanology and Geothermal Research*, **167**, 212–238.

MAENO, F., IMAMURA, F. & TANIGUCHI, H. 2006. Numerical simulation of tsunami generated by caldera collapse during the 7.3 ka Kikai eruption, Japan. *Earth Planets and Space*, **58**, 1013–1024.

MAENO, F., NAGAI, M., NAKADA, S., BURDEN, R., ENGWELL, S., SUZUKI, Y. & KANEKO, T. 2014. Constraining tephra dispersion and deposition from three subplinian explosions in 2011 at Shinmoedake volcano, Kyushu, Japan. *Bulletin of Volcanology*, **76**, 823, http://doi.org/10.1007/s00445-014-0823-9

MAHONY, S. H., WALLACE, L. M., MIYOSHI, M., VILLAMOR, P., SPARKS, R. S. J. & HASENAKA, T. 2011. Volcano-tectonic interactions during rapid plate-boundary evolution in the Kyushu region, SW Japan. *Geological Society of America Bulletin*, **123**, 2201–2223.

MATSUMOTO, A. & NAKAGAWA, M. 2010. Formation and evolution silicic magma plumbing system: petrology of the volcanic rocks of Usu volcano, Hokkaido, Japan. *Journal of Volcanology and Geothermal Research*, **196**, 185–207.

MATSUMOTO, S., SHIMIZU, H., ONISHI, M. & UEHIRA, K. 2012. Seismic refection survey of the crustal structure beneath Unzen volcano, Kyushu, Japan. *Earth, Planets and Space*, **64**, 405–414.

MINAKAMI, T., ISHIKAWA, T. & YAGI, K. 1951. The 1944 eruption of Volcano Usu in Hokkaido, Japan. *Bulletin Volcanologique*, **11**, 45–160.

MIYAGI, I., ITOH, J. I., HOANG, N. & MORISHITA, Y. 2012. Magma systems of the Kutcharo and Mashu volcanoes (NE Hokkaido, Japan): petrogenesis of the medium-K trend and the excess volatile problem. *Journal of Volcanology and Geothermal Research*, **231–232**, 50–60.

MIYAJI, N. 1988. History of younger Fuji Volcano. *Journal of Geological Society of Japan*, **94**, 433–452 [in Japanese with English abstract].

MIYAJI, N. 2007. Eruptive history, eruption rate and scale of eruptions for the Fuji Volcano during the last 11 000 years, Fuji volcano. *In*: ARAMAKI, S., FUJII, T., NAKADA, S. & MIYAJI, N. (eds) *Fuji Volcano*. Yamanashi Institute of Environmental Sciences, Fuji-Yoshida, 79–95 [in Japanese with English abstract].

MIYAJI, N. & KOYAMA, M. 2007. Recent studies on the 1707 (Hoei) eruption of Fuji volcano. *In*: ARAMAKI, S., FUJII, T., NAKADA, S. & MIYAJI, N. (eds) *Fuji Volcano*. Yamanashi Institute of Environmental Sciences, Fuji-Yoshida, 339–348 [in Japanese with English abstract].

MIYAJI, N., KAN'NO, A., KANAMARU, T. & MANNEN, K. 2011. High-resolution reconstruction of the Hoei eruption (AD 1707) of Fuji volcano, Japan. *Journal of Volcanology and Geothermal Research*, **207**, 113–129.

MIYASHIRO, A. 1974. Volcanic rock series in island arcs and active continental margins. *American Journal of Science*, **274**, 321–355.

MIYAZAKI, T. 1984. Features of historical eruptions at Miyake-jima Volcano. *Bulletin of Volcanological Society of Japan*, **29**, S1–S15 [in Japanese with English abstract].

MORRIS, P. A. 1995. Slab melting as an explanation of Quaternary volcanism and aseismicity in southwest Japan. *Geology*, **23**, 395–398.

NAGAOKA, S. 1988. The late Quaternary tephra layers from the caldera volcanoes in and around Kagoshima bay, southern Kyushu, Japan. *Geographical Reports of Tokyo Metropolitan University*, **23**, 49–122.

NAGAOKA, S. & OKUNO, M. 2011. Tephrochronology and eruptive history of Kirishima volcano in southern Japan. *Quaternary International*, **246**, 260–269.

NAGAOKA, S., OKUNO, M. & ARAI, F. 2001. Tephrostratigraphy and eruptive history of the Aira caldera volcano during 100–30 ka, Kyushu, Japan. *Journal of Geological Society of Japan*, **107**, 432–450 [in Japanese with English abstract].

NAKADA, S. 2000. Hazards from pyroclastic flows and surges. *In*: SIGURDSSON, H. ET AL. (eds) *Encyclopedia of Volcano*. Academic Press, San Diego, 945–955.

NAKADA, S. 2001. Sequence of 2000 eruption of Usu volcano, Japan. *Bulletin of Earthquake Research Institute, University of Tokyo*, **76**, 203–314 [in Japanese with English abstract].

NAKADA, S. & FUJII, T. 1993. Preliminary report on the activity at Unzen Volcano (Japan), November 1990-November 1991: dacite lava domes and pyroclastic flows. *Journal of Volcanology and Geothermal Research*, **54**, 319–333.

NAKADA, S. & MOTOMURA, Y. 1999. Petrology of the 1991–1995 eruption at Unzen: effusion pulsation and groundmass crystallization. *Journal of Volcanology and Geothermal Research*, **89**, 173–196.

NAKADA, S. & WILSON, L. (eds) 2008. Special Issue: Scientific drilling at Mount Unzen. *Journal of Volcanology and Geothermal Research*, **175**, 240.

NAKADA, S., EICHELBERGER, J. C. & SHIMIZU, H. (eds) 1999. Unzen eruption; magma ascent and dome growth. *Journal of Volcanology and Geothermal Research*, **89**, 1–125.

NAKADA, S., NAGAI, M., KANEKO, K., NOZAWA, A. & SUZUKI-KAMATA, K. 2005*a*. Chronology and products of the 2000 eruption of Miyakejima Volcano, Japan. *Bulletin of Volcanology*, **67**, 205–218.

NAKADA, S., UTO, K., SAKUMA, S., EICHELBERGER, J. C. & SHIMIZU, H. 2005*b*. Scientific results of conduit drilling in the Unzen Scientifc Drilling Project (USDP). *Scientific Drilling*, **1**, 18–22.

NAKADA, S., NAGAI, M., KANEKO, T., SUZUKI, Y. & MAENO, F. 2013. The outline of the 2011 eruption at Shinmoe-dake (Kirishima), Japan. *Earth, Planets and Space*, **65**, 475–488.

NAKAGAWA, M., SHIMOTORI, H. & YOSHIDA, T. 1986. Aoso-Osore volcanic zone -The volcanic front of the Northeast Honshu arc, Japan-. *Journal of Mineralogy, Petrology and Economic Geology*, **81**, 471–478 [in Japanese with English abstract].

NAKAMURA, H., IWAMORI, H. & KIMURA, J.-I. 2008. Geochemical evidence for enhanced fluid flux due to overlapping subducting plates. *Nature Geoscience*, **1**, 380–384.

NAKAMURA, K. 1964. Volcano-stratigraphic study of Oshima volcano, Izu. *Bulletin of Earthquake Research Institute, University of Tokyo*, **42**, 649–728.

NAKAMURA, K. 1984. Distribution of flank craters of Miyake-jima Volcano and the nature of ambient crustal stress field. *Bulletin of Volcanological Society of Japan*, **29**, S16–S23 [in Japanese with English abstract].

NAKAMURA, M. 1995. Continuous mixing of crystal mush and replenished magma in the ongoing Unzen eruption. *Geology*, **23**, 807–810.

NAKANO, S. & YAMAMOTO, T. 1991. Chemical variations of magmas at Izu-Oshima volcano, Japan: plagioclase-controlled and differentiated magmas. *Bulletin of Volcanology*, **53**, 112–120.

NAKAO, S., MORITA, Y. *ET AL*. 2013. Volume change of the magma reservoir relating to the 2011 Kirishima Shinmoe-dake eruption - Charging, discharging and recharging process inferred from GPS measurements. *Earth, Planets and Space*, **65**, 505–515.

NEDO (NEW ENERGY DEVELOPMENT ORGANIZATION) 1994. *FY1993 Geothermal Development Promotion Survey in the Western region of Aso Volcano*. New Energy Development Organization, Tokyo [in Japanese].

NEWHALL, C. G. & SELF, S. 1982. The volcanic explosivity index (VEI) an estimate of explosive magnitude for historical volcanism. *Journal of Geophysical Research*, **87**, 1231–1238.

NOHDA, S., TATSUMI, Y. & OTOFUJI, Y. 1988. Asthenospheric injection and back-arc opening. *Chemical Geology*, **68**, 317–327.

OBA, Y. & KATSUI, Y. 1983. Petrology of the felsic volcanic rocks from Usu Volcano, Hokkaido, Japan. *Journal of Mineralogy, Petrology and Economic Geology*, **78**, 123–131 [in Japanese with English abstract].

OHKI, J., WATANABE, N., SHUTO, K. & ITAYA, T. 1993. Shifting of the volcanic fronts during Early Miocene ages in the Northeast Japan arc. *Island Arc*, **2**, 87–93.

OKINO, K., KASUGA, S. & OHARA, Y. 1998. A new scenario of the Parecce Vela basin genesis. *Marine Geophysical Research*, **20**, 21–40.

OKUNO, M. 2002. Chronology of tephra layers in southern Kyushu, SW Japan, for the last 30 000 years. *Quaternary Research*, **41**, 225–236 [in Japanese with English abstract].

OKUNO, M., NAKAMURA, T. & KOBAYASHI, T. 1998. AMS ^{14}C dating of historic eruptions of the Kirishima, Sakurajima and Kaimondake volcanoes, southern Kyushu, Japan. *Radiocarbon*, **40**, 825–832.

OKUNO, M., FUKUSHIMA, D. & KOBAYASHI, T. 2000. Tephrochronology in Southern Kyushu, SW Japan: tephra layers for the past 100 000 years. *Journal of the Society of Human History*, **12**, 9–23 [in Japanese].

ONO, K. & WATANABE, K. 1983. Aso caldera. *Monthly Chikyu*, **5**, 73–82 [in Japanese].

ONO, K. & WATANABE, K. 1985. *Geological Map of Aso Volcano 1:50 000*. Geological Survey of Japan, Tsukuba.

ONO, K., SOYA, T. & HOSONO, T. 1982. *Geology of the Satsuma-Io-Jima District. Quadrangle Series, Scale 1:50 000*. Geological Survey of Japan, Tsukuba [in Japanese with English abstract].

ONO, K., WATANABE, K. & KOMAZAWA, M. 1993. Structure of the Aso caldera implied by gravitational data. *Monthly Chikyu*, **15**, 686–690 [in Japanese].

OSHIMA, H. & UI, T. 2003. The 2000 eruption of Usu volcano. *In*: *Reports on Volcanic Activities and Volcanological Studies in Japan for the Period from 1999 to 2002*. Bulletin of the Volcanological Society of Japan, **484**, 22–31.

OTOFUJI, Y. & MATSUDA, T. 1987. Amount of clockwise rotation of Southwest Japan - fan shape opening of the southwestern part of the Japan Sea. *Earth and Planetary Science Letters*, **85**, 289–301.

OTOFUJI, Y., MATSUDA, T. & NOHDA, S. 1985. Opening mode of the Japan Sea inferred from paleomagnetism of the Japan arc. *Nature*, **317**, 603–604.

OTOFUJI, Y., KAMBARA, A., MATSUDA, T. & NOHDA, S. 1994. Counterclockwise rotation of Northeast Japan: paleomagnetic evidence for regional extent and timing of rotation. *Earth and Planetary Science Letters*, **121**, 503–518.

OZAWA, S., NISHIMURA, T., SUITO, H., KOBAYASHI, T., TOBITA, M. & IMAKIIRE, T. 2011. Coseismic and postseismic slip of the 2011 magnitude-9 Tohoku-Oki earthquake. *Nature*, **475**, 373–376.

OZAWA, T. & KOZONO, T. 2013. Temporal variation of the Shinmoe-dake crater in the 2011 eruption revealed by spaceborne SAR observations. *Earth Planets and Space*, **65**, 527–537.

SAITO, G., KAZAHAYA, K., SHINOHARA, H., STIMAC, J. & KAWANABE, Y. 2001. Variation of volatile concentration in a magma system of Satsuma–Iwojima volcano deduced from melt inclusion analyses. *Journal of Volcanology and Geothermal Research*, **108**, 11–31.

SAITO, G., STIMAC, J., KAWANABE, Y. & GOFF, F. 2002. Mafic inclusions in rhyolites of Satsuma–Iwojima volcano: evidence for mafic–silicic magma interaction. *Earth, Planets and Space*, **54**, 303–325.

SAITO, G., UTO, K., KAZAHAYA, K., SHINOHARA, H., KAWANABE, Y. & SATOH, H. 2005. Petrological characteristics and volatile content of magma from the 2000 eruption of Miyakejima Volcano, Japan. *Bulletin of Volcanology*, **67**, 268–280.

SAKUYAMA, M. 1977. Lateral variation of phenocryst assemblages in volcanic rocks of the Japanese islands. *Nature*, **269**, 134.

SATO, H. 1994. The relationship between late Cenozoic tectonic events and stress field and basin development in northeast Japan. *Journal of Geophysical Research*, **99**, 22 261–22 274.

SATO, H. & AMANO, K. 1991. Relationship between tectonics, volcanism, sedimentation and basin development, Late Cenozoic, central part of Northern Honshu, Japan. *Sedimentary Geology*, **74**, 323–343.

SHIMODA, G., TATSUMI, Y., NOHDA, S., ISHIZAKA, K. & JAHN, B. M. 1998. Setouchi high-Mg andesites revised: geochemical evidence for melting of subducting sediments. *Earth and Planetary Science Letters*, **160**, 479–492.

SHINJO, R., CHUNG, S.-L., KATO, Y. & KIMURA, M. 1999. Geochemical and Sr-Nd isotopic characteristics of volcanic rocks from the Okinawa Trough and Ryukyu Arc: implications for the evolution of a young, intracontinental back arc basin. *Journal of Geophysical Research*, **104**, 10 591–10 608.

SHINJO, R., WOODHEAD, J. D. & HERGT, J. M. 2000. Geochemical variation within the northern Ryukyu arc: magma source compositions and geodynamic implications. *Contributions to Mineralogy and Petrology*, **140**, 263–282.

SHINJOE, H. 1997. Origin of the granodiorite in the forearc region of southwest Japan: melting of the Shimanto accretionary prism. *Chemical Geology*, **134**, 237–255.

SHIONO, K., MIKUMO, T. & ISHIKAWA, Y. 1980. Tectonics of the Kyushu-Ryuku arc as evidenced from seismicity and focal mechanism of shallow to intermediate-depth earthquakes. *Journal of Physics of the Earth*, **28**, 17–43.

SHUTO, K., ISHIMOTO, H. *ET AL*. 2006. Geochemical secular variation of magma source during Early to Middle Miocene time in the Niigata area, NE Japan: asthenospheric mantle upwelling during back-arc basin opening. *Lithos*, **86**, 1–33.

SIBUET, J.-C., HSU, S.-K., SHYU, C.-T. & LIN, C.-S. 1995. Structural and kinematic evolutions of the Okinawa Trough backarc basin. *In*: TAYLOR, B. (ed.) *Backarc Basins: Tectonics and Magmatism*. New York, Plenum Press, 347–379.

SIEBELT, L., SIMIKIN, T. & KIMBERLY, P. 2010. *Volcanoes of the World*. 3rd ed. Smithsonian Institution, University of California Press, California.

SOYA, T., SAKAGUCHI, K. ET AL. 1987. The 1986 eruption and products of Izu–Oshima volcano. *Bulletin of Geological Survey of Japan*, **38**, 609–630 [in Japanese with English abstract].

SOYA, T., KATSUI, Y., NIIDA, K., SAKAI, I. & TOMIYA, A. 2007. *Geological Map of Usu Volcano, 1:25 000. Geological Map of Volcanoes 2*. Geological Survey of Japan, AIST.

SUDO, M., UTO, K., MIKI, D., ISHIHARA, K. & TATSUMI, Y. 2000. K-Ar dating of volcanic rocks along the Aira caldera rim: volcanic history before explosive Aira pyroclastic eruption. *Annuals of Disaster Prevention Research Institute, Kyoto University*, **43**, B-1, 15–35 [in Japanese with English abstract].

SUDO, M., UTO, K., MIKI, D. & ISHIHARA, K. 2001. K-Ar dating of volcanic rocks along the Aira caldera rim: Part 2. Volcanic history of western and northwestern area of caldera and Sakurajima volcano. *Annuals of Disaster Prevention Research Institute, Kyoto University*, **44**, B-1, 305–316 [in Japanese with English abstract].

SUZUKI, Y., NAGAI, M. ET AL. 2013*a*. Precursory activity and evolution of the 2011 eruption of Shinmoe-dake in Kirishima volcano – insights from ash samples. *Earth, Planets and Space*, **65**, 591–607.

SUZUKI, Y., YASUDA, A., HOKANISHI, N., KANEKO, T., NAKADA, S. & FUJII, T. 2013*b*. Syneruptive deep magma transfer and shallow magma remobilization during the 2011 eruption of Shimoe-dake, Japan. Constraints from melt inclusions and phase equilibria experiments. *Journal of Volcanology and Geothermal Research*, **257**, 184–204.

TAKADA, A., ISHIZUKA, Y., NAKANO, S., YAMAMOTO, T., KOBAYASHI, M. & SUZUKI, Y. 2007. Characteristic and evolution inferred from eruptive fissures of Fuji volcano, Japan. *In*: ARAMAKI, S., FUJII, T., NAKADA, S. & MIYAJI, N. (eds) *Fuji Volcano*. Yamanashi Institute of Environmental Sciences, Fuji-Yoshida, 183–202 [in Japanese with English abstract].

TAKAHASHI, M., HOSHI, H. & YAMAMOTO, T. 1999. Miocene counterclockwise rotation of the Abukuma Mountains, Northeast Japan. *Tectonophysics*, **306**, 19–31.

TAKAHASHI, M., MATSUDA, F., YASUI, M., CHIBA, T. & MIYAJI, N. 2007. Geology of the Aokigahara lava flow field in Fuji Volcano and its implications in the process of Jogan Eruption. *In*: ARAMAKI, S., FUJII, T., NAKADA, S. & MIYAJI, N. (eds) *Fuji Volcano*. Yamanashi Institute of Environmental Sciences, Fuji-Yoshida, 303–338 [in Japanese with English abstract].

TAKARADA, S., HOSHIZUMI, H. ET AL. 2007. C1: Unzen Volcanoa and new lava dome climb. *Field Trip Guidebook, Cities on Volcanoes 5 Conference*, November 2007, Shimabara.

TAMAKI, K. 1988. Geological structure of the Japan Sea and its tectonic implications. *Bulletin of Geological Survey of Japan*, **39**, 269–365.

TAMURA, Y., TATSUMI, Y., ZHAO, D., KIDO, Y. & SHUKUNO, H. 2002. Hot fingers in the mantle wedge: new insights into magma genesis in subduction zone. *Earth and Planetary Science Letters*, **197**, 105–116.

TANAKA, H. K. M., NAKANO, T. ET AL. 2007. Imaging the conduit size of the dome with cosmic-ray muons: the structure beneath Showa-Shinzan lava dome, Japan. *Geophysical Research Letters*, **34**, http://doi.org/10.1029/2007GL031389

TATSUMI, Y. 2001. Geochemical modeling of partial melting of subducting sediments and subsequent melt-mantle interaction: generation of high-Mg andesites in the Setouchi volcanic belt, southwest Japan. *Geology*, **29**, 323–326.

TATSUMI, Y., OTOFUJI, Y., MATSUDA, T. & NOHDA, S. 1989. Opening of the Sea of Japan back-arc basin by asthenospheric injection. *Tectonophysics*, **166**, 317–329.

TOGASHI, S. & TAKAHASHI, M. 2007. Geochemistry of rocks from the Fuji volcano, Japan; constraints for evolution of magmas. *In*: ARAMAKI, S., FUJII, T., NAKADA, S. & MIYAJI, N. (eds) *Fuji Volcano*. Yamanashi Institute of Environmental Sciences, Fuji-Yoshida, 219–231 [in Japanese with English abstract].

TOMIYA, A. & TAKAHASHI, E. 1995. Reconstruction of an evolving magma chamber beneath Usu volcano since the 1663 eruption. *Journal of Petrology*, **36**, 617–636.

TOMIYA, A. & TAKAHASHI, E. 2005. Evolution of the magma chamber beneath Usu volcano since 1663: a natural laboratory for observing changing phenocryst compositions and textures. *Journal of Petrology*, **46**, 2395–2426.

TOSHA, T. & HAMANO, Y. 1988. Paleomagnetism of Tertiary rocks from the Oga Peninsula and the rotation of Northeast Japan. *Tectonics*, **7**, 653–662.

TSUKUI, M., NIIHORI, K., KAWANABE, Y. & SUZUKI, Y. 2001. Stratigraphy and formation of Miyakejima volcano. *Journal of Geography*, **110**, 156–167 [in Japanese with English abstract].

TSUYA, H. 1968. *Geology of Volcano Mt. Fuji. Explanatory text of geologic map 1:50 000 scale*. Geological Survey of Japan, Tsukuba.

UENO, T. 2001. Emplacement mechanism of the Ito pyroclastic flow inferred from the lateral and vertical variations in ash components. *Bulletin of Volcanological Society of Japan*, **46**, 257–268 [in Japanese with English abstract].

UHIRA, K., BABA, T., MORI, H., KATAYAMA, H. & HAMADA, N. 2005. Earthquake swarms preceding the 2000 eruption of Miyakejima volcano, Japan. *Bulletin of Volcanology*, **67**, 219–230.

UI, T. 1973. Exceptionally far-reaching, thin pyroclastic flow in Southern Kyushu, Japan. *Bulletin of Volcanological Society of Japan*, **18**, 153–168 [in Japanese with English Abstract].

UI, T., METSUGI, H. & SUZUKI, K. 1984. Flow lineation of Koya low-aspect ratio ignimbrite, South Kyusyu, Japan. *Progress report of the US–Japan Cooperative Science Program*. Ministry of Education, Science and Culture, Japan, 9–12.

UI, T., NAKAGAWA, M., INABA, C. & YOSHIMOTO, M. & GEOLOGICAL PARTY, JOINT RESEARCH GROUP FOR THE USU 2000 ERUPTION, 2002. Sequence of 2000 eruption, Usu volcano. *Bulletin of Volcanological Society of Japan*, **47**, 105–117 [in Japanese with English abstract].

UKAWA, M. 2005. Deep low-frequency earthquake swarm in the mid-crust beneath Mount Fuji (Japan) in 2000 and 2001. *Bulletin of Volcanology*, **68**, 47–56.

UKAWA, M., FUJITA, E., YAMAMOTO, E., OKADA, Y. & KIKUCHI, M. 2000. The 2000 Miyakejima eruption: crustal deformation and earthquakes observed by NIED Miyakejima observation network. *Earth, Planets and Space*, **52**, xix–xxxvi.

UMAKOSHI, K., SHIMIZU, H. & MATSUWO, N. 2001. Volcano-tectonic seismicity at Unzen Volcano. *Journal of Volcanology and Geothermal Research*, **112**, 117–131.

UTO, K., MIKI, D., HOANG, N., SUDO, M., FUKUSHIMA, D. & ISHIHARA, K. 2005. Temporal evolution of magma composition in Sakurajima volcano, southwest Japan. *Annuals of Disaster Prevention Research Institute, Kyoto University*, **48B**, 341–347 [in Japanese with English abstract].

WALKER, G. P. L., MCBROOME, L. A. & CARESS, M. E. 1984. Products of the Koya Eruption from the Kikai Caldera, Japan. *Progress report of the US–Japan Cooperative Science Program*. Ministry of Education, Science and Culture, Japan, 4–8.

WATANABE, H. 1984. Gradual bubble growth in dacite magma as a possible cause of the 1977–1978 long-lived activity of Usu Volcano. *Journal of Volcanology and Geothermal Research*, **20**, 133–144.

WATANABE, K. & HOSHIZUMI, H. 1995. *Geological map of Unzen volcano, scale 25 000:1*. Geological Survey of Japan, Tsukuba.

YAMAJI, A. 1990. Rapid intra-arc rifting in Miocene northeast Japan. *Tectonics*, **9**, 365–378.

YAMAMOTO, T. 2006. Pyroclastic density current from the caldera-forming eruption of Izu-Oshima volcano, Japan: restudy of the Sashikiji 2 Member based on stratigraphy, lithofacies, and eruption age. *Bulletin of Volcanological Society of Japan*, **51**, 257–271 [in Japanese with English abstract].

YAMAMOTO, T. 2009. Sedimentary processes caused by felsic caldera-forming volcanism in the Late Miocene to Early Pliocene intra-arc Aizu basin, NE Japan arc. *Sedimentary Geology*, **220**, 337–348.

YAMAMOTO, T. 2011. Origin of the sequential Shirakawa ignimbrite magmas from the Aizu caldera cluster, northeast Japan: evidence for renewal of magma system involving a crustal hot zone. *Journal of Volcanology and Geothermal Research*, **204**, 91–106.

YAMAMOTO, T. & HOANG, N. 2009. Synchronous Japan Sea opening Miocene fore-arc volcanism in the Abukuma Mountains, NE Japan: an advancing hot asthenosphere flow v. Pacific slab melting. *Lithos*, **112**, 575–590.

YAMAMOTO, T., TAKADA, A., ISHIZUKA, Y. & NAKANO, S. 2005*a*. Chronology of the products of Fuji volcano based on new radiometric carbon ages. *Bulletin of Volcanological Society of Japan*, **50**, 53–70 [in Japanese with English abstract].

YAMAMOTO, T., TAKADA, A., ISHIZUKA, Y., MIYAJI, N. & TAJIMA, Y. 2005*b*. Basaltic pyroclastic flows of Fuji volcano, Japan: characteristics of the deposits and their origin. *Bulletin of Volcanology*, **67**, 622–633.

Yamamoto, T., Ishizuka, Y. & Takada, A. 2007. Surface and subsurface geology at the southwestern foot of Fuji volcano, Japan: new stratigraphy and chemical variations of the products. *In*: Aramaki, S., Fujii, T., Nakada, S. & Miyaji, N. (eds) *Fuji Volcano*. Yamanashi Institute of Environmental Sciences, Fuji-Yoshida, 97–118 [in Japanese with English abstract].

Yamaoka, K., Kawamura, M., Kimata, F., Fujii, N. & Kudo, T. 2005. Dike intrusion associated with the 2000 eruption of Miyakejima Volcano, Japan. *Bulletin of Volcanology*, **67**, 231–242.

Yasui, M., Takahashi, M., Ishihara, K. & Miki, D. 2007. Eruption style and its temporal variation through the 1914–1915 eruption of Sakurajima volcano, southern Kyushu, Japan. *Bulletin of Volcanological Society of Japan*, **52**, 161–186 [in Japanese with English abstract].

Yokoyama, I., Yamashita, H., Watanabe, H. & Okada, H. M. 1981. Geophysical characteristics of dacite volcanism – the 1977–1978 eruption of Usu volcano. *Journal of Volcanology and Geothermal Research*, **9**, 335–358.

Yokoyama, S. 2000. The northern limits of the distribution of the Ito ignimbrite, south Kyushu, Japan. *Bulletin of Volcanological Society of Japan*, **45**, 209–216 [in Japanese with English abstract].

Yoshida, T., Ohguchi, T. & Abe, T. 1995. Structure and evolution of source area of the Cenozoic volcanic rocks in Northeast Honshu arc, Japan. *Memoirs of the Geological Society of Japan*, **44**, 263–308 [in Japanese with English Abstract].

Yoshida, T., Nakajima, J. *et al.* 2005. Evolution of Late Cenozoic magmatism in the NE Honshu Arc and its relation to the crust-mantle structures. *Quaternary Research (Daiyonki-Kenkyu)*, **44**, 195–216 [in Japanese with English Abstract].

Zielinski, G. A., Mayewski, P. A., Meeker, L. D., Whitlow, S. & Twickler, M. S. 1996. A 110 000-Yr record of explosive volcanism from the GISP2 (Greenland) Ice Core. *Quaternary Research*, **45**, 109–118.

6 Neogene–Quaternary sedimentary successions

MAKOTO ITO, KOJI KAMEO, YASUFUMI SATOGUCHI, FUJIO MASUDA, YOSHIHISA HIROKI, OSAMU TAKANO, TAKESHI NAKAJIMA & NORIYUKI SUZUKI

Overview (MI and KK)

Neogene and Quaternary sedimentary successions are widely distributed across the Japanese islands, commonly underlying or distributing nearby heavily populated areas where the infrastructure of Japanese society is fully developed. Furthermore, some Neogene sedimentary successions contain oil and gas and have played an important role in hydrocarbon production in Japan. The Neogene and Quaternary sedimentary basins formed as a response to the development of the Japanese arc-trench system as a result of the opening of the Sea of Japan (*c.* 15 Ma), and spatial and temporal variations in the sedimentary successions are considered to have been controlled by the interaction between glacial–interglacial sea-level changes and active tectonic movements of the arc-trench system, which has been controlled mainly by subduction of the Pacific and Philippine Sea plates beneath the Japanese islands.

This chapter mainly focuses on the Neogene and Quaternary sedimentary successions formed in NE and SW Japan (Fig. 6.1), and we describe their lithofacies organization and tectonic setting in the forearc, intra-arc and back-arc regions of the arc-trench system. Of particular interest within these young sedimentary basins is the common presence of volcanic ash (tephra) beds, which provide a high-resolution chronostratigraphic framework allowing both local and regional correlations of the sedimentary successions detailed reconstruction of Neogene and Quaternary palaeoenvironmental changes in Japan.

Tephrostratigraphy of the Pliocene–Pleistocene successions (YS)

Tephras are important key beds for both local and regional stratigraphy and chronology of sedimentary successions; they can provide time horizons within heterogeneous lithofacies assemblages, and permit detailed chronostratigraphic correlation of sedimentary successions between separated outcrops within a sedimentary basin and those formed in separated sedimentary basins. The potential role of tephras in the correlation of stratigraphic successions within the Japanese islands was first recognized by the discovery of the widespread nature of the Late Pleistocene Aira-Tn volcanic ash (Machida & Arai 1976). This tephra is distributed over almost the whole of Honshu Island as a result of an extraordinarily large-scale explosive eruption which occurred at *c.* 29 ka. Subsequently, many Pliocene and Pleistocene tephras have been investigated (e.g. Machida *et al.* 1980; Yoshikawa *et al.* 1996), and widespread tephras have been identified and used for establishing regional correlation of different types of Pliocene–Pleistocene stratigraphic successions across the Japanese islands (e.g. Satoguchi *et al.* 1999). Recently, a standardized stratigraphic model of the Japanese Pliocene and Pleistocene successions has been proposed on the basis of detailed correlation and age frameworks of many widespread tephras (Satoguchi & Nagahashi 2012). This model is crucial for detailed reconstruction of Neogene and Quaternary palaeoenvironmental changes in Japan.

Widespread tephras

Although the formal use of the term 'widespread tephra' has not been defined, it refers to a tephra which is widely distributed for a few hundred to a thousand kilometres across separate sedimentary basins in the Japanese islands. Detailed identification of a 'widespread tephra' bed or bed sets has been conducted on the basis of distinctive features such as: (1) lithofacies and colours; (2) chemical and mineralogical composition; (3) shape and refractive index of volcanic glass; (4) stratigraphic position relative to some other already identified tephra beds; and (5) stratigraphic position of the tephra bed and/or bed sets relative to defined boundaries identified by the other types of stratigraphic study. Use of a combination of several distinctive tephras is more robust for both local and regional correlation rather than relying only on evidence from a single tephra bed (e.g. Satoguchi & Hattori 2008).

The Pliocene–Middle Pleistocene sedimentary successions which contain widespread tephras are mainly developed in Honshu and Kyushu (Fig. 6.2), and their distributions are here tentatively divided into 12 areas: Miyazaki; Osaka; Lake Biwa; Ise Bay; Kakegawa; Himi; Kanazawa–Toyama; Nagaoka; the Boso Peninsula; Choshi; Fukushima–Sendai; and the Oga Peninsula (Fig. 6.3). Groups and formations within these sedimentary successions comprise marine and/or fluvio-lacustrine clastic deposits and are intercalated with many tephra beds, which represent distal facies of parent volcanic activities. Each tephra bed has its own local name in each sedimentary basin (Fig. 6.2), and in some cases has been grouped together under a unified name as a widespread tephra, typically with a name indicating the source volcano and/or pyroclastic flow deposits. Fifty-six widespread tephras have been identified from sedimentary successions of age 0.4–5.3 Ma in Honshu and Kyushu, although the locations of the volcanic sources have only been clarified for 17 of them. Most such tephras in Japan have been produced by huge eruptions and are distributed towards the east of their source volcanoes as a result of the influence of the prevailing westerly winds. Given the lack of a proven source area in many cases, most of the Pliocene–Middle Pleistocene widespread tephras have been named by combining the names of two areas where they were described in detail in an early stage of their study (Satoguchi & Nagahashi 2012). Some of the tephras have been mapped well enough to allow the reconstruction of isopach maps and eruptive magnitudes (e.g. Nagahashi *et al.* 2000).

Stratigraphy, chronology and volcanism

The tephrostratigraphy of sedimentary successions formed during the period between 0.4 and 5.3 Ma has been established on the

From: Moreno, T., Wallis, S., Kojima, T. & Gibbons, W. (eds) 2016. *The Geology of Japan*. Geological Society, London, 309–337.

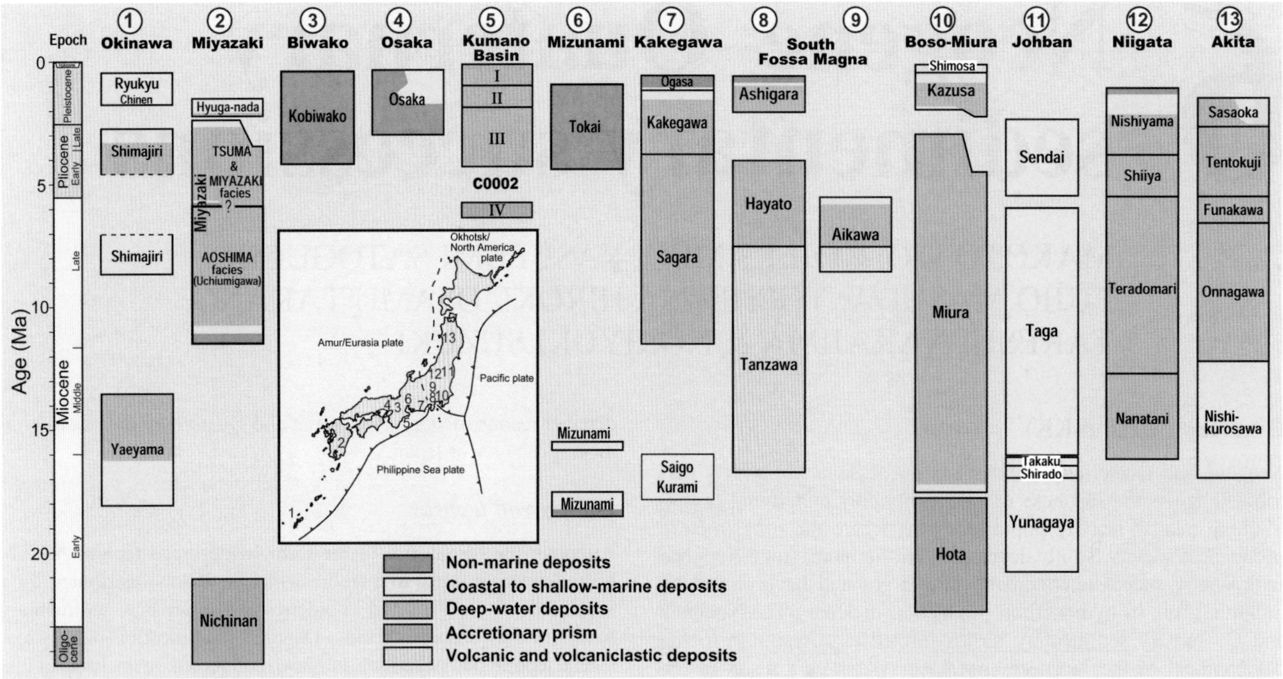

Fig. 6.1. Chronostratigraphic correlation of the Neogene and Quaternary successions described in the chapter and some additional successions from NE Japan (Joban), collision zone (South Fossa Magna) and Okinawa. See inset map for locations of each succession (Ikebe *et al.* 1975; Kato 1985; Suzuki & Tsukawaki 1989; Yoshikawa & Yamazaki 1998; Aoike 1999; Arima *et al.* 1999; Nakamura *et al.* 1999*a*, *b*; Kawagata *et al.* 2002; Suto *et al.* 2005; Chiyonobu *et al.* 2009; Expedition 315 Scientists 2009; Satoguchi 2009; Yao *et al.* 2009). Niigata and Akita areas use formation names and the other areas are illustrated by group names. Plate-tectonic framework of the Japanese islands in the inset map is from Nakajima & Hasegawa (2010).

basis of the 56 recognized widespread tephras (Fig. 6.3; Satoguchi & Nagahashi 2012). The depositional ages of the tephras are basically determined by magnetostratigraphy and biostratigraphy. The latest version of the tephrostratigraphy in Figure 6.3 has reduced discrepancies between tephra correlations and local chronology (Satoguchi *et al.* 1999), and by now most of the major Pliocene–Pleistocene

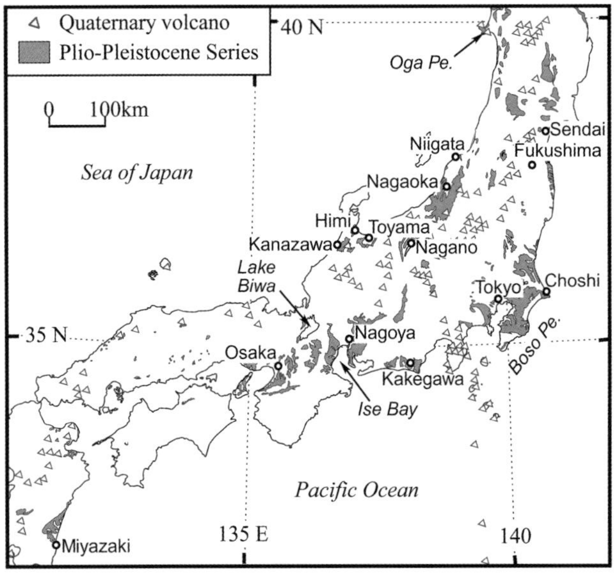

Fig. 6.2. Distribution of Pliocene and Pleistocene sedimentary successions and Quaternary volcanoes. Simplified from the Japan Association for Quaternary Research (1987) and Shimozuru *et al.* (1995).

formations contain tephra beds that can allow correlation between the 12 areas previously listed. Despite this success, however, there are still many tephras that have been described in detail but have not yet been correlated over a wide area. For example, while more than 300 tephras have been described from the Kazusa Group on the Boso Peninsula, only 28 of them have been correlated with tephras elsewhere. Furthermore, some areas have few widespread tephras useful for the regional correlation. This is the case in the Miyazaki Group in the Miyazaki area, where most of the tephras also have some distinctive features useful only for local correlation and their features are quite similar to some of the widespread tephras already identified elsewhere. In such a case, the tephras can only be used for the reconstruction of local volcanic activities. Tephras formed at around the magnetostratigraphic chronozone boundaries are of especial importance. Such examples include the Ss-Az (under the Brunhes–Matuyama), Ss-Pnk (lower part of the Jaramillo Subchronozone), Eb-Fukuda (above the Olduvai Subchronozone), Ass-Tmd2 (under the Matuyama–Gauss), Sr-Ity (above the Gauss–Gilbert) and Sk-Ya5 (above the Cochiti Subchronozone) tephras. These tephras have been described in many areas and their distribution patterns in Japan and volcanic sources have been well studied (Machida 1999; Nagahashi *et al.* 2000; Tamura *et al.* 2008).

Even if the eruptive centres of the widespread tephras have not been identified, the degree of explosive activities of the source volcanoes can be detected on the basis of thickness and volumes of tephras preserved in sedimentary successions. Consequently, explosive volcanism is considered to have been extremely active in Japan since at least Pliocene times. Because these tephras each provide a specific record of the nature and timing of volcanic eruptions, they can be used to reveal spatial and temporal variations in explosive volcanism over the last 5 Ma or more. The distributions, source volcanoes and eruption volumes of most tephras that formed after Pleistocene times have been particularly well documented (e.g.

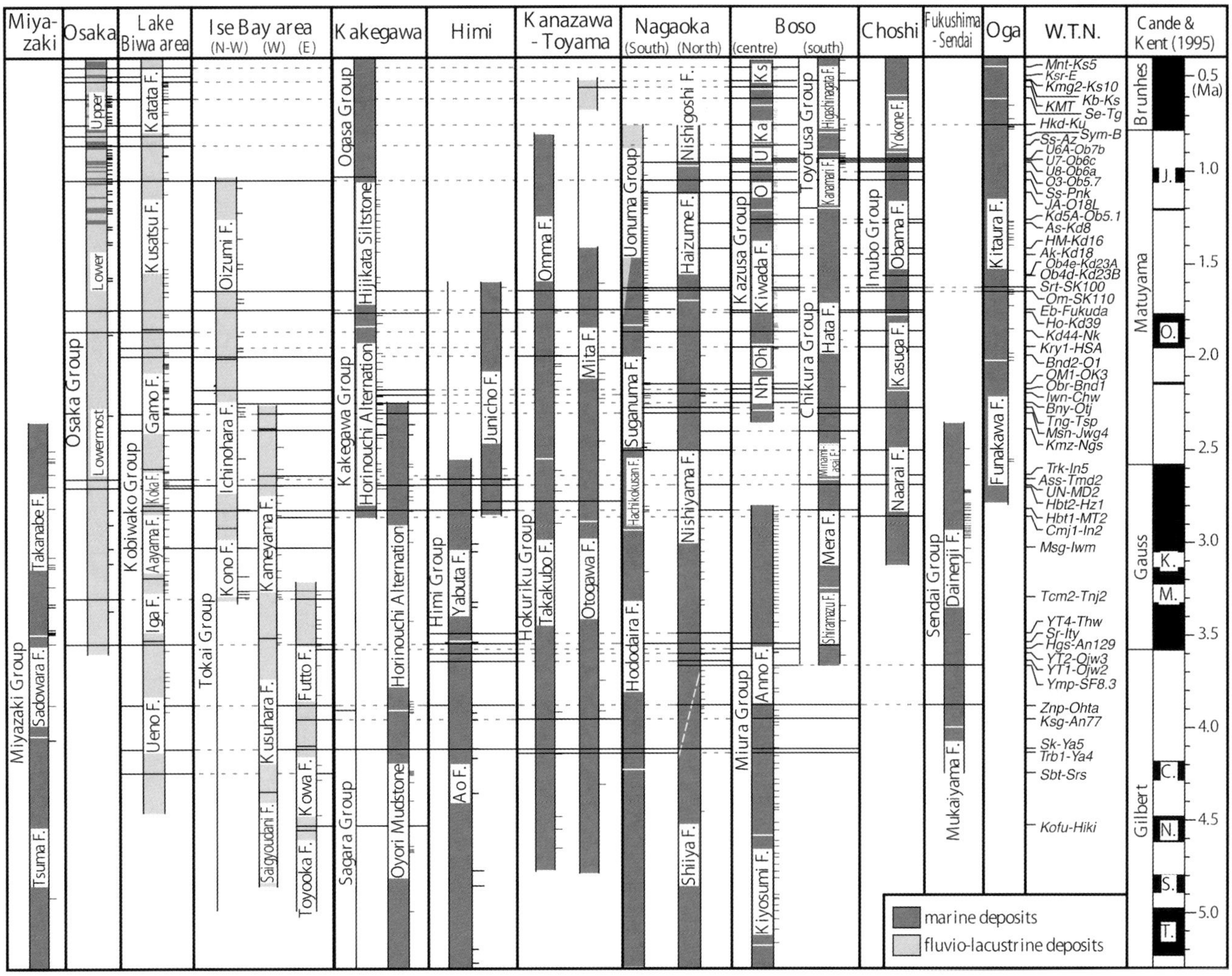

Fig. 6.3. Tephrostratigraphy of the Pliocene–Middle Pleistocene on the magnetostratigraphic time scale. Simplified from Satoguchi & Nagahashi (2012). Abbreviations: L, Lake; Fu-Se, Fukushima–Sendai; WTN, widespread tephra name. Solid lines indicate correlation of each tephra between the basins, and dashed lines indicate the lack of the widespread tephra between the basins.

Machida 1999). In the case of those tephras whose source volcanoes are well identified, it is clear that the major focus of volcanism in Japan changed from the Chubu Mountains area of central Honshu and moved both SW to Kyushu Island and NE to the Tohoku region around 1.4 Ma ago (Satoguchi & Nagahashi 2012). Although the origin of such regional variation in active volcanism responsible for the formation of the widespread tephras remains controversial, Kimura *et al.* (2010) and Mahony *et al.* (2011) suggested that volcano-tectonic changes occurred in SW Japan at *c.* 4 and 2 Ma, respectively.

Forearc basin successions (MI and KK)

Active sedimentation in the Neogene–Quaternary forearc basins began during the period 15–10 Ma in SE Japan, although thinner blankets of sedimentary successions formed in forearc regions in NE Japan (Fig. 6.1). The development of forearc basins in SW Japan is interpreted to have been induced by subduction of the Philippine Sea Plate beneath the Eurasia Plate along the Nankai Trough, superimposed by the effect of collision of the Izu–Bonin Arc with the Honshu Arc after the opening of the Sea of Japan by *c.* 15 Ma (e.g. Masuda 1984; Taira 2001; Takahashi 2006). The forearc basin successions developed over intensely deformed accretionary prisms and/or older forearc-basin-fill successions, and the basins are interpreted to have developed as a response to the uplift of accretionary prisms along the Nankai Trough (Taira 2001). Three major areas are available for detailed description and interpretation of lithofacies organization and sequence stratigraphy of the forearc basin successions using onshore outcrop belts in SW Japan; these are the Boso–Miura, Kakegawa and Miyazaki areas, respectively (Fig. 6.1).

Boso–Miura area

The forearc basin successions in the Boso–Miura area are exposed on the central and northern parts of the Boso and Miura peninsulas (Fig. 6.4). These successions are defined as the Miura and Kazusa groups, in ascending order, and the Kazusa Group is in turn unconformably overlain by the Shimosa Group, which represents an infill of the palaeo-Tokyo Bay, a relict of the Kazusa Group forearc basin (i.e. Kazusa forearc basin: Ito & Katsura 1992) (Fig. 6.4).

Miura Group

The Miura Group is up to 4100 m in maximum thickness and was formed during the period 16–3.0 Ma (Fig. 6.4). Although lithostratigraphic classification of the lowest part of the group remains

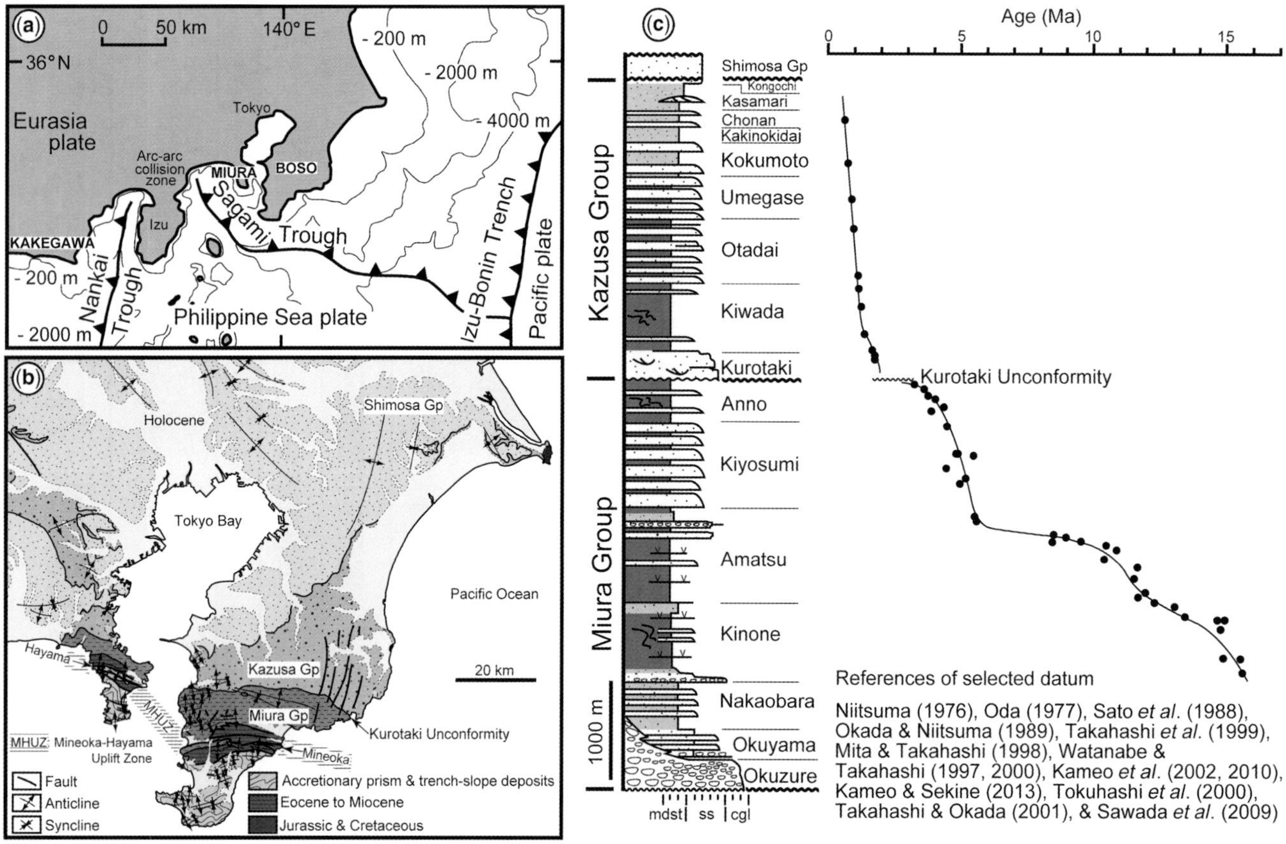

Fig. 6.4. (**a**) Plate-tectonic framework of the Boso–Miura and Kakegawa areas. Modified form Ito (1998). (**b**) Distribution and tectonic framework of the Miura, Kazusa and Shimosa groups. Simplified from Kakimi *et al.* (1982). (**c**) A composite stratigraphic section and ages of the Miura, Kazusa and Shimosa groups. Lithological data of the lower Miura Group are from Saito (1992). See Figure 6.7 for the other explanations.

controversial (e.g. Nakajima *et al.* 1981; Mitsunashi 1990; Saito 1992), the group is represented by an overall transgressive–regressive cycle and shows east- to SE-directed palaeocurrents (e.g. Ito *et al.* 2002; Saito & Ito 2002). The Miura Group sedimentary basin is interpreted to have formed as a response to the uplift of a trench-slope break to the south, which is represented by the Mineoka and Hayama groups in the Mineoka–Hayama Uplift Zone (e.g. Takahashi 2004) (Fig. 6.4). The lower part of the group (up to 1100 m thick) (the Okuzure, Okuyama and Nakaobara formations) consists of breccias and conglomerates, which intertongue with interbedded sandstones, siltstones, sandy siltstones and conglomerates, and fine upwards to siltstones intercalated with turbiditic sandstones, minor volcanic ash beds and slumped deposits (the Kinone Formation) (Fig. 6.4). The lower part represents the initial sedimentation in shallow-marine and slope environments (*c.* up to 1000 m in palaeowater depth: Kitazato 1997) and is interpreted to have formed in a strike-slip basin which developed in response to dextral strike-slip movement of the trench-slope break (Saito *et al.* 1991). The lower part fines further upwards to the middle part, which is represented by siltstones intercalated with thin- to medium-bedded turbiditic sandstones and many volcanic ash beds (the Amatsu Formation) (Fig. 6.4). The middle part is up to 980 m thick and shows a palaeowater depth of up to 2000 m (Kitazato 1997). Locally, the middle part contains shelf and coastal deposits formed in response to ephemeral forced regressions and subsequent transgressions (Ito *et al.* 2002). The middle part coarsens upwards to the upper part, which is represented by siltstones and interbedded turbiditic sandstones and their related sediment-gravity-flow deposits (Tokuhashi 1979; Saito & Ito 2002; Ishihara & Tokuhashi 2005; Kase *et al.* 2013) and shows a palaeowater depth of 1000–2000 m (Kitazato 1997). The turbidite-dominated upper part is considered to have formed by submarine-fan (the Kiyosumi Formation, *c.* 850 m thick) and slope-apron (the Anno Formation, *c.* 400 m thick) systems, in ascending order. Abrupt replacement of the siltstone-dominated middle part by the turbidite-dominated upper part occurred at *c.* 5 Ma, and an average sedimentation rate also abruptly increased in the upper part (Masuda 1984; Takahashi 2006) (Fig. 6.4). Active shedding of siliciclastic sediments into the forearc basin from the western hinterlands is interpreted to have started in response to collision of the Tanzawa Block in the north of the Izu Peninsula with the Honshu Arc at *c.* 5 Ma (Ito & Masuda 1988; Niitsuma 1989) (Fig. 6.1), although four depositional sequences defined within the Kiyosumi Formation submarine-fan succession are interpreted to have formed in response to the oxygen isotopic sea-level index (Saito & Ito 2002).

The Miura Group is unconformably overlain by the Kazusa Group, a younger forearc-basin-fill succession in this area. The boundary is defined by the Kurotaki Unconformity (Fig. 6.4) and has been interpreted to be a sedimentary signal of distinct change in relative motion of the Philippine Sea and Pacific plates (Masuda 1984; Ito & Masuda 1988; Niitsuma 1989; Takahashi 2006). However, this unconformity is not necessarily ubiquitous in the sedimentary successions of forearc basins in SW Japan, including offshore forearc basins such as the Kumano forearc basin (Saffer *et al.* 2010) (Fig. 6.1), and this interpretation still remains controversial.

Kazusa Group

The Kazusa Group formed in the Kazusa forearc basin, which is interpreted to have developed in response to the WNW-directed subduction of the Pacific Plate beneath the Eurasia Plate (later the Okhotsk Plate: Seno *et al.* 1994) superimposed by the NW-directed subduction of the Philippine Sea Plate (Katsura 1984; Watanabe *et al.* 1987; Ito & Masuda 1988). This group is well exposed on the Boso Peninsula and age-equivalent coastal and shallower-marine deposits are observed on the Miura Peninsula. The group on the Boso Peninsula is up to 3000 m in total thickness and developed during the period 2.4–0.45 Ma (Fig. 6.4). The sedimentary successions of the Kazusa Group represents a series of marine environments ranging from deep-sea basin plain, through submarine-fan and slope to shallow seas (e.g. Katsura 1984; Ito & Katsura 1992) (Fig. 6.5). In general, the lower part of the group is characterized by deep-water basin-plain, submarine-fan and lower-slope deposits in the northeastern area, with the deepest part of the basin of up to *c.* 1500 m in palaeowater depth (Kitazato 1997), and by upper-slope and outer-shelf deposits in the southwestern area (Ito & Katsura 1992) (Fig. 6.5). Everywhere, the upper part of the group is represented by upper-slope and outer-shelf deposits, in association with minor inner-shelf and coastal deposits in the southwestern area. These lithofacies successions of the group show an overall transgressive–regressive cycle with duration of about 1.95 myr (i.e. a third-order depositional sequence in the sense of Vail *et al.* 1991), and indicate east- and NE-directed palaeocurrents (Ito & Katsura 1992). Although the average sedimentation rate of the entire Kazusa Group is up to 1.5 m ka^{-1}, the lowest sedimentation rate is defined in the middle part of the Kiwada Formation; this horizon is interpreted to document a maximum transgression in the third-order depositional sequence (Ito & Katsura 1992).

On the basis of the mapping and lateral termination patterns of time lines as defined by volcanic ash beds, together with lateral and vertical discontinuity of lithofacies associations, 17 higher-frequency depositional sequences were identified over the entire Kazusa Group (Fig. 6.5) and their development is interpreted to have been controlled primarily by glacioeustatic sea-level changes superimposed on the forearc tectonics (Ito & Katsura 1992; Pickering *et al.* 1999; Tsuji *et al.* 2005). In a deep-water environment, although active shedding of coarse-grained terrigenous sediments was dominant mainly during falling and lowstand stages, these sediments were also supplied by turbidity currents and their related sediment-gravity flows even during transgressive and highstand stages; this likely reflects a narrow shelf, steep slope and high sediment supply in the Kazusa forearc basin (Ito 1998). In a shallow-sea environment, thick sand bodies formed during transgressive stages as sand-ridge deposits, which are considered to have been controlled by the intrusion of a strong warm-water palaeo-Kuroshio Current into the Kazusa forearc basin in response to glacioeustatic sea-level rises (Nakayama & Masuda 1987; Ito 1992; Horikawa & Ito 2004) (Fig. 6.5). Correlation of depositional sequences of the Kazusa Group with those of the Kakegawa and Osaka groups has been investigated on the basis of tephrochronology and tentative correlation of sequence boundaries with the oxygen isotopic sea-level index (Masuda & Ito 1999; Urabe 1999) (Fig. 6.3).

The Kazusa Group is unconformably overlain by the Middle–Upper Pleistocene Shimosa Group, which developed in the palaeo-Tokyo Bay and now constitutes uplands (20–50 m in altitude) in the Kanto Plain, the largest plain in Japan and centred on the Tokyo area (Figs 6.1 & 6.4).

Shimosa Group

The Shimosa Group is interpreted to have developed in response to the uplift of the eastern margin of the Kazusa forearc basin since Middle Pleistocene times (*c.* 0.45–0.08 Ma) (i.e. Kashima–Boso Uplift Zone: Kaizuka 1974) (Fig. 6.6). The bay expanded and contracted several times in response to Middle and Late Pleistocene glacioeustatic sea-level changes (Machida *et al.* 1980; Watanabe *et al.* 1987) (Fig. 6.6), and had a maximum palaeowater depth of 50–120 m (e.g. O'Hara 1973). The Shimosa Group is represented by coastal and shallow-marine deposits, together with a local association of fluvial deposits (Okazaki & Masuda 1992), and consists mainly of six unconformity-bounded lithostratigraphic units (from 1–2 m to as thick as 70 m) (Tokuhashi & Endo 1984) (Fig. 6.6). Each unit is represented by a transgressive–regressive cycle, and an incised-valley-fill deposit locally defines its basal part (e.g. Murakoshi & Masuda 1992; Ito & O'Hara 1994; Nishikawa & Ito 2000). Consequently, each unit is interpreted to be a depositional sequence, which formed mainly during transgressive and highstand stages, and each sequence boundary can be correlated with falling stages of the oxygen isotopic sea-level index (i.e. 100 ka Milankovitch cyclicity) on the basis of the age framework of the Shimosa Group (Fig. 6.6). Because the palaeo-Tokyo Bay was tectonically active, higher-frequency depositional sequences with durations shorter than the Milankovitch cyclicity have also been identified in the highstand coastal deposits (Ito *et al.* 1999).

The transgressive–regressive coastal and shallow-marine successions formed in the palaeo-Tokyo Bay are interpreted to have developed in response to retrogradation and subsequent progradation of barrier island systems along the Pacific margin (i.e. eastern margin) of the bay (Masuda 1992; Okazaki & Masuda 1995; Nishikawa & Ito 2000) (Fig. 6.6). Deltaic systems also developed in some other parts of the margin during highstand stages, where siliciclastic sediments were supplied from the hinterland hills and mountains (e.g. Ito & O'Hara 1994). Before the development of the barrier island systems, locally developed incised valleys were infilled during lowstand and early transgressive stages under the influence of tidal and wave processes. Wave and tidal climates in the palaeo-Tokyo Bay were reconstructed from wave-ripple and tidal-dune deposits (Masuda & Makino 1987; Masuda *et al.* 1988), and are estimated to have been *c.* 1.8 m in mean wave height and *c.* 1.5 m in mean tidal range, which are quite similar to those of present-day Tokyo Bay.

Kakegawa area

A forearc basin in the Kakegawa area was initiated at *c.* 10 Ma and was infilled by the Sagara and Kakegawa groups in ascending order (Figs 6.1 & 6.7). The physiography of the basin for the Sagara Group is interpreted to have changed at *c.* 3.8 Ma, in response to the uplift of a trench-slope break in the eastern margin which produced the north–south-trending Megami Anticline. The Kakegawa Group was subsequently formed in the reorganized basin (e.g. Tsukawaki 1994). Although the switching of depocentres occurred at *c.* 3.8 Ma, no distinct break in sedimentation was documented between the Sagara and Kakegawa groups in the southern offshore area; continuous sedimentation was recorded in the Tamari Formation (Kameo 1998), except for an unconformable contact with basement rocks in the NW basin margin and at the base of submarine-channel-fill deposits near the Megami Anticline (e.g. Tsukawaki 1994; Kameo 1998). The Kakegawa Group is, in turn, unconformably overlain by the Ogasa Group (Fig. 6.7), which is represented by gravelly submarine-channel-fill deposits and overlying gravelly fan-delta deposits (e.g. Muto & Blum 1989).

Sagara Group

The Sagara Group onlaps onto basement rocks older than 15 Ma (Fig. 6.1) and consists of an overall transgressive–regressive cycle

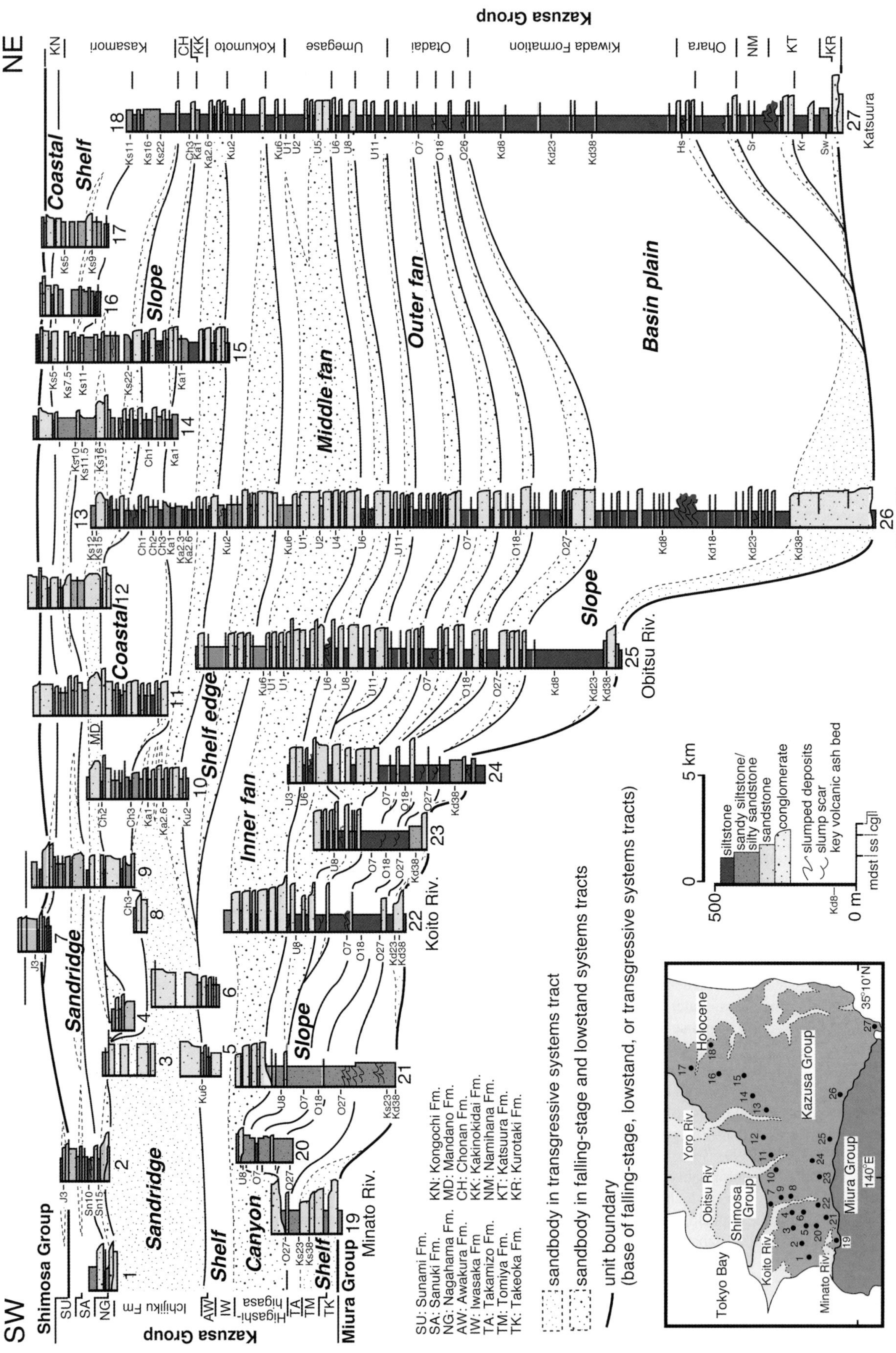

Fig. 6.5. Sequence-stratigraphic classification of the Kazusa Group. Slightly modified from Ito & Katsura (1992).

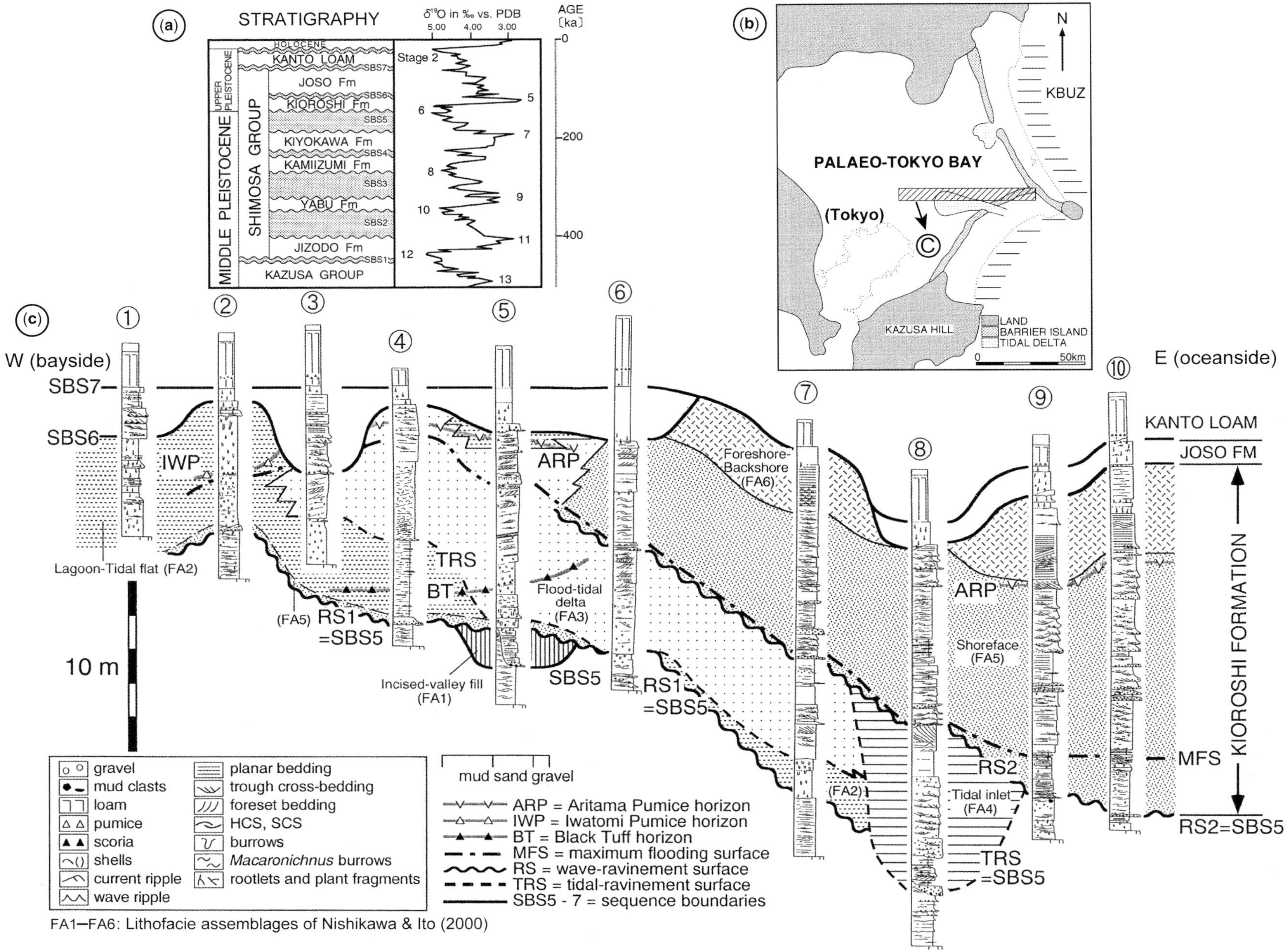

Fig. 6.6. (**a**) Stratigraphic classification of the Shimosa Group and correlation between unconformable boundaries with the oxygen isotopic sea-level index. (**b**) Palaeogeographic map of the palaeo-Tokyo Bay and reconstruction of a barrier-island system of the Kioroshi Formation. (**c**) Stratigraphic cross-section of the Kioroshi Formation, indicating major depositional environments and sequence-stratigraphic classification. See (b) for the location of this cross-section. Slightly modified from Nishikawa & Ito (2000).

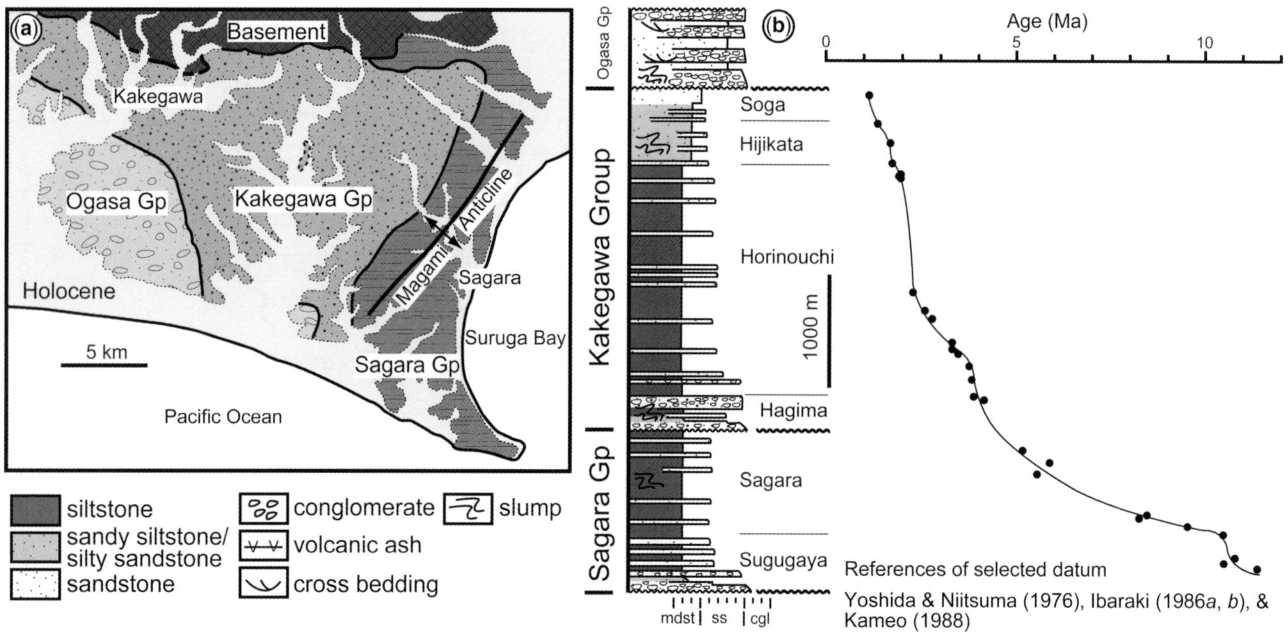

Fig. 6.7. (**a**) Geological sketch map of the Kakegawa area. Simplified from Nobuhara & Takatori (1999). (**b**) A composite stratigraphic section and ages of the Sagara, Kakegawa and Ogasa groups. Lithological data are from Ishibashi (1989), Tsukawaki (1994) and Sakai & Masuda (1996).

of up to 1400 m, which is characterized mainly by interbedded turbiditic sandstones and hemipelagic siltstones, with local associations of slumped deposits and coarse-grained sediment-gravity-flow deposits (e.g. Tsukawaki 1994) (Fig. 6.7). Furthermore, sandy siltstones with local intercalations of turbiditic sandstones and slumped deposits developed along the basin margin (the Lower Tamari Formation: Kameo 1998). Lithofacies features of the group are interpreted to have developed in slope and submarine-fan environments, which were also associated with submarine channels in the basal part of the group (Tsukawaki 1994). Finer-grained, muddy deposits are dominant in the southern downslope area and their palaeowater depth is estimated at 600–1000 m (Kitazato 1997).

Kakegawa Group

The Kakegawa Group is up to 3000 m in maximum thickness and formed during the period 3.8–0.9 Ma (Fig. 6.7). The group was deposited in various sedimentary environments ranging from fluvial and coastal, through shallow-marine and slope to submarine-fan and basin-plain environments. As for the Sagara Group, it also shows an overall transgressive–regressive cycle (e.g. Ujiié 1962; Aoshima 1978; Ishibashi 1989; Tsukawaki 1994) which has been interpreted to be equivalent to a third-order depositional sequence (Sakai & Masuda 1996) (Fig. 6.8). In general, SE-directed palaeocurrents are dominant throughout the group, and fluvial and coastal deposits developed along the northern and NW basin margins. The deepest parts of the basin in the SE show palaeowater depth in the range 300–600 m (Kitazato 1997). Locally, some turbiditic sandstones in the lower part of the group document a possible reflection of turbidity currents on the trench-slope break in the eastern basin margin (Sakai 2000).

On the basis of the termination patterns of time lines represented by volcanic ash beds and lateral and vertical discontinuities of lithofacies assemblages, the entire Kakegawa Group is classified into lowstand, transgressive and highstand systems tracts (in the sense of Vail *et al.* 1991) in a third-order depositional sequence (Fig. 6.8). Because fluvial, coastal, shallow-marine and slope deposits are dominant in the transgressive and highstand systems tracts, higher-frequency depositional sequences and their equivalent parasequences with 40 ka Milankovitch cyclicity have also been defined in the two systems tracts, and are tentatively correlated with the oxygen isotopic sea-level index (Sakai & Masuda 1996). Furthermore, the third-order transgressive deposits are characterized by active shoreface erosion, bypassing of coarse-grained sediments in shelf and upper-slope environments and development of thicker turbidite packages in a lower-slope environment. In contrast, the third-order highstand deposits are generally coarser than those in the transgressive deposits and document storing of coarser-grained sediments in shelf and upper-slope environments with local association of submarine channels (Ishibashi 1989; Sakai & Masuda 1996).

Miyazaki area

A forearc basin succession in the Miyazaki area is represented by the Miyazaki Group which is up to 3000 m thick (e.g. Kino *et al.* 1984) and shows spatial variations in lithofacies and diagenetic features from south to north in the basin (Fig. 6.9). These variations have informally been defined as the Aoshima (south), Miyazaki (central) and Tsuma (north) facies (Shuto 1961). The Aoshima facies has been overprinted by a higher degree of diagenesis than the other two facies and is represented by interbedded turbiditic sandstones and hemipelagic siltstones except for the lower part, which is characterized by coastal and shallow-marine deposits (Nakamura *et al.* 1999*a*). The Miyazaki facies is also represented by interbedded turbiditic sandstones and hemipelagic siltstones formed mainly in shelf and slope environments with local associations of coarse-grained fluvial, coastal and submarine-canyon-fill deposits (Takashimizu 2009). The Tsuma facies is represented by siltstones, sandy siltstones and silty sandstones of coastal, shelf and slope deposits, which are associated locally with interbedded turbiditic sandstones and hemipelagic siltstones, slumped deposits and coarse-grained submarine-canyon-fill deposits (Endo & Suzuki 1986; Majima *et al.* 2003). Overall, each facies shows an overall transgressive–regressive cycle (Fig. 6.9).

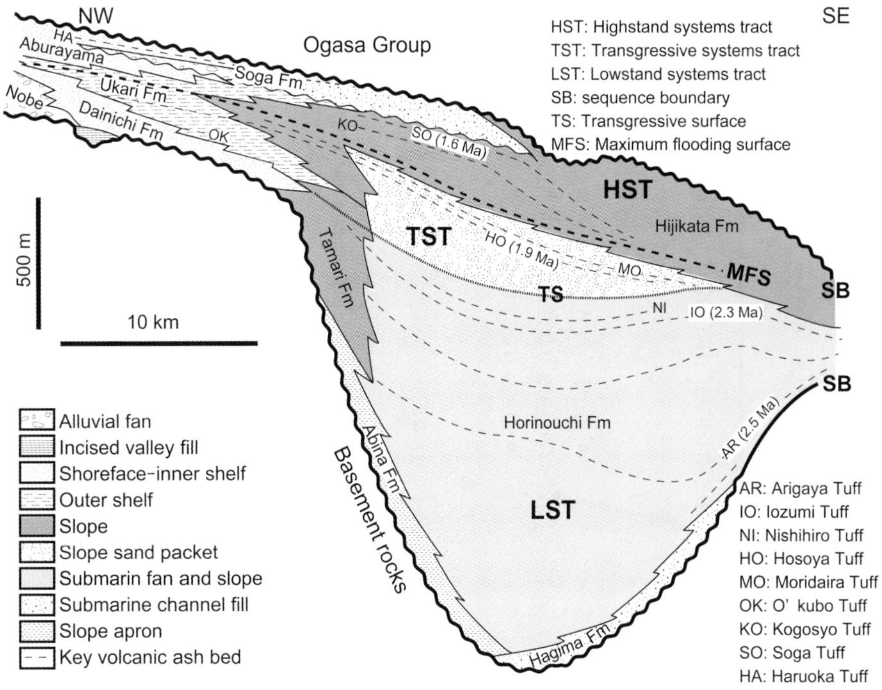

Fig. 6.8. Schematic stratigraphic cross-section of the Kakegawa Group, illustrating depositional environments and sequence-stratigraphic classification. Slightly modified from Sakai & Masuda (1996).

Although temporal relationships between the three facies and their lithostratigraphic and chronostratigraphic classification and correlation still remain controversial, and the Aoshima facies has been more formally defined as the Uchiumigawa Group (e.g. Kino *et al.* 1984; Endo & Suzuki 1986; Suzuki 1987; Nakamura *et al.* 1999*a*; Oda *et al.* 2011), the initial sedimentation of the Aoshima and Miyazaki facies is interpreted to have occurred at *c.* 10.5 and 5.6 Ma, respectively, and the age of the uppermost part of the Aoshima facies is considered to be equivalent to that of the lowest part of the Miyazaki facies (Nakamura *et al.* 1999*a*) (Fig. 6.9). The initiation of the Tsuma facies was slightly later than that of the Miyazaki facies and is considered to have been at *c.* 5.4 Ma (Endo & Suzuki 1986). The base of the submarine-canyon-fill deposits in the uppermost part of the Miyazaki Group (i.e. the Hisamine Member of the Takanabe Formation: Endo & Suzuki 1986) is interpreted to indicate a distinct erosional hiatus formed at 2.3–2.2 Ma (Oda *et al.* 2011). This erosional hiatus is called the Hisamine Unconformity and is correlated with the Kurotaki Unconformity in the Boso–Miura area (Oda *et al.* 2011). On the basis of the recognition of this unconformity, Oda *et al.* (2011) separated the uppermost part of the Tsuma facies from the Miyazaki Group as the Hyuga-nada Group (Fig. 6.9), and correlated this newly defined group with the lower Kazusa Group in the Boso–Miura area (Fig. 6.1).

Although the origin of northwards migration of the depocentre of the Miyazaki Group is not yet clearly understood, the tectonic framework of the forearc basin is interpreted to have been controlled by fluctuation of the convergent direction of the Philippine Sea Plate relative to the Eurasia Plate since 6 Ma (e.g. Kamata & Kodama 1994). Furthermore, anticlockwise rotation of about 30° is considered to have commenced in and around the forearc basin at *c.* 3 Ma, as a response to rifting of the southern Kyushu and the northern Okinawa Trough (Kodama *et al.* 1995; Torii & Oda 2001). Although detailed classification of depositional sequences and correlation of high-frequency depositional cyclicities with the oxygen isotope sea-level index have not yet been investigated in the Miyazaki Group, some faunal features of the group document palaeo-oceanographic fluctuation, which is considered to have responded to glacial–interglacial global environmental changes (e.g. Nakamura *et al.* 1999*a*; Iwatani *et al.* 2009; Chiyonobu *et al.* 2012).

Intra-arc basin successions (FM and YH)

This section describes four major intra-arc basin successions in central Japan, namely the Pliocene–Pleistocene Kobiwako, Osaka and Tokai groups and the Miocene Mizunami Group. The thickness of these groups varies from 200 to 3000 m in response to variations in tectonic regimes in each sedimentary basin, and they were deposited mainly in the Kinki and Tokai districts in the central part of SW Japan (Fig. 6.10).

Kobiwako and Osaka groups

Stratigraphy and major lithology

The Pliocene–Pleistocene Kobiwako Group, which is distributed around Lake Biwa and the adjacent Ohmi Basin to the SE (Fig. 6.10), consists mainly of thick lacustrine mudstones alternating with alluvial sandstones and conglomerates which are intercalated with more than 100 volcanic ash (tephra) beds and up to 1500 m in total thickness (Yokoyama 1969; Kawabe 1989). The depocentre for these sediments is referred to as the Kobiwako Basin, and the strata display a homoclinal structure. The intercalated tephra beds enable detailed correlation of the Kobiwako Group with strata formed in the other basins far from the Kobiwako Basin (Kurokawa *et al.* 2004; Satoguchi & Nagahashi 2012). The incipient deposition of the group began at 5–6 Ma (Takemura 2002) or 4.5 Ma (Satoguchi & Nagahashi 2012), and deposition has continued to the present day in Lake Biwa.

The Pliocene–Pleistocene Osaka Group, 1000–3000 m in thickness, is distributed across Osaka Bay, the Osaka Plain and Kyoto (Yamashiro) and Nara basins (Fig. 6.10). The lower part of the group consists of alternating gravels, sands and muds of alluvial and lacustrine in origin, and the upper part is represented by alternations of marine bay muds and terrestrial fluvial sands

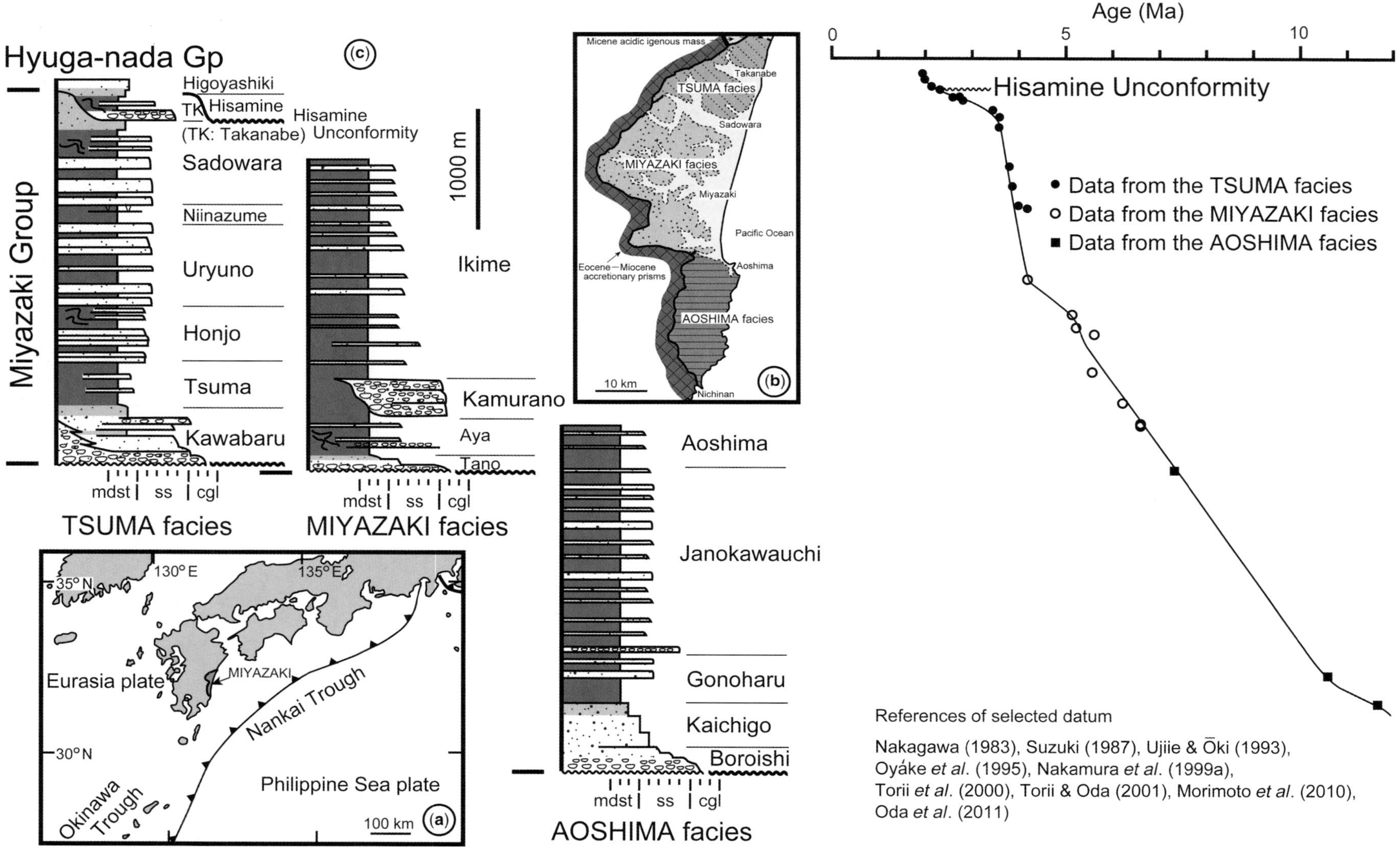

Fig. 6.9. (**a**) Plate-tectonic framework of the Miyazaki area. Slightly modified form Oda *et al.* (2011). (**b**) Distribution of the Aoshima, Miyazaki and Tsuma facies of the Miyazaki Group. Simplified from Suzuki (1987) and Nakamura *et al.* (1999*a*, *b*). (**c**) A composite stratigraphic section and ages of the Miyazaki and Hyuga-nada groups. Lithological data are from Endo & Suzuki (1986), Nakamura *et al.* (1999*a*, *b*) and Takashimizu (2009). See Figure 6.7 for the other explanations.

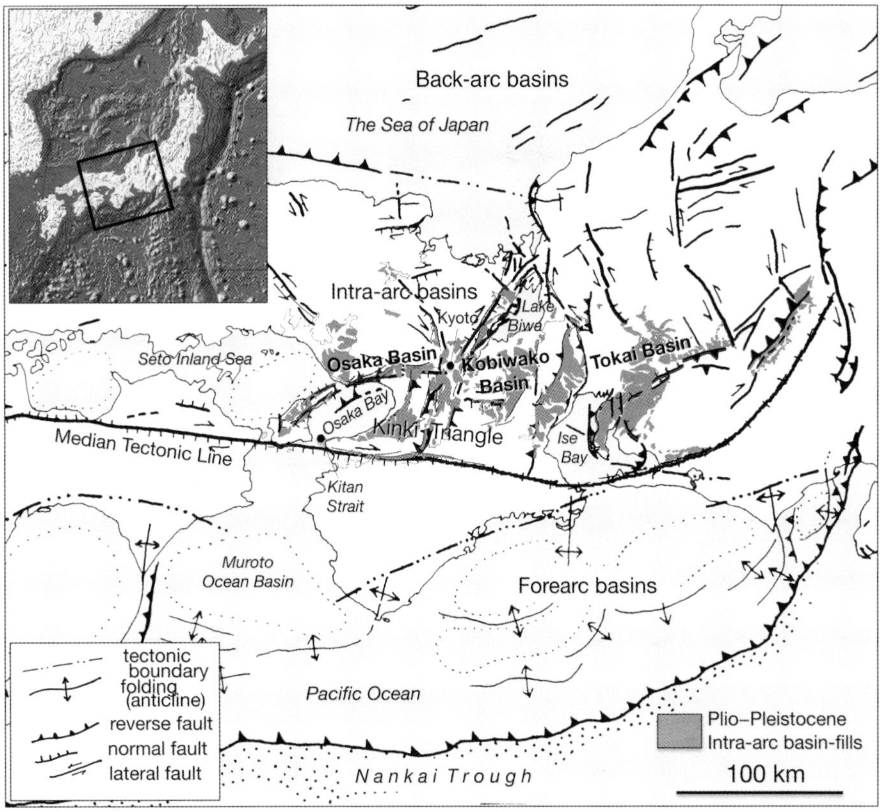

Fig. 6.10. Geological sketch map showing the distribution of the Pliocene–Pleistocene intra-arc basin-fill successions in the Kinki and Tokai districts, central Japan. The distribution and structures of basin-fill successions are partially from Ota *et al.* (2004).

and gravels (Itihara 1993; Kitada *et al.* 2008), which are considered to have been controlled by Quaternary glacioeustatic sea-level changes (Masuda & Ito 1999). Deposition of the Osaka Group began at 3.2–3.5 Ma as documented by two deep-borehole cores, the Higashinada (Biswas *et al.* 1999) and Kankuu cores (Kitada *et al.* 2008), in the central part of the sedimentary basin. The Osaka Group is unconformably overlain by uppermost Pleistocene–Holocene deposits (known as Chuseki-so in Japanese) which comprise, in ascending order, alluvial gravels, estuarine sands and muds, marine clays and deltaic and coastal sands (Masuda 2007) (Fig. 6.11).

Tectonics of basin formation

Two tectonic models have been proposed for the origin of the Osaka and Kobiwako groups sedimentary basins: (1) tectonic construction in the 'Kinki Triangle' area (Huzita 1962); and (2) east–west compression of the 'Rokko Movements' beginning at 0.5–0.6 Ma (Ikebe & Huzita 1966; Huzita 1968). The movements are interpreted to have been induced by basement folding under a phase of east–west-aligned compressional stress which started at *c.* 2 Ma and subsequently changed to block faulting at *c.* 0.5 Ma, resulting in a rapid increase in the topographic contrast between rising mountains and subsiding sedimentary basins.

In the Kobiwako Basin, three major gravel horizons dated at 3, 1.5 and 0.5 Ma have been reported as providing signals of the changing tectonic regimes (Yokoyama 1983; Uemura 2001; Takemura 2002). According to our re-examination of the gravel facies tracts, which were first described by Hayashi & Kawabe (1993), these gravel horizons are not laterally continuous and change locally into muddy deposits. The gravel horizons therefore cannot be used as basin-wide time lines, and are considered as not necessarily documenting tectonic events in the basin. The timing and modes of tectonic movements in this basin therefore remain controversial.

Each of the sedimentary-fill successions of the Osaka and Kobiwako basins shows an overall geometry quite similar to that formed in a half-graben under tensional tectonics. That is, in longitudinal cross-sections the successions thicken towards the major fault with large displacements along the basin-margin mountains, while a simple homoclinal structure has developed on the opposite side of the basin margin. In contrast, in transverse cross-sections the successions have symmetrical cross-sectional geometry with respect to the marginal faults on both sides in association with horst-and-graben structures in the basin centres, with homoclinal structures dipping towards both margins. Uemura (2001) interpreted the basement structure beneath Lake Biwa as showing a geometry quite similar to that of the basin–range system. In our interpretation, the sedimentary basins of the Osaka and Kobiwako groups are considered to have initially formed under a tensional tectonic regime, which was subsequently converted into a compressional tectonic regime in response to variations in relative plate motion.

Topography of basement surface

The basement to both the Osaka and Kobiwako groups comprises Cretaceous granitic rocks and Mesozoic–Palaeozoic sedimentary rocks. The basement topography beneath the *c.* 1000 m thick deposits of the Kobiwako Group in Lake Biwa has been investigated by high-resolution seismic reflection surveys (Horie 1991; Ikawa 1991). Similar seismic surveys were also carried out in the Kyoto Basin and the northeastern part of the Osaka Basin (Kansai Geo-informatics Council 2002) (Fig. 6.11) and in Osaka Bay (Iwabuchi *et al.* 2000) (Fig. 6.12). These surveys showed that (1) there are ridge- and valley-like basement topographies with a relief of 500–1000 m; (2) this relief is comparable to that of the modern valley–ridge system of the surrounding land; and (3) the basement topography is continuous with that of modern valleys and ridges on land (Uemura & Taishi 1990; Masuda *et al.* 2010). These findings

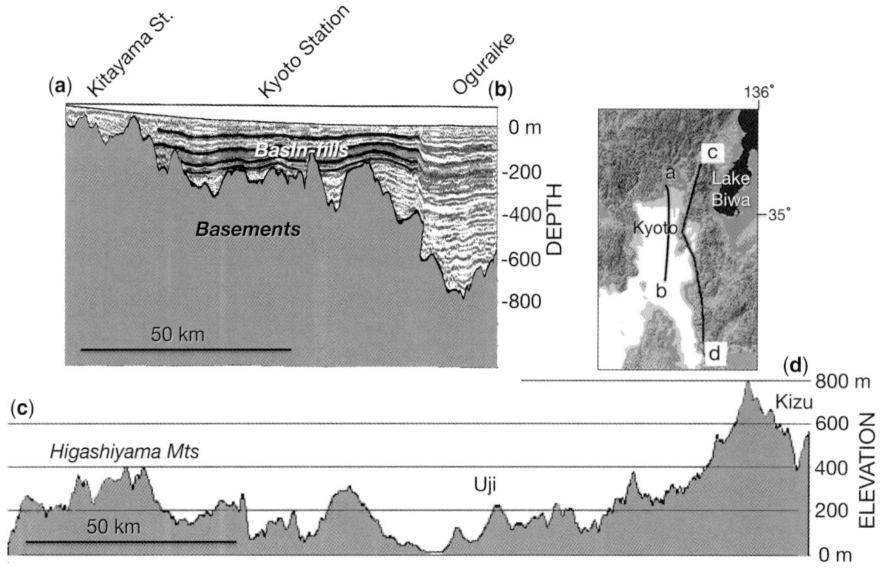

Fig. 6.11. Surface and basement topography in a profile across the Kyoto Basin. Geological sections are from Kansai Geo-informatics Council (2002).

show that the original topography of the basin before the deposition of the basin-fill successions was characterized by valleys and ridges similar to the present-day topography of the surrounding areas. Each basin has large boundary faults, including the active Biwako-seigan fault system (West Coast Fault of Lake Biwa) in the Kobiwako Basin and the Arima–Takatsuki tectonic line and/or Osaka-wan Fault in the main Osaka Basin (Fig. 6.12). These faults are therefore considered to have played an important role in the formation of the basins. In contrast, the origin of the undulating basement surface of the basins (which originally had peneplain topography before the Rokko Movements) and of the Rokko Movements themselves remain controversial.

Lake formation by fault movement

Masuda *et al.* (2010) conducted a facies analysis of the lowermost deposits (*c.* 1.7 Ma) in the Karasuma Deep Drilling Core in Lake Biwa and reported that palaeo-Lake Biwa was a shallow and narrow lake. The palaeolake is considered to have been created by floodwaters backing up behind natural dams formed by repeated basement uplift events as a response to seismic movement of the basin boundary faults. Active fault movements were therefore the main mechanism for the initiation of naturally dammed lakes in intra-arc basins.

There were three or four such palaeolakes in the Kobiwako Basin during the depositional period of the Kobiwako Group. It was previously believed that the positions of the palaeolakes moved from south to north in the basin along with depocentre migration of the group (Yokoyama 1984). These lakes were more likely formed by the damming of rivers than by differential subsidence of the basin, however. Thick lacustrine deposits have also been reported from a deep drilling core in the southern Osaka Basin (Kitada *et al.* 2008). The development of the thick lacustrine deposits is considered to have been controlled by fault movement of the Median Tectonic Line (MTL), one of the most significant faults in the Japanese islands, which may have formed a natural dam responsible for the

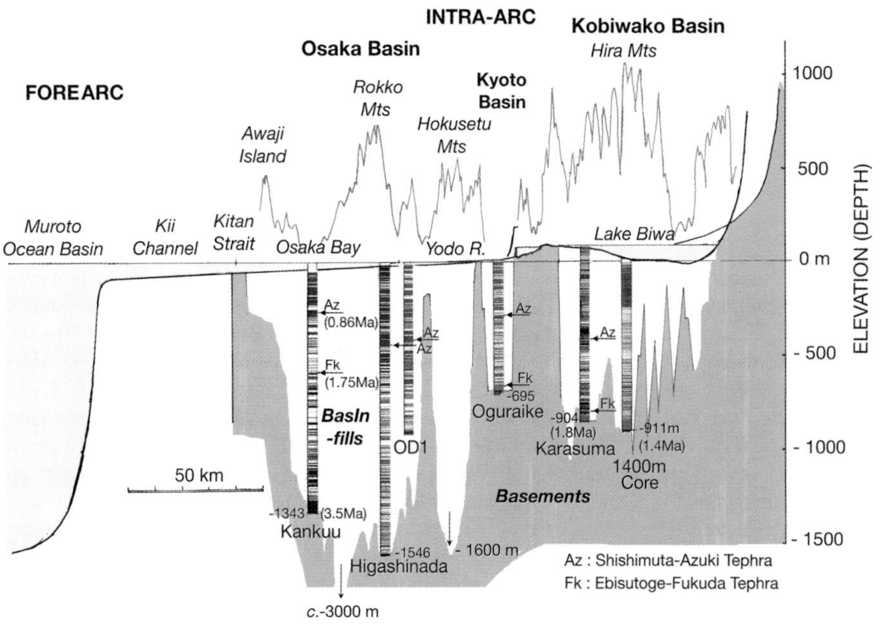

Fig. 6.12. Cross-section of the Kobiwako and Osaka basins and stratigraphic columns from the Karasuma core (Karasuma Deep Core Investigation Group 1999), Oguraike core (Kansai Geo-informatics Council 2002), OD1 core (Ikebe *et al.* 1970), Higasinada core (Kansai Geo-informatics Council 1999) and Kankuu core (Kitada *et al.* 2008).

deposition of the Osaka Group at the narrow Kitan Strait. Behind this natural dam, a lake comparable to modern Lake Biwa in area and depth occupied the southern part of the Osaka Basin during the period 3.5–2.0 Ma, and is referred to as the palaeo-Osaka Lake (Yokoyama 1983).

Mizunami Group

Distribution of the Miocene sedimentary rocks (First Setouchi Series)

The inland zone of SW Japan on the north side of the MTL is called the Setouchi province, and is where Miocene sedimentary basins developed (Shibata & Itoigawa 1980) (Fig. 6.13). The sedimentary rocks in these basins are referred to as the First Setouchi Series (e.g. Ikebe 1957) and rest upon Cretaceous granites and the Ryoke metamorphic complex. The southern limit of the Setouchi province almost coincides with the MTL, which is considered to have been activated as a normal fault and played an important role in the development of half-grabens in some areas far west from the Setouchi province (Takagi *et al.* 1992; Kusuhashi & Yamaji 2001). The crustal movement along the MTL is therefore also thought to have controlled the development of the Miocene sedimentary basins in the Ise–Tokai district.

The Setouchi province of the Ise–Tokai district is further subdivided into two areas: the Iga–Owari area to the north and the Ise–Mikawa area to the south (Huzita 1962) (Fig. 6.13). Fluvial and shallow-marine deposits are dominant in the Iga–Owari area, and the sedimentary successions in this area are 200–400 m in thickness (Yamashita *et al.* 1988). The sedimentary successions of the Ise–Mikawa area, as much as 1000–1500 m in thickness (Yamashita *et al.* 1988), consist mainly of marine sediments and exhibit sedimentary environments deeper than those represented by the Iga–Owari exposures.

Local stratigraphic names have been assigned for the Miocene sedimentary successions. In the Iga–Owari area these include: the Mizunami Group in Gifu; Chikusa Group in Aichi; Ayukawa Group in Shiga; and Suzuka and Awa groups in Mie. In the Ise–Mikawa area they include: the Shitara and Okazaki groups in Aichi; Morozaki Group in the Chita Peninsula, Aichi; Ichishi Group in Mie; and Yamakasu Group in Nara. The following section focuses mainly on the Mizunami Group, which has been studied by many researchers with regard to its litho- and biostratigraphy, palaeontology and sedimentology.

Stratigraphy and sedimentary facies of the Mizunami Group. The Mizunami Group is well exposed along the Toki River in Mizunami and Toki cities of Gifu Prefecture, and is up to 170 m thick in Mizunami city. The Mizunami Group unconformably overlies Mesozoic and Palaeozoic sedimentary rocks of the Mino Zone and Cretaceous granites and quartz-porphyry, and is unconformably overlain by the Pliocene–Pleistocene Seto Group. The Mizunami Group consists of four formations, namely the Toki (lignite-bearing), Hongo, Akeyo and Oidawara formations in ascending order (Itoigawa 1974*a*) (Fig. 6.14). The Toki Formation is a lignite-bearing succession and is distributed only in the northern part of Mizunami city. The Hongo Formation is characterized by a fining-upwards sedimentary succession from conglomerates, through coarse- and fine-grained sandstones to pumiceous sandstones intercalated with mudstones which contain abundant plant fossils and non-marine molluscan fossils (Itoigawa 1974*b*). This fining-upwards succession is considered to have developed in response to retreat of an alluvial fan system, and reflects a change in depositional environments from proximal to distal parts of the fan. The boundary between the Hongo Formation and the overlying Akeyo Formation is defined by a distinct erosional surface, considered to be a ravinement surface (Hiroki & Matsumoto 1999). Ravinement deposits on this surface are characterized by poorly sorted, coarse-grained sandstones with pebbles and rip-up clasts. Medium-grained sandstones, which contain swaley and hummocky cross-stratifications and ripple lamination, indicating a shoreface environment, overlie the ravinement deposits. The lowest part of the Akeyo Formation is assigned as the Tsukiyoshi Member, and consists of massive muddy sandstones and sandy mudstones deposited in an offshore–shoreface transition zone. The Tsukiyoshi Member contains abundant marine molluscan fossils and *Ophiomorpha*-type burrows and is conformably overlain by the Togari

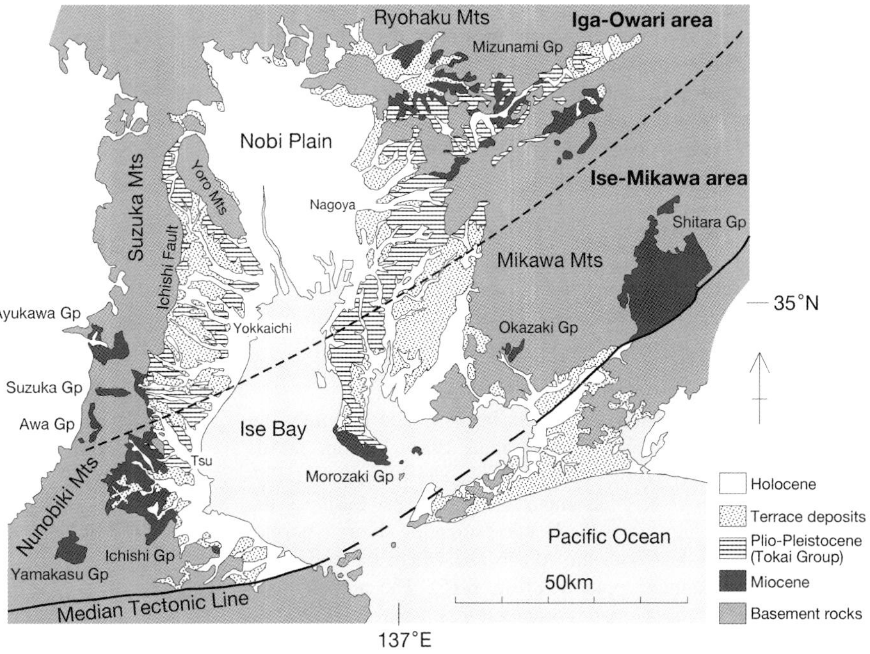

Fig. 6.13. Distribution of the Miocene and Pliocene–Pleistocene strata in the Ise–Tokai district, central Japan. Illustration based on Geological Survey of Japan AIST (2012).

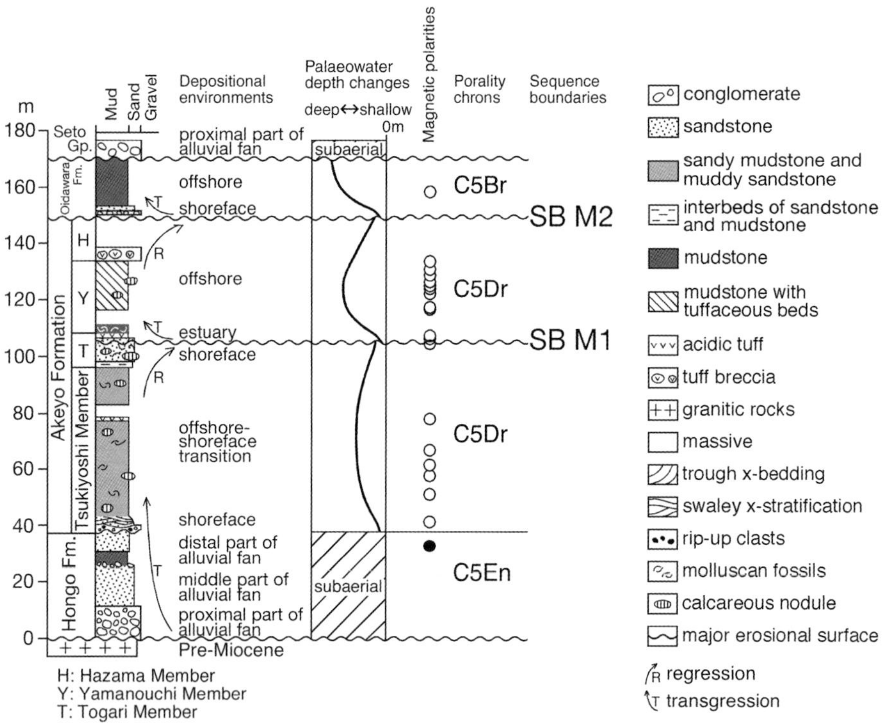

Fig. 6.14. Schematic columnar section of the Mizunami Group, showing vertical changes in lithofacies, regressive–transgressive patterns, depositional environment, palaeowater depth and the stratigraphic positions of sequence boundaries, magnetic polarities and polarity chron assignments. The illustration was drawn based on Hiroki & Matsumoto (2003) and Irizuki & Hosoyama (2006).

Member which consists mainly of intensely bioturbated, yellowish-grey shoreface sandstones with a massive appearance. The uppermost part of the Togari Member is represented by tidal deposits up to 2 m thick with an unconformable basal contact. These tidal deposits comprise interbeds of fine-grained tuffs and medium-grained sandstones with abundant *Ophiomorpha*-type burrows. The interbeds are interpreted as tidal bundles formed in an estuary. The tidal deposits, in turn, are covered by offshore deposits of the Yamanouchi Member, indicating transgression. The Yamanouchi Member consists of dark bluish-grey sandy mudstones with intense bioturbation and abundant molluscan fossils.

The Oidawara Formation unconformably overlies the Akeyo Formation (Kitamura 1959; Itoigawa 1960), and consists of homogeneous diatomaceous massive mudstones formed in an offshore environment. The unconformable basal surface of the Oidawara Formation represents another ravinement surface; the ravinement deposits, which are defined as the Nataki Conglomerate (Itoigawa 1974*a*; Itoigawa *et al.* 1974), contain coral and various molluscan fossils, showing a wide range of palaeowater depth from 0 to 200 m.

Two major unconformities are recognized in this succession at the uppermost part of the Togari Member of the Akeyo Formation and at the base of the Oidawara Formation, and have been interpreted as sequence boundaries (SB M1 and SB M2 in Fig. 6.14) formed by relative sea-level falls (Hiroki & Matsumoto 1999).

Records of Miocene glacioeustasy

The age of the Mizunami Group has been studied by analysing foraminiferal fossils (Saito 1963; Ibaraki 1981), diatom fossils (Koizumi 1981; Yanagisawa 1993) and magnetostratigraphy (Hayashida 1986; Hiroki & Matsumoto 1999). On the basis of the magnetostratigraphic studies the Hongo Formation shows a normal magnetic polarity, and both the Akeyo and Oidawara formations show reversed magnetic polarities (Fig. 6.15). The normal magnetic polarity of the Hongo Formation has been correlated to Chron C5En (Irizuki & Hosoyama 2006); the reversed magnetic polarities of the Akeyo and Oidawara formations have been correlated to Chron C5Dr and Chron C5Br, respectively (Irizuki & Hosoyama 2006).

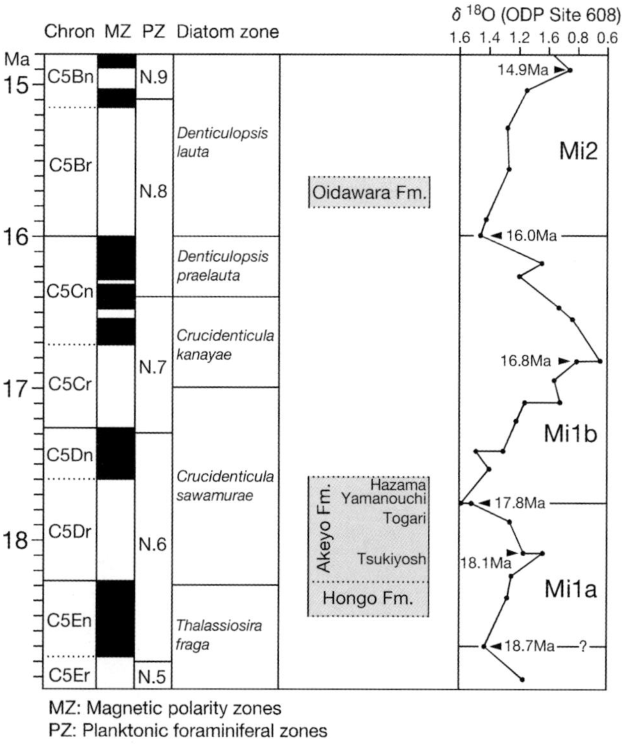

Fig. 6.15. Chronostratigraphy of the Mizunami Group and the Miocene marine oxygen isotope fluctuation of ODP Site 608. Modified from Irizuki & Hosoyama (2006). The age assignments for the marine oxygen isotope fluctuation are based on Takahashi (2004).

Marine oxygen isotope events (positive excursion of $\delta^{18}O$) with the growths of the Antarctic ice sheet have been identified in Miocene sedimentary cores (Wright & Miller 1992; Miller *et al.* 1996). The oxygen isotope events exhibit excellent correlations with SB M1 and SB M2 of the Mizunami Group (Irizuki & Hosoyama 2006) (Fig. 6.15). The positive shift of oxygen isotope at 17.8 Ma between Mi1a and Mi1b oxygen isotope zones and SB M1 occurred simultaneously in Chron C5Dr. Sequence Boundary M2 exhibits a time gap of 15.8–17.6 Ma in which the positive shift at 16.0 Ma between Mi1b and Mi2 oxygen isotope zones occurred. The synchronous occurrences of the positive shifts of $\delta^{18}O$ and the sequence boundaries of the Mizunami Group indicate that the relative sea-level falls that formed SB M1 and SB M2 were primarily caused by glacioeustatic sea-level changes.

Tokai Group

Distribution of the Tokai Group

The Tokai Group represents the Pliocene–Pleistocene succession in the Tokai district, and constitutes the hills around Ise Bay (Ishida & Yokoyama 1969). The group contains abundant freshwater molluscan fossils, plant fossils and proboscidea fossils, and consists of alluvial-fan, fluvial, swamp and lake deposits which developed in the Tokai Basin (Yoshida 1990; Nakayama 1991; Tado Collaborate Research Group 1998) (Fig. 6.1). The southern limit of the Tokai Basin is the MTL and the western limit is the Ichishi Fault, which defines the eastern margin of the Suzuka and Nunobiki mountains. The Tokai Basin is surrounded by the Mikawa Mountains to the east and the Ryohaku Mountains to the north. The maximum extent of the Tokai Basin is *c.* 100 km both east–west and north–south, and the total distribution area is *c.* 6000 km^2 (Yoshida 1990). The Tokai Group equivalent sedimentary successions have local names within the Tokai Basin. The Seto Group is distributed mainly in Nagoya and Seto (Makiyama 1950), and the Tokoname and Age groups are in the Chita Peninsula (Oze 1929) and the northern part of Mie Prefecture (Ogawa 1919), respectively (Fig. 6.13). The thickness of the sedimentary successions is 200 m for the Seto Group, 700 m for the Tokoname Group and 1800 m for the Age Group (Nakayama & Yoshikawa 1990; Yoshikawa *et al.* 1991; Yoshida 1988).

Stratigraphy and ages of the Tokai Group

Because the Tokai Group consists mainly of non-marine deposits, lateral facies changes are remarkable even within each distribution area. The Tokai Group contains many volcanic ash beds, allowing detailed correlations of heterolithic sedimentary successions formed in different areas within the Tokai Basin. The ages of the sedimentary successions have been estimated by the fission-track dating of some selected volcanic ash beds. The Kaminoma volcanic ash bed in the lower part of the Tokoname Group, for example, is estimated to have been deposited at 5.3 Ma (Makinouchi *et al.* 1983), and the Terakawa volcanic ash bed in the middle part of the Age Group in the Kameyama area of Mie Prefecture was deposited at 3.7 Ma (Yoshida 1987). Furthermore, the Ichinohara volcanic ash bed in the lower part of the Age Group and the Rokkoku volcanic ash bed in the upper part of the Age Group in the Inabe area of Mie Prefecture were deposited at 2.9 and 1.4 Ma, respectively (Yokoyama *et al.* 1980). The sedimentation in the Tokai Basin therefore began at *c.* 6 Ma, and continued in response to basin subsidence until 1 Ma. Inundation of seawater into the basin started during the interglacial periods in Middle–Late Pleistocene times, and marine environments became dominant in the basin. Consequently, the area of the Tokai Basin was subsequently reduced to that of the present Ise Bay.

Back-arc basin successions (OT and TN)

The northwestern side of the SW and NE Japan arcs forms a large back-arc marginal sea, the Sea of Japan, which developed between latest Oligocene (*c.* 24 Ma) and Early Miocene (*c.* 17 Ma) times as a result of back-arc opening (e.g. Otofuji *et al.* 1985). The Sea of Japan and its margins comprise several basins (Fig. 6.16), reflecting a complex back-arc opening history. The Japan Basin is the largest such basin and occupies the northern half of the Sea of Japan. The southern half is segmented into the Tsushima (Ulleung) Basin, Yamato Basin, San'in Basin, Hokuriku Basin, Toyama Trough and Japan Sea East Margin Basins. Among these basins, the Tsushima (Ulleung) Basin, Toyama Trough and Japan Sea East Margin Basins are filled with a huge amount of siliciclastic and volcaniclastic sediments as a response to the interplay between active subsidence and large sediment supply from the uplifted Japanese islands and Korean Peninsula. This section focuses mainly on the Japan Sea East Margin Basins, because these basins have accumulated thick sediments and have been well studied in terms of stratigraphy, depositional systems and tectonics.

The Japan Sea East Margin Basins were generated as a series of NE–SW-trending rift basins at *c.* 17–16 Ma (Suzuki 1989; Yamaji 1989, 1990; Takano 2002*a*, *b*), which corresponded to the final phase of the back-arc opening. These rifts together form a composite basin consisting of several large basins, such as the Niigata Basin in

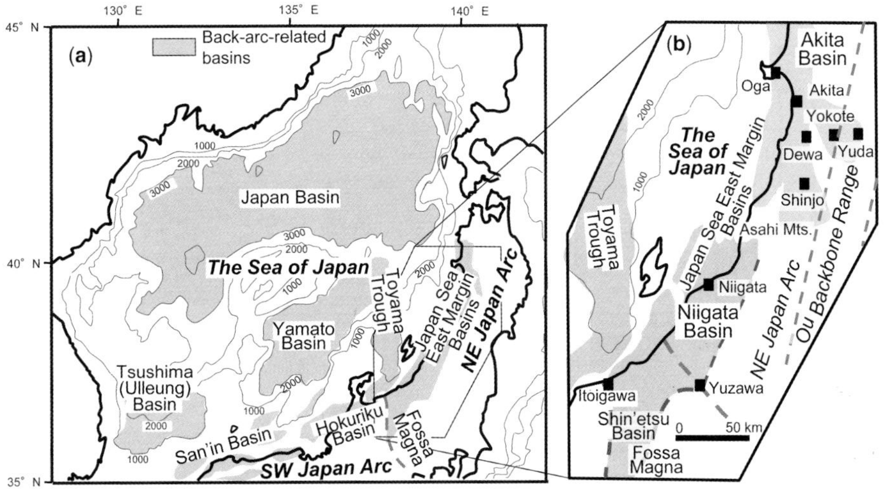

Fig. 6.16. Maps showing the locations of (**a**) major back-arc basins of the NE and SW Japan arcs, and (**b**) the Japan Sea East Margin Basins including the Niigata and Akita basins.

the south and the Akita Basin in the north (Fig. 6.16), both of which yield hydrocarbon resources. Recent studies suggest that a boundary between the NE Japan microplate and the Eurasian plate can be traced through the Japan Sea East Margin Basins, and that the eastern margin of the Sea of Japan is represented by an active fold and fault zone under a compressional stress field (e.g. Nakamura 1983; Okamura *et al.* 1995).

Niigata Basin

Tectonostratigraphy and basin evolution

The Niigata Basin constitutes the southern half of the Japan Sea East Margin Basins (Fig. 6.16) and extends continuously southwestward to the Shin'etsu Basin in the Northern Fossa Magna, which is a large graben zone along the junction between the NE and SW Japan arcs. The Niigata Basin has been filled with thick siliciclastic and volcaniclastic sediments and volcanic rocks during Miocene–Holocene times, and the sediments are composed of submarine-fan turbidites, slope to basin-floor shales, shelf siltstones, delta to shallow-marine sandstones, fluvial alternating beds of sandstone and shale, and fluvial to alluvial-fan conglomerates (Kobayashi & Tateishi 1992; Tateishi *et al.* 1997; Takano 2002*a*, *b*). The Niigata Basin can be categorized as an inverted rift basin; since the basin initially originated as a rift basin and was subsequently converted to a compressive basin since Late Miocene times. Accordingly, the basin evolution can be divided into four tectonic stages (Fig. 6.17) on the basis of the temporal changes in subsidence patterns, major depositional systems and their stacking patterns (Takano 2002*a*, *b*), as detailed in the following.

Synrift to early post-rift successions (Stage I: 16.5–12.5 Ma)

The Stage I sediments represent synrift to early post-rift successions, which filled the bottom part of the rift basins and are considered to have originally developed as a wide rift mode (*sensu* Ingersoll & Busby 1995) in and around the Niigata Basin. Representative lithostratigraphic units of the Stage I sediments include the Middle Miocene Tsugawa and Nanatani formations. The Stage I sediments are dominated by slope to basin-floor systems (the Nanatani Formation), which consist of massive or bedded mudstone facies (Fig. 6.17). The slope to basin-floor systems overlies alluvial-fan, fan-delta and nearshore systems and pyroclastic deposits (the Tsugawa Formation) developed at the base of the Stage I sediments, in particular along

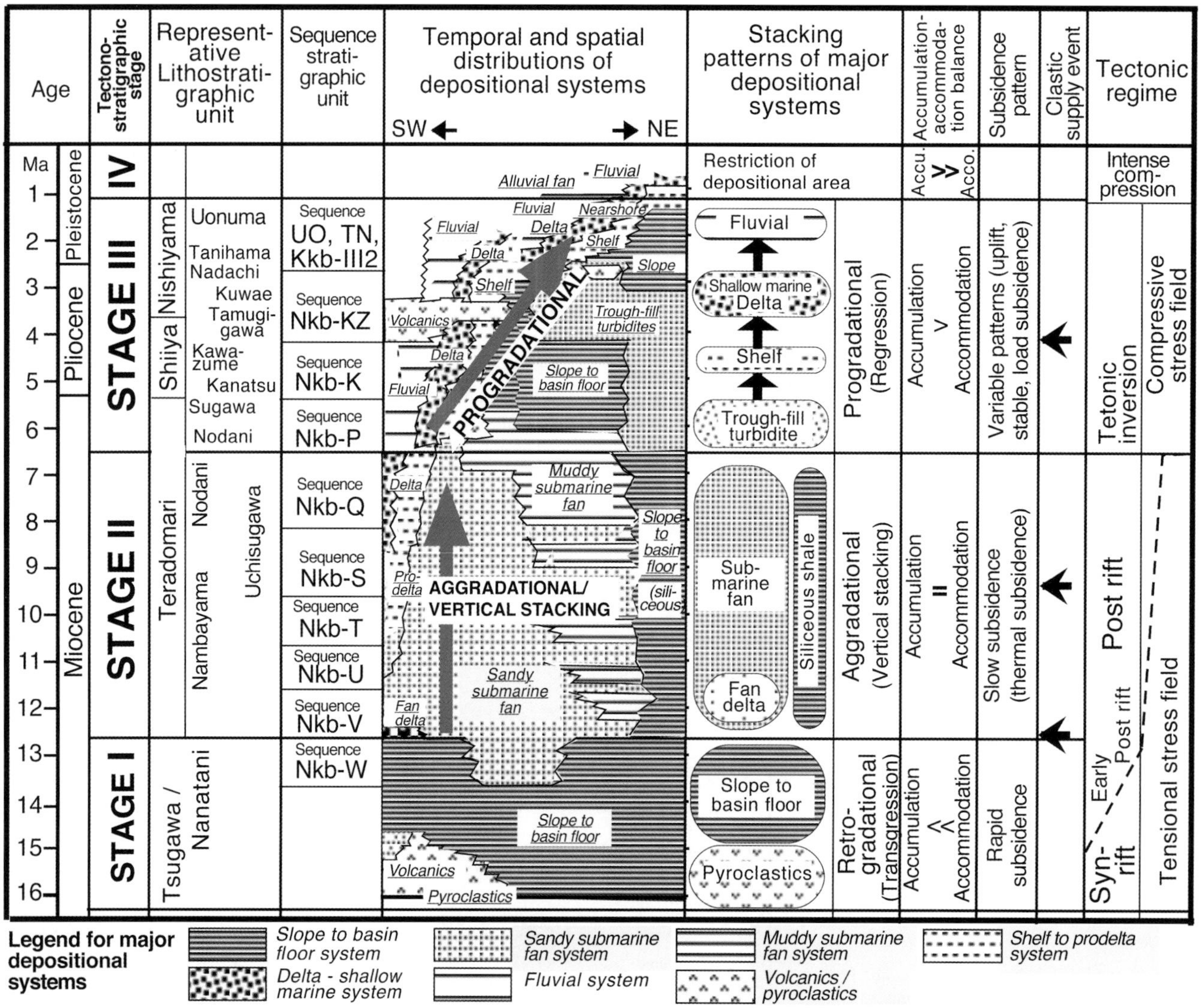

Fig. 6.17. Tectonostratigraphic stage division chart of the Niigata Basin showing classification of lithostratigraphic and sequence stratigraphic units, major depositional systems, stacking patterns, subsidence pattern and tectonic regimes for each stage. After Takano (2002*a*, *b*; 2011).

the basin margin. This basal succession suggests that soon after the deposition of coarse siliciclastic and pyroclastic sediments, an overall transgression occurred and most of the basin was inundated and abruptly changed to a middle to lower bathyal marine environment. Because of rapid subsidence in the synrift and early post-rift stage, the rate of creation of accommodation space is considered to have exceeded that of sediment accumulation during Stage I. The latter resulted in abrupt deepening of the basin, which was documented by a retrogradational stacking of the basal successions at the basin margin (Fig. 6.17). Another important reason for the predominance of muddy strata in the Stage I sediments was likely the reduction of coarse clastic sediment supply to the basin as a response to the transgression and weak tectonic activity in the provenance area (Takano 2002*a*, *b*).

Late post-rift successions (Stage II: 12.5–6.5 Ma)
The Stage II sediments represent late post-rift successions. Because of an exponentially decreasing subsidence rate during Stage II, the rate of creation of accommodation space is considered to have been balanced with that of sediment accumulation, having resulted in an aggradational stacking pattern of depositional systems (Fig. 6.17; Takano 2002*a*, *b*). The representative lithostratigraphic units of the Stage II sediments include the uppermost Middle–Upper Miocene Nambayama Formation in the southern part, the Teradomari Formation in the middle part and the Uchisugawa Formation in the northern part of the Niigata Basin. In the southern part, submarine-fan systems (defined as the 'Nambayama submarine fan') (Endo & Tateishi 1990) were widely developed as a response to the activation of coarse clastic sediment supply from the Hida and Kanto mountains (Figs 6.17 & 6.18) (Takano 2002*a*, *b*). The Stage II sediments in the south to central parts of the basin can be divided into five third-order depositional sequences: Sequence Nkb-V, -U, -T, -S and -Q, in ascending order. Each depositional sequence consists of coarser turbidite successions in the lower to middle sections and mudstone-dominated turbidite successions in the upper section (Takano 2011). In the central part of the basin, finer-grained and thin-bedded turbidites are dominant, and are interbedded with mudstones with high total organic carbon (TOC) contents, which have a high potential as petroleum source rocks (the Teradomari Formation). In the northern part of the basin, the Stage II sediments consist mostly of siliceous shale (the Uchisugawa Formation), which reflects a minor clastic sediment supply into the basin, and diatom-blooming conditions established under a confined, stagnant setting in the Sea of Japan.

Basin inversion and compressive successions (Stage III: 6.5–1 Ma)
The Stage III sediments represent basin inversion and compressive stress field successions. The representative lithostratigraphic units of the Stage III sediments include: the Pliocene Shiiya Formation, the Pliocene–Pleistocene Nishiyama Formation and the Pleistocene Uonuma Group in the central part of the basin; the upper Nodani Formation and the Sugawa, Kawazume, Tamugigawa, Nadachi and Tanihama formations in the southern part; and the Kanatsu, Taira and Kuwae formations in the northern part. The Stage III sediments are characterized by large-scale northwards progradation of depositional systems, consisting of slope to basin-floor, submarine-fan, shelf, nearshore, delta, fan-delta, fluvial and alluvial-fan systems (Figs 6.17 & 6.18). In Stage III, tectonic subsidence ceased as a response to tectonic inversion under a compressive stress field. The sediment accumulation rate increased and is considered to have exceeded that of creation of accommodation because large amounts of coarse clastic sediments were supplied from the uplifted provenance areas (Takano 2002*a*, *b*). This condition caused large-scale progradation, and the basin was filled up rapidly from south to north. The basin physiography during Stage III was strongly affected by syndepositional intra-basinal folding due to the compressive stress field, which caused local depressions for trough-fill turbidite systems at the bottom part of the large-scale progradational successions (the Tamugigawa and Shiiya formations; Takano *et al.* 2005; Figs 6.17 & 6.18). During the latter half of Stage III (2.5–1 Ma) a thick prograding wedge of the Pleistocene Uonuma Group, comprising shallow-marine to alluvial-fan successions, developed in the Uonuma area in the middle central part of the basin (Kazaoka *et al.* 1986). Based on sedimentary successions and palaeontological data, the Stage III sediments can be divided into four third-order depositional sequences (in ascending order): Nkb-P, -K, -KZ and -TN (or UO, Kkb-III.2) (Takano *et al.* 2001, 2011). High-frequency depositional sequences were also locally defined, in particular in shallow-marine to fluvial successions in the Uonuma Group, and are considered to have formed in response to glacioeustatic sea-level changes (Takano 1995).

Intense compressive successions (Stage IV: 1 Ma–present)
The distribution of the Stage IV sediments is restricted mainly to the present-day alluvial plain and offshore areas as a result of uplifting events of the present intra-basinal mountainous and hilly areas related to folding and faulting under the strong compressional tectonic regime. In Stage IV, north- or NW-directed progradation of shallow-marine, coastal and fluvial systems have continued, and these deposits have formed the present coastal plain and surrounding coastal terraces.

Akita Basin

Tectonostratigraphy and basin evolution
The Akita Basin constitutes the northern half of the Japan Sea East Margin Basins (Fig. 6.16) and is separated from the Niigata Basin by a structural high called the Asahi Mountains. The Akita Basin has been further subdivided into the coastal basin (basin centre) and Yokote and Shinjo inland basins by an intervening ridge, which has uplifted to form the present-day Dewa Hills (Fig. 6.16). Similar to the Niigata Basin, the Akita Basin has been filled with thick sedimentary successions of Miocene–Pleistocene siliciclastic and volcanic rocks, which are composed of slope to basin-floor mudstones and siliceous shale, submarine-fan turbidites, shelf sandstones, deltaic sandstones and fluvial conglomerates (Sato *et al.* 1991, 2009, 2012; Nakajima *et al.* 2000, 2006; Moriya *et al.* 2008; Kano *et al.* 2011; Nakajima 2013). The Akita Basin is also classified as an inverted rift basin. However, the Ou Backbone Range in the eastern margin of the Akita Basin inverted earlier (at the end of Middle Miocene times) than the basin centre, and records a complex uplift–subsidence history in shallow-marine and terrestrial environments (Nakajima *et al.* 2006) (Fig. 6.19). In contrast, the basin centre was filled mainly with siliceous shales and mudstones, which almost excluded intercalations of coarse clastic sediments during the post-rift stages and which were overlain by progradational successions formed with turbidites and shelf sandstones during the subsequent compression stages after Pliocene times (Fig. 6.19). As detailed below, the basin history can be divided into six tectonic stages (Fig. 6.19) on the basis of temporal changes in basin subsidence patterns (Nakajima *et al.* 2006; Nakajima 2013).

Synrift successions (Stage I: 17–13.5 Ma)
The Stage I sedimentary and volcanic rocks represent synrift successions, which filled ridges and grabens formed in the Akita Basin. Representative lithostratigraphic units of Stage I include the Middle Miocene Nishikurosawa, Aosawa, Sugota and Oishi formations from the west to east. However, the lithology of the Stage I rocks varies significantly according to the host basin structures of the

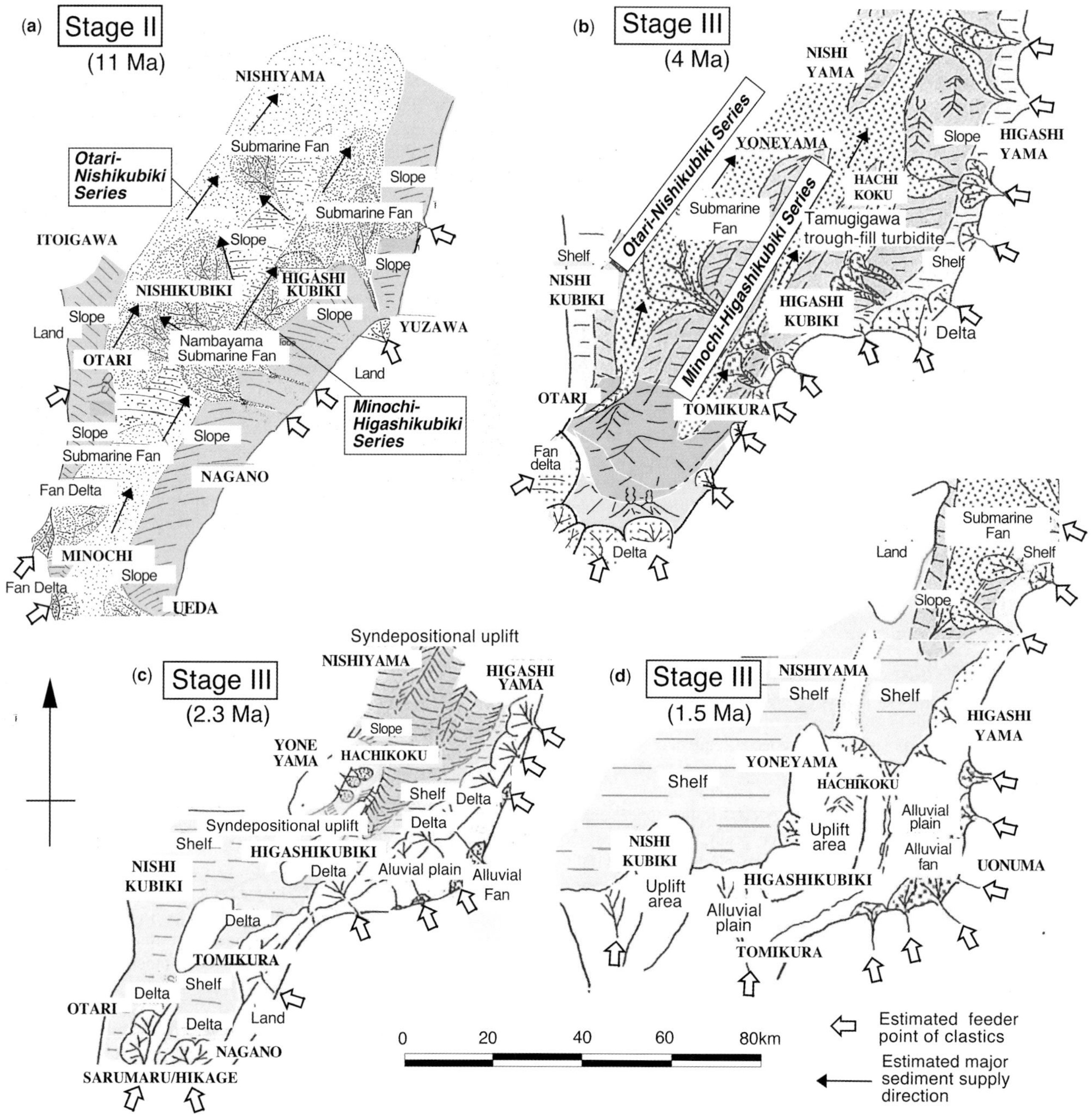

Fig. 6.18. Restored palaeogeographic maps of the middle to southern parts of the Niigata Basin for Stages II and III – (**a**) 11; (**b**) 4; (**c**) 2.3; and (**d**) 1.5 Ma – showing temporal and spatial variations in major depositional systems and sediment supply routes. After Takano (2002*a*, *b*).

two component rifts: the Aosawa rift along the Akita coast and the Kuroko rift in the Ou Backbone Range (Yoshida 2009), which were formed in the Akita Basin during Stage I (Fig. 6.19). The Aosawa rift is a large-scale graben formed by pull-apart tectonism, and was filled with thick piles (*c.* 2000 m) of graben-fill basalt lavas (the Aosawa Formation) (Yagi *et al.* 2001). The Kuroko rift is composed of half-grabens which were filled with thick piles (*c.* 1000 m) of felsic volcanic and pyroclastic rocks interbedded with mudstones (the Oishi Formation) (Yamada & Yoshida 2004; Nakajima *et al.* 2006). In contrast the Nishikurosawa and Sugota formations, which are only 30–400 m thick, are distributed on the Oga and Dewa ridges, respectively, with unconformable contacts with the underlying pre-rift successions; they consist of shallow-marine sandstones, calcareous sandstones and conglomerates which were deposited on a shelf environment (Sato *et al.* 1991; Kano *et al.* 2011) (Fig. 6.19). These lithofacies and thickness contrasts are likely attributed to differential subsidence between ridges and grabens during the rifting under extensional tectonics (Ingersoll & Busby 1995). Volcanic activity in Stage I is classified as the back-arc basin stage based on the chemical composition of volcanic products (Yoshida 2009). The Kuroko ore deposits were formed at the end of Stage I at *c.* 13.5 Ma, which may have been caused by the tectonic conversion from a back-arc to an island-arc setting (Yamada & Yoshida 2004; Yoshida 2009).

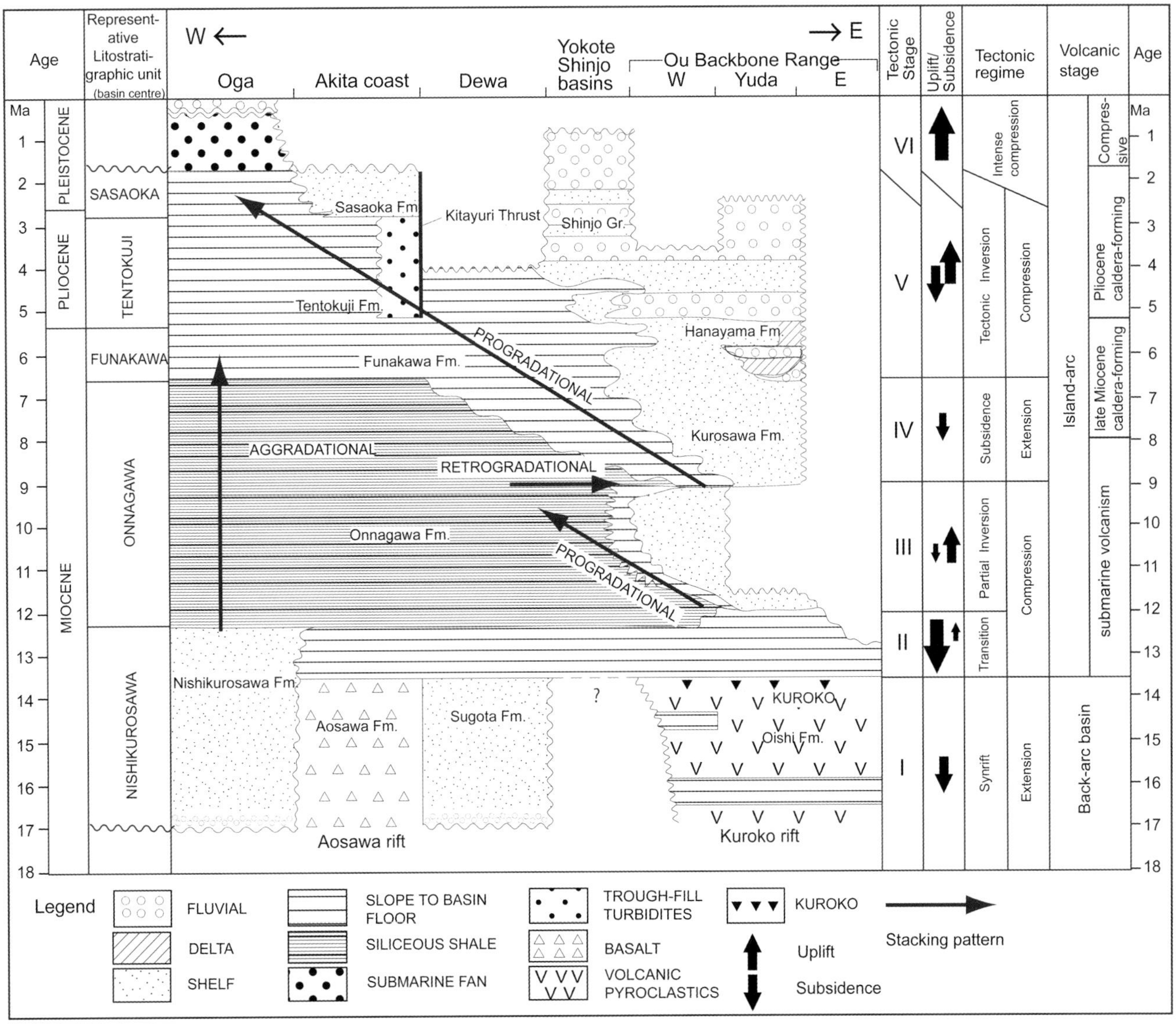

Fig. 6.19. Tectonostratigraphic stage division chart of the Akita Basin showing classification of lithostratigraphic units, major depositional systems, stacking patterns, uplift and subsidence patterns, tectonic regimes and volcanic regimes. Compiled from Sato *et al.* (1991, 2009, 2012), Nakajima *et al.* (2000, 2006), Moriya *et al.* (2008), Yoshida (2009) and Kano *et al.* (2011).

Post-rift successions (stages II, III and IV: 13.5–6.5 Ma)

Stage II (post-rift transition stage: 13.5–12 Ma) was characterized by the cessation of activities of the Aosawa and Kuroko rifts, and by marine transgression associated with rapid subsidence of the Dewa and Oga ridges. The Dewa ridge subsided rapidly from inner sublittoral to middle bathyal environments at 13.5 Ma, followed by rapid subsidence of the Oga ridge from upper bathyal to lower bathyal environments at 12.3 Ma (Sato *et al.* 1991; Kano *et al.* 2011). This subsidence mode cannot solely be attributed to a post-rift thermal subsidence, because the subsidence rate far exceeded the estimated rate (*c.* 70 m Ma^{-1}) of the thermal subsidence (Nakajima *et al.* 2006) and the timing of rapid subsidence was out of phase between the Dewa and Oga ridges. In terms of volcanism, Stage II was represented by a major change from back-arc basin stage to island-arc stage (Yoshida 2009). These changes in tectonic and volcanic styles suggest a stress change from extension to compression (Nakajima *et al.* 2006; Nakajima 2013).

Stage III (partial inversion stage: 12–9 Ma) was represented by the temporal uplift and associated unconformity in the Ou Backbone Range. The eastern sector of the Ou Backbone Range emerged by *c.* 12 Ma and an unconformity was formed at *c.* 10 Ma within an intermountain basin called the Yuda Basin (Nakajima *et al.* 2006) (Fig. 6.19). The uplift of the eastern sector of the Backbone Range may be attributed to a subsequent inversion of a half-graben formed during Stage I (Nakajima *et al.* 2006). The western sector of the Backbone Range also uplifted from middle bathyal to shelf environments, although the bounding normal fault does not seem to have been inverted. Volcanic activity was reduced around this time interval (10–8 Ma) and seems to have been linked to an increase in horizontal compression stress (Nakajima *et al.* 2006; Yoshida 2009). However, the basin centre and the Dewa Hills are considered to have remained bathyal environments in association with continuous deposition of siliceous shales (Fig. 6.19). The eastern margin of the Sea of Japan did not document any deformation or unconformity during this stage (Okamura *et al.* 1995). These observations suggest attenuation of compressional stress and its related deformation towards the Sea of Japan (Nakajima *et al.* 2006).

Stage IV (subsidence stage: 9–6.5 Ma) was represented by regional slow subsidence and increased felsic volcanism in association with intense caldera formation (Nakajima *et al.* 2006; Yoshida 2009). The subsidence resumed and shallow-marine sandstones were deposited in the Yuda Basin, while the eastern sector of the Backbone Range remained as a sediment source (Nakajima *et al.* 2006) (Fig. 6.19). Regional subsidence resumed with increased volcanic activity under weak extensional stress (Nakajima *et al.* 2006).

The representative lithostratigraphic units of the post-rift stages (Stage II, III, IV) include the latest Middle–Late Miocene Onnagawa Formation in both the basin centre and Dewa Hills, and the Kurosawa Formation in the Backbone Range. The Onnagawa Formation consists of siliceous shales and hard mudstones, which yield high TOC contents and constitute major hydrocarbon source rocks (Sato *et al.* 2009). The Kurosawa Formation in the Backbone Range consists of shallow-marine sandstones deposited in a shelf environment. These sedimentary successions show aggradational stacking patterns in the basin centre, possibly reflecting low supply of terrigeneous sediments and slow subsidence in the basin floor, and also show westwards progradation from the Backbone Range to the Dewa Hills as a local response to excess sediment supply over the rate of creation of accommodation space. Furthermore, a rapid retrogradation occurred at *c.* 9 Ma in the Ou Backbone Range because of the resumption of subsidence (Nakajima 2013) (Fig. 6.19).

Basin inversion and compressive successions (Stage V: 6.5–3–2 Ma)

Stage V represents the differentiation of uplifted and subsided areas because of basin inversion under a compressive stress field (Nakajima *et al.* 2006). The representative lithostratigraphic units of the Stage V sediments include: the Late Miocene Funakawa Formation, the Pliocene Tentokuji Formation and the Early Pleistocene Sasaoka Formation in the basin centre; the Shinjo Group in the Shinjo Basin; and the Hanayama Formation in the Yuda Basin. The Dewa Hills and the Backbone Range uplifted during Stage V. These sedimentary successions show basin-scale westwards-directed progradational stacking patterns including upwards-shallowing cycles, which consist of slope to basin-floor, trough-fill-turbidite, shelf, nearshore, delta and fluvial systems, and likely reflect an increased accumulation rate caused by large amount of sediment supply from the uplifted Ou Backbone Range and the Dewa Hills in response to the increase in compressional stress (Fig. 6.19). The activity of the Kitayuri Thrust extending north–south along the Akita coast started in this stage and resulted in the deposition of trough-fill turbidite system (Katsurane Facies) on the footwall trough of the thrust as a response to syndepositional faulting and folding under compressional stress (Sato *et al.* 2012; Fig. 6.19). Four third-order depositional sequences consisting of shallow-marine to fluvial successions (i. e. the Shinjo Group) developed in the Shinjo Basin, which represent gradual retreat of marine environments from the Shinjo Basin in response to the successive uplift of the Dewa Hills (Moriya *et al.* 2008). Four third-order depositional sequences superimposed by high-frequency depositional sequences, which consist of shallow-marine, deltaic and fluvial successions, developed in the Yuda Basin by 4 Ma, indicating marine incursion in the centre of the Backbone Range until *c.* 4 Ma (Nakajima *et al.* 2000) (Fig. 6.19).

Intense compressive successions (Stage VI: 3–2 Ma–present)

Stage VI represents crustal deformation associated with the uplift and emergence of all present land areas because of increased compressive stress (Sato 1994; Nakajima *et al.* 2006). The Akita coastal plain emerged at 1.7 Ma, resulting in westwards shift of a sedimentary basin and submarine-fan deposition in Oga, followed by gradual fill of the basin with coarse sediments and by emergence of the basin-fill successions (Sato *et al.* 2012) (Fig. 6.19).

Neogene hydrocarbon resources in Japan (NS)

History of petroleum exploration

Petroleum exploration in Japan began in the 1870s, with the first oil-deposit discovery being made in the Neogene Niigata sedimentary basin in the late nineteenth century. An aggressive exploration effort has been pursued to date leading to the discovery of approximately 100 oil and gas deposits, although many of these have already been exhausted. Some oil and gas deposits have also been found in the Palaeogene sedimentary basins. At present, Japan has two offshore and approximately 15 onshore producing oil and gas fields. Representative producing oil and gas fields in the Neogene sedimentary basins are the Yurihara, Ayukawa and Yabase oil and gas fields in Akita and Minami Nagaoka, Katakai and Iwafune-oki oil and gas fields in Niigata (Fig. 6.20). The Yufutsu oil and gas field in Hokkaido is one of the largest producing fields in Japan and located in a Palaeogene sedimentary basin (Fig. 6.20). The total domestic annual production of oil and gas in the period between 2007 and 2011 ranged from 832 to 961 × 10^3 kL and 3.3–3.7 × 10^9 m^3, respectively, corresponding to 0.4% and 3.5% of the nation's annual oil and gas demands, respectively (METI 2011).

Distribution of oil and gas fields in Japan

Japan is located on the NW margin of the Pacific Ocean, and most of the oil- and gas-producing Neogene sedimentary basins developed in association with the subduction of the Pacific Plate. The major Japanese oil and gas fields are located in the back-arc basins near the Sea of Japan along northern Honshu and Hokkaido (Fig. 6.20), and hydrocarbons are mainly derived from Middle Miocene source rocks (e.g. Taguchi 1976). The geological ages of reservoir rocks vary by location and range from Miocene to Pliocene. Very few oil and gas deposits have been found in the Neogene fore-arc basins along the Pacific side of Japan, these including the Yufutsu, Iwaki-oki and Sagara oil and gas fields which are located within Palaeogene sedimentary basins (Fig. 6.20). Another type of gas resource, known as 'gas dissolved-in-water deposits', is distributed across Chiba, Miyazaki and Niigata, with the largest such deposit occurring in Mobara, Chiba Prefecture (Fig. 6.4). The carbon and hydrogen isotopic compositions of methane in these gas dissolved-in-water deposits indicate a microbial origin (Igari & Sakata 1989; Kaneko *et al.* 2002).

Source rock distribution and quality

Formation of the Neogene petroleum source rocks was closely related to the tectonic and palaeo-oceanographic evolution of the Sea of Japan. Primary production in the palaeo-Sea of Japan significantly increased during Middle–Late Miocene times, resulting in the formation of diatomaceous and organic-rich sediments. These organic-rich sedimentary rocks are the major source of oil and gas in the Japanese Neogene sedimentary basins. The total organic carbon (TOC) concentration of the Middle–Late Miocene marine mudstones is in the range 0.1–4.0%, with most being *c.* 0.4–0.6% (Sampei *et al.* 2000). Rocks with comparatively higher TOC (i.e. >1.5%) are mainly diatomaceous siliceous shale (Takayama & Kato 1995). Mature siliceous shale and organic-rich clastic shale of the Onnagawa Formation (Akita Basin), the Kusanagi Formation (Yamagata Basin) and the Nanatani and Teradomari formations (Niigata Basin) are widely distributed along NE Honshu near the Sea of Japan (Figs 6.17 & 6.19), which is consistent with the distribution of the Neogene oil and gas deposits (Fig. 6.20). The potential source rocks in the Neogene sedimentary basins contain Type II to II/III kerogen (Sato 1976; Sekiguchi *et al.* 1984;

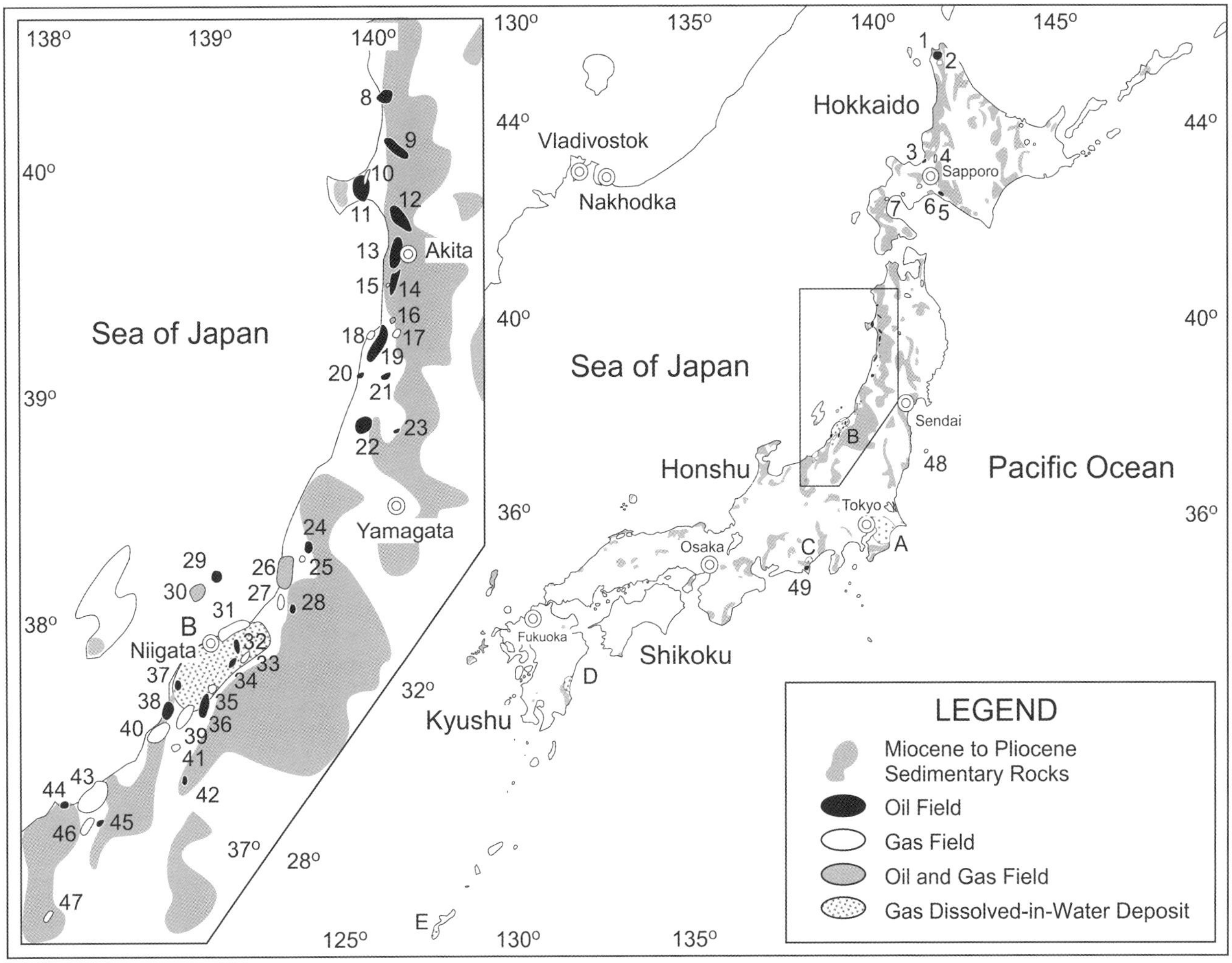

Fig. 6.20. Distribution of oil and gas fields in Japan. (**a**) Hokkaido Region: 1, Wakkanai, Koetoi, Yuuchi, Masuhoro and Menashi; 2, Toyotomi; 3, Ishikari, Atsuta and Barato; 4, Minenobu and Kanazawa; 5, Furaoi, Karumai and Biratori; 6, Yufutsu (producing); and 7, Oshamanbe (producing). (**b**) Akita Region: 8, Hachimori; 9, Minami Noshiro, Sakaki and Hibiki; 10, Mino and Katanishi; 11, Sarukawa (producing), Hashimoto, Fukumezawa, Katanishi, Fukukawa and Nishi Oogata; 12, Urayama, Toyokawa, Kurokawa, Michikawa, Nigorikawa and Asahikawa; 13, Yabase (producing) and Tsuchizaki-oki; 14, Kinshoji, Niida, Toyoiwa, Katsurane, Hanekawa and Katte; 15, Shimohama; 16, Ayukawa (producing); 17, Yurihara (producing); 18, Nishioguni; and 19, Innai, Katsurazaka, Odaki and Kamihama. (**c**) Yamagata Region: 20, Fukura; 21, Chokaizan; 22, Narahashi, Ishinazaka, Shinbori, Sagoshi and Amarume (producing); and 23, Mogami. (**d**) Niigata Region: 24, Echigokurokawa; 25, Hirakida; 26, Shintainai and Nakajo (producing); 27, Shiunji (producing) and Seiro; 28, Shibata; 29, Aga-oki Kita; 30, Aga-oki and Iwafune-oki (producing); 31, Higashi Niigata (producing) and Matsuzaki gas deposits; 32, Minami Aga oil deposit; 33, Minamisuibara and Kuwayama; 34, Niitsu; 35, Higashi Sanjo and Nagasawa; 36, Oomo, Mitsuke (producing) and Higashiyama; 37, Yoshida; 38, Nishiyama, Miyagawa and Torigo; 39, Ooguchi, Nishi Nagaoka, Fujikawa and Sekihara gas deposits; 40, Myouhouji, Yoshii (producing) and Hihashi Kashiwazaki; 41, Minami Nagaoka (producing) and Katakai (producing); 42, Tamugiyama; 43, Kubiki, Kuroi and Meiji; 44, Gotsu; 45, Maki; 46, Umaya and Bessh; and 47, Odagir. (**e**) Other regions: 48, Iwaki-oki; and 49, Sagara. (**f**) Gas dissolved-in-water deposits: A, Mobara; B, Niigata; C, Shizuoka; D, Okinawa.

Waseda & Omokawa 1990). Kerogen in the Onnagawa siliceous shale from the Akita Basin tends to be rich in organic sulphur. The Onnagawa kerogen from the central part of the Akita Basin is often classified as Type II-S kerogen, similar to that in the Middle Miocene Monterey shale of California, USA (Suzuki *et al.* 1995). Although the Miocene siliceous mudstones are also distributed in western Hokkaido, they are generally immature and prior to the stage of oil generation (Omokawa *et al.* 1990; Ogura & Kamon 1992).

Correlation of oil and source rock

The oil and gas found in the Niigata Basin are mainly derived from the marine source rocks of the Nanatani and Teradomari formations. The Niigata oils are classified into two types on the basis of the relative amount of a biomarker called oleanane, which is derived from terrestrial angiosperm plants (Hirai *et al.* 1995). Oils from the Niitsu, Aga-oki and Higashi Niigata oil and gas fields in the northern Niigata Basin are poor in oleanane compared to those from the Kubiki, Minami Nagaoka and Fujikawa oil and gas fields in the southern Niigata Basin. This indicates a significant contribution of terrestrial organic matter to marine oils in the southern Niigata Basin. Oils from the Akita and Yamagata basins are comparatively poor in terrestrial biomarkers, similar to those from the northern part of the Niigata Basin (Sakata *et al.* 1988). Oils from the Akita and Yamagata basins are mainly derived from contemporary formations equivalent to the Onnagawa siliceous shale. Crude oils from the Akita and Yamagata basins tend to be rich in sulphur compared to those from the Niigata Basin (Kato *et al.* 1997). Neogene oils from Hokkaido are mainly derived from marine organic matter, with a little contribution from terrestrial organic matter (Omokawa *et al.* 1990).

Reservoirs and seals

Active volcanism on the ocean floor as a response to the expansion of the Sea of Japan occurred during Early–Middle Miocene times. The contemporary Neogene sedimentary basins on the Sea of Japan side of Japan therefore contain significant amounts of lava and pyroclastic rocks. Porous tuffaceous sandstones and tuffs act as good reservoir rocks and characterize the Japanese oil and gas fields (Katahira 1974; Komatsu *et al.* 1984). The Early–Middle Miocene rhyolite–dacite volcanic rocks of the Tsugawa and Nanatani formations in the Niigata Basin and the basaltic volcanic rocks of the Nishikurosawa Formation are known as important reservoir rocks. Typical oil and gas deposits with volcanic and volcaniclastic reservoir rocks are the Minami Nagaoka, Katakai, Yoshii and Higashi Kashiwazaki oil and gas fields in the Niigata Basin and the Sarukawa, Ayukawa and Yurihara oil and gas fields in the Akita Basin (JNGA & JOPDA 1982, 1992; Araki & Kato 1993; Okubo *et al.* 1996; Yamada & Uchida 1997; Inaba 2001). Turbidite sandstones with volcaniclastic deposits in the Late Miocene–Pliocene successions are also important major reservoirs (Abe 1978; Moriya *et al.* 2007). Anticlinal folds, which formed under the inversion tectonic regime in the sedimentary basins, are the dominant trap type. Stratigraphic traps are also occasionally found in some oil and gas fields (Akiba *et al.* 1987; Saito *et al.* 2008).

Timing of expulsion and accumulation

The maturity levels of the Neogene oils, including condensate oils, are highly variable. These oils show a wide range of vitrinite reflectance (R_o) of 0.5–1.2%, as estimated from biomarkers and aromatic maturity parameters (Suzuki *et al.* 1987; Kato & Nishita 2010). Akita crude oils are characterized by low maturity levels, whereas Niigata crude oils tend to be more mature (Sakata *et al.* 1990). The comparatively low maturity of oils in the Akita Basin is partly due to their siliceous source rocks, containing Type II-S kerogen. Natural gas condensate is abundant in the Niigata Basin as compared to the Akita Basin. Various oil maturity parameters show that the maturity levels of condensates are generally higher than those of crude oils, suggesting a thermogenic origin of the condensate. The maturity level of oil is closely related to the timing of oil expulsion from the source rock, because in many cases the maturation of the oil occurred in the source rocks. Maturation in the reservoir is suggested for some condensates however, such as those from the Yoshii and Katakai oil and gas deposits in the Niigata Basin (Suzuki 1990; Waseda 1993). The expulsion and primary migration of oil from the Middle Miocene source rocks primarily occurred during the Pliocene–Pleistocene epochs under a compressional tectonic regime in the back-arc basins (Aiba 1980; Sekiguchi *et al.* 1984; Waseda & Omokawa 1990).

Undiscovered resources

While petroleum exploration in Japan has concentrated on the shallow-water shelf areas on the Sea of Japan side, many unexplored offshore areas at water depth >200 m remain to be fully investigated around Japan. Three-dimensional (3D) seismic surveys have been performed covering a large part of the shelf areas where thick Palaeogene–Neogene sedimentary successions are developed. Sedimentary basins that can host unconventional oil and gas deposits, such as coalbed methane, shale gas/oil and tight gas/oil, have attracted recent exploration interest. Test production of shale oil from Middle Miocene siliceous shale has been successfully performed in the Akita Basin. Shale gas resources generally occur within low-permeability and highly mature shale. The shale gas potential of the Neogene sedimentary basins may not be high, because most of the Neogene shale rocks are in the moderately mature stage before the stage where dry gas is generated. The shale gas potential of the Palaeogene and older sedimentary rocks distributed on the Pacific side of Japan is currently under investigation.

Appendix

English to Kanji and Hiragana translations for geological and place names

Aga-oki	阿賀沖	あがおき
Age	奄芸	あげ
Aichi	愛知	あいち
Aira	姶良	あいら
Akeyo	明世	あけよ
Akita	秋田	あきた
Amatsu	天津	あまつ
Anno	安野	あんの
Aosawa	青沢	あおさわ
Aoshima	青島	あおしま
Arima	有馬	ありま
Asahi	朝日	あさひ
Awa	阿波	あわ
Ayukawa	鮎川	あゆかわ
Biwa	琵琶	びわ
Bonin	ボニン	ぼにん
Boso	房総	ぼうそう
Chiba	千葉	ちば
Chikusa	千種	ちくさ
Chita	知多	ちた
Choshi	銚子	ちょうし
Chubu	中部	ちゅうぶ
Dewa	出羽	でわ
Fujikawa	藤川	ふじかわ
Fukuda	福田	ふくだ
Fukushima	福島	ふくしま
Funakawa	船川	ふなかわ
Gifu	岐阜	ぎふ
Hanayama	花山	はなやま
Hayama	葉山	はやま
Hida	飛騨	ひだ
Higashi Kashiwazaki	東柏崎	ひがしかしわざき
Higashi Niigata	東新潟	ひがしにいがた
Higashinada	東灘	ひがしなだ
Himi	氷見	ひみ
Hisamine	久峰	ひさみね
Hokkaido	北海道	ほっかいどう
Hokuriku	北陸	ほくりく
Hongo	本郷	ほんごう
Honshu	本州	ほんしゅう
Hyuga-nada	日向灘	ひゅうがなだ
Ichinohara	市之原	いちのはら
Ichishi	一志	いちし
Iga	伊賀	いが
Inabe	員弁	いなべ
Ise	伊勢	いせ
Itoigawa	糸魚川	いといがわ
Iwafune-oki	岩船沖	いわふねおき
Iwaki-oki	磐城沖	いわきおき
Izu	伊豆	いず

(*Continued*)

English to Kanji and Hiragana translations for geological and place names (Continued)

Joban	常磐	じょうばん
Kakegawa	掛川	かけがわ
Kameyama	亀山	かめやま
Kaminoma	上野間	かみのま
Kanatsu	金津	かなつ
Kanazawa	金沢	かなざわ
Kankuu	関空	かんくう
Kanto	関東	かんとう
Karasuma	烏丸	からすま
Kashima	鹿島	かしま
Katakai	片貝	かたかい
Katsurane	桂根	かつらね
Kawazume	川詰	かわづめ
Kazusa	上総	かずさ
Kinki	近畿	きんき
Kinone	木ノ根	きのね
Kioroshi	木下	きおろし
Kitan	紀淡	きたん
Kitayuri	北由利	きたゆり
Kiwada	黄和田	きわだ
Kiyosumi	清澄	きよすみ
Kobiwako	古琵琶湖	こびわこ
Kubiki	頚城	くびき
Kumano	熊野	くまの
Kuroko	黒鉱	くろこう
Kurosawa	黒沢	くろさわ
Kuroshio	黒潮	くろしお
Kurotaki	黒滝	くろたき
Kusanagi	草薙	くさなぎ
Kuwae	鍬江	くわえ
Kyoto	京都	きょうと
Kyushu	九州	きゅうしゅう
Megami	女神	めがみ
Mie	三重	みえ
Mikawa	三河	みかわ
Minami Nagaoka	南長岡	みなみながおか
Mineoka	嶺岡	みねおか
Mino	美濃	みの
Miura	三浦	みうら
Miyazaki	宮崎	みやざき
Mizunami	瑞浪	みずなみ
Mobara	茂原	もばら
Morozaki	師崎	もろざき
Nadachi	名立	なだち
Nagaoka	長岡	ながおか
Nagoya	名古屋	なごや
Nakaobara	中尾原	なかおばら
Nambayama	難波山	なんばやま
Nanatani	七谷	ななたに
Nankai	南海	なんかい
Nara	奈良	なら
Nataki	名滝	なたき
Niigata	新潟	にいがた
Niitsu	新津	にいつ
Nishikurosawa	西黒沢	にしくろさわ
Nishiyama	西山	にしやま
Nodani	能生谷	のうだに
Nunobiki	布引	ぬのびき
Oga	男鹿	おが
Ogasa	小笠	おがさ
Oguraike	巨椋池	おぐらいけ

(Continued)

English to Kanji and Hiragana translations for geological and place names (Continued)

Ohmi	近江	おうみ
Oidawara	生俵	おいだわら
Oishi	大石	おおいし
Okazaki	岡崎	おかざき
Okinawa	沖縄	おきなわ
Okuyama	奥山	おくやま
Okuzure	大崩	おおくずれ
Onnagawa	女川	おんながわ
Osaka	大阪	おおさか
Ou	奥羽	おうう
Owari	尾張	おわり
Rokko	六甲	ろっこう
Rokkoku	六石	ろっこく
Ryohaku	両白	りょうはく
Sagara	相良	さがら
San’in	山陰	さんいん
Sarukawa	申川	さるかわ
Sasaoka	笹岡	ささおか
Sendai	仙台	せんだい
Seto	瀬戸	せと
Setouchi	瀬戸内	せとうち
Shiga	滋賀	しが
Shiiya	椎谷	しいや
Shimosa	下総	しもうさ
Shin’etsu	信越	しんえつ
Shinjo	新庄	しんじょう
Shitara	設楽	したら
Sugawa	須川	すがわ
Sugota	須郷田	すごうた
Suzuka	鈴鹿	すずか
Taira	平	たいら
Takanabe	高鍋	たかなべ
Takatsuki	高槻	たかつき
Tamari	満水	たまり
Tamugigawa	田麦川	たむぎがわ
Tanihama	谷浜	たにはま
Tanzawa	丹沢	たんざわ
Tentokuji	天徳寺	てんとくじ
Teradomari	寺泊	てらどまり
Terakawa	寺川	てらかわ
Togari	戸狩	とがり
Tohoku	東北	とうほく
Tokai	東海	とうかい
Toki	土岐	とき
Tokoname	常滑	とこなめ
Tokyo	東京	とうきょう
Toyama	富山	とやま
Tsugawa	津川	つがわ
Tsukiyoshi	月吉	つきよし
Tsuma	妻	つま
Tsushima	対馬	つしま
Uchisugawa	内須川	うちすがわ
Uchiumigawa	内海川	うちうみがわ
Uonuma	魚沼	うおぬま
Yabase	八橋	やばせ
Yamagata	山形	やまがた
Yamakasu	山粕	やまかす
Yamanouchi	山野内	やまのうち
Yamashiro	山城	やましろ
Yamato	大和	やまと
Yokote	横手	よこて
Yoshii	吉井	よしい

(Continued)

English to Kanji and Hiragana translations for geological and place names (Continued)

Yuda	湯田	ゆだ
Yufutsu	勇払	ゆうふつ
Yurihara	由利原	ゆりはら

References

Abe, M. 1978. The sedimentary history of Minami-Aga reservoir sandstones. *Journal of the Japanese Association for Petroleum Technologists*, **43**, 398–406 [in Japanese with English abstract].

Aiba, J. 1980. Timing of hydrocarbon migration in the main oil and gas fields in the Northeast Japan. *Journal of the Japanese Association for Petroleum Technologists*, **45**, 373–382 [in Japanese with English abstract].

Akiba, F., Sasaki, E. & Saga, H. 1987. Note on sandstone development of the Shiiya-Nishiyama stages around the Shiunji gas field, Kitakanbara plain. *Journal of the Japanese Association for Petroleum Technology*, **52**, 48–57 [in Japanese with English abstract].

Aoike, K. 1999. Tectonic evolution of the Izu Collision zone. *Research Report of the Kanagawa Prefectural Museum Natural History*, **9**, 113–151 [in Japanese with English abstract].

Aoshima, M. 1978. Depositional environment of the Plio-Pleistocene Kakegawa Group, Japan–A comparative study of the fossil and the recent foraminifera. *Journal of the Faculty of Science, University of Tokyo, Series 2*, **19**, 401–441.

Araki, N. & Kato, S. 1993. A discovery of the Ayukawa oil and gas field, Akita Prefecture. *Journal of the Japanese Association for Petroleum Technology*, **55**, 191–127 [in Japanese with English abstract].

Arima, M., Aoike, K. & Kawate, S. 1999. Tectonic evolution of the Tanzawa Mountainland. *Research Report of the Kanagawa Prefectural Museum Natural History*, **9**, 57–77 [in Japanese with English abstract].

Biswas, D. K., Hyodo, M. *et al.* 1999. Magnetostratigraphy of Plio-Pleistocene sediments in a 1700-m core from Osaka Bay, southwestern Japan and short geomagnetic events in the middle Matuyama and early Brunhes chrons. *Palaeogeography, Palaeoclimatology, Palaeoecology*, **148**, 233–248.

Cande, S. C. & Kent, D. V. 1995. Revised calibration of the geomagnetic polarity timescale for the Late Cretaceous and Cenozoic. *Journal of Geophysical Research, B4*, **100**, 6093–6095.

Chiyonobu, S., Saruwatari, H., Sato, T., Kabamoto, J. & Iryu, Y. 2009. Geologic ages of the Chinen Formation on the Katsuren Peninsula, Okinawa-jima based on calcareous nannofossil biostratigraphy. *Journal of the Geological Society of Japan*, **115**, 528–539 [in Japanese with English abstract].

Chiyonobu, S., Morimoto, J., Torii, M. & Oda, M. 2012. Pliocene/Pleistocene boundary and paleoceanographic significance of the upper Miyazaki Group, southern Kyushu, Southwest Japan, based on calcareous nannofossil and planktonic foraminiferal assemblages. *Journal of the Geological Society of Japan*, **118**, 109–116 [in Japanese with English abstract].

Endo, H. & Suzuki, Y. 1986. *Geology of the Tsuma and Takanabe District. With Geological Sheet Map at 1:50 000*. Geological Survey of Japan, Ibaraki [in Japanese with English abstract].

Endo, M. & Tateishi, M. 1990. Reconstruction of the Miocene Nambayama submarine fan, northern Fossa Magna, central Japan. *Journal of the Geological Society of Japan*, **96**, 193–209.

Expedition 315 Scientists 2009. Expedition 315 Site C0002. *In*: M. Kinoshita, H. Tobin, *et al.* (eds) *Proceedings IODP 314/315/316*. Integrated Ocean Drilling Program Management International, Inc., Washington, DC, http://doi.org/10.2204/iodp.proc.314315316.124.2009

Geological Survey of Japan AIST (ed.) 2012. *Seamless Digital Geological Map of Japan 1: 200 000. July 3, 2012 Version. Research Information Database DB084*. Geological Survey of Japan, National Institute of Advanced Industrial Science and Technology, Ibaraki.

Hayashi, T. & Kawabe, T. 1993. Kobiwako Group, terrace deposits and Chusekiso. *In*: Itihara, M. (ed.) *The Osaka Group*. Sogensya, Osaka, 158–168 [in Japanese].

Hayashida, A. 1986. Timing of rotational motion of Southwest Japan inferred from palaeomagnetism of the Setouchi Miocene Series. *Journal of Geomagnetism and Geoelectricity*, **38**, 295–310.

Hirai, A., Okada, R., Wakamatsuya, S., Miyamoto, Y. & Hachinohe, K. 1995. Organic geochemical relationship between pooled oils and distribution of their source rocks in the Niigata basin, Japan. *Journal of the Japanese Association for Petroleum Technology*, **60**, 87–97 [in Japanese with English abstract]

Hiroki, Y. & Matsumoto, R. 1999. Magnetostratigraphic correlation of Miocene regression-and-transgression boundaries in central Honshu, Japan. *Journal of the Geological Society of Japan*, **105**, 87–107.

Hiroki, Y. & Matsumoto, R. 2003. Correlation of Miocene (18–12 Ma) sequence boundaries in central Japan to major Antarctic glaciation events. *Sedimentary Geology*, **157**, 303–315.

Horie, S. (ed.) 1991. *Die Geschichte des Biwa-Sees in Japan*. Universitatsverlag Wagner, Innsbruck [in German].

Horikawa, K. & Ito, M. 2004. Long-term ENSO-like events represented in the Middle Pleistocene shelf successions, Boso Peninsula, Japan. *Palaeogeography, Palaeoclimatology, Palaeoecology*, **203**, 239–251.

Huzita, K. 1962. Tectonic development of the Median zone (Setouchi) of Southwest Japan, since the Miocene, with special reference to the characteristic structure of Central Kinki Area. *Journal of Geoscience, Osaka City University*, **6**, 103–144.

Huzita, K. 1968. Rokko movements and its appearance. *The Quaternary Research*, **7**, 248–260 [in Japanese with English abstract].

Ibaraki, M. 1981. Mizunami area, Gifu Prefecture. *In*: Tsuchi, R. (ed.) *Fundamental Data on Japanese Neogene Bio- and Chronostratigraphy – Supplement*. Publication of IGCP Project No. 114, National Working Group of Japan, Shizuoka University, Shizuoka [in Japanese].

Ibaraki, M. 1986*a*. Planktonic foraminiferal datum levels recognized in the Neogene sequence of the Kakegawa area, and their relationship with the lithostratigraphy. *Journal of the Geological Society of Japan*, **92**, 119–134 [in Japanese with English abstract].

Ibaraki, M. 1986*b*. Neogene planktonic foraminiferal biostratigraphy of the Kakegawa area on the Pacific coast of central Japan. *Reports of the Faculty of Science, Shizuoka University*, **20**, 39–173 [in Japanese with English abstract].

Igari, S. & Sakata, S. 1989. Origin of natural gas of dissolved-in-water type in Japan inferred from chemical and isotopic compositions: occurrence of dissolved gas of thermogenic origin. *Geochemical Journal*, **23**, 139–142.

Ikawa, T. 1991. Die kanozoischen Sedimente des Biwa-Sees – dargestellt mit Hilfe der Reflexionsseismik. *In*: Horie, S. (ed.) *Die Geschichte des Biwa-Sees in Japan*. Universitatsverlag Wagner, Innsbruck, 41–57 [in German].

Ikebe, N. 1957. Cenozoic sedimentary basins in Japan. *Cenozoic Research*, **24–25**, 1–10 [in Japanese].

Ikebe, N. & Huzita, K. 1966. The Rokko movements, the Pliocene–Pleistocene movements in Japan. *Quaternaria*, **8**, 277–287.

Ikebe, N., Iwatsu, J. & Takenaka, J. 1970. Quaternary geology of Osaka with special reference to land subsidence. *Journal of Geoscience, Osaka City University*, **13**, 39–98.

Ikebe, N., Chiji, M. & Morozumi, Y. 1975. *Lepidocyclina* horizon in the Miocene Kumano Group in reference to planktonic foraminiferal biostratigraphy. *Bulletin of the Osaka Museum of Natural History*, **29**, 81–89.

Inaba, M. 2001. Basalt reservoir in the Yurihara oil and gas field, northeast, Japan. *Journal of the Japanese Association for Petroleum Technology*, **66**, 56–67 [in Japanese with English abstract].

Ingersoll, R. V. & Busby, C. J. 1995. Tectonics of sedimentary basins. *In*: Busby, C. J. & Ingersoll, R. V. (eds) *Tectonics of Sedimentary Basins*. Blackwell Science, Cambridge, MS, 1–51.

Irizuki, T. & Hosoyama, M. 2006. Shukunohora member and Oidawara formation, mizunami group –marine strata deposited in tropical shallow marine to maximum flooding. Geological Society of Japan (ed.) *Japanese Regional Geological Feature, Chubu Region*. Asakura Publishing, Tokyo, 370–371 [in Japanese].

Ishibashi, M. 1989. Sea-level controlled shallow-marine systems in the Plio-Pleistocene Kakegawa Group, Shizuoka, central Honshu, Japan: comparison of transgressive and regressive phases. *In*: Taira, A. & Masuda, F. (eds) *Sedimentary Facies in the Active Plate Margin*. Terrapub, Tokyo, 345–363.

Ishida, S. & Yokoyama, T. 1969. Tephrochronology, paleogeography and tectonic development of Plio-Pleistocene in Kinki and Tokai districts, Japan. The research of younger Cenozoic strata in Kinki province, part 10. *Quaternary Research*, **8**, 31–43 [in Japanese with English abstract].

Ishihara, Y. & Tokuhashi, S. 2005. Depositional process of the Pliocene Anno Formation, the uppermost part of the Awa Group, Boso Peninsula, central Japan: A case study on a turbidite depositional system of a forearc-basin filling. *Journal of the Geological Society of Japan*, **111**, 269–285 [in Japanese with English abstract].

Itihara, M. (ed.) 1993. *The Osaka Group*. Sogensha, Osaka [in Japanese].

Itoigawa, J. 1960. Paleoecological studies of the Miocene Mizunami group, central Japan. *Journal of Earth Sciences, Nagoya University*, **8**, 246–300.

Itoigawa, J. 1974*a*. Geology of the Mizunami Group. *Bulletin of Mizunami Fossil Museum*, **1**, 9–42 [in Japanese].

Itoigawa, J. 1974*b*. Paleoenvironment, paleogeography and depositional history of the Mizunami Group. *Bulletin of Mizunami Fossil Museum*, **1**, 365–367 [in Japanese].

Itoigawa, J., Shibata, H. & Nishimoto, H. 1974. Molluscan fossils of the Mizunami Group. *Bulletin of Mizunami Fossil Museum*, **1**, 43–203 [in Japanese].

Ito, M. 1992. High-frequency depositional sequences of the upper part of the Kazusa Group, a middle Pleistocene forearc basin fill in Boso Peninsula, Japan. *Sedimentary Geology*, **76**, 155–175.

Ito, M. 1998. Contemporaneity of component units of the lowstand systems tract: an example from the Pleistocene Kazusa forearc basin, Boso Peninsula, Japan. *Geology*, **26**, 939–942.

Ito, M. & Katsura, Y. 1992. Inferred glacio-eustatic control for high-frequency depositional sequences of the Plio-Pleistocene Kazusa Group, a forearc basin fill in Boso Peninsula, Japan. *Sedimentary Geology*, **80**, 67–75.

Ito, M. & Masuda, F. 1988. Late Cenozoic deep-sea fan-delta sedimentation in an arc-arc collision zone, central Honshu, Japan: sedimentary response to varying plate-tectonic regime. *In*: Nemec, W. & Steel, R. S. (eds) *Fan Deltas: Sedimentology and Tectonic Settings*. Blackie, Glasgow, 400–418.

Ito, M. & O'Hara, S. 1994. Diachronous evolution of systems tracts in a depositional sequence from the middle Pleistocene palaeo-Tokyo Bay, Japan. *Sedimentology*, **41**, 677–697.

Ito, M., Nishikawa, T. & Sugimoto, H. 1999. Tectonic control of high-frequency depositional sequences with duration shorter than Milankovitch cyclicity: an example from the Pleistocene paleo-Tokyo Bay, Japan. *Geology*, **27**, 763–766.

Ito, M., Saito, T. & Someya, H. 2002. Tectonic control of facies architecture in falling-stage deposits in a forearc basin: upper Miocene Senhata Formation, Boso Peninsula, Japan. *Journal of Sedimentary Research*, **72**, 491–499.

Iwabuchi, Y., Nishikawa, H. *et al.* 2000. Basement and active structures revealed by the seismic reflection survey in Osaka Bay. *Report of Hydrographic Research*, **36**, 1–23 [in Japanese with English abstract].

Iwatani, H., Murai, K., Irizuki, T., Hayashi, H. & Tanaka, Y. 2009. Discovery of the oldest fossil of *Argonauta hians* in Japan, from the middle Pliocene Sadowara Formation, southwest Japan, and its depositional age. *Journal of the Geological Society of Japan*, **115**, 548–551 [in Japanese with English abstract].

Japan Association for Quaternary Research (ed.) 1987. *Quaternary Maps of Japan*. University of Tokyo Press, Tokyo. [in Japanese].

JNGA and JOPDA (Japan National Gas Association and Japan Offshore Petroleum Development Association) 1982. *Oil and Gas Resources in Japan*, new edition. Shirahashi Press, Co., Tokyo [in Japanese].

JNGA and JOPDA (Japan National Gas Association and Japan Offshore Petroleum Development Association) 1992. *Oil and Gas Resources in Japan*, revised edition. Shirahaishi Press, Co., Tokyo [in Japanese].

Kaizuka, S. 1974. Tectonic setting and the Quaternary crustal movements in Kanto district. *In*: Kakimi, T. & Suzuki, Y. (eds) *Earthquake and Crustal Movements in Kanto District*. Ratesu, Tokyo, 99–118 [in Japanese].

Kakimi, T., Yamazaki, H., Samukawa, A., Sugiyama, Y., Shimokawa, K. & Oka, S. 1982. *Neotectonic Map Sheet 8*. Geological Survey of Japan, Ibaraki.

Kamata, H. & Kodama, K. 1994. Tectonics of an arc-arc junction: an example from Kyushu Island at the junction of the Southwest Japan Arc and the Ryukyu Arc. *Tectonophysics*, **233**, 69–81.

Kameo, K. 1998. Upper Neogene and Quaternary stratigraphy in the Kakegawa district based on the calcareous nannofossil datum planes, with reference to the stratigraphic position of the Tamari Formation. *Journal of the Geological Society of Japan*, **104**, 672–686 [in Japanese with English abstract].

Kameo, K. & Sekine, T. 2013. Calcareous nannofossil biostratigraphy and the geologic age of the Anno Formation, the Awa Group in the Boso Peninsula, central Japan. *Journal of the Geological Society of Japan* [in Japanese with English abstract].

Kameo, K., Mita, I. & Fujioka, M. 2002. Calcareous nannofossil biostratigraphy of the Amatsu Formation (Middle Miocene to Lower Pliocene), Awa Group, distributed in the central part of the Boso Peninsula, central Japan. *Journal of the Geological Society of Japan*, **108**, 813–828 [in Japanese with English abstract].

Kameo, K., Shindo, R. & Takayama, T. 2010. Calcareous nannofossil biostratigraphy and geologic age of the Kiyosumi Formation of the Awa Group, Boso Peninsula, central Japan: age determination based on size variations of *Reticulofenestra* specimens. *Journal of the Geological Society of Japan*, **116**, 563–574 [in Japanese with English abstract].

Kaneko, N., Maekawa, T. & Igari, S. 2002. Generation of archaeal methane and its accumulation mechanism into interstitial water. *Journal of the Japanese Association for Petroleum Technology*, **67**, 97–110 [in Japanese with English abstract].

Kano, K., Ohguchi, T. *et al.* 2011. *Geology of the Toga and Funakawa District, Quadrangle Series, 1:50 000*. Geological Survey of Japan, AIST [in Japanese with English abstract].

Kansai Geo-informatics Council 1999. *New Kansai-jiban, Kobe and Hanshin Regions*. KG-NET, Osaka [in Japanese].

Kansai Geo-informatics Council 2002. *New Kansai-jiban, Kyoto Basin*. KG-NET, Osaka [in Japanese].

Karasuma Deep Core Investigation Group (ed.) 1999. The Karasuma Deep Core Investigation on south-east coast of Lake Biwa, central Japan. *Research Report of the Lake Biwa Museum*, **12**, 1–169 [in Japanese with English abstract].

Kase, Y., Sato, M., Nishida, N. & Ito, M. 2013. Geometry and microstructures of muddy turbidites in the Plio–Pleistocene deep-water successions on the Boso Peninsula, central Japan. *Journal of the Sedimentological Society of Japan*, **72**, 31–37 [in Japansese].

Katahira, T. 1974. Hydrocarbon deposits found in the Green Tuff in the Niigata sedimentary basin – Petroleum geology of the Neogene Tertiary in the Chuetsu and the Kaetsu regions, Niigata, Japan. *Journal of the Japanese Association for Petroleum Technologists*, **39**, 337–356 [in Japanese with English abstract].

Kato, S. & Nishita, H. 2010. Methylphenanthrene maturity parameters of crude oils from oil and gas fields in Niigata Prefecture. *Journal of the Japanese Association for Petroleum Technology*, **75**, 286–295 [in Japanese with English abstract].

Kato, S., Kajiwara, Y. & Nakano, T. 1997. Sulfur content and sulfur isotopic composition of crude oils from the oil-field region of northeastern Japan. *Journal of the Japanese Association for Petroleum Technologists*, **62**, 142–150 [in Japanese with English abstract].

Kato, T. 1985. Stratigraphy of Nichinan Group in southeastern Kyushu, Japan. *Contributions from the Institute of Geology and Paleontology, Tohoku University*, **87**, 1–23 [in Japanese with English abstract].

Katsura, Y. 1984. Depositional environments of the Plio-Pleistocene Kazusa Group, Boso Peninsula, Japan. *Science Report, Institute of Geoscience, University of Tsukuba, Section C, Geological Sciences*, **5**, 69–104.

Kawabe, T. 1989. Stratigraphy of the lower part of the Kobiwako Group around the Ueno basin, Kinki district, Japan. *Journal of Geoscience, Osaka City University*, **32**, 39–90.

Kawagata, S., Tanaka, Y., Motoyama, I., Noda, H., Yamazato, N. & Segawa, T. 2002. Some new calcareous nannofossil ages of the Neogene marine Shimajiri Group in Kume-jima Island, Ryukyu Islands, Japan. *Journal of the Geological Society of Japan*, **108**, 540–543.

Kazaoka, O., Tateishi, M. & Kobayashi, I. 1986. Stratigraphy and facies of the Uonuma Group in the Uonuma district, Niigata Prefecture, central Japan. *Journal of the Geological Society of Japan*, **92**, 829–853 [in Japanese with English abstract].

Kimura, J., Stern, R. J. & Yoshida, T. 2010. Reinitiation of subduction and magmatic responses in SW Japan during Neogene time. *Geological Society of America Bulletin*, **117**, 969–986.

Kino, Y., Kageyama, K., Okumura, K., Endo, H., Fukuta, O. & Yokoyama, S. 1984. *Geology of the Miyazaki District. Quadrangle Series, Scale 1:50 000*. Geological Survey of Japan, Ibaraki [in Japanese with English abstract].

Kitada, N., Inoue, N., Takemura, K., Masuda, F., Hayashida, A., Tabata, T. & Emura, T. 2008. Stratigraphy around Kansai Airport and its properties: reconstruction of the Osaka Group (Pliocene to Pleistocene) in the southern Osaka Basin, Japan. 33rd International Geological Congress (Oslo), August, 2008.

Kitamura, S. 1959. Stratigraphy of Shukunohora Member (Hiyoshi Member). *Journal of the Geological Society of Japan*, **65**, 705–706 [in Japanese].

Kitazato, H. 1997. Palaeogeographic changes in central Honshu, Japan, during the late Cenozoic in relation to the collision of the Izu–Ogasawara Arc with the Honshu Arc. *Island Arc*, **6**, 144–157.

Kobayashi, I. & Tateishi, M. 1992. Neogene stratigraphy and paleogeography in the Niigata region, central Japan. *In*: Kobayashi, I., Tateishi, M., Takayasu, K., Matoba, M. & Akiyama, M. (eds) *The Neogene in the Eastern Margin of the Paleo Sea of Japan -Stratigraphy, Paleogeography, Paleoenvironment*. Memoir of The Geological Society of Japan, Tokyo, **37**, 53–70 [in Japanese with English abstract].

Kodama, K., Tashiro, H. & Takeuchi, T. 1995. Quaternary counterclockwise rotation of south Kyushu, southwest Japan. *Geology*, **23**, 823–826.

Koizumi, I. 1981. Mizunami area, Gifu Prefecture. *In*: Tsuchi, R. (ed.) *Fundamental Data on Japanese Neogene Bio- and Chronostratigraphy – Supplement*. Publ. IGCP Project No. 114, National Working Group of Japan, Shizuoka University, Shizuoka, 68–69 [in Japanese].

Komatsu, N., Fujita, Y. & Sato, O. 1984. Cenozoic volcanic rocks as potential hydrocarbon reservoirs. *Proceedings of 11th World Petroleum Congress*. London, 1983, Chichester, UK, John Wiley, **2**, 411–420.

Kurokawa, K., Ohashi, A., Higuchi, Y. & Satoguchi, Y. 2004. Correlation of the late Pliocene Mushono-Shiraiwa-Tephra beds in the Kobiwako and Kakegawa Groups to the Kyp-NA11-Jwg4 Tephra Beds in the Niigata region, central Japan. *Memoirs of Faculty of Education and Human Sciences (Natural Sciences), Niigata University*, **6**, 107–120.

Kusuhashi, N. & Yamaji, A. 2001. Miocene tectonics of SW Japan as inferred from the Kuma Group, Shikoku. *Journal of the Geological Society of Japan*, **107**, 26–40 [in Japanese with English abstract].

Machida, H. 1999. The stratigraphy, chronology and distribution of distal marker-tephras in and around Japan. *Global and Planetary Change*, **21**, 71–94.

Machida, H. & Arai, F. 1976. The discovery and significance of the very widespread tephra: the Aira-Tn ash. *Kagaku*, **46**, 339–347 [in Japanese].

Machida, H., Arai, F. & Sugihara, S. 1980. Tephrochronological study on the middle Pleistocene deposits in the Kanto and Kinki districts, Japan. *Quaternary Research*, **19**, 233–261 [in Japanese with English abstract].

Mahony, S. H., Wallace, L. M., Miyoshi, M., Villamor, P., Sparks, R. S. J. & Hasenaka, T. 2011. Volcano-tectonic interactions during rapid plate-boundary evolution in the Kyushu region, SW Japan. *Geological Society of America Bulletin*, **123**, 2201–2223.

Majima, R., Ikeda, K., Wada, H. & Kato, K. 2003. An outer-shelf cold-seep assemblage in forearc basin fill, Pliocene Takanabe Formation, Kyushu Island, Japan. *Paleontological Research*, **7**, 297–311.

Makinouchi, T., Danhara, T. & Isoda, K. 1983. Fission-track ages of the Tokai Group and associate formations in the east coast areas of Ise Bay and their significance in geohistory. *Journal of the Geological Society of Japan*, **89**, 257–270 [in Japanese with English abstract].

Makiyama, J. 1950. *Japanese Regional Geological Feature, Chubu Region*. Asakura Publishing, Tokyo [in Japanese].

Masuda, F. 1984. Sedimentary basins in arc-trench system as a high-sensitive recorder of oceanic plate motion. *Mining Geology*, **34**, 1–20 [in Japanese with English abstract].

Masuda, F. 1992. Barrier islands in Paleo-Tokyo Bay. *Chishitsu News*, **458**, 16–27 [in Japanese].

Masuda, F. 2007. Formation of depositional sequences and landforms controlled by relative sea-level change: the result of the Holocene to Upper Pleistocene study in Japan. *Transactions, Japanese Geomorphological Union*, **28**, 365–379 [in Japanese with English abstract].

Masuda, F. & Ito, M. 1999. Contribution to sequence stratigraphy from the Quaternary studies in Japan. *Quaternary Research*, **38**, 184–193.

Masuda, F. & Makino, Y. 1987. Paleo-wave conditions reconstructed from ripples in the Pleistocene Paleo-Tokyo Bay deposits. *Journal of Geography*, **96**, 23–45 [in Japanese with English abstract].

Masuda, F., Nakayama, N. & Ikehara, K. 1988. A cross-stratification produced by tidal current during nine days in the Pleistocene at Uchijiku in Kitaura, Ibaraki. *Environmental Studies, Tsukuba*, **11**, 91–105 [in Japanese].

Masuda, F., Saitoh, Y. & Satoguchi, Y. 2010. Depositional environments and a paleogeographic position for the Pleistocene basal part of the Karasuma Deep Drilling core from Lake Biwa, central Japan. *Quaternary Research*, **49**, 121–131.

METI (Ministry of Economy, Trade and Industry) 2011. *Yearbook of Mineral Resources and Petroleum Products Statistics*. Petroleum/Non-Metallic Minerals/Coke and Metallic Minerals, Tokyo.

Miller, K. G., Mountain, G. S. & Leg 150 Shipboard Party 1996. Drilling and dating New Jersey Oligocene-Miocene sequences: ice volume, global sea level, and EXXON records. *Science*, **271**, 1092–1095.

Mita, I. & Takahashi, M. 1998. Calcareous nannofossil biostratigraphy of the Middle Miocene Kinone and lower Amatsu Formations in the Boso Peninsula, central Japan. *Journal of the Geological Society of Japan*, **104**, 877–890 [in Japanese with English abstract].

Mitsunashi, T. 1990. Synsedimentary tectonics of the southern part of the Kanto sedimentary basin, central Japan. *In*: Mitsunashi, T., Suzuki, Y., Yamauchi, S., Kodama, K. & Komuro, H. (eds) *Development of Sedimentary Basin and Its Relating Folding*. Memoir of the Geological Society of Japan, Tokyo, **34**, 1–9 [in Japanese with English abstract].

Morimoto, J., Oda, M., Torii, M., Schiyonobu, S., Shibuya, H. & Domitsu, H. 2010. Integrated stratigraphy of the Middle to Late Pliocene upper Miyazaki Group, southern Kyushu, Southwest Japan. *Stratigraphy*, **7**, 25 32.

Moriya, S., Yamane, T., Saito, Y., Kato, S. & Nakayama, K. 2007. Hydrocarbon migration modeling through turbidite sands in the Iwafuneoki field, offshore Japan. *Journal of the Japanese Association for Petroleum Technology*, **72**, 89–97 [in Japanese with English abstract].

Moriya, S., Chinzei, K., Nakajima, T. & Danhara, T. 2008. Uplift of the Dewa Hills recorded in the Pliocene paleogeographic change of the western Shinjo Basin, Yamagata Prefecture. *Journal of the Geological Society of Japan*, **114**, 389–404 [in Japanese with English abstract].

Murakoshi, N. & Masuda, F. 1992. Estuarine, barrier-island to strand-plain sequence and related ravinement surface developed during the last interglacial in the Paleo-Tokyo Bay, Japan. *Sedimentary Geology*, **80**, 167–184.

Muto, T. & Blum, P. 1989. An illustration of a sea level control model from a subsiding coastal fan system: pleistocene Ogasayama Formation, central Japan. *Journal of Geology*, **97**, 451–463.

Nagahashi, Y., Satoguchi, Y. & Yoshikawa, S. 2000. Correlation and stratigraphic eruption age of the pyroclastic flow deposits and wide spread volcanic ashes intercalated in the Pliocene-Pleistocene strata, central Japan. *Journal of the Geological Society Japan*, **106**, 51–69 [in Japanese with English abstract].

Nakagawa, H. 1983. Outline of Cenozoic history of the Ryukyu Islands, southwest Japan. *Memoir of the Geological Society of Japan*, **22**, 67–79 [in Japanese with English abstract].

Nakajima, J. & Hasegawa, A. 2010. Cause of M ~ 7 intraslab earthquakes beneath the Tokyo metropolitan area, Japan: possible evidence for a vertical tear at the easternmost portion of the Philippine Sea slab. *Journal of Geophysical Research*, **115**, B04301.

Nakajima, T. 2013. Late Cenozoic tectonic events and intra-arc basin development in Northeast Japan. *In*: Itoh, Y. (ed.) *Mechanism of Sedimentary Basin Formation – Multidisciplinary Approach on Active Plate Margins*. InTech, Rijeka, 153–189.

Nakajima, T., Makimoto, H., Hirayama, J. & Tokuhashi, S. 1981. *Geology of Kamogawa District. Quadrangle Series, Scale 1:50 000*. Geological Survey of Japan, Ibaraki [in Japanese with English abstract].

Nakajima, T., Danhara, T. & Chinzei, K. 2000. Late Cenozoic basin development of the Yuda Basin in the axial part of the Ou Backbone Range, northeast Japan. *Journal of the Geological Society of Japan*, **106**, 93–111 [in Japanese with English abstract].

Nakajima, T., Danhara, T., Iwano, H. & Chinzei, K. 2006. Uplift of the Ou Backbone Range in Northeast Japan at around 10 Ma and its implication for the tectonic evolution of the eastern margin of Asia. *Palaeogeography, Palaeoclimatology, Palaeoecology*, **241**, 28–48.

Nakamura, K. 1983. Possible nascent trench along the eastern Japan Sea as the convergent boundary between Eurasian and North American Plates. *Bulletin of Earthquake Research Institute*, **58**, 711–722 [in Japanese with English abstract].

Nakamura, Y., Ozawa, T. & Nobuhara, T. 1999*a*. Stratigraphy and molluscan fauna of the upper Miocene to lower Pliocene Miyazaki Group in the Aoshima area, Miyazaki Prefecture, southwest Japan. *Journal of the Geological Society of Japan*, **105**, 45–60 [in Japanese with English abstract].

Nakamura, Y., Kameo, K., Asahara, Y. & Ozawa, T. 1999*b*. Stratigraphy and geologic age of the Neogene Shimajiri Group in Kumejima Island,

Ryukyu Islands, southwestern Japan. *Journal of the Geological Society of Japan*, **105**, 757–770.

NAKAYAMA, K. 1991. Depositional process of the Neogene Seto Porcelain Clay Formation in the northern part of Seto City, central Japan. *Journal of the Geological Society of Japan*, **97**, 945–958 [in Japanese with English abstract]

NAKAYAMA, K. & YOSHIKAWA, S. 1990. Magnetostratigraphy of the late Cenozoic Tokai Group in central Japan. *Journal of the Geological Society of Japan*, **96**, 967–976 [in Japanese with English abstract]

NAKAYAMA, N. & MASUDA, F. 1987. Ocean current-controlled sedimentary facies of the Pleistocene Ichijiku Formation, Kazusa Group, Boso Peninsula, Japan. *Journal of the Geological Society of Japan*, **93**, 833–845 [in Japanese with English abstract].

NIITSUMA, N. 1976. Magnetic stratigraphy in the Boso Peninsula. *Journal of the Geological Society of Japan*, **82**, 163–181 [in Japanese with English abstract].

NIITSUMA, N. 1989. Collision tectonics in the southern Fossa Magna, central Japan. *Modern Geology*, **14**, 3–18.

NISHIKAWA, T. & ITO, M. 2000. Late Pleistocene barrier-island development reconstructed from genetic classification and timing of erosional surfaces, paleo-Tokyo Bay, Japan. *Sedimentary Geology*, **137**, 25–42.

NOBUHARA, T. & TAKATORI, R. 1999. Occurrence of Calyptogena sp. (Bivalvia: Vesicomyidea) from the Pliocene Horinouchi Formation, Shizuoka Prefecture, central Japan. *Journal of the Geological Society of Japan*, **105**, 140–150 [in Japanese with English abstract].

O'HARA, S. 1973. Molluscan fossils from the Higashiyatsu Formation. *College of Arts and Sciences, Chiba University, Journal*, **B-6**, 67–87.

ODA, M. 1977. Planktonic foraminiferal biostratigraphy of the Late Cenozoic sedimentary sequence, central Honshu, Japan. *Science Reports, Tohoku University, 2nd Series, Geology*, **48**, 1–76.

ODA, M., CHIYONOBU, S. *ET AL.* 2011. Integrated magnetobiochronology of the Pliocene–Pleistocene Miyazaki succession, southern Kyushu, southwest Japan: implications for an Early Pleistocene hiatus and defining the base of the Gelasian (P/P boundary type section) in Japan. *Journal of Asian Earth Sciences*, **40**, 84–97.

OGAWA, T. 1919. *About the Tertiary in Ise*. Shimazu Hyo-hon Jiho, Kyoto, **6** [in Japanese].

OGURA, M. & KAMON, M. 1992. The subsurface structures and hydrocarbon potentials in the Tenpoku and Haboro area, the northern Hokkaido, Japan. *Journal of the Japanese Association for Petroleum Technology*, **57**, 32–44 [in Japanese with English abstract]

OKADA, M. & NIITSUMA, N. 1989. Detailed paleomagnetic records during the Brunhes- Matuyama geomagnetic reversal, and a direct determination of depth lag for magnetization in marine sediments. *Physics of the Earth and Planetary Interiors*, **56**, 133–150.

OKAMURA, Y., WATANABE, M., MORIJIRI, R. & SATOH, M. 1995. Rifting and basin inversion in the eastern margin of the Japan Sea. *Island Arc*, **4**, 166–181.

OKAZAKI, H. & MASUDA, F. 1992. Depositional systems of the Late Pleistocene sediments in Paleo-Tokyo Bay area. *Journal of the Geological Society of Japan*, **98**, 235–258 [in Japanese with English abstract].

OKAZAKI, H. & MASUDA, F. 1995. Sequence stratigraphy of the late Pleistocene Palaeo-Tokyo Bay: barrier islands and associated tidal delta and inlet. *In*: FLEMMING, B. W. & BARTHOLOMā, A. (eds) *Tidal Signatures in Modern and Ancient Sediments*. International Association of Sedimentologists, Special Publications, Oxford, **24**, 275–288.

OKUBO, S., HOSHI, K., KATO, K. & SUZAKI, T. 1996. The dolerite reservoir and its alteration in the Ayukawa oil and gas field, Akita Prefecture, Japan. *Journal of the Japanese Association for Petroleum Technology*, **61**, 61–70 [in Japanese with English abstract].

OMOKAWA, M., KONDO, K. & WASEDA, A. 1990. Geochemical study and geological background for the oil and gas fields in northern Ishikari Area, Hokkaido. *Journal of the Japanese Association for Petroleum Technology*, **55**, 23–36 [in Japanese with English abstract].

OTA, Y., NARUSE, T., TANAKA, S. & OKADA, A. 2004. *Kinki, Chugoku, and Shikoku: Geography of Japan*. Tokyo University Press, Tokyo [in Japanese].

OTOFUJI, Y., MATSUDA, T. & NOHDA, S. 1985. Paleomagnetic evidence from the Miocene counter-clockwise rotation of Northeast Japan. *Earth Planetary Science Letters*, **75**, 265–277.

OYAKE, T., ODA, M. & TORII, M. 1995. Planktic foraminiferal biostratigraphy of the lower part of the Miyazaki Group. *102th Annual Meeting, Geological Society of Japan*, Abstract, Hiroshima, **96** [in Japanese].

OZE, T. 1929. Geography and geology of Chita Peninsula. *Journal of Geography*, **41**, 338–345 [in Japanese with English abstract].

PICKERING, K. T., COUTER, C., OBA, T., TAIRA, A., SCHAAF, M. & PLATZMAN, E. 1999. Glacio-eustatic control on deep-marine clastic forearc sedimentation, Pliocene–mid-Pleistocene (*c.* 1180–600 ka) Kazusa Group, SE Japan. *Journal of the Geological Society, London*, **156**, 125–136.

SAFFER, D., MCNEILL, L., BYRNE, T., ARAKI, E., TOCZKO, S., EGUCHI, N.,TAKAHASHI, K. & THE EXPEDITION 319 SCIENTISTS 2010. NanTroSEIZE Stage 2: NanTroSEIZE Riser/Riserless Observatory. *Proceedings of the IODP*, **319**, http://doi.org/10.2204/iodp.proc.319.2010

SAITO, S. 1992. Stratigraphy of Cenozoic strata in the southern terminus area of Boso Peninsula, Central Japan. *Contribution of Institute of Geology and Paleontology, Tohoku University*, **93**, 1–37 [in Japanese with English abstract].

SAITO, S., SAKAI, T., ODA, MO., HASEGAWA, S. & TANAKA, Y. 1991. The Miura Group in the southern Boso Peninsula–Uplifted forearc basin. *Earth Monthly*, **13**, 15–19 [in Japanese].

SAITO, T. 1963. Planktonic foraminifera from the Neogene Tertiary of Japan. *Fossils*, **5**, 8–19 [in Japanese with English abstract].

SAITO, T. & ITO, M. 2002. Deposition of sheet-like turbidite packets and migration of channel-overbank systems on a sandy submarine fan: an example from the Late Miocene–Early Pliocene forearc basin, Boso Peninsula, Japan. *Sedimentary Geology*, **149**, 265–277.

SAITO, Y., YAMAMOTO, T., YAMANE, T., MORIYA, S., NISHIMURA, M. & TAKANO, O. 2008. Exploration case study of sandstone pinch-out stratigraphic traps offshore Kitakanbara – Challenge to discovery of 'the second Iwafuneoki oil and gas field'. *Journal of the Japanese Association for Petroleum Technology*, **73**, 38–36 [in Japanese with English abstract].

SAKAI, T. 2000. Reflected turbidites in the Plio-Pleistocene Kakegawa Group, central Japan. *Journal of the Geological Society of Japan*, **106**, 87–90 [in Japanese with English abstract].

SAKAI, T. & MASUDA, F. 1996. Sequence stratigraphy of the upper part of the Plio-Pleistocene Kakegawa Group, western Shizuoka, Japan. *Journal of Sedimentary Research*, **66**, 778–787.

SAKATA, S., KANEKO, N. & SUZUKI, N. 1988. A biomarker study of petroleum from the Neogene Tertiary sedimentary basins in Northeast Japan. *Geochemical Journal*, **22**, 89–105.

SAKATA, S., KANEKO, N. & SUZUKI, N. 1990. Biomarkers in petroleum from sedimentary basins in northeast Japan. *Journal of the Japanese Association for Petroleum Technology*, **55**, 48–53 [in Japanese with English abstract].

SAMPEI, Y., SUZUKI, N. *ET AL.* 2000. Organic carbon contents of Neogene mudstones in Japan. *Geoscience Report Shimane University*, **19**, 77–95 [in Japanese with English abstract].

SATO, H. 1994. The relationship between late Cenozoic tectonic events and stress field and basin development in northeast Japan. *Journal of Geophysical Research*, **99**, 261–274.

SATO, S. 1976. Organo geochemical study on kerogen of sedimentary rocks in Japan. *Science Reports of the Tohoku University*, Third Series (Geology), **12**, 85–113.

SATO, T., TAKAYAMA, T., KATO, M., KUDO, T. & KAMEO, K. 1988. Calcareous microfossil biostratigraphy of the uppermost Cenozoic Formations distributed in the coast of the Japan Sea–Part 4: conclusion. *Journal of the Japanese Association for Petroleum Technology*, **53**, 475–491 [in Japanese with English abstract].

SATO, T., BABA, K., OHGUCHI, T. & TAKAYAMA, T. 1991. Discovery of Early Miocene calcareous nannofossils from Japan Sea side, Northern Honshu, Japan, with reference to palaeoenvironment in the Daijima and Nishikurosawa ages. *Journal of the Japanese Association of Petroleum Technology*, **56**, 263–279 [in Japanese with English abstract].

SATO, T., YAMASAKI, M. & CHIYONOBU, S. 2009. Geology of Akita Prefecture. *Daichi*, **50**, 70–83 [in Japanese].

SATO, T., SATO, N., YAMASAKI, M., OGAWA, Y. & KANEKO, M. 2012. Late Neogene to Quaternary paleoenvironmental changes in the Akita area, Northeast Japan. *Journal of the Geological Society of Japan*, **118**, 62–73 [in Japanese with English abstract].

SATOGUCHI, Y. 2009. Iga–Ohmi Basin. *In*: YAO, A., YOSHIKAWA, S. *ET AL.* (eds) *Regional Geology of Japan 5, Kinki District*, Asakura Publishing, Tokyo, 253–258 [in Japanese].

SATOGUCHI, Y. & HATTORI, N. 2008. Correlation of volcanic ash beds of the upper Kobiwako Group with those of the upper Kazusa Group in the Middle Pleistocene, central Japan. *Quaternary Research (Daiyonki Kenkyu)*, **47**, 15–27 [in Japanese with English abstract].

SATOGUCHI, Y. & NAGAHASHI, Y. 2012. Tephrostratigraphy of the Pliocene to Middle Pleistocene Series in Honshu and Kyushu Islands, Japan. *Island Arc*, **21**, 149–169.

SATOGUCHI, Y., NAGAHASHI, Y., KUROKAWA, K. & YOSHIKAWA, S. 1999. Tephrostratigraphy of the Pliocene to Lower Pleistocene formations in central Honshu, Japan. *Earth Science*, **53**, 275–290 [in Japanese with English abstract].

SAWADA, T., SHINDO, R., MOTOYAMA, I. & KAMEO, K. 2009. Geology and radiolarian biostratigraphy of the Miocene and Pliocene series exposed along the Koito-gawa River, Boso Peninsula, Japan. *Journal of the Geological Society of Japan*, **115**, 206–222 [in Japanese with English abstract].

SEKIGUCHI, K., OMOKAWA, M., HIRAI, A. & MIYAMOTO, Y. 1984. Geochemical study of oil and gas accumulation in 'Green Tuff' reservoir in Nagaoka to Kashiwazaki region. *Journal of the Japanese Association for Petroleum Technology*, **49**, 56–64 [in Japanese with English abstract].

SENO, T., SAKURAI, T. & STEIN, S. 1994. Can the Okhotsk plate be discriminated from the North American plate? *Journal of Geophysical Research*, **101**, 11305–11315.

SHIBATA, H. & ITOIGAWA, J. 1980. Miocene paleogeography of the Setouchi province, Japan. *Bulletin of Mizunami Fossil Museum*, **7**, 1–49 [in Japanese with English abstract].

SHIMOZURU, D., ARAKMAKI, S. & IDA, Y. (eds) 1995. *Encyclopedia of Volcanoes*. Asakura Shoten, Tokyo [in Japanese]

SHUTO, T. 1961. Palaeontological study of the Miyazaki Group: A general account of the faunas. *Memoirs of the Faculty of Science, Kyushu University, Series D, Geology*, **10**, 73–206.

SUTO, I., YANAGISAWA, Y. & OGASAWARA, K. 2005. Tertiary geology and chronostratigraphy of the Joban area and its environs, northeastern Japan. *Bulletin of Geological Survey of Japan*, **56**, 375–409 [in Japanese with English abstract].

SUZUKI, A. & TSUKAWAKI, S. 1989. Geological age and sedimentary facies of the Yaeyama Group, Yonaguni-jima, Okinawa Prefecture. *96th Annual Meeting Geological Society of Japan, Abstract*, Mito, 170 [in Japanese].

SUZUKI, H. 1987. Stratigraphy of the Miyazki Group in the southeastern part of Miyazaki Prefecture, Kyushu, Japan. *Contributions from the Institute of Geology and Paleontology, Tohoku University*, **90**, 1–24 [in Japanese with English abstract].

SUZUKI, N. 1990. Application of sterane epimerization to evaluation of Yoshii gas and condensate reservoir, Niigata basin, Japan. *American Association of Petroleum Geologists Bulletin*, **74**, 1571–1589.

SUZUKI, N., SAKATA, S. & KANEKO, N. 1987. Biomarker maturation levels and primary migration stage of the Neogene Tertiary crude oils and condensates in the Niigata sedimentary basin. *Journal of the Japanese Association for Petroleum Technology*, **52**, 499–510 [in Japanese with English abstract].

SUZUKI, N., SAMPEI, Y. & MATSUBAYASHI, H. 1995. Organic geochemical difference between source rocks from Akita and Niigata oil field, Neogene Tertiary, Japan. *Journal of the Japanese Association for Petroleum Technology*, **60**, 62–75 [in Japanese with English abstract]

SUZUKI, U. 1989. Geology of Neogene Basins in the eastern part of the Sea of Japan. *In*: KITAMURA, N., OTSUKI, K. & OHGUCHI, T. (eds) *Cenozoic Geotectonics of Northeast Honshu Arc*. Memoir of the Geological Society of Japan, Tokyo, **32**, 143–183 [in Japanese with English abstract].

TADO COLLABORATE RESEARCH GROUP 1998. Palaeoenvironments of the Plio-Pleistocene Tokai Group in the northern part of Mie Prefecture, central Japan. *Earth Science*, **52**, 115–135 [in Japanese with English abstract].

TAGUCHI, K. 1976. Geochemical relationships between Japanese Tertiary oils and their source rocks. *Proceedings of 9th World Petroleum Congress*. Tokyo, 1975. Applied Science Publishers, Barking, Essex, England, **2**, 193–194.

TAIRA, A. 2001. Tectonic evolution of the Japanese Island arc system. *Annual Reviews of Earth and Planetary Sciences*, **29**, 109–134.

TAKAGI, H., TAKESHITA, T., SHIBATA, K., UCHIUMI, S. & INOUE, M. 1992. Middle Miocene normal faulting along the Tobe Thrust in western Shikoku. *Journal of the Geological Society of Japan*, **98**, 1069–1072 [in Japanese].

TAKAHASHI, M. 2004. Oxygen isotope fluctuation pattern between 19 and 8 Ma re-calculated on the basis of Cande and Kent's (1995) geomagnetic polarity time scale. *Journal of the Japanese Association for Petroleum Technology*, **69**, 83–93 [in Japanese with English abstract].

TAKAHASHI, M. 2006. Tectonic development of the Japanese Islands controlled by Philippine Sea plate motion. *Journal of Geography*, **115**, 116–123 [in Japanese].

TAKAHASHI, M. & OKADA, T. 2001. K-Ar age of the Kn-1 Tuff in the Miocene marine sequence in the Boso Peninsula, central Japan. *Journal of the Japanese Association for Petroleum Technology*, **66**, 396–403 [in Japanese with English abstract].

TAKAHASHI, M., MITA, I. & OKADA, T. 1999. K-Ar age of the Am-4 Tuff related to the CN5a/CN5b boundary on the Miocene marine sequence in the Boso Peninsula, central Japan. *Journal of the Japanese Association for Petroleum Technology*, **64**, 282–287 [in Japanese with English abstract].

TAKANO, O. 1995. Delta to shelf systems and depositional sequences of the Pliocene Higashigawa and Naradate Formations, deposited during the prograding stage of the Northern Fossa Magna Basin, central Japan. *In*: SAITO, Y., HOYANAGI, K. & ITO, M. (eds) *Sequence Stratigraphy –Toward a New Dynamic Stratigraphy*. Memoir of the Geological Society of Japan, Tokyo, **45**, 170–188 [in Japanese with English abstract].

TAKANO, O. 2002*a*. Changes in depositional systems and sequences in response to basin evolution in a rifted and inverted basin: an example from the Neogene Niigata-Shin'etsu basin, Northern Fossa Magna, central Japan. *Sedimentary Geology*, **152**, 79–97.

TAKANO, O. 2002*b*. Tectonostratigraphy and changes in depositional architecture through rifting and basin inversion in the Neogene Niigata-Shin'etsu basin, Northern Fossa Magna, central Japan: implications for tectonic history of the Japan Sea marginal regions. *In*: TATEISHI, M. & KURITA, H. (eds) *Development of Tertiary Sedimentary Basins Around Japan Sea (East Sea)*. Niigata University, 157–181.

TAKANO, O. 2011. Upper Neogene sequence stratigraphy and submarine-fan turbidite sequences in the Kubiki area in the Niigata–Shin'etsu basin, Northern Fossa Magna, central Japan. *Journal of the Geological Society of Japan*, **117**, 238–258 [in Japanese with English abstract].

TAKANO, O., MORIYA, S., NISHIMURA, M., AKIBA, F., ABE, M. & YANAGIMOTO, Y. 2001. Sequence stratigraphy and characteristics of depositional systems of the Upper Miocene to Lower Pleistocene in the Kitakambara area, Niigata Basin, central Japan. *Journal of the Geological Society of Japan*, **107**, 585–604.

TAKANO, O., TATEISHI, M. & ENDO, M. 2005. Tectonic controls of a backarc trough-fill turbidite system: the Pliocene Tamugigawa Formation in the Niigata–Shin'etsu inverted rift basin, Northern Fossa Magna, central Japan. *Sedimentary Geology*, **176**, 247–279.

TAKASHIMIZU, Y. 2009. Ancient submarine-canyon system developed within fore-arc basin fill, Mio–Pliocene 'Miyazaki facies', Miyazaki Group, western Japan. *Journal of the Geological Society of Japan*, **115**, 559–577 [in Japanese with English abstract].

TAKAYAMA, M. & KATO, S. 1995. Lithology and total organic carbon of the Onnagawa Formation in the subsurface of the Akita-Yamagata Sedimentary Basin. *Journal of the Japanese Association for Petroleum Technology*, **60**, 39–48 [in Japanese with English abstract]

TAKEMURA, K. 2002. Environmental changes in Lake Biwa. *Earth Environment*, **7**, 59–75 [in Japanese].

TAMURA, I., YAMAZAKI, H. & MIZUNO, K. 2008. Characteristics for the recognition of Pliocene and early Pleistocene marker tephras in central Japan. *Quaternary International*, **178**, 85–99.

TATEISHI, M., TAKANO, O., TAKASHIMA, T. & KUROKAWA, K. 1997. Depositional system and provenance of Cenozoic coarse sediments in Northern Fossa Magna. *Journal of the Japanese Association for Petroleum Technology*, **62**, 35–44 [in Japanese with English abstract].

TOKUHASHI, S. 1979. Three dimensional analysis of a large sandy-flysch body, Mio-Pliocene Kiyosumi Formation, Boso Peninsula, Japan. *Memoir of the Faculty of Science, Kyoto University, Series of Geology & Mineralogy*, **46**, 1–60.

TOKUHASHI, S. & ENDO, H. 1984. *Geology of the Anesaski District. Quadrangle Series, Scale 1:50 000*. Geological Survey of Japan, Ibaraki [in Japanese with English abstract].

TOKUHASHI, S., DANHARA, T. & IWANO, H. 2000. Fission track ages of eight tuffs in the upper part of the Awa Group, Boso Peninsula, central Japan. *Journal of the Geological Society of Japan*, **106**, 560–573 [in Japanese with English abstract].

TORII, M. & ODA, M. 2001. Correlation of the Izaku pyroclastic flow deposit in Kagoshima Prefecture with tuff bed intercalated in the Miyazaki Group: the eruption age of the Izaku-Hisamine tephra based on integrated stratigraphy of the Miyazaki Group and its significance. *Journal of the Geological Society of Japan*, **107**, 379–391 [in Japanese with English abstract].

TORII, M., ODA, M. & ITAYA, T. 2000. K-Ar dating of tuff beds intercalated in the Miyazaki Group, Kyushu, Japan. *Bulletin of the Volcanlogical Society of Japan*, **45**, 131–148 [in Japanese with English abstract].

TSUJI, T., MIYATA, Y., OKADA, M., MITA, I., NAKAGAWA, H., SATO, Y. & NAKAMIZU, M. 2005. High-resolution chronology of the lower Pleistocene Otadai and Umegase Formations of the Kazusa Group, Boso Peninsula, central Japan–Chronostratigraphy of the JNOC TR-3 cores based on oxygen isotope, magnetostratigrahy and calcareous nannofossil. *Journal of the Geological Society of Japan*, **111**, 1–120 [in Japanese with English abstract].

TSUKAWAKI, S. 1994. Depositional Environments of the Sagara and Kakegawa Groups (Middle Miocene–Early Pleistocene), and the evolution of the sedimentary basin, central Japan. *Science Reports of the Tohoku University, Sendai, Second Series (Geology)*, **63**, 1–38.

UEMURA, Y. 2001. *A Study of Comparative Tectonic Geomorphology: Tectonic Landform of Plate Boundary Zones and Quaternary Crustal Movement.* Kokon-shoin, Tokyo [in Japanese].

UEMURA, Y. & TAISHI, H. 1990. Active tectonics of the bottom of Lake Biwa and development of its Lake Basin, southwest Japan. *Geographical Review of Japan*, **63**, 722–740 [in Japanese with English abstract].

UJIIÉ, H. 1962. Geology of the Sagra-Kakgawa sedimentary basin in central Japan. *Tokyo Kyoiku Daigaku, Science Report, Section C*, **8**, 1–65.

UJIIÉ, K. & OKI, K. 1993. On the stratigraphic and structural relationship between the Miyazaki and the Aoshima Facies of the upper Neogene Miyazaki Group, Kyushu, Japan. *Reports of Faculty of Science, Kagoshima University*, **26**, 67–84 [in Japanese with English abstract].

URABE, A. 1999. Correlation of relative sea-level changes in Plio-Pleistocene forearc basin-fill successions of the Osaka, Kakegawa, Kazusa and Uonuma Groups, Japan. *Earth Science*, **53**, 247–257 [in Japanese with English abstract].

VAIL, P. R., AUDEMARD, F., BOWMAN, S. A., EISNER, P. N. & PERETS-CRUZ, C. 1991. The stratigraphic signatures of tectonics, eustasy and sedimentology–an overview. *In*: EINSELE, G., RICKEN, W. & SEILACHER, A. (eds) *Cycles and Events in Stratigraphy*. Springer-Verlag, Berlin, 617–659.

WASEDA, A. 1993. Effect of maturity on carbon and hydrogen isotopes of crude oils in Northeast Japan. *Journal of the Japanese Association for Petroleum Technology*, **58**, 199–208 [in Japanese with English abstract].

WASEDA, A. & OMOKAWA, M. 1990. Generation, migration and accumulation of hydrocarbons in the Yurihara oil and gas field. *Journal of the Japanese Association for Petroleum Technology*, **55**, 233–244 [in Japanese with English abstract].

WATANABE, K., MASUDA, F., KATSURA, Y. & OKAZAKI, H. 1987. Environmental changes in Kanto, Japan (2). *Education of Earth Sciences*, **40**, 79–90 [in Japanese with English abstract].

WATANABE, M. & TAKAHASHI, M. 1997. Diatom biostratigraphy of the Middle Miocene Kinone and lower Amatsu Formations in the Boso Peninsula, central Japan. *Journal of the Japanese Association for Petroleum Technology*, **62**, 215–225 [in Japanese with English abstract].

WATANABE, M. & TAKAHASHI, M. 2000. Diatom biostratigraphy of the middle Miocene marine sequence of the Kawadani section in the Kamogawa area, Boso Peninsula, central Japan. *Journal of the Geological Society of Japan*, **106**, 489–500 [in Japanese with English abstract].

WRIGHT, J. D. & MILLER, K. G. 1992. Miocene stable isotope stratigraphy, Site 747, kerguelen Plateau. *In*: WISE, S. W. JR, ET AL. (eds) *Proceedings of the Ocean Drilling Program, Scientific Results*, College Station, Texas, **120**, 855–866.

YAGI, M., HASENAKA, T. ET AL. 2001. Transition of magmatic composition reflecting an evolution of rifting activity – a case study of the Akita–Yamagata basin in Early to Middle Miocene, Northeast Honshu, Japan. *Japanese Magazine of Mineralogical and Petrological Sciences*, **30**, 265–287 [in Japanese with English abstract].

YAMADA, R. & YOSHIDA, T. 2004. Volcanic sequences related to Kuroko mineralization in the Hokuroku district, Northeast Japan. *Resource Geology*, **54**, 399–412.

YAMADA, Y. & UCHIDA, T. 1997. Characteristics of hydrothermal alteration and secondary porosityies in volcanic rock reservoirs, the Katakai gas field. *Journal of the Japanese Association for Petroleum Technology*, **62**, 311–320 [in Japanese with English abstract].

YAMAJI, A. 1989. Geology of Atsumi area and Early Miocene rifting in the Uetsu district, northeast Japan. *Memoir of the Geological Society of Japan*, **32**, 305–320 [in Japanese with English abstract].

YAMAJI, A. 1990. Rapid intra-arc rifting in Miocene northeast Japan. *Tectonics*, **9**, 365–378.

YAMASHITA, K. Y. & ITOIGAWA, J. 1988. *Geology of Japan 5, Chubu Region II.* Kyoritsu Shuppan, Tokyo [in Japanese].

YANAGISAWA, Y. 1993. Mediaria magna Yanagisawa, sp. nov., a new fossil raphid diatom species useful for Middle Miocene diatom biostratigraphy. *Transactions and Proceedings of Palaeontological Society of Japan, New Series*, **174**, 411–425.

YAO, A., YOSHIKAWA, S. ET AL. (eds) 2009. *Regional Geology of Japan 5, Kinki District*. Asakura Publishing, Tokyo [in Japanese].

YOKOYAMA, T. 1969. Tephrochronology and paleogeography of thc Plio-Pleistocene in the eastern Setouchi Geologic Province, Southwest Japan. *Memoirs of Faculty of Science, Kyoto University, Series of Geology and Mineralogy*, **36**, 19–85.

YOKOYAMA, T. 1983. *Natural History of Southwest Japan.* Sanwashobo, Kyoto [in Japanese].

YOKOYAMA, T. 1984. Stratigraphy of the Quaternary system around Lake Biwa and geohistory of the ancient Lake Biwa. *In*: HORIE, S. (ed.) *Lake Biwa*. Dr. W. Junk Pub., Dordrecht, 43–128.

YOKOYAMA, T., MATSUDA, T. & TAKEMURA, K. 1980. Fission-track ages of volcanic ashes in the Tokai Group, central Japan. *Quaternary Research*, **19**, 301–309 [in Japanese with English abstract].

YOSHIDA, F. 1987. *Geology of the Tsu-Tobu District. With geological sheet map at 1:50 000.* Geological Survey of Japan, Ibaraki [in Japanese with English abstract].

YOSHIDA, F. 1988. Plio-Pleistocene Tokai Group between the Suzuka and Yoro Mountains, central Japan. *Earth Science*, **42**, 1–16 [in Japanese with English abstract].

YOSHIDA, F. 1990. Stratigraphy of the Tokai Group and paleogeography of the Tokai sedimentary basin in the Tokai region, central Japan. *Bulletin of Geological Survey of Japan*, **41**, 303–340 [in Japanese with English abstract].

YOSHIDA, K. & NIITSUMA, N. 1976. Magnetostratigraphy of the Sagara-Kakegawa area, Shizuoka Prefecture. *83th Annual Meeting, The Geological Society of Japan*, Matsumoto, Abstract, 167 [in Japanese].

YOSHIDA, T. 2009. Late Cenozoic Magmatism in the northeast Honshu Arc, Japan. *Earth Science*, **63**, 269–288 [in Japanese with English abstract].

YOSHIKAWA, S. & YAMAZAKI, H. 1998. *Changes in Paloe-Lake Biwa and the development of Lake Biwa.* Urban Kubota, Osaka, **37**, 2–11 [in Japanese].

YOSHIKAWA, S., YOSHIDA, F. & SUGAWA, E. 1991. Volcanic ash layers of the Tokai Group and their correlation. *Earth Science*, **45**, 453–467 [in Japanese with English abstract].

YOSHIKAWA, S., SATOGUCHI, Y. & NAGAHASHI, Y. 1996. A widespread volcanic ash bed in the horizon close to the Pliocene-Pleistocene boundary: Fukuda-Tsujimatagawa-Kd38 volcanic ash bed occurring in central Japan. *Journal of the Geological Society of Japan*, **102**, 258–270 [in Japanese with English abstract].

7 Deep seismic structure

AKIRA HASEGAWA, JUNICHI NAKAJIMA & DAPENG ZHAO

Subducting slab structure and related seismic activity (AH)

Japan is located in a region of subduction zones in which four tectonic plates converge (Fig. 7.1), with two oceanic plates subducting beneath two continental plates. Frequent earthquakes are associated with this plate convergence, including several major earthquakes that have caused extensive damage to inhabited areas. The recent *M* 9.0 11 March 2011 great Tohoku-oki earthquake, which occurred along the plate interface east of Tohoku, NE Japan, is an example of a significantly destructive earthquake in the region. This event was the greatest earthquake in the modern history of Japan, the world's fourth-largest earthquake to occur during the era of instrumental seismology, and it caused severe damage to eastern Japan, resulting in *c.* 20 000 dead and missing.

As shown in Figure 7.1, NE Japan lies on the southernmost portion of the North American (NA) plate, whereas SW Japan is located at the eastern edge of the Eurasian Plate. These two plates converge along the Itoigawa–Shizuoka Tectonic Line in the south and along the eastern margin of the Japan Sea in the north. The Philippine Sea (PHS) Plate subducts beneath SW Japan northwestwards at a rate of 3–5 cm a^{-1} along the Sagami Trough in the east, along the Nankai Trough in the west and along the Ryukyu Trench in the SW (Seno *et al.* 1993, 1996; Wei & Seno 1998; Heki & Miyazaki 2001; Miyazaki & Heki 2001). The Pacific (PAC) plate subducts west-northwestwards at a rate of 8–9 cm a^{-1} along the Kuril Trench in the north and along the Japan Trench in the centre beneath NE Japan, and in the south at a rate of approximately 6 cm a^{-1} along the Izu–Bonin Trench beneath the PHS Plate (DeMets *et al.* 1994).

This complex circumstance of tectonic plate convergence causes not only extremely high seismic activity, mainly along the plate interfaces (interplate earthquakes), within the subducted plates (intraslab earthquakes) and in the shallow crust of the overriding plates (inland crustal earthquakes), but also leads to a complex structure of subducting slabs beneath the Japanese islands. At the present time, approximately one-tenth of earthquakes in the world occur in and around the Japanese islands. Figure 7.2 shows the epicentre distribution of such earthquakes of magnitude 4 or greater ($M > 4$) for the period of 1995–2005.

The Japanese islands are covered by a dense (station separation of *c.* 20 km) nationwide seismic network, known as the Kiban seismic network, which provides a large volume of high-quality data. The spatial resolution of data that can be used for imaging and determination of deep seismic structures beneath the region has therefore rapidly increased. Data from the Kiban network, which covers the entirety of the Japanese islands, have been used in a number of studies on seismic activity, earthquake source processes and seismic structures. These studies have revealed the detailed structures of the subducting oceanic plates and the mantle wedge and crust above these plates, contributing to a deeper understanding of the generation mechanism of earthquakes in the Japanese subduction zones, particularly with respect to their shallow portions down to *c.* 200 km. As a result, the Japanese islands have what is now the best-understood deep seismic structure in subduction zones around the world.

In this section we briefly introduce the subducting slab structure beneath the islands, recently obtained from the high-quality nationwide seismic network data, and its implications for the genesis of subduction-zone earthquakes.

Thickness of the oceanic PAC and PHS plates

The thickness of an oceanic plate is thought to grow with age, which has been confirmed by analysing surface waves (e.g. Kanamori & Press 1970) although the resolution is not very good. Recently, Kawakatsu *et al.* (2009) showed that the plate thickness increases with age as expected from a thermal model based on higher-resolution P- and S-receiver function analyses of teleseismic body waves observed at ocean-bottom borehole broadband seismographs (Fig. 7.3). They detected sharp lithosphere–asthenosphere boundaries at the base of the oceanic PAC and PHS plates, beneath the Pacific Ocean east of the Japanese mainland and beneath the Philippine Sea south of the mainland, respectively. Depths to these boundaries range from *c.* 55 to *c.* 83 km and are age dependent, as expected. Observed S-wave velocity reductions at the boundaries are very large, which requires a partially molten asthenosphere. By using S-to-P converted seismic waves along three margins of oceanic plates with ages in the range *c.* 130–10 Ma, Kumar & Kawakatsu (2011) also confirmed that the thickness of the oceanic plate increases with the plate age, suggesting the evolution of the oceanic plate is mainly governed by temperature. The young PHS Plate, with an age of 70–15 Ma subducting beneath SW Japan, is relatively thin and its thickness ranges from *c.* 30 km to 60 km depending on the location. In contrast, the old PAC Plate with an age of *c.* 140–120 Ma subducting beneath NE Japan and beneath the PHS Plate is thicker, its thickness being *c.* 80 km. A 80–90 km thickness for the PAC Plate was also estimated from a seismic tomography study (Zhao *et al.* 1994).

Configuration of the subducting PAC Plate

The subduction of the 80–90 km thick PAC Plate beneath the Japanese islands forms a deep planar seismic zone within the plate known as the Wadati–Benioff Zone, which extends to a depth of more than 600 km. In the depth range of *c.* 50–180 km, these intraslab earthquakes within the PAC Plate form a double-planed deep seismic zone beneath Hokkaido, Tohoku and Kanto in NE Japan (Tsumura 1973; Hasegawa *et al.* 1978*a*; Suzuki *et al.* 1983).

Using converted ScS-to-P waves (ScSp phase) at the plate interface beneath Tohoku, Hasegawa *et al.* (1978*b*) determined that the upper interface of the subducting PAC Plate lies immediately above the upper seismic plane of the double-planed deep seismic zone. Subsequently, Matsuzawa *et al.* (1986, 1990) confirmed this observation. They detected converted S-to-P and P-to-S waves (SP and PS phases) at the plate interface in seismograms from

From: Moreno, T., Wallis, S., Kojima, T. & Gibbons, W. (eds) 2016. *The Geology of Japan*. Geological Society, London, 339–370.

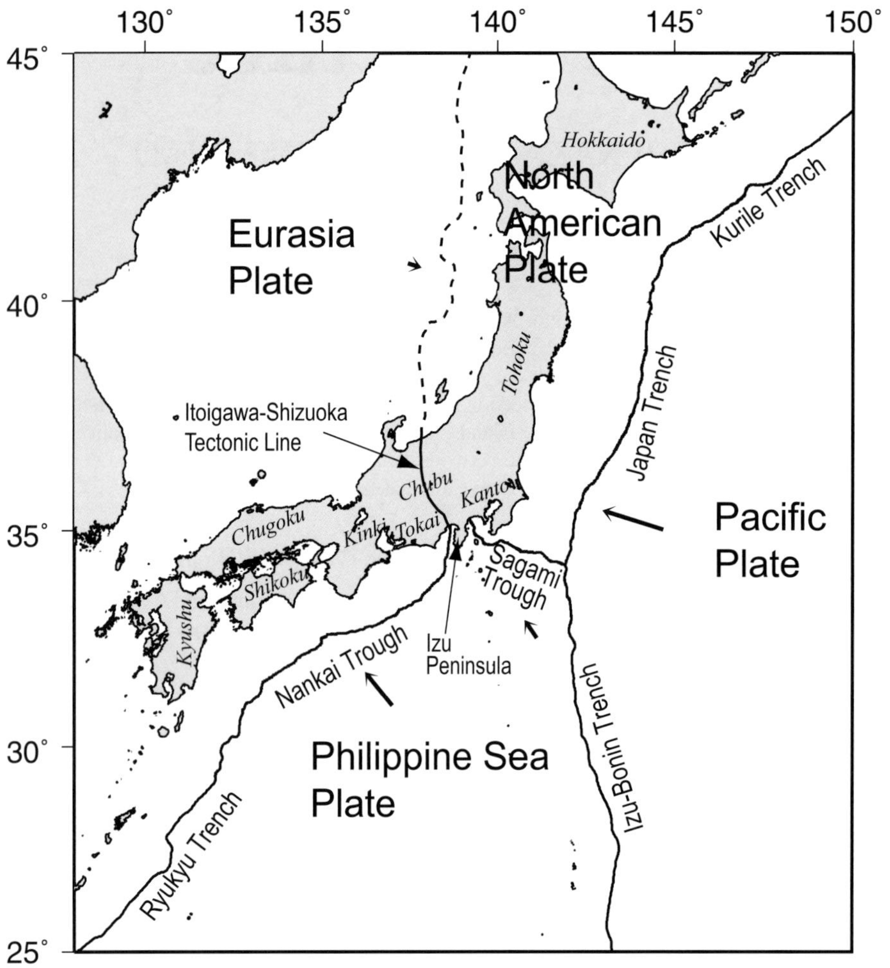

Fig. 7.1. Tectonic setting of Japan.

intermediate-depth intraslab events and located the upper interface of the subducting plate using the arrival times of these waves. Zhao *et al.* (1997*a*) located the upper plate interface for the whole area of Tohoku by inverting arrival time data of the SP waves. These studies revealed that, at depths greater than *c.* 50 km, most of the events in the upper seismic plane of the double seismic zone occur within the crust of the subducting plate, which indicates that the upper envelope of intermediate-depth intraslab seismicity coincides approximately with the upper surface of the subducting PAC Plate.

Adopting the upper envelope of intraslab seismicity as the upper surface of the subducting plate, as indicated by the above-mentioned observations, Nakajima *et al.* (2009*a*) and Kita *et al.* (2010) updated previous studies (e.g. Hasegawa *et al.* 1994; Zhao *et al.* 1997*a*) and estimated the configuration of the upper surface of the subducting PAC Plate down to a depth of *c.* 500 km beneath the entirety of Japan. In their estimation of the upper surface of the plate at depths shallower than *c.* 50 km, they used the relocated hypocentres of interplate earthquakes such as low-angle thrust-type and small repeating earthquakes-seismic events that are thought to occur at the plate interface. The geometry of the upper surface of the subducting PAC Plate beneath the Japanese islands estimated in this manner is shown as isodepth contours in Figure 7.4

At depths greater than *c.* 150 km, intraslab earthquakes tend to occur not in the crust but rather in the mantle portion of the subducted slab as will be described in section 'Internal structure of the slab and intermediate-depth intraslab earthquakes'. This means that the upper envelope of these earthquakes does not represent the upper plate interface, but instead is probably located slightly below the upper plate interface. This difference is not very large however (perhaps *c.* 10 km).

Configuration of the subducting PHS Plate

A number of studies have estimated the configuration of the PHS slab subducting beneath SW Japan by assuming that the upper envelope of the distribution of intraslab earthquake hypocentres represents the upper surface of the subducting plate (e.g. Yamazaki & Ooida 1985; Ishida 1992; Noguchi 1996), as in the case of the PAC slab subducting beneath NE Japan. The assumption used for these estimations is roughly supported by the following observations of ScSp waves and seismic-guided waves.

ScSp waves converted at the upper surface of the PHS Plate subducting beneath SW Japan were also detected and used to estimate the location of the plate interface (Nakanishi 1980). In addition, Fukao *et al.* (1983) and later Hori *et al.* (1985) detected seismic-guided waves propagating within the seismic low-velocity crust of the PHS Plate, indicating that the crust of the subducting plate in this region remains untransformed down to depths of at least 60 km. Comparing the obtained results with the hypocentre locations of intraslab earthquakes, these studies showed that many of the intraslab events in the PHS slab occur in the upper portion near the upper surface of the slab.

Although a number of studies have been conducted as described above, the configuration of the subducting PHS Plate beneath SW

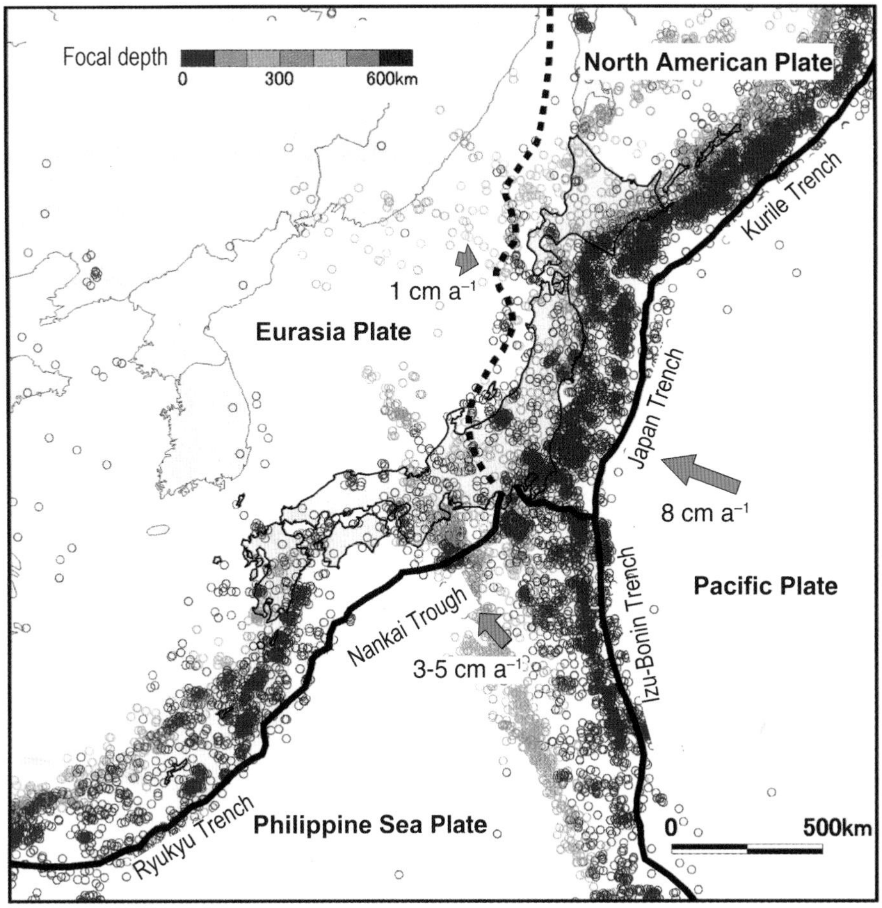

Fig. 7.2. Epicentre distribution of $M > 4$ earthquakes located by USGS for the period of 1995–2005 (Headquarters for Earthquake Research Promotion, Ministry of Education, Culture, Sports, Science and Technology, http://www.jishin.go.jp/). The focal depths are indicated by the colour scale at the top.

Japan has been poorly understood until recently. This is mainly because the seismic activity associated with the subduction of this plate is very limited. Interplate earthquakes on the upper surface of the subducting PHS Plate seldom occur, which indicates that reliable estimation of the upper interface of the PHS Plate is very difficult at its shallower portion. Moreover, seismic wide-angle reflection and refraction surveys performed along lines crossing the arc in the Tokai, Kinki and Shikoku areas indicate that simply adopting the upper envelope of intraslab seismicity as the upper plate interface is not a reliable approach (Kodaira *et al.* 2000, 2004; Kurashimo *et al.* 2002). These surveys detected the seismic low-velocity crust of the subducting plate and showed that the upper plate interface in this region is several kilometres shallower than the upper limit of intraslab seismicity.

Subsequently, Hirose *et al.* (2008*a*) detected this oceanic crust of the subducted PHS slab as a layer with low S-wave velocity (V_s) and high V_p/V_s located immediately above a region of intraslab seismicity by applying the double-difference tomography method of Zhang & Thurber (2003) to arrival time data obtained by the nationwide seismic network. They found that this low-V_s, high-V_p/V_s oceanic crust of several kilometres in thickness is continuously distributed in the entire area from Tokai to Kyushu in SW Japan. Hirose *et al.* (2008*a*) used this low-V_s, high-V_p/V_s layer to estimate the configuration of the upper surface of the PHS slab at depths ranging from 20 to 60 km for the entire area of SW Japan from Tokai to Kyushu, as shown by isodepth contours in Figure 7.4. In the figure, the shallower portion of the upper surface of the subducting plate beneath the Pacific Ocean (isodepth contour of 10 km) was determined by offshore seismic reflection and refraction surveys (Baba *et al.* 2002), and the deeper portions (60–200 km) beneath Chubu and Chugoku–Kyushu were determined by the seismic-tomography-derived upper envelope of the seismic high-velocity slab (Nakajima & Hasegawa 2007*a*).

For the configuration of the subducting PHS Plate beneath Kanto, a number of different models have been proposed based primarily on hypocentre distributions, focal mechanisms, later phase analyses, seismic tomography studies and seismic reflection and refraction surveys (e.g. Iidaka *et al.* 1990; Ishida 1992; Sekiguchi 2001; Kodaira *et al.* 2004; Sato *et al.* 2005; Hori 2006; Wu *et al.* 2007). As in the area from Tokai to Kyushu of SW Japan, Hirose *et al.* (2008*b*) clearly detected a several-kilometre-thick, low-V_s, high-V_p/V_s oceanic crust that is shallowly inclined along the subduction direction of the PHS slab by applying double-difference tomography to earthquake arrival time data for the Kanto area. Based on this observation, they delineated the upper surface of the subducting PHS Plate beneath Kanto at depths down to *c.* 90 km, which is shown as 10–90 km isodepth contours in the Kanto area of Figure 7.4

The PHS Plate below the region north of the Izu Peninsula between Kanto and Tokai has long been a subject of controversy, because this area is characterized by a large fan-shaped seismicity gap in intraslab earthquakes and by a sparse distribution of Quaternary volcanoes (see Fig. 7.4). Some researchers have suggested that the PHS slab tears in this area of the Izu collision zone, splitting into the Kanto slab in the east and the Tokai slab in the west (e.g. Ishida 1992; Mazzotti *et al.* 1999). In contrast, other researchers have suggested that an aseismic portion of the PHS slab exists beneath this region (e.g. Iidaka *et al.* 1990; Sekiguchi 2001; Matsubara *et al.*

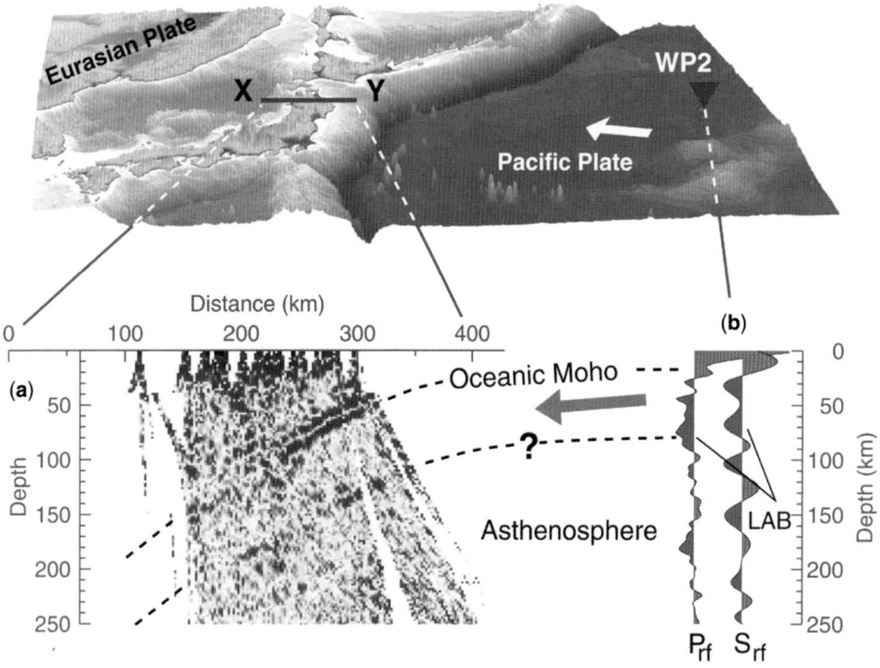

Fig. 7.3. (**a**) P-to-S receiver function image of the subducting PAC Plate beneath Tohoku along the profile XY; and (**b**) P-to-S and S-to-P receiver functions for an ocean-bottom borehole seismic station WP2 (Kawakatsu *et al.* 2009). Red and blue colours denote velocity increase and decrease from shallow to deep, respectively. Locations of the profile and the station are shown by a red line and by an inverted red triangle in the topographic map at the top.

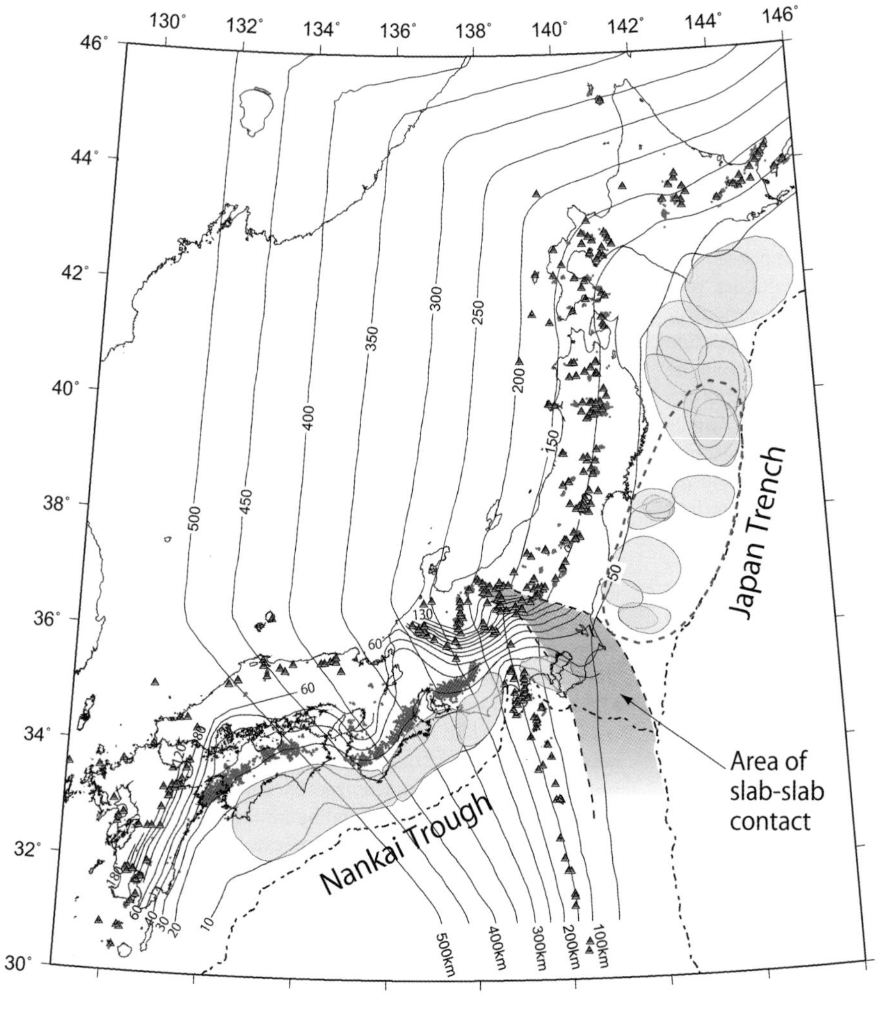

Fig. 7.4. Map showing isodepth contours of the upper surfaces of the PAC and PHS plates subducting beneath the Japanese islands (Baba *et al.* 2002; Nakajima & Hasegawa 2007*a*; Hirose *et al.* 2008*a*; Nakajima *et al.* 2009*a*; Kita *et al.* 2010). The contact zone between the PAC Plate and the overlying PHS Plate is shaded in grey and is enclosed by two broken curves. The source area of the 1923 Kanto earthquake (Wald & Somerville 1995) and those of the forthcoming Tokai, Tonankai and Nankai earthquakes (Headquarters for Earthquake Research Promotion, Ministry of Education, Culture, Sports, Science and Technology, http://www.jishin.go.jp/) are shown by light blue ellipses. The source areas of *M* 7 interplate earthquakes on the upper surface of the PAC Plate for the past 80 years (Umino *et al.* 1990) are also shown by light blue ellipses. The red triangles denote Quaternary volcanoes. Deep low-frequency earthquakes (Obara 2002) occurring on the plate interface in SW Japan are indicated by dots. The source area of the 2011 *M* 9.0 Tohoku-oki earthquake is enclosed by a red broken curve.

2008), although the nature of this aseismic slab was not known until recently.

A seismic tomography study of central Japan recently conducted has shown that a seismic high-velocity zone extends continuously down to a depth of *c.* 140 km, even in the region north of the Izu Peninsula (Nakajima *et al.* 2009*a*). This observation suggests that the PHS Plate is present as a single slab from Kanto to Tokai without disruption, with the subduction of an aseismic slab below the region north of the Izu Peninsula. Nakajima *et al.* (2009*a*) further detected an aseismic portion of the subducting PHS slab down to a depth of *c.* 140 km beneath Kanto as a seismic high-velocity layer. Connecting this slab to that north of the Izu Peninsula we obtain a continuous slab geometry from Kanto to Tokai, as shown by the 10–130 km isodepth contours beneath the region between Kanto and Tokai in Figure 7.4

Slab–slab contact zone beneath Kanto

In the region south of the Sagami Trough, the PAC Plate subducts beneath the PHS Plate west-northwestwards along the Izu–Bonin Trench, bringing the two plates into contact at shallow depths. In the region north of the Sagami Trough, the PHS Plate in turn subducts northwestwards beneath the NA plate. After the subduction of the PHS Plate beneath Kanto, this contact zone is expected to deepen and migrate northwestwards due to the westwards dip of the immediately underlying PAC Plate. Investigations of the spatial distribution of slip vectors for interplate earthquakes and seismic tomographic imaging have shown that this contact zone between the bottom of the PHS slab and the top of the PAC slab underlies a wide area beneath the Kanto Plain (Nakajima *et al.* 2009*a*; Uchida *et al.* 2009).

The northeastern end (or updip end) of this slab–slab contact zone corresponds to the border between the overlying PHS and NA plates, the former subducting northwestwards beneath the latter to the SW of this updip end. Consequently, the updip end can be estimated from the lateral variation of interplate earthquake slip vectors. Uchida *et al.* (2009) used a large number of focal mechanism solutions for interplate events that occurred along the interfaces between the PAC and PHS plates and between the PAC and NA plates based on the catalogue of the F-net moment tensor solution database (Broadband Seismic Network Laboratory, NIED 2008) to identify this northeastern end with a higher resolution than in previous estimates.

The southwestern end, or downdip end, of the slab–slab contact zone can be estimated from the precise location and configuration of the upper surface of the PAC slab and the lower surface of the PHS slab. Nakajima *et al.* (2009*a*) reported an accurate location for the upper surface of the subducting PAC slab using relocated hypocentre distributions, low-angle thrust-type focal mechanisms and a high-velocity anomaly imaged by seismic tomography. The location of the lower surface of the PHS slab was estimated using the thickness of the seismic high-velocity anomaly imaged by seismic tomography and the locations of intraslab earthquakes. The configuration of the PHS slab directly above the contact zone in the Kanto region estimated in this manner is wedge shaped. Its maximum thickness is *c.* 60 km, approximately twice the estimated thickness for the area from Tokai to Shikoku in SW Japan (e.g. Ishida 1992). These slab thickness estimates, which vary from region to region, are approximately consistent with those expected based on the age of the PHS slab, which is >48 Ma in the Kanto area and 28–15 Ma in the area from Tokai to Shikoku (Seno & Maruyama 1984).

The contact zone between the PAC slab and the PHS slab estimated in this manner is shown by two broken curves in Figure 7.4. The figure shows that the contact zone is present over a wide area beneath Kanto, with a correlation between the lateral extent of the contact zone and the location of the Kanto Plain, the largest plain in the Japanese islands, which suggests that the slab–slab contact might play a part in the subsidence of the Kanto plain. Further discussion on a geological timescale is however required to relate the slab–slab contact zone to the formation of the Kanto plain (Nakajima *et al.* 2009*a*). Figure 7.4 also shows that the isodepth contours delineating the upper surface of the PHS slab fluctuate significantly in the north–south direction over a broad area extending from Kanto to Shikoku, including the area north of the Izu Collision Zone. This undulatory configuration of the subducting PHS slab is inferred to be related to along-arc contraction and buckling deformation caused by the subduction of this slab into the mantle; the descending PHS slab becomes spatially constrained with increasing depth, primarily due to the westwards dip of the immediately underlying PAC slab.

Interplate earthquakes and anomalous deepening of their downdip limit in the slab–slab contact zone

As shown in Figure 7.2, many shallow earthquakes in NE Japan occur beneath the Pacific Ocean between the trench axis and the Pacific coast. These shallow earthquakes occur primarily on the plate interface between the subducting PAC Plate and the overriding plate. Large destructive earthquakes have also frequently occurred along this plate interface. This is also clear from the distribution of the source areas of large interplate earthquakes of *M* 7 or greater, as shown by the light blue ellipses in Figure 7.4. The 2011 *M* 9 Tohoku-oki earthquake, which is the greatest earthquake in the modern history of Japan, is one typical example; it ruptured approximately two-thirds of the megathrust zone east of the entire NE Japan Arc, as shown in Figure 7.4. In contrast to the region beneath the Pacific Ocean east of NE Japan, few shallow earthquakes occur in the region beneath the Pacific Ocean between the Nankai Trough and the Pacific coast in SW Japan, as shown in Figure 7.2. However, large destructive earthquakes of *M* 8 or greater have repeatedly occurred along the plate interface in this region.

The downdip limit of interplate earthquakes in the Japanese subduction zones estimated from the hypocentre distribution of the low-angle, thrust-type earthquakes and from the distribution of fault planes of large interplate events is known to be subparallel to the isodepth contour of the subducting oceanic plate. However, the depth of the downdip limit is different between the two subduction zones; it is *c.* 50 km for the PAC Plate in NE Japan and *c.* 25 km for the PHS Plate in SW Japan. This large difference in the depth of the downdip limit between that for the PAC Plate subducting beneath NE Japan and that for the PHS Plate subducting beneath SW Japan is due to the difference in age and consequently to the difference in thermal structure between the two oceanic plates.

The broad slab–slab contact zone existing in depths in the range 0–140 km beneath Kanto hinders heating of both the PAC and PHS slabs at this location by the hot mantle wedge. This causes an anomalously deep downdip limit of interplate earthquake activity in the contact zone beneath Kanto, because the downdip limit of this earthquake activity is controlled by temperature. This downdip limit of interplate events locally deepens for both the PAC and PHS plates beneath Kanto. As shown in Figure 7.5, interplate earthquakes along the upper surface of the subducting PHS Plate occur down to a depth of *c.* 55 km immediately above the slab–slab contact zone, whereas interplate earthquakes along the upper surface of the PAC Plate take place down to a depth of *c.* 80 km within the slab–slab contact zone (Nakajima *et al.* 2009*a*; Uchida *et al.* 2009). These depths of the downdip limit of interplate events are

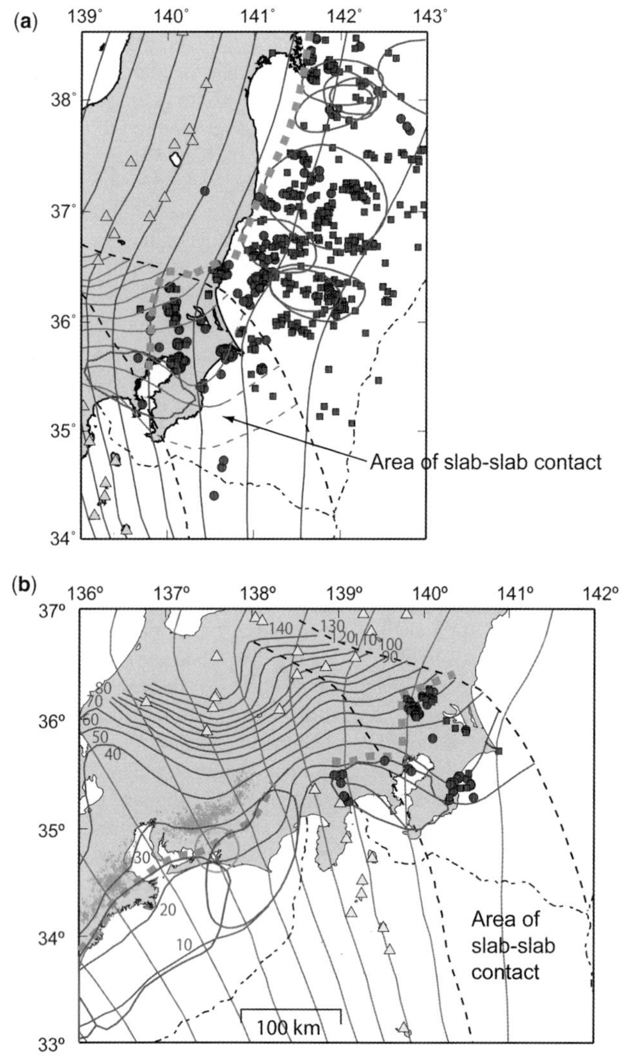

Fig. 7.5. Distribution of interplate earthquakes on the upper surfaces of the (**a**) PAC and (**b**) PHS plates (Nakajima *et al.* 2009a). Red circles and squares indicate the epicentres of low-angle, thrust-type events and small repeating earthquakes (Uchida *et al.* 2003), respectively. Source areas of large interplate events are enclosed by red curves. Pink dashed lines show the downdip limit of interplate earthquakes. The orange crosses in (b) denote deep low-frequency earthquakes. The slab–slab contact area is enclosed by two broken curves.

significantly greater than the depths of *c.* 25 and 50 km for the PHS and PAC plates, respectively, observed in other regions of the Japanese islands, which indicates an anomalously low-temperature condition in the slab–slab contact zone beneath Kanto.

The slab–slab contact beneath Kanto also causes a locally deep cut-off depth of shallow crustal inland earthquakes at this location. The cut-off depth of shallow seismicity within the continental NA plate is as deep as 20–30 km directly above the slab contact zone beneath Kanto, contrasting with depths of 10–15 km in other inland areas of the Japanese islands (Omuralieva *et al.* 2012). Heat flow is also locally low directly above the slab–slab contact zone beneath Kanto (Tanaka *et al.* 2004). This supports the interpretation that a locally low geothermal gradient there is caused by slab–slab contact, with the subduction of the two cold oceanic slabs yielding a dual cooling effect in addition to the broad contact zone hindering the heating of the two oceanic slabs by the hot mantle wedge.

Internal structure of the slab and intermediate-depth intraslab earthquakes

As described in the previous section, intermediate-depth intraslab earthquakes beneath NE Japan form a double-planed deep seismic zone at depths of *c.* 50–180 km. Earthquakes in the upper seismic plane occur primarily in the crust of the slab, whereas those in the lower seismic plane occur in the middle of the mantle of the slab.

At the depths of such intermediate-depth earthquakes, the lithostatic pressure and so normal stress on the fault become too high to cause ordinary brittle shear faulting. The generation of intermediate-depth intraslab earthquakes therefore requires some special weakening mechanism. One possible mechanism that decreases the strength of the fault for intermediate-depth earthquakes is dehydration embrittlement (Raleigh & Paterson 1965). If this is the mechanism that generates intermediate-depth earthquakes, these events will occur in the portion of the slab that contains hydrous minerals, and dehydration is expected to occur. Using experimentally derived phase diagrams of rocks, Yamasaki & Seno (2003) estimated the locations at which dehydration decompositions of metamorphosed crust and serpentinized mantle of the slab occur and compared these diagrams with the observed locations of intermediate-depth intraslab earthquakes. Their results for NE Japan revealed that estimated dehydration loci of the crust and mantle correspond roughly to the locations of the upper and lower planes of the double seismic zone (Fig. 7.6), which supports the dehydration embrittlement model for the genesis of intermediate-depth intraslab earthquakes. This further explains why the double seismic zone is formed in such a depth range.

Recent studies on the detailed hypocentre distribution and seismic velocity structure within the subducting slab using high-quality data obtained by the dense nationwide seismic network have provided further evidence supporting the dehydration embrittlement hypothesis. Precise relocations of intermediate-depth earthquakes revealed the existence of a pronounced seismic belt in the upper plane of the double seismic zone, which runs nearly parallel to the *c.* 80 km isodepth contour of the upper surface of the PAC slab beneath Hokkaido and Tohoku (Kita *et al.* 2006). This pronounced belt-like seismicity, referred to here as the upper-plane seismic belt, for Tohoku is indicated by the pink zone in Figure 7.7c. An across-arc vertical cross-section of earthquakes (Fig. 7.7a) also reveals that there exists a concentration of intraslab earthquakes at depths of 70–90 km, which corresponds to the seismic belt indicated in Figure 7.7c

Since many uncertainties exist in the estimation of temperature in the slab, direct comparison with mineralogical research results does not allow us to verify whether the phase transformation

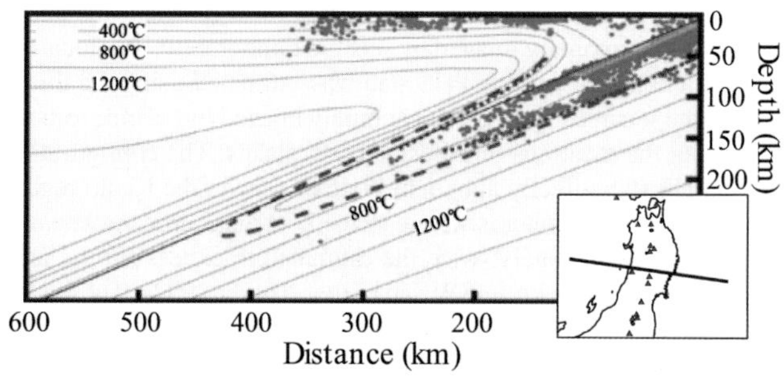

Fig. 7.6. Across-arc vertical cross-section of earthquakes and dehydration loci within the PAC slab beneath central Tohoku (Yamasaki & Seno 2003). Dehydration loci of metamorphosed crust and those of serpentinized mantle are shown by a green line and red broken lines, respectively. Blue dots are earthquakes. Thermal structure is shown by isothermal contours.

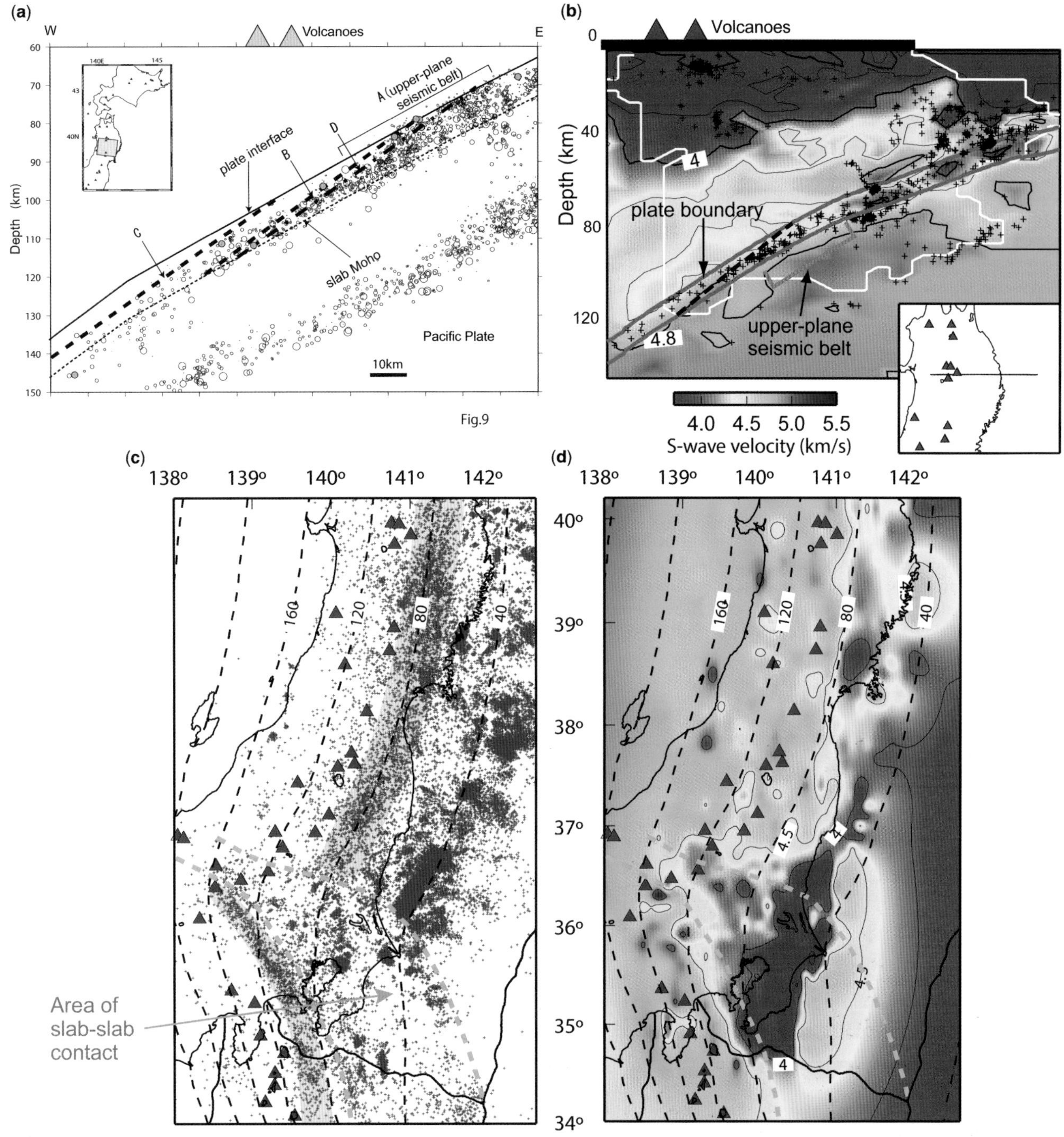

Fig. 7.7. (**a**) Across-arc vertical cross-section of intraslab earthquakes in central Tohoku (Kita *et al.* 2006). Earthquakes are shown by open circles. The estimated jadeite lawsonite blueschist-lawsonite amphibole eclogite facies boundary (B) and the lawsonite amphibole eclogite-eclogite facies boundary (C) of the MORB (Hacker *et al.* 2003) are represented by thick broken lines. The facies boundary (D) based on Omori *et al.* (2004) is also shown by a dashed-dotted line. (**b**) Across-arc vertical cross-section of S-wave velocity (indicated by the colour scale at the bottom of the figure) in central Tohoku (Tsuji *et al.* 2008). The upper interface of the PAC Plate is indicated by a solid line, and the upper-plane seismic belt is denoted by a pink hook symbol. Contours with derivative weighted sum of 500 are shown by white lines. The facies boundary B in (a) is shown as a broken line. (**c**) Distribution of earthquakes within the crust of the PAC slab. Earthquakes occurring at 0–10 km below the upper-plate surface are indicated as blue dots, and the upper-plane seismic belt is shown in pink. The contact zone between the PHS and PAC slabs beneath Kanto is delineated by two broken green curves. The broken black curves and red triangles show the isodepth contours of the upper surface of the PAC slab and the location of active volcanoes, respectively. (**d**) S-wave velocity (colour scale at the top of the figure) distribution in the crust of the PAC slab on a curved plane 5 km below the upper-plate surface (Nakajima *et al.* 2009*b*). The broken black curves denote the isodepth contours of the upper surface of the PAC slab. The slab–slab contact zone is enclosed by two green curves.

accompanying dehydration in the slab crust actually occurs at the depths of this upper-plane seismic belt. However, we can make at least a rough estimation. As two examples, facies boundaries estimated in this manner based on the phase diagram of mid-oceanic ridge basalt (MORB) by Hacker *et al.* (2003) and the slab geotherm by Peacock & Wang (1999), and that based on the MORB phase diagram by Omori *et al.* (2004) and the geotherm by van Keken *et al.* (2002), are shown by broken lines and by a dashed-dotted line, respectively, in the vertical cross-section of Figure 7.7a. The figure shows that the concentration of earthquakes forming the upper-plane seismic belt is found around the metamorphic facies boundary in shallower areas (B or D in Fig. 7.7a), which approximately agrees with the prediction from the dehydration embrittlement hypothesis. Although further validation of the location of the facies boundary is required, this observation suggests that the upper-plane seismic belt is formed with intraslab earthquake activity caused by dehydration-related embrittlement (Kita *et al.* 2006).

A phase transformation of crustal rock is expected to cause an increase in seismic wave velocity. This means that we can verify from seismic tomography whether the upper-plane seismic belt is caused by a phase transformation. An across-arc vertical cross-section of S-wave velocity obtained by seismic tomography, which is shown in Figure 7.7b, indicates that low-seismic-velocity slab crust persists down to, but not below, the depth of this upper-plane seismic belt, suggesting that a phase transformation and eclogite formation takes place in the slab crust at this depth (Tsuji *et al.* 2008).

The upper-plane seismic belt and the seismic low velocity persisting down to its depth in the slab crust correlate well with the slab–slab contact zone beneath Kanto mentioned in the previous subsection. As described above, the depth at which the phase transformation of crustal rock occurs is thought to be dependent on the temperature within the slab (e.g. Hacker *et al.* 2003). If this is the case, the locally low-temperature conditions in the PAC slab immediately beneath the slab–slab contact zone in Kanto should cause a delay in phase transformation. As expected, the upper-plane seismic belt in Kanto is not parallel to the *c.* 80 km isodepth contour, but is oblique to this contour and deepens locally towards the north from a depth of *c.* 100–140 km along a trend nearly parallel to the downdip edge of the slab–slab contact zone (Hasegawa *et al.* 2007), as shown in Figure 7.7c

A seismic tomography study further revealed that the depth range of the low-velocity layer in the slab crust deepens locally in the zone of contact with the overlying PHS slab, where the obliquely trending upper-plane seismic belt is observed (Fig. 7.7d). These observations indicate that low-temperature conditions associated with the slab–slab contact beneath Kanto cause delayed onset of eclogite-forming phase transformations in the slab crust in this region. The observations also show that the location and areal extent of the contact zone estimated from earthquake slip vectors, hypocentre distribution and seismic tomography, as described in the previous section, is accurate.

Seismic tomography studies also revealed the detailed structure of the mantle portion of the subducted PAC Plate. Zhang *et al.* (2004) first detected prominent P-wave low-velocity anomalies along the lower plane of the double seismic zone within the PAC Plate beneath central Tohoku by double-difference tomography. Subsequently, Shelly *et al.* (2006) detected a similar P-wave low-velocity layer along the lower plane of the double seismic zone beneath southern Tohoku. Figure 7.8 shows another example obtained for Hokkaido, showing across-arc vertical cross-sections of P- and S-wave velocities beneath Hokkaido imaged by double-difference seismic tomography (Nakajima *et al.* 2009*c*). A distinct seismic low-velocity layer *c.* 10 km in thickness is imaged at the

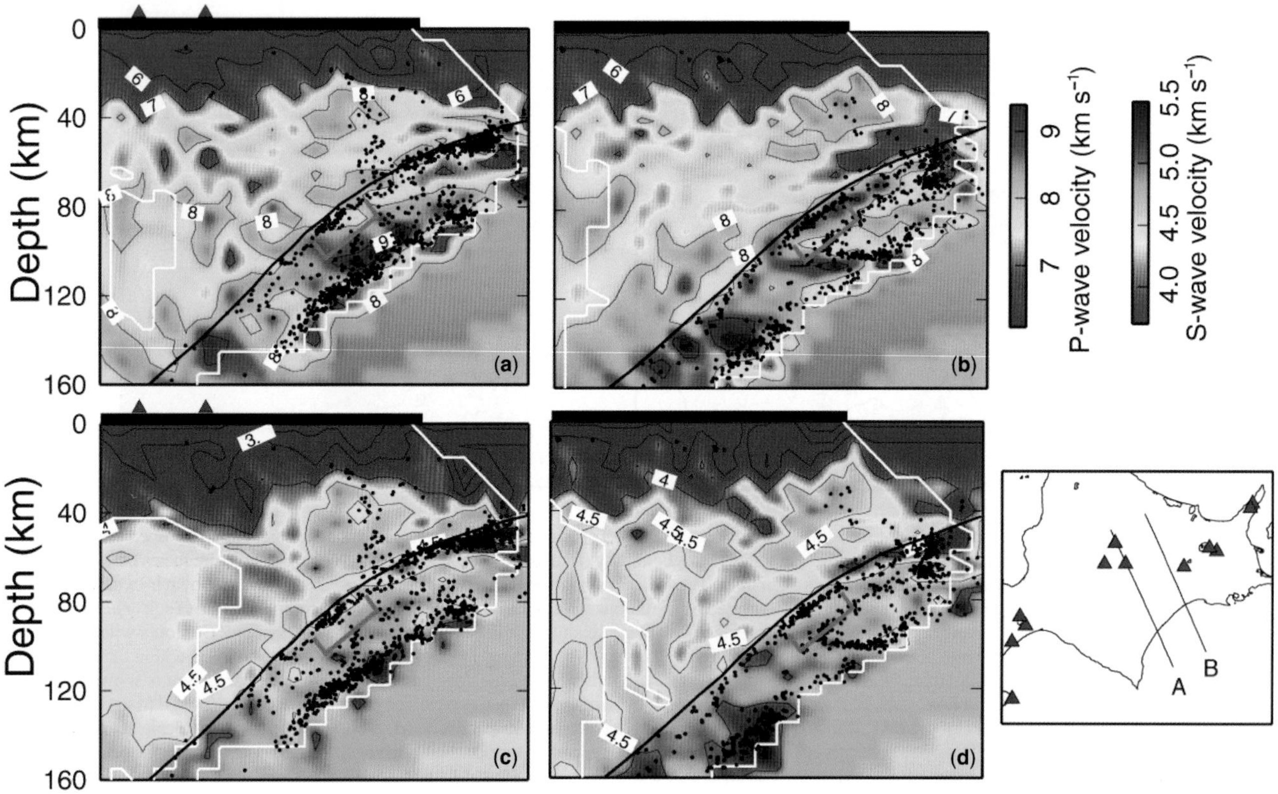

Fig.7.8. Across-arc vertical cross-sections of (**a**) and (**b**) P-wave velocity and (**c**) and (**d**) S-wave velocity along the two lines indicated in the inset map (Nakajima *et al.* 2009*c*). The P-wave and S-wave velocities are shown by the colour scales on the right. Contours with derivative weighted sum of 500 are shown by white lines. The solid lines denote the location of the upper surface of the PAC Plate. The solid lines and red triangles at the top of the figure indicate the land area and active volcanoes, respectively. Grey hook symbols denote the upper-plane seismic belt in this region.

uppermost part of the subducting PAC Plate. This low-velocity layer corresponds to the hydrated oceanic crust, which gradually disappears at depths of 70–80 km, suggesting the breakdown of hydrous minerals at this depth range as in the case for Tohoku shown in Figure 7.7b. Furthermore, Figure 7.8 reveals the existence of prominent P-wave low-velocity anomalies in the mantle of the PAC slab along the lower plane of the double seismic zone and immediately above the aftershock area of the 1993 *M* 7.8 Kushiro-oki earthquake. Un-metamorphosed peridotite has a much higher P-wave velocity than the abovementioned velocities. In contrast, in regions between the upper and lower planes where seismicity is not so high, seismic velocities are relatively high and are comparable to those of un-metamorphosed peridotite.

The prominent P-wave low-velocity anomalies both in the lower seismic plane and immediately above the aftershock area may suggest the existence of hydrated minerals in those areas. However, it is still debated whether water is supplied to such a great depth within the oceanic plate before its subduction. Seismic observations show that hydrothermal serpentinization of the oceanic plate along faults due to the plate bending at the trench outer rise region is limited down to several kilometres below the oceanic crust (e.g. Ranero *et al.* 2003). Numerical modelling shows that fluid infiltration due to the plate bending decreases with increasing depth and is limited to *c.* 5 km below the crust (Faccenda *et al.* 2009). Taking these results into consideration, Reynard *et al.* (2010) gave an alternative interpretation for the low-velocity anomalies in the lower seismic plane. They pointed out that the low-velocity anomalies can be explained by seismic anisotropy of anhydrous deformed peridotite and that lower plane earthquakes might be caused by shear instability along anisotropic planar shear zones without requiring the presence of water there. In contrast, Seno & Yamanaka (1996) proposed the possibility that hydration of the middle to deep portion of the oceanic plate occurs when the plate passes over plumes and superplumes. Recently, Hirano *et al.* (2013) detected small volcanoes on the outer rise system of the PAC Plate. These petit-spot volcanoes, erupted along lithospheric fractures in response to the plate flexure during subduction, provide a possibility of ubiquitous small-volume melts under the PAC Plate in offshore Japan and might also contribute to the hydration of the deep portion of the oceanic plate. If hydration actually occurs at such a great depth within the PAC Plate, lower plane earthquakes also occur primarily in regions containing hydrous minerals.

Seismic tomography, precise earthquake hypocentre relocations and focal mechanism studies conducted using data acquired by a dense nationwide seismic network that was recently constructed in Japan have shown that the PAC Plate subducts continuously beneath eastern Japan down to the bottom of the upper mantle. Results also show that the PHS Plate subducting beneath SW Japan is continuous throughout the entire region from Kanto to Kyushu without disruption or splitting, even beneath the area to the north of the Izu Peninsula which is a collision zone between the two arc-type crusts. The PHS Plate has an extremely undulatory geometry during subduction, which is potentially caused by contractive deformation in an along-arc direction due to the restricted space within the mantle with increasing depth. Furthermore, the estimated geometry of the PAC and PHS slabs shows a broad contact zone between the two slabs located directly below the Kanto plain. This slab–slab contact hinders heating of both the PAC and PHS slabs by the hot mantle wedge and causes a local reduction in temperature, resulting in anomalously deep downdip limits for interplate, intraslab and shallow inland earthquakes beneath Kanto. Precise earthquake hypocentre relocations also showed that earthquakes in the crust of the subducted PAC slab form a seismic belt (upper-plane seismic belt) running parallel to the *c.* 80 km isodepth contour of the upper plate interface, which is located near the estimated dehydration loci of the metamorphosed crust. Seismic tomography studies revealed the existence of seismic low-velocity oceanic crust which persists down to, but not below, the depth of this upper-plane seismic belt. This confirms that the estimated dehydration loci are located at the depth of the upper-plane seismic belt. The upper-plane seismic belt and the seismic low velocity persisting down to its depth in the crust of the PAC slab correlate well with the slab–slab contact area beneath Kanto. Both the upper-plane seismic belt and the downdip end of the seismic low-velocity layer in the crust of the PAC slab deepen locally in the zone of contact with the overlying PHS slab beneath Kanto. Seismic tomographic studies also revealed the detailed structure of the PAC slab mantle and detected prominent P-wave low-velocity anomalies along the lower plane of the double seismic zone. These observations support the hypothesis that intermediate-depth intraslab earthquakes occur primarily in regions containing hydrous minerals, although there remains the possibility that the low-velocity anomalies in the lower-plane seismicity are caused by seismic anisotropy of anhydrous deformed peridotite.

Mantle wedge structure, arc magmatism and shallow crustal earthquakes (JN)

In most subduction zones, the volcanic front is formed subparallel to the trench axis. The subducting slab below the volcanic front has been considered to be located at a roughly constant depth of 100–120 km (Tatsumi 1986), and this has been used to model the genesis of magmatism as a strong constraint on locations of dehydration reactions in the subducting slab. However, the compilation of new data on volcano-specific subduction parameters indicates that the depths to the slab surface below the volcanic front vary among subduction zones across ranges as wide as 60–130 km (England *et al.* 2004) and 72–173 km (Syracuse & Abers 2006). The new estimations suggest that volcanism is controlled by a process that depends critically on the slab geometry, the incoming plate temperatures and convection in the mantle wedge.

The mantle wedge structures beneath the Japanese islands have been intensively investigated at local and regional scales by seismic velocity and attenuation tomography, shear-wave splitting and P-wave azimuthal anisotropy analyses. Although seismological observations represent only present-day snapshots of ongoing mantle dynamics, quantitative interpretations of the results in combination with petrological, geochemical and geological observations and geodynamic modelling have provided important clues for understanding magmatism beneath the Japanese islands.

This section presents the upper mantle and crustal structures and reviews the contribution of recent seismological observations for the understanding of fluid flow and magmatism beneath eastern Japan (Hokkaido, Tohoku and Izu–Bonin) and western Japan (Chubu, Kii, Chugoku and Kyushu and Ryukyu). On the basis of the mantle wedge structures, seismic activity in the continental crust is briefly summarized in 'Crustal deformation and shallow crustal seismicity'. Other geophysical approaches such as magneto telluric surveying, terrestrial heat flow measurements and gravity measurements have definitely provided additional constraints on the interpretation of ongoing dynamics, but we focus mainly on seismological observations that have been progressively enhanced in the last decade by virtue of high-quality seismic data recorded at the dense Kiban seismograph network in Japan.

Eastern Japan: Hokkaido, Tohoku and Izu–Bonin

Northern Japan is located at a typical subduction zone where the PAC Plate is subducting beneath the continental plate at a rate of *c.* 10 cm a^{-1}. As described in the previous section, the subduction of the PAC Plate has governed the tectonic framework in northern Japan, where the volcanic front is running through the middle of the arc and intermediate-depth earthquakes occur in the PAC slab (Figs 7.2 & 7.4). An interesting observation is that active volcanoes are not distributed continuously along the volcanic front but form somewhat isolated volcanic clusters striking transverse to the arc (Tamura *et al.* 2002). There are strong contrasts, for example, in terrestrial heat flow (e.g. Tanaka *et al.* 2004) and $^3He/^4He$ ratios (e.g. Sano & Nakajima 2008) between the forearc and back-arc sides. The along-arc and across-arc variations in these surface manifestations suggest the important role of mantle wedge processes on present-day differentiation and arc volcanism.

As summarized in the previous section, dehydration reactions of hydrous minerals are expected to occur intensively at depths of 70–90 km in the subducting PAC slab where the upper-plane seismic belt is observed (Fig. 7.7a). The fluids released by dehydration reactions at such depths may migrate into the overlying mantle wedge unless the plate interface is completely sealed, forming a thin hydrous layer composed of serpentine or chlorite immediately above the plate interface (Iwamori 1998) (Fig. 7.9). This hydrated layer is dragged with the underlying PAC slab and acts as a secondary source of fluids. The hydrous layer has been detected as a low-velocity layer at depths of 70–120 km (Figs 7.7b, d & 7.9) (e.g. Kawakatsu & Watada 2007; Tsuji *et al.* 2008). The depths for breakdown of hydrous minerals in this layer are estimated to be from 100–130 km (Cagnioncle *et al.* 2007) to 150–200 km (Iwamori 1998), depending on thermal structures derived in geodynamic modelling. In any case, a large amount of fluids can be released to the mantle wedge at depths of <200 km.

The most important structure for the mantle wedge is the existence of an inclined low-velocity and high-attenuation zone subparallel to the downdip direction of the PAC slab (e.g. Zhao *et al.* 1992, 2012; Tsumura *et al.* 2000; Nakajima *et al.* 2001, 2013) (Fig. 7.10a). As shown in 'Deep structure of Japan subduction zone', this inclined low-velocity zone in the back-arc mantle is commonly observed from eastern Hokkaido to northern Izu–Bonin through Tohoku, and extends continuously from >200 km depths in the back-arc side to 40 km depth beneath the volcanic front, suggesting a sheet-like distribution of the inclined low-velocity zone (Wang & Zhao 2005; Zhao *et al.* 2012). A low-velocity and high-attenuation zone is continuously distributed along the volcanic front at a depth of 40 km immediately below the continental Moho (Fig. 7.10b). Numerical simulation of secondary mantle convection associated with slab subduction has demonstrated the development of an inclined low-velocity zone subparallel to the downdip direction (e.g. Eberle *et al.* 2002), and the inclined low-velocity and high-attenuation zone is interpreted as mantle upwelling flow generated by the slab subduction.

Seismic anisotropy above the PAC slab shows a strong contrast between the forearc and back-arc mantle. Trench-normal anisotropy is observed in the back-arc mantle wedge, and trench-parallel anisotropy is observed in the forearc side (Fig. 7.11) (e.g. Nakajima & Hasegawa 2004; Ishise & Oda 2005; Nakajima *et al.* 2006; Wang

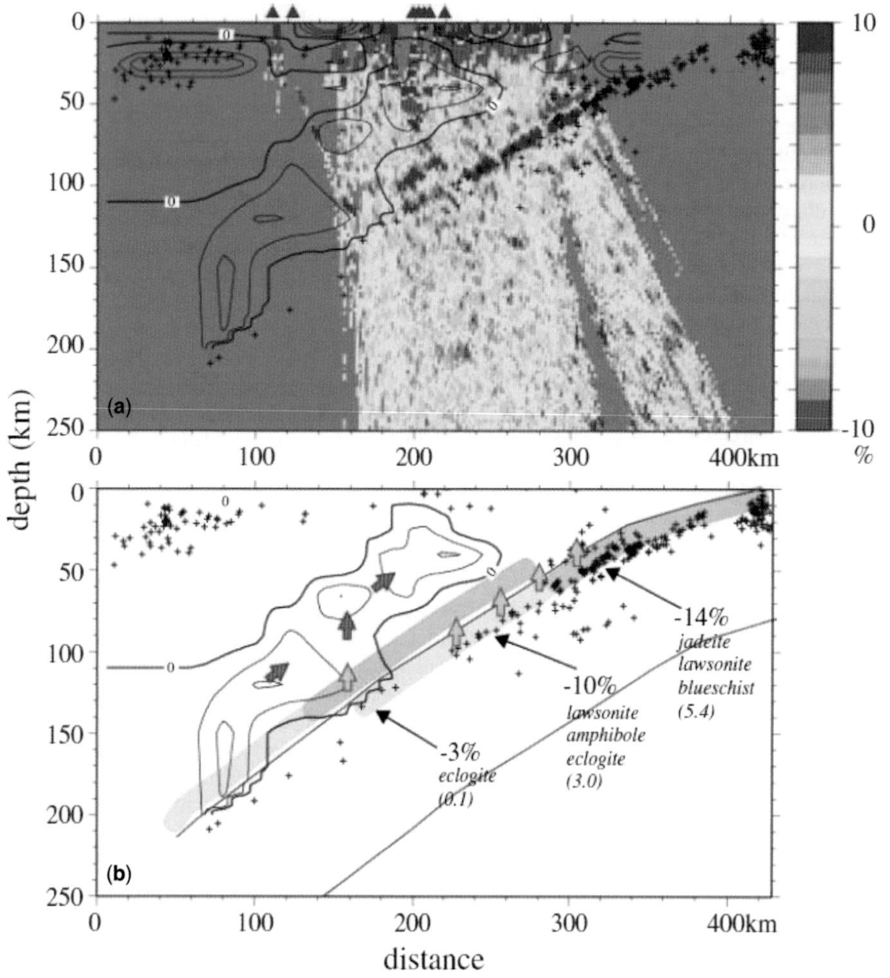

Fig. 7.9. Integrated seismic images beneath the Tohoku subduction zone and a schematic model for the water transportation (Kawakatsu & Watada 2007). (**a**) The dip-corrected reflectivity image compared with S-wave velocity structure plotted as contours with 3% velocity-change intervals. Plus signs indicate earthquakes. Red triangles are active volcanoes. (**b**) The pathway of water transportation postulated by Iwamori (1998) is schematically illustrated by light blue and red colours for non-magmatic and magmatic paths, respectively. Negative percentages give the S-wave velocities of fully hydrated oceanic crust relative to those of the slab mantle.

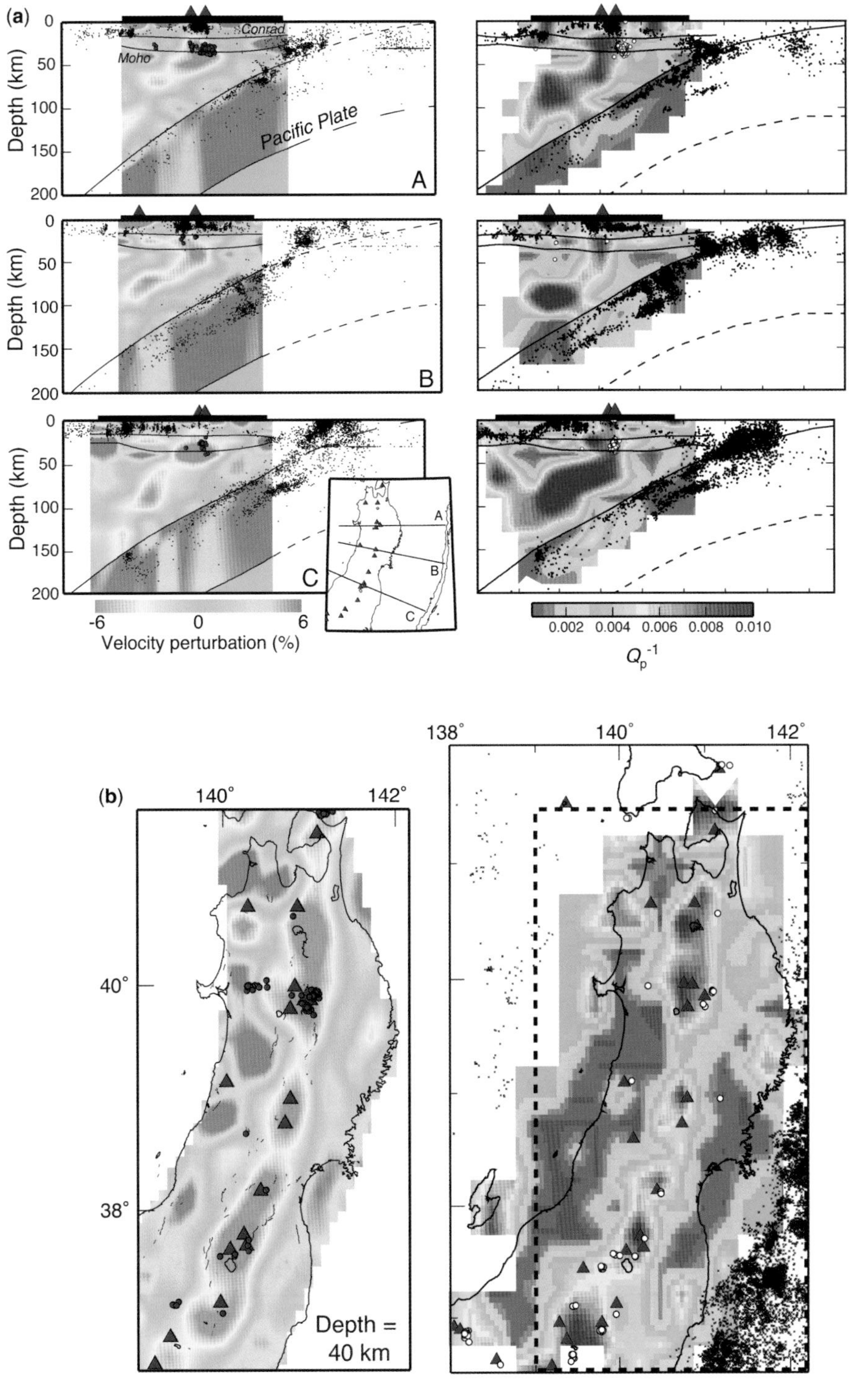

Fig. 7.10. (**a**) Across-arc vertical cross-sections of (left) S-wave velocity and (right) P-wave attenuation structures along lines as shown in the insert map (Nakajima *et al.* 2001, 2013). Triangles and bars on the top represent active volcanoes and the land area, respectively. Black dots and red or white circles show earthquakes and deep, low-frequency earthquakes, respectively. (**b**) (left) S-wave velocity and (right) P-wave attenuation structure at a depth of 40 km. A dashed rectangle indicates an area shown in the S-wave velocity images.

& Zhao 2008; Huang *et al.* 2011*b*). Katayama (2009) proposed dislocation creep of olivine immediately above the plate interface as a cause of the trench-normal anisotropy. The trench-parallel anisotropy observed in the forearc side is attributable to anisotropy in the subducting crust or B-type olive in the mantle wedge or crustal anisotropy in the continental plate. Seismic anisotropy is an important clue to understand the deformation in the mantle wedge that cannot be constrained by the seismic velocity.

Petrological models have estimated temperatures of the mantle wedge as high as 1200–1400°C (Tatsumi *et al.* 1983), which is sufficiently high to generate magmas with the assistance of fluids liberated from the PAC slab. For a calibration, Nakajima & Hasegawa (2003) showed, through a combined analysis of seismic attenuation structures with experimental data, that temperatures in the inclined low-velocity zone are 1000–1200°C. The temperatures are lower than the estimation by Tatsumi *et al.* (1983) but still higher than the wet solidus of peridotite, suggesting that partial melting occurs within the upwelling flow. On the basis of this estimated thermal structure, Nakajima *et al.* (2005) calculated aspect ratios and melt volume fractions from the reduction of V_p and V_s and showed

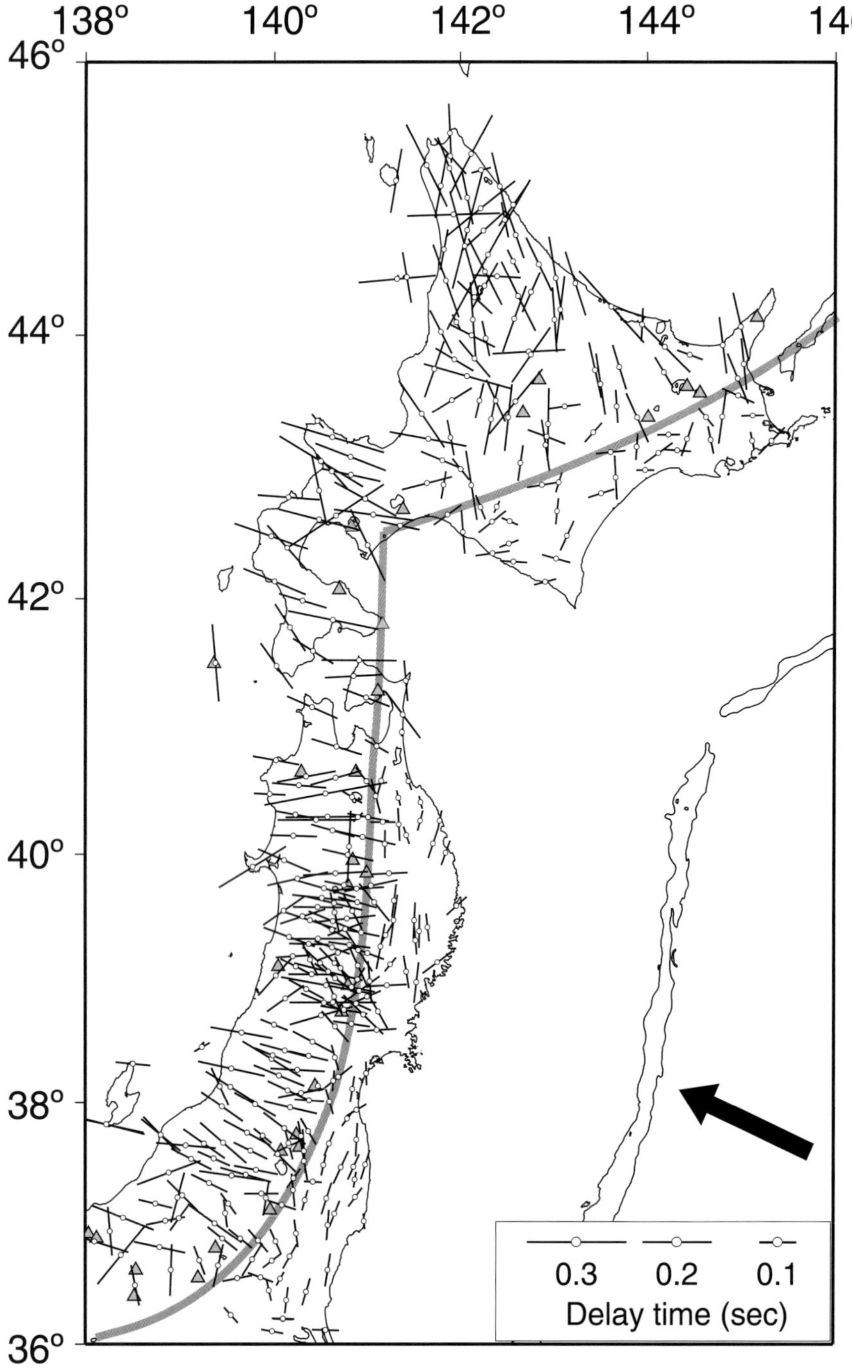

Fig. 7.11. Distributions of the average fast direction and delay times for each station plotted on the station location as black bars. Length of the bar is normalized to the averaged delay time (Nakajima & Hasegawa 2004; Nakajima *et al.* 2006).

that melt-filled pores with aspect ratios of 0.1–0.01 and fraction of *c.* 0.1–5 vol% exist in the mantle wedge. These observations strongly suggest that the low-velocity and high-attenuation zone in the mantle wedge is caused by partial melt and acts as a deep source of arc magmas, as recently confirmed by Nakajima *et al.* (2013). Based on the comprehensive interpretation of seismological observations in combination with petrological and mineralogical aspects and numerical simulations, fluid transportation paths have been inferred (Fig. 7.12a) (Hasegawa & Nakajima 2004; Hasegawa *et al.* 2005).

The fluids liberated from the hydrous layer above the PAC slab migrate upwards to the overlying mantle wedge and react chemically with the mantle wedge rock. The fluids lower the solidus of the mantle wedge rock (e.g. Arcay *et al.* 2006), and generate melts in the mantle wedge. Consequently, the fluids are incorporated into the melts and eventually transported to the uppermost mantle below the Moho by the upwelling flow. These melts migrate further upwards due to buoyancy, penetrating the arc crust and ultimately reaching the subsurface. The model for arc volcanism considered here suggests that the location of the volcanic front is governed by the locations where inclined sheet-like upwelling flows reach the arc Moho. The upwelling flow may be first accumulated below the Moho due to the density contrast, and later the melt may segregate from the accumulated materials as a result of chemical and physical differentiation resultant buoyancy. Finally the melt may migrate upwards subvertically with further differentiation on the way to the surface.

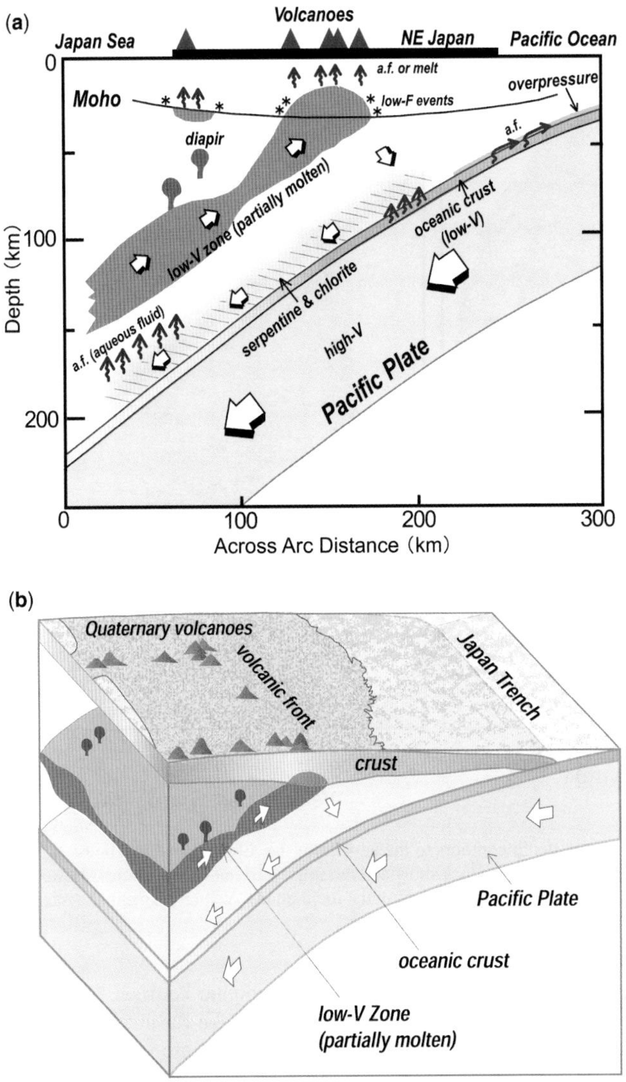

Fig. 7.12. Schematic cartoons of (**a**) 2D view and (**b**) 3D view of the crust and upper mantle beneath NE Japan that show inferred transportation paths of aqueous fluids (Hasegawa & Nakajima 2004; Hasegawa *et al.* 2009).

The model for arc magmatism can work well in the two-dimensional scheme, but it is known that Quaternary volcanoes are grouped into a number of clusters with spatial gaps in between (e.g. Kondo *et al.* 1998; Tamura *et al.* 2002) (Fig. 7.13a). High-resolution seismic tomography (Nakajima & Hasegawa 2004) has revealed the existence of an along-arc variation in velocity reduction rates in the mantle wedge, spatially correlated with volcanic clusters defined by Tamura *et al.* (2002). In the updated velocity model, areas of marked low velocity are distributed in the mantle wedge periodically every *c.* 80 km along the strike of the arc (Fig. 7.13c). These marked low-velocity areas are correlated spatially with topography and deep low-frequency earthquakes around volcanoes (Fig. 7.13b). These observations suggest that mantle upwelling with higher melt contents or higher temperatures has been sporadically developed and caused clustering of Quaternary volcanoes at the surface as a result of repeated supply of magmas to the arc crust through the upwelling flow, as schematically shown in Figure 7.12b. The gaps of Quaternary volcanoes are probably due to insufficient melt contents in the upwelling flow to allow for segregation. This suggests that spacing of Quaternary volcanoes is determined by the along-arc variation in melt contents in the upwelling flow of the mantle wedge (e.g. Nakajima *et al.* 2013).

As well as the Quaternary volcanism on the land area of eastern Japan, alkali basalt volcanoes have also been reported offshore to the east of the Japan Trench where volcanism was not expected (Hirano *et al.* 2006). The volcanoes discovered here are small and younger than 1 million years old and have been named 'petit-spot' after both the size and location far from known ridges and hotspots (Fig. 7.14). Hirano *et al.* (2013) reported that petit-spot submarine volcanoes are also present off the trench-outer slope of central Chile. Small-scale low-velocity zones are observed below the incoming plate in some areas of the circum-Pacific region, some of which are interpreted as a high-temperature anomaly (e.g. Obayashi *et al.* 2006) and as a consequence of the thermal instability caused by the collapsing of the old slabs down to lower mantle depths (e.g. Forte *et al.* 2007; Zhao *et al.* 2012). The upwelling flow below the incoming plate can occasionally cause petit-spot volcanism in response to plate flexure (Hirano *et al.* 2006, 2013). Chemical analysis of rocks at petit-spot volcanoes will provide information on the composition of source materials in the asthenosphere below the oceanic plates.

Western Japan: Chubu

A remarkable feature of the volcanism in the Chubu district in central Japan is that the volcanic front is deflected towards the west (Fig. 7.4). This deflection is considered to be associated with the subduction of the PHS Plate immediately above the PAC slab (Iwamori 2000). Seismic tomography studies have shown that the low-velocity crust in the PAC slab is preserved to depths of 110–140 km beneath Kanto, which is 40–70 km deeper than its depth extent observed in NE Japan (Fig. 7.7d). This deeper preservation of the low-velocity subducting crust suggests the shifts of dehydration reactions to greater depths as a result of the lower-temperature environment in the PAC slab due to the contact with the overlying PHS slab, as already described in the previous section. This observation confirms the numerical simulation by Iwamori (2000) and strongly suggests that the thermal recovery of the subducting PAC Plate is slow due to the overlapping PHS Plate, which shifts the dehydration reactions to greater depths along the PAC Plate.

Quaternary volcanoes are intensively distributed in the northern part of Chubu district (Fig. 7.4). The PHS slab is subducting seismically to a depth of 80 km to the south of the region with volcanic clusters, but it becomes aseismic at deeper depths with a steeper dip angle (Fig. 7.15a). The aseismic PHS slab is imaged as a high-velocity zone down to a depth of *c.* 140 km, and Quaternary volcanoes lie in the region below which the PHS slab is subducting at a steeper dip. The subduction of the PHS slab at a steep dip efficiently induces an upwelling flow and results in a low-velocity zone in the mantle wedge as shown in Figure 7.15b. The upwelling flow works as a carrier of fluids to the uppermost mantle. Geochemical analysis of volcanic rocks has showed that aqueous fluids from both the PAC and PHS slabs have contributed to the volcanism in central Japan (Fig. 7.15c) (Nakamura *et al.* 2008). These observations indicate the importance of the dual roles of the PHS and PAC slabs in the genesis of magmatism in central Japan. Seismic anisotropy observed in the northern part of central Japan (e.g. Hiramatsu *et al.* 1998) supports the development of the mantle upwelling flow subparallel to the maximum-dip direction of the PHS slab, as discussed in Nakajima & Hasegawa (2007*a*). A low-velocity zone imaged below the PHS slab (Fig. 7.15a) is also interpreted as a small-scale upwelling flow, but no volcanism occurs in offshore regions at present. The sub-slab upwelling flow would eventually cause offshore small

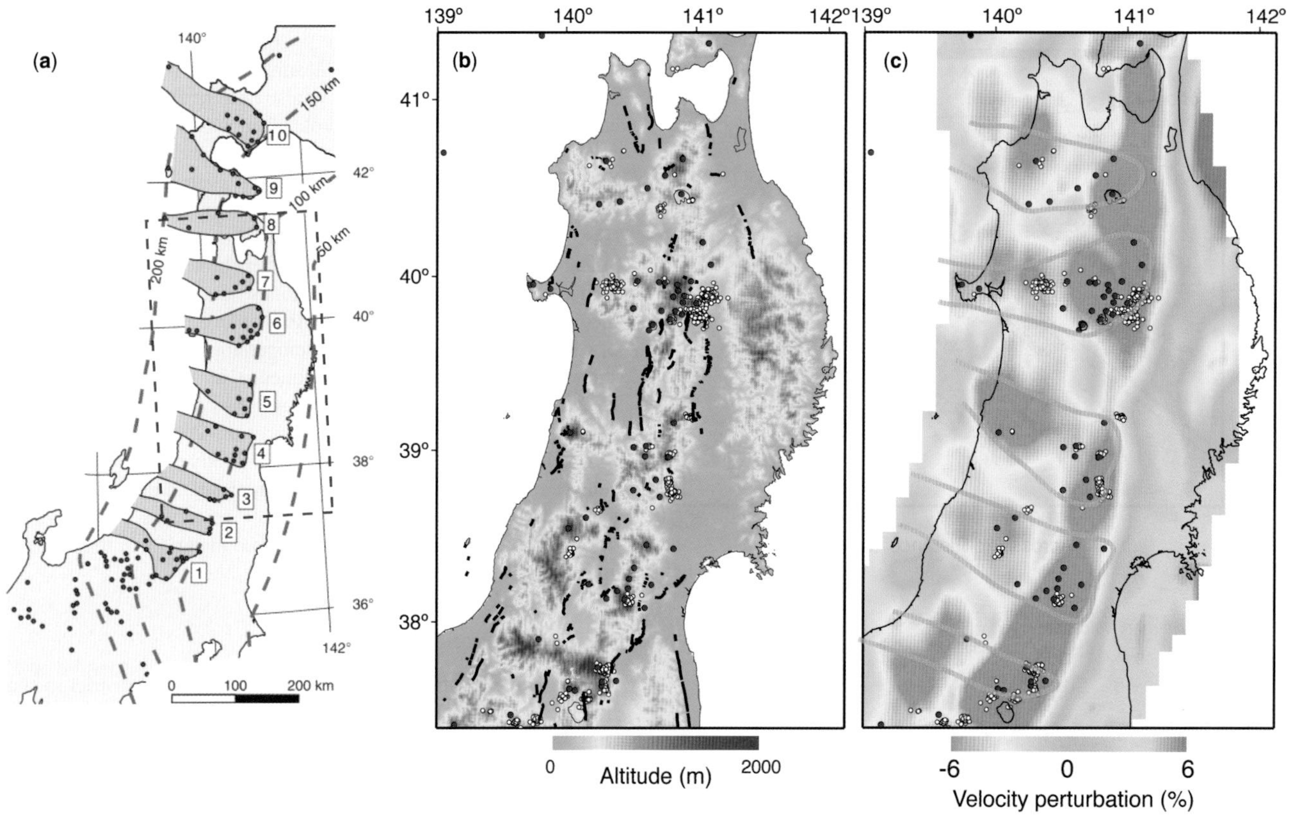

Fig. 7.13. (**a**) Hot fingers in the mantle wedge (Tamura *et al.* 2002). Dashed lines show depth contours to the surface of the dipping seismic zones. A red dashed rectangle indicates an area shown in (b) and (c). (**b**) Topography map of NE Japan. Black lines show active faults. Red and white circles show Quaternary volcanoes and deep low-frequency earthquakes, respectively. (**c**) S-wave velocity perturbations along the inclined low-velocity zone in the mantle wedge of NE Japan (Hasegawa & Nakajima 2004). Grey lines show the positions of hot fingers shown in (a).

volcanoes such as petit-spot (Hirano *et al.* 2006) when the physical conditions required for volcano formation are satisfied.

Western Japan: Chugoku

The temporal and spatial variations in the late Cenozoic volcanic activity in the Chugoku district have been examined in terms of experimental petrology, geographic distribution of volcanoes and geochemical approaches (e.g. Kimura *et al.* 2003). Briefly,

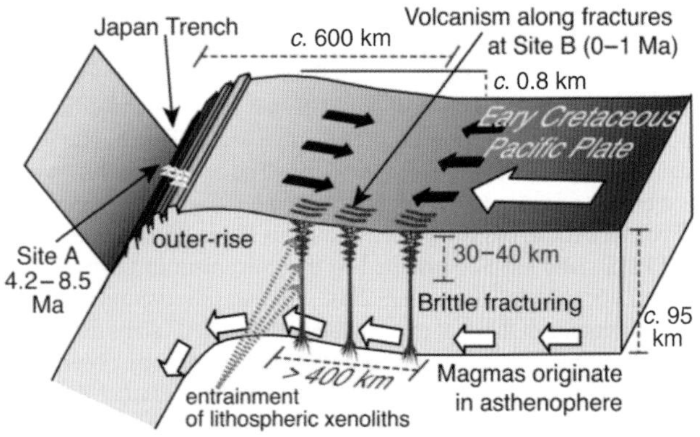

Fig. 7.14. A conceptual model of petit-spot volcanism (Hirano *et al.* 2006). Magmas from the asthenosphere escape to the shallow depths because of the extensional environment of the lower lithosphere and migrate up through the brittle compressed upper lithosphere by exploiting fractures created by flexure of the plate.

distinctive volcanism occurred in the forearc between 17 and 12 Ma, resulting from the opening of the Shikoku Basin and the subsequent opening of the Japan Sea. Persistent rear-arc volcanism began at *c.* 25 Ma while forearc volcanism ceased at *c.* 12 Ma, probably due to the subduction of the PHS slab. Volcanism that occurred in the Quaternary period has been discussed in terms of adakite volcanism due to the melting of the PHS slab (e.g. Kimura *et al.* 2003) or alkalic basalt volcanism resulting from a wet plume in a more primitive or enriched upper mantle (e.g. Iwamori 1992).

Deep mantle structure in the Chugoku district has been inferred from local, regional and teleseismic tomography (e.g. Honda & Nakanishi 2003; Matsubara *et al.* 2008). These studies revealed the subduction of an aseismic PHS slab beneath the Chugoku district, and heterogeneous mantle structures possibly acting as a deep source of magmatism. Figure 7.16 shows that a large low-velocity zone (4–6% velocity reduction) exists in the upper mantle, which originates from depths of *c.* 300 km below the PHS Plate (Nakajima & Hasegawa 2007*b*). This low-velocity zone appears to reach the uppermost mantle beneath rear-arc volcanism, passing through the north of the leading edge of the subducting PHS slab. Recent teleseismic tomography revealed the existence of a high-velocity and hence aseismic slab further to the north beyond the coastline of the Japan Sea side in western Chugoku, but the high-velocity anomaly is very weak beneath Quaternary volcanoes (Huang *et al.* 2013; Zhao *et al.* 2012). The upwelling flow may therefore disrupt the slab beneath northern Chugoku, and not avulse the northern edge of the slab.

The reduction of P-wave velocity of 4–6% corresponds to an increase in temperature of 400–600°C (e.g. Takei 2002; Wiens

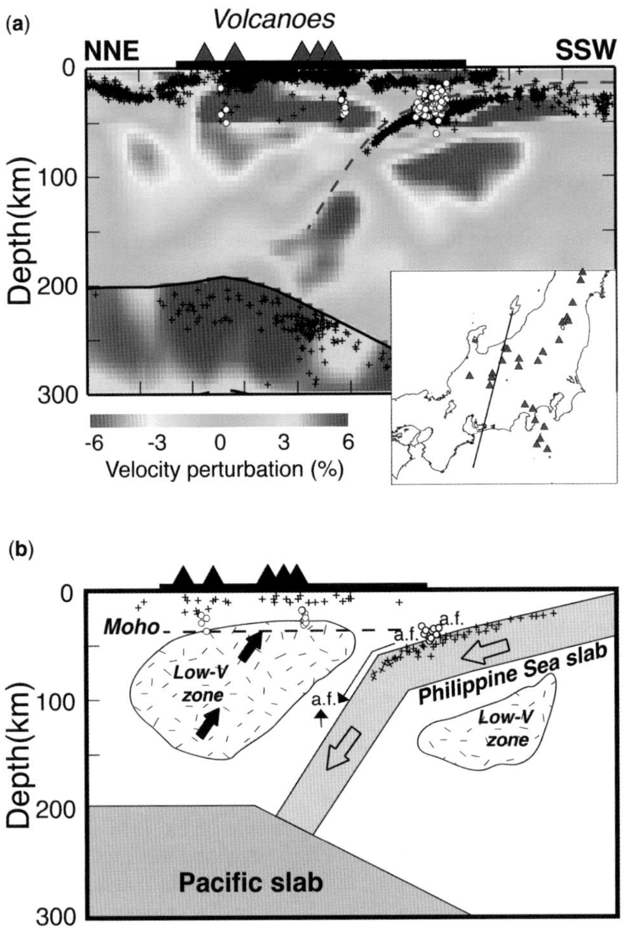

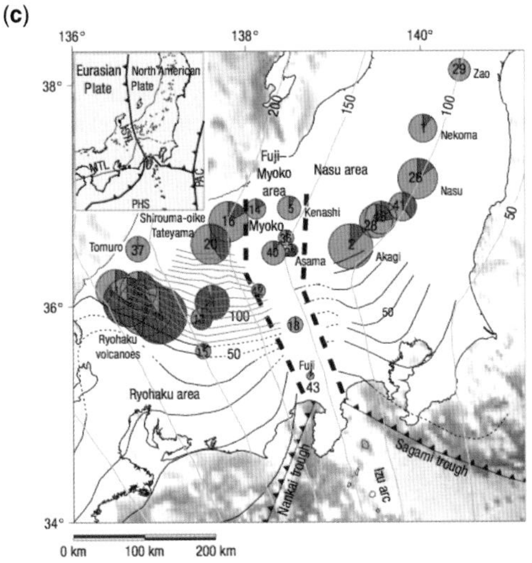

Fig. 7.15. (a) Vertical cross-sections of S-wave velocity perturbations along a line in the inset map (Nakajima & Hasegawa 2007a). Crosses and white circles denote earthquakes and low-frequency earthquakes, respectively. Red triangles and a bar on the top indicate active volcanoes and the land area, respectively. A red broken curve shows the upper boundary of the PHS slab. (b) A schematic illustration of the interpretation of the velocity images in (a). (c) Spatial distribution of fluid flux from the PAC and PHS slabs. The red and green parts represent the proportion of PHS and PAC fluids, respectively. The radius of each circle is proportional to the weight proportion of the fluid added to the mantle of the melting region.

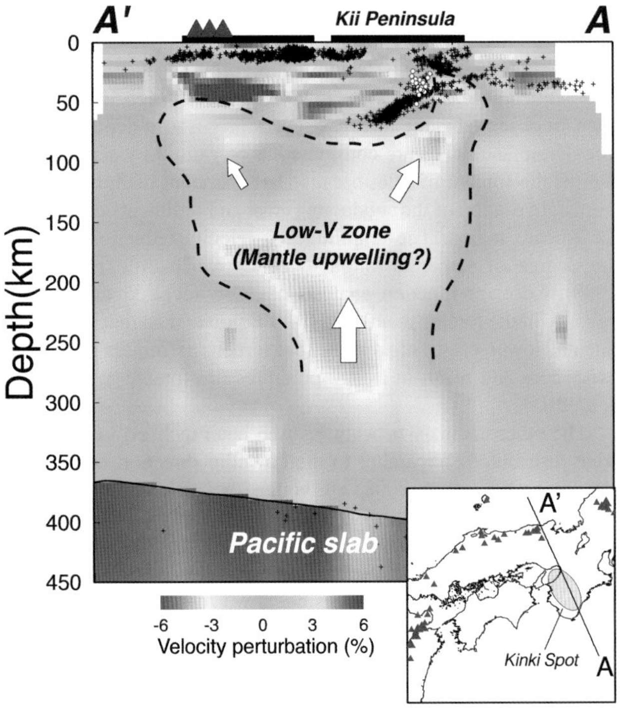

Fig. 7.16. Across-arc vertical cross-section of S-wave velocity perturbations along line A–A′ in the inset map (Nakajima & Hasegawa 2007*b*). Crosses and white circles show earthquakes and deep low-frequency earthquakes in a 10 km wide zone along each profile. The solid curve denotes the upper boundary of the subducted PAC slab. White arrows indicate interpretative flow directions. A grey ellipse in the insert map denotes the location of the 'Kinki Spot' defined by Sano & Wakita (1985).

et al. 2008). Since such a high-temperature anomaly is unlikely, Nakajima & Hasegawa (2007*b*) interpreted the low-velocity zone as caused by higher fluid content with a combined effect of chemical composition and thermal anomaly (e.g. Shito & Shibutani 2004). This interpretation does not contradict the petrological model that volcanism has been caused by a plume with abundant volatiles rather than by only a plume of anomalously high temperature (Iwamori 1992). This low-velocity zone corresponds to upwelling asthenosphere and probably acts as a primary source of magmas. The upwelling asthenosphere can explain high $^3He/^4He$ ratios (we define high $^3He/^4He$ as $^3He/^4He > 4R_a$, where R_a is the atmospheric $^3He/^4He$ ratio) observed in the Chugoku district (Sano *et al.* 2006), as well as high terrestrial heat flow (e.g. Tanaka *et al.* 2004). The origins of the upwelling flow have been related to a remnant of the upwelling due to the opening of the Shikoku Basin in the period 30–15 Ma, or by mantle convection associated with the opening of the Japan Sea back-arc basin in the early–middle Miocene (Nakajima & Hasegawa 2007*b*).

Western Japan: Kyushu and Ryukyu

It is apparent that the age of the PHS slab changes abruptly across the Kyushu–Palau Ridge, south of which the older (>50 Ma) PHS Plate is subducting beneath south Kyushu to Ryukyu (e.g. Okino *et al.* 1994). As a result of the subduction of the old PHS slab, a well-defined volcanic front is formed with active volcanism in Kyushu and Ryukyu. The PHS slab is seismic down to a depth of *c.* 200 km in southern Kyushu, and the aseismic PHS slab exists beyond the seismic portion down to mantle transition zone depths (Abdelwahed & Zhao 2007; Zhao *et al.* 2012).

The mantle wedge structure in Kyushu is characterized by two marked anomalies. One is low-velocity with high V_p/V_s anomalies, as observed in the forearc of central Kyushu at depths of <60 km. These anomalies are considered to be caused by serpentinization of the forearc mantle, because large amount of fluids derived from dehydration of the subducting crust and sediments can hydrate the forearc region. The amplitude of velocity reduction suggests the degree of serpentinization to be up to 20–30% (Xia *et al.* 2008). Receiver function analysis has also revealed that S-wave velocity in the forearc mantle is 20–30% slower than that in the continental lower crust, suggesting that serpentine minerals or fluids exist there and result in a high-pore-pressure forearc mantle (Abe *et al.* 2011).

The other anomaly beneath Kyushu is an inclined low-velocity zone distributed subparallel to the downdip direction of the PHS slab (e.g. Xia *et al.* 2008). Although the inclined low-velocity zone is limited to depths of *c.* 100 km, deduced from the analyses of local seismic tomography (e.g. Wang & Zhao 2006; Matsubara *et al.* 2008), recent regional tomography clearly shows that the inclined low-velocity zone is visible down to depths of *c.* 300 km in southern Kyushu (Zhao *et al.* 2012). Resistivity models have demonstrated a somewhat-inclined conductor that extends from a depth of >100 km to the lower crust just beneath the volcanic front (Hata *et al.* 2012), which is interpreted as regions of fluid released from the slab or partial melting of the mantle (Fig. 7.17). The inclined low-velocity and low-resistivity zone developed in the mantle wedge plays a crucial role in the generation of arc magmas in Kyushu.

A unique tectonic setting in Kyushu is an opening of the Okinawa Trough, a back-arc basin formed by extension within the continental lithosphere behind the Ryukyu Trench. As a result, northern Kyushu is crossed by a major tectonic structure, known as the Beppu–Shimabara Graben, representing a northeastern continuation of the Okinawa Trough. In the back-arc of Kyushu, a low-velocity zone is observed in the upper mantle and is considered to correspond to a mantle upwelling at this northeastern edge of the extending Okinawa Trough (Sadeghi *et al.* 2000). A low-velocity anomaly is also observed in the depth range of 20–120 km beneath Unzen Volcano (Wang & Zhao 2006).

Regional and local seismic tomography is limited in the Ryukyu subduction zone but has revealed prominent seismic velocity structures associated with the subduction of the PHS slab and the opening of the Okinawa Trough. Nakamura *et al.* (2003) showed that prominent P- and S-wave low-velocity zones exist along the Okinawa Trough at *c.* 50 km depth, where V_p/V_s values are high. The low-velocity and high V_p/V_s anomalies may reflect mantle upwelling flow from depths greater than 50 km along the Okinawa Trough. Teleseismic tomographic study has shown that low-velocity areas are distributed in the uppermost mantle along the volcanic front as well as along the Okinawa Trough, with a total length of 1200 km from north Taiwan to south Kyushu (Wang *et al.* 2008). The low-velocity zone below the volcanic front is interpreted to be caused by fluids liberated by the dehydration process of the subducting PHS slab.

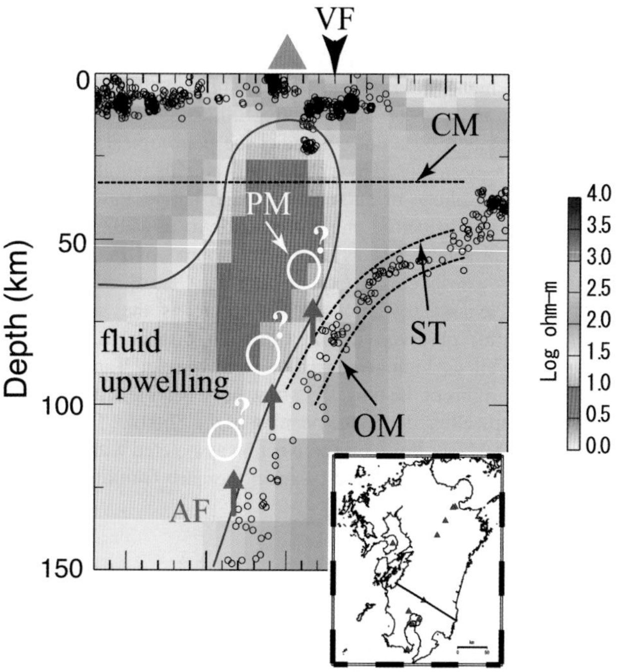

Fig. 7.17. 2D resistivity model together with interpretations of fluid upwelling along a line in the insert map (Hata *et al.* 2012). AF, aqueous fluids released from the subducting slab; PM, aqueous fluids released from the partial melting; ST, slab top; OM, oceanic Moho; CM, continental Moho; VF, volcanic front.

Kii Peninsula

The Kii Peninsula is known as an anomalous region characterized by exceptionally high $^3He/^4He$ ($>4R_a$) ratios (e.g. Sano & Wakita 1985), deeper intraslab seismicity down to a depth of *c.* 80 km and serpentinization of the subducting mantle of the PHS slab (Seno *et al.* 2001). Sano & Wakita (1985) suggested that high $^3He/^4He$ ratios may be indicative of renewed or incipient magmatism due to the subducting young PHS slab. We briefly summarize the seismological features in this particular region, located in the forearc side of SW Japan.

Whereas the subducting mantle of the PHS slab typically shows a high-velocity anomaly, a low-velocity anomaly is observed in the subducting mantle of the PHS slab beneath the Kii Peninsula (Seno *et al.* 2001). This low-velocity anomaly appears to originate from the large low-velocity anomaly below the PHS slab (Fig. 7.16) and has V_p/V_s values of 1.75–1.90 (Nakajima & Hasegawa 2007*b*). Since velocity reductions of *c.* 6% and moderate–high V_p/V_s values suggest the existence of fluid-filled pores (Takei 2002), the low-velocity anomaly intersecting the PHS slab is interpreted as a part of the hydrated mantle upwelling widely distributed in the upper mantle of SW Japan. Anomalous slab structures beneath the Kii Peninsula have also been highlighted by P-wave anisotropy analysis (Ishise *et al.* 2009) and receiver function analysis (Shiomi & Park 2008).

The Kinki Spot, known as an area with anomalously high $^3He/^4He$ ratios (a grey ellipse in the inset map of Fig. 7.16) (Sano & Wakita 1985), correlates in space with the region where the mantle upwelling intersects the PHS slab. These observations suggest that the ^3He carried by the mantle upwelling below the PHS slab contributes to anomalously high $^3He/^4He$ ratios observed in the Kii Peninsula (Nakajima & Hasegawa 2007*b*; Sano & Nakajima 2008). A long-term swarm activity in the upper crust (e.g. Kato *et al.* 2010) is probably related to locally concentrated fluids. High terrestrial heat flow observed in the Kii Peninsula (e.g. Tanaka *et al.* 2004) may also be explained by the existence of the upwelling flow from the upper mantle. Although these models assumed a continuous slab model, Ide *et al.* (2010) proposed a model of the split PHS slab at the west of the Kii peninsula and attributed high $^3He/^4He$ ratios to the split of the slab, in which they explained high $^3He/^4He$ observed only in the Kii Peninsula (at the eastern side of the slab tear) in terms of the wake of the tear moving to the west.

Deep, low-frequency earthquakes around the continental Moho

It is known that deep, low-frequency (LF) earthquakes are often observed around volcanic areas in Japan. These LF earthquakes have been frequently observed at depths of *c.* 30 km, accompanied by impulsive body waves with a predominant frequency of 1–2 Hz (e.g. Ukawa & Obara 1993; Hasegawa & Yamamoto 1994). Since LF earthquakes occur immediately above low-velocity areas interpreted as partial melting materials, their occurrences are discussed in terms of aqueous fluids that are released to surroundings as a result of the solidification of magmas.

The dense nationwide seismograph network in Japan has detected LF earthquakes even in non-volcanic areas at depths of *c.* 30 km (pink crosses in Fig. 7.18a) (e.g. Ohmi & Obara 2002; Katsumata & Kamaya 2003; Aso *et al.* 2013). Ohmi & Obara (2002) detected five LF earthquakes in the source area of the 2000 Tottori earthquake (M_w 6.7) before the main shock. Their result showed that a single-force source mechanism is more preferable than a double-couple source mechanism, which suggests the transport of fluid such as water or magma. Recently, Aso & Ide (2013) revealed that focal mechanisms for many LF earthquakes are dominated by a compensated linear vector dipole with symmetry axes parallel to the lineation of hypocentre distributions and the direction of the minimum principal axis of regional stress, and concluded that LF earthquakes potentially represent the oscillation of magma induced by fluid movement. Tomographic studies for the whole area of the Japanese islands has shown that LF earthquakes in non-volcanic areas are also distributed in low-velocity zones of the lower crust (e.g. Omuralieva *et al.* 2012), suggesting that LF earthquakes are one of the manifestations of the widespread existence of abundant fluids in the lower crust that are carried by the upwelling flow from the slab to the crust.

Crustal deformation and shallow crustal seismicity

Through an analysis of dense GPS network data, zones of high strain rate have been observed in the Japanese islands (e.g. Sagiya *et al.* 2000; Miura *et al.* 2004). The high-strain-rate zones undergo contraction a few times larger than that in surrounding regions. Historically, many large earthquakes have occurred along these high-strain-rate zones, suggesting that the present-day deformation field has persisted for at least the last several hundred years. Local concentrations of arc crust deformation are interpreted to be caused by abundant fluid in the crustal depths that are originally supplied from the subducting oceanic plates (Hasegawa *et al.* 1991, 2000, 2005; Iio & Kobayashi 2002; Iio *et al.* 2002). The presence of abundant and widespread fluids in the lower crust is suggested by bright S-wave reflectors (e.g. Ito 1993; Hori *et al.* 2004) as well as by low-velocity zones (e.g. Nakajima & Hasegawa 2007*c*).

There are many observations that moderate to large crustal earthquakes in the Japanese islands occurred in the upper crust underlain

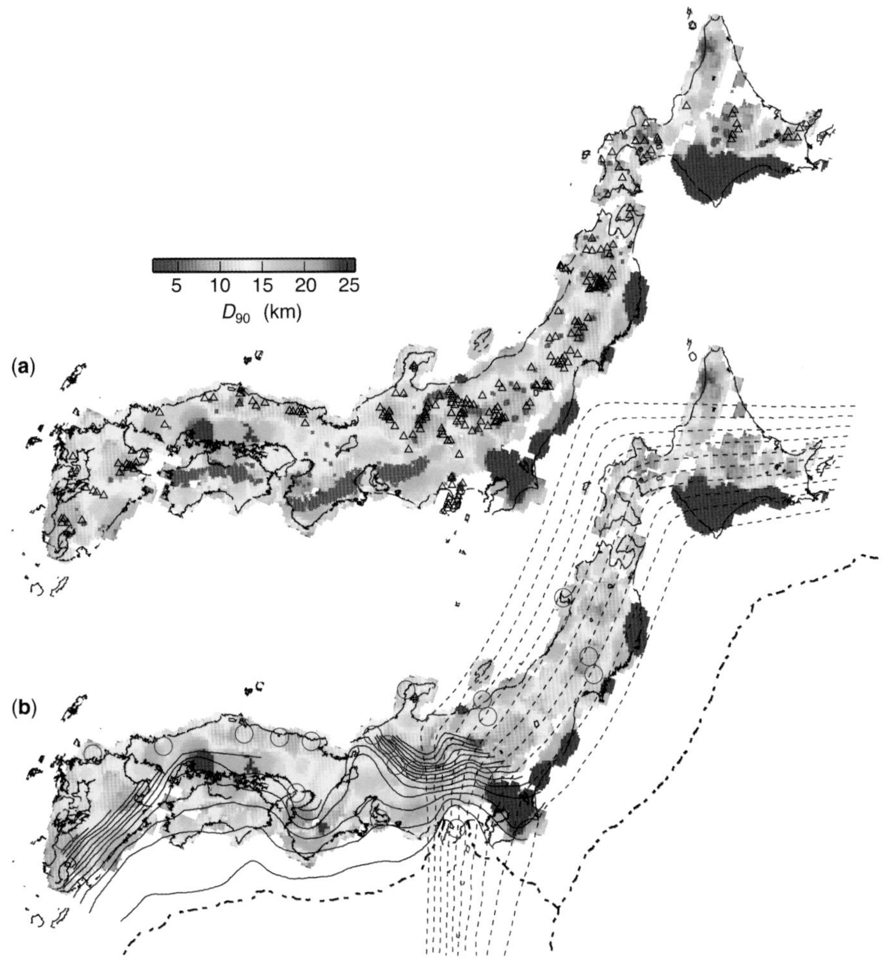

Fig. 7.18. (**a**) Distributions of D_{90} with Quaternary volcanoes (triangles) and deep, low-frequency earthquakes (magenta crosses) (Omuralieva *et al.* 2012). (**b**) Distributions of D_{90} with the geometries of the PAC slab (broken lines) and PHS slab (solid lines) and large earthquakes ($M > 6.5$) (red circles) in the period 1926–2010.

by the low-velocity lower crust where aqueous fluids may exist (e.g. Zhao *et al.* 1996; Nakajima & Hasegawa 2008; Hasegawa *et al.* 2009). Zhao *et al.* (2010) showed that hypocentres of 164 crustal earthquakes (M 5.7–8.0) during 1885–2008 are distributed in the upper crust, below which low-velocity zones are widely observed in the lower crust and uppermost mantle. In addition, fine-scale seismic tomography has revealed that low-velocity zones are distributed in the lower crust beneath active faults that are characterized by the linear alignment of seismicity (Nugraha *et al.* 2013). Omuralieva *et al.* (2012) estimated the spatial distribution of the cut-off depth (D_{90}) of shallow seismicity over the whole of Japan using relocated hypocentres during the last decade. D_{90} is defined as a depth above which 90% of shallow crustal earthquakes occur. The estimated D_{90} values show a considerable lateral variation, ranging from *c.* 5 km to *c.* 40 km with shallower D_{90} in volcanic areas, the Kinki district and the middle of Shikoku district and deeper D_{90} along the coastal area of the Pacific ocean (Fig. 7.18). An important observation is that large crustal earthquakes tend to occur around areas with shallower D_{90} areas (Fig. 7.18b).

Aqueous fluids in the lower crust continue to rise to the subsurface and result in fluid-rich areas in some part of the brittle upper crust. The upper crust deforms elastically under the regional stress field but ductile deformation may be partly involved in areas with abundant fluids and elevated temperatures, resulting in the thinning of the seismogenic layer and shallow D_{90}. The local and inhomogeneous deformation enhances stress concentration, resulting in very active seismicity in areas where a large amount of fluids exits in the lower crust (e.g. Ito 1993; Hasegawa *et al.* 2005). Seismic activity and the occurrence of large crustal earthquakes in the upper crust may be one of the manifestations of abundant fluids in the crust.

Seismological observations have revealed three-dimensional mantle wedge structures beneath the Japanese islands and provided new insights for the understanding of magmatism associated with the subduction of the PAC and PHS plates. An inclined sheet-like low-velocity zone that is distributed in the mantle wedge subparallel to the slab from Hokkaido to Izu and from Kyushu to Ryukyu probably controls the evolution of the arc systems and acts as a deep source of arc magmas. A large upwelling in the upper mantle beneath SW Japan is likely to control magmatism in the northern part of Chugoku district. A branch of this upwelling flow appears to penetrate into the PHS slab beneath the Kii Peninsula, probably resulting in anomalous geophysical and geochemical signatures observed in that area. Partial melts that are conveyed through the mantle upwelling to the continental crust may enhance ductile deformation in the lower crust and result in stress concentration in the upper crust immediately above. Crustal earthquakes and deep low-frequency earthquakes are considered to occur as a result of these mechanical processes. Volcanoes are formed above regions where large amounts of melts are successively supplied by the mantle upwelling flow to the crustal levels. The observations described in this section suggest that both volcanisms and seismogenesis are closely associated with fluids that are supplied from the subducting oceanic plate. Shallow crustal earthquakes occur as a consequence of stress concentration to faults by inhomogeneous inelastic deformation in the crust.

Stagnant slab and deep mantle processes (DZ)

As mentioned in the preceding sections, beneath the Japanese islands earthquakes occur actively in the crust and in the subducting PAC and PHS slabs. Tomographic studies using local-earthquake arrival-time data can determine the detailed 3D P- and S-wave velocity structure down to *c.* 200 km depth under Japan, including the crust, upper-mantle wedge and the upper portion of the PAC and PHS slabs, but cannot determine the deeper 3D structure for the lower portion of the slabs and the mantle below the slabs. Zhao *et al.* (1994) proposed a joint inversion of local-earthquake arrival times and relative travel-time residuals of teleseismic events in order to determine the deep 3D velocity structure of the Japan subduction zone. This joint inversion approach preserves the advantages of the separate approaches of local-earthquake tomography (Aki & Lee 1976) and teleseismic tomography (Aki *et al.* 1977) and overcomes their drawbacks. Moreover, the horizontally propagating local rays and the vertically travelling teleseismic rays cross each other in the shallow portion of the model, which can improve the resolution there.

Because of the narrowness of the Japanese islands where the dense seismic network is installed, however, the resolved areas of the deep structure by the local and teleseismic joint inversion (Zhao *et al.* 1994) are spatially limited. To understand the deep 3D structure under the Pacific Ocean and the Japan Sea, we have to rely on the models of global tomography or regional tomography under East Asia, although those models have a lower resolution than that of the local tomography models under Japan.

In this section, we first introduce the recent global and regional tomography images which show (1) that the PAC slab becomes stagnant in the mantle transition zone under the Korean Peninsula and East China, and (2) a big mantle wedge has formed above the stagnant slab, causing active tectonics and intraplate volcanoes in East Asia. High-resolution images of the deeper portion of the Japan subduction zone are provided, which show that the PHS slab has subducted aseismically down to the mantle transition zone and the inclined mantle-wedge low-velocity (V) zone under Tohoku has extended towards the eastern margin of the Japan Sea. A metastable olivine wedge with lower seismic velocity is detected within the subducting PAC slab at the mantle transition zone depths, which may be related to the occurrence of deep earthquakes in the slab. Finally, we summarize our current knowledge and understanding of the deep structure and dynamics under the Japan islands and East Asia.

Stagnant slab and big mantle wedge

Figure 7.19 shows two vertical cross-sections of P-wave tomography down to the core–mantle boundary under East Asia from a global tomography model (Zhao 2004). In the upper mantle, the subducting PAC slab is imaged clearly and earthquakes occur down to about 600 km depth within the slab. Under East China, the PAC slab becomes stagnant in the mantle transition zone, and the slab extends westwards to the Beijing area. Prominent low-V zones exist in the upper mantle above the stagnant slab, and the active intraplate volcanoes in NE China (e.g. Changbai, Wudalianchi and Datong) are located above the low-V zones (Fig. 7.19). In the lower mantle, high-V anomalies are visible under the stagnant slab. Similar features were also found in other global tomographic models (e.g. Bijwaard *et al.* 1998; Fukao *et al.* 2001), receiver functions (e.g. Gu *et al.* 2012) and waveform modelling (e.g. Tajima *et al.* 2009; Li *et al.* 2013). These results suggest that the subducting PAC slab meets strong resistance when it encounters the 660 km discontinuity. The slab bends horizontally and accumulates there for a long period (*c.* 100–140 myr), and then finally collapses to sink as blobs onto the core–mantle boundary as a result of very large gravitational instability induced by phase transitions (Maruyama 1994; Zhao 2004).

Higher-resolution regional tomography has been determined for the crust and mantle under East Asia (e.g. Huang & Zhao 2006; Wei *et al.* 2012). Figure 7.20 shows nine east–west vertical cross-sections of the tomography down to 1300 km depth under East Asia. The stagnant PAC slab is more clearly visible than in the global tomography model (Fig. 7.19). In the northern vertical cross-sections

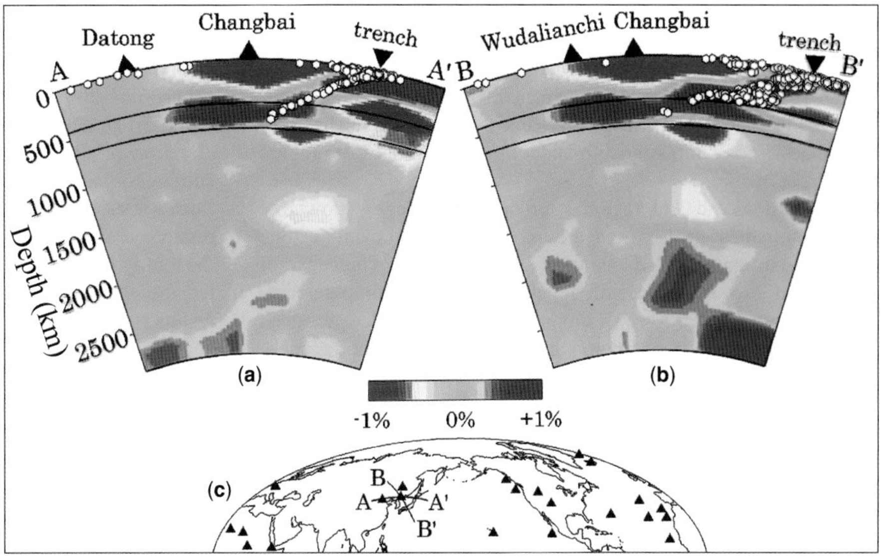

Fig. 7.19. (**a**, **b**) Vertical cross-sections of whole-mantle P-wave tomography (Zhao 2004). Red and blue colours denote slow and fast velocity perturbations (in %), respectively, from the 1D iasp91 Earth model (Kennett & Engdahl 1991). The velocity perturbation scale is shown below the cross-sections. White dots denote earthquakes within 150 km width of the profiles. The two lines show the 410 and 660 km discontinuities. Solid triangles show the active intraplate volcanoes. The reverse triangles show the location of the Japan Trench. (**c**) Map showing the locations of the two cross-sections in (a, b). Triangles denote hotspots or intraplate volcanoes on Earth.

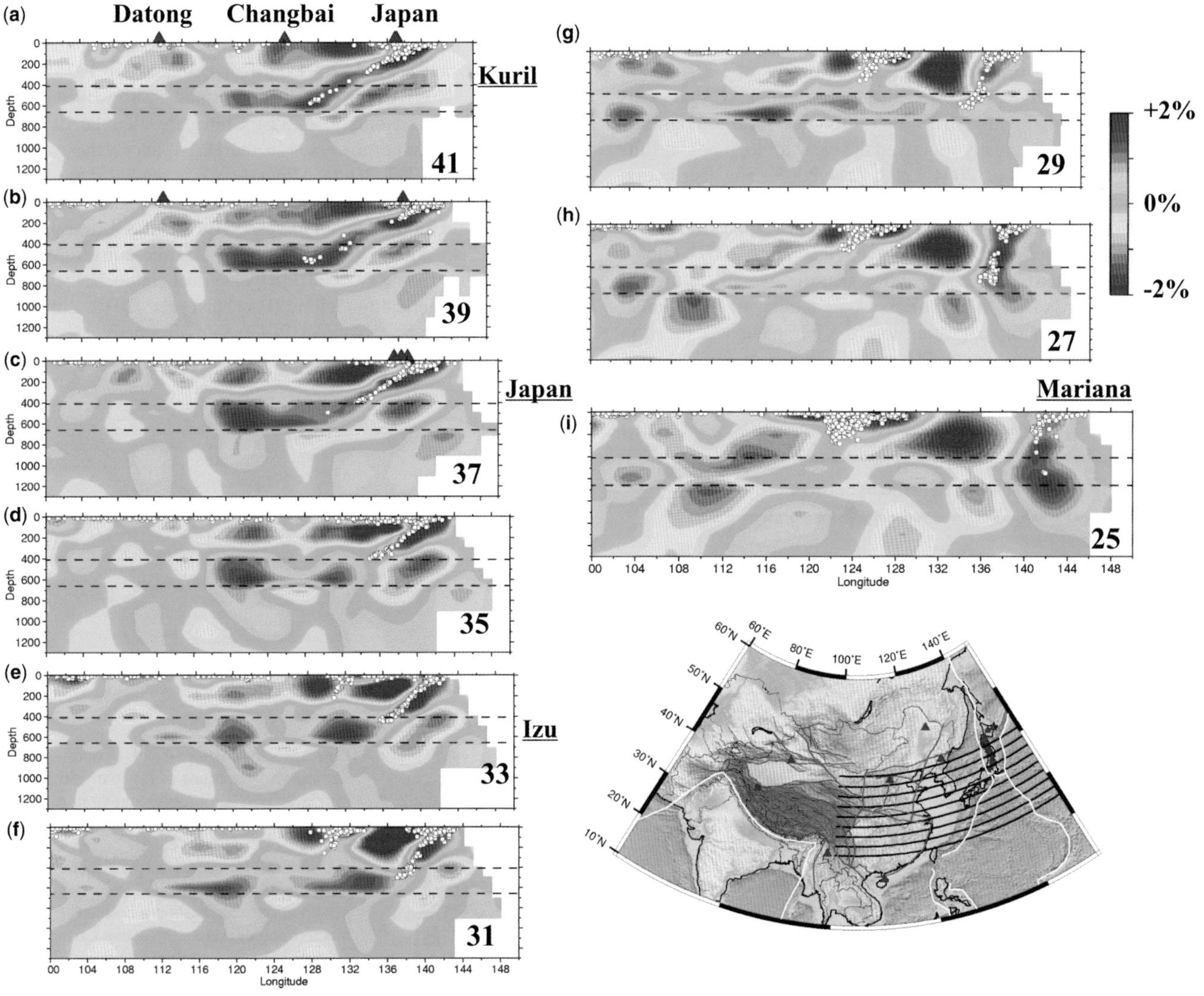

Fig. 7.20. (**a–i**) East–west vertical cross-sections of P-wave tomography along the profiles shown on the insert map. The latitude of each cross-section is shown on the right. Red and blue colours denote low and high velocities, respectively. The velocity perturbation scale is shown on the right. The white dots show earthquakes which occurred within 100 km from each profile. The two dashed lines denote the 410 and 660 km discontinuities. Modified from Huang & Zhao (2006).

(Fig. 7.20a, b), the entire stagnant slab is located in the mantle transition zone. Toward the south, however, part of the slab material starts to drop down to the lower mantle (Fig. 7.20c–e). Under the Mariana region, the PAC slab has penetrated directly to the lower mantle (Fig. 7.20h, i). It is also observable that the PHS slab has subducted down to *c.* 500 km depth under the Ryukyu back-arc region (Fig. 7.20f, h).

Both the global and regional tomographic images show that low-*V* anomalies exist in and around the mantle transition zone below the subducting PAC slab (Figs 7.19 & 7.20). The cause of the sub-slab low-*V* zones is still not clear. Results of computer simulations suggest that the low-*V* anomalies might have been produced in the Pacific super-plume region and carried towards the Japanese islands by the horizontal flow associated with the plate movement, where they were dragged down by the subducting PAC slab (Honda *et al.* 2007).

High-resolution local tomography of the crust and upper mantle beneath the Changbai intraplate volcano was determined using a large quantity of arrival-time data from local earthquakes and teleseismic events that were recorded by many permanent seismic stations in North China and some portable stations around the Changbai Volcano (Zhao *et al.* 2009) (Fig. 7.21). The result shows a clear low-*V* anomaly extending down to 410 km depth under the Changbai Volcano. A high-*V* anomaly is visible in the mantle transition zone and deep earthquakes occur at depths of 500–600 km within the high-*V* zone, suggesting that the stagnant PAC slab exists under the Changbai region. This high-resolution local tomography under Changbai has greatly improved the early tomographic results in this region (Zhao *et al.* 2004; Lei & Zhao 2005). The distribution of seismic stations is very sparse in Wudalianchi and other intraplate volcanic areas in NE Asia, so the tomographic images in those areas have a lower resolution (e.g. Duan *et al.* 2009). Global tomography (Zhao 2004, 2009) shows that the overall features of mantle structure under those volcanoes are similar to those under Changbai (Fig. 7.19), suggesting that the intraplate volcanoes in NE Asia have the same origin (Zhao 2012; Wei *et al.* 2012).

The stress regime in the subducting PAC slab has been investigated using focal-mechanism solutions of intermediate-depth and deep earthquakes under the Japan Sea and the East Asian margin (Zhao *et al.* 2009). The results show that compressional stress axes of almost all deep earthquakes are nearly parallel with the downdip direction of the slab, indicating that the PAC slab is under the compressional stress regime in the depth range of 200–600 km. Such a stress regime in the slab is caused by the slab meeting strong resistance at the 660 km discontinuity, this being consistent with the tomographic results (Figs 7.20 & 7.21). Ichiki *et al.* (2006) studied the electric-conductivity structure under NE China, and their results suggest that the asthenosphere under East China is both hot and wet, being associated with the deep dehydration of the stagnant PAC slab.

A big mantle wedge (BMW) model was proposed to explain the intraplate magmatism and mantle dynamics under East Asia based on the multiscale tomographic images (Zhao *et al.* 2004, 2007). It is considered that the upper mantle above the stagnant slab has formed a BMW under East Asia. Corner flow in the BMW and fluids from deep slab dehydration and/or brought down from the shallow mantle wedge by convection cause upwelling of hot asthenospheric materials, resulting in thinning and fracturing of the lithosphere under East China. The formation of active intraplate volcanism (such as the Changbai and Wudalianchi volcanoes), strong intraplate earthquakes (such as the *M* 7.8 Tangshan earthquake in 1976), reactivation of the North China craton and a boundary in surface topography and gravity anomaly in East China are all related to the structure and dynamic processes in the BMW above the stagnant slab (Zhao *et al.* 2007, 2011). Mineral physics studies demonstrate that deep slab dehydration in the mantle transition zone is possible (e.g. Inoue *et al.* 2004; Ohtani *et al.* 2004; Ohtani & Litasov 2006; Ohtani & Zhao 2009). Recent results of seismic anisotropy (e.g. Liu *et al.* 2008; Huang *et al.* 2011*a*), electrical conductivity (Ichiki *et al.* 2006), numerical modelling (Faccenna *et al.* 2010; Zhu *et al.* 2010; Kameyama & Nishioka 2012) and geochemical analysis (Chen *et al.* 2007; Zou *et al.* 2008; Kuritani *et al.* 2009; Sakuyama

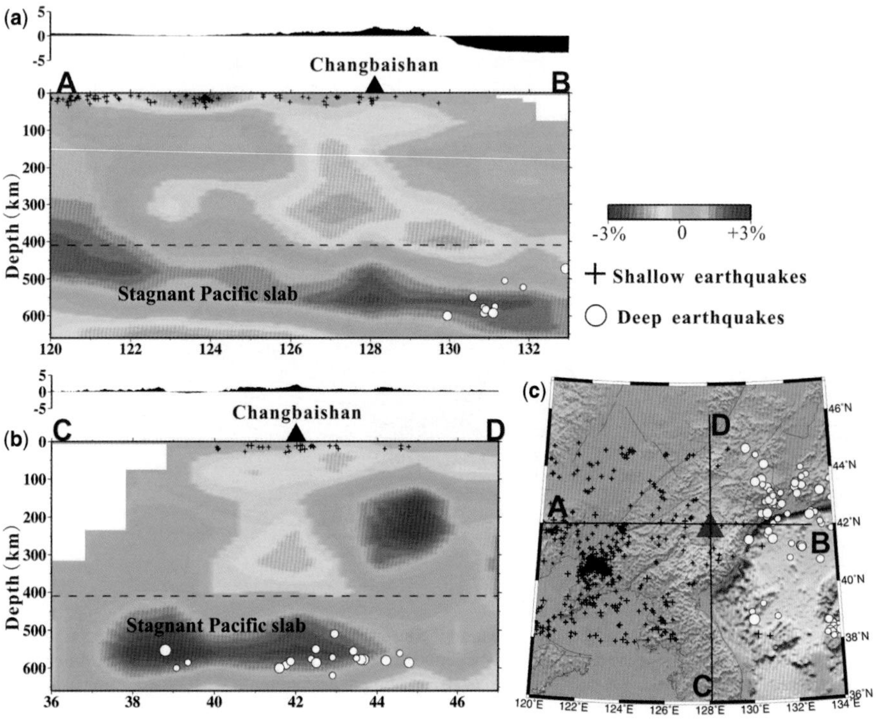

Fig. 7.21. (**a**) East–west and (**b**) north–south vertical cross-sections of P-wave velocity tomography along profiles shown in (c) (Zhao *et al.* 2009). Red and blue colours denote slow and fast velocity anomalies (in %) from the 1D iasp91 velocity model (Kennett & Engdahl 1991). The velocity perturbation scale is shown beside (a). The black triangles denote the active Changbai volcano. The dashed lines show the 410 km discontinuity. (**c**) Map showing the locations of the two cross-sections in (a, b). The red triangle denotes the Changbai volcano. Black crosses and white dots show the shallow (0–30 km) and deep (450–600 km) earthquakes, respectively.

et al. 2014) all suggest a hot and wet upper mantle above the stagnant PAC slab and its close relationship to the intraplate volcanism and active tectonics in East Asia, therefore supporting the BMW model.

Deep structure of Japan subduction zone

The deep structure of the Japan subduction zone has been investigated using a large number of high-quality arrival-time data from local and distant earthquakes recorded by the seismic networks on the Japanese islands (Zhao *et al.* 1994, 2012; Abdelwahed & Zhao 2007; Huang *et al.* 2013). Figures 7.22–7.25 show vertical cross-sections of P-wave tomography down to 700 km depth under Japan. Beneath Tohoku the inclined low-*V* zone in the mantle wedge is found to extend westwards beneath the Japan Sea; it is not just confined beneath Honshu Island (Fig. 7.23). Results of extensive synthetic tests indicate that this is a reliable feature (Zhao *et al.* 2012). However, it is still unclear whether this mantle-wedge low-*V* zone extends further westwards under the entire Japan Sea or not. The global and regional tomographic models show broad low-*V* zones in the upper mantle under the Japan Sea and East China (Figs 7.19 & 7.20). These large-scale models have a lower resolution however, so it is unclear whether the BMW low-*V* zones are continuous from the volcanic front in Honshu to East China or if they exist intermittently beneath Honshu, Japan Sea and East China. The spatial extent and detailed structure of the BMW low-*V* zones are an important issue related to the deep slab dehydration, mantle-wedge corner flow and origin of back-arc and intraplate magmatism. To clarify this issue, a portable ocean-bottom seismometer (OBS) network should be installed in the Japan Sea to determine a high-resolution upper-mantle structure under the Japan Sea and East Asia.

Beneath northern Kyushu, a significant low-*V* zone is visible in the crust and upper-mantle wedge above the PHS slab (Fig. 7.25). The low-*V* zone extends down to 300 km depth dipping toward the west subparallel to the PHS slab, which is considered to be associated with the shallow and deep dehydration of the PHS slab and convective circulation process in the mantle wedge, similar to the processes occurring in the mantle wedge above the PAC slab. These processes may have led to the formation and opening of the Okinawa Trough with its northern end reaching the Unzen Volcano in western Kyushu.

Earthquakes occur within the PHS slab down to a depth of *c.* 50 km under Chugoku and *c.* 200 km under Kyushu, whereas a

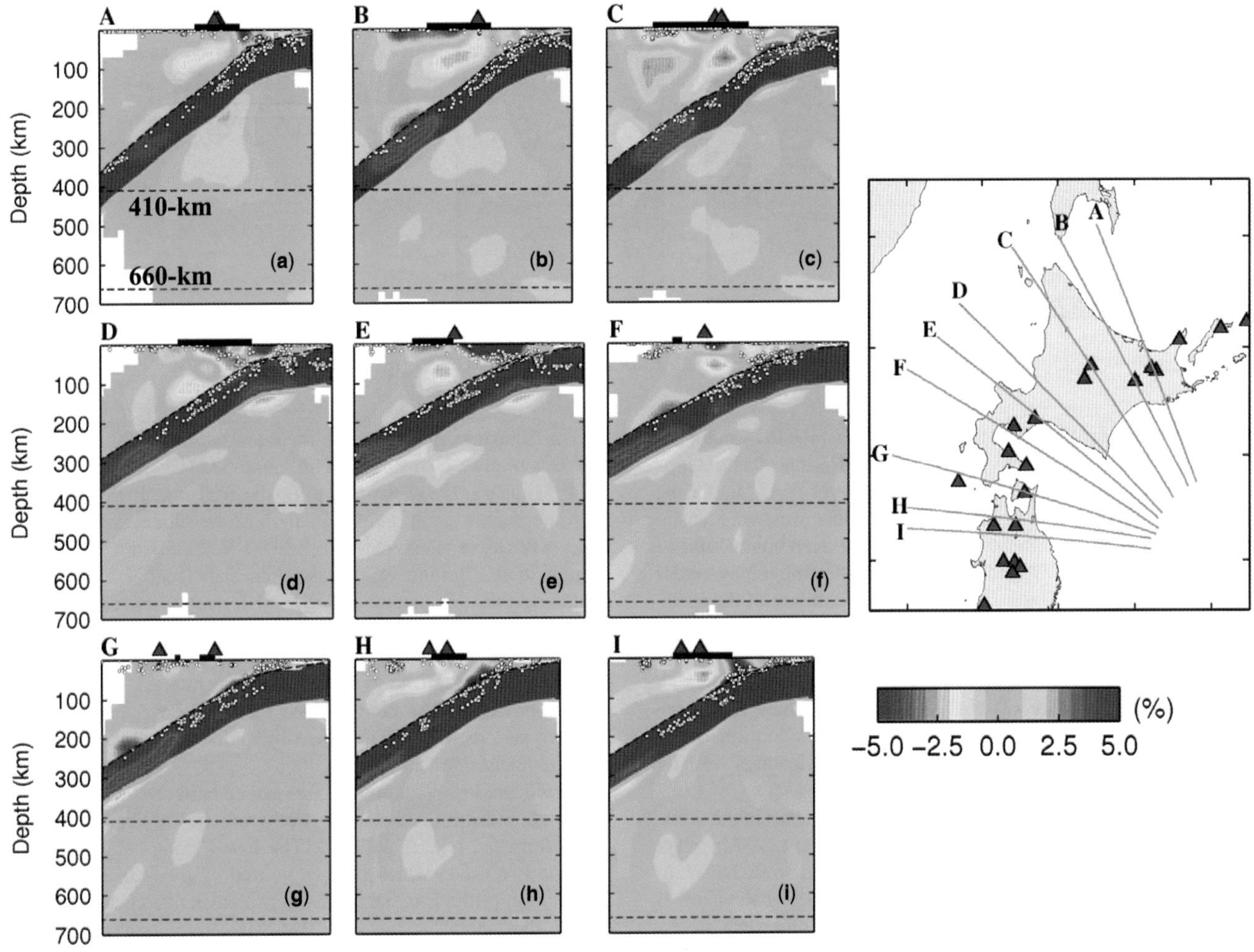

Fig. 7.22. (**a–i**) Vertical cross-sections of P-wave tomography along the profiles shown on the inset map (Zhao *et al.* 2012). Red and blue colours denote low and high velocities, respectively. The velocity perturbation (in %) scale is shown below the inset map. The red triangles denote the active arc volcanoes. The black bar atop each cross-section denotes the land area. The two dashed lines in each cross-section represent the 410 and 660 km discontinuities. The white dots denote seismicity that occurred within a 20 km width of each profile. Note that the subducting PAC slab was introduced into the starting model for the tomographic inversion, because the general geometry and structure of the PAS slab under the Japan islands have been well constrained by many previous studies using seismicity, receiver functions and reflected and converted waves at the slab boundary. By introducing the PAC slab into the model, ray paths and travel times can be calculated more precisely and the final 3D velocity model fits the data better. For details, see Zhao *et al.* (2012).

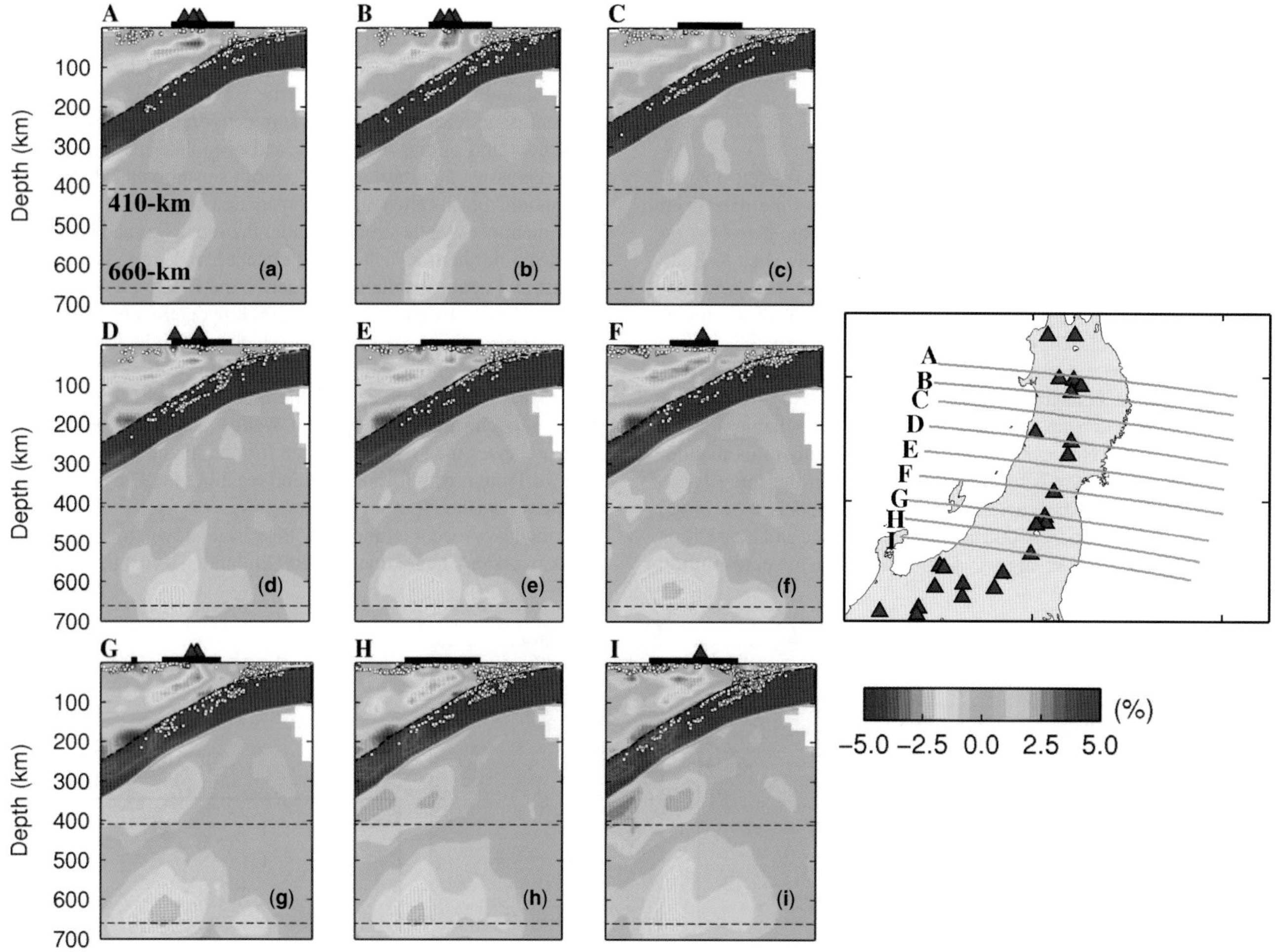

Fig. 7.23. **(a–i)** As for Figure 7.22 but for vertical cross-sections beneath Tohoku as shown on the inset map. After Zhao *et al.* (2012).

high-*V* zone is revealed below the seismic zone in the PHS slab. The latter represents the aseismic portion of the PHS slab that extends down to *c.* 430 km depth with a dipping angle of *c.* 45° under the Japan Sea and down to *c.* 460 km depth under western Kyushu (Figs 7.25 & 7.26). As mentioned above, the deep subduction of the PHS slab is also imaged by the regional tomography under East Asia (Fig. 7.20). Figure 7.26 shows the geometry of the aseismic PHS slab under SW Japan estimated from teleseismic tomography using data recorded by dense seismic networks in both western Japan and South Korea (Huang *et al.* 2013). The aseismic PHS slab appears to be discontinuous and intermittent. Detailed resolution analyses and synthetic tests were conducted, which confirmed that these new results of the PHS slab are reliable features (Huang *et al.* 2013).

The PHS Plate is one of the marginal sea complexes in the western Pacific, and started to subduct northwestwards at *c.* 40 Ma when the Pacific Plate changed its direction of motion from NNW to WNW (Karig 1971; Seno & Maruyama 1984). Along the Nankai Trough off SW Japan, the PHS Plate is composed of several blocks with ages increasing from the east to west: the Izu–Bonin Arc and back-arc (2–0 Ma); Shikoku Basin (30–15 Ma); Kyushu–Palau Ridge; and Amami Plateau (49–40 Ma) (Hilde & Lee 1984; Hall *et al.* 1995). The subduction rate of the PHS Plate also increases from 28 mm a^{-1} at the Sagami Trough to 50 mm a^{-1} off southern Kyushu (Seno *et al.* 1993; Bird 2003). It has been inferred that the subduction of the PHS Plate occurred intermittently throughout the late Cenozoic, depending on the rates and modes of destruction and formation of plates in the Philippine Sea (Uyeda & Miyashiro 1974; Hirahara 1981). Considering the aseismic PHS slab from the tomographic result, the total length of the subducting PHS slab under SW Japan is estimated to be 800–900 km from the Nankai Trough to the bottom of the aseismic slab (Figs 7.25 & 7.26). Assuming that this part of the PHS slab started its subduction at 40–15 Ma, then the tomographic results suggest its average subduction rate is 20–60 mm a^{-1}. This value is roughly consistent with the estimates (28–50 mm a^{-1}) by the abovementioned studies using other approaches. The convergence rate may have varied during different periods of the PHS Plate subduction (Uyeda & Miyashiro 1974; Hirahara 1981).

Significant low-*V* anomalies are imaged between the subducting PHS slab and the PAC slab at depths of 100–500 km under SW Japan (Figs 7.24, 7.25 & 7.27). The low-*V* zones are connected with the PAC slab, which may be caused by fluids from the deep dehydration of the PAC slab as well as convective circulation process in the mantle wedge. The PAC slab has a large subduction rate of 8–9 cm a^{-1} at the Japan Trench; the slab therefore has a lower temperature which allows deep slab dehydration at 400–500 km depth, similar to that occurring in the subducting PAC slab under the Tonga Arc and the Lau back-arc spreading centre (Zhao *et al.* 1997*b*; Conder & Wiens 2006). Water transported to the deep upper mantle and the mantle transition zone is stored in some hydrous minerals such as phase A, phase E, superhydrous

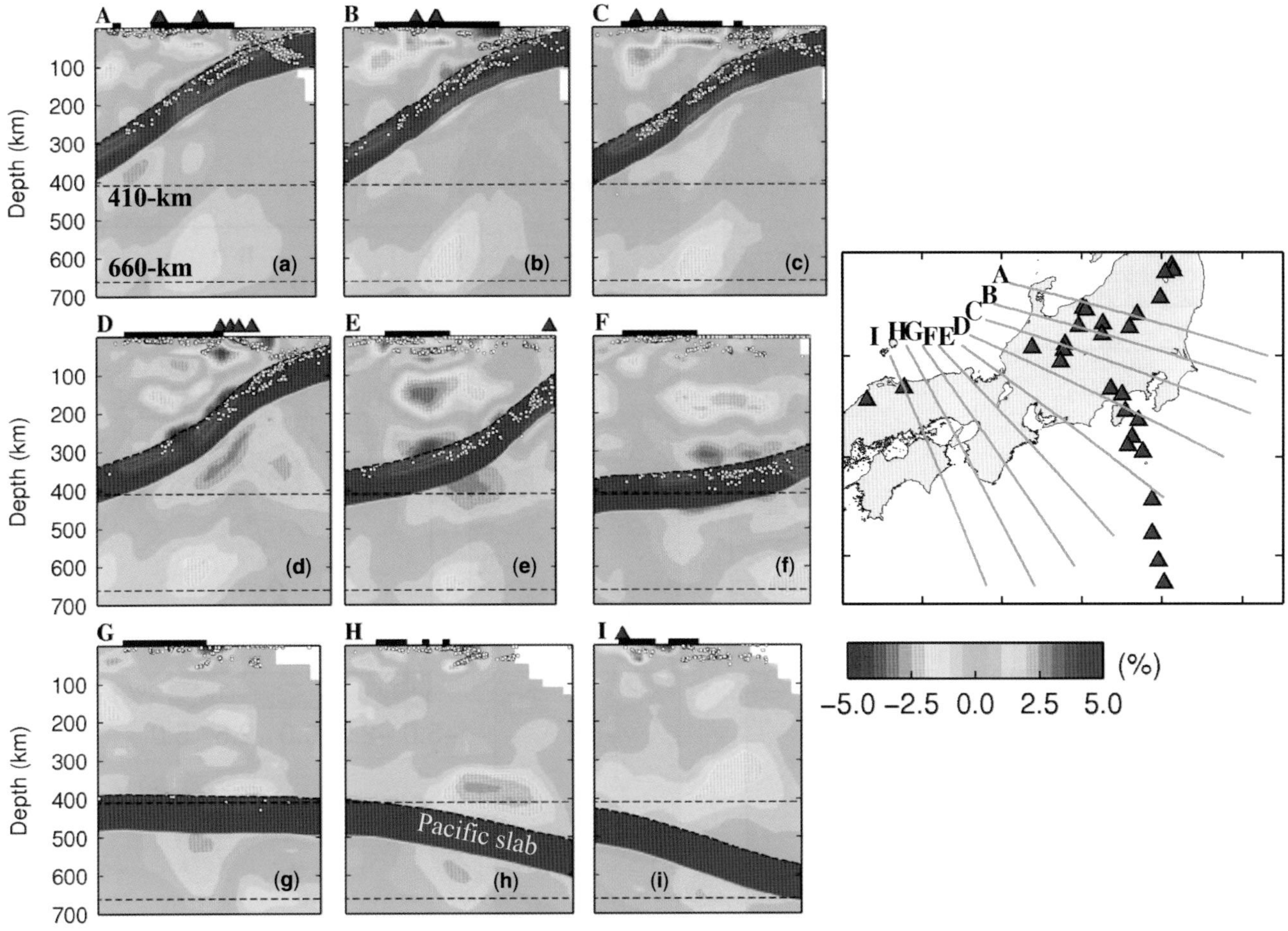

Fig. 7.24. (a–i) As for Figure 7.22 but for different vertical cross-sections as shown on the inset map. After Zhao *et al.* (2012).

phase B, phase D and nominally anhydrous minerals such as wadsleyite and ringwoodite (Inoue *et al.* 2004; Komabayashi *et al.* 2004; Ohtani *et al.* 2004). Dehydration reactions proceed by decomposition of the hydrous minerals due to the temperature increase of the stagnant slab. The decrease of the maximum water solubility in wadsleyite and ringwoodite may also cause slab dehydration (Inoue *et al.* 2004; Komabayashi *et al.* 2004; Ohtani *et al.* 2004; Ohtani & Zhao 2009).

The deep low-*V* zones at depths of 100–500 km between the PAC and PHS slabs should have a higher temperature due to the corner flow in the mantle wedge above the PAC slab. The PHS slab is heated from below and so its brittleness is lost at a shallower depth; intermediate-depth earthquakes therefore do not occur within the slab below 50 km depth under Chugoku and below 200 km under Kyushu. The heating by the hot low-*V* zones from below may also reduce the stiffness of the PHS slab, so that the slab suddenly bends where the intraslab seismicity stops (Fig. 7.25).

It is generally considered that corner flow in the mantle wedge is driven by the traction of the subducting slab on the overlying mantle. Because both the PAC and PHS slabs exist under SW Japan, it is possible that the wedge flow below the PHS slab induced by the PAC slab provides additional heat. Fluids from slab dehydration may change the rheology of the interface between the slab and the wedge mantle, and so may influence the effective coupling of the mantle wedge with the slab. This issue should be investigated in future studies by combining petrologic, geochemical and numerical modelling approaches.

A dipping low-*V* zone is visible at depths of 80–300 km above the PAC slab and beneath the active volcanoes in the Izu–Bonin Arc (Fig. 7.27), which represents the upwelling flow in the central part of the mantle wedge similar to that under Hokkaido and Tohoku (Figs 7.10, 7.12, 7.22 & 7.23). Another low-*V* anomaly is visible at 80–450 km depth directly beneath the PHS slab and directly above the PAC slab (Fig. 7.27), which may reflect an upwelling flow caused by the deep dehydration of the PAC slab and corner flow in the BMW. This low-*V* zone may interact with the PHS slab from below and, in doing so, may have changed the PHS slab geometry and affected the cut-off depth of intraslab seismicity in the PHS slab.

In the mantle below the PAC slab under Tohoku and central Japan low-*V* anomalies are visible down to 700 km depth, the bottom of the local tomographic model (Figs 7.23 & 7.24). Results of synthetic tests indicate that this is a reliable feature (Zhao *et al.* 2012). Being driven by gravitational force, the subducting PAC slab continues to sink down to the 660 km discontinuity where the sudden increase in viscosity will not allow direct slab penetration into the lower mantle under some regions (e.g. Turcotte & Schubert 2001). The PAC slab therefore becomes stagnant in the mantle transition zone and will finally collapse down to the lower mantle as a result of very large gravitational instability from phase transitions (Fig. 7.19). This will cause turbulences and thermal instability in the mantle transition zone and the lower mantle, and mantle upwelling occurs as a consequence of the thermal instability at a thermal boundary layer (Albers & Christensen 1996). The thermal plume

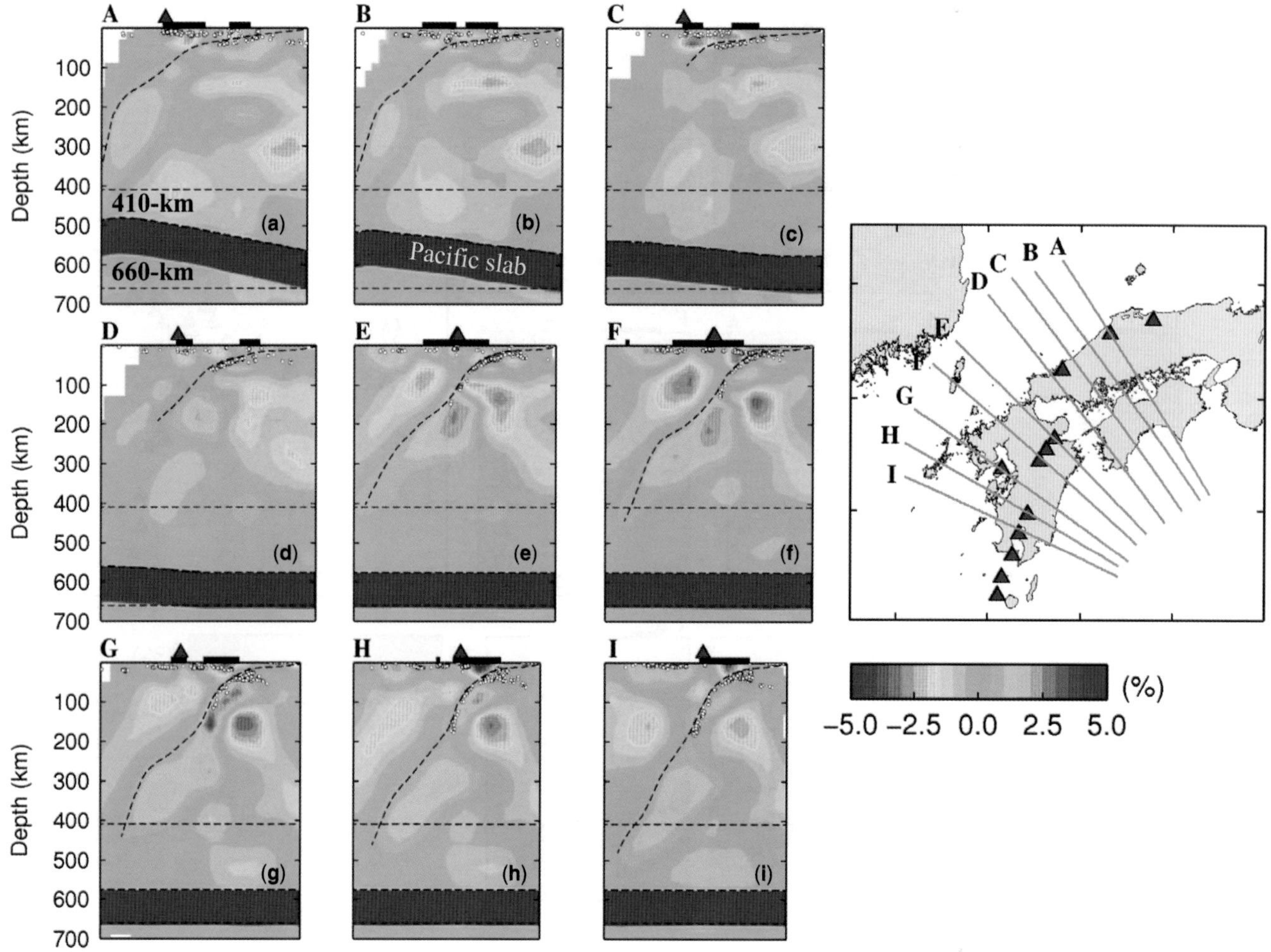

Fig. 7.25. (**a–i**) As for Figure 7.22 but for vertical cross-sections beneath western Japan as shown on the inset map. The estimated upper boundary of the subducting Philippine Sea slab is shown by a dashed line in each cross-section. After Zhao *et al.* (2012).

model (Morgan 1971) proposes that the thermal disturbance at the core–mantle boundary is responsible for the generation of mantle plumes. The low-*V* anomalies in the mantle below the PAC slab under Japan (Figs 7.23 & 7.24) may be a consequence of the thermal instability caused by the collapsing of the PAC slab materials down to the lower mantle (Fig. 7.19). The deep-subduction-related hot mantle upwelling has also been revealed in other regions. For example, recent geophysical and geochemical studies show that a young mantle plume exists under the Hainan hotspot in southern China, and the Hainan plume is caused by the thermal instability associated with the Indian plate's deep subduction in the west and the PAC and PHS slabs' deep subduction in the east down to the lower mantle (Wang *et al.* 2012; Wei *et al.* 2012; Zhao 2012). An alternative explanation for the sub-slab low-*V* zone is that it was a hot mantle upwelling dragged down by the subducting PAC slab (Honda *et al.* 2007), as mentioned above.

Metastable olivine wedge in the Pacific slab

Although deep earthquakes occupy only a few percent of global seismic activity, estimated from either the number of earthquakes or moment release, they can provide direct information on the thermal, thermodynamic and mechanical properties inside the subducting slab (Kirby *et al.* 1996). The mechanism of deep earthquakes is however debated, and several models have been proposed including dehydration embrittlement (Raleigh & Paterson 1965), transformational faulting of olivine (Green & Burnley 1989; Green & Houston 1995; Green 2003) and adiabatic shear instability (Hobbs & Ord 1988; Karato *et al.* 2001). The second model (i.e. transformational faulting of olivine) seems more popular than the other hypotheses because it was recognized early that the deepest earthquakes occur in the mantle transition zone (Green & Houston 1995).

Using travel-time residuals from deep earthquakes, Iidaka and Suetsugu (1992) showed the presence of a metastable olivine wedge (MOW) with a depth of *c.* 550 km inside the subducting PAC slab beneath the Izu–Bonin region, and they suggested that the occurrence of deep earthquakes is related to the MOW in the PAC slab. Koper *et al.* (1998) investigated this issue for the subducting slab in the Tonga region, but they failed to detect a MOW in the Tonga slab. Kaneshima *et al.* (2007) and Kubo *et al.* (2009) revealed a MOW of 5% low velocity relative to the iasp91 Earth model (Kennett & Engdahl 1991) inside the subducting Mariana slab.

Jiang *et al.* (2008) and Jiang & Zhao (2011) studied the fine structure of the subducting PAC slab under the Japan Sea using a large number of high-quality arrival-time data from deep earthquakes under the Japan Sea and the East Asian margin recorded by the Japanese seismic network. They detected a low-*V* finger within the PAC slab in the mantle transition zone under the Japan Sea, which is interpreted to be a MOW (Fig. 7.28). They carefully relocated the deep

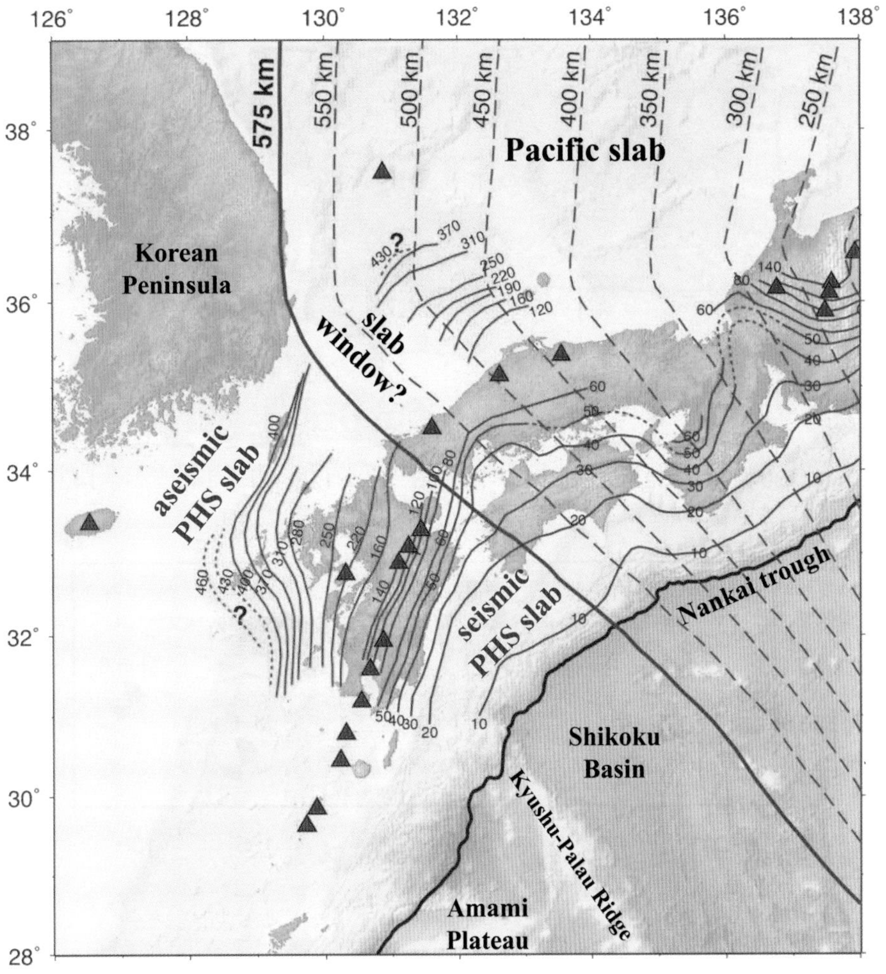

Fig. 7.26. Depth contours to the upper boundary of the subducting Pacific slab (the blue dashed lines) and that of the subducting Philippine Sea (PHS) slab. The bold blue line denotes the 575 km depth contour of the Pacific slab which becomes stagnant in the mantle transition zone to the west of this contour line, according to Zhao (2004). The grey lines denote the seismic parts of the PHS slab estimated from the seismicity in the PHS slab and local-earthquake tomography (Nakajima *et al.* 2009*a*). The red lines show the upper boundary of the aseismic PHS slab estimated from teleseismic tomography (Huang *et al.* 2013). Red triangles denote active volcanoes. After Huang *et al.* (2013).

earthquakes using the final slab model and found that all the deep earthquakes are located within the MOW or along its edges (Fig. 7.28), suggesting that the occurrence of deep earthquakes at >400 km depth is related to the fine structural heterogeneity and phase changes in the subducting slab, as suggested by earlier studies (e.g. Kirby 1991; Green & Houston 1995).

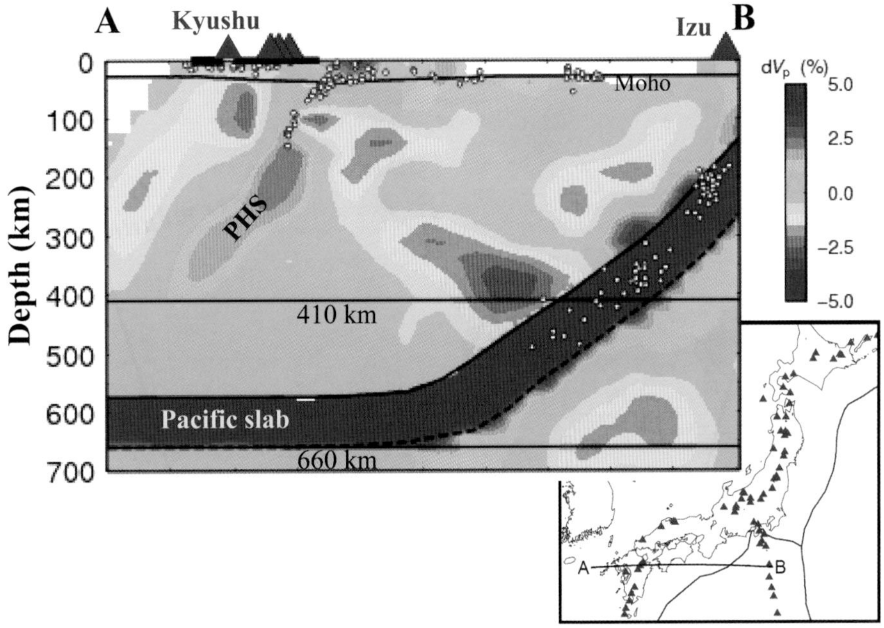

Fig. 7.27. A vertical cross-section of P-wave tomography from 0 to 700 km depth along the profile as shown on the inset map (Zhao 2012). Red and blue colours denote slow and fast velocities, respectively. The velocity perturbation scale is shown on the right. The red triangles and the thick line on the top denote active volcanoes and the land area, respectively. Earthquakes within a 30 km width from the cross-section are represented by white dots.

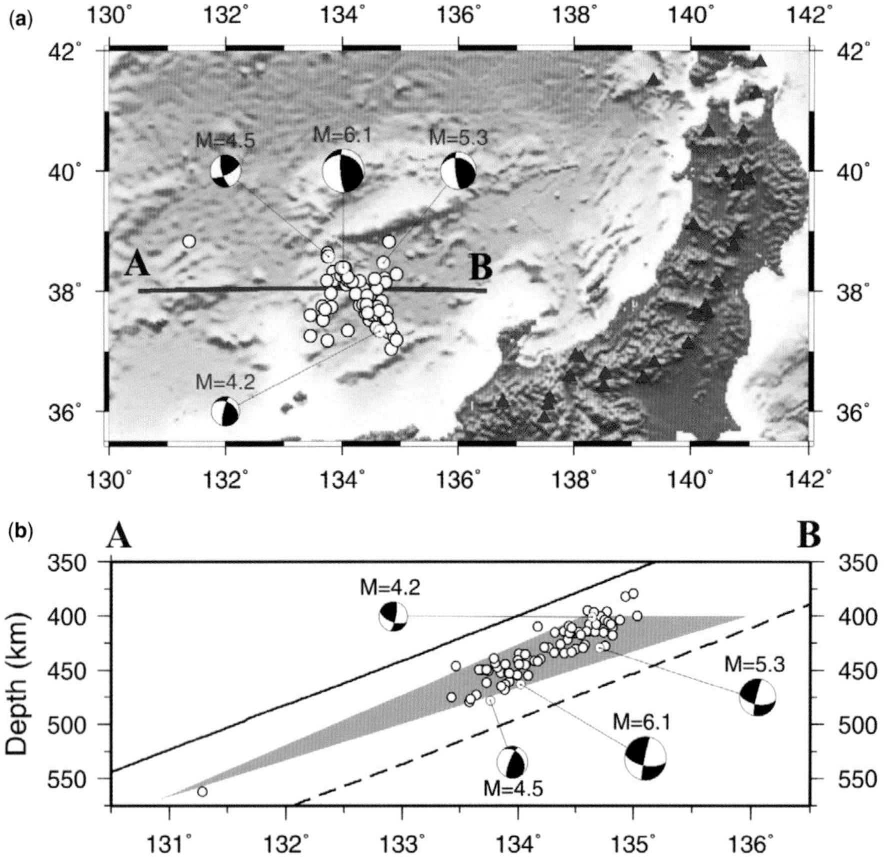

Fig. 7.28. (**a**) Epicentre locations of 78 well-located deep earthquakes under the Japan Sea. Focal mechanism solutions (lower-hemisphere projection) of four earthquakes are also shown, whose sizes are proportional to their magnitudes shown above each beach ball. The black portions in each beach ball denote the compressional area. The red triangles represent the active arc volcanoes. (**b**) Vertical cross-section of the well-located 78 earthquakes and the focal mechanism solutions of four earthquakes in the optimal slab model along line AB. The solid and dashed lines denote the upper and lower boundaries of the Pacific slab. The grey triangle in the slab represents the optimal MOW model. After Jiang & Zhao (2011).

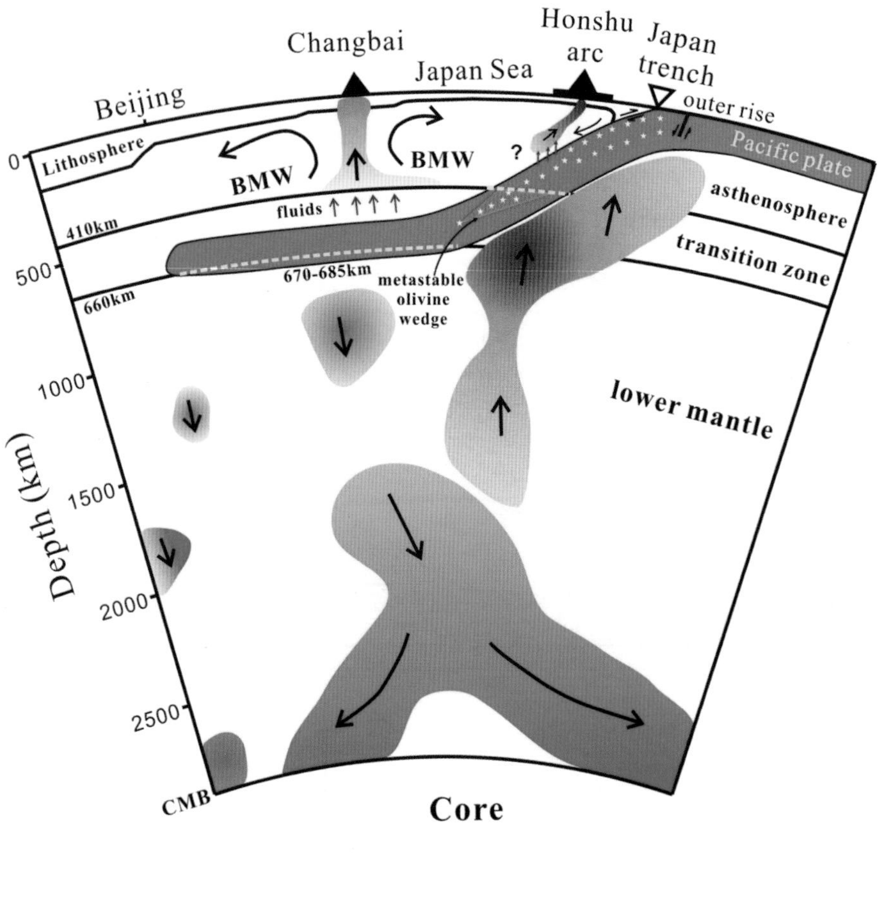

Fig. 7.29. A cartoon showing the main features of mantle structure and dynamics under the Japan islands and East Asia (see text for details). After Zhao *et al.* (2011).

The MOW is a tiny feature in the subducting slab, and the resolution of current tomography under the Japan Sea is still too low to image such a small feature because of the lack of seismic stations in the Japan Sea (e.g. Abdelwahed & Zhao 2007; Zhao *et al.* 2012). Seismologists have therefore adopted forward-modelling approaches to investigate the MOW structure so far. In future studies however, it would be ideal to detect the MOW by tomographic imaging when a dense OBS network is installed in the Japan Sea above the deep earthquakes.

Our current understanding of the main features of the crust and mantle structure and dynamics under the Japan islands and East Asia is summarized in a cartoon shown in Figure 7.29. The western PAC Plate has an age of *c.* 130 Ma near the Japan Trench and has a thickness of *c.* 85 km (Zhao *et al.* 1994, 2012; Kawakatsu *et al.* 2009). Before the plate subduction at the Japan Trench, normal-faulting type earthquakes occur in the outer-rise portion because of the downwards bending of the oceanic lithosphere (e.g. Obana *et al.* 2012). Some of the outer-rise earthquakes are very large and the faulting may even cut through the lithosphere (e.g. the 1933 *M* 7.8 Sanriku-oki earthquake; Kanamori 1971), so that seawater may enter the deep portion of the lithosphere (e.g. Wickham & Taylor 1985; Faccenda *et al.* 2009; Zhao & Tian 2013). Strong coupling occurs between the subducting oceanic plate and the overlying continental lithosphere down to a depth of *c.* 50 km, and large and small thrust-type earthquakes occur actively in the interplate megathrust zone beneath the forearc region (e.g. Kanamori 1986), such as the 2011 Tohoku-oki earthquake (M_w 9.0). Earthquakes within the subducting slab form a Wadati–Benioff deep seismic zone from the Japan Trench to a depth of *c.* 600 km. The Wadati–Benioff zone consists of two planes from the Japan Trench to *c.* 180 km depth (Hasegawa *et al.* 1978*a*, *b*, 2009; Gamage *et al.* 2009). The upper-plane earthquakes occur within the subducting oceanic crust, while those in the lower plane occur in the central portion of the slab; the distance between the two planes is 30–40 km (Matsuzawa *et al.* 1986). Earthquakes in the two planes are caused by different stress regimes, which may be explained by slab bending or unbending (Hasegawa *et al.* 2009). A metastable olivine wedge exists in the depth range of 400–600 km within the subducting PAC slab under the Japan Sea, which may be related to the occurrence of deep earthquakes (Jiang *et al.* 2008; Jiang & Zhao 2011).

Under the volcanic arc, corner flow in the mantle wedge and fluids supplied continuously from the slab dehydration produce arc magmas and active arc volcanoes (Hasegawa *et al.* 1991; Zhao *et al.* 1992; Hasegawa & Nakajima 2004). The arc magma and fluids from slab dehydration can affect the generation of all types of earthquakes in the forearc, arc and back-arc areas (Zhao *et al.* 2002, 2010; Hasegawa *et al.* 2005, 2009). The upper-mantle structure beneath the Japan Sea is unclear, and so the westwards extent of the inclined upwelling flow in the mantle wedge is still unknown.

Under the Korean Peninsula and East China, the subducting PAC slab becomes stagnant in the lower part of the mantle transition zone because the slab meets strong resistance at the boundary between the upper and lower mantle due to the large increase in viscosity at the boundary. The '660 km' discontinuity has shifted to 670–685 km depth under the stagnant slab due to the lower temperature of the slab (e.g. Helffrich 2000; Ai *et al.* 2003; Yamada & Zhao 2007). A BMW has formed in the upper mantle and the upper part of the mantle transition zone above the subducting PAC slab and the stagnant slab (Zhao *et al.* 2004, 2007). Dehydration of the stagnant slab and corner flow in the BMW cause upwelling of asthenospheric materials, which lead to the formation of intraplate volcanoes, lithospheric thinning in East China and reactivation of the North China Craton, expressed in the frequent occurrence of large crustal earthquakes there (e.g. Huang & Zhao 2004, 2009). The western edge of the stagnant slab coincides with a boundary in surface topography, gravity anomaly and also lithospheric thickness in East China (Huang & Zhao 2006), which may be controlled by the processes in the BMW (Zhao *et al.* 2007, 2011).

Pieces of high-*V* anomalies are imaged in the lower mantle down to the core–mantle boundary, suggesting that the stagnant slab finally collapses down to the bottom of the mantle (Maruyama 1994). As a result of this collapsing of the cold slab materials, hot mantle materials rise from the lower mantle in a return flow and reach the upper mantle beneath the subducting slab to cause the slab–plume interactions (Zhao *et al.* 2011).

Future efforts are needed to deploy a dense OBS network in the Japan Sea in order to clarify the relationship between the flow in the BMW and the flow in the regular (small) mantle wedge under the Japan islands, and to understand the mechanism of deep earthquakes and the fine structure of the PAC and PHS slabs.

Appendix

English to Kanji and Hiragana translations for geological and place names

Amami	奄美	あまみ
Beppu	別府	べっぷ
Bonin	ボニン	ぼにん
Chubu	中部	ちゅうぶ
Chugoku	中国	ちゅうごく
Hokkaido	北海道	ほっかいどう
Honshu	本州	ほんしゅう
Itoigawa	糸魚川	いといがわ
Izu	伊豆	いず
Kanto	関東	かんとう
Kii	紀伊	きい
Kinki	近畿	きんき
Kushiro-oki	釧路沖	くしろおき
Kyushu	九州	きゅうしゅう
Nankai	南海	なんかい
Okinawa	沖縄	おきなわ
Ryukyu	琉球	りゅうきゅう
Sagami	相模	さがみ
Sanriku-oki	三陸沖	さんりくおき
Shikoku	四国	しこく
Shimabara	島原	しまばら
Shizuoka	静岡	しずおか
Tohoku	東北	とうほく
Tohoku-oki	東北沖	とうほくおき
Tokai	東海	とうかい
Tonankai	東南海	とうなんかい
Tottori	鳥取	とっとり
Unzen	雲仙	うんぜん

References

Abdelwahed, M. F. & Zhao, D. 2007. Deep structure of the Japan subduction zone. *Physics of the Earth and Planetary Interiors*, **162**, 32–52.

Abe, Y., Ohkura, T., Hirahara, K. & Shibutani, T. 2011. Water transportation through the Philippine Sea slab subducting beneath the central Kyushu region, Japan, as derived from receiver function analyses. *Geophysical Research Letters*, **38**, L23305, http://doi.org/10.1029/2011GL049688

Ai, Y., Zheng, T., Xu, W., He, Y. & Dong, D. 2003. A complex 660 km discontinuity beneath northeast China. *Earth and Planetary Science Letters*, **212**, 63–71.

Aki, K. & Lee, W. 1976. Determination of three-dimensional velocity anomalies under a seismic array using first P arrival times from local earthquakes, 1. A homogeneous initial model. *Journal of Geophysical Research*, **81**, 4381–4399.

Aki, K., Christoffersson, A. & Husebye, E. 1977. Determination of the three-dimensional seismic structure of the lithosphere. *Journal of Geophysical Research*, **82**, 277–296.

Albers, M. & Christensen, U. 1996. The excess temperature of plumes rising from the core-mantle boundary. *Geophysical Research Letters*, **23**, 3567–3570.

Arcay, D., Doin, M.-P., Tric, E., Bousquet, R. & de Capitan, C. 2006. Overriding plate thinning in subduction zones: localized convection induced by slab dehydration. *Geochemistry, Geophysics, Geosystems*, **7**, Q02007, http://doi.org/10.1029/2005GC001061

Aso, N. & Ide, S. 2013. Focal mechanisms of deep low-frequency earthquakes in eastern Shimane in western Japan. *Journal of Geophysical Research*, **119**, 364–377, http://doi.org/10.1002/2013JB010681

Aso, N., Ohta, K. & Ide, S. 2013. Tectonic, volcanic, and semi-volcanic deep low-frequency earthquakes in western Japan. *Tectonophysics*, **600**, 27–40.

Baba, T., Tanioka, Y., Cummins, P. R. & Uhira, K. 2002. The slip distribution of the 1946 Nankai earthquake estimated from tsunami inversion using a new plate model. *Physics of the Earth and Planetary Interiors*, **132**, 59–73.

Bijwaard, H., Spakman, W. & Engdahl, E. 1998. Closing the gap between regional and global travel time tomography. *Journal of Geophysical Research*, **103**, 30 055–30 078.

Bird, P. 2003. An updated digital model of plate boundaries. *Geochemistry, Geophysics, Geosystems*, **4**, 1027.

Broadband Seismic Network Laboratory, 2008. *National Research Institute for Earth Science and Disaster Prevention*, http://www.fnet.bosai.go.jp/top.php.

Cagnioncle, A. M., Parmentier, E. M. & Elkins-Tanton, L. T. 2007. Effect of solid flow above a subducting slab on water distribution and melting at convergent plate boundaries. *Journal of Geophysical Research*, **112**, B09402, http://doi.org/10.1029/2007JB004934

Chen, Y., Zhang, Y., Graham, D., Su, S. & Deng, J. 2007. Geochemistry of Cenozoic basalts and mantle xenoliths in Northeast China. *Lithos*, **96**, 108–126.

Conder, J. & Wiens, D. 2006. Seismic structure beneath the Tonga arc and Lau back-arc basin determined from joint Vp, Vp/Vs tomography. *Geochemistry, Geophysics, Geosystems*, **7**, Q03018.

DeMets, C., Gordon, R. G., Argus, D. F. & Stein, S. 1994. Effect of recent revisions to the geomagnetic reversal time scale on estimates of current plate motions. *Geophysical Research Letters*, **21**, 2192–2194.

Duan, Y., Zhao, D., Zhang, X. & Xia, S. 2009. Seismic structure and origin of active intraplate volcanoes in Northeast Asia. *Tectonophysics*, **470**, 257–266.

Eberle, M. A., Grasset, O. & Sotin, C. 2002. A numerical study of the interaction between the mantle wedge, subducting slab, and overriding plate. *Physics of the Earth and Planetary Interiors*, **134**, 191–202.

England, P., Engdahl, R. & Thatcher, W. 2004. Systematic variation in the depths of slabs beneath arc volcanoes. *Geophysical Journal International*, **156**, 377–408.

Faccenda, M., Gerya, T. V. & Burlini, L. 2009. Deep slab hydration induced by bending-related variations in tectonic pressure. *Nature Geoscience*, **2**, 790–793, http://doi.org/10.1038/ngeo656

Faccenna, C., Becker, T., Lallemand, S., Lagabrielle, Y., Funiciello, F. & Piromallo, C. 2010. Subduction-triggered magmatic pulses: a new class of plumes? *Earth and Planetary Science Letters*, **299**, 54–68.

Forte, A. M., Mitrovica, J. X., Moucha, R., Simmons, N. A. & Grand, S. P. 2007. Descent of the ancient Farallon slab drives localized mantle flow below the New Madrid seismic zone. *Geophysical Research Letters*, **34**, L04308, http://doi.org/10.1029/2006GL027895

Fukao, Y., Hori, S. & Ukawa, M. 1983. A seismological constraint on the depth of the basalt–eclogite transition in a subducting oceanic crust. *Nature*, **303**, 413–415.

Fukao, Y., Widiyantoro, S. & Obayashi, M. 2001. Stagnant slabs in the upper and lower mantle transition region. *Reviews of Geophysics*, **39**, 291–323.

Gamage, S., Umino, N., Hasegawa, A. & Kirby, S. 2009. Offshore double-planed shallow seismic zone in the NE Japan forearc region revealed by sP depth phases recorded by regional networks. *Geophysical Journal International*, **178**, 195–214.

Green, H. 2003. Tiny triggers deep down. *Nature*, **424**, 893–894.

Green, H. & Burnley, P. 1989. A new self-organizing mechanism for deep-focus earthquakes. *Nature*, **341**, 733–737.

Green, H. W. & Houston, H. 1995. The mechanisms of deep earthquakes. *Annual Review of Earth and Planetary Sciences*, **23**, 169–213.

Gu, Y. J., Okeler, A. & Schultz, R. 2012. Tracking slabs beneath northwestern Pacific subduction zones. *Earth and Planetary Science Letters*, **331–332**, 269–280.

Hacker, B. R., Abers, G. A. & Peacock, S. M. 2003. Subduction factory 1. Theoretical mineralogy, densities, seismic wave speeds, and H_2O contents. *Journal of Geophysical Research*, **108**, 2029, http://doi.org/10.1029/2001JB001127

Hall, R., Ali, J., Anderson, C. & Baker, S. 1995. Origin and motion history of the Philippine Sea plate. *Tectonophysics*, **251**, 229–250.

Hasegawa, A. & Nakajima, J. 2004. Geophysical constraints on slab subduction and arc magmatism. *In*: Sparks, R. S. J. & Hawkesworth, C. J. (eds) *State of the Planet: Frontiers and Challenges in Geophysics*. AGU, Washington, Geophysical Monograph, **150**, 81–94.

Hasegawa, A. & Yamamoto, A. 1994. Deep, low-frequency microearthquakes in or around seismic low-velocity zones beneath active volcanoes in northeastern Japan. *Tectonophysics*, **233**, 233–252.

Hasegawa, A., Umino, N. & Takagi, A. 1978*a*. Double-planed structure of the deep seismic zone in the northeastern Japan arc. *Tectonophysics*, **47**, 43–58.

Hasegawa, A., Umino, N. & Takagi, A. 1978*b*. Double-planed deep seismic zone and upper mantle structure in the northeastern Japan arc. *Geophysical Journal of the Royal Astronomical Society*, **54**, 281–296.

Hasegawa, A., Zhao, D., Hori, S., Yamamoto, A. & Horiuchi, S. 1991. Deep structure of the northeastern Japan arc and its relationship to seismic and volcanic activity. *Nature*, **352**, 683–689.

Hasegawa, A., Horiuchi, S. & Umino, N. 1994. Seismic structure of the northeastern Japan convergent margin: a synthesis. *Journal of Geophysical Research*, **99**, 22 295–22 311.

Hasegawa, A., Yamamoto, A., Umino, N., Miura, S., Horiuchi, S., Zhao, D. & Sato, H. 2000. Seismic activity and deformation process of the crust within the overriding plate in the northeastern Japan subduction zone. *Tectonophysics*, **319**, 225–239.

Hasegawa, A., Nakajima, J., Umino, N. & Miura, S. 2005. Deep structure of the northeastern Japan arc and its implications for crustal deformation and shallow seismic activity. *Tectonophysics*, **403/1–4**, 59–75.

Hasegawa, A., Nakajima, J., Kita, S., Okada, T., Matsuzawa, T., Kirby, S. 2007. Anomalous deepening of a belt of intraslab earthquakes in the Pacific slab crust under Kanto, central Japan: possible anomalous thermal shielding, dehydration reactions, and seismicity caused by shallower cold slab material. *Geophysical Research Letters*, **34**, L09305, http://doi.org/10.1029/2007GL029616

Hasegawa, A., Nakajima, J., Uchida, N., Okada, T., Zhao, D., Matsuzawa, T. & Umino, N. 2009. Plate subduction, and generation of earthquakes and magmas in Japan as inferred from seismic observation: an overview. *Gondwana Research*, **16**, 370–400.

Hata, M., Oshiman, N., Yoshimura, R., Tanaka, Y. & Uyeshima, M. 2012. Fluid upwelling beneath arc volcanoes above the subducting Philippine Sea Plate: evidence from regional electrical resistivity structure. *Journal of Geophysical Research*, **117**, B07203, http://doi.org/10.1029/2011JB009109

Heki, K. & Miyazaki, S. 2001. Plate convergence and long-term crustal deformation in Central Japan. *Geophysical Research Letters*, **28**, 2313–2316.

Helffrich, G. 2000. Topography of the transition zone seismic discontinuities. *Review of Geophysics*, **38**, 141–158.

Hilde, T. & Lee, C. 1984. Origin and evolution of the west Philippine basin: a new interpretation. *Tectonophysics*, **102**, 85–104.

Hirahara, K. 1981. Three-dimensional seismic structure beneath southwest Japan: the subducting Philippine Sea plate. *Tectonophysics*, **79**, 1–44.

Hiramatsu, Y., Ando, M., Tsukuda, T. & Ooida, T. 1998. Three-dimensional image of the anisotropic bodies beneath central Honshu, Japan. *Geophysical Journal International*, **135**, 801–816.

Hirano, N., Takahashi, E. et al. 2006. Volcanism in response to plate flexure. *Science*, **313**, 1426–1428.

Hirano, N., Abe, N., Morishita, T., Tamura, A. & Arai, S. 2013. Petit-spot lava fields off the central Chile trench induced by plate flexure. *Geochemical Journal*, **47**, 249–257.

Hirose, F., Nakajima, J. & Hasegawa, A. 2008*a*. Three-dimensional seismic velocity structure and configuration of the Philippine Sea slab in southwestern Japan estimated by double-difference tomography. *Journal of Geophysical Research*, **13**, B09315, http://doi.org/10.1029/2007JB005274

Hirose, F., Nakajima, J. & Hasegawa, A. 2008*b*. Three-dimensional velocity structure and configuration of the PHS slab beneath Kanto district, central Japan, estimated by double-difference tomography. *Journal of the Seismological Society of Japan*, **60**, 123–138 [in Japanese with English abstract].

Hobbs, B. & Ord, A. 1988. Plastic instabilities: implication for the origin of intermediate and deep focus earthquakes. *Journal of Geophysical Research*, **93**, 10521–10540.

Honda, S. & Nakanishi, I. 2003. Seismic tomography of the uppermost mantle beneath southwestern Japan: seismological constraints on modeling subduction and magmatism for the Philippine Sea slab. *Earth, Planets and Space*, **55**, 443–462.

Honda, S., Morishige, M. & Orihashi, Y. 2007. Sinking hot anomaly trapped at the 410 km discontinuity near the Honshu subduction zone, Japan. *Earth and Planetary Science Letters*, **261**, 565–577.

Hori, S. 2006. Seismic activity associated with the subducting motion of the Philippine Sea plate beneath the Kanto district, Japan. *Tectonophysics*, **417**, 85–100.

Hori, S., Inoue, H., Fukao, Y. & Ukawa, M. 1985. Seismic detection of the untransformed 'basaltic' oceanic crust subducting into the mantle. *Geophysical Journal of the Royal Astronomical Society*, **83**, 169–197.

Hori, S., Umino, N., Kono, T. & Hasegawa, A. 2004. Distinct S-wave reflectors (bright spots) extensively distributed in the crust and upper mantle beneath the northeastern Japan arc. *Zisin*, **56**, 435–446 [in Japanese with English abstract].

Huang, J. & Zhao, D. 2004. Crustal heterogeneity and seismotectonics of the region around Beijing, China. *Tectonophysics*, **385**, 159–180.

Huang, J. & Zhao, D. 2006. High-resolution mantle tomography of China and surrounding regions. *Journal of Geophysical Research*, **111**, B09305.

Huang, J. & Zhao, D. 2009. Seismic imaging of the crust and upper mantle under Beijing and surrounding regions. *Physics of the Earth and Planetary Interiors*, **173**, 330–348.

Huang, Z., Wang, L., Zhao, D., Mi, N. & Xu, M. 2011*a*. Seismic anisotropy and mantle dynamics beneath China. *Earth and Planetary Science Letters*, **306**, 105–117.

Huang, Z., Zhao, D. & Wang, L. 2011*b*. Frequency dependent shear-wave splitting and multilayer anisotropy in northeast Japan. *Geophysical Research Letters*, **38**, L08302, http://doi.org/10.1029/2011GL046804

Huang, Z., Zhao, D., Hasegawa, A., Umino, N., Park, J. & Kang, I. 2013. Aseismic deep subduction of the Philippine Sea plate and slab window. *Journal of Asian Earth Sciences*, **75**, 82–94.

Ichiki, M., Baba, K., Obayashi, M. & Utada, H. 2006. Water content and geotherm in the upper mantle above the stagnant slab: interpretation of electrical conductivity and seismic P-wave velocity models. *Physics of the Earth and Planetary Interiors*, **155**, 1–15.

Ide, S., Shiomi, K., Mochizuki, K., Tonegawa, T. & Kimura, G. 2010. Split Philippine Sea plate beneath Japan. *Geophysical Research Letters*, **37**, L21304, http://doi.org/10.1029/2010GL044585

Iidaka, T. & Suetsugu, D. 1992. Seismological evidence for metastable olivine inside a subducting slab. *Nature*, **356**, 593–595.

Iidaka, T., Mizoue, M. et al. 1990. The upper boundary of the Philippine Sea plate beneath the western Kanto region estimated from SP converted waves. *Tectonophysics*, **79**, 321–326.

Iio, Y. & Kobayashi, Y. 2002. A physical understanding of large interplate earthquakes. *Earth, Planets and Space*, **54**, 1001–1004.

Iio, Y., Sagiya, T., Kobayashi, Y. & Shiozaki, I. 2002. Water-weakened lower crust and its role in the concentrated deformation in the Japanese Islands. *Earth and Planetary Science Letters*, **203**, 245–253.

Inoue, T., Tanimoto, Y., Irifune, T., Suzuki, T., Fukui, H. & Ohtaka, O. 2004. Thermal expansion of wadsleyite, ringwoodite, hydrous wadsleyite and hydrous ringwoodite. *Physics of the Earth and Planetary Interiors*, **143**, 279–290.

Ishida, M. 1992. Geometry and relative motion of the Philippine Sea plate and Pacific plate beneath the Kanto-Tokai district, Japan. *Journal of Geophysical Research*, **97**, 453–489.

Ishise, M. & Oda, H. 2005. Three-dimensional structure of P-wave anisotropy beneath the Tohoku district, northeast Japan. *Journal of Geophysical Research*, **110**, B07304, http://doi.org/10.1029/2004JB003599

Ishise, M., Koketsu, K. & Miyake, H. 2009. Slab segmentation revealed by anisotropic P-wave tomography. *Geophysical Research Letters*, **36**, L08308, http://doi.org/10.1029/2009GL037749

Ito, K. 1993. Cutoff depth of seismicity and large earthquakes near active volcanoes in Japan. *Tectonophysics*, **217**, 11–21.

Iwamori, H. 1992. Degree of melting and source composition of Cenozoic basalts in southwest Japan: evidence for mantle upwelling by flux melting. *Journal of Geophysical Research*, **97**, 10 983–10 995.

Iwamori, H. 1998. Transportation of H_2O and melting in subduction zones. *Earth and Planetary Science Letters*, **160**, 65–80.

Iwamori, H. 2000. Deep subduction of H_2O and deflection of volcanic chain towards backarc near triple junction due to lower temperature. *Earth and Planetary Science Letters*, **181**, 41–46.

Jiang, G. & Zhao, D. 2011. Metastable olivine wedge in the subducting Pacific slab and its relation to deep earthquakes. *Journal of Asian Earth Sciences*, **42**, 1411–1423.

Jiang, G., Zhao, D. & Zhang, G. 2008. Seismic evidence for a metastable olivine wedge in the subducting Pacific slab under Japan Sea. *Earth and Planetary Science Letters*, **270**, 300–307.

Kameyama, M. & Nishioka, R. 2012. Generation of ascending flows in the Big Mantle Wedge (BMW) beneath northeast Asia induced by retreat and stagnation of subducted slab. *Geophysical Research Letters*, **39**, L10309.

Kanamori, H. 1971. Great earthquakes at island arcs and the lithosphere. *Tectonophysics*, **12**, 187–198.

Kanamori, H. 1986. Rupture process of subduction-zone earthquakes. *Annual Review of Earth and Planetary Sciences*, **14**, 293–322.

Kanamori, H. & Press, F. 1970. How thick is the lithosphere? *Nature*, **226**, 330–331, http://doi.org/10.1038/226330a0

Kaneshima, S., Okamoto, T. & Takenaka, H. 2007. Evidence for a metastable olivine wedge inside the subducted Mariana slab. *Earth and Planetary Science Letters*, **258**, 219–227.

Karato, S., Riedel, M. & Yuen, D. 2001. Rheological structure and deformation of subducted slabs in the mantle transition zone: implications for mantle circulation and deep earthquakes. *Physics of the Earth and Planetary Interiors*, **127**, 83–108.

Karig, D. 1971. Origin and development of marginal basins in the western Pacific. *Journal of Geophysical Research*, **76**, 2542–2561.

Katayama, I. 2009. Thin anisotropic layer in the mantle wedge beneath northeast Japan. *Geology*, **37**, 211–214.

Kato, A., Sakai, S., Iidaka, T., Iwasaki, T. & Hirata, N. 2010. Non-volcanic seismic swarms triggered by circulating fluids and pressure fluctuations above a solidified diorite intrusion. *Geophysical Research Letters*, **37**, L15302, http://doi.org/10.1029/2010GL043887

Katsumata, A. & Kamaya, N. 2003. Low-frequency continuous tremor around the Moho discontinuity away from volcanoes in the southwest Japan. *Geophysical Research Letters*, **30**, http://doi.org/10.1029/2002GL015981

Kawakatsu, H. & Watada, S. 2007. Seismic evidence for deep water transportation in the mantle. *Science*, **316**, 1468–1471.

Kawakatsu, H., Kumar, P., Takei, Y., Shinohara, M., Kanazawa, T., Araki, E. & Suyehiro, K. 2009. Seismic evidence for sharp lithosphere-asthenosphere boundaries of oceanic plates. *Science*, **324**, 499–502.

Kennett, B. & Engdahl, E. 1991. Travel times for global earthquake location and phase identification. *Geophysical Journal International*, **105**, 426–465.

Kimura, J., Nagao, T. et al. 2003. Late Cenozoic volcanic activity in the Chugoku area, southwest Japan arc during back arc basin opening and re-initiation of subduction. *Island Arc*, **12**, 22–45.

Kirby, S. 1991. Mantle phase changes and deep-earthquake faulting in subducting lithosphere. *Science*, **252**, 216–224.

Kirby, S., Stein, S., Okal, E. & Rubie, D. 1996. Metastable mantle transformations and deep earthquakes in subducting oceanic lithosphere. *Review of Geophysics*, **34**, 261–306.

Kita, S., Okada, T., Nakajima, J., Matsuzawa, T. & Hasegawa, A. 2006. Existence of a seismic belt in the upper plane of the double seismic zone extending in the along-arc direction at depths of 70–100 km beneath NE Japan. *Geophysical Research Letters*, **33**, http://doi.org/10.1029/2006GL028239

Kita, S., Okada, T., Hasegawa, A., Nakajima, J. & Matsuzawa, T. 2010. Anomalous deepening of a seismic belt in the upper-plane of the double seismic zone in the Pacific slab beneath the Hokkaido corner: possible evidence for thermal shielding caused by subducted forearc crust materials. *Earth and Planetary Science Letters*, **290**, 415–426.

KODAIRA, S., TAKAHASHI, N., PARK, J., MOCHIZUKI, K., SHINOHARA, M. & KIMURA, S. 2000. Western Nankai Trough seismogenic zone: results from a wide-angle ocean bottom seismic survey. *Journal of Geophysical Research*, **105**, 5887–5905.

KODAIRA, S., IIDAKA, T., KATO, A., PARK, J., IWASAKI, T. & KANEDA, Y. 2004. High pore fluid pressure may cause silent slip in the Nankai Trough. *Science*, **304**, 1295–1298.

KOMABAYASHI, T., OMORI, S. & MARUYAMA, S. 2004. Petrogenetic grid in the system MgO-SiO2-H2O up to 30 GPa, 1600 C: applications to hydrous peridotite subducting into the Earth's deep interior. *Journal of Geophysical Research*, **109**, B03206.

KONDO, H., KANEKO, K., TANAKA, K. 1998. Characterization of spatial and temporal distribution of volcanoes since 14 Ma in the Northeast Japan arc. *Bulletin of the Volcanological Society of Japan*, **43**, 173–180.

KOPER, K., WIENS, D., DORMAN, L., HILDEBRAND, J. & WEBB, S. 1998. Modeling the Tonga slab: can travel time data resolve a metastable olivine wedge? *Journal of Geophysical Research*, **103**, 30 079–30 100.

KUBO, T., KANESHIMA, S., TORII, Y. & YOSHIOKA, S. 2009. Seismological and experimental constraints on metastable phase transformations and rheology of the Mariana slab. *Earth and Planetary Science Letters*, **287**, 12–23.

KUMAR, P. & KAWAKATSU, H. 2011. Imaging the seismic lithosphere-asthenosphere boundary of the oceanic plate. *Geochemistry, Geophysics, Geosystems*, **12**, Q01006, http://doi.org/10.1029/2010GC003358

KURASHIMO, E., TOKUNAGA, M. ET AL. 2002. Geometry of the subducting Philippine Sea plate and the crustal and upper mantle structure beneath eastern Shikoku island revealed by seismic refraction/wide-angle reflection profiling. *Zisin*, **54**, 489–505 [in Japanese with English abstract].

KURITANI, T., KIMURA, J. ET AL. 2009. Intraplate magmatism related to deceleration of upwelling asthenospheric mantle: implications from the Changbaishan shield basalts, northeast China. *Lithos*, **112**, 247–258.

LEI, J. & ZHAO, D. 2005. P-wave tomography and origin of the Changbai intraplate volcano in Northeast Asia. *Tectonophysics*, **397**, 281–295.

LI, J., WANG, X., WANG, X. & YUEN, D. 2013. P and SH velocity structure in the upper mantle beneath Northeast China: evidence for a stagnant slab in hydrous mantle transition zone. *Earth and Planetary Science Letters*, **367**, 71–81.

LIU, K., GAO, S., GAO, Y. & WU, J. 2008. Shear wave splitting and mantle flow associated with the deflected Pacific slab beneath northeast Asia. *Journal of Geophysical Research*, **113**, B01305.

MARUYAMA, S. 1994. Plume tectonics. *Journal of the Geological Society of Japan*, **100**, 24–49.

MATSUBARA, M., OBARA, K. & KASAHARA, K. 2008. Three-dimensional P- and S-wave velocity structures beneath the Japan Islands obtained by high-density seismic stations by seismic tomography. *Tectonophysics*, **454**, 86–103.

MATSUZAWA, T., UMINO, N., HASEGAWA, A. & TAKAGI, A. 1986. Upper mantle velocity structure estimated from PS-converted wave beneath the northeastern Japan arc. *Geophysical Journal of the Royal Astronomical Society*, **86**, 767–787.

MATSUZAWA, T., KONO, T., HASEGAWA, A. & TAKAGI, A. 1990. Subducting plate boundary beneath the northeastern Japan arc estimated from SP converted waves. *Tectonophysics*, **181**, 123–133.

MAZZOTTI, S., HENRY, P., PICHON, X. L. & SAGIYA, T. 1999. Strain partitioning in the zone of transition from Nankai subduction to Izu–Bonin collision (central Japan): implications for an extensional tear within the subducting slab. *Earth and Planetary Science Letters*, **172**, 1–10.

MIURA, M., SATO, T., HASEGAWA, A., SUWA, Y., TACHIBANA, K. & YUI, S. 2004. Strain concentration zone along the volcanic front derived by GPS observations in NE Japan arc. *Earth, Planets and Space*, **56**, 1347–1355.

MIYAZAKI, S. & HEKI, K. 2001. Crustal velocity field of Southwest Japan: subduction and arc–arc collision. *Journal of Geophysical Research*, **106**, 4305–4326.

MORGAN, W. 1971. Convective plumes in the lower mantle. *Nature*, **230**, 42–43.

NAKAJIMA, J. & HASEGAWA, A. 2003. Estimation of thermal structure in the mantle wedge of northeastern Japan from seismic attenuation data. *Geophysical Research Letters*, **30**, http://doi.org/10.1029/2003GL017185

NAKAJIMA, J. & HASEGAWA, A. 2004. Shear-wave polarization anisotropy and subduction-induced flow in the mantle wedge of northeastern Japan. *Earth and Planetary Science Letters*, **225**, 365–377.

NAKAJIMA, J. & HASEGAWA, A. 2007*a*. Subduction of the Philippine Sea plate beneath southwestern Japan: slab geometry and its relationship to arc magmatism. *Journal of Geophysical Research*, **112**, B08306, http://doi.org/10.1029/2006JB004770

NAKAJIMA, J. & HASEGAWA, A. 2007*b*. Tomographic evidence for the mantle upwelling beneath southwestern Japan and its implications for arc magmatism. *Earth and Planetary Science Letters*, **254**, 90–105.

NAKAJIMA, J. & HASEGAWA, A. 2007*c*. Deep crustal structure along the Niigata-Kobe Tectonic Zone, Japan: its origin and segmentation. *Earth Planets and Space*, **59**, e5–e8.

NAKAJIMA, J. & HASEGAWA, A. 2008. Existence of low-velocity zones under the source areas of the 2004 Niigata-Chuetsu and 2007 Niigata-Chuetsu-Oki earthquakes inferred from travel-time tomography. *Earth, Planets and Space*, **60**, 1127–1130.

NAKAJIMA, J., MATSUZAWA, T., HASEGAWA, A. & ZHAO, D. 2001. Three-dimensional structure of Vp, Vs, and Vp/Vs beneath northeastern Japan: implications for arc magmatism and fluids. *Journal of Geophysical Research*, **106**, 843–857.

NAKAJIMA, J., TAKEI, Y. & HASEGAWA, A. 2005. Quantitative analysis of the inclined low-velocity zone in the mantle wedge of northeastern Japan: a systematic change of melt-filled pore shapes with depth and its implications for melt migration. *Earth and Planetary Science Letters*, **234**, 59–70.

NAKAJIMA, J., SHIMIZU, J., HORI, S. & HASEGAWA, A. 2006. Shear-wave splitting beneath the southwestern Kurile arc and northeastern Japan arc: a new insight into mantle return flow. *Geophysical Research Letters*, **33**, L05305, http://doi.org/10.1029/2005GL025053

NAKAJIMA, J., HIROSE, F. & HASEGAWA, A. 2009*a*. Seismotectonics beneath the Tokyo metropolitan area, Japan: effect of slab-slab contact and overlap on seismicity. *Journal of Geophysical Research*, **114**, B08309, http://doi.org/10.1029/2008JB006101

NAKAJIMA, J., TSUJI, Y. & HASEGAWA, A. 2009*b*. Seismic evidence for thermally-controlled dehydration reaction in subducting oceanic crust. *Geophysical Research Letters*, **36**, L03303, http://doi.org/10.1029/2008GL036865

NAKAJIMA, J., TSUJI, Y., HASEGAWA, A., KITA, S., OKADA, T. & MATSUZAWA, T. 2009*c*. Tomographic imaging of hydrated crust and mantle in the subducting Pacific slab beneath Hokkaido, Japan: evidence for dehydration embrittlement as a cause of intraslab earthquakes. *Gondwana Research*, **16**, 470–481.

NAKAJIMA, J., HADA, S. ET AL. 2013. Seismic attenuation beneath northeastern Japan: constraints on mantle dynamics and arc magmatism. *Journal of Geophysical Research: Solid Earth*, **118**, 5838–5855, http://doi.org/10.1002/2013JB010388

NAKAMURA, H., IWAMORI, H. & KIMURA, J. 2008. Geochemical evidence for enhanced fluid flux due to overlapping subducting plates. *Nature Geoscience*, **1**, 380–384.

NAKAMURA, M., YOSHIDA, Y., ZHAO, D., KATAO, H. & NISHIMURA, S. 2003. Three-dimensional P- and S-wave velocity structures beneath the Ryukyu arc. *Tectonophysics*, **369**, 121–143.

NAKANISHI, I. 1980. Precursors to ScS phases and dipping interface in the upper mantle beneath southwestern Japan. *Tectonophysics*, **69**, 1–35.

NOGUCHI, S. 1996. Geometry of the Philippine Sea slab and the convergent tectonics in the Tokai district, Japan. *Zisin*, **49**, 295–325 [in Japanese with English abstract].

NUGRAHA, A. D., OHMI, S., MORI, J. & SHIBUTANI, T. 2013. High resolution seismic velocity structure around the Yamazaki fault zone of southwest Japan as revealed from travel-time tomography. *Earth, Planets and Space*, **65**, 871–881.

OBANA, K., FUJIE, G. ET AL. 2012. Normal-faulting earthquakes beneath the outer slope of the Japan Trench after the 2011 Tohoku earthquake: implications for the stress regime in the incoming Pacific plate. *Geophysical Research Letters*, **39**, L00G24.

OBARA, K. 2002. Nonvolcanic deep tremor associated with subduction in southwest Japan. *Science*, **296**, 1679–1681, http://doi.org/10.1126/science.1070378

OBAYASHI, M., SUGIOKA, S., YOSHIMITSU, J. & FUKAO, Y. 2006. High temperature anomalies oceanward of subducting slabs at the 410-km discontinuity. *Earth Planetary Science Letters*, **243**, 149–158.

OHMI, S. & OBARA, K. 2002. Deep low-frequency earthquakes beneath the focal region of the Mw 6.7 2000 Western Tottori earthquake. *Geophysical Research Letters*, **29**, 1807, http://doi.org/10.1029/2001GL014469

OHTANI, E. & LITASOV, K. 2006. Effect of water on mantle phase transitions. *Reviews in Mineralogy and Geochemistry*, **62**, 397–420.

OHTANI, E. & ZHAO, D. 2009. The role of water in the deep upper mantle and transition zone: dehydration of stagnant slabs and its effects on the big mantle wedge. *Russian Geology and Geophysics*, **50**, 1073–1078.

OHTANI, E., LITASOV, K., HOSOYA, T., KUBO, T. & KONDO, T. 2004. Water transport into the deep mantle and formation of a hydrous transition zone. *Physics of the Earth and Planetary Interiors*, **143**, 255–269.

OKINO, K., SHIMAKAWA, Y. & NAGAOKA, S. 1994. Evolution of the Shikoku Basin. *Journal of Geomagnetism and Geoelectricity*, **46**, 463–279.

OMORI, S., KOMABAYASHI, T. & MARUYAMA, S. 2004. Dehydration and earthquakes in the subducting slab: empirical link in intermediate and deep seismic zones. *Physics of the Earth and Planetary Interiors*, **146**, 297–311.

OMURALIEVA, A. M., HASEGAWA, A., MATSUZAWA, T., NAKAJIMA, J. & OKADA, T. 2012. Lateral variation of the cutoff depth of shallow earthquakes beneath the Japan Islands and its implications for seismogenesis. *Tectonophysics*, **518–521**, 93–105.

PEACOCK, S. M. & WANG, K. 1999. Seismic consequences of warm v. cool subduction metamorphism: examples from southwest and northeast Japan. *Science*, **286**, 937–939.

RALEIGH, C. B. & PATERSON, M. S. 1965. Experimental deformation of serpentinite and its tectonic implications. *Journal of Geophysical Research*, **70**, 3965–3985.

RANERO, C. R., MORGAN, J. P., MCINTOSH, K. & REICHERT, C. 2003. Bending related faulting and mantle serpentinization at the Middle America trench. *Nature*, **425**, 367–373, http://doi.org/10.1038/nature01961

REYNARD, B., NAKAJIMA, J. & KAWAKATSU, H. 2010. Earthquakes and plastic deformation of anhydrous slab mantle in double Wadati-Benioff zones. *Geophysical Research Letters*, **37**, L24309, http://doi.org/10.1029/2010GL045494

SADEGHI, H., SUZUKI, S. & TAKENAKA, H. 2000. Tomographic low-velocity anomalies in the uppermost mantle around the northeastern edge of Okinawa Trough, the backarc of Kyushu. *Geophysical Research Letters*, **27**, 277–280.

SAGIYA, T., MIYAZAKI, S. & TADA, T. 2000. Continuous GPS array and present day crustal deformation of Japan. *Pure and Applied Geophysics*, **157**, 2303–2322.

SAKUYAMA, T., NAGAOKA, S. ET AL. 2014. Melting of the uppermost metasomatized asthenosphere triggered by fluid fluxing from ancient subducted sediment: constraints from the Quaternary basalt lavas at Chugaryeong volcano, Korea. *Journal of Petrology*, **55**, 499–528.

SANO, Y. & NAKAJIMA, J. 2008. Geographical distribution of 3Hc/ 4Hc ratios and seismic tomography in Japan. *Geochemical Journal*, **42**, 51–60.

SANO, Y. & WAKITA, H. 1985. Geographical distribution of 3He/4He ratios in Japan: implications for arc tectonics and incipient magmatism. *Journal of Geophysical Research*, **90**, 8729–8741.

SANO, Y., TAKAHARA, N. & SENO, T. 2006. Geographical distribution of 3He/4He in the Chugoku district, southwestern Japan. *Pure and Applied Geophysics*, **163**, 745–757.

SATO, H., HIRATA, N. ET AL. 2005. Earthquake source fault beneath Tokyo. *Science*, **309**, 462–464.

SEKIGUCHI, S. 2001. A new configuration and a seismic slab of the descending Philippine Sea plate revealed by seismic tomography. *Tectonophysics*, **341**, 19–32.

SENO, T. & MARUYAMA, S. 1984. Paleogeographic reconstruction and origin of the Philippine Sea. *Tectonophysics*, **102**, 53–84.

SENO, T. & YAMANAKA, Y. 1996. Double seismic zones, compressional deep trench-outer rise events and superplumes. *In*: BEBOUT, G. E., SCHOLL, D. W., KIRBY, S. H. & PLATT, J. P. (eds) *Subduction Top to Bottom*. AGU, Washington, Geophysical Monograph **96**, 347–355.

SENO, T., STEIN, S. & GRIPP, A. E. 1993. A model for the motion of the Philippine Sea plate consistent with NUVEL-1 and geological data. *Journal of Geophysical Research*, **98**, 17 941–17 948.

SENO, T., SAKURAI, T. & STEIN, S. 1996. Can the Okhotsk plate be discriminated from the North American plate? *Journal of Geophysical Research*, **101**, 11 305–11 315.

SENO, T., ZHAO, D., KOBAYASHI, Y. & NAKAMURA, M. 2001. Dehydration of serpentinized slab mantle: seismic evidence from southwest Japan. *Earth, Planets and Space*, **53**, 861–871.

SHELLY, D. R., BEROZA, G. C., ZHANG, H., THURBER, C. H. & IDE, S. 2006. High-resolution subduction zone seismicity and velocity structure beneath Ibaraki Prefecture, Japan. *Journal of Geophysical Research*, **111**, B06311, http://doi.org/10.1029/2005JB004081

SHIOMI, K. & PARK, J. 2008. Structural features of the subducting slab beneath the Kii Peninsula, central Japan: seismic evidence of slab segmentation, dehydration, and anisotropy. *Journal of Geophysical Research*, **113**, B10318, http://doi.org/10.1029/2007JB005535

SHITO, A. & SHIBUTANI, T. 2004. Nature of heterogeneity of the upper mantle beneath the northern Philippine Sea as inferred from attenuation ad velocity tomography. *Physics of the Earth and Planetary Interiors*, **140**, 331–341.

SUZUKI, S., MOTOYA, Y. & SASATANI, T. 1983. Double seismic zone beneath the middle of Hokkaido, Japan, in the southwestern side of the Kurile arc. *Tectonophysics*, **96**, 59–76.

SYRACUSE, E. M. & ABERS, G. A. 2006. Global compilation of variations in slab depth beneath arc volcanoes and implications. *Geochemistry, Geophysics, Geosystems*, **7**, Q05017, http://doi.org/10.1029/2005GC001045

TAJIMA, F., KATAYAMA, I. & NAKAGAWA, T. 2009. Variable seismic structure near the 660 km discontinuity associated with stagnant slabs and geochemical implications. *Physics of the Earth and Planetary Interiors*, **172**, 183–198.

TAKEI, Y. 2002. Effect of pore geometry on Vp/Vs: from equilibrium geometry to crack. *Journal of Geophysical Research*, **107**, 2043, http://doi.org/10.1029/2001JB000522

TAMURA, Y., TATSUMI, Y., ZHAO, D., KIDO, Y. & SHUKUNO, H. 2002. Hot fingers in the mantle wedge: new insights into magma genesis in subduction zones. *Earth and Planetary Science Letters*, **197**, 105–116.

TANAKA, A., YAMANO, M., YANO, Y. & SASADA, M. 2004. Geothermal gradient and heat flow data in and around Japan (I): appraisal of heat flow from geothermal gradient data. *Earth, Planets and Space*, **56**, 1191–1194.

TATSUMI, Y. 1986. Formation of the volcanic front in subduction zones. *Geophysical Research Letters*, **13**, 717–720.

TATSUMI, Y., SAKUYAMA, M., FUKUYAMA, H. & KUSHIRO, I. 1983. Generation of arc basalt magmas and thermal structure of the mantle wedge in subduction zones. *Journal of Geophysical Research*, **88**, 5815–5825.

TSUJI, Y., NAKAJIMA, J. & HASEGAWA, A. 2008. Tomographic evidence for hydrated oceanic crust of the Pacific slab beneath northeastern Japan: implications for water transportation in subduction zones. *Geophysical Research Letters*, **35**, L14308, http://doi.org/10.1029/2008GL034461

TSUMURA, K. 1973. Microearthquake activity in the Kanto District. *Publications for the 50th Anniversary of the Great Kanto Earthquake, 1923 (in Japanese with English Abstract)*. Earthquake Research Institute, University of Tokyo, 67–87.

TSUMURA, N., MATSUMOTO, S., HORIUCHI, S. & HASEGAWA, A. 2000. Three-dimensional attenuation structure beneath the northeastern Japan arc estimated from spectra of small earthquakes. *Tectonophysics*, **319**, 241–260.

TURCOTTE, D. & SCHUBERT, G. 2001. *Geodynamics*, 2nd edn. Cambridge University Press, Cambridge.

UCHIDA, N., MATSUZAWA, T., IGARASHI, T. & HASEGAWA, A. 2003. Interplate quasistatic slip off Sanriku, NE Japan, estimated from repeating earthquakes. *Geophysical Research Letters*, **30**, http://doi.org/10.1029/2003 GL017452

UCHIDA, N., HASEGAWA, A., NAKAJIMA, J. & MATSUZAWA, T. 2009. What controls interplate coupling? Evidence for abrupt change in coupling across a border between two overlying plates in the NE Japan subduction zone. *Earth and Planetary Science Letters*, **283**, 111–121, http://doi.org/10.1016/j.epsl.2009.04.003

UKAWA, M. & OBARA, K. 1993. Low frequency earthquakes around Moho beneath the volcanic front in the Kanto district, central Japan. *Bulletin of the Volcanological Society of Japan*, **38**, 187–197.

UMINO, N., HASEGAWA, A. & TAKAGI, A. 1990. The relationship between seismicity patterns and fracture zones beneath northeastern Japan. *Tohoku Geophysical Journal*, **33**, 149–162.

UYEDA, S. & MIYASHIRO, A. 1974. Plate tectonics and the Japanese Islands: a synthesis. *Geological Society of America Bulletin*, **85**, 1159–1170.

VAN KEKEN, P. E., KIEFER, B. & PEACOCK, S. M. 2002. High-resolution models of subduction zones: implications for mineral dehydration reactions and the transport of water into the deep mantle. *Geochemistry, Geophysics, Geosystems*, **3**, 10, http://doi.org/10.1029/2001GC000256

WALD, D. J. & SOMERVILLE, P. G. 1995. Variable-slip rupture model of the great 1923 Kanto, Japan Earthquake: geodetic and body-waveform analysis. *Bulletin of the Seismological Society of America*, **85**, 159–177.

WANG, X., LI, Z. ET AL. 2012. Temperature, pressure, and composition of the mantle source region of Late Cenozoic basalts in Hainan Island, SE

Asia: a consequence of a young thermal mantle plume close to subduction zones? *Journal of Petrology*, **53**, 177–233.

Wang, Z. & Zhao, D. 2005. Seismic imaging of the entire arc of Tohoku and Hokkaido in Japan using P-wave, S-wave and sP depth-phase data. *Physics of the Earth and Planetary Interiors*, **152**, 144–162.

Wang, Z. & Zhao, D. 2006. Vp and Vs tomography of Kyushu, Japan: new insight into arc magmatism and forearc seismotectonics. *Physics of the Earth and Planetary Interiors*, **157**, 269–285.

Wang, J. & Zhao, D. 2008. P-wave anisotropic tomography beneath northeast Japan. *Physics of the Earth and Planetary Interiors*, **170**, 115–133.

Wang, Z., Huang, R., Huang, J. & He, Z. 2008. P-wave velocity and gradient images beneath the Okinawa Trough. *Tectonophysics*, **455**, 1–13.

Wei, D. & Seno, T. 1998. Determination of the Amurian plate motion, in mantle dynamics and plate interactions in East Asia. *Geodynamics Series*, **27**, 337–346.

Wei, W., Xu, J., Zhao, D. & Shi, Y. 2012. East Asia mantle tomography: new insight into plate subduction and intraplate volcanism. *Journal of Asian Earth Sciences*, **60**, 88–103.

Wickham, S. & Taylor, H. 1985. Stable isotopic evidence for large-scale seawater infiltration in a regional metamorphic terrane; the Trois Seigneurs Massif, Pyrenees, France. *Contributions to Mineralogy and Petrology*, **91**, 122–137.

Wiens, D. A., Conder, J. A. & Faul, U. H. 2008. The seismic structure and dynamics of the mantle wedge. *Annual Review of Earth and Planetary Sciences*, **36**, 421–455.

Wu, F., Okaya, D., Sato, H. & Hirata, N. 2007. Interaction between two subducting plates under Tokyo and its possible effects on seismic hazards. *Geophysical Research Letters*, **34**, L18301, http://doi.org/10.1029/2007GL030763

Xia, S., Zhao, D. & Qiu, X. 2008. Tomographic evidence for the subducting oceanic crust and forearc mantle serpentinization under Kyushu, Japan. *Tectonophysics*, **449**, 85–96.

Yamada, A. & Zhao, D. 2007. Mapping the 660 km discontinuity under Japan Islands using mantle reflected waves. *Earth Science Frontiers*, **14**, 47–56.

Yamasaki, T. & Seno, T. 2003. Double seismic zone and dehydration embrittlement of the subducting slab. *Journal of Geophysical Research*, **108**, http://doi.org/10.1029/2002JB001918

Yamazaki, F. & Ooida, T. 1985. Configuration of subducted Philippine Sea plate beneath the Chubu district, central Japan (in Japanese with English abstract). *Zisin*, **38**, 193–201.

Zhang, H. & Thurber, C. H. 2003. Double-difference tomography: the method and its application to the Hayward Fault, California. *Bulletin of the Seismological Society of America*, **93**, 1875–1889.

Zhang, H., Thurber, C. H., Shelly, D. R., Ide, S., Beroza, G. C. & Hasegawa, A. 2004. High resolution subducting-slab structure beneath northern Honshu, Japan, revealed by double-difference tomography. *Geology*, **32**, 361–364.

Zhao, D. 2004. Global tomographic images of mantle plumes and subducting slabs: insight into deep Earth dynamics. *Physics of the Earth and Planetary Interiors*, **146**, 3–34.

Zhao, D. 2009. Multiscale seismic tomography and mantle dynamics. *Gondwana Research*, **15**, 297–323.

Zhao, D. 2012. Tomography and dynamics of Western-Pacific subduction zones. *Monographs on Environment, Earth and Planets*, **1**, 1–70.

Zhao, D. & Tian, Y. 2013. Changbai intraplate volcanism and deep earthquakes in East Asia: a possible link? *Geophysical Journal International*, **195**, 706–724.

Zhao, D., Hasegawa, A. & Horiuchi, S. 1992. Tomographic imaging of P and S wave velocity structure beneath northeastern Japan. *Journal of Geophysical Research*, **97**, 19 909–19 928.

Zhao, D., Hasegawa, A. & Kanamori, H. 1994. Deep structure of Japan subduction zone as derived from local, regional and teleseismic events. *Journal of Geophysical Research*, **99**, 22 313–22 329.

Zhao, D., Kanamori, H., Negishi, H. & Wiens, D. 1996. Tomography of the source area of the 1995 Kobe earthquake: evidence for fluids at the hypocenter? *Science*, **274**, 1891–1894.

Zhao, D., Matsuzawa, T. & Hasegawa, A. 1997*a*. Morphology of the subducting slab boundary and its relationship to the interplate seismic coupling. *Physics of the Earth and Planetary Interiors*, **102**, 89–104.

Zhao, D., Xu, Y., Wiens, D., Dorman, L., Hildebrand, J. & Webb, S. 1997*b*. Depth extent of the Lau back-arc spreading center and its relation to subduction processes. *Science*, **278**, 254–257.

Zhao, D., Mishra, O. P. & Sanda, R. 2002. Influence of fluids and magma on earthquakes: seismological evidence. *Physics of the Earth and Planetary Interiors*, **132**, 249–267.

Zhao, D., Lei, J. & Tang, R. 2004. Origin of the Changbai intraplate volcanism in Northeast China: evidence from seismic tomography. *Chinese Science Bulletin*, **49**, 1401–1408.

Zhao, D., Maruyama, S. & Omori, S. 2007. Mantle dynamics of western Pacific to East Asia: new insight from seismic tomography and mineral physics. *Gondwana Research*, **11**, 120–131.

Zhao, D., Tian, Y., Lei, J., Liu, L. & Zheng, S. 2009. Seismic image and origin of the Changbai intraplate volcano in East Asia: role of big mantle wedge above the stagnant Pacific slab. *Physics of the Earth and Planetary Interiors*, **173**, 197–206.

Zhao, D., Santosh, M. & Yamada, A. 2010. Dissecting large earthquakes in Japan: role of arc magma and fluids. *Island Arc*, **19**, 4–16.

Zhao, D., Yu, S. & Ohtani, E. 2011. East Asia: seismotectonics, magmatism and mantle dynamics. *Journal of Asian Earth Sciences*, **40**, 689–709.

Zhao, D., Yanada, T., Hasegawa, A., Umino, N. & Wei, W. 2012. Imaging the subducting slabs and mantle upwelling under the Japan Islands. *Geophysical Journal International*, **190**, 816–828.

Zhu, G., Shi, Y. & Tackley, P. 2010. Subduction of the western Pacific plate underneath Northeast China: implications of numerical studies. *Physics of the Earth and Planetary Interiors*, **178**, 92–99.

Zou, H., Fan, Q. & Yao, Y. 2008. U-Th systematics of dispersed young volcanoes in NE China: asthenosphere upwelling caused by piling up and upward thickening of stagnant Pacific slab. *Chemical Geology*, **255**, 134–142.

8 Crustal earthquakes

SHINJI TODA

Chapter 7 introduced the plate configuration in and around the Japanese islands, particularly with respect to the deep seismic structure and associated seismicity. This chapter overviews the shallow seismicity in Japan, with a particular focus on inland active faulting from the viewpoint of earthquake geology. This is followed by a consideration of the long-term earthquake forecasting and seismic hazard assessment – research that has been accelerated since the 1995 Kobe earthquake – emphasizing several of the more important current issues such as the presence of hidden faults, the effects of fault segmentation, sequential rupture and seismic cycles, as well as time dependency due to stress perturbation.

In this chapter, to describe the size of earthquake, we use M for magnitude determined by the Japan Meteorological Agency (JMA), often called JMA magnitude (M_{JMA}), whereas M_w refers to moment magnitude (Hanks & Kanamori 1979). M is systematically larger than M_w for large inland earthquakes. For example, the 1995 Kobe (Hyogo-ken–Nanbu) earthquake is $M = 7.3$ but $M_w = 6.9$, and the 1891 Nobi earthquake is $M = 8.0$ but $M_w = 7.4$.

Seismicity in and around Japan (instrumental and historical earthquakes)

Historical destructive earthquakes in Japan

Japan is one of the few countries with an unusually long record of large earthquakes. The oldest, although unreliable, historical account of an earthquake, 'jishin' in Japanese, was written in AD 416 within the *Nihonshoki* (*The Chronicles of Japan*), which is the second oldest book of classical Japanese history. The oldest record of widespread shaking, tsunami and problems with hot springs induced by one of the Nankai Trough earthquakes is mentioned in AD 684. Vast amounts of historical documents of shaking, damage, tsunami and related phenomena were recovered and compiled by Kinkichi Musha in the early 1940s. Following this, Professor Tatsuo Usami took over the compilation and published a book entitled *Materials for Comprehensive List of Destructive Earthquakes in Japan*, the first edition of which appeared in 1975 (Fig. 8.1). The Usami catalogue has been pervasively used and updated about every ten years, with the most recent edition published in 2013 with four other co-authors (Usami *et al.* 2013). The catalogue includes not only the summary of documentation for each earthquake but also estimated date (time), epicentre location and magnitude which is inferred from the area of assigned JMA intensities (isoseismal) larger than 5 or 6 (see Table 8.1 for the JMA intensity scale compared with the Modified Mercalli intensity scale). Recent editions also include numerous data from 'earthquake archaeology' (e.g. Sangawa 2010) and palaeoseismology to estimate plausible source faults corresponding to particular earthquakes. Hypocentres of large earthquakes that occurred in AD 1885–1922 determined by Utsu (1982) are also incorporated into the Usami catalogue, and a list of instrumentally determined hypocentres since August 1923 has been provided by the JMA.

The largest historical earthquake, the 2011 Tohoku earthquake of $M_w = 9.0$, ironically occurred after all this work on destructive events had been compiled. Before 2011, the largest earthquake estimated from historical documents was the 1707 $M = 8.6$ Hoei event, which ruptured the entire Nankai Trough (Fig. 8.2) and generated a huge tsunami that travelled from Kanto to Kyushu (up to 26 m high in Kochi, Shikoku Island). The Hoei earthquake was one of the megathrust earthquakes that have repeatedly been recorded along the Nankai Trough every *c.* 180 years on average (90–262 years) since the Hakuo earthquake in 684 (e.g. Ando 1975; Ishibashi & Satake 1998). The reason why eight cycles (nine megathrust events) of large earthquakes have been revealed along the Nankai Trough is because there has been a longer historical record kept in SW Japan, especially of events involving Nara and Kyoto where the ancient imperial court of Japan was located. The latest of these events, namely the 1944 Tonankai ($M = 7.9$) and 1946 Nankai ($M = 8.0$) megathrust earthquakes, occurred with a two-year time gap and produced not only strong ground motion of JMA intensity ≥ 6 along the coastal regions, but also devastating tsunamis up to 8–10 m high along the SE coast of the Kii Peninsula in 1944 and 5–6 m high along the entire coast of southern Shikoku and the Kii Peninsula in 1946. Despite the destructive power of these tsunamis, however, they were relatively small compared to the other historical Nankai earthquakes. It is interesting to note that the time gap between the 1854 Ansei–Tokai ($M = 8.3$) and 1854 Ansei–Nankai ($M = 8.3$) events was only 32 hours. A significant difference between the 1944 Tonankai and 1854 Ansei–Tokai earthquake was the involvement of rupture of segment E in Figure 8.2, which in turn led to the 'Tokai seismic gap' hypothesis of Ishibashi (Ishibashi 1976), subsequent intensive monitoring by JMA and the 'Large-Scale Earthquake Countermeasures Law' enforced since 1978.

The historical record of large earthquakes along the other subduction zones is relatively short and their recurrent behaviours are not well understood. In the Sagami Trough south of Tokyo (formerly Edo), the most recent 1923 Taisho Kanto ($M = 7.9$, Fig. 8.1) earthquake struck the cities of Tokyo, Yokohama, Yokosuka and Chiba Prefecture and killed 105 000 people (Takemura 2003). The earthquake, caused by a megathrust on the interface between the subducting Philippine Sea Plate and overlying Eurasian (or North American) Plate (Fig. 8.1), brought a coseismic 1.5 m uplift of the southern tip of the Boso Peninsula and triggered a tsunami of several metres height along the coastal regions from northern Izu Peninsula (maximum 12 m at Atami) to the southern Boso Peninsula. The historical 1703 Genroku Kanto earthquake ($M = 8.3$) is thought to have shared the same source fault on the subduction interface of the 1923 Taisho Kanto earthquake (Matsuda *et al.* 1978; Shishikura 2003). However, the Genroku Kanto earthquake additionally involved rupture of an adjacent fault segment and so produced larger and more widespread tsunami generation affecting an area from the entire Izu Peninsula to

From: Moreno, T., Wallis, S., Kojima, T. & Gibbons, W. (eds) 2016. *The Geology of Japan*. Geological Society, London, 371–408.

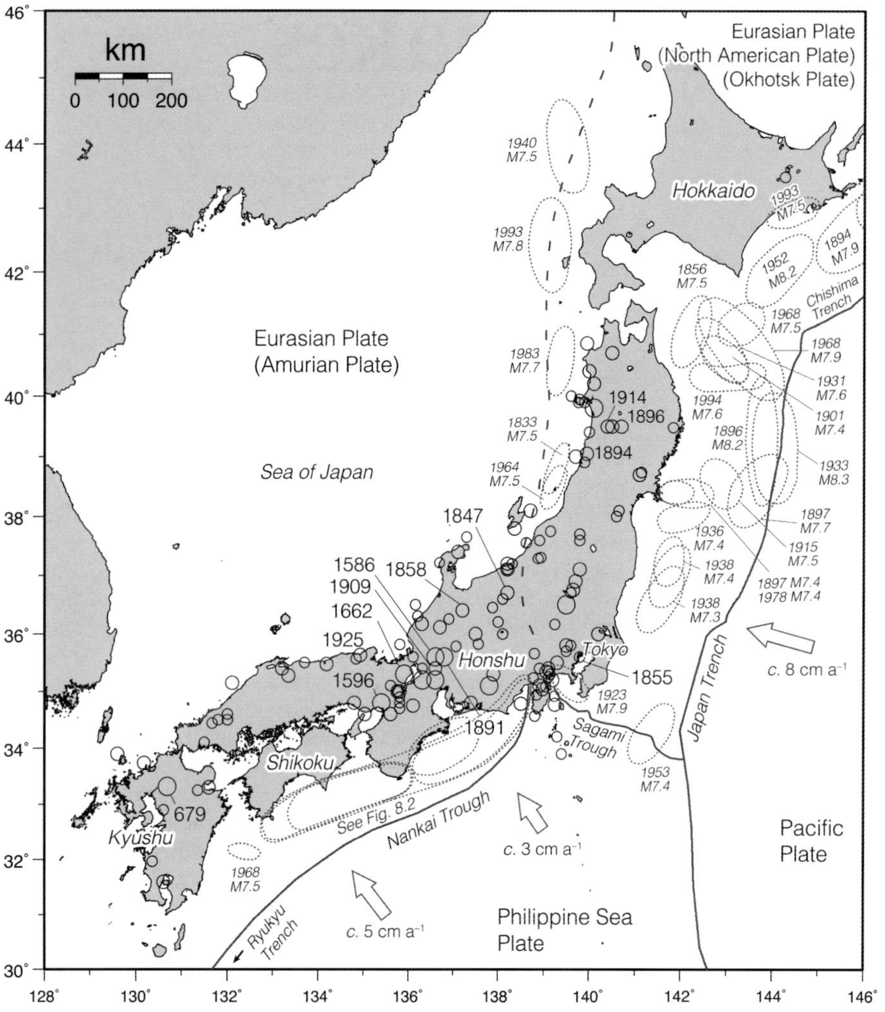

Fig. 8.1. Historical destructive earthquakes and plate configuration in and around the Japanese islands (HERP 2005*a*, *b*, *c*). Inland historical earthquakes are from Usami *et al.* (2013). Large oceanic earthquakes and their sources since 1800s are from HERP (2005*a*, *b*, *c*).

east of the Boso Peninsula. Coseismic uplift at the southern Boso Peninsula is estimated to have been as great as 6 m (Shishikura 2003). Similar large megathrust earthquakes might have been recorded before the Edo period (the Tokugawa shogunate) began in AD 1603 (e.g. the 1293 earthquake estimated by Shimazaki *et al.* 2011). Matsuda *et al.* (1978) classified two types of large subduction earthquakes along the Sagami Trough, Taisho-type and Genroku-type, based on the crustal deformation associated with two recent earthquakes and uplifted marine terraces. Shishikura (2003) further studied 16 uplifted marine terraces including the

Table 8.1. *JMA intensity scale and equivalent ground motion parameters after Stein* et al. *(2006)*

Observations used to assess shaking intensity	JMA intensity scale	Estimated peak ground acceleration at 5–8 Hz (g)	Estimated peak ground velocity at 5–8 Hz (m s^{-1})	Modified Mercalli Intensity (MMI)
Felt by most people in the building; some people are frightened	3	*c.* 0.02	*c.* 0.02	V
Many people are frightened; some people try to escape from danger; most sleeping people awoken	4	*c.* 0.07	*c.* 0.07	VI
Occasionally, less earthquake-resistant houses suffer damage to walls and pillars	5 lower	*c.* 0.26	*c.* 0.26	VII, VIII
Occasionally, less earthquake-resistant houses suffer heavy damage to walls and pillars, and lean	5 upper			
Occasionally, less earthquake-resistant houses collapse and even walls and pillars of highly earthquake-resistant houses are damaged	6 lower	*c.* 0.95	*c.* 0.93	IX, X
Many less earthquake-resistant houses collapse; in some cases, even walls and pillars of highly earthquake-resistant houses are heavily damaged	6 upper			
Occasionally, even highly earthquake-resistant houses are severely damaged and lean	7	*c.* 3.4	*c.* 3.3	XI, XII

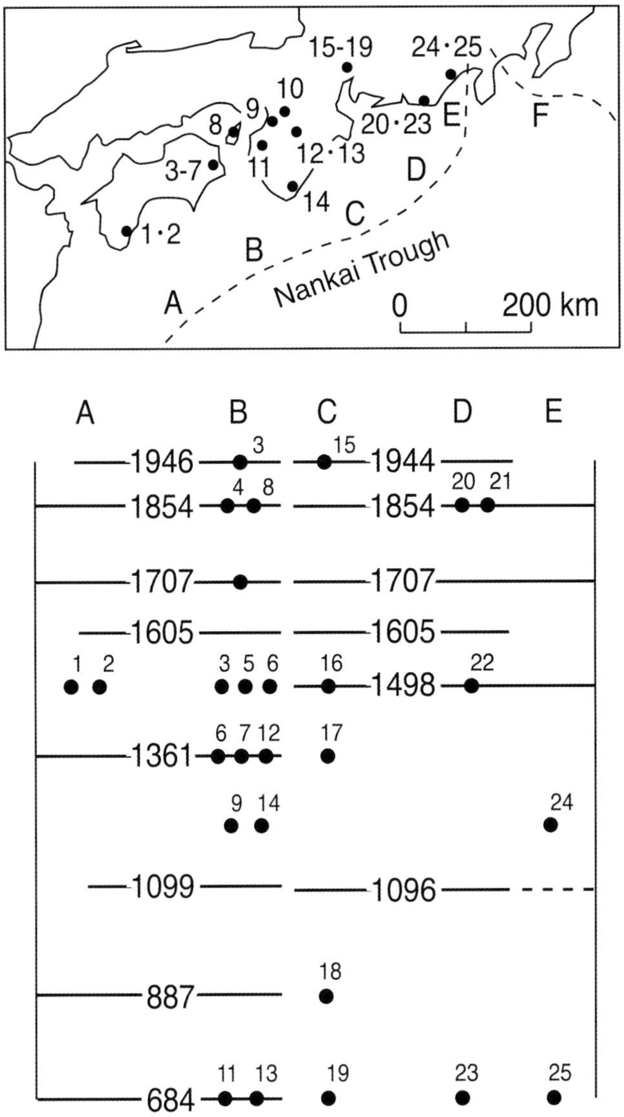

Fig. 8.2. History of megathrust earthquakes along the Nankai Trough, Japan (Sangawa 2010). The subduction zone along the Nankai Trough is divided into five segments, A–E. The horizontal lines with number in lower panel denote the rupture extents and occurrence years of large earthquakes. Dots with number indicate archaeological sites with exposed significant liquefactions at specific ruins and their calendar years. Recent archaeo-seismological data filled the gaps in long recurrence intervals originally estimated by only historic accounts, revealing the Nankai subduction earthquake cycles to be more regular and frequent.

local subsidence associated with the Genroku-type earthquakes during the past 7000 years and concluded that the Taisho-type and Genroku-type earthquakes have occurred about every 400 years and every 2000–2700 years, respectively.

The records of subduction earthquakes in the Tohoku district are more limited, due to its rural location far from ancient Japanese central governments. However, due to the fast convergence rate between the Pacific and Eurasian (or North American) plates (*c.* 8 cm a^{-1}), numerous large earthquakes have occurred along the Japan Trench over the past 200 years (Fig. 8.1). Among these, the relatively short recurrence intervals of $M = 7.3$–7.4 earthquakes off Miyagi Prefecture occurring in 1793 ($M = 8.2$), 1835, 1861, 1897, 1936 and 1978 during the past 200 years seemed to support a common belief that stresses were being relieved by these quasi-periodic repetitions, a conclusion that unfortunately made the much larger 2011 Tohoku earthquake all the more unexpected.

In general, there are three types of large earthquakes occurring in and around the Japanese archipelago: (1) subduction megathrusts; (2) intraslab; and (3) shallow inland crustal earthquakes (Fig. 8.3). Subduction megathrust earthquakes occur, from north to south, along the Kuril (Chishima) Trench, Japan Trench, Sagami Trough, Nankai Trough, and Ryukyu Trench. Large intra-oceanic-plate earthquakes are divided into two types: intraslab and outer-rise earthquakes. The former is represented by the 1994 $M = 8.2$ Hokkaido-toho-oki earthquake, while a typical example of the latter is the 1933 $M = 8.3$ Showa–Sanriku earthquake which devastated the Sanriku coast with a large tsunami (Fig. 8.1). Shallow-crust earthquakes, those with hypocentres shallower than 20 km, occur beneath inland and shallow coastal regions.

The largest recorded inland earthquake is the 1891 $M = 8.0$ Nobi (Mino–Owari) earthquake, central Japan, which killed 7273 people (Usami *et al.* 2013). The Nobi earthquake brought Japanese geologist Bunjiro Koto to the idea of 'earthquake faulting hypothesis', which contributed substantially to the elastic-rebound theory presented after the 1906 San Francisco earthquake (Reid 1910). The Nobi earthquake occurred soon after the first seismometer was installed and first seismological society had been established in Japan in the 1880s. Since inland crustal earthquakes typically bring strong but locally restricted ground motion due to their shallow focus, historical records of such inland shocks are more incomplete compared to those recording the subduction megathrust earthquakes. The oldest inland earthquake documented is the $M = 6.5$–7.5 Chikushi earthquake in AD 679 which occurred in northern Kyushu (Usami *et al.* 2013) and is inferred to have been associated with the most recent activity of the Minou Fault (Chida *et al.* 1994).

More than 40 destructive earthquakes were documented in the Nara (AD 710–794) and Heian (Kyoto) (AD 794–1192) periods (Usami *et al.* 2013). Among them, in AD 869 the Jogan earthquake and tsunami devastated the ancient city of Sendai (Tagajo) which had been established as a regional military centre during the Nara period to exert control over the region. This Jogan earthquake might have been the predecessor of the 2011 Tohoku-oki earthquake (e.g. Sugawara *et al.* 2013).

Earthquake records during the period between AD 1192 and 1603 are highly limited because of the warring states. However, both the 1586 Tensho and 1596 Keicho earthquakes, which struck the Nagano–Nagoya and Osaka–Kyoto regions respectively in the Japanese middle ages, are also thought to have been $M \approx 8$ class inland earthquakes (details of these shocks are described in 'Dense fault networks in Osaka and Kyoto').

During the Edo (Tokugawa) period, under the stable control of the Tokugawa shogunate lasting from 1603 to 1868, numerous destructive earthquakes were documented including the 1605, 1707, 1854 $M \approx 8$ earthquakes along the Nankai Trough (Fig. 8.2) and 1703 Genroku earthquake along the Sagami Trough. Another candidate for a previous 2011 Tohoku-oki type event is the 1611 Keicho–Sanriku earthquake that again inundated the Sendai plain and killed more than 1800 people. Tsunami deposits corresponding to the 1611 earthquake were not only recovered from Sendai but also from Aomori, *c.* 400 km north of Sendai (Tanigawa *et al.* 2014), suggesting that the source location and dimension might have been different from the 2011 event. In the 1600–1700s, the documented number of destructive earthquakes increased in the Tohoku district. Although this is partly due to the expanded populations in NE Honshu and improved administration by the Tokugawa shogonate, a temporal clustering of earthquakes, possibly sequential ruptures involving nearby faults (discussed in the next section), could also be argued for.

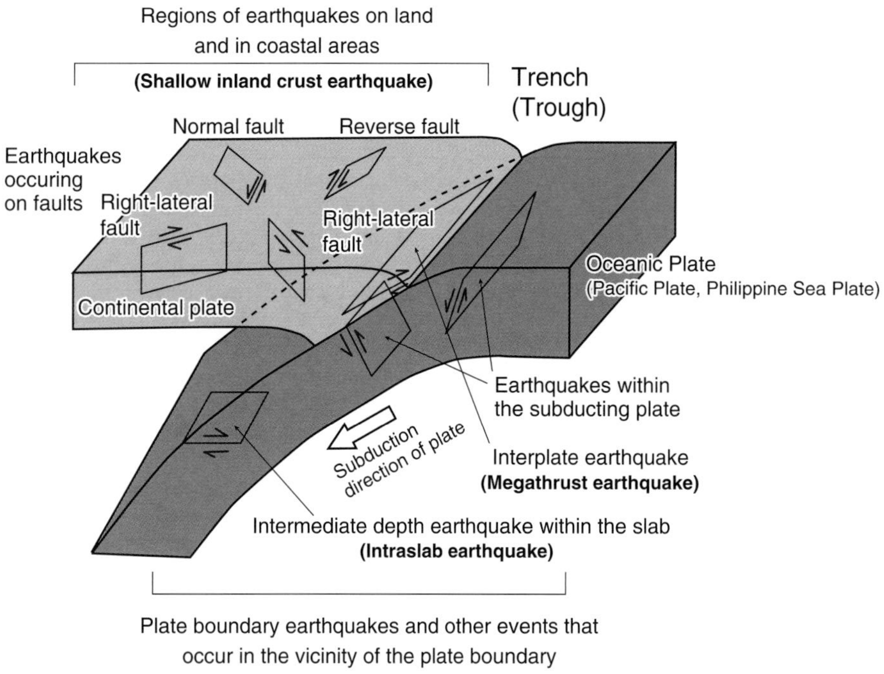

Fig. 8.3. Schematic illustration of seismic sources of large earthquakes in and around the Japanese islands (HERP 2005*a*, *b*, *c*).

The larger earthquakes other than megathrust earthquakes in the Edo period include: the 1662 Kanbun earthquake of $M = 7.3$–7.6 (Fig. 8.1); 1771 Yaeyama earthquake of $M = 7.4$ and its triggered tsunami which struck the Ryukyu islands; 1847 Zenkoji, Nagano earthquake of $M = 7.4$; 1858 Hietsu earthquake of $M = 7.0$–7.1; and the 1855 Ansei Edo earthquake of $M = 6.9$–7.3. This latter event, occurring on 11 November 1855, killed more than 7000 people in Edo (Tokyo) and is the last devastating earthquake with a source located just beneath Tokyo (e.g. Bakun 2005). It is also interesting to note that the 1855 Ansei Edo shock occurred one year after the 1854 $M = 8.4$ Ansei Tokai earthquake. In the early Meiji period of Japanese modernization, starting from AD 1868, the occurrence and recording of large earthquakes continued. In the mid-Meiji period of the late 1800s, European scientists such as John Milne (1850–1913) and James A. Ewing (1855–1935) were invited by the Japanese government as professors to the Imperial University. They led the early era of seismology in Japan, established the Seismological Society of Japan in 1880 and, in 1883, invented the first seismometer to be introduced to a meteorological observatory. Utsu (1982) later compiled the earthquakes which occurred in the Meiji and Taisho (AD 1912–1926) periods. The 1891 $M = 8.0$ Nobi, 1894 $M = 7.0$ Shonai, 1896 $M = 7.2$ Rikuu, 1909 $M = 6.8$ Anekawa, 1914 $M = 7.1$ Akita-Senboku and 1925 $M = 6.8$ Kita–Tajima earthquakes were destructive inland earthquakes that occurred during this period, as well as the 1896 $M \approx 8.3$ Meiji–Sanriku earthquake and the 1923 Taisho Kanto earthquake of $M = 7.9$ (Fig. 8.1).

Instrumentally recorded destructive crustal earthquakes near and over M7 include: 1927 Kita–Tango earthquake of $M = 7.3$; 1930 Kita–Izu earthquake of $M = 7.3$; 1943 Tottori earthquake of $M = 7.2$; 1945 Mikawa earthquake of $M = 6.8$; 1948 Fukui earthquake of $M = 7.1$; 1961 Kita-mino earthquake of $M = 7.0$; 1978 Izu-Oshima-kinkai earthquake of $M = 7.0$; 1984 Nagano-ken–seibu earthquake of $M = 6.8$; 1995 Kobe (Hyogo-ken–Nanbu) earthquake of $M = 7.3$; 2000 Tottori-ken–seibu earthquake of $M = 7.3$; 2004 Niigata-ken–Chuetsu earthquake of $M = 6.8$; 2005 Fukuoka-ken–seiho-oki earthquake of $M = 7.0$; 2007 Noto–Hanto earthquake of $M = 6.9$; 2007 Niigata-ken–Chuetsu-oki earthquake of $M = 6.8$; 2008 Iwate–Miyagi-nairiku earthquake of $M = 7.2$; and 2011 Fukushima–Hamadori earthquake of $M = 7.0$ (Fig. 8.1). A series of large destructive earthquakes occurred in the eastern margin of Japan Sea; the 1940 Shakotan-hanto-oki earthquake of $M = 7.5$, 1964 Niigata earthquake of $M = 7.5$, 1983 Nihonkai-chubu (Japan Sea) earthquake of $M = 7.8$ and 1993 Hokkaido-nansei-oki earthquake of $M = 7.9$ might also be categorized as crustal earthquakes along a newly proposed plate boundary. Even though damage and casualties are not always proportional to the size of earthquake (they are related to a combination of exposure and vulnerability), a few of these $M \geq 7$ earthquakes killed more than 1000 people. The worst case was the 1995 Kobe earthquake which took the lives of 6434 people, mostly due to collapse of buildings.

Seismotectonics and stress field in and around Japan

The driving force to produce the numerous large earthquakes in the Japanese islands, the Japan Trench, Sagami Trough and Nankai Trough derives from plate convergence between the subducting Pacific Plate and the subducting and colliding Philippine Plate beneath the Eurasian Plate (Fig. 8.1). The 1983 Nihonkai-chubu earthquake ($M = 7.8$) occurred in the eastern Japan Sea, and GPS geodesy suggests that the traditionally defined Eurasian Plate can be better divided into Amurian and North American (or Okhotsk) plates along the eastern margin of the Japan Sea as far as the Itoigawa–Shizuoka Tectonic Line (Heki *et al.* 1999; Sagiya *et al.* 2000; Miyazaki & Heki 2001). In NE Japan, east–west compressional stress is sustained by the subduction of Pacific Plate subducting beneath Honshu at a rate of 8.0–8.5 cm a^{-1} (Fig. 8.1). In central Honshu, together with the stress transfer from the Pacific Plate, collision of the Zenisu and Izu–Bonin ridges on the Philippine Sea Plate widely influences the regional stress field. Due to this collision, north–south compression is dominant in and around the Izu Peninsula, deforming the Median Tectonic Line (MTL) and its associated rocks, inducing late Quaternary uplift of the Kanto Mountains and activating strike-slip faults on the Izu Peninsula. In SW Japan, the Philippine Sea Plate is being subducted northwestwards beneath the overriding Eurasian (or Amurian) Plate along the Nankai Trough at a rate of 6.3–6.8 cm a^{-1} (Miyazaki & Heki 2001).

The current crustal stress field in Japan has been investigated in several ways. An inexpensive and effective method is to use focal mechanisms of moderate to large earthquakes. In the 1970s and 1980s, regional stress fields from a limited number of available focal mechanisms were examined. The estimated regional stress fields are roughly consistent with those approximated from fault type and slip sense of active faulting (e.g. Okada & Ando 1979; Wesnousky *et al.* 1982). From this, the azimuth of maximum horizontal compression, S_{Hmax}, is estimated from average focal mechanism P axes and a very limited number of *in situ* stress measurements (Tsukahara & Kobayashi 1991; Tsukahara & Ikeda 1991). The data obtained from this shows that a NW-trending S_{Hmax} is present across most of Honshu except in the southern Shikoku, Boso, Kanto, Izu and Tokai regions where subduction of the Philippine Sea Plate is possibly perturbing the stress field. A recently developed inversion method using the centroid moment tensor (CMT) of seismic events combined with Bayesian statistical theory allows us to estimate the 3D pattern of complex tectonic stress fields in and around the Japanese islands (Terakawa & Matsu'ura 2010; Fig. 8.4). These stress patterns show the Kuril–Japan–Nankai Arc is basically under east–west compression, but the direction of intermediate principal stress changes from north–south (reverse faulting type) in NE Japan to vertical (strike-slip faulting type) in SW Japan. The stress pattern of the Ryukyu and Izu–Bonin Arc regions defines a trench-perpendicular tension (normal faulting type). In addition, the map shows several local stress patterns associated with the motion of the Kuril forearc sliver, the Izu collision with Honshu and the rift-zone opening in northern Kyushu (Fig. 8.4).

Geodetic data are another source of information when discussing current stress field in Japan. Geodetic measurements not only provide strain rate from the detected deformation during the observation periods, but also offer a clue to infer regional stress field. Hashimoto (1990) and Ishikawa & Hashimoto (1999) used data from the first-order triangulation stations from the Meiji era (AD 1883) and estimated strain rate in Hokkaido, Honshu, Shikoku and Kyushu islands. Their calculated rates of strain are greater than those derived from seismic moment release rates or displacements of active faults (e.g. Wesnousky *et al.* 1982). In their map, a north–south- to NE–SW-directed extension is currently prevailing in the Tohoku district, whereas a NW–SE compression is predominant in the southern Kanto to Chubu district and a principal axis defining east–west compression is estimated for SW Japan except for an area of major north–south extension in the central Kyushu district. Recent permanent GPS networks monitored since the mid-1990s enable us to model the short-term deformation of the Japanese islands. The first comprehensive study using several years of data from about 1000 permanent GPS stations was performed by Sagiya *et al.* (2000). In addition to the regions of plate boundaries and active volcanoes, large strain rates along the Japan Sea coast and in the northern Chubu and Kinki districts were recognized. They referred to this newly discovered area of high strain rate as the Niigata–Kobe Tectonic Zone (NKTZ) and claimed that the NKTZ may be related to a hypothetical boundary between the Eurasian (or Amurian) and the Okhotsk (or North America) plates. Recently, Townend & Zoback (2006) compared GPS strain with the axis of maximum horizontal compressive stress obtained from stress tensor inversion with earthquake focal mechanisms in central Japan. They concluded that both geodetic and seismic estimates agree only after the effects of

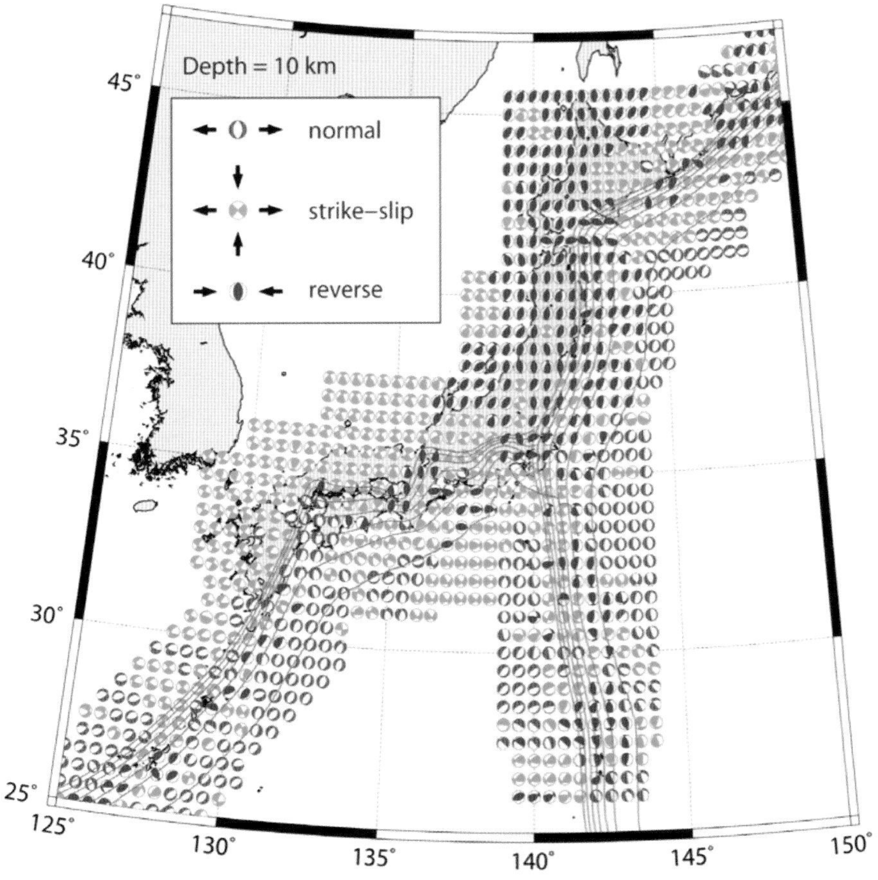

Fig. 8.4. Estimated stress pattern at 10 km depth using an inversion method of centroid moment tensor of earthquakes (Terakawa & Matsu'ura 2010). The stress pattern is represented by the lower-hemisphere projection of focal mechanisms of potential seismic events.

interseismic strain accumulation on geodetic observations have been accounted for using an elastic dislocation model of subduction thrust locking. Residual deformation might be explained by long-term inelastic deformation associated with mountain-building processes.

In addition to seismological and geodetic estimates, geologists and geomorphologists have made attempts to estimate the long-term regional stress field in the Quaternary period to the present using the distribution of active faults, their geometry, fault style, slip information, deformed strata associated with fault-related folds and the alignment of volcanic dyke intrusions. Wesnousky *et al.* (1982) was the first comprehensive study to elucidate the geological contraction rate, strain rate and moment release rates in active faulting in Honshu and the Shikoku islands using active fault data compiled by the Research Group for Active Faults of Japan (1980). Their calculated east–west crustal shortening rates are 0.4–4.4 mm a^{-1}, whereas the strain rates are $1–25 \times 10^{-9}$ a^{-1}. Their map shows that only the Izu region is under north–south contraction, to the northwards impingement of the Izu Peninsula into Honshu. However, the north–south shortening rate cannot be fully calculated from the data along the active faults alone, due to active folding (fault-related fold) of Neogene–Pleistocene strata up to several kilometres thick. Wesnousky *et al.* (1982) indeed calculated a contraction rate of 0.6 mm a^{-1} in Tohoku district, which is one order-of-magnitude smaller than those from the other districts in Honshu. In another study, Sato (1989) focused on the geological cross-sections of the late Cenozoic strata across the NE Honshu Arc, taking the active folds into consideration. He found significant contraction along the regions facing the eastern margin of the Japan Sea and a more moderate rate along the volcanic front of the Tohoku Arc. He concluded that the average strain rate of shortening in the central part of NE Honshu is on the order of 10^{-8} a^{-1}, which is much greater than that from Wesnousky *et al.* (1982). Okada & Ikeda (2012) largely updated the analyses of Sato (1989), compiling the profiles of numerous seismic reflections, gravity anomaly data and balanced cross-sections. The total amount of shortening across the Tohoku region from Pliocene to late Quaternary times is estimated from these data to be 10–15 km, which is equivalent to an average strain rate of 10^{-7} a^{-1}.

Nohara *et al.* (2000) measured the horizontal strain rate caused by fault activity during the past several thousand years using the active fault database. The pattern of strain rate resembles that from the GPS observations of Sagiya *et al.* (2000). However, the geological strain rate of 10^{-8} a^{-1} is much smaller than the geodetic strain on the order of 10^{-7} a^{-1}. It indicates that geodetically estimated strain is transient, reflecting fast interseismic deformation due to the plate coupling as was revealed by the large east–west stretching associated with the 2011 Tohoku earthquake of $M_w = 9.0$.

Recent well-detected seismicity by rapid progress in seismic networks

The history and development of the seismic observation networks in Japan are well summarized in Okada *et al.* (2004). Before the 1995 Kobe earthquake, several high-sensitivity seismic observation networks had been operated in Japan by various organizations independently, with only limited data exchanges. The Japan Meteorological Agency (JMA) operated a nationwide network consisting of 188 stations in order to detect small to large earthquakes occurring in and around Japan. University groups maintained 274 local networks for microearthquake observation as well as dense arrays at specific sites of research interest. The National Research Institute for Earth Science and Disaster Prevention (NIED) operated a network consisting of 89 highly sensitive stations in the Kanto–Tokai region, central Japan. Only 28 broadband stations were in operation, three of these working on a real-time basis.

Since the Kobe earthquake, the density of the seismic networks in Japan has changed dramatically. There is now a new high-sensitivity seismograph network (Hi-net) consisting of around 800 stations, along with a broadband seismograph network (F-net) comprising around 70 stations. A nationwide strong-motion seismograph network (K-net), consisting of more than 1000 observation stations, has been deployed since 1996. In addition, a network of around 700 stations with an uphole/downhole pair of strong-motion seismographs is called KiK-net. For offshore seismic observation, the NIED and the Japan Agency for Marine-Earth Science and Technology (JAMSTEC) have been respectively deploying submarine-cabled real-time seafloor observatory networks, called 'Seafloor Observation Network for Earthquakes and Tsunamis along the Japan Trench' (S-net) along the Japan Trench and 'Dense Ocean-floor Network System for Earthquakes and Tsunamis' (DONET) along the Nankai Trough.

High-sensitivity data from Hi-net and pre-existing seismic networks operated by various institutions are shared by university groups in real time using satellite communication for their research work. The graphic displays of current and recent seismicity are openly available from the NIED website (e.g. 24-hour to one-month seismicity detected by Hi-net: http://www.hinet.bosai.go.jp/hypomap/)

The detection capability for earthquakes in and around the Japanese islands has been improving year by year. For earthquake forecasting and prediction research, it is highly important for seismologists to detect earthquakes as small as, and as many as, possible. Nanjo *et al.* (2010) carefully examined the smallest magnitude fully detected, that is, minimum magnitude of completeness (M_c) of the earthquake catalogues in and around Japan based on the Gutenberg-Richter frequency-magnitude law. It was found that presently $M_c = 1.0$ might be typical in the mainland (Fig. 8.5), although to have a complete catalogue they recommend using $M_c \geq 1.9$ earthquakes. Iwata (2013) found M_c values of 2.5–3.0 during 2006–2010 if daily variation in detection capability is considered.

One of the recent progresses in earthquake detection is 'waveform matching technique' which enables us to pick up missing

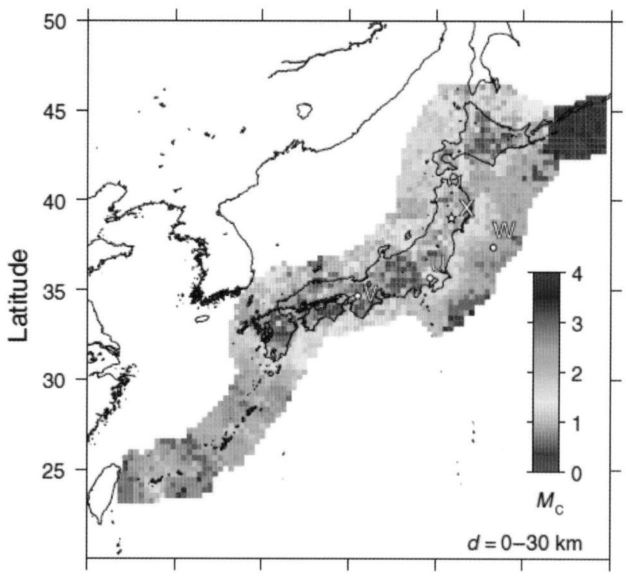

Fig. 8.5. Smallest magnitude completely detected by Japan's seismic networks (M_c) for 2008 at depth of 0–30 km (Nanjo *et al.* 2010).

and/or smaller events using the cross-correlation of the continuous seismic signal (e.g. of Hi-net) with pre-determined template events. The technique has been used in particular for detecting low-frequency earthquakes from non-volcanic tremors (e.g. Shelly *et al.* 2007), recovering missing earthquakes hidden in coda waves during aftershock sequence (e.g. Enescu *et al.* 2009; Lengline *et al.* 2012), foreshock activity (e.g. Kato *et al.* 2012) and remotely triggered earthquakes after a large earthquake (e.g. Kato *et al.* 2013; Shimojo *et al.* 2014).

Microseismicity delineates plate configurations in and around the Japanese islands and spatio-temporal clusters of occurrence of earthquakes. Figure 8.6 presents the 5-year seismicity that occurred during 2006–2010. The subducting Pacific Plate slab is visible as an area of dense seismicity due to the small–moderately sized earthquakes constantly occurring on the plate interface with the overriding inland plate, and the double zone of seismicity (e.g. Hasegawa *et al.* 1978; see Chapter 7 for details) in the slab. In contrast, the seismicity associated with the subduction of the Philippine Sea Plate along the Nankai Trough is low, except in the Hyuga-nada region of east Kyushu. Regarding crustal earthquakes mostly shallower than 20 km, these can be divided into two types of activity: one is represented by continuous stable-background seismicity including frequent volcanic swarms; the other is a cluster of aftershocks from recent large earthquakes. The former is typical for example in SW Kyushu, San'in, Wakayama, Tamba, Takayama and Izu (Fig. 8.6). Regarding the latter, if epicentres of large earthquakes that occurred in late 1900s are compared with the shallow earthquake clusters they are seen to be spatially correlated. Since Figure 8.6 includes the occurrence times of the 2007 Noto–Hanto earthquake ($M = 6.9$), 2007 Chuetsu-oki earthquake ($M = 6.8$), 2008 Iwate–Miyagi-nairiku earthquake ($M = 7.2$), dense earthquake clusters are significantly seen on and nearby their epicentres. However, despite these and the other recent large shocks such as the 2000 Tottori-ken–Seibu earthquake ($M = 7.3$), 2004 Niigata-ken–Chuetsu earthquake ($M = 6.8$) and 2005 Fukuoka-ken–seiho-oki earthquake ($M = 7.0$), even aftershocks of the 1983 Nihonkai-chubu earthquake ($M = 7.8$) and 1993 Hokkaido–nansei-oki earthquake ($M = 7.9$) are visible. These long-lasting aftershocks of the large crustal earthquakes are attributed to their low strain accumulation relative to the subduction zones. The high detection level of seismicity in and around Japan demonstrates that aftershock duration is proportional to fault recurrence interval and therefore inversely proportional to the fault stressing rate (Dieterich 1994; Stein & Liu 2010).

Palaeoseismology and earthquake geology in Japan

Modern instrumental records of earthquakes cover only a little more than a century since the 1890s. Historic accounts in Japan regionally allow us to extend the records of large earthquakes up to a little more

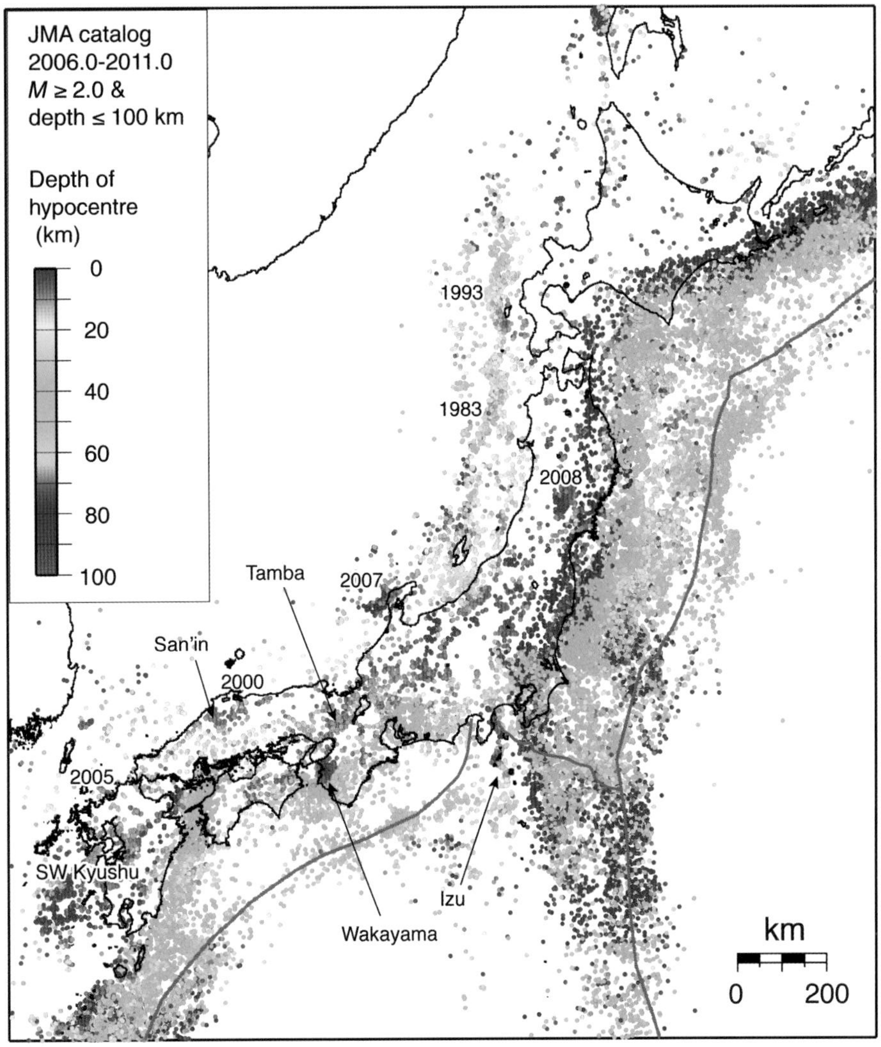

Fig. 8.6. Five-year seismicity in and around the Japanese islands based on the JMA catalogue.

than a thousand years, as outlined in the previous section. However, compared to the typical repeat time of large earthquakes on an active fault of several hundred to several tens of thousands of years, this is too short to understand the fault behaviours and forecast earthquakes. Furthermore, only a limited number of historical accounts are available for the regions (except in and around the ancient capitals of the Kyoto–Nara area) over the last several hundreds of years. However, sedimentary layers and geomorphic features can also play a role as seismograms to preserve evidence for strong shaking, ground breaks and landscape changes associated with large earthquakes. Although these geological and geomorphic records are incomplete and have poor constraints in time and space, they greatly help in understanding faulting processes and the recurrence of earthquakes that are unavailable in the brevity of historic documents. We have named this discipline 'palaeoseismology' (e.g. McCalpin 2009), and the individual earthquakes revealed are referred to as 'palaeoearthquakes.' In Japan, on-fault palaeoseismology is available for the study of inland crustal earthquakes and off-fault palaeoseismology for the effects of subduction earthquakes. Approaches in off-fault palaeoseismology are well described in McCalpin (2009). Due to such frequent megathrust events in and around the Japanese islands, particularly after the 2004 $M = 9.2$ Sumatra–Andaman earthquake and 2011 $M_w = 9.0$ Tohoku earthquake, the importance of studies of uplifted marine terraces and tsunami sediments have been more widely appreciated not only by the scientists but also by the general public. Off-fault paleoseismological studies associated with the subduction megathrust earthquakes in and around Japan have been well reviewed by Sawai *et al.* (2012) and Goto *et al.* (2012). In this chapter, we only focus on the on-fault palaeoseismology in Japan.

History of earthquake geology for Japan's inland active faults

As mentioned in the previous section, the 1891 $M = 8.0$ Nobi (Mino–Owari) earthquake, central Japan, that killed 7273 people has become one of the two key earthquakes (together with the 1906 San Francisco earthquake) that led to an understanding of earthquake mechanisms. An 80 km long surface rupture with a maximum 8 m left-lateral slip along the Nukumi, Neodani and Umehara faults occurred during the Nobi earthquake (Koto 1893). The impressive photograph of a 5 m scarp (Fig. 8.7a), recorded by Prof. Koto, occurred at a transpressional step between left-lateral fault sections at the centre of the rupture zone.

Most of the historically destructive $M \geq 6.5$ earthquakes within inland Japan are thought to have been accompanied by surface faulting (surface rupture, Fig. 8.7) regardless of their surface dimensions. Matsuda (1981) pointed out that about 80% of the inland damaging earthquakes during 1850–1980 (130 years) occurred either along mapped active faults or very close to them. Takemura (1998) also found that all events of $M \geq 6.8$ were accompanied by surface faulting. It is therefore certain that all the major active faults have the potential to be the sources of future destructive earthquakes.

Despite the evidence from the 1891 Nobi and 1906 San Francisco earthquakes, which led to Reid's elastic rebound theory, for around another half a century many geologists in Japan did not believe that faulting is the cause of an earthquake (Matsuda 2008). Several geological studies at this time nevertheless observed cumulative offset of geological boundaries and Holocene marine terraces by faulting, and by the 1960s and early 1970s seismologists in Japan were convinced that all earthquakes are indeed a product of faulting. Seismic waveforms, ground deformation and other characteristics associated with the instrumentally recorded 1923 Taisho Kanto, 1927 Kita-tango, 1943 Tottori and 1948 Fukui earthquakes were all numerically modelled by dislocation in elastic half-space (e.g. Kanamori 1972). Intriguingly, numerous active strike-slip faults were discovered in the 1960s for the first time (e.g. Atera Fault, central Honshu by Sugimura & Matsuda 1965, Fig. 8.8a). Matsuda (2008) retrospectively commented that several major active fault systems such as the MTL, Itoigawa–Shizuoka Tectonic Line, Atotugawa Fault and Atera Fault had already been confirmed as important strike-slip faults back in the 1960s.

Fig. 8.7. Significant examples of surface rupture. (**a**) The 6 m high fault scarp that emerged during the 1891 $M = 8.0$ Nobi earthquake (Koto 1893). (**b**) A coseismic scarp across a rice paddy resulting from the 1995 Kobe earthquake. (**c**) Normal faulting scarp produced by the 11 April 2011 Fukushima-ken–Hamadori earthquake of $M = 7.0$ (Toda & Tsutsumi 2013).

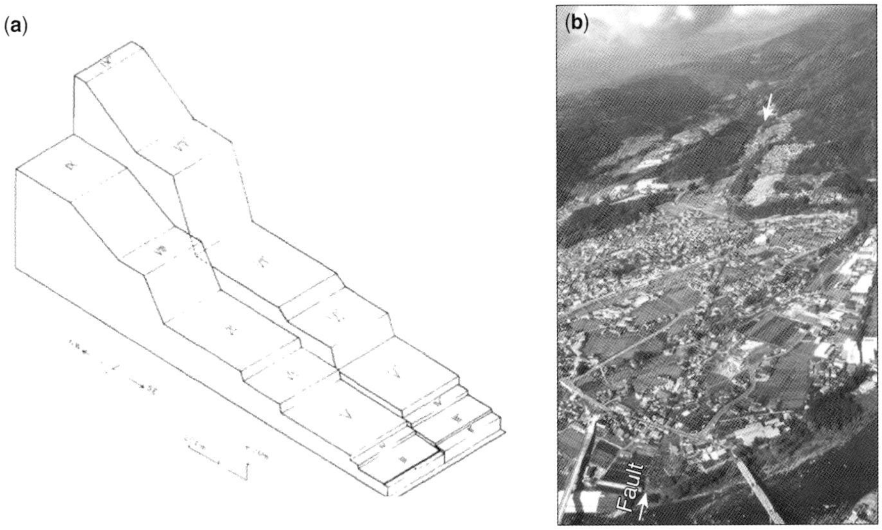

Fig. 8.8. Cumulative left-lateral offsets of rising flights of terraces across the Atera fault. (**a**) Schematic diagram showing the cumulative displacements (Sugimura & Matsuda 1965). (**b**) Photograph of the offset river terraces at Sakashita along the Kiso River.

The surveying of active faults was incorporated into the Earthquake Prediction Research Project, authorized by the government after the 1964 Niigata earthquake, and started in 1965. It became a great opportunity for geomorphologists and geologists to become directly involved in earthquake research. The Research Group for Active Faults, composed of numerous geomorphologists and a few geologists and seismologists, was inaugurated in 1975 in response to the Blueprint project. Together with the increased availability of aerial photos across the Japanese islands and a pervasive radiocarbon-dating database, active fault research in Japan progressed rapidly in the 1970s. One of the landmarks associated with seismic hazard is the empirical equations expressing the relationship between earthquake magnitude M and active fault length L in kilometres (Matsuda 1975, Fig. 8.9):

$$\log L = 0.6\,M - 2.9.$$

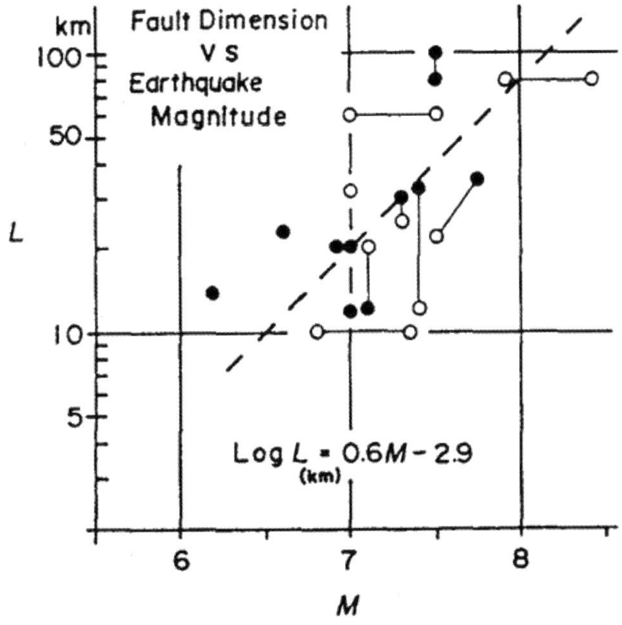

Fig. 8.9. Empirical relationship between earthquake magnitude (M) and length of the surface rupture or subsurface fault (L, km) (Matsuda 1975).

The equation allows earthquake engineers to take the size of the expected earthquake and the maximum acceleration into account with attenuation equations. The building boom associated with rapid construction of nuclear power plants in the 1970s provided a major boost to the research progress. Several other important empirical equations between earthquake magnitude and other fault parameters were later proposed around the world (e.g. Wells & Coppersmith 1994; Stirling *et al.* 2002) including Japan (e.g. Takemura 1998); the Japanese still use Matsuda's equation for conventional size estimates of future earthquakes, however.

The first full-scale catalogue of active faults in Japan appeared in 1980, a voluminous book of 437 pages entitled *Active Faults in Japan: Sheet Maps and Inventories*, published by the Research Group for Active Faults. The demand for this first publicly available active fault map was greater than expected by the authors. A fully revised edition of the catalogue was subsequently published in 1991, by now displaying over 2000 active fault traces at a scale of about 1:200 000 (Fig. 8.10). Based on the vast amounts of papers and reports published up until that time, the book offers a detailed catalogue for each of the mapped faults, including fault length, slip sense, amount of displacement and age and type of displaced topographic and geological features such as piercing point and line. A large amount of offshore active fault data is also included in this 1991 edition, based on seismic-reflection data and offshore topographic interpretation.

From the viewpoint of earthquake occurrences associated with active faults in Japan, Matsuda (1991) proposed the definition of a 'seismogenic fault', focusing on the length and distribution of the known active faults in the 1980 catalogue. A seismogenic fault is defined using the following criteria: (1) an independent fault longer than 10 km without any other faults within a distance of 5 km; (2) a group of discontinuous faults separated by gaps of $\leq$5 km; (3) a group of faults in a 5 km zone of the similar trend; and (4) a secondary or branching fault with a midpoint further than 5 km from a main fault. These criteria are based on the assumption that the surface rupture of the historical earthquakes emerges within a fault zone width of 5 km, which is also consistent with the world-wide study by Wesnousky (1988) who observed that earthquake rupture fronts do not propagate across gaps greater than 5 km. The '5 km rule', in particular for strike-slip faults, has been further confirmed by the updated database of Wesnousky (2006, 2008).

As well as the mapping effort made to identify active faults in Japan during the 1970–1980s, the earthquake potential from a

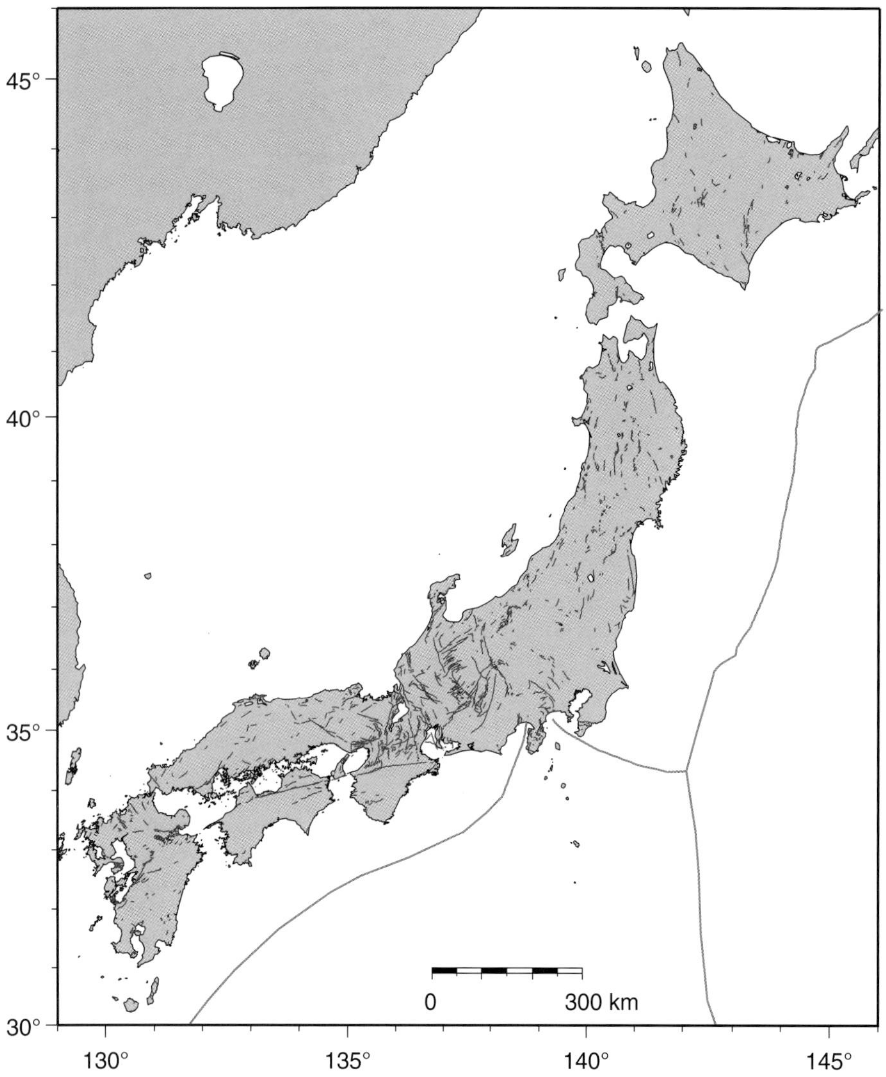

Fig. 8.10. Active faults in Japan (Research Group for Active Faults of Japan 1991). See Figure 8.24 for the 110 major active faults among them.

given active fault has been evaluated by modelling (Matsuda 1975; Schwartz & Coppersmith 1984). Even though the length of an active fault and a seismogenic fault as defined by Matsuda can be determined from the map, one of the palaeoseismic parameters hardly addressed by these studies is the temporal information of earthquake occurrence in the past. The timing of a palaeo-surface-rupturing earthquake is commonly constrained by the ages of deformed sedimentary units and overlying undeformed strata, which we call an 'event horizon'. Because natural exposures of deformed late Quaternary sediments are highly limited, we need intentional excavation of a palaeoseismic trench to expose the fault-related strata. To meet this need, a methodology of trench excavation surveying was imported to Japan from the US where intensive excavations across the San Andreas Fault had been already performed by Sieh (1978). Starting from the first trench excavation survey across the Shikano Fault that emerged during the 1943 Tottori earthquake (Okada *et al.* 1981), numerous full-blown and costly surveys were carried out across several major faults and historical earthquake rupture zones in the 1980s and early 1990s by groups of universities, the Geological Survey of Japan, the Central Research Institute of Electric Power Industries and others (e.g. the Tanna Fault Trenching Research Group 1983, Fig. 8.11). Identifying the timing of the most recent surface-rupturing earthquake is crucial for earthquake prediction, and an excellent compilation of historical earthquake data (Usami *et al.* 2013) has helped to estimate the most recent large earthquake which occurred on active faults, combining field data with radiocarbon ages (extracted from the trench wall) associated with the most recent movement of the fault. As a result of this work 'precaution faults', namely those expected to generate future destructive earthquakes, were identified from among the many major active faults in Japan (e.g. Matsuda 1977). The Nojima Fault, which caused the 1995 Kobe earthquake, was indeed one of the precaution faults identified by Matsuda (1977) at that time. Despite all this work, and unlike in California (e.g. Working Group on California Earthquake Probabilities 1988), the palaeoseismic data available on active faults in Japan was still not sufficient in the early 1990s to allow the probabilistic estimates of large earthquakes.

On 17 January 1995, the Kobe (Hyogo-ken–Nambu) earthquake of $M = 7.3$ ($M_w = 6.9$) devastated the northern part of Awaji Island and the city of Kobe, and killed 6434 people. It was only four years since the 1991 catalogue had been published. Because the Nojima Fault (one of the listed active faults in the 1991 book) caused the Kobe earthquake, not only scientists but also the general public have become aware of the importance of an active fault as a potential source of a destructive inland earthquake. The earthquake triggered immediate action from the Japanese government. A Special Measures Law on Earthquake Disaster Prevention was enacted to shore up existing measures against earthquake damage. Within

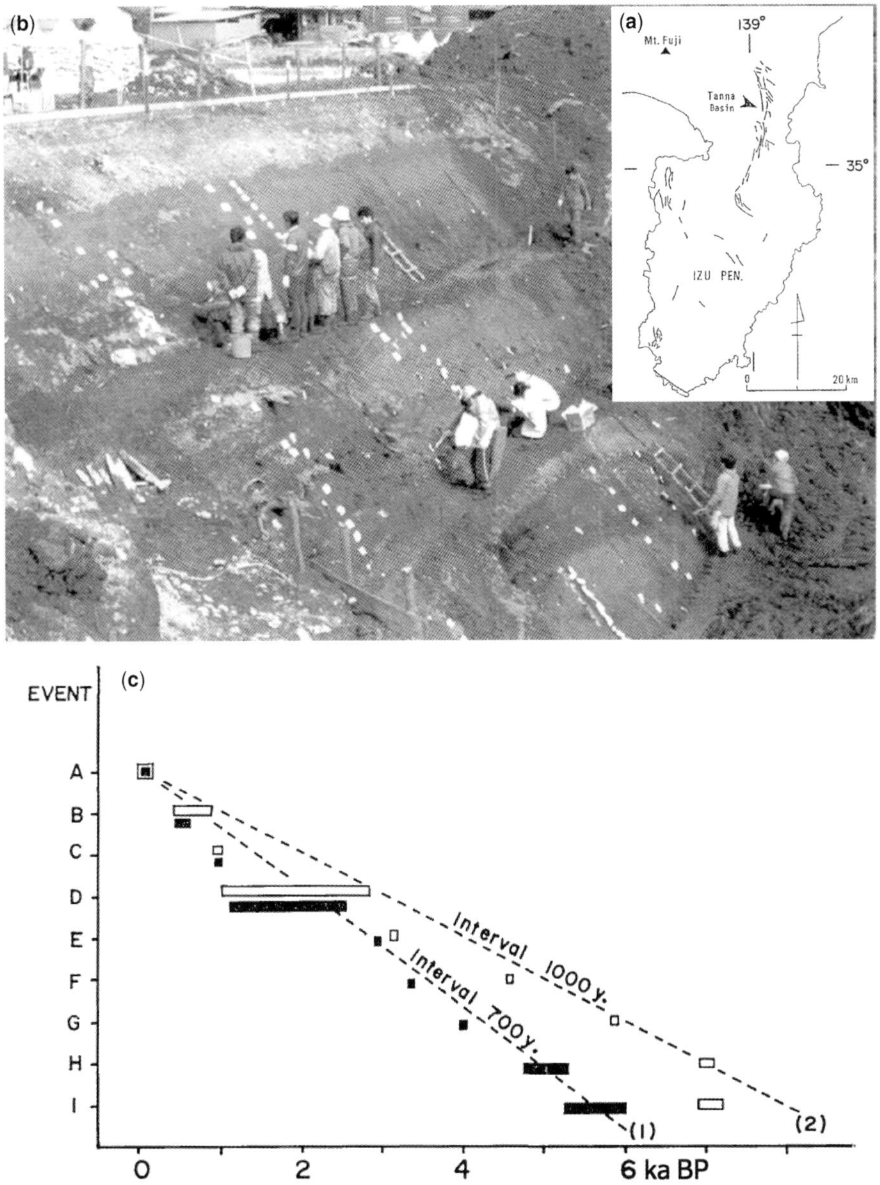

Fig. 8.11. Palaeoseismic trench study across the Tanna Fault performed in the early 1980s (Tanna Fault Trenching Research Group 1983). (**a**) Major active faults in the Izu Peninsula; the arrow indicates the trench site. (**b**) Photograph of one of the trench walls. (**c**) Diagram showing estimated ages of surface-rupturing events. Solid and open bars indicate time ranges of events inferred from two different interpretations.

this organization, the Earthquake Research Committee is responsible for collecting and organizing the results of seismic surveys, conducting research on earthquakes and analysing and evaluating this information (based at the Headquarters for Earthquake Research Promotion or HERP from 1995). The HERP produced the *Fundamental Survey and Observation Plan for Earthquake Research*, referred to the KIBAN project, on 29 August 1997. The KIBAN project consists of the following core items: (1) seismic observation; (2) observation of strong motion; (3) observation of crustal movement; and (4) survey of active faults in inland and coastal regions. The 1995 Kobe earthquake changed the direction of Japanese seismic hazard research from 'earthquake prediction' to the long-term probabilistic estimates of large earthquakes (seismic hazard map), the better understanding of earthquake phenomena and the mechanisms of strong ground motion.

To evaluate long-term seismic hazard, a huge amount of surveying including drilling, trench excavation, seismic reflection and other geophysical explorations have been deployed across more than 100 major active faults selected by HERP. In addition, re-evaluation of the fault landforms (palaeoseismic landforms) using more aerial photos by various groups has led to the publication of the *1:25 000-scale Active Fault Map in Urban Areas* (Geospatial Information Authority of Japan 2014), the *Digital Active Fault Map of Japan* (Nakata & Imaizumi 2002) and *Atlas of Quaternary Thrust Faults in Japan* (Ikeda *et al.* 2002). The most up-to-date data are compiled within the Active Fault Database of Japan operated by the Geological Survey of Japan, AIST (https://gbank.gsj.jp/activefault/index_e_gmap.html), including information from the studies preceding the Kobe earthquake.

Examples of major active fault behaviour in inland Japan

Even given Japan's relatively long record of historical documents on destructive earthquakes spanning over 1000 years, only a very few examples of large events have been repeated on the same active fault. Tremendous efforts to reveal the history of active faulting in Japan have been made, particularly since the Kobe earthquake. However the number of active faults already found or inferred is over 2000, so that research work has inevitably had to concentrate on a relatively limited number of these fractures in order to understand long-term faulting behaviours of the most potentially hazardous zones, rather than disperse and dilute our research budget and efforts. In this context the following text introduces several representative active faults in Japan which have already produced (or will in the future) $M = 7.5$–8 class destructive earthquakes.

North–south-striking reverse faults due to east–west crustal contraction in the Tohoku district

Short-term east–west contraction rates based on GPS data are more than 10 times faster than those estimated from geological shortening (Nohara *et al.* 2000). As mentioned in the section 'Inland earthquake hazard time dependency due to stress perturbation' the 2011 Tohoku-oki earthquake of $M_w = 9.0$ has released at least 200 years of geodetic shortening to cancel out the contradiction between geodetic and geological estimates discussed in the previous section. Permanent plastic contraction has been accommodated by the north–south-trending reverse faults since *c.* 3 Ma in NE Honshu (Sato 1994). Such significant horizontal east–west shortening across the Tohoku region is a product of the reactivation of densely populated normal faults as reverse faults. These faults were originally developed as normal faults in the rift zone associated with the opening of the Japan Sea during Miocene times (e.g. Okamura *et al.* 1995). Topographically, such north–south-striking reverse faults have developed 20–50 km wavelength basin-and-ranges centred on the Ou backbone range (Sekiryo Mountain, Fig. 8.12). Along the arc, distributions of these reverse faults are complemented by the active volcanoes not only at the volcanic front but also along the coastal region facing the Japan Sea, which can be explained by a 'hot mantle finger' model (Tamura *et al.* 2002).

Historically numerous destructive inland earthquakes have struck the Tohoku region. Among them, the largest thrust-faulting earthquakes ever recorded are the 31 August 1896 Rikuu earthquake of $M = 7.2$ and the most recent 14 June 2008 Iwate–Miyagi Nairiku earthquake of $M = 7.2$ (Fig. 8.12). The former occurred 5 years after the Nobi earthquake and killed 209 people lived in the villages along the eastern margin of the Yokote Basin in Akita Prefecture. A surface rupture of length *c.* 30 km and with up to 3 m vertical ground separation appeared along the Senya Fault (Yamasaki 1896; Research Group for the Senya Fault 1986) which uplifted the Mahiru Range, a part of the Ou backbone range. The other ground break, the *c.* 15 km long Kawafune Fault, was found *c.* 10 km east of the epicentre as a west-dipping back-thrust fault along

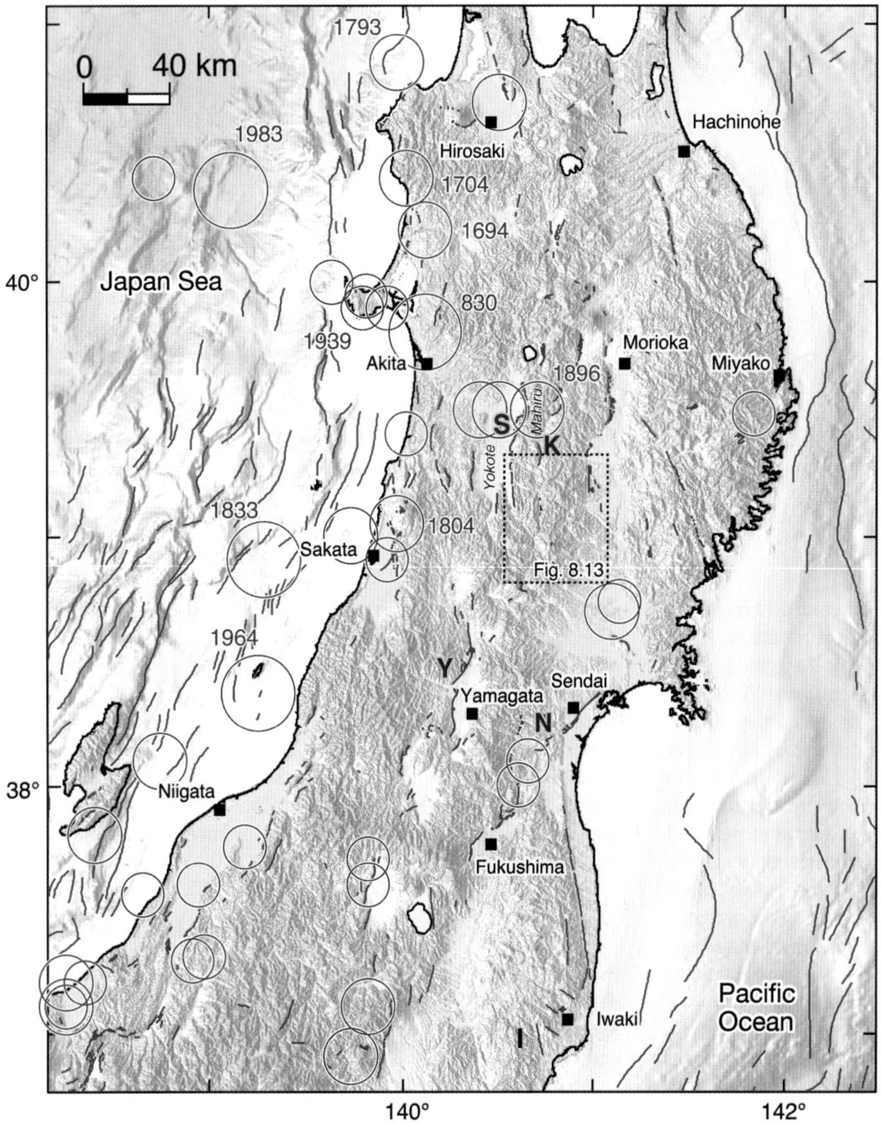

Fig. 8.12. Active faults in the Tohoku district (Research Group for Active Faults of Japan 1991) and destructive historical earthquakes of $M \geq 6.5$ (Usami *et al.* 2013).

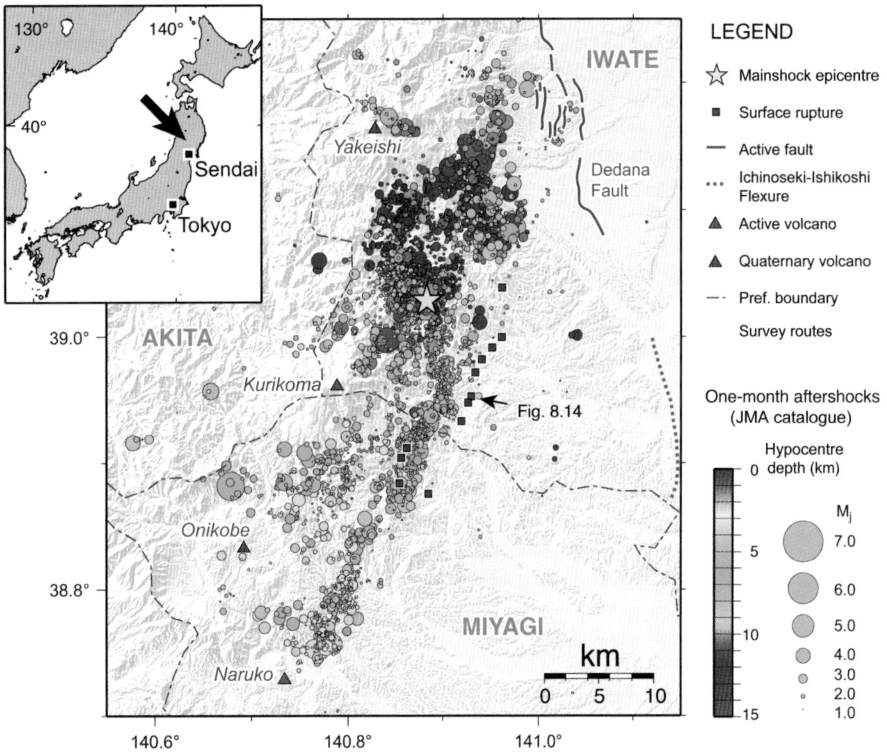

Fig. 8.13. Sporadic distribution of surface ruptures (red squares) associated with the 2008 $M = 7.3$ Iwate–Miyagi-nairiku earthquake on the distribution of observed aftershocks (Toda *et al.* 2010). The geometry of the source faults is complex (e.g. Takada *et al.* 2009), but they dip overall to the WNW.

the eastern margin of the Mahiru Range (Research Group for the Senya Fault 1986; Oyama *et al.* 1991). Seismic reflection surveys across the Senya Fault revealed shallow fault branching to the range-front fault and a basinwards-migrated fault, ruptured during the 1896 earthquake (Sato *et al.* 2002) as pointed out by Ikeda (1983). Palaeoseismic trenches across the Senya Fault exposed near-surface flattening of the fault dip and horizontal shortening, referred to as 'bulldozing' in the paper (Research Group for the Senya Fault 1986). The penultimate surface-rupturing earthquake is estimated to have occurred at *c.* 3500 a BP (Research Group for the Senya Fault 1986), which might also be a representative recurrence interval as supported by the vertical slip rate of 1.0 mm a^{-1} and coseismic vertical displacement at the Rikuu earthquake (HERP 2005*a*).

The $M = 7.2$ ($M_w = 6.9$) Iwate–Miyagi-nairiku earthquake struck mountainous regions east of Mt Kurikoma which is one of the active volcanoes in the Ou backbone range (Fig. 8.13). The earthquake killed 23 persons, including 6 missing, due mostly to the vast number of landslides. The earthquake occurred along the southern structural extension of the Kitakami-teichi-seien Fault Zone (Dedana Fault in Fig. 8.13) where no active traces were previously mapped. More than 13 fault-rupture observations suggest that the estimated total length of the tectonic ground breakages reaches *c.* 20 km, even though their locations are separate rather than continuous along the entire trend (Toda *et al.* 2010). However, contractional features such as thrust fault exposures, flexure, tilting and buckling deformations predominate on the rupture zone, which is consistent with reverse faulting under the WNW–ESE compressional stress field in northern Honshu. Amounts of vertical offset and horizontal shortening measured using cultural piercing points are mostly smaller than 50 cm. However, near the southern end of the rupture zone, fault structure and slip sense become complex and measured offsets are exceptionally large; an east–west-striking *c.* 1 km long continuous rupture involving 4–8 m of dextral and 2–4 m of vertical offsets were found. The mapped zone of the surface ruptures approximately defines the central part of the surface projection of a *c.* 40 km long west-dipping source fault and associated aftershock zone. The rupture process and subsurface faults might be complex in this fault system, and Takada *et al.* (2009) has modelled five source faults inverted from Synthetic Aperture Radar (SAR) pixel matching technique. Post-seismic trenches exposed the cumulative bedrock faulting and revealed evidence for Holocene palaeoseismic events at several locations where only offsets of <50 cm were measured (Fig. 8.14; Maruyama *et al.* 2009). The dates of palaeoseismic events are not well correlated between multiple sites, which may suggest that discontinuous surface faulting was a result of seismogenic faults rather than a direct connection from the subsurface faulting.

All the local prefectural offices in Tohoku are located nearby major active faults (Fig. 8.12). One of the typical range front faults in the Tohoku region is the Yamagata-bonchi Fault Zone which is a 60 km long west-dipping reverse fault located along the western margin of the Yamagata Basin in Yamagata Prefecture. From the surface fault traces and geomorphic features, the fault zone is divided into northern and southern sections. In the southern section, the Quaternary sediments (up to 400 m thick) thicken towards the west and are bounded by the fault zone. This suggests that the footwall of the fault has been continuously subsiding, with vertical movement being predominant over crustal shortening. In the northern section, however, the thickness of the Quaternary sediments decreases rapidly and there are several subparallel fault traces, suggesting that significant fault branching and front fault migration have occurred. Palaeoseismic trenches across the northern section exposed evidence for seven surface-rupturing earthquakes over the past 10 ka, with a recurrence interval of *c.* 1500 years (Toda *et al.* 2008). In contrast, trench walls across the southern section of the fault zone exposed evidence for only one event in the past 10 ka. Together with the major fault discontinuity at the centre of the fault zone, the Yamagata-bonchi Fault is therefore clearly divided into two 30 km long segments with distinct behaviours, both of which have probably produced (and will continue to generate) $M \approx 7.3$ earthquakes. The average recurrence interval of *c.* 1500

Fig. 8.14. Photograph showing deformed strata as evidence for much larger offset caused by a larger earthquake (penultimate earthquake) prior to the 2008 Iwate–Miyagi-nairiku earthquake.

years and the long elapsed time since the most recent event on the northern section allow us to calculate 3–14% of 30-year conditional earthquake probabilities. Since the time range of the event on the southern segment overlaps one of the events recovered from the northern segment, we cannot rule out the near future likelihood of the infrequent worst scenario of 60 km long simultaneous multiple ruptures that might produce a $M \approx 7.5$ earthquake.

The city of Sendai, the largest city in Tohoku and with over a million people, is built on one of these major reverse-fault systems known as the Nagamachi–Rifu Fault Zone. A NE–SW-striking active fault trace of length *c.* 40 km bisects Sendai city, with flights of river terraces younger than *c.* 200 ka showing significant offset (Nakata *et al.* 1976). Amounts of cumulative vertical separation and estimated ages of the river terraces provide a vertical slip rate of 0.5–0.7 mm a^{-1}. Poorly constrained ages of palaeoseismic events recovered from trenches do not allow us to directly estimate the recurrence interval of large earthquakes, although an estimate of about 3000 years of recurrence interval can be inferred from an ideal coseismic displacement corresponding to $M = 7.0$–7.5 from the fault length and estimated slip rate (HERP 2002).

As discussed in the following section, one of the features of large inland earthquakes recorded in the Tohoku district is temporal clustering possibly influenced by the seismic cycle of subduction megathrust earthquakes. In this context the recent events of the 2003 Miyagi-ken–hokubu ($M = 6.5$), 2004 Niigata-ken–Chuetsu ($M = 6.8$), 2007 Noto–hanto ($M = 6.9$), 2007 Chuetsu-oki ($M = 6.8$), and 2008 Iwate–Miyagi-nairiku ($M = 7.2$) earthquakes preceded the 2011 $M = 9.0$ Tohoku-oki earthquake. Another characteristic is high palaeoseismicity in the eastern margin of the Japan Sea and the coastal regions. As introduced in the previous section, the late Quaternary contraction rate in the region is highest in NE Honshu. Densely populated active thrust faults and fault-related folds, which are reactivated normal faults originally generated during Miocene rifting, caused destructive large earthquakes in 1694 ($M = 7.0$), 1704 ($M = 7.0$), 1793 ($M = 6.9$–7.1), 1804 ($M = 7.0$), 1810 ($M = 6.5$), 1833 ($M \approx 7.5$), 1894 ($M = 7.0$), 1939 ($M = 6.8$, $M = 6.7$), 1964 ($M = 7.5$) and 1983 ($M = 7.7$) (Fig. 8.12). Among them, the 1833 and 1983 Nihonkai-chubu earthquakes triggered widespread tsunamis that reached down as far as the Japan Sea coast in SW Japan. One of the symbolic changes in landform associated with a coastal active fault occurred during the 10 July 1804 Kisakata earthquake of $M = 7.0$ when a scenic coast with a shoal of tiny islands, the subject of a famous seventeenth century poem, was coseismically uplifted by up to 2 m (Hirano *et al.* 1979). A stretch of the coastal zone of length *c.* 25 km emerged from the sea to form new land.

The Itoigawa–Shizuoka Tectonic Line fault system, central Honshu
The Itoigawa-Shizuoka Tectonic Line (ISTL) active fault system in central Japan is a complex 150 km long fault system consisting of three distinct segments: north-trending east-dipping reverse faults; NW-trending left-lateral strike-slip faults; and north-trending west-dipping reverse faults (Fig. 8.15, Okumura *et al.* 1994; Shimokawa *et al.* 1995). The ISTL is a part of the geologically defined western boundary of the 'Fossa Magna' (Naumann 1887), a wide and deep sedimentary graben separating SW Japan from NE Japan in the Miocene period. The ISTL originally comprised mostly normal faults, but these have now been reactivated as thrust and strike-slip faults accommodating east–west contraction in late Quaternary times. The ISTL is one of the most active faults in Japan, with slip rates of up to 10 mm a^{-1} on the Gofukuji Fault along the central strike-slip section. Despite such a high rate of late Quaternary slip, a number of palaeoseismic studies completed in the 1980s suggested that the recurrence interval of large earthquakes is 3500–5000 years, with the most recent large earthquake likely corresponding

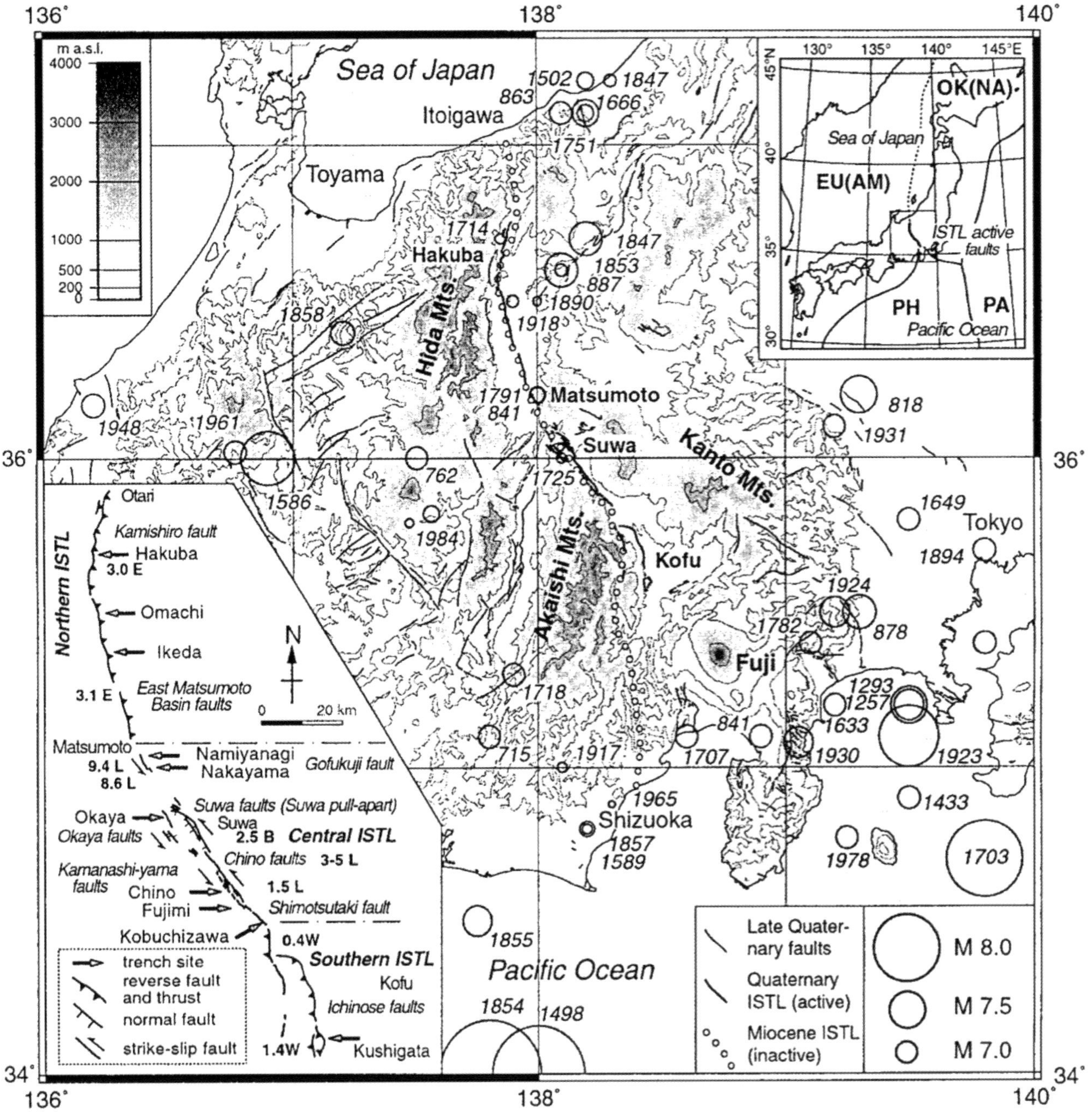

Fig. 8.15. Late Quaternary faults and large historic earthquakes in and around central Japan (Okumura 2001). Outline of the Itoigawa–Shizuoka Tectonic Line (ISTL) active fault system based on Shimokawa *et al.* (1995) is in the inset lower left. Numbers indicate average slip rate of the faults in m ka^{-1}. E, east-side-up vertical component; W, west-side-up vertical component; L, left-lateral; B, subsidence of Suwa pull-apart basin.

to historically documented earthquakes occurring in AD 841 (e.g. Research Group for the Ito-Shizu Tectonic Line Active Faults 1988). This implies that the elapsed time since the last earthquake is short relative to the average recurrence interval, so that the next large earthquake is unlikely to occur in the near future. Such a recurrence interval would however produce an unrealistic 35–50 m coseismic slip, taking the faster slip rate of several millimetres per year into account. Intensive trenching studies by the Geological Survey of Japan in the 1990s resolved the contradiction (e.g. Okumura *et al.* 1994) by excavating three trenches across the Gofukuji Fault, one of the middle segments of the ISTL. It was discovered that the recurrence interval was *c.* 1000 years and that the most recent surface faulting would have occurred either in AD 841 or 762. Okumura *et al.* (1998) also revealed that the recurrence interval of the northern ISTL is 1000–1800 years and the last event might have occurred together with ruptures on the middle ISTL either in AD 841 or 762, implying that the next surface-rupturing earthquake is highly likely to occur very soon. The ISTL has therefore been prioritized as an active fault in need of evaluation by the Headquarters for Earthquake Research Promotion (HERP) in 1996. The HERP completed its evaluation and warned of a near-future earthquake of $M \approx 8.3$ involving rupture along the entire ISTL. At that time however, even though they mentioned the potentially different behaviours of the northern and southern ISTL, the data were insufficient to forecast the details of the next large event. In 2014, the 30-year probability of a *M*8 class earthquake on the ISTL was calculated to be 14%, based on the elapsed time of 12 ka and its shorter 1000-year recurrence time at the Gofukuji Fault (HERP 2014).

Since the official warning by the government in 1996, continuous efforts by earthquake geologists have been performed. The ISTL is now the most-excavated fault system in Japan, not only in the evaluation of palaeoseismicity but also in terms of fault segmentation studies. Natural outcrops, palaeoseismic trenches, drilling and Geo-slicer samples at more than 40 sites have exposed palaeoseismic events during the past 10 ka (Fig. 8.16). Maruyama *et al.* (2010) compiled all the palaeoseismic data and redefined ten behavioural segments with various recurrence intervals. The recurrence intervals of surface-rupturing earthquakes on the northern, central and southern segments are 1300–2000 years, 1100–1500 years and 3000–6000 years, respectively. The slip rate of the southern segment is significantly slower than the others and their recurrence intervals are more than double (Toda *et al.* 2000; Miura *et al.* 2002). The excavations across the southern ISTL did not expose any evidence for rupture having occurred either in AD 762 or 841. These results suggest that rupture of the entire 150 km, producing a huge earthquake, is unlikely to occur.

Recent instrumental seismicity along the ISTL has been low except for several distinct clusters of small earthquakes near the Gofukuji Fault in the central segment. Detailed studies of focal mechanisms of small earthquakes along the ISTL (Fig. 8.17, Kuwahara *et al.* 2010) have revealed that the strike-slip earthquakes occurred in the middle ISTL and thrust-faulting earthquakes occurred in the southern ISTL, which is consistent with the sense of active faults. However, the observed focal mechanisms in the northern ISTL indicate a mixture of both strike-slip and reverse events. The southern part of the northern ISTL is in particular dominated by strike-slip earthquakes, which is indeed consistent with the recent discovery of strike-slip offset near Matsumoto by detailed geomorphic studies using airborne light detection and ranging (LiDAR) (Kondo *et al.* 2008).

A significant increase in seismicity along the ISTL has been observed since the 2011 Tohoku earthquake (Toda *et al.* 2011a). A moderate-sized earthquake of $M = 5.5$ occurred near the Gofukuji Fault on 30 June 2011 about four months after the Tohoku earthquake, killing one person in Matsumoto. Computed Coulomb failure stress increase on the central ISTL due to the Tohoku earthquake is up to 0.5 bar (Toda *et al.* 2011b), and continuing related seismicity is estimated to last more than a decade (Toda *et al.* 2011a).

The Atera fault system, central Honshu and the 1568 Tensho earthquake of $M \approx 7.8$

The Atera fault system defines a topographical boundary between the elevated northern Atera Range and the lower SW Mino Heights. The main rivers (Kiso, Tsukechi, Kashimo and Hida) that cross the

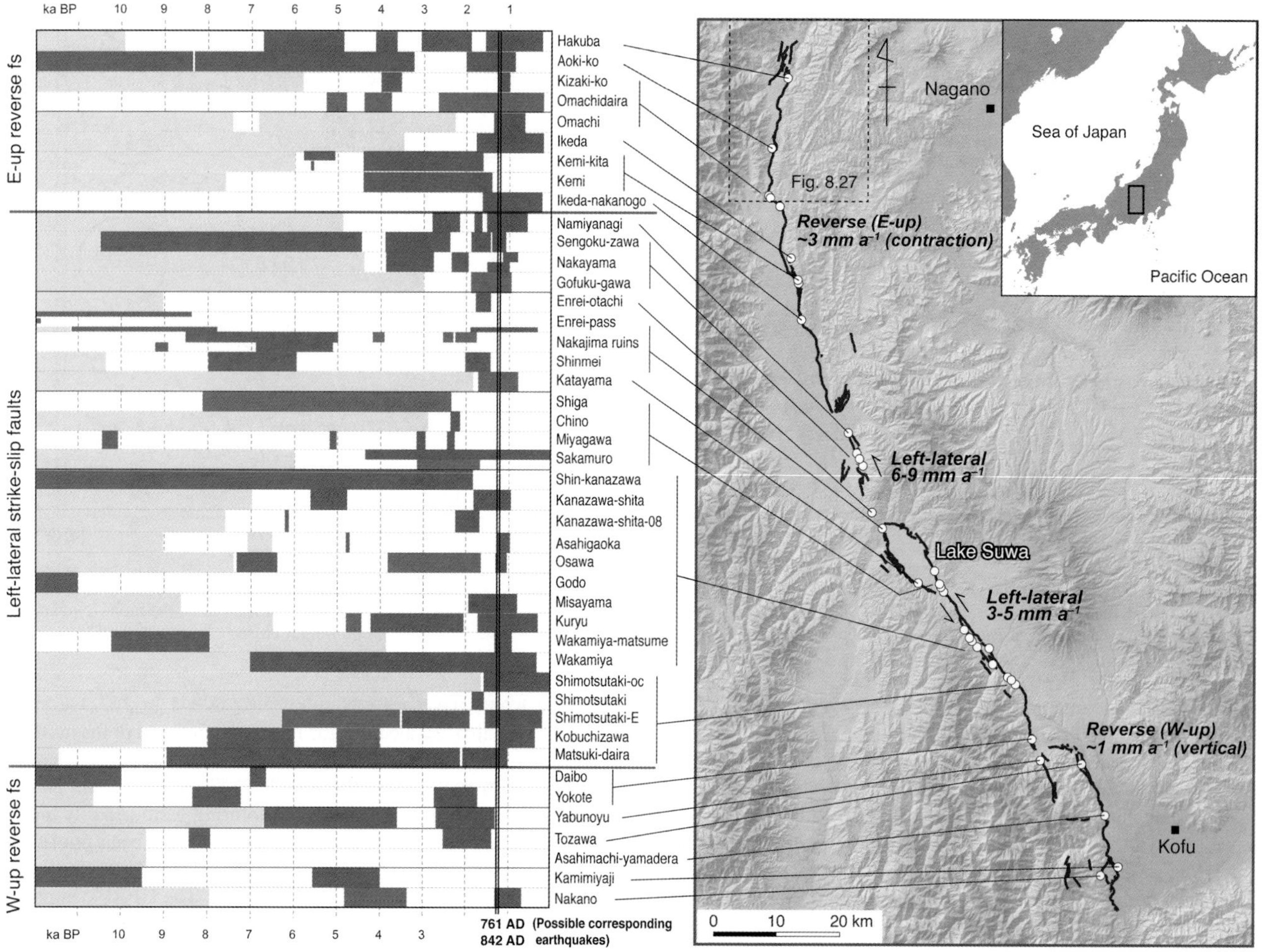

Fig. 8.16. Space–time diagram for large-magnitude surface-rupturing earthquakes on the ISTL during the past 12 ka (Maruyama *et al.* 2010). Horizontal dark grey bar indicates a range of palaeo-earthquake occurrence. Horizontal light grey bar denotes a time range across which no datable strata were recovered.

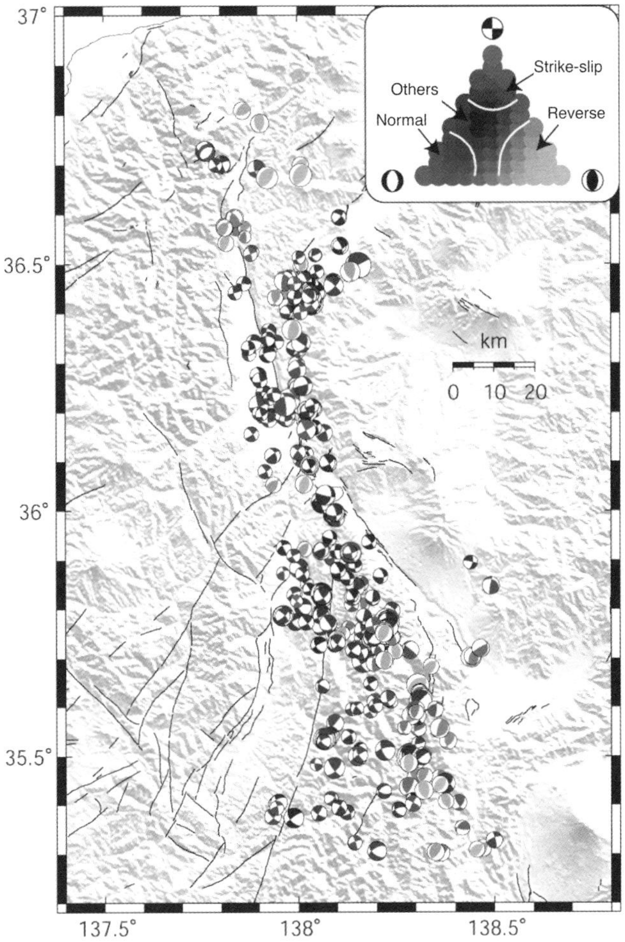

Fig. 8.17. Focal mechanisms of small-to-moderate earthquakes obtained in and around the ISTL (Kuwahara *et al.* 2010). Earthquakes caused by strike-slip faults extend north further than that expected from geological structure.

fault system show conspicuous inflections due to a 5–15 km left-lateral offset by accumulated faulting. Late Quaternary fluvial terraces and fans along these rivers are also clearly offset. Among them, all terraces in the town of Sakashita, where the Kiso River flows, have been displaced such that the greater the displacement, the older the terrace (Fig. 8.8; Sugimura & Matsuda 1965). About 1–5 mm a^{-1} of left-lateral slip and 1 mm a^{-1} of NE-up vertical slip have been estimated from the terraces (e.g. Tsukuda *et al.* 1993).

The Atera fault system bisects the upper Cretaceous Nohi Rhyolitic rocks. Along the northern to central parts of the fault system, these rhyolites juxtapose the Mino sedimentary complex and an intrusive granite. Restoring the original geological structure by the offset displayed by the Nohi Rhyolites, a left-lateral offset of *c.* 7 km can be estimated (Collaborative Research Group for the Nohi Rhyolite 1973; Yamada 1981).

The Atera fault system consists of eight geometric segments (Tsukuda *et al.* 1993). In the 1970s, the fault system was not highlighted as especially dangerous because of the lack of evidence for a surface rupturing earthquake corresponding to any historical earthquake. However, subsequent palaeoseismic trenching surveys were performed across the Atera fault system in the 1980s which, along with a road-cut exposure found in early 1990s (Toda *et al.* 1994) and the discovery of new historical documents (Awata *et al.* 1986; Usami *et al.* 2013), suggest that the Atera fault system ruptured during the 1586 Tensho earthquake of $M \approx 7.8$.

The 29 November 1586 Tensho earthquake is one of the largest and most enigmatic inland earthquakes ever documented in Japan. Since this earthquake occurred during a state of anarchy (the age of warlords, Sengoku period), historical documents are highly limited. Isoseismal areas of JMA intensity 6 are however estimated to have extended from south of the Noto Peninsula to the north (current Toyama Prefecture) and around the Atera Fault (current Gifu Prefecture), Nagoya and Lake Biwa to the west (Fig. 8.18). There were massive landslides in the northern part of the disaster area, one of which completely eradicated a large fortress on the hill. Many other local large fortresses also collapsed in response to the earthquake. Several reliable accounts described submerged small islands along the Kiso-gawa River and a tsunami in the Ise Bay that killed several thousand people. They are interpreted as massive liquefactions occurring on the Nobi Plain, where the city of Nagoya is now located. Huge sand blows and sand dykes corresponding to the historical documents were found in several archaeological sites in the plain (Kanaori *et al.* 1993). In Kyoto, several temples suffered minor damage and there exist historical documents describing how the aftershocks lasted for the following year (Usami *et al.* 2013). In addition, one of the missionaries of the Society of Jesus (Luis Frois) recorded the damage caused by the earthquake.

The epicentre of the Tensho earthquake is estimated to have been located at the centre of the isoseismals (Fig. 8.18). However, the historical records and palaeoseismic data suggest that the Atera Fault is only one of the sources of the huge inland earthquake. Sugiyama *et al.* (1991) found the most recent events on the Miboro Fault occurred after the eleventh century judging from trench excavations, so this fault may have been responsible for the 1586 Tensho earthquake. Since then, numerous archaeological exposures of palaeoliquefactions in the Nobi Plain and drilling surveys (Kanaori *et al.* 1993; Sugai 2011) have revealed that the Yoro Fault was also involved in the Tensho earthquake and induced the massive damage in the current Nagoya and Gifu areas. A recent study of historical documents argued that the Tensho earthquake would have been separated into an $M \approx 7.0$ earthquake on 16 January on the Miboro Fault and the subsequent $M = 7.8–8.0$ earthquake on 18 January in the Nobi Plain, and then opposed to include the Atera fault as one of the sources (Matsu'ura 2011). The exact sources of the Tensho earthquake are therefore still controversial, but it is more likely that this great inland shock was a product of sequential multiple fault ruptures (doublet or triplet) occurring across the dense fault network.

Dense fault networks in Osaka and Kyoto

Tectonic forces associated with the subduction of the Pacific Plate are transmitted even further to the Inner Zone of SW Japan through the northern Fossa Magna region (Huzita 1980). Even though active faults in the Kinki district are distributed densely and can lack obvious order, it appears that several crustal blocks or microplates are bounded by systematic fault clusters associated with major fault zones (Huzita 1962; Kanaori 1990; Kanaori *et al.* 1993). The majority of destructive earthquakes have occurred along these major fault zones (microplate boundaries of Kanaori *et al.* 1991).

Rupture of the 1995 Kobe earthquake started beneath the Akashi Strait between the city of Kobe and northern edge of the Awaji Island (Figs 8.19 & 8.20), and then propagated bilaterally to the NE and SW. Slip beneath the city of Kobe occurred mostly deeper than 5 km (Sekiguchi *et al.* 2002*a*), whereas significant right-lateral movement with a reverse component of up to 2.5 m slip broke surface (Fig. 8.7b) along the 11 km trace of the previously mapped Nojima Fault (Fig. 8.20, Awata *et al.* 1996). Based on the long-term slip rate

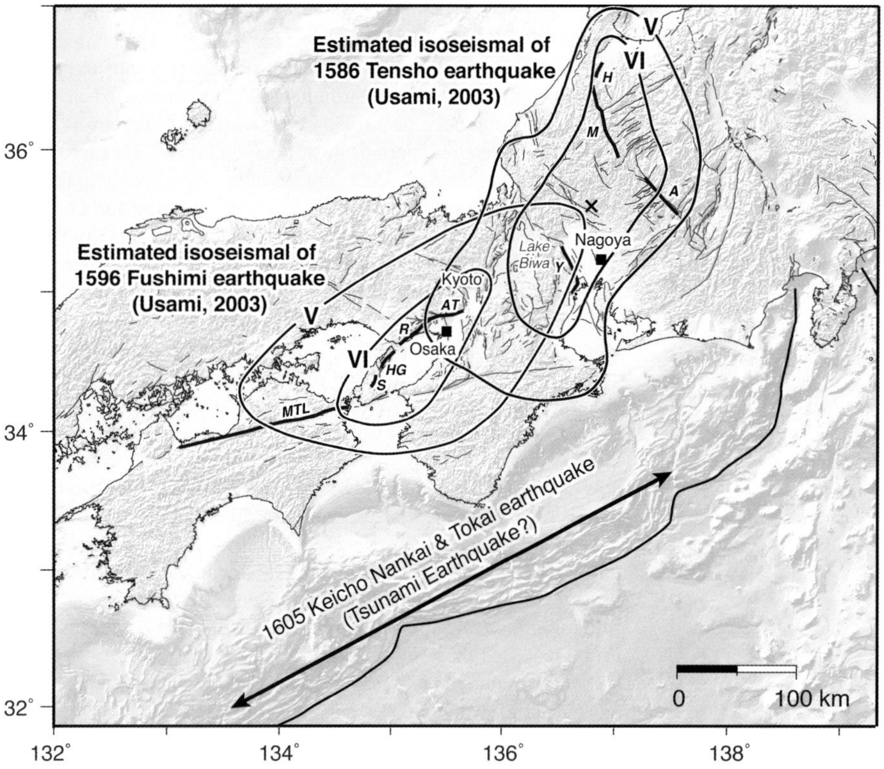

Fig. 8.18. Isoseismal of the JMA intensities V and VI at the 1586 Tensho earthquake and the 1596 Fushimi earthquake (Usami 2003) and their candidates of the source faults (H, Horinji Fault; M, Miboro Fault; A, Atera Fault; Y, Yoro Fault; AT, Arima–Takatsuki Tectonic Line active fault system; R, Rokko fault system; HG, Higashiura Fault; S, Senzan Fault). The locations of active faults are from Research Group for Active Faults of Japan (1991).

of the Nojima Fault (Mizuno *et al.* 1990) and the amount of measured slip at the Kobe earthquake, the recurrence interval of such surface-rupturing earthquakes is estimated to be 1300–2500 years (Awata *et al.* 1996).

The penultimate large crustal earthquake in this region, the Fushimi earthquake, occurred on 5 September 1596 (Figs 8.18 & 8.19). The size of the Fushimi earthquake is estimated to be $M \approx 7.5$ and devastated southern Kyoto, being especially notorious for destroying a castle and a large Buddha statue of the Shogunate (chancellor) Hideyoshi Toyotomi. Evidence for palaeoseismic events possibly corresponding to the 1596 earthquake was found in several trench walls across the Arima–Takatsuki Tectonic Line (ATTL) NE of Kobe, the Senzan Fault and the Higashiura Fault in northern Awaji Island (Fig. 8.18; Sugiyama 1998; Yoshioka *et al.* 1997). Lin *et al.* (1998) also found evidence for the surface rupture having occurred about 400 years ago on one of the segments of the Rokko fault system, where deeper extension of the fault slipped during the 1995 Kobe earthquake. Prior to the 5 September Fushimi

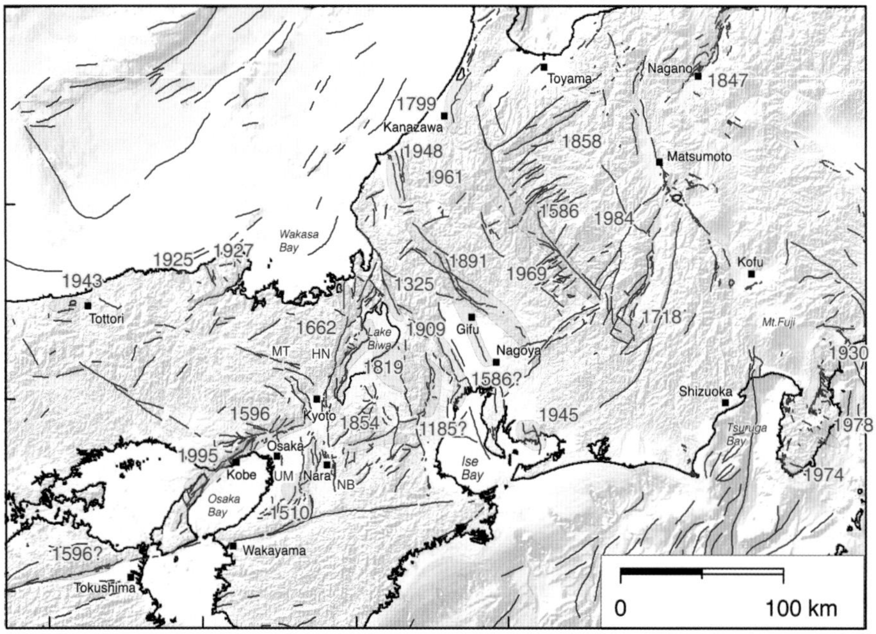

Fig. 8.19. Confirmed and inferred rupture zones (shaded) of large inland earthquakes in the Kinki and Chubu district, central Japan during the past 1000 years (Tsukuda 2002). Number denotes occurrence year.

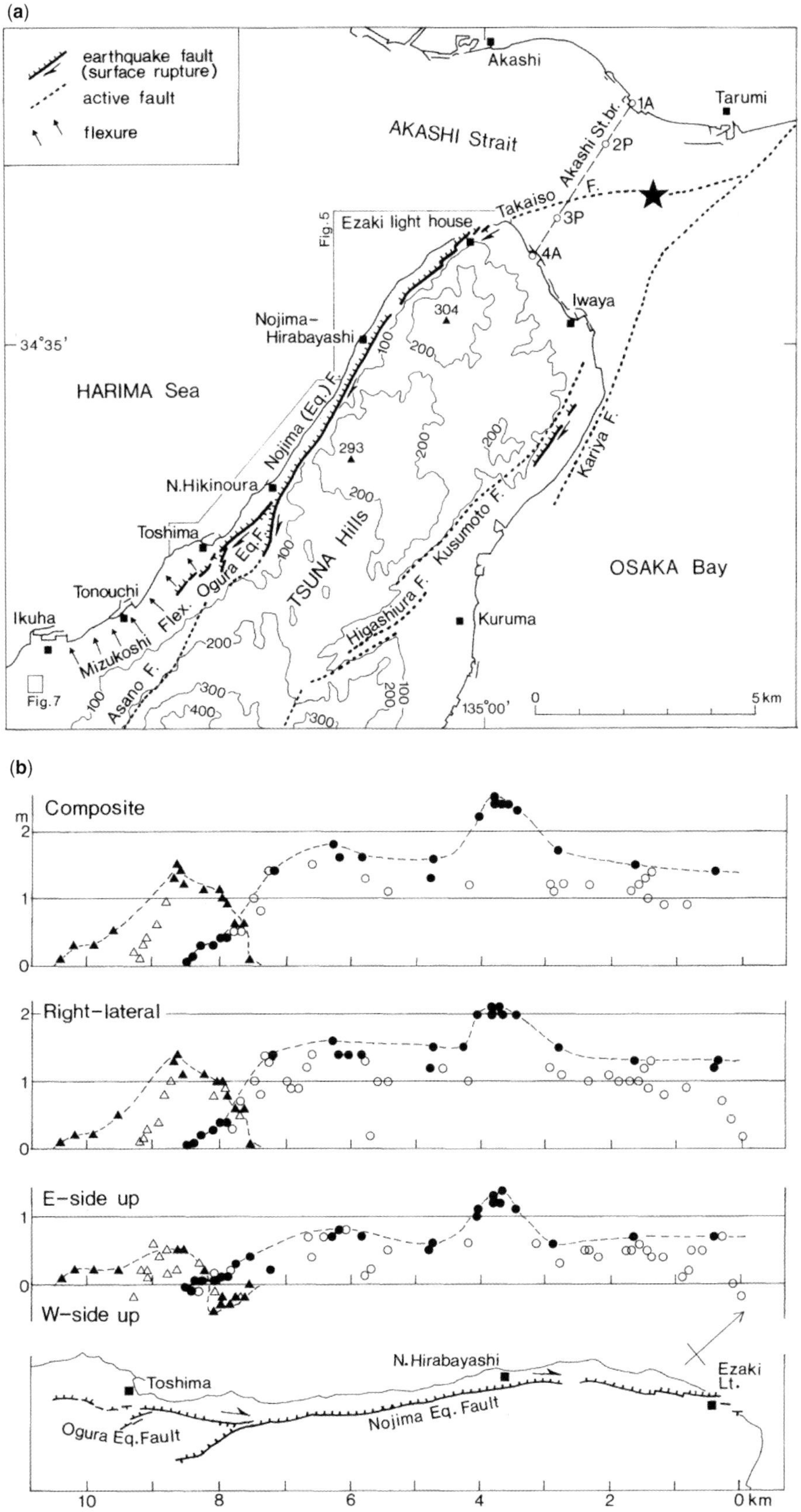

Fig. 8.20. Surface rupture associated with the 1995 *M* 7.3 Kobe earthquake (Awata *et al.* 1996). (**a**) Distribution of the ruptures along the Nojima Fault and Ogura Fault on the Awaji Island. The solid star is the epicentre of the earthquake. (**b**) Displacements along the Nojima and Ogura faults are denoted as open circles and open triangles, respectively. Solid symbols indicate the total displacements of the fault strand.

earthquake, a few large earthquakes had occurred along the MTL in Shikoku, beginning with an $M \approx 7.0$ earthquake on 1 September (Usami *et al.* 2013). The Fushimi earthquake might therefore have been the last event of a delayed multiple sequential rupture involving a number of active faults. Sugiyama (1998) and Matsuda (1998) suggest that recurrence interval of the Rokko–Awaji Fault Zone is c. 400 years with various rupture patterns. Unlike in the 1596 Fushimi earthquake, the rupture in 1995 stopped beneath the city of Kobe. Hashimoto (1995) and Toda *et al.* (1998) have calculated the static Coulomb stress change caused by the 1995 Kobe earthquake, and concluded that the ATTL was brought closer to failure by several bars.

Most of the 8.9 million population of Osaka Prefecture live in the Osaka Basin which is bounded by the ATTL to the north, the Ikoma Fault to the east and the uplifted Izumi Mountains associated with the MTL to the south. In particular, the east-dipping Uemachi Fault underlies the city centre of Osaka as a blind-thrust fault. The central to southern part of the Uemachi Fault is identifiable due to uplifted terraces on the hanging wall of the fault. Several seismic reflection surveys (Fig. 8.21; Sugiyama *et al.* 2003; Iwata *et al.* 2013) and drilled cores have revealed that the fault length is c. 50 km, with a vertical slip rate of 0.3 mm a^{-1} (Iwata *et al.* 2013). Timing of the most recent event is still unknown, with the most reliable data suggesting that the last earthquake occurred sometime after 9 ka BP (Sugiyama *et al.* 2003). The anticipated size of the earthquake from the Uemachi Fault is $M \approx 7.5$, which will cause widespread severe shaking in the Osaka Basin which is underlain by over 2 km of Quaternary sediments. Numerous strong ground motion simulations were demonstrated based on scenario earthquakes of the Uemachi Fault (e.g. Sekiguchi *et al.* 2002*b*). The city of Osaka has revealed the anticipated intensity map to the public, showing that the areas likely to suffer severe shaking of JMA intensities 6+ and 7, roughly equal to the Modified Mercalli Intensity (MMI) scale XI to XII, occupy more than half of the city.

The ancient capitals of Japan, Kyoto and Nara are located on basins bounded by active faults (Fig. 8.19). Kyoto, where the imperial palace was located from AD 794 to 1868, lies on a sedimentary sequence up to 700 m thick and bounded by the Hanaore Fault to the east and the Mitoke–Nishiyama Fault to the west. Among many historical earthquakes documented in Kyoto, probably the 16 June 1662 Kanbun earthquake was the most damaging (Fig. 8.19). The widely distributed devastated areas, suggesting $M \approx 7.8$, occurred across northern Kyoto and included villages along the western coast of Lake Biwa (Usami *et al.* 2013). Based on recent palaeoseismic studies (Okada 1984; Komatsubara *et al.* 1999; Ishimura *et al.* 2010), the earthquake is also thought to be a multiple-rupture event involving the Kumagawa Fault, a left-lateral strike-slip fault, and conjugate right-lateral Mikata Fault NW of Lake Biwa. Since a significant area of the western coast of Lake Biwa was submerged into the lake by the earthquake (Sangawa & Tsukuda 1986), the northern part of the Hanaore Fault was also involved in the rupture process, although with some time delay (T. Komatsubara, 2010 pers. comm.). The older capital of Nara is bounded to the east by the Nara-bonchi-toen Fault Zone which is interpreted as some kind of southern extension of the Hanaore fault system. The vertical slip rate and recurrence interval of the fault are estimated to be 0.6 mm a^{-1} and about 5000 years, respectively (HERP 2001). No severely damaging earthquake larger than *M*6.5 has been recorded beneath the Nara Basin, although the Nankai Trough megathrust earthquakes have inflicted significant damage on the city.

MTL active fault system

The MTL active fault system is one of the fastest-moving and longest-active strike-slip faults in Japan. As for the ISTL, the MTL is a major geological boundary, in this case separating the low-pressure–high-temperature metamorphic rocks and granites of the Ryoke Belt to the north from the high-pressure–low-temperature metamorphic rocks of the Sanbagawa Belt to the south. The MTL has been accommodating the right-lateral shear component associated with the oblique subduction of the Philippine Sea Plate at a northern edge of a forearc sliver (Miyazaki & Heki 2001; Ikeda *et al.* 2009; Fig. 8.22). The right-lateral slip began in late Pliocene–early Pleistocene times (Sangawa 1978), with surface traces lying a little north of the geological boundary (e.g. Mizuno *et al.* 1993). The maximum slip rate of the MTL is c. 10 mm a^{-1} on eastern Shikoku Island (Okada 1973; Tsutsumi & Okada 1996), decreasing to 1–2 mm a^{-1} in Wakayama in the Kii Peninsula

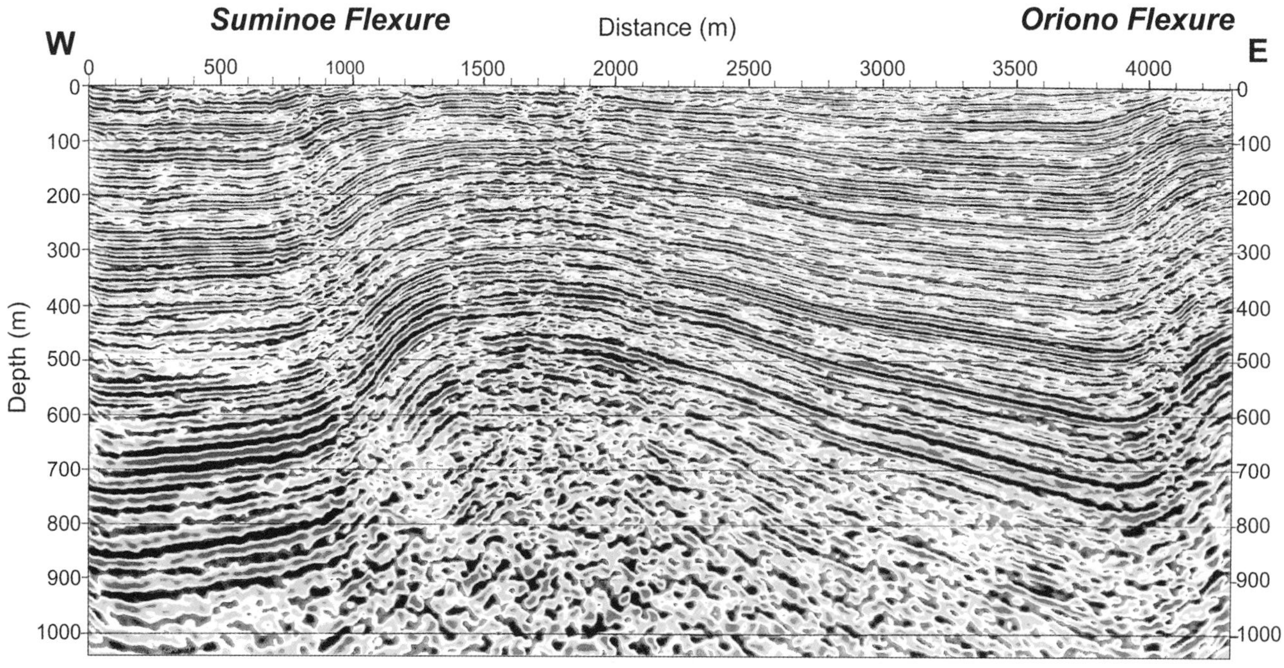

Fig. 8.21. Depth-converted P-wave seismic reflection profile across two flexures of the Uemachi Fault in south Osaka (Sugiyama *et al.* 2003). Cumulative vertical separations are seen in fault-related fold of Plio-Pleistocene non-marine and marine sediments.

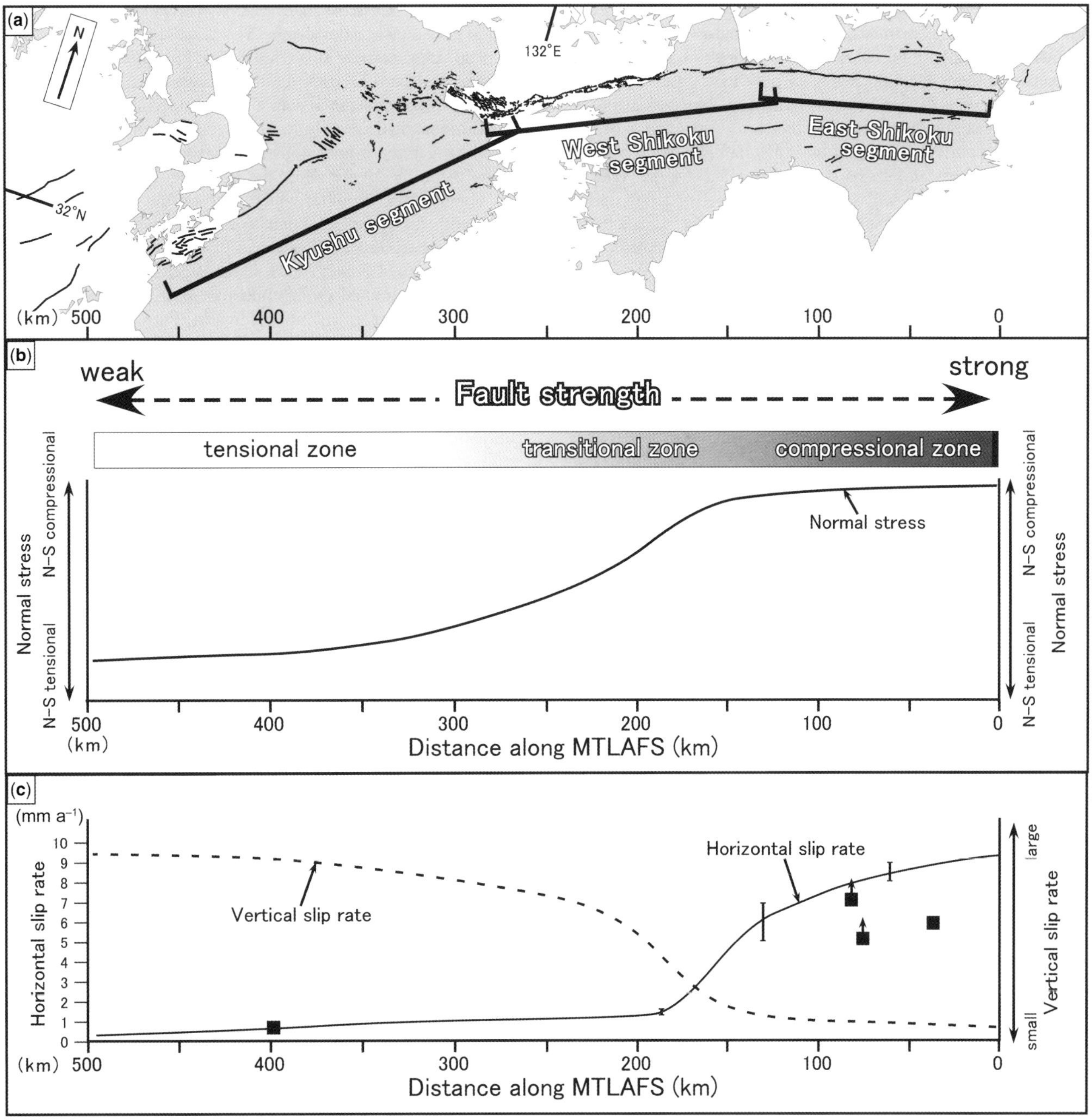

Fig. 8.22. (**a**) Segmentation model for the MTL active fault system based on fault distribution and stress condition. (**b**) Schematic stress conditions and fault strength profile along the MTL. (**c**) Horizontal slip rates and estimated trend of vertical slip rates along the MTL. (Ikeda *et al.* 2009)

(Okada & Sangawa 1978; Mizuno *et al.* 1993). Although the MTL as a geological boundary continues further east to the Kanto region no evidence for Quaternary activity has been found further east of the Kongo Fault (Takehisa 1973), which is accommodating the right-lateral displacement as a north–south-striking reverse fault.

The MTL changes its structure, slip style and stress conditions from transpression on eastern Shikoku Island to transtension on Kyushu Island (Fig. 8.22, Ikeda *et al.* 2009). Ikeda *et al.* (2009) have argued that the change of stress conditions along the fault system is caused by the anticlockwise rotation of the Nankai forearc sliver in response to the relative motion between the Philippine Sea and Eurasian tectonic plates and back-arc spreading in the Okinawa Trough. They identify three main segments with contrasting stress: East Shikoku, West Shikoku and Kyushu (Fig. 8.22). The East Shikoku segment, where the right-lateral slip rate is 5–9 mm a^{-1}, is characterized by straight fault traces with minor transpressional steps. In contrast, the West Shikoku segment is composed of several small fault segments connected by transtensional gaps and pull-apart grabens. The horizontal slip rate decreases to 1 mm a^{-1} in western Shikoku, but the vertical movements associated with a group of normal faults in Beppu Bay become significant. Beyond the Beppu–Aso volcanic zones (eastern part of Beppu–Shimabara graben), which includes numerous normal faults, the Futagawa–Hinagu fault system renews the right-lateral movement with a slip rate of 0.8 mm a^{-1}. The southwestern end of the Futagawa–Hinagu fault system has numerous branches and may represent the continuation to the Okinawa Trough in the Ryukyu subduction system. It is interesting to note that these MTL fault traces, stress conditions and the movement of the Nankai forearc sliver are quite similar to those along the North Anatolian fault system during the current rotation of the Anatolia Plate block.

The most recent surface-breaking earthquake on the MTL is thought to be a series of sequential ruptures which occurred during 1–5 September 1596 (Nakanishi 2002; Okada 2012; Usami *et al.* 2013). The first event appears to have occurred on 1 September in western Shikoku, followed by a second which occurred in Beppu Bay at around 16:00 hours on 4 September. The largest rupture occurred in eastern Shikoku as part of the 1596 Fushimi earthquake on 5 September (summarized in Okada 2012). Recent palaeoseismic trenches at several locations along the entire MTL in Shikoku exposed evidence for the last movement as having occurred after the sixteenth century (Morino & Okada 2002). The largest earthquake that has been recorded in eastern Shikoku, the Fushimi event in 1596 (see above), is inferred to have involved up to 7 m of right-lateral slip (Tsutsumi & Goto 2006), as measured from the geomorphic and artificial features. Extensive palaeoseismic data obtained from the MTL in Shikoku has also yielded evidence for a penultimate event that occurred about 2 ka BP (Morino & Okada 2002). It suggests that the recurrence interval of surface-rupturing earthquakes on the MTL is longer than 1000 years, but involves large coseismic slip so that large magnitudes would be expected when the fault moves. Tsutsumi & Goto (2006) concluded that the entire MTL Shikoku segments would produce an earthquake as large as $M_w = 7.8$, which would be far greater than the 1891 Nobi earthquake ($M_w = 7.4$, $M = 8.0$).

Long-term earthquake forecasting and seismic hazard assessment

Probabilistic seismic hazard map

Before the Kobe earthquake there were several attempts to make a national seismic hazard assessment in Japan based on active fault data (e.g. Wesnousky *et al.* 1984), with some movement toward the probabilistic seismic hazard estimates following the methodology applied in California (Working Group on California Earthquake Probabilities 1988). Since the earthquake, the Headquarters for Earthquake Research Promotion (HERP) has accelerated the effort by revisiting historical documents, re-evaluating seismicity and compiling vast amounts of recent palaeoseismic data.

The HERP, a special governmental organization attached to the Prime Minister's office (currently belonging to the Ministry of Education, Culture, Sports, Science and Technology, MEXT) was established in 1995 and has been releasing probabilistic national seismic hazard maps for Japan every year since 2005 (Fig. 8.23), compiling thousands of instrumental, historical and palaeoseismological data and numerical models of plausible future earthquake sources and their shaking intensities.

Seismic hazard maps are traditionally based on epicentre densities, which is often called 'seismic zoning'. However, if historical, palaeoseismological and geological data as well as instrumental data are sufficiently available, these data can be used directly for making probabilistic seismic hazard maps that incorporate information on fault location, dimension, geometry, largest earthquake from such dimension, slip rates and recurrence intervals (e.g. Working Group on California Earthquake Probabilities 1995). Two types of statistical models of earthquake occurrence are commonly used to estimate earthquake probability: a stationary Poisson model and a conditional quasi-periodic model. The former is used for active faults or active seismic regions where only the frequency of large earthquakes is available. The latter considers fluctuations of repeat times, namely as a coefficient of variation in recurrence time, and elapsed time since the most recent event, which is time-dependent and compatible with the strain accumulation process. Several probabilistic density functions such as normal, log-normal, Weibull or gamma distributions are used for the renewal process. Among them, the HERP adopt the Brownian Passage Time function (Matthews *et al.* 2002) which is so far the best function due to its ability to account for the stress perturbations, and is also adopted for the probabilistic seismic hazards in California (HERP 2005*a*, *b*, *c*; Field *et al.* 2009). Time-dependent seismic hazard considering conditional probabilities on most major active faults provides us with more realistic estimates (Fujiwara *et al.* 2009). These outputs are available to the general public and researchers on the portal web named as 'J-SHIS', a browser-based GIS (http://www.j-shis.bosai.go.jp/map/?lang=en) to enhance utilization of the seismic hazard maps, as well as providing the fundamental information of fault locations, subsurface and surface geological conditions (site amplifications), exposed population and other models. Users can choose many options to see the information as preferred.

The national seismic hazard map for Japan is a result of a probabilistic seismic hazard analysis (PSHA). It shows the probability of exceedance of the Japan Meteorological Agency (JMA) intensity 6 (roughly equal to MMI IX-X) in Japan for a 30-year period (Fig. 8.23a). PSHA has two essential ingredients: seismic-scale source characterization (SSC) and ground-motion characterization (GMC) (Hanks *et al.* 2012). For any site of interest, SSC and GMC are combined in the hazard integral, which integrates over all magnitudes M and distance R to determine the exceedance rate of chosen ground-motion values on abscissa and their rates of exceedance on the ordinate. The hazard curves for many sites are then synthesized into a map showing hazard values for a fixed ground-motion level or ground-motion values for a fixed hazard level (Hanks *et al.* 2012).

In addition to the PSHA-based map for Japan, since 2005 the HERP has also produced detailed seismic hazard maps based on scenario earthquakes for the rupture of a seismic source fault, officially called *Seismic Hazard Maps for Specific Seismic Source Faults*. This is prepared for prediction of strong ground motions by considering the characteristics of an earthquake of interest when incorporating three-dimensional subsurface structure (Fig. 8.23b). This reflects the lessons learnt from the Kobe earthquake, in which strong shaking with significant damage was largely affected by the local characteristics of the source fault and subsurface geological structure. To simulate strong ground motion from a particular source fault, the standard method assigns the amount of slip, size and location of asperity, rupture propagation parameters and so on, together called the 'recipe' (Irikura & Miyake 2001; Irikura 2004). When providing palaeoseismic data to seismic engineers, possible rupture patterns for the next earthquake, slip distribution and plausible locations of asperities provide valuable additional information (Somerville *et al.* 1999). These maps are useful in the formulations of disaster prevention measures and emergency restoration programs for lifelines, such as water supplies and gas facilities, in cases where extensive damage is likely (HERP 2005*a*, *b*, *c*).

Three main features of seismic hazard assessment

As mentioned in the previous section, the 1995 Kobe (Hyogo-ken–Nambu) earthquake demonstrated three key features for seismic hazard assessment associated with active fault research. The first is

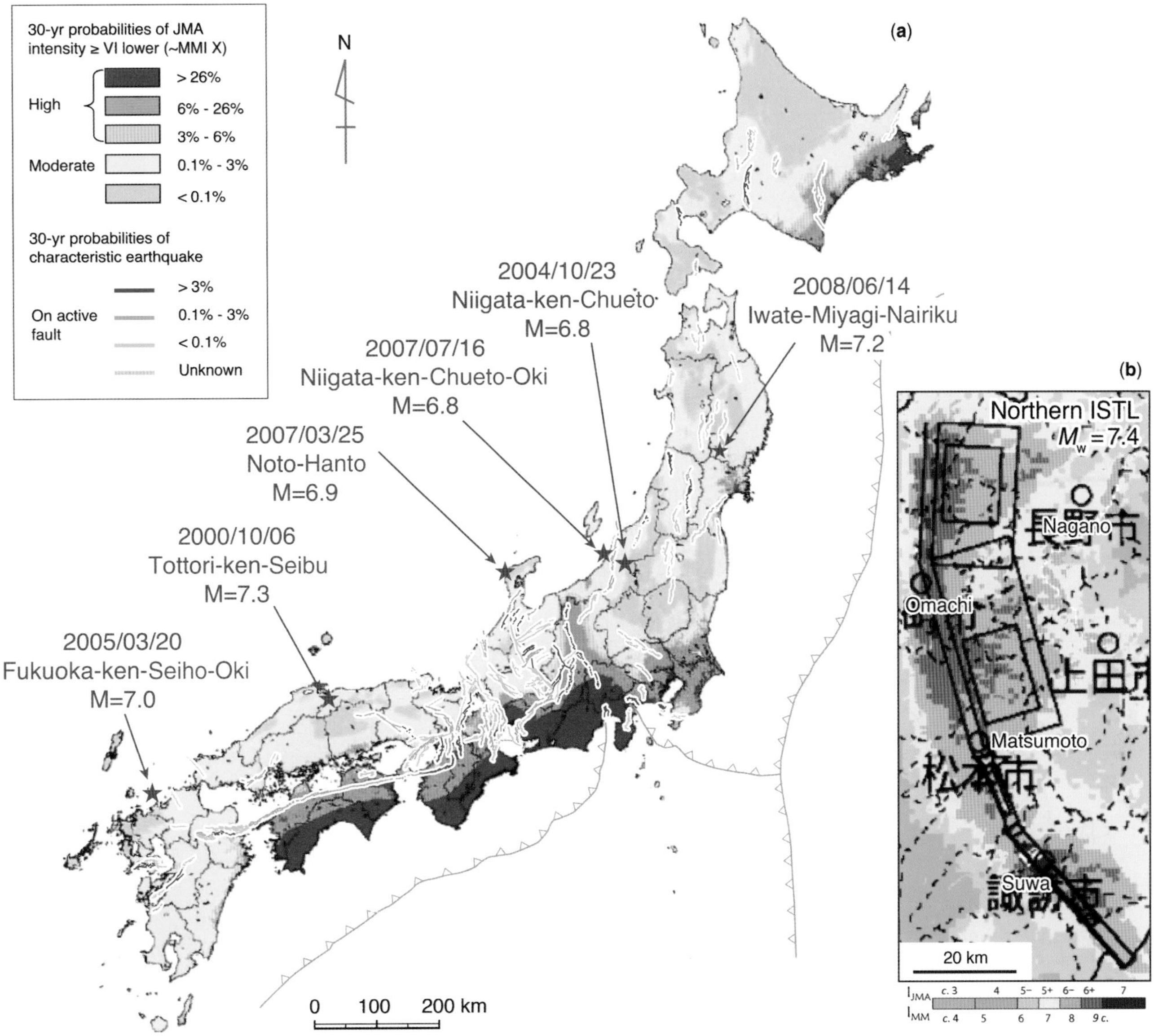

Fig. 8.23. (**a**) National seismic hazard map showing probabilities of ground motions ≥ JMA intensity 6 for the next 30 years (HERP 2005*a*, *b*, *c*). Destructive inland earthquakes since the 1995 Kobe earthquake and 30-year probabilities of characteristic earthquakes (largest earthquakes) on major active faults are also shown. Because locations of $M \approx 7$ epicentres cannot be directly compared with strong ground motion prediction, such a plot can be misleading to people. (**b**) An example Seismic Hazard Map for Specific Seismic Source Faults: forecast intensity distribution of an earthquake of $M_w = 7.4$ on the northern ISTL (HERP 2005*a*, *b*, *c*).

that the surface rupture occurred along the Nojima Fault, one of the mapped faults of inland Japan. From 1995, it gave us more confidence to progress with the excavation of large numbers of palaeoseismic trenches across mapped faults. The second feature is that seismogenic faulting under the city of Kobe did not reach the Earth's surface, which had implications for subsequent earthquakes. Most of the destructive inland shocks of $M \approx 7$ since then have indeed been associated with such blind faults. The third feature is the importance of fault geometry, which controls rupture initiation, propagation and termination; for example, the 1995 epicentre was located in a transtensional jog between the Nojima Fault on Awaji Island and the Rokko Fault beneath the city of Kobe. Recent studies have confirmed that a 5 km separation of fault spacing works well to estimate the maximum extent of rupture. The following subsections will discuss the second and third features.

Misunderstanding the seismic hazard map and issues with hidden and class C faults

Several large inland earthquakes in Japan since the 1995 Kobe earthquake have struck the lower-probability areas mapped by the HERP in 2005 (Figs 8.23 & 8.24). In particular, after the 2011 Tohoku earthquake, Geller (2011) and Stein *et al.* (2011) criticized the hazard map, writing: 'The regions assessed as most dangerous are the zones of three hypothetical 'scenario earthquakes' (Tokai, Tonankai, and Nankai). However, since 1979, earthquakes that have caused 10 or more fatalities in Japan actually occurred in places assigned a relatively low probability. This discrepancy – the latest in a string of negative results for the characteristic model and its cousin the seismic-gap model – might appear to strongly suggest that the hazard map and the methods used to produce it are flawed and should be discarded.' In addition, despite the 2010 version of the HERP

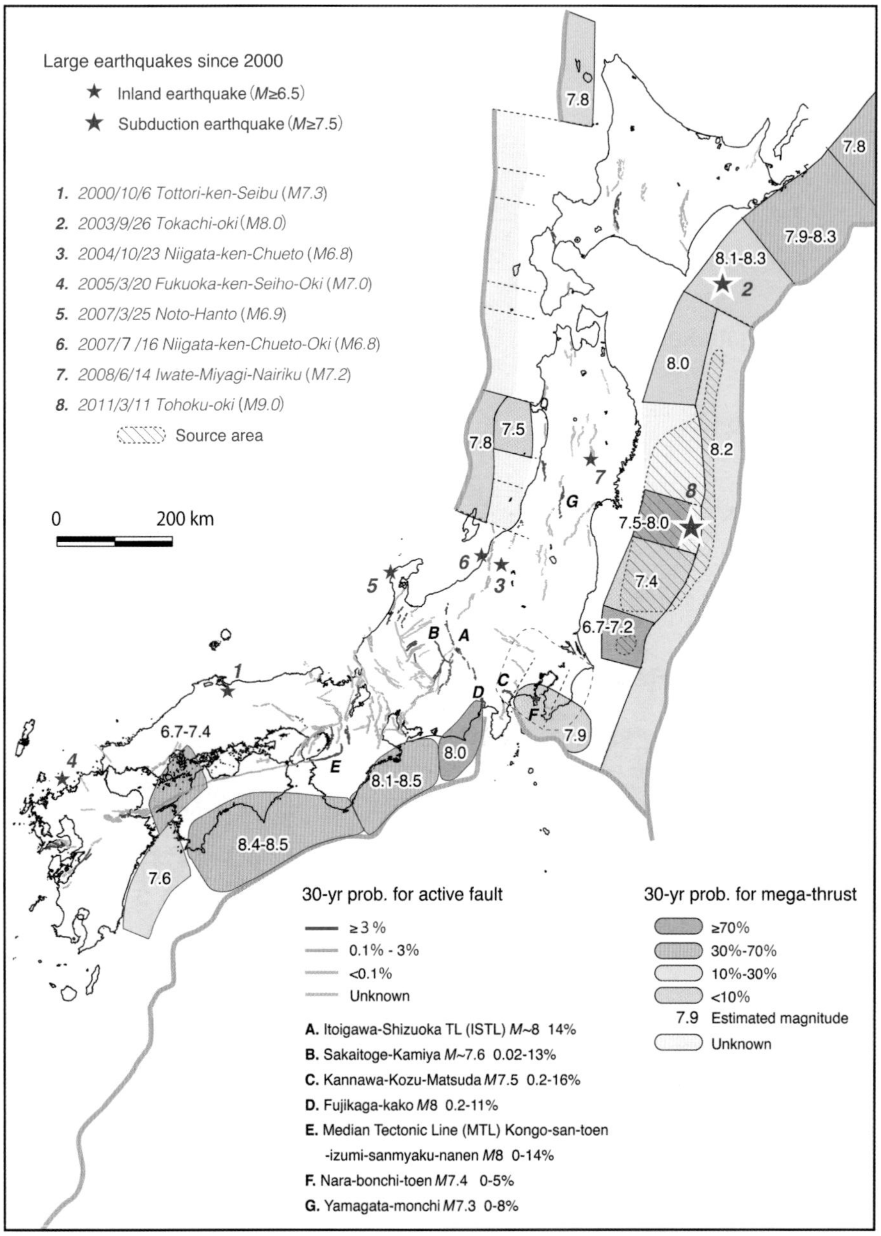

Fig. 8.24. Destructive inland earthquakes since the 1995 Kobe earthquake and 30-year probabilities of characteristic earthquakes (largest earthquakes) on major active faults and subduction interfaces.

hazard map predicting a less than a 0.1% probability of shaking with JMA intensity 6-lower in the next 30 years at Fukushima (once in the next 30 ka), such shaking occurred within 2 years (Stein *et al.* 2012).

Why does the hazard map appear to have failed? Events since its publication have indeed reminded us of the important message that low seismic hazard does not mean no seismic hazard. Several unforeseen large earthquakes occurring in the low probability regions do not mean a failure of probabilistic seismic hazard analysis (PSHA). The most meaningful tests will come from observations collected over long periods, where 'long' means times some multiple of the reciprocal of the hazard levels of interest (Hanks *et al.* 2012). Thirty years of such data is not long enough to prove that the map is wrong.

More importantly, a simple spatial comparison with the epicentre of large shocks superimposed on the hazard map in Figure 8.23a and the figures falsifying the hazard map in Geller (2011) are inappropriate. Ground motion hazard estimates are different from earthquake-occurrence observations. In Figure 8.23a, areas assigned high probabilities of large shaking depend highly on the megathrust earthquakes along the Nankai and Sagami subduction zones. Just one $M \approx 8$ near the coasts would cause a huge area to suffer strong shaking of JMA intensity 6 if it occurred during the 30-year prediction period. In contrast, shallow local $M \approx 7$ shocks within inland Japan result in much more limited areas of strong ground motion. Using an average occurrence rate of $M \geq 6.8$ once every 6 years in Inner Honshu, only 2–5% (3000–7500 km^2) of the area of Inner Honshu will experience JMA 6 intensity shaking during the 30-year period (Hanks *et al.* 2012). A substantial issue in avoiding misunderstanding of the hazard map based on PHSA is the need to separate the earthquake occurrence from ground motion hazard. The HERP has started to make three versions of the hazard map, categorizing all the sources into repeated subduction megathrust earthquakes (category I), unknown sources of subduction earthquakes (category II) and inland active faults (category III, Fig. 8.25). To view the forecast hazard of $M \approx 7$-class inland shocks, a PSHA for

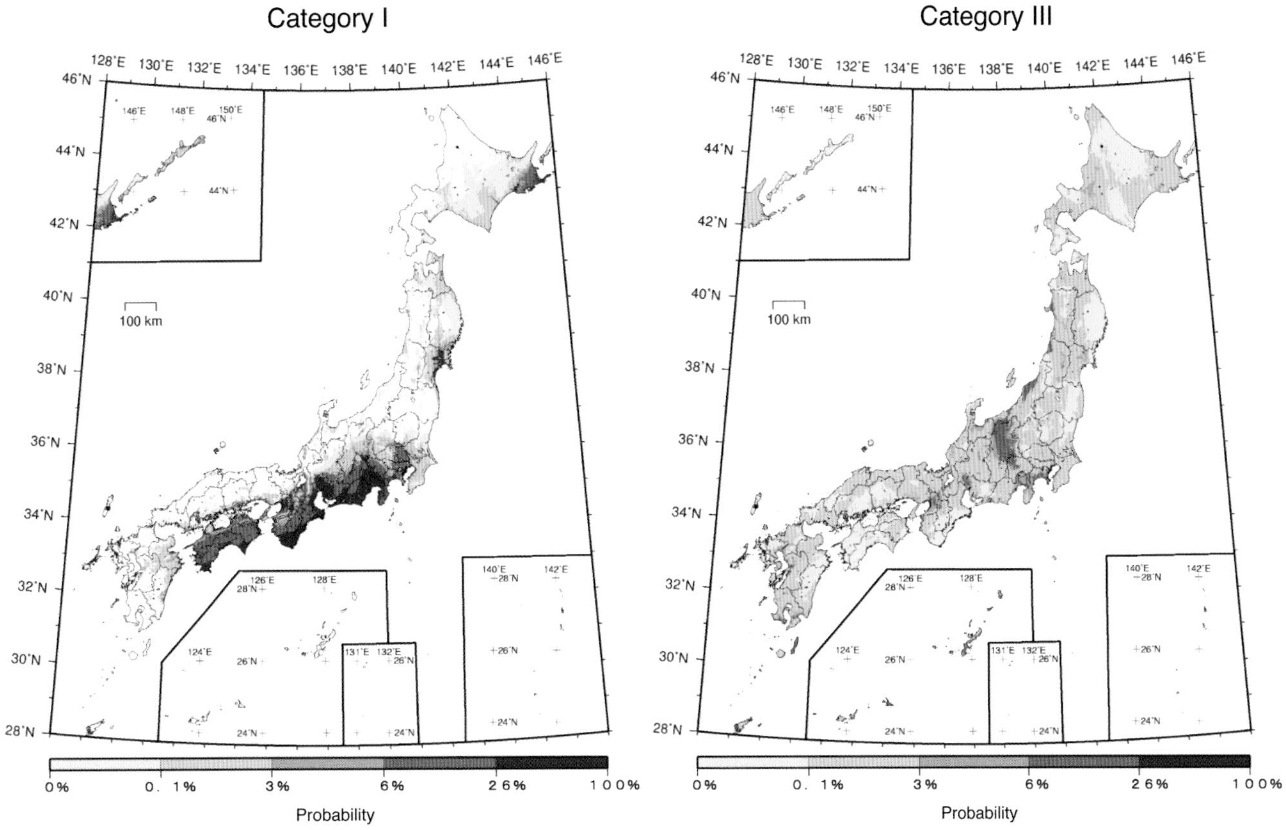

Fig. 8.25. Thirty-year probabilities of strong ground motion of JMA intensity ≥ 6 caused by seismic sources of Category I (subduction-zone earthquakes with specified seismic source faults) and Category III (shallow earthquakes inland and in sea area) (HERP 2010).

category III is appropriate and avoids confusing regional differences of the hazard with the extreme high probabilities due to the subduction earthquakes. This approach would appear to be very effective in correcting misleading impressions acquired by the general public, although these categorized maps are still hidden in the HERP technical report as one of many figures and are unfortunately easy to miss.

Even if we exclude the significant effects of the subduction earthquakes, such as on the category III map, there are still at least 110 major active faults affecting the inland earthquake hazard. As already mentioned, however, palaeoseismic trenching studies and other approaches to extract active fault parameters largely rely on surface-rupturing processes. In other words, if no surface rupture had taken place, no large earthquake would have been identified. With this in mind, Toda (2013) re-examined the surface-rupturing earthquakes since 1923 when the Japan Meteorological Agency (JMA) officially started their recording catalogue (Fig. 8.26; Table 8.2). Only 20% of $M \geq 6.5$ and 44% of $M \geq 7.0$ shallow earthquakes left the surface breaks that correspond to their source fault dimension. It can therefore be concluded that the number of potential destructive earthquakes of *M*6–7 estimated from the major active faults is likely to be significantly underestimated. Numerical calculations with active fault data and their assigned probabilities in the HERP report also largely underestimate the number of observed earthquakes. Both of these independent analyses suggest that there are far more minor active faults hidden beneath the Japanese islands.

Asada (1991) has already pointed out the so-called 'class C fault issue' which raises a question about hidden active faults with slip rates of <0.01 mm a^{-1}. He claimed that 'since the recent destructive inland earthquakes are occupied equally by all the classes (1 mm/yr $\leq$ class A $<$ 10 mm/yr, 0.1 mm/yr $\leq$ class B $<$ 1 mm/yr, 0.01 mm/yr $\leq$ class C $<$ 0.1 mm/yr), the number of class C faults must be 100 times of the class A faults'. In the catalogue of the Research Group for Active Faults of Japan (1991) however, the estimated numbers of class A, B and C faults are 103, 884 and 660 respectively. This suggests that the majority of the slowly moving class C faults are hidden, possibly due to fast rates of erosion and sedimentation associated with Japan's climate and surface processes.

From the viewpoint of elastic rebound on different scales, it is useful to consider the efficiency of stress release associated with inland earthquakes depending on the fault orientation. Simple elastic half-space models qualitatively demonstrate that highly dipping reverse-faulting earthquakes such as the 2004 Niigata-ken–Chuetsu earthquake (associated with tectonic inversion) are unfavourably oriented for the recent stress field and therefore inefficient at releasing regional differential stress. The models also imply that numerous moderate-sized earthquakes due to minor faults cannot compensate for one large earthquake caused by a mature fault system. It explains why the inland deformation zone sustains high continuous seismicity, whereas the outer zone facing the Pacific Ocean is significantly influenced by the seismic cycles of subduction megathrust events. Regarding the probabilistic estimates of large earthquakes, the former would be appropriate for more Poisson forecasting taking bulk deformation into account, whereas the latter would be better evaluated from time-dependent conditional probabilities.

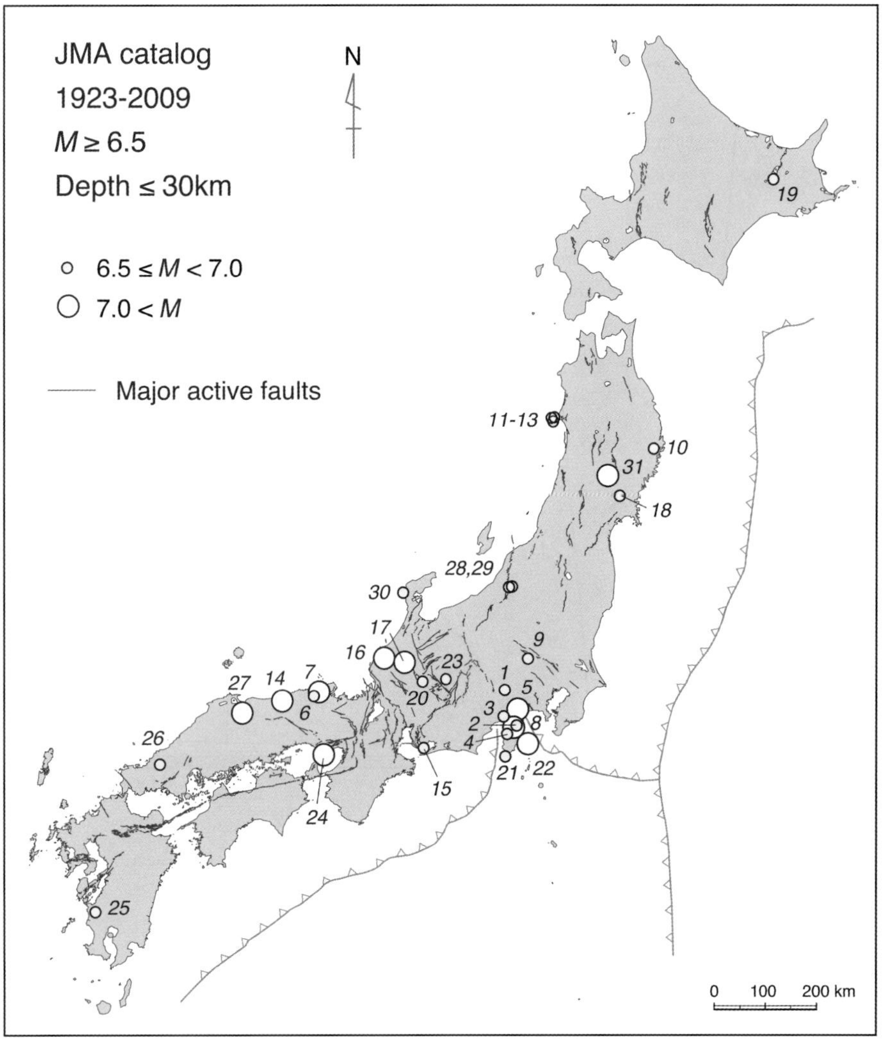

Fig. 8.26. Distribution of inland earthquakes which have occurred in Japan since 1923. Identification numbers are listed in Table 8.2.

Fault segmentation, sequential rupture and complex seismic cycle in active faulting

The size and location of the 11 March 2011 Tohoku earthquake surprised many earthquake scientists in Japan. The Tohoku earthquake broke five segments (sections), yielding a magnitude 9.0 earthquake (Fig. 8.24). Such segmentation along the Japan Trench is based on the 'characteristic earthquake model' (CEM, Schwartz & Coppersmith 1984), in which each segment has repeatedly produced a characteristic size of rupture length, displacement and thus magnitude within a certain interval. The CEM does not consider any significant change in size of the repeating earthquakes. The HERP's estimate was therefore that the largest future earthquakes along the different segments of the Japan Trench were expected to have magnitudes between 7 and 8 with different inter-event times (Fig. 8.24). Although the time-predictable model applied to the Nankai earthquake sequence (Shimazaki & Nakata 1980) implicitly includes a fault-scaling law, it was forecast that the seismic moment of sequential multiple rupture events involving adjacent segments would be a simple summation of the moments assigned to the segments. In other words, maximum displacement on each fault segment (subduction compartment) is limited and characteristic. The maximum slip at the Tohoku earthquake unfortunately exceeded 50 m (e.g. Simons *et al.* 2011), which forced us to reconsider the CEM and the traditional segmentation model.

Proper assessment of multiple segment rupture for inland earthquakes, which determines the size of the earthquake, is also a serious issue for design-and-construction decisions. As already noted, the HERP basically follows the definition of 'seismogenic active fault' provided by Matsuda (1991) in which the seismogenic faults are chosen and connected within a gap smaller than 5 km. However, the strategy only calculates the maximum size of large earthquakes and their frequency, ignoring the more frequent segment-based smaller ruptures. As well as the advanced palaeoseismic studies along the San Andreas Fault in California (e.g. Scharer *et al.* 2014), numerous palaeoseismic excavations across the ISTL (Fig. 8.16) already show that it is more likely that frequent $M \approx 7$ shocks have occurred, along with complex combinations of multiple segment ruptures.

The 22 November 2014 Nagano-ken–hokubu earthquake ($M = 6.7$, $M_w = 6.2$) also demonstrated a clear difficulty in estimating size of an earthquake. The quake struck northern Nagano, central Japan, destroying more than 100 houses and injuring 46 people. More than 9 km of complex surface faulting occurred on the previously mapped NNW-trending Kamishiro Fault, one of the segments of the 150 km long ISTL active fault system (Fig. 8.27; Okada *et al.* 2015). The surface rupture lies *c.* 5 km west of the epicentre, juxtaposing a basin with a NNW-trending mountain range. Although the free face of the bedrock fault was not observed, there were many

Table 8.2. *Large inland crustal earthquakes occurring in Japan between 1923 and 2009 (*M ≥ *6.5, depth* ≤ *30 km) with associated coseismic surface ruptures (modified from Toda 2013)*

ID	Year	Month	Day	Lon. (°)	Lat. (°)	Location	*M*	M_w	L_{sub} (km)	Surface rupture main surface fault	Slip type	Length (km)	Rank	Remarks
1	1923	9	1	138.793	35.661	Yamanashi-ken–Tobu	6.8	–	–	–	–	–	–	Aftershock of Kanto earthquake
2	1923	9	1	139.042	35.086	Izu Peninsula	6.6	–	–	–	–	–	–	Aftershock of Kanto earthquake
3	1923	9	1	138.769	35.233	Fujinomiya	6.6	–	–	–	–	–	–	Aftershock of Kanto earthquake
4	1923	9	2	138.843	34.937	Izu Peninsula	6.5	–	–	–	–	–	–	Aftershock of Kanto earthquake
5	1924	1	15	139.055	35.341	Tanzawa	7.3	–	–	–	–	–		Aftershock of Kanto earthquake
6	1925	5	23	134.835	35.563	Kita–Tajima	6.8	–	–	Tai?	V	–	–	
7	1927	3	7	134.931	35.632	Kita–Tango	7.3	7.0	35	Gomura	LL	22	1	
8	1930	11	24	138.974	35.043	Kita–Izu	7.3	6.9	24	Tanna	LL	32	1	
9	1931	9	21	139.248	36.158	Nishi–Saitama	6.9	6.5	20	–	–	–	–	
10	1931	11	4	141.839	39.481	Iwate	6.5	–	–	–	–	–	–	
11	1939	5	1	139.786	39.946	Oga	6.8	6.8	16	Sarukawa?	–	–	–	Most of the source is offshore
12	1939	5	1	139.600	39.997	Oga	6.7	6.8	16	–	–	–	–	Most of the source is offshore
13	1939	5	2	139.766	39.899	Oga	6.6	–	–	–	–	–		Most of the source is offshore
14	1943	9	10	134.184	35.473	Tottori	7.2	7.0	33	Shikano	RL	12	1	
15	1945	1	13	137.114	34.703	Mikawa	6.8	6.6	12	Fukozu	R	20	1	
16	1948	6	28	136.291	36.172	Fukui	7.1	6.8	30	Fukui	RL	25	2	
17	1961	8	19	136.700	36.112	Kita–Mino	7.0	6.4	12	–	–	–	–	
18	1962	4	30	141.138	38.740	Miyagi-ken–Hokubu	6.5	6.2	12	–	–	–	–	
19	1967	11	4	144.267	43.483	Kussharo	6.5	–	–	–	–	–	–	
20	1969	9	9	137.067	35.783	Gifu-ken–Chubu	6.6	6.3	23	–	–	–	–	
21	1974	5	9	138.800	34.567	Izu–Hanto–Oki	6.9	6.5	18	Irozaki	RL	6	2	Half of the source might be offshore
22	1978	1	14	139.250	34.767	Izuoshima–Kinkai	7.0	6.8	15	Inatori–Omineyama	RL	4	3	Most of the source is offshore
23	1984	9	14	137.557	35.825	Nagano-ken–Seibu	6.8	6.3	12	–	RL	–	3	
24	1995	1	17	135.035	34.598	Hyogo-ken–Nanbu (Kobe)	7.3	6.9	55	Nojima	RL	17	1	Source length in Awaji Island is 21 km
25	1997	3	26	130.359	31.973	Kagoshima-ken–Hokuseibu	6.6	6.1	15	–	LL	–	–	
26	1997	6	25	131.666	34.441	Yamaguchi-ken–Hokubu	6.6	5.9	17	–	RL	–	–	
27	2000	10	6	133.349	35.274	Tottori-ken–Seibu	7.3	6.7	33	–	LL	*c.* 4	3	
28	2004	10	23	138.867	37.292	Niigata-ken–Chuetsu	6.8	6.6	24	Obiro	R	*c.* 1	3	
29	2004	10	23	138.930	37.306	Niigata-ken–Chuetsu	6.5	6.3	20	–	R	–	–	Aftershock of #28 earthquake
30	2007	3	25	136.686	37.221	Noto–Hanto	6.9	6.6	21	–	R	–	–	Half of the source is offshore
31	2008	6	14	138.499	34.786	Iwate–Miyagi-nairiku	7.2	6.9	40	–	R	20	2	Extremely discontinuous traces

M, Japan Meteorological Agency magnitude (JMA magnitude); M_w, moment magnitude; L_{sub}, subsurface length of each source fault.
Slip type: V, vertical separation; LL, left-lateral slip; RL, right-lateral slip; R, reverse slip.

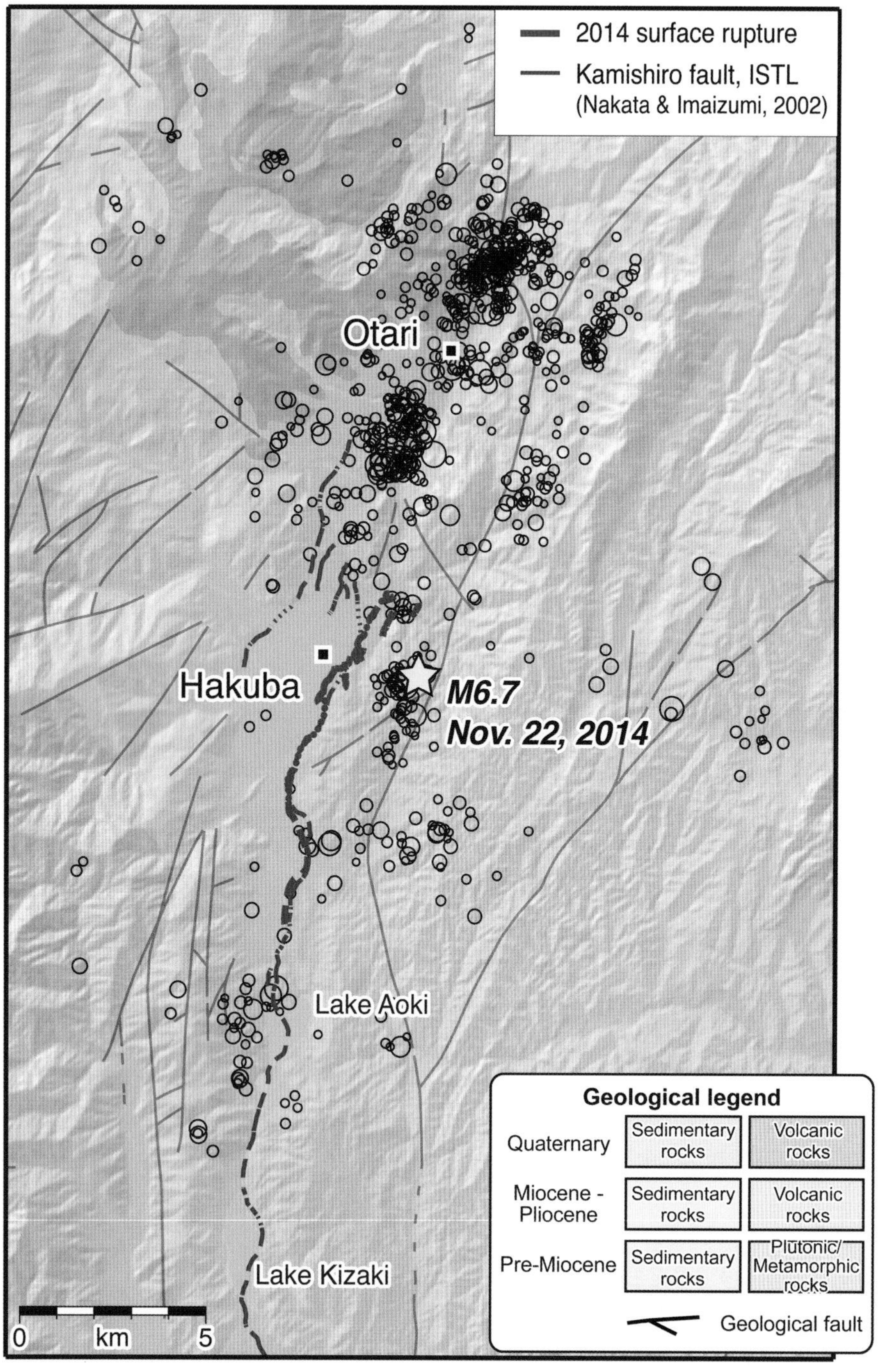

Fig. 8.27. Nagano-ken–hokubu earthquake (map). Surface rupture associated with the 22 November 2014 Nagano-ken–hokubu earthquake of $M=6.7$ ($M_w=6.2$) along the Kamishiro Fault (Okada *et al.* 2015). Aftershocks which occurred during the following 1.5 months are superimposed on the map. See the location of the Kamishiro Fault on the ISTL in Figures 8.15 and 8.16.

features that warped and buckled the ground surface, corresponding to vertical and contractional deformation associated with a shallow east-dipping thrust fault (east side up), accommodating NW–SE compression. The maximum vertical displacement and contraction measured were 80 and 50 cm, respectively (Fig. 8.28). The 2014 earthquake is the first surface-rupturing earthquake to strike on one of the 110 major active faults evaluated by the HERP since 1995. Numerous palaeoseismic data, in particular on the most active central ISTL, allowed the HERP to forecast a 14% 30-year probability of an $M=8.3$ earthquake if the entire ISTL ruptured (see the section 'The Itoigawa–Shizuoka Tectonic Line fault system, central Honshu'). Probabilities of more frequent $M\approx7$ ruptures had been under consideration. One of the rupture scenarios was a $M_w=7.4$ event on the northern half of the ISTL, involving the Kamishiro Fault (Figs 8.15 & 8.23b). Even a single 26 km long rupture specifically on the Kamishiro fault, as defined by its geomorphological continuity, would have produced a larger event of $M=7.2$ ($M_w=6.7$). In addition, the observed 80 cm vertical displacement is much smaller than the expected 3–4 m slip from a 1000–1500 year elapsed time and a vertical slip rate of 3 mm a^{-1} of the Kamishiro fault. Given the evidence so far available, it can therefore be concluded that either the 2014 earthquake was not a characteristic event or the palaeoseismic data are extremely incomplete.

Fig. 8.28. Fault surface ruptures associated with the 2014 Nagano-ken–Hokubu earthquake. (**a**) Vertical displacement of 40 cm which occurred in 2014 amplified the pre-existing scarp of the Kamishiro Fault. (**b**) Maximum 80 cm vertical separation of the paved road across the fault.

Inland earthquake hazard time dependency due to stress perturbation

One of the characteristic features associated with earthquake occurrence is clustering in space and time. Aftershocks and swarms typically display such clustering behaviours. Analysing the aftershocks of the 1891 Nobi earthquake, Omori (1894) found that temporal decay of aftershocks follows the simple equation:

$$n(t) = K(t + c)^{-1}$$

where K and c are constants and t is the time elapsed from the mainshock. Utsu (1961) extended Omori's law, giving it more flexibility to fit more observation data:

$$n(t) = K(t + c)^{-p}$$

where p is also a constant, explaining how fast or slow aftershock activity decays. These empirical equations are now explained by earthquake physics of rate- and state-dependent friction law (Dieterich 1994). In rate- and state-dependent friction, a sudden stress step causes accelerations of micro-slip on the nucleation sources of earthquakes and then, subsequent to the mainshock, produces a burst of earthquakes known as aftershocks. In addition to normal aftershocks, such a mechanism enables us to explain highly sensitive

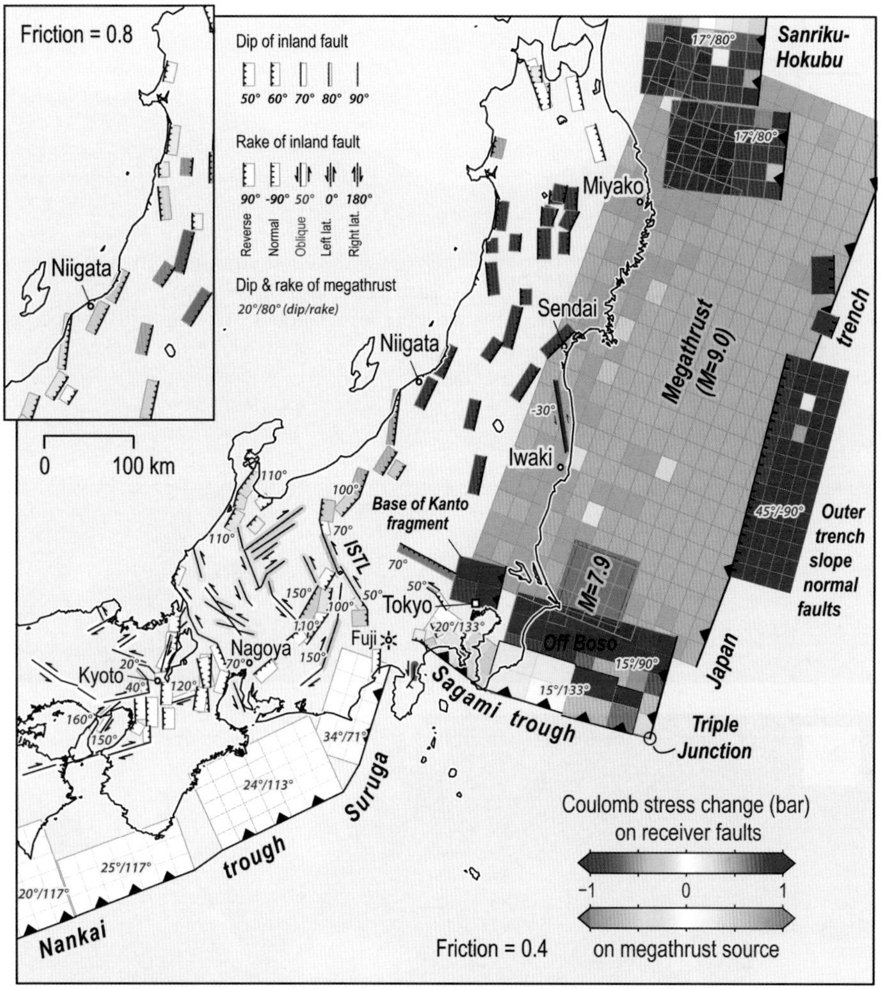

Fig. 8.29. Coulomb stress imparted by the 2011 $M_w = 9.0$ Tohoku mainshock and its largest aftershock of $M = 7.9$ to surrounding active faults, resolved in their inferred rake directions (oblique rakes are labelled). Top and bottom depths of most of the active faults are set to 0 and 15 km. Modified from Toda *et al.* 2011*b*.

seismic responses to a small stress change off the mainshock rupture zone. These post-mainshock earthquakes are often called 'off-fault aftershocks' (King *et al.* 1994) and they also show Omori-like temporal decay as well as traditional aftershocks. Since the late 1980s, numerous observations of Omori-like temporal behaviours in off-fault aftershocks have been reported (e.g. Toda *et al.* 1998; Stein 1999), aiding calculation of the near-future earthquake probability of a large earthquake on a fault stress loaded by a nearby mainshock (e.g. Stein *et al.* 1997). More importantly, in the Dieterich law, the influencing period of stress transfer, namely aftershock duration, is inversely proportional to the fault stressing rate, a hypothesis supported by observation data (Stein & Liu 2010). It suggests that high seismic hazard in the areas next to those with recent large earthquakes is sustained during subsequent years to decades, and even for more than a century. Some of the recent seismicity affecting inland Japan represents aftershocks of large destructive earthquakes that have occurred in recent decades (e.g. Ishibe *et al.* 2011).

The 11 March 2011 $M_w = 9.0$ great Tohoku earthquake has dramatically changed the stress field (Fig. 8.29) and therefore seismicity in eastern Honshu (Asano *et al.* 2011; Ishibe *et al.* 2011; Toda *et al.* 2011a). The Tohoku earthquake ruptured a 500 km long and 200 km wide megathrust of the interface between the Pacific Plate and the overriding Eurasian (North American) Plate, with a maximum displacement of *c.* 50 m (e.g. Simons *et al.* 2011). It also caused a significant eastwards movement of the Pacific coast of up to 5.3 m (Ozawa *et al.* 2011, 2012) that changed the inland region from a zone of secular interseismic east–west contraction to a zone of significant east–west stretching. Widespread off-fault aftershocks as far as *c.* 450 km from the locus of high seismic slip (Toda *et al.* 2011a) and their complex fault plane solutions (Asano *et al.* 2011; Kato *et al.* 2011) bear out the extensive coseismic stress change. Large offshore aftershocks caused by normal faulting explicitly demonstrate extraordinary stress perturbation, possibly suggesting a more than *c.* 30° rotation of the principal stress axes (Hasegawa *et al.* 2011; Yoshida *et al.* 2012). Such significant coseismic stress change brought with it a widespread seismicity rate increase across central Japan, extending west to the Japan Sea and south to the Izu islands. Broadly, there have been strong increases in seismicity rate across a region extending up to 300 km from the distal edges of the $M = 9$ rupture surface, and 425 km from the locus of high ($\geq$15 m) seismic slip (Toda *et al.* 2011*a*). The seismicity rate jump observed along the ISTL, described in the previous section, is one of the more remarkable examples of this.

Among the induced inland seismicity associated with the Tohoku earthquake, one of the most significant features was the highly active normal-faulting aftershocks that partly extended westwards onshore to the area of southern Fukushima and northern Ibaraki prefectures (e.g. Kato *et al.* 2011; Imanishi *et al.* 2012, Fig. 8.30). The unusually high seismicity started immediately after the Tohoku earthquake and has continued for longer than four years (at least until now) as an extensive massive seismic swarm including four $M \geq 6$ earthquakes in the area shown in Figure 8.30. This triggered swarm zone corresponds to the southern margin of the Abukuma Mountains, composed of Cretaceous granite and schist (Abukuma granites and metamorphic complex, Geological Survey of Japan, AIST 2014) along a boundary of low hills underlain by Tertiary sedimentary rocks to the west. No major active fault has been mapped in the area, but several isolated fault strands shorter than 20 km were identified. These fault strands were interpreted to be potentially active without discernible slip sense (Research Group on Active Faults of Japan 1991; Nakata & Imaizumi 2002). On 11 April (one month after the Tohoku mainshock) the largest normal faulting event of $M = 7.0$ ($M_w = 6.6$), officially named the Fukushima-ken–Hamadori earthquake (referred to here as the Iwaki earthquake), occurred at the centre of the swarm zone where it triggered numerous rock falls and landslides, caused structural damage and killed four people. The Iwaki earthquake involved multiple traces of surface rupture on mountain slopes and lowlands NE of the epicentre. Two *c.* 15 km long subparallel ruptures appeared along the previously mapped Itozawa and Yunodake faults (Mizoguchi *et al.* 2012; Toda & Tsutsumi 2013). These surface breaks had previously been the site of active geomorphic features along reactivated geological boundaries originally developed as a half-graben in the Neogene Period. A trench excavation revealed evidence for a penultimate surface-deforming earthquake sometime between 12 620 and 17 410 years BP on the fault, suggesting a long recurrence interval for such a forearc inland normal fault, and could also have been linked to a past giant megathrust earthquake along the Japan Trench (Toda & Tsutsumi 2013).

The Iwaki earthquake is one of the more remarkable examples of how a megathrust earthquake controls the timing of inland active faulting. However, there are numerous papers that have already discussed the possibility that the seismic cycle of subduction megathrust earthquakes influences inland seismicity in SW Japan (Utsu 1974; Shimazaki 1976; Hori & Oike 1996; Pollitz & Sacks 1997; Hori & Oike 1999; Shikakura *et al.* 2014) and northern Honshu (Shimazaki 1978; Seno 1979; Rydelek & Sacks 1990). Based on historical earthquake data since 868, an approximate 100-year wavelength temporal fluctuation of seismicity in SW Japan, particularly in the Kinki region (Osaka–Kyoto–Nara region), is clearly evident. Stacking all the seismicity as a function of occurrence times of the great Nankai Trough earthquakes in 887, 1096, 1361, 1498, 1605, 1707, 1854 and 1944 (Fig. 8.2), Hori & Oike (1996, 1999) found that the activity of inland earthquakes in SW Japan increases during the period from about 50 years before to 10 years after the occurrence of great megathrust earthquakes along the Nankai Trough (Fig. 8.31). Shikakura *et al.* (2014) also demonstrated that the temporal evolution of Coulomb stress accumulation on active faults in the Kinki region and found the inland reverse-faulting activity mostly increases before Nankai Trough interplate earthquakes and decreases after the earthquakes, whereas strike-slip activity is mostly suppressed before interplate earthquake and increases thereafter. If these papers are correct, seismicity across inland SW Japan will be more active during the next few decades until the next Nankai Trough earthquake occurs.

Independently of the temporal correlation between the inland earthquakes and the great interplate earthquakes in Japan, a significant dormancy of large earthquake occurrence has been observed in the San Francisco Bay area since the 1906 $M_w = 7.8$ San Francisco earthquake which ruptured a 470 km long segment along the San Andreas Fault (Thatcher *et al.* 1997). The term ‘stress shadow’ was proposed to explain how the 1906 earthquake inhibited

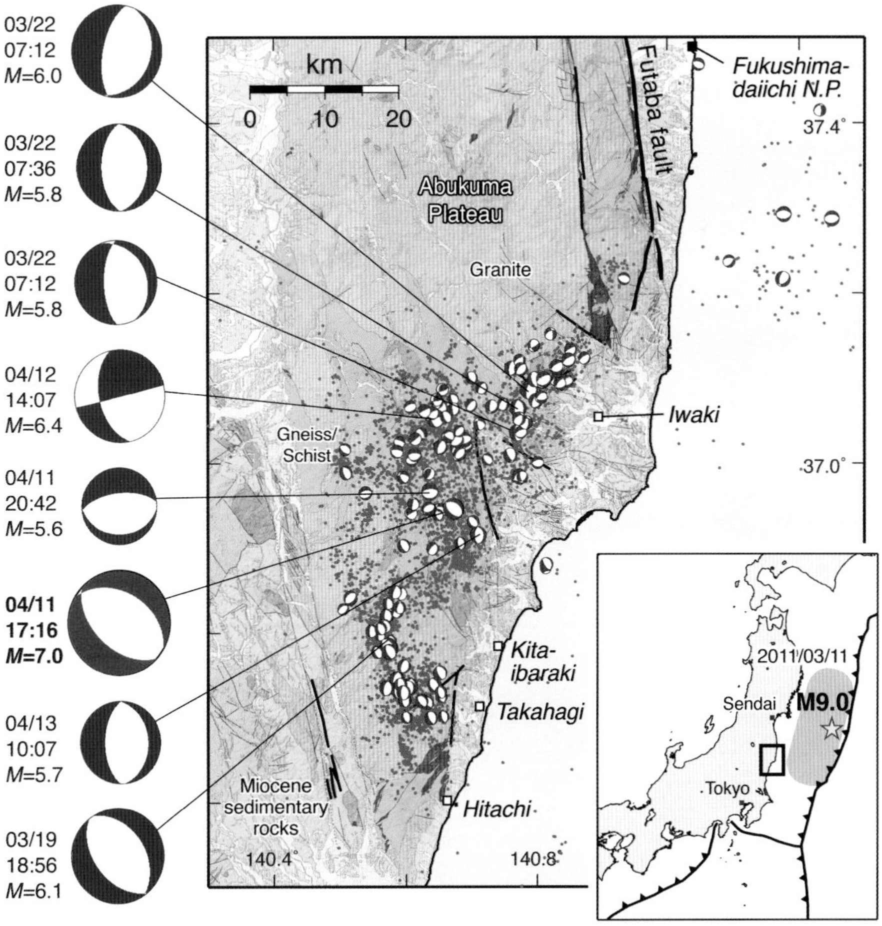

Fig. 8.30. Focal mechanisms of large shallow earthquakes in the SE Fukushima and northern Ibaraki areas after the Tohoku–Oki ($M_w = 9$) mainshock (F-net catalogue from NIED), including the M_w 6.6 11 April 2011 Iwaki earthquake. Small dots indicate earthquakes that occurred after the Tohoku–Oki mainshock (depth ≤20 km, 2011/03/11–2011/08/10). The location of Figure 8.2 is indicated by a dotted rectangular box. The geological map behind the earthquake plots is from the Seamless Digital Geological Map of Japan (Geological Survey of Japan 2014) and shows that swarm activity has been occurring in and around the boundary between Mesozoic granite/metamorphic rocks and Neogene sedimentary rocks. Thick black lines denote mapped active faults (Research Group for Active Faults of Japan 1991). The inset map shows the relationship between the $M_w = 9$ source fault (grey) and the Fukushima region (box).

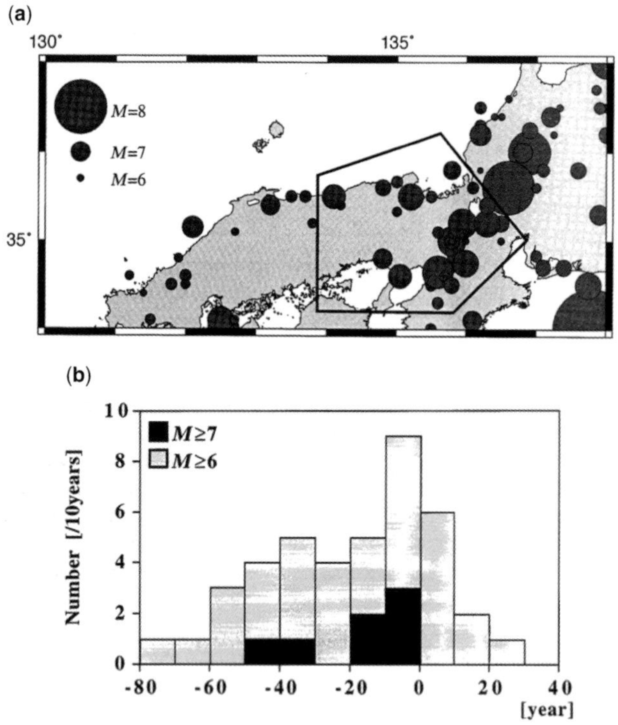

Fig. 8.31. (**a**) Epicentres of destructive earthquakes from 868 to 1995. The pentagon indicates the Kinki region. (**b**) Histogram of the time series of large inland earthquakes in the Kinki region before and after the great Nankai Trough subduction earthquakes (Hori & Oike 1996). The occurrence time of the Nankai Trough earthquakes is set to 0 at the x axis.

subsequent earthquakes due to the significant coseismic stress decrease around the source fault (Harris & Simpson 1996; Jaume & Sykes 1996). The above papers also discussed the inland seismic cycles synchronized with the interplate occurrence in Japan, implicitly suggesting the effect of stress shadow during the dormant period. The long seismic quiescence in the Tokyo metropolitan region after 1923 (Okada 1994) can also be explained by a stress shadow, in this case following the 1923 $M=7.8$ Taisho Kanto earthquake (Nyst *et al.* 2005; Toda 2013). Seismic cycles with megathrust earthquakes intervals of 200–400 years along the Sagami Trough (Shishikura 2003) also influence the seismic hazard in and around the Tokyo metropolitan area. We must seek better ways of incorporating the effects of such temporal fluctuations into the probabilistic seismic hazard.

Appendix

English to Kanji and Hiragana translations for geological and place names

Abukuma	阿武隈	あぶくま
Akashi	明石	あかし
Akita	秋田	あきた
Aomori	青森	あおもり
Arima	有馬	ありま
Aso	阿蘇	あそ
Atami	熱海	あたみ
Atera	阿寺	あてら
Atotsugawa	跡津川	あとつがわ

(*Continued*)

English to Kanji and Hiragana translations for geological and place names (Continued)

Awaji	淡路	あわじ
Beppu	別府	べっぷ
Biwa	琵琶	びわ
Bonin	ボニン	ぼにん
Boso	房総	ぼうそう
Chiba	千葉	ちば
Chishima	千島	ちしま
Chubu	中部	ちゅうぶ
Dedana	出店	でだな
Edo	江戸	えど
Fukui	福井	ふくい
Fukushima	福島	ふくしま
Futagawa	布田川	ふたがわ
Gifu	岐阜	ぎふ
Gofukuji	牛伏寺	ごふくじ
Hanaore	花折	はなおれ
Heian	平安	へいあん
Hida	飛騨	ひだ
Higashiura	東浦	ひがしうら
Hinagu	日奈久	ひなぐ
Honshu	本州	ほんしゅう
Horinji	法輪寺	ほうりんじ
Hyuga-nada	日向灘	ひゅうがなだ
Ibaraki	茨城	いばらき
Ikoma	生駒	いこま
Ise	伊勢	いせ
Itoigawa	糸魚川	いといがわ
Itozawa	井戸沢	いとざわ
Iwaki	いわき	
Izu	伊豆	いず
Izumi	和泉	いずみ
Kamishiro	神城	かみしろ
Kanto	関東	かんとう
Kashimo	加子母	かしも
Kawafune	川舟	かわふね
Kii	紀伊	きい
Kinki	近畿	きんき
Kiso	木曽	きそ
Kitakami-teichi-seien	北上低地西縁	きたかみていちせいえん
Kobe	神戸	こうべ
Kochi	高知	こうち
Kumagawa	熊川	くまがわ
Kurikoma	栗駒	くりこま
Kyoto	京都	きょうと
Kyushu	九州	きゅうしゅう
Mahiru	真昼	まひる
Matsumoto	松本	まつもと
Meiji	明治	めいじ
Miboro	御母衣	みぼろ
Mikata	三方	みかた
Mikawa	三河	みかわ
Mino	美濃	みの
Minou	水縄	みのう
Mitoke	三峠	みとけ
Miyagi	宮城	みやぎ
Nagamachi	長町	ながまち
Nagano	長野	ながの
Nagoya	名古屋	なごや
Nankai	南海	なんかい
Nara	奈良	なら
Nara-bonchi-toen	奈良盆地東縁	ならぼんちとうえん

(*Continued*)

English to Kanji and Hiragana translations for geological and place names (Continued)

Neodani	根尾谷	ねおだに
Nihonshoki	日本書紀	にほんしょき
Niigata	新潟	にいがた
Nishiyama	西山	にしやま
Nobi	濃尾	のうび
Nohi	濃飛	のうひ
Nojima	野島	のじま
Noto	能登	のと
Nukumi	温見	ぬくみ
Ogura	小倉	おぐら
Okinawa	沖縄	おきなわ
Osaka	大阪	おおさか
Ou	奥羽	おうう
Owari	尾張	おわり
Rifu	利府	りふ
Rokko	六甲	ろっこう
Ryoke	領家	りょうけ
Ryukyu	琉球	りゅうきゅう
Sagami	相模	さがみ
Sakashita	坂下	さかした
San'in	山陰	さんいん
Sanbagawa	三波川	さんばがわ
Sanriku	三陸	さんりく
Sendai	仙台	せんだい
Sengoku	戦国	せんごく
Senya	千屋	せんや
Senzan	先山	せんざん
Shikano	鹿野	しかの
Shikoku	四国	しこく
Shimabara	島原	しまばら
Shizuoka	静岡	しずおか
Suwa	諏訪	すわ
Tagajo	多賀城	たがじょう
Taisho	大正	たいしょう
Takatsuki	高槻	たかつき
Takayama	高山	たかやま
Tanba (Tamba)	丹波	たんば
Tanna	丹那	たんな
Tohoku	東北	とうほく
Tokai	東海	とうかい
Tokugawa	徳川	とくがわ
Tokyo	東京	とうきょう
Tottori	鳥取	とっとり
Toyama	富山	とやま
Tsukechi	付知	つけち
Uemachi	上町	うえまち
Umehara	梅原	うめはら
Wakayama	和歌山	わかやま
Yamagata	山形	やまがた
Yamagata-bonchi	山形盆地	やまがたぼんち
Yokohama	横浜	よこはま
Yokosuka	横須賀	よこすか
Yokote	横手	よこて
Yoro	養老	ようろう
Yunodake	湯ノ岳	ゆのだけ
Zenisu	銭洲	ぜにす

References

ANDO, M. 1975. Source mechanisms and tectonic significance of historical earthquakes along the Nankai trough. *Tectonophysics*, **27**, 119–140.

ASADA, T. 1991. Some questions on active faults. *Active Fault Research*, **9**, 1–3 [in Japanese].

ASANO, Y., SAITO, T. *ET AL.* 2011. Spatial distribution and focal mechanisms of aftershocks of the 2011 off the Pacific coast of Tohoku earthquake. *Earth, Planets and Space*, **63**, 669–673.

AWATA, Y., TSUKUDA, E., YAMAZAKI, H. & MIZUNO, K. 1986. Recent event history of the Atera fault (Atera-danso no saikin no katsudoushi). *Abstracts of Atera Fault Symposium*, **2**, 1–11 [in Japanese].

AWATA, Y., MIZUNO, K. *ET AL.* 1996. Surface fault ruptures on the northwest coast of Awaji Island associated with the Hyogo-ken Nanbu earthquake of 1995, Japan. *Zishin*, **49**, 113–124 [in Japanese with English abstract].

BAKUN, W. H. 2005. Magnitude and location of historical earthquakes in Japan and implications for the 1855 Ansei Edo earthquake. *Journal of Geophysical Research*, **110**, B02304, http://doi.org/10.1029/2004JB003329

CHIDA, N., MATSUMURA, K., SANGAWA, A. & MATSUDA, T. 1994. Recent activity of Mihoh fault system, central Kyushu, Japan. *Quaternary Research*, **33**, 261–267 [in Japanese].

COLLABORATIVE RESEARCH GROUP FOR THE NOHI RHYOLITE 1973. Stratigraphy and the volcanic history of the Nohi rhyolite in its eastern marginal part. *Earth Science*, **27**, 161–179 [in Japanese].

DIETERICH, J. 1994. A constitutive law for rate of earthquake production and its application to earthquake clustering. *Journal of Geophysical Research*, **99**, 2601–2618.

ENESCU, B., MORI, J., MIYAZAWA, M. & KANO, Y. 2009. Omori-Utsu law c-value associated with recent moderate earthquakes in Japan. *Bulletin of the Seismological Society of America*, **99**, 884–891, http://doi.org/10.1785/0120080211

FIELD, E. H., DAWSON, T. E. *ET AL.* 2009. Uniform California Earthquake Rupture Forecast, Version 2 (UCERF 2). *Bulletin of the Seismological Society of America*, **99**, 2053–2107.

FUJIWARA, H., MORIKAWA, N., ISHIKAWA, Y., OKUMURA, T., MIYAKOSHI, J., NOJIMA, N. & FUKUSHIMA, Y. 2009. Statistical comparison of National Probabilistic Seismic Hazard Maps and frequency of recorded JMA seismic intensities from the K-NET strong-motion observation network in Japan during 1997–2006. *Seismological Research Letters*, **80**, 458–464.

GELLER, R. J. 2011. Shake-up time for Japanese seismology. *Nature*, **472**, 407–409.

GEOLOGICAL SURVEY OF JAPAN, AIST 2014. Seamless Digital Geological Map of Japan (1:200,000), https://gbank.gsj.jp/seamless/index_en.html [last accessed 6 August 2015].

GEOSPATIAL INFORMATION AUTHORITY OF JAPAN 2014. An introduction of the 1:25,000 scale Active Fault Maps in Urban Areas, http://www.gsi.go.jp/bousaichiri/active_fault.html [in Japanese] [last accessed 6 August 2015].

GOTO, K., CHAGUE-GOFF, C., GOFF, J. & JAFFE, B. 2012. The future of tsunami research following the 2011 Tohoku-oki event. *Sedimentary Geology*, **282**, 1–13.

HANKS, T. C. & KANAMORI, H. 1979. Moment magnitude scale. *Journal of Geophysical Research*, **84**, 2348–2350.

HANKS, T. C., BEROZA, G. C. & TODA, T. 2012. Have recent earthquakes exposed flaws in or misunderstandings of probabilistic seismic hazard analysis? *Seismological Research Letters*, **83**, 759–764.

HARRIS, R. A. & SIMPSON, R. W. 1996. In the shadow of 1857 - the effect of the great Ft. Teton earthquake on subsequent earthquakes in southern California. *Geophysical Research Letters*, **23**, 229–232.

HASEGAWA, A., UMINO, N. & TAKAGI, A. 1978. Double-planed structure of the deep seismic zone in the northeastern Japan arc. *Tectonophysics*, **47**, 43–58.

HASEGAWA, A., YOSHIDA, K. & OKADA, T. 2011. Nearly complete stress drop in the 2011 Mw 9.0 off the Pacific coast of Tohoku Earthquake. *Earth, Planets and Space*, **63**, 703–707, http://doi.org/10.5047/eps.2011.06.007

HASHIMOTO, M. 1990. Horizontal strain rates in the Japanese islands during interseismic period deduced from geodetic survey (part I): Honshu, Shikoku and Kyushu. *Zisin*, **43**, 13–26 [in Japanese with English abstract].

HASHIMOTO, M. 1995. Static stress changes associated with the Kobe earthquake: calculation of changes in Coulomb failure function and comparison with seismicity change. *Zisin*, **48**, 521–530 [in Japanese with English abstract].

HEADQUARTERS FOR EARTHQUAKE RESEARCH PROMOTION 2001. Evaluation of the Kyoto-bonchi-Nara-bonchi fault zone, http://www.jishin.go.jp/main/

chousa/01jul_keina/index.htm [in Japanese] [last accessed 6 August 2015].

Headquarters for Earthquake Research Promotion 2002. Evaluation of the Nagamachi-Rifu fault zone, http://www.jishin.go.jp/main/chousa/02feb_rifu/index.htm [in Japanese] [last accessed 6 August 2015].

Headquarters for Earthquake Research Promotion 2005*a*. Seismic Activity in Japan, http://www.hp1039.jishin.go.jp/eqchreng/eqchrfrm.htm [last accessed 6 August 2015].

Headquarters for Earthquake Research Promotion 2005*b*. The Promotion of Earthquake Research, http://www.jishin.go.jp/main/suihon-e/sogo.pdf [last accessed 6 August 2015].

Headquarters for Earthquake Research Promotion 2005*c*. National seismic hazard maps for Japan 2005, Earthquake Research Committee (K. Tsumura, chair), Headquarters for Earthquake Research Promotion, http://www.jishin.go.jp/main/index-e.html [last accessed 6 August 2015].

Headquarters for Earthquake Research Promotion 2010. National seismic hazard maps for Japan 2010, Earthquake Research Committee (K. Abe, chair), Headquarters for Earthquake Research Promotion, www.jishin.go.jp/main/chousa/10_yosokuchizu/index.htm [in Japanese] [last accessed 6 August 2015].

Headquarters for Earthquake Research Promotion 2014. National seismic hazard maps for Japan 2014, Earthquake Research Committee (Y. Honkura, chair), Headquarters for Earthquake Research Promotion, http://www.jishin.go.jp/evaluation/seismic_hazard_map/shm_report/shm_report_2014/ [in Japanese] [last accessed 9 November 2015].

Heki, K., Miyazaki, S. *et al.* 1999. The Amurian plate motion and current plate kinematics in eastern Asia. *Journal of Geophysical Research*, **104**, 29147–29155.

Hirano, S., Nakata, T. & Imaizumi, T. 1979. Crustal deformation associated with the Kisakata earthquake of 1804 on the Japan Sea Coast, northeast Japan. *Quaternary Research*, **18**, 17–30 [in Japanese with English abstract].

Hori, T. & Oike, K. 1996. A statistical model of temporal variation of seismicity in the inner zone of southwest Japan related to the great interplate earthquakes along the Nainkai Trough. *Journal of Physics of the Earth*, **44**, 349–356.

Hori, T. & Oike, K. 1999. A physical mechanism for temporal variation in seismicity in the inner zone of southwest Japan related to the great interplate earthquakes along the Nankai Trough. *Tectonophysics*, **308**, 83–98.

Huzita, K. 1962. Tectonic development of the median zone (Setouchi) of southwest Japan since the miocene, with special reference to the characteristic structure of Central Kinki Area. *Journal of Geosciences, Osaka City University*, **6**, 103–144.

Huzita, K. 1980. Role of the median tectonic line in the Quaternary tectonics of the Japanese Islands. *Memoirs of the Geological Society of Japan*, **18**, 129–153 [in Japanese with English abstract].

Ikeda, M., Toda, S., Kobayashi, S., Ohno, Y., Nishizaka, N. & Ohno, I. 2009. Tectonic model and fault segmentation of the Median Tectonic Line active fault system on Shikoku, Japan. *Tectonics*, **28**, TC5006, http://doi.org/10.1029/2008TC002349

Ikeda, Y. 1983. Thrust-front migration and its mechanisms: evolution of intraplate thrust-fault systems. *Bulletin of the Department of Geography, University of Tokyo*, **15**, 125–159.

Ikeda, Y., Imaizumi, T., Togo, M., Hirakawa, K., Miyauchi, T. & Sato, H. 2002. *Atlas of Quaternary Thrust Faults in Japan*. University of Tokyo Press, Tokyo [in Japanese].

Imanishi, K., Ando, R. & Kuwahara, Y. 2012. Unusual shallow normal-faulting earthquake sequence in compressional northeast Japan activated after the 2011 off the Pacific coast of Tohoku earthquake. *Geophysical Research Letters*, **39**, L09306, http://doi.org/10.1029/2012GL051491

Irikura, K. 2004. Recipe for predicting strong ground motion from future large earthquake. *Annuals of Disaster Prevention Research Institute, Kyoto University*, **47**, A317–A322 [in Japanese].

Irikura, K. & Miyake, H. 2001. Prediction of strong ground motion for scenario earthquakes. *Journal of Geography*, **110**, 849–875 [in Japanese with English abstract].

Ishibashi, K. 1976. Re-examination of a great earthquake expected in the Tokai district: possibility of the 'Suruga Bay earthquake'. *Abstracts, Seismological Society of Japan*, **2**, 30–34 [in Japanese].

Ishibashi, K. & Satake, K. 1998. Problems on forecasting great earthquakes in the subduction zones around Japan by means of paleoseismology. *Zisin*, **50**(Suppl.), 1–21 [in Japanese with English abstract].

Ishibe, T., Shimazaki, K., Tsuruoka, H., Yamanaka, Y. & Satake, K. 2011. Correlation between Coulomb stress changes imparted by large historical strike-slip earthquakes and current seismicity in Japan. *Earth Planets Space*, **63**, 301–314.

Ishikawa, N. & Hashimoto, M. 1999. Average horizontal crustal strain rates in Japan during interseismic period deduced from geodetic surveys (part 2). *Zisin*, **52**, 299–315 [in Japanese with English abstract].

Ishimura, D., Kato, S., Okada, A. & Takemura, K. 2010. Late Pleistocene episodic subsidence events recorded in drilled cores from the east coast of Lake Mikata, western Japan, suggesting Mikata fault zone activity. *Journal of Geography*, **119**, 775–793 [in Japanese with English abstract].

Iwata, T. 2013. Estimation of completeness magnitude considering daily variation in earthquake detection capability. *Geophysics Journal International*, **194**, 1909–1919.

Iwata, T., Hashimoto, M., Takemura, K., Kimura, H., Kusumoto, N., Ito, Y. & Toda, S. 2013. 2010–2012 report of the comprehensive study of the Uemachi fault zone, Ministry of Education. *Culture, Sports, Science and Technology (MEXT)*, 66–163 [in Japanese].

Jaume, S. C. & Sykes, L. R. 1996. Evolution of moderate seismicity in the San Francisco Bay region, 1850 to 1993: seismicity changes related to the occurrence of large and great earthquakes. *Journal of Geophysical Research*, **101**, 765–789.

Kanamori, H. 1972. Determination of effective tectonic stress associated with earthquake faulting. The Tottori earthquake of 1943. *Physics of the Earth and Planetary Interiors*, **5**, 426–434.

Kanaori, Y. 1990. Late Mesozoic-Cenozoic strike-slip and block rotation in the inner belt of southwest Japan, *Tectonophysics*, **177**, 381–399.

Kanaori, Y., Kawakami, S. & K.& Yairi, 1991. Space-time distribution patterns of destructive earthquakes in the inner belt of central Japan: activity intervals and locations of earthquakes. *Engineering Geology*, **31**, 209–230.

Kanaori, Y., Kawakami, S., Yairi, K. & Hattori, T. 1993. Liquefaction and flow age at archaeological sites in the inner belt of central Japan: tectonic and hazard implications. *Engineering Geology*, **35**, 65–80.

Kato, A., Sakai, S. & Obara, K. 2011. A normal-faulting seismic sequence triggered by the 2011 off the Pacific coast of Tohoku Earthquake: wholesale stress regime changes in the upper plate. *Earth, Planets and Space (EPS)*, **63**, 745–748.

Kato, A., Obara, K., Igarashi, T., Tsuruoka, H., Nakagawa, S. & Hirata, N. 2012. Propagation of slow slip leading up to the 2011 Mw 9.0 Tohoku-Oki earthquake. *Science*, **335**, 705–708, http://doi.org/10.1126/science.1215141

Kato, A., Fukuda, J. & Obara, K. 2013. Response of seismicity to static and dynamic stress changes induced by the 2011 M9.0 Tohoku-Oki earthquake. *Geophysical Research Letters*, **40**, 3572–3578, http://doi.org/10.1002/grl.50699

King, G. C. P., Stein, R. S. & Lin, J. 1994. Static stress changes and the triggering of earthquakes. *Bulletin of the Seismological Society of America*, **84**, 935–953.

Komatsubara, T., Mizuno, K., Sangawa, A. & Yamazaki, H. 1999. *Trenching Study of the Mikata Fault and Reconstruction of Crustal Movements During the 1662 Kambun Earthquake by Historical Records*. Geological Survey of Japan Interim Report, no. EQ/99/3 [in Japanese with English abstract].

Kondo, H., Toda, S., Okumura, K., Takada, K. & Chiba, T. 2008. A fault scarp in an urban area identified by LiDAR survey: a case study on the Itoigawa-Shizuoka Tectonic Line, central Japan. *Geomorphology*, **101**, 731–739, http://doi.org/10.1016/j.geomorph.2008.02.012

Koto, B. 1893. On the cause of the great earthquake in Central Japan, 1891. *Journal of the College of Science, Imperial University of Japan*, **5**, 196–353.

Kuwahara, Y., Imanishi, K. & Cho, I. 2010. Seismic observation along the Itoigawa-Shizuoka Tectonic Line active fault system, 2009 report of the comprehensive study of the Itoigawa-Shizuoka Tectonic Line active fault system. *Ministry of Education, Culture, Sports, Science and Technology (MEXT)*, 139–154 [in Japanese].

Lengline, O., Enescu, B., Peng, Z. & Shiomi, K. 2012. Decay and expansion of the early aftershock activity following the 2011, Mw 9.0 Tohoku earthquake. *Geophysical Research Letters*, **39**, L18309, http://doi.org/10.1029/2012GL052797

Lin, A., Maruyama, T. & Miyata, T. 1998. Paleoseismic events and the 1596 Keicho-Fushimi large earthquake produced by a slip on the

Gosukebashi fault at the eastern Rokko Mountains, Japan. *Island Arc*, **7**, 621–636.

Maruyama, T., Toda, S. *et al.* 2009. Paleoseismological trench investigation of the surface rupture associated with the 2008 Iwate-Miyagi Nairiku earthquake (Mj 7.2) NE Japan. *Annual Report on Active Fault and Paleoearthquake Researches, AIST*, **9**, 19–54, [In Japanese with English abstract].

Maruyama, T., Toda, S., Okumura, K., Miura, D., Sasaki, T., Haraguchi, T. & Tsuji, Y. 2010. Historical and paleoseismic investigations along the Itoigawa-Shizuoka Tectonic Line active fault system, 2009 report of the comprehensive study of the Itoigawa-Shizuoka Tectonic Line active fault system. *Ministry of Education, Culture, Sports, Science and Technology (MEXT)*, 230–254 [in Japanese].

Matsuda, T. 1975. Magnitude and recurrence interval of earthquakes from a fault. *Journal of the Seismological Society of Japan (Zisin)*, **28**, 269–283 [in Japanese with English abstract].

Matsuda, T. 1977. Estimation of future destructive earthquakes from active faults on land in Japan. *Journal of Physics of the Earth*, **25**(Suppl.), s251–s260.

Matsuda, T. 1981. Active faults and damaging earthquakes in Japan: macroseismic zoning and precaution fault zones. *In*: D. W.Simpson, & P. G. Richarss, (eds) *Earthquake Prediction – An International Review*. AGU, Washington, Maurice Ewing Series, **4**, 279–289.

Matsuda, T. 1991. Seismic zoning map of Japanese Islands, with maximum magnitudes derived from active fault data. *Bulletin of the Earthquake Research Institute, University of Tokyo*, **65**, 289–319 [in Japanese with English abstract].

Matsuda, T. 1998. Present state of long-term prediction of earthquakes based on active fault data in Japan: An example for the Itoigawa-Shizuoka Tectonic active fault system. *Journal of the Seismological Society of Japan (Zisin)*, **50**, 23–33 [in Japanese with English abstract].

Matsuda, T. 2008. A review of the history of active fault research in Japan. *Active Fault Research*, **28**, 15–22 [in Japanese with English abstract].

Matsuda, T., Ota, Y., Ando, M. & Yonekura, N. 1978. Fault mechanism and recurrence time of major earthquakes in southern Kanto district, Japan, as deduced from coastal terrace data. *Bulletin of the Geological Society of America*, **89**, 1610–1618.

Matsu'ura, R. 2011. What is reliable and physically plausible about 1586 Tensho earthquake from historical materials? – The latest research on historical seismicity. *Active Fault Research*, **35**, 29–1639 [in Japanese with English abstract].

Matthews, M. V., Ellsworth, W. L. & Reasenberg, P. A. 2002. A Brownian model for recurrent earthquakes. *Bulletin of the Seismological Society of America*, **92**, 2233–2250.

McCalpin, J. P. (ed.) 2009. *'Paleoseismology'*, 2nd edn. Academic Press International, Geophysics Series, Burlington, MA, USA, **95**.

Miura, D., Hataya, R., Shintaro, A., Miyakoshi, K., Nikaido, M., Tachibana, T. & Takase, S. 2002. Recent faulting history at the Ichinose fault group, southern part of the Itoigawa-Shizuoka Tectonic Line, central Japan. *Journal of the Seismological Society of Japan (Zisin)*, 2nd Series, **55**, 33–45 [in Japanese with English abstract].

Miyazaki, S. & Heki, K. 2001. Crustal velocity field of southwest Japan: subduction and arc-arc collision. *Journal of Geophysical Research*, **106**, 4305–4326.

Mizoguchi, K., Uehara, S. & Ueta, K. 2012. Surface fault ruptures and slip distributions of the Mw 6.6 11 April 2011 Hamadoori, Fukushima Prefecture, northeast Japan, earthquake. *Bulletin of the Seismological Society of America*, **102**, 1949–1956.

Mizuno, K., Hattori, H., Sangawa, A. & Takayashi, H. 1990. *Geology of the Akashi district with geological sheet map at 1:50,000*. Geological Survey of Japan, Tsukuba, Japan [in Japanese with English abstract 5p.].

Mizuno, K., Okada, A., Sangawa, A. & Shimizu, F. 1993. *Explanatory text of strip map of the Median Tectonic Line Active Fault System in Shikoku, Japan, scale 1:25,000. Tectonic map series (8)*. Geological Survey of Japan, Tsukuba, Japan [in Japanese with English abstract].

Morino, M. & Okada, A. 2002. Faulting history of the Median Tectonic Line active fault system in Shikoku, based on re-examination of trench survey results. *Active Fault and Paleoearthquake Research, AIST*, **2**, 153–182 [In Japanese with English abstract].

Nakanishi, I. 2002. Historical material showing disasters due to the September 1, 1596 earthquake in the northwestern Shikoku Island, Japan. *Journal of the Seismological Society of Japan (Zisin)*, **55**, 311–316 [in Japanese with English abstract].

Nakata, T. & Imaizumi, T. (ed.) 2002. *Digital Active Fault Map of Japan*. University of Tokyo Press, Tokyo [in Japanese].

Nakata, T., Otsuki, K. & Imaizumi, T. 1976. Quaternary crustal movements along Nagamachi-Rifu dislocation line in the vicinity of Sendai, northern Japan, Tohoku Chiri. *Quarterly Journal of Geography*, **28**, 111–120 [in Japanese with English abstract].

Nanjo, K. Z., Ishibe, T., Tsuruoka, H., Schorlemmer, D., Ishigaki, Y. & Hirata, N. 2010. Analysis of the completeness magnitude and seismic network coverage of Japan. *Bulletin of the Seismological Society of America*, **100**, 3261–3268.

Naumann, E. 1887. The physical geography of Japan, with remarks on the people. *Proceedings of the Royal Geographical Society and Monthly Record of Geography*. **9**, 86–102.

Nohara, T., Koriya, Y. & Imaizumi, T. 2000. An estimation of the crustal strain rate using the active fault GIS data. *Active Fault Research*, **19**, 23–32 [in Japanese with English abstract].

Nyst, M., Hamada, N., Pollitz, F. F. & Thatcher, W. 2005. The stress triggering role of the 1923 Kanto earthquake. *AGU Fall Meeting 2005*, San Francisco, CA, USA, Abstract.

Okada, A. 1973. On the Quaternary faulting along the Median Tectonic Line. *In*: Sugiyama, R. (ed.) *Median Tectonic Line*. Tokyo University Press, Tokyo, 49–86 [in Japanese with English abstract].

Okada, A. 1984. Processes of Mikatagoko Lowlands formation and crustal movements. *In*: *Torihama Shell-mound' – Survey of Low-wetland Remains during the Early Jomon Period 4*. Fukui Prefectural Wakasa History and Folklore Museum, Obama, 9–42 [in Japanese].

Okada, A. 2012. Research on Quaternary faulting history and long-term seismic evaluation of the Median Tectonic Line (MTL) fault zone in southwest Japan. *Quaternary Research*, **51**, 131–150 [in Japanese with English abstract].

Okada, A. & Ando, M. 1979. Active faults and earthquakes in Japan. *Kagaku (Science in Japanese)*, **49**, 158–169 [in Japanese].

Okada, A. & Sangawa, A. 1978. Fault morphology and Quaternary faulting along the Median Tectonic Line in the southern part of the Izumi Range. *Geographical Review of Japan*, **50**, 385–405 [in Japanese with English abstract].

Okada, A., Ando, M. & Tsukuda, T. 1981. Trenches, late Holocene displacement and seismicity of the Shikano fault associated with the 1943 Tottori earthquake. *Bulletin of the Disaster Prevention Research Institute*, **24**(B-1), 105–126 [in Japanese with English abstract].

Okada, S. & Ikeda, Y. 2012. Quantifying crustal extension and shortening in the back-arc region of northeast Japan. *Journal of Geophysical Research*, **117**, B01404, http://doi.org/10.1029/2011JB008355

Okada, S., Ishimura, D., Niwa, Y. & Toda, S. 2015. The first surface-rupturing earthquake in 20 years on a HERP active fault is not characteristic: the 2014 Mw 6.2 Nagano event along the Northern Itoigawa–Shizuoka tectonic line. *Seismological Research Letters*, **86**, 1287–1300, http://doi.org/10.1785/0220150052

Okada, Y. 1994. Seismotectonics in the Tokyo metropolitan area, central Japan. *Proceedings of the 9th Joint Meeting of the UJNR Panel on Earthquake Prediction Technology*, Kyoto, Japan, **9**, 47–59.

Okada, Y., Kasahara, K., Hori, S., Obara, K., Sekiguchi, S., Fujiwara, H. & Yamamoto, A. 2004. Recent progress of seismic observation networks in Japan: Hi-net, F-net, K-NET and KiK-net. *Earth, Planets and Space*, **56**, xv–xxviii.

Okumura, K. 2001. Paleoseismology of the Itoigawa-Shizuoka tectonic line in central Japan. *Journal of Seismology*, **5**, 411–431.

Okumura, K., Shimokawa, K., Yamazaki, H. & Tsukuda, E. 1994. Recent surface faulting events along the middle section of the Itoigawa-Shizuoka tectonic line - Trenching survey of the Gofukuji fault near Matsumoto, central Japan. *Journal of the Seismological Society of Japan (Zisin), 2nd Series*, **46**, 425–438 [in Japanese with English abstract].

Okumura, K., Imura, R., Imaizumi, T., Togo, M., Sawa, H., Mizuno, K. & Kariya, Y. 1998. Recent surface faulting events along the northern part of the Itoigawa-Shizuoka Tectonic Line - Trenching survey of the Kamishiro fault and east Matsumoto Basin faults, central Japan. *Journal of the Seismological Society of Japan (Zisin)*, **50**, 35–51 [in Japanese with English abstract].

Okamura, Y., Watanabe, M., Morijiri, R. & Satoh, M. 1995. Rifting and basin inversion in the eastern margin of the Japan Sea. *Island Arc*, **4**, 166–181.

Omori, F. 1894. On the aftershocks of earthquake. *Journal of the College of Science, Imperial University of Tokyo*, **7**, 111–200.

Oyama, T., Sone, K. & Ueta, K. 1991. *Survey of Faults under Alluvial Deposits – (3) Trench Log Survey of Kawafune Fault*. Central Research Institute of Electric Power Industry (CRIEPI) Research Report, U91032 [in Japanese with English abstract].

Ozawa, S., Nishimura, T., Suito, H., Kobayashi, T., Tobita, M. & Imakiire, T. 2011. Coseismic and postseismic slip of the 2011 magnitude-9 Tohoku-Oki earthquake. *Nature*, **475**, 373–376, http://doi.org/10.1038/nature10227

Ozawa, S., Nishimura, T., Munekane, H., Suito, H., Kobayashi, T., Tobita, M. & Imakiire, T. 2012. Preceding, coseismic, and postseismic slips of the 2011 Tohoku earthquake, Japan. *Journal of Geophysical Research*, **117**, B07404, http://doi.org/10.1029/2011JB009120

Pollitz, F. F. & Sacks, I. S. 1997. The 1995 Kobe, Japan, earthquake: a long-delayed aftershock of the offshore 1944 Tonankai and 1946 Nankaido earthquakes. *Bulletin of the Seismological Society of America*, **87**, 1–10.

Reid, H. F. 1910. The mechanism of the earthquake. *In*: Lawson, A. C. (chair) *The California Earthquake of April 18, 1906, Report of the State Earthquake Investigation Commission*. Carnegie Institution, Washington, DC, **2**, 1–192.

Research Group for Active Faults of Japan 1980. *Active Faults in Japan, Sheet Maps and Inventories*. University of Tokyo Press, Tokyo [in Japanese].

Research Group for Active Faults of Japan 1991. *Active Faults in Japan, Sheet Maps and Inventories*, rev. edn. University of Tokyo Press, Tokyo [in Japanese].

Research Group for Ito-Shizu Tectonic Line Active Faults 1988. Late Quaternary activities in the central part of Itoshizu tectonic line – excavation study at Wakamiya and Osawa faults, Nagano prefecture, central Japan. *Bulletin of the Earthquake Research Institute, University of Tokyo*, **63**, 349–408 [in Japanese with English abstract].

Research Group for the Senya Fault 1986. Holocene activities and near-surface features of the Senya fault, Akita Prefecture, Japan: Excavation study at Komori, Senhata-cho. *Bulletin of the Earthquake Research Institute, University of Tokyo*, **61**, 339–402 [in Japanese with English abstract].

Rydelek, P. A. & Sacks, I. S. 1990. Asthenospheric viscosity and stress diffusion: a mechanism to explain correlated earthquakes and surface deformations in NE Japan. *Geophysics Journal International*, **100**, 39–58.

Sagiya, T., Miyazaki, S. & Tada, T. 2000. Continuous GPS array and present-day crustal deformation of Japan. *Pure and Applied Geophysics*, **157**, 2302–2322.

Sangawa, A. 1978. Geomorphic development of the Izumi and Sanuki Ranges and relating crustal movement. *The Science Reports of the Tohoku University, 7th Series (Geography)*, **28**, 313–338.

Sangawa, A. 2010. A study of paleoearthquakes at archeological sites – a new interdisciplinary area between paleoseismology and archeology. *Synthesiology, National Institute of Advanced Industrial Science and Technology (AIST)*, **2**, 84–94.

Sangawa, A. & Tsukuda, E. 1986. Active faults on the west coast of Biwa lake and submergence of coastal zone by the earthquake of 1662 (Kanbun 2). *Seismological Society of Japan, Programme and Abstracts*, **2**, 136 [in Japanese].

Sato, H. 1989. Degree of deformation of late Cenozoic strata in the Northeast Honshu Arc. *Memoirs of the Geological Society of Japan*, **32**, 257–268 [in Japanese with English abstract].

Sato, H. 1994. The relationship between late Cenozoic tectonic events and stress field and basin development in northeast Japan. *Journal of Geophysical Research*, **99**, 22261–22274.

Sato, H., Hirata, N., Iwasaki, T., Matsubara, M. & Ikawa, T. 2002. Deep seismic reflection profiling across the Ou Backbone range, northern Honshu Island, Japan. *Tectonophysics*, **355**, 41–52.

Sawai, Y., Namegaya, Y., Okamura, Y., Satake, K. & Shishikura, M. 2012. Challenges of anticipating the 2011 Tohoku earthquake and tsunami using coastal geology. *Geophysical Research Letters*, **39**, L21309, http://doi.org/10.1029/2012GL053692

Scharer, K., Weldon, R., Streig, A. & Fumal, T. 2014. Paleoearthquakes at Frazier Mountain, California delimit extent and frequency of past San Andreas Fault ruptures along 1857 trace. *Geophysical Research Letters*, **41**, 4527–4534.

Schwartz, D. P. & Coppersmith, K. J. 1984. Fault behavior and characteristic earthquakes: examples from the Wasatch and San Andreas faults. *Journal of Geophysical Research*, **89**, 5681–5698.

Sekiguchi, H., Irikura, K. & Iwata, T. 2002*a*. Source inversion for estimating continuous slip distribution on the fault -Introduction of Green's functions convolved with a correction function to give moving dislocation effects in sub faults. *Geophysics Journal International*, **150**, 377–391.

Sekiguchi, H., Kase, Y., Horikawa, H., Satake, K., Sugiyama, Y. & Pitarka, A. 2002*b*. Ground motion prediction in Osaka Plain based on dynamic rupture scenario and three-dimensional subsurface structure model. *Annual Report on Active Fault and Paleoearthquake Researches, AIST*, **2**, 341–357 [in Japanese with English abstract].

Seno, T. 1979. Intraplate seismicity in Tohoku and Hokkaido and large interplate earthquakes: a possibility of a large interplate earthquake off the southern Sanriku coast, northern Japan. *Journal of Physics of the Earth*, **27**, 21–51.

Shelly, D. R., Beroza, G. C. & Ide, S. 2007. Non-volcanic tremor and low-frequency earthquake swarms. *Nature*, **446**, 305–307, http://doi.org/10.1038/nature05666

Shikakura, Y., Fukahata, Y. & Hirahara, K. 2014. Long-term changes in the Coulomb failure function on inland active faults in southwest Japan due to east-west compression and interplate earthquakes. *Journal of Geophysical Research*, **119**, 502–518.

Shimazaki, K. 1976. Intra-plate seismicity and inter-plate earthquakes: historical activity in southwest Japan. *Tectonophysics*, **33**, 33–42.

Shimazaki, K. 1978. Correlation between intraplate seismicity and interplate earthquakes in Tohoku, northeast Japan. *Bulletin of the Seismological Society of America*, **68**, 181–192.

Shimazaki, K. & Nakata, T. 1980. Time-predictable recurrence model for large earthquakes. *Geophysical Research Letters*, **7**, 279–282.

Shimazaki, K., Kim, H. Y.Chiba, T. & Satake, K. 2011. Geological evidence of recurrent great Kanto earthquakes at the Miura Peninsula, Japan. *Journal of Geophysical Research*, **116**, B12408, http://doi.org/10.1029/2011JB008639

Shimojo, K., Enescu, B., Yagi, Y. & Takeda, T. 2014. Fluid-driven seismicity activation in northern Nagano region after the 2011 M9.0 Tohoku-oki earthquake. *Geophysical Research Letters*, **41**, 7524–7531, http://doi.org/10.1002/2014GL061763

Shimokawa, K., Mizuno, K., Imura, R., Okumura, K., Sugiyama, Y. & Yamazaki, H. 1995. *Strip map of the Itoigawa-Shizuoka tectonic line active fault system, Tectonic Map Series 11*. Geological Survey of Japan, Tsukuba, Japan.

Shishikura, M. 2003. Cycle of interplate earthquakes along the Sagami Trough deduced from tectonic geomorphology. *Bulletin of the Earthquake Research Institute, University of Tokyo*, **78**, 245–254 [in Japanese with English abstract].

Sieh, K. 1978. Pre-historic large earthquakes produced by slip on the San Andreas fault at Pallett Creek, California. *Journal of Geophysical Research*, **83**, 3907–3939.

Simons, M., Minson, S. E. *et al.* 2011. The 2011 magnitude 9.0 Tohoku-Oki earthquake: mosaicking the megathrust from seconds to centuries. *Science*, **332**, 1421–1425.

Somerville, P., Irikura, K. *et al.* 1999. Characterizing crustal earthquake slip models for prediction of strong ground motion. *Seismological Research Letters*, **70**, 59–80.

Stein, R. S. 1999. The role of stress transfer in earthquake occurrence. *Nature*, **402**, 605–609.

Stein, R. S., Barka, A. A. & Dieterich, J. H. 1997. Progressive failure on the north Anatolian fault since 1939 by earthquake stress triggering. *Geophysics Journal International*, **128**, 594–604.

Stein, R. S., Toda, S., Parsons, T. & Grunewald, E. 2006. A new probabilistic seismic hazard assessment for greater Tokyo. *Philosophical Transactions of the Royal Society A*, **364**, 1965–1988, http://doi.org/10.1098/rsta.2006.1808

Stein, S. & Liu, M. 2010. Long aftershock sequences within continents and implications for earthquake hazard assessment. *Nature*, **462**, 87–89.

Stein, S., Geller, R. J. & Liu, M. 2011. Bad assumptions or bad luck: why earthquake hazard maps need objective testing. *Seismological Research Letters*, **82**, 623–626.

Stein, S., Geller, R. J. & Liu, M. 2012. Why earthquake hazard maps often fail and what to do about it. *Tectonophysics*, **562–563**, 1–25.

Stirling, M., Rhoades, D. & Berryman, K. 2002. Comparison of earthquake scaling relations derived from data of the instrumental and pre instrumental era. *Bulletin of the Seismological Society of America*, **92**, 812–820.

SUGAI, T. 2011. Geomorphological and geological records indicates that activity on the Yoro fault system caused the AD 1586 Tensho earthquake. *Active Fault Research*, **35**, 15–28 [in Japanese with English abstract].

SUGAWARA, D., IMAMURA, F., GOTO, K., MATSUMOTO, H. & MINOURA, K. 2013. The 2011 Tohoku-oki earthquake tsunami: similarities and differences to the 860 Jogan tsunami on the Sendai Plain. *Pure and Applied Geophysics*, **170**, 831–843.

SUGIMURA, A. & MATSUDA, T. 1965. Atera fault and its displacement vectors. *Bulletin of the Geological Society of America*, **76**, 509–522.

SUGIYAMA, Y. 1998. Present state and prospect of active fault research and paleoseismology by the Geological Survey of Japan. *Chishitsu News*, **523**, 12–20 [in Japanese].

SUGIYAMA, Y., AWATA, Y. & TSUKUDA, E. 1991. Holocene activity of the Miboro fault system, central Japan and its implication for the Tensho earthquake of 1586 – verification by excavation survey. *Journal of the Seismological Society of Japan (Zisin)*, **44**, 283–295 [in Japanese with English abstract].

SUGIYAMA, Y., MIZUNO, K. *ET AL.* 2003. Study of blind thrust faults underlying Tokyo and Osaka urban areas using a combination of high-resolution seismic reflection profiling and continuous coring. *Annals of Geophysics*, **46**, 1071–1085.

TAKADA, Y., KOBAYASHI, T., FURUYA, M. & MURAKAMI, M. 2009. Coseismic displacement due to the 2008 Iwate-Miyagi Nairiku earthquake detected by ALOS/PALSAR: preliminary results. *Earth, Planets and Space*, **61**, e9–e12.

TAKEHISA, Y. 1973. Crustal movement suggested by fault scarplets along the eastern foot of the Kongo-Katsuragi Range. *Annual Report of Studies in Humanities and Social Sciences, Faculty of Letters, Nara Women's University*, **16**, 145–167 [in Japanese with English abstract].

TAKEMURA, M. 1998. Scaling law for Japanese intraplate earthquakes in special relations to the surface faults and the damages. *Journal of the Seismological Society of Japan (Zisin)*, **51**, 211–228 [in Japanese with English abstract].

TAKEMURA, M. 2003. *The Great Kanto Earthquake, Knowing Ground Shaking in Greater Tokyo Area, Tokyo Japan*. Kajima Institute Publishing, Tokyo [in Japanese].

TAMURA, Y., TATSUMI, Y., ZHAO, D., KIDO, Y. & SHUKUNO, H. 2002. Hot fingers in the mantle wedge: new insights into magma genesis in subduction zones. *Earth and Planetary Science Letters*, **197**, 105–116.

TANIGAWA, K., SAWAI, Y., SHISHIKURA, M., NAMEGAYA, Y. & MATSUMOTO, D. 2014. Geological evidence for an unusually large tsunami on the Pacific coast of Aomori, northern Japan. *Journal of Quaternary Science*, **29**, 200–208.

TANNA FAULT TRENCHING RESEARCH GROUP 1983. Trenching study for Tanna fault, Izu, at Myoga, Shizuoka Prefecture, Japan. *Bulletin of the Earthquake Research Institute, University of Tokyo*, **53**, 797–830 [in Japanese with English abstract].

TERAKAWA, T. & MATSU'URA, M. 2010. The 3-D tectonic stress fields in and around Japan inverted from centroid moment tensor data of seismic events. *Tectonics*, **29**, TC6008, http://doi.org/10.1029/2009TC002626

THATCHER, W., MARSHALL, G. & LISOWSKI, M. 1997. Resolution of fault slip along the 470-km-long rupture of the great 1906 San Francisco earthquake and its implications. *Journal of Geophysical Research*, **102**, 5353–5367.

TODA, S. 2013. Current issues and a prospective view to the next step for long-term crustal earthquake forecast in Japan. *Joural of the Geological Society of Japan*, **119**, 105–123 [in Japanese with English abstract].

TODA, S. & TSUTSUMI, H. 2013. Simultaneous Reactivation of Two, Subparallel, Inland Normal Faults during the Mw 6.6 11 April 2011 Iwaki Earthquake Triggered by the Mw 9.0 Tohoku-oki, Japan, Earthquake. *Bulletin of the Seismological Society of America*, **103**, 1584–1602, http://doi.org/10.1785/0120120281

TODA, S., INOUE, D., TAKASE, N., KUBOUCHI, A. & TOMIOKA, N. 1994. The latest activity of the Atera fault: a possibility of the 1586 Tensho earthquake. *Journal of the Seismological Society of Japan (Zisin), 2 Series*, **47**, 73–77 [in Japanese].

TODA, S., STEIN, R. S., REASENBERG, P. A. & DIETERICH, J. H. 1998. Stress transferred by the Mw = 6.9 Kobe, Japan, shock: effect on aftershocks and future earthquake probabilities. *Journal of Geophysical Research*, **103**, 24,543–24,565.

TODA, S., MIURA, D., MIYAKOSHI, K. & INOUE, D. 2000. Recent surface faulting events along the southern part of the Itoigawa-Shizuoka tectonic line. *Journal of the Seismological Society of Japan (Zisin), 2nd Series*, **52**, 445–468 [in Japanese with English abstract].

TODA, S., YOSHIOKA, K., OMATA, M., KORIYA, Y. & IWASAKI, T. 2008. Holocene paleoseismic history and possible segmentation on the Yomagata-bonchi fault zone, in northern Honshu, Japan. *Active Fault Research*, **29**, 35–57 [in Japanese with English abstract].

TODA, S., MARUYAMA, T., YOSHIMI, M., KANEDA, Y., AWATA, Y., YOSHIOKA, T. & ANDO, R. 2010. Surface rupture associated with the 2008 Iwate-Miyagi Nairiku, Japan earthquake and its implications to the rupture process and evaluation of active faults. *Journal of the Seismological Society of Japan (Zisin), 2 Series*, **62**, 153–178 [in Japanese with English abstract].

TODA, S., STEIN, R. S. & LIN, J. 2011*a*. Widespread seismicity excitation throughout central Japan following the 2011 M = 9.0 Tohoku earthquake and its interpretation by Coulomb stress transfer. *Geophysical Research Letters*, **38**, L00G03, http://doi.org/10.1029/2011GL047834

TODA, S., LIN, J. & STEIN, R. S. 2011*b*. Using the 2011 Mw 9.0 off the Pacific coast of Tohoku Earthquake to test the Coulomb stress triggering hypothesis and to calculate faults brought closer to failure. *Earth, Planets and Space*, **63**, 725–730.

TOWNEND, J. & ZOBACK, M. D. 2006. Stress, strain, and mountain building in central Japan. *Journal of Geophysical Research*, **111**, B03411, http://doi.org/10.1029/2005JB003759

TSUKAHARA, H. & IKEDA, R. 1991. Crustal stress orientation pattern in the central part of Honshu, Japan: stress provinces and their origins. *Journal of the Geological Society of Japan*, **97**, 461–474 [in Japanese with English abstract].

TSUKAHARA, H. & KOBAYASHI, Y. 1991. Crustal stress in the central and western parts of Honshu, Japan. *Journal of the Seismological Society of Japan (Zisin)*, **44**, 221–231 [in Japanese with English abstract].

TSUKUDA, E. 2002. Rupturing history of active faults during the last 1000 years in the central Japan. *In*: FUJINAWA, Y. & YOSHIDA, A. (eds) *Seismotectonics in Convergent Plate Boundary*. Terra Scientific Publication Company, Tokyo, 209–218.

TSUKUDA, E., AWATA, Y., YAMAZAKI, H., SUGIYAMA, Y., SHIMOKAWA, K. & MIZUNO, K. 1993. *Explanatory text of the strip map of the Atera fault system, scale 1:25,000*. Tectonic map series (7) Geological Survey of Japan [in Japanese with English abstract].

TSUTSUMI, H. & GOTO, H. 2006. Surface offsets associated with the most recent earthquakes along the Median Tectonic Line active fault system in Shikoku, southwest Japan. *Journal of the Seismological Society of Japan (Zisin)*, **59**, 117–132 [in Japanese with English abstract]

TSUTSUMI, H. & OKADA, A. 1996. Segmentation and Holocene surface faulting on the Median Tectonic Line, southwest Japan. *Journal of Geophysical Research*, **101**, 5855–5871, http://doi.org/10.1029/95JB01913

USAMI, T. 2003. *Materials for Comprehensive List of Destructive Earthquakes in Japan*. University of Tokyo Press, Tokyo [in Japanese].

USAMI, T., ISHII, H., IMAMURA, T., TAKEMURA, M. & MATSU'URA, R. S. 2013. *Materials for Comprehensive List of Destructive Earthquakes in Japan, 599–2012*. University of Tokyo Press, Tokyo [in Japanese].

UTSU, T. 1961. A statistical study on the occurrence of aftershocks. *Geophysics Magazine*, **30**, 521–605.

UTSU, T. 1974. Correlation between great earthquakes along the Nankai trough and destructive earthquakes in western Japan. *Report of the Coordinating Committee for Earthquake Prediction, Japan*, **12**, 120–122.

UTSU, T. 1982. Catalog of large earthquakes in the region of Japan from 1885 through 1980. *Bulletin of Earthquake Research Institute*, **57**, 401–463 [in Japanese with English abstract].

WELLS, D. L. & COPPERSMITH, K. J. 1994. New empirical relationships among magnitude, rupture length, rupture area, and surface displacement. *Bulletin of the Seismological Society of America*, **84**, 974–1002.

WESNOUSKY, S. G . 1988. Seismological and structural evolution of strike-slip faults. *Nature*, **335**, 340–343.

WESNOUSKY, S. G. 2006. Predicting the endpoints of earthquake ruptures. *Nature*, **444**, 358–360.

WESNOUSKY, S. G. 2008. Displacement and geometrical characteristics of earthquake surface ruptures: issues and implications for seismic-hazard analysis and the process of earthquake rupture. *Bulletin of the Seismological Society of America*, **98**, 1609–1632.

Wesnousky, S. G., Scholz, C. & Shimazaki, K. 1982. Deformation of an island arc: rates of moment release and crustal shortening in intraplate Japan determined from seismicity and Quaternary fault data. *Journal of Geophysical Research*, **87**, 6829–6852.

Wesnousky, S. G., Scholz, C. H., Shimazaki, K. & Matsuda, T. 1984. Integration of geological and seismological data for the analysis of seismic hazard – a case study of Japan. *Bulletin of the Seismological Society of America*, **74**, 687–708.

Working Group on California Earthquake Probabilities (WGCEP) 1988. Probabilities of large earthquakes occurring in California on the San Andreas fault. US Geological Survey, Open-File Report, **88–398**, 62.

Working Group on California Earthquake Probabilities (WGCEP) 1995. Seismic hazards in southern California: probable earthquakes, 1994–2024. *Bulletin of the Seismological Society of America*, **85**, 379–439.

Yamada, N. 1981. The Atera fault – its geological background. *GekkanChikyu*, **3**, 237–243 [in Japanese].

Yamasaki, N. 1896. Preliminary report of the Rikuu earthquake. *Report of the Imperial Earthquake Investigation Commission*, **11**, 50–74 [in Japanese].

Yoshida, K., Hasegawa, A., Okada, T., Iinuma, T., Ito, Y. & Asano, Y. 2012. Stress before and ater the 2011 great Tohoku-oki earthquake and induced earthquakes in inland areas of eastern Japan. *Geophysical Research Letters*, **39**, L03302, http://doi.org/10.1029/2011 GL049729

Yoshioka, T., Mizuno, K. & Sakakibara, N. 1997. Last faulting event of the Senzan fault in the central Awaji Island, southwestern Japan. *Active Fault Research*, **16**, 87–94 [in Japanese with English abstract].

9 Coastal geology and oceanography

YOSHIKI SAITO, KEN IKEHARA & TORU TAMURA

Coastal geology (YS, TT)

The Japanese islands comprise 59% mountains, 6% volcanoes and 35% hills and (mostly coastal) lowlands (Yonekura 2001). Quaternary strata excluding volcanoes occupy around 25% of the geological map of Japan, and constitute hills and lowland plains. Holocene strata are distributed mainly in the lowlands and occupy 13–15% of Japan (Murata & Kano 1995; D. Kawabata 2013, pers. comm.). These extensive lowlands comprise four depositional environments, namely delta, fan delta, strand plain and barrier/estuary systems, all of which have been impacted by Holocene sea-level changes. Most of Holocene Japanese coastal plains record a hydro-isostatically controlled relatively stable to falling sea level over the last 6–7 ka (Ota *et al.* 1981, 1987*a*, 1990; Nakada *et al.* 1991; Okuno *et al.* 2014). This hydro-isostatic effect varies spatially, being more strongly developed in the central rather than marginal parts of the islands (Yokoyama *et al.* 1996; Nakada *et al.* 1998; Nakada & Okuno 2011; Okuno *et al.* 2014). Local tectonics have also strongly impacted relative sea levels since middle Holocene times, ranging from +30 m in the southern tip of the uplifting Boso Peninsula, SE of Tokyo, to −20 m in the subsiding Echigo Plain (Ota *et al.* 1987*a*, *b*; Shishikura 2001; Urabe *et al.* 2004; Tanabe *et al.* 2009). The overall scenario of falling sea levels over the last 6–7 ka has encouraged delta progradation and shoreline migration seawards, resulting in the formation of wide coastal plains and the infilling of lagoons and estuaries (referred to here as barrier and estuary systems). The barrier and estuary system lowlands, dominant during maximum Holocene transgression, have evolved into delta and strand plain systems following the infilling and abandonment of lagoons.

In this first part of the chapter, after a brief overview of previous work we describe in more detail the four Holocene coastal depositional systems. This is followed by a consideration of Japanese tsunami sediments, which provide a valuable geological record of past earthquakes that can be used in disaster mitigation for tsunamis in the future. We then focus on the striking history of recent subsidence that has occurred in some of the Holocene coastal plains. Finally we provide a brief overview of the famous site of Lake Suigetsu, which is one of the auxiliary stratotypes for the base of the Holocene Series/Epoch in Global Stratotype Section and Point (GSSP) and also offers a standard radiocarbon calibration dataset.

Early work

Research on the lowlands of Japan was first described in the geological map of Hokkaido (Lyman 1876), which separated New Alluvium (Holocene) from Old Alluvium (Pleistocene), and this was followed by studies on the Tokyo Lowland (Naumann 1879; Brauns 1881; Suzuki 1888). After the 1923 Great Kanto Earthquake, which induced major devastation in Tokyo where more than 100 000 people were killed, the Reconstruction Bureau produced systematic research on lowlands in Tokyo and Yokohama with 797 newly collected boreholes and 1508 compiled boreholes, and revealed the stratigraphy and buried valleys in the lowlands (Construction Department, Reconstruction Bureau 1929). The topography, stratigraphy and formation of the lowlands were discussed from the viewpoint of post-glacial sea-level changes in the 1950s (Sugimura 1950; Sugimura & Naruse 1954, 1955). Research on the lowlands progressed largely by the analysis of several tens of thousands of borehole cores (e.g. Kuwano *et al.* 1971; Kaizuka *et al.* 1977), particularly during the high-growth economy of the 1950s–1960s, and Holocene basement maps for construction were published in major cities and lowlands in Japan in the 1960s–1970s.

As for Quaternary research in Japan, Quaternary Maps of Japan: Landforms Geology and Tectonics (1: 1 000 000) were published in 1987 (Japan Association for Quaternary Research 1987), and Holocene and Last Interglacial Strata and shoreline maps of Japan were published by Ota *et al.* (1987*b*) and Ota *et al.* (1992), respectively.

Architecture of incised-valley fills and coastal strata

Strata infilling incised valleys formed during the Last Glacial Maximum (LGM) and constituting lowlands are called ‘Chuseki-so’ in Japanese, translated from the word alluvium. The stratotype in Japan occurs in the Tokyo Lowland and consists of a lower unit (Nanagochi Formation) which comprises fluvial sediments overlain by brackish estuarine sediments, and an upper unit (Yurakucho Formation) which comprises marine sediments overlain by present-day fluvial sediments (Otuka 1934; Aoki 1969; Tokyo Institute of Civil Engineering 1969; Kuwano *et al.* 1971). As post-glacial transgression is closely linked with these formations; the late Pleistocene transgression is named the Nanagochi Transgression (Endo *et al.* 1982, 1983) and the early Holocene transgression is named the Yurakucho Transgression (Kobayashi 1957) or Jomon Transgression (from the Jomon period, an archaeological period in Japan). Ota *et al.* (1981, 1987*a*, *b*, 1990) reviewed Holocene sea-level changes in the Japanese islands and recently Tanabe *et al.* (2010*a*, *c*), Tanigawa *et al.* (2013) and Tanabe & Ishihara (2013) showed more high-resolution sea-level changes.

The coastal lowlands in Japan are characterized by relatively coarse-grained sediments and thick incised-valley fills which developed over the last 6–7 ka. The coarse sediments are a product of steep-gradient and large-sediment-delivery mountainous rivers. The thickness of incised-valley fills formed since the LGM beneath the coastal plains is >30 m in most of the coastal plains (Ota *et al.* 1987*a*, *b*), >70 m in the Tokyo Lowland (Kaizuka *et al.* 1977; Tanabe *et al.* 2008*b*) and >150 m in the subsiding Echigo Plain

From: Moreno, T., Wallis, S., Kojima, T. & Gibbons, W. (eds) 2016. *The Geology of Japan*. Geological Society, London, 409–430.

(Niigata Plain) (Tanabe *et al.* 2013). The lowest sea levels during the LGM in Japan were *c.* 130 m below the present sea level (Saito *et al.* 1989; Nakada *et al.* 1991; Saito 2005), allowing excavation of these deep-incised valleys across the coastal plains and exposed shelf areas. The shelf width around the Japanese islands is relatively narrow, ranging from <1 km in Toyama Bay to *c.* 60 km in Sendai Bay. High-resolution sequence analysis from shelves to coastal zones using dated sediment cores and seismic data since the LGM has been conducted around the Japanese islands (e.g. Saito 1991*b*, 1994; Nishida & Ikehara 2013).

Japanese stratotype since the LGM: the Tokyo Lowland

The stratigraphy of the Chuseki-so incised-valley fills and coastal strata was summarized in the 1970s–1980s by Iseki (1975, 1983), Matsuda (1974) and Kaizuka *et al.* (1977) as follows (in ascending order): fluvial basal gravel beds (BG); fluvial to brackish lower sand and mud (LS); brackish to marine middle sand (MS); marine middle mud (MM; or upper mud: UM); marine upper sand (US); and uppermost terrestrial mud (TM). In the stratotype Tokyo Lowland, the Nanagochi and Yurakucho formations consist of BG to LS, and MS to TM, respectively (Aoki 1969; Matsuda 1974). The sedimentary environments of the Chuseki-so are interpreted as mostly braided river systems for BG, meandering river to estuarine systems for LS and MS to the lowest part of MM, and deltaic systems for MM, US and TM (e.g. Saito 1995; Tanabe *et al.* 2008*a*, 2010*b*). The Holocene maximum transgression boundary between these essentially estuarine (Nanagochi-type) and deltaic (Yurakucho-type) systems is recognized at *c.* 7 ka (6–8 ka), depending on the differences between relative sea-level changes and sediment supply in each region (e.g. Tamura *et al.* 2008*b*; Ogami *et al.* 2009; Sato & Masuda 2010; Tanabe *et al.* 2010*b*; Saegusa et al. 2011). The formation boundary was initially interpreted as an unconformity correlated with the Younger Dryas sea-level fall (Matsuda 1974; Kaizuka *et al.* 1977), but later redefined and interpreted as a diachronous facies boundary (Saito 2006; Tanabe *et al.* 2010*b*; Tanabe 2013) or as an unconformity indicating the base of the Holocene basal gravel bed, named HBG (Endo *et al.* 1983, 2013). The transgressive sediments demonstrate complicated facies distributions which include estuarine transgressive spit systems in the Tokyo Lowland (Tanabe *et al.* 2012). Although the Tokyo Lowland is a stratotype for strata formed since the LGM in Japan, its detailed stratigraphy is still under debate (Endo *et al.* 2013; Tanabe 2013).

Coastal systems

The Holocene development and classification of coastal plains in Japan have been mainly studied using geomorphological approaches (Oya 1977; Umitsu 1981, 1994; Moriwaki 1982), although here we approach the subject from the viewpoint of depositional systems. Coastal systems constituting coastal plains have been grouped into deltaic, strand plain, and barrier and estuary systems (e.g. Saito 1987, 1995, 2006; Masuda & Saito 1999). These three coastal environments are closely linked to each other, and together comprise a composite system that has evolved over the last 6–7 ka. In central Japan, braided rivers flowing on alluvial fans reach the shoreline directly and form fan deltas along the coast, forming a fourth depositional system.

These four depositional systems are closely linked to coastal oceanography and sediment discharge from rivers. Most of the coasts facing the Pacific Ocean, the Japan Sea and the Sea of Okhotsk are wave- or storm-dominated with mean wave heights of 0.8–1.3 m which can exceed 5 m during storms. In contrast, more sheltered areas such as the Seto Inland Sea (Seto–Nai–Kai) and enclosed bays (e.g. Tokyo Bay) show less than 0.5 m in mean wave height. Mean maximum tidal range is 1.0–2.0 m for coasts facing the Pacific Ocean, <0.5 m in the Japan Sea, 1.0–1.5 m in the Sea of Okhotsk (micro-tidal), *c.* 2 m in Tokyo Bay and Ise Bay, >2 m in the Seto Inland Sea (micro- to meso-tidal) and >4 m in Ariake Bay of Kyushu (macro-tidal). In summary, the Japanese coasts are wave- or storm-dominated coasts facing open seas, or tide-wave-dominated coasts in enclosed bays except for the tide-dominated Ariake Bay.

The four depositional systems in the Japanese islands each have their characteristic distributions, resulting from sediment delivery and coastal oceanography (Fig. 9.1). Most of the delta systems are located in enclosed bay areas which are relatively less influenced by waves. Fan delta systems occur mostly in central Japan, where very steep rivers deliver coarse-grained sediments to the coasts. Barriers and estuary systems are found along the coasts facing wave- or storm-dominated open seas. Some of the lagoons or central basins of an estuary system have been filled by sediments supplied from bay-head deltas (or lagoonal or estuarine deltas) and/or supplied by longshore currents through inlets. In such cases, bay-head rivers directly discharge into open seas at present, forming a delta or strand plain. It can be hard to discern previous systems by looking only at present-day morphology.

Another important factor controlling depositional systems in Japan is relative sea-level change. Stable and falling sea level for the last 6–7 ka after a rapid sea-level rise in early Holocene times (Ota *et al.* 1990; Hori & Saito 2007) has exposed wide coastal plains along the Japanese islands and changed coastal depositional systems from transgressive to regressive. Some of the barriers and estuaries, which are typical transgressive systems, have evolved into other systems after the infilling of central basins and lagoons. Most of the large rivers in Japan currently empty directly into open seas and so deliver sediments to open coasts. Japanese coastal geomorphology and depositional systems were therefore very different during the middle Holocene (6–8 ka), when marine inundation was at its maximum, from those of the present. In the following sections we describe in more detail each of the main four coastal depositional systems seen around Japan today.

Delta systems. These are well developed in enclosed bays such as Tokyo Bay, Ise Bay and the Seto Inland Sea. Regarding the former, the Edo, Naka and Arakawa rivers have together formed a wide delta plain comprising the Tokyo, Nakagawa and Arakawa lowlands at the head of Tokyo Bay (Kaizuka *et al.* 1977; Endo *et al.* 1995; Tanabe & Ishihara 2013). Within this setting the Tama, Tsurumi (Matsushima 1987) and Obitsu river deltas (Saito 1991*a*, 1995) are distributed across the middle part of Tokyo Bay. Other delta systems in central Japan feature those draining into Ise Bay, notably those of the combined Kiso, Ibi and Nagara rivers (Nobi Plain: Umitsu 1992; Yamaguchi *et al.* 2006*b*; Saegusa *et al.* 2011), the Kumozu River (Funabiki *et al.* 2010) and Yahagi River (Okazaki Plain: Fujimoto *et al.* 2009; Sato & Masuda 2010) around Ise Bay. Further SW, other notable examples of deltas include those of the Ota (Hiroshima Plain; Shirogami 1985), Goto (Takamatsu Plain; Kawamura 2000), Shigenobu (Matsuyama Plain; Kawamura 2009) in the Seto Inland Sea, Chikugo (Chikushi Plain; Shimoyama *et al.* 1994), Midorikawa and Shirakawa (Kumamoto Plain; Hase *et al.* 2006) rivers in Ariake Bay.

A typical cross section and internal structure of one of these delta systems is shown in Figure 9.2. The figure indicates a geological cross-section across the Kiso River delta (Nobi Plain) with 1 ka interval chronostratigraphic lines based on more than 100 radiocarbon dates (Ogami *et al.* 2009, 2015). Sediment facies are identified by sedimentological and diatom analyses (Yamaguchi *et al.* 2003, 2005, 2006*a*, *b*; Ogami *et al.* 2009; Saegusa *et al.* 2011).

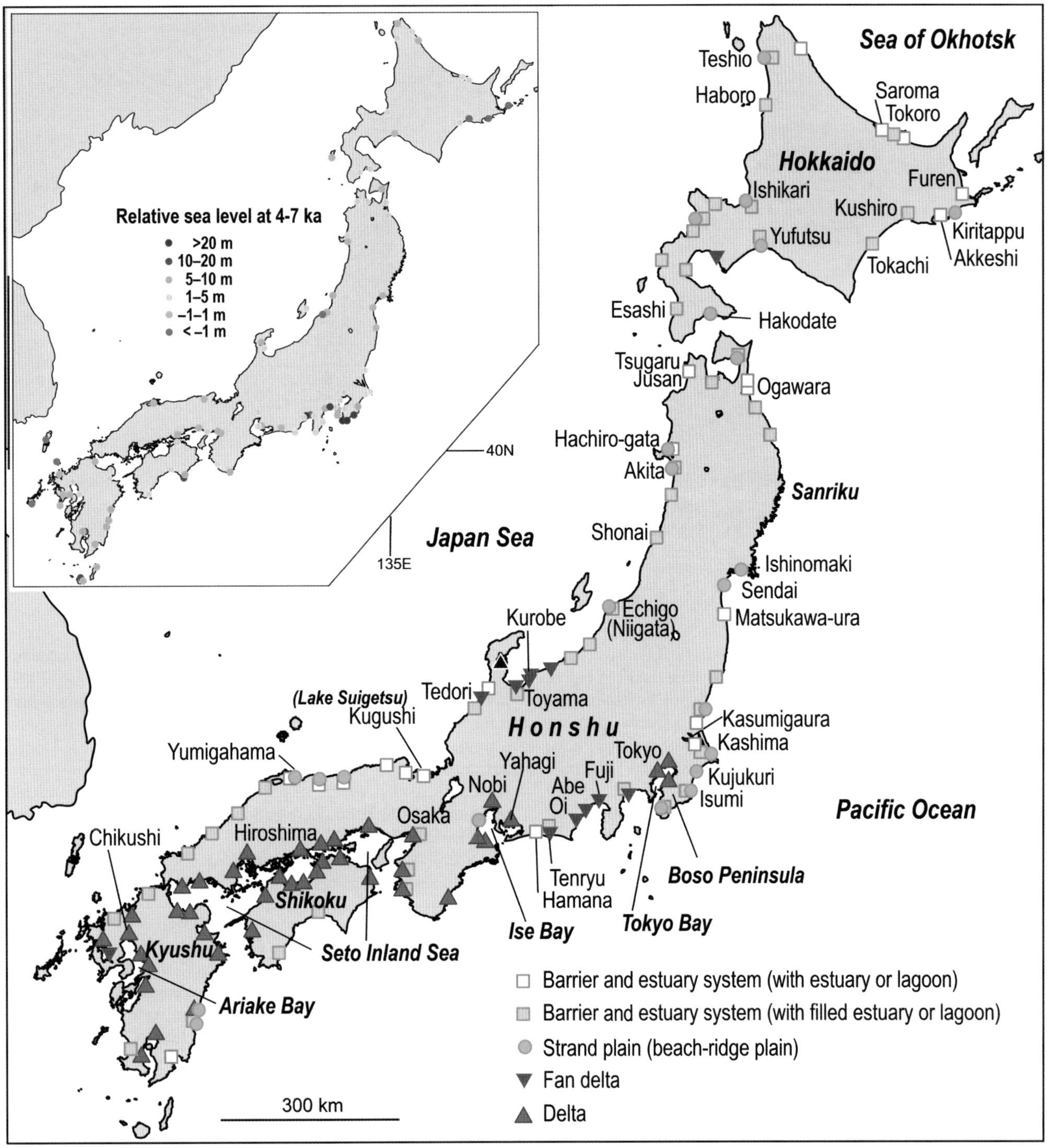

Fig. 9.1. The distribution of four types of coastal systems in Japan. Note that fan delta systems are found mainly only in central Japan, except volcano areas. Delta systems are well developed in bays and the Seto Inland Sea. Most coastal systems facing open oceans indicate a wave-dominated system. Shore-normal overlapped symbols show coastal evolution since the middle Holocene to the present. Coastal systems at the middle Holocene were dominated by estuary or barrier systems. These middle Holocene systems are shown as overlapped symbols. Inserted figure shows the middle Holocene sea levels (modified from Ota *et al.* 1987*a*; Sato *et al.* 2001; Tanigawa *et al.* 2013).

Since the delta was initiated at *c.* 7.3 ka (K-Ah tephra), it has prograded more than 30 km into Ise Bay. Spatial changes of the shoreline during the Holocene epoch have also been illustrated by Saegusa *et al.* (2011). Deltaic successions in Japan mostly show a coarsening-upwards (shallowing-upwards) succession for the subaqueous part. Deltaic sediments in the tide-dominated Ariake Bay are characterized by more muddy sediment facies (Shimoyama *et al.* 1994), particularly for prodelta sediment (known as the Ariake Clay Formation) and for intertidal mudflat sediment.

Delta progradation has not been constant since middle Holocene times. Ono (2004) and Kawase (1998) showed a rapid shoreline migration seawards during 3–2 ka for the Nobi Plain (Kiso River delta) and Okazaki Plain (Yahagi River delta), respectively, caused by relative sea-level fall during 3–2.5 ka (the so-called Yayoi

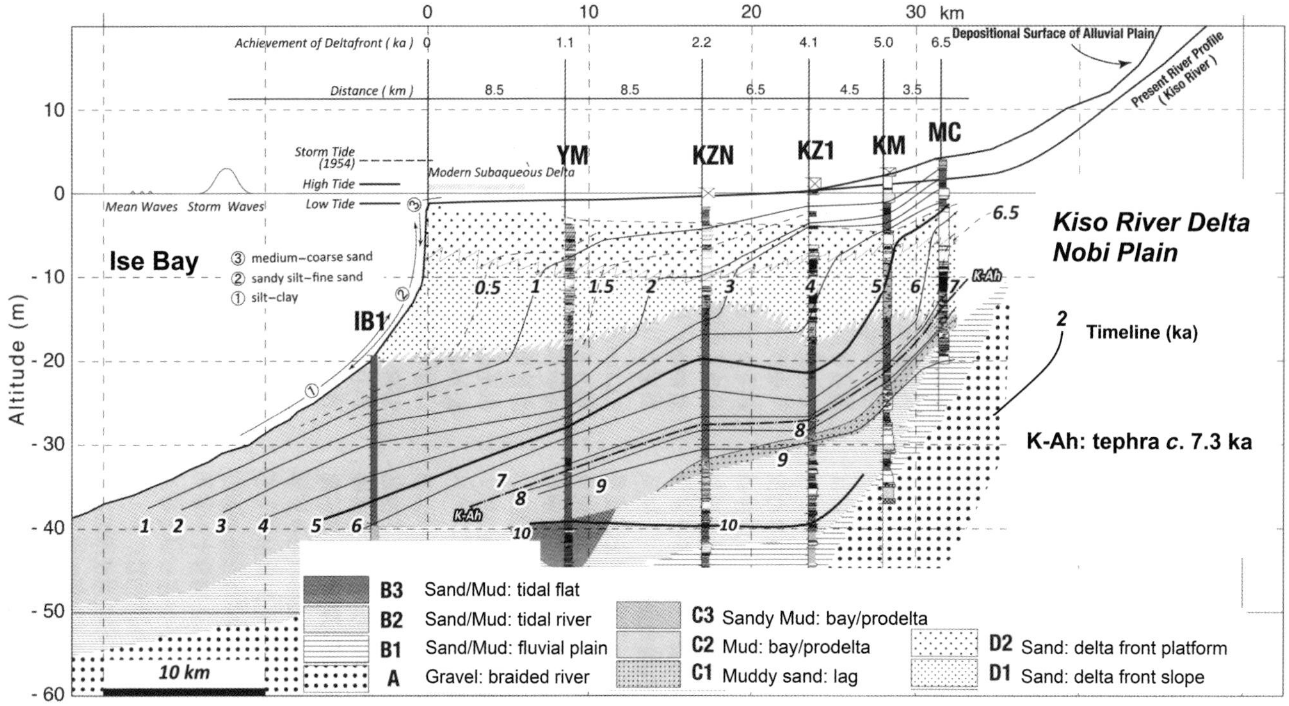

Fig. 9.2. Cross-section of the Kiso River delta system in the Nobi Plain (modified after Ogami *et al.* 2009).

Regression: Furukawa 1972). Tanabe & Ishihara (2013) also demonstrate a sea-level fall from 3.2–2.8 ka down to more than 1 m below the present sea level in the Tokyo and Nakagawa lowlands, also correlative with the Yayoi Regression. Sediment delivery variations due to climate changes and human activities (deforestation) have also impacted shoreline migration (Kawase 1998; Ono *et al.* 2001; Ono 2004). Deforestation for iron sand mining (Kanna-nagashi) in the San'in region of western Honshu during *c.* 1600–1950 yielded a huge amount of sediments, resulting in a rapid progradation of deltas (e.g. Hii River, Hayashi 1991; Takahashi River, Sadakata 2009) and in the development of beach-ridge plains (Yumigahama and Hino River; Sadakata 1991; Kagohara & Imaizumi 2003).

Fan delta systems. These depositional systems are found around central Japan, where steep mountainous rivers deliver very coarse-grained sediments to the coasts. The Sagami, Sakawa, Fuji, Abe, Ooi and Tenryu rivers form fan deltas on the Pacific coast, and the Kurobe, Katakai, Hayatsuki, Joganji and Tedori rivers form fan deltas on the Japan Sea coast (Fig. 9.1).

These fan deltas are divided into two groups: fan delta throughout the Holocene epoch and evolving fan delta during middle–late Holocene times. An example of the former fan delta is the Kurobe River delta, which consists of an aggradational fan formed during the post-glacial sea-level rise and a progradational margin formed during a stable sea level for the last 6–7 ka (Fujii 1992). Early Holocene fan sediments including standing tree stumps crop out at present in erosional nearshore to shelf areas along canyons (Fujii *et al.* 1986). The Fuji River delta also shows aggradational gravelly sediments. Examples of an evolving fan delta are provided by those of the Tenryu, Sagami and Joganji rivers. These deltas were located at the bay head of barriers or estuaries formed during early–middle Holocene times. After the initial infilling of lagoons or coastal wetlands by the progradation of the bay-head fan delta, the current steep-gradient fan delta system was established. In the case of the Tenryu fan delta system, sediment facies beneath surficial gravel beds show an upwards-fining sequence with very high accumulation rates (10–8 ka) overlain by a muddy sequence with low accumulation rates (8–4 ka) (Kobayashi 1964; Nagasawa & Hori 2009), interpreted as having been deposited in back-barrier or estuary environments. The Sagami, Sakawa and Joganji river deltas also show a similar internal architecture with fine-grained sediments. Modern fan delta systems in central Japan have relatively narrow shelves (Fig. 9.1) so that coarse materials supplied by rivers are largely transported to the deep sea (e.g. Soh *et al.* 1995; Saito 2000, 2011; Yoshikawa & Nemoto 2010, 2014).

Other fan deltas are found in Hokkaido and Kyushu, closely linked to active volcanoes.

Strand plain systems. These systems are well developed along the coastal plains facing both the Pacific Ocean and the Sea of Japan, and are characterized by a series of beach ridges. The widest strand plain in Japan is that of Kujukuri in the Boso Peninsula, central Japan, which has a 60 km long shoreline and 10 km wide plain at its maximum, and is flanked by a low plateau and hills <100 m high composed of Pleistocene sedimentary rocks. A series of arcuate beach ridges formed during the last 6 ka is well developed on the Kujukuri strand plain, and has been divided into three groups (Moriwaki 1979). The subsurface geology of the plain is shown in Figure 9.3 (Tamura *et al.* 2008*a*), indicating a regressive beach–shoreface system consisting of (in ascending order): basal shoreface; lower shoreface; upper shoreface; foreshore–backshore; and dunes. Relative sea levels estimated from foreshore sediments show a fall for the last 6 ka. High-resolution analyses of borehole cores and ground-penetrating radar (GPR) surveys reveal step-like lowering of relative sea levels, caused by tectonic uplift of the Boso Peninsula (Tamura *et al.* 2007, 2008*a*, 2010). Relative sea-level falls in the southern Kujukuri strand plain for the last 6 ka are 5–8 m, which are larger than the 1–3 m typically found elsewhere in Japan (Fig. 9.1).

The strand plain system is characterized not only by the above-mentioned sedimentary facies but also by a sharp erosional surface (wave ravinement surface) bounding underlying strata. Due to shoreface erosion during a rise of sea level after the LGM into the

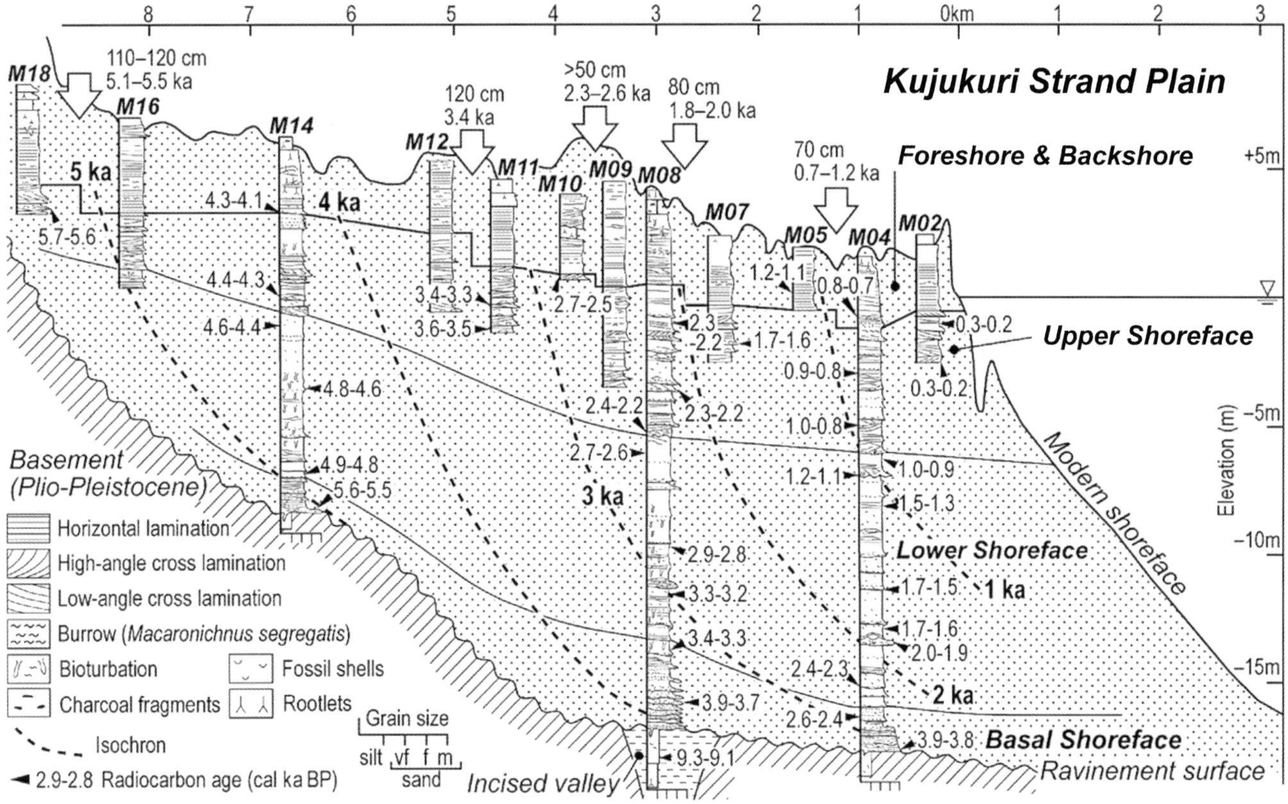

Fig. 9.3. Cross-section of the Kujukuri strand plain system (after Tamura *et al.* 2008*a*). High-resolution ^{14}C dating analyses allow temporal development of a prograding beach–shoreface system for the last 6000 years. Step-like migration of foreshore sediments indicates an abrupt uplift caused by earthquakes.

middle Holocene, the preservation potential of transgressive sediments is generally very low.

The sediment sources of a strand plain system are usually twofold, namely river discharge and longshore drift. In the case of the Kujukuri strand system, most of the sediments have been supplied by coastal erosion from cliffs to the north and south. The Sendai and Ishinomaki coastal plains are examples of a strand plain system (Matsumoto 1981), these occurring in northern Honshu and receiving sediments from the Abukuma, Natori and Nanakita rivers for the Sendai Plain and the Kitakami River for the Ishinomaki Plain. A deltaic morphology, which is characterized by seawards-outbuilding subaqueous and subaerial morphology at the river mouth and river mouth bars, is however very limited, occurring only at the river mouth area even though most of the sediment of the plain comes from the rivers (Saito 1989*a*). As a regressive beach–shoreface system is dominant in these coastal plains, this system is therefore classified here as a strand plain (Tamura & Masuda 2005; Tamura *et al.* 2006, 2008*b*). The major difference between the Kujukuri and Sendai strand plains is the abundance of river-derived fine-grained sediments in the latter, these being overlain by shoreface sediments and distributed across present-day inner shelf areas (Saito 1989*a*, 1991*b*).

Relatively wide and long beach-ridge plains are also found in other parts of Japan, for example at Ishikari and Yufutsu in Hokkaido (Uesugi & Endo 1973; Matsushita 1979; Moriwaki 1982) and Hachirogata and Echigo (or Niigata) on the Japan Sea coast of NE Japan (Matsumoto 1984; Tanaka *et al.* 1996). The Ishikari River emptied into the Japan Sea after the infilling of lagoons at *c.* 6 ka and formed a *c.* 5 km wide beach-ridge plain (classified here as a strand plain system evolved from an earlier barrier system) The Hachirogata and Echigo coastal plains are regarded as parts of barrier systems because they still display lagoons or wide wetlands behind the beach-ridge plains.

Barrier and estuary systems. Barrier and estuary systems comprise the dominant depositional coastal environment in Japan. Lagoons or central basins of such systems have been mostly filled over the last 6 ka and remain as coastal ponds/lakes. Active barrier or estuary systems including tidal deltas and inlets are presently abundant in the eastern parts of Hokkaido, where the present sea level is at its highest of the Holocene, forming the coastal brackish lakes of Furen-ko, Akkeshi-ko, Notori-ko and Saroma-ko ('ko' means lake in Japanese). There are similar brackish lakes in Honshu, such as Ogawara-ko, Tsugaru Jusan-ko, Hinuma, Hamana-ko and Nakaumi, and there were even more before recent engineering works in places such as Kasumigaura, Hachiro-gata and Kahoku-gata.

Typical examples of coastal plains of this system are provided by the Kushiro marsh, the Sarobetsu Plain and the Tokoro Plain in Hokkaido, and Palaeo-Kinu Bay (Tone River lowland and Lake Kasumigaura: Fig. 9.4), Echigo Plain, Shonai Plain and Toyama (Imizu) Plain, Fukui Plain and Tsuruga Plain in Honshu (e.g. Saito *et al.* 1991; Fujii 1992; Kamoi & Yasui 2004; Urabe *et al.* 2004, 2006).

The barrier and estuary system consists of three parts, namely sandy-gravelly barrier complex (estuary mouth sand body), muddy lagoon (muddy central basin) and sandy lagoonal deltas (sandy bay-head deltas). These systems are typically well developed during transgression (Dalrymple *et al.* 1992). Related to the rise of sea level after the LGM, incised valleys formed during the LGM were inundated by marine transgression and became drowned. Coastal

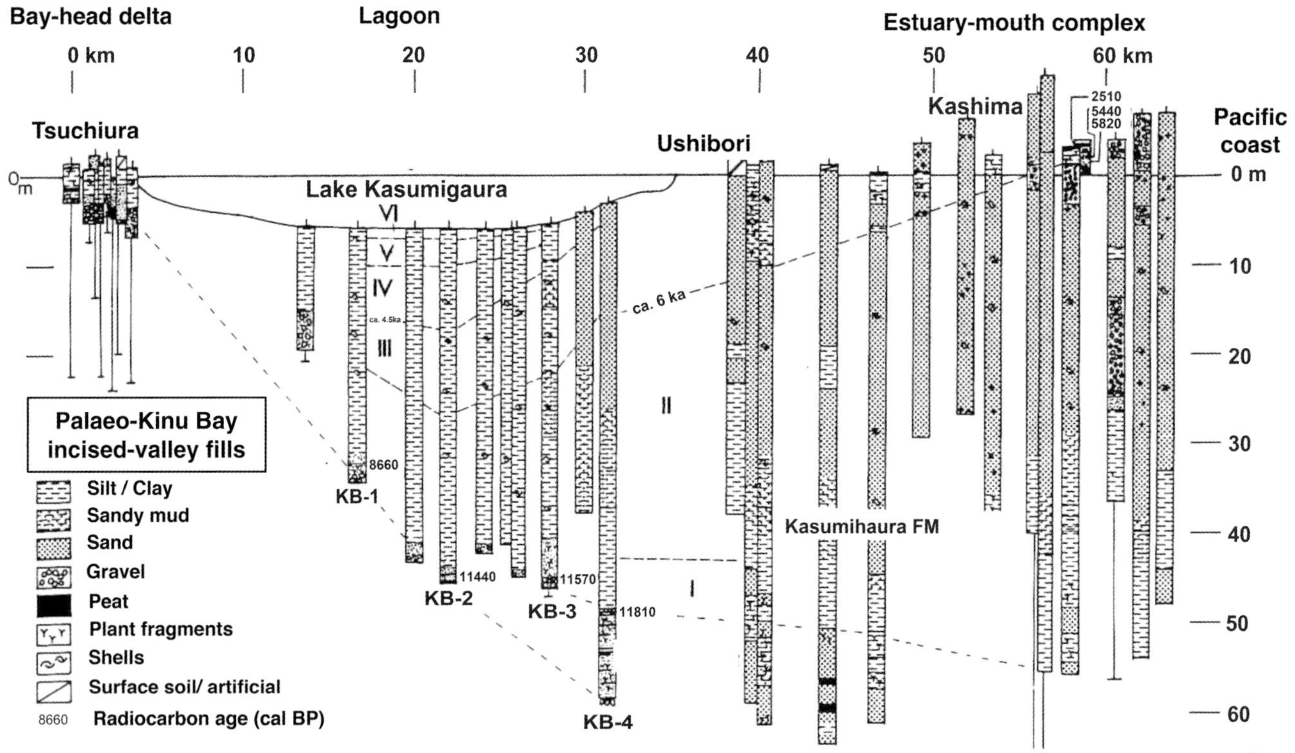

Fig. 9.4. Cross-section of the palaeo-Kinu Bay estuary system (modified after Suzuki & Saito 1987; Saito 1995). A lagoon, Lake Kasumigaura, subsists as a coastal lake. Dashed time lines are based on ^{14}C ages and *c.* 30 tephra-intercalated Holocene sediments. Only recent 500 years (stage VI) show brackish to freshwater environments. Sediment facies change from sand and gravel in an estuary mouth (Kashima) to mud in a lagoon (Kasumigaura) during an active estuary system from early Holocene transgression to *c.* 4.5 ka. Most sediments were delivered from coastal erosion through the estuary mouth (Kashima area). Sediment thickness distribution also shows a thinning trend lagoonwards.

erosion was dominant at headlands in the areas facing open seas and eroded materials were transported alongshore, producing estuary-mouth sand bodies. This is the reason why most of the lowlands facing open seas in the Japanese islands show an estuary system. Shore-parallel lagoon and barriers are rare, although examples include a barrier island in Shunkunitai and the lagoon of Furen-ko in eastern Hokkaido.

Lagoons and central basins have been infilled by bay-head delta progradation and/or sediment supply through bay mouths during the stable or relative fall of sea level after the maximum transgression. This has resulted in a change into fluvial plains or marshy wetlands, although in some cases they are still preserved as coastal ponds/lakes. Currently we can see different stages of the development of an estuary system in the form of active estuaries (e.g. Furen-ko, Akkeshi-ko), marshy wetlands (e.g. Kushiro wetland) and fluvial plains (e.g. Ishikari Plain, Echigo Plain, Tenryu fan). Finally, bay-head rivers empty directly into open seas and form new regressive systems such as the Ishikari River which has formed a beach-ridge plain at the coast. These systems show a compound structure consisting of an early–middle Holocene estuary (barrier) system inside and a late Holocene delta or strand plain system outside. Figure 9.1 also shows examples of a former estuary or barrier system in middle Holocene times.

Tsunami deposits

The first systematic study of tsunami deposits and their potential impacts in Japan was conducted by the group of Tohoku University (Kon'no *et al.* 1960, 1961) based on the 1960 Chilean examples. They subsequently surveyed inundation areas, wave heights, deposits and damage for the Sanriku coast of north Honshu, discussed sedimentary processes and the relationship with topography, and classified seven kinds of sediments. Tsunami records in sediments for the last 300 years were investigated first in Japan by Minoura *et al.* (1987) using sediment cores taken from coastal lagoon Tsugaru Jusan-ko, based on a tsunami deposit study of the 1983 Sea of Japan earthquake. Following their study, research on past tsunamis has been widely carried out from the 1990s onwards (e.g. Abe *et al.* 1990; Minoura & Nakaya 1991; Nishimura & Miyaji 1995, 1998; Fujiwara *et al.* 1997; Okamura *et al.* 1997; Minoura *et al.* 2001).

Tsunami deposits are well preserved at depressions between beach ridges on strand plains (e.g. Minoura & Nakaya 1991; Nanayama *et al.* 2003, 2007; Sawai *et al.* 2007, 2012; Komatsubara *et al.* 2008), coastal lakes/lagoons and wetlands (Minoura *et al.* 1987, 1994; Minoura & Nakata 1994; Okamura *et al.* 1997; Sawai 2002; Nanayama *et al.* 2003; Soeda & Nanayama 2005; Sawai *et al.* 2008*a*, 2009; Okamura & Matsuoka 2012*a*, *b*), and enclosed bay bottoms (Fujiwara *et al.* 1997, 2000; Fujiwara & Kamataki 2007). Tsunami boulders are also recognized on shore (Goto *et al.* 2010).

The distribution of locations of past tsunami sediments reported before 2011 are summarized by Fujiwara (2004) and Goto *et al.* (2012). Most of the tsunami sediments are located along the Pacific coasts, indicating that subduction-zone earthquakes are major causes of tsunamis along the Kuril and Japan trenches and the Nankai Trough. Other areas are along the Japan Sea coast from central Japan towards the north, where plate-boundary-zone earthquakes between the North American and Eurasia plates also occur frequently. There are fewer reports from the Japan Sea coast of the western part of Japan. Most of the triggers of these tsunamis have been earthquakes, although other causes include volcanic eruption and collapse. For example, the 1640 Hokkaido Komagatake eruption was accompanied by a tsunami-generating mountain collapse (Nishimura & Miyaji 1995, 1998). The eruption of the Kikai Caldera on Kyushu 7300 years ago also led to the deposition of tsunami sediments 300 km north of the source caldera (Fujiwara *et al.* 2010).

These analyses of tsunami sediments, combined with numerical simulation of tsunami inundation in eastern Hokkaido, have revealed that unusually large earthquakes occurred about every 500 years. New insight into multi-segment earthquakes, which had not occurred for the past 200 years in this area (Nanayama *et al.* 2003), has been obtained in this way. Nanayama *et al.* (2007) and Sawai *et al.* (2009) worked on detailed recurrence intervals in eastern Hokkaido, and high-resolution analysis of sediment cores also showed vertical movement associated with tsunami sediments, pre-seismic subsidence and post-seismic uplift (Sawai *et al.* 2004).

After the 2011 Tohoku-oki earthquake and tsunami, a tremendous amount of tsunami studies have been conducted and reported. The Sendai coastal plain was thought to be a relatively less tsunami-prone area in comparison with the Sanriku coast on the north side of Sendai, which suffered heavy damage by tsunamis in 1896, 1933 and 1960. Abe *et al.* (1990) and Minoura & Nakaya (1991) had however demonstrated the tsunami caused by the Jogan earthquake of AD 869 using tsunami sediments and historical records in and around Sendai (Fig. 9.5). The inundation areas of the 2011 Tohoku and 869 Jogan tsunamis show a very similar distribution (e.g. Sawai *et al.* 2012; Sugawara *et al.* 2012). Furthermore, Sawai *et al.* (2007, 2008*b*), Shishikura *et al.* (2007) and Namegaya *et al.* (2010) revealed the spatial distribution of Jogan tsunami sediments with numerical simulations of tsunami inundation before the 2011 tsunami, although this scientific information was not fully used for disaster mitigation before the tsunami struck in 2011. The geological survey and identification of past tsunami sediments will continue to provide crucially important information not only for science, but also for society.

Subsidence

Land subsidence is pronounced in major bays and was recognized first in the Tokyo Lowland in 1915; accumulated subsidence has now reached *c.* 4.5 m (Fig. 9.6). Areas below the mean spring high tide level, which is called 'Zero-Meter Region' in Japan, currently occupy a combined space of *c.* 580 km^2 and are inhabited by four million people within the three major bays of Japan in total (Tokyo, Ise and Osaka bays; Panel on Storm Surge Control Measures in Areas Below Sea Level 2006). These lowlands have already been hit by flooding disasters caused by high tides, typhoons and river overflow.

The rate of subsidence in the Tokyo Lowland was 2–3 cm a^{-1} for 1914–1917, *c.* 5 cm a^{-1} for 1918–1922 and reached 17.77 cm a^{-1} in 1924. The total subsidence from 1923 to 1944 was *c.* 2 m. After World War II, continuous subsidence at a rate of more than 10 cm a^{-1} was recorded from 1955 to 1973 with a maximum rate of 23.89 cm a^{-1} in 1968. It then decreased from around 1973 and subsidence has almost stopped since around 1979 to the present. The maximum accumulated subsidence in the Tokyo Lowland is 4.5156 m at Minamisuna 2 chome, Koto-ku over the period 1918–1997 (Endo *et al.* 2001).

The reason for this subsidence was not clear in the early twentieth century. However, as its distribution almost coincides with that of the Holocene lowlands (Chuseki-so), the subsidence was regarded as the result of groundwater extraction and compaction of the Chuseki-so. A survey using monitoring wells was initiated in 1933 and the continuous monitoring of groundwater level started in 1952. The Japanese Government restricted groundwater withdrawal by implementing the Industrial Water Law (1961) for industrial use, and the Law Controlling Pumping of Groundwater for Use in Buildings (1963) for air conditioning use. As a result of these regulations, groundwater level rose rapidly from around 1965 although subsidence continued. Finally, in 1972 the Tokyo Metropolitan Government bought mining rights in the Tokyo Lowland for extracting natural gas dissolved in water. After the cessation of natural gas mining from deep brackish water aquifers, subsidence rates decreased immediately (Endo *et al.* 2001; Kawashima *et al.* 2008; Kaneko & Toyoda 2011).

Monitoring data from wells of different depths has revealed the compaction of both Chuseki-so and underlying Pleistocene sediments equally during 1955–1973. During 1974–1998 however, compaction only occurred for the Holocene Chuseki-so sediments with rebound occurring in the Pleistocene sediments (Endo *et al.* 2001).

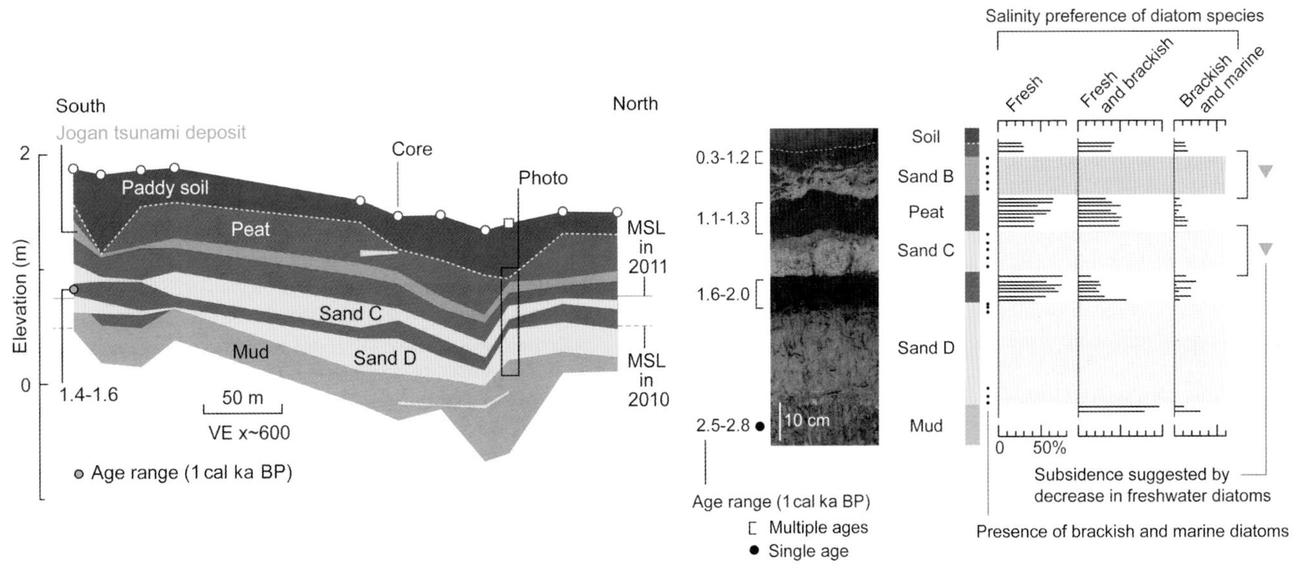

Fig. 9.5. Tsunami sediments from the Sendai coastal plain. Orange layer shows the 869 Jogan tsunami deposit. Three tsunami deposits are recognized for the last 3000 years. After Sawai *et al.* (2012). *Note*: VE x, vertical exaggeration.

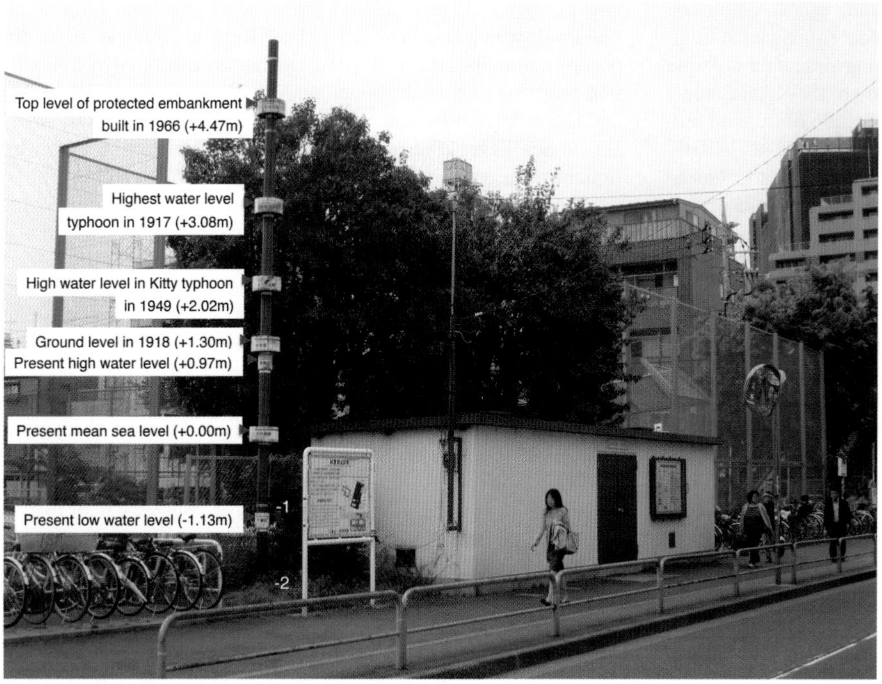

Fig. 9.6. Pole with panels showing present sea levels and former extreme high water levels during storm surges at the subsiding Tokyo Lowland. Minamisuna-machi monitoring well site, Minamisuna 3-chome park. The pole was constructed in March 2000.

The Ministry of Environment of Japan issues an annual report on subsidence in Japan. Recent data show that the number of areas with a subsidence rate of more than 2 cm a^{-1} was 7 (total only 2 km^2 for more than 1 km^2 subsiding area) in 2012, 14 (total 5920 km^2) in 2011 and 6 (total 6 km^2) in 2010. The higher number in 2011 is the result of tectonic subsidence due to the 2011 Tohoku-oki earthquake. Current issues associated with the subsidence problem include the rise of groundwater level and its impacts on artificial structures underground, and subsidence in Pleistocene uplands.

Lake Suigetsu

Lake Suigetsu (35° 35′ N, 135° 53′ E, 0 m above present sea level) is a tectonically subsiding lake located on the Japan Sea coast of the western part of Honshu and one of five lakes, named Mikata-go-ko, in that area. Lake Kugushi is the only natural barrier–lagoon system (Fig. 9.1) of five lakes. Lake Suigetsu is 34 m deep and covers an area of *c.* 4.3 km^2 with a diameter of *c.* 2 km. The lake was originally a freshwater lake, but became an artificially brackish-water lake in the middle of the seventeenth century. Lake sediments are characterized by well-preserved varves with seasonally alternating materials. The sediment core taken from Lake Suigetsu in 2006 (named ‘SG06’) is 73.5 m long, which is estimated to fully cover the last glacial since the last interglacial with varves occurring between *c.* 70 and 10 ka (Nakagawa *et al.* 2012; Staff *et al.* 2013). Lake Suigetsu provides one of the global auxiliary stratotypes for the base of the Holocene Series/Epoch (Walker *et al.* 2009), and the SG06 core is additionally the main data source for radiocarbon calibration beyond 11.2 kyr BP (IntCal13, Ramsey *et al.* 2012; Reimer *et al.* 2013). This core will be one of the global standards for palaeoclimate study (e.g. Kossler *et al.* 2011; Demske *et al.* 2013; Schlolaut *et al.* 2014) and chronology (e.g. Smith *et al.* 2011, 2013).

Oceanography, marine sedimentology and palaeoceanography (KI)

The Japanese islands occur along the northwestern North Pacific margin, with a strong westerly oceanic flow known as the Kuroshio Current prevailing offshore from the southern coastline where storm-dominated shelves predominate. The meeting of subtropical and subarctic waters in this region forms polar fronts, and the East Asian monsoon system has a major influence on the oceanography of Japanese waters. In addition to the open ocean environments, marginal seas such as the Okhotsk Sea, the Japan Sea and the East China Sea have their own unique oceanic environments. This section provides an overview of the Quaternary sedimentology and palaeoceanography around Japan, a complex oceanic setting that has been controlled by tectonic movements, eustatic sea-level changes, and orbital- and millennial-scale climate changes.

General submarine topography

The Japanese islands are located where the Pacific and Philippine Sea plates are subducting at the Kuril and Japan trenches, and the Nankai Trough and Ryukyu Trench, respectively. The oceanography of the region is therefore dominated by the classic trench–forearc–arc–back-arc system typical of subduction zones. As a result of frequent active tectonic movements within this island-arc setting, narrow shelves and steep slopes typify the submarine topography around the archipelago. The water depth at the edge of the Japanese shelf normally lies at 120–170 m (average 140 ± 10 m), although there is considerable spatial variation which reflects the local effects of tectonic movements (Iwabuchi & Kato 1988). Closer to the trenches and troughs, water depths in the forearc basins range from several hundred to a few thousand metres. Further offshore still, steeper trench slopes occur between the outer ridges or outer highs of the forearc basins and the trench floor. Numerous submarine canyons and channels are found on these slopes (e.g. Shimamura 1989), and some of them form submarine fans on the basin and trench floors.

On the back-arc side of the island arcs, basins and/or rifts include the East China, Japan and Okhotsk seas which have developed along the western NW Pacific margin (Fig. 9.7). The Japan Sea and the southern part of the Okhotsk Sea constitute a back-arc basin opening in relation to collision of the Indian landmass with the Asian

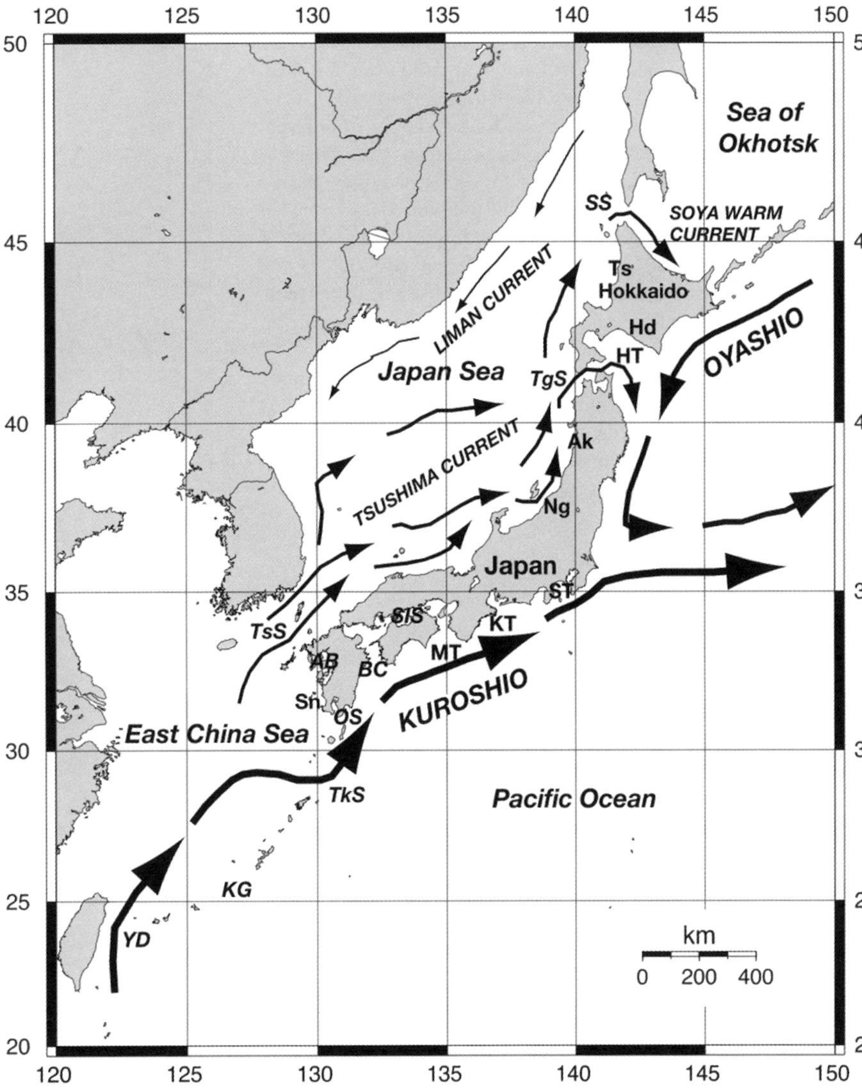

Fig. 9.7. Outline of surface currents around the Japanese islands. YD, Yanaguni Depression; KG, Kerama Gap; TkS, Tokara Strait; OS, Osumi Strait; BC, Bungo Channel; TsS, Tsushima Strait; TgS, Tsugaru Strait; SS, Soya Strait; AB, Ariake Bay; SIS, Seto Inland Sea; MT, Muroto Trough; KT, Kumano Trough; ST, Sagami Trough; HT, Hidaka Trough; Sn, Satsunan; Ng, Niigata; Ak, Akita; Hd, Hidaka; Ts, Teshio.

continent (e.g. Jolivet *et al.* 1994; Honza *et al.* 2004). In addition there are active back-arc rift basins east of the Izu–Ogasawara Ridge along the Japan Trench, and a back-arc rift basin (Okinawa Trough) north of the Ryukyu islands along the Ryukyu Trench.

Modern oceanography

The surface current system around the Japanese islands is summarized in Figure 9.7. The strong Kuroshio Current prevails along the southern coast of Japan, and acts as a dominant influence on present oceanographic conditions in western Japan and the Ryukyu islands. This current flows into the East China Sea through the Yonaguni Depression which lies between Taiwan and Yonaguni Island, and flows out to the NW Pacific through the Tokara Strait between Amami–Oshima and Tanegashima islands (Fig. 9.7). It then flows parallel to the coast of Kyushu, Shikoku and southern Honshu with a straight or meandering path, continuing eastwards to leave the Japanese islands east of Kanto. A branch of the Kuroshio Current, known as the Tsushima Warm Current, flows into the Japan Sea through the Tsushima Strait between Kyushu and Korea, and flows out mainly through the Tsugaru Strait into the NW Pacific (Fig. 9.7). A more minor component of this current flows northward past Hokkaido and escapes out to the Okhotsk Sea through the Soya Strait, turning into the Soya Warm Current which flows southeastwards along the Hokkaido coast (Fig. 9.7). The portion of the flow that remains trapped within the Japan Sea flows westwards and changes into the Liman Cold Current, which flows back southwestwards along the Russian coast of the Japan Sea. The boundary between the Tsushima and Liman currents forms the polar front, which is located in the central Japan Sea (Fig. 9.7).

Another important Pacific circulation system is provided by the Oyashio Current, a cold, southwestwards-flowing surface water mass with a high nutrient content. In the open Pacific, the polar front is marked by the boundary between this flow and the warmer Kurushio Current (Fig. 9.7). The cold Oyashio waters mix with the warm and saline Kurushio Current east of Honshu, producing eddies which lead to the upwelling of nutrient-rich subsurface water which maintains a high primary productivity in this region.

Surface water cooling by the East Asian winter monsoon has an important effect on the intermediate and deep waters in the Japan and Okhotsk seas. Brine rejection during the formation of seasonal sea ice contributes to increased water density and consequent sinking of the denser surface waters. This process, resulting in replenished intermediate and deep waters with high oxygen concentrations, favours oxygenated seafloor conditions.

The coastal area generally lies under the strong influence of wave action, so that wave-dominated coastal environments prevail around the Japanese islands; tide-dominated conditions are only found in the Seto Inland Sea and in some of the larger bays (Fig. 9.1).

Two major sources of terrigenous sediment supply from rivers are of especial importance in the seas around Japan. One relates to the flood events produced by summer monsoon activities and/or strong typhoons in early summer to late autumn. Heavy rainfall triggers landslides during these events, leading to the transport of large quantities of sand and mud downstream and out to the ocean. The other important terrigenous sediment source is transportation by snow meltwater discharge in spring, a process occurring mainly in northern Japan and along the Japan Sea.

Eolian dust (*Kosa*) is another significant source of terrigenous material deposited in the Japanese offshore waters. Strong winds in central China re-suspend desert dust which becomes transported eastwards by the jetstream across the East China Sea, Korea, Japan Sea and Japanese islands, and further out into the North Pacific.

Surface sediment distribution

Surface sediment distribution around the Japanese islands has been controlled by late Quaternary sea-level changes, tectonic movements and oceanic conditions such as currents, waves and water mass structure. An outline of sediment distribution has been provided by Hoshino (1958), Japan Association for Quaternary Research (1987) and Ikehara (1993), and the geochemical characteristics of surface sediments have been described by Ohta *et al.* (2004, 2007, 2010, 2013) and Imai *et al.* (2010).

Johnson & Baldwin (1986) classified continental shelves into six categories: (1) storm-dominated shelf with palimpsest and/or relict sediments; (2) tide-dominated shelf with palimpsest sediments; (3) storm-dominated shelf with modern sediments; (4) storm-dominated shelf with palimpsest and/or modern sediments; (5) storm-dominated graded shelf with modern sediments; and (6) ocean-current-dominated shelf with palimpsest and/or modern sediments. Saito (1989*b*) noted that categories (1) and (2) are absent from the Japanese shelf, and added a new category: (7) tide-dominated shelf with palimpsest and/or modern sediments.

Storm-dominated shelf is the most common oceanographic setting along the Japanese coast, with sediment grain size becoming finer away from the shore. Under these conditions well-sorted sandy sediments on the shoreface slope give way to fine-grained sediments further offshore. The fairweather wave base recorded in bottom sediments, close to the maximum depth of closure, is generally located at *c.* 20 m water depth; the storm wave base, which is the boundary between inner and outer shelf, is located at a depth of *c.* 50–80 m (Saito 1989*b*). In general, offshore mud is deposited below the storm wave base. The Akita (Ikehara *et al.* 1994*a*) and Satsunan (Fig. 9.8; Ikehara 2013) shelves offer typical examples of this kind of graded storm-dominated shelf.

A *tide-dominated shelf* setting has developed in sheltered areas such as the Seto Inland Sea and Ariake Bay (Inouchi 1982; Kamada 1985; Ikehara & Kinoshita 1989; Ikehara 1998). In the case of Ariake Bay, wide muddy and sandy tidal flats have developed under a tidal range that can exceed 5 or 6 m under maximum spring tides. There has been extensive deposition of offshore tidal sands in the straits of the Seto Inland Sea (Inouchi 1982; Ikehara & Kinoshita 1989; Ikehara 1998). Topographically accelerated tidal currents in the narrow straits have eroded the sea bottom, forming erosional scours (cauldrons) and transporting sands downstream to deposit them as sandbanks in areas of decreased current velocity. Bedload partition therefore occurs in offshore tidal sand systems (Johnson *et al.* 1982; Harris *et al.* 1995). Large subaqueous dunes are found on the surfaces of these sandbanks (Ikehara & Kinoshita 1989; Ikehara 1998).

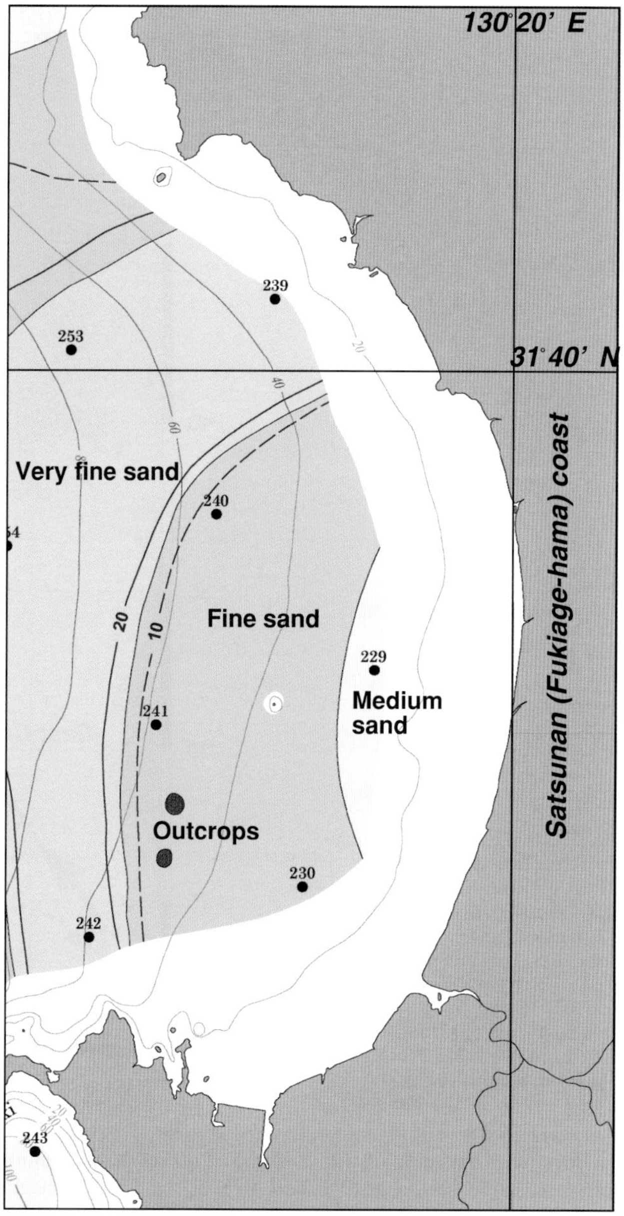

Fig. 9.8. Surface sediment distribution off Satsunan coast, a typical example of graded shelf (modified from Ikehara 2013).

Ocean-current-dominated shelf conditions occur along the paths of the Kuroshio and Tsushima warm currents (Ikehara 1988; Nishida & Ikehara 2013). This depositional system has similarities to the offshore tidal sand system in terms of grain size and bedform changes from the straits to the peripheral areas (Ikehara 1988). Late Quaternary sea-level rise has influenced current conditions (velocity, direction and water depth) in both tidal and ocean currents, changing patterns of bottom erosion and sediment transport (Ikehara 1988; Nishida & Ikehara 2013).

A large amount of fine-grained sediment supplied from rivers has formed extensive offshore muddy shelf areas, with the Niigata and Teshio shelves providing typical examples. On the Niigata shelf silty sediments, mainly supplied from the Shinano and Agano rivers, occur across an offshore area with water depth >35 m

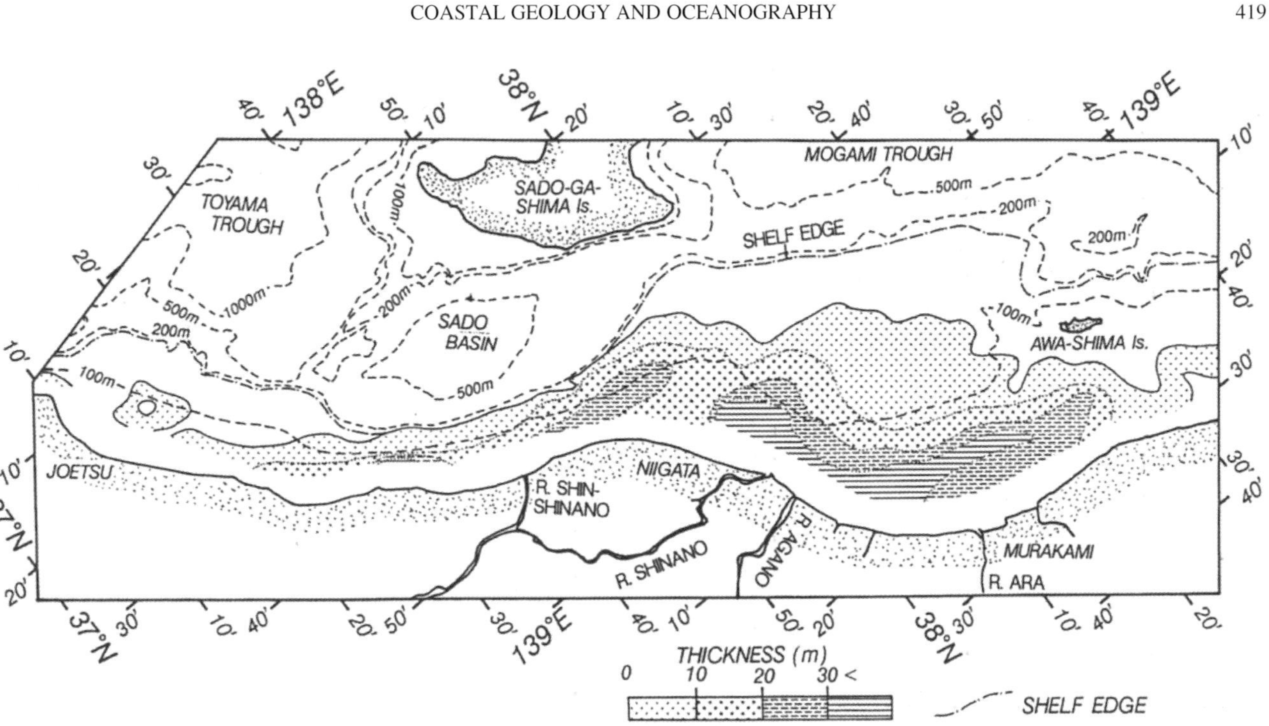

Fig. 9.9. Thickness distribution of fine-grained sediment accumulation on the Niigata shelf (after Ikehara *et al.* 1994b).

(Fig. 9.9; Ikehara *et al.* 1994*b*). In the case of the Teshio shelf, terrigenous fine-grained particles supplied from the Teshio River have formed the offshore mud. The occurrence of freshwater diatoms in the muddy sediments (Sagayama 2006) confirms the large contribution of river discharges to the formation of these offshore muddy shelves. The important influence of flood events to these shelf deposits, even from relatively small mountainous rivers, is exemplified by the 2004 Hokkaido Hidaka flood event which deposited up to 1 m of mud on the inner shelf off the mouth of the Saru River (Yamashita 2004), although most of the flood mud had disappeared one year later (Katayama *et al.* 2007).

Bioclastic material is sometimes a major component of surface sediments on the Japanese shelves, whether dominated by storms, tides and/or ocean currents. These cool water carbonates, comprising biogenic carbonate remains such as those of bivalves, bryozoans and rodoliths, are concentrated by the winnowing effect of strong currents and wave action which remove finer clastic materials (Kamada & Kondo 1983; Ikehara 2004; Nishida & Ikehara 2013). Coral reefs develop only around islands of the Ryukyu islands, where tropical carbonate deposits are found on the shelves and upper slopes with their distribution influenced by tidal currents (Ujiie *et al.* 1983; Tsuji 1993; Ujiie & Oshiro 1993; Tsuji *et al.* 1994).

The deposition of slope and basin floor sediments has been controlled by the mode of sediment supply, sea-level changes, climate changes and tectonics. Omura *et al.* (2012), for example, identified a sea-level control on basin floor turbidite deposition in the Kumano forearc basin, offshore between central Honshu and the Nankai Trough. In this area the formation of a large inner bay (Ise Bay) as a result of sea-level rise has trapped terrigenous materials from rivers, thus reducing sandy turbidite deposition on the Kumano Basin floor. In contrast, Nakajima & Itaki (2007) have emphasized the influence of climate change on offshore turbidite deposition along the Toyama Deep-Sea Channel in the central Japan Sea. Increased precipitation resulting from inflow of the Tsushima Warm Current to the Japan Sea during the last deglaciation facilitated river transport of clastic materials produced under a dry climate during the glacial period, to form turbidites on the offshore basin floor. Nakajima (2006) pointed out that there is more than 700 km of hyperpycnal transport from river mouths through the Toyama Deep-Sea Channel to the Japan Basin floor. This suggests a highly efficient long-distance transport mechanism involving confined flows through the channel-like topography.

Slope failures and subaqueous debris-flow generation provide another important depositional mechanism in the Japanese basins. Evidence of line-source gravitational movements from slope to basin floor is found in seismic reflection records and cores (Katayama *et al.* 1988; Ikehara *et al.* 1990; Noda *et al.* 2013; Strasser *et al.* 2013; Fink *et al.* 2014). Tectonic movement is a major triggering mechanism for slope failures, with gas-hydrate-related fluid migration being an additional cause of such failures in the Hidaka Trough (Morita *et al.* 2011, 2012; Noda *et al.* 2013). Large sliding masses associated with vertically oriented structures, which were formed in relation to fluid migration linked to decomposition of gas-hydrate in sediment, have been recognized in the seismic reflection records (Morita *et al.* 2011, 2012).

Earthquake-related sediment transport and redeposition

The position of the Japanese archipelago along highly active convergent plate margins results in frequent large and destructive earthquakes. Even over just the last few decades there have been several earthquakes with magnitude >7, such as the 1989 Nihonkai-chubu, the 1994 Hokkaido-nansei-oki, the 1994 Sanriku-haruka-oki, the 2003 Tokachi-oki, the 2004 Kii-hanto-oki and the 2011 Tohoku-oki events. As the classical papers by Heezen & Ewing (1952) and Heezen *et al.* (1954) on the generation of turbidity currents during the 1929 Grand Bank earthquake have demonstrated, strong ground motion associated with large earthquakes is a triggering mechanism for submarine slope failure and subsequent turbidity current production. The generation of turbidity currents by this mechanism has been observed along the western margin of the Sagami Trough during the Izu-toho-oki earthquake swarms (Iwase *et al.* 1997, 1998; Kinoshita *et al.* 2006), and the formation of highly turbid bottom water was reported in a small slope basin on the Nankai Trough accretionary prism as a consequence of the 2004 Kii-hanto-oki earthquake (Ashi *et al.* 2012). Similarly, an abnormally high particulate flux was observed in a sediment trap in the Japan Trench just after the 1994 Sanriku-haruka-oki earthquake (Itou *et al.* 2000).

Turbidite formed in response to earthquake events has been called 'seismo-turbidite' (Mutti *et al.* 1984), and is thought to have potential as a palaeoseismological tool (Adams 1990; Goldfinger 2011). Deposition of seismo-turbidites has been reported for several earthquakes such as the 1983 Nihonkai–Chubu (Nakajima & Kanai 1995), the 1993 Hokkaido-nansei-oki (Shimokawa & Ikehara 2002; Abdeldayem *et al.* 2004; Ikehara & Usami 2007) and the 2011 Tohoku-oki (Ikehara *et al.* 2011*a*; Arai *et al.* 2013) events. However, turbidity currents generated by earthquakes do not always form a turbidite bed. Using evidence from western Sagami Bay off the coast of central Japan, Ikehara *et al.* (2012*a*) pointed out that in this area earthquakes of magnitude *c.* 6 have insufficient energy to generate clearly defined sandy turbidites. Taira & Murakami (1984) and Fujioka *et al.* (1991) provide pioneer papers in the study of turbidite palaeoseismology in Japan, with subsequent studies on this subject having been conducted along the Nankai Trough (Fig. 9.10: Ikehara 1999, 2001; Iwai *et al.* 2004), Japan Trench (Ikehara *et al.* 2012*b*), Kuril Trench (Noda *et al.* 2004, 2008), Ryukyu Trench (Ujiie *et al.* 1997) and eastern margin of Japan Sea (Nakajima & Kanai 1995, 2000; Ikehara 2000; Shimokawa & Ikehara 2002; Ikehara *et al.* 2014*a*).

Large tsunamis associated with huge earthquakes may be another possible mechanism to form turbidites, even in offshore areas. Sugawara & Goto (2012) calculated the friction velocity of the tsunami wave for the 2011 Tohoku-oki tsunami, and suggested that it had enough potential to move fine-grained sea bottom sediments in the area shallower than 450 m. Resuspended sediments under these conditions might be capable of forming the turbid water masses and turbidity currents necessary to form turbidites. Tsunami-related

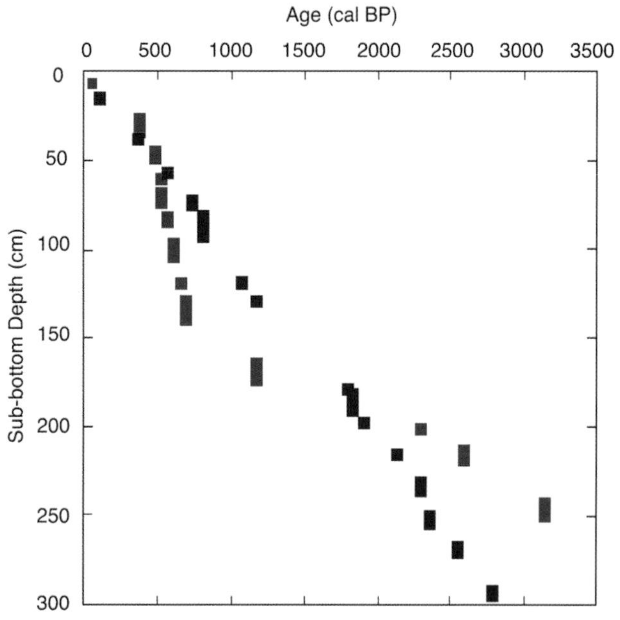

Fig. 9.10. Turbidite recurrence intervals in two slope basins along the eastern Nankai Trough (core data after Ikehara 2001, dark blue; GH97-310 core and red; GH97-311 core). Height of bars indicates the thickness of turbidites.

sediment erosion, movement and redeposition might have occurred across a wide area along the Tohoku coast in 2011 (Arai *et al.* 2013; Ikehara *et al.* 2014*b*), but more detailed information is needed to clarify our understanding of tsunami behaviour in shallow and deep water and its influence on the sea bottom environments.

Strong ground motion by large earthquakes produces sediment deformation. In this context, vein structures in Miocene forearc basin deposits in the Miura Group are thought to have formed in response to earthquakes (Hanamura & Ogawa 1993); although there is as yet no report of the occurrence of such veins in modern Japanese forearc and slope basin deposits, it nevertheless remains a possible proxy for large earthquakes. Sakaguchi *et al.* (2011) considers another kind of sediment deformation, namely mud brecciation, viewing this to be a result of surface rupture during strong earthquake ground motion near a large spray fault on the Nankai accretionary prism slope. Tectonic displacements during the 2011 Tohoku-oki earthquake reached the Japan Trench axis (Kodaira *et al.* 2012) where *c.* 50 m bathymetrical changes were observed near the epicentre (Fujiwara *et al.* 2011). Strasser *et al.* (2013) observed a rotational slump on the lowermost trench slope, and Kawamura *et al.* (2012) suggested that submarine landslides along the Japan Trench have the potential for tsunami generation. Similar bathymetric characteristics of submarine landslides are found on the slopes of the Nankai Trough.

Late Quaternary palaeoceanography

The Japanese islands are located in a mid-latitude position where warm subtropical water encounters cold subarctic water; the polar front is created between these two contrasting water masses. Global glacial–interglacial cycles may have influenced the expansion and retreat of these two water masses, changing the position of the polar front over time. Furthermore, eustatic sea-level fluctuations during Quaternary time have changed environmental conditions (width, depth, tide, wind stress, water temperature, salinity, water circulation and so on) in both shallow and deep waters in the open Pacific Ocean as well as its marginal seas.

The circulation of surface and intermediate-depth waters in the NW Pacific changed between the LGM and the Holocene epoch (e.g. Ahagon *et al.* 2003; Ohkushi *et al.* 2003; Harada *et al.* 2004; Ikehara *et al.* 2006; Minoshima *et al.* 2007; Shibahara *et al.* 2007; Sagawa & Ikehara 2008; Inagaki *et al.* 2009). A cold water mass expanded to the south of the Japanese islands in the modern Kuroshio area (Chinzei *et al.* 1987; Oda & Takemoto 1992; Oba & Yasuda 1992; Oba *et al.* 2006). Southwards migration of this subarctic cold water mass during the LGM to the last deglaciation resulted in a larger marine reservoir correction than that seen in subtropical warm water masses at present (Ikehara *et al.* 2011*b*). This indicates weakening of the Kuroshio Current during the LGM. Radiolarian assemblages of the deglacial northwestern North Pacific suggest that the source of intermediate-depth water was the Bering Sea and was different from the modern source (i.e. the Okhotsk Sea) (Ohkushi *et al.* 2003). Because deep-water circulation in the North Pacific during the LGM is thought to be different from modern circulation (Matsumoto *et al.* 2002), the mode of vertical water circulation in the glacial North Pacific might be different between the LGM and Holocene times.

The modern Kuroshio Current flows into the East China Sea through the Yonaguni Depression. There are two opposite visions of the glacial Kuroshio Current pathway during the LGM. Using planktonic foraminiferal evidence Ujiie & Ujiie (1999) and Ujiie *et al.* (2003) speculated that at the LGM the current flowed to the east of the Ryukyu islands. In contrast, using a numerical modelling approach, Kao *et al.* (2006) argued for a significant change of the

LGM Kuroshio Current pathway during the period when sea level was lower than 80 m: flowing into the East China Sea through the Yonaguni Depression, but then back out to the Pacific through the Kerama Gap located SW of Okinawa Island (Fig. 9.7). However, Kawahata & Ohshima (2004) reported the occurrence of tropical tree pollen in the northern Okinawa Trough sediments, and consequently suggested the entry here of the Kuroshio Current during the LGM. Recently, Lee *et al.* (2013) showed no significant difference in glacial sea-surface temperature and salinity between the Okinawa Trough (East China Sea) and the Ryukyu forearc (Pacific), and so suggested that the Kuroshio surface water flowed into the East China Sea even during the LGM. The decrease of surface water temperature both inside and outside of the Ryukyu islands was 3–4°C, this being almost the same as that (3°C) in the subtropical North Pacific on the Amami Plateau east of Okinawa (Sagawa *et al.* 2011). Evidence presented by Kawamura *et al.* (2007) indicated a change in bottom water conditions within the central Okinawa Trough at *c.* 8 ka, suggesting a possible difference in the structure of the oceanic water mass between glacial and interglacial times. A three-dimensional reconstruction of the water masses in the glacial Okinawa Trough is necessary in order to understand the LGM Kuroshio Current behaviour.

Freshwater and sediment discharges from Chinese rivers provide another important influence on geological oceanography in the East China Sea. Wahyudi & Minagawa (1997) and Ujiie *et al.* (2001) showed that there has been a gradual decrease of organic C/N ratio in Okinawa Trough sediments since the LGM. Similar decreases have been observed in lignin phenol content and the ratio of C28 profiles (Ujiie *et al.* 2001). These changes are interpreted to be a consequence of the retreat of the continental shoreline during post-glacial sea-level rise. East Asian summer monsoon fluctuation is also recorded in the northern East China Sea sediments. Kubota *et al.* (2010) reconstructed the surface water temperature and salinity from Mg/Ca and oxygen isotope ratios of planktonic foraminifera in the northern East China Sea. They recognized millennial-scale variations between warmer, more saline surface water and cooler, less saline surface water during the early deglacial period and the Holocene epoch, and interpreted high-salinity events as related to millennial-scale dryer conditions over the Changjiang River drainage basin due to a weaker summer monsoon. Furthermore, they found that the ages of the low-salinity events correlated with North Atlantic ice-rafting events. In addition, Ijiri *et al.* (2005) identified a freshwater supply to the northern East China Sea during the warm (interstadial) periods of Dansgaard–Oeschger (D-O) cycles observed in Greenland ice cores (Dansgaard *et al.* 1993), using alkenone-based surface water temperatures and oxygen isotope ratios of planktonic foraminifera. Both results suggest a teleconnection between North Atlantic climate and the East Asian summer monsoon.

Both orbital and millennial timescale palaeoenvironmental changes have been reported from the Japan Sea and Okhotsk Sea. Late Neogene and Quaternary sediments of the Japan Sea are characterized by the presence of darker- and lighter-coloured layers (Fig. 9.11; e.g. Tada *et al.* 1992, 1999; Stax & Stein 1994; Tada 1994). Thick laminated dark layers with high total sulphur/organic carbon ratios deposited during the LGM are thought to represent euxinic deep water conditions resulting from sea-level lowering to below −90 m, and consequent development of low-salinity surface water and a strong density stratification of the water column (Oba *et al.* 1991; Matsui *et al.* 1998; Tada *et al.* 1999). Tada *et al.* (1999) and Tada (2004) suggested that distinct alternations of thin dark and light layers were deposited in association with D-O cycles, with dark layers corresponding to warm interstadials and light layers corresponding to cold stadials. Moreover, thin dark layers were deposited when surface waters were less saline due to the expansion of low-salinity East China Sea shelf water, reflecting large freshwater discharge from the Changjiang River during high precipitation under a strong summer monsoon (Tada *et al.* 1999; Tada 2004). Enhanced deposition of organic carbon formed thin dark layers, especially at intermediate depths where an oxygen minimum zone developed (Watanabe *et al.* 2007). Usami *et al.* (2013) suggest, however, that the faunal fluctuations in benthic foraminifera were related to Northern Hemisphere insolation changes. Wang & Oba (1998) pointed out that high-latitude warming during interstadials could result in a large meander in the path of the prevailing westerlies. This meander could in turn heat up the East Asian landmass, leading to large differences in atmospheric pressure between land and ocean and enhancing the summer monsoon. The Japan Sea dark layer

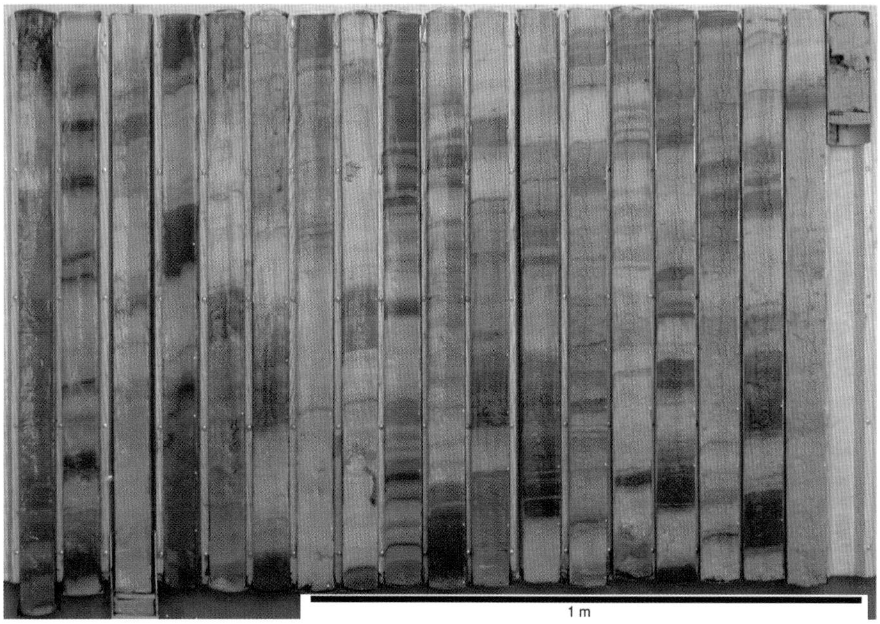

Fig. 9.11. Quaternary Japan Sea sediments. Alternation of dark- and light-coloured layers reflecting orbital- and millennial-timescale palaeoceanographic changes is a characteristic feature of the Quaternary Japan Sea sediments.

therefore also suggests the presence of climate teleconnection between North Atlantic and East Asia.

Surface water conditions in the Japan Sea have had an important influence on palaeoclimate and vegetation in the Japan Sea region (Yasuda 1982; Ikehara & Oshima 2009). Marine pollen data from the Japan Sea appears to have been influenced by orbital-scale changes in global climate and millennial-scale fluctuations in the East Asian summer monsoon (Ikehara & Oshima 2009). Pollen assemblages indicate overall cooling from the Last Interglacial to the LGM, a cold climate during LGM, gradual warming from 12 ka and rapid warming after 10 ka. These stages are concordant with published palaeoceanographic changes in the Japan Sea (e.g. Oba *et al.* 1991). During mid-marine isotope stage (MIS) 5 to MIS 3, high levels of precipitation are inferred from the abundance of *Cryptomeria*. In contrast, a cool and dry climate is interpreted for the light layers, based on an abundance of subarctic conifer pollen.

In addition to the summer monsoon, the East Asian climate system has a winter monsoon which is typically characterized by cold, dry northwesterlies that blow over the Japan Sea. Cooling of surface water by the winter monsoon promotes the formation of sea ice and deep water in the northern Japan Sea. Modern observations suggest a close relationship between intensification of this monsoon and both sea ice formation (Talley *et al.* 2003) and sea-surface temperature decrease in the northeastern Japan Sea (Minobe *et al.* 2004). Records of ice-rafted debris and alkenone-based surface water temperatures in the northeastern Japan Sea suggest close relationships with Northern Hemisphere summer insolation (Ikehara & Fujine 2012). This indicates that the intensity of the winter monsoon was enhanced during periods of low summer insolation. The ice-rafted debris peaks in some cases are coeval with North Atlantic Heinrich events, suggesting a possible climate linkage between the North Atlantic and East Asia; in contrast however, other peaks occurred even during periods of high summer insolation. Some of the peaks can be correlated with minima in reconstructed surface water temperature in the northeastern Japan Sea and maxima in oxygen isotope ratios of Chinese stalagmites (Wang *et al.* 2001, 2008; Yuan *et al.* 2004). These results suggest that extremely cold winters occur even during periods of enhanced summer monsoon intensity, and indicate that solar insolation in summer is not the only factor that controls variations in the intensity of the winter monsoon on millennial–centennial timescales (Ikehara & Fujine 2012).

Orbital- and millennial-timescale palaeoenvironmental variations are also found in the Okhotsk Sea. During the LGM to early deglaciation before 15–16 ka, expanded sea-ice coverage and longer seasonal sea-ice cover may have resulted in a reduced primary productivity in the surface water (Okazaki *et al.* 2005; Itaki *et al.* 2008). Existence at this time of well-ventilated intermediate and deep waters has been inferred from the radiolarian assemblages, suggesting deeper ventilation than during the Holocene epoch (Itaki *et al.* 2008). During 15–9 ka, there was an enhancement in surface water primary productivity (Itaki *et al.* 2008) and meltwater pulses 1A and 1B are probably recorded by two prominent $CaCO_3$ and total organic carbon peaks. High total sulphur content and lack of a deep-water radiolarian fauna at this time suggest oxygen-depleted conditions in deep water, although at intermediate depths the radiolarian assemblages indicate better-ventilated conditions. High diatom productivity occurs in the Holocene epoch after 9–8 ka, possibly due to the influence of the East Sakhalin Cold Current (Okazaki *et al.* 2005; Itaki *et al.* 2008). The influence of the Soya Warm Current is discernible after 12 ka (Okazaki *et al.* 2005). On a millennial timescale the strength of the Tsushima Warm Current in the Japan Sea may have been greater during the mid-Holocene climate warming/optimum, but weaker during Neoglacial times when the position of the Aleutian Low in winter moved southwards (Itaki & Ikehara 2004; Shimada *et al.* 2004; Itaki *et al.* 2008). Harada *et al.* (2006) reported an alkenone-based surface water temperature fluctuation in the southwestern Okhotsk Sea, with repeated abrupt changes on a centennial timescale coinciding with changes from warm interstadials to cold stadials. The periodicities of the alkenone temperature, which reveal that a *c.* 8 ky cycle is prominent during 90–10 ky BP and a 4–5 ky cycle is characteristic during 40–30 ky BP, are both similar and dissimilar to those in the Polar Circulation Index recorded in Greenland ice cores (Mayewski *et al.* 1997). This implies that the surface water temperature is controlled mainly by the atmosphere–ocean circulation system in the Northern Hemisphere, although the relationship between the surface water temperature in the Okhotsk Sea and the climate in Greenland is not linear. Expansion of seasonal sea ice is a characteristic feature in the modern Okhotsk Sea. Sakamoto *et al.* (2005) demonstrated a pattern of ice-raft debris occurrence during the past 100 ka which suggests that sea ice expanded over the whole Okhotsk Sea during the glacial periods, although it is unlikely that sea-ice cover was perennial. Sea-ice cover was at a minimum during 50–40 ka and during the Holocene epoch. In contrast, there have been 13 abrupt ice-raft debris peaks over the past 100 ka, most likely recording a strengthening of polar atmospheric circulation.

Appendix

English to Kanji and Hiragana translations for geological and place names

Abe	安倍	あべ
Abukuma	阿武隈	あぶくま
Agano	阿賀野	あがの
Akita	秋田	あきた
Akkeshi-ko	厚岸湖	あっけしこ
Amami-Oshima	奄美大島	あまみおおしま
Arakawa	荒川	あらかわ
Ariake	有明	ありあけ
Boso	房総	ぼうそう
Bungo	豊後	ぶんご
Chikugo	筑後	ちくご
Chikushi	筑紫	ちくし
Chuseki-so	沖積層	ちゅうせきそう
Echigo	越後	えちご
Edo	江戸	えど
Fuji	富士	ふじ
Fukui	福井	ふくい
Furen-ko	風蓮湖	ふうれんこ
Goto	郷東	ごうとう
Hachirogata	八郎潟	はちろうがた
Hamana-ko	浜名湖	はまなこ
Hayatsuki	早月	はやつき
Hidaka	日高	ひだか
Hii	斐伊	ひい
Hino	日野	ひの
Hinuma	涸沼	ひぬま
Hokkaido	北海道	ほっかいどう
Honshu	本州	ほんしゅう
Ibi	揖斐	いび
Imizu	射水	いみず
Ise	伊勢	いせ
Ishikari	石狩	いしかり
Ishinomaki	石巻	いしのまき
Izu	伊豆	いず
Jogan	貞観	じょうがん

(*Continued*)

English to Kanji and Hiragana translations for geological and place names (Continued)

Joganji	常願寺	じょうがんじ
Jomon	縄文	じょうもん
Kahoku-gata	河北潟	かほくがた
Kanto	関東	かんとう
Kashima	鹿島	かしま
Kasumigaura	霞ヶ浦	かすみがうら
Katakai	片貝	かたかい
Kerama	慶良間	けらま
Kii	紀伊	きい
Kikai	鬼界	きかい
Kinu	鬼怒	きぬ
Kiso	木曽	きそ
Kitakami	北上	きたかみ
Komagatake	駒ヶ岳	こまがたけ
Koto-ku	江東区	こうとうく
Kugushi	久々子	くぐし
Kujukuri	九十九里	くじゅうくり
Kumamoto	熊本	くまもと
Kumano	熊野	くまの
Kumozu	雲出	くもず
Kurobe	黒部	くろべ
Kuroshio	黒潮	くろしお
Kushiro	釧路	くしろ
Kyushu	九州	きゅうしゅう
Matsuyama	松山	まつやま
Midorikawa	緑川	みどりかわ
Mikata-go-ko	三方五湖	みかたごこ
Minamisuna	南砂	みなみすな
Miura	三浦	みうら
Muroto	室戸	むろと
Nagara	長良	ながら
Naka	那珂	なか
Nakagawa	那珂川	なかがわ
Nakaumi	中海	なかうみ
Nanagochi	七号地	ななごうち
Nanakita	七北田	ななきた
Nankai	南海	なんかい
Natori	名取	なとり
Nihonkai	日本海	にほんかい
Niigata	新潟	にいがた
Nobi	濃尾	のうび
Notori-ko	能取湖	のとりこ
Obitsu	小櫃	おびつ
Ogasawara	小笠原	おがさわら
Ogawara-ko	小川原湖	おがわらこ
Okazaki	岡崎	おかざき
Okinawa	沖縄	おきなわ
Ooi	大井	おおい
Osaka	大阪	おおさか
Osumi	大隅	おおすみ
Ota	太田	おおた
Oyashio	親潮	おやしお
Ryukyu	琉球	りゅうきゅう
Sagami	相模	さがみ
Sakawa	酒匂	さかわ
San'in	山陰	さんいん
Sanriku	三陸	さんりく
Sarobetsu	サロベツ	さろべつ
Saroma-ko	サロマ湖	さろまこ
Saru	沙流	さる
Satsunan	薩南	さつなん
Sendai	仙台	せんだい

(*Continued*)

English to Kanji and Hiragana translations for geological and place names (Continued)

Seto	瀬戸	せと
Shigenobu	重信	しげのぶ
Shikoku	四国	しこく
Shinano	信濃	しなの
Shirakawa	白川	しらかわ
Shonai	庄内	しょうない
Shunkunitai	春国岱	しゅんくにたい
Soya	宗谷	そうや
Suigetsu	水月	すいげつ
Takahashi	高梁	たかはし
Takamatsu	高松	たかまつ
Tama	多摩	たま
Tanegashima	種子島	たねがしま
Tedori	手取	てどり
Tenryu	天竜	てんりゅう
Teshio	天塩	てしお
Tohoku	東北	とうほく
Tokara	トカラ (吐噶喇)	とから
Tokoro	常呂	ところ
Tokyo	東京	とうきょう
Tone	利根	とね
Toyama	富山	とやま
Tshushima	対馬	つしま
Tsugaru	津軽	つがる
Tsugaru Jusan-ko	津軽十三湖	つがるじゅうさんこ
Tsuruga	敦賀	つるが
Tsurumi	鶴見	つるみ
Yahagi	矢作	やはぎ
Yayoi	弥生	やよい
Yokohama	横浜	よこはま
Yonaguni	与那国	よなぐに
Yufutsu	勇払	ゆうふつ
Yumigahama	弓ヶ浜	ゆみがはま
Yurakucho	有楽町	ゆうらくちょう

References

Abdeldayem, A. L., Ikehara, K. & Yamazaki, T. 2004. Flow path of the 1993 Hokkaido-Nansei-oki earthquake seismoturbidite, southern margin of the Japan sea north basin, inferred from anisotropy of magnetic susceptibility. *Geophysical Journal International*, **157**, 15–24.

Abe, F., Sugeno, Y. & Chigama, A. 1990. Estimation of the height of the Sanriku Jogan earthquake-tsunami (A.D. 869) in the Sendai Plain. *Zishin (Journal of the Seismological Society of Japan, 2nd series)*, **43**, 513–525 [in Japanese with English abstract].

Adams, J. 1990. Paleoseismicity of the Cascadia subduction zone: evidence from turbidites off Pregon-Washington Margin. *Tectonics*, **9**, 569–583.

Ahagon, N., Ohkushi, K., Uchida, M. & Mishima, T. 2003. Mid-depth circulation in the northwest Pacific during the last deglaciation: evidence from foraminiferal radiocarbon ages. *Geophysical Research Letters*, **30**, 2097, http://doi.org/10.1029/2003GL018287

Aoki, S. 1969. The Quaternary deposits in the Tokyo lowland. *Proceedings of the Coastal Plains (Kaigan Heiya) Symposium of the 76th Annual Meeting of the Geological Society of Japan*, 13 October 1969, Niigata, 15–20 [in Japanese].

Arai, K., Naruse, H. *et al.* 2013. Tsunami-generated turbidity current of the 2011 Tohoku-oki earthquake. *Geology*, **41**, 1195–1198, http://doi.org/10.1130/G34777.1

Ashi, J., Ikehara, K., Kinoshita, M. & KY04-11 & KH-10-3 Shipboard Scientists 2012. Settling of earthquake-induced turbidity on the accretionary prism slope of the central Nankai subduction zone. *In*: Yamada, Y., Kawamura, K. *et al.* (eds) *Submarine Mass Movements and Their Consequences*. Springer, Dordrecht, 561–571.

Brauns, D. 1881. Geology of the Environs of Tokio. *Memoirs of the Science Department, Tokio Daigaku (University of Tokyo)*, **4**, 82.

Chinzei, K., Fujioka, K. *et al.* 1987. Postglacial environmental change of the Pacific Ocean off the coasts of central Japan. *Marine Micropaleontology*, **11**, 273–291.

Construction Department, Reconstruction Bureau of Japan 1929. *Report of geological survey in Tokyo and Yokohama*. Reconstruction Bureau of Japan, Tokyo [in Japanese].

Dalrymple, R. W., Zaitlin, B. A. & Boyd, R. 1992. Estuarine facies models-Conceptual basis and stratigraphic implications. *Journal of Sedimentary Petrology*, **62**, 1130–1146.

Dansgaard, W., Johnsen, S. J. *et al.* 1993. Evidence for general instability of past climate from a 250-kyr ice-core record. *Nature*, **364**, 218–220.

Demske, D., Tarasov, P. & Nakagawa, T. 2013. Atlas of pollen, spores and further nonpollen palynomorphs recorded in the glacialeinterglacial late Quaternary sediments of Lake Suigetsu, central Japan. *Quaternary International*, **290–291**, 164–238.

Endo, K., Sekimoto, K. & Takano, T. 1982. Holocene stratigraphy and paleoenvironments in the Kanto Plain, in relation to the Jomon Transgression. *Proceedings of Institute of Natural Sciences, Nihon University*, **17**, 1–16 [in Japanese with English abstract].

Endo, K., Sekimoto, K., Takano, T., Suzuki, M. & Hirai, Y. 1983. ‘Chuseki-so’ in Kanto Plain. *Urban Kubota*, **21**, 26–43 [in Japanese].

Endo, K., Makinouchi, T., Tsubota, K. & Iwao, Y. 1995. Formation processes of ‘Chuseki-so’ in Japan. *Tsuchi to Kiso*, **43**, 8–12 [in Japanese].

Endo, K., Ishiwata, S., Hori, S. & Nakao, Y. 2013. Tokyo Lowland and Chuseki-so: formation of the soft ground and Jomon Transgression. *Journal of Geography*, **122**, 968–991 [in Japanese with English abstract].

Endo, T., Kawashima, S. & Kawai, M. 2001. Historical review of development of land subsidence and its cease in Shitamachi Lowland, Tokyo. *Journal of the Japan Society of Engineering Geology*, **42**, 74–87 [in Japanese with English abstract].

Fink, H. G., Strasser, M. *et al.* & R/V Sonne SO219A Cruise Participants 2014. Evidence for mass transport deposits at the IODP JFAST-site in the Japan Trench. *In*: Krastel, S., Behrmann, J.-H. *et al.* (eds) *Submarine Mass Movements and their Consequences*. Springer, Dordrecht, 33–43.

Fujii, S. 1992. Toyama Plain. *Urban Kubota*, **31**, 38–47 [in Japanese].

Fujii, S., Nasu, N. *et al.* 1986. Submerged forest off Nyuzen, Kurobegawa alluvial fan, Toyama Bay, Central Japan. *Boreas*, **15**, 265–277.

Fujimoto, K., Kawase, K., Ishizuka, S., Shichi, K., Ohira, A. & Adachi, H. 2009. Sediment and carbon storages in the Yahagi River Delta during the Holocene, central Japan. *Quaternary Science Reviews*, **28**, 1472–1480.

Fujioka, K., Gamo, T., Kinoshita, M., Koga, K., Kaiho, K., Yamano, M. & Tokuyama, H. 1991. Seismo-volcanic catastrophe happened in western Sagami Bay, central Japan: Results on R/V Tansei Maru KT89-11 cruise. *Journal of Physics of Earth*, **39**, 267–297.

Fujiwara, O. 2004. Sedimentological and paleontological characteristics of tsunami deposits. *Memoirs of the Geological Society of Japan*, **58**, 35–44 [in Japanese with English abstract].

Fujiwara, O. & Kamataki, T. 2007. Identification of tsunami deposits considering the tsunami waveform: an example of subaqueous tsunami deposits in Holocene shallow bay on southern Boso Peninsula, central Japan. *Sedimentary Geology*, **200**, 295–313.

Fujiwara, O., Masuda, F., Sakai, T., Fuse, K. & Saito, A. 1997. Tsunami deposits in Holocene bay-floor mud and the uplift history of the Boso and Miura peninsulas. *Quaternary Research (Daiyonki–Kenkyu)*, **36**, 73–86 [in Japanese with English abstract].

Fujiwara, O., Masuda, F., Sakai, T., Irizuki, T. & Fuse, K. 2000. Tsunami deposits in Holocene bay mud in southern Kanto region Pacific coast of central Japan. *Sedimentary Geology*, **135**, 219–230.

Fujiwara, O., Machida, J. & Shiochi, J. 2010. Tsunami deposit from the 7300 cal BP Akahoya eruption preserved in the Yokoo midden, north Kyushu, West Japan. *Quaternary Research (Daiyonki–Kenkyu)*, **49**, 23–33 [in Japanese with English abstract].

Fujiwara, T., Kodaira, S., No, T., Kaiho, Y., Takahashi, N. & Kaneda, Y. 2011. Tohoku-oki Earthquake: displacement reaching the trench axis. *Science*, **334**, 1240, http://doi.org/10.1126/science.1211554

Funabiki, A., Haruyama, S. & Dinh, H. T. 2010. Holocene evolution of the Kumozu River delta, Mie Prefecture, central Japan. *Quaternary Research (Daiyonki–Kenkyu)*, **49**, 201–218.

Furukawa, H. 1972. Alluvial deposits of the Nohbi Plain, Central Japan. *Memoirs of the Geological Society of Japan*, **7**, 39–59 [in Japanese with English abstract].

Goldfinger, C. 2011. Submarine paleoseismology based on turbidite records. *Annual Review of Marine Science*, **3**, 35–66.

Goto, K., Kawana, T. & Imamura, F. 2010. Historical and geological evidence of boulders deposited by tsunamis, southern Ryukyu Islands, Japan. *Earth-Science Reviews*, **102**, 77–99.

Goto, K., Nishimura, Y., Sugawara, D. & Fujino, S. 2012. The Japanese tsunami deposit researches. *Journal of the Geological Society of Japan*, **118**, 431–436 [in Japanese with English abstract].

Hanamura, Y. & Ogawa, Y. 1993. Layer-parallel faults, duplexes, imbricated thrusts and vein structures of the Miura Group: keys to understanding the Izu forearc-arc sediment accretion to the Honshu forearc. *Island Arc*, **2**, 126–141.

Harada, N., Ahagon, N., Uchida, M. & Murayama, M. 2004. Northward and southward migrations of frontal zones during the past 40 ka in the Kuroshio-Oyashio transition area. *Geochemistry, Geophysics, Geosystems*, **5**, Q09004, http://doi.org/10.1029/2004GC000740

Harada, N., Ahagon, N., Sakamoto, T., Uchida, M., Ikehara, M. & Shibata, Y. 2006. Rapid fluctuation of alkenone temperature in the southwestern Okhotsk Sea during the past 120 ky. *Global and Planetary Change*, **53**, 29–46.

Harris, P. T., Pattiaratchi, C. B., Collins, M. B. & Dalrymple, R. W. 1995. What is a bedload parting? *In*: Flemming, B. W. & Bartholoma, A. (eds) *Tidal Signatures in Modern and Ancient Sediments*. International Association of Sedimentologists, Special Publication, Blackwell, Oxford, **24**, 3–18.

Hase, Y., Hiraki, K., Nakahama, K., Iwauchi, A., Matsushima, Y., Okuno, M. & Nakamura, T. 2006. Environmental changes based on analyses of sediments, pollen/diatom, and radiocarbon dating in late Pleistocene to Holocene of the Kumamoto Plain and the southern part of Ariake Bay in central Kyushu, Japan. *Memoirs of the Geological Society of Japan*, **59**, 141–155 [in Japanese with English abstract].

Hayashi, M. 1991. Geomorphic development of the Izumo Plain, Western Japan. *Geographical Review of Japan*, **64A**, 26–46 [in Japanese with English abstract].

Heezen, B. C. & Ewing, M. 1952. Turbidity currents and submarine slumps, and the 1929 Grand Banks earthquake. *American Journal of Science*, **250**, 849–873.

Heezen, B. C., Ericson, D. B. & Ewing, M. 1954. Further evidence for a turbidity current following the 1929 Grand Banks Earthquake. *Deep-Sea Research*, **1**, 193–202.

Honza, E., Tokuyama, H. & Soh, W. 2004. Formation of the Japan and Kuril Basins in the Late Tertiary. *In*: Clift, P., Wang, P., Kuhnt, W. & Hayes, D. (eds) *Continent-Ocean Interactions within East Asian Marginal Seas*. AGU, Washington, 87–108.

Hori, K. & Saito, Y. 2007. An early Holocene sea-level jump and delta initiation. *Geophysical Research Letters*, **34**, L18401.

Hoshino, M. 1958. The shelf sediments in the adjacent seas of Japan. *Monograph of the Association for the Geological Collaboration*, **7**, 41 [in Japanese with English abstract].

Ijiri, A., Wang, L., Oba, T., Kawahata, H., Huang, C.-Y. & Huang, C.-Y. 2005. Paleoenvironmental changes in the northern area of the East China Sea during the past 42 000 years. *Palaeogeography, Palaeoclimatology, Palaeoecology*, **219**, 239–261.

Ikehara, K. 1988. Ocean current generated sedimentary facies in the Osumi Strait, south of Kyushu, Japan. *Progress in Oceanography*, **21**, 515–524.

Ikehara, K. 1993. Modern sedimentation in the shelf to basin areas around Southwest Japan, with special reference to the relationship between sedimentation and oceanographic conditions. Bulletin of Geological Survey of Japan, **44**, 283–349.

Ikehara, K. 1998. Sequence stratigraphy of tidal sand bodies in the Bungo Channel, southwest Japan. *Sedimentary Geology*, **122**, 217–232.

Ikehara, K. 1999. Recurrence interval of deep-sea turbidites and its importance for paleoseismicity analysis: an example from a piston core analysis from Kumano Trough, southwest Japan forearc. *Journal of Sedimentological Society of Japan*, **49**, 13–21 [in Japanese with English abstract].

Ikehara, K. 2000. Large earthquakes recorded as deep-sea turbidites in the Rishiri Trough, northernmost Hokkaido. *Quaternary Research (Daiyonki–Kenkyu)*, **39**, 569–574 [in Japanese with English abstract].

IKEHARA, K. 2001. Recurrence interval of large earthquakes along the eastern Nankai Trough inferred from deep-sea turbidites. *Journal of Geography*, **110**, 471–478 [in Japanese with English abstract].

IKEHARA, K. 2004. Current-swept cool-water carbonate to clastic facies around the Soya Strait, northernmost Japan. *Abstracts of Tidalite-2004*, 2–5 August, Copenhagen, 93–94.

IKEHARA, K. 2013. *Sedimentological Map Offshore of Cape Noma Misaki*. Marine Geology Map Series, 79. Geological Survey of Japan, AIST, Tsukuba, CD.

IKEHARA, K. & FUJINE, K. 2012. Fluctuations in the Late Quaternary East Asian winter monsoon recorded in sediment records of surface water cooling in the northern Japan Sea. *Journal of Quaternary Science*, **27**, 866–872.

IKEHARA, K. & KINOSHITA, Y. 1989. Bedforms and their migration patterns in the southern Bungo Strait, Japan. *In*: TAIRA, A. & MASUDA, F. (eds) *Sedimentary Facies in the Active Plate Margin*. Terra Publishing, Tokyo, 261–273.

IKEHARA, K. & OSHIMA, H. 2009. Orbital- and millennial-scale fluctuations in late Quaternary marine pollen records from the Japan Sea. *Journal of Quaternary Science*, **24**, 866–879.

IKEHARA, K. & USAMI, K. 2007. Sedimentary processes of deep-sea turbidites caused by the 1993 Hokkaido-Nansei-oki Earthquake. *Quaternary Research (Daiyonki–Kenkyu)*, **46**, 477–490 [in Japanese with English abstract].

IKEHARA, K., SATO, M. & YAMAMOTO, H. 1990. Sedimentation in the Oki Trough, southern Japan Sea, as revealed by high resolution seismic records (3.5 kHz echograms). *Journal of the Geological Society of Japan*, **96**, 37–49 [in Japanese with English abstract].

IKEHARA, K., NAKAJIMA, T. & KATAYAMA, H. 1994*a*. *Sedimentological Map West of Akita*. Marine Geology Map Series, 41. Geological Survey of Japan, Tsukuba [in Japanese with English abstract].

IKEHARA, K., KATAYAMA, H. & NAKAJIMA, T. 1994*b*. *Sedimentological Map of the Vicinity of Awashima*. Marine Geology Map Series, 42. Geological Survey of Japan, Tsukuba [in Japanese with English abstract].

IKEHARA, K., OHKUSHI, K., SHIBAHARA, A. & HOSHIBA, M. 2006. Changes of bottom water conditions at intermediate depths of the Oyashio region, NW Pacific over the past 20 000 yrs. *Global and Planetary Change*, **53**, 78–91.

IKEHARA, K., USAMI, K., JENKINS, R. & ASHI, J. 2011*a*. Occurrence and lithology of seismo-turbidites by the 2011 off the Pacific Coast of Tohoku Earthquake. *Abstract of IGCP the 5th International Symposium Submarine Mass Movements and Their Consequences*, 24–26 October, Kyoto, 74.

IKEHARA, K., DANHARA, T., YAMASHITA, T., TANAHASHI, M., MORITA, S. & OHKUSHI, K. 2011*b*. Paleoceanographic control on a large marine reservoir effect offshore of Tokai, south of Japan, NW Pacific, during the last glacial maximum-deglaciation. *Quaternary International*, **246**, 213–221.

IKEHARA, K., ASHI, J., MACHIYAMA, H. & SHIRAI, M. 2012*a*. Submarine slope response to earthquake shaking within western Sagami Bay, central Japan. *In*: YAMADA, Y., KAWAMURA, K. *ET AL*. (eds) *Submarine Mass Movements and their Consequences*. Springer, Dordrecht, 539 547.

IKEHARA, K., KANAMATSU, T. *ET AL*. 2012*b*. Past 'earthquake/tsunami' event deposits found in the Japan Trench: Results from the Sonne SO219A and Mirai MR12-E01 cruises. *AGU 2012 Fall Meeting Abstract*, 3–7 December, San Francisco, NH41C-01.

IKEHARA, K., ITAKI, T., TUZINO, T. & HOYANAGI, K. 2014*a*. Deep-sea turbidite evidence on the recurrence of large earthquakes off Shakotan Peninsula, northeastern Japan Sea. *In*: KRASTEL, S., BEHRMANN, J.-H. *ET AL*. (eds) *Submarine Mass Movements and their Consequences*. Springer, Dordrecht, 639–647.

IKEHARA, K., IRINO, T., USAMI, K., JENKINS, R., OMURA, A. & ASHI, J. 2014*b*. Possible submarine tsunami deposits on the outer shelf of Sendai Bay, Japan resulting from the 2011 earthquake and tsunami off the Pacific coast of Tohoku. *Marine Geology*, **358**, 120–127.

IMAI, N., TERASHIMA, S. *ET AL*. 2010. *Geochemical Map of Sea and Land of Japan*. Geological Survey of Japan, AIST, Tsukuba [in Japanese with English abstract].

INAGAKI, M., YAMAMOTO, M., IGARASHI, Y. & IKEHARA, K. 2009. Biomarker records from core GH02-1030 off Tokachi in the northwestern Pacific over the last 23 000 years: environmental changes during the last deglaciation. *Journal of Oceanography*, **24**, 866–879.

INOUCHI, Y. 1982. Distribution of bottom sediments in the Seto Inland Sea – the influence of the tidal currents on the distribution of bottom sediments. *Journal of the Geological Society of Japan*, **88**, 665–681 [in Japanese with English abstract].

ISEKI, H. 1975. On the basal gravel bed of the Recent deposits. *Journal of Geography*, **84**, 246–264 [in Japanese with English abstract].

ISEKI, H. 1983. *Chuseki-heiya (Alluvial Coastal Plains)*. University of Tokyo Press, Tokyo [in Japanese].

ITAKI, T. & IKEHARA, K. 2004. Middle to late Holocene changes of the Okhotsk Sea intermediate water and their relation to atmospheric circulation. *Geophysical Research Letters*, **31**, L24309, http://doi.org/10.1029/2004GL021384

ITAKI, T., KHIM, B.-K. & IKEHARA, K. 2008. Last glacial-Holocene water structure in the southwestern Okhotsk Sea inferred from radiolarian assemblages. *Marine Micropaleontology*, **67**, 191–215.

ITOU, M., MATSUMURA, I. & NORIKI, S. 2000. A large flux of particulate matter in the deep Japan Trench observed just after the 1994 Sanriku-Oki earthquake. *Deep-Sea Research I*, **47**, 1987–1998.

IWABUCHI, Y. & KATO, S. 1988. Some features of the continental shelf around the Japanese islands on compilation of the Quaternary map. *Quaternary Research (Daiyonki–Kenkyu)*, **26**, 217–225 [in Japanese with English abstract].

IWAI, M., FUJIWARA, O. *ET AL*. 2004. Holocene seismoturbidites from the Tosabae Trough, a landward slope basin of Nankai Trough off Muroto: Core KR9705P1. *Memoirs of the Geological Society of Japan*, **58**, 137–152 [in Japanese with English abstract].

IWASE, R., MOMMA, H., KAWAGUCHI, K., FUJIWARA, N., SUZUKI, S. & MITSUZAWA, K. 1997. Turbidity currents on the deep seafloor triggered by the earthquake swarm in the east off Izu Peninsula in March 1997 – observation by the long-term deep seafloor observatory off Hatsushima Island in Sagami Bay. *JAMSTEC Journal of Deep Sea Research*, **13**, 433–442.

IWASE, R., MITSUZAWA, K. & MOMMA, H. 1998. Mudflow associated with the swarm earthquakes east off Izu Peninsula in April 1998, observed by the long-term deep sea floor observatory off Hatsushima Island in Sagami Bay. *JAMSTEC Journal of Deep Sea Research*, **14**, 301–317 [in Japanese with English abstract].

JAPAN ASSOCIATION FOR QUATERNARY RESEARCH 1987. *Quaternary maps of Japan*. University of Tokyo Press, Tokyo [in Japanese].

JOHNSON, H. D. & BALDWIN, C. T. 1986. Shallow siliciclastic seas. *In*: READING, H. G. (ed.) *Sedimentary Environments and Facies*. 2nd edn. Blackwell, London, 229–282.

JOHNSON, M. A., KENYON, N. H., BELDERSON, R. H. & STRIDE, A. H. 1982. Sand transport. *In*: STRIDE, A. H. (ed.) *Offshore Tidal Sand, Processes and Deposits*. Chapman and Hall, London, 58–94.

JOLIVET, L., TAMAKI, K. & FOURNIER, M. 1994. Japan Sea, opening history and mechanism: a synthesis. *Journal of Geophysical Research*, **99**, B11, 22 237–22 259.

KAGOHARA, K. & IMAIZUMI, T. 2003. Holocene landform developments and coastal changes of the Yumigahama Peninsula, Southwest Japan. *Bulletin of the Faculty of Education & Human Sciences, University of Yamanashi*, **5**, 1–22 [in Japanese with English abstract].

KAIZUKA, S., NARUSE, Y. & MATSUDA, I. 1977. Recent formations and their basal topography in and around Tokyo Bay, Central Japan. *Quaternary Research*, **8**, 32–50.

KAMADA, Y. 1985. Geology, Ariake Bay. *In*: COASTAL OCEANOGRAPHY RESEARCH COMMITTEE, OCEANOGRAPHICAL SOCIETY OF JAPAN (ed.) *Coastal Oceanography of Japanese Islands*. Tokai University Press, Tokyo, 815–830 [in Japanese].

KAMADA, Y. & KONDO, H. 1983. Particle-size distribution and content of calcium carbonate in marine sediments in the northern part of the Goto-nada Sea, Nagasaki Prefecture, Japan. *Science Bulletin of Faculty of Education, Nagasaki University*, **34**, 35–51 [in Japanese with English abstract].

KAMOI, Y. & YASUI, S. 2004. Formation process of the Echigo Plain that traces it with several Holocene paleogeographical maps. *Tsuchi-to-Kiso*, **52**, 8–10 [in Japanese].

KANEKO, S. & TOYODA, T. 2011. Long-term urbanization and land subsidence in Asian megacities: in indicators system approach. *In*: TANIGUCHI, M. (ed.) *Groundwater and Subsurface Environments: Human Impacts in Asian Coastal Cities*. Springer, Tokyo, 249–270.

KAO, S. J., WU, C.-R., HSIN, Y. C. & DAI, M. 2006. Effects of sea level change in the upstream Kuroshio Current through the Okinawa Trough. *Geophysical Research Letters*, **33**, L16604, http://doi.org/10.1029/2006GL026822

KATAYAMA, H., IKEHARA, K., YAMAMOTO, H. & SATO, M. 1988. Occurrence of debris flow deposits from the Oki Trough, southern part of the Japan

Sea. *Journal of the Geological Society of Japan*, **94**, 633–636 [in Japanese].

Katayama, H., Ikehara, K., Suga, K., Sagayama, T., Irino, T., TuZino, T. & Inoue, T. 2007. Distribution of surface sediments after the 2003 flood on the shelf off Hidaka, southern Hokkaido. Bulletin of Geological Survey of Japan, **58**, 189–199 [in Japanese with English abstract].

Kawahata, H. & Ohshima, H. 2004. Vegetation and environmental record in the northern East China Sea during the late Pleistocene. *Global and Planetary Change*, **41**, 251–273.

Kawamura, K. 2000. Stratigraphy of Late Quaternary sediments and depositional environment in the Takamatsu Plain, Kagawa Prefecture, Japan. *Quaternary Research (Daiyonki–Kenkyu)*, **39**, 489–504 [in Japanese with English abstract].

Kawamura, K. 2009. Stratigraphy of the Pleistocene and Holocene sediments associated with correlation of some tephra beds in the Matsuyama Plain, Ehime Prefecture, southwest Japan. *Quaternary Research (Daiyonki–Kenkyu)*, **48**, 379–394 [in Japanese with English abstract].

Kawamura, K., Ikehara, K. & Fujioka, K. 2007. Abrupt change of sedimentary environment in ca. 7300 years ago recorded in the piston cores from the central Okinawa Trough. *Journal of the Geological Society of Japan*, **113**, 184–192 [in Japanese with English abstract].

Kawamura, K., Sasaki, T., Kanamatsu, T., Sakaguchi, A. & Ogawa, Y. 2012. Large submarine landslides in the Japan Trench: a new scenario for additional tsunami generation. *Geophysical Research Letters*, **39**, L05308, http://doi.org/10.1029/2011GL050661

Kawase, K. 1998. Late Holocene paleoenvironmental changes in the Yahagi River lowlands, central Japan. *Geographical Review of Japan*, **71A**, 411–435 [in Japanese with English abstract].

Kawashima, S., Kawai, M. & Endo, T. 2008. *Chronology of Land Subsidence and its Survey, Countermeasures in the Tokyo Lowland*. 2008 Annual Report of the Institute of Civil Engineering of Tokyo Metropolitan Government [in Japanese].

Kinoshita, M., Kasaya, T., Goto, T., Asakawa, K., Iwase, R. & Mitsuzawa, K. 2006. Seafloor landslides off Hatsushima, western Sagami Bay induced by east off Izu Peninsula earthquake. *Journal of Japanese Landslide Society*, **43**, 41–43 [in Japanese].

Kobayashi, K. 1957. Stratigraphic situation of the non-Ceramic Cultures and its related problems. *Cenozoic Research*, **23**, 486–498 [in Japanese].

Kobayashi, K. 1964. *The Geology of Hamamatsu City*. Report of Geological Survey. Hamamatsu City [in Japanese with English abstract].

Kodaira, S., No, T. *et al.* 2012. Coseismic fault rupture at the trench axis during the 2011 Tohoku-oki earthquake. *Nature Geoscience*, **5**, 646–650, http://doi.org/10.1038/ngeo1547

Komatsubara, J., Fujiwara, O., Takada, K., Sawai, Y., Aung, T. T. & Kamataki, T. 2008. Historical tsunamis and storms recorded in a coastal lowland, Shizuoka Prefecture, along the Pacific Coast of Japan. *Sedimentology*, **55**, 1703–1716 [in Japanese with English abstract].

Kon'no, E., Kitamura, N., Kotaka, T. & Kataoka, J. 1960. Erosion and deposition due to the tsunami caused by the Chile earthquake on May 24, 1960. *Annals of the Tohoku Geographical Association*, **12**, 120–127.

Kon'no, E., Iwai, J. *et al.* 1961. Geological observations of the Sanriku coastal region damaged by the tsunami due to the Chile earthquake in 1960. *Contributions from the Institute of Geology and Paleontology, Tohoku University*, **52**, 40 [in Japanese with English abstract].

Kossler, A., Tarasov, P. *et al.* Suigetsu 2006 project members 2011. Onset and termination of the late-glacial climate reversal in the high-resolution diatom and sedimentary records from the annually laminated SG06 core from Lake Suigetsu, Japan. *Palaeogeography, Palaeoclimatology, Palaeoecology*, **306**, 103–115.

Kubota, Y., Kimoto, K., Tada, R., Oda, H., Yokoyama, Y. & Matsuzaki, H. 2010. Variations of East Asian summer monsoon since the last deglaciation based on Mg/Ca and oxygen isotope of planktonic foraminifera in the northern East China Sea. *Paleoceanography*, **25**, PA4205, http://doi.org/10.1029/2009PA001891

Kuwano, Y., Shibasaki, T. & Aoki, S. 1971. Significance of buried valleys and other topographies in elucidating the Late Quaternary geohistory of Japanese coastal plains. *Quaternaria*, **14**, 217–236.

Lee, K. Y., Lee, H. J., Park, J.-H., Chang, Y.-P., Ikehara, K., Itaki, T. & Kwon, H. K. 2013. Stability of the Kuroshio path with respect to glacial sea level lowering. *Geophysical Research Letters*, **40**, 392–396.

Lyman, B. S. 1876. *A geological sketch map of the Island of Yesso, Japan*. Geological Survey of Hokkaido, Sapporo.

Masuda, F. & Saito, Y. 1999. Temporal variations in depositional rates within a Holocene sequence in Japan. *In*: Saito, Y., Ikehara, K. & Katayama, H. (eds) *Land-Sea Link in Asia 'Prof. Kenneth O. Emery Commemorative International Workshop'*. Proceedings of an International Workshop on Sediment Transport and Storage in Coastal Sea-Ocean System. STA (JISTEC) & Geological Survey of Japan, Tsukuba, 421–426.

Matsuda, I. 1974. Distribution of the recent deposits and buried landforms in the Kanto Lowland, Central Japan. *Geographical Reports of Tokyo Metropolitan University*, **9**, 1–36.

Matsui, H., Tada, R. & Oba, T. 1998. Low salinity isolation event in the Japan Sea in response to eustatic sea-level drop during LGM: reconstruction based on salinity-balance model. *Quaternary Research (Daiyonki–Kenkyu)*, **37**, 221–233 [in Japanese with English abstract].

Matsumoto, H. 1981. Sea-level changes during Holocene and geomorphic developments of the Sendai coastal plain, northeast Japan. *Geographical Review of Japan*, **54**, 72–85 [in Japanese with English abstract].

Matsumoto, H. 1984. Beach ridges on Holocene coastal plains on northeast Japan. *Geographical Review of Japan*, **57**, 720–738 [in Japanese with English abstract].

Matsumoto, K., Oba, T., Lynch-Stieglitz, J. & Yamamoto, H. 2002. Interior hydrography and circulation of the glacial Pacific Ocean. *Quaternary Science Review*, **21**, 1693–1704.

Matsushima, Y. 1987. Holocene relative sea-1evel changes in the Tama and Tsurumi River lowlands. *In*: Matsushima, Y. (ed.) *The Geological Studies on the Alluvial Deposits in the Tama and Tsurumi River Lowlands, Kawasaki*. Kawasaki City Museum, Kawasaki, 125–132 [in Japanese].

Matsushita, K. 1979. Buried landforms and the Upper Pleistocene-Holocene deposits of the Ishikari coastal plain. *Quaternary Research (Daiyonki–Kenkyu)*, **18**, 69–78 [in Japanese with English abstract].

Mayewski, P. A., Meeker, L. D., Twickler, M. S., Whitlow, S. I., Yang, Q., Lyons, W. B. & Prentice, M. 1997. Major features and forcing of high latitude northern hemisphere atmospheric circulation over the last 110 000 years. *Journal of Geophysical Research*, **102**, 26 345–26 366.

Minobe, S., Sako, A. & Nakamura, M. 2004. Interannual to interdecadal variability in the Japan Sea based on a new gridded upper water temperature dataset. *Journal of Physical Oceanography*, **34**, 2382–2397.

Minoshima, K., Kawahata, H. & Ikehara, K. 2007. Changes in biological production in the mixed water region (MWR) of the northwestern North Pacific during the last 27 kyr. *Palaeogeography, Palaeoclimatology, Palaeoecology*, **254**, 430–447.

Minoura, K. & Nakata, T. 1994. Discovery of an ancient tsunami deposit in coastal sequences of southwest Japan: verification of a large historic tsunami. *Island Arc*, **3**, 66–72.

Minoura, K. & Nakaya, S. 1991. Traces of tsunami preserved in intertidal lacustrine and marsh deposits: some examples from northeast Japan. *Journal of Geology*, **99**, 265–287.

Minoura, K., Nakaya, S. & Sato, Y. 1987. Traces of tsunami recorded in lake deposits Examples from Jusan, Shiura-mura, Aomori. *Zishin (Journal of the Seismological Society of Japan, 2nd series)*, **40**, 183–196 [in Japanese with English abstract].

Minoura, K., Nakaya, S. & Uchida, M. 1994. Tsunami deposits in a lacustrine sequence of the Sanriku coast, northeast Japan. *Sedimentary Geology*, **89**, 25–31.

Minoura, K., Imamura, F., Sugawara, D., Kono, Y. & Iwashita, T. 2001. The 869 Jogan tsunami deposit and recurrence interval of large-scale tsunami on the Pacific coast of northeast Japan. *Journal of Natural Disaster Science*, **23**, 83–88.

Morita, S., Nakajima, T. & Hanamura, Y. 2011. Submarine slump sediments and related dewatering structures: observations of 3D seismic data obtained for the continental slope off Shimokita Peninsula, NE Japan. *Journal of the Geological Society of Japan*, **117**, 95–98 [in Japanese with English abstract].

Morita, S., Nakajima, T. & Hanamura, Y. 2012. Possible ground instability factor implied by slumping and dewatering structures in high-methane-flux continental slope. *In*: Yamada, Y., Kawamura, K. *et al.* (eds) *Submarine Mass Movements and their Consequences*. Springer, Dordrecht, 311–320.

Moriwaki, H. 1979. Landform evolution of the Kujukuri coastal plain, central Japan. *Quaternary Research (Daiyonki–Kenkyu)*, **18**, 1–16 [in Japanese with English abstract].

MORIWAKI, H. 1982. Geomorphic development of Holocene coastal plains in Japan. *Geographical Reports of Tokyo Metropolitan University*, **17**, 1–42.

MURATA, Y. & KANO, K. 1995. The areas of the geological units comprising the Japanese Islands, calculated by using the Geological Map of Japan 1:1 000 000, 3rd Edition, CD-ROM Version. *Chishitsu News*, **493**, 26–29 [in Japanese].

MUTTI, E., RICCI LUCCHI, F., SEGURET, M. & ZANZUCCHI, G. 1984. Seismoturbidites: a new group of resedimented deposits. *In*: CITA, M., B. RICCI LUCCHI, F. (eds) *Seismicity and Sedimentation*. Elsevier, Amsterdam, 103–116.

NAGASAWA, S. & HORI, K. 2009. Sedimentary facies and accumulation rates of core sediments obtained from the Tenryu River alluvial fan. *Transactions, Japanese Geomorphological Union*, **30**, 305–316 [in Japanese with English abstract].

NAKADA, M. & OKUNO, J. 2011. Glacio-hydro isostasy. *Transactions, Japanese Geomorphological Union*, **32**, 327–331 [in Japanese with English abstract].

NAKADA, M., YONEKURA, N. & LAMBECK, K. 1991. Late Pleistocene and Holocene sea-level changes in Japan: implications for tectonic histories and mantle rheology. *Palaeogeography, Palaeoclimatology, Palaeoeeology*, **85**, 107–122.

NAKADA, M., OKUNO, J., YOKOYAMA, Y., NAGAOKA, S., TAKANO, S. & MAEDA, Y. 1998. Mid-Holocene underwater Jomon sites along the west coast of Kyushu, Japan, hydro-isostasy and asthenospheric viscosity. *Quaternary Research (Daiyonki–Kenkyu)*, **37**, 315–323 [in Japanese with English abstract].

NAKAGAWA, T., GOTANDA, K. *ET AL*. 2012. SG06, a fully continuous and varved sediment core from Lake Suigetsu, Japan: stratigraphy and potential for improving the radiocarbon calibration model and understanding of late quaternary climate changes. *Quaternary Science Reviews*, **36**, 164–176.

NAKAJIMA, T. 2006. Hyperpycnites deposited 700 km away from river mouths in the central Japan Sea. *Journal of Sedimentary Research*, **76**, 59–72.

NAKAJIMA, T. & ITAKI, T. 2007. Late Quaternary terrestrial climate variability recorded in deep-sea turbidites along the Toyama Deep-sea Channel, central Japan Sea. *Palaeogeography, Palaeoclimatology, Palaeoecology*, **247**, 162–179.

NAKAJIMA, T. & KANAI, Y. 1995. Recurrence intervals of earthquakes inferred from turbidites near the epicenter area of the 1983 Japan Sea Earthquake. *Zisin*, **48**, 223–228 [in Japanese with English abstract].

NAKAJIMA, T. & KANAI, Y. 2000. Sedimentary features of seismoturbidites triggered by the 1983 and older historical earthquakes in the eastern margin of the Japan Sea. *Sedimentary Geology*, **135**, 1–19.

NAMEGAYA, Y., SATAKE, K. & YAMAKI, S. 2010. Numerical simulation of AD 869 Jogan tsunami in Ishinomaki and Sendai plains and Ukedo river-mouth lowland. *Annual Report of Active Fault and Paleoearthquake Research*, **10**, 1–21. Active Fault and Earthquake Research Center, AIST [in Japanese with English abstract].

NANAYAMA, F., SATAKE, K., FURUKAWA, R., SHIMOKAWA, K., ATWATER, B. F., SHIGENO, K. & YAMAKI, S. 2003. Unusually large earthquakes inferred from tsunami deposits along the Kuril trench. *Nature*, **424**, 660–663.

NANAYAMA, F., FURUKAWA, R., SHIGENO, K., MAKINO, A., SOEDA, Y. & IGARASHI, Y. 2007. Nine unusually large tsunami deposits from the past 4000 years at Kiritappu marsh along the southern Kuril Trench. *Sedimentary Geology*, **200**, 275–294.

NAUMANN, E. 1879. Uber die Ebene von Yedo. Eine geographisch-geologische Studie. *Petermanns Geographische Mittheilungen (Gotha)*, **25** (IV), 121–135.

NISHIDA, N. & IKEHARA, K. 2013. Holocene evolution of depositional processes off southwest Japan: response to the Tsushima Warm Current and sea-level rise. *Sedimentary Geology*, **290**, 138–148.

NISHIMURA, Y. & MIYAJI, N. 1995. Tsunami deposits from the 1993 southwest Hokkaido earthquake and 1640 Hokkaido Komagatake eruption, northern japan. *Pure and Applied Geophysics*, **144**, 719–733.

NISHIMURA, Y. & MIYAJI, N. 1998. On height distribution of Tsunami caused by the 1640 eruption of Hokkaido-Komagatake, northern Japan. *Bulletin of the Volcanological Society of Japan*, **43**, 239–242.

NODA, A., TSUJINO, T., FURUKAWA, R. & YOSHIMOTO, N. 2004. Character, provenance, and recurrence intervals of Holocene turbidites in the Kushiro Submarine Canyon, eastern Hokkaido forearc. *Memoirs of the Geological Society of Japan*, **58**, 123–135.

NODA, A., TUZINO, T., KANAI, Y., FURUKAWA, R. & UCHIDA, J. 2008. Paleoseismicity along the southern Kuril Trench deduced from submarine-fan turbidites. *Marine Geology*, **254**, 73–90.

NODA, A., TUZINO, T., JOSHIMA, M. & GOTO, S. 2013. Mass transport-dominated sedimentation in a foreland basin, the Hidaka Trough, northern Japan. *Geochemistry, Geophysics, Geosystems*, **14**, 2638–3660, http://doi.org/10.1002/ggge.20169

OBA, T. & YASUDA, H. 1992. Paleoenvironmental change of the Kuroshio region since the last glacial age. *Quaternary Research (Daiyonki–Kenkyu)*, **31**, 329–339 [in Japanese with English abstract].

OBA, T., KATO, M., KITAZATO, H., KOIZUMI, I., OMURA, A., SAKAI, T. & TAKAYAMA, T. 1991. Paleoenvironmental changes in the Japan Sea during the last 85 000 years. *Paleoceanography*, **6**, 499–518.

OBA, T., IRINO, T., YAMAMOTO, M., MURAYAMA, M., TAKAMURA, A. & AOKI, K. 2006. Paleoceanographic change off central Japan since the last 144 000 years based on high-resolution oxygen and carbon isotope records. *Global and Planetary Change*, **53**, 5–20.

ODA, M. & TAKEMOTO, A. 1992. Planktonic foraminifera and paleoceanography in the domain of the Kuroshio Current around Japan during the last 20 000 years. *Quaternary Research (Daiyonki–Kenkyu)*, **31**, 341–357 [in Japanese with English abstract].

OGAMI, T., SUGAI, T., FUJIWARA, O., YAMAGUCHI, M. & SASAO, E. 2009. Development of the Kiso River delta during the last 10 000 years based on analyses of sedimentary cores and ^{14}C datings. *Journal of Geography*, **118**, 665–685 [in Japanese with English abstract].

OGAMI, T., SUGAI, T. & FUJIWARA, O. 2015. Dynamic particle segregation and accumulation processes in time and space revealed in a modern river-dominated delta: a spatiotemporal record of the Kiso River delta, central Japan. *Geomorphology*, **235**, 27–39.

OHKUSHI, K., ITAKI, T. & NEMOTO, N. 2003. Last glacial-Holocene change in intermediate-water ventilation in the Northwest Pacific. *Quaternary Science Review*, **22**, 1477–1484.

OHTA, A., IMAI, N., TERASHIMA, S., TACHIBANA, Y., IKEHARA, K. & NAKAJIMA, T. 2004. Geochemical mapping in Hokuriku, Japan: influence of surface geology, mineral occurrence and mass movement from terrestrial to marine environments. *Applied Geochemistry*, **19**, 1453–1469.

OHTA, A., IMAI, N. *ET AL*. 2007. Elemental distribution of coastal sea and stream sediments in the island-arc region of Japan and mass transfer processes from terrestrial to marine environments. *Applied Geochemistry*, **22**, 2872–2891.

OHTA, A., IMAI, N., TERASHIMA, S., TACHIBANA, Y., IKEHARA, K., KATAYAMA, H. & NODA, A. 2010. Factors controlling regional spatial distribution of 53 elements in coastal sea sediments in northern Japan: comparison of geochemical data derived from stream and marine sediments. *Applied Geochemistry*, **25**, 357–376.

OHTA, A., IMAI, N., TERASHIMA, S., TACHIBANA, Y. & IKEHARA, K. 2013. Regional spatial distribution of multiple elements in the surface sediments of the eastern Tsushima Strait (southwestern Sea of Japan). *Applied Geochemistry*, **37**, 43–56.

OKAMURA, M. & MATSUOKA, H. 2012*a*. Recurrence of the Nankai Earthquakes revealed by tsunami sediments. *Kagaku*, **82**, 182–194.

OKAMURA, M. & MATSUOKA, H. 2012*b*. Mega-earthquake recurrences recorded in lacustrine deposits along the Nankai Trough. *AGU Fall Meeting Abstract*, 3–7 December, 2012, San Francisco, NH11A-1540.

OKAMURA, M., KURIMOTO, T. & MATSUOKA, H. 1997. Coastal and lacustrine deposits as a monitor of tectonic movements. *Chikyu Monthly*, **24**, 698–703.

OKAZAKI, Y., TAKAHASHI, K. *ET AL*. 2005. Late Quaternary paleoceanographic changes in the southwestern Okhotsk Sea: evidence from geochemical, radiolarian, and diatom records. *Deep-Sea Research II*, **52**, 2332–2350.

OKUNO, J., NAKADA, M., ISHII, M. & MIURA, H. 2014. Vertical tectonic crustal movements along the Japanese coastlines inferred from late Quaternary and recent relative sea-level changes. *Quaternary Science Reviews*, **91**, 42–61.

OMURA, A., IKEHARA, K., SUGAI, T., SHIRAI, M. & ASHI, J. 2012. Determination of the origin and processes of deposition of deep-sea sediments from the composition of contained organic matter: an example from two forearc basins on the landward flank of the Nankai Trough, Japan. *Sedimentary Geology*, **249–250**, 10–25.

ONO, E. 2004. Factors affecting late Holocene marine regression in the Nobi Plain, Central Japan. *Geographical Review of Japan*, **77**, 77–98 [in Japanese with English abstract].

ONO, E., UMITSU, M. & KAWASE, K. 2001. Landform evolution of shallow buried valleys in the alluvial lowland in the Northern part of the Nobi Plain,

central Japan. *Quaternary Research (Daiyonki–Kenkyu)*, **40**, 345–352 [in Japanese with English abstract].
OTA, Y., MASTUSHIMA, Y. & MORIWAKI, H. 1981. *Atlas of Holocene sea level records in Japan*. Japanese Working Group of Project 61, Holocene sea level project. IGCP, 195p.
OTA, Y., MASTUSHIMA, Y., UMITSU, M. & KAWANA, T. 1987*a*. *Middle Holocene shoreline map of Japan*. Japanese Working Group for IGCP Project 200. Yokohama National University.
OTA, Y., MASTUSHIMA, Y. & UMITSU, M. 1987*b*. *Atlas of Late Quaternary sea level records in Japan. Volume I, Review papers and Holocene*. Japanese Working Group of IGCP Project 200. Yokohama National University.
OTA, Y., UMITSU, M. & MATSUSHIMA, Y. 1990. Recent Japanese research on relative sea level changes in the Holocene and related problems: review of studies between 1980 and 1988. *Quaternary Research (Daiyonki–Kenkyu)*, **29**, 31–48 [in Japanese with English abstract].
OTA, Y., KOIKE, K., OMURA, A. & MIYAUCHI, T. 1992. *Last Interglacial Shoreline Map of Japan*. Contribution to IGCP project 274 and to the Western Pacific Subcommission of the Commission on the Quaternary Shoreline, INQUA. Yokohama National University.
OTUKA, Y. 1934. Physiography of Tokyo during late Quaternary. *Proceedings of the Imperial Academy*, **10**, 274–277.
OYA, M. 1977. Comparative study of the fluvial plain based on the geomorphological land classification. *Geographical Review of Japan*, **50**, 1–31 [in Japanese with English abstract].
PANEL ON STORM SURGE CONTROL MEASURES IN AREAS BELOW SEA LEVEL 2006. *Future Storm Surge Control Measures in Areas below Sea Level (Recommendation)*. Ministry of Land, Infrastructure, Transport and Tourism. Tokyo.
RAMSEY, C. B., STAFF, R. A. *ET AL*. 2012. A complete terrestrial radiocarbon record for 11.2 to 52.8 kyr B.P. *Science*, **338**, 370–374.
REIMER, P. J., BARD, E. *ET AL*. 2013. IntCal13 and Marine13 radiocarbon age calibration curves 0–50 000 years cal BP. *Radiocarbon*, **55**, 1869–1887.
SADAKATA, N. 1991. The influence of iron sand mining (Kanna-nagashi) on the formation of Sotohama beach ridges in the Yumigahama Peninsula of South-western Japan. *Geographical Review of Japan*, **64A**, 759–778 [in Japanese with English abstract].
SADAKATA, N. 2009. Landform transformation by human activities in the historical time of Japan. *In*: JAQUA (ed.) *Digital Book: Progress in Quaternary Research in Japan. The Japan Association for Quaternary Research*. Japan Association for Quaternary Research (JAQUA), Tokyo, 4-4-4-26 [in Japanese].
SAEGUSA, Y., SUGAI, T., OGAMI, T., KASHIMA, K. & SASAO, E. 2011. Reconstruction of Holocene environmental changes in the Kiso-Ibi-Nagara compound river delta, Nobi Plain, central Japan, by diatom analyses of drilling cores. *Quaternary International*, **230**, 67–77.
SAGAWA, T. & IKEHARA, K. 2008. Intermediate water ventilation change in the subarctic northwest Pacific during the last deglaciation. *Geophysical Research Letters*, **35**, L24702, http://doi.org/10.1029/2008GL035133
SAGAWA, T., YOKOYAMA, Y., IKEHARA, M. & KUWAE, M. 2011. Vertical thermal structure history in the western subtropical North Pacific since the Last Glacial Maximum. *Geophysical Research Letters*, **38**, L00F02, http://doi.org/10.1029/2010GL045827
SAGAYAMA, T. 2006. Distribution of fresh water diatom valves in the surface sediments off the Tokachi-gawa, Teshi-gawa and Ishikari-gawa, Hokkaido, Japan. *Journal of the Geological Society of Japan*, **112**, 594–607 [in Japanese with English abstract].
SAITO, Y. 1987. Depositional model of Holocene coastal sediments. *Earth Monthly (Gekkan-Chikyu)*, **9**, 533–541 [in Japanese].
SAITO, Y. 1989*a*. Modern storm deposits in the inner shelf and their recurrence intervals, Sendai Bay, Northeast Japan. *In*: TAIRA, A. & MASUDA, F. (eds) *Sedimentary Facies in the Active Plate Margin*. Terra Publishing, Tokyo, 331–344.
SAITO, Y. 1989*b*. Classification of shelf sediments and their sedimentary facies in the storm-dominated shelf: a review. *Journal of Geography*, **98**, 350–365 [in Japanese with English abstract].
SAITO, Y. 1991*a*. Morphology and sediments of the Obitsu Delta in Tokyo Bay, Japan. *Journal of the Sedimentological Society of Japan*, **35**, 41–48.
SAITO, Y. 1991*b*. Sequence stratigraphy on the shelf and upper slope in response to the latest Pleistocene-Holocene sea-level changes off Sendai, northeast Japan. *In*: MACDONALD, D. I. M. (ed.) *Sedimentation, Tectonics and Eustasy: Sea-Level Changes at Active Margins*. Blackwell Scientific Publications, Oxford, International Association of Sedimentologists, Special Publication **12**, 133–150.
SAITO, Y. 1994. Shelf sequence and characteristic bounding surfaces in a wave-dominated setting: latest Pleistocene-Holocene examples from Northeast Japan. *Marine Geology*, **120**, 105–127.
SAITO, Y. 1995. High-resolution sequence stratigraphy of an incised-valley fill in a wave- and fluvial-dominated setting: latest Pleistocene-Holocene examples from Kanto Plain of central Japan. *Memoirs of the Geological Society of Japan*, **45**, 76–100.
SAITO, Y. 2000. Research trend of sediment transport from land to sea. *Gekkan Kaiyo (Kaiyo Monthly)*, **32**, 197–201 [in Japanese].
SAITO, Y. 2005. Continental shelves and their characteristics. *In*: KOIKE, K., TAMURA, T., CHINZEI, K. & MIYAGI, T. (eds) *Tohoku, Geomorphology of Japan*. University of Tokyo Press, Tokyo, **3**, 80–85 [in Japanese].
SAITO, Y. 2006. Fascinating Chuseki-so research and unsolved problems. *Memoirs of the Geological Society of Japan*, **59**, 205–212 [in Japanese with English abstract].
SAITO, Y. 2011. Delta-front morphodynamics of the Kurobe river fan-delta, central Japan. *In*: SHAO, X., WANG, Z. & WANG, G. (eds) *Proceedings of the 7th IAHR Symposium of River, Coastal and Estuarine Morphodynamics (RCEM2011)*. 6–8 September 2011, Beijing, Tsinghua University Press, Beijing, 506–511.
SAITO, Y., MATSUMOTO, E. & KASHIMA, K. 1989. Sea level of the last glaciation maximum based on in-place sediments on the shelf off Sendai, Northeast Japan. *Quaternary Research (Daiyonki–Kenkyu)*, **28**, 111–119.
SAITO, Y., INOUCHI, Y. & YOKOTA, S. 1991. Coastal lagoon evolution influenced by Holocene sea-level changes, Lake Kasumigaura, central Japan. *Memoirs of the Geological Society of Japan*, **36**, 103–118 [in Japanese with English abstract].
SAKAGUCHI, A., KIMURA, G., STRASSER, M., SCREATON, E. J., CUREWITZ, D. & MURAYAMA, M. 2011. Episodic seafloor mud brecciation due to great subduction zone earthquakes. *Geology*, **39**, 919–922, http://doi.org/10.1130/G32043.1
SAKAMOTO, T., IKEHARA, M., AOKI, K., IIJIMA, K., KIMURA, N., NAKATSUKA, T. & WAKATSUCHI, M. 2005. Ice-rafted debris (IRD)-based sea-ice expansion events during the past 100 kyrs in the Okhotsk Sea. *Deep-Sea Research II*, **52**, 2275–2301.
SATO, H., OKUNO, J., NAKADA, M. & MAEDA, Y. 2001. Holocene uplift derived from relative sea-level records along the coast of western Kobe, Japan. *Quaternary Science Reviews*, **20**, 1459–1474.
SATO, T. & MASUDA, F. 2010. Temporal changes of a delta: example from the Holocene Yahagi delta, central Japan. *Estuarine, Coastal and Shelf Science*, **86**, 415–428.
SAWAI, Y. 2002. Evidence for 17th century tsunamis generated on the Kuril-Kamchatka subduction zone, Lake Tokotan, Hokkaido, Japan. *Journal of Asian Earth Science*, **20**, 903–911.
SAWAI, Y., SATAKE, K. *ET AL*. 2004. Transient uplift after a 17th-century earthquake along the Kuril subduction zone. *Science*, **306**, 1918–1920.
SAWAI, Y., SHISHIMURA, M. *ET AL*. 2007. A study on paleotsunami using handy geoslicer in Sendai Plain (Sendai, Natori, Iwanuma, Watari, and Yamamoto), Miyagi, Japan. *Annual Report of Active Fault and Paleoearthquake Research*, **7**, 47–80. Active Fault and Earthquake Research Center, AIST [in Japanese with English abstract].
SAWAI, Y., FUJII, Y. *ET AL*. 2008*a*. Marine inundations of the past 1500 years and evidence of tsunamis at Suijin-numa, a coastal lake facing the Japan Trench. *The Holocene*, **18**, 517–528.
SAWAI, Y., SHISHIMURA, M. & KOMATSUBARA, J. 2008*b*. A study on paleotsunami using handy corer in Sendai Plain (Sendai City, Natori City, Iwanuma City, Watari Town, Yamamoto Town), Miyagi, Japan. *Annual Report of Active Fault and Paleoearthquake Research*, **8**, 17–70. Active Fault and Earthquake Research Center, AIST [in Japanese with English abstract].
SAWAI, Y., KAMATAKI, T. *ET AL*. 2009. Aperiodic recurrence of geologically recorded tsunamis during the past 5500 years in eastern Hokkaido, Japan. *Journal of Geophysical Research*, **114**, B01319.
SAWAI, Y., NAMEGAYA, Y., OKAMURA, Y., SATAKE, K. & SHISHIKURA, M. 2012. Challenges of anticipating the 2011 Tohoku earthquake and tsunami using coastal geology. *Geophysical Research Letters*, **39**, L21309.
SCHLOLAUT, G., BRAUER, A. *ET AL*. & SUIGETSU 2006 PROJECT MEMBERS 2014. Event layers in the Japanese Lake Suigetsu 'SG06' sediment core: description, interpretation and climatic implications. *Quaternary Science Reviews*, **83**, 157–170.

Shibahara, A., Ohkushi, K., Kennet, J. P. & Ikehara, K. 2007. Late Quaternary changes in intermediate water oxygenation and oxygen minimum zone, northern Japan: a benthic foraminiferal perspective. *Paleoceanography*, **22**, PA3213, http://doi.org/10.1029/2005PA001234

Shimada, C., Ikehara, K., Tanimura, Y. & Hasegawa, S. 2004. Millennial-scale variability of Holocene hydrography in the southwestern Okhostk Sea: diatom evidence. *The Holocene*, **14**, 641–650.

Shimamura, K. 1989. How do the submarine canyons grow? The chart of submarine canyon systems around the Japanese Islands. *Journal of the Geological Society of Japan*, **95**, 769–780 [in Japanese with English abstract].

Shimokawa, K. & Ikehara, K. 2002. Sedimentary records of past earthquakes. *In*: Otake, M., Taira, A. & Ota, Y. (eds) *Active Faults and Seismo-Tectonics of the Eastern Margin of the Japan Sea*. University of Tokyo Press, Tokyo, 95–108 [in Japanese].

Shimoyama, S., Matsumoto, N., Yumura, H., Takemura, K., Iwao, Y., Miura, N. & Tohno, I. 1994. Quaternary geology of the low-land along the north coast of Ariake bay, west Japan. *Memoirs of the Faculty of Science, Kyushu University. Series D, Earth and Planetary Sciences*, **18**, 103–129 [in Japanese with English abstract].

Shirogami, H. 1985. The structure of alluvial formations deduced from FeS2 contents of cored sediments taken from Hiroshima Plain. *Geographical Review of Japan*, **58A**, 631–644 [in Japanese with English abstract].

Shishikura, M. 2001. *Crustal Movements in the Boso Peninsula from the Analysis of Height Distribution of the Highest Holocene Paleo-Shoreline*. Annual Report on Active Fault and Paleoearthquake Research, 1. Active Fault Research Center, AIST, **1**, 273–285 [in Japanese with English abstract].

Shishikura, M., Sawai, Y. *et al.* 2007. *Age and Distribution of Tsunami Deposit in the Ishinomaki Plain, Northern Japan*. Annual Report of Active Fault and Paleoearthquake Research, Active Fault Research Center, **7**, 31–46 [in Japanese with English abstract].

Smith, V. C., Mark, D. F. *et al.* 2011. Toward establishing precise $^{40}Ar/^{39}Ar$ chronologies for Late Pleistocene palaeoclimate archives: an example from Suigetsu SG06, Japan. *Quaternary Science Reviews*, **30**, 2845–2850.

Smith, V. C., Staff, R. A. *et al.* & Suigetsu 2006 Project Members 2013. Identification and correlation of visible tephras in the Lake Suigetsu SG06 sedimentary archive, Japan: chronostratigraphic markers for synchronizing of east Asian/west Pacific palaeoclimatic records across the last 150 ka. *Quaternary Science Reviews*, **67**, 121–137.

Soeda, Y. & Nanayama, F. 2005. Drastic changes of the paleoenvironments identified in the Holocene sediments cored from Lake Harutori-ko, Kushiro City on the relation with the great earthquake tsunamis along the Kuril Subduction Zone. *Journal of Geography*, **114**, 626–630 [in Japanese with English abstract].

Soh, W., Tanake, T. & Taira, A. 1995. Geomorphology and sedimentary processes of a modern slope-type fan (Fujigawa fan delta), Suruga Trough, Japan. *Marine Geology*, **98**, 79–95.

Staff, R. A., Schlolaut, G. *et al.* 2013. Integration of the old and new Lake Suigetsu (Japan) terrestrial radiocarbon calibration data sets. *Radiocarbon*, **55**, 2049–2058.

Stax, R. & Stein, R. 1994. Quaternary organic carbon cycles in the Japan Sea (ODP-Site 798) and their paleoceanographic implications. *Palaeogeography, Palaeoclimatology, Palaeoecology*, **108**, 509–521.

Strasser, M., Kolling, M. *et al.* 2013. A slump in the trench: tracking the impact of the 2011 Tohoku-oki earthquake. *Geology*, **41**, 935–938, http://doi.org/10.1130/g34477.1

Sugawara, D. & Goto, K. 2012. Numerical modeling of the 2011 Tohoku-oki tsunami in the offshore and onshore of Sendai Plain, Japan. *Sedimentary Geology*, **104**, 127–153.

Sugawara, D., Goto, K., Imamura, F., Matsumoto, H. & Minoura, K. 2012. Assessing the magnitude of the 869 Jogan tsunami using sedimentary deposits: prediction and consequence of the 2011 Tohoku-oki tsunami. *Sedimentary Geology*, **282**, 14–26.

Sugimura, A. 1950. On the submarine terraces along the coast of the Kanto region and the others. *Geographical Review of Japan*, **23**, 10–16 [in Japanese, English abstract].

Sugimura, A. & Naruse, A. 1954. Changes in sea level, seismic upheavals, and coastal terraces in the southern Kanto region, Japan (I). *Japanese Journal of Geology and Geography, Transactions*, **24**, 101–113.

Sugimura, A. & Naruse, A. 1955. Changes in sea level, seismic upheavals, and coastal terraces in the southern Kanto region, Japan (II). *Japanese Journal of Geology and Geography, Transactions*, **26**, 165–176.

Suzuki, T. 1888. *Geological Map of Tokyo, 1:20 000*. Geological Map of Tokyo (Tokio), Tokyo.

Suzuki, T. & Saito, Y. 1987. Heavy mineral composition and provenance of Holocene marine sediments in Lake Kasumigaura, Ibaraki, Japan. Bulletin of the Geological Survey of Japan, **38**, 139–164 [in Japanese with English abstract].

Tada, R. 1994. Paleoceanographic evolution of the Japan Sea. *Palaeogeography, Palaeoclimatology, Palaeoecology*, **108**, 487–508.

Tada, R. 2004. Onset and evolution of millennial-scale variability in the Asian monsoon and its impact on paleoceanography of the Japan Sea. *In*: Clift, P., Wang, P., Kuhnt, W. & Hayes, D. (eds) *Continent-Ocean Interactions within East Asian Marginal Seas*. AGU, Washington, 283–298.

Tada, R., Koizumi, I., Cramp, A. & Rahman, A. 1992. Correlation of dark and light layers, and origin of their cyclicity in the Quaternary sediments from the Japan Sea. *Proceedings of Ocean Drilling Program, Scientific Results*, **127–128**, 577–601.

Tada, R., Irino, T. & Koizumi, I. 1999. Land-ocean linkage over orbital and millennial timescales recorded in late Quaternary sediments of the Japan Sea. *Paleoceanography*, **14**, 236–247.

Taira, A. & Murakami, H. 1984. Past Nankai and Tokai earthquakes detected using the turbidite: Possibility of turbidite paleoseismology. *Programme and Abstracts, Seismological Society of Japan*, **2**, 195 [in Japanese].

Talley, L. D., Lobanov, V., Ponomarev, V., Salyuk, A., Tishchenko, P., Zhabin, I. & Riser, S. 2003. Deep convection and brine rejection in the Japan Sea. *Geophysical Research Letters*, **30**, 1159, http://doi.org/10.1029/2002GL016451

Tamura, T. & Masuda, F. 2005. Bed thickness characteristics of inner shelf storm deposits associated with a transgressive to regressive Holocene wave-dominated shelf, Sendai coastal plain, Japan. *Sedimentology*, **52**, 1375–1395.

Tamura, T., Saito, Y. & Masuda, F. 2006. Stratigraphy and sedimentology of marine deposits in a Holocene strand plain: examples from the Sendai and Kujukurihama coastal plains. *Memoirs of the Geological Society of Japan*, **59**, 83–92 [in Japanese with English abstract].

Tamura, T., Nanayama, F., Saito, Y., Furakami, F., Nakashima, R. & Watanabe, K. 2007. Intra-shoreface erosion in response to rapid sea-level fall: depositional record of a tectonically-uplifted strand plain, Pacific coast of Japan. *Sedimentology*, **54**, 1149–1162.

Tamura, T., Murakami, F., Nanayama, F., Watanabe, K. & Saito, Y. 2008*a*. Ground-penetrating radar profiles of Holocene raised-beach deposits in the Kujukuri strand plain, Pacific coast of eastern Japan. *Marine Geology*, **248**, 11–27.

Tamura, T., Saito, Y. & Masuda, F. 2008*b*. Variations in depositional architecture of Holocene to modern prograding shorefaces along the Pacific coast of eastern Japan. *In*: Hampson, G. J., Steel, R. J., Burgess, P. & Dalrymple, R. W. (eds) *Recent Advances in Models of Siliciclastic Shallow-Marine Stratigraphy*. SEPM Special Publication, **90**, 189–203.

Tamura, T., Murakami, F. & Watanabe, K. 2010. Holocene beach deposits for assessing coastal uplift of the northeastern Boso Peninsula, Pacific coast of Japan. *Quaternary Research*, **74**, 227–234.

Tanabe, S. 2013. Paleogeography of the Tokyo and Nakagawa Lowlands since the Last Glacial Maximum. *Journal of Geography*, **122**, 949–967 [in Japanese with English abstract].

Tanabe, S. & Ishihara, Y. 2013. Evolution of the Uppermost Alluvium in the Tokyo and Nakagawa Lowlands, Kanto Plain, central Japan. *Journal of the Geological Society of Japan*, **119**, 350–367 [in Japanese with English abstract].

Tanabe, S., Ishihara, Y. & Nakashima, R. 2008*a*. Sequence stratigraphy and paleogeography of the Alluvium under the northern area of the Tokyo Lowland, central Japan. Bulletin of the Geological Survey of Japan, **59**, 509–547 [in Japanese with English abstract].

Tanabe, S., Nakanishi, T., Kimura, K., Hachinohe, S. & Nakayama, T. 2008*b*. Basal topography of the Alluvium under the northern area of the Tokyo Lowland and Nakagawa Lowland, central Japan. Bulletin of the Geological Survey of Japan, **59**, 497–508 [in Japanese with English abstract].

Tanabe, S., Tateishi, M. & Shibata, Y. 2009. The sea-level record of the last deglacial in the Shinano River incised-valley fill, Echigo Plain, central Japan. *Marine Geology*, **266**, 223–231.

Tanabe, S., Nakanishi, T. & Yasui, S. 2010*a*. Relative sea-level change in and around the Younger Dryas inferred from the late Quaternary incised-valley fills along the Japan Sea. *Quaternary Science Reviews*, **29**, 3956–3971.

TANABE, S., ISHIHARA, Y. & NAKANISHI, T. 2010*b*. Stratigraphy and physical property of the Alluvium under the Tokyo and Nakagawa lowlands, Kanto Plain, central Japan: implications for the Alluvium subdivision. *Journal of the Geological Society of Japan*, **116**, 85–98 [in Japanese with English abstract].

TANABE, S., NAKANISHI, T., NAKASHIMA, R., ISHIHARA, Y., UCHIDA, M. & SHIBATA, Y. 2010*c*. Sediment accumulation pattern of the muddy Alluvium in the Nakagawa incised valley, Saitama Prefecture, central Japan. *Journal of the Geological Society of Japan*, **116**, 252–269 [in Japanese with English abstract].

TANABE, S., NAKASHIMA, R., UCHIDA, M. & SHIBATA, Y. 2012. Depositional process of an estuary mouth shoal identified in the Alluvium of the Tokyo Lowland along the Tokyo Bay, central Japan. *Journal of the Geological Society of Japan*, **118**, 1–19 [in Japanese with English abstract].

TANABE, S., NAKANISHI, T., MATSUSHIMA, H. & HONG, W. 2013. Sediment accumulation patterns in a tectonically subsiding incised valley: insight from Echigo Plain, central Japan. *Marine Geology*, **336**, 33–43.

TANAKA, H., HASEGAWA, T., KIMURA, S., OKAMOTO, I. & SAKAI, Y. 1996. The geohistory of the formation of the Niigata sand dunes. *Quaternary Research (Daiyonki–Kenkyu)*, **35**, 207–218 [in Japanese with English abstract].

TANIGAWA, K., HYODO, M. & SATO, H. 2013. Holocene relative sea-level change and rate of sea-level rise from coastal deposits in the Toyooka basin, western Japan. *The Holocene*, **23**, 1039–1051.

TOKYO INSTITUTE OF CIVIL ENGINEERING 1969. Jiban Chishitsu Zu (Soil mechanic properties of ground of the Tokyo city). Tokyo Institute of Civil Engineering [in Japanese].

TSUJI, Y. 1993. Tide influenced high energy environments and rhodolith-associated carbonate deposition on the outer shelf and slopes off Miyako Islands, southern Ryukyu Island arc, Japan. *Marine Geology*, **113**, 255–271.

TSUJI, Y., HONDA, N., MATSUDA, H. & SUNOUCHI, H. 1994. Research on Quaternary carbonates in the Ryukyu Island arc, southern Japan, and the implications on petroleum geology. *Carbonate Rocks: Hydrocarbon Exploration & Reservoir Characterization*. Special Publication of the Technological Research Center, Japan National Oil Corporation, Chiba, **3**, 13–26.

UESUGI, Y. & ENDO, K. 1973. On the topographies and soils of the Ishikari coastal plain. *Quaternary Research (Daiyonki–Kenkyu)*, **12**, 115–124 [in Japanese with English abstract].

UJIIE, H. & OSHIRO, Y. 1993. Surface sediments of coral seas, west of Miyako Island and their environs, Ryukyu Island Arc, Japan. *Report of the Technology Research Center, JNOC*, **24**, 79–92.

UJIIE, H. & UJIIE, Y. 1999. Late Quaternary course changes of the Kuroshio Current in the Ryukyu Arc region, northwestern Pacific Ocean. *Marine Micropaleontology*, **37**, 23–40.

UJIIE, H., YAMAMOTO, S., OKITSU, M. & NAGANO, K. 1983. Sedimentological aspects of Nakagusuku Bay, Okinawa, subtropical Japan. *Galaxea*, **2**, 95–117.

UJIIE, H., NAKAMURA, T., MIYAMOTO, Y., PARK, J.-O., HYUN, S. & OYAKAWA, T. 1997. Holocene turbidite cores from the southern Ryukyu Trench slope: suggestions of periodic earthquakes. *Journal of the Geological Society of Japan*, **103**, 590–603.

UJIIE, H., HATAKEYAMA, Y., GU, X. X., YAMAMOTO, S., ISHIWATARI, R. & MAEDA, L. 2001. Upward decrease of organic C/N ratios in the Okinawa Trough cores: proxy for tracing the post-glacial retreat of the continental shore line. *Palaeogeography, Palaeoclimatology, Palaeoecology*, **165**, 129–140.

UJIIE, Y., UJIIE, H., TAIRA, A., NAKAMURA, T. & OGURI, K. 2003. Spatial and temporal variability of surface water in the Kuroshio source region, Pacific Ocean, over the past 21 000 years: evidence from planktonic foraminifera. *Marine Micropaleontology*, **49**, 335–364 [in Japanese with English abstract].

UMITSU, M. 1981. Geomorphic evolution of the alluvial lowlands in Japan. *Geographical Review of Japan*, **54**, 142–160 [in Japanese with English abstract].

UMITSU, M. 1992. Holocene deltaic sequence in the Kiso river delta, Central Japan. *Journal of Sedimentological Society of Japan*, **36**, 47–56 [in Japanese with English abstract].

UMITSU, M. 1994. *Late Quaternary Environment and Landform Evolution of Riverine Coastal Lowlands*. Kokon-Shoin, Tokyo [in Japanese].

URABE, A., TAKAHAMA, N. & YABE, H. 2004. Identification and characterization of a subsided barrier island in the Holocene alluvial plain, Niigata, central Japan. *Quaternary International*, **115–116**, 93–104.

URABE, A., YOSHIDA, M. & TAKAHAMA, N. 2006. Development process of barrier-lagoon system in the Holocene sediments of the Echigo Plain, central Japan. *Memoirs of the Geological Society of Japan*, **59**, 111–127 [in Japanese with English abstract].

USAMI, K., OHI, T., HASEGAWA, S. & IKEHARA, K. 2013. Foraminiferal records of bottom-water oxygenation and surface-water productivity in the southern Japan Sea during 160–15 ka: associations with insolation changes. *Marine Micropaleontology*, **101**, 10–27.

WAHYUDI, & MINAGAWA, M. 1997. Response of benthic foraminifera to organic carbon accumulation rates in the Okinawa Trough. *Journal of Oceanography*, **53**, 411–420.

WALKER, M., JOHNSEN, S. *ET AL.* 2009. Formal definition and dating of the GSSP (Global Stratotype Section and Point) for the base of the Holocene using the Greenland NGRIP ice core, and selected auxiliary records. *Journal of Quaternary Science*, **24**, 3–17.

WANG, L. & OBA, T. 1998. Tele-connections between East Asian monsoon and the high-latitude climate: a comparison between the GISP2 ice core record and the high resolution marine records from the Japan and the South China Seas. *Quaternary Research (Daiyonki–Kenkyu)*, **37**, 211–219.

WANG, Y. J., CHENG, H., EDWARDS, R. L., AN, Z. S., WU, J. Y., SHEN, C.-C. & DORALE, J. A. 2001. A high-resolution absolute-dated Late Pleistocene monsoon record from Hulu cave, China. *Science*, **294**, 2345–2348.

WANG, Y. J., CHENG, H. *ET AL.* 2008. Millennial- and orbital-scale changes in the East Asian monsoon over the past 224 000 years. *Nature*, **451**, 1090–1093.

WATANABE, S., TADA, R., IKEHARA, K., FUJINE, K. & KIDO, Y. 2007. Sediment fabrics, oxygenation history, and circulation modes of Japan Sea during the Late Quaternary. *Palaeogeography, Palaeoclimatology, Palaeoecology*, **247**, 50–64.

YAMAGUCHI, M., SUGAI, T., FUJIWARA, O., OHMORI, H., KAMATAKI, T. & SUGIYAMA, Y. 2003. Depositional process of the Holocene Nobi Plain, Central Japan, reconstructed from drilling core analysis. *Quaternary Research (Daiyonki–Kenkyu)*, **42**, 335–346 [in Japanese with English abstract].

YAMAGUCHI, M., SUGAI, T., FUJIWARA, O., OHMORI, H., KAMATAKI, T. & SUGIYAMA, Y. 2005. Depositional process and landform of the Kiso River delta, reconstructed from grain size distributions and accumulation rate of sediment cores. *Quaternary Research (Daiyonki–Kenkyu)*, **44**, 37–44 [in Japanese with English abstract].

YAMAGUCHI, M., SUGAI, T., OGAMI, T., FUJIWARA, O. & OHMORI, H. 2006*a*. Three-dimensional structures of the Latest Pleistocene to Holocene sequence at Nobi Plain, Central Japan. *Journal of Geography*, **115**, 41–50 [in Japanese with English abstract].

YAMAGUCHI, M., SUGAI, T., FUJIWARA, O., OGAMI, T. & OHMORI, H. 2006*b*. Stratigraphy of Upper Holocene deposits and landform evolution in the Kiso River delta system, Central Japan. *Quaternary Research (Daiyonki–Kenkyu)*, **45**, 451–462 [in Japanese with English abstract].

YAMASHITA, K. 2004. *Behavior of fine-grained sediment in coastal area.* Report of surveys on the 2011 Hokkaido flood disaster, Committee on Hydroscience and Hydraulic Engineering, Japan Society of Civil Engineering

YASUDA, Y. 1982. Pollen analytical study of the sediment from the Lake Mikata in Fukui Prefecture, central Japan: especially on the fluctuation of precipitation since the last glacial age on the side of Sea of Japan. *Quaternary Research (Daiyonki–Kenkyu)*, **21**, 255–271 [in Japanese with English abstract].

YOKOYAMA, Y., NAKADA, M. *ET AL.* 1996. Holocene sea-level change and hydro-isostasy along the west coast of Kyushu, Japan. *Palaeogeography, Palaeoclimatology, Palaeoecology*, **123**, 29–47.

YONEKURA, N. 2001. Characteristics of plains. *In*: YONEKURA, N., KAIZUKA, S., NOGAMI, M. & CHINZEI, K. (eds) *Introduction to Japanese Geomorphology, Regional Geomorphology of the Japanese Islands*. University of Tokyo Press, Tokyo, **1**, 200–207 [in Japanese].

YOSHIKAWA, S. & NEMOTO, K. 2010. Seasonal variations of sediment transport to a canyon and coastal erosion along the Shimizu coast, Suruga Bay, Japan. *Marine Geology*, **271**, 165–176.

YOSHIKAWA, S. & NEMOTO, K. 2014. The role of summer monsoon-typhoons in the formation of nearshore coarse-grained ripples, depression, and sand-ridge systems along the Shimizu coast, Suruga Bay facing the Pacific Ocean, Japan. *Marine Geology*, **353**, 84–98.

YUAN, D., CHENG, H. *ET AL.* 2004. Timing, duration, and transitions of the last interglacial Asian monsoon. *Science*, **304**, 575–578.

10 Mineral and hydrocarbon resources

YASUSHI WATANABE, TETSUICHI TAKAGI, NOBUYUKI KANEKO & YUICHIRO SUZUKI

The Japanese islands host a variety of mineral and hydrocarbon resources, although the produced amounts and reserves of most of these resources are relatively small compared to the Japanese domestic demands (except for a few commodities such as iodine and limestone). Nevertheless, Japan has historically been a major producer of Au, Ag, Cu, coal and sulphur, and was once described as Zipangu (gold country) in 'Divisement dou monde' by Marco Polo in the thirteenth century. The variety of the mineral resource commodities of Japan is attributed to the complex tectonic and geological history of the island arc. The accretion of allochthonous and autochthonous complexes during late Palaeozoic–Palaeogene time to the Eurasian continent provided Cu ores in massive sulphide deposits and Mn ores in stratiform deposits, in addition to limestone and chert. Continental arc magmatism during Cretaceous–Palaeogene times was responsible for formation of base-metal- and precious metal-bearing mineral occurrences, as well as non-metallic resources such as feldspar, pyrophyllite and sericite, all associated with causative granitoids. Island-arc volcanism from the Miocene to the present time was associated with numerous base-metal- and precious metal-bearing epithermal vein and Kuroko massive sulphide deposits, as well as economic amounts of clay. Deformation associated with the compressional tectonic regime in the southern Kurile arc and northern Izu–Bonin arc since the Late Miocene facilitated hydrocarbon and iodine resource accumulation. This chapter describes the distribution, occurrence and geneses of the major mineral and hydrocarbon resources in Japan.

Metallic mineral deposits of Japan (YW)

The first recorded domestic metal production in Japan is reflected by iron tools and bronze ritual utensils found in third–seventh century archeological sites. In the seventh century Ag was discovered on Tsushima Island in Kyushu, and Cu and Hg were produced from several places in Japan. By the beginning of the eighth century Cu and Sb, and later Sn and Pb in the tenth century, were being produced from the Nagato and Kurameki mines in southwestern Japan in order to manufacture coins (Saito *et al.* 2002). In the fifteenth and sixteenth centuries, the Ashikaga shogunate and local generals encouraged production of Au, Ag and Cu which were mainly exported to China. In the seventeenth century, major base metal deposits such as those at Ashio in central Honshu (discovered 1610), Osarizawa in northern Honshu (discovered 1666) and Besshi in Shikoku (discovered 1690) were developed. After the Meiji Restoration in 1868 major metallic deposits, including those of Kosaka and Ani in northern Honshu, Sado Island in the Sea of Japan and Ikuno in central Honshu, were disposed from the government to private companies. The production of metals in Japan peaked in 1917, fell in the ensuing economic depression, rose again before and during World War II, but then ceased temporarily in 1945.

In the post-war period, metal production revived and more than 200 metallic mines were in operation, peaking in 1970 during the rapid growth period of the Japanese economy. However, the introduction of a floating exchange rate for the yen in 1967 and the subsequent increase in the value of the currency diminished the competitive power of Japan's domestic mining industry and led to extensive mine closures (Ishihara 1992). Only a few Au–Ag mines (e.g. Hishikari) exploit metallic minerals in Japan at present.

Metallogenic setting

The Japanese islands have been situated at the active margin of the Eurasian continent since late Palaeozoic times, and consequently have a long history of plate subduction, arc magmatism and terrane accretion as has been detailed in previous chapters. In the Early Miocene the Japanese islands were separated from the main continent due to back-arc spreading, resulting in the formation of the Sea of Japan. The metallogenetic setting of the Japanese islands is divided into five stages: (1) pre-accretionary Palaeozoic continental margin; (2) Late Palaeozoic–Palaeogene terrane accretion; (3) Triassic–Palaeogene active continental margin arc magmatism; (4) mineralization associated with Neogene back-arc rifting; and (5) the modern island-arc system. Such a complex evolutionary history of the Japanese archipelago is reflected in its metallogeny, with a wide variety of mineral deposits that repeatedly overprint each other in space and time. The following discussion reviews the tectonomagmatic settings in which the various types of mineral deposits developed.

Pre-accretionary continental margin setting: Cu–Zn massive sulphide mineralization

The pre-accretionary stage of the Japanese islands involves the evolution of fragmented continental blocks (Hida, Oki, Southern Kitakami and Hitachi belts; Fig. 10.1) which have their own geological history and characteristics. The Hida and Oki blocks consist mainly of metamorphic rocks such as psammitic and pelitic biotite gneiss with a wide variety of protoliths (granite, basalt, sandstone, mudstone and limestone) of Permo-Carboniferous age (Wakita 2013). The Southern Kitakami Belt is regarded as an eastern extension of the Yangtze Block of China, on the basis of the stratigraphic characteristics that include middle–late Palaeozoic continental shelf facies sedimentary rocks (Isozaki & Maruyama 1991). The Hitachi Terrane is a small block (15 km × 5 km) composed of 510–500 Ma volcanic and plutonic arc rocks, as well as continental shelf sedimentary rocks (Tagiri *et al.* 2011). Isozaki *et al.* (2010) regarded all these pre-accretionary belts, blocks and terranes to have been situated along the southeastern China continental margin during Palaeo-Pacific ocean plate subduction. The only important metallic

From: Moreno, T., Wallis, S., Kojima, T. & Gibbons, W. (eds) 2016. *The Geology of Japan*. Geological Society, London, 431–455.

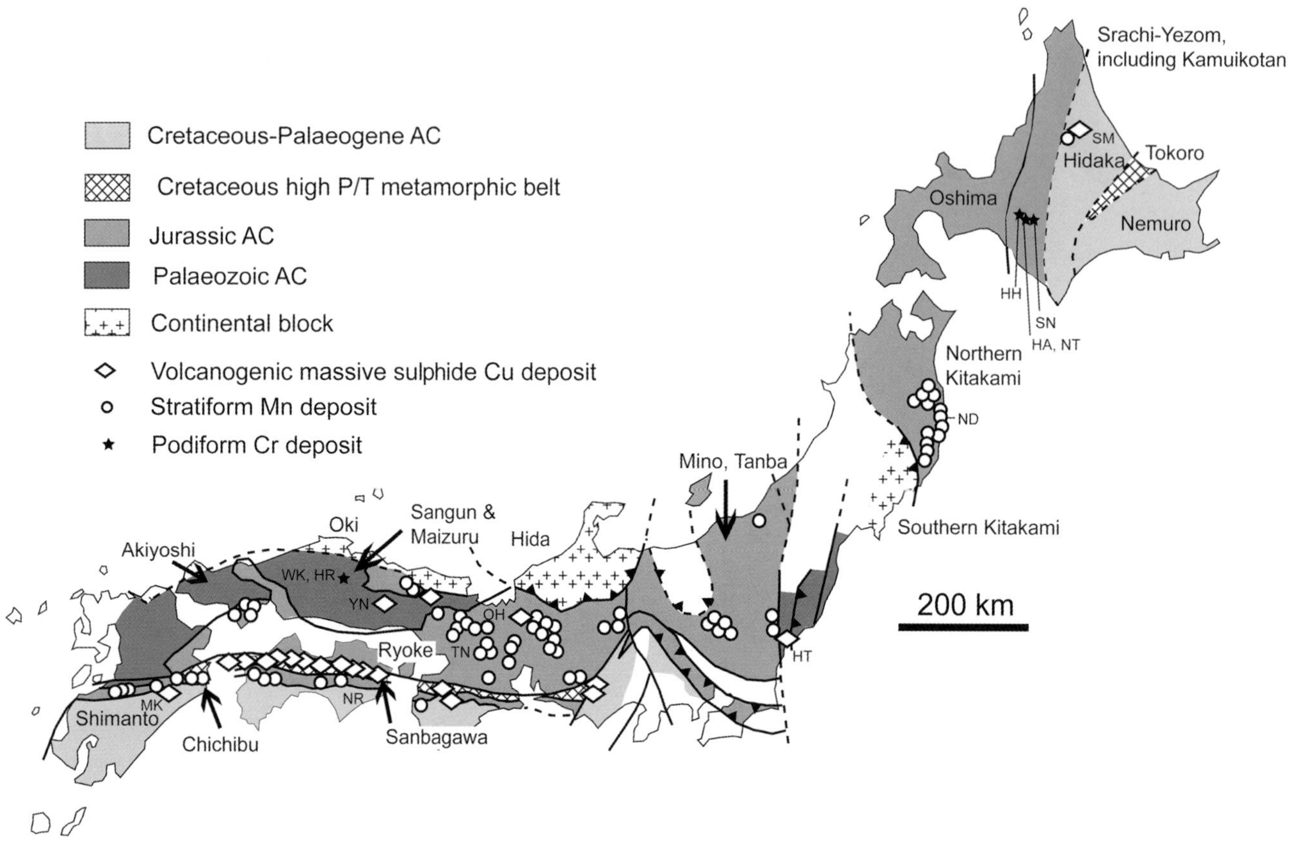

Fig. 10.1. Division of the allochthonous and autochthonous terranes and complexes of the Japanese islands with distribution of volcanogenic massive sulphide Cu, stratiform Mn and podiform Cr deposits, modified from Sato & Kase (1996). Abbreviations of deposit names: HA, Hatta (Cr); HH, Hattahachiman (Cr); HR, Hirose (Cr); HT, Hitachi (Cu); MK, Makimine (Mn); ND, Noda–Tamagawa (Mn); NR, Nirou (Mn); NT, Nitto (Cr); OH, Ohono (Mn); SM, Shimokawa (Cu); SN, Shinnitto (Cr); TN, Tonoda (Mn); YN, Yanahara (Cu); WK, Wakamatsu (Cr).

deposit produced during this stage is the Hitachi Cu–Zn massive sulphide deposit in the Hitachi Terrane exposed in the Abukuma Plateau of northern Honshu. This deposit produced 3.48 Mt of ores at 1.35% Cu and is hosted by an acidic schist unit underlying basic to intermediate schist of the Akazawa Formation (Mariko & Kato 1994). An Re–Os isochron age of 533 $\pm$ 12 Ma is reported from one of the orebodies of the deposit (Nozaki *et al.* 2012). The orebodies are stratabound, and the ores are massive, banded and/or disseminated in texture. Major ore minerals are chalcopyrite, sphalerite, pyrite and pyrrhotite $\pm$ galena (Kase & Yamamoto 1985). The massive sulphide mineralization is inferred to have formed in a submarine back-arc volcanic setting, based on the occurrence of bimodal felsic and tholeiitic basaltic rocks (Mariko & Kato 1994).

Accretionary setting: Cr, Mn and Cu mineralization

A number of mineral deposits are associated with the ophiolite sequences within the Japanese islands accreted terranes described previously in Chapters 1–3 (Fig. 10.1). The following sections describe the three most important types of these deposits, which include: Cr ore in ultramafic rocks in southwestern Japan and Hokkaido; widespread manganiferous charts; and cupriferous massive sulphide deposits that are most commonly associated with the Sanbagawa Belt.

Podiform-type Cr mineralization in ultramafic complex
Chromium metallogenic provinces in the Sangun Belt in SW Japan and in the Kamuikotan Belt in Hokkaido (Fig. 10.1) have yielded 70% and 30%, respectively, of the Cr ores that have been mined in the country (Hirano 2009). In the Sangun Belt, the host rocks form ultramafic blocks several kilometres in diameter and consist of harzburgite, dunite and subordinate chromitite, except for the Ochiai–Hokubo complex which is composed of layered dunite-wehrlite and lherzolite-harzburgite (Matsumoto & Arai 2001). These complexes are contact metamorphosed by younger granitic intrusions (Arai 1975). An Re–Os isochron age of 322 $\pm$ 12 Ma is reported for chromespinel in chromitite from one of these ultramafic units, known as the Tari–Misaka ultramafic complex (Matsumoto & Suzuki 2002). In the northern margin of this latter complex there are two major lenticular chromite deposits (Hirose and Wakamatsu deposits: Table 10.1; Fig. 10.1) that have each produced more than 200 000 t of chromitite ores (Hirano *et al.* 1978). The origin of these podiform chromitite orebodies is related to the intrusion of Ti-rich magma into harzburgite in which orthopyroxene was selectively dissolved, resulting in the formation of a secondary magma enriched in Si and Cr. Mixing of this secondary magma with later, more primitive magma caused oversaturation with spinel, resulting in the formation of podiform chromian spinel (Arai & Yurimoto 1994; Matsumoto & Arai 2001).

The Kamuikotan Belt is composed of serpentine mélange and sedimentary mélange that include various blocks of low-temperature, high-pressure metamorphic rocks emplaced within the Cretaceous Sorachi–Yezo accretionary complex. Serpentinite derived from peridotite contains several podiform chromitite deposits that together have yielded a total of 0.4 Mt of chromitite ore. The major deposits, such as Hatta, Hattahachiman, Nitto and Shinnitto

Table 10.1. *Major metallogenic provinces in Japan*

Metallogenic provinces	Tectonic setting	Geological setting	Metallic element	Principal ore type	Representative deposits	Age
Pre-accretion continental margin stage						
Hitachi terrane	Back-arc	Bimodal volcanism of felsic and mafic schist	Cu, Zn	Metamorphosed massive sulphide	Hitachi	Cambrian
Accretionary stage						
Kamuikotan Belt	Serpentine mélange	Ultramafic complexes	Cr	Podiform chromitite	Hattahachiman, Hatta, Nitto, Shinnitto	Mesozoic
Sangun Belt	Serpentine mélange	Ultramafic complex	Cr	Podiform chromitite	Wakamatsu, Hirose	Palaeozoic
Mino–Tamba complex	Accretionary complex	Pelagic sedimentary units	Mn	Stratiform manganese	Ohori, Tonoda	Permian– Jurassic
Northern Kitakami complex	Accretionary complex	Pelagic sedimentary unit	Mn	Metamorphosed stratiform manganese	Noda–Tamagawa	Jurassic
Shimanto complex	Accretionary complex	Pelagic sedimentary unit and basalt	Mn, Cu	Stratiform manganese and massive sulphide	Nirou (Mn)	Jurassic
Sanbagawa Belt	Accretionary complex	Greenschist facies basic rocks	Cu, Zn	Volcanogenic massive sulphide	Besshi, Sazare, Shirataki, Okuki, Iimori, Makimine	Jurassic
Hidaka complex	Accretionary complex	Basalt	Cu	Volcanogenic massive sulphide	Shimokawa	Palaeogene
Continental arc stage						
San'yo Belt	Magmatic arc	Ilmenite-series granitoids	W, Cu	Hydrothermal vein, skarn	Kiwada (W, Cu), Kawayama (Cu, Pb, Zn), Kuga (W, Cu), Fujigatani (W, Cu), Yoshioka (Cu, Ag, Au), Obie (Cu, Ag), Kaneuchi (W), Otani (W)	Cretaceous
San'in Belt	Magmatic arc	Magnetite-series granitoids	Mo	Hydrothermal vein, skarn	Higashiyama (Mo), Seikyu (Mo), Daito (Mo), Komaki (Mo, W), Tsumo (Cu, Zn, W), Sasagatani (Cu, Ag), Bandojima (Pb, Ag, Zn), Nakatatsu (Zn, Pb, Ag, Cu), Omodani (Ag, Cu, Pb, Au), Kamioka (Zn, Pb, Ag)	Palaeogene
Abukuma Belt	Magmatic arc	Mixture of magnetite and ilmenite-series granitoids	Au, Ag, Cu, W	Skarn	Yakuki	Cretaceous
Kitakami Belt	Magmatic arc	Magnetite-series granitoids	Fe, Au, Ag, Cu	Skarn, hydrothermal vein	Ogayu (Au, Ag, Cu), Omine (Cu, Au, Ag), Kamaishi (Fe, Cu), Akagane (Cu, Au, Fe, Ag). Oya (Au, Ag), Kohoku (Au, Cu, Ag)	Cretaceous
Back-arc rifting stage						
Sado–Hokuriku Belt	Continental rift	Felsic volcanic rocks	Au, Ag	Epithermal vein	Sado (Au, Ag)	Early Miocene
Island-arc stage						
Outer Zone of SW Japan Arc	Magmatic island arc	Ilmenite-series granitoids and felsic volcanic rocks	Sn W. Sb, Cu, Hg, Au	Skarn, epithermal vein, dissemination	Hoei (Cu, Sn), Ichinokawa (Sb), Kishu (Cu), Myoho (Cu), Yamato (Hg), Kando (Sb)	Middle Miocene
Kurile Arc	Magmatic island arc	Bimodal volcanism of basalt and rhyolite	Au, Ag, Cu, Hg	Epithermal vein, dissemination	Konomai (Au, Ag), Kitami (Cu, Pb, Zn), Kitano-o (Au), Itomuka (Hg)	Middle– Late Miocene
NE Japan Arc	Magmatic island arc	Bimodal volcanism of basalt and rhyolite-dacite, unimodal andesite-dacite volcanism	Au, Ag, Cu, Pb, Zn, Mn, Fe	Epithermal vein, dissemination, Kuroko, skarn, sedimentary	Teine (Au, Cu), Oe–Inakuraishi (Mn, Pb, Zn), Toyoha (Pb, Zn, Ag, In), Chitose (Au), Jokoku (Mn, Zn, Ag), Nurukawa (Cu, Au, Ag, Pb, Zn), Hosokura (Pb, Zn, Ag), Furotobe (Cu, Pb, Zn), Kosaka (Cu, Pb, Zn), Hanawa (Zn, Cu, Pb), Osarizawa (Cu, Zn, Au), Fukazawa (Zn, Cu, Pb), Shakanai (Cu, Zn, Pb), Hanaoka (Cu, Zn, Pb), Takatama (Au, Ag), Ashio (Cu, Ag, Au), Chichibu (Au, Ag, Cu), Kucchan (Fe), Gunma (Fe)	Middle Miocene–Holocene
Izu–Ogawawara Arc	Magmatic island arc	Intermediate and bimodal basalt-rhyolite volcanism	Au, Ag, Cu, Pb, Zn	Epithermal vein, dissemination, Kuroko	Mochikoshi (Au, Ag), Seigoshi (Au, Ag), Takara (Cu, Pb, Zn)	Miocene– Pleistocene
SW Japan Arc	Magmatic island arc	Intermediate volcanism, partly adakitic, with felsic volcanism	Au, Ag, Sb	Epithermal vein	Takeno (Au, Ag), Nakase (Au, Ag, Sb), Iwami (Ag, Au), Shinko–Hoshino (Au, Ag), Bajo (Au, Ag), Taio (Au, Ag)	Pliocene– Pleistocene
Ryukyu Arc	Magmatic island arc	Intermediate volcanism with felsic volcanism	Au, Ag	Epithermal vein	Okuchi (Au, Ag), Hishikari (Au, Ag), Yamagano (Au, Ag), Kushikino (Au, Ag)	Pliocene– Pleistocene

(Fig. 10.1), are all present in a unit known as the Mukawa serpentinite. In the Nitto deposit, compact chromitite orebodies are present within an area of 350 m × 100 m as local lenses that are a few metres thick (Bamba 1984).

Stratiform Mn deposits

Numerous bedded Mn and Fe–Mn deposits occur in accretionary complexes such as Mino–Tamba, Northern Kitakami–Oshima and Shimanto (Fig. 10.1), in deposits which are typically less than 100 000 t in size. However, more than 1000 of these Mn and Fe–Mn deposits were extensively mined from the 1930s to 1960s (Nakamura 1990; Sato & Kase 1996). The major occurrences are the Ohono and Tonoda deposits in the Mino–Tamba complex, Noda–Tamagawa in the Northern Kitakami complex and Nirau in the Shimanto complex (Fig. 10.1; Table 10.1).

The Mn deposits are hosted in chert or chert-basalt complexes in the Jurassic accretionary complexes such as Northern Kitakami, Ashio, Mino–Tamba and Chichibu. The occurrences of Mn deposits are rare in Cretaceous–Palaeogene complexes, where basaltic rocks are much less common, and the close association of chert with Mn ores indicates deposition on the deep-sea floor. These deposits, particularly those in the Sanbagawa Belt, are overprinted by regional low-temperature, high-pressure metamorphism. Those in the Mino–Tamba and Northern Kitakami complexes are, in addition, contact-metamorphosed by Late Cretaceous granitic intrusions (Nakamura 1990).

Nakamura (1990) classified the bedded Mn deposits into the following three types: (1) basalt-free, chert-hosted Mn deposits characterized by dominant rhodochrosite; (2) basalt-hosted bedded Fe–Mn deposits composed mainly of very fine-grained bementite and hematite; and (3) basalt-associated and chert-hosted Mn–Fe deposits composed dominantly of braunite, with rhodochrosite and minor bementite. The first deposit type occurs extensively in the Mino–Tamba and Northern Kitakami–Oshima complexes, whereas the second and third types occur mainly in the Chichibu and Shimanto complexes (Nakamura 1990).

Manganese in the Noda–Tamagawa deposit, hosted in a basalt-free chert-hosted deposit type in the Northern Kitakami complex, precipitated initially as manganese oxyhydroxides that are analogous to modern ferromanganese nodules forming in the deep ocean due to the change of oxidation state (Komuro *et al.* 2005). In contrast, the Fe–Mn ores in the Kunimiyama deposit and Mn ores in the Ananai deposit, both hosted in Early Permian chert in the Chichibu complex, are regarded as precipitates produced by hydrothermal activity at a mid-ocean ridge (Kato *et al.* 2005; Fujinaga & Kato 2005) and on an oceanic island (Fujinaga *et al.* 2006), respectively.

Cu-rich volcanogenic massive sulphide deposits

In Japan, Cu-dominant volcanogenic massive sulphide deposits in accretionary terranes occur mainly in the Sanbagawa Belt, although some deposits are known in weakly or non-metamorphosed Mino–Tamba, Hidaka and Shimanto complexes (Fig. 10.1). These massive sulphide deposits, also known as Besshi-type deposits, share the following common characteristics: (1) they are generally stratabound and associated with submarine mafic volcanic rocks or their metamorphic equivalents; (2) they formed on the seafloor above stockwork veins that represent the conduits for hydrothermal solutions; and (3) they are massive orebodies, consisting of pyrite with lesser amount of chalcopyrite. The major gangue minerals are quartz and chlorite. Minor amounts of sphalerite are common but galena is absent or very rare, in contrast to the Kuroko deposits which comprise zinc- and lead-rich massive sulphides (Sato & Kase 1996).

More than 100 massive sulphide Cu deposits are known in the Sanbagawa Belt, including the Besshi deposit in northern Shikoku which produced more than 0.7 Mt of Cu and the Iimori deposit in central Honshu which yielded 2.7 Mt at 1.30% Cu (Nozaki *et al.* 2010). The deposits comprise thin layers of massive sulphide ores. The largest orebody in the Besshi deposit is 3 m in average thickness and extends for 1.8 km along-strike and 2.7 km downdip from the surface. The orebodies at Besshi are hosted in pelitic schist interbeded with mafic schist, and are highly deformed and overprinted by later metamorphism. The main Cu minerals are bornite and chalcopyrite (Kase 2003). The normal mid-ocean-ridge basalt (N-MORB) geochemical characteristics of the mafic schist at Besshi suggest that the deposit formed by hydrothermal activity at a mid-ocean ridge (Nozaki *et al.* 2006). An Re–Os age of 156.8 ± 3.6 Ma for the pyrite from the Iimori deposit indicates that the deposit formed at least 20–30 million years before accretion of its host rocks to the Eurasian continent (Nozaki *et al.* 2010).

A few massive sulphide deposits, including the Shimokawa Cu deposit (685 Mt at 2.3% Cu; 0.16 Mt Cu) are present in the Cretaceous–Palaeogene Hidaka complex (Miyashita & Watanabe 1988; Nakayama 2003). The massive sulphide orebodies at Shimokawa are hosted in mafic rocks underlying thick terrigenous sedimentary rocks. These orebodies are composed of pyrite, chalcopyrite, pyrrhotite, sphalerite and magnetite (Mariko 1988*a*), and occur in dolerite or at the base of pillow lavas along a major regional fault (Miyashita & Watanabe 1988; Nakayama 2003). Such associations suggest that the deposit formed at a spreading centre and were subsequently overlain by terrigenous sediments (Mariko 1984; Miyashita & Watanabe 1988; Nakayama 2003).

Nozaki *et al.* (2011) reported Re–Os sulphide ages of 89 and 48 Ma for the Makimine deposit in the Shimanto Complex and the Shimokawa deposit in the Hidaka Complex, respectively (Fig. 10.1). They interpreted the Cretaceous and Palaeogene Cu-dominant massive sulphide mineralization to have been associated with the Kula–Pacific Ridge, which migrated northeastwards during Late Cretaceous–Eocene time along the eastern margin of the Asian continent.

Continental arc setting: skarns and vein deposits

The major felsic arc plutonism of Japan can be divided into Triassic–Jurassic, Early–middle Cretaceous and Late Cretaceous–Palaeogene stages (Ishihara & Sasaki 1991). The granitoids generated within the arc are classified into magnetite series and ilmenite series on the basis of magmatic oxidation state, which controlled the concentrations of different metal species and the formation of different metallogenic belts (Sn–W and Cu–Mo) in the Japanese islands (Ishihara 1977).

Plutonism during the first major magmatic cycle (Triassic–Jurassic) is represented by the Funatsu I-type magnetite-series granitoids in central Japan. These are granodiorite in average composition, are not accompanied by volcanism and have almost no related mineralization (Ishihara & Sasaki 1991). The subsequent (Early–middle Cretaceous and Late Cretaceous–Palaeogene) cycles of felsic magmatism and associated mineralization are described in more detail in the following sections.

Cretaceous arc mineralization: Kitakami and Abukuma

Outcrops representing this phase of arc activity are present in northern Honshu as the Kitakami granitic belt and in the Abukuma granitic and metamorphic belt (Fig. 10.2), with co-magmatic volcanic rocks occurring only along the eastern margin of the Kitakami Belt. I-type, magnetite-series granitoids are predominant in the Kitakami Belt, and the magnetite content of these granitoids

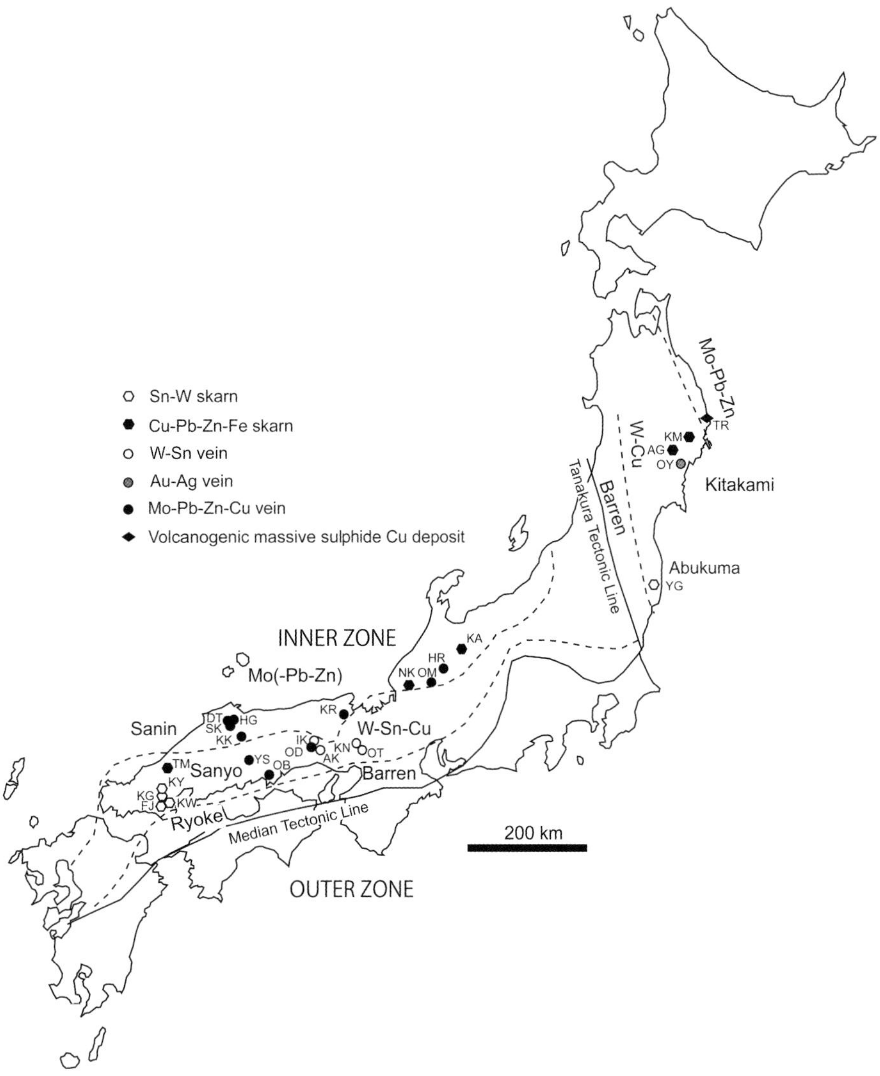

Fig. 10.2. Distribution of metallic mineral deposits associated with Late Cretaceous–Palaeogene granitoids, modified from Ishihara & Sasaki (1991). Abbreviations of deposit names: AG, Akagane (Cu–Au–Fe–Ag); AK, Akenobe (Cu–Pb–Zn–Sn); DT, Daito (Mo); FJ, Fujigatani (W–Cu); HG, Higashiyama (Mo); HR: Hirase (Mo); IK, Ikuno (Ag–Cu–Pb–Zn–Sn); KA, Kamioka (Pb–Zn); KG, Kuga (W–Cu); KK, Komaki (Mo); KM, Kamaishi (Fe–Cu); KN, Kaneuchi (W); KR, Komori (Cu); KW, Kiwada (W); KY, Kawayama (Cu–Pb–Zn–W); NK, Nakatatsu (Zn–Pb–Ag–Cu); OB, Obie (Cu–Ag); OD, Ohmidani (Ag–Au); OM, Omodani (Ag–Cu–Pb); OT, Otani (W); OY, Oya (Au–Ag); SK, Seikyu (Mo); TM, Tsumo (W–Cu–Zn); TR, Taro (Cu–Pb–Zn); YG, Yakuki (Au–Ag–Cu–W); YS, Yoshioka (Cu–Ag–Au).

decreases westwards towards the Tanakura Tectonic Line. In contrast, I-type, ilmenite-series rocks of average granodioritic composition are predominant across most of the Abukuma Belt (Ishihara 1990).

The Cretaceous granitoids in the Kitakami Belt are associated with several skarn and intrusion-hosted vein deposits. The ages of the mineralization are identical to those of the granitoids, indicating that these deposits are genetically related to the magmatism (e.g. Ishihara *et al.* 1989). Metallic mineralization in the Kitakami Belt is associated with small stocks and occurs mostly in the intrusive rocks (Mo) or in the sedimentary wall rocks (Au, W, Cu and Fe). The Mo-bearing granitoids are associated with coeval volcanic rocks, suggesting that they formed at a higher level than deposits containing the other metals. The latter are characterized by scheelite, rather than wolframite as the main W-bearing phase, which reflects the calcic nature of the associated granitoids (Ishihara & Murakami 2004). Ore deposit types are mostly skarn and veins containing Fe, Cu, W and Mo, with the exception of the Pb–Zn-rich Kuroko-type Taro massive sulphide deposit. The ore metals are asymmetrically zoned from west to east from W–Cu to Mo–Pb–Zn provinces in the Kitakami Belt (Fig. 10.2). Auriferous quartz vein deposits are associated with granitic intrusions in the W–Cu province, but are separated from W and Cu deposits. These deposits are mostly exogranitic, and are hosted in sedimentary rocks and ultramafic rocks in addition to the Early Cretaceous plutonic rocks and coeval volcanic rocks (Ishihara & Murakami 2004).

The most representative skarn deposit in this belt is the Kamaishi Cu–Fe deposit (58 Mt at 50–64% Fe, 0.14 Mt Cu), which is associated with the high Sr/Y series Ganidake granodiorite stock (Ishihara & Murakami 2004) intruding Carboniferous and Permian limestones. The deposit is composed of Fe, Fe–Cu and Cu orebodies, in which chalcopyrite tends to occur in and near the limestone whereas magnetite is more abundant in the granodiorite (Ishihara & Murakami 2004). The Cu mineralization in the Tengumori orebody of this deposit is divided into an early calc-silicate stage, ore stage and late calcite stage, with major sulphide minerals being chalcopyrite, pyrrhotite, cubanite and pyrite (Haruna *et al.* 1990). The representative Au deposit is Oya, which produced 18.9 t of Au and 6.3 t of Ag. The veins of the Oya deposit contain native Au, arsenopyrite, pyrrhotite, chalcopyrite, tellurobismuthite and scheelite with gangue quartz and rare K-feldspar (Ishihara & Murakami 2004).

In the Abukuma Belt, only a small number of metallic deposits are present. The Yakuki deposit is a Cu–Fe–W skarn that has yielded 0.06 Mt Cu, 0.3557 Mt Fe and 2491 t WO_3. Skarn minerals are mostly Fe-rich andradite and hedenbergite with minor amounts of epidote, diopside, wollastonite, babingtonite, quartz and calcite.

The ore minerals are chalcopyrite, cubanite, sphalerite, magnetite, pyrite, pyrrhotite and scheelite. The associated intrusive rocks are magnetite-free ilmenite-series granodiorite and granite (Ishihara & Chappell 2008).

Late Cretaceous–Palaeogene arc mineralization: Ryoke, San'yo and San'in

The final phase of felsic arc magmatism produced the largest volume of granitic and coeval volcanic rocks, which are exposed in the Inner Zone of SW Japan to the west of the Tanakura Tectonic Line. In the Inner Zone batholith, the Ryoke granitic and metamorphic belt and the San'yo and San'in volcanic and granitic belts are distributed to the north of the Median Tectonic Line (Fig. 10.2). Many skarn and intrusion-related vein deposits containing W, Mo, Sn, Cu, Pb, Zn and F occur in these volcanic and plutonic belts. The ore elements show a distinct metal zoning from W–Sn–Cu to Mo $\pm$ Pb–Zn) provinces from the south to north in the Inner Zone (Ishihara & Sasaki 1991; Fig. 10.2), and the ages of mineralization are the same as those of the associated granitoids (Ishihara *et al.* 1988*b*).

The Ryoke Belt is typified by I-type, ilmenite-series granodiorite, although locally S-type, ilmenite-series, two-mica monzogranites are present. The San'yo Belt is similarly dominated by I-type, ilmenite-series rocks, mostly granodiorite and monzogranite, whereas the plutons of the San'in Belt are I-type, magnetite-series granodiorite and monzogranite. The concentration of sulphur in the magnetite-series granitoids is generally low because more oxidizing conditions favoured its fractionation into the aqueous phase and consequent transport out of the silicate melt. In contrast, sulphur contents of the ilmenite-series granitoids are commonly higher although variable, particularly in younger intrusions. The reduced-sulphur species of the ilmenite series remains as pyrrhotite in the silicate melts during crystallization (Ishihara *et al.* 1988*a*). The sulphur isotope values of sulphides in the ores are negative in the San'yo Belt and positive in the San'in Belt, reflecting a sedimentary and magmatic origin, respectively (Ishihara & Sasaki 2002). The formation of the sulphide-bearing ilmenite-series granitoids is ascribed to subduction of Farallon–Izanagi and Kura–Pacific spreading ridges which produced high geothermal gradients in the upper mantle and lower crust, resulting in the melting and mixing of crustal sedimentary rocks.

In the San'yo Belt, there is a large number of skarn, vein-hosted and pegmatite-hosted metallic mineral deposits associated with the ilmenite-series granitoids. The mineralization ages of these deposits range from 102 Ma to 60 Ma, with the majority being in the range 90–80 Ma. In this belt, W mineralization is older than base- and precious-metal mineralization, and tends to become younger in age to the east (Watanabe *et al.* 1998). The major metallic deposits in the belt are the Tsumo W–Cu–Zn skarn, Kiwada W skarn, Kawayama Cu–Pb–Zn skarn, Kuga Cu–W skarn, Fujigatani W–Cu skarn, Yoshioka hydrothermal Cu vein, Obie hydrothermal Cu vein, Kaneuchi hydrothermal W vein, and Otani W–Cu–Sn xenothermal vein (Fig. 10.2; Watanabe *et al.* 1998).

The Kiwada deposit produced a total of 4500 t WO_3 and comprises eleven orebodies, the largest of which has yielded more than 68 250 t of ore at 5.5% WO_3. These orebodies replace limestone blocks in the Jurassic Kuga Formation. The ore minerals of the deposit are composed mainly of scheelite, chalcopyrite, pyrrhotite, arsenopyrite, sphalerite and galena, and occur in and surrounding quartz veins cutting hedenbergite-garnet skarn (Nagahara & Shima 1992). The skarn mineral assemblage at Kiwada is a reduced type, which is consistent with the reduced nature of the host granitoids (Shibue *et al.* 1994). The Ag veins at Ohmidani, Yabu and Tada and the Co–Ni veins at Natsume and Oya in the San'yo Belt are related to the Late Cretaceous granitoids, as are polymetallic Cu–Zn–Pb–Sn veins at Akenobe (16.8 Mt at 1.035% Cu, 1.64% Zn and 0.4% Pb) (Watanabe *et al.* 1998). This polymetallic vein mineralization at Akenobe grades transitionally into later and marginal Au–Ag mineralization. The major ore minerals of the polymetallic stage are sphalerite, chalcopyrite, cassiterite, pyrrhotite, pyrite and galena, and those of the Au–Ag stage are pearcite-polybasite, Ag-bearing tenntantite-tetrahedrite, canfieldite, electrum and native Ag (Furuno *et al.* 1992).

In the San'in Belt the mineralization ages of the metallic deposits range from 76 to 37 Ma, with the majority being in the range 66–56 Ma which is younger than the ores of the San'yo Belt. An initial phase of W–Mo mineralization (76–64 Ma) in the San'in Belt was followed by solely Mo mineralization (63–37 Ma) (Watanabe *et al.* 1998). The large San'in deposits are characterized by hydrothermal Mo veins, such as the Higashiyama, Seikyu, Hirase, Daito and Komaki deposits, and polymetallic Ag–Pb–Zn–Mo skarns, such as the Kamioka and Nakatatsu deposits. Other important deposits include the Omodani felsic volcanic rock-hosted Pb–Zn vein deposit and the Komori granite-hosted Cu vein deposit (Watanabe *et al.* 1998; Table 10.2).

The Daito, Seikyu and Higashiyama Mo deposits comprise gently dipping veins consisting mainly of quartz, molybdenite and K-feldspar, with minor amounts of pyrite, chalcopyrite, sphalerite and galena and alteration halos of andalusite, biotite, K-feldspar and muscovite. There are about 40 veins distributed in the mining areas, which are located in the marginal zones of the Rengeji granodiorite. The Komaki Mo deposit is located in a two-mica granite close to a coarse-grained granodiorite, and consists of pipe-like and quartz-molybdenite vein orebodies (Ishihara 1971).

Kamioka is a skarn deposit that replaced limestone in the Hida metamorphic rocks. The deposit produced 0.9 Mt of ore with average grades of 30 g/t Ag, 0.7% Pb and 5.0% Zn, and is composed of the Mozumi, Maruyama and Tochibora lenticular orebodies that are associated with thin 62–57 Ma felsic porphyritic dykes (Sakurai *et al.* 1993). Major ore minerals are chalcopyrite, pyrrhotite, sphalerite and galena, with gangue clinopyroxene, actinolite, garnet, epidote, quartz and calcite (Mariko 1988*b*). The precipitation of sulphides was triggered by the reduction of meteoric water-dominated hydrothermal fluids (Shimazaki & Kusakabe 1990) reacting with host rocks at *c.* 360°C (Kato 1999; Takeno *et al.* 1999).

Back-arc rift setting: epithermal Au–Ag mineralization

Early Miocene back-arc spreading separated the NE Japan Arc from the Eurasian continent, with a *c.* 60° anticlockwise rotation of the arc between 21 and 18 Ma (Hoshi & Takahashi 1999). The only metallic mineral deposit that is clearly related to this event is the Sado epithermal Au–Ag deposit that has yielded 72.7 t Au and 2278 t Ag. The 24.4–22.1 Ma age (Ministry of International Trade & Industry 1987) of the Sado deposit is slightly older than the back-arc spreading and other epithermal Au deposits in the NE Japan Arc, suggesting that the deposit formed at the beginning of or just before the spreading. Similar epithermal Au–Ag deposits associated with felsic volcanism are distributed along the Sea of Japan from Sado to Hokuriku and through the Noto Peninsula, although geochronological data have not yet been obtained from these deposits.

Island-arc setting: epithermal base and precious metal mineralization

The Japanese islands currently include five different island arcs: Kurile, NE, Izu–Ogasawara, SW and Ryukyu (Fig. 10.3). These

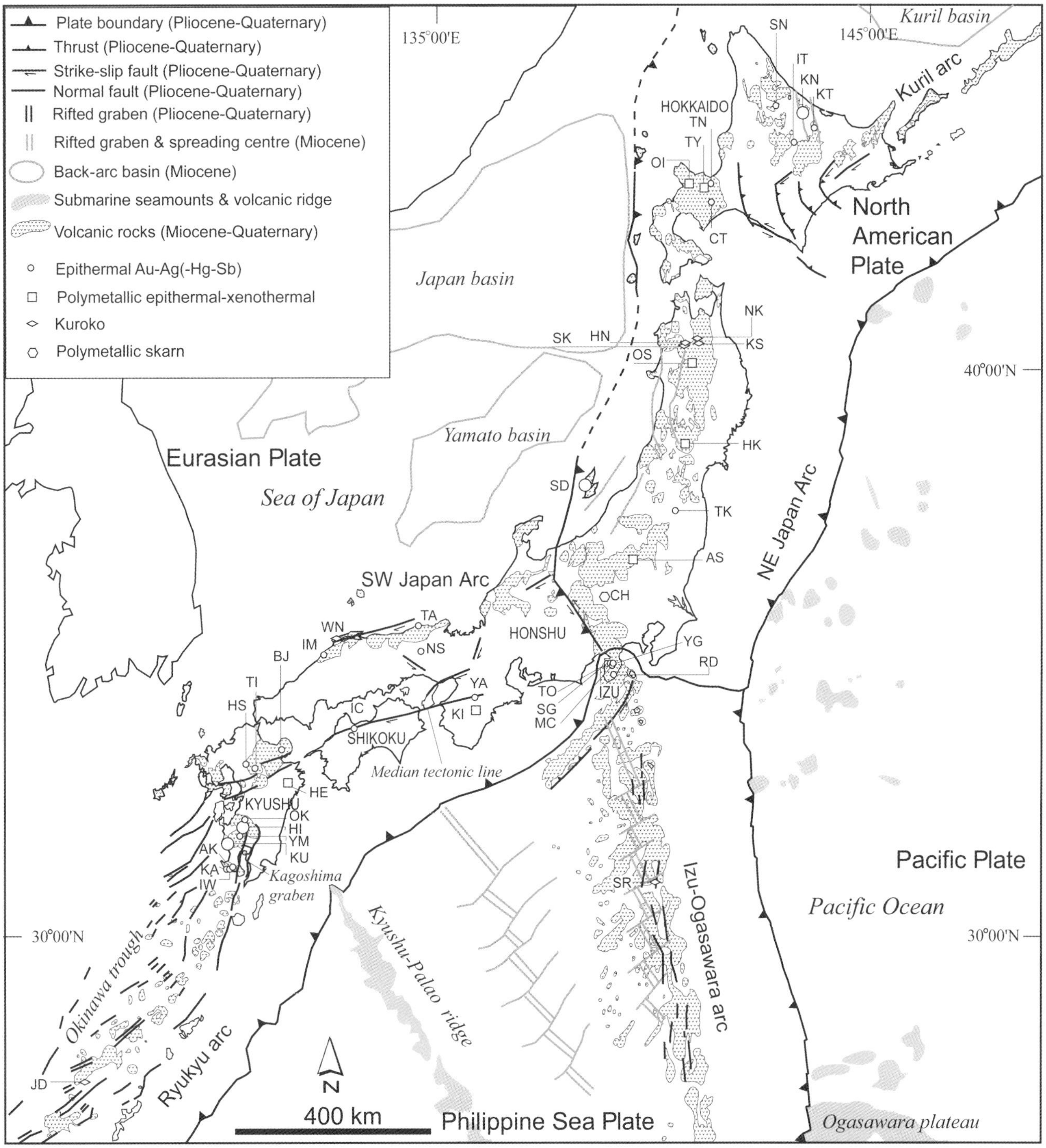

Fig. 10.3. Distribution of metallic mineral deposits formed in the late Cenozoic–Quaternary period, modified from Garwin *et al.* (2005). Abbreviations of deposit names: AK, Akeshi (Au–Ag); AS, Ashio (Ag–Cu); BJ, Bajo (Au); CH, Chichibu (Cu–Au–Ag); CT, Chitose (Au–Ag); HE, Hoei (Cu–Sn); HI, Hishikari (Au–Ag); HK, Hosokura (Pb–Zn–Ag); HN, Hanaoka (Cu–Pb–Zn); HS, Hoshino (Au–Ag); IC, Ichinokawa (Sb); IM, Iwami (Ag–Au); IT, Ikutawara (Au–Ag); IW, Iwato (Au–Ag); KA, Kasuga (Au–Ag); KI, Kishu (Cu); KN, Konomai (Au–Ag); KS, Kosaka (Cu–Pb–Zn); KT, Kitano-o (Au–Ag); KU, Kushikino (Au–Ag); MC, Mochikoshi (Au–Ag); NK, Nurukawa (Cu–Au–Ag–Pb); NS, Nakase (Au–Ag–Sb); OI, Oe–Inakuraishi (Mn–Pb–Zn); OK, Okuchi (Au–Ag); OS, Osarizawa (Cu–Zn–Au); RD, Rendaiji (Au–Ag); SD, Sado (Au–Ag); SG, Seigoshi (Au–Ag); SK, Shakanai (Cu–Pb–Zn); SN, Sanru (Au–Ag); SR, Sunrise (Cu–Pb–Zn); TA, Takeno (Au–Ag); TI, Taio (Au–Ag); TK, Takatama (Au–Ag); TN, Teine (Au–Cu); TO, Toi (Au–Ag); TY, Toyoha (Pb–Zn–Ag–In); WN, Wanibuchi (Cu); YA, Yamato (Hg); YG, Yugashima; YM, Yamagano (Au–Ag).

individual arcs have each undergone a different volcanic evolution, resulting in a number of types of mineralization. Nevertheless, for the most part the ore deposits of this stage are dominated by epithermal base-metal and precious-metal mineralization, in addition to Kuroko massive sulphide mineralization, all of which formed at shallow levels of the upper crust.

Kurile Arc

The Miocene–Holocene Kurile magmatic arc extends for *c.* 2200 km from the northeastern Kamchatka peninsula to southwestern Hokkaido, where it connects to the Aleutian Arc and NE Japan Arc, respectively. The southwestern portion of the Kurile Arc is associated with the Kurile back-arc basin, which formed before the Middle Miocene, due to NE-trending rifting with northwesterly spreading axes (Baranov *et al.* 2002). The volcanic rocks of the southwestern Kurile Arc change from Middle Miocene andesites to Middle–Late Miocene bimodal basalts and rhyolites, although there was also a period of 12–8 Ma basalt-only volcanism. The andesitic volcanism migrated southwards during the Middle Miocene (Watanabe 1995), followed by Middle–Late Miocene bimodal and basalt-only volcanism which occurred mainly in a north–south-trending graben normal to the arc trend (Watanabe 1995). Since Pliocene times, bimodal volcanism in the back-arc has been absent and andesitic volcanic activity at the volcanic front has become dominant. This Pliocene–Pleistocene activity has been associated with the formation of calderas such as Kussharo and Mashu, which are as much as several to tens of kilometres in diameter and have erupted large amounts of felsic ignimbrite (Ikeda 1991).

More than 40 low-sulphidation epithermal gold and mercury deposits and prospects are distributed along the southwestern portion of the Kurile Arc. They are associated mainly with rhyolite intrusions and domes of the Miocene bimodal assemblage and typically occur as gold-bearing quartz-adularia veins in ENE-trending strike-slip faults, although some of them are disseminated in the host rocks (Watanabe 1996). The age of the epithermal gold mineralization ranges from 14 to 4 Ma, with a 12–8 Ma gap corresponding to the period of the basalt-only volcanism. A small number of young (3–1.5 Ma) low-sulphidation epithermal gold deposits are also located near the calderas along the present andesitic volcanic front (Yahata *et al.* 1999). Representative deposits in northeastern Hokkaido include Konomai (73.2 t Au at about 6.4 g/t and 1243 t Ag), Sanru (6.7 t Au at about 7.4 g/t and 46.4 t Ag) and Itomuka (3086 t Hg) (Watanabe 1995). The sulphur isotope values of these epithermal deposits are mostly negative, suggesting the involvement of ilmenite-series magmas generated in or interacting with the continental crust (Ishihara & Matsueda 1997).

NE Japan Arc

The NE Japan Arc extends for *c.* 1600 km from western Hokkaido to the Chichibu area, where it connects to the Izu–Ogasawara Arc. The arc has a slight concave configuration towards the Pacific Ocean side of Japan. Presently the arc is bounded on the southwestern side of Japan by a major fault zone (Itoigawa–Shizuoka Tectonic Line), which also marks the boundary between the North American Plate and Eurasian Plate (Uyeda 1991). The Pacific Plate has been subducting beneath the NE Japan Arc continuously since the Early Miocene, resulting in arc volcanism. This volcanism is divided into Early–Middle Miocene rift-related activity, including bimodal basalt and rhyolite volcanism in the back-arc, and subsequent subduction-related intermediate to felsic activity during Late Miocene, Pliocene and Quaternary times (Tsuchiya 1990; Nakajima *et al.* 1995).

The tectonic regime of the NE Japan Arc since the Miocene has changed from Early–Middle Miocene extension, characterized by arc-parallel normal faulting with rifted basins, to Late Pliocene–Quaternary shortening with arc-parallel thrusting and folding. The less-active transitional period between these two regimes, from Late Miocene to Early Pliocene, lacked significant tectonic deformation (Sato 1994; Watanabe 2002). The Miocene rifting and related bimodal volcanism is best developed in the middle part of the NE Japan Arc, whereas it is not clear or mixed with andesite-dacite magmas at the northern and southern margins of the arc where the Okhotsk continental block and the Izu–Ogasawara Arc have collided with the NE Japan Arc (Kimura *et al.* 1983; Amano 1991). In the Chichibu area at the southern margin of the NE Japan Arc, Miocene magnetite-series I type granitic stocks crop out. These stocks are partly reduced due to their assimilation of sedimentary country rocks (Ishihara *et al.* 1987).

Metallic mineral deposits of the NE Japan Arc include Middle Miocene Cu–Pb–Zn (–Ag–Au) Kuroko-type deposits and Late Miocene–Pleistocene Cu–Pb–Zn and Au–Ag epithermal deposits (Watanabe 2002). About 70 Kuroko deposits, including massive gypsum and barite deposits, have been discovered in the 800 km long NE Japan Arc (Sato 1974). These deposits are associated with rhyodacite submarine volcanism of the Middle Miocene back-arc bimodal assemblage. Despite the wide distribution of the Kuroko deposits in the NE Japan Arc, economic deposits are limited to the middle part of the segment. In particular, productive Kuroko deposits cluster in submarine calderas in the Hokuroku district. Representative deposits in the Hokuroku district include Hanaoka (0.96 Mt Cu, 1.3 Mt Zn, 0.3 Mt Pb), Shakanai (0.1 Mt Cu, 0.3 Mt Zn, 0.1 Mt Pb), Kosaka (0.6 Mt Cu, 1.7 Mt Zn, 0.4 Mt Pb) (Ohmoto *et al.* 1983) and gold-rich Nurukawa (6.8 t Au and 123 t Ag). An Re–Os isochron age of 14.3 $\pm$ 0.5 Ma has been reported for the Kuroko deposits in the Hokuroku district (Terakado 2001*a*). A small number of Early or Middle Miocene epithermal Au–Ag deposits are also distributed in the NE Japan Arc, and are related to felsic volcanism (Watanabe 2002).

Late Miocene–Pleistocene Cu–Pb–Zn–Ag epithermal (-xenothermal) and Au–Ag epithermal deposits are distributed throughout the NE Japan Arc, and are associated with calc-alkaline andesite-dacite volcanism in a terrestrial environment. Miocene examples include the Takatama low-sulphidation Au–Ag deposit (28.7 t Au, 280 t Ag), the Ashio xenothermal Cu-Ag deposit (0.6 Mt Cu), the Hosokura Pb–Zn–Ag deposit and the Chichibu Au–Ag–Cu deposit. During the Pliocene–Pleistocene, a number of epithermal Au–Ag and base-metal deposits formed in the northern part of the arc. They are subdivided into low- or high-sulphidation types, with the high-sulphidation deposits being restricted in size (Watanabe 2002). Representative examples include the Chitose Au–Ag (18 t Ag, 83 t Ag), Teine Au–Cu (9 t Au, 7600 t Cu) and Toyoha Ag–Pb–Zn–In deposits (1.8 Mt Zn, 0.5 Mt Pb, 3000 t Ag, 3000 t In). In the Pleistocene and Quaternary, sedimentary goethite-hematite deposits formed in the lower parts of andesite volcanoes where advanced argillic alteration is extensively developed, with examples including the Gunma and Kucchan deposits.

Izu–Ogasawara Arc

The north-trending Izu–Ogasawara Arc extends for *c.* 1200 km from the Izu Peninsula of Honshu Island to the Iwo-jima islands at 25° N and 142° E in the Pacific Ocean (Fig. 10.3). This magmatic arc is situated along the eastern margin of the Philippine Sea Plate due to northwestwards subduction of the Pacific Plate. It connects to the NE Japan Arc in the north and the Mariana Arc in the south. Rift-related grabens are located in the back-arc of the central portion of the Izu–Ogasawara Arc (Tamaki 1985). The arc probably reached its present position by the Middle Miocene and has collided with and accreted to the Japan Arc at its northern end in the Izu Peninsula (Takahashi & Saito 1997). Middle Miocene–Holocene Izu–Ogasawara Arc volcanism is bimodal (basalt and rhyolite-dacite; Taylor 1992), whereas calc-alkaline andesite and dacite characterized the back-arc from 12.5 to 2.9 Ma (Ishizuka *et al.* 1998). The tectonic regime of the Izu–Ogasawara Arc has been extensional, characterized by several stages of normal faulting since the

Oligocene, except along its northern margin where the arc has collided with the NE Japan Arc. The accretion of the Izu Block to the NE Japan Arc at *c.* 1 Ma (Amano 1991) formed NW-trending right-lateral and NNE-trending left-lateral strike-slip faults that localize epithermal Au deposits.

Metallic mineral deposits along the Izu–Ogasawara Arc include Kuroko-type massive sulphides and epithermal deposits (Fig. 10.3). Kuroko-style submarine hydrothermal mineralization is presently recognized at Myojin Knoll, Myojinsho and in the Sumisu rift of the Izu–Bonin Arc (Glasby *et al.* 2000). The gold-rich Cu–Zn–Pb Sunrise deposit at Myojin Knoll is estimated to have 9 Mt of mineralized material (Iizasa *et al.* 1999). A few Miocene-age Kuroko deposits, represented by the Takara Cu–Pb–Zn deposit (Kinoshita 1920), also occur in the accreted blocks of the Izu–Ogasawara Arc to the NE Japan Arc (Amano 1991). Low-sulphidation epithermal Au–Ag deposits are located in the Izu Peninsula at the northern end of the Izu–Ogasawara Arc. These include Seigoshi (16.0 Au, 511 Ag), Toi (12.1 t Au; 94 t Ag) and Mochikoshi (4.9 t Au, 104 t Ag). The deposits consist of Au-bearing adularia-quartz veins oriented subparallel to regional strike-slip faults. Mineralization ages vary from 2.5 to 1.4 Ma, which corresponds to a period of bimodal volcanism of basalt and rhyolite-dacite as well as andesitic volcanism (Garwin *et al.* 2005).

SW Japan Arc

The Pacific Plate continued to subduct beneath the SW Japan Arc until the Middle Miocene, causing submarine felsic volcanism along the Sea of Japan margin. However, the eastwards sweep of the Izu–Ogasawara Arc from latest Oligocene to Middle Miocene along the SW Japan Arc caused the subducting oceanic plate to switch from the Pacific Plate to the young Philippine Sea Plate (Shikoku Basin). The new subduction of this young, hot plate caused melting of crustal sediments along the SW Japan forearc, resulting in the generation of ilmenite-series granitoids and related volcanism during the Middle Miocene. The abundance of sedimentary enclaves in the ilmenite-series granitoids and the negative sulphur isotope ratios for the whole rock granitoids (−8.3 to − 7.0 per mil for Osumi pluton) indicate the contribution of sedimentary sulphur (Ishihara *et al.* 1999). The negative sulphur isotope values of the Sn-, W-, Sb-, Hg- and base-metal-rich Kishu epithermal vein deposits in central Honshu, which have yielded 7.0 Mt at 1.05% Cu, reflect sulphur sourced from these sediment-enriched ilmenite-series magmas (Ishihara *et al.* 1999).

Subsequent adakitic volcanism along the present volcanic front far from the Nankai Trough started in the Pliocene (Kimura *et al.* 2003). Since the latest Pliocene, east-trending right-lateral strike-slip faulting along the Median Tectonic Line and other tectonic lines resulted from the oblique subduction of the buoyant Philippine Sea Plate beneath the SW Japan Arc (Ito *et al.* 2002). Widespread rift-related basaltic magmatism took place during the Pliocene and Pleistocene in northern Kyushu at the western end of the SW Japan Arc. This rifting is related to spreading along the Okinawa Trough located in the Ryukyu back-arc (Kamata & Kodama 1999).

Such changes in the tectonomagmatic setting resulted in different styles and ages of mineralization in the SW Japan Arc (Sato 1974). The 18.4 ± 0.6 Ma Re–Os age obtained from the Wanibuchi Kuroko deposit (Terakado 2001*b*) in the SW Japan back-arc is older than the 15.4–12.4 Ma age range (Sawai & Itaya 1993) for the Kuroko mineralization in the NE Japan Arc, suggesting that a Kuroko-favourable rift setting developed slightly earlier in the SW Japan Arc than in the NE Japan Arc.

Subsequent Middle Miocene subaerial felsic volcanism in the SW Japan Arc was associated with epithermal Au–Ag mineralization, as demonstrated by the Takeno deposit of Hyogo Prefecture in central Honshu. The forearc magmatism included the generation of ilmenite-series granitoid plutons that are associated with Sn–W and Sb mineral occurrences.

The youngest mineral deposits include those of an epithermal gold province along the western margin of the SW Japan Arc (northern Kyushu), where back-arc rifting along the Okinawa Trough has propagated into the arc. This province contains about 20 low-sulphidation Au–Ag deposits of Pliocene and Pleistocene age, associated with intermediate to felsic volcanism in and surrounding the Beppu–Shimabara Graben. They include the Taio low-sulphidation epithermal Au–Ag deposit (36.0 t Au, 137 t Ag) (Watanabe 2005).

Ryukyu Arc

The Ryukyu Arc extends for *c.* 1200 km from Kyushu to Taiwan and comprises, from east to west, the Ryukyu Trench, a row of islands, an active volcanic belt and the Okinawa Trough. The Ryukyu Arc is related to westwards subduction of the Philippine Sea Plate beneath the Eurasian Plate at a velocity of 6–7 cm a^{-1} (Shinjo 1999).

The basement rocks of the Ryukyu Arc consist of Permian–Cretaceous sedimentary or serpentinite mélange, which includes blocks of limestone and metamorphic rocks, and Cretaceous–Palaeogene flysch-type sedimentary rocks. The basement rocks are overlain by late Cenozoic sedimentary rocks. Middle–Late Miocene ilmenite-series and magnetite-series granitoids intruded both the forearc and back-arc in the northern part of the arc, respectively. This intrusive activity was followed by Pliocene and Quaternary calc-alkaline andesitic volcanism, which is associated with rhyolite and dacite in the back-arc. Volcanic and subvolcanic rocks of the Ryukyu Arc are calcic and low in initial $^{87}Sr/^{86}Sr$ ratios (0.70474–0.70486), suggesting that the most primitive felsic to intermediate magmas are related to Au mineralization (Ishihara *et al.* 1990). Palaeomagnetic data (Kodama *et al.* 1995) indicate that the forearc part of Kyushu has rotated anticlockwise with respect to the back-arc part since 2 Ma, resulting in east-trending extension along the Kagoshima Graben at the northern end of the Ryukyu Arc. Back-arc spreading started at the Okinawa Trough from the latest Pliocene (Sibuet *et al.* 1998), characterized by bimodal basaltic and rhyolitic volcanism (Shinjo & Kato 2000) and by seafloor hydrothermal activity that deposited Kuroko ores in a caldera (Marumo & Hattori 1999).

The southern Kyushu epithermal gold province in the northern end of the Ryukyu Arc is dominated by Pliocene and Pleistocene low-sulphidation deposits, represented by the Hishikari (260 t Au, 208 t Ag), Kushikino (55.9 t Au, 477 t Ag), Yamagano, (28.4 t Au, 28.3 t Ag) and Okuchi (22.2 t Au, 17.0 t Ag) deposits as well as a small number of high-sulphidation deposits that include Akeshi (8.9 t Au, 4.7 t Ag), Kasuga (8.8 t Au, 5.0 t Ag) and Iwato (8.1 t Au, 13.7 t Ag) (Izawa & Watanabe 2001). These deposits cluster on the western side of the Kagoshima Graben (Izawa & Urashima 1987). The epithermal Au deposits in the province first formed during the Pliocene, with both high- and low-sulphidation deposits developing in the west of the province (Kushikino, 3.7–3.4 Ma; Kasuga. 5.5–5.3 Ma; Iwato, 4.7–4.2 Ma and Akeshi, 3.7 Ma). During the Pleistocene, low-sulphidation deposits (Okuchi, 1.6–1.2 Ma; Hishikari, 1.1–0.7 Ma) formed further east in the province. This eastwards expansion of the metallogenic province followed an eastwards shift of the Ryukyu volcanic front (Izawa & Watanabe 2001) due to steepening of the subduction angle of the Philippine Sea Plate (Watanabe 2005).

Non-metallic mineral resources in Japan (TT)

Non-metallic mineral resources characteristic of active continental margins are abundantly distributed across the Japanese archipelago.

Many of these continue to be exploited, and support the economic development of Japan in the twenty-first century. The geology and development of these industrial mineral deposits are briefly reviewed in this section, using resource data mainly from 2012 compilations (MIC 2012).

Limestone

Limestone is the most abundant non-metallic mineral resource in Japan, with more than 200 mines currently in operation with an annual production of 140 Mt (Limestone Association of Japan 2012). The limestone masses mostly originate from coral reefs of middle–late Palaeozoic and Mesozoic age, and occur generally as irregularly shaped blocks or fragments that are as much as 20 km in diameter within Permian–Cretaceous accretionary complexes. These limestones are typically characterized by a high purity of calcite composition rather than dolomite or dolomitic limestone, both of which are relatively scarce. Some of the limestones have been altered to marble by thermal metamorphism during granitic intrusion. Akiyoshi-dai in western Honshu Island is the largest occurrence of limestone-marble in Japan. Approximately 50% of the quarried limestone is used for cement production, with the remainder being utilized as rock aggregates, paper fillers and in steel refineries.

Silica

Silica recovered from siliceous rock types is the second-most-abundant non-metallic mineral resource in Japan. In terms of their occurrences and uses, siliceous resources have commonly been divided into the following three types: silica stone, silica sand and diatomite.

Silica stone

High-grade silica stone occurs as pegmatite and vein quartz, hydrothermal porous silica stone, chert and siliceous schist and is used mainly for glass, refractory materials, construction materials and refining of steel. The hydrothermal porous silica stone originates from strongly silicified volcanic rocks that have been subjected to acid leaching in or near volcanic fumaroles (Hamasaki 2009). Medium- to low-grade silica stone ($SiO_2 < 90$ wt%) is produced from chert, sandstone and hydrothermally altered massive volcanic rocks, and used mainly for cement and rock aggregates. The annual production of silica stone is *c.* 5 Mt, with medium- to low-grade materials accounting for *c.* 90% of the total production.

Silica sand

Silica sand in Japan is produced mostly from Tertiary sediments. The largest deposits occur as thick arkosic sand beds intercalated in Pliocene lacustrine sediments in the Seto and Tajimi region, north of Nagoya (e.g. Sudo & Naito 2000). These arkosic sands can easily be separated into silica sand and kaolin clay by elutriation, and these two components have been given the local names of Gaerome (frog eyes = large quartz crystals) silica sand and Gaerome clay, respectively. Both of these materials are used entirely for ceramics. Another source of silica sand is provided by siliceous sandstone beds in Miocene sedimentary rocks. The overall annual production of silica sand in Japan is *c.* 1 Mt, with the Gaerome silica sand accounting for 65% of this amount.

Diatomite

Medium- to small-scale diatomite deposits are ubiquitously distributed across the Japanese archipelago, occurring mostly as Tertiary marine and lacustrine sediments. Around 130 000 t of diatomite are annually supplied for filter, construction materials, refractory materials and agricultural uses and as fillers in paper and rubber manufacturing.

Feldspar

The annual production of feldspathic materials in Japan is *c.* 650 000 t, with these resources being divided into pegmatitic and metasomatic feldspar, aplite feldspar and saprolite (Sudo 2001).

Pegmatitic and metasomatic feldspar

Pegmatitic feldspar is obtained from two large-scale pegmatites within Cretaceous granitic rocks at Kanamaru in Yamagata Prefecture and Umanotani–Shiroyama in Shimane Prefecture. The two pegmatites occur as irregularly shaped orebodies of several tens to hundreds of metres in diameter, and K-feldspar is more dominant than plagioclase (Ishihara *et al.* 1998; Kihara *et al.* 2005). Metasomatic feldspar occurs as small albitized masses within granitic batholiths. This type of feldspar is used for ceramics such as sanitary wares, tiles and glass wool due to its purity and whiteness.

Aplite feldspar

These aplitic deposits comprise equal amounts of quartz, plagioclase and K-feldspar, without any mafic minerals, and occur as lenticular-shaped bodies or dykes of several to ten of metres width. They are used mainly as glazes for pottery and tiles.

Saprolite (Saba)

In the context of feldspar resources, saprolite defines a kind of weathered granite whose feldspar phenocrysts are extractable by sieving, acid washing and magnetic separation (Sudo 2000*a*). The local name of the saprolite is Saba (or Soukei), and improvements in mineral processing techniques have resulted in Saba feldspar output currently reaching *c.* 85% of total feldspar production in the country. The distribution of Saba is restricted to zones along active faults in central Honshu and is therefore formed by a combination of chemical weathering and weak cataclastic deformation (Sudo 2000*b*). The main uses of Saba feldspar are in ceramics, construction materials and glass.

Clay resources

Clay resources are indispensable materials for ceramics, foundry industries and civil engineering. Due to their voluminous nature and corresponding transport costs, more than one-half of the current demand is still being supplied from *c.* 100 local mines. Clay resources in Japan can be divided into sedimentary clay, hydrothermal clay and bentonite in terms of their origins. Their distribution is shown in Figure 10.4

Sedimentary clay

The main component of sedimentary clay is generally kaolin-group minerals such as kaolinite and halloysite, but it commonly includes various impurities. In practice, sedimentary clay has therefore been classified not according to its constituent minerals but more on its Seger Cone (SK) number, which is an indicator of refractoriness; clay with SK > 26 is defined as 'refractory' as opposed to 'miscellaneous' clay.

Production of refractory clay in Japan is confined to the Gaerome and Kibushi clays produced during the separation of Gaerome silica sand mentioned above. Gaerome clay is a kaolinite-rich, viscous, white- to grey-coloured clayey material produced from Pliocene arkosic sand beds during elutriation processes. Kibushi (ligneous) clay is produced from thin and dark-coloured clayey beds, less than several metres thick, which are often intercalated with the arkosic sand beds. Kibushi clay consists mainly of ligneous materials and fine kaolinite, and needs to be refined by elutriation to remove

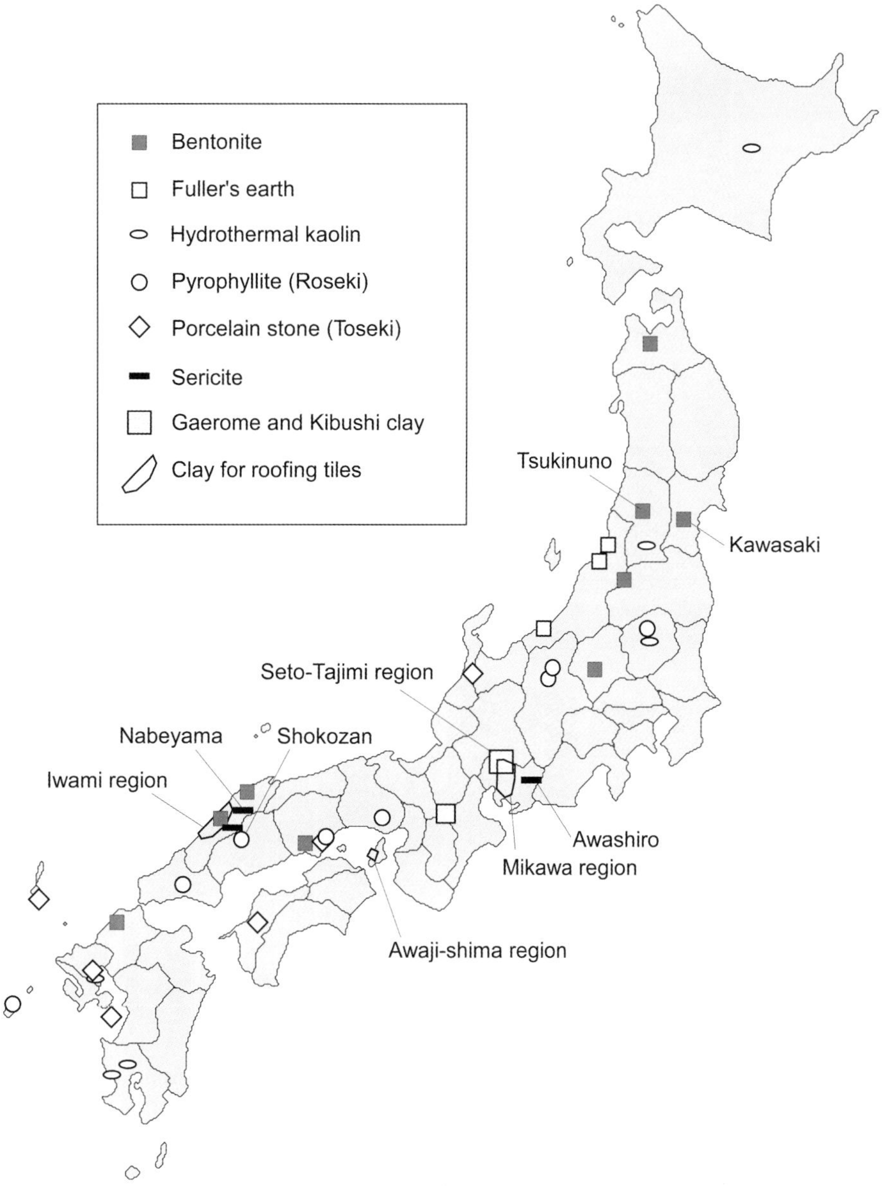

Fig. 10.4. Distribution of clay resources in Japan. Northern Territories and Nansei islands are omitted from the map due to the limitation of space.

impurities. The viscosity and whiteness of Kibushi clay are greater than those of Gaerome clay (Sudo 2000*c*). Kibushi clay is therefore useful as an additive to Gaerome clay for quality adjustment. Gaerome and Kibushi clays have been supplied mostly for the ceramics industries and their annual production is *c.* 170 000 t, although this is in decline due to the urbanization of mining areas in the Nagoya area.

Miscellaneous clay is produced from fluvial sediments and marsh/paddy fields and is widely used for roofing tiles, bricks and plant pots. The clay for roofing tiles is mined mostly in the Mikawa, Iwami and Awaji-shima regions (Sudo 1999), and annual production exceeds 1 Mt.

Hydrothermal clay

Hydrothermal clay is a characteristic resource of the active continental margins in Asia, including southern China, South Korea, Vietnam and Japan, in contrast to clay resources in North America and Europe that have mostly been supplied from sedimentary rocks or weathered igneous rocks. In Japan, the major hydrothermal clay resources are kaolin, pyrophyllite, porcelain stone and sericite.

Hydrothermal kaolin deposits occur as irregular-shaped orebodies, several tens to hundreds of metres in diameter, and are hosted in Neogene rhyolitic to andesitic volcanic rocks (e.g. Kawano & Tomita 1990). Kaolin ores are generally soft and fragile, consisting of kaolinite, dickite, halloysite and subordinate smectite. The demand of domestic kaolin dropped drastically during the 1980s to 1990s due to its replacement by imported kaolin, and no mines are operating as of 2014.

Pyrophyllite (Roseki, in Japanese) deposits in Japan have been formed by the intense hydrothermal alteration of silicic volcanic rocks to produce grey- to yellow-green-coloured, soft and massive ores showing a waxy surface. The Roseki deposits of western Honshu are hosted in Late Cretaceous pyroclastic rocks (Fig. 10.5a), and tend to occur as large-scale lenticular-shaped orebodies several tens to hundreds of metres in diameter (e.g. Hida *et al.* 1996). In contrast, those in central Honshu are smaller scale and hosted within Neogene volcanic rocks (Sudo 2009). Roseki deposits commonly show an

Fig. 10.5. (**a**) Takinotani quarry of Yano–Shokozan pyrophyllite mine, Hiroshima Prefecture. (**b**) Corundum-diaspore zone, Yano–Shokozan pyrophyllite mine. The dark spots are corundum and diaspore aggregates. (**c**) Underground quarry of Awashiro sericite mine, Aichi Prefecture. (**d**) Unstiffening of sericite ores by water immersion, Nabeyama sericite mine, Shimane Prefecture. (**e**) Kawasaki bentonite mine, Miyagi Prefecture. (**f**) Underground quarry of the Tsukinuno bentonite mine, Yamagata Prefecture. The dark layer is a bentonite bed.

alteration zoning from the central to peripheral parts as follows: corundum-diaspore zone (Fig. 10.5b); pyrophyllite zone; kaolinite zone; quartz-sericite zone; and weakly altered zone. The main uses of high-grade, high alumina Roseki ores are in glass fibre and refractories, whereas medium- to low-grade ores are used in cement, tiles and as a carrier of agricultural chemicals. Although the annual production as of 1990 was *c.* 1 Mt, it has since decreased significantly to less than 100 000 t over the past decade due to the increase of imported products.

Porcelain stone (Toseki, in Japanese) deposits were formed by the weak hydrothermal alteration of silicic volcanic rocks, producing grey to white ores that consist of quartz, sericite and subordinate kaolin. Toseki deposits occur as silicic dykes in western Kyushu (Hamasaki & Sudo 1999) and as irregular-shaped massive orebodies in central Honshu. The annual production of Toseki ores is *c.* 100 000 t and they are used mostly for chinaware and insulators.

Sericite ores are hydrothermally altered igneous rocks composed mainly of sericite (illite) grains several tens of microns in diameter. Two small sericite deposits produced by the alteration of Palaeogene granodiorite and Neogene andesitic dykes have been exploited in western and central Honshu, respectively (Kitagawa *et al.* 1982; Okamura 1999; Fig. 10.5c). The ores are softened by initial water immersion (Fig. 10.5d), and then sericite grains are separated and refined into sericite crystals of high-aspect ratios using hydrocyclones. The sericite crystals are used for cosmetics, paint as nano-composite and plastic filler using the properties of platy sericite crystals. The annual production is *c.* 1000 t.

Bentonite

Bentonite is a clay composed mainly of montmorillonite, and is formed by diagenesis or low-temperature hydrothermal alteration of rhyolitic to andesitic vitric tuff. In Japan, there are abundant bentonite resources in Neogene strata, with the larger-scale deposits being located mainly in northern Honshu (Fig. 10.5e). Bentonite ores generally consist of montmorillonite, opal-CT and quartz, with subordinate zeolite and calcite. These ores are classified into Na- and Ca-bentonite based on the interlayer cation chemistry of the montmorillonite. The Na-bentonite ore has a greater water absorption swelling capability than Ca-bentonite. In Japanese deposits, Ca-bentonite tends to alter to Na-bentonite with increasing depth (e.g. Takagi *et al.* 2005), with the large-scale and deep-seated Tsukinuno deposit in Yamagata Prefecture providing a good example of pure Na-bentonite formed by this alteration process (Fig. 10.5f). The main uses of bentonites are as bonding agents for foundry sand, clay grouting for civil engineering, drilling mud, cat litter and the carrier of agricultural chemicals. The annual production is *c.* 300 000 t.

Fuller's earth is a kind of bentonite altered by groundwater near the surface and involving the partial replacement of the montmorillonite Ca or Na interlayer by H. A number of small- to medium-scale deposits of Fuller's earth are distributed across Niigata Prefecture, and the annual production is less than 20 000 t. Fuller's earth is generally treated by heated sulphuric acid, and is then converted to activated clay having more than double the surface area of the original clay. Activated clay is used in oil refineries, as filter materials, silica powder and in ceramics.

Rock aggregates and building stones

Rock aggregates are indispensable for all infrastructural engineering projects and therefore abundantly utilized all over Japan, with about 1400 quarries being operative across the country in 2013 (ANRE 2014). Although riverbed gravel was the major source of rock aggregates through the 1960s, during the following two decades such workings were replaced by mountain/sea gravel and crushed rocks due to the imposition of regulations for the protection of river environments. The sources of crushed rocks are mainly sandstone, shale and limestone of Palaeozoic–Mesozoic age and Tertiary rhyolite, andesite and basalt. The annual production is *c.* 400 Mt.

Building stones are important for tombstones, building walls and floor materials, with about 40% of these rocks being granitic and the remainder mainly andesite and tuffs (Ishihara 1993). As of 2013, 940 quarries were operating in Japan with an annual production of 1.5 Mt, although the stone industry in Japan is in decline due to the combined effects of economic stagnation and competition from abroad.

Hydrocarbon and iodine resources in Japan (NK)

Because the Japanese islands are positioned along a highly active subducting plate margin, they do not have a geotectonic setting suitable for the large-scale accumulation of conventional hydrocarbons. Although some oil and gas is produced, the amount is too small to satisfy domestic consumption. The history of production in Japanese oil and gas fields is described in Figure 10.6 and Table 10.2. There are currently about 700 active wells. Crude oil production in 2010 was about 8.7×10^5 kL whereas that of natural gas was 3.43×10^9 m^3, totalling 0.3% and 3.3% of domestic demand, respectively. Since nuclear power plant operation was suspended as a result of the Great East Japan Earthquake in 2011, demand for imported liquefied natural gas (LNG) for power generation has been growing and the domestic share of natural gas has fallen.

Conventional oil and gas fields are distributed only in Neogene back-arc basins and Cretaceous–Neogene forearc basins in NE Japan (Fig. 10.7). A total isopach map and schematic cross-sections of sedimentary basins surrounding Japan were produced by Sumii *et al.* (1992). Exploration targets have been changing from shallow to deep, from land to submarine and from anticlinal to stratigraphic and fractured basement types of reservoirs. At the end of 2010, recoverable reserves of conventional crude oil and natural gas on land were estimated to be 1.0×10^7 kL and 4.9×10^{10} m^3, respectively, whereas in the ocean the estimates are 9.5×10^5 kL and 1.2×10^9 m^3, respectively (Japan Natural Gas Association 2012).

Because Japan is poor in domestic hydrocarbon resources there is much interest in unconventional natural gas deposits, advancing studies of the development of dissolved methane in formation waters and investigating submarine methane hydrate sources. The brine associated with some Japanese deposits includes iodine at high concentrations, and this iodine supplies 30% of the world's demand (Fig. 10.8).

Conventional hydrocarbons

Back-arc basins in NE Japan

The 'Green Tuff Basin' is the general term for an assembly of back-arc basins running from NE to SW Japan, including the Fossa Magna area, which are filled with Neogene volcanic and clastic sediments after the opening of the Sea of Japan. The name 'Green Tuff' came from the greenish colour of the altered mafic to felsic submarine volcanic ejecta in the area. The oil and gas fields distributed in this basin are restricted to NE Japan (Fig. 10.7). Their source rocks are the siliceous sediments of the Onnagawa Formation that originated from diatoms in the Akita Basin, and the shales of the Nanatani and Teradomari formations which contain abundant terrigenous organic matter in the Niigata Basin. The source rocks were deposited under anaerobic conditions in the sedimentary basins, which were formed in an extensional regime across the arc after the opening of the Sea of Japan. Quaternary compression of the basin promoted burial and maturation of the source rock, forming traps and causing the accumulation of hydrocarbons.

Akita Basin. Oil fields are located adjacent to the Sea of Japan, where the fold axes tend to run north–south. The reservoir horizons are the Upper Onnagawa, Funakawa and Lower Tentokuji formations. The source rock is well-laminated hard shale of the Onnagawa Formation that has been diagenetically altered from diatoms (see Chapter 6 for geology and stratigraphy). The kerogen-rich material is amorphous and a favourable oil source (Hirai 1982), with a total organic carbon content of >1%. The thermal gradient in the Akita Basin is 4–5°C/100 m, and the porosity and permeability of the reservoir are 15–40% and 1–800 md, respectively (Japan Natural Gas Association & Japan Offshore Petroleum Development Association 1992).

The Yabase oil field located in Akita city is the largest oil field in Japan. It has an anticlinal trap running in a NNE–SSW direction, which is 13 km long and 0.6 km wide. Its source rock is the Onnagawa Formation, and its hydrocarbon generating ('kitchen') area is a zone of subsidence in the west. The reservoirs are the Onnagawa,

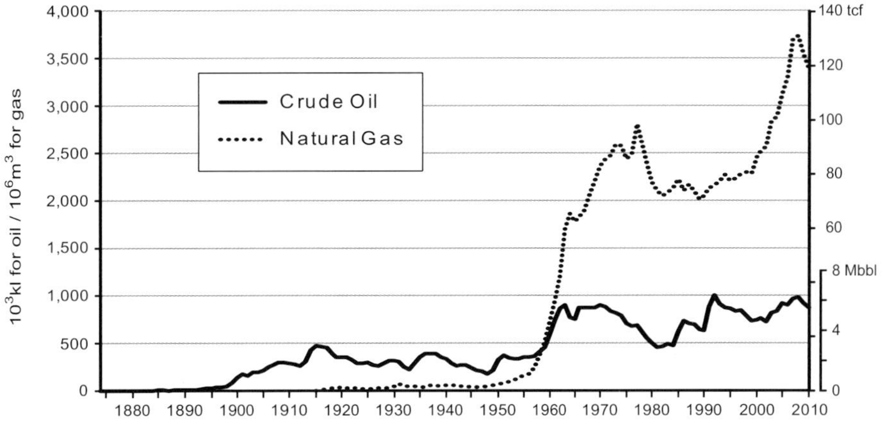

Fig. 10.6. Japanese hydrocarbon production from 1874 to 2010.

Table 10.2. *Oil and gas field production in Japan*

No.*	Field name	Type of fluid†	Year of discovery	Cumulative production	
				Oil (KL)	Gas ($\times 10^3$ m^3)‡
Tenpoku Basin					
	Wakkanai	O	1914	1148	–
	Koetoi	O	1922	5905	1426
	Yuchi	O	1941	898	–
	Masuhoro	O	1922	7207	93
	Menashi	O	1915	2712	174
	Kita–Toyotomi	O	1938	916	21
	Toyotomi	G	1925	–	163 237
Ishikari Basin					
	Atsuta	O	1931	4766	20 655
	Ishikari	O	1903	155 168	
	Minami–Kanazawa	G	1985	–	–
	Minenobe	G	1987	–	–
	Barato	O	1957	84 708	8742
Hidaka Basin					
	Fureoi	O	1887	121 598	22 283
	Karumai	O	1907		
	Biratori	O	1956	1327	–
1	Yufutsu	O&G	1989	2 546 768	4 355 055
Oshima Basin					
	Oshamanbe	W	1954	–	9315
Akita Basin					
	Hachimori	O	1935	117 000	12 733
	Minami–Noshiro	O	1969	8410	4748
	Sakaki	O	1940	3976	–
	Hibiki	O	1918	?	–
	Yoshino	O	1972	–	23 354
2	Sarukawa	O	1958	2 433 775	237 805
	Hashimoto	O	1962	46 985	3072
	Fukumezawa	O	1964	723 806	85 586
	Katanishi	G	1963	–	2540
	Fukukawa	O	1968	241 148	30 071
	Nishi–Ogata	O	1968	472 055	48 385
	Urayama	O	1915	1 280 000	36 000
	3	Toyokawa	O	1913	
4	Kurokawa	O	1912	1 221 000	100 000
	Michikawa	O	1913	147 560	366
	Nigorikawa	O	1908	287 406	17 836
	Asahikawa	O	1904	337 817	
	Hamada	O	1951	69 673	8212
	Kita–Akita	O	1957	25 949	9157
	Shin–Akita	O	1957	744	–
5	Yabase	O	1933	5 664 056	1 347 177
	Tuchizaki–oki	O	1959	183 920	18 427
	Hodono	O	1924	295	–
	Kinshoji	O	1924	5247	21
	Niida	O	1914	24 104	4284
	Toyoiwa	O	1920	9642	264
	Ktsurane	O	1903	129 028	7816
	Hagawa	O	1919	51 170	3047
	Shimohama	G	1968	–	23 454
	Katte	O	1921	127 502	13 116
	Uchimichikawa	O	1921	32 127	–
6	Ayukawa	O&G	1989	1 224 866	461 707
7	Yurihara	O&G	1976		
	Chokai	W	1954	–	?
8	Innai	O	1922	1 222 744	42 230
	Katsurazaka	O	1937	125 545	7528
	Nishi–Oguni	G	1964	–	2357
	Kisakata & Konoura	W	1951	–	?
	Kotaki	O	1938	60 777	2768
	Kamihama	O	1941	56 983	11 309
	Fukura	O	1958	41 113	8306
	Chokaisan	O	1934	68 404	1564
	Narahashi	O	1947	105 625	–
	Sagoshi	O	1960	1 451	744
	Ishinazaka	O	1944	212 386	14
	Niibori	O	1960	106 747	60 041

(Continued)

Table 10.2. *Oil and gas field production in Japan (Continued)*

No.*	Field name	Type of fluid†	Year of discovery	Cumulative production Oil (KL)	Gas ($\times 10^3$ m^3)‡
	Amarume	O	1960	706 782	301 780
	Higashi–Amarume	O	1964		
	Mogami	O	1953	14 301	903
	Sakata	G	?	–	9
Niigata Basin					
9	Iwafune–oki	O&G	1983	4 848 196	2 787 626
	Echigo–Kurokawa	O	19c	24 190	58
	Hirakida	G	1966	17 922	183 706
10	Shintainai	G	1967	496 506	2 624 080
11	Nakajo	O&G, W	1957	801 224	5 478 175
12	Shiunji	G	1962	61 907	1 029 340
	Seiro	G	1965	–	14 749
	Kajikawa	O	1949	5907	–
	Shibata	O	1945	21 660	1973
	Toyosaka	G	1963	–	716
	Maki	G	1959	–	5374
13	Aga–oki–kita	O	1981	1 155 000	167 000
14	Aga–oki	O&G	1972	1 433 669	4 072 907
15	Higashi–Niigata	G	1959	3 296 670	9 416 207
16	Matsuzaki	G	1966	383 912	1 227 000
17	Niigata	W	1926	–	3 917 340
	Kita–Aga	G	1957	–	61 082
18	Minami–Aga	O	1964	2 789 191	849 639
	Minami–Suibara	G	1968	8280	51 877
	Kuwayama	G	1968	80 981	465 455
	Minami–Kuwayama	O	2003	90 158	10 676
19	Niitsu	O	1874	3 026 636	130 078
	Minami–Niitsu	O	1979	799	47
	Higashi–Sanjo	G	1960	–	4917
	Omo	O	1916	222 748	81 279
20	Mitsuke	O	1958	1 828 701	656 169
	Okuchi	W	1906	351	8022
21	Higashiyama	O	1888	1 232 001	3452
	Tamugiyama	O	1956	53 054	17 838
	Yoshida	O	1969	14 177	3401
	Nishi–Nagaoka	W	1960	304	41 727
	Shin–Nishi–Nagaoka	G	1986	31	16
22	Fujikawa	G	1964	276 000	1 286 333
	Torigoe	O	1906	85 563	3688
	Kumoide	G	1962	112 923	699 970
	Sekihara	G	1960	10	378 545
23	Minami–Nagaoka	G	1979	2 379 287	16 677 885
23	Katakai	G	1960	743 628	7 497 136
	Ojiya	O	1900	9580	2591
	Teradomari	O	1904	149	16 008
24	Amaze	O	1888	33 857	
	Ishiji	O	1899	4307	
	Miyagawa	O	1895	312 011	3630
25	Nishiyama	O	1896	2 961 258	814 064
	Myohoji	G	18c	–	?
26	Yoshii	G	1968	2 185 730	11 416 052
26	Higashi–Kashiwazaki	G	1970	1 681 280	7 934 496
27	Kubiki	O&G	1954	2 418 804	4 221 834
	Naoetsu	O	1963	8439	2432
	Gotsu	O	1934	10 001	22 198
	Nadachi	O	1879	11 176	–
	Maki	O	1900	149 984	691
	Bessho	G	1967	14 677	160 443
	Odagiri	G	1944	–	1430
Suwa Basin					
	Suwako	W	1900	–	?
Joban Basin					
28	Iwaki–oki	G	1973	71 000	5 600 000
Kanto Basin					
29	Minami–Kanto	W	1865	–	23 326 449
Tokai Basin					
	Yaizu	W	?	–	7366

(Continued)

Table 10.2. *Oil and gas field production in Japan* (*Continued*)

No.*	Field name	Type of fluid†	Year of discovery	Cumulative production Oil (KL)	Cumulative production Gas ($\times 10^3$ m^3)‡
30	Sagara	O	1904	2517	–
Miyazaki Basin					
	Sadowara	W	1973	–	137 180
	Miyazaki & Nichinan	W	1920's	–	16 477
Shimajiri Basin					
	Okinawa–Honto–Nanbu	W	1950's	–	–

*Numbers as in Figure 10.7.
†O: oil; G: gas; W: water (dissolved gas).
‡Under standard conditions.

Funakawa and Lower Tentokuji formations of depths 350–1750 m, and oil in the shallower levels has been biodegraded.

Recently, the main producing oil and gas fields in the Akita Basin are the Yurihara and Ayukawa oil and gas fields, whose reservoirs are in the 'Green Tuff' of the Nishikurosawa Formation, the siliceous shale of the Onnagawa Formation and the sandstone of the Funakawa Formation. Exploration for shale oil in the rocks of the Onnagawa Formation is planned.

Niigata Basin. Crude oil was being obtained from the Niigata area as far back as the year AD 668. The Amaze oil field was the site of the world's first production from a submarine accumulation in 1888. Modern development started after cable tool drilling was introduced in 1891. The main fluid produced was crude oil from turbidite sandstone in the early days, but is now natural gas originating in deep 'Green Tuff'. The source rock is argillaceous sediment of the Nanatani and Lower Teradomari formations with Type II and Type II–III kerogen, respectively (Sekiguchi *et al.* 1984). The reservoirs are turbidite sandstone of the Siiya and Nishiyama formations, both tuffaceous, with porosities of 25–35% and permeabilities of 10–1000 md; the 'Green Tuff' of the Nanatani Formation is a fracture-type reservoir, with porosity of 10–30% and permeability of 0.5–100 md. There are also non-reservoirs with low permeability (Japan Natural Gas Association & Japan Offshore Petroleum Development Association 1992). In other areas within the Niigata Basin, such as the Minaminagaoka–Katakai and Yoshii–Higashikashiwazaki gas fields, oil migrated into the 'Green Tuff' reservoir, where it matured and thermodegraded to wet gas

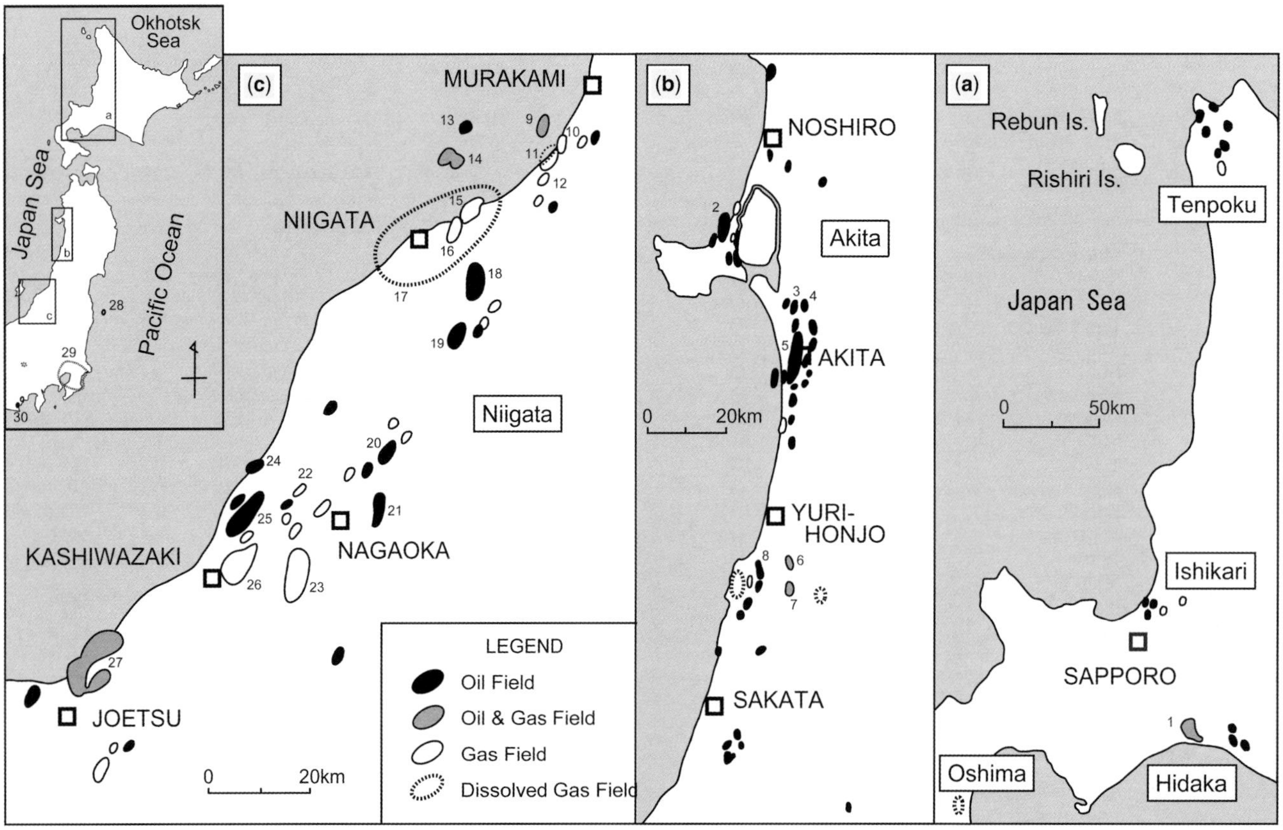

Fig. 10.7. Japanese petroleum provinces and fields. (**a**) Tenpoku and Ishikari–Hidaka basins; (**b**) Akita Basin; and (**c**) Niigata Basin. Other oil and gas fields in NE Japan are illustrated on the index map. Fields' numbers are the same as in Table 10.2.

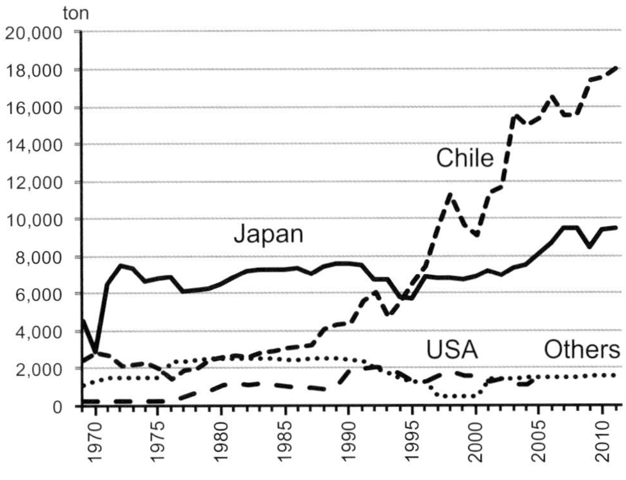

Fig. 10.8. World iodine production. Japan's recent increase includes recycling.

containing several percent CO_2 (Waseda 1993). The main reservoir of the Minaminagaoka–Katakai gas field is deeper than 4000 m, and its temperature is above 150°C. After the Nishiyama stage of basin formation in the Pliocene, the Niigata Basin evolved under a compressional stress field across the arc into a hydrocarbon generative 'kitchen' area during subsidence and accumulation in traps (Ogura 1983; Sekiguchi *et al.* 1984). The structural trend of the area is generally north–south to NNE–SSW, which is parallel to the coastline and mountain ranges.

Forearc basins in NE Japan

The source rocks for hydrocarbons in the forearc basins in NE Japan are Palaeogene non-marine sediments containing coal, and are distributed along the coast adjacent to the Pacific Ocean. They contain Type III kerogen. In the Palaeogene, these were forearc basins of the Eurasia continent. The Tenpoku, Ishikari–Hidaka, Kitakami–Sanriku and Joban basins are distributed from north to south and are described in the following sections.

Tenpoku and Ishikari–Hidaka basins in Hokkaido. Small oil and gas fields are distributed in the Tenpoku Basin, and their source rock is Palaeogene coal and coaly shale. The source rock of the oil fields in the Ishikari Basin is Miocene marine shale, whereas the sources of the eastern gas fields are Palaeogene coal and coaly shale (Waseda & Nishita 1998). The Yufutsu oil and gas field in the Hidaka Basin has reservoirs in the shallower Miocene Kawabata Formation and in deeper fracture-bearing granite and conglomerate. Their condensate is too waxy to flow at low temperatures. Because a pipeline was connected to Sapporo city in 1996, the amount of gas production has increased.

Sanriku and Joban basins off NE Japan. Submarine accumulations of hydrocarbons were discovered off Hachinohe in the Sanriku Basin, but have not yet been developed. The Iwaki-oki gas field in the Joban Basin has a reservoir of Oligocene–Miocene sandstone with an anticlinal trap in a horst (Funayama 1993), but production from it stopped in 2002.

East China Sea

The potential that hydrocarbons could be contained within the Neogene sediments below the continental shelf of the East China Sea was pointed out by an ECAFE (Economic Commission for Asia and the Far East) survey (Emery *et al.* 1969). Joint exploration between Japan and Korea was conducted off the western shore of Kyushu Island, but did not succeed. As China has developed some gas fields close to the intermediate line of the East China Sea, there might also be similar potential on the Japanese side.

Unconventional hydrocarbons

There are two unconventional hydrocarbon sources in Japan: water-dissolved gas and methane hydrate. Water-dissolved gas deposits supply methane by degassing from porewater pumped from below. If the water is a brine, it can contain iodine at high concentrations. Bottom simulating reflectance, which identifies the basements of submarine hydrates, can now demonstrate distributions of this potential resource surrounding Japan; exploratory drilling has confirmed their existence.

Water-dissolved iodine and gas deposits

Water-dissolved natural gas is in the form of methane, and is released from porewater by uplift and depressurization. The deposits are classified into two groups, namely those hosted in marine and non-marine sediments. Large deposits occur in marine sediments, and the depths of the reservoirs lie between several hundreds of metres and 2000 m. Dissolved methane typically originates from anaerobic microbes, but thermogenic gas might also be mixed at some concentration. The deposits are restricted to relatively young (post-Miocene) sediments, have experienced rapid uplift and have avoided being recharged by meteoric water, so that the fossil water remains. To prevent land subsidence, the total volume of uplifted water needs to be controlled so that gas production rate remains stable. There are currently about 500 producing wells.

Minami–Kanto gas field. Dissolved gas exists beneath part of the Kanto Plain, including downtown Tokyo. Development of these resources is however restricted in urban areas to prevent land subsidence resulting from the discharge of large volumes of water. However, pumping water up for hot springs is not prohibited because the amount is small, although gas explosions sometimes occur. The Kujukuri area facing the Pacific Ocean in Chiba Prefecture is the main gas-producing area.

There are gas-bearing aquifers in unconsolidated middle Miocene turbidite sandstones, with the main reservoirs being the Kiwada, Otadai and Umegase formations of the Kazusa Group. Production behaviour from these reservoirs can be divided into two types: normal and Mobara. The normal type has a steady gas/water ratio <4, whereas in Mobara-type production the gas/water ratio rises to >4 and, in extreme cases, to 40 after several years of development. This occurs when production decreases the reservoir pressure, inducing a continuous flow of gas from the siltstone to sandstone (Akibayashi & Zhou 1986; Tazaki 1988). The shallow reservoirs surrounding Mobara and Otaki show Mobara-type behaviour.

Gas production has remained at *c.* 4.5×10^8 m^3 a^{-1} because of restrictions on the volume of water that can be discharged. The ratio of reserves to yearly production is of the order several hundred. The natural gas produced is dry gas, composed mainly of methane with small amounts of nitrogen and CO_2 and trace amounts of C_{2+} hydrocarbons. Carbon and hydrogen isotope compositions have revealed that the methane originated from CO_2-reductive/H_2-philic methanogens (Kaneko *et al.* 2002).

Iodine occurs as I- in brine at concentrations >100 ppm, which is 2000 times more concentrated than seawater. The iodine was deposited as organic matter, and the brine contains large amounts of ammonium and humic substances. Japan has been the world's largest iodine producer since 1974, although recently its share has dropped to one-third of the world's production because Chile's output has been rising (Fig. 10.8). The amount produced has been steady for the same reason as dissolved gas production has remained constant, but recycling has allowed for some recent increase. The Minami–Kanto gas field supplies 90% of the iodine produced in Japan.

Niigata area. Conventional types of hydrocarbon accumulation are distributed across the Niigata area. The Higashi–Niigata and Nakajo oil and gas fields near the coastline have shallow reservoirs of microbial dissolved gas and iodine, whereas thermogenic oil and gas reside in a deep trap defined by an anticline and the pinching-out of a sandstone. The chemical and isotopic compositions of the gas from medium-depth show mixed gas characteristics (Kaneko *et al.* 2002; Waseda *et al.* 2011). The Niigata gas field, including the Higashi–Niigata gas field, is distributed over a wide area; development in urban areas has however been stopped because of the problem of land subsidence. All of the brine remaining after iodine extraction is reinjected underground.

Miyazaki and Okinawa areas. Neogene forearc basins have developed on the accreted basement rocks of Cretaceous–Palaeogene sandstone/siltstone sequences in both of these areas of SW Japan. Microbial gas and iodine are abundant in the Lower Pliocene rocks at Sadowara in the Miyazaki area and in the Upper Miocene rocks in the southern part of Okinawa-jima. These deposits include thermogenic gas dissolved in brine in the fractured basement and the pores of the overlying Miocene basal conglomerate and sandstone cover succession (Kato *et al.* 2011, 2012).

Methane hydrate

Gas hydrate has long been suspected to exist in the seas adjacent to Japan, and indeed its presence has now been confirmed (http://www.mh21japan.gr.jp/english/mh21-1/2-2/). The Nankai Trough is the most promising area and research into its development has been promoted by the government. The total amount of methane gas as hydrate in the eastern part of the Nankai Trough has been estimated to be 1.1×10^{12} m^3, increasing to *c.* 5.7×10^{11} m^3 in the more concentrated zones (Fujii *et al.* 2008). Whereas natural gas consisting of hydrates of microbial origin can occur in most submarine environments (Waseda & Uchida 2004), hydrates of thermogenic gas were discovered at the sea bottom off Joetsu in Niigata, an oil- and gas-producing area (Matsumoto *et al.* 2009).

Coal resources (YS)

The major coal fields in Japan are mainly distributed in Hokkaido and Kyushu, apart from Joban in central Honshu (Fig. 10.9). The coal beds of these fields formed in the Palaeogene except for those in the Tenpoku field of Neogene age. Small-scale coal beds present in the Chugoku and Shikoku districts are Mesozoic in age (Fig. 10.10). The majority of Japanese coal was produced from Kyushu before World War II, but coal production in Hokkaido later became competitive in the 1960s and has been dominant since the 1970s. Annual total production in Japan reached more than 50 Mt in 1940s and 1960s, although it drastically decreased after 1970 (Fig. 10.11). The last major coal mine was the Taiheiyo Mine in the Kushiro coal field which was operative until 2002; only small-scale operations are now conducted at a few open pits. In contrast

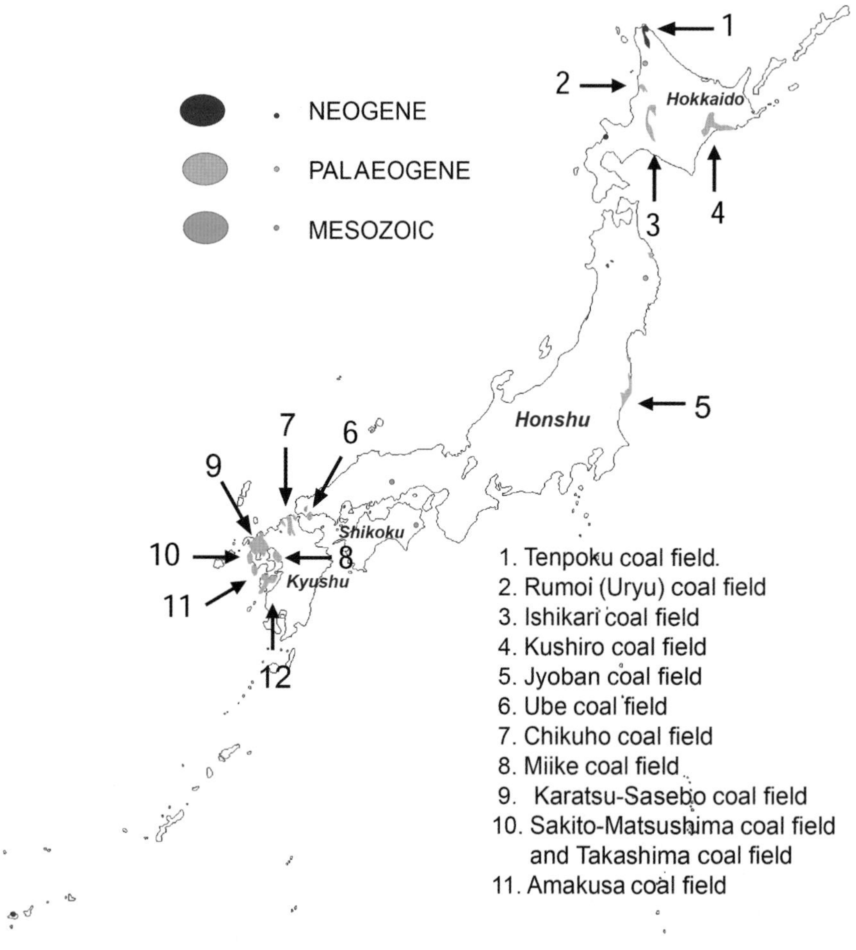

Fig. 10.9. Distribution of the major coal fields in Japan.

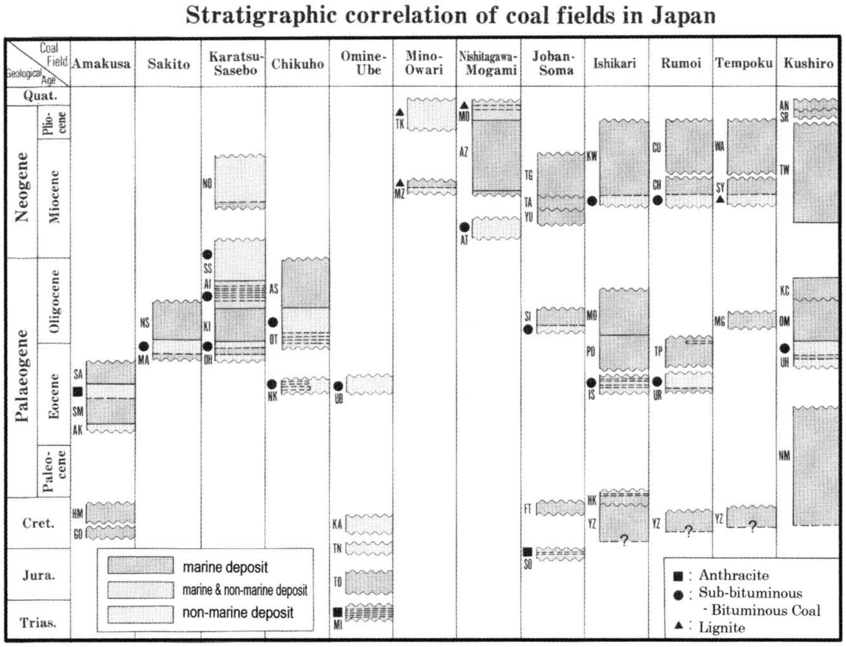

Amakusa Coal-field
SA: Sakasegawa Group
SM: Shimojima Group
AK: Akasaki Group
HM: Himenoura Group
GO: Goshonoura Group

Sakito Coal-field
NS: Nishisonoki Group
MA: Matsushima Group

Karatsu-Sasebo Coal-field
NO: Nojima Group
SS: Sasebo Group
AI: Ainoura Group
KI: Kijima Group
OH: Ohchi Group

Chikuho Coal-field
AS: Ashiya Group
OT: Ohtsuji Group
NK: Noukata Group

Omine-Ube Coal-field
UB: Ube Group
KA: Kanmon Group
TN: Toyonishi Group
TO: Toyora Group
MI: Mine Group

Joban-Soma Coal-field
TG: Taga Group
TA: Takaku Group
YU: Yunagaya Group
SI: Shiramizu Group
FT: Futaba Group
SO: Soma Group

Ishikari Coal-field
KW: Minenobu Formation
Oiwake Formation
Iwamizawa Formation
Kawabata Formation
Takinoue Formation
Asahi Formation
MO: Momijiyama Formation
PO: PoronaiFormation
IS: Ishikari Group
HK: Hakobuchi Group
YZ: Yezo Group

Rumoi Coal-field
CU: Mochikubetsu Formation
Enbetsu Formation
Chepotsunaizawa Formation
CI: Kotanbetsu Formation
Chikubetsu Formation
Haboro Formation
TP: Tappu Formation
UR: Uryu Formation
YZ: Yezo Group

Tempoku Coal-field
WA: Yuchi Formation
Koetoi Formation
Wakkanai Formation
SY: Masuporo Formation
Onishibetsu Formation
Sohya Formation
MG: Magaribuchi Formation
YZ: Yezo Group

Kushiro Coal-field
AN: Akan Group
SR: Shiranuka Formation
TW: Atsunai Formation
Chokubetsu Formation
Okkobezawa Formation
Tokiwa Formation
KC: Kamicharo Formation
OM: Ombetsu Group
UR: Urahoro Group
NM: Nemuro Group

Fig. 10.10. Stratigraphic correlation of coal-bearing horizons of major coal fields in Japan.

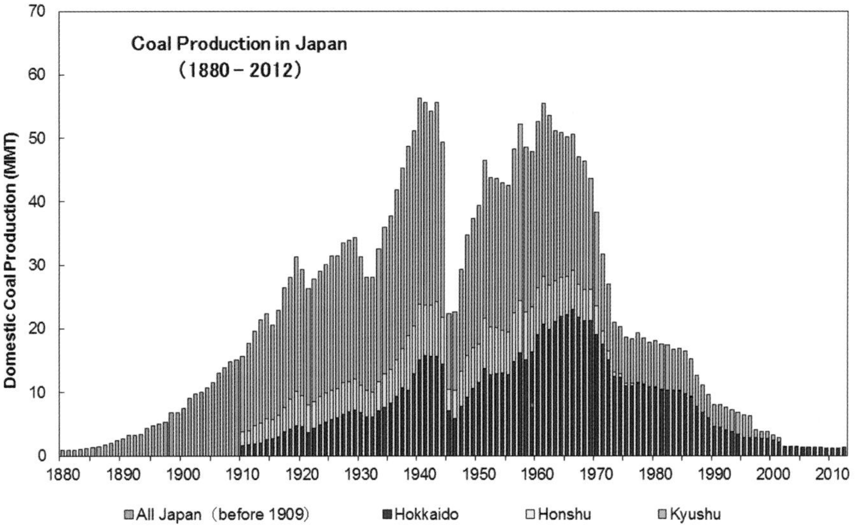

Fig. 10.11. Coal production in Japan. The production from 1880 to 1910 is not subdivided into the regions.

to this decline in domestic coal mining in Japan, the annual Japanese demand for coal has steadily increased to more than 175 Mt in 2013, mainly from the steel industry and for coal power generation.

Coal fields in Kyushu

Several coal fields, such as those of Chikuho, Fukuoka, Karatsu, Sasebo, Miike, Sakito–Matsushima, Takashima and Amakusa, are distributed across NW Kyushu (Fig. 10.9). Of these, the Karatsu and Sasebo fields have proven to be most continuous; the Sakito–Matsushima and Takashima fields are also relatively extensive.

The coal beds are hosted within non-marine horizons of Eocene and Oligocene sedimentary successions (Fig. 10.10). The major lithostratigraphic host units of the coal beds are the Shimojima Group in the Amakusa field, the Matsushima Group in the Sakito field, the Ochi, Ainuma and Sasebo Groups in the Karatsu-Sasebo field, the Ohmuta Group in the Miike field and the Nogata and Ohtsuji Groups in the Chikuho field.

The coals produced in the Chikuho field are >8100 kcal kg^{-1} in calorific power and are classified as bituminous. Those in the Karatsu–Sasebo field are classified as bituminous to sub-bituminous coal and those in the Miike and Sakido–Matsushima fields are bituminous coal, although some of the coals in the Miike field are high in sulphur content which is interpreted to result from ocean water contamination. The coal in the Amakusa field is anthracitic in quality, although the coal bed is <1 m in thickness.

Coal fields in Hokkaido

There are four major coal fields in Hokkaido. The largest of these is the Ishikari field (Fig. 10.9), where coal beds are hosted in several formations of the Eocene Ishikari Group (Fig. 10.10). Rocks of the Ishihari Group unconformably overlie the marine sediments of the Hakobuchi Group and underlie the mudstone of the Eocene Poronai Formation. The sedimentary sequence of the Ishikari Group is represented by a cyclic assemblage of non-marine, brackish, shallow marine and lacustrine deposits. The coal beds are heavily deformed by west-vergent, north-trending fold-and-thrust faults.

The Kushiro coal field extends from the Kushiro Plain to beneath the shallow sea region of the Pacific Ocean, and is hosted within the Upper Eocene Urahoro Group (Fig. 10.10). The coals in the Ishikari field and the Kushiro field are classified into bituminous and sub-bituminous, respectively.

Coal fields in Honshu

The Joban coal field extends in a north–south direction, with the coal beds in this field occurring in the Eocene Shiramizu Group and being sub-bituminous in grade. The Ube field, also in Honshu, is the eastern continuation of the Chikuho field in Kyushu, and yields poor-quality sub-bituminous coals.

Appendix

English to Kanji and Hiragana translations for geological and place names

Abukuma	阿武隈	あぶくま
Aichi	愛知	あいち
Akagane	赤金	あかがね
Akazawa	赤沢	あかざわ
Akenobe	明延	あけのべ

(Continued)

English to Kanji and Hiragana translations for geological and place names (Continued)

Akeshi	赤石	あけし
Akita	秋田	あきた
Akiyoshi-dai	秋吉台	あきよしだい
Amakusa	天草	あまくさ
Amaze	尼瀬	あまぜ
Ananai	穴内	あなない
Ani	阿仁	あに
Ashio	足尾	あしお
Awaji-shima	淡路島	あわじしま
Awashiro	粟代	あわしろ
Ayukawa	鮎川	あゆかわ
Bajo	馬上	ばじょう
Bandojima	坂東島	ばんどうじま
Beppu	別府	べっぷ
Besshi	別子	べっし
Bonin	ボニン	ぼにん
Chiba	千葉	ちば
Chichibu	秩父	ちちぶ
Chikuho	筑豊	ちくほう
Chitose	千歳	ちとせ
Daito	大東	だいとう
Fujigatani	藤ヶ谷	ふじがたに
Fukazawa	深沢	ふかざわ
Fukuoka	福岡	ふくおか
Funakawa	船川	ふなかわ
Funatsu	船津	ふなつ
Furutobe	古遠部	ふるとおべ
Ganidake	蟹岳	がにだけ
Gunma	群馬	ぐんま
Hachinohe	八戸	はちのへ
Hakobuchi	函淵	はこぶち
Hanaoka	花岡	はなおか
Hanawa	花輪	はなわ
Hatta	八田	はった
Hattahachiman	八田八幡	はったはちまん
Hida	飛騨	ひだ
Hidaka	日高	ひだか
Higashi-Niigata	東新潟	ひがしにいがた
Higashikashwazaki	東柏崎	ひがしかしわざき
Higashiyama	東山	ひがしやま
Hirase	平瀬	ひらせ
Hirose	広瀬	ひろせ
Hiroshima	広島	ひろしま
Hishikari	菱刈	ひしかり
Hitachi	日立	ひたち
Hoei	豊栄	ほうえい
Hokkaido	北海道	ほっかいどう
Hokubo	北房	ほくぼう
Hokuriku	北陸	ほくりく
Hokuroku	北麓	ほくろく
Honshu	本州	ほんしゅう
Hoshino	星野	ほしの
Hosokura	細倉	ほそくら
Hyogo	兵庫	ひょうご
Ichinokawa	市ノ川	いちのかわ
Iimori	飯盛	いいもり
Ikuno	生野	いくの
Ikutawara	生田原	いくたわら
Inakuraishi	稲倉石	いなくらいし
Ishikari	石狩	いしかり
Itoigawa	糸魚川	いといがわ
Itomuka	イトムカ	いとむか

(Continued)

English to Kanji and Hiragana translations for geological and place names (Continued)

Iwaki-oki	磐城沖	いわきおき
Iwami	石見	いわみ
Iwato	岩戸	いわと
Iwo-jima (Ioto)	硫黄島	いおうじま
Izu	伊豆	いず
Joban	常磐	じょうばん
Joetsu	上越	じょうえつ
Jokoku	上国	じょうこく
Kagoshima	鹿児島	かごしま
Kamaishi	釜石	かまいし
Kamioka	神岡	かみおか
Kamuikotan	神居古潭	かむいこたん
Kanamaru	金丸	かなまる
Kaneuchi	鐘打	かねうち
Kanto	関東	かんとう
Karatsu	唐津	からつ
Kasuga	春日	かすが
Katakai	片貝	かたかい
Kawabata	川端	かわばた
Kawasaki	川崎	かわさき
Kawayama	河山	かわやま
Kazusa	上総	かずさ
Kishu	紀州	きしゅう
Kitakami	北上	きたかみ
Kitami	北見	きたみ
Kitano-o	北の王	きたのおう
Kiwada*	黄和田	きわだ
Kiwada†	喜和田	きわだ
Kohoku	興北	こうほく
Komaki	小馬木	こまき
Komori	河守	こうもり
Konomai	鴻之舞	こうのまい
Kosaka	小坂	こさか
Kucchan	倶知安	くっちゃん
Kuga	玖珂	くが
Kujukuri	九十九里	くじゅうくり
Kunimiyama	国見山	くにみやま
Kushikino	串木野	くしきの
Kushiro	釧路	くしろ
Kussharo	屈斜路	くっしゃろ
Kyushu	九州	きゅうしゅう
Makimine	槇峰	まきみね
Maruyama	円山	まるやま
Mashu	摩周	ましゅう
Matsushima	松島	まつしま
Miike	三池	みいけ
Mikawa	三河	みかわ
Minami-Kanto	南関東	みなみかんとう
Minaminagaoka	南長岡	みなみながおか
Mino	美濃	みの
Misaka	三坂	みさか
Miyagi	宮城	みやぎ
Miyazaki	宮崎	みやざき
Mobara	茂原	もばら
Mochikoshi	持越	もちこし
Mozumi	茂住	もずみ
Mukawa	鵡川	むかわ
Myoho	妙法	みょうほう
Myojin	明神	みょうじん
Myojinsho	明神礁	みょうじんしょう
Nabeyama	鍋山	なべやま
Nagato	長門	ながと

(Continued)

English to Kanji and Hiragana translations for geological and place names (Continued)

Nagoya	名古屋	なごや
Nakajo	中条	なかじょう
Nakase	中瀬	なかせ
Nakatatsu	中竜	なかたつ
Nanatani	七谷	ななたに
Nankai	南海	なんかい
Nansei	南西	なんせい
Natsume	夏梅	なつめ
Niigata	新潟	にいがた
Nirou	韮生	にろう
Nishikurosawa	西黒沢	にしくろさわ
Nishiyama	西山	にしやま
Nitto	日東	にっとう
Noda	野田	のだ
Nogata	直方	のおがた
Noto	能登	のと
Nurukawa	温川	ぬるかわ
Obie	帯江	おびえ
Ochi	相知	おうち
Ochiai	落合	おちあい
Oe	大江	おおえ
Ogasawara	小笠原	おがさわら
Ogayu	大ヶ生	おおがゆ
Ohmidani	大見谷	おおみだに
Ohmuta	大牟田	おおむた
Ohori	大堀	おおほり
Ohtsuji	大辻	おおつじ
Oki	隠岐	おき
Okinawa	沖縄	おきなわ
Okinawa-jima	沖縄島	おきなわじま
Okuchi	大口	おおくち
Okuki	大久喜	おおくき
Omine	大峰	おおみね
Omodani	面谷	おもだに
Onnagawa	女川	おんながわ
Osarizawa	尾去沢	おさりざわ
Oshima	渡島	おしま
Osumi	大隅	おおすみ
Otadai	大田代	おおただい
Otaki	大多喜	おおたき
Otani	大谷	おおたに
Oya‡	大谷	おおや
Oya§	大屋	おおや
Poronai	幌内	ぽろない
Rendaiji	蓮台寺	れんだいじ
Rengeji	蓮花寺	れんげじ
Ryoke	領家	りょうけ
Ryukyu	琉球	りゅうきゅう
Sado	佐渡	さど
Sadowara	佐土原	さどわら
Sakito	崎戸	さきと
San'in	山陰	さんいん
San'yo	山陽	さんよう
Sanbagawa	三波川	さんばがわ
Sangun	三郡	さんぐん
Sanriku	三陸	さんりく
Sanru	珊瑠	さんる
Sapporo	札幌	さっぽろ
Sasagatani	笹ヶ谷	ささがたに
Sasebo	佐世保	させぼ
Sazare	佐々連	さざれ
Seigoshi	清越	せいごし

(Continued)

English to Kanji and Hiragana translations for geological and place names (Continued)

Seikyu	清久	せいきゅう
Seto	瀬戸	せと
Shakanai	釈迦内	しゃかない
Shiiya	椎谷	しいや
Shikoku	四国	しこく
Shimabara	島原	しまばら
Shimane	島根	しまね
Shimanto	四万十	しまんと
Shimojima	下島	しもじま
Shimokawa	下川	しもかわ
Shinnitto	新日東	しんにっとう
Shiramizu	白水	しらみず
Shirataki	白滝	しらたき
Shiroyama	城山	しろやま
Shizuoka	静岡	しずおか
Shokozan	勝光山	しょうこうざん
Sorachi	空知	そらち
Sumisu	須美寿	すみす
Tada	多田	ただ
Taio	鯛生	たいお
Tajimi	多治見	たじみ
Takara	宝	たから
Takashima	高島	たかしま
Takatama	高玉	たかたま
Takeno	竹野	たけの
Takinotani	滝の谷	たきのたに
Tamagawa	玉川	たまがわ
Tanakura	棚倉	たなくら
Tanba (Tamba)	丹波	たんば
Tari	多里	たり
Taro	田老	たろう
Teine	手稲	ていね
Tengumori	天狗森	てんぐもり
Tenpoku	天北	てんぽく
Tentokuji	天徳寺	てんとくじ
Teradomari	寺泊	てらどまり
Tochibora	栃洞	とちぼら
Toi	土肥	とい
Tonoda	殿田	とのだ
Toyoha	豊羽	とよは
Tsukinuno	月布	つきぬの
Tsumo	都茂	つも
Tsushima	対馬	つしま
Ube	宇部	うべ
Umanotani	馬谷	うまのたに
Umegase	梅ヶ瀬	うめがせ
Urahoro	浦幌	うらほろ
Wakamatsu	若松	わかまつ
Wanibuchi	鰐淵	わにぶち
Yabase	八橋	やばせ
Yabu	養父	やぶ
Yakuki	八茎	やくき
Yamagano	山ヶ野	やまがの
Yamagata	山形	やまがた
Yamato	大和	やまと
Yanahara	柵原	やなはら
Yano	矢野	やの
Yezo	蝦夷	えぞ
Yoshii	吉井	よしい
Yoshioka	吉岡	よしおか
Yufutsu	勇払	ゆうふつ
Yugashima	湯ヶ島	ゆがしま

(*Continued*)

English to Kanji and Hiragana translations for geological and place names (Continued)

Yurihara	由利原	ゆりはら
Zomeki	蔵目喜	ぞうめき

*For 'Kiwada gas field'.
†For 'Kiwada tangsten skarn'.
‡For 'Oya in the Abukuma Belt'.
§For 'Oya in the San'yo Belt'.

References

Akibayashi, S. & Zhou, P. 1986. A numerical model for simulating production performance in the Mobara type water-dissolved natural gas field. *Journal of Japanese Association for Petroleum Technology*, **51**, 487–491 [in Japanese with English abstract].

Amano, K. 1991. Multiple collision tectonics of the south Fossa Magna in central Japan. *Modern Geology*, **15**, 315–329.

ANRE (Agency of Natural Resources and Energy) 2014. Statistic data on the operative situation of rock aggregate and building stone industries in Heisei 25. http://www.enecho.meti.go.jp/category/resources_and_fuel/mineral_resources/situation/004/pdf/h25-141202.pdf [in Japanese].

Arai, S. 1975. Contact metamorphosed dunite-harzburgite complex in the Chugoku district, Western Japan. *Contributions to Mineralogy and Petrology*, **52**, 1–16.

Arai, S. & Yurimoto, H. 1994. Podiform chromitites of the Tari-Misaka ultramafic complex, Southwestern Japan, as mantle-melt interation products. *Economic Geology*, **89**, 1279–1288.

Bamba, T. 1984. A model illustrating the formation process of the podiform chroite deposits in some Alpine orogenic terrains. *In*: Wauschkuhn, A., Kluth, C. & Zimmermann, R. A. (eds) *Syngenesis and Epigenesis in the Formation of Mineral Deposits*. Springer-Verlag, Berlin, 507–518.

Baranov, B., Wong, H. K., Dozorova, K., Karp, B., Lüdmann, T. & Karnaukh, V. 2002. Opening geometry of the Kurile Basin (Okhotsk Sea) as inferred from structural data. *Island Arc*, **11**, 206–219.

Emery, K. O., Hayashi, Y. *et al.* 1969. Geological structure and some water characteristics of the East China Sea and the Yellow Sea. *CCOP Technical Bulletin*, **2**, 3–43.

Fujii, T., Saeki, T. *et al.* 2008. Resource assessment of methane hydrate in the Eastern Nankai Trough, Japan. *Proceedings of 2008 Offshore Technology Conference*, OTC19310, 5–8 May, Houston.

Fujinaga, K. & Kato, Y. 2005. Radiolarian age of red chert from the Kunimiyama ferromanganese deposit in the Northern Chichibu belt, central Shikoku, Japan. *Resource Geology*, **55**, 353–356.

Fujinaga, K., Nozaki, T., Nishiuchi, T., Kuwahara, K. & Kato, Y. 2006. Geochemistry and origin of Ananai stratiform manganese deposit in the Northern Chichibu belt, central Shikoku, Japan. *Resource Geology*, **56**, 399–414.

Funayama, M. 1993. Iwaki-oki gas field. *In*: *Explolation and Development of Hydrocarbon in Japan*. Japanese Association for Petroleum Technology, Tokyo, 18–19 [in Japanese].

Furuno, M., Ito, K. & Mariko, T. 1992. Polymetallic and gold-silver mineralizations in and around the Akenobe ore deposits, southwest Japan. *Mining Geology*, **42**, 33–46 [in Japanese with English abstract].

Garwin, S., Hall, R. & Watanabe, Y. 2005. Tectonic setting, geology, and gold and copper mineralization in Cenozoic magmatic arcs of southeast Asia and the west Pacific. *Society of Economic Geologists*, 100th Anniversary Volume, 891–930.

Glasby, G. P., Iizasa, K., Yuasa, M. & Usui, A. 2000. Submarine hydrothermal mineralization on the Izu-Bonin arc, south of Japan: an overview. *Marine Georesources and Geotechnology*, **18**, 141–176.

Hamasaki, S. 2009. Quaternary volcanic-related acid hydrothermal alteration in the Ugusu silica stone deposit, western Izu peninsula, central Japan: geology and alteration. *Resource Geology*, **59**, 153–169.

Hamasaki, S. & Sudo, S. 1999. Potterystone of Amakusa area, Kumamoto prefecture. *Chishitsu-News*, **538**, 38–47 [in Japanese].

Haruna, M., Ueno, H. & Ohmoto, H. 1990. Development of skarn-type ores at the Tengumori copper deposit of the Kamaishi Mine, Iwate Prefecture, northeastern Japan. *Mining Geology*, **40**, 223–244.

Hida, T., Ishiyama, D., Mizuta, T. & Ishikawa, Y. 1996. Geologic characteristics and formation environments of the Yano-Shokozan pyrophyllite deposit, Hiroshima prefecture, Japan: volcanic successions and hydrothermal alterations processes. *Nendo-Kagaku*, **36**, 62–72 [in Japanese with English abstract].

Hirai, A. 1982. Organic facies in the Akita-Niigata oil field. *Researches in Organic Geochemistry*, **3**, 46–49 [in Japanese].

Hirano, H. 2009. Review on chromium, its resources and uses. *Chishitsu News*, **664**, 37–42 [in Japanese].

Hirano, H., Higashimoto, S. & Kamitani, M. 1978. Geology and chromite deposits of the Tari district, Tottori prefecture. *Bulletin of the Geological Survey of Japan*, **29**, 61–71.

Hoshi, H. & Takahashi, M. 1999. Miocene counterclockwise rotation of Northeast Japan: a review and new model. *Bulletin of the Geological Survey of Japan*, **50**, 3–16.

Iizasa, K., Fiske, R. S. et al. 1999. A Kuroko-type polymetallic sulfide deposit in a submarine silicic caldera. *Science*, **283**, 975–977.

Ikeda, Y. 1991. Geochemistry and magmatic evolution of Pliocene-early Pleistocene pyroclastic flow deposits in central Hokkaido, Japan. *Journal of the Geological Society of Japan*, **97**, 645–666.

Ishihara, S. 1971. *Major Molybdenum Deposits and Related Granitic Rocks*. Report of the Geological Survey of Japan, No. 239 [in Japanese with English abstract].

Ishihara, S. 1977. The magnetite-series and ilmenite-series granitic rocks. *Mining Geology*, **27**, 293–305.

Ishihara, S. 1990. The Inner Zone batholith v. the Outer Zone batholith in Japan: evaluation from their magnetic susceptibility. *The University Museum, University of Tokyo, Nature and Culture, Tokyo*, **2**, 21–34.

Ishihara, S. 1992. Resource Geology: the Year 2000. *Resource Geology Special Issue*, **13**, 1–4 [in Japanese with English abstract].

Ishihara, S. 1993. Stone mining in Japan. *Resource Geology*, **43**, 387–396.

Ishihara, S. & Chappell, B. W. 2008. Chemical compositions of the late Cretaceous granitoids across the central part of the Abukuma Highland, Japan-Revised. *Bulletin of the Geological Survey of Japan*, **59**, 151–170.

Ishihara, S. & Matsueda, H. 1997. Genesis of two contrasting metallogenic provinces in the back-arc basins of Hokkaido, Japan. *Proceedings of the 30th International Geological Congress*. 12 August 1996, Beijing, **9**, 3–13.

Ishihara, S. & Murakami, H. 2004. Granitoid types related to Cretaceous plutonic Au-quartz vein and Cu-Fe skarn deposits, Kitakami Mountains, Japan. *Resource Geology*, **54**, 281–298.

Ishihara, S. & Sasaki, A. 1991. Ore deposits related to granitic magmatism in Japan: a magmatic viewpoint. *Episodes*, **14**, 286–292.

Ishihara, S. & Sasaki, A. 2002. Paired sulfur isotopic belts: Late Cretaceous-Paleogene ore deposits of Southwest Japan. *Bulletin of the Geological Survey of Japan*, **53**, 461–477.

Ishihara, S., Terashima, S. & Tsukimura, K. 1987. Spatial distribution of magnetic susceptibility and ore elements, and cause of local reduction on magnetite-seires granitoids and related ore deposits at Chichibu, central Japan. *Mining Geology*, **37**, 15–28.

Ishihara, S., Sasaki, A. & Terashima, S. 1988*a*. Sulfur in granitoids and its role for mineralization. *Proceedings of the Seventh Quadrennnial IAGOD Symposium*. 18–22 August 1986, Luleå, 573–581.

Ishihara, S., Shibata, K. & Utsumi, S. 1988*b*. K-Ar ages of ore deposits related to Cretaceous-Paleogene granitoids-Summary in 1987. *Bulletin of the Geological Survey of Japan*, **39**, 81–94.

Ishihara, S., Shibara, K. & Uchiumi, S. 1989. K-Ar age of molybdenum mineralization in the east-central Kitakami Mountains, Northern Honshu, Japan: comparison with the Re-Os age. *Geochemical Journal*, **23**, 85–89.

Ishihara, S., Shibata, K. & Terashima, S. 1990. Alkalinity and initial $^{87}Sr/^{86}Sr$ ratios of igneous rocks related to late Cenozoic gold mineralization of the Ryukyu arc, Japan. *CCOP Technical Bulletin*, **21**, 1–16.

Ishihara, S., Hamano, K. & Ikegami, A. 1998. Isotopic evaluation on the genesis of the Kanamaru pegmatite deposit, Niigata prefecture, Japan. *Resource Geology*, **48**, 1–6 [in Japanese with English abstract].

Ishihara, S., Yamamoto, M. & Sasaki, A. 1999. Sulfur and carbon contents and $\delta^{34}S$ ratio of Miocene ilmenite-series granitoids: Osumi and Shibisan plutons, Kyushu, SW Japan. *Bulletin of the Geological Survey of Japan*, **50**, 671–682.

Ishizuka, O., Uto, K., Yuasa, M. & Hochstaedter, A. G. 1998. K-Ar ages from seamount chains in the back-arc region of the Izu-Ogasawara arc. *Island Arc*, **7**, 408–421.

Isozaki, Y. & Maruyama, S. 1991. Studies on orogeny based on plate tectonics in Japan and a new geotectonic subdivision of the Japanese islands. *Journal of Geography*, **100**, 697–761 [in Japanese with English abstract].

Isozaki, Y., Aoki, K., Nakama, T. & Yanai, S. 2010. New insight into a subduction-related orogen: a reappraisal of the geotectonic framework and evolution of the Japanese Islands. *Gondwana Research*, **18**, 82–105.

Ito, Y., Tsutsumi, H., Yamamoto, H. & Arato, H. 2002. Active right-lateral strike-slip fault zone along the southern margin of the Japan Sea. *Tectonophysics*, **351**, 301–314.

Izawa, E. & Urashima, Y. 1987. Geologic and tectonic setting of the epithermal gold and geothermal areas in Kyushu. *In*: Urashima, Y. (ed.) *Gold Deposits and Geothermal Fields in Kyushu*. Society of Mining Geologists of Japan, Tokyo, Guidebook, 2, 1–12.

Izawa, E. & Watanabe, K. 2001. Overview of epithermal gold mineralization in Kyushu, Japan. *SEG Guidebook Series*, **34**, 11–15.

Japan Natural Gas Association 2012. Annual report of natural gas data in 2011 [in Japanese].

Japan Natural Gas Association & Japan Offshore Petroleum Development Association 1992. *Oil and Natural Gas Resources in Japan (revised)*. Japan Natural Gas Association, Tokya & Japan Offshore Petroleum Development Association, Tokyo [in Japanese].

Kamata, H. & Kodama, K. 1999. Volcanic history and tectonics of the Southwest Japan Arc. *Island Arc*, **8**, 393–403.

Kaneko, N., Maekawa, T. & Igari, S. 2002. Generation of archaeal methane and its accumulation mechanism into interstitial water. *Journal of Japanese Association for Petroleum Technology*, **67**, 97–110 [in Japanese with English Abstract]

Kase, K. 2003. *Genesis and Classification of Besshi-type Cu Deposits*. Resource and Environmental Geology, Society of Resource Geology, Tokyo, 87–94 [in Japanese].

Kase, K. & Yamamoto, M. 1985. Geochemical study of conformable massive sulfide deposits of the Hitachi Mine, Ibaraki Prefecture, Japan. *Mining Geology*, **35**, 17–29.

Kato, S., Waseda, A. & Iwano, H. 2011. Geochemistry of natural gas and formation water from water-dissolved gas field in Miyazaki Prefecture. *Journal of Japanese Association for Petroleum Technology*, **76**, 244–253 [in Japanese with English abstract].

Kato, S., Honda, T. & Omiiya, K. 2012. Petroleum geology of the Nanjo R1 exploratory well, Okinawa Prefecture. *Journal of Japanese Association for Petroleum Technology*, **77**, 86–95 [in Japanese with English abstract].

Kato, Y. 1999. Genesis of the Kamioka skarn deposits: an important role of clinopyroxene skarn and graphite-bearing limestone in precipitating sulfide ore. *Resource Geology*, **49**, 213–222.

Kato, Y., Fujinaga, K., Nozaki, T., Osawa, H., Nakamura, K. & Ono, R. 2005. Rare earth, major and trace elements in the Kunimiyama ferromanganese deposit in the Northern Chichibu belt, central Shikoku, Japan. *Resource Geology*, **55**, 291–299.

Kawano, M. & Tomita, K. 1990. Mineralogical properties and formation process of kaolinite in the Iriki kaolin deposit, Kagoshima prefecture, Japan. *Nendo-Kagaku*, **30**, 229–239 [in Japanese with English abstract].

Kihara, S., Hoshino, K., Watanabe, M., Nishido, H. & Ishihara, S. 2005. K-Ar ages of granitic magmatism and related pegmatite formation at the Umanotani-Shiroyama mine, Shimane prefecture, SW Japan and their bearings on cooling history. *Resource Geology*, **55**, 123–129 [in Japanese with English abstract].

Kimura, G., Miyashita, S. & Miyasaka, S. 1983. Collision tectonics in Hokkaido and Sakhalin. *In*: Hashimoto, M. & Uyeda, S. (eds) *Accretion Tectonics in the Circum Pacific Regions*. Terrapub, Tokyo, 123–134.

Kimura, J.-I., Kunikiyo, T. et al. 2003. Late Cenozoic volcanic activity in the Chugoku area, southwest Japan arc during back-arc basin opening and reinitiation of subduction. *Island Arc*, **12**, 22–45.

Kinoshita, K. 1920. Mineral deposits of the Takara mine. *Journal of Geological Society of Japan*, **75**, 127–140 [in Japanese].

Kitagawa, R., Kakitani, S. & Funaki, A. 1982. Mica clay minerals occurring in the sericite deposits at the Mitoya district, Iishi-gun, Shimane prefecture. *Nendo-Kagaku*, **22**, 54–67 [in Japanese with English abstract].

Kodama, K., Tashiro, H. & Takeuchi, T. 1995. Quaternary counterclockwise rotation of south Kyushu, southwest Japan. *Geology*, **23**, 823–826.

KOMURO, K., YAMAGUCHI, K. & KAJIWARA, Y. 2005. Chemistry and sulfur isotopes in a chert-dominant sequence around the stratiform manganese deposit of the Noda-Tamagawa Mine, Northern Kitakami Terrane, Northeast Japan: implication for paleoceanograhic environmental setting. *Resource Geology*, **55**, 337–351.

LIMESTONE ASSOCIATION OF JAPAN 2012. History of production and shipping of limestone in Japan. http://www.limestone.gr.jp/doc/toukei/pdf/toukei2012.pdf [in Japanese]

MARIKO, T. 1984. Sub-sea hydrothermal alteration of basalt, diabase and sedimentary rocks in the Shimokawa copper mining area, Hokkaido, Japan. *Mining Geology*, **34**, 307–321.

MARIKO, T. 1988*a*. Ores and ore minerals from the volcanogenic massive sulfide deposits of the Shimokawa Mine, Hokkaido, Japan. *Mining Geology*, **38**, 233–246.

MARIKO, T. 1988*b*. Compositional zoning and 'chalcopyrite disease' in sphalerite contained in the Pb-Zn-quartz-calcite ore from the Mozumi deposit of the Kamioka mine, Gifu Prefecture, Japan. *Mining Geology*, **38**, 393–400 [in Japanese with English abstract].

MARIKO, T. & KATO, Y. 1994. Host rock geochemistry and tectonic setting of some volcanogenic massive sulfide deposits in Japan: examples of the Shimokawa and the Hitachi ore deposit. *Resource Geology*, **44**, 353–367.

MARUMO, K. & HATTORI, K. 1999. Seafloor hydrothermal clay alteration at Jade in the back-arc Okinawa Trough: mineralogy, geochemistry and isotope characteristics. *Geochimica et Cosmochimica Acta*, **63**, 2785–2804.

MATSUMOTO, I. & ARAI, S. 2001. Morphological and chemical variations of chromian spinel in dunite-harzburgite complexes from the Sangun zone 8SW Japan): implications for mantle/melt reaction and chromitite formation processes. *Mineralogy and Petrology*, **73**, 305–323.

MATSUMOTO, I. & SUZUKI, K. 2002. Re-Os age and isotopic constraints on genesis of Tari-Misaka ultramafic complex, of the Sangun zone, Southwest Japan. *Abstract of 4th International Workshop on Orogenic lherzolite and Mantle Processes*, Samani, Hokkaido.

MATSUMOTO, R., OKUDA, Y. *ET AL.* 2009. Formation and collapse of gas hydrate deposits in high methane flux area of the Joetsu basin, eastern margin of Japan Sea. *Journal of Geography*, **118**, 43–71 [in Japanese with English abstract].

MIC (MINISTRY OF INTERNAL AFFAIRS AND COMMUNICATIONS), STATISTICS BUREAU 2012. Economy census and activity survey in Heisei 24. http://www.stat.go.jp/index.htm [in Japanese].

MINISTRY OF INTERNATIONAL TRADE AND INDUSTRY 1987. *Report of Regional Geological Structure Survey in the Sado Area During the Fiscal Year Showa 61*. Ministry of International Trade and Industry, Tokyo [in Japanese].

MIYASHITA, S. & WATANABE, Y. 1988. Genetic environments of the greenstones in the Hidaka Zone, Hokkaido, with special reference to the metallogeny of massive sulfide, and bedded iron and manganese ores. *In*: BAMBA, T. & TOGARI, K. (eds) *Wall Rock Alteration and Ore Genesis*. Society of Mining Geologists of Japan, Tokyo, Mining Geology Special Issue, **12**, 93–104.

NAGAHARA, M. & SHIMA, H. 1992. Kiwada tungsten mine, Iwakuni City, Yamaguchi Prefecture. *Chishitsu News*, **460**, 13–20 [in Japanese with English abstract].

NAKAJIMA, S., SHUTO, K., KAGAMI, H., OHKI, J. & ITAYA, T. 1995. Across-arc chemical and isotopic variation of Late Miocene to Pliocene volcanic rocks from the Northeast Japan arc. *Memoirs of the Geological Society of Japan*, **44**, 197–226 [in Japanese with English abstract].

NAKAMURA, T. 1990. Pre-Cretaceous strata-bound ore deposits. *In*: ICHIKAWA, K., MIZUTANI, S., HARA, I., HADA, S. & YAO, A. (eds) *Pre-Cretaceous Terranes of Japan*. IGCP Project No. 224, Osaka City University Pub., Osaka, 381–399.

NAKAYAMA, K. 2003. Internal structure of the Shimokawa greenstone-argillaceous sediment complex-reconstruction of the paleo-sedimented spreading center. *Shigen-Chishitsu*, **53**, 81–94 [in Japanese with English abstract].

NOZAKI, T., NAKAMURA, K., AWAJI, S. & KATO, Y. 2006. Whole-rock geochemistry of basic schists from the Beshi area, central Shikoku: implications for the tectonic setting of the Besshi sulfide deposit. *Resource Geology*, **56**, 423–432.

NOZAKI, T., KATO, Y. & SUZUKI, K. 2010. Re-Os geochronology of the Iimori Besshi-type massive sulfide deposit in the sanbagawa metamorphic belt, Japan. *Geochimica et Cosmochimica Acta*, **74**, 4322–4331.

NOZAKI, T., KATO, Y., SUZUKI, K., TAKAYA, Y. & NAKAYAMA, K. 2011. Re-Os ages of Besshi-type massive sulfide deposits associated with in-situ basalt as a new age constraint for ridge subduction. *21st VM Goldschmidt Conference, Mineralogical Magazine*, August, 2011, Prague, **75**, 1553.

NOZAKI, T., KATO, Y., SUZUKI, K. & KASE, K. 2012. Re-Os geochronology of the Fudotaki and Fujimi deposits in the Hitachi mine, Ibaraki Prefecture: the oldest ore deposit in the Japanese Island. *Abstracts with Programs, The Society of Resource Geology*, 27–29 June 212, Tokyo, Japan, O-31, 57.

OGURA, N. 1983. Estimation of thermal maturation of sedimentary rocks in the Niigata basin, northeast Japan Part 1: estimation of paleotemperature gradients. *Journal of Japanese Association for Petroleum Technology*, **48**, 227–238 [in Japanese with English abstract].

OHMOTO, H., TANIMURA, S., DATE, J. & TAKAHASHI, T. 1983. Geologic setting of the Kuroko deposits, Japan. *Economic Geology Monograph*, **5**, 9–54.

OKAMURA, Y. 1999. Sericite deposit of Awashiro mine, Aichi Prefecture: the characteristics and utility of sericite. *Chishitsu-News*, **540**, 49–53 [in Japanese].

SAITO, T., TAKAHASHI, T. & NISHIKAWA, Y. 2002. *Physicochemical Research on Old Coins*. IMES Discussion Paper Series, Institute for Monetary and Economic Studies, Tokyo, 2002-J-30.

SAKURAI, W., OKADA, Y. & MIZUYACHI, O. 1993. On the exploration of the Atotsugawa district in the Kamioka mining area, central Japan. *Resource Geology*, **43**, 79–91 [in Japanese with English abstract].

SATO, H. 1994. The relationship between late Cenozoic tectonic events and stress field and basin development in northeast Japan. *Journal of Geophysical Research*, **99**, 22261–22274.

SATO, K. & KASE, K. 1996. Pre-accretionary mineralization of Japan. *Island Arc*, **5**, 216–228.

SATO, T. 1974. Distribution and geological setting of the Kuroko deposits. *Mining Geology Special Issue*, **6**, 1–9.

SAWAI, O. & ITAYA, T. 1993. K-Ar ages of kuroko-type deposits in the Shakotan-Toya district, southwest Hokkaido, Japan. *Resource Geology*, **43**, 165–172 [in Japanese with English abstract].

SEKIGUCHI, K., OMOKAWA, M., HIRAI, A. & MIYAMOTO, Y. 1984. Geochemical study of oil and gas accumulation in 'Green Tuff' reservoir in Nagaoka to Kashiwazaki region. *Journal of Japanese Association for Petroleum Technology*, **49**, 56–64 [in Japanese with English abstract].

SHIBUE, Y., FUJIOKA, K. & ISHII, T. 1994. Chemical compositions of skarn minerals from the Kiwada Tungsten deposit, Yamaguchi Prefecture. *Hyogo University of Teacher Education Journal*, **14**, 25–34.

SHIMAZAKI, H. & KUSAKABE, M. 1990. Oxygen isotope study of the Kamioka Zn-Pb skarn deposits, central Japan. *Mineralium Deposita*, **25**, 221–229.

SHINJO, R. 1999. Geochemistry of high Mg andesites and the tectonic evolution of the Okinawa Trough-Ryukyu arc system. *Chemical Geology*, **157**, 69–88.

SHINJO, R. & KATO, Y. 2000. Geochemical constraints on the origin of bimodal magmatism at the Okinawa Trough, an incipient back-arc basin. *Lithos*, **54**, 117–137.

SIBUET, J., DEFFONTAINES, B., HSU, S., THAREAU, N., LE FORMAL, J. P., LIU, C. S. & ACT PARTY 1998. Okinawa trough back-arc basin: early tectonic and magmatic evolution. *Journal of Geophysical Research*, **103**, 30245–30267.

SUDO, S. 1999. Roofing tiles: their type, form and origin. *Chishitsu-News*, **536**, 39–50 [in Japanese].

SUDO, S. 2000*a*. Weathered granite and feldspar resources in Mino-Mikawa kogen area, Central Japan. *Chishitsu-News*, **554**, 39–43 [in Japanese].

SUDO, S. 2000*b*. Feldspar resources of Nagiso district, Nagano prefecture, Central Japan: Origin and occurrence of Ohira choseki. *Chishitsu-News*, **555**, 12–17 [in Japanese].

SUDO, S. 2000*c*. Industrial minerals '99 of Tokai area, Central Japan. *Chishitsu-News*, **552**, 23–29 [in Japanese].

SUDO, S. 2001. Feldspar and feldspathic resources in Japan. *Chishitsu-News*, **559**, 50–58 [in Japanese].

SUDO, S. 2009. Roseki ore deposits in North Fossa Magna area. Roseki ore deposit of Shinyo mine: appearance of heat source rock body. *Chishitsu-News*, **660**, 16–31 [in Japanese].

SUDO, S. & NAITO, K. 2000. 'Setoyaki' pottery and industrial minerals of Seto-city and surrounding area. *Chishitsu-News*, **552**, 30–41 [in Japanese].

SUMII, T., WATANABE, Y., SUZUKI, Y., KODAMA, K. & TANAHASHI, M. 1992. Fuel Resources Map of Japan. *Geological Atlas of Japan*. 2nd edn. Sheet 10. Asakura Publishing Co. Ltd., Tokyo.

TAGIRI, M., DUNKLEY, D. J., ADACHI, T., HIROI, Y. & FANNING, C. M. 2011. Shrimp dating of magmatism in the Hitachi metamorphic terrance, Abukuma Belt, Japan: evidence for a Cambrian volcanic arc. *Island Arc*, **20**, 259–279.

TAKAGI, T., KOH, S-M., SONG, M-S., ITOH, M. & MOGI, K. 2005. Geology and properties of the Kawasaki and Dobuyama bentonite deposits of Zao region in northeastern Japan. *Clay Minerals*, **40**, 333–350.

TAKAHASHI, M. & SAITO, M. 1997. Miocene intra-arc bending at an arc-arc collision zone, central Japan. *Island Arc*, **6**, 168–182.

TAKENO, N., SAWAKI, T., MURAKAMI, H. & MIYAKE, K. 1999. Fluid inclusion study of skarns in the Maruyama deposit, the Kamioka Mine, Central Japan. *Resource Geology*, **49**, 233–242.

TAMAKI, K. 1985. Two modes of back-arc spreading. *Geology*, **13**, 475–478.

TAYLOR, B. 1992. Rifting and the volcanic-tectonic evolution of the Izu-Bonin-Mariana arc. *In*: TAYLOR, B., FUJIOKA, K., JANECEK, T. R. & LANGMUIR, C. (eds) *Proceedings of the Oceanic Drilling Program, Scientific Results*, **126**. Texas A&M University, College Station, TX, 627–651.

TAZAKI, Y. 1988. Natural gas deposits of dissolved-in-water type and sand/silt system. *Journal of Japanese Association for Petroleum Technology*, **53**, 256–264 [in Japanese with English Abstract].

TERAKADO, Y. 2001*a*. Re-Os dating of the Kuroko ore deposits from the Hokuroku district, Akita Prefecture, Northeast Japan. *Journal of the Geological Society of Japan*, **107**, 354–357.

TERAKADO, Y. 2001*b*. Re-Os dating of the Kuroko ores from the Wanibuchi Mine, Shimane Prefecture, southwestern Japan. *Geochemical Journal*, **35**, 169–174.

TSUCHIYA, N. 1990. Middle Miocene back-arc rift magmatism of basalt in the NE Japan arc. *Bulletin of the Geological Survey of Japan*, **41**, 473–505.

UYEDA, S. 1991. The Japanese island arc and the subduction process. *Episode*, **14**, 190–198.

WAKITA, K. 2013. Geology and tectonics of Japanese islands: a review: The key to understanding the geology of Asia. *Journal of Asian Earth Sciences*, **72**, 75–87.

WASEDA, A. 1993. Effect of maturity on carbon and hydrogen isotopes of crude oils in Northeast Japan. *Journal of Japanese Association for Petroleum Technology*, **58**, 199–208 [in Japanese with English Abstract].

WASEDA, A. & NISHITA, H. 1998. Geochemical characteristics of terrigenous- and marine-sourced oils in Hokkaido, Japan. *Organic Geochemistry*, **28**, 27–41.

WASEDA, A. & UCHIDA, T. 2004. The geochemical context of gas hydrate in the eaatern Nankai Trough. *Resource Geology*, **54**, 69–78.

WASEDA, A., IWANO, H. & ASARI, Y. 2011. Migration and accumulation of hydrocarbons in the Kitakanbara area, Niigata, Japan, based on gas carbon isotope compositions. *Journal of Japanese Association for Petroleum Technology*, **76**, 43–51 [in Japanese with English abstract].

WATANABE, M., NISHIDO, H., HOSHINO, K., HAYASAKA, Y. & IMOTO, N. 1998. Metallogenic epochs in the Inner Zone of southwest Japan. *Ore Geology Review*, **12**, 267–288.

WATANABE, Y. 1995. A tectonic model for epithermal Au mineralization in NE Hokkaido, Japan. *Resource Geology Special Issue*, **18**, 257–269.

WATANABE, Y. 1996. Genesis of vein-hosting fractures in the Kitami region, Hokkaido, Japan. *Resource Geology*, **46**, 151–166.

WATANABE, Y. 2002. Late Cenozoic metallogeny of southwest Hokkaido, Japan. *Resource Geology*, **52**, 191–210.

WATANABE, Y. 2005. Late Cenozoic evolution of epithermal gold metallogenic provinces in Kyushu, Japan. *Mineralium Deposita*, **40**, 307–327.

YAHATA, M., KUBOTA, Y., KUROSAWA, K. & YAMAMOTO, K. 1999. Evolution in space and time of epithermal mineralization in northeastern Hokkaido, Japan. *Shigen-Chishitsu*, **49**, 191–202 [in Japanese with English abstract].

11 Engineering geology

M. CHIGIRA, Y. KANAORI, Y. WAKIZAKA,
H. YOSHIDA & Y. MIYATA

Engineering geology in Japan started as a practical science with a focus on exploration for natural resources in the middle of the eighteenth century before shifting its scope to the construction sector and environmental protection, a scope which has persisted to the present day. Engineering geology has contributed to the construction of infrastructure such as railways, roads, dams and power plants, the mitigation of natural disasters induced by mass movements and to protection of the environment from contamination by general waste as well as by radioactive waste.

The Japanese islands are tectonically active and have a complicated geology; adverse geological conditions have frequently had enormous influence on engineering geological projects. Many engineering geology projects in Japan are affected by wide shear zones, deep weathering, pressurized groundwater and landslides induced by excavation or dam impoundment. The methods and techniques of geological investigations, which are major fields within engineering geology, have been closely adapted to the geological conditions encountered in Japan. The many active faults in Japan, which have the potential to generate earthquakes, must be appropriately evaluated to secure the safety of constructions and the public. In addition to the geological and geomorphological setting of the Japanese islands, the region is located in an area of monsoon and tropical typhoons and receives large amounts of rain and snow, resulting in frequent landslides. Earthquakes induce liquefaction in densely populated areas on flat lands, as typified by the 2011 Tohoku-oki earthquake which inspired a renewed discussion on the possibility of the underground disposal of high-level radioactive waste, another important topic in the field of engineering geology.

This chapter covers aspects of engineering geology related to the construction of infrastructure, active faults, landslides, liquefaction and the disposal of radioactive waste.

Engineering geology in the construction of infrastructure (YW)

Infrastructure projects in Japan have encountered many difficult geological conditions such as swelling rocks associated with mud volcanism, high-temperature ground resulting from volcanic activity, frequent earthquakes, the complicated structure of accretional complexes, deep weathering and weak volcanic materials. The following sections provide examples of such geological conditions, selected in the context of tunnel, bridge and dam construction. Their locations are shown in Figure 11.1.

Tunnel construction in swelling rocks

Nabetachiyama Tunnel provides a good example of the challenges faced by engineers excavating the swelling rocks of Japan. This is a 9116 m railway tunnel on the Hokuetsu Express Hokuhoku Line, which crosses Nabetachiyama and is located between the cities of Tokamachi and Joetsu, Niigata Prefecture. As a result of the considerable problems encountered during construction, it took 21 years to complete.

The tunnel is located in Higashi-Kubiki hill, 300–400 m above sea level, and is excavated through Miocene–Pliocene rocks of the Teradomari and Nishiyama formations (Akita & Sato 1993). The Nishiyama Formation, which caused the problem of swelling rocks in the middle section of the tunnel, is mainly composed of crushed mudstone with thin layers of tuff and tuffaceous sandstone (Fig. 11.2). The formation described by Akita & Sato (1993) corresponds to the late Miocene Sugawa Formation reported by Takeuchi *et al.* (2000).

The rock encountered in the Nabetachiyama Tunnel exhibits the following characteristics comparable with other mudstones of a similar age: the constituent grains are very fine; the plasticity index is high; and the ratio of uniaxial compressive strength to overburden pressure is very small (uniaxial strength: 0.196–0.294 MPa). The ground contains combustible gases such as methane with a maximum pressure of 1.568 MPa. The gas permeability of the ground is poor, so it was difficult to degas when the tunnel was cut. The earth pressure reaches more than 2.94 MPa due to the load of overburden (*c.* 150 m). The rock on the cutting face was squeezed out into the cavern, which was an extreme phenomena not observed in other tunnels with swelling rocks in Japan (Akita & Sato 1993).

According to Kogure & Kimura (1995), the construction chronology of the 3327 m long middle section of the tunnel, in which the swelling rocks are distributed, was as follows. The middle section is composed of single-track parts (west end side, 131 m long; middle to east side, 2576 m long) and a double track part (west side, 680 m long). The double-track part at the bottom of the inclined shaft was initially constructed by the top-heading method. In the single-track part, collapse of the cutting face and replacement of timbers frequently occurred due to the high swelling pressure. This forced a change to the short-bench cut method, using a circular shape. Given that the amount of thrusting back of the cutting face was larger than the amount of excavation, various methods (e.g. the New Austrian Tunnelling Method (NATM) with an advancing centre drift, the mini-bench cut method and using face-reinforcement bolts) were all attempted to overcome this problem. However, these methods had no effect. Construction was stopped in 1982 after 8 years after work, leaving 645 m of tunnel uncompleted. Japanese National Railways, which had been an entrepreneur of this difficult construction project, was dissolved by the law for business reconstruction.

Construction of the remaining part of the tunnel was resumed in August 1985 by Hokuetsu Express Co. Ltd. Initially, the short-bench cut and drift advancing short-bench cut methods proved effective. However, deformation of the supports became critical when excavation had progressed 266 m, and so a centre drift excavated from the other side of the cutting face was advanced prior to the main tunnel. The centre drift acted as an anchor to control the thrusting back of the cutting face when the main tunnel was excavated. A tunnel boring machine (TBM) facilitated high-speed excavation, but extreme squeezing occurred when the TBM had excavated another 65 m. The TBM was withdrawn by 100 m, and the ground was further consolidated with cement. From then on, excavation and grouting

From: Moreno, T., Wallis, S., Kojima, T. & Gibbons, W. (eds) 2016. *The Geology of Japan*. Geological Society, London, 457–483.

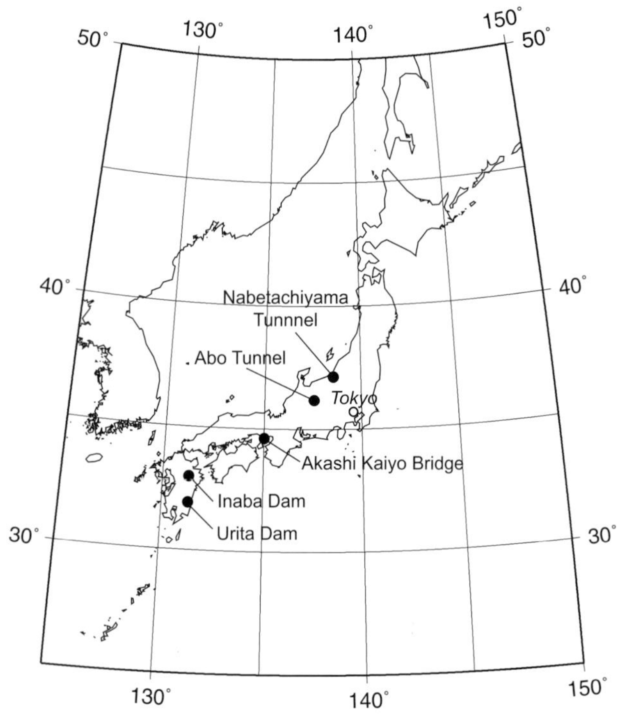

Fig. 11.1. Locality map of the case studies of infrastructure construction.

were alternated. The drift was finished in April 1992 and the main tunnel was finished in March 1995.

Recently, Shinya & Tanaka (2005) reported that mud volcanoes were present in the upper part of the middle section of the Nabetachiyama Tunnel. Tanaka & Ishihara (2009) subsequently attributed the swelling bedrock encountered in the tunnel to mud volcanism. They concluded that the controlling mechanism generating the swelling ground was as follows. Fluids derived from extreme pore-water pressure from deep underground formed a mud chamber due to degassing under the decreased pressure. Mudstone located in the upper part then became crushed by hydraulic pressure and separated into a muddy matrix and fragments. When the muddy matrix, which was saturated with very salty water (a strong brine), came into contact with much less salty water (a weak brine), an electric double layer formed. Consequently, the mudstone converted into mud through the repulsive force of the electric double layer. Ca-smectite changed to Na-smectite through ion exchange with the uplifted strong brine, inducing the rapid deterioration of the ground. Finally, the ratio of uniaxial compressive strength to overburden pressure decreased. Methane with a maximum pressure of 1.568 MPa was derived from degassing, and this high pressure caused squeezing mud. Whereas Akita & Sato (1993) had originally reported that the mudstone was crushed to fine grains by folding, Tanaka & Ishihara (2009) subsequently argued that the crushing was caused by the hydraulic pressure of mud volcanism.

Tunnel construction in high-temperature ground

The 4370 m long Abo Tunnel on the Abo-Toge Expressway (National Highway No. 158) provides a classic example of a tunnel in high-temperature ground around an active volcano in Japan (Matsushita 1992). The tunnel is excavated through: the central part of the Norikura Volcanic Zone in central Honshu; the Yakedake and Akandana volcanoes which are located north of the tunnel; and the Norikuradake Volcano which is situated south of the tunnel. All of these volcanoes are active. The distance between the tunnel entrance on the Nagano Prefecture (east) side and Yakedake Volcano is only 3 km. The geology around the tunnel mostly comprises dacitic to andesitic lavas and pyroclastic rocks from Akandana Volcano, mélanges belonging to the Mino Terrane (Hirayu Complex: Fig. 11.3), aplite dykes and granodiorite porphyry. The Hirayu Complex mélanges consist of a muddy matrix and blocks of sandstone, chert, altered basalt and limestone (Harayama 1990), with radiolaria-bearing mudstones indicating a middle Jurassic age (Otsuka 1988).

The 300–1200 m long section inside the eastern entrance of the tunnel (Nakanoyu side) is a high-temperature zone in which the temperature of the ground is >50°C, with maximum temperatures reaching 75°C. The high-temperature zone comprises cracked alternating sandstone and slate, and these rocks are rather soft. Geological investigation based only on drilling was believed to be insufficient for solving the geological problems associated with this high-temperature ground. Exploratory tunnels were therefore excavated to investigate the geology, practicality of excavation and construction method for the main tunnel. Excavation of the exploratory tunnel on the Nakanoyu side began in 1980, and that on the Hirayu side began in 1983.

Prior to excavation of the exploratory tunnel on the Nakanoyu side, advanced boring was performed to investigate the geology, ground temperature, groundwater and possible presence of poisonous volcanic gases. The exploratory tunnel was excavated using the NATM method. Heat-resistant gunpowder was used in the section (distance from the entrance on the Nakanoyu side was 500–775 m) where the ground temperature was higher than 70°C, whereas ordinary gunpowder was used in the other section. Before excavating the tunnel, exploratory borings were always drilled to examine the ground conditions and groundwater and monitor poisonous volcanic gases. The ground temperature drastically increased near the entrance of the Nakanoyu side, with a maximum temperature of 75°C being recorded 650 m from the entrance and a temperature of 50°C at 1200 m from the entrance. The work conditions were very poor due to the high temperature, so ventilation needed to be installed using a blastpipe. The concentration of hydrogen sulphide, sulphur dioxide, hydrogen chloride, carbon dioxide and methane was measured in boreholes from advanced and exploratory borings as a countermeasure against poisonous volcanic gases. However, only minute amounts of hydrogen sulphide were detected in these exploratory boreholes, such that working conditions were not affected by poisonous volcanic gases.

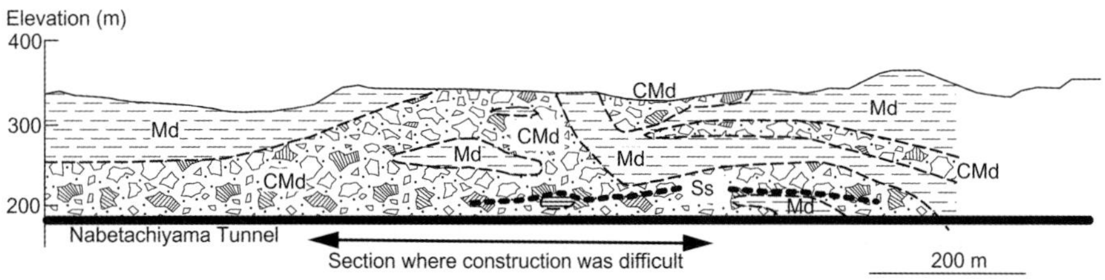

Fig. 11.2. Geological profile of section where construction of Nabetachiyama Tunnel was difficult. Md, mudstone; CMd, crushed mudstone; Ss, sandstone. Modified from Suzuki *et al.* (2009).

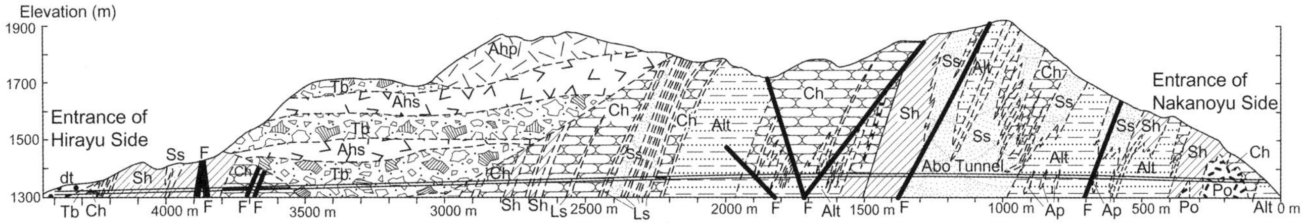

Fig. 11.3. Geological section along Abo Tunnel. Ahp, hornblende-pyroxene andesitic lava; Ahs, hornblende andesitic lava; Alt, alternating sandstone and shale; Ap, aplite; Ch, chert; dt, detritus; F, fault; Ls, limestone; Po, porphyrite; Sh, shale; Ss, sandstone; Tb, tuff breccia. Modified from Takayama National Highway Work Office (1999).

Based on the results of many different trial examinations, Portland blast-furnace slag cement B was found to be the best type of cement in terms of durability at high temperature. This cement was therefore used for the final lining and shotcrete in the high-temperature section.

Excavation of the main tunnel at the Nakanoyu side began in 1991, and the exploratory tunnel was changed to a working drift for construction of the main tunnel. Just before construction of the tunnel was completed, a steam explosion occurred near the entrance of the tunnel at the Nakanoyu side on 11 February 1995. The geology at the explosion site was a pyroclastic dyke, and the chemical components of the explosion gas were similar to those of gas from the nearby Yakedake Volcano. Miyake & Ossaka (1998) consequently proposed that the origin of the explosion was volcanic gas rising from the magma chamber inside Yakedake to reach the 1 km depth of the explosion site. Four people who were working on the construction of the Abo–Toge Expressway died in this explosion. The construction of the Abo Tunnel was finally completed in April 1995.

Suspension bridge construction in earthquake-prone areas

The Akashi Kaikyo (Channel) suspension bridge is part of the Kobe–Awaji–Naruto Expressway, which is one of three legs of the Honshu–Shikoku Bridge Expressway linking Maiko, Kobe and Matsuho, Awaji Island. Its total length is 3911 m including a centre span of 1991 m, the longest in the world. Construction of the bridge began in 1986, and service started in 1998. It is situated close to the epicentre of the 1995 Hyogo-ken-nambu ('Kobe') earthquake, so has become a textbook example of the challenges involved in bridge building across seismically active areas.

A narrow basin, oriented NW–SE and 400 m wide, crosses the central part of the Akashi Channel below the bridge. The depth of the basin is 110 m from the surface of the sea, and both sides of the basin form steep slopes with a relative height of 50 m. Several flat underwater terraces are located between the basin and coast at depths of 10, 30 and 50 m (Yamagata *et al.* 1993). Yamagata *et al.* (1993) described the geology of the Akashi Channel as comprising a basement of Cretaceous granite overlain by the Miocene Kobe Group, Pliocene–Pleistocene Akashi Formation, diluvium and alluvium (Fig. 11.4). The upper part of the granite is weathered and altered. The Kobe Group consists mainly of alternating sandstone and mudstone beds on the Kobe side, and coarse-grained arkosic sandstones on the Awaji Island side. The Akashi Formation is a sand and gravel layer, 40 m thick at 2P (Kobe-side pier: shown on Fig. 11.4), and with a gravel size ranging from fine to over 200 mm. Several faults running north–south and NE–SW belonging to the Rokko Fault System were detected in this area by acoustic exploration and were taken account of when the route of the bridge was designed.

The bedrock at 1A (Kobe-side abutment) and 3P (Awaji Island-side pier) comprises the Kobe Group, whereas that at 2P belongs to the Akashi Formation and that at 4A (Awaji Island-side abutment) is granite. Sampling for soil tests (such as a triaxial compressive testing) was carried out using a triple tube sampler with a 360 mm diameter for the Akashi Formation, while the same type of sampler with a 116 mm diameter was used for the Kobe Group (Yamagata 1990).

The foundations of the main piers (2P and 3P) were excavated down to 60 m below the mean sea level of Tokyo Bay (Tokyo Peil or TP) as illustrated in Figure 11.4. Cylindrical steel caissons (watertight retaining structures with a diameter of 78–80 m and a height of 62–65 m) were then sunk into each of the excavated holes at 2P and 3P. After the caissons were sunk, concrete was poured into them. The weight of the caisson was *c.* 196 MN, so the total weights of the caisson and concrete were 7760.1 MN for 2P and 6964.7 MN for 3P. The displacements of the foundations were then measured. Displacement of the seafloor at TP −60 m in 2P was 86.133 mm just after the pouring of concrete (Yamagata *et al.* 1993).

On 17 January 1995 the $M_J = 7.3$ ($M_w = 6.9$) Hyogo-ken-nambu earthquake occurred, the epicentre of which was very close to the Akashi Kaikyo Bridge. The seismogenic fault was part of the Rokko Fault System and a surface fault rupture appeared on Awaji Island. When the earthquake occurred erection of the towers of the bridge had been completed, all cable strands were in place and cable stretching work was in progress (Tada *et al.* 1995). After the earthquake, visual inspection and displacement measurement of the bridge and a geological survey of Akashi Channel were carried

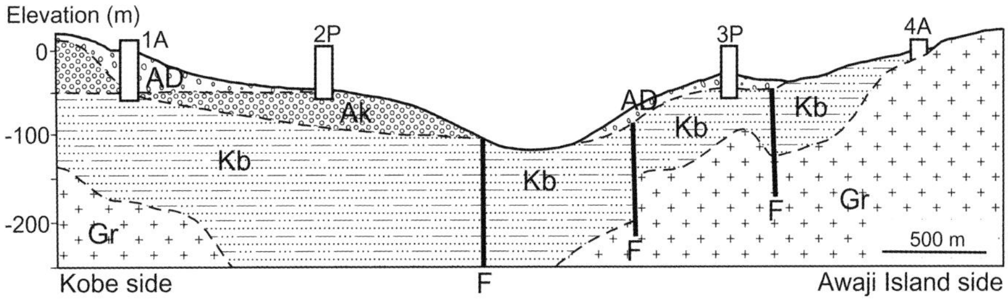

Fig. 11.4. Geological section along Akasi Kaikyo Bridge. AD, alluvium and diluvium; Ak, Akashi Formation; F, fault; Gr, granite; Kb, Kobe Group. Modified from Yamagata (1990).

out. The visual inspection showed no damage to the bridge. The geological survey, performed by the Maritime Safety Agency, indicated no new fault on the seafloor of Akashi Channel. According to the displacement measurement, the anchorage on the Awaji Island side and foundation of 3P moved 1.4 and 1.3 m, respectively, relative to the Kobe side anchorage. The original centre span length of 1990 m was therefore changed to 1990.8 m (Tada *et al.* 1995). The foundation of 2P sunk 20 mm as a result of the earthquake (Kurihara 1998).

Dam construction in mélange bedrock

Urita Dam exemplifies the challenges faced by dam construction in accretionary complex mélanges, a rock type common to many parts of Japan. Construction of the dam in Miyazaki Prefecture, SE Kyushu, began in 1984 and was completed in 1998. It is a concrete gravity dam with a height of 42.0 m, crest length of 160.4 m and a dam body volume of 100 200 m^3. Its main purpose is flood control.

The geology of the dam foundation comprises Cretaceous Shimanto Supergroup and late Miocene–early Pleistocene Miyazaki Group sandstones. The Shimanto Supergroup consists of blocks of altered basalt, sandstone, alternating sandstone and shale, shale and a matrix of shale and pebbly shale. These rocks were initially considered to be Palaeogene–Neogene olistostromes; after excavation of the dam body foundation however, the rocks were reinterpreted as accretionary prism tectonic mélange based on the extreme folding and discontinuity of the strata (Matsuura & Okazaki 1999). Radiolaria from shale indicated they were Cretaceous in age. The complicated distribution of lithologies within the mélange bedrock proved difficult to estimate and predict. Figure 11.5 compares a geological map of the excavated surface for the dam body based on the drilled cores with another geological map based on observation of the excavated surface. Although there are broad similarities between the maps, notably in the elongated strike direction of the various lithologies, the details differ, especially with regard to the distributions of small rock bodies of sandstone and alternating sandstone and shale. This difference demonstrated that the distribution of the smaller rock mélange clasts with a size of 10 m or less cannot be accurately estimated by drilling at 20 m intervals (Miyazaki Prefecture 1998).

Rocks of the Shimanto Supergroup are divided into massive lithologies, such as altered basalt and sandstone, and non-massive rocks, such as alternating sandstone and shale, shale and pebbly shale. Cleavage was strongly developed in the non-massive rocks, and bedding planes were easily exfoliated. In contrast, the Miyazaki Group is dominated by massive sandstone, and this difference provided the initial basis for rock classification from an engineering perspective. Massive rocks were classified based on a combination of subclasses such as hardness, interval of joints, state of joints and weathering of rocks, while non-massive rocks were classified based on a combination of these subclasses plus the interval of separated cleavages and brilliance of cleavage planes (Matsuura & Okazaki 1999). The brilliance of cleavage planes was divided into three classes: strong, medium and weak. In the strong class cleavage planes showed strong brilliance, and most formed a very smooth plane like a slickenside. In the medium class most cleavage planes indicated weak brilliance, although some indicated strong brilliance. In the weak class most cleavage planes showed weak brilliance (Miyazaki Prefecture 1998). Applying this classification allowed recognition of rock classes ranging from strongest (CM) to weakest (D), as shown in Table 11.1. As with the geological outcrop pattern, it nevertheless proved difficult to estimate the exact distribution of each rock mass class. In particular, estimation of the rock mass class in cleavage-bearing shale based on drilled cores was partially mistaken. In the drilled cores, cleavages were separated by the mechanical effect of drilling, leading to misinterpretation of lithologies as belonging to weaker classes than was actually the case. Direct observation of excavated drill core locations in rocks previously classified during drilling as CL–CM, for example, showed that *in situ* cleavage planes were much better adhered than in the drill core so that the rock mass class could be evaluated as CM (Miyazaki Prefecture 1998). Another specific area of difficulty was presented by the mudstone matrix of the mélange, which was typically strongly cleaved and required care during drilling.

Dam construction in non-consolidated pyroclastic rocks

Inaba Dam in Oita Prefecture, eastern Kyushu, is a typical dam constructed upon a weak volcanic foundation of pyroclastic rocks NE of the Aso Caldera, the largest active volcano in Japan. It is a concrete

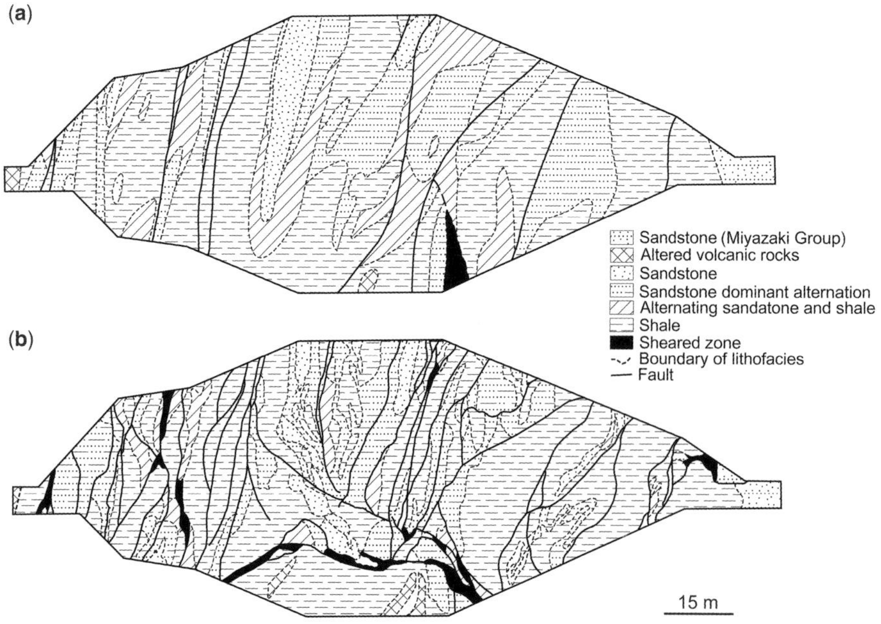

Fig. 11.5. Geological map of excavated surface of Urita Dam. (**a**) Map based on observation of drilled cores. (**b**) Map based on observation of excavated surface. Rocks belonging to Shimanto Group except for sandstone of Miyazaki Group. Modified from Matsuura & Okazaki (1999).

Table 11.1. *Characteristics of each rock mass class for non-massive rocks of Urita Dam. Translated from Matsuura & Okazaki (1999)*

Class	Characteristics	Note
CM	Rock pieces are fresh and hard, and are not exfoliated. Cleavage is exfoliated when rock is strongly struck by a hammer. Cracks are fresh to partly oxidized. Interval of cracks is longer than 15 cm. Cracks are not softened by weathering. Cleavages are almost stuck each other. Brilliance of cleavage is weak.	Rock mass is CL–CM when cleavage is rather developed, is easily exfoliated and brilliance of cleavage is strong. Rock mass is CL–CM when cleavage is easily exfoliated by weathering.
CL–CM	Intermediate between CM and CL.	
CL	Rock piece is totally fresh and medium hard. Rock pieces are partly fine fragment. Cracks are rare, although rich exfoliation. Interval of separated cleavages is less than 5 cm. Cleavages are easily exfoliated, and indicate strong brilliance. Rock pieces become soft.	CL–D or D depending on degree of hardness affected by weathering.
CL–D	Rock mass shows intermediate state between CL and D. Cleavages are extremely developed. Rock mass is a shear zone with fine-grained fragments and thin layers of clay. Tightness of rock mass is good.	A shear zone with many clay thin layers and a weathered shear zone are D.
D	Cleavages are extremely developed. Rock mass is mainly composed of fine-grained, soft fragments and clay.	Weathering is developed. A part where composed of concentration of weak zones, a fault and a shear zone.

gravity dam with a height of 56 m, crest length of 233.5 m and a dam body volume of 223 000 m^3. Construction of the dam began in 2003, and the pouring of concrete was completed in 2007.

Using the same classification outlined in the previous example, the rock mass at the dam site was classified as CH, CM, CL and D in descending order of shear strength. The rock-mass class correlated with the degree of welding. Strongly welded pyroclastic flows were grouped in the CH and CM classes, moderately to weakly welded pyroclastic flows were evaluated as CL class, and non-welded pyroclastic flows and the soft deposits between them were classified as D class (Fig. 11.6).

The foundation rock of the riverbed was sufficiently strong to allow construction of a gravity concrete dam because the local volcanic deposit (Imaichi Pyroclastic Flow) was moderately to strongly welded and therefore belonged to class CM–CH. However, the geology lying below both left and right abutments included several weak layers between different welded pyroclastic flows and so led to the rock mass being classified as CL and D.

As a result of these volcanostratigraphic complexities, construction of the gravity concrete dam proved very difficult at both abutments. Box-type diaphragm walls and concrete displacement by tunnelling were considered as countermeasures for these weak foundations, but the predicted long construction period, high cost and binding between the countermeasures and foundation presented serious problems. Finally, an inclined artificial abutment was determined to provide an acceptable solution (Hasegawa 2008) and was constructed. The left artificial abutment had the following specifications: a height of 37.5 m, stream-axis length of 44.9 m, dam-axis width of 8 m and a slope of 1:1. A similar approach was adopted for the right abutment.

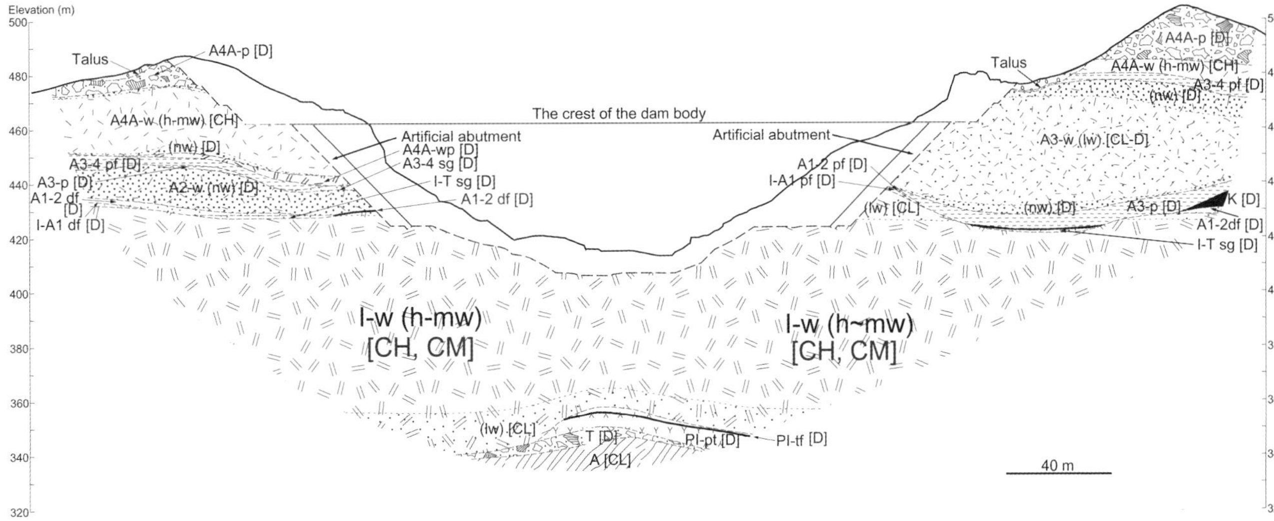

Fig. 11.6. Geological section along dam axis of Inaba Dam. Capital letters and numbers indicate stratigraphy: A4A, A3, A2, I: Aso-4A, Aso-3, Aso2, Imaichi pyroclastic flows; A3–4, A1–2, I–A1, I–T: deposits between Aso-3 and Aso-4, Aso-1 and Aso-2, Imaichi pyroclastic flow and Takashomachi pyroclastic flow; K, Miyagi pyroclastic flow; PI, pre-Imaichi pyroclastic flow; T, Tuff; A, Asaji metamorphic rocks. Lower-case letters represent lithofacies, p, pumiceous tuff breccia; w, welded tuff; wp, fine pumice tuff; pf, pyroclastic fall; sg, sand and gravel; df, debris flow; tf, tuff; pt, pumice tuff. Lower-case letters in parentheses describe the degree of welding, hw–mw, highly to moderately welded; lw, low welded; nw, non-welded. Rock mass classes CH, CM, CL and D are also indicated in this figure. Modified from Hasegawa (2008).

Rock permeability at the site posed additional problems. The strongly welded pyroclastic flows had very high permeability (>100 Lu) because of the presence of open cooling joints. In addition, the non-welded pyroclastic flows and intervening weaker deposits had very low resistance to hydraulic fracturing because they were not consolidated. In the context of this high permeability and complex geology, grouting tests confirmed that cutting water off with a grout curtain was going to be difficult (Sakemi 2010). Consequently, the use of reservoir surface facing was selected to cut off the water. Reservoir surface facing consists of three methods: concrete facing, earth blanket and asphalt facing. Concrete facing was adopted for steep slopes with strongly welded pyroclastic flows and unconsolidated flows and deposits. Asphalt facing was used in parts with non-consolidated flows and deposits as the asphalt can deform as these non-consolidated foundations deform. The earth blanket method was adopted for riverbed areas exposing the strongly welded Imaichi Pyroclastic Flow. Before constructing the earth blanket, the opened cooling joints were filled with cemented sand and gravel (Sakemi 2010). Finally, after constructing the dam body and cutting off the water, the first impoundment was carried out from 24 February to 29 May in 2010. The results of the first impoundment indicated no problems for the dam body and reservoir (Tagawa 2011).

Faults from an engineering geology perspective (YK)

When selecting sites in Japan for large structures such as high dams, nuclear power stations and routes for highways or tunnels, it is important to investigate active faults at the site from the early planning stages (Kanaori 1997; Kanaori *et al.* 2000). This section refers to faults that can be recognized by fracture zones and are included on geological maps as 'geological faults', as well as those specifically identified or presumed from tectonic relief to be 'active faults'.

Active faults are usually defined as faults that have been active in the Quaternary Period and which may remain active in the future (Research Group for Active Faults of Japan 1980, 1991) although, as discussed later, definitions are being revised. Doke *et al.* (2012) indicates that, based on statistical analysis of the start of motion of active faults in Japan, the number of active faults initiating motion began to increase at *c.* 3 Ma, the number significantly increased after 1.5 Ma, and the number peaked at *c.* 0.5 Ma.

Next in the early stages of site or route selection, the static and dynamic effects of fracture zones associated with faults in the basement rock beneath the structure are investigated in detail using geological and geophysical techniques. To evaluate static effects, the fracture zone associated with the fault and the surrounding process zone are characterized and physical properties are measured. For dynamic effects, investigations include whether a fault in the basement rock is active. If so, the magnitude of earthquakes likely to be generated is estimated from the length of the active fault (Matsuda 1975), and this is used as basic data for earthquake-proof design of the structure. In particular, the Nuclear Regulation Authority demands detailed evaluation of active faults for nuclear power stations.

Distribution maps of faults

Distribution maps of geological and active faults throughout the Japanese islands have been published. These maps can be used in site and route selection at early planning stages.

Geological faults

Kosaka *et al.* (2010) lists more than 3000 on-land faults of 10 km or longer, based on 1:200 000 and 1:50 000 scale geological mapping published by the Geological Survey of Japan. Faults of 10 km or longer are of interest for the following reasons:

(1) They have multiple lines of evidence by which the fault can be verified, improving certainty of their existence.
(2) They reflect, to some degree, geological or structural scale.
(3) Their movement will result in a $M = 6.5$ or greater earthquake.
(4) They will likely have a significant influence on large underground facilities and surrounding underground geological environments.

In addition to these faults, the Kosaka *et al.* (2010) map duplicates the active faults illustrated in the publication 'Active Faults in Japan' (Research Group for Active Faults of Japan 1991).

Active faults

There are more than 2000 active faults 10 km or longer in Japan. Typical distribution maps showing active faults include, in chronological order, 'Active Fault Map of Japan 1:2 000 000' (Kakimi *et al.* 1978), 'Active Faults in Japan' (Research Group for Active Fault of Japan 1980), 'Active Faults in Japan (revised edition)' (Research Group for Active Fault of Japan 1991), 'Active Fault Map of Japan 1:2 000 000 (Working Group for Compilation of 1:2 000 000 Active Faults Map of Japan 2001) and 'Digital Active Fault Map of Japan' (Nakata & Imaizumi 2002). A database of currently active faults is also provided by the Active Fault and Earthquake Research Center (2012).

Among these maps, 'Active Faults in Japan (revised edition)' (Research Group for Active Fault of Japan 1991) is frequently referred to for regional distribution of active faults when siting large structures and selecting highway or tunnel routes. In this map, active faults are drawn on 1:200 000 topographical maps published by the Geospatial Information Authority of Japan, and are categorized according to certainty of existence, as follows:

- Certainty level I: Existence of an active fault is assured.
- Certainty level II: Existence of an active fault is assumed, for example because location and slip sense are indicated, but there are insufficient data for a Certainty level I classification.
- Certainty level III: Existence of a lineament that may be an active fault but with unknown slip-sense, the possibility of lineament creation by other causes (e.g. scarp erosion by a river or sea) or a lineament that likely formed by erosional topography along an inactive fault.

The degree of activity is tentatively defined using the average slip rate S (m ka^{-1}) as an indicator. Active faults are categorized according to slip rate into the following three classes:

- Class A: $1.0 \leq S < 10$;
- Class B: $0.1 \leq S < 1.0$; and
- Class C: $0.01 \leq S < 0.1$.

Fault process zone

Faults accompanied by fracture zones commonly occur in the basement rock of large structures. The fracture zone has static and dynamic effects on the basement rock so that it is surrounded by a 'process zone', which is involved in fault generation and faulting. This process zone is characterized by a high density of joints, secondary shear fractures and microcracks even in the host rock (Vermilye & Scholz 1998; Scholz 2002). The term 'damage zone' has recently been used instead of 'process zone' (Kim *et al.* 2004). Figure 11.7 provides a schematic illustration showing fracture zones accompanying a fault and a process zone surrounding them. Microcrack analysis of the process zone is required when evaluating effects of faulting and permeability in detail in the basement rock around a fault. In particular, for geological disposal of nuclear

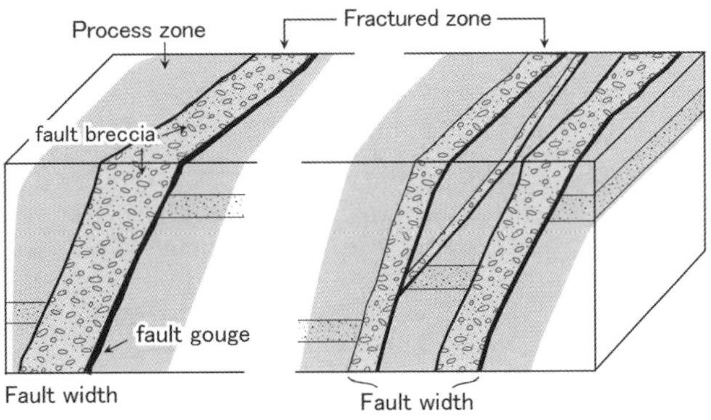

Fig. 11.7. Schematic diagram of a fracture zone accompanying a fault and a process zone around the fault (modified after Kanaori 2001).

wastes, the process zone and fracture zone must be characterized to evaluate long-term circulation of groundwater and diffusion of nuclides to ensure long-term stability (Yoshida *et al.* 2009).

Fault density

Small-scale faults commonly occur in basement rock and may be expected even if we select a structure site in an area without large faults. Figure 11.8 shows the density of small faults in basement rock of different formational ages. The fault density is estimated from cut surfaces and test pits in basement rock at sites such as large dams, tunnels and nuclear power stations (Ogata 1976). Fault density in Palaeozoic or Mesozoic (*c.* 400–100 Ma) basement rock ranges from 0.06 to 0.23 m^{-1}, whereas that for younger rocks decreases so that the number of faults in Pliocene basement rock for example is approximately one order of magnitude less. As fault spacing is equal to the reciprocal of the fault density, small-scale faults typically occur at a spacing of 5–15 m in Palaeozoic or Mesozoic basement rock.

Fault rock classification

Fault rocks are mainly classified according to their degree of consolidation and texture (Fig. 11.9), and show a wide range of different physical properties that are of direct interest to the engineering geologists. Fault gouge and breccias are categorized as unconsolidated fault rocks. Because unconsolidated fault rock is less cohesive and weak, engineering-geology zones that include this kind of lithology are usually referred to as fracture zones and are given special attention during site investigations (Kanaori 1997). In contrast, consolidated fault rocks such as mylonite (which is foliated) and cataclasite (an unfoliated but cohesive fault breccia) may in some cases be stronger than the host rock. Because rocks weather near the surface, it can be difficult to distinguish weathered cataclasite from unconsolidated fault rocks such as fault gouge and breccias. Zones including cataclasite are appropriately called cataclasite zones, not fracture zones. These differences in fault-rock types are primarily related to its formational depth (Sibson 1977). Fault gouge and breccias form near the surface, cataclasite is generated in deeper, predominantly brittle failures, while mylonite forms by predominantly ductile flow at crustal depths within and below the brittle–ductile transition.

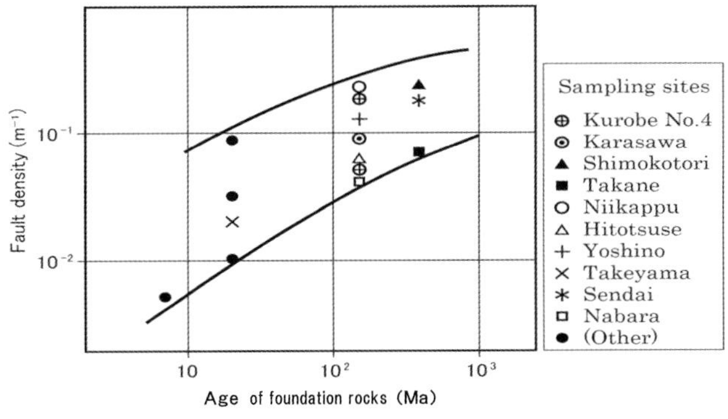

Fig. 11.8. Fault density v. geological ages (Ogata 1976).

Relationship between width and length of fault fracture and process zones

Fault-rock zones containing fractured and cataclasite lithologies are easily identified at outcrop scale. In contrast, it can be difficult to trace the length of long faults running across unexposed ground. The fault length L (m) can be estimated from its width W (m) or the one-side width P (m) of a process zone by the following relationships (Fig. 11.10):

$$\log L = 0.68 - 1.74 \log W.$$

The relationship between P and L is defined:

$$\log P = 0.18 - 2.0 \log L.$$

Both equations indicate that longer faults have wider fracture and process zones, and are used to estimate L from the width W of the fault in the basement rock. The ratios of width W of the fracture zone and the one-side width P of the process zone with respect to the fault length L are approximately 1:1000 and 1:100, respectively.

Case study on a fault process zone: Usukidani Fault

Trending NE–SW, the 30 km long Usukidani Fault is situated near the boundary between Shimane and Hiroshima prefectures in SW Japan. The fault cuts through rhyolitic tuff of the Cretaceous age. Figure 11.11 shows outcrops within the process zone accompanying the Usukidani Fault. The joint spacing tends to increase with increasing distance from the fault centre. Figure 11.12 shows the changes in width of the fault gouge, cataclastic and process zones along the fault strike. The total width of the fault gouge and cataclastic zones ranges from 3 to 5 m. Outside of these zones, the fault process zone is *c.* 200 m wide at maximum. The process zone becomes narrower towards each terminus of the fault, and has a spindle shape overall. In summary, the Usukidani Fault is accompanied by fault gouge and cataclastic zones of 3–5 m in total width, outside of which is a process zone with a maximum width of 200 m. The process zone narrows towards each terminus, creating a spindle shape with symmetry about the fault strike.

Seismogenic faults and interlocking

Some faults are isolated and others form en echelon arrays. A fault branches or incorporates others to form a set of faults, which together can be referred to as a fault zone or a fault system. Among multiple intermittently distributed faults, a unit of faults generating an earthquake is referred to as a seismogenic fault. The criteria for determining a seismogenic fault was proposed by Matsuda (1990): (1) an isolated fault 10 km or longer and without other faults within 5 km; (2) a set of faults with a similar strike within 5 km of each another; (3) a fault with a different strike and a centre more than 5 km from the main fault; or (4) multiple faults with a similar strike aligned in a straight line at a spacing of less than 5 km.

These criteria are also referred to as the '5 km rule', because the 5 km distance is important. It is not clear whether faults are

Cohesion	Nature of matrix		Random fabric		Foliated		Percent of matrix
Incohesive			Fault breccia (visible fragments >30% of rock mass				
Incohesive			Fault gouge (visible fragments <30% of rock mass		Foliated gouge		
Cohesive	Nature of matrix	Glass/devitrified glass	Pseudotachylyte				
Cohesive	Nature of matrix	Tectonic reduction in grain size dominates grain growth by recrystallization and neomineralization	Crush breccia (fragments >0.5 cm) Fine crush breccia (0.1<fragments <0.5 cm) Crush microbreccia (fragments <0.1 cm)				0–10
Cohesive	Nature of matrix	Tectonic reduction in grain size dominates grain growth by recrystallization and neomineralization	Cataclasite series	Protocataclasite	Mylonite series	Protomylonite	10–50
Cohesive	Nature of matrix	Tectonic reduction in grain size dominates grain growth by recrystallization and neomineralization	Cataclasite series	Cataclasite	Mylonite series	Mylonite (Phyllonite series)	50–90
Cohesive	Nature of matrix	Tectonic reduction in grain size dominates grain growth by recrystallization and neomineralization	Cataclasite series	Ultracataclasite	Mylonite series	Ultramylonite (Phyllonite series)	90–100
Cohesive		Grain growth pronounced			Blastomylonite		

Fig. 11.9. Textural classification of fault rocks (Sibson 1977). Cited from Scholz (2002).

dynamically a set of those which belong to a seismogenic fault, because it is defined geometrically. Yamaguchi & Kanaori (2011) checked whether the four-part 5 km rule definition is realized using a two-dimensional faulting simulation incorporating a fault process zone (Fukushima *et al.* 2010). The simulation shows that only the first of the given criteria is realized, and the other three are not.

The Headquarters for Earthquake Research Promotion creates long-term predictions of inland earthquake magnitudes by applying the 5 km rule to active faults. However, when obtaining basic data for countermeasures to inland earthquakes, it is necessary for regional committees to re-examine seismogenic faults when an individual fault moves independently or constituent faults move at the same time.

Nuclear power stations and active faults

Earthquake-resistant design of a nuclear power station requires a detailed evaluation of active faults at and around the site. The 'Guidelines for Earthquake-proof Design of Generation Reactor Facilities', published in 1978, is often used during such evaluations. According to these guidelines, faults that have moved in the past 50 ka are regarded as active faults of interest. In 2006, the guidelines were revised so that any fault activity in the past 120 to 130 ka at or near a nuclear power site must be investigated.

The Headquarters for Earthquake Research Promotion (2010) gives a revised definition of active faults as 'faults that have moved repeatedly in the most recent geological age', which is roughly the past 400 ka. Furthermore, in June 2012 the Nuclear Regulation

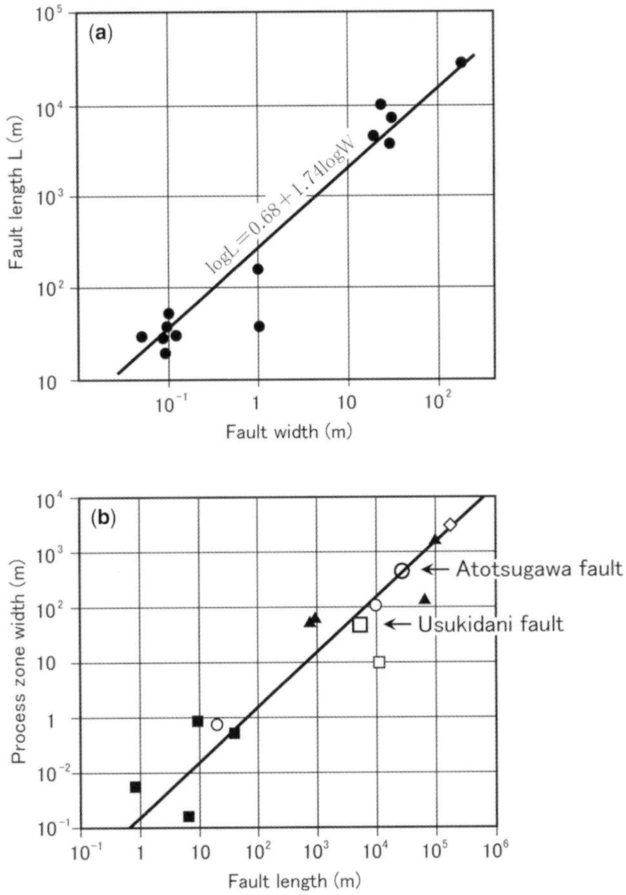

Fig. 11.10. Relationship between width and length of fault fracture and process zones. (**a**) Length of the fault v. fault width (Ogata 1976). (**b**) Width of the process zone wake v. fault length. Data for two faults in Japan from Vermilye & Scholz (1998) and Scholz (2002).

Authority began consideration of defining active faults as faults active during the late Quaternary, a period considered to be the past *c.* 400 ka. Following these revisions of the definition period, active faults in and around nuclear power stations in Japan are being reinvestigated for activity during the revised age.

As mentioned in the section of fault density, small faults commonly occur in basement rock. For example, spacing of small faults ranges from 5 to 15 m in Mesozoic basement rock. A few faults probably occur in Mesozoic basement rock in and around nuclear power stations. Judgment is required whether such faults are active over the intended period. However, it is difficult to check whether faults have been active in the past 400 ka where cover strata was removed during construction, and there is currently no dating method to measure the age without cover strata.

Summary

Checking whether faults are active is a top priority in Japanese engineering geology. Distribution maps of active and geological faults at the scale of 1:200 000 are checked at the early stages of site selection for large structures and route selection for highways and tunnels. The existence of faults has static and dynamic effects. As examples of static effects, deterioration of basement rock reduces their strength and groundwater passes through fracture zones with relatively high permeability. Dynamic effects include direct displacement and seismic motion by faulting. In particular, strict earthquake-resistant design for nuclear power stations requires checking whether faults in basement rock have been active in the indicated age. Definitions of the active period are now being discussed and the Nuclear Regulation Authority, established after the 2011 Tohoku-oki earthquake, has been investigating faults in and around nuclear power plants for activity over longer periods.

Landslides and related phenomena (MC)

Japan is one of the more landslide-susceptible countries in the world because of the high rates of tectonic activity, the monsoonal climate and the complicated geology. In addition, 70% of Japan's land area is mountainous. Each year, Japan experiences landslides induced by

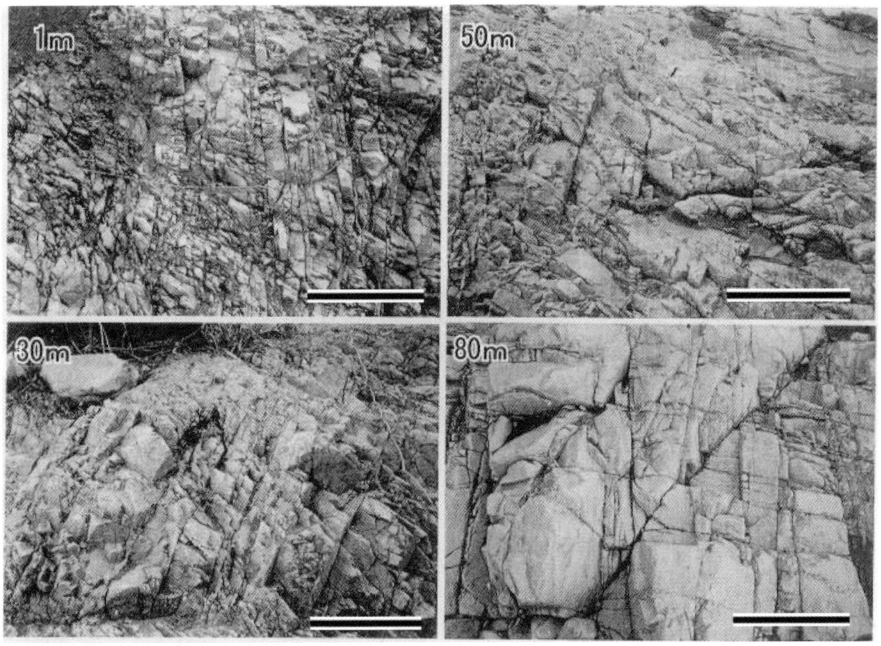

Fig. 11.11. Photographs of joints formed in the process zone accompanying the Usukidani fault in SW Japan (from Shiraishi 2000). The figure in the upper left of each photograph is the distance from the fault centre, as measured perpendicular to the fault strike. The number of joints decreases with the increasing distance. Scale bar: 0.5 m.

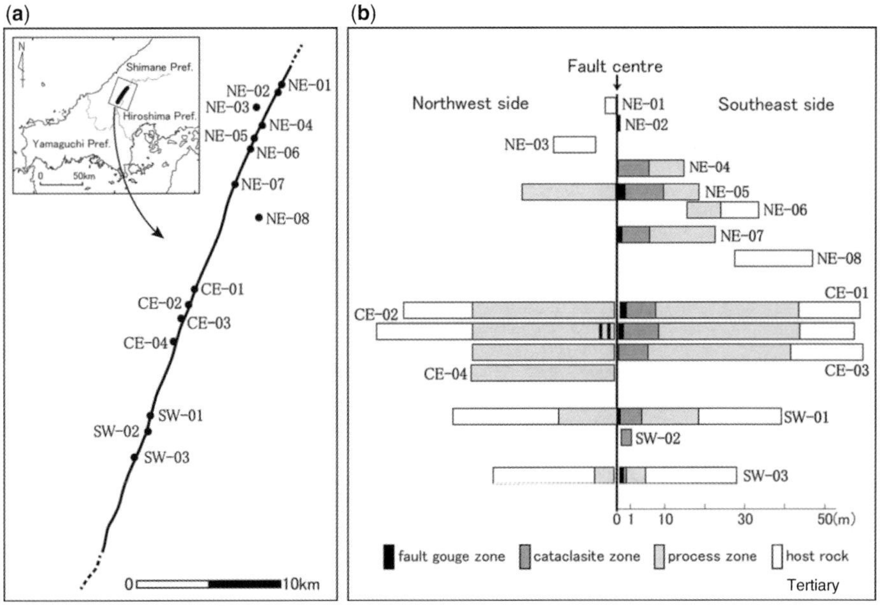

Fig. 11.12. Width distributions of the fault gouge, cataclastic and process zones of the Usukidani Fault along its strike. (**a**) Measurement points along the fault strike and locality map of the fault (insert). (**b**) Measurement values of the width (modified after Shiraishi 2000).

rainstorms, earthquakes and snowmelt, alongside other complicating factors. This section refers to typical mass-movement types along with their classification in Japan.

Classification of landslides in Japan

Scientific study of landslides in Japan started in the early 1900s. Wakimizu (1919) first classified landslides into two categories with respect to their velocity: 'landslide proper' is characterized by sudden and rapid movement, while 'land creep' is characterized by slow and continuous movement. Nakamura (1934) also classified landslides into two categories: 'Jikuzure' and 'Jisuberi', which describe rapid and slow-moving landslides with and without disintegration, respectively. Miyabe (1935) further classified landslides into four groups: shallow landslides or avalanches, debris flows, slow moving landslides and rock falls.

Compared to North American and European classifications (e.g. Varnes 1978; Cruden & Varnes 1996), the Japanese classification of landslides gives greater importance to velocity as a crucial factor in determining the risk to those in the path of the landslide. Today, a classification that distinguishes landslides, slope failures, debris flows and creep is most commonly used in Japan. Using this system of classification, 'landslides' are typically defined as slow-moving slides of rock or debris with little displacement and a master sliding surface where most of the sliding material remains in its source area. In contrast, 'slope failure' describes rapid failure events during which most of the moving material moves out of its source area. In this sense, slope failures may be compared to the 'disintegrating landslides' of Keefer (1984). Dramis & Sorriso-Valvo (1994) defined deep-seated gravitational slope deformation (DGSD) as a process of large-scale mass movement with or without a master sliding surface and with little displacement. The process of DGSD includes most landslides (or 'Jisuberi') using the Japanese system of classification. Zischinsky (1966) defined 'Sackung' as large-scale gravitational slope deformation without a through-going master sliding surface, which can develop into 'Gleitung' when a master sliding surface is made. Oyagi (2004) further proposed that 'Jisuberi' can be used as an all-encompassing term equivalent to 'landslide' in English.

Koide (1955) proposed another widely used classification of landslides based on landslide-susceptible bedrock types in Japan, comprising: Tertiary type, involving incompetent Tertiary sedimentary strata; hot-spring-volcanic type, which occurs in highly altered rocks; and fracture-zone type, which occurs in fault zones and highly broken metamorphic rocks. Koide (1955) assumed the existence of a clayey shear zone, which is the zone of sliding during the shearing of rocks. Indeed, many landslides in Japan have occurred in association with these rock features. Although this system of classification is still widely used in Japan, the word 'Tertiary' has been generally abandoned as a chronological term in geology.

Landslides including deep-seated gravitational deformations and rock avalanches have been mapped by the National Research Institute for Earth Science and Disaster Prevention for most of the Japanese islands (Fig. 11.13). This mapping is based on the interpretation of aerial photographs, the results of which are distributed via the Internet and as printed 1:50 000 (locally 1:25 000) maps.

Earthquake-induced landslides

Earthquakes induce various types of landslides ranging from small rockfalls on steep slopes that are not strongly controlled by geology to large geologically controlled catastrophic landslides. Recent earthquakes in Japan have induced many large, catastrophic landslides in pyroclastic and sedimentary rocks. Some earthquakes have also caused significant changes to groundwater systems, including the diversion of groundwater to the surface which itself can induce landslides (Morimoto *et al.* 1967; Toda *et al.* 1995).

Pyroclastic rock slopes

Table 11.2 lists earthquakes that have caused landslides on slopes formed of pyroclastic rocks in Japan. Four earthquakes (the 1949 Imaichi earthquake, 1968 Tokachi-oki earthquake, 1978 Izu–Oshima-kinkai earthquake and 2011 Tohoku-oki earthquake) induced catastrophic landslides with long run-outs of Quaternary pyroclastic fall deposits (Figs 11.14 & 11.15). These landslides were shallow (mostly <5 m thick) but highly mobile, with apparent friction angles commonly $<10°$ (Chigira *et al.* 2012b). Sliding surfaces were formed in layers of palaeosol, weathered ash or pumice, with abundant halloysite (Morimoto 1951; Inoue *et al.* 1970; Chigira 1982; Chigira *et al.* 2012b). The 2008 Iwate–Miyagi-nairiku earthquake hit an area of mixed pyroclastic rocks of Neogene–Quaternary age; however, these slopes were not mantled by younger pyroclastic

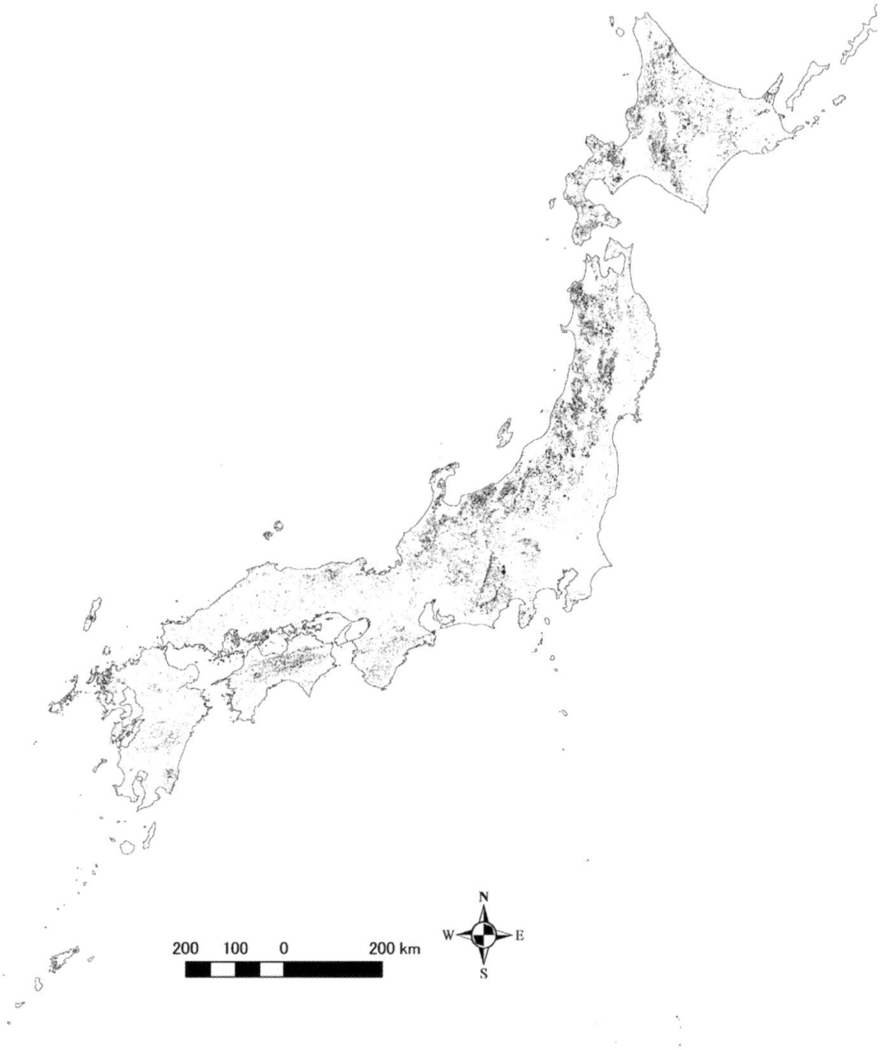

Fig. 11.13. Map of landside distribution in Japan, produced by the National Research Institute for Earth Science and Disaster Prevention.

Table 11.2. *Earthquakes that have induced landslides of pyroclastic fall deposits, with corresponding seismic intensities and antecedent rainfall. M_j: Magnitude of the Japanese Meteorological Agency (JMA)*

Earthquake	Date	Magnitude (M_j)	Seismic intensity at the sites of landslide (JMA)	Observatory	Preceding rainfall (mm)			Number of landslides	Sliding surface depth (m)	Slope-parallel bedding	Undercut
					10 days	30 days	60 days				
1949 Imaichi	26 Dec.	6.2 (8:17) 6.4 (8:24)	5–6 (Imaichi)	Utsunomiya	22.5	80.8	255	88*	3–5*	○	○
1968 off Tokachi	16 May	7.9	5 (Hachinohe)	Hachinohe	181	292	307	152[†]	<3[†]	○	○
1978 Izu–Oshima-Kinkai	14 Jan.	7.0	5–6	Inatori	12	172	334	7[‡] (Material distrbution limited)	2–6[‡]	○	○
2008 Iwate–Miyagi inland	14 June	7.2	+5 to +6	Komanoyu	89	284.5	388	>100	–	–	Various
2011 Tohoku	11 March	9.0	+6 (Shirakawa) −6 (Nakagawa)	Shirakawa	12.5	83.5	93.5	<10	3–9	○	○

*Morimoto (1951)
[†]Inoue *et al.* (1970)
[‡]Chigira (1982)
Note: Circles indicate 'yes'.

Fig. 11.14. Oblique photograph of a long-run-out landslide induced by the 2011 Tohoku earthquake (courtesy of the Ministry of Land, Infrastructure, Transportation and Tourism).

fall deposits, meaning that long-run-out landslides did not occur. The 1923 Kanto earthquake and the 1984 Nagano-ken–Seibu earthquake, which are not included in Table 11.2, triggered the Nebukawa Landslide (1 million m^3 in volume) and the Ontake-san Landslide (36 million m^3 in volume) respectively, involving the formation of sliding surfaces in a weathered pumice layer (Okuda *et al.* 1985; Kamai 1990). The Ontake-san Landslide had a velocity of 71–95 km/hour and a run-out distance of 10 km (Okuda *et al.* 1985).

Comparing the five earthquakes listed in Table 11.2, the 1978 Izu–Oshima-kinkai earthquake and the 2011 Tohoku-oki earthquake induced far fewer landslides than the other three events. The 1978 Izu–Oshima-kinkai earthquake, for example, triggered only seven long-run-out landslides on slopes formed of pyroclastic fall deposits, due to the limited distribution of these deposits close to the earthquake epicentre (Chigira 1982). Pyroclastics that slid during the 2011 Tohoku-oki earthquake are, in comparison, widely distributed, but fewer than 10 long-run-out landslides were induced. This is probably attributable to lower amounts of rainfall before the event (Table 11.2; Chigira *et al.* 2012*b*).

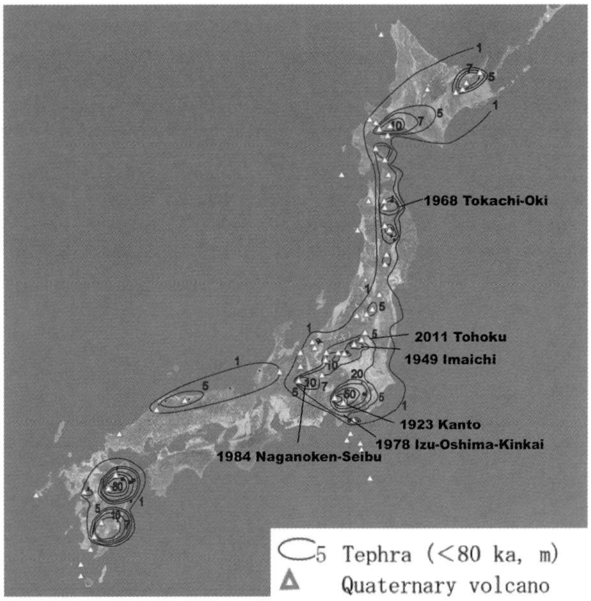

Fig. 11.15. Distribution of volcanic ash soils and associated long-run-out landslides induced by earthquakes (modified from Chigira 1982, with data by T. Suzuki, fig. 15).

Sedimentary-rock slopes

The 2004 Mid Niigata Prefecture earthquake occurred in areas formed of sedimentary rocks of Neogene–Quaternary age, where many slow-moving landslides have occurred. Before this earthquake, it had been supposed by many researchers that earthquake tremors would not trigger catastrophic reactivation of existing slow-moving landslides. However, more than 100 deep-seated large landslides were induced during the 2004 earthquake (Figs 11.16 & 11.17) (Chigira & Yagi 2006). Many of these landslides were reactivated, and had originally collided with the opposite slopes and become undercut by river incision. Two new (i.e. not reactivated) landslides in siltstone also occurred during this event, with sliding surfaces forming in tuff beds *c.* 5 cm thick. The occurrence of these slides is attributed to the weak, weathered tuff, and to the fact that the beds had been previously undercut and destabilized by human excavation for road construction (Chigira & Yagi 2006).

The 2004 Mid Niigata Prefecture earthquake was followed by two further earthquakes in 2007 (Table 11.3) of similar magnitude (6.9 and 6.8 on the Japan Meteorological Agency Scale) that affected geologically and topographically similar areas as the 2004 event within 180 km from the 2004 epicentre. However, these later events triggered comparatively few large landslides, which may again be attributable to antecedent rainfall (Fig. 11.18); more than 150 mm of rainfall fell during the 10 days preceding the 2004 earthquake, while the 2007 earthquakes were preceded by less than 50 mm of rainfall. As a result, the 2004 event caused the mobilization of saturated valley-bottom sediments probably via partial liquefaction in many locations (Chigira & Yagi 2006).

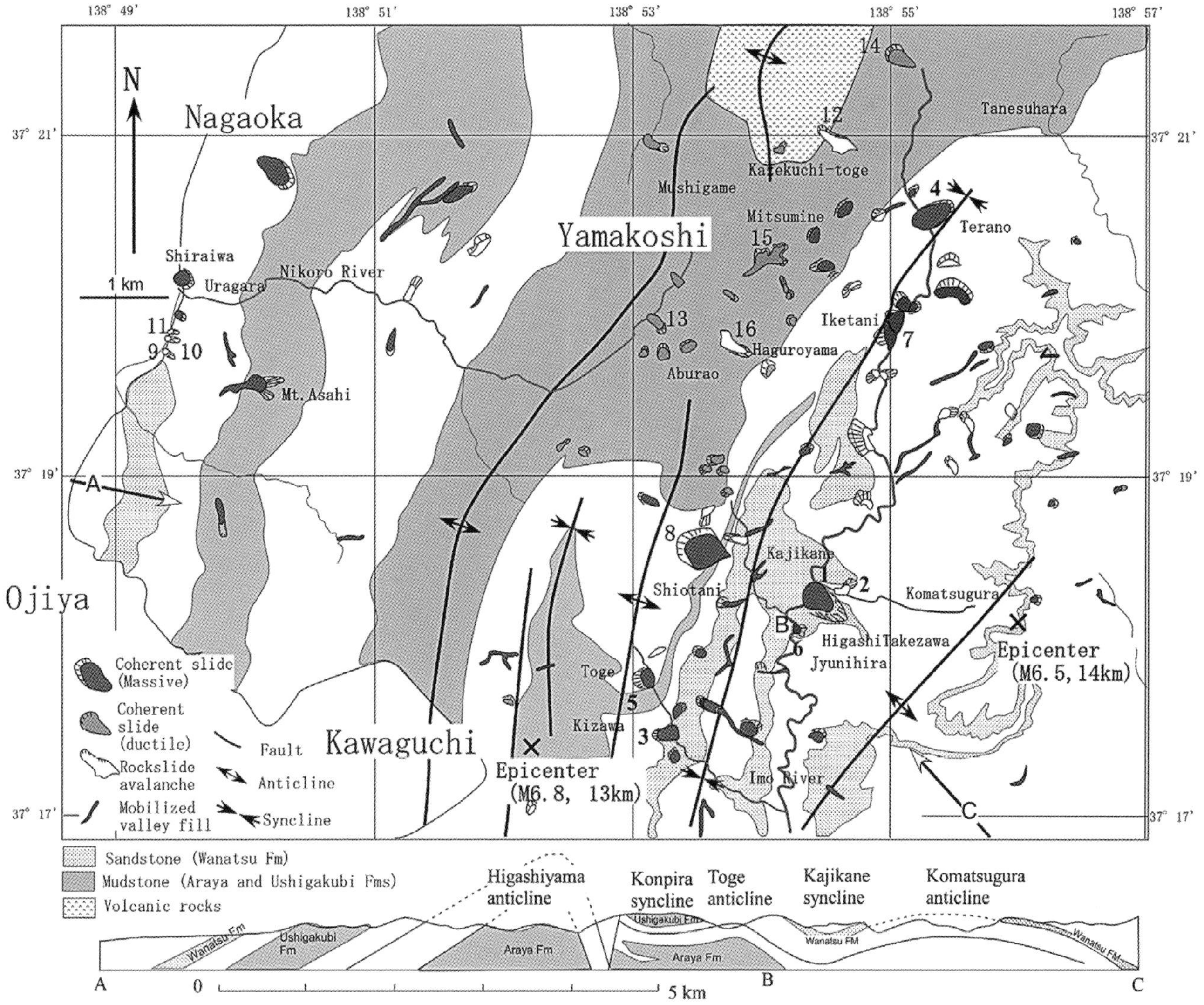

Fig. 11.16. Geology and distribution of major landslides induced by the 2004 Mid Niigata Prefecture earthquake (from Chigira & Yagi 2006).

Rainstorm-induced landslides

Historic records show maximum rainfall amounts of up to 1317 mm/day and maximum rates of 187 mm/hour in the period between 1976 and 2012. In addition, climate change has increased the frequency of heavy and intense rainfall events in Japan. A full review of the influence of rainfall on landslide initiation is beyond the scope of this chapter. Generally speaking, rainfall events totalling 400 mm or more or with an hourly rainfall amount of 50 mm or more are sufficient to induce landslides in Japan.

Rainfall events that have triggered landslides in Japan have mainly been associated with the Baiu Front in June–July and typhoons in August–September (Table 11.4). Interestingly, deep-seated large landslides (commonly greater than 100 000 m^3 in volume) and shallow landslides are generally caused by different rainfall patterns. In this respect, one rainfall event can induce numerous shallow landslides without any larger deep-seated landslides, while the opposite may also be true. These differences are assumed to be related to the control of geology on landslide occurrence. For example, numerous shallow landslides can occur during a single rainfall event on slopes with particular surface layers originating from weathering (e.g. granitoids, pyroclastics and volcanic ash). In contrast, deep-seated catastrophic landslides are more typically associated with accretionary complexes, as described below.

In response to the recognized importance of antecedent rainfall, and as an early warning system, the Japan Meteorological Agency began issuing alarms in 2000 based on soil water content. This alarm system is based on the amount of water in conceptual underground storage tanks and is calculated hourly (in 25 km^2 grid cells) using AMeDAS (Automated Meteorological Data Acquisition System) and radar data. An alarm is issued when soil water content exceeds the record level for the previous 10 years.

Shallow landslides and debris avalanches

Granitoid slopes. Granitoid slopes are sometimes believed to be very stable and not prone to shallow landslides, especially in previously glaciated regions such as northern Europe and northern North America where weathered materials have been largely eroded away (Durgin 1977). However, in humid and temperate or tropical regions, very thick weathering profiles develop that are prone to landslides. For example, in Japan, numerous shallow landslides (less than a few metres in depth) have occurred on moderately weathered (not heavily weathered) granite slopes at many sites

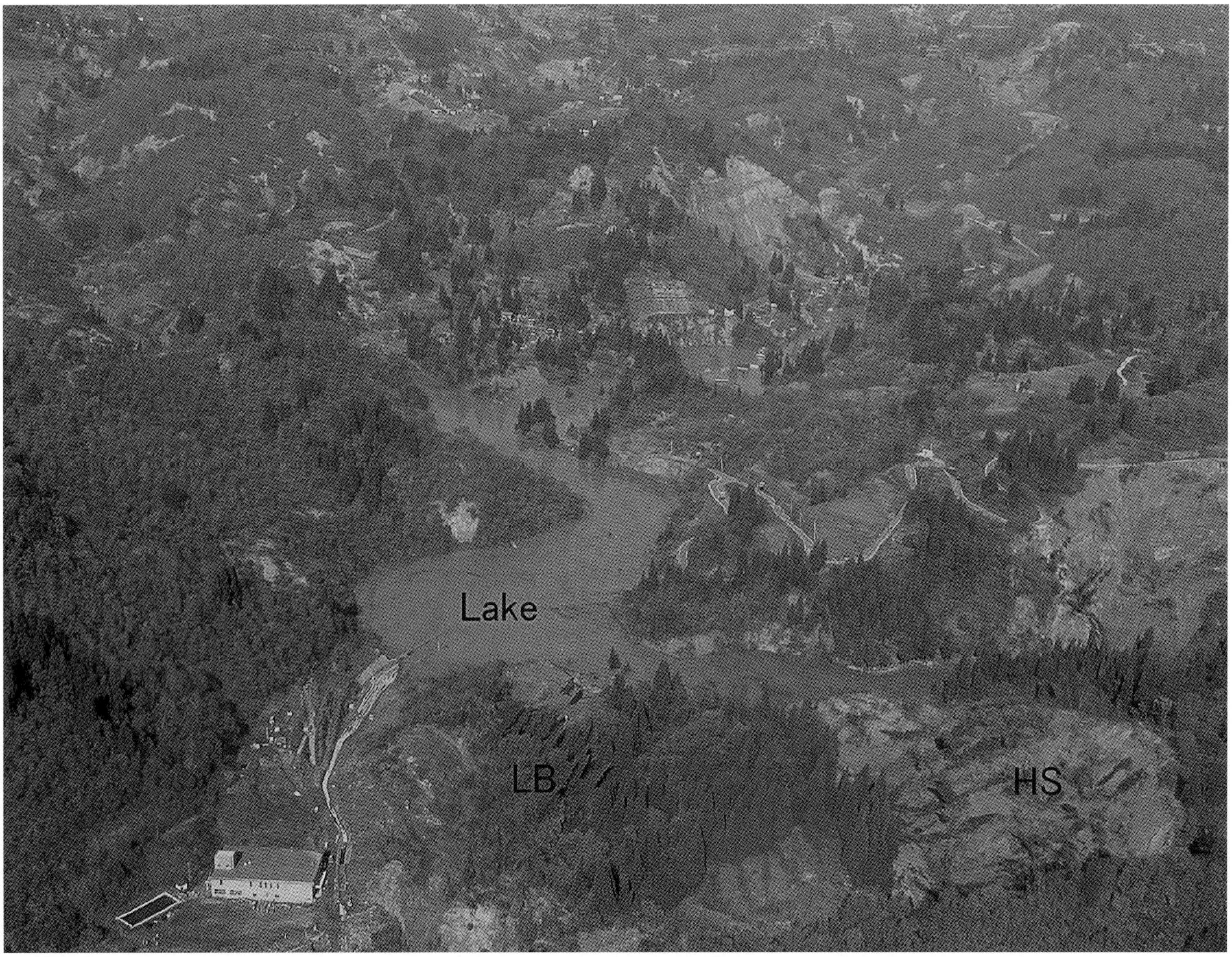

Fig. 11.17. The Higashi Takezawa Landslide, one of the largest landslides induced by the 2004 Mid Niigata Prefecture earthquake. LB: landslide body; HS: head scarp. The lake which formed behind the landslide dam is also labelled.

(Fig. 11.19). The slip surfaces of these landslides are commonly along the base of extensively loosened granite (Iida & Okunishi 1979; Onda 1992), although the loosening mechanism is unknown. Many of these landslides have occurred in association with heavy rain, but have not occurred on common granodiorite slopes without micro-sheeting, described in the following paragraph. Durgin (1977) characterized such moderately weathered granite as 'decomposed granite' (not saprolite), noting that this material is vulnerable to shallow landslides, whereas saprolite is more often subject to deep rotational slides rather than shallow landslides. According to Chigira & Ito (1999) and Suzuki *et al.* (2002), moderately weathered granite is easily loosened to a depth of 50–70 cm within 12 years of a slope being artificially cut, and is characterized by an abrupt weathering front.

Certain types of granite, and some granodiorite, have clearly recognizable microscopic sheets (micro-sheeting) to a depth of 50 m as a result of unloading, which is another primary cause of landslides on granitoid slopes (Hashikawa & Miyahara 1977; Hashikawa 1978;

Table 11.3. *Earthquakes of similar magnitude and intensity that have affected areas of Neogene sedimentary rock*

Earthquake	Date	Magnitude (M_j)	Seismic intensity (JMA)	Fault type	Earthquake fault	Rock types	Age	Landslide number	References
Mid Niigata Prefecture EQ in 2004	23 Oct.	6.8	<6 to 7	Reverse	Yes (obscure)	Sedimentary rocks	Neogene and younger	More than 100 large landslides	Chigira & Yagi (2006)
Noto Peninsula EQ in 2007	25 March	6.9	<6 to >6	Reverse	No	Sedimentary and volcanic rocks	Neogene and younger	A few large landslides	
Off Mid Niigata Prefecture EQ in 2007	16 July	6.8	>6	Reverse	No	Sedimentary rocks	Neogene and younger	A few large landslides	Chigira *et al.* (2012*a*, *b*)

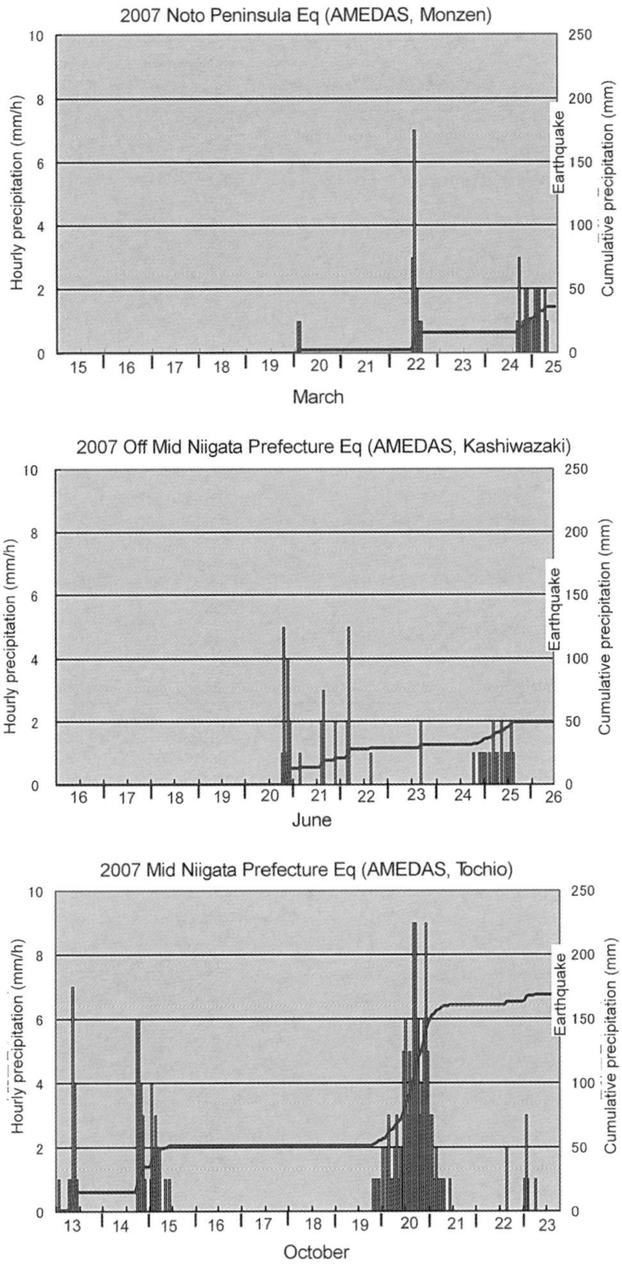

Fig. 11.18. Antecedent rainfall for the three earthquakes affecting areas of Neogene sedimentary rock, of which only the 2004 Mid Niigata Prefecture earthquake resulted in numerous landslides (also see Fig. 11.8).

Chigira 2001). For example, a rainstorm in June 1999 generated numerous shallow landslides in weathered granite in Hiroshima Prefecture, Japan (Chigira 2001), where micro-sheeting formed low-angle micro-joints dipping gently downslope. Micro-sheeted granite is easily disintegrated by the development and interconnection of small cracks, which is accelerated by wetting and drying cycles, temperature fluctuations and creep movement. These types of disintegration can proceed with an abrupt front and the loosened layer can slide. The rate of granite disintegration in this way is poorly constrained but, based on observations of granite micro-sheeting during the course of engineering construction projects, it is estimated to be in the order of years to decades.

Ignimbrite slopes. Pyroclastic rocks such as ash and tuff are easily weathered by hydration, ion exchange, dissolution and clay mineralization (Petit *et al.* 1990). Pyroclastic flow deposits (i.e. ignimbrite) are the most common type of pyroclastic rock in Japan and are frequently subject to landslides during heavy rainfall (Fig. 11.20). Shirasu, a typical unwelded ignimbrite dated at 30 ka, has been subject to shallow landslides on many occasions in Kagoshima Prefecture, southern Japan, resulting in numerous casualties (Yokota 1999). These events have often been in association with the weathering of newly exposed Shirasu via hydration of volcanic glass and clay mineralization with a well-defined weathering front (Chigira & Yokoyama 2005). The migration of the weathering front and deterioration within the weathering zone proceed over a timescale of years, with a landslide recurrence interval on the order of tens to a few hundred years (Shimokawa 1984).

Compared to the unwelded type, weakly welded ignimbrite or vapour-phase crystallized ignimbrite is not so easily weathered, but does have a characteristic weathering profile. Such profiles were recognized as the principal cause of landslides during the heavy rains of 1999 in Fukushima Prefecture, NE Japan (Chigira *et al.* 2002). In this region, the ignimbrite (the Quaternary Shirakawa Pyroclastic Flow) is generally separated into thin plates oriented parallel to the ground surface at the base of heavily weathered sections beneath the topsoil. This so-called foliated zone is *c.* 30–100 cm thick and each plate ranges from a few centimetres to 5 cm in thickness. The tuff in the foliated zone and the overlying soil are much softer than the underlying tuff, although few clay minerals have been detected in the rocks from this zone or in the overlying soil. According to this structure of the weathering zone, rainwater infiltrates the soil and the foliated zone along fractures but cannot penetrate the underlying, less-weathered and unweathered rocks due to a scarcity of cracks. Consequently, the soil and rock of the foliated zone quickly become saturated, leading to sliding. Slopes formed of weakly welded ignimbrite are prone to landslides via this mechanism, while those formed of strongly welded ignimbrite are comparatively stable except for the occasional toppling of rock columns divided by columnar joints. The degree of welding in such deposits is therefore a critical factor that must be considered in landslide hazard assessments.

Volcanic ash slopes. Young deposits of volcanic ash are susceptible to landsliding in heavy rainstorms, partly because they cover slope surfaces in parallel-bedded layers. For example, a rainstorm in 2012 in northern Kyushu induced many shallow landslides and debris flows on the Mount Aso Caldera (Fig. 11.21), one of the largest calderas in the world. During this event, over 400 mm of rainfall fell in four hours, causing widely distributed landslides with slip surfaces a few metres deep. Slip surfaces apparently formed along the boundary between brownish, less permeable, heavily weathered ash layers and overlying black volcanic ash. This area was also affected by intense rainstorms in 1990 and 2001, both of which induced similar landslides. Miyabuchi *et al.* (2004) reported that the sliding layers during the 2001 event were formed of volcanic ash with an age younger than 3 ka.

Deep-seated and catastrophic landslides

Many deep-seated and catastrophic landslides have been induced in recent years in the accretionary complexes of the Outer Zone of SW Japan by rainstorms that produced 1000 mm of precipitation or more. This includes the Miyagawa (in 2004), Kisawa (in 2004), Mimikawa (in 2005) and Kii Mountains (in 2011) disasters (Table 11.2; Fig. 11.22). Landslides occurred in the Kii Mountains at more than 70 different locations, resulting in 56 fatalities (Chigira *et al.* 2012*a*). These landslides, induced by extreme rainfall, caused

Table 11.4. Recent rainstorm events in Japan

Year	Date	Trigger (T: typhoon)	Place (Prefecture)	Geology	Large landslide	Many shallow landslides
1998	26–31 Aug	Rain (Front)	Fukushima	Vapour-phase crystallized ignimbrite	–	○
1999	29 June	Rain (Baiu Front)	Hiroshima	Granite	–	○
2000	28–29 July	Rain (Front)	Rumoi (Hokkaido)	Soft sedimentary rocks	–	○
	11–12 Sept	Rain (Front+T14)	Tokai (Aichi)	Granite	–	○
2003	20 July	Rain (Baiu Front)	Minamata (Kumamoto), Hishikari (Kagoshima)	Andesite lava and pyroclastics	○	○
	9–10 Aug	Rain (T10)	Hidaka (Hokkaido)	Sandstone and conglomerate	–	○
	9–10 Aug	Rain (T10)	Ditto	Mélange	–	○
2004	13 July	Rain (Baiu Front)	Nagaoka (Niigata)	Weak mudstone	–	○
	13 July	Rain (Baiu Front)	Fukui	Volcanic rocks	–	○
	28–29 Sept	Rain (T21)	Miyagawa (Mie)	Accretional complex (hard sedimentary rocks)	○	–
	1 Aug	Rain (T10)	Kisawa (Tokushima)	Accretional complex (Greeenstone and hard sedimentary rocks)	○	–
	29 Sept	Rain (T21)	Ehime–Kagawa	Heavily weathered hard sandstone and mudstone	–	○
	29 Sept	Rain (T21)	Saijo (Ehime)	Schist	○	○
2005	6 Sept	Rain (T14)	Mimikawa (Miyazaki)	Accretional complex (hard sedimentary rocks)	○	–
2006	19 July	Rain (Baiu Front)	Okaya (Nagano)	Loam	–	○
2009	21 July	Rain (Baiu Front)	Hofu (Yamaguchi)	Granite	–	○
2010	16 July	Rain (Baiu Front)	Shobara (Hiroshima)	Soil	–	○
2011	4 Sept.	Rain (T12)	Kii Mountains (Nara, Wakayama)	Accretional complex (hard sedimentary rocks)	○	–
2012	12 July	Rain (Baiu Front)	Aso (Kumamoto)	Volcanic ash	–	○

Note: Circles indicate 'yes'.

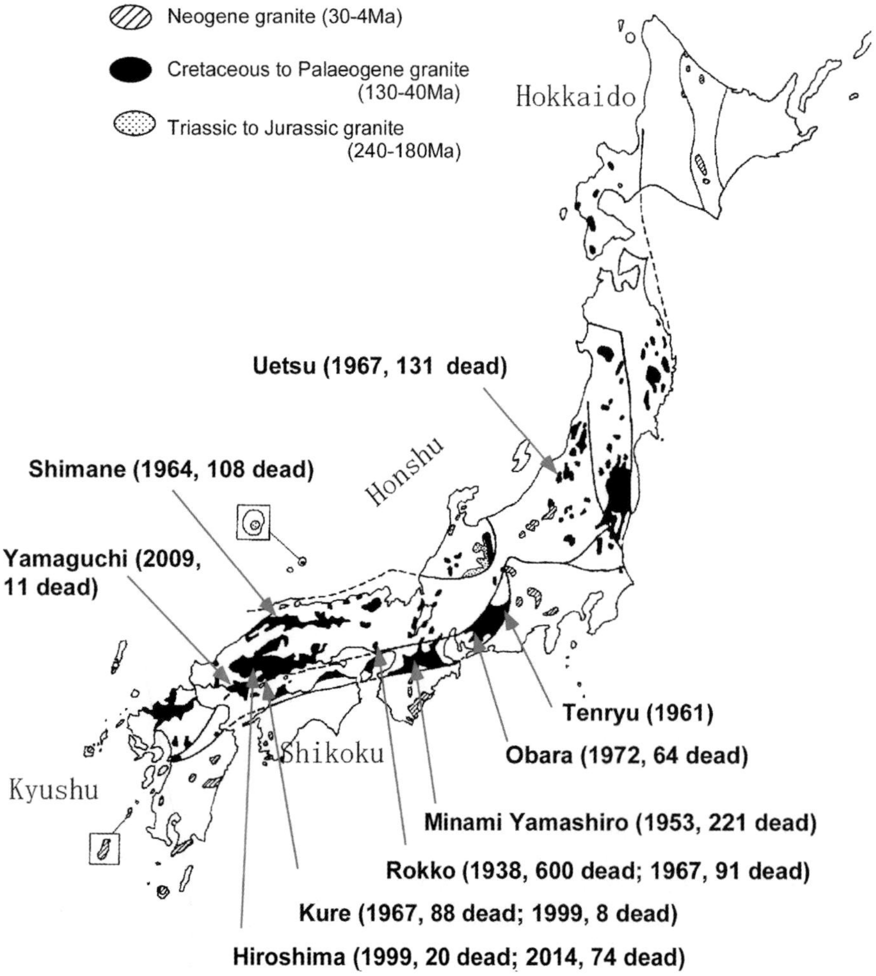

Fig. 11.19. Rainstorm disasters affecting areas of granitic rock (modified from Chigira 2001, fig. 1).

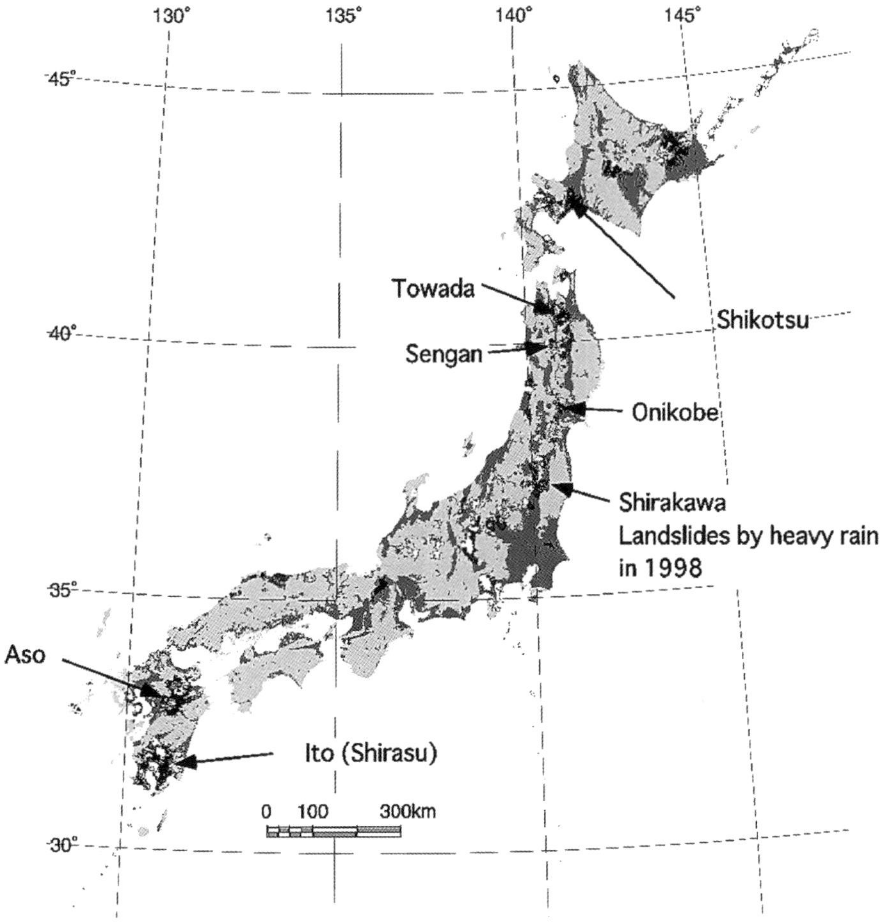

Fig. 11.20. Distribution of Quaternary pyroclastic flow deposits (black areas are from Chigira & Yokoyama 2005).

direct damage to houses, initiated tsunami waves by sliding into swollen rivers and created landslide dams, five of which persisted for more than six months before being breached. Ten of the landslide sites in the Kii Mountains had digital elevation models (at 1 m resolution) from before the event. By comparing the topography of these areas before and after the landslides, Chigira *et al.* (2012*a*) concluded that they had been preceded by gravitational slope deformation as evidenced by the presence of scarplets along

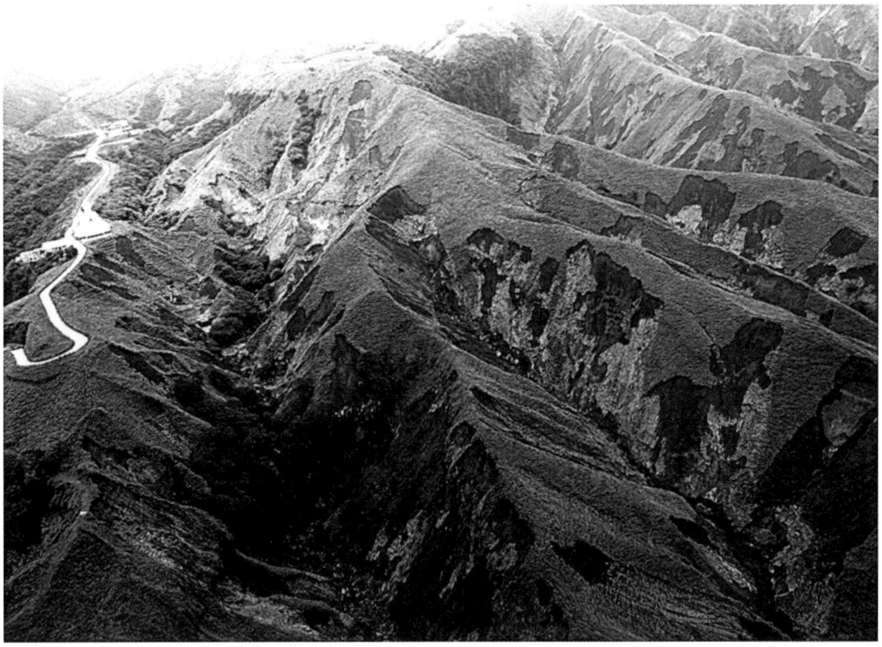

Fig. 11.21. Shallow landslides of volcanic ash at Aso during the 2012 northern Kyushu rainstorm.

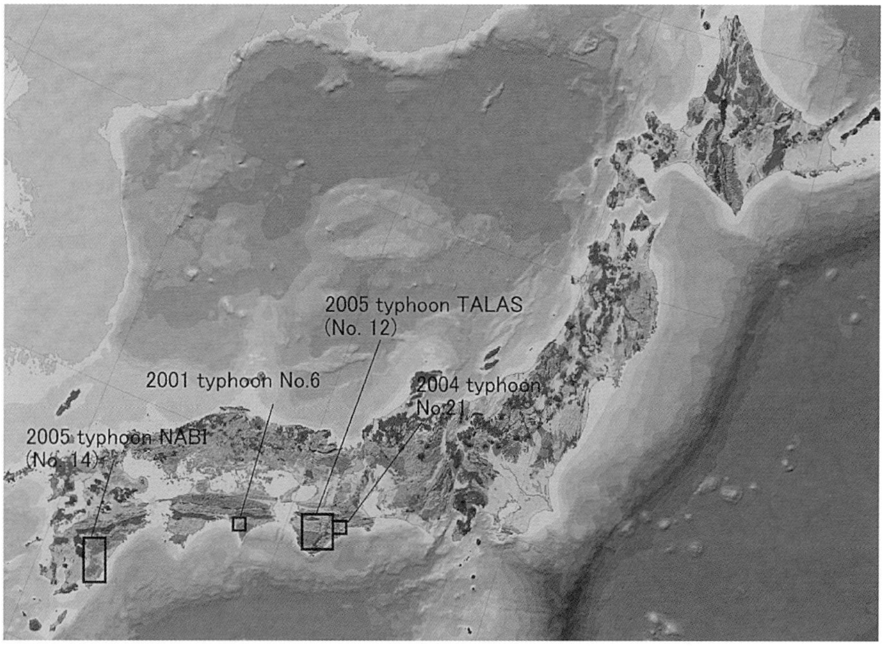

Fig. 11.22. Deep-seated catastrophic landslide disasters induced by rainstorms in the Outer Zone of SW Japan.

the future landslide crowns (Fig. 11.23). The sliding surfaces of most of the landslides were also found to occur along undulating shear surfaces (Fig. 11.23b) that were formed during accretion. Importantly, the presence of features such as scarplets could be used as criteria in predicting catastrophic landslides in the future.

Slow and continuous landslides and gravitational slope deformation

Slow and continuous landslides

There are many deep-seated and slow-moving landslides (which move repeatedly) in Japan, particularly along the coastline of the Japan Sea on Honshu Island. These areas are formed of Neogene sedimentary or volcanic rocks and are also subject to heavy snowfall. Many landslides here occur in mudstone, which is classified as a geotechnically weak rock with poor diagenesis, typically of the Tertiary type as classified by Koide (1955). Many of the landslides have sliding surfaces at the base of the mudstone weathering profile, or in intercalated weathered acid tuff. The development of sliding surfaces may be related to the formation of smectite by weathering, which is a weak clay mineral commonly found in samples taken from the sliding zone of landslides. Weathering of chlorite, a common constituent of mudstone, can form smectite or vermiculite (Chigira 1990; Tsushima & Oho 1999) under the acid-oxidizing

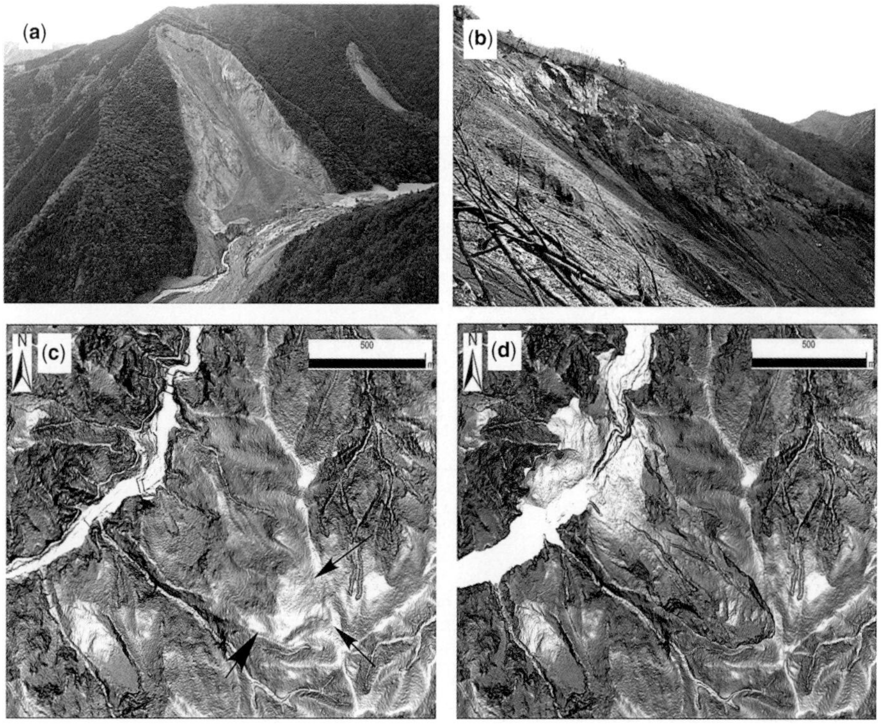

Fig. 11.23. The Akatani Landslide, one of the deep-seated catastrophic landslides induced by Typhoon Talas, 2011. (**a**) Oblique view; (**b**) glossy shear surfaces that formed during accretion and became sliding surfaces; (**c**) slope that existed prior to the landslide (arrows indicate scarplets); and (**d**) slope after the landslide.

conditions induced by pyrite oxidation. Acid tuff is also commonly altered to form bentonite with abundant smectite (Shuzui 2001), which is a major component of sliding zones. Many landslides in this region are also reactivated by snowfall or snowmelt (Matsuura *et al.* 2008).

Deep-seated gravitational slope deformation

Deep-seated gravitational slope deformation has been described using various terms since the 1960s including Sackung, sagging, mass rock creep, rock creep and rock flow (Zischinsky 1966; Radbruch-Hall *et al.* 1976; Varnes 1978; Chigira 1992; Dramis & Sorriso-Valvo 1994; Crosta 1996). Chigira (1992) defined mass rock creep as a mass movement that accompanies long-term gravitational deformation of rocks without a well-defined master sliding surface, and noted that mass rock creep would develop to a landslide in the strictest sense with a master sliding surface as proposed by Zischinsky (1966). Dramis & Sorriso-Valvo (1994) and Agliardi *et al.* (2001) defined deep-seated gravitational deformation as large mass movements with minor displacement, irrespective of whether a master continuous sliding surface is present. The small displacement implicit in this definition is consistent with the definition of 'Jisuberi', a term also used to describe a landslide in Japanese.

Deep-seated gravitational slope deformation is particularly common in densely foliated argillaceous rocks such as slate and pelitic schist (Zischinsky 1966; Chigira 1992; Agliardi *et al.* 2009; Yamasaki & Chigira 2010). Deformation of foliated rocks causes bending (i.e. flexural toppling) (De Freitas & Watters 1973; Cruden 1989), buckling and sliding (Chigira 1992; Tommasi *et al.* 2009). Bending and buckling are easily recognized by identifying a change in foliation attitude. In comparison, sliding is not easily identified, primarily because it does not produce a significant change in foliation attitude and because sliding-related shear zones or non-tectonic faults (Lettis *et al.* 1998) are not well characterized. Two types of gravitational deformation, buckling and sliding, are more unstable than flexural toppling, and can sometimes transform into catastrophic rock avalanches. The 2008 Shiaolin Landslide in Taiwan, for example, had been preceded by a buckling type of gravitational deformation (Tsou *et al.* 2011), and many landslides induced by Typhoon Talas in Japan in 2011 were preceded by sliding along undulating shear surfaces (Chigira *et al.* 2012*a*).

Landslide mitigation in Japan

The Japanese islands have suffered from slope movement hazards for a long period of time, and in response to this risk Japan has established four Acts to mitigate landslide disasters (Table 11.5). Three of these Acts have specifically targeted debris flows, slow-moving landslides, the failure of slag heaps and rapid slope failure of steep slopes, and have set out 'hard' engineering countermeasures. The fourth Act, established in 2000, was the first to focus on 'soft' countermeasures such as evacuation and early warning systems, as well as land-use regulations. This 'soft' approach followed two rain-induced landslide disasters in 1998 (Fukushima) and 1999 (Hiroshima) in which numerous shallow landslides and debris flows affected residential areas and welfare facilities. This Act recognized that a countermeasures approach based solely on hard engineering solutions is not enough to mitigate landslide hazards, because increasing numbers of people live in susceptible areas. Today, every prefectural government in Japan is undertaking investigations to identify landslide-susceptible areas so that appropriate evacuation procedures and warning systems can be developed, and many engineering geologists have been contributing to this practice.

Seismic liquefaction of sandy ground (YM)

Severe structural damage caused by seismic-related ground liquefaction presents a major engineering challenge in Japan. A recent reminder of the problem was provided by the 2011 Tohoku-oki earthquake and its aftershocks, which produced extensive sand liquefaction and associated phenomena at thousands of localities in the Kanto Plain. The problem is typically associated with Quaternary sediments which amplify the seismic waves transmitted from the basin floor crust to increase ground motions. Near-surface unconsolidated sand beds below the water table become liquefied by the seismic motions, consequently exhibiting a characteristic readiness to flow and cause various forms of damage to constructed facilities such as subsidence, tilting and deformation.

Deposits likely to undergo liquefaction

Sand deposits most likely to exhibit liquefaction during an earthquake are those which are water-saturated (below the water table) and with lower permeability but high porosity. Clay or muddy silt horizons are less likely to liquefy under cyclic seismic stress, mainly because of the cohesive strength exerted by clay minerals. If the liquefied layer is capped by less permeable material such as clay, a water layer or film can be formed (Kokusho 2000). In such cases the water mass may be erupted through the overlying bed to emerge on the ground surface as sand volcanoes (Fig. 11.24), providing a superficial indicator of the underground liquefaction.

Figure 11.25 shows the sedimentary deposit types associated with 3332 liquefaction sites and the number percentage of 496 785 grids in the Kanto area following the 2011 Tohoku-oki earthquake (MLIT & the Japanese Geotechnical Society 2011). The figure demonstrates how liquefaction problems are most frequently associated with artificially filled ground. In many cases this phenomenon is a result of mid-twentieth-century coastal post-war restoration, which involved extensive coastal dredging in the Tokyo, Osaka and Nagoya bay areas, for example. The dredged bottom sand was mostly used for land reclamation, resulting in filled ground with loosely packed sand lying beneath the water table. Artificial islands, such as Port Island in Kobe, were constructed using so-called 'masa' sand, which is a coarse-grained sandy material derived from weathered granite. These sands have proven especially vulnerable to seismic liquefaction, as demonstrated in the Kobe Port area during the 1995 Hyogo-ken–Nambu ('Kobe') earthquake (M_w 7.2).

Table 11.5. *Landslide-related Acts in Japan*

Year	Act	Target
1897	Erosion Control Act	Debris flow
1958	Landslide Prevention Act	Slow-moving landslide and failure of slag heaps
1969	Act on Prevention of Disasters Caused by Steep Slope Failure	Rapid slope failure (shallow)
2000	Act on Sediment Disaster Countermeasures for Sediment Disaster Prone Areas	Promotion of soft measures such as evacuation warning systems as well as land usage regulation

Fig. 11.24. Sand volcanoes at Uminonakamichi–Kaihin Park formed by seismic liquefaction of the 2005 Fukuoka earthquake.

Most of the large cities in Japan are located on coastal alluvial plains which provide another sedimentary setting likely to undergo liquefaction during earthquakes. The sea level around the Japanese islands during the last 6000 years has remained at the present level. River estuaries filled with muddy deposits during Holocene transgression to highstand have been covered by sandy deltaic deposits, commonly derived from mountainous provenance areas with high rainfall. These near-surface sandy deposits are commonly several metres thick and widespread across the Japanese coastal plains. Their high porosity, resulting from low overburden pressure and water saturation, makes them an important focus for liquefaction and infrastructure damage as amplified seismic waves travel across the low-lying plains.

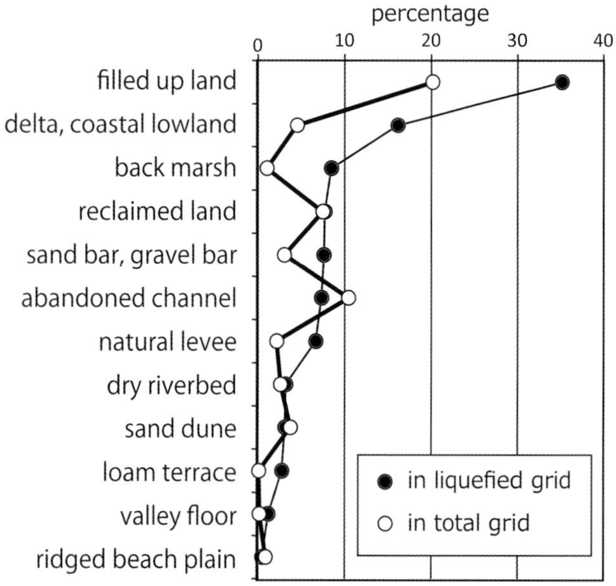

Fig. 11.25. The deposit types of the 3332 liquefaction sites (black circle) and the number percentage of 496 785 grids (white circle) in the Kanto region (after MLIT & the Japanese Geotechnical Society 2011).

A third important type of sedimentary unit that is highly prone to liquefaction is that associated with abandoned river channel deposits. Many meander bends of natural rivers in Japan have been filled for flood control, producing artificial ground that is at risk from seismic liquefaction (Fig. 11.25). The risk is enhanced by the fact that these filled channels are located at low altitudes with a high water table and are mostly composed of sandy sediments. A famous example of the involvement of river terrace deposits in seismic damage resulted from the 1964 Niigata earthquake (M_w 7.6) on the west coast of northern Honshu. The pattern of liquefaction-related subsidence and the eruption of sand volcanoes closely followed the locations of former river channels. The damage inflicted upon the city of Niigata, with thousands of houses damaged or destroyed and the spectacular collapse of the Showa Bridge, was important in attracting attention to the disaster potential of sand liquefaction.

Liquefaction in historical records

Wakamatsu (1993) examined about 1000 earthquakes in Japan between the year 416 and 2008, using ancient documents and literatures. She found almost 16 500 locations indicating liquefaction-related phenomena, such as the eruption of sand, mud and water and floating of underground facilities. Among these 150 sites were damaged by more than two earthquakes, indicating repeated liquefaction of the same sediment body. The records show 11 such repetitions in the Nobi and Niigata plains, 10 in the Akita and Noshiro plains, 9 in the Osaka Plain, 8 in the Kanto Plain and 6 in the Tokachi and Kushiro plains and the Kyoto Basin. Such observations appear to contradict the prediction that post-liquefaction sand is likely to have had its volume reduced by dewatering, and therefore changed into more densely packed and resistant materials unlikely to be remobilized during subsequent shocks. Instead, the data suggest that, in some cases, the sand layer in these Japanese low-lying areas is neither homogeneous nor completely liquefied by the first event, so that some risk remains of future seismic re-liquefaction. A recent example is again provided by the liquefaction sites induced by the 2011 Tohoku-oki earthquake (M_w 9.0), some of which are very close to sites previously mobilized during the 1987 Chiba-ken–Toho-oki earthquake (M_w 6.7) (Fig. 11.26).

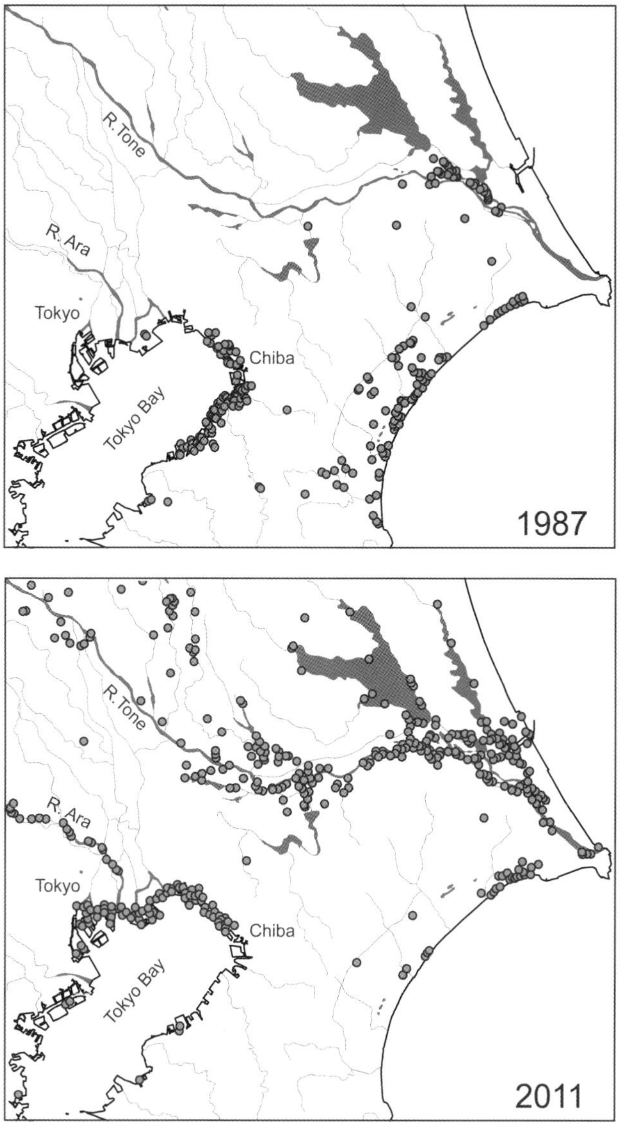

Fig. 11.26. Liquefaction sites in the Kanto area by the 1987 Eastward Offshore Chiba Earthquake and by the 2011 Great East Japan Earthquake. Liquefaction damage is mostly concentrated within the filled ground around Tokyo Bay, the old channels of the Tone river and the coastal plain facing the Pacific Ocean. Data from MLIT & the Japanese Geotechnical Society (2011) and Chiba Prefecture Headquarters for Earthquake Disaster Prevention (1989).

Liquefaction in the geological record

The occurrence of underground liquefaction can be inferred from erupted sand and water or other ground-surface deformation. However, such observations at the surface do not reveal exactly the depth to liquefaction or how the underground sand or water behaved. Although the shallow subsurface structures of sand volcanoes have been observed by trench surveys, little is known of the deeper structures. On the other hand, there are records of liquefaction preserved in ancient sedimentary sequences. Most sand dykes are believed to provide evidence for palaeo-liquefaction, and in some cases they can be linked to particular archaeological remains relating to earthquake events recorded in ancient documents (e.g. Sangawa 1999).

Conspicuous examples of deeper deformation structures were reported by Mitani (2006) from middle Pleistocene fluvial deposits in northern Chiba Prefecture (Fig. 11.27), especially at the interface between sandy channel deposits and overlying marsh muddy deposits. These beds are intercalated within highstand shallow-marine strata, and can be traced laterally more than 10 km along the former bed of the River Tone. Deformation structures such as convoluted or mixed or faded laminations are observed in the fluvial sandy deposits, sand dykes and sills intruded into the overlying mud interval, or sand doming into the sand interval of the marsh facies. The convolutions with wavelength <1 m have pointed ridges involving pumice fragments and rounded troughs with small-scale sags. Vertical sand dykes and lenticular sand sills cut through the overlying mud, both involving mud clasts originated from thin layers in the underling fluvial sand-dominated part. The strike direction of the dykes is mostly consistent within each outcrop. Upwards-intruded sand produces thick sand layers with dome- or mushroom-shaped structures instead of dykes, often formed above the crest of the convolute lamination. Stratigraphic examination of these structures revealed that the liquefaction occurred in the sandy deposits of 2.5–4 m depth beneath the flat marsh ground, and that vertical movement rather than lateral flow of sand was predominant. A lack of evidence for eruption through to the ground surface or surface displacement serves as a reminder that liquefied sand loses its strength, but does not necessarily deform or flow unless an external force is applied. The external force needed to induce convolution and intrusion structures may be applied by the gravitational force due to the density stratification in a sand bed and loading by the overlying mud layer, which confined the increased pore pressure within the liquefied sand bed.

Nuclear waste disposal (HY)

Deep geological disposal of vitrified high-level radioactive waste (HLW) is being considered in Japan. HLW wastes are produced during the recycling of spent nuclear fuel and are contained in vitrified form by stainless steel canisters (each *c.* 40 cm in diameter and 1.2 m in height). As of the end of 2011, a total amount of HLW equivalent to 26 000 canisters had been accumulated. Appropriate disposal sites are located at a depth of more than 300 m below the ground surface (e.g. JNC 2000), in a suitable geological environment with long-term stability and host rocks that function as barriers to radionuclide migration. Both geological stability and host-rock barrier functions are crucial factors that must be demonstrated to build confidence among stakeholders, including the public, that nuclear waste will be isolated safely from human beings by deep geological disposal.

Long-term stability

Long-term stability of the geological environment is a key factor that must be considered when siting an HLW repository in Japan. In the orogenic field in particular, assessments of the long-term safety and performance of a repository must be based on a sound understanding of the potential impacts on deep geological environments of volcanic activity, faulting and uplifting. To help develop this understanding and hence define transparent siting factors (factors that indicate whether a site may be suitable to host a repository), the nature of basic geological conditions, volcanic activity, geothermal gradient, locations of active faults and uplift rates throughout Japan have been compiled (Fig. 11.28; Committee for Geological Stability Research 2011). Generic criteria for recognizing areas within the Japanese archipelago that show long-term stability have therefore been determined.

In the Japanese islands volcanoes are generally distributed parallel to a 'volcanic front', which is caused by magmatism above plates being subducted under the island arc. Areas where the geothermal gradient is higher than the background average of *c.* 3°C per 100

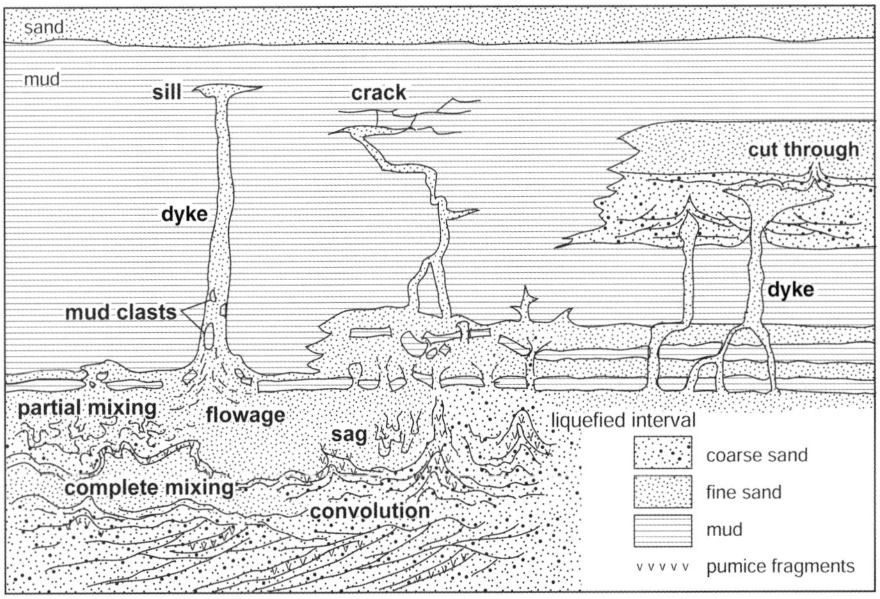

Fig. 11.27. Summarized liquefaction-induced deformation structures in the Pleistocene fluvial deposits (after Mitani 2006).

m also correspond closely to volcanic activity. Temperature is an important siting factor that influences the long-term performance of a repository system, since the degradation of synthetic barrier materials is temperature dependent.

Uplift is also a basic siting factor because it impacts upon the long-term physical isolation of a deep repository. Uplift is caused by the heterogeneous stress distribution due to the tectonic movement of lithospheric plates in and around the Japanese islands.

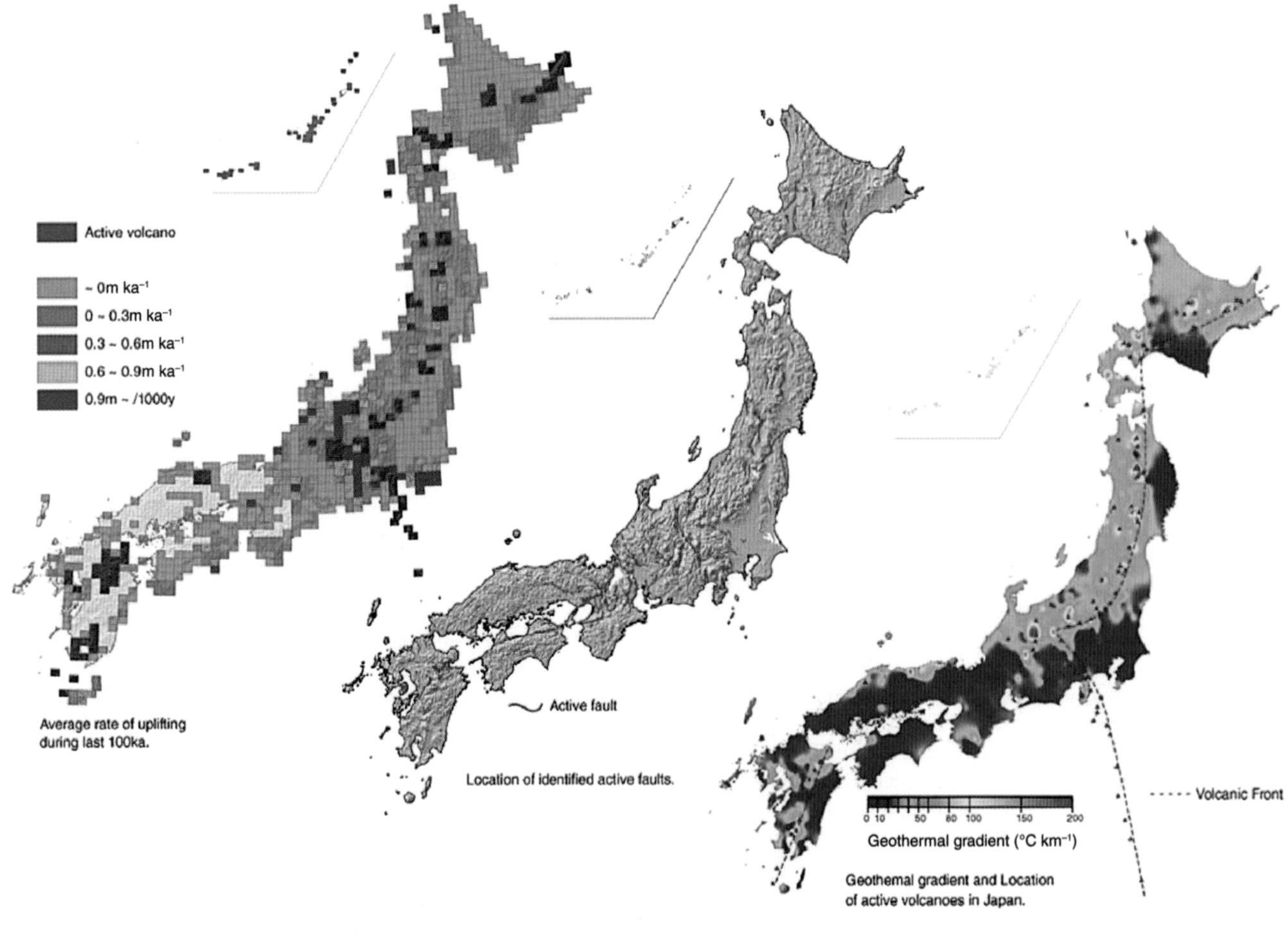

Fig. 11.28. Geological conditions of Japan (modified from Committee for Geological Stability Research 2011).

Due to the time required for the radionuclides within HLW to undergo radioactive decay, it is estimated that vitrified wastes will need to be isolated for more than 100 ka. A deep geological repository should not be located within an area where rapid uplift of more than 3 mm a^{-1} (i.e. 300 m/100 ka) is taking place.

When siting a deep geological repository it is also important to avoid potentially adverse impacts from active fault movements and earthquakes, especially damage to the barrier system of the repository. In order to ensure that active faults do not occur in the vicinity of a proposed repository site, detailed site and fault characterization is carried out when a candidate site has been selected.

The general characteristics of Japanese geology demonstrate that not all locations in Japan are sufficiently stable in the long-term to be suitable for the site of a final deep geological repository. In particular, an area must be avoided if it shows active volcanic activity and high geothermal gradients, active faulting and/or rapid uplift. Although there are many such areas in Japan as shown in Figure 11.28, there are also parts of the country that are clearly stable, and these provide potentially suitable sites for selection as a deep geological repository for HLW.

Barrier function of the geological environment

Bedrock geology has a major influence on solute transport and significantly affects the ability of subsurface rock masses to behave as natural barriers. Since the 1980s, one aspect of engineering geology studies in Japan has been the investigation of natural barrier functions within abandoned mines (e.g. Kamaishi Mine and Tono Mine) and Underground Research Laboratories (URL, at Horonobe and Mizunami).

Orogenic granitic rocks have been studied in Kamaishi Mine and the URL at Mizunami. The former is an abandoned iron mine located in the Kitakami area of northern Japan. The Kamaishi study was conducted as a prototype *in situ* study (JNC 1999) that preceded the construction of a URL in granitic rock in Mizunami, Gifu Prefecture, central Japan. These underground studies identified features of fracture systems in orogenic granites that influence hydrogeology (groundwater flow) and geochemistry, and that would affect nuclide migration around any repository constructed in these kinds of rocks. The processes by which fractures evolve were also determined. It was shown that the overall geometries and characteristics of fluid-conducting fracture systems in plutons of the orogenic field are in fact broadly stable (Fig. 11.29; Yoshida *et al.* 2013). Knowledge of the characteristics of fluid flow in fractured granitic rocks is important for developing a conceptual model for the barrier functions of granitic rocks in the orogenic field of Japan. This conceptual model can then be used in assessments of the safety and performance of a final deep geological repository. Another Japanese URL is the Horonobe Underground Facility located in Hokkaido, the northernmost large island of Japan. The URL at Horonobe is constructed in Tertiary sedimentary rock, and studies are being conducted in order to develop methodologies for characterizing such lithologies and to understand their generic geological features. At both URLs, engineering studies as well as geoscientific studies have been carried out. These engineering studies include investigations of: rock mechanics during deep underground excavation; influences of excavation on groundwater hydrology and geochemistry; and interactions between engineered materials and the host rock. Understanding these processes in the long term is a challenge that must be met in order to develop confidence among stakeholders in the feasibility of nuclear waste disposal.

Tono Uranium Mine (Tsukiyoshi uranium deposits) is a disused mine excavated within Tertiary sedimentary rocks and located in the Tono area of Gifu Prefecture, central Japan. This mine was used for a 'natural analogue' study (e.g. Yoshida 1994). A natural analogue is a natural phenomenon that has some similarities to one or more aspects of a deep geological repository. Consequently, a natural analogue can be used to help understand the processes that would influence radionuclide migration (i.e. retardation and sorption) and/or the long-term barrier function of the geological environment (e.g. Miller *et al.* 2000). It is particularly important to evaluate the long-term processes that cannot be investigated fully in simple, short-term laboratory experiments. The Tono Mine was excavated within the sandstone-type Tsukiyoshi uranium deposit, which formed near the base of the Tertiary sedimentary rock sequence *c.* 10 Ma ago. Studies of the uranium deposit were used to qualitatively evaluate the long-term processes by which uranium-series nuclides migrated and accumulated in the long-term, due to redox potential changes in sedimentary rock (Yoshida 1994). Such natural phenomena suggest that commonly identified mineral constituents of many kinds of sedimentary rock have a capacity to retain migrating radionuclides.

Another important analogue of long-term natural barrier functioning is redox front formation along groundwater-conducting fractures, combined with solute diffusion from the fracture surfaces into the rock matrices, known as 'matrix diffusion' (e.g. Miller *et al.* 2000). These processes have been investigated in order to estimate quantitatively the significance of radionuclide retardation processes in fractured rocks that might be considered to host a repository. Redox reactions along a conductive fracture extending downwards from the land surface occur as downwards-flowing oxygenated or

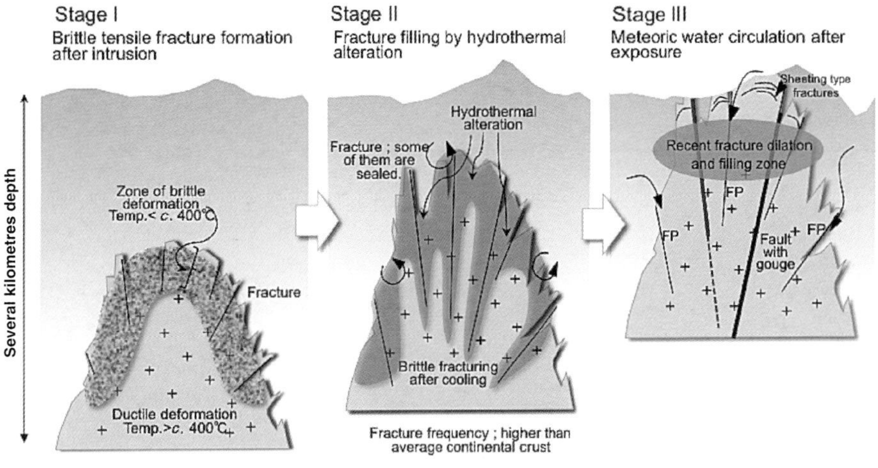

Fig. 11.29. Schematic model of fracture formation during the evolution of a granite pluton within the Japanese orogenic field (modified from Yoshida *et al.* 2013).

acid-reducing porewater diffuses into the wall rock from the fracture surfaces (Yamamoto *et al.* 2013). Mineralogical evidence for the redox reactions that occur indicates the depth to which solutes penetrate from the fracture surfaces into the rock matrix. Mineralogical evidence for redox reactions and associated analyses of trace element distributions can therefore be treated as an analogue of 'matrix diffusion' involving radionuclides and used as a basis for estimating radionuclide retardation. Natural analogues are effective tools to demonstrate realistic long-term processes of radionuclide migration and retardation in deep geological environments.

Conclusions

On the subject of engineering geology in Japan, we have reviewed engineering geological issues related to construction projects of infrastructures, landslides, liquefaction and disposal of radioactive waste, with special reference to faults which provide one of the most difficult problems both from earthquake generation and from degraded rock qualities. Our forerunners have been struggling with adverse geological conditions, created during a long geological history under active tectonic settings, by: identifying and locating particular geological features such as fault shear zones, high-temperature zones, mud diapirs and weak rocks; siting and designing structures; and employing particular techniques for construction.

New findings in geological science and new investigation techniques have yielded new insights into engineering geology. The finding of accretion complexes and mélanges was epoch making in engineering geology in Japan; disrupted and foliated sedimentary rocks, which have been commonly encountered, can now be appropriately interpreted from the geological viewpoint. Many thrust faults of accretionary complexes have been found, which can become an embryo of sliding surfaces of deep-seated gravitational slope deformation and catastrophic rockslides. Such new viewpoints have been improving research and techniques in landslide hazard mitigation. New understanding of mud diapirs has led to a re-interpretation of the bulging ground encountered during a very difficult tunnel construction. Airborne laser scanner techniques have provided new theories for the prediction of potential landslide sites. Liquefaction of sandy ground during an earthquake, which has been studied from engineering sectors, can be newly interpreted from the viewpoint of geological materials and liquefaction structures recorded in geological stratum. One of the most important challenges in the engineering geology of Japan is the geological disposal of high-level radioactive wastes. We must evaluate the long-term safety of a geological repository in such tectonically active islands regardless of the energy policies of the Japanese government.

Appendix

English to Kanji and Hiragana translations for geological and place names

Abo	安房	あぼう
Abo-toge	安房峠	あぼうとうげ
Akandana	アカンダナ	あかんだな
Akashi	明石	あかし
Akashi Kaikyo	明石海峡	あかしかいきょう
Akatani	赤谷	あかたに
Akita	秋田	あきた
Asaji	朝地	あさじ
Aso	阿蘇	あそ

(Continued)

English to Kanji and Hiragana translations for geological and place names (Continued)

Awaji	淡路	あわじ
Chiba	千葉	ちば
Fukuoka	福岡	ふくおか
Fukushima	福島	ふくしま
Gifu	岐阜	ぎふ
Higashi Takezawa	東竹沢	ひがしたけざわ
Higashi-Kubiki	東頸城	ひがしくびき
Hirayu	平湯	ひらゆ
Hiroshima	広島	ひろしま
Hokkaido	北海道	ほっかいどう
Honshu	本州	ほんしゅう
Horonobe	幌延	ほろのべ
Hyogo	兵庫	ひょうご
Imaichi	今市	いまいち
Inaba	稲葉	いなば
Iwate	岩手	いわて
Izu-Oshima-kinkai	伊豆大島近海	いずおおしまきんかい
Joetsu	上越	じょうえつ
Kagoshima	鹿児島	かごしま
Kamaishi	釜石	かまいし
Kanto	関東	かんとう
Kii	紀伊	きい
Kisawa	木沢	きさわ
Kitakami	北上	きたかみ
Kobe	神戸	こうべ
Kushiro	釧路	くしろ
Kyoto	京都	きょうと
Kyushu	九州	きゅうしゅう
Maiko	舞子	まいこ
Matsuho	松帆	まつほ
Mimikawa	耳川	みみかわ
Mino	美濃	みの
Miyagawa	宮川	みやがわ
Miyagi	宮城	みやぎ
Miyazaki	宮崎	みやざき
Mizunami	瑞浪	みずなみ
Nabetachiyama	鍋立山	なべたちやま
Nagano	長野	ながの
Naganoken-seibu	長野県西部	ながのけんせいぶ
Nagoya	名古屋	なごや
Nakanoyu	中の湯	なかのゆ
Naruto	鳴門	なると
Nebukawa	根府川	ねぶかわ
Niigata	新潟	にいがた
Nishiyama	西山	にしやま
Nobi	濃尾	のうび
Norikura	乗鞍	のりくら
Norikuradake	乗鞍岳	のりくらだけ
Noshiro	能代	のしろ
Oita	大分	おおいた
Ontake-san	御嶽山	おんたけさん
Osaka	大阪	おおさか
Rokko	六甲	ろっこう
Shikoku	四国	しこく
Shimane	島根	しまね
Shimanto	四万十	しまんと
Shirakawa	白河	しらかわ
Sugawa	須川	すがわ
Takashomachi	鷹匠町	たかしょうまち
Teradomari	寺泊	てらどまり
Tohoku	東北	とうほく
Tokachi-oki	十勝沖	とかちおき

(Continued)

English to Kanji and Hiragana translations for geological and place names (Continued)

Tokamachi	十日町	とおかまち
Tokyo	東京	とうきょう
Tone	利根	とね
Tono	東濃	とうのう
Tsukiyoshi	月吉	つきよし
Uminonakamichi-Kaihin Park	海の中道海浜公園	うみのなかみちかいひんこうえん
Urita	瓜田	うりた
Usukidani	臼杵谷	うすきだに
Yakedake	焼岳	やけだけ

References

Active Fault and Earthquake Research Center 2012. Active Fault Database of Japan. https://gbank.gsj.jp/activefault/index_cyber.html

Agliardi, F., Crosta, G. & Zanchi, A. 2001. Structural constraints on deep-seated slope deformation kinematics. *Engineering Geology*, **59**, 83–102.

Agliardi, F., Zanchi, A. & Crosta, G. B. 2009. Tectonic v. gravitational morphostructures in the central Eastern Alps (Italy): constraints on the recent evolution of the mountain range. *Tectonophysics*, **474**, 250–270.

Akita, K. & Sato, K. 1993. Tunnel through expansive mudstone zone (Mabetachiyama Tunnel). *Tsuchi-to-Kiso*, **47**, 41–53 [in Japanese].

Chiba Prefecture Headquarters for Earthquake Disaster Prevention (eds) 1989. *The 1987 Offshore Chiba Earthquake — Disaster Records*. Chiba Prefecture, Chiba [in Japanese].

Chigira, M. 1982. Dry debris flow of pyroclastic fall deposits triggered by the 1978 Izu-Oshima-Kinkai earthquake: the 'collapsing' landslide at Nanamawari, Mitaka-Iriya, southern Izu Peninsula. *Journal of Natural Disaster Science*, **4**, 1–32.

Chigira, M. 1990. A mechanism of chemical weathering of mudstone in a mountainous area. *Engineering Geology*, **29**, 119–138.

Chigira, M. 1992. Long-term gravitational deformation of rocks by mass rock creep. *Engineering Geology*, **32**, 157–184.

Chigira, M. 2001. Micro-sheeting of granite and its relationship with landsliding specifically after the heavy rainstorm in June 1999, Hiroshima Prefecture, Japan. *Engineering Geology*, **59**, 219–231.

Chigira, M. & Ito, E. 1999. Characteristic weathering profiles as basic causes of shallow landslides. *In*: Yagi, N., Yamagami, T. & Jiang, J.-C. (eds) *Slope Stability Engineering*. Balkema, Rotterdam, 1145–1150.

Chigira, M. & Yagi, H. 2006. Geological and geomorphological characteristics of landslides triggered by the 2004 Mid Niigta prefecture earthquake in Japan. *Engineering Geology*, **82**, 202–221.

Chigira, M. & Yokoyama, O. 2005. Weathering profile of non-welded ignimbrite and the water infiltration behavior within it in relation to the generation of shallow landslides. *Engineering Geology*, **78**, 187–207.

Chigira, M., Nakamoto, M. & Nakata, E. 2002. Weathering mechanisms and their effects on the landsliding of ignimbrite subject to vapor-phase crystallization in the Shirakawa pyroclastic flow, northern Japan. *Engineering Geology*, **66**, 111–125.

Chigira, M., Matsushi, Y., Tsou, C.-Y., Hiraishi, N., Matsuzawa, M. & Matsuura, S. 2012*a*. Deep-seated catastrophic landslides induced by typhoon 1112 (Talas). *Annuals of Disaster Prevention Research Institute, Kyoto University*, **55**, 193–211 [in Japanese with English abstract].

Chigira, M., Nakasuji, A., Fujiwara, S. & Sakagami, M. 2012*b*. Soil slide-avalanches of pyroclastic fall deposits induced by the 2011 off the Pacific Coast of Tohoku Earthquake. *Journal of Japan Society of Engineering Geology*, **52**, 222–230 [in Japanese with English abstract].

Committee for Geological Stability Research 2011. *Geological Leaflet 4, Japanese Island-arc and Geosphere Stability*. The Geological Society of Japan, Tokyo [in Japanese].

Crosta, G. 1996. Landslide, spreading, deep seated gravitational deformation: analysis, examples, problems and proposals. *Geografia Fisica e Dinamica Quaternaria*, **19**, 297–313.

Cruden, D. M. 1989. Limits to common toppling. *Canadian Geotechnical Journal*, **26**, 737–742.

Cruden, D. M. & Varnes, D. J. 1996. Landslide types and processes. *In*: Turner, A. K. & Schuster, R. L. (eds) *Special Report 247: Landslides, Investigation and Mitigation*. TRB, National Research Council, Washington, DC, 36–75.

De Freitas, M. H. & Watters, R. J. 1973. Some field examples of toppling failure. *Geotechnique*, **23–4**, 495–514.

Doke, R., Tanikawa, R., Yasue, K., Nakayasu, A., Niizato, T., Umeda, K. & Tanaka, T. 2012. Spatial patterns of initiation ages of active faulting in the Japanese Islands. *Active Fault Research*, **37**, 1–16.

Dramis, F. & Sorriso-Valvo, M. 1994. Deep-seated gravitational slope deformations, related landslides and tectonics. *Engineering Geology*, **38**, 231–243.

Durgin, P. B. 1977. Landslides and the weathering of granitic rocks. *Reviews in Engineering Geology*, **3**, 127–131.

Fukushima, K., Kanaori, Y. & Miura, F. 2010. Influence of fault process zone on ground shaking of inland earthquakes: verification of Mj = 7.3 Western Tottori Prefecture and Mj = 7.0 West off Fukuoka Prefecture Earthquakes, Southwest Japan. *Engineering Geology*, **116**, 157–165.

Harayama, S. 1990. *Geology of the Kamikochi district with Geological Sheet Map at 1:50 000*. Geological Survey of Japan, Tsukuba, 175 [in Japanese with English abstract].

Hasegawa, K. 2008. Adoption of inclined artificial abutment in the Inaba Dam. *Engineering for Dams*, **260**, 81–90 [in Japanese].

Hashikawa, K. 1978. Weathering structure of the grnitic rocks in the dissected pediment. *Journal of Japan Society of Engineering Geology*, **19**, 45–59 [in Japanese with English abstract].

Hashikawa, K. & Miyahara, K. 1977. Structure of the weathered granitic rocks and in its engineering significance. *Jounal of Japan Society of Engineering Geology*, **15**, 47–57. [in Japanese with English abstract].

Headquarters for Earthquake Research Promotion 2010. *Report on Research Technique of Active Faults*. Temporary Edition, The Ministry of Education, Culture, Sports, Science and Technology, Tokyo [in Japanese].

Iida, T. & Okunishi, K. 1979. On the slope development caused by the surface landslides. *Geographical Review of Japan*, **52**, 426–438 [in Japanese with English abstract].

Inoue, Y., Honsho, S., Matsushima, M. & Esashi, Y. 1970. *Geological and Soil Mechanical Studies on the Slides Occurred During the 1968 Toakachioki Earthquake in Southeastern Area of Aomori Prefecture*. Central Research Institute of Electric Power Industry, Report no. 69086 [in Japanese with English abstract].

JNC (Japan Nuclear Fuel Cycle Development Institute) 1999. *Proceedings of International Workshop for the Kamaishi In-situ Experiments*. Kamaishi, March 1999, JNC TN, 7400, 99-007, 275p [in Japanese].

JNC (Japan Nuclear Fuel Cycle Development Institute) 2000. *Project to Establish the Scientific and Technical Basis for HLW Disposal in Japan*. Project Overview Report, JNC TN, 1410, 2000–001 [in Japanese].

Kakimi, T., Kinugasa, Y. & Kato, T. 1978. *Active Fault Map of Japan 1:2 000 000*. Geological Survey of Japan, Tsukuba [in Japanese].

Kamai, T. 1990. Failure mechanism of deep-seated landslides caused by the 1923 Kanto earthquake, Japan. *Proceedings of the Sixth International Conference and Field Workshop on Landslides*, 12 September 1990, Milan, Italy, 187–198.

Kanaori, Y. 1997. *Earthquake Proof Design and Active Faults*. Elsevier, Amsterdam.

Kanaori, Y. 2001. How far does the influence of faulting extend. *Journal of the Japan Society of Engineering Geology*, **41**, 323–332 [in Japanese with English abstract].

Kanaori, Y., Tanaka, T. & Chigira, M. 2000. *Engineering Geological Advances in Japan for the New Millennium*. Elsevier, Amsterdam.

Keefer, D. 1984. Landslides caused by earthquakes. *Geological Society of America Bulletin*, **95**, 406–421.

Kim, Y.-S., Peacock, D. C. P. & Sanderson, D. J. 2004. Fault damage zone. *Journal of Structural Geology*, **26**, 503–517.

Kogure, M. & Kimura, H. 1995. Completion of excavation of heavily swelled tunnel: Nabetachiyama Tunnel of Hokuetsu-Hokusen. *Tunnels and Underground*, **26–27**, 7–12 [in Japanese].

Koide, H. 1955. *Landslides in Japan*. Toyo Keizai Shinpo, Tokyo [in Japanese].

Kokusho, T. 2000. Emergence of water film in layered sand due to seismic liquefaction and its effect on soil stability. *Journal of Japan Society of Engineering Geology*, **41**, 77–86 [in Japanese with English abstract].

KOSAKA, K., KANAORI, Y., CHIGIRA, M. & YOSHIDA, S. 2010. *Fault Map of Japan*. Baifukan Publishing Co., Tokyo [in Japanese].

KURIHARA, T. 1998. Effects of the Kobe Earthquake on the Akashi Kaikyo Bridge. *Bridge and Foundation Engineering*, **98-8**, 94–97 [in Japanese].

LETTIS, W. R., KELSON, K. I., HANSON, K. L. & ANGELL, M. A. 1998. Is a fault by any other name? Differentiating tectonic from nontectonic faults. *Proceedings of the 8th International Conference of IAGE*, 21 September, Vancouver, Canada, 609–629.

MATSUDA, T. 1975. Magnitude and recurrence interval of earthquakes from a fault. *Zisin 2 (Journal of the Seismological Society of Japan, Second Series)*, **28**, 269–283 [in Japanese with English abstract].

MATSUDA, T. 1990. Seismic zoning map of Japanese Islands, with maximum magnitudes derived from active fault data. *Bulletin of the Earthquake Research Institute, University of Tokyo*, **65-1**, 289–319 [in Japanese with English abstract].

MATSUSHITA, T. 1992. An exploratory tunnel was opened by conquest of high temperature ground, high water pressure and low velocity ground, Abo Tunnel of national highway 158. *Tunnels and Underground*, **23-3**, 7–14 [in Japanese].

MATSUURA, Y. & OKAZAKI, K. 1999. On the design and construction of Urita Dam. *Dam Nippon*, **657**, 19–48 [in Japanese].

MATSUURA, S., ASANO, S. & OKAMOTO, T. 2008. Relationship between rain and/or meltwater, pore-water pressure and displacement of a reactivated landslide. *Engineering Geology*, **101**, 49–49.

MILLER, W., ALEXANDER, R., CHAPMAN, N., MCKINLEY, I. & SMELLIE, J. (eds) 2000. *Geological Disposal of Radioactive Wastes and Natural Analogue*. Waste Management Series, **2**. Elsevier Science Ltd, Oxford.

MITANI, Y. 2006. Earthquake-induced liquefaction and fluidization in middle Pleistocene in the northern area of Chiba Prefecture, Japan. *Earth Science (Chikyu Kagaku)*, **60**, 241–252 [in Japanese with English abstract].

MIYABE, N. 1935. A study on landslide. *Bulletin of the Earthquake Research Institute, Tokyo University*, **13**, 85–113.

MIYABUCHI, Y., DAIMARU, H. & KOMATSU, Y. 2004. Landslides and lahars triggered by the rainstorm of June 29, 2001, at Aso Volcano, southwestern Japan. *Transactions, Japanese Geomorphological Union*, **25**, 23–43 [in Japanese with English abstract].

MIYAKE, Y. & OSSAKA, J. 1998. Steam explosion of February 11th, 1995 at Nakanoyu Hot Spring, Nagano Prefecture, Central Japan. *Bulletin of the Volcanological Society of Japan*, **43**, 113–121 [in Japanese with English abstract].

MIYAZAKI PREFECTURE 1998. *Construction Record of Urita Dam*. Miyazaki Prefecture, Miyazaki [in Japanese].

MLIT (MINISTRY OF LAND, INFRASTRUCTURE AND TRANSPORT) & THE JAPANESE GEOTECHNICAL SOCIETY 2011. Survey report on the ground liquefaction in the Kanto region by the 2011 Great Tohoku earthquake, Tokyo. The Japanese Geotechnical Society, Tokyo [in Japanese].

MORIMOTO, R. 1951. Geology of Imaichi District with special reference to the earthquakes of Dec. 26th, 1949. (II). *Bulletin of the Earthquake Research Institute*, **29**, 349–358.

MORIMOTO, R., NAKAMURA, K., TSUNEISHI, Y., OSSAKA, J. & TSUNODA, N. 1967. Landslides in the epicentral area of the Matsushiro earthquake swarm – their relation to the earthquake fault. *Bulletin of the Earthquake Research Institute*, **45**, 241–263.

NAKAMURA, K. 1934. *Landslides*. Iwanami Shoten, Tokyo [in Japanese].

NAKATA, T. & IMAIZUMI, T. 2002. *Digital Active Fault Map of Japan*. Tokyo Press, Tokyo [in Japanese].

OGATA, S. 1976. Activity evaluation of fault in the basememt Terrain: characteristics of its fracture thickness and filled materials. *Journal of Japan Society of Engineering Geology*, **3**, 118–121 [in Japanese with English abstract].

OKUDA, S., OKUNISHI, K., SUWA, H., YOKOYAMA, K. & YOSHIOKA, R. 1985. Restoration of motion of debris avalanche at Mt. Ontake in 1984 and some discussions on its moving state. *Bulletin of the Disaster Prevention Research Institute, Kyoto University*, **28**, 491–504 [in Japanese with English abstract].

ONDA, Y. 1992. Influence of water storage capacity in the regolith zone on hydrological characteristics, slope processes, and slope form. *Zeitschrift für Geomorphologie*, **36**, 165–178.

OTSUKA, T. 1988. Paleozoic-Mesozoic sedimentary complex in the eastern Mino Terrane, central Japan. *Journal of Geosciences, Osaka City University*, **31**, 63–122 [in Japanese with English abstract].

OYAGI, N. 2004. Definition and classification landslides. *In*: *The Committee on Landslide Terminology Based on Geology and Geomorphology* (ed.) *Landslides – Terminology in Geological and Geomorphological Studies*. Japan Landslide Society, Tokyo, 3–15 [in Japanese].

PETIT, J. C., MEA, G. D., DRAN, J. C., MAGONTHIER, M. C., MANDO, P. A. & PACCAGELLA, A. 1990. Hydrated-layer formation during dissolution of complex silicate glasses and minerals. *Geochimica et Cosmochimica Acta*, **54**, 1941–1955.

RADBRUCH-HALL, D. H., VARNES, D. J. & SAVAGE, W. Z. 1976. Gravitational spreading of steep-sided ridges ('sackung') in western United States. *Bulletin of the International Association of Engineering Geology*, **14**, 23–25.

RESEARCH GROUP FOR ACTIVE FAULTS OF JAPAN 1980. *Active Faults in Japan: Sheet Maps and Inventories*. University of Tokyo Press, Tokyo [in Japanese].

RESEARCH GROUP FOR ACTIVE FAULTS OF JAPAN 1991. *Active Faults in Japan: Sheet Maps and Inventories*. Revised edition. University of Tokyo Press, Tokyo [in Japanese].

SAKEMI, Y. 2010. Adoption of an asphalt facing method in the Inaba Dam. *Engineering for Dams*, **283**, 48–55 [in Japanese].

SANGAWA, S. 1999. Paleoliquefaction features at archaeological sites in Japan. *Journal of Geography*, **108**, 391–398 [in Japanese].

SCHOLZ, C. H. 2002. *The Mechanics of Earthquakes and Faulting*. 2nd edition. Cambridge University Press, Cambridge.

SHIMOKAWA, E. 1984. Natural recovery process of vegetation on landslide scars and landslide periodicity in forested drainage basins. *Proceedings of the Symposium on Effects of Forest Land Use on Erosion and Slope Stability*. Environment and Policy Institute, East-West Center, University of Hawaii, 99–107.

SHINYA, T. & TANAKA, K. 2005. Mud volcanoes in Matsudai, Tokamachi City, Niigata Prefecture. *Journal of Japan Society for Natural Disaster Science*, **24**, 49–58 [in Japanese with English abstract].

SHIRAISHI, K. 2000. Characterization of a process zone accompanied with an intermediate fault: a case study from the Usukidani fault, southwestern Shimane prefecture. Master's Thesis, Yamaguchi University, Yamaguchi [in Japanese with English Abstract].

SHUZUI, H. 2001. Process of slip-surface development and formation of slop-surface clay in landslides in Tertiary volcanic rocks, Japan. *Engineering Geology*, **61**, 199–219.

SIBSON, R. H. 1977. Fault rocks and fault mechanism. *Journal of the Geological Society, London*, **133**, 191–213.

SUZUKI, K., ITO, E. & CHIGIRA, M. 2002. Loosening process of surface area in weathered granite and infiltration of rainwater to excavated slope: evaluation using geophysical exploration and observed field data. *Journal of the Japan Society of Engineering Geology*, **43**, 270–283 [in Japanese with English abstract].

SUZUKI, K., TOKUYASU, S. & TANAKA, K. 2009. Underground structure of mud volcanoes in Tokamachi City, Niigata Prefecture determined by electromagnetic exploration, and geographical and geological surveys. *Journal of Geography*, **118**, 373–389 [in Japanese with English abstract].

TADA, K., JIN, H., KITAGAWA, M., NITTA, A. & TORIUMI, R. 1995. Effect of the Southern Hyogo Earthquake on the Akashi Kaikyo Bridge. *Structural Engineering International*, **3/95**, 179–181.

TAGAWA, T. 2011. Analyses of behavior and observation plans for the first impoundment in the Inaba Dam. *Engineering for Dams*, **295**, 78–97 [in Japanese].

TAKAYAMA NATIONAL HIGHWAY WORK OFFICE 1999. Construction Record of Abo-Toge Expressway. Takayama National Highway Work Office, Takayama [in Japanese].

TAKEUCHI, K., YOSHIKAWA, T. & KAMAI, T. 2000. *Geology of the Matsunoyama Onsen District with Geological Sheet Map at 1:50,000*. Geological Survey of Japan, Tsukuba [in Japanese with English abstract].

TANAKA, K. & ISHIHARA, T. 2009. Mud volcanism near the Nabetachiyama Tunnel and the formation mechanism of swelling rock mass. *Journal of Geography*, **118**, 499–510 [in Japanese with English abstract].

TODA, S., TANAKA, K., CHIGIRA, M., MIYAKAWA, K. & HASEGAWA, T. 1995. Coseismic behavior of groundwater by the 1995 Hyogo-ken nanbu earthquake. *Zishin*, **48**, 547–553 [in Japanese with English abstract].

TOMMASI, P., VERRUCCI, L., CAMPEDEL, P., VERONESE, L., PETTINELLI, E. & RIBACCHI, R. 2009. Buckling of high natural slopes: the case of Lavini di Marco (Trento-Italy). *Engineering Geology*, **109**, 93–108.

Tsou, C.-Y., Feng, Z.-Y. & Chigira, M. 2011. Catastrophic landslide induced by Typhoon Morakot, Shiaolin, Taiwan. *Geomorphology*, **127**, 166–178.

Tsushima, A. & Oho, Y. 1999. Mineralogical weathering of mudstone in western Shimane Peninsula, Japan. *Journal of Geological Society of Japan*, **105**, 852–865 [in Japanese with English abstract].

Varnes, D. J. 1978. Slope movement types and processes. *In*: Schuster, R. L. & Krizek, R. J. (eds) . *Landslides: Analysis and Control*. TRB, Washington, DC, **176**, 11–33.

Vermilye, J. M. & Scholz, C. H. 1998. The process zone: a microstructural view of fault growth. *Journal of Geophysical Research*, **103**, 12 223–12 237.

Wakamatsu, K. 1993. *History of soil liquefaction in Japan and assessment of liquefaction potential based on geomorphology*. PhD thesis, Waseda University, Tokyo, Japan [in Japanese].

Wakimizu, T. 1919. Landslide causes and types. *Journal of the Japan Society of Civil Engineers*, **5**, 19–32 [in Japanese].

Working Group for Compilation of 1:2 000 000 Active Faults Map of Japan 2001. 1:2 000 000 active fault map of Japan. *Active Fault Research*, **19**, 3–12 [in Japanese].

Yamagata, M. 1990. Geological investigations on Honshu-Sikoku Bridge Expressway: an example of Akashi Kaikyo Bridge. *The Foundation Engineering & Equipment*, **18–10**, 107–115 [in Japanese].

Yamagata, M., Okada, R., Nasu, S. & Abe, M. 1993. Monitoring and analysis of seabed-layer deformation at the Akashi Kaikyo Bridge tower foundation. *Tsuchi to Kiso*, **41–2**, 33–36 [in Japanese].

Yamaguchi, Y. & Kanaori, Y. 2011. Successive movement of active faults by the fault motion simulation incorporating a fault process zone: case study of the Western Chugoku District of SW Japan. *Proceedings of Research Meeting of Japanese Society of Engineering Geology*, Sapporo, October 2011, 177–178.

Yamamoto, K., Yoshida, H., Akagawa, F., Nishimoto, S. & Metcalfe, R. 2013. Redox front penetration in the fractured Toki Granite, central Japan: an analogue for redox reactions and redox buffering in fractured crystalline host rocks for repositories of long-lived radioactive waste. *Applied Geochemistry*, **35**, 75–87.

Yamasaki, S. & Chigira, M. 2010. Weathering mechanisms and their effects on landsliding in pelitic schist. *Earth Surface Processes and Landforms*, **36**(4), 481–494, http://doi.org/10.1002/esp.2067

Yokota, S. 1999. Weathering distribution in a steep slope of soft pyroclastic rocks as an indicator of slope instability. *Engineering Geology*, **55**, 57–68.

Yoshida, H. 1994. Relation between U-series nuclide migration and microstructural properties of sedimentary rocks. *Applied Geochemistry*, **9**, 479–490.

Yoshida, H., Oshima, A., Yoshimura, K., Nagatomo, A. & Nishimoto, S. 2009. Fracture characterization along the fault – a case study of 'damaged zone' analysis on Atera Fault, central Japan. *Journal of the Japan Society of Engineering Geology*, **50**, 16–28 [in Japanese with English abstract].

Yoshida, H., Metcalfe, R., Ishibashi, M. & Minami, M. 2013. Long-term stability of fracture systems and their behaviour as flow-paths in uplifting granitic rocks from the Japanese orogenic field. *Geofluids*, **13**(1), 45–55, http://doi.org/10.1111/gfl.12008

Zischinsky, U. 1966. On the deformation of high slopes. *Proceedings of the 1st Conference of International Society of Rock Mechanics*, Lisbon, 25 September 1966, 179–185.

12 Field geotraverse, geoparks and geomuseums

WES GIBBONS, TERESA MORENO & TOMOKO KOJIMA

The main part of this chapter describes a geological journey westwards from central to SW Japan and is specifically aimed at the overseas visitor new to Japanese field geology. This is followed by a brief overview of the Japanese geopark system and museums with geological exhibits. There are currently over 30 geoparks in Japan, providing one of the most dense networks in the world of sites of special geological interest.

Field geotraverse: plate collision, strike-slip and subduction in western Japan (WG, TM)

The field geotraverse route runs from central Japan to Kyushu Island in the far SW, covering a distance of over 2000 km (Fig. 12.1). It begins in the Izu Collision Zone in Central Honshu and includes visits to Mount Fuji and the Japanese Alps before heading west to Nagoya and the fault-active 'Kinki Triangle'. Continuing past Kyoto, Osaka and Kobe to Okayama, the route deviates south to cross the island of Shikoku and its classical paired metamorphic belt separated by the Median Tectonic Line (MTL), one of the most conspicuous strike-slip terrane boundaries in the world. Emerging from the metamorphic mountains of central Shikoku, the geotraverse reaches the Pacific coast, offering the opportunity to examine coastal exposures of subduction mélanges and sediments in the Shimanto accretionary complex. Back in Honshu, the journey continues past Hiroshima then heads SW to the island of Kyushu, visiting spectacular examples of active arc volcanoes built on young crust accreted above the subducting Philippine Plate.

Previous chapters in this book have provided the detailed geological background for this journey, so we have kept our reference list very short. The chapter does not attempt the kind of description suited to research-level fieldtrips but instead aims at an overview for both professional and amateur geologists interested in Japanese geology, including foreign student groups keen to visit what is one of the most geologically and culturally fascinating countries in the world. While researching this chapter a colleague drew our attention to a short book on geological travel in Japan written in both English and Japanese, ambitiously aimed at a wide audience (Kato *et al.* 2008). The book is lavishly illustrated and, although lacking a specific itinerary, features several of the classic geological localities described in this chapter. We hope our geotraverse transmits a similar enthusiasm for geology in the field.

The route would be ideally suited to a two-week stay in Japan, but we deliberately offer considerable flexibility depending on the time available (see travel logistics section below); a rewarding journey requiring just one week is perfectly feasible. Luggage transport is minimized by organizing a route whereby the majority of day excursions can be made by radiating out from the urban bases of Matsumoto (in Honshu), Kochi (in Shikoku) and Kumamoto (in Kyushu), all attractive and interesting cities in their own right (Fig. 12.1). Although it is possible to make the entire journey by road, given the challenges of driving such long distances in sometimes densely populated areas and with road signs that can be exclusively in Japanese, the route is designed primarily for non-Japanese visitors using a Japan Rail Pass. For those keen to drive, then cars can be hired on arrival at Matsumoto, Kochi and Kumamoto rail stations, but this is not essential as the route can be made using public transport alone.

We realize that our approach is unusual: in most countries around the world, attempting a major geotraverse using public transport would be so limiting as to be impractical. The inevitable loss of mobility and flexibility is, however, considerably lessened in Japan which has an excellent rail system, commonly linked to buses. Luggage lockers are commonplace, and hotels are always available within easy and safe walking distance from major rail stations. Trains are comfortable and (with some planning) easy to use for the foreign visitor, providing stress-free travel through attractive and interesting landscapes and making the travelling hours part of the experience. Furthermore, the Japan Rail Pass is relatively inexpensive (given the distances travelled) and includes both free seat reservation and access to many of the superb Shinkansen 'bullet train' high-speed lines. In particular, the Shinkansen connection between Shin–Osaka and Kagoshima which opened in spring 2011 has greatly facilitated travel to the far SW.

When to visit and where to arrive

Japan has a highly seasonal climate, and without doubt the best times of the year to visit are either spring (April or May, always specifically avoiding the busy Golden Week national holiday period in latest April/early May) or autumn (October) after the typhoon season has abated. If visiting in winter or early spring the Japanese Alps area around Matsumoto may be blocked by snow. In summer the high humidity, abundant monsoonal rainfall and frequent thunderstorms are likely to interfere with travel plans, especially in Shikoku and Kyushu.

Any of the three main international airports in Japan (Tokyo, Nagoya or Osaka) can be used as arrival and departure points. If plans include the use of domestic flights, then Nagoya (Central Japan International Airport: Chūbu Centrair) has the distinct advantage of housing both domestic and international terminals, removing

From: Moreno, T., Wallis, S., Kojima, T. & Gibbons, W. (eds) 2016. *The Geology of Japan*. Geological Society, London, 485–508.

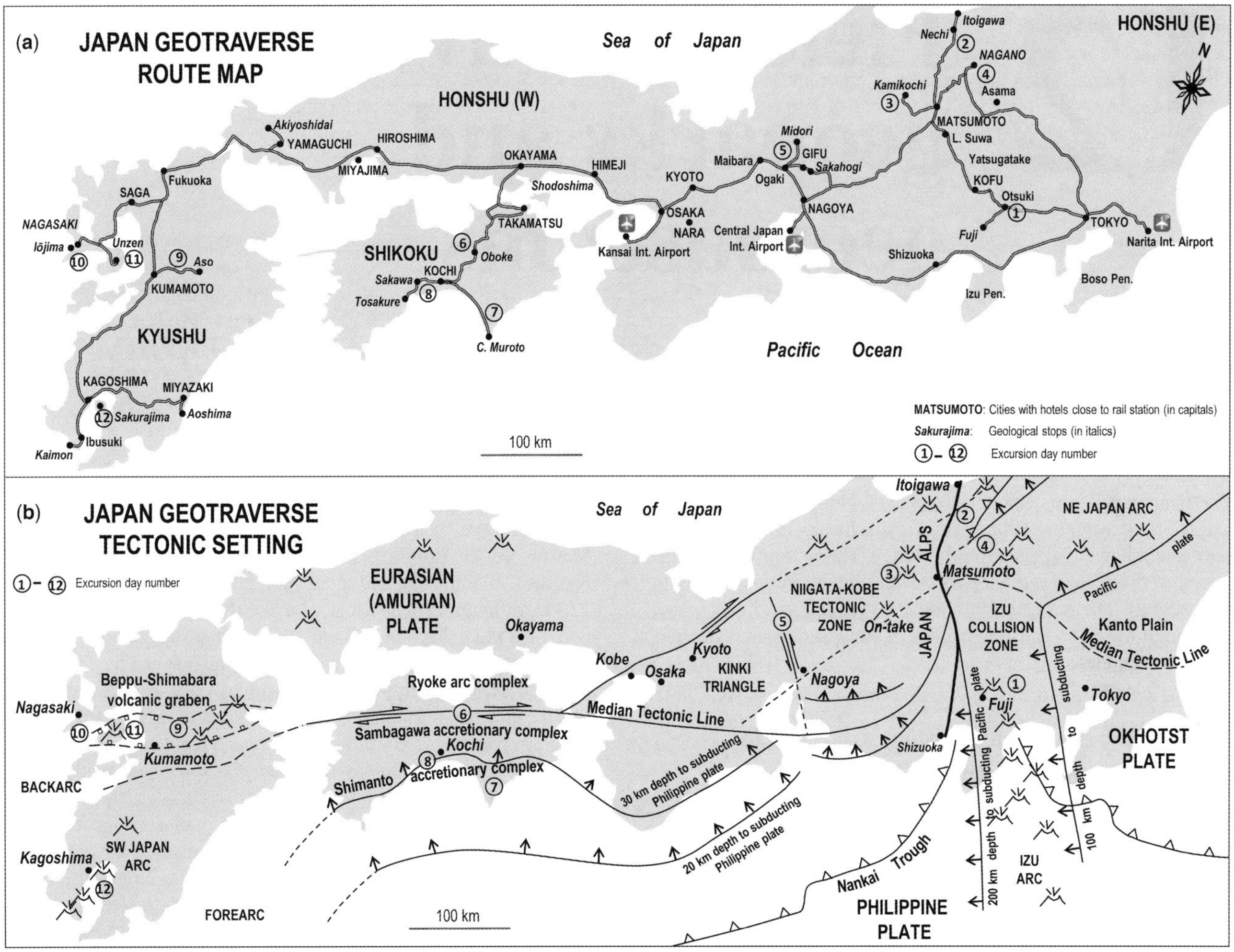

Fig. 12.1. (**a**) Geotraverse route map and (**b**) plate tectonic setting.

the need for tedious bus transfers on busy city highways. Osaka Kansai International (like Centrair) is an attractive modern airport built off the coast of Honshu, and has the different advantage of its own JR rail station (with direct connection to the popular tourist destination of Kyoto) so that rail pass vouchers can be exchanged upon arrival and the journey begun immediately. It is also the closest to Kagoshima of the three international airports, facilitating the journey back from Kyushu. If arriving in Tokyo, busy Narita International Airport has a direct JR service to Shinjuku (1.5 hours) on the west side of the megacity. The starting point for the geotraverse is either Shinjuku if arriving at Tokyo, or Matsumoto if arriving at Nagoya or Osaka (Fig. 12.1).

Whereas it is perfectly possible to complete the geotraverse without speaking any Japanese, resorting to such universal basics as sign language and menu photos, visitors should not expect English to be widely spoken (it is not). It is both courteous as well as undeniably useful to make the effort to master at least some basic language skills before arrival.

Travel logistics in Japan

The JR rail pass is available only to overseas visitors arriving on a temporary visitor's visa. It must be bought (as a voucher) before arrival in Japan (where the voucher is exchanged for the real thing), and gives unlimited travel on JR trains for a period of 7, 14 or 21 consecutive days. Whereas the geotraverse route is designed for a visit of 2 weeks, those with time only for a 7-day pass can decide to limit their journey to Central Honshu (based in Matsumoto) or Shikoku–Kyushu (based in Kochi and Kumamoto) or make a more ambitious 'limited stop highlights' trip from Tokyo to Kagoshima, taking in Mount Fuji (Excursion 1; overnight in Matsumoto), Sakahogi in the Japanese Rhine Valley (Excursion 5; overnight in Himeji), Oboke Gorge (Excursion 6; overnight in Kochi), Shimanto Belt mélanges at Tosa-Kure (Excursion 8; overnight in Okayama) and both Aso (Excursion 9) and Sakurajima (Excursion 12) volcanoes on Kyushu (two nights in Kumamoto) (Fig. 12.1). The luxury of a full 21-day pass allows enough time to visit all locations plus extra sightseeing trips, and offers the greatest flexibility. We recommend extra nights in Kochi, Takamatsu, Kumamoto, Nagasaki, Sakurajima and Miyajima.

A JR timetable of mainline rail services (in English) is provided on request when vouchers are exchanged for the rail pass on arrival in Japan. However, by that time one should already have planned the route and have a list ready to reserve seats at least on the first few trains to be used. Information about train services within Japan can easily be obtained online and in English from the excellent Hyperdia.com website (remember to deselect the Nozomi train services from the search parameters, as rail pass holders cannot travel on these trains). This website provides timetables, train names and numbers and stations in route, even for local lines (although note that sometimes it does not give all the options actually available for complex journeys involving several changes, in which case break the journey down to individual stages). Armed with this information it becomes an easy matter to gather the information necessary before reserving seats at the larger stations prior to travel.

Seat reservation on mainline trains is always recommended, particularly at weekends and is free with the rail pass. Although seats can commonly be reserved just minutes before the train arrives (depending on the queue at the booking office) it is always better to book as far in advance as possible because some trains fill up surprisingly early, making it necessary to find a vacant seat in the non-reserved coaches. Further, although Shinkansen services are almost invariably on time, this is not necessarily true of other train lines; avoid changing non-Shinkansen trains with only a few minutes to spare, especially if switching from JR to other train companies (where tickets need to be bought) and in the larger stations where transfers can be complicated.

There is not much storage space on Japanese trains (or, for that matter, in Japanese hotels), so the geotraveller is urged to excel in the art of carrying as little as possible. Ideally this means just one airline-hand-luggage-style backpack per person and ensuring that nothing is carried that is not going to be used. Geological 'equipment' can be limited to a compass, camera and hand lens, and heavy boots are not necessary. Clothes can be washed and dried in machines provided for this purpose in many business-style hotels, and cheap umbrellas are usually sold or loaned out from hotels. Our approach of using city bases from which to radiate will help minimize luggage transport. It will not, however, reduce the inconvenience of train travel with large bags. Many stations are equipped with coin-operated lockers where luggage can be stored for the day.

With regard to hotel accommodation, there is always plenty of choice to be found using the usual web search engines. In most places it will be possible to find Ryokan accommodation, where traditional Japanese hospitality, eating and sleeping arrangements can be experienced, although these are expensive and are best reserved for memorable nights in special places such as Unzen village, Sakurajima and Miyajima (Fig. 12.1a). A simpler approach, more suited to following this geotraverse, is to use business hotels. Our personal favourite is the Toyoko Inn chain (but there are several others) which has friendly, inexpensive hotels with small, clean rooms, free wired internet access from your own laptop and reduced prices and early check-in if you join their membership scheme. There are Toyoko Inns conveniently close to the JR rail stations along the route in Matsumoto, Nagoya, Gifu, Nara, Himeji, Okayama, Takamatsu, Kumamoto, Kagoshima, Nagasaki and Hiroshima (Fig. 12.1b). An exception is Kochi in Shikoku, where alternatives include Comfort, 7 Days and Super hotels. Such accommodation can be booked online in English, cancelled at the last moment without exacting a fee and includes breakfast in the price. It is recommended to book these hotels as early as possible, especially for Saturday nights and preferably months before arrival in Japan. Finally, digital geological maps may be accessed on the Geological Survey of Japan/AIST website (http://riodb02.ibase.aist.go.jp/db084/kihon_main_e.html).

Excursion 1: Mount Fuji

Most visitors to Japan will want to view the world-famous symmetric profile of Mount Fuji, the volcanic peak of which rises to 3776 m and forms the highest mountain in the country; for obvious reasons, the trip is only worth it in good weather (Fig. 12.2a). The visit can be made during a day trip from Tokyo (Shinjuku) to Matsumoto, a day return from Matsumoto (in which case luggage need not be carried) or as a longer stay using hotels near Mount Fuji (at Kawaguchiko) and exploring the area in more detail by bus or hire car. There are various ways to reach the Mount Fuji area by train or bus. Here we use the JR Azusa Limited Express (LEX) train route running between Tokyo and Matsumoto, changing trains at Otsuki where an easy platform switch is made to the private Fuji Kyuko train line (JR pass not valid so a ticket must be purchased) which takes around an hour to reach Kawaguchiko, a touristic town built on the northern slopes of the volcano. With current (2015) timetables, an 08:30 start in Tokyo Shinjuku will allow over 3 hours at Fuji, and arrival in Matsumoto before 18:00 (although consider making an even earlier start if a stop at Lake Suwa is to be included on the same day).

If travelling from Tokyo and the weather is poor that day, then avoid Mount Fuji and instead travel to Matsumoto northwards via a stop in Nagano (see Excursion 4). This alternative route also offers an opportunity to consider a visit to the famous and scenic riverside exposures of Sanbagawa schists at Nagatoro in Chichibu Geopark (2 hours north of Tokyo via Kumagaya, partly by non-JR Chichibu railway train) before moving on to Nagano. Nagatoro, claimed by some to be the 'Birthplace of Japanese Geology', exposes metabasaltic, serpentinic and metasedimentary schists with a wide variety of mineral assemblages that include piemontite and stilpnomelane.

Mount Fuji is a conical composite stratovolcano, the products of which were erupted over an older andesitic edifice of Middle Pleistocene age (Komitake Volcano). The earliest products of Fuji (Kofuji) were mostly pyroclastic falls and lahars produced 80–11 ka ago, these building a volcano subsequently covered by voluminous, mainly olivine basalt lavas and pyroclastic rocks which have built up the current symmetric cone (Shinfuji). Over the last 2000 years extensive Strombolian activity from NW–SE-aligned scoria cones has produced abundant *aa* flows of pyroxene olivine basalt, altering the landscape particularly in the NW side of the volcano (see map in http://www.eri.u-tokyo.ac.jp/VRC/vrc/others/fujigeol.html). The most recent eruption of the volcano started in the morning of 16 December 1707 when a Plinian ash plume was emitted from the SE side of the volcano, drifting eastwards over Edo (former Tokyo) to cause darkness at midday. The event, referred to as the Hoei eruption and which occurred 7 weeks after an M 8.4 Tokai (Tokai = southern Central Honshu) Earthquake, did not cease until 1 January. The tephra emitted during this recent eruption, beginning as dacitic and ending as basaltic in composition, was deposited widely over land and ocean east of the volcano. Although no human victims were reported as a direct result of the eruption there was widespread hardship as villages and farmland were destroyed, leading to deaths by starvation and disease. The date of the next Fuji eruption is unknown, although a series of low-frequency earthquakes at the turn of this century indicated possible magma recharging and the volcano continues to be very closely monitored. Another major Tokai earthquake is overdue (Mogi 2004) and could provide the trigger for a future eruption.

Kawaguchiko lies mainly on the oldest olivine basalt flows of the Shinfuji eruptions and is built on the southern shores of Lake Kawaguchi, the eastern remnant of a much larger lake subsequently partially infilled by recent eruptions. Pick up a map from the tourist office by the rail station and walk north for 15 min to lakeside exposures of the old basalts (Tatamiiwa Lava Rock). Just east of this locality a funicular ropeway leads to the viewpoint of Fujimidai on Mount Tenjo (1104 m), which provides some of the best views of Mount Fuji (you can also walk up but it is hard work, better to walk back down; Fig. 12.2a). The smoothly symmetric tephra-covered upper slopes of the volcano contrast with the topographically more irregular lower slopes, especially on the lower northwestern side which is pockmarked with Strombolian basaltic

Fig. 12.2. (**a**) Symmetric profile of Mount Fuji (3776 m) as seen from the forests of Mount Tenjo (Excursion 1), less than 100 km west of Central Tokyo. (**b**) Peaks exceed 3000 m above sea level in the Hida Mountain Range of the Japanese Alps. Plate collision has induced rapid uplift of 2000 m since Middle Pleistocene times (Kamikochi, Excursion 3). (**c**) The steaming summit of Yakedake ('burning peak', 2455 m), the most active volcano in the Hida Mountains, viewed from the Kappa Bridge, Kamikochi (Excursion 3). In recent times (since 1907) there have been repeated phreatic eruptions, leading to deaths on more than one occasion. (**d**) Porphyritic co-magmatic mafic enclave (left) in the Quaternary Takidani Granodiorite, claimed to be the youngest exposed granitoid pluton on Earth (Harayama 1992; Excursion 3). (**e, f**) Damage to the Zenkoji Temple during the 1854 earthquake at Nagano, with ground movements causing the heavy bronze bell to fall and indent (inset) the wooden support pillar and forcefully twist vertical entrance columns (Excursion 4). (**g**) Surface rupture of the Neodani Fault at Midori during the 1891 Nobi Earthquake produced a fault scarp with offsets of 6 m (vertical) and 8 m (left-lateral). The scarp is still clearly visible on either side of the road (note edge of Earthquake Museum building lower right) (Excursion 5). (**h**) Sakura (cherry blossom) springtime in Oboke Gorge, Central Shikoku, where the banks of the Yoshino River expose Sambagawa belt HP/LT metasediments (Excursion 6).

vents. There are also fine views from Fujimidai west across Lake Kawaguchi, blocked and reduced in size by extensive lava outpourings from these Strombolian vents. If staying overnight in Kawaguchiko bus tours may be taken across this young basaltic area to the west, allowing visits to some of the 100 or so lava tube caves discovered in this area such as the nearly 400 m long Saiko Komori Ana (Bat Cave). From the Fujimidai viewpoint a woodland path leads back down to the town and station.

Return from Kawaguchiko to Otsuki. The onward Azusa LEX trains bound for Matsumoto (usually changing at Kofu) leave the narrow valley confines of Otsuki to pass westwards through a long tunnel and emerge in the southern part of the Fossa Magna (FM) basin. There are extensive views to the left across this basin, initiated in Miocene times as a back-arc rift during the rotation of Japan away from the Asian mainland and marine transgression from the Pacific to the Sea of Japan. Later uplift of the central zone separated this southern FM from a northern FM marine basin (Excursion 2). Across the southern basin here the Akaishi Mountains (Southern Japanese Alps, with some peaks exceeding 3000 m) provide a grand backdrop to the scenery.

Beyond the town of Kofu the railway skirts the SW edge of the huge but now inactive Pleistocene Yatsugatake volcanic chain, the mountains of which rise to the NE between the active volcanoes of Fuji and Asama (Fig. 12.1a). Continuing NW towards Chino the route curves into the Itoigawa Shizuoka Tectonic Line (ISTL) which, in its middle segment, comprises a 50 km long zone of strike-slip faults where surface-rupturing earthquakes are well documented (Okumura 2001). Beyond Chino, with the ISTL prominently expressed by the abrupt rise of the hills to the west, the train traverses the northeastern urbanized corner of Lake Suwa (Fig. 12.1a). This lake lies in a strike-slip pull-apart basin formed during left-lateral movement between NW–SE-oriented faults along the ISTL. To visit the lake alight from the train at Kami-Suwa, an area famous for hot spring activity. There is a thermal footbath in the station itself and the Lake Suwa Geyser Centre (the geyser erupts roughly hourly to a height of 40–50 m) is located just 10 minutes walk north of the station at the lakeside. Continuing on the train journey, beyond the lake is another long tunnel which provides access to the Matsumoto Basin, a prominent elongate (5–15 km × 60 km) intermontane basin overlooked by the Hida Mountains of the Northern Japanese Alps.

The elevated location of Matsumoto city, which has the highest airport in Japan (658 m asl), ensures a cold winter with daytime temperatures usually well below 10°C. This elevation is a consequence of uplift during the collision of the Izu–Bonin Arc with central Honshu, with Matsumoto positioned immediately north of the resulting syntaxial bend ('Kanto Syntaxis') and above the ISTL which separates east and west Japan (Fig. 12.1b). Quaternary rise of the adjacent Hida Mountains by >1 km has shed thick alluvial fan sediments eastwards to infill the Matsumoto Basin, producing the flat terrain of the city and covering the ISTL. The western and eastern margins of this basin are defined by active faults, the most prominent of which is the subvertical East Matsumoto Basin Fault, with the last major earthquake event in the area thought to have taken place in the ninth century (AD 841; Okumura 2001).

Excursion 2: Itoigawa–Shizuoku Tectonic Line in Itoigawa Geopark

This journey runs north from Matsumoto towards Itoigawa, following the northern part of the tectonic boundary that separates eastern and western Japan. Unless a car is hired (follow Routes 147 to Omachi (大町) then 148), the excursion requires taking several trains with a leisurely start (after 10:00) on the Azusa 3 train (buy food beforehand). It is the scenic nature of the journey itself combined with a visit to Itoigawa Geopark which provides the day's highlights. The geopark visit includes a rare (and easily accessible) artificially enhanced exposure of ITSL fault rocks at Nechi, some 100 km north of Matsumoto. An optional visit can be made to the Fossa Magna Museum in Igoitawa, where there are excellent displays of locally discovered jade. With current timetables (2015) it is possible to visit both the ITSL exposures at Nechi and (briefly) the Fossa Magna Museum using public transport but only by visiting the museum first, arriving back in Matsumoto just before 20:00.

The journey north from Matsumoto runs to the NW corner of the Matsumoto Basin, following the ISTL and passing the fault-generated lakes of Kizaki and Aoki at the foot of the Hida Mountains. This northern part of the Fossa Magna is bounded by two major fault systems: the ISTL to the west and the Western Nagano Basin Fault Zone (WNBFZ) to the east (Takeda *et al.* 2004). Both of these tectonic boundaries include reverse faults capable of causing destructive earthquakes, with the ISTL thrusting to the west and the WNBFZ thrusting east (Okumura 2001). The structural complexity of this part of Japan is further compounded by a 100 km wide NE–SW-aligned tectonic zone of anomalously high strain rate running through Niigata and Kobe, within which crustal deformation is being concentrated (Fig. 12.1b). This zone, responding to regional stresses, has been attributed to lithospheric weakening due to fluid infiltration above the subducting Phillipine and Pacific plates (Nakajima & Hasegawa 2007).

North of Lake Aoki, the route enters the narrow tectonic depression of the Kamishiro Basin which, drained northwards by the Hime River, provides road and rail access north to the coast. The LEX Azusa train terminates in Minami-Otari (75 minutes from Matsumoto), a location deep in the Hime Valley. A quick change of trains here (follow other passengers up over the footbridge and don't delay) allows a memorable journey on a single car diesel train which descends the river valley towards Itoigawa on the Sea of Japan shoreline (show your rail pass when you leave the train). This is one of Japan's classic train journeys, following a line (finally opened in 1957 after having been built with great engineering skill) which curves through tunnels and across bridges as it tries to avoid the worst of the landslide hazard zones in the Himekawa Gorge. Initially the valley follows the ISTL, but beyond Kitaotari it curves left to enter the serpentinite-rich metamorphic terrane outcropping west of the tectonic boundary. At Hiraiwa the route crosses into the Itoigawa Geopark, the oldest such park in Japan. Established in 1990, it now comprises a group of 24 areas of special geological interest. This area is famous for yielding jade deposits, especially along the 'Jade Gorge' valley of the Kotaki River which joins the Hime River at Kotaki. The jadeitic minerals occur within high-pressure–low-temperature (HP-LT) meta-ophiolitic lithologies which include abundant serpentinite mélange.

Continuing north from Kotaki, a major attraction within Itoigawa Geopark is an artificially excavated exposure of ISTL fault rocks easily accessible by signposted path 5 minutes north of Nechi rail station (45 minutes by Oito Line train from Minamiotari, or parking just north of trail on R148). Allow at least 90 minutes for the return visit along this trail (if a museum visit is planned do this first, staying on the train to Himekawa or Itoigawa then backtrack to Nechi). From Nechi station walk north to Nechi River and cross the main road bridge (R148) to locate the signposted trail leading east: 'Major Fault. Fossa Magna Park'. Progressing along the 'Fossa Magna Promenade' trail, a series of well-presented information panels (mounted on polished granitoid gneiss pedestals, displaying excellent ductile shear textures) describe various aspects of the local and regional geology. The first of these shows the plate boundary setting of the ISTL and epicentres of destructive earthquakes recorded in the region. The nearest such event on record occurred south of Otari in 1714 (M 6.3) but note also the 1725 M6.3 event at Lake Suwa (Excursion 1), the 1847 M7.4 event at Nagano (Excursion 4) and the several earthquakes recorded around Matsumoto.

The trail passes weathered exposures of Permian metagabbro, with a display panel placing this rock in context within a terrane map of Japan. The gabbro has been assigned to the Maizuru Terrane, a pre-Jurassic tectonic unit that includes Palaeozoic ophiolites, crops out in both western and eastern Honshu on either side of the ISTL and is thought to continue north into Russia. The next display panel focuses on Heinrich Edmund Naumann's identification of the Fossa Magna, the sediments within which are known to reach a thickness of >4000 m adjacent to the ISTL here in the northern part of the basin. This thickness, added to the 3000 m peaks of the Hida Mountains, records the enormous amount of displacement that has taken place across the ISTL. Since late Pliocene times the Northern Fossa Magna has been under NW–SE-directed compression however, folding the basin fill and inducing thrust propagation westwards in the west and eastwards in the east (Takeda *et al.* 2004).

Naumann (1854–1927) arrived in Japan in 1875 as geological advisor to the Meiji Government, and founded the Geological Survey of Japan in 1878. He mapped extensively across Japan over a 10 year period, and taught at one of the institutes that were later to merge and form Tokyo University. In addition to his fame for recognizing the Fossa Magna, his interest in Japanese Neogene palaeontology led him to conclude that the archipelago had once been connected to the Asian Mainland. He gained considerable further publicity by demonstrating that elephants had once been native to Japan, under tropical conditions in late Pliocene times. In 1973 the Fossa Magna Museum, dedicated to Naumann, was opened in Itoigawa.

The next two panels illustrate the Neogene opening of the Sea of Japan during the clockwise rotation of the Japanese landmass, the corresponding evolution of the Fossa Magna over the last 16 Ma and recent seismicity and volcanicity in the area. Shortly beyond the second panel, steps lead (right) down to the trail highlight: the superbly self-explanatory excavated site revealing ISTL fault rocks. The tectonic boundary is expressed here as a zone of fault breccias and gouges up to 1.5 m wide separating Palaeozoic metagabbro from Miocene andesite. The varied colouration of these high-level cataclastic fault rocks testifies to the involvement of different protoliths.

Return to the main path and continue east to pass several more information panels. The first of these provides colour photographs of fault displacement exposures at four other places along the ISTL. Next are panels on volcanic centres in the Fossa Magna area and hydrocarbon resources within the Neogene Niigata oil and gas fields. These are followed by explanations of the local landslide hazard in the rivers draining north into the Sea of Japan, and the detailed local geology and geomorphology of the area between Nechi and the coast at Itoigawa. The trail reaches a rock shelter at an exposure of jointed giant pillow lava, the form of which is somewhat obscured by the presence of the shed. More typical exposures of Miocene basaltic pillow lavas within the Fossa Magna are to be found by continuing to the trail end and following the signs to descend (right) past a picnic site to the river. Return by the same route and be sure not to miss the last train back to Matsumoto.

The Fossa Magna Museum is most easily visited by road transport as it lies in the hills 3.5 km south of Itoigawa rail station and 2.8 km east of Himekawa rail station. If using rail transport either take a taxi from Itoigawa station or walk from Himekawa station (one stop before Itoigawa), the latter option offering an attractive walk but with current timetables (2015) leaving only just under 1 hour in the museum. To start the 40 minute walk from tiny Himekawa station, turn left along the main road (route 148) and walk south to traffic lights. Here turn left, cross the rail and continue by road and track across rice fields to an irrigation canal which you follow (right) to a road. Turn left on this road to climb up the valley side: stay left at a road junction to reach the top to meet another road where a path leads straight on alongside a small pond on the left. At the end of this path turn right into a road which climbs past a sports stadium to a T-junction. Turn left and the museum is a few minutes along this road on the right. Return by the same route.

Excursion 3: Northern Japanese Alps: Pleistocene plutons and Holocene volcanoes

The majestic Alpine scenery of the Hida Mountain Range which rises abruptly west of the Matsumoto Basin can only be visited by non-JR public transport. Despite this, if time is available and the weather is good (with roads not blocked by snow or landslides) then a visit to the famously scenic valley of Kamikochi (where private cars are banned) offers an unforgettable day trip into the Japanese Alps. Private cars must be left in the official Hirayu (平湯) or Sawando (沢渡) car parks 50 min from Matsumoto (from where taxis or shuttle buses run into Kamikochi). Using public transport from Matsumoto is easy and takes 2 hours, involving a local train ride then efficient transfer to a coach that tunnels into the high mountains. Visit the tourist office in Matsumoto Station for maps, timetables and instructions regarding how to buy combined tickets, and remember to book a seat on the return coach once you have arrived. The excursion provides the opportunity to see one of the youngest exposed granitoid intrusions in the world, located in a deep valley at the foot of the steaming andesitic Yakedake volcano which last erupted in 1995.

The Hida mountain range has been uplifting episodically since late Miocene times (Harayama 1992), but has risen particularly rapidly (*c.* 2000 m) over the last 500 ka. The range includes a chain of active andesitic peaks (Norikura Volcanic Zone) erupting above the subducting slab of the Pacific plate (located some 200–250 km below) and contributing heat to the actively deforming Niigata–Kobe Tectonic Zone (Fig. 12.1b). Immediately north of Yakedake is a volcanic-plutonic complex including the Takidani granodiorite pluton which intrudes Plio-Pleistocene andesite (see map and cross-section in Harayama 1992). Intrusion, cooling, uplift and erosion over the last 2 Ma have revealed a >2000 m thick eastwards-tilted granodioritic sheet which has locally contact-metamorphosed its envelope of welded tuff to hornblende hornfels facies. The pluton shows a textural and geochemical zoning downwards from hornblende-bearing porphyritic biotite granite to equigranular biotite-hornblende granodiorite in the deeper exposed levels (Harayama 1992), these being partly covered in the Kamikochi area by the recent products of Yakedake Volcano.

Pick up a map from the tourist office by the bus station then walk north to the Kappa Bridge for views north into the pale mountains, exposing the pluton and its country rock (Fig. 12.2b), and west to the volcanic peak of Yakedake (2455 m; Fig. 12.2c). Visiting exposures of these rocks on the mountain ridges requires an arduous climb through bear country, requiring an early (and well-equipped) start. Day visitors will have to content themselves with exposures of the pluton at lower levels in the valley such as those seen around the Weston Memorial, a wall plaque commemorating the English climber Walter Weston. From the Kappa Bridge walk downstream, staying on the right bank of the Azusa River. The medium-grained fresh granodiorite exposed here contains co-magmatic porphyritic mafic enclaves (Fig. 12.2d) and provides a good idea of the nature of the lower levels of this strikingly young intrusion. Reaching the peak of Yakedake, currently the most active of all the volcanoes in the Hida Mountains, takes up to 4 hours; once again an early start is essential, and the return trek is too long for the average day visitor. Instead, cross the bridge downstream from the Weston Memorial and continue walking SW on the nature trail parallel to the river to reach Taisho Lake immediately below the volcano. The lake was created by the damming effect of muddy debris flows produced during an eruption of Yakedake in June–July 1915.

Excursion 4: Zenkoji Temple (Nagano), then SW to Kiso Valley and Nōbi Plain

A visit to the famously earthquake-damaged Zenkoji Temple is only a half-day excursion from Matsumoto as Nagano is less than an hour away by train, and from the station it is an easy 1.5 km walk north through the city. This excursion therefore allows a few morning hours for sightseeing in Matsumoto. Returning to Matsumoto from Nagano, the scenic train journey continues SW to Nagoya and Gifu (Fig. 12.1).

The train route to Nagano runs NNE of Matsumoto to the Western Nagano Basin Fault Zone described in Excursion 2. A severe (*M* 7.4) and highly destructive earthquake occurred along this fault zone in Nagano on 8 May 1847. Surface rupture due to SE-directed thrusting occurred across an area of *c.* 40 km and nearly 10 000 people were reported killed. Most of the damage occurred on the hanging wall (NW side) of the fault, which rose 2–3 m above the footwall (Hattori *et al.* 2004). The heavy death toll was due to the earthquake coinciding with a special event in Zenkoji Temple which had attracted large numbers of people to the city. Surviving evidence of the earthquake can be viewed at the temple in the form of displacement and damage to the building (Fig. 12.2e, f) and stone garden lanterns. The earthquake was quickly followed by widespread fires in the built area immediately in front of the temple, and it is reported that over 80 aftershocks occurred during the following night (Piccardi & Masse 2007).

Return to Matsumoto by the same route and continue SW to Nagoya. The initial part of the journey passes through Shiojiri and climbs away from the Matsumoto Basin. Around 25 minutes from Matsumoto, just after Narai Station, the train route tunnels through the mountain watershed to enter the Kiso River drainage basin, one of seven river systems identified as of especial importance to Japan. While only covering 17% of national land area, these seven drainage basins concentrate around 50% of the total population and industrial activities. In the case of the Kiso River System, this involves several rivers draining south from the mountains and into the Nobi Plain, an important rice-growing area. The upper part of the Kiso River runs through a scenic valley separating the Central Alps (Kiso Mountains) to the east from the huge On-take stratovolcano (Fig. 12.1b) at the southern end of the Norikura volcanic zone. It is the second highest volcano in Japan, and its continued activity is evidenced by the phreatic eruption on 27 September 2014 that killed over 50 excursionists out enjoying the autumn scenery.

The rail route continues following the river, now cutting deeply into Ryoke granites; the MTL lies to the east (Fig. 12.1b). Scenic exposures of these pale granites in Nezame-no–Toko Gorge can be enjoyed from the train 7–10 minutes after Kisofukushima Station. Legend has it that Urashima Taro, the Japanese equivalent of Rip Van Winkle, woke up from his 300 year sleep on these granites.

Nagoya, the fourth largest city in Japan, lies on a low plateau of westerly tilted Plio-Quaternary sediments on the east side of the Nōbi Plain and at the SE apex of the Kinki Triangle. This is a seismically active and topographically subdued area north of the MTL and defined by conjugate fault zones running NW–SE and NE–SW (Fig. 12.1b; Nakajima & Hasegawa 2007). Around 20 million people live within the Kinki District, which includes the large cities of Nagoya, Kyoto, Osaka and Kobe. The earthquake that hit Kobe in 1995, the second-most damaging seismic event in the twentieth century (after the Great Kanto Earthquake of 1923), produced right-lateral NE–SW fault displacement (Fig. 12.1b). Just over 100 years earlier it had been the turn of Nagoya to be damaged by a severe seismic event, the great Nobi Earthquake of 1891 which, in contrast, showed left-lateral NW–SE movement (Fig. 12.1b).

The train journey from Nagoya to Gifu crosses the Nobi Plain, produced by the confluence of the Ibi, Kiso and Nagara rivers, with the alluvial ground surface here lying just 2–3 m above sea level. This area was devastated by the Nobi Earthquake, the severity of the event claiming worldwide attention and recorded as 'The Great Earthquake in Japan' in the 1894 edition of The Geographical Journal. Stay overnight in Gifu.

Excursion 5: The Neodani Fault and Japanese Rhine Valley

This excursion visits two particularly interesting geological sites en route from Central Honshu to Shikoku Island (Fig. 12.1a). The first site, at Midori in Gifu Prefecture, preserves one of the most spectacular inland fault-related landforms in Japan; the second examines famous exposures of Mesozoic oceanic cherts along the Kiso River at Sakahogi, along a stretch of water that has been compared scenically to the German Rhine Valley. Note that when planning this journey with Hyperdia.com, it is better to refer to timetable information separately for each journey segment. The trip from Gifu to Midori involves a change from JR trains at Ogaki to the non-JR Tarumi Railway Company (better to allow at least 10 minutes for this as tickets have to be bought; validate the ticket on entry into the train). Returning from Midori to Ogaki, the onward journey to Sakahogi involves a change of trains at Gifu. With an early start (08.10 train) and careful planning luggage can be left stored in Gifu for the day, still leaving enough time to journey on to Okayama in the evening.

After changing trains in Ogaki, the single coach Tarumi Railway train crosses the Ibi River and turns north for a delightful 30 km journey upstream along a tributary valley to Midori, situated on the NE boundary of the Kinki Triangle. The peaceful, rural scene at Midori belies the enormous violence inflicted by surface rupture during the Nobi Earthquake. From Midori station walk a short distance west and turn left into the main road for a few 100 m south to locate the resulting fault scarp, over which a pyramidal museum has been built (Fig. 12.2g).

The *M* 8 Nobi Earthquake occurred early in the morning of 28 October 1891 and was the largest inland shallow earthquake ever recorded in Japan, producing over 7000 deaths and tremendous damage to property and transport systems. One of the most prominent surface-rupturing faults which appeared during the earthquake was the Neodani Fault, showing offset across a zone running 38 km NW–SE, some 60 km from Nagoya. Here at Midori, this fault has a vertical offset of 6 m and left-lateral offset of 8 m (Fig. 12.2g), and has been excavated and enclosed within the museum (open 09:00–17:00, closed on Mondays). The uplifted (NE) side of the fault exposes cataclastic fault rocks derived from Mino Terrane Mesozoic basalts and sediments juxtaposed against Quaternary river deposits. Individual cobbles within the coarse river deposits have been rotated vertically against the fault plane, which is covered with alluvial sediments deposited after the earthquake. Outside the museum the scarp is well preserved, cutting across the agricultural valley, and a viewpoint with explanation panels is provided. Good views over the valley and fault scarp can be obtained by climbing steps (right) from the road south of the museum. Return to Ogaki by the same route.

The journey from Midori to Sakahogi, changing trains at Ogaki and Gifu, takes 2 hours 30 minutes. Sakahogi is a small town within a 13 km valley that runs from the meeting point of the Kiso and Hida rivers to Inuyama city in Aichi Prefecture. There are extensive riverbank exposures of Mesozoic oceanic cherts around 500 m SSW of Sakahogi train station, where the Nakasendo Highway 21 runs alongside the river. From the station walk south for 150 m to cross this highway at traffic lights, and immediately climb to the river flood protection embankment. Follow this embankment west (downstream) for 5 min, passing a watchtower (left) to a picnic site (right) with wooden roof shelter, beyond which descend the sloping ramp access (left) to a concrete level running along the wall. Here, double back left and follow this level for 50 m to old railed steps which provide access to some of the best chert exposures via a small path crossing the riverside undergrowth. A visit to this site should never be considered if the river is in flood stage during the rainy season.

The Sakahogi cherts were deposited in the deep Palaeo-Pacific (Panthalassa) Ocean over a period lasting *c.* 30 Ma during middle–late Triassic times in equatorial latitudes, and subsequently drifted north to become incorporated into the Jurassic accretionary complex of the Mino Terrane. Most of the sequence comprises rhythmically bedded red and green cherts separated by thin siliceous claystone seams, although there are also white, purple, yellow and brown cherts depending on the content of impurities such as Fe, Mn, clay minerals and organics (Fig. 12.3a). The overall dip is steep, with NNW–SSE-aligned strike and WSW-directed younging, except around the hinges of steeply plunging folds. The cherts record deep oceanic radiolarian blooms of biogenic silica deposited relatively rapidly against a background slow rainout of atmospheric aeolian dust, which produced the thin siliceous clays. The siliceous claystone seams are typically of the order 10 times thinner than the chert beds, but may have taken around 100 times longer to be deposited. An added curiosity here is the fact that anomalous concentrations of platinum group elements, nickel-rich magnetite and microspherules in the lower part of a relatively thick (4–5 cm) claystone layer ('Sakahogi ejecta layer') discovered within this succession have been attributed to the global reach of the 215 Ma Manicouagan extra-terrestrial bolide impact event in Canada (Onoue *et al.* 2012). There is however no evidence of radiolarian mass extinction across the impact event horizon at Sakahogi, unlike in coeval marine faunas and terrestrial tetrapods and floras in North America, suggesting the pelagic marine realm survived relatively unscathed. Stay overnight in Gifu, Himeji or Okayama.

Excursion 6: Traverse through Shikoku to Kochi: crossing the Median Tectonic Line

In this excursion starting from Okayama the westwards journey across Honshu is interrupted by a north–south traverse across Shikoku, the smallest of the four major Japanese islands and the only one without active volcanoes. Shikoku is instead geologically famous for the MTL, one of the world's most prominent strike-slip terrane boundaries (Fig. 12.1b). The MTL slices east–west across the island, separating the HT-LP Ryoke Belt from the HP-LT Sanbagawa Belt, further south of which is the Shimanto subduction accretionary complex of mélanges and oceanic sediments. This excursion leaves the Shinkansen rail track at Okayama, arrives in Shikoku over a series of spectacular bridges, then traverses the Ryoke Belt to enter the remote, mountainous centre of the island. After crossing the MTL, the highlight of the day is an examination of Sanbagawa metasediments in the deep chasm of the Oboke Gorge before continuing south to Kochi (Fig. 12.1).

The journey south from Okayama initially crosses the Sanyo Belt, characterized by Late Cretaceous continental arc granitoid and ignimbritic rocks associated with mineral deposits (notably tungsten and copper) and thermally metamorphosed Jurassic accretionary complex basement (Ishihara 2003). The route runs SW across flat expanses of Holocene sediments extensively covered by rice cultivation, before tunnelling through hills of the Mesozoic Sanyo rocks to arrive at Kojima on the northern coast of the Seto Inland Sea. Shikoku is reached via an impressive 18 km long series of high bridges. Completed in 1988, this chain of six suspension structures spans the 9.4 km strait between Kojima on Honshu and Sakaide on Shikoku, with the double railway line running beneath a road expressway. This crossing brings us south from the Sanyo Belt into the Ryoke Belt which, although also dominated by Late Cretaceous plutonic granitoid rocks, represents deeper levels of the continental magmatic arc. Here, the Jurassic basement has commonly been metamorphosed to sillimanite grade, there is a lack of important metallic mineralization, the granitoid rocks are more commonly foliated and there is locally abundant mingling between mafic magmas injected into hot, solidifying granitoid intrusions (Ishihara 2003).

The route across Shikoku initially crosses the flat coastal plain of Quaternary sediments covering Ryoke Belt granitic plutonic basement overlain by Miocene volcanic outliers, which rise from the low-lying landscape to form prominent inselbergs. Arriving in the ancient shrine town of Kotohira, one of these Miocene volcanic outliers rises to the west to form a long NW–SE-oriented ridge capping the granitic basement. A stop here to make the climb up this ridge to the temple (allow 2 hours for the return walk; there are luggage lockers in the station) will be rewarded by expansive views over the surrounding geomorphology. While exposures are poor and deeply weathered, the Ryoke Cretaceous granites (and locally gabbro) are present everywhere in paving and kerb stones, pillars and sculptures.

South of Kotohira rise high hills marking the outcrop of the Izumi Group, a thick sequence of latest Cretaceous marine sediments deposited in a narrow, left-lateral strike-slip controlled basin developed along the MTL. The train route south tunnels through this Izumi Group outcrop which rises above to nearly 1000 m. The railway emerges from these hills around 15 minutes from Kotohira, curving into the deep valley of the Yoshino River (the longest river in Shikoku) which defines the position of the MTL, the dominant fault in the most prominent strike-slip fault system in Japan (http://www.glgarcs.net/intro/sum_landgeo.html; Fig. 12.1b). The train makes a 180° curve across this valley, crossing the river and providing expansive views both east and west along the strike of this famous terrane boundary which here defines the Izumi–Sanbagawa boundary, to arrive in Awaikeda. This is a valley town built in the middle of the central segment of the now right-lateral fault system (left-lateral in Late Cretaceous–early Palaeogene times). To the east of Awaikeda the MTL is transpressional, whereas westwards the fault system makes a gradual transition into a transtensional stress state.

Leaving Awaikeda the route turns south, still following the Yoshino River, to enter the most scenically spectacular part of the journey. Vertiginous glimpses of the HP-LT metasediments and metabasites belonging to the Besshi and underlying Oboke nappe complexes of the Sanbagawa Belt (Wallis *et al.* 2004) are interrupted by numerous tunnels as the railway and Highway 32 weave up the valley to reach Oboke (Fig. 12.2h). In the river gorge here are abundant exposures of thick psammitic, chlorite-bearing low-grade (<300°C) but highly deformed metasediments locally including metaconglomeratic (psephitic) schists and belonging to the lower exposed part of the Sanbagawa sequence (the Oboke Nappe

Fig. 12.3. (**a**) Triassic radiolarian cherts exposed on the banks of the Kiso River in the 'Japanese Rhine Valley' (Excursion 5). (b, c) Clasts of Mesozoic oceanic materials in chaotic units produced by subduction underthrusting within the older (Cretaceous) part of the Shimanto accretionary complex: (**b**) Mesozoic pillow lava in coastal exposures of the Tei Mélange (Nishibun, Excursion 7); and (**c**) disrupted sediments in shaley matrix within the Kure Mélange (Tosa-Kure, Excursion 8). (**d**) Miocene turbidites exposed at the tip of the Muroto Peninsula (Excursion 7). Originally deposited in oceanic (?trench-slope) environments, these sediments were later added to the Shimanto accretionary complex by continued Neogene subduction. (**e**) The fiercely degassing 'First crater' of Aso Volcano, the current focus of volcanic activity inside the giant Aso Caldera, can only be approached during relatively quiescent periods when explosions are less likely and when winds carry away toxic gases (Excursion 9). (**f**) Sand volcanoes preserved on a tessellated sandstone pavement exposing the Oligocene Funazu Fm on Iōjima within the Takashima Coalfield (Excursion 10). (**g**) The steaming, geothermal field at Unzen tourist village provides scenes reminiscent of Yellowstone Park, USA (Excursion 11). (**h**) Sakurajima is one of the world's most active volcanoes, frequently producing explosions which can deposit large quantities of ash over the nearby city of Kagoshima as was the case with this eruption at midday on 25 September 2012. View from Yogan Nagisa footbaths (Excursion 12).

Complex; see map in Wallis *et al.* 2004). Excellent exposures of Sanbagawa metasediments can be examined 1 km downstream from the railway station and, combined with a visit to the mineral and rock collection in the Lapis Oboke Museum, should occupy you for 2.5–3 hours in total.

Leaving the station (lockers available) continue ahead up steps then turn right to climb to a red bridge across the Yoshino River, passing a prominent ornamental slab of Sanbagawa greenschist on the right. Cross the bridge, turn right ('Oboke 1 km') and follow the left side of the road to pass a wall displaying a collection of fluvial boulders of typical Sanbagawa metabasic and metasedimentary rocks, then cross the road at traffic lights and continue to the Lapis Oboke Museum (¥500) which provides a collection of mythical monsters followed (on the first floor) by an excellent mineral collection and finally local geological maps and specimens of local HP-LT rocks, including garnet riebeckite schists, albite-porphroblastic ('prasinitic') metabasites and metagabbros, as well as the famous local metaconglomerate. You can descend the building complex by following signs to the 'promenade', which lead through the concrete substructure and out into the rocky gorge to exposures of strongly foliated, quartz-veined, fine-grained psammitic schists with south-dipping metamorphic foliation on the southern limb of a prominent antiform.

Back on the road continue to another red bridge beyond which is a shop/restaurant complex selling tickets for boat trips in the gorge below. Enter the shop and turn right to descend steps signposted to the boat tour, passing toilets and an underground car park to emerge again on the west side of the gorge. At the point where the path to the boat turns sharp left to descend to the jetty, continue straight on but then immediately descend a few steps on the left to reach a flat concrete platform surrounded by good exposures of the metaconglomerate. The clasts include granitoid and felsic volcanic rocks which can exceed 5 cm in size, so are very obvious despite their deformed state. There are further excellent exposures of this unusual lithology along the route down towards the boat jetty. To make the return journey, continue up the path back in the direction of the red bridge, pass under the bridge to reach a hotel, then look for a short flight of steps up to the right which leads back to the road to join the outward route before returning to the station.

The rail journey through Central Shikoku from Oboke continues south then curves west, still following the Yoshino Valley and Oboke Nappe outcrop as far as Osugi. From here the route turns south again, initially tunnelling through the narrow (<5 km) outcrop of the Mikabu Belt (mostly mafic volcanic rocks belonging to a Cretaceous ophiolite) to cross the Northern Chichibu and Kurosegawa belts (Jurassic–Early Cretaceous accretionary complex and mélanges including pre-Jurassic basement) which crop out south of the Sanbagawa Belt across the entire length of Shikoku. The railway finally emerges from tunnelling through the Kurosegawa outcrop and, leaving the mountains of central Shikoku behind, turns SW to Tosayamada. The final part of the journey crosses estuarine coastal plain sediments (which provide the only large flat agricultural area in southern Shikoku) to reach Kochi, the capital of Kochi Prefecture, situated some 30 km above the subducting Philippine Plate (Fig. 12.1b).

The area around Kochi marks the southern limit of the older accretionary mélanges and metamorphic rocks. The younger accretionary mélanges and sediments of the Shimanto Belt appear further south, coastal exposures of which can be visited on daytrips to the SW (Tosa-Kure) or SE (Muroto Peninsula). Whereas the mélanges exposed at Tosa-Kure are reachable by JR train (Excursion 8; a journey which offers the added opportunity of a visit to Sakawa Geology Museum), the Muroto Peninsula can only be reached by hire car or non-JR train and coach (Excursion 7). Stay overnight in Kochi.

Excursion 7: SE from Kochi: Shimanto belt mélanges, turbidites and gabbros

This long day excursion follows the coast SE from Kochi to the tip of the Muroto Peninsula (Fig. 12.1). The first half of the day is spent examining a beach cliff section at Nishibun where there are good exposures of disrupted oceanic basalts and deep-water sediments within a mélange in the northern (Cretaceous) part of the Shimanto Belt. The afternoon stop visits Muroto Geopark in the southern (Palaeogene–Neogene) Shimanto Belt, where there is a fine coastal section of Miocene turbidites intruded and contact metamorphosed by the Muroto-misaki Gabbro.

Arrival at Nishibun by public transport from Kochi requires taking a JR train to Gomen then changing trains to the Tosa–Kuroshio Gomen Nahari Line (no JR rail pass), which takes 25 minutes to arrive at Nishibun station. The onward journey to the Muroto Peninsula takes another 40 minutes by train to Nahari, followed by an hour in non-JR bus to Muroto-misaki. Although the geology and scenery are excellent, we view this day (especially the afternoon in Muroto) as one of the most optional for those with limited time; it is quite an expedition and involves non-JR transport, the extra cost of which is currently (2015) around 5500 Yen return per person.

The journey from Gomen to Nishibun crosses the wide coastal plain SE of Kochi from which parallel ranges of forested hills rise, marking more competent units within the accretionary complex that forms southern Shikoku. Reaching the coast at the port of Akaoka, the railway continues to the eastern margin of the coastal plain. It tunnels through the Shimanto Belt outcrop to emerge at Nishibun, now on an elevated line that forms part of the coastal tsunami defence system. The Pacific coast of Shikoku is hit by major tsunamis every 100–150 years, the last being on 21 December 1946 and causing around 1300 deaths.

From the rail station pass under the railway and follow the coastal cycle path west, initially at beach level then climbing to run alongside a busy road. Twenty-five minutes from the station turn left to descend a road signposted to Nishibun Fishing Port (西分漁港). A few minutes after reaching the port, turn left past exposures of disrupted, originally finely bedded dark marine sediments to reach the NE corner of the harbour and climb the breakwater to access the beach. Easily accessible and continuous cliff exposures can be followed from here 600 m eastwards, then on across the beach back to the rail station (allow an hour for this traverse). Basaltic pillow lavas emerge from the beach below the ocean end of the harbour wall, providing a photogenic view across the beach and cliff section.

The cliff section here exposes rocks of the Tei Mélange, an oceanic subduction unit within the older (Cretaceous) part of the Shimanto Belt (see Chapter 2d). It comprises clasts of oceanic volcanic rocks and sediments of Jurassic–Cenomanian age, lying within a shaley matrix dated by radiolaria as Coniacian–Santonian in age (Taira *et al.* 1980). Although within a mélange unit, many of the exposures at Nishibun are characterized by a semi-coherent character in which original aspects of the deep oceanic stratigraphy have been preserved. The first rocks encountered in the cliffs after crossing the harbour wall are purple basalts (locally mixed with limestone) and pale-weathering, vertically bedded deep-marine sediments. The next few 100 m of the cliff section exposes variously coloured radiolarian cherts and pelagic mudstones originally deposited on the basaltic basement. Although extensively disrupted by faulting, subvertical bedding is generally well preserved and both cleavage–bedding relationships and pillow lava form indicate a younging direction to the north. This is the typical situation within the Shimanto Belt, with individual tectono-stratigraphic units each internally younging northwards progressively scraped off the subducting plate during the construction of an accretionary complex which overall youngs southwards.

The best exposures of bedded cherts occur 200 m east of the harbour wall, and strongly recall the 'Rhine Valley' exposures at Sakahogi seen on Excursion 5 with the same colourations, steep dip and steeply plunging isoclinal folds. The difference in age between these sequences however provides a reminder of the long history of oceanic plate subduction that has taken place beneath what is now the Japanese archipelago. Beyond these exposures there is a prominent steep fault zone strongly displacing and disrupting the chert outcrop, followed by more northwards-younging purple pillow lavas (Fig. 12.3b). The degree of structural disruption increases towards the end of the beach sequence where there are grey metasedimentary mélanges similar to those previously observed in Nishibun Harbour, typical of many stratigraphically chaotic outcrops within the northern Shimanto Belt.

Continue eastwards and return to Nishibun station via the cycle path. Take the next train onwards to Nahari (or Aki) and transfer to the Kochi Tobu bus line, which runs on to Muroto (the same bus stops outside both Nahari and Aki station; it is slightly cheaper to go by rail to Nahari). Bus times from Aki (安芸駅) and Nahari (奈半利駅) to Muroto-misaki (室戸岬) are posted at the bus stops outside the station. The town of Aki has given its name to the fault boundary (Aki Tectonic Line) separating the Cretaceous mélanges from the younger (Eocene–early Miocene in age), deformed but generally more coherent accreted turbidite sequence of the Muroto Peninsula. Beyond Nahari, the bus enters the Muroto Geopark area, making slow progress as it follows the fishing communities confined within the narrow coastal platform between former seacliff and the open ocean. Well-documented exposures of the Shimanto Belt sedimentary sequence in this geopark include the classical Palaeogene trace fossil localities at Hane and Gyodo points (Nara & Ikari 2011). Around an hour from Nahari, the bus finally rounds the Cape to stop outside the Muroto Geopark Information Office where leaflets in English are available. Allow 2.5–3 hours for the field visit. Bus times back from Muroto-misaki are posted at the stop opposite.

Walk west a short distance from the Information Office, past the imposing statue of Shintaro Nakaoka built on a pedestal of coarse-grained gabbroic boulders, to a small car park (left) which is the starting point of an informative geotrail similar to that seen at Nechi in the Itoigawa Geopark (Excursion 2). Before following this trail, however, locate the information panel in the far corner of the car park to view the contact between steeply dipping Shimanto Group turbidites (left) and the base of the Muroto-misaki gabbroic sill (right). This gabbro is up to 220 m thick and was emplaced horizontally into the Shimanto sediments before being later tilted steeply to the NW. A detailed description and interpretation of the magmatic evolution of this intrusion has been provided by Hoshide & Obata (2009, 2012).

The igneous stratigraphy of the sill evolves from a lower chilled margin of glassy olivine basalt up through fine- to medium-grained olivine gabbro. Some 40 m up into the intrusion the gabbros become more heterogeneous in texture, containing feldspathic and pegmatitic layers and pods. To visit this interesting lower zone (see map in Hoshide & Obata 2012), descend to beach level from the car park and clamber around to the right (west), passing from the lower chilled margin and textually homogeneous gabbros above to the overlying pegmatitic zone. The origin of the layering and pegmatites has been attributed to the upwards expulsion of residual H_2O-rich fluids as the sill crystallized. Infiltration of these fluids into the newly crystallizing gabbroic mass above caused local remelting and crystal-melt segregation, producing podiform pegmatites commonly preserved with a mafic (picritic) base and felsic top, presumably caused by gravitational segregation of crystal mushes. There are excellent examples of these pale feldspathic segregations forming pods and layers with bulbous, diapiric upper surfaces, and remobilized protrusions demonstrating their gravitational instability within the hot, recently crystallized intrusion. These heterogeneous, pegmatitic olivine gabbros continue upwards to over 100 m from the basal chilled margin and are overlain by 75 m of textually more homogeneous coarse gabbro generally lacking olivine and representing the youngest part of the crystallizing sill.

Return to the car park and follow the geotrail to the information panel exposures of turbidites on the beach *c.* 75 m beneath the gabbro sill. These represent the youngest sediments exposed on the Peninsula and were likely deposited in early Miocene oceanic trench-slope environments before being added to the accretionary complex. The exposures preserve well-bedded, steeply dipping turbiditic sandstones and shales (Fig. 12.3c), extensively faulted and folded about NNE-trending axes. Occasional high-energy sedimentary events are recorded by mudclast conglomerates (there are good exposures of these down to the right of the prominent outcrop featured in the display panel). Moving on, there are extensive exposures of this deformed and disrupted marine sedimentary sequence across the coastal platform to the right of the geotrail, which continues NNE parallel to the coast and back up through the gabbro outcrop.

The geotrail highlights several exposures of the gabbro partially coated by pale calcareous deposits originally produced in the intertidal zone by the serpulid tubeworm *Pomatoleios kraussii*. The presence of these fossils above sea level, successively uplifted during major earthquake events, provides a graphic reminder of the tectonically active and exposed setting of the Muroto Peninsula. Further north along the geotrail is the prominent uplifted sea stack (with wave-cut platform) of Bishago Iwa, comprising olivine gabbro within the upper zone of the steeply dipping sill intrusion. In a mirror image to the exposures examined earlier, the gabbro grain size decreases from SE to NW towards the upper chilled margin. Continuing north, this chilled margin is well exposed alongside the geotrail, with a basaltic-doleritic contact against pale, hornfelsed metasediments. A 14 Ma Rb–Sr isotopic date from felsic rocks produced by local contact-melting of these sediments indicates middle Miocene intrusion after early Miocene sediment deposition. Return by the same route and overnight in Kochi.

Excursion 8: SW from Kochi: Shimanto Subduction Mélange and Sakawa Geological Museum

This excursion can be made entirely using JR train and combines a visit to coastal exposures of Shimanto Belt accretionary mélanges on the Pacific coast at Tosa-Kure with the excellent geological museum at Sakawa. Tosa-Kure, an attractive fishing village (with fish market and Buddhist temple; allow around 3 hours to enjoy the geology and location) lies some 50 minutes by LEX train from Kochi. Avoid arriving in Tosa-Kure at highest tide. With current (2016) timetables, only an early start allows visits to both Tosa-Kure and Sakawa, retrieving luggage left in Kochi, and then returning north to Okayama all in one day (Fig. 12.1a).

From Tosa-Kure rail station entrance (11 m above sea level) walk 200 m east then, forking right, 300 m SE through the village, curving south past the temple to arrive at the beachfront. Turn right and follow the beach promenade south for 400 m to cross a road bridge and reach Highway 25. Continue along this road, following the coast south for another 300 m to where Highway 25 disappears into a tunnel. Here take the old, overgrown, tarmac lane on the left and follow it around a rocky headland above the port. This headland exposes the Kure Mélange, a good example of one of several Late Cretaceous chaotic units within the Northern Belt of the Shimanto Complex

(Taira *et al.* 1980; Mukoyoshi *et al.* 2006, 2007; see Chapter 2d). These mélanges occur sandwiched between coherent sedimentary sequences and have been interpreted as tectonic in origin, produced during subduction underthrusting and accretion of oceanic sediments and basaltic crust and later exhumed by out-of-sequence thrusts slicing through the accretionary prism (Mukoyoshi *et al.* 2006; Kitamura & Kimura 2012). The Kure Mélange crops out over a zone *c.* 1 km wide between the coherent packages of the Nonokawa (south) and Shimotsui (north) formations (see maps in Mukoyoshi *et al.* 2006, 2007). The village of Tosa-Kure lies on the Upper Cretaceous Shimotsui Formation sandstones, exposures of which occur on the headland 800 m to the north but are currently (2015) out of bounds due to landslide hazard.

In this excursion we describe a visit to four coastal cliff exposures of the Kure Mélange. The first of these (Locality 1) can be examined by following the overgrown tarmac lane around the headland to a house entrance (right), in front of which the beach (left) is accessed down a series of steps. Following the beach around to the left (north), exposures of sandstone-shale mélange with a subvertical foliation (Fig. 12.3d) are immediately encountered. Shearing disruption of the original marine sequence is extreme, with sandstone clasts lying within a shale matrix and little or no preservation of coherent bedded structure. Moving further north there are remnants of thicker (>1 m) sandstone units within the mélange. Exposure is briefly lost across a small beach before exposures of mélange resume, with good examples of a disrupted thick grey sandstone unit preserving its coarse clastic texture within individual clasts. Beyond this, a zone of sandstone-shale mélange containing large basaltic blocks is entered. The largest of the green-red weathering metabasic clasts is >20 m thick and forms a small headland, beyond which access further north is difficult.

The wide range of textures observable in these mélange exposures reflects a long history of progressive deformation from initial pure shear to increasingly simple shear deeper within the accretionary prism. Early disruption of sandstone bodies by flattening and boudinage is further accentuated by cataclasis, granular flow, pervasive pressure solution effects and increasingly ductile behaviour as the mélange rock material moves through the seismogenic zone, with temperatures rising from <100°C to >300°C (Hyndman *et al.* 1997; Kitamura & Kimura 2012). Deformation is strongly partitioned into the shale matrix which develops a pervasive scaly foliation, whereas the more competent sandstone blocks commonly preserve original sedimentary textures.

Climbing back up the steps, turn left and continue following the coastline SW, initially rejoining Highway 25 then branching left on a minor road to reach the end of the concrete beach wall where steps lead to the beach and rocky promontory (Locality 2). Here the exposures are dominated by massive grey sandstones similar to some of the coarser clasts seen at Locality 1 but with little obvious evidence for mélange disruption. Thin shaley beds within this sandstone-dominated sequence are more deformed however, once again demonstrating strong strain partitioning.

Walk back 100 m along the concrete beach wall and turn left to branch SW along a narrow lane which passes between low-lying fields (right) and a wooded isolated hill (left), marking the outcrop of the massive sandstone unit viewed at Locality 2. Some 200 m further SW there is hidden access (left) to a wide beach with views east towards Cape Otsuzaki and cliff exposures both left (north) and right (south). Turn left and walk to the northern beach corner (Locality 3) to view exposures of steeply dipping and faulted grey metasedimentary mélange lying against the other side of the same massive sandstone unit viewed at Locality 2 (Mukoyoshi *et al.* 2006, 2007). Overviewing these exposures at the three localities usefully illuminates how tectonic mélanges demonstrate stratal disruption at different scales, from individual outcrop scale to map scale, destroying stratigraphic continuity as the deformation progresses.

Finally, if the tide state allows, walk south across the beach to examine further mélange exposures on the far side. Here the mélange is darker and more argillaceous in character, recording the disruption of a marine mudstone sequence with only minor, thin-bedded arenaceous layers. In contrast, several locally derived boulders of coarse, matrix-supported mudclast conglomerates, although not *in situ*, indicate the presence of massive, high-energy sedimentary slump units in the Shimanto succession. Further east, the Kure Mélange outcrop is terminated by a north–south fault so that coherent Upper Cretaceous turbidites of the Nonokawa Formation form Cape Otsuzaki (Mukoyoshi *et al.* 2006). Return to Tosa-Kure village by the same route.

The return train from Tosa-Kure stops in Sakawa where a visit to the geology museum is recommended (15 minute walk from station, open until 17:00, closed Mondays). From Sakawa train station entrance walk 100 m to the main road, turn left and follow it 400 m to cross the railway and river, where it merges into Route 33. The museum is signposted (in English) to the right at traffic lights 200 m further on. The museum is not especially large but it houses an interesting and high-quality range of material; the palaeontological collections and rock garden highlighting local Palaeozoic and Mesozoic accretionary strata are especially good. There is a scale model of the Sakawa Basin and photographic records of flooding in Kochi, and foreign visitors are given a museum guide (in English) which includes a simple geological cross-section through Shikoku. Sakawa is just over 3 hours train journey from Okayama or Takamatsu (Fig. 12.1a).

While on Shikoku, those with enough time and an interest in the plutonic geology of the Ryoke Belt can consider adding a day visit by Uchinomi Ferry from Takamatsu to Kusakabe Port on the island of Shodoshima. The journey is best made on a non-public holiday weekday and JR passes are not valid; bicycles, cars or taxis can be hired to ride the 9 km SE from Kusakabe to the Tanoura Peninsula where there is an open-air museum for the classic Japanese film *24 Eyes* (entrance fee). On the beach behind the film set turn left to traverse from granitoid to gabbroid rocks of the Tanoura Gabbroic Complex (Ishihara 2003), with intermediate (grey granodioritic/dioritic) lithologies representing various stages of magma mingling and mixing. The more basic rocks tend to be finer grained, commonly chilling against the more granitoid fractionates to produce pillow-like contacts. Heterogeneous, grey tonalitic and paler granodioritic hybrid rocks show extensive back-veining into synmagmatic diorites, gabbros and finer-grained mafic dykes, which in places progressively disaggregate into blocky enclaves. Return to Takamatsu by the same route, with views northwards to inland hills exposing thick Miocene basalts erupted over the Cretaceous granite basement, and passing (left) coastal and road exposures of Miocene white brecciated pumiceous blocky ash flow tuff and flow-banded rhyolite. Stay overnight in Okayama.

Excursion 9: Aso Volcano

The morning Sakura Shinkansen service from Okayama to Kyushu takes 2 hours 30 minutes to Kumamoto where transfer to a LEX train allows an afternoon excursion into the active caldera of the Aso Volcanic Complex. Current timetables (2016) allow arrival in Aso rail station just before 13:00 (weekends and holidays best avoided). The route east from Kumamoto follows the Shira River, running over Holocene and Pleistocene valley deposits to Tateno where the train reverses direction to zig-zag up the steep valley side. The train continues east but then turns north to curve around the northwestern perimeter of the volcano via Akamizu to reach Aso, a town built inside what is one of the largest, most active and most populated volcanic calderas in the world. The basement beneath the Aso area, present at depths <500 m, include Cretaceous granitoid lithologies intruding older metamorphic rocks analogous to those exposed north of the MTL in Shikoku and Honshu. In contrast to Shikoku, where modern plate movements are predominantly strike-slip, Aso lies 140 km above the Benioff Zone produced by the subduction of the Phillipine Plate at the Nankai–Ryukyu Trench some 275 km to the SE, and is the northernmost of five giant caldera complexes running south through Kyushu (Fig. 12.1). Earthquakes around the Aso caldera rim have included several recorded events reaching M 6–7 (in 1894, 1895, 1898 and 1975).

Andesitic pre-caldera volcanic activity began in this area in Late Pliocene times (3 Ma), followed by four major, caldera-forming Pleistocene eruptions (370–270, 170, 100 and 80–70 ka) from a zoned and recharging magma chamber. Magma evolution has followed a tholeiitic to calc-alkaline trend and produced basaltic to rhyolitic lavas and pyroclastic deposits (Hunter 1998; Kaneko *et al.* 2007; Miyabuchi 2011 with maps and references therein). The last of these four eruptions was the largest, producing *c.* 600 km^3 of pyroclastic flow and ash deposits which covered much of Kyushu and other parts of Japan. Post-caldera activity has been concentrated into a cluster of 15 volcanic cones erupting basaltic to andesitic and dacitic materials and forming a series of peaks in the centre of the caldera. Substantial deposits of scoria, ash and pumice were produced during this continuing post-caldera phase. At the present time only one of the central cones (Nakadake) is active (Fig. 12.3e). Emerging from Aso rail station (where there are luggage lockers) the central volcanic complex and the steaming vent of Nakadake can be seen to the south. There is a tourist office (left) and bus stop (right) from where infrequent afternoon buses (in 2015 leaving at 13:10, 13:40, 14:30 and 15:20; pay on the bus) take 40 minutes to reach the cable car station at the foot of Nakadake.

On the climb towards the central volcanic mountains, the bus passes the 80 m high symmetric basaltic cone of Komezuka (right), a monogenetic Strombolian vent *c.* 3300 years old. The bus then makes a brief stop at the viewpoint of Kusasenri 6 km east of the active Nakadake vent (1506 m), and the enormous scale of the Aso Caldera becomes apparent. Eboshidake (1337 m) rises to the south across the green pastureland of Kusasenri, one of the four other volcanic peaks in this central part of the caldera. The other three peaks are Kishimadake (1321 m), rising just 1 km NE from Kusasenri, and Takadake (1592 m) and Nekodake (1433 m), both to the east of Nakadake. The bus continues past Aso Volcano Museum to stop at the car park in front of a round building housing shops and the cable car station.

Nakadake is an extremely active vent and has been so throughout historical time, having erupted over 60 times in the twentieth century alone. It is common for the vent to be closed off to visitors either because of eruptions or volcanic gas emissions. However, when the volcano is quiet and the wind favourable, the active crater can be accessed either by the cable car (upstairs in the tourist centre) or by walking up the road to the south of the cable car. The elongate summit area has been active for at least 25 ka and comprises seven small craters of which only the northernmost (known as the 'First Crater') is currently active. The steaming First Crater lies 300 m north of the cable car station and can be visited when atmospheric SO_2 concentrations are <5 ppm. Information on current SO_2 levels is provided by a series of warning lights (red >5 ppm; orange 2–5 ppm; blue 0.2–2 ppm; green <0.02 ppm). In response to several deaths due to inhalation of toxic volcanic gas (these occurring both at the crater edge and 250 m away from the crater), approaching the crater is these days prohibited for those with asthma or other chronic respiratory diseases such as pulmonary emphysema or for those with heart disease.

During quiescent periods there is a fuming hyperacidic lake (Yudamari) at the crater bottom with water temperatures typically around 65°C, a pH below 1 and highly sulphurous lacustrine sediments. Degassing takes place both through the lake and from the fumarole field at the south side of the crater, where surface temperatures can exceed 300°C. In times of increased volcanic activity, the lake becomes more active and progressively dries up as phreatomagmatic explosions and strombolian and ash eruptions take place. There are several concrete shelters built around the crater providing protection from sudden episodes of volcanic bomb showers. Mafic scoria-fall and ash-fall deposits resulting from eruptive episodes can be seen draping the crater rim. Walk SSE from the First Crater to view similar volcanic deposits around the rims of the adjacent pair of craters (Second and Third), which have been inactive since a major eruption in 1933. The walk back down the access road to the tourist centre passes a cluster of similarly inactive small craters (Fourth to Seventh) at the southern end of the Naka-dake summit area. Return by the same route and overnight in Kumamoto (we recommend staying in the city centre near the Ginza-dori arcade, 6 tram stops from the rail station).

Excursion 10: Nagasaki and Iōjima

The town of Nagasaki can be visited on a day trip from Kumamoto (2 hours 30 minutes via Shin–Tosu) or as an overnight stop. It is an attractive journey, following the coastal plain of the Ariake Sea (the largest bay in Kyushu) and the eastern slopes of the Tara Mountains which were built from Miocene–Pleistocene mafic volcanic rocks.

Nagasaki lies in a southwards-draining alluvial valley flanked by Neogene felsic pyroclastic flows overlain by mafic lavas. Beneath these volcanic rocks are Palaeogene arenaceous sediments of the Takashima coal field, themselves lying on a basement of Sanbagawa metamorphic rocks which forms much of the Nomo Peninsula to the SE of the city. Just 40 km to the east rises the volcanic peak of Mount Unzen (Excursion 11). A day or more can easily be spent in Nagasaki which has much to offer the tourist, including a visit to the nearby island of Iōjima which is the type locality for the youngest sediments in the Takashima coal field.

Close to the train station, the nearby 'Site of the Martyrdom of the 26 Saints of Japan' provides one of the few areas where bedrock is exposed in the city. Leave the station with the JR reservation office on the right and climb steps to cross the main road on an elevated pedestrian walkway. On the far side turn left along the main road for 120 m to locate a blue and white signpost for the site where 26 Christians were crucified in 1597. Climbing from here, follow a wall constructed from local Neogene basalts for another 120 m to enter the hilltop cruxificion site (left), from where there are views across the city and up the river valley. The hypocentre of the August 1945 atomic bomb, which exploded 500 m above ground level, lies 2.5 km NNW up the valley; the hilltop location marks the edge of the central zone of extreme destruction, within which all buildings were either seriously damaged or obliterated. On the west side of the hilltop there are pale exposures of the Neogene porphyritic felsic pyroclastic rocks which underlie much of the city. Just before the entrance to the museum, an ornamental rock wall displaying river cobbles provides good examples of the Sanbagawa basement lithologies (greenschist metabasites and metaquartzites) and Palaeogene sandstones, and there are further exposures of the young ignimbrites. Return to the train station.

Regular ferries to Iōjima, a small island just south of Nagasaki in the northern part of the Takashima coal field, run from the port 1 km south of the rail station. Turn right out of the station, walk south to cross traffic lights at a major road junction then dog-leg right then left along a paved back route which continues south towards the port to another main road. Turn right to pass Starbucks then follow the south side of the huge 'you me Saito' department store to locate the port. Ferries to Iōjima run every 1–2 hours and take 20 minutes, sailing south then southwest down the estuary, leaving the Neogene volcanic sequence to reach the underlying Palaeogene sediments.

Palaeogene rocks are common in NW Kyushu, typically comprising shallow-marine to brackish siliciclastic sediments intercalated with carbonaceous deposits that form several coal fields. Iōjima forms an elongate island (1 × 2 km) oriented NNW–SSE and formed entirely of Eocene–Oligocene sediments belonging to the Iōjima Group (>550 m thick), the youngest stratigraphic unit within the Takashima coal field (see map in Yamaguchi *et al.* 2008). The Iōjima Group comprises breccias, conglomerates, sandstones, shales and coal-rich seams, the latter having been mined in the generally poorly exposed lower part of the sequence on adjacent Okinoshima (Magome Formation, middle Eocene in age). The Magome Formation coals are overlain by lower Oligocene Funazu Formation sandstones and conglomerates, easily accessible coastal exposures of which occur 1 km NW of the harbour. Turn right from the ferry harbour, pass the popular local hot baths (Nagasaki Onsen Yasuragi Ioujima) and follow the coast road to reach the prominent north-facing beach ('Costa del Sol'). In the cliffs behind the beach are exposures of Funazu Formation sandstones and matrix-supported breccias and conglomerates, containing vein-quartz pebbles and other detritus derived from the underlying metamorphic basement.

Continue past the beach to reach an artificial isthmus leading out to the small island of Kojima. Steps lead down to exposures of a tessellated sandstone pavement near sea level on the west side of the concrete isthmus. Erosion of the regular system of vertical joints that gives rise to the tessellated texture produces concave slabs of grey-brown sandstone, the central area of which is sometimes interrupted by excellent examples of sand volcanoes produced by the extrusion of liquefied sediments from isolated pipes (Fig. 12.3f). Follow the walkway clockwise around the west side of Kojima to reach further instructive exposures revealing evidence of soft-sediment deformation, probably induced by seismic activity soon after deposition. Cross-beds within the sandstone locally show overturning due to liquefied slumping down the lee-slope of individual dunes. The sediments have been tilted to the SSE and cut by several normal faults. Although the dominant lithology of this sequence is brown sandstone, there are layers and lenses of conglomerate and, towards the northern tip of the island, a poorly exposed silty mudstone layer provides an exception to the dominantly high-energy sedimentary environment recorded by the Iōjima Formation. Returning along the east side of the island there are fine views towards Nagasaki and the Nomo Peninsula.

Finally, if time allows, a 2 km return walk to exposures of the breccio-conglomerates at the base of the Magome Formation can be made by returning to the harbour and walking westwards, following the channel separating the islands of Iōjima and Okinoshima (Yamaguchi *et al.* 2008). Cross over to Okinoshima (with exposures of Magome Formation sandstones in the cliffs opposite to the right) and continue west to locate a tunnel, on the other side of which are extensive cliff exposures and beach boulders of NE-dipping middle Eocene coarse breccias and conglomerates packed with basement clasts (the 40 m thick Dezaki Conglomerate; Yamaguchi *et al.* 2008). Stay overnight in Kumamoto, Nagasaki or Unzen village.

Excursion 11: Unzen Volcanic Complex

The volcanic peaks produced by recent eruptions at Unzen rise from the Shimabara Peninsula halfway between Kumamoto and Nagasaki (Fig. 12.1); they can be visited on a long day trip from either city or by staying overnight in Unzen itself, a village famous for its active geothermal phenomena. Those with limited time and lacking a hire car should consider this the most optional of the three volcanoes visited in Kyushu (Excursions 9, 11 and 12) because the journey by public transport to this currently quiescent volcano is time consuming. Buses (according to the 2015 timetable) run to Unzen village from opposite the JR train stations of both Nagasaki (Ken-ei bus terminal; Stop 3; 1800 Yen one way; 1 hour 40 minutes) and Isahaya (below the shopping arcade a few minutes' walk directly ahead from the rail station; 1350 Yen; 1 hour 20 minutes). Once in Unzen village allow 2 hours to visit the volcanic summit area (by minibus and cable car) and a further 2 hours locally to study the geological displays on the first floor of the visitor centre and enjoy a walk through the steaming, bubbling and roaring *jigoku* vents scattered across the adjacent thermal field and emerging from the porphyritic dacitic bedrock (see the geopark website listed in Table 12.1). The sulphurous scene is reminiscent of Yellowstone Park, except for the unsightly network of pipes bringing thermal waters to the Onsen hotels, and the monument to seventeenth-century Christians martyred in the scalding waters.

To reach the summit area of the modern volcano from Unzen village it is necessary to book a minibus seat with Heisei Kanko Taxi (currently 3 times a day from the tourist information centre at 09:00, 11:00 and 14:00). The journey takes 20 minutes each way and includes a 1 hour stop at Nita Pass, allowing a cable car ride up to a viewpoint over the current volcanic peaks, the youngest of which rose as a dacite dome in the eruption of 1990–95. During the minibus ascent to Nita Pass there are views north across to this lava dome (Mount Heisei Shinzan), which currently forms the highest peak. For maps and descriptions of Unzen geology and eruptive history see Hoshizumi *et al.* (1999) and Nakada *et al.* (1999).

Unzen Volcano is unlike those of Aso (Excursion 9) and Sakurajima (Excursion 12) because it lies in a back-arc position 70 km west of the volcanic front produced by Philippine Plate subduction (Fig. 12.1b). Its origin is linked to lithospheric extension producing a rift zone running east across Kyushu from Beppu to Shimabara and out towards the intercontinental Okinawa back-arc basin. The east–west rift at Unzen is *c.* 35 km long, currently subsiding at a rate of 2–3 mm a^{-1}, and is infilled to a depth of 1 km below sea level with young volcanic products derived from hornblende-bearing silicic andesite-dacite magmas. Volcanic activity in the area commenced with basalt-andesite eruptions in Pliocene times, evolving to produce a major Middle Pleistocene andesite-dacite eruptive centre (Older Unzen Volcano); more recent eruptions have built the current cluster of eruptive peaks (Younger Unzen Volcano). Outcrops of the older volcano occur mostly to the west of the modern younger volcano, including those around Unzen village, and comprise mostly thick, hydrothermally altered porphyritic lava flows, domes and pyroclastic deposits that now form the basement to the modern volcano.

Arrival at the Nita Pass parking area below the cable car offers views across the modern composite volcanic edifice. Immediately to the south rises the lava dome of Nodake Volcano (1147 m), the oldest eruptive centre of the Younger Unzen Volcano. Andesitic-dacitic lavas, pyroclastic flows and debris avalanches began to be erupted from here around 100 ka ago and were directed mostly towards the SE. To the north rises the main andesitic volcanic edifice of Myokendake Volcano which has two peaks, Myokendake (1333 m) and Kunimidake (1347 m), comprising dissected 20–30 ka lava domes. On the east side of this volcano is a 1.5 km diameter horseshoe-shaped scar where the youngest volcanic peaks have been constructed. Ascent of the cable car from Nita Pass provides access to Myokendake peak which offers a grand panorama across the scar eastwards to these peaks, the highest of which comprise Fugendake (1359 m) and the Heisei Shinzan lava dome (1483 m) 700 m further east.

Volcanic activity in historic times has included relatively minor eruptions from the Myokendake scar of basaltic andesite in 1662 and dacite in 1792. The 1792 event was accompanied by fumarolic activity and tremors which included a M 6.4 earthquake and induced the collapse one month later (21 May) of a major dacitic lava dome built just east of the scar around 5000 years earlier. The human consequences of this collapse were catastrophic because the event occurred so close to the coast. The debris avalanche hit Shimbara city then entered the sea to create a mega-tsunami (reportedly in the order of 100 m high) which in turn devastated the far side of the bay. With a final death toll of around 15 000, in human terms this remains the most disastrous volcanic event on record in Japan.

Dormant for nearly 200 years, renewed earthquake activity beginning west of the volcano heralded an eruption of ash 600 m east of Fugendake in November 1990. By February 1991 ash eruptions were taking place from three vents, becoming violently phreatomagmatic in April–May as the ascending magma interacted with the aquifer beneath the volcano. Arrival of the magma in late May exposed a rising porphyritic dacite dome, the surface of which quickly broke into large blocks as magma was fed from below so that successive lobes emerged and froze in and around the crater area. Continuous discharge of increasingly degassed, viscous dacite was accompanied at first by several Vulcanian explosions, the largest of which (on 11 June 1991) threw pumice fragments up to 45 cm in size into populated areas lying 5 km to the NE of the crater, causing considerable damage.

Over 9000 pyroclastic flow events were triggered during dome growth from late May 1991 onwards, mostly channelled eastwards down the Mizunashi River valley. On 3 June 1991 an ash-cloud surge from one of the larger of these flows, induced by east-directed dome collapse, killed 43 people 3.5 km down this valley. The victims, who had passed beyond the police control line, were mostly press photographers but included the three volcanologists Harry Glicken, Katia Krafft and Maurice Krafft. Two years later another person fell victim to a similar pyroclastic ash-cloud surge after having crossed into the no-go area to visit his damaged house. By the end of the eruption, around 2500 houses had been burned or destroyed by pyroclastic flows and lahars, the latter proving most hazardous during the monsoon period, with the worst time being during the summer of 1991 when *c.* 11 000 people had to be evacuated from their homes.

The eruption came to an end with the extrusion of a lava spine which emerged from the infilled crater during October–December 1994, and by February the next year dome growth was undetectable and earthquake and rockfall activity had virtually disappeared. Today the Myokendake viewpoint overlooks the devastated Mizunashi Valley, partially filled in its upper levels by pyroclastic flows and in its lower part by debris flows which form a wide delta feeding into the Ariake Sea. Viewed from here, the vertiginous gradients from volcanic peaks to sea level, and the obvious potential instability of the young lava dome, underline the vulnerability of the populated coastline during any future activity of this dangerous volcano.

Excursion 12: Kagoshima area and Sakurajima Volcano

The coastal city of Kagoshima is located on the SW margin of the Aira Caldera, an eruptive centre *c.* 20 km in diameter lying in the middle of a 120 km long volcanic structure known as the Kagoshima Graben. Most of the Aira Caldera is flooded by the sea and, unlike Aso, is topographically inconspicuous at ground level. Instead it is the smoking andesitic stratovolcano of Sakurajima, built on the southern rim of the Aira Caldera, which immediately commands attention upon arrival in the city and gives Kagoshima its nickname of 'Naples of the East'. Sakurajima is one of Japan's most active volcanoes, currently rising to a height of 1117 m asl, and frequently depositing ash over the adjacent city (Fig. 12.3h).

On a clear day an excellent way to gain an impression of the geomorphological and volcanic setting of the city is to ride the red Ferris wheel built on top of the Amu shopping centre which is attached to Kagoshima-Chūō Station, the Kyushu Shinkansen terminus. Turn right from the ticket barrier, follow signs to the Amu centre (left) and take the lift to the 6th floor. The spectacular view from the wheel is dominated by Sakurajima, to the north of which the sea covers the Aira Caldera, beyond which rises the volcanic complex of Mount Kirishima (1700 m) some 50 km NNE of the city. Sakurajima lies astride the Kagoshima Graben which continues south as a marine channel leading to the partially submerged Ata Caldera, and out to the Kikai Caldera in the open Pacific Ocean (http://www.eri.u-tokyo.ac.jp/fmaeno/kikai/Fig/Fig2-1.jpg). Thick pyroclastic flow deposits produced by major eruptions from these calderas cover much of the surrounding area. In the case of the Aira Caldera, a massive outflow event *c.* 25 ka ago followed a Plinian pumice eruption and produced the Ito pyroclastic flow. This catastrophic outflow was directed mainly towards the peninsula to the SE of the caldera, where it covered an area of over 500 km^2 with more than 8 m of pyroclastic flow deposits. Soft, easily eroded ignimbritic deposits from this eruption are known as Shirasu and mantle the landscape to form a plateau later dissected by Holocene fluvial erosion. Much of the city of Kagoshima lies on the low ground beneath the fluvially dissected Shirasu Plateau, a geomorphology readily appreciated from the Ferris wheel. The basement to the younger volcanic deposits comprises Shimanto Belt accretionary rocks of Cretaceous age intruded by Miocene granites, exposed both east and west of the graben.

To visit Sakurajima take a JR train from Kagoshima-Chūō to Kagoshima station (or board a tram outside Kagoshima-Chūō) and walk 700 m south to Kagoshima ferry terminal (follow the tram one stop, turn left and walk five minutes to locate the covered overhead walkway leading to the ferry terminal). Frequent, inexpensive ferries (pay fare upon arrival) ply the 3 km wide strait between city and the western flank of the volcano. As the ferry approaches the volcano note the steep, cliff-like vegetated hill rising to a plateau immediately behind the port. This is an inlier of the Shirasu Plateau, forming a small remnant of the Aira Caldera ash deposits which, in most places, were later overprinted by the growth of Sakurajima Volcano.

There is an information desk immediately on arrival with bus timetables and leaflets in English, from which a Sakurajima map can be obtained. Descend the escalator to the tourist and local bus stops at street level (there are also cars and bicycles for hire opposite). One tourist bus does a twice-daily full circuit of the volcano (36 km, 2 hours, currently starting at Kagoshima-Chūō station and later leaving Sakurajima Port at 09:40 and 14:20), whereas another ('Sakurajima Island View') follows a shorter (1 hour) circular route 8 times a day and offers an inexpensive daily bus pass allowing for hop-on-hop-off stops en route.

Our recommended route is either to hire a car or buy an Island View Bus Pass and take the bus from the port for the short (3 minutes) ride to the visitor centre. Alighting here, walk to the nearby coast to locate the Yogan Nagisa footbaths and enjoy a cleansing and contemplative view of the volcanic peaks (Fig. 12.3h). From here, walk south along the attractive coastal trail through the Taisho lava field, with continuous grand views of the volcano and straits currently separating Sakurajima from Kagoshima. Abundant spreads of grey andesitic ash underfoot are deposited when easterly winds send eruptive plumes across to the city, causing considerable inconvenience and expense.

The Taisho eruption of 1914 was preceded by several days of earthquake activity, prompting the evacuation of almost the entire island population as well as many people from Kagoshima. The explosive eruption finally began on the morning of 12 January when groups of vents opened on both the western and eastern sides of the volcano, throwing material 10 km high to form a Plinian column. This collapsed to form pyroclastic flows, resulting in extensive ash deposition especially to the east of the volcano. A major earthquake estimated as *M* 7 and accompanied by a tsunami occurred during the evening of the same day, causing extensive damage in Kagoshima and 35 deaths. On the evening of the following day, following another major earthquake, effusion of andesitic lava began with flows directed both west and southeast. By 1 February the lava flows entering the sea in the SE connected Sakurajima with the mainland, whereas those here in the west overwhelmed an islet known as Karasujima which had lain 500 m offshore and is now marked by the Karasujima Observatory. The eruption left nearly half the village on Sakurajima destroyed by lava flows, produced extensive ash damage to buildings and agriculture and prompted one of the first recorded large-scale government relief operations in response to a volcanic eruption. Since the Taisho events of 1914–15 the volcano has erupted again in 1935, 1946 and 1955, since when it has been semi-continuously active (>7500 eruptions) with explosive Vulcanian events providing an ever-present threat to the surrounding area (Fig. 12.3h). Shock waves associated with these Vulcanian eruptions and travelling at several 100 m s^{-1} are capable of breaking windows in Kagoshima. Such explosions likely owe their origin to a build-up of gas pressure finally fracturing the confining volcanic vent plug and so producing a violent expansion during sudden outgassing of water-saturated magma (Ishihara 1985; Iguchi *et al.* 2008).

Follow the route hugging the coast south then SE for 2 km through the blocky lava flows, which vary from more slowly cooled massive to chilled clinkery textures. The lava initially erupted from two vents high on the mountainside, followed by successive effusions from several other vents which opened lower on the slopes *c.* 3 km from the present coastline. The eruption continued for around two weeks. The coastal path curves to reach a road and the Karasujima Observatory, from which a panoramic view back across the Taisho lava field to the inlier of Shirasu ignimbrite behind the port can be enjoyed.

Walk a further 300 m SE along the road past the observatory area to locate the artificial channel dug to release water and debris flow material from the Hikinohira River. These engineering works, started here in 1995, form part of an extensive network of 'Sabo' lahar warning and control facilities which began to be built around the volcano after a series of particularly severe debris-flow events in the 1970s. The sudden appearance of rapidly moving debris flows along the normally dry river channels radiating from the volcanic peaks claimed many lives prior to the implementation of the Sabo system. Along some of the rivers, such events can occur as frequently as 20 times per year and represent a serious hazard to the local population. Real-time information from an array of monitoring sensors is fed to the Sakurajima International Volcanic Sabo Center (situated near the Nojiri River 2 km further SE). The problem is also being ameliorated by reforestation programmes to try to stabilize the damaged land lying closer to the vents.

Continue south along the road for a few minutes to locate (right) the Akamizu View Park, known principally for its andesitic sculpture 'Portrait of a Shout' by Hiroshi Oonari, commemorating an all-night concert here in August 2004. Rejoin the Island View bus which climbs to Yunohira Observatory (373 m), the closest reachable point to the active crater. From here on a clear day there are tremendous views north into the Aira Caldera and beyond to the peaks of Mt Kirishima. Return to the port.

For those wishing to spend more time on Sakurajima but lacking a hire car, we recommend a visit to the SE corner of the peninsula where the currently active Showa Crater may be viewed (at a relatively safe distance of 3 km) from the Arimura Lava Observatory, a 20 minute ride on local bus No. 4 (ask for a timetable at the tourist information desk). The return journey from Arimura allows the option of a stop in the Furusato Kanko Hotel (2 km east) to indulge in the volcanic waters of their justly famous open-air coastal thermal bath shrine (mixed bathing, wearing a hired white yukata). The hotel offers a free shuttle back to the port.

Those able to spend longer in the Kagoshima area can make rewarding day excursions south to Kaimondake Volcano (the 'Fuji of the Satsuma Peninsula') and/or east to Aoshima. Kaimondake (924 m) is the most prominent volcano of the Ibusuki volcanic field which lies on the west side of the Ata Caldera. The scenic rail journey by small diesel train passes slowly through rural Japan, and provides an opportunity to visit the famous sand baths at Ibusuki. For those wishing to face the challenge of the tough climb up the volcano for views over the Satsuma Peninsula, allow at least 6 hours return trip from Kaimon Station and avoid walking after sunset or in wet conditions.

The journey to visit the late Miocene forearc sediments exposed on the shoreline at Aoshima (3 hours east from Kagoshima via Minamimiyazaki; time your arrival with lowish tide) passes around the northern margin of the Aira Caldera with expansive views of Sakurajima Volcano, and crosses the southern slopes of the active volcanic field of Mount Kirishima. Shinmoedake Volcano (a location in the James Bond film 'You Only Live Twice') is currently the most active within the cluster of vents on Kirishima, erupting in 2011 and causing considerable damage which included disruption to the rail system. Upon arrival in Aoshima, walk left then curve right to cross the main road and pass an array of tourist shops to reach a causeway across to the island, famous for its palm trees, Buddhist temple and coastal geology. Extensive exposures at low tide show alternations of grey turbiditic sandstones and hemipelagic siltstones belonging to the Uchiumigawa Group (see Chapter 6), the regularity of which (the 'Devil's Washboard') has been attributed to earthquake triggering of turbidite flows down the slope of the forearc basin deposited above the subducting Philippine Plate.

Return journey

The train journey back across Kyushu and Honshu from Matsumoto to the international airports can be made in one day (Kansai: 4 hours 30 minutes; Chūbu Centrair: 5 hours 30 minutes; Narita: 8 hours), or there are regular flights from both Kumamoto and Kagoshima airports to Osaka (Itami), Tokyo (Haneda) or Nagoya (Chūbu Centrair). Using a JR rail pass, interesting options for breaking the journey include a visit to the famous Akiyoshi-dai limestone karst and cave system in Yamaguchi Prefecture (western Honshu) or the beautiful granitic island of Miyajima (Itsukushima) in the Seto Inland Sea near Hiroshima.

Akiyoshi-do is the largest cave in Japan, exposing reef limestones deposited on an oceanic basement in Permo-Carboniferous times. Returning from Kyushu to the Japanese 'mainland' of western Honshu, leave the Shinkansen at Shin–Yamaguchi for a local connection to Yamaguchi from where a blue and white JR bus (free with rail pass; one currently leaves at 12:15 according to the 2015 timetable) takes just under an hour to reach the tourist centre at Akiyoshi-do. The 20 m high cave entrance leads to a 750 m underground walkway, following a river and passing terraced rimstone pools and impressive displays of stalactitic phenomena. Emerge from the lift at the back of the cave and turn left to climb to viewpoints over the extensive grassy karstic plateau. Allow 2 hours for the visit then return to Shin–Yamaguchi to reconnect with the Shinkansen system.

Miyajima is justifiably one of the most popular tourist sites in Japan, and is reached by JR Sanyo line train from Hiroshima to Miyajimaguchi then by a short ride on a JR ferry (free with rail pass). Although the scenic setting is outstanding, the geology is less compelling being entirely composed of Cretaceous Ryoke Belt granites comparable to those on northern Shikoku (Kotohira, Excursion 6) and Shodoshima (Excursion 8), and mostly covered by thick forest. In places however there are good streamside exposures of these coarse-grained, texturally homogeneous granitoid rocks, such as along the path up Mount Misen just behind the Daisho-in Temple. While a visit to Miyajima cannot perhaps be justified on geological grounds alone, the island is steeped in ancient Japanese culture and has a certain magic. The Daisho-in Temple offers a peaceful place to contemplate the immensity of the geological journey just accomplished.

Japanese geoparks and geomuseums (TK, TM, WG)

It is difficult to escape geology in Japan. The incredibly active plate tectonic setting of the Japanese archipelago means that geology has become woven into the fabric of society, affecting local culture, customs and mindset of the Japanese people. It is therefore not surprising to find that Japan has one of the most organized and rapidly burgeoning geopark systems in the world, with over 50 museums specifically dedicated to geology and around another 50 museums with some kind of geological exhibit.

The initiative to develop a network of geoparks in Japan was begun in 2007 with the establishment of the Japanese Geopark Liason Council. By 2008, following the efforts of the Japanese Geopark Committee, seven sites had been successfully evaluated as National Geoparks. The following year the Japanese Geoparks Network (JGN) was formally instigated as a non-profit organization providing support and promoting rapid expansion, so that by December 2013 the JNG regular membership comprised seven Global Geoparks and no less than 29 National Geoparks, combined with a further 16 Aspiring Geoparks listed as associate members (http://geopark.jp/en/pdf/JGNbook_EG.pdf). Each certified geopark must be revaluated every 4 years, and a JGN National Conference is held in one of the geoparks annually.

These geoparks are heavily committed to education schemes designed to increase Earth Science awareness in local people, particularly schoolchildren. The sites also cater for foreign visitors, with information boards and booklets including explanations in English, and typically offer an enthusiastic welcome to visitors from abroad. In this section of the book we therefore provide a

Fig. 12.4. Map showing the approximate location of each of the geoparks and geomuseums listed in Tables 12.1 and 12.2.

Table 12.1. *Geoparks in Japan*

Map	Geopark (English name)	Website	Geology	Japanese name
1	Shirataki	http://geopark.engaru.jp/	Shirataki Mine in volcanogenic massive sulphide deposit	白滝ジオパーク
4A	Mikasa	http://www.city.mikasa.hokkaido.jp/geopark	Ishikari coal field geology	三笠ジオパーク
9	Mount Apoi	http://www.apoi-geopark.jp/english/	Horoman Peridotite Complex	アポイ岳ジオパーク
10A	Toya Caldera, Usu Volcano	http://www.toya-usu-geopark.org/?pagename=english	Basaltic-andesitic stratvolcano and caldera lake	洞爺湖有珠山ジオパーク
12	Happo Shirakami	http://www.geopark.jp/geopark/happou/	Mount Shirakami, ultramafic, granitic, volcanic and sedimentary rocks	八峰白神ジオパーク
13	Oga Peninsula–Ogata	http://www.oga-ogata-geo.jp	Palaeogene–Pleistocene volcanic and sedimentary succession	男鹿半島・大潟ジオパーク
15A	Sanriku	http://sanriku-geo.com/	Memorial to 2011 tsunami disaster zone	三陸ジオパーク
16	Yuzawa	http://www.yuzawageopark.com	Geothermal activity and metalliferous mining history	ゆざわジオパーク
18	Sado Island	http://sado-geopark.com	Neogene volcanic rocks	佐渡ジオパーク
19A	Mount Bandai	http://bandaisan-geo.com	Recently active Mount Bandai Volcano	磐梯山ジオパーク
21A	Itoigawa	http://geo-itoigawa.com/eng/	24 geosites including Fossa Magna Museum and Kotakigawa Jade Gorge	糸魚川世界ジオパーク
25	North Ibaraki	http://www.ibaraki-geopark.com/	Geosites include Palaeozoic basement, ammonites and Miocene volcano	茨城県北ジオパーク
27	Shimonita	http://www.shimonita-geopark.jp/english/index.html	Geosites include Atokura Klippe, Median Tectonic Line and Myogi Volcano	下仁田ジオパーク
28	Chichibu	http://www.chichibu-geo.com/	Famous copper mine	ジオパーク秩父
30A	Hakusan Tedorigawa	http://hakusan-geo.main.jp/	45 geosites including Hakusan Volcano	白山手取川ジオパーク
31	Choshi	http://www.choshi-geopark.jp	Permian, Cretaceous, Neogene sediments, fossils and volcanic rocks	銚子ジオパーク
32A	Dinosaur Valley Fukui Katsuyama	http://www.city.katsuyama.fukui.jp/geopark/	Tetori Group (Jurassic–Cretaceous) siliciclastic sediments, dinosaurs	恐竜渓谷ふくい勝山ジオパーク
34A	Southern Alps MTL Area	http://minamialps-mtl-geo.jp	Dramatic tectonic uplift in Minami Alps	南アルプスジオパーク（中央構造線エリア）
38A	Hakone	http://www.hakone-geopark.jp	Geothermal zone in Fuji–Hakone–Izu National Park	箱根ジオパーク
39A	Izu–Oshima	http://www.izu-oshima.or.jp/geopark/index.html	Modern volcano built on Plio-Pleistocene precursors	伊豆大島ジオパーク
40	Izu Penninsula	http://izugeopark.org/en/	A peninsula created out of the ongoing collision between two island arcs	伊豆半島ジオパーク
45A	San'in Kaigan	http://sanin-geo.jp/en/index_english.html	Geosites include Tottori Sand Dunes and Genbudo basalt cave	山陰海岸ジオパーク
48	Oki Islands	http://www.oki-geopark.jp/english-top.htm	Geosites include Neogene–Pleistocene volcanic rocks and sediments	隠岐世界ジオパーク
56	Muroto	http://www.muroto-geo.jp/en/index.php	Young accretionary complex intruded by Miocene gabbro	室戸世界ジオパーク
57A	Shikoku Seiyo	http://www.seiyo1400.jp/geopark	Geosites include Shikoku karst	四国西予ジオパーク
61	Oita Himeshima	http://www.himeshima.jp/geopark	Volcanic geology of Himeshima Island including obsidian sites	おおいた姫島ジオパーク
62	Oita Bungo–Ono	http://bungo-ohno.com	Catastrophic pyroclastic flow (Aso-4).	おおいた豊後大野ジオパーク
63A	Aso	http://aso-geopark.jp/	Degassing crater in one of the world's largest active volcanic caldera	阿蘇ジオパーク
64A	Shimabara Peninsula	http://www.unzen-geopark.jp/en-top	Various sites around one of Japan's most hazardous stratovolcanoes	島原半島世界ジオパーク
65A	Amakusa Goshoura	http://www.city.amakusa.kumamoto.jp/amakusa-geopark/	Fossiliferous Cretaceous strata with dinosaurs	天草御所浦ジオパーク
67	Kirishima	http://www.mct.ne.jp/users/kiri-geopark/	Kirishima Volcanic Complex includes active Shinmoedake Volcano	霧島ジオパーク
68	Sakurajima–Kinkowan	http://www.kagoshima-yokanavi.jp/shizen/geopark/geo01.html	Ongoing eruption history of Japan's most active volcano	桜島・錦江湾ジオパーク

Table 12.2. *Geomuseums in Japan*

Map	Geomuseum (English name)	Website	Address	Japanese name
2	Numata Fossil Museum	http://numata-kaseki.sakura.ne.jp/	2-7-49 Minami 1jo, Numata-cho, Uryu-gun, Hokkaido 078-2202	沼田町化石館
3	Ashoro Museum of Paleontology	http://www.museum.ashoro.hokkaido.jp/	1-29-25 Konan, Ashoro-cho, Ashoro-gun, Hokkaido 089-3727	足寄動物化石博物館
4B	Mikasa City Museum	http://www.city.mikasa.hokkaido.jp/museum/category/310.html	1-212-1 Nishiki-cho, Mikasa City, Hokkaido 068-2111	三笠市立博物館
5	Hokkaido University Museum	http://www.museum.hokudai.ac.jp/english/index.html	8 Kita 10jo-nishi, Kita-ku, Sapporo City, Hokkaido 060-0810	北海道大学総合博物館
6	Hobetsu Museum	http://www.town.mukawa.lg.jp/1908.htm	80-6 Hobetsu, Mukawa-cho, Yuhutsu-gun, Hokkaido 054-0211	穂別博物館
7	Churui Naumann Elephant Museum	http://www.makubetsu.jp/kankobussankyokai/english/nauman/index.html	384-1 Shirogane-machi, Churui, Nakagawa-gun, Hokkaido 089-1701	忠類ナウマン象記念館
8	Hidaka Mountains Museum	hhttp://www.town.hidaka.hokkaido.jp/hmc/	1-297-12 Honcho-Higashi, Saru-gun, Hokkaido 055-2301	日高山脈博物館
10B	Toyako Volcano Science Museum	http://www.toyako-vc.jp/en/volcano/	1142-5 Toyako Onsen, Toyako Town, Abuta-gun, Hokkaido 049-5721	洞爺湖火山科学館
10C	Mimatsu Masao Volcano Museum	http://www005.upp.so-net.ne.jp/usuvolcano/eco77/mmuseum.htm	1184-12 Showa Shinzan, Sobetsu-cho, Usu-gun, Hkkaido 052-0102	三松正夫記念館
11	Towada Science Museum	http://aohakubutukan.ninja-web.net/i/area06-6.html	Towada-kohan Yasumiya Sanbashimae, Towada City, Aomori 018-5501	十和田科学博物館
14	Akita Mining Museum	http://kuroko.mus.akita-u.ac.jp/indexE.html	28-2 Osawa, Tegata, Akita City, Akita 010-8502	秋田大学国際資源学部附属鉱業博物館
15	Kuji Amber Museum	http://www.kuji.co.jp/museum/	19-156-133, Kokuji-cho, Kuji City, Iwate 028-0071	久慈琥珀博物館
17	Tohoku University Museum	http://www.museum.tohoku.ac.jp/english/index.html	6-3 Aoba, Aramaki, Aoba-ku, Sendai 980-8578	東北大学総合学術博物館
19B	Mount Bandai Eruption Museum	http://www.bandaimuse.jp/	11093-36 Kengamine, Hibara, Kitashiobara, Yama-gun, Fukushima	磐梯山噴火記念館
20A	Iwaki City Ammonite Center	http://www.ammonite-center.jp/	147-2 Tsurubo, Ohisa-machi, Iwaki City, Fukushima 979-0338	いわき市アンモナイトセンター
20B	Iwaki City Coal and Fossil Museum	http://www.sekitankasekikan.or.jp/	3-1 Mukaida, Jobanyumoto-machi, Iwaki, Fukushima 972-8321	いわき市石炭・化石館「ほるる」
21B	Fossa Magna Museum	http://www.city.itoigawa.lg.jp/dd.aspx?menuid=4586	1313 Ichinomiya, Itoigawa City, Niigata 941-0056	糸魚川市フォッサマグナミュージアム
22A	Asama Volcano Museum	http://www.asamaen.tsumagoi.gunma.jp/	Kitakaruizawa, Naganohara-machi, Agatsuma-gun, Gunma 377-1405	浅間火山博物館
22B	Gunma Museum of Natural History	http://www.gmnh.pref.gunma.jp/en/	1674-1, Kamikuroiwa, Tomioka City, Gunma 370-2345	群馬県立自然史博物館
22C	Kanna Town Dinosur Center	http://www.dino-nakasato.org/index.html	51-2 Kagahara, Kanna-machi, Tano-gun, Gunma 370-1602	神流町恐竜センター
23	Konoha Fossil Museum	http://www.konohaisi.jp/	472 Nakashiobara, Nasushiobara City, Tochigi 329-2924	木の葉化石園
24	Kuzuu Fossil Museum	http://www.city.sano.lg.jp/kuzuufossil/	1-11-15 Kuzuu-higashi, Sano City, Tochigi 327-0501	葛生化石館
26A	Geological Museum	https://www.gsj.jp/Muse/	1-1-1 Tsukuba Central no.7 Higashi, Tsukuba, Ibaraki 305-8567	地質標本館
26B	Ibaraki Nature Museum	https://www.nat.museum.ibk.ed.jp	700 Osaki, Bando, Ibaraki, 306-0622	茨城県自然博物館

(*Continued*)

Table 12.2. *Geomuseums in Japan* (*Continued*)

Map	Geomuseum (English name)	Website	Address	Japanese name
29	Tateyama Caldera Sabo Museum	http://www.tatecal.or.jp/top.htm	68 Bunazaka, Tateyama-machi, Nakaniikawa-gun, Toyama 930-1405	立山カルデラ砂防博物館
30B	Hakusan Dinosaur Park Shiramine	http://www.city.hakusan.lg.jp/kyouiku/hakusanrokubunsitukyouiku/kyouryupark.html	4-99-1 Kuwajima, Hakusan City, Ishikawa 920-2502	白山恐竜パーク白峰
32B	Fukui Museum of Dinosaurs	http://www.dinosaur.pref.fukui.jp/en/	51-11 Terao, Muroko, Katsuyama City, Fukui 911-8601	福井県立恐竜博物館
33A	Lake Nojiri elephant Museum	http://www.avis.ne.jp/~nojiriko/	287-5 Nojiri, Shinano-machi, Kamiminochi-gun, Nagano 389-1303	野尻湖ナウマンゾウ博物館
33B	Shiga Fossil Museum	http://youkoso.city.matsumoto.nagano.jp/wordpress_indexp_148-htm	85-1 Nanaarashi, Matsumoto City, Nagano 390-1701	松本市四賀化石館
33C	Togakushi Fossil Museum	http://www.avis.ne.jp/~kaseki/index.htm	3400 Togakushitochiwara, Nagano City, Nagano 381-4104	戸隠地質化石博物館
34B	Oshika Median Tectonic Line Museum	http://www.osk.janis.or.jp/~mtl-muse/	988 Ohkawara, Oshika-mura, Shimoina-gun, Nagano 399-3502	大鹿村中央構造線博物館
34C	Anan-cho Fossil Museum	http://www.town.anan.nagano.jp/tyomin/cat11/cat154/000114.html	3905 Tomikusa, Anan-cho, Shimoina-gun, Nagano 399-1505	阿南町化石館
34D	Kashiwagi Museum	http://www.kmuseum.com/	5513 Kitayama, Chino City, Nagano 391-0301	柏木博物館
34E	Koken Earth Science Museum	http://www.koken-boring.co.jp/jwlbox/index.html	Ikoinomori Park 4688 Kitaono, Shiojiri City, Nagano 399-0651	ミュージアム鉱研 地球の宝石箱
35A	Saitama Museum of Natural history	http://www.shizen.spec.ed.jp/?page_id=164	1417-1 Nagatoro, Nagatoro-machi, Saitama, 369-1305	埼玉県立自然の博物館
35B	Ogano Fosiil Museum	http://www.town.ogano.lg.jp/menyu/kankou/midokoro/kasekikan/kasekikan.html	453 Shimoogano, Ogano-machi, Saitama 368-0101	おがの化石館
36	National Museum of Nature and Science	http://www.kahaku.go.jp/english/	7-20 Ueno Park, Taito-ku, Tokyo 110-8718	国立科学博物館
37	Chiba Natural History Museum	http://www.chiba-muse.or.jp/NATURAL/english/index-e.htm	955-2 Aoba-cho, Chuo-ku, Chiba 260-8682	千葉県立中央博物館
38B	Kanagawa Prefectural Museum of Natural History	http://nh.kanagawa-museum.jp/index_en.html	499 Iryuda, Odawara, Kanagawa 250-0031	神奈川県立生命の星・地球博物館
39B	Izu-Oshima Museum of Volcanoes	http://www.izu-oshima.or.jp/work/look/kazan.html	617 Kandayashiki, Ohshima-machi, Tokyo 100-0101	伊豆大島火山博物館
41	Fukui City Natural History Museum	http://www.nature.museum.city.fukui.fukui.jp/	147 Asuwakami-cho, Fukui City, Fukui 918-8006	福井市自然史博物館
42A	Kinshozan Fossil Museum	http://www.city.ogaki.lg.jp/0000000664.html	4527-19 Akasaka-cho, Ohgaki City, Gifu 503-2213	金生山化石館
42B	Mizunami Fossil Museum	http://www.city.mizunami.lg.jp/docs/2014092922650/	1-47 Yamanouchi, Akiyo-cho, Mizunami City, Gifu 509-6132	瑞浪市化石博物館
42C	Nakatsugawa Mineral Museum	http://mineral.n-muse.jp	6639-15 Naegi, Nakatsugawa City, Gifu 508-0101	中津川市鉱物博物館
42D	Stone Museum	http://www.hakusekikan.co.jp/	5263-7 Hirukawa, Nakatsugawa City, Gifu 509-8301	博石館

42E	Hichiso Precambrian Museum	http://www.hichiso.jp/ishi	1160 Nakaaso, Hichiso, Kamo-gun, Gifu 509-0403	日本最古の石博物館
43A	Kiseki Museum of World Stone	http://www.kiseki-jp.com/	3670 Yamamiya, Fujinomiya City, Shizuoka 418-0111	奇石博物館
43B	Tokai Univ. Natural History Museum	http://www.sizen.muse-tokai.jp/	2389 Miho, Sihmizu-ku, Shizuoka City, Shizuoka 424-8620	東海大学自然史博物館
44	Toyohashi Natural History Museum	http://www.toyohaku.gr.jp/sizensi/English/English-top.htm	1-238 Oana, Oiwa, Toyohashi, Aichi, 441-3147	豊橋市自然史博物館
46A	Lake Biwa Museum	http://www.lbm.go.jp/english/index.html	1091 Oroshimo, Kusatsu, Shiga 525-0001	琵琶湖博物館
46B	Masutomi Geology Museum	http://www.masutomi.or.jp/	394 Nakademizu-cho, Kamigyo-ku, Kyoto City, Kyoto 602-8012	益富地学会館
47A	Ikuno Ginzan	http://www.ikuno-ginzan.co.jp/	33-5 Kono, Ikuno-cho, Asago City, Hyogo 679-3324	生野銀山
47B	Genbudo Museum	http://www3.ocn.ne.jp/~genbudo/	1362 Akaishi, Toyooka City, Hyogo 668-0801	玄武洞ミュージアム
49	The Museum of Nature and Human Activities	http://www.hitohaku.jp/english/	6 Yayoigaoka, Sanda, Hyogo 669-1546	兵庫県立人と自然の博物館
50	Nojima Fault Preservation Museum	http://www.nojima-danso.co.jp/nojimafaultpreservationmuseum.php	177 Ogura, Awaji City, Hyogo 656-1736	野島断層保存館
51	Osaka Natural History Museum	http://www.mus-nh.city.osaka.jp/	1-23 Nagai Park, Higashi-Sumiyoshi-ku, Osaka, 546-0034	大阪市立自然史博物館
52	Kurashiki Natural History Museum	http://www2.city.kurashiki.okayama.jp/musnat	2-6-1 Chuo, Kurashiki City, Okayama, 710-0046	倉敷市立自然史博物館
53	Wakayama Natural History Museum	http://www.shizenhaku.wakayama-c.ed.jp/	370-1 Funoo, Kainan City, Wakayama 642-0001	和歌山県立自然博物館
54	Stone Museum, Mure Town	http://www.takamatsu-webmuseum.jp/ishi/	1810 Mure, Mure-cho, Takamatsu City, Kagawa 761-0121	高松市石の民俗資料館
55	Sakawa Geology Museum	http://www.town.sakawa.lg.jp/chishitsukan/index.htm	Ko 360 Sakawa-cho, Takaoka-gun, Kochi 789-1201	佐川町立佐川地質館
57B	Shirokawa Geology Museum	http://www.seiyo1400.jp/c/geopoint/西予市立城川地質館/	2080 Kubono, Shirokawa-cho, Seiyo City, Ehime 797-1703	西予市立城川地質館
58	Shimane Mt. Sanbe Sahimel Museum	http://nature-sanbe.jp/sahimel/	1121-8 Tane, Sanbe-cho, Ohda City, Shimane 694-0003	島根県立三瓶自然館サヒメル
59	Okuizumo Tane Natural History Mus.	http://tanemuseum.jp/	236-1 Sajiro, Okuizumo-cho, Nita-gun, Shimane 699-1434	奥出雲多根自然博物館
60A	Akiyoshi-dai Natural History Museum	http://www.c-able.ne.jp/~akihaku/amnh/	1237-938, Akiyoshi, Shuho-cho, Mine City, Yamaguchi, 754-0511	美祢市立秋吉台科学博物館
60B	Mine Fossil Museum	http://www.c-able.ne.jp/~naganobo/mmhfmfm/mfm_index.html	315-12 Higashibun, Ohmine-cho, Mine City, Yamaguchi 759-2292	美祢市化石館
63B	Aso Volcano Museum	http://www.asomuse.jp/	1930-1, Akamizu, Aso-shi, Kumamoto 869-2232	阿蘇火山博物館
64B	Mt. Unzen Disaster Memorial Hall	http://www.udmh.or.jp/english.html	1-1 Heisei-machi Shimabara-city Nagasaki 855-0879	雲仙岳災害記念館
65B	Goshoura Cretaceous Museum	http://gcmuseum.ec-net.jp/	4310-5, Goshoura, Goshoura-machi, Amakusa City, Kumamoto 866-0313	御所浦白亜紀資料館
66A	Kumamoto City Museum	http://webkoukai-server.kumamoto-kmm.ed.jp/web/index.shtml	3-2 Furukyo-machi, Chuoh-ku, Kumamoto City, Kumamoto 860-0007	熊本市立熊本博物館
66B	Mifune Dinosaur Museum	http://www.mifunemuseum.jp/	995-3 Mifune, Mifune-cho, Kamimashiki-gun, Kumamoto 861-3207	御船町恐竜博物館

simple location map (Fig. 12.4), and list relevant websites and brief descriptions of the geoparks currently formally certified (Table 12.1). More detailed descriptions of localities in several of these geoparks have already been provided in the preceding geotraverse.

Figure 12.4 also shows the location of most of the museums in Japanese with geological exhibits, and Table 12.2 lists their addresses and websites. The oldest and most visited of these, established in the late nineteenth century, is in the National Museum of Nature and Science, Tokyo. Other major museums may be linked to important institutions such as the Geological Museum (attached to the Geological Survey of Japan) or the Mineral Industry Museum (attached to Akita University). Many museums, in contrast, are funded by local government, such as the Fukui Prefectural Museum of Dinosaurs or the Fossa Magna Museum in Itoigawa (see geotraverse). A majority of such geomuseums came into being after 1980 during a time of increased public awareness and economic well-being, and they typically have highly active educational outreach programmes. In many cases the museums are also linked to activities taking place within nearby geoparks and they are keen to promote themselves through geotourism, hazard awareness and exploring the links between geology, Japanese history and culture. Even in smaller, more local geomuseums, such as that at Sakawa in Shikoku (also visited in the geotraverse), it is common to find superb, well-catalogued collections and sometimes good-quality informative material written in English. We encourage foreign visitors to make the effort and seek out such places, and enrich their experience of Japanese geology.

We wish to thank the two referees for their comments. In particular we acknowledge the enthusiastic help of Theodore Brown of the Geopark Promotion Office in Itoigawa (where the excellent Fossa Magna Museum may be found) who provided background information and maps of the Japanese Geopark network, Tony Reedman (British Geological Survey) for bringing helpfully relevant material to our attention, and our fellow editor Simon Wallis who made several suggestions for improvements. TM and TK also gratefully acknowledge the generous support of the Invitation Fellowship Program for Research of the Japan Society for the Promotion of Science (No. 11019).

Appendix

English to Kanji and Hiragana translations for geological and place names

Aichi	愛知	あいち
Aira	姶良	あいら
Akaishi	赤石	あかいし
Akamizu*	赤水	あかみず
Akaoka	赤岡	あかおか
Akita	秋田	あきた
Akiyoshi	秋吉	あきよし
Akiyoshi-dai	秋吉台	あきよしだい
Akiyoshi-do	秋芳洞	あきよしどう
Aoki	青木	あおき
Aoshima	青島	あおしま
Ariake	有明	ありあけ
Arimura	有村	ありむら
Asama	浅間	あさま
Aso	阿蘇	あそ
Ata	阿多	あた
Awaikeda	阿波池田	あわいけだ
Azusa[†]	梓	あずさ
Beppu	別府	べっぷ
Besshi	別子	べっし
Bishago Iwa	ビシャゴ岩	びしゃごいわ
Bonin	ボニン	ぼにん

(*Continued*)

English to Kanji and Hiragana translations for geological and place names (Continued)

Chichibu	秩父	ちちぶ
Chino	茅野	ちの
Daisho-in	大聖院	だいしょういん
Dezaki	出崎	でざき
Eboshidake	烏帽子岳	えぼしだけ
Fugendake	普賢岳	ふげんだけ
Fuji	富士	ふじ
Fujimidai	富士見台	ふじみだい
Fukui	福井	ふくい
Funazu	船津	ふなづ
Gifu	岐阜	ぎふ
Ginza-dori	銀座通	ぎんざどおり
Gomen	後免	ごめん
Gyodo	行当	ぎょうど
Hane	羽根	はね
Heisei Shinzan	平成新山	へいせいしんざん
Hida	飛騨	ひだ
Hikinohira	引ノ平	ひきのひら
Hime	姫	ひめ
Himeji	姫路	ひめじ
Himekawa	姫川	ひめかわ
Hiraiwa	平岩	ひらいわ
Hirayu	平湯	ひらゆ
Hiroshima	広島	ひろしま
Honshu	本州	ほんしゅう
Ibi	揖斐	いび
Ibusuki	指宿	いぶすき
Inuyama	犬山	いぬやま
Iōjima	伊王島	いおうじま
Isahaya	諫早	いさはや
Ito	入戸	いと
Itoigawa	糸魚川	いといがわ
Itsukushima	厳島	いつくしま
Izu	伊豆	いず
Izumi	和泉	いずみ
jigoku	地獄	じごく
Kagoshima	鹿児島	かごしま
Kagoshima-Chūō	鹿児島中央	かごしまちゅうおう
Kaimon	開聞	かいもん
Kaimondake	開聞岳	かいもんだけ
Kamikochi	上高地	かみこうち
Kamishiro	神城	かみしろ
Kamisuwa	上諏訪	かみすわ
Kanto	関東	かんとう
Kappa	河童	かっぱ
Karasujima	烏島	からすじま
Kawaguchiko	河口湖	かわぐちこ
Kinki	近畿	きんき
Kirishima	霧島	きりしま
Kishimadake	杵島岳	きしまだけ
Kiso	木曽	きそ
Kiso-Fukushima	木曽福島	きそふくしま
Kitaotori	北小谷	きたおたり
Kizaki	木崎	きざき
Kobe	神戸	こうべ
Kochi	高知	こうち
Kofu	甲府	こうふ
Kofuji	古富士	こふじ
Kojima[‡]	児島	こじま
Kojima[§]	小島	こじま
Komezuka	米塚	こめづか
Komitake	小御岳	こみたけ
Kotaki	小滝	こたき
Kotohira	琴平	ことひら
Kumagaya	熊谷	くまがや
Kumamoto	熊本	くまもと
Kunimidake	国見岳	くにみだけ
Kure	久礼	くれ
Kurosegawa	黒瀬川	くろせがわ

(*Continued*)

English to Kanji and Hiragana translations for geological and place names (Continued)

Kusakabe	草壁	くさかべ
Kusasenri	草千里	くさせんり
Kyoto	京都	きょうと
Kyushu	九州	きゅうしゅう
Magome	馬込	まごめ
Maizuru	舞鶴	まいづる
Matsumoto	松本	まつもと
Midori	水鳥	みどり
Mikabu	御荷鉾	みかぶ
Minamimiyazaki	南宮崎	みなみみやざき
Minamiotari	南小谷	みなみおたり
Mino	美濃	みの
Misen	弥山	みせん
Miyajima	宮島	みやじま
Miyajimaguchi	宮島口	みやじまぐち
Mizunashi	水無	みずなし
Muroto	室戸	むろと
Muroto-misaki	室戸岬	むろとみさき
Myokendake	妙見岳	みょうけんだけ
Nagano	長野	ながの
Nagara	長良	ながら
Nagasaki	長崎	ながさき
Nagatoro	長瀞	ながとろ
Nagoya	名古屋	なごや
Nahari	奈半利	なはり
Nakadake	中岳	なかだけ
Nakasendo	中山道	なかせんどう
Nankai	南海	なんかい
Nara	奈良	なら
Narai	奈良井	ならい
Narita	成田	なりた
Nechi	根知	ねち
Nekodake	根子岳	ねこだけ
Neodani	根尾谷	ねおだに
Nezame-no-Toko	寝覚の床	ねざめのとこ
Niigata	新潟	にいがた
Nishibun	西分	にしぶん
Nita	仁田	にた
Nobi	濃尾	のうび
Nodake	野岳	のだけ
Nojiri	野尻	のじり
Nomo	野母	のも
Nonokawa	野々川	ののかわ
Norikura	乗鞍	のりくら
Oboke	大歩危	おおぼけ
Ogaki	大垣	おおがき
Oito	大糸	おおいと
Okayama	岡山	おかやま
Okinawa	沖縄	おきなわ
Okinoshima	沖之島	おきのしま
Omachi	大町	おおまち
Ontake	御嶽	おんたけ
Osaka	大阪	おおさか
Osugi	大杉	おおすぎ
Otari	小谷	おたり
Otsuki	大月	おおつき
Otsuzaki	大津崎	おおつざき
Ryoke	領家	りょうけ
Ryukyu	琉球	りゅうきゅう
Saiko Komori Ana	西湖コウモリ穴	さいここうもりあな
Sakahogi	坂祝	さかほぎ
Sakaide	坂出	さかいで
Sakawa	佐川	さかわ
Sakurajima	桜島	さくらじま
Sanbagawa (Sambagawa)	三波川	さんばがわ
Sanyo	山陽	さんよう
Satsuma	薩摩	さつま
Sawando	沢渡	さわんど

(Continued)

English to Kanji and Hiragana translations for geological and place names (Continued)

Seto	瀬戸	せと
Shikoku	四国	しこく
Shimabara	島原	しまばら
Shimanto	四万十	しまんと
Shimotsui	下津井	しもつい
Shin-Osaka	新大阪	しんおおさか
Shin-Tosu	新鳥栖	しんとす
Shin-Yamaguchi	新山口	しんやまぐち
Shinfuji	新富士	しんふじ
Shinjuku	新宿	しんじゅく
Shinmoedake	新燃岳	しんもえだけ
Shiojiri	塩尻	しおじり
Shira	白	しら
Shizuoka	静岡	しずおか
Shodoshima	小豆島	しょうどしま
Showa	昭和	しょうわ
Suwa	諏訪	すわ
Taisho	大正	たいしょう
Takadake	高岳	たかだけ
Takamatsu	高松	たかまつ
Takashima	高島	たかしま
Takidani	滝谷	たきだに
Tanoura	田ノ浦	たのうら
Tara	多良	たら
Tarumi	樽見	たるみ
Tatamiiwa	畳岩	たたみいわ
Tateno	立野	たての
Tei	手結	てい
Tenjo	天上	てんじょう
Tokai	東海	とうかい
Tokyo	東京	とうきょう
Tosa-Kure	土佐久礼	とさくれ
Tosayamada	土佐山田	とさやまだ
Uchinomi	内海	うちのみ
Uchiumigawa	内海川	うちうみがわ
Unzen	雲仙	うんぜん
Yakedake	焼岳	やけだけ
Yamaguchi	山口	やまぐち
Yatsugatake	八ヶ岳	やつがたけ
Yogan Nagisa	溶岩なぎさ	ようがんなぎさ
Yoshino	吉野	よしの
Yudamari	湯だまり	ゆだまり
Yunohira	湯之平	ゆのひら
Zenkoji	善光寺	ぜんこうじ

*For both 'Akamizu near Aso' and 'Akamizu near Sakurajima'.
†For 'Azusa River'.
‡For 'Kojima along the coast of Seto Inland Sea'.
§For 'Kojima beside Iōjima'.

References

HARAYAMA, S. 1992. Youngest exposed granitoid pluton on Earth: cooling and rapid uplift of the Pliocene-Quaternary Takidani Granodiorite in the Japan Alps, central Japan. *Geology*, **20**, 657–660.

HATTORI, H., KOBAYASHI, K., KIKUCHI, T., ENOMOTO, T., IWATATE, T. & SHIMA, H. 2004. Distribution of seismic intensity due to strong ground motion estimated from damage of temples in the 1847 Zenkoji earthquake. *13th World Conference on Earthquake Engineering*, Vancouver, BC, 1–6 August 2004, Paper No. 2057. www.iitk.ac.in/nicee/wcee/article/ 13_2057.pdf

HOSHIDE, T. & OBATA, M. 2009. Zoning and resorption of plagioclase in a layered gabbro, as a petrographic indicator of magmatic differentiation. *Earth and Environmental Science Transactions of the Royal Society of Edinburgh*, **100**, 235–249.

HOSHIDE, T. & OBATA, M. 2012. Amphibole-bearing multiphase solid inclusions in olivine and plagioclase from a layered gabbro: origin of the trapped melts. *Journal of Petrology*, **53**, 1–31.

HOSHIZUMI, H., UTO, K. & WATANABE, K. 1999. Geology and eruptive history of Unzen volcano, Shimabara Peninsula, Kyushu, SW Japan. *Journal of Volcanology and Geothermal Research*, **89**, 81–94.

HUNTER, A. G. 1998. Intracrustal controls on the coexistence of tholeiitic and calc-alkaline magma series at Aso Volcano, SW Japan. *Journal of Petrology*, **39**, 1255–1284.

HYNDMAN, R. D., YAMANO, M. & OLESKEVICH, D. A. 1997. The seismogenic zone of subduction thrust faults. *Island Arc*, **6**, 244–260.

IGUCHI, M., YAKIWARA, H., TAMEGURI, T., HENDRASTO, M. & HIRABAYASHI, J. 2008. Mechanism of explosive eruption revealed by geophysical observations at the Sakurajima, Suwanosejima and Semeru volcanoes. *Journal of Volcanology and Geothermal Research*, **178**, 1–9.

ISHIHARA, K. 1985. Dynamic analysis of volcanic explosion. *Journal of Geodynamics*, **3**, 327–349.

ISHIHARA, S. 2003. Chemical contrast of the Late Cretaceous granitoids of the Sanyo and Ryoke Belts, SW Japan: Okayama-Kagawa transect. *Bulletin of the Geological Survey of Japan*, **54**, 95–116.

KANEKO, K., KAMATA, H., KOYAGUCHI, T., YOSHIKAWA, M. & FURUKAWA, K. 2007. Repeated large-scale eruptions from a single compositionally stratified magma chamber: an example from Aso volcano, Southwest Japan. *Journal of Volcanology and Geothermal Research*, **167**, 160–180.

KATO, H., WAKITA, K. & WILSON, T. H. 2008. *Geologic travel in Japan*. Aichi Publishing Company, Tokyo.

KITAMURA, Y. & KIMURA, G. 2012. Dynamic role of tectonic mélange during interseismic process of plate boundary mega earthquakes. *Tectonophysics*, **568–569**, 39–52.

MIYABUCHI, Y. 2011. Post-caldera explosive activity inferred from improved 67–30 ka tephrostratigraphy at Aso Volcano, Japan. *Journal of Volcanology and Geothermal Research*, **205**, 94–113.

MOGI, K. 2004. Two grave issues concerning the expected Tokai earthquake. *Earth, Planets and Space*, **56**, li–lxvi.

MUKOYOSHI, H., SAKAGUCHI, A., OTSUKI, K., HIRONO, T. & SOH, W. 2006. Co-seismic frictional melting along an out-of-sequence thrust in the Shimanto accretionary complex. Implications on the tsunamigenic potential of splay faults in modern subduction zones. *Earth and Planetary Science Letters*, **245**, 330–343.

MUKOYOSHI, H., HIRONO, T., SEKINE, K., TSUCHIDA, N. & SOH, W. 2007. Cathodoluminescence and fluid inclusion analyses of mineral veins within major thrusts in the Shimanto accretionary complex: evidence of hydraulic fracturing during thrusting. *Earth Planets and Space*, **59**, 937–942.

NAKADA, S., SHIMIZU, H. & OHTA, K. 1999. Overview of 1990–1995 eruptions at Unzen Volcano. *Journal of Volcanology and Geothermal Research*, **89**, 1–22.

NAKAJIMA, J. & HASEGAWA, A. 2007. Deep crustal structure along the Niigata-Kobe Tectonic Zone, Japan: its origin and segmentation. *Earth, Planets and Space*, **59**, e5–e8.

NARA, M. & IKARI, Y. 2011. Deep-sea bivalvian highways: an ethological interpretation of branched Protovirgularia of the Palaeogene Muroto-Hanto Group, southwestern Japan. *Palaeogeography, Palaeoclimatology, Palaeoecology*, **305**(1–4), 250–255, http://doi.org/10.1016/j.palaeo.2011.03.005

OKUMURA, K. 2001. Paleoseismology of the Itoigawa-Shizuoka tectonic line in central Japan. *Journal of Seismology*, **5**, 411–431.

ONOUE, T., SATO, H. ET AL. 2012. Deep-sea record of impact apparently unrelated to mass extinction in the Late Triassic. *Proceedings of the National Academy of Sciences*, **109**, 19 134–19 139.

PICCARDI, L. & MASSE, W. B. 2007. *Myth and Geology*. Geological Society, London, Special Publication no. 273.

TAIRA, A., OKAMURA, M., KATTO, J., TASHIRO, M., SAITO, Y., KODAMA, K., HASHIMOTO, M., TIBA, T. & AOKI, T. 1980. Lithofacies and geologic age relationship within melange zone of northern Shimanto belt (Cretaceous), Kochi Prefecture, Japan. *In*: TAIRA, A. & KATTO, J. (eds) *Geology and Paleontology of the Shimanto Belt: Kochi, Japan*. Rinya-Kosaikan Press, Kochi, 179–214.

TAKEDA, T., SATO, H., IWASAKI, T., MATSUTA, N., SAKAI, S., IIDAKA, T. & KATO, A. 2004. Crustal structure in the northern Fossa Magna region, central Japan, modelled from refraction/wide-angle r reflection data. *Earth, Planets and Space*, **56**, 1293–1299.

YAMAGUCHI, T., TANAKA, Y. & NISHI, H. 2008. Calcareous nannofossil and planktic foraminiferal biostratigraphy of the Paleogene Iojima Group in the Takashima Coafield, Nagasaki Prefecture, southwest Japan. *Paleontological Research*, **12**, 223–236.

WALLIS, S., MORIYAMA, Y. & TAGAMI, T. 2004. Exhumation rates and age of metamorphism in the Sanbagawa belt: new constraints from zircon fission track analysis. *Journal of Metamorphic Geology*, **22**, 17–24.

Index

Page numbers in *italic* refer to Figures. Page numbers in **bold** refer to Tables.